10TH EDITION

&TOPLEY &WILSON'S

MICROBIOLOGY & MICROBIAL INFECTIONS

MEDICAL MYCOLOGY

10TH EDITION

TOPLEY & WILSON'S
MICROBIOLOGY & MICROBIAL INFECTIONS

Topley & Wilson's Microbiology and Microbial Infections has grown from one to eight volumes since first published in 1929, reflecting the ever-increasing breadth and depth of knowledge in each of the areas covered. This tenth edition continues the tradition of providing the most comprehensive reference to microorganisms and the resulting infectious diseases currently available. It forms a unique resource, with each volume including examples of the best writing and research in the fields of virology, bacteriology, medical mycology, parasitology, and immunology from around the globe.

www.topleyandwilson.com

VIROLOGY Volumes 1 and 2

Edited by Brian W.J. Mahy and Volker ter Meulen
Volume 1 ISBN 0 340 88561 0; Volume 2 ISBN 0 340 88562 9; 2 volume set ISBN 0 340 88563 7

BACTERIOLOGY Volumes 1 and 2

Edited by S. Peter Borriello, Patrick R Murray, and Guido Funke
Volume 1 ISBN 0 340 88564 5; Volume 2 ISBN 0 340 88565 3; 2 volume set ISBN 0 340 88566 1

MEDICAL MYCOLOGY

Edited by William G. Merz and Roderick J. Hay
ISBN 0 340 88567 X

PARASITOLOGY

Edited by F.E.G. Cox, Derek Wakelin, Stephen H. Gillespie, and Dickson D. Despommier
ISBN 0 340 88568 8

IMMUNOLOGY

Edited by Stephan H.E. Kaufmann and Michael W. Steward
ISBN 0 340 88569 6

Cumulative index

ISBN 0 340 88570 X

8 volume set plus CD-ROM

ISBN 0 340 80912 4

CD-ROM only

ISBN 0 340 88560 2

For a full list of contents, please see the *Complete table of contents* on page 863

10TH EDITION

& TOPLEY & WILSON'S

MICROBIOLOGY & MICROBIAL INFECTIONS

MEDICAL
MYCOLOGY

EDITED BY

William G. Merz PhD
Professor of Pathology, Dermatology, Epidemiology, and Molecular Microbiology and Immunology
The Johns Hopkins University, Baltimore, Maryland, USA

Roderick J. Hay DM FRCP FRCPATH
Dean, Faculty of Medicine and Health Sciences
Queens University Belfast, Belfast, Northern Ireland, UK

Hodder Arnold

A MEMBER OF THE HODDER HEADLINE GROUP

ASM
PRESS

First published in Great Britain in 1929
Second edition 1936
Third edition 1946
Fourth edition 1955
Fifth edition 1964
Sixth edition 1975
Seventh edition 1983 and 1984
Eight edition 1990
Ninth edition 1998
This tenth edition published in 2005 by
Hodder Arnold, an imprint of Hodder Education and a member of the Hodder Headline Group,
338 Euston Road, London NW1 3BH

http://www.hoddereducation.com

Distributed in the United States of America by ASM Press, the book publishing division of the American Society for Microbiology, 1752 N Street, N.W. Washington, D.C. 20036, USA

Hodder Headline's policy is to use papers that are natural, renewable and recyclable products and made from wood grown in sustainable forests. The logging and manufacturing processes are expected to conform to the environmental regulations of the country of origin.

Whilst the advice and information in this book are believed to be true and accurate at the date of going to press, neither the author[s] nor the publisher can accept any legal responsibility or liability for any errors or omissions that may be made. In particular (but without limiting the generality of the preceding disclaimer) every effort has been made to check drug dosages; however it is still possible that errors have been missed. Furthermore, dosage schedules are constantly being revised and new side-effects recognized. For these reasons the reader is strongly urged to consult the drug companies' printed instructions before administering any of the drugs recommended in this book.

British Library Cataloguing in Publication Data
A catalogue record for this book is available from the British Library

Library of Congress Cataloging-in-Publication Data
A catalog record for this book is available from the Library of Congress

This volume only ISBN-10 0 340 885 67X   ISBN-13 978 0 340 885 673
Complete set and CD-ROM ISBN-10   0 340 80912 4   ISBN-13   978 0 340 80912 9
Indian edition ISBN-10   0 340 88559 9   ISBN-13   978 0 340 88559 8

1 2 3 4 5 6 7 8 9 10

Commissioning Editor: Serena Bureau / Joanna Koster
Development Editor: Layla Vandenberg
Project Editor: Zelah Pengilley
Production Controller: Deborah Smith
Index: Merrall-Ross International Ltd.
Cover Designer: Sarah Rees

Cover image: Candida albicans fungus, SEM. Stem Jems / Science Photo Library

Typeset in [TYPE SIZE AND NAME] by Lucid to complete
Printed and bound in Italy

What do you think about this book? Or any other Hodder Arnold title? Please send your comments to www.hoddereducation.com

Contents

Contributors

L. Ajello PHD†
Formerly Adjunct Professor
Department of Ophthalmology
Emory University Eye Center
Atlanta, GA, USA

Roberto Arenas MD
Professor of Dermatology and
Medical Mycology
Hospital General 'Dr. Manuel Gea González'
University of Mexico
México D.F.

Sevtap Arikan MD
Professor of Microbiology
Head of Mycology Laboratory
Hacettepe University Faculty of Medicine
Department of Microbiology and Clinical Microbiology
Ankara Turkey

Sarath N. Arseculeratne MBBS (CEY) DIP BACT
(MANCH) DPHIL (OXON)
Department of Microbiology, Faculty of Medicine
University of Peradeniya
Peradeniya Sri Lanka

Gil Benard
Medical Researcher, Laboratory of Dermatology and
Immunodeficiencies
Medical School of the University of São Paulo,
São Paulo, Brazil

James P. Burnie DSC MD PHD FRCPATH MRCP MSC
Professor of Medical Microbiology
University of Manchester; and
Chief Executive Officer, NeuTec Pharma plc
Manchester, UK

Colin K. Campbell BSC MSC PHD
Mycology Reference Laboratory and National
Collection of Pathogenic Fungi
Health Protection Agency
Specialist and Reference Microbiology
Division, HPA South West Laboratory
Bristol, UK

Francis W. Chandler DVM PHD
Professor Emeritus
Department of Pathology
Medical College of Georgia
Augusta, GA, USA

Chester R. Cooper, Jr PHD
Associate Professor
Department of Biological Sciences
Youngstown State University
Youngstown, OH, USA

Gary M. Cox MD
Associate Professor of Medicine
Duke University Medical Center
Durham, NC, USA

Vicente Crespo Erchiga MD PHD
Head, Department of Dermatology
Hospital Reginal Universitario Carlos Haya
Málaga, Spain

Melanie T. Cushion
Professor, Department of Internal Medicine
University of Cincinnati College of Medicine and the
Cincinnati Veterans Affairs Medical Center
Cincinnati, OH, USA

G. Sybren de Hoog PHD
Senior researcher
Centralbureau voor Schimmelcultures,
Utrecht, The Netherlands; and
Professor, Institue for Biodiversity and
Ecosystem Dynamics, University of Amsterdam,
Amsterdam, The Netherlands

David W. Denning FRCP FRCPATH
University of Manchester and
Wythenshawe Hospital
Manchester, UK

Arthur F. Di Salvo MD
Former South Carolina State Public
Health Laboratory Director
Reno, NV, USA

Edouard Drouhet MD†
Formerly Professor of Mycology,
Pasteur Institute, Mycology Unit,
Paris, France

Daniel Elad DVM PHD
Head, Division of Bacteriological and Mycological
Laboratories
Kimron Veterinary Institute
Bet Dagan, Israel

Boni Elewski MD
Professor of Dermatology
Department of Dermatology
University of Alabama at Birmingham
Birmingham, AL, USA

David H. Ellis BSC (Hons) MSC PHD FASM
FRCPA (Hon)
Associate Professor, Mycology Unit
Women's and Children's Hospital
North Adelaide, Australia

Philippe Esterre DVM PHD MSC
Head of Plasmodium Chemoresistance
Reference Center (CNRCP)
Institut Pasteur de Guyane
French Guiana, South America

Paul L. Fidel Jr PHD
Carl Baldridge Research Professor,
Department of Microbiology, Immunology, and
Parasitology
Louisiana State University Health
Sciences Center
New Orleans, LA, USA

Marcello Franco MD PHD
Chairman, Department of Pathology
Federal University of São Paulo
Paulista School of Medicine
São Paulo, Brazil

P. Geraldine MSC MPHIL PHD
Department of Animal Science
Bharathidasan University
Tiruchirapalli, India

Norman L. Goodman
Professor Emeritus
Department of Pathology
College of Medicine, University of Kentucky
Lexington, KY, USA

John R. Graybill
Professor of Medicine
Division of Infectious Diseases

University of Texas Health Science Center
San Antonio, TX, USA

Eveline Guého PHD
Researcher INSERM
Formerly Pasteur Institute, Mycology Unit
Paris, France

Roderick J. Hay DM FRCP FRCPATH
Dean, Faculty of Medicine and Health Sciences,
Queens University Belfast,
Belfast

Henrik Elvang Jensen DVM DR VET SCI PHD
Diplomate, European College of Veterinary
Pathologists
The Royal Veterinary and Agricultural University
Copenhagen, Denmark

Elizabeth M. Johnson BSC PHD
Director, Mycology Reference Laboratory and
National Collection of Pathogenic Fungi
Health Protection Agency, Specialist and Reference
Microbiology Division, HPA, South West Laboratory
Bristol, UK

Bruce S. Klein
Professor of Pediatrics
Internal Medicine and Medical Microbiology and
Immunology
University of Wisconsin Medical School
Madison, WI, USA

Paul F. Lehmann PHD
Professor, Department of Medical Microbiology and
Immunology
Medical College of Ohio
Toledo, OH, USA

Janine R. Maenza MD
Clinical Assistant Professor of Medicine
Division of Allergy and Infectious Diseases
Department of Medicine, University of Washington,
Seattle, WA, USA

Ruth Matthews MSC MD PHD FRCPATH
Professor of Infectious Diseases
University of Manchester; and
Research & Development Director
NeuTec Pharma plc
Manchester, UK

Leonel Mendoza MSC PHD
Associate Professor, Medical Technology Program
Microbiology and Molecular Genetics
Michigan State University
East Lansing, MI, USA

William G. Merz PHD
Professor of Pathology,
Dermatology, Epidemiology, and Molecular
Microbiology and Immunology
The Johns Hopkins University
Baltimore, MD, USA

Thomas G. Mitchell PHD
Associate Professor
Department of Molecular Genetics and Microbiology
Duke University Medical Center
Durham, NC, USA

Caroline B. Moore MSC PHD SRCS CBIOL
Senior Clinical Scientist
Regional Mycology Centre
Department of Microbiology, Hope Hospital
Salford, UK

Arvind A. Padhye PHD
Guest Researcher, Mycotic Diseases Branch,
Division of Bacterial and Mycotic Diseases,
Centers for Disease Control and Prevention
Atlanta, GA, USA

Demosthenes Pappagianis MD PHD
Professor of Medical Microbiology and
Immunology, School of Medicine
University of California
Davis, CA, USA

Peter G. Pappas MD
Professor of Medicine
University of Alabama at Birmingham School of
Medicine
Birmingham, AL, USA

Thomas F. Patterson MD FACP
Professor of Medicine
Director, San Antonio Center for Medical Mycology
The University of Texas Health Science Center
at San Antonio
San Antonio, TX, USA

John R. Perfect MD
Professor of Medicine
Duke University Medical Center
Durham, NC, USA

R. Scott Pore PHD
Professor of Microbiology, Immunology
and Cell Biology
West Virginia University School of Medicine
Morgantown, WV, USA

Roger Pradinaud
Centre Hospitalier Andree Rosemon
Service de Dermatologie
Cayenne, Cedex, France

Sanjay G. Revankar MD
Associate Professor of Medicine
Division of Infectious Diseases
Dallas VA Medical Center; and
Department of Medicine, UT Southwestern
Dallas, TX, USA

John H. Rex MD FACP
Vice President and Medical Director for Infection
Astra Zeneca Pharmaceuticals, Cheshire, UK; and
Adjunct Professor of Medicine,
University of Texas Medical School,
Houston, TX, USA

Malcolm D. Richardson BSC PHD FRCPATH FIBIOL
Associate Professor in Medical Mycology
Department of Bacteriology and Immunology
Haartman Institute, University of Helsinki
Helsinki, Finland

Glenn D. Roberts PHD
Consultant, Division of Clinical Microbiology
Department of Laboratery Medicine and Pathology
Mayo Clinic; and Professor of Microbiology and of
Laboratory Medicine, Mayo Clinic College of Medicine
Rochester, MN, USA

Esther Segal PHD
Professor of Microbiology and Mycology,
Head, Department of Human Microbiology
Sackler School of Medicine, Tel-Aviv University
Tel Aviv, Israel

Lynne Sigler MSC
Professor and Curator,
University of Alberta Microfungus Collection &
Herbarium, Devonian Botanic Garden,
Edmonton, AB, Canada

Richard C. Summerbell
Centraalbureau voor Schimmelcultures,
Utrecht, The Netherlands

Sinésio Talhari PHD
Chief, Department of Tropical Dermatology
Institute of Tropical Medicine
School of Medicine
University of Amazonas
Manaus, Brazil

Ram P. Tewari DVM MPH PhD
Professor of Microbiology
Department of Cell Biology and Neurosciences
Rutgers University
Piscataway, NJ, USA

Philip A. Thomas MD PhD MAMS FIMSA
Professor and Head of Ocular Microbiology
Institute of Ophthalmology, Joseph Eye Hospital
Tiruchirapalli, Tamilnadu, India

Nongnuch Vanittanakom Dr rer nat
Professor of Medical Microbiology
Department of Microbiology
Faculty of Medicine, Chiang Mai University
Chiang Mai, Thailand

Maria Anna Viviani MD
Associate Professor of Hygiene
Laboratory of Medical Mycology
Institute of Hygiene and Preventive Medicine
School of Medicine, Universita degli Studi di Milano
Milano, Italy

L. Joseph Wheat MD
President and Director
MiraVista Diagnostics &
MiraBella Technologies
Indianapolis, IN, USA

Jeff Weeks MD
University of Alabama at Birmingham
Department of Dermatology
Birmingham, AL, USA

Karen L. Wozniak MS PhD
Department of Medicine
Section of Infectious Diseases
Boston University Medical Center
Boston, MA, USA

Preface

During the last few decades there has been a rise in the global incidence and prevalence of fungal diseases afflicting humans. At the beginning of the twentieth century, most of the prevailing fungal infections were thought to be primarily superficial, whereas systemic mycoses were considered to be rare and exotic. Nowadays, opportunistic fungi, formerly considered to be saprophytic, have been found to possess latent capabilities to cause life-threatening infections in certain individuals such as those with primary or secondary immunodeficiencies. We also have recognized that our endemic mycoses are not rare pathogens causing life-threatening disease, but are common causing a wide spectrum of disease, even in a normal host. Increased interest in academic, industrial, and government scientists in the pathogenic molds and yeasts has stimulated research that has resulted in a significant expansion of our knowledge of the biology, clinical expression, and management of fungal diseases.

The introductory chapters of this volume are devoted to the history and taxonomy of the fungal pathogens, diagnostic techniques used in medical mycology, and approaches to their management as well. Thereafter, the etiological agents are presented on the basis of the range of diseases that they cause. The mycoses have been grouped by the principle sites primarily targeted by the invading pathogens and the host response they elicit.

Although it is often said that progress in the biology, diagnosis, and treatment of fungal infections proceeds at a slow pace, we are very pleased to have had significant new information to incorporate in this revised version. Innovative research and applications of new technologies in medical mycology are appearing regularly we even now have a new class of antifungal agent all of which make it difficult to keep any text current. We have attempted to add new information and make this revision as up to date as possible. While not altering the content or general structure of this edition, our intention has been to make it easier for the reader to explore this text and to facilitate the location of specific fungi or clinical infections.

In these days of specialization, our goal has been to produce a text that will provide an important resource for all members of the medical mycology community and associated disciplines including research scientists, clinical laboratory scientists, infectious disease and other clinicians, and epidemiologists. In addition, and perhaps most importantly, we hope this release will be a powerful resource and inspiration for students interested in studying the relationship of the fungi and the infections they cause.

We are deeply grateful to all the authors who contributed so much of their time, efforts, and expertise to make this volume reflect the clinical, microbiological, and public health importance of the mycoses and their etiological agents and the challenge that they pose. We also are most appreciative of the individuals at Hodder Headline who have brought these efforts to fruition.

It will be extremely interesting to observe what exciting new concepts the molecular era will bring to the discipline before the next edition. There is already evidence discussed in the chapter on Rhinosporidiosis that suggests that the causative organism should more correctly be classified as a Protozoan.

William G. Merz
Roderick J. Hay
Belfast and Baltimore
May 2005

Abbreviations

AB	asteroid bodies
ABC	avidin–biotin enzyme complex
ABCD	amphotericin B colloidal dispersion
ABG	arterial blood gas
ABLC	amphotericin B lipid complex
ABPA	allergic bronchopulmonary aspergillosis
AD	atopic dermatitis
ADCC	antibody-dependent cellular cytotoxicity
AdoMet	*S*-adenosylmethionine
AFLP	amplified fragment length polymorphism
AFS	allergic fungal sinusitis
Ag2	antigen 2
AIDS	acquired immunodeficiency syndrome
ALAT	alanine aminotransferase
ALP	alkaline protease
Als1p	agglutin-like sequence-1
AMB	amphotericin B
APAAP	alkaline phosphatase-antialkaline phosphatase
APEX	arrayed primer extension
ARDS	acute respiratory distress syndrome
ASWS	alkali-soluble water-soluble
AZT	azidothymidine
BA	Bayesian analysis
BAD1	Blastomyces adhesin 1
BAL	bronchoalveolar lavage
BALF	bronchoalveolar lavage fluid
BCG	bacille Calmette–Guérin
BCP	bromocresol purple
BHI	brain–heart infusion
BSS	balanced salt solution
CAPD	continuous ambulatory peritoneal dialysis
CARE	candidal DNA repetitive elements
CCPA	chronic cavitary pulmonary aspergillosis
CDC	Centers for Disease Control and Prevention
CF	complement fixation
CFA	culture filtrate antigen
CFPA	chronic fibrosing pulmonary aspergillosis
CFTR	cystic fibrosis transmembrane conductance regulator
CFU	colony-forming units
CGD	chronic granulomatous disease

CHEF	contour-clamped homogeneous electrical field, or contour-clamped homogeneous electric field electrophoresis
CIE	counterimmunoelectrophoresis
CMA	corn meal agar
CMC	chronic mucocutaneous candidiasis
CMI	cell-mediated immunity
CMV	cytomegalovirus
CNPA	chronic necrotizing pulmonary aspergillosis
CNS	central nervous system
ConA	concanavalin A
COPD	chronic obstructive pulmonary disease
CR3	complement receptor 3
CRP	confluent and reticulate papillomatosis of Gougerot–Carteaud
CRS	chronic rhinosinusitis
CSA	*Coccidioides*-specific antigen
CSF	cerebrospinal fluid; or colony-stimulating factor
CT	computed tomography
CTL	cytotoxic T lymphocytes
Cu/Zn SOD	copper and zinc-containing superoxide dismutase
DAPI	4′,6-diamidino-2-phenylindole
DC	disseminated candidiasis
DD	double diffusion
DFA	direct fluorescent antibody
DFMO	α-difluoromethylornithine
2-DIE	two-dimensional immunoelectrophoresis
DMSO	dimethyl sulfoxide
DHFR	dihydrofolate reductase
DSP	deep-seated phaeohyphomycosis
DTH	delayed-type hypersensitivity
DTM	dermatophyte test medium
ECM	extracellular materials; or extracellular matrix
EDB	electron-dense bodiess
EF3	elongation factor 3
EIA	enzyme immunoassay
ELISA	enzyme-linked immunosorbent assay
EM	electron microscope
EORTC	European Organization for Research in the Treatment of Cancer
ERG	electroretinograms

EST	expressed sequence tag		**Int1p**	integrin-like protein-1
			IPO	immunoperoxidase
5-FC	5-fluorocytosine		**ISHAM**	International Society for Human and Animal Mycology
5-FU	5-fluorouracil			
5-FUTP	5-fluorouracil triphosphate		**ITS**	internal transcribed spacer
FDA	Food and Drug Administration (USA)		**ITS1**	internal transcribed spacer 1
FITC	fluorescein isothiocyanate		**ITS2**	internal transcribed spacer 2
FNA	fine-needle aspiration		**IV**	intravenous
G-CSF	granulocyte colony stimulating factor		**JMVM**	Journal of Medical and Veterinary Mycology
G6PDH	glucose-6-phosphate dehydrogenase			
GAPDH	glyceraldehyde-3-phosphate-dehydrogenase		**KOH**	potassium hydroxide
G+C	guanine+cytosine content		**KTR**	killer toxin receptor
GC-MS	gas chromatography-mass spectrometry			
GF	Gridley's fungus		**LA**	latex agglutination
GI	gastrointestinal		**LAF**	laminar air flow
GLC	gas–liquid chromatography		**L-AMB**	liposomal amphotericin B
GM	granulocyte-macrophage		**LASIK**	laser-in-situ keratomileusis
GM-CSF	granulocyte–macrophage colony-stimulating factor		**LAT**	latex particle agglutination
			LB	laminated (multilamellar) body
GMS	Gomori's methenamine silver nitrate		**LD50**	median lethal dose
GPA	glucose peptone agar		**LDH**	lactic dehydrogenase
GPI	glycosylphosphatidylinositol		**LMI**	leukocyte migration inhibition
GVHD	graft-versus-host disease		**LPR**	lymphoproliferative responses
			LSU	large subunit
H&E	hematoxylin–eosin			
HAART	highly active antiretroviral therapy		**MAIDS**	murine AIDS model
hBD-1	β-defensin 1		**M-CSF**	macrophage colony-stimulating factor
HEPA	high-efficacy particulate air		**MEC**	minimal effective concentration
HES	hematoxylin–eosin–saffron		**MEM**	minimum essential medium
HIR	humoral immune responses		**MF**	*Malassezia* folliculitis
HIV	human immunodeficiency virus		**MFC**	minimum fungicidal concentration
HPLC	high-pressure liquid chromatography		**MFS**	major facilitator superfamily
HRP	horseradish peroxidase		**MHC**	major histocompatibility complex
HRT	human rectal tumor		**MIC**	minimal inhibitory concentration
HSP	heat shock protein		**MIF**	migration inhibition factor
HSP90	heat shock protein 90		**ML**	maximum likelihood
HSV-2	herpes simplex virus-2		**MLC**	minimum lethal concentration
HWP-1	hyphal wall protein-1		**MLST**	multilocus sequence typing
			MOPS	3-[*N*-morpholino]-propanesulfonic acid
ICU	intensive care unit		**MP**	maximum parsimony
Id	idiotypic		**MPS**	mononuclear phagocytic system
ID	immunodiffusion		**MRI**	magnetic resonance imaging
IDCF	immunodiffusion complement fixation		**MSR**	MSG-related proteins
IDTP	immunodiffusion tube precipitin		**mtDNA**	mitochondrial DNA
IEL	intraepithelial lymphocytes		**MTT**	3-4,5-dimethyl-2-thiazolyl-2,5-diphenyl-2*H*-tetrazolium bromide
IFA	incomplete Freund's adjuvant; or indirect fluorescent antibody assay			
IFN-γ	interferon-gamma		**NAO**	nucleus-associated organelle
IGS	intergenic spacer		**NASBA**	nucleic acid sequence based amplification
IL	interleukin		**NCCLS**	National Committee for Clinical Laboratory Standards
IL-3	interleukin-3			
IL-4	interleukin-4		**NJ**	neighbor-joining
IL-10	interleukin-10		**NK**	natural killer
IL-12	interleukin-12			
IMA	inhibitory mold agar		**ODC**	ornithine decarboxylase
indels	insertions or deletions		**OI**	opportunistic infection

OLT	orthotopic liver transplantation		**SBHI**	Sabouraud's brain–heart infusion
OPC	oropharyngeal candidiasis		**SCID**	severe combined immunodeficiency disease
OPN	osteopontin		**SCIDS**	severe combined immunodeficiency disease syndrome
pABA	*para*-aminobenzoic acid		**SD**	seborrheic dermatitis
PAP	peroxidase–antiperoxidase		**SDA**	Sabouraud's dextrose agar
PAS	periodic acid–Schiff		**SDS-PAGE**	sodium dodecylsulfate-polyacrylamide gel electrophoresis
PBL	peripheral blood leukocyte		**SEM**	scanning electron microscopy
PBMC	peripheral blood mononuclear cells		**SGA**	Sabouraud's glucose (dextrose) agar
PBS	phosphate buffered saline		**SI**	stimulation indices
PCMS	*p*-chloromercuriphenylsulfonic acid		**SIB**	single-step immunoblot
PCP	*Pneumocystis* pneumonia		**SIDS**	sudden infant death syndrome
PCR	polymerase chain reaction		**SNP**	single nucleotide polymorphisms
PCR-EIA	polymerese chain reaction-enzyme immunoassay		**SP**	surfactant protein
			SP-A	surfactant protein A
PCR-REA	polymerase chain reaction-restriction enzyme amplification		**SP-D**	surfactant protein D
			SSAT	spermidine/spermine-N^1-acetyltransferase
PCR-SSCP	polymerase chain reaction single-strand conformation polymorphism		**SSCP**	single-strand conformational polymorphism
			SSU	small subunit
PDA	potato dextrose agar			
PFGE	pulsed-field gel electrophoresis			
PHMB	polyhexamethylene biguanide		**TABM**	T-cell-derived antigen-binding molecule
PIM	prototheca isolation medium		**TCA**	tricarboxylic acid
PML	polymorphonuclear leukocyte		**TCR**	T-cell receptor
PMN	polymorphonuclear neutrophil		**TEM**	transmission electron microscopy
PRA	proline-rich antigen		**TGF-β**	transforming growth factor beta
PUO	pyrexia of unknown origin		**Th1**	T helper 1
PV	pityriasis versicolor		**Th2**	T helper 2
			TNF	tumor necrosis factor
RAPD	random amplified polymorphic DNA		**TNF-α**	tumor necrosis factor alpha
RAST	radioallergoabsorbent test		**TP**	tube precipitin
rDNA	ribosomal DNA		**TRITC**	tetra-methylrhodamine isothiocyanate isomer R
RE	restriction enzymes		**TS**	thymidylate synthase
REA	restriction endonuclease analysis			
RFLP	restriction fragment length polymorphism		**UCS**	upstream conserved sequence
RGD	arginine-glycine-asparagine		**UPGMA**	unweighted pair-group method with arithmetic mean
RIA	radioimmunoassay			
RMVM	Review of Medical and Veterinary Mycology		**UPRTase**	uracil phosphoribosyltransferase
RPLA	reverse passive latex agglutination		**UTI**	urinary tract infection
rRNA	ribosomal RNA		**UV**	ultraviolet
RT-PCR	reverse transcriptase polymerase chain reaction			
RVVC	recurrent VVC		**VP**	ventriculoperitoneal
			VVC	vulvovaginal candidiasis
SALT	skin-associated lymphoid tissue			
SAM:SMT	*S*-adenosyl-Lmethionine:sterol C-24 methyl-transferase		**WCE**	Wolff–Chaikoff effect
SAM	*S*-adenosyl-L-methionine		**XTT**	2,3-bis-2-methoxy-4-nitro-5-sulphophenyl-5-[phenylaminocarbonyl]-2*H*-tetrazo-lium hydroxide
SAP	secretory aspartyl proteases			

PART I

BACKGROUND AND BASIC INFORMATION

Historical introduction: evolution of knowledge of the fungi and mycoses from Hippocrates to the twenty-first century

EDOUARD DROUHET

Since the beginnings of medical mycology at the turn of the century, with the discovery of the principal agents of human and animal mycoses (ringworm, aspergillosis, candidosis, cryptococcosis, histoplasmosis, etc.), there has been continual progress in our knowledge of pathogenic fungi that cause diseases and their role in the pathology of infectious diseases.

The pathogenic fungi and the mycoses, which have emerged in recent years in the immunocompromised host, have emphasized the importance of medical mycology, within the discipline of medical microbiology. This has been made possible by the pioneers in medical mycology, and by enthusiastic medical mycologists who founded the International Society of Human and Animal Mycology (ISHAM) in 1954. This now has more than 1 000 members and a score of affiliated medical mycological societies throughout the world. The expansion, over the past 50 years, of the spectrum of fungal infections, the clinical and pathological aspects of the mycoses, the ecology and epidemiology of the classical pathogenic and opportunistic fungi, as defined in 1962 at the International Symposium on Opportunistic Infections, has been accompanied by an increased understanding of pathogenic fungi and the mycoses. The physiology, biochemistry, immunology, morphogenesis (studied by transmission and scanning electron microscopy), sexuality, genetics, and molecular biology of fungi pathogenic to humans and animals have been studied in depth.

Many of the systemic mycoses have been known since the end of the nineteenth century, but they were considered to be extremely rare, usually discovered postmortem, and observed only in compromised patients. In recent decades, since the introduction of antimicrobials in 1945, of corticosteroids in 1950, of immunosuppressive therapy in 1960, of catheters, prosthetic devices, and organ transplantation in the 1960s and 1970s, and with the emergence of the acquired immunodeficiency syndrome (AIDS) and the increasing use of bone marrow grafting in the 1980s, these deep mycoses have started to develop new clinical aspects, occurring with high frequency and severity. Many of the opportunistic mycoses are now nosocomial, hospital-acquired, invasive infections caused by a new spectrum of opportunistic pathogenic fungi (Bodey 1988; Weber and Rutala 1988; Jarvis 1995).

FUNGI IN NATURE AND PATHOLOGY: CLASSIFICATION

Among the more than 72 000 described species of fungi that are widespread in nature (soil, plants, water, air), more than 300 are now recognized as true pathogens or potential pathogens responsible for mycoses in humans and animals. The word fungus is derived from the Latin *fungus*, which came into use in the fifteenth century. The Latin was derived from the Greek *sphongirs* or sponge. The term 'mycosis' was derived from the Greek word *mykes* for mushroom.

The Fungi represent one of the five kingdoms of organisms classified, according to Whittaker (1969) and Margulis and Schwartz (1988), into three levels of organization:

1 the prokaryotes (kingdom Monera, including bacteria, actinomycetes, blue–green bacteria)
2 the unicellular eukaryotes (kingdom Protoctista), protozoa, nucleated algae, oomycetes)
3 the eukaryotes (kingdom Fungi, Plantae or Animalia).

The actinomycetes, which are true bacteria, have traditionally been studied by medical mycologists because, like fungi, they have mycelial forms and cause similar diseases.

The boundaries of these kingdoms and their components are still not well delineated. In the past, the causative agents for key mycoses, such as histoplasmosis (*Histoplasma capsulatum* var. *capsulatum*), which was discovered in 1906 by Darling, and coccidioidomycosis (*Coccidioides immitis*), which has been known since 1892 (Posadas 1892; Wernicke 1892; Rixford and Gilchrist 1896), were described by protozoologists as 'protozoans' because of their invasive unicellular tissue forms. They were recognized as fungi many years later – *H. capsulatum* var. *capsulatum* in 1934 by DeMonbreun, and *C. immitis* in 1900 by Ophüls and Moffitt – when they were isolated and cultured in their mycelial, saprophytic form.

Even now, revolutionary changes may occur, as with *Pneumocystic carinii*, which was reclassified as a fungus by molecular taxonomists because of its ribosomal RNA (rRNA) sequences (Edman et al. 1988). The study of the 5S rRNA of *P. carinii* showed that its phylogenetic position is closely associated with the Rhizopoda (ameba), Myxomycota, and Zygomycota (Watanabe et al. 1989); the most recent genetic studies in biomolecular taxonomy reinforced the close relationship of this organism to fungi rather than protozoa (Cushion et al. 1994) (see also Chapter 37, *Pneumocystis* pneumonia).

Two algae that lack chlorophyll and have long been included in the kingdom Protoctista are dealt with by medical mycologists. The first of these species was described as *Prototheca zopfii* by Krüger in 1894 and the second as *Prototheca wickerhamii* by Tubaki and Sonida in 1954; they have been recognized, since the first proven infection in humans found by Davies et al. in 1964, as new pathogens (Ravisse et al. 1993).

Oomycetes are also within the province of medical mycologists. *Pythium insidiosum* (de Cock et al. 1987) was first described by Bridges and Emmons (1961) as *Hyphomyces destruens* in infected horses; it is now an emerging agent of animal (cattle, dogs, and cats) and of human infections in Australia, Japan, Latin America, Thailand, and the USA (Mendoza et al. 1996).

The principal fungi causing the mycoses are responsible for:

• superficial and cutaneous mycoses
• deep, systemic mycoses caused by the opportunistic yeasts or filamentous fungi or highly pathogenic dimorphic fungi that are introduced primarily into the respiratory tract (Figures 1.1 and 1.2) and
• subcutaneous mycoses.

This clinical classification is not rigorous because superficial fungal infections may become systemic mycoses that present as mucocutaneous manifestations.

MILESTONES OF MEDICAL MYCOLOGY

In two fascinating books, one devoted to the history of mycology (Ainsworth 1976) and the other to the history of medical and veterinary mycology (Ainsworth 1986), Ainsworth, the author and former Director of the Commonwealth Mycological Institute (at Kew in the UK), gives a straightforward account of the main views held about fungi in the past three millennia and the development of the study of fungi as a branch of the science of medical microbiology over the last 250 years.

As emphasized by Ajello (1975a), the evolution of medical mycology can be compared to the great network of roads that linked the farthest reaches of the Roman empire to the Eternal City, where at designated distances between cities, stone pillars were erected to mark the miles. The milestones of medical mycology were designated by Ainsworth (Ainsworth 1976, 1986; Ainsworth and Stockdale 1984), Ajello (1975a), Emmons et al. (1977), Seeliger (1985, 1988), Rippon (1988), and numerous other authors in their introductions to books and articles (Baldwin 1981; Huntington 1985; Drouhet 1991; Kwon-Chung and Bennett 1992).

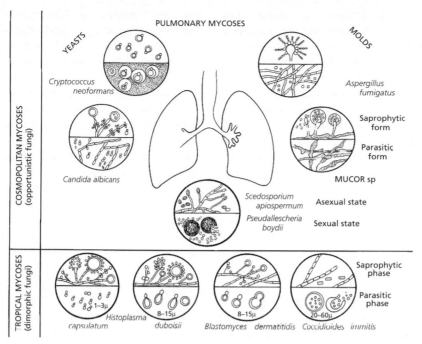

Figure 1.1 *Principal fungi responsible for pulmonary and deep mycoses: saprophytic form (upper hemisphere) and parasitic form (lower hemisphere). For* Scedosporium apiospermum *(asexual state), the sexual stage is in the lower hemisphere (*Pseudallescheria boydii*).*

The earliest known record of a mycotic infection is that in the Atharva Veda in India (about 2000–1000 BC), of mycetoma of the foot which was described under the name 'foot anthill.'

As a result of their visibility, the superficial mycoses of the mucosa or tegument have been known since antiquity. Hippocrates (460–377 BC) and Gallen (AD 130–200) described *aphthae albae* (in English called thrush, and in French *muguet* because of the whiteness of the infected tongue), whereas in 1847, Robin identified the yeast *Oidium (Candida) albicans* as the agent of thrush in infants (Robin 1853).

Celsus (first century AD), following the nomenclature of Hippocrates, dealt, in *De re medicina*, not only with

thrush but described favus, the inflammatory lesions of some ringworm infections – kerion of Celsus. The first dermatophytic fungal infections were recognized only about 1840 by Remak, Schoenlein, and Gruby (Seeliger 1985, 1988).

The centrifugal growth of fungi in a circular form on the skin led to various names such as herpes (Greek: dandruff), now known as *herpes circinatus*. The Romans associated the lesions with insects and named the disease tinea, meaning any small insect larva. This name has remained in English clinical terminology (tinea capitis, tinea corporis, tinea pedis, etc.), and is translated in French as *teigne* and in popular English as ringworm. *Favus* (Latin: honeycomb), a specific chronic scalp

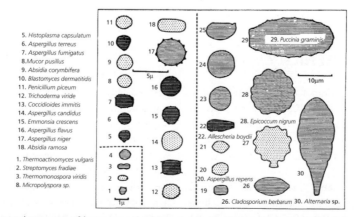

5. *Histoplasma capsulatum*
6. *Aspergillus terreus*
7. *Aspergillus fumigatus*
8. *Mucor pusillus*
9. *Absidia corymbifera*
10. *Blastomyces dermatitidis*
11. *Penicillium piceum*
12. *Trichoderma viride*
13. *Coccidioides immitis*
14. *Aspergillus candidus*
15. *Emmonsia crescens*
16. *Aspergillus flavus*
17. *Aspergillus niger*
18. *Absidia ramosa*

1. *Thermoactinomyces vulgaris*
2. *Streptomyces fradiae*
3. *Thermomonospora viridis*
8. *Micropolyspora sp.*

29. *Puccinia graminis*
28. *Epicoccum nigrum*
22. *Allescheria boydii*
20. *Aspergillus repens*
26. *Cladosporium berbarum* 30. *Alternaria sp.*

Figure 1.2 *Spores in allergies and mycoses of humans and other animals. Outlines of the airborne spores of Fungi and Actinomycetes causing allergies and mycoses. Directly pathogenic species – dark shading; allergenic species – light shading; species isolated from animal lungs – dotted. (Redrawn from Austwick 1966)*

mycosis, characterized by yellowish cup-shaped crusts (scutula) was the first human mycosis to be recognized as caused by a dermatophytic fungus by Remak and Schoenlein (Schoenlein 1839; Remak 1845); this discovery has been described by Seeliger (1985). Tinea imbricata, a tropical dermatophyte disease, easily recognized by the polycyclic, concentrically arranged rings of scales which are scattered over most of the body, was described by travellers in the Pacific Islands of Oceania. Dampier (1729) named it 'Tokelau disease' after the Islands where the disorder was prevalent. This disease was recognized also in southeast Asia, and it has been suggested that it may be of anthropological interest, as it was possibly introduced to the western coasts of South and Central America by pre-Colombian voyagers from Polynesia (Rippon 1988). Only in 1896 did Blanchard describe the agent as *Trichophyton concentricum* and later isolated cultures with difficulty.

The invention of the microscope during the last decade of the sixteenth century enabled seventeenth century microscopists, such as Hooke of England with his *Micrographia* London (1665), who presented the first illustrations of microfungi such as *Mucor*, followed by the Dutchman van Leeuwenhoek (1632–1723) and the Italian Malpighi (1628–94), in *Anatomia plantarum pars altera*, London, 1679, to give the first descriptions of some common microfungi. van Leeuwenhoek is credited with being the first person to have observed yeasts microscopically. In 1680 he sent descriptions and drawings of yeast cells to the Royal Society of London, 4 years after making public his observations on bacteria.

Aided by the binomial system of the Swedish physician and botanist Linnaeus (1707–78), descriptive mycology grew gradually, but steadily. Micheli (1679–1737), a Florentine botanist of the Public Gardens in Florence, was the first to give generic names to fungi that are now classified among the Hyphomycetes. His monograph *Nova plantarum genera* (published in Florence in 1729) was not only a historical milestone but also the starting point for mycology as a science. Familiar genera such as *Botrytis* and *Mucor* were described for the first time.

CONCEPT OF PATHOGENICITY

It required more than 200 years for a proposal of the concept of infectivity to be made. Fracastora of Verona (1546) postulated the existence of a *contagium vivum* as the cause of infectious diseases and foresaw a relationship between microfungi in nature and fungal diseases in humans and animals.

The first milestone in the medical mycological highway is attributed by most historians of medical mycology to the Italian, A. Bassi (1773–1856) (Ajello 1975a, 1993; Ainsworth 1976, 1986). In 1807, after he had studied 'physics, chemistry, natural history, some

branches of medicine and the principles of mathematics,' Bassi began to study a devastating disease of the silkworm (*Bombyx mori*), the muscardine, and proved that it was caused by a fungus subsequently named, in his honor, *Beauveria bassiana*.

Bassi's revelation of the role played by a fungus in an animal disease inspired others to investigate the etiology of human diseases. Schoenlein and his assistant Remak succeeded in discovering, in 1837, the agent of the human favus, which was named *Achorion (Trichophyton) schoenleinii* by Remak (1845) in honor of his mentor Schoenlein (1793–1864), but the studies of Gruby (1810–98) proved to have a greater impact among the physicians at that time through his communications to the Academy of Sciences of Paris (1841–44) about the fungal origin of favus. For the first time, Gruby tried to reproduce a human infection by inoculating a colleague in the arm and also his own body with the fungus. He fulfilled Koch's postulates for the criteria of the etiology of infection 40 years before Koch had formulated them. In 1841, Gruby also described the in vivo appearance, in thrush, of the yeast named *Oidium (Candida) albicans* by Robin (Figure 1.3) in his doctoral thesis 6 years later (Robin 1847). Robin (1821–85), elected to the chair of histology at the University of Paris, made other important contributions to mycology, describing for the first time the dermatophyte *Trichophyton ('Microsporon') mentagrophytes* in his book published in 1853: *Histoire naturelle des végétaux parasites qui croissent sur l'homme et des animaux*. Gruby described an important agent of tinea capitis which he named *Microsporum audouinii* in honor of the great Parisian naturalist, V. Audouin (1797–1841), of the Paris Natural History Museum, known for his work on muscaridine disease. This is considered the second important historical milestone of medical mycology (Zakon and Benedek 1944). His microscopic descriptions (after his death more than 15 000 microscopic preparations and 2 000 photomicrographs were found among his possessions), interpreting host–parasite relationships, prepared the work of Sabouraud (Rook 1978). The role of fungi in these early concepts was preceded by the work of dermatologists such as W. Wilson (1817–84), J. Hoog (1817–99), T. Fox (1836–79), F. van Hebra (1816–80) of Vienna, and E. Bazin (1807–78); the last named wrote a thesis entitled 'La contagion et l'infection' (1839) and became a physician at the St Louis Hospital in Paris, publishing a monograph on ringworm in 1853.

Over several decades, pathogenicity was associated with the concept of spontaneous generation, and finally Pasteur succeeded in winning the fight against this concept of spontaneous generation of microorganisms – a theory that had been well established since antiquity. Pasteur (1822–95; Figure 1.4), unanimously recognized as the founder of microbiology, in the famous memoirs about the 'organised corpuscles existing in the atmo-

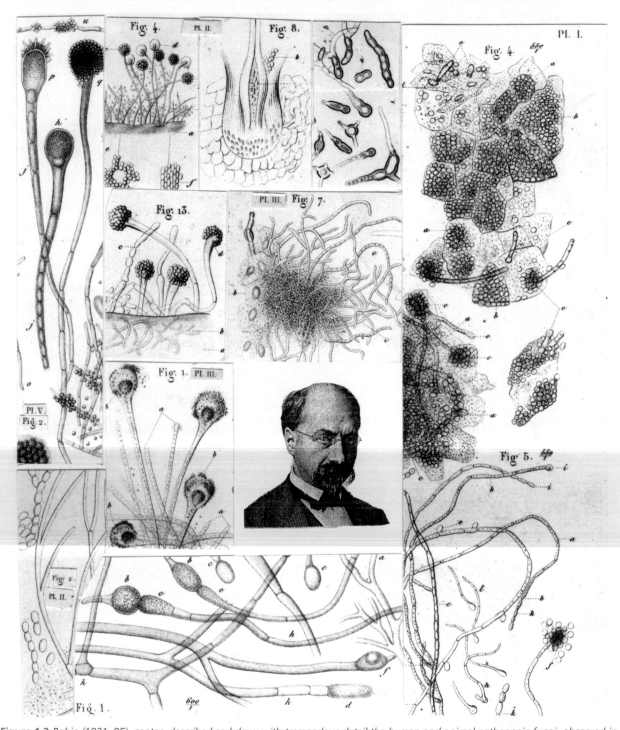

Figure 1.3 *Robin (1821–85), center, described and drew with tremendous detail the human and animal pathogenic fungi, observed in vivo ('d'après nature') on skin, oral, genital, digestive mucosae, ears, lungs, etc., in his book/atlas* Natural history of Plant's parasites of living humans and animals *(Robin 1853: 15 plates, 100 figures all detailed).* Oidium albicans *(for the first time the name of the thrush agent is specified) is represented as: clusters of yeast cells on epithelial cells and filaments which become entangled in depth; dermatophytes of tinea capitis with follicle invaded by spores on the inside and outside of hair; molds such as aspergilli in the auditory canal, lungs of ducks, or aerial sacs of other birds; and even mucorales in a bovine uterus.*

sphere,' reported his studies of air microorganisms collected in Paris and at 2 000 metres in Chamonix, near Mont Blanc. From his drawings, it is easy to recognize yeasts, yeast-like fungi (*Trichosporon*, *Geotrichum*), and filamentous fungi (*Penicillium*, *Mucor*, etc.) in the aerial biota, which later led to the science of aeromycology. His student, Miguel, made detailed studies for a doctoral thesis in 1883 about microorganisms present in the air, not only in the Paris Observatory, but also in dwellings, houses, and hospitals. The interest

Figure 1.4 *Pasteur (1822–95), center, has drawn the 'organized corpuscles from the atmosphere' in the studies which came out against the theory of spontaneous generation (Pasteur 1861), illustrating the 'air spora' formed by filamentous and yeast-like fungi collected from the air of Paris and Chamonix in swan-neck flasks (at 2000 metres; aquarelle by Clair-Guyot, on the right).*

of Pasteur and his disciples in mycology was considerable. He revealed the role of yeasts in enzymatic fermentation, and he was the first to describe dimorphism in such filamentous fungi as *Mucor* spp.,

which transformed into a yeast-like form, and to improve culture media.

Jules Raulin (1836–96), van Tieghem (1839–1914), and Duclaux (1840–1904) were the first three students to work with Pasteur in his laboratory at the Ecole Normale in Paris in the early 1860s. Raulin's study was designed to elucidate the mineral nutrition of *Aspergillus niger* by quantitative measurement of the growth obtained in recognition of the importance of trace elements for the culture of fungi. The controversy between Pasteur and Leibig, about the size of inoculum necessary for the growth of yeasts, led to the bios of Wildiers, and the growth factors known as vitamins.

van Tieghem and Le Monnier (1873) introduced the microscopic examination of hanging drop cultures of fungi (Figure 1.5) – the 'van Tieghem cell;' they described the genus *Absidia* in 1876 and the species *Mucor circinelloides* in 1875. In 1887, the German Petri invented the 'Petri dish' – the covered glass plate in which the solid support of gelatine or agar allowed the culture of bacteria and fungi; this led to the greater development of bacteriology and mycology (Petri 1887).

Sabouraud (1864–1938), a young dermatologist at the St Louis Hospital, started his studies on dermatophytes in 1890 with the famous 'Great Course of Microbiology' inaugurated by Roux 2 years after the foundation of the Pasteur Institute (1888), at the end of Pasteur's historical career. This explains Sabouraud's studies on the culture of dermatophytes. Duclaux (1886) described an enzyme (zymase) in a dermatophyte (*Trichophyton tonsurans*) and discredited the opinion held by some of his predecessors and contemporaries that all dermatophytoses were caused by a single fungal species, a mutable *Aspergillus* or *Penicillium* species. Pure cultures were needed for the identification and definition of fungus species on solid taxonomic grounds, and to reveal the existence of a multiplicity of pathogenic species; this was realized by Sabouraud.

Other milestones in the history of medical mycology continued in the 1880s, 1890s, and 1900s, with the discovery of the principal agents of systemic and subcutaneous mycoses and with the evolution of the diseases that they provoked.

HISTORY OF THE PRINCIPAL FUNGI RESPONSIBLE FOR MYCOSES

Dermatophytes and dermatophytoses: essential milestones in the history of medical mycology

The history of the tineas and their agents, the dermatophytes, started with Sabouraud in his monograph, *Les Teignes* in 1910. The history was continued by Ainsworth (1976, 1986), Ajello (1974, 1975a), Emmons et al. (1963, 1970, 1977), Seeliger (1985, 1988) and later Rippon (1988). They outlined the evolution of our knowledge of this important group of mycoses and their etiological agents.

The work of Sabouraud is one of the most important milestones in the history of medical mycology. He truly revolutionized our concepts about dermatophytes and significantly contributed to the development of medical mycology. He studied medicine in Paris, was a man of multiple talents and an excellent dermatologist. He was also a talented sculptor. He trained in microbiology at the Pasteur Institute under Roux and so knew pure culture techniques. His studies, which started in 1892, culminated in 1910 in the publication of his classic book *Les teignes* (Sabouraud 1910), the most comprehensive account of all the then-known dermatophytes, to which mycologists still refer. He published 1 000 scientific papers and 12 huge volumes. His work was acknowledged at the dermatology hospital, the St Louis Hospital, by the creation of a laboratory for dermatophyte infections and a Children's School for the treatment and education of children suffering from tinea capitis, a severe, chronic disease that was difficult to treat at that period. He used pure culture techniques and new media to determine the physiological and morphological characteristics and parasitological aspects of the development of dermatophytes in vivo, and observed variations and pleomorphism in dermatophytes. Sabouraud laid down a solid foundation for modern medical mycology, and consequently the journal of the ISHAM was initially named *Sabouraudia* (1961). His medium for the identification of the dermatophytes still bears his name and is universally used.

Sabouraud developed a classification system, based on the in vitro and in vivo characteristics of dermatophytes, which remains valid for the three genera *Epidermophyton*, *Microsporum*, and *Trichophyton*, but not for the genus *Achorion*. The last was invalidated by Emmons (1934; Figure 1.6), based on the botanical rules of nomenclature and taxonomy. Many of the common dermatophytoses and dermatophytes in humans were first differentiated in Sabouraud's time: in 1902 Bodin described *T. violaceum*; Castellani and Chalmers (1910) described *T. rubrum* (as *E. rubrum*); Sabouraud (1910) proposed the name *E. inguinale* for the fungus respon-

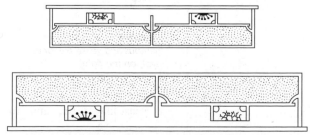

Figure 1.5 *The 'van Tieghem cell.' (From van Tieghem and Le Monnier 1873)*

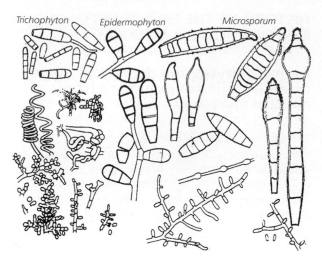

Figure 1.6 *Drawings from Emmons (1934): dermatophytes. Natural groupings based on the form of conidia and accessory organs.*

sible for tinea cruris (now *E. floccosum*); the Hebra's eczema marginatum described in 1860; and also *T. persicolor* (1902) now classified in the genus *Microsporum*. Animal dermatophyte infections included ringworm of cattle (*T. verrucosum* Bodin 1902), of horses (*T. equinum* Matruchot and Dassonville 1898), of cats, dogs, and humans (*Microsporum canis* Bodin 1902), and of fowl (*Microsporum gallinae* as *E. gallinae* by Megnin in 1881). Other dermatophytes were discovered, such as *M. ferrugineum* Ota 1921, long considered to be a *Trichophyton* species, until typical echinulate macroconidia were obtained by Vanbreuseghem et al. (1970). This mold is found in Japan and the Far East and is responsible for tinea capitis.

In the years after Sabouraud, various tentative modifications of his generic classification were suggested, especially by Langeron (1874–1950) of the Faculty of Medicine of Paris, Vanbreuseghem (1909–93), who developed the field of medical mycology in Belgium after studying with Langeron, and Rivalier (1892–1979), who became the Director of Sabouraud's Laboratory at the St Louis Hospital. Many mycologists adopted Sabouraud's classification, which was simplified by Emmons (1934; see Figure 1.6). In 1954, Conant who had worked with Sabouraud (1936–37) also modified his classification (Conant 1936a, b, 1937a).

Langeron and Milochevitch (1930) emphasized the importance of natural media to induce the characteristic sporulation of dermatophytes. Studies on the nutrition of dermatophytes, as related to morphogenesis in chemically defined synthetic or semisynthetic media, were carried out after 1950, essentially by Georg (Georg 1950; Georg and Camp 1957), Drouhet and Mariat (1952a, 1953), Drouhet (1952), and Silva and Benham (1952). The need was demonstrated for vitamins and amino acids as essential or complementary growth factors for certain groups of dermatophytes which were strictly

adapted to human or animal hosts. The capacity for sporulation was lost in media without the addition of exogenous growth factors. These biochemical and physiological studies led to routine nutritional tests for the specific identification of dermatophytes (Georg and Camp 1957).

The introduction by Vanbreuseghem (1952) of the hair-baiting technique, for the specific purpose of isolating dermatophytes from soil, revealed the worldwide occurrence of dermatophyte-like fungi and geophilic dermatophytes, as studied and reviewed by Ajello (1974, 1975a) and Rippon (1988).

The next major development in the taxonomy of dermatophytes came in 1959 with the description of the teleomorph, the perfect or sexual stage of *Trichophyton (Keratinomyces) ajelloi* by Dawson and Gentles (1959) under the name *Arthroderma uncinatum*.

Nannizzi (1877–1961) of the University of Sienna (Italy) had, in 1927, described the sexual state of *Microsporum gypseum* as *Gymnoascus gypseum* (Figure 1.7). In 1960, just before his death, Griffin reisolated the perfect stage of *M. gypseum* from Australian soil. Independently, in 1961, Stockdale (1927–89), of England's Commonwealth Mycological Institute, discovered several teleomorphic stages of this dermatophyte and proposed the new genus *Nannizzia* to accommodate the teleomorphs of the *Microsporum* spp. (Stockdale 1961, 1963, 1967; Hasegawa and Usui 1975). Nannizzi's remarkable breakthrough was not recognized by some medical mycologists at that time because the substrates used were erroneously considered not to be sterile. Since then, numerous imperfect or anamorphic species have been found to have one or more ascigerous states obtained in pure culture with cycloheximide [the introduction of cycloheximide (Acti-Dione) in the selective isolation of pathogenic fungi to humans by Georg et al. (1954) contributed to the discovery of numerous dermatophytes in their sexual states]. The seminal work of

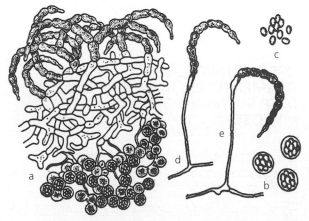

Figure 1.7 Gymnoascus gypseus Nannizzi, *teleomorphic state of* Microsporum gypseum *(Nannizzi 1927). Stockdale described several teleomorph states for* Microsporum gypseum *and created the genus* Nannizzia *in honor of Nannizzi (1877–1961).*

Nannizzi was unanimously recognized (Rippon 1985; Ajello 1993). The perfect state was discovered for common dermatophytes such as *Microsporum canis*, named *Nannizzia otae* in honor of the famous Japanese mycologist Ota (Hasegawa and Usui 1975). All of the teleomorphs of dermatophytes and dermatophyte-like fungi were classified into two genera until 1986, when Weitzman et al. proposed treating *Nannizzia* as a later synonym of the genus *Arthroderma*. Currently, the use of physiological characteristics, mating types, mitochondrial DNA, scanning and transmission electron microscopy, and antigenic analysis, based on monoclonal antibodies, in addition to classic methods of descriptive mycology, has given the taxonomy of dermatophytes and other pathogenic fungi a firm scientific basis (Tanaka et al. 1992) (see also Chapter 3, Fungal structure and morphology).

The combination of improved hygiene and the revolutionary application of griseofulvin in 1958 in the treatment of dermatophytoses decreased the prevalence of tinea capitis in occidental regions, although this disease remains prevalent in some tropical areas. On the other hand, although unknown in Sabouraud's era, tinea pedis caused by *Trichophyton rubrum* of Asian origin has now spread worldwide.

Dermatophyte-like infections

Ringworm-like nondermatophyte infections (skin, nails) caused by pheoid (black) fungi of tropical origin, such as *Hendersonula (Nattrassia) toruloidea* Nattrass 1930, *Scytalidium hyalinum* Campbell and Mulder 1977, have been described in Europe (Gentles and Evans 1970; Hay and Moore 1984), as well as some tropical superficial mycoses known for numerous years as hair colonizers: white piedra caused by *Trichosporon beigelii* (1887), black piedra caused by *Piedraia hortai* (1913) and the skin lesion, tinea nigra, caused by *Exophiala (Cladosporium) werneckii* (1921). Since their discovery, these superficial dermatomycoses represent a rare group of cutaneous mycoses.

OPPORTUNISTIC YEASTS AND FILAMENTOUS FUNGI AS AGENTS OF DEEP SYSTEMIC OR VISCERAL MYCOSES

Opportunistic fungi and opportunistic fungal infections: predisposing factors

These new terms have been introduced into medical mycology since the International Symposium on opportunistic infections in 1962, as emphasized by Ajello, Baker, Emmons, Segretain, Utz, and other participants at that historical meeting. These mycoses have increased progressively and intensively as a result of the favorable factors for opportunistic fungal infections which have evolved over the past decades (Emmons 1962).

INTRINSIC FACTORS DEPENDENT ON THE HOST

Apart from physiological factors resulting from age, pregnancy, immunological and/or endocrinological conditions, the pathological factors are most important. For a long time it has been known that normally saprophytic fungi may invade patients with severe, debilitating, primary diseases such as diabetes, malignant blood diseases, Hodgkin's disease, and various neoplasms. However, before the advent of antibiotic and corticosteroid treatments, opportunistic mycoses were found only in patients in the final stages of serious microbial infections, severe diabetes, other endocrinopathies or malignant diseases (Emmons et al. 1963, 1970, 1977; Drouhet 1972; Warnock and Richardson 1982, 1991; Armstrong 1981). At present, these secondary opportunistic fungal infections are extremely common in immunocompromised hosts, particularly those with granulocytopenia below 1 000 cells/ml. *Candida* spp. infections are seen as sequels to antibiotic treatment. Sometimes fungal infections are diagnosed before the primary disorder and the mycoses become indicative 'signal' diseases such as cryptococcosis. The occurrence of opportunistic fungal infections has risen progressively (Drouhet 1972; Fraser et al. 1979; Bullock and Deepe 1989). Invasive fungal infections have been reported in recent years in 26 percent of patients who are chronically and intensively immunosuppressed as a result of underlying disease or drug therapy. The incidence of aspergillosis increased from 1.91 to 4.8 per million persons (+ 158 percent). *Cryptococcus neoformans* infections increased from 1.3 per million to 2.3 per million (+ 78 percent) between 1970 and 1976 in the USA. However, the incidence rate became substantially higher after 1980 as a result of the opportunistic infections, cryptococcosis, candidiasis, and histoplasmosis, which occur in AIDS (Graybill 1988; Dupont et al. 1994; Stevens et al. 1994).

Alterations of the immune system

The pathogenesis of fungal infections in the immunocompromised host has recently been reviewed and discussed in the general context of microbial infections. Nonspecific immune mechanisms, including cellular and humoral ones, and specific cell-mediated and humoral immunity play important but unequal roles in host resistance to fungal infections (Warnock and Richardson 1982, 1991; Murphy et al. 1994).

Functional defects in chemotaxis

Adherence, or killing (degranulation), phagocytic disorders caused by low cell numbers, granulopenia, and disorders of leukocyte metabolism are often correlated with fungal infections such as aspergillosis, candidiasis, and other opportunistic mycoses (Waldorf 1986; Walsh et al. 1994).

Acquired immunodeficiency syndrome

Severe fungal opportunistic infections have been reported in the USA, Europe, Africa, and other countries (Chandler 1985; Drouhet and Dupont 1990; Dupont et al. 1994). Usually candidiasis, particularly thrush and esophagitis (Gottlieb et al. 1981; Klein et al. 1984; Tavitian et al. 1986; Dupont and Drouhet 1988), cryptococcosis (Kovacs et al. 1985; Dupont 1989; Drouhet and Dupont 1990) and histoplasmosis (Graybill 1988; Mackenzie 1992), are observed because natural resistance to these infections relies on cell-mediated immunity. Other opportunistic infections in the immunocompromised host, such as disseminated aspergillosis, were very rare in acquired immunodeficiency syndrome (AIDS) in the 1980s, because their development correlates with the absence or reduced function of polymorphonuclear leukocytes. The mycosis with the highest incidence in AIDS, between 50 and 90 percent, is oral and esophageal candidiasis. In high-risk patients, the presence of unexplained oral candidiasis predicts the development of serious opportunistic infections more than 50 percent of the time. Cryptococcosis is the most dramatic fungal infection in AIDS. In the USA, out of 13 834 AIDS patients registered by the Centers for Disease Control (CDC) in October 1985, 904 cases of cryptococcosis (6.5 percent) were diagnosed. In France during the years 1985–95, most of the 1 300 diagnosed cases of cryptococcosis developed in AIDS patients. Although histoplasmosis may occur in immunocompetent patients, we wonder, as do other authors, whether it should not be considered one of the major deep infections in those patients with AIDS who have resided in an endemic area.

A new candidate for systemic mycosis has appeared in AIDS – infection caused by *Penicillium marneffei*, an Asian tropical fungus described in 1956 by Capponi et al. which was responsible for a mycosis of the reticuloendothelial system in Vietnamese bamboo rats, and for an accidental human infection (Capponi et al. 1956; Segretain 1959). Since 1987, human cases of disseminated, pulmonary, cutaneous penicilliosis caused by *P. marneffei* have been observed in England, France, Italy, and the USA, in AIDS patients who travelled in southeast Asia or autochthonous patients with AIDS living in China, Hong Kong, and Thailand. The 44 cases reported in patients positive for the human immunodeficiency virus (HIV) traveling or living in southeast Asia during 1985–93 increased to 117 cases in the following 2 years and to more than 400 cases in 1995 (Drouhet 1993a; Supparatpinyo et al. 1994; Drouhet and Dupont 1995) (see also Chapter 29, Penicilliosis).

The pattern of mycotic infection in AIDS patients has been radically altered by the introduction of combination antiretroviral chemotherapy.

EXTRINSIC, IATROGENIC FACTORS

These include effects caused by either drug treatment or breach of the integument resulting from medical (vascular catheters for drug administration or parenteral nutrition and other catheters) and surgical (organ transplants, abdominal surgery, etc.) devices or interventions, which allow the penetration of opportunistic fungi.

Antibiotics

The role of antibiotics in the pathogenesis of infections with *Candida* spp. is based not only on yeast selection in the alimentary canal and mucous membranes (imbalance between the bacterial and yeast biota), but also on the action of antibiotics on the immunological defence systems (Drouhet 1972). The latter includes decreased antibody production, suppression of phagocytosis, and metabolic alteration of the host cells. Recent data indicate that antibiotics acting against bacteria augment the adhesion of *C. albicans* to the surface of epithelial cells by mannan receptors taking the unoccupied place of the bacteria that disappeared, thereby facilitating the transition of *C. albicans* from the yeast form to its filamentous aggressive form (Kennedy 1988; Kennedy et al. 1992). This was demonstrated both in vivo and in vitro in a few hours and may explain the remarkable in vivo activity of azoles, such as ketoconazole, which arrest mycelial but not yeast cell development at very low concentrations (Borgers and Vanden Bossche 1982).

Immunosuppressants

Corticosteroids, immunosuppressants (such as azathioprine, cyclophosphamide, cyclosporin, etc.) and cytostatic agents favor the development of opportunistic fungal infections. The granulopenia induced by these drugs is responsible for susceptibility to the invasive forms of deep mycoses (Odds 1988). In the 1960s and 1970s, major renal transplant units reported fungal fatalities of up to 15 percent and up to 30 percent in heart transplant recipients (Warnock and Richardson 1991).

Intravenous drug abuse

Heroin addicts are a population at risk for certain infectious diseases, such as bacterial and fungal endocarditis, and hepatitis B. Since 1980, a new pathological septicemic syndrome of *C. albicans* infection, characterized by fever followed immediately by cutaneous disseminated lesions of folliculitis, pustulosis, subcutaneous nodules, and ocular metastasis, and several weeks later by osteoarticular involvement, was observed in the brown heroin addict population. This was observed not only in France (Drouhet et al. 1981; Drouhet and Dupont 1983; Dupont and Drouhet 1985) where more than 200 cases occurred, but also in Italy and principally in Spain (> 890 cases) and even in Australia. Although crude heroin or Iranian or Asian brown heroin does not contain *C. albicans*, the yeast is introduced by the syringes used by the addicts, which are contaminated from the spoon in which the heroin is dissolved in lemon juice. All patients have oral and digestive tract candidiasis as the primary source of this contamination. Alterations of T- and null-lymphocyte frequencies in the peripheral blood of opiate addicts were observed; these result from the opiate receptor sites on T lymphocytes. An immunosuppressive of the cellular immunity contaminant of brown heroin may also have been the origin of this new pathology, which is unknown in the USA where white or purified heroin is used.

Desferrioxamine

Therapy for iron or aluminum overload in dialyzed patients, with desferrioxamine, has become an important predisposing factor, since 1986, for disseminated severe zygomycosis caused by *Rhizopus* spp. (Kwon-Chung and Bennett 1992).

Candida spp. and candidoses

What is extraordinary in the history of *C. albicans*, and the 10 major pathogenic thermophilic *Candida* spp. adapted to warm-blooded animals including humans, is that these unicellular fungi, which have been recognized in the past 130 years, are capable of showing up multiple new conditions in all branches of medicine. The notion of the 'compromised' patient was developed by the famous French pediatricians Trousseau (1868) and Parrot (1869a, b) who revealed that athrepsy in newborn infants may lead to disseminated thrush of the digestive and pulmonary tracts (Figure 1.8a). The first description by Parrot (1869a, b, 1877) of esophageal, gastric, intestinal, laryngeal, tracheal, and pulmonary thrush, initiated by oropharyngeal thrush, was histologically proven to be a clinical entity (Figure 1.8b). In the first edition of his monumental book *Candida and candidosis* published in 1979, Odds analyzed 2 265 references, but in the second edition (1988) no fewer than 5 796 references were cited. This emphasized the many developments that have occurred in the molecular, medical, and genetic aspects of *C. albicans* and candidiasis (Kwon-Chung et al. 1985; Whelan 1987; Lott et al. 1992; Odds 1994).

Thrush, an important opportunistic marker in AIDS patients (Gottlieb et al. 1981; Klein et al. 1984; Tavitian et al. 1986), is a clinical manifestation caused exclusively by *C. albicans*. The other forms of candidiasis can be produced by various other *Candida* spp. The superficial manifestations of candidiases are conventionally sepa-

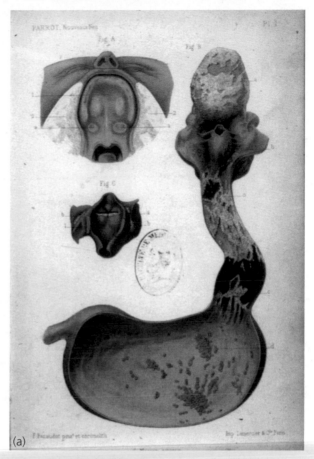

(a)

Figure 1.8 Drawings of Parrot (1869a) showing **(a)** the clinical aspects of generalised thrush in sick newborn infants with white patches on the tongue, esophagus, gastric mucosa, larynx (even bronchi and lungs, not shown here). (Continued over)

rated from deep-seated, disseminated or systemic candidiasis, which now principally appears in iatrogenic conditions, but the boundary is difficult to establish between the two forms (Bodey and Fainstein 1988, 1993).

Chronic mucocutaneous candidiasis (CMC), characterized by recurrent and persistent infections of the mucosa, nails, and skin by *C. albicans,* can be observed in patients with cellular immune abnormalities (Kirkpatrick et al. 1971). It has been shown that some of these abnormalities may be primarily induced by the mannan antigen of *C. albicans* (Durandy et al. 1987). The elimination of this immunosuppressive antigen by intensive antifungal therapy (amphotericin B, ketoconazole) restores the impaired cell-mediated immunity to *C. albicans* and leads to the disappearance of severe, chronic skin lesions (Drouhet and Dupont 1980, 1983; Drouhet et al. 1985).

The incidence of candidiasis has increased annually since the introduction of antibiotic therapy, immunosuppressive therapy, and other conditions. Odds (1988), basing estimates on official statistics and publications in four countries over a period of 30–35 years starting in

Figure 1.8 *Drawings of Parrot (1869a) showing* **(b)** *Histological aspect of* Candida albicans *yeast cells on the epithelial digestive mucosa with deep penetration by filaments ('aggressive form') into submucosal and muscular strata (explaining septicemia). The famous pediatricians Trousseau and Parrot at that time had described systemic candidiasis (Continued).*

1950, showed that, in the USA, the number of deaths attributed to candidiasis has increased roughly threefold between the late 1960s and the late 1970s; in France there has been a twofold increase between 1973 and 1982. During the 1980s and 1990s, the incidence of fungemia caused by *Candida* spp. acquired in hospitals increased more than threefold. Finally, adherence of *Candida* spp. to host surfaces was shown to be an important feature in the pathogenesis of candidiasis, and biomolecular studies about adherence accordingly developed rapidly (Kennedy et al. 1992; Robert et al. 1992) (see also Chapter 30, Candidiasis).

Lipophilic yeasts and malasseziasis

The history of lipophilic yeasts grouped in the genera *Malassezia* Baillon 1889, and *Pityrosporum* Sabouraud 1904 are now unified in a single genus *Malassezia* (Guého and Meyer 1988). Their clinical manifestations – pityriasis versicolor, pityriasis capitis, and seborrheic dermatitis – are among the most interesting in medical mycology. Eichstedt (1846) and Sluyter (1856) termed the disease pityriasis versicolor as a result of the yellow–brown or achromic macules (*pityron* = bran or scale forming in Greek). In 1853, Robin observed the spherical budding cells grouped in clusters associated with hyphae and named the agent *Microsporon furfur*. Rivolta (1873) considered it to belong to *Cryptococcus*, but Malassez (1974) clearly described the yeast-like cells from scalp scales and dandruff (pityriasis simplex capitis) as spores. Bizzozero, in 1884, observed the morphological diversity in the epidermal scales, and named the organisms with spherical cells *Saccharomyces sphaericus* and those with oval cells *S. ovalis*. In 1889, Baillon in his *Traité de botanique* created the generic name *Malassezia* and named the agent of pityriasis versicolor *Malassezia furfur*. Matakieff (1899) drew the sculpting of the cell wall, showing spiral grooves and branching hyphae in the scales of pityriasis versicolor (Figure 1.9a). These aspects were confirmed more recently using electron microscopy (Guillot et al. 1995) (Figure 1.9b). Sabouraud (1904) individualized clearly the 'scaling dermatoses' such as pityriasis capitis (dandruff) and seborrheic dermatitis (Figure 1.10) – a dermatologically complex subject of controversy (see Hay and Midgley 1988). He also created the genus *Pityrosporum*, with *P. malassezi* being distinct for him from the dimorphic agent of pityriasis versicolor. Attempts to elucidate the role of yeasts in these scaling dermatoses were hindered by difficulties in culturing the lipophilic organism and determining its exact growth requirements, until studies by Benham (1939) and Gordon (1951). The first culture of this yeast was obtained in the presence of lipids by Ota and Huang (1933). The polymorphism of *M. furfur* led Gordon to separate *P. orbiculare*, considered to be the etiologic agent of pityriasis versi-

color, and *P. ovale* related to pityriasis capitis and seborrheic dermatitis as Sabouraud had claimed in 1904. There followed a long series of studies favoring this proposition, but these were contradicted by others (reviewed by Midgley 1989). In 1925, Weidman isolated a yeast from a severe exfoliative dermatitis of an Indian rhinoceros (*Rhinoceros unicornis*), which he cultured without the addition of lipids; he named it *P. pachydermatis*. Later, it was described as *P. canis*, by Gustafson in 1955. This species is largely observed in dogs. Studies on the ultrastructure and genomic structure confirm the validity of three species: *M. furfur* (*M. furfur* biovar. *orbiculare* and *M. furfur* biovar. *ovale*), *M. pachydermatis*, and *M. sympodialis* (Guého et al. 1987, 1994; Guého and Meyer 1988; Simmons and Guého 1990). This new classification has recently been extended further. An experimental model of Malassezia infection reproduced seborrheic dermatitis (Drouhet et al. 1980, 1988b) with follicular and cutaneous invasion in guinea pigs and mice, as shown by Sabouraud in 1904 (Figure 1.10). The therapeutic activity of topical azoles, and particularly of oral ketoconazole, led to the cure of pityriasis capitis, seborrheic dermatitis, and pityriasis versicolor. Clinical studies have shown the important role of lipophilic yeasts in folliculitis after antibacterial and corticosteroid therapy, and in AIDS (Drouhet and Dupont 1990; Hay 1991) as well as in septicemia and systemic malassezia infection in preterm newborns receiving parenteral lipid hyperalimentation (Dankner et al. 1987; Guého et al. 1987) (see also Chapter 12, Superficial diseases caused by *Malassezia* species).

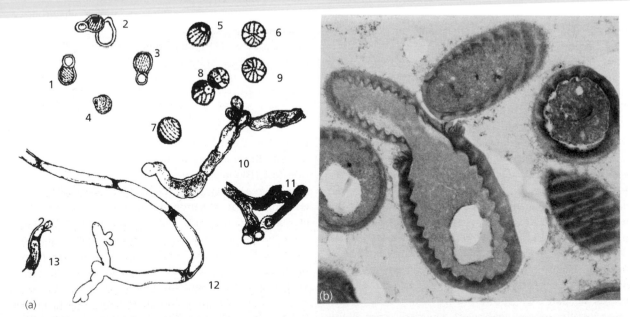

(a)   (b)

Figure 1.9 (a) *From Matakieff (1899): sketches 1–9 took into account the sculpturing of the innermost layer of the yeast's cell wall; sketches 10–12 illustrated budding and branching hyphae in scales of pityriasis versicolor; sketch 13 showed the purported formation of endospores.* **(b)** *Ultrastructure of the serrated cell wall of the yeast* Malassezia *showing growths on the plasma membrane as Matakieff had observed almost 100 years ago with an ordinary microscope. (From Guillot et al. 1995.)*

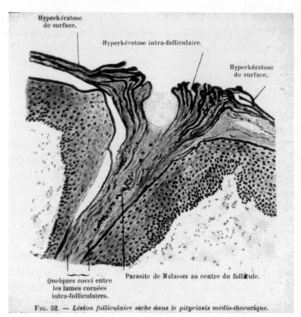

Figure 1.10 *Sabouraud's drawings of the follicular lesions with Malassez's 'spores' in the center of the follicle in pityriasis of the medial thoracic region (Sabouraud 1904).*

Cryptococcus neoformans and cryptococcosis

In 1894, the Italian microbiologist Sanfelice isolated an encapsulated yeast-like fungus, from peach juice, which he named *Saccharomyces neoformans* because of the pseudotumoral, sarcoma-like, lymphadenitic lesions observed in animals after intraperitoneal inoculation (Sanfelice 1895). At the same time, in Germany, a pathologist, Busse (1894, 1895), and a surgeon, Buschke (1895), separately reported the same fungus from sarcoma-like lesions of the tibia, and cutaneous and other lesions in a 31-year-old woman; they drew different aspects of the capsulated yeast including pseudomycelial or moniliform cells (Figure 1.11) in addition to histopathological cellular reactions (Figure 1.12). In 1896, in France, Curtis described a similar case using the term human saccharomycosis for the observed disease and he called the organism *Saccharomyces subcutaneous tumefaciens*. This second human case of cryptococcosis was retrospectively considered to be *Cryptococcus neoformans* var *gattii*, serotype B. When, in 1901, the eminent French mycologist Vuillemin (1861–1932) from Nancy examined several cultures, he did not find ascospores characteristic of the genus *Saccharomyces* and named it *Cryptococcus* – a name used earlier in 1833 by Kutzing. Therefore, he reclassified the yeast isolate of Busse and of Curtis into the genus *Cryptococcus* and the isolate of Sanfelice (1894, 1895) as *C. neoformans*. Subsequently, several cases were reported, especially in Europe, but the disease was confused with blastomycosis, so it was known as European blastomycosis. In 1905 Von Hansemann was the first to observe this

fungus in meningitis. In 1916, Stoddard and Cutler described the clinical aspects of the disease. Ignoring the European literature, the authors mistook the mucoid capsule of the yeast for histolysis of host tissue, and therefore named the fungus *Torula histolytica*. Unfortunately, this name became popular in the Anglo-Saxon literature, and the disease was known for a long time as torulosis or torula meningitis, until the terms of Vuillemin (1901) – *C. neoformans* and cryptococcosis – became widely used in medical mycology. In 1935, Benham clearly differentiated cryptococcosis from the 'blastomycoses,' especially from blastomycosis (caused by *Blastomyces dermatitidis*), by studying numerous strains of cryptococci. She showed that only one species was responsible for cryptococcosis, although serological differences existed among the strains (Benham 1935, 1950). Benham is considered to be one of the most remarkable women pioneer scientists in medical mycology (O'Hearn 1985).

An important step in the history of this fungus was the discovery in 1951 by Emmons of *C. neoformans* in barnyard soil and in 1955 the frequent saprobic association with nests and droppings of pigeons (Emmons 1955). Although it was known that *C. neoformans* was isolated from a peach by Sanfelice in 1894, during the following 55 years isolation of this fungus was only obtained from humans and lower animals, so an endogenous source of infection was erroneously assumed. Since these discoveries, the epidemiology of cryptococcosis has been intensively studied and the role of other plant sources such as *Eucalyptus* species acknowledged.

Another important step originated from the studies of the capsular polysaccharides of *C. neoformans*. In 1950, Drouhet et al. showed that the principal capsular polysaccharide of *C. neoformans* was a glucuronoxylomannan, a virulence factor that inhibits the migration of leukocytes (Drouhet and Segretain 1951). Since then numerous studies, particularly by Bulmer and Sans (1967, 1968), showed the interrelationships of the polysaccharide capsule, phagocytosis and the mechanism of host defence against *C. neoformans*. The quantity of capsular polysaccharide in the host may be so great that it can be found as a free, circulating antigen in body fluids and easily detected by a latex agglutination test (Bloomfield et al. 1961). Evans, in 1950, identified three serotypes (A, B, and C), and in 1951 confirmed that glucuronomannan is the predominant capsular polysaccharide, but also found the presence of galactose (Evans and Kessel 1951), to which a fourth serotype D was added by Wilson et al. (1968), on the basis of antigenic differences in capsular polysaccharides. The historical Sanfelice's strain (CBS 132 = ATCC 32045) belongs to serotype D, Busse's strain (CBS 879 = ATCC 4189) to serotype A, and Curtis' strain (CBS 1622 = ATCC 2344) to serotype B.

The observation that pigeon droppings were a nutrient source led Staib (1963), from the Koch Institute in

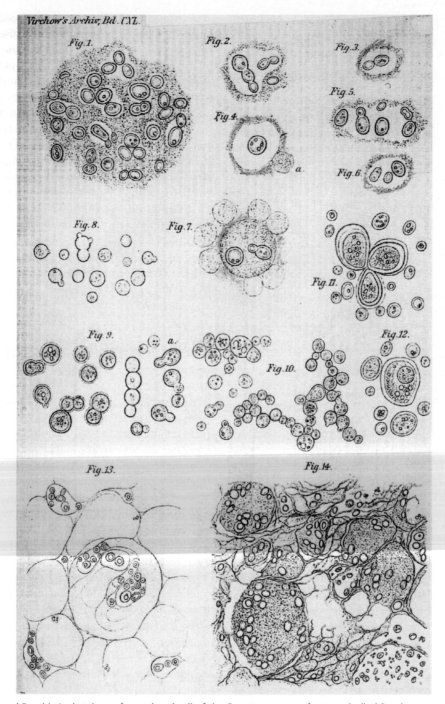

Figure 1.11 *Busse and Buschke's drawings of capsulated cell of the* Cryptococcus neoformans *(called* Saccharomyces *and the disease saccharomycosis hominis in their publication), showing a moniliform yeast form converted into a 'gelatinous cyst' (Busse 1895).*

Berlin, to new concepts for the identification of the fungus. Staib reported the brown color effect specific for *C. neoformans* produced by nigerseed of the composite *Guizotia abyssinica*. The enzyme responsible for the pigment formed was a phenoloxidase involved in virulence, according to the studies of Kwon-Chung, Polacheck and Rhodes in the 1970s. This work has now been extended with the recognition of melanin as a key virulence determinant.

Another valuable addition to the knowledge about this pathogenic yeast was the discovery of its sexual reproduction by Kwon-Chung (1975), who proposed the genus *Filobasidiella* to accommodate this basidiomycetous yeast species. Kwon-Chung (1976) showed that there are two varieties of *F. neoformans*: *F. neoformans* var. *neoformans*, corresponding to the asexual state *C. neoformans* var. *neoformans*, and *F. neoformans* var. *bacillispora*, corresponding to the asexual state *C. neoformans* var. *gattii*. This last variety was isolated by Gatti (Gatti and Eeckels 1970) in Zaire from an African child with meningitis. It was described as *C. neoformans* var. *gattii* by Vanbreuse-

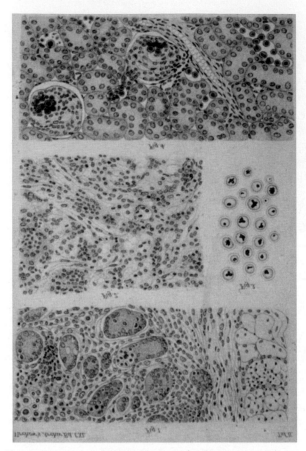

Figure 1.12 *Histopathological lesions of cryptococcosis with stained yeasts undergoing granulomatous cellular reactions and yeast cells infiltrating the tissue with no cellular reaction. Drawings by Busse (1894).*

ghem and Takashio (1970). Further studies showed that important biochemical, epidemiologic, and pathological differences existed among the four serotypes and the two sexual forms of this basidiomycete fungus (Bennett et al. 1977, 1978; Kwon-Chung et al. 1992, 1994).

In order to evaluate antifungal compounds in a reproducible way, Van Cutsem et al. (1986) established a model in the guinea pig of progressive, disseminated, cryptococcal infections including skin manifestations, as observed in patients with systemic cryptococcosis, particularly in AIDS patients showing a dermotropism of *C. neoformans*.

Another important event in the history of cryptococcosis is that this disease, which was very rare in the past, essentially developed only in people with low host resistance, malignant diseases of the reticuloendothelial system (Hodgkin's disease, lymphoma), or endocrine problems, and was named the 'sleeping giant'by Ajello in 1970. Within a few years cryptococcosis became the 'awakening giant' (Kaufman and Blumer 1978), probably as a result of the intensive immunosuppressive treatments of patients with leukaemia and those undergoing organ transplantations. The incidence in 1978

ranged from 200 to 400 cases of cerebral meningitis per year in the USA, to 15 000 subclinical respiratory infections in New York City alone. A few years after the emergence of AIDS, the frequency and the severity of cryptococcosis increased considerably. In addition, the disease responded poorly to the classic treatment (amphotericin B + 5-fluorocytosine), which necessitated the introduction of new triazole molecules (fluconazole, itraconazole). Cryptococcosis, now a principal life-threatening opportunistic mycosis, was considered to be the 'mycosis of the future' (Drouhet and Dupont 1990). In the USA, among 13 834 AIDS patients, registered by the CDC in October 1985, 904 cases of cryptococcosis (6.5 percent) were diagnosed. In France during the period 1985–88, among 283 diagnosed cases of cryptococcosis, 204 (72.08 percent) had developed in AIDS patients (Drouhet and Dupont 1990) and in 1994 more than 1 300 cases were reported (see also Chapter 32, Cryptococcosis).

Aspergillus spp. and aspergillosis

Micheli (1729) described the genus *Aspergillus* and derived its name from the Latin word *aspergillum* (to sprinkle) which referred to the perforated globe used to sprinkle holy water during Catholic religious ceremonies *'aspergus te.'* Link delineated several distinct species in 1809 such as *A. candidus*, *A. flavus*, and *A. glaucus*, which he had isolated from decaying vegetation.

The first reports of infections with *Aspergillus* spp. were those of Mayer and Emmert in 1815, who described an infection in the lung of a European jay (*Garrulus glandarius*), but principally those of the German Fresenius, in 1850, who observed an air sac infection in a captive bird, the Great bustard (*Otis tarda*) at the Frankfurt Zoo. Fresenius named the isolate *Aspergillus fumigatus* and first used the term aspergillosis.

The first cases in humans with detailed microscopic drawings of an *Aspergillus* sp. in vivo were described by Virchow in 1856 (Figure 1.13). Virchow noted that the organism he had observed (probably *A. fumigatus*) in human disease was closely related to the 'parasitic' vegetable structures growing in living animals reported by Bennett in 1844 and by Sluyter in 1847, except for the association of the pulmonary form with 'pigeon feeder's' disease by Virchow and later, in 1890, by Dieulafoy, Chantemesse, and Widal. In the following years, very few cases of the disease were reported. In 1897, Renon published his thesis showing the association of aspergillosis with certain occupations including pigeon stuffers ('*gaveurs de pigeons*') and wig cleaners. They concluded that moldy grain was the source of the infectious conidia. Renon also showed that some patients had tuberculosis because their primary disease antedated the aspergillosis. Between 1897 and 1938, the aspergillosis mentioned was not the localized fungus ball ('bronch-

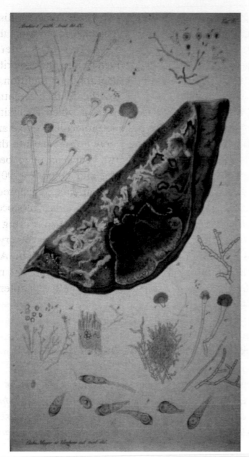

Figure 1.13 *Virchow's drawings of the first case of human pulmonary aspergillosis.* Aspergillus fumigatus *can be recognized from the drawings.*

iectosiant aspergilloma'), which was defined by the French surgeon Devé only in 1938 as a rare colonization of congenital bronchial cysts (see also Chapter 34, Aspergillosis).

The growing importance of aspergillosis was recognized only after the remarkable description, by Monod (surgeon), Pesle (physician) and Segretain (mycologist) (Monod et al. 1954), of the development of bronchopulmonary aspergilloma (fungus ball) in the pulmonary cavities of tuberculous patients who had been treated with powerful antituberculous medications. Radiological evidence demonstrated that *Aspergillus* spp., most frequently *A. fumigatus*, colonized preformed cavities in 12 percent of healed tuberculous patients along the radiolucent crest. Aspergilloma is accompanied by the appearance of anti-*A. fumigatus* precipitating antibodies, studied particularly since the 1960s in the UK (Pepys 1969; Longbottom 1986; Longbottom and Austwick 1986), and in France (Biguet et al. 1964; Drouhet et al. 1972). This localized form was observed particularly in Europe (Segretain 1962) where the incidence of tuberculosis was high, but it is now observed worldwide. At the same time, allergic bronchopulmonary aspergillosis was described by Hinson et al. (1952). In addition,

progressive invasive aspergillosis in the immunocompromised host has became more and more frequent and severe. Invasive aspergillosis has become a major problem in immunosuppressed or myelosuppressed patients, or recipients of organ transplants. It has raised new, epidemiologic, clinical, immunological, and therapeutic problems. The First International Symposium on *Aspergillus* spp. and aspergillosis (Vanden Bossche et al. 1988) presented a summation of the then current knowledge of this mycosis which has become one of the most important in fungal diseases.

Scedosporium spp. and pseudallescheriasis

The history of the taxonomy of the fungi and of the clinical and epidemiologic aspects of the mycosis, known under various names such as monosporiosis, allescheriosis, petriellidiosis, pseudallescheriasis, or scedosporiosis, is fascinating.

The history of *Monosporium apiospermum,* described in 1911 by the famous Italian taxonomic mycologist Saccardo (1845–1920), started with a white grain mycetoma of the foot of an Italian patient. In 1922, Shear (1865–1956) in Texas (USA) described and named an ascospore-producing fungus isolated from a mycetoma of the foot as *Allescheria boydii*, in honor of the clinician studying this American case. For 30 years, the two fungi were accepted as different agents of mycetoma, until 1944 when Emmons reported that *M. apiospermum* was the anamorph or asexual state of *A. boydii*. The names of both the teleomorphic and anamorphic states have undergone several successive changes: the genus *Allescheria* was reclassified, in 1970, by Malloch as *Petriellidium*. McGinnis et al., in 1982, recognized the priority of the genus *Pseudallescheria* Negroni et Fisher 1944 and the sexual telemorph became *Pseudallescheria boydii* (Shear) McGinnis, Padhye et Ajello 1972. The anamorph *Monosporium apiospermum* was invalidated by Hughes in 1958 and *Scedosporium apiospermum* (Saccardo) Castellani et Chambers 1919 is now the valid name of the anamorph.

After the first isolation from the soil by Ajello and Zeidberg (1951), numerous studies established its presence as a saprophyte in nature, explaining the source of post-traumatic infections and bronchopneumopathies.

The first clinical infection, other than a mycetoma, was observed by Benham and Georg (1948), who described granulomatous meningitis caused by *P. boydii* in a patient with no underlying disease, after spinal anesthesia a month before the onset of symptoms. The first reported cases of pulmonary infection, resulting from *P. boydii* infections, were those of Creitz and Harris (1955) and Drouhet (1955), followed by numerous reports carefully reviewed by Rippon (1988) and Dupont et al. (1991).

In 1984, Malloch and Salkin described a new species of *Scedosporium, S. inflatum*. The fungus was isolated from a bone biopsy from a 6-year-old boy. Previously, a fungal osteoarthritis of the knee with severe joint destruction, caused by an atypical *S. apiospermum* that was highly virulent for laboratory mice, had later been identified as *S. inflatum* (Drouhet et al. 1991). Since then, more than a dozen cases caused by this species have been described in young boys as a result of trauma to the knee or the foot, in patients who have undergone renal transplantation or in drug abusers (endocarditis and hip arthritis). In 1991, Guého and de Hoog, based on genomic taxonomy (DNA/DNA reassociation, G + C percentage) and the cultural aspects of numerous strains of *Sepedonium* and *Pseudallescheria*, found that *S. inflatum* was identical to *Lomentospora prolificans* Hennebert et Desai 1974, a soil fungus. They proposed the name, *Scedosporium prolificans* (Malloch et Salkin) Guého et de Hoog 1991 (Figure 1.14).

Mucor spp., Mucorales and zygomycosis (in part)

The nonseptate filamentous fungi with sporangia and sexual reproduction by zygospores, now belonging to the

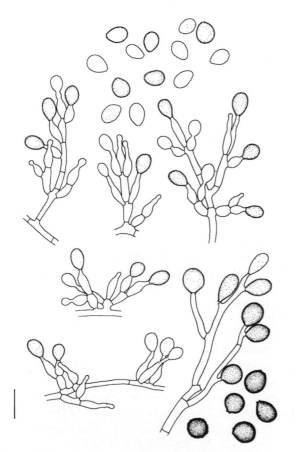

Figure 1.14 Scedosporium prolificans *(Hennebert and Desai) Guého and de Hoog comb. nov. (1991): a new emerging fungus.*

class Zygomycetes and the order Mucorales, were recognized by the first microscopists, namely Hooke (1636–1703), Malpighi (1628–84) and van Leewenhoek (1637–1723). They illustrated these fungi in 1665, 1669 and 1680, respectively, but the genera related to pathogenic fungi were defined and described later as *Mucor* Micheli 1729, *Rhizopus* Corda 1838, *Absidia* van Tieghem 1876, *Rhizomucor* Lucet et Costantin 1899, *Cunninghamella* Matruchot 1903 and, more recently, *Saksenaea* Saksena 1953 and *Apophysomyces* (Misra et al., 1979), which are all new emerging opportunistic pathogens in medical mycology.

The earliest reports of authentic zygomycosis were described 150 years ago, but the clinical descriptions were often inaccurate and incomplete as was the identification of the isolates, necessitating accommodation in genera other than those indicated in the old literature. The first authentic human pulmonary case of zygomycosis was recognized by the German Kurchenmeister in 1855, in a cancerous lung. Based on the drawings showing sporangia and nonseptate hyphae, the isolated fungus was called a *Mucor*. After two other doubtful pulmonary cases reported by Fürbinger in 1876, Paltauf published, in 1885 in *Virchow's Arch Pathol Anat*, the first observation of disseminated infection that had started in the larynx and pharynx. The classic primary localizations of rhinocerebral zygomycosis were first observed more than 50 years later (Gregory et al. 1943). Paltauf called the etiologic agent *Mucor corymbifera*, in spite of the absence of cultures for this case. In 1884, Lichtheim isolated from nature two species of *Mucor*, which caused fatal mycoses in rabbits: *Mucor corymbifera* Cohn 1884 and *M. rhizopodiformis* Cohn apud Lichtheim 1884. After a new case of pulmonary mucormycosis described by the Belgian Herla in 1895, Lindt described and found *M. pusillus*, in 1896, as well as *M. racemosus* Fresenius 1850 in animal mycoses.

Since 1943, when Gregory et al. described three cases of rhinocerebral zygomycosis and delineated this syndrome, 40 cases of rhinocerebral zygomycosis are noted per year in the USA. Around 1960, zygomycosis became a progressively increasing mycosis in compromised hosts with ketoacidotic diabetes mellitus (the rhinocerebral severe form) and in immunocompromised hosts; since 1984, the pulmonary form has been observed in 18 percent of patients with malignant leukemias, lymphoma, and severe neutropenia treated with corticosteroids. Gastrointestinal and esophageal zygomycosis is observed in 20 percent of kwashiorkor, malnourished children eating moldy food, and cutaneous or subcutaneous forms in 3 percent of cases from various origins (burns, local trauma, etc.). Zygomycosis became the fourth most common opportunistic mycosis in the compromised host after candidiasis, aspergillosis, and cryptococcosis. Since 1960, organ transplant recipients, particularly recipients of renal transplants who have been treated with stronger and stronger immunosuppressive drugs and who do or do not have associated diabetes, have caused a considerable increase in the number of cases of severe zygomycosis.

Since the 1980s, nosocomial zygomycosis, presenting as primary skin and wound infections caused by *Rhizopus* spp., *Cunninghamella bertholletiae*, *Rhizomucor pusillus*, *Absidia corymbifera*, and now *Saksenaea vasiformis* and *Apophysomyces elegans*, is often reported. Infections have been contracted from contaminated elastic bandages, infected burns, intravenous catheters, surgical wounds, ostomy sites or intravenous drug abuse whether or nor associated with AIDS. Another important condition is desferrioxamine therapy for iron or aluminum overload, which was responsible, between 1986 and 1990, for 24 lethal disseminated cases of zygomycosis caused particularly by *R. arrhizus (oryzae)*, *R. rhizopodiformis*, and *C. bertholletiae*. (For further information on zygomycosis, see also Chapter 33, Systemic zygomycosis.)

McGinnis et al. (1993) reviewed 10 cases of rapidly fatal necrotizing fasciitis (Figure 1.15) caused by *Apophysomyces elegans* after trauma or burns in healthy patients or from various other predisposing conditions.

Figure 1.15 *Necrotizing fasciitis caused by* Apophysomyces elegans. *Back of patient showing extensive deep débridement in the exposed back muscles and necrotic margins. (From McGinnis et al. 1993)*

Another emerging organism with coenocytic hyphae and asexual spores produced in a sporangium is the oomycete *Pythium insidiosum* (de Cock et al. 1987) (synonyms *P. destruens* Shipton 1987, *Hyphomyces destruens* Bridges et Emmons 1961); this has been responsible for animal diseases and, since 1987, for severe human orbitofacial infections and arteritis among patients in Thailand and the USA with α- or β-thalassemia (Rinaldi et al. 1989). (For further information about *Pythium insidiosum*, see also Chapter 22, Pythiosis.)

SYSTEMIC MYCOSES CAUSED BY DIMORPHIC FUNGI

Fungal dimorphism

In medical mycology the term dimorphism has been employed for the pathogenic, highly virulent, dimorphic fungi with a saprophytic, vegetative, mycelial form in nature or under ordinary laboratory conditions, and a pathogenic, unicellular, yeast-like, or spherule form in human and animal tissues. Under this medical definition, three important groups of dimorphic fungi are included, based on their morphology in vivo (see Figure 1.1).

The first group consists of yeast-like dimorphic fungi (Figures 1.16 and 1.17) such as:

- *Histoplasma capsulatum* var. *capsulatum* (dimorphism proven by DeMonbreun 1934) and *H. capsulatum* var. *duboisii*, respective agents of small-form or classic histoplasmosis and African or large-form histoplasmosis
- *Blastomyces dermatitidis*, agent of blastomycosis
- *Paracoccidioides brasiliensis*, agent of paracoccidioidomycosis
- *Sporothrix schenckii*, agent of sporotrichosis.

To this group *Penicillium marneffei* has been added. It is a redoubtable fungus emerging since 1988 in AIDS patients in southeast Asia which has been reviewed (Drouhet 1993a; Drouhet and Dupont 1995). It is the only known species, among more than 200 *Penicillium* spp., that is transformed in tissues into a unicellular form which multiplies not by budding, but by schizogeny. (For further information about *Penicillium marneffei*, see also Chapter 29, Penicilliosis).

The second group consists of dimorphic fungi forming spherules in vivo:

- the agent of coccidioidomycosis, *Coccidioides immitis*, was first considered to be a protozoan but proved to be a dimorphic mold only in 1900 (Ophüls and Moffitt 1900) (see also Chapter 26, Coccidioidomycosis)
- the agent of rhinosporidiosis thought to be a mycosis by Seeber (1900), but the agent *Rhinosporidium seeberi* has yet to be cultured (see also Chapter 24, Rhinosporidiosis).

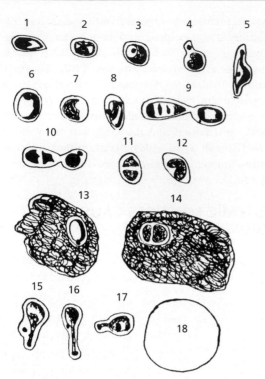

Figure 1.16 *Darling's drawing of intrahistiocytic aspects of* Histoplasma capsulatum *var* capsulatum *which he considered to be a protozoan, in spite of visible budding yeast cells.*

- the agents of adiaspiromycosis, a rare pulmonary human mycosis, but a frequent pulmonary mycosis in rodents caused by *Emmonsia crescens* and *E. parva* (see also Chapter 38, Adiaspiromycosis and other infections caused by *Emmonsia* species).

Besides these pathogenic dimorphic fungi with yeast-like or spherical unicellular forms in vivo, there are other dimorphic or rather polymorphic fungi such as *Candida albicans* with unicellular budding cells and hyphae in tissues or as *Exophiala (Wangiella) dermatitidis, Phialophora verrucosa, Fonsecaea pedrosoi*, which are agents of chromoblastomycosis when, in vivo, they form muriform cells. When this parasitic form is absent and subcutaneous, even deep organs such as the brain are invaded by hyphae; the disease is then referred to as a pheohyphomycosis (Ajello 1975b). Their mycological, clinical, and epidemiologic aspects are different from the systemic mycoses caused by the dimorphic fungi.

Pasteur's experimental conditions for M → Y transformation of *Mucor* spp., such as CO_2 high tension, were found for most pathogenic thermal dimorphic fungi for in vitro transformation. The Y form of *Sporothrix schenckii* is obtained at 37°C in a 5 percent CO_2 atmosphere in a simple chemical defined medium with an inorganic $[(NH_4)_2SO_4]$ or organic (arginine) nitrogen source, glucose, biotin (role in CO_2 fixation into the cell)

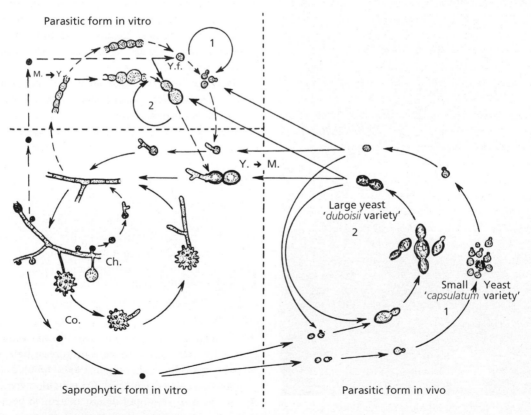

Figure 1.17 *Diagrammatic representation of the cycle in vitro (saprophytic and parasitic forms) and in vivo of* Histoplasma capsulatum *var* capsulatum *(1) and the* duboisii *var of* H. capsulatum *(2). (Ch., chlamydoconidia; Co. conidium; M.→Y., mycelial–yeast conversion; Y., yeast; Y.f., yeast form; Y.→M., yeast–mycelial conversion.) (Redrawn from Mariat 1964)*

and pyrimidine as growth factors, minerals, and oligoelements (Drouhet and Mariat 1952b). Conant (1941) obtained the difficult transformation of *H. capsulatum* var. *capsulatum* on a complex blood agar medium in sealed paraffinized tubes which provided an increased CO_2 tension. A higher content of CO_2 (15–20 percent) was necessary for *H. capsulatum* var. *farciminosum* on blood agar at pH 7.4 by Bullen (1949). Pine (1954) also found a stimulating effect of CO_2 and, in a more complex, chemically defined, nutritional agar medium in sealed tubes, obtained the transformation and conservation of cultures of the *capsulatum* and *duboisii* varieties of *H. capsulatum* for the production of antisera and vaccines (Pine and Drouhet 1963). Spherule formation in *Coccidioides immitis* is stimulated by 10 percent CO_2 in a synthetic medium.

This category of mycoses includes infections caused by species of *Blastomyces, Coccidioides, Histoplasma*, and *Paracoccidioides*. Their fascinating history began at the turn of the century and is still evolving today.

Histoplasmosis

Histoplasmosis was discovered in 1905 by Darling (1872–1925), an American pathologist and parasitologist working at the Canal Zone Hospital in Panama. There he observed three fatal cases of an intracellular infection of the reticuloendothelial system involving lymphatic tissues and all the principal organs of the body. Darling considered the observed microorganism to be a protozoan and named it *Histoplasma capsulatum* (Figure 1.16). However, da Rocha-Lima, a Brazilian studying in Hamburg, suggested that *H. capsulatum* was a fungus in 1912–13, because of its similarity to the tissue forms of the agent of epizootic lymphangitis of horses, an intracellular yeast then known as *Cryptococcus farciminosum* (Rivolta 1873) but now known as *H. capsulatum* var. *farciminosum*.

In 1934, DeMonbreun, a pathologist at Vanderbilt University in Nashville, TN, obtained a culture of the etiologic agent of histoplasmosis from a child diagnosed antemortem by Dodd and Tompkins (DeMonbreun 1934). He established the dimorphic nature of the fungus (Figure 1.17) by studying its in vitro characteristics at 25 and 37°C, and reproduced the disease in experimental animals. Independently, in 1934, Hausman and Schenken also obtained the mycelial form of *H. capsulatum* but erroneously considered it to belong to the genus *Sepedonium*. The culture of *H. capsulatum*, originally grown by DeMonbreun, served as the source of histoplasmin (comparable to coccidioidin from *C. immitis* used by Smith since 1943 to detect coccidioidomycosis), which was used by Christie and Peterson in 1945 in a skin-test survey. They showed that histoplasmosis capsulati was responsible for pulmonary calcifications in infants, young World War II recruits and young control subjects, who were negative for both tuberculin and coccidioidin. Parsons and Zarafonetis reviewed 71 cases of histoplasmosis, in 1945, collected during 1905–45, and found that the disease was rare but uniformly fatal. However, since the work of Christie, Peterson, Palmer, and others, benign asymptomatic histoplasmosis has been considered a widely prevalent disease.

Emmons (1949) first isolated the *capsulatum* variety of *H. capsulatum* from soil around a rodent burrow and under a chicken coop. Positive cultures were obtained only after 156 negative specimens – a feat that demonstrated that perseverance is necessary for successful completion of such a study. Hundreds more isolations from suspected sites and patients in many parts of the world have supplied information of epidemiologic importance. Soil contaminated by the excreta of chickens, pigeons, and bats in the USA, Central and South America, and South Africa, and by the nocturnal fruit-eating oil bird (*Steatornis caripensis*) in caves in Colombia, Peru, Trinidad, and Venezuela, as shown by Ajello and Zeidberg (1951) and Emmons (1958), and air and water, as shown by Furcolow (1958), have proved to be sources of benign pulmonary outbreaks of histoplasmosis. The first epidemiologic study on a large scale (since the study in 1945 by Furcolow) led to the rapid development of investigations of all phases of the disease. Since the first complement fixation serological test by Tenenberg for diagnosis in 1947, numerous other tests have been developed for antigen and antibody detection (Wheat et al. 1986). Clinical aspects of histoplasmosis have been described, ranging from pulmonary and splenic calcifications (Schwarz et al. 1955) to the disseminated form, which is now observed in immunosuppressed patients and particularly in AIDS patients (Graybill 1988). Epidemiologic and clinical studies have revealed that histoplasmosis has a worldwide distribution with the densest endemic areas chiefly in the USA, where 40 000 000 people have had the disease and 500 000 new infections occur every year (Rippon 1988). Parallel with these studies, there has been intensive study of the biological aspects of dimorphism (Rippon 1980; Szaniszlo et al. 1983), the regulation of the metabolism of the yeast (Y) and mycelial (M) forms at a molecular level, differences in cell wall composition accompanying morphogenesis (San-Blas and San-Blas 1985, 1993) and the role of temperature for the regulation of the Y/M form (Howard 1967). Another important contribution was the discovery by Kwon-Chung, in 1972, of the sexual state of *H. capsulatum*: *Emmonsiella capsulata* (1972), now known as *Ajellomyces capsulatus* (Kwon-Chung 1972) McGinnis et Katz 1979.

More historical details on this fascinating disease were discussed at two national conferences held in the USA in 1952 and 1962, organized by Furcolow and Ajello, and in books such as *Histoplasmosis* by Sweany in 1960.

A new form of histoplasmosis was discovered as a geographically localized form in central Africa, particularly Senegal, Nigeria, and Zaire. It is characterized by a parasitic tissue-form comprising large, oval yeast cells in giant cells (Figure 1.17) and by a distinct clinical disease. The cutaneous, subcutaneous and bone lesions predominate over pulmonary and visceral disease. The first case was reported by Catanei and Kervran in 1945 at the Pasteur Institute in Algiers. Other fatal cases were observed by Kervran and Aretas in 1947 and by Duncan, who suggested that the agent was either a species distinct from *H. capsulatum* or a variant (Duncan 1947, 1958). This agent was considered to be a distinct species by Vanbreuseghem in 1952, who named it *H. duboisii* in honor of Dubois, Director of the Institute of Tropical Medicine (Antwerp), where the strain from a Belgian patient infected in Africa was studied (Dubois et al. 1952; Vanbreuseghem 1953). The disease was called African or large-form histoplasmosis, although histoplasmosis is also endemic in Africa. For that reason an alternative term, histoplasmosis duboisii, was proposed (Ajello 1983) to describe the disease caused by *H. capsulatum* var. *duboisii* (Drouhet 1957). Subsequently, this organism was confirmed by Kwon-Chung (1972, 1975) who showed that the sexual states of both varieties are identical. This new form of histoplasmosis, also endemic in Madagascar, now has a small but important place in the pathology of Africa and is an imported mycosis in Europe, particularly in Belgium and France (Drouhet 1989).

More than 200 published cases have been reported among Africans and Europeans, with 30 percent developing severe systemic infections. In animals, the spontaneous infection has been observed only in monkeys imported from Africa into France, England, and the USA. The natural history of this particular form is not well known, but, recently, Guignani et al. (1994) reported the isolation of *H. capsulatum* var. *duboisii* from soil admixed with bat guano from a sandstone bat cave in a rural area of Nigeria.

The third form of histoplasmosis (see also Chapter 27, Histoplasmosis), epizootic lymphangitis of horses and mules, was first described by Rivolta (1873), who named the agent *Cryptococcus farciminosum*. This fungus was renamed *H. farciminosum* by Redaelli and Ciferri (1934). It is now considered to be a variety of *H. capsulatum*: *H. capsulatum* var. *farciminosum* (Rivolta) Weeks, Padhye et Ajello 1985, responsible for the disease. It is particularly prevalent among equines in Egypt, India, and Russia.

Blastomycosis

In 1894, Gilchrist observed the first case of a verrucous, granulomatous dermatitis produced by a yeast-like fungus. In 1896, it was named *Blastomyces dermatitidis*, when together with Stokes he published the second case (Gilchrist and Stokes 1896). This disease, once called Gilchrist's disease and the Chicago disease (15 cases were described there in 1901), was considered to be endemic in North America until 1953, when cases were observed in Tunisia (Vermeil et al. 1954), and later in Algeria, Central Africa, Saudi Arabia, India, Israel, Morocco, Mozambique, and South Africa, indicating extensive geographical distribution. The natural habitat of *B. dermatitidis* remains an enigma in spite of various isolates from nature. The discovery by McDonough and Lewis, in 1967, of the sexual state of *B. dermatitidis*, named *Ajellomyces dermatitidis*, is an important event in the history of this disease. (See also Chapter 25, Blastomycosis.)

Paracoccidioidomycosis

This Latin American mycosis was first described by Lutz (1908), in a Brazilian patient with severe granulomatous naso-oropharyngeal lesions and cervical adenopathy. Lutz described and cultured this dimorphic fungus which he considered, in tissue, to resemble the parasitic form of coccidioidomycosis, and called the disease a pseudo-coccidioidal mycosis. The Italian microbiologist, Splendore, described the fungus and its disease in several studies between 1908 and 1912, and Almeida named the fungus *Paracoccidioides brasiliensis* in 1930. Consequently, the disease was called the Lutz–Splendore–Almeida disease. In 1942, Conant and Howell considered this fungus to be similar to *B. dermatitidis* and named it *B. brasiliensis*, and the disease South American blastomycosis. However, later investigators restored the names *P. brasiliensis* and paracoccidioidomycosis (Figure 1.18). The ecology of *P. brasiliensis* remains an unsolved puzzle (Restrepo 1985), but the disease responds well to azole antifungals (Restrepo 1990). (See also Chapter 28, Paracoccidioidomycosis.)

Coccidioidomycosis

Now recognized as a disease of major public health importance in the USA and especially Mexico, coccidioidomycosis was first described, in 1892, by Posadas in Argentina. Posadas was a student of the famous pathologist Wernicke. He considered this granulomatous disease to be a neoplasia. Two years later, Rixford and Gilchrist (1896) observed a new case in California, considered that the agent was a protozoan, and named it *Coccidioides immitis*. It was later identified by Ophüls and Moffitt (1900) as a fungus with a dimorphic life cycle (Figure 1.19). The dimorphism of *C. immitis* was critically studied and reviewed by Cole et al. (1993). The disease is considered to be rare in Argentina (only a few cases in 1929, 27 in 1967); however, it is most common in southern California. In 1915, Dickson of Stanford University reported 40 cases of a primary pulmonary infection with mild and

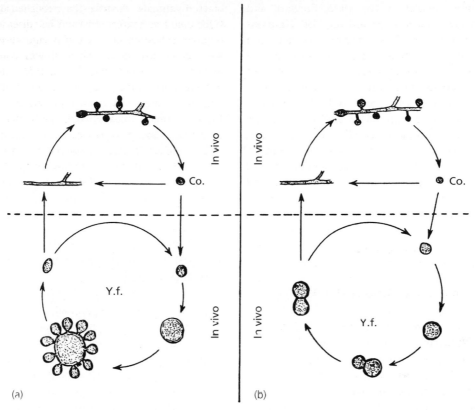

Figure 1.18 *Diagrammatic representation of cycles in vitro and in vivo of* **(a)** Paracoccidioides brasiliensis *and* **(b)** B. dermatitidis *(Co., conidium; Y.f., yeast form).*

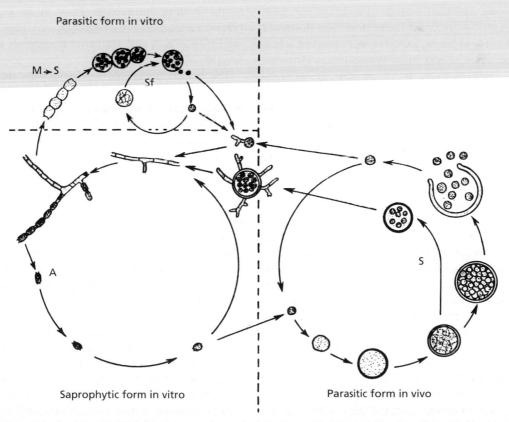

Figure 1.19 *Diagrammatic representation of cycle in vitro (saprophytic and parasitic forms) and in vivo of* Coccidioides immitis *(A, arthroconidias; M→S, mycelium–spherule conversion; S, spherule; Sf, spherule form). (From Mariat 1964)*

fatal disseminated evolution. In 1932, Stewart and Meyer isolated *C. immitis* from soil; in 1942 Emmons isolated it from soil as well as from rodents, and as a result of the extensive studies of Dickson, Gifford, and Smith, it was established that San Joaquin Valley fever was coccidioidomycosis. The work of C.E. Smith (1904–67) on the epidemiology, diagnosis, and prophylaxis of coccidioidomycosis was considerable and has been the subject of a voluminous literature. The work of Smith, Huppert, Cole, Pappagianis, Levine, and others contributed largely to the understanding of the immunology of this disease. Ajello edited the proceedings of the second symposium on coccidioidomycosis in 1967, and of the third symposium in 1977, followed by the fourth symposium in 1984 and several books (Fiese 1958; Ajello 1977). (See also Chapter 26, Coccidioidomycosis.)

Emmonsia spp. and adiaspiromycosis

Studying the ecology and natural history of *C. immitis* in the desert of southern Arizona, Emmons and Ashburn (1942) revealed that many rodents had mycotic pulmonary lesions and described the mycelial fungus obtained in culture as *Haplosporangium parvum* because of its resemblance to the phycomycete *H. bisporale*. Emmons described the adiaspores in the tissues and the later work showed that the conidia of the new filamentous fungus grew in vivo to form uninucleate spherules (< 40 mm), with thick laminated walls and no endospores. this cystic pulmonary disease had been described earlier by Kirschenblatt in 1939. He had observed adiaspores in the lungs of rats from Transcaucasia (Russia), and named the organism *Rhinosporidium pulmonale*. The organism was not isolated and was therefore not given a Latin name. In 1947, Dowding reported a fungus similar in culture to *H. parvum*, but with adiaspores attaining a diameter of 270 mm in rodent lungs from western Canada and the northern USA. Jellison, in numerous publications between 1950 and 1959, reported the widespread distribution of this mycosis in rodent lungs and other small mammals in Africa, Ecuador, Finland, Japan, Korea, Norway, Sweden, the USA, and Yugoslavia, in animals preserved in formalin since 1845, and in France, in certain specimens of animals preserved for 60 years in the Museum of Natural History of Paris (Jellison et al. 1961; Jellison 1969).

The natural habitat of the *Emmonsia (Haplosporangium)* spp. is soil. In 1954, Menges and Haberman reported the isolation of *H. parvum* from soil in Missouri, but Emmons found the strain to be different from isolates of this species. In 1959, Ciferri and Montemartini isolated *E. parva* from soil in Italy. They renamed this fungus and placed it in the new genus *Emmonsia*. In 1960, Emmons and Jellison, after studying several hundred dimorphic fungi in cultures and in pulmonary tissues of naturally, or experimentally,

infected animals, described a new species *E. crescens* which could be segregated from *E. parva* on the basis of its larger adiaspores, which also enlarge without multiplication in host tissue. Since the first human case of pulmonary adiaspiromycosis described in France by Doby-Dubois (1964), numerous other chronic benign pulmonary cases have been reported in Brazil, Czechoslovakia, France, India, Russia, and Venezuela. All these cases were caused by *E. crescens*, except for a case in 1993 in an AIDS patient observed in Colombia which was caused by *E. parva* (Echavarria et al. 1993). (See also Chapter 38, Adiaspiromycosis and other infections caused by *Emmonsia* species.)

Infection caused by *Penicillium marneffei* – the new emerging systemic mycosis in AIDS

Penicillium marneffei (Capponi et al. 1956; Segretain 1959), the only dimorphic pathogenic *Penicillium* sp. among 223 described species of *Penicillium*, was first isolated from bamboo rats (*Rhizomys sinensis*), native to Burma, China, and Vietnam (Capponi et al. 1956). These rats had been kept in captivity for experimental infections at the Pasteur Institute of Indochina at Dalat, and they developed a fatal reticuloendothelial mycosis. The fungus isolated from a rat was used to infect a laboratory mouse and both fungus and mouse were sent to the Pasteur Institute in Paris to be studied. The fungus was identified by Segretain as a new species, *P. marneffei*, in honor of Marneffe, Director of the Pasteur Institute of Indochina.

The typical feature of this pathogenic fungus is its thermal dimorphism: unicellular, oval or elongated cells, which multiply by schizogeny in vivo in histiocytes and macrophages or in vitro at 37°C. At 25°C, *P. marneffei* grows as a mold with hyphae and conidia, producing a characteristic diffusible red pigment.

Segretain (1959) pricked his finger accidentally, during experimental studies, with a needle used to inoculate hamsters with *P. marneffei*. He developed a small nodule at the site of inoculation 9 days later, followed by lymphangitis and axillary lymph node hypertrophy. Antifungal sensitivity studies by Drouhet (in Segretain 1959) demonstrated high in vitro sensitivity of this fungus to nystatin. An intensive treatment with oral nystatin (20 million units – 10–13 times the usual daily dose) for 30 days cleared the infection. This accidental infection emphasized the possibility of human pathogenicity.

The first natural human infection was described only 17 years later by DiSalvo et al. (1973) in the USA, in a 64-year-old minister who had traveled in southeast Asia. *P. marneffei* infection was diagnosed when the patient underwent a splenectomy for management of Hodgkin's disease. A second imported pulmonary case was reported in the USA by Pautler et al. (1984), in a

59-year-old man who had travelled extensively in the Far East.

Since 1984, cases of *Penicillium* infection have been reported in southern China by Deng and Connor in 1985, and in Thailand by Jaynetra et al. in 1984, among healthy or immunocompromised patients, as a systemic mycosis involving the entire reticuloendothelial system, and mimicking histoplasmosis. Since 1988, numerous cases have been reported among AIDS patients from Australia, France, Italy, the Netherlands, the UK, and the USA, who had travelled in southeast Asia, and in an African patient diagnosed in France who had acquired a laboratory infection in Paris. The clinical, epidemiologic, mycological and biological aspects of 44 cases in patients negative for the human immunodeficiency virus (HIV) and 44 cases in HIV-positive patients were analysed and compared (Drouhet 1993a). In the following 2 years an increasing number of AIDS patients from Thailand and Hong Kong (Supparatpinyo et al. 1994) have been infected with *P. marneffei*. The number of cases reported is now close to 400 (Drouhet and Dupont 1995).

Epidemiologic studies since 1986 have shown that wild bamboo rats in China, Thailand and Vietnam – *Cannomys badius*, *R. pruinosus*, *R. sinensis*, and *R. sumatrensis* – are commonly infected by *P. marneffei* (Deng et al. 1988; Ajello et al. 1995). Penicilliosis can be considered to be a 'new emerging systemic mycosis' for individuals with HIV infection living or traveling in southeast Asia. This constitutes a dangerous risk for contracting penicilliosis. This disease must be included in the clinical definition of AIDS as other systemic mycoses already are (candidiasis, coccidioidomycosis, cryptococcosis, and histoplasmosis) in defined endemic areas (Drouhet 1993a). (See also Chapter 29, Penicilliosis.)

SUBCUTANEOUS MYCOSES AND OTHER DEEP MYCOSES

Mycetoma

The term mycetoma encompasses all pseudotumors produced by fungi. It was first used by Vandyke Carter in 1874 after he had established the fungal etiology of this disease. The division of this disease into separate categories dependent on a fungal or actinomycotic etiology was first suggested by Pinoy (1913). At present, according to the literature, including comprehensive and extensive reviews by Mariat et al. (1977), who surveyed 854 cases identified between 1940 and 1960, and the classic manuals of medical mycology by Emmons et al. (1977) and Rippon (1988), the single clinical entity mycetoma may be evoked by 33 well-recognized agents, 23 species of true fungi (i.e. eukaryotic microorganisms) and 10 actinomycetes (i.e. prokaryotic organisms). These figures, as well as the number of cases, have increased

according to the last review of Kwon-Chung and Bennett (1992).

Many authors from all over the world have studied this disease. In particular several groups of medical mycologists have contributed: Abbott in 1956 in clinical studies of 1 231 Sudanese mycetomas, Murray on the immunology of mycetoma, Hay and Mackenzie on ultrastructure and histopathology, Mahgoub in 1976 in Sudan (epidemiology, immunology, and therapy of mycetoma), Latapi and Lavalle in Mexico, Mackinnon in Uruguay, Borelli in Venezuela, da Silva Lacaz in Brazil (da Silva Lacaz 1983), Ajello, Conant, Emmons, Gordon, and Rippon in the USA, Avram et al. in Romania, Vanbreuseghem in Belgium, and many others. In France, the first historical studies on actinomycotic and fungal agents of mycetoma by Vincent in 1894, Brumpt in 1906, Laveran in 1906, Pinoy in 1913, and Langeron in 1928, have been developed intensively at the Pasteur Institute since 1956 by Segretain and Mariat, in association with the histopathologists Destombes, Camain, and Ravisse; there was also help from French and African physicians, surgeons, and hygienists. All these data have been reviewed (Mariat et al. 1977). Several discoveries have helped to clarify the extremely confusing host–parasite relationship of this syndrome with its 23 fungal agents. In addition to the discovery of new agents of mycetoma (Figure 1.20), i.e. *Leptosphaeria senegalensis* (Segretain et al. 1959), *Leptosphaeria tompkinsii* (Segretain et al. 1974), *Neotestudina rosatii* (Segretain and Destombes 1961), there have been new data on the epidemiology of mycetomas, such as the discovery by Segretain of these two fungi in African soil and on the thorns of acacia trees, the discovery of the sulfurophilic fungus *Madurella mycetomatis* in termitaria (Segretain and Destombes 1969), the classification of mycetomas according to the microscopic aspects of actinomycotic and fungal 'grains' in tissues (Figure 1.20). There are many questions to be answered and work to be done on the eumycetomas, particularly on their treatment which is ineffective in most cases, and which necessitates amputation of legs and feet. Two International Symposia (Venezuela 1977; Mexico 1987) have been dedicated to the mycetomas. (See also Chapter 20, Eumycetomas.)

Pheohyphomycoses

As of 1994, 101 species of pheoid fungi had been isolated from human and animal infections and grouped by Ajello (1975b) under the umbrella term pheohyphomycoses. Taxonomically, the fungi responsible for pheohyphomycosis include hyphomycetes in the Dematiaceae, Spheropsidales and various Ascomycetes. Clinically, there is a wide spectrum of infections which vary from superficial colonization of the skin and hair or the cornea of the eye, to cyst formation, to cutaneous, subcutaneous, cerebral, and systemic opportunistic infec-

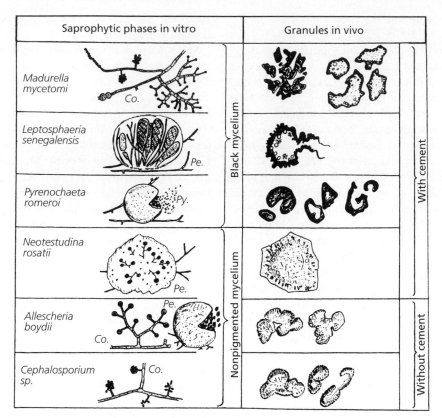

Figure 1.20 *Diagrammatic representation of the principal agents of fungal mycetomas in vitro and in vivo (Co., conidium; Pe., perithecium; Py., pycnidium). (Redrawn from Mariat 1964)*

tions; these have all been well detailed by Ajello (1975b, 1981); McGinnis (1983); Rippon (1988); and Matsumoto et al. (1994). (See also Chapter 38, Adiospiromycosis and other infections caused by *Emmonsia* species.)

Sporotrichosis

As pointed out by Mariat (1968), sporotrichosis can be considered a historical 'model of medical mycology.' The first case of sporotrichosis was reported by Schenck, from the Johns Hopkins Hospital in Baltimore in 1898. The fungus was isolated by Schenck and classified in the genus *Sporotrichum* by the plant mycologist E.F. Smith. The second case, also reported from the USA (Chicago) by Hektoen and Perkins in 1900, was observed in a boy who had struck his finger with a hammer. He developed an inoculation chancre with lymphocutaneous lesions. The authors named the fungus *Sporothrix schenckii*. It is a dimorphic fungus (Figure 1.21) with a yeast form obtained in a chemically defined medium under 5 percent carbon dioxide (Drouhet and Mariat 1952b). The lesions cured spontaneously and 65 years later the patient still had a positive sporotrichin skin test. In Europe, the disease was initially described in 1903 by de Beurmann and Ramond. The fungus isolated was studied in 1910 by the French botanist, Matruchot, who considered it to be a species different from *S. schenckii* because of the presence of a black pigment. It was named *Sporotrichum beurmannii*. The identity of the American and French species was established in 1921 by Davis and, in 1962, Carmichael determined that *Sporothrix schenckii* is the correct binomial for the fungus. Between 1906 and 1912, de Beurmann and Gougerot identified at least 10 more cases and tabulated more than 200 additional cases in their famous book, *Les sporotrichoses* (de Beurmann and Gougerot 1912). They were the first to suggest that sporotrichosis could be considered as an opportunistic infection, which has recently been confirmed by the occurrence of sporotrichosis in AIDS. Following the suggestion of Sabouraud, in 1903 de Beurmann and Ramond were the first to employ iodides in the chemotherapy of sporotrichosis with remarkable success (de Beurmann and Gougerot 1912). This therapy still remains the treatment of choice for sporotrichosis. Over a period of 50 years, potassium iodide remains the only oral chemotherapeutic agent for deep mycoses. To date several hundred cases have been reported all over the world, particularly in Europe. Curiously, since 1940, sporotrichosis has virtually disappeared from Europe. (See also Chapter 19, Sporotrichosis.)

Intensive studies on the ecology and epidemiology of sporotrichosis, reviewed by Mariat (1968) and Rippon (1988), established that *S. schenckii* can be isolated from soil, timber, house plants, straw, bird nests, and animal burrows, etc. The most endemic areas are Latin America

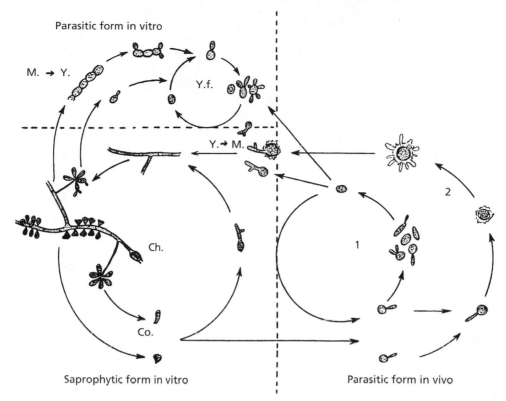

Parasitic form in vitro

M. → Y.

Y.f.

Y. → M.

Ch.

Co.

2

1

Saprophytic form in vitro

Parasitic form in vivo

Figure 1.21 *Diagrammatic representation of the cycle of* Sporothrix schenckii *in vitro (mycelial and yeast forms) and in vivo (Ch., chlamydoconidium; Co., conidium; M.→Y., mycelial–yeast conversion; Y.f., yeast form; Y.→M., yeast–mycelial conversion; 1, 'cigar bodies;' 2, asteroid bodies). (From Mariat 1964)*

and South Africa. *S. schenckii* is phylogenetically related to the plant fungus *Ceratocystis stenoceras* (Mariat 1977).

Chromoblastomycosis and its agents

The term chromoblastomycosis was coined by the Brazilian mycologists and dermatologists Terra, Torres, Fonseca, and Area Leao in 1922, to define a clinical entity presenting with verrucous dermatitis caused by various pheoid fungi. The attempt by Moore and Almeida, in 1935, to replace the term chromoblastomycosis with chromomycosis, including cases of similarly pigmented hyphal fungi, was disputed by Ajello (1974) in favor of the term pheohyphomycosis for cutaneous, subcutaneous, cerebral, and systemic opportunistic infections, caused by a larger spectrum of pheoid fungi. Historically, although chromoblastomycosis is essentially a tropical and subtropical disease, the first authentic case was reported in a patient from Boston (USA) by Lane and Medlar in 1915. The fungus was named *Phialophora verrucosa* Medlar 1915. Pedroso and Gomes (1920) reported four cases from Brazil of the disease called locally 'formiguiero.' One of the patients had been observed since 1911 and is considered to be the first known case of chromoblastomycosis. The etiologic agent was considered to be different from *P. verrucosa* and was named *Hormodendron pedrosoi* by Brumpt in 1920.

The valid name is *Fonsecaea pedrosoi* (Brumpt) Negroni 1936. This species is the most frequent agent of chromoblastomycosis in Latin America. The first case outside the Americas was described in 1927 by Montpellier and Catanei from the Pasteur Institute of Algiers. The responsible fungus was described as *Hormodendron algeriensis* (now *Fonsecaea pedrosoi*). In Madagascar, Brygoo and Segretain (1960), during the 4-year period 1955–59, reported a high prevalence of 129 cases (a case rate of one per 32 500 population). Numerous other cases were found in Central and South America, Africa, Asia, and Australia, where intensive mycological studies were performed for identification and taxonomic classification of the dimorphic agents (Figure 1.22) of chromoblastomycosis. (See also Chapter 18, Chromoblastomycosis.)

Five principal agents of chromoblastomycosis have been recognized since 1915:

1 *Phialophora verrucosa* Medlar 1915
2 *Fonsecaea pedrosoi* (Brumpt) Negroni 1936
3 *Fonsecaea compacta* Carrion 1940
4 *Cladophialophora carrionii* Trejos 1954, genetically related to *P. verrucosa* and *F. pedrosoi*
5 *Rhinocladiella aquaspersa* (Borelli) Schnell, McGinnis et Borelli (1983).

Among numerous historical, mycological, and epidemiologic studies on chromoblastomycosis and its responsible fungi, special mention is necessary for the

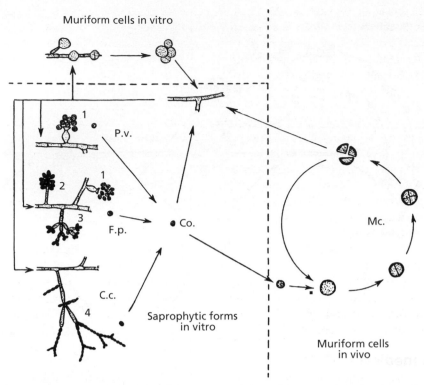

Figure 1.22 *Diagrammatic representation of the cycles of three agents of chromoblastomycosis in vitro (saprophytic and tissue forms) and in vivo (C.c., Cladophialophora carrionii; Co., conidium; F.p., Fonsecae pedrosoi; Mc., muriform cells; P.v., Phialophora verrucosa; 1, phialide form; 2, acrotheca form; 3, short cladosporium form; 4, long cladosporium form). (Redrawn from Mariat 1964)*

studies of Carrion between 1935 and 1971 and Silva between 1947 and 1971 at the School of Tropical Medicine, San Juan, Puerto Rico and Columbia University, New York, where new species were identified and the parasitic form obtained in vitro. The first evidence that the etiologic agents exist in nature was obtained by Conant (1937b) at Duke University (Durham, NC). He demonstrated that *P. verrucosa* was a cause of 'blueing' of lumber, followed by the discovery of the presence of *F. pedrosoi* and *C. carrionii* in timber, soil, wood or pulps, and air, and in numerous infections in animals with characteristic muriform cells in tissues. The first immunological studies in chromoblastomycosis were performed by Conant and Martin (1937). More recently, in Venezuela, Zeppenfeldt et al. (1994) isolated *C. carrionii* from the medullary tissue of two members of the family *Cactaceae* with 'muriform cells' similar to those in patients with chromoblastomycosis.

HISTORICAL DEVELOPMENTS IN MEDICAL MYCOLOGY

The International Society for Human and Animal Mycology

Medical mycology was recognized as an individual full branch of microbiology only in 1954 with the creation of the International Society for Human and Animal

Mycology (ISHAM), at the Thirteenth International Botanical Congress in Paris. Until that time, sessions on the mycoses were admitted (sometimes with difficulty) at dermatological and botanical congresses. The idea had long been encouraged by some mycopathologists, such as those participating in the Fifth Congress on Microbiology in Rio de Janeiro (1950), but nothing was decided until the Sixth International Congress on Microbiology in Rome in 1953, when Redaelli and Ciferri invited a small group of mycologists composed of Vanbreuseghem (Belgium), Ainsworth (the UK), Segretain, Drouhet, Mariat (France), Lodder (the Netherlands), Seeliger (Germany), Paldrok (Sweden), and several others to found the International Society for Human and Animal Mycology (ISHAM), which was formalized in 1954, in Paris. A president was elected (Redaelli, Milan), as well as a very active general secretary (Vanbreuseghem, Anvers) and four vice-presidents: Segretain (Paris), Emmons (Bethesda), Ainsworth (Exeter), and Negroni (Buenos Aires). Mycologists from a host of other countries have joined ISHAM, which, as a result of the value of scientific contributions from its members, has encouraged the steady development of medical mycology as an individual branch of medical microbiology, alongside bacteriology, parasitology, and virology.

Once ISHAM was established, some founders and charter members claimed the right to form National Societies of Medical Mycology for intensive education, training, and development of research in this field in their own countries. Foundation of the Société Française de Mycologie Médicale in 1956 was followed by the Japanese Society for Medical Mycology (in 1957), the Deutschsprachige Mykologische Gesellschaft (in 1960), the British Society for Mycopathology (in 1961), the Société Belge de Mycologie Humaine et Animale, the Medical Society of the Americas (in 1961) and another 18 other national societies all affiliated to ISHAM. The ISHAM congresses to date have been held in Paris (1954), Lisbon (1958), Montréal (1962), New Orleans (1967), Paris (1971), Tokyo (1975), Tel Aviv (1979), Palmerston (1982), Atlanta (1985), Barcelona (1988), Montréal (1991), Adelaide (1994), Salsomaggiore, Italy (1997), Buenos Aires (2000), and San Antonio (2003).

International and National Societies have contributed through their symposia, workshops, reference committees, and publications to the development of basic and applied research, education, and training. Their historical evolution is recorded in several publications (Ainsworth 1986; Drouhet 1991; Espinel-Ingroff 1996).

Literature on medical mycology

The literature of medical and veterinary mycology directed towards teaching and research has increased considerably since 1945, at a rate estimated to be about equal to that of scientific literature in general. The *Review of Medical and Veterinary Mycology* (*RMVM*), the quarterly journal of abstracts published by the Commonwealth Mycological Institute, presents complete information on this literature. In 1966, Ainsworth, analyzing the pattern of the medical and veterinary literature reviewed since the foundation in 1943 of the *RMVM*, estimated that the increase in the number of abstracted papers was linear and had doubled over a 12-year period. Since these data, the increasing number of papers published concern principally the immunology and chemotherapy of the mycoses. The presidential address of Mackenzie (1992) at the Eleventh Congress of ISHAM held in Montréal (Canada) gave a full analysis of the evolution over the last five decades, with topics of interest ranging from laboratory studies largely directed at fungal biology (the 1960s), pathology, etiology/epidemiology (up to the 1970s) to antifungal/chemotherapy (dominant since the 1980s), immunology, host–parasite relations, and, more recently, molecular biology.

Several publications are devoted exclusively, or in large part, to medical mycology:

● The *Journal of Medical and Veterinary Mycology* (*JMVM*), formerly named *Sabouraudia*, in 1961, when this journal was founded as the official journal of ISHAM. The four annual numbers of *Sabouraudia* were increased to six in 1985 when the journal changed its name to *JMVM* with nearly 150 papers

annually. Two large volumes were published of the Proceedings of the last Congresses of ISHAM (Montréal, 1992; Adelaide, 1994) [*J Med Vet Mycol* 1992, **30** (Suppl 1): 1–331; *J Med Vet Mycol* 1994, **32** (Suppl 1): 1–420].

● *Mycopathologia*, founded in 1938 by Ciferri and Redaelli under the name *Mycopathologia et Mycopathologia Applicata*, one of the oldest journals of mycology, devoted to the fungi of plant pathology, mycotoxicoses, applied industrial mycology, and human and animal mycoses.

● *Mycoses*, a bimonthly journal since 1988, formerly named *Mykosen*, official publication of the German-speaking Mycological Society.

● The *Journal de Mycologie Médicale* (*Journal of Medical Mycology*), founded in 1991 (formerly, since 1956, the *Bulletin de la Société Française de Mycologie Médicale*), an international journal (in English and French) for mycological studies related exclusively to human and animal mycology.

● The *Japanese Journal of Medical Mycology*, founded in 1956.

Besides these publications devoted to medical mycology, there are other general medical journals that publish, only occasionally, articles on various aspects of human mycoses, such as *Journal of Infectious Diseases, Reviews of Infectious Diseases, Infection and Immunity, Clinical Infectious Diseases, American Journal of Medicine*, etc., or about fungi of medical interest, such as *Antonie van Leeuwenhoek, Mycotaxon, Experimental Mycology*, and national or regional journals of some Societies of Medical Mycology.

During the period up to World War I (1914), the representative mycopathological textbooks and monographs included those on dermatophytes, such as Morris's *Ringworm in the light of recent research* (London), published in 1898, although the main texts were the monumental book of Sabouraud, *Les teignes*, published in 1910, and the book on sporotrichoses by de Beurmann and Gougerot (1912). Robin's *The vegetals* (1853), in which the mycoses in human and lower animals are developed, and two French monographs on aspergillosis by the veterinarian Lucet and the medical pathologist Renon, published in 1897, complete this pioneering period. The truly educational period for mycopathologists began with the famous *Manual of tropical medicine* (first published in 1910; subsequent editions in 1913 and 1919) by Castellani, considered to be one of the first modern medical mycologists, and Chalmers, and the *Précis de parasitologie* by Brumpt (first edition in 1910, sixth edition in 1949). These handbooks were followed by Dodge's *Medical mycology* (1935), an extensively referenced book with 900 pages and 4 000 references. The breakthrough in medical mycology teaching occurred in 1944, when Conant and his colleagues Martin, Smith, Baker, and

Callaway at Duke University, Durham, NC, published the *Manual of medical mycology* (Conant et al., 1944, 1954, 1971). This book set the pattern for subsequent textbooks such as those of Emmons, Kwon-Chung, and Rippon, the latter from the Pritzker School of Medicine, University of Chicago, entitled *Medical mycology*. From the first edition of 1974 to the third edition of 1988, this book gave the most significant changes occurring in recent years, but it also gives historical data for all pathogenic fungi and actinomycetes. This book still represents a powerful tool which will inspire researchers and clinicians for many years to come.

Other general textbooks, manuals, monographs and proceedings (such as Langeron 1945, 1952; Nickerson 1947; Lodder and Kreger-van Rij 1952; Littman and Zimmerman 1956; Baker 1971; McGinnis 1980; Howard 1983; Rose and Harrison 1987; etc.) have been published and the reader can also refer to Ainsworth (1986) and Drouhet (1991).

Education in medical mycology

In medicine and health sciences, medical mycological education was carried out with minimal mycological input at the start; but, progressively, postgraduate courses have been developed by well-trained mycologists. The training of mycopathologists and education in medical mycology started in the second half of the nineteenth century in Europe (mainly in France, Great Britain, Italy, Germany, Belgium, the Netherlands), North America (the USA, Canada), Latin America (Argentina, Brazil, Uruguay, Colombia, Venezuela, Mexico) and in the second half of the twentieth century in Asia, Africa, and Australia. This evolution has largely been described and analyzed for all countries by Ainsworth (1986); Drouhet (1991) and more recently for the USA by Espinel-Ingroff (1996). Ajello (1975a) pointed out that the history of medical mycology can well be divided into the periods BC and AC – before Conant and after Conant. He cited the work of Bill Conant and his colleagues at Duke University as a milestone in ending the previous perfunctory treatment of medical mycology which had prevailed before his era.

Another important course on medical mycology was initiated in 1947 by Ajello, Director of the Division of Mycotic Diseases at the Centers for Disease Control in Atlanta, Georgia. This month-long course was directed primarily at staff members of state and city health departments and federal agencies, and later also to international institutions (Ajello et al. 1975). In addition to the 4-week course with Georg, Kaplan, Padhye, Kaufman, Standard, and Ajello, courses in veterinary and medical mycology, mycoserology, exoantigen procedures for pathogenic fungi, and public health mycology were organized.

Other federal mycological laboratories, including the National Institutes of Health (Bethesda, MD), where Emmons (1900–85) was appointed in 1936, contributed to the development of modern medical mycology. Emmons' book, *Medical mycology* (Emmons et al. 1963, 1970, 1977), followed by that of Kwon-Chung and Bennett (1992), contributed significantly to the education of mycopathologists. A great number of other medical mycological courses are now available in the USA, Salkin (1988). Twenty-four universities or colleges in the USA provide graduate research programs in medical mycology leading to a doctoral degree.

In the UK, since the end of the nineteenth century, research in medical mycology was dominated by interest in the dermatophytes (Fox 1863; Adamson 1895). New developments were realized after World War II. Duncan (1884–1958), at the London School of Hygiene and Tropical Medicine, was the first Reader in Medical Mycology at London University. Ainsworth, who was interested in medical and veterinary mycology, was one of the founding members of ISHAM (1953–54). He contributed to the creation of the Medical Mycology Committee of the Medical Research Council (1943–69) and developed, with the late Phyllis Stockdale, the *Review of Medical and Veterinary Mycology* (*RMVM*), a joint publication of the Bureau of Hygiene and Tropical Medicine, the Imperial Bureau of Animal Health at Weybridge, and the Imperial Mycological Institute. Important centers for medical mycology at Leeds, Glasgow, and Bristol were developed. A history of important developments in medical mycology in the UK was outlined by Ainsworth, who was author of numerous books dedicated to fungi and to the history of medical and veterinary medicine (1986), general mycology (1976) and animal mycoses (Ainsworth and Austwick 1973). Consequently, Ainsworth made a substantial contribution to bettering knowledge of this branch of microbiology. At the Commonwealth Mycological Institute, Kew, Stockdale rediscovered the sexual state of *Microsporum gypseum*, a starting point for significant further research in this field. In 1958 at Glasgow University, Gentles discovered the oral activity of griseofulvin on tinea. Stockdale, editor of the *RMVM* for many years, and Gentles, Odds and Evans, editors of *Sabouraudia* (later *JMVM*), contributed a great deal to education in medical mycology. A Mycological Diagnostic Laboratory was set up in 1959 in the Department of Microbiology, Queens University, Belfast.

In France, Sabouraud (1864–1938), trained at the Pasteur Institute under Roux (1891) and at the Saint Louis Hospital, the center of French dermatology at that time, was the father of modern medical mycology even though he did not do much university teaching. His 'Laboratoire des Teignes' became a shrine for dermatologists and mycopathologists from all over the world. The first regular university course on medical mycology was organized, at the Faculty of Medicine in Paris, by

Langeron (1874–1950); it took place in the Department of Parasitology and was directed by Brumpt (1877–1951), who had been interested over many years in the human mycoses (particularly mycetomas and black piedra). Numerous French and other distinguished mycologists such as Talice (Uruguay), Negroni (Argentina), Ota (Japan), and Milochevitch (Yugoslavia), followed his teaching that was devoted, particularly, to yeasts and dermatophytes. In 1887, Pasteur (1822–95), at the end of his long career as father of modern microbiology, founded the Institute that bears his name, independent of university and state institutions. He assigned Roux to organize a course for the study of all microorganisms responsible for infections in humans and animals, which became the well-known annual Great Course of the Pasteur Institute. Each microorganism was treated by a specialized microbiologist. The pathogenic fungi of humans and animals were well covered in this course. Four lectures were given by Magrou (1883–1951), botanist and physician, head of the Department of Mycology and Plant Physiology (created by Pinoy in 1912), and a fifth lecture was given by Sabouraud every year until his death in 1938. Three mycologists of the Pasteur Institute, trained as microbiologists – Segretain, Mariat, and Drouhet – but oriented their research work and teaching principally towards medical mycology after 1946. After the death of Magrou in 1951, an annual postgraduate course in medical mycology was set up in 1953. Twenty-four participants – graduates in medicine, veterinary sciences, pharmacy, and science – particularly interested in teaching and research are enrolled annually: two-thirds from France and one-third from other countries, principally Europe, Latin America, and Africa. The course covers all aspects of fungi and actinomycetes that cause diseases in humans and animals and provides extensive practical laboratory work full time. After theoretical and practical examinations, a diploma is awarded by the Institute Pasteur. This course has trained more than 1 000 medical mycologists. There are now more than 15 university centers in the principal cities of France, in which advanced medical mycological teaching is given (Drouhet 1991).

At the Centraalbureau voor Schimmelcultures, Baarn, in the Netherlands, courses on medical and veterinary mycology were organized by de Vries since 1953; now these are organized by de Hoog and provide for an international group of students. Another important course in tropical mycology was given every year at the Institut de Médecine Tropicale Prince Leopold, Antwerp; it was started by Vanbreuseghem.

HISTORICAL EVOLUTION OF THE DIAGNOSTIC MYCOLOGY LABORATORY

The evolution of laboratory diagnoses from the classic conventional methods to the modern, rapid techniques in general followed a similar course to that in microbiology and other new and developing disciplines (immunology, electron microscopy, molecular biology). Their progress contributed, to a great extent, to the development of medical mycology.

The discovery of the microscope (first mentioned by John Faber around 1625) was followed by its application to the study of fungi in nature (Hooke 1665; van Leeuwenhoek 1673; Malpighi 1679; Micheli 1729; Pasteur 1861) and only later, at the turn of the century, to the study of fungi involved in human and animal infections. Phase contrast microscopy, invented by the Dutch physicist F. Zernicke (recipient of a Nobel Prize in 1932), prepared the way for the introduction in the 1950s of transmission electron microscopy (TEM). This was introduced in medical mycology only in the 1960s (Garrisson 1983). It was followed by the apperance of scanning electron microscopy (SEM); both electron microscopic techniques have had a broad application to fungal studies since the 1970s (Cole 1986). Fluorescent microscopy using nonspecific staining procedures (calcofluor white) has been used since the 1980s; specific immunolabeling, which has developed since 1956 as a result of Coon's labeling of protein with fluorescent substances, has become an important tool for the immunological diagnosis of mycoses. Fluorescent microscopy was successfully applied to medical mycology by Kaplan and Kaufman (1961), Drouhet et al. (1972), and others, and was reviewed by Kaufman and Reiss (1986), Longbottom (1986), and Drouhet (1988). Immunoperoxidase (IPO), using an anti-IgG serum label attached to the peroxidase, rather than to fluorescein isothiocyanate, has proved to be as valuable as the immunofluorescent method.

Microscopic detection of fungi in clinical preparations (KOH/NaOH, lactophenol cotton blue, India ink) is made on smears or histological sections stained first with classic stains used in bacteriology and parasitology, such as Gram's stain (1884), May–Grünwald–Giemsa (Wright's) stain, which are based on methylene blue eosinates (1902), hematoxylin–eosin–saffran (HES) (1911) and mucicarmine. Later, two fungus-specific stains superseded all others, namely:

1 the periodic acid–Schiff (PAS) stain based on the same Schiff reagent as that used in Gram's stain, introduced in 1947 by Hotchkiss–MacMannus with Kligman and Mescon's modification (1950), and
2 Gomori's (1946) and Grocott's (1955) silver methenamine stain (GMS).

These two stains that color the cell wall distinctly red (PAS) or black (GMS), are used universally by histopathologists (Baker 1957, 1971; Chandler et al. 1980).

Culture of fungi evolved after the creation of Raulin's first chemically defined liquid medium under standardized conditions, from cultures on shallow porcelain dishes to glass flasks or tubes sealed or plugged with cotton wool to prevent contamination (Pasteur 1861; Roberts 1874). The use of solid media, such as gelatin, used in 1852 by Vittadini for culture of *Botrytis (Beauveria) bassiana*, was followed by use of agar poured onto glass dishes – a simple technique described by the German bacteriologist Petri (1887), one of Koch's assistants. The agar Petri dish technique, which allowed isolation and identification by morphogenesis, was adapted for bacteriology and mycology. In his early work, Sabouraud (1896, 1910) grew dermatophytes on a variety of solid media and noted that their colony characteristics and microscopic features differed considerably on the various media. Sabouraud's glucose medium became the most widely used in medical mycology, not only for dermatophytes, but also for the isolation and routine subculture of most fungal pathogens (Odds 1991). Later, it was modified by the addition of thermostable antibacterial antibiotics (chloramphenicol, 1947; gentamicin, 1969) and cycloheximide (Acti-Dione) for the selective isolation and identification of pathogenic fungi (Georg et al. 1954). A new culture medium called the dermatophyte test medium (DTM), based on a modification of Sabouraud's medium by the addition of a pH indicator and cycloheximide which inhibits saprophytic mold, led to a simplified and rapid detection of dermatophytes. Many natural media (potato dextrose, potato–carrot–bile, and malt extract agars) or chemically defined media with essential carbon and nitrogen compounds, vitamins or growth factors such as oleic acid for lipophilic yeasts have been described for particular groups or single species of yeasts and molds (Saint-Germain and Summerbell 1996).

Important improvements allowed for the rapid isolation of fungi from blood (lysis–centrifugation system, non-radiometric CO_2 detection, after the 1980s), rapid yeast identification procedures, germ tubes for *Candida albicans*, assimilation tests using C and N compounds (up to 32 carbon sources) by a scaling down of old techniques (described by Beijering in the 1800s, by Wickerham in 1948, and transformed and developed as commercial kits on microplate systems, including rapid enzyme identification, e.g. phenol oxidase developed by Staib 1962, urease, etc.). All these new procedures have been reviewed (Pincus et al. 1988; Roberts et al. 1992; Salkin et al. 1994).

The immunoidentification of fungal pathogens by exoantigens, soluble immunogenic macromolecules produced by fungi, has been intensively developed and standardized by Kaufman and Standard since 1978. The earliest attempt to identify fungi by their soluble antigens was made by Manyck and Sourek in 1966, when they used a single-diffusion tube precipitin test and inoculated cultures of *B. dermatitidis*, *C. immitis*, *H. capsulatum*, and *P. brasiliensis* on the surface of serum agar; their test required up to 21 days to complete, however. Extensive studies carried out since 1983, by Kaufman et al., using standardized rapid procedures

with specific reference antigens, allow the identification of most dimorphic fungi, hyaline and pheoid fungi in 2–5 days.

Since 1990, DNA probes have been developed for the rapid identification of those medically important fungi (*H. capsulatum*, *B. dermatitidis*, *C. immitis*), for which conventional techniques are time-consuming. Probe assay results are available within 2 hours compared with the 72 hours for exoantigen tests (Roberts et al. 1992; Mitchell et al. 1994).

IMMUNODIAGNOSIS OF INVASIVE, DEEP MYCOSES

The need for sensitive, rapid, routine, standardized tests for the immunodiagnosis of these mycoses has become increasingly important, and the detection of circulating fungal antigens and serum antibodies has been emphasized in recent reviews (de Repentigny and Reiss 1984; Longbottom and Austwick 1986; Drouhet 1988, 1993b; Buckley et al. 1992; de Repentigny et al. 1994). A new discipline, molecular immunology in medical mycology, was developed (Reiss 1986).

Fungal antibodies

Over the last decades, important progress has been made in the detection of specific serum antibodies as a result of improved antigen preparation and the techniques of immunodiffusion in gels and immunoelectroprecipitation.

These methods were developed at the Pasteur Institute: in 1946, Oudin's gel precipitation method using simple and double diffusion (DD) in tubes, followed in 1948 by Ouchterlony's modification of his practical DD plate technique carried out in Petri dishes. Important improvements were achieved by Grabar and Williams in 1952 with the introduction of immunoelectrophoresis, which combined gel electrophoresis with immunodiffusion. In 1959, Bussard described counter-immunoelectrophoresis (CIE), a highly sensitive and rapid method based on the migration of antigens towards the anode and antibodies towards the cathode in a gel (agarose) at pH 8.2. The two-dimensional immunoelectrophoresis (2DIE) technique was described by Ressler in 1960 and developed by Laurell in 1965 under the name crossed immunoelectrophoresis. In mycology this method made a significant contribution to the identification and characterization of the complex antigens of the fungi and to the antigenic interrelationships of antigenic preparations of species and genera. In 1973, Axelsen first applied 2DIE to the antigenic mosaic of *C. albicans* which is made up of at least 78 antigens, including several major ones. Numerous variants have been reviewed (Axelsen 1973; Longbottom 1986).

Specific enzyme staining techniques applied to immunodiffusion and immunoelectrophoresis of pathogenic fungi started with the work of Uriel in 1963 at the Pasteur Institute. The French school of Biguet showed that the principal precipitating antigen–antibody complexes have enzymatic activities of diagnostic interest, such as the activities of chymotrypsin and catalase for *A. fumigatus,* of chymotrypsin and malic dehydrogenase for *A. flavus*, of malic dehydrogenase activity for *Candida* spp., and of catalase and glucuronidase activities for *H. capsulatum.*

The first enzyme immunoassay (EIA) techniques were described independently by Avrameas and Guilbert in 1971 at the Pasteur Institute and by Engvall and Perlman who gave the name ELISA to enzyme-linked immunosorbent assay. Subsequently, this test was rapidly and extensively applied to the immunology of pathogenic fungi and their respective mycoses. These tests provide an objective end point; they can be automated and are simple to carry out. In addition, they have a sensitivity and specificity comparable to radioimmunoassay (RIA), which is capable of detecting nanogram amounts of antigens and antibodies. Various adaptations of EIA, particularly the enzyme-linked immunoelectrotransfer blot or Western blotting technique developed by Towbin in 1979 and Burnett in 1981, have been reported and compared with other immunological tests (Mackenzie 1983, 1996; de Camargo et al. 1984; de Repentigny and Reiss 1984; Longbottom 1986; Drouhet 1988; Guesdon et al. 1988; de Repentigny et al. 1994).

CIRCULATING FUNGAL ANTIGENS

Research on fungal circulating antigens received considerable attention, principally because of the evolution of systemic opportunistic mycoses in immunocompromised or AIDS patients.

Since the First International Symposium on Fungal Antigens held in 1986 at the Pasteur Institute in Paris, during which many problems in isolation, purification and detection of principal fungal antigens were analysed, new tests for the detection of circulating antigens have been developed for aspergillosis, candidiasis and cryptococcosis; these tests are rapid, routine and commercially available, compared with other reference tests (Drouhet et al. 1988a; Drouhet 1989, 1993b; Buckley et al. 1992; de Repentigny et al. 1994).

Historically, the first circulating antigen was the polysaccharide galactoxylomannan from *C. neoformans*, which was detected by Neil and colleagues in infected animals (Neil and Kapros 1950) and later in humans (Neil et al. 1951). In 1961, Bloomfield et al. developed a simple rapid reactive test for antigen detection in body fluids by latex particle agglutination (LAT), using 0.8 mm particles coated with a polyclonal rabbit anti-*C. neoformans* IgG. This was an extremely sensitive and reliable test when standard conditions were used. The monoclonal antibodies, first reported in 1975 by Kohler and Mildstein, were successfully applied by Dromer et al. (1987) for *C. neoformans* galactoxylomannan antigen detection. Monoclonal antibodies against the mannan of *C. albicans* and galactomannan of *A. fumigatus*, prepared by Stynen (Diagnostics Pasteur 1989), were successfully used in invasive candidiasis (Georges et al. 1991) and aspergillosis (Dupont et al. 1990), respectively; these have high specificity but less sensitivity than the tests for cryptococcosis.

Numerous circulating antigen tests have been performed not only by LAT, but also by EIA and RIA, using monoclonal and polyclonal antisera from cell wall polysaccharides, heat labile glycoproteins (*C. albicans*), cytoplasmic thermostable glycoprotein and 48 kDa protein (*C. albicans*), 18, 32, 92 kDa proteins (*A. fumigatus*) or yeast or spherule forms of dimorphic fungi (*H. capsulatum, C. immitis*). Most of these rapid tests are now commercially available (Drouhet 1993b; Dromer 1993).

SENSITIVITY TESTING OF ANTIFUNGAL AGENTS

Although in bacteriology the necessity of laboratory sensitivity followed soon after the discovery of antibiotics, in 1945–50, by disk sensitivity and minimum inhibitory concentration (MIC) testing under standardized conditions (microdilution techniques), in medical mycology the use of in vitro tests and the assay of drugs in body fluids only started in the 1970s with flucytosine (5-FC); this resulted from the emergence of primary or secondary resistant strains and a need to monitor body fluid levels. The tests have also become available for the azoles since the 1980s. Disk as well as diffusion and microplate automated techniques became easier to carry out and were standardized for the in vitro evaluation of antifungal agents against yeasts and sporulating pathogenic fungi (Drouhet et al. 1986), but not for filamentous nonsporulating fungi or dimorphic fungi (Galgiani et al. 1992). Physicochemical methods were used mainly for assays of drugs in body fluids. The methodology, development, and standardization for these tests have been reviewed (Speller 1980; Drouhet and Dupont 1986; Drouhet et al. 1986; Iwata and Vanden Bossche 1986; Galgiani et al. 1992; Van Cutsem et al. 1994; Vanden Bossche et al. 1994).

FINAL CONCLUSIONS

Since the beginnings of the concerted interest in medical mycology which began at the turn of the last century, with the discovery of the principal agents of human and animal mycoses, there has been constant progress in our knowledge of the pathogenic fungi that cause mycotic diseases.

The opportunistic pathogenic fungi and their resultant mycoses, which have emerged in the last few decades under new environmental and host conditions, have reinforced the place of medical mycology as a key component of microbiology. This became possible as a result of the work of the pioneers of medical mycology: enthusiastic medical mycologists, who founded the ISHAM in Paris in 1954, which, with its affiliates, now has more than 1 000 members in medical mycology throughout the world.

The evolution over the last 50 years in the spectrum of fungal infections, their clinical and pathological features, and the ecology and epidemiology of pathogenic and opportunistic fungi has been accompanied by an increased understanding of the fungal cell. Our knowledge of the physiology, biochemistry, immunology, morphogenesis, ultrastructure, sexuality, genetics, and molecular biology of many pathogenic fungi in humans and animals has advanced enormously, but it still needs further intensive investigation.

The need for education and training in medical mycology, as with all closely related basic and applied scientific disciplines, requires further development, with the objective of resolving the burning issues of fungal disease, such as mycoses in high-risk individuals and immunocompromised patients, immunological defence, the cost-effective management of life-threatening mycoses, and the rapid and early detection of infection.

ACKNOWLEDGMENTS

Sadly Edouard Drouhet died during the production of this chapter. His original material has been revised by Roderick J. Hay.

REFERENCES

Adamson, H.G. 1895. Observations on the parasites of ringworm. *Br J Dermatol*, **7**, 202–11.

Ainsworth, G.C. 1976. *Introduction to the history of mycology*. Cambridge: Cambridge University Press.

Ainsworth, G.C. 1986. *Introduction to the history of medical and veterinary mycology*. Cambridge: Cambridge University Press.

Ainsworth, G.C. and Austwick, P.K.C. 1973. *Fungal diseases of animals*. Farnham Royal: Commonwealth Agricultural Bureau.

Ainsworth, G.C. and Stockdale, P.M. 1984. Biographical notices of deceased medical and veterinary mycologists. *Rev Med Vet Mycol*, **19**, 1–13.

Ajello, L. 1970. The medical mycological iceberg. *Proceedings of the International Symposium on Mycoses. Pan American Health Organization Scientific Publication no. 205*. Washington, DC: PAHO, 3–12.

Ajello, L. 1974. Natural history of the dermatophytes and related fungi. *Mycopathol Mycol Appl*, **53**, 93–110.

Ajello, L. 1975a. Milestones in the history of medical mycology. The dermatophytes. *Recent Advances in Medical and Veterinary Mycology: Proceedings of the 6th International Congress of ISHAM*. Tokyo: University of Tokyo Press, 3–11.

Ajello, L. 1975b. Phaeohyphomycosis: definition and etiology. *Mycoses. Pan American Health Organization Science Publication, no. 304*, PAHO, Washington DC, 126–30.

Ajello, L. 1977. *Coccidioidomycosis. Current clinical and diagnostic status*. Miami, FL: Symposia Specialists.

Ajello, L. 1981. The gamut of human infections caused by dematiaceous fungi. *Jpn J Med Mycol*, **22**, 1–5.

Ajello, L. 1983. Histoplasmosis – a dual entity: histoplasmosis capsulati and histoplasmosis duboisii. *Igiene Mod*, **79**, 3–30.

Ajello, L. 1993. Contributo italiano alla storia della micologia. In: Ajello, L., Farina, C., et al. (eds), *Fondamenti de micologia clinica*. Milan: AMCLI, 13–28.

Ajello, L. and Zeidberg, D. 1951. Isolation of *Histoplasma capsulatum* and *Allescheria boydii* from soil. *Science*, **113**, 662–3.

Ajello, L., Georg, L.K., et al. 1975. *Laboratory manual for medical mycology*, 2nd edition. Atlanta, GA: Centers for Disease Control.

Ajello, L., Padhye, A.A., et al. 1995. Occurrence of *Penicillium marneffei* infections among wild bamboo rats in Thailand. *Mycopathologia*, **131**, 1–8.

Armstrong, D. 1981. Fungal infections in the compromised host. In: Rubin, R.H. and Young, L.S. (eds), *Clinical approach to infection in the compromised host*. New York: Plenum, 67–84.

Austwick, P.C. 1966. The role of spores in the allergies and mycoses of man and animals. *Proceedings of the Eighteenth Symposium, Colston Research Society, University of Bristol*. London: Butterworths, 321.

Axelsen, N.H. 1973. Quantitative immunoelectrophoretic methods as tools for a polyvalent approach to standardization in the immunochemistry of *Candida albicans*. *Infect Immun*, **7**, 949.

Baker, R.D. 1957. The diagnosis of fungus diseases in biopsy. *J Chron Dis*, **5**, 552–70.

Baker, R.D. 1971. *The pathologic anatomy of mycoses*. Berlin: Springer-Verlag.

Baldwin, R.S. 1981. *The fungus fighters. Two women scientists and their discovery*. Ithaca, NY: Cornell University Press.

Benham, R.W. 1935. Cryptococci – their identification by morphology and by serology. *J Infect Dis*, **57**, 255–74.

Benham, R.W. 1939. The cultural characteristics of *Pityrosporum ovale* – a lipophilic fungus. *J Invest Dermatol*, **2**, 187–202.

Benham, R.W. 1950. Cryptococcosis and blastomycosis. *Ann N Y Acad Sci*, **50**, 1299–314.

Benham, R.W. and Georg, L.K. 1948. *Allescheria boydii*, causative agent in a case of meningitis. *J Invest Dermatol*, **10**, 99–110.

Bennett, J.E., Kwon-Chung, K.J. and Howard, D.H. 1977. Epidemiologic differences among serotypes of *Cryptococcus neoformans*. *Am J Epidemiol*, **105**, 582–6.

Bennett, J.E., Kwon-Chung, K.J. and Theodore, T.S. 1978. Biochemical differences between serotypes of *Cryptococcus neoformans*. *Sabouraudia*, **16**, 167–74.

Biguet, J., Van Tran, K., et al. 1964. Analyse immunoélectrophorétique d'extraits cellulaires et de milieux de culture d'*Aspergillus fumigatus* par des immunosérums expérimentaux et des sérums de malades atteints d'aspergillome bronchopulmonaire. *Ann Inst Pasteur, Paris*, **107**, 72.

Bloomfield, N., Gordon, M.A. and Elmendorf, D.F.J. 1961. Dectection of *Cryptococcus neoformans* antigen in body fluids by latex particle agglutination. *Proc Soc Exp Biol Med*, **114**, 64–7.

Bodey, G.P. 1988. The emergence of fungi as major hospital pathogens. *J Hosp Infect*, **11**, Suppl A, 411–26.

Bodey, G.P. and Fainstein, V. 1988. *Candidiasis*. New York: Raven Press.

Bodey, G.P. and Fainstein, V. 1993. *Candidiasis*, 2nd edition. New York: Raven Press.

Bodin, E. 1902. *Les champignons parasites de l'homme*. Paris.

Borgers, M. and Vanden Bossche, H. 1982. The mode of action of antifungal drugs. In: Levine, H.B. (ed.), *Ketoconazole in the management of fungal disease*. Sydney: ADIS Press, 25–47.

Bridges, C.H. and Emmons, C.W. 1961. A phycomycosis of horse caused by *Hyphomyces destruens*. *J Am Vet Med Assoc*, **138**, 579–89.

Brumpt, E. 1906. Les mycétomes. *Arch Pathol*, **10**, 489–572.

Brumpt, E. 1910. *Précis de parasitologie*. Paris: Masson.

Brumpt, E. 1949. *Précis de parasitologie*, 6th edition. Paris: Masson.

Brygoo, E.R. and Segretain, G. 1960. Etude clinique, épidémiologique et mycologique de la chromoblastomycose à Madagascar. *Bull Soc Exot*, **53**, 443–75.

Buckley, H.R., Richardson, M.D., et al. 1992. Immunodiagnosis of invasive fungal infection. *J Med Vet Mycol*, **30**, Suppl 1, 249–60.

Bullen, J.J. 1949. The yeast-like form of *Cryptococcus farciminosus* (Rivolta) (*Histoplasma farcininosum*). *J Pathol Bacteriol*, **61**, 117–20.

Bullock, N.E. and Deepe, G.S. 1989. Medical mycology in crisis. *J Lab Clin Med*, **102**, 685–93.

Bulmer, G.S. and Sans, M.D. 1967. *Cryptococcus neoformans*. II. Phagocytosis by human leucocytes. *J Bacteriol*, **94**, 1480–3.

Bulmer, G.S. and Sans, M.D. 1968. *Cryptococcus neoformans*. III. Inhibition of phagocytosis. *J Bacteriol*, **95**, 5–8.

Buschke, A. 1895. Uber eine durch Coccidien hervorgerufene Krankheit des Menschen. *Dtsch Med Wochenschr*, **21**, 14.

Busse, O. 1894. Uber parasitare Zellenschlüsse und ihre Züchtung. *Zentralbl Bakteriol*, **167**, 175–80.

Busse, O. 1895. Uber Saccharomycosis Hominis. *Virchows Arch A*, **140**, 23–46.

Capponi, M., Sureau, P. and Segretain, G. 1956. Penicilliose de *Rhizomys sinensis*. *Bull Soc Pathol Exot*, **49**, 418–21.

Castellani, A. and Chalmers, A.I. 1910. *Manual of tropical medicine*. New York: William Wood & Co..

Castellani, A. and Chalmers, A.I. 1913. *Manual of tropical medicine*, 2nd edition. New York: William Wood & Co..

Castellani, A. and Chalmers, A.I. 1919. *Manual of tropical medicine*, 3rd edition. New York: William Wood & Co..

Chandler, F.W. 1985. Pathology of the mycoses in patients with the acquired immunodeficiency syndrome (AIDS). *Curr Topics Med Mycol*, **1**, 1–23.

Chandler, F.W., Kaplan, W. and Ajello, L. 1980. *Color atlas and text of the histopathology of mycotic diseases*. Chicago: Year Book Medical Publisher, 333.

Ciferri, R. and Montemartini, A. 1959. Taxonomy of *Haplosporangium parvum*. *Mycopathol Mycol Appl*, **10**, 303–16.

Cole, G.T. 1986. Preparation of microfungi for scanning electron microscopy. In: Aldrich, H.C. and Todd, W.J. (eds), *Ultrastructure techniques for microorganisms*. New York: Plenum Press, 128–38.

Cole, G.T., Kruse, D., et al. 1993. Factors regulating morphogenesis in *Coccidioides immitis*. In: Vanden Bossche, H., Odds, F.C. and Kerridge, D. (eds), *Dimorphic fungi in biology in medicine*. New York: Plenum Press, 191–212.

Conant, N.F. 1936a. Studies on the genus *Microsporum*, I Cultural studies (1937). *Arch Dermatol Syphilol Chicago*, **33**, 665–83.

Conant, N.F. 1936b. Studies on the genus *Microsporum*, II Biometric studies. *Arch Dermatol Syphilol Chicago*, **34**, 79–89.

Conant, N.F. 1937a. Studies on the genus *Microsporum*, III Taxonomic studies. *Arch Dermatol Syphilol Chicago*, **35**, 791–808.

Conant, N.F. 1937b. The occurrence of a human pathogenic fungus as a saprophyte in nature. *Mycologia*, **29**, 597–8.

Conant, N.F. 1941. A cultural study of the life-cycle of *Histoplasma capsulatum* Darling 1906. *J Bacteriol*, **41**, 563–80.

Conant, N.F. and Howell, A. Jr 1942. The similarity of the fungi causing South American blastomycosis (paracoccidioidal granuloma) and North American blastomycosis (Gilchrist's disease). *J Invest Dermatol*, **5**, 353.

Conant, N.F. and Martin, D.S. 1937. The morphologic and serologic relationships of various fungi causing dermatitis verrucosa (chromoblastomycosis). *Am J Trop Med*, **17**, 553–78.

Conant, N.F., Martin, D.S., et al. 1944. *Manual of clinical mycology*. London: W.B. Saunders.

Conant, N.F., Smith, D.T., et al. 1954. *Manual of clinical mycology*, 2nd edition. London: W.B. Saunders.

Conant, N.F., Martin, D.S., et al. 1971. *Manual of clinical mycology*, 3rd edition. London: W.B. Saunders.

Creitz, J. and Harris, H.W. 1955. Isolation of *Allescheria boydii* from sputum. *Am Rev Tuberc*, **71**, 126–30.

Curtis, F. 1896. Contribution à l'étude de la saccharomycose humaine. *Ann Inst Pasteur*, **10**, 449–68.

Cushion, M.T., Harmsen, A., et al. 1994. Recent advances in the biology of *Pneumocystis carinii*. *J Med Vet Mycol*, **32**, Suppl 1, 217–28.

Dampier, W. 1729. *A new voyage around the world*. London, 228–9.

Dankner, W.M., Spector, S.A., et al. 1987. *Malassezia furfur* in neonates and adults: complication of hyperalimentation. *Rev Infect Dis*, **9**, 743–53.

Darling, S.T.A. 1906. A protozoan general infection producing pseudotubercles in the lungs and focal necroses in the liver, spleen and lymph mode. *JAMA*, **46**, 1283–5.

da Silva Lacaz, C. 1983. Historia de micologia médica no Brasil. *Ann Bras Derm*, **58**, 265–70.

Davies, R.R., Spencer, H. and Wakelin, P.O. 1964. A case of human protothecosis. *Trans R Soc Trop Med Hyg*, **58**, 448–51.

Davis, D.J. 1921. The identity of American and French sporotrichoses. *Univ Wis Stud Sci*, **2**, 104–30.

Dawson, C.O. and Gentles, J.C. 1959. The perfect stage of *Keratinomyces ajelloi*. *Nature (Lond)*, **183**, 1345–6.

de Beurmann, L. and Gougerot, H. 1912. *Les sporotrichoses*. Paris: Alcan.

de Camargo, Z.P., Guesdon, J.C. and Drouhet, E. 1984. Enzyme-linked immunoabsorbent assay (ELISA) in paracoccidioidomycosis. Comparison with counterimmuno electrophoresis and erythroimmunoassay. *Mycopathologia*, **88**, 31–7.

de Cock, A.W.A.M., Mendoza, L, et al. 1987. *Pythium insidiosum* n sp, the etiologic agent of pythiosis. *J Clin Microbiol*, **33**, 317–23.

DeMonbreun, W.A. 1934. The cultivation and cultural characteristics of Darling's *Histoplasma capsulatum*. *Am J Trop Med*, **14**, 93–125.

Deng, Z., Ribas, J.L., et al. 1988. Infections caused by *Penicillium marneffei* in China and Southeast Asia: review of eighteen published cases and report of four more Chinese cases. *Rev Infect Dis*, **10**, 640.

de Repentigny, L. and Reiss, E. 1984. Current trends on immunodiagnosis of candidiasis and aspergillosis. *Rev Infect Dis*, **6**, 301–12.

de Repentigny, L., Kaufman, L., et al. 1994. Immunodiagnosis of invasive fungal infections. *J Med Vet Mycol*, **32**, 239–52.

DiSalvo, A.F., Fickling, A.M. and Ajello, L. 1973. Infection caused by *Penicillium marneffei*: description of first natural infection in man. *Am J Clin Pathol*, **60**, 259–63.

Doby-Dubois, M. 1964. Premier cas human d'adiaspiromycose par *Emmonsia crescens*. *Bull Soc Pathol Exot*, **57**, 204–44.

Dodge, C.W. 1935. *Medical mycology*. St Louis, MO: CV Mosby Co.

Dromer, F. 1993. The detection of fungal antigenaemia. In: Spencer, R.C., Wright, E.P. and Newsom, S.W.B. (eds), *Rapid methods and automation in microbiology and immunology*. Andover: Intercept Ltd, 13–22.

Dromer, F., Salamero, J., et al. 1987. Production, characterisation and antibody specificity of a mouse monoclonal antibody reactive with *Cryptococcus neoformans* capsular polysaccharide. *Infect Immunol*, **55**, 742–8.

Drouhet, E. 1952. II – Action des acides aminés sur la morphogénèse et la croissance des dermatophytes. *Ann Inst Pasteur*, **82**, 348–56.

Drouhet, E. 1955. The status of fungus diseases in France. In *Therapy of fungus diseases, an international symposium, University of California at Los Angeles*. Boston, MA: Little, Brown & Co., 43–53.

Drouhet, E. 1957. Quelques aspects biologiques et mycologiques des histoplasmoses. *Pathol Biol (Paris)*, **33**, 439–91.

Drouhet, E. 1972. Champignons opportunistes et mycoses iatrogènes. *Bull Inst Pasteur*, **70**, 391–464.

Drouhet, E. 1988. Overview on fungal antigens. In: Drouhet, E., Cole, G., et al. (eds), *Fungal antigens, First International Symposium on Fungal Antigens (Paris)*. New York: Plenum Press, 1–36.

Drouhet, E. 1989. African histoplasmosis. In: Hay, R. (ed.), *Tropical fungal infections. Baillère's clinical tropical medicine and communicable diseases*, vol. 4, no. 1. London: Baillière Tindall, 221–47.

Drouhet, E. 1991. Education and training in medical mycology with an introduction to medical mycology. In: Arora, D.K., Ajello, L. and Mukerji, K.G. (eds), *Handbook of applied mycology humans, animals, and insects*, 2nd edition. New York: Marcel Dekker Inc., 1–50.

Drouhet, E. 1993a. Penicilliosis due to *Penicillium marneffei*: a new emerging systemic mycosis in AIDS patients travelling or living in Southeast Asia. Review of 44 cases reported in HIV infected patients during last 5 years compared to 44 cases HIV negative patients reported over 20 years. *J Mycol Med (Paris)*, **3**, 195–224.

Drouhet, E. 1993b. Rapid tests for immunodiagnosis of invasive opportunistic mycoses. *Rev Iberoam Micol*, , Suppl 1, S53–67.

Drouhet, E. and Dupont, B. 1980. Chronic mucocutaneous candidiasis and the other superficial and systemic mycoses successfully treated with ketoconazole. *Rev Infect Dis*, **2**, 606–19.

Drouhet, E. and Dupont, B. 1983. Laboratory and clinical assessment of ketoconazole in deep seated mycoses. *Am J Med*, **74**, Suppl 1B, 30–47.

Drouhet, E. and Dupont, B. 1986. Evaluation of systemic antifungal agents in body fluids by sensitive bioassays. Clinical application in monitoring deep mycoses, in vitro and in vivo. In: Iwata, K. and Vanden Bossche, H. (eds), *Evaluation of antifungal antigens*. New York: Elsevier Science, 174–81.

Drouhet, E. and Dupont, B. 1990. Mycoses in AIDS patients: an overview. In: Vanden Bossche, H., Mackenzie, D.W.R., et al. (eds), *Mycoses in AIDS patients*. New York: Plenum Press, 27–54.

Drouhet, E. and Dupont, B. 1995. Infection due *Penicillium marneffei*: systemic mycosis with cutaneous manifestations associated to AIDS. *J Mycol Med (Paris)*, **5**, Suppl 1, 21–34.

Drouhet, E. and Mariat, F. 1952a. Recherche sur la nutrition des dermatophytes. I – Etude des besoins vitaminiques. *Ann Inst Pasteur*, **82**, 337–47.

Drouhet, E. and Mariat, F. 1952b. Etude des facteurs déterminant le développement de la phase levure de *Sporotrichum schenckii*. *Ann Inst Pasteur*, **83**, 506–14.

Drouhet, E. and Mariat, F. 1953. Nutrition et facteurs de croissance des dermatophytes. Symposium Nutrition and Growth Factors, Istituto Superiore di Sanità, Roma, 113–40.

Drouhet, E. and Segretain, G. 1951. Inhibition de la migration leucocytaire in vitro par un polyoside capsulaire du *Torulopsis neoformans*. *Ann Inst Pasteur*, **81**, 674–6.

Drouhet, E., Segretain, G. and Aubert, J.P. 1950. Polyoside capsulaire d'un champignon pathogène, *Torulopsis neoformans*. Relation avec la virulence. *Ann Inst Pasteur*, **79**, 891–901.

Drouhet, E., Camey, L. and Segretain, G. 1972. Valeur de l'immunoprécipitation et de l'immunofluorescence indirect dans les aspergilloses bronchopulmonaires. *Ann Inst Pasteur*, **123**, 379–95.

Drouhet, E., Dompmartin, D., et al. 1980. Dermatite expérimentale à *Pityrosporum ovale* et (ou) *Pityrosporum orbiculare* chez le cobaye et la souris. *Sabouraudia*, **18**, 149–56.

Drouhet, E., Dupont, B., et al. 1981. Une nouvelle pathologie: candidose folliculaire et nodulaire avec des localisations ostéoarticulaires et oculaires au cours des septicémies à *Candida albicans* chez les héroïnomanes. Mono et polythérapie antifongiques. *Bull Soc Fr Mycol Med*, **10**, 179–84.

Drouhet, E., Dupont, B. and Dikeacou, T. 1985. Antifungal agents and immunity. In: Pulverer, G. and Jeliaszewicz, J. (eds), *Chemotherapy and immunity*. Stuttgart: Gustav Fischer Verlag, 1–38.

Drouhet, E., Dupont, B., et al. 1986. Disc agar diffusion and microplate automatized technics for *in vitro* evaluation of antifungal agents on yeasts and sporulated pathogenic fungi *in vitro* and *in vivo*. In: Iwata, K. and Vanden Bossche, H. (eds), *Evaluation of antifungal antigens*. Amsterdam: Elsevier Science, 202–13.

Drouhet, E., Ravisse, P., et al. 1988a. Pouvoir pathogène expérimental des levures lipophiles. Nouvelles données expérimentales. *Bull Soc Fr Mycol Med*, **17**, 255–66.

Drouhet, E., Ravisse, P., et al. 1988b. Etude mycologique, ultrstructurale et expérimentale sur Penicillium marneffei isolé d'une pénicilliose disséminée chez un SIDA. *Bull Soc Fr Mycol Med*, **17**, 77–82.

Drouhet, E., Dupont, B. and Ravisse, P. 1991. Experimental scedosporiosis with a high virulent strain of *Scedosporium inflatum* isolated from a knee arthritis. *J Mycol Med*, **1**, 16–20.

Dubois, A., Janssens, P.G. and Brutsaert, P. 1952. Un cas d'histoplasmose africaine, avec une note mycologique sur *Histoplasma duboisii* n. sp., par R Vanbreuseghem. *Ann Soc Belg Med Trop*, **32**, 569–84.

Duclaux, E. 1886. Sur le *Trichophyton tonsurans*. *C R Soc Biol Paris*, **38**, 14–16.

Duncan, J.T. 1947. A unique form of *Histoplasma*. *Trans R Soc Trop Med Hyg*, **40**, 364–5.

Duncan, J.T. 1958. Tropical African histoplasmosis. *Trans R Soc Trop Med Hyg*, **52**, 468–74.

Dupont, B. 1989. Cryptococcosis. In: Hay, R. (ed.), *Tropical fungal infections. Baillière's clinical tropical and communicable disease*, vol. 4, no. 1. . London: Baillière Tindall, 119–52.

Dupont, B. and Drouhet, E. 1985. Cutaneous, ocular, and osteoarticular candidiasis in heroin addicts: new clinical and therapeutic aspects in 38 patients. *J Infect Dis*, **152**, 577–91.

Dupont, B. and Drouhet, E. 1988. Fluconazole in the management of oropharyngeal candidosis in a predominantly HIV antibody-positive group of patients. *J Med Vet Mycol*, **26**, 67–71.

Dupont, B., Improvisi, L. and Provost, F. 1990. Detection de galactomannane dans les aspergilloses invasives humaines et animales avec un test au latex. *Bull Soc Fr Mycol Méd*, **19**, 35–41.

Dupont, B., Improvisi, L. and Ronin, O. 1991. Aspects épidémiologiques et cliniques des infections à *Scedosporium* et à *Pseudallescheria*. *J Mycol Med*, **1**, 33–42.

Dupont, B., Denning, D.W., et al. 1994. Mycoses in AIDS patients. *J Med Vet Mycol*, **32**, Suppl 1, 65–78.

Durandy, A., Fischer, A., et al. 1987. Mannan-specific and mannan-induced T-cell suppressive activity in patients with chronic mucocutaneous and candidiasis. *J Clin Immunol*, **7**, 400–9.

Echavarria, E., Cano, E.L. and Restrepo, A. 1993. Disseminated adiaspiromycosis in a patient with AIDS. *J Med Vet Mycol*, **31**, 91–7.

Edman, J.C., Kovacs, J.A., et al. 1988. Ribosomal RNA sequences shows *Pneumocystis carinii* to be a member of the fungi. *Nature (Lond)*, **334**, 519–22.

Eichstedt, C. 1846. Pilzbildung in der Pityriasis versicolor. *Neue Not Geb Nature-u-Heilk*, **853**, 270–1.

Emmons, C.W. 1934. Dermatophytes: natural groupings based on the form of spores and accessory organs. *Arch Dermatol Syphilol*, **30**, 337–62.

Emmons, C.W. 1942. Isolation of *Coccidioides immitis* from soil and rodents. *Public Health Rep*, **57**, 109–11.

Emmons, C.W. 1949. Isolation of *Histoplasma capsulatum* from soil. *Public Health Rep*, **64**, 892–6.

Emmons, C.W. 1955. Saprophytic sources of *Cryptococcus neoformans* associated with the pigeon (*Columbia livia*). *Am J Hyg*, **62**, 227–32.

Emmons, C.W. 1958. Association of bats with histoplasmosis. *Public Health Rep*, **73**, 590–5.

Emmons, C.W. 1962. Natural occurrence of opportunistic fungi. *Lab Invest*, **11**, 1026–32.

Emmons, C.W. and Ashburn, L.L. 1942. The isolation of *Haplosporangium parvum* n. sp. and *Coccidioides immitis* from wild rodents. Their relationship to coccidioidomycosis. *Public Health Rep*, **57**, 1715–27.

Emmons, C.W. and Jellison, W.L. 1960. *Emmonsia crescens* sp. n. and adiaspiromycosis (haplomycosis in mammals). *Ann N Y Acad Sci*, **89**, 91–101.

Emmons, C.W., Binford, C.H. and Utz, J.P. (eds), 1963. *Medical mycology*, 1st edition. Philadelphia: Lea & Febiger.

Emmons, C.W., Binford, C.H. and Utz, J.P. (eds), 1970. *Medical mycology*, 2nd edition. Philadelphia: Lea & Febiger.

Emmons, C.W., Binford, C.H. and Utz, J.P. (eds), 1977. *Medical mycology*, 3rd edition. Philadelphia: Lea & Febiger.

Espinel-Ingroff, A. 1996. History of medical mycology in the United States. *Clin Microbiol Rev*, **9**, 235–72.

Evans, E.E. 1950. The antigenic composition of *Cryptococcus neoformans*. I. A serologic classification by means of the capsular agglutinations. *J Immunol*, **64**, 423–30.

Evans, E.E. and Kessel, J.F. 1951. The antigenic composition of *Cryptococcus neoformans*. II. Serologic studies with the capsular polysaccharide. *J Immunol*, **67**, 109–14.

Fiese, M.F. 1958. *Coccidioidomycosis*. Springfield, IL: Charles C Thomas.

Fox, T.W. 1863. *Skin diseases of parasitic origin*, London

Fraser, D.W., Ward, J.I., et al. 1979. Aspergillosis and other systemic mycoses. The growing problem. *JAMA*, **242**, 1631–5.

Furcolow, M.L. 1958. Recent studies on the epidemiology of histoplasmosis. *Ann N Y Acad Sci*, **72**, 127–64.

Galgiani, J.N., Rinaldi, M.G., et al. 1992. Standardization of antifungal susceptibility testing. *J Med Vet Mycol*, **30**, Suppl 1, 213–24.

Garrisson, R.G. 1983. Ultrastructural cytology of pathogenic fungi. In: Howard, D.H. (ed.), *Fungi pathogenic for humans and animals. Part A, Biology*. New York: Marcel Dekker, 229–322.

Gatti, F. and Eeckels, H. 1970. An atypical strain of *Cryptococcus neoformans* (Sanfelice) Vuillemin 1894. Part 1. Description of the disease and of the strain. *Ann Soc Belge Méd Trop*, **50**, 689–94.

Gentles, J.C. 1958. Experimental ringworm in guinea pigs, oral treatment with griseofulvin. *Nature (Lond)*, **182**, 476–7

Gentles, J.C. and Evans, E.G.V. 1970. Infection of the feet and nails with *Hendersonula toruloidea*. *Sabouraudia*, **8**, 72–5.

Georg, L.K. 1950. The relation of nutrition to the growth and morphology of *Trichophyton faviforme*. *Mycologia*, **42**, 693–716.

Georg, L.K. and Camp, L.B. 1957. Routine nutritional tests for the identification of dermatophytes. *J Bacteriol*, **74**, 477–90.

Georg, L.K., Ajello, L. and Papageorge, C. 1954. Use of cycloheximide in the selective isolation of fungi pathogenic to man. *J Lab Clin Med*, **44**, 422–8.

Georges, E., Garrigues, M.L., et al. 1991. Diagnostic des candidoses systémiques par un test au latex: résultat d'une année de suivi sérologique chez les malades à risques. *J Mycol Méd*, **1**, 25–8.

Gilchrist, T.C. and Stokes, W.R. 1896. A case of pseudolupus caused by *Blastomyces*. *J Exp Med*, **3**, 53–78.

Gomori, G. 1946. A new histochemical test for glycogen and mucin. *Am J Clin Pathol*, **10**, 177–9.

Gordon, M. 1951. Lipophilic yeast-like organism associated with tinea versicolor. *J Invest Dermatol*, **17**, 267–72.

Gottlieb, M.S., Schrodd, R., et al. 1981. *Pneumocystis carinii* pneumonia and mucosal candidiasis in previously healthy homosexual men. Evidence of a new acquired cellular immunodeficiency. *N Engl J Med*, **A305**, 1425.

Graybill, J.R. 1988. Histoplasmosis and AIDS. *J Infect Dis*, **158**, 623–6.

Gregory, J.E., Golden, A., et al. 1943. Mucormycosis of the central nervous system: report of three cases. *Bull Johns Hopkins Hosp*, **73**, 405–19.

Grocott, R.C. 1955. A strain for fungi in tissue and smears using Gomori's methenamine-silver nitrate technic. *Am J Clin Pathol*, **25**, 975–9.

Gruby, D. 1841. Sur les mycodermes qui constituent la teigne faveuse. *C R Acad Sci*, **13**, 309–12.

Guého, E. and de Hoog, G.S. 1991. Taxonomy of the medical species of *Pseudallescheria* and *Scedosporium*. *J Mycol Méd*, **118**, 329.

Guého, E. and Meyer, S.A. 1988. A reevaluation of the genus *Malassezia* by means of genome comparison. *Antonie van Leewenhoeck*, **54**, 245–51.

Guého, E., Simmons, R.B., et al. 1987. Association of *Malassezia pachydermatis* with systemic infections of humans. *J Clin Microbiol*, **25**, 1789–90.

Guého, E., Faergemann, J., et al. 1994. *Malassezia* and *Trichosporon*: two emerging pathogenic basidiomycetous yeast-like fungi. *J Med Vet Mycol*, **30**, Suppl, 367–78.

Guesdon, J.L., de Camargo, P.Z., et al. 1988. Enzyme-linked immunosorbent assay and related sensitive assays for antigen and antibody detection. In: Drouhet, E., Cole, G., et al. (eds), *Fungal antigens. First International Symposium on Fungal Antigens*. New York: Plenum Press, 32–9.

Guignani, H.C., Muotol-Drafor, F.A., et al. 1994. A natural focus of *Histoplasma capsulatum* var *duboisii* in a bat cave. *Mycopathologia*, **127**, 151–7.

Guillot, J., Guého, E. and Prévost, M.C. 1995. Ultrastructural features of the dimorphic yeast *Malassezia furfur*. *J Mycol Méd*, **5**, 86–91.

Hasegawa, A. and Usui, K. 1975. *Nannizzia otae* sp. nov., the perfect state of *Microsporum canis* Bodin. *Jpn J Med Mycol*, **6**, 148–53.

Hay, R.J. 1991. Clinical manifestations and management of superficial fungal infection in the compromised patient. In: Warnock, D.U. and Richardson, M.D. (eds), *Fungal infection in the compromised patient*, 2nd edition. Chichester: John Wiley & Sons, 232–45.

Hay, R.J. and Midgley, G. 1988. Pathogenic mechanisms of *Pityrosporum* infection, seborrhoeic dermatitis and dandruff. A fungal disease. In Shuster, S. and Blatchford, N. (eds), *Royal Society of Medicine International Congress and Symposium Series*, no. 132. London: Royal Society of Medicine Services Ltd.

Hay, R.J. and Moore, M.K. 1984. Clinical features of superficial fungal infections caused by *Hendersonula toruloidea* and *Scytalidium hyalinum*. *Br J Dermatol*, **110**, 677–83.

Hinson, K.F.W., Moon, A.J. and Plummer, N.S. 1952. Bronchopulmonary aspergillosis: a review and a report of eight new cases. *Thorax*, **7**, 317–33.

Hooke, R. 1665. *Micrographia*. London.

Howard, D.H. 1967. Effect of temperature on the intracellular growth of *Histoplasma capsulatum*. *J Bacteriol*, **89**, 518–23.

Howard, D.H. 1983. *Fungi pathogenic for humans and animals. Part A Biology*. New York: Marcel Dekker.

Huntington, R.W. 1985. Four great coccidioidomycologists: William Ophuls (1871-1933), Myrnie Gifford (1892-1966), Charles Edward Smith (1904-1967) and William A Winn (1903-1967). *Sabouraudia*, **23**, 361–70.

Iwata, K. and Vanden Bossche, G. 1986. *In vitro* and *in vivo* evaluation of antifungal agents. *Second Topic Symposium on Medical Mycology*. Amsterdam: Elsevier.

Jarvis, W.R. 1995. Epidemiology of nosocomial fungal infections with emphasis on *Candida* species. *Clin Infect Dis*, **20**, 1526–30.

Jellison, W.L. 1969. *Adiaspiromycosis (= Haplomycosis)*. Missoula: Montana Mountain Press.

Jellison, W.L., Drouhet, E., et al. 1961. Adiaspiromycose (haplomycose) chez les mammifères sauvages en France. *Ann Inst Pasteur*, **100**, 747–52.

Kaplan, W. and Kaufman, L. 1961. The application of fluorescent antibody techniques to medical mycology: a review. *Sabouraudia*, **1**, 137–44.

Kaufman, L. and Blumer, S. 1978. Cryptococcosis: the awakening giant. *Proceedings of the 4th International Conference on the Mycoses. The Black and White Yeasts, Brasilia*. Pan American Health Organization Scientific Publication no. 353. Washington, DC: PAHO, 176–82.

Kaufman, L. and Reiss, E. 1986. Serodiagnosis of fungal diseases. In: Rose, N.R. and Friedman, H. (eds), *Manual of clinical immunology*, 3rd edition. Washington, DC: American Society for Microbiology, 212–9.

Kaufman, L., Standard, P.G. and Padhye, A.A. 1983. Exoantigen tests for the identification of fungal cultures. *Mycopathologia*, **82**, 3.

Kennedy, M.J. 1988. Adhesion and association mechanisms of *Candida albicans*. In: McGinnis, M.R. (ed.), *Current topics in medical mycology*. New York: Springer-Verlag, 73–169.

Kennedy, M.J., Calderone, R.A. and Cutler, J.E. 1992. Molecular basis of *Candida albicans* adhesion. *J Med Vet Mycol*, **30**, Suppl 1, 95–112.

Kirkpatrick, C.H., Rich, R.R. and Bennett, J.E. 1971. Chronic mucocutaneous candidiasis: model building in cellular immunity. *Ann Intern Med*, **75**, 955–78.

Klein, R.S., Harris, C.A., et al. 1984. Oral candidiasis in high risk patients as the initial manifestation of the acquired immunodeficiency syndrome. *N Engl J Med*, **311**, 354–8.

Kligman, A.M. and Mescon, H. 1950. The periodic acid-Schiff stain for the demonstration of fungi in animal tissues. *J Bacteriol*, **60**, 415.

Kovacs, J.A., Kovacs, A.A., et al. 1985. Cryptococcosis in the acquired immunodeficiency syndrome. *Ann Intern Med*, **103**, 533–8.

Kruger, W. 1894. Kurze characteristick einiger niederer einen neuuer pilz-typus repräsentiert duch die gattung *Prototheca moriformis* et *P. zopfii*. *Hedwigt*, **33**, 241–51.

Kwon-Chung, K.J. 1972. *Emmonsiella capsulata*: perfect state of *Histoplasma capsulatum*. *Science*, **177**, 368–9.

Kwon-Chung, K.J. 1975. A new genus *Filobasidiella* the perfect state of *Cryptococcus neoformans*. *Mycologia*, **67**, 1197–200.

Kwon-Chung, K.J. 1976. A new species of *Filobasidiella*, the sexual state of *Cryptococcus neoformans* B and C serotypes. *Mycologia*, **68**, 821–33.

Kwon-Chung, K.J. and Bennett, J.E. 1992. *Medical mycology*. Philadelphia: Lea & Febiger.

Kwon-Chung, K.J., Lehman, D., et al. 1985. Genetic evidence for the role of extracellular proteinase in virulence of *Candida albicans*. *Infect Immun*, **49**, 571–5.

Kwon-Chung, K.J., Kozel, T.R., et al. 1992. Recent advances in biology and immunology of *Cryptococcus neoformans*. *J Med Vet Mycol*, **30**, Suppl 1, 133–42.

Kwon-Chung, K.J., Pfeiffer, T., et al. 1994. Molecular biology of *Cryptococcus neoformans* and therapy of cryptococcosis. *J Med Vet Mycol*, **32**, Suppl 1, 407–16.

Langeron, M. 1945. *Précis de mycologie, mycologie générale, mycologie médicale*. Paris: Masson.

Langeron, M. 1952. *Précis de mycologie, mycologie générale, mycologie médicale*, 2nd edition. Vanbreuseghem, R. (ed.). Paris: Masson.

Langeron, M. and Milochevitch, S. 1930. Morphologie des dermatophytes sur milieux naturels et milieux à base de polysaccharides. Essai de classification (deuxième mémoire). *Ann Parasitol Hum Comp*, **8**, 465–508.

Littman, M.L. and Zimmerman, L.E. 1956. *Cryptococcosis*. New York: Grune & Stratton.

Lodder, J. and Kreger-van Rij, N.J.W. 1952, 1970. *The yeasts, a taxonomic study*. Amsterdam: North Holland Publishing Co.

Longbottom, J.M. 1986. Applications of immunological methods in mycology. In: Weir, D.M. (ed.), *Handbook of experimental immunology, 4, Applications of immunological methods in biomedical sciences*. Oxford: Blackwell, 12.4–12.30.

Longbottom, J.M. and Austwick, P.K.C. 1986. Fungal antigens. In: Weir, D.M. (ed.), *Handbook of experimental immunology, 1, Immunochemistry*. Oxford: Blackwell, 7.1–7.

Lott, T.J., Magee, P.T., et al. 1992. The molecular genetics of *Candida albicans*. *J Med Vet Mycol*, **30**, Suppl 1, 77–86.

Lutz. 1908. Uma mycose pseudococcidioidica localizad ne bocco no Brasil contribuicao ao conhecimento ds hyphablastomycoses americanas. *Brasil Med*, **22**, 141–144.

Mackenzie, D.W.R. 1983. Serodiagnosis. In: Howard, D.H. (ed.), *Fungi pathogenic for humans and animals. Part B, Pathogenicity and detection I*. New York: Marcel Dekker, 121–218.

Mackenzie, D.W.R. 1992. Presidential address. Proceedings of the XI Congress of the ISHAM. *J Med Vet Mycol*, **30**, Suppl 1, 1–7.

Mackenzie, D.W.R. 1996. Diagnosis of fungal disease. In: Kibbler, C.C., Mackenzie, D.W.R. and Odds, F.C. (eds), *Principles and practice of clinical mycology*. Chichester: John Wiley & Sons Ltd, 23–34.

Malassez, L. 1974. Note sur le champignon du pityriasis simple. *Arch Physiol Norm Pathol (ser 2)*, **1**, 451–64.

Malloch, D. and Salkin, I. 1984. A new species of *Scedosporium* associated with osteomyelitis in humans. *Mycotaxon*, **21**, 247–55.

Malpighi, M. 1679. *Anatome plantarum pars altera*. London, 64–7.

Margulis, L. and Schwartz, K.V. 1988. *Five Kingdoms*, 2nd edition. New York: Freeman & Co., 1–376.

Mariat, F. 1964. Saprophytic and parasitic morphologic of pathogenic fungi. In: Smith, H. (ed.), *Microbial behavior, in vivo and in vitro*. Cambridge: Cambridge University Press, 85–111.

Mariat, F. 1968. The epidemiology of sporotrichosis. In: Wolstenholme, G.E.W. (ed.), *Systemic mycoses*. London: A. Churchill, 144–59.

Mariat, F. 1977. Taxonomic problems related to the fungal complex *Sporothrix schenckii/Ceratocystis* spp.. In: Iwato, K. (ed.), *Recent advances in medical and veterinary mycology*. Baltimore/London: University Park Press, 265–77.

Mariat, F., Destombes, P. and Segretain, G. 1977. The mycetomas: clinical features, pathology, etiology and epidemiology. *Contrib Microbiol Immunol*, **4**, 1–39.

Matakieff, E. 1899. Le pityriasis versicolor et son parasite. Thesis, University of Nancy.

Matsumoto, T., Ajello, L., et al. 1994. Developments in hyalohyphomycosis and phaeohyphomycosis. *J Med Vet Mycol*, **32**, Suppl 1, 329–50.

McDonough, E.S. and Lewis, A. 1967. *Blastomycoses dermatitidis*: production of the sexual stage. *Science*, **156**, 528–9.

McGinnis, M.R. 1980. *Laboratory handbook of medical mycology*. New York: Academic Press.

McGinnis, M.R. 1983. Chromoblastomycosis and phaeohyphomycosis: new concepts, diagnosis and mycology. *J Am Acad Dermatol*, **8**, 1–6.

McGinnis, M.R., Padhye, A.A. and Ajello, L. 1982. *Pseudallescheria* Negroni et Fischer 1943, and its later synonym *Petriellidium* Malloch 1970. *Mycotaxon*, **14**, 94–102.

McGinnis, M.R., Midez, J., et al. 1993. Necrotizing fasciitis caused by *Apophysomyces elegans*. *J Mycol Med*, **3**, 175–9.

Mendoza, L., Ajello, L. and McGinnis, M.R. 1996. Infections caused by the oomycetous pathogen *Phythium insidiosum*. *J Mycol Med*, **6**, 151–64.

Micheli, P.A. 1729. *Nova plantarum genera juxta tournefortü methodum disposita*. Florence, 234.

Midgley, G. 1989. The diversity of *Pityrosporum* (*Malassezia*) yeasts in vivo and in vitro. *Mycopathologia*, **106**, 143–53.

Misra, P.C., Srivastava, K.J. and Lata, K. 1979. *Apophysomyces*, a new genus of Mucorales. *Mycotaxon*, **8**, 377–82.

Mitchell, T.G., Sandin, R.L., et al. 1994. Molecular mycology: DNA probes and applications of PCR technology. *J Med Vet Mycol*, **32**, Suppl 1, 351–66.

Monod, O., Pesle, G. and Segretain, G. 1954. Sur une forme nouvelle d'aspergillose pulmonaire: l'aspergillome bronchiectasiant. *Bull Acad Nat Med (Paris)*, **135**, 508.

Murphy, J.W., Wu-Hsieh, B.A. and Singer-Vermes, L.M. 1994. Cytokines in the host response to mycotic agents. *J Med Vet Mycol*, **32**, Suppl 1, 203–10.

Nannizzi, A. 1927. Ricerche sull'origine saprofitica dei funghi delle tigne. II *Gymnoascus gypseum* sp.n, forma ascofora del *Sabouradites* (*Achorion*) *gypseum* (Bodin) Ota et Langeron (Nota preventiva). *Atti R Acad Fissiocritici Siena Ser X*, **2**, 89.

Neil, J.M. and Kapros, C.E. 1950. Serological tests on soluble antigens from mice infected with *Cryptococcus neoformans* and *Sporotrichum schenkii*. *Proc Soc Exp Biol Med*, **72**, 557.

Neil, J.M., Sugg, J.Y. and McCuley, D.W. 1951. Serologically reactive material in spinal fluid, blood and urine from a human case of cryptococcosis (torulosis). *Proc Soc Exp Biol Med*, **77**, 775–8.

Nickerson, W.J. 1947. *Biology of pathogenic fungi*. Waltham, MA: Chronica Botanica Co.

Odds, F.C. 1988. *Candida and candidosis. A review and bibliography*, 2nd edition. London: Baillière Tindall.

Odds, F.C. 1991. Sabouraud('s) agar. *J Med Vet Mycol*, **29**, 355–9.

Odds, F.C. 1994. Presidential address. *Candida albicans*, the life and times of a pathogenic yeast. *J Med Vet Mycol*, **32**, Suppl 1, 1–8.

O'Hearn, E.M. 1985. *Profiles of pioneers, women scientists*. Washington, DC: Acropolis Books.

Ophüls, W. and Moffitt, H.C. 1900. A new pathogenic mould (formerly described as a protozoan): *Coccidioides immitis*. Preliminary report. *Phila Med J*, **5**, 1472–900.

Ota, M. and Huang, P. 1933. Sur les champignons du genre *Pityrosporum* Sabouraud. *Ann Parasitol*, **11**, 49–59.

Parrot, J. 1869a. Du muguet gastrique et de quelques autres localisations de ce parasite. *Arch Physiol Norm Pathol*, **2**, 504–17.

Parrot, J. 1869b. Note sur un cas de muguet du gros intestin. *Arch Physiol Norm Pathol 1870*, **3**, 621–5.

Parrot, J. 1877. *Clinique des nouveaux-nés. L'athrépsie*. Paris: Masson.

Pasteur, L. 1861. Mémoire sur les corpuscules organisés qui existent dans l'atmosphère. Examen de la doctrine des générations spontanées. *Ann Sci Nat Paris*, **16**, 4th edn sér, 5–98.

Pautler, K.B., Padhye, A. and Ajello, L. 1984. Imported penicillosis marneffei in the United States: report of second human infection. *Sabouraudia*, **22**, 433–8.

Pedroso, A. and Gomes, J.M.Y. 1920. Four cases of dermatitis verrucosa produced by *Phialophora verrucosa*. *Ann Paulistas Med Chir*, **9**, 53.

Pepys, J. 1969. Hypersensitivity diseases of the lungs due to the fungi and organic dusts. *Monogr Allergy*, **4**, 1–147.

Petri, R.J. 1887. Eine kleine Modification des Koch'schen Plattenverfahrens. *Zentrabl Bakteriol*, **1**, 279–80.

Pincus, D.H., Salkin, I.F. and McGinnis, M.R. 1988. Rapid methods in medical mycology. *Lab Med*, **19**, 315–20.

Pine, L. 1954. Studies on the growth of *Histoplasma capsulatum*. I. Growth of the yeast phase in liquid media. *J Bacteriol*, **68**, 671–9.

Pine, L. and Drouhet, E. 1963. Sur l'obtention et la conservation de la phase levure d'*Histoplasma capsulatum* et d'*H. duboisii* en milieu chimiquement défini. *Ann Inst Pasteur*, **105**, 798–804.

Pinoy, E. 1913. Actinomycoses and mycetomes. *Bull Inst Pasteur*, **11**, 929–38.

Posadas, A. 1892. Un neuvo caso de micosis fungoidea con psorospermias. *Circulo Med Argent*, **5**, 585–97.

Ravisse, P., de Bièvre, C., et al. 1993. Protothécose cutanée traitée avec succès par l'itraconazole: nouveau cas et revue générale des protothécoses humaines. *J Mycol Med*, **3**, 84–94.

Redaelli, P. and Ciferri, R. 1934. Affinité entre les agents d'histoplasmose humaine, du farcin equin et d'une mycose spontanée des muridées. *Bull Sez Ital Internaz Microbiol*, **6**, 376–9.

Reiss, E. 1986. *Molecular immunology of mycotic and actinomycotic infections*. 1–423, New York: Elsevier, 1–423, .

Remak, R. 1845. *Diagnostische und pathogenetische Untersuchungen in der Klinik des Geh. Raths Dr Schönlein auf dessen Veranlassung angestellt und mit Benutzung anderweitiger Beobachtungen veröffentlicht*. Berlin, 242 pp.

Renon, R. 1897. *Etude sur l'aspergillose chez les animaux et chez l'homme*. Paris: Masson.

Restrepo, M.A. 1985. The ecology of *Paracoccidioides brasiliensis*: a puzzle still unsolved. *Sabouraudia*, **23**, 323–34.

Restrepo, M.A. 1990. Paracoccidioidomycosis (south American blastomycosis). In: Jacobs, P.H. and Nahl, L. (eds), *Antifungal drug therapy*. New York: Marcel Dekker, 163–9.

Rinaldi, M.G., Seidenfeld, S.M. et al. 1989. *Pythium insidiosum* as an agent of devastating orbital-facial mycosis: mycological and management aspects. *Abst Am Soc Microbiol Ann Meeting*, F 11, New Orleans.

Rippon, J.W. 1980. Dimorphism in pathogenic fungi. *CRC Crit Rev Microbiol*, **8**, 49–79.

Rippon, J.W. 1985. The changing epidemiology and emerging patterns of dermatophyte species. *Curr Topics Med Mycol*, **1**, 208–34.

Rippon, J.W. 1988. *Medical mycology. The pathogenic fungi and the pathogenic actinomycetes*, 3rd edition. Philadelphia: W.B. Saunders.

Rivolta, S. 1873. *Dei parasiti vegetali come introduzione allo studio delle malattie parassitaire e delle alterzione dell' ailmento degli animali domestici*. Torino: G. Speirani e F, 246–52.

Rixford, E. and Gilchrist, T.C. 1896. Two cases of protozoan (coccidioidal) infection of the skin and other organs. *Johns Hopkins Hosp Rep*, **1**, 209–69.

Robert, R., Senet, J., et al. 1992. Molecular basis of the interactions between *Candida albicans* fibrinogen and platelets. *J Mycol Med*, **2**, 19–25.

Roberts, G.D., Pfaller, M.A., et al. 1992. Developments in the diagnostic mycology laboratory. *J Med Vet Mycol*, **30**, Suppl 1, 241–8.

Roberts, W. 1874. Studies on biogenesis. *PhilosTrans R Soc Lond*, **164**, 457–77.

Robin, C. 1847. Des végétaux qui croissent sur l'homme et sur les animauz vivants. Thesis, Faculty of Science, Paris.

Robin, C. 1853. *Histoire naturelle des végétaux parasites qui croissent sur l'homme et sur les animaux vivants*. Paris: Baillière.

Rook, A. 1978. Early concepts of the host-parasite relationship in mycology: the discovery of dermatophyte. *Int J Dermatol*, **17**, 666–77.

Rose, A.H. and Harrison, J.S. 1987. *The yeasts, vol. 1, Biology of yeasts*, 2nd edition. London: Academic Press, 1–423.

Sabouraud, R. 1896. Recherche des milieux de culture propres à la différenciation des espèces trichophytiques à grosse spore. In *Les trichophyties humaines*. Paris: Masson et Cie, 49–55.

Sabouraud, R. 1904. *Maladies du cuir chevelu. II. Les maladies desquamatives*. Paris: Masson et Cie.

Sabouraud, R. 1910. *Les teignes (= Maladies du cuir chevelu III. Les maladies cryptogamiques)*. Paris: Masson et Cie.

Saint-Germain, G. and Summerbell, R. 1996. *Identifying filamentous fungi. A clinical laboratory handbook*. Belmont, CA: Star Publishing Co..

Salkin, I. 1988. Medical mycology in the United States. In Torres-Rodriguez, J.M. and Prous, J.R. (eds), *Proceedings of the Congress of the International Society of Human and Animal Mycology (ISHAM), Barcelona*.

Salkin, I., McGinnis, M.R., et al. 1994. Current priorities for the clinical mycology laboratory. *J Med Vet Mycol*, **32**, Suppl 1, 309–20.

San-Blas, F. and San-Blas, G. 1985. *Paracoccidioides brasiliensis*. In: Szanizlo, P.J. (ed.), *Fungal dimorphism*. New York: Plenum Press, 93–120.

San-Blas, F. and San-Blas, G. 1993. Biochemical and physiological aspects in the dimorphism of *Paracoccidioides brasiliensis*. In: Vanden Bossche, H., Odds, F., et al. (eds), *Dimorphic fungi in biology and medicine*. New York: Plenum Press, 219–24.

Sanfelice, F. 1894. Contributo all morfologia e biologia dei blastomycete che si sviluppano nei suchi di alcuni frutte. *Ann Igiene*, **4**, 463–95.

Sanfelice, F. 1895. Ueber einen neuen Pathogenen Blastomyceten, welcher innerthalb der Gewebe unter Bildung kalkartig aussenhender Massendegeneriert. *Zentbl Bakteriol Parasitenkd Infektionskr Hyg*, **18**, 521–6.

Schoenlein, J.L. 1839. Zur pathogenie der impetigines. *Arch Anat Physiol Wiss Med (Muller's)*, **82**.

Schwarz, J., Silverman, F.N., et al. 1955. The relation of splenic calcifications to histoplasmosis. *N Engl J Med*, **252**, 887–91.

Seeber, G.R. 1900. Un neuvo esporozuario parasito del hombre. Dos casos encontrades en polops nasales. Thesis, Fac Med Univ Nat de Buenos Aires.

Seeliger, H.P.R. 1985. The discovery of *Achorion schoenleinii*. Facts and 'stories'. *Mykosen*, **28**, 161–82.

Seeliger, H.P.R. 1988. The beginnings of medical mycology. In: Tümbay, E. (ed.), *FEMS Symposium on dermatophytes and dermatophytosis in man and animals. Proceedings*. Izmir, Turkey: Izmir Bilgehim, 1–15.

Segretain, G. 1959. *Penicillium marneffei* n. sp. agent d'une mycose du système réticulo-endothélial. *Mycopathol Mycol Appl*, **11**, 327–53.

Segretain, G. 1962. Pulmonary aspergillosis. *Lab Invest*, **11**, Part 2, 1046–52.

Segretain, G. and Destombes, P. 1961. Description d'un nouvel agent de maduromycose. *C R Acad Sci (Paris)*, **253**, 257–9.

Segretain, G. and Destombes, P. 1969. Recherche sur les mycétomes à *Madurella grisea* et *Pyrenochaeta romeroi*. *Sabouraudia*, **7**, 51–61.

Segretain, G., Baylet, R., et al. 1959. *Leptosphaeria senegalensis* n. sp. agent de mycétome à grains noirs. *C R Acad Sci (Paris)*, **248**, 3730–2.

Segretain, G., André, M., et al. 1974. *Leptosphaeria tompkinsii*, agent de mycétomes au Senegal. *Bull Soc Fr Mycol Med*, **3**, 71–4.

Silva, M. and Benham, R.W. 1952. Nutritional studies of the dermatophytes with special reference to *Trichophyton megninii* Blanchard 1896 and *Trichophyton gallinae* (Megnin 1881) comb nov. *J Invert Dermatol*, **18**, 453–72.

Simmons, R.B. and Guého, E. 1990. A new species of *Malassezia*. *Mycol Res*, **94**, 1146–9.

Sluyter, T. 1856. De Vegetbilibus organismi animalis parasitis ac de novo epiphyto in pityriasi versicolore obvio. Inaugural Dissertation, Berlin. Fide Virchow.

Speller, D.C.E. 1980. *Antifungal chemotherapy*. Chichester: John Wiley & Sons.

Staib, F. 1962. *Cryptococcus neoformans* and *Guizotia abyssinica* (Syn. *G. oleifera* D.C.). *Z Hyg*, **148**, 466–75.

Staib, F. 1963. New concepts in the occurrence and identification of *Cryptococcus neoformans*. *Mycopathol Mycol Appl*, **19**, 43–5.

Stevens, D.A., Domer, J.E., et al. 1994. Immunomodulation in mycoses. *J Med Vet Mycol*, **32**, Suppl 1, 253–66.

Stewart, R.A. and Meyer, K.F. 1932. Isolation of *Coccidioides immitis* (Stiles) from soil. *Proc Soc Exp Biol Med*, **29**, 937–8.

Stockdale, P.M. 1961. *Nannizzia incurvata* gen. nov., sp. nov., a perfect state of *Microsporum gypseum* (Bodin) Guiart et Grigorakis. *Sabouraudia*, **1**, 41–8.

Stockdale, P.M. 1963. The *Microsporum gypseum* complex [*Nannizzia incurvata* Stockd, *N. gypsea* (Nann.). comb. nov., *N. fulva* sp. nov]. *Sabouraudia*, **3**, 114–26.

Stockdale, P.M. 1967. *Nannizzia persicolor* sp. nov., the perfect state of *Trichophyton persicolor*. *Sabouraudia*, **5**, 355–9.

Stoddard, Y.L. and Cutler, E. 1916. *Torula* infection in man. (*Torula histolytica*, n sp.). *Monograph Rockefeller Institute for Medical Research no. 6*, 1–98.

Supparatpinyo, K., Khamwan, C., et al. 1994. Disseminated *Penicillium marneffei* infection in Southeast Asia. *Lancet*, **344**, 110–13.

Sweany, H.C. (ed.) 1960. *Histoplasmosis*. Springfield, IL: Charles C Thomas.

Szaniszlo, P.J., Jacobs, C.W. and Geis, P.A. 1983. Dimorphism: morphological and biochemical aspects. In: Howard, D.H. (ed.), *Fungi pathogenic for humans and animals. Part A, Biology*. New York: Marcel Dekker, 323–436.

Tanaka, S., Summerbell, R.C., et al. 1992. Advances in dermatophytes and dermatophytosis. *J Med Vet Mycol*, **30**, Suppl 1, 29–39.

Tavitian, A., Raufman, J.P. and Rosenthal, L.E. 1986. Oral candidiasis as a marker for esophageal candidiasis in the acquired immunodeficiency syndrome. *Ann Intern Med*, **104**, 54–5.

Trousseau, A. 1868. *Lectures on clinical medicine delivered at the Hotel Dieu, Paris (translated from the end of 1868 by JR Cormack) [Lecture XXI Thrush London (New Sydentrans Soc)]*, **2**, 618–30.

Vanbreuseghem, R. 1952. Technique biologique pour l'isolement des dermatophytes du sol. *Ann Soc Belg Med Trop*, **32**, 173–8.

Vanbreuseghem, R. 1953. *Histoplasma duboisii* and African histoplasmosis. *Mycologia*, **45**, 803–16.

Vanbreuseghem, R. and Takashio, M. 1970. An atypical strain of *Cryptococcus neoformans* (Sanfelice) Vuillemin 1894. Part II. *Cryptococcus neoformans* var. *gattii* var. nov. *Ann Soc Belg Med Trop*, **50**, 695–702.

Vanbreuseghem, R., de Vroey, C. and Takashio, M. 1970. Production of macroconidia by *Microsporum ferrugineum* Ota 1922. *Sabouraudia*, **7**, 252–6.

Van Cutsem, J., Fransen, J., et al. 1986. Experimental cryptococcosis: dissemination of *Cryptococcus neoformans* and dermotropism in guinea-pigs. *Mykosen*, **29**, 561–75.

Van Cutsem, J., Kurata, H. and Matsuoka, H. 1994. Antifungal drug susceptibility testing. *J Med Vet Mycol*, **32**, Suppl 1, 267–76.

Vanden Bossche, H., Mackenzie, D.W.R. and Cauwenbergh, G. 1988. *Aspergillus* and aspergillosis. *Proceedings of the Second International Symposium Topics in Mycology*. New York: Plenum Press, 31.

Vanden Bossche, H., Warnock, D.W. and Dupont, B. 1994. Mechanisms and clinical impact of antifungal drug resistance. *J Med Vet Mycol*, **32**, Suppl 1, 189–202.

van Leeuwenhoek, A. 1673. A specimen of some observations made by microscope. *Philos Trans*, **8**, 94, 60–37.

van Tieghem, P.E.L. and Le Monnier, G. 1873. Recherches sur les Mucorinées. *Ann Sci Nat Paris sér 6*, **17**, 261–399.

Vermeil, C., Gordeef, A., et al. 1954. Sur un cas tunisien de mycose généralisée mortelle. *Ann Inst Pasteur*, **86**, 636–46.

Virchow, R. 1856. Beitrage zur lehre von dem beim menschen vorkom menden planzlichen parasiten. *Arch Pathol Anat Physiol (Virchow's)*, **9**, 557–93.

Vuillemin, P. 1901. Les blastomycètes pathogènes. *Rev Gen Sci Pures Appl*, **12**, 732–51.

Waldorf, A.R. 1986. Host–parasite relationships in opportunistic mycoses. *Crit Rev Microbiol*, **13**, 133–72.

Walsh, T.J., de Pauw, B. and Anaissie, E. 1994. Recent advances in the epidemiology, prevention and treatment of invasive fungal infections in neutropenic patients. *J Med Vet Mycol*, **32**, Suppl 1, 33–52.

Warnock, D.W. and Richardson, M.D. 1982. *Fungal infection in the compromised patient*, 1st edition. Chichester: John Wiley & Sons.

Warnock, D.W. and Richardson, M.D. 1991. *Fungal infection in the compromised patient*, 2nd edition. Chichester: John Wiley & Sons.

Watanabe, J.L., Hori, H., et al. 1989. Phylogenetic association of *Pneumocystis carinii* with the Rhizopoda/Myxomycota/Zygomycota groups indicated by comparison of 5S ribosomal RNA sequences. *Mol Biochem Parasitol*, **32**, 163–8.

Weber, D.J. and Rutala, W.A. 1988. Epidemiology of nosocomial fungal infections. *Curr Top Med Mycol*, **2**, 305–30.

Weitzman, I., McGinnis, M.R., et al. 1986. The genus *Arthroderma* and its later synonym *Nannizzia*. *Mycotaxon*, **25**, 505–18.

Wernicke, R. 1892. Ueber einen Protozoenbefund bei Mycosis Fongoides. *Zentbl Bakteriol*, **12**, 859–61.

Wheat, L.J., Kohler, R.B. and Tewari, R.P. 1986. Diagnosis of disseminated histoplasmosis by detection of *Histoplasma capsulatum* antigen in serum and urine specimens. *N Engl J Med*, **314**, 82–8.

Whelan, W.L. 1987. The genetics of medically important fungi. *CRC Crit Rev Microbiol*, **14**, 99–170.

Whittaker, R.H. 1969. New concepts of kingdoms of organisms. *Science*, **163**, 150–60.

Wilson, D.E., Bennett, J.E. and Bailey, J.W. 1968. Serologic grouping of *Cryptococcus neoformans*. *Proc Soc Exp Biol Med*, **127**, 820–3.

Zakon, S.J. and Benedek, T. 1944. David Gruby and the centenary of medical mycology 1841-1941. *Bull Hist Med*, **16**, 155–68.

Zeppenfeldt, G., Richard-Yegres, N., et al. 1994. *Cladosporium carrionii*: hongo dimorfico en cactaceas dela zona endémica para la cromomicosis en Venezuela. *Rev Iberoam Micol*, **11**, 61–3.

Kingdom Fungi: fungal phylogeny and systematics

THOMAS G. MITCHELL

Taxonomy is often regarded as the dullest of subjects, fit only for mindless ordering and sometimes denigrated within science as mere 'stamp collecting.' If systems of classification were neutral hat racks for hanging the facts of the world, this disdain might be justified. But classifications both reflect and direct our thinking. The way we order represents the ways we think. Historical changes in classification are the fossilized indicators of conceptual revolutions. (Gould 1996)

ORIGIN OF THE FUNGI

The evolutionary origins of biology began with the prokaryotes, the Archaea, and Eubacteria. The Eukaryota, or eukaryotes, are believed to have evolved from prokaryotes about 1.6–2.1 billion years ago with the emergence of protists (Javaux et al. 2001). The protists include protozoa, certain algae, and fungal-like organisms, the Oomycota. Eukaryotic cells are distinguished by a cytoskeleton and intracellular organelles, which are usually enclosed by membranes. The endomembranes of eukaryotic cells include the nuclear membrane, endoplasmic reticulum, Golgi apparatus, peroxisomes, and the plasma membrane. Most of the DNA is organized into chromosomes and contained with the nucleus, but DNA is also found in the mitochondria (in fungi and animals) as well as in extrachromosomal forms. The cytoskeleton includes major proteins that evolved early and have been conserved throughout the evolution of eukaryotes, such as tubulin (which forms microtubules) and actin (microfilaments).

Two major lineages arose early, one led to the algae and terrestrial plants, and the other gave rise to fungi and animals (Figure 2.1). The common origins of animals and fungi are supported by phylogenetic analyses of the nucleotide sequences of several conserved genes, such as the small subunit (SSU) 18S and large subunit (LSU) 26S of ribosomal DNA (rDNA), elongation factor, tubulin, and actin (Bruns et al. 1991; Baldauf and Palmer 1993; Wainright et al. 1993), although not by sequences of the RNA polymerase II gene (Sidow and Thomas 1994). Based on phylogenetic studies, molecular clock estimations of the rate of mutation, and limited fossil data, the time when animals and fungi diverged has been estimated at approximately 1.5 billion years ago (Heckman et al. 2001; Hedges et al. 2004).

Most fungi grow vegetatively as yeasts or molds, and fungal cells may have more than one nucleus. Similar to other eukaryotes, fungi typically contain mitochondria, a Golgi apparatus, and endoplasmic reticulum. They also contain vacuoles, but lack chloroplasts and pseudopodia. Most have cell walls composed of chitin and β-glucans, as well as other polysaccharides, proteins, and lipids.

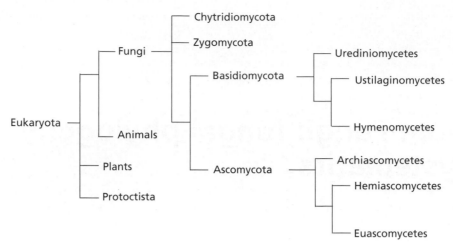

Figure 2.1 *Major kingdoms of the Eukaryota, the four phyla of the Fungi, and the three classes of the Basidiomycota and the Ascomycota.*

Collectively, fungi exhibit a rich diversity, inhabit every ecological niche on the planet, and manifest a variety of lifestyles. Most fungi are chemotrophic and secrete catabolic enzymes that digest vicinal substrates, which are absorbed, as are minerals, salts, and other small molecules, by passive or active transport mechanisms. Fungi distinctively utilize α-aminoadipic acid in the biosynthesis of lysine. As saprobic fungi proliferate in aquatic and terrestrial locales, they degrade organic substrates for their nutrition, and in the process recycle the organic detritus of plants and animals. Other fungi are exploited by industry for the production of secondary metabolites or pharmacological compounds. Some serve science as model eukaryotes for the study of cellular, molecular, and genetic processes (e.g. *Saccharomyces cerevisiae*, *Neurospora crassa*). Many provide gustatory pleasure, either directly (edible mushrooms, truffles, morels) or indirectly (bread, cheese, alcoholic beverages). Certain groups of fungi are intriguingly specialized, and have the ability to entrap nematodes, form symbiotic relationships (lichens), nourish or protect plants (mycorrhizae and endophytes, respectively), or produce disease in plants or animals. More than 80 000 species of fungi have been described, but the total number of extant fungal species has been estimated to be about 1 500 000 (de Hoog et al. 2000; Kirk et al. 2001). Compared to the enormous kingdom of fungi, the number of infectious, or mycotic, species is minuscule. Nevertheless, the medical fungi exert a profound, global impact on human health. Their numbers continue to increase, but no one knows how many human pathogenic fungi exist. To date, approximately 400 clinically relevant species have been described (de Hoog et al. 2000; Kirk et al. 2001; Howard 2003).

CONCEPTS AND TERMS

Systematics aims to describe and organize biological diversity, usually by determining the phylogenetic relationships among organisms. Systematicists study the relationships of organisms and the processes by which they evolved and are maintained. Classification is the assignment of organisms to defined groups. Systematics entails nomenclature and taxonomy, the science of naming and classifying organisms. Because of their historical association with plants, the fungi are named according to the rules established by the International Code of Botanical Nomenclature. However, reflecting their affinity with animals, the higher fungal groups are called Phyla, not Divisions. Major taxonomic groups of fungi are denoted by the following suffixes: Phylum, -mycota; Subphylum, -mycotina; Class, -mycetes; Subclass, -mycetidae; Order, -ales; Suborder, -ineae; Family, -aceae; and Subfamily, -oideae. The basic taxon is the species. However, the species name is often the antepenultimate rank of classification, just above the variety and individual (Kirk et al. 2001).

Each species is identified by its genus and species name, followed if necessary by a varietal or subspecies designation. A full descriptor of a given species will include the binomial in Italic font followed in non-Italic font by the surname(s) of the author(s) who first described and named the species and the year of its publication, for example, *Cryptococcus neoformans* (Sanfelice) Vuillemin (1901). If a species name is modified, the names of subsequent authors may be added. Whenever possible, the original isolate from which the species was described becomes its permanent reference or the type strain for that species, and it is deposited in an established culture collection (e.g. American Type Culture Collection, Centraalbureau voor Schimmelcultures). If a species is inadvertently described more than once, its name reverts to the original description (with sanctioned exceptions for names with longstanding familiarity).

There are several definitions of a fungal species (Taylor et al. 2000; Burnett 2003). Members of a species are quite similar, if not identical, in their morphology, mode of reproduction, physiology and temperature range, biochemical structure, ecology, and genomic sequence. Species have been historically demarcated on the basis of morphological and other phenotypic similarities. Biological species are defined by mating compatibility, the abilities to engage in sexual reproduction and successfully interbreed, thus sharing a common gene pool with other members of the species (Burnett 2003). Other criteria have been used to define certain species, including both ecological isolation in nature and the host specificity of pathogenic fungi. Species may also be identified by phylogenetic analyses, using molecular markers as characters to determine evolutionary relationships. Phylogenetic criteria define a species as the smallest recognizable population of individuals or lineage united by shared derived characters (synapomorphy) (Burnett 2003; Vilgalys 2004). Molecular studies have discovered situations where organisms were morphologically indistinguishable, similar in numerous other phenotypes, and occupied the same habitat (sympatric), but they were not members of the same species. One such cryptic species is *Candida dubliniensis*, which was long misidentified as *Candida albicans* (Sullivan and Coleman 1998; Sullivan et al. 2004). Another example is the discovery that the common cause of coccidioidomycosis, *Coccidioides posadasii*, was for decades subsumed by the phenotypically similar species, *Coccidioides immitis* (Fisher et al. 2002).

An important route to speciation occurs when organisms become geographically and reproductively isolated (allopatry). Over time, mutations accrue, selective pressure or genetic drift may impact the organisms, and they become sufficiently divergent that mating with the ancestral population is no longer possible. Advances in fungal genomics are beginning to discover some of the genetic mechanisms that may drive speciation, including mitotic recombination (parasexuality), gene duplications, and chromosomal rearrangements and translocations (Lynch 2002; Wong et al. 2002; Dujon et al. 2004).

The sexually reproductive life cycle of a fungal species is its teleomorph, which indicates that the species has the ability to undergo meiosis and produce meiospores. The anamorph refers to mitotic or asexual reproduction. If a teleomorphic species produces multiple anamorphs, they may be called synanamorphs. Since many fungal species lack a known teleomorph, or were described before their sexual reproductive mode was recognized, their names were based on their anamorphic or asexual reproductive morphology and ontogeny. This method of classification relies on morphology and other phenotypes, and it may or may not reflect evolutionary relationships. Many fungi have both a teleomorphic and one or more anamorphic species names, reflecting their separate description at different times and places. The teleomorph, anamorph, and any synanamorphs of a fungal species comprise its holomorph.

Furthermore, the reproductive structures of many fungal species are indistinguishable. Mating between fungal isolates usually constitutes evidence of species identity. However, isolates with identical anamorphs may not be recognized as different species if they are incapable of sexual reproduction. Consequently, an anamorphic 'species' may represent more than one teleomorphic species, and vice versa. The dermatophytes offer good examples: the geophilic anamorph *Trichophyton terrestre* may represent three different teleomorphs, *Arthroderma quadrifidum*, *Arthroderma insingulare*, or *Arthroderma lenticulare*. Similarly, the teleomorphic species *Arthroderma uncinatum* has two anamorphs, *Trichophyton ajelloi* and *Trichophyton phaseoliforme* (Summerbell and Kane 1997).

The apparent predominance of anamorphy among mycotic fungi may be advantageous for pathogenicity. Asexual fungal species tend to consist of one to several distinct clones or lineages, and sexual reproduction is limited or absent. Evolution occurs by various genomic mechanisms, as noted above, and the gradual accumulation of spontaneous mutations (Taylor et al. 1999b; Taylor et al. 2000; Xu and Mitchell 2002). The environmental growth conditions will subsequently promote the positive selection, survival, and proliferation of clones with favorable mutations. By contrast, sexual reproduction offers the opportunity to more rapidly exchange genetic information. However, in sexually reproducing populations, favorable combinations of mutations that arise in some lineages may be lost through subsequent recombination. Many fungi exhibit a mixture of both clonal reproduction, which can be favored under uniform growth conditions, and sexual recombination, which can be advantageous in changing environments. Perhaps as commensal and pathogenic fungi become more entrenched on the mammalian host, their need for genetic variation is minimized, and anamorphy becomes an important survival mechanism. The dermatophytes exemplify support for this concept. Except for *Trichophyton mentagrophytes* and *Microsporum canis*, the anthropophilic and zoophilic species exist only as anamorphs (e.g. *Trichophyton rubrum*, *Trichophyton tonsurans*, *Trichophyton verrucosum*, *Trichophyton equinum*, *Microsporum ferrugineum*, *Epidermophyton floccosum*), but most of the saprobic, geophilic species have teleomorphic states (e.g. *Microsporum cookei*, *Microsporum gypseum*, *Microsporum persicolor*) (Howard et al. 2003). The evolution to anthropophilism can also be associated with the loss of asexual sporulation. Routine cultures of most geophilic species regularly manifest both macroconidia and microconidia, but many human pathogens tend to produce only, or predominantly, microconidia (*Trichophyton soudanense*, *T. rubrum*, *T. tonsurans*, *Trichophyton yaoundei*), macroconidia (*E. floccosum*), or neither (*Microsporum audouinii, M. ferrugineum, Trichophyton concentricum, Trichophyton schoenleinii, T. verrucosum, Trichophyton violaceum*).

Phylogenetics is the taxonomical classification of organisms that reflects their evolutionary relationships. Using molecular or phenotypic characters, a phylogenetic tree depicts the ancestor–descendant relationships among organisms. Morphology – macroscopic, microscopic and ultrastructural – provided the original basis for classifying fungi, as well as other organisms, and proposing phylogenetic relationships. Molecular methods include comparative analyses of the nucleotide sequences of RNA and DNA and the amino acid sequences of proteins (Hillis et al. 1996a). The most comprehensive phylogenies have been constructed from comparisons of the highly conserved nucleotide sequences of ribosomal RNA, especially alignable regions of the small (18S) and large (26S) subunits (SSU and LSU) of rDNA.

A cluster of taxa on a branch of a phylogenetic tree or clade is monophyletic if all organisms in that group are known to have developed from a common ancestor, and all descendants of that ancestor are included in the group. A group is paraphyletic if all the members have a common ancestor but the group does not include all the descendants of the common ancestor. Taxonomic groups that contain organisms but not their common ancestor are called polyphyletic. Phylogenetic classifications aim to group species such that every group is descended from a single common ancestor, and the elimination of groups that are found to be polyphyletic often stimulates major revisions of the classification.

PHENOTYPES AND MOLECULAR MARKERS TO IDENTIFY FUNGI

As summarized in Table 2.1, traditional methods of classifying fungi utilize phenotypic differences in vegetative and reproductive morphology (e.g. colonial and micromorphology), physiology (e.g. patterns of assimilation and fermentation, temperature, and pH range of growth), the presence of structural macromolecules (e.g. cell wall polysaccharides, isoprene units of coenzyme Q), and sexual mating. Since most medical fungi do not evince a teleomorph, their identification and classification have rested on phenotypic characteristics. Definitive phenotypic features are often difficult to discern or highly variable. Thus, for many pathogens, the species boundaries are loosely defined. For example, any elliptical, fermentative, nonpigmented, asexual yeast producing multilateral buds and pseudohyphae is likely to be placed in the genus *Candida*, which is large, multifarious, and polyphyletic (Meyer et al. 1998; Diezmann et al. 2004).

Molecular markers are definitive and more stable than most phenotypic attributes, and they are helping to resolve these issues. There is continuing progress in the biotechnology of rapid, high-throughput methods to genotype strains and the development of innovative statistical approaches and software to analyze the data. Some of the more common DNA-based methods are listed in Table 2.2, discussed in Chapter 4, Laboratory diagnosis, and detailed in recent reviews (Bruns et al. 1991; Taylor et al. 1999a; Mitchell and Xu 2003; Taylor and Fisher 2003).

Table 2.1 *General phenotypic approaches to identify and characterize fungi*

Phenotype category	Observations (examples)
Morphology	Teleomorph: Sexual reproduction (zygospore, ascocarp, basidiocarp) Anamorph: Mitospore ontogeny and morphology (sporangia, conidiogenesis) Vegetative yeast or hyphal morphology (septate, dematiaceous, Diazonium Blue B) Colony texture, pigment, size Septal pore type Dimorphism
Structural macromolecules	Cell wall polysaccharides (chitin, glucans, mannans) Coenzyme Q isoprene units Isozymes Antigens
Metabolism	Assimilation of particular compounds as sole nitrogen or carbon sources Fermentation of various carbon compounds Metabolic pathways Vitamin requirements (biotin, thiamin) Production of enzymes (proteases, phospholipases, urease, catalase, etc.) Secondary metabolites (mycotoxins)
Growth parameters	Optimum and tolerable range of temperature, pH, concentration of salts, oxygen requirements, etc.
Susceptibility to inhibitors	Inorganic compounds (chlorine salts) Fungicides, detergents, organic compounds (canavanine, benomyl, etc.) Drugs (antifungal antibiotics, cycloheximide) Killer yeasts
Pathogenicity	Host range Virulence factors (adhesins, secretory enzymes, etc.)

Table 2.2 *General DNA-based methods in medical mycology*

Method	Application		
	Strain identification, molecular epidemiology, population genetics	Species identification	Phylogenetics and systematics
Electrophoretic karyotype	✓		
RFLP	✓	✓	
Southern hybridization	✓	✓	
RAPD, PCR fingerprint, AFLP	✓	✓	
Microsatellites	✓	✓	
Microarrays	✓	✓	
SNP, MLST, DNA sequencing	✓	✓	✓

Abbreviations: AFLP, amplified fragment length polymorphism; MLST, multilocus sequence typing; RAPD, random amplified polymorphic DNA; RFLP, restriction fragment length polymorphism; SNP, single nucleotide polymorphism, detected by high-throughput, automated system. Microarray technology is currently only theoretical.
Other methods: mol% G+C; DNA–DNA hybridization.
References: Bruns et al. 1991; Mitchell and Xu 2003; Taylor and Fisher 2003.

A molecular marker is defined as any identifiable region of the genome. Some markers are better at discriminating individual strains, discrete species or higher taxonomic groups. For certain studies, it is desirable to use markers in specific genes, and for others, markers in noncoding, usually anonymous portions of the genome are preferable because they are assumed to be neutral or unaffected by selective pressure. Molecular markers for fingerprinting medical fungi include allozymes, electrophoretic karyotypes, DNA hybridization probes, PCR-based genotypes, endonuclease restriction fragment length polymorphisms (RFLP), and DNA sequencing (Table 2.2).

The comparison of nucleotide sequences among species has revolutionized phylogenetic studies. The most reliable DNA markers identify nucleotide sequence polymorphisms, such as single nucleotide polymorphisms (SNP), insertions or deletions (indels), and length variations in tandem oligonucleotide repeats (microsatellites). Such markers can be detected by an array of molecular techniques, such as PCR amplicon length polymorphisms, differences in the lengths of RFLP, single-strand conformational polymorphisms (SSCP), or direct DNA sequencing of homologous genomic regions of the strains or species under comparison [multilocus sequence typing (MLST)]. These methods and their applications to fungal phylogenetics have been described in several reviews (Berbee and Taylor 1999; Taylor et al. 2000; Mitchell and Xu 2003; Taylor and Fisher 2003). Some methods are more suitable for population and phylogenetic studies of closely related taxa and strains of a species, such as isozymes (see below), amplified fragment length polymorphisms (AFLP), randomly amplified polymorphism DNA (RAPD), and hybridization of nucleotide probes to electrophoretic blots of digested genomic DNA (Bakkeren et al. 2000; Soll 2000; Mitchell and Xu 2003).

Protein polymorphisms, such as isozymes, provide excellent markers for population genetics, but they are not widely applicable for large phylogenetic investigations. They usually reflect nucleotide sequence differences resulting in the substitution of one or more amino acids, which alter the chromatographic mobility of the protein to enable detection (Brandt et al. 1993; Murphy et al. 1996; Mitchell and Xu 2003).

In summary, these markers and the attendant analytical software have greatly innovated the methods to:

- resolve taxonomic uncertainties
- investigate the transmission of strains in local outbreaks
- accurately identify fungi in clinical specimens
- recognize strains with clinically important phenotypes (e.g. virulence factors, resistance to antifungal drugs)
- analyze the genetic structure of populations of a pathogen
- trace the evolution of mycotic species
- validate the medical relevance of strains used in basic research, comparative genomics, and the development of new antifungal antibiotics and diagnostic tests.

PHYLOGENETIC METHODS

The sequences of any two or more related taxa can be compared to detect phylogenetic relationships. The underlying assumptions are that with the passage of time, there is an increase in the number of mutations, divergence, and phylogenetic distance between organisms and their genomic sequences. The 'molecular clock' hypothesis assumes that molecules evolve in direct proportion to time; consequently, the differences between homologous DNA sequences (or proteins) can be used to estimate the elapsed time since the two DNA sequences (or their taxa) last shared a common ancestor (Hillis et al. 1996b; Hedges and Kumar 2004). However, the observed rate at which nucleotides change is not constant. Some sequences exhibit less variation over time than expected because their functionality is essential and lethal mutations are lost. This problem is usually obviated by analyzing sequences of noncoding DNA. However, after sufficient time, some nucleotides may undergo further mutation(s) and revert to the ancestral sequence. Most algorithms to construct phylogenetic trees consider such variables and produce a consensus or single 'best' tree with branches of optimal length and likelihood. The common approaches of phylogenetic analysis include Neighbor-Joining (NJ), parsimony, Maximum Likelihood (ML), and Bayesian methods (Xu and Mitchell 2002; Baldauf 2003b; Felsenstein 2003; Holder and Lewis 2003).

The usual phylogenetic methods use algorithms that are based on genetic distances, optimality criteria, such as parsimony or ML, or Bayesian probabilities (Swofford et al. 1996). The first step involves the alignment of nucleotide (or protein) sequences of the taxa being compared. The most popular distance methods are NJ and Unweighted Pair-Group Method with Arithmetic Mean (UPGMA), which estimate phylogenies based on the nucleotide differences among the taxa. Maximum Parsimony (MP) or ML apply optimality criteria to infer phylogenies that minimize the number of evolutionary steps required to explain the data. The selection of one or more methods is governed by the number and diversity of the taxa, as well as the sequences that are compared. Other considerations are the types of assumptions associated with a particular data set (e.g. constant or varying rate) and the required computational power, as algorithms such as ML are more demanding than others.

The products of these methods are phylogenetic trees, which estimate the evolutionary history of a group of sequences. A phylogenetic tree is often rooted with an outgroup of some evolutionary distance from the taxa being analyzed, which anchors the study taxa and provides structure for the tree. Two kinds of diagrams are commonly generated with rooted trees – cladograms and phylograms. A cladogram presents only the nodes and branches. In a phylogram, branch length is proportional to the evolutionary distance, providing a schematic depiction of the number of character state differences (e.g. nucleotide substitutions) between the taxa on that branch. There are several excellent reviews of phylogenetic methods and theory (Hillis et al. 1996a; Swofford et al. 1996; Felsenstein 2003; Holder and Lewis 2003; Hall 2004).

Distance methods approximate a phylogenetic tree by estimating the relative number of nucleotide differences between each pair of taxa in a multiple alignment. This method usually assumes that the rate of mutation for the aligned sequences is neutral, independent, and constant, and that the occurrence of multiple changes of the same nucleotide(s) can be estimated. However, corrections for unequal substitution rates and other variables are commonly introduced. UPGMA is the simplest clustering algorithm, which starts with the two most similar sequences, and then sequentially adds taxa of increasing genetic distance. UPGMA also assumes that all taxa are equally distant from the root (i.e. assumes a molecular clock) and thus yields symmetrical trees. NJ is a similar distance method but does not assume a molecular clock. NJ initially calculates the net divergence of each taxon from all the others, which is the sum of the individual differences from the taxon. Next, the program selects the pair of taxa that are least diverged and calculates their distances from the node that connects them. This node is substituted for the pair, and the process continues, reducing the matrix with each step. One of the accessible programs for constructing phylogenies is PAUP* (Swofford 2002; Hall 2004). UPGMA and NJ programs are the least computationally demanding of the common methods and run very quickly on most computers. However, their speed sacrifices accuracy, as neither method attempts to optimize a tree to determine the best estimate. By contrast, MP, ML, and Bayesian inferences perform iterative rounds of optimization to find the best tree.

An excellent test of the statistical support for each branch of the tree is bootstrapping, which can be appliedto all methods (Felsenstein 1985). There is some debate about the interpretation of bootstrap values in phylogenetic analyses. For parsimony analysis, bootstrap values >70 percent are usually considered acceptable, but for the other methods, values >90 percent are considered significant support for a clade. Bayesian probabilities often achieve >95 percent support for branches.

Unlike distance methods, MP seeks a tree(s) that requires the minimum number of nucleotide (or other characters) changes. Therefore, MP only needs to analyze informative sites, namely, nucleotide changes that occur in two or more sequences. The algorithm then determines which one or more of the possible tree(s) requires the fewest number of changes.

The ML approach is usually applied to a particular tree or evolutionary model and computes the probability of generating that tree from the dataset. ML seeks to infer the tree with highest probability from the observed sequence alignment.

A Bayesian analysis (BA) undertakes a succession of iterative phylogenetic analyses, each of which employs the optimized parameters of the previous analysis, until a best tree or set of trees is obtained. BA resembles ML, but ML searches for the best tree that maximizes the probability of observing the data for that tree, and BA seeks the tree(s) that maximizes the posterior probability of the tree for that data. BA, which is commonly employed with the MrBayes program, using the Metropolis-coupled Markov Chain Monte Carlo sampling method, is more robust and exhaustive and will determine the best set of trees (Huelsenbeck and Ronquist 2001). These programs do their work by analyzing the alignments, sequentially and repetitively comparing tree configurations to identify the tree or trees with the highest probability (Huelsenbeck et al. 2001; Holder and Lewis 2003; Hall 2004).

In addition to genomic sequences, other molecular characters and phenotypes are helpful in analyzing phylogenies. Certain cell wall polysaccharides are associated with specific fungi, and the chemical or immunological detection of signature motifs can be used to identify fungi within those groups. Similarly, the capacity to produce secondary metabolites and metabolic byproducts (e.g. arabinitol) under the appropriate conditions have sufficient specificity in clinical settings to be useful in diagnostic and prognostic tests, and these properties may be significantly correlated with certain taxonomic lineages. These structural and metabolic compounds include coenzyme Q, which is a mitochondrial electron carrier with a varying number of isoprene units. The length of the isoprene chain is often the same within a monophyletic group (Yamada et al. 1977). As noted below, pathogenicity is a polyphyletic trait, as it evolved, and was lost, multiple times throughout the history of fungi. Many of the attributes required for virulence by certain fungal species are absent in other pathogens, as well as present in nonpathogens. There are many examples of such properties, including dimorphism, secretory enzymes that catabolize host substrates, and cell wall ligands that promote adhesion to host cell receptors. At least one phenotype seems to be universally required if not necessarily sufficient for systemic fungal infection, growth at 37°C.

GENOMICS

The new disciplines of genomics and bioinformatics offer panoptic approaches to integrate molecular and phenotypic data to elucidate fungal classification and evolution. The number of complete genome sequences of mycotic, phytopathogenic, and model fungi is increasing (www.genome.gov/11008243). These databases permit the comparison of not only genes but entire genomes, and there are multiple ways to exploit these resources. Following completion of the genome sequence of *S. cerevisiae*, sequences of 13 related hemiascomycetous yeasts were compared to evaluate the conservation of chromosome maps and the similarity and synteny among genes (Gaillardin et al. 2000; Malpertuy et al. 2000; Souciet et al. 2000). Data from this ongoing multilaboratory project are regularly updated at the Génolevures ('yeast gene') web site: cbi.labri.fr/Genolevures/. A subsequent comparison of the genomic sequences of *Candida glabrata*, *Debaryomyces hansenii*, *Kluyveromyces lactis*, *S. cerevisiae*, and *Yarrowia lipolytica*, which manifest different reproductive modes, led to the discovery of many new genes and gene families and specific mechanisms whereby these yeast lineages have evolved (Dujon et al. 2004). Comparative genomics approaches are being used to identify phyla and other taxon- or pathogen-specific

genes, as well as test the validity of phylogenetic methods (Hardison 2003).

CLASSIFICATION

It has been difficult to resolve the ancestry of fungi and other eukaryotes, and analyses of various taxa and genes have generated inconsistent phylogenies. Several issues contribute to the problem of parsing these ancient lineages:

- there is a dearth of fossil data
- extant ancestral species have saturated their available mutable nucleotides, even back-mutating to their original sequences
- the results vary depending upon the selected taxa, genes, methods, and outgroups (Baldauf 2003a).

Eventually these problems will be surmounted by increasing the number of taxa, using multigene datasets, and refining the phylogenetic methods.

Lacking a consensus about the classification of the deep branches of the eukaryotic domain, a neutral approach has been to place the diverse ancestral phyla within the Kingdom Protoctista, which contains early eukaryotes that do not qualify as plants, animals, or fungi (Figure 2.1). Some of the 18–20 protoctistan phyla were previously thought to be fungi, such as the Acrasiomycota (cellular slime molds), Myxomycota (plasmodial slime molds), and Oomycota (water molds). The Protoctista also includes the protozoa and algae. Table 2.3 lists the higher taxa containing pathogenic fungi, including for comparison a few prominent taxa that do not contain mammalian pathogens. At least two human pathogens are included among these 'pseudofungi,' *Rhinosporidium seeberi* in the Mesomycetozoa (Herr et al. 1999; Mendoza et al. 2002) and *Pythium insidiosum* in the Oomycota (Cooke et al. 2000; Tyler 2001).

The true fungi are contained within the Kingdom Eumycota, which is comprised of four phyla, the Chytridiomycota, Zygomycota, Ascomycota, and Basidiomycota (Figure 2.1). Multiple phylogenetic analyses of molecular, morphological and other phenotypic data support the view that the Kingdom Fungi is monophyletic and originated with the Chytridiomycota (Sugiyama 1998). Fungi were classically defined by the limits of mating, and three phyla – Zygomycota, Ascomycota, and Basidiomycota – are named according to the product of sexual reproduction. For asexual fungi, the morphology and ontogeny of their mitosporic reproductive structures became the basis for the artificial form-taxa within the Class Deuteromycetes or Fungi Imperfecti. This alternative approach groups anamorphs according to their morphology, such as yeasts (blastomycetes) and molds (hyphomycetes). However, these morphology-based schemata do not represent phylogenetic relationships and thus do not reflect reliable biological or evolutionary information. Using available SSU rDNA sequences, the phylogram in Figure 2.2

Table 2.3 *Higher taxa of mycotic fungi and selected pseudofungi*

Kingdom	Phylum	Class	Order	Mycotic family or species
Protoctista	Acrasiomycota			
	Hyphocytriomycota			
	Labyrinthulomycota			
	Myxomycota	Dictyosteliomycetes		
		Myxomycetes	Physarales	
	Mesomycetozoa	Mesomycetozoea	Dermocystida	*Rhinosporidium seeberi*
			Ichthyophonida	
	Oomycota	Peronosporomycetes	Peronosporales	
			Pythiales	*Pythium insidiosum*
			Saprolegniales	
Eumycota	Chytridiomycota			*Basidiobolus ranarum*
	Zygomycota	Trichomyetes		
		Zygomycetes	Entomophthorales	*Ancylistaceae*
				Basidiobolaceae
			Mortierellales	*Mortierellaceae*
			Mucorales	*Cunninghamellaceae*
				Mucoraceae
				Saksenaeaceae
				Syncephalastraceae
				Thamnidiaceae
	Ascomycota	Archiascomycetes	Neolectomycetes	*Neolecta*
			Pneumocystidales	*Pneumocystidaceae*
			Taphrinales	*Taphrinaceae*
		Schizosaccharomycetes	Schizosaccharomycetales	*Schizosaccharomycetaeae*
		Hemiascomycetes	Saccharomycetales	*Ascoideaceae*
				Candidaceae
				Dipodascaceae
				Endomycetaceae
				Eremotheciaceae
				Lipomycetaceae
				Metschnikowiaceae
				Saccharomycetaceae
		Euascomycetes	Chaetothyriales	*Herpotrichiellaceae*
			Clavicipitales	*Clavicipitaceae*
			Dothideales	*Botryosphaeriaceae*
				Didymosphaeriaceae
				Dothioraceae
				Lophiostomataceae
				Mycosphaerellaceae
				Piedraiaceae
			Eurotiales	*Eremomycetaceae*
				Monascaceae
				Pseudeurotiaceae
				Thermoascaceae
				Trichocomaceae
			Hypocreales	*Hypocreaceae*
			Lectiales	*Leotiaceae*
			Microascales	*Microascaceae*
			Onygenales	*Arthrodermataceae*
				Gymnoascaceae
				Onygenaceae
			Ophiostomatales	*Ophiostomataceae*
			Pezizales	*Ascodesmiaceae*
			Phyllachorales	*Phyllachoraceae*
			Pleosporales	*Leptosphaeriaceae*
				Pleosporaceae
			Polystigmatales	*Polystigmataceae*
			Sordariales	*Chaetomiaceae*
				Coniochaetaceae
				Lasiosphaeriaceae
				Magnaporthaceae
				Myxotrichiaceae
				Sordariaceae

(Continued over)

Table 2.3 *Higher taxa of mycotic fungi and selected pseudofungi (Continued)*

Kingdom	Phylum	Class	Order	Mycotic family or species
	Basidiomycota	Hymenomycetes	Agaricales	*Coprinaceae*
			Stereales	*Bjerkanderaceae*
				Corticiaceae
				Schizophyllaceae
			Tremellales	*Filobasidiaceae*
		Urediniomycetes	Sporidiales	*Sporidiobolaceae*
		Ustilaginomycetes	Malasseziales	
			Microstromatales	
			Tilletiales	
			Trichosporonales	
			Ustilaginales	

References: de Hoog et al. 2000; Mendoza et al. 2002.

illustrates the relationships among many of the medical fungi.

As noted above, traditional classifications of the fungi relied on morphological characters and sexual reproduction (Guarro et al. 1999; de Hoog ct al. 2000). These approaches continue to serve well, especially for the identification of taxa. Of course, clinical mycology has more pragmatic goals: to obtain a rapid and accurate identification of a fungal isolate from a clinical specimen. Most human pathogenic fungi are anamorphic, and with rare exceptions (the homothallic ascomycete, *Piedraia hortai*), they are asexual when causing an infection. Therefore, traditional and commercial methods have combined morphological and physiological phenotypic characters to develop keys for the timely identification of fungal isolates from clinical specimens. Expedient shortcuts utilizing physiological tests are especially useful for the rapid identification of yeasts (Freydiere et al. 2001).

Beyond the bedside, clinical mycology benefits from an authentic phylogeny of pathogenic fungi. Knowledge of the lineages and nearest relatives of the mycotic fungi is useful because closely related species share similar genes, macromolecules, and biochemical processes. This information is invaluable in selecting comparative taxa to investigate the specificity of new diagnostic tests and antifungal antibiotics, as well as directing rational studies of pathogenesis, virulence factors, and mechanisms of drug resistance.

Modern systematics uses molecular data to assign asexual species to their likely positions within the taxonomic framework based on their most closely related taxa, as determined by molecular and phenotypic similarities. The ultimate goal – to create a unified taxonomy that reflects the phylogenetic evolution of the fungi – has been facilitated by advances in biotechnology, sequencing, and computer programs to analyze the data. As molecular evolution is the foundation of biology, the aim is to assign asexual species to their proper phylogenetic place in the classification, and it is feasible with sufficient molecular data and analyses. However, molecular data are lacking for many pathogenic anamorphic fungi (and countless saprobic anamorphs). Such species are usually classified with phenotypically similar fungi (Seifert and Gams 2001).

The classification of fungi will continue to be a work in progress for many years. New species are constantly being discovered. The phenotypic and genotypic characterization of each new species allows it to be placed on the mycological tree, which will often lead to refinement and adjustment of various branches of the phylogeny. This ongoing process will further elucidate the relationships among fungi and their evolution. Species, families, and orders will continue to be created, conflated, and reorganized. Thus revision is constant and inexorable, understandably but unfortunately causing some confusion, as familiar fungal names and taxonomic associations are periodically supplanted. This perpetual change is often accompanied by controversy and reasoned disagreement based on conflicting data or differing interpretations of the data.

The protocols for naming new species and higher taxa are well established, and mycologists follow the rules of Botanical Nomenclature. However, there is not a recognized mycological governing body to arbitrate controversies and decide controversial issues. Lacking an official consensus, the classification presented here is an amalgam of several respected sources, including recent publications (de Hoog et al. 2000; Kirk et al. 2001; Howard 2003) and useful web sites, such as Myconet (www.umu.se/myconet/Myconet.html), the Fungal Tree of Life (ocid.nacse.org/research/aftol/), the Index Fungorum (www.indexfungorum.org/Names/Names.asp), Deep Hypha (ocid.nacse.org/research/deephyphae/index.php?id=projects), and the Tree of Life Web Project (tolweb.org/tree?group=Fungi&contgroup=Eukaryotes).

PHYLUM CHYTRIDIOMYCOTA

Members of this phylum are organized into a single class with five orders. The Chytridiomycetes are characterized by motile asexual zoospores, and more than 900 species have been described (de Hoog et al. 2000; Kirk et al. 2001). The Orders Neocallimasticales and Blastocladiales

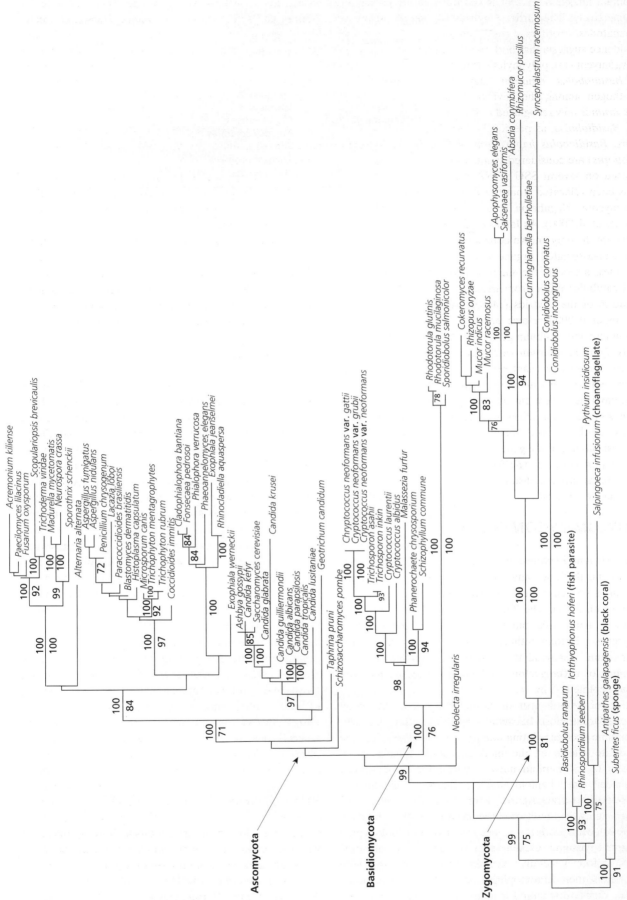

Figure 2.2 Phylogram of representative mycotic fungi. Single most likely tree based on a ML analysis of the nuclear SSU rDNA sequences. Nodes with statistically significant support by bootstrap (≥70 percent) or Bayesian (≥95 percent) analysis are indicated by the values below or above the nodes, respectively.

contain intestinal anaerobes and invertebrate pathogens, respectively. The Order Chytridiales has pathogens of nematodes, rotifers, and amphibians. Phylogenetic evidence suggests that both the Chytridiomycota and the Zygomycota are polyphyletic (Sugiyama 1998).

Basidiobolus ranarum may be the sole human pathogen among the chytridiomycetes (Table 2.3). *B. ranarum* is now recognized as the only pathogenic species of *Basidiobolus*, as previously described mycotic species (viz. *Basidiobolus haptosporus* and *Basidiobolus meristosporus*) are considered synonymous (Ribes et al. 2000). Based on several SSU rDNA phylogenies, *B. ranarum* has been relocated from the Zygomycota to the Chytridiomycota (Nagahama et al. 1995; Jensen et al. 1998; James et al. 2000). As James et al. observed, *Basidiobolus* spp. are the only known nonzoosporic fungi that possess a nucleus-associated organelle (NAO) containing microtubules, although the number and arrangement of the microtubules differ from those of the centriolar microtubules of the chytrids (McKerracher and Heath 1985; James et al. 2000). However, phylogenetic analyses of the α-tubulin and β-tubulin genes provided strong support for classifying *B. ranarum* within the Zygomycetes (Keeling 2003). These discrepant results may be explained by the apparently faster rate of evolution of the SSU rDNA of the chytrids and zygomycetes compared with the tubulin genes (James et al. 2000; Keeling 2003). Another phylogeny of these basal fungi analyzed sequences of the largest subunit of the gene encoding the DNA-dependent RNA polymerase II (*RPB1*), and these results indicated that *B. ranarum* is distantly related to the zygomycetes (Tanabe et al. 2004).

PHYLUM ZYGOMYCOTA

The Phylum Zygomycota has two classes, the Zygomycetes and the Trichomycetes, and together they contain about 1 100 described species (de Hoog et al. 2000; Kirk et al. 2001). The Trichomycetes are obligate parasites of arthropods, but two orders of the Class Zygomycetes, Entomophthorales and Mucorales, have important mammalian pathogens. The vegetative hyphae of zygomycetous molds are coenocytic or sparsely septated. The zygomycetes are defined by the mode of sexual reproduction, which is initiated by hyphal anastomosis between compatible mating types of the same species. Mating compatibility is determined by opposite alleles at the mating type locus. Following fusion of the haploid nuclei, meiosis, and one mitotic division, one haploid nucleus usually survives to develop into a zygospore. When exposed to the appropriate growth conditions, the zygospore of Mucorales species will produce a sporangiophore and a sporangium, yielding asexual sporangiospores that germinate to form vegetative haploid hyphae. Most members of the Order Entomophthorales reproduce asexually with forcibly discharged conidia.

The zygomycetous pathogens are listed in Table 2.4. Within the Entomophthorales, the two classically recognized pathogenic genera, *Basidiobolus* and *Conidiobolus*, cause primary subcutaneous infections and are responsible for subcutaneous phycomycosis and rhinoentomophthoromycosis, respectively (see Chapter 17, Subcutaneous zygomycoses). However, as noted above, molecular phylogenies have indicated that *B. ranarum* may be more closely related to nonflagellate chytridiomycetes (Nagahama et al. 1995; Jensen et al. 1998; James et al. 2000).

Most of the pathogens in this phylum are members of the Order Mucorales and agents of the more invasive infection, mucormycosis, whose name refers to the order and not the genus, *Mucor*, which is a rare cause of disease. However, as elaborated in Chapter 33, Systemic zygomycosis, most clinical mycologists favor the term 'zygomycosis,' referring to an infection caused by any member of the Class Zygomycetes. The most common agent of invasive zygomycosis in humans is *Rhizopus arrhizus* (*Rhizopus oryzae*) (Ribes et al. 2000).

PHYLUM ASCOMYCOTA

The Ascomycota is the largest fungal phylum, and it embraces 90 percent of the nearly 400 pathogenic fungi (Table 2.5). Their evolution has been estimated to date from 900 to 500 million years ago (Berbee and Taylor 1993; Taylor et al. 1999; Hedges et al. 2004). Approximately 33 000 ascomycetous species have been described, and they are unified by sexual reproduction that culminates in the production of asci and ascospores (de Hoog et al. 2000; Kirk et al. 2001). In addition, most of the ≈16 000 anamorphic species of fungi, which were first identified according to their conidial mitospores, have been placed within the existing ascomycetous classification based on their molecular and phenotypic affinities. The availability of nucleotide sequences has enabled asexual fungi to be placed with confidence among their closest relatives. The Ascomycota is also home to most of the lichenized fungi.

During vegetative growth, ascomycetous fungi produce either budding or fission yeast cells or branching substrate (vegetative) and aerial (asexually reproductive) hyphae. Hyphal septa possess simple pores that permit cytoplasmic continuity between cells. Some species, including many important human pathogens, are dimorphic, and the morphogenetic conversion to hyphae or yeasts (or spherules) is triggered by environmental signals such as temperature, available nutrients, or carbon dioxide concentration. The Phylum Ascomycota has three classes, the Archiascomycetes, the Hemiascomycetes (yeasts), and the large Euascomycetes (molds). According to SSU rDNA sequence analyses, the Archiascomycetes contains taxa that arose before the latter two classes, which are monophyletic lineages (Bruns et al. 1992; Berbee and Taylor 1993; Nishida and

Sugiyama 1994). Members of the Archiascomycetes are highly diverse, encompassing mycelial species, fission yeasts, and unicellular forms. The genera include many nonpathogens, such as *Protomyces* and *Saitoella*, as well as *Taphrina*, *Schizosaccharomyces*, and *Pneumocystis* (Table 2.5). Evidence from molecular and biochemical studies indicates that some of these taxa are monophyletic, but the status of others is questionable (Sjamsuridzal et al. 1997). Figure 2.3 depicts an rDNA phylogeny of representative ascomycetes.

Class Hemiascomycetes

The Hemiascomycetes (yeasts) comprise one large order, the Saccharomycetales, which was established in 1960 by V.I. Kudrjavzev (Kurtzman and Fell 1998; Kirk et al. 2001). The monumental compendium of yeasts, edited by Kurtzman and Fell, lists 42 teleomorphic and 15 anamorphic genera of ascomycetous yeasts (including several genera of Archiascomycetes). The other indispensable resource, written by Barnett et al. (2000), describes 678 species of ascomycetous and basidiomycetous yeasts. Members of the Saccharomycetes are characterized by:

- absent or only rudimentary hyphae
- vegetative cells that reproduce by budding or fission
- cell walls that lack chitin

- asci that are formed singly or in chains (Kurtzman and Fell 1998).

Phylogenetic analyses of SSU rDNA and RNA polymerase II (*RPB2*) gene sequences support a single evolutionary origin of the Saccharomycetales (Berbee and Taylor 1992; Liu et al. 1999). Perhaps the largest phylogenetic analysis of the Saccharomycetes entailed more than 500 species and a comparison of LSU rDNA sequences (Kurtzman and Robnett 1998).

The Saccharomycetales contains many yeasts of medical importance, such as species of *Candida*, which are the most prevalent pathogenic fungi (Table 2.5; Chapters 10, Management of superficial infections; 15, Superficial candidiasis; and 30, Candidiasis) (Calderone 2002). As mentioned earlier, *Candida* is a large and broadly defined anamorphic genus with 163 species (Meyer et al. 1998). They include *C. albicans*, *C. glabrata*, *Candida parapsilosis*, and *Candida tropicalis*. In a recent analysis of the medical yeasts and related hemiascomycetes, DNA sequences of six nuclear genes were analyzed using ML and Bayesian phylogenetic methods to produce the phylogram in Figure 2.4, which depicts the relationships of the most common mycotic yeasts with rare opportunists and strict saprobes (Diezmann et al. 2004). During the translation of mRNA to polypeptides, several species of *Candida* exhibit alternative codon usage; the codon CUG is translated as serine instead of leucine, as in most other

Table 2.4 *Mycotic taxa of Zygomycota*

Class	Order	Family	Species
Zygomycetes	Entomophthorales	*Ancylistaceae*	*Conidiobolus coronatus*
			Conidiobolus incongruus
			Conidiobolus lamprauges
		Basidiobolaceae	*Basidiobolus*
	Mortierellales	*Mortierellaceae*	*Mortierella polycephala*
			Mortierella wolfii
	Mucorales	*Cunninghamellaceae*	*Cunninghamella bertholletiae*
		Mucoraceae	*Absidia coerulea*
			Absidia corymbifera
			Apophysomyces elegans
			Chlamydoabsidia padenii
			Mucor amphibiorum
			Mucor circinelloides
			Mucor hiemalis
			Mucor indicus
			Mucor racemosus
			Mucor ramosissimus
			Rhizomucor miehei
			Rhizomucor pusillus
			Rhizomucor variabilis
			Rhizopus arrhizus (*oryzae*)
			Rhizopus azygosporus
			Rhizopus microsporus
			Rhizopus schipperae
			Rhizopus stolonifer
		Saksenaeaceae	*Saksenaea vasiformis*
		Syncephalastraceae	*Syncephalastrum racemosum*
		Thamnidiaceae	*Cokeromyces recuvatus*

References: Kurtzman and Fell 1998; Guarro et al. 1999; de Hoog et al. 2000.

Table 2.5 *Mycotic taxa of the Phylum Ascomycota*

Class	Order	Family	Teleomorph	Anamorph
Archiascomycetes	Pneumocystidales	Pneumocystidaceae	Pneumocystis jiroveci	
	Schizosaccharomycetales	Schizosaccharomycetaeae	Schizosaccharomyces pombe[a]	
	Taphrinales	Taphrinaceae	Taphrina[a]	
Hemiascomycetes	Saccharomycetales	Ascoideaceae	Yarrowia lipolytica	Candida lipolytica
		Candidaceae		Arxula
				Blastobotrys
				Candida albicans
				Candida castelli
				Candida glabrata
				Candida haemulonii
				Candida inconspicua
				Candida intermedia
				Candida maltosa
				Candida norvegica
				Candida parapsilosis
				Candida rugosa
				Candida tropicalis
				Candida viswanathii
				Candida zeylanoides
		Dipodascaceae		Geotrichum
			Galactomyces geotrichum	Geotrichum candidum
			Dipodascus (Blastoschizomyces) capitatus	Geotrichum capitatum
				Geotrichum clavatum
		Endomycetaceae	Stephanoascus ciferrii	Candida ciferrii
			Debaryomyces hansenii	Candida famata
			Pichia guilliermondii	Candida guilliermondii
			Pichia fermentans	Candida lambica
			Pichia norvegensis	Candida norvegensis
			Pichia (Hansenula) anomala	Candida pelliculosa
			Pichia jadinii	Candida utilis
		Eremotheciaceae	Eremothecium gossypii	Ashbya gossypii[a]
		Lipomycetaceae	Kluyveromyces marxianus	Candida kefyr
		Metschnikowiaceae	Clavispora lusitaniae	Candida lusitaniae
			Clavispora opuntiae[a]	
			Metschnikowia pulcherrima	Candida pulcherrima
		Saccharomycetaceae	Dekkera	Brettanomyces
			Debaryomyces	
			Zygoascus hellenicus	Candida hellenica
			Issatchenkia orientalis	Candida krusei
			Arxiozyma telluris	Candida pintolopesii
			Saccharomyces cerevisiae	
Euascomycetes	Chaetothyriales	Herpotrichiellaceae		Anthopsis deltoidea
				Cladophialophora arxii
				Cladophialophora bantiana
				Cladophialophora boppii
				Cladophialophora carrionii
				Cladophialophora devriesii
				Cladophialophora modesta
				Exophiala bergeri
				Exophiala castellanii (Exophiala mansonii)
				Wangiella dermatitidis (Exophiala dermatitidis)
				Exophiala jeanselmei
				Exophiala lecanii-corni
				Exophiala moniliae
				Exophiala pisciphila
				Exophiala salmonis
				Exophiala spinifera
				Fonsecaea compacta
				Fonsecaea pedrosoi
				Phaeoannellomyces (Exophiala) elegans
			Capronia semiimmersa	Phialophora americana

(Continued over)

Table 2.5 *Mycotic taxa of the Phylum Ascomycota (Continued)*

Class	Order	Family	Teleomorph	Anamorph
				Phialophora bubakii
				Phialophora europaea
				Phialophora reptans
				Phialophora repens
				Phialophora richardsiae
				Phialophora verrucosa
				Ramichloridium obovoideum (R. mackenziei)
				Ramichloridium schulzeri
				Rhinocladiella aquaspersa
				Rhinocladiella atrovirens
				Sarcinomyces phaeomuriformis
				Veronaea botryosa
				Xylohypha (Cladophialophora) emmonsii
	Clavicipitales			*Beauveria bassiana*
				Engyodontium album
				Paecilomyces fumosoroseus
				Paecilomyces javanicus
	Dothideales	*Botryosphaeriaceae*	*Botryosphaeria rhodina*	*Lasiodiplodia theobromae*
			Botryosphaeria subglobosa	*Sphaeropsis subglobosa*
		Didymosphaeriaceae	*Neotestudina rosatii*	
		Dothioraceae	*Discosphaerina fulvida*	*Aureobasidium pullulans*
				Cyphellophora laciniata
				Cyphellophora pluriseptata
			Sydowia polyspora	*Hormonema dematioides*
				Hortaea (Phaeoannellomyces) werneckii
				Scytalidium dimidiatum (Nattrassia mangiferae)
				Scytalidium hyalinum
				Pseudomicrodochium suttonii
		Lophiostomataceae		*Madurella grisea*
				Madurella mycetomatis
				Pseudochaetosphaeronema larense
				Pyrenochaeta mackinnonii
				Pyrenochaeta romeroi
				Pyrenochaeta unguis-hominis
				Tetraploa aristata
		Mycosphaerellaceae		*Cladosporium cladosporioides*
				Cladosporium elatum
			Mycosphaerella tassiana	*Cladosporium herbarum*
				Cladosporium oxysporum
				Cladosporium sphaerospermum
		Mytilinidiaceae		*Taeniolella exilis*
				Taeniolella stilbaspora
		Piedraiaceae	*Piedraia hortai*	
	Eurotiales	*Eremomycetaceae*	*Eremomyces langeronii*	*Arthrographis kalrae*
		Monascaceae	*Monascus ruber*	*Basipetospora rubra*
		Pseudeurotiaceae	*Pseudeurotium ovale*	*Sporothrix*
		Thermoascaceae	*Byssochlamys*	
			Thermoascus crustaceus	*Paecilomyces crustaceus*
		Trichocomaceae	*Petromyces alliaceus*	*Aspergillus alliaceus*
				Aspergillus avenaceus
				Aspergillus caesiellus
				Aspergillus candidus
				Aspergillus carneus
			Eurotium chevalieri	*Aspergillus chevalieri*
				Aspergillus clavato-nanicus
				Aspergillus clavatus
				Aspergillus conicus
				Aspergillus deflectus
			Neosartorya fischeri	*Aspergillus fischerianus*
			Fennellia flavipes	*Aspergillus flavipes*
				Aspergillus flavus

(Continued over)

Table 2.5 *Mycotic taxa of the Phylum Ascomycota (Continued)*

Class	Order	Family	Teleomorph	Anamorph
				Aspergillus fumigatus
				Aspergillus granulosus
			Eurotium herbariorum	*Aspergillus glaucus*
				Aspergillus janus
				Aspergillus japonicus
			Emericella nidulans	*Aspergillus nidulans*
			Emericella echinulata	*Aspergillus nidulans*
				Aspergillus niger
			Fennellia nivea	*Aspergillus niveus*
				Aspergillus ochraceus
				Aspergillus oryzae
			Eurotium repens	*Aspergillus reptans*
				Aspergillus restrictus
				Aspergillus sclerotiorum
				Aspergillus sydowii
				Aspergillus tamarii
				Aspergillus terreus
			Emericella quadrilineata	*Aspergillus tetrazonus*
			Neosartorya pseudofischeri	*Aspergillus thermomutatus*
			Neosartorya spinosa	*Aspergillus spinosus*
			Emericella unguis	*Aspergillus unguis*
				Aspergillus ustus
				Aspergillus varians
				Aspergillus versicolor
			Eurotium amstelodami	*Aspergillus vitis (Aspergillus hollandicus)*
				Paecilomyces lilacinus
				Paecilomyces marquandii
				Paecilomyces puntonii
				Paecilomyces variotii
				Paecilomyces viridis
				Penicillium aurantiogriseum
				Penicillium brevicompactum
				Penicillium chrysogenum
				Penicillium citrinum
				Penicillium commune
				Penicillium decumbens
				Penicillium expansum
				Penicillium griseofulvum
				Penicillium marneffei
				Penicillium piceum
				Penicillium purpurogenum
				Penicillium rugulosum
				Penicillium spinulosum
				Penicillium verruculosum
				Polypaecilum insolutum
	Hypocreales	*Hypocreaceae*	*Neocosmospora vasinfecta*	*Acremonium*
				Acremonium atrogriseum
				Acremonium blochii
				Acremonium curvulum
				Acremonium falciforme
				Acremonium hyalinulum
				Acremonium kiliense
				Acremonium potronii
				Acremonium recifei
				Acremonium roseogriseum
				Acremonium strictum
				Acremonium spinosum
				Cylindrocarpon cyanescens
			Nectria radicicola	*Cylindrocarpon destructans*
				Cylindrocarpon lichenicola
				Fusarium antophilum
			Cosmospora episphaerica	*Fusarium aquaeductuum*
				Fusarium chlamydosporum
				Fusarium dimerum
				Fusarium incarnatum

(Continued over)

Table 2.5 *Mycotic taxa of the Phylum Ascomycota (Continued)*

Class	Order	Family	Teleomorph	Anamorph
				Fusarium napiforme
			Gibberella nygamai	*Fusarium nygamai*
				Fusarium oxysporum
			Gibberella fujikuroi	*Fusarium proliferatum*
			Gibberella fujikuroi var. subglutinans	*Fusarium subglutinans*
			Nectria haematococca var. breviconia	*Fusarium solani*
			Gibberella moniliformis	*Fusarium verticillioides*
				Metarhizium anisopliae
				Trichoderma harzianum
			Hypocrea koningii	*Trichoderma konigii*
				Trichoderma longibrachiatum
			Hypocrea pseudokoningii	*Trichoderma pseudokoningii*
			Hypocrea rufa	*Trichoderma viride*
	Leotiales	Leotiaceae		*Ochroconis constricta*
				Ochroconis gallopavum
				Ochroconis humicola
				Ochroconis tshawytschae
				Pleurophoma cava
				Pleurophomopsis lignicola
				Pleurophoma pleurospora
	Microascales	Microascaceae	*Petriella setifera*	*Graphium*
			Pseudallescheria boydii	*Graphium eumorphum*
				Graphium putredinis
			Petriella setifera	*Scedosporium*
			Pseudallescheria boydii	*Scedosporium apiospermum*
				Scedosporium prolificans (S. inflatum)
			Microascus cinereus	*Scopulariopsis cinereus*
			Microascus cirrosus	*Scopulariopsis paisii*
			Microascus manginii	*Scopulariopsis candida*
				Scopulariopsis acremonium
				Scopulariopsis asperula
			Microascus brevicaulis	*Scopulariopsis brevicaulis (S. koningii)*
				Scopulariopsis brumptii
				Scopulariopsis candida
				Scopulariopsis flava
				Scopulariopsis fusca
	Onygenales	Arthrodermataceae		*Epidermophyton floccosum*
				Keratinomyces ceretanicus
			Arthroderma corniculatum	*Microsporum*
			Arthroderma borellii	*Microsporum amazonicum*[a]
				Microsporum audouinii
			Arthroderma otae	*Microsporum canis*
			Arthroderma cajetani	*Microsporum cookei*
				Microsporum equinum
				Microsporum ferrugineum
			Arthroderma fulvum	*Microsporum fulvum*
			Arthroderma grubyi	*Microsporum gallinae*
			Arthroderma gypseum	*Microsporum gypseum*
			Arthroderma incurvatum	*Microsporum gypseum*
			Arthroderma obtusum	*Microsporum nanum*
			Arthroderma persicolor	*Microsporum persicolor*
				Microsporum praecox
			Arthroderma racemosum	*Microsporum racemosum*
			Arthroderma grubyi	*Microsporum vanbreuseghemii*
			Arthroderma uncinatum	*Trichophyton ajelloi*
				Trichophyton concentricum
				Trichophyton duboisii
				Trichophyton equinum
			Arthroderma gloriae	*Trichophyton gloriae*[a]
				Trichophyton gourvilii
			Arthroderma vanbreuseghemii	*Trichophyton mentagrophytes*

(Continued over)

Table 2.5 *Mycotic taxa of the Phylum Ascomycota (Continued)*

Class	Order	Family	Teleomorph	Anamorph
				var. *interdigitale*
				Trichophyton kanei
				Trichophyton krajdenii
				Trichophyton megninii
			Arthroderma benhamiae	*Trichophyton mentagrophytes*
				Trichophyton phaseoliforme[a]
				Trichophyton raubitschekii
				Trichophyton rubrum
				Trichophyton schoenleinii
			Arthroderma simii	*Trichophyton simii*
				Trichophyton soudanense
			Arthroderma insingulare	*Trichophyton terrestre*[a]
			Arthroderma lenticulare	*Trichophyton terrestre*[a]
			Arthroderma quadrifidum	*Trichophyton terrestre*[a]
				Trichophyton thuringiense
				Trichophyton tonsurans
			Arthroderma gertleri	*Trichophyton vanbreuseghemii*
				Trichophyton verrucosum
				Trichophyton violaceum
				Trichophyton yaoundei
		Gymnoascaceae	*Gymnoascella dankaliensis*	
			Gymnoascella hyalinospora	
			Narasimhella hyalinospora	
				Malbranchea gypsea
				Malbranchea pulchella
			Arachnomyces nodosetosus	*Onychocola canadensis*
		Onygenaceae	*Ajellomyces dermatitidis*	*Blastomyces dermatitidis*
			Aphanoascus fulvescens	*Chrysosporium*
			Arthroderma tuberculatum	*Chrysosporium*
			Nannizziopsis vriesii	*Chrysosporium*
				Chrysosporium inops
			Aphanoascus keratinophilus	*Chrysosporium keratinophilum*[a]
				Chrysosporium pannicola
			Uncinocarpus queenslandicus	*Chrysosporium queenslandicum*
				Chrysosporium tropicum
			Uncinocarpus orissi	*Chrysosporium zonatum*
				Coccidioides immitis
				Coccidioides posadasii
			Ajellomyces crescens	*Emmonsia crescens*
				Emmonsia parva
				Emmonsia pasteuriana
			Ajellomyces capsulatus	*Histoplasma capsulatum*
			Ajellomyces capsulatus	*Histoplasma capsulatum* var. *duboisii*
				Histoplasma capsulatum var. *farciminosum*
				Lacazia loboi
			Uncinocarpus reesii	*Malbranchea*[a]
			Neoarachnotheca keratinophila	*Myriodontium keratinophilum*
				Paracoccidioides brasiliensis
	Ophiostomatales	*Ophiostomataceae*	*Ophiostoma stenoceras*	*Sporothrix*
				Sporothrix schenckii
	Pezizales	*Ascodesmiaceae*		*Cephaliophora irregularis*
	Phyllachorales	*Phyllachoraceae*	*Plectosphaerella cucumerina*	*Plectosporium tabacinum*
	Pleosporales	*Leptosphaeriaceae*	*Leptosphaeria coniothyrium*	*Coniothyrium fuckelii*
			Leptosphaeria senegalensis	
			Leptosphaeria thompkinsii	
				Microsphaeropsis olivacea
		Pleosporaceae		*Alternaria alternata*
				Alternaria chartarum
				Alternaria chlamydospora
				Alternaria dianthicola
			Lewia infectoria	*Alternaria infectoria*
				Alternaria longipes
				Alternaria tenuissima

(Continued over)

Table 2.5 *Mycotic taxa of the Phylum Ascomycota (Continued)*

Class	Order	Family	Teleomorph	Anamorph
			Cochliobolus australiensis	*Bipolaris australiensis*
			Cochliobolus hawaiiensis	*Bipolaris hawaiiensis*
				Bipolaris papendorfii
			Cochliobolus spiciferus	*Bipolaris spicifera*
				Botryomyces caespitosus
				Corynespora cassiicola
				Curvularia brachyspora
				Curvularia clavata
			Cochliobolus geniculatus	*Curvularia geniculata*
			Cochliobolus lunata	*Curvularia lunata*
			Cochliobolus pallescens	*Curvularia pallescens*
				Curvularia senegalensis
			Cochliobolus verruculosus	*Curvularia verruculosa*
				Dichotomophthoropsis nymphaearum
				Dissitimurus exedrus
				Drechslera biseptata
				Exserohilum longirostratum
				Exserohilum mcginnisii
			Setosphaeria rostrata	*Exserohilum rostratum*
				Mycocentrospora acerina
				Papulaspora equi
				Phaeosclera dematioides
				Phaeotrichoconis crotalariae
				Phoma dennissiii var. oculo-hominis
				Phoma eupyrena
				Phoma glomerata
				Phoma herbarum
				Phoma hibernica
				Phoma minutella
				Phoma minutispora
				Phoma sorghina
				Polycytella hominis
				Ulocladium botrytis
				Ulocladium chartarum
	Polystigmatales	*Polystigmataceae*		*Colletotrichum coccodes*
				Colletotrichum crassipes
				Colletotrichum dematium
			Glomerella cingulata	*Colletotrichum gloeosporioides*
			Glomerella tucumanensis	*Colletotrichum graminicola*
	Sordariales	*Chaetomiaceae*	*Thielavia terrestris*	*Acremonium alabamense*
			Ascotricha chartarum	*Dicyma ampullifera*
			Chaetomium atrobrunneum	
			Chaetomium funicola	
			Chaetomium globosum	
			Chaetomium murorum	
			Chaetomium perlucidum	
			Chaetomium (Achaetomium) strumarium	
			Corynascus heterothallicus	*Myceliophthora thermophila*
				Myceliophthora lutea
				Phaeoisaria dematidis
				Staphylotrichum coccosporum
		Coniochaetaceae		*Acrophialophora fusispora*
			Coniochaeta ligniaria	*Lecythophora hoffmannii*
				Lecythophora mutabilis
				Phaeoacremonium inflatipes
				Phaeoacremonium parasiticum
				Phaeoacremonium rubrigenum
				Phialemonium curvatum
				Phialemonium obovatum
		Lasiosphaeriaceae	*Arnium leporinum*	
				Arthrinium phaeospermum
				Nigrospora sphaerica
		Magnaporthaceae	*Omnidemptus* sp.	*Mycoleptodiscus indicus*
		Myxotrichaceae	*Myxotrichum deflexum*[a]	

(Continued over)

Table 2.5 *Mycotic taxa of the Phylum Ascomycota (Continued)*

Class	Order	Family	Teleomorph	Anamorph
				Geomyces pannorum
				Oidiodendron cerealis
				Ovadendron sulphureo-ochraceum
		Sordariaceae	*Neurospora sitophila*	*Chrysonilia sitophila*
				Thermomyces lanuginosus (*Humicola lanuginosa*)
		Xylariaceae		*Nodulisporium*

References: Summerbell & Kane 1997; Kurtzman & Fell, 1998; Guarro et al. 1999; de Hoog et al. 2000; Machouart-Dubach et al. 2002; Revankar et al. 2002; Barron et al. 2003; Schell 2003; Sigler 2003a, b; Summerbell 2003; Cano et al. 2004; Sutton et al. 2004.
a) Questionable human pathogen.

eukaryotes (Massey et al. 2003). However, *Candida* is not a monophyletic genus, and species vary in codon usage as well as the number of isoprene units in coenzyme Q and other characters (Diezmann et al. 2004).

Unlike molds, which are identified primarily by observing their colonial and microscopic morphology, yeasts vary little in morphology and are routinely speciated according to a battery of physiological criteria (Freydiere et al. 2001; Hazen and Howell 2003). However, a commercial method has been recently developed for the automated sequencing of the D2 region of LSU rDNA to identify both yeast and mold cultures within 24 hours (Hall et al. 2003, 2004).

Class Euascomycetes

This large class of molds holds most of the Ascomycota, including most of the anamorphic taxa (see Table 2.5). The phylogenies of numerous groups of Euascomycetes have been analyzed using a number of gene sequences (e.g. SSU and LSU rDNA, *RPB1*, *EF2*) and phylogenetic approaches (e.g. MP, NJ, Bayesian) (Prillinger et al. 2002; Reeb et al. 2004). The number of orders and families remains in flux, and the classification of the Euascomycetes will likely be mutable for many years as there is a steady and natural accrual of new information (Kirk et al. 2001; Prillinger et al. 2002; Eriksson et al. 2004). Some of the more medically relevant groups are described below.

ORDER ONYGENALES

Many notable pathogens are members of the Order Onygenales, which was originally defined using detailed morphological observations and other phenotypic data (Currah 1985). This order is noted for the production of enclosed ascomata, either cleistothecia or gymnothecia. The anamorphs produce distinctive thallic conidia, such as macroconidia, microconidia, chlamydospores, and arthroconidia. Two medically prominent families are the *Arthrodermataceae*, which contains the dermatophytes and related keratinophilic saprobes, and the *Onygenaceae*, which includes the endemic, dimorphic pathogens (Table 2.5). The more famous onygenalean genera are *Blastomyces*, *Coccidioides*, *Epidermophyton*, *Histoplasma*,

Microsporum, *Paracoccidioides*, and *Trichophyton*. The *Arthrodermataceae*, *Onygenaceae*, and *Gymnoascaceae* were subjected to a masterful phylogenetic investigation that involved a comparison of sequences of both the SSU and LSU rDNA using NJ and ML methods (Sugiyama et al. 2002). From this analysis of 80 taxa, four major clades emerged. One clade included saprobes and systemic pathogens, while the *Arthrodermataceae* were ensconced in a different clade. Two pathobiological phenotypes, keratinolysis and dimorphism, are likely apomorphic, deriving independently by convergent evolution, as they occurred in taxa in different clades. For example, since *C. immitis* was in a different clade from the *Ajellomyces* spp., pathogenicity and dimorphism are apparently apomorphic (Sugiyama et al. 2002).

Initially described in the 1840s, the dermatophytes are among the earliest documented human pathogens (Howard et al. 2003). The many species of *Microsporum* and *Trichophyton* are closely related and similar in their morphology, physiology, degradative enzymes (e.g. keratinases, elastases), and antigens (Kane et al. 1997). However, some species exhibit marked differences in ecology and host range. As mentioned above, the evolution of dermatophytes toward zoophilism and anthropophilism correlates with the loss of both sexual and asexual sporulation. Most of the clinical species are resolutely anamorphic and defined by subtle phenotypic features. After decades of taxonomic revisions, several molecular phylogenies are finally resolving the phylogeny of the dermatophytes, but classifying dermatophytes remains a dynamic enterprise (Gräser et al. 1999; Makimura et al. 1999).

ORDER EUROTIALES

In the Order Eurotiales, sexual reproduction results in perithecia or cleistothecia, and the anamorphs produce a variety of morphologically distinct conidiophores and phialoconidia. As indicated in Table 2.5, the family *Trichocomaceae* includes *Penicillium* and *Aspergillus* (Berbee et al. 1995; LoBuglio and Taylor 1995; Summerbell 2003). The Order Ophiostomatales contains the dimorphic *Sporothrix schenckii* and a number of closely related phytopathogens (de Beer et al. 2003). Numerous dematiaceous pathogens are

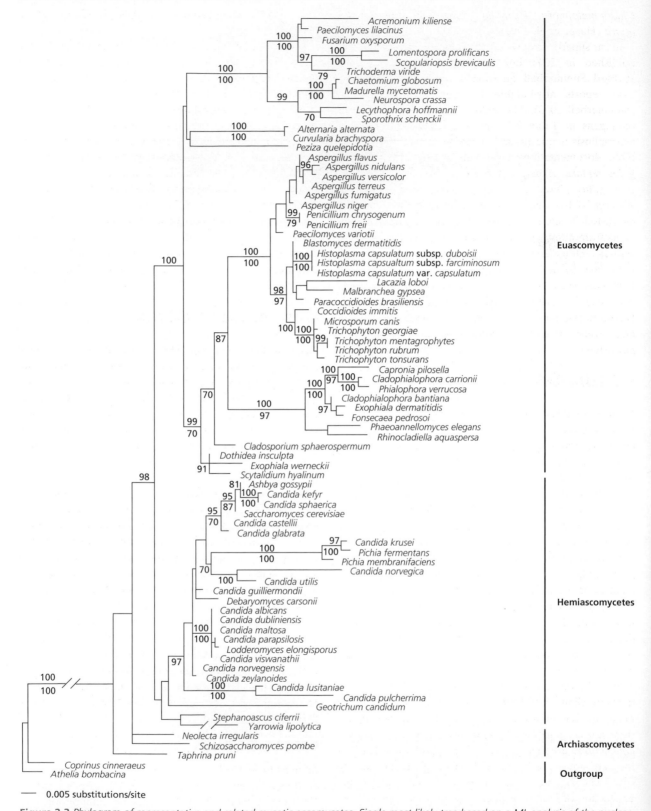

Figure 2.3 *Phylogram of representative and related mycotic ascomycetes. Single most likely tree based on a ML analysis of the nuclear SSU rDNA sequences. Nodes with statistically significant support by bootstrap (≥70 percent) or Bayesian (≥95 percent) analysis are indicated by the values below or above the nodes, respectively. Vertical lines on the right denote the three classes of the Ascomycota.*

classified in the Order Dothideales (Spatafora et al. 1995; de Hoog et al. 1999; Sterflinger et al. 1999). Although the exact phylogenetic placement of these

taxa is the subject of ongoing investigations, many familial relationships have been resolved; for example, the family *Herpotrichiellaceae* contains species of

Cladophialophora, *Exophila*, *Fonsecaea*, and *Phialophora* (Haase et al. 1999).

In a superb review of the Eurotiales, which was published in 2003 but apparently written in 2000, Richard Summerbell questioned the validity of many case reports attributable to species of *Aspergillus* (Summerbell 2003). The following species are listed as pathogens in Table 2.5, but their status as agents of aspergillosis is questionable because of the lack of definitive documentation: *Aspergillus glaucus* group, *Aspergillus reptans*, *Aspergillus restrictus*, *Aspergillus unguis*, *Aspergillus janus*. Of course, many other species of *Aspergillus* have been recovered from patients but are not listed because they were isolated only as contaminants from nonsterile clinical specimens (e.g. skin, sputa, sinuses). Other *Aspergillus* spp. are listed even though they have to date only caused infection in nonhuman mammals, such as *Aspergillus rugulovalvus* (*Emericella rugulosa*) in cattle and *Aspergillus deflectus* in dogs. Nevertheless, they have been cited in numerous texts and reviews. Exclusively nonmammalian pathogens are not listed.

PHYLUM BASIDIOMYCOTA

Each of the three classes of the Phylum Basidiomycota – Urediniomycetes, Ustilaginomycetes, and Hymenomycetes – contains species of yeasts and molds, as well as species that are capable of infecting mammalian hosts. Similar to the Ascomycota, the basidiomycetes are ubiquitous. Many are phytopathogens, such as the rusts and smuts, which are found in the Orders Uredinales and Ustilaginales, respectively. Approximately 30 000 basidiomycetous species have been described, yet only a dozen or so cause human infection (Table 2.6). (Toxigenic fungi, such as poisonous mushrooms, are not included in Table 2.6.)

The basidiomycetes are characterized by the sexual production of basidia and basidiospores. Sexual reproduction is initiated by the fusion of compatible mating partners, and a dikaryon is often formed in which each cell contains two haploid nuclei, one from each partner. This dikaryophase may be protracted, and the hyphae develop clamp connections to ensure that each cell maintains the requisite two nuclei. The clamp connections are accompanied by complex septal pores. Clamp connections are only produced by basidiomycetes, but they are not made by every member of the Basidiomycota. The signature feature of basidiomycetes is the production of basidia, which culminates the sexual cycle. The basidium is the site of karyogamy, which is followed by meiosis and mitosis, typically yielding four haploid basidiospores. The basidiospores germinate to form vegetative septate hyphae or budding yeast cells. Similar to the Ascomycota, much phenotypic and molecular evidence suggests that the Basidiomycota is monophyletic (Oberwinkler 1987; Swann and Taylor

1993, 1995b; McLaughlin et al. 1995; Prillinger et al. 2002).

Most described basidiomycetes are members of the Class Hymenomycetes, and they share the feature of distinctive, perforated, dolipore septa (Moore 1978). The largest subdivision of the Hymenomycetes is the Homobasidiomycetes, which houses mushrooms, puffballs, and other higher fungi. The classification of the higher taxa of Hymenomycetes is not completely resolved (Binder and Hibbett 2002). Another significant clade of the Hymenomycetes is the Tremellales, which contains many of the basidiomycetous yeasts (Swann and Taylor 1995a, c; Fell et al. 2000). The genera of greatest clinical importance are *Filobasidiella* (*Cryptococcus*) (Sivakumaran et al. 2003), *Malassezia* (Kano et al. 1999) *Rhodosporidium* (*Rhodotorula*), *Sporidiobolus* (*Sporobolomyces*), and *Trichosporon* (Guého et al. 1992; Sugita et al. 2002).

CLINICAL APPLICATIONS

As mentioned earlier, DNA sequences have been exploited not only to determine phylogenetic relationships but to develop molecular methods to identify cultures and detect fungal DNA in clinical material, to genotype species and strains for molecular epidemiology, and to analyze the population structure of pathogens (Taylor et al. 1999a; Burnett 2003; Xu and Mitchell 2003).

Phylogenetic studies have greatly abetted efforts to develop more accurate methods to identify species (and strains) of many pathogenic fungi. Signature DNA sequences of the target species are identified, often within the rDNA gene cluster, and specificity can be evaluated by including the closest taxa based on phylogenies of the target species (Iwen et al. 2002). PCR methods are rapid, definitive, specific, and considerably more accurate and reliable than conventional methods of identification. Increasing reports confirm the utility of PCR methods to identify isolates of dermatophytes (Kanbe et al. 2003; Shin et al. 2003), dimorphic pathogens (Walsh et al. 2003), and opportunistic molds (Kanbe et al. 2002; de Aguirre et al. 2004) and yeasts (Elie et al. 1998; Ahmad et al. 2002; Luo and Mitchell 2002; Selvarangan et al. 2003). A new commercial method, originally developed for the identification of bacteria, has proven successful in identifying fungal cultures (Hall et al. 2003, 2004). After amplifying the D2 region of the LSU rDNA of the isolate, the amplicon is then purified, cycle-sequenced, and the nucleotide sequence is compared with a database of D2 sequences of >1 000 mycotic fungi to identify the closest taxon. Multiple isolates of the same species usually exhibit <1.00 percent variation (Hall et al. 2003, 2004).

The natural next step is to develop methods to recognize and diagnose mycotic infections directly from clinical specimens, and many DNA probes and PCR

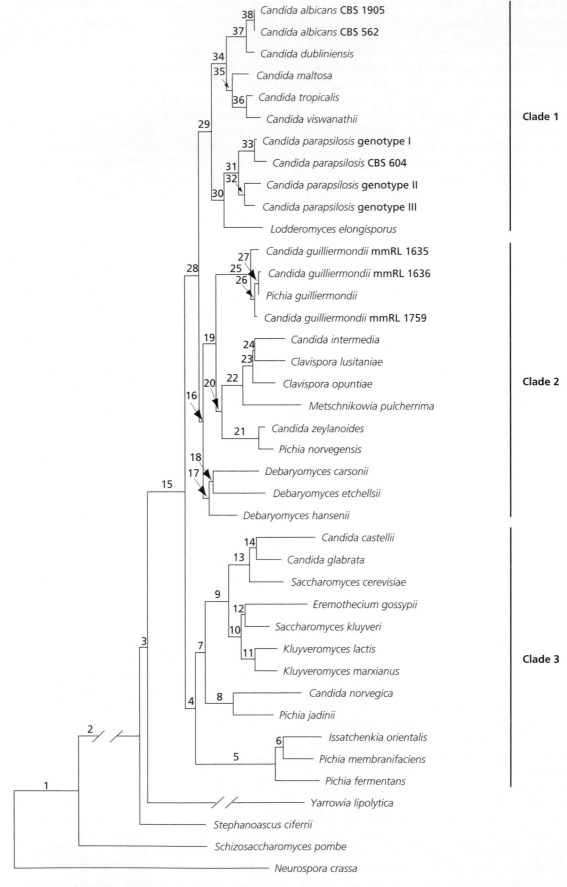

—— 0.05 substitutions/site

Table 2.6 *Mycotic taxa of the Phylum Basidiomycota*

Class	Order	Family	Teleomorph	Anamorph
Hymenomycetes	Agaricales	*Coprinaceae*	*Coprinus cinereus*	*Hormographiella aspergillata*
				Hormographiella verticillata
	Stereales	*Bjerkanderaceae*	*Bjerkandera adusta*	
		Corticiaceae	*Phanerochaete chrysosporium*	*Sporotrichum pruinosum*
		Schizophyllaceae	*Schizophyllum commune*	
	Tremellales	*Filobasidiaceae*		*Cryptococcus albidus*
				Cryptococcus ater
				Cryptococcus curvatus
				Cryptococcus laurentii
				Cryptococcus macerans
			Filobasidiella bacillispora	*Cryptococcus neoformans* var. *gattii*
			Filobasidiella neoformans	*Cryptococcus neoformans* var. *grubii*
			Filobasidiella neoformans	*Cryptococcus neoformans* var. *neoformans*
			Filobasidium unigutttulatum	*Cryptococcus uniguttulatus*
	Trichosporonales			*Cryptococcus humicola* complex
				Trichosporon asahii
				Trichosporon asteroides
				Trichosporon cutaneum
				Trichosporon inkin
				Trichosporon loubieri
				Trichosporon mucoides
				Trichosporon ovoides
Urediniomycetes	Sporidiales	*Sporidiobolaceae*	*Rhodosporidium diobovatum*	*Rhodotorula glutinis*
			Rhodosporidium sphaerocarpum	*Rhodotorula glutinis*
			Rhodosporidium toruloides	*Rhodotorula glutinis*
				Rhodotorula minuta
				Rhodotorula mucilaginosa (*R. rubra*)
			Sporidiobolus salmonicolor	*Sporobolomyces salmonicolor*
			Sporidiobolus johnstonii	*Sporobolomyces holsaticus*
				Sporobolomyces roseus
Ustilaginomycetes	Malasseziales			*Malassezia furfur*
				Malassezia globosa
				Malassezia obtusa
				Malassezia pachydermatis
				Malassezia restricta
				Malassezia slooffiae
				Malassezia sympodialis
				Moniliella suaveolens[a]
	Microstromatales			*Fugomyces* (*Cerinosterus*) *cyanescens*
	Tilletiales			*Tilletiopsis minor*

References: Kurtzman and Fell 1998; Guarro et al. 1999; de Hoog et al. 2000; Fell et al. 2000; Takashima et al. 2001.
a) Questionable human pathogen.

Figure 2.4 *Combined ML analysis of six genes (ACT1, EF2, RPB1, RPB2, SSU and LSU rDNA) for 38 taxa of Hemiascomycetes and two outgroup species, an Archiascomycetes (Schizosaccharomyces pombe) and a Euascomycetes (Neurospora crassa). The following nodes are supported by heterogeneous Bayesian posterior probabilities ≥95 percent as calculated in the combined analysis: 2, 4, 5, 7–11, 13, 15, 16, 21–23, 25–38. Branch lengths leading to Yarrowia lipolytica (0.19157), and the outgroup taxa (0.31967) were shortened to fit the figure. Vertical lines on right indicate the three major clades (Diezmann et al. 2004).*

tests have been developed to identify specific fungal DNA in clinical specimens (Yeo and Wong 2002). These topics are amplified in Chapters 4, Laboratory diagnosis; and 5, Mycoserology and molecular diagnosis.

ACKNOWLEDGMENTS

This review was supported by grants from the NIH, AI 25783, AI 28836 and AI 44975. Invaluable comments were generously provided by Stephanie Diezmann and Rytas J. Vilgalys. I am most grateful to Stephanie Diezmann for generously creating the figures.

REFERENCES

Ahmad, S., Khan, Z., et al. 2002. Seminested PCR for diagnosis of candidemia: comparison with culture, antigen detection, and biochemical methods for species identification. *J Clin Microbiol*, **40**, 2483–9.

Bakkeren, G., Kronstad, J.W. and Levesque, C.A. 2000. Comparison of AFLP fingerprints and ITS sequences as phylogenetic markers in Ustilaginomycetes. *Mycologia*, **92**, 510–21.

Baldauf, S.L. 2003a. The deep roots of eukaryotes. *Science*, **300**, 1703–6.

Baldauf, S.L. 2003b. Phylogeny for the faint of heart: a tutorial. *Trends Genet*, **19**, 345–51.

Baldauf, S.L. and Palmer, J.D. 1993. Animals and fungi are each other's closest relatives: congruent evidence from multiple proteins. *Proc Natl Acad Sci U S A*, **90**, 11558–62.

Barnett, J.A., Payne, R.W. and Yarrow, D. 2000. *Yeasts: characteristics and identification*, 3rd edition. Cambridge, UK: Cambridge University Press.

Barron, M.A., Sutton, D.A., et al. 2003. Invasive mycotic infections caused by *Chaetomium perlucidum*, a new agent of cerebral phaeohyphomycosis. *J Clin Microbiol*, **41**, 5302–7.

Berbee, M.L. and Taylor, J.W. 1992. Detecting morphological convergence in true fungi, using 18S rRNA gene sequence data. *Biosystems*, **28**, 117–25.

Berbee, M.L. and Taylor, J.W. 1993. Dating the evolutionary radiations of the true fungi. *Can J Bot*, **71**, 1114–27.

Berbee, M.L. and Taylor, J.W. 1999. Fungal phylogeny. In: Oliver, R.P. and Schweizer, M. (eds), *Molecular fungal biology*. Cambridge, UK: Cambridge University Press, 21–77.

Berbee, M.L., Yoshimura, A., et al. 1995. Is *Penicillium* monophyletic? An evaluation of phylogeny in the family Trichocomaceae from 18S, 5.8S and ITS ribosomal DNA sequence data. *Mycologia*, **87**, 210–22.

Binder, M. and Hibbett, D.S. 2002. Higher-level phylogenetic relationships of Homobasidiomycetes (mushroom-forming fungi) inferred from four rDNA regions. *Mol Phylogenet Evol*, **22**, 76–90.

Brandt, M.E., Bragg, S.L. and Pinner, R.W. 1993. Multilocus enzyme typing of *Cryptococcus neoformans*. *J Clin Microbiol*, **31**, 2819–23.

Bruns, T.D., White, T.J. and Taylor, J.W. 1991. Fungal molecular systematics. *Annu Rev Ecol Systematics*, **22**, 525–64.

Bruns, T.D., Vilgalys, R.J., et al. 1992. Evolutionary relationships within the fungi: analyses of nuclear small subunit rRNA sequences. *Mol Phylogenet Evol*, **1**, 231–41.

Burnett, J. 2003. *Fungal populations and species*. Oxford, UK: Oxford University Press.

Calderone, R.A. (ed.) 2002. *Candida and Candidiasis*. Washington, DC: ASM Press.

Cano, J., Guarro, J. and Gené, J. 2004. Molecular and morphological identification of *Colletotrichum* species of clinical interest. *J Clin Microbiol*, **42**, 2450–4.

Cooke, D.E., Drenth, A., et al. 2000. A molecular phylogeny of *Phytophthora* and related oomycetes. *Fungal Genet Biol*, **30**, 17–32.

Currah, R.S. 1985. Taxonomy of the Onygenales: Arthrodermataceae, Gymnoascaceae, Myxotrichaceae and Onygenaceae. *Mycotaxon*, **24**, 1–216.

de Aguirre, L., Hurst, S.F., et al. 2004. Rapid differentiation of *Aspergillus* species from other medically important opportunistic molds and yeasts by PCR-enzyme immunoassay. *J Clin Microbiol*, **42**, 3495–504.

de Beer, Z.W., Harrington, T.C., et al. 2003. Phylogeny of the *Ophiostoma stenoceras Sporothix schenckii* complex. *Mycologia*, **95**, 434–41.

de Hoog, G.S., Zalar, P., et al. 1999. Relationships of dothideaceous black yeasts and meristematic fungi based on 5.8S and ITS2 rDNA sequence comparison. *Stud Mycol*, **43**, 31–7.

de Hoog, G.S., Guarro, J., et al. 2000. *Atlas of clinical fungi*, 2nd edition. Utrecht, The Netherlands: Centraalbureau voor Schimmelcultures.

Diezmann, S., Cox, C., et al. 2004. Phylogeny and evoluation of *Candida* and related taxa: a multilocus analysis. *J Clin Microbiol*, **42**, 5624–35.

Dujon, B., Sherman, D.R., et al. 2004. Genome evolution in yeasts. *Nature*, **430**, 35–44.

Elie, C.M., Lott, T.J., et al. 1998. Rapid identification of *Candida* species with species-specific DNA probes. *J Clin Microbiol*, **36**, 3260–5.

Eriksson, O.E., Baral, H.-O. et al. (eds). *Outline of Ascomycota –2004*. Myconet: http://www.umu.se/myconet/M10a.html 10, 1-99. Ref Type: Generic.

Fell, J.W., Boekhout, T., et al. 2000. Biodiversity and systematics of basidiomycetous yeasts as determined by large-subunit rDNA D1/D2 domain sequence analysis. *Int J Syst Evol Microbiol*, **50**, 1351–71.

Felsenstein, J. 1985. Confidence limits on phylogenies: an approach using the bootstrap. *Evolution*, **39**, 783–91.

Felsenstein, J. 2003. *Inferring phylogenies*. Sunderland, MA: Sinauer Associates, Inc.

Fisher, M.C., Koenig, G.L., et al. 2002. Molecular and phenotypic description of *Coccidioides posadasii* sp. nov., previously recognized as the non-California population of *Coccidioides immitis*. *Mycologia*, **94**, 73–84.

Freydiere, A.-M., Guinet, R. and Boiron, P. 2001. Yeast identification in the clinical microbiology laboratory: phenotypical methods. *Med Mycol*, **39**, 9–33.

Gaillardin, C., Duchateau-Nguyen, G., et al. 2000. Genomic exploration of the hemiascomycetous yeasts: 21. Comparative functional classification of genes. *FEBS Lett*, **487**, 134–49.

Gould, S.J. 1996. *Full house. The spread of excellence from Plato to Darwin*. New York: Crown Publishers.

Gräser, Y., el Fari, M., et al. 1999. Phylogeny and taxonomy of the family Arthrodermataceae (dermatophytes) using sequence analysis of the ribosomal ITS region. *Med Mycol*, **37**, 105–14.

Guarro, J., Gené, J. and Stchigel, A.M. 1999. Developments in fungal taxonomy. *Clin Microbiol Rev*, **12**, 454–500.

Guého, E., Smith, M.T., et al. 1992. Contributions to a revision of the genus *Trichosporon*. *Antonie van Leeuwenhoek Int J Microbiol*, **61**, 289–316.

Haase, G., Sonntag, L., et al. 1999. Phylogenetic inference by SSU-gene analysis of members of the Herpotrichiellaceae with special reference to human pathogenic species. *Stud Mycol*, **43**, 80–97.

Hall, B.G. 2004. *Phylogenetic trees made easy. A how-to manual*, 2nd edition. Sunderland, MA: Sinauer Associates.

Hall, L., Wohlfiel, S.L. and Roberts, G.D. 2003. Experience with the MicroSeq D2 large-subunit ribosomal DNA sequencing kit for identification of commonly encountered, clinically important yeast species. *J Clin Microbiol*, **41**, 5099–102.

Hall, L., Wohlfiel, S.L. and Roberts, G.D. 2004. Experience with the MicroSeq D2 large-subunit ribosomal DNA sequencing kit for identification of filamentous fungi encountered in the clinical laboratory. *J Clin Microbiol*, **42**, 622–6.

Hardison, R.C. 2003. Comparative genomics. *PLoS Biol*, **1**, 156–60.

Hazen, K.C. and Howell, S.A. 2003. *Candida, Cryptococcus*, and other yeasts of medical importance. In: Murray, P.R., Barron, E.J., et al.

(eds), *Manual of clinical microbiology*, 8th edition. Washington, DC: ASM Press, 1693–711.

Heckman, D.S., Geiser, D.M., et al. 2001. Molecular evidence for the early colonization of land by fungi and plants. *Science*, **293**, 1129–33.

Hedges, S.B. and Kumar, S. 2004. Precision of molecular time estimates. *Trends Genet*, **20**, 242–7.

Hedges, S.B., Blair, J.E., et al. 2004. A molecular timescale of eukaryote evolution and the rise of complex multicellular life. *BMC Evol Biol*, **4**, 2.

Herr, R.A., Ajello, L., et al. 1999. Phylogenetic analysis of *Rhinosporidium seeberi*'s 18S small-subunit ribosomal DNA groups this pathogen among members of the protoctistan mesomycetozoa clade. *J Clin Microbiol*, **37**, 2750–4.

Hillis, D.M., Mable, B.K. and Moritz, C. 1996a. *Applications of molecular systematics: the state of the field and a look to the future*, 3rd edition. Sunderland, MA: Sinauer Associates, Inc, 515-543.

Hillis, D.M., Moritz, C. and Mable, B.K. (eds) 1996b. *Molecular systematics*, 2nd edition. Sunderland, MA: Sinauer Associates, Inc.

Holder, M.T. and Lewis, P.O. 2003. Phylogeny estimation: traditional and Bayesian approaches. *Nat Rev Genet*, **4**, 275–84.

Howard, D.H. (ed.) 2003. *Pathogenic fungi in humans and animals*, 2nd edition. New York: Marcell Dekker.

Howard, D.H., Weitzman, I. and Padhye, A.A. 2003. Onygenales: Arthrodermataceae. In: Howard, D.H. (ed.), *Pathogenic fungi in humans and animals*. New York: Marcel Dekker, 141–94.

Huelsenbeck, J.P. and Ronquist, F. 2001. MRBAYES: Bayesian inference of phylogenetic trees. *Bioinformatics*, **17**, 754–5.

Huelsenbeck, J.P., Ronquist, F., et al. 2001. Bayesian inference of phylogeny and its impact on evolutionary biology. *Science*, **294**, 2310–14.

Iwen, P.C., Hinrichs, S.H. and Rupp, M.E. 2002. Utilization of the internal transcribed spacer regions as molecular targets to detect and identify human fungal pathogens. *Med Mycol*, **40**, 87–109.

James, T.Y., Porter, D., et al. 2000. Molecular phylogenetics of the Chytridiomycota supports the utility of ultrastructural data in chytrid systematics. *Can J Bot*, **78**, 336–50.

Javaux, E.J., Knoll, A.H. and Walter, M.R. 2001. Morphological and ecological complexity in early eukaryotic ecosystems. *Nature*, **412**, 66–9.

Jensen, A.B., Gargas, A., et al. 1998. Relationships of the insect-pathogenic order Entomophthorales (Zygomycota, Fungi) based on phylogenetic analyses of nuclear small subunit ribosomal DNA sequences (SSU rDNA). *Fungal Genet Biol*, **24**, 325–34.

Kanbe, T., Yamaki, K. and Kikuchi, A. 2002. Identification of the pathogenic Aspergillus species by nested PCR using a mixture of specific primers to DNA topoisomerase II gene. *Microbiol Immunol*, **46**, 841–8.

Kanbe, T., Suzuki, Y., et al. 2003. Species-identification of dermatophytes *Trichophyton*, *Microsporum* and *Epidermophyton* by PCR and PCR-RFLP targeting of the DNA topoisomerase II genes. *J Dermatol Sci*, **33**, 41–54.

Kane, J., Summerbell, R.C., et al. 1997. *Laboratory handbook of Dermatophytes*. Belmont, CA: Star.

Kano, R., Aizawa, T., et al. 1999. Chitin synthase 2 gene sequence of *Malassezia* species. *Microbiol Immunol*, **43**, 813–15.

Keeling, P.J. 2003. Congruent evidence from α-tubulin and β-tubulin gene phylogenies for a zygomycete origin of microsporidia. *Fungal Genet Biol*, **38**, 298–309.

Kirk, P.M., Cannon, P.F., et al. (eds) 2001. *Ainsworth & Bisby's dictionary of the fungi*, 9th edition. Oxford, UK: CAB International.

Kurtzman, C.P. and Fell, J.W. (eds) 1998. *The yeasts. A taxonomic study*, 4th edition. Amsterdam: Elsevier.

Kurtzman, C.P. and Robnett, C.J. 1998. Identification and phylogeny of ascomycetous yeasts from analysis of nuclear large subunit (26S) ribosomal DNA partial sequences. *Antonie Van Leeuwenhoek*, **73**, 331–71.

Liu, Y.J.J., Whelen, S. and Hall, B.D. 1999. Phylogenetic relationships among ascomycetes: evidence from an RNA polymerase II subunit. *Mol Biol Evol*, **16**, 1799–808.

LoBuglio, K.F. and Taylor, J.W. 1995. Phylogeny and PCR identification of the human pathogenic fungus *Penicillium marneffei*. *J Clin Microbiol*, **33**, 85–9.

Luo, G. and Mitchell, T.G. 2002. Rapid identification of pathogenic fungi directly from cultures by using multiplex PCR. *J Clin Microbiol*, **40**, 2860–5.

Lynch, M. 2002. Genomics. Gene duplication and evolution. *Science*, **297**, 945–7.

Machouart-Dubach, M., Lacroix, C., et al. 2002. Nucleotide structure of the *Scytalidium hyalinum* and *Scytalidium dimidiatum* 18S subunit ribosomal RNA gene: evidence for the insertion of a group IE intron in the rDNA gene of *S. dimidiatum*. *FEMS Microbiol Lett*, **208**, 187–96.

Makimura, K., Tamura, Y., et al. 1999. Phylogenetic classification and species identification of dermatophyte strains based on DNA sequences of nuclear ribosomal internal transcribed spacer 1 regions. *J Clin Microbiol*, **37**, 920–4.

Malpertuy, A., Tekaia, F., et al. 2000. Genomic exploration of the hemiascomycetous yeasts: 19. Ascomycetes-specific genes. *FEBS Letters*, **487**, 113–21.

Massey, S.E., Moura, G., et al. 2003. Comparative evolutionary genomics unveils the molecular mechanism of reassignment of the CTG codon in *Candida* spp. *Genome Res*, **13**, 544–57.

McKerracher, L.J. and Heath, I.B. 1985. The structure and cycle of the nucleus-associated organelle in 2 species of *Basidiobolus*. *Mycologia*, **77**, 412–17.

McLaughlin, D.J., Frieders, E.M. and Lu, H.S. 1995. A microscopist's view of heterobasidiomycete phylogeny. *Stud Mycol*, **38**, 91–109.

Mendoza, L., Taylor, J.W. and Ajello, L. 2002. The class mesomycetozoea: a heterogeneous group of microorganisms at the animal–fungal boundary. *Annu Rev Microbiol*, **56**, 315–44.

Meyer, S.A., Payne, R.W. and Yarrow, D. 1998. *Candida* Berkhout. In: Kurtzman, C.P. and Fell, J.W. (eds), *The yeasts. A taxonomic study*, 4th edition. Amsterdam: Elsevier, 454–573.

Mitchell, T.G. and Xu, J. 2003. Molecular methods to identify pathogenic fungi. In: Howard, D.H. (ed.), *Pathogenic fungi in humans and animals*, 2nd edition. New York: Marcel Dekker, 677–702.

Moore, R.T. 1978. Taxonomic significance of septal ultrastructure with particular reference to jelly fungi. *Mycologia*, **70**, 1007–24.

Murphy, R.W., Sites, J.W. Jr, et al. 1996. Proteins: isozyme electrophoresis. In: Hillis, D.M., Mortiz, C. and Mable, B.K. (eds), *Molecular systematics*, 2nd edition. Sunderland, Massachusetts: Sinauer Associates, 51–120.

Nagahama, T., Sato, H., et al. 1995. Phylogenetic divergence of the entomophthoralean fungi: evidence from nuclear 18S ribosomal RNA gene sequences. *Mycologia*, **87**, 203–9.

Nishida, H. and Sugiyama, J. 1994. Archiascomycetes: detection of a major new lineage within the ascomycota. *Mycoscience*, **35**, 361–6.

Oberwinkler, F. 1987. Heterobasidiomycetes with ontogenetic yeast stages – systematic and phylogenetic aspects. In: de Hoog, G.S., Smith, M.T. and Weijman, A.C.M. (eds), *The expanding realm of yeast-like fungi*. Amsterdam: Elsevier Science, 61–74.

Prillinger, H., Lopandic, K., et al. 2002. Phylogeny and systematics of the fungi with special reference to the Ascomycota and Basidiomycota. *Chem Immunol*, **81**, 207–95.

Reeb, V., Lutzoni, F.M. and Roux, C. 2004. Contribution of *RPB2* to multilocus phylogenetic studies of the euascomycetes (Pezizomycotina, Fungi) with special emphasis on the lichen-forming Acarosporaceae and evolution of polyspory. *Mol Phylogenet Evol*, **32**, 1036–60.

Revankar, S.G., Patterson, J.E., et al. 2002. Disseminated phaeohyphomycosis: review of an emerging mycosis. *Clin Infect Dis*, **34**, 467–76.

Ribes, J.A., Vanover-Sams, C.L. and Baker, D.J. 2000. Zygomycetes in human disease. *Clin Microbiol Rev*, **13**, 236–301.

3

Fungal structure and morphology

PAUL F. LEHMANN

The species of the fungi that are associated with infections in humans represent a number of very different groups within the Kingdom Fungi. Three phyla of the true fungi, the Ascomycota, Basidiomycota, and Zygomycota, include infectious fungi. The remaining phylum, the Chytridiomycota, includes the order Neocallimastigales within which several genera of anaerobic fungi have been described. Anaerobic fungi have been found in the gut lumen of several herbivores but they have not been associated with human disease.

Additional organisms resemble fungi and are studied by medical mycologists. These include a pathogenic member of the genus *Pythium* (*Pythium insidiosum*), but this is a genus that is no longer included within the Kingdom Fungi being assigned, with other oomycetes, to the Stramenopiles or Kingdom Chromista (Hawksworth et al. 1995; Van de Peer et al. 2000). Another group of species, those in the *Pneumocystis carinii* complex, have traditionally been studied by parasitologists on account of their morphological similarity to some of the parasitic protozoa; however, comparative analysis of the sequences of the ribosomal RNA (rRNA) and of a number of different genes has placed this complex of species among the fungi (Cailliez et al. 1996). It is likely that several other organisms that have been treated as parasitic protozoa will be found to be fairly close relatives of fungi. Such appears to be the case for *Encephalitozoon caniculi*, a microsporidian that has emerged as a problem in immunosuppressed patients including those with the acquired immunodeficiency syndrome (AIDS) (Vivarès et al. 2002).

Most recently, the class Mesomycetozoea has been defined within the Kingdom Protozoa. This includes a number of fungus-like organisms which, based on rRNA sequences, have evolved as a lineage which is distinct from the eukaryotic kingdoms represented by fungi, animals, and plants (Mendoza et al. 2002). The Mesomycetozoea include a number of fish parasites and the human pathogen, *Rhinosporidium seeberi*, the etiologic agent of rhinosporidiosis. Relative to the fungal pathogens from the Kingdom Fungi that are the traditional study of medical mycologists, these organisms are poorly understood and, unless directly noted, will not be the subject for further discussion within this chapter.

In spite of their diversity, fungi share many features. At the simplest level, they are nonphotosynthetic, eukaryotic organisms that possess a cell wall lying external to the plasma membrane. This membrane, also known as the plasmalemma, represents the outermost cellular membrane. It is the cell wall that imparts a shape to the fungal cell and prevents it from bursting when the fungus is placed in environmental settings with low osmotic pressure. If the wall is removed, by enzymatic digestion, the cell takes on a spherical form and becomes osmotically fragile. Differential patterns of cell wall synthesis play a central role in determining both the shape of the fungal cell and the patterns of fungal growth that will be discussed herein.

INTRACELLULAR COMPONENTS

Being eukaryotic cells, all true fungi possess a nuclear pore complex within a nuclear membrane. Within the nucleus there is a nucleolus and linear chromosomes. The cytoplasm contains a cytoskeleton in which the microfilaments, which are composed largely of actin, and the tubulin-containing microtubules provide the major components (Gow and Gadd 1995). The cytoplasm also contains organelles including mitochondria, although these are not found in the anaerobic fungi, the Golgi apparatus, vacuoles, often several types of vesicle and 80S ribosomes. The plasmalemma typically contains large quantities of sterols, usually ergosterol. It is their preferential binding to ergosterol rather than cholesterol, the sterol that is dominant in human cell membranes, which explains the selectivity of the polyene antifungal agents, nystatin and amphotericin B. The pathway for synthesis of ergosterol is the target of two groups of antifungal agents. The azole antifungals, which include fluconazole, itraconazole, ketoconazole, miconazole, and voriconazole, inhibit the cytochrome P450-dependent lanosterol demethylase, while the allylamines, which include terbinafine and naftidine, inhibit squalene expoxidase.

BASIC GROWTH PATTERNS: MOLDS AND YEASTS

Fungi, not being photosynthetic, require environmental sources of fixed carbon for their growth, and two major strategies for exploiting substrata have been developed. Filamentous growth and apical extension of the filament, in association with the formation of lateral branches, are exemplified by hyphae and mycelium, while single cell division is associated with yeasts (Figure 3.1). Apical growth, in association with a somewhat rigid wall, allows a hypha to generate forward pressure at its tip and this, combined with the release of degradative enzymes, enables it to force its way through tissues. Branching of the hyphae leads to the production of mycelium, a large structure composed of interconnected hyphal filaments.

In contrast to the filamentous form of growth that leads to the production of a mycelium, the yeast form of growth results in the physical separation of parental and progeny cells. To exploit new sources of nutrients, yeast cells spread to more distant sites by being carried passively in moving fluids, such as in plasma, or in fluid films such as those that bathe tissue surfaces.

In addition to the basic pattern of growth by molds and yeasts, several fungi or fungus-like organisms have unusual morphologies and methods of growth. These include *Pneumocystis* where clusters of multinucleate cysts and trophozoite (single-celled) stages are seen in infected persons. In both *Coccidioides* and *Rhinosporidium*, parasitic development is characterized by the

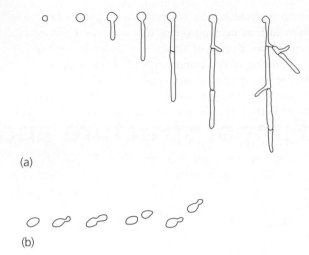

(a)

(b)

Figure 3.1 *Growth forms of fungi:* **(a)** *molds – the ungerminated conidium swells, then germinates to produce an apically growing hypha which develops septa and branches. Growth continues at the apices of the original hypha and branches.* **(b)** *Budding yeast cell: a common growth pattern where reproduction occurs by separation of the emerging bud from the mother cell. Both mother and daughter cells continue to bud.*

production of spherical bodies that undergo endosporulation and with the subsequently-released endospores swelling to form new spherical structures. In *Coccidioides* spp., the spherical structure (= a spherule) undergoes segmentation and fungal cell wall material divides the contents into sections in which the endospores mature. Rupture of the spherule leads to release of endospores. By contrast, the fungus-like *R. seeberi* produces a large spherical sporangium in which there is no segmentation and the endospores appear to be released through a pore in the sporangial wall (Mendoza et al. 1999). A further form of growth is seen for the parasitic stage of some etiologic agents of chromoblastomycosis: these produce sclerotic bodies showing planar division. Figures showing these forms can be seen in the appropriate chapters covering these diseases.

Cell wall synthesis

The actual shape of the fungal cell is determined by the cell wall. This lies exterior to the cell membrane and is composed primarily of carbohydrates, some of which are covalently linked to proteins. The cell walls of distantly related fungi can show substantial differences in their chemical composition, and so it is remarkable that hyphae of quite distantly related species may appear very similar both in morphology and in their pattern of growth. The carbohydrate components appear to be synthesized at the surface of the plasmalemma and include materials that are crystalline, such as α-glucans and chitin, which is the $\beta(1,4)$-linked polymer of *N*-acetyl-D-glucosamine. The crystalline materials are thought to provide strength and rigidity to the wall.

Other polysaccharidic components include matrix materials such as mannoproteins that may allow the wall to have some degree of flexibility. Evidence for covalent crosslinking of cell components has been found and this would be expected to strengthen the walls. Those glucan synthetases that generate β-glucans within the cell wall provide the target for the echinocandin antifungal agents such as caspofungin acetate.

The expanding hyphal tip has, at its apex, a cell wall that shows significant plasticity. Here the wall is quite thin and also is thought to be composed of materials that are more loosely associated with each other than that seen in the more mature hyphal walls where the crosslinking of the various materials appears to have been completed. For example, in contrast to the situation with mature walls, the newly synthesized chitin chains found in the tip appear not to be fully crystallized. Models for fungal cell wall synthesis have been reviewed (Sentandreu et al. 1994; Sietsma and Wessels 1994; Wessels 1994).

Hydrophobic proteins (hydrophobins) are often produced by aerial hyphae and conidia. The assembly of these molecules results in a superficial hydrophobic rodlet layer. Our understanding of these hydrophobic molecules is far from complete (Wessels 1994).

Mold growth pattern

APICAL EXTENSION OF HYPHAE

During hyphal extension, there is directed flow of precursors into the hyphal tip. Many vesicles are detectable at hyphal tips and these are considered to be the vehicles for both the enzymes and possibly some of the precursors that are required for cell wall synthesis (Heath 1994; Sentandreu et al. 1994; Sietsma and Wessels 1994; Wessels 1994; Gow and Gadd 1995). For example, some of the vesicles, termed 'chitosomes,' may carry chitin synthase in its zymogenic form. Chitosomes, when activated and incubated with UDP-N-acetyl-D-glucosamine, have been shown to cause the synthesis of chitin microfibrils (Bracker et al. 1976). A cluster of vesicles near growing hyphal tips can be observed by light microscopy and it appears as a distinct structure that has been named Spitzenkörper. This structure has been suggested to represent a distribution center for the vesicles migrating to the growing tips of hyphae. The Spitzenkörper is not observed when hyphal growth ceases and experimentally induced changes to its position relative to the growing tip have been linked to changes in the shape of the hyphal wall (Bartnicki-Garcia et al. 1995).

An understanding of the basic mechanisms that are involved both in maintaining the uniform cylindrical shape of a growing hypha and in producing apical extension of the hyphal tip is beginning to emerge. Multiple gene products are involved: indeed a large-scale genetic screen for mutations in *Neurospora crassa* that were associated with defects in polarized growth and hyphal morphogenesis obtained >100 complementation groups and 21 different morphological phenotypes (Seiler and Plamann 2003). Among the genes involved were those regulating the formation of the cytoskeleton or were part of the secretory pathway. Genes encoding members of the cell wall and outer membrane biosynthetic pathways and also encoding components of the signal transduction pathways, including interactive kinases, were also identified, indicating that the process of morphogenesis is complex.

The large number of morphological phenotypes and the abundance of complementation groups involved in morphogenesis enable the creation of many possible models for the control of hyphal growth. Clearly the three-dimensional form of the cytoskeleton is important; indeed, the shape may be the predominant factor that controls the shape of the cell (Heath 1994; Wessels 1994). However, a simple model is inadequate. For example, Heath has pointed out that apical growth cannot simply be explained by use of a model where turgor pressure causes tip expansion by pushing the cell contents outwards at the site where the hyphal cell wall is weakest, i.e. at the hyphal tip (Heath 1994). Among the phenomena that are not explained by that model are: the capability of cell wall mutants, for example, mutants of *Aspergillus niger* that lack chitin, to produce relatively normal looking hyphae in media of high osmotic strength, and the poor correlation between changes in internal turgor pressure and the hyphal growth rate.

As the hypha grows forward, materials are incorporated into the wall at the apex of its hyphal tip. The newly synthesized wall pushes aside the previously formed wall components that soon become part of the mature walls. In actively growing fungi, the hyphal contents do not have a uniform metabolic activity throughout the length of the hypha. Maximal activity is found at the growing tip and it is in this region that most of the absorption of nutrients appears to occur and where hydrolytic enzymes, such as lipases, polysaccharidases and proteases, are secreted. These enzymes may remain in the hyphal wall or they can be released into the surrounding substrate.

The factors controlling polar growth exhibited by hyphae have been examined for a number of different species and the subject has been reviewed fairly recently (Momany 2002). The reader is invited to obtain references to several important papers in the original literature from that review as they are beyond the scope of this introductory chapter. In spite of the difference in typical growth habit of yeasts and molds, studies have demonstrated a similarity between the central protein machinery required for the polarized growth exhibited by hyphae of filamentous fungi and the protein machinery used in maintaining polarized growth associated with the more filamentous growth morphology of baker's yeast, *Saccharomyces cerevisiae*. This form of growth can be seen when yeast is grown on certain media. For example, a Rho-GTPase, Cdc42, drives polar growth in yeast and the homologous protein appears to perform a similar function in the ascomycetes, *Ashbya gossypii* and *Penicillium marneffei*. The function of Cdc42 appears to be to act as a target for additional proteins that themselves act as a scaffold for the cytoskeletal components (Johnson 1999). Polar growth appears dependent on the state of the plasma membrane where it has been suggested that rafts of polarity-inducing proteins may be localized. Post-translational lipid modification of some of these proteins via *N*-myristoylation (*Aspergillus nidulans*) or geranylgeranylation (*S. cerevisiae*) has been suggested as a mechanism whereby the polarity proteins develop affinity for the plasma membrane but this alone does not explain the control and maintenance of a polarized form of growth. Here, it is likely that vesicle trafficking is organized so that vesicles containing wall precursors are targeted to a single area, a process that is complex and certainly involves the correct deposition of both formins and septins. This is discussed further in the next section.

DEVELOPMENT OF MYCELIUM

Although the mature hyphal wall becomes thickened and rigid, it may still undergo modifications at a later stage. When, at some distance behind the growing hyphal apex, a small area of wall is rendered elastic, a lateral hypha can emerge as a branch (Figure 3.2). Typically, this new hypha grows apically and itself may give rise to lateral branches. The whole pattern leads to the development of an interconnected mycelium.

Cross-walls, known as septa, are often formed. Septa have a variety of structures with certain types being restricted to specific major divisions of the fungi, for example, septa containing dolipores are restricted to the Basidiomycota (Figure 3.3). The different types of septa have different effectiveness in providing barriers to the movement of nutrients, vesicles, organelles, and nuclei. Those with large pores appear to provide few restrictions. This contrasts with the control provided by septa with small or complex pores.

Septum production has been studied fairly extensively for *Aspergillus nidulans* and requires, after a conidium has germinated, that the hypha reaches a certain minimum size before a primary septum is formed. The septum develops from a cylindrical band of material that includes actin. This ring constricts so that fungal components become separated after mitosis has occurred. Septal formation is linked to mitotic events: the linkage retained as the hypha continues to grow and form new septa being at some distance behind the hyphal tip. Several proteins are involved in septal formation:

- the actins that form a ring appear to interact with the microtubules of the mitotic spindle

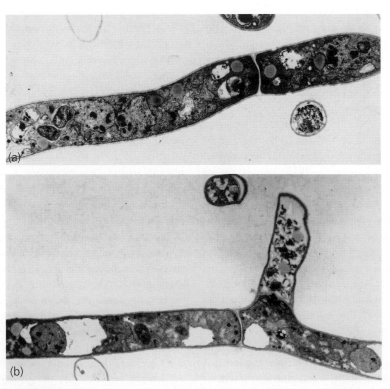

Figure 3.2 Coccidioides immitis *hyphae:* **(a)** *showing septum and intracellular organelles (magnification ×4 765);* **(b)** *showing formation of a branch (magnification ×6 800). Courtesy of K.R. Seshan*

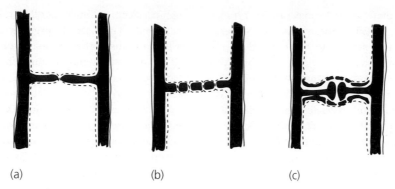

Figure 3.3 *Septa: there are several types. The diagramatic representations show a hyphal cross-wall developing from inward growth of the inner cell wall layers with* **(a)** *a simple pore,* **(b)** *multiple pores, and* **(c)** *a complex dolipore. In the last, the doughnut-shaped septal pore can be covered by a perforated cap.*

- the septins that belong to a conserved family of proteins found in eukaryotes where they act as morphogenetic scaffolds important in polarized growth and cell division
- the formins that seem to be needed for the correct organization of the actin cytoskeleton during cell division and the concurrent constriction of the ring of septal material.

One formin, SepA, has been detected both in the hyphal tip and at sites where septation is occurring; it appears to be required for proper deposition of the septin, AspB (Sharpless and Harris 2002). Though deposited on both sides of the constricting septum, the AspB deposited on the basal surface is not retained and disappears. By contrast, AspB is retained on the apical side of a septum and it may act as an signal or marker for cell polarity that can be exploited by the fungal cell (Harris 2001).

Some fungi, notably many species within the Zygomycota, produce few or no septa within a mycelium. Such a mycelium is essentially a coenocytic organism. There is little understanding of the molecular mechanisms accounting for polarized growth and branching of hyphae in these organisms.

The yeast pattern of growth

The pattern of growth that follows cell divisions, which result in the physical separation of daughter and parental cells, is quite different from that found for mycelial growth. Most yeasts of medical importance produce daughter cells by budding (see Figure 3.1). These species include *Cryptococcus neoformans*, the species of *Candida*, and the parasitic forms of several dimorphic pathogens including *Histoplasma capsulatum*, *Blastomyces dermatitidis*, and *Sporothrix schenckii*. Typically, in a budding yeast, cell wall synthesis occurs over the whole surface of the bud and is not directed at a single pole. After mitosis and the migration of a nucleus into the bud, the separation of the daughter cell from the parent cell involves both the formation of a cross-

wall at the site of the bud and the subsequent cleavage of this wall as the daughter cell dissociates. The wall formed at the bud site has a different structure from the rest of the yeast's cell wall and is termed the 'bud scar.' A great deal is known about the control of the growth cycle and budding in brewer's yeast, *S. cerevisiae*, which therefore acts as a model for budding yeast growth. Many of the features are conserved among species, for example, the production of a chitin-rich ring in the bud scar representing the site on the mother cell from which the daughter yeasts are cleft. But there can be significant differences in the complement of enzymes expressed by different species. Given the differences in cell wall composition associated with different species, and especially when basidiomycetous yeast species such as *Cryptococcus* are compared to ascomycetous species, it is clear that a thorough understanding of the mechanism of budding growth will require its investigation in several species.

A second form of cell division is characterized by the fission yeasts. Here the parental cell elongates and, after mitosis has occurred, the nuclei become separated by a septum that divides the elongated cell. The two daughter yeasts then dissociate after cleavage of the septum. This pattern of division is seen among few species of infectious fungi. It can be observed in the parasitic form of *Penicillium marneffei*, and also occurs in an incomplete fashion during the development of the muriform cells that are formed in vivo by the agents of chromoblastomycosis. The muriform cells are composed of cells that have reproduced by elongation and have formed a dividing septum. However, fission is delayed and further septal formation in the progeny cells can lead to somewhat irregular multicellular structures.

The basis of maintaining the growth patterns seen for budding and fission yeasts have been subject to extensive research and detailed discussion is beyond the scope of this introduction. While signal transduction via MAP kinases and other pathways such as TOR, which acts as a nutrient sensor and controls translation, are essential for allowing of growth to proceed (Crespo and

Hall 2002), it is clear that there also have to be intracellular structures that direct synthesis in an ordered manner. These structures appear to be supplied in part by polarized actin cables that, as is also seen in mycelium, are produced in a process that is dependent of formins. The actin cables act as tracks along which myosin V motors transport vesicles, organelles, and vacuolar components into the bud. The subject has been reviewed for *S. cerevisiae* by Bretscher (2003).

DIMORPHISM

Though a certain plasticity in morphology is often seen for fungi grown under different environmental conditions, several fungi associated with infections of humans can switch between the two basic growth patterns which are typified by molds and yeasts (Table 3.1). Indeed, the basic growth pattern of a fungus found during an infection can be quite different from that seen when it grows as a saprophyte in an external habitat such as soil. For example, *H. capsulatum* produces mycelium at environmental temperatures of less than 30°C, but reproduces as a budding yeast when growing intracellularly in patients with histoplasmosis. The capacity to switch back and forth between very distinct forms of growth is probably quite usual for fungi. Several organisms show unusual morphologies when placed at high temperature, for example, the conidia of *Aspergillus*

Table 3.1 *Dimorphism among selected infectious fungi*

Species	Parasitic form[a]
Agents of chromoblastomycosis[b]	Muriform cells
Blastomyces dermatitidis	Yeasts
Coccidioides immitis	Spherules
Emmonsia parva	Adiaspores[c]
Histoplasma capsulatum	Yeasts
Paracoccidioides brasiliensis	Yeasts
Penicillium marneffei	Yeasts
Sporothrix schenckii	Yeasts
Mixed growth patterns in vivo	
Malassezia furfur	Yeasts and short hyphae[d]
Candida spp.[e]	Yeasts, pseudohyphae, hyphae and chlamydoconidia[f]

a) Characteristic form seen during active infection.
b) In several examples of chromoblastomycosis a mixture of structures may be seen. In some cases, the dominant morphology may depend in part on the host immune reactivity.
c) These represent swollen spores and do not replicate.
d) This yeast species, which requires lipid for growth, generally does not produce filaments when cultured on conventional media.
e) There is substantial variability among the species. Some *Candida* spp., for example *C. (Torulopsis) glabrata*, grow only in the yeast form. Some others produce a mixture of yeasts and pseudohyphae, yet do not appear to produce true hyphae.
f) Among *Candida* spp., only *C. albicans* is known to produce chlamydoconidia in vivo. Structures that probably represent these conidia have been detected in experimentally infected mice (see text).

niger can form spherical swollen cells in vitro (Anderson and Smith 1972). Such swelling is also seen in lung tissues containing the adiaspores of *Emmonsia parva*.

The reversible modification of growth pattern in response to environmental changes results from modifications affecting both the components of the cell wall and the pattern in which cellular components are deposited. This capacity to switch the pattern of growth in response to an environmental change has been termed 'dimorphism.' The same term has also been applied to describe situations where a multiplicity of growth forms can be produced by making adjustments to the conditions under which a species is cultivated. The signal for a switch in morphology from the saprophytic form to the parasitic form can, in many cases, be shown to follow an elevation in environmental temperature. However, adequate nutrients must also be present: for example, the switch to the parasitic yeast form by *H. capsulatum*, which is achieved by transferring a culture from room temperature to 37°C, is inhibited if cysteine is not included within the culture medium. Nutritional changes and elevations in P_{CO_2} can be required to induce changes in the growth patterns of *Coccidioides immitis*. For *Candida albicans*, several seemingly independent environmental stimuli may enhance a switch between yeast and filamentous growth patterns (Berman and Sudbery 2002).

A number of species are known that produce a variety of growth patterns even while causing infection. *Malassezia furfur*, an obligatorily lipophilic yeast, produces both budding yeast and short hyphae on the skin surface. *C. albicans* and a number of other *Candida* spp. grow both as budding yeasts and in filamentous forms. Indeed, in a single lesion caused by *C. albicans*, one may see histological evidence for yeast cells, hyphae, and pseudohyphae (Figure 3.4). In addition to its growth in filamentous forms and as a yeast, *C. albicans* is also known to produce chlamydoconidia. These spherical conidia, having thick cell walls, are easily seen when the species is cultivated on yeast morphology agars such as rice extract agar. Although similar conidia have been seen in sputum from an AIDS patient, and can be detected by use of differential cell wall staining and electron microscopy in submucosal tissues of immunosuppressed mice after experimental intragastric infection with *C. albicans* (Chabasse et al. 1988; Cole et al. 1991), it is remarkable that chlamydospores are hardly ever reported from clinical specimens. The *C. albicans* chlamydoconidium may be a resting structure and may not actually represent a replicative form.

With the recent sequencing of the *C. albicans* genome (www-sequence.stanford.edu/group/candida/search.html), it has been possible to identify many genes that appear to be homologous to ones previously investigated in *S. cerevisiae*. The expression of the genes involved in morphogenesis can change quite substantially during dimorphic growth and this is discussed in a recent review

(a)

(b)

(c)

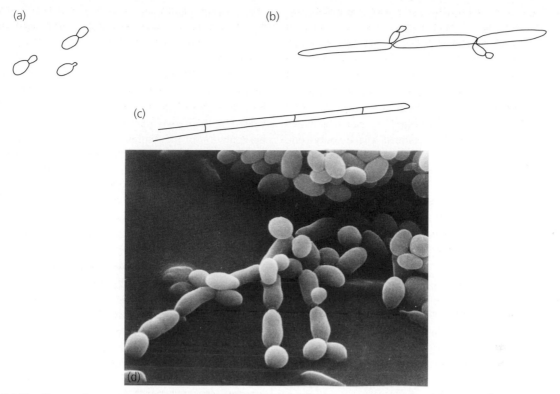

(d)

Figure 3.4 *The diagramatic representation shows morphological types of the yeast* Candida albicans. **(a)** *Yeast cells,* **(b)** *pseudohyphae and yeast cells;* **(c)** *a hypha;* **(d)** *budding yeast cells and pseudohyphae of* Candida kefyr *(magnification ×2 520). Courtesy of G.T. Cole*

which includes photomicrographs indicating how septin is deposited in a different fashion in the various morphological forms of *C. albicans*, i.e. budding yeast, hyphae, and pseudohyphae (Berman and Sudbery 2002). The pseudohyphal rather than the hyphal form of *C. albicans* appears to show most similarity with the 'filamentous' form of *S. cerevisiae*. At least four major signal transduction pathways appear involved in controlling the switch from yeast to filamentous forms of *C. albicans*.

Typically, *Candida* spp. grow as yeasts in media containing abundant glucose and oxygen, and isolates of many of these species can produce filamentous extensions when placed in conditions where the $P\text{CO}_2$ is raised and the glucose concentration is reduced. Such conditions are associated with growth under coverslips on cornmeal agar and in media used to test for germ-tube production by *C. albicans*. Other yeast species, such as *Candida glabrata*, a species which is more closely related to *S. cerevisiae* than to *C. albicans*, show no evidence for this capacity and, until recently, were widely considered to be species having no capacity to form filaments. However, as already alluded to above, research with *S. cerevisiae* has elucidated nutrient conditions that can induce polar cell growth causing production of pseudohyphae (Gimeno et al. 1993). Thus, even in species where there has been little evidence of dimorphism, there may exist an unexpected flexibility in growth pattern. Indeed, the haploid pathogenic basidiomycetous

yeast species, *Cryptococcus neoformans*, can be induced to grow in a filamentous fashion when cultivated under conditions that differ from the standard glucose-rich media used in Sabouraud's agar or other media that are typically used when cultivating fungi in the clinical laboratory (Wickes et al. 1996). Before this discovery, a filamentous morphology had only been known for the teleomorph produced when two haploid strains mate.

Although it seems likely that a capacity to produce filamentous growth patterns will be found for many other yeasts that previously have not been considered dimorphic, it remains to be proven that the filamentous growth, seen only on certain media, plays any significant role in the establishment and pathogenesis of infections caused by either *Cryptococcus neoformans* or *S. cerevisiae*. By contrast, the yeast and filamentous forms of *C. albicans* behave so differently that it is likely that dimorphism is essential for maximal virulence. For example, they show different capacities for binding to epithelial cells and to several host proteins including complement components: such variability likely enables the fungus to colonize, grow, and become released from a variety of niches. The report that a mutant lacking mitochondrial NADH dehydrogenase complex I fails to develop filaments, though it grows normally as a yeast, suggests that energy demands for yeast and filamentous growth could be rather different with the hyphal form requiring relatively high production of ATP and reducing power (McDonough et al. 2002). While differences in the metabolic state of cells might contribute to the induction of morphological variability, including white–opaque switching (next section), environmental cues are important for inducing the morphological switch as culture media containing very similar nutrients can support both types of growth. The morphological flexibility of *C. albicans* is so notable in that it is extremely rare to find clinical lesions where only one morphological form exists and, indeed in biofilms formed on plastic appliances, both forms are observed (Chandra et al. 2001).

INSTABILITY IN GROWTH PATTERN: 'PHENOTYPIC SWITCHING'

In addition to the changes in growth patterns associated with dimorphism, an additional form of morphological flexibility has been described whereby, at least in culture, certain species undergo 'phenotypic switching.' This process, first described for certain strains of *C. albicans*, differs from dimorphism in that the changes in growth form appear spontaneously, usually at a rate of between 1:1 000 and 1:10 000 progeny cells, (Slutsky et al. 1985). The different patterns of growth result in the generation of differently shaped colonies when these are cultivated on agar media. Most progeny colonies from a single colony type retain the growth pattern of the parental colony; however, some colonies develop where the pattern appears to have switched to one of a number of alternative phenotypes (Figure 3.5). It appears that phenotypic switching increases the capability of *C. albicans* to adapt to different environments and environmental stresses for switches of metabolic activity may also be found. Features of the switching process, including its relevance to pathogenesis, have been reviewed (Soll 1993).

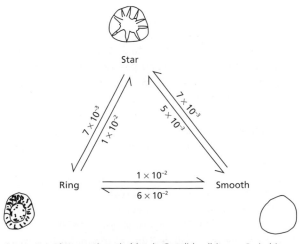

Figure 3.5 *Phenotypic switching in* Candida albicans. *Switching rates from one colony phenotype to another are examples taken from data showing switching between several distinctive colony phenotypes that were reported by Slutsky et al. (1985).*

A second phenotypic switching system, the white–opaque system, has been described and the regulation of genes expressed in each form has been investigated by using a microarray of *C. albicans* genes (Lan et al. 2002). There were substantial changes not only in the expression of genes that are likely to interact directly with the host, but also in the general metabolic activities of the yeasts in these two phases. In addition, there was an upregulation of the expression of mating type-associated genes in the opaque phenotype. This latter property correlates well with the recent discovery that the opaque cell type can undergo efficient mating (Miller and Johnson 2002), a feature that is not seen for the regular 'white' phenotype which is characterized by having a heterozygous mating type locus (*MTLa/α*). In contrast, the mating type locus found in the opaque phenotype is homozygous (*MTLa/a* or *MTLα/α*). This novel finding suggests a process by which gene flow can occur fairly easily in populations of *C. albicans*.

Phenotypic switching is now reported for several other fungal pathogens of humans including *Candida parapsilosis* and *Candida tropicalis*, which are both closely related to *C. albicans*, and for more distant species including *C. glabrata*, a species that is more closely related to *Saccharomyces*, and the basidiomycetous yeast *Cryptococcus neoformans* (cited in Soll et al. 2003). It has been clearly linked to pathogenicity in *C. albicans* and it is very likely to found in several other organisms, including plant pathogens and non-pathogens, which can exist in highly variable environmental niches where the switching may be coordinated by changes in the activity of global regulators controlling the expression of families of genes. Further insights into the signaling process leading to phenotypic switching could identify useful targets of antifungal chemotherapy.

STRUCTURES USED IN IDENTIFICATION OF FUNGI

Although there have been substantial advances in the use of molecular characteristics, such as antigens and nucleic acid sequences, for the identification of fungal pathogens, the morphological features of fungi still play a dominant role in allowing microbiologists to identify filamentous fungi. By far the most important taxonomic characters are those associated with the morphology of spores and fungal structures on or within which the spores are produced. With molds, spore-bearing hyphae typically grow aerially in a manner that raises the spores away from the vegetative hyphae. A substantial number of differential characteristics are associated with these specialized hyphae and the spores they bear, and these allow isolates to be identified both to genus and, very often, to species level. The vegetative hyphae, by contrast, tend to be fairly simple and show few characteristics that allow different species to be discriminated. However, certain hyphal features, including pigmentation, size, and the type and abundance of the septa, may be helpful in eliminating particular species as choices in the identification of an unknown mold.

Many fungal propagules may be produced in an asexual process whereby their production does not require prior fusion of nuclei and a subsequent meiosis. Sporulating colonies that bear such conidia are known as anamorphs and, because more than one form of conidium may be produced by some fungi, there are examples where a single species can produce more than one anamorph. The sporulating structures that are associated with sexual spores that are produced after meiosis has occurred are named teleomorphs. The terms anamorph and teleomorph are also known as the imperfect and perfect states, and both are separately classified and named for a genus and species. As a single organism may produce both anamorphs and teleomorphs, many fungi have more than one name. It is important to realize that the different names do not describe different organisms; rather they describe different reproductive types. One sees a parallel among humans where a change of name does not alter the identity of the person.

The reason that more than one name is used for a fungus is largely based on practicality. Most fungi are heterothallic, i.e. they have mating types that ensure outbreeding, and for these species the production of teleomorphs requires the mating of different strains. As most clinical isolates involve a single strain, it would be necessary to employ a battery of tester strains were we to try to identify each anamorph via the production of its teleomorph. Not only is this inconvenient compared with naming a fungus on the basis of its anamorph, but conditions for obtaining teleomorphs have not been found for many important pathogenic species, for example *Aspergillus fumigatus* and *Coccidioides immitis*. Even when a teleomorph is known, the anamorphic form may have been studied for many decades before the discovery of the perfect state. Such situations exist for many important pathogens that are primarily known by the name given to their anamorphs. Such species include *B. dermatitidis*, *H. capsulatum*, *Cryptococcus neoformans*, and many species of dermatophytes within the genera *Trichophyton* and *Microsporum*. Imposition of nomenclature based on teleomorphs, for these examples *Ajellomyces dermatitidis*, *Ajellomyces capsulatus*, *Filobasidiella neoformans*, and species of *Arthroderma*, would be very confusing for most physicians and clinical microbiologists.

The major groups of anamorphic sporulation structures associated with infectious fungi are illustrated in Figures 3.6, 3.7 and 3.8. There are two major groups of asexual propagules. Those propagules that are produced as naked, asexual, nonmotile propagules, and which usually fall away from the parent hypha or yeast, are known as conidia. The specialized hypha on which they are formed is termed a 'conidiophore.' There are two basic ways that conidia are produced. Blastoconidia differentiate after ballooning out of a fertile hypha and

are characterized by developing a conidial wall that is produced *de novo* (Figure 3.6). By contrast, those conidia that are produced via thallic conidiogenesis (Figure 3.7) develop from a preexisting hypha where modification of the cell wall and the cell contents leads to the development of the conidium (Cole and Samson 1979).

The second major form of asexual sporulation involves the production of sporangiospores (Figure 3.8). These asexual spores are only produced by fungi within the phylum Zygomycota. The most common form of sporangiospore production is associated with the multispored sporangium which has the appearance of a swollen vesicle produced on the end of a specialized hypha, the sporangiophore. Within the sporangium the spores are formed after nuclear divisions and cytoplasmic cleavage have taken place. After this, new spore walls are synthesized and these surround the spore initials (Cole and Samson 1979). Once the sporangiospores are mature, the sporangial wall breaks open to allow their release. In species that produce sporangiola, such as those within the genus *Cunninghamella*, the spores are pushed out of the swollen ends of the branches produced by the sporangiophore and the spore wall becomes fused to the sporangial wall. At first sight the sporangiola can resemble conidiophores that bear blastoconidia, but their pattern of spore development is quite different (Cole and Samson 1979).

Current taxonomic studies are increasingly becoming dependent on molecular genetic analysis, especially with the introduction of several allied organisms including agents of the Mesomycetozoea that were discussed in the introduction to this chapter. However, within the true fungi, the study of teleomorphs has been most useful in clarifying the systematic position of many anamorphic fungi and yeasts. For any one species, one type of spore is produced by the teleomorph. Species from within the phyla Ascomycota, Basidiomycota, and Zygomycota produce, respectively, ascospores, basidiospores, and zygospores.

Spores produced by teleomorphs appear rarely to be involved with human infections although their infectious potential has seldom been studied. However, there are certain situations where the sexually derived spores appear important. In black piedra, the asci and ascospores of *Piedraia hortae* are found in the nodules that develop on infected hairs. Also, basidiospores are the most likely source of infection in the few cases of invasive disease that have been reported for the mushrooms, *Schizophyllum commune* and *Coprinus cinereus*.

CONIDIAL STRUCTURE AND GERMINATION

Asexual propagules produced by the infectious fungi are the form of the fungus responsible for establishing most human infections, especially those infections that follow pulmonary exposure. Fungal conidia also play an

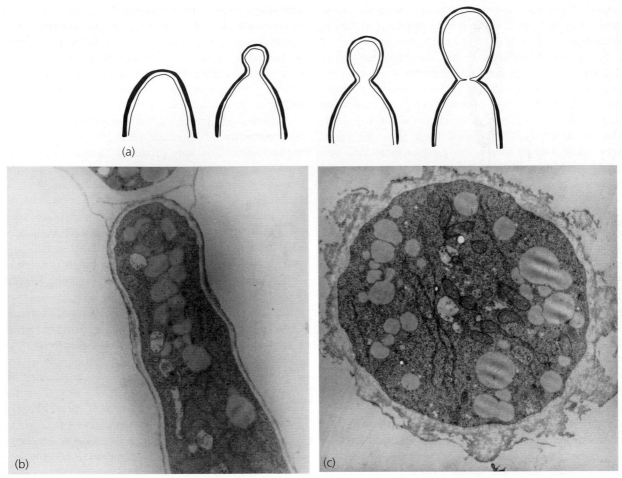

Figure 3.6 (a) *Conidiospore development by blastic conidiogenesis where the conidium is pushed out of a conidiophore and its cell forms newly synthesized cell wall material. The inward growth of the inner wall seals the area of emergence.* **(b)** Scopulariopsis brevicaulis *conidiophore (magnification ×11 000) and* **(c)** *mature conidium (magnification ×14 350). Courtesy of G.T. Cole*

important role as allergens and may carry mycotoxins. Although the sexual reproductive process has been studied extensively because of its importance in identification, the biological activities of the fungal propagules produced by pathogens of humans have not received the same scrutiny.

It is clear that the conidium's cell wall can show quite different surface features to the walls of vegetative cells. In many cases it can be quite hydrophobic. The components responsible for the hydrophobicity appear in the form of sheets of rodlets that have been named 'rodlet fascicles.' These, depending on the species and strain, can be extracted from the conidial surface by use of water and detergents (Cole and Samson 1984; Sietsma and Wessels 1994). The hydrophobic rodlet material resembles the hydrophobins that are found in mushroom fruiting bodies, but the latter require more powerful solubilization techniques, such as the use of trifluoroacetic acid (Lugones et al. 1996).

Conidia germinate by producing germ tubes that represent hyphal outgrowths. Regardless of conidium type, the germination process usually requires the presence of CO_2, oxygen, and water for activation to occur. In many cases, exogenous low molecular weight nutrients may also be required. When conidial germination inhibitors are present in the conidia, they can be leached into the surrounding liquid and their dilution allows the conidium to begin to germinate. Germination is associated with the swelling of the conidium where water is taken up and the wall begins to expand. At the same time, there is a marked increase in metabolic activity that is observable before the emergence of the germ tube. When cell wall deposition becomes polarized, the germ tube forms. Though much research has been done on factors influencing the germination of spores, there have been relatively few studies on the molecular controls of the process in different fungi. Spores often contain large quantities of trehalose, a disaccharide that is hydrolyzed to glucose by spore-associated trehalases. These may require activation during germination or, if in a separate cellular compartment such as the cell wall, may be brought into contact with trehalose when the spore swells (D'Enfert 1997).

Germination signals can vary for different species and for different conidial types produced by a single species. For example, the arthroconidia of *Trichophyton*

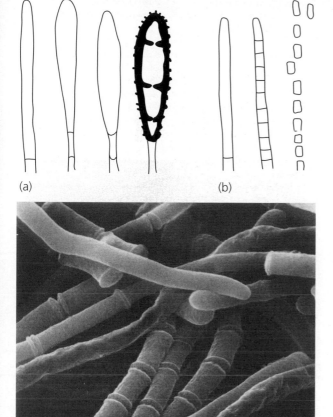

Figure 3.7 *Conidiospore development by thallic conidiogenesis where a conidium forms after remodeling of existing hyphal walls.* **(a)** *Development of an aleuriconidium after swelling, septation, and ornamentation of a hypha.* **(b)** *Development of arthroconidia after formation of cross-walls that allow disarticulation of the hypha.* **(c)** *Hyphae and arthroconidia formed by a species of* Trichosporon *(magnification ×1 680). Courtesy of G.T. Cole*

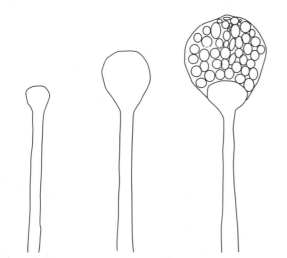

Figure 3.8 *Sporangiospore development: spores form after division of nuclei and cleavage of the cytoplasm which accumulates within the swollen tip of the sporangiophore. Mature sporangiospores are formed within a sac (the sporangium) and are released when this ruptures.*

mentagrophytes germinate after being 'activated' by either a short exposure to elevated temperature (45°C, 10 minutes), or by soaking in water at room temperature for 24 hours. After this pretreatment, the conidia germinate in the absence of exogenous nutrients, producing germ tubes when held in water at 37°C. By contrast, the microconidia of *T. mentagrophytes* do not germinate in the absence of exogenous amino acids or dipeptides. Such differences in germination requirements may well explain the poor infectivity of microconidia when compared to arthroconidia produced by the same dermatophyte (Hashimoto 1991).

FEATURES OF FUNGI ASSOCIATED WITH PATHOGENICITY AND VIRULENCE

Much has been written concerning those characteristics of pathogens that contribute to their virulence. Certain properties are necessary for effective pathogenicity, for example, tolerance of body temperatures and ability to grow using the nutrients available in tissues. But a successful pathogen must also be able to establish an infection, a process that not only involves ensuring that the host is exposed to infectious propagules but also that, once exposed, the propagules can survive and grow in the face of a mobilized host defense. In the case of the pathogens that infect via the lungs, spore size can be of great importance. For example, the conidia of *Aspergillus fumigatus* are small enough to enter the alveoli, a feature that is not the case for all aspergilli.

There is a significant body of work demonstrating the importance of the antiphagocytic capsule to the virulence of *Cryptococcus neoformans*. But other features, such as the production of phenoloxidase (a laccase), can play a role in virulence when this species invades the central nervous system. The features associated with virulence are considered in the chapters concerning the different fungi.

Normal cellular processes such as the signal-transduction cascades, exemplified by the MAP kinases, have been found to be necessary for normal development and virulence in *C. albicans* and *Cryptococcus neoformans*. The specificity of pathway selection and activation during host–parasite interactions is an area of active research and it seems likely that at least some of these signals induce the production of virulence factors by pathogens (Lengeler et al. 2000). However, this may not turn out to be the case for many of those opportunistic fungi that are saprophytes and are very unusually seen as pathogens in humans or animals. In a similar fashion to the situation with signal transduction pathways, certain steps in metabolic pathways may need to be present for effective pathogenicity. Auxotrophic mutants can have greatly diminished virulence, a feature reported for *Aspergillus nidulans* (Purnell 1973), *Cryptococcus neoformans* (Perfect et al. 1993), and *C. albicans* (Guo and Bhattacharjee 2003). While such 'normal' processes may not play a central role in host–parasite interactions, they might well provide useful targets for antifungal drugs.

Evidence of fungal components being important in virulence has been sought by using molecular genetic methods. Typically, a protein that is thought to be important in virulence may be inactivated, usually by generating a mutant. The most efficient way to generate targeted mutations where reversion does not occur has been to disrupt the gene encoding the target protein. After gene disruption and demonstration that the protein is no longer expressed, the significance of the protein may be estimated by comparing the virulence of the mutant to the original wild-type. Should the mutant show reduced virulence, it is then essential to show that it regains virulence when a functional target gene is reinserted. What has become clear is that most fungi appear to exploit numerous components for virulence and that simple disruption of one gene may not have much obvious effect when there are several genes encoding a particular activity, for example with serine proteinases or chitin synthetases. In some cases, gene disruption shows significant effects on certain stages in the infectious process: for example phospholipase B1 mutants of *C. albicans* retain some pathogenicity but have reduced invasive capacity (Ghannoum 2000). Details of current knowledge on fungal virulence factors are beyond the scope of this introductory chapter.

REFERENCES

Anderson, J.G. and Smith, J.E. 1972. The effects of elevated temperature on spore swelling and germination in *Aspergillus niger*. *Can J Microbiol*, **18**, 289–97.

Bartnicki-Garcia, S., Bartnicki, D.D., et al. 1995. Evidence that Spitzenkörper behaviour determines the shape of a fungal hypha: a test of the hyphoid model. *Exp Mycol*, **19**, 153–9.

Berman, J. and Sudbery, P.E. 2002. *Candida albicans*: a molecular revolution built on lessons from budding yeast. *Nature Rev Genet*, **3**, 918–30.

Bracker, C., Ruiz-Herrera, J. and Bartnicki-Garcia, S. 1976. Structure and transformation of chitin synthetase particles (chitosomes) during microfibril synthesis in vitro. *Proc Natl Acad Sci U S A*, **73**, 4570–4.

Bretscher, A. 2003. Polarized growth and organelle segregation in yeast: the tracks, motors, and receptors. *J Cell Biol*, **160**, 811–16.

Cailliez, J.C., Seguy, N., et al. 1996. *Pneumocystis carinii*: an atypical fungal micro-organism. *J Med Vet Mycol*, **34**, 227–39.

Chabasse, D., Bouchara, J.P., et al. 1988. Chlamydospores de *Candida albicans* observées in vivo chez un patient atteint de SIDA. *Ann Biol Clin*, **46**, 817–18.

Chandra, J., Kuhn, D.M., et al. 2001. Biofilm formation by the fungal pathogen *Candida albicans*: development, architecture, and drug resistance. *J Bacteriol*, **183**, 5385–94.

Cole, G.T. and Samson, R.A. 1979. *Patterns of development of conidial fungi*. London: Pitman.

Cole, G.T. and Samson, R.A. 1984. The conidia. In: Al-Doory, Y. and Domson, J.F. (eds), *Mould allergy*. Philadelphia, PA: Lea & Febiger, 66–103.

Cole, G.T., Seshan, K.R., et al. 1991. Chlamydospore-like cells of *Candida albicans* in the gastrointestinal tract of infected, immunocompromised mice. *Can J Microbiol*, **37**, 637–46.

Crespo, J.L. and Hall, M.N. 2002. Elucidating TOR signaling and rapamycin action: lessons from *Saccharomyces cerevisiae*. *Microbiol Mol Biol Rev*, **66**, 579–91.

D'Enfert, C. 1997. Fungal spore germination: insights from the molecular genetics of *Aspergillus nidulans* and *Neurospora crassa*. *Fungal Genet Biol*, **21**, 163–92.

Ghannoum, M.A. 2000. Potential role for phospholipases in virulence and fungal pathogenicity. *Clin Microbiol Rev*, **13**, 122–43.

Gimeno, C.J., Ljungdahl, P.O., et al. 1993. Characterization of *Saccharomyces cerevisiae* pseudohyphal growth. In: Vanden Bossche, H., Odds, F.C. and Kerrigge, D. (eds), *Dimorphic fungi in biology and medicine*. New York: Plenum Press, 83–103.

Gow, N.A.R. and Gadd, G.M. 1995. *The growing fungus*. London: Chapman & Hall.

Guo, S. and Bhattacharjee, J.K. 2003. Molecular characterization of the *Candida albicans LYS5* gene and site-directed mutational analysis of the PPTase (Lys5p) domains for lysine biosynthesis. *FEMS Microbiol Lett*, **224**, 261–7.

Harris, S.D. 2001. Septum formation in *Aspergillus nidulans*. *Curr Opin Microbiol*, **4**, 736–9.

Hashimoto, T. 1991. Infectious propagules of dermatophytes. In: Cole, G.T. and Hoch, H.C. (eds), *The fungal spore and disease initiation in plants and animals*. New York: Plenum Press, 181–202.

Hawksworth, D.L., Kirk, P.M., et al. 1995. *Ainsworth and Bisby's dictionary of the fungi*, 8th edition. Wallingford: CAB International.

Heath, I.B. 1994. The cytoskeleton in hyphal growth, organelle movements, and mitosis. In: Wessels, J.G.H. and Meinhardt, F. (eds), *The Mycota. I Growth, differentiation and sexuality*. Berlin: Springer-Verlag, 43–65.

Johnson, D.I. 1999. Cdc42: an essential Rho-type GTPase controlling eukaryotic cell polarity. *Microbiol Mol Biol Rev*, **63**, 54–105.

Lan, C.Y., Newport, G., et al. 2002. Metabolic specialization associated with phenotypic switching in *Candida albicans*. *Proc Natl Acad Sci U S A*, **99**, 14907–12.

Lengeler, K.B., Davidson, R.C., et al. 2000. Signal transduction cascades regulating fungal development and virulence. *Microbiol Mol Biol Rev*, **64**, 746–85.

Lugones, L.G., Bosscher, J.S., et al. 1996. An abundant hydrophobin (ABH1) forms hydrophobic rodlet layers in *Agaricus bisporus* fruiting bodies. *Microbiology*, **142**, 1321–9.

McDonough, J.A., Bhattacherjee, V., et al. 2002. Involvement of *Candida albicans* NADH dehydrogenase complex I in filamentation. *Fungal Genet Biol*, **36**, 117–27.

Mendoza, L., Herr, R.A., et al. 1999. In vitro studies on the mechanisms of endospore release by *Rhinosporidium seeberi*. *Mycopathologia*, **148**, 9–15.

Mendoza, L., Taylor, J.W. and Ajello, L. 2002. The class Mesomycetozoea: a heterogeneous group of microorganisms at the animal-fungal boundary. *Annu Rev Microbiol*, **56**, 315–44.

Miller, M.G. and Johnson, A.D. 2002. White-opaque switching in *Candida albicans* is controlled by mating-type locus homeodomain proteins and allows efficient mating. *Cell*, **110**, 293–302.

Momany, M. 2002. Polarity in filamentous fungi: establishment, maintenance and new axes. *Curr Opin Microbiol*, **5**, 580–5.

Perfect, J.R., Toffaletti, D.L. and Rude, T.H. 1993. The gene encoding phosphoribosylaminoimidazole carboxylase (*ADE2*) is essential for growth of *Cryptococcus neoformans* in cerebrospinal fluid. *Infect Immun*, **61**, 4446–51.

Purnell, D.M. 1973. The effects of specific auxotrophic mutations on the virulence of *Aspergillus nidulans* for mice. *Mycopathol Mycol Appl*, **50**, 195–203.

Seiler, S. and Plamann, M. 2003. The genetic basis of cellular morphogenesis in the filamentous fungus *Neurospora crassa*. *Mol Cell Biol*, **14**, 4352–64.

Sentandreu, R., Mormeneo, S. and Ruiz-Herrera, J. 1994. Biogenesis of the fungal cell wall. In: Wessels, J.G.H. and Meinhardt, F. (eds), *The Mycota. I Growth, differentiation and sexuality*. Berlin: Springer Verlag, 111–24.

Sharpless, K.E. and Harris, S.D. 2002. Functional characterization and localization of the *Aspergillus nidulans* formin SepA. *Mol Cell Biol*, **13**, 469–79.

Sietsma, J.H. and Wessels, J.G.H. 1994. Apical wall biogenesis. In: Wessels, J.G.H. and Meinhardt, F. (eds), *The Mycota I. Growth, differentiation and sexuality*. Berlin: Springer Verlag, 125–41.

Slutsky, B., Buffo, J. and Soll, D.R. 1985. High-frequency switching of colony morphology in *Candida albicans*. *Science*, **230**, 666–9.

Soll, D.R. 1993. Switching and the regulation of gene transcription in *Candida albicans*. In: Vanden Bossche, H., Odds, F.C. and Kerridge, D. (eds), *Dimorphic fungi in biology and medicine*. New York: Plenum Press, 72–82.

Soll, D.R., Lockhart, S.R. and Zhao, R. 2003. Relationship between switching and mating in *Candida albicans*. *Eukaryot Cell*, **2**, 390–7.

Van de Peer, W., Baldauf, S.L., et al. 2000. An updated and comprehensive rRNA phylogeny of (Crown) eukaryotes based on rate calibrated evolutionary distances. *J Mol Evol*, **51**, 565–76.

Vivarès, C.P., Gouy, M., et al. 2002. Functional and evolutionary analysis of a eukaryotic parasitic genome. *Curr Opin Microbiol*, **5**, 499–505.

Wessels, J.G.H. 1994. Developmental regulation of fungal cell wall formation. *Annu Rev Phytopathol*, **32**, 423–37.

Wickes, B.L., Mayorga, M.E., et al. 1996. Dimorphism and haploid fruiting in *Cryptococcus neoformans:* association with the alpha mating type. *Proc Natl Acad Sci U S A*, **93**, 7327–31.

4

Laboratory diagnosis

GLENN D. ROBERTS AND NORMAN L. GOODMAN

GENERAL STATEMENT

The diagnosis of a fungal disease is a complicated event requiring the cooperation and collaboration of many people with diverse expertise. Most fungal diseases require a laboratory diagnosis, so it is imperative that the laboratory is in close communication with the physician. This is especially important with the dramatic changes in potential hosts for fungal diseases in this era of immunosuppression and transplantation. As the host has changed, so has the range of potentially 'pathogenic' fungi. It is now well documented that many fungi in the environment may cause disease in severely compromised human hosts. The physician and laboratorian must thus be prepared to consider any fungal isolate as being etiologically significant and take appropriate measures to incriminate or disprove its role in each patient's case. The most commonly used methods for the laboratory diagnosis of fungal infections are, in general, old and time-honored. The recovery of the etiologic agent from clinical specimens is still considered to be the 'gold standard.' Methods used for identification of the fungi are a mixture of traditional and newer commercially available systems, e.g. yeast identification kits. Molecular methods are being developed for the detection and identification of fungi at a very fast rate; however, no comparative studies have been done to determine which are optimal and no standards for testing have been developed to date. Molecular testing is the future for many areas of clinical mycology but it will be some time before routine

testing will be standard practice within the laboratory. A section at the end of this chapter will present a preview of what has been done and some insight in to what will be useful in the future.

SPECIMEN SELECTION

The laboratory diagnosis of a fungal disease starts with careful collection of the appropriate clinical specimen. The specimen must contain the viable etiologic agent, if it is to be recovered and identified. The anatomic site in which the organism is present must be carefully selected and the specimen collected in such a manner that it will allow the fungus to remain viable in its 'natural' state, without contamination (Merz and Roberts 2003). Sites from which to collect specimens for the most common mycoses are given in Table 4.1.

When the appropriate site is selected, the specimen should be placed in a suitable container and promptly sent to the laboratory.

SPECIMEN COLLECTION

Specimens that are inappropriately collected, stored, or not processed on time may cause misdiagnoses, delays in appropriate therapy and increased costs for the patient.

The following are guidelines for specimen collection:

- Collect the specimen from an active lesion; old 'burned-out' lesions often do not contain viable organisms.

Table 4.1 Selection of clinical specimens for fungal culture: order of preference for site selection[a]

Specimen	Blastomycosis	Coccidioidomycosis	Histoplasmosis	Paracoccidioidomycosis	Candidiasis	Cryptococcosis	Aspergillosis	Zygomycosis	Hyalohyphomycosis	Dermatophytosis	Chromoblastomycosis	Pheohyphomycosis	Sporotrichosis	Mycetoma
Lower respiratory tract	1	1	1	2	x	1	1	1	3	–	–	–	3	–
Blood	–	6	2	–	1	2	–	–	2	–	–	–	x	–
Bone	4	x	–	x	–	x	x	x	–	–	–	–	x	–
Bone marrow	x	x	3	x	x	x	–	–	x	–	–	–	–	–
Brain	x	x	–	x	x	x	x	3	x	–	x	2	–	–
Cerebrospinal fluid	x	x	x	–	x	5	–	–	–	–	–	–	x	–
Eye	–	–	x	–	x	–	x	x	5	–	–	–	–	–
Nose–nasal sinus	x	x	–	x	x	–	2	2	4	–	–	–	4	–
Prostate	x	x	–	–	x	–	–	–	–	–	–	–	–	–
Mucous membrane	3	2	5	3	4	x	x	x	x	–	x	–	x	–
Subcutaneous tissue	–	x	–	–	x	–	–	x	–	–	2	1	2	1
Joints	x	x	–	–	x	–	x	–	x	–	–	–	–	–
Urine	5	5	4	–	2	3	x	–	–	–	–	–	–	–
Skin	2	3	x	1	x	x	x	4	1	1	1	–	1	2
Hair and nails	–	–	–	–	x	–	x	–	x	2	–	–	–	–
Multiple systemic sites during disseminated infection	6	4	6	4	3	4	3	5	6	–	x	–	x	–

a) Predominant sites for recovery of organism are ranked in order of importance (based on most common clinical presentations). x indicates other sites from which organisms have been recovered.

- Collect the specimen under aseptic conditions.
- Collect a sufficient amount of the specimen.
- Collect specimens before instituting therapy.
- Use sterile collection devices and containers.
- Label the specimens appropriately; all clinical specimens should be considered as potential biohazards and should be handled with care using universal precautions.

Respiratory tract specimens

The most common specimen for the diagnosis of fungal pneumonia and chronic pulmonary disease is sputum. The best specimen can be obtained when the patient first awakens from sleep, after the oral cavity has been cleansed. Bronchial alveolar lavage (BAL) and bronchial brushings are usually superior to sputum, provided that double- or triple-lumen bronchoscopes are used to prevent upper respiratory microbial contamination. Well-collected lower respiratory specimens are very useful for direct microscopic observation and culture for fungi.

Sterile body fluids (cerebrospinal, peritoneal, pericardial, synovial and vitreous humor)

These provide excellent specimens because there are usually no contaminating organisms present. Sterile fluids should be obtained under aseptic conditions using a sterile syringe. The specimen should be transferred to a sterile, closed tube for transport to the laboratory. Small amounts of fluid can be transported in the syringe, after the needle has been removed and the syringe re-capped.

Tissue

Tissue from the site of active disease process is the ideal specimen for diagnosing a mycosis and it should be collected under strict aseptic conditions. It should be placed into a sterile, closed container of the appropriate size. An adequate volume of sterile saline, without preservative, should be added to the container to keep the tissue moist. If the tissue cannot be processed immediately, it may be refrigerated, but not frozen, before it is examined microscopically and cultured.

Exudates/pus

Specimens from closed lesions and abscesses are usually of high quality. The outer surface of the site should be thoroughly cleansed and decontaminated. The specimen should then be aspirated with a syringe and needle and transferred to a sterile, closed container. Swabs should never be used to collect a specimen from a closed lesion

that has been surgically opened. Drainage from open lesions should be collected only after the lesion has been thoroughly cleansed. Care should be taken to collect any granules that may be present in specimens from open or closed lesions. If swabs must be used to collect a specimen from an open lesion, a maximum amount of material should be collected on the swab and kept moist *en route* to the laboratory.

External eye

Corneal scrapings should ideally be placed directly on to the surface of fungal culture media and on to the center of a clean, glass microscope slide to be used for direct microscopic examination. Fluid from an eye should be aspirated with a sterile syringe and needle and, if only a small quantity is present, transported to the laboratory in the syringe. Quantities greater than 0.5 ml should be transferred to a sterile tube for transport to the laboratory (Jones et al. 1981).

Urine

Urine specimens (50–200 ml) should be placed in a sterile, screw-capped container. Urine specimens should be freshly collected. Specimens from catheter bags are unacceptable. Urine may be refrigerated if it cannot be cultured promptly.

Blood

Blood should be collected under strict aseptic conditions. The specimen may be collected directly into a culture system or with a syringe, and then transferred to an appropriate culture medium. Strict aseptic conditions must be maintained in all situations (see section on Blood cultures below for more details).

Stool

Under very special circumstances, fecal cultures for fungi are appropriate, e.g. profoundly immunosuppressed bone marrow transplant recipients. The specimen should be collected in a sterile, screw-capped container or on two rectal swabs.

Hair

Infected hair should be drawn from the scalp with forceps and placed into a clean, dry container. Clipped hair is not acceptable because the root is the prime specimen. It is not necessary to use a sterile container. Hair for culture should be free of topical medications, conditioners or dressings.

Skin

Specimens of skin may be obtained by scraping or biopsy. A site on the active border of the lesions should be used. As skin is usually contaminated with other microorganisms, the surface of the selected site should be cleansed with 70 percent isopropyl alcohol. The epidermis should be scraped with a scalpel or other instrument and the scrapings placed in a clean, dry container, such as a Petri dish, specimen cup or envelope. Biopsied specimens should be placed in a sterile container.

Nails

Nails should be cleansed with 70 percent alcohol. A portion of the infected nail should be clipped and the excess keratin scraped from the nail bed. The clippings and scrapings should then be placed in a clean, dry container.

SPECIMEN TRANSPORT AND STORAGE

For best results, all clinical specimens should be microscopically examined and cultured as soon as possible. Except for blood and corneal scrapings, it is not necessary to transport the specimen in a transport medium. Blood should be placed in the preferred culture system or the lysis centrifugation system (Isolator, Carter Wallace Wampole Laboratories, Cranbury, NJ, USA). Corneal scrapings should be inoculated onto the medium when collected. Specimens should not be frozen before culture. Specimens not likely to contain contaminating microorganisms, e.g. spinal fluid, and those that could contain dermatophytes should not be refrigerated. Transported specimens should never be allowed to dry out. Sputum specimens may be transported without preservation if only dimorphic fungi are to be recovered (Hariri et al. 1982).

SPECIMEN PROCESSING

When specimens reach the laboratory, they must be appropriately processed to ensure viability of the etiologic agent and to minimize the chances of contamination. Specimens from normally sterile sites, e.g. spinal and peritoneal fluids, need no special processing, but may be cultured directly onto primary recovery media. If the volume is adequate, specimens should be concentrated by centrifugation membrane filtration, and the concentrated specimen aseptically transferred to primary recovery media for culture and on to a clean microscope slide for direct microscopic examination. Aseptically collected tissue should be homogenized in a tissue grinder, mortar pestle or 'stomacher.' A portion of tissue

suspected to contain a zygomycete should be finely minced with a pair of scissors or scalpel.

A highly viscous specimen, such as sputum, should be liquefied before culture, but it should not be excessively diluted. It should be concentrated by centrifugation if it is too dilute.

CULTURE

Most of the mycotic diseases are diagnosed by recovering and identifying the etiologic agent in the clinical laboratory. Few fungal diseases can be diagnosed clinically. The recovery of fungi from a clinical specimen begins with the selection of appropriate primary culture media that must include one that will support the growth of any potentially pathogenic fungus in the specimen and others that will selectively inhibit contaminating microorganisms. These media must be selected with great care; however, a number of options are available.

Media

Fungi may be recovered on most bacteriological media if given enough time to grow. There are, however, well-established fungal culture media, that are optimal for the recovery of fungi from clinical specimens. The following are some of the most commonly used media for recovering fungi from clinical specimens: Sabouraud's dextrose agar; Emmons' modified Sabouraud's agar; brain–heart infusion (BHI) agar (with or without 5 percent sheep red blood cells), Sabouraud's brain–heart infusion (SBHI) agar; inhibitory mold agar (IMA); mycosel or mycobiotic agars; potato flake agar; and yeast extract phosphate agar.

Combinations of antibacterial antibiotics and cycloheximide may be added to these media to make them selective. Suggested uses for these media are presented in Table 4.2 and, as indicated, a battery of different media should be used to culture specimens when more than one organism is expected to be recovered.

As in all areas of microbiology, the key to rapid and accurate identification of an organism is a pure culture. Isolated colonies are needed for morphological or biochemical identification. It is, therefore, necessary to spread specimens suspected of containing multiple organisms on to a large surface area. Petri dishes are generally optimal but others, such as tissue culture flasks or large bottles, are adequate. Standard, microbiological, culture tubes should not be used because colonies cannot be easily isolated from fluid specimens such as sputum and urine.

Incubation conditions

Fungi commonly associated with disease will grow well at temperatures between 25°C and 30°C. It is generally accepted that 30°C is the optimal temperature for

Table 4.2 *Media useful for the recovery of fungi from clinical specimens*

Specimen	Inhibitory media (bacteria and/or saprobic fungi)[a] with 5% sheep blood	Inhibitory media (bacteria and/or saprobic fungi)[b]	Yeast extract–phosphate agar
Lower respiratory tract	x	x	x
Secretions (e.g. sputum)	x	x	x
Tissue (lung or pleura)	x	x	–
Bone marrow	x	x	–
CSF	x	x	–
Eye	–	x	–
Nares/nasal sinuses	x	x	–
Skin/mucous membrane lesions	x	x	–
Tissue biopsy (not contaminated)	x	x	–
Contamination suspected	x	x	–
Urine	x	x	–

a) Includes brain–heart infusion (BHI) agar.
b) Includes Sabouraud's dextrose agar; Emmons' modification of Sabouraud's dextrose agar; BHI agar; Sabouraud's brain–heart infusion (SBHI) agar; inhibitory mold agar (IMA); mycosel or mycobiotic agars and potato flake agar.

recovery of fungi from clinical specimens, although 'room' temperature (25–30°C) is adequate. Care should be taken to provide a humid environment, primarily to prevent plated media from drying during the incubation period of 4–6 weeks. Adequate moisture can usually be achieved by placing containers of water within the incubator or placing plates in a sealed oxygen-permeable plastic bag. Other specific incubator environments are necessary for special procedures; e.g. conversion of dimorphic fungi from the mold to the yeast form requires a temperature of 35–37°C and a moist environment. Higher temperatures may be needed to test for temperature-tolerance studies, which will differentiate some fungi. It should be noted that growth at 37°C is not a requirement for pathogenic fungi that may infect anatomic sites having a lower temperature.

Blood cultures

Fungemia has become a more common event, particularly in immunosuppressed patients and those who receive broad-spectrum antibiotics, or in those with intravascular devices in place. It is common for nosocomial fungal infections to develop in these groups of patients. Further, fungemia caused by dimorphic fungi has increased over the past few years. Laboratories must be sufficiently proficient to offer a reliable blood culture system that will recover the etiologic agent in the shortest possible time.

A number of variables must be carefully considered in the provision of optimal recovery of fungi from blood. Blood should be collected at intervals as described for the detection of bacteremia (Geha and Roberts 1994). The volume of blood cultured is critical in ensuring optimal recovery with, in general, the larger the volume collected, the greater the chance for recovery. It is necessary to obtain at least 10 ml of blood in a 1:10 ratio

of blood to broth in bottles, or 10 ml inoculated on to the surface of several different media after lysis-centrifugation. Automated systems may allow for use of a larger amount of blood for inoculation.

Blood culture bottles must be vented, either transiently or permanently, and the temperature of incubation should range from 25 to 30°C, with the higher temperature being preferable. The duration of incubation for most isolates of yeasts is one week, although the range of time for recovery of dimorphic fungi may extend to 4–6 weeks. Using lysis-centrifugation (with an Isolator), 3 weeks of incubation seems to be adequate. The average recovery time for yeasts and molds using the lysis-centrifugation method is 3.8 and 10.5 days, respectively (Telenti and Roberts 1989).

Media used for the recovery of fungi from blood vary with the system used in the laboratory. For the most part, media such as those used in commercially available bottles will recover yeast species adequately. The lysis-centrifugation method allows the user to select the solid media used. Yeasts are often recovered by employing sheep blood or chocolate agars. The addition of BHI, IMA and Sabouraud's dextrose agars allows for the recovery of all fungal pathogens within 2 weeks of incubation.

As a rule, broth cultures require about twice as long for recovery as the lysis-centrifugation method. *Cryptococcus neoformans* may not be recovered or the time for its recovery is significantly delayed by commercially available systems. Furthermore, the dimorphic fungi are frequently not recovered using broth media or commercial systems. However, a Myco/F/Lytic culture medium is reported potentially to offer an alternative method for the recovery of dimorphic fungi from blood (Waite and Woods 1998; Vetter et al. 2001). Other studies have not shown the Myco/F/Lytic culture medium to be superior to previously reported methods (Fuller et al. 2001).

The detection of fungemia has improved remarkably over the years and recovery time is clinically relevant. The lysis-centrifugation system is optimal when considering the recovery of all agents of fungemia; when only yeasts are considered, however, broth media or commercially available systems are probably satisfactory. The lysis-centrifugation method is preferred for the recovery of *Histoplasma capsulatum* and *Cryptococcus neoformans*; the system also has the added benefit of providing quantitative cultures.

DIRECT MICROSCOPIC EXAMINATION

A definitive diagnosis of most fungal diseases requires recovery and identification of the fungus from cultures. This often requires a period of several days to weeks for the organisms to produce enough growth and sporulation for identification. A tentative or preliminary diagnosis can often be made by microscopically detecting fungal elements present in the clinical specimen. A well-trained person, with knowledge of fungal morphology, can often provide a definitive diagnosis by this method or at least provide a tentative diagnosis days or weeks before culture confirmation.

The direct microscopic examination of clinical specimens may be as simple as placing a drop of liquid specimen on to a clean microscope slide and examining it with a light microscope, or it may involve more complex procedures, including staining of tissue. The sensitivity of direct examination depends on several factors, including quality of specimen, specimen type, quality of the microscope and, not least, the expertise of the microscopist in recognizing fungal elements and their diagnostic morphology.

If the specimen is transparent, it may be examined without treatment. If it is opaque, it must be cleared to reveal fungal elements, or the elements must be differentially stained within the specimen.

Treatment with potassium hydroxide (KOH) will clear the specimen of cells or debris Calcofluor white (Hageage and Harrington 1984); Gram's, Giemsa, Wright, periodic acid–Schiff (PAS), methenamine silver, and Papanicolaou's stains (Chandler et al. 1989; Conner et al. 1997) are satisfactory for use in detecting fungi in clinical specimens. These procedures may be performed in different areas of a laboratory, e.g. cytology, surgical pathology, hematology or microbiology; but it must be remembered that the above factors affecting sensitivity and specificity apply to any laboratory. The morphology of fungi appears essentially to be the same in other clinical specimens as it does in tissue, so it is possible to become very proficient in reading and interpreting results from a variety of specimens. Methods commonly used for detecting fungal elements in clinical specimens are presented in Table 4.3.

Fungi demonstrate enormous diversity in morphological features. Environmental factors play a significant role in determining how a fungus grows, so morphological diversity of fungi therefore should be expected in diseased tissue. Despite this, there is reasonable stability in the general morphology of the groups of fungi found in clinical specimens. Yeast forms are readily identifiable, but hyphae may be pleomorphic, depending on the source and host response. The microscopist must view a sufficient number of fields to ensure observation of the true morphological form of a fungus. Direct microscopic examination of clinical specimens can be very helpful in providing an early diagnosis for the clinician. The characteristics of the fungal elements commonly found in tissue and other clinical specimens are given in Table 4.4.

Examples of yeasts, pseudohyphae, septate and non-septate hyphae in tissue are shown in Figures 4.1, 4.2, 4.3 and 4.4.

METHODS OF IDENTIFICATION

When a fungus is recovered from a clinical specimen, a decision must be made as to its importance as a cause of disease. It is very helpful to know if the isolate is a 'classic pathogen' or one of the opportunistic fungi asso-

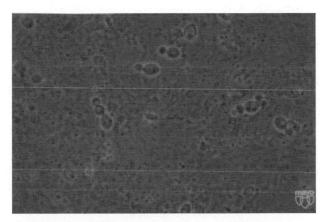

Figure 4.1 *Wet preparation of* Histoplasma capsulatum *var.* capsulatum *yeasts in bronchial lavage.*

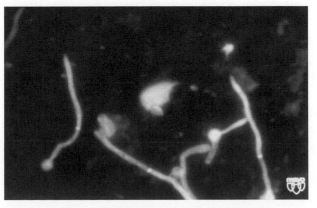

Figure 4.2 *Wet preparation of pseudohyphae of* Candida albicans *in scraping of oral lesion.*

Table 4.3 *Selected methods useful for direct microscopic detection of fungal elements in clinical specimens*

Method	Use	Time required	Advantages	Disadvantages
Calcofluor white	Detecting fungi	1 min	Can be mixed with KOH; fungi brightly fluoresce	Requires fluorescence microscope and special filters. Background sometimes fluoresces but fungi exhibit more intense fluorescence
Giemsa stain	Examining bone marrow and peripheral blood smears	15 min	Detects intracellular *H. capsulatum*	Detection is usually limited to *H. capsulatum*
Gram's stain	Detecting bacteria	3 min	Commonly performed on most clinical specimens submitted for bacteriology and detects most fungi present	Some fungi stain well; however, others (e.g. *Cryptococcus* spp.) stain weakly in some instances and exhibit only stippling. Common GS artifacts appear as yeast cells
India ink	Detecting *C. neoformans* in CSF and other body fluids	1 min	When positive in CSF, diagnostic of cryptococcosis	Negative in many cases of meningitis (non-AIDS)
Potassium hydroxide (KOH)	Clearing of specimens to make fungi more readily visible	Variable, depending on specimen	Rapid detection of fungal elements	Experience required; background artifacts are often confusing
Methylene blue	Detecting fungi in skin scrapings	2 min	Usually added to KOH; provides contrast for detection of fungal elements	Background staining of cells makes reading difficult
Methenamine silver stain	Detecting fungi in histological sections	1 hour	Best stain for detecting fungal elements	Requires specialized staining method not usually available in microbiology laboratories
Papanicolaou stain	Examining secretions for presence of malignant cells	30 min	Cytotechnologist can detect fungal elements	Requires specialized staining and a reader familiar with this stain
PAS stain	Detecting fungi	20 min; 5 min additional if counterstain is employed	Stains fungal elements well; hyphae and yeasts can be readily distinguished	*B. dermatitidis* appears pleomorphic. PAS-positive artifacts can appear as yeast cells
Wright stain	Examining bone marrow and peripheral blood smears	7 min	Detects intracellular *H. capsulatum*	Detection is usually limited to *H. capsulatum*

PAS, periodic acid–Schiff.

Figure 4.3 *Preparation of non-septate hyphae in bronchial washings stained with calcofluor white.*

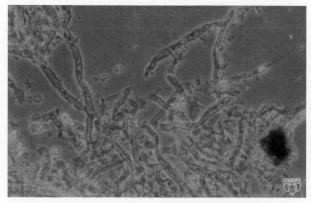

Figure 4.4 *Wet preparation of septate hyaline hyphae in bronchial washings.*

Table 4.4 *Characteristics of fungal elements seen in direct examination of clinical specimens*

Morphology of fungal elements	Organism(s)	Diameter range (μm)	Characteristic features
Yeast form	*Blastomyces dermatitidis*	8–15	Cells usually large and spherical, doubly refractile; buds usually single and attached by broad base, small forms of 2–5 μm may be seen
	Cryptococcus neoformans	2–15	Cells vary in size; usually spherical but may be football shaped; buds usually single and 'pinched off'; capsule may or may not be evident; rarely, pseudohyphae forms with or without a capsule may be seen
	Histoplasma capsulatum var. *capsulatum*	2–5	Small; oval to round budding cells; often found clustered within histocytes; difficult to detect when present in small numbers
	Paracoccidioides brasiliensis	5–60	Cells usually large and surrounded by smaller buds around periphery ('mariner's wheel appearance'); smaller cells (2–5 μm) that resemble *H. capsulatum* may be present; buds have 'pinched-off' appearance
	Sporothrix schenckii	2–6	Small; oval to round to cigar shaped; single or multiple buds present; often not seen in clinical specimens
Yeast form and pseudohyphae or true hyphae	*Candida* spp.	3–4 (yeast forms); 5–10 (pseudohyphae)	Cells usually exhibit single budding; pseudohyphae, when present, are constricted at ends and remain attached like links of sausage; true hyphae, when present, have parallel walls and are septate
Yeast form and hyphae in ski Yeast form in tissues	*Malassezia furfur*	3–8 (yeast forms); 2.5–4 (hyphae) 2.5–4.0 (yeast)	Short, curved hyphal elements usually present along with round yeast cells that retain their spherical shapes in compacted clusters; when found in skin exhibit monopolar budding, small broad base at point of attachment of buds
Spherules	*Coccidioides immitis*	10–200	Spherules vary in size; some contain endospores, others are empty. Adjacent spherules may resemble *B. dermatitidis*; endospores may resemble *H. capsulatum* but show no evidence of budding. Spherules may produce multiple germ tubes if direct preparation is kept in moist chamber for ⩾24 hours; hyphae may be found in cavitary lesions
Sporangia	*Rhinosporidium seeberi*	6–300	Large, thick-walled sporangia containing sporangiospores; mature sporangia are larger than spherules of *C. immitis*
Wide, non-septate hyphae	Zygomycetes: *Mucor* spp; *Rhizopus* spp. and other genera	10–30	Hyphae are large, ribbon-like, often fractured or twisted. Occasionally, septae may be present, branching usually at right angles. Smaller hyphae overlap those of *Aspergillus* spp., particularly *A. flavus*
Hyaline septate hyphae	Dermatophytes: skin, hair and nails	3–15	Hyaline septate hyphae commonly seen; chains of arthroconidia may be present
	Aspergillus spp.	3–12	Hyphae are septate and exhibit dichotomous, 45° angle branching; cannot be distinguished from other hyaline molds; larger hyphae may resemble those of zygomycetes
	Geotrichum spp.	4–12	Hyphae and rectangular arthroconidia are present and are sometimes rounded. Irregular forms may be present
	Trichosporon spp.	2–4 by 8	Hyphae and rectangular arthroconidia are present and are sometimes rounded. Blastoconidia may be produced, but are difficult to observe
	Scedosporium aprospermum (cases other than mycetoma)	3–12	Hyphae are septate and are impossible to distinguish from those of other hyaline molds, e.g. *Aspergillus* spp.
	Fusarium spp.	3–12	Hyphae are septate and are impossible to distinguish from those of other hyaline molds, e.g. *Aspergillus* spp.

(Continued over)

Table 4.4 *Characteristics of fungal elements seen in direct examination of clinical specimens (Continued)*

Morphology of fungal elements	Organism(s)	Diameter range (μm)	Characteristic features
Pheoid septate hyphae	*Bipolaris* spp. *Curvularia* spp., *Exserohilum* spp., *Exophiala* spp., *Phialophora* spp., *Wangiella dermatitidis*, *Cladophialophora bantiana*	2–6	Pheoid polymorphous hyphae are seen; budding cells with single septa and chains of swollen rounded, sometimes budding, cells may be present. Occasionally, aggregates may be present when infection is caused by *Phialophora* and *Exophiala* spp.
	Phaeoannellomyces elegans	1.5–5	Usually, large numbers of frequently branched hyphae are present along with budding cells
Muriform cells	*Cladophialophora carrionii*, *Fonsecaea compacta*, *Fonsecaea pedrosoi*, *Phialophora verrucosa*, *Rhinocladiella aquaspersa*	5–20	Pheoid, round to pleomorphic, thick-walled cells with transverse septa. Commonly, cells contain two fission plates that form a tetrad of cells. Occasionally, branched septate hyphae may be found in addition to muriform cells
Granules	*Acremonium falciforme, A. kiliense, A. recifei*	200–300	White, soft granules without cement-like matrix
	Curvularia geniculata, C. lunata	500–1 000	Black, hard grains with cement-like matrix at periphery
	Aspergillus nidulans	65–160	White, soft granules without cement-like matrix
	Exophiala jeanselmei	200–300	Black, soft granules, vacuolated, without cement-like matrix, made of dark hyphae and swollen cells
	Fusarium spp.	200–500	White, soft granules without cement-like matrix
	F. moniliforme, F. solani	300–600	
	Leptosphaeria spp.	400–600	Black, hard granules; cement-like matrix; periphery composed of polygonal swollen cells and center composed of hyphal network
	L. senegalensis, L. tompkinsii	500–1 000	
	Madurella grisea	350–500	Black, soft granules without cement-like matrix; periphery composed of polygonal swollen cells and center composed of hyphal network
	M. mycetomatis	200–900	Black to brown hard granules of two types: (1) rust brown, compact, and filled with cement-like matrix; (2) deep brown, filled with numerous vesicles, 6–14 μm in diameter, cement-like matrix in periphery and central area of light-colored hyphae
	Neotestudina rosatti	300–600	White, soft granules with cement-like matrix at periphery
	Scedosporium aprospermum	200–300	White, soft granules composed of hyphae and swollen cells at periphery in cement-like matrix
	Pyrenochaeta romeroi	300–600	Black, soft granules composed of polygonal swollen cells at periphery; center is network of hyphae; no cement-like matrix

ciated with infection of immunocompromised hosts. A rapid identification of the fungus is thus very important for diagnosis and early management of fungal diseases (Kwon-Chung and Bennett 1992).

Morphology

The standard method for identifying moulds is by observing their colonial and microscopic morphological features and following taxonomic keys to make an identification. This usually entails observing, on standardized media, the rate of growth, colonial morphology and color of the colony, and microscopic morphology including: formation of conidia; conidial arrangement;

conidial morphology and pigmentation; and hyphal morphology and pigmentation. There is considerable variation in the characteristics of molds and their identification is difficult for an untrained laboratory worker.

Yeasts can be identified by morphological and biochemical methods. Details on the identification of yeasts can be found in Chapters 30, Candidiasis and 32, Cryptococcosis. There are many commercial systems available for identifying yeasts on the basis of biochemical reactions. Most of these systems are very good for the identification of the clinically important yeasts, provided that their morphology is considered when interpreting the biochemical results. Most systems,

Table 4.5 *Commonly used media for determining morphology of fungi*

Organism	Media
Aspergillus spp.	Czapek's Dox agar
Pigmented molds	Cornmeal agar, Sabouraud's dextrose agar, Lactrimel agar
Dermatophytes	Cornmeal agar, potato dextrose agar
Hyaline molds	Cornmeal agar, Sabouraud's dextrose agar (Emmons' modification), potato flake agar, inhibitory mold agar
Yeasts	Cornmeal agar
Zygomycetes	Cornmeal agar, potato flake agar, Water agar, Sabouraud's dextrose agar

however, have serious flaws in identifying some yeasts without morphological information. It is necessary to use standardized media to be able to observe consistent morphological features that are stable enough to be used in taxonomic keys. Some of the more commonly used media are listed in Table 4.5.

Incubation requirements for morphological media are generally the same as for primary isolation. When a fungus appears mature enough to have formed conidia, it is necessary to examine the microscopic structures. Looking at colonies under a stereoscopic microscope is sometimes helpful for observing large fruiting structures and conidial arrangements. Culture dishes should not be opened during observation.

More detailed observation requires removal of a small representative sample of a portion of the colony, and its transfer to a drop of lactophenol cotton blue or aniline blue placed on a clean microscope slide.

The colony segment is carefully torn apart, using inoculation wires or picks; a clean coverslip is then added and the specimen is observed under the low- and high-power objectives of a microscope. If the fungal colony is soft, a portion is placed in the dye on a microscope slide and the specimen is covered with a glass coverslip, which is pressed down with a pencil eraser to flatten the specimen. An alternative and often-superior method for observing the morphology of molds is the 'Scotch tape' technique. A strip of clear Scotch tape is pressed firmly on the colony; it is then placed in a drop of lactophenol cotton blue or aniline blue dye on a clean microscope slide. The tape holds hyphae and reproductive structures in place and serves as a coverslip.

Most morphological structures can be observed adequately by transmitted light microscopy, but some unusual structures, e.g. annellides, may require more sophisticated microscopy, such as phase contrast or a high-resolution digital photograph that can be enlarged. A detailed description of the morphological characteristics of fungi is given in Chapter 3, Fungal structure and morphology.

Identification of dimorphic fungi using nucleic acid probes

The definitive identification of the dimorphic fungi *Histoplasma capsulatum*, *Blastomyces dermatitidis* and *Coccidioides immitis* has traditionally been made on the basis of the microscopic morphology of both mold and parasitic (yeast or spherule) forms. In vitro conversion of the mould form to its corresponding yeast form for *H. capsulatum* and *B. dermatiditis* is often difficult and requires an extended period of time. Conversion of *C. immitis* requires specialized media and conditions; clinical laboratories do not undertake it. Laboratories with little experience have found that it is impractical to use the in vitro conversion process.

The exoantigen test (Kaufman and Standard 1978) was considered for many years as the method of choice for the definitive identification of dimorphic fungi. The method relied on the principle that soluble antigens are produced by fungi when grown in a broth medium and that they can be extracted and concentrated and subsequently reacted with serum known to contain antibodies specific for a particular species. This immunodiagnostic method utilized the immunodiffusion test as a means of detecting the species-specific antigens produced by these dimorphic fungi.

The exoantigen test has been replaced by specific nucleic acid probes that provide a more rapid and specific identification. The identification of dimorphic fungi using nucleic acid probes has changed the laboratory aspects of clinical mycology more than any other single contribution made during the past few years. This technology allows for the rapid (one hour) identification of the organisms, whether found in mixed culture or not. The probes are commercially available from Gen-Probe Incorporated (San Diego, CA) and are genus and species specific.

Fungal cells are lysed and the ribosomal RNA is hybridized to a DNA probe which has been labeled with an acridinium ester (Stockman et al. 1993). In short, the hybrids are detected by the chemiluminescence of the attached probe. Chemiluminescence is quantitated using a luminometer and the limits for positivity are well defined. The method is simple and rapid. It is recommended for laboratories that deal with dimorphic fungi with some frequency. Financial considerations must be made before incorporating these probes into a clinical laboratory's diagnostic routine; however, the time saved for patients awaiting a diagnosis is dramatically shortened and appropriate therapy can be instituted promptly.

FUNGAL SEROLOGICAL TESTING

Serological procedures, together with clinical and patient history data, are often useful in obtaining an early diagnosis of the life-threatening systemic fungal diseases (Kaufman et al. 1997). Most of the currently available, standardized, serological procedures detect antibodies to specified fungi; they are, therefore, of limited value in diagnosing disease in immunocompromised patients. This deficiency is being addressed by the rapid development of tests for fungal antigens.

The use of serological tests to measure antibodies is most useful in the diagnosis of acute, systemic, fungal diseases. Agglutination, precipitation (immunodiffusion), complement fixation, and enzyme immunosorbent assays are widely available commercially for use in the diagnosis of aspergillosis, blastomycosis, coccidioidomycosis, histoplasmosis and paracoccidioidomycosis. The most widely used is the assay for the detection of *Histoplasma* antigen (Durkin et al. 1997). This test has proven useful for the detection of antigenemia in patients having disseminated disease (Wheat et al. 1986), particularly those with acquired immunodeficiency syndrome (AIDS) (Wheat et al. 1992). Serological tests have been developed commercially for detecting cryptococcal and candida antigens. The latex agglutination test for cryptococcal capsular antigen is one of the most reliable fungal serological tests for use in diagnosing cryptococcal meningitis. The role of tests for antigens of *Candida* spp. in diagnosing candidemia is unclear. Considerable effort is being directed towards developing tests to detect fungal antigens, in the serum and other body fluids, as a means of diagnosing the increasing number of opportunistic fungal diseases in immunocompromised patients (Yeo and Wong 2002). The galactomannan EIA assay offers promise of being an important test for invasive aspergillosis (Wheat 2003). A listing of the currently commercially available serological tests and their uses can be found in Table 4.6. A more detailed description of fungal immunoserology can be found in Chapter 5, Mycoserology and molecular diagnosis.

LABORATORY SAFETY

Biosafety in the clinical laboratory is of the utmost importance to the laboratory worker and to those who work in adjacent areas. The fungi produce conidia that are easily aerosolized and may contaminate the laboratory, air-handling systems, and, in some instances, other cultures and specimens. Comprehensive safety guidelines should be available to laboratory workers to minimize the risk of acquiring infection in this setting. The following are guiding principles which should be remembered and practised by laboratory personnel:

- All clinical specimens should be considered as infectious and universal precautions must be used to protect laboratory personnel.

- All fungi should be considered as potentially pathogenic. Some laboratory personnel may be immunocompromised and highly susceptible to infection.

- Even fungi that are not endemic to a certain locale must be taken into consideration when making a diagnosis. Cultures of *Blastomyces dermatitidis*, *Coccidioides immitis*, *Histoplasma capsulatum*, *Paracoccidioides brasiliensis* and *Penicillium marneffei* may be encountered in any laboratory as a result of the ease of travel today.

- All molds must be handled within a biological safety cabinet to prevent aerosolization of their conidia within the laboratory.

- When laboratory-teaching rounds are done, cultures of molds should not be passed around from person to person; they should remain on the bench to prevent inadvert breakage of a culture dish or tube.

- When flaming loops or wires, excess inoculum should be removed before flaming. Flaming should be used to decontaminate the entire wire or loop, not only the tip.

- Culture dishes can be used safely in the laboratory; however, lids should be removed only inside a biological safety cabinet. Lids should be taped entirely or in two places to prevent inadvertent opening.

- Biological safety cabinets must be certified annually to ensure proper protection to the laboratory worker.

- Work surfaces must be wiped with a disinfectant after a spill and before leaving for the day. A number of suitable disinfectants are available; however, hypochlorite is cost-effective and is suitable.

- A laboratory safety manual must be prepared that includes procedures for treatment of laboratory spills, an evacuation plan, and overall laboratory decontamination in the event of biohazardous spills.

MOLECULAR TESTING

When one reviews the literature relating to molecular methods used in fungal diagnosis, a plethora of reports are found. The number of reports is so numerous that it is difficult to determine which are of practical value in the clinical laboratory (Sandhu et al. 1995; Chen et al. 2002; Erjavec and Verweij 2002; Stevens 2002; Yeo and Wong 2002). Many of the studies have been validated only by the reporting investigator; many used only a small number of samples for the validation, and most lacked strong clinical correlation. However, these studies represent the beginning of a molecular era that will evolve over the next several years and these tests will find their particular niches within the laboratory. Most clinical mycology laboratories will not replace all traditional testing with molecular methods due to the extremely large number of fungi that are capable of causing human disease. To develop testing that will detect each etiologic agent would be neither practical nor cost-

Table 4.6 *Commonly used fungal serological tests*

Disease	Test antigen(s)	Interpretation
Aspergillosis	CF (aspergillin)	Titer ⩾1:32 suggestive of infection
	Aspergillus fumigatus, A. niger and *A. flavus*	Higher titers have a stronger correlation with disease
	ID, aspergillin EIA, galactomannan assay uses a monoclonal antibody	Multiple precipitin bands increase the suspicion of active disease. Positive in 95% of cases with fungus ball; 50% with bronchopulmonary disease. May be positive in invasive disease, rarely. Galactomannan antigen is detected in some patients, more evaluations are needed to determine efficacy. Piperacillin-tazo bactam cause false positive reactions
Blastomycosis	CF, yeast form of *Blastomyces dermatitidis*	Titers of ⩾1:8 are significant. Higher titers may be more indicative of disease. Cross-reactions occur in patients with histoplasmosis and coccidioidomycosis may occur. High percentage of cases are negative
	ID, yeast culture filtrate	Specific band for A antigen. Many culture-proved cases are negative
Candidiasis	ID, *Candida albicans* sonicate	Most cases have bands; however, high percentage of normal populations have antibody to *C. albicans*
	LA, yeast cytosol or mannan	⩾1:4 or 1:8 are considered positive; difficult to interpret; must have clinical correlation
Coccidioidomycosis	CF, coccidioidin	Titer ⩾1:2 significant. Higher titer parallels severity of disease. May cross-react with patients with histoplasmosis and blastomycosis. May not be positive early in disease
	ID, coccidioidin	Specific F antigen; correlates with CF results
	TP, coccidioidin	Detects IgM during first 3 weeks of disease
	LA, coccidioidin	Correlates well with CF
	EIA, coccidioidin	A positive relative EIA value indicates disease. Separate test for IgG and IgM. Positive IgM indicates early disease
Cryptococcosis	LA, anticryptococcal capsular polysaccharide	Detects cryptococcal capsular polysaccharide in cerebrospinal fluid or serum. Any titer significant. Rheumatoid factor causes false positive; 95% of *Cryptococcus* meningitis cases are positive *neoformans*
	EIA, antibody to cryptococcal capsular polysaccharide	Appears to be more sensitive than LA test. Use as quantitative test not confirmed
Histoplasmosis	CF, histoplasmin (mycelial form)	⩾1:8 significant. Higher titers parallel more severe disease. May be influenced by skin test
	CF, yeast cells (yeast form)	⩾1:8 significant. Higher titers parallel more severe disease. Usually higher titer than mycelial form. May cross-react with blastomycosis and coccidioidomycosis
	ID Histoplasmin	Specific bands to H and M antigens. 'M' band usually appears first in disease and persists. 'H' band is associated with active disease and often disappears during therapy. 1.0–2.0 indicates a weak positive; suggest repeat; 2.1–4.0, positive; 4.1–10.0, moderate positive; >10.0, highly positive (all values are EIA units).
	Histoplasma Immunosorbent Assay[a]	–
Sporotrichosis	LA, Yeasts of *Sporothrix schenckii*	Titer ⩾1:80 usually indicates active disease. Usually negative with systemic sporotrichosis

CF, complement fixation; EIA, enzyme immunoassay; ID, immunodiffusion; LA, latex agglutination; TP, tube precipitin.
a) Testing performed only by MiraVista Laboratories, Indianiapolis, IN.

effective; but, testing for selected fungi is certainly possible and work is already well underway to provide this to the clinical laboratory. However, amplification with universal fungal primers followed by sequencing of the amplicon offers the promise of being able to detect a variety of fungi in clinical specimens.

Molecular assays will allow the laboratory to detect and/or identify the fungi present in clinical specimens whether it is done by nucleic acid amplification technology or by specific probes or nucleic acid sequencing. This is an exciting time since molecular technology is in its infancy in routine clinical laboratories and many new developments lie ahead. A brief summary of what has been done to date will be presented; however, it will not be exhaustive. In this way, the reader will have a general appreciation for what has been currently accomplished and, perhaps, what might find its way into the routine clinical laboratory. Before choosing a specific molecular testing method, a thorough comparison of useful methods should be completed so that the best can be verified and validated prior to being used. Further, standards must be set for test performance analytical and clinical sensitivity; specificity, and reproducibility within the laboratory. We must develop a focused and standard approach that is optimal for patient diagnosis. Hopefully this will occur, but it may not be realistic with the financial and personnel constraints imposed on clinical laboratories currently. Molecular testing should be embraced with great enthusiasm and excitement since it has the potential to provide for a more accurate and sensitive diagnosis of fungal infections in patients whom we serve.

Direct detection of the nucleic acid of fungi in clinical specimens

Immunocompromised patients and fungal infections are being seen at a much higher rate than ever before. Traditional cultures often require an extended time before the etiologic agent can be recovered and identified. During this, the disease may have progressed to a point where therapy may be ineffective; molecular detection should allow for monitoring patients on a periodic basis for the presence of circulating nucleic acid of the infecting organism. Further, if nucleic acid can be detected in blood or urine, this would dramatically reduce the need for invasive procedures such as biopsy or bronchoalveolar lavage. Being able to start therapy at a much earlier time during the clinical course should significantly affect the survival rate of the patient with a life-threatening fungal infection.

One of the simplest approaches used has been in situ hybridization using specific nucleic acid probes for the identification of organisms in patient specimens. This method lacks amplification and is less sensitive than other assays but is useful for identifying fungi that can be seen in tissue and other clinical specimens. Assays for *Aspergillus*, *Candida* and many other fungi are used primarily in pathology laboratories and are helpful when a morphological identification cannot be made (Hayden et al. 2001, 2002, 2003; Procop 2002).

Amplification assays using the polymerase chain reaction (PCR) or similar methods allow for the detection of small amounts of target DNA in clinical specimens. Specific primers with or without specific probes have been used with some success. Recent reviews by Yeo and Wong (2002) and by Chen et al. (2002), summarize specific targets used and detection methods. Assays have been developed to detect DNA of *Candida* (Chryssanthou et al. 1999; Loeffler et al. 2000a; Wahyiningshi et al. 2000; Ahmad et al. 2002), *Aspergillus* (Lass-Florl et al. 2000; Buchheidt et al. 2001, 2002;

Ferns et al. 2002), *Fusarium* (Hue et al. 1999), *Cryptococcus* (Sandhu et al. 1995; Hendolin et al. 2000), *Histoplasma* (Bialek et al. 2002; Rickerts et al. 2002), *Blastomyces* (Sandhu et al. 1995; Bialek et al. 2003), *Coccidioides* (Sandhu et al. 1995), *Paracoccidioides* (Gomes et al. 2000), *Pneumocystis jirovecii* (Lindsley et al. 2001; Nevez et al. 2002), *Penicillium marneffei* (Vanittanakom et al. 2002), and dermatophytes (Turin et al. 2000).

More recent technology has utilized real time PCR using the Light Cycler offered by Roche Molecular Systems (Mannheim, Germany). Fluorescent resonance energy transfer is used to detect amplified products and the system allows for the quantification of the target DNA. Results are available within one hour of testing and sensitivity is exquisite. Loeffler has used this technology to develop assays for *Aspergillus* (Loeffler et al. 2000b), *Candida* (Loeffler et al. 2000a), and the zygomycetes (Loeffler et al. 2001). A recent report described the use of real time PCR for the identification of culture isolates of *Histoplasma capsulatum* (Martagon-Villamil et al. 2003). This study also reported the detection of *H. capsulatum* DNA in a few clinical specimens; however, these are preliminary data and clinical trials will be needed to confirm the utility of the test as a diagnostic tool. Further, the MagNApure (Roche Molecular Systems, Mannheim, Germany) automated DNA extraction system was evaluated for use with the fungi (Loeffler et al. 2002). A combination of the Light Cycler and MagNApure appears to offer great promise for the detection of fungal DNA in clinical specimens in a very rapid manner. The Light Cycler may also be used to identify organisms to the species level using specific primers and probes.

Traditionally, the identification of clinically important fungi has been made based on macroscopic and microscopic morphological features. Considering the large number of fungi in the environment that are capable of causing human disease, it is difficult to believe that molecular methods will replace conventional methods anytime soon. However, a limited number of fungi may be identified to genus and species using PCR and specific probes. Yeasts in blood cultures have been identified and most were species of *Candida* (Chang et al. 2001). *Aspergillus* (Henry et al. 2000), zygomycetes (Voigt et al. 1999), dermatophytes (Mochizuki et al. 2001; Ninet et al. 2003), and several filamentous fungi may be identified using amplification and probes. Nucleic acid sequencing has been use with great success for the identification of fungi in culture (Turenne et al. 1999). Currently, a comprehensive database of fungal sequences is available in the MicroSeq 500 microbial identification system (Applied Biosystems, Foster City, CA). This database is not yet complete, but when used with GenBank and other public databases, many fungi may be more easily be identified when compared to traditional identification methods.

The use of molecular methods is important to investigate the epidemiology and environmental sources of fungi that infect immunocompromised patients and, in some instances, immunocompetent patients. Epidemiologic typing can determine whether or not organisms share the same DNA profile and this can be related to environmental isolates to determine the point source. Most of the studies have been related to isolates of *Candida* (Stephan et al. 2002), *Cryptococcus* (Anonymous 2002), *Aspergillus* and *Fusarium* (Raad et al. 2002). The use of molecular tools has allowed for the reduction of hospital-acquired infections and their spread.

Overall, molecular methods have and will continue to have a major impact on the diagnosis and appropriate treatment of fungal infection. Analytical parameters of these methods need to be standardized to optimize sensitivity and specificity and comparative studies need to be performed to determine which are best to use in the laboratory. Ideally, tests should be as simple as possible to perform so that most clinical laboratories can use them. The most important element, clinical correlation, must be established for methods. If these criteria are met, most of the newly developed molecular-based tests will be available to all of the patients whom we serve. One needs to remember that all of us should be in the field to improve patient care and that the needs of the patient come first.

REFERENCES

Ahmad, S., Khan, Z., et al. 2002. Seminested PCR for diagnosis of candidemia: comparison with culture, antigen detection and biochemical methods for species identification. *J Clin Microbiol*, **40**, 7, 2483–9.

Anonymous 2002. European Confederation of Medical Mycology (ECMM) prospective survey of cryptococcosis: report from Italy. *Med Mycol*, **40**, 5, 507–17.

Bialek, R., Feucht, A., et al. 2002. Evaluation of two nested PCR assays for detection of *Histoplasma capsulatum* DNA in human tissue. *J Clin Microbiol*, **40**, 5, 1644–7.

Bialek, R., Cirera, A.C., et al. 2003. Nested PCR assays for detection of *Blastomyces dermatitidis* DNA in paraffin-embedded canine tissue. *J Clin Microbiol*, **41**, 1, 205–8.

Buchheidt, D., Spiess, B., et al. 2001. Systemic infections with *Aspergillus* species in patients with hematological malignancies: current serologic and molecular diagnostic approaches. *Onkologie*, **24**, 6, 531–6.

Buchheidt, D., Baust, C., et al. 2002. Clinical evaluation of a polymerase chain reaction assay to detect *Aspergillus*. *Br J Haematol*, **116**, 4, 803–11.

Chandler, F.W., Kaplan, W., et al. 1989. *A colour atlas and textbook of the histopathology of mycotic diseases*. London: Wolfe Medical.

Chang, H.C., Leaw, S.N., et al. 2001. Rapid identification of yeasts in positive blood cultures by a multiplex PCR method. *J Clin Microbiol*, **39**, 10, 3466–71.

Chen, S.C., Halliday, C.L., et al. 2002. A review of nucleic acid-based diagnostic tests for systemic mycoses with an emphasis on polymerase chain reaction-based assays. *Med Mycol*, **40**, 333–57.

Chryssanthou, E., Klingspor, L., et al. 1999. PCR and other non-culture methods for diagnosis of invasive *Candida* infections in allogeneic bone marrow and solid organ transplant recipients. *Mycoses*, **42**, 4, 239–47.

Conner, D.H., Chandler, F.W., et al. 1997. *Pathology of infectious diseases*. Standforn, CT, Appleton and Lange.

Durkin, M.M., Connolly, P.A., et al. 1997. Comparison of radioimmunoassay and enzyme-linked immunoassay methods for detection of *Histoplasma capsulatum* antigen. *J Clin Microbiol*, **35**, 2252–5.

Erjavec, Z. and Verweij, P.E. 2002. Recent progress in the diagnosis of fungal infections in the immunocompromised host. *Drug Resistance Updates*, **5**, 3–10.

Ferns, R.B., Fletcher, H., et al. 2002. The prospective evaluation of a nested polymerase chain reaction assay for the early detection of *Aspergillus* infection in patients with leukemia or undergoing allograft treatment. *Br J Haematol*, **119**, 3, 720–5.

Fuller, D.D., Davis, T.E.J., et al. 2001. Evaluation of BACTEC MYCO/ F Lytic medium for recovery of mycobacteria, fungi, and bacteria from blood. *J Clin Microbiol*, **39**, 8, 2933–6.

Geha, D.J. and Roberts, G.D. 1994. Laboratory detection of fungemia. *Clin Lab Med*, **14**, 83–97.

Gomes, G.M., Cisalpino, P.S., et al. 2000. PCR for diagnosis of paracoccidioidomycosis. *J Clin Microbiol*, **38**, 9, 3478–80.

Hageage, G.J. and Harrington, B.J. 1984. Use of calcofluor white in clinical mycology. *Lab Med*, **15**, 109–12.

Hariri, A.R., Hempel, H.O., et al. 1982. Effects of time lapse between sputum collection and culturing on isolation of clinically significant fungi. *J Clin Microbiol*, **15**, 425–8.

Hayden, R.T., Qian, X., et al. 2001. In situ hybridization for the identification of yeastlike organisms in tissue section. *Diagn Mol Pathol*, **10**, 1, 15–23.

Hayden, R.T., Qian, X., et al. 2002. In situ hybridization for the identification of filamentous fungi in tissue section. *Diagn Mol Pathol*, **11**, 2, 119–26.

Hayden, R.T., Isotalo, P.A., et al. 2003. In situ hybridization for the differentiation of *Aspergillus*, *Fusarium* and *Pseudallescheria* species in tissue section. *Diagn Mol Pathol*, **12**, 1, 21–6.

Hendolin, P.H., Paulin, L., et al. 2000. Panfungal PCR and multiplex liquid hybridization for detection of fungi in tissue specimens. *J Clin Microbiol*, **38**, 11, 4186–92.

Henry, T., Iwen, P.C., et al. 2000. Identification of *Aspergillus* species using internal transcribed spacer regions 1 and 2. *J Clin Microbiol*, **38**, 4, 1510–15.

Hue, F., Huerre, M., et al. 1999. Specific detection of *Fusarium* species in blood and tissues by a PCR technique. *J Clin Microbiol*, **37**, 8, 2434–8.

Jones, D.B., Liesegang, T.J., et al. 1981. *Cumitech 13*. Washington, DC: American Society for Microbiology.

Kaufman, L. and Standard, P. 1978. Improved version of the exoantigen test for identification of *Histoplasma capsulatum*. *J Clin Microbiol*, **8**, 42–5.

Kaufman, L., Kovacs, J.A., et al. 1997. *Clinical immunomycology*. Washington, DC: American Society for Microbiology.

Kwon-Chung, J. and Bennett, J. 1992. *Medical mycology*. Philadelphia, PA: Lea and Febiger.

Lass-Florl, C., Rath, P.M., et al. 2000. *Aspergillus terreus* infections in haematological malignancies: molecular epidemiology suggest association with in-hospital plants. *J Hosp Infect*, **46**, 1, 31–5.

Lindsley, M.D., Hurst, S.F., et al. 2001. Rapid identification of dimorphic and yeast-like fungal pathogens using specific DNA probes. *J Clin Microbiol*, **39**, 10, 3505–11.

Loeffler, J., Hagmeyer, L., et al. 2000a. Rapid detection of point mutations by fluorescence resonance energy transfer and probe melting curves in *Candida* species. *Clin Chem*, **46**, 5, 631–365.

Loeffler, J., Henke, N., et al. 2000b. Quantification of fungal DNA by using fluorescence resonance energy transfer and the light cycler system. *J Clin Microbiol*, **38**, 2, 586–90.

Loeffler, J., Hebart, H., et al. 2001. Nucleic acid sequence-based amplification of Aspergillus RNA in blood samples. *J Clin Microbiol*, **39**, 4, 1626–9.

Loeffler, J., Schmidt, K., et al. 2002. Automated extraction of genomic DNA from medically important yeast species and filamentous fungi by using the MagNA Pure LC system. *J Clin Microbiol*, **40**, 6, 2240–3.

Martagon-Villamil, J., Shrestha, N., et al. 2003. Identification of *Histoplasma capsulatum* from culture extracts by real-time PCR. *J Clin Microbiol*, **41**, 3, 1295–8.

Merz, W.G. and Roberts, G.D. 2003. *Detection and recovery of fungi from clinical specimens*. Washington, DC: American Society for Microbiology.

Mochizuki, T., Sugita, Y., et al. 2001. Advances in molecular biology of dermatophytes. *Nippon Ishinkin Gakkai Zasshi*, **42**, 2, 81–6.

Nevez, G., Totet, A., et al. 2002. *Pneumocystis jiroveci* detection using the polymerase chain reaction in patients with sacdoidosis. *J Mycol Med*, **12**, 183–6.

Ninet, B., Jan, I., et al. 2003. Identification of dernatophyte species by 28S ribosomal DNA sequencing with a commercial kit. *J Clin Microbiol*, **41**, 2, 826–30.

Procop, G.W. 2002. In situ hybridization for the detection of infectious agents. *Clin Microbiol Newsl*, **24**, 16, 121–4.

Raad, I., Tarrand, J., et al. 2002. Epidemiology, molecular mycology and environmental sources of *Fusarium* infections in patients with cancer. *Infect Control Hosp Epidemiol*, **23**, 9, 532–7.

Rickerts, V., Bialek, R., et al. 2002. Rapid PCR-based diagnosis of disseminated histoplasmosis in an AIDS patient. *Eur J Clin Microbiol Infect Dis*, **21**, 11, 821–3.

Sandhu, G.S., Kline, B.C., et al. 1995. Molecular probes for diagnosis of fungal infections. *J Clin Microbiol*, **33**, 2913–19.

Stephan, F., Bah, M.S., et al. 2002. Molecular diversity and routes of colonization of *Candida albicans* in a surgical intensive care unity, as studied using microsatellite markers. *Clin Infect Dis*, **35**, 12, 1477–83.

Stevens, D.A. 2002. Diagnosis of fungal infections: current status. *J Antimicrob Chemother*, **49(S1)**, 11–19.

Stockman, L., Clark, K.A., et al. 1993. Evaluation of commercially available acridinium ester-labeled chemiluminescent DNA probes for culture identification of *Blastomyces dermatitidis*, *Coccidioides immitis*, *Cryptococcus neoformans* and *Histoplasma capsulatum*. *J Clin Microbiol*, **31**, 845–50.

Telenti, A. and Roberts, G.D. 1989. Laboratory detection of fungemia. *J Clin Microbiol*, **8**, 825–31.

Turenne, C., Sanche, S.E., et al. 1999. Rapid identification of fungi by using the ITS2 genetic region and an automated fluorescent capillary electrophoresis system. *J Clin Microbiol*, **37**, 6, 1846–51.

Turin, L., Riva, F., et al. 2000. Fast, simple and highly sensitive double-rounded polymerase chain reaction assay to detect medically relevant fungi in dermatological specimens. *Eur J Clin Invest*, **30**, 511–18.

Vanittanakom, N., Vanittanakom, P., et al. 2002. Rapid identification of *Penicillium marneffei* by PCR-based detection of specific sequences on the rRNA gene. *J Clin Microbiol*, **40**, 5, 1739–42.

Vetter, E., Torgerson, C., et al. 2001. Comparison of the BACTEC MYCO/F Lytic bottle to the isolator tube, BACTEC Plus Aerobic F/bottle, and BACTEC Anaerobic Lytic/10 bottle and comparison of the BACTEC Plus Aerobic F/bottle to the Isolator tube for recovery of bacteria, mycobacteria, and fungi from blood. *J Clin Microbiol*, **39**, 12, 4380–6.

Voigt, K., Cigelnik, E., et al. 1999. Phylogeny and PCR identification of clinically important zygomycetes based on nuclear ribosomal-DNA sequence data. *J Clin Microbiol*, **37**, 3957–64.

Wahyiningshi, R., Freisleben, H.J., et al. 2000. Simple and rapid detection of *Candida albicans* DNA in serum by PCR for diagnosis of invasive candidiasis. *J Clin Microbiol*, **38**, 8, 3016–21.

Waite, T.R. and Woods, G.L. 1998. Evaluation of Myco/Flytic medium for recovery of mycobacteria and fungi from blood. *J Clin Microbiol*, **36**, 1176–9.

Wheat, L.J. 2003. Rapid diagnosis of invasive aspergillosis by antigen detection. *Transpl Infect Dis*, **5**, 158–66.

Wheat, L.J., Kohler, R.B., et al. 1986. Diagnosis of disseminated histoplasmosis by detection of *Histoplasma capsulatum* antigen in serum and urine specimens. *N Engl J Med*, **314**, 83–88.

Wheat, L.J., Connolly-Stingfiekd, P., et al. 1992. Diagnosis of histoplasmosis in patients with the acquired immunodeficiency syndrome by detection of *Histoplasma capsulatum* polysaccharide antigen in bronchoalveolar lavage fluid. *Am Rev Respir Dis*, **145**, 1421–4.

Yeo, S.F. and Wong, B. 2002. Current status of nonculture methods for diagnosis of invasive fungal infections. *Clin Microbiol Rev*, **15**, 3, 465–84.

5

Mycoserology and molecular diagnosis

RUTH MATTHEWS AND JAMES P. BURNIE

This chapter describes the application of serology and the polymerase chain reaction (PCR) to the diagnosis of systemic fungal infections. Culture-confirmation is still the cornerstone of diagnosis, and has become even more desirable as new treatment options and antifungal sensitivity testing become available. But the problem remains that positive cultures can be difficult to obtain – the yield being low, growth often slow, and biopsying a deep infected site may be difficult or impossible. Therefore, new, nonculture-based tests have been developed, usually based on detection of circulating fungal antigens or DNA, in an attempt to make diagnosis more rapid and reliable.

Traditionally the rationale behind this strategy has been that by confirming the diagnosis more quickly, antifungal treatment will be initiated earlier thereby reducing mortality (Horn et al. 1985) – confirmation being particularly important when the only treatment option was conventional amphotericin B. Now that safer antifungals are available, there is greater use of antifungal prophylaxis in high-risk groups and of empiric therapy in symptomatic patients – but these approaches are not without problems. Paralleling the increased use of fluconazole in the early 1990s, fluconazole-resistant non-*albicans* species of *Candida* have become commoner (Kontoyiannis and Lewis 2002). Amphotericin B resistance is much rarer but well documented. Greater empiric use of antifungal drugs inevitably exposes a larger number of patients, some inappropriately, to the risk of drug-induced toxicity and substantially increases pharmaceutical costs. Therefore the goal should remain to develop reliable diagnostic tests to provide an early diagnosis, preferably before beginning treatment.

Some of these assays are commercially available, whereas others are still under research and development. Inevitably the latter have not been widely assessed outside the originating research laboratory, although they may take advantage of the latest technical advances. Assessing the diagnostic value of a serology test raises questions such as those given in Table 5.1. For the referring clinician to get the most out of serodiagnostic assays, it is important that they provide a full clinical history, ascertain the frequency with which sera should be examined, and, where possible, avoid repeated freeze-thawing of samples.

CANDIDIASIS

Clinical diagnosis of invasive (systemic and disseminated) candidiasis is hampered by a lack of specific features in most cases. Confirmation of the diagnosis from blood cultures tends to be slow (growth taking 2–5 days) and insensitive. Blood cultures are negative in about half of necropsy-proven cases (de Repentigny and Reiss 1984). Recovery from blood cultures can be improved using lysis-centrifugation or the BACTEC high-blood-volume fungal medium (Telenti and Roberts 1989; Wilson et al. 1993), but the sensitivity of even the most advanced blood culture systems becomes progressively poorer when fewer than three deep-tissue sites are infected (Berenguer et al. 1993). In chronic, hepatosplenic candidiasis, only five out of 60 cases had positive blood cultures (Thaler et al. 1988; Kauffman et al. 1991).

A single positive blood culture or culture from a normally sterile organ, body cavity, or fluid, in a

Table 5.1 *Assessment of serodiagnostic tests for systemic fungal infections*

Question	Example
How frequently should sera be examined?	Mannanemia is transient so frequent sampling is essential for detecting DC
How sensitive is the assay in different forms of the infection?	Acute DC versus hepatosplenic candidiasis; ABPA versus aspergillosis
How sensitive is the assay at different stages of the infection?	ID tests for dimorphic fungi usually miss early primary infections when IgM predominates
Is serology likely to make the diagnosis before other diagnostic procedures?	Cryptococcal RPLA tests, when direct microscopy is negative
Is sensitivity affected by the underlying condition?	Neutropenia; AIDS
Can it be used to distinguish between systemic infection and colonization?	Generally true of candidal antigen detection, but not candidal antibody assays
Can it be used prognostically?	Cryptococcal RPLA tests; serial measurement of candidal antibodies
Can it be used to monitor response to treatment and to detect relapses?	Cryptococcal RPLA tests; *Aspergillus* galactomannan tests; some candidal antigen detection systems; *Histoplasma* antigen detection in AIDS
Is the antigen detected stable to heat or protease treatment or freeze-thawing?	Protease pretreatment is essential for immune complex dissociation before mannan detection
Is the assay species specific?	Assays for *C. albicans* mannan miss *C. krusei*
Can the assay determine intrinsic (species dependent) or acquired antifungal resistance?	Potentially PCR versus resistance genes; species-specific targets, e.g. *Candida* mannan
Known causes of false positives?	*T. beigelii* with cryptococcal RPLA tests
Known causes of false negatives?	Prozone effect with cryptococcal RPLA tests
Reagent costs and hands-on time	Relatively high for PCR
Trained staff, special equipment, and facilities	GLC; becoming less of a problem with PCR

ABPA, allergic bronchopulmonary aspergillosis; AIDS, acquired immune deficiency syndrome; DC, disseminated candidiasis; EIA, enzyme immunoassay; GLC, gas–liquid chromatography; ID, immunodiffusion; IgM, immunoglobulin M; PCR, polymerase chain reaction; RPLA, reverse passive latex agglutination.

septic patient, is in practice enough to start therapy, but care should be taken to avoid contamination. For example, preferably, blood cultures should be taken percutaneously rather than through an intravenous line – which may have become colonized. In the latter case, simply removing the infected line may resolve the infection, provided the *Candida* has not already spread, via the bloodstream, to set up a second focus of infection.

Serology and PCR are particularly useful in helping to support the diagnosis in suspected cases in whom culture-confirmation cannot be obtained. They can help distinguish disseminated candidiasis (DC) from transient candidemia or heavy colonization. They can also be used to screen high-risk patients or monitor response to treatment in known cases. Development of these assays is complicated by the diversity of candidiasis: a test designed to detect circulating candidal antigen in acute DC will not necessarily detect heptosplenic candidiasis or localized infections such as candidal peritonitis or endophthalmitis. Sensitivity may also vary between different patient populations (such as neutropenic compared to non-neutropenic). Another increasingly important challenge is the detection of a wide range of species, not just the commonest, *Candida albicans*, and ideally their speciation, as an indication of their probable antifungal sensitivity pattern.

Candida antibodies

Several kits for measuring antibodies to *Candida* spp. are commercially available (Gutierrez and Maroto 1993; Ruchel 1993), but these are intrinsically flawed because antibody often occurs in superficially infected individuals, antibody takes time to develop, and patients with worsening candidiasis often fail to make or sustain a detectable antibody response (Matthews et al. 1984, 1987). An important exception is endocarditis caused by *Candida* spp. Here the antibody is raised and this may be of diagnostic value (Dee and Rytel 1977). In invasive candidiasis, the development of a good antibody response has been used as a marker of recovery, whereas lack of antibody is strongly indicative of a bad prognosis. This lack of antibody may reflect antibody consumption or host immunosuppression.

Cand-Tec latex agglutination assay

The first of the commercially available antigen-based diagnostics was the Cand-Tec test developed by Gentry et al. (1983) and marketed by Ramco Laboratories Inc., Houston, TX. It is an example of reverse passive latex agglutination (RPLA), latex particles being sensitized with rabbit antibodies raised against heat-killed blastoconidia.

The target antigen has not been fully characterized but is proteinaceous, clearly distinct from mannan, and could not be found in the culture supernatant of *Candida* spp. by Ruchel (1989a). It is heat labile at 57°C.

Considerable variation in the success rate of this particular test is reported in the literature, with sensitivities ranging from 33 percent (Burnie and Williams 1985) to 71 percent (Fung et al. 1986), but more commonly around 44 to 48 percent (Kahn and Jones 1986; Lemieux et al. 1990; Herent et al. 1992). This probably reflects differences in the patient population being examined and the threshold titer regarded as significant. Burnie (1991a) found that the test gave a better bimodal distribution between colonized and systemically infected patients when assessed using postoperative patients with DC rather than neutropenic patients. This could possibly be related to the higher frequency with which neutropenic patients produce the immunoglobulin IgM over IgG, which may lead to less free antigen being present (Matthews et al. 1987) or to a need for the antigen being measured to be processed by leukocytes (Ruchel 1993). In the instructions that come with the kit, a titer of 1:4 is regarded as positive. Both Burnie and Williams (1985) and Ruchel (1989b) commonly observed weakly positive titers (1:4) in neutropenic patients with DC. Although a low threshold is necessary in neutropenic patients, in order to retain sensitivity, a higher cutoff point of 1:8 is probably more accurate in non-neutropenic patients to raise specificity.

Candidiasis caused by species such as *Candida krusei*, *Candida lusitaniae*, and *Candida parapsilosis* is not readily detected by Cand-Tec (Ness et al. 1989; Ruchel 1993). Fungal infections caused by *Aspergillus* spp. (Ness et al. 1989; Phillips et al. 1990; Ruchel 1993) or *Cryptococcus neoformans* (Lemieux et al. 1990) can give a positive Cand-Tec result. False positives occur as a result of rheumatoid factor and this needs to be excluded. Burnie and Williams (1985) found 12 percent of rheumatoid factor-positive sera gave false-positive reaction with titers as high as 1:8. The importance of frequent, preferably daily, serum sampling to raise sensitivity was also emphasized. The test proved helpful in monitoring response to therapy, titers falling as the patient recovered and rising during relapse.

In conclusion, Cand-Tec does have a limited, but sometimes useful, role to play in the diagnosis and management of DC, particularly if care is taken to exclude rheumatoid factor-positive patients, daily sera are available, and titers are interpreted in the context of whether or not the patient is neutropenic. Its greatest asset is that it is quick and easy to perform.

Mannan

The first of the circulating antigens for *Candida* spp. to be fully characterized was the cell wall mannoprotein or mannan, consisting of the carbohydrate homopolymer mannose plus small amounts of protein and phosphate. This antigen is stable and resistant to boiling, protease treatment, and acidic pH. In the serum, mannan is found in immune complexes and the first step in its detection is the dissociation of these complexes by protease digestion, exposure to extremes of pH (such as NaOH at 56°C for 2 hours), boiling, or heat extraction.

Some of the first assays developed for its detection were enzyme-linked immunoassays (EIA) based on indirect inhibition (Segal et al. 1979; Meckstroth et al. 1981). These were superseded by the double antibody sandwich EIAs, developed by several different groups, reviewed by Reiss and Morrison (1993). These had the advantage that absorbance units could be directly converted into concentration (ng/ml). Specificity was consistently high, probably helped by the fact that sera were first heat treated to dissociate immune complexes, but sensitivity varied from 23 percent (Lemieux et al. 1990) to 90 percent (Fujita et al. 1986). Factors that probably influenced sensitivity included the quality of the antibody probe, the infecting species of *Candida*, the patient's underlying condition, frequency of serum sampling and time of collection relative to disease severity.

The relative species specificity of mannan antibodies is well documented. Nakamura et al. (1991), using an avidin-biotin-amplified EIA with an overall sensitivity of 84 percent for DC, found the lower limits of mannan detection were 1.0–2.8 ng/ml for *C. albicans*, *Candida tropicalis*, and *C. parapsilosis*, 6.7 ng/ml for *Candida guilliermondii*, 20 ng/ml for *Candida glabrata*, and more than 50 ng/ml for *C. krusei*. By including antibodies to the mannan of *C. krusei* as well as *C. albicans*, Fujita and Hashimoto (1992) identified a case of DC caused by *C. krusei* which was only detectable with the homologous antibody. They noted a further six cases of *C. glabrata* DC in whom serum mannan could not be detected by either antibody.

Therefore, incorporation of antibodies to the mannan of several species of *Candida* improves the performance of these assays. The greatest limitation, however, is probably that the circulation of mannan is only transient (Kahn and Jones 1986; Herent et al. 1992). In the rabbit, the serum half-life of mannan was only 2 hours (Kappe and Muller 1991). As a result of the fast clearance of mannan from the blood, tests for mannan must be performed frequently. Studies in which a large number of sera have been examined per patient, or samples have been collected when DC was most severe, are likely to give higher sensitivities.

Sensitivities reported for RPLA tests for mannan have varied from 0 to 81 percent. The highest sensitivity was obtained by Bailey et al. (1985) who used their own RPLA assay with polyvalent rabbit antisera raised against heat-killed *Candida* spp. and dissociated immune complexes with protease and heat. Sensitivity was 81 percent (17 of 21 patients) for DC, 60 percent (three of five) for candidemia and 40 percent (four of 10) for esophagitis. Five of six cases of DC caused by *C. tropicalis* were also detected. It was negative in patients with endophthalmitis or superficial candidiasis. One false positive occurred in 86 controls. Mannan was, however, reliably detected only late on in the course of infection. Sera examined during the penultimate week before death were less than 50 percent likely to be positive. Another disappointing feature was the lack of correlation between antigen titer and severity of infection.

There are two commercially available kits for the detection of mannan by RPLA: one uses a polyclonal antibody probe (the LA-Candida Antigen Detection System, Immuno-Mycologics, Inc., Norman, OK) and the other a monoclonal antibody (Pastorex *Candida*, BioRad, Marnes-la-Coquette, France). Using the former, Fung et al. (1986) failed to detect mannan in 24 patients who had positive cultures for *C. albicans*, six of whom had DC. These results were inferior to Cand-Tec which had a sensitivity of 71 percent and a specificity of 98 percent, using a titer of 1:8 or greater as the criterion for dissemination. Disappointing results were also obtained by Phillips et al. (1990). With the Pastorex test, Herent et al. (1992) detected mannan in 10 of 19 patients with DC (52.6 percent) compared with nine positives using the Cand-Tec test. Specificity was 100 percent for Pastorex and 98 percent for Cand-Tec. Ruchel (1993) detected mannan in about half the patients with DC tested with Pastorex.

The RPLA test may be intrinsically less sensitive than EIA. Fujita and Hashimoto (1992) compared RPLA and a sandwich EIA for *C. albicans* and *C. krusei* mannan in parallel with the same antibody reagents and clinical specimens. The sensitivity of the EIA was 74 percent whereas that of the RPLA was only 38 percent. A sandwich EIA is commercially available (ICON *Candida*, Hybritech Inc., Torrey Pines, CA) and has been assessed by Pfaller et al. (1993). Of 14 patients developing DC, the test was positive in 12 (86 percent) but, notably, antigenemia preceded diagnosis by culture or biopsy in just five cases. Only a single serum was available from the two patients who gave false negatives. The authors found the test easy to perform and interpret. Girmenia et al. (1997), using dot immunobinding, found that 7.1 percent of patients with transient or catheter-related candidemia had detectable mannan compared with 76.5 percent (13 of 17) patients with persistent candidemia.

In conclusion, the commercially available RPLA tests do not, on the basis of currently published reports, significantly outperform Cand-Tec, although the sensitivity of EIA may be better, and these are becoming commercially available (Erjavec and Verweij 2002). They may miss cases of DC caused by *C. glabrata* and *C. krusei*. Use of a combination of tests to detect both mannan antigen and anti-mannan antibody may improve sensitivity, since Sendib et al. (1999) found the presence of serum mannan to be inversely related to the presence of anti-mannan antibody – detecting both raised sensitivity to 80 percent, with specificity being 93 percent.

Another cell wall component, glucan, has been found circulating in patients with candidemia or invasive aspergillosis (Obayashi et al. 1995; Mori and Matsumura 1999). It is a characteristic component of fungal cell walls (except zygomycetes) but lacking in human cells, prokaryotes, and viruses. An assay for its detection is available commercially (Fungitec G-test, Seikagaku Corp., Tokyo, Japan). The test depends on the activation, by glucan, of factor G, a horseshoe-crab coagulation factor. A broad range of fungi induce reactivity, with high sensitivity (90 percent) and specificity (100 percent) being reported in patients undergoing treatment for hematological malignancies (Obayashi et al. 1995), but negative results have been observed in pulmonary cryptococcosis.

Protease

Both a circulating secreted acid protease antigen and specific antibodies have been described in patients infected with *C. albicans* (MacDonald and Odds 1980; Ruchel et al. 1988). The protease is thought to facilitate epidermal invasion. An assay detecting urinary protease in a rabbit model has been reviewed by Reiss and Morrison (1993), but there are few human data (Ruchel et al. 1988).

It is now known that there are at least seven genes encoding a family of secreted proteases in *C. albicans*, of which six are differentially expressed and one has not yet been successfully expressed (Hube et al. 1994). Different species of *Candida* secrete antigenically distinct proteases and some species, such as *C. krusei*, do not appear to secrete proteases. Na and Song (1999) found that an inhibition enzyme-linked immunosorbent assay (ELISA) for the detection of secretory aspartyl proteinase, using a specific monoclonal antibody, gave a sensitivity of 94 percent and specificity of 96 percent when applied to retrospectively selected sera from patients with invasive candidiasis.

Enolase and heat-shock protein 90

The application of immunoblotting to the serology of invasive candidiasis created the opportunity to examine the antibody responses of infected patients to a range of *Candida* antigens in one assay. This led to the identification of two immunodominant antigens which have been used as diagnostic markers: the 47 kDa antigen (Matthews et al. 1984), now identified as the carboxy fragment of heat shock protein 90 (Hsp90) (Matthews and Burnie 1989), and the 48 kDa antigen (Strockbine et al. 1984), subsequently identified as enolase (Sundstrom and Aliaga 1992; Franklyn and Warmington 1993).

Enolase is an abundant cytoplasmic enzyme involved in the glycolytic pathway of yeasts. As it is produced by all *Candida* spp., species other than *C. albicans* should also be detected by an enolase assay, although *C. glabrata* may be an exception (Walsh et al. 1991). To detect serum enolase, a liposomal sandwich-type immunoassay was developed and made available commercially (Directigen, Becton Dickinson Inc., Philadelphia, PA). Evaluating the assay prospectively in oncology patients, Walsh et al. (1991) obtained an overall sensitivity of 75 percent and specificity of 96 percent. Many patients who were positive on almost daily sampling would have been negative if sampling had occurred weekly. In a prospective comparative assessment with Pastorex, Gutierrex and Maroto (1993) found the Directigen test to be positive in eight of 10 patients (13 of 20 sera) with candidemia, with a specificity of 97.1 percent, whereas the Pastorex test remained negative. The Directigen test never became widely used and it was subsequently withdrawn by the manufacturer.

Heat-shock proteins, also known as stress proteins, are highly conserved, ubiquitous proteins which have been categorized into families, such as Hsp90, on the basis of their molecular mass and sequence homologies. They act as immunodominant antigens in a range of bacterial, parasitic, and fungal infections. Both murine monoclonal (Matthews et al. 1991a) and human recombinant antibodies (Matthews et al. 1995) to Hsp90 were protective in mouse models of the infection and an antibody to Hsp90 is currently being assessed as a therapeutic agent. The 47 kDa antigen can be isolated from the sera of systemically infected patients by affinity chromatography (Matthews et al. 1987) and was used as a diagnostic marker in a dot immunobinding assay (Matthews and Burnie 1988).

Once *C. albicans* Hsp90 had been sequenced it was possible to map the epitopes recognized by infected patients' sera. Both a conserved immunodominant epitope, LKVIRK, and a species-specific epitope recognized by about half the patients, DEPAGE, were identified (Matthews et al. 1991b). A rabbit antiserum raised against LKVIRK, again in the dot immunobinding assay, picked up 70 percent of patients with DC caused by *C. albicans*, with a further 18 percent giving weak positives and 12 percent false negatives. Two patients with DC caused by *C. glabrata* gave negative results but a patient with *C. parapsilosis* was positive. For 12 patients, multiple sera were available and here there was a good correlation with clinical outcome, titers falling in those who recovered and rising in those who died. Each of four blood culture-negative patients with severe focal infections requiring amphotericin B were positive. Colonized patients were negative or only weakly positive. All 404 uninfected control patients were negative in this series, but false positives have been observed with some other infections, possibly as a result of circulating microbial Hsp90. One possibility would be to use antibody against the conserved LKVIRK epitope to capture serum Hsp90 and a range of species-specific epitopes, such as DEPAGE in the case of *C. albicans*, for speciation.

Metabolites

Most species of *Candida*, with the exception of *C. glabrata* and *C. krusei*, produce the metabolite D-arabinitol (Bernard et al. 1981). It can be measured by gas–liquid chromatography (GLC) but, as this is a cumbersome technique, its availability as a diagnostic procedure is limited. As arabinitol is excreted by the kidney, it accumulates in patients with renal insufficiency, so an arabinitol/creatinine ratio is calculated to be independent of renal function.

An enzymatic fluorometric method has been developed as an alternative to GLC (Soyama and Ono 1985). It depends upon the oxidation of arabinitol by arabinitol dehydrogenase, with concomitant reduction of NAD, generating NADH. The initial rate of NAD reduction is proportional to the serum arabinitol content and can be measured by a recording spectrofluorometer. A potential problem is that mannitol is also reduced by arabinitol dehydrogenase producing NADH.

Several preliminary assessments have been published of the kit made commercially available by Nacalai Tesque Co. Ltd (Kyoto, Japan). Fujita and Hashimoto (1992), in a series of 58 patients with DC, reported a sensitivity of 50 percent and a specificity of 91 percent. This compared with a sensitivity of 74 percent for an EIA test for mannan. The false positives obtained (13 of 109 patients) were attributed to heavy colonization with *Candida* spp. causing a rise in serum arabinitol or interference with the fluorometric assay by mannitol. De Repentigny et al. (1985) likewise found that an EIA for mannan, combined with blood cultures, gave superior results to GLC measurement of arabinitol. In that series the low sensitivity, 13 percent, of arabinitol/creatinine ratios was associated with a significant increase in arabinitol concentrations in high-risk patients without candidiasis compared with normal blood donors. One cause of these false positives could be steroid therapy, which has been shown to raise arabinitol levels in rabbits.

Urinary samples can be used to monitor the ratio of D-arabinitol (fungal origin) to L-arabinitol (human origin) and this was found to be positive 3–31 days (mean 12 days) before blood cultures become positive or empiric therapy initiated, in immunocompromised children with systemic candidosis (Christensson et al. 1997).

PCR

The application of PCR to fungal diagnosis has been reviewed by Walsh and Chanock (1998), Buchheidt et al. (2000), and Erjavec and Verweij (2002). Early assays were directed against a defined target gene present in a single species or genus, but now panfungal PCR primers are being used increasingly to detect a broad range of fungi, including yeasts and molds. Many of these assays are now achieving very high levels of sensitivity in-house but head-to-head comparisons and standardized, inter-laboratory assessments are as yet rare.

The first target for the development of PCR-based diagnostics for invasive candidiasis was the gene *L1A1* encoding cytochrome P450 lanosterol 14α-demethylase which catalyzes an essential step in the conversion of lanosterol to ergosterol. As ergosterol is not a component of human or bacterial cells, it is believed to be fungus specific. The enzyme is inhibited by the imidazoles. Therefore, antifungal resistance related to alteration in the sequence of this gene might potentially be used to pick up resistant yeasts. Buchman et al. (1990) carried out an assessment with 13 clinical specimens, including urine, sputa, wound fluid, and two whole bloods, all from sites that were culture positive for *C. albicans*, and in each case they were PCR positive. Among 17 additional culture-negative specimens, PCR suggested the presence of a yeast in one blood and one urine specimen. Results were available within 6 hours and it was estimated, by serial dilution of *C. albicans* mixed with fresh blood, that as few as 12 yeast cells per 0.1 ml could be detected.

Chryssanthou et al. (1994) developed a nested PCR against the same target to increase sensitivity. They also avoided the cumbersome sample preparation required for whole blood by detecting naked DNA in serum. The primers used by Buchman et al. (1990) were used for the outer pair, and an inner pair of primers was selected which gave a PCR product of the expected size. This fragment was only amplified for *C. albicans*, other faint bands appearing with *C. lusitaniae* and *C. tropicalis*. They applied it to serum and tissue samples from infected mice and the sera of patients with DC and candidemia. The nested PCR detected as little as 1 pg *C. albicans* DNA, serially diluted in water, compared with 10 pg using the outer primer pair alone. It was more sensitive than culture in mouse experiments and in seven blood culture-negative, necropsy-positive patients, five of whom gave PCR-positive sera before death (usually taken within a week of death but in one case 84 days before and therefore of doubtful significance). However, in 11 patients with proven or, in one case, suspected candidemia, 15 of 17 blood cultures were positive, compared with only 11 of 17 sera positive by PCR. These discrepancies might result partly because blood cultures and sera for PCR were not taken simultaneously.

Buchman's group followed up their earlier work on *C. albicans* by sequencing homologous genes from *C. glabrata* and *C. krusei*, and using a published sequence for *C. tropicalis* to develop a nested PCR (Burgener-Kairuz et al. 1994). The first PCR, with genus-specific primers

from a conserved region, detected a wide range of yeast species when tested with purified DNA, including *C. albicans*, *C. glabrata*, *C. krusei*, *C. parapsilosis*, *C. tropicalis*, and *Cryptococcus neoformans*. The second nested PCR with species-specific primers had a detection level 10 times higher for the homologous species. Southern blots were used to confirm the nature of the PCR product and the sensitivity was increased slightly further. The sensitivity and specificity of the nested PCR on 80 culture-positive clinical specimens were 71 percent and 95 percent for *C. albicans* and 100 percent and 97 percent for *C. glabrata*, respectively. Clinical specimens often contained large quantities of leukocytes, cell debris, and bacteria, all potential sources of strong inhibitors of the PCR which could be causes of false negatives.

In a prospective study, by Morace et al. (1999), in febrile patients with hematological malignancies, the P450 lanosterol 14α-demethylase gene was amplified and the resulting amplicon digested with restriction enzymes (RE) in order to speciate the *Candida*. The polymerase chain reaction-restriction enzyme amplification (PCR-REA) was much more sensitive (92.8 percent) and faster (24–36 hours) than blood culture systems, but there was some question over the method's specificity. Posteraro et al. (2000) used the same primers to develop a PCR-reverse crossblot hybridization assay. Chryssanthou et al. (1999) undertook a prospective study in high-risk patients comparing a nested PCR against lanosterol 14α-demethylase, with urinary D/L-arabinitol ratios and three commercially available tests for the detection, respectively, of serum glucan, mannan, and the Cand-Tec assay. No single test was sufficient for diagnosis and the correlation between some of the tests was poor. This study illustrated the difficulty in evaluating these diagnostic tests because of the lack of a definitive culture-confirmed diagnosis in many cases. In a comparison of a PCR assay (targeting a rRNA gene), the β-glucan and the Cand-Tec test, Sakai et al. (2000) showed a significant correlation between the PCR and β-glucan assays and that PCR was significantly more sensitive than the β-glucan test.

Hsp90 is an alternative target, a potential source of both species-specific and conserved DNA sequences. By selecting a species-specific primer, Crampin and Matthews (1993) obtained the expected 317 bp product only with *C. albicans* and, less intensely, *C. parapsilosis*. In a prospective study on a 100 clinical specimens, 23 percent were positive by routine culture, 31 percent by extended culture, and 37 percent by PCR. The identity of the PCR product was confirmed by hybridization with an internal probe and also by obtaining the predicted restriction enzyme digest pattern. This assay was sensitive down to 50 pg of DNA (5 pg after Southern blotting) and 100 yeast cells. Generally, specimens that were culture negative or grew bacteria or germ-tube-negative yeasts gave negative PCR results. The exceptions were four urine samples from female patients and seven high vaginal swabs from women with vaginal discharge. Five culture-positive swabs gave negative PCR results, possibly as a result of a sampling error because plates for culture were inoculated first. Whole blood seeded with *C. albicans* gave positive results only when inhibitors were removed by the glass adhesion method. Some of the other yeast species, particularly *C. glabrata* and *C. parapsilosis*, gave bands about the same size as *C. albicans*, but in each case these were accompanied by ectopic bands not seen with *C. albicans*.

Another target is the actin gene (Kan 1993). Primers were selected which amplified a 158 bp fragment and the product was confirmed by hybridization against an internal probe. This gave a lower limit of detection estimated as 10 yeast cells for clinical specimens. The assay was assessed in both experimentally infected mice and patients with candidemia. In the mouse model of candidemia, pooled plasma samples were culture and PCR positive, whereas in the thigh-abscess model they were culture and PCR negative. In the 14 blood culture-positive patients, six of whom had *C. albicans*, 11 (79 percent) gave a positive PCR. This was performed on sera rather than on whole blood. The greatest difficulty was experienced in detecting *C. krusei* and *C. parapsilosis*. Sera from 12 patients with superficial candidiasis were PCR negative as were 17 normal volunteers.

Flahaut et al. (1998) targeted the *C. albicans* secreted aspartic proteinase (*SAP*) genes, a multigene family. A single primer pair was designed to detect six of these *SAP* genes, thereby increasing the sensitivity of the test. Detection of the PCR product by EIA was found to be as sensitive as Southern blotting with a *SAP*-labeled probe, and the sensitivity and specificity, when applied to clinical samples, were 100 percent and 98 percent, respectively.

Others have deliberately selected multicopy targets such as mitochondrial or ribosomal DNA in an attempt to maximize sensitivity. Miyakawa et al. (1992), targeting a species-specific mitochondrial DNA fragment 1.8 kbp long, could detect as few as 10 yeast cells in human urine but as many as 105 cells were required in human blood before detection. They attributed this to the loss of yeast cells during precipitation from the blood and the use of too long a target. They then developed a means of coprecipitating *C. albicans* with carrier *C. tropicalis* cells, aided by a monoclonal antibody which recognized both species, and selected a shortened sequence of 120 bp. This raised the sensitivity of detection of *C. albicans* seeded into blood to 3 cells/0.1 ml (Miyakawa et al. 1993). No clinical specimens were examined.

Panfungal PCR assays directed against rDNA are increasingly being used because not only are they present in multiple copies but, as a result of slow divergence of rDNA over evolutionary time, regions of sequence conservation across entire kingdoms can be

identified. Separating the multicopy rRNA gene subunits are noncoding intergenic transcriber spacers (ITS). While rRNA genes are highly conserved, the ITS regions are divergent both in sequence and size. Consequently, panfungal primers can be used to amplify the fungal DNA and then species-specific ITS DNA probes can be used to speciate the amplicon.

Using such panfungal primers, directed against a region conserved throughout the kingdom Fungi, Hopfer et al. (1993) amplified a 310 bp product from fungi, while obtaining negative results from human and bacterial DNA. This PCR detected 10–15 yeast cells, but sensitivity and specificity with clinical samples were not determined. *Hae*III digests were used to characterize the fungus into one of five groups: *Aspergillus*, *Candida*, *Cryptococcus*, and *Trichosporon*, zygomycetes, and the dimorphic fungi. Holmes et al. (1994), also targeting rDNA, co-amplified using both conserved primers, suitable for the detection of many fungal species, and *C. albicans*-specific primers, facilitating rapid speciation. A 105 bp fragment was amplified from all the yeast species tested, whereas a second 684 bp fragment was amplified only from the *C. albicans* DNA. The level of sensitivity was 15 yeast cells/ml using spiked blood specimens. Makimura et al. (1994), using primers from conserved fungal rDNA sequences, applied this to a limited number of clinical specimens, including blood and cerebrospinal fluid, and also blood samples from mice systemically infected with *C. albicans*. Their preliminary results were encouraging, suggesting PCR to be more sensitive than conventional blood cultures.

Again targeting fungal rDNA, Fujita et al. (1995) developed a microtitration plate enzyme immunoassay to detect PCR-amplified DNA from *Candida* spp. Species-specific probes for *C. albicans*, *C. krusei*, *C. parapsilosis*, and *C. tropicalis* were labeled with the digoxigenin, avoiding the use of radioisotopes. The PCR product was hybridized to both digoxigenin- and biotin-labeled probes and detected in an EIA by capture on streptavidin-coated microtiter plates. The assay detected as few as 2 cells/0.2 ml of blood seeded with *C. albicans*. Although no clinical samples were examined, this approach holds great promise. It combines the use of universal fungal primers against a multicopy target to broaden detection capabilities with identification of the fungus, using species-specific probes, in an EIA format that is particularly suitable to clinical laboratories; this has the potential for automation unlike Southern blotting. Ahmad et al. (2002) applied a seminested PCR, using panfungal primers against rDNA followed by the same species-specific primers from these four *Candida* spp., to detect and speciate from the sera of patients with candidemia: the results were in full agreement with blood culture results with respect to both positivity and species identity, and more sensitive than culture. Others have targeted ITS regions using species-specific primers in a multiplex PCR to speciate multiple fungal pathogens in a single PCR reaction, from either fungal colonies (Fujita et al. 2001; Luo and Mitchell 2002) or positive blood culture broths (Chang et al. 2001). Luo and Mitchell (2002) found that DNA from only about half the molds (*Aspergillus fumigatus*) examined could be amplified directly from the mycelial fragments whereas DNA from every yeast colony (*Candida* spp. and *C. neoformans*) could be readily amplified. Rigby et al. (2002) have applied fluorescence in situ hybridization with peptide nucleic acid probes (DNA probe mimics) to rapid identification of *C. albicans* directly from positive blood culture bottles.

Shin et al. (1997) successfully applied PCR-EIA to identify *Candida* spp. in BacT/Alert blood culture bottles in 7 hours – 2.5 days sooner than by conventional culture. Alternatively, the variability in the length of the ITS2 region can be used to speciate fungal isolates. These size differences cannot be reliably detected by agarose gel electrophoresis and instead automated capillary electrophoresis has been used (Turenne et al. 1999; Chen et al. 2000).

Burnie et al. (1997) applied a similar semiquantitative PCR-EIA to sera from infected patients, this time digoxigenylating the amplicon during the PCR process. The assay was positive (optical density >0.1) in 28 cases of culture-confirmed DC. It was negative in 15 control patients and 13 of 16 patients with local infections. The remaining three, each of whom received amphotericin B, were positive. A universal, rather than yeast-specific, method of extracting the DNA was applied to the sera.

Einsele et al. (1997), targeting a consensus sequence from a variety of fungal pathogens present in 18S rRNA, achieved 100 percent sensitivity when two blood samples were examined in 21 bone marrow transplant patients with documented invasive fungal infections. Specificity was 98 percent. PCR remained persistently positive in patients failing to respond to antifungal therapy but cleared in those who did respond. Van Burik et al. (1998) developed a panfungal PCR assay optimized to detect *C. albicans* and *A. fumigatus* from whole-blood specimens, and obtained positive results in four patients with candidemia and three with pulmonary aspergillosis (one of these occurring a week prior to culture-confirmation by bronchoalveolar lavage). Wahyuningsih et al. (2000) successfully used the QIAamp blood kit (Qiagen, Hilden, Germany) to extract *C. albicans* DNA from human serum samples. They used fungus-specific universal primers against ITS regions corresponding to the 5.8S and 28S rRNA genes for amplification and then hybridized this to a biotinylated probe directed against a *C. albicans*-specific ITS2 region located between these two genes, on a streptavidin-coated plate. All seven patients with culture-confirmed invasive candidiasis were positive by PCR as were a further three high-risk patients.

PCR can be performed with either serum or whole blood. Bougnoux et al. (1999) formally compared serum with whole blood samples, in rabbits with candidemia, applying a nested PCR with five species-specific inner pairs of primers directed against ITS. Their results suggested that serum was the sample of choice.

Hendolin et al. (2000) applied a panfungal PCR to the detection of fungi in tissue specimens. Using broad-range fungal primers, they PCR amplified the fungal ITS region (ITS1-5.8SrRNA-ITS2) and then captured this with species-specific probes against eight fungal pathogens (*Candida* species, *Cryptococcus neoformans*, and *Aspergillus* spp.), species identification therefore relying on both specific hybridization and an assessment of PCR product length. A detection level of 0.1–1 pg of purified DNA was achieved. Of the 20 specimens, PCR was positive for 19 (95 percent), of which 10 (53 percent) were hybridization positive – compared to 12 (60 percent) positive by direct microscopy and seven (35 percent) by culture.

Reiss et al. (1998) applied TaqMan (Perkin-Elmer, Foster City, CA) PCR directed against the ITS2 region to blood cultures, to obtain species identification in 5 hours. TaqMan combines, in one step, PCR, hybridization with a fluorogenic probe, and fluorescent signal generation. Guiver et al. (2001) applied TaqMan PCR to the rapid identification of *Candida* species within 4 hours, compared to a mean of 3.5 days for conventional phenotypic tests. Alternatively, real-time species identification and quantification of fungal load in clinical specimens can be achieved with a LightCycler (Roche Diagnostics, Mannheim, Germany), which combines rapid thermocycling with glass capillaries with online fluorescence detection of the PCR amplicon (Loeffler et al. 2000).

In conclusion, PCR has the potential to offer a high level of sensitivity and specificity within hours of specimen collection. Care must be taken to avoid laboratory contamination, but automated sample preparation and use of real-time PCR is helping to standardize these procedures and reduce the risk of false positives. PCR is likely to be used increasingly, not only to confirm the diagnosis but to speciate the fungus.

ASPERGILLOSIS

The key to understanding the serodiagnosis of infection with *Aspergillus* spp. is an appreciation of the heterogeneity of the diseases that these fungi can cause and the resultant variations in host response. The three main conditions are:

1 an aspergilloma occurring in a pre-existing lung lesion
2 allergic bronchopulmonary aspergillosis (ABPA)
3 invasive aspergillosis.

In all three conditions, the prime organ involved is the lung, which means that diagnosis based on culture is dependent on obtaining adequate bronchial secretions. For aspergilloma, recovery rates vary from 31 to 91 percent (Schonheyder 1987). In ABPA, a positive sputum and the demonstration of hyphae of *Aspergillus* spp. are some of the features necessary to prove the diagnosis; detection of antibody is also essential (Wardlaw and Geddes 1992).

In invasive aspergillosis, lung involvement occurs in 90 percent of cases. The classic clinical presentation is that of unremitting fever, usually higher than 38°C, in a neutropenic patient, but 10 percent of cases occur in non-neutropenic patients in whom steroid therapy is often a complicating factor (Young et al. 1970). Dyspnea and a nonproductive cough are often absent during the early stage of the disease, but pleuritic chest pain and a pleural rub occur in 15–61 percent of patients (Young et al. 1970; Meyer et al. 1973). No sputum is produced by 50 percent of patients and two-thirds of those who do produce sputum are culture negative (Fisher et al. 1981). Positive cultures can occur in the absence of significant clinical disease and this may reflect the aspiration of conidia or possibly hyphal fragments, which transiently contaminate respiratory secretions. Tregar et al. (1985) reviewed 89 patients with positive sputum cultures. In nine cases there were two or more positives and eight of these patients were subsequently proved by histology to have invasive disease. Histological evidence of mycelial invasion from tissue biopsy is often not possible because of the clinical condition of the patient and often the presence of thrombocytopenia. Computed tomography can be used to detect lesions typical of invasive aspergillosis and has a significant input on management (Caillot et al. 1997). Blood cultures are occasionally of value in endocarditis caused by *Aspergillus* spp., but in invasive aspergillosis a positive blood culture is normally the result of contamination (Pennington 1980).

The failure of conventional techniques based on culture has led to numerous attempts to develop alternative assays. These can be grouped into antibody detection, antigen detection, and the detection of the DNA of *Aspergillus* spp. by PCR. *A. fumigatus* is the dominant species, but clinically significant numbers of infections can be caused by other species such as *Aspergillus niger* and *Aspergillus flavus*. Tests which specifically target one species may miss infection caused by a different species. Furthermore, *A. fumigatus* is a diverse microorganism phenotypically (Burnie et al. 1989); therefore strain-specific antigens must be avoided in developing antigen-based diagnostics. *Aspergillus* spp. can colonize patients without causing disease, so very sensitive tests may need to be quantified.

Allergic bronchopulmonary aspergillosis

Clinical diagnosis is dependent on the typical chest x-ray changes of fleeting alveolar infiltrates caused by eosinophilic pneumonia, or a combination of mucus plugging with segmental, lobar, or whole lung collapse. Detection of antibody is essential to the diagnosis of the disease. This includes IgE-mediated immediate-type hypersensitivity to *A. fumigatus* as defined by a positive immediate skin-prick test, a positive radioallergosorbent test (RAST), a high total IgE level (i.e. >2 500 ng/ml), and a raised IgG (Wardlaw and Geddes 1992). These can be demonstrated by radioimmunoassay and EIA, although the latter has been criticized for its lack of reproducibility (Richardson et al. 1982; Richardson and Warnock 1984). The antibody response has been further characterized by self-crossed radioimmunoelectrophoresis (Longbottom 1986). Antigens were subdivided into those with strong precipitin reactions and weak IgE and those that were poorly precipitating but showed strong IgE binding. Of these last antigens, three reacted with more than 50 percent of sera and were regarded as the major allergens.

Piechura et al. (1983) analyzed the cell sap and culture filtrate from three strains of *A. fumigatus*. Two-dimensional electrophoresis followed by gel filtration identified a major component of apparent molecular mass 150 kDa (CS2). On immunoelectrophoresis this produced a major precipitin arc with aspergilloma sera or sera of ABPA patients. On SDS-PAGE, following reduction, it produced three major bands at molecular masses of 41, 53, and 81 kDa. Kurup et al. (1986) prepared culture filtrate antigens from *A. fumigatus* by isoelectric focusing with a pH gradient of 4–6.5. A fraction with an isoelectric point of 6.5 had three components of apparent molecular masses 20, 40, and 80 kDa on two-dimensional electrophoresis. This fraction reacted in an EIA, showing high levels of IgG and IgE reactivity in patients with ABPA and of IgG in those with aspergilloma.

The advent of immunoblotting has helped with detailed dissection of the antibody response, but many anomalies arose regarding the apparent molecular masses of immunodominant antigens. These may reflect, in part, the different conditions under which the gels were run. With a gradient gel of 11–13 percent acrylamide, Brouwer (1988) identified antigens at molecular masses of 36, 50, and 100–120 kDa, the last being estimated by extrapolation. Baur et al. (1989), analyzing 28 patients at varying stages in the disease, found that IgE levels increased as the disease progressed. Immunodominant bands at molecular masses of 40, 51, and 80 kDa were observed, similar in size to the same antigens described by Piechura et al. (1983). Immunodominant antigens at molecular masses of 23 kDa and 17 kDa were thought to be identical to Ag3 (Longbottom 1986) and Ag40 (Wallenbeck et al. 1984). Ag40, along with Ag10, was isolated from a complex mixture of more than 50 precipitating antigens and identified as the major allergens in six patients with ABPA and in 80 percent of 25 patients with asthma caused by *Aspergillus* spp. (Wallenbeck 1991). Samuelsen et al. (1991), looking at the IgE of patients exposed to *A. fumigatus*, found that bands at molecular masses of 20, 31, 44, 50, 53, 77, and 92 kDa reacted with more than 50 percent of sera. Others have described immunodominant antigens at 92 kDa, a 55 kDa glycoprotein (Teshima et al. 1993), and a protease at 32 kDa which was subsequently cloned and identified as a monoglycosylated alkaline protease with elastase activity (Moser et al. 1993, 1994). Leser et al. (1992) examined sera from 43 cases of ABPA and showed specific IgG and IgE against 12 different bands varying from 15 to 97 kDa. The IgG response broadly reflected the IgE with strong antibody responses to high-molecular-mass bands (between 66 and 97 kDa) as well as bands at 15, 35, and 40 kDa. Arruda et al. (1992) identified major antigens of 18 and 45 kDa. Of the 24 patients with ABPA, 11 had IgE to the 18 kDa antigen. Kobayashi and Miyoshi (1993), examining 10 aspergilloma patients and three ABPA patients, demonstrated IgG binding to antigens of greater than 30 kDa in all sera tested, and IgE binding to the 18 kDa antigen in seven of eight sera with a positive precipitin reaction by counterimmunoelectrophoresis (CIE). When immunoblotting sera from patients with cystic fibrosis and ABPA, Knutsen et al. (1994) observed heterogeneous IgE, IgG, and IgA responses to 11 major bands at 14, 18, 20, 34, 47, 50, 55, 61, 70, 81, and 93 kDa. IgE was more frequently detected against the bands at 47, 50, 61, 70, and 81 kDa.

Some of these bands are likely to be heat-shock proteins. Immunodominant antigens at 88, 84, 51, and 40 kDa were demonstrated by immunoblotting sera from 35 patients with ABPA (Burnie and Matthews 1991). The antigens at 88, 51, and 40 kDa were then shown to react with monoclonal antibodies against Hsp90 of *C. albicans* and the oomycete *Achlya ambisexualis*. Screening an *A. fumigatus* cDNA library with pooled sera from ABPA patients produced a clone which, on sequencing, was homologous with *Candida*, *Saccharomyces*, and human Hsp90 (Kumar et al. 1992). Expressed as a fusion protein, it reacted with the IgE and IgG of patients with ABPA.

Aspergilloma

A pulmonary aspergilloma can often be diagnosed on chest x-ray, or by computed tomography, by the characteristic sign of an opacity surrounded by a crescent-shaped translucent area. Levels of antibody are high and multiple lines can be detected by immunodiffusion (ID) or CIE against either mycelial or culture filtrates (Kurup and Kumar 1991). In patients with ABPA, the same levels of *A. fumigatus*-specific antibodies occur in both subclasses IgG1 and IgG2. The sera of patients with aspergilloma react to the carbohydrate and glycoprotein antigens of *A. fumigatus* with more IgG1 than IgG2 (Kurup et al. 1990).

Immunoblotting has been used to analyze the antibody response. Matthews et al. (1985) demonstrated bands ranging from 33 to 88 kDa with an immunodominant antigen at 40 kDa. Latgè et al. (1991) identified immunodominant antigens of 18, 28, 30, 38, 45–48, 94, 110, and 125 kDa with sera from 75 patients, but only a 35 kDa antigen reacted specifically with sera from the aspergilloma patients. The 18 kDa antigen was purified to homogeneity and related to a similar antigen found in the urine of patients with invasive aspergillosis (Haynes et al. 1990). It was subsequently shown to have ribonuclease activity (Latgè et al. 1993). Burnie and Matthews (1991) demonstrated immunodominant antigens at 40, 84, and 88 kDa. The 88 kDa antigen reacted with monoclonal antibodies against Hsp90 of *Candida* and *Achlya* spp., as did the extract from an aspergilloma removed from a patient. This gave a prominent band at 50–52 kDa. Kobayashi et al. (1993) confirmed the immunodominance of a band at 88 kDa but suggested that it was not Hsp90. Moutaouakil et al. (1993) biochemically characterized a 33 kDa serine protease but failed to demonstrate specific antibodies against it in aspergilloma patients.

Increasingly, recombinant proteins are being used as the basis of assays for the detection of specific, circulating antibodies to *A. fumigatus*. These are often highly specific and, in the case of aspergilloma, highly sensitive: 100 percent sensitivity was obtained by Holdom et al. (2000) using recombinant Cu, Zn superoxide dismutase, by Weig et al. (2001) using recombinant mitogillin protein, and by Chan et al. (2002) using recombinant galactomannoprotein.

Invasive aspergillosis

ANTIBODY DETECTION

Early investigations into the serodiagnosis of invasive aspergillosis concentrated on antibody detection. Coleman and Kaufman (1972) found precipitins in 82 percent of proven cases of invasive aspergillosis and in 83 percent of suspected cases. Positive precipitins in invasive disease were also observed by Meyer et al. (1973), Warnock (1977), and Kurup and Fink (1978).

Several groups working independently with different assays have found a correlation between rising antibody titers and recovery. Gold et al. (1980) developed a passive hemagglutination test based on sheep red cells treated with concanavalin A and sensitized with a partially purified antigen to *Aspergillus* spp. Sequential testing in 55 cancer patients, with sera available in the 2 weeks before diagnosis of invasive aspergillosis was made, demonstrated that 15 seroconverted. Recovery was associated with an increased antibody titer. Holmberg et al. (1980) developed an EIA with a concentrated culture filtrate precipitated by 75 percent saturated ammonium sulfate. Examining material from 10 cases of invasive aspergillosis, they correlated a serial rise in EIA titer with recovery. Matthews et al. (1985) found, by immunoblotting serial sera, that antibody to an immunodominant 40 kDa antigen was notably absent or fell in titer in 12 fatal cases of invasive aspergillosis but persisted in two survivors. A solid-phase radioimmunoassay has also been developed for measurement of antibody (Marier et al. 1979).

There have been a number of studies describing immunodominant antigens. Antigens at 41, 54, and 71 kDa were described by de Repentigny et al. (1991) using a rabbit model. Fratamico and Buckley (1991), examining 172 sera from 38 patients with invasive aspergillosis, found 90 percent of sera detected a 58 kDa antigen on immunoblots. This antigen was abundant in mycelial extracts and comprised about 50 percent of the Coomassie blue-stained protein. It contained carbohydrate, staining with periodic acid–Schiff agent and binding to the lectin concanavalin A. Monoclonal antibodies raised against it reacted with a similar antigen in *C. albicans*, shown by localized cell surface immunofluorescence. They also recognized an *Aspergillus* antigen with a molecular mass of more than 89 kDa (Fratamico et al. 1991). Burnie (1995) immunoblotted sera from 55 cases of proven and 20 cases of suspected invasive aspergillosis and demonstrated immunodominant bands at 88, 84, 63, and 40 kDa. Possession of either IgM or IgG against the bands at 88, 84, and 40 kDa correlated with survival. The band at 88 kDa was heat inducible and reacted with monoclonal antibodies specific to the Hsp90 homologs of *A. ambisexualis* and *C. albicans* (Burnie and Matthews 1991). Reactivity was neutralized by cross-absorption with Hsp90.

Hamilton et al. (1995) identified a Cu, Zn superoxide dismutase which, by immunoblot, gave a sensitivity of 25 percent in cases of invasive aspergillosis (Holdom et al. 2000). Using recombinant mitogillin protein, in an ELISA, Weig et al. (2001) obtained a sensitivity of 62.2 percent and specificity of 95.4 percent (in normal blood donors), with a cutoff value of mean ±2 SD. Increasing the stringency by using a cutoff value of mean ±5 SD improved specificity by lowered sensitivity. Using recombinant galactomannoprotein, Chan et al. (2002), with a cutoff value of mean ±5 SD, obtained a sensitivity of 33.3 percent (five of 15 sera from 15 patients with invasive aspergillosis positive) and specificity of 100 percent.

ANTIGEN DETECTION

In cases of suspected invasive aspergillosis, antigen detection is generally preferred because the at-risk groups are likely to be immunocompromised and poor antibody producers. Antigen detection stemmed from the initial description of circulating galactomannan in the serum of a patient with invasive aspergillosis (Reiss and Lehmann 1979). The importance of this antigen has been confirmed in a rabbit model (Dupont et al. 1987) and in patients using radioimmunnoassay, EIA, and latex agglutination (LA) (Shaffer et al. 1979; Dupont et al. 1990).

In clinical studies, the radioimmunoassay had a sensitivity of 70–80 percent, a specificity of 90 percent, a positive predictive value of 82 percent, and a negative predictive value of 85 percent (Talbot et al. 1987). Antigen was detected before invasive aspergillosis was suspected in 30 percent and before laboratory confirmation in 46 percent of patients. The main limitations were the use of radioactive material and the absence of antigen in about 25 percent of cases.

Sabetta et al. (1985) devised a competitive EIA that detected an undefined carbohydrate antigen in five of six immunosuppressed, infected rabbits and 11 of 19 patients with invasive aspergillosis. In experimentally infected rabbits, de Repentigny et al. (1987) detected circulating antigen which bound to concanavalin A with an apparent molecular mass of 50–100 kDa. The size of the galactomannan antigen was then substantiated by a monoclonal antibody which reacted with the immunodominant oligogalactoside side chains of *A. fumigatus* galactomannan recognizing 80, 62, and 49 kDa components of a hyphal homogenate (Ste-Marie et al. 1990).

Rogers et al. (1990) described two inhibition EIAs using a high-titer human serum to *Aspergillus* and a rat IgM monoclonal antibody to galactomannan, respectively. The results were encouraging in that 15 of 16 patients with invasive disease had detectable antigen at some point during their illness, but only 13 percent of sera and 20 percent of urine specimens were positive with the human serum. This improved to 18 percent for sera and 44 percent for urine samples when the monoclonal antibody against galactomannan was employed.

A commercially available RPLA test, Pastorex *Aspergillus* (BioRad, Marnes-la-Coquette, France) utilized a rat monoclonal EB-A2 specific to the galactomannan of *Aspergillus*. It reliably detected galactomannan in the sera of experimentally infected guinea pigs and showed no crossreactivity in animals with invasive candidiasis or cryptococcosis (Van Cutsem et al. 1990). Sensitivities of up to 95 percent were reported in patients with proven invasive aspergillosis (Dupont et al. 1990; Haynes and Rogers 1994), though stored sera were liable to become negative (Warnock et al. 1991; Knight and Mackenzie 1992). Hopwood et al. (1995), examining patients undergoing liver and bone marrow transplants, found that four of eight patients subsequently diagnosed as having invasive aspergillosis were Pastorex positive, but only for eight of the 187 sera examined from these four patients. In a prospective study, Verweij et al. (1995b) obtained a positive Pastorex test in eight patients who were subsequently shown to have invasive aspergillosis. The maximum titer was 1 in 8. The initial serum sample was negative in five of the cases. Storage at −20°C for longer than 6 months resulted in a loss of reactivity in six of the 13 sera tested. A further problem was the crossreactions observed, with no defined cause, reported in five patients by Hopwood et al. (1995) and also in a serum contaminated by *Penicillium chrysogenum* (Kappe and Schulze-Berge 1993).

Styner et al. (1995) demonstrated that the limit of detection was lowered 10-fold in an EIA employing the galactomannan monoclonal antibody as both captor and detector. Comparison with LA demonstrated improvement in both sensitivity and specificity (Verweij et al. 1995c). There is now a commercially available sandwich ELISA (Platalia Aspergillus, Bio-Rad Laboratories, Marnes La Coquette, France) with a reported sensitivity of 67–100 percent and specificity of 81–98 percent. Most importantly, this assay is sensitive enough to give positives at an early stage of the infection. Antigen titers increase in the terminal stages of the disease and serial assessment of serum titers is useful in monitoring response to treatment (Verweij et al. 1996, 1997). Prospective studies in patients with hematological malignancies have shown high sensitivity (>89 percent) and specificity (>92 percent) (Maertens et al. 1999, 2001; Sulahian et al. 2001). Sera were collected twice a week and a positive result reported when galactomannan was detected in two consecutive samples: increasing antigen titers in consecutive samples was a particularly strong indication of infection. These studies suggested that the cutoff value recommended by the manufacturer could be lowered. There is evidence that the test can be reliably applied to bronchoalveolar lavage and CSF specimens (Verweij et al. 1995a; Viscoli et al. 2002). The great majority of galactomannan in the CSF appeared to be produced intrathecally. The usefulness of monitoring response to treatment may vary with the drug used, since there is some data from a rabbit model of pulmonary aspergillosis to suggest that during treatment with the echinocandin caspofungin circulating antigen failed to decrease despite clinical improvement (Petraitiene et al. 2002).

Urine from recipients of bone marrow transplants with invasive aspergillosis contained not only galactomannan with a molecular weight of over 45 kDa but also several protein antigens between 11 and 44 kDa (Haynes et al. 1990). Other antigens described include an 80 kDa antigen (Phillips and Radigan 1989) and antigens at around 88, 40, 27, and 20 kDa (Yu et al. 1990). Burnie (1991b) developed a RPLA test detecting *Aspergillus* antigen with a rabbit polyclonal antiserum raised against a crushed extract of *A. fumigatus*. Antigen extracted from high-titer patient sera, by affinity chromatography using the same rabbit antiserum, reacted with monoclonal antibodies to *A. ambisexualis* and *C. albicans* Hsp90, showing that *Aspergillus* Hsp90 circulated during infection (Burnie and Matthews 1991).

PCR

In invasive aspergillosis, targets for PCR have been limited copy genes, for example the 18 kDa IgE-binding protein (Reddy et al. 1993) and alkaline protease (Tang et al. 1993), or multicopy genes such as the 26S intergenic spacer gene of rDNA (Spreadbury et al. 1993), the 18S rRNA genes (Yamakami et al. 1996), mitochondrial DNA (Jones et al. 1998), or panfungal targets such as the small subunit rRNA gene (van Burik et al. 1998). Specimens obtained at bronchoscopy have been tried, but the sensitivity can be low and false positives have been a persistent problem (Brètagne et al. 1995; Verweij et al. 1995a; Bart-Delabesse et al. 1996). This problem may be independent of the target sequence; it may be caused by colonization of the patient's respiratory tract or by contamination during processing.

Yamakami et al. (1996) developed an aspergillus-specific nested PCR, based on sequences derived from the 18S rRNA gene, which detected *Aspergillus* DNA in the sera of experimentally infected mice (sensitivity 71 percent) and 20 patients with invasive aspergillosis (sensitivity 70 percent). The same gene was targeted by Einsele et al. (1997), who detected circulating DNA in all 13 patients with invasive aspergillosis. Golbang et al. (1999) modified this assay to convert it from a nested PCR to a semiquantitative PCR-EIA, incorporating the digoxigenin label during the PCR and binding the amplicon to a streptavidin plate with a biotinylated capture probe. This assay was used to quantify the level of circulating DNA in invasive aspergillosis. The sensitivity was 91 percent, 30 of the 33 culture-confirmed cases of invasive aspergillosis giving positive optical densities (>0.1). It was negative in all 20 control sera. Where sequential samples were available, levels fell in survivors and gradually rose in fatal cases. PCR has also been successfully applied to cerebrospinal fluid specimens in cases of aspergillus meningitis (Kami et al. 1999; Verweij et al. 1999) and to ocular samples in fungal endophthalmitis (Jaeger et al. 2000). Comparative studies of the galactomannan ELISA with PCR have yielded conflicting results as to which is superior (Bretagne et al. 1998; Kawamura et al. 1999; Becker et al. 2000). Both systems have shown the ability to detect the infection early (Verweij et al. 1997; Hebart et al. 2000), opening up the possibility of prospective screening of high-risk patients and prompter treatment.

Real-time PCR greatly reduces the risk of false-positive results because the reaction tubes do not need to be opened following amplification, thus avoiding environmental contamination with the amplicon, the main source of contamination. Using the LightCycler, and a fully automated nucleic acid extraction technique to further reduce the risk of cross-contamination, Costa et al. (2002) compared the performance of real-time PCR with an ELISA for detecting galactomannan. They found serum to be preferable to white cells as a source of *A. fumigatus* DNA. The galactomannan assay was more frequently positive than the PCR assay (52 percent versus 45 percent). It may be that *Aspergillus* releases large amounts of galactomannan, which is relatively easily detected compared to the tiny quantities of circulating fungal DNA, despite amplification by PCR. Even before death, the DNA concentration was never greater than 30 fg/ml of serum. More studies on the origins and kinetics of fungal DNA release would help resolve these issues.

CRYPTOCOCCOSIS

This systemic infection is caused by the encapsulated basidiomycetous yeast *Cryptococcus neoformans*. It can cause meningitis in non-immunocompromised patients, but the great majority of cases now occur in patients with the acquired immunodeficiency syndrome (AIDS), in whom it may present as meningitis, pneumonia, or fever resulting from an underlying cryptococcemia. Although the principles of laboratory diagnosis are the same, the lack of sufficient capsular material produced by some isolates from AIDS patients occasionally causes problems for standard LA tests.

Latex agglutination

Several latex agglutination tests (RPLA) are commercially available for the rapid detection of the capsular polysaccharide antigen. The capsular antigen is produced in large quantities, has a serum half-life of 24 hours (Kappe and Muller 1991), and is stable against heat or pronase treatment, allowing dissociation from bound antibodies. These RPLA tests, which use antibody-coated latex particles, are easy to perform and highly sensitive, and specific for the diagnosis of meningeal and disseminated cryptococcosis (Prevost and Newell 1978; Kauffman et al. 1981; Bhattacharjee et al. 1984). Although *C. neoformans* is usually easy to isolate from clinical material, it may take up to 10 days to appear. The RPLA test can give the diagnosis in 30 minutes. Titres of 1:100 or more can be demonstrated in the CSF and, particularly in AIDS patients, in serum. Antigen titers can help monitor response to therapy and give an indication of the extent of infection and, indirectly, of the prognosis. Antibody detection adds little (Hay 1992).

The sensitivity of RPLA tests is high, well above 90 percent. Reports as to the lower limits of detection with the Cryptococcal Antigen Latex Agglutination System (CALAS, Meridian Diagnostics Inc., Cincinnati, OH) and the Crypto-LA test (International Biological Labs Inc., Cranbury, NJ) are in the range 2.5–10 ng/ml of capsular polysaccharide for serotypes A–D (Temstet et al. 1992) and 7.6–61 ng/ml, the highest level being for serotype D (Gade et al. 1991). False positives can occur. A heat-stable cell wall component of *Trichosporon beigelii* (syn. *Trichosporon cutaneum*) crossreacts with cryptococcal polysaccharide (Melcher et al. 1991). Disseminated infections caused by *T. beigelii* can occur in immunosuppressed patients and they are a well-documented cause of positive results with cryptococcal RPLA tests (McManus et al. 1985). Other rare causes of false positives are malignancy (Hopfer et al. 1982) and septicemia with DF-2 (Westerink et al. 1987). Technical false positives have been reported as a result of the introduction of talc from latex gloves into CSF samples, following immersion of a platinum wire inoculating loop into a CSF sample for culture before testing, and after repeated washing of ring slides with certain disinfectants and soaps (Heelan et al. 1991; Blevins et al. 1995).

A prozone-like effect, probably the result of immune complex formation, can give rise to false negatives. This can be avoided by diluting out the specimen (Stamm and Polt 1980) or treating with pronase. Pronase considerably reduces the occurrence of false reactions by cleaving antibodies present in immune complexes, which can otherwise mask the detection of antigen, and destroying rheumatoid factor (Sadamoto et al. 1993). Pronase pretreatment is a recommended modification of many of the commercially available RPLA tests; it increases sensitivity for most sera and some (but not all) CSF samples, and improves the end-point readings by making the agglutination reactions stronger (Gray and Roberts 1988; J.R. Hamilton et al. 1991). Kohno et al. (1993) obtained significantly higher rates of detection of serum cryptococcal antigen using the Eiken test (Eiken Co., Tokyo) in conjunction with pronase pretreatment.

One of the commercially available LA tests (Pastorex *Cryptococcus*, BioRad, Marnes-La-Coquette, France) has a murine monoclonal antibody to cryptococcal polysaccharide (Dromer et al. 1987), sensitizing the latex particles instead of the more commonly used rabbit polyclonal antisera. Temstet et al. (1992) compared this with two commercially available RPLA tests using polyclonal immune sera, CALAS and Crypto-LA. Clinical specimens from 87 AIDS patients, 40 of whom had culture-proven cryptococcosis, were retrospectively tested in a blind study. Pastorex performed well during the initial screening, concordance among the three tests being 97 percent. The predominant serotypes were A and D, only one patient having serotype B and none having serotype C, which reflected the prevalent serotypes. However, the threshold for detection of capsular polysaccharide with Pastorex was at least a log poorer for serotypes B, C, and D, compared with CALAS and Crypto-LA. Pronase treatment enhanced both the sensitivity and specificity of Pastorex and it was recommended for all samples except urine. Interestingly pronase treatment did not resolve the prozone effect in one case which was corrected by sample dilution. The antigen titers obtained with Pastorex and CALAS after pronase treatment agreed within one dilution for 67 percent of CSF samples and 57 percent of sera. It was emphasized, as previously observed (J.R. Hamilton et al. 1991), that kits for the detection of cryptococcal antigen cannot be used interchangeably to monitor antigen titres in patients.

Occasionally false negatives occur in AIDS patients as a result of a capsular-deficient variant of *C. neoformans*, resulting in low concentrations of the antigen. Both serum and CSF specimens are usually positive in AIDS patients with cryptococcal meningitis, definitive diagnosis then being made by culture of the CSF (Panther and Sande 1990). Routine screening for serum cryptococcal antigen does not, however, appear to predict patients who subsequently develop cryptococcal meningitis (Nelson et al. 1990). Localized pulmonary cryptococcosisis is likely to give negative RPLA results with serum samples (Prevost and Newell 1978; Kauffman et al. 1981), although Baughman et al. (1992) successfully screened specimens of bronchoalveolar lavage fluid with the CALAS test to diagnose cryptococcal pneumonia.

EIA

An EIA (Premier, Meridian Diagnostics), using a polyclonal capture system to bind capsular polysaccharide and visualized with a peroxidase-labelled monoclonal antibody, was compared with the CALAS RPLA by Gade et al. (1991). They obtained close agreement (98 percent). Overall sensitivity for the EIA was 99 percent and specificity 97 percent, there being 120 EIA and latex agglutination (LA) positives out of the 475 specimens screened. Detection limits for serotypes A–D were 0.63, 0.63, 7.8, and 62 ng/ml, respectively. Knight (1992) compared this with the RPLA test of Immuno-Mycologics and reported 92 percent agreement, after examination of 54 samples, mainly sera and CSF, from patients with and without cryptococcosis. There was some indication, but only in a couple of patients, that the EIA gave fewer false positives and improved the diagnosis of genuine early cases. The EIA test was easy to perform and had a similar cost. As a screening test it took no longer than the RPLA, although determining the titer of first time positives took longer.

Sekhon et al. (1993) compared the Premier EIA with two commercially available RPLA tests, Immuno-Mycologics and Crypto-LA, and a non-commercial latex agglutination test. Of 143 sera and CSF from proven and suspected cases, 115 were negative by EIA and RPLA. The sensitivity of the EIA test was 85.2 percent and specificity 97 percent, compared with 100 percent sensitivity and specificity for the LA tests. Out of 26 true-positive specimens, 23 were EIA positive but three RPLA positive specimens were negative by EIA. In the great majority of cases, however, the EIA yielded higher titers than the RPLA, especially for specimens from AIDS patients, observations similar to those made by Gade et al. (1991). They found the EIA, unlike RPLA, did not require pretreatment of specimens to prevent false-positive reactions.

An EIA has also been described by Casadevall et al. (1992), who used monoclonal antibodies, of different isotypes, both to capture and to detect cryptococcal polysaccharide. Immuno-Mycologics have produced a biotin-amplified sandwich EIA.

Noncapsular antigens

Chaturvedi et al. (2001) have cloned and compared the Cu, Zn superoxide dismutase gene from three varieties of *C. neoformans*, with variable host predilection, in order to provide insight into its role in pathogenesis. A.J. Hamilton et al. (1991) have produced a range of murine monoclonal antibodies which recognized culture filtrate antigens (exoantigens) from both encapsulated and non-encapsulated isolates of *C. neoformans*. This suggests that they did not recognize capsular polysaccharide material, and immunofluorescence data showed reactivity with the cytoplasm and cell membranes only. Some of these monoclonal antibodies crossreacted with *T. beigelii*. One monoclonal, which recognized neither *T. beigelii* nor *C. neoformans* var. *gattii*, recognized a species-specific epitope on a 34–38 kDa glycoprotein produced as an exoantigen by *C. neoformans* var. *neoformans* (Hamilton et al. 1992). Another monoclonal antibody recognized a 115 kDa exoantigen which has been shown to react strongly with the sera of patients with cryptococcosis (Hamilton and Goodley 1993). Monoclonal antibodies such as these may prove useful in the serodiagnosis of cryptococcosis in AIDS patients infected with isolates lacking polysaccharide capsules.

PCR

PCR has been successfully applied to the detection of *C. neoformans* in clinical specimens from confirmed cases (Rappelli et al. 1998; Posteraro et al. 2000), but whether it can outperform the exisiting, highly successful assays for this infection remains to be determined. Bialek et al. (2002c) have developed nested and real-time PCR assays, both targeting the 18S rRNA gene, to a murine model of cryptococcal meningitis.

HISTOPLASMOSIS

This is a systemic fungal infection caused by the dimorphic fungus *Histoplasma capsulatum*. The fungus is found in the soil and endemic to extensive parts of the Americas, Africa, and Asia. There are two varieties: the small-celled form (var. *capsulatum*), which occurs globally, and the large-celled form (var. *duboisii*), confined to Africa and Madagascar. Histoplasmosis is a common and serious complication of AIDS, occurring with highest frequency in the endemic areas of the USA (Wheat et al. 1991). Infection usually begins in the lungs, but a progressive disseminated form of the disease often develops in AIDS patients. Clinical diagnosis may be hampered because the severity of the systemic disease overshadows the respiratory component and radiologically the chest may appear normal initially. Most of these patients present with fever and weight loss.

In exposed communities, up to 70 percent of the population have positive histoplasmin skin tests. The histoplasmin skin test is similar to the Mantoux test. It indicates exposure to the organism, and is used mainly for epidemiologic purposes. Many patients with disseminated infections are anergic and give negative skin tests whereas, in a sensitized individual, skin testing can induce antibodies which then complicate the serological diagnosis (Hay 1992; Kaufman 1992).

Laboratory diagnosis is required to confirm clinical and radiological signs of a systemic mycosis and to establish the etiology. Histopathological examination and culture of bone marrow biopsies and blood cultures are common means of obtaining a definitive diagnosis. But isolation and identification of the *capsulatum* variety of *H. capsulatum* from clinical material can take several weeks and requires conversion of the mycelial isolate to the yeast form or identification by commercially available exoantigen tests or DNA probes (Kaufman 1992; Huffnagle and Gander 1993). Therefore, serology is still the best method available for screening suspected cases of histoplasmosis.

Antibody assays

Proper interpretation of serology is helped by taking a full clinical history, including occupation, travel, residence, and whether the patient has undergone histoplasmin skin testing. The laboratory, which must be familiar with the tests' capabilities and limitations, can then select the appropriate battery of tests. Complement fixation (CF), ID, and LA tests for the detection of antibody are available commercially.

The histoplasmin LA test (available commercially from Immuno-Mycologics) detects IgM, which is produced in the early stages of infection. It is therefore only positive early in the course of a primary infection, becoming negative within the first few months of a chronic infection.

The ID test is based on the principles of double diffusion. The patient's serum and soluble antigen are placed in separate wells cut in a suitable diffusion medium (such as agarose or Cleargel) and allowed to diffuse outward. Visible lines of precipitate form where the antigen and antibody have combined in the proper relative concentrations (equivalence). Using the mycelial-form culture filtrate of *H. capsulatum* var. *capsulatum* as

a source of H and M antigens ('histoplasmin'), both H and M precipitins are detected, the M precipitin band appearing nearer the antigen well. Positive control sera, containing specific antibodies against H and M antigens, are placed adjacent to each patient serum in order to determine the specificity of any patient reactions through the formation of an 'identity' or 'partial identity' reaction where the two precipitin bands interact.

The ID test detects IgG antibodies and is negative in the first 3–6 weeks of a primary infection when IgM predominates. It usually becomes positive within about 1 month of infection, but can remain completely negative in patients with confirmed histoplasmosis (Helner 1958). The M band is present in 70 percent of active cases (Kaufman 1992). Provided a skin test has not been performed recently (Kaufman 1973), it is an indicator of active infection or recent past infection, because M precipitins persist some months after recovery. The H band occurs less frequently than the M band, and the absence of an H band does not exclude active histoplasmosis. In active cases, 2 percent are H+ and 10 percent are H+ and M+ (Kaufman 1992). An H band usually indicates active clinical disease and, if both M and H precipitins are present, it is highly suggestive of active histoplasmosis regardless of other serological results. Demonstration of M and H precipitins in the CSF is indicative of meningeal histoplasmosis (Kaufman 1992).

The CF test is based on the principle that complement (C′) is irreversibly taken up by certain classes of antibody when they react with antigen, the presence of free C′ being indicated by lysis of sheep red blood cells sensitized with antibody because C′ is required for this process. If complement-fixing antibodies are present, there is no hemolysis. It is essential to determine serum anticomplementary activity. This is when the patient's serum fixes or destroys C′ in the absence of added antigen and can be caused by a number of factors. Other essential prerequisites include determination of the optimal antigen dilution with known case sera (not hyperimmune animal control sera), standardization of all reagents, and inclusion of positive and negative control sera.

The CF test is more sensitive than ID. Both histoplasmin (H and M), derived from the mycelial form of *H. capsulatum* var. *capsulatum*, and yeast form antigens are available commercially for use in CF tests. Leland et al. (1991) compared reagents from Immuno-Mycologics with those obtained from Meridian Diagnostics and found the former to be more reactive. Measurement of yeast antibody with the CF test has a higher sensitivity (90 percent) than histoplasmin antibody (80 percent) (Kaufman 1992). CF titers of more than 32, or a fourfold rise in titer, give good presumptive evidence of histoplasmosis. Lower titers are nonspecific, particularly with the yeast antibody. Skin testing may interfere with histoplasmin antibody, which can be induced or significantly raised in histoplasmin-sensitized individuals after one histoplasmin skin test. Crossreactions can occur with sera from patients with blastomycosis, coccidioidomycosis, and other mycoses, and, with yeast-form antigen, they can occur in patients with leishmaniasis (Kaufman 1992). Lack of an antibody response should not exclude the diagnosis, particularly if only one specimen has been tested and the clinical picture is highly suggestive. CSF titers are usually 1:8 or greater in cases of meningeal histoplasmosis (Plouffe and Fass 1980).

Antibody assays are less reliable in immunocompromised patients and disseminated histoplasmosis. Torres et al. (1993) found that neither CF nor an EIA was useful for the diagnosis of disseminated histoplasmosis in AIDS patients. Kaufman (1992) found that the histoplasmin antigens used in CF and ID tests more readily detected histoplasmosis in AIDS patients than yeast form antigens.

Detection of antigens

A radioimmunoassay for detection of *H. capsulatum* var. *capsulatum* polysaccharide in urine or serum has been used successfully in the rapid diagnosis of disseminated (Wheat et al. 1989) and pulmonary (Wheat et al. 1992) histoplasmosis in AIDS patients. The antigen was found in the urine of 92 percent and serum of 88 percent of such AIDS patients. It could also be detected in the bronchoalveolar lavage fluid of 70 percent of cases, and was not detectable in negative controls. Wheat et al. (1991) found the assay particularly useful in the detection of relapses in AIDS patients with histoplasmosis. Antigen levels fell in response to antifungal therapy and an increase of more than 2 units in antigen levels during maintenance therapy was strongly indicative of relapse. They suggested routine measurement of antigen levels in both serum and urine every 1–3 months and in the event of clinical evidence of relapse. But in a comparative assessment of itraconazole versus fluconazole for the treatment of disseminated histoplasmosis in patients with AIDS, they found clearance of fungal blood cultures a better measure of antifungal effect than clearance of antigen (Wheat et al. 2002).

Other histoplasma antigens are now being characterized. Hamilton et al. (1990a) have immunological data suggesting that the M antigen of *H. capsulatum* var. *capsulatum* is a catalase. They have also produced monoclonal antibodies that distinguish between the varieties *capsulatum* and *duboisii*. These could be useful for epidemiology and serodiagnosis in areas where both these forms of histoplasmosis occur (Hamilton et al. 1990b). Torres et al. (1993) identified antibodies against four different antigens of *Histoplasma* spp. (91, 83, 70, and 38 kDa) by immunoblotting sera from patients with histoplasmosis. Gomez et al. (1992, 1995) have identified two heat-shock proteins, Hsp60 (HIS-62) and Hsp70

(HIS-80), mediating protective immunity against murine histoplasmosis through induction of cellular immunity. Maresca et al. (1994) have correlated the virulence of a strain with the time required to transform to the yeast form and to express heat shock genes. Detection of Hsp60 or Hsp70 in serum or urine could have diagnostic value for histoplasmosis analogous to the detection of Hsp90 in DC.

PCR

The detection of *H. capsulatum* in the soil currently involves a slow (6–8 weeks) and expensive assay requiring inoculation of mice. Therefore the development by Reid and Schafer (1999) of a 2-day PCR method, using fungal-specific primers followed by nested primers against an ITS region of a rRNA gene, should be a significant advantage. Bialek et al. (2002a) have developed a nested PCR assay to unambiguously identify *H. capsulatum* in tissue sections.

BLASTOMYCOSIS

This is caused by the dimorphic fungus, *Blastomyces dermatitidis*, endemic to parts of North America, Africa, India, and the Middle East. Diagnosis is usually made by the identification of the yeast in body fluids or tissues (such as sputa, bronchial lavage, or biopsy specimens) by potassium hydroxide smears, cytology, histology, or culture (Lemos et al. 2000; Martynowicz and Prakash 2002). As for histoplasmosis, serology can provide an indication of the likely diagnosis earlier than culture or histology. Serodiagnosis has improved following better purification of the A antigen of *B. dermatitidis*. Sensitivity varies, depending on the assay format, from 40 percent to 88 percent, but specificity is consistently very high (98–100 percent) (Kaufman 1992). Commercial sources include Immuno-Mycologics, Meridian Diagnostics, and Gibson Laboratories, Lexington, KY (formerly Nolan Scott Biological Laboratories, Atlanta, GA).

The widely used CF and ID tests, with purified A antigen, are only positive, respectively, in 40 percent and 65 percent of confirmed cases (Turner et al. 1986). Therefore a negative result in no way excludes active blastomycosis. The ID test is totally specific, an A band being presumptive evidence of active or recent infection. Sera from patients with other infections, such as histoplasmosis and coccidioidomycosis, may produce bands against the antigen to *B. dermatitidis*, but these do not form identity reactions with the A bands produced by the positive control serum.

More sensitive assay systems are now being developed. An indirect EIA, using purified A antigen, gave a sensitivity of 80 percent and a specificity of 98 percent (Turner et al. 1986). Titers of 1:32 or greater were indicative of active blastomycosis while lower titers were considered suggestive but could be the result of crossreactivity with other dimorphic fungi, particularly the *capsulatum* variety of *H. capsulatum*. Lo and Notenboom (1990), using commercial A antigen captured with rabbit antibody to A antigen with a sandwich EIA, succeeded in avoiding crossreactivity with sera from patients with histoplasmosis or coccidioidomycosis. They detected seven of eight active cases of blastomycosis, compared with three positive cases by ID or CF. The false-negative result was from a patient with AIDS. A cell wall-derived radiolabelled 120 kDa protein has been used to detect antibody in infected patients by Klein and Jones (1990).

COCCIDIOIDOMYCOSIS

This systemic mycosis, caused by the dimorphic fungus *Coccidioides immitis*, generally takes the form of an acute, mild respiratory disease but can become chronic or disseminated, and occurs as an opportunistic infection in AIDS patients. The fungus is endemic only to certain semidesert-like regions between southwest USA and Argentina. Both clinical diagnosis and laboratory diagnosis, by demonstration of the organism histologically or by culture, are sometimes difficult. Often positive serology is the first evidence of the diagnosis, indicating the need for more definitive techniques which may require biopsy.

Serodiagnosis is generally based on the detection of antibodies to two main antigens to *C. immitis*: the tube precipitin (TP) and CF antigens found in mycelial-form culture filtrate (coccidioidin). Antibody to the former was thought to be mainly IgM, although recent data suggest that IgG responses are often largely directed against TP (Gade et al. 1992). Antibody to CF antigens is mainly IgG (Pappagianis and Zimmer 1990). The early IgM response is useful in the diagnosis of acute primary coccidioidomycosis. Later, IgG is produced, usually outlasting the IgM, and persisting in chronic coccidioidomycosis.

Partial characterization of both TP and CF antigens has occurred. Both 120 kDa and 110 kDa components of TP have been identified as a glycoprotein co-localized within cytoplasmic vesicles and the wall of the spherule (Cole et al. 1991). Deglycosylation of these antigens resulted in significant loss of reactivity with patient IgM. Although mannose and glucose were the predominant monosaccharides, a 3-*O*-methylmannose sugar is also present which appears to be unique to this systemic fungal pathogen and at least partially responsible for the reactivity of IgM with the 120 kDa antigen. One of the CF antigens, which participates in the ID reaction, appears to be a chitinase (Johnson and Pappagianis 1992).

A LA test is commercially available (Immuno-Mycologics). It uses TP-coated latex particles to detect agglutinating antibodies. These are primarily IgM, but high titers of IgG can also cause agglutination (Martins et al. 1995a). It is a sensitive and simple rapid screening test, useful in detecting early acute disease, but it has a high rate of false positives so confirmation by ID is essential (Drutz and Cantanzaro 1978). Rheumatoid factor is one cause of false positives and should be excluded. It can be removed by IgG-absorbent diluents and columns that leave the IgM intact (Martins et al. 1995b).

The ID test can be used to detect antibodies to TP or CF antigens. The immunodiffusion tube precipitin (IDTP) test, thought to measure IgM, is a sensitive and specific indicator of early acute disease (Pappagianis and Zimmer 1990; Kaufman 1992). IgG is detectable by immunodiffusion complement fixation (IDCF) or CF tests and is useful in diagnosing later stages of infection. It can also be used to detect past infection. Quantification of antibody titers provides an indicator of disease progression, falling titers being a measure of response to therapy. Changing or high CF titers are diagnostic (Kaufman 1992). Lower CF titers may reflect crossreactivity and such results should be confirmed by IDCF. The IDCF test can be used as a qualitative screen or it may be quantified. CF and IDCF can be applied to serum or CSF samples.

An EIA for the detection of IgM and IgG against a purified mixture of TP and CF is now commercially available from Meridian Diagnostics (Gade et al. 1992). In assessing this assay, Martins et al. (1995a) compared the IgM EIA with LA, and the IgG EIA with ID and CF tests. It proved to be a reliable assay for the detection of antibodies in serum or CSF against both TP and CF antigens. The best overall correlation was achieved when EIA was compared with the combined results of the traditional assays; there was an agreement of 96.7 percent, a specificity of 98.5 percent and a sensitivity of 94.8 percent. EIA had several advantages. It used specific anti-IgG and anti-IgM conjugates to give a definitive IgG or IgM result. Results were determined objectively with a spectrophotometer. Unlike the CF test, indeterminate results caused by anticomplementary activity were not a problem, and there was no need to heat inactivate the specimen or for an overnight incubation. The EIA could be run in less than 2 hours, allowing results to be reported the same day the sample was received. A disadvantage of the EIA was that because the same serum dilution was used for both IgG and IgM testing, there was no IgG absorbent in the diluent; samples positive for specific IgM must therefore be tested for rheumatoid factor.

PARACOCCIDIOIDOMYCOSIS

This systemic mycosis, caused by the dimorphic fungus *Paracoccidioides brasiliensis*, is endemic to Latin America. It primarily involves the lungs but may disseminate, causing secondary lesions. Definitive diagnosis can only be achieved using laboratory procedures such as microscopy, histology, and culture.

Many different tests have been used for the sero-diagnosis of paracoccidioidomycosis. A comparative assessment including ID, CIE, and CF tests was reported by Del Negro et al. (1991). The ID and CIE tests with yeast culture filtrate were the most sensitive (91.3 percent and 95.6 percent, respectively) and highly specific (100 percent), significantly outperforming the CF test. A positive ID result is, therefore, taken as presumptive evidence of current or recent infection. The ID test is commercially available for research use from Immuno-Mycologics. Most frequently a single E band (also called band 1) is produced, but sometimes two additional bands are present. The greatest limitation of the test is seen with specimens from patients with early primary infections. During the first 4–8 weeks, IgM antibodies predominate and these react poorly in gel diffusion tests, because of their large size. In disseminated disease, CF antibody titers are high, usually falling with successful therapy and rising with relapses, but not uncommonly titers of 8–32 persist even after successful therapy (Restrepo et al. 1978). Do Valle et al. (2001) found immunoblotting gave better results than double immunodiffusion, raising sensitivity to 100 percent. Costa et al. (2000) improved the diagnostic coverage of ID from 93–95 percent, using crude antigens, to 100 percent in an enzyme-linked ID assay capture test using the gp43 antigen described below. Mamoni et al. (2001) developed a capture ELISA to detect specific IgE, the levels of which correlated with clinical severity, falling with clinical improvement. In conclusion, measurement of antibody is useful for both diagnosis and prognosis.

Detection of circulating *P. brasiliensis* antigens could facilitate early diagnosis or confirm a preliminary diagnosis when antibody assays are inconclusive. Several techniques with different sensitivities have been used (Brummer et al. 1993; Mendes-Giannini et al. 1993). Mendes-Giannini et al. (1989) identified a circulating 43 kDa glycoprotein, gp43, in the sera of both acute and chronic cases. This shared epitopes with Fava-Netto's (1990) cytoplasmic polysaccharide antigen (FNPA), another antigen widely used diagnostically (Rodrigues and Travassos 1994). The gp43 antigen has now been cloned and expressed in *Escherichia coli* (Diniz et al. 2002). Taborda and Camargo (1993, 1994) have developed both a passive hemagglutination assay and a dot immunobinding assay, based on detection of antibody to gp43, for the diagnosis of paracoccidioidomycosis.

Two heat-shock proteins, Hsp60 and Hsp87, have been characterized. The 87 kDa antigen was detected in infected patients' sera, its levels correlating well with response to treatment and clinical cure. Direct amino acid sequencing revealed substantial homology with heat-shock proteins and immunological crossreactivity with a monoclonal antibody against *H. capsulatum* Hsp80 (Diez et al. 2002). A monoclonal antibody to this 87 kDa antigen demonstrated that the level of expression was higher in the yeast than the mycelial phase and that the yeasts expressed the antigen in tissues from infected patients. The Hsp60 gene from *P. brasiliensis* has been cloned and characterized; both the recombinant protein and Hsp60 extracted from the yeast reacted with sera from infected patients (Izacc et al. 2001). Immunoblots of the recombinant protein indicated reactivity with 72 of 75 sera from paracoccidioidomycosis patients (Cunha et al. 2002); there was also some reactivity with sera from healthy controls (9.5 percent) and patients with histoplasmosis (11.5 percent).

Bialek et al. (2000b) developed a nested PCR directed against an outer-membrane protein gene unique to *P. brasiliensis*, and successfully used it to detect the fungus in tissue samples from a mouse model of the infection. Gomes et al. (2000) have a PCR assay based on the gene sequence of the gp43 antigen. Others have sequenced rDNA genes and intergenic spacers to derive primers capable of discriminating *P. brasiliensis* from other human fungal pathogens by PCR (Motoyama et al. 2000). Sano et al. (2001) have applied nested PCRs based on both the gp43 gene and the 5.8S ribosomal RNA gene to infected murine tissues. These early developments may well evolve into PCR-based assays useful in the diagnosis of this infection.

REFERENCES

Ahmad, S., Khan, Z., et al. 2002. Seminested PCR for diagnosis of candidemia: comparison with culture, antigen detection and biochemical methods for species identification. *J Clin Microbiol*, **40**, 2483–9.

Arruda, L.K., Platts-Mills, T.A.E., et al. 1992. *Aspergillus fumigatus*: Identification of 16, 18 and 45 kDa antigens recognised by human IgG and IgE antibodies and murine monoclonal antibodies. *J Allergy Clin Immunol*, **89**, 1166–76.

Bailey, J.W., Sada, E., et al. 1985. Diagnosis of systemic candidiasis by latex agglutination for serum antigen. *J Clin Microbiol*, **21**, 749–52.

Bart-Delabesse, E., Marmorat-Khuong, A., et al. 1996. Detection of Aspergillus DNA in bronchoalveolar lavage fluid of AIDS patients by the polymerase chain reaction. *Eur J Microbiol Infect Dis*, **15**, 24–5.

Baughman, R.P., Rhodes, J.C., et al. 1992. Detection of cryptococcal antigen in bronchoalveolar lavage fluid: a prospective study of diagnostic utility. *Am Rev Respir Dis*, **145**, 1226–9.

Baur, X., Weiss, W., et al. 1989. Immunoprint pattern in patients with allergic bronchopulmonary aspergillosis in different stages. *J Allergy Clin Immunol*, **83**, 839–44.

Becker, M.J., de Marie, S., et al. 2000. Quantitative galactomannan detection is superior to PCR in diagnosing and monitoring invasive pulmonary aspergillosis in an experimental rat model. *J Clin Microbiol*, **38**, 1434–8.

Berenguer, J., Buck, M., et al. 1993. Lysis-centrifugation blood cultures in the detection of tissue-proven invasive candidiasis: disseminated versus single organ infections. *Diagnostic Microbiol Infect Dis*, **17**, 103–9.

Bernard, E.M., Christiansen, K.J., et al. 1981. Rate of arabinitol production by pathogenic yeast species. *J Clin Microbiol*, **14**, 189–94.

Bhattacharjee, A.K., Bennett, J.E. and Glaudemans, C.P.J. 1984. Capsular polysaccharides of *Cryptococcus neoformans*. *Rev Infect Dis*, **6**, 619–24.

Bialek, R., Ernst, F., et al. 2002a. Comparison of staining methods and a nested PCR assay to detect *Histoplasma capsulatum* in tissue sections. *Am J Clin Pathol*, **117**, 597–603.

Bialek, R., Ibricevic, A., et al. 2000b. Detection of *Paracoccidioides brasiliensis* in tissue samples by a nested PCR assay. *J Clin Microbiol*, **38**, 2940–2.

Bialek, R., Weiss, M., et al. 2002c. Detection of *Cryptococcus neoformans* DNA in tissue samples by nested and real-time PCR assays. *Clin Diagn Lab Immunol*, **9**, 461–9.

Blevins, L.B., Fenn, J., et al. 1995. False-positive cryptococcal antigen latex agglutination caused by disinfectants and soaps. *J Clin Microbiol*, **33**, 1674–5.

Bougnoux, M.-E., Dupont, C., et al. 1999. Serum is more suitable than whole blood for diagnosis of systemic candidiasis by nested PCR. *J Clin Microbiol*, **37**, 925–30.

Brètagne, S., Costa, J.M., et al. 1995. Detection of *Aspergillus* species DNA in bronchoalveolar lavage samples by competitive PCR. *J Clin Microbiol*, **33**, 1164–8.

Bretagne, S., Costa, J.M., et al. 1998. Comparison of serum galactomannan antigen detection and competitive polymerase chain reaction for diagnosing invasive aspergillosis. *Clin Infect Dis*, **26**, 1407–12.

Brouwer, J. 1988. Detection of antibodies against *Aspergillus fumigatus*: comparison between double immunodiffusion, ELISA and immunoblot analysis. *Int Arch Allergy Appl Immunol*, **85**, 225–49.

Brummer, E., Castaneda, E. and Restrepo, A. 1993. Paracoccidioidomycosis: an update. *J Clin Microbiol*, **6**, 89–117.

Buchheidt, D., Skladny, H., et al. 2000. Systemic infections with *Candida* sp. and *Aspergillus* sp. in immunocompromised patients with hematological malignancies: current serological and molecular diagnostic methods. *Chemotherapy*, **46**, 219–28.

Buchman, T.G., Rossier, M., et al. 1990. Detection of surgical pathogens by in vitro DNA amplification. I. Rapid identification of *Candida albicans* by in vitro amplification of a fungus-specific gene. *Surgery*, **108**, 338–47.

Burgener-Kairuz, P., Zuber, J.P., et al. 1994. Rapid detection and identification of *Candida albicans* and *Torulopsis (Candida) glabrata* in clinical specimens by species-specific nested PCR amplification of a cytochrome P-450 lanosterol-demethylase (L1A1) gene fragment. *J Clin Microbiol*, **32**, 1902–7.

Burnie, J.P. 1991a. Developments in the serological diagnosis of opportunistic fungal infections. *J Antimicrob Chemother*, **28**, Suppl A, 23–33.

Burnie, J.P. 1991b. Antigen detection in invasive aspergillosis. *J Immunol Methods*, **143**, 187–95.

Burnie, J.P. 1995. Hsps in aspergillosis. In: Matthews, R.C. and Burnie, J.P. (eds), *Heat shock proteins in fungal infections*. Austin, TX: R.G. Landes.

Burnie, J.P. and Matthews, R.C. 1991. Heat shock protein 88 and *Aspergillus* infection. *J Clin Microbiol*, **29**, 2099–106.

Burnie, J.P. and Williams, D. 1985. Evaluation of the Ramco latex agglutination test in the early diagnosis of systemic candidiasis. *Eur J Clin Microbiol*, **4**, 98–101.

Burnie, J.P., Matthews, R.C., et al. 1989. Immunoblot fingerprinting *Aspergillus fumigatus*. *J Immunol Methods*, **118**, 179–86.

Burnie, J.P., Golbang, N. and Matthews, R.C. 1997. Semiquantitative PCR-EIA for diagnosis of disseminated candidiasis. *Eur J Clin Microbiol Infect Dis*, **16**, 346–50.

Caillot, D., Casasnovas, O., et al. 1997. Improved management of invasive aspergillosis in neutropenic patients using early thoracic computed tomographic scan and surgery. *J Clin Oncol*, **15**, 139–47.

Casadevall, A., Mukherjee, J. and Scharff, M.D. 1992. Monoclonal antibody based ELISAs for cryptococcal polysaccharide. *J Immunol Methods*, **154**, 27–35.

Chan, C., Woo, P.C.Y., et al. 2002. Detection of antibodies specific to an antigenic cell wall galactomannoprotein for serodiagnosis of *Aspergillus fumigatus* aspergillosis. *J Clin Microbiol*, **40**, 2041–5.

Chang, H.C., Leaw, S.N., et al. 2001. Rapid identification of yeasts in positive blood cultures by a multiplex PCR method. *J Clin Microbiol*, **39**, 3466–71.

Chaturvedi, S., Hamilton, A.J., et al. 2001. Molecular cloning, phylogenetic analysis and three-dimensional modeling of Cu, Zn superoxide dismutase (CnSOD1) from three varieties of *Cryptococcus neoformans*. *Gene*, **268**, 41–51.

Chen, Y.C., Eisner, J.D., et al. 2000. Identification of medically important yeasts using PCR-based detection of DNA sequence polymophisms in the internal transcribed spacer region 2 of the rRNA genes. *J Clin Microbiol*, **38**, 2302–10.

Christensson, B., Wiebe, T., et al. 1997. Diagnosis of invasive candidiasis in neutropenic children with cancer by determination of D-arabinitol/L-arabinitol ratios in urine. *J Clin Microbiol*, **35**, 636–40.

Chryssanthou, E., Anderson, B., et al. 1994. Detection of *Candida albicans* DNA in serum by polymerase chain reaction. *Scand J Infect Dis*, **26**, 479–85.

Chryssanthou, E., Klingspor, L., et al. 1999. PCR and other non-culture methods for diagnosis of invasive *Candida* infections in allogenic bone marrow and solid organ transplant recipients. *Mycoses*, **42**, 239–47.

Cole, G.T., Kruse, D. and Seshan, K.R. 1991. Antigen complex of *Coccidioides immitis* which elicits a precipitin antibody response in patients. *Infect Immun*, **59**, 2434–6.

Coleman, R.M. and Kaufman, L. 1972. Use of immunodiffusion test in the serodiagnosis of aspergillosis. *Appl Microbiol*, **23**, 301–8.

Costa, C., Costa, J.-M., et al. 2002. Real-time PCR coupled with automated DNA extraction and detection of galactomannan antigen in serum by enzyme-linked immunosorbent assay for diagnosis of invasive aspergillosis. *J Clin Microbiol*, **40**, 2224–7.

Costa, E.M.R., Da Silva Lacaz, C., et al. 2000. Conventional versus molecular diagnostic tests. *Med Mycol*, **38**, S1, 139–45.

Crampin, A.C.C. and Matthews, R.C. 1993. Application of the polymerase chain reaction to the diagnosis of candidosis by amplification of an HSP90 gene fragment. *J Med Microbiol*, **26**, 233–8.

Cunha, D.A., Zancope-Oliveira, R.M., et al. 2002. Heterologous expression, purification and immunological reactivity of a recombinant HSP60 from *Paracoccidioides brasiliensis*. *Clin Diagn Lab Immunol*, **9**, 374–7.

Dee, T.H. and Rytel, M.W. 1977. Detection of *Candida* serum precipitins by counterimmunoelectrophoresis: an adjunct in determining significant candidiasis. *J Clin Microbiol*, **5**, 453–7.

Del Negro, G.M., Garcia, N.M., et al. 1991. The sensitivity, specificity and efficiency values of some serological tests used in the diagnosis of paracoccidioidomycosis. *Rev Inst Med Trop Sao Paulo*, **33**, 277–80.

de Repentigny, L. and Reiss, E. 1984. Current trends in immunodiagnosis of candidiasis and aspergillosis. *Rev Infect Dis*, **6**, 301–2.

de Repentigny, L., Marr, L.D., et al. 1985. Comparison of enzyme immunoassay and gas-liquid chromatography for the rapid diagnosis of invasive candidiasis in cancer patients. *J Clin Microbiol*, **21**, 972–9.

de Repentigny, L., Boushira, M., et al. 1987. Detection of galactomannan antigenemia by enzyme immunoassay in experimental invasive aspergillosis. *J Clin Microbiol*, **25**, 863–7.

de Repentigny, L., Kilanowski, E., et al. 1991. Immunoblot analyses of the serologic response to *Aspergillus fumigatus* antigens in experimental invasive aspergillosis. *J Infect Dis*, **163**, 1305–11.

Diez, S., Gomez, B.L., et al. 2002. *Paracoccidioides brasiliensis* 87-kilodalton antigen, a heat shock protein useful in diagnosis: characterization, purification and detection in biopsy material via immunohistochemistry. *J Clin Microbiol*, **40**, 359–65.

Diniz, S.N., Carvalho, K.C., et al. 2002. Expression in bacteria of the gene encoding the gp43 antigen of *Paracoccidioides brasiliensis*: immunological reactivity of the recombinant fusion proteins. *Clin Diagn Lab Immunol*, **9**, 1200–4.

Do Valle, A.C., Costa, R.L., et al. 2001. Interpretation and clinical correlation of serological tests in paracoccidioidomycosis. *Med Mycol*, **39**, 373–7.

Dromer, F., Salamero, J., et al. 1987. Production, characterization and antibody specificity of a mouse monoclonal antibody reactive with *Cryptococcus neoformans* capsular polysaccharide. *Infect Immune*, **55**, 742–8.

Drutz, D.J. and Cantanzaro, A. 1978. Coccidioidomycosis. Part 1. *Am Rev Respir Dis*, **117**, 559–85.

Dupont, B., Huber, M., et al. 1987. Galactomannan antigenemia and antigenuria in aspergillosis studies in patients and experimentally infected rabbits. *J Infect Dis*, **155**, 1–11.

Dupont, B., Improvisi, L. and Provost, F. 1990. Detection de galactomannane dans les aspergillosis invasives humaines et animales avec un test au latex. *Bull Soc Fr Mycol Med*, **19**, 35–42.

Einsele, H., Hebart, H., et al. 1997. Detection and identification of fungal pathogens in blood by using molecular probes. *J Clin Microbiol*, **35**, 1353–60.

Erjavec, Z. and Verweij, P.E. 2002. Recent progress in the diagnosis of fungal infections in the immunocompromised host. *Drug Resist Updates*, **5**, 3–10.

Fava-Netto, C. 1990. Antigeno polissacaridico do *Paracoccidioides brasiliensis. Interciencia (Venezuela)*, **15**, 209–11.

Fisher, B.D., Armstrong, D., et al. 1981. Invasive aspergillosis. Progress in early diagnosis and treatment. *Am J Med*, **71**, 751–77.

Flahaut, M., Sangland, D., et al. 1998. Rapid detection of *Candida albicans* in clinical samples by DNA amplification of common regions from *C. albicans*-secreted aspartic proteinase genes. *J Clin Microbiol*, **36**, 395–401.

Franklyn, K.M. and Warmington, J.R. 1993. Cloning and nucleotide sequence analysis of the *Candida albicans* enolase gene. *FEMS Microbiol Lett*, **111**, 101–8.

Fratamico, P.M. and Buckley, H.R. 1991. Identification and characterization of an immunodominant 58-kilodalton antigen of *Aspergillus fumigatus* recognised by sera of patients with invasive aspergillosis. *Infect Immun*, **59**, 309–15.

Fratamico, P.M., Long, W.K. and Buckley, H.R. 1991. Production and characterization of monoclonal antibodies to a 58-kilodalton antigen of *Aspergillus fumigatus. Infect Immun*, **59**, 316–22.

Fujita, S.I. and Hashimoto, T. 1992. Detection of serum *Candida* antigens by enzyme-linked immunosorbent assay and a latex agglutination test with anti-*Candida albicans* and anti-*Candida krusei* antibodies. *J Clin Microbiol*, **30**, 3132–7.

Fujita, S.I., Matsubara, F. and Matsuda, T. 1986. Enzyme-linked immunosorbent assay measurement of fluctuations in antibody titer and antigenemia in cancer patients with and without candidiasis. *J Clin Microbiol*, **23**, 568–75.

Fujita, S.I., Lasker, B.A., et al. 1995. Microtitration plate enzyme immunoassay to detect PCR-amplified DNA from *Candida* species in blood. *J Clin Microbiol*, **33**, 962, 67.

Fujita, S.I., Senda, Y., et al. 2001. Multiplex PCR using internal transcribed spacer 1 and 2 regions for rapid detection and identification of yeast strains. *J Clin Microbiol*, **39**, 3617–22.

Fung, J.C., Donta, S.T. and Tilton, R.C. 1986. Candida detection system (CAND-TEC) to differentiate between *Candida albicans* colonization and disease. *J Clin Microbiol*, **24**, 542–7.

Gade, W., Hinnefeld, S.W., et al. 1991. Comparison of the PREMIER cryptococcal antigen enzyme immunoassay and the latex agglutination assay for detection of cryptococcal antigens. *J Clin Microbiol*, **29**, 1616–19.

Gade, W., Ledman, D.W. and Wethington, R. 1992. Serological responses to various *Coccidioides* antigen preparations in a new enzyme immunoassay. *J Clin Microbiol*, **30**, 1907–12.

Gentry, L.O., Wilkinson, I.D., et al. 1983. Latex agglutination test for detection of *Candida* antigen in patients with disseminated disease. *Eur J Clin Microbiol*, **2**, 122–8.

Girmenia, C., Marino, P., et al. 1997. Assessment of detection of *Candida* mannoproteinemia as a method to differentiate central venous cather-related candidemia from invasive disease. *J Clin Microbiol*, **35**, 903–6.

Golbang, N., Burnie, J.P. and Matthews, R.C. 1999. A polymerase chain reaction enzyme immunoassay for diagnosing infection caused by *Aspergillus fumigatus. J Clin Pathol*, **52**, 419–23.

Gold, J.W.M., Fisher, B., et al. 1980. Diagnosis of invasive aspergillosis by passive hemagglutination assay of antibody. *J Infect Dis*, **142**, 87–94.

Gomes, G.M., Cisalpino, P.S., et al. 2000. PCR for diagnosis of paracoccidioidomycosis. *J Clin Microbiol*, **38**, 3470–80.

Gomez, F.J., Gomez, A.M. and Deepe, G.S. Jr 1992. An 80 kilodalton antigen from *Histoplasma capsulatum* that has homology to heat shock protein 70 induces cell-mediated immune responses and protection in mice. *Infect Immun*, **60**, 2565–71.

Gomez, F.J., Allendoerfer, R. and Deepe, G.S. Jr 1995. Vaccination with recombinant heat shock protein 60 from *Histoplasma capsulatum* protects mice against pulmonary histoplasmosis. *Infect Immun*, **63**, 2587–95.

Gray, L.D. and Roberts, G.D. 1988. Experience with the use of pronase to eliminate interference factors in the latex agglutination test for cryptococcal antigen. *J Clin Microbiol*, **26**, 2450–1.

Guiver, M., Levi, K. and Oppenheim, B.A. 2001. Rapid identification of candida species by TaqMan PCR. *J Clin Path*, **54**, 362–6.

Gutierrez, J. and Maroto, C. 1993. Circulating *Candida* antigens and antibodies: useful markers of candidemia. *J Clin Microbiol*, **31**, 2550–2.

Hamilton, A.J. and Goodley, J. 1993. Purification of the 115-kd exoantigen of *Cryptococcus neoformans* and its recognition by immune sera. *J Clin Microbiol*, **31**, 335–9.

Hamilton, A.J., Bartholomew, M.A., et al. 1990a. Evidence that the M antigen of *Histoplasma capsulatum* var. *capsulatum* is a catalase which exhibits cross-reactivity with other dimorphic fungi. *J Med Vet Mycol*, **28**, 479–85.

Hamilton, A.J., Bartholomew, M.A., et al. 1990b. Preparation of monoclonal antibodies that differentiate between *Histoplasma capsulatum* variant *capsulatum* and *H. capsulatum* variant *duboisii. Trans R Soc Trop Med Hyg*, **84**, 425–8.

Hamilton, A.J., Bartholomew, M.A., et al. 1991. Production of species-specific murine monoclonal antibodies against *Cryptococcus neoformans* which recognise a noncapsular exoantigen. *J Clin Microbiol*, **29**, 980–4.

Hamilton, A.J., Jeavons, L., et al. 1992. A 34- to 38-kilodalton *Cryptococcus neoformans* glycoprotein produced as an exoantigen bearing a glycosylated species-specific epitope. *Infect Immun*, **60**, 143–9.

Hamilton, A.J., Holdom, M.D. and Hay, R.J. 1995. Specific recognition of purified Cu, Zn superoxide dismutase from *Aspergillus fumigatus* by immune human sera. *J Clin Microbiol*, **33**, 495–6.

Hamilton, J.R., Noble, A., et al. 1991. Performance of *Cryptococcus* antigen latex agglutination kits on serum and cerebrospinal fluid specimens of AIDS patients before and after pronase treatment. *J Clin Microbiol*, **29**, 333–9.

Hay, R.J. 1992. Laboratory techniques in the investigation of fungal infections. *Genitourin Med*, **68**, 409–12.

Haynes, K. and Rogers, T.R. 1994. Retrospective evaluation of a latex agglutination test for diagnosis of invasive aspergillosis in immunocompromised patients. *Eur J Clin Microbiol Infect Dis*, **13**, 670–4.

Haynes, K.A., Latgè, J.P. and Rogers, T.R. 1990. Detection of *Aspergillus* antigens associated with invasive infection. *J Clin Microbiol*, **28**, 2040–4.

Hebart, H., Loffler, J., et al. 2000. Early detection of aspergillus infection after allogeneic stem cell transplantation by polymerase chain reaction screening. *J Infect Dis*, **181**, 1713–19.

Heelan, J.S., Corpus, L. and Kessimian, N. 1991. False-positive reactions in the latex agglutination test for *Cryptococcus neoformans* antigen. *J Clin Microbiol*, **29**, 1260–1.

Helner, D.C. 1958. Diagnosis of histoplasmosis using precipitin reactions in agar gel. *Pediatrics*, **22**, 616–27.

Hendolin, P.H., Paulin, L., et al. 2000. Panfungal PCR and multiplex liquid hybridization for detection of fungi in tissue specimens. *J Clin Microbiol*, **38**, 4186–92.

Herent, P., Stynen, D., et al. 1992. Retrospective evaluation of two latex agglutination tests for detection of circulating antigens during invasive candidosis. *J Clin Microbiol*, **30**, 2158–64.

Holdom, M.D., Lechenne, B., et al. 2000. Production and characterization of recombinant *Aspergillus fumigatus* C, Zn superoxide dismutase and its recognition by immune human sera. *J Clin Microbiol*, **38**, 558–62.

Holmberg, K., Berdischewsky, M. and Young, L.S. 1980. Serologic immunodiagnosis of invasive aspergillosis. *J Infect Dis*, **141**, 656–64.

Holmes, A., Cannon, R.D., et al. 1994. Detection of *Candida albicans* and other yeasts in blood by PCR. *J Clin Microbiol*, **32**, 228–31.

Hopfer, R.L., Perry, E.V. and Fainstein, V. 1982. Diagnostic value of cryptococcal antigen in the cerebrospinal fluid of patients with malignant disease. *J Infect Dis*, **145**, 915.

Hopfer, R.L., Waldon, P., et al. 1993. Detection and differentiation of fungi in clinical specimens using polymerase chain reaction (PCR) amplification and restriction enzyme analysis. *J Med Vet Mycol*, **31**, 65–75.

Hopwood, V., Johnson, E.M., et al. 1995. Use of the pastorex aspergillus antigen latex agglutination test for the diagnosis of invasive aspergillosis. *J Clin Pathol*, **48**, 210–13.

Horn, R., Wong, B., et al. 1985. Fungemia in a cancer hospital: changing frequency, earlier onset and results of therapy. *Rev Infect Dis*, **7**, 646–55.

Hube, B., Monod, M., et al. 1994. Expression of seven members of the gene family encoding secretory aspartyl proteinasis in *Candida albicans*. *Mol Microbiol*, **14**, 87–99.

Huffnagle, K.E. and Gander, R.M. 1993. Evaluation of Gen-probe's *Histoplasma capsulatum* and *Cryptococcus neoformans* accuProbes. *J Clin Microbiol*, **31**, 419–21.

Izacc, S.M., Gomez, F.J., et al. 2001. Molecular cloning, characterization and expression of the heat shock protein 60 gene from the human pathogenic fungus *Paracoccidioides brasiliensis*. *Med Mycol*, **39**, 445–55.

Jaeger, E.E., Carroll, N.M., et al. 2000. Rapid detection and identification of *Candida*, *Aspergillus* and *Fusarium* species in ocular samples using nested PCR. *J Clin Microbiol*, **38**, 2902–8.

Johnson, S.M. and Pappagianis, D. 1992. The coccidioidal complement fixation and immunodiffusion-complement fixation antigen is a chitinase. *Infect Immun*, **60**, 2588–92.

Jones, M.E., Fox, A.J., et al. 1998. PCR-ELISA for the early diagnosis of invasive pulmonary aspergillus infection in neutropenic patients. *J Clin Pathol*, **51**, 652–6.

Kahn, F.W. and Jones, J.M. 1986. Latex agglutination tests for detection of *Candida* antigens in sera of patients with invasive candidiasis. *J Infect Dis*, **153**, 579–85.

Kami, M., Ogawa, S., et al. 1999. Early diagnosis of central nervous system aspergillosis using polymerase chain reaction, latex agglutination test, and enzyme-linked immunosorbent assay. *Br J Haematol*, **106**, 2, 536–7.

Kan, V.L. 1993. Polymerase chain reaction for the diagnosis of candidemia. *J Infect Dis*, **168**, 779–83.

Kappe, R. and Muller, J. 1991. Rapid clearance of *Candida albicans* mannan antigens by liver and spleen in contrast to prolonged circulation of *Cryptococcus neoformans* antigens. *J Clin Microbiol*, **29**, 1665–9.

Kappe, R. and Schulze-Berge, A. 1993. New cause for false-positive results with the Pastorex *Aspergillus* antigen latex agglutination test. *J Clin Microbiol*, **31**, 2489–90.

Kauffman, C.A., Bergman, A.G., et al. 1981. Detection of cryptococcal antigen: comparison of two latex agglutination tests. *Am J Clin Pathol*, **75**, 106–9.

Kauffman, C.A., Bradley, S.F., et al. 1991. Hepatosplenic candidiasis: successful treatment with fluconazole. *Am J Med*, **91**, 137–41.

Kaufman, L. 1973. Value of immunodiffusion tests in the diagnosis of systemic mycotic diseases. *Ann Clin Lab Sci*, **3**, 141–6.

Kaufman, L. 1992. Laboratory methods for the diagnosis and confirmation of systemic mycoses. *Clin Infect Dis*, **14**, Suppl 1, S23–9.

Kawamura, S., Maesaki, S., et al. 1999. Comparison between PCR and detection of antigen in sera for diagnosis of pulmonary aspergillosis. *J Clin Microbiol*, **37**, 218–20.

Klein, B.S. and Jones, J.M. 1990. Isolation, purification, and radiolabeling of a novel 120-kd surface protein on *Blastomyces dermatitidis* yeasts to detect antibody in infected patients. *J Clin Invest*, **85**, 152–61.

Knight, F. and Mackenzie, D.W.R. 1992. *Aspergillus* antigen latex test for diagnosis of invasive aspergillosis. *Lancet*, **339**, 188.

Knight, F.R. 1992. New enzyme immunoassay for detecting cryptococcal antigen. *J Clin Pathol*, **45**, 836–7.

Knutsen, A.P., Mueller, K.R., et al. 1994. Serum anti-*Aspergillus fumigatus* antibodies by immunoblot and ELISA in cystic fibrosis with allergic bronchopulmonary aspergillosis. *J Allergy Clin Immunol*, **93**, 926–31.

Kobayashi, M. and Miyoshi, I. 1993. Immunoblot analysis of *Aspergillus fumigatus* antigen with human antibodies and lectin probes. *Int Med*, **32**, 99–105.

Kobayashi, H., Depbeaupuis, J.P., et al. 1993. An 88-kilodalton antigen secreted by *Aspergillus fumigatus*. *Infect Immun*, **61**, 4767–71.

Kohno, S., Yasuoka, A., et al. 1993. High detection rates of cryptococcal antigen in pulmonary cryptococcosis by Eiken latex agglutination test with pronase pretreatment. *Mycopathologia*, **123**, 75–9.

Kontoyiannis, D.P. and Lewis, R.E. 2002. Antifungal drug resistance of pathogenic fungi. *Lancet*, **359**, 1135–44.

Kumar, A., Reddy, L.V., et al. 1992. Isolation and characterization of a recombinant heat shock protein of *Aspergillus fumigatus*. *J Allergy Clin Immunol*, **91**, 1024–9.

Kurup, V.P. and Fink, J.N. 1978. Evaluation of methods to detect antibodies against *Aspergillus fumigatus*. *Am J Clin Pathol*, **69**, 414–17.

Kurup, V.P. and Kumar, A. 1991. Immunodiagnosis of aspergillosis. *Clin Microbiol Rev*, **4**, 439–56.

Kurup, V.P., John, K.V., et al. 1986. A partially purified glycoprotein antigen from *Aspergillus fumigatus*. *Int Arch Allergy Appl Immunol*, **79**, 263–9.

Kurup, V.P., Resnick, A., et al. 1990. Antibody isotype responses in *Aspergillus*-induced diseases. *J Lab Clin Med*, **115**, 298–303.

Latgè, J.P., Moutaouakil, M., et al. 1991. The 18-kilodalton antigen secreted by *Aspergillus fumigatus*. *Infect Immun*, **59**, 2586–94.

Latgè, J.P., Debeaupuis, J.P., et al. 1993. Cell wall antigens in *Aspergillus fumigatus*. *Arch Med Res*, **24**, 269–74.

Leland, D.S., Zimmerman, S.E., et al. 1991. Variability in commercial histoplasma complement fixation antigens. *J Clin Microbiol*, **29**, 1723–4.

Lemieux, C., St-Germain, G., et al. 1990. Collaborative evaluation of antigen detection by a commercial latex agglutination test and enzyme immunoassay in the diagnosis of invasive candidiasis. *J Clin Microbiol*, **28**, 249–53.

Lemos, L.B., Guo, M. and Baliga, M. 2000. Blastomycosis: organ involvement and etiologic diagnosis. A review of 123 patients from Mississippi. *Ann Diagn Pathol*, **4**, 391–406.

Leser, C., Kauffman, H.F., et al. 1992. Specific serum immunopatterns in clinical phases of allergic bronchopulmonary aspergillosis. *J Allergy Clin Immunol*, **90**, 589–99.

Lo, C. and Notenboom, R.H. 1990. A new enzyme immunoassay specific for blastomycosis. *Am Rev Respir Dis*, **141**, 84–8.

Loeffler, J., Henke, N., et al. 2000. Quantification of fungal DNA by using fluorescence resonance energy transfer and the light cycler system. *J Clin Microbiol*, **38**, 586–90.

Longbottom, J.L. 1986. Antigens and allergens of *Aspergillus fumigatus* II. Their further identification and partial characterization of a major allergen (Ag3). *J Allergy Clin Immunol*, **78**, 18–24.

Luo, G. and Mitchell, T.G. 2002. Rapid identification of pathogenic fungi directly from cultures by using multiplex PCR. *J Clin Microbiol*, **40**, 2860–5.

MacDonald, F. and Odds, F.C. 1980. Inducible proteinase of *Candida albicans* in diagnostic serology and in the pathogenesis of systemic candidosis. *J Med Microbiol*, **13**, 423–35.

Maertens, J., Verhaegen, J., et al. 1999. Autopsy-controlled prospective evaluation of serial screening for circulating galactomanna by a sandwich enzyme-linked immunosorbent assay for haematological patients at risk for invasive aspergillosis. *J Clin Microbiol*, **37**, 3223–8.

Maertens, J., Verhargen, J., et al. 2001. Screening for circulating galactomannan as a non-invasive diagnostic tool for invasive aspergillosis in prolonged neutropenic patients and stem cell transplantation recipients: a prospective validation. *Blood*, **97**, 1604–10.

Makimura, K., Murayama, S. and Yamaguchi, H. 1994. Detection of a wide range of medically important fungi by the polymerase chain reaction. *J Med Microbiol*, **40**, 358–64.

Mamoni, R.L., Rossi, C.L., et al. 2001. Capture enzyme-linked immunosorbent assay to detect specific immunoglobulin E in sera of patients with paracoccidioidomycosis. *Am J Trop Med Hyg*, **65**, 237–41.

Maresca, B., Carratu, L. and Kobayashi, G. 1994. Morphological transition in the human fungal pathogen *Histoplasma capsulatum*. *Trends Microbiol*, **2**, 110–14.

Marier, R., Smith, W., et al. 1979. A solid-phase radioimmunoassay for the measurement of antibody to *Aspergillus* in invasive aspergillosis. *J Infect Dis*, **140**, 771–9.

Martins, T.B., Jaskowski, T.D., et al. 1995a. Comparison of commercially available enzyme immunoassay with traditional serological tests for detection of antibodies to *Coccidioides immitis*. *J Clin Microbiol*, **33**, 940–3.

Martins, T.B., Jaskowski, T.D., et al. 1995b. An evaluation of the effectiveness of three immunoglobulin G (IgG) removal procedures for routine IgM serological testing. *Clin Diagn Lab Immunol*, **2**, 98–103.

Martynowicz, M.A. and Prakash, U.B. 2002. Pulmonary blastomycosis: an appraisal of diagnostic techniques. *Chest*, **121**, 768–73.

Matthews, R.C. and Burnie, J.P. 1988. Diagnosis of systemic candidiasis by an enzyme-linked dot immunobinding assay for a circulating 47-kilodalton antigen. *J Clin Microbiol*, **26**, 459–63.

Matthews, R.C. and Burnie, J.P. 1989. Cloning of a DNA sequence encoding a major fragment of the 47 kilodalton stress protein homologue of *Candida albicans*. *FEMS Microbiol Lett*, **51**, 25–30.

Matthews, R.C., Burnie, J.P. and Tabaqchali, S. 1984. Immunoblot analysis of the serological response in systemic candidosis. *Lancet*, **2**, 1415–18.

Matthews, R., Burnie, J.P., et al. 1985. Immunoblot analysis of serological response in invasive aspergillosis. *J Clin Pathol*, **38**, 1300–3.

Matthews, R.C., Burnie, J.P. and Tabaqchali, S. 1987. Isolation of immunodominant antigens from sera of patients with systemic candidiasis and characterization of serological response to *Candida albicans*. *J Clin Microbiol*, **25**, 230–7.

Matthews, R.C., Burnie, J.P., et al. 1991a. Autoantibody to heat shock protein 90 can mediate protection against systemic candidosis. *Immunology*, **74**, 20–4.

Matthews, R.C., Burnie, J.P. and Lee, W. 1991b. The application of epitope mapping in the development of a new serological test for systemic candidosis. *J Immunol Methods*, **143**, 73–9.

Matthews, R.C., Hodgetts, S. and Burnie, J.P. 1995. Preliminary assessment of a human recombinant antibody fragment to hsp90 in murine invasive candidiasis. *J Infect Dis*, **171**, 1668–71.

McManus, E.J., Bozdech, M.J. and Jones, J.M. 1985. Role of the latex agglutination test for cryptococcal antigen in diagnosing disseminated infection with *Trichosporon beigelii*. *J Infect Dis*, **151**, 1167–9.

Meckstroth, K.L., Reiss, E., et al. 1981. Detection of antibodies and antigenemia in leukemic patients with candidiasis by enzyme-linked immunosorbent assay. *J Infect Dis*, **144**, 24–32.

Melcher, G.P., Reed, K.D., et al. 1991. Demonstration of a cell wall antigen cross-reacting with cryptococcal polysaccharide in experimental disseminated trichosporonosis. *J Clin Microbiol*, **29**, 192–6.

Mendes-Giannini, M.J., Bueno, J.P., et al. 1989. Detection of a 43,000-molecular-weight glycoprotein in the sera of patients with paracoccidiodomycosis. *J Clin Microbiol*, **27**, 2842–5.

Mendes-Giannini, M.J., del Negro, G.B. and Siqueira, A.M. 1993. Serodiagnosis. In: Franco, M., da Silva Lacaz, C.S., et al. (eds), *Paracoccidiodomycosis*. Boca Raton, FL: CRC Press, 345–63.

Meyer, R.D., Young, L.S., et al. 1973. Aspergillosis complicating neoplastic disease. *Am J Med*, **54**, 6–15.

Miyakawa, Y., Mabuchi, T., et al. 1992. Isolation and characterization of a species-specific DNA fragment for detection of *Candida albicans* by polymerase chain reaction. *J Clin Microbiol*, **30**, 894–900.

Miyakawa, Y., Mabuchi, T. and Fukazawa, Y. 1993. New methods for detection of *Candida albicans* in human blood by polymerase chain reaction. *J Clin Microbiol*, **31**, 3344–7.

Morace, G., Pagano, L., et al. 1999. PCR-restriction enzyme analysis for detection of *Candida* DNA in blood from febrile patients with hemotological malignancies. *J Clin Microbiol*, **37**, 1871–5.

Mori, T. and Matsumura, M. 1999. Clinical evaluation of diagnostic methods using plasma and/or serum for three mycoses: aspergillosis, candidosis and pnemocystosis. *Nippon Ishinkin Gakki Zasshi*, **40**, 223–30.

Moser, M., Crameri, R., et al. 1993. Diagnostic value of recombinant Aspergillus fumigatus allergen 1/a for skin testing and serology. *J Allergy Clin Immunol*, **93**, 1–11.

Moser, M., Menz, G., et al. 1994. Recombinant expression and antigenic properties of a 32-kilodalton extracellular alkaline protease, representing a possible virulence factor from *Aspergillus fumigatus*. *Infect Immun*, **62**, 936–42.

Motoyama, A.B., Venancio, E.J., et al. 2000. Molecular identification of *Paracoccidioides brasiliensis* by PCR amplification of ribosomal DNA. *J Clin Microbiol*, **38**, 3106–9.

Moutaouakil, M.M., Monod, M., et al. 1993. Identification of the 33 kDa alkaline protease of *Aspergillus fumigatus* in vitro and in vivo. *J Med Microbiol*, **39**, 393–9.

Na, B.K. and Song, C.Y. 1999. Use of monoclonal antibody in diagnosis of candidiasis caused by *Candida albicans*: detection of circulating aspartyl proteinase antigen. *Clin Diagn Lab Immunol*, **6**, 924–9.

Nakamura, A., Ishikawa, N. and Suzuki, H. 1991. Diagnosis of invasive candidiasis by detection of mannan antigen by using the avidin-biotin enzyme immunoassay. *J Clin Microbiol*, **29**, 2363–7.

Nelson, M.R., Bower, M., et al. 1990. The value of serum cryptococcal antigen in the diagnosis of cryptococcal infection in patients infected with the human immunodeficiency virus. *J Infect*, **21**, 175–81.

Ness, M.J., Vaughan, W.P. and Woods, G.L. 1989. *Candida* antigen latex test for detection of invasive candidiasis in immunocompromised patients. *J Infect Dis*, **159**, 495–502.

Obayashi, T., Toshida, M., et al. 1995. Plasma $(1{\rightarrow}3)$-β-D-glucan measurement in diagnosis of invasive deep mycosis and fungal febrile episodes. *Lancet*, **345**, 17–20.

Panther, L.A. and Sande, M.A. 1990. Cryptococcal meningitis in the acquired immunodeficiency syndrome. *Semin Respir Infect*, **5**, 138–45.

Pappagianis, D. and Zimmer, B.L. 1990. Serology of coccidioidomycosis. *Clin Microbiol Rev*, **3**, 247–68.

Pennington, J.E. 1980. Aspergillus lung disease. *Med Clin North Am*, **64**, 475–90.

Petraitiene, T.F., Petraitis, V., et al. 2002. Antifungal efficacy of caspofungin (MK-0991) in experimental pulmonary aspergillosis in persistently neutropenic rabbits: pharmacokinetics, drug disposition and relationship to galactomannan antigenemia. *Antimicrob Agents Chemother*, **46**, 12–23.

Pfaller, M.A., Cabezudo, I., et al. 1993. Value of the Hybritech ICON *Candida* assay in the diagnosis of invasive candidiasis in high risk patients. *Diagn Microbiol Infect Dis*, **16**, 53–60.

Phillips, R. and Radigan, G. 1989. Antigenemia in a rabbit model of invasive aspergillosis. *J Infect Dis*, **159**, 1147–50.

Phillips, P., Dowd, A., et al. 1990. Nonvalue of antigen detection immunoassays for diagnosis of candidemia. *J Clin Microbiol*, **28**, 2320–6.

Piechura, J.E., Huang, C.J., et al. 1983. Antigens of *Aspergillus fumigatus* II. Electrophoretic and clinical studies. *Immunology*, **49**, 657–65.

Plouffe, J.F. and Fass, R.J. 1980. Histoplasma meningitis: diagnostic value of cerebrospinal fluid serology. *Ann Intern Med*, **92**, 189–91.

Posteraro, B., Sanguinetti, M., et al. 2000. Reverse cross blot hybridization assay for rapid detection of PCR-amplified DNA from *Candida* species. *J Clin Microbiol*, **38**, 1609–14.

Prevost, E. and Newell, R. 1978. Commercial cryptococcal latex kit: clinical evaluation in a medical center hospital. *J Clin Microbiol*, **8**, 529–33.

Rappelli, P., Are, R., et al. 1998. Development of a nested PCR for detection of *Cryptococcus neoformans* in cerebrospinal fluid. *J Clin Microbiol*, **36**, 11, 3438–40.

Reddy, L.V., Kumar, A. and Kurup, V.P. 1993. Specific amplification of *Aspergillus fumigatus* DNA by polymerase chain reaction. *Mol Cell Probes*, **7**, 121–6.

Reid, T.M. and Schafer, M.P. 1999. Direct detection of *Histoplasma capsulatum* in soil suspensions by two-stage PCR. *Mol Cell Probes*, **13**, 4, 269–73.

Reiss, E. and Lehmann, P.F. 1979. Galactomannan antingenemia in invasive aspergillosis. *Infect Immun*, **25**, 357–65.

Reiss, E. and Morrison, C.J. 1993. Nonculture methods for diagnosis of disseminated candidiasis. *Clin Microbiol Rev*, **6**, 311–23.

Reiss, E., Taneka, K., et al. 1998. Molecular diagnosis and epidemiology of fungal infections. *Med Mycol*, **36**, Suppl 1, 249–57.

Restrepo, A., Restrepo, M., et al. 1978. Immune responses in paracoccidioidomycosis. A controlled study of 16 patients before and after therapy. *Sabouraudia*, **16**, 151–63.

Richardson, M.D. and Warnock, D.W. 1984. Antigen and antibody attachment in ELISA for *Aspergillus fumigatus* IgG antibodies. *J Immunol Methods*, **66**, 119–32.

Richardson, M.D., Stubbins, J.M. and Warnock, D.W. 1982. Rapid enzyme-linked immunosorbent assay (ELISA) for *Aspergillus fumigatus* antibodies. *J Clin Pathol*, **35**, 1134–7.

Rigby, S., Procop, G.W., et al. 2002. Fluorescence in situ hybridization with peptide nucleic acid probes for rapid identification of *Candida albicans* directly from blood culture bottles. *J Clin Microbiol*, **40**, 2182–6.

Rodrigues, E.G. and Travassos, L.R. 1994. Nature of the reactive epitopes in *Paracoccidioides brasiliensis* polysaccharide antigen. *J Med Vet Mycol*, **B32**, 77–81.

Rogers, T.R., Haynes, K.A. and Barnes, R.A. 1990. Value of antigen detection in predicting invasive pulmonary aspergillosis. *Lancet*, **336**, 1210–13.

Ruchel, R., Boning-Stutzer, B. and Mari, A. 1988. A synoptical approach to the diagnosis of candidosis, relying on serological antigen and antibody tests, on culture, and on evaluation of clinical data. *Mycoses*, **31**, 87–106.

Ruchel, R. 1989a. Identification of certain false-positive results in the Cand-Tec test for candidal antigen. *Mycoses*, **32**, 627–30.

Ruchel, R. 1989b. Candidosis: diagnostic tools in the laboratory. *Mycoses*, **32**, Suppl 2, 18–22.

Ruchel, R. 1993. Diagnosis of invasive mycoses in severely immunosuppressed patients. *Ann Hematol*, **67**, 1–11.

Sabetta, J.R., Miniter, P. and Andriole, V.T. 1985. The diagnosis of invasive aspergillosis by an enzyme-linked immunosorbent assay for circulating antigen. *J Infect Dis*, **152**, 946–53.

Sadamoto, S., Ikeda, R., et al. 1993. Evidence for interference by immune complexes in the serodiagnosis of cryptococcosis. *Microbiol Immunol*, **37**, 129–33.

Sakai, T., Ikegami, K., et al. 2000. Rapid, sensitive and simple detection of candida deep mycosis by amplification of 18S ribosomal RNA gene;

comparison with assay of serum beta-D-glucan level in clinical samples. *Tohoku J Exp Med*, **190**, 119–28.

Samuelsen, H., Karlsson-Borg, A., et al. 1991. Purification of a 20 kD allergen from *Aspergillus fumigatus*. *Allergy*, **20**, 115–24.

Sano, A., Yokoyama, K., et al. 2001. Detection of gp43 and ITS1-5.8S-ITS2 ribosomal RNA genes of *Paracoccidioides brasiliensis* in paraffin-embedded tissue. *Nippon Ishinkin Gakkai Zasshi*, **42**, 23–7.

Schonheyder, H. 1987. Pathogenetic and serological aspects of pulmonary aspergillosis. *Scand J Infect Dis*, **151**, 1–62.

Segal, E., Berg, R.A., et al. 1979. Detection of *Candida* antigen in sera of patients with candidiasis by an enzyme-linked immunosorbent assay-inhibition technique. *J Clin Microbiol*, **10**, 116–18.

Sekhon, A.S., Garg, A.K., et al. 1993. Evaluation of a commercial enzyme immunoassay for the detection of cryptococcal antigen. *Mycoses*, **36**, 31–4.

Sendib, B., Tabouret, M., et al. 1999. New enzyme immunoassays for sensitive detection of circulating *Candida albicans* mannan and antimannan antibodies: useful combined test for diagnosis of systemic candidiasis. *J Clin Microbiol*, **37**, 1510–17.

Shaffer, P.J., Kobayashi, G.S., et al. 1979. Demonstration of antigenemia in patients with invasive aspergillosis by solid phase (protein A-rich *Staphylococcus aureus*) radioimmunoassay. *Am J Med*, **67**, 627–30.

Shin, J.-H., Nolte, F.S. and Morrison, C.J. 1997. Rapid identification of *Candida* species in blood cultures by a clinically useful PCR method. *J Clin Microbiol*, **35**, 1454–9.

Soyama, K. and Ono, E. 1985. Enzymatic fluorometric method for the determination of D-arabinitol in serum by initial rate analysis. *Clin Chim Acta*, **149**, 149–54.

Spreadbury, C., Holden, D. and Cohen, J. 1993. Detection of *Aspergillus fumigatus* by PCR. *J Clin Microbiol*, **31**, 615–21.

Stamm, A.M. and Polt, S.S. 1980. False-negative cryptococcal antigen test. *JAMA*, **244**, 1359.

Ste-Marie, L., Senechal, S., et al. 1990. Production and characterization of monoclonal antibodies to cell wall antigens of *Aspergillus fumigatus*. *Infect Immun*, **58**, 2105–14.

Strockbine, N.A., Largen, M.T., et al. 1984. Identification and molecular weight characterization of antigens from *Candida albicans* that are recognized by human sera. *Infect Immunol*, **43**, 715–21.

Styner, D., Goris, A., et al. 1995. A new sensitive sandwich enzyme-linked immunosorbent assay to detect galactofuran in patients with invasive aspergillosis. *J Clin Microbiol*, **33**, 497–500.

Sulahian, A., Boutboul, F., et al. 2001. Value of antigen detection using an enzyme immunoassay in the diagnosis and predication of invasive aspergillosis in two adults and pediatric hematology units during a 4-year prospective study. *Cancer*, **91**, 311–18.

Sundstrom, P. and Aliaga, G.R. 1992. Molecular cloning of cDNA and analysis of protein secondary structure of *Candida albicans* enolase an abundant immunodominant glycolytic enzyme. *J Bacteriol*, **174**, 6789–99.

Taborda, C.P. and Camargo, Z.P. 1993. Diagnosis of paracoccidioidomycosis by passive haemagglutination assay using a purified and specific antigen – gp43. *J Med Vet Mycol*, **31**, 155–60.

Taborda, C.P. and Camargo, Z.P. 1994. Diagnosis of paracoccidiodomycosis by dot immunobinding assay for antibody detection using the purified and specific antigen gp-43. *J Clin Microbiol*, **32**, 554–6.

Talbot, G.H., Weiner, M.H., et al. 1987. Serodiagnosis of invasive aspergillosis in patients with meatologic malignancy: validation of the *Aspergillos fumigatus* antigen radioimmunoassay. *J Infect Dis*, **155**, 12–27.

Tang, C.M., Holden, D.W., et al. 1993. The detection of *Aspergillus* spp. by the polymerase chain reaction and its evaluation in bronchoalveolar lavage fluid. *Am Rev Respir Dis*, **148**, 1313–17.

Telenti, A. and Roberts, G.D. 1989. Fungal blood cultures. *Eur J Clin Microbiol Infect Dis*, **8**, 825–31.

Temstet, A., Roux, P., et al. 1992. Evaluation of a monoclonal antibody-based latex agglutination test for diagnosis of cryptococcosis:

comparison with two tests using polyclonal antibodies. *J Clin Microbiol*, **30**, 2544–30.

Teshima, R., Ikebuchi, H., et al. 1993. Isolation and characterization of a major allergenic component (gp55) of *Aspergillus fumigatus*. *J Allergy Clin Immunol*, **92**, 698–705.

Thaler, M.B., Pastakia, T.H., et al. 1988. Hepatic candidiasis in cancer patients: the evolving picture of the syndrome. *Ann Intern Med*, **108**, 88–100.

Torres, M., Diaz, H., et al. 1993. Evaluation of enzyme linked immunosorbent-assay and western blot for diagnosis of histoplasmosis. *Rev Invest Clin*, **45**, 155–60.

Tregar, T.R., Visscher, D.W., et al. 1985. Diagnosis of pulmonary infection caused by *Aspergillus* usefulness of respiratory cultures. *J Infect Dis*, **152**, 572–6.

Turenne, C.Y., Sanche, S.E., et al. 1999. Rapid identification of fungi using the ITS2 genetic region and an automated fluorescent capillary electrophoresis system. *J Clin Microbiol*, **37**, 1846–51.

Turner, S., Kaufman, L. and Jalbert, M. 1986. Diagnostic assessment of an enzyme-linked immunosorbent assay for human and canine blastomycosis. *J Clin Microbiol*, **23**, 294–7.

van Burik, J., Myerson, D., et al. 1998. Parfungal PCR assay for detection of fungal infection in human blood specimens. *J Clin Microbiol*, **36**, 1169–75.

Van Cutsem, J., Meulemans, L., et al. 1990. Detection of circulating galactomannan by Pastorex *Aspergillus* in experimental invasive aspergillosis. *Mycoses*, **33**, 61–9.

Verweij, P.E., Latge, J.P., et al. 1995a. Comparison of antigen detection and PCR assay using bronchoalveolar lavage fluid for diagnosing invasive pulmonary aspergillosis in patients receiving treatment for hematological malignancies. *J Clin Microbiol*, **33**, 3150–3.

Verweij, P.E., Rijs, A.J.M.M., et al. 1995b. Clinical evaluation and reproducibility of the Pastorex *Aspergillis* antigen latex agglutination test for diagnosing invasive aspergillosis. *J Clin Pathol*, **48**, 474–6.

Verweij, P.E., Stynen, D., et al. 1995c. Sandwich enzyme-linked immunosorbent assay compared with Pastorex latex agglutination test for diagnosing invasive aspergillosis in immunocompromised patients. *J Clin Microbiol*, **33**, 1912–14.

Verweij, P.E., Donnelly, J.P., et al. 1996. Prospects for early diagnosis of invasive aspergillosis in the immunocompromised patient. *Rev Med Microbiol*, **7**, 105–13.

Verweij, P.E., Dompeling, E.C., et al. 1997. Serial monitoring of *Aspergillus* antigen in the early diagnosis of invasive aspergillosis. Preliminary investigations with two examples. *Infection*, **25**, 86–9.

Verweij, P.E., Brinkman, K., et al. 1999. *Aspergillus* meningitis: diagnosis by non-culture-based microbiological methods and management. *J Clin Microbiol*, **37**, 1186–9.

Viscoli, C., Machetti, M., et al. 2002. *Aspergillus* galactomannan antigen in cerebrospinal fluid of bone marrow transplant recipients with probable cerebral aspergillosis. *J Infect Dis*, **40**, 1496–9.

Wahyuningsih, R., Freisleben, H.-J., et al. 2000. Simple and rapid detection of *Candida albicans* DNA in serum by PCR for diagnosis of invasive candidiasis. *J Clin Microbiol*, **38**, 3016–21.

Wallenbeck, I. 1991. *Aspergillus fumigatus* specific IgE and IgG antibodies for diagnosis of aspergillus-related lung diseases. *Allergy*, **46**, 372–8.

Wallenbeck, I., Aukrust, L. and Einarsson, R. 1984. Antigenic variability of different strains of *Aspergillus fumigatus*. *Int Arch Allergy Appl Immunol*, **73**, 166–72.

Walsh, T.J. and Chanock, S.J. 1998. Diagnosis of invasive fungal infections: advances in nonculture systems. *Curr Clin Top Infect Dis*, **18**, 101–53.

Walsh, T.J., Hathorn, J.W., et al. 1991. Detection of circulating *Candida* enolase by immunoassay in patients with cancer and invasive candidiasis. *N Engl J Med*, **324**, 1026–31.

Wardlaw, A. and Geddes, D.M. 1992. Allergic bronchopulmonary aspergillosis: a review. *J R Soc Med*, **85**, 747–51.

Warnock, D.W. 1977. Detection of *Aspergillus fumigatus* precipitins: a comparison of counter immunoelectrophoresis and double diffusion. *J Clin Pathol*, **30**, 388–9.

Warnock, D.W., Foot, A.B.M., et al. 1991. *Aspergillus* antigen latex test for diagnosis of invasive aspergillosis. *Lancet*, **338**, 1023–4.

Weig, M., Frosch, K., et al. 2001. Use of recombinant mitogillin for improving serodiagnosis of *Aspergillus fumigatus* – associated diseases. *J Clin Microbiol*, **39**, 1721–30.

Westerink, M.A.L., Amsterdam, D., et al. 1987. Septicemia due to DF-2: cause of a false-positive cryptococcal latex agglutination result. *Am J Med*, **83**, 155–8.

Wheat, L.J., Connolly-Stringfield, P., et al. 1989. *Histoplasma capsulatum* polysaccharide antigen detection in diagnosis and management of disseminated histoplasmosis in patients with acquired immunodeficiency syndrome. *Am J Med*, **87**, 396–400.

Wheat, L.J., Connolly-Stringfield, P., et al. 1991. Histoplasmosis relapse in patients with AIDS: detection using *Histoplasma capsulatum* variety *capsulatum* antigen levels. *Ann Intern Med*, **115**, 936–41.

Wheat, L.J., Connolly-Stringfield, P., et al. 1992. Diagnosis of histoplasmosis in patients with the acquired immunodeficiency syndrome by detection of *Histoplasma capsulatum* polysaccharide antigen in bronchoalveolar lavage fluid. *Am Rev Respir Dis*, **145**, 1421–4.

Wheat, L.J., Connolly, P., et al. 2002. Antigen clearance during treatment of disseminated histoplasmosis with itraconazole versus fluconazole in patients with AIDS. *Antimicrob Agents Chemother*, **46**, 248–50.

Wilson, M.L., Davis, T.E., et al. 1993. Controlled comparison of the BACTEC high-blood-volume fungal medium, BACTEC plus 26 aerobic blood culture bottle, and 10-milliliter isolator blood culture system for detection of fungemia and bacteremia. *J Clin Microbiol*, **31**, 865–7.

Yamakami, Y., Hashimoto, A., et al. 1996. PCR detection of DNA specific for *Aspergillus* species in serum of patients with invasive aspergillosis. *J Clin Microbiol*, **34**, 2464–8.

Young, R.C., Bennett, J.E., et al. 1970. Aspergillosis: the spectrum of disease in 98 patients. *Medicine (Baltimore)*, **49**, 147–73.

Yu, B., Yoshihito, N. and Armstrong, D. 1990. Use of immunoblotting to detect *Aspergillus fumigatus* antigen in sera and urines of rats with experimental invasive aspergillosis. *J Clin Microbiol*, **28**, 1575–9.

6

Histopathological diagnosis of mycotic diseases

HENRIK ELVANG JENSEN AND FRANCIS W. CHANDLER

INTRODUCTION

Ideally, diagnosis of a fungal pathogen is made by observation of typical clinical symptoms, by demonstration of the presence of a fungus within lesions accompanied by a host reaction, and by subsequent isolation in culture of the infectious agent. Unfortunately, this has been achieved comparatively rarely in certain deep mycoses. Moreover, patients with fungal infections are frequently also suffering from other diseases, which may mask the fungal-related symptoms. As a result of the nonspecific nature of the clinical observations including radiological findings in fungal infections, a diagnosis usually depends on several laboratory approaches apart from mycological culture and histopathology. These include detection of:

- fungal antigens in serum, spinal fluid, bronchoalveolar lavage specimens, and urine as, for example, *Histoplasma* antigen, cryptococcal antigen, and galactomannan in cases of aspergillosis
- antifungal antibodies useful in, for example, histoplasmosis
- fungal metabolites such as D-arabinitol in *Candida* infections
- cell wall markers like $(1\rightarrow3)$-β-D-glucan which is present in the wall of almost all fungi with the exception of zygomycetes (Reiss et al. 2000).

In recent years, molecular-based techniques such as the polymerase chain reaction (PCR) have been developed for the non-culture based diagnosis of fungal infections in blood and other clinical specimens. Due to the risk of false-negative and false-positive results in both culture- and nonculture-based diagnostic tests for mycotic infections, clinical features should always complement histopathology in the diagnosis of fungal infections.

Many of the nonculture-based techniques are not available for many of the unusual mycotic pathogens. Isolation of fungi from tissues is often also problematic. Culture of fungi is often a slow process, and results may not be available for several days or weeks. Moreover, when a mycosis is not suspected (often they are mistaken for neoplasms or inflammation of unknown etiology), the whole biopsy specimen is often fixed for histopathological examination, and thus portions are not available for culture (Jensen 1994). When this occurs, histopathological and immunohistological techniques may be the only means of establishing an etiologic diagnosis short of repeating the biopsy procedure. A systematic approach must then be taken to arrive at the most accurate diagnosis possible by examining fixed, paraffin-embedded tissue sections. Although the morphology of fungal elements within tissues may provide a tentative diagnosis, it is not always possible to establish a clearcut diagnosis based on morphological details because of morphological similarities among tissue forms of several fungal genera and the presence of sparse or atypical fungal elements (Chandler et al. 1980; Jensen et al. 1996).

Histopathological studies can contribute in a variety of ways to the diagnosis of diseases caused by fungi and related pathogens. A generic or specific etiologic diagnosis may be made in 24 hours or less by the unequivocal demonstration and identification of characteristic fungal elements in tissue sections (Baker 1971; Anthony 1973; Schwarz 1982). Direct microscopic examination is the only reliable way of diagnosing certain fungal diseases, such as lobomycosis, rhinosporidiosis, and *Pneumocystis* infections, for which isolation techniques do not currently exist. The etiologic significance of a culture isolate can usually be determined by careful histopathological evaluation (Chandler and Watts 1987). Microscopic demonstration of tissue invasion and host reaction resolve the clinical dilemma of whether a fungal isolate is truly pathogenic, or merely a superficial colonizer, a component of the normal mycobiota (e.g. *Candida albicans*), or an environmental contaminant (e.g. a species of *Aspergillus* or a member of the class Zygomycetes) (Hoog and de Guého 1985). Microscopic evaluation of the inflammatory reaction and the distribution of fungal elements in a tissue section can also help to determine whether the disease is an invasive infection or a purely allergic reaction (e.g. invasive versus allergic bronchopulmonary aspergillosis); it can sometimes be used to assess the efficacy of antifungal therapy in pre- and post-treatment biopsy specimens (Chandler and Watts 1987). Finally, histopathological studies may sometimes confirm the presence of multiple infections, e.g. in acquired immunodeficiency syndrome (AIDS), or exclude a fungal disease from the clinical differential diagnosis altogether by revealing another process that accounts for the clinical and radiological findings. Accurate diagnosis may not only give the benefit of optimal therapy, but is also essential for the study of the pathological and epidemiologic aspects of different mycoses (Jensen 1994).

HISTOLOGICAL STAINS FOR DEMONSTRATING FUNGI

A number of traditional histochemical stains can be used to detect fungi and related pathogens in tissue sections. These stains, along with their applications and limitations, are summarized in Table 6.1. Details of the staining procedures are not included here, but can be found in standard manuals and textbooks (Elias 1982; Bancroft and Stevens 1996) of histochemistry and histological techniques.

Hematoxylin and eosin (H&E) is a versatile stain which enables pathologists to evaluate the host's inflammatory response, including the Splendore–Hoeppli (asteroid bodies) reaction, and to determine whether a fungus is hyaline (colorless) or pheoid (naturally pigmented–dematiaceous) (Matsumoto et al. 1994). It is the histological stain of choice to demonstrate the hematoxylinophilic or amphophilic nuclei of yeast-form cells that are multinucleated, particularly those of *Blastomyces dermatitidis*, *Loboa loboi*, and *Paracoccidioides brasiliensis*. Some fungi, e.g. the aspergilli and zygomycetes, are hematoxylinophilic and readily

delineated with H&E, but many fungal agents are not stained or stain poorly. Even in the instances of poor staining, careful examination often reveals the outlines of unstained fungal elements, which suggest the existence of a fungal infection. Moreover, pathologists should always pay special attention to and carefully search for fungal elements when a granulomatous or pyogranulomatous inflammation is present (Jensen 1994). The H&E procedure has limitations, however. With this stain, it is sometimes difficult to distinguish fungal elements from tissue components, e.g. thin blood vessels within brain and lung tissues, and, when fungi are sparse, they are easily overlooked (Jensen et al. 1997). As a result of these limitations, special stains for fungi are usually needed to adequately reveal the presence and morphology of fungal pathogens in tissue sections (Chandler and Watts 1987).

Special histological stains such as Gomori's methenamine silver (GMS), the periodic acid–Schiff (PAS) reaction, and Gridley's fungus (GF) procedures are useful to demonstrate fungi in tissue sections. The staining reactions of these procedures are based on the principle that adjacent hydroxyl groups of complex polysaccharides in fungal cell walls are oxidized to aldehydes by chromic acid or periodic acid. In the GMS procedure, the aldehydes reduce the methenamine silver nitrate complex; fungal cell walls stain brownish–black because reduced silver is deposited wherever aldehydes are located. In the PAS and GF procedures, the aldehydes react with Schiff's reagent, coloring fungi magenta. All three procedures stain the cell walls of fungal elements, but tend to obscure the internal details of these cells. Moreover, hydroxyl groups in tissue elements such as basement membranes will also be stained by these techniques. The GMS stain is the best procedure for detecting fungi in tissue sections because it provides better contrast for screening and, in most instances, it will stain old and nonviable organisms that may be refractory to the PAS and GF procedures. Fungi stained with the last-mentioned procedures also tend to fade with prolonged storage. When properly done, the GMS procedure usually eliminates nonspecific background staining of normal tissue components and necrotic debris, whereas the PAS and GF procedures do not. The GMS stain is also more sensitive because it demonstrates certain nonfungal pathogens, e.g. *Actinomyces israelii* and related species, *Nocardia* spp., *Mycobacterium* spp., nonfilamentous bacteria with polysaccharide capsules, and free-living soil amoebae, algal cells, spores of microsporidians, and cytoplasmic granular inclusion bodies of cytomegalovirus (Chandler and Watts 1987). Such an all-purpose, low-specificity stain is extremely helpful for the rapid screening of specimens from immunocompromised patients, who are often infected by multiple pathogens.

When performing the GMS stain, the histotechnologist should vary the staining time according to the etiologic agent under consideration, guided by a

Table 6.1 *Stains for demonstrating fungi and related pathogens in tissue sections*

Stains	Applications	Limitations
Hematoxylin and eosin (H&E)	Demonstrates inflammatory response Stains some fungi Allows determination of innate color of pheoid (pigmented) fungi Demonstrates Splendore–Hoeppli material Hematoxylin stains nuclei of most yeast-like cells	Does not stain many fungi Inadequate to screen for sparse fungal elements Does not stain individual filaments of *Actinomyces*, *Nocardia* and *Streptomyces* spp.
Special stains for fungi	All three stains are excellent for detecting fungi	Do not allow determination of the innate color of a fungus (GMS stains fungi black–brown; GF and PAS stain fungi red–purple, which mask innate color)
Gomori's methenamine silver (GMS) Gridley's fungus (GF) Periodic acid–Schiff (PAS)	The GMS is better than GF or PAS for screening; it stains old and nonviable fungal elements better than the other two stains, and it also demonstrates filaments of *Actinomyces* spp. and *Nocardia* spp. and elements of *Pneumocystis carinii*	GMS may overstain fungi and obscure internal details GMS and GF do not allow proper study of host response GF and PAS do not stain filaments of the actinomycetes
GMS with H&E counterstain	Permits study of host response and is excellent for detecting fungi and most actinomycetes The stain of choice, if only one slide is available for histopathological evaluation Excellent for photomicrography	Does not allow determination of the innate color of phaeoid (pigmented) fungi
Mucin stains Mayer's mucicarmine Southgate's mucicarmine Alcian blue	Stain mucopolysaccharide capsular material of *Cryptococcus neoformans* Permit differentiation of *C. neoformans*, from other fungi of similar size and form	Not specific for *Cryptococcus neoformans* (may stain *Rhinosporidium seeberi* and some cells of *Blastomyces dermatitidis*)
Modified Gram's stains Brown and Brenn Brown–Hopps MacCallum–Goodpasture	Demonstrate gram-positive filaments of *Actinomyces* spp., *Nocardia* spp. and other actinomycetes Useful for detecting bacterial copathogens Demonstrate gram-positive and gram-negative bacteria of botryomycosis	Do not selectively stain most fungi (those that do stain are gram positive)
Modified acid-fast stains Modified Ziehl–Neelsen Fite–Faraco Kinyoun's	Useful to detect *Nocardia asteroides*, *N. brasiliensis*, and *N. otitidiscaviarum* Valuable to detect *Mycobacterium tuberculosis* and other mycobacteria	*Nocardia* spp. are weakly acid-fast and non-alcohol-fast in tissue sections Fungal cell walls and the agents of actinomycosis are not acid-fast
Melanin stain Modified Fontana–Masson	Stains cell walls of *Cryptococcus neoformans* and other *Cryptococcus* spp. Accentuates melanin or melanin-like substances in weakly pigmented agents of pheohyphomycosis	May stain cell walls of *Sporothrix schenckii* and immature spherules of *Coccidioides immitis*
Whitening agents Calcofluor White M2R, Blankophor, Uvitex 2B and others	Stain cell walls of most fungi	Require a fluorescence microscope Do not consistently stain fungal elements May not stain degenerated fungi

concurrently stained positive control slide. To accomplish this, the control slide should be periodically removed from the silver nitrate bath and examined under a light microscope for optimal staining. Optimal staining time in the silver nitrate solution is usually increased when old, degenerated, and nonviable fungal elements are suspected, such as those of *Coccidioides immitis* and *Histoplasma capsulatum* var. *capsulatum* in the caseated centers of residual granulomas. The depth of staining should never be so intense that it obscures the morphological details of fungal elements; preferably, some internal (cytoplasmic and nuclear) details should remain visible. When a tissue section is overstained with GMS, the contrast between fungal cells and background tissue elements is obscured. Blood vessels with darkened basement membranes can mimic fungal hyphae, particularly if the vessels are empty, branched, and of approximately the same caliber as the suspected fungus. Erythrocytes, leukocytes, and naked nuclei can be overstained and mimic yeast cells, whereas darkened reticular and elastic fibers can be mistaken for filamentous fungi or bacteria. In caseated granulomas, calcific bodies that stain with H&E and PAS procedures often simulate yeast cells, particularly those of *Cryptococcus neoformans* and *H. capsulatum* var. *capsulatum*. As the chromic acid used as an oxidant in the GMS and GF procedures dissolves calcium, calcific bodies are unstained when these procedures are used to confirm the presence of a fungal pathogen (Chandler and Watts 1987).

A major disadvantage of special stains is that they mask the innate color of fungal elements, making it impossible to determine whether a fungus is hyaline or pheoid. Such a determination may be crucial to the diagnosis of a mycosis caused by pigmented fungi, e.g. pheohyphomycosis, chromoblastomycosis, and black-grain eumycotic mycetomas (Chandler and Watts 1987). As a result of masking, duplicate H&E-stained or unstained tissue sections should always be examined to look for brown pigmentation of fungal cell walls. Another limitation of the special stains for fungi is that they do not allow adequate study of the host reaction to fungal invasion. To avoid this limitation, H&E can be used as the counterstain for the GMS procedure. This combination of stains (GMS–H&E) readily colors fungal elements brownish–black while staining background tissue components as expected. Thus, it is possible simultaneously to detect the fungus and to evaluate the host's inflammatory response and its relationship to the fungus. Tissue sections stained with GMS–H&E are also optimal for photomicrography.

When only a single unstained section from a suspected lesion is available for examination, GMS–H&E is the best stain combination for attempting to make a diagnosis. If desired, the section can be stained initially with H&E, checked for the presence of pheoid fungi, and then restained with GMS–H&E. For retrospective studies of unexplained lesions, tissue sections previously stained with Giemsa, modified Gram's, PAS, and GF procedures can be decolorized in acid alcohol after the coverslip has been removed and then restained with GMS.

Stains for mucin, such as Alcian blue and Mayer's or Southgate's mucicarmine procedures, readily demonstrate the mucopolysaccharide capsule of *Cryptococcus neoformans*. This staining reaction is a convenient diagnostic marker and usually differentiates typical cryptococci from nonencapsulated yeast-form pathogens of similar size and appearance. In rare cases, however, capsule-deficient cryptococci may not stain positively for mucin. In addition, mucin stains are not specific for *C. neoformans*, because the cell walls of *Blastomyces dermatitidis*, *Paracoccidioides brasiliensis*, and *Rhinosporidium seeberi* are also variably colored with these stains. As the last three fungi are nonencapsulated and morphologically distinctive, ordinarily they would not be mistaken for *C. neoformans*. Mucin stains are not required for their identification.

The cell walls of cryptococci contain silver-reducing substances (melanin-like substances derived from dihydroxyphenylalanine), which react positively with a modified Fontana–Masson stain for melanin (Kwon-Chung et al. 1981; Wheeler and Bell 1987; Dixon and Polak 1991). As this stain reaction does not depend on the presence of a mucinous capsule, it can be used routinely to demonstrate capsule-deficient cryptococci (so-called 'dry' variants) (Ro et al. 1987). Capsule-deficient cryptococci stained with conventional procedures can easily be confused with nonencapsulated yeasts of similar size and morphology. When the agents of pheohyphomycosis are either nonpigmented or lightly pigmented in tissue sections, stains for melanin can also be used to accentuate and confirm the presence of melanin in their cell walls (Wood and Russel-Bell 1983; Chandler and Watts 1987). Studies have shown that the pigment in many of these fungi is a particular type of melanin derived from dihydroxynaphthalene, which differs from the melanin produced by *C. neoformans* (Dixon et al. 1991).

Modified Gram's stains, such as the Brown and Brenn and MacCallum–Goodpasture procedures, demonstrate the delicate, branched, gram-positive filaments of the agents of actinomycosis, nocardiosis, streptomycosis, dermatophilosis, and actinomycotic mycetomas. The GMS procedure can also be used to demonstrate these agents, but they often do not stain intensely or uniformly. The H&E, PAS, and Gridley procedures do not stain the actinomycetes (although entire granules in actinomycosis and actinomycotic mycetomas stain well, individual filaments within the granules are unstained). Modified Gram's stains are also needed to detect bacteria other than actinomycetes that may complicate a mycotic or actinomycotic infection, or that might be the primary cause of disease, e.g. botryomycosis (Chandler and Watts 1987). Some fungi, especially the yeast forms of *Candida* spp. and the conidia of *Aspergillus* spp. are usually gram-positive. Gram's stains are not, however, recommended for screening or for definitive diagnosis of mycoses caused by these agents.

As the *Nocardia* spp. are weakly acid-fast and non-alcohol-fast in tissue sections, these filamentous bacteria can be distinguished from the agents of actinomycosis with modified acid-fast stains that use a weak aqueous solution consisting of 0.5 or 1.0 percent sulfuric acid for decolorization; the nocardiae are weakly acid-fast, whereas the agents of actinomycosis are not acid-fast. The cytoplasm of certain fungi with yeast-like tissue forms, especially *B. dermatitidis* and *H. capsulatum* var. *capsulatum*, is also variably acid-fast (Wages and Wear 1982). The acid-fast properties of fungi are, however, inconsistent and should not be depended on for diagnosis. The cell walls of fungi are not acid-fast.

When H&E-stained tissue sections that contain fungi are examined under ultraviolet light, some of the fungal elements usually autofluoresce (Graham 1983). Although this property may help delineate sparse or poorly stained fungi, fungal autofluorescence is inconsistent and should not be used as a substitute for the more reliable special stains such as GMS or for immuno-histological procedures. Some fungi in paraffin-embedded or fresh-frozen tissue sections also stain nonspecifically with optical brightening histochemistry, i.e. whitening agents that have a diaminostilbene backbone structure, such as Calcofluor White M2R, Blankophor, and Uvitex 2B, which fluoresce under ultraviolet light (Monheit et al. 1984, 1986; Jensen 1994). Most of these fluorescent brighteners have an affinity for certain glycosidic linkages of the polysaccharides of the fungal cell wall. Tissue sections are usually stained for one minute with a 0.1 percent solution of the whitening agent, counterstained with Evan's blue, and examined with a fluorescence microscope with an excitation band of 500 nm. This rapid and simple procedure has been used in the intraoperative examination of fresh-frozen tissues for fungi, scrapings, swabs, and body fluids. Whitening agents, however, do not always highlight fungi, and these compounds should only be used for definitive diagnosis when positive. There is no advantage in using whitening agents compared with the rapid methenamine silver techniques for the detection of fungi in tissue sections (Shimono and Harman 1986).

HISTOPATHOLOGICAL IDENTIFICATION OF FUNGI

As a result of their large size, polysaccharide content, and morphological diversity, fungi can be detected readily and may also to a certain level be identified in histological sections by conventional light microscopy (Schwarz 1982; Salfelder 1990). Fungal elements in tissue sections are either hyaline or pheoid, and usually appear as yeast-like cells, hyphae, endosporulating spherules, granules, or a combination of these forms. These four broad morphological categories form a useful

nontaxonomic system to approach the histopathological differential diagnosis of fungal diseases. Once a pathogen is correctly recognized as belonging to one of the four categories, it can in many cases be accurately identified to the level of genus or even to the species level by further consideration of its defining morphological and tinctorial features. For example, within the broad category of yeast-like pathogens, further defining features would include:

- size and shape of individual cells
- thickness of the cell wall
- number and shape of blastoconidia (buds)
- type of attachment of buds to parent cells
- presence or absence of septations
- pigmentation
- encapsulation
- number of nuclei
- presence of pseudohyphae, hyphae or arthroconidia.

Thus, a yeast-like pathogen with a thick cell wall, single broad-based bud, and multiple nuclei can only be *B. dermatitidis*. An encapsulated yeast-like pathogen can only be a *Cryptococcus* sp. A pathogen that forms budding yeast-like cells, pseudohyphae, and hyphae, but not arthroconidia, and is not pheoid (naturally pigmented), can only be a *Candida* sp. The system is less reliable for the category of pathogens with hyphal tissue forms, many of which appear quite similar. Moreover, the appearances of hyphae in sections are often affected by steric orientation, age of the hyphae, type of infected tissue, and the host response (Jensen and Schønheyder 1989). The morphological and tinctorial features of fungi that occur in tissue as yeast-like cells, large spherules, and nonpigmented hyphae are summarized in Tables 6.2, 6.3 and 6.4.

In the practice of surgical pathology, a specific etiologic or generic diagnosis of a fungal infection often has important prognostic and therapeutic implications. Certain fungal diseases are caused by agents that can be specifically identified in tissue sections because they are morphologically distinctive. In this group, a single species or variety of fungi is usually responsible for each particular disease, and if sufficient numbers of typical organisms are demonstrated, an etiologic diagnosis can be made. This group of diseases includes adiaspiromycosis, blastomycosis, coccidioidomycosis, cryptococcosis, small and large form histoplasmosis, lobomycosis, paracoccidioidomycosis, infection due to *Penicillium marneffei*, rhinosporidiosis, sporotrichosis, and pneumocystosis. Other fungal diseases are caused by any of several members of a genus that, in tissues, are all morphologically similar to each other. Typical forms of these fungi also have a distinctive morphology but can only be identified to the genus level. Diseases caused by fungi in this group, which can be diagnosed generically, include candidosis and trichosporonosis. Table 6.4 clearly shows that agents of aspergillosis, fusariosis and scedosporiosis most often cannot be separated in tissue

Table 6.2 *Morphological features of fungi that occur as yeast-like cells in tissue*

Feature	Histoplasma capsulatum var. capsulatum	Histoplasma capsulatum var. duboisii	Blastomyces dermatitidis	Cryptococcus neoformans	Penicillium marneffei	Paracoccidioides brasiliensis	Loboa loboi	Sporothrix schenckii var. schenckii[d]	Candida spp.	Candida glabrata	Pneumocystis carinii
Size (µm)	2–4	6–12	7–15[a]	2–20	2.5–5	5–60	5–12	2–10	3–6	2–5	2–10
Shape	Spherical or oval	Oval	Spherical	Pleomorphic	Spherical, oval or elongated	Spherical	Spherical	Pleomorphic[e]	Spherical or oval	Spherical or oval	Spherical, oval or crescentic
Number of buds	Single	Single	Single	Single and rarely multiple	None	Multiple; 'steering wheel' forms	Multiple; chains	Single and rarely multiple	Single; chains	Single	None
Attachment of buds	Narrow	Narrow	Very broad	Narrow	NA[c]	Narrow	Narrow; tubular	Narrow	Narrow	Narrow	NA[c]
Schizogony	–	–	–	–	+	–	–	–	–	–	+
Thickness of cell wall	Thin	Thick	Thick	Thin	Thin	Variable	Thick	Thin	Thin	Thin	Thick in 'cysts'; thin in 'trophozoites'
Pseudohyphae and/or hyphae	Rare	Absent	Rare	Rare	Absent	Rare	Absent	Rare	Characteristic	Absent	Absent
Number of nuclei	Single	Single	Multiple	Single	Single	Multiple	Multiple	Single	Single	Single	Multiple in 'cysts'; single in 'trophozoites'
Mucicarmine reaction	–	–	±	+[b]	–	±	–	–	–	–	–

a) Microforms, 2–4 µm in diameter, also may occur in tissue.
b) Some strains may be capsule deficient and noncarminophilic.
c) NA, not applicable.
d) The cells of *S. schenckii* var. *luriei* are larger (10–20 µm) compared with the var. *schenckii*, and typical 'eyeglass' forms are often seen.
e) Elongated, cigar-shaped cells characteristic but infrequent.

Table 6.3 *Morphological features of fungi that occur as large spherules in tissue*

Feature	Coccidioides immitis	Rhinosporidium seeberi	Chrysosporium parvum var. crescens
External diameter of spherule (μm)	20–200	100–350	200–400
Thickness of spherule wall (μm)	1–2	3–5	20–70
Diameter of endospores	2–5	6–10	None
Hyphae or arthroconidia	Rare	None	None
Special stain reactions			
GMS	+	+	+
Mucicarmine	–	+	–

sections (Jensen et al. 1996). Moreover, aspergillosis, fusariosis and scedosporiosis may in some cases also be confused with zygomycosis and candidosis, respectively (Jensen et al. 1997). In especially chronic lesions with atypical fungal forms, similar problems do also apply to a number of other fungi that belong to different genera but appear morphologically similar to each other in tissue sections. Although the etiologic agent in some lesions cannot be specifically or generically identified, the disease can still be named, i.e. chromoblastomycosis, dermatophytosis, hyalohyphomycosis, pheohyphomycosis, and zygomycosis.

In organs with access to air, especially within the upper parts of the respiratory tract, but also within the lungs, conidial heads are sometimes formed. The morphology of such frutification bodies may sometimes allow an identification of the infecting agent to the species level (Salfelder 1990). Heavy deposits of birefringent calcium oxalate crystals in tissues may also be used to sustain the identification of *Aspergillus niger* (Salfelder 1990). In other cases, *A. niger* infections may result in the developing of generalized oxalosis (Salfelder 1990).

Mycetomas constitute special cases. By studying the granules – the diagnostic hallmark of the mycetomas – in appropriately stained sections of infected tissue, one can easily determine whether the etiologic agent is an actinomycete (branched, filamentous bacterium) or a eumycete (true fungus), and, if a eumycete, whether it is hyaline (white grained) or pheoid (black grained). The size, shape, color, and microscopic architecture of the granules are correlated with the etiologic agent, and each species, with some exceptions, forms its own distinctive type of granule (Chandler and Watts 1987). Thus, the gross and microscopic appearances of the granules provide insight into the identity of the organisms involved. Definitive identification of the etiologic agents should, however, be based, whenever possible, on microbiological culture.

Finally, in some cases, intact fungal elements or fragments may be detected in tissue sections but cannot be definitively or even presumptively identified. In other cases, the morphology of fungi may be atypical or bizarre as a result of changes induced by hypoxia, necrosis or antifungal therapy. Although the disease cannot be named in such instances, the pathologist can nevertheless often conclude that a fungal infection is present and apply a panel of specific antibodies using different immunohistochemical techniques (Jensen et al. 1996).

The histomorphological features of the common pathogenic fungi and related microorganisms are summarized in Table 6.5 and illustrated in Figures 6.1–6.62 at the end of the chapter. To facilitate differential diagnosis, the diseases cited in the table are grouped according to the morphological similarities of their etiologic agents in tissue. As diseases caused by the actinomycetes and the protothecae traditionally fall within the province of medical mycology, they are included in Table 6.5. Fungal infections that involve the hair exclusively, e.g. black piedra and white piedra, are not included in Table 6.5 because histological methods are not required for their diagnosis.

Although certain inflammatory patterns in an otherwise healthy person may suggest that a fungal infection

Table 6.4 *Morphological features of fungi that occur as hyaline hyphae in tissue*

Feature	Aspergillus spp.	Fusarium spp.	Scedosporium spp.; Pseudallescheria boydii	Rhizopus spp. and other zygomycetes
Width (μm)	3–6	3–8	2–5	6–25
Contours	Parallel	Parallel	Parallel	Irregular
Pattern of branching	Dichotomous	Dichotomous or right angle	Dichotomous and/or haphazard	Haphazard
Orientation of branches	Parallel or radial	Random and parallel	Random and parallel	Random
Frequency of septation	Frequent	Frequent	Frequent	Absent or infrequent
Angioinvasiveness	Yes	Yes	Yes	Yes

Table 6.5 *Histomorphological features of the common pathogenic fungi and related pathogens*

Etiologic agent(s)	Mycosis	Typical morphology
Pathogens that occur in tissue as yeast-like cells		
Blastomyces dermatitidis	Blastomycosis	Spherical, multinucleate, yeast-like cells, 7–15 µm diameter, with thick walls and single, broad-based buds
Cryptococcus neoformans	Cryptococcosis	Pleomorphic yeast-like cells, 2–20 µm diameter, with mucin-positive capsules and single or, rarely, multiple narrow-based buds; some strains are capsule-deficient and their capsular material may not be detected with mucin stain
Histoplasma capsulatum var. *capsulatum*	Histoplasmosis (classic, small form, histoplasmosis capsulati)	Spherical to oval, uninucleate, yeast-like cells, 2–4 µm diameter; often clustered because of growth within mononuclear phagocytes
Histoplasma capsulatum var. *duboisii*	African histoplasmosis (large form, histoplasmosis duboisii)	Spherical to oval uninucleate, yeast-like cells, 6–12 µm diameter, which have thick walls and bud by a narrow base, creating typical 'hour-glass' or 'figure-of-eight' forms
Loboa loboi	Lobomycosis	Spherical, budding, yeast-like cells, 5–12 µm diameter, that form chains of cells, each connected by a narrow tube-like isthmus; secondary budding sometimes present
Paracoccidioides brasiliensis	Paracoccidioidomycosis	Large, spherical, yeast-like cells 5–60 µm diameter, with multiple buds attached by narrow necks ('steering wheel' forms)
Penicillium marneffei	Infection due to *Penicillium marneffei* (penicilliosis marneffei)	Spherical to oval yeast-like cells, 2.5–5 µm diameter, with a single transverse septum; short hyphal forms and elongated, curved 'sausage' forms with one or more septa may be formed in necrotic and cavitary lesions
Sporothrix schenckii var. *schenckii*	Sporotrichosis	Pleomorphic, spherical to oval and, at times, 'cigar'-shaped yeast forms, 2–10 µm diameter, which produce single and, rarely, multiple buds
Sporothrix schenckii var. *luriei*	Sporotrichosis due to *S. schenckii* var. *luriei*	Hyaline, large (10–20 µm) thick-walled yeast forms together with the typical 'eyeglass' configuration of incompletely separated cells
Candida glabrata	Candidosis	Spherical to oval yeast-like cells 2–5 µm diameter
Pneumocystis carinii	Pneumocystosis	Thin-walled spherical, oval and crescentic trophozoites, 2–10 µm diameter; cysts are thick-walled, 4–6 µm diameter
Pathogens that occur in tissue as yeast-like cells and mycelial elements		
Candida albicans, C. guilliermondii, C. krusei, C. parapsilosis, C. pseudotropicalis, C. tropicalis and others	Candidosis	Oval, budding, yeast-like cells (blastoconidia), 2–6 µm diameter; pseudohyphae; and septate hyphae
Geotrichum candidum	Geotrichosis	Septate, infrequently branched hyphae, 3–6 µm wide; spherical yeast-like cells; and rectangular or oval arthroconidia, 4–10 µm wide, with rounded or squared ends
Trichosporon beigelii, T. capitatum (*Blastoschizomyces capitatus*)	Trichosporonosis	Pleomorphic yeast-like cells, 3–8 µm diameter; septate hyphae; and, rarely, arthroconidia
Pathogens that occur in tissue as large spherules		
Chrysosporium parvum var. *crescens*	Adiaspiromycosis	Large adiaconidia, 200–400 µm diameter, with thick (20–70 µm) walls; budding and endosporulation do not occur
Coccidioides immitis	Coccidioidomycosis	Spherical, thick-walled, endosporulating spherules, 20–200 µm diameter; mature spherules contain small, 2–5 µm diameter, uninucleate endospores; septate hyphae and chains of arthroconidia sometimes occur in necrotic nodules

(Continued over)

Table 6.5 *Histomorphological features of the common pathogenic fungi and related pathogens (Continued)*

Etiologic agent(s)	Mycosis	Typical morphology
Prototheca wickerhamii, P. zopfii	Protothecosis	Spherical, oval or polyhedral sporangia, 2–25 μm diameter, that, when mature, contain 2–20 sporangiospores
Rhinosporidium seeberi	Rhinosporidiosis	Large sporangia, 100–350 μm diameter, with thin walls (3–5 μm) that enclose numerous sporangiospores, 6–10 μm diameter
Pathogens that occur in tissue as hyaline hyphae		
Absidia corymbifera, Basidiobolus ranarum, Conidiobolus coronatus, Mucor ramosissimus, Rhizomucor pusillus, Rhizopus arrhizus, and others	Zygomycosis	Broad, thin-walled, aseptate or infrequently septate hyphae, 6–25 μm wide, with non-parallel contours and random branches. Sporangia sometimes formed in cavitary lesions with access to air
Aspergillus fumigatus, A. flavus, A. niger, and other *Aspergillus* spp.	Aspergillosis	Septate, dichotomously branched hyphae of uniform width (3–6 μm); conidial heads sometimes formed in cavitary lesions with access to air. Calcium oxalate crystals may be present in case of *A. niger*
Pathogenic members of the genera *Acremonium, Beauvaria, Cylindrocarpon, Fusarium, Paecilomyces, Phialomonium, Scedosporium, Scopulariopsis,* and others	Hyalohyphomycosis	Septate, branched, hyaline hyphae, 2–8 μm wide, which usually have regular contours and dichotomous random branches
Malassezia furfur	Pityriasis versicolor	Short, curved, and bent hyaline hyphae, 2–4 μm wide; and clusters of spherical to oval, thick-walled phialoconidia, 3–8 μm diameter
Pathogenic members of the genera *Epidermophyton, Microsporum,* and *Trichophyton*	Dermatophytosis	Branched, septate hyphae that break up into chains of arthroconidia
Pseudallescheria boydii (anamorph of *Scedosporium apiospermum*)	Pseudallescheriosis (scedosporidiosis)	Septate, randomly branched hyphae, 2–5 μm wide; conidia of scedosporium type may be formed in cavitary lesions
Pigmented (pheoid) nonmycetomatous pathogens		
Phaeoannellomyces werneckii, Stenella araguata	Tinea nigra	Pheoid, branched, septate hyphae, 1–3 μm wide; and elongated budding cells, 1–5 μm diameter
Cladophialophora carrionii, Fonsecaea compacta, F. pedrosoi, Phialophora verrucosa, and *Rhinocladiella aquaspersa*	Chromoblastomycosis	Large, 6–12 μm diameter, spherical to polyhedral, thick-walled, dark-brown, muriform cells (sclerotic bodies) with septations along one or two planes; pheoid hyphae sometimes present
Bipolaris spicifera, Cladophialophora bantiana, Exophiala jeanselmei, Phialophora parasitica, and other pheoid fungi	Pheohyphomycosis	Pheoid (brown) hyphae, 2–6 μm wide, which may be branched and are often constricted at their frequent and prominent septations; pheoid yeast forms, and chlamydoconidia sometimes present
Mycetomatous pathogens		
Acremonium spp., *Aspergillus nidulans, Curvularia geniculata, Exophiala jeanselmei, Leptosphaeria senegalensis, Madurella grisea, M. mycetomatis, Neotestudina rosatii, Pseudallescheria boydii, Pyrenochaeta romeroi,* and others	Mycetoma (eumycotic) or eumycetoma	Granules, 0.2 mm to several millimeters in diameter, composed of broad (2–6 μm), hyaline or phaeoid, septate hyphae that often branch, and form chlamydoconidia
Actinomadura madurae, A. pelletieri, Nocardia spp., *Nocardiopsis dassonvillei, Streptomyces somaliensis,* and others	Mycetoma (actinomycotic) or actinomycetoma	Granules, 0.1 mm to several millimeters diameter, composed of delicate gram-positive filaments, 1 μm wide, which are often branched, and beaded
Filamentous bacterial pathogens		
Actinomyces israelii, A. meyeri, A. naeslundii, A. odontolyticus, A. viscosus, Arachnia propionica, Rothia dentocariosa, and others	Actinomycosis	Organized aggregates (granules) composed of delicate, branched, gram-positive filaments, 1 μm wide, that are often beaded; entire granules are 30–3000 μm diameter
Nocardia asteroides, N. brasiliensis, and *N. otitidiscaviarum*	Nocardiosis	Delicate, branched, individual filaments, 1 μm wide, that are gram-positive, weakly acid-fast, and often beaded

exists, there are no absolute histological criteria that permit an etiologic diagnosis based only on host response. In addition, inflammatory patterns are often atypical and unreliable in immunodeficient patients. A fungus must first be detected and then identified in tissue sections before a definitive histopathological diagnosis can be made. Knowledge of the patient's travel history may help resolve differential diagnostic problems because the geographic distribution of some pathogenic fungi is limited (Warnock et al. 1998).

DIAGNOSTIC APPLICATION OF IMMUNOHISTOCHEMICAL TECHNIQUES

The prerequisite for all immunohistochemical staining systems is a primary antibody properly characterized especially in terms of specificity under optimal conditions (Jensen 1994). The technique used for obtaining an immunohistochemical diagnosis of mycoses may be either direct (conjugated primary antibodies) or indirect (conjugated secondary antibodies or tertiary enzyme complexes). In accordance with other immunohistochemical staining systems, the reaction complexes may be vizualized with the help of fluorochromes [e.g. fluorescein isothiocyanate (FITC) and tetra-methylrhodamine isothiocyanate isomer R (TRITC)], gold-silver complexes, or complexes of enzymes [e.g. peroxidase–antiperoxidase (PAP) techniques, and alkaline phosphatase-antialkaline phosphatase (APAAP) techniques] (El Nageeb and Hay 1981; Kaufman 1992; Jensen 1994). Also, avidin–biotin enzyme complex (ABC) methods may be used with horseradish peroxidase (HRP) and galactosidase (Jensen et al. 1996). The major advantage of using enzyme immunohistochemistry compared with immunofluorescence techniques is that permanent sections are provided, specialized microscopic equipment is not needed, and pathological reactions may be assessed simultaneously during the evaluation of the immunoreactivity (Jensen 1994; Jensen et al. 1996). Application of immunohistochemical techniques may, in several cases, be the only means of establishing an accurate etiologic diagnosis in fixed tissue sections because of morphological similarities among the tissue forms of several fungal genera (e.g. aspergillosis, fusariosis, and scedosporiosis), when atypical forms of a fungus are present, or when fungal elements are sparse (Jensen et al. 1996). A range of different forms of fungal elements are frequently observed within lesions, especially when more than one organ is studied, and it must be ascertained whether these elements belong to a single or more taxa. In such cases, dual immunostaining techniques are useful tools for obtaining a reliable diagnosis (Jensen et al. 1996). In the case of *Cryptococcus neoformans*, the combination of reactions obtained by different primary antibodies is used to differentiate between serotypes (Krockenberger et al. 2001)

Apart from tissue sections, immunohistochemical techniques can also be used to identify fungi in smears of lesional exudates, bronchial washings, bronchoalveolar lavage fluid, blood, bone marrow, cerebrospinal fluid, and in sputum specimens that have been enzymatically or chemically digested.

An important limitation to the widespread application of immunohistochemical techniques and their use in the routine diagnosis of mycoses lies in the fact that sensitive and specific reagents are usually derived from multiple heterologously adsorbed polyclonal antisera, which are not commercially available (Jensen et al. 1996). However, in recent years more specific monoclonal antibodies have been developed, some of which can be obtained through companies offering diagnostic reagents (Jensen et al. 1997). Moreover, in addition the development of specific DNA probes for in situ hybridization techniques may facilitate the access to specific fungal identification (Jensen et al. 1996). In Table 6.6, fungi and related pathogens, which can be identified immunohistochemically, are listed.

Before immunostaining, localization and presumptive identification of fungal elements in GMS-stained tissue sections enable the pathologist to narrow the etiologic

Table 6.6 *Fungal and related pathogens that can be identified immunohistochemically in deparaffinized sections of formalin-fixed tissue*

Fungal and related pathogens
Actinomycetes
Actinomyces israelii
Actinomyces naeslundii
Actinomyces viscosus
Arachnia propionica
Fungi
Aspergillus to the genus level or to the species level, i.e. *A.fumigatus, A. flavus,* and *A. niger*
Blastomyces dermatitidis
Candida to the genus level or to the species level, i.e. *C. albicans, C. glabrata, C. tropicalis*
Coccidioides immitis
Cryptococcus neoformans to the genus level or to the level of varieties and serotypes, i.e. *C. neoformans* var. *neoformans* (A, D, and AD) and *C. neoformans* var. *gatti* (B and C), respectively
Fusarium spp.
Geotrichum candidum
Histoplasma capsulatum (tissue forms of capsulatum and duboisii varieties)
Paracoccidioides brasiliensis
Penicillium marneffei
Pneumocystis carinii
Pseudallescheria boydii and *Scedosporium* spp.
Sporothrix schenckii var. *schenckii*
Trichosporon beigelii and *Trichosporon capitatum* (*Blastoschizomyces capitatus*)
Zygomycotic fungi (Order Mucorales) to the level of genus, i.e. *Absidia, Rhizopus,* and *Rhizomucor*
Algae
Prototheca wickerhamii
Prototheca zopfii

Table 6.7 *Special terminology*

Term	Definition
Anamorph	A somatic or reproductive structure of the asexual cycle
Adiaconidium	Asexual conidium (spore) that enlarges after formation in vitro or implantation in vivo; adiaconidia do not reproduce in vivo
Arthroconidium	Asexual conidium (spore) formed by mycelial disarticulation
Asteroid body	Radially oriented Splendore–Hoeppli material that sometimes surrounds the yeast-like cells of *Sporothrix schenckii* or other fungi in tissue
Bud (blastoconidium)	Variety of conidium produced by lateral outgrowth from a parent cell; buds may be single or multiple
Budding	Asexual reproductive process characteristic of unicellular fungi or conidia in which a lateral outgrowth from the parent cell is pinched off to form a new cell
Capsule	Hyaline (colorless) mucopolysaccharide coat external to the wall of a fungal cell or conidium
Chlamydoconidium	Thick-walled, rounded, resistant conidium formed by direct differentiation of the mycelium
Coenocytic	Nonseptate
Conidiophore	Specialized hypha that produces and bears conidia
Conidium (plural, conidia)	Asexual propagule formed on, but easily detached from, a hypha, conidiophore, or other parent cell
Dematiaceous	Naturally pigmented, usually brown or black; term considered to be obsolete; replacement by the term pheoid advocated
Dichotomous	Equal branching of hyphae
Dimorphic	Having two forms; growth as hyphae in vitro at 25°C and as budding yeasts, muriform cells, or spherules in infected tissues or in vitro at 37°C on special media
Endospore	Asexual spore formed within a closed structure such as a spherule
Endosporulation	Process of producing endospores
'Fruiting body'	Imprecise term for conidia-bearing organs produced by fungi
Germ tube	Tube-like process, produced by a germinating conidium, which eventually develops into a hypha
Granule (grain)	Mass of organized mycelium that may be embedded in a cement-like matrix; formed in actinomycosis and in mycetomas caused by fungi and aerobic actinomycetes
Hyaline	Colorless
Hypha (plural, hyphae)	Filament that forms the thallus or body of most fungi
Intercalary	Between two fungal cells
Muriform	Having horizontal and vertical septations or cross-walls
Mycelium (pl. mycelia)	Mass of intertwined and branched hyphae
Perfect state	Developmental state of a fungus when sexual reproduction takes place and sexual spores are produced
Pheoid	Dark, dusky; term applied to fungal cell walls darkened by melanins
Phialide	A conidiogenous cell that successively produces and extrudes conidia from within
Pseudohypha	Short hyphal-like filament produced by successive yeast buds that elongate and fail to separate
Schizogony	Process of multiplication of yeasts by the splitting and separation of a septum that permits adjacent cells to be set free
Septate	Having cross-walls
Septum (pl. septa)	Cross-wall of a mycelial filament or conidium
Spherule	Closed, thick-walled, spherical structure within which asexual endospores are produced by progressive cytoplasmic cleavage
Splendore–Hoeppli material	Eosinophilic, refractile, homogeneous and often radially oriented material sometimes found around fungi in tissue sections; represents a localized antigen–antibody reaction in a hypersensitized host
Sporangium (pl. sporangia)	A closed structure within which asexual spores (sporangiospores) are produced by cytoplasmic cleavage
Sterigma (pl. sterigmata)	Short or elongate specialized projection of a conidiophore on which conidia are developed (see Phialide)
Yeast	Spherical to oval unicellular fungus that reproduces by a budding or fission process

possibilities, so that the most appropriate panel of immunoreagents can be employed. Depending on the nature of reactive fungal epitopes, pretreatment with enzymes (e.g. trypsin and pepsin) and addition of a substance such as dimethyl sulfoxide may enhance the reactivity of tests on formalin-fixed tissues. A number of appropriate control procedures are essential in the evaluation of immunohistochemical diagnostic assays, e.g. replacement of antifungal antibodies by antibodies raised against irrelevant antigens and including both positive and negative control sections with each batch of slides (Jensen et al. 1996).

As the somatic and cell wall antigens of most pathogenic fungi survive formalin fixation and processing into paraffin quite well, immunohistochemical and in situ techniques can be performed retrospectively on archival material. Unstained sections, paraffin blocks, or formalin-fixed 'wet' tissue can be sent to specialized reference centers for immunohistopathology. If the number of sections available is limited, those already stained with H&E, Giesma or tissue Gram's stains can be decolorized in acid–alcohol and reused for immunohistochemistry. If immunofluorescence is used, as much as possible of the eosin should be removed because this stain autofluoresces. In cases where the antigenic epitope is a polysaccharide, sections previously stained with GMS, PAS, or GF procedures are not suitable for immunohistochemistry, because oxidation of the cell wall polysaccharides irreparably alters the antigenicity of the fungal cells.

SPECIAL TERMINOLOGY

As medical mycology and the histopathology of mycotic diseases are highly specialized disciplines, the following selected terms, used to describe the appearance of fungi and related pathogens in tissue sections, are defined as shown in Table 6.7.

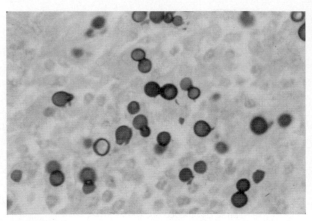

Figure 6.2 *Pulmonary blastomycosis in an immunocompromised patient. Numerous yeast-like cells of* Blastomyces dermatitidis *fill alveolar spaces. Note the diagnostic broad-based budding, thick cell walls, and variation in size of fungal cells. GMS; magnification × 360*

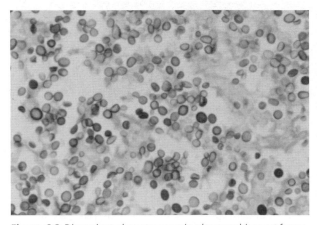

Figure 6.3 *Disseminated cryptococcosis: pleomorphic yeast forms of* Cryptococcus neoformans *are predominantly extracellular, bud by a relatively narrow base, range from 2 to 20 μm in diameter, and are surrounded by unstained (optically clear) capsules. A multiple-budding cell is seen at left center. GMS; magnification × 225*

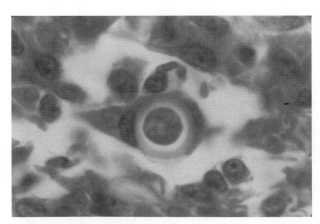

Figure 6.1 *Pulmonary blastomycosis: a multinucleated giant cell contains a single yeast form of* Blastomyces dermatitidis. *In tissue, the cells of this fungus are yeast-like, spherical to oval, 8–15 μm in diameter, and have thick 'doubly contoured' walls and single buds attached to parent cells by a broad base. Note that the fungal cell is multinucleate. H&E; magnification × 1 125*

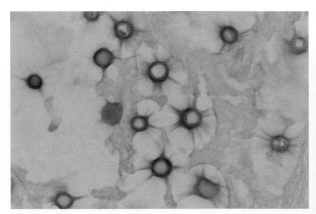

Figure 6.4 *Cerebromeningeal cryptococcosis: yeast-like cells of* Cryptococcus neoformans *distend the leptomeninges, elicit little or no inflammation, and are surrounded by wide, carminophilic capsules that have a crenated or spinous appearance because of uneven shrinkage during tissue processing. Mayer's mucicarmine; magnification × 563*

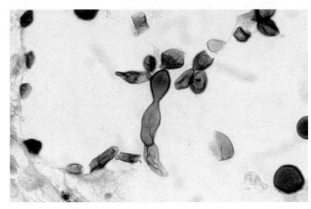

Figure 6.5 *Hepatic cryptococcosis: the cells of* Cryptococcus neoformans *may sometimes form hyphae and pseudohyphae that may be confused with elements of candidosis. GMS; magnification* × *360*

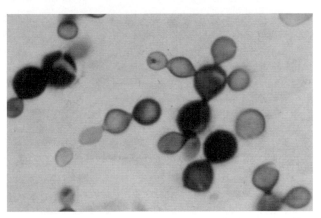

Figure 6.8 *Lobomycosis: spherical to oval and budding yeast-like cells of* Loboa loboi, *5–12 μm in diameter, are attached in chains. Budding cells are often connected by a narrow, tube-like isthmus. GMS; magnification* × *1 125*

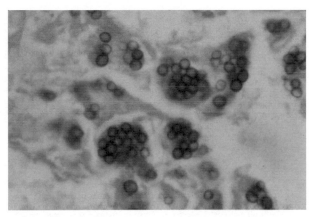

Figure 6.6 *Disseminated histoplasmosis involving the adrenal gland. Numerous spherical to oval yeast forms of* Histoplasma capsulatum *var.* capsulatum, *2–4 μm in diameter, have single buds attached to parent cells by a relatively narrow base, and are clustered within histiocytes. GMS; magnification* × *1 125*

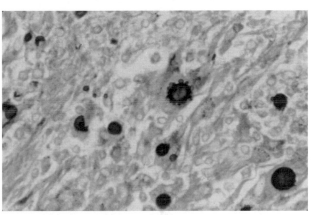

Figure 6.9 *Pulmonary paracoccidioidomycosis. A granuloma contains diagnostic, multiple-budding ('steering wheel') yeast forms of* Paracoccidioides brasiliensis *(right center), as well as nonbudding and single-budding cells that range from 10 to 60 μm in diameter. Note the small detached blastoconidia. GMS; magnification* × *225*

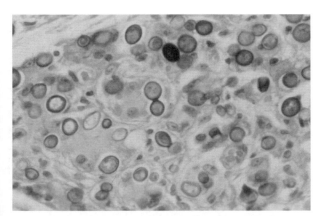

Figure 6.7 *Cutaneous lesion of African histoplasmosis. Histiocytes and multinucleated giant cells contain numerous yeast-like cells of* Histoplasma capsulatum *var.* duboisii, *8–15 μm in diameter. Note the classic 'double-cell' yeast form (top center) with narrow-based bud and thick cell walls. GMS–H&E; magnification* × *360*

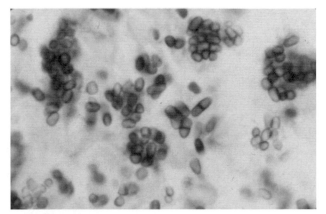

Figure 6.10 *Disseminated* Penicillium marneffei: *hepatic granuloma contains individual and clustered, spherical to oval yeast-like cells of* Penicillium marneffei, *2.5–5.0 μm in maximum diameter. Note the diagnostic yeast forms with single, wide, transverse septa. GMS; magnification* × *1 125*

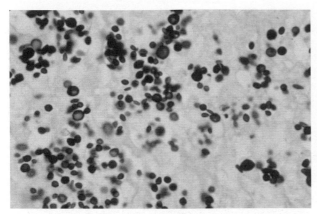

Figure 6.11 *Disseminated sporotrichosis: pleomorphic spherical and elongated (cigar-shaped) yeast-like cells of* Sporothrix schenckii *are located in a testicular granuloma. GMS; magnification × 1 125*

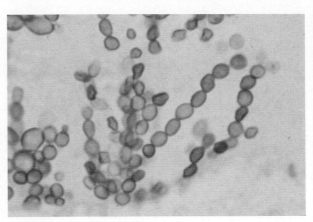

Figure 6.14 *Disseminated candidosis: numerous yeast-like cells and mycelial elements of* Candida albicans *occupy a nodular infarct in the lung of a granulocytopenic patient. GMS; magnification × 563*

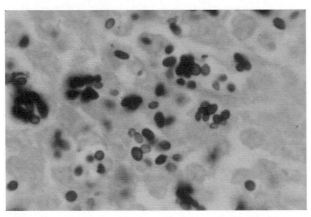

Figure 6.12 *Disseminated candidosis. Individual and clustered yeast-like cells of* Candida glabrata *are spherical to oval, 2–5 µm in diameter, have single buds attached to parent cells by a relatively narrow base, and resemble those of* Histoplasma capsulatum var. capsulatum. *GMS; magnification × 1 125*

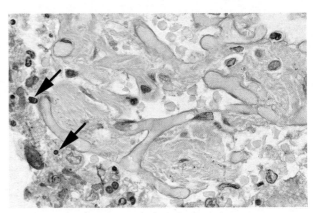

Figure 6.15 *Pulmonary candidosis: capillaries may sometimes be mistaken for fungal hyphae. However, by paying special attention, endothelial cells usually can be identified. In the present case, only a few PAS-positive* Candida glabrata *cells were demonstrated (arrows). PAS; magnification × 360*

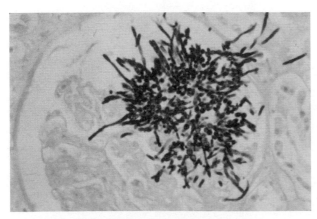

Figure 6.13 *Renal candidosis caused by* Candida albicans. *Blastoconidia (budding yeast-like cells), pseudohyphae and true hyphae proliferate in a radial pattern, and efface the normal renal architecture. GMS; magnification × 225*

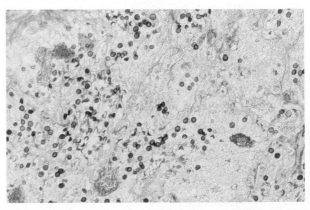

Figure 6.16 *Pulmonary pneumocystosis.* Pneumocystis carinii – *'cysts' and 'trophozoites' are seen as round, distorted, or crescent structures from 3 to 8 µm in diameter. GMS; magnification × 360*

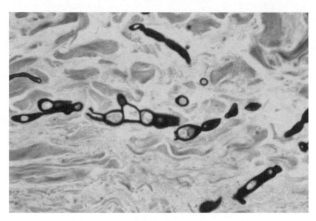

Figure 6.17 *Cutaneous geotrichosis: individual hyphae of* Geotrichum candidum *have non-parallel contours and prominent septa, and branch at acute angles. Spherical yeast-like cells and rectangular arthroconidia, 4–10 µm wide, with rounded or squared ends are also produced by this fungus. GMS; magnification × 360*

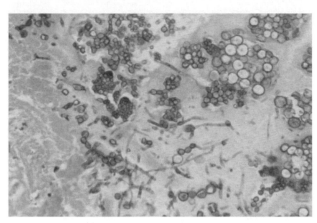

Figure 6.20 *Miliary pulmonary coccidioidomycosis: necrotic nodule contains characteristic endosporulating spherules, immature spherules, hyphae and arthroconidia (lower centre) of* Coccidioides immitis. *GMS–H&E; magnification × 113*

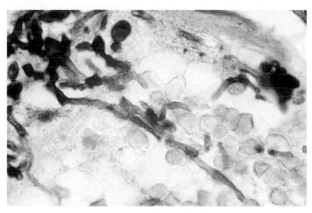

Figure 6.18 *Disseminated trichosporonosis: a splenic infarct contains radiating, septate hyphae and pleomorphic yeast-like cells of* Trichosporon beigelii. *GMS; magnification × 360*

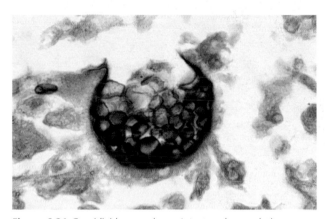

Figure 6.21 *Coccidioidomycosis: an intact endosporulating spherule of* Coccidioides immitis *is located in a pulmonary granuloma. GMS–H&E; magnification × 563*

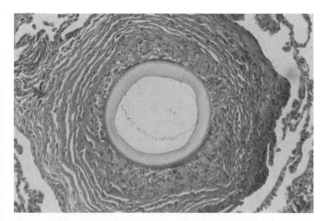

Figure 6.19 *Adiaspiromycosis: a large, spherical, thick-walled adiaconidium of* Chrysosporium parvum var. crescens *is surrounded by a fibrogranulomatous reaction in a lung biopsy specimen. The interior of the conidium is empty. Budding and endosporulation do not occur. H&E; magnification × 56*

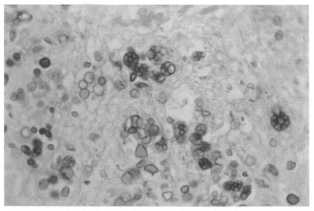

Figure 6.22 *Protothecal olecranon bursitis: a necrotizing granuloma contains numerous single and endosporulating sporangia of* Prototheca wickerhamii. *Note the characteristic 'morula' forms. Gridley; magnification × 225*

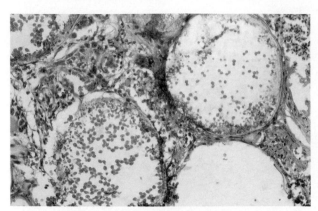

Figure 6.23 *Rhinosporidiosis: two mature sporangia of* Rhinosporidium seeberi *with thin walls are located in the inflamed stroma of a nasal polyp. Note typical maturation and zonation of sporangiospores. H&E; magnification × 113*

Figure 6.26 *Disseminated zygomycosis involving the myocardium. When hyphae with uniform wide, parallel walls, and dichotomous branching are formed, the identification may be confused with aspergillosis. PAS; magnification × 563*

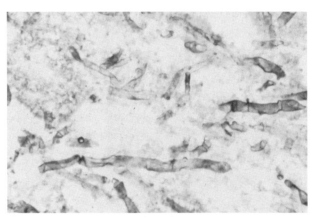

Figure 6.24 *Disseminated zygomycosis involving the myocardium. Typical hyphae of the mucoraceous zygomycetes are 6–25 μm wide, thin walled, infrequently septate, and have nonparallel contours and random branches. Hyphae of the zygomycetes are weakly stained with GMS procedure. GMS; magnification × 360*

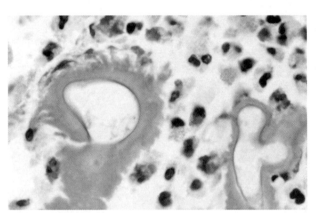

Figure 6.27 *Zygomycotic enteritis. Around hyphae, huge amounts of eosinophilic, homogenous and radially radiating Splendore-Hoeppli material has formed. H&E; magnification × 1 125*

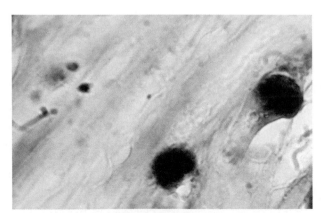

Figure 6.25 *Pulmonary zygomycosis caused by* Absidia corymbifera. *In tissues with access to air, sporangia are sometimes produced. If sectioned optimally, the sporangia may be useful for the identification of the fungus. H&E; magnification × 360*

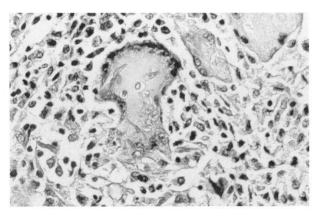

Figure 6.28 *Chronic zygomycotic lymphadenitis. Within the giant cells only sparse fungal elements are seen. The identification of elements as belonging to the zygomycetes was confirmed by immunohistochemistry. (Compare with Figure 6.59.) H&E; magnification × 360*

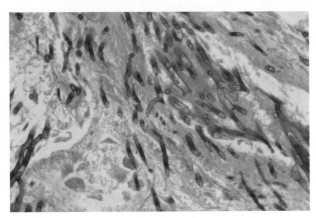

Figure 6.29 *Invasive pulmonary aspergillosis: hyphae of the Aspergillus spp. are septate, uniform in width (3–6 μm), have parallel contours, and branch progressively and dichotomously. The aspergilli are readily demonstrated with GMS and other special histological stains for fungi. GMS; magnification × 360*

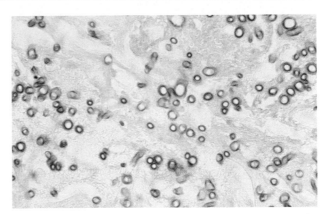

Figure 6.32 *Pulmonary aspergillosis. In cross-section, hyphae of Aspergillus spp. may be confused with yeast cells. However, the elements may easily be identified as Aspergillus spp. by immunohistochemistry. (Compare with Figure 6.55.) GMS; magnification × 563*

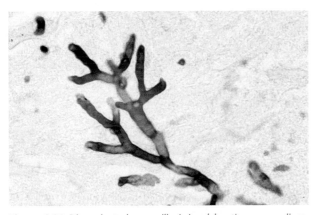

Figure 6.30 *Disseminated aspergillosis involving the myocardium. The hyphae present are typical of Aspergillus spp. They are frequently septate, have uniform widths (3-6 μm), have parallel contours, and branch dichotomously. GMS; magnification × 563*

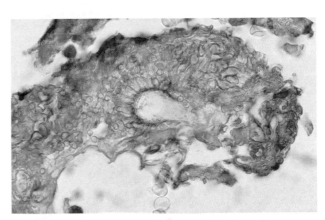

Figure 6.33 *Pulmonary aspergillosis caused by* Aspergillus fumigatus. *In tissues with access to air, conidial heads are sometimes produced. If the conidial heads are cut optimally they are useful for the identification of the fungus. H&E; magnification × 563*

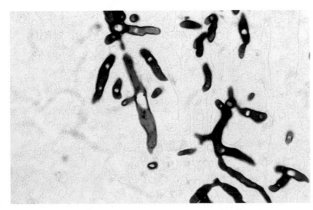

Figure 6.31 *Cerebral aspergillosis caused by* Aspergillus fumigatus. *Hyphae of* Aspergillus *spp. may adapt to highly bizarre forms in chronic lesions, which may be confused with hyphae of zygomycotic fungi. GMS; magnification × 360*

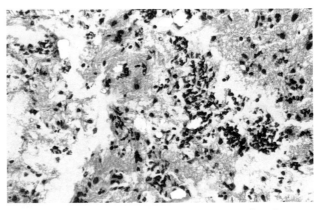

Figure 6.34 *Cerebral aspergillosis caused by* Aspergillus niger. *Deposits of birefringent calcium oxalate crystals are sometimes observed in lesions caused by* Aspergillus niger. *H&E and polarized light; magnification × 225*

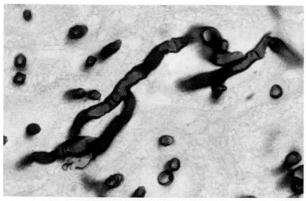

Figure 6.35 *Renal fusariosis caused by* Fusarium solani. *Hyphae in cross-sections and longitudinal sections are seen. Hyphae are septate, 3–8 μm wide and branch at dichotomous angles. Often fusariosis cannot be differentiated from aspergillosis and scedosporiosis. GMS; magnification × 563*

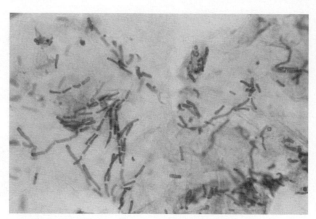

Figure 6.38 *Pityriasis versicolor: in skin scrapings,* Malassezia spp. *appear as clusters of single and budding yeast-like cells, 3–8 μm in diameter, mixed with short, narrow and sometimes curved hyphal fragments, 2.5–4 μm wide, that are oriented end to end. PAS; magnification × 225*

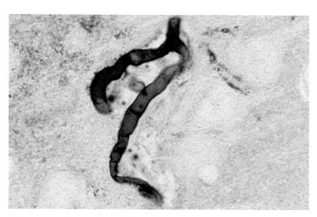

Figure 6.36 *Cerebral scedosporiosis caused by* Scedosporium apiospermum. *Hyphae of* Scedosporium spp. *are septate, 3–8 μm wide and branch at dichotomous angles. Often scedosporiosis cannot be differentiated from aspergillosis and fusariosis. GMS; magnification × 563*

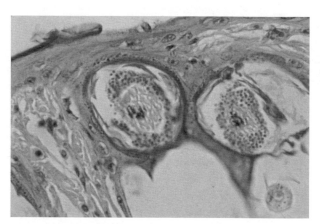

Figure 6.39 *Dermatophytosis caused by* Microsporum canis. *Hematoxylinophilic arthroconidia within two hair follicles surround partially degenerated hairs. H&E; magnification × 360*

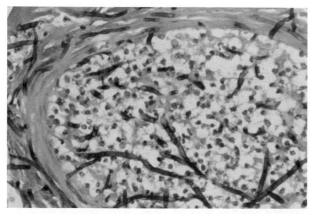

Figure 6.37 *Cutaneous hyalohyphomycosis caused by* Fusarium moniliforme *in a burned patient. Hyphae of the* Fusarium spp. *are septate, 3–8 μm wide and usually branch at right angles. Hyphal branches are often constricted at their origins from parent hyphae. GMS–H&E; magnification × 225*

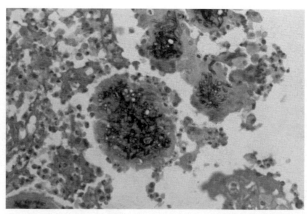

Figure 6.40 *Deep dermatophytic granuloma (pseudo-mycetoma) caused by* Microsporum canis. *The deep dermis contains two compact aggregates (pseudogranules) of distorted dermatophytic hyphae ensheathed by eosinophilic, Splendore–Hoeppli material. GMS–H&E; magnification × 225*

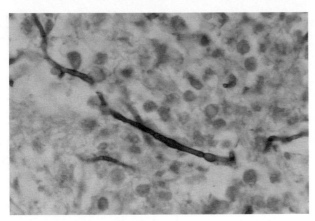

Figure 6.41 *Disseminated hyalohyphomycosis: a cerebral infarct contains narrow (2–5 μm), septate, randomly branched hyphae of* Pseudallescheria boydii. *GMS; magnification × 563*

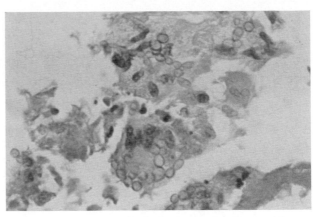

Figure 6.44 *Systemic (cerebral) pheohyphomycosis caused by* Cladophialophora bantiana. *Multinucleate giant cells in the wall of a cystic granuloma contain numerous pheoid yeast forms and hyphae that are constricted at their prominent septations. H&E; magnification × 563*

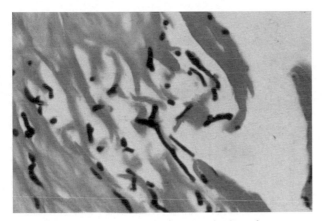

Figure 6.42 *Tinea nigra. branched, septate hyphae of* Phaeoannellomyces werneckii, *1–3 μm wide, are present in the hyperkeratotic layer of the epidermis. In H&E-stained tissue sections, the fungal elements are distinctly pheoid (naturally pigmented). GMS–H&E; magnification × 563*

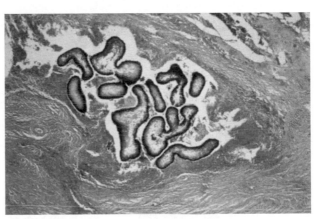

Figure 6.45 *Eumycotic black grain mycetoma caused by* Curvularia geniculata. *Several pheoid, lobulated grains (granules) are embedded in an abscess which is encapsulated by dense fibrous connective tissue. H&E; magnification × 45*

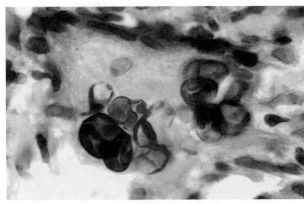

Figure 6.43 *Chromoblastomycosis caused by* Phialophora verrucosa. *Phaeoid muriform cells (sclerotic bodies) that divide by septation in one or more planes are located within multinucleated giant cells in a dermal granuloma. H&E; magnification × 1125*

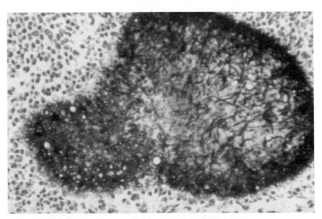

Figure 6.46 *Eumycotic white grain mycetoma caused by* Aspergillus nidulans. *The grain (granule) is composed of compact, radially oriented hyphae and vesicular chlamydoconidia. GMS–H&E; magnification × 113*

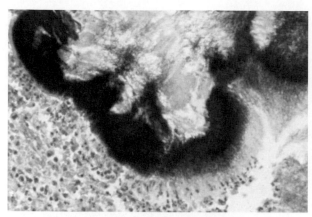

Figure 6.47 *Actinomycotic mycetoma caused by* Actinomadura pelletieri. *A deeply hematoxylinophilic, lobulated granule is embedded in a subcutaneous abscess. The granule is composed of delicate, compact, wavy and randomly oriented bacterial filaments, about 1 μm wide, that are often branched and beaded. H&E; magnification × 50*

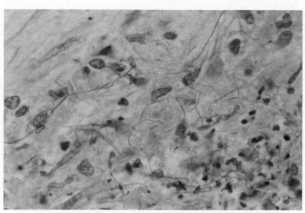

Figure 6.50 *Disseminated nocardiosis involving the cerebrum. Delicate, weakly acid-fast filaments of* Nocardia asteroides, *about 1 μm in width, branch at predominantly right angles. Modified Ziehl–Neelsen; magnification × 563*

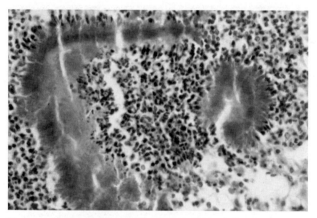

Figure 6.48 *Thoracic actinomycosis caused by* Actinomyces israelii. *A pulmonary abscess contains two granules that are bordered by intensely eosinophilic, club-like, Splendore–Hoeppli material. Individual filaments within the granules are not visible with the H&E stain. H&E; magnification × 225*

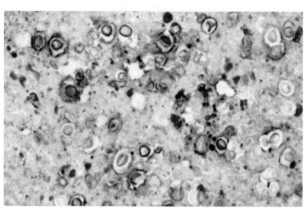

Figure 6.51 *Cerebral cryptococcosis. By the application of a serotype (A + D) specific primary monoclonal antibody in an avidin–biotin–horseradish peroxidase (HRP) technique, the etiology was found to be* Cryptococcus neoformans *var. gatti. HRP; magnification × 563*

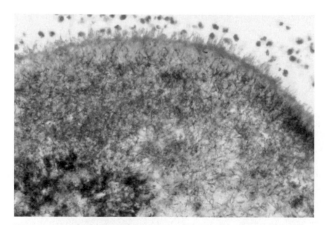

Figure 6.49 *Abdominal actinomycosis: the granule is composed of randomly oriented, gram-positive, branched and beaded bacterial filaments, about 1 μm in width, embedded in a matrix or ground substance of uncertain composition. The peripheral Splendore–Hoeppli material is gram negative. Brown and Brenn; magnification × 225*

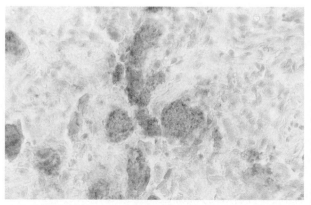

Figure 6.52 *Pulmonary pneumocystosis.* Pneumocystis carinii *within alveolar space was demonstrated immunohistochemically by the use of a specific monoclonal antibody in an APAAP technique. APAAP; magnification × 563*

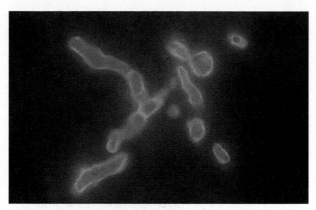

Figure 6.53 *Pulmonary aspergillosis caused by* Aspergillus fumigatus. *The hyphae were stained by an indirect immunofluorescence (IIF) technique using FITC as fluorochrome. IIF; magnification × 1 125*

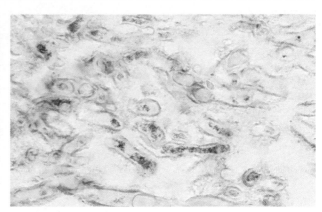

Figure 6.56 *Pulmonary aspergillosis. Hyphae of* Aspergillus fumigatus *may also be identified by in situ hybridization. In the present case, an alkaline phosphatase-conjugated DNA probe hybridizing with genomic DNA and mRNA responsible for the 33 kDa protease was used. In the technique, Fast red was the substrate. Magnification × 1 125*

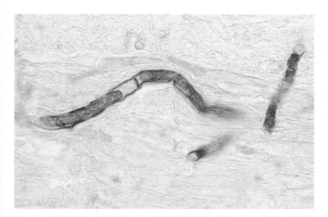

Figure 6.54 *Myocardial aspergillosis. The hyphae were identified immunohistochemically by using a primary genus specific monoclonal antibody raised towards galactomannan. APAAP technique; magnification × 1 125*

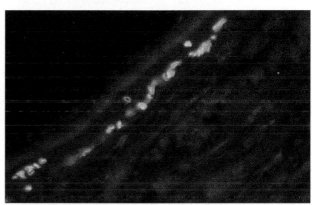

Figure 6.57 *Corneal fusariosis. The hyphae within a corneal biopsy were identified by the application of heterologously adsorbed polyclonal antibodies in an IIF technique using FITC as fluorochrome. The primary antibody was raised against somatic antigens of* Fusarium solani. *IIF; magnification × 1 125*

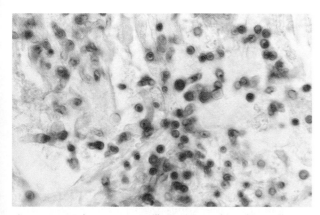

Figure 6.55 *Pulmonary aspergillosis. Immunohistochemically, cross-sectioned hyphae of* Aspergillus spp. *may be identified by using a genus specific primary monoclonal antibody in an APAAP technique. (Compare with Figure 6.32.) APAAP; magnification × 563*

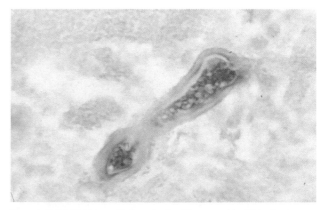

Figure 6.58 *Zygomycotic lymphadenitis caused by* Rhizopus arrhizus. *When using a monoclonal antibody raised somatic antigens as the primary antibody in an APAAP technique, the hyphal walls remain unstained. APAAP; magnification × 1 125*

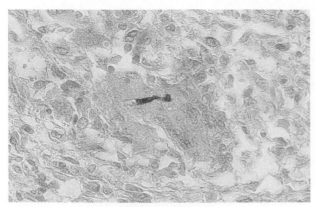

Figure 6.59 *Chronic zygomycotic lymphadenitis. A single fungal fragment within a giant cell may be identified immunohistochemically by using a panel of specific primary antibodies. (Compare with Figure 6.28.) In the present case the agent was identified as a zygomycotic fungus due to the reactivity with a monoclonal antibody raised towards somatic antigens of Rhizopus arrhizus. The primary monoclonal antibody was used in a PAP technique. PAP; magnification × 360*

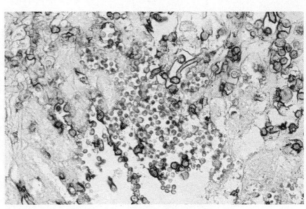

Figure 6.61 *Concomitant intestinal infection due to* Candida albicans *and* Candida glabrata. *The elements of* Candida glabrata *were immunostained bluish by an indirect ABC method using galactosidase as an enzyme and 5-bromo-4-chloro-indolyl-galactosidase as a substrate. The* Candida albicans *elements were stained brownish by a PAP technique using 3,3́-diaminobenzidine tetrahydrochloride as substrate. ABC and PAP; magnification × 360*

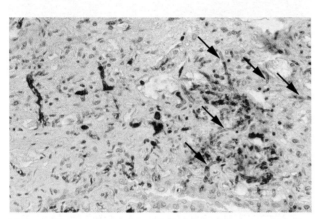

Figure 6.60 *Concomitant enteral candidosis and zygomycosis. The* Candida *hyphae (arrows) were immunostained brown (substrate = 3-amino-9-ethylcarbazole) by a genus-specific polyclonal rabbit antibody, whereas the zygomycotic hyphae were immunostained green (substrate = dioctyl sulfosuccinate sodium and 3,3́,5,5́-tetramethylbenzidine) by a murine monoclonal antibody. Both antibodies were used in a PAP technique. PAP; magnification × 360*

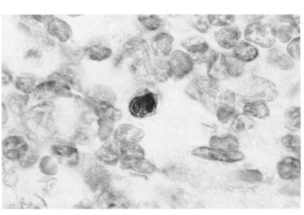

Figure 6.62 *Protothecal lymphadenitis. In tissues the cells of* Prototheca zopfii *can be identified immunohistochemically by using primary heterologously adsorbed polyclonal antibodies in a PAP technique. PAP; magnification × 563*

REFERENCES

Anthony, P.P. 1973. A guide to the histological identification of fungi in tissues. *J Clin Pathol*, **26**, 828–31.

Bancroft, J.D. and Stevens, A. 1996. *Theory and practice of histopathological techniques*. New York: Churchill Livingstone.

Baker, R.D. (ed.) 1971. *Human infection with fungi, actinomycetes and algae*. Berlin: Springer-Verlag.

Chandler, F.W. and Watts, J.C. 1987. *Pathologic diagnosis of fungal infections*. Chicago, IL: ASCP Press.

Chandler, F.W., Kaplan, W. and Ajello, L. 1980. *Colour atlas and text of the histopathology of mycotic diseases*. Chicago IL: Year Book Medical Publishers.

Dixon, D.M. and Polak, A. 1991. The medically important dematiaceous fungi and their identification. *Mycoses*, **34**, 1–18.

Dixon, D.M., Szaniszlo, P.J. and Polak, A. 1991. Dihydroxynaphthalene (DHN) melanin and its relationship with virulence in the early stages of phaeohyphomycosis. In: Cole, G.T. and Hoch, H.C. (eds), *The fungal spore and disease initiation in plants and animals*. New York: Plenum Press, 297–318.

El Nageeb, S. and Hay, R.J. 1981. Immunoperoxidase staining in the recognition of *Aspergillus* infections. *Histopathology*, **5**, 437–44.

Elias, J.M. 1982. *Principles and techniques in diagnostic histopathology*. Park Ridge, NJ: Noyes Publications.

Graham, A.R. 1983. Fungal autofluorescence with ultraviolet illumination. *Am J Clin Pathol*, **79**, 231–4.

Hoog, G.S. and de Guého, E. 1985. A plea for the preservation of opportunistic fungal isolates. *Diagn Microbiol Infect Dis*, **3**, 369–72.

Jensen, H.E. 1994. Systemic bovine aspergillosis and zygomycosis in Denmark with reference to pathogenesis, pathology, and diagnosis. *APMIS*, **42(Suppl 102)**, 1–48.

Jensen, H.E. and Schønheyder, H. 1989. Immunofluorescence staining of hyphae in the histopathological diagnosis of mycoses in cattle. *J Med Vet Mycol*, **27**, 33–44.

Jensen, H.E., Schønheyder, H., et al. 1996. Diagnosis of systemic mycoses by specific immunohistochemical tests. *APMIS*, **104**, 241–258.

Jensen, H.E., Salonen, J. and Ekfors, T.O. 1997. The use of immunohistochemistry to improve sensitivity and specificity in the diagnosis of systemic mycoses in patients with haematological malignancies. *J Pathol*, **181**, 100–5.

Kaufman, L. 1992. Immunohistochemical diagnosis of systemic mycoses: an update. *Eur J Epidemiol*, **8**, 377–82.

Krockenberger, M.B. and Canfield, P.J. 2001. An immunohistochemical method that differentiates *Cryptococcus neoformans* varieties and serotypes in formalin-fixed paraffin-embedded tissues. *Med Mycol*, **39**, 523–33.

Kwon-Chung, K.J., Hill, W.B. and Bennett, J.E. 1981. New, special stain for histopathological diagnosis of cryptococcosis. *J Clin Microbiol*, **13**, 383–7.

Matsumoto, T., Ajello, L., et al. 1994. Developments in hyalohyphomycosis and phaeohyphomycosis. *J Med Vet Mycol*, **32**, Suppl 1, 329–49.

Monheit, J.E., Cowan, D.F. and Moore, D.G. 1984. Rapid detection of fungi in tissues using calcofluor white and fluorescence microscopy. *Arch Pathol Lab Med*, **108**, 616–18.

Monheit, J.E., Brown, G., et al. 1986. Calcofluor white detection of fungi in cytopathology. *Am J Clin Pathol*, **85**, 222–5.

Reiss, E., Obayashi, T., et al. 2000. Non-culture based diagnostic tests for mycotic infections. *Med Mycol*, **38(Suppl 1)**, 147–59.

Ro, J.Y., Lee, S.S. and Ayala, A.G. 1987. Advantage of Fontana-Masson stain in capsule-deficient cryptococcal infection. *Arch Pathol Lab Med*, **111**, 53–7.

Salfelder, K. 1990. *Atlas of fungal pathology*. Lancaster, UK: Kluwer Academic Publishers.

Schwarz, J. 1982. The diagnosis of deep mycoses by morphologic methods. *Hum Pathol*, **13**, 519–33.

Shimono, L.H. and Harman, B. 1986. A simple and reliable rapid methenamine silver stain for *Pneumocystis carinii* and fungi. *Arch Pathol Lab Med*, **110**, 855–6.

Warnock, D.W., Dupont, B., et al. 1998. Imported mycoses in Europe. *Med Mycol*, **36** (Suppl 1). 87–94.

Wages, D.S. and Wear, D.J. 1982. Acid-fastness of fungi in blastomycosis and histoplasmosis. *ArchPathol Lab Med*, **106**, 440–1.

Wheeler, M.H. and Bell, A.A. 1987. Melanins and their importance in pathogenic fungi. *Curr Topics Med Mycol*, **2**, 338–7.

Wood, C. and Russel-Bell, B. 1983. Characterization of pigmented fungi by melanin staining. *Am J Dermatopathol*, **5**, 77–81.

PART II

THERAPEUTIC AGENTS AND VACCINES

7

Antifungal agents and antifungal susceptibility testing

SANJAY G. REVANKAR, JOHN R. GRAYBILL, AND THOMAS F. PATTERSON

In the late nineteenth century, the appreciation of systemic infections by *Coccidioides immitis* and *Histoplasma capsulatum* launched an effort to find effective treatment for these rare, debilitating, and lethal diseases. Unfortunately, broad-spectrum agents were not discovered until the 1950s, in the forms of amphotericin B, hamycin, filipin, and nystatin. As a result of their great toxicity, only hamycin and nystatin survived as topical drugs, and amphotericin B as a systemic drug. As fungal infections were rather esoteric, relatively few physicians used amphotericin B. In more recent decades, the increasing use of antibacterial drugs, immunosuppressive agents, and prosthetic implanted devices, and the appearance of the acquired immunodeficiency syndrome (AIDS) provided settings for a dramatic increase in fungal infections, and a subsequent increase in the development of systemic antifungal drugs.

Flucytosine was introduced in the 1960s. Although rapidly active as an antimetabolite, the spectrum was limited to *Candida* spp. and *Cryptococcus neoformans*, and the drug now is used only in combination with amphotericin B (Francis and Walsh 1992).

A new era in antifungal therapy was introduced with the azole antifungals, developed with the aim of reducing toxicity and improving efficacy against fungal pathogens. The first of these drugs was clotrimazole. When given parenterally, clotrimazole was soon found to activate the hepatic cytochrome enzymes responsible for its degradation, a sort of suicide effect, which limited its use as a systemic antifungal agent. Presently, clotrimazole is used only for topical therapy of mucosal

candidiasis and for dermatophyte infections. The second of the imidazole drugs was miconazole, which was used intravenously for several years. Unfortunately, the carrier vehicle caused severe histamine-mediated reactions, and miconazole today is used almost exclusively topically (Stevens 1977; Heel et al. 1980).

The major breakthrough in antifungal drug development was ketoconazole, which was released in the late 1970s. Ketoconazole has a broad antifungal spectrum, is absorbed when given orally, and is much better tolerated than amphotericin B or flucytosine (Dismukes et al. 1983; Saag et al. 1985; Sugar et al. 1987; Graybill 1989, 1990; Bodey 1992; Como and Dismukes 1994). However, its lack of oral absorption in seriously ill patients, its poor central nervous system penetration, and its lack of activity against molds limited its clinical use. Thus, in the late 1980s, ketoconazole was supplemented and, in some countries supplanted by, the more recently developed azoles fluconazole and itraconazole (Como and Dismukes 1994). These triazole antifungals are very well tolerated and achieve significant activity in systemic disease. Modifications of these molecules are aimed at further improvement of their clinical properties.

During this period of azole expansion, multiple efforts to modify the amphotericin B molecule to improve tolerance were initiated, but they all failed because of toxicity. The most recent efforts involved incorporating amphotericin B or other toxic antifungals such as nystatin into lipid vehicles, which reduce the dose-limiting toxicity of those compounds. Despite continuing advances, none of the drugs presently available is absolutely fungicidal, either in vivo or in vitro, so that further advances in antifungal drugs are still needed.

SYSTEMIC ANTIFUNGAL DRUGS IN CURRENT USE

Polyenes

MECHANISM OF ACTION

The polyenes are macrolide antibiotics with unsaturated diene bonds. All the polyenes rapidly bind to sterols, preferentially to ergosterol, the primary fungal cell membrane sterol, and less avidly to cholesterol, the primary mammalian cell membrane sterol (Table 7.1). The consequences of this binding include disruption of osmotic integrity of the membrane, with leakage of intracellular potassium and magnesium (Hamilton-Miller 1974; Hammond 1977; Kuroda et al. 1978; Brajtburg et al. 1980, 1986; Vertut-Doi et al. 1994; Wasan et al. 1994). The polyenes also disrupt the function of oxidative enzymes in target cells (Wilson et al. 1991). Of the polyenes in use, amphotericin B is the only one with a toxicity that is sufficiently limited to permit intravenous administration. However, tolerance is also limited by toxic effects on renal cells, with secondary renal tubular acidosis and hypokalemia. Ultimately there is glomerular failure, suppression of erythropoietin synthesis, and anemia (MacGregor et al. 1978). Other systemic symptoms include chills, fever, and nausea. The spectrum of antifungal activity of polyenes is similar and broad; few fungal species are resistant to these compounds. Resistance is most commonly mediated by substitution of ergosterol by other sterols in fungal cell membranes (Woods et al. 1974). Resistance tends to be species

dependent and emerges uncommonly and slowly in isolates from patients treated with amphotericin B (Woods et al. 1974; Pappagianis et al. 1979; Dick et al. 1980; Guého et al. 1994; Matsumoto et al. 1994). These include *Candida lusitaniae*, *Fusarium* spp., *Trichosporon beigellii*, *Pseudallescheria boydii*, and *Scedosporium prolificans*. Some other species of *Candida* may develop lesser degrees of resistance.

Despite the very broad range of amphotericin B susceptibility, there has been a study suggesting that higher minimum inhibitory concentration (MIC) of *Candida* species are correlated with clinical failure (Vanden Bossche et al. 1994b). Nguyen et al. (1998) have reviewed in vitro responses and outcome in 105 patients treated with amphotericin B. They found that isolates with an MIC at 24 hours $\geqslant 1$ µg/ml correlated with more clinical failures than those with MIC < 1 µg/ml. They also found that the minimum lethal concentration (MLC) was even more useful, correlating well with failure ($P<0.03$). In their study an MLC >1 µg/ml was associated with 78 percent failure in 18 patients, versus only 22 percent failures in 87 patients with isolates having MLC $\leqslant 1$ µg/ml. Their method utilized antibiotic medium 3, which is now recommended for amphotericin B testing.

NYSTATIN

This drug is used topically, as an oral suspension at a concentration of 100 000 units/ml for mucosal candidiasis, where it is marginally effective. It is also used in ophthalmic preparations and other ointments.

AMPHOTERICIN B

The most widely used form of amphotericin B is commercially available as a deoxycholate and is infused

Table 7.1 *Antifungal mechanisms of action*

Target	Chemical class	Antifungal agent
DNA/RNA synthesis	Pyrimidine	Flucytosine
Membrane barrier function	Polyenes	Amphotericin B, nystatin
Ergosterol synthesis		
Squalene epoxidase	Allylamines	Terbinafine, naftifine
14α-Demethylase	Azoles	
	Imidazoles	Clotrimazole, econazole, ketoconazole, miconazole
	Triazoles	Fluconazole, itraconazole, voriconazole, posaconazole
Δ^{14}-Reductase/Δ^{7}–Δ^{8}-isomerase	Morpholines	Amorolfine
Mitosis (sliding of microtubules)		Griseofulvin
1,3-β-Glucan synthesis	Echinocandins	Caspofungin, anidulafungin, micafungin
Metabolite uptake	Pyridone	Ciclopiroxolamine
Chitin synthase	Nikkomycin	Nikkomycins K, Z, T
Cell wall	Pradimicin	BMS-181184
Protein synthesis (elongation factor 2)	Sordarins	GM-237354

Adapted from Vanden Bossche et al. (1994a).

intravenously at doses ranging from 0.3 to 1.5 mg/kg over 1–4 hours (Gallis et al. 1990). The optimal time of infusion has been controversial. Two studies have addressed this, with conflicting conclusions. Oldfield et al. (1990) examined 128 infusions in 12 patients. They were randomized to 1 versus 4 h. Fever, rigors, and need for meperidine (pethidine) were similar in both groups, though the onset of infusion-related toxicity occurred more quickly in 1-hour infusion recipients. Conversely, Ellis et al. (1992) found that chills, meperidine dose, and maximum 7-day total pulse rise were all worse in patients receiving 1-hour infusions. Infusion-related intolerance includes fever, chills, rigors, and thrombophlebitis, which may occur in more than 70 percent of patients; hypotension occurs rarely. These side effects may be blocked by premedication with antihistamines, corticosteroids, and acetaminophen (paracetamol) (Goodwin et al. 1995). However, the efficacy of these measures remains controversial (Goodwin et al. 1995). Nephrotoxicity is dose limiting for patients who have received in excess of 5–10 g. To minimize nephrotoxicity, 1 liter of physiological saline is usually infused before amphotericin B is given (Heidemann et al. 1983). Dosing is often interrupted if serum creatinine rises above 2.5 mg/dl. Potassium wasting occurs initially as kidney tubule function is disrupted. Hypokalemia is a common toxicity of amphotericin B administration, which occurs before the development of renal insufficiency. As glomerular function later declines, patients develop typical metabolic acidosis of uremia, with potassium retention. Rapid administration to these patients may cause leaking of potassium into the plasma and, with poor clearance from the kidneys, acute hyperkalemia may develop. Although amphotericin B is highly nephrotoxic, only 10–20 percent of the dose is excreted by the kidneys. Therefore, anephric patients may receive identical doses to individuals with normal renal function.

LIPID VEHICLES FOR POLYENES

Three licensed lipid preparations are available to encase amphotericin B in lipid vehicles that bypass the kidneys, the primary site of nephrotoxicity, but permit access to phagocytic cells (Table 7.2). Nephrotoxicity is markedly reduced with each of these lipid preparations, but patients still encounter infusion-related side effects and hypokalemia (Chopra et al. 1991; Collette et al. 1991; de Marie et al. 1994; Jangnegt et al. 1995). Acute dyspnea

has also been reported in a few patients (Arning et al. 1995). Although these lipid preparations are less toxic than amphotericin B deoxycholate, it appears that on a milligram per milligram basis they are less potent (Collette et al. 1991; Meunier et al. 1991; Schmitt 1993; de Marie et al. 1994; Jangnegt et al. 1995). Doses of 3–5 mg/kg of liposomal forms of amphotericin B are well tolerated and commonly used (Collette et al. 1991; Meunier et al. 1991; Schmitt 1993; de Marie et al. 1994; Jangnegt et al. 1995). These preparations are all efficacious (Chopra et al. 1991; Ringden et al. 1991; Mills et al. 1994; Viviani et al. 1994). Comparative studies with amphotericin B deoxycholate are now emerging. The first such report, comparing amphotericin B lipid complex (ABLC) 5 mg/kg/day with amphotericin B 0.6 mg/kg/day for candidiasis, found comparable responses, with less renal toxicity in the ABLC-treated patients (Anaissie et al. 1995b).

All are effective antifungal agents for a variety of mycoses, including *Candida* species, *Aspergillus*, *Fusarium*, *Cryptococcus neoformans*, and others (Anaissie and Ramphal 1992; Jangnegt et al. 1995; Tollemar and Ringden 1995; Walsh et al. 1997; White et al. 1997; Graybill et al. 1998b). These agents have also been used for empirical therapy for potential mycoses in the febrile neutropenic patient. Here it appears that AmBisome is superior to amphotericin B, whereas Amphotec is only equal (Walsh et al. 1998; White et al. 1998). Anecdotal data are accumulating that high doses (⩾5 mg/kg/day) of lipid amphotericin B preparations may be more efficacious than amphotericin B deoxycholate for the treatment of invasive zygomycosis, and some would suggest that these agents be considered as first line therapy for these refractory infections (Gonzalez et al. 2002; Mondy et al. 2002). The very high costs of lipid preparations (up to US$1 000 per day) may preclude their widespread use for indications such as empirical therapy, though pharmacoeconomic analyses point out that this may be offset by reducing the costs associated with renal dysfunction and other toxicities attributed to standard amphotericin B (Graybill 1996; Hann and Prentice 2001; Ostrosky-Zeichner et al. 2003).

Flucytosine

This is the only antifungal compound of its kind. Flucytosine is activated by deamination within fungal cells to

Table 7.2 *Characteristics of lipid-associated amphotericin B*

Drug	Form	Percentage amphotericin	Dose (mg/kg)
Amphotericin B lipid complex	Ribbons	33	Up to 5
Amphotericin B colloidal dispersion	Disks	50 (molar)	Up to 7
Amphotericin B liposomes	Small unilamellar vesicles (SUV)	10	Up to 5
Amphotericin B + Intralipid	Micelles	30	Up to 1.5

the antimetabolite 5-fluorouracil. Flucytosine is highly water soluble and can be given orally or intravenously, although the intravenous preparation is not widely available (Francis and Walsh 1992). Flucytosine is nontoxic in in vitro assays to mammalian cells, but more than one-third of patients treated with this drug experience bone marrow depression and gastrointestinal intolerance. The toxicity of flucytosine in vivo is thought to be mediated by bacterial conversion intraluminally, in the gut, of flucytosine to 5-fluorouracil, a compound with significant human toxicity (Harris et al. 1986). Flucytosine has a narrow spectrum aimed largely at yeasts, including *Candida* spp. and *Cryptococcus neoformans* (Tassel and Madoff 1968). Some activity has been reported against *Aspergillus* spp. and some agents of chromoblastomycosis (Vandevelde et al. 1972). Flucytosine has been used as monotherapy for treatment of cryptococcosis, but resistance emerged rapidly and only 10 (43 percent) of 23 evaluable patients had a response to therapy (Hospenthal and Bennett 1998). Flucytosine is subject to development of resistance by mutation of cytosine permease, deaminase, and other mechanisms (Francis and Walsh 1992). Thus, flucytosine is used clinically only in combination with amphotericin B and, more recently, fluconazole (Larsen et al. 1994). Flucytosine is excreted largely unchanged by the kidneys. Renal failure, commonly induced by amphotericin B, causes an increase in plasma concentrations of flucytosine which potentiates its toxicity. Some clinicians have argued that serum concentrations should be measured on all patients receiving flucytosine, but others have found that monitoring leukocyte counts and creatinine, and adjusting doses based on these parameters, is sufficient to minimize intolerance. This has best been demonstrated in a study of cryptococcosis, wherein 2 weeks of amphotericin B was compared with and without flucytosine. If dosing was adjusted for renal and bone marrow function, toxicity was not sufficient to remove or exclude many patients from the study (Van der Horst et al. 1997).

Azoles

All of the azole antifungal agents act by inhibiting fungal cytochrome 14α-demethylase (Table 7.1) (Vanden Bossche 1985). Inhibition of this enzyme blocks synthesis of ergosterol for fungal cell membranes. Consequently, the integrity of membrane-associated oxidative enzymes is impaired, with accumulation of phospholipids within the cell and ultimately cell death. The action of antifungal azoles takes several generations of fungi to become apparent and is consequently much slower than that of the polyenes.

Differences among antifungal azoles are primarily in their pharmacokinetics, and in the relative avidity of the drug for fungal versus mammalian enzymes (Tables 7.3, 7.4, 7.5, and 7.6). In distinction to the drugs developed initially, fluconazole and itraconazole have a vastly increased affinity for fungal membranes over mammalian cell membranes, and are better tolerated than ketoconazole (Table 7.5) (Como and Dismukes 1994). Pharmacokinetics distinguish two general classes of azole antifungals (Table 7.3). Fluconazole is the only representative of the first group available for therapeutic use at the present time. It is relatively water soluble, is absorbed extremely well, distributed to all tissues, and is excreted largely unchanged by the kidneys (Humphrey et al. 1985). There are relatively few side effects with fluconazole as compared with ketoconazole and itraconazole, the other azoles commonly used (Table 7.6) (Lazar and Wilner 1990; Como and Dismukes 1994). However, fluconazole is less potent than other azoles against many fungi and its spectrum of activity is confined largely to yeasts (Table 7.4). Fluconazole is 10–100 times less potent than itraconazole against many species, including *Penicillium marneffei*, *Aspergillus* spp., and some *Candida* spp. (Drouhet 1993; Supparatpinyo et al. 1993). Fluconazole is remarkably well tolerated even when given at doses as high as 2000 mg/day (Anaissie et al. 1995a).

Table 7.3 *Azole antifungals*

Parameter	Ketoconazole	Itraconazole	Fluconazole	Voriconazole	Posaconazole
C_{max} (μg/ml) (200 mg orally)	2–4	0.2–0.4	8–12	2.7	0.3
Intravenous	No	Yes	Yes	Yes	No
Oral absorption					
Acid	+++	++	0	0	–[a]
Lipid meal	+	+++	0	–	++
Clearance	Hepatic	Hepatic	Renal	Hepatic	Hepatic
Linear	Yes	No	Yes	No	Yes
$t_{1/2}$ (h)	7–10	34–42	22–31	6	24
Penetration (% serum)					
Urine	<5	<1	>70	<2	–
Cerebrospinal fluid	3–7	<1	>70	38–68	–

Graybill (1994); Hoffman et al. (2000).
a) – indicates information is not known.

Table 7.4 *Spectrum of activity of azole compounds*

Organism	Ketoconazole	Itraconazole	Fluconazole	Voriconazole	Posaconazole
Opportunistic yeasts					
C. albicans	++	+++	++++	++++	++++
C. tropicalis	++	++	++	+++	+++
C. krusei	+	++	0	+++	+++
C. glabrata	+	++	+	+++	+++
C. parapsilosis	++	+++	++++	++++	++++
Cryptococcus neoformans	+	++	+++	+++	+++
Opportunistic molds					
Aspergillus spp.	0	+++	0	++++	++++
Fusarium spp.	0	±	0	++	+
Pseudallescheria boydii	+	++	0	+++	++
Zygomycetes	0	0	0	0	++
Agents of phaeohyphomycosis	+	+++	+	+++	+++
Dimorphic molds					
H. capsulatum	++	++++	+++	++++	++++
Blastomyces dermatitidis	++	+++	+	++++	++++
Coccidioides immitis	++	+++	+++	++++	++++
Sporothrix schenckii	+	++++	++	++	+++
Paracoccidioides brasiliensis	+++	++++	++	+++	
Penicillium marneffei	+	++++	+	+++	

Ketoconazole and itraconazole share several pharma-cokinetic parameters (Brass et al. 1982; Heykants et al. 1982, 1989; Daneshmend et al. 1984; Hardin et al. 1988; Barone et al. 1993) (Tables 7.3, 7.4, and 7.5). They are both absorbed optimally in the gastric acid environment, and are cleared by the liver via cytochrome P450-mediated degradation (Table 7.3). In achlorhydric patients, absorption of these drugs is markedly decreased (Chin et al. 1995). Itraconazole has recently been reformulated in a solution with β-hydroxy-cyclodextrin. This allows intravenous administration and also improves oral absorption of the drug by >60 percent. Further, the pH-associated absorption is no longer an effect, and the drug is best absorbed when taken in a fasting state (Reynes et al. 1997; Vandewoude et al. 1997; Barone et al. 1998). Itraconazole solution is as effective as fluconazole for treatment of esophagitis and thrush, and topical effect may allow treatment of mucosal candidiasis even when patients are taking rifampin. Coadministration of drugs such as rifampin (rifampicin), rifabutin, or carbamazepine, which induce hepatic drug-metabolizing enzymes, causes itraconazole or ketoconazole serum concentrations to fall, and may be associated with treatment failure (Tucker et al. 1992; Como and Dismukes 1994). Alternatively, itraconazole and ketoconazole may cause elevations of serum concentrations of other drugs by competing for the same site that is degraded by the same hepatic enzymes. Such

Table 7.5 *Adverse reactions to azole compounds*

Intolerance	Ketoconazole	Itraconazole	Fluconazole	Voriconazole	Posaconazole
Nausea, vomiting	++++	+	+	+	+
Rash	±	±	+	++	±
Hepatic	+++	+	+	++	+
Endocrine					
Decreased testosterone	+++	0	0	0	±
Decreased cortisol	++	±	0	0	0
Miscellaneous					
Hypokalemia	0	+	0	0	0
Hypertension	0	+	0	0	0
Edema	0	+	0	0	0
Headache	+	+	+	+	+
Altered mentation	+	+	+	0	0
Dizziness	+	+	+	0	+
Visual disturbances	0	0	0	+++	0

Table 7.6 *Important drug interactions with azole compounds*

Interaction	Ketoconazole	Itraconazole	Fluconazole	Voriconazole
Increased azole clearance				
Rifampin	++++	++++	++	++++
Rifabutin		+++	+	++++
Phenytoin	+++	+++	0	++
Isoniazid	+++	0	0	
Increased levels of other drug				
Phenytoin	++	++	+	++
Carbamazepine	++	++	+	+++
Warfarin	++	++	+	++
Ciclosporin	+++	+++	+	+++
Terfenadine	+++	++	+	+++
Astemizole	++	++	?	+++
Sulfonylureas	+	+	+	+
Digoxin	+	+		0
Increased level of azole				
Clarithromycin			+	0

drugs include ciclosporin, digoxin, coumadin (warfarin), terfenadine, and astemizole, and may include oral sulfonylureas (Kramer et al. 1990; Crane and Shih 1993; Pohjola et al. 1993; Sachs et al. 1993; Como and Dismukes 1994) (Table 7.6).

Itraconazole is in general more potent than ketoconazole or fluconazole, has a broader spectrum, and may be used to treat infections with *Aspergillus* spp. (Table 7.4) (Denning et al. 1989). Itraconazole is much better tolerated than ketoconazole. Ketoconazole binds more avidly to mammalian cells than itraconazole, and causes hepatotoxicity in up to 5 percent of patients (Sugar et al. 1987), whereas hepatotoxicity is rare with itraconazole. Ketoconazole also suppresses synthesis of testosterone and, at high concentrations, cortisol (Table 7.5) (Pont et al. 1982, 1984). Itraconazole in high doses causes edema, hypokalemia, and hypertension (Sharkey et al. 1991). All three azole compounds, when given at high doses for a prolonged time, may cause alopecia.

The doses of all three azole drugs are 200–400 mg/day, although higher doses of fluconazole appear well tolerated and may be more effective. The half-life of ketoconazole is less than 12 hours, and only at doses above 400 mg/day is there an accumulation from one day to the next. As a result of slow clearance, with more than a 24-hour half-life, there is accumulation over 2 weeks for itraconazole and fluconazole to reach a steady state (Hardin et al. 1988; Como and Dismukes 1994). Some recommend using 300 mg twice daily for 3 days as a 'loading dose'(for itraconazole, and an initial dose twice that of a daily dose for fluconazole)(Wheat et al. 1995). As yet, there is no clear correlation between clinical response and serum concentration, but serum concentration of itraconazole should be measured several days after starting treatment, especially if the patient is seriously ill; this indicates that the drug has been absorbed.

In neutropenic patients fluconazole has been used at 400 mg/day for empirical therapy or prophylaxis of fungal infections. Inherently resistant species that are not *Candida albicans*, such as *Candida glabrata* and *Candida krusei*, may emerge in that setting (Goodman et al. 1992; Rex et al. 1995b). In patients infected with the human immunodeficiency virus (HIV) with severe immune deficiency, fluconazole is the mainstay of treatment for thrush. However, long courses and low-dose exposure to fluconazole are associated with emergence of resistance to *C. albicans* (Heinic et al. 1993; Redding et al. 1994; Rex et al. 1995b). The antifungal spectrum of the antifungal azoles is broadest for itraconazole, which is effective against *Aspergillus* spp., *P. marneffei*, and many *Candida* spp. resistant to fluconazole (Denning et al. 1992; Supparatpinyo et al. 1992, 1993; Barchiesi et al. 1994a). In non-HIV-infected patients with candidemia, there has been a change from *C. albicans* to non-*albicans Candida* spp., such as *C. glabrata* and *Candida tropicalis*. These are more resistant to fluconazole. However, against fluconazole-resistant *Candida* spp. isolates, the potency of itraconazole is somewhat less than against fluconazole-susceptible isolates (Barchiesi et al. 1994a).

At present there are no indications that azoles have cumulative toxicity. Fluconazole has been used at 400–800 mg/day for treatment of fungal meningitis for years, and for shorter times up to 2 000 mg/day have been used with no major adverse sequelae (Berry et al. 1992; Galgiani et al. 1993). As fluconazole is excreted largely unchanged by the kidneys, it accumulates in renal failure. Other antifungal azoles are degraded by hepatic cytochrome P450 enzymes and have a number of important drug interactions (Table 7.6). Maximal chronic dosing of itraconazole tolerance is probably in the range of 600–800 mg/day, and is mediated by gastrointestinal intolerance and a curious phenomenon of hypotension, edema, and hypokalemia (Sharkey et al. 1991). No endocrine or renal cause for this phenomenon has been identified. Ketoconazole toxicity is mediated by gastrointestinal intolerance, cortisol suppression, and testosterone suppression at levels of 1 200 mg/day or more (Pont et al. 1984).

Voriconazole is the first of the second generation triazole antifungals, which includes posaconazole and ravuconazole (Johnson and Kauffman 2003). These agents are potent, broad-spectrum agents which were developed specifically to target molds, especially *Aspergillus*, and yeasts other than *C. albicans*, that may exhibit resistance to the earlier triazoles, fluconazole, and itraconazole (Sheehan et al. 1999).

Voriconazole is active in both oral and parenteral formulations and has been extensively evaluated in clinical trials for a wide range of fungal pathogens. In vitro, voriconazole demonstrates fungicidal activity against many strains of *Aspergillus*, including *Aspergillus fumigatus* (Espinel-Ingroff et al. 2001). It also demonstrates activity against *Aspergillus terreus*, which may be resistant to amphotericin B, although MICs to voriconazole may be increased (Walsh et al. 2003). The broad spectrum of activity extends to other molds such as *Fusarium* and *Scedosporium*, and includes agents of pheohyphyomycosis, such as *Bipolaris*, *Scedosporium*, *Alterneria*, and others, but not the zygomycetes (Pfaller et al. 2002). Notably, excellent in vitro and animal model data support its potent activity against *C. krusei* as well as *C. glabrata*, although susceptibility may be less for the latter, particularly when the strains are fluconazole resistant (Pfaller et al. 2003a). Except for *C. krusei*, generally a linear relationship exists between fluconazole and voriconazole, with MICs being approximately 100-fold lower for voriconazole, although breakpoints have yet to be established.

Clinical studies have documented superior efficacy and better tolerance of voriconazole as compared to standard amphotericin B for the treatment of invasive aspergillosis so that voriconazole should be considered for primary therapy for most patients with invasive aspergillosis (Denning et al. 2002; Herbrecht et al. 2002). Improved outcomes were also demonstrated in patients with extensive infection, including central nervous system infection (Troke et al. 2003). For rare and refractory molds, such as *Fusarium* and *Scedosporium*, results have also been favorable with successful outcomes reported in 30–40 percent of patients failing other antifungal therapies (Walsh et al. 2002a; Perfect et al. 2003). Clinical data are less extensive for serious *Candida* infection, although good efficacy and safety was demonstrated in patients with *Candida* esophagitis (Ally et al. 2001). Additional clinical data for candidemia are in progress. In an open labeled trial comparing voriconazole to liposomal amphotericin B for fever in patients with neutropenia, voriconazole was less successful at achieving overall success, but importantly was more

effective at reducing the clinically relevant endpoint of breakthrough fungal infections, particularly *Aspergillus* (Walsh et al. 2002b).

The safety of voriconazole has been favorable in clinical trials of patients with serious fungal infections (Herbrecht et al. 2002; Walsh et al. 2002a). The most common serious adverse event has been a transient and reversible visual toxicity that has not been associated with pathology and is rarely dose limiting (Johnson and Kauffman 2003). This effect occurs after oral or intravenous administration and likely correlates with higher serum levels and occurs in approximately 30 percent of patients, generally 30 minutes after administration, and lasts for a mean of 30 minutes. Other less common events include liver toxicity in 10–15 percent, skin rash, nausea, and vomiting.

Voriconazole achieves CSF levels of approximately 50 percent that of serum but little (<5 percent) active drug is excreted in the urine. A loading dose of 6 mg/kg every 12 hours for two doses is recommended for initial therapy intravenously, followed by 4 mg/kg twice daily for invasive aspergillosis and molds for which the drug is indicated in the USA. Oral therapy is recommended at 200 mg twice daily. The drug is metabolized through the cytochrome P450 enzyme system so that drug interactions and variability in patient metabolism can occur (Groll et al. 2003). Metabolism occurs most extensively through CYP2C19, which demonstrates genetic variability, so that some patients such as non-Indian Asians may have higher drug levels. CYP3A4 and other isoenzymes generally contribute less to its metabolism. Drug interactions are common including those with immunosuppressive agents. Dose reductions of ciclosporin by half and tacrolimus by one-third are advised along with close follow-up of the immunosuppressive drug level and for clinical toxicity. Sirolimus is contraindicated because of the dramatic increase in those levels that can occur. Importantly, the immunosuppressive drug dosage will need to be increased after voriconazole is discontinued.

A major issue is whether the azoles and amphotericin B are antagonistic when used together. There is some suggestion, in animal models, that this may be the case in aspergillosis and candidiasis (Polak et al. 1982; Schaffner and Frick 1985). Sugar argues that fluconazole is water soluble, can pass readily into the fungal cell, and may act at a site distant from the cell membrane site of amphotericin B (Sugar 1995). Thus there is no antagonism, and there may even be an additive effect of fluconazole and amphotericin B. In a large study comparing fluconazole alone versus fluconazole with amphotericin B for treatment of candidemia, no clinical antagonism was observed, and there was a trend actually favoring the combination (Rex et al. 2003). Conversely, the lipid-soluble triazoles may concentrate in the cell membrane and act at a site or in a way that antagonizes polyenes, though clinical data are sparse.

Echinocandins

A unique mechanism of action sets the echinocandins apart from other antifungals. Echinocandins are semisynthetic cyclic glycopeptides that inhibit fungal 1,3-β-glucan synthase, which is necessary for the production of 1,3-β-glucan, a major component in many fungal cell walls (Hector 1993; Kurtz et al. 1994a; Abruzzo et al. 1995; Bartizal et al. 1995). Along with chitin and mannans and mannoproteins, glucans make up the major components of the fungal cell wall. The cell wall is responsible for maintaining integrity of the cell shape, for ion and small molecule exchange between the cell and its milieu, and for presentation of surface molecules. 1,3-β-Glucan synthase is an attractive target because it is not present in mammalian cells. This class of drugs appears to have no nephrotoxicity and minimal hepatotoxicity. However, oral absorption is poor and parenteral administration is required. In vitro studies and animal model studies have confirmed great potency in candidiasis and aspergillosis (Abruzzo et al. 1995). *Cryptococcus neoformans*, which contains mostly 1,6-β-glucans in its cell walls, is resistant.

A stop-and-start phase of initial development ultimately led to the first derivative, cilofungin (Lilly). Cilofungin failed in phase I clinical trials because it required a vehicle for aqueous solubilization which turned out to be nephrotoxic. Also limiting cilofungin was the perceived narrow antifungal spectrum, which was largely limited to *Candida*. This class of drugs was at one time close to being discarded because of narrow spectrum and very poor oral absorption. However, persistent efforts on the part of several pharmaceutical companies have led to the licensing of caspofungin (Merck), and late phase II trials with two other contenders, micafungin (Fujisawa) and anidulafungin (Vicuron). The rapid development of echinocandins and their introduction into clinical practice has been nothing short of spectacular. Indeed, the advent of echinocandins is one major reason why Ostrosky-Zeichner et al. were recently able to sound the death knell for amphotericin B desoxycholate (Ostrosky-Zeichner et al. 2003).

The action of echinocandins is very rapid, occurring in minutes, and is noncompetitive (Uzun et al. 1997; Karlowsky et al. 2001). Echinocandins bind to the fungal cell sites where cell wall synthesis is most active. In yeast cells, such as *Candida* spp., echinocandins bind to the cell in many places, and diffusely affect cell wall synthesis. As a result, the fungal cell collapses, resembling an erythrocyte, and may die. Against *Candida*, echinocandins have a very prolonged post-antifungal effect, and are considered fungicidal in vitro (Ernst et al. 2000). In contrast to yeasts, the activity of echinocandins in filamentous fungi is concentrated at the hyphal tip and the budding sites (Kurtz and Rex 2001; Bowman et al. 2002). Supravital stains have shown that echinocandins are more or less fungistatic against *Aspergillus* (Kurtz

et al. 1994b). Part of the fungal mycelium is destroyed, but not all of it.

The antifungal spectrum of echinocandins appears to be similar for caspofungin, anidulafungin, and micafungin. These agents are broadly effective in vitro and in animal models against *Candida* spp., including fluconazole-susceptible or resistant isolates (Nelson et al. 1996; Graybill et al. 1997; Zhanel et al. 1997; Petraitiene et al. 1999; Abruzzo et al. 2000). It does not appear at present that the higher MIC values reported for caspofungin against *Candida parapsilosis* are reflected in clinical failures (Mora-Duarte et al. 2002). There is more varying activity against molds, where a new test, the minimal effective concentration (MEC), appears more useful (Oakley et al. 1998). Most *Aspergillus* spp. and some other molds, such as certain species of phaeohyphomycetes, appear to be susceptible, while zygomycetes and *Trichosporon beigelii* are resistant (Espinel-Ingroff 1998; Del Poeta et al. 1996). *Coccidioides immitis* is susceptible, while there are conflicting reports of *H. capsulatum* (Graybill et al. 1998a; Kohler et al. 2000; Gonzalez et al. 2001). The in vitro antifungal spectrum has been most widely explored for caspofungin, and we are assuming that the other echinocandins are similar. The limited data in vitro and in animals have acted to delay clinical exploration of these compounds in mycoses other than *Candida* and *Aspergillus* infections. There is a recent report of a patient with breakthrough *T. beigelii* infection while receiving caspofungin, which suggests resistance (Goodman et al. 2002). The full range of activity of echinocandins has yet to be elucidated. Because the target is unique to fungi (with little toxicity to mammalian cells), the echinocandins are thought to be excellent candidates to replace the more toxic conventional amphotericin B – at least against a few major pathogens.

All three echinocandins now in clinical use or trials share characteristics of water solubility and poor oral absorption. Thus all drugs must be administered intravenously. Although there have been concerns of histamine-like reactions following intravenous infusion, these have been rare. There are some differences among the drugs, but it is not yet clear whether these will impact on clinical use (Table 7.7). The slower clearance of anidulafungin, as well as its clearance entirely by chemical, not metabolic, degradation, may allow for less frequent dosing. The lower protein binding of anidulafungin (86 versus >95 percent for caspofungin and micafungin) may also influence the ultimate determination of in vitro breakpoints.

It is thought that the peak serum concentration/MIC ratio determines effect of echinocandins. This is similar to amphotericin B, but in contrast to fluconazole, where the AUC/MIC ratio appears more important (Andes 2003). With very few serious adverse reactions to the echinocandins (though data are much more extensive with caspofungin), the maximum tolerated dose and the

Table 7.7 *Characteristics of echinocandins*

Property	Caspofungin 40 mg dose	Micafungin 50 mg dose	Anidulafungin 50 mg dose
Protein binding (%)	97	>99	86
AUC (μg/h/ml)	56		102
C_{max} (μg/ml)	6		4.1
Terminal $t_{1/2}$ (h)	11	15	28
Clearance	Hepatic, metabolism	Hepatic, metabolism	Chemical degradation
Drug interaction	? Ciclosporin	None	None
Urine excretion (%)	<5	<5	<5

Azuma et al. (1998); Stone et al. (2002); Dowell et al. (2003).

therapeutic ratio have not been determined (Arathoon et al. 2002; Mora-Duarte et al. 2002; Letscher-Bru and Herbrecht 2003). Caspofungin and anidulafungin have adverse effects similar to fluconazole (Villanueva et al. 2002; Krause et al. 2003).

The major clinical uses of echinocandins are being defined. All three drugs are effective against *Candida* mucosal infection, with responses similar to fluconazole. A dose of 50 mg/day appears to be effective, given for up to 2 weeks for patients with esophagitis (Kohno et al. 2001; Arathoon et al. 2002; Krause et al. 2003). The only large study with systemic candidiasis was caspofungin versus amphotericin B, and found that both drugs were similarly effective, between 65 and 75 percent, but that caspofungin had far fewer adverse reactions (Mora-Duarte et al. 2002). Caspofungin is also effective in mucosal infection caused by fluconazole-resistant *Candida* (Kartsonis et al. 2003). For fluconazole-resistant *Candida* in the bloodstream, caspofungin is now considered the drug of choice (Walsh 2002).

Studies of aspergillosis are generally unpublished and open label, salvage trials (Maertens et al. 2000). Caspofungin is as effective as a variety of other alternatives. Data with micafungin and anidulafungin are spotty. Micafungin has been compared with fluconazole for antifungal prophylaxis. Results are equivalent to fluconazole, but unfortunately not quite superior to fluconazole against *Aspergillus*. Animal data suggest that there may be additive effects of micafungin and a triazole against *Aspergillus* (Petraitis et al. 2003). Clinically, there are only a few collections of anecdotes, and these are not helpful.

Overall, clinical experience has been restricted to some comparative trials with mucosal candidiasis, one large study with disseminated *Candida*, and unpublished data on prophylaxis and aspergillosis. Beyond candidemia, the roles of the echinocandins have yet to be defined.

FUTURE ANTIFUNGAL DRUGS

New triazoles are under consideration, including posaconazole (Schering, NJ, USA). These agents are much more lipid soluble than fluconazole, have an increased

fungal spectrum against fluconazole-resistant *C. albicans*, other *Candida* spp., and molds (Atkinson et al. 1994; Hoffman et al. 2000), and are excreted primarily by hepatic degradation. Hepatic metabolism places them at potentially similar risks of drug reaction as ketoconazole and fluconazole. Although these drugs have increased activity against fluconazole-resistant *Candida* spp., the MIC is higher than for fluconazole-sensitive isolates.

Other antifungal targets are being explored. These include elongation factor 2, chitin synthases, topoisomerase, translation, myristoylation, and steps specific to certain fungi, such as phenyloxidase and capsule synthesis for *Cryptococcus neoformans*. Agents aimed at these new targets have not yet reached clinical trials. Finally, there are numerous in vivo animal studies examining combined antifungal drugs and immune modulators, such as phagocyte colony-stimulating factors, interferons, and interleukins (Rex et al. 1991; Uchida et al. 1992; Ausiello et al. 1993; Brummer and Stevens 1994; Brummer et al. 1994; Joly et al. 1994; Nassar et al. 1994; Roilides et al. 1994, 1995; Sewell et al. 1994; Vitt et al. 1994). The potential clinical value of these cytokine/antifungal combinations has not been clearly defined.

ANTIFUNGAL SUSCEPTIBILITY TESTING

The combined effects of increasing numbers of fungal infections, growing numbers of patients at risk, the increasing rate of fungal resistance, and the expanded antifungal armamentarium have led to an increased recognition of the need for standardized laboratory testing for antifungal drug susceptibility. However, numerous problems have been encountered in the quest for a simple, reproducible, and inexpensive testing system applicable to all clinically important fungi. These include multiple variables affecting outcome, such as pH, inoculum size, medium, and time and temperature of incubation, which produce varying results; unclear in vitro to in vivo correlation; and differences in testing yeasts versus filamentous fungi (Rex et al. 1993). Several excellent reviews have been written on the subject, which detail the difficulties in developing in vitro antifungal susceptibility testing and insights into controversial issues (Stevens 1984; Drutz 1987; Galgiani 1987, 1993; Espinel-Ingroff and Shadomy 1989; Pfaller and Rinaldi 1993; Rex et al. 1993, 2001; Sheehan et al. 1993; Pfaller et al. 1997; Pfaller and Yu 2001; Rex and Pfaller 2002).

Early studies

Only recently has antifungal drug susceptibility testing become an important issue. Initially, amphotericin B was the only agent available for systemic infection, so testing did not appear to have practical clinical significance. Even after several antifungal agents became available, only a few laboratories were performing antifungal testing, usually less than 10–20 isolates per year (Rex et al. 1993). Studies in the 1980s revealed that between-laboratory susceptibility results varied by as much as 512-fold, and standard methods were not available for susceptibility testing (Rex et al. 1993). It was noted, however, that the rank order of isolates was relatively consistent among the different laboratories, suggesting a definite need for standardized methods.

Many techniques have been used to determine antifungal susceptibility, including measurement of germ tubes, uptake of metabolites, flow cytometry, agar-based methods, and broth dilution (Rex et al. 1993). Most are considered impractical for large-scale use. Agar techniques are appealing because of their ease and low cost, but suffer from a wide variation of results, including factors such as inoculum size, temperature, time of incubation, and ability of poorly soluble drugs, such as ketoconazole and itraconazole, to diffuse through agar (Rex et al. 1993). Broth dilution methods are widely used, and have been standardized by the National Committee for Clinical Laboratory Standards (NCCLS) for testing of *Candida*, *Cryptococcus* (document M-27A), and a variety of clinically important filamentous fungi (document M-38A) (National Committee for Clinical Laboratory Standards 1997, 2002).

Factors affecting antifungal susceptibility testing

END-POINT DETERMINATION

This aspect of antifungal testing is an important source of variability among laboratories in the testing of azoles and occasionally 5-flucytosine. These agents are generally fungistatic and usually tend not to have distinct end points on MIC testing, and thereby introduce subjective interpretations of susceptibility results. By contrast, amphotericin B, which is fungicidal, generally produces distinct end points. Azoles in particular may also produce 'trailing end points,' with growth evident at all concentrations tested (Rex et al. 1993). A solution to

quantify end-point determination has been proposed by the NCCLS method, which uses an 80 percent decrease in turbidity as compared with control to establish an end point. A study by Pfaller et al. (1995) demonstrated that agitation of a microbroth well or spectrophotometric measurement led to a more definitive end point. More recently, studies have suggested that a 50 percent end point may lead to fewer trailing end points and may better predict clinical outcome (St Germain 2001; Arthington-Skaggs et al. 2002). Isolates that are susceptible at 24 hours but resistant at 48 hours due to trailing appear to revert to susceptible at lower media pH, and actually behave as if they are susceptible in vivo (Revankar et al. 1998b; Rex et al. 1998; Marr et al. 1999). Further clinical studies are needed with these isolates.

For testing echinocandins, a unique end point termed the MEC has been proposed that addresses the difficulties in assessing response to the drug, particularly for filamentous fungi (Oakley et al. 1998).

INOCULUM SIZE

It has been shown that increasing the size of inoculum can drastically increase MICs for most drugs tested, and interlaboratory consistency is improved by using smaller inocula (Rex et al. 1993). The NCCLS method for yeasts suggests an inoculum of 0.5–2.5×10^3 c.f.u./ml, prepared by matching turbidity at 530 nm of a 0.5 McFarland standard (National Committee for Clinical Laboratory Standards 1997). For filamentous fungi, NCCLS recommends an inoculum of 0.4–5×10^4 c.f.u./ml, by spectrophotometric measurement, with varying optical densities depending on the species tested (National Committee for Clinical Laboratory Standards 2002).

Hemocytometer count is also a reasonable technique for preparing an inoculum, although with results that are less consistent in between-laboratory tests (Rex et al. 1993). In comparing a wide variety of filamentous fungi, this method may be more reliable, though it is unclear if this would significantly affect the results of susceptibility testing compared with the standardized method (Aberkane et al. 2002).

INCUBATION

Both time and temperature can have significant effects on MIC determinations. MICs generally tend to increase with longer incubation periods (Rex et al. 1993). Temperature variations are not as predictable, but are felt to be least apparent at 35°C (Rex et al. 1993). The NCCLS methods recommend incubation at 35°C for 48 hours for *Candida* spp., 72 hours for *Cryptococcus* spp., and 24–72 hours for molds, depending on the species (National Committee for Laboratory Standards 1997, 2002).

MEDIA

Different media can give rise to widely varying results of antifungal testing, due to interactions between the drugs and the testing media. This can be especially seen with relatively undefined media. The pH can also be important, with a lower pH of the test medium associated with higher MICs of most antifungal agents. For this reason, the use of a defined, synthetic medium, RPMI-1640 (Sigma Chemical Company, St Louis, MO, USA), pH 7.0 with 3-[N-morpholino]-propanesulfonic acid (MOPS) (Sigma) is advocated by NCCLS for both yeasts and filamentous fungi (National Committee for Clinical Laboratory Standards 1997, 2002).

Testing of filamentous fungi

Until recently, no standardized method for antifungal susceptibility testing of filamentous fungi was available. This area of susceptibility testing presents its own, unique problems, namely how to quantify inhibition of growth for organisms that change morphological forms. For example, *Aspergillus* spp. have conidia, which are small sphere-like forms that are easily quantifiable, but germinate to hyphal forms that are not. Many studies have used germination of hyphal growth as an indicator in susceptibility testing. Denning et al. (1992) found a wide range of in vitro susceptibility results while reviewing studies by different investigators, although within their own laboratory they achieved consistent results. In response to these issues, the NCCLS developed M-38A, a standardized broth dilution method for susceptibility testing of filamentous fungi (National Committee for Clinical Laboratory Standards 2002). The primary differences with the yeast testing method are variable incubation times for different species and a no growth or 100 percent inhibition end point for amphotericin B, itraconazole, and the newer azoles. Other methods have also been studied, including E-test, automated blood culture systems, and colorimetric methods, which in general have had good correlation with the broth method, particularly the E-test (Espinel-Ingroff 2001; Meletiadis et al. 2001; Rath et al. 2002; Pfaller et al. 2003b). In addition, the concept of minimum fungicidal concentration (MFC) may be important for in vitro–in vivo correlation, and a proposed definition is the lowest dilution with <3 c.f.u. on subculture, corresponding to 99–99.5 percent killing based on the NCCLS standardized inoculum (Espinel-Ingroff et al. 2002).

METHODS OF SUSCEPTIBILITY TESTING

Macrobroth dilution

This is the currently approved reference method for antifungal susceptibility testing of yeasts and filamentous fungi. Numerous studies have shown good correlation between laboratories using this methodology: amphotericin B, 90 percent; ketoconazole, 75 percent, 5-flucytosine, 85 percent; fluconazole, 88 percent (Rex et al. 1993). However, the NCCLS broth technique is insensitive for detecting resistance to amphotericin B (Rex et al. 1993). The current recommendation uses a macrobroth technique with 1 ml of medium, but microbroth dilution using 0.2 ml in a 96-well microtiter plate may be more practical for rapid, widespread use, and appears to produce similar results (Barchiesi et al. 1994b; Pfaller et al. 1994a; Sewell et al. 1994; Hacek et al. 1995). It is considered an acceptable alternative to the standardized method by the NCCLS.

Colorimetric

ALAMAR BLUE

Alamar Blue (Alamar Biosciences, Inc., Sacramento, CA) is a novel colorimetric indicator that changes color from blue to red when reduced in the presence of microbial growth (Pfaller and Barry 1994). It has shown excellent correlation with bacterial reference methods and several studies have shown promise in antifungal susceptibility testing (Pfaller and Barry 1994; Pfaller et al. 1994a, b; Tiballi et al. 1995; To et al. 1995). In general, studies have confirmed that results agree closely with the NCCLS macrobroth method at 24 hours and less so at 48 hours (Pfaller and Barry 1994; Pfaller et al. 1994b; To et al. 1995). Some combinations, such as fluconazole with *C. albicans*, *C. glabrata*, and *C. tropicalis*, had relatively poor correlations (11–65 percent) in one study (To et al. 1995). A commercial system, Sensititre YeastOne (TREK Diagnostics Systems Inc., Westlake, OH) which uses Alamar Blue, was recently approved by the FDA. Studies using this system have shown good correlation with the NCCLS method, though some drug-species combinations have less correlation than others (Espinel-Ingroff et al. 1999; Morace et al. 2002).

TETRAZOLIUM SALTS

A tetrazolium salt, 2,3-bis-2-methoxy-4-nitro-5-sulphophenyl-5-[phenylaminocarbonyl]-2H-tetrazo-lium hydroxide (XTT), has been used in eukaryotic cell drug assays (Tellier et al. 1992). It is a member of a class of compounds that produce a colored formazan crystal when reduced, which can be detected using a spectrophotometer (Tellier et al. 1992). A study by Tellier et al. (1992) used this agent to test various yeast species, and distinct MICs were observed, although this method was not directly compared to the NCCLS reference method.

Another tetrazolium derivative, 3-4,5-dimethyl-2-thiazolyl-2,5-diphenyl-2H-tetrazolium bromide (MTT), was used to determine antifungal susceptibility in a study by Jahn et al. (1995). It demonstrated reasonable correlation to NCCLS macrobroth methods, but few isolates were tested.

E-test

This method of susceptibility testing has been used successfully with bacteria. It involves use of a plastic strip impregnated with a defined gradient of the antimicrobial agent to be tested. The strip is then placed on agar seeded with the test organism. The graded zone of inhibition, if present, is read from a scale printed on the strip, corresponding to the particular concentration of drug. Multiple studies have shown that good correlation is achieved with the NCCLS method with amphotericin B, azoles, and caspofungin, though occasional problems were noted. A study by Colombo et al. (1995) comparing the E-test and NCCLS macrobroth methods showed reasonable MIC correlations, with 71 percent for ketoconazole, 80 percent for fluconazole, and 84 percent for itraconazole. However, the E-test results were usually one- to twofold less than for the NCCLS method, with *C. tropicalis* showing the least concordance (Colombo et al. 1995). Another study by Sewell et al. (1994) demonstrated similar findings, with poor correlations against NCCLS methodology for *C. tropicalis* and *C. glabrata* at 24 and 48 hours. *C. albicans* testing was in good agreement with NCCLS at 24 hours, but not at 48 hours (Sewell et al. 1994).

Other methods

Alternative testing methods that may be easier to perform and interpret than the NCCLS method continue to be evaluated and show promise. These include use of flow cytometry (results within 6 hours), automated blood culture systems, disk diffusion, and sterol quantitation (Ramani and Chaturvedi 2000; Arthington-Skaggs et al. 2002; Rath et al. 2002; Matar et al. 2003). In addition, testing of *Candida* spp. in the presence of biofilms has been studied to better evaluate the effect of antifungal drugs in this setting (Ramage et al. 2001). Agar dilution screening for *Candida* and *Cryptococcus* has also shown promising results compared with the NCCLS macrobroth method, and may be useful to rapidly detect clinical resistant isolates (Patterson et al. 1996; Kirkpatrick et al. 1998; Nelson and Cartwright 2003).

IN VIVO–IN VITRO CORRELATION

Many studies have shown a disparity between in vitro MICs and in vivo efficacy for antifungal agents tested. However, standardized methods have improved correlation in more recent studies. There are interpretive breakpoint guidelines for antifungal susceptibility testing of fluconazole and itraconazole to *Candida* spp. (Rex et al. 1997). These were primarily based on studies of the correlation between in vitro testing and clinical

response in oropharyngeal candidiasis in HIV-infected patients (Laguna et al. 1997; Revankar et al. 1998a). Correlations between increased MICs of *C. albicans* to fluconazole and poor clinical response in oropharyngeal candidiasis have been shown (Redding et al. 1994; Revankar et al. 1998a). Studies of systemic fungal infections have also shown correlation with in vitro testing (Aller et al. 2000; Kovacicova et al. 2000; Lee et al. 2000). There are also clinical responses when in vitro testing reveals the organism to be resistant. In a study by Rex et al. (1995a), bloodstream isolates from *Candida* spp. revealed an inverse correlation between in vitro resistance to fluconazole at MIC of 32 µg/ml or more and response to therapy at 400 mg/day, suggesting that in this setting other factors may be more important in determining clinical outcome than in vitro susceptibility testing. However, the number of patients with resistant organisms was very small. Increased MICs to amphotericin B to over 0.8 µg/ml were associated with higher mortality in oncology patients in a study by Powderly et al. (1988). In a study by Denning et al., in vitro testing conditions for itraconazole were adjusted to produce good correlation to clinical response with six isolates of *A. fumigatus*, of which two were resistant (Denning et al. 1997). When it occurs, a lack of correlation between in vitro susceptibility and in vivo efficacy may result from host factors or differences in drug pharmacokinetics that alter its availability at the site of infection. In summary, clinical correlation of in vitro activity of antifungal agents is a complex issue that will require ongoing study. There is emerging evidence, however, that in vitro testing will be useful to the clinician.

REFERENCES

Aberkane, A., Cuenca-Estrella, M., et al. 2002. Comparative evaluation of two different methods of inoculum preparation for antifungal susceptibility testing of filamentous fungi. *J Antimicrob Chemother*, **50**, 719–22.

Abruzzo, G.K., Flattery, A.M., et al. 1995. Evaluation of water-soluble pneumocandin analogs L-733560, L-705589 and L-731373 with mouse models of disseminated aspergillosis, candidiasis, and cryptococcosis. *Antimicrob Agents Chemother*, **39**, 1077–81.

Abruzzo, G.K., Gill, C.J., et al. 2000. Efficacy of the echinocandin caspofungin against disseminated aspergillosis and candidiasis in cyclophosphamide-induced immunosuppressed mice. *Antimicrob Agents Chemother*, **44**, 2310–18.

Aller, A.I., Martin-Mazuelos, E., et al. 2000. Correlation of fluconazole MICs with clinical outcome in cryptococcal infection. *Antimicrob Agents Chemother*, **44**, 1544–8.

Ally, R., Schurmann, D., et al. 2001. A randomized, double-blind, double-dummy, multicenter trial of voriconazole and fluconazole in the treatment of esophageal candidiasis in immunocompromised patients. *Clin Infect Dis*, **33**, 1447–54.

Anaissie, E.J. and Ramphal, R. 1992. Efficacy and safety of amphotericin B lipid complex (ABLC) in the treatment of patients with life-threatening fusariosis. *Thirty Fourth Interscience Conference on Antimicrobial Agents and Chemotherapy*, M87 (abstract).

Anaissie, E.J., Kontoyiannis, D.P., et al. 1995a. Safety, plasma concentrations, and efficacy of high-dose fluconazole in invasive mold infections. *J Infect Dis*, **172**, 599–602.

Anaissie, E.J., White, M., et al. 1995b. Amphotericin B lipid complex (ABLC) versus amphotericin B (AMB) for treatment of hematogenous and disseminated candidiasis. *ICAAC*, **35**, 330, (abstract).

Andes, D. 2003. In vivo pharmacodynamics of antifungal drugs in treatment of candidiasis. *Antimicrob Agents Chemother*, **47**, 1179–86.

Arathoon, E.G., Gotuzzo, E., et al. 2002. Randomized, double-blind, multicenter study of caspofungin versus amphotericin B for treatment of oropharyngeal and esophageal candidiasis. *Antimicrob Agents Chemother*, **46**, 451–7.

Arning, M., Heer-Sonderhoff, A.H., et al. 1995. Pulmonary toxicity during infusion of liposomal amphotericin B in two patients with acute leukemia. *Eur J Clin Microbiol Infect Dis*, **14**, 41–3.

Arthington-Skaggs, B.A., Lee-Yang, W., et al. 2002. Comparison of visual and spectrophotometric methods of broth microdilution MIC end point determination and evaluation of a sterol quantitation method for in vitro susceptibility testing of fluconazole and itraconazole against trailing and non-trailing *Candida* isolates. *Antimicrob Agents Chemother*, **46**, 2477–81.

Atkinson, B.A., Bocanegra, R., et al. 1994. Treatment of disseminated *Torulopsis glabrata* infection with DO870 and amphotericin B. *Antimicrob Agents Chemother*, **38**, 1604–7.

Ausiello, C.M., Urbani, F., et al. 1993. Cytokine gene expression in human peripheral blood mononuclear cells stimulated by mannoprotein constituents from *Candida albicans*. *Infect Immun*, **61**, 4105–11.

Azuma, J., Yamamoto, I., et al. 1998. Phase I study of FK463, a new antifungal, in healthy male volunteers. *Thirty Eighth Interscience Conference on Antimicrobial Agents and Chemotherapy*, **38**, 146, (abstract).

Barchiesi, F., Colombo, A.L., et al. 1994a. In vitro activity of itraconazole against fluconazole-susceptible and -resistant *Candida albicans* isolates from oral cavities of patients infected with human immunodeficiency virus. *Antimicrob Agents Chemother*, **38**, 1530–3.

Barchiesi, F., Colombo, A.L., et al. 1994b. Comparative study of broth macrodilution and microdilution techniques for in vitro antifungal susceptibility testing of yeasts by using the National Committee for Clinical Laboratory Standards' proposed standard. *J Clin Microbiol*, **32**, 2494–500.

Barone, J.A., Koh, J.G., et al. 1993. Food interaction and steady-state pharmacokinetics of itraconazole capsules in healthy male volunteers. *Antimicrob Agents Chemother*, **37**, 778–84.

Barone, J.A., Moskovitz, B.L., et al. 1998. Enhanced bioavailability of itraconazole in hydroxyprolyl-beta-cyclodextrin solution versus capsules in healthy volunteers. *Antimicrob Agents Chemother*, **42**, 1862–5.

Bartizal, K., Scott, T., et al. 1995. In vitro evaluation of the pneumocandin antifungal agent L-733560, a new water-soluble hybrid of L-705589 and L-731373. *Antimicrob Agents Chemother*, **39**, 1070–6.

Berry, A.J., Rinaldi, M.G. and Graybill, J.R. 1992. High-dose fluconazole as salvage therapy for cryptococcal meningitis in patients with AIDS. *Antimicrob Agents Chemother*, **36**, 690–2.

Bodey, G.P. 1992. Azole antifungal agents. *Clin Infect Dis*, **19**, Suppl 1, S161–9.

Bowman, J.C., Hicks, P.S., et al. 2002. The antifungal echinocandin caspofungin acetate kills growing cells of *Aspergillus fumigatus* in vitro. *Antimicrob Agents Chemother*, **46**, 3001–12.

Brajtburg, J., Medoff, G., et al. 1980. Influence of extracellular K^+ or Mg^{2+} on the stages of the antifungal effects of amphotericin B and filipin. *Antimicrob Agents Chemother*, **18**, 593–7.

Brajtburg, J., Elberg, S., et al. 1986. Effects of serum lipoproteins on damage to erythrocytes and *Candida albicans* cells by polyene antibiotics. *J Infect Dis*, **153**, 623–6.

Brass, C., Galgiani, J.N., et al. 1982. Disposition of ketoconazole, an oral antifungal, in humans. *Antimicrob Agents Chemother*, **21**, 151–8.

Brummer, E. and Stevens, D.A. 1994. Macrophage colony-stimulating factor induction of enhanced macrophage anticryptococcal activity: synergy with fluconazole for killing. *J Infect Dis*, **170**, 173–9.

Brummer, E., Nassar, F. and Stevens, D.A. 1994. Effect of macrophage colony-stimulating factor on anticryptococcal activity of bronchoalveolar macrophages: synergy with fluconazole for killing. *Antimicrob Agents Chemother*, **38**, 2158–61.

Chin, T.W.F., Loeb, M. and Fong, I.W. 1995. Effects of an acidic beverage (Coca-Cola) on absorption of ketoconazole. *Antimicrob Agents Chemother*, **39**, 1671–8.

Chopra, R., Blair, S., et al. 1991. Liposomal amphotericin B (AmBisome) in the treatment of fungal infections in neutropenic patients. *J Antimicrob Chemother*, **28**, Suppl B, 93–104.

Collette, N., van der Auwera, P., et al. 1991. Tissue distribution and bioactivity of amphotericin B administered in liposomes to cancer patients. *J Antimicrob Chemother*, **27**, 535–48.

Colombo, A.L., Barchiesi, F., et al. 1995. Comparison of Etest and National Committee for Clinical Laboratory Standards broth macrodilution method for azole antifungal susceptibility testing. *J Clin Microbiol*, **33**, 535–40.

Como, J.A. and Dismukes, W.E. 1994. Oral azole drugs as systemic antifungal therapy. *N Engl J Med*, **330**, 263–72.

Crane, J.K. and Shih, H.-T. 1993. Syncope and cardiac arrhythmia due to an interaction between itraconazole and terfenadine. *Am J Med*, **95**, 445–7.

Daneshmend, T.K., Warnock, D.W. and Ene, M.D. 1984. Influence of food on the pharmacokinetics of ketoconazole. *Antimicrob Agents Chemother*, **25**, 1–3.

Del Poeta, M., Schell, W.A. and Perfect, J.R. 1996. In vitro antifungal activity of L-743872 against a variety of moulds. *Thirty Sixth Interscience Conference on Antimicrobial Agents and Chemotherapy*, **36**, 105, abstract.

de Marie, S., Janknegt, R. and Bakker-Woudenberg, I.A.J.M. 1994. Clinical use of liposomal and lipid-complexed amphotericin B. *J Antimicrob Chemother*, **33**, 907–16.

Denning, D.W., Tucker, R.M., et al. 1989. Treatment of invasive aspergillosis with itraconazole. *Am J Med*, **86**, 791–800.

Denning, D.W., Hanson, L.H., et al. 1992. In vitro susceptibility and synergy studies of *Aspergillus* species to conventional and new agents. *Diagn Microbiol Infect Dis*, **15**, 21–34.

Denning, D.W., Radford, S.A., et al. 1997. Correlation between in-vitro susceptibility testing to itraconazole and in-vivo outcome of *Aspergillus fumigatus* infection. *J Antimicrob Chemother*, **40**, 401–14.

Denning, D.W., Ribaud, P., et al. 2002. Efficacy and safety of voriconazole in the treatment of acute invasive aspergillosis. *Clin Infect Dis*, **34**, 563–71.

Dick, J.D., Merz, W.G. and Saral, R. 1980. Incidence of polyene resistant yeasts recovered from clinical specimens. *Antimicrob Agents Chemother*, **18**, 158–63.

Dismukes, W.E., Stamm, A.M., et al. 1983. Treatment of systemic mycoses with ketoconazole: emphasis on toxicity and clinical response in 52 patients. *Ann Intern Med*, **98**, 13–20.

Dowell, J.A., Pu, F., et al. 2003. A clinical mass balance study of anidulafungin showing complete fecal elimination. *Forty Third Interscience Conference on Antimicrobial Agents and Chemotherapy*, **43**, 1576, abstract.

Drouhet, E. 1993. Penicilliosis due to *Penicillium marneffei*: a new emerging systemic mycosis in AIDS patients traveling or living in Southeast Asia. Review of 44 cases reported in HIV infected patients during last 5 years compared to 44 cases HIV negative patients reported over 20 years. *J Mycol Méd*, **3**, 195–224.

Drutz, D.J. 1987. In vitro antifungal susceptibility testing and measurement of levels of antifungal agents. *Rev Infect Dis*, **9**, 392–7.

Ellis, M.E., Hokail, A.A., et al. 1992. Double-blind randomized study of the effect of infusion rates on toxicity of amphotericin B. *Antimicrob Agents Chemother*, **36**, 172–9.

Ernst, E.J., Klepser, M.E. and Pfaller, M. 2000. Postantifungal effects of echinocandin, azole, and polyene antifungal agents against *Candida albicans* and *Cryptococcus neoformans*. *Antimicrob Agents Chemother*, **44**, 1108–11.

Espinel-Ingroff, A. 1998. Comparison of in vitro activities of the new triazole SCH56592 and the echinocandins MK-0991 (L-743,872) and LY303366 against opportunistic filamentous and dimorphic fungi and yeasts. *J Clin Microbiol*, **36**, 2950–6.

Espinel-Ingroff, A. 2001. Comparison of the E-test with the NCCLS M38-P method for antifungal susceptibility testing of common and emerging pathogenic filamentous fungi. *J Clin Microbiol*, **39**, 1360–7.

Espinel-Ingroff, A. and Shadomy, S. 1989. In vitro and in vivo evaluation of antifungal agents. *Eur J Clin Microbiol Infect Dis*, **8**, 352–61.

Espinel-Ingroff, A., Pfaller, M., et al. 1999. Multicenter comparison of the Sensititre YeastOne colorimetric antifungal panel with the National Committee for Clinical Laboratory Standards M27-A reference method for testing clinical isolates of common and emerging *Candida* spp., *Cryptococcus* spp., and other yeasts and yeast-like organisms. *J Clin Microbiol*, **37**, 591–5.

Espinel-Ingroff, A., Boyle, K., et al. 2001. In vitro antifungal activities of voriconazole and reference agents as determined by NCCLS methods: review of the literature. *Mycopathologia*, **150**, 3, 101–15.

Espinel-Ingroff, A., Fothergill, A., et al. 2002. Testing conditions for determination of minimum fungicidal concentrations of new and established antifungal agents for Aspergillus spp.: NCCLS collaborative study. *J Clin Microbiol*, **40**, 3204–8.

Francis, P. and Walsh, T.J. 1992. Evolving role of flucytosine in immunocompromised patients: new insights into safety, pharmacokinetics, and antifungal therapy. *Clin Infect Dis*, **15**, 1003–18.

Galgiani, J.N. 1987. Antifungal susceptibility tests. *Antimicrob Agents Chemother*, **31**, 1867–70.

Galgiani, J.N. 1993. Susceptibility testing of fungi: current status of the standardization process. *Antimicrob Agents Chemother*, **37**, 2517–21.

Galgiani, J.N., Catanzaro, A., et al. 1993. Fluconazole therapy for coccidioidal meningitis. *Ann Intern Med*, **119**, 28–35.

Gallis, H.A., Drew, R.H. and Pickard, W.W. 1990. Amphotericin B: 30 years of clinical experience. *Rev Infect Dis*, **12**, 308–29.

Gonzalez, G., Tijerina, R., et al. 2001. Correlation between antifungal susceptibilities of *Coccidioides immitis* (CI) in vitro and antifungal treatment with caspofungin in a mouse model. *Antimicrob Agents Chemother*, **45**, 1854–9.

Gonzalez, C.E., Rinaldi, M.G. and Sugar, A.M. 2002. Zygomycosis. *Infect Dis Clin North Am*, **16**, 895–914.

Goodman, D., Pamer, E., et al. 2002. Breakthrough trichosporonosis in a bone marrow transplant recipient receiving caspofungin acetate. *Clin Infect Dis*, **35**, E35–6.

Goodman, J.L., Winston, D.J., et al. 1992. A controlled trial of fluconazole to prevent fungal infections in patients undergoing bone marrow transplantation. *N Engl J Med*, **326**, 845–51.

Goodwin, S.D., Cleary, J.D., et al. 1995. Pretreatment regimens for adverse events related to infusion of amphotericin B. *Clin Infect Dis*, **20**, 755–61.

Graybill, J.R. 1989. Azole therapy in systemic fungal infections. In: Holmberg, K. and Meyer, R.D. (eds), *Diagnosis and therapy of systemic fungal infections*. New York: Raven Press, 133–44.

Graybill, J.R. 1990. Systemic azole antifungal drugs – into the 's. In: Armstrong, D., Ryley, J.F. and Ryley, J.F. (eds), *Chemotherapy of fungal diseases*. New York: Springer-Verlag, 455–82.

Graybill, J.R. 1994. Is there a correlation between serum antifungal drug concentration and clinical outcome? *J Infect*, **28**, Suppl 1, 17–24.

Graybill, J.R. 1996. Lipid formulations for amphotericin B: does the emperor need new clothes? *Ann Intern Med*, **126**, 921–3.

Graybill, J.R., Bocanegra, R., et al. 1997. Treatment of murine *Candida krusei* or *Candida glabrata* infection with L-743,872. *Antimicrob Agents Chemother*, **41**, 1937–9.

Graybill, J.R., Najvar, L.C., et al. 1998a. Treatment of histoplasmosis with MK-991 (L-743,872). *Antimicrob Agents Chemother*, **42**, 151–3.

Graybill, J.R., Vazquez, J., et al. 1998b. Itraconazole oral solution: a novel and effective treatment for oropharyngeal candidiasis in HIV/AIDS patients. *Am J Med*, **104**, 33–9.

Groll, A.H., Gea-Banacloche, J.C., et al. 2003. Clinical pharmacology of antifungal compounds. *Infect Dis Clin North Am*, **17**, 1, 159–191, ix.

Guého, E., Faergemann, J., et al. 1994. *Malassezia* and *Trichosporon*: two emerging pathogenic basidiomycetous yeast-like fungi. *J Med Vet Mycol*, **32**, Suppl 1, 367–78.

Hacek, D.M., Noskin, G.A., et al. 1995. Initial use of a broth microdilution method suitable for in vitro testing of fungal isolates in a clinical microbiology laboratory. *J Clin Microbiol*, **33**, 1884–9.

Hamilton-Miller, J.M.T. 1974. Fungal sterols and the mode of action of the polyene antibiotics. *Adv Appl Microbiol*, **17**, 109–33.

Hammond, S.M. 1977. Biological activity of polyene antibiotics. *Prog Med Chem*, **14**, 105–79.

Hann, I.M. and Prentice, H.G. 2001. Lipid-based amphotericin B: a review of the last 10 years of use. *Int J Antimicrob Agents*, **17**, 161–9.

Hardin, T.C., Graybill, J.R., et al. 1988. Pharmacokinetics of itraconazole following oral administration to normal volunteers. *Antimicrob Agents Chemother*, **32**, 1310–13.

Harris, B.E., Manning, B.W., et al. 1986. Conversion of 5-fluorocytosine to 5-fluorouracil by intestinal microflora. *Antimicrob Agents Chemother*, **29**, 44–8.

Hector, R.F. 1993. Compounds active against cell walls of medically important fungi. *Clin Microbiol Rev*, **6**, 1–21.

Heel, R.C., Brogden, R.N., et al. 1980. Miconazole: a preliminary review of its therapeutic efficacy in systemic fungal infections. *Drugs*, **19**, 7–30.

Heidemann, H.Th., Gerkens, J.F., et al. 1983. Amphotericin B nephrotoxicity in humans decreased by salt repletion. *Am J Med*, **75**, 476–81.

Heinic, G.S., Stevens, D.A., et al. 1993. Fluconazole-resistant *Candida* in AIDS patients. *Oral Surg Oral Med Oral Pathol*, **76**, 711–15.

Herbrecht, R., Denning, D.W., et al. 2002. Voriconazole versus amphotericin B for primary therapy of invasive aspergillosis. *N Engl J Med*, **347**, 6, 408–15.

Heykants, J., van Peer, A. and Van De Velde, V. 1989. The clinical pharmacokinetics of itraconazole: an overview. *Mycoses*, **32**, Suppl 1, 67–87.

Heykants, J.M., Michiels, W. and Meuldermans, W. 1982. The parmacokinetics of itraconazole in animals and man: an overview. In: Fromtling, R.A. (ed.), *Recent trends in the disovery, development and evaluation of antifungal agents*. Barcelona, Spain: JR Prous Science Publishers, 223–49.

Hoffman, H.L., Ernst, E.J. and Klepser, M.E. 2000. Novel triazole antifungal agents. *Exp Opin Invest Drugs*, **9**, 593–605.

Hospenthal, D.R. and Bennett, J.E. 1998. Flucytosine monotherapy for cryptococcosis. *Clin Infect Dis*, **27**, 260–4.

Humphrey, M.J., Jevons, S. and Tarbit, M.H. 1985. Pharmacokinetic evaluation of UK-49,858, a metabolically stable triazole antifungal drug, in animals and humans. *Antimicrob Agents Chemother*, **28**, 648–53.

Jahn, B., Martin, E., et al. 1995. Susceptibility testing of *Candida albicans* and *Aspergillus* species by a simple microtiter menadione-augmented 3-(4, 5-dimethyl-2-thiazolyl)-2,5-diphenyl-2H-tetrazolium bromide assay. *J Clin Microbiol*, **33**, 661–7.

Jangnegt, R., de Marie, S., et al. 1995. Liposomal and lipid formulations of amphotericin B. *Clin Pharmacokinet*, **23**, 279–91.

Johnson, L.B. and Kauffman, C.A. 2003. Voriconazole: a new triazole antifungal agent. *Clin Infect Dis*, **36**, 630–7.

Joly, V., Saint-Julien, L., et al. 1994. In vivo activity of interferon-gamma in combination with amphotericin B in the treatment of experimental cryptococcosis. *J Infect Dis*, **170**, 1331–4.

Karlowsky, J.A., Harding, G.A.J., et al. 2001. In vitro kill curves of a new semisynthetic echinocandin, LY-303366, against fluconazole-sensitive and -resistant *Candida* species. *Antimicrob Agents Chemother*, **41**, 2576–8.

Kartsonis, N., DiNubile, M.J., et al. 2003. Efficacy of caspofungin in the treatment of esophageal candidiasis resistant to fluconazole. *J Acquir Immune Defic Syndr*, **31**, 183–7.

Kirkpatrick, W.R., McAtee, R.K., et al. 1998. Comparative evaluation of National Committee for Clinical Laboratory Standards broth macrodilution and agar dilution screening methods for testing fluconazole susceptibility of *Cryptococcus neoformans. J Clin Microbiol*, **36**, 1330–2.

Kohler, S., Wheat, L.J., et al. 2000. Comparison of the echinocandin caspofungin with amphotericin B for treatment of histoplasmosis following pulmonary challenge in a murine model. *Antimicrob Agents Chemother*, **44**, 1850–4.

Kohno, S., Masoka, T. and Yamaguchi, H. 2001. A multicenter, open-label clinical study of FK463 in patients with deep mycosis in Japan. *Forty First Interscience Conference on Antimicrobial Agents and Chemotherapy*, **41**, 384, abstract.

Kovacicova, G., Krupova, Y., et al. 2000. Antifungal susceptibility of 262 bloodstream yeast isolates from a mixed cancer and non-cancer patient population: is there a correlation between in-vitro resistance to fluconazole and the outcome of fungemia? *J Infect Chemother*, **6**, 216–21.

Kramer, M.R., Marshall, S.E., et al. 1990. Cyclosporine and itraconazole interaction in heart and lung transplant recipients. *Ann Intern Med*, **113**, 327–9.

Krause, D., Schranz, J. and Birmingham, W. 2003. Safety results from a phase 3, randomized, double-blind, double-dummy study of anidulafungin (ANID) vs fluconazole (FLU) in patients with esophageal candidiasis. *Forty Third Interscience Conference on Antimicrobial Agents and Chemotherapy*, **43**, 136, abstract.

Kuroda, S., Uno, J. and Arai, T. 1978. Target substances of some antifungal agents in the cell membrane. *Antimicrob Agents Chemother*, **13**, 454–9.

Kurtz, M.B. and Rex, J.H. 2001. Glucan synthase inhibitors as antifungal agents. *Adv Protein Chem*, **56**, 423–75.

Kurtz, M.B., Douglas, C., et al. 1994a. Increased antifungal activity of L-733,560, a water-soluble, semisynthetic pneumocandin, is due to enhanced inhibition of cell wall synthesis. *Antimicrob Agents Chemother*, **38**, 2750–7.

Kurtz, M.B., Heath, I.B., et al. 1994b. Morphological effects of lipopeptides against *Aspergillus fumigatus* correlate with activities against (13)-β-D-glucan synthase. *Antimicrob Agents Chemother*, **38**, 1480–9.

Laguna, F., Rodriguez-Tudela, J.L., et al. 1997. Patterns of fluconazole susceptibility in isolates from human immunodeficiency virus-infected patients with oropharyngeal candidiasis due to *Candida albicans. Clin Infect Dis*, **24**, 124–30.

Larsen, R.A., Bozzette, S.A., et al. 1994. Fluconazole combined with flucytosine for treatment of cryptococcal meningitis in patients with AIDS. *Clin Infect Dis*, **19**, 741–5.

Lazar, J.D. and Wilner, K.D. 1990. Drug interactions with fluconazole. *Rev Infect Dis*, **12**, Suppl 1, S327–33.

Lee, S.C., Fung, C.P., et al. 2000. Clinical correlates of antifungal macrodilution susceptibility test results for non-AIDS patients with severe *Candida* infections treated with fluconazole. *Antimicrob Agents Chemother*, **44**, 2715–18.

Letscher-Bru, V. and Herbrecht, R. 2003. Caspofungin: the first representative of a new antifungal class. *J Antimicrob Chemother*, **51**, 513–21.

MacGregor, R.R., Bennett, J.E. and Ersley, A.J. 1978. Erythropoietin concentration in amphotericin B induced anemia. *Antimicrob Agents Chemother*, **14**, 270–3.

Maertens, J., Raad, I., et al. 2000. Multicenter, noncomparative study to evaluate the safety and efficacy of caspofungin (CAS) in adults with invasive aspergillosis (IA) refractory (R) or intolerant (I) to standard therapy (ST). *Fortieth Interscience Conference on Antimicrobial Agents and Chemotherapy*, **40**, 371, abstract.

Marr, K.A., Rustad, T.R., et al. 1999. The trailing end point phenotype in antifungal susceptibility testing is pH dependent. *Antimicrob Agents Chemother*, **43**, 1383–6.

Matar, M.J., Ostrosky-Zeichner, L., et al. 2003. Correlation between E-test, disk diffusion, and microdilution methods for antifungal susceptibility testing of fluconazole and voriconazole. *Antimicrob Agents Chemother*, **47**, 1647–51.

Matsumoto, T., Ajello, L., et al. 1994. Developments on hyalohyphomycosis and phaeohyphomycosis. *J Med Vet Mycol*, **32**, Suppl 1, 329–49.

Meletiadis, J., Mouton, J.W., et al. 2001. Colorimetric assay for antifungal susceptibility testing of *Aspergillus* species. *J Clin Microbiol*, **39**, 3402–8.

Meunier, F., Prentice, H.G. and Ringden, O. 1991. Liposomal amphotericin B (AmBisome): safety data from a phase II/III clinical trial. *J Antimicrob Chemother*, **28**, Suppl B, 83–91.

Mills, W., Chopra, R., et al. 1994. Liposomal amphotericin B in the treatment of fungal infections in neutropenic patients: a single-centre experience of 133 episodes in 116 patients. *Br J Haematol*, **86**, 754–60.

Mondy, K.E., Haughey, B., et al. 2002. Rhinocerebral mucormycosis in the era of lipid-based amphotericin B: case report and literature review. *Pharmacotherapy*, **22**, 519–26.

Morace, G., Amato, G., et al. 2002. Multicenter comparative evaluation of six commercial systems and the National Committee for Clinical Laboratory Standards M27-A broth microdilution method for fluconazole susceptibility testing of *Candida* species. *J Clin Microbiol*, **40**, 2953–8.

Mora-Duarte, J., Betts, R., et al. 2002. Comparison of caspofungin and amphotericin B for invasive candidiasis. *N Engl J Med*, **347**, 2020–9.

Nassar, F., Brummer, E. and Stevens, D.A. 1994. Effect of in vivo macrophage colony-stimulating factor on fungistasis of bronchoalveolar and peritoneal macrophages against *Cryptococcus neoformans. Antimicrob Agents Chemother*, **38**, 2162–4.

National Committee for Clinical Laboratory Standards. 1997. *Reference method for broth dilution antifungal susceptibility testing for yeasts: proposed standard M27-A*. Wayne, PA: National Commitee for Laboratory Standards.

National Committee for Clinical Laboratory Standards. 2002. *Reference method for broth dilution antifungal susceptibility testing of filamentous fungi: approved standard M38-A*. Wayne, PA: National Commitee for Laboratory Standards.

Nelson, P.W., Lozano-Chiu, M. and Rex, J.H. 1996. In vitro activity of L-743872 against putatively amphotericin B (AmB) and fluconazole (Flu)-resistant *Candida* isolates. *Thirty Sixth Interscience Conference on Antimicrobial Agents and Chemotherapy*, **36**, 104, abstract.

Nelson, S.M. and Cartwright, C.P. 2003. Detection of fluconazole-resistant isolates of *Candida glabrata* by using an agar screen assay. *J Clin Microbiol*, **41**, 2141–3.

Nguyen, M.H., Clancy, C.J., et al. 1998. Do in vitro susceptibility data predict the microbiologic response to amphotericin B? Results of a prospective study of patients with Candida fungemia. *J Infect Dis*, **177**, 425–30.

Oakley, K.L., Moore, C.B. and Denning, D.W. 1998. In vitro activity of the echinocandins antifungal agent LY303,366 in comparison with itraconazole and amphotericin B against *Aspergillus* spp.. *Antimicrob Agents Chemother*, **42**, 2726–30.

Oldfield, E.C., Garst, P.D., et al. 1990. Randomized, double-blind trial of 1- versus 4-hour amphotericin B infusion durations. *Antimicrob Agents Chemother*, **34**, 1402–6.

Ostrosky-Zeichner, L., Marr, K.A., et al. 2003. Amphotericin B: Time for a new gold standard. *Clin Infect Dis*, **37**, 415–25.

Pappagianis, D., Collins, M.S., et al. 1979. Development of resistance to amphotericin B in Candidalusitaniae infecting a human. *Antimicrob Agents Chemother*, **16**, 123–6.

Patterson, T.F., Kirkpatrick, W.R., et al. 1996. Comparative evaluation of macrodilution and chromogenic agar screening for determining fluconazole susceptibility of *Candida albicans. J Clin Microbiol*, **34**, 3237–9.

Perfect, J.R., Marr, K.A., et al. 2003. Voriconazole treatment for less-common, emerging, or refractory fungal infections. *Clin Infect Dis*, **36**, 9, 1122–31.

Petraitiene, R., Petraitis, V., et al. 1999. Antifungal activity of LY303366, a novel echinocandin B, in experimental disseminated candidiasis in rabbits. *Antimicrob Agents Chemother*, **43**, 2148–55.

Petraitis, V., Petraitiene, R., et al. 2003. Combination therapy in treatment of experimental pulmonary aspergillosis: synergistic interaction between an antifungal triazole and an echinocandin. *J Infect Dis*, **187**, 1834–43.

Pfaller, M.A. and Barry, A.L. 1994. Evaluation of a novel colorimetric broth microdilution method for antifungal susceptibility testing of yeast isolates. *J Clin Microbiol*, **32**, 1992–6.

Pfaller, M.A. and Rinaldi, M.G. 1993. Antifungal susceptibility testing: current state of technology, limitations, and standardization. *Infect Dis Clin North Am*, **7**, 435–44.

Pfaller, M.A. and Yu, W.L. 2001. Antifungal susceptibility testing: new technology and clinical applications. *Infect Dis Clin North Am*, **15**, 1227–61.

Pfaller, M.A., Bale, M., et al. 1994a. Multicenter comparison of a colorimetric microdilution broth method with the macrodilution method for in vitro susceptibility testing of yeast isolates. *Diagn Microbiol Infect Dis*, **19**, 9–13.

Pfaller, M.A., Grant, C., et al. 1994b. Comparative evaluation of alternative methods for broth dilution susceptibility testing of fluconazole against *Candida albicans*. *J Clin Microbiol*, **32**, 506–9.

Pfaller, M.A., Messer, S.A. and Coffmann, S. 1995. Comparison of visual and spectrophotometric methods of MIC endpoint determinations by using broth microdilution methods to test five antifungal agents, including the new triazole D0870. *J Clin Microbiol*, **33**, 1094–7.

Pfaller, M.A., Rex, J.H. and Rinaldi, M.G. 1997. Antifungal susceptibility testing: technical advances and potential clinical applications. *Clin Infect Dis*, **24**, 776–84.

Pfaller, M.A., Messer, S.A., et al. 2002. Antifungal activities of posaconazole, ravuconazole, and voriconazole compared to those of itraconazole and amphotericin B against 239 clinical isolates of *Aspergillus* spp. and other filamentous fungi: report from SENTRY Antimicrobial Surveillance Program, 2000. *Antimicrob Agents Chemother*, **46**, 1032–7.

Pfaller, M.A., Diekema, D.J., et al. 2003a. In vitro activities of voriconazole, posaconazole and four licensed systemic antifungal agents against *Candida* species infrequently isolated from blood. *J Clin Microbiol*, **41**, 78–83.

Pfaller, M.A., Messer, S., et al. 2003b. In vitro susceptibility testing of filamentous fungi: comparison of E-test and reference M38-A microdilution methods for determining posaconazole MICs. *Diagn Microbiol Infect Dis*, **45**, 241–4.

Pohjola, S.P., Viitasalo, M., et al. 1993. Itraconazole prevents terfenadine metabolism and increases risk of torsades de pointes ventricular tachycardia. *Eur J Clin Pharmacol*, **45**, 191–3.

Polak, A.M., Scholer, H.J. and Wall, M. 1982. Combination therapy of experimental candidiasis, cryptococcosis, and aspergillosis in mice. *Chemotherapy*, **28**, 461–79.

Pont, A., Williams, P.L., et al. 1982. Ketoconazole blocks adrenal steroid synthesis. *Ann Intern Med*, **97**, 370–2.

Pont, A., Graybill, J.R., et al. 1984. Effect of high dose ketoconazole on adrenal and testicular function. *Arch Intern Med*, **144**, 2150–3.

Powderly, W.G., Kobayashi, G.S., et al. 1988. Amphotericin B resistant yeast infection in severely immunocompromised patients. *Am J Med*, **84**, 826–32.

Ramage, G., Vande Walle, K., et al. 2001. Standardized method for in vitro susceptibility testing of *Candida albicans* biofilms. *Antimicrob Agents Chemother*, **45**, 2475–9.

Ramani, R. and Chaturvedi, V. 2000. Flow cytometry antifungal susceptibility testing of pathogenic yeasts other than *Candida albicans* and comparison with the NCCLS broth microdilution test. *Antimicrob Agents Chemother*, **44**, 2752–8.

Rath, P.M., Freise, J.M. and Ansorg, R. 2002. Susceptibility testing of *Aspergillus* spp. by means of an automated blood culture system. *J Antimicrob Chemother*, **50**, 115–17.

Redding, S., Smith, J., et al. 1994. Resistance of *Candida albicans* to fluconazole during treatment of oropharyngeal candidiasis in a patient with AIDS: documentation by in vitro susceptibility testing and DNA subtype analysis. *Clin Infect Dis*, **18**, 240–2.

Revankar, S.G., Dib, O.P., et al. 1998a. Clinical evaluation and microbiology of oropharyngeal infection due to fluconazole-resistant *Candida* in human immunodeficiency virus-infected patients. *Clin Infect Dis*, **26**, 960–3.

Revankar, S.G., Kirkpatrick, W.R., et al. 1998b. Interpretation of trailing endpoints in antifungal susceptibility testing by the National Committee for Clinical Laboratory Standards method. *J Clin Microbiol*, **36**, 153–6.

Rex, J.H. and Pfaller, M.A. 2002. Has antifungal susceptibility testing come of age? *Clin Infect Dis*, **35**, 982–9.

Rex, J.H., Bennett, J.E., et al. 1991. In vivo interferon-gamma therapy augments the in vitro ability of chronic granulomatous disease neutrophils to damage *Aspergillus* hyphae. *J Infect Dis*, **163**, 849–52.

Rex, J.H., Pfaller, M.A., et al. 1993. Antifungal susceptibility testing. *Clin Microbiol Rev*, **6**, 367–81.

Rex, J.H., Pfaller, M.A., et al. 1995a. Antifungal susceptibility testing of isolates from a randomized, multicenter trial of fluconazole versus amphotericin B as treatment of nonneutropenic patients with candidemia. *Antimicrob Agents Chemother*, **39**, 40–4.

Rex, J.H., Rinaldi, M.G. and Pfaller, M.A. 1995b. Resistance of *Candida* species to fluconazole. *Antimicrob Agents Chemother*, **39**, 1–8.

Rex, J.H., Pfaller, M.A., et al. 1997. Development of interpretive breakpoints for antifungal susceptibility testing: conceptual framework and analysis of in vitro-in viva correlation data for fluconazole, itraconazole, and candida infections. Subcommittee on Antifungal Susceptibility Testing of the National Committee for Clinical Laboratory Standards. *Clin Infect Dis*, **24**, 235–47.

Rex, J.H., Nelson, P.W., et al. 1998. Optimizing the correlation between results of testing in vitro and therapeutic outcome in vivo for fluconazole by testing critical isolates in a murine model of invasive candidiasis. *Antimicrob Agents Chemother*, **42**, 129–34.

Rex, J.H., Pfaller, M.A., et al. 2001. Antifungal susceptibility testing: practical aspects and current challenges. *Clin Microbiol Rev*, **14**, 643–58.

Rex, J.H., Pappas, P.G., et al. 2003. A randomized and blinded multicenter trial of high-dose fluconazole plus placebo versus fluconazole plus amphotericin B as therapy for candidemia and its consequences in nonneutropenic subjects. *Clin Infect Dis*, **36**, 1221–8.

Reynes, J., Bazin, C., et al. 1997. Pharmacokinetics of itraconazole (oral solution) in two groups of human immunodeficiency virus-infected adults with oral candidiasis. *Antimicrob Agents Chemother*, **41**, 2554–8.

Ringden, O., Meunier, F., et al. 1991. Efficacy of amphotericin B encapsulated in liposomes (AmBisome) in the treatment of invasive fungal infections in immunocompromised patients. *J Antimicrob Chemother*, **28**, Suppl B, 73–82.

Roilides, E., Holmes, A., et al. 1994. Antifungal activity of elutriated human monocytes against *Aspergillus fumigatus* hyphae: enhancement by granulocyte-macrophage colony-stimulating factor and interferon-gamma. *J Infect Dis*, **170**, 894–9.

Roilides, E., Holmes, A., et al. 1995. Effects of granulocyte colony-stimulating factor and interferon-gamma on antifungal activity of human polymorphonuclear neutrophils against pseudohyphae of different medically important *Candida* species. *J Leukocyte Biol*, **57**, 651–6.

Saag, M., Bradsher, R.W. and Chapman, S.W. 1985. Treatment of blastomycosis and histoplasmosis with ketoconazole. *Ann Intern Med*, **103**, 861–72.

Sachs, M.K., Blanchard, L.M. and Green, P.J. 1993. Interaction of itraconazole and digoxin. *Clin Infect Dis*, **16**, 400–3.

St Germain, G. 2001. Impact of endpoint definition on the outcome of antifungal susceptibility tests with *Canidida* species: 24- versus 48-hr incubation and 50 percent versus 80 percent reduction in growth. *Mycoses*, **44**, 37–45.

Schaffner, A. and Frick, P.G. 1985. The effect of ketoconazole on amphotericin B in a model of disseminated aspergillosis. *J Infect Dis*, **151**, 902–10.

Schmitt, H.J. 1993. New methods of delivery of amphotericin B. *Clin Infect Dis*, **17**, Suppl 2, S501–6.

Sewell, D.L., Pfaller, M.A. and Barry, A.L. 1994. Comparison of broth macrodilution, broth microdilution and E test antifungal susceptibility tests for fluconazole. *J Clin Microbiol*, **32**, 2099–102.

Sharkey, P.K., Rinaldi, M.G., et al. 1991. High dose itraconazole in the treatment of severe mycoses. *Antimicrob Agents Chemother*, **35**, 707–13.

Sheehan, D.J., Espinel-Ingroff, A., et al. 1993. Antifungal susceptibility testing of yeasts: a brief overview. *Clin Infect Dis*, **17**, Suppl 2, S494–500.

Sheehan, D.J., Hitchcock, C.A., et al. 1999. Current and emerging azole antifungal agents. *Clin Microbiol Rev*, **12**, 40–79.

Stevens, D.A. 1977. Miconazole in the treatment of systemic fungal infections. *Am Rev Respir Dis*, **116**, 801–5.

Stevens, D.A. 1984. Antifungal drug susceptibility testing: a critical review. *Mycopathologia*, **87**, 137–40.

Stone, J.A., Holland, S.D., et al. 2002. Single- and multiple-dose pharmacokinetics of caspofungin in healthy men. *Antimicrob Agents Chemother*, **46**, 739–45.

Sugar, A.M. 1995. Use of amphotericin B with azole antifungal drugs: what are we doing? *Antimicrob Agents Chemother*, **39**, 1907–12.

Sugar, A.M., Alsip, S.G., et al. 1987. Pharmacology and toxicity of high-dose ketoconazole. *Antimicrob Agents Chemother*, **31**, 1874–8.

Supparatpinyo, K., Chiewchanvit, S., et al. 1992. *Penicillium marneffei* infection in patients infected with human immunodeficiency virus. *Clin Infect Dis*, **14**, 871–4.

Supparatpinyo, K., Nelson, K.E., et al. 1993. Response to antifungal therapy by human immunodeficiency virus-infected patients with disseminated *Penicillium marneffei* infections and in vitro susceptibilities of isolates from clinical specimens. *Antimicrob Agents Chemother*, **37**, 2407–11.

Tassel, D. and Madoff, M.A. 1968. Treatment of *Candida* sepsis and *Cryptococcus* meningitis with 5-fluorocytosine. *JAMA*, **206**, 830–2.

Tellier, R., Krajden, M., et al. 1992. Innovative endpoint determination system for antifungal susceptibility testing of yeasts. *Antimicrob Agents Chemother*, **36**, 1619–25.

Tiballi, R.N., He, X., et al. 1995. Use of a colorimetric system for yeast susceptibility testing. *J Clin Microbiol*, **33**, 915–17.

To, W.K., Fothergill, A.W. and Rinaldi, M.G. 1995. Comparative evaluation of macrodiluton and Alamar colorimetric microdiluton broth methods for antifungal susceptibility testing of yeast isolates. *J Clin Microbiol*, **33**, 2660–4.

Tollemar, J. and Ringden, O. 1995. Lipid formulations of amphotericin B: less toxicity, but at what cost? *Drug Saf*, **13**, 207–18.

Troke, P.F., Schwartz, S., et al. 2003. Voriconazole (VRC) therapy (Rx) in 86 patients (pts) with CNS aspergillosis (CNSA): a retrospective analysis. *Forty Third Interscience Conference on Antimicrobial Agents and Chemotherapy*, **43**, abstract M1755, 476.

Tucker, R.M., Denning, D.W., et al. 1992. Interaction of azoles with rifampin, phenytoin, and carbamazepine, in vitro and clinical observations. *Clin Infect Dis*, **14**, 165–74.

Uchida, K., Yamamoto, Y., et al. 1992. Granulocyte-colony stimulating factor facilitates the restoration of resistance to opportunistic fungi in leukopenic mice. *J Med Vet Mycol*, **30**, 293–300.

Uzun, O., Kocagoz, S., et al. 1997. In vitro activity of a new echinocandin, LY303366, compared with those of amphotericin B and fluconazole against clinical yeast isolates. *Antimicrob Agents Chemother*, **41**, 1156–7.

Vanden Bossche, H. 1985. Biochemical targets for antifungal azole derivatives: hypothesis on the mode of action. *Curr Topics Med Mycol*, **1**, 3132–51.

Vanden Bossche, H., Marichal, P. and Odds, F. 1994a. Molecular mechanisms of drug resistance in fungi. *Trends Microbiol*, **2**, 393–400.

Vanden Bossche, H., Warnock, D.W., et al. 1994b. Mechanisms and clinical impact of antifungal drug resistance. *J Med Vet Mycol*, **32**, Suppl 1, 189–202.

Van der Horst, C., Saag, M., et al. 1997. Treatment of cryptococcal meningitis associated with the acquired immunodeficiency syndrome. *N Engl J Med*, **337**, 15–21.

Vandevelde, A., Mauceri, A.A. and Johnson, J.E. 1972. 5-Fluorocytosine in the treatment of mycotic infections. *Ann Intern Med*, **77**, 43–51.

Vandewoude, K., Vogelaers, D., et al. 1997. Concentrations in plasma and safety of 7 days of intravenous itraconazole followed by 2 weeks of oral itraconazole solution in patients in intensive care units. *Antimicrob Agents Chemother*, **41**, 2714–18.

Vertut-Doi, A., Ohnishi, S. and Bolard, J. 1994. The endocytic process in CHO cells, a toxic pathway of the polyene antibiotic amphotericin B. *Antimicrob Agents Chemother*, **38**, 2373–9.

Villanueva, A., Gotuzzo, E., et al. 2002. A randomized double-blind study of caspofungin versus fluconazole for the treatment of esophageal candidiasis. *Am J Med*, **113**, 294–9.

Vitt, C.R., Fidler, J.M., et al. 1994. Antifungal activity of recombinant human macrophage colony-stimulating factor in models of acute and chronic candidiasis in the rat. *J Infect Dis*, **169**, 369–74.

Viviani, M.A., Rizzardini, G., et al. 1994. Lipid-based amphotericin B in the treatment of cryptococcosis. *Infection*, **22**, 77–82.

Walsh, T.J. 2002. Echinocandins – an advance in the primary treatment of invasive candidiasis. *N Engl J Med*, **347**, 2070–2.

Walsh, T.J., Hiemenez, J.W., et al. 1997. Amphotericin B lipid complex in patients with invasive fungal infections: analysis of safety and efficacy in 556 cases. *13th Congress of the International Society for Human and Animal Mycology*, Abstract P571, 221, abstract.

Walsh, T.J., Bodensteiner, D., et al. 1998. A randomized, double-blind trial of AmBisome (liposomal amphotericin B) versus amphotericin B in the empirical treatment of persistently febrile neutropenic patients. *Thirty Seventh Interscience Conference on Antimicrobial Agents and Chemotherapy*, **37**, LM–87, abstract.

Walsh, T.J., Lutsar, I., et al. 2002a. Voriconazole in the treatment of aspergillosis, scedosporiosis and other invasive fungal infections in children. *Pediatr Infect Dis J*, **21**, 240–8.

Walsh, T.J., Pappas, P., et al. 2002b. Voriconazole compared with liposomal amphotericin B for empirical antifungal therapy in patients with neutropenia and persistent fever. *N Engl J Med*, **346**, 225–34.

Walsh, T.J., Petraitis, V., et al. 2003. Experimental pulmonary aspergillosis due to Aspergillus terreus: pathogenesis and treatment of an emerging fungal pathogen resistant to amphotericin B. *J Infect Dis*, **188**, 2, 305–19.

Wasan, K.M., Rosenblum, M.G., et al. 1994. Influence of lipoproteins on renal cytotoxicity and antifungal activity of amphotericin B. *Antimicrob Agents Chemother*, **38**, 223–7.

Wheat, J., Hafner, R., et al. 1995. Itraconazole treatment of disseminated histoplasmosis in patients with the acquired immunodeficiency syndrome. *Am J Med*, **98**, 336–42.

White, M.H., Anaissie, E.J., et al. 1997. Amphotericin B colloidal dispersion vs amphotericin B as therapy for invasive aspergillosis. *Clin Infect Dis*, **24**, 635–42.

White, M.H., Bowden, R.A., et al. 1998. Randomized double-blind clinical trial of amphotericin B colloidal dispersion vs amphotericin B in the empirical treatment of fever and neutropenia. *Clin Infect Dis*, **27**, 296–302.

Wilson, E., Thorson, L. and Speert, D.P. 1991. Enhancement of macrophage superoxide anion production by amphotericin B. *Antimicrob Agents Chemother*, **35**, 796–800.

Woods, R.A., Bard, M., et al. 1974. Resistance to polyene antibiotics and correlated sterol changes in two isolates of *Candida tropicalis* from a patient with amphotericin B resistant fungemia. *J Infect Dis*, **129**, 53–8.

Zhanel, G., Karlowsky, J.A., et al. 1997. In vitro activity of a new semisynthetic echinocandin, LY-303366, against systemic isolates of *Candida* species, *Cryptococcus neoformans*, *Blastomyces dermatitidis*, and *Aspergillus* species. *Antimicrob Agents Chemother*, **41**, 863–5.

8

Principles of antifungal therapy

PETER G. PAPPAS

Any discussion of antifungal therapy must begin with a clear understanding of the terminology that describes the strategies for prevention and treatment of invasive fungal infections. Moreover, a basic understanding of the problems inherent to clinical trial design and the challenges in interpretation of clinical trial data involving antifungals is essential for the appropriate clinical application of these therapies. This chapter provides a brief overview of these two important areas.

PROPHYLACTIC AND PREEMPTIVE THERAPY

The terms prophylaxis, targeted prophylaxis, and preemptive are often used interchangeably in conjunction with antifungal therapy. While there is some overlap in meaning, these terms represent different approaches to therapy. Prophylaxis generally refers to the broad use of antifungal therapy in a heterogeneous group of patients who are at variable risk of developing superficial or invasive fungal infection. By definition, prophylactic antifungal therapy is administered to patients who are considered to be at risk for fungal infection, but who have no symptoms of infection at the time that the antifungal agent is initiated. Prophylaxis may be systemic (e.g. oral or parenteral fluconazole) or topical (e.g. oral nystatin). Virtually any population can be given prophylactic antifungal therapy, but it is typically administered to high-risk patient populations including selected medical and surgical intensive care unit (ICU) patients (Eggimann et al. 1999; Garbino et al. 2002), allogeneic hematopoietic stem cell transplant recipients (Winston et al. 2003), and selected solid organ transplant recipients (Winston et al. 1999). The chief aim of antifungal prophylaxis is the prevention of colonization and disease due to target organisms.

Targeted prophylaxis implies a measure of selectivity, that is, utilizing specific risk factors or patient characteristics in determining the population to whom prophylaxis is administered. It is assumed that the target population is uniformly at higher risk of developing fungal infection than patients with similar underlying conditions or severity of illness, and this high-risk population is selected on the basis of certain known risk factors. For example, targeted prophylaxis using fluconazole in the ICU is often administered to patients who have multiple risk factors for invasive candidiasis such as the presence of a central venous catheter, systemic antibacterial therapy, endotracheal intubation, renal failure, and recent abdominal surgery (Garbino et al. 2002). Targeted antifungal prophylaxis with voriconazole is often given to prevent invasive aspergillosis and other mold infections among allogeneic stem cell transplant recipients at risk for graft-versus-host disease.

Preemptive antifungal therapy is administered to persons who are not only at risk but also have markers of early infection, for example, colonization with a fungal pathogen. The group that has been best described in conjunction with preemptive therapy is the liver transplant population who meet criteria for very high risk of invasive fungal infection including prolonged intraoperative time, preexisting renal failure, early colonization with *Candida* spp., retransplant for early graft failure, and choledochojejunostomy anastomosis (Collins et al. 1994).

The optimal use of any form of antifungal prophylaxis or preemptive therapy must utilize a practical and validated assessment of risk. Studies to better define specific risk factors for invasive fungal infection in a variety of clinical settings are essential, recognizing that risk varies substantially between study populations, and that each patient group is characterized by unique conditions that place them at a greater or lesser risk for fungal colonization and disease.

EMPIRIC THERAPY

Empiric antifungal therapy refers to the use of these agents among patients with findings and/or symptoms of suspected invasive fungal disease. The use of empiric antifungal therapy has been studied most extensively in persistently febrile and neutropenic patients (Walsh et al. 1999, 2002; Wingard et al. 2000). The primary goal of empiric antifungal therapy in this setting is to prevent breakthrough fungal infections and to treat baseline infections due to molds and other important fungi in this uniquely susceptible host. However, all neutropenic patients are not alike, and the risk of invasive fungal infection is directly related to the underlying condition(s) and the duration and depth of neutropenia. For instance, induction therapy for acute myelogenous leukemia is associated with a very high rate of invasive fungal infection and associated high mortality, especially due to invasive mold infections. By contrast, myeloablative therapy for most solid tumors usually leads to shorter periods of neutropenia and far less risk of invasive fungal infection. The appropriate use of empiric antifungal therapy in the setting of persistent fever and neutropenia requires that the patient have persistent fever despite a reasonable course (usually 96 hours or more) of broad-spectrum antibacterial therapy, and that there is no other obvious explanation for the clinical deterioration of the patient.

The administration of empiric antifungal therapy to persistently febrile non-neutropenic ICU patients with presumed invasive candidiasis has become common practice in developed countries throughout the world, yet studies that address the appropriateness of this practice and the best approach to therapy are lacking. Several epidemiologic studies have identified multiple risk factors associated with nosocomial candidiasis (Blumberg et al. 2001; Trick et al. 2002; Pappas et al. 2003). Despite the recognition of these risk factors, this group of patients has proven particularly difficult to enroll into clinical trials, and thus, the use of empiric antifungal therapy in the ICU patients with unexplained fever, leukocytosis, hypotension, or other evidence of clinical deterioration is the most common reason to prescribe parenteral antifungal therapy. It remains one of the most poorly understood and understudied areas in the discipline of antimicrobial therapy.

SPECIFIC THERAPY

Specific therapy refers to therapy directed at a specific pathogen which has been detected by culture, histopathology, and/or serology, or in the absence of laboratory evidence, then clinical/radiographic evidence strongly suggestive of invasive fungal disease. These include such findings as hepatosplenic bull's eye lesions suggesting chronic disseminated candidiasis, the halo or air crescent sign on chest CT suggesting invasive aspergillosis, and disseminated cutaneous lesions consistent with invasive candidiasis. Specific therapy is often based on presumptive evidence of infection, but it is nonetheless targeted towards the organism(s) most likely responsible for the clinical picture.

In practice and clinical trials, few infections have received as much attention as candidemia in the non-neutropenic host because of its prevalence and the general familiarity of most clinicians with this entity. Our current practices towards the treatment of candidemia generally reflect the results of large and well-done published comparative clinical trials in this arena (Pappas et al. 2004). Similarly, the treatment of central nervous system (CNS) cryptococcosis has become more uniform because of data generated from therapeutic trials among patients with and without the acquired immunodeficiency syndrome (AIDS) who have CNS cryptococcosis (Saag et al. 2000). By contrast, very few large clinical trials have been done among patients with less common invasive mycoses such as aspergillosis, histoplasmosis, blastomycosis, coccidioidomycosis, and sporotrichosis. Thus, the 'gold standard' of treatment for these mycoses is often based upon data generated from smaller comparative studies without significant power to discern differences in therapeutic outcome.

CLINICAL TRIAL DESIGN FOR ANTIFUNGAL COMPOUNDS

The design and implementation of the clinical trials involving antimycotic agents has been a challenge to clinicians since the availability of amphotericin B in 1958. The key problems encountered in conducting clinical trials of antifungal agents are not dissimilar to those facing the development of other antimicrobial agents; however, these issues have been recently examined in the context of systemic antifungals because of the availability of a growing number of these agents, and the growing difficulties encountered in studying them in a randomized and controlled fashion. The key challenges to the clinical investigator include:

- slow patient accrual because of restrictive eligibility criteria for uncommon diseases
- establishment of meaningful clinical end points
- the absence of validated surrogate markers of success or failure (e.g. serological studies)

- the unwillingness of many investigators to perform double-blinded clinical trials
- the need for large numbers of study centers to facilitate completion of these trials in a reasonable timeframe (Rex et al. 2001).

Patient accrual is an obstacle faced by most clinical trials of anti-infective agents, thus it is not unique to trials of antifungal agents. With the exception of candidemia and possibly CNS cryptococcosis, invasive mycoses are relatively uncommon. Yet, despite the low frequency of most invasive mycoses, the relative lack of effective agents has served as an incentive to place patients into some of these trials for less common mycoses such as histoplasmosis and blastomycosis. By contrast, randomized trials for the treatment of candidemia in non-neutropenic patients have enrolled slowly despite the relatively high incidence of candidemia in the nosocomial setting, reflecting, in part, fairly restrictive entry criteria and an inability to establish a diagnosis consistently within a short timeframe. For example, a recently published candidemia treatment trial enrolled only about 10 percent of all patients with candidemia at study sites (Pappas et al. 2003; Rex et al. 2003). Most of these potentially eligible patients were excluded due to prior antifungal therapy, abnormal laboratory values, age, comorbid conditions, and other considerations. Patient accrual into studies for therapy of invasive aspergillosis has been even more challenging. In a recent open-label trial comparing voriconazole to amphotericin B and other licensed antifungal therapy for the primary treatment of invasive aspergillosis, it required more than 5 years and almost 100 centers to enroll almost 300 eligible patients, again reflecting the effect of restrictive inclusion criteria and the difficulty in establishing a firm diagnosis of an uncommon disorder on patient accrual (Herbrecht et al. 2002).

The reluctance of many investigators to perform nonrandomized trials and the unwillingness of regulatory agencies to accept data from uncontrolled trials for purposes of licensing are other important considerations in the conduct of these trials. Double-blinded studies are difficult to perform, particularly when one of the study arms includes an amphotericin B (AmB) formulation. However, several recent studies have been successfully completed that have successfully blinded both conventional AmB and lipid formulations of AmB (Mora-Duarte et al. 2002; Rex et al. 2003). A randomized and double-blind study comparing AmB in combination with high-dose fluconazole (800 mg/day) to high-dose fluconazole alone for non-neutropenic patients with candidemia, demonstrated that a double-blinded placebo-controlled trial involving AmB can be performed successfully and without undue patient risk (Rex et al. 2003). Similarly, a randomized and double-blinded study comparing AmB to liposomal AmB for empiric antifungal therapy in persistently febrile and neutropenic patients demonstrated the ability to perform a double-blind trial with these agents (Walsh et al. 1999).

Study end points have generated considerable controversy in antifungal trials, relying almost exclusively on mycological (culture), clinical and/or radiographic parameters. Among trials comparing agents for treatment of candidemia, rendering the blood culture negative in conjunction with resolution of clinical signs and symptoms has become a standard end point for a successful outcome. However, clinical end points, including resolution of fever, hypotension, and/or other clinical signs of candidemia, are often difficult to discern due to concomitant illness and other comorbidities that are common among these patients. Thus, in most candidemia studies, an element of investigator bias is introduced in what at first glance might appear to be a fairly 'hard' end point. Similarly, end points for studies of CNS cryptococcosis have utilized a composite end point of cerebrospinal fluid culture negativity at important time intervals such as 2 and 10 weeks, together with the somewhat subjective measurement of clinical improvement based on neurological status and other clinical parameters subject to investigator bias. In both of these examples, the ability to use a surrogate marker of efficacy, such as a serological assay, might have helped improve objectivity and lessen reliance on subjective clinical measurements.

Controversy has surrounded the use of the five-component composite end point utilized in several of the comparative trials of empiric antifungal therapy in persistently febrile and neutropenic patients (Walsh et al. 1999, 2002). The five components include:

1 survival at 7 days after study drug completion
2 no discontinuation of study drug due to toxicity or lack of efficacy
3 prevention of emergence of incident fungal infection
4 successful treatment of baseline fungal infection
5 resolution of fever during antifungal therapy.

To be considered a success in these studies, all five end points must be met. Not surprisingly, these studies have had rates of success that vary between 26 and 50 percent when these criteria are applied stringently. Again, the use of a validated surrogate marker might obviate the need for such complex end points, simplifying outcome assessment by removing such vague and questionable end points as 'fever resolution,' which is prone to much subjective interpretation by investigators.

By necessity, most large antifungal trials require large numbers of participating sites to provide adequate patient accrual, adding another degree of complexity in terms of quality assurance, consistent practice patterns, and variations in interpretation of the protocol.

Despite these important considerations, much progress has been made in the design and implementation of clinical trials of antifungal agents in the last two decades. It will be important in the design and implementation of future studies to consider each of the challenges, and to strive for more efficient patient selection and more reproducible and timely results.

REFERENCES

Blumberg, H.M., Jarvis, W.R., et al. 2001. Risk factors for candidal bloodstream infections in surgical intensive care unit patients: the NEMIS prospective multicenter study. The National Epidemiology of Mycosis Survey. *Clin Infect Dis*, **33**, 177–86.

Collins, L.A., Samore, M.H., et al. 1994. Risk factors for invasive fungal infections complicating orthotopic liver transplantation. *J Infect Dis*, **170**, 644–52.

Eggimann, P., Francioli, P., et al. 1999. Fluconazole prophylaxis prevents intra-abdominal candidiasis in high-risk surgical patients. *Crit Care Med*, **27**, 1066–72.

Garbino, J., Lew, D.P., et al. 2002. Prevention of severe *Candida* infections in nonneutropenic, high-risk, critically ill patients: a randomized, double-blind, placebo-controlled trial in patients treated by selective digestive decontamination. *Intensive Care Med*, **28**, 1708–17.

Herbrecht, R., Denning, D.W., et al. 2002. Voriconazole versus amphotericin B for primary therapy of invasive aspergillosis. *N Engl J Med*, **347**, 408–15.

Mora-Duarte, J., Betts, R., et al. 2002. Comparison of caspofungin and amphotericin B for invasive candidiasis. *N Engl J Med*, **347**, 2020–9.

Pappas, P.G., Rex, J.H., et al. 2003. A prospective observational study of candidemia: epidemiology, therapy, and influences on mortality in hospitalized adult and pediatric patients. *Clin Infect Dis*, **37**, 634–43.

Pappas, P.G., Rex, J.H., et al. 2004. Guidelines for treatment of candidiasis. *Clin Infect Dis*, **38**, 161–89.

Rex, J.H., Walsh, T.J., et al. 2001. Need for alternative trial designs and evaluation strategies for therapeutic studies of invasive mycoses. *Clin Infect Dis*, **33**, 95–106.

Rex, J.H., Pappas, P.G., et al. 2003. A randomized and blinded multicenter trial of high-dose fluconazole plus placebo versus fluconazole plus amphotericin B as therapy for candidemia and its consequences in nonneutropenic subjects. *Clin Infect Dis*, **36**, 1221–8.

Saag, M.S., Graybill, R.J., et al. 2000. Practice guidelines for the management of cryptococcal disease. *Clin Infect Dis*, **30**, 710–18.

Trick, W.E., Fridkin, S.K., et al. 2002. Secular trend of hospitalized-acquired candidemia among intensive care unit patients in the United States during 1989–1999. *Clin Infect Dis*, **35**, 627–30.

Walsh, T.J., Finberg, R.W., et al. 1999. Liposomal amphotericin B for empirical therapy in patients with persistent fever and neutropenia. National Institute of Allergy and Infectious Diseases Mycoses Study Group. *N Engl J Med*, **340**, 764–71.

Walsh, T.J., Pappas, P.G., et al. 2002. Voriconazole compared with liposomal amphotericin B for empirical antifungal therapy in patients with neutropenia and persistent fever. *N Engl J Med*, **346**, 225–34.

Wingard, J.R., White, M.H., et al. 2000. A randomized, double-blind comparative trial evaluating the safety of liposomal amphotericin B versus amphotericin B lipid complex in the empirical treatment of febrile neutropenia. L Amph/ABLC Collaborative Study Group. *Clin Infect Dis*, **31**, 1155–63.

Winston, D.J., Pakrasi, A. and Busuttil, R.W. 1999. Prophylactic fluconazole in liver transplant recipients: a randomized, double-blind, placebo-controlled trial. *Ann Intern Med*, **131**, 729–37.

Winston, D.J., Maziarz, R.T., et al. 2003. Intravenous and oral itraconazole versus intravenous and oral fluconazole for long-term antifungal prophylaxis in allogeneic hematopoietic stem-cell transplant recipients. *Ann Intern Med*, **138**, 705–13.

9

Resistance to antifungal agents

SEVTAP ARIKAN AND JOHN H. REX

INTRODUCTION

The field of antifungal therapy is now enjoying its most dynamic era. The increased interest in antifungal agents is attributable to several factors. First, the fungi, which were once known only as the common causes of superficial infections, now have a well-established role in opportunistic invasive infections. The increase in the incidence of invasive mycoses originates from the increase in the number of patients who are at high risk of developing these infections. The use of more intensive chemotherapeutic regimens in patients with cancer, the appearance of acquired immunodeficiency syndrome (AIDS), and the increasing practice of organ and bone marrow transplantation all resulted in emergence of a particular host population who are highly immunocompromised and predisposed to develop invasive mycoses. The most significant aspects of these infections are that they are difficult to treat and associated with high rates of mortality (Anaissie 1992; McNeil et al. 2001; Kullberg and Lashof 2002). Second, more efficacious and less toxic antifungal agents appeared in clinical practice. Novel compounds with distinctive modes of antifungal activity and novel derivatives of the currently available agents were developed. As a consequence, we now have a diverse spectrum of antifungal agents in contrast to the limited number of such compounds available in the past (Table 9.1) (Arikan and Rex 2001, 2002). Third, due both to the increase in invasive infections and the related more common use of antifungal therapy, resistance to antifungal agents has become an issue, with both intrinsic (primary) and acquired (secondary) resistance being realized (Alexander and Perfect 1997; Bille 2000; Canuto and Rodero 2002; Sanglard and Odds 2002).

This chapter reviews:

- the definitions of resistance
- methods for in vitro antifungal susceptibility testing
- the correlation between in vitro resistance and clinical response in fungal infections
- the modes of action and mechanisms of resistance for currently available and novel antifungal drugs.

DEFINITIONS OF RESISTANCE

Meaning of resistance

Resistance, by definition, denotes the absence of activity of a drug against a particular microorganism. However, due to the multifactorial feature of the issue, the term is imprecise when used alone. It is important to distinguish in vitro, molecular, and clinical resistance (White et al. 1998). In vitro resistance is a minimum inhibitory concentration (MIC) value (µg/ml) of a drug against a particular strain that is above a predefined limit for that drug. Molecular resistance, on the other hand, is demonstration of resistance at genetic level and may include detection of point mutations, gene conversions, gene amplifications leading to overexpression, or mitotic recombinations. The best way to explore molecular mechanisms of resistance is to use matched sets of susceptible and resistant clinical isolates from the same strain. Clinical resistance is the ultimate issue and means

Table 9.1 *Classification of antifungal agents according to their chemical structure and modes of action*

Chemical structure	Antifungal agent(s)	Mode of action
Allylamines	Terbinafine, naftifine, butenafine	Inhibition of ergosterol synthesis
Azoles	Imidazoles: ketoconazole, clotrimazole, econazole, miconazole, oxiconazole, sulconazole Trizoles: fluconazole, itraconazole, voriconazole, posaconazole, ravuconazole	Inhibition of ergosterol synthesis
Echinocandins	Caspofungin, anidulafungin, micafungin	Inhibition of glucan synthesis
Fluorinated pyrimidine	Flucytosine	Nucleic acid inhibitor
Morpholine	Amorolfine	Inhibition of ergosterol synthesis
Peptide-nucleoside	Nikkomycin	Inhibition of chitin synthesis
Polyenes	Amphotericin B Lipid amphotericin B formulations Amphotericin B lipid complex (ABLC) Amphotericin B colloidal dispersion (ABCD) Liposomal amphotericin B (L-AMB) Nystatin Liposomal nystatin Pimaricin Filipin	Disruption of fungal cytoplasmic membrane
Tetrahydrofuran derivatives	Sordarins Azasordarins	Inhibition of fungal protein synthesis
Other	Griseofulvin	Antimitotic activity (spindle disruption)

clinical failure in response to treatment with an anti-fungal agent. The most striking point is that detection of resistance in vitro does not always mean clinical resistance. This follows from the multifactorial nature of the clinical response. Not only the in vitro activity of the drug but also its pharmacokinetic and pharmacodynamic properties, the immune status of the host, virulence factors of the infecting fungus, and existence of indwelling catheters determine the ultimate clinical outcome (Ghannoum and Rice 1999; Rex and Pfaller 2002). The major topic of interest is, of course, clinical resistance and we use in vitro susceptibility tests and molecular methods in an effort to predict clinical response.

Origin of resistance

The sources for the emergence of resistant fungal strains are diverse. Primary (intrinsic) resistance denotes natural resistance of a particular genus or species. The most typical example for this type of resistance is the resistance of *Candida krusei* (Collin et al. 1999) and *Aspergillus* to fluconazole (Edlind et al. 2001). Secondary (acquired) resistance, on the other hand, is acquisition of resistance or selection of a resistant strain following the pressure of antifungal therapy. Secondary resistance may emerge in various ways. One of the commonly observed types of secondary resistance is replacement with a more resistant species. Fluconazole prophylaxis or therapy, for instance, may lead to selection of a fluconazole-resistant *Candida glabrata* (Collin et al. 1999) or *C. krusei* strain (Bignardi et al. 1991; Casanovas et al. 1992). Antifungal therapy may also result in replacement with a more resistant strain. For

example, following prolonged fluconazole therapy, a previously fluconazole-susceptible strain of *Candida albicans* may become fluconazole-resistant (Revankar et al. 1996; Marr et al. 1997; Franz et al. 1999). Temporary (epigenetic) resistance is another type of secondary resistance. This mostly originates from transient gene expressions under the pressure of antifungal therapy. Such a strain quickly reverts to a susceptible phenotype once the drug pressure is eliminated (Calvet et al. 1997). Finally, a single strain may include a resistant subpopulation which may again be selected under the pressure of antifungal therapy. This entity originates from genomic instability within a single strain and is known as 'population bottleneck' (Schoofs et al. 1997; White et al. 1998). Beyond these forms of resistance, alterations in cell type, such as yeast–hypha transitions or phenotypic switches (Ha and White 1999), as well as existence of putative virulence factors (Fekete-Forgacs et al. 2000) may be associated with resistance or decreased susceptibility. However, related data are limited.

DETECTION OF RESISTANCE IN VITRO

In vitro antifungal susceptibility testing emerged as an area of very active research since the 1990s. The increase in the invasive fungal infections and availability of an increased number of antifungal agents than ever before led to a great demand for standardization of antifungal susceptibility testing methods. The first standardized procedure for *Candida* and *Cryptococcus neoformans* was proposed in 1992 as NCCLS M27-P, which was followed by M27-T in 1995, M27-A in 1997, and finally M27-A2 in 2002 (National Committee for

Clinical Laboratory Standards 2002b). This remarkable progress led to the development of a reference method for some filamentous fungi as well; the NCCLS M38-P document was published in 1998 for testing *Aspergillus*, *Fusarium*, *Rhizopus*, *Pseudallescheria*, and mycelial phase of *Sporothrix schenckii*. The M38-P document has also been revised recently and the M38-A document is now also available (National Committee for Clinical Laboratory Standards 2002a). These testing procedures are based on a microdilution format procedure and have proven valuable in many settings. Some uncertainties still remain and clinical data are required for clarification of some in vitro findings. Some modifications of the reference microdilution method and assays other than microdilution (e.g. disk diffusion, Etest, and flow cytometry) have also been investigated and appear useful (Ramani and Chaturvedi 2000; Cuenca-Estrella and Rodriguez-Tudela 2001; Rex et al. 2001; Vandenbossche et al. 2002). Nevertheless, the reference microdilution method remains as the standard procedure and in vitro–in vivo correlation studies are mostly based on microdilution data.

The challenges for the reference method in terms of detection of resistance

Despite its being a great step in the field of mycology, the reference susceptibility testing method still provides only limited data. For yeasts, the MIC breakpoints to be used in detection of resistant isolates have as yet been defined only for fluconazole, itraconazole, and flucytosine and only against *Candida* (Table 9.2). For molds, an MIC breakpoint was proposed for detection of itraconazole resistance in *Aspergillus* (>8 µg/ml) (Espinel-Ingroff et al. 2001). The situation is also quite limited for amphotericin B against both yeasts and molds. The method frequently fails to detect amphotericin B-resistant isolates. Although some modifications of the reference method [antibiotic medium 3 instead of RPMI 1640 medium (Rex et al. 1995; Lozano-Chiu et al. 1998a, b); Etest instead of microdilution (Wanger et al. 1995)] have led to easier discrimination of the resistant isolates in some investigators' hands, work by others has failed

to support the finding, particularly about the use of antibiotic medium 3 (Nguyen et al. 1998) and significant variability is the rule rather than the exception (Lozano-Chiu et al. 1997). Detection of unusually high MICs, however, may suggest putative resistance for any fungus–drug combination with undefined MIC breakpoints.

Resistant fungi

The overall data obtained by the reference method suggest that resistance is more likely for particular fungal genera or species against specific antifungal agents. This finding emphasizes the significance of identification of a fungal isolate to species level. The fungi which have been observed to display significant rates of antifungal resistance are summarized in Table 9.3. For some of these, resistance is primary and includes all strains classified in that genus (e.g. fluconazole resistance in *C. krusei*). For others, resistance is detected at variable rates for the strains of a species. Apart from those noted in Table 9.3, strain-to-strain variations in antifungal susceptibility profiles are always possible even for a genus known to be susceptible to an antifungal agent. The typical example for this entity is *C. albicans*, a species well known to be susceptible to azoles. As noted previously, fluconazole resistance may develop in a previously susceptible *C. albicans* strain following antifungal prophylaxis or therapy with fluconazole (Marr et al. 1997).

Another topic that merits discussion under this heading is cross-resistance. Cross-resistance is now more significant than ever before due to the availability of novel derivatives of the azoles. Does, for instance, a fluconazole-resistant *C. glabrata* isolate remain resistant to itraconazole or voriconazole as well? Available data suggest that cross-resistance between old and novel azoles is likely but not universal (Cartledge et al. 1997; Nguyen and Yu 1998; Goldman et al. 2000; Muller et al. 2000; Mosquera and Denning 2002). The same is true for polyenes. Some amphotericin B-resistant *Candida* isolates remain susceptible to nystatin and liposomal nystatin (Arikan et al. 2002a). These data indicate that

Table 9.2 *MIC interpretive guidelines for susceptibility testing of* Candida *(National Committee for Clinical Laboratory Standards 2002b)*

Antifungal agent	Susceptible (S)	Susceptible-dose dependent (S-DD)	Intermediate (I)	Resistant (R)
Fluconazole[a]	⩽8	16-32[b]	–	⩾64
Itraconazole	⩽0.125	0.25-0.5[c]	–	⩾1
Flucytosine	⩽4	–	8-16[d]	⩾32

a) *C. krusei* is intrinsically resistant to fluconazole and these guidelines should not be used for assessment of fluconazole susceptibility of *C. krusei*.
b) Fluconazole doses of at least 400 mg/day may be required in adults with normal renal function to achieve adequate blood levels for treating infections due to these isolates.
c) Plasma itraconazole concentrations of >0.5 µg/ml may be required for optimal response in treating infections due to these isolates.
d) Isolates with uncertain susceptibility profile.

Table 9.3 *Fungal genera/species for which a significant incidence of resistance to antifungal agents has been observed*[a]

Fungus	Antifungal agent	Type of resistance	Reference(s)
Yeasts			
Candida glabrata	Azoles	Primary/secondary	Yamaguchi et al. (1989); Collin et al. (1999); Arikan et al. (2002b); Pfaller et al. (2002)
Candida guilliermondii	Amphotericin B	Primary	Dick et al. (1985); Collin et al. (1999); Krcmery and Barnes (2002)
Candida krusei	Azoles	Primary	Rex et al. (1998); Collin et al. (1999); Pfaller et al. (2001)
	Amphotericin B	Primary	Rex et al. (1998); Arikan et al. (2002a); Krcmery and Barnes (2002)
Candida lusitaniae	Amphotericin B	Primary/secondary	Pietrucha-Dilanchian et al. (2001); Arikan et al. (2002a); Ellis (2002); McClenny et al. (2002)
Cryptococcus neoformans	Flucytosine	Primary/secondary	Block et al. (1973); Brandt et al. (2001)
Trichosporon spp.	Amphotericin B	Primary	Walsh et al. (1990); Arikan and Hascelik (2002); Toriumi et al. (2002)
Molds			
Aspergillus spp.	Fluconazole	Primary	Abrahamsen et al. (1992); Edlind et al. (2001)
Aspergillus terreus	Amphotericin B	Primary	Lass-Florl et al. (1998); Sutton et al. (1999)
Fusarium solani	Amphotericin B	Primary	Anaissie et al. (1992); Arikan et al. (1999); Ellis (2002)
Scedosporium apiospermum	Amphotericin B	Primary	Walsh et al. (1995); Espinel-Ingroff (1998); Ellis (2002)
Scedosporium prolificans	Amphotericin B	Primary	Berenguer et al. (1997); Maertens et al. (2000); Ellis (2002)

a) This table covers the fungi that are relatively commonly reported to be resistant to antifungal agents. Fungi for which resistance was detected less frequently and/or in a few strains were not included. It should be remembered that strain-to-strain variations in susceptibility are possible.

class-based in vitro susceptibility tests are not adequate for definitive determination of the susceptibility profile of an isolate.

CORRELATION BETWEEN IN VITRO AND CLINICAL RESISTANCE

A recent meta-analysis of the available in vitro–in vivo correlation studies found that the percentage of clinical success was 91 percent for infections due to isolates susceptible to the corresponding antifungal agent while it was 48 percent for those where the isolate were resistant (Rex and Pfaller 2002). These success rates were shown to be similar to the success rates for therapy of susceptible/resistant bacterial isolates, and the general concept of a 90/60 rule was proposed in which susceptible isolates responded about 90 percent of the time and resistant isolates responded about 60 percent of the time. This analysis makes it clear that in vitro susceptibility tests can predict clinical resistance. However, several other factors, such as the immune status of the host and severity of the infection in the interim, influence the clinical outcome and should be considered.

Noteworthy, in vitro–in vivo correlation studies for amphotericin B are not included in this meta-analysis. Available data are limited and remain ambiguous in this respect. Although a correlation between in vitro and clinical resistance could be demonstrated in some *Candida* (Nguyen et al. 1998) and *Aspergillus* (Lass-Florl et al. 1998) infections by using a proposed MIC breakpoint value, further data to support these findings, unfortunately, are not available.

MECHANISMS OF ANTIFUNGAL RESISTANCE

The cellular and molecular bases of resistance are tightly related to the modes of antifungal activity. The modes of action of currently available and novel antifungal agents are summarized in Table 9.1.

Polyenes

This group of antifungal agents includes amphotericin B, nystatin, lipid formulations of amphotericin B and nystatin, pimaricin, and filipin (Table 9.1). The most significant agent in terms of clinical implications of resistance is amphotericin B and most of our knowledge about polyene resistance is provided by clinical and investigational data for amphotericin B.

Amphotericin B has been in clinical use since the 1960s and remains a mainstay of systemic antifungal therapy. The use of amphotericin B was further augmented by the development of its lipid formulations [amphotericin B lipid complex (ABLC), amphotericin B colloidal dispersion (ABCD), and liposomal amphotericin B (L-AMB)] in the 1990s. These lipid formulations are at least as efficacious as the parent drug and definitely less nephrotoxic than the conventional formulation, amphotericin B deoxycholate. Their acquisition cost, however, is remarkably high (Arikan and Rex 2001, 2002).

As noted previously, the major challenge with in vitro amphotericin B susceptibility testing is the frequent inability of the reference method in detection of resistant isolates. Unusually high MICs of amphotericin B have been remarkable for isolates of *Candida guilliermondii*, *C. krusei*, *Candida lusitaniae*, *Trichosporon* spp., *Aspergillus terreus*, *Scedosporium apiospermum*, *Scedosporium prolificans*, and *Fusarium solani* (Table 9.3). Although some differences have been detected between the in vitro activity of amphotericin B deoxycholate and its lipid formulations (Pahls and Schaffner 1994; Carrillo-Munoz et al. 1999; Oakley et al. 1999), based on the fact that the compound responsible for the antifungal activity is identical in all formulations, most authorities suggest that testing the in vitro activity of lipid formulations is not required. Importantly, in vivo resistance to amphotericin B is not common and, when present, usually originates from the impaired immune status of the host. Development of secondary resistance to amphotericin B is also not a major problem, despite its being in clinical use for more than 40 years. This is probably due to the fungicidal nature of the agent.

The antifungal activity of amphotericin B and other polyenes is via a specific interaction between the agent and ergosterol, the bulk sterol found in fungal cell membrane. Ergosterol is an essential element to maintain fluidity and integrity of the membrane and proper functioning of the membrane-bound enzymes. Amphotericin B intercalates into the membrane and generates channels and pores. Through these pores, several cellular components, particularly potassium and magnesium ions, leak and destroy the proton gradient within the membrane, leading to death of the fungal cell (Vanden Bossche et al. 1994a; Ellis 2002). In addition, it has previously been reported that amphotericin B may be cytotoxic to *Candida* via its prooxidative activity (Sokol-Anderson et al. 1986). However, and contrary to this finding, more recent data suggest that amphotericin B may act as an antioxidant (Osaka et al. 1997) and this question remains unresolved. The affinity of amphotericin B is high for ergosterol and less for 3-hydroxy or 3-oxo sterols. This low affinity for sterols such as fecosterol and episterol may be significant in emergence of resistance to amphotericin B (Ghannoum and Rice 1999).

Any quantitative or qualitative change in sterol content of the fungal cell influencing the amount or the availability of ergosterol may result in resistance to amphotericin B (Table 9.4). The three mechanisms of

Table 9.4 *Possible mechanisms of amphotericin B resistance*

Mechanism
Reduced ergosterol content of the fungal cytoplasmic membrane (defective *ERG2* or *ERG3* genes)
Alterations in sterol content of the membrane (predominancy of sterols with reduced affinity, such as fecosterol and episterol)
Alterations in sterol to phospholipid ratio in the membrane
Reorientation or masking of ergosterol in the membrane
Stationary growth phase of the fungus
Increase in the levels of alkali-soluble and -insoluble glucans in the cell wall
Production of melanin
Previous exposure to azoles

quantitative ergosterol changes are decrease in amount of ergosterol due to inhibition of its synthesis (Merz and Sanford 1979; Dick et al. 1980; Peyron et al. 2002), replacement of ergosterol with episterol, fecosterol or other sterols (Woods et al. 1974; Kelly et al. 1994; Peyron et al. 2002), and alterations in the ratio of sterols to phospholipids. A fourth mechanism that may lead to resistance is a qualitative change in ergosterol; its reorientation or masking in the membrane. This in turn results in sterically or thermodynamically less favored binding with amphotericin B (Ghannoum and Rice 1999). A fifth mechanism related to amphotericin B resistance is related to the growth phase of the fungal cell. As expected, during exponential growth, breakdown and resynthesis of the cell wall occurs at a high rate, providing enhanced access of amphotericin B to the cell membrane. By contrast, at stationary growth phase, breakdown and synthesis of the cell wall occurs at a much lower rate, leading to relative resistance to amphotericin B (Gale et al. 1975).

There are two newly proposed mechanisms associated with amphotericin B resistance. The first is related to the cell wall. Alterations in the cell wall components of mycelia, specifically an increase in the levels of alkali-soluble and insoluble glucans, have been postulated to lead to amphotericin B resistance in *Aspergillus flavus* (Seo et al. 1999). The second is related to the production of melanin. Killing assays demonstrated that melanized isolates of *Cryptococcus neoformans* and *Histoplasma capsulatum* were less susceptible to amphotericin B compared to the nonmelanized strains. Of interest, the reference M27-A method failed to detect this melanin-associated difference in amphotericin B susceptibility (van Duin et al. 2002).

Previous exposure to an azole compound may also be responsible for amphotericin B resistance. The key issue here is the depletion of ergosterol, the target molecule of amphotericin B, by the previous antifungal activity of the azole. This kind of interaction of an azole with amphotericin B is one of the major examples for antagonism of the action of one antifungal by that of another (Schaffner and Böhler 1993; Martin et al. 1994; Sugar and Liu 1998). However, the related clinical implications are complex and the details are beyond the scope of this chapter.

MOLECULAR ASPECTS OF AMPHOTERICIN B RESISTANCE

Only a few studies have focused on the genetic basis of amphotericin B resistance in clinical isolates. Several enzymes take part in the chain of reactions starting from squalene and ending up with synthesis of ergosterol. In this ergosterol synthesis pathway, the mutations in two genes encoding two enzymes have been found to be associated with amphotericin B resistance. C-8 sterol isomerase catalyzes the production of episterol from fecosterol and is regulated by the *ERG2* gene. C-5 sterol desaturase, on the other hand, is one of the enzymes in conversion of episterol to ergosterol and is encoded by the *ERG3* gene. Clinical strains of amphotericin B-resistant *C. albicans* with reduced ergosterol content and defective *ERG2* or *ERG3* genes have been reported (Broughton et al. 1991; Haynes et al. 1996; Nolte et al. 1997). The stability of these mutations remains unknown. Sterol analysis of an amphotericin B-resistant mutant isolate of *C. neoformans* has also suggested an alteration in the *ERG2* gene (Kelly et al. 1994).

Azoles

This group of antifungal agents includes imidazoles and triazoles (Table 9.1). Fluconazole is the most widely used azole compound. Its favorable clinical efficacy and safety, as well its availability in both oral and parenteral formulations, led to widespread use of fluconazole for both treatment and prophylaxis. With such broad usage, it is perhaps not surprising that fluconazole-resistant isolates were soon noted and this led to efforts to develop novel compounds with favorable activity against fluconazole-resistant organisms.

The antifungal activity of the azoles is via inhibition of lanosterol (14-alpha) demethylase, an enzyme which functions in synthesis of ergosterol from lanosterol (Sanati et al. 1997). There are data suggesting that the azoles primarily target the heme protein which cocatalyzes cytochrome P450-dependent 14-alpha-demethylation of lanosterol (Hitchcock et al. 1990). The inhibition of demethylation results in depletion of ergosterol and accumulation of 14-alpha-methylated sterols such as lanosterol, 4,14-dimethylzymosterol, and 24-methylene-dihydrolanosterol. Finally, the integrity of the membrane and its functions, such as nutrient transport and chitin synthesis, are lost, leading to inhibition of fungal growth. The mechanism of antifungal activity may vary depending on the genus of the fungus. For example, fluconazole and itraconazole affect the reduction of obtusifolione to obtusifoliol as well as inhibiting 14-alpha-demethylase in *C. neoformans* and both mechanisms result in accumulation of methylated sterol precursors (Vanden Bossche et al. 1993; Ghannoum et al. 1994).

The clinical effect of inhibition of sterol biosynthesis on human enzymes merits discussion. This adverse effect is mostly observed for ketoconazole. Azoles block the synthesis of mammalian cholesterol as well at the stage of 14-alpha demethylation. However, it has been demonstrated by Hitchcock et al. (1995) that the 50 percent inhibitory concentration of voriconazole needed to inhibit 14-alpha-demethylase of rat liver cholesterol is much higher than that required to inhibit fungal demethylase (7.4 μM versus 0.03 μM).

As shown in Table 9.5, azole resistance may originate from qualitative or quantitative changes in the target enzyme lanosterol demethylase, reduction in the intercellular concentration of the drug or a combination of these mechanisms (Lupetti et al. 2002; Sanglard and Odds 2002). Qualitative alterations in the enzyme frequently originate from the mutations in the encoding gene and result in reduced binding affinity of the enzyme for azoles. Interestingly, species-dependent intrinsic qualitative differences may also be observed. Compared to that of *C. albicans*, lanosterol demethylase of *C. krusei* displays reduced susceptibility to inhibition by fluconazole (Orozco et al. 1998). Quantitative modifications, on the other hand, are due to increased copy number of the enzyme and lead to increased ergosterol synthesis due to its overexpression. Azole resistance may sometimes develop due to its unfavorably low intercellular concentration. This may be observed for two reasons. First, the azole may penetrate poorly across the membrane due to the alterations in sterol and/or phospholipid composition of the membrane and related reduced permeability. Second, and more importantly, the azole can readily get access to its target initially but is then pumped out by overexpressed efflux systems (Ghannoum and Rice 1999). There are no reports suggesting the relevance of modification of azoles as a mechanism of resistance.

Table 9.5 *Possible mechanisms of azole resistance*

Mechanism
Alteration in lanosterol demethylase
Overexpression of lanosterol demethylase
Reduction in intercellular concentration of the azole
1 Overexpression of energy-dependent efflux systems
i   ATP-binding cassette (ABC) superfamily proteins
ii Major facilitator superfamily proteins
2 Changes in sterol and/or phospholipid composition of fungal cell membrane leading to decreased permeability

MOLECULAR ASPECTS OF AZOLE RESISTANCE

Lanosterol demethylase is encoded by the *ERG11* gene and the active site pocket of the enzyme is located on top of the heme cofactor. This gene has previously been referred to as *ERG16* and *CYP51A1* in *C. albicans*. Various genetic alterations in *ERG11*, including point mutations, overexpression, gene amplification leading to overexpression, gene conversion, and mitotic recombination have been observed in *C. albicans* (White et al. 1998) and appear related to azole resistance. In addition, alterations in other *ERG* genes encoding enzymes that function in ergosterol biosynthetic pathway may also be associated with azole resistance. Among these are *ERG3* (Kelly et al. 1997) and *ERG5* (White et al. 1998).

Point mutations in *ERG11*

Point mutations in *ERG11* associated with azole resistance have been reported. While some of them have been identified in clinical isolates, others were developed in laboratory strains. Seven different point mutations associated with azole resistance have so far been defined. However, since matched sets of susceptible and resistant isolates from the same strain have not been tested, it is hard to know whether these mutations are the actual causes of resistance (Loeffler et al. 1997). Moreover, recent data suggest that the resistance mechanisms identified in matched sets of susceptible and resistant isolates frequently remain insufficient to explain resistance in unmatched clinical isolates (White et al. 2002). A point mutation leading to replacement of arginine with lysine at amino acid 467 (abbreviated as *R467K*) has been found to be associated with azole resistance in a clinical strain of *C. albicans* when matched strains were tested (White 1997b). Another point mutation was constructed in the laboratory in a *C. albicans* strain, leading to the replacement of threonine with alanine at position 315 (*T315A*) and this mutation has also been studied in *Saccharomyces cerevisiae*. Importantly, the *T315A* mutation was located in the active site pocket of the enzyme and led to reduction in azole binding and fluconazole resistance (Lamb et al. 1997). *D116E* and *E266D* are the other two of the most common point mutations in *ERG11* of *C. albicans*. However, these mutations are frequently observed and are not necessarily associated with resistance (White et al. 2002). Geber et al. (1995) demonstrated that the *ERG11* deletion mutant of *C. glabrata* was resistant to fluconazole (at 100 µg/ml) and itraconazole (at 16 µg/ml) as well as amphotericin B (at 2 µg/ml).

Overexpression of *ERG11*

Although overexpression of the *ERG11* gene has been shown in clinical isolates of *C. albicans* resistant to azoles, the contribution of this phenomenon alone to development of resistance remains unknown (Sanglard et al. 1995; Albertson et al. 1996; White 1997a; Sanglard et al. 1998). Other alterations, such as the *R467K* mutation and the overexpression of the genes encoding efflux pump systems, have always been present in the resistant isolates with an overexpressed *ERG11* gene. Importantly, recent data from a collection of unmatched clinical isolates of *C. albicans* suggest that overexpression of *ERG11* is not correlated with azole resistance (White et al. 2002). Overexpression of *ERG11* may originate from an increase in the number of gene copies and *ERG11* gene amplification is another postulated mechanism for azole resistance. In a clinical isolate of resistant *C. glabrata*, *ERG11* gene amplification, increase in *ERG11* mRNA levels, and an associated increase in lanosterol demethylase have been detected. The gene amplification observed in this isolate was linked to a chromosome duplication. In the same isolate, increased drug efflux and change in *ERG7* activity was also present (Vanden Bossche et al. 1992, 1994a, b; Marichal et al. 1997). Overexpression of *ERG11* in *C. albicans*, however, could not be linked to *ERG11* amplification by some authorities (Sanglard et al. 1995; White 1997a).

Efflux pumps

There are two types of efflux pumps that contribute to azole resistance; ATP binding cassette (ABC) transporters and major facilitator superfamily (MFS) proteins. The ABC system uses ATP as the energy source while the MFS proteins use proton motive force of the membrane (the H^+ gradient across the membrane). Efflux pump systems have been found to be associated with fluconazole and itraconazole resistance in *C. krusei* (Marichal et al. 1995; Venkateswarlu et al. 1996) and *C. glabrata* (Sanglard et al. 1999), itraconazole resistance in *Aspergillus fumigatus* (Denning et al. 1997), and secondary fluconazole resistance in *C. albicans* (Marr et al. 1997).

The ABC transporters are encoded by several genes. The major ABC genes proven to be correlated with azole resistance are the *CDR* genes. *CDR* genes appear to function in efflux of various azoles. Overexpression of *CDR1* (Sanglard et al. 1995; White 1997a; Maebashi et al. 2001; White et al. 2002) and *CDR2* (Maebashi et al. 2001; White et al. 2002) have been correlated with azole resistance in *C. albicans* strains. Similar to the data demonstrated for *C. albicans*, the *CgCDR1* gene was found to be involved in azole resistance in *C. glabrata* (Sanglard et al. 1999). The *ABC1* gene, on the other hand, appears to be potentially involved in azole resistance of *C. krusei* (Katiyar and Edlind 2001).

The data concerning the genetic regulation of MFS proteins are more limited than those for the ABC system. *MDR1* (*BEN-R*) is the major MFS gene cloned from *C. albicans* and appears to be specific for fluconazole. Current data suggest a correlation between *MDR1* overexpression and azole resistance in *C. albicans* (Sanglard et al. 1995; Albertson et al. 1996; White 1997a) and *C. dublininensis* (Wirsching et al. 2001). No correlation could be found between overexpression of *FLU1*, which is an efflux pump gene related to *MDR1*, and azole resistance (White et al. 2002).

The overexpression of *CDR* and *MDR1* genes has so far been demonstrated by the increase in their mRNA levels. Increased mRNA levels of CDR were found to be associated with development of rapid, transient fluconazole resistance in *C. albicans* (Marr et al. 1998). The definitive causes of the increase in mRNA levels, however, remain yet unknown (Lyons and White 2000). It may result from gene amplification, increased transcription from endogenous copies of the gene, or increase in half-life of mRNA and its slower degradation (Borst 1991).

Decreased membrane permeability

Changes in sterol and/or phospholipid composition of the fungal cell membrane have been postulated by some investigators to lead to decreased membrane permeability and a reduction in the amount of the drug taken up by the fungal cell. The most significant alteration in sterol composition was found to be the depletion of ergosterol and its replacement with methlyfecosterol and other methylated sterols in *C. albicans* (Kohli et al. 2002). However, other workers reported that no correlation between the sterol composition and azole resistance could be demonstrated in *C. albicans* (Ghannoum and Rice 1999). A recent study by Kohli et al. (2002) demonstrated that changes in status of the membrane lipid phase, asymmetrical distribution of phosphatidylethanolamine and its increased exposition to the outer monolayer could contribute to azole resistance in *C. albicans*.

Cross-resistance in azoles

CDR overexpression and the *R467K* point mutation in *ERG11* appear to be associated with azole cross-resistance. Due to its specificity for fluconazole, *MDR1* overexpression does not lead to cross-resistance to other azoles (Sanglard et al. 1995, 1996). Whether other point mutations and mechanisms result in cross-resistance awaits further analysis.

Current status

Azole resistance is a gradual process and involves several alterations, particularly in strains under continuous antifungal drug pressure (Marr et al. 1997; Franz et al. 1998; Lopez-Ribot et al. 1998). The data obtained by analysis of ergosterol biosynthetic pathway in a particular genus or species may not be relevant to others, including the closely related ones. The contribution of some mechanisms alone remains unknown and available

data suggest that further mechanisms may have a significant role in development of resistance. One of the novel mechanisms reported to be associated with fluconazole resistance in *C. albicans* is sequestration of fluconazole in vesicular vacuoles (Maebashi et al. 2002). Importantly, simultaneous contribution of several mechanisms to fluconazole resistance in *C. albicans* strains was verified by Perea et al. (2001). In this study, overexpression of efflux pump genes, amino acid substitutions in lanosterol demethylase due to mutations in *ERG11* genes, and overexpression of *ERG11* were detected in 85, 65, and 35 percent, respectively. Overall, multiple mechanisms of resistance were found to be active in 75 percent of the studied isolates.

Allylamines

This group of antifungal agents (Table 9.1) yield antifungal activity via inhibition of squalene epoxidase enzyme (encoded by the *ERG1* gene) which converts squalene to 2,3-oxidosqualene in the ergosterol biosynthetic pathway. This in turn leads to accumulation of squalene and a resulting increase in membrane permeability and disruption of cellular organization. Terbinafine is the most commonly used allylamine derivative in clinical practice. While it has proven to be efficacious in cutaneous mycoses, subcutaneous mycoses (including sporotrichosis, chromoblastomycosis, and maduromycosis), refractory and disseminated aspergillosis, infections due to some azole-resistant *Candida* strains, pheohyphomycosis, and cutaneous leishmaniasis (Ghannoum and Elewski 1999; Perez 1999), dermatophytosis is the most common clinical indication for terbinafine use (Gupta and Shear 1997).

Cases of onychomycosis that failed to respond to terbinafine have been seen particularly with *Trichophyton rubrum* infection. However, in vitro susceptibility data suggested that clinical failure in these cases was not due to development of resistance to terbinafine and host-related factors might have contributed (Bradley et al. 1999). More recently, primary resistance to terbinafine was observed in a clinical isolate of *T. rubrum* from a patient with onychomycosis, suggesting that terbinafine-resistant dermatophytes are rare, but do exist (Leidich et al. 2001).

MOLECULAR ASPECTS OF ALLYLAMINE RESISTANCE

Due to the very rare detection of terbinafine-resistant dermatophytes, no data are available on genetic analysis of terbinafine resistance in this particular group of fungi. However, parallel data for other fungal genera suggest that efflux pumps may play a role in terbinafine resistance. Vanden Bossche et al. (1998) reported that overexpression of *MDR1* may be related to resistance to fluconazole and terbinafine while that of *CDR1* may lead to cross-resistance to fluconazole, itraconazole, ketoconazole, and terbinafine in *S. cerevisiae*. Also, *ERG* upregulation, particularly of *ERG11*, has been observed in vitro following treatment of *C. albicans* cultures with azoles and terbinafine (Henry et al. 2000).

Flucytosine

Flucytosine (5-fluorocytosine) is a fluorinated pyrimidine that exerts antifungal activity via inhibition of nucleic acid and protein synthesis. Flucytosine is taken up into the cell by fungal cytosine permease and then is deaminated to 5-fluorouracil (5-FU) via action of cytosine deaminase. 5-FU is initially converted to 5-fluorodeoxy-uridine monophosphate and 5-fluorouridylic acid. The latter reaction is catalyzed by uracil phosphoribosyl-transferase (UPRTase). Further phosphorylation results in production of 5-fluorouracil triphosphate (5-FUTP). There are two key metabolites of flucytosine that provide antifungal activity. The first, 5-fluorodeoxyuridine monophosphate, inhibits DNA synthesis via inhibition of thymidylate synthase. The second, 5-FUTP, is incorporated into RNA and disrupts protein synthesis (Polak 1977).

Flucytosine is primarily indicated for treatment of cryptococcal meningitis in combination with amphotericin B and/or fluconazole due to its favorable interaction with amphotericin B in this particular setting (Dismukes et al. 1987; Rodero et al. 2000). Primary resistance to flucytosine has been reported in *C. albicans*, non-*albicans Candida*, *C. neoformans*, and *Aspergillus*, while secondary resistance has been observed in *C. albicans* and *C. neoformans* (Vanden Bossche et al. 1994b, 1998; Barchiesi et al. 2000; Cuenca-Estrella et al. 2001). The key precaution to prevent development of secondary resistance to flucytosine is to avoid flucytosine monotherapy (Vermes et al. 2000). Resistance to flucytosine may develop due to loss of permease activity, a defect in cytosine deaminase activity or decrease in the activity of UPRTase enzyme. The defect in cytosine deaminase usually results in primary resistance, while a decrease in UPRTase activity leads to secondary resistance in *C. albicans* and *C. neoformans* (Whelan 1987; Vanden Bossche et al. 1994b). Loss of permease activity, on the other hand, has been found to be associated with flucytosine resistance in *C. glabrata* and *S. cerevisiae*, but does not seem to be significant in *C. albicans* and *C. neoformans* (Jund and Lacroute 1970; Whelan 1987).

MOLECULAR ASPECTS OF FLUCYTOSINE RESISTANCE

Genetic basis of flucytosine resistance has been explored in several fungi, particularly in *C. albicans* and *C. neoformans*. Two resistance genes, *FCY1* and *FCY2*, have been defined for flucytosine (Alexander and Perfect 1997). Since the fungus is diploid, it has two alleles for each gene and the sequential differences between the two alleles lead to *FCY* homozygotes and heterozygotes. Accordingly, a strain may be homozygous sensitive (*FCY/FCY*), homozygous resistant (*fcy/fcy*), or heterozygous (*fcy/FCY*). The heterozygotes are slightly resistant to flucytosine but they occur at significantly high ratio. Also, they carry preexisting resistance determinants which may lead to clinical treatment failures following development of homozygosity. Resistance in *fcy1/fcy1* strains was found to be associated with decreased UPRTase activity, whereas resistance in *fcy2/fcy2* strains was associated with decreased cytosine deaminase activity (Whelan 1987).

Echinocandins

Echinocandins (Table 9.1) are lipopeptides which exert antifungal activity via noncompetitive inhibition of synthesis of 1,3-beta-D-glucan, the essential cell wall homopolysaccharide in several pathogenic fungi. Glucan synthase is a heteromeric enzyme complex composed of two subunits; the large integral membrane protein referred to as FKS1 or FKS2 (encoded by *FKS1* and *FKS2* genes which are highly homologous), and the small subunit more loosely associated with the membrane and referred to as *RHO1*. *FKS1p* constitutes the majority of the activity in vegetatively growing cells while the *FKS2* gene product is required for sporulation. Although some investigators have postulated that inhibition of glucan synthesis by echinocandins is via inhibition of glucan synthase enzyme, definitive evidence to support this theory is lacking. According to these authorities, echinocandins affect the function of the Fksp component from either *FKS* gene (Kurtz and Douglas 1997).

Echinocandins are active against *Candida* and *Aspergillus* spp. Based on their distinct mode of action, activity of echinocandins against azole-resistant strains is one of their most significant features (Cuenca-Estrella et al. 2000). Caspofungin is as yet the only licensed novel echinocandin in the USA and is indicated for salvage therapy in refractory invasive aspergillosis (Groll and Walsh 2001). *C. neoformans* and *Fusarium* spp. are intrinsically resistant to echinocandins. The data obtained by Feldmesser et al. (2000) suggest that the limited action of the echinocandins against *C. neoformans* may originate from their reduced activity against glucan synthase of this particular organism or from yet undefined mechanisms. Interestingly, the production of melanin has recently been demonstrated to lead to reduced susceptibility to caspofungin in *C. neoformans* and *H. capsulatum* (van Duin et al. 2002). Echinocandin-resistant mutants of *S. cerevisiae* and *C. albicans* have been studied. The results obtained by Kurtz et al. (1996) suggested that these mutants may have alterations in glucan synthase activity. The use of caspofungin has been limited to date and reports of secondary resistance are not available.

MOLECULAR ASPECTS OF ECHINOCANDIN RESISTANCE

Although genetic studies in *S. cerevisiae* have shown that mutations in the *FKS1* and *FKS2* genes result in caspofungin resistance, no data have been available about the influence of gene overexpression on caspofungin resistance. *FKS1* mutations conferred high-level resistance to echinocandins. Mutations in another cell wall synthesis gene, *GNS1* (encodes an enzyme that functions in fatty acid elongation), on the other hand, were shown to result in low level resistance (Kurtz and Douglas 1997). In addition and recently, a novel mechanism of echinocandin resistance has been proposed. Osherov et al. (2002) demonstrated that overexpression of Sbe2p (encodes a Golgi protein involved in the transport of cell wall components) under the regulated control of the *GAL1* promoter results in caspofungin resistance in *S. cerevisiae.*

Griseofulvin

Griseofulvin has remained a first-line drug for treatment of dermatophytosis for may years. However, following the emergence of more effective and less toxic alternatives, itaconazole and terbinafine, the clinical use of griseofulvin is now limited (Zaias et al. 1996). Griseofulvin exerts antifungal activity via inhibition of fungal mitosis by disrupting the mitotic spindle through interaction with polymerized microtubules (Gull and Trinci 1973). Cases refractory to griseofulvin therapy have been reported (Lukacs et al. 1994). The clinical success rate in dermatophytic onychomycosis may be as low as 30 percent (Roberts 1994). The lack of a standardized method for susceptibility testing of dermatophytes and the absence of a defined griseofulvin MIC breakpoint limit our ability to estimate the rate of in vitro resistance to griseofulvin. Griseofulvin yielded higher MICs compared to terbinafine and itraconazole when tested against clinical *Trichophyton rubrum* isolates (Korting et al. 1995; Jessup et al. 2000) and was found to be less active than voriconazole against most dermatophytes (Wildfeuer et al. 1998). The in vitro activity of griseofulvin against *Trichophyton mentagrophytes* was less when compared to that against *T. rubrum* (Korting et al. 1995). Laboratory mutants resistant to griseofulvin have been observed following exposure to increasing gradients of the drug, mutagenic treatment or UV radiation (Aytoun et al. 1960; Iwata 1991; Fachin et al. 1996). Data on griseofulvin resistance are thus limited and the molecular basis of resistance mechanisms remains uncertain.

CONCLUSION AND FUTURE DIRECTIONS

Antifungal resistance is a complex, gradual, and multifactorial issue. Areas of significant uncertainty persist and molecular assays to detect antifungal resistance are not simple. Based on the dynamic interaction between the host, the infecting fungus, and the antifungal agent, antifungal resistance is not the only factor influencing clinical outcome. The immune status of the host plays a very important role in this regard. Proper dosing strategies and restricted and well-defined indications for antifungal prophylaxis appear as the future directions to avoid development of secondary resistance.

REFERENCES

Abrahamsen, T.G., Widing, E., et al. 1992. Disseminated fungal disease resistant to fluconazole treatment in a child with leukemia. *Scand J Infect Dis*, **24**, 391–3.

Albertson, G.D., Niimi, M., et al. 1996. Multiple efflux mechanisms are involved in *Candida albicans* fluconazole resistance. *Antimicrob Agents Chemother*, **40**, 2835–41.

Alexander, B.D. and Perfect, J.R. 1997. Antifungal resistance trends towards the year 2000: Implications for therapy and new approaches. *Drugs*, **54**, 657–78.

Anaissie, E. 1992. Opportunistic mycoses in the immunocompromised host: experience at a cancer center and review. *Clin Infect Dis*, **14**, Suppl 1, S43–53.

Anaissie, E.J., Hachem, R., et al. 1992. Lack of activity of amphotericin B in systemic murine fusarial infection. *J Infect Dis*, **165**, 1155–7.

Arikan, S. and Hascelik, G. 2002. Comparison of NCCLS microdilution method and Etest in antifungal susceptibility testing of clinical *Trichosporon asahii* isolates. *Diagn Microbiol Infect Dis*, **43**, 107–11.

Arikan, S. and Rex, J.H. 2001. Lipid-based antifungal agents: current status. *Curr Pharm Design*, **7**, 393–415.

Arikan, S. and Rex, J.H. 2002. New agents for the treatment of systemic fungal infections – current status. *Expert Opin Emerg Drugs*, **7**, 3–32.

Arikan, S., Lozano-Chiu, M., et al. 1999. Microdilution susceptibility testing of amphotericin B, itraconazole, and voriconazole against clinical isolates of *Aspergillus* and *Fusarium* species. *J Clin Microbiol*, **37**, 3946–51.

Arikan, S., Ostrosky-Zeichner, L., et al. 2002a. In vitro activity of nystatin compared with those of liposomal nystatin, amphotericin B, and fluconazole against clinical *Candida* isolates. *J Clin Microbiol*, **40**, 1406–12.

Arikan, S., Sancak, B. and Hascelik, G. 2002b. In vitro activity of caspofungin compared to that of amphotericin B, fluconazole, and itraconazole against *Candida* species. *4th European Congress of Chemotherapy*, Paris, France, abst. no. PM144.

Aytoun, R.S.C., Campbell, A.H., et al. 1960. Mycological aspects of action of griseofulvin against dermatophytes. *Arch Dermatol*, **81**, 650–6.

Barchiesi, F., Arzeni, D., et al. 2000. Primary resistance to flucytosine among clinical isolates of *Candida* spp.. *J Antimicrob Chemother*, **45**, 408–9.

Berenguer, J., Rodriguez-Tudela, J.L., et al. 1997. Deep infections caused by *Scedosporium prolificans*. A report on 16 cases in Spain and a review of the literature. *Scedosporium prolificans* Spanish Study Group. *Medicine (Baltimore)*, **76**, 256–65.

Bignardi, G.E., Savage, M.A., et al. 1991. Fluconazole and *Candida krusei* infections (letter). *J Hosp Infect*, **18**, 326–7.

Bille, J. 2000. Mechanisms and clinical significance of antifungal resistance. *Int J Antimicrob Agents*, **16**, 331–3.

Block, E.R., Jennings, A.E. and Bennett, J.E. 1973. 5-Fluorocytosine resistance in *Cryptococcus neoformans*. *Antimicrob Agents Chemother*, **3**, 649–56.

Borst, P. 1991. Genetic mechanisms of drug resistance. A review. *Acta Oncol*, **30**, 87–105.

Bradley, M.C., Leidich, S., et al. 1999. Antifungal susceptibilities and genetic relatedness of serial *Trichophyton rubrum* isolates from patients with onychomycosis of the toenail. *Mycoses*, **42**, 105–10.

Brandt, M.E., Pfaller, M.A., et al. 2001. Trends in antifungal drug susceptibility of *Cryptococcus neoformans* isolates in the United States: 1992 to 1994 and 1996 to 1998. *Antimicrob Agents Chemother*, **45**, 3065–9.

Broughton, M.C., Bard, M. and Lees, N.D. 1991. Polyene resistance in ergosterol producing strains of *Candida albicans*. *Mycoses*, **34**, 75–83.

Calvet, H.M., Yeaman, M.R. and Filler, S.G. 1997. Reversible fluconazole resistance in *Candida albicans*: a potential in vitro model. *Antimicrob Agents Chemother*, **41**, 535–9.

Canuto, M.M. and Rodero, F.G. 2002. Antifungal drug resistance to azoles and polyenes. *Lancet Infect Dis*, **2**, 550–63.

Carrillo-Munoz, A.J., Quindos, G., et al. 1999. In vitro antifungal activity of liposomal nystatin in comparison with nystatin amphotericin B cholesteryl sulphate, liposomal amphotericin B, amphotericin B lipid complex, amphotericin B desoxycholate, fluconazole and itraconazole. *J Antimicrob Chemother*, **44**, 397–401.

Cartledge, J.D., Midgley, J. and Gazzard, B.G. 1997. Clinically significant azole cross-resistance in *Candida* isolates from HIV-positive patients with oral candidosis. *AIDS*, **11**, 1839–44.

Casanovas, R.O., Caillot, D., et al. 1992. Prophylactic fluconazole and *Candida krusei* infections. *N Engl J Med*, **326**, 891–2.

Collin, B., Clancy, C.J. and Nguyen, M.H. 1999. Antifungal resistance in non-*albicans Candida* species. *Drug Resist Update*, **2**, 9–14.

Cuenca-Estrella, M. and Rodriguez-Tudela, J.L. 2001. Present status of the detection of antifungal resistance: the perspective from both sides of the ocean. *Clin Microbiol Infect*, **7**, 46–53.

Cuenca-Estrella, M., Mellado, E., et al. 2000. Susceptibility of fluconazole-resistant clinical isolates of *Candida* spp. to echinocandin LY303366, itraconazole and amphotericin B. *J Antimicrob Chemother*, **46**, 475–7.

Cuenca-Estrella, M., Diaz-Guerra, T.M., et al. 2001. Flucytosine primary resistance in *Candida* species and *Cryptococcus neoformans*. *Eur J Clin Microbiol Infect Dis*, **20**, 276–9.

Denning, D.W., Venkateswarlu, K., et al. 1997. Itraconazole resistance in *Aspergillus fumigatus*. *Antimicrob Agents Chemother*, **41**, 1364–8.

Dick, J., Merz, W. and Saral, R. 1980. Incidence of polyene-resistant yeasts recovered from clinical specimens. *Antimicrob Agents Chemother*, **18**, 158–63.

Dick, J.D., Rosengard, B.R., et al. 1985. Fatal disseminated candidiasis due to amphotericin B-resistant *Candida guilliermondii*. *Ann Intern Med*, **102**, 67–8.

Dismukes, W.E., Cloud, G., et al. 1987. Treatment of cryptococcal meningitis with combination amphotericin B and flucytosine for four as compared with six weeks. *N Engl J Med*, **317**, 334–41.

Edlind, T.D., Henry, K.W., et al. 2001. *Aspergillus fumigatus* CYP51 sequence: potential basis for fluconazole resistance. *Med Mycol*, **39**, 299–302.

Ellis, D. 2002. Amphotericin B: spectrum and resistance. *J Antimicrob Chemother*, **49**, 7–10.

Espinel-Ingroff, A. 1998. In vitro activities of the new triazole voriconazole (UK-109,496) against opportunistic filamentous and dimorphic fungi and common and emerging yeast pathogens. *J Clin Microbiol*, **36**, 198–202.

Espinel-Ingroff, A., Bartlett, M., et al. 2001. Optimal susceptibility testing conditions for detection of azole resistance in *Aspergillus* spp.: NCCLS collaborative evaluation. *Antimicrob Agents Chemother*, **45**, 1828–35.

Fachin, A., Maffei, C.M. and Martinez-Rossi, N.M. 1996. In vitro susceptibility of *Trichophyton rubrum* isolates to griseofulvin and tioconazole. Induction and isolation of a resistant mutant to both antimycotic drugs. Mutant of *Trichophyton rubrum* resistant to griseofulvin and tioconazole. *Mycopathologia*, **135**, 141–3.

Fekete-Forgacs, K., Gyure, L. and Lenkey, B. 2000. Changes of virulence factors accompanying the phenomenon of induced fluconazole resistance in *Candida albicans*. *Mycoses*, **43**, 273–9.

Feldmesser, M., Kress, Y., et al. 2000. The effect of the echinocandin analogue caspofungin on cell wall glucan synthesis by *Cryptococcus neoformans*. *J Infect Dis*, **182**, 1791–5.

Franz, R., Kelly, S.L., et al. 1998. Multiple molecular mechanisms contribute to a stepwise development of fluconazole resistance in clinical *Candida albicans* strains. *Antimicrob Agents Chemother*, **42**, 3065–72.

Franz, R., Ruhnke, M. and Morschhauser, J. 1999. Molecular aspects of fluconazole resistance development in *Candida albicans*. *Mycoses*, **42**, 453–8.

Gale, E.F., Johnson, A.M., et al. 1975. Factors affecting the changes in amphotericin B sensitivity of *Candida albicans* during growth. *J Gen Microbiol*, **87**, 20–36.

Geber, A., Hitchcock, C.A., et al. 1995. Deletion of the *Candida glabrata ERG3* and *ERG11* genes: effect on cell viability, cell growth, sterol composition, and antifungal susceptibility. *Antimicrob Agents Chemother*, **39**, 2708–17.

Ghannoum, M.A. and Elewski, B. 1999. Successful treatment of fluconazole-resistant oropharyngeal candidiasis by a combination of fluconazole and terbinafine. *Clin Diagn Lab Immunol*, **6**, 921–3.

Ghannoum, M.A. and Rice, L.B. 1999. Antifungal agents: Mode of action, mechanisms of resistance, and correlation of these mechanisms with bacterial resistance. *Clin Microbiol Rev*, **12**, 501–17.

Ghannoum, M., Spellberg, B.J., et al. 1994. Sterol composition of *Cryptoccoccus neoformans* in the presence and absence of fluconazole. *Antimicrob Agents Chemother*, **38**, 2029–33.

Goldman, M., Cloud, G.A., et al. 2000. Does long-term itraconazole prophylaxis result in in vitro azole resistance in mucosal *Candida albicans* isolates from persons with advanced human immunodeficiency virus infection? *Antimicrob Agents Chemother*, **44**, 1585–7.

Groll, A.H. and Walsh, T.J. 2001. Caspofungin: pharmacology, safety and therapeutic potential in superficial and invasive fungal infections. *Expert Opin Investig Drugs*, **10**, 1545–58.

Gull, K. and Trinci, A.P.J. 1973. Griseofulvin inhibits fungal mitosis. *Nature*, **244**, 292–4.

Gupta, A.K. and Shear, N.H. 1997. Terbinafine: an update. *J Am Acad Dermatol*, **37**, 979–88.

Ha, K.C. and White, T.C. 1999. Effects of azole antifungal drugs on the transition from yeast cells to hyphae in susceptible and resistant isolates of the pathogenic yeast *Candida albicans*. *Antimicrob Agents Chemother*, **43**, 763–8.

Haynes, M.P., Chong, P.L., et al. 1996. Fluorescence studies on the molecular action of amphotericin B on susceptible and resistant fungal cells. *Biochemistry*, **35**, 7983–92.

Henry, K.W., Nickels, J.T. and Edlind, T.D. 2000. Upregulation of *ERG* genes in *Candida* species by azoles and other sterol biosynthesis inhibitors. *Antimicrob Agents Chemother*, **44**, 2693–700.

Hitchcock, C., Dickinson, K., et al. 1990. Interaction of azole antifungal antibiotics with cytochrome P450-dependent 14 alpha-sterol demethylase purified from *Candida albicans*. *J Biochem*, **266**, 475–80.

Hitchcock, C.A., Pye, G.W., et al. 1995. UK 109,496: a novel wide spectrum triazole derivative for the treatment of fungal infections: antifungal activity and selectivity in vitro. In *35th Interscience Conference on Antimicrobial Agents and Chemotherapy*. Washington, DC: American Society for Microbiology, abst. no. 2739.

Iwata, K. 1991. Drug resistance in human pathogenic fungi. *Eur J Epidemiol*, **8**, 407–21.

Jessup, C.J., Warner, J., et al. 2000. Antifungal susceptibility testing of dermatophytes: establishing a medium for inducing conidial growth and evaluation of susceptibility of clinical isolates. *J Clin Microbiol*, **38**, 341–4.

Jund, R. and Lacroute, F. 1970. Genetic and physiological aspects of resistance to 5-fluoropyrimidines in *Saccharomyces cerevisiae*. *J Bacteriol*, **102**, 607–15.

Katiyar, S.K. and Edlind, T.D. 2001. Identification and expression of multidrug resistance-related ABC transporter genes in *Candida krusei*. *Med Mycol*, **39**, 109–16.

Kelly, S.L., Lamb, D.C., et al. 1994. Resistance to amphotericin B associated with defective sterol delta 8→7 isomerase in a *Cryptococcus neoformans* strain from an AIDS patient. *FEMS Microbiol Lett*, **122**, 39–42.

Kelly, S.L., Lamb, D.C., et al. 1997. Resistance to fluconazole and cross-resistance to amphotericin B in *Candida albicans* from AIDS patients caused by defective sterol delta (5,6) desaturation. *FEBS Lett*, **400**, 80–2.

Kohli, A., Smriti, et al. 2002. In vitro low-level resistance to azoles in *Candida albicans* is associated with changes in membrane lipid fluidity and asymmetry. *Antimicrob Agents Chemother*, **46**, 1046–52.

Korting, H.C., Ollert, M., et al. 1995. Results of German multicenter study of antimicrobial susceptibilities of *Trichophyton rubrum* and *Trichophyton mentagrophytes* strains causing tinea unguium. *Antimicrob Agents Chemother*, **39**, 1206–8.

Krcmery, V. and Barnes, A.J. 2002. Non-*albicans Candida* spp. causing fungaemia: pathogenicity and antifungal resistance. *J Hosp Infect*, **50**, 243–60.

Kullberg, B.J. and Lashof, A. 2002. Epidemiology of opportunistic invasive mycoses. *Eur J Med Res*, **7**, 183–91.

Kurtz, M.B. and Douglas, C.M. 1997. Lipopeptide inhibitors of fungal glucan synthase. *J Med Vet Mycol*, **35**, 79–86.

Kurtz, M.B., Abruzzo, G., et al. 1996. Characterization of echinocandin-resistant mutants of *Candida albicans*: genetic, biochemical, and virulence studies. *Infect Immun*, **64**, 3244–51.

Lamb, D.C., Kelly, D.E., et al. 1997. The mutation T315A in *Candida albicans* sterol 14 alpha-demethylase causes reduced enzyme activity and fluconazole resistance through reduced affinity. *J Biol Chem*, **272**, 5682–8.

Lass-Florl, C., Kofler, G., et al. 1998. In-vitro testing of susceptibility to amphotericin B is a reliable predictor of clinical outcome in invasive aspergillosis. *J Antimicrob Chemother*, **42**, 497–502.

Leidich, S.D., Isham, N., et al. 2001. Primary resistance to terbinafine in a clinical isolate of the dermatophyte *Trichophyton rubrum*. In: *41st Interscience Conference on Antimicrobial Agents and Chemotherapy*. Washington, DC: ASM, abst. no. J-104.

Loeffler, J., Kelly, S.L., et al. 1997. Molecular analysis of cyp51 from fluconazole-resistant *Candida albicans* strains. *FEMS Microbiol Lett*, **151**, 263–8.

Lopez-Ribot, J.L., McAtee, R.K., et al. 1998. Distinct patterns of gene expression associated with development of fluconazole resistance in serial *Candida albicans* isolates from human immunodeficiency virus-infected patients with oropharyngeal candidiasis. *Antimicrob Agents Chemother*, **42**, 2932–7.

Lozano-Chiu, M., Nelson, P.W., et al. 1997. Lot to lot variability of antibiotic medium 3 when used for susceptibility testing of *Candida* isolates to amphotericin B. *J Clin Microbiol*, **35**, 270–2.

Lozano-Chiu, M., Arikan, S. et al. 1998a. Reliability of Antibiotic Medium 3 (AM3) agar and E-test for detection of amphotericin B (amB)-resistant isolates of *Candida* spp.: results of a collaborative two-center study. *38th Interscience Conference on Antimicrobial Agents and Chemotherapy*. Washington, DC: American Society for Microbiology, abst. no. J-18, p. 455.

Lozano-Chiu, M., Arikan, S. et al. 1998b. A two-center study of Antibiotic Medium 3 (AM3) broth for detection of amphotericin B (amB)-resistant isolates of *Candida* species (CAND) and *Cryptococcus neoformans* (CNEO). *38th Interscience Conference on Antimicrobial Agents and Chemotherapy*. Washington, DC: American Society for Microbiology, abst. no. J-19b, p. 456.

Lukacs, A., Korting, H.C. and Lindner, A. 1994. Successful treatment of griseofulvin-resistant tinea capitis in infants. *Mycoses*, **37**, 451–3.

Lupetti, A., Danesi, R., et al. 2002. Molecular basis of resistance to azole antifungals. *Trends Mol Med*, **8**, 76–81.

Lyons, C.N. and White, T.C. 2000. Transcriptional analyses of antifungal drug resistance in *Candida albicans*. *Antimicrob Agents Chemother*, **44**, 2296–303.

Maebashi, K., Niimi, M., et al. 2001. Mechanisms of fluconazole resistance in *Candida albicans* isolates from Japanese AIDS patients. *J Antimicrob Chemother*, **47**, 527–36.

Maebashi, K., Kudoh, M., et al. 2002. A novel mechanism of fluconazole resistance associated with fluconazole sequestration in *Candida albicans* isolates from a myelofibrosis patient. *Microbiol Immunol*, **46**, 317–26.

Maertens, J., Lagrou, K., et al. 2000. Disseminated infection by *Scedosporium prolificans*: an emerging fatality among hematology patients. Case report and review. *Ann Hematol*, **79**, 340–4.

Marichal, P., Gorrens, J., et al. 1995. Origin of differences in susceptibility of *Candida krusei* to azole antifungal agents. *Mycoses*, **38**, 111–17.

Marichal, P., Vanden Bossche, H. and Odds, F.C. 1997. Molecular and biological characterization of an azole-resistant *Candida glabrata* isolate. *Antimicrob Agents Chemother*, **41**, 2229–37.

Marr, K.A., White, T.C., et al. 1997. Development of fluconazole resistance in *Candida albicans* causing disseminated infection in a patient undergoing marrow transplantation. *Clin Infect Dis*, **25**, 908–10.

Marr, K.A., Lyons, C.N., et al. 1998. Rapid, transient fluconazole resistance in *Candida albicans* is associated with increased mRNA levels of CDR. *Antimicrob Agents Chemother*, **42**, 2584–9.

Martin, E., Maier, F. and Bhakdi, S. 1994. Antagonistic effects of fluconazole and 5-fluorocytosine on candidacidal action of amphotericin B in human serum. *Antimicrob Agents Chemother*, **38**, 1331–8.

McClenny, N.B., Fei, H.H., et al. 2002. Change in colony morphology of *Candida lusitaniae* in association with development of amphotericin B resistance. *Antimicrob Agents Chemother*, **46**, 1325–8.

McNeil, M.M., Nash, S.L., et al. 2001. Trends in mortality due to invasive mycotic diseases in the United States, 1980–1997. *Clin Infect Dis*, **33**, 641–7.

Merz, W.G. and Sanford, G.R. 1979. Isolation and characterization of a polyene-resistant variant of *Candida tropicalis*. *J Clin Microbiol*, **9**, 677–80.

Mosquera, J. and Denning, D.W. 2002. Azole cross-resistance in *Aspergillus fumigatus*. *Antimicrob Agents Chemother*, **46**, 556–7.

Muller, F.M.C., Weig, M., et al. 2000. Azole cross-resistance to ketoconazole, fluconazole, itraconazole and voriconazole in clinical *Candida albicans* isolates from HIV-infected children with oropharyngeal candidosis. *J Antimicrob Chemother*, **46**, 338–41.

National Committee for Clinical Laboratory Standards, 2002a. *Reference method for broth dilution antifungal susceptibility testing of filamentous fungi; Approved standard NCCLS document M38-A*. Wayne, PA: National Committee for Clinical Laboratory Standards.

National Committee for Clinical Laboratory Standards, 2002b. *Reference method for broth dilution antifungal susceptibility testing of yeasts; Approved standard NCCLS document M27-A2*. Wayne, PA: National Committee for Clinical Laboratory Standards.

Nguyen, M.H. and Yu, C.Y. 1998. Voriconazole against fluconazole-susceptible and resistant *Candida* isolates: in-vitro efficacy compared with that of itraconazole and ketoconazole. *J Antimicrob Chemother*, **42**, 253–6.

Nguyen, M.H., Clancy, C.J., et al. 1998. Do in vitro susceptibility data predict the microbiologic response to amphotericin B? Results of a prospective study of patients with *Candida* fungemia. *J Infect Dis*, **177**, 425–30.

Nolte, F.S., Parkinson, T., et al. 1997. Isolation and characterization of fluconazole- and amphotericin B-resistant *Candida albicans* from blood of two patients with leukemia. *Antimicrob Agents Chemother*, **44**, 196–9.

Oakley, K.L., Moore, C.B. and Denning, D.W. 1999. Comparison of in vitro activity of liposomal nystatin against *Aspergillus* species with those of nystatin, amphotericin B (AB) deoxycholate, AB colloidal dispersion, liposomal AB, AB lipid complex, and itraconazole. *Antimicrob Agents Chemother*, **43**, 1264–6.

Orozco, A.S., Higginbotham, L.M., et al. 1998. Mechanism of fluconazole resistance in *Candida krusei*. *Antimicrob Agents Chemother*, **42**, 2645–9.

Osaka, K., Ritov, V.B., et al. 1997. Amphotericin B protects *cis*-parinaric acid against perosyl radical-induced oxidation: amphotericin B as an antioxidant. *Antimicrob Agents Chemother*, **41**, 743–7.

Osherov, N., May, G.S., et al. 2002. Overexpression of Sbe2p, a golgi protein, results in resistance to caspofungin in *Saccharomyces cerevisiae*. *Antimicrob Agents Chemother*, **46**, 2462–9.

Pahls, S. and Schaffner, A. 1994. Comparison of the activity of free and liposomal amphotericin B in vitro and in a model of systemic and localized murine candidiasis. *J Infect Dis*, **169**, 1057–61.

Perea, S., Lopez-Ribot, J.L., et al. 2001. Prevalence of molecular mechanisms of resistance to azole antifungal agents in *Candida albicans* strains displaying high-level fluconazole resistance isolated from human immunodeficiency virus-infected patients. *Antimicrob Agents Chemother*, **45**, 2676–84.

Perez, A. 1999. Terbinafine: broad new spectrum of indications in several subcutaneous and systemic and parasitic diseases. *Mycoses*, **42**, 111–14.

Peyron, F., Favel, A., et al. 2002. Sterol and fatty acid composition of *Candida lusitaniae* clinical isolates. *Antimicrob Agents Chemother*, **46**, 531–3.

Pfaller, M.A., Messer, S.A., et al. 2001. In vitro activities of posaconazole (Sch 56592) compared with those of itraconazole and fluconazole against 3,685 clinical isolates of *Candida* spp. and *Cryptococcus neoformans*. *Antimicrob Agents Chemother*, **45**, 2862–4.

Pfaller, M.A., Diekema, D.J., et al. 2002. Trends in antifungal susceptibility of *Candida* spp. isolated from pediatric and adult patients with bloodstream infections: SENTRY Antimicrobial Surveillance Program, 1997 to 2000. *J Clin Microbiol*, **40**, 852–6.

Pietrucha-Dilanchian, P., Lewis, R.E., et al. 2001. *Candida lusitaniae* catheter-related sepsis. *Ann Pharmacother*, **35**, 1570–4.

Polak, A. 1977. 5-Fluorocytosine – current status with special references to mode of action and drug resistance. *Contrib Microbiol Immunol*, **4**, 158–67.

Ramani, R. and Chaturvedi, V. 2000. Flow cytometry antifungal susceptibility testing of pathogenic yeasts other than *Candida albicans* and comparison with the NCCLS broth microdilution test. *Antimicrob Agents Chemother*, **44**, 2752–8.

Revankar, S.G., Kirkpatrick, W.R., et al. 1996. Detection and significance of fluconazole resistance in oropharyngeal candidiasis in human immunodeficiency virus-infected patients. *J Infect Dis*, **174**, 821–7.

Rex, J.H. and Pfaller, M.A. 2002. Has antifungal susceptibility testing come of age? *Clin Infect Dis*, **35**, 982–9.

Rex, J.H., Cooper, C.R. Jr, et al. 1995. Detection of amphotericin B-resistant *Candida* isolates in a broth-based system. *Antimicrob Agents Chemother*, **39**, 906–9.

Rex, J.H., Lozano-Chiu, M. et al. 1998. Susceptibility testing of current *Candida* bloodstream isolates from Mycoses Study Group (MSG) collaborative study #34: isolates of *C. krusei* are often resistant to both fluconazole and amphotericin B. *36th Annual Meeting of the Infectious Diseases Society of America, Denver, CO*, abst. no. 324.

Rex, J.H., Pfaller, M.A., et al. 2001. Antifungal susceptibility testing: practical aspects and current challenges. *Clin Microbiol Rev*, **14**, 643–58.

Roberts, D.T. 1994. Oral therapeutic agents in fungal nail disease. *J Am Acad Dermatol*, **31**, S78–81.

Rodero, L., Cordoba, S., et al. 2000. In vitro susceptibility studies of *Cryptococcus neoformans* isolated from patients with no clinical response to amphotericin B therapy. *J Antimicrob Chemother*, **45**, 239–42.

Sanati, H., Belanger, P., et al. 1997. A new triazole, voriconazole (UK-109,496), blocks sterol biosynthesis in *Candida albicans* and *Candida krusei*. *Antimicrob Agents Chemother*, **41**, 2492–6.

Sanglard, D. and Odds, F.C. 2002. Resistance of *Candida* species to antifungal agents: molecular mechanisms and clinical consequences. *Lancet Infect Dis*, **2**, 73–85.

Sanglard, D., Kuchler, K., et al. 1995. Mechanisms of resistance to azole antifungal agents in *Candida albicans* isolates from AIDS patients involve specific multidrug transporters. *Antimicrob Agents Chemother*, **39**, 2378–86.

Sanglard, D., Ischer, F., et al. 1996. Susceptibilities of *Candida albicans* multidrug transporter mutants to various antifungal agents and other metabolic inhibitors. *Antimicrob Agents Chemother*, **40**, 2300–5.

Sanglard, D., Ischer, F., et al. 1998. Multiple resistance mechanisms to azole antifungals in yeast clinical isolates. *Drug Resist Update*, **1**, 255–65.

Sanglard, D., Ischer, F., et al. 1999. The ATP binding cassette transporter gene CgCDR1 from *Candida glabrata* is involved in the resistance of clinical isolates to azole antifungal agents. *Antimicrob Agents Chemother*, **43**, 2753–65.

Schaffner, A. and Böhler, A. 1993. Amphotericin B refractory aspergillosis after itraconazole: evidence for significant antagonism. *Mycoses*, **36**, 421–4.

Schoofs, A., Odds, F.C., et al. 1997. Isolation of *Candida* species on media with and without added fluconazole reveals high variability in relative growth susceptibility phenotypes. *Antimicrob Agents Chemother*, **41**, 1625–35.

Seo, K., Akiyoshi, H. and Ohnishi, Y. 1999. Alteration of cell wall composition leads to amphotericin B resistance in *Aspergillus flavus*. *Microbiol Immunol*, **43**, 1017–25.

Sokol-Anderson, M.L., Brajtburg, J. and Medoff, G. 1986. Amphotericin B-induced oxidative damage and killing of *Candida albicans*. *J Infect Dis*, **154**, 76–83.

Sugar, A.M. and Liu, X.P. 1998. Interactions of itraconazole with amphotericin B in the treatment of murine invasive candidiasis. *J Infect Dis*, **177**, 1660–3.

Sutton, D.A., Sanche, S.E., et al. 1999. In vitro amphotericin B resistance in clinical isolates of *Aspergillus terreus*, with a head-to-head comparison to voriconazole. *J Clin Microbiol*, **37**, 2343–5.

Toriumi, Y., Sugita, T., et al. 2002. Antifungal pharmacodynamic characteristics of amphotericin B against *Trichosporon asahii*, using time-kill methodology. *Microbiol Immunol*, **46**, 89–93.

van Duin, D., Casadevall, A. and Nosanchuk, J.D. 2002. Melanization of *Cryptococcus neoformans* and *Histoplasma capsulatum* reduces their susceptibilities to amphotericin B and caspofungin. *Antimicrob Agents Chemother*, **46**, 3394–400.

Vanden Bossche, H., Marichal, P., et al. 1992. Characterization of an azole-resistant *Candida glabrata* isolate. *Antimicrob Agents Chemother*, **36**, 2602–10.

Vanden Bossche, H., Marichal, H.P., et al. 1993. Effects of itraconazole on cytochrome P-450-dependent 14-alpha demethylation and reduction of 3-ketosteroids in *Cryptoccocus neoformans*. *Antimicrob Agents Chemother*, **37**, 2101–5.

Vanden Bossche, H., Marichal, P. and Odds, F.C. 1994a. Molecular mechanisms of drug resistance in fungi. *Trends Microbiol*, **2**, 393–400.

Vanden Bossche, H., Warnock, D.W., et al. 1994b. Mechanisms and clinical impact of antifungal drug resistance. *J Med Vet Mycol*, **32**, Suppl. 1, 189–202.

Vanden Bossche, H., Dromer, F., et al. 1998. Antifungal drug resistance in pathogenic fungi. *Med Mycol*, **36**, 119–28.

Vanden Bossche, I., Vaneechoutte, M., et al. 2002. Susceptibility testing of fluconazole by the NCCLS broth macrodilution method, E-test, and disk diffusion for application in the routine laboratory. *J Clin Microbiol*, **40**, 918–21.

Venkateswarlu, K., Denning, D.W., et al. 1996. Reduced accumulation of drug in *Candida krusei* accounts for itraconazole resistance. *Antimicrob Agents Chemother*, **40**, 2443–6.

Vermes, A., Guchelaar, H.J. and Dankert, J. 2000. Flucytosine: a review of its pharmacology, clinical indications, pharmacokinetics, toxicity and drug interactions. *J Antimicrob Chemother*, **46**, 171–9.

Walsh, T.J., Melcher, G.P., et al. 1990. *Trichosporon beigelii*, an emerging pathogen resistant to amphotericin B. *J Clin Microbiol*, **28**, 1616–22.

Walsh, T.J., Peter, J., et al. 1995. Activities of amphotericin B and antifungal azoles alone and in combination against *Pseudallescheria boydii*. *Antimicrob Agents Chemother*, **39**, 1361–4.

Wanger, A., Mills, K., et al. 1995. Comparison of Etest and National Committee for Clinical Laboratory Standards broth macrodilution method for antifungal susceptibility testing: enhanced ability to detect amphotericin B-resistant *Candida* isolates. *Antimicrob Agents Chemother*, **39**, 2520–2.

Whelan, W.L. 1987. The genetic basis of resistance to 5-fluorocytosine in *Candida* species and *Cryptococcus neoformans*. *Crit Rev Microbiol*, **15**, 45–56.

White, T.C. 1997a. Increased mRNA levels of ERG16, CDR and MDR1 correlate with increases in azole resistance in *Candida albicans* isolates from a patient infected with human immunodeficiency virus. *Antimicrob Agents Chemother*, **41**, 1482–7.

White, T.C. 1997b. The presence of an R467K amino acid substitution and loss of allelic variation correlate with an azole-resistant lanosterol 14a demethylase in *Candida albicans*. *Antimicrob Agents Chemother*, **41**, 1488–94.

White, T.C., Marr, K.A. and Bowden, R.A. 1998. Clinical, cellular, and molecular factors that contribute to antifungal drug resistance. *Clin Microbiol Rev*, **11**, 382–402.

White, T.C., Holleman, S., et al. 2002. Resistance mechanisms in clinical isolates of *Candida albicans*. *Antimicrob Agents Chemother*, **46**, 1704–13.

Wildfeuer, A., Seidl, H.P., et al. 1998. In vitro evaluation of voriconazole against clinical isolates of yeasts, moulds and dermatophytes in comparison with itraconazole, ketoconazole, amphotericin B and griseofulvin. *Mycoses*, **41**, 309–19.

Wirsching, S., Moran, G.P., et al. 2001. MDR1-mediated drug resistance in *Candida dubliniensis*. *Antimicrob Agents Chemother*, **45**, 3416–21.

Woods, R.A., Bard, M., et al. 1974. Resistance to polyene antibiotics and correlated sterol changes in two isolates of *Candida tropicalis* from a patient with an amphotericin B-resistant funguria. *J Infect Dis*, **129**, 53–8.

Yamaguchi, H., Uchida, K., et al. 1989. In vitro activity of fluconazole, a novel bistriazole antifungal agent. *Jpn J Antibiot*, **42**, 1–16.

Zaias, N., Glick, B. and Rebell, G. 1996. Diagnosing and treating onychomycosis (see comments). *J Fam Pract*, **42**, 513–18.

10

Management of superficial infections

JEFF WEEKS AND BONI ELEWSKI

Successful treatment of superficial fungal infections requires knowledge of the various antifungal agents and familiarity with measures that can be taken to reduce the chance of spread or reinfection. There are multiple topical and systemic antifungal agents available to treat these infections. In order to choose the most effective medication, the clinician needs a basic understanding of the drugs available as well as their mechanisms of action, indications, dosages, interactions, and adverse effects. Many conditions, such as tinea corporis and tinea cruris, usually respond well to topical agents. Others, such as tinea capitis and onychomycosis, generally require systemic therapy. The newer systemic agents including fluconazole, itraconazole, and terbinafine offer greater efficacy, broader spectrums of activity and shortened treatment duration. However, they are not without potential adverse effects and drug interactions and their use requires careful patient selection and monitoring.

This chapter discusses the management of a variety of superficial cutaneous infections including:

- superficial mycoses
- dermatophytoses
- disease caused by nondermatophyte molds
- cutaneous candidiasis.

A summary of the topical and systemic pharmaceutical agents commonly used for these conditions is presented. This includes a discussion of the indications, dosages, mechanisms of action, drug interactions, and adverse effects of selected antimycotics. Also discussed are measures that should be taken to reduce the chance of spread or recurrence of the infection.

ANTIFUNGAL AGENTS

There are multiple topical and systemic antimycotics that are effective in the management of superficial fungal infections. The most widely used today fall into three main classes: the polyenes, azoles, and allylamines/benzylamine. Others that are not included in these classes include griseofulvin, ciclopirox, tolnaftate, and selenium sulfide.

Polyenes

The polyenes were the first specific antifungals (Phillips and Rosen 2001). Members of this group include topical/oral nystatin and topical/intravenous (IV) amphotericin B (see Table 10.1). Although nystatin can be orally dosed, it is not absorbed. The polyenes bind fungal cell wall sterols, thus altering membrane permeability and allowing leakage of intracellular contents. Both agents are active against *Candida* species when topically applied but are ineffective against dermatophytes. Although still available from the manufacturer upon special request, topical amphotericin is rarely used today (Phillips and Rosen 2001). The usage of IV amphotericin is generally limited to systemic mycoses, and is beyond the scope of this chapter.

Azoles

The azole family can be subdivided into the topical/oral imidazoles (see Table 10.1) and the newer oral/IV triazoles (fluconazole, itraconazole, and voriconazole). Compared to the polyenes, both subsets have extended

Table 10.1 *Selected topical antifungals*

Class	MOA	Active against	Dosing frequency
Polyenes			
Nystatin	Bind to fungal cell wall sterols, increasing permeability	*Candida*	b.i.d.–t.i.d.
Amphotericin B			b.i.d.–q.i.d.
Imidazoles			
Clotrimazole	Block lanosterol 14α-methylase thus inhibiting fungal cell wall ergosterol synthesis	*Candida*, dermatophytes, *Malassezia*; econazole has some antibacterial activity	b.i.d.
Econazole			q.d.–b.i.d.
Ketoconazole			q.d.–b.i.d.
Miconazole			b.i.d.
Oxiconazole			q.d.–b.i.d.
Sulconazole			q.d.–b.i.d.
Allylamines			
Butenafine	Inhibit squalene epoxidase thus inhibiting fungal cell wall ergosterol synthesis. Squalene accumulation within cell is toxic	*Candida*, dermatophytes	q.d.
Naftifine			q.d.
Terbinafine			q.d.–b.i.d.
Substituted pyridone			
Ciclopirox	Alters active membrane transport, cell membrane integrity and cell respiratory processes	*Candida*, dermatophytes, *Malassezia*, and some bacteria. Some anti-inflammatory activity	b.i.d.
Others			
Tolnaftate	Inhibits squalene epoxidase	Dermatophytes, *Malassezia*	b.i.d.
Selenium sulfide	Cytostatic effect on keratinocytes. Fungicidal against *Malassezia*	*Malassezia*	q.d.

spectrums of antifungal activity. Not only effective against *Candida*, they also inhibit dermatophytes and *Malassezia* (De Doncker et al. 1996; Odom et al. 1996; Elewski et al. 1997; Havu et al. 1997; Scher 1999). Additionally, topical econazole appears to exhibit some antibacterial activity (Brooks 1996; Rupke 2000). Their fungistatic properties are due to inhibition of the cytochrome P450-dependent enzyme, lanosterol 14α-demethylase, an important step in the synthesis of ergosterol. Ergosterol is vital for fungal cell membrane integrity.

The inhibition of the cytochrome P450 enzyme system can lead to significant drug interactions when the systemic azoles are taken with other drugs metabolized by these enzymes, such as astemizole, some benzodiazepines, cisapride, cyclosporine, warfarin, and the statins. Medications such as antacids, proton-pump inhibitors, and H$_2$-blockers may decrease the systemic absorption of ketoconazole and itraconazole by raising the gastric pH. These drugs are better absorbed when taken with food or an acidic beverage. By contrast, fluconazole can be taken without regard to food.

Following oral administration and absorption, both fluconazole and itraconazole achieve excellent tissue distribution. Fluconazole is hydrophilic and is widely distributed throughout the body. It has been detected in the skin and nails within 2–3 hours of oral administration (Haneke 1990, 1992; Hay 1993; Scher 1999; Moossavi et al. 2001). Therapeutic levels remain in the stratum corneum for at least one week after treatment (Wildfeuer et al. 1994; Lesher 1999). In contrast to fluconazole, itraconazole is highly lipophilic. It also has a strong affinity for keratin, resulting in persistently high concentrations is the stratum corneum and in nails. Significant drug levels have been found in the stratum corneum from the back, palms and face for 2, 3, and 4 weeks, respectively, after stopping therapy (Cauwenbergh et al. 1988; Lesher 1999). A level far exceeding the minimum inhibitory concentration (MIC) of common dermatophytes was detected in nails 90 days after completing a 7-day course (Meinhof 1993; Scher 1999). This preferential persistence in keratin permits dosage flexibility such as once weekly administration of fluconazole as well as 'pulse' dosing of itraconazole.

Adverse effects of the topical azole antifungal agents are generally limited to local irritation, stinging, burning, and contact dermatitis. However, gastrointestinal adverse events are not uncommon with the oral agents. Indeed, approximately 10 percent of patients taking itraconazole or ketoconazole may report nausea or vomiting (Moossavi et al. 2001), which can be improved by dosing with meals. Gastrointestinal symptoms are also seen with fluconazole, although less frequently. Six percent of the patients taking fluconazole in a study reported in 1998 experienced headache, which was the most frequently reported adverse effect (Scher et al. 1998). Itraconazole has recently been associated with a negative inotropic effect on the heart and congestive heart failure. It is contraindicated in patients with known ventricular dysfunction (Ahmad et al. 2001). The most worrisome adverse effect seen with systemic azole therapy is hepatotoxicity. The incidence of ketoconazole-induced hepatotoxicity ranges from 1:10 000 to 1:15 000 and has been fatal in at least seven people. The hepatotoxicity seems to be idiosyncratic and is not related to the dose or length of therapy (Knight et al. 1991; Moossavi et al. 2001). This, combined with the significant drug interactions, has effectively ended frequent systemic use of ketoconazole for cutaneous fungal infections. Serum transaminases may also be elevated with itraconazole (Tucker et al. 1990) and less frequently with fluconazole (Scher et al. 1998). Monitoring of hepatic enzymes is recommended in patients taking itraconazole continuously for longer than one month and in those with preexisting hepatic function abnormalities (Mehta 2001; Moossavi et al. 2001) but routine monitoring is generally not required with fluconazole.

Allylamines/butenafine

A third group of antifungals is composed of the allylamines (topical naftifine and topical/oral terbinafine) and a single topical benzylamine (butenafine) (see Table 10.1). Like the azoles, these agents have broad antifungal spectrums, and they have the additional advantage of fungicidal activity. This activity is due to their inhibition of squalene epoxidase, an enzyme needed for the production of ergosterol. As a result, squalene accumulates, producing a toxic effect on the organism and eventual cell death (Ryder 1992). It is still unclear if the in vitro fungicidal activity is clinically relevant.

Unlike the azole family, oral terbinafine does not affect the cytochrome P450 3A4 isoenzyme, so drug interactions are less problematic. Cimetidine may increase the levels of terbinafine by decreasing its metabolism and clearance. Conversely, rifampin induces hepatic metabolism of terbinafine and causes decreased blood levels. The combination of cyclosporine and terbinafine may increase the clearance of cyclosporine. Terbinafine does inhibit cytochrome P450 2D6, which may lead to toxic levels of tricyclic antidepressants and thioridazine.

Terbinafine is well absorbed after oral administration and absorption is not affected by administration with food. Delivery to the stratum corneum occurs rapidly. Levels were detected in peripheral nail clippings of healthy volunteers after 7 days of 250 mg/day dosing (Gupta et al. 1997; Scher 1999). The concentrations in the stratum corneum of patients taking the 250 mg/day dose for 1 or 2 weeks were far greater than the MICs of most dermatophytes (10–100 and 100–1 000, respectively) (Faergemann et al. 1994; Lesher 1999). Similarly,

the drug level in nails 90 days after completing one week of therapy was much higher than the MICs (Gupta et al. 1997; Scher 1999).

As with other oral antifungals, gastrointestinal side effects are among the most commonly reported with terbinafine. Hepatic adverse effects range from transient elevations of serum transaminases in 3.3–7 percent of patients to severe cholestatic or hepatocellular hepatotoxicity (Boldewijn et al. 1996; Moossavi et al. 2001). This is thought to be an idiosyncratic, nondose-related reaction. Rarely, severe cutaneous events have been reported with terbinafine including toxic epidermal necrolysis, erythema multiforme, Stevens–Johnson syndrome, and acute generalized exanthematous pustulosis (Moossavi et al. 2001). The topical agents are well tolerated. Adverse effects such as allergic or irritant contact dermatitis, itching, burning or stinging have rarely been reported (Phillips and Rosen 2001).

Griseofulvin

Griseofulvin was the first widely used systemic antifungal medication. Its spectrum of activity is limited to dermatophytes and it is ineffective against *Candida*, *Malassezia*, bacteria, deep fungal infections, and nondermatophyte molds such as *Scopulariopsis* and *Scytalidium* (Moossavi et al. 2001). Griseofulvin's fungistatic activity is due to interference in microtubule spindle formation in actively growing fungi. Poor absorption from the gastrointestinal tract is counteracted by microsizing or ultramicrosizing griseofulvin particles. The ultramicrosized particles are better absorbed and require lower dosages. Absorption is also enhanced by administration with a fatty meal. Drug interactions are generally due to griseofulvin's induction of the cytochrome P450 enzyme system with resultant decreased blood levels of other drugs metabolized by this system. Examples include warfarin, oral contraceptives, and salicylates. Headache is the most common adverse effect and this usually resolves with continued therapy. Although rare, hepatotoxicity, neutropenia, and leukopenia have been reported.

Miscellaneous antimycotic agents

CICLOPIROX

Ciclopirox is a broad-spectrum topical antifungal which is effective against dermatophytes, yeasts, fungal saprophytes, and some gram-positive and gram-negative bacteria (Jue et al. 1985) (see Table 10.1). In addition, it has some anti-inflammatory activity and may be especially useful in treating superficial inflammatory mycoses (Hanel et al. 1991). It acts by altering active membrane transport, cell membrane integrity, and cell respiratory processes (Abrams et al. 1991; Phillips and Rosen 2001).

It is available as a cream, gel, lotion, nail lacquer, and shampoo.

TOLNAFTATE

Tolnaftate is a topical antifungal with a narrow spectrum of activity (see Table 10.1). It is effective in superficial infections caused by dermatophytes and *Malassezia*, but is not active against *Candida* or bacteria (Gupta et al. 1998; Weinstein and Berman 2002). It inhibits squalene epoxidase, thus interfering with fungal cell wall sterol synthesis (Phillips and Rosen 2001).

SELENIUM SULFIDE

Selenium sulfide is a liquid topical antifungal agent used in the treatment of tinea versicolor and seborrheic dermatitis (Sanchez and Torres 1984; Faergemann 2000) (see Table 10.1). It has some in vivo fungicidal activity against *Malassezia* (Phillips and Rosen 2001). It also appears to have a cytostatic effect on keratinocytes, which may explain its effectiveness as an adjutant in the treatment of tinea capitis (Allen et al. 1982; Givens et al. 1995).

Choice of agent

Topical antifungal agents can be used as sole therapy in most superficial fungal infections. They offer the advantages of very few adverse effects and negligible systemic absorption. For these reasons, they are generally considered first-line therapy except in tinea capitis and onychomycosis, which do not respond well to topical therapy and require treatment with an oral agent. Other instances in which systemic antifungal therapy may be required include:

- failure of topical agents
- patients immunocompromised by disease or therapy
- extensive fungal disease.

THERAPY OF SELECTED SUPERFICIAL INFECTIONS

Malassezioses

TINEA VERSICOLOR

Tinea versicolor usually responds well to topical therapy. Selenium sulfide 2.5 percent, ketoconazole 2 percent, or zinc pyrithione 1 percent shampoos/lotions can be applied to the affected areas and left in place for 5–10 minutes once a day for 2 weeks. Alternatively, the azole antifungal creams or ciclopirox 0.1 percent solution are also generally effective when used twice daily (Corte et al. 1989).

Failure of the above regimens or extensive disease may force trial of systemic medications (see Table 10.2).

Table 10.2 *Systemic antifungals for tinea versicolor*

Drug	Dosage
Fluconazole	400 mg once 400 mg once and repeat in 7 days
Itraconazole	200 mg/day for 7 days
Ketoconazole	200 mg/day for 5–7 days 400 mg once and repeat in 7 days

Ketoconazole 200 mg/day for 5–7 days has been proven effective. One study showed 17 of 23 patients treated with fluconazole in a 400 mg one-time dose were clear of lesions 3 weeks after treatment (Faergemann 1992). Another study found itraconazole 200 mg/day for 7 days to be safe and effective (Hickman 1996). Oral terbinafine does not appear to be effective in treating tinea versicolor (Lesher 1999).

As many as 60–80 percent of patients successfully treated with oral or topical therapy will have a recurrence of tinea versicolor within one year (Rupke 2000). To reduce this rate, appropriate prophylaxis may be necessary. A single dose of oral ketoconazole, fluconazole, or itraconazole could be dosed once or twice a month. Topical antifungal agents may also be used once or twice monthly. In either case, however, compliance may be an issue.

SEBORRHEIC DERMATITIS

Although the role of *Malassezia* in the etiology of seborrheic dermatitis is unclear, agents effective against these organisms are also generally effective in controlling this condition. Ketoconazole 2 percent shampoo has been successful when applied at least twice a week (Peter and Richarz-Barthauer 1995). Ciclopirox 1 percent, selenium sulfide 2.5 percent shampoo (Shuster 1984) and zinc pyrithione shampoo (Marks et al. 1985) have also worked well. Antifungal creams offer another therapeutic option. Due to its anti-inflammatory activity, ciclopirox may be particularly useful in seborrheic dermatitis (Hanel et al. 1991). Nonantifungal treatments for seborrheic dermatitis on the face include low-potency topical corticosteroids such as desonide, hydrocortisone, or alclometasone. More potent corticosteroid lotions, solutions, or foams may be used on the scalp. The topical immunomodulating agent, pimecrolimus may prove to be effective (Ling 2001; Crutchfield 2002).

PITYROSPORAL FOLLICULITIS

Pityrosporal folliculitis usually responds well to topical treatment with an azole cream or shampoo. Selenium sulfide shampoo/lotion may also be used (Ford et al. 1982). When patients do not respond to topical therapy, oral fluconazole (100–200 mg/day for up to 3 weeks), ketoconazole (200 mg/day for 7–10 days), or itraconazole (200 mg/day for 5–7days) may prove effective (Ford et al. 1982; Rhie et al. 2000). Itraconazole may be the most effective option because its lipophilic properties permit penetration into the pilosebaceous unit.

Other superficial mycoses

TINEA NIGRA

Topical azole antifungals such as ketoconazole, econazole, or miconazole are normally sufficient for clearing tinea nigra (Marks et al. 1980; Burke 1993; Hughes et al. 1993). Ciclopirox and terbinafine have also been effective when used topically (Sayegh-Carreno et al. 1989; Shannon et al. 1999). Whitfield's ointment is an old medication that has also been used successfully. Some authors have suggested scraping away the visibly affected area before topical application (Sayegh-Carreno 1989). Oral therapy is rarely required.

PIEDRAS

Both black and white piedra may be treated by shaving or clipping off the infected hairs. This is not always a cosmetically acceptable therapy, however. Ketoconazole 2 percent shampoo, ciclopirox solution, selenium sulfide, and zinc pyrithione have been used with varying degrees of success. A small study reported complete cure in 11 of 12 patients with white piedra treated with 100 mg of itraconazole taken by mouth once daily for 8 weeks (Khandpur and Reddy 2002).

Dermatophytoses

TINEA CORPORIS AND TINEA CRURIS

Most topical antimycotics are effective in treating uncomplicated cases of tinea corporis and tinea cruris when used as directed by the manufacturer (Drake et al. 1996a; Brodell and Elewski 1997). See Table 10.1 for a list of these medications, as well as dosing frequencies. Nystatin is ineffective against dermatophytes and would therefore not be a good choice in treating tinea corporis (Lesher 1996). In intertriginous areas, tolnaftate should not be used, as it is not effective against *Candida* (Demis 1999). When bacterial superinfection is also suspected, topical antifungals with some antibacterial activity, such as econazole or ciclopirox, may be particularly useful (Drake et al. 1996a). All topical agents should be applied to the lesion and an approximately 2.5 cm border of surrounding normal skin. Ideally, application should be continued for 1–2 weeks after all signs of infection have cleared in order to minimize recurrence. More specific dosing suggestions can be found in the manufacturer's recommendations.

Indications for systemic therapy include failure of topical antimycotic agents, extensive disease, primary or secondary immune deficiency, and inflammatory infections such as Majocchi's granuloma. In addition, severely obese persons with tinea cruris may require systemic

therapy. Suitable oral medications include griseofulvin, fluconazole, itraconazole, and terbinafine (Lambert et al. 1989; Lesher 1999) (see Table 10.3).

Due to the ubiquity of the pathogens, reinfection and recurrence are common. Simple strategies to minimize recurrence in patients with intertriginous dermatophytoses such as tinea cruris include keeping the area dry in order to make the environment less conducive to fungal growth. In addition, patients should avoid tight-fitting, nonabsorbent clothing and minimize exposure to wet clothing such as bathing suits. Affected areas should be dried thoroughly after bathing and hand-held hair dryers work well for this purpose. Other suggestions might include application of absorbent antifungal powders and weight reduction.

TINEA BARBAE

Although some cases of inflammatory tinea barbae may resolve spontaneously, alopecia and scar formation can occur. The less common, noninflammatory sycosiform variant rarely heals without therapy. Since the hair follicles are infected, systemic therapy is required (Drake et al. 1996a). Griseofulvin, fluconazole, itraconazole, and oral terbinafine have all been used successfully (see Table 10.4). In most cases, therapy should be continued until all lesions have cleared and new hair growth is occurring.

Adjunctive therapies may be of some benefit. Warm compresses may soften crusts and help in the removal of debris. Ketoconazole 2 percent or selenium sulfide 2.5 percent shampoos also help remove debris and may provide some antifungal activity.

Since most cases of tinea barbae are acquired from zoophilic sources, avoidance, identification, and treatment of infected animals are important.

TINEA PEDIS AND MANUUM

Interdigital tinea pedis and tinea manuum with no associated onychomycosis in immune competent patients generally respond well to topical antifungals. See Table 10.1 for a list of suitable agents along with their dosing frequencies. Antifungals with antibacterial

Table 10.3 *Systemic antifungals for tinea corporis and tinea cruris*

Medication	Dose
Fluconazole	50–100 mg daily or 150 mg once weekly for 2–3 weeks[a]
Itraconazole	100 mg daily for 2 weeks or 200 mg daily for 7 days[a]
Terbinafine	250 mg daily for 1–2 weeks[a]
Griseofulvin	Microsize 20 mg/kg/day for 2–4 weeks Ultramicrosize 15 mg/kg/day for 2–4 weeks

a) Lesher (1999)

Table 10.4 *Systemic antifungals for tinea barbae*

Drug	Dose
Griseofulvin	Microsize 20 mg/kg/day Ultramicrosize 15 mg/kg/day Continue for at least 2 weeks after resolution. Six weeks or longer are generally required
Fluconazole	150–300 mg once a week for 4–6 weeks or 200–400 mg daily for up to 4 weeks
Itraconazole	400 mg/day for 1 week. May require a second 1-week pulse 3 weeks after the first pulse is completed or 200 mg/day for 2–4 weeks
Terbinafine	250 mg/day for 2–4 weeks

Table 10.5 *Systemic antifungals for tinea pedis*

Drug	Dosage
Fluconazole	150–300 mg once a week for up to 8 weeks (Lesher 1999)
Itraconazole	100 mg/day for 2 weeks or 400 mg/day for 1 week
Terbinafine	250 mg/day for 1–2 weeks

activity, such as econazole and ciclopirox, are good initial choices as interdigital disease is often complicated by secondary bacterial infections (Rupke 2000). Activity against *Candida* may also be important and drugs such as tolnaftate, which lack this attribute, are not first-line (Drake et al. 1996b). Generally, topical antifungal agents should be used for about 4 weeks or as directed by the manufacturer. Topical terbinafine and butenafine may be effective in as little as one week of therapy (Villars and Jones 1989; Bergstresser et al. 1993; Savin et al. 1994, 1997; Elewski et al. 1995). Emollients containing lactic acid or urea may be useful adjuncts to the antifungals when extensive scaling or hyperkeratosis is present.

Systemic antifungals are generally required in moccasin and inflammatory tinea pedis. Other patients who may require oral therapy include those with concomitant onychomycosis, peripheral vascular disease, or compromised immune status. Griseofulvin may be ineffective in tinea pedis (Moossavi et al. 2001) but terbinafine, itraconazole, and fluconazole have all been used successfully (Lesher 1999) (see Table 10.5). When

onychomycosis is present, the nail infection should be treated to prevent recurrence of tinea pedis or tinea manuum. See Table 10.6 for medications and dosages used in treating onychomycosis.

As with other dermatophytoses, efforts to minimize exposure to the infective fungi help to prevent initial or recurrent infections. Protective footwear should be worn in public facilities such as locker rooms and hotel bathrooms. Taking measures to eliminate the warm, moist conditions in which dermatophytes thrive can also reduce the recurrence rate. For example, patients should be instructed to dry feet thoroughly after bathing, and to apply an antifungal powder to both the feet and inside shoes. The use of cotton socks wicks moisture away from the feet. Prevention and effective treatment of tinea pedis is likely to prevent tinea manuum and tinea corporis.

TINEA UNGUIUM (ONYCHOMYCOSIS)

Topical therapy of onychomycosis is often unsuccessful. Ciclopirox 8 percent in nail lacquer applied daily to infected nails for 48 weeks produced a complete cure in only 5.5 percent of patients in one study and 8.5 percent in another (Dermik Laboratories 2000; Mehta 2001). Evaluations of other topical agents are ongoing.

Due to the low efficacy of topical medications, oral antifungals are generally required (Drake et al. 1996b). Fluconazole, itraconazole, and terbinafine are options. See Table 10.6 for dosage and monitoring guidelines. Continuous terbinafine therapy has been shown to be more effective when compared to both pulsed and continuous itraconazole (Brautigam et al. 1995; De

Table 10.6 *Systemic antifungals for onychomycosis*

Drug	Dosage	Duration	Monitoring
Fluconazole	150–300 mg once a week	3 months for fingernails, 6 months for toenails. Some authors recommend 6–9 months	Generally none recommended
Itraconazole	Intermittent (pulse) dosing – 200 mg b.i.d.	One week per month for 3 consecutive months for toenails, 2 months for fingernails	Baseline LFTs. Monitor LFTs every 4–6 weeks for continuous dosing of more than 1-month duration (Rodgers and Bassler 2001)
	Continuous dosing – 200 mg q.d.	8 weeks for fingernails, 12 weeks for toenails	
Terbinafine	250 mg/day	6 weeks for fingernails, 12 weeks for toenails	Baseline liver function tests (LFTs) and complete blood count (CBC). Repeat monthly (Rodgers and Bassler 2001)

Backer et al. 1998; Evans and Sigurgeirsson 1999). A large North American study showed terbinafine to have the greatest in vitro activity against dermatophytes and nondermatophyte molds when compared to fluconazole and itraconazole. Itraconazole had the greatest activity against *Aspergillus* spp. and *Candida parapsilosis*. Fluconazole was the most active of the three against *Candida albicans* (Ghannoum et al. 2000). Although its use appears to be growing, there are few data available comparing the clinical efficacy of fluconazole to the other agents. Griseofulvin is used infrequently due to its low cure rates, high recurrence rates, and the need for protracted treatment courses. Likewise, oral ketoconazole is rarely used due to its hepatotoxicity.

In addition to systemic antifungal therapy, certain adjunct measures may provide additional benefits and help recurrent infections. Occasionally, a nidus of infection (dermatophytoma) may form under the nail, which responds poorly to current therapy and may require nail avulsion. Topical antifungals may also be useful, especially when patients have severe tinea pedis. The patient should be instructed to wear protective footwear in public showers and facilities and to avoid sharing manicure equipment and other fomites. Feet should be kept as dry as possible. Patients should be advised to wear cotton socks and to change them regularly. Antifungal powders help keep feet dry and may provide some therapeutic benefit. Finally, early recognition and effective treatment of tinea pedis will help prevent tinea unguium.

TINEA CAPITIS

Topical treatment of tinea capitis is considered ineffective as sole therapy and systemic therapy is required for penetration of hair follicles (Elewski 1999). See Table 10.7 for medications, dosages, and duration of therapy. Griseofulvin has been the gold standard and remains an effective alternative. In the USA, the microsized preparation is available in a 125 mg/5 ml suspension, 250 mg tablets and 500 mg tablets. The ultramicrosized product

has enhanced absorption and is available in tablets ranging from 125 mg to 330 mg in strength. Availability may be limited in some countries. Griseofulvin's disadvantages include its prolonged treatment course and relatively high dosage when compared to the newer antifungals. The bitter taste of the suspension makes compliance a problem. Nausea, vomiting, and headache are not uncommon adverse effects that may reduce compliance, as does the bitter taste of the liquid formulation. Perhaps as a result of these disadvantages, fluconazole, itraconazole, and terbinafine have become more commonly used to treat tinea capitis. The pleasant-tasting fluconazole suspension is especially appealing for children. It is available in 10 mg/ml and 40 mg/ml strengths. In addition, tablet strengths range from 50 to 200 mg. Precise dosing of itraconazole is more difficult. The liquid preparation should not be considered a substitute for the capsule due to the adverse effects associated with its cyclodextrin-containing vehicle. The inability to divide the 100 mg capsule makes accurate pediatric dosing problematic. Unlike fluconazole, itraconazole needs to be taken with food or a cola beverage to enhance absorption. Terbinafine may be taken with or without food. It is available only in 250 mg tablets that can be divided for administration to children. In tinea capitis caused by *Microsporum canis*, terbinafine may prove less effective than the other systemic alternatives (Mock et al. 1998). Tinea capitis caused by this organism generally requires longer treatment courses with any selected antimycotic (Hamm et al. 1999).

In addition to systemic therapy, use of topical medications may help minimize spread of the infection. Selenium sulfide 1 and 2.5 percent (Allen et al. 1982; Givens et al. 1995) and ketoconazole 2 percent shampoos (al-Fouzan et al. 1993; Greer 2000) are thought to reduce shedding of infected hairs and to stabilize infectious particles. Although there are fewer data on ciclopirox shampoo, it is likely to be equally efficacious. Patients should be instructed to use one of these several times a week. When feasible, family members should also use

Table 10.7 *Systemic antifungals for tinea capitis*

Drug	Pediatric dose	Adult dose
Griseofulvin	Microsized 20–25 mg/kg/day in single or divided doses for 8–12 weeks Ultramicrosized 15 mg/kg/day for 8–12 weeks	Same
Fluconazole	6 mg/kg/day for 3–6 weeks	200 mg/day for 3–6 weeks
Itraconazole	5 mg/kg/day for 4 weeks Most require 100 mg/day Very small children: 100 mg every other day *M. canis* may require 6–8 weeks	5 mg/kg/day Usually 200 mg/day for 4 weeks
Terbinafine	Weight 10–20 kg: 62.5 mg/day (1/4 tablet) Weight 20–40 kg: 125 mg/day (1/2 tablet) Weight > 40 kg: 250 mg/day (1 tablet) For 2–4 weeks *M. canis* may require 6–8 weeks	250 mg/day for 2–4 weeks

these shampoos, because they may be asymptomatic carriers (Elewski 2000). Other preventative measures include discarding fomites such as combs, brushes, and other hair accessories. When a zoophilic pathogen such as *M. canis* is isolated, an attempt to identify and treat the infected animal could prove beneficial.

Nondermatophyte molds

Treatment of onychomycosis caused by nondermatophyte molds such as *Scytalidium* spp., *Acremonium* spp., *Fusarium* spp. and *Scopulariopsis* spp. is difficult. Topical antifungals are ineffective, as are systemic ketoconazole and griseofulvin (Gupta and Elewski 1996). A study published in 2000 showed that terbinafine had the highest in vitro antifungal activity against *Acremonium*, *Fusarium*, *Scopulariopsis*, and *Scytalidium* spp. when compared to fluconazole, itraconazole, and griseofulvin (Ghannoum et al. 2000). However, the clinical relevance is uncertain. Terbinafine and itraconazole have also shown some clinical efficacy against *Scopulariopsis brevicaulis* (Gupta et al. 2001). Avulsion may be the best option.

Superficial candidiasis

CANDIDAL PARONYCHIA

Chronic paronychia may be associated with *Candida* infection in some patients. However, one study has shown that treatment of chronic paronychia with a topical steroid (methylprednisolone aceponate) was more effective than systemic antifungal therapy with itraconazole or terbinafine (Tosti et al. 2002). Nevertheless, both topical and oral antifungals continue to be used with varying degrees of success. Topical imidazoles and polyene antifungal solutions, gels, or lotions are applied frequently over a prolonged course of therapy. Thymol 4 percent solution in acetone, isopropyl alcohol, or chloroform and sulfacetamide lotion have also been used (Hay 1999). Oral fluconazole and itraconazole are alternatives, but their effectiveness has not yet been established (Hay 1999). Patients should also be instructed to keep their fingers warm and dry.

CANDIDAL ONYCHOMYCOSIS

Systemic therapy is often required for microscopy, biopsy or culture-proven candidal onychomycosis. Itraconazole has been used successfully in three pulsed doses for toenails and two pulses for fingernails (Gupta et al. 2000). Weekly dosing of 150–300 mg of fluconazole should also be effective although few trials have been conducted. Terbinafine has unpredictable in vitro activity against *C. albicans* (Ryder et al. 1998) and has not shown consistently good results clinically (Rex et al. 2000).

CANDIDAL INTERTRIGO AND INTERDIGITAL CANDIDIASIS

In contrast to onychomycosis, topical therapy is usually sufficient for candidal intertrigo (including diaper dermatitis) and interdigital candidiasis. Topical allylamines, butenafine, or imidazoles are applied once or twice daily (Drake et al. 1996c). As in any intertriginous condition, it is important for patients to keep the area dry. This can be done by using a hair dryer and by applying an antifungal powder.

OROPHARYNGEAL CANDIDIASIS (THRUSH)

For immunocompetent patients, topical treatment of oropharyngeal candidiasis with nystatin or clotrimazole is generally effective. Nystatin suspension or lozenges are given four to five times a day in doses ranging from 100 000 to 500 000 units. Clotrimazole 10 mg troches are given five times a day. Immunocompromised patients may require systemic antifungals. Fluconazole or itraconazole are given in 100–200 mg daily doses for 5–10 days. The development of resistance to these drugs can be a problem, especially in patients with acquired immunodeficiency syndrome (AIDS). Generally, when patients fail to respond clinically, the medication dose is doubled. If they still do not respond, it is necessary to check the identity and MIC value of the organism and to change drugs (Hay 1999).

VAGINAL CANDIDIASIS

Topical antifungals are usually effective in treating vaginal candidiasis. Imidazole (miconazole, terconazole, etc.) or nystatin vaginal creams, tablets or suppositories are used for from 1 to 7 days. Alternatively, oral therapy with fluconazole (100 mg/day for 5–7 days or 150 mg once) (Hay 1999), or itraconazole (200 mg/day for 3 days or 200 mg b.i.d. for 1 day) (Sobel et al. 1998) is also effective.

CHRONIC MUCOCUTANEOUS CANDIDIASIS

Systemic therapy with fluconazole, ketoconazole or itraconazole is usually necessary for patients with chronic mucocutaneous candidiasis. Dosing strategies similar to those used for oropharyngeal candidiasis in immune suppressed patients are employed. Once remissions are induced, maintenance therapy is not used in order to reduce the development of drug resistance (Hay 1999).

CONCLUSION

Successful management of superficial fungal infections requires not only knowledge of the various antimycotic agents but also familiarity with adjunctive measures that can help reduce the chance of spread or reinfection. This chapter has been a discussion of these measures along with a summary of the indications, dosages, mechanisms of action, adverse effects, and drug interactions of the most commonly used topical and systemic antifungal drugs. Some superficial infections respond well to therapy with topical medications alone. Others such as tinea capitis require systemic therapy. The newer oral agents such as fluconazole, itraconazole, and terbinafine offer greater efficacy, broader spectrums of activity and shortened treatment duration. Awareness of their limitations, potential adverse effects and drug interactions is essential for proper patient selection and monitoring.

REFERENCES

Abrams, B.B., Hanel, H. and Hoehler, T. 1991. Ciclopirox olamine: a hydroxypyridone antifungal agent. *Clin Dermatol*, **9**, 4, 471–7.

Ahmad, S., Singer, S. and Leissa, B. 2001. Congestive heart failure associated with itraconazole. *Lancet*, **357**, 1766–77.

al-Fouzan, A., Nanda, A. and Kubec, K. 1993. Dermatophytosis of children in Kuwait: a prospective survey. *Int J Dermatol*, **32**, 798–801.

Allen, H.B., Honig, P.J., et al. 1982. Selenium sulfide: adjunctive therapy for tinea capitis. *Pediatrics*, **69**, 1, 81–3.

Bergstresser, P., Elewski, B.E., et al. 1993. Topical terbinafine and clotrimazole in interdigital tinea pedis: a multicenter comparison of cure and relapse rates with 1- and 4-week regimens. *J Am Acad Dermatol*, **28**, 648–51.

Boldewijn, O., Ottervanger, J.P., et al. 1996. Hepatitis attributed to the use of terbinafine. *Ned Tijdschr Geneeskd*, **140**, 669–72.

Brautigam, M., Nolting, S., et al. 1995. Randomised double blind comparison of terbinafine and itraconazole for treatment of toenail tinea infection. Seventh Lamisil German Onychomycosis Study Group. (Published erratum appears in 1995 *BMJ* **311**, 919–922). *BMJ*, **311**, 1350.

Brodell, R.T. and Elewski, B.E. 1997. Superficial fungal infections: errors to avoid in diagnosis and treatment. *Postgrad Med*, **101**, 279–87.

Brooks, K.E. 1996. Tinea pedis: diagnosis and treatment. *Clin Pediatr Med Surg*, **13**, 31–46.

Burke, W.A. 1993. Tinea nigra: treatment with topical ketoconazole. *Cutis*, **52**, 4, 209–11.

Cauwenbergh, G., Degreef, H., et al. 1988. Pharmacokinetic profile of orally administered itraconazole in human skin. *J Am Acad Dermatol*, **18**, 263–8.

Corte, M., Jung, K., et al. 1989. Topical application of a 0.1% ciclopirox olamine solution for the treatment of pityriasis versicolor. *Mycoses*, **32**, 200–3.

Crutchfield, C.E. 2002. Pimecrolimus: a new treatment for seborrheic dermatitis. *Cutis*, **70**, 4, 207–8.

De Backer, M., De Vroey, C., et al. 1998. Twelve weeks of continuous oral therapy for toenail onychomycosis caused by dermatophytes: a double-blind comparative trial of terbinafine 250 mg/day versus itraconazole 200 mg/day. *J Am Acad Dermatol*, **38**, S57–63.

De Doncker, P., Decroix, J., et al. 1996. Antifungal pulse therapy for onychomycosis: a pharmacokinetic and pharmacodynamic investigation of monthly cycles of 1-week pulse therapy with itraconazole. *Arch Dermatol*, **132**, 34–41.

Demis, D.J. 1999. Tinea corporis. In: Demis, D.J. (ed.), *Clinical dermatology*, 26th edition. Philadelphia, PA: Lippincott Williams and Wilkins, Unit 17–6, 1.

Dermik Laboratories. 2000. *Ciclopirox topical solution 8% package insert*. Berwyn, PA: Dermik Laboratories.

Drake, L.A., Dinehart, S.M., et al. 1996a. Guidelines of care for superficial mycotic infections of the skin: tinea corporis, tinea cruris, tinea faciei, tinea manuum, and tinea pedis. Guideline/Outcomes Committee, American Academy of Dermatology. *J Am Acad Dermatol*, **34**, 282–6.

Drake, L.A., Dinehart, S.M., et al. 1996b. Guidelines of care for superficial mycotic infections of the skin: onychomycosis. Guideline/

Outcomes Committee, American Academy of Dermatology. *J Am Acad Dermatol*, **34**, 116–21.

Drake, L.A., Dinehart, S.M., et al. 1996c. Guidelines of care for superficial mycotic infections of the skin: mucocutaneous candidiasis. Guideline/Outcomes Committee, American Academy of Dermatology. *J Am Acad Dermatol*, **34**, 110–15.

Elewski, B.E. 1999. Treatment of tinea capitis: beyond griseofulvin. *J Am Acad Dermatol*, **40**, S27–30.

Elewski, B.E. 2000. Tinea capitis: a current perspective. *J Am Acad Dermatol*, **42**, 1–20.

Elewski, B.E., Bergstresser, P.R., et al. 1995. Long-term outcome of patients with interdigital tinea pedis treated with terbinafine or clotrimazole. *J Am Acad Dermatol*, **32**, 290–2.

Elewski, B.E., Scher, R.K., et al. 1997. Double-blind randomized comparison of itraconazole capsules vs. placebo in the treatment of toenail onychomycosis. *Cutis*, **59**, 4, 217–20.

Evans, E. and Sigurgeirsson, B. 1999. Double blind, randomised study of continuous terbinafine compared with intermittent itraconazole in treatment of toenail onychomycosis The LION Study Group. *BMJ*, **318**, 1031–5.

Faergemann, J. 1992. Treatment of pityriasis versicolor with a single dose of fluconazole. *Acta Dermatol Venereol*, **72**, 74–5.

Faergemann, J. 2000. Management of seborrheic dermatitis and pityriasis versicolor. *Am J Clin Dermatol*, **1**, 2, 75–80.

Faergemann, J., Zehender, H. and Millerioux, L. 1994. Levels of terbinafine in plasma stratum corneum, dermis-epidermis (without stratum corneum), sebum, hair and nails during and after 250 mg terbinafine orally once daily for 7 and 14 days. *Clin Exp Dermatol*, **19**, 121–5.

Ford, G.P., Ive, F.A. and Midgley, G. 1982. *Pityrosporum* folliculitis and ketoconazole. *Br J Dermatol*, **107**, 691–5.

Ghannoum, M., Hajjeh, R., et al. 2000. A large-scale North American study of fungal isolates from nails: the frequency of onychomycosis, fungal distribution, and antifungal susceptibility patterns. *J Am Acad Dermatol*, **43**, 641–8.

Givens, T.G., Murray, M.M. and Baker, R.C. 1995. Comparison of 1% and 2.5% selenium sulfide in the treatment of tinea capitis. *Arch Pediatr Adolesc Med*, **149**, 7, 808–11.

Greer, D.L. 2000. Successful treatment of tinea capitis with 2% ketoconazole shampoo. *Int J Dermatol*, **39**, 302–4.

Gupta, A.K. and Elewski, B.E. 1996. Nondermatophyte causes of onychomycosis and superficial mycoses. *Curr Topics Med Mycol*, **7**, 87–97.

Gupta, A.K., Scher, R.K. and De Doncker, P. 1997. Current management of onychomycosis: an overview. *Dermatol Clin*, **15**, 121–35.

Gupta, A.K., Einarson, T.R., et al. 1998. An overview of topical antifungal therapy in dermatomycoses; a North American perspective. *Drugs*, **55**, 645–74.

Gupta, A.K., De Doncker, P. and Haneke, E. 2000. Itraconazole pulse therapy for the treatment of *Candida* onychomycosis. *J Eur Acad Dermatol Venereol*, **15**, 2, 112–15.

Gupta, A.K., Gregurek-Novak, T., et al. 2001. Itraconazole and terbinafine treatment of some nondermatophyte molds causing onychomycosis of the toes and a review of the literature. *J Cut Med Surg*, **5**, 3, 206–10.

Hamm, H., Schwinn, A., et al. 1999. Short duration treatment with terbinafine for tinea capitis caused by *Trichophyton* or *Microsporum* species. The Study Group. *Br J Dermatol*, **140**, 3, 480–2.

Haneke, E. 1990. Fluconazole levels in human epidermis and blister fluid. *Br J Dermatol*, **123**, 273–7.

Haneke, E. 1992. Pharmacokinetic evaluation of fluconazole in plasma, epidermis and blister fluid. *Int J Dermatol*, **31**, 3–5.

Hanel, H., Smith-Kurtz, E. and Pastowsky, S. 1991. Therapy of seborrheic eczema with an antifungal agent with an antiphlogistic effect. *Mycoses*, **34**, 1, 91–3.

Havu, V., Brandt, H., et al. 1997. A double-blind randomized study comparing itraconazole pulse therapy with continuous dosing for the treatment of toe-nail onychomycosis. *Br J Dermatol*, **136**, 230–4.

Hay, R.J. 1993. Onychomycosis: agents of choice. *Dermatol Clin*, **11**, 161–9.

Hay, R.J. 1999. The management of superficial candidiasis. *J Am Acad Dermatol*, **40**, 6 Pt 2, S35–42.

Hickman, J. 1996. A double-blind, randomized, placebo-controlled evaluation of short-term treatment with oral itraconazole in patients with tinea versicolor. *J Am Acad Dermatol*, **34**, 785–7.

Hughes, J., Moore, M. and Pembroke, A. 1993. Tinea nigra palmaris. *Clin Exp Dermatol*, **18**, 5, 481–2.

Jue, S.G., Dawson, G.W. and Brogden, R.N. 1985. Ciclopirox olamine 1% cream. A preliminary review of its antimicrobial activity and therapeutic use. *Drugs*, **29**, 4, 330–41.

Khandpur, S. and Reddy, B. 2002. Itraconazole therapy of white piedra affecting scalp hair. *J Am Acad Dermatol*, **47**, 415–18.

Knight, T.E., Shikuma, C.Y. and Knight, J. 1991. Ketoconazole-induced fulminant hepatitis necessitating liver transplantation. *J Am Acad Dermatol*, **25**, 398–400.

Lambert, D.R., Siegle, R.J. and Camisa, C. 1989. Griseofulvin and ketoconazole in the treatment of dermatophyte infections. *Int J Dermatol*, **28**, 300–4.

Lesher, J.L. Jr 1996. Recent developments in antifungal therapy. *Dermatol Clin*, **14**, 163–9.

Lesher, J.L. Jr 1999. Oral therapy of common superficial fungal infections of the skin. *J Am Acad Dermatol*, **40**, 31–4.

Ling, M.R. 2001. Topical tacrolimus and pimecrolimus: future directions. *Semin Cut Med Surg*, **20**, 268–74.

Marks, J. Jr, King, R. and Davis, B. 1980. Treatment of tinea nigra palmaris with miconazole. *Arch Dermatol*, **116**, 3, 321–2.

Marks, R., Pears, A.D. and Walker, A.P. 1985. The effects of a shampoo containing zinc pyrithione on the control of dandruff. *Br J Dermatol*, **112**, 415–22.

Meinhof, W. 1993. Kinetics and spectrum of activity of oral antifungals; the therapeutic implications. *J Am Acad Dermatol*, **29**, 37–41.

Mehta M. (ed.) 2001. *PDR electronic library*. Montvale, NJ: Medical Economics.

Mock, M., Monod, M., et al. 1998. Tinea capitis dermatophytes: susceptibility to antifungal drugs tested in vitro and in vivo. *Dermatology*, **197**, 4, 361–7.

Moossavi, M., Bagheri, B. and Scher, R. 2001. Systemic antifungal therapy. *Dermatol Clin*, **19**, 35–52.

Odom, R., Daniel III, C.R. and Aly, R. 1996. A double-blind randomized comparison of itraconazole capsules and placebo in the treatment of onychomycosis of the toenail. *J Am Acad Dermatol*, **35**, 110–11.

Peter, R.U. and Richarz-Barthauer, U. 1995. Successful treatment and prophylaxis of scalp seborrhoeic dermatitis and dandruff with 2% ketoconazole shampoo: results of a multicentre, double-blind, placebo-controlled trial. *Br J Dermatol*, **132**, 3, 441–5.

Phillips, R.M. and Rosen, T. 2001. Topical antifungal agents. In: Wolverton, S. (ed.), *Comprehensive dermatologic drug therapy*. Philadelphia: W.B. Saunders, 497–523.

Rex, J.H., Walsh, T.J., et al. 2000. Practice guidelines for the treatment of candidiasis. *Clin Infect Dis*, **30**, 4, 552–78.

Rhie, S., Turcios, R., et al. 2000. Clinical features and treatment of Malassezia folliculitis with fluconazole in orthotopic heart transplant recipients. *J Heart Lung Transplant*, **19**, 2, 215–19.

Rodgers, P. and Bassler, M. 2001. Treating onychomycosis. *Am Fam Physician*, **63**, 663-672, 677–8.

Rupke, S.J. 2000. Fungal skin disorders. *Prim Care*, **27**, 2, 407–21.

Ryder, N.S. 1992. Terbinafine: mode of action and properties of the squalene epoxidase inhibition. *Br J Dermatol*, **126**, 2–7.

Ryder, N.S., Wagner, S. and Leitner, I. 1998. In vitro activities of terbinafine against cutaneous isolates of Candida albicans and other pathogenic yeasts. *Antimicrob Agents Chemother*, **42**, 5, 1057–61.

Sanchez, J.L. and Torres, V.M. 1984. Double-blind efficacy study of selenium sulfide in tinea versicolor. *J Am Acad Dermatol*, **11**, 235–8.

Savin, R., Atton, A.V., et al. 1994. Efficacy of terbenafine 1% cream in the treatment of moccasin-type tinea pedis: results of placebo-controlled multicenter trials. *J Am Acad Dermatol*, **30**, 4, 663–7.

Savin, R., De Villez, R.L., et al. 1997. One-week therapy with twice-daily butenafine 1% cream versus vehicle in the treatment of tinea pedis: a multicenter, double-blind trial. *J Am Acad Dermatol*, **36**, 2 Pt 1, 515–19.

Sayegh-Carreno, R., Abramovits-Ackerman, W. and Giron, G.P. 1989. Therapy of tinia nigra plantaris. *Int J Dermatol*, **28**, 1, 46–8.

Scher, R.K. 1999. Onychomycosis: therapeutic update. *J Am Acad Dermatol*, **40**, S21–26.

Scher, R.K., Breneman, D., et al. 1998. Once-weekly fluconazole (150, 300, or 450 mg) in the treatment of distal subungual onychomycosis of the toenail. *J Am Acad Dermatol 38*, **38**, 77–86.

Shannon, P., Ramos-Caro, F., et al. 1999. Treatment of tinea nigra with terbinafine. *Cutis*, **64**, 3, 199–201.

Shuster, S. 1984. The aetiology of dandruff and the mode of action of therapeutic agents. *Br J Dermatol*, **111**, 235–42.

Sobel, J.D., Faro, S., et al. 1998. Vulvovaginal candidiasis: epidemiologic diagnostic, and therapeutic considerations. *Am J Obstet Gynecol*, **178**, 2, 203–11.

Tosti, A., Piraccini, D.M., et al. 2002. Topical steroids versus systemic antifungals in the treatment of chronic paronychia: an open, randomized double-blind and double dummy study. *J Am Acad Dermatol*, **47**, 1, 73–6.

Tucker, R.M., Haq, Y., et al. 1990. Adverse events associated with itraconazole in 189 patients on chronic therapy. *J Antimicrob Chemother*, **26**, 4, 561–6.

Villars, V. and Jones, T.C. 1989. Clinical efficacy and tolerability of terbinafine (Lamisil) – a new topical and systemic fungicidal drug for treatment of dermatomycoses. *Clin Exp Dermatol*, **14**, 124–7.

Weinstein, A. and Berman, B. 2002. Topical treatment of common superficial tinea infections. *Am Fam Physician*, **65**, 2095–102.

Wildfeuer, A., Faergemann, J., et al. 1994. Bioavailability of fluconazole in the skin after oral medication. *Mycoses*, **37**, 127–30.

PART III

SUPERFICIAL AND OCULAR FUNGAL INFECTIONS

11

White piedra, black piedra, and tinea nigra

G. SYBREN DE HOOG AND EVELINE GUÉHO

Superficial mycoses are defined as the development of fungal growth on epithelial tissues such as human hair, skin, or nails, without noticeable invasion of living tissue and without apparently provoking an immune response by the host. (The term 'pityriasis' is often preferred rather than 'tinea' because the latter term should properly be reserved for dermatophyte infections.) Ecologically, such mycoses are comparable to epiphytic or epilithic growth, i.e. superficial colonization of plants, wood, or stones, almost without making use of the supporting substratum. In the case of colonizers of human skin, utilization of products excreted by the host seems likely.

Two ecological categories can be distinguished in superficial fungi. Growth on the human body may be entirely coincidental, the fungus occupying a natural niche outside the human body. This is the case with *Hortaea werneckii* (*Phaeoannellomyces werneckii*), the causative agent of human tinea nigra, which is a halophilic fungus living in salt pans at subtropical coasts (Zalar et al. 1999). Alternatively, the mammalian body may be the natural habitat of the fungus. This is the case, for example, with members of the ascomycete genus *Piedraia*, which are unique in the fungal kingdom in that they complete their entire lifecycle, including production of ascospores, on hairs of living humans.

ETIOLOGIC AGENTS

White piedra is caused by species of *Trichosporon* Behrend, a genus of basidiomycetous arthroconidial yeasts. *Trichosporon* spp. are also known as agents of cutaneous and of systemic infections. Black piedra is caused by members of the ascomycete genus *Piedraia* da Fonseca and de Area Leão, which are exclusively known as agents of this particular disorder. Tinea nigra is caused by the single species, *Hortaea werneckii* (Horta) Nishimura and Miyaji [*Phaeoannellomyces werneckii* (Horta) McGinnis and Schell].

ECOLOGY

Commensal fungi assimilate compounds that are produced by the human body, without invasion of living tissue (de Hoog et al. 2000). They may also decompose dying or dead host remains without the need for the host itself to be present. As they are frequently found on humans and may form macroscopically visible colonies, the fungi under consideration have frequently been referred to as opportunistic pathogens. However, infections caused by most of the species treated in this chapter are rare, even in immunocompromised patients.

When the target compound is an integral part of the host, e.g. keratin assimilated by dermatophytes, adjacent living tissue can also be affected, and a gradual evolution towards true pathogenicity may be expected (Rippon 1985). In the presence of excreted target compounds, the fungi remain strictly superficial. This is the case with the agents of black piedra and of tinea nigra, and with members of the lipophilic genus *Malassezia* Baillon.

WHITE PIEDRA

Clinical presentation

White piedra is a non-inflammatory, non-invasive fungal growth on the outside of the hair shaft. Soft nodules are formed, which can easily be pulled off the hair. The nodules are white, pale greenish, or yellowish and are composed of compacted fungal elements (Figure 11.1). At the outside the elements produce rounded arthroconidia-like cells. The hairs are not invaded, but they may break if the fungi have been present for long periods.

White piedra is caused by several *Trichosporon* species. The infections may affect hairs on the head, in the axilla, or on the crural area. Human white piedra of the scalp occurs with a low incidence in temperate and subtropical climates (Kaplan 1959a; Gold et al. 1984). Hairs of the beard, eyelashes, eyebrows, and scalp may be affected. In the nineteenth century, the disorder may have been quite common in temperate climates, judging from the fact that patients harbored such infections for several years (Guého et al. 1992a). White piedra has also been observed on the fur of other mammals such as the horse and lower primates worldwide (Kaplan 1959a).

White piedra of the groin does not depend on personal hygiene (Gold et al. 1984; Kalter et al. 1986). At present, genital white piedra is surprisingly common in the Americas (Fischman et al. 1980; Benson et al. 1983; Kalter et al. 1986), Africa (Thérizol-Ferly et al. 1994a) and Europe (Stenderup et al. 1986). Smith et al. (1973) suspected that the disorder might be under-diagnosed, as a result of the site of infection. It is frequently accompanied by intertrigo (Thérizol-Ferly et al. 1994a), but it may also be strictly harmless (Benson et al. 1983; Ellner et al. 1991). Prevalence varies with socioeconomic background (Ellner et al. 1991), gender (Thérizol-Ferly et al. 1994a), sexual behavior (Torssander et al. 1985; Stenderup et al. 1986), and with race (Avram et al. 1987), although marked regional differences can be noted. The fungus may also colonize adjacent cotton fibers in underwear (de Almeida et al. 1990). Perhaps the warm and moist microclimate of groin and axilla is a major predisposing factor. Axillary white piedra has not been investigated in detail; it probably occurs with a comparable frequency.

Microbiology and identification

TRICHOSPORON ASAHII AKAGI

Colonies on Sabouraud's glucose agar (SGA) are dry, pustular, with a broad, deeply fissured, marginal zone and with a white, farinose covering at the center. Budding cells and lateral conidia are absent. Arthroconidia are regular and barrel shaped (Figure 11.2). Appressoria are absent. Guého et al. (1992a) found that documented strains from superficial mycoses mainly originated from animal white piedra; numerous other strains came from human cutaneous and deep infections.

TRICHOSPORON CUTANEUM (DE BEURMANN ET AL.) OTA

Colonies on SGA are moist and shiny, with a broad, fissured, marginal zone without farinose covering. Budding cells are abundant in primary cultures, but hyphae predominate after repeated transfer. Arthroconidia are cylindrical to ellipsoidal, and lateral conidia are present. The species is known only from a small collection of strains, among which is one originating from axillary white piedra.

TRICHOSPORON INKIN (OHO EX OTA) DO CARMO SOUSA AND VAN UDEN

Colonies on SGA are restricted, cerebriform, with no marginal zone, often cracking the agar medium. Budding cells and lateral conidia are absent. Arthroconidia are long and cylindrical, and appressoria are present (Figure 11.3). Sarcinae may be present in tissue; in vitro this is reproduced on media with high sugar content. The species is limited to hairs in the groin area.

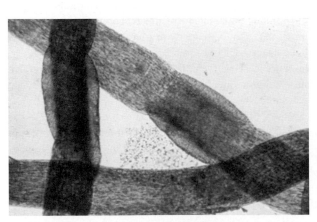

Figure 11.1 *Nodules of white piedra caused by* Trichosporon inkin *on pubic hair. Magnification* × *940*

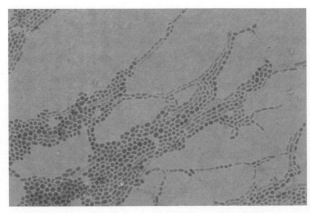

Figure 11.2 *Regular arthroconidia of* Trichosporon asahii. *Magnification* × *940*

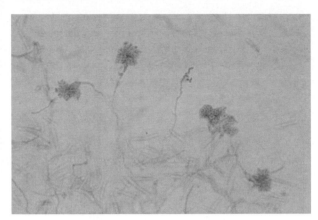

Figure 11.3 *Arthroconidia and appressoria of* Trichosporon inkin. *Magnification × 940*

TRICHOSPORON MUCOIDES GUÉHO AND M.TH. SMITH

Colonies on SGA are moist and shiny, and elevated with deep, narrow, radial fissures. Arthroconidia are rectangular. Short lateral branches terminating with clavate conidia are present (Figure 11.4). The taxon is the prevalent species in groin infections of humans in Africa and is frequently isolated from other superficial locations in Europe.

TRICHOSPORON OVOIDES BEHREND

Colonies are white, farinose, and irregularly wrinkled at the center, with a flat marginal zone. Budding cells and lateral conidia are absent. Arthroconidia are cylindrical and appressoria are present. The species is physiologically indistinguishable from *T. asahii*. A strain originating from scalp white piedra, CBS 7556, was used as the neotype strain for the type species of the genus *Trichosporon* (Guého et al. 1992a).

DISCUSSION OF THE GENUS *TRICHOSPORON*

The genus *Trichosporon* has been revised using molecular characteristics (Guého et al. 1992b). A large

Figure 11.4 *Arthroconidia and lateral blastoconidia of* Trichosporon mucoides. *Magnification × 940*

number of species was recognized, several of which may be encountered as causative agents of opportunistic mycoses. Since that time several new species have been introduced (Sugita et al. 1997, 2001; Sugita and Nakase 1998; Middelhoven et al. 1999, 2000, 2001), but none of these is an agent of white piedra. Overviews of clinically relevant species were published by Guého et al. (1994) and Middelhoven (2002). It should be noted that the species involved in animal mycoses are different to those found in the environment. Thus, infections by *Trichosporon* spp. are probably transmitted through human or animal vectors.

Members of the genus have multilamellar cell walls in common and contain more or less developed dolipore septa, with or without vesicular or tubular parenthesomes (Guého et al. 1992b). They are therefore classified in the Basidiomycetes and are related to *Filobasidiella* (*Cryptococcus*), a position that has been confirmed by partial 26S rRNA sequencing (Guého et al. 1993) and antigenic similarity (Seeliger and Schröter 1963; McManus and Jones 1985). Nevertheless, the phylogenetic distance to remaining heterobasidiomycetes has been judged large enough to erect a separate order, the Trichosporonales (Fell et al. 2000).

Trichosporon spp. do not generally form budding cells but cubic or rectangular arthroconidia. Identification of the five species has been summarized by Guého et al. (1994) (Table 11.1; see Figure 11.1) and Thérizol-Ferly et al. (1994b). Serological recognition of the three most prevalent species was demonstrated by Douchet et al. (1994).

Therapy

The disorder can be controlled by shaving and local application of clotrimazole cream or 5 percent ammoniated mercury ointment (Kwon-Chung and Bennett 1992). All clinically significant *Trichosporon* spp. show high in vitro susceptibility to amphotericin B, clotrimazole, ketoconazole, and itraconazole (Kalter et al. 1986; Guého et al. 1994). Remissions of the disorder are, however, frequent (Walzman and Leeming 1989).

BLACK PIEDRA

Clinical presentation

Black piedra is an asymptomatic but macroscopically visible colonization of the shaft of scalp hair, caused by species of the ascomycete genus *Piedraia*. The species form hard, black nodules, consisting of dense fungal stromata, which are thicker (up to 150 μm) at one end. The nodules cannot be pulled off the hair shaft. The hairs may break after prolonged fungal growth. The disorder is restricted to humid tropical areas (Fischman

Table 11.1 *Key characteristics of* Trichosporon *spp. isolated from white piedra nodules using the API 20C Aux Identification System* (bioMérieux)[a]

Species	Con	Glu	Gly	2-Ket	Ara	Oxy	Ado	Xyl	Gal	Ino	Sor	α-Met	N-Ace	Cel	Lac	Mal	Suc	Tre	Mel	Raf
T. ovoides[b]	−	+	−	+	−	+	−	−	+	+	−	+	+	+	+	+	+	−	v	−
T. inkin	−	+	−	+	−	+	−	−	+	+	−	+	+	+	+	+	+	+	v	−
T. asahii	−	+	−	+	+	+	−	−	+	+	−	+	+	+	+	+	+	s	−	−
T. cutaneum	−	+	v	+	+	+	−	−	+	+	−	−	+	v	+	+	+	v	v	v
T. mucoides	−	+	+	+	+	+	s	s	+	+	+	+	+	+	+	+	+	s	v	+

Ado, adonitol; Ara, L-arabinose; Cel, cellobiose; Con, control; Gal, galactose; Glu, glucose; Gly, glycerol; Ino, inositol; Lac, lactose; Mal, maltose; Mel, melezitose; N-Ace, N-acetyl-D-glucosamine; Oxy, D-oxylose; Raf, raffinose; Sor, sorbitol; Suc, sucrose; Tre, trehalose; Xyl, xylitol; α-Met, α-methyl-D-glucoside; 2-Ket, 2-keto-D-gluconate.
a) After a 4-day incubation at 25°C: −, no growth; +, growth in 24–48 h; s, growth in 72–96 h; v, variable growth. The five species are further separated by absence of growth at 37°C for *T. cutaneum* and no colonial margin for *T. inkin* (see Guého et al. 1994).
b) *T. ovoides*, responsible for the scalp hair white piedra, is the type species of the genus (Guého et al. 1992a).

et al. 1980), particularly South America (Coimbra and Santos 1989), Asia (Adam et al. 1977) and Africa, and may be associated with hair care involving the application of lipid-containing substances (Simons 1954). Black piedra commonly occurs on the pelts of primates living in tropical forests (Kaplan 1959b; Ajello 1964).

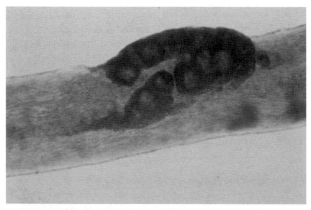

Figure 11.5 *Black piedra caused by* Piedraia hortae *on Asian scalp hair. The nodule corresponds to pseudoparenchymatous ascomata. Magnification × 940*

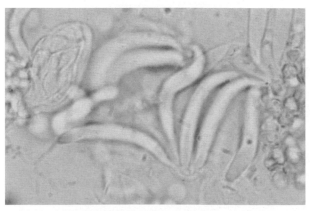

Figure 11.6 *Spindle-shaped ascospores of* Piedraia hortae *tapering towards both ends to form whip-like extensions. Magnification × 2 350*

Microbiology and identification

PIEDRAIA HORTAE (BRUMPT) DA FONSECA AND DE AREA LEÃO

Colonies on SGA containing glycerin grow slowly and are folded, velvety, and dark-brown to black; most remain sterile. Ascostromata on hairs are compact, elongated, up to 1 mm long and 0.3 mm wide. Black, spherical cavities (Figure 11.5) are formed by lysis, each containing a few broadly ellipsoidal asci, usually with eight ascospores (Figure 11.6). Ascospores are hyaline, one-celled, reniform (30–45 × 5.5–10 μm) with whip-like appendages at both ends.

DISCUSSION OF THE GENUS PIEDRAIA

A related species which does not produce ascospore appendages is *Piedraia quintanilhae* van Uden, reported from chimpanzees in Central Africa (Takashio and de Vroey 1975). It is not known to occur on humans.

Therapy

Treatment of black piedra is notoriously difficult, although topical application of azole antifungals or selenium sulfide has been used with variable degrees of success.

TINEA NIGRA

Clinical presentation

Brown to black lesions are formed, mainly on the skin of the palm (Severo et al. 1994), but occasionally on the sole of the foot. Pigmentation is more intense near the border. Only the outer, dead layers of the skin (stratum corneum) are affected; no invasion of living tissue occurs. The fungus is present in the form of melanized, ellipsoidal cells, which often have a thick median crosswall.

Microbiology and identification

HORTAEA WERNECKII (HORTA) NISHIMURA AND MIYAJI

One of the synonyms is *Phaeoannellomyces werneckii* (Horta) McGinnis and Schell. Colonies on oatmeal agar are restricted, smooth, and slimy with an oily, glistening, olivaceous–black color. After several transfers the colonies may become velvety as a result of the production of aerial hyphae. Budding cells are subhyaline, 0–1 septate, and produce daughter cells by annellidic conidiogenesis from their poles. Hyphae are up to 6 μm wide, becoming thick-walled, olivaceous–black, and densely septate at maturation. Conidia are produced laterally from hyphae or from the poles of budding cells by annellidic conidiogenesis. Annellated zones are conspicuous and 1–2 μm wide, with clearly visible annellations. Conidia are initially hyaline and one-celled but soon become olivaceous and measure 7–9.5 × 3.5–4.5 μm. After liberation they inflate and develop transverse and occasionally oblique septa; liberated cells are converted into budding cells or chlamydospore-like cells (Figure 11.7).

DISCUSSION OF THE GENUS *HORTAEA*

The genus *Hortaea* contains two species. In addition to *H. werneckii*, which is halophilic (Zalar et al. 1999), *H. acidophila* has been introduced (Hölker et al. 2004), which is an extremophile able to grow in very acidic substrates. *H. werneckii* is recognized by resistance to 10 percent sodium chloride, production of extracellular DNAase (de Hoog and Gerrits van den Ende 1992), and by the presence of over 1 μm wide annellated zones with relatively widely spaced annellations (Miyaji and Nishimura 1985). Conidia soon become dark, swell, and develop transverse and oblique septa. Hyphae become dark and thick-walled with age and develop further transverse septa. The old hyphal cells are frequently wider than long, and finally produce extracellular granular material. Yeast-like strains without hyphae are recognized morphologically by bipolar annellated zones.

Phylogenetic position

The species was found to be phylogenetically related to *Aureobasidium pullulans* (De Bary) Arnaud (Haase et al. 1995; Spatafora et al. 1995), which may be an anamorph member of the ascomycete family *Dothioraceae* (order Dothideales). These findings are consistent with the similarities in karyology and thallus maturation (Takeo and de Hoog 1991), coenzyme Q systems (Yamada et al. 1989) and the absence of Woronin bodies at septal pores (Mittag 1993). The family contains plant-parasitic fungi and osmo- and acidophilic species living in the phyllosphere; numerous species occur on inert surfaces such as stone (Sterflinger et al. 1999). No human pathogenic species are known. Phylogenetically they are not related to the human pathogenic black yeasts classified in *Exophiala* Carmichael (Masclaux et al. 1995), which are anamorph members of the ascomycete family *Herpotrichiellaceae* (order Chaetothyriales).

Ecology

Hortaea werneckii is halophilic, its natural niche containing significantly elevated concentrations of salt, e.g. beach soil, salt pans, sea water, or salted fish (Zalar et al. 1999). The finding of this habitat confirms speculations of de Cock (1994) and Uijthof et al. (1994) based on mitochondrial DNA and molecular fingerprinting, respectively. Human cases are primarily reported from coastal areas in the subtropics (Rippon 1988). Hyperhydrotic humans may acquire the colonization of their skin. Primary adhesion may be explained by hydrophobic interaction of fungal cells with human skin; the fungus may subsequently thrive on the human body by the production of extracellular polysaccharides and feed by the assimilation of lipidic excretion products (Göttlich et al. 1995).

Similar behaviour may be adduced for the equally saprophytic fungus, *Stenella araguata* Sydow (*Cladosporium castellanii* Borelli and Marcano). This species was reported as an etiologic agent of cases of tinea nigra in Venezuela (di Prisco and Borelli 1973; Marcano and Hutton 1973; Reyes and Borelli 1974).

Therapy

The fungus can easily be removed by keratinolytic compounds such as Whitfield's ointment (Rippon 1988).

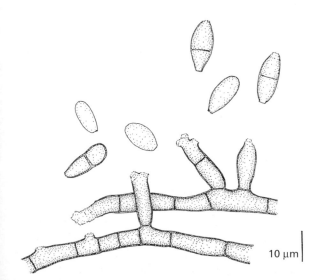

10 μm

Figure 11.7 Hortaea werneckii: *morphology of 1-month-old culture on oatmeal agar.*

REFERENCES

Adam, B.A., Soo-Hoo, T.S. and Chong, K.C. 1977. Black piedra in west Malaysia. *Australas J Dermatol*, **18**, 45–7.

Ajello, L. 1964. Survey of three shrew pelts for mycotic infections. *Mycologia*, **56**, 455–8.

Avram, A., Buot, G., et al. 1987. Etude clinique et mycologique concernant 11 cas de trichosporie noueuse (piedra blanche) génito-pubienne. *Ann Derm Venereol (Stockh)*, **114**, 819–27.

Benson, P.M., Lapins, N.A. and Odom, R.B. 1983. White piedra. *Arch Dermatol*, **119**, 602–4.

Coimbra, C.E.A. and Santos, R.V. 1989. Black piedra among the Zoro indians from Amazonia (Brazil). *Mycopathologia*, **107**, 57–60.

de Almeida, H.L., Rivitti, E.A. and Jaeger, R.G. 1990. White piedra: ultrastructure and a new micro-ecological aspect. *Mycoses*, **33**, 491–7.

de Cock, A.W.A.M. 1994. Population biology of *Hortaea werneckii* based on restriction patterns of mitochondrial DNA. *Antonie van Leeuwenhoek*, **65**, 21–8.

de Hoog, G.S. and Gerrits van den Ende, A.H.G. 1992. Nutritional pattern and eco-physiology of *Hortaea werneckii*, agent of human tinea nigra. *Antonie van Leeuwenhoek*, **62**, 321–9.

de Hoog, G.S., Guarro, J., et al. 2000. *Atlas of clinical fungi*, 2nd edition. Reus: Centraalbureau voor Schimmelcultures, Baarn/Universitat Rovira i Virgili.

di Prisco, J. and Borelli, D. 1973. Tinea nigra por *Cladosporium* species. *Castellania*, **1**, 97–100.

Douchet, C., Thérizol-Ferly, M., et al. 1994. White piedra and *Trichosporon* species in Africa. III. Identification of *Trichosporon* species by slide agglutination test. *Mycoses*, **37**, 261–4.

Ellner, K., McBride, M.E., et al. 1991. Prevalence of *Trichosporon beigelii*. Colonization of normal perigenital skin. *J Med Vet Mycol*, **29**, 99–103.

Fell, J.W., Boekhout, T., et al. 2000. Biodiversity and systematics of basidiomycetous yeasts as determined by large-subunit rDNA D1/D2 domain sequence analysis. *Int J Syst Evol Microbiol*, **50**, 1351–71.

Fischman, O., Pires de Camargo, Z. and Meireles, M.C.A. 1980. Genital white piedra: an emerging fungal disease? *Pan American Health Organization, Scientific Publication no. 396*. Washington, DC: PAHO, 70–6.

Göttlich, E., de Hoog, G.S., et al. 1995. Cell-surface hydrophobicity and lipolysis as essential factors in human tinea nigra. *Mycoses*, **38**, 489–94.

Gold, I., Sommer, B., et al. 1984. White piedra, a frequently misdiagnosed infection of hair. *Int J Dermatol*, **23**, 621–3.

Guého, E., de Hoog, G.X., et al. 1992a. Neotypification of the genus *Trichosporon*. *Antonie van Leeuwenhoek*, **61**, 285–8.

Guého, E., Smith, M.T., et al. 1992b. Contributions to a revision of the genus *Trichosporon*. *Antonie van Leeuwenhoek*, **61**, 289–316.

Guého, E., Improvisi, L., et al. 1993. Phylogenetic relationships of *Cryptococcus neoformans* and some related basidiomycetous yeasts determined from partial large subunit rRNA sequences. *Antonie van Leeuwenhoek*, **63**, 175–89.

Guého, E., Improvisi, L., et al. 1994. *Trichosporon* on humans: a practical account. *Mycoses*, **37**, 3–10.

Haase, G., Sonntag, L., et al. 1995. Phylogenetic analysis of ten black yeast species using nuclear small subunit rRNA gene sequences. *Antonie van Leeuwenhoek*, **68**, 19–33.

Hölker, U., Bend, J., et al. 2004. *Hortaea acidophila*, a new acidophilic black yeast from lignite. *Antonie van Leeuwenhoek*, **in press.**

Kalter, D.C., Tschen, J.A., et al. 1986. Genital white piedra: epidemiology, microbiology, and therapy. *J Am Acad Dermatol*, **14**, 982–93.

Kaplan, W. 1959a. Piedra in lower animals. A case report of white piedra in a monkey and a review of the literature. *J Am Vet Med Assoc*, **134**, 113–17.

Kaplan, W. 1959b. The occurrence of black piedra in primate pelts. *Trop Geogr Med*, **11**, 115–26.

Kwon-Chung, K.J. and Bennett, J.W. 1992. *Medical mycology*. Philadelphia: Lea & Febiger.

Marcano, C. and Hutton, B. 1973. Tinea nigra plantaris por *Cladosporium* sp, segondo caso. *Castellania*, **1**, 129–31.

Masclaux, F., Guého, E., et al. 1995. Phylogenetic relationships of human-pathogenic *Cladosporium (Xylohypha)* species inferred from partial LS rRNA sequences. *J Med Vet Mycol*, **33**, 327–38.

McManus, E.J. and Jones, J.M. 1985. Detection of a *Trichosporon beigelii* capsular polysaccharide in serum from a patient with disseminated *Trichosporon* infection. *J Clin Microbiol*, **21**, 681–5.

Middelhoven, W.J. 2002. Identification of clinically relevant *Trichosporon* species. *Mycoses*, **46**, 7–11.

Middelhoven, W.J., Scorzetti, G. and Fell, J.W. 1999. *Trichosporon guehoae* sp. nov., an anamorphic basidiomycetous yeast. *Can J Microbiol*, **45**, 686–90.

Middelhoven, W.J., Scorzetti, G. and Fell, J.W. 2000. *Trichosporon veenhuisii* sp. nov. an alkane-assimilating anamorphic basidiomycetous yeast. *Int J Syst Evol Microbiol*, **50**, 381–7.

Middelhoven, W.J., Scorzetti, G. and Fell, J.W. 2001. *Trichosporon porosum* comb. nov., an anamorphic basidiomycetous yeast inhabiting soil, related to the *loubieri/laibachii* group of species that assimilate hemicelluloses and phenolic compounds. *FEMS Yeast Res*, **1**, 15–22.

Mittag, H. 1993. The fine structure of *Hortaea werneckii*. *Mycoses*, **36**, 343–50.

Miyaji, M. and Nishimura, K. 1985. Conidial ontogenesis of pathogenic black yeasts and their pathogenicity for mice. *Proc Ind Acad Sci Plant Sci*, **94**, 437–51.

Reyes, O. and Borelli, D. 1974. Caso de tiña negra por cepa peculiar de *Cladosporium castellanii*. *Rev Dermatol Venez*, **13**, 20–8.

Rippon, J.W. 1985. The changing epidemiology and emerging patterns in dermatophyte species. *Curr Topics Med Mycol*, **1**, 208–34.

Rippon, J.W. 1988. *Medical mycology*, 3rd edition. Philadelphia: W.B. Saunders.

Seeliger, H.P.R. and Schröter, R. 1963. A serological study on the antigenic relationships of the form genus *Trichosporon*. *Sabouraudia*, **2**, 248–63.

Severo, L.C., Bassanesi, M.C. and Londero, A.T. 1994. Tinea nigra: report of four cases observed in Rio Grande do Sul (Brazil) and a review of Brazilian literature. *Mycopathologia*, **126**, 157–62.

Simons, R.D.G.P. 1954. Piedra and *Piedraia*. In: Simons, R.D.G.P. (ed.), *Handbook of tropical dermatology*, vol. 2. Amsterdam: Elsevier, 236–48.

Smith, J.D., Murtishaw, W.A. and McBride, M.E. 1973. White piedra (trichosporonosis). *Arch Dermatol*, **107**, 439–42.

Spatafora, J.W., Mitchell, T.G. and Vilgalys, R. 1995. Analysis of genes coding for small-subunit rRNA sequences in studying phylogenetics of dematiaceous fungal pathogens. *J Clin Microbiol*, **33**, 1322–6.

Stenderup, A., Schoenheyder, H., et al. 1986. White piedra and *Trichosporon beigelii* carriage in homosexual men. *J Med Vet Mycol*, **24**, 399–406.

Sterflinger, K., de Hoog, G.S. and Haase, G. 1999. Phylogeny and ecology of meristematic ascomycetes. *Stud Mycol*, **43**, 5–22.

Sugita, T. and Nakase, T. 1998. *Trichosporon japonicum* sp. nov. isolated from the air. *Int J Syst Bacteriol*, **48**, 1425–9.

Sugita, T., Makimura, K., et al. 1997. Partial sequences of large subunit ribosomal DNA of a new yeast species, *Trichosporon domesticum* and related species. *Microbiol Immunol*, **41**, 571–3.

Sugita, T., Takashima, M., et al. 2001. Two new yeasts, *Trichosporon debeurmannianum* sp. nov. and *Trichosporon dermatis* sp. nov., transferred from the *Cryptococcus humicola* complex. *Int J Syst Evol Microbiol*, **51**, 1221–8.

Takashio, M. and de Vroey, C. 1975. Piedra noire chez des chimpanzes du Zaire. *Sabouraudia*, **13**, 58–62.

Takeo, K. and de Hoog, G.S. 1991. Karyology and hyphal characters as taxonomic criteria in ascomycetous black yeasts and related fungi. *Antonie van Leeuwenhoek*, **60**, 35–42.

Thérizol-Ferly, M., Kombila, M., et al. 1994a. White piedra and *Trichosporon* species in equatorial Africa. I History and clinical aspects: an analysis of 449 superficial inguinal specimens. *Mycoses*, **37**, 249–53.

Thérizol-Ferly, M., Kombila, M., et al. 1994b. White piedra and *Trichosporon* species in equatorial Africa. II Clinical and mycological associations: an analysis of 449 superficial inguinal specimens. *Mycoses*, **37**, 255–60.

Torssander, J., Carlsson, B. and von Krogh, G. 1985. *Trichosporon beigelii*: Increased occurrence in homosexual men. *Mykosen*, **28**, 355–6.

Uijthof, J.M.J., Cock, A.W.A.M., et al. 1994. Polymerase chain reaction-mediated genotyping of Hortaeawerneckii, causative agent of tinea nigra. *Mycoses*, **37**, 307–12.

Walzman, M. and Leeming, J.G. 1989. White piedra and *Trichosporon beigelii*: the incidence in patients attending a clinic in genitourinary medicine. *Genitourin Med*, **65**, 331–4.

Yamada, Y., Sugihara, K., et al. 1989. Coenzyme Q systems in ascomycetous black yeasts. *Antonie van Leeuwenhoek*, **56**, 349–56.

Zalar, P., de Hoog, G.S. and Gunde-Cimerman, N. 1999. Ecology of halotolerant dothideaceous black yeasts. *Stud Mycol*, **43**, 38–48.

12

Superficial diseases caused by *Malassezia* species

VICENTE CRESPO ERCHIGA AND EVELINE GUÉHO

INTRODUCTION

Definition and history

The genus *Malassezia* comprises opportunistic lipophilic yeasts with a natural habitat on the skin of humans and warm-blooded animals. Their ability to cause diseases in humans was recognized more than 150 years ago, when Eichsted in 1846 stated that a fungus was responsible for the lesions of pityriasis versicolor (PV). However, the impossibility at this early stage of isolating and culturing the pathogenic agent brought a long-lasting nomenclatural and clinical controversy that was detailed by Slooff in 1970. As a short introductory history, Baillon created the genus *Malassezia* in 1889, with the species *Malassezia furfur* as the agent of PV. In the lesions, yeasts and filaments (= mycelium or hyphae) were observed, but in the absence of growth on isolation there was no certainty that both features were produced by a single microorganism. The genus *Pityrosporum* was thus created to accommodate similar yeasts observed, without any filaments, in the scales of the human scalp (*Pityrosporum ovale* Castellani and Chalmers 1913), rhinoceros skin (*Pityrosporum pachydermatis* Weidman 1925), and dog-ear (*Pityrosporum canis* Gustafson 1955). The species *Pityrosporum orbiculare* (Gordon 1951b) had been also described because of the morphological similarity of yeasts present in PV scales with those of the two former *Pityrosporum*

species. The two species described in humans were found to require complex media to grow whereas the two zoophilic species were less demanding for lipids, growing on enriched media such as Sabouraud dextrose agar (also mycosel agar). The lipophilicity, which excluded the use of any yeast identification system, either conventional or commercial, led mycologists, in the absence of clear key characteristics, to limit this group of fungi to two species (Yarrow and Ahearn 1984): a human species *M. furfur*, and an animal species *Malassezia pachydermatis*, the generic name *Malassezia* having a nomenclatural priority to *Pityrosporum*. In 1990, Simmons and Guého described a third species *Malassezia sympodialis* on the basis of genome differences, but in the absence of well-defined differences, mycologists, and dermatologists as well, up to 1996 ignored this new species.

Phylogeny and taxonomy

The genus *Malassezia* was recognized as having basidiomycete characteristics, even although no teleomorph could be found for any of the species. These features included a multilayered cell wall, repetitive monopolar budding, and a capacity to hydrolyze urea and to give red staining with diazonium Blue B. These characteristics, combined with the lipophilicity, were the basis for a study involving the collection of as many fresh strains as possible from humans and animals, coupled with the various isolates still available in international fungal collections, and for performing a taxonomic revision of the genus. The access to new molecular techniques, such as ribosomal RNA sequencing, allowed investigators to identify seven genetic species (Guillot and Guého 1995). In addition to the already known species *M. furfur*, *M. pachydermatis*, and *M. sympodialis*, the species *Malassezia globosa*, *Malassezia obtusa*, *Malassezia restricta*, and *Malassezia slooffiae* were thus added to the genus, all species having morphological and physiological key characteristics (Guého et al. 1996).

Then, the genus *Malassezia* with its seven species was confirmed as belonging phylogenetically to the Basidiomycota. These species form a branch, individualized as the order of Malasseziales, in the class of Ustilaginomycetes. However, and surprisingly, this branch is more closely related to plant pathogens such as *Exobasidium* spp. and *Tilletia* spp. (Begerow et al. 2000; Fell et al. 2000) than to the human and animal pathogens *Cryptococcus neoformans* and *Trichosporon* spp. classified as Tremellales and Trichosporonales, respectively (Sugita et al. 2002; Boekhout and Guého 2003).

The seven species can be easily identified in routine of medical laboratories, by combining their morphological and physiological characteristics (Guillot et al. 1996; Mayser et al. 1997; Guého et al. 1998). In cases of doubt, sequencing removes any ambiguity (Duarte et al. 2002). Furthermore, owing to the great reliability of sequences, several techniques for molecular identification of *Malassezia* yeasts can be developed (see section on Molecular methods of identification below).

These different techniques of identification have led already to a number of studies that have investigated the relationships of different species with both diseased and healthy skin. From some of these works (Crespo Erchiga et al. 1999a, b, 2000, 2002; Nakabayashi et al. 2000; Aspiroz et al. 2002), it now appears clear that *M. globosa* is the main species associated with PV. This is the only cutaneous disease in which the involvement of *Malassezia* is undisputed. Nevertheless, this species can also be found in normal skin, alone, or mixed with *M. sympodialis*, *M. restricta*, or *M. furfur* (Sugita et al. 2001). In the remaining dermatological disorders associated with *Malassezia* yeasts, their role is controversial. In seborrheic dermatitis, atopic dermatitis, and folliculitis, several studies have focused on the immunological aspects that might explain the pathogenic mechanisms (Ashbee et al. 1994a, b; Devos and van der Valk 2000; Kesavan et al. 2000; Koyama et al. 2000; Tengvall Linder et al. 2000; Faergemann et al. 2001; Ashbee and Evans 2002; Faergemann 2002). In other diseases, such as confluent and reticulate papillomatosis, neonatal pustulosis, psoriasis, blepharitis, otitis, and onychomycosis, the presence and/or significance of *Malassezia* is still a matter of discussion.

CHARACTERISTICS OF *MALASSEZIA* SPP.

Identification in vivo and isolation

The cutaneous pathologies caused by *Malassezia* yeasts are sufficiently characteristic to be identified readily by dermatologists. Usually, because they are not equipped to perform cultures, dermatologists do not isolate yeasts from lesions, but often they confirm their diagnosis with a direct microscopic examination of material from lesions. Indeed, the yeasts are among the smallest seen in tissues, but the repetitive unipolar budding leaves a very typical prominent bud scar on the mother cells. The most satisfactory method for visualizing this phenomenon can be performed when ink (Parker Permanent Blue/Black Quink) is incorporated with the classical 20–30 percent potassium hydroxide in equal proportions (Cohen 1954). *Malassezia* elements (yeasts and/or filaments) take up the stain immediately and appear blue against a clear background (Figure 12.1). A more selective staining method is the calcofluor white stain which can only be visualized if the laboratory has access to a fluorescence microscope. This technique provides an excellent view of seborrheic dermatitis and healthy skin samples (Figure 12.2). Indeed, the appearance of the fungus in tissues, including the yeast shape, the presence or absence of filaments, and the frequency of the two features, provides information about the disease itself. When just one species *M. furfur* (Yarrow and Ahearn 1984) was recognized in humans (alias *P. ovale* or *P. orbiculare*), and another, *M. pachydermatis* (alias *P. pachydermatis* and *P. canis*), in animals, further identification from cultures was not necessary. The interest in *Malassezia*-related pathology has resurrected recognition that several lipid-dependent *Malassezia* species are present in humans (and animals), and those that can be isolated from healthy or diseased skin.

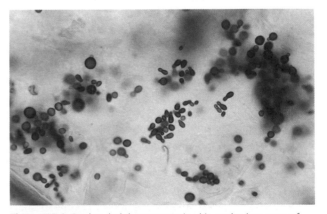

Figure 12.1 *Oval and globose yeasts in skin scales in a case of Gougerot–Carteaud disease. KOH+Parker ink, ×1 000*

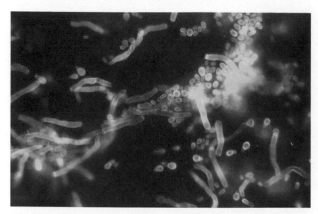

Figure 12.2 Malassezia *oval yeasts and pseudo-hyphae in healthy human skin from the paranasal area. Calcofluor white, ×1 000*

For years the lipophilic fungus *Malassezia* was isolated from skin scales, nails, or other infected materials on regular Sabouraud agar (2 percent glucose, 1 percent peptone, 2 percent agar) or 1 percent olive oil Sabouraud agar, both media supplemented with 0.5 percent chloramphenicol and 0.5 percent cycloheximide to eliminate, as much as possible, contamination by bacteria and molds, respectively. The regular Sabouraud medium was sufficient to isolate the lipophilic, but nonlipid-dependent, species *M. pachydermatis*, whereas the same medium enriched with 1 percent emulsioned olive oil seemed to be very convenient for isolation of the lipid-dependent species *M. furfur* (Guého and Meyer 1989). The taxonomic revision showed that olive oil could be toxic for a few species, so this medium, which is too selective, is no longer used. Two media are now widely employed to isolate and grow all *Malassezia* species: the Dixon medium described by van Abbe in 1964 or its modified formula (mDixon): 3.6 percent malt extract, 0.6 percent peptone, 2.0 percent desiccated ox-bile (Oxoid), 1.0 percent Tween 40, 0.2 percent glycerol, 0.2 percent oleic acid, and 1.2 percent agar, and the medium described by Leeming and Notman in 1987. Because *Malassezia* yeasts are a part of the cutaneous flora of warm-blooded animals, the addition of chloramphenicol and cycloheximide is recommended for all media. The advantage of the former medium (mDixon) over the latter (Leeming and Notman) is that its dark coloration makes easier counting of colonies, particularly when several species are mixed, and a better recognition of their morphologies. This is the reason that the present authors favor the mDixon rather the Leeming and Notman formula in their work. Moreover, before the taxonomic revision, cultures were almost systematically incubated at 37°C. As for the olive oil, the taxonomic revision has shown that several species are inhibited above 37°C. So all cultures are now performed between 30 and 35°C, according to the facilities of the laboratory, but never at 37°C or above, except to characterize the maximum temperature of growth.

Identification in vitro

The seven species of *Malassezia* were clearly separated genetically (Guillot and Guého 1995). However, they are easily identified with routine techniques in any microbiological laboratory on the basis of their characteristics, including their location on the body, colony and yeast morphology, maximum growth temperatures, β-glucosidase (splitting of esculin), and catalase activities, and requirements for growth, in other words growth with Tweens 20, 40, 60, 80, and cremophor EL as sole source of complex lipid. In the lead taxonomic paper (Guého et al. 1996), growth with Tweens at various concentrations was evaluated in liquid medium. Maximum temperatures of growth and catalase activity were fixed for all strains included in the study. Then a simplified method of identification was deduced from these basic results with the diffusion of Tweens in Sabouraud agar plates (Guillot et al. 1996). The identification scheme was improved by Mayser et al. (1997) who suggested the addition of cremophor EL to the diffusion plates and the characterization of the β-glucosidase activity using the esculin agar tubes. The complete identification system was finally summarized by Guého et al. in 1998 (see also Figure 12.3).

M. furfur forms thick, convex or umbonate colonies, 4–5 mm in diameter, with a surface that is smooth to rough and a cream color (Figure 12.4). The texture is soft and the cells are easy to emulsify. This is not the case for all species and, for eliciting this characteristic, it is important to prepare suspensions for all experiments. The species is morphologically heterogeneous with globose, oval, or cylindrical yeast cells (usually not in the same culture), 4.0–5.0 μm and up to 6.0 μm long, some strains being able to produce filaments, spontaneously or under particular culture conditions (Figures 12.2 and 12.5). These are now considered to be pseudohyphae since they have no dolipores, which interrupt the true filaments in Basidiomycetes (Guillot et al. 1995). This species is genetically heterogeneous with high ribosomal RNA similarity, but with two karyotypes (Boekhout et al. 1998). By contrast, *M. furfur* is physiologically homogeneous. It can be routinely identified by combining all its characteristics: ability to grow above 37°C, a strong catalase reaction, an absence or very weak β-glucosidase, and an equal growth in the presence of Tweens 20, 40, 60, 80, and cremophor EL as sole lipid (Figure 12.3). In fact, *M. furfur* is weakly lipid dependent since any lipid supplement is sufficient for its growth. Mayser et al. (1997) demonstrated this as castor oil and ricinoleic acid were sufficient for growth of *M. furfur*. Murai et al. (2002) showed also that *M. furfur* was the least dependent species with a unique positive effect on the growth with glycine. Specific characteristics are quite stable for all the species. However, atypical variants may occur, such as the *M. furfur* dog isolate,

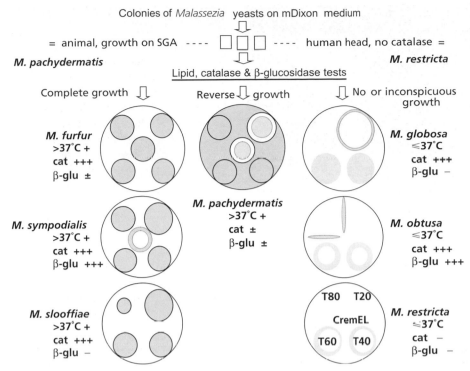

Figure 12.3 *Scheme of Malassezia* spp. *identification. Each culture isolated on mDixon agar is tested for its catalase (cat) and β-glucosidase (β glu) activities; its growth at 37°C is given; then its capacity to grow with Tweens 20, 40, 60, 80, and Cremophor EL is evaluated using Sabouraud agar plates: complete growth (circled gray color); inconspicuous growth (dotted-circled gray color); only precipitate (not circled light gray color).*

confirmed specifically by rDNA sequencing, but unable to grow with cremophor EL (Duarte et al. 2002). Normal isolates of the species are also regularly, but not abundantly, isolated from various animals (Crespo et al. 1999, 2002).

M. pachydermatis is historically the second species described in the genus. It is different from all the other species since it is able to grow on rich conventional laboratory media (Sabouraud or mycosel agars). The species is lipophilic but can grow in the presence of simple lipids present in the peptone of such media. *M. pachydermatis* grows also on mDixon agar and generates

thick, cream-colored, and convex colonies with a mat surface and usually a brittle texture which makes it difficult to emulsify (Figure 12.6). The cells are small, 2.5–4.0 μm, oval to short, and cylindrical with a very broad base to the bud (Figure 12.7). Pink isolates (Midgley 1989) have been reported and a genotype, isolated from dogs, and characterized by smaller colonies, was shown to be lipid dependent in primary culture (Bond and Anthony 1995; Guillot et al. 1997). Duarte et al. (2002) isolated also an atypical lipid-dependent strain of *M. pachydermatis* that remained dependent on lipids even after transfers. Therefore, all epidemiologic surveys of

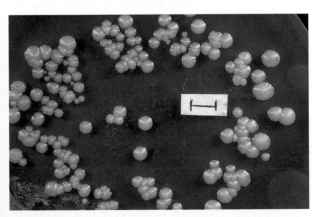

Figure 12.4 *Colonies of* M. furfur, *7 days old, on mDixon medium*

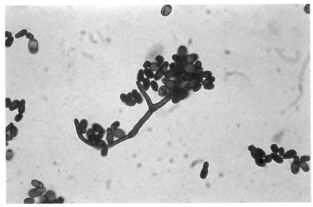

Figure 12.5 *Yeast cells and pseudohyphae of* M. furfur. *Periodic acid Schiff stain,* × 1 000

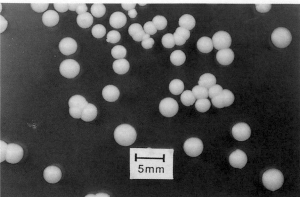

Figure 12.6 *Colonies of* M. pachydermatis, *7 days old, on mDixon medium*

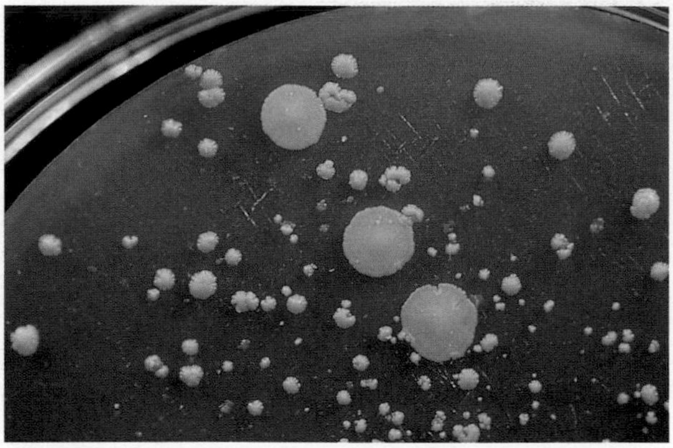

Figure 12.8 *Mixed colonies of* M. sympodialis *(a few) and* M. restricta *(numerous), 7 days old, on mDixon medium*

Malassezia yeasts, from animal or human sources, should utilize directly lipid-supplemented media. The taxonomic revision maintained a single genetic species (Guillot and Guého 1995) but many of its characteristics show that it may be in the process of speciation, a possible adaptation in host specificity (Midreuil et al. 1999). All isolates grow at 37°C, but differences occur in catalase and β-glucosidase expression, and the Tweens 20, 40, 60, 80, and cremophor EL growth tests. These different compounds, particularly Tween 20 and cremophor EL, may be more or less inhibitory, a secondary growth appearing around the wells after dilution of the substrate by diffusion (Guého and Guillot 1999, and Figure 12.3). This species is rare in humans, although it has been found to cause epidemics of sepsis, usually in infants as a complication of prematurity (Mickelsen et al. 1988). In humans its isolation from skin may be considered as accidental. By contrast, the species is likely to be responsible for seborrheic dermatitis and otitis in warm-blooded animals, especially dogs (Guillot and Bond 1999).

M. sympodialis is an easy species to identify. It forms extended, up to 5–6 μm in diameter, flat colonies, often with a slight central elevation. These are cream to buff in color with a smooth, shiny surface, and a homogeneous texture, and are very easy to emulsify (Figure 12.8). The yeasts are small, 2.5–5.0 μm, and ovoid with a narrow bud base (Figure 12.9). As a result of sympodial budding at the same site, yeasts may remain somewhat grouped as 'clover leaves'. The species correlates with the former *M. furfur* serovar A (Cunningham et al. 1990). It grows at 37°C and it is easily characterized in routine by a strong β-glucosidase activity, which deeply darkens the esculin medium in 24 hours at 37°C, and the assimilation of the four Tweens, but only very weakly of cremophor EL (Figures 12.3 and 12.10). Primary isolates may develop a ring of tiny colonies at some distance from the cremophor source, as also seen in a few *M. slooffiae* isolates, but this latter species does not split esculin.

Sugita et al. in 2002 isolated and described a new species, *Malassezia dermatis*, from patients with atopic dermatitis. Analysis of ribosomal DNA 26S and internal transcribed spacers (ITS)s sequences showed this species to be closely related to *M. sympodialis*. The two species differ by only 1.2 percent rDNA base divergence, a difference acceptable as occurring within a single species (Kurtzman and Blanz 1998), and their colony

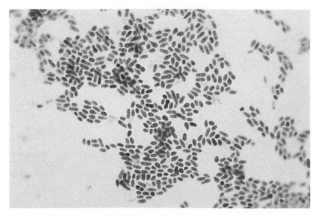

Figure 12.7 *Yeast cells of* M. pachydermatis. *Lactophenol Cotton Blue,* ×1 000

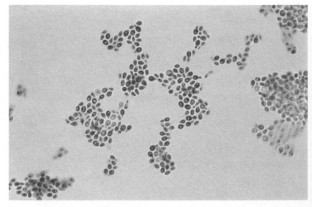

Figure 12.9 *Yeast cells of* M. sympodialis. *Lactophenol cotton blue,* ×1 000

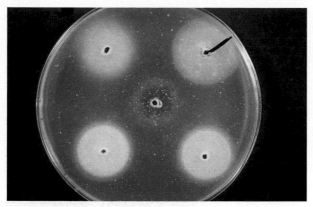

Figure 12.10 M. sympodialis *showing assimilation of the four Tweens (right top, Tween 20 with a black mark, and clockwise, Tweens 40, 60, and 80), and in the center, a ring of tiny colonies around Cremophor EL*

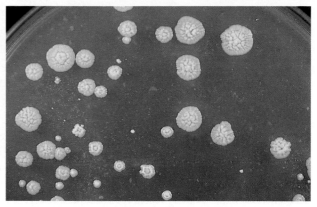

Figure 12.11 *Typical wrinkled colonies of* M. globosa, *7 days old, on mDixon medium*

characteristics, but they cannot be separated with the basic laboratory tests (Guillot et al. 1996). *M. dermatis*, as with *M. sympodialis*, has strong catalase activity and assimilates the four Tweens. However, esculin and cremophor EL are missing in the description of this new species. Similarly, several horse *Malassezia* isolates were found with a rDNA sequence analysis that would suggest that they represent two new species (Nell et al. 2002). No description was given in the paper, although *Malassezia equi* was suggested as the name of one of these new species. As with *M. dermatis*, these zoophilic entities are very closely related to *M. sympodialis* phylogenetically. More studies are in process to better characterize and possibly validate these different entities. *M. nana*, another new species isolated from animals, also resembles *M. sympodialis* but does not split esculin (Hirai et al. 2004). The species *sensu stricto* was found to be associated with otitis externa in two cats (Crespo et al. 2000a).

M. globosa corresponds morphologically to the original description of *P. orbiculare* given by Gordon (1951a). This specific name was not preserved, because it was obvious that most data published in the past as *P. orbiculare* corresponded to one or the other lipid-dependent species. Immunologically, *M. globosa* correlates with the former serovar B (Cunningham et al. 1990). The species produces recognizable colonies, 4 mm in diameter, rough with a deeply folded surface, cream to buff in color, and a very brittle texture; they are particularly difficult to emulsify (Figure 12.11). The cells are spherical, reaching in some strains 6–8 μm in diameter with buds formed on a narrow base (Figure 12.12). In PV skin scales, these yeasts are regularly mixed with pseudofilaments, which are usually still present in primary cultures as germinative tubes, but unfortunately they disappear after several transfers. The species does not grow at 37°C, or very poorly, does not split esculin, and does not grow with any individual lipid supplements (Figure 12.3). Fresh isolates may produce tiny colonies

in the presence of Tweens, in particular Tween 20. Whitish disks, frequently present with the different species, around Tweens 40 and 60 correspond to precipitates (Guillot et al. 1996). *M. globosa* is very often associated with another species, especially *M. sympodialis*, but also *M. furfur*, *M. slooffiae*, or *M. restricta* (Crespo Erchiga et al. 1999b). It has been isolated occasionally from horses and domestic ruminants (Crespo et al. 2002).

M. obtusa is a species that may be confused with *M. furfur* morphologically but not physiologically. The species grows slowly, so colonies remain small in diameter with a sticky texture. By contrast, cylindrical cells are among the biggest in the genus, up to 10 μm long when the bud is still attached to the mother cell. The species does not grow at 37°C, or with any of the five lipids used as identification criteria as with *M. globosa* and *M. restricta*, but expresses strong β-glucosidase activity (Figure 12.3). It is a very rare species that so far is known to occur only in healthy skin. In animals it has been isolated, with *M. furfur*, from a case of canine otitis (Crespo et al. 2000b), and from healthy horses and goats (Crespo et al. 2002).

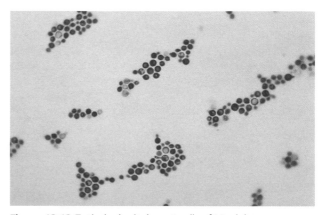

Figure 12.12 *Typical spherical yeast cells of* M. globosa. *Lactophenol Cotton Blue, × 1 000*

M. restricta might correspond to the former *P. ovale*, but it is difficult to correlate with the old literature, a task that is even greater than for *M. globosa*. The species was named 'restricta' because of its very poor performances. It grows very slowly, giving irregular, small (2 mm average in diameter) and cream-colored colonies with a hard texture; these are very difficult to emulsify (see Figure 12.8). Cells are spherical to oval, 2.0–4.0 μm, with buds formed on a relatively narrow base (Figure 12.13). Its micromorphology resembles that of *M. sympodialis*, but hopefully not its colonial macromorphology. This new species could be correlated with serovar C, the third serotype characterized by Cunningham et al. (1990) among lipid-dependent *Malassezia* yeasts. At this point it is important to point out that serotyping, as did biotyping proposed by Aspiroz et al. (1999), recognized only the three species *M. sympodialis*, *M. globosa*, and *M. restricta*, and ignored the others, in particular *M. furfur* and *M. slooffiae*. Hopefully for its identification, *M. restricta* is the only lipid-dependent species lacking catalase activity. This test is thus recommended as a first step, particularly when very restricted numbers of *Malassezia* colonies are obtained from the head. Furthermore, this species does not grow at 37°C, does not split esculin, and does not grow in the presence of different Tweens or cremophor EL (Figure 12.3). The species, as with *M. obtusa*, has also been isolated for the first time from domestic animals (Crespo et al. 2002). Because of the great difficulty encountered with this species, its presence in humans, and possibly in animals, is likely to have been underestimated. Furthermore, such difficulties in isolation, obtaining sufficient cultures, and maintaining *M. restricta* may be a barrier to understanding the dermatological disorders such as seborrheic dermatitis and atopic dermatitis, and consequently their treatment.

M. slooffiae, one of the four new species, is less frequently isolated from humans but appears to be regularly associated with the skin of pigs. It may be morphologically misidentified as *M. furfur*. It has colonies with

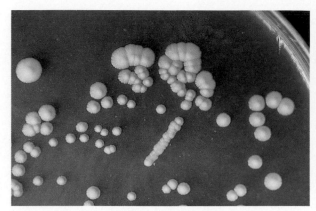

Figure 12.14 *Typical colonies of* M. slooffiae *with folded margins, 7 days old, on mDixon medium*

finely folded margins, that are 3 mm in diameter, cream to buff in color, and have a brittle texture (Figure 12.14). The cells are small and shortly cylindrical, 1.5–3.5 μm long, but never globose or elongate as are some strains of *M. furfur* (Figure 12.15). However, physiologically, there is no ambiguity in clearly identifying organisms as *M. slooffiae*. The species has catalase activity and grows at up to 40°C as does *M. furfur*, but it does not split esculin at all, grows weakly with Tween 80, and does not grow with cremophor EL (Figure 12.3). The species is found principally on healthy skin, or mixed with other species such as *M. furfur*, *M. sympodialis*, *M. globosa*, or *M. restricta*, but always at a lower frequency. It is unlikely to be a major pathogen, at least on humans.

Maintenance of *Malassezia* in collections

Several reasons explain why the taxonomy of the genus *Malassezia* remained for so long a matter of controversy. Primarily, the lipophilicity was not understood and for years the different entities were known only from the literature, without any material deposited in fungal collections. When cultural isolation became possible,

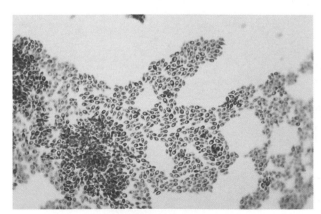

Figure 12.13 *Yeast cells of* M. restricta. *Lactophenol Cotton Blue,* ×1 000

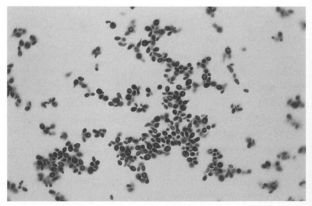

Figure 12.15 *Cylindrical yeast cells of* M. slooffiae. *Lactophenol cotton blue,* ×1 000

isolates started to be deposited. However, with time, only the less demanding species, essentially *M. furfur* and *M. pachydermatis*, have survived. In 1989, Guého and Meyer published the reevaluation of the genus *Malassezia*, based for the first time on genome comparisons but using lyophilized strains or strains maintained for years in the Dutch CBS fungal collection (Centralbureau voor Schimmelcultures). Furthermore, the lipid-dependent strains were grown on olive-oil-supplemented medium. The authors limited, as described by Yarrow and Ahearn (1984), the genus to only two species. In fact, only *M. furfur*, *M. pachydermatis*, *M. sympodialis*, and *M. slooffiae* survive lyophilization. For successful taxonomic revision (Guého et al. 1996), the work compared strains maintained in collections housing a large number of fresh isolates. All cultures were performed on mDixon which allows a better maintenance of all species. They are also never incubated or preserved at room temperature, but at 34°C with transfers every few months. Indeed, at room temperature, even with *M. pachydermatis*, losses are significant. The collection was then preserved in parallel with mDixon freeze-dried cultures, stored at –80°C. Crespo et al. (2000c) confirmed the importance of storage conditions used to preserve *Malassezia* yeasts.

ECOLOGY OF *MALASSEZIA* SPP.

The *Malassezia* yeasts have an absolute requirement for lipids to grow and survive. Furthermore, they are mesophilic, in other words their optimal temperature of life is between 30 and 35°C. These two physiological peculiarities explain why they are known only from the skin of warm-blooded animals. In the laboratory they can be obtained and maintained in culture only if specific nutrients and growth temperature are provided. Veterinarians have investigated the role of *M. pachydermatis* in causing ear infections in dogs for a long time (Guillot et al. 1998; Guillot and Bond 1999). Recently, some cases have been reported where *M. furfur* and *M. obtusa* were isolated as agents of otitis externa in dogs, and *M. sympodialis* in a similar condition in cats (Crespo et al. 1999, 2000a, b). Many studies had already demonstrated that *Malassezia* lipid-dependent spp. also colonize the skin of many other animals: monkeys, pigs, rhinoceros, bears, and birds (Midgley and Clayton 1969; Dufait 1985; Guillot et al. 1994, 1998).

The presence of *Malassezia* spp. in healthy human skin had been detected in earlier work during the second half of the nineteenth century. The frequency and density of colonization are related to the subject age and to sebaceous gland activity in the studied area (Marcon and Powell 1992).

Since the new descriptions, some studies have focused on their precise distribution in normal human skin. Crespo Erchiga et al. (1999a, b, 2002) with three successive studies made up of nearly 300 adults, showed that

M. sympodialis appears as the predominant species in healthy skin, especially of the trunk, where it can be recovered in great numbers from more than 50 percent of individuals. It can also be found on the face, though at a lower percentage, together with *M. globosa*. This latter species can be found alone in about 10 percent of subjects. In a new ongoing study by the same authors (unpublished data), *M. restricta* seems to be predominant in the scalp, where it is isolated in 20 percent of adults, though always associated with a certain degree of scaling. Aspiroz et al. (1999) also found *M. restricta* associated particularly with the scalp skin, *M. sympodialis* with the back, while *M. globosa* was evenly distributed on scalp, forehead, and trunk. These data are similar to those published by Midgley in the UK in 2000, although this author reports a higher percentage of *M. globosa* in the scalp (45–50 percent). In these different surveys, the number of *M. furfur sensu stricto* isolates is very low (Midgley 2000; Crespo Erchiga et al. 2002), and even absent (Aspiroz et al. 1999).

In a Japanese study of 35 normal subjects, only *M. globosa* (51 percent) and *M. sympodialis* (26 percent) were recovered from the trunk. Sixty percent of scalp and face samples remained negative in culture, whereas *M. globosa*, *M. furfur*, and *M. sympodialis* were isolated from a few cases (Nakabayashi et al. 2000). In Russia, *M. sympodialis* was found to be the most common species in the skin of 32 healthy individuals, whilst *M. globosa* was encountered much less frequently (Arzumanyan 2001). In Canada, *M. sympodialis* was also the commonest species in 20 healthy control subjects (Gupta et al. 2001a). In this study, *M. globosa* was equally recovered from the scalp, forehead, and trunk, but less from the arms and legs, whereas *M. restricta* and *M. slooffiae* were recovered more frequently from the scalp and forehead than from the lower body. Differences encountered among these various studies may be explained by the different sampling techniques, culture media (Leeming-Notman by Gupta et al. 2001a, and mDixon by Midgley 2000, Nakabayashi 2002 and Crespo Erchiga et al. 2002), and possibly also by ethnic and geographical factors.

Finally, Sugita et al. in 2001, using a nested polymerase chain reaction (PCR) technique, detected *Malassezia* DNA in 78 percent of the scalp and nape samples of 18 healthy subjects. Among these isolates, *M. restricta*, *M. sympodialis*, *M. globosa*, and *M. furfur* were found at frequencies ranging from 61 to 11 percent. These results confirm that at least four *Malassezia* species, *M. sympodialis*, *M. restricta*, *M. globosa*, and *M. furfur*, are common inhabitants of the human skin.

SUPERFICIAL DISEASES ASSOCIATED WITH *MALASSEZIA* SPP.

The yeasts of the genus *Malassezia* are associated in a variable way with a number of human diseases, basically affecting the skin, although in the last two decades a number of opportunistic systemic infections, most in premature neonates and related to the administration of lipids through intravenous catheters, have been described (Mickelsen et al. 1988; Midgley 2000; Crespo Erchiga and Delgado 2002).

Pityriasis versicolor

Pityriasis versicolor (PV) is a chronic, benign cutaneous disorder, usually asymptomatic, and characterized by slightly scaly patches of variable color (either pink, brown, or white) localized in the upper trunk. It has a worldwide distribution, though it is more frequent in tropical climates, where it has been reported in 30–40 percent of the population. The incidence in temperate climates is much lower (1–4 percent). It affects mainly young adults of both sexes although in tropical zones it is also very common in infancy and even in babies. The onset of the disease seems to be related mainly to local factors such as high temperature and humidity. The appearance on the trunk in the classic form of PV is probably related to occlusion produced by clothing. The condition appears generally in otherwise healthy people, and its incidence is not higher in those with acquired immunodeficiency syndrome (AIDS).

The diagnosis is usually performed by direct observation of the typical yeast and pseudo-hyphae features present in PV skin scales, using the reactive KOH-Parker ink (Figure 12.16). The yeasts are spherical, as are those of *M. globosa*, in the majority of cases, and occasionally oval, as are those of *M. furfur*. Indeed, in the taxonomic revision, it appeared that only *M. globosa* and *M. furfur* were able to filament more or less spontaneously (Guého et al. 1996).

Historically, Gordon was the first to assume that the rounded forms of *Malassezia* yeasts, he referred to *P. orbiculare*, were the etiologic agents of PV (Gordon 1951a, b). Following the description of new taxa, a few epidemiologic studies have dealt with this question.

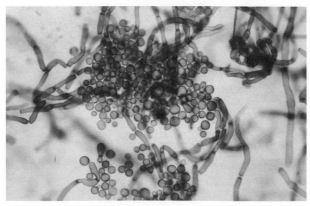

Figure 12.16 *Typical globose yeasts and pseudohyphae in PV skin scales. KOH+Parker ink, ×1000*

In Japan, Nakabayashi et al. (2000) found *M. globosa* present in 55 percent of PV scale samples of 22 patients, whilst all other species were below 10 percent. They suggested that this species plays a pathogenic role in the disease, as proposed previously by Crespo Erchiga et al. (1999a).

In a study carried out in Canada, Gupta et al. (2001b) found *M. sympodialis* (59 percent), *M. globosa* (25 percent), and *M. furfur* (11 percent) in 111 PV cases. In this study, a high number of cases with positive microscopy failed to grow in culture, and the authors suggested that this was consistent with the standard sampling practice of scraping the older (central) parts of PV lesions rather than the margins.

In Southern Spain, Crespo Erchiga et al. (2000) performed three successive epidemiologic studies, comparing the isolation of *Malassezia* spp. from PV, seborrheic dermatitis, and healthy skin from the same patients, with normal subjects. In the lesions, direct microscopy always showed the typical mixture of globose yeasts and filaments. As a result of this study, the authors did not find that the central parts of clinical lesion were less suitable for culture than margins. In contrast to other dermatomycoses (such as tinea), *Malassezia* yeasts and pseudomycelium can be distributed in similar amounts all over the affected skin, while scraping margins could lead to a higher recovery of surrounding commensal species, such as *M. sympodialis*. In the last group of this survey, consisting of 96 adult patients, *M. globosa* was isolated from 97 percent of cases, alone in 60 percent of them, and associated with *M. sympodialis* in 29 percent, or *M. slooffiae* in 7 percent. These two species were also found in similar percentages in the clinically uninvolved skin of the trunk, whereas *M. globosa* was not isolated at these sites. However, on the forehead, a small number of *M. globosa* colonies were recovered in 12 percent of cases. The presence of this species, in its yeast phase, in seborrheic dermatitis and healthy skin, indicates that local factors (humidity, sweat, heat, and seborrhea), together with some degree of idiosyncratic individual predisposition are responsible for the transformation to the mycelial form and the development of clinical lesions. In 2002, Crespo Erchiga et al. presented new results obtained with a case study of 210 PV patients. These new results were consistent with previously published data that thus confirmed the role of *M. globosa* in the etiology of PV, at least in the clinical form seen in temperate climates.

Three other studies support this conclusion. In Spain (Zaragoza), Aspiroz et al. (2002), with a survey of 79 PV patients, recovered *M. globosa* in 90% and *M. sympodialis* in 41 percent. In Japan, Nakabayashi (2002), using PCR identification of *Malassezia* species involved in PV, found *M. globosa* (97 percent) as the most commonly detected species, followed by *M. restricta* (79 percent) and *M. sympodialis* (68 percent). Finally, in Greece,

Gaitanis et al. (2002), using a DNA-PCR procedure, directly applicable to pathological skin scales, detected and identified only *M. globosa* from PV scales, an identification also confirmed by positive culture and biochemical tests.

An interesting point, already discussed by earlier authors, is whether the species associated with PV can vary depending on clinical and geographical distribution. Ingham and Cunningham in 1993 affirmed that this discussion had started at the beginning of the century. Castellani (1925), for instance, thought that pityriasis seen in tropical zones was a separate entity, and Panja (1927) divided the disorder into two diseases, named pityriasis versicolor (due to *M. furfur*) and pityriasis flava (due to *M. tropica*). In temperate zones, lesions affect the upper trunk principally and, although there are variations under direct microscopy, they regularly show globose cells. In tropical regions, by contrast, lesions preferentially affect the face and limbs and oval or cylindrical cells are found. Midgley (2000) concluded that the species responsible for PV could vary with the body site and different regions of the globe. According to Crespo Erchiga's experience, only 1 percent of PV cases displays, under direct scale examination, oval or cylindrical yeasts (Figure 12.17). The lesions showing this kind of picture are usually of an atypical distribution, affecting the face or limbs, and seem to be found mostly in patients from tropical regions. In Guého's experience, the number of PV cases for which the responsibility of *M. furfur* can be proved increases significantly when material taken from lesions of PV is incubated at 37°C (Guého and Meyer 1989). Indeed, *M. furfur* is a regular skin inhabitant and this species is responsible for all deep infections due to a lipid-dependent *Malassezia* species whatever the geographical origin of the patient (Boekhout et al. 1998; Theelen et al. 2001). However, this species has been unexpectedly rare in most recent investigations using either the Dixon or Leeming and Notman media. *M. furfur* may also be

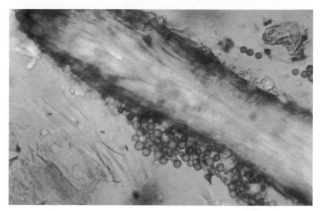

Figure 12.18 *Direct microscopy of PV scales, with* M. globosa *yeasts surrounding a villous hair and penetrating the follicle. KOH+Parker ink, ×1 000*

considered to be a regular but sparsely distributed inhabitant of human skin, taking advantage of an ecological niche only when the other species are eliminated or not present.

The immunological responses of PV patients to *Malassezia* yeasts have been investigated with equivocal results. Wu and Chen (1985) found an increase in humoral *Malassezia* antibodies in patients, compared to control subjects, while other studies did not find any difference between the two groups (Faergemann 1983; Midgley and Hay 1988; Ashbee et al. 1994a). Conflicting results have also been documented in investigations of cellular immunity.

The treatment of PV is simple, with rapid disappearance of lesions after treatment with different antifungal drugs, given either orally or topically, although the achromic areas can persist for weeks and even months. Topically, the most common substances used are selenium sulfide, propyleneglycol, and antifungal azoles. In extensive cases, the short-term use of systemic drugs, such as ketoconazole or itraconazole, is recommended. However, in spite of good initial results, the rates of recurrence after treatment are very high, about 60 percent after 1 year and 80 percent after 2 years, probably because underlying conditions and local factors remain unmodified. The maintenance of yeasts in the deep portions of the pilosebaceous follicles, frequently observed in KOH (Figure 12.18), may also explain such a high rate of recurrence. Prophylactic itraconazole treatment, 200 mg 1 day per month for 6 consecutive months, showed mycological cures in 88 percent of cases at the end point (Faergemann et al. 2002).

Seborrheic dermatitis

Seborrheic dermatitis (SD) is a chronic, relapsing skin disorder, characterized by the appearance of greasy, scaly, reddish patches localized in sebum-rich areas, such as the scalp, eyebrows, paranasal and middle thoracic

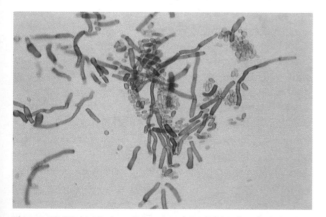

Figure 12.17 *Atypical oval yeasts and pseudohyphae in PV scales. The patient was a child with hypochromic lesions on the face. KOH+Parker ink, ×1 000*

regions. Dandruff, which affects 5–10 percent of the adult population, is considered to be the mildest or initial form of the disease. AIDS patients have an increased incidence of SD, suggesting that the immune system plays an important role in the pathogenesis of the disease. The incidence is also higher in patients with neurological disorders such as Parkinson's disease, multiple sclerosis, and mood depression. The rare cases of *Malassezia* blepharitis can also be related to this skin disorder (Thygeson and Vaughan 1954; Ninomiya et al. 2002).

The relationship between *Malassezia* yeasts and SD is still a matter of controversy. Sabouraud (1932), during the first half of the past century, considered the yeast *Pityrosporum malassezii* to be the causative agent of the disease. In a critical review of the literature, Shuster (1984) described *P. ovale* as the etiologic agent. Since then, numerous studies have shown that a relationship exists between the severity of SD and the density of *Malassezia* cells present in dandruff. The taxonomic revision has generated several studies, aimed at tracing the different distribution of species in SD lesions. In Japan, Nakabayashi et al. (2000) isolated *M. furfur* (35 percent) and *M. globosa* (22 percent) from SD facial lesions of 42 patients, and less from normal subjects. In Canada, during a study using contact plates from 28 SD patients, Gupta et al. (2001a), found decreasing proportions of *M. globosa*, *M. sympodialis*, and *M. slooffiae*. In Spain, Crespo Erchiga et al. (1999a) concluded, in a survey of 75 SD patients, that *M. restricta* (43 percent) and *M. globosa* (34 percent) were the predominant species on the scalp and face of SD patients. Indeed, *M. sympodialis* (19 percent), *M. slooffiae* (2.5 percent), and *M. furfur* (0.8 percent) were much less prominent in this clinical feature. In more than 50 percent of lesional skin specimens, more than one species was obtained in culture. By contrast, healthy skin of the same area (forehead) remained sterile in 90 percent of samples, whereas 62 percent of the healthy trunk skin (shoulders) of the same patients yielded *M. sympodialis* predominantly. This striking difference might, of course, reflect the use of shampoos containing antifungal additives.

Crespo Erchiga and Delgado Florencio (2002) in a new study of 100 SD patients confirmed that *M. restricta* (65 percent) and *M. globosa* (64 percent) were the predominant species in SD lesions, followed by *M. sympodialis* (27 percent), mainly in lesions located on the trunk, and *M. furfur* (4 percent). In the healthy skin of the forehead, *M. sympodialis* was found in 20 percent of patients and *M. globosa* in 10 percent, whilst on the shoulders *M. sympodialis* was present in 58 percent of individuals and *M. globosa* in 7 percent. The same year, Gemmer et al. in the USA (Cincinnati), in a survey of 70 subjects with SD, using a direct fluorescent nested rDNA PCR of scalp samples, demonstrated that *M. restricta* (50–72 percent, depending on the flaking levels) and *M. globosa* (33–45 percent) were the predominant

species in dandruff, whilst *M. sympodialis* was detected in a very small percentage (7–8 percent), and *M. furfur* not at all.

The increased incidence of SD in inmunosuppressed patients suggests that immune mechanisms are important in the pathogenesis of this disease. Both cellular and humoral immune responses have been investigated with conflicting results. Some studies of humoral *Malassezia* antibodies, compared with normal subjects, have shown increased titres in patients (Ashbee et al. 1994b; Silva et al. 1997a), whereas another did not show significant differences (Parry and Sharpe 1998). Lately, Midgley (2000) has shown that 72.5 percent of SD patients and 32 percent of PV patients had precipitating antibodies against *M. globosa*, in contrast to only 20 percent of normal control subjects. The results of testing cellular immunity are also contradictory. Meuber et al. (1996) found a reduced lymphocyte stimulation reaction when SD patients' cells were stimulated with *P. ovale* extract. By contrast, Ashbee et al. (1994b) showed an increase in both lymphocyte transformation responses and leukocyte migration inhibition. Faergemann et al. (2001) have reported increased numbers of NK1+ and CD16+ cells and in complement staining that suggest that an irritant nonimmunogenic stimulation of the immune system is important in the pathogenesis of the disease. The results of interleukin (IL) assays (increase in the production of inflammatory IL and in the regulatory IL for both Th1 and Th2 cells) showed similarities to the immune response described for *Candida* infections.

Kesavan et al. (2000) compared the inmunomodulatory capacity of normal and lipid-depleted *Malassezia* yeasts (using *M. sympodialis*, *M. globosa*, and *M. restricta*). They demonstrated that the extraction of lipids reversed their capacity to reduce the levels of proinflammatory cytokines. The lipid microfibrillar layer of *Malassezia* cells may prevent them from inducing inflammation and explain their normal commensal status on human skin. The absence or alteration of this lipid layer may provide an explanation for the inflammatory nature of the lesions.

Treatment options include selenium sulfide-, pyrithione zinc-, piroctone olamine-, or ketoconazole-containing shampoos, topical terbinafine solution or ketoconazole cream as well as topical corticosteroids. The corticosteroids, even the mildest ones such as hydrocortisone acetate, provide a rapid improvement of skin and scalp lesions, and relapse may be delayed by combining corticosteroids with ketoconazole. These favorable treatment responses suggest that *Malassezia* yeasts, especially *M. restricta* and *M. globosa*, are important in the etiology of SD and dandruff.

This skin disorder is also known from animals. Historically, the species *M. pachydermatis* was isolated from an exfoliative dermatitis of an Indian rhinoceros (Weidman 1925), and now the species is regularly reported to cause SD in various animals, dogs and cats particularly, possibly because pets are more often investigated than other animals (Guillot et al. 1998; Guillot and Bond 1999). Bond and Lloyd (2002) compared IgG responses of various healthy and infected dogs, using Western immunoblotting. They found differences in the immunoreactivity according to the type of dog.

Atopic dermatitis

Atopic dermatitis (AD), or atopic eczema, is a common chronic inflammatory skin disease of unknown etiology. Clemmensen and Hjorth suggested for the first time in 1983 a role for *Malassezia* yeasts in this disease. Furthermore, they found a significant effect of oral ketoconazole in the treatment of adult patients with eczematous lesions on the head and neck. Since then, many reports highlighted the importance of such yeasts in AD. Faergemann has published comprehensive reviews of this disorder (Faergemann 1999, 2002). He concluded that *Malassezia* yeasts may play a major role as allergens in some patients, especially those with a head and neck eczema distribution, and suggested that treatment of such cases should include antifungal therapy. Tengvall Linder et al. (2000) found *P. orbiculare*-specific IgE in serum from 13/15 AD patients, and a positive patch test reaction in 8 of them, whereas SD patients and healthy controls were radioallergoabsorbent test (RAST) and patch test negative. They concluded that *Malassezia* yeasts can trigger an eczematous reaction in sensitized patients. Devos and van der Valk in 2000 have also shown that *P. ovale*-specific IgE is strongly related to a head and neck AD localization. Koyama et al. (2000) have described a *M. globosa* glycoprotein, named Malg46b, as a possible major antigen for IgE antibodies in these patients. Using a nested PCR technique, Sugita et al. (2001) compared the cutaneous colonization by *Malassezia* species in AD patients and healthy subjects. *M. restricta* and *M. globosa* were detected in approximately 90 percent of the patients, and *M. sympodialis* and *M. furfur* in about 40 percent (species mixture in a certain number of cases). In control subjects, *M. restricta*, *M. sympodialis*, *M. globosa*, and *M. furfur* were isolated at a frequency ranging from 61 to 11 percent. They concluded that the four species are common inhabitants of the skin of both AD patients and healthy subjects, but with an increased prevalence of *M. restricta* and *M. globosa* in AD patients. A new species, closely related to *M. sympodialis*, was isolated during this AD study from three patients, and was described as *Malassezia dermatis* by Sugita et al. (2002). However, its rarity excludes this species as being solely responsible for the disease. In 2003, Sugita et al., using the rRNA intergenic spacer (IGS) 1 region, divided *M. globosa* into four major groups and suggested that one of these genotypes may play a significant role in AD. Just after that, the same team (Sugita et al., 2003b) described *Malassezia*

japonica, another new species that may be implicated in AD. This species is able to assimilate Tween 40 and Tween 60, but not Tween 20 and Tween 80; however, this is not clearly defined as yet.

Further studies are needed to clarify the immunological as well as the mycological aspects of this dermatological condition, including a clear identification of etiologic species and of their antigens. Indeed, studies using *M. restricta* and *M. globosa*, the two major species recognized in AD, are still very rare (Koyama et al. 2000). When old names are cited, the connection of these organisms to the present species is not evident. *P. orbiculare* and *P. ovale* in such studies probably correspond to *M. sympodialis* in most cases (Mayser and Gross 2000), or to *M. furfur*.

Malassezia folliculitis

Mallassezia folliculitis (MF) is a benign disorder, characterized by follicular papules and pustules, localized predominantly on the back and chest. The condition is probably common, but often misdiagnosed as acne. The pruritus, associated with the absence of comedones and facial lesions, may facilitate the distinction between the two diseases. MF is more frequent in tropical countries, or during the summer in temperate regions. Occlusion seems to play an important role in this disorder, which has also been associated with antibiotic treatment (tetracyclines), corticosteroids, and immunosuppression (Alves et al. 2000). A minor epidemic outbreak has been reported in an intensive care unit (Archer-Dubon et al. 1999), and, during a 4-month period, 11 cases were described (out of 198 patients) in heart transplant recipients receiving immunosuppressive treatment (Rhie et al. 2000).

Direct microscopy and histopathology of pustules show the presence of abundant *Malassezia* yeasts in the pilosebaceous follicles. Back et al. (1985) observed round yeasts in most cases. At that time they identified yeasts obtained in culture as *P. orbiculare*. Such results suggest that *M. globosa*, as in PV, could be the predominant MF causative agent. Rhie et al. (2000) referred some of their cases as either being due to *M. furfur* or *M. pachydermatis*, but these authors did not apply the new *Malassezia* taxonomy. However, if hair follicle colonization by *Malassezia* yeasts is not abnormal, the diagnosis of MF is confirmed indirectly by the positive response to the antifungal therapy. Topical treatment may be effective in most cases, whilst others may respond to oral ketoconazole, itraconazole, or fluconazole.

Confluent and reticulate papillomatosis of Gougerot–Carteaud

The role of *Malassezia* yeasts in this rare cutaneous disorder, characterized by confluent hyperkeratotic papulae, grayish-brown pigmented, and located preferentially on the trunk, is controversial. The diagnosis of confluent and reticulate papillomatosis of Gougerot–Carteaud (CRP), as that of MF, is based both on the observation (direct microscopy and histopathology) of lipophilic yeasts in the stratum corneum, and on a favorable response to topical or systemic antifungal treatments. In 1969, Roberts and Lachapelle suggested that CRP could represent a peculiar host reaction to heavy colonization by *P. orbiculare*. However, in the series revised by Nordby and Mitchell in 1986, only 5 of 19 patients responded to topical treatment with selenium sulfide. Other reports described complete recovery of CRP using retinoids (Baalbaki et al. 1993), minocycline (Shimizu and Han-Yaku 1997) or azithromycin (Raja Babu et al. 2000). Furthermore, two cases were also successfully treated with topical vitamin D analogs, either calcipotriol (Gülec and Seçkin 1999) or tacalcitol (Ginarte et al. 2002).

It is possible that this rare dermatological disorder represents more than one pathological condition, with *Malassezia* yeasts involved in only a few cases, perhaps those with a clinical pattern similar to that of PV. It could also be a basic disorder of keratinization, determined by an abnormal host response to *Malassezia* yeasts, or to bacteria or their products.

Neonatal *Malassezia* pustulosis

Since 1991, a few cases, characterized by non-follicular papulopustules located on face and neck of newborn infants, have been reported (Aractingi et al. 1991; Rapelanoro et al. 1996; Pont et al. 1998). The diagnosis, made in most of these cases by direct microscopy and histopathology, shows typical *Malassezia* yeasts. Moreover, the lesions cleared up after topical ketoconazole therapy. Rapelanoro et al. assumed that this condition could have been misdiagnosed as neonatal acne. Two more studies (Niamba et al. 1998; Bernier et al. 2002) have suggested that *M. sympodialis* could be the triggering agent for this rash.

Otitis

In 2000, a case of malignant otitis externa caused by *M. sympodialis* has been described in Australia (Chai et al. 2000). The patient, a man with a poorly controlled type II diabetes mellitus, had developed a severe infection, which started around the external auditory canal, and progressed with marked destruction of the temporal bone and an adjacent large soft tissue mass, below the skull base. The infection showed a total resolution after therapy with intravenous amphotericin B and fluconazole.

In animals, especially dogs, the role of the species *M. pachydermatis* in otitis is now well recognized by veterinary teams (Guillot et al. 1998; Guillot and Bond 1999; Crespo et al. 2000a, b).

Onychomycosis

A few onychomycosis cases, involving *Malassezia* spp., have been reported (Crozier and Wise 1993; Silva et al. 1997b; Escobar et al. 1999). However, the presence of *Malassezia* yeasts in nail material, a well-known though infrequent finding, is probably caused by subungual colonization by members of the cutaneous flora which makes the relevance of such cases uncertain. *Malassezia* yeasts have been observed occasionally in patients with distal fingernail onycholysis, associated with *Candida* spp. (Crespo Erchiga, personal observation). *M. furfur* was identified from an AIDS patient with an infected nail (Guého, unpublished result), and in a series of patients with onychomycosis in Brazil (Silva et al. 1997b). This species, which is also able to form hair nodules (Lopes et al. 1994, with identification confirmed afterwards by E. Guého), seems to be the only *Malassezia* species able to survive in such unusual conditions. The role of *Malassezia* yeasts in psoriasis (nails or scalp) is similarly questionable (Elewski 1990).

MOLECULAR METHODS OF IDENTIFICATION

The lipophilic yeasts of the genus *Malassezia* remained for years a complex and indeterminate group with a taxonomy based more on clinical and epidemiologic features than mycological characteristics, as demonstrated by the four former species *P. ovale*, *P. pachydermatis*, *P. orbiculare* and *P. canis*. The absence of a sexual cycle, appropriate classification of yeast morphology, and the absolute growth requirements for complex media were the reasons for this situation. Access to molecular methodology, in particular large subunit (LSU) ribosomal RNA sequencing, has led to a solution of these problems (Guillot and Guého 1995; Guého et al. 1996). Once the taxa were defined genetically, it was possible to define key characteristics applicable for routine use in any microbiological laboratory (Guého et al. 1998). However, such methodologies are time consuming (minimum 4 days after growth of cultures), and demand a high standard of practice in laboratory work. More and more laboratories are now equipped with PCR and are able to develop reproducible, precise and, above all, rapid molecular methods of identification. However, the molecular method must be chosen depending on the purpose of the study, either identification of species, typing of strains for epidemiologic surveys, or in vivo diagnosis from hosts. Indeed, the unique and ideal technique for every application does not exist as yet.

Species identification

RIBOSOMAL DNA (OR RNA) SEQUENCING

The taxonomic revision of the genus *Malassezia* was based on sequencing of the large subunit of ribosomal RNA D1/D2 domains which allowed investigators to define the seven species (Guillot and Guého 1995; Guého et al. 1996). These papers showed that this variable rRNA was a good alternative to nuclear DNA/DNA reassociation for yeast taxonomic work. Indeed, the former conventional characters of the nDNA guanine+cytosine content (GC %) and nDNA/DNA reassociation have now been used to validate each specific sequence. The DNA methodology, which is time consuming and fastidious yet still does not provide sufficient information, has now been abandoned in favor of rRNA or rDNA sequencing. Because most yeast mycologists have adopted the same rDNA region, recognition of an unusual yeast is becoming comparatively easy. Furthermore, with a sequence library extending day after day, a new species may be instantly positioned among the other related species (Nell et al. 2002), and in the classification of the Fungi (Sugita et al. 2002). In common with these latter authors, other mycologists have completed LSU rDNA sequencing with the use of rDNA ITSs. However, at the species level, one must keep in mind that strains of the same species may display about a 1.0 percent sequence divergence in the rRNA or rDNA D1/D2 regions (Kurtzman and Blanz 1998), and even more with other regions (Makimura et al. 2000). However, in 1997, Guillot et al., using the former rRNA D1/D2 sequences, recognized within the sole species *M. pachydermatis* seven sequence types (sequevars), with up to a 2 percent base substitution, but they did not divide the species. By contrast Sugita et al. (2002) described a new species *M. dermatis* which differs from *M. sympodialis* by only a 1.2 percent base substitution in their 26S rDNA D1/D2 and internal transcribed spacer 1 (ITS1)/internal transcribed spacer 2 (ITS2) region sequences.

PCR-RESTRICTION ENDONUCLEASE ANALYSIS

PCR-restriction endonuclease analysis (REA) with PCR of the 26S rRDA sequence allows one to separate the seven species (Guillot et al. 2000). The PCR products are digested with only three restriction endonucleases, *Ban*I, *Hae*II, and *Msp*I, which generate a unique pattern for each species. Moreover, this technique is very cheap and rapid, and detects and characterizes mixtures of several species. Gupta et al. (2000) have also assayed a similar analysis but the patterns generated were useful in the identification of only five species out of seven.

PULSED FIELD GEL ELECTROPHORESIS

Pulsed field gel electrophoresis (PFGE) shows that all *Malassezia* species display different karyotypes that do not vary intraspecifically, except in *M. furfur* (Boekhout et al. 1998). Senczek et al. (1999) applied PFGE, coupled to phenotypic characteristics, successfully to identify 220 *Malassezia* isolates, mainly from dogs. However, this technique, which has confirmed the seven genetic *Malassezia* species, does not appear to be the best choice for rapid species identification.

Molecular typing of *Malassezia* species

RNA AND DNA SEQUENCES OF IGS1 AND ITS1 REGIONS

rRNA sequences of the IGS1 region allowed the characterization of four genotypes within the species *M. globosa* with short sequence repeats CT(n) and GT(n). The authors (Sugita et al. 2003a) suggested also that one of these genotypes could be related to AD.

DNA sequences of ITS1 region may be used to identify (see above) or type species. Makimura et al. in 2000 using ITS1 sequencing were able to separate the seven *Malassezia* species and to recognize 22 ITS1-identical, individual groups. Among the 46 clinical isolates, they identified only *M. furfur* (11 strains from PV, four from SD, and four from AD), *M. sympodialis* (seven from PV, three from SD, one from AD and 11 from healthy controls), and *M. slooffiae* (three from chronic otitis and two from healthy controls).

RESTRICTION FRAGMENT LENGTH POLYMORPHISM

Restriction fragment length polymorphism (RFLP) of nDNA detects more or less specific variability according to the choice of the region amplified and the restriction enzyme battery. This technique allows investigators to trace sources of *M. pachydermatis* in bloodstream infections. One isolate was correlated in this way to a nurse's pet dog (Chang et al. 1998).

RANDOM AMPLIFIED POLYMORPHIC DNA

Random amplified polymorphic DNA (RAPD) implies a genomic DNA PCR using oligonucleotides as primers. In contrast to specific sequences of an rRNA- or rDNA-determined region, the RAPD markers are amplified by using arbitrary sequences. Boekhout et al. (1998) applied this methodology to the genus *Malassezia*, and found genetic variations in all species, particularly with four out of the 20 decamer primers tested.

AMPLIFIED FRAGMENT LENGTH POLYMORPHISM

Amplified fragment length polymorphism (AFLP) is a powerful method for fingerprinting genomic DNA, as well as generating a large number of dominant markers for genotypic analysis (Mitchell and Xu 2003). The technique, which combines the strategy of enzymatic digestion and, usually, two PCR steps, is able to generate numerous fragments, when varying the restriction enzymes and the selective primers. In addition, the fragments are stable and highly reproducible. Like RAPD markers, AFLP markers are amplified partly by arbitrary primers, but with greater reproducibility and fidelity. However, this sophisticated technique, which generates complex patterns, is more appropriate for solving an unusual epidemiologic question. Theelen et al. (2001) found better typing resolution of *Malassezia* yeasts, in particular within *M. furfur*, with AFLP than with RAPD. Using AFLP, the seven species were separated and four genotypes could be distinguished within *M. furfur*, the genotype four containing only isolates from deep human sources.

MULTILOCUS ENZYME ELECTROPHORESIS

Multilocus enzyme electrophoresis does not use PCR technology but is among the most cost-effective methods for investigating genetic variations within a species at the molecular level. It was applied to *M. pachydermatis* (Midreuil et al. 1999), which was confirmed to be genetically heterogeneous. A total of 27 electrophoretic types were identified that could be divided into five groups with different host specificity. Unfortunately, traces of lipids used to grow lipid-dependent species were present, a problem that could not be resolved in this study.

In vivo identification of *Malassezia* species

Malassezia yeasts, particularly *M. globosa* and *M. restricta*, have fastidious culture conditions and exceedingly different growth rates. These difficulties have stimulated investigators to devise molecular techniques that are able to recognize *Malassezia* species directly from hosts. In 2001, Sugita et al. proposed the first method. They examined variations in cutaneous colonization by *Malassezia* species in AD patients and healthy subjects. DNA was extracted directly from dressings and amplified in a nested PCR assay. The diversity of *Malassezia* species found in AD was greater than in healthy subjects. In 2002, two teams described similar methods. Gaitanis et al. extracted *Malassezia* DNA directly from pathological skin scales by single and nested PCR. These authors amplified the rDNA ITS genes and then analyzed the PCR products by the technique of RFLP. The same analyses were obtained from positive cultures obtained from the same site and identified using basic biochemical tests. They obtained a good correlation with the two approaches and recognized the seven species, which were present as single or multiple isolates in the same specimen. The second paper (Gemmer et al.) investigated only the scales of dandruff from two groups of subjects, differentiated by their rate of composite adherent scalp flaking scores. Their technique involved fluorescent nested PCR of ITSs, coupled to a terminal fragment length polymorphism analysis. In this work, *M. restricta* and *M. globosa* were found to be the predominant species present in both groups, while the authors eliminated *M. furfur* as the causal organism for dandruff in their patients.

REFERENCES

Alves, E.V., Martins, J.E., et al. 2000. *Pityrosporum* folliculitis: renal transplantation case report. *J Dermatol*, **27**, 49–51.

Aractingi, S., Cdranel, S., et al. 1991. Pustulose néonatale induite par *Malassezia furfur*. *Ann Dermatol Venereol*, **118**, 856–8.

Archer-Dubon, C., Icaza-Chivez, M.E., et al. 1999. An epidemic outbreak of *Malassezia* folliculitis in three adult patients in an intensive care unit: a previously unrecognized nosocomial infection. *Int J Dermatol*, **38**, 453–6.

Arzumanyan, V.G. 2001. The yeast *Malassezia* on the skin of healthy individuals and patients with atopic dermatitis. *Vestn Ross Akad Med Nauk*, **2**, 29–31.

Ashbee, H.R. and Evans, E.G. 2002. Immunology of diseases associated with *Malassezia* species. *Clin Microbiol Rev*, **15**, 21–57.

Ashbee, H.R., Ingham, E., et al. 1994a. Cell-mediated immune response to *Malassezia furfur* serovars A, B and C in patients with pityriasis versicolor, seborrheic dermatitis and controls. *Exp Dermatol*, **3**, 106–12.

Ashbee, H.R., Fruin, A., et al. 1994b. Humoral immunity to *Malassezia furfur* serovars A, B and C in patients with pityriasis versicolor, seborrheic dermatitis and controls. *Exp Dermatol*, **3**, 227–33.

Aspiroz, C., Moreno, L.A., et al. 1999. Differentiation of three biotypes of *Malassezia* species on normal human skin. Correspondence with *M. globosa*, *M. sympodialis* and *M. restricta*. *Mycopathologia*, **145**, 69–74.

Aspiroz, C., Ara, M., et al. 2002. Isolation of *Malassezia globosa* and *M. sympodialis* from patients with pityriasis versicolor in Spain. *Mycopathologia*, **154**, 111–17.

Baalbaki, S.A., Malak, J.A. and Al-Khars, M.A. 1993. Confluent and reticulated papillomatosis: treatment with etretinate. *Arch Dermatol*, **129**, 961–3.

Back, O., Faergemann, J. and Hornqvist, R. 1985. *Pityrosporum* folliculitis: a common disease of the young and middle aged. *J Am Acad Dermatol*, **12**, 56–61.

Baillon, H. 1889. *Traité de botanique médicale cryptogamique*. Paris: Octave Doin, 234–5.

Begerow, D., Bauer, R. and Boekhout, T. 2000. Phylogenetic placements of ustilaginomycetous anamorphs as deduced from nuclear LSU rDNA sequences. *Mycol Res*, **104**, 53–60.

Bernier, V., Weill, F.X., et al. 2002. Skin colonization by *Malassezia* species in neonates. A prospective study and relationship with neonatal cephalic pustulosis. *Arch dermatol*, **138**, 215–18.

Boekhout, T. and Guého, E. 2003. Basidiomycetous yeasts. In: Howard, D.H. (ed.), *Pathogenic fungi in humans and animals*, 2nd edition. New York: Marcel Dekker, Inc., 535–64.

Boekhout, T., Kamp, M. and Guého, E. 1998. Molecular typing of *Malassezia* species with PFGE and RAPD. *Med Mycol*, **36**, 365–72.

Bond, R. and Anthony, R.M. 1995. Characterization of markedly lipid-dependent *Malassezia pachydermatis* isolates from dogs. *J Appl Microbiol*, **78**, 537–42.

Bond, R. and Lloyd, D.H. 2002. Immunoglobulin G responses to *Malassezia pachydermatis* in healthy dogs and dogs with *Malassezia* dermatitis. *Vet Rec*, **150**, 509–11.

Castellani, A. 1925. Notes on three new yeast-like organisms and a new bacillus, with remarks on the clinical conditions from which they have been isolated: furunculosis blastomycetia, macroglossia blastomycetia stomatitis cryptobacillus. *J Trop Med Hyg*, **28**, 217–23.

Castellani, A. and Chalmers, A.J. 1913. *Manual of tropical medecine*, 2nd edition. London: Baillière, Tindall and Cox, 1747.

Chai, F.C., Auret, K., et al. 2000. Malignant otitis externa caused by *Malassezia sympodialis*. *Head Neck*, **22**, 87–9.

Chang, H.J., Miller, H.L., et al. 1998. An epidemic of *Malassezia pachydermatis* with colonization of health care workers' pet dogs. *N Engl J Med*, **338**, 706–11.

Clemmensen, O.J. and Hjorth, N. 1983. Treatment of dermatitis of the head and neck with ketoconazole in patients with type I sensitivity to *Pityrosporum orbiculare*. *Semin Dermatol*, **2**, 26–9.

Cohen, M.M. 1954. A simple procedure for staining tinea versicolor (*M. furfur*) with fountain pen ink. *J Invest Dermatol*, **22**, 9–10.

Crespo, M.J., Abarca, M.L. and Cabañes, F.J. 1999. Isolation of *Malassezia furfur* from a cat. *J Clin Microbiol*, **37**, 1573–4.

Crespo, M.J., Abarca, M.L. and Cabañes, F.J. 2000a. Otitis externa associated with *Malassezia sympodialis* in two cats. *J Clin Microbiol*, **38**, 1263–6.

Crespo, M.J., Abarca, M.L. and Cabañes, F.J. 2000b. A typical lipid-dependent *Malassezia* species isolated from dogs with otitis externa. *J Clin Microbiol*, **38**, 2383–5.

Crespo, M.J., Abarca, M.L. and Cabañes, F.J. 2000c. Evaluation of different preservation and storage methods for *Malassezia* spp.. *J Clin Microbiol*, **38**, 3872–5.

Crespo, M.J., Abarca, M.L. and Cabañes, F.J. 2002. Occurrence of *Malassezia* spp. in horses and domestic ruminants. *Mycoses*, **45**, 333–7.

Crespo Erchiga, V. and Delgado Florencio, V. 2002. *Malassezia* species in skin diseases. *Curr Opin Infect Dis*, **15**, 133–42.

Crespo Erchiga, V., Ojeda Martos, A., et al. 1999a. Aislamiento e identificación de *Malassezia* spp. en pitiriasis versicolor, dermatitis seborreica y piel sana. *Rev Ioberoam Micol*, **16**, S16–21.

Crespo Erchiga, V., Ojeda Martos, A., et al. 1999b. Mycology of pityriasis versicolor. *J Mycol Méd*, **9**, 143–8.

Crespo Erchiga, V., Ojeda Martos, A., et al. 2000. *Malassezia globosa* as the causative agent of pityriasis versicolor. *Br J Dermatol*, **143**, 799–803.

Crespo Erchiga, V., De Toro, I., et al. 2002. Pityriasis versicolor. Etude épidémiologique sur 210 cas dans le Sud de l'Espagne. *Proc Cong Soc Fr Mycol Méd (Paris)*P9.

Crozier, W.J. and Wise, K.A. 1993. Onychomycosis due to *Pityrosporum*. *Aust J Dermatol*, **34**, 109–12.

Cunningham, A.C., Leeming, J.P., et al. 1990. Differentiation of three serovars of *Malassezia furfur*. *J Appl Bacteriol*, **68**, 439–46.

Devos, S.A. and van der Valk, P.G. 2000. The relevance of skin prick tests for *Pityrosporum ovale* in patients with head and neck dermatitis. *Allergy*, **55**, 1056–8.

Duarte, E.R., Lachance, M.-A. and Hamdan, J.S. 2002. Identification of atypical strains of *Malassezia* spp. from cattle and dog. *Can J Microbiol*, **48**, 749–52.

Dufait, R. 1985. Présence de *Malassezia pachydermatis* (syn *Pityrosporum canis*) sur les poils et les plumes des animaux domestiques. *Bull Soc Fr Mycol Méd*, **14**, 19–22.

Eichstedt, E. 1846. Pilzbildung in der pityriasis versicolor. *Froriep Neue Notiz Natur Heilk*, **39**, 270.

Elewski, B. 1990. Does *Pityrosporum ovale* have a role in psoriasis? *Arch Dermatol*, **126**, 1111–12.

Escobar, M.L., Carmona-Fonseca, J. and Santamaria, L. 1999. Onicomicosis por *Malassezia*. *Rev Iberoam Micol*, **16**, 225–9.

Faergemann, J. 1983. Antibodies to *Pityrosporum orbiculare* in patients with tinea versicolor and controls of various ages. *J Invest Dermatol*, **80**, 133–5.

Faergemann, J. 1999. *Pityrosporum* species as a cause of allergy and infection. *Allergy*, **54**, 413–19.

Faergemann, J. 2002. Atopic dermatitis and fungi. *Clin Microbiol Rev*, **15**, 545–63.

Faergemann, J., Bergbrant, I.M., et al. 2001. Seborrhoeic dermatitis and *Pityrosporum* (*Malassezia*) folliculitis: characterization of inflammatory cells and mediators in the skin by immunohistochemistry. *Br J Dermatol*, **144**, 549–56.

Faergemann, J., Gupta, A.K., et al. 2002. Efficacy of itraconazole in prophylactic treatment of pityriasis (tinea) versicolor. *Arch Dermatol*, **138**, 69–73.

Fell, J.W., Boekhout, T., et al. 2000. Biodiversity and systematics of basidiomycetous yeasts as determined by large-subunit rDNA D1/D2 domain sequence analysis. *Int J Syst Evol Microbiol*, **50**, 1351–71.

Gaitanis, G., Velegraki, A., et al. 2002. Identification of *Malassezia* species from skin scales by PCR-RFLP. *Clin Microbiol Infect*, **8**, 162–73.

Gemmer, C.M., DeAngelis, Y., et al. 2002. Fast, noninvasive method for molecular detection and differentiation of *Malassezia* yeast species on human skin and application of the method to dandruff microbiology. *J Clin Microbiol*, **40**, 3350–7.

Ginarte, M., Fabeiro, J.M. and Toribio, J. 2002. Confluent and reticulate papillomatosis (Gougerot-Carteaud) successfully treated with tacalcitol. *J Dermatol Treat*, **13**, 27–30.

Gordon, M.A. 1951a. Lipophilic yeast organism associated with tinea versicolor. *J Invest Dermatol*, **17**, 267–72.

Gordon, M.A. 1951b. The lipophilic mycoflora of the skin. I. In vitro culture of *Pityrosporum orbiculare* n.sp.. *Mycologia*, **43**, 524–35.

Guého, E. and Guillot, J. 1999. Comments on *Malassezia* species from dogs and cats. *Mycoses*, **42**, 673–4.

Guého, E. and Meyer, S.A. 1989. A reevaluation of the genus *Malassezia* by means of genome comparison. *Antonie van Leeuwenhoek*, **55**, 245–51.

Guého, E., Midgley, G. and Guillot, J. 1996. The genus *Malassezia* with description of four new species. *Antonie van Leeuwenhoek*, **69**, 337–55.

Guého, E., Boekhout, T., et al. 1998. The role of *Malassezia* species in the ecology of human skin and as pathogen. *Med Mycol*, **36**, Suppl 1, 220–9.

Guillot, J. and Bond, R. 1999. *Malassezia pachydermatis*: a review. *Med Mycol*, **37**, 295–306.

Guillot, J. and Guého, E. 1995. The diversity of *Malassezia* yeasts confirmed by rRNA sequence and nuclear DNA comparisons. *Antonie van Leeuwenhoek*, **67**, 297–314.

Guillot, J., Chermette, R. and Guého, E. 1994. Prévalence du genre *Malassezia* chez les mammifères. *J Mycol Méd*, **4**, 72–9.

Guillot, J., Guého, E. and Prévost, M.C. 1995. Ultrastructural features of the dimorphic yeast *Malassezia furfur*. *J Mycol Méd*, **5**, 86–91.

Guillot, J., Guého, E., et al. 1996. Identification of *Malassezia* species. A practical approach. *J Mycol Méd*, **6**, 103–10.

Guillot, J., Guého, E., et al. 1997. Epidemiological analysis of *Malassezia pachydermatis* isolates by partial sequencing of the large subunit ribosomal RNA. *Res Vet Sci*, **62**, 22–5.

Guillot, J., Guého, E., et al. 1998. Importance des levures du genre *Malassezia* en dermatologie vétérinaire. *Point Vét*, **29**, 21–31.

Guillot, J., Deville, M., et al. 2000. A single PCR-restriction endonuclease analysis for rapid identification of *Malassezia* species. *Lett Appl Microbiol*, **31**, 400–3.

Gülec, A.T. and Seçkin, D. 1999. Confluent and reticulate papillomatosis: treatment with topical calcipotriol. *Br J Dermatol*, **141**, 1150–1.

Gupta, A.K., Kohli, Y. and Summerbell, R.C. 2000. Molecular differentiation of seven *Malassezia* species. *J Clin Microbiol*, **38**, 1869–75.

Gupta, A.K., Kohli, Y., et al. 2001a. Quantitative culture of *Malassezia* species from different body sites of individuals with or without dermatoses. *Med Mycol*, **39**, 243–51.

Gupta, A.K., Kohli, Y., et al. 2001b. Epidemiology of *Malassezia* yeasts associated with pityriasis versicolor in Ontario, Can. *Med Mycol*, **39**, 199–206.

Gustafson, B.A. 1955. Otitis externa in the dog. A bacteriological and experimental study. Thesis, Royal Veterinary College of Sweden, Stockholm.

Hirai, A., Kano, R., et al. 2004. *Malassezia nana* sp. nov., a novel lipid-dependent yeast species isolated from animals. *Int J Syst Evol Microbiol*, **54**, 623–7.

Ingham, E. and Cunningham, A.C. 1993. *Malassezia furfur. J Med Vet Mycol*, **31**, 265–88.

Kesavan, S., Holland, K.T. and Ingham, E. 2000. The effects of lipid extraction on the immunomodulatory activity of *Malassezia* species in vitro. *Med Mycol*, **38**, 239–47.

Koyama, T., Kanbe, T., et al. 2000. Isolation and characterization of a major antigenic component of *Malassezia globosa* to IgE antibodies in sera of patients with atopic dermatitis. *Microbiol Immunol*, **44**, 373–9.

Kurtzman, C.P. and Blanz, P.A. 1998. Ribosomal RNA/DNA sequence comparisons for assessing phylogenetic relationships. In: Kurtzman, C.P. and Fell, J.W. (eds), *The yeasts – a taxonomic study*, 4th revised edition. Amsterdam: Elsevier, 69–74.

Leeming, J.P. and Notman, F.H. 1987. Improved methods for isolation and enumeration of *Malassezia furfur* from human skin. *J Clin Microbiol*, **25**, 2017–19.

Lopes, J.O., Alves, S.H., et al. 1994. Nodular infection of the hair caused by *Malassezia furfur. Mycopathologia*, **125**, 149–52.

Makimura, K., Tamura, Y., et al. 2000. Species identification and strain typing of *Malassezia* species stock strains and clinical isolates based on the DNA sequences of nuclear ribosomal internal transcribed spacer 1 regions. *J Med Microbiol*, **49**, 29–35.

Marcon, M.J. and Powell, D.A. 1992. Human infections due to *Malassezia* spp.. *Clin Microbiol Rev*, **5**, 101–19.

Mayser, P. and Gross, A. 2000. IgE antibodies to *Malassezia furfur*, *M. sympodialis* and *Pityrosporum orbiculare* in patients with atopic dermatitis, seborrheic eczema or pityriasis versicolor, and identification of respective allergens. *Acta Derm Venereol*, **80**, 357–61.

Mayser, P., Haze, P., et al. 1997. Differentiation of *Malassezia* species: selectivity of cremophor EL, castor oil and ricinoleic acid for *M. furfur. Br J Dermatol*, **137**, 208–13.

Meuber, K., Kröger, S., et al. 1996. Effects of *Pityrosporum ovale* on proliferation, immunoglobulin (IgA,G,M) synthesis and cytokine (IL-2, IL-10, IFN-gamma) production of peripheral blood mononuclear cells from patients with seborrhoeic dermatitis. *Arch Dermatol Res*, **288**, 532–6.

Mickelsen, P.A., Viano-Paulson, M.C., et al. 1988. Clinical and microbiological features of infection with *Malassezia pachydermatis* in high-risk infants. *J Infect Dis*, **157**, 1163–8.

Midgley, G. 1989. The diversity of *Pityrosporum (Malassezia)* yeasts in vivo and in vitro. *Mycopathologia*, **106**, 143–53.

Midgley, G. 2000. The lipophilic yeasts: state of the art and prospects. *Med Mycol*, **38**, Suppl 1, 9–16.

Midgley, G. and Clayton, Y.M. 1969. The yeast flora of birds and mammals in captivity. *Antonie van Leeuwenhoek*, **35**, Suppl E, 23–4.

Midgley, G. and Hay, R.J. 1988. Serological responses to *Pityrosporum (Malassezia)* in seborrhoeic dermatitis demonstrated by ELISA and western blotting. *Bull Soc Fr Mycol Méd*, **17**, 267–76.

Midreuil, F., Guillot, J., et al. 1999. Genetic diversity in the yeast species *Malassezia pachydermatis* analysed by multilocus enzyme electrophoresis. *Int J Syst Bacteriol*, **49**, 1287–94.

Mitchell, T.G. and Xu, J. 2003. Molecular methods to identify pathogenic fungi. In: Howard, D.H. (ed.), *Pathogenic fungi in humans and animals*, 2nd edition. New York: Marcel Dekker, Inc., 677–702.

Murai, T., Nakamura, Y., et al. 2002. Differentiation of *Malassezia furfur* and *Malassezia sympodialis* by glycine utilization. *Mycoses*, **45**, 180–3.

Nakabayashi, A. 2002. Identification of *Malassezia*-associated dermatoses. *Jap J Med Mycol*, **43**, 65–8.

Nakabayashi, A., Sei, Y. and Guillot, J. 2000. Identification of *Malassezia* species isolated from patients with seborrhoeic dermatitis, atopic dermatitis, pityriasis versicolor and normal subjects. *Med Mycol*, **38**, 337–41.

Nell, A., James, S.A., et al. 2002. Identification and distribution of a novel *Malassezia* species yeast on normal equine skin. *Vet Rec*, **150**, 395–8.

Niamba, P., Weill, F.X., et al. 1998. Is common neonatal cephalic pustulosis (neonatal acne) triggered by *Malassezia sympodialis? Arch Dermatol*, **134**, 995–8.

Ninomiya, J., Nakabayashi, A., et al. 2002. A case of seborrhoeic blepharitis; treatment with itraconazole. *Nippon Ishinkin Gakkai zasshi*, **43**, 189–91.

Nordby, C.A. and Mitchell, A.J. 1986. Confluent and reticulated papillomatosis responsive to selenium sulfide. *Int J Dermatol*, **25**, 194–9.

Panja, G. 1927. The *Malassezia* of the skin, their cultivation, morphology and species. *Trans 7th Congr Far East Assoc Trop Med*, **2**, 305–6.

Parry, M.E. and Sharpe, G.R. 1998. Seborrhoeic dermatitis is not caused by an altered immune response to *Malassezia* yeasts. *Br J Dermatol*, **139**, 254–63.

Pont, V., Grau, C., et al. 1998. Pustulosis neonatal por *Malassezia furfur. Actas Dermosifiliogr*, **89**, 199–205.

Raja Babu, K.K., Snehal, S. and Sudha Vani, D. 2000. Confluent and reticulate papillomatosis: successful treatment with azithromycin. *Br J Dermatol*, **142**, 1252–3.

Rapelanoro, R., Mortureux, P., et al. 1996. Neonatal *Malassezia furfur* pustulosis. *Arch Dermatol*, **132**, 190–3.

Rhie, S., Turcios, R., et al. 2000. Clinical features and treatment of *Malassezia* folliculitis with fluconazole in orthotopic heart transplant recipients. *J Heart Lung Transplant*, **19**, 215–19.

Roberts, S. and Lachapelle, J. 1969. Confluent and reticulate papillomatosis (Gougerot-Carteaud) and *Pityrosporum orbiculare. Br J Dermatol*, **81**, 841–5.

Sabouraud, R. 1932. *Diagnostic et traitement des affections du cuir chevelu*. Paris: Masson et Cie.

Senczek, D., Siesenop, U. and Böhm, K.H. 1999. Characterization of *Malassezia* species by means of phenotypic characteristics and detection of electrophoretic karyotypes by pulsed-field gel electrophoresis (PFGE). *Mycoses*, **42**, 409–14.

Shimizu, S. and Han-Yaku, H. 1997. Confluent and reticulated papillomatosis responsive to minocycline. *Dermatology*, **194**, 59–61.

Shuster, S. 1984. The etiology of dandruff and the mode of action of therapeutic agents. *Br J Dermatol*, **111**, 235–42.

Silva, V., Fishman, O. and Camargo, Z.P. 1997a. Humoral immune response to *Malassezia furfur* in patients with pityriasis versicolor and seborrhoeic dermatitis. *Mycopathologia*, **139**, 79–85.

Silva, V., Moreno, G.A., et al. 1997b. Isolation of *Malassezia furfur* from patients with onychomycosis. *J Med Vet Mycol*, **35**, 73–4.

Simmons, R.B. and Guého, E. 1990. A new species of *Malassezia. Mycol Res*, **94**, 1146–9.

Slooff, W.C. 1970. Genus 6 *Pityrosporum* Sabouraud. In: Lodder, J. (ed.), *The yeasts: a taxonomic study*, 2nd edition. Amsterdam: North-Holland Publishing Co., 1167–86.

Sugita, T., Suto, H., et al. 2001. Molecular analysis of *Malassezia* microflora on the skin of atopic dermatitis patients and healthy subjects. *J Clin Microbiol*, **39**, 3486–90.

Sugita, T., Takashima, M., et al. 2002. New yeast species, *Malassezia dermatis*, isolated from patients with atopic dermatitis. *J Clin Microbiol*, **40**, 1363–7.

Sugita, T., Kodama, M., et al. 2003a. Sequence diversity of the intergenic spacer region of the rRNA of *Malassezia globosa* colonizing the skin of patients with atopic dermatitis and healthy individuals. *J Clin Microbiol*, **43**, 3022–7.

Sugita, T., Takashima, M., et al. 2003b. Description of a new yeast species, *Malassezia japonica*, and its detection in patients with atopic dermatitis and healthy subjects. *J Clin Microbiol*, **41**, 4695–9.

Tengvall Linder, M., Johansson, C., et al. 2000. Positive atopy patch test reactions to *Pityrosporum orbiculare* in atopic dermatitis patients. *Clin Exp Allergy*, **30**, 122–31.

Theelen, B., Silvestri, M., et al. 2001. Identification and typing of *Malassezia* yeasts using amplified length polymorphism (AFLP[Tm]), random amplified polymorphic DNA (RAPD) and denaturing gradient gel electrophoresis (DGGE). *FEMS Yeast Res*, **1**, 79–86.

Thygeson, P. and Vaughan, D.G. 1954. Seborrhoeic blepharitis. *Trans Am Ophthalmol Soc*, **52**, 173–88.

van Abbe, N.J. 1964. The investigation of dandruff. *J Soc Cosmetic Chem*, **15**, 609–30.

Weidman, F.D. 1925. Exfoliative dermatitis in the Indian rhinoceros (*Rhinoceros unicornis*), with description of a new species: *Pityrosporum pachydermatis*. In Fox, H. (ed.) *Rep Lab Museum Comp Pathol Zoo Soc Philadelphia*. Philadelphia, 36–45.

Wu, Y. and Chen, K.T. 1985. Humoral immunity in patients with tinea versicolor. *J Dermatol*, **12**, 161–6.

Yarrow, D. and Ahearn, D.G. 1984. Genus 7 *Malassezia* Baillon. In: Kreger-van Rij, N.J.W. (ed.), *The yeasts: a taxonomic study*, 3rd edition. Amsterdam: Elsevier Science Publ BV, 882–5.

13

The dermatophytes

ARVIND A. PADHYE AND RICHARD C. SUMMERBELL

HISTORY

Historically, Agostino Bassi (1835–36) was the first to elucidate the microbial nature of a deadly disease of silkworms (*Bombyx mori*). Through meticulous studies and animal experiments carried out over a period of 25 years, he established unequivocally that a mold, now known as *Beauveria bassiana*, was the cause of a devastating disease of the commercially valuable silkworm. This was the first microorganism to be recognized as causing disease, long before the classic work of Robert Koch and Louis Pasteur. The discovery that a fungus could cause dermatophytosis was probably made by Robert Remak, who observed unusual microscopic structures in favic lesions. According to Seeliger's (1985) historical account, Remak claimed that he did not at first recognize the structures he saw as being fungal. Instead, he credited this recognition to his mentor, Professor Johann L. Schönlein, who described the fungal etiology in 1839. Remak (1845) later, however, cultured the causal fungus, established that it was infectious, and described the etiologic agent, in the context of its attendant disease symptoms, as *Achorion schoenleinii*.

Although Remak has priority for the discovery of the first fungus causing human disease, the real founder of medical mycology, based on his discoveries from 1841 to 1844, was the Parisian physician, David Gruby (Gruby 1841a, b, 1843, 1844; Zakon and Benedek 1944). Unaware of the observations of Remak and Schönlein,

Gruby described the clinical and microscopic features of the causal agent of favus, and established the contagious nature of the disease (Gruby 1841a, b). He also described ectothrix hair invasion in beards and scalp hair of other patients, and named the etiologic agent involved *Microsporum audouinii*. The generic name he established referred in its etymology to the small spores that surrounded the hair shaft (Gruby 1843). Gruby (1844) also described a fungus he saw causing endothrix hair invasion as *Herpes (Trichophyton) tonsurans*.

In the early 1890s, Raimond Sabouraud, a French dermatologist, convincingly established the 'plurality' of the ringworm fungi and integrated the mycological and clinical aspects of ringworm. Based on the advances made by his contemporaries in medicine and veterinary science, as well as on his personal observations, Sabouraud wrote and published his monumental *Les teignes* in 1910. He classified dermatophytes into four genera: *Achorion*, *Epidermophyton*, *Microsporum*, and *Trichophyton*. This taxonomy was based on the clinical aspects of the disease that they caused, combined with their cultural and microscopic characters.

Emmons (1934) modernized the taxonomic scheme of Sabouraud. He critically reviewed the myriad dermatophyte species that had been described by Sabouraud and later workers in the European school, and discarded genera and species based on extrinsic factors such as their clinical manifestations and the host from which they were isolated. In addition, he rejected the attribution of undue taxonomic import to trivial and highly variable characters, such as colony texture, chlamydospores, nodular organs, pigmentation, racquet hyphae, and spiral appendages. He discarded the genus *Achorion* and redefined the remaining three anamorphic genera – *Epidermophyton*, *Microsporum*, and *Trichophyton* – based on conidial morphology and other relatively reliably formed microscopic structures. Two decades later, nutritional, physiological, and morphological studies (Georg 1952; Silva and Benham 1952; Benham 1953; Swartz and Georg 1955; Ajello and Georg 1957; Georg and Camp 1957; Padhye et al. 1980) simplified the identification of the dermatophytes and led to a reduction in the number of species recognized.

The discovery of the teleomorphs (sexual or perfect states) of *Trichophyton ajelloi*, *Trichophyton terrestre*, *Microsporum gypseum*, and *Microsporum nanum* (Dawson and Gentles 1961; Stockdale 1961), using the hair-bait technique (Vanbreuseghem 1952), led to rapid discoveries of the teleomorphs of several other dermatophyte species and related keratinophilic taxa. The successful oral treatment of experimental *Microsporum canis* infections in guinea pigs with griseofulvin, by Gentles in 1958, opened new doors to the therapy of the dermatophytoses, especially ringworm of the scalp.

ETIOLOGIC AGENTS

The dermatophytes are closely-related keratinophilic fungi causing dermatophytosis (ringworm or tinea) by virtue, in part, of their ability to degrade keratin and thus to invade the skin, hair, and nails. The three anamorphs (asexual, conidial or imperfect state) of genera *Epidermophyton*, *Microsporum*, and *Trichophyton* are recognized as belonging to the family *Arthrodermataceae* of the order Onygenales, a group also containing other human pathogens such as *Histoplasma capsulatum* and *Coccidioides immitis*. Formerly, these asexual states were placed in a separate group known as the Deuteromycota or Fungi Imperfecti, but molecular studies and official changes in the guidelines for fungal nomenclature have allowed the abandonment of this artificial taxonomic category.

The dermatophytes capable of reproducing sexually are considered to be members of the genus *Arthroderma* (Weitzman et al. 1986) in the family *Arthrodermataceae*. The currently recognized dermatophytes and related nonpathogenic dermatophytoids are listed in Table 13.1 along with the names of their corresponding teleomorphs.

Physiologically, the dermatophytes facilitate distinction with special growth media by virtue of their high tolerance for cycloheximide and their ability to use proteins as a sole carbon source, releasing excess ammonium ion in a way that causes the pH of protein-containing media to rise into the alkaline range.

Table 13.1 *Dermatophytes and their nonpathogenic relatives: teleomorph and anamorph states*

Teleomorph (sexual or perfect state)	Anamorph (asexual or imperfect state)
Arthroderma spp.	*Microsporum* and *Trichophyton* spp.
A. benhamiae	*T. mentagrophytes* complex in part
A. cajetani	*M. cookei*
A. fulvum	*M. fulvum*
A. grubyi	*M. vanbreuseghemii*
A. gypseum	*M. gypseum* complex in part
A. incurvatum	*M. gypseum* complex in part
A. otae	*M. canis*
A. obtusum	*M. nanum*
A. persicolor	*M. persicolor*
A. racemosum	*M. racemosum*
A. simii	*T. simii*
A. vanbreuseghemii	*T. mentagrophytes* complex in part

Relatively few other fungi except some closely-related nonpathogenic members of the *Arthrodermataceae* combine these attributes, and among those that do, e.g. some *Exophiala* species in the 'black yeast' family *Herpotrichiellaceae*, most do not resemble dermatophytes in culture.

Morphological identification of the dermatophytes today essentially follows the classification scheme of Emmons: it is predominantly based on the microscopic characters of the anamorph and on the macromorphology of the colony on Sabouraud's glucose (dextrose) agar (SGA). The addition of 1 percent yeast extract to SGA, or use of other media such as cornmeal glucose agar or potato glucose agar may be necessary to stimulate conidiation. In addition, nutritional and other physiological tests may be required for species identification, especially of the *Trichophyton* spp.

Colonies should be examined for the color of the surface and underside, presence of diffusible pigment, surface texture, topography, and rate of growth. With regard to microscopic morphology, the types of conidia present, and their shapes and sizes, are essential criteria for identification; the presence of other structures such as pectinate (comb-like), coiled or antler-like hyphae, reflexive branches, chlamydospores or nodular organs may be helpful.

Many dermatophytes, especially those primarily associated with nonhuman hosts, regularly produce two kinds of conidia, large multicellular macroconidia and smaller unicellular microconidia. The presence or absence of these conidial types and the appearance of the walls of macroconidia (rough or smooth) are of generic significance. Species identification is largely based on the morphology and arrangement of the conidia.

Epidermophyton (Sabouraud 1907)

This genus is characterized by the presence of numerous, broadly clavate, nearly smooth-walled macroconidia; microconidia are absent. There is only a single recognized species, *E. floccosum*. It is anthropophilic, that is, specifically adapted to cause ongoing human infection. This species attacks skin, nails, and, very rarely, hair (Sherna et al. 1993).

Microsporum (Gruby 1843)

Members of this genus usually produce both macroconidia and microconidia. The essential distinguishing feature is the presence of macroconidia that have rough walls, with textures ranging from spiny to warty. The macroconidia vary in shape from egg-shaped (obovate) to cylindrofusiform; they may have thin to thick cell walls and 1–15 septa depending upon the species. The roughness of the cell walls may not be readily apparent in some isolates or species, and may require special media to induce its formation (Kane et al. 1997). Similarly, special media may be required for conidial production. Microconidia are typically clavate (club-shaped). In microscopic examination, one should also note the formation and arrangement of both types of conidia. Members of this genus regularly attack skin and hair but only occasionally nails.

Trichophyton (Malmsten 1845)

Smooth-walled macroconidia and microconidia are produced by typical members of this genus. The macroconidia may range in shape from elongate to pencil-shaped, clavate, fusiform, or cylindrofusiform. They are multiseptate and may be thin or thick walled. Microconidia are usually produced in greater abundance than macroconidia and their shape varies from pyriform, clavate, spherical to elongate. They are borne along the hyphae singly or in clusters and are sessile or borne on short stalks. Their arrangement on fertile hyphae is one of the important factors in their identification. Members of this genus attack skin, hair, and nails. For many *Trichophyton* spp., special media may be required to stimulate conidial formation. Some species produce few or no conidia regardless of the growth conditions used.

Table 13.2 summarizes the essential features of the dermatophyte genera; Table 13.3 summarizes the recognized species, along with some nonpathogenic relatives, and Tables 13.4 and 13.5, (p. 226) present short synopses of the characteristic features of the species.

ECOLOGY

Dermatophytes have been grouped into geophilic, zoophilic and anthropophilic species based on the ecological niches in which they naturally occur. For pathogens, this classification is based on their host preferences as manifested at the population level (Georg 1960; Ajello 1962). That is to say, sporadic occurrences of pathogenic species on atypical hosts are ignored. Table 13.6, (p. 227) summarizes classification of dermatophyte species according to ecology and host preference.

Geophilic species

Geophilic species are considered ancestral to the pathogenic dermatophytes (Chemel 1980; Ozegovic 1980; Rippon 1988; Weitzman and Summerbell 1995). The natural habitat of these species is the soil, where they are associated with decomposing keratinous material, e.g. hair, feathers, horns, hooves, nails, etc. Carnivore dung predominantly containing animal hair is another prominent habitat; the inoculum presumably comes from underlying soil. Exposure to the soil is the main source of infection for humans and lower animals (Alsop and Prior 1961; Chemel and Buchvald 1970; Ajello 1974; Rippon 1985). Transmission of geophilic species from lower animals to humans or from human to human is rare. *M. gypseum*, one of the best examples of a geophilic dermatophyte, was reported as the causal agent of two unusual outbreaks in association with soil

Table 13.2 *Characteristics of the genera of the dermatophytes*

Genus	Macroconidia	Microconidia	Clinical
Microsporum	Rough walls; egg-shaped, fusiform; usually present and more numerous than microconidia	Usually present; pyriform or clavate; typically borne singly along the hypha	Members may attack skin, hair, rarely nails. *M. persicolor* does not invade hair
Trichophyton	Smooth-walled; pencil-shaped, clavate, fusiform to cylindrofusiform; often absent or less numerous than microconidia	Usually more numerous than macroconidia but may also be scarce or absent; clavate, pyriform or spherical, in clusters or singly along the hypha	Members may attack skin, hair or nails. *T. concentricum* does not attack hair
Epidermophyton	Smooth-walled; broadly clavate; singly or in banana-like clusters	Absent	Attacks skin and nails; generally does not invade hair

Table 13.3 *Dermatophytes and their nonpathogenic relatives: species of the genera* Epidermophyton, Microsporum, *and* Trichophyton

Epidermophyton	Microsporum	Trichophyton
E. floccosum	M. audouinii	T. concentricum
	M. canis	T. equinum
	M. cookei	T. erinacei
	M. ferrugineum	T. interdigitale
	M. fulvum	T. megninii
	M. gypseum	T. mentagrophytes
	M. gallinae	T. rubrum
	M. nanum	T. schoenleinii
	M. persicolor	T. simii
	M. praecox	T. soudanense
	M. racemosum	T. tonsurans
	M.	T. verrucosum
	vanbreuseghemii	T. violaceum

(Alsop and Prior 1961; Rippon 1985). One such outbreak occurred in a cucumber greenhouse among workers handling soil enriched with horn and hoof fertilizer. Secondary infections occurred in children and adults in close contact with the infected greenhouse workers (Alsop and Prior 1961).

Zoophilic species

Zoophilic species have gradually evolved from soil to parasitize animals (Chemel 1980; Ozegovic 1980; Rippon 1988). They have lost their ability to survive over the long term in nonsterile soil, though their mating to form teleomorphic structures cannot occur on the host and must therefore occur on shed hairs of the host, along with other such keratinous materials, on soil. These fungi are primarily animal parasites or may be carried by apparently healthy, wild animals (Feurman et al. 1975; Chemel 1980; Ozegovic 1980; Pier et al. 1994). Human infections are acquired either by direct contact with an infected animal or indirectly by contact with fomites or other inanimate objects associated with keratinous material from the animal (Pier et al. 1994; Thomas et al. 1994; Weitzman and Summerbell 1995). Human infections by *M. canis*, an important zoophile, are usually acquired from cats and dogs (Rippon 1985; Pier et al. 1994).

Anthropophilic species

Anthropophilic dermatophyte species appear generally to have evolved from zoophilic species, although recent molecular studies indicate that *E. floccosum*, possibly derived from within the *M. gypseum* complex, may be an exception (Gräser et al. 2000a). Humans are the normal hosts for this group of species and transmission may occur by direct contact or indirectly by fomites (Weitzman and Summerbell 1995). Human-to-animal transmission of infection by an anthropophilic species is rare but has been documented in the literature (Kaplan and Gump 1958; Mayr 1989).

The evolution from saprobic life in soil to an almost exclusive existence as a colonizer of keratinized tissue of humans generally corresponds to a decrease or loss of conidial formation (Rippon 1985; Rippon 1988), as well as an inability to reproduce sexually (Rippon 1988; Tanaka et al. 1992; Summerbell 2002). This loss of sexual reproduction (Kwon-Chung and Bennett 1992) is observed in almost all of the anthropophilic dermatophytes, although a small number of apparently anthropophilic members of the *Trichophyton mentagrophytes* complex (isolates with velutinous morphology traditionally associated with the name *Trichophyton interdigitale*) retain the ability to mate as *Arthroderma benhamiae* or *Arthroderma vanbreuseghemii* in laboratory tests (Watanabe and Hironaga, 1981; Vasilyev and Bogomolova, 1985; Hejtmánek and Hejtmánkova 1989).

Certain zoophilic and anthropophilic species may represent an intermediate stage in the evolution of dermatophytes towards loss of the ability to reproduce sexually. *Arthroderma otae* (anamorph *M. canis*) exists predominantly as one mating type but is still capable of reproducing sexually when paired with the rare + mating type (Weitzman and Padhye 1978). Many anthropophilic and a few zoophilic species are apparently incapable of reproducing sexually but appear to exist exclusively as one mating type; they may be induced to form rudimentary sexual organs (abortive gymnothecia) when paired with appropriate tester strains of *Arthroderma simii* (Stockdale 1968; Young 1968; Kwon-Chung and Bennett 1992). An example of such a species is *Trichophyton rubrum*, which consists entirely of strains of (−) mating type.

In addition to a decrease or loss of conidial production and the ability to reproduce sexually, anthropophilic fungi generally tend to produce chronic infections that are less likely to resolve spontaneously (Rippon 1988). The distinction between a geophilic and zoophilic species may not always be obvious and may even be controversial. A typical geophilic species is *M. gypseum*. Its natural habitat is soil and it is not always associated with animals (Weitzman and Summerbell 1995). *M. gypseum* can be isolated readily from garden soil by using Vanbreuseghem's hair-baiting technique (Vanbreuseghem 1952). In addition, it can not only survive and flourish in the soil environment, but also propagates by producing numerous macroconidia in nature (Gordon 1953). It is also a common sporadic animal pathogen, and may cause limited outbreaks, but it is not known to cause enduring, chronic, directly transmissible infections in populations of any animal species. Although *M. nanum* causes ringworm mainly in pigs, its partially geophilic nature has been well established by its isolation from soil as well as by the demonstration that its characteristic macroconidia could be seen directly in soil suspensions (Ajello et al. 1964). If it

Table 13.4 *Characteristic of pathogenic* Epidermophyton *and* Microsporum *species in culture*

Species	Colony on SGA	Microscopic morphology	Key tests, features or comments
Epidermophyton floccosum	Surface olive, khaki or yellow-brown, flat or radially folded, white tufts with age; suede-like texture; reverse yellow-brown, slow growth rate	Numerous macroconidia, 20–40 × 6–8 μm; broadly clavate, 0–4 septa, single or in clusters	No microconidia
Microsporum audouinii	Surface white, gray or tan; flat or velvety; reverse salmon-pink or rust; moderate growth	Conidia few or absent; macroconidia bizarre, fusiform, elongated, thick, rough walled, septa irregular in intervals; microconidia clavate, borne singly along sides of hyphae, pectinate hyphae, terminal or intercalary chlamydospores	Brownish discoloration of rice grains and poor growth; does not perforate hair in vitro
M. canis (= *M. canis* var. *distortum*)	Surface white to buff, woolly, reverse yellow, orange to orange-brown; nonpigmented and dysgonic variants occur; growth rapid	Numerous macroconidia, fusiform with thick, rough walls, 18–125 × 5–25 μm, knobbed apex, up to 15 septa; few microconidia; in *M. canis* var. *distortum* many distorted thick, rough-walled macroconidia	Heavy sporulation on rice grains; perforates hair in vitro
M. cookei	Surface yellowish, reddish tan or lavender, powdery or granular; reverse deep wine-red; moderate growth	Numerous macroconidia, mostly ellipsoid, thick walled, some moderately thin walled resembling *M. gypseum*; abundant microconidia	Must be distinguished from *M. gypseum* and *M. racemosum*
M. ferrugineum	Surface yellowish to rust; heaped, folded, waxy; very slow growth	Irregular hyphae lacking conidia; coarse hyphae with prominent septa ('bamboo hyphae')	Does not perforate hair in vitro
M. gallinae	Surface white to pinkish, satiny, slightly folded; reverse strawberry red diffusible pigment; rapid growth	Macroconidia elongate, curved with blunt tips, smooth to slightly roughened at tips, 2–10 cells, 6–8 × 15–50 μm; microconidia present	Characteristic diffusible red pigment; does not perforate hair in vitro
M. gypseum complex (*M. gypseum*, *M. fulvum*)	Surface tan, rosy-buff, cinnamon, powdery to floccose; reverse buff or reddish brown; rapid growth	Numerous ellipsoidal to fusiform macroconidia, 25–60 × 7–15 μm, thin, rough walls; up to 6 septa; microconidia abundant	Differentiation of *M. gypseum* from *M. fulvum* by inducing the teleomorph on special media
M. nanum	Surface cream to tan; powdery; reverse reddish brown; moderate growth	Abundant ovoid or egg-shaped 1–3 celled macroconidia, 10–30 × 6–13 μm; microconidia rare to moderate	Colony resembles *M. gypseum* but with slower growth rate; macroconidia slightly resemble conidia of *Trichothecium roseum*
M. persicolor	Surface yellowish buff to peach; powdery or downy; reverse pink to reddish brown	Abundant spherical microconidia; usually in clusters, some clavate to pear-shaped; macroconidia clavate, smooth, thin walls, rough walls with age or on special media; spiral hyphae often present	Uncommon pathogen; perforates hair in vitro, growth at 37°C minimal; rosy colonies on peptone agar; no alkalinity on BCP milk solids glucose agar in contrast to similar-looking *T. mentagrophytes*
M. praecox	Surface pale yellow to yellow, powdery to velvety; pale yellow to orange on reverse; moderate growth	Abundant long, fusiform macroconidia, 40–90 × 7–17 μm, thin, smooth to spiny macroconidia, 2–8 septa; microconidia absent	Rare pathogen; does not perforate hair in vitro

(Continued over)

Table 13.4 *Characteristic of pathogenic* Epidermophyton and Microsporum species in culture (*Continued*)

Species	Colony on SGA	Microscopic morphology	Key tests, features or comments
M. racemosum	Surface beige-cream, granular; flat; wine red reverse; rapid growth	Abundant macroconidia, fusiform to ellipsoidal, 41–77 × 9–10 μm; resembles *M. gypseum*; numerous clavate, stalked microconidia in grape-like clusters (racemes)	Must be differentiated from *M. gypseum*, *M. cookei*; rare pathogen; perforates hair in vitro
M. vanbreuseghemii	Surface yellowish, cream or pink; powdery to downy; buff to pale yellow reverse	Numerous cylindrofusiform macroconidia, 44–87 × 11 μm; up to 12 septa, thick walls smooth to spiny; numerous microconidia	Rare pathogen; perforates hair in vitro

were a true geophile, however, it would be distributed in many environments lacking its specific animal population host, but this has never been demonstrated for *M. nanum*. Visualization of the characteristic macroconidia in the environment offers proof that a capability for saprobic existence has been retained, in that these characteristic conidia are not formed in infected animals or humans (Ajello 1974; De Vroey 1985). Summerbell (2002) has pointed out that the ability for sexual reproduction, a process that can only occur in the context of saprobic growth, can be correlated with numerous other characters indicating an alternatingly soil-borne and host-borne existence for most zoophilic dermatophytes.

Some mycologists consider *Microsporum persicolor* and *T. simii* to be geophilic (Padhye and Thirumalachar 1967; Matsumoto and Ajello 1987; Kwon-Chung and Bennett 1992), whereas others have treated them as zoophilic (Rippon 1988; Tanaka et al. 1992). Both species have been isolated repeatedly from soil and from the hair of apparently healthy animals harboring these fungi without any ringworm lesions (Gugnani et al. 1967; Contet-Audonneau and Percebois 1986). *M. persicolor* is most frequently isolated from rodents, notably bank voles (Ozegovic 1980) and can cause severe human infections (Rippon 1988). The isolation of *M. persicolor* from the cutaneous lesion of a woman and from her garden soil (both isolates being of the same mating type) (Contet-Audonneau and Percebois 1986) could be taken to suggest geophilism, though recent contact of infected voles with the garden in question can by no means be excluded. Similarly, *A. simii* (anamorph *T. simii*) causes sporadic ringworm in dogs, monkeys, poultry, and humans in India (Gugnani et al. 1967). Macroconidia associated with *M. persicolor* and *T. simii* have not been observed in the direct examination of soil samples.

EPIDEMIOLOGY

Species identification of the dermatophytes and knowledge of their host preference and ecology play an important role in epidemiology, public health issues, and infection control. Of special concern are those fungi capable of producing family or institutional outbreaks. Knowledge of the geographically endemic dermatophytes such as the African *Trichophyton soudanense* is also relevant to epidemiology, as it allows pinpointing exposure during travel, residence, or contact with a person from an endemic area. Geographical distribution will be covered elsewhere in this chapter (see Geographical distribution).

Geophilic fungi such as *M. gypseum* are usually transmitted from a soil source and can be secondarily transmitted by animals to humans (Kaplan et al. 1957).

Infections caused by *M. canis* can involve a variety of animals as hosts, but cats and dogs are the principal carriers (Rebell and Taplin 1974; De Vroey 1985; Rippon 1985, 1988; Kwon Chung and Bennett 1992). Fomites play an important role in the acquisition of infections (De Vroey 1985; Thomas et al. 1994). Wakimoto et al. (1985) reported an outbreak of ringworm caused by *M. canis* in Japan involving a kitten, fomites, and person-to-person spread of the infection. However, institutional outbreaks of infections in adults in a chronic healthcare facility (Shah et al. 1988), and in neonates (Mossovitch et al. 1986; Snider et al. 1993), have been reported with no clearly demonstrated link to an infected animal. Transmission in these cases resulted from person-to-person spread and fomites. *Trichophyton verrucosum* and zoophilic forms of *T. mentagrophytes* are typically encountered in rural areas and are acquired by humans in these locations mostly from herd animals, usually cattle (Georg 1956, 1960; Rippon 1985). Tinea capitis and tinea corporis are the most common manifestations in children and tinea corporis, while tinea barbae occurs in adult males (Rippon 1985). Fomites play an important role in zoophile infections, as noted in an outbreak caused by the mouse dermatophyte, traditionally called *T. mentagrophytes* var. *quinckeanum* (Blank 1957; Georg 1960) but recently renamed *T. mentagrophytes sensu stricto* (i.e. 'in the strict sense of the name') (Gräser et al. 1999c). Granular varieties within the *T. mentagrophytes* complex, formerly referred to as *T. mentagrophytes* var. *mentagrophytes*, have rodent

Table 13.5 *Characteristics of common pathogenic* Trichophyton *species in culture*

Species	Colony on SGA	Microscopic morphology	Key tests, features, or comments
T. concentricum	Surface cream, amber or brown; heaped, folded; glabrous to velvety; very slow growth	Conidia absent, only sterile hyphae, antler-like hyphae lacking terminal swelling may be present	50% isolates stimulated by thiamine; geographically limited
T. equinum	Surface white, fluffy, dome shaped; becoming buff and folded; reverse lemon yellow to reddish brown, growth moderate	Abundant microconidia, variable, spherical, pyriform to elongated clavate forms, borne usually singly along hyphae, occasionally in small clumps; macroconidia rare, clavate, smooth walled	Requires nicotinic acid; autotrophic strains only known from Australia and New Zealand; some isolates perforate hair in vitro
T. erinacei	Surface white, powdery, flat; yellow pigment on reverse; rapid growth	Numerous clavate to pyriform microconidia borne singly along the hyphae and in regular clusters; macroconidia clavate to cylindrical, smooth, thin walled, rare	Perforates hair in vitro; urease test often slow
T. interdigitale anthropophilic subtype	White to cream-colored surface, downy to fluffy, reverse tan	Microconidia few, moderate or heavy, pyriform, singly along hyphae	Perforates hair in vitro
T. megninii	Surface pinkish, velvety; radial folding, reverse wine red; moderate growth rate	Abundant pyriform to clavate microconidia; macroconidia pencil-shaped, smooth, rare	Requires L-histidine; urease positive; in urea indole broth, may be negative in Christensen urea broth in 7 days
T. mentagrophytes complex, zoophilic types	Surface creamy tan to pink; powdery to granular, flat; reverse buff to red or reddish-brown; rapid growth	Abundant spherical microconidia in grape-like clusters; macroconidia smooth walled, clavate to pencil-shaped, spirals often present	No nutritional requirement; urease positive; perforates hair in vitro
T. rubrum	Surface typically white, velvety to fluffy, occasionally powdery to granular; reverse wine red, rarely yellow or brown diffusible pigment; moderate to slow growth; dysgenic, lavender heaped varieties rare	Microconidia scanty to numerous, pyriform borne singly along the hyphae; macroconidia rare in fluffy isolates, numerous pencil-shaped to clavate macroconidia in granular to powdery colonies, 15–30 × 4–6 μm	Urease negative except in some flat, heavily conidial isolates of the afro-asiatic variant, in vitro hair perforation negative
T. schoenleinii	Surface glabrous, waxy, heaped, cerebriform becoming velvety, white, gray to tan; very slow growth, cracks agar	Conidia usually absent; microconidia in old, velvety subcultures; antler-like hyphae (favic chandeliers) with swollen nail-head tips; chlamydospores and hyphal swellings frequent	Autotrophic for vitamins, equal growth at 25–37°C differentiates it from T. verrucosum
T. simii	Surface white, cream or buff; powdery, granular, flat; reverse straw to salmon; rapid growth	Abundant clavate to fusiform to cylindrical, thin-walled macroconidia, 30–80 × 6–11 μm, fragmenting or with intercalary chlamydospores; microconidia pyriform, clavate, singly along the hyphae, spirals may be present	Resembles T. mentagrophytes but produces an unusually high number of macroconidia in isolates that have been transferred several times
T. soudanense	Surface flat, yellow to wine red, with star-like margin in primary isolates	Macroconidia absent, microconidia clavate, scarce or absent; reflexive branches are seen along larger hyphae within radial striations	May or may not appear to have vitamin requirements

(Continued over)

Table 13.5 *Characteristics of common pathogenic* Trichophyton *species in culture* (*Continued*)

Species	Colony on SGA	Microscopic morphology	Key tests, features, or comments
T. tonsurans	Surface white, cream, pale yellow to sulphur yellow, tan, pink; powder, suede-like or velvety; flat or raised and folded; mahogany reddish-brown reverse; slow growth	Microconidia numerous, clavate or elongate, some swollen, variable in size and shape, borne singly along hyphae, often on relatively long, empty 'matchstick' stalks and often attached to open branched clusters of hyphae; macroconidia rare, cylindrical, smooth, thin walled, tips often bent on one side	Growth stimulated by thiamine; urease positive; in vitro hair perforation variable; major incitant of endothrix type of ringworm
T. verrucosum	Surface cream to tan, flat, velvety, discoid; white, heaped glabrous to downy; yellow-ochre, glabrous, convoluted; very slow growth	Conidia usually absent; microconidia clavate; macroconidia 'rat-tail', irregular; chlamydospores in chains characteristic ('string of pearls') best at 37°C	About 16% of isolates require thiamine and inositol; growth stimulated at 37°C (faster than at 25°C)
T. violaceum	Surface lavender to deep purple, glabrous, heaped, very slow growth; white glabrous variants may occur	Conidia typically absent except after extended cultivation on thiamine-rich medium; irregular hyphae and chlamydospores present	Growth and sporulation stimulated by thiamine

Gräser et al. (2000c) concluded that *T. megninii* is synonymous with *T. rubrum* and that *T. gourvilii*, *T. soudanense*, and *T. yaoundei* were synonymous with *T. violaceum*; Summerbell (2003), however, concluded that *T. megninii* was a distinct species and that *T. gourvilii* was a synonym of *T. soudanense*, while *T. yaoundei* was a synonym of *T. violaceum*.

or rabbit-family (lagomorph) population hosts, and inoculum from such sources have caused infections in numerous laboratory workers (Georg 1960; Hironaga et al. 1981; Sewell 1995). Zoophilic members of the *T. mentagrophytes* complex (including isolates mating as *A. benhamiae* as well as isolates mating as *A. vanbreuseghemii*) have been isolated from many mammalian species (Ozegovic 1980; Rippon 1985), but chronic population carriage is only known from rodents, lagomorphs, and hedgehogs to date.

Table 13.6 *Classification of dermatophytes based on ecology and host preference*

Geophilic	Zoophilic	Anthropophilic
M. cookei	*M. canis*	*E. floccosum*
M. gypseum complex	*M. gallinae*	*M. audouinii*
M. nanum	*T. equinum*	*M. ferrugineum*
M. persicolor[a]	*T. erinacei*	*T. concentricum*
M. praecox	*T. mentagrophytes* complex, in part	*T. interdigitale*
M. racemosum	*T. verrucosum*	*T. megninii*
M. vanbreuseghemii		*T. rubrum*
T. simii[a]		*T. schoenleinii*
		T. soudanense
		T. tonsurans
		T. violaceum

a) Considered as zoophilic by Rippon (1988) and Weitzman and Summerbell (1995)

Some examples of less common infections contracted from other animals include *Microsporum gallinae* from fowl (Rippon 1985), *M. nanum* from pigs (Ajello et al. 1964; Morganti et al. 1976), *Trichophyton equinum* from horses (Georg et al. 1957; Zuckerman et al. 1992), *T. mentagrophytes* var. *quinckeanum* from mouse (Blank et al. 1961), *Trichophyton erinacei* (= *T. mentagrophytes* var. *erinacei*) from hedgehogs (Morris and English 1973; Philpot and Bowen 1992), *M. persicolor* from field voles (English et al. 1978; Onsberg 1978), and *T. simii* from poultry, dogs, and monkeys (Gugnani et al. 1967; Tewari 1969). Infections acquired from animals are usually inflammatory, but are more likely to resolve spontaneously than are infections caused by anthropophilic dermatophyte species (Rippon 1988; Kwon-Chung and Bennett 1992).

Infections caused by anthropophilic fungi are mostly acquired by direct contact with infected humans; fomites also play an important role and infection may even be acquired after aerosolization of infectious arthroconidia. *M. audouinii* and *T. tonsurans* have been isolated from the air (Friedman et al. 1960; Arnow et al. 1991) and several other dermatophyte species belonging to all three ecological groupings have been reported from house dust (Shimmura 1985). *M. audouinii* is an example of the extent of contagion and the potentially rapid spread of ectothrix tinea capitis. It was responsible for an epidemic in Europe in the nineteenth century that was later exported to North America and ended in the mid-1950s (Rippon 1985). It was essentially eliminated by griseofulvin, but was replaced by a *T. tonsurans* epidemic that swept into the south and southwest of the USA and Canada, apparently derived from an ongoing epidemic in Puerto Rico and other Hispanic countries (Rippon 1985). Tinea capitis, in the form caused by *T. tonsurans*, is an endothrix type of scalp infection. It may persist subclinically in the scalp, resulting in long-term carriers who may shed viable propagules for decades (Weitzman and Summerbell 1995). In adults, it is more frequently manifested as tinea corporis and uncommonly as tinea manuum and tinea unguium (Weitzman and Summerbell 1995).

T. tonsurans has also been responsible for several institutional outbreaks (Mackenzie 1961; Kane et al. 1988; Arnow et al. 1991). One such outbreak of nosocomial tinea corporis occurred in a pediatric ward. The infection was spread by the index patient to the staff and a visitor. *T. tonsurans* was isolated from the air, soiled linen and the backs of chairs (Arnow et al. 1991). Another nosocomial outbreak was reported in a nursing home for elderly people and persisted for 9 months despite remedial and sanitary measures (Kane et al. 1988). The index patient, who was immobile, infected a staff member who, in turn, may have spread the infection to other immobile patients who in turn infected other staff. *T. tonsurans* was found in several environmental samples. A third outbreak occurred in a residential school for girls where clinical infections were found in 21 children, mostly as tinea capitis and less frequently as tinea corporis or tinea unguium (Mackenzie 1961). The fungus was isolated from such inanimate sites as the air, combs, brushes, bedding, floors, and a curtain. Tinea corporis among adolescents and young adults, caused by *T. tonsurans*, has been reported from contact sports such as wrestling (Stiller et al. 1992; Beller and Gessner 1994).

Tinea capitis is most commonly a dermatophytosis of childhood, as is tinea corporis caused by *M. canis*. Other forms of tinea are seen most commonly in adolescents and adults, and are most commonly caused by anthropophilic dermatophytes. Tinea cruris (groin), tinea pedis (feet), and tinea unguium (nails) are frequently caused by *T. rubrum*, *T. interdigitale* (a name recently redefined after molecular studies by Gräser et al. (1999c) and used here in its modern sense), and *E. floccosum*, although this may vary with geographical location. Fomites play an important role in transmission, but host factors such as immunological status and local factors such as trauma, excessive moisture or occlusive clothing may constitute risk factors when combined with exposure to the fungus.

Outbreaks of tinea cruris involving *T. rubrum* and *E. floccosum* have been attributed to sources of contagion such as toenail clippings, bedding, toilet seats, and saunas, as well as to exchange of clothing (Neves and Xavier 1964; Lundell 1974; Rippon 1988). Ringworm

caused by other anthropophilic species such as *T. violaceum* and *T. schoenleinii* can be transmitted from person to person by clothing, cotton caps, pillows, combs, and towels (Padhye 1962; Kwon-Chung and Bennett 1992). Infection by *Trichophyton concentricum* is transmitted from mother to child after birth by contact (Rippon 1988), but fomites may also play a role in transmission (De Vroey 1985). Contagion either by direct contact or from fomites has been essentially established in tinea capitis, tinea corporis, and tinea barbae. With regard to tinea pedis, exogenous exposure to the fungus in showers, swimming pools, locker rooms, etc., plays an important role in the acquisition of clinical disease, though host susceptibility factors may also be involved (Gentles 1957; Gentles and Holmes 1957; Gip 1967; Zaias et al. 1996). Viable propagules of dermatophytes are shed from the feet of people with tinea pedis and these fungi have been isolated from areas where people walk bare footed and from towels, shoes, and socks (Ajello and Getz 1954; Gentles 1957; Gip 1967; Fujuhiro 1994). Epidemiological study showed a threefold increase of tinea pedis when communal showers were installed in a coal mine (Gentles and Holmes 1957).

The dermatophyte propagules most commonly associated with contagion in all anthropophilic dermatophyte infections and most zoophile infections are the substrate arthroconidia found within desquamated epithelium and hairs; these conidia can survive for an extended period of time in the environment (McPherson 1957; Mackenzie 1961; Kane et al. 1988; Rippon 1988; Kwon-Chung and Bennett 1992).

GEOGRAPHICAL DISTRIBUTION

The geographical distribution of the anthropophilic and zoophilic dermatophytes, rather than being static within a city, state, or country, is dynamic. It tends to be affected by periods of social change featuring changing immigration patterns, health habits, standards of living, or propensities for travel. For example, before 1900, tinea capitis was rare in North America, and was caused mostly by *M. canis* transmitted by animal contact (Rippon 1985). From 1900 to the mid-1950s a 'gray patch' ectothrix type of ringworm in children, caused by *M. audouinii*, spread over the USA and Canada. As mentioned above, this epidemic was in turn displaced by 'black dot' endothrix ringworm caused by *T. tonsurans* (Rippon 1985, 1992; Babel 1990), which was for the most part imported from Latin America and the Caribbean but went on to become the current predominant cause of tinea capitis in North America (Rippon 1985; Gupta and Summerbell 2000). *T. tonsurans* is now the second most commonly isolated dermatophyte in the USA; only *T. rubrum* is more common. *M. audouinii*, although still cultured sporadically, usually in connection with recent migrants from Africa, has virtually disappeared in the

USA (Sinski and Kelley 1991). In some southern European and middle eastern countries, *M. canis* is the most frequently isolated dermatophyte and threatens to become a serious epidemiological problem as a result of carriage by feral cats (Lunder and Lunder 1992). The geographical distribution of the geophilic, zoophilic, and anthropophilic dermatophytes is shown in Tables 13.7, 13.8 and 13.9. Dermatophytes endemic in certain geographical areas may be isolated outside such locations as a result of travel and immigration. *Trichophyton violaceum* (= *T. soudanense*), for example, which is widespread in Africa, has been reported among African immigrants in several European countries and in North America (Kwon-Chung and Bennett 1992; Viguie et al. 1992). For a more comprehensive review of the geographical distribution of the dermatophytes, readers should refer to publications by Ajello (1960); Sinski and Flouras (1984); Rippon (1988) and Kwon-Chung and Bennett (1992).

MOLECULAR TAXONOMY AND PHYLOGENETIC RELATIONSHIPS

In recent years, dermatophytes have been subject to extensive molecular systematic study. It has been confirmed that all sexual and asexual dermatophytes are, indeed, closely related to one another, and that all are members of the family *Arthrodermataceae*. (Anamorphic species and genera are now correctly said to belong to teleomorphic families in mycology when this information is unambiguously known, e.g. through molecular study.) Though further study is needed before firm conclusions can be drawn, it appears there may be four genera rather than three within the traditional grouping of dermatophytes and dermatophytoids. (The latter term refers to nonpathogenic species in genera containing dermatophytes.) The studies of Y. Gräser and colleagues (Gräser et al. 1999b, 2000a) have found that most of the nonpathogenic *Trichophyton* spp., e.g. *T. ajelloi* and *T. terrestre*, form a clade separate from the pathogenic species. *Microsporum* appears to be a distinct genus, but *Epidermophyton*, while supported as separate by some

studies (Kano et al. 1999, 2002), is apparently contained within *Microsporum* in other studies (Gräser et al. 1999b, 2000a, b). Molecular mycologists tend to find it inconsistent to recognize the anamorphs in a clade as a separate genus, but not at the same time to recognize the generic status of the corresponding teleomorphs. Application of this viewpoint in the dermatophytes would tend to revive the generic name *Nannizzia* for teleomorphs in the *Microsporum* clade. No formal proposal to this effect, however, has yet been made.

At the species level, the separate status of all known sexual species within the dermatophytes has been upheld wherever studied. It has been found, however, that many asexual species appear to have arisen relatively recently in evolutionary time, and each of them except *E. floccosum* is closely related to a known sexual species. This has caused the validity of some species to be questioned. Numerous taxonomic challenges are posed by unusual population genetics patterns that can be seen within the asexual dermatophytes and their

Table 13.7 *Geographical distribution of geophilic dermatophytes*

Geophilic	Distribution
M. cookei	Cosmopolitan
M. gypseum and *M. fulvum*	Cosmopolitan
M. nanum	Cosmopolitan
M. persicolor	Cosmopolitan
M. praecox	Western Europe, North America
M. racemosum	Romania, Venezuela, USA
M. vanbreuseghemii	Russia, Africa, India, USA
T. simii	India

Table 13.8 *Geographical distribution of zoophilic dermatophytes*

Species	Geographical distribution in major animal hosts
M. canis	Cosmopolitan
M. canis var. *distortum*	Australia, New Zealand, USA
M. canis var. *equinum*	Africa, Australia, Europe, New Zealand, North and South America
M. gallinae	Cosmopolitan
T. equinum	Cosmopolitan
T. erinacei	Europe, New Zealand
T. mentagrophytes complex (zoophilic types)	Cosmopolitan; Australia, Canada, Eastern Europe, Italy
T. verrucosum	Cosmopolitan

Table 13.9 *Endemic geographical distribution of anthropophilic dermatophytes, excluding isolates in travelers and recent immigrants*

Species	Major geographical regions
E. floccosum	Cosmopolitan
M. audouinii	Parts of Africa
M. ferrugineum	Far East, West Africa, Eastern Europe
T. concentricum	Parts of Asia, Oceania, isolated areas inhabited by indigenous peoples in parts of Latin America
T. megninii	Portugal, Sardinia, Corsica, and parts of Africa
T. interdigitale	Cosmopolitan
T. rubrum	Cosmopolitan
T. schoenleinii	Parts of Africa and Asia
T. tonsurans	Cosmopolitan
T. violaceum	North and east Africa, west Asia

immediate sexual relatives. In some cases, sequencing studies have found interbreeding strains within a sexual species to be apparently more distantly related to one another than some of them are to certain morphologically and physiologically distinct asexual species. Moreover, in such cases, the genetically varied individual sexual species and the genetically restricted asexual species to which they appear to have given rise tend to infect different population hosts. For example, based on ribosomal internal transcribed spacer (ITS) sequencing, oligonucleotide polymerase chain reaction (PCR) typing, and amplified fragment length polymorphism (AFLP) typing, CBS 495.86, the feline-derived, (+) mating type, taxonomic cotype strain of *Arthroderma otae*, the teleomorph of *M. canis*, is more closely related to the anthropophilic, asexual species *M. audouinii* than it is to the (−) *A. otae* mating type strain, CBS 496.86, that serves as its complementary cotype (Gräser et al. 1999c). While it is not difficult to conceive that one clone of the (+) mating type of the cat dermatophyte may have transformed itself within a moderately short evolutionary time frame into a morphologically and physiologically distinct organism with a uniquely strong genetic adaptation for causing ectothrix infection of human preadolescent scalps, it is somewhat controversial to interpret this event as a speciation process. Sequences of noncoding and housekeeping parts of the genome, e.g. the ribosomal sequence and associated spacer regions, tend to have changed only slightly between at least some *A. otae*/*M. canis* strains and *M. audouinii*. At the same time, it is clear that in both interbreeding status and general population genetics, *M. canis* has a strong potential unity as a 'biological species,' and also has a common habitat leading to stabilizing selection of pathogenicity factors, morphology, and other aspects of phenotype. None of these factors is shared by *M. audouinii*, the connection of which to *M. canis* is purely historical. In view of this population genetics divergence, Summerbell (2002) has recently proposed recognizing *M. audouinii* and all other 'unifactorial asexual radiate lineages' within the dermatophytes, that is, all single-mating-type, asexual dermatophyte lineages that have undergone a population-host shift in respect to their most closely related sexual relative, as being distinct at the species level.

Some long-recognized asexual dermatophyte species have been formally proposed as synonymous with closely related species as a result of molecular phylogenetic studies. For example, *T. equinum* has been proposed as a synonym of *T. tonsurans* by Gräser et al. (1999c) based on ITS sequence and other molecular similarities. The rejection of this synonymy has subsequently been suggested, however, since the former dermatophyte, with equine population hosts, is of the (+) mating type, while the latter, with human hosts, is of the (−) mating type (Summerbell 2003). This evidence of long-term evolutionary separation was not acknowledged in the study originally proposing the synonymy of these species. Existing data suggest separate, though relatively evolutionarily recent, origins from a common ancestor rather than conspecificity for *T. equinum* and *T. tonsurans*.

Trichophyton gourvilii, *T. soudanense*, and *T. yaoundei* have been suggested to be synonyms of *T. violaceum* (Gräser et al. 2000c). Summerbell (2003), citing the apparently premature synonymization of *T. tonsurans* and *T. equinum* based on similar techniques, has called for additional confirmatory molecular genetic study before the morphologically and geographically distinct cluster containing *T. soudanense* and *T. gourvilii* is synonymized with *T. violaceum*.

Population genetics studies within other traditionally defined dermatophyte species have revealed highly divergent patterns. *T. rubrum*, for example, has been found to be characterized by a high degree of overall genetic uniformity suggestive of recent subclonal derivation from a preexisting clonal lineage that also gave rise to *T. violaceum* and *T. soudanense* (Gräser et al. 1999b, 2000c; Summerbell et al. 1999). *T. mentagrophytes*, on the other hand, has been found to subtend an unexpected degree of biodiversity. While it was long known to contain anamorphs of two biological species, *A. benhamiae* and *A. vanbreuseghemii*, and to be closely related to the *T. simii* anamorphs of *A. simii*, *T. mentagrophytes* as traditionally conceived by Emmons (1934) has recently been found also to phylogenetically encompass asexual radiates such as *T. schoenleinii*, *T. concentricum*, *T. verrucosum*, and the lineages giving rise to *T. equinum*, *T. tonsurans*, *T. soudanense*, *T. violaceum*, *T. megninii*, and *T. rubrum*. Gräser et al. (1999c) have proposed a formal nomenclatural neotypification that limits application of the name *T. mentagrophytes* to the group of isolates formerly referred to as *T. mentagrophytes* var. *quinckeanum*, the mouse favus dermatophyte. These isolates are capable of mating successfully with mating type testers of *A. benhamiae* (Weitzman and Padhye 1976); this teleomorphic species, however, is conceived as sufficiently phylogenetically distinct to be accepted as a taxon separate from *T. mentagrophytes sensu stricto* (Probst et al. 2002). This viewpoint is based on the common knowledge that in many biological groups, the ability to manifest interfertility under artificial conditions may be retained as a symplesiomorphic (inherited ancestral-type) character among closely related species that never interbreed in nature. The apparently clonal (Mochizuki et al. 1990, 1996) *T. mentagrophytes*-like lineage that is contagious among humans has been referred to a neotypified *T. interdigitale* (Gräser et al. 1999c). A few isolates compatible with this species retain an ability to mate with testers of the normally zoophilic *A. vanbreuseghemii*; the exact status of *A. vanbreuseghemii* vis-à-vis *T. interdigitale* is not yet settled. Certain other *T. mentagrophytes*-like lineages that are capable of interbreeding, at least in vitro, with members of the established biological species, have also been conceived as separate species, e.g. the hedgehog dermatophyte, *T. erinacei*. In general, unraveling the complexity of these phylogenetic and tokogenetic relationships among dermatophyte lineages is a process that will require additional study.

CLINICAL FORMS

Infections caused by dermatophytes (ringworm, tineas) are clinically classified on the basis of the location of the lesions on the body. Although different body sites may be affected, each focus of infection is generally the result of local inoculation. The invading dermatophyte grows in a centrifugal manner, forming irregular rings with inflammatory borders with some clearing in the central area of the lesion. The name ringworm was based on the worm-like appearance of the lesions with irregular, inflammatory borders. The infection is named according to body site after the Latin word *tinea*. The word tinea itself comes from the clothes moth and the similarity of its effects on wool garments to the shape of the dermatophyte skin lesions (Ajello 1974). The clinical manifestations are classified as follows: tinea pedis (feet); tinea cruris (groin); tinea corporis (glabrous skin); tinea barbae (ringworm of the beard and moustache); tinea capitis (scalp, eyebrows, eyelashes); tinea manuum (hand); tinea unguium (nails); tinea favosa (favus); and tinea imbricate (ringworm caused by *T. concentricum*). The clinical manifestations and their major etiologic agents are described here briefly. More detailed information may be obtained by referring to texts by Rippon (1988); Kwon-Chung and Bennett (1992) and Kane et al. (1997).

Tinea pedis ('athlete's foot')

Ringworm of the feet involves infection of the interdigital webs and soles. The most common clinical manifestation is the intertriginous form associated with maceration, scaling, fissuring, and erythema, which presents with itching and burning. The other clinical form, which is commonly noted, involves a chronic, squamous, hyperkeratotic type of infection of the soles and heels extending up the sides of the foot (moccasin foot). This is an acute condition characterized by the formation of vesicles, and inflammation plus occasional pustules and bullae are most commonly caused by granular, zoophilic members of the *T. mentagrophytes* complex. The common agents of tinea pedis in cases of chronic infection are *E. floccosum*, *T. interdigitale*, and *T. rubrum*.

Tinea cruris ('jock itch')

This infection of the inguinal area involves the groin, perianal, and perineal areas, often involving the upper thighs. It is frequently caused in adults by *T. rubrum* and *E. floccosum*. Lesions are erythematous and scaly, exhibiting raised, inflamed borders often with vesicles. They are symptomatically associated with itching and burning. They are usually bilateral, extending down the inner thighs and over the proximal waist area and buttocks.

Tinea corporis

This ringworm of the upper parts of the body usually involves the shoulders, axilla, chest, and back. It may also involve the face, the legs, or the dorsa of hands or feet. The lesions are well marginated with raised erythematous, vesicular borders. The infection may be mild to severe. The annular, scaly patches may coalesce to form large areas of chronic infection when *T. rubrum* is the etiologic agent. Zoophilic dermatophytes frequently cause this infection.

Tinea barbae

This infection may be mild and superficial but, more often, the lesions are inflammatory and pustular, especially when caused by zoophilic dermatophytes such as *T. mentagrophytes* and *T. erinacei*. Erythematous patches are often scaly, and show lusterless hairs and folliculitis.

Tinea capitis

This ringworm involves infection of the scalp hairs and intervening skin. Infection may range from mild, with slight erythema and a few, patchy scaly areas, to severe, with inflammatory lesions involving folliculitis or kerion formation, often with secondary bacterial infection resulting in scarring and alopecia. Hair invasion occurs in two patterns, ectothrix and endothrix. In the ectothrix type of invasion, the invading fungus infects the hair shaft at mid-follicle and forms a sheath of hyphae and arthroconidia 2–3 µm in diameter surrounding the hair shaft. The infected hair become lusterless and brittle, and the hair filaments break off at the level of the scalp to give an appearance of partial alopecia. The best known agents of ectothrix scalp ringworm are *M. audouinii* (seen in Africa as a variant classically indicated by the synonymous name *M. langeronii*), *M. canis*, *M. ferrugineum*, and *M. gypseum*. *T. verrucosum* and *T.megninii* also occasionally cause ectothrix colonization of the hair shaft. In endothrix hair invasion, the hyphae invade the hair follicle and then grow into the hair shaft, forming numerous arthroconidia within it and severely damaging the hair in the process. The infected hairs become grayish-white and break off easily at the level of the scalp to give a 'black dot' appearance. The predominant cause of endothrix tinea capitis in the USA at present is *T. tonsurans*. *T. violaceum* and *T. soudanense* are prevalent in other parts of the world. Lesions may coalesce to form large patches of alopecia.

Tinea manuum

This infection involves the palmar and interdigital areas of the hands. Lesions present as diffuse hyperkeratotic areas. Most infections are caused by *T. rubrum*.

Tinea unguium

Invasion of the nail plate by a dermatophyte is referred to as tinea unguium. Distal subungual infection is one of the common patterns of nail invasion and generally involves the nail bed and the underside of the distal portion of the nail. A mild inflammation produces hyperkeratosis and focal parakeratosis (Elewski et al. 1995). Subungual hyperkeratosis causes lifting of the nail plate and its detachment from the nail bed. *T. rubrum*, *T. interdigitale*, and *E. floccosum* are the main causes of tinea unguium in the USA (Elewski et al. 1995) as well as most other parts of the world.

Tinea favosa

Favus (meaning honeycomb) is now an exceedingly rare disease except in a few redoubts in central Africa and central Asia. It is usually caused by *T. schoenleinii*. Infections can be severe and chronic and involve pus formation in the hair follicles, a process producing cup-shaped crusts called scutula. Longstanding favus infections may result in alopecia with scarring.

Tinea imbricata

This chronic infection is caused by *T. concentricum*. Seven different clinical patterns of infection are distinguished: imbricate, featuring concentric rings of overlapping scales, lamellar, lichenified, plaque-like, annular, palmar/plantar, and ungual (Hay et al. 1984; Rippon 1988).

Dermatophyte pseudomycetoma

A distinct but uncommon dermatophyte infection is the invasion of subcutaneous tissue. In such infections, fungal elements form yellowish-white pseudogranules lacking any of the cementing material found in some true mycetomas. Granulomatous dermatitis is rare and is generally observed when a hair follicle ruptures and individual fungal hyphal elements escape into the dermis. This condition, also known as Majocchi's dermatophyte granuloma, results from instances of trauma to the infected skin. For example, a puncture wound may

introduce the dermatophyte into deep tissue and facilitate invasion. Loose or compactly arranged hyphal aggregates of varying sizes embedded in eosinophil-rich Splendore–Hoeppli material may be erroneously referred to as granules or grains, by analogy with structures seen in mycetoma, and the lesions may indeed be referred to as mycetoma. However, the fungal aggregates may result from hair follicles with mycelial growth rupturing, permitting the mycelial filaments and fungal cells formed therein to escape into the surrounding dermis and continue to develop, eliciting a strong Splendore–Hoeppli reaction and granulomatous tissue response. Based on their loosely arranged mycelial aggregation surrounded by striking and abundant Splendore–Hoeppli reaction material, they should more appropriately be called pseudogranules and the lesion should be called a dermatophytic pseudomycetoma (Ajello et al. 1980; Chandler and Watts 1987). The most common cause of dermatophytic pseudomycetoma is *M. canis*. Other species that have been reported are *M. ferrugineum* and *M. audouinii* (Kwon-Chung and Bennett 1992).

IMMUNOLOGY

Dermatophytes colonize the keratinized tissue of the stratum corneum. The degree of inflammation produced in the lesions depends primarily on the species of the causal agent and, to some extent, on the immunological competence of the host. Infections by anthropophilic fungi, such as *T. rubrum*, often elicit less inflammatory response than is seen with infections caused by zoophilic and geophilic dermatophytes such as *T. verrucosum* or *T. mentagrophytes*.

Invasion by dermatophytes may be favored by environmental conditions. Warm, moist conditions are known to favor infection, as was seen in swampy areas of Vietnam featuring a high incidence of dermatophytosis in combat troops (Blank et al. 1969). Occlusion of sites exposed to dermatophyte inoculum appears to increase susceptibility. In experimental infections, this factor played a major role, apparently because it increased hydration of the underlying skin and retention of CO_2, helping dermatophyte growth (King et al. 1978).

Medical conditions such as collagen vascular disease, Cushing's disease, diabetes mellitus, hematological malignancy, atopy, and old age may play a significant role in predisposing patients to chronic dermatophytic infections, as may systemic corticosteroid therapy. Susceptibility to chronic dermatophytosis by *T. concentricum* may occur as an inherited trait (Hay 1992). An inherited autosomal recessive trait has been shown to predispose to infection by *T. concentricum* (Serjeantson and Lawrence 1977; Ravine et al. 1980). Similarly, susceptibility to chronic *T. rubrum* infection is likely to be mediated by inheritance of an autosomal dominant allele (Zaias et al. 1996).

Although a host develops a variety of antibodies as a response to a dermatophyte infection, including antibodies derived from immunoglobulins IgM, IgG, IgA, and IgE, it has been accepted that IgE plays a key role in the suppression of cell-mediated immunity (CMI), perhaps through modulating histamine activity. The variability seen among patients in humoral immunity responses to dermatophytosis has been attributed to a lack of standardized antigens and to differences in the sensitivity and specificity of the methods used (Matsumoto et al. 1996). It is known that CMI is the cornerstone of host defense in regard to dermatophytosis. The development of CMI, which is correlated with delayed-type hypersensitivity, is usually associated with clinical cure and with exclusion of the dermatophyte elements from the stratum corneum (Dahl 1993; Jones 1993). Delayed hypersensitivity responses to intradermal injections of trichophytin are commonly observed in the normal population (Wood and Cruickshank 1962; Palmer and Reed 1974; Grossman et al. 1975). These responses are probably caused by earlier exposures to dermatophytoses, or by crossreactivity with one or more related environmental organisms.

The immune response to dermatophytic infections has been investigated in human infections as well as in various experimental studies using calves, guinea pigs, rabbits, and rats. Experimental dermatophyte infections in animals appear to result in CMI to the antigens of the infecting dermatophyte (Cruickshank et al. 1960; Krebs et al. 1977; Green and Balish 1980). In guinea pigs with experimental infections caused by a member of the *T. mentagrophytes* complex, maximal erythema in the infected skin occurs at the time the infected animals developed CMI to trichophytin (Krebs et al. 1977). Similar findings have been noted in cattle experimentally infected with *T. verrucosum*. The delayed hypersensitivity reactions in the skin corresponded to clearance of infection by means of an increase in the desquamation rate of the stratum corneum (Lepper 1974; Lepper and Anger 1976; Wagner and Sohnle 1995). A live vaccine (LTF 130) against *T. verrucosum* ringworm in cattle was developed and successfully used to reduce infections in cattle herds in the former Soviet Union and some countries in eastern Europe (Segal 1989). The use of vaccine in humans, however, has not been effective.

TREATMENT

The increase in mycotic infections stimulated the pharmaceutical industry to develop new drugs to meet this challenge. The newer antifungal agents available to treat the dermatophytoses include orally active triazoles (fluconazole and itraconazole), the allylamines (naftifine and terbinafine), and the morpholines (amorolfine). These antifungals can be added to others currently in use, such as orally administered griseofulvin and itraconazole as well as topical antifungals such as haloprogin, the thiocarbonates (tolnaftate), hydroxypyridones (ciclopirox olamine), and numerous other imidazoles (Smith 1993; Gupta et al. 1994a).

Decisions regarding the best regimen for therapy must consider the extent, location, and clinical type of infection, the etiologic agent, the spectrum of activity and pharmacokinetics of the antifungal, any pharmacodynamic drug–drug interactions that may occur (Schafer-Korting 1993; Bickers 1994; Brodell and Elewski 1995), side effects (Hay 1993), cost (Einarson et al. 1994), and the overall risk/benefit analysis. The interested reader is referred to several reviews in which antifungal agents for the dermatophytoses are discussed along with their modes of action, dosages, therapeutic indications and contraindications, and drug–drug reactions (Elewski 1993; Stiller et al. 1993; Degreef and DeDoncker 1994; Gupta et al. 1994a, b; Brodell and Elewski 1995).

Oral antifungals are usually prescribed for tinea capitis, tinea unguium, chronic dry forms of tinea cruris, invasive infections, and infections resistant to topical therapy or with extensive involvement. These require extensive treatment before a clinical cure can be achieved. Generally, successful therapy necessitates compliance and good personal hygiene.

Tinea capitis

Oral griseofulvin has been the mainstay of therapy for tinea capitis since the 1950s. It is the most frequently prescribed therapy, usually requiring a 6-week course (Frieden and Howard 1994). No topical therapy has been shown to be as effective as oral griseofulvin, although topical applications of various preparations may be used as adjuncts to impede shedding of viable fungi from infected scalps (Del Palacio et al. 2000). Ectothrix infections caused by *M. canis* and other species of *Microsporum* usually respond best to griseofulvin (Jacobs 1990; Degreef and DeDoncker 1994); however, Jacobs (1990) considered ketoconazole as the first choice for treating endothrix infections caused by *T. tonsurans*.

Other studies indicate better results with oral ketoconazole (Gan et al. 1987; Tanz et al. 1988). Generally, oral ketoconazole is considered as an effective alternative when there is resistance, intolerance, or nonresponse to griseofulvin (Elewski 1994; Frieden and Howard 1994). Rare side effects, such as hepatotoxicity with long-term use of ketoconazole, limit its application to second-line treatment. Legendre and Escola-Macre (1990) and Elewski (1994) found itraconazole to be an effective alternative to either griseofulvin or ketoconazole. Gupta and colleagues (Gupta et al. 1997, 1998; Gupta and Adam 1998) investigated the efficacy of pulse therapy with itraconazole and terbinafine in children with *T. tonsurans* infections reporting 100 percent cure rate when one to three pulses of itraconazole were used and 92 percent cure rate with terbinafine. Gupta et al. stated that, given the pharmacokinetics of both drugs, they found the pulse-dosing format to be a reasonable option. The pulse regimens allowed the physician to tailor therapy to the patient's response. Terbinafine, given orally, was found to be a safe and effective drug for treating tinea capitis (Haroon et al. 1992; Villars and Jones 1992).

Adjunct therapy may include selenium sulfide shampoos (Frieden and Howard 1994) or ketoconazole shampoo and topical econazole nitrate (Elewski 1994) to decrease inoculum shedding, as well as oral antibiotics to treat secondary infection and prednisone to diminish severe kerion (Frieden and Howard 1994). In addition to treatment with antifungals, general sanitation measures are necessary to prevent spread and recurrence of infection (Weitzman and Summerbell 1995). However, fomites and asymptomatic carriage contribute to spread and reinfection (Frieden and Howard 1994; Howard and Frieden 1995). Hebert et al. (1985) recommended treating not only the patient, but also all family members including adults, and sanitizing the home environment.

Tinea corporis

Tinea corporis, when caused by geophilic or zoophilic dermatophyte, usually resolves spontaneously within a few months. A variety of topical agents are usually applied to speed up healing of local and uncomplicated lesions (Smith 1993; Gupta et al. 1994a; Weitzman and Summerbell 1995). Topical imidazoles appeared to be highly effective with cure rates of up to 80 percent (Del Palacio et al. 2000). Amorolfine and allylamines are another alternative (Del Palacio-Hernanz et al. 1989; Del Palacio et al. 1991). Topical therapy is preferable because there is less potential for side effects. However, systemic therapy is also indicated for extensive lesions resulting from the anthropophilic dermatophyte species, for invasive lesions, and for chronic dry manifestations of tinea corporis caused by *T. rubrum*. For many years, oral griseofulvin and ketoconazole have been used as alternatives to treat recalcitrant or widespread lesions. However, griseofulvin-resistant mutants have rarely been reported (Artis et al. 1981). It has also been noted that one-third of patients do not respond to treatment with griseofulvin (Jacobs 1990). The therapeutic use of ketoconazole is limited by the rare side effects of this drug, such as liver toxicity and depressed adrenal activity and testosterone secretion (Kwon-Chung and Bennett 1992). Ketoconazole also has drug–drug interactions that may be problematical (Bickers 1994; Brodell and Elewski 1995). Itraconazole and terbinafine, taken orally, were reported to be generally safe and effective therapies against most of the common dermatophytes (Degreef

and DeDoncker 1994; Gupta et al. 1994b; Weitzman and Summerbell 1995; Del Palacio-Hernanz et al. 1990; Del Palacio et al. 1993; Faergemann et al. 1997; Nozickova et al. 1998).

Tinea cruris

A variety of topical medications, applied as creams, lotions, or powders, may be used for uncomplicated tinea cruris (Smith 1993; Gupta et al. 1994a; Weitzman and Summerbell 1995). However, the dry recalcitrant type of infections caused by *T. rubrum* may require systemic therapy such as griseofulvin, itraconazole, fluconazole, or terbinafine (Gupta et al. 1994b). Relapses are more common with griseofulvin. Butenafine is effective in tinea cruris with 2 weeks of treatment, with a cure rate around 70 percent (Lesher et al. 1997). Fluconazole (150 mg once weekly) for 4–5 weeks has been effective in the management of tinea cruris and tinea corporis with a 74 percent clinical cure rate (Faergemann et al. 1997). Terbinafine was reported to be superior to griseofulvin in several studies (Degreef and DeDoncker 1994). Currently, itraconazole may be given as a dose of 400 mg/day given as two daily doses of 200 mg for 1 week as an effective treatment for tinea cruris infection (Del Palacio et al. 2000).

Tinea unguium

Tinea unguium, or dermatophyte infection of the nails, has generally been found to be difficult to cure. This has especially been noted with infections of the toenails, except in cases of superficial white onychomycosis. Tinea unguium was generally regarded as an incurable disease until the advent of systemic oral antifungal drugs. Previous treatment consisted of mechanical or chemical avulsion of the nail followed by topical therapy. Recurrence was, however, common. The use of older topical antifungal agents applied directly to the nail has given variable and disappointing results (Hay 1993; Hay et al. 1994) as a result of the inability of these agents to penetrate through the nail plate down to the nail bed where the fungus resides. Newer topically active antifungals, such as 28 percent tioconazole, amorolfine, and 8 percent ciclopirox, have produced higher remission rates than did older topical formulations, but overall cure rates have also been disappointing (Hay 1993). The use of the newer agents as an adjunct to oral antifungal therapy with griseofulvin improved recovery rates significantly (Hay 1993).

Griseofulvin was the first orally active, systemic antifungal agent applicable for the treatment of onychomycosis (fungal infection of the nail), but its use was limited to dermatophytes and required long-term therapy of 6–9 months for fingernails and 12–18 months for toenails (Roberts 1994). This was because it did not diffuse into the nail bed but only reached the site of infection via uptake of newly produced nail keratin. In addition, without removal of the nail, overall cure rates of toenail infections were low, around 30–50 percent, and relapse was common (Roberts 1994).

Newer, orally active systemic medications have recently been shown to be effective when administered for short periods (up to 3 months) or when used intermittently. Both itraconazole and terbinafine reach the nail via incorporation into the matrix and remain unchanged in the nails at a therapeutic level for at least 6 months after termination of therapy in patients treated for 12 weeks (Hay 1993; Hay et al. 1994; Del Palacio et al. 1994, 1999). Either continuous therapy with itraconazole (200 mg/day) for 3 months or pulse therapy (200 mg twice a month) for 3–4 months may be used (Heikkila and Stubb 1997). The pharmacokinetic comparison of continuous and intermittent itraconazole pulse therapy showed that intermittent therapy resulted in relatively high itraconazole plasma concentrations but relatively low drug exposure. This resulted in a lower nail concentration of itraconazole in intermittent therapy than that obtained with continuous itraconazole therapy (Havu et al. 1999). However, this characteristic of intermittent therapy did not result in a reduced cure rate, as the concentration of itraconazole in the nail remained within the therapeutic range. The total itraconazole administered in pulse therapy was about half of what was given in a continuous therapy. This reduction in the amount of itraconazole administered reduced side effects and improved cost-effectiveness, in addition to providing increased convenience for the patient (Van Doorslaer et al. 1996; Nolting et al. 1998).

Nail removal plays a limited role in the therapy of onychomycosis (Hay 1993). It may be accomplished by surgery combined with the use of systemic antifungals before avulsion. Alternatively, it may be done chemically. The best known chemical procedure entails the use of 40 percent urea paste formulation under occlusion, sometimes with the incorporation of 1 percent bifonazole into the paste.

Although dermatophytes play a major role as causative agents of onychomycosis, nondermatophytic molds and yeasts account for a significant number of cases. Laboratory tests are needed to confirm the species involved and to rule out bacterial or nonfungal etiology; this will also ensure selection of the most effective treatment for each patient, and increase the probability of a complete cure (Elewski et al. 1995).

LABORATORY DIAGNOSIS

Direct microscopic examination

The direct microscopic examination of a properly collected specimen is one of the most rapid and effective methods of detecting a fungal infection. This highly effective screening technique will provide useful information regarding the etiologic agent, e.g. whether it is a mold or a yeast, and whether it causes ectothrix, endothrix, or favic hair invasion. In a few cases, direct specimen microscopy may suggest the genus of the causative agent.

The choice of method for the direct examination will depend upon the type of specimen screened and the laboratory's resources. The typical specimens submitted for the diagnosis of a dermatophyte infection are skin scrapings, hair stubs, and nail clippings or scrapings. The solutions usually chosen for the examination of these specimens are potassium or sodium hydroxide preparations, most commonly potassium hydroxide (10–20 percent KOH), often in combination with the fluorescent stain calcofluor white. In some laboratories and procedures, the periodic acid–Schiff (PAS) stain is used.

POTASSIUM HYDROXIDE

Potassium hydroxide is the most widely used preparation for the direct examination of clinical specimens for the presence of fungi. Basically, a drop of 10–20 percent KOH is placed on a slide, a small amount of specimen is added to the drop, a coverslip is placed over it and the preparation is gently heated (short of boiling). The KOH softens and clears the specimen for easier detection of hyphae by digesting any proteinaceous debris and disrupting the keratin's cellular sheets, thereby rendering the more biochemically resistant fungus more visible as highly refractile, hyaline, septate, branched, or unbranched hyphae and arthroconidia. In hairs, fungal elements may appear as arthroconidia on the outside (ectothrix invasion) of the hair shaft, or on the inside (endothrix), or they may appear as hyphae co-occurring with bubbles and tunnels (favic invasion). The reader is referred to several texts for descriptions and photographs of dermatophytes in invaded keratinized tissue (Fragner 1987; Rippon 1988; Weitzman et al. 1988; Kwon-Chung and Bennett 1992; Weitzman et al. 1995; Weitzman and Summerbell 1995; Kane et al. 1997).

Through the years, several modifications of the basic 10 percent KOH preparation have been made for more rapid detection. One modification is the incorporation of Parker superchrome blue-black ink in the KOH solution for selective staining of the fungus (Rippon 1988). Other modifications of the basic method include the addition of 36 percent dimethyl sulfoxide (DMSO) to 20 percent KOH to aid in the preparation and clearing of the specimen without heating (Rebell and Taplin 1974; Rippon 1988), and the addition of 5–10 percent glycerine to the KOH preparation (usually 10–25 percent for nails) to delay dehydration, crystallization of the KOH, and concomitant degradation of the fungal structures (Rebell and Taplin 1974; Weitzman and

Summerbell 1995). The slides are examined under the microscope with reduced light by lowering the condenser and adjusting the condenser's diaphragm.

Fungal hyphae must be differentiated from a variety of hypha-like artifacts such as cotton wool and synthetic fibers and from the so-called 'mosaic fungus.' This last artifact, which is more difficult to distinguish, consists of cholesterol crystals deposited around the periphery of the epidermal cells. It can be recognized by its regularity of outline, abrupt changes in width, presence of re-entrant angles in the flat crystalline structures, and lack of internal organelles (Rippon 1988). In experienced hands, the KOH preparation is one of the most useful and inexpensive diagnostic procedures in medical mycology.

CALCOFLUOR WHITE

Calcofluor white is a whitening agent used in the textile and paper industry. It binds to chitin and cellulose in fungus cell walls and fluoresces on excitation by long-wave ultraviolet rays or short-wave visible light. The use of this technique requires a fluorescent microscope with a proper ultraviolet source and a filter combination that may be too expensive for many laboratories. This method has, however, the advantage of allowing relatively easy fungal detection with reduced search time and a lower degree of technical experience (Elder and Roberts 1986). Calcofluor white can be combined with KOH for rapid clearance of the specimens. Fungal hyphae must be differentiated from textile fibers that may also fluoresce. Several publications are available describing calcofluor white application and technique (Hageage and Harrington 1984; Elder and Roberts 1986; Aslanzadeh and Roberts 1991; Harrington and Hageage 1991). Although background elements may also fluoresce, the fungal components are generally brighter and readily recognizable.

PERIODIC ACID–SCHIFF (PAS) STAIN

The PAS stain is based on the reaction of fungal cell wall polysaccharide with the PAS reagents, resulting in the fungus developing a red-violet fuchsin color in the tissue. This stain has also been applied successfully to histopathological analysis of nail plate clippings where onychomycosis was suspected (Suarez et al. 1991). One must be aware that a negative direct microscopic examination does not rule out the existence of a fungal infection.

Culture

ISOLATION MEDIA

Culture is a necessary adjunct to direct microscopic examination. It is only by culture that we can achieve definitive identification of the etiologic agent. In many instances, the choice of therapy may depend upon the specific identification of the invasive mold. This is especially important in nail and skin infections caused by nondermatophytic filamentous fungi, which are often resistant to the therapy used for dermatophyte infections (Summerbell et al. 1989).

A primary medium for the isolation of dermatophytes should be selective against bacteria and non-dermatophyte saprobic molds. Two such media that are commonly used are Sabouraud glucose agar (SGA) (a glucose peptone agar) with cycloheximide and chloramphenicol, and the dermatophyte test medium (DTM). Both media are available commercially and contain cycloheximide to inhibit saprobic contaminating molds. DTM incorporates gentamicin and chloramphenicol to inhibit bacteria and a phenol red indicator that changes color from yellow to red when the medium becomes alkaline as the result of growth of dermatophytes (Rebell and Taplin 1974). However, nonpathogenic fungi can also turn this medium red (Merz et al. 1970). Some systemically infecting fungi such as *Coccidioides immitis* (Salkin 1973) can resemble a dermatophyte on this medium, causing a serious biohazard situation. In addition, some dermatophytes, for example, *M. canis*, may give a false-negative reaction (Moriello and Deboer 1991). Therefore, DTM is a good screening medium for the isolation of dermatophytes but not a specific indicator of a dermatophyte. It also has the disadvantage of not allowing visualization of pigmentation on the reverse of the colony – a characteristic often used in identification. Clearly, for maximum yield of pathogenic fungi, it is important also to use a medium devoid of cycloheximide, especially for the culture of specimens from nails, soles and palms where nondermatophytic fungi sensitive to cycloheximide may be the causal agents (Summerbell et al. 1989; Weitzman and Kane 1991; Kane et al. 1997).

Nondermatophytic fungal elements may often resemble dermatophyte elements in direct microscopic examination, and hence culture on cycloheximide-free medium is necessary for their isolation and identification. The simple, commercially available, but effective, cycloheximide-free medium is SGA. If used without antibacterial agents, it allows bacterial growth from skin and nail specimens, and this may prevent the outgrowth of the fungi causing the infection. If used with chloramphenicol or with combined chloramphenicol and gentamicin, it can be very effective in the isolation of causal nondermatophytes; incidental contaminating fungi, of course, will also grow, and dermatophytes may do so as well. Other media containing antibacterial agents and lacking cycloheximide are inhibitory mold agar (BBL Microbiology Systems, Cockeysville, MD, USA) and Littman oxgall agar (Difco Laboratories Detroit, MI, USA) with added gentamicin and chloramphenicol (Summerbell et al. 1989; Weitzman and Kane 1991). Most of these media are available commercially. Specialized isolation media include: a Casamino acids/erythritol/albumin medium for the isolation of dermatophytes from lesions heavily contaminated with bacteria or with cycloheximide-tolerant *Candida albicans* (Fischer and Kane 1974); and a bromcresol purple (BCP)/casein yeast extract agar designed for the isolation and recognition of *T. verrucosum* in specimens from rural areas (Kane and Smitka 1978). The addition of 0.1 percent yeast extract or thiamine to SGA may increase the isolation rate of dermatophytes, especially of *T. verrucosum* (Carmichael and Kraus 1959; Aly 1994).

Identification

Identification of a dermatophyte is based on its gross colonial morphology on SGA and on its microscopic morphology. These criteria, however, may not be sufficient, especially for *Trichophyton* spp. that resemble each other or are variable in appearance or atypical. In addition, some isolates do not sporulate readily and special sporulation media that may not be readily available are necessary to obtain sporulation. Thus, additional physiological and biochemical tests may be necessary in conjunction with gross and microscopic morphology for species identification.

COLONY CHARACTERISTICS

Gross colony characters observed on SGA include:

- color of the surface, i.e. of the aerial mycelium
- color of the reverse or obverse faces of the colony
- production of a diffusible pigment
- texture of the surface (glabrous or waxy, powdery, granular, suede-like, velvety, downy, or fluffy)
- topography (flat, cerebriform, crateriform)
- growth rate.

These features are best noted when cultures are grown on Petri dishes rather than in tubes.

MICROSCOPIC MORPHOLOGY

Microscopic morphology may be studied in teased mounts, slide cultures (Riddell 1950; Larone 1995), or pressure-sensitive tape preparations (Rebell and Taplin 1974; St Germain and Summerbell 1996); they are examined for micro- and macroconidia and the other structures described in Tables 13.4 and 13.5. These preparations may be mounted in lactophenol cotton blue (Poirrer's blue) or lactophenol aniline blue (St Germain and Summerbell 1996). Specialized media that may be necessary to stimulate sporulation include corn meal, potato flake, potato glucose, Sabouraud's glucose + 3–5 percent NaCl (Kane and Fischer 1975), lactrimel (Kaminski 1985) and autoclaved polished rice grains (McGinnis 1980). The formulas for these media are found in several textbooks (McGinnis 1980; Rippon 1988; Kwon-Chung and Bennett 1992; de Hoog et al. 2000).

PHYSIOLOGICAL TESTS

In vitro hair perforation test

This test (Ajello and Georg 1957; Padhye et al. 1980; Clayton and Midgley 1989) originally devised to distinguish atypical isolates of *T. mentagrophytes* and *T. rubrum* (Ajello and Georg 1957), may also be used to distinguish *M. equinum* from *M. canis* (Padhye et al. 1980). *T. mentagrophytes* and *M. canis* perforate hair whereas *T. rubrum* and *M. equinum* do not. This test is also helpful in identifying other dermatophyte species (Padhye et al. 1980).

Production of urease or urea hydrolysis

The ability to hydrolyze urea (Philpot 1967; Clayton and Midgley 1989), in either an agar or a broth medium, aids in the distinction of *T. rubrum* (urease-negative) and *T. mentagrophytes* (urea-positive). The dermatophyte *Trichophyton raubitschekii*, considered by some mycologists to be a variant of *T. rubrum* (Kwon-Chung and Bennett 1992; Summerbell et al. 1999; Gräser et al. 2000a, c) or as a distinct species by others, is urease-positive (Kane et al. 1981). *T. megninii* [considered by some (Gräser et al. 1999a, 2000c) to be synonymous with *T. rubrum*] is urease positive in a urea-indole broth (Sequeira et al. 1991), but may be negative on Christensen's urea medium, whereas *T. rubrum*, apart from the relatively uncommon afro-asiatic isolates sometimes separated as *T. raubitschekii*, is urease-negative in both media (Rosenthal and Sokolsky 1965; Sequeira et al. 1991). Both Christensen's urea agar and broth media may be used. A broth medium is, however, preferred over the agar medium because it is more sensitive (Kane and Fischer 1971). The test is considered negative if there is no color change from straw to reddish purple within 7 days at 23 30°C. Cultures to be tested must be pure, because the presence of bacteria can produce a false-positive reaction. When in doubt about the purity of a culture, incubate the test culture on a blood agar medium to detect the presence of bacteria, before carrying out the urease test.

Nutritional requirements

A series of vitamin and amino acid test agars that is available commercially as trichophyton agars 1–7 (Difco, Remel, Lenexa, KA, USA) is used to distinguish some *Trichophyton* spp. by demonstrating their requirements for special growth factors (Georg and Camp 1957; Clayton and Midgley 1989). Trichophyton agar no. 1 is a vitamin-free casein basal medium to which various growth factors have been added, namely, inositol for Trichophyton agar no. 2, inositol and thiamine for no. 3, thiamine for no. 4, and nicotinic acid for no. 5. Ammonium nitrate basal medium makes up Trichophyton no. 6, and the same medium with added histidine makes up medium no. 7. The special nutritional requirements are indicated in descriptions of the *Trichophyton* spp. in Table 13.5. Care must be taken to avoid vitamin and amino acid carryover from the transferring agar and to avoid use of large inocula.

Additional useful tests for identification may include examination of the type of growth and change of indicator color occurring on bromcresol milk solids glucose agar (Padhye et al. 1973; Kane et al. 1987; Summerbell et al. 1988; Weitzman et al. 1995), a test for growth on autoclaved polished rice grains, and tests for temperature tolerance and high-temperature enhancement of growth (Clayton and Midgley 1989; Weitzman et al. 1995). Details regarding the preparation of these media, performance of these tests, and their results are discussed by Kane et al. 1997, Clayton and Midgley 1989, Ellis et al. 1992, St Germain and Summerbell 1996, and Weitzman et al. 1995.

Histopathology

Dermatophyte infections are generally restricted to the horny epidermal layers of the skin. A toxic reaction in the epidermis is often the first response to the presence of the fungus in the stratum corneum. It presents as an eczemiform pattern or as a nonspecific, subacute, chronic dermatitis. If the biopsy tissue is from a dry scaly area, hyperkeratosis and parakeratosis are seen. Special fungal stains demonstrate mycelium in the horny layer of the skin. In tinea capitis, there is no specific histological picture for dermatophytic infections. Fungal hyphal elements are seen sparsely in the stratum corneum penetrating through the squames. Hyphae extend down into the hair shaft and penetrate hair lying parallel to them. Hyphal tips growing down within the shaft reach to the edge of living keratinizing cells and form 'Adamson's fringe.' The overall pattern in ringworm of the scalp is that of a subacute or chronic dermatitis. The newly keratinized material of the growing hair extending through the aperture of the follicle carries, either within it (endothrix) or surrounding the cuticular surface (ectothrix), hyphal elements of the invading fungus which break up into arthroconidia. In tinea favosa, mycelium is present in the horny layer of the scalp, within and around the hairs, and in the scutulum. The scutula consist of mycelial masses, scales, sebum, and other debris, cementing together to form a cup-shaped crust (Rippon 1988; Weitzman and Summerbell 1995). Development of a hypersensitive kerion reaction on the scalp is accompanied by infiltration of lymphocytes, plasma cells, eosinophils, and neutrophils into the dermis (Kwon-Chung and Bennett 1992).

In the most common, distal-lateral subungual type of tinea unguium, fungi are readily seen in infected nail sections stained with PAS. Filamentous hyphae and arthroconidia are generally aligned horizontally between lamellae of the nail and restricted to the lowermost portion of the nail. There is little or no inflammation in the underlying tissue. The etiologic agent is usually *T. rubrum*, though *E. floccosum* can also be involved. By contrast, in superficial white onychomycosis, infection is restricted to the surface of the nail and is characterized by irregular hyphae that are flattened with frond-like branches that spread out between keratin lamellae. The etiologic agent is generally *T. interdigitale*, but *Fusarium oxysporum*, *Acremonium* spp., and *Scopulariopsis* spp. are among numerous other organisms that may also be involved.

A rare but distinctive clinical manifestation of dermatophytosis is invasion of the subcutaneous tissue with formation of yellowish-white pseudogranules. These structures measure 80–500 µm in diameter and consist of septate hyphae, frequently vesiculate, surrounded by a thick, radiating coat of Splendore–Hoeppli eosinophilic material. Most of the causal agents are *Microsporum* spp. or *T. rubrum* and patients are often north Africans, Australian aboriginals, and African-Americans (Chandler et al. 1980; Kwon-Chung and Bennett 1992).

PREVENTION AND CONTROL

The prevention and control of dermatophytic infection depend on the area of the body involved, the causal agent, and the source of infection. In connection with scalp ringworm caused by *M. audouinii*, all contacts with infected individuals should be carefully examined using Wood's light for fluorescent hair. In *M. canis* infections, likely pet animal sources should similarly be examined along with other persons who may have been in contact with them. In chronic care and scholastic residential institutions where *M. canis* may spread widely from a single index patient who has contacted an infected animal, all inmates (e.g. patients, students) and staff should be examined. In case of nonfluorescent tinea capitis caused by *T. tonsurans* and *T. violaceum*, scalp examination should be done more carefully for spotty alopecia and scaly lesions. Hair from suspicious areas should be cultured regularly. The hairbrush method (Mackenzie 1963) may be helpful in collecting scalp hair from children. In populations and geographical areas with an ongoing tinea capitis problem, routine scalp examinations of young school children should be done on a regular basis to avoid large-scale outbreaks. Good hygiene and avoidance of sharing headgear, combs, and hairbrushes should be stressed to students. Barbershop instruments should be disinfected after use. All children or adults infected with ringworm of the scalp should be treated promptly to prevent further transmission of their infections.

Nosocomial spread of dermatophytosis, although rare, has been reported in recent years (Mossovitch et al. 1986; Kane et al. 1988; Shah et al. 1988; Arnow et al.

1991; Snider et al. 1993). In one such outbreak, caused by *T. tonsurans*, the infection was transmitted from an infected child to hospital personnel (Arnow et al. 1991). Two other outbreaks resulted from transmission of *M. canis* to neonates by nursing personnel (Mossovitch et al. 1986; Snider et al. 1993). Such nosocomial outbreaks should be investigated promptly to avoid further dissemination. Involved personnel and direct contacts should be screened using Wood's light, and scrapings from suspected areas of skin and scalp should be cultured. Until the source of infection in a nursery is identified and treated, protective clothing should be worn by healthcare workers to avoid direct contact.

Tinea corporis and tinea cruris caused by anthropophilic species can be transmitted by contaminated clothing, bedding, towels, etc., and these should be washed and disinfected before being used by other people. Individuals with tinea infection of the glabrous skin should also avoid contact sports such as wrestling. A number of recent dermatophytic outbreaks among high school and college wrestlers have been reported in recent years (Stiller et al. 1992; Beller and Gessner 1994).

When zoophilic species such as *M. canis*, *T. mentagrophytes*, and *T. verrucosum* are detected as etiologic agents, infected animal reservoir sources should be sought out and treated. Infections caused by *T. verrucosum* are difficult to detect with the use of Wood's light because infected hairs do not fluoresce, and the shed, infected scales have been shown to survive for years on fomites. Good hygiene and sanitation, and use of fungicidal sprays and washes, are effective methods for prevention of such infections (Weitzman and Summerbell 1995).

The spread of tinea pedis infections may be controlled by good foot hygiene, use of antifungal powders, avoiding excessive moisture, and occlusion of feet by wearing sandals or ventilated footwear, and avoiding walking barefooted in swimming pool areas, community baths, and shower areas. Prompt treatment of individuals with tinea pedis should be undertaken to avoid spread of infection.

REFERENCES

Ajello, L. 1960. Geographic distribution and prevalence of the dermatophytes. *Ann N Y Acad Sci*, **89**, 30–8.

Ajello, L. 1962. Present day concepts in the dermatophytes. *Mycopathol Mycol Appl*, **17**, 315–24.

Ajello, L. 1974. Natural history of the dermatophytes and related fungi. *Mycopathol Mycol Appl*, **53**, 93–110.

Ajello, L. and Georg, L.K. 1957. In vitro hair cultures for differentiating between atypical isolates of *Trichophyton mentagrophytes* and *Trichophyton rubrum*. *Mycopathol Mycol Appl*, **8**, 3–17.

Ajello, L. and Getz, M.E. 1954. Recovery of dermatophytes from shoes and shower stall. *J Invest Dermatol*, **22**, 17–21.

Ajello, L., Varsavsky, E., et al. 1964. The natural history of *Microsporum nanum*. *Mycologia*, **36**, 873–84.

Ajello, L., Kaplan, W. and Chandler, F.W. 1980. Dermatophyte mycetomas: fact or fiction? *Fifth International Conference on the Mycoses – Superficial, Cutaneous and Subcutaneous Infections, Pan American Health Organization Scientific Publication no. 396.* Washington, DC: PAHO, 135–40.

Alsop, J. and Prior, A.P. 1961. Ringworm infection in a cucumber greenhouse. *Br Med J*, **1**, 1081–3.

Aly, R. 1994. Culture media for growing dermatophytes. *J Am Acad Dermatol*, **31**, S107–8.

Arnow, P.M., Houchins, S.G. and Pugliase, G. 1991. An outbreak of tinea corporis in hospital personnel caused by a patient with *Trichophyton tonsurans* infection. *Pediatr Infect Dis J*, **10**, 355–9.

Artis, W.M., Odle, B.M. and Jones, H.E. 1981. Griseofulvin-resistant dermatophytosis correlates with in vitro resistance. *Arch Dermatol*, **117**, 16–19.

Aslanzadeh, J. and Roberts, G.D. 1991. Direct microscopic examination of clinical specimens for the laboratory diagnosis of fungal infections. *Clin Microbiol Newsl*, **13**, 185–92.

Babel, D.E. 1990. Dermatophytosis of the scalp: incidence, immune response, and epidemiology. *Mycopathologia*, **109**, 69–73.

Bassi, A. 1935–36. *Del male del segno calcinacio o moscardino, malattia che affligge: i bachi da seta e sul modo di liberarne le bagattaje anche le peu infeste.* Lodi: Tipografia Orcesi.

Beller, M. and Gessner, B.D. 1994. An outbreak of tinea corporis gladiatorum on a high school wrestling team. *J Am Acad Dermatol*, **31**, 197–201.

Benham, R.W. 1953. Nutritional studies of the dermatophytes: effect on growth and morphology, with special reference to the production of macroconidia. *Trans N Y Acad Sci*, **15**, 102–6.

Bickers, D.R. 1994. Antifungal therapy: potential interactions with other classes of drugs. *J Am Acad Dermatol*, **31**, S87–90.

Blank, F. 1957. Favus in mice. *Can J Microbiol*, **3**, 885–96.

Blank, F., Leclerc, G. and Telner, P. 1961. Clinical manifestations of mouse favus in man. *Arch Dermatol*, **83**, 587–90.

Blank, H., Taplin, D. and Zaias, N. 1969. Cutaneous *Trichophyton mentagrophytes* infections in Vietnam. *Arch Dermatol*, **99**, 135–44.

Brodell, R.T. and Elewski, B.E. 1995. Clinical pearl: systemic antifungal drugs and drug interactions. *J Am Acad Dermatol*, **33**, 259–60.

Carmichael, J.W. and Kraus, H.J. 1959. The cattle ringworm fungus, *Trichophyton verrucosum* in Alberta. *Alberta Med Bull*, **24**, 137–9.

Chandler, F.W. and Watts, J.C. 1987. *Pathologic diagnosis of fungal infections.* Chicago, IL: American Society of Clinical Pathologists.

Chandler, F.W., Kaplan, W. and Ajello, L. 1980. *A colour atlas and textbook of the histopathology of mycotic diseases.* London: Wolfe Medical Publications.

Chemel, L. 1980. Zoophilic dermatophytes and infections in man. *Med Mycol*, **8**, Suppl, 61–6.

Chemel, L. and Buchvald, J. 1970. Ecology and transmission of *Microsporum gypseum* from soil to man. *Sabouraudia*, **8**, 149–56.

Clayton, Y.M. and Midgley, G. 1989. Identification of agents of superficial mycoses. In: Evans, E.G.V. and Richardson, M.D. (eds), *Medical mycology – a practical approach.* Oxford: IRL Press at Oxford University Press, 65, 95.

Contet-Audonneau, N. and Percebois, G. 1986. *Microsporum persicolor* isolement du sol. *Bull Soc Fr Mycol*, **15**, 193–6.

Cruickshank, C.N.D., Trotter, M.D. and Wood, S.R. 1960. Studies on trichophytin sensitivity. *J Invest Dermatol*, **35**, 219–23.

Dahl, M.V. 1993. Suppression of immunity and inflammation by products produced by dermatophytes. *J Am Acad Dermatol*, **28**, S19–28.

Dawson, C.O. and Gentles, J.C. 1961. The perfect states of *Keratinomyces ajelloi* Vanbreuseghem, *Trichophyton terrestre* Durie and Frey and *Microsporum nanum* Fuentes. *Sabouraudia*, **1**, 49–57.

De Vroey, C. 1985. Epidemiology of ringworm (dermatophytosis). *Semin Dermatol*, **4**, 185–200.

Degreef, H.J. and DeDoncker, P.R.G. 1994. Current therapy of dermatophytosis. *J Am Acad Dermatol*, **31**, S25–30.

de Hoog, G.S, Guarro, J., et al. 2000. *Atlas of clinical fungi*, 2nd edition. Reus: Centraalbureau voor Schimmelcultures, Baarn/Universitat Rovira i Virgili.

Del Palacio, A., Lopez, S., et al. 1991. A randomized comparative study: amorolfine (cream 0.125, 0.25 and 0.5) in dermatomycoses. *J Dermatol Treat*, **1**, 299–303.

Del Palacio, A., Van Cutsem, J., et al. 1993. Estudio doble ciego randomizado comparative con itraconazol y griseofulvina en tinea corporis y tinea cruris. *Rev Iberoam Micol*, **10**, 51–8.

Del Palacio, A., Cuetara, M.S. and Castejon, A. 1994. Avances en el tratamiento de onicomicosis y dermatofitosis. *Medicine*, 6th edition. 24–36.

Del Palacio, A., Garau, M., et al. 1999. Tratamiento antifungico: ultimos avances en dermatologia. *Rev Iberoam Micol*, **16**, 86–91.

Del Palacio, A., Garau, M., et al. 2000. Trends in the treatment of dermatophytosis. In: Kushwaha, R.K.S. and Guarro, J. (eds), *Biology of dermatophytes and other keratinophilic fungi*. Bilbao, Spain: Revista Iberoamericana Micologia, 148–58.

Del Palacio-Hernanz, A., Lopez-Gomez, S., et al. 1989. A clinical double-blind trial comparing amorolfine cream 0.5% (RO-14-4767) with bifonazole cream 1% in the treatment of dermatomycosis. *Clin Exp Dermatol*, **14**, 141–4.

Del Palacio-Hernanz, A., Lopez-Gomez, S., et al. 1990. A comparative double-blind study of terbinafine (Lamisil) and griseofulvin in tinea corporis and tinea cruris. *Clin Exp Dermatol*, **15**, 210–16.

Einarson, T.R., Arikian, S.R. and Shear, N.H. 1994. Cost-effectiveness analysis for onychomycosis therapy in Canada from a government perspective. *Br J Dermatol*, **130**, Suppl 43, 32–4.

Elder, B.L. and Roberts, G.D. 1986. Rapid methods for the diagnosis of fungal infections. *Lab Med*, **17**, 591–6.

Elewski, B.E. 1993. Mechanisms of action of systemic antifungal agents. *J Am Acad Dermatol*, **28**, S28–34.

Elewski, B.E. 1994. Tinea capitis: itraconazole in *Trichophyton tonsurans* infection. *J Am Acad Dermatol*, **31**, S65–7.

Elewski, B.E., Rinaldi, M.G. and Weitzman, I. 1995. *Diagnosis and treatment of onychomycosis – a clinician's handbook*. Titusville, FL: SynerMed.

Ellis, D., Davis, S., et al. 1992. *Descriptions of medical QAP fungi*. Underdale, Australia: Gillingham Printers.

Emmons, C.W. 1934. Dermatophytes: natural groupings based on the form of the spores and accessory organs. *Arch Dermatol Syphilol*, **30**, 337–62.

English, M.P., Kapica, L. and Maciejewska, J. 1978. On the occurrence of *Microsporum persicolor* in Montreal, Canada. *Mycopathologia*, **64**, 35–7.

Faergemann, J., Mork, N.J. and Haglund, A. 1997. A multicentre (double-blind) comparative study to assess the safety and efficacy of fluconazole and griseofulvin in the treatment of tinea corporis and tinea cruris. *Br J Dermatol*, **136**, 575–7.

Feurman, E., Alteras, I., et al. 1975. Saprophytic occurrence of *Trichophyton mentagrophytes* and *Microsporum gypseum* in the coats of healthy laboratory animals. *Mycopathologia*, **55**, 13–15.

Fischer, J.B. and Kane, J. 1974. The laboratory diagnosis of dermatophytosis complicated by *Candida albicans*. *Can J Microbiol*, **20**, 167–82.

Fragner, P. 1987. Microscopic diagnosis of onychomycosis. *Ceska Mykol*, **41**, 153–61.

Frieden, I.J. and Howard, R. 1994. Tinea capitis: epidemiology, diagnosis, treatment and control. *J Am Acad Dermatol*, **31**, S42–6.

Friedman, L., Derbes, V.L., et al. 1960. The isolation of dermatophytes from the air. *J Invest Dermatol*, **35**, 3–5.

Fujuhiro, M. 1994. Study of dermatophytes isolated in a hospital environment and nosocomial infection cases of tinea pedis. *Jpn J Med Mycol*, **35**, 25–32.

Gan, V.N., Petruska, M. and Ginsburg, C.M. 1987. Epidemiology and treatment of tinea capitis: ketoconazole vs. griseofulvin. *Pediatr Infect Dis J*, **6**, 46–9.

Gentles, J.C. 1957. Athlete's foot fungus on the floors of community bathing places. *Br Med J*, **1**, 746–8.

Gentles, J.C. 1958. Experimental ringworm in guinea pigs: oral treatment with griseofulvin. *Nature (Lond)*, **182**, 476.

Gentles, J.C. and Holmes, J.G. 1957. Foot ringworm in coal miners. *Br J Ind Med*, **14**, 22–9.

Georg, L.K. 1952. Cultural and nutritional studies of *Trichophyton gallinae* and *Trichophyton megninii*. *Mycologia*, **44**, 470–92.

Georg, L.K. 1956. The role of animals as vectors of human fungus diseases. *Trans N Y Acad Sci Ser II*, **18**, 639–47.

Georg, L.K. 1960. Epidemiology of the dermatophytes: sources of infection, modes of transmission and epidemicity. *Ann N Y Acad Sci*, **89**, 69–77.

Georg, L.K. and Camp, L.B. 1957. Routine nutritional tests for the identification of dermatophytes. *J Bacteriol*, **74**, 113–21.

Georg, L.K., Kaplan, W. and Camp, L.B. 1957. Equine ringworm with special reference to *Trichophyton equinum*. *Am J Vet Res*, **18**, 798–810.

Gip, L. 1967. Estimation of incidence of dermatophytes on floor areas after barefoot walking with washed and unwashed feet. *Acta Dermatol Venereol*, **47**, 89–93.

Gordon, M.A. 1953. The occurrence of the dermatophyte *Microsporum gypseum* as a saprophyte in soil. *J Invest Dermatol*, **20**, 201–6.

Gräser, Y., El Fari, M., et al. 1999a. Phylogeny and taxonomy of the family *Arthrodermataceae* (dermatophytes) using sequence analysis of the ribosomal 17S region. *Med Mycol*, **37**, 105–14.

Gräser, Y., Kühnisch, J. and Presber, W. 1999b. Molecular markers reveal exclusively clonal reproduction in *Trichophyton rubrum*. *J Clin Microbiol*, **37**, 3713–17.

Gräser, Y., Kuijpers, A.F.A., et al. 1999c. Molecular taxonomy of Trichophyton mentagrophytes and T. tonsurans. *Med Mycol*, **37**, 315–30.

Gräser, Y., de Hoog, G.S. and Kuijpers, A.F.A. 2000a. Recent advances in the molecular taxonomy of dermatophytes. In: Kushwaha, R.K.S. and Guarro, J. (eds), *Biology of dermatophytes and other keratinophilic fungi*. Bilbao, Spain: Revista Iberoamericana de Micologia, 17–21.

Gräser, Y., Kuijpers, A.F.A., et al. 2000b. Molecular and conventional taxonomy of the *Microsporum canis* complex. *Med Mycol*, **38**, 143–53.

Gräser, Y., Kuijpers, A.F.A., et al. 2000c. Molecular taxonomy of the *Trichophyton rubrum* complex. *J Clin Microbiol*, **38**, 3329–36.

Green, F. and Balish, E. 1980. *Trichophyton mentagrophytes* dermatophytosis in germfree guinea pigs. *J Invest Dermatol*, **75**, 476–80.

Grossman, J., Baum, J., et al. 1975. The effect of aging and acute illness on delayed hypersensitivity. *J Allergy Clin Immunol*, **55**, 268–75.

Gruby, D. 1841a. Memoire sur une vegetation qui constituent la vraie teigne. *C R Acad Sci*, **13**, 72–5.

Gruby, D. 1841b. Sur les mycodermes qui constituent la teigne faveuse. *C R Acad Sci*, **13**, 309–12.

Gruby, D. 1843. Recherches sur la nature, le siège et le développement du porrigo décalvans ou phytoalopécie. *C R Acad Sci*, **17**, 301–2.

Gruby, D. 1844. Recherches sur les cryptogams qui constituent la maladie contagieuse du cuir chevelu decrite sous le nom de teigne tondante (Mahon), *Herpes tonsurans* (Cazenave). *C R Acad Sci*, **18**, 583–5.

Gugnani, H.C., Shrivastav, J.B. and Gupta, N.P. 1967. Occurrence of *Arthroderma simii* in soil and on hair of small mammals. *Sabouraudia*, **6**, 77–80.

Gupta, A.K. and Adam, P. 1998. Terbinafine pulse therapy is effective in tinea capitis. *Pediatr Dermatol*, **15**, 56–8.

Gupta, A.K. and Summerbell, R.C. 2000. Tinea capitis. *Med Mycol*, **38**, 255–87.

Gupta, A.K., Sauder, D.N. and Shear, N.H. 1994a. Antifungal agents: an overview, Part I. *J Am Acad Dermatol*, **30**, 677–98.

Gupta, A.K., Sauder, D.N. and Shear, N.H. 1994b. Antifungal agents: an overview, Part II. *J Am Acad Dermatol*, **30**, 911–33.

Gupta, A.K., Alexis, M.E., et al. 1997. Itraconazole pulse therapy is effective in the treatment of tinea capitis in children: an open multicentre study. *Br J Dermatol*, **137**, 251–4.

Gupta, A.K., Adam, P. and DeDoncker, P. 1998. Itraconazole pulse therapy for tinea capitis: a novel treatment schedule. *Pediatr Dermatol*, **15**, 225–8.

Hageage, G.J. and Harrington, B.J. 1984. The use of calcofluor white in clinical mycology. *Lab Med*, **15**, 109–12.

Haroon, T.S., Hussain, I., et al. 1992. A randomized double-blind comparative study of terbinafine vs. griseofulvin in tinea capitis. *J Dermatol Treat*, **3**, Suppl, 25–7.

Harrington, B.J. and Hageage, J.R. 1991. Calcofluor white: tips for improving its use. *Clin Microbiol Newsl*, **13**, 3–5.

Havu, V., Brandt, H., et al. 1999. Continuous and intermittent itraconazole dosing schedules for the treatment of onychomycosis: a pharmacokinetic comparison. *Br J Dermatol*, **140**, 96–101.

Hay, R.J. 1992. Genetic susceptibility to dermatophytosis. *Eur J Epidemiol*, **8**, 346–9.

Hay, R.J. 1993. Onychomycosis, agents of choice. *Dermatol Clin*, **11**, 161–9.

Hay, R.J., Reid, S., et al. 1984. Endemic tinea imbricata – a study of Goodenough island, Papua New Guinea. *Trans R Soc Trop Med Hyg*, **78**, 246–51.

Hay, R.J., Baran, R., et al. 1994. Diseases of the nails and their management, 2nd edition. London: Blackwell Scientific, 97–121.

Hebert, A.A., Head, E.S. and MacDonald, E.M. 1985. Tinea capitis caused by *Trichophyton tonsurans*. *Pediatr Dermatol*, **2**, 219–23.

Heikkila, H. and Stubb, S. 1997. Long term results of patients with onychomycosis treated with itraconazole. *Acta Dermatol Venereol*, **77**, 70–1.

Hejtmánek, M. and Hejtmánkova, N. 1989. Hybridization and sexual stimulation in *Trichophyton mentagrophytes*. *Folia Microbiol*, **34**, 77–9.

Hironaga, M., Fujigaki, T. and Watanabe, S. 1981. *Trichophyton mentagrophytes* skin infections in laboratory animals as cause of zoonosis. *Mycopathologia*, **73**, 101–4.

Howard, R. and Frieden, I.J. 1995. Tinea capitis: new perspectives on an old disease. *Semin Dermatol*, **14**, 2–8.

Jacobs, P.H. 1990. Treatment of fungal skin infections: state of the art. *J Am Acad Dermatol*, **23**, 549–51.

Jones, H.E. 1993. Immune response and host resistance of humans to dermatophyte infections. *J Am Acad Dermatol*, **28**, S12–18.

Kaminski, G.W. 1985. The routine use of modified Borelli's lactritmel (MBLA). *Mycopathologia*, **91**, 57–9.

Kane, J. and Fischer, J.B. 1971. The differentiation of *Trichophyton rubrum* and *T. mentagrophytes* by use of Christensen's urea broth. *Can J Microbiol*, **17**, 911–13.

Kane, J. and Fischer, J.B. 1975. The influence of sodium chloride on the growth and morphology of dermatophytes and some other keratinophilic fungi. *Can J Microbiol*, **21**, 742–9.

Kane, J. and Smitka, C.M. 1978. Early detection and identification of *Trichophyton verrucosum*. *J Clin Microbiol*, **8**, 740–7.

Kane, J., Salkin, I.F., et al. 1981. *Trichophyton raubitschekii* sp. nov.. *Mycotaxon*, **13**, 259–66.

Kane, J., Sigler, L. and Summerbell, R.C. 1987. Improved procedures for differentiating Microsporum persicolor from *Trichophyton mentagrophytes*. *J Clin Microbiol*, **25**, 2449–52.

Kane, J., Leavitt, E., et al. 1988. An outbreak of *Trichophyton tonsurans* dermatophytosis in a chronic care institution for the elderly. *Eur J Epidemiol*, **4**, 144–9.

Kane, J., Summerbell, R., et al. 1997. *Laboratory handbook of dermatophytes*. Belmont, CA: Star Publishing Company.

Kano, R., Nakamura, Y., et al. 1999. Phylogenetic relation of *Epidermophyton floccosum* to the species of *Microsporum* and *Trichophyton* in chitin synthase 1 (*CHS1*) gene sequences. *Mycopathologia*, **146**, 111–13.

Kano, R., Nakamura, Y., et al. 2002. Chitin synthase 1 and 2 genes of dermatophytes. *Stud Mycol*, **47**, 49–55.

Kaplan, W. and Gump, R.H. 1958. Ringworm in a dog caused by *Trichophyton rubrum*. *Vet Med*, **53**, 139–42.

Kaplan, W., Hopping, J.L. and Georg, L.K. 1957. Ringworm in horses caused by the dermatophyte *Microsporum gypseum*. *J Am Vet Med Assoc*, **131**, 329–32.

King, R.D., Cunico, R.L., et al. 1978. The effect of occlusion on carbon dioxide emission from human skin. *Acta Dermatol Venereol*, **58**, 135–8.

Krebs, S., Greenberg, J. and Jesrani, K. 1977. Temporal correlation of lymphocyte blastogenesis, skin test responses, and erythema during dermatophyte infection. *Clin Exp Immunol*, **27**, 526–30.

Kwon-Chung, K.J. and Bennett, J.E. 1992. *Medical mycology*. Philadelphia: Lea & Febiger.

Larone, D.H. 1995. *Medically important fungi*, 3rd edition. Washington, DC: American Society for Microbiology.

Legendre, R. and Escola-Macre, J. 1990. Itraconazole in the treatment of tinea capitis. *J Am Acad Dermatol*, **23**, 559–60.

Lepper, A.W.D. 1974. Experimental bovine *Trichophyton verrucosum* infection. The cellular responses in primary lesions of the skin resulting from surface or intradermal inoculation. *Res Vet Sci*, **16**, 287–98.

Lepper, A.W.D. and Anger, H.S. 1976. Experimental bovine *Trichophyton verrucosum* infection. Comparison of the rate of epidermal cell proliferation and keratinization in non-infected and reinoculated cattle. *Res Vet Sci*, **20**, 117–21.

Lesher, J.L., Babel, D.E., et al. 1997. Butenafine 1% cream in the treatment of tinea cruris: a multicenter vehicle controlled, double-blind trial. *J Am Acad Dermatol*, **36**, S20–24.

Lundell, E. 1974. *Epidermophyton floccosum* endemic in the sauna of a trade school hostel. *Mykosen*, **17**, 219–20.

Lunder, M. and Lunder, M. 1992. Is *Microsporum canis* infection about to become a serious dermatological problem? *Dermatology*, **184**, 87–9.

Mackenzie, D.W.R. 1961. The extra-human occurrence of *Trichophyton tonsurans* var. *sulfureum* in a residential school. *Sabouraudia*, **1**, 58–64.

Mackenzie, D.W.R. 1963. 'Hairbrush diagnosis' in detection and eradication of non-fluorescent scalp ringworm. *Br Med J*, **2**, 363–5.

Matsumoto, T. and Ajello, L. 1987. Current taxonomic concepts pertaining to the dermatophytes and related fungi. *Int J Dermatol*, **26**, 491–9.

Matsumoto, T., Kibbler, C.C., et al. 1996. Principles and practice of clinical mycology. New York: John Wiley & Son, 103–29.

Mayr, A. 1989. Infections which humans in the household transmit to dogs and cats. *Zentbl Bakteriol Mikrobiol Hyg [B]*, **187**, 508–26.

McGinnis, M.R. 1980. *Laboratory handbook of medical mycology*. New York: Academic Press.

McPherson, E.A. 1957. The influence of physical factors on dermatophytosis in domestic animals. *Vet Rec*, **69**, 1010–13.

Merz, W.G., Berger, C.L. and Silva-Hutner, M. 1970. Media with pH indicators for the isolation of dermatophytes. *Arch Dermatol*, **102**, 545–7.

Mochizuki, T., Takada, K., et al. 1990. Taxonomy of *Trichophyton interdigitale* (*Trichophyton mentagrophytes* var. *interdigitale*) by restriction enzyme analysis of mitochondrial DNA. *J Med Vet Mycol*, **28**, 191–6.

Mochizuki, T., Watanabe, S. and Uehara, M. 1996. Genetic homogeneity of *Trichophyton mentagrophytes* var. *interdigitale* isolated from geographically distant regions. *J Med Vet Mycol*, **34**, 139–43.

Morganti, L., Bianchedi, M., et al. 1976. First European report of swine infection by *Microsporum nanum*. *Mycopathologia*, **59**, 179–82.

Moriello, K.A. and Deboer, D.J. 1991. Fungal flora of haircoat of cats with and without dermatophytosis. *J Med Vet Mycol*, **29**, 285–92.

Morris, P. and English, M.P. 1973. Transmission and course of *Trichophyton erinacei* infections in British hedgehogs. *Sabouraudia*, **11**, 42–7.

Mossovitch, M., Mossovitch, B. and Alkan, M. 1986. Nosocomial dermatophytosis caused by *Microsporum canis* in a newborn department. *Infect Control*, **7**, 593–5.

Neves, H. and Xavier, N.C. 1964. The transmission of tinea cruris. *Br J Dermatol*, **76**, 429–36.

Nolting, S.K., Sanchez-Carazo, J., et al. 1998. Oral treatment schedules for onychomycosis: a study of patient preference. *Int J Dermatol*, **37**, 454–6.

Nozickova, M., Koudelkova, V., et al. 1998. A comparison of the efficacy of oral fluconazole 150 mg/day versus 50 mg/day, in the treatment of tinea corporis, tinea cruris, tinea pedis and cutaneous candidosis. *Int J Dermatol*, **37**, 701–8.

Onsberg, P. 1978. Human infections with *Microsporum persicolor* in Denmark. *Br J Dermatol*, **99**, 531–6.

Ozegovic, L. 1980. Wild animals as reservoirs of human pathogenic dermatophytes. *Zentbl Bakteriol Parasitenkd Infekt Hyg [A]*, **Suppl 8**, 369–80.

Padhye, A.A. 1962. Isolation of *Trichophyton violaceum* from extra human source in a remand home for boys, Poona. *Ind Practnr*, **15**, 333–6.

Padhye, A.A. and Thirumalachar, M.J. 1967. Isolation of *Trichophyton simii* and *Cryptococcus neoformans* from soil in India. *Hindustan Antibiot Bull*, **9**, 155–7.

Padhye, A.A., Koblenzer, P.J., et al. 1973. *Microsporum persicolor* infection in the United States. *Arch Dermatol*, **108**, 561–2.

Padhye, A.A., Young, C.N. and Ajello, L. 1980. Hair perforation as the diagnostic criterion in the identification of *Epidermophyton*, *Microsporum* and *Trichophyton* species. *Fifth International Conference on the Mycoses – Superficial, Cutaneous and Subcutaneous Infections. Pan American Health Organization Scientific Publication no. 396.* Washington, DC: PAHO, 115–20.

Palmer, D.L. and Reed, W.P. 1974. Delayed hypersensitivity skin testing I. Response rates in a hospitalized population. *J Infect Dis*, **130**, 132–7.

Philpot, C. 1967. The differentiation of *Trichophyton mentagrophytes* from *Trichophyton rubrum* by a simple urease test. *Sabouraudia*, **5**, 189–93.

Philpot, C.N. and Bowen, R.G. 1992. Hazards from hedgehogs: two case reports with a survey of the epidemiology of hedgehog ringworm. *Clin Exp Dermatol*, **17**, 156–8.

Pier, A.C., Smith, J.M.B., et al. 1994. Animal ringworm – its etiology, public health significance and control. *J Med Vet Mycol*, **32**, Suppl 1, 133–50.

Probst, S., de Hoog, G.S. and Gräser, Y. 2002. Development of DNA markers to explore host shifts in dermatophytes. *Stud Mycol*, **47**, 57–74.

Ravine, D., Turner, K.J. and Alpers, M.P. 1980. Genetic inheritance of susceptibility to tinea imbricata. *J Med Genet*, **17**, 342–8.

Rebell, G. and Taplin, D. 1974. *Dermatophytes, their recognition and Identification.* Coral Gables, FL: University of Miami Press.

Remak, R. 1845. *Diagnostische und pathogentische unterschungen in der klinik des Herrn Geh. Raths Dr. Schoenlein auf dessen Veranlassung angestell und mit Benutzung anderweitiger Beobachtungen veroffenlicht.* Berlin: A Hirschwald.

Riddell, R.W. 1950. Permanent stained mycological preparations obtained by slide culture. *Mycologia*, **42**, 265–70.

Rippon, J.W. 1985. The changing epidemiology and emerging patterns of dermatophyte species. *Curr Topics Med Mycol*, **1**, 208–34.

Rippon, J.W. 1988. *Medical mycology. The pathogenic fungi and the pathogenic actinomycetes*, 3rd edition. Philadelphia: W.B. Saunders.

Rippon, J.W. 1992. Forty-four years of dermatophytes in the Chicago clinic (1944–1988). *Mycopathologia*, **119**, 25–8.

Roberts, D.T. 1994. Oral therapeutic agents in fungal nail disease. *J Am Acad Dermatol*, **30**, 663–7.

Rosenthal, S.A. and Sokolsky, H. 1965. Enzymatic studies with pathogenic fungi. *Dermatol Int*, **4**, 72–9.

Sabouraud, R. 1910. *Les teignes.* Paris: Masson.

St-Germain, G. and Summerbell, R.C. 1996. *Identifying filamentous fungi: a clinical laboratory handbook.* Belmont CA: Star Publishing Co.

Salkin, I.F. 1973. Dermatophyte test medium: evaluation with non-dermatophyte pathogens. *Appl Microbiol*, **26**, 134–7.

Schafer-Korting, M. 1993. Pharmacokinetic optimization of oral antifungal therapy. *Clin Pharmacokinet*, **25**, 329–41.

Schoenlein, J.L. 1839. Zur Pathogenic der Impetigines. *Arch Anat Physiol Wiss Med*, 82.

Seeliger, H.P.R. 1985. The discovery of *Achorion schoenleinii* facts and 'stories'. *Mykosen*, **28**, 161–82.

Segal, E. 1989. Vaccines for the management of dermatophyte and superficial yeast infection. *Clin Microbiol Rev*, **8**, 317–35.

Sequeira, H., Cabrita, J., et al. 1991. Contributions to our knowledge of *Trichophyton megninii. J Med Vet Mycol*, **29**, 417–18.

Serjeantson, S. and Lawrence, G. 1977. Autosomal recessive inheritance of susceptibility to tinea imbricate. *Lancet*, **1**, 13–15.

Sewell, D.L. 1995. Laboratory associated infections and biosafety. *Clin Microbiol Rev*, **8**, 389–405.

Shah, P.C., Krajden, S., et al. 1988. Tinea corporis caused by *Microsporum canis*: report of nosocomial outbreak. *Eur J Epidemiol*, **4**, 33–8.

Sherna, F., Farella, V., et al. 1993. Epidemiology of the dermatophytes in the Florence area of Italy: 1985–1990. *Mycopathologia*, **122**, 153–62.

Shimmura, Y. 1985. Isolation of dermatophytes from human cases of dermatophytosis and from house dust. *Jpn J Med Mycol*, **26**, 74–80.

Silva, M. and Benham, R.W. 1952. Nutritional studies of the dermatophytes with special reference to *Trichophyton megninii* Blanchard 1890 and *Trichophyton gallinae* (Megnin 1881) comb nov. *J Invest Dermatol*, **18**, 453–72.

Sinski, J.T. and Flouras, K. 1984. A survey of dermatophytes isolated from human patients in the United States from 1979 to 1981 with chronological listings of world-wide incidence of five dermatophytes often isolated in the United States. *Mycopathologia*, **85**, 97–120.

Sinski, J.T. and Kelley, L.M. 1991. A survey of dermatophytes from human patients in the United States from 1985 to 1987. *Mycopathologia*, **114**, 87–9.

Smith, E.B. 1993. Topical antifungal drugs in the treatment of tinea pedis, tinea cruris and tinea corporis. *J Am Acad Dermatol*, **28**, S24–8.

Snider, R., Landers, S. and Levy, M.L. 1993. The ringworm riddle: an outbreak of *Microsporum canis* in a nursery. *Pediatr Infect Dis J*, **12**, 145–8.

Stiller, M.J., Klein, W.P., et al. 1992. Tinea corporis gladiatorum: an epidemic of *Trichophyton tonsurans* in student wrestlers. *J Am Acad Dermatol*, **27**, 632–3.

Stiller, M.J., Sangueza, O.P. and Shupak, J.L. 1993. Systemic drugs in the treatment of dermatophytoses. *Int J Dermatol*, **32**, 16–21.

Stockdale, P.M. 1961. *Nannizzia incurvata* gen. nov., sp. nov., a perfect state of *Microsporum gypseum* (Bodin) Guiart et Grigorakis. *Sabouraudia*, **1**, 41–8.

Stockdale, P.M. 1968. Sexual stimulation between *Arthroderma simii* Stockd., Mackenzie and Austwick and related species. *Sabouraudia*, **6**, 176–81.

Suarez, S.M., Silvers, D.N., et al. 1991. Histologic evaluation of nail clippings for diagnosing onychomycosis. *Arch Dermatol*, **127**, 1517–19.

Summerbell, R.C. 2002. What is the evolutionary and taxonomic status of asexual lineages in the dermatophytes? *Stud Mycol*, **47**, 97–101.

Summerbell, R.C. 2003. *Trichophyton, Microsporum, Epidermophyton*, and agents of superficial mycosis. In: Murray, P.R., Baron, E.J., et al. (eds), *Manual of clinical microbiology*, 8th edition. Washington, DC: ASM Press, 1798–819.

Summerbell, R.C., Rosenthal, S.A. and Kane, J. 1988. Rapid method of differentiation of *Trichophyton rubrum, Trichophyton mentagrophytes*, and related dermatophyte species. *J Clin Microbiol*, **26**, 2279–82.

Summerbell, R.C., Kane, J. and Krajden, S. 1989. Onychomycosis, tinea pedis, and tinea manuum caused by non-dermatophytic filamentous fungi. *Mycoses*, **32**, 609–19.

Summerbell, R.C., Haugland, R.A. and Gupta, A.K. 1999. rRNA gene internal transcribed spacer 1 and 2 sequences of asexual, anthropophilic dermatophytes related to *Trichophyton rubrum. J Clin Microbiol*, **37**, 4005–11.

Swartz, H.E. and Georg, L.K. 1955. The nutrition of *Trichophyton tonsurans. Mycologia*, **47**, 475–93.

Tanaka, S., Summerbell, R.C., et al. 1992. Advances in dermatophytes and dermatophytosis. *J Med Vet Mycol*, **30**, Suppl 1, 29–39.

Tanz, R.R., Hebert, A.A. and Esterly, N.B. 1988. Treating tinea capitis: should ketoconazole replace griseofulvin? *J Pediatr*, **112**, 987–91.

Tewari, R.P. 1969. *Trichophyton simii* infections in chickens, dogs and man in India. *Mycopathol Mycol Appl*, **39**, 293–8.

Thomas, P., Korting, H.C., et al. 1994. *Microsporum canis* infection in a 5-year-old boy: transmission from the interior of a second hand car. *Mycoses*, **37**, 141–2.

Vanbreuseghem, R. 1952. Technique biologique pour l'isolement des dermatophytes du sol. *Ann Soc Belg Trop*, **32**, 173–8.

Van Doorslaer, E.K.A., Tormans, G. and Gupta, A.K. 1996. Economic evaluation of antifungal agents in the treatment of toenail onychomycosis in Germany. *Dermatol*, **193**, 239–44.

Vasilyev, O.D. and Bogomolova, T.S. 1985. Tipy sparivaniya i sovershennaya forma shtammov griba *Trichophyton mentagrophytes* (Robin) Blanchard, vydelennykh ot bol'nikh dermatofitiyami. *Mikol Fitopatol*, **19**, 309–17.

Viguie, C., Ancelle, T., et al. 1992. Epidemiological survey on *Trichophyton soudanense* tinea at school. *J Mycol Med*, **2**, 160–3.

Villars, V. and Jones, T.C. 1992. Special features of the clinical use of oral terbinafine in the treatment of fungal diseases. *Br J Dermatol*, **126**, Suppl 39, 61–9.

Wagner, D.K. and Sohnle, P.G. 1995. Cutaneous defenses against dermatophytes and yeasts. *Clin Microbiol Rev*, **8**, 317–35.

Wakimoto, A., Sei, Y., et al. 1985. Group contagion of tinea corporis caused by *Microsporum canis* among junior high school students. *Jpn J Med Mycol*, **26**, 109–11.

Watanabe, S. and Hironaga, M. 1981. Differences or similarities of the clinical lesions produced by '+' and '–' types members of the 'mentagrophytes' complex in Japan. In: Vanbreuseghem, R. and de Vroey, C. (eds), *Sexuality and pathogenicity of fungi*. Paris, New York: Masson, 83–96.

Weitzman, I. and Kane, J. 1991. Dermatophytes and agents of superficial mycoses. In: Balows, A., Hausler, E.J., et al. (eds), *Manual of clinical microbiology*, 5th edition. Washington, DC: American Society for Microbiology, 601–16.

Weitzman, I. and Padhye, A.A. 1976. Is *Arthroderma simii* the perfect state of *Trichophyton quinckeanum*? *Sabouraudia*, **14**, 65–74.

Weitzman, I. and Padhye, A.A. 1978. Mating behaviour of *Nannizzia otae* (*Microsporum canis*). *Mycopathologia*, **64**, 17–22.

Weitzman, I. and Summerbell, R.C. 1995. The dermatophytes. *Clin Microbiol Rev*, **8**, 24059.

Weitzman, I., McGinnis, M.R., et al. 1986. The genus *Arthroderma* and its later synonym *Nannizzia*. *Mycotaxon*, **25**, 505–18.

Weitzman, I., Rosenthal, S., et al. 1988. Diagnostic procedures for mycotic and parasitic diseases. 7th edition. Washington, DC: American Public Health Association, Inc., 33–97.

Weitzman, I., Kane, J. and Summerbell, R.C. 1995. Trichophyton, microsporum, epidermophyton and agents of superficial mycoses. In: Murray, P.R., Baron, E.J., et al. (eds), *Manual of clinical microbiology*, 6th edition. Washington, DC: American Society for Microbiology, 791–808.

Wood, S.R. and Cruickshank, C.N.D. 1962. The relationship between trichophytin sensitivity and fungal infection. *Br J Dermatol*, **74**, 329–35.

Young, C.N. 1968. Pseudo-cleistothecia in *Trichophyton rubrum*. *Sabouraudia*, **6**, 160–2.

Zaias, N., Tosti, A., et al. 1996. Autosomal dominant pattern of distal subungual onychomycosis caused by *Trichophyton rubrum*. *J Am Acad Dermatol*, **34**, 302–4.

Zakon, S.J. and Benedek, T. 1944. David Gruby and the centenary of medical mycology – 1841–1941. *Bull Hist Med*, **16**, 155–68.

Zukerman, I., Yeruham, I. and Hadani, A. 1992. An outbreak of ringworm (trichophytosis) in horses accompanied by human infection. *Israel J Vet Med*, **47**, 34–6.

14

Dermatomycotic molds

COLIN K. CAMPBELL AND ELIZABETH M. JOHNSON

Although the dermatophytes show considerable adaptation to their life as epidermal parasites, it is clear that most other fungi are unable to invade keratinized tissue. In those cases where nondermatophyte molds are recognized as superficial pathogens, the affected site is remote from living tissue, as in the extremities of the nails, or there is structural and presumably biochemical abnormality of the keratin as a result of trauma or preexisting disease. Very few molds have been able to invade the epidermis of unaltered skin in the way dermatophytes can, and only the tropical ascomycete, *Piedraia hortai*, invades hair. This chapter describes the characteristics of mold infections of skin and nail, and of otitis externa, together with the distinguishing features of species commonly involved.

SKIN AND NAIL INFECTIONS

Clinical features

The main clinical features, etiological aspects, and therapy of nail infections have been reviewed (Midgley et al. 1994; Denning et al. 1995). Estimates of the importance of molds, compared with dermatophyte fungi in skin and nail infections, vary considerably. Summerbell et al. (1989) reported that less than 10 percent of isolations from infected nails were nondermatophytes. Of these, half were yeasts (mostly *Candida* spp.) and half were various mold species. Certain classes of patient show a raised incidence of nondermatophyte

molds in nails. In particular, elderly or immunocompromised individuals and patients with keratin abnormalities may be affected. Such infections are more often reported from toenails than fingernails. English and Atkinson (1972), using the efficient sampling method of pulverized nails, estimated that 18 percent of elderly chiropody patients seen and 68 percent of those with positive direct microscopy and culture, were infected by molds.

Two main types of nail infection have been recognized: the common distal and lateral subungual onychomycosis and the less common superficial white onychomycosis. Both are typically caused by dermatophyte fungi such as *Trichophyton rubrum* and *Trichophyton mentagrophytes* var. *interdigitale* in addition to a number of mold species. In the case of the molds, infection is often confined to a single nail. In the distal subungual form, the fungus invades underneath the free end of the nail plate, eliciting a hyperkeratotic response in the nailbed epithelium. This results in a build-up of amorphous keratinized tissue, greatly thickening the nail and, in time, lifting the nail plate to an angle of 30° or more from its original position.

Although the hard nail plate may be invaded, the softer amorphous material is the major site of invasion by the fungal mycelium. A variety of yeasts and bacteria may also grow in this material and it is difficult to attribute sole pathogenic role to the presence of the mold. Many affected nails become discolored, possibly resulting from the mixture of organisms present and from the opacity of the amorphous keratin itself.

In superficial white onychomycosis, the fungus infects the top surface of the hard nail plate, and slowly erodes the keratin to give a white powdery area. The disease seldom progresses beyond the superficial layers of the nail plate and there is no response by the underlying nail bed epithelium.

Diagnosis

As most nondermatophyte species invading keratinized tissue are also common saprophytes (a few are plant parasites), some authors have stressed the importance of assessing the significance of isolation from diseased sites. This is done by the demonstration of hyphae in tissue and by growth of a mold in pure culture from most tissue samples. Growth of one or only a few colonies from perhaps 10 or 20 samples cannot be interpreted, especially in the absence of direct observation of hyphae in the tissue, because they may have grown from dormant spores contaminating the skin or nail. Unfortunately not all reports of mold infection have made this distinction. Zaias et al. (1969) reported isolation of mold species from 183 abnormal toenails and demonstrated that KOH-negative nails could yield a profusion of saprophytic molds in culture.

If there is ready access to a fluorescence microscope, then use of the optical brightener calcofluor white can enhance the detection of fungal elements in skin, nail, and hair specimens. For skin and hair specimens, calcofluor white can be used in equal proportion to KOH and placed over the specimen on a microscope slide, covered with a coverslip and left to digest for at least 20 minutes. During this time the slides should be protected from light. It is important that nail samples are presoftened before the addition of calcofluor white or it is unable to penetrate the tissue.

Microscopic examination will reveal the presence of any fungal hyphae. It is a feature of dermatophyte infections that the production of the asexual conidia is suppressed in the infectious state, with direct microscopy of the tissue revealing only vegetative hyphae, some of which may show fragmentation into arthroconidia. Thus, chains of rectangular spores are typical of dermatophyte infection but are not usually seen in mold-infected nails; this is therefore an important feature to note as it may help in the assessment of significance of a subsequent mold isolate. Some molds, such as *Scopulariopsis brevicaulis* and *Scytalidium dimidiatum*, may be distinguished from dermatophytes on direct examination. Mold hyphae may produce fronding which is not usually seen with dermatophyte hyphae and occasionally, when there is an air pocket in the nail, sporing structures may be visible within the nail preparation (Figure 14.1). Molds usually infect nails damaged either due to trauma, disease, or underlying dermatophyte infection and sometimes form fungal balls within the nail plate.

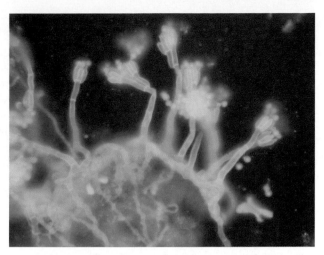

Figure 14.1 *Calcofluor staining of a nail specimen infected with a* Penicillium *sp. demonstrating the production of sporing structures produced in an air pocket within the nail*

The common use of cycloheximide in agar culture media for selection of dermatophyte fungi may mask the presence of infectious molds in skin and nail because this drug is inhibitory to most saprophytic molds. Many laboratories now use a cycloheximide-free medium for isolation from nails in addition to the routine medium. It should be remembered, however, that mold infections of the skin do occur (see sections on Infections by *Scytalidium* spp., Infections by *Aphanoascus fulvescens* (*Anixiopsis stercoraria*) , and Infections by *Onychocola canadensis*), and these will also require a less selective medium.

Isolation of a mold is not considered significant if direct microscopy was negative; a repeated attempt at isolation of a dermatophyte should also be considered if chains of arthroconidia were observed on direct microscopy as these are more indicative of a dermatophyte infection. If a mold is isolated from a specimen from which a dermatophyte is recovered, the mold is likely to be an incidental contaminant. In a study of onychomycosis, Ellis et al. (1997) reported the presence of a mold or yeast in addition to a dermatophyte in 64 percent of 118 patients. Significantly, only three of these patients (2.5 percent) yielded the same yeast or mold from two or more consecutive specimens.

Etiology

The relative incidence of the mold species reported in nails varies from one study to another, but it is clear that *Scopulariopsis brevicaulis* (see section on *Scopulariopsis brevicaulis* Bainier below) is the mold most readily acknowledged as capable of causing nail dystrophy. It is one of the few molds that appears to be able to infect otherwise healthy nails and can sometimes be identified on microscopy of an infected nail because of its propensity to produce conidia in vivo. Microscopy reveals broad, septate hyphae, which are easily distinguished from those

Table 14.1 *Other molds rarely isolated from cases of onychomycosis*

Species	References
Aspergillus sclerotiorum	Feuilharde de Chauvin and de Bièvre 1985; Garcia-Martos et al. 2001
Aspergillus ustus	Walshe and English 1966
Aspergillus unguis	Schönborn and Schmoranzer 1970
Chaetomium globosum	Costa et al. 1988; Naidu et al. 1991; Hattori et al. 2000
Curvularia lunata	Barde and Singh 1983
Exophiala jeanselmei	Boisseau-Garsaud et al. 2002
Geomyces pannorum	Schönborn and Schmoranzer 1970
Geotrichum candidum	Restrepo and de Uribe 1976
Gymnascella dankaliensis	Summerbell et al. 1989
Lasiodiplodia theobromae	Restrepo et al. 1976; Vélez and Díaz 1985
Myxotrichum deflexum	de Vroey 1976
Ophiostoma stenoceras	Summerbell et al. 1993
Penicillium spp.	Walshe and English 1966; McAleer 1981; Vélez and Díaz 1985
Pyrenochaeta unguis-hominis	Punithalingham and English 1975
Polypaecilum insolitum	Piontelli and Toro 1988

of dermatophytes. A few reports exist of onychomycoses caused by other *Scopulariopsis* spp., including *Scopulariopsis acremonium*, *Scopulariopsis asperula*, *Scopulariopsis fusca*, and *Scopulariopsis koningii* (Krempl-Lamprecht 1970; Schönborn and Schmoranzer 1970), and *Scopulariopsis flava* (Piontelli and Toro 1988).

Next in order of frequency to *S. brevicaulis* are species of *Aspergillus*, *Acremonium*, and *Fusarium*. The species causing deep aspergillosis have all been described from nails, although the *Aspergillus versicolor* group (including *Aspergillus sydowii*), the *Aspergillus glaucus* group, and *Aspergillus candidus* (see sections on *Aspergillus versicolor* (Vuill.) Tiraboschi, *Aspergillus glaucus* group, and *Aspergillus candidus*) are probably more frequent (Walshe and English 1966; Zaias 1972). As with onychomycosis caused by *Scopulariopsis* spp., a small number of reports involve well-documented cases of infection by related but little-known species. These include *Aspergillus hollandicus*, a member of the *A. glaucus* group (Grigoriu and Grigoriu 1975), *Aspergillus restrictus*, *Aspergillus unguis* (Schönborn and Schmoranzer 1970), and *Aspergillus sclerotiorum* (Feuilharde de Chauvin and de Bièvre 1985; Garcia-Martos et al. 2001). Reports of onychomycosis caused by *Fusarium* and *Acremonium* (*Cephalosporium*) spp. usually have not carried the identification beyond genus level (Walshe and English 1966; Zaias 1972; McAleer 1981; Vélez and Díaz 1985), although experience suggests that *Fusarium solani*, *Fusarium oxysporum*, and *Acremonium strictum* are the most common (see sections on *Fusarium solani* (Mart.) Sacc., *Fusarium oxysporum* Schlecht, and *Acremonium strictum* W. Gams, respectively). There have been reports in neutropenic patients of disseminated infections with *Fusarium* spp. arising from cutaneous lesions associated with infected nails (Merz et al. 1988; Girmenia et al. 1992). *Paecilomyces variotii* and *Paecilomyces lilacinus* are occasionally implicated in nail infection (see Arenas et al. 1998 for a review of onycho-

mycosis due to *P. variotii*). Onychomycosis due to *Alternaria* spp., in particular *Alternaria alternata*, is rare, but there are several reports suggesting their involvement (Wadhwani and Srivastava 1985; Singh et al. 1990; Glowacka et al. 1998; Romano et al. 2001)

In patients of tropical origin, onychomycoses can result from *Scytalidium hyalinum* and *Scytalidium dimidiatum*. These, together with the mold recently described from nails as a new genus and species, *Onychocola canadensis*, are discussed in detail below. Other molds are occasionally reported, some of which may have been implicated without a careful appraisal of their true significance. Those with good evidence of infection are summarized in Table 14.1.

INFECTIONS BY *SCYTALIDIUM* SPP.

Gentles and Evans (1970) were the first to report human infection by *S. dimidiatum* (under its earlier name, *Hendersonula toruloidea*; see section on *Scytalidium dimidiatum* (Penz.) Sutton and Dyko (pycnidial synanamorph *Nattrassia mangiferae* (Sydow and Sydow) Sutton and Dyko)). First described in 1933 as a cause of dieback and wilt disease in stone-fruit trees in Egypt, it had become well recognized as an important pathogen of a wide range of tropical crop plants. Over a period of 7 years, Gentles and Evans collected eight cases of a condition clinically indistinguishable from tinea pedis, in which this species was the only isolate recovered, sometimes repeatedly. In four cases, the skin of the palms or fingernails was also involved. Most patients were middle-aged male immigrants to the UK from tropical countries. Subsequent reports from other UK laboratories (Campbell et al. 1973; Moore 1978) followed and established this fungus as a frequent cause of infection of both hands and feet in people who originated from tropical countries. More recently, reports have appeared of its occurrence as a human pathogen in Canada

(Summerbell et al. 1989), Nigeria (Gugnani and Oyeka 1989), Sweden (Rollman and Johansson 1987), Spain (Revilla et al. 1992), Thailand (Kotrajaras et al. 1988), and the USA (Greer and Gutierrez 1987; Frankel and Rippon 1989; Elewski and Greer 1991).

Unusually for nondermatophyte pathogens, *S. dimidiatum* and *S. hyalinum* (see section on *Scytalidium hyalinum* Campbell and Mulder below) are able to invade the skin as well as nails, mimicking tinea manuum and tinea pedis caused by the dermatophyte *T. rubrum* (Hay and Moore 1984). The microscopic appearance of the infectious hyphae in skin shows, however, great variation in width along any one length and some hyphae are twisted and contorted, unlike the appearance of typical dermatophyte hyphae (Campbell et al. 1973).

During an investigation into the occurrence of *S. dimidiatum* infections, several cases of essentially the same disease entity were encountered in which the isolated fungus was a white arthroconidiate mold. Campbell and Mulder (1977) described eight such cases and named the organism *Scytalidium hyalinum*. All occurred in immigrants from the Caribbean area or west Africa. Moore (1986) later described the same species in 11 other British immigrants, also originating from these areas. Pointing to the many parallels in the clinical presentation, conidiogenesis, and antigenic composition of these two species, Moore suggested that *S. hyalinum* could merely be a nonpigmented variant of *S. dimidiatum*. This is now supported by DNA comparisons (Roeijmans et al. 1997).

A small survey carried out on hospital attendees in Tobago indicated that infections by this fungus may be common in some tropical areas (Allison et al. 1984). Of 45 patient samples, seven showed microscopic and/or cultural evidence of *S. hyalinum*, and three others yielded *S. dimidiatum*. With the figures added for positive microscopy but failed culture, almost half the total population studied would appear to have had some form of *Scytalidium* spp. infection. None of these people was aware of the infection. Possibly patients referred to clinics in the UK have been motivated by severity of the lesions. Reports of *S. hyalinum* infections have also come from Canada (Summerbell et al. 1989), the USA (Elewski and Greer 1991), Australia (Maslen and Hogg 1992), and Nigeria (Gugnani and Oyeka 1989).

INFECTIONS BY *APHANOASCUS FULVESCENS* (*ANIXIOPSIS STERCORARIA*)

Although rare (it is the subject of only a handful of case reports), this keratinophilic soil fungus (see section on *Aphanoascus fulvescens* (Cooke) Apinis below) stands out as unusual among nondermatophyte molds in that it apparently invades skin rather than nail; it is, however, unwise to rule out the possibility of nail infection. Ring-worm-like lesions with epidermal scaling have been described on thighs (Rippon et al. 1970), arms (Albala et al. 1982), neck (Marín and Campos 1984), and scalp (Guého et al. 1985), and similar infections have occurred in animals (Rippon et al. 1970; Vanbreuseghem and de Vroey 1980). In most of these cases, the anamorphic state – a *Chrysosporium* sp. close to *Chrysosporium keratinophilum* – was isolated, and the *Aphanoascus* state developed on prolonged incubation. Other *Chrysosporium* spp. have also been associated with skin lesions (de Hoog et al. 2000).

INFECTIONS BY *ONYCHOCOLA CANADENSIS*

Onychocola canadensis (see section on *Onychocola canadensis* Sigler below) is a slow-growing, arthroconidiate hyphomycete which was first reported in Canada from nail and skin infections (Sigler and Congly 1990), although evidence of skin infection was not supported by positive microscopy. Since then the organism has been reported from a number of other countries, including France (Contet-Audonneau et al. 1997), Spain (Llovo et al. 2002), New Zealand (Sigler et al. 1994), and the UK (Campbell et al. 1997) where it is referred to the Mycology Reference Laboratory for identification several times a year. There has also been a further more extensive report of cases in Canada and a review of the literature (Gupta et al. 1998).

Microscopy of the subungual keratin from infected nails shows a picture distinct from that seen with dermatophytes. There are septate, hyaline, branching hyphae of varying widths reminiscent of those seen from infections with *S. dimidiatum* and *S. hyalinum*. The round-to-barrel-shaped arthroconidia are more rounded than the cylindrical arthroconidia seen in dermatophyte infections. There are also unusual thick-walled, light-brown hyphae which may be a feature of chronic infections as similar hyphae are only produced on prolonged culture in vitro.

The environmental source of *O. canadensis* has not been established, but it is likely to be a soil saprophyte in common with other nondermatophyte agents of onychomycosis. In vitro studies have shown that it is unable to break down the keratin of hair, but weak cellulolytic activity has been demonstrated. As elderly individuals are most often affected, it appears that altered keratin may be necessary for infection. A description of its teleomorph, *Arachnomyces nodosetosus* sp. nov., was published by Sigler et al. in 1994.

OTITIS EXTERNA

Fungal infections of the ear are common conditions in which there is infection, usually of the external auditory canal, with fungi that are either spread by airborne conidia or derived from direct invasion of the epithelium through the external canal (Kaur et al. 2000; Ozcan et al. 2003). Examples of the former include *Aspergillus* spp., and of the latter, dermatophytes. Sometimes cutaneous pathogens cause this condition, but their presence is likely to have followed airborne contamination by spores, e.g. *S. brevicaulis* (Besbes 2002). Rarely, otitis externa may follow dissemination of a systemic fungal infection such as coccidioidomycosis (Harvey et al. 1978). As such, the pathogenetic role of each organism has to be carefully weighed against the pathological response, as in some cases it is likely that the isolation of a fungus is merely a reflection of temporary carriage of the organism in the external canal. Growth of the fungus by itself can cause loss of hearing due simply to obstruction of the lumen. *Aspergillus niger*, for instance, is a common cause of this form of hearing loss (Loh et al. 1998). The main organisms isolated are *Aspergillus* and *Candida* spp., a whole range of different environmental fungi including *Scedosporium apiospermum* have been implicated (Milne et al. 1986). Rarely, infection may result in an acute and severe destructive process with subsequent invasion of adjacent tissue, malignant otitis externa (Bellini et al. 2003). This is a rare event and mainly associated with profound immunosuppression; the commonest cause is bacterial infection with *Pseudomonas*, but fungi such as *Aspergillus* spp. are occasional causes (Bellini et al. 2003). Other even rarer causes have included *Malassezia pachydermatis* (Chai et al. 2000) and *Absidia corymbifera* (Paterson et al. 2000).

Factors predisposing to infection of the external auditory canal range from existing disease of the external canal such as seborrhoeic dermatitis, eczema, or psoriasis, to trauma and climatic conditions. The latter seem to be particularly important as overgrowth of organisms such as *A. niger* is more common in tropical and humid conditions (Schuster et al. 2002). It is also apparent that in many cases there is mixed growth of different organisms from the canal.

The pathogenesis of fungal otitis externa is not always clear. As stated previously, growth of fungi may simply reflect colonization with growth exacerbated by climatic conditions or other factors such as the administration of antibiotics. However, the excessive growth of fungi may physically obstruct the narrow lumen. Otherwise, invasion is almost always confined to the epithelium except in malignant otitis externa (Vennewald et al. 2003). In addition, in some cases of *A. niger* infection local deposition of oxalate crystals may contribute to underlying tissue damage (Landry and Parkins 1993).

The main clinical features of fungal otitis externa are irritation and discharge from the external auditory canal (Kurnatowski and Filipiak 2001). There may also be hearing impairment due to the production of wax, scaling, or fungal growth. On inspection, the ear canal is usually reddened and may have an inflammatory exudate. If *A. niger* is present, the fungal growth may appear as a dark mat and black fruiting heads are seen together with accompanying oozing and scaling. The malignant form of otitis externa shows ulceration and necrosis with bleeding. There may be pain, often indicating deeper extension into the mastoid ear cells.

Fungi may also be found deeper in the middle ear in association with chronic otitis media. Once again their pathogenic role is not clear and many are contaminants. The organisms isolated from this site are similar to those found in otitis externa (Vennewald et al. 2003).

These infections are managed through a variety of different techniques such as local toilet including debridement and intensive cleansing (Kurnatowski and Filipiak 2001). Topical antifungals such as ciclopirox olamine (del Palacio et al. 2002), 1 percent bifonazole (Falser 1984), clotrimazole, nystatin, or the application of gauze packs soaked in a 1 percent azole solution have also been used. These are repeatedly changed. Oral or intravenous therapy is indicated for necrotizing infections.

The external ear itself may also be affected by facial infection, the commonest of which is *T. rubrum*. However, rarer infections such as chromoblastomycosis and lobomycosis may also affect the external ear. *Rhinosporidium* may also involve this area.

DESCRIPTIONS OF MOLD FUNGI CAUSING ONYCHOMYCOSIS AND DERMATOMYCOSIS

Acremonium strictum W. Gams

Colonies are moderately fast growing, attaining a diameter of 50 mm in 1 week at 30°C. The surface is flat to slightly folded and may be covered in a loose network of aerial hyphae, or may be wet and almost glabrous. The predominant color of the aerial growth is pale salmon pink, although the more floccose strains may be white. Microscopically there is an abundance of small (3.5–5×1–2 μm) conidia, cylindrical to ovoid in shape, which are produced sequentially from the phialidic tips of long, slender, lateral hyphae (Figure 14.2). These arise from bundles of aerial hyphae. The conidiophores can be viewed in situ on the colony using a binocular dissecting microscope, when each phialide can be seen to carry a droplet-like mass of conidia, held together by surface tension.

Alternaria alternata (Fr.) Keissl

The genus *Alternaria* is characterized by producing branching chains of multicellular conidia with dark pigmented walls and septa arranged longitudinally, transversely, and obliquely. The group are mostly pathogens of plant leaves, but *A. alternata* seems to be mainly saprophytic and is a common laboratory contaminant.

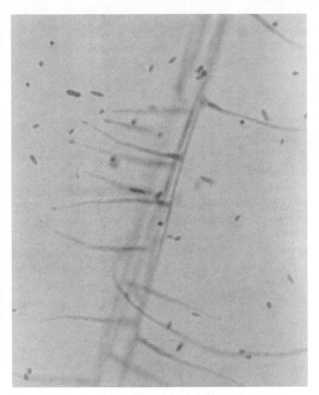

Figure 14.2 Acremonium strictum: *slide culture preparation showing delicate lateral phialides with terminal conidia. In this mount all but the last-formed conidia have dispersed into the mounting fluid. In life they are aggregated in round heads on the phialide apices.*

On normal media it grows as a cottony white colony which slowly develops black pigment in the center. On less rich media the colonies are black to dark brown, suede-like or powdery, and abundant conidia are produced. Conidia are typically clavate, about 25–50 µm long, with a short apical beak which produces the successive conidium in the chain.

Aphanoascus fulvescens (Cooke) Apinis

Colonies are of medium growth rate, flat and felt-like, white to cream in color. On prolonged incubation, a ring of pale purple–brown ascocarps develops near the center. On microscopy there are many club-shaped aleurioconidia, of similar shape to the microconidia of dermatophytes but much larger (mostly 10–15×4–6 µm) and frequently slightly curved. This is the conidial state (*Chrysosporium* spp.). The *Aphanoascus* teleomorph is a smooth-walled ascocarp, up to 500 µm diameter, with pale-brown, rough-walled, ovoid ascospores in round asci.

Aspergillus versicolor (Vuill.) Tiraboschi

This is an *Aspergillus* sp. of moderate growth rate, reaching 10–20 mm diameter in 1 week. A. *versicolor* is so called because it can appear in a variety of colonial colors, ranging from shades of green, light brown, yellow, and grayish pink. The conidial color is, however, dull gray–green and this dominates in cultures that spore well. Microscopically, the distinguishing features are the small head, consisting of an ovoid vesicle that bears supporting cells (metulae) below the layer of phialides (i.e. biseriate heads). There may also be reduced heads resembling those of *Penicillium* spp. A. *sydowii*, often reported from nails, is microscopically identical but has a blue–green colony color. Some mycologists place it as a variant of A. *versicolor* (Figure 14.3).

Aspergillus glaucus group

The name A. *glaucus* is no longer used as a single species but represents an assemblage of closely related aspergilli separable mostly on the basis of their sexual ascosporic states. The colony is generally slow growing on normal media, but this can often be accelerated on special media with low water activity. On glucose peptone agar, the surface is granular and colored mid-green (conidia), with areas of bright-yellow ascocarps. Microscopically, the heads are small relative to the wide stalks, and are composed of small phialides mounted directly upon the subspherical vesicle. In many strains aberrant heads, consisting of secondary conidiophores radiating out from the vesicle, may be seen.

Aspergillus candidus Link

Colonies are of moderate growth rate. Young colonies often appear wet or glabrous but develop into dry colonies, varying from dense and granular to loose and floccose. The conidia are unpigmented so even heavily sporing cultures are white or pale cream. Microscopically, there may be large spherical heads, but reduced heads reminiscent of *Penicillium* spp. may be present. Unlike the dry conidia of most *Aspergillus* spp., the conidia of A. *candidus* readily become wetted when touched with a needle.

Fusarium solani (Mart.) Sacc.

The colony is fast growing, reaching a diameter of 30–40 mm in a week, and producing a uniform, densely floccose, aerial growth. Many strains remain white or dull cream–buff, although others develop shades of bluish–purple. Microscopically, the conidia are ovoid to kidney-shaped, 4–8×3–4 µm and are produced on long hypha-like phialides.

Fusarium oxysporum Schlecht

This species is similar to *F. solani*, but colonies are salmon pink to violet, and the conidia are produced on

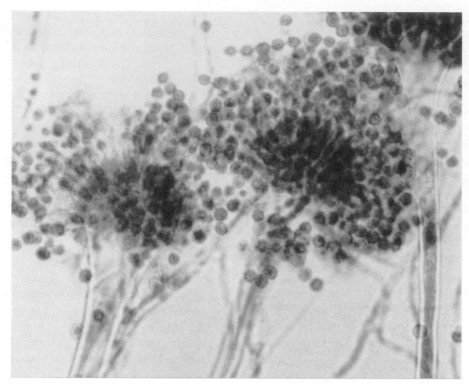

Figure 14.3 Aspergillus versicolor: *sporing heads showing the typical small vesicles with 2 rows of cells below the conidia*

short lateral phialides. Both species may produce smaller numbers of sickle-shaped multicellular macroconidia.

Onychocola canadensis Sigler

Unlike species of *Scytalidium*, *O. canadensis* is resistant to cycloheximide at 0.4 mg/l so it will grow on routine isolation media but shows enhanced growth on carbohydrate-rich medium. Colonies are restricted, attaining a diameter of 14–20 mm in 6 weeks at 25°C on glucose peptone agar, and may cause cracking of the agar. Typically, after 2 weeks at 25°C, colonies are more than 1 cm in diameter, yellowish–white to pale gray with a darker brownish–gray reverse. The surface is domed with abundant aerial mycelium and a cottony, velvety, or woolly texture. *O. canadensis* will grow at 37°C but is less vigorous.

Microscopy of older cultures may reveal abundant broad, pheoid hyphae up to 6.5 μm in width with few septa and with darker-brown wall thickenings; these are similar to the hyphae observed in infected nail specimens. Sporulation is slow, conidia are formed in chains from poorly differentiated, septate, hyaline hyphae 1.5–3 μm wide. Formation of conidia may be either intercalary along the length of the hyphae or in basipetal succession on lateral chains. Conidia (arthroconidia) are broadly ellipsoidal, cylindrical, or irregular in shape, and either unicellular or with a single septum, often persisting in long, branched chains. Release of arthroconidia is by dissolution (rhexolysis) of adjacent

thin-walled cells or by schizolysis of adjacent conidia; the remnants of wall material may be visible on the end walls of the detached conidium (Figure 14.4).

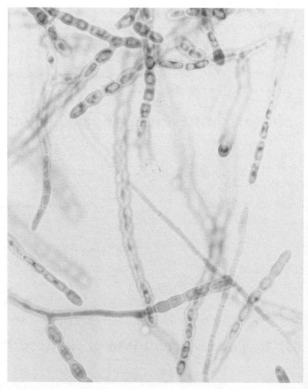

Figure 14.4 Onychocola canadensis: *thick-walled arthroconidia in sparsely branching chains removed from a 4-week-old culture*

Paecilomyces variotii and P. lilacinus

The genus is a group of saprophytes and insect parasites closely related to *Aspergillus* and *Penicillium*. The conidia are produced in unbranching chains from phialides. In some species the phialides are grouped in whorls as in *Penicillium*, but single phialides are common, as are irregular branching systems of phialides. In contrast to many *Aspergillus* spp. and all *Penicillium* spp., the conidia are not colored green. *P. variotii* produces colonies which are flat, fast growing, and with a finely powdery surface. The conidia are variable in size as they continue to grow in size as the chain of spores develops, but are generally ovoid to lozenge-shaped, 3–7 µm long (Figure 14.5). The other species in nails, *P. lilacinus*, produces domed, cottony colonies, which become flattened with age. The aerial growth is a pale lilac purple color. Microscopically the phialides are narrow and tightly compacted together in irregular *Penicillium* like clusters. Conidia are smaller than in *P. variotii*, mostly about 3 µm long.

Scopulariopsis brevicaulis Bainier

S. brevicaulis is an annelidic hyphomycete which will grow on cycloheximide-containing medium and forms spreading colonies, often with a raised and furrowed central area. It has a powdery surface as a result of the production of abundant conidia, and it is grayish–white at first, developing a cinnamon-colored central area on production of conidia.

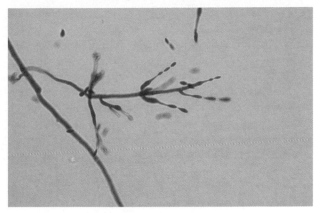

Figure 14.5 Paecilomyces variotii *showing slender phialides producing unbranched chains of oat-shaped cells*

Microscopy reveals an abundance of lemon-shaped, roughened conidia with truncated bases, which are produced in basipetal chains from the tips of annellidic conidiogenous cells. The annellides, which increase in length with age as each successive spore leaves a collar of wall material, are produced singly or in penicillate heads (Figure 14.6).

Scytalidium dimidiatum (Penz.) Sutton and Dyko (pycnidial synanamorph Nattrassia mangiferae (Sydow and Sydow) Sutton and Dyko)

In cultures of *S. dimidiatum*, two types of growth may be produced. These were described in detail by Campbell

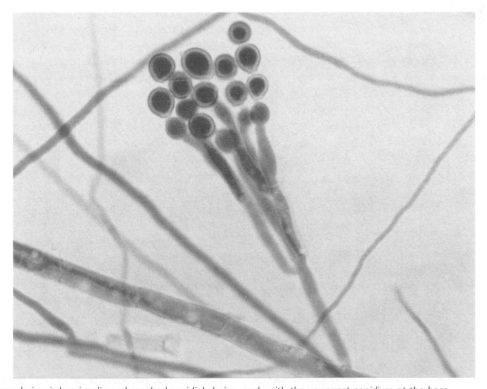

Figure 14.6 Scopulariopsis brevicaulis: *unbranched conidial chains, each with the youngest conidium at the base*

(1974). Type A is fast growing, filling a 9 cm Petri dish in 2–4 days with a dark-gray, floccose, aerial mycelium which becomes brownish–black after a week. A few strains show concentric zones of growth. Microscopy of this type shows a mixture of wide, brown-pigmented hyphae and narrow hyaline hyphae. Both hyphal forms are much fragmented into long-cylindrical to barrel-shaped arthroconidia, 2.5–10×2.5–7 µm, many with two cells (Figure 14.7). Type B is much slower growing, reaching a diameter of 2 cm in 1 week, with little aerial mycelium; the colour is gray–olive, with an irregular submerged margin. Scanty brown arthroconidia are seen on microscopy. In addition to the spores, this type also features coiled hyphae, and older hyphae may bear irregular thickenings on the outer surface of the walls. Type A was associated with immigrants from the Caribbean area, and type B with patients of Indo-Pakistani origin. In a larger series of cases, Moore (1988) found the same geographical split between the fast- (type 1) and slow- (type 3) growing strains. In this study several intermediate colonial forms (type 2) were also found.

On prolonged incubation, strains of type A may develop the pycnidial state, referred to the taxon *N. mangiferae* (Sydow and Sydow) Dyko and Sutton or its earlier synonym *H. toruloidea* Nattrass. This takes the form of hard, carbonaceous stromata, each with several multilocular pycnidia embedded inside. Each connects to the surface via an ostiole through which the pycnidioconidia are released. The pycnidioconidia are ovoid with pointed ends, 10–14×4–6 µm, and become biseptate, with the central cell darker than the end cells.

Scytalidium hyalinum Campbell and Mulder

S. hyalinum grows rapidly on glucose peptone agar or other common media at 28–32°C to produce a white, effuse, cottony, aerial mycelium. With age, a yellowish color may develop and the mycelium collapses somewhat. Microscopy shows an abundance of arthroconidia formed in chains by the disintegration of the aerial hyphae. The spores are cylindrical to ellipsoidal, mostly aseptate but some with two cells, thin-walled, smooth and hyaline, separated by septal splitting. Attempts to induce pycnidia or other conidial states have failed.

TREATMENT OF ONYCHOMYCOSIS AND DERMATOMYCOSIS CAUSED BY MOLDS

Treatment of onychomycosis due to molds is difficult and there is no consistent treatment of proven efficacy. Organisms such as *S. brevicaulis* are not susceptible in vitro to any of the agents licensed for superficial fungal infections. However, in vivo–in vitro correlation has not been established for molds and, despite high MICs in vitro, there are reports of successful treatment. It is possible that a significant number of mold infections only arise due to trauma to the nail by an underlying dermatophyte infection. In these situations it may be that the presence of the dermatophyte is masked by the overgrowth of mold. However, treatment of the underlying dermatophyte can lead to successful resolution of the mold infection. It was reported by Ellis et al. (1997) in their study on onychomycosis that the presence of a mold had no influence on the outcome of treatment with terbinafine.

Treatment of nail infections usually entails oral treatment with a systemically active agent such as terbinafine or itraconazole. Topical application of an antifungal agent, such as terbinafine, tioconazole, or amorolfine, as a paint or lacquer directly onto the nail surface, is unlikely to have the penetrative power to reach all infected tissue, although it may be useful in cases of superficial white onychomycosis. There has also been a report of successful treatment of a *S. hyalinum* onychomycosis with frequent application of amorolfine nail lacquer (Downs et al. 1999), and it may lead to higher rates of resolution when used in combination with an oral agent or following nail avulsion.

Another factor that may have some bearing on outcome of treatment of a mold infection is that oral treatment with itraconazole can lead to high tissue concentrations, particularly in keratinized tissue. Levels in skin and nail tissue may far exceed serum levels; thus

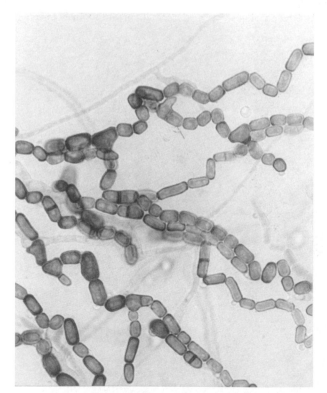

Figure 14.7 Scytalidium dimidiatum: *brown-walled arthroconidia formed by the irregularly timed formation of cross-walls results in some spores with two cells*

the drug may have activity against molds demonstrating relative insusceptibility in vitro. The accumulation and persistence of this drug in keratinized tissue also allows for the use of this agent as pulse therapy. The regimen suggested for pulse therapy is 200 mg twice daily for 1 week followed by a 3-week break. Two pulses may be sufficient for fingernail infections whereas three pulses are usually required to eradicate toenail infection.

A prospective, comparative parallel group, single-blind, randomized study investigated the efficacy of systemic fluconazole, griseofulvin, itraconazole, ketoconazole, and terbinafine in the treatment of 59 patients with toenail onychomycosis due to *S. brevicaulis*. Standard dosage regimens were employed and patients were evaluated for clinical and mycological cure at 12 months after the start of treatment. Terbinafine and itraconazole produced the best response rates with high levels of both clinical and mycological cure (Gupta and Gregurek-Novak 2001). A second group evaluated the efficacy of systemic itraconazole or terbinafine, topical terbinafine following nail plate avulsion, and ciclopirox lacquer in the treatment of patients with onychomycosis due to a variety of molds. Response rates were highest with *Aspergillus* spp. infections (100 percent), *Acremonium* spp. (71.4 percent), *S. brevicaulis* (69.2 percent), but only 40 percent of 26 patients with onychomycosis due to *Fusarium* spp. responded (Tosti et al. 2000).

An initial treatment with any of the systemically active antifungal agents licensed for dermatophyte infections may be beneficial in secondary mold infections, but itraconazole and terbinafine appear to be associated with the most consistent success rates. In extreme cases, it may be necessary to perform surgical removal of the nail plate or avulse the nail with a chemical application such as urease paste. However, following such treatment the nail tissue regrowth may not appear entirely normal.

REFERENCES

Albala, F., Moreda, A. and Lopez, G. 1982. Isolement de *Anixiopsis stercoraria* sur lesions dermiques humaines en Zaragoza (Espagne). *Bull Soc Fr Mycol Méd*, **11**, 287–90.

Allison, V.Y., Hay, R.J. and Campbell, C.K. 1984. *Hendersonula toruloidea* and *Scytalidium hyalinum* infections in Tobago. *Br J Dermatol*, **111**, 371–2.

Arenas, R., Arce, M., et al. 1998. Onychomycosis due to *Paecilomyces variotii*: case report and review. *J Mycol Med*, **8**, 32–3.

Barde, A.K. and Singh, S.M. 1983. A case of onychomycosis caused by *Curvularia lunata*. *Mykosen*, **26**, 311–16.

Besbes, M., Makni, F., et al. 2002. Otomycosis due to *Scopulariopsis brevicaulis*. *Rev Laryngol Otol Rhinol*, **123**, 77–8.

Bellini, C., Antonini, P., et al. 2003. Malignant otitis externa due to *Aspergillus niger*. *Scand J Infect Dis*, **35**, 284–8.

Boisseau-Garsaud, A.M., Desbois, N., et al. 2002. Onychomycosis due to *Exophiala jeanselmei*. *Dermatology*, **204**, 150–2.

Campbell, C.K. 1974. Studies on *Hendersonula toruloidea* isolated from human skin and nail. *Sabouraudia*, **12**, 150–6.

Campbell, C.K. and Mulder, J.L. 1977. Skin and nail infection by *Scytalidium hyalinum* sp. nov.. *Sabouraudia*, **15**, 161–6.

Campbell, C.K., Kurwa, A., et al. 1973. Fungal infection of skin and nail by *Hendersonula toruloidea*. *Br J Dermatol*, **89**, 45–52.

Campbell, C.K., Johnson, E.M. and Warnock, D.W. 1997. Nail infection caused by *Onychocola canadensis*: report of the first four British cases. *J Med Vet Mycol*, **35**, 423–5.

Chai, F.C., Auret, K., et al. 2000. Malignant otitis externa caused by *Malassezia sympodialis*. *Head Neck*, **22**, 87–9.

Contet-Audonneau, N., Schmutz, J.L., et al. 1997. A new agent of onychomycosis in the elderly: *Onychocola canadensis*. *Eur J Dermatol*, **7**, 115–17.

Costa, A.R., Porto, E., et al. 1988. Cutaneous and ungueal phaeohyphomycosis caused by species of *Chaetomium* Kunze (1817) ex Fresenius 1829. *J Med Vet Mycol*, **26**, 261–8.

de Hoog, G.S., Guarro, J., et al. 2000. *Atlas of clinical fungi*, 2nd edn. Baarn/Rus: Centraalbureau voor Schimmelcultures, 545-559.

del Palacio, A., Cuetara, M.S., et al. 2002. Randomized prospective comparative study: short-term treatment with ciclopiroxolamine (cream and solution) versus boric acid in the treatment of otomycosis. *Mycoses*, **45**, 317–28.

Denning, D.W., Evans, E.G.V., et al. 1995. Fungal nail disease: a guide to good practice (report of a working group of the British Society for Medical Mycology). *Br Med J*, **311**, 1277–81.

de Vroey, C. 1976. Sur quelques ascomycètes isolés de lesions cutanées chez l'homme. *Bull Soc Fr Mycol Méd*, **5**, 161–3.

Downs, A.M.R., Lear, J.T. and Archer, C.B. 1999. *Scytalidium hyalinum* onychomycosis successfully treated with 5% amorolfine nail laquer. *Br J Dermatol*, **140**, 555.

Elewski, B.E. and Greer, D.L. 1991. *Hendersonula toruloidea* and *Scytalidium hyalinum*. *Arch Dermatol*, **127**, 1041–4.

Ellis, D.H., Watson, A.B., et al. 1997. Non-dermatophytes in onychomycosis of the toenails. *Br J Dermatol*, **136**, 490–3.

English, M.P. and Atkinson, R. 1972. Onychomycosis in elderly chiropody patients. *Br J Dermatol*, **91**, 67–72.

Falser, N. 1984. Fungal infection of the ear. Etiology and therapy with bifonazole cream or solution. *Dermatologica*, **169**, Suppl 1, 135–40.

Feuilharde de Chauvin, M. and de Bièvre, C. 1985. Onyxis et perionyxis à *Aspergillus sclerotiorum*. *Bull Soc Fr Mycol Méd*, **14**, 77–9.

Frankel, D.H. and Rippon, J.W. 1989. *Hendersonula toruloidea* infection in man. Index cases in the non-endemic North American host, and a review of the literature. *Mycopathologia*, **105**, 175–86.

Garcia-Martos, P., Guarro, J., et al. 2001. Onychomycosis caused by *Aspergillus sclerotiorum*. *J Mycol Méd*, **11**, 222–4.

Gentles, J.C. and Evans, E.G.V. 1970. Infection of the feet and nails with *Hendersonula toruloidea*. *Sabouraudia*, **8**, 72–5.

Girmenia, C., Arcese, W., et al. 1992. Onychomycosis as a possible origin of disseminated *Fusarium solani* infection in a patient with severe aplastic anaemia. *Clin Infect Dis*, **14**, 1167.

Glowacka, A., Wasowska-Krolikowska, K., et al. 1998. Childhood onychomycosis: alternariosis of all ten fingernails. *Cutis*, **62**, 3, 125–8.

Greer, D.L. and Gutierrez, M.M. 1987. Tinea pedis caused by *Hendersonula toruloidea*. *J Am Acad Dermatol*, **16**, 1111–15.

Grigoriu, D. and Grigoriu, A. 1975. Les onychomycoses. *Rev Med Suisse Romande*, **95**, 839–49.

Guého, E., Villard, J. and Guinet, R. 1985. A new human case of *Anixiopsis stercoraria* mycosis; discussion of its taxonomy and pathogenicity. *Mykosen*, **28**, 430–6.

Gugnani, H.C. and Oyeka, C.A. 1989. Foot infections due to *Hendersonula toruloidea* and *Scytalidium hyalinum* in coal miners. *J Med Vet Mycol*, **27**, 169–79.

Gupta, A.K. and Gregurek-Novak, T. 2001. Efficacy of itraconazole, terbinafine, fluconazole, griseofulvin and ketoconazole in the treatment of *Scopulariopsis brevicaulis* causing onychomycosis of the toes. *Dermatology*, **202**, 235–8.

Gupta, A.K., Horgan-Bell, C.B. and Summerbell, R.C. 1998. Onychomycosis associated with *Onychocola canadensis*: ten case reports and a review of the literature. *J Am Acad Dermatol*, **39**, 410–17.

Harvey, R.P., Pappagianis, D., et al. 1978. Otomycosis due to coccidioidomycosis. *Arch Intern Med*, **138**, 1434–5.

Hattori, N., Adachi, M., et al. 2000. Onychomycosis due to *Chaetomium globosum* successfully treated with itraconazole. *Mycoses*, **43**, 89–92.

Hay, R.J. and Moore, M.K. 1984. Clinical features of fungal infections caused by *Hendersonula toruloidea* and *Scytalidium hyalinum*. *Br J Dermatol*, **110**, 677–83.

Kaur, R., Mittal, N., et al. 2000. Otomycosis: a clinicomycologic study. *Ear Nose Throat J*, **79**, 606–9.

Kotrajaras, R., Chongsathien, S., et al. 1988. *Hendersonula toruloidea* infection in Thailand. *Int J Dermatol*, **27**, 391–5.

Krempl-Lamprecht, L. 1970. *Scopulariopsis* – Arten bei Onychomykosen. *Proceedings of the 2nd International Symposium on Medical Mycolology, Poznan, 1967*. Krakow: Drukarnia Narodowa, 45–8.

Kurnatowski, P. and Filipiak, A. 2001. Otomycosis: prevalence, clinical symptoms, therapeutic procedure. *Mycoses*, **44**, 472–9.

Landry, M.M. and Parkins, C.W. 1993. Calcium oxalate crystal deposition in necrotizing otomycosis caused by *Aspergillus niger*. *Mod Pathol*, **6**, 493–6.

Loh, K.S., Tan, K.K., et al. 1998. Otitis externa -- the clinical pattern in a tertiary institution in Singapore. *Ann Acad Med Singapore*, **27**, 215–18.

Llovo, J., Prieto, E., et al. 2002. Onychomycosis due to *Onychocola canadensis*: report of the first two Spanish cases. *Med Mycol*, **40**, 209–12.

Marín, G. and Campos, R. 1984. Dermatofitosis por *Aphanoascus fulvescens*. *Sabouraudia*, **22**, 311–14.

Maslen, M.M. and Hogg, G.G. 1992. *Scytalidium hyalinum* isolated from the toe nail of an Australian patient. *Australas J Dermatol*, **33**, 165–8.

McAleer, R. 1981. Fungal infections of the nails in Western Australia. *Mycopathologia*, **73**, 115–20.

Merz, W.G., Karp, J.E., et al. 1988. Diagnosis and successful treatment of fusariosis in the compromised host. *J Infect Dis*, **158**, 1046–55.

Midgley, G., Moore, M.K., et al. 1994. Mycology of nail disorders. *J Am Acad Dermatol*, **31**, 568–74.

Milne, L.J.R., McKerrow, W.S., et al. 1986. Pseudallescheriasis in northern Britain. *J Med Vet Mycol*, **24**, 377–82.

Moore, M.K. 1978. Skin and nail infections by non-dermatophyte filamentous fungi. *Mykosen*, Suppl 1, 128–32.

Moore, M.K. 1986. *Hendersonula toruloidea* and *Scytalidium hyalinum* infections in London, England. *J Med Vet Mycol*, **24**, 219–30.

Moore, M.K. 1988. Morphological and physiological studies of *Hendersonula toruloidea* Nattrass cultured from human skin and nail samples. *J Med Vet Mycol*, **26**, 25–39.

Naidu, J., Singh, S.M. and Pouranik, M. 1991. Onychomycosis caused by *Chaetomium globosum* Kunze. *Mycopathologia*, **113**, 31–4.

Ozcan, K.M., Ozcan, M., et al. 2003. Otomycosis in Turkey: predisposing factors, aetiology and therapy. *J Laryngol Otol*, **117**, 39–42.

Paterson, P.J., Marshall, S.R., et al. 2000. Fatal invasive cerebral *Absidia corymbifera* infection following bone marrow transplantation. *Bone Marrow Transplant*, **26**, 701–3.

Piontelli, E. and Toro, M.A. 1988. Un raro caso de hialohifomicosis en uñas por *Polypaecilum insolitum* G Smith. *Bol Micol*, **4**, 155–9.

Punithalingham, E. and English, M.P. 1975. *Pyrenochaeta unguium-hominis* sp. nov. on human toe nails. *Trans Br Mycol Soc*, **64**, 539–41.

Restrepo, A. and de Uribe, L. 1976. Isolation of fungi belonging to the genera *Geotrichum* and *Trichosporon* from human dermal lesions. *Mycopathologia*, **59**, 3–9.

Restrepo, A., Arango, M., et al. 1976. The isolation of *Botryodiplodia theobromae* from a nail lesion. *Sabouraudia*, **14**, 1–4.

Revilla, T., Moore, M.K., et al. 1992. Infeccion por *Scytalidium dimidiatum* diagnosticada en España. *Rev Iberoam Micol*, **9**, 1–3.

Rippon, J.W., Lee, F.C. and McMillen, S. 1970. Dermatophyte infection caused by *Aphanoascus fulvescens*. *Arch Dermatol*, **102**, 552–5.

Roeijmans, H.J., de Hoog, G.S., et al. 1997. Molecular taxonomy and GC/MS of metabolites of *Scytalidium hyalinum* and *Nattrassia mangiferae* (*Hendersonula toruloidea*). *J Med Vet Mycol*, **35**, 181–8.

Rollman, O. and Johansson, S. 1987. *Hendersonula toruloidea* infection: successful response of onychomycosis to nail avulsion and topical cyclopiroxolamine. *Acta Derm Venereol (Stockh)*, **67**, 506–10.

Romano, C., Paccagnini, E. and Difonzo, E.M. 2001. Onychomycosis caused by *Alternaria* spp. in Tuscany, Italy from 1985 to 1999. *Mycoses*, **44**, 73–6.

Schönborn, C. and Schmoranzer, H. 1970. Untersuchungen uber Schimmelpilzinfektionen der Zehennagel. *Mykosen*, **13**, 253–72.

Schuster, E., Dunn-Coleman, N., et al. 2002. On the safety of *Aspergillus niger* — a review. *Appl Microbiol Biotechnol*, **59**, 426–35.

Sigler, L. and Congly, H.A. 1990. Toenail infection caused by *Onychocola canadensis* gen et sp. nov.. *J Med Vet Mycol*, **28**, 405–17.

Sigler, L., Abbott, S.P. and Woodgyer, A.J. 1994. New records of skin and nail infection due to *Onychocola canadensis* and description of its teleomorph *Arachnomyces nodosetosus* sp nov. *J Med Vet Mycol*, **32**, 275.

Singh, S.M., Naidu, J. and Pouranik, M. 1990. Ungual and cutaneous phaeohyphomycosis caused by *Alternaria alternata* and *Alternaria chlamydospora*. *J Med Vet Mycol*, **28**, 275.

Summerbell, R.C., Kane, J. and Krajden, S.Y. 1989. Onychomycosis, tinea pedis and tinea manuum caused by non-dermatophytic filamentous fungi. *Mycoses*, **32**, 609–19.

Summerbell, R.C., Kane, J., et al. 1993. Medically important *Sporothrix* species and related ophiostomatoid fungi, *Ceratocystis* and *Ophiostoma*. In: Wingfield, M.J., Seifert, K.A. and Webber, J.F. (eds), *Taxonomy, ecology and pathogenicity*. St Paul MI: APS, 185–92.

Tosti, A., Piraccini, B.M. and Lorenzi, S. 2000. Onychomycosis caused by nondermatophytic molds: clinical features and response to treatment of 59 cases. *J Am Acad Dermatol*, **42**, 217–24.

Vanbreuseghem, R. and de Vroey, C. 1980. Dermatophytic infection by *Anixiopsis stercoraria* in a wild boar (*Sus scrofa*). *Mykosen*, **23**, 183–7.

Vélez, H. and Díaz, F. 1985. Onychomycosis due to saprophytic fungi. *Mycopathologia*, **91**, 87–92.

Vennewald, I., Schonlebe, J. and Klemm, E. 2003. Mycological and histological investigations in humans with middle ear infections. *Mycoses*, **46**, 12–18.

Wadhwani, K. and Srivastava, A.K. 1985. Some cases of onychomycosis in North India in different working environments. *Mycopathologia*, **92**, 149–55.

Walshe, M.M. and English, M.P. 1966. Fungi in nails. *Br J Dermatol*, **78**, 198–207.

Zaias, N. 1972. Onychomycosis. *Arch Dermatol*, **105**, 263–74.

Zaias, N., Oertel, I. and Elliott, D.F. 1969. Fungi in toe nails. *J Invest Dermatol*, **53**, 140–2.

15

Superficial candidiasis

PAUL L. FIDEL Jr AND KAREN L. WOZNIAK

Superficial candidiasis includes infections of cutaneous or mucocutaneous tissues. *Candida albicans* is the causative agent in the majority of infections. Other *Candida* species that can cause the superficial infections include *C. tropicalis*, *C. glabrata*, *C. parapsilosis*, *C. stellatoidea*, *C. dubliniensis*, and *C. krusei* (Cann 1992). Superficial candidiasis can be acute, chronic, or recurrent. *Candida* spp. and *C. albicans*, in particular, are ubiquitous dimorphic fungal organisms that are part of the normal microflora of healthy individuals. However, they are also opportunistic pathogens that can quickly transform from harmless mucosal commensals to a highly pathogenic organism of the same tissues with significant morbidity and even mortality under the appropriate conditions (Fidel 1999).

Superficial candidiasis can include skin, oropharyngeal, gastrointestinal, vaginal, and conjunctival tissue infections. Cutaneous candidiasis usually occurs as erythematous patches on the thigh and buttocks, or as superficial infections of the hands in individuals who routinely wear plastic gloves. Mucocutaneous infections may be chronic and/or recurrent and one particular form is classified as the disorder known as chronic mucocutaneous candidiasis (CMC). These infections affect a variety of mucosal tissues, skin, and nails. Oropharyngeal candidiasis (OPC) involves infections of the buccal mucosa, gingiva, and tongue. The infections can be atrophic with erythematous lesions or pseudomembranous (thrush) with characteristic white lesions.

Esophagitis is the most common form of gastrointestinal candidiasis. The second most common site of gastrointestinal candidiasis is the stomach. Vaginitis involves infections of the vaginal lumen and often the vulva as well. An estimated 75 percent of all women will experience at least one episode of acute vulvovaginal candidiasis (VVC). A separate population of women (5–10 percent) will experience recurrent VVC (RVVC) (Sobel 1988). Conjunctivitis caused by *C. albicans* infects the mucous membranes of the eye, while keratitis affects the cornea. The common superficial *Candida* infections are summarized below:

> Cutaneous
> Mucocutaneous
> > Oropharyngeal
> > Gasatrointestinal
> > > Stomach
> > > Esophageal
> Vaginal
> Conjunctival/keratitis

Host defense mechanisms that protect against *C. albicans* infections include innate resistance and acquired immunity. The relative contributions of these two systems, however, differ depending upon the site at which *C. albicans* enters the host. Innate host defenses (i.e. polymorphonuclear leukocytes, macrophages) are heavily involved in protecting the systemic circulation, antibodies (i.e. IgA, IgG) appear to play a role at both systemic and mucosal sites, and cell-mediated immunity (CMI) (i.e. T cells, relevant cytokines) protects primarily mucosal tissues. With regard to mucosal tissues, while historically there was no distinction between different anatomic sites, there is now increasing evidence that 'not all mucosal tissues are created equal' relative to protective host defenses. Accordingly, current data show roles for several types of host responses at any one site, albeit at different levels of priority with respect to their protective capacity. This review covers the epidemiology of each superficial infection, properties of *C. albicans* as a commensal of the gastrointestinal and genitourinary tracts, properties of *C. albicans* as a pathogen causing superficial infections, and host defenses critical to protection against and susceptibility to superficial candidiasis at the various anatomic sites.

EPIDEMIOLOGY

Cutaneous/mucocutaneous candidiasis

Cutaneous infections affect the skin, whereas mucocutaneous infections affect the mucous membranes. Cutaneous candidiasis can be acute or chronic (reviewed in Odds 1988a) (Kirkpatrick 1989) and are characterized by erythema, edema, and a creamy exudate. Acute infections usually occur as a diaper rash in infants or between folds of skin in obese individuals or on the hands of individuals who routinely wear plastic gloves. The major symptoms are burning and itching. If left untreated, infections usually persist as long as the tissue remains moist. Chronic infections usually involve the feet and present with a thick layer of infected stratum corneum overlaying the epidermis.

Chronic mucocutaneous candidiasis is a heterogenous description that applies to infections involving chronic or recurrent infections of the skin, nails, and mucous membranes. The disease usually appears in the first 3 years of life, and affects males and females equally (Kirkpatrick 1984). Four types of CMC exist, including familial CMC, CMC with endocrinopathy, CMC with familial susceptibility and endocrinopathy, and CMC with onset after 10 years of age (reviewed in Ruhnke 2002). Patients usually develop oropharyngeal candidiasis followed by perianal candidiasis. Additionally, lesions appear on the scalp, torso, and extremities, and nail infections occur (Kirkpatrick 1984). Several disorders are associated with CMC, including hypoendrocrine function, autoimmune diseases, immunosuppression, hepatitis, diabetes, thymoma, dental enamel dysplasia, vitiligo, alopecia, and infectious diseases (Filler and Edwards 1993; Kirkpatrick 1994). Symptoms of CMC include disfiguring lesions on the skin and nails and dysphagia, due to ulcerative lesions on the buccal mucosa and esophagus (Kirkpatrick 1984).

Oropharyngeal candidiasis

Oropharyngeal candidiasis (OPC) involves infections of the hard and soft palate, tongue, buccal mucosa, and floor of the mouth. It presents as reddened patches (erythematous) or white curd-like lesions (pseudomembranous). Chewing and swallowing can be difficult under these conditions. Infections can be acute or recurrent, and are common in immunocompromised patients, especially those infected with the human immunodeficiency virus (HIV). OPC is also a common manifestation of CMC (reviewed in Odds 1988a) and occurs in patients with lymphoma, those undergoing steroid therapy, and in transplant recipients. Although OPC will occur under several immunocompromising conditions, it appears to be much more common in HIV-infected persons than under any other condition. In fact, OPC is often one of the first clinical signs of underlying HIV infection and will occur in 50–95 percent of all HIV-positive persons sometime during their progression to full-blown acquired immunodeficiency syndrome (AIDS) (Rabeneck et al. 1993). Thus, it is possible that a link between HIV and OPC is present that enhances susceptibility to OPC. Interestingly, highly active antiretroviral therapy (HAART) has reduced the incidence of OPC (Palella et al. 1998). This is postulated to be due both to increased immune responsiveness as well as the action of the protease inhibitors in HAART on the secretory aspartyl proteases (SAP), important virulence factors of *C. albicans* (Cassone et al. 1999; Gruber et al. 1999; Calderone and Fonzi 2001).

Gastrointestinal candidiasis

For the purposes of the discussion at hand, gastrointestinal (GI) candidiasis will be defined as infections of the esophagus, stomach, and small and large intestines. GI candidiasis is not easily recognized clinically but does occur in cancer patients who are heavily immunosuppressed and patients receiving antibiotic prophylaxis (Anaissie and Bodey 1990; Samonis et al. 1994). Esophagitis is the most common form of gastrointestinal candidiasis that occurs in later stages of immunosuppression and is considered an AIDS-defining illness (Macher 1988). It is extremely painful and can promote wasting because of the difficulty with swallowing. White plaques are usually observed similar to thrush. Complications include bleeding and perforation. Diagnosis is made on appearance of ulcers and erosions on the esophagus, and may need to be confirmed by endoscopy and biopsy of the affected site (Wheeler et al. 1987). The second most common site of GI candidiasis is the stomach where symptoms of gastritis occur. GI candidiasis is sometimes difficult to diagnose due to the lack of specific symptoms and the lack of the ability to differentiate between infectious and commensal *C. albicans*.

Vaginal candidiasis

Vulvovaginal candidiasis (VVC) affects a significant number of women predominantly in their reproductive years (Sobel 1988, 1992; Kent 1991). An estimated 75 percent of all women will experience an episode of acute VVC in their lifetime with another 5–10 percent succumbing to RVVC (Sobel 1988, 1992). Vaginitis involves infections of the vaginal lumen and often the vulva as well. Symptoms include burning, itching, soreness, an abnormal discharge, and dyspareunia. Signs include vaginal and vulvar erythema and edema. Acute VVC has several known predisposing factors including antibiotic and oral contraceptive usage, hormone replacement therapy, pregnancy, and uncontrolled diabetes mellitus (Sobel 1988, 1992; Kent 1991). RVVC is multi-factorial in etiology, but is usually defined as idiopathic with no known predisposing factors in the majority of those affected (Sobel 1988, 1992). In women with RVVC, antifungal therapy is highly effective for individual symptomatic attacks, but does not prevent subsequent recurrence. There is little evidence that resistance to antifungal drugs plays a role in the pathogenesis of RVVC (Lynch et al. 1996). Instead, susceptibility to RVVC is postulated to be immune-based in that these otherwise healthy women experience repeated symptomatic episodes as a result of some immunological dysfunction or deficiency. Furthermore, it is postulated that recurrences are a result of relapse rather than re-infection since strain types of *C. albicans* tend to remain the same in women for multiple recurrences over several years (Vazquez et al. 1994). This is consistent with the fact that antifungal drugs are static rather than cidal and, as such, do not completely eliminate the organisms.

Conjunctivitis and keratitis

Conjunctivitis caused by *C. albicans* infects the mucous membranes of the eye. The eyelid and surrounding soft tissue can be swollen by adjacent tissue involved in the mycotic infection (Klotz et al. 2000). *Candida* spp. have been found in cases of ulcerative blepharitis in patients with skin atopy (Huber-Spitzy et al. 1992). Symptoms of *Candida* conjunctivitis include conjunctival erythema, cheesy discharge, and corneal ulceration. However, *Candida* can be isolated from the conjunctiva and eyelids of normal healthy individuals and remain asymptomatic (commensal) (reviewed in Klotz et al. 2000).

Keratitis caused by *Candida* spp. affects the cornea and causes ulcerative lesions on the eye (Klotz et al. 2000). The predisposing factor most often associated with *Candida* keratitis is chronic dry eye, but it can also be found in cases of trauma to the eye, abuse of corneal anesthetics, and following corneal transplant (reviewed in Klotz et al. 2000). Additionally, keratitis can be caused by wearing hard and soft extended-wear contact lenses, due to the fact that *Candida* has the ability to adhere to contact lenses and form a biofilm (Butrus and Klotz 1986; Elder et al. 1995). *Candida* keratitis presents as a small, demarcated ulcer with underlying opacity, and a granular infiltrate in the corneal epithelium and stroma can be observed. Diagnosis can be made upon microscopic examination of corneal scrapings.

General diagnosis

Together with compatible signs and symptoms, a definitive diagnosis of cutaneous and mucocutaneous *C. albicans* infection involves the demonstration of hyphae detected by microscopic analysis of tissue scraping (KOH wet mount preparation) or tissue invasion by histopathological examination. A swab culture is used to isolate the organism. Colonies are speciated in the laboratory using a variety of tests, such as germ tube formation, appearance (color) of colonies on CHROMagar Candida (Odds and Bernaert 1994; Sullivan and Coleman 1998), commercial API ID 32C or API 20C AUX (Richardson and Carlson 2002), polymerase chain reaction (PCR) (Janusz et al. 1988; Okhravi et al. 1998; Guiver et al. 2001), contour-clamped homogeneous electric field electrophoresis (CHEF) (Vazquez et al. 1992), DNA fingerprinting (Lockhart et al. 1997, 2001; Steffan et al. 1997), as well as other biochemical tests (Sood et al. 2000; Freydiere et al. 2002; Richardson and Carlson 2002).

Treatment

Nystatin, a topical polyene antifungal related chemically to amphotericin B, can be used for treatment of cutaneous candidiasis. Azoles (i.e. fluconazole, ketoconazole, butaconazole, itraconazole, etc.) are recommended for all other forms of superficial candidiasis. Antifungal drugs are highly effective for individual attacks or acute infections, but recurrence remains a problem. In fact, recurrent attacks can occur within days of cessation of treatment. Thus, in the appropriate clinical setting, maintenance therapy is encouraged. Table 15.1 summarizes the symptoms, diagnosis, and treatment for the various forms of superficial candidiasis.

C. ALBICANS AS A COMMENSAL OF THE GASTROINTESTINAL AND GENITOURINARY TRACTS

As stated previously, *C. albicans* is a commensal organism of the gastrointestinal and reproductive tracts. In this colonized state, individuals are considered asymptomatic carriers. It is widely recognized that as a commensal, *C. albicans* exists exclusively in the blastospore form, although it is rarely detectable in wet

Table 15.1 *Symptoms, diagnosis, and treatment of superficial candidiasis*

Infection	Signs/symptoms	Diagnosis	Treatment
Cutaneous	Erythema, edema, creamy exudates, burning, itching	Erythematous patches in folds of skin, KOH stain showing hyphae	Nystatin
Oropharyngeal	Reddened erythematous or white curd-like lesions, difficulty swallowing	Erythematous or pseudomembranous lesions (can be scraped), KOH stain showing hyphae	Azoles
Gastrointestinal–esophageal–stomach	Difficulty swallowing, bleeding, perforation, gastritis	Ulcers and erosions on esophagus, biopsy of lesion, KOH stain showing hyphae	Azoles
Vaginal	Burning, itching, soreness, white cottage cheese-like discharge, vuvlar edema, pruritis	KOH stain showing hyphae, culture positive, signs and symptoms	Azoles
Conjunctival/keratitis	Conjunctival edema, cheesy discharge, corneal ulceration	Corneal scraping, KOH stain showing hyphae	Topical/injected azoles

mount preparations of tissue scrapings. Instead, asymptomatic colonization is usually detected by swab culture. The most common sites showing detectable asymptomatic colonization are the rectum, oral cavity, and vagina. Although many *Candida* spp. are considered normal flora, *C. albicans* is the most common species recovered. In the oral cavity, the approximate rate of yeast carriage is 50–60 percent (Glick and Siegel 1999), but it has been reported to be as low as 2 percent and as high as 69 percent (Odds 1988b). In recent studies from our laboratory, we found that HIV-negative individuals had an oral colonization rate between 45 and 50 percent, with *C. albicans* present in 85 percent of isolates recovered. In contrast, HIV-positive individuals had an asymptomatic colonization rate in the oral cavity as high as 76 percent, with *C. albicans* present in 83 percent of isolates recovered (Leigh et al. 1998; Wozniak et al. 2002a). The distribution of *C. albicans* in the oral cavity includes the tongue, palate, and buccal mucosa. It can also be recovered from diverse areas of the gastrointestinal tract, including oropharynx, stomach, jejunum, illeum, and colon. In the stomach, *Candida* colonization ranges from 47 percent (when stomach pH is less than 3) to 70 percent (when pH is greater than 3), and species found include *C. albicans*, *C. krusei*, and *C. tropicalis* (Bernhardt and Knoke 1997). Additionally, at pH greater than 3, *C. glabrata* has been isolated from the stomach (Bernhardt and Knoke 1997). Rectal swabs were positive for *C. albicans* in 20–25 percent of healthy individuals (Odds 1988b; Soll et al. 1991). A similar proportion of normal healthy women (i.e. 20–25 percent) carry *C. albicans* in the vagina (Sobel 1988), although one study reported carriage in 40 percent of women (Soll et al. 1991), and two recent studies reported asymptomatic colonization of 13 percent (Fidel et al. 2003) and 11 percent (Bauters et al. 2002), respectively. A study of HIV-positive women reported that 37 percent were asymptomatically colonized at the vaginal mucosa, compared to 21 percent of HIV-negative high-risk women from the same study (Schuman et al. 1998b). The recovery of yeast increases during pregnancy and during select stages of the menstrual cycle, in particular when reproductive hormones are elevated (luteal phase). In contrast to colonization with *C. albicans* at mucosal sites, *C. albicans* is not considered normal flora of the skin. *C. albicans* is also infrequently isolated from water sources, soil, air, and plants, and those isolates recovered from food and beverages are considered contaminants from human and animal sources (Hasenclever et al. 1961).

Colonization with *C. albicans* at mucosal tissues is considered to occur at an early age, the organism having been acquired either by traveling through the birth canal, nursing, or from food sources. Aside from direct culture of mucosal surfaces, long-term colonization with *C. albicans* is assumed to be responsible for acquired immune responsiveness. Anti-*Candida* antibodies, circulating IgG, and mucosal IgA and IgG, can be detected in most healthy individuals (Witkin et al. 1988, 1989; Regulez et al. 1994; Wozniak et al. 2002a). Moreover, greater than 80 percent of healthy persons have positive cutaneous skin test reactivity to *Candida* antigen, and peripheral blood lymphocytes from more than 90 percent of healthy individuals proliferate in vitro to *Candida* antigen (Kirkpatrick et al. 1971; Mathur et al. 1977; Odds 1988a; Fidel et al. 1993). It is presumed that these acquired host responses in conjunction with innate resistance (i.e. polymorphonuclear leukocytes and macrophages) play a significant role in restricting *C. albicans* to mucosal surfaces in an asymptomatic commensal state.

C. ALBICANS AS A PATHOGEN OF THE GASTROINTESTINAL AND GENITOURINARY TRACTS

There are several scenarios in which *C. albicans* can convert from the harmless commensal to a serious pathogen. These include a lack of protective host defenses, expression of virulence factors by the organism, or environmental conditions that allow the yeast to overwhelm the protective host defenses.

C. albicans has several virulence factors that promote adherence and ultimately disease. Dimorphism itself is considered a virulence factor of C. albicans (reviewed in Calderone and Fonzi 2001). There are several environmental factors that allow C. albicans blastospores to transform into the more pathogenic hyphal form, the most common of which are temperature and pH. Acidic pH and temperatures below 35°C usually favor the blastospore form, whereas alkaline pH (> 6.5) and temperatures 37°C or higher favor hyphal formation. Although both forms can exist at any mucosal site, it is the hyphal form of C. albicans that is often observed in infected tissue. Hyphae appear to adhere and invade tissue by elaborating proteolytic enzymes. A secondary effect of dimorphism that can be associated with virulence is antigenic variation (De Bernardis et al. 1994). Antigens present on the blastospore and hyphal surface may be different and thus one or the other form may evade immune surveillance.

Aside from dimorphism, there are other components of Candida that are considered to be virulence factors. One virulence factor is the presence of integrin analogs on C. albicans that enhance adhesion to epithelial cells (Cann 1992). Interestingly, compared to non-albicans species, C. albicans has been reported to express the highest levels of the integrin analog. Another virulence factor is hyphal wall protein-1 (HWP-1), which is an adhesin that is important in attachment to human buccal epithelial cells through the action of transglutaminase (Staab et al. 2000). Other adhesins that are considered to be virulence factors include agglutin-like sequence-1 (Als1p), Als5p, and integrin-like protein-1 (Int1p) (reviewed in Calderone and Fonzi 2001). Phenotype switching, an in vitro phenomenon which involves phenotypic changes in the absence of genotypic changes, represents yet another possible virulence factor (Soll 1992, 2002). If it occurs in vivo, as predicted, it would affect both adherence and immunological escape. Also, SAPs, which are known to be involved in invasion of tissues and tissue damage, are thought to be virulence factors for Candida (reviewed in Calderone and Fonzi 2001; Hube and Naglik 2002) (De Bernardis et al. 1990, 1999). This seems especially true at the oral and vaginal mucosa where isolates recovered from infections have been shown to express high levels of specific SAPs (Hube et al. 1997; De Bernardis et al. 1999; Schaller et al. 1999). Additional virulence factors are the secretion of hydrolases which enhance adherence to mucosal tissues, and heat-shock proteins that bind to serum proteins and inhibit proper folding of the serum proteins as well as interactions with other proteins (reviewed in Ross et al. 1990; Matthews et al. 1998; Calderone and Fonzi 2001). More recently, other components, such as spindle assembly checkpoint component CaMad2p (Bai et al. 2002), copper and zinc-containing superoxide dismutase (Cu/Zn SOD) (Hwang et al. 2002), the transcription factor Spt3 (Laprade et al. 2002), and the

two-component response regulator SSK1 (Calera et al. 2000) have been suggested as possible virulence factors for Candida. Finally, antimycotic drug resistance of some Candida spp. allows them to escape therapeutic treatment regimens.

Regardless of the level of tissue penetration that is usually quite shallow as a superficial infection, there can be large numbers of organisms associated with the tissue. It is generally considered that as a result of some change in the host, the organism grows in high enough numbers to overcome the host defenses, penetrates the tissue and begins to cause the symptoms associated with the infection. It is unclear, however, at what point in the pathway the organism converts into the hyphal form. It is postulated that the conversion begins prior to penetration and that the extension on the germ tube initiates the penetration into the tissue. Other proponents suggest that the yeast initially invades the tissue and the hyphal form occurs as a consequence of exposure to serum components as the immune cells infiltrate the tissue in an attempt to eliminate the organism. The latter is supported by the high incidence of mucosal infections caused by C. glabrata, an organism that does not form hyphae (Fidel et al. 1999b). Nevertheless, it is clear that both the host and the organism contribute significantly to C. albicans as a pathogen.

HOST RESPONSE AGAINST C. ALBICANS

Innate resistance

The polymorphonuclear leukocyte (PMNL) (neutrophil) is the first line of defense against C. albicans. The role of macrophages in defense is less clear although they readily phagocytize the yeast form of Candida. Anti-Candida activities by PMNL or macrophages can be oxidative (i.e. hydrogen peroxide, nitric oxide, superoxide dismutase) or non-oxidative (i.e. lactoferrin, lysozyme, calprotectin, defensins) (reviewed in Fidel 1999). Phagocytosis is usually required for oxidative mechanisms although there is some evidence that oxidative products can be secreted by PMNL that affect nearby fungal hyphal forms too large to be phagocytosed (Levitz and Farrell 1990).

PMNL are relatively effective as a first line of defense, providing that organisms gain access to the vasculature or that the PMNL attach to hyphae. However, PMNL function can be greatly enhanced by cytokines such as tumor necrosis factor-alpha (TNF-α), interferon-gamma (IFN-γ), interleukin (IL)-2, IL-8, and colony-stimulating factors (CSF) (Djeu 1993). Oxidative mechanisms are enhanced in the presence of serum as well. There are also both classical and alternate complement pathways that become activated by C. albicans which promote increased adherence to phagocytes

(Heidenreich and Dierich 1985; Edwards et al. 1986; Kozel et al. 1987).

The functional mechanisms by which monocytes and macrophages affect *C. albicans* are less well defined (reviewed in Fidel 1999). In general, once phagocytosed in vitro, killing is at least as effective as PMNLs and involves oxidative mechanisms and candidicidal molecules. Activation of macrophages with cytokines appears to increase the killing activity, although the effects appear short-lived at best. Killing of both blastospores and hyphae have been demonstrated, but the rate and efficiency depends on the source of the cells. For example, pulmonary macrophages appear to be particularly efficient against *C. albicans*.

A role for dendritic cells against fungal infections has also been suggested. Immature myeloid dendritic cells can phagocytose both yeast and hyphae of *C. albicans* (Romani et al. 2002). Recognition of yeast forms is through the mannose-fucose and complement receptor 3 (CR3) receptors, while recognition of hyphae is through the FcγR and CR3 receptors (Romani et al. 2002). Mature dendritic cells are less phagocytic than immature cells, but they have increased expression of major histocompatibility complex (MHC) class II, which leads to more efficient antigen presentation (Banchereau and Steinman 1998; Pulendran et al. 2001; Reis e Sousa 2001). Dendritic cells are also capable of directing the acquired immune response to *Candida*, depending on the form phagocytosed and the receptors used, and the type of response induced can affect whether a protective or nonprotective response is generated (Romani et al. 2002).

The activity of natural killer (NK) cells against *C. albicans* has been controversial. While a limited number of studies showed inhibition of *C. albicans* growth by murine leukocyte suspensions containing NK cells (Brummer et al. 1986; Morrison et al. 1987; Steele et al. 1999a), the general consensus is that NK cells are not effective against *Candida* (Djeu and Blanchard 1987; Arancia et al. 1995). Suffice it to say that, although inhibition of *C. albicans* can be mediated by cells from lymphoid tissue, the primary and secondary effector cells are innate, not specific, and consist of PMNLs and macrophages, respectively.

A final type of cellular innate resistance mechanism has only recently been discovered and involves epithelial cells lining the surface of mucosal tissues. Oral and vaginal epithelial cells from mice, macaques, and humans have the ability to inhibit the growth of *Candida* in vitro (Steele et al. 1999a, b; Barousse et al. 2001). The inhibition requires contact between viable epithelial cells and *Candida*, with no role for soluble factors and does not involve phagocytosis (Steele et al. 1999a, 2000). While the mechanism is not fully elucidated, the inhibition appears to be mediated by a carbohydrate moiety on viable epithelial cells (Steele et al. 2001) and is static rather than cidal (Nomanbhoy et al. 2002).

Humoral immunity

C. albicans antigens readily induce the production of immunoglobulins as evidenced by the presence of *Candida*-specific IgG in serum and IgA in mucosal secretions of healthy individuals. Thus, it might be presumed that humoral immune mechanisms are important host defenses in both the systemic circulation and at mucosal sites. However, a role for antibody-mediated immunity at mucosal sites has been controversial. A major argument against a role for humoral immunity comes from clinical observations wherein individuals with congenital or acquired B-cell abnormalities are not more susceptible to mucosal (or systemic) candidiasis than immunocompetent individuals (Rogers and Balish 1980). Other observations suggestive of a lack of a role for antibody include the following:

- levels of *Candida*-specific IgA were higher in the saliva of HIV-positive patients than in HIV-negative controls (Coogan et al. 1994; Challacombe and Sweet 1997; Millon et al. 2001)
- levels of *Candida*-specific IgA, sIgA, IgG, or subclasses did not differ in saliva of HIV-positive OPC-positive patients compared to HIV-negative OPC-negative controls (Wozniak et al. 2002a)
- women with RVVC had normal levels of *Candida*-specific IgG in serum and IgA in vaginal secretions (Kirkpatrick et al. 1971; Mathur et al. 1977; Fidel et al. 1997)
- mice protected from a secondary vaginal challenge did not have detectable levels of *Candida*-specific antibodies in vaginal secretions (Wozniak et al. 2002b)
- in vitro studies have provided little evidence that serum antibody and complement can kill *C. albicans* (Rogers and Balish 1980) despite increased phagocytosis of the organism in the presence of complement.

Despite the vast information showing little role for *Candida*-specific antibodies or at least little evidence of any deficiency in such naturally induced antibodies, there are considerable data derived from animals that support a role for *Candida*-specific IgA and IgG antibodies in protection (Cassone et al. 1995; Polonelli et al. 1996; De Bernardis et al. 1997; Han et al. 1998, 2000). These contradictory findings may be reconciled by taking into consideration a theory proposed by Casadevall (1995). He suggested that a pool of natural antibodies may consist of a mixture of protective, nonprotective, and indifferent antibodies. The type that predominates in the pool of antibodies dictates the outcome (protection versus infection). If this theory is correct, then 'protective' antibodies should be present under conditions of 'immunity' and should be identifiable and able to be isolated. Unfortunately, the lack of protection observed in the presence of normal levels of

Candida-specific antibodies suggests that 'protective' antibodies, if they exist, are in the minority of most pools examined. On the other hand, protection in animal models using 'protective antibodies' with known specificities and of a specific class and subclass, provides considerable hope for the use of these antibodies for vaccine development.

Cell-mediated immunity (CMI)

In contrast to the controversial issues regarding the role of humoral immunity against mucosal candidiasis, the role of CMI is somewhat better defined, at least as far as clinical correlations are concerned. Clinical observations support an important role for CMI in protection of mucosal tissues against invasion by *C. albicans*. The strongest evidence is that individuals with reduced or absent CMI (AIDS patients, transplantation recipients, lymphoma patients treated with T-cell immunosuppressive therapy, and patients on corticosteroid therapy) have a high incidence of mucosal candidiasis (i.e. oropharyngeal, esophageal, mucocutaneous) (reviewed in Fidel and Sobel 1994). In addition, there is a direct causal relationship between reduced *Candida*-specific CMI in the peripheral circulation and the incidence of individuals with CMC (Kirkpatrick et al. 1970, 1971). The role of T cells, however, at each respective mucosal site (i.e. oral mucosa, vaginal mucosa, GI tract) is poorly understood as are the relative roles of cells migrating from the systemic circulation into the mucosal sites versus those resident in the mucosal tissue.

The mechanisms associated with CMI against *C. albicans* appear to be dominated by CD4$^+$ T cells. Protection often correlates with delayed-type hypersensitivity (DTH), but a direct role for DTH has been difficult to assess. There have been reports where the presence of DTH did not correlate with protection (Fidel et al. 1994; Garner and Domer 1994). Nevertheless, *Candida*-specific DTH develops in vivo in the immunocompetent individual. DTH is demonstrable at an early age and likely results from early exposure to *C. albicans*. Thus, a pool of *Candida*-specific memory T cells continues to circulate in the peripheral circulation. The response is presumably initiated at the mucosal site where antigen-presenting cells such as Langerhans cells or dendritic cells process and present the antigen to locally associated T cells or transport it to the draining lymph nodes where T cells can be primed. Upon migration of the *Candida*-specific T cells to the site of antigen deposition and contact with additional processed antigen, cytokine secretion by the T cells activates tissue macrophages and signals other phagocytic cells to infiltrate the site of the insult (mononuclear infiltrate), in an attempt to eliminate or reduce the organism. The current dogma is that Th1-type responses are associated with resistance against intracellular pathogens, while Th2-type responses are associated with resistance to extracellular pathogens. However, for *C. albicans*, and in fact most medically important fungi, Th1-type responses have been associated with resistance to infection, with Th2-type responses showing an association with susceptibility to infection (Magee and Cox 1995; Romani et al. 1996; Cenci et al. 1999; Tachibana et al. 1999; Cain and Deepe 2000; Wuthrich et al. 2000).

In contrast to the role CD4$^+$ cells play in protection from *C. albicans* infections at mucosal tissues, a role for CD8$^+$ cells as an acquired host defense is unknown. To date, there has been no evidence that CD8$^+$ cytotoxic T lymphocytes (CTL) are directly involved in host defense against *C. albicans*. However, gnotobiotic mice deficient in CD8$^+$ T cells are susceptible to gastric candidiasis (Balish et al. 1996; Jones-Carson et al. 2000). Additionally, CD8$^+$ T cells have a role in suppression of CMI in candidiasis. There are numerous reports showing that CD8$^+$ T cells specific for *C. albicans* cell wall antigens, both glycoprotein and mannan, suppress DTH and lymphocyte proliferation in infected mice (Domer et al. 1989; Garner et al. 1990; Garner and Domer 1994). Suppressor or regulatory cells also appear to hinder the ability to clear the infection (Garner and Domer 1994). It appears that the downregulation of DTH includes an antagonistic function of IL-12 p40, and that the production of CD8$^+$ downregulatory T cells is dependent on the presence of IL-10, IL-4, and IFN-γ in the absence of biologically active IL-12 (Li et al. 1998; Wang et al. 1998). Additionally, naïve CD8$^+$ T cells stimulated with IL-2 inhibit the growth of *C. albicans* (Beno et al. 1995). This was first shown in cells from uninfected mice, but more recently has also been suggested clinically in persons who have had a recent episode of OPC (Colon et al. 1998). The mechanism appears to be non-MHC restricted through the action of excreted enzymes (Mathews and Beno 1993; Beno et al. 1995) and may in fact be a form of innate immunity.

HOST DEFENSE AGAINST SUPERFICIAL CANDIDIASIS AT ANATOMIC SITES

Cutaneous/mucocutaneous candidiasis

Host defense against cutaneous or mucocutaneous candidiasis involves both innate and adaptive host defenses. Innate defenses involve PMNL and monocytes/macrophages. These cells can be resident in the respective tissues or recruited to the site by chemotactic factors. The adaptive host defenses are predominantly T cells, most probably of the CD4$^+$ Th1-type cells. In cases of CMC, in vitro *Candida*-specific responses by peripheral blood mononuclear cells (proliferation and cytokine production) are reduced compared to those without CMC, suggesting some immunological defect in CMI (Kirkpatrick et al. 1971; Ruhnke 2002). In addition, delayed skin test reactivity to *Candida* antigen is usually absent (Kirkpatrick et al. 1970; Ruhnke 2002). These reduced responses can be restricted to *Candida* or can be part of an overall reduced responsiveness to several antigens (Kirkpatrick et al. 1970, 1971). The results of studies involving experimental cutaneous disease also suggest a protective role for CMI (Wilson and Sohnle 1986). In a more recent report, in fact, the authors have described the costimulatory isoform of B7 (B7-2) as being required on antigen-presenting cells to induce a protective Th1-type response (Gaspari et al. 1998). By contrast, there is little evidence that humoral immunity has a role in protection in CMC. *Candida*-specific antibodies are present in sera and mucosal secretions of those with or without CMC (Lehner et al. 1972; Ruhnke 2002). However, aside from immunoreactivity for detection of antibodies, no studies have assessed the actual function of the antibodies present. Serum factors have been suggested as a possible source of the immunological defect(s) in patients with CMC (Valdimarsson et al. 1973; Twomey et al. 1975), in that it has been reported that *Candida*-specific carbohydrate antigens can be present and persist in the serum of those with CMC that inhibit peripheral blood lymphocyte responses (Fischer et al. 1978).

Interestingly, the immune responses against cutaneous and mucocutaneous candidiasis may involve an immune network in the epidermis termed skin-associated lymphoid tissue (SALT) as well as those infiltrating from the periphery (Singl et al. 1999). SALT includes epidermal Langerhan's cells, dermal dendritic cells, dermal T lymphocytes consisting primarily of α/β T-cell receptor (TCR)-positive cells and small numbers of γ/δ TCR$^+$ cells, epidermal T cells with considerably more γ/δ TCR$^+$ cells, keratinocytes, and microvascular endothelial cells (Bos et al. 1990).

Gastrointestinal candidiasis

Host defense against GI candidiasis has been studied primarily through the use of immunodeficient mice. Studies using both congenitally deficient and severe combined immunodeficient (SCID) mice have shown that T cells are critical for effective protection against GI candidiasis (Cantorna and Balish 1990). SCID mice were shown to be more susceptible to GI tract colonization, but were not susceptible to intravenous challenge or dissemination from the GI tract. Mice with congenital T-cell (*nude/nude*) and phagoctyic cell (*beige/beige*) deficiencies (*nu/nu, bg/bg*) developed persistent GI candidiasis, while those with only phagocytic cell deficiencies (*nu/+, bg/bg*) cleared *C. albicans* from the GI tract effectively. The *nu/+* mice (wild-type) responded to *Candida* antigens with IL-2 production and DTH, while *nu/nu* mice did not, suggesting the correlation between a Th1-type response and clearance of the organism. An

important question, however, is what role circulating and mucosal T cells actually play in the protection. It has been shown that the mucosal tissue of the GI tract has intraepithelial lymphocytes (IEL), Peyer's patches (organized lymphoid tissue), T cells, and antigen-presenting cells that function in many types of GI infections. Much less is known, however, about the function of any of these cell populations against *C. albicans*, although Bistoni and colleagues have begun to address this issue. It was shown that Peyer's patch lymphocytes from colonized mice which went on to clear the fungus from the GI tract produced Th1, but not Th2-type cytokines in response to *Candida* antigen (Bistoni et al. 1993). Additionally, GI candidiasis appears to conform to the Th1/Th2-type immune response pattern seen in cutaneous infection and experimental systemic infections, namely that Th1-type responses are associated with resistance against infection, while Th2-type responses promote susceptibility (Magee and Cox 1995; Romani et al. 1996; Cenci et al. 1999; Tachibana et al. 1999; Cain and Deepe 2000; Wuthrich et al. 2000). The role of antibodies against GI candidiasis has not been evaluated experimentally except through early studies that showed no protection from *Candida* antibody-containing sera (Balish and Filutowicz 1991; Greenfield 1992).

Host response to esophageal candidiasis has not been studied clinically or in animal models. However, clinical observations show susceptibility to infection almost exclusively during severe T cell immunocompromised conditions (i.e. late-stage AIDS) (Macher 1988; Hoover et al. 1993; Martinez et al. 2000), suggesting that protective host defenses are intact for longer periods of time in the esophagus compared to the oral cavity, although this could also be due to less of a commensal presence of *Candida* in the esophagus. Additionally, animal models have been developed for study of esophageal candidiasis (De Repentigny et al. 2000; Walsh et al. 2000), but have been used for evaluation of chemotherapeutic treatment or fungal pathogenesis, and studies on host defense against esophageal candidiasis are still sorely needed.

Oropharyngeal candidiasis

Oropharyngeal candidiasis (OPC) occurs primarily in individuals with HIV (Nielsen et al. 1994; Sangeorzan et al. 1994; Schuman et al. 1998a; Greenspan et al. 2000). Host defense against OPC has only recently begun to be studied in greater detail due primarily to the HIV epidemic. Animal models have not been of great value in understanding the natural host defense, as the disease resolves too quickly with variable and inconsistent fungal titers in immunocompetent animals (reviewed in Samaranayake and Samaranayake 2001). On the other hand, information has accumulated from humans who are immunocompromised. In general, most data to date suggest that CD4+ Th1-type cells are critical for host defense against infection. Clinically, OPC is most common in HIV-positive persons when CD4+ cell numbers drop below 200 cells/μl (Rabeneck et al. 1993; Nielsen et al. 1994; Schuman et al. 1998a; Greenspan et al. 2000). In vitro immune analyses using peripheral blood mononuclear cells (PBMC) show that cells from most individuals respond to *Candida* antigens with Th1-type cytokines. Thus, it is generally considered that susceptibility to OPC is enhanced under reduced CD4+ T cells due to either a lack of Th1-type responses and/or a shift to Th2-type responses. Some early studies in HIV-positive individuals suggested that reduced Th1-type cytokine responses were responsible for OPC (Quinti et al. 1991). However, a study evaluating PBMC reactivity in HIV-negative and HIV-positive persons with or without symptomatic OPC stratified by CD4+ T-cell numbers showed little to no appreciable differences in *Candida*-specific proliferation or cytokine production between the two groups (Leigh et al. 2001). In another study, no demonstrable deficiencies in clonal responses of the PBMC from such patients to a variety of antigenic peptides were shown (Kunkl et al. 1998). These results suggested that the *Candida*-specific T cells themselves were not becoming defective with immunosuppression, but that a threshold number of CD4+ T cells was required to protect the oral cavity against infection by this commensal organism. Below this threshold number of cells, local immune mechanisms must function exclusively for protection. The prevalence of OPC may then depend on the status of the local immune mechanisms. Indeed, some individuals with < 200 CD4 cells/μl never have OPC while others have recurrent bouts of OPC.

Local immunity has only now begun to be studied clinically. In support of the Th1/Th2 dichotomy concept, it was recently reported that HIV-negative individuals had Th1/Th0 cytokines in their saliva, whereas HIV-positive individuals had primarily Th2-type cytokines, which were in higher concentration in those patients with OPC (Leigh et al. 1998). Interestingly, the Th2-type profile was the result of reduced Th1-type cytokines rather than increased Th2-type cytokines. Lymphocytes have also been examined in the OPC lesions and both CD4+ and CD8+ cells have been identified (Romagnoli et al. 1997). Other investigators have suggested that only CD8+ cells were present. Data from our laboratory supports the latter observation, and, in fact, we have shown an accumulation of CD8+ T cells at a considerable distance from the *Candida* located superficially at the outer epithelium (Myers et al. 2003). This cellular accumulation is not seen in those without OPC. This may suggest a role for CD8+ T cells against infection, with a potential problem in cell trafficking leading to OPC. Interestingly, a murine AIDS model (MAIDS) showed a rate of 30 percent recurrent OPC in inoculated

mice with a predominance of CD8+ T cells recruited into the tissues (Deslauriers et al. 1997).

Studies using experimental models support a role for a combination of T cells and innate cells against OPC. Both CD4+ and CD8+ T cells were shown to be recruited along with macrophages into the mucosal tissue, and intraepithelial CD4+ T cells persisted after resolution of the infection (Allen et al. 1994; Chakir et al. 1994). Moreover, there was a time-dependent recruitment of γ/δ TCR+ cells that correlated with the resolution of the disease (Chakir et al. 1994; Elahi et al. 2000). Analysis of cytokines showed a role for both Th1- and Th2-type cytokines in resistance to infection depending on the strain of mice used (Elahi et al. 2000). A recent study examined cytokines in oral tissues from infected mice and found increases in IL-6, IFN-γ, and TNF-α in mice recovering from oral infection (Farah et al. 2002). However, no IL-2, 4, or 10 was detected during infection (Farah et al. 2002). As with GI candidiasis, however, little is known concerning the specific roles of mucosal versus systemic T cells in the mucosa versus circulating T cells in protection against infection. In a recent study in vivo cellular depletions were used to create immunocompromised mice. CD4+ T cells, PMNL, and macrophages all appeared to be important in resistance to disease (Farah et al. 2001).

Humoral immunity, on the other hand, again does not appear to play a role in protection against or susceptibility to OPC. Early studies of *Candida*-specific antibodies in saliva of HIV-positive persons with or without OPC yielded similar or elevated levels of IgA or IgG, although in one study a reduced affinity of *Candida*-specific IgA antibodies was shown in AIDS patients (Coogan et al. 1994). There is no evidence to date, however, that a deficiency in *Candida*-specific antibodies is present in HIV-positive persons that could account for the increased prevalence of OPC (Wray et al. 1990; Millon et al. 2001). A recent comprehensive analysis of *Candida*-specific IgA and IgG in the saliva of a large number of HIV-negative and HIV-positive persons stratified by OPC status as well as CD4+ T-cell numbers, which included subclass analysis, supports the earlier findings (Wozniak et al. 2002a).

Other anti-OPC mechanisms involve innate cellular defenses. PMNL appear to play a role, as neutropenic patients are susceptible to OPC (Anaissie and Bodey 1990). However, in-depth studies with PMNL in the oral cavity have not been conducted. Epithelial cells represent another cell type with potential innate function. As stated earlier, epithelial cells have been shown to inhibit the growth of *Candida* spp. in vitro by a static mechanism (Steele et al. 1999a, 2000; Nomanbhoy et al. 2002). While both oral and vaginal epithelial cells have anti-*Candida* activity via cell contact through a putative carbohydrate moiety (Steele et al. 2001), oral epithelial cells have a much increased capacity to inhibit the growth of *Candida* (up to 80 percent versus 40 percent for vaginal cells) (Steele et al. 2000). Analysis of oral epithelial cells in HIV-positive persons showed significantly reduced activity when the cells were derived from patients with OPC compared to that from patients without OPC, potentially contributing to the susceptibility to infection (Steele et al. 2000). Additionally, epithelial cells produce both cytokines and chemokines in response to *Candida*, which may contribute to the innate and/or adaptive immune response (Steele and Fidel 2002). Thus, several lines of defense may be important for protection against OPC, many of which do not become evident until CD4$^+$ cells are reduced below the protective threshold.

Vaginal candidiasis

In contrast to the fairly well documented role for CMI by CD4$^+$ T cells against oral, gastrointestinal, and cutaneous candidiasis, the role for CMI against vaginal candidiasis has not been clear and more recently has been challenged. Thus, despite intense efforts to date, the natural protective host defenses against VVC remain poorly understood.

Studies of host defense against VVC began with clinical observations in which women with RVVC were examined for systemic *Candida*-specific responsiveness by culture of their PBMC in vitro or through skin-testing in search of DTH responses. The results over time conflicted with the results of some studies wherein it was suggested that there was a deficiency in *Candida*-specific CMI, or an antigen nonspecific deficiency, or no deficiency at all (reviewed in Fidel and Sobel 1996). Thus, these studies did little to identify an immunological deficiency, if any, in women with RVVC.

In subsequent studies, animal models have been employed to attempt to understand the natural host defense mechanisms against vaginal *C. albicans* infection. Such studies have included rat, mouse, and macaque models of vaginal candidiasis (reviewed in Fidel and Sobel 1999). All are dependent on a state of pseudoestrus for the infection to become established. The rat model has been particularly helpful in suggesting a potential role of antibodies against *C. albicans* at the vaginal mucosa. The mouse model, on the other hand, has been used primarily to study the role of CMI (reviewed in Sobel et al. 1998). The macaque model is relatively new and to date has been described with only limited immunological data (Steele et al. 1999b). Mice and rats, in contrast to macaques and humans, are not normally colonized with *C. albicans* and do not have any preexisting *Candida*-specific immunity.

Mice inoculated under pseudoestrus conditions acquire a persistent vaginal infection together with systemic Th1-type responses, as evidenced by Th1-type cytokine production in response to *Candida* antigen by lymph node cells draining the vaginal tissue, and

Candida-specific DTH. However, this systemic Th1-type CMI or that preinduced by systemic immunization with *Candida* antigen in adjuvant did not protect mice against vaginal candidiasis (Fidel et al. 1994). Partial protection against vaginitis was achieved, however, in animals given a second inoculation following the spontaneous resolution of a primary infection in the absence of estrogen (Fidel et al. 1995b). Interestingly, suppression of *Candida*-specific systemic CMI or systemic T-cell depletion had no effect on the protection, suggesting that systemic CMI had a limited role at the vaginal mucosa in protection against vaginal candidiasis. In support of this suggestion as well, mice, either resistant or susceptible to systemic *C. albicans* infection (Hector et al. 1982), were shown to be equally susceptible to vaginitis and could all be partially protected against a second vaginal infection (Fidel et al. 1995a). These data were supported by those gathered in another study in which six different strains of mice representing four haplotypes were used (Black et al. 1999a). Finally, adoptive transfer of *Candida*-sensitized T cells into T-cell deficient (nude) mice had no effect on the course of vaginitis (Black et al. 1999b). Based on these results, it was postulated that some form of locally acquired mucosal immunity, T-cell and/or antibody-mediated, was responsible for protecting mice against vaginal *C. albicans* infection and that the vaginal mucosa had a significant level of immunological independence with little influence from systemic sources.

The independent compartmentalization concept correlates well with a clinical study showing that RVVC patients had normal levels of *Candida*-specific Th1-type CMI in the peripheral circulation (Fidel et al. 1993). It should be mentioned, however, that others have obtained data to the contrary (Corrigan et al. 1998; Carvalho et al. 2002). Nevertheless, the lack of identifiable deficiencies in systemic CMI in women with RVVC correlates with two relevant clinical observations; women with RVVC are not susceptible to CMC or other forms of cutaneous candidiasis (Sobel 1988), and women with CMC are generally not susceptible to RVVC (Odds 1988a). Thus, the immune deficiency, if any, in RVVC patients is presumed to be localized to the vagina and does not occur at the systemic level. These data were the first evidence to suggest some level of immunological independence regarding CMI host defense mechanisms at the vaginal mucosa against *C. albicans*. Further support for these observations comes from studies with HIV-infected women, who, despite the higher incidence of OPC under conditions of immunosuppression, do not get VVC any more frequently than HIV-negative women (Clark et al. 1995; White 1996; Schuman et al. 1998b; Leigh et al. 2001). Furthermore, peripheral blood lymphocyte proliferation and Th1/Th2 cytokine production in response to *Candida* antigens was relatively similar in HIV-positive women with or without symptomatic VVC (Leigh et al. 2001), similar to RVVC patients.

Subsequent studies focused on the presence of T-cell subpopulations in the vaginal mucosa. It had been established that the vaginal mucosa, while not composed of organized lymphoid tissues as the Peyer's patches in the GI tract, is immunologically competent for both CMI and T-dependent antibody production with the presence of both T cells and MHC class II$^+$ cells. To this end, we and others reported that vaginal lymphocytes are phenotypically distinct from those in the peripheral circulation (Nandi and Allison 1991; Fidel et al. 1996). Despite the fact that CD4$^+$ α/β TCR$^+$ cells predominate in the vaginal tissue, a higher percentage of γ/δ TCR$^+$ cells are present with few if any CD8$^+$ cells, compared to the systemic circulation. Interestingly, human vaginal lymphocytes are also phenotypically distinct in comparison to those in the peripheral circulation, but differ from mice in the types and relative proportions of cells; the human vagina has increased CD4$^+$ and γ/δ T cells but also equal numbers of CD8$^+$ T cells (Hladik et al. 1999; Fidel et al. 2001).

Studies to examine changes in vaginal T cells during experimental vaginal Candida infection have all shown little evidence for modulation of α/β or γ/δ vaginal T cells during either primary or secondary experimental vaginal C. albicans infections, and no evidence for systemic T-cell infiltration during a vaginal Candida infection (Fidel et al. 1999a; Wormley et al. 2001a). These results are consistent with those from immunodeficient or knockout mice (Cantorna et al. 1990; Black et al. 1999b), and correlate with the systemic immune studies in immunocompetent mice (Fidel et al. 1994, 1995a, b, c). This lack of responsiveness to Candida infection in mice is in direct contrast, however, to a murine model of a genital Chlamydia trachomatis or herpes simplex virus-2 (HSV-2) infection where CD4$^+$ T cells were shown to infiltrate readily into the vaginal mucosa in response to infection (Kelly and Rank 1997; Parr and Parr 1997). Taken as a whole, these data emphasize the uniqueness of the host vaginal response(s) against C. albicans. Interestingly, a study in which a dual infection with Candida and Chlamydia was studied, showed a complete independence of 'responsiveness' in the upper genital tract (Chlamydia) and 'lack of responsiveness' in the lower tract (Candida) (Kelly et al. 2001), illustrating further the complexity of immunity in the genital tract. Together these data suggested that instead of a simple lack of a role for systemic or local CMI against C. albicans at the vaginal mucosa, there may be some form of immunoregulation or tolerance that prohibits a more profound CMI response.

There are several pieces of evidence, both experimentally and clinically, that support immunoregulation against CMI at the vaginal mucosa. First, experimental vaginitis in δ-chain TCR knockout mice has been found to be less severe, suggestive of a tolerance role for vaginal γ/δ T cells (Wormley et al. 2001b). Evaluation of Th cytokines in the vaginal tissue homogenates from mice showed high constitutive concentrations of transforming growth factor-beta (TGF-β) in naïve mice that were increased further in estrogen-treated and/or vaginally infected mice; these same mice had low levels of other Th1- or Th2-type cytokines (Taylor et al. 2000). Clinically, TGF-β was found in fairly high concentrations in human vaginal secretions with modulations upward over the course of the menstrual cycle (Fidel et al. 2003). In addition, although Th1/Th2 cytokines are detectable in vaginal secretions of women with RVVC, there are no discernible patterns that would account for susceptibility to RVVC when comparisons were made to control women (Fidel et al. 1997). In the mouse model of infection, T cells with appropriate homing receptors for infiltration into the vaginal mucosa are reduced/lost during a vaginal Candida infection despite the upregulation of reciprocal adhesion molecules on the vaginal endothelium (Wormley et al. 2001b). Further supporting immunoregulation, a T-cell-derived antigen-binding molecule (TABM) has been identified in serum of RVVC patients, and to a lesser extent in control women, that can become associated with TGF-β and inhibit Candida-specific peripheral blood leukocyte (PBL) proliferation and cytokine production (Little et al. 2000). Finally, despite some preliminary evidence that the intravaginal challenge of normal healthy women with Candida antigen stimulated Th1-type cytokine secretion in vaginal lavages 16–18 hours post-challenge (Fidel et al. 1997), the results of a more recent study in which responsiveness to the antigen in a large number of women was tested during different stages of the menstrual cycle showed no evidence of immune responsiveness (Fidel et al. 2003). Thus, there is convincing evidence that immunoregulation or tolerance is functioning at the vaginal mucosa and that it inhibits the ability of Candida-specific systemic CMI to reach the vaginal mucosa.

The rat model, used primarily to study humoral immunity, was also used to evaluate CMI (De Bernardis et al. 2000). The progression of infection, however, is quite different in the rat than the mouse, in that spontaneous resolution occurs within a 3-week period in the rat. Consistent with this property is the fact that T cells could be demonstrated infiltrating the vaginal mucosa following infection, and Th1-type cytokines were produced and secreted into vaginal secretions. In addition, in contrast to data obtained from the mouse model of vaginitis, a recent study in the rat model showed that passive transfer of both CD4$^+$ and CD8$^+$ T cells from previously infected rats accelerated clearance of Candida in the vagina of naïve animals (Santoni et al. 2002). Also, the CD4$^+$ T cells were more effective at clearance than the CD8$^+$ T cells, suggesting that, in the rat model, vaginal lymphocytes, especially CD4$^+$ T cells, are important in clearance of Candida from the vagina (Santoni et al. 2002). Thus, the rat, in contrast to the mouse and human, may exemplify what would occur in response to infection in the absence of immunoregulation.

The role of humoral immunity in recovery from vaginitis is equally controversial. Immunoglobulins are clearly present in vaginal secretions, including *Candida*-specific antibodies (IgG and IgA) (Witkin et al. 1988, 1989; Regulez et al. 1994). Interestingly, though, B cells are a minor constituent of the mouse, rat, or human vaginal mucosa (Parr and Parr 1991). In contrast to other mucosal sites, IgG antibodies predominate over IgA antibodies, at least on the basis of total immunoglobulin (Fidel et al. 1997; Kutteh et al. 1998). Despite that, clinical studies show that serum IgG and vaginal IgG and IgA are similar in women with RVVC (symptomatic or in remission) and healthy control women with no history of RVVC (Kirkpatrick et al. 1971; Mathur et al. 1977; Fidel et al. 1997). Experimentally, similar results were observed in infected and uninfected rhesus macaques, including the predominance of vaginal IgG over IgA immunoglobulins (Steele et al. 1999b). In addition, in a mouse model of vaginal infection, *Candida*-specific IgA and IgG were not detectable in lavage fluids of animals protected against secondary challenge, suggesting little or no protective role for *Candida*-specific antibodies against the murine vaginal infection (Wozniak et al. 2002b). While these experimental and clinical observations largely fail to support a role for antibody against candidiasis, there are studies in rodents that indicate a protective role for humoral immunity in experimental vaginal candidiasis. Pollonelli and colleagues showed that rats given an intravaginal immunization using antibodies that neutralized the anti-*Candida* activity of a yeast killer toxin were protected against a vaginal infection (Polonelli et al. 1994). The protection was found to be associated with rising vaginal titers of anti-idiotypic IgA antibodies that could transfer protection passively to unimmunized rats. Interestingly, killer toxin anti-idiotypic-like antibodies were also detected in women with symptomatic vaginitis (Polonelli et al. 1996). Similarly, Cassone and colleagues (Cassone et al. 1995; De Bernardis et al. 1997) have shown that a primary *C. albicans* vaginal infection in estrogen-treated rats induced anti-mannan and anti-aspartyl IgA antibodies in vaginal secretions, and that those antibodies appeared to protect the same rats against a secondary vaginal infection. Moreover, recipient rats treated intravaginally with vaginal fluids from primary-infected rats were protected. In a follow-up study, vaginal lymphocyte analysis during primary or secondary vaginal *C. albicans* infections in rats revealed an increase in B cells in the tissue (De Bernardis et al. 2000). Successful protection against vaginitis in the presence of these IgA antibodies could reflect 'protective' antibodies based on the Casadevall concept of nonprotective, protective, and indifferent antibodies (Casadevall 1995). In support of this, Cutler and colleagues have shown that two different 'protective' antibodies specific for mannan (IgM and

IgG3) protect mice against *Candida* vaginitis when given either locally or systemically (Han et al. 1998, 2000). Taking this into account, one might speculate that when antibodies are detected clinically that do not appear to have a protective effect, indifferent or unprotective antibodies may predominate in the antibody pool. If, indeed, there is some precedent for this, it appears to be the norm rather than the exception.

Innate immunity appears to play a significant role in protection against vaginitis as well. This was initially predicted based on a leukocytic infiltrate of predominantly PMNL that is often observed in vaginal lavage fluid of infected animals. These leukocytic cells are usually associated with or attached to the hyphae and/or sheets of epithelial cells. On the other hand, when this infiltrate is present during infection it does not correlate with lower vaginal fungal burden (Fidel et al. 1999a; Saavedra et al. 1999). Furthermore, this leukocytic infiltrate is generally not observed in clinical cases of *Candida* vaginitis. Although PMNL and macrophages are potential candidates for anti-*Candida* innate resistance and are present at or near the vaginal mucosa, Balish and colleagues (Cantorna et al. 1990) showed that animals with the beige mutation, which are immunodeficient in phagocytic cells (*bg/bg*), were not more susceptible to a natural *C. albicans* infection under non-estrogenized conditions. Two more recent studies addressed the specific role of PMNL during experimental vaginitis. The results indicated that depletion of PMNL under estrogen- or non-estrogen-treated conditions had no effect on vaginal fungal burden (Black et al. 1998; Fidel et al. 1999a), suggesting that PMNL do not play a significant role against *C. albicans* in the vagina despite their presence in the vaginal lumen during some experimental infections. Perhaps this reflects their inability to function against *C. albicans* in the vaginal microenvironment compared to their ability to function well in blood or tissue culture. It might reflect immunoregulatory activity as well. Alternatively, the presence of PMNL during an infection might not represent a response to the yeast, but simply reflect their normal presence during the diestrus stage of the menstrual cycle (every 2 days in mice) when they are deployed to phagocytize the apoptotic squamous epithelial cells from the estrus (2-day cycle) stage.

As mentioned above, vaginal epithelial cells represent a potential innate anti-*Candida* host defense mechanism. This has been shown for both mouse, macaque, and human cells (Steele et al. 1999b, 2000; Barousse et al. 2001). As described earlier, the anti-*Candida* activity mediated by a putative carbohydrate moiety inhibits rather than kills the *Candida* (Nomanbhoy et al. 2002), but vaginal cells are not as effective as oral epithelial cells (Steele et al. 2000). If these cells are important as an innate mechanism, less activity by vaginal cells may reflect a higher prevalence of vaginal versus oral candidiasis when both tissues are colonized

Table 15.2 *Host defenses against superficial* Candida *infections*

Host defense mechanism	Site of infection			
	Cutaneous	GI[a]	Oral	Vaginal
Innate				
PMNL[b]	+[d]	− −	+/?	− −/?
Macrophage	+	− −	− −	− −
NK cells[c]	− −	− −	− −	− −
Epithelial cells	?	?	++	+
Humoral				
IgM	− −	− −	− −	+/−?
IgG	− −	− −	− −	+/−?
IgA	− −	− −	− −/?	+/−?
Cell-mediated				
CD4[+] T cells (Th1)	+++	+++	+++	−/?
CD8[+] T cells	− −	+/?	++	− −

a) Gastrointestinal.
b) Polymorphonuclear leukocytes.
c) Natural killer cells.
d) Symbols: + to +++ indicates level of activity from weakly positive (+) to strongly positive (+++); +/−? indicates variable activity, if any; +/? indicates probaby activity, but unknown at present; − −/? indicates probably negative activity, but unknown at present; − − indicates negative activity; ? indicates unknown.

(Steele et al. 2000). Interestingly, a recent study in humans showed that while the vaginal epithelial cell-mediated anti-*Candida* activity was not different at the various stages of the menstrual cycle, it was significantly reduced in women with a history of RVVC (Barousse et al. 2001). Thus, reduced epithelial cell anti-*Candida* activity may represent, in part, a local immune deficiency associated with RVVC.

Table 15.2 summarizes the various host defenses against each form of superficial candidiasis with what is known and what is still questionable.

IMMUNOTHERAPY

In addition to antifungal drugs available, new treatment strategies are being explored, especially in the case of azole-resistance (Walker et al. 2000; Mathema et al. 2001; Ferris et al. 2002) and the appearance of infections by non-*albicans* species that are intrinsically azole resistant (Sobel 1998; Fidel et al. 1999b; Abu-Elteen 2001). One goal of studying host defense is to design possible immunotherapeutic strategies to treat or prevent infection. For example, antibodies directed at SAPs, manno-proteins, mannan, and mannoprotein extract have been proposed as potential treatments and vaccine candidates (reviewed in Magliani et al. 2002). Intravaginal and intranasal immunization using mannoprotein MP65 and Sap2 have both proved to be protective in a rat model of vaginal candidiasis (De Bernardis et al. 2002) and a mannan-specific antibody was protective against a murine vaginal infection (Han et al. 2000). Additionally, anti-idiotypic (Id) vaccination and anti-Id therapy against the yeast killer toxin receptor (KTR) have been described for both systemic and mucosal *Candida* infection (reviewed in Magliani et al. 2002). Monoclonal and recombinant rat and mouse antibodies to the KTR (mKAb and rKAb) have been produced and can kill *C. albicans* in vitro (Magliani et al. 1997; Polonelli et al. 1997). In humans, *Streptococcus gordonii*, which is a commensal organism of the vaginal mucosa, has been engineered to secrete rKAb, and this has been shown to be as effective as fluconazole treatment against vaginitis in the rat model (Beninati et al. 2000). Recently, protease inhibitors used for treatment of HIV have been shown to reduce the incidence/occurrence of OPC in HIV-positive individuals (Cassone and Cauda 2002). This is thought to be mediated by inhibiting SAPs, which are a major virulence factor and are important in invasion and tissue damage (reviewed in Hube and Naglik 2002) (De Bernardis et al. 1990, 1999). In a clinical study that examined OPC following treatment with protease inhibitors, a reduction in OPC cases as well as a reduction of the presence of SAPs in the saliva of treated patients was observed (Cassone et al. 2002). Furthermore, in the case of vaginitis, overcoming immunoregulation by gene therapy, administration of cytokines, or use of antibodies against immunoregulatory cytokines may be potential methods of immunotherapy. Alternative treatments for candidiasis are promising and may be used to treat antifungal drug-resistant infections.

CONCLUSION

In summary, superficial candidiasis includes several types of infections, including cutaneous, oropharyngeal, gastrointestinal, vaginal, and ocular. Infections at each site are accompanied by unique signs and symptoms and affect different groups of individuals. Protective host defenses against infection at each anatomic site are independent/compartmentalized, and quite distinct, with some sites revealing mechanisms consistent with the dogma of CD4[+] Th1-type CMI as the main protective mechanism, and other sites, particularly the vaginal mucosa, challenging this dogma. In addition, innate immunity appears to be playing more of a role than previously thought, while humoral immunity is still considered controversial. Finally, immunotherapeutic strategies developed as a result of the research efforts appear promising, with more strategies predicted as additional immune mechanisms against superficial candidiasis are uncovered.

REFERENCES

Abu-Elteen, K.H. 2001. Increased incidence of vulvovaginal candidiasis caused by *Candida glabrata* in Jordan. *Jpn J Inf Dis*, **54**, 103–7.

Allen, C.M., Saffer, A., et al. 1994. Comparison of a lesion-inducing isolate and a non-lesional isolate of *Candida albicans* in an immunosuppressed rat model of oral candidiasis. *J Oral Pathol Med*, **23**, 133–9.

Anaissie, E.J. and Bodey, G.P. 1990. Fungal infections in patients with cancer. *Pharmacotherapy*, **10**, 164S–9S.

Arancia, G., Molinari, A., et al. 1995. Noninhibitory binding of human interleukin-2-activated natural killer cells to the germ tube forms of *Candida albicans*. *Inf Immun*, **63**, 280–8.

Bai, C., Ramanan, N., et al. 2002. Spindle assembly checkpoint component CaMad2p is indispensable for *Candida albicans* survival and virulence in mice. *Mol Microbiol*, **45**, 31–44.

Balish, E. and Filutowicz, H. 1991. Serum antibody response of gnotobiotic athymic and euthymic mice following alimentary tract colonization and infection with *Candida albicans*. *Can J Microbiol*, **37**, 204–10.

Balish, E., Vazquez-Torres, A., et al. 1996. Importance of B2-microglobulin in murine resistance to mucosal and systemic candidiasis. *Inf Immun*, **64**, 5092–7.

Banchereau, J. and Steinman, R.M. 1998. Dendritic cells and the control of immunity. *Nature*, **392**, 245–52.

Barousse, M.M., Steele, C., et al. 2001. Growth inhibition of *Candida albicans* by human vaginal epithelial cells. *J Infect Dis*, **184**, 1489–93.

Bauters, T.G.M., Dhont, M.A., et al. 2002. Prevalence of vulvovaginal candidiasis and susceptibility to fluconazole in women. *Am J Obstet Gynecol*, **187**, 569–74.

Beninati, C., Oggioni, M.R., et al. 2000. Therapy of mucosal candidiasis by expression of an anti-idiotype in human commensal bacteria. *Nature Biotechnol*, **18**, 1060–4.

Beno, D.W.A., Stover, A.G., et al. 1995. Growth inhibition of *Candida albicans* hyphae by CD8+ lymphocytes. *J Immunol*, **154**, 5273–81.

Bernhardt, H. and Knoke, M. 1997. Mycological aspects of gastrointestinal microflora. *Scand J Gastroenterol*, **222**, 102–6.

Bistoni, F., Cenci, E., et al. 1993. Mucosal and systemic T helper cell function after intragastric colonization of adult mice with *Candida albicans*. *J Infect Dis*, **168**, 1449–57.

Black, C.A., Eyers, F.M., et al. 1998. Acute neutropenia decreases inflammation associated with murine vaginal candidiasis but has no effect on the course of infection. *Infect Immun*, **66**, 1273–5.

Black, C.A., Eyers, F.M., et al. 1999a. Major histocompatibility haplotype does not impact the course of experimentally induced murine vaginal candidiasis. *Lab Anim Sci*, **49**, 668–72.

Black, C.A., Eyers, F.M., et al. 1999b. Increased severity of *Candida* vaginitis in BALB/c *nu/nu* mice versus the parent strain is not abrogated by adoptive transfer of T cell enriched lymphocytes. *J Reprod Immunol*, **45**, 1–18.

Bos, J.D., Teunissen, M.B., et al. 1990. T-cell receptor gamma delta bearing cells in normal human skin. *J Invest Dermatol*, **94**, 37–42.

Brummer, E., McEwen, J.G. and Stevens, D.A. 1986. Fungicidal activity of murine inflammatory polymorphonuclear neutrophils: comparison with murine peripheral blood PMN. *Clin Exp Immunol*, **66**, 681–90.

Butrus, S.I. and Klotz, S.A. 1986. Blocking *Candida* adherence to contact lenses. *Curr Eye Res*, **5**, 745–50.

Cain, J.A. and Deepe, G.S. Jr 2000. Interleukin-12 neutralization alters lung inflammation and leukocyte expression of CD80, CD86, and major histocompatibility complex class II in mice infected with *Histoplasma capsulatum*. *Inf Immun*, **68**, 2069–76.

Calderone, R.A. and Fonzi, W.A. 2001. Virulence factors of *Candida albicans*. *Trends Microbiol*, **9**, 327–35.

Calera, J.A., Zhao, X.J. and Calderone, R. 2000. Defective hyphal development and avirulence caused by a deletion of the *SSK1* response regulator gene in *Candida albicans*. *Infect Immun*, **68**, 518–25.

Cann, C. 1992. Candidiasis (moniliasis, thrush, *Candida* paronychia, *Candida* endocarditis, bronchomycosis, mycotic vulvovaginitis candiosis). In: Kwon-Chung, K.J. and Bennett, J.E. (eds), *Medical mycology*. Philadelphia/London: Lea & Febiger, 280–336.

Cantorna, M.T. and Balish, E. 1990. Mucosal and systemic candidiasis in congenitally immunodeficient mice. *Infect Immun*, **58**, 1093–100.

Cantorna, M.T., Mook, D. and Balish, E. 1990. Resistance of congenitally immunodeficient gnotobiotic mice to vaginal candidiasis. *Infect Immun*, **58**, 3813–15.

Carvalho, L.P., Bacellar, O., et al. 2002. Downregulation of IFN-gamma production in patients with recurrent vaginal candidiasis. *J Allergy Clin Immmunol*, **109**, 102–5.

Casadevall, A. 1995. Antibody immunity and invasive fungal infections. *Infect Immun*, **63**, 4211–18.

Cassone, A. and Cauda, R. 2002. Response: HIV proteinase inhibitors: do they really work against *Candida* in a clinical setting? *Trends Microbiol*, **10**, 177–8.

Cassone, A., Boccanera, M., et al. 1995. Rats clearing a vaginal infection by *Candida albicans* acquire specific antibody-mediated resistance to vaginal infection. *Infect Immun*, **63**, 2619–24.

Cassone, A., De Bernardis, F., et al. 1999. *In vitro* and *in vivo* anticandidal activity of human immunodeficiency virus protease inhibitors. *J Infect Dis*, **180**, 448–53.

Cassone, A., Tacconelli, E., et al. 2002. Antiretroviral therapy with protease inhibitors has an early immune reconstitution-independent beneficial effect on *Candida* virulence and oral candidiasis in human immunodeficiency virus-infected subjects. *J Infect Dis*, **185**, 188–95.

Cenci, E., Mencacci, A., et al. 1999. Interleukin-4 causes susceptibility to invasive pulmonary aspergillosis through suppression of protective type I responses. *J Infect Dis*, **180**, 1957–68.

Chakir, J., Cote, L., et al. 1994. Differential pattern of infection and immune response during experimental oral candidiasis in BALB/c and DBA/2 (H-2d) mice. *Oral Microbiol Immunol*, **9**, 88–94.

Challacombe, S.J. and Sweet, S.P. 1997. Salivary and mucosal immune responses to HIV and its co-pathogens. *Oral Dis*, **3**, Suppl 1, S79–84.

Clark, R.A., Blakley, S.A., et al. 1995. Predictors of HIV progression in women. *J Acquired Immune Defic Synd Hum Retrovirol*, **9**, 43–50.

Colon, M., Toledo, N., et al. 1998. Anti-fungal and cytokine producing activities of CD8+ T lymphocytes from HIV-1 infected individuals. *Boletih-Asociacion Medica de Puerto Rico*, **90**, 21–6.

Coogan, M.M., Sweet, S.P. and Challacombe, S.J. 1994. Immunoglobulin A (IgA), IgA1 and IgA2 antibodies to *Candida albicans* in whole and parotid saliva in human immunodeficiency virus infection and AIDS. *Inf Immun*, **62**, 892–6.

Corrigan, E.M., Clancy, R.L., et al. 1998. Cellular immunity in recurrent vulvovaginal candidiasis. *Clin Exp Immunol*, **111**, 574–8.

De Bernardis, F., Agatensi, L., et al. 1990. Evidence for a role for secreted aspartate proteinase of *Candida albicans* in vulvovaginal candidiasis. *J Infect Dis*, **161**, 1276–83.

De Bernardis, F., Molinari, A., et al. 1994. Modulation of cell surface-associated mannoprotein antigen expression in experimental candidal vaginitis. *Infect Immun*, **62**, 509–19.

De Bernardis, F., Boccanera, M., et al. 1997. Protective role of antimannan and anti-aspartyl proteinase antibodies in an experimental model of *Candida albicans* vaginitis in rats. *Infect Immun*, **65**, 3399–405.

De Bernardis, F., Arancia, S., et al. 1999. Evidence that members of the secretory aspartyl proteinase gene family, in particular SAP2, are virulence factors for *Candida* vaginitis. *J Infect Dis*, **179**, 201–8.

De Bernardis, F., Santoni, G., et al. 2000. Local anticandidal immune responses in a rat model of vaginal infection by and protection against *Candida albicans*. *Infect Immun*, **68**, 3297–304.

De Bernardis, F., Boccanera, M., et al. 2002. Intravaginal and intranasal immunizations are equally effective in inducing vaginal antibodies and conferring protection against vaginal candidiasis. *Infect Immun*, **70**, 2725–9.

De Repentigny, L., Aumont, F., et al. 2000. Characterization of binding of *Candida albicans* to small intestinal mucin and its role in adherence to mucosal epithelial cells. *Infect Immun*, **68**, 3172–9.

Deslauriers, N., Cote, L., et al. 1997. Oral carriage of *Candida albicans* in murine AIDS. *Infect Immun*, **65**, 661–7.

Djeu, J.Y. 1993. Modulators of immune response to fungi. In: Murphy, J.W., Friedman, H. and Bendinelli, M. (eds), *Fungal infections and immune responses*. New York: Plenum Press, 521–32.

Djeu, J.Y. and Blanchard, D.K. 1987. Regulation of human polymorphonuclear neutrophil (PMN) activity against *Candida albicans* by large granular lymphocytes via release of a PMN-activating factor. *J Immunol*, **139**, 2761–7.

Domer, J.E., Garner, R.E. and Befidi Menguc, R.N. 1989. Mannan as an antigen in cell-mediated immunity (CMI) assays and as a modulator of mannan-specific CMI. *Infect Immun*, **57**, 693–700.

Edwards, J.E. Jr, Gaither, T.A., et al. 1986. Expression of specific binding sites on *Candida* with functional and antigenic characteristics of human complement receptors. *J Immunol*, **137**, 3577–83.

Elahi, S., Pang, G., et al. 2000. Cellular and cytokine correlates of mucosal protection in murine model of oral candidiasis. *Infect Immun*, **68**, 5771–7.

Elder, M., Stapleton, F., et al. 1995. Biofilm-related infections in ophthalmology. *Eye*, **9**, 102–9.

Farah, C.S., Elahi, S., et al. 2001. T cells augment monocyte and neutrophil function in host resistance against oropharyngeal candidiasis. *Infect Immun*, **69**, 6110–18.

Farah, C.S., Gotjamanos, T., et al. 2002. Cytokines in the oral mucosa of mice infected with *Candida albicans*. *Oral Microbiol Immunol*, **17**, 375–8.

Ferris, D.G., Nyirjesy, P., et al. 2002. Over-the-counter antifungal drug misuse associated with patient-diagnosed vulvovaginal candidiasis. *Obstet Gynecol*, **99**, 419–25.

Fidel, P.L. Jr 1999. *Candida albicans*: from commensal to pathogen. In: Tannock, G.W. (ed.), *Medical importance of the normal microflora*. London: Chapman and Hall, 441–76.

Fidel, P.L. Jr and Sobel, J.D. 1994. The role of cell-mediated immunity in candidiasis. *Trends Microbiol*, **2**, 202–6.

Fidel, P.L. Jr and Sobel, J.D. 1996. Immunopathogenesis of recurrent vulvovaginal candidiasis. *Clin Microbiol Rev*, **9**, 335–48.

Fidel, P.L. Jr and Sobel, J.D. 1999. Murine models of *Candida* vaginal infections. In: Zak, O. and Sande, M. (eds), *Experimental models in antimicrobial chemotherapy*. London: Academic Press, 741–8.

Fidel, P.L. Jr, Lynch, M.E., et al. 1993. Systemic cell-mediated immune reactivity in women with recurrent vulvovaginal candidiasis (RVVC). *J Infect Dis*, **168**, 1458–65.

Fidel, P.L. Jr, Lynch, M.E. and Sobel, J.D. 1994. Effects of preinduced Candida-specific systemic cell-mediated immunity on experimental vaginal candidiasis. *Infect Immun*, **62**, 1032–8.

Fidel, P.L. Jr, Cutright, J.L. and Sobel, J.D. 1995a. Effects of systemic cell-mediated immunity on vaginal candidiasis in mice resistant and susceptible to *Candida albicans* infections. *Inf Immun*, **63**, 4191–4.

Fidel, P.L. Jr, Lynch, M.E., et al. 1995b. Mice immunized by primary vaginal *C. albicans* infection develop acquired vaginal mucosal immunity. *Infect Immun*, **63**, 547–53.

Fidel, P.L. Jr, Lynch, M.E. and Sobel, J.D. 1995c. Circulating CD4 and CD8 T cells have little impact on host defense against experimental vaginal candidiasis. *Infect Immun*, **63**, 2403–8.

Fidel, P.L. Jr, Wolf, N.A. and KuKuruga, M.A. 1996. T lymphocytes in the murine vaginal mucosa are phenotypically distinct from those in the periphery. *Infect Immun*, **64**, 3793–9.

Fidel, P.L. Jr, Ginsburg, K.A., et al. 1997. Vaginal-associated immunity in women with recurrent vulvovaginal candidiasis: evidence for vaginal Th1-type responses following intravaginal challenge with *Candida* antigen. *J Infect Dis*, **176**, 728–39.

Fidel, P.L. Jr, Luo, W., et al. 1999a. Analysis of vaginal cell populations during experimental vaginal candidiasis. *Infect Immun*, **67**, 3135–40.

Fidel, P.L. Jr, Vazquez, J.A. and Sobel, J.D. 1999b. *Candida glabrata*: review of epidemiology, pathogenesis, and clinical disease with comparison to *C. albicans*. *Clin Microbiol Rev*, **12**, 80–96.

Fidel, P.L. Jr, Wormley, F.L. Jr, et al. 2001. Analysis of the CD4 protein on human vaginal CD4+ T cells. *Am J Reprod Immunol*, **45**, 200–4.

Fidel, P.L. Jr, Barousse, M. and Espinosa, T. 2003. Local immune responsiveness following intravaginal challenge with *Candida* antigen in adult women at different stages of the menstrual cycle. *Med Mycol*, **41**, 97–109.

Filler, S.G. and Edwards, J. Jr 1993. Chronic mucocutaneous candidiasis. In: Murphy, J.W., Friedman, H. and Bendinelli, M (eds), *Fungal Infections and Immune Responses*. New York: Plenum Press, 117–33.

Fischer, A., Ballet, J.J. and Griscelli, C. 1978. Specific inhibition of in vitro *Candida*-induced lymphocyte proliferation by polysaccharidic antigens present in the serum of patients with chronic mucocutaneous candidiasis. *J Clin Invest*, **62**, 1005–13.

Freydiere, A.M., Parant, F., et al. 2002. Identification of *Candida glabrata* by a 30-second trehalase test. *J Clin Microbiol*, **40**, 3602–5.

Garner, R.E. and Domer, J.E. 1994. Lack of effect of *Candida albicans* mannan on development of protective immune responses in experimental murine candidiasis. *Infect Immun*, **62**, 738–41.

Garner, R.E., Childress, A.M., et al. 1990. Characterization of *Candida albicans* mannan-induced, mannan- specific delayed hypersensitivity suppressor cells. *Infect Immun*, **58**, 2613–20.

Gaspari, A.A., Burns, R., et al. 1998. CD86 (B7-2), but not CD80 (B7-1), expression in the epidermis of transgenic mice enhances the immunogenicity of primary cutaneous *Candida albicans* infections. *Infect Immun*, **66**, 4440–9.

Glick, M. and Siegel, M.A. 1999. Viral and fungal infections of the oral cavity in immunocompetent patients. *Infect Dis Clin North Am*, **13**, 817–31.

Greenfield, R.A. 1992. Host defense system interactions with *Candida*. *J Med Vet Mycol*, **30**, 89–104.

Greenspan, D., Komaroff, E., et al. 2000. Oral mucosal lesions and HIV viral load in the Women's Interagency HIV study (WIHS). *J Acquired Immune Defic Syndr*, **25**, 44–50.

Gruber, A., Speth, C., et al. 1999. Human immunodeficiency virus type 1 protease inhibitor attenuates *Candida albicans* virulence properties in vitro. *Immunopharmacology*, **41**, 227–34.

Guiver, M., Levi, K. and Oppenheim, B.A. 2001. Rapid identification of *Candida* species by TaqMan PCR. *J Clin Pathol*, **54**, 362–6.

Han, Y., Morrison, R.P. and Cutler, J.E. 1998. A vaccine and monoclonal antibodies that enhance mouse resistance to *Candida albicans* vaginal infection. *Infect Immun*, **66**, 5771–6.

Han, Y., Riesselman, M.H. and Cutler, J.E. 2000. Protection against candidiasis by an immunoglobulin G3 (IgG3) monoclonal antibody specific for the same mannotriose as an IgM protective antibody. *Infect Immun*, **68**, 1649–54.

Hasenclever, H.F., Mitchell, W.O. and Loewe, J. 1961. Antigenic studies of *Candida*. II. Antigenic relation of *Candida albicans* Group A and Group B to *Candida stellatoidea* and *Candida tropicalis*. *J Bacteriol*, **82**, 570–3.

Hector, R.F., Domer, J.E. and Carrow, E.W. 1982. Immune responses to *Candida albicans* in genetically distinct mice. *Infect Immun*, **38**, 1020–8.

Heidenreich, F. and Dierich, M.P. 1985. *Candida albicans* and *Candida stellatoidea*, in contrast to other *Candida* species, bind iC3b and C3d but not C3b. *Infect Immun*, **50**, 598–600.

Hladik, F., Lentz, G., et al. 1999. Coexpression of CCR5 and IL-2 in human genital but not blood cells: implications for the ontogeny of the CCR5+ Th1 phenotype. *J Immunol*, **163**, 2306–13.

Hoover, D.R., Saah, A.J., et al. 1993. Clinical manifestations of AIDS in the era of pneumocystis prophylaxis. Multicenter AIDS Cohort Study. *N Engl J Med*, **329**, 1922–6.

Hube, B. and Naglik, J.R. 2002. Extracellular hydrolases. In: Calderone, R (ed.), *Candida and candidiasis*. Washington, DC: ASM Press, 107–22.

Hube, B., Sanglard, D., et al. 1997. Disruption of each of the secreted aspartyl proteinase genes SAP1, SAP2, and SAP3 of *Candida albicans* attenuates virulence. *Infect Immun*, **65**, 3529–38.

Huber-Spitzy, V., Bohler-Sommeregger, K., et al. 1992. Ulcerative blepharitis in atopic patients – is *Candida* species the causative agent? *Br J Ophthalmol*, **76**, 272–4.

Hwang, C.S., Rhie, G.E., et al. 2002. Copper- and zinc-containing superoxide dismutase (Cu/ZnSOD) is required for the protection of *Candida albicans* against oxidative stresses and the expression of its full virulence. *Microbiology*, **148**, 3705–13.

Janusz, M.J., Austen, K.F. and Czop, J.K. 1988. Phagocytosis of heat-killed blastospores of *Candida albicans* by human monocyte beta-glucan receptors. *Immunology*, **65**, 181–5.

Jones-Carson, J., Vazquez-Torres, A., et al. 2000. Disparate requirement for T cells in resistance to mucosal and acute systemic candidiasis. *Infect Immun*, **68**, 2363–5.

Kelly, K.A. and Rank, R.G. 1997. Identification of homing receptors that mediate the recruitment of CD4 T cells to the genital tract following intravaginal infection with *Chlamydia trachomatis*. *Infect Immun*, **65**, 5198–208.

Kelly, K.A., Gray, H.L., et al. 2001. *Chlamydia trachomatis* infection does not enhance local cellular immunity against concurrent *Candida albicans* vaginal infection. *Infect Immun*, **69**, 3451–4.

Kent, H.L. 1991. Epidemiology of vaginitis. *Am J Obstet Gynecol*, **165**, 1168–75.

Kirkpatrick, C.H. 1984. Host factors in defense against fungal infections. *Am J Med*, **77**, 1–12.

Kirkpatrick, C.H. 1989. Chronic mucocutaneous candidiasis. *Eur J Clin Microbiol Infect Dis*, **8**, 448–56.

Kirkpatrick, C.H. 1994. Chronic mucocutaneous candidiasis. *J Am Acad Dermatol*, **31**, S14–17.

Kirkpatrick, C.H., Chandler, J.W. and Schimke, R.N. 1970. Chronic mucocutaneous moniliasis with impaired delayed hypersensitivity. *Clin Exp Immunol*, **6**, 375–85.

Kirkpatrick, C.H., Rich, R.R. and Bennett, J.E. 1971. Chronic mucocutaneous candidiasis: model building in cellular immunity. *Ann Intern Med*, **74**, 955–78.

Klotz, S.A., Penn, C.C., et al. 2000. Fungal and parasitic infections of the eye. *Clin Microbiol Rev*, **13**, 662–85.

Kozel, T.R., Brown, R.R. and Pfrommer, G.S.T. 1987. Activation and binding of C3 by *Candida albicans*. *Infect Immun*, **55**, 1890–4.

Kunkl, A., Mortara, L., et al. 1998. Recognition of antigenic clusters of *Candida albicans* by T lymphocytes from human immunodeficiency virus-infected persons. *J Infect Dis*, **178**, 488–96.

Kutteh, W.H., Moldoveanu, Z. and Mestecky, J. 1998. Mucosal immunity in the female reproductive tract: correlation of immunoglobulins, cytokines, and reproductive hormones in human cervical mucus around the time of ovulation. *AIDS Res Hum Retroviruses*, **14**, 51–5.

Laprade, L., Boyartchuk, V.L., et al. 2002. Spt3 plays opposite roles in filamentous growth in *Saccharomyces cerevisiae* and *Candida albicans* and is required for *C. albicans* virulence. *Genetics*, **161**, 509–19.

Lehner, T., Wilton, J.M. and Ivanyi, L. 1972. Immunodeficiencies in chronic mucocutaneous candidiasis. *Immunology*, **22**, 775–87.

Leigh, J.E., Steele, C., et al. 1998. Th1/Th2 cytokine expression in saliva of HIV-positive and HIV-negative individuals: a pilot study in HIV-positive individuals with oropharyngeal candidiasis. *J Acquired Immune Defic Syndr Hum Retrovirol*, **19**, 373–80.

Leigh, J.E., Barousse, M., et al. 2001. *Candida*-specific systemic cell-mediated immune reactivities in HIV-infected persons with and without mucosal candidiaisis. *J Infect Dis*, **183**, 277–85.

Levitz, S.M. and Farrell, T.P. 1990. Human neutrophil degranulation stimulated by *Aspergillus fumigatus*. *J Leukoc Biol*, **47**, 170–5.

Li, S.P., Lee, S.I. and Domer, J.E. 1998. Alterations in frequency of interleukin-2 (IL-2)-, gamma interferon-, or IL-4-secreting splenocytes induced by *Candida albicans* mannan and/or monophosphoryl lipid A. *Infect Immun*, **66**, 1392–9.

Little, C.H., Georgiou, G.M., et al. 2000. Measurement of T-cell-derived antigen binding molecules and immunoglobulin G specific to *Candida albicans* mannan in sera of patients with recurrent vulvovaginal candidiasis. *Infect Immun*, **68**, 3840–7.

Lockhart, S.R., Joly, S., et al. 1997. Development and verification of fingerprinting probes for *Candida glabrata*. *Microbiology*, **143**, 3733–46.

Lockhart, S.R., Pujol, C., et al. 2001. Development and use of complex probes for DNA fingerprinting the infectious fungi. *Med Mycol*, **39**, 1–8.

Lynch, M.E., Sobel, J.D. and Fidel, P.L. Jr 1996. Role of antifungal drug resistance in the pathogenesis of recurrent vulvovaginal candidasis. *J Med Vet Mycol*, **34**, 337–9.

Macher, A.M. 1988. The pathology of AIDS. *Public Health Rep*, **103**, 246–54.

Magee, D.M. and Cox, R.A. 1995. Roles of gamma interferon and interleukin-4 in genetically determined resistance to *Coccidioides immitis*. *Inf Immun*, **63**, 3514–19.

Magliani, W., Conti, S., et al. 1997. Therapeutic potential of antiidiotype single chain antibodies with yeast killer toxin activity. *Nat Biotechnol*, **15**, 155–8.

Magliani, W., Conti, S., et al. 2002. New immunotherapeutic strategies to control vaginal candidiasis. *Trends Mol Med*, **8**, 121–6.

Martinez, A.C., Tobal, F.G., et al. 2000. Risk factors for esophageal candidiasis. *Eur J Clin Microbiol Infect Dis*, **19**, 96–100.

Mathema, B., Cross, E., et al. 2001. Prevalence of vaginal colonization by drug-resistant *Candida* species in college-age women with previous exposure to over-the-counter azole antifungals. *Clin Infect Dis*, **33**, E23–7.

Mathews, H.L. and Beno, D.W.A. 1993. Quantitative measurement of lymphocyte mediated growth inhibition of *Candida albicans*. *J Immunol Methods*, **164**, 155–7.

Mathur, S., Virella, G., et al. 1977. Humoral immunity in vaginal candidiasis. *Infect Immun*, **15**, 287–94.

Matthews, R., Maresca, B., et al. 1998. Stress proteins in fungal diseases. *Med Mycol*, **36**, 45–51.

Millon, L., Drobacheff, C., et al. 2001. Longitudinal study of anti-*Candida albicans* mucosal immunity against aspartic proteinases in HIV-infected patients. *J Acquired Immune Defic Syndr*, **26**, 137–44.

Morrison, C.J., Brummer, E. and Stevens, D.A. 1987. Effect of a local immune reaction on peripheral blood polymorphonuclear neutrophil microbicidal function: studies with fungal targets. *Cell Immunol*, **110**, 176–82.

Myers, T.A., Leigh, J.E., et al. 2003. Immunohistochemical evaluation of T cells in oral lesions from human immunodeficiency virus-positive persons with oropharyngeal candidiasis. *Infect Immun*, **71**, 956–63.

Nandi, D. and Allison, J.P. 1991. Phenotypic analysis and gamma/delta-T cell receptor repertoire of murine T cells associated with the vaginal epithelium. *J Immunol*, **147**, 1773–8.

Nielsen, H., Bentsen, K.D., et al. 1994. Oral candidiasis and immune status of HIV-infected patients. *J Oral Pathol Med*, **23**, 140–3.

Nomanbhoy, F., Steele, C., et al. 2002. Vaginal and oral epithelial cell anti-*Candida* activity. *Infect Immun*, **70**, 7081–8.

Odds, F.C. 1988a. Chronic mucocutaneous candidiosis. In *Candida and candidosis*. London: Baillière Tindall, 104–110.

Odds, F.C. 1988b. Ecology and epidemiology of candidiasis. In *Candida and candidosis*. London: Baillière Tindall, 89.

Odds, F.C. and Bernaert, R. 1994. CHROMagar Candida, a new differential isolation medium for presumptive identification of clinically important Candida species. J Clin Microbiol, 32, 1923–9.

Okhravi, N., Adamson, P., et al. 1998. Polymerase chain reaction and restriction fragment length polymorphism mediated detection and speciation of Candida spp. causing intraocular infection. Invest Ophthalmol Vis Sci, 39, 859–66.

Palella, F.J. Jr, Delaney, K.M., et al. 1998. Declining morbidity and mortality among patients with advanced human immunodeficiency virus infection. HIV Outpatient Study Investigators. N Engl J Med, 338, 853–60.

Parr, M.B. and Parr, E.L. 1991. Langerhans cells and T lymphocyte subsets in the murine vagina and cervix. Biol Reprod, 44, 491–8.

Parr, M.B. and Parr, E.L. 1997. Protective immunity against HSV-2 in the mouse vagina. J Reprod Immunol, 36, 77–92.

Polonelli, L., De Bernardis, F., et al. 1994. Idiotypic intravaginal vaccination to protect against candidal vaginitis by secretory, yeast killer toxin-like anti-idiotypic antibodies. J Immunol, 152, 3175–82.

Polonelli, L., De Bernardis, F., et al. 1996. Human natural yeast killer toxin-like candidicidal antibodies. J Immunol, 156, 1880–5.

Polonelli, L., Seguy, N., et al. 1997. Monoclonal yeast killer toxin-like candidacidal anti-idiotypic antibodies. Clin Diag Lab Immunol, 4, 142–6.

Pulendran, B., Palucka, K. and Banchereau, J. 2001. Sensing pathogens and tuning immune responses. Science, 293, 253–6.

Quinti, I., Palma, C., et al. 1991. Proliferative and cytotoxic responses to mannoproteins of Candida albicans by peripheral blood lymphocyes of HIV-infected subjects. Clin Exp Immunol, 85, 1–8.

Rabeneck, L., Crane, M.M., et al. 1993. A simple clinical staging system that predicts progression to AIDS using CD4 count, oral thrush, and night sweats. J Gen Intern Med, 8, 5–9.

Regulez, P., Garcia Fernandez, J.F., et al. 1994. Detection of anti-Candida albicans IgE antibodies in vaginal washes from patients with acute vulvovaginal candidiasis. Gynecol Obstet Invest, 37, 110–14.

Reis e Sousa, C. 2001. Dendritic cells as sensors of infection. Cell, 14, 495–8.

Richardson, M.D. and Carlson, P. 2002. Culture- and non-culture based diagnostics for Candida species. In: Calderone, R. (ed.), Candida and candidiasis. Washington, DC: ASM Press, 387–94.

Rogers, T.J. and Balish, E. 1980. Immunity to Candida albicans. Microbiol Rev, 44, 660–82.

Romagnoli, P., Pimpinelli, N., et al. 1997. Immunocompetent cells in oral candidiasis of HIV-infected patients: an immunohistochemical and electron microscopical study. Oral Dis, 3, 99–105.

Romani, L., Puccetti, P. and Bistoni, F. 1996. Biological role of Th cell subsets in candidiasis. In: Romagnani, S. (ed.), Th1 and Th2 cells in health and disease. Farmington, CT: Karger, 114–37.

Romani, L., Bistoni, F. and Puccetti, P. 2002. Fungi, dendritic cells and receptors: a host perspective of fungal virulence. Trends Microbiol, 10, 508–14.

Ross, I.K., De Bernardis, F., et al. 1990. The secreted aspartate proteinase of Candida albicans: physiology of secretion and virulence of a proteinase-deficient mutant. J Gen Microbiol, 136, 687–94.

Ruhnke, M. 2002. Skin and mucous membrane infections. In: Calderone, R. (ed.), Candida and candidiasis. Washington, DC: ASM Press, 307–25.

Saavedra, M., Taylor, B., et al. 1999. Local production of chemokines during experimental vaginal candidiasis. Infect Immun, 67, 5820–9.

Samaranayake, Y.H. and Samaranayake, L.P. 2001. Experimental oral candidiasis in animal models. Clin Microbiol Rev, 14, 398–429.

Samonis, G., Gikas, A., et al. 1994. Prospective study of the impact of broad-spectrum antibiotics on the yeast flora of the human gut. Eur J Clin Microbiol Infect Dis, 13, 665–7.

Sangeorzan, J.A., Bradley, S.F., et al. 1994. Epidemiology of oral candidiasis in HIV infected patients: colonization, infection, treatment, and emergence of fluconazole resistance. Am J Med, 97, 339–46.

Santoni, G., Boccanera, M., et al. 2002. Immune cell-mediated protection against vaginal candidiasis: evidence for a major role of vaginal CD4(+) T cells and possible participation of other local lymphocyte effectors. Infect Immun, 70, 4791–7.

Schaller, M., Hube, B., et al. 1999. In vivo expression and localization of Candida albicans secreted aspartyl proteinases during oral candidiasis in HIV-infected patients. J Invest Dermatol, 112, 383–6.

Schuman, P., Ohmit, S., et al. 1998a. Oral lesions among women living with or at risk for HIV infecion. Am J Med, 104, 559–63.

Schuman, P., Sobel, J.D., et al. 1998b. Mucosal candidal colonization and candidiasis in women with or at risk for human immunodeficiency virus infection. Clin Infect Dis, 27, 1161–7.

Singl, G., Maurer, D., et al. 1999. The epidermis: an immunologic microenvironment. In: Freedberg, I.M., Fitzpatrick, T.B., et al. (eds), Dermatology in general medicine. New York: McGraw-Hill, 343–70.

Sobel, J.D. 1988. Pathogenesis and epidemiology of vulvovaginal candidiasis. Ann NY Acad Sci, 544, 547–57.

Sobel, J.D. 1992. Pathogenesis and treatment of recurrent vulvovaginal candidiasis. Clin Infect Dis, 14, S148–53.

Sobel, J.D. 1998. Vulvovaginitis due to Candida glabrata: an emerging problem. Mycoses, 41, Suppl 2, 18–22.

Sobel, J.D., Hasegawa, A., et al. 1998. Selected animal models: vaginal candidiosis, Pneumocystis pneumonia, dermatophytosis and trichosporonosis. Med Mycol, 36, 129–36.

Soll, D.R. 1992. High-frequency switching in Candida albicans. Clin Microbiol Rev, 5, 183–203.

Soll, D.R. 2002. Phenotypic switching. In: Calderone, R. (ed.), Candida and candidiasis. Washington, DC: ASM Press, 123–42.

Soll, D.R., Galask, R., et al. 1991. Genetic dissimilarity of commensal strains of Candida spp. carried in different anatomical locations of the same healthy women. J Clin Microbiol, 29, 1702–10.

Sood, P., Mishra, B., et al. 2000. Comparison of Vitek Yeast Biochemical Card with conventional methods for speciation of Candida. Indian J Pathol Microbiol, 43, 143–5.

Staab, J.F., Bradway, S.D., et al. 2000. Adhesive and mammalian transglutaminase substrate properties of Candida albicans Hwp1. Science, 283, 1535–8.

Steele, C. and Fidel, P.L. Jr 2002. Cytokine and chemokine production by human oral and vaginal epithelial cells in response to Candida albicans. Infect Immun, 70, 577–83.

Steele, C., Ozenci, H., et al. 1999a. Growth inhibition of Candida albicans by vaginal cells from naive mice. Med Mycol, 37, 251–60.

Steele, C., Ratterree, M. and Fidel, P.L. Jr 1999b. Differential susceptibility to experimental vaginal candidiasis in macaques. J Infect Dis, 180, 802–10.

Steele, C., Leigh, J.E., et al. 2000. Growth inhibition of Candida by human oral epithelial cells. J Infect Dis, 182, 1479–85.

Steele, C., Leigh, J.E., et al. 2001. Potential role for a carbohydrate moiety in anti-Candida activity of human oral epithelial cells. Infect Immun, 69, 7091–9.

Steffan, P., Boikov, D., et al. 1997. Identification of Candida species by randomly amplified polymorphic DNA fingerprinting of colony lysates. J Clin Microbiol, 35, 2031–9.

Sullivan, D.J. and Coleman, D.C. 1998. Candida dubliniensis: characteristics and identification. J Clin Microbiol, 36, 329–34.

Tachibana, T., Matsuyama, T. and Mitsuyama, M. 1999. Involvement of CD4+ T cells and macrophages in acquired protection against infection with Sporothrix schenckii in mice. Med Mycol, 37, 397–404.

Taylor, B.N., Saavedra, M. and Fidel, P.L. Jr 2000. Local Th1/Th2 cytokine production during experimental vaginal candidiasis. Med Mycol, 38, 419–31.

Twomey, J., Waddell, C.C., et al. 1975. Chronic mucocutaneous candidiasis with macrophage dysfunction, a plasma inhibitor, and coexistent aplastic anemia. J Lab Clin Med, 85, 968–77.

Valdimarsson, H., Higgs, J.M., et al. 1973. Immune abnormalities associated with chronic mucocutaneous candidiasis. Cell Immunol, 6, 348–61.

Vazquez, J.A., Donabedian, S., et al. 1992. Comparison of restriction enzyme analysis versus pulse-field gradient electrophoresis as a typing system for Torulopsis glabrata and Candida species other than C. albicans. J Clin Micro, 1, 2021–30.

Vazquez, J.A., Sobel, J.D., et al. 1994. Karyotyping of *Candida albicans* isolates obtained longitudinally in women with recurrent vulvovaginal candidiasis. *J Infect Dis*, **170**, 1566–9.

Walker, P.P., Reynolds, M.T., et al. 2000. Vaginal yeasts in the era of 'over the counter' antifungals. *Sex Transm Infect*, **76**, 437–8.

Walsh, T.J., Gonzalez, C.E., et al. 2000. Correlation between in vitro and in vivo antifungal activities in experimental fluconazole-resistant oropharyngeal and esophageal candidiasis. *J Clin Microbiol*, **38**, 2369–73.

Wang, Y.Y., Li, S.K.P., et al. 1998. Cytokine involvement in immunomodulatory activity affected by *Candida albicans* mannan. *Infect Immun*, **66**, 1384–91.

Wheeler, R.R., Peacock, J.E., et al. 1987. Esophagitis in the immunocompromised host: role of esophagoscopy in diagnosis. *Rev Infect Dis*, **9**, 88–96.

White, M.H. 1996. Is vulvovaginal candidiasis an AIDS-related illness. *Clin Infect Dis*, **22**, Suppl 2, S124–7.

Wilson, B.D. and Sohnle, P.G. 1986. Participation of neutrophils and delayed hypersensitivity in the clearance of experimental cutaneous candidiasis in mice. *Am J Pathol*, **123**, 241–9.

Witkin, S.S., Jeremias, J. and Ledger, W.J. 1988. A localized vaginal allergic response in women with recurrent vaginitis. *J Allergy Clin Immmunol*, **81**, 412–16.

Witkin, S.S., Jeremias, J. and Ledger, W.J. 1989. Vaginal eosinophils and IgE antibodies to *Candida albicans* in women with recurrent vaginitis. *J Med Vet Mycol*, **27**, 57–8.

Wormley, F.L. Jr, Chaiban, J. and Fidel, P.L. Jr 2001a. Cell adhesion molecule and lymphocyte activation marker expression during experimental vaginal candidiasis. *Infect Immun*, **69**, 5072–9.

Wormley Jr, F.L., Steele, C., et al. 2001b. Resistance of TCR δ chain knock-out mice to experimental *Candida* vaginitis. *Infect Immun*, **69**, 7162–4.

Wozniak, K.L., Leigh, J.E., et al. 2002a. A comprehensive study of *Candida*-specific antibodies in saliva of HIV-infected persons with oropharyngeal candidiasis. *J Infect Dis*, **185**, 1269–76.

Wozniak, K.L., Wormley, F.L. Jr and Fidel, P.L. Jr 2002b. *Candida*-specific antibodies during experimental vaginal candidiasis in mice. *Infect Immun*, **70**, 5790–9.

Wray, D., Felix, D.H. and Cumming, C.G. 1990. Alteration of humoral responses to *Candida* in HIV infection. *Br Dent J*, **168**, 326–9.

Wuthrich, M., Finkel-Jiminez, B.E. and Klein, B.S. 2000. Interleukin 12 as an adjuvant to WI-1 adhesin immunization augments delayed-type hypersensitivity, shifts the subclass distribution of immunoglobulin G antibodies, and enhances protective immunity to *Blastomyces dermatitidis* infection. *Infect Immun*, **68**, 7172–4.

Oculomycosis

PHILIP A. THOMAS AND P. GERALDINE

Opportunistic fungal infections that occur in the eye and its associated structures are collectively termed 'oculomycoses.' Until about 30 years ago, instances of oculomycosis were sufficiently infrequent to justify single case reports in the ophthalmic literature if the etiology of the disorder could be conclusively established by culture (Wilson and Ajello 1998). However, such infections are being increasingly recognized as important causes of morbidity and blindness; certain types of oculomycoses may even be life threatening (Yohai et al. 1994; Levin et al. 1996). The cornea is the site most frequently affected (Srinivasan et al. 1991), but the orbit, lids, lacrimal apparatus, conjunctiva, sclera, and intraocular structures may also be involved.

Fungal infections of the eye affect the orbit and the eyeball (globe) itself; infections of the globe manifest as external ocular or intraocular infections.

ETIOLOGIC AGENTS OF OCULOMYCOSES

By 1998, some 105 species in 56 genera of fungi had been reported as causes of mycotic keratitis and other oculomycoses (Wilson and Ajello 1998). Since the publication of the previous edition of this book, other species and genera of fungi have been reported as

ophthalmic pathogens, including *Chrysosporium parvum* (Wagoner et al. 1999), *Metarhizium anisopliae* var. *anisopliae* (de Garcia et al. 1997), *Phaeoisaria clematidis* (Guarro et al. 2000), and *Sarcopodium oculorum* (Guarro et al. 2002). Unfortunately, it may be difficult to distinguish between true ophthalmic pathogens and organisms from the environment that are inadvertently introduced into specimens during or after collection (D'Mellow et al. 1991). Hence, certain criteria need to be satisfied before a fungal strain isolated from an ocular specimen can be considered significant; this is especially important when reporting fungal genera or species that have not been previously implicated in ocular infections.

McGinnis (1980) applied strict criteria to evaluate more than 300 reports pertaining to human fungal infections published in the literature from the late 1940s to the beginning of 1979. A report was considered to be acceptable when:

- an adequate clinical history was presented that suggested a mycotic infection
- the fungus was seen in the clinical specimens
- the morphology of the fungus in the clinical specimens was compatible with the reported etiologic agent
- there was adequate evidence that the fungus was identified properly.

The assessment included reports on 60 species in 30 genera of fungi isolated from ophthalmic infections, principally keratitis, but only reports pertaining to 32 species in 19 genera of fungi satisfied the strict criteria to be considered acceptable. In this chapter, an attempt has been made to apply these and similar criteria (Thomas 2003) to assess the significance of the fungal genera and species that have been reported as pathogens in different parts of the eye (Tables 16.1, 16.2 p. 278, 16.3, p. 280, and 16.4, p. 281). In vitro susceptibility data pertaining to various species of fungi that have been implicated as causes of ocular infection, and derived from studies performed by a method approved by the National Committee for Clinical Laboratory Standards (NCCLS) are also presented (Sutton et al. 1998). It should be noted that the clinical relevance of these data to the therapy of ocular fungal infections, especially when topical antifungals are used, is uncertain. It is also not clear whether these data can be used as a guide to therapy.

There appear to be variations in the genera and species of fungi reported as causes of oculomycoses, depending on the geographical location and the time period. Up to 1962, most published reports on mycotic keratitis had cited species of *Aspergillus* as the causative organisms (Gingrich 1962); since that time, however, species of *Fusarium* have been reported as frequent causes of mycotic keratitis (Nelson et al. 1994; Thomas 1994, 2003). In the first half of a 9-year study on microbial keratitis in South Florida, nine strains of *Candida albicans* were isolated, but only one in the second half (Liesegang and Forster 1980). In the period from 1969 to 1977, *Fusarium solani* was the commonest fungal isolate from patients with keratitis in South Florida (Liesegang and Forster 1980) whereas from 1982 to 1992, *Fusarium oxysporum* was the commonest isolate (Rosa et al. 1994). Although *Fusarium solani* is apparently the commonest cause of mycotic keratitis in many parts of the world (Rosa et al. 1994; Srinivasan et al. 1997), species of *Aspergillus* have predominated in many reports from the Indian subcontinent (Srinivasan et al. 1991; Upadhyay et al. 1991; Dunlop et al. 1994; Thomas 1994). *Candida albicans* is a frequent cause of keratitis in temperate regions, including the northern United States (Thygeson and Okumoto 1974; Chin et al. 1975; Tanure et al. 2000), but is infrequently reported from tropical and subtropical regions (Rosa et al. 1994; Thomas 1994; Srinivasan et al. 1997). The reasons for these variations are not clear.

Hyaline filamentous fungi

Species of *Fusarium*, widespread saprobic or parasitic organisms that grow on decaying vegetation and various parts of wild and cultivated plants, also cause important plant diseases, particularly in major crops (Cuero 1980).

These fungi are believed to account for at least one-third of all reported cases of mycotic keratitis in tropical and subtropical zones (Srinivasan et al. 1997; Gopinathan et al. 2002; Leck et al. 2002); they have also been reported to cause intraocular infections (Pflugfelder et al. 1988; Patel et al. 1994; Glasgow et al. 1996; Goldblum et al. 2000; Sponsel et al. 2002; Verma and Tuft 2002) and scleritis (Moriarty et al. 1993). *Fusarium solani* is the species most frequently isolated from ocular infections, other species reported being *Fusarium dimerum*, *Fusarium episphaeria*, *Fusarium moniliforme*, *Fusarium nivale*, and *Fusarium oxysporum* (Rebell and Forster 1980; Gonawardena et al. 1994; Rosa et al. 1994; Gopinathan et al. 2002; Vismer et al. 2002). In ocular samples, *Fusarium* spp. appear as septate, hyaline, branching hyphae (2–4 µm in diameter). Adventitious sporulation (the formation of conidia in tissues) may also occur; the conidia of *Fusarium* spp. are generally seen to be larger than those of *Paecilomyces* spp. (Liu et al. 1998). In culture, species of *Fusarium* exhibit distinctive, septate, large, banana-shaped macroconidia, with apical and basal cells (Figure 16.1, p. 282), occurring on sporodochia. This sporodochial type often mutates, with an associated loss of virulence, giving rise to either a mycelial type or a pionnotal type (Toussoun and Nelson 1976). Species of *Fusarium* are notoriously resistant to antifungal agents in vitro. At one reference laboratory, only 25 percent of isolates of *Fusarium solani* were found susceptible to amphotericin B in vitro, while none of the isolates tested was susceptible to flucytosine, fluconazole, itraconazole, ketoconazole, or miconazole, when tested by the NCCLS method (Sutton et al. 1998).

Fungi of the genus *Aspergillus* abound in the environment worldwide, thriving on various substrates, including corn, decaying vegetation, and soil. These fungi are also found commonly in hospital air (Leenders et al. 1999); an outbreak of ocular aspergillosis following cataract surgery was traced to ongoing hospital construction, with probable wide dispersal of conidia of *Aspergillus* spp. (Tabbara and al-Jabarti 1998). *Aspergillus* spp. are important causes of keratitis following occupational trauma (Upadhyay et al. 1991; Dunlop et al. 1994; Gopinathan et al. 2002) or surgery (Heidemann et al. 1995; Sridhar et al. 2000a) in both tropical and temperate zones of the world. Species of *Aspergillus* have also been implicated in dacryocystitis (Levin et al. 1996; Kristinsson and Sigurdsson 1998), exogenous endophthalmitis (Tabbara and al-Jabarti 1998), endogenous endophthalmitis (Weishaar et al. 1998), orbital lesions (Kronish et al. 1996; Levin et al. 1996; Johnson et al. 1999), and scleritis (Bernauer et al. 1998). While *Aspergillus fumigatus* is the most commonly isolated species, *Aspergillus flavus*, *Aspergillus terreus*, and *Aspergillus sydowii* have also been reported as etiologic agents in ocular infections (Rippon 1988; Sutton et al. 1998). In ocular samples, species of *Aspergillus* manifest as

Table 16.1 *Etiologic agents of and types of ocular mycoses: hyaline molds*

Genera and species	Types of ocular lesions[a]	Articles in which cited	Acceptable[b]	Uncertain[c]
Acremonium				
A. atrogriseum	CO	Read et al. 2000	a	. . .
A. curvum	CO	Wilson and Ajello 1998	. . .	1
	E	Wilson and Ajello 1998	. . .	1
A. kiliense	CO	Weissgold et al. 1998	a	. . .
	E	Paiva et al. 1960	b	
A. potronii	CO	Forster et al. 1975a	b	. . .
A. recifei	CO	Simonsz 1983	. . .	3
Acremonium sp.	CO	Cho and Lee 2002	. . .	3
Arthrographis				
A. kalrae	CO	Perlman and Binns 1997	a	. . .
Aspergillus				
A. candidus	E, optic nerve	McCormick et al. 1975	. . .	2
A. clavatus	CO	Wilson and Ajello 1998	. . .	1
A. fischerianus	C, CO, E	Coriglione et al. 1990	a	. . .
A. flavipes	CO	Jones 1975	. . .	2
A. flavus	CO	Albesi and Zapater 1975	b	. . .
	E	Lance et al. 1988	a	. . .
	O	Miloshev et al. 1966	a	. . .
	S	Carlson et al. 1992	a	. . .
A. fumigatus	C	Perry et al. 1998	. . .	3
	CO	Singh et al. 1989	a	. . .
	S	Rodriguez-Arres et al. 1995	. . .	3
	E	Tabbara and al-Jabarti 1998	a	. . .
	O	Kronish et al. 1996	a	. . .
A. janus	CO	Neuhann 1976	. . .	2
A. glaucus	CO, E	Wilson and Ajello 1998	. . .	1
A. nidulans	CO	Koul and Pratap 1975	. . .	3
	E	Wilson and Ajello 1998	. . .	1
A. niger	CO	Abboud and Hanna 1970	b	. . .
	E	Brar et al. 2002	a	. . .
A. oryzae	CO	Wilson and Ajello 1998	. . .	1
	F	Wilson and Ajello 1998	. . .	1
	O	Miloshev et al. 1966	b	. . .
	S	Stenson et al. 1982	. . .	3
A. terreus	CO	Singh et al. 1990	a	. . .
	E	Kalina and Campbell 1991	a	. . .
A. wentii	CO	Jones 1975	. . .	2
Beauveria				
B. bassiana	CO	Kisla et al. 2000	a	. . .
Blastomyces				
B. dermatitidis	CO	Rodrigues and Laibson 1973	a	. . .
	E	Li et al. 1998	a	. . .
	EL	Bartley 1995	a	. . .
	O	Li et al. 1998	a	. . .
	U	Gottlieb et al. 1995	a	. . .
Cephaliophora				
C. irregularis	CO	Mathews and Kuriakose 1995	a	. . .
Chrysonilia				
C. sitophila	CO	Rosa et al. 1994	. . .	3
	E	Theodore et al. 1962	. . .	2
Chrysosporium				
C. parvum	CO	Wagoner et al. 1999	. . .	4

(Continued over)

Table 16.1 *Etiologic agents of and types of ocular mycoses: hyaline molds (Continued)*

Genera and species	Types of ocular lesions[a]	Articles in which cited	Acceptable[b]	Uncertain[c]
Coccidioides				
C. immitis	C	Maguire et al. 1994	a	...
	CO	Wilson and Ajello 1998	...	1
	E	Cutler et al. 1978	a	...
	EL	Irvine 1968	a	...
	O	Wilson and Ajello 1998	...	1
	S	Wilson and Ajello 1998	...	1
	U	Bell and Font 1972	a	...
Cylindrocarpon				
C. lichenicola (C. tonkinense)	CO	Laverde et al. 1973	b	...
Engyodontium				
E. alba	CO	McDonnell et al. 1984–85	a	...
Fonsecaea				
F. pedrosoi	C	Wilson and Ajello 1998	...	1
	CO	Barton et al. 1997	a	...
Fusarium				
F. aquaeductum	E	Wilson and Ajello 1998	...	1
F. dimerum	CO	Zapater et al. 1972, 1976	b	...
	CO	Greer et al. 1973	b	...
	E	Wilson and Ajello 1998	...	1
F. verticilloides (F. moniliforme)	CO	Anderson et al. 1959	b	...
	CO	Polenghi and Lasagni 1976	b	...
	E	Wilson and Ajello 1998	...	1
F. nivale	C	Wilson and Ajello 1998	...	1
	CO	Perz 1966	...	3
F. oxysporum	CO	Mikami and Stemmermann 1958	b	...
F. solani	CO	Gugnani et al. 1976	b	...
	E	Goldblum et al. 2000	a	...
F. subglutinans	CO	Wilson and Ajello 1998	...	1
F. ventricosum	CO	Wilson and Ajello 1998	...	1
Histoplasma				
H. capsulatum var. capsulatum	E	Goldstein and Buettner 1983; Pulido et al. 1990	a	...
	O	Wilson and Ajello 1998	...	1
	U	Macher et al. 1985	a	...
H. capsulatum var. duboisii	O	Bansal et al. 1977	a	...
	O	Ajayi et al. 1986	...	3
Metarhizium				
M. anisopliae	CO	de Garcia et al. 1997	a	...
Microsporum				
Microsporum sp.	CO	Cho and Lee 2002	...	3
M. canis	EL	Wilson and Ajello 1998	...	1
Ovadendron				
O. ochraceum	E	Wilson and Ajello 1998	...	1
O. sulphureo-ochraceum	CO	Wilson and Ajello 1998	...	1
	E	Lee et al. 1995	a	...
	S	Wilson and Ajello 1998	...	1
Paecilomyces				
P. farinosus	CO	Gonawardena et al. 1994	a	...
P. lilacinus	CO	Gordon and Norton 1985	a	...
	E	Mosier et al. 1977	b	...
	LS	Henig et al. 1973	...	2
	O	Agrawal et al. 1979	a	...

(Continued over)

Table 16.1 *Etiologic agents of and types of ocular mycoses: hyaline molds (Continued)*

Genera and species	Types of ocular lesions[a]	Articles in which cited	Acceptable[b]	Uncertain[c]
P. variotii+[d]	CO	Wilson and Ajello 1998	. . .	1
P. viridis	E	Rodrigues and MacLeod 1975	b	. . .
Paracoccidioides				
P. brasiliensis	CO	Silva et al. 1988	a	. . .
	EL	Belfort et al. 1975	a	. . .
	U	Silva et al. 1988	a	. . .
Penicillium				
P. citrinum	CO	Gugnani et al. 1976	b	. . .
P. expansum	CO	Gugnani et al. 1976	b	. . .
Pythium				
P. insidiosum	CO	Virgile et al. 1993	a	. . .
	O	Mendonza et al. 1996	a	. . .
Rhinosporidium				
R. seeberi	C	Engzell and Jones, 1973; Rolon 1974	a	. . .
	EL	Kapoor et al. 1976	a	. . .
	LS	Sengupta et al. 1975	a	. . .
	S	Wilson and Ajello 1998		1
Rhizoctonia				
Rhizoctonia sp.	CO	Srivastava et al. 1977	b	. . .
Sarcopodium				
S. oculorum	CO	Guarro et al. 2002	a	. . .
Scedosporium				
S. apiospermum (reported as Pseudallescheria boydii)	CO	Ernest and Rippon 1966	b	. . .
	O	Anderson et al. 1984	a	. . .
	S	Taravella et al. 1997	a	. . .
	E	Ksiazek et al. 1994	a	. . .
S. prolificans	C	Arthur et al. 2001	. . .	3[e]
	S	Moriarty et al. 1993	. . .	3
Scopulariopsis				
S. brevicaulis	CO	Ragge et al. 1990	a	. . .
Sporothrix				
S. schenckii	C	Wilson and Ajello 1998	. . .	1
	CO	Wilson and Ajello 1998	. . .	1
	E	Kurosawa et al. 1988	a	. . .
	EL	Wilson and Ajello 1998	. . .	3
	O	Streeten et al. 1974	a	. . .
	S	Wilson and Ajello 1998	. . .	1
	U	Font and Jakobiec 1976	a	. . .
Trichophyton				
Trichophyton sp.	EL	Velazquez et al. 2002	. . .	3
Tritirachium				
T. oryzae	CO	Rodrigues et al. 1975	b	. . .
Tubercularia				
T. vulgaris	E	Wilson and Ajello 1998	. . .	1
Verticillium				
Verticillium sp.	CO	Shin et al. 2002	. . .	3
V. searrae	CO	Wilson and Ajello 1998	. . .	1
Volutella				
V. cinerescens	E	Foster et al. 1958	b	. . .

a) C, conjunctiva; CO, cornea; E, endophthalmitis; EL, eyelid; LS, lacrimal system; O, orbit and optic nerve; S, sclera; U, uvea.

b) a, microscopy and culture positive, reliable identification; b, deemed accepted by McGinnis (1980).

c) 1, not listed in Medline; 2, deemed questionable by McGinnis (1980); 3, details of identification inadequate; 4, morphology in tissue not consistent with the fungus isolated.

d) *Paecilomyces variotii+*, initially identified as *Paecilomyces viridis*.

e) Subsequently identified as *Acrophialophora fusispora*.

Table 16.2 *Etiologic agents and types of ocular mycoses: pheoid (dematiaceous) molds*

Genera and species	Types of ocular lesions[a]	Articles in which cited	Acceptable[b]	Uncertain[c]
Alternaria				
A. alternata	CO	Ando and Takatori 1987	a	...
	E	Wilson and Ajello 1998	...	1
A. infectoria	CO	Forster et al. 1975b	...	2
Alternaria sp.		Azar et al. 1975	...	2
Aureobasidium				
A. pullulans	CO	Jones and Christensen 1974	...	2
Bipolaris				
B. hawaiiensis	CO	Anandi et al. 1988	a	...
	E	Pavan and Margo 1993	...	3
	O	Maskin et al. 1989	a	...
B. spicifera[d]	CO	Zapater et al. 1975	b	...
Cladorrhinum				
C. bulbillosum	CO[e]	Chopin et al. 1997	a	...
Cladosporium				
C. cladosporioides	CO	Polack et al. 1976	...	2
Colletotrichum				
C. capsici	CO	Upadhyay et al. 1991	...	3
C. coccodes	CO	Liesegang and Forster 1980	...	3
C. dematium	CO	Liao et al. 1983; Fernandez et al. 2002	...	3
C. graminicola	CO	Ritterband et al. 1997	a	...
C. gloeosporioides	CO	Yamamoto et al. 2001	a	...
Colletotrichum state of *Glomerella cingulata*	CO	Shukla et al. 1983	a	...
Curvularia				
C. brachyspora	CO	Marcus et al. 1992	a	...
C. geniculata	CO	Nityananda et al. 1964; Warren 1964	b	...
C. lunata	CO	Nityananda et al. 1962	b	...
	E	Kaushik et al. 2001	a	...
C. pallescens	CO	Wilson and Ajello 1998	...	1
C. senegalensis	CO	Forster et al. 1975b	b	...
C. verruculosa	CO	Forster et al. 1975b	...	2
Dichotomophthoropsis				
D. nymphearum	CO	Wright et al. 1990	a	...
D. portulacae	CO	Wilson and Ajello 1998	...	1
Doratomyces				
D. stemonitis	CO	Wilson and Ajello 1998	...	1
Exophiala				
E. dermatitidis	CO, E	Levenson et al. 1984[f]	a	...
	CO, E	Benaoudia et al. 1999	...	3
E. jeanselmei	CO	Laverde et al. 1973	b	...
	E	Hammer et al. 1983	...	3
Exserohilum				
E. rostratum[g]	CO	Jones 1975	b	...
E. longirostratum	CO	Bouchon et al. 1994	a	...
Fonsecaea				
F. pedrosoi	C	Wilson and Ajello 1998	...	1
	CO	Barton et al. 1997	a	...
Lasiodiplodia				
L. theobromae	CO	Rebell and Forster 1976	b	...
	E	Borderie et al. 1997	a	...
Lecythophora				
L. mutabilis	CO	Ho et al. 1991	...	3

(Continued over)

Table 16.2 *Etiologic agents and types of ocular mycoses: pheoid (dematiaceous) molds (Continued)*

Genera and species	Types of ocular lesions[a]	Articles in which cited	Acceptable[b]	Uncertain[c]
Microsphaeropsis				
M. olivacea	CO, E	Shah et al. 2001	. . .	3
Phaeoisaria				
P. clematidis	CO	Guarro et al. 2000	a	. . .
Phaeotrichoconis				
P. crotalariae	CO	Shukla et al. 1989	a	. . .
Phialophora				
P. bubakii	CO	Wilson and Ajello 1998	. . .	1
P. richardsiae	LS	Pitrak et al. 1988	a	. . .
P. verrucosa	CO	Wilson et al. 1966	b	. . .
Phoma				
P. oculo-hominis	CO	Punithalingam 1976	. . .	2
Phoma sp.	CO	Rishi and Font 2003	a	. . .
Sphaeropsis				
S. subglobosa	CO, E	Kirkness et al. 1991	a	. . .
Tetraploa				
T. aristata	CO	Newmark and Polack 1970	b	. . .

a) C, conjunctiva; CO, cornea; E, endophthalmitis; EL, eyelid; LS, lacrimal system; O, orbit and optic nerve, S, sclera; U, uvea.
b) a, microscopy and culture positive, reliable identification; b, deemed accepted by McGinnis (1980).
c) 1, not listed in Medline; 2, deemed questionable by McGinnis (1980); 3, details of identification inadequate.
d) Identified as *Drechslera spicifera*.
e) Isolated from a Percheron cross horse.
f) Identified as *Wangiella dermatitidis*; inoculation into animal cornea reproduced disease.
g) Identified as *Drechslera rostrata*.

septate, hyaline, branching hyphae, 3–6 µm wide, which exhibit parallel walls and radiate from a single point (Figure 16.2, p. 282), and which are smaller than the hyphae of zygomycetes (Liu et al. 1998); dichotomous (45°) branching may occur (Figure 16.2), and although widely believed to be pathognomonic of *Aspergillus* spp., may be mimicked by *Pseudallescheria boydii* (anamorph *Scedosporium apiospermum*) in ocular samples (McGuire et al. 1991). Identification of the genus *Aspergillus* in culture depends on the morphology of its asexual reproductive structures; the conidiophore, with its swollen terminal end, is surrounded by flask-shaped sterigmata, each of which produces long chains of coccoid conidia that radiate out from the terminal end. Identification of the species involved is important since different species may exhibit differing susceptibilities to antifungals. For example, in vitro, 95 percent of isolates of *Aspergillus fumigatus* were found susceptible to amphotericin B, 97 percent to itraconazole, 92 percent to ketoconazole, 50 percent to miconazole, and none to fluconazole or flucytosine; by contrast, 67 percent of isolates of *Aspergillus flavus* were found to be susceptible to amphotericin B, 98 percent to itraconazole, 100 percent to ketoconazole and miconazole and none to fluconazole or flucytosine, while only 13 percent of strains of *Aspergillus terreus* were reported susceptible to amphotericin B (Sutton et al. 1998). These data suggest that different species of *Aspergillus* show varying degrees of susceptibility to amphotericin B but are generally susceptible to the imidazoles and triazoles (except fluconazole).

Scedosporium apiospermum (teleomorph *Pseudallescheria boydii*), which is found in soil, sewage, and polluted water, has been reported to cause severe ocular infection following trauma during agricultural work or by polluted water, or following immunosuppression (Taravella et al. 1997). There is unequivocal evidence to implicate *Scedosporium apiospermum* as a cause of endogenous or exogenous endophthalmitis (McGuire et al. 1991; Pfeifer et al. 1991; McKelvie et al. 2001), keratitis (Bloom et al. 1992; Sridhar et al. 2000b; Tanure et al. 2000; Diaz-Valle et al. 2002; Wu et al. 2002; Saracli et al. 2003), orbital infections (Anderson et al. 1984; Nunery et al. 1985; Jones et al. 1999), and scleritis (Moriarty et al. 1993; Taravella et al. 1997). Another species, *Scedosporium prolificans*, which was first described as a human pathogen in 1984, has been reported to cause sclerokeratitis (Sullivan et al. 1994; Kumar et al. 1997). In infected tissue, *Scedosporium apiospermum* and *Scedosporium prolificans* resemble other hyaline filamentous fungi in having septate, hyaline, branching hyphae, 2–4 µm wide. In culture, *Scedosporium apiospermum* forms white to pale or dark gray, floccose colonies on glucose peptone agar after one week (Figure 16.3a, p. 282). Microscopic morphology is characterized by abundant yellow to pale brown oval conidia (with a scar at the base) formed from single or branched long slender annellides (Figure 16.3b, p. 282),

Table 16.3 *Etiologic agents of, and types of, ocular mycoses: yeasts*

Genera and species	Types of ocular lesions[a]	Articles in which cited	Acceptable[b]	Uncertain[c]
Candida				
C. albicans	C	Gumbel et al. 1990	. . .	3
	CO	Wilhelmus and Robinson 1991	a	. . .
	E	Stone et al. 1975	a	. . .
	LS	Purgason et al. 1992	a	. . .
	S	Garg et al. 2003	a	. . .
	U	Elliott et al. 1979	a	. . .
C. famata	CO	Wilson and Ajello 1998	. . .	1
	E	Rao et al. 1991	a	. . .
C. glabrata	CO	Djalilian et al. 2001	. . .	3
	E	Chapman et al. 1998	a	. . .
	O	Wilson and Ajello 1998	. . .	1
	U	Larsen 1973	. . .	3
C. guilliermondii	C	Wilson and Ajello 1998	. . .	1
	CO	Ainbinder et al. 1998	a	. . .
C. krusei	CO	Tandon et al. 1984	a	. . .
	E	McQuillen et al. 1992	. . .	3
C. lusitaniae	U	Wilson and Ajello 1998	. . .	1
C. parapsilosis	CO	Rhem et al. 1996	a	. . .
	E	Fekrat et al. 1995	a	. . .
	L	Wilson and Ajello 1998	. . .	1
	S	Wilson and Ajello 1998	. . .	1
	U	Wilson and Ajello 1998	. . .	1
C. tropicalis	CO	Rosa et al. 1994	. . .	3
	E	Behrens-Baumann et al. 1991	. . .	3
	U	Wilson and Ajello 1998	. . .	1
Cryptococcus				
C. laurentii	CO	Ritterband et al. 1998	a	. . .
	E	Custis et al. 1995	. . .	3
C. neoformans	C	Waddell et al. 2000	a	. . .
	CO	Tanure et al. 2000	. . .	3
	E	Charles et al. 1992	a	. . .
	EL	Coccia et al. 1999	. . .	3
	O	Cohen and Glasgow 1993	a	. . .
	S	Limathe et al. 2002	. . .	3
	U	Muccioli et al. 1995	a	. . .
Geotrichum				
G. candidum	C	Wilson and Ajello 1998	. . .	1
	CO	Wilson and Ajello 1998	. . .	1
Malassezia				
M. furfur	LS	Wilson and Ajello 1998	. . .	. . .
	EL	Toth et al. 1996	. . .	3
	CO	Roodhooft et al. 1998	. . .	3

(Continued over)

Table 16.3 *Etiologic agents of, and types of, ocular mycoses: yeasts (Continued)*

Genera and species	Types of ocular lesions[a]	Articles in which cited	Acceptable[b]	Uncertain[c]
M. pachydermatis	LS	Wilson and Ajello 1998	. . .	. . .
Rhodotorula				
R. glutinis	CO	Guerra et al. 1992	. . .	3
R. minuta	E	Pinna et al. 2001	. . .	3
R. rubra	C	Segal et al. 1975	. . .	2
	CO	Romano et al. 1973	. . .	2
	E	Merkur and Hodge 2002	. . .	3
	LS	Muralidhar and Sulthana 1995	a	. . .
Rhodotorula sp.	CO	Panda et al. 1999	a	. . .
Torulopsis				
T. magnoliae	E	Rosenfeld et al. 1994	. . .	3
	O	Wilson and Ajello 1998	. . .	1
Trichosporon				
T. asahii	U	Wilson and Ajello 1998	. . .	1
T. beigelii	E	Sheikh et al. 1974	b	. . .

a) C, conjunctiva; CO, cornea; E, endophthalmitis; EL, eyelid; LS, lacrimal system; O, orbit and optic nerve; S, sclera; U, uvea.
b) a, microscopy and culture positive, reliable identification; b, deemed accepted by McGinnis (1980).
c) 1, not listed in Medline; 2, deemed questionable by McGinnis (1980); 3, details of identification inadequate.

and sometimes aggregated into bundles (*Graphium* state); ascocarps are sometimes present. *Scedosporium prolificans* grows more slowly, forming oval conidia (frequently in groups) from short annellides, with inflated bases and narrow tapering tips. The correct identification of the species, with confirmation by DNA sequencing if warranted (Guarro and Gené 2002), is important in view of differing susceptibilities to different antifungals. In vitro susceptibility testing by the NCCLS method revealed that all isolates of *Scedosporium apiospermum* were susceptible to miconazole, 95 percent to ketoconazole, 77 percent to fluconazole, and 58 percent to amphotericin B; by contrast, two of nine isolates of *Scedosporium prolificans* were found to be susceptible in vitro to miconazole, whereas all isolates tested were resistant to amphotericin B, flucytosine, itraconazole,

Table 16.4 *Etiologic agents of and types of ocular mycoses: zygomycetes*

Genera and species	Types of ocular lesions[a]	Articles in which cited	Acceptable[b]	Uncertain[c]
Absidia				
A. corymbifera	CO	Marshall et al. 1997	a	. . .
	O	Gebhard et al. 1995	a	. . .
Apophysomyces				
A. elegans	O	Fairley et al. 2000	a	. . .
Chlamydoabsidia				
C. padenii	CO	Wilson and Ajello 1998	. . .	1
Mucor				
M. ramosissimus	O	Bullock et al. 1974	b	. . .
Rhizopus				
R. arrhizus (oryzae)	O	Kronish et al. 1996	a	. . .
R. stolonifer	O	Wilson and Ajello 1998	. . .	1
	EL	Champion and Johnson 1969	. . .	2
	EL	Coetzee and de Bruin 1974	. . .	2

a) CO, cornea; EL, eyelid; O, orbit and optic nerve.
b) a, microscopy and culture positive, reliable identification; b, deemed accepted by McGinnis (1980).
c) 1, not listed in Medline; 2, deemed questionable by McGinnis (1980); 3, details of identification inadequate.

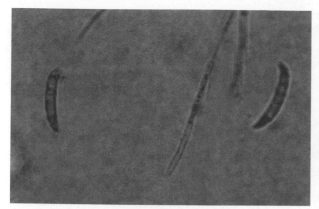

Figure 16.1 *Photomicrograph of large, multicelled, banana-shaped macroconidia of* Fusarium solani *in culture. Lactophenol cotton blue stain; magnification ×400*

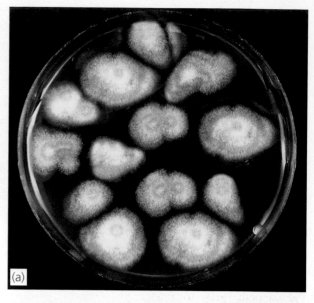

(a)

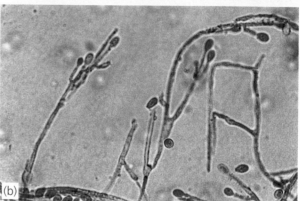

(b)

Figure 16.3 (a) *Macroscopic appearance of* Scedosporium apiospermum *(7-day-old growth) on Sabouraud glucose neopeptone agar.* **(b)** *Microscopic morphology of* Scedosporium apiospermum *(7-day-old culture, Sabouraud glucose neopeptone agar). Abundant pale-brown, oval conidia (with scar at the base) formed from long slender annellides. Lactophenol cotton blue stain; magnification ×400*

ketoconazole, and fluconazole (Sutton et al. 1998). Again, the relevance of these data to the therapy of oculomycoses is uncertain.

Species of *Paecilomyces* are found worldwide in soil and decaying vegetation. They may also contaminate sterile solutions and culture media, since they are resistant to most of the common sterilizing procedures (Castro et al. 1990). Many of the documented infections by *Paecilomyces* spp. in the eye and associated structures have followed surgical procedures (Castro et al. 1990). The species implicated in human infections are *Paecilomyces lilacinus*, *Paecilomyces variotii*, and *Paecilomyces marquandii* (Castro et al. 1990; Sutton et al. 1998). Documented ocular infections include endophthalmitis (Pettit et al. 1980; Okhravi et al. 1997), intralenticular infection (D'Mellow et al. 1991), keratitis (Kozarsky et al. 1984; Gordon and Norton 1985; Starr 1987; Legeais et al. 1994), and orbital infection (Agrawal et al. 1979). Keratitis due to *Paecilomyces lilacinus* frequently fails to respond to medical therapy, necessitating surgery, and endophthalmitis due to this fungus

does not respond to amphotericin B (Scott et al. 2002). In infected tissue, species of *Paecilomyces* exhibit septate, hyaline branching hyphae (2–4 μm wide), which are indistinguishable from those of other agents of hyalohyphomycosis. However, abundant adventitious sporulation frequently occurs, and has been reported from samples of patients with keratitis; adventitious conidia are subglobose to very short ellipsoidal, and this feature may help in identification and selection of appropriate antifungal therapy (Liu et al. 1998). Culture is essential to identify the causal agent. In culture at 25–30°C, *Paecilomyces lilacinus* grows rapidly; the colonies are initially white in color but gradually take on a lilac–lavender hue. Culture mounts reveal flask-shaped phialides with a swollen basal portion tapering in a long distinct neck; the conidiogenous cells are borne singly, in whorls or in penicillate heads. Accurate species identification is important when dealing with *Paecilomyces*

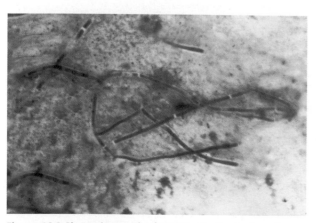

Figure 16.2 *Photomicrograph of septate, regular hyphae of* Aspergillus flavus *in corneal scrape material; dichotomous branching seen. Lactophenol cotton blue stain; magnification ×400*

infection since *Paecilomyces lilacinus* and *Paecilomyces marquandii* are highly resistant to polyene antifungals (amphotericin B, natamycin) and to flucytosine and susceptible to imidazoles (clotrimazole, ketoconazole, econazole, miconazole), whereas *Paecilomyces variotii* is almost universally susceptible to amphotericin B and flucytosine (Castro et al. 1990; Sutton et al. 1998).

Species of *Acremonium* are ubiquitous, being found in abundance in the soil and air (Guarro et al. 1997). *Acremonium atrogriseum*, *Acremonium kiliense*, and *Acremonium potronii* have been reported as causes of endophthalmitis (Weissgold et al. 1996; Guarro et al. 1997) and keratitis (Fincher et al. 1991; Mino de Kasper et al. 1991; Vajpayee et al. 1993; Chander and Sharma 1994; Read et al. 2000). In infected tissue, septate, hyaline, branching hyphae (2–4 µm wide) are seen; adventitious sporulation may occur (Liu et al. 1998). In culture on glucose peptone agar at 30°C, colonies are flat, smooth, gray to orange, and rapidly growing, attaining a diameter of 50 mm in 1 week; conidiophores are long, straight and slightly tapering, while conidia are ellipsoidal and accumulated in slime balls (Campbell et al. 1996).

Arthrographis kalrae is an ascomycetous filamentous fungus that is widespread in decaying plant material and in the soil. McAleer et al. (1988) reported the isolation of this fungus in two corneal scrapings taken from a patient with a corneal ulcer. Perlman and Binns (1997) reported the isolation of this fungus from corneal lesions in a contact lens wearer, although the patient was believed to have acquired the infection while not wearing the lens; although there was no response to topical natamycin 5 percent, the keratitis ultimately resolved with hourly topical 0.15 percent amphotericin B and daily oral ketoconazole 100 mg. Interestingly, this isolate of *Arthrographis kalrae* was reported to be susceptible to ketoconazole, itraconazole, and amphotericin B, but resistant to natamycin. Reported minimal inhibitory concentrations for *Arthrographis kalrae* are 0.5 µg/ml amphotericin B, 0.06 µg/ml itraconazole, and 0.5 µg/ml ketoconazole (Sutton et al. 1998). Colonies of *Arthrographis kalrae* on potato flakes agar at 25°C are rather slow-growing. Microscopy reveals septate hyaline hyphae; arthroconidia formed from short, branched or unbranched conidiophores are one-celled, hyaline, and rectangular (Sutton et al. 1998).

Fungi of the genus *Cylindrocarpon* are soil fungi and are rarely associated with human disease. *Cylindrocarpon lichenicola*, previously known as *Cylindrocarpon tonkinensis* or *Cylindrocarpon tonkinense*, has a wide geographical distribution, but occurs frequently in tropical climates, where it has a wide host range for woody post-harvest fruit invasion (de Hoog and Guarro 1995). In experimental rabbit eyes, *Cylindrocarpon tonkinense* was found to produce as severe a corneal infection as did *Fusarium solani* (Ishibashi et al. 1986). A literature survey on human keratitis due to

Cylindrocarpon lichenicola revealed only three cases (Laverde et al. 1973; Matsumoto et al. 1979; Mangiaterra et al. 2001). The histological appearance of the fungus in corneal tissue is very similar to that seen in infection by species of *Aspergillus* or *Fusarium*, with septate branching hyphae and globular structures associated with the hyphae being noted. In fact, *Cylindrocarpon* spp. appear closely related both morphologically and taxonomically to *Fusarium* spp., with both sharing teleomorphs in the genus *Nectria* (Samuels and Brayford 1993; James et al. 1997). Because of morphological similarities, separation of the genus *Cylindrocarpon* from the genus *Fusarium* by cultural characteristics may become problematic at times (Guarro and Gené 1992). *Cylindrocarpon lichenicola* closely resembles *Fusarium solani*, but differs from it microscopically by forming macroconidia which are predominantly straight rather than curved, by having apical cells that are rounded rather than tapering, by having basal cells with truncate and offset rather than attenuated pedicels (foot cells) (Figure 16.4), by lacking microconidia, by having pigmented chlamydoconidia, and by the formation of a brown color rather than cream color on reverse of the Sabouraud glucose neopeptone agar (Iwen et al. 2000).

Dematiaceous fungi

The primary factor unifying the dematiaceous fungi is the dark pigmentation of the hyphae (McGinnis et al. 1986). This cell coloring results in colonies that are olive to black and in a brown-to-olive-to-black color in the cell walls of their vegetative cells, conidia or both. The dematiaceous fungi reported as causes of ocular infections, especially keratitis, include *Alternaria alternata*, *Aureobasidium pullulans*, *Bipolaris spicifera*, and *Bipolaris hawaiiensis*, *Curvularia lunata*, *Curvularia geniculata*, *Curvularia brachyspora*, *Curvularia senegalensis*, *Dichotomophthoropsis nymphaearum*, *Exophiala*

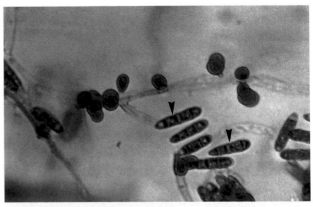

Figure 16.4 *Microscopic morphology of* Cylindrocarpon lichenicola *(7-day-old culture, Sabouraud glucose neopeptone agar). Multicelled macroconidia are predominantly straight, with rounded apical cells and truncate basal cells. Lactophenol cotton blue stain; magnification ×400*

jeanselmei var. *jeanselmei*, *Exophiala* (formerly *Wangiella*) *dermatitidis*, *Exserohilum rostratum*, *Exserohilum longirostratum*, *Lecytophora* (formerly *Phialophora*) *mutabilis*, *Lecytophora hoffmannii*, *Phaeoisaria clematidis*, *Phialophora verrucosa*, *Phoma oculo-hominis*, *Phoma* spp., and *Tetraploa aristata* (Newmark and Polack 1970; Jones and Christensen 1974; Forster et al. 1975b; Zapater et al. 1975; Punithalingam 1976; Liesegang and Forster 1980; Levenson et al. 1984; McGinnis et al. 1986; Douer et al. 1987; Wright et al. 1990; Ho et al. 1991; Stern and Buttross 1991; Marcus et al. 1992; Bouchon et al. 1994; Chang et al. 1994; Sutton et al. 1998; Guarro et al. 1999, 2000; Garg et al. 2000; Kaushik et al. 2001; Rishi and Font 2003). However, the validity of some of the earlier reports has been questioned (McGinnis 1980). In various studies, dematiaceous fungi are reported to be the third most important cause of fungal keratitis (behind *Aspergillus* and *Fusarium*) (Sundaram et al. 1989; Panda et al. 1997; Srinivasan et al. 1997; Garg et al. 2000; Gopinathan et al. 2002; Leck et al. 2002). Dematiaceous fungi may also cause infections of the orbit (Jay et al. 1988; Maskin et al. 1989), or intraocular infections (Kaushik et al. 2001). For diagnosis, direct microscopy of wet preparations of clinical material in 10 percent potassium hydroxide wet mounts permits the demonstration of the characteristic brown-pigmented septate fungal hyphae (Garg et al. 2000); lactophenol cotton blue stain can also be used (Figure 16.5) (Thomas et al. 1991b). Histology is also helpful in the diagnosis, although these fungi may be difficult to identify in hematoxylin and eosin-stained sections due to variability in the amount of pigment in the walls; to overcome this difficulty, the microscope condenser is lowered to make the hyphae more refractile, or the sections are stained by melanin-specific stains such as Fontana–Masson (Chandler 1991). Culture is required to confirm the etiologic diagnosis and to initiate appropriate treatment. Plates of Sabouraud

glucose neopeptone agar or other appropriate media are inoculated with the material and incubated at 25–30°C for up to 3 weeks. In general, the colonies tend to be light-brown to olive-gray to black in color. The different genera and species are identified by the microscopic morphology of the conidia (McGinnis et al. 1986; Sutton et al. 1998).

The form class Coelomycetes comprises asexual fungi that produce conidia within fruiting bodies named conidiomata. These structures may be spherical (pycnidia), with conidiogenous cells lining the inner cavity wall, or cup-shaped (acervuli), with the conidiogenous cells forming a palisade on the surface of the conidiomata. Coelomycetous fungi implicated in ocular infections include *Lasiodiplodia theobromae*, species of *Colletotrichum*, and *Phoma* spp.

Lasiodiplodia theobromae is an important cause of rot in corn, yams, citrus, bananas, and other plants, mainly in tropical regions (Adisa and Fajola 1983; Nwufo and Fajola 1988; Sutton et al. 1998). This fungus was initially implicated as an etiologic agent of human keratitis in two patients in India (Puttana 1967). Subsequently, reports from Colombia, France, Ghana, other parts of India, the Philippines, the southern USA, and Sri Lanka have confirmed the pathogenic potential of this fungus for the human cornea (Laverde et al. 1973; Valenton et al. 1975; Rebell and Forster 1976; Liesegang and Forster 1980; Slomovic et al. 1985; Thomas et al. 1991a, 1995; Gonawardena et al. 1994; Borderie et al. 1997; Garg et al. 2000; Gopinathan et al. 2002; Leck et al. 2002). This fungus causes severe keratitis in experimental animals and in humans (Puttana 1967; Rebell and Forster 1976; Thomas et al. 1991a, 1995; Borderie et al. 1997), perhaps due to its ability to secrete collagenases (Rebell and Forster 1976) and amylolytic and cellulolytic enzymes (Adisa and Fajola 1983; Nwufo and Fajola 1988). This fungus has been found to form intrahyphal hyphae and thickened cell walls in parasitized human corneal tissue (Figure 16.6a), and also in vitro, in the presence of an antifungal (Thomas et al. 1991a, 1995). Such intrahyphal hyphae have been demonstrated in experimental *Fusarium solani* keratitis in rabbits (Kiryu et al. 1991), and may represent a virulence factor for these fungi. In infected corneal tissues, *Lasiodiplodia theobromae* exhibits brown, highly bulged, septate hyphae. In culture on glucose peptone agar, the colony is rapidly growing (90 mm in one week), floccose, and gray to brown–black in color, the reverse being black (Figure 16.6b). Macroscopically visible fruiting bodies (pycnidia) are formed in culture. The pycnidial stromata, which are only formed on cornmeal or other nutritionally deficient media, are up to 5 mm in diameter, consisting of several pycnidia, each with a wide ostiole to permit release of the conidia; the conidia (Figure 16.6c) are 20–30 µm × 10–15 µm, initially colorless, ellipsoidal, and nonseptate, later becoming dark brown and septate, with longitudinal striations and

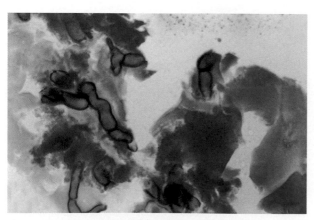

Figure 16.5 *Photomicrograph of pigmented fungal hyphae seen in the corneal scrape material of a patient with keratitis due to* Curvularia geniculata. *Lactophenol cotton blue stain; magnification ×400*

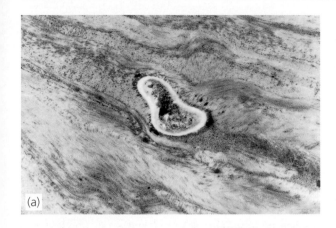

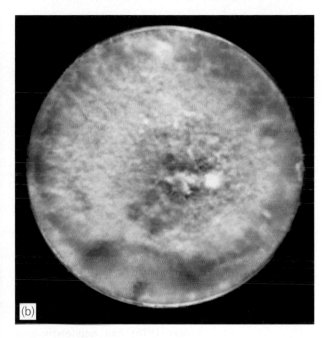

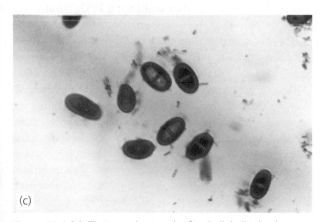

Figure 16.6 (a) *Electron micrograph of* Lasiodiplodia theobromae *fungal cell showing thickened cell wall in corneal scrape material (×6 050).* **(b)** *Macroscopic appearance of* Lasiodiplodia theobromae *(7-day-old growth) on Sabouraud glucose neopeptone agar.* **(c)** *Microscopic morphology of* Lasiodiplodia theobromae *(10-day-old culture, Sabouraud glucose neopeptone agar). Oval conidia, dark-brown, septate with longitudinal striations, truncate bases. Lactophenol cotton blue stain; magnification ×400*

truncate bases (Campbell et al. 1996). In vitro, *Lasiodiplodia theobromae* is resistant to many antifungals; minimal inhibitory concentrations exceeding 100 µg/ml for miconazole, ketoconazole, itraconazole, and natamycin, and ⩽10 µg/ml for econazole against ocular isolates of this fungus have been reported (Thomas et al. 1995).

The complex form genus *Colletotrichum* Corda comprises several hundred species, which are anamorphs of the genus *Glomerella* Spauld & H. Schrenk; they are placed in the order Phyllachorales, although there is evidence linking them to the Sordariales (Uecker 1994). *Colletotrichum* spp. are usually pathogenic on plants, in which they cause anthracnosis, necrosis, leaf spot, and fruit rot, and are believed to be only rarely pathogenic to humans. Although a search of the Medline database through PubMed yielded four reports describing 12 patients with *Colletotrichum* keratitis (Shukla et al. 1983; Ritterband et al. 1997; Yamamoto et al. 2001; Fernandez et al. 2002), a more detailed manual search through the results of various studies on mycotic keratitis reveals that species of *Colletotrichum* have been previously implicated as etiologic agents in as many as 20 patients with keratitis (Kaliamurthy et al. 2004). The species reported include *Colletotrichum gloeosporioides* in seven patients (Shukla et al. 1983; Matsuzaki et al. 1988; Yamamoto et al. 2001; Fernandez et al. 2002), *Colletotrichum coccodes* (reported as *Colletotrichum atramentarium*) in four (Liesegang and Forster 1980; Rosa et al. 1994), *Colletotrichum dematium* in three (Liao et al. 1983; Fernandez et al. 2002), *Colletotrichum graminicola* in one (Ritterband et al. 1997), *Colletotrichum capsici* in one (Upadhyay et al. 1991), and unidentified species of *Colletotrichum* in four (Fernandez et al. 2002). Antecedent ocular trauma appears to have been the principal risk factor for keratitis in most of these patients, but insulin-dependent diabetes mellitus, prolonged use of topical corticosteroids, and prior use of antivirals were also reported as risk factors. These patients received different antifungal drugs, principally topical natamycin alone (seven patients), amphotericin B alone or in combination with an azole or flucytosine (four patients), combined therapy with natamycin, intraocular amphotericin B, and oral azole (two patients), and natamycin followed by topical amphotericin B therapy. In most of the patients, the corneal lesions resolved completely with good visual recovery following initiation of medical therapy, possibly because the lesions were relatively superficial in nature. Microbiological investigations such as culture (Figure 16.7a) are essential to confirm the diagnosis of *Colletotrichum* keratitis and to ensure prompt initiation of specific antifungal therapy.

Colletotrichum dematium can be clearly distinguished from other species of *Colletotrichum* by the occurrence of falcate macroconidia (Figure 16.7b) (de Hoog and Guarro 1995). The gross appearance of these

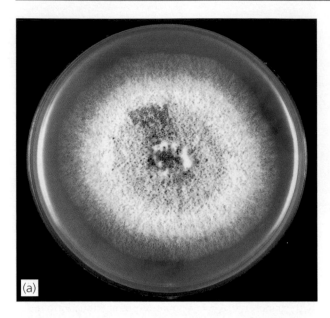

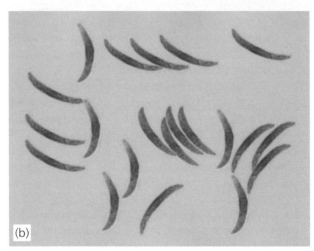

Figure 16.7 (a) *Macroscopic appearance of* Colletotrichum dematium *(7-day-old culture, Sabouraud glucose neopeptone agar).* **(b)** *Microscopic morphology of a* Colletotrichum *sp. (7-day-old culture, Sabouraud glucose neopeptone agar). Falcate macroconidia. Lactophenol cotton blue stain; magnification ×400.* **(c)** *Microscopic morphology of a* Colletotrichum *sp. (14-day-old culture, Sabouraud glucose-neopeptone agar). Brown-colored setae and sclerotia. Lactophenol cotton blue stain; magnification ×400*

macroconidia is very similar to that of the macroconidia of species of *Fusarium*. This may lead an inexperienced observer to mistakenly identify an isolate of *Colletotrichum dematium* as a *Fusarium* sp., particularly if the characteristic sclerotia and setae are not visible when the young culture is examined. To overcome this drawback, it is advisable to incubate all mycology culture plates for an additional 2 weeks after growth occurs, particularly if such falcate macroconidia have been observed (Kaliamurthy et al. 2004). It is important to remember that the macroconidia of *Colletotrichum dematium* are aseptate (*Fusarium* macroconidia are septate), and appear thinner than the macroconidia of species of *Fusarium*. Sclerotia and setae are found in abundance in older cultures of *Colletotrichum dematium* (Figure 16.7c); such structures are not observed in cultures of *Fusarium*.

Yeast and yeast-like fungi

Most yeast infections in corneal ulcers and other ocular infections are due to various *Candida* spp., predominantly *Candida albicans* and *Candida parapsilosis*. There is conclusive evidence to implicate species of *Candida* in dacryocystitis (Purgason et al. 1992), infectious crystalline keratopathy (Wilhelmus and Robinson 1991), intraocular lesions (Okhravi et al. 1998; Hidalgo et al. 2000; Jaeger et al. 2000), and keratitis (Rosa et al. 1994; Tanure et al. 2000); there is less convincing evidence for their implication in blepharitis, conjunctivitis, dacryocanaliculitis, and orbital lesions (McGinnis 1980; Behrens-Baumann 1999). The presence of small budding yeast cells (3–4 μm in size) and pseudohyphae in a corneal scraping is almost diagnostic for *Candida*. On microscopic examination of samples collected from the conjunctiva and lid margins, species of *Candida* can be easily distinguished from *Malassezia* spp. by the narrow attached base of the bud, the lack of symmetry of the yeast, and the off-axis position of the bud in *Candida*. Although both the yeast and the hyphal forms may be seen in corneal scrapings from *Candida* infections, the mycelial form is considered to be the invasive form (O'Day and Burd 1994). In culture, species of *Candida* form smooth, creamy-white colonies, resembling staphylococci in the very early stages of growth. The organisms grow well on blood agar, and do not require special conditions for isolation. It is important to identify isolates of *Candida* down to the species level, since different species exhibit varying susceptibilities to antifungal agents (Sutton et al. 1998). For example, 96–100 percent of isolates of *Candida parapsilosis* and 81–100 percent of isolates of *Candida albicans*, when tested by the recently standardized NCCLS broth microdilution method, were found susceptible to amphotericin B, flucytosine, fluconazole, itraconazole, ketoconazole, and miconazole; by contrast, only 59 and 77 percent of isolates of *Candida tropicalis*, respectively, and 85 and 52 percent strains of *Candida glabrata*, respectively, were found to be susceptible to fluconazole and itraconazole (Sutton et al. 1998). However, the relevance of such in vitro susceptibility test results to the clinical outcome of ocular *Candida* infections remains unclear.

The yeast *Cryptococcus neoformans* var. *neoformans* is typically 5–10 μm in diameter (range 2–20 μm). Acute inflammation and necrosis may obscure the organism, but a useful cytological feature is the appearance of teardrop-shaped, narrow-based budding. The identification of this yeast as the etiologic agent in ocular infections still rests primarily on its cytomorphology (Klotz et al. 2000), with confirmation, as necessary, by special stains such as methenamine silver, which stain the organisms, and mucicarmine, periodic acid–Schiff, and alcian blue, which stain the capsule. These yeasts may be mistaken for *Histoplasma capsulatum* var. *capsulatum* when engulfed by histiocytes or when nonencapsulated (Powers 1998). There is clear evidence to show that *Cryptococcus neoformans* var. *neoformans* can cause blepharitis (Doorenbos-Bot et al. 1990; Coccia et al. 1999), chorioretinitis (Morinelli et al. 1993), endophthalmitis (Morinelli et al. 1993), keratitis (Liesegang and Forster 1980; Tanure et al. 2000), and solitary subretinal lesions (Hester et al. 1992). Recently, *Cryptococcus laurentii* was reported as the etiologic agent of keratitis, in conjunction with *Fusarium solani*, in one eye of a person wearing gas-permeable contact lenses (Ritterband et al. 1998). In vitro, 94–100 percent of strains have been found susceptible to fluconazole, itraconazole, amphotericin B, and ketoconazole (Sutton et al. 1998).

Zygomycetous fungi

Zygomycetes are filamentous fungi which exhibit aseptate or sparsely septate, hyaline hyphae. There are two orders, the Mucorales and the Entomophthorales, containing organisms causing human disease, although most human infections are caused by the Mucorales (Ribes et al. 2000). Ocular zygomycosis is commonly linked to *Rhizopus* spp., particularly *Rhizopus arrhizus*, but species of *Absidia*, *Apophysomyces*, *Cunninghamella*, *Mucor*, *Rhizomucor*, *Saksenaea*, and *Syncephalastrum* have also been implicated (Kaufman et al. 1988; Yohai et al. 1994; Marshall et al. 1997; Fairley et al. 2000; Ribes et al. 2000). Ocular infections caused by this group of fungi include rhino-orbital-cerebral zygomycosis (Anand et al. 1992; Yohai et al. 1994), where the spores are inhaled, and keratitis (Marshall et al. 1997), where the infection manifests after trauma. A *Rhizopus* sp. has also been reported as a cause of scleritis (Locher et al. 1998), but the evidence is not convincing. The detection of fungi belonging to the Mucorales in clinical material by direct microscopy is more significant than their isolation in culture since, even if cultured,

contamination always has to be excluded (Richardson and Shankland 1999; Ribes et al. 2000). In clinical material, the hyphae are broad and aseptate or sparsely septate with right-angled (90°) branching (Figure 16.8); they do not possess parallel walls, nor do they radiate from a single point in tissues (Liu et al. 1998). Although the hyphae stain poorly with periodic acid–Schiff, they stain well with hematoxylin–eosin and Gomori methenamine silver stains. Cresyl-fast violet stains zygomycete walls brick red, and other fungi blue or purple (Richardson and Shankland 1999). In tissue sections, the hyphae are seen in the midst of associated prominent inflammation, necrosis, and invasion of blood vessels (Ribes et al. 2000). In vitro, 91 percent of strains of *R. arrhizus*, 100 percent of strains of *Absidia corymbifera*, and 50 percent of strains of *Apophysomyces elegans* were found susceptible to amphotericin B, while most strains were resistant to flucytosine, fluconazole, and itraconazole (Sutton et al. 1998). In practice also, amphotericin B is the drug of choice in the treatment of ocular zygomycosis.

Thermally dimorphic fungi

Blastomycosis, a granulomatous inflammatory disease caused by *Blastomyces dermatitidis*, usually occurs following inhalation of airborne spores that arise from the soil, and thus the primary infection is usually pulmonary; however, characteristic cutaneous lesions are often the presenting symptom. Eyelid lesions are believed to be the most common ocular manifestation of disseminated infection due to *Blastomyces dermatitidis* (Barr and Gamel 1986; Slack et al. 1992); however, a recent survey of 79 patients revealed just one with eyelid lesions (Bartley 1995). Other reported ocular lesions due to *Blastomyces dermatitidis* include conjunctivitis (Slack et al. 1992), orbital cellulitis (Vida and Moel 1974; Li et al. 1998), keratitis (Rodrigues and Laibson 1973),

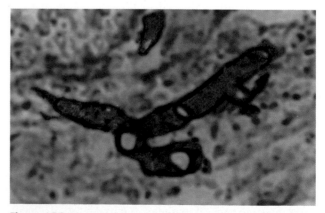

Figure 16.8 *Large, aseptate, irregular, bizzare hyphae of a zygomycetous fungus in tissue section of necrotic orbital material from a patient with rhino-orbital zygomycosis. Hematoxylin and eosin stain; magnification × 1 200*

and intraocular lesions such as anterior uveitis, iris nodules, or yellow-white posterior fundus lesions (Safneck et al. 1990), or endophthalmitis (Safneck et al. 1990; Li et al. 1998). Potassium hydroxide mounts of clinical material reveal characteristic yeasts with broad-based buds. The tissue reaction is a combination of suppurative and epithelioid cell granulomas with giant cells. Yeast-like spherical cells of *Blastomyces dermatitidis* are generally larger (8–20 μm) than those of cryptococci, and have refractile, double-contoured walls; broad-based budding is a useful criterion to differentiate *Blastomyces dermatitidis* from other dimorphic yeasts (Powers 1998). In culture at 26°C, the mycelial phase is noted (round or oval microconidia borne on conidiophores along sides of hyphae or terminally on hyphal branches), while at 37°C, on enriched media, the yeast phase is produced (large, oval, multinucleate yeasts, budding on a broad base, often in chains, with refractile cell walls). In vitro, most strains of *Blastomyces dermatitidis* are reported to be susceptible to amphotericin B, fluconazole, itraconazole, and ketoconazole (Sutton et al. 1998).

Coccidioidomycosis is caused by the thermally dimorphic fungal pathogen *Coccidioides immitis*. The occurrence of ocular lesions in coccidioidomycosis has been reported infrequently (Rodenbiker and Ganley 1980). Anterior segment manifestations are usually a mild hypersensitivity response, occurring as a chronic granulomatous iridocyclitis (Moorthy et al. 1994; Cunningham et al. 1998); however, phlyctenular conjunctivitis, episcleritis, scleritis, and keratoconjunctivitis may occur in association with underlying pulmonary infection, and lid granulomata and inflammation in disseminated disease. The manifestations in the posterior part of the eye range from asymptomatic focal chorioretinitis to a fulminating granulomatous process involving the entire eye. In potassium hydroxide mounts of pus or necrotic material from infected ocular tissue, large multinucleate thick-walled cells, the spherules, are seen, usually within giant cells (Rippon 1988); at maturity, these spherules are filled with spores, which escape by rupture of the cell wall. In vitro, more than 90 percent of strains of *Coccioides immitis* have been found susceptible to amphotericin B, miconazole, ketoconazole, and itraconazole, and 86 percent to fluconazole (Sutton et al. 1998).

The classical form of histoplasmosis is caused by the dimorphic fungus *Histoplasma capsulatum* var. *capsulatum*, while a variant form, known as African or large-celled histoplasmosis, is caused by *Histoplasma capsulatum* var. *duboisii*. The mode of infection is presumably by inhalation or entry through broken skin surfaces, manifesting as acute pulmonary infections or cutaneous and subcutaneous disease; dissemination may occur in immunocompromised patients (Gonzales et al. 2000). *Histoplasma capsulatum* var. *capsulatum* has been isolated from the eyes of patients suffering from the 'presumed histoplasmosis syndrome,' suggesting that this

fungus is the etiologic agent (Roth 1977; Khalil 1982). *Histoplasma capsulatum* var. *capsulatum* has been reported to cause endogenous endophthalmitis (Carroll and Franklin 1981; Goldstein and Buettner 1983; Gonzales et al. 2000), exogenous endophthalmitis after cataract extraction (Pulido et al. 1990), and panophthalmitis (Schwarz et al. 1977). Anterior segment lesions due to *Histoplasma capsulatum* var. *capsulatum* are less common but have been reported (Feman and Tilford 1985; Font et al. 1995). In patients with the acquired immunodeficiency syndrome (AIDS), *Histoplasma capsulatum* var. *capsulatum* has been reported to cause endophthalmitis (Gonzales et al. 2000) and lesions of the choroid, retina, and optic nerve (Specht et al. 1991; Yau et al. 1996). *Histoplasma capsulatum* var. *duboisii* has been reported to cause orbital lesions (Bansal et al. 1977; Ajayi et al. 1986). Organisms may be missed by direct microscopic examination of wet preparations of necrotic tissue, hence all material should be examined as stained smears (Powers 1998). Culture is done on media at 26°C (for mycelial phase) and enriched media at 37°C (for yeast phase). Conversion of the mycelial phase to the yeast phase must be demonstrated by culturing the mycelial phase on appropriate media at 37°C, or by inoculating the organism into a mouse, and reisolating it from the liver and spleen. In vitro, 100 percent of strains are susceptible to amphotericin B, itraconazole, and ketoconazole, 82 percent to fluconazole, and 50 percent to flucytosine (Sutton et al. 1998).

Paracoccidioides brasiliensis causes a severe, usually chronic disease involving the skin, lungs, and lymphoid organs. Ocular involvement usually represents reactivated disease, and commonly presents as a chronic papular or ulcerating lesion of the eyelid in a male over 30 years of age engaged in agriculture, and coming from a region endemic for paracoccidioidomycosis (Silva et al. 1988). Although ocular paracoccidioidomycosis, which is usually unilateral, rarely occurs in the absence of lesions in other parts of the body, palpebral lesions (Burnier and Sant'Anna 1997) and granulomatous uveitis (Dantas et al. 1990) have sometimes been reported as the first indications of the disseminated disease. In an experimental model of ocular paracoccidioidomycosis in guinea pigs obtained by the intracardiac inoculation of yeast forms of *Paracoccidioides brasiliensis*, ocular involvement was observed in 80 percent of the infected animals; the uvea, ciliary body, choroid, iris, lids, and the conjunctiva were the structures most commonly affected (Kamegasawa et al. 1988). In the initial stages, ocular paracoccidioidomycosis needs to be differentiated from hordeolum, bacterial blepharitis, trachoma, leishmaniasis, sporotrichosis, lupus erythematosus, tuberculosis, and secondary syphilis (Silva et al. 1988). Spherical yeast-like cells with multiple buds attached by narrow necks ('steering wheel forms') can be seen in 10 percent potassium hydroxide mounts of necrotic tissue material or in tissue sections. In culture, *Paracoccidioides brasiliensis* exhibits dimorphism, which is temperature- and media-dependent.

Sporotrichosis, which is caused by the thermally dimorphic fungus *Sporothrix schenckii*, occurs worldwide in both tropical and temperate regions, and manifests as nodular lesions in the cutaneous and subcutaneous tissues that suppurate, ulcerate, and drain. Up to 1990, 17 episodes of endophthalmitis due to this fungus (Witherspoon et al. 1990) and orbital sporotrichosis (Streeten et al. 1974) had been reported. Since then, granulomatous uveitis progressing to scleral perforation and endophthalmitis due to use of corticosteroids (Cartwright et al. 1993), scleritis (Brunette and Stulting 1992), uveitis (Vieira-Dias et al. 1997), and a retinal granuloma associated with disseminated lesions (Curi et al. 2003) have been reported. In tissues or in culture at 37°C, the fungus appears as small, spherical, oval or elongated budding yeast cells with irregular-stained cytoplasm, or is 'cigar-shaped.' In culture at 26°C, the mycelial phase is recovered. In tissues, 'asteroid bodies' may occur, consisting of a central spherical or oval basophilic cell (3–5 μm diameter), surrounded by a thick, radiate, eosinophilic substance (which is believed to be a mass of antigen–antibody complexes); the asteroid body is characteristic of sporotrichosis only if the fungal cell within is typical of *Sporothrix schenckii*, since similar bodies may be observed in other mycoses (Rippon 1988).

Organisms of uncertain taxonomic classification

Rhinosporidiosis is caused by *Rhinosporidium seeberi*, an endosporulating microorganism traditionally considered to be a fungus but now of uncertain taxonomic classification (Pe'er et al. 1996). Rhinosporidiosis appears to be endemic in the Indian subcontinent (Chitravel et al. 1980; Krishnan et al. 1986; Moses et al. 1990; Shrestha et al. 1998); the prevalence of the condition in a south Indian village was reported to be 470/100 000 population (Chitravel et al. 1980). However, a significant number of cases has also been reported from other countries (Jimenez et al. 1984; Reidy et al. 1997). Farmers or their families, particularly children or young (up to 30 years of age) adults, and males, appear to be the most frequently affected; however, a female predominance is sometimes noted (Chitravel et al. 1980). The mode of transmission, while not definitely known, is believed to be by frequent exposure to water contaminated with spores of *Rhinosporidium seeberi* (Moses et al. 1990); in conjunctival rhinosporidiosis, the infection may arise by accidental injury to the eye by soil dust containing the spores (Jimenez et al. 1984). Lesions of rhinosporidiosis manifest as polypoid or papillomatous, very friable, proliferative outgrowths principally in the nasal cavity; ocular lesions may account for 13 percent of all lesions, the ratio of nasal to ocular lesions

being 1.4:1 (Owor and Wamukota 1978). The principal ocular lesion in rhinosporidiosis is conjunctival, but involvement of the limbus, sclera, lacrimal sac (Krishnan et al. 1986; Kalavathy et al. 1998), and peripheral cornea (Bhomaj et al. 2001) have also been reported. Two distinct phases, namely, trophic and endosporulating, have been noted in the life cycle of this organism in tissues (Savino and Margo 1983), and the diagnosis of rhinosporidiosis is based on detecting these cyst-like structures in the affected tissues (Figure 16.9). The formation of the cyst wall appears to be a continuous morphological and biochemical spectrum throughout the cytological maturation of the organism (Sunba and al-Ali 1989); variations in this pattern have been noted to occur, probably as a protective mechanism, in concurrent rhinosporidiosis and *Papillomavirus* infection.

Pythium insidiosum, a cosmopolitan fungus-like aquatic organism, is found predominantly in swampy environments, where water lilies, various vegetables, and especially certain grasses support the asexual phase of its life cycle; motile zoospores, which appear to be chemotactically attracted to plant leaves or human/horse hairs, are the likely infective particles (Sutton et al. 1998). Originally considered to be an oomycetous member of the kingdom Fungi and later reclassified in the kingdom Protoctista (Imwidthaya 1994; Sutton et al. 1998), *Pythium insidiosum* is now considered to belong to the kingdom Stramenopila, organisms that are related to algae (Sutton et al. 1998). This organism causes diseases in plants and animals (horses, cattle, dogs, cats, or fishes), particularly in tropical and subtropical parts of the world (Imwidthaya 1994; Thianprasit et al. 1996). Virgile et al. (1993) presented what they believed to be the first case of a human corneal ulcer due to *Pythium insidiosum*. Subsequently, *Pythium insidiosum* has been reported as a cause of keratitis in Thailand (Imwidthaya 1994, 1995), where it has also been reported to cause subcutaneous lesions in thalassemic patients (Thianprasit

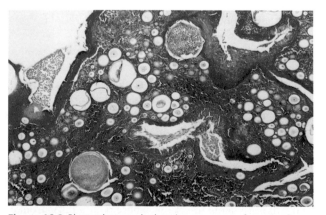

Figure 16.9 *Photomicrograph showing presence of sporangia (cysts) of* Rhinosporidium seeberi *in stroma of lacrimal sac. The cysts possess a sharply defined wall, with the largest sporangium revealing maturing endospores. Hematoxylin and eosin stain; magnification ×400*

et al. 1996), and in the temperate climate of New Zealand (Murdoch and Parr 1997), endophthalmitis (Srimuang et al. 1996), and orbital cellulitis with deep facial tissue involvement (Mendonza et al. 1996). Colonies on potato flakes agar at 25°C (optimal growth is at 33–36°C) are white to yellowish-white with short aerial hyphae; these are rapid-growing, and addition of human blood appears to stimulate colonial growth (Sutton et al. 1998). The most significant diagnostic microscopic morphological feature, biflagellate/motile asexual zoospores, may be induced by placing small pieces of *Pythium insidiosum* grown on agar into water and incubating at 37°C in the dark; after 1 hour incubation, encysted zoospores may be seen and after 24 hours, numerous zoosporangia may be observed (Sutton et al. 1998). In vitro susceptibility data are currently not available. Serological diagnosis of infection due to *Pythium insidiosum* by using sera from patients with keratoconjunctivitis or corneal ulcers with endophthalmitis has been reported (Srimuang et al. 1996). Badenoch et al. (2001) described a patient with keratitis in which DNA-based technology yielded a diagnosis of *Pythium insidiosum* keratitis.

Pneumocystis carinii was originally considered to be a protozoon, based on its morphology and response to antifungal drugs (Cailliez et al. 1996). However, subsequent to analysis of its nucleic acids, it has been reclassified as a member of the kingdom Fungi (Cailliez et al. 1996). It has been implicated as a cause of choroiditis (Dugel et al. 1990; Foster et al. 1991; Shami et al. 1991) and orbital infection (Friedberg et al. 1992) in patients with AIDS. In tissues, granulomatous inflammation mixed with foamy material containing *Pneumocystis carinii* is seen; round cysts with thickened walls, containing the crescent-shaped trophozoites, are demonstrated by the methenamine silver or Giemsa stains (Friedberg et al. 1992). No continuous in vitro culture system has been described for this organism. However, animals can be infected.

FUNGAL INFECTIONS OF THE ORBIT

Fungal infections of the orbit rarely occur spontaneously. Fungi usually gain access to the orbital space by direct extension from adjacent tissues (sinuses, teeth, lacrimal sac, lids); traumatic implantation of foreign bodies contaminated with fungi, or hematogenous seeding of fungi from a distant focus may also occur. Spread of infection from the sinuses to the orbit is believed to occur in 67–85 percent of orbital infections (Goodwin 1985; Parfrey 1986). In children, orbital complications usually accompany concurrent maxillary and ethmoid sinus infection, whereas in adults, frontal, maxillary, and ethmoidal sinus disease are about equally responsible for orbital complications (Hornblass et al. 1984). Due to the intimate relationship between the orbit and paranasal sinuses, virulent fungal pathogens causing sinusitis can devastate orbital structures by contiguous spread through the wafer-thin walls of the sinuses (Maskin et al. 1989). In fact, the variable presentations of orbital fungal infections parallel the presentations of paranasal sinus mycoses (Maskin et al. 1989; Klapper et al. 1997). Just as the distinctions between the different types of paranasal sinus mycoses are not clear because of closely associated patterns of clinical behaviour and pathological reactions (Levin et al. 1996; de Shazo et al. 1997), so also, the distinctions between the different types of orbital fungal infections, especially the chronic varieties, may not always be clear-cut.

Orbital fungal infections are usually demarcated into the invasive category, comprising the acute/fulminant (typified by rhinocerebral zygomycosis) and chronic/indolent (granulomatous) varieties (typified by orbital aspergillosis), and the noninvasive category, comprising the 'fungus ball' and orbital manifestations of allergic fungal rhinosinusitis (Klapper et al. 1997).

Since rhinocerebral zygomycosis and sino-orbital aspergillosis appear to be the principal types of orbital fungal infections, these are dealt with in greater detail.

Acute/fulminant orbital infections

RHINO-ORBITAL-CEREBRAL ZYGOMYCOSIS (RHINOCEREBRAL MUCORMYCOSIS)

Rhino-orbital-cerebral zygomycosis is the prototype of the acute/fulminant variety of invasive fungal orbital infection, usually running an acute course in an immunocompromised host (rarely in non-immunocompromised individuals), with angioinvasion and marked tissue necrosis being key features. The infection usually begins in the paranasal sinuses and then spreads to involve the orbit, face, palate, or brain (Anand et al. 1992). *Rhizopus* spp., especially *Rhizopus arrhizus*, are the most frequent causes (Ribes et al. 2000), less common causes being *Absidia corymbifera*, *Apophysomyces elegans* (Radner et al. 1995; Brown et al. 1998; Fairley et al. 2000), *Cunninghamella* spp. (Chetchotisakd et al. 1991), and *Saksenaea vasiformis* (Ajello et al. 1976; Kaufman et al. 1988). The major mode of disease transmission for the zygomycetes is presumed to be by inhalation of fungal conidia from environmental sources (Yohai et al. 1994; Ribes et al. 2000).

Being opportunistic pathogens, the Mucorales require a breakdown in immune defense mechanisms, especially those disease processes leading to neutropenia or neutrophil dysfunction, to cause disease. Neutrophil dysfunction induced by ketoacidosis underlies the majority of cases of human zygomycosis (Ribes et al. 2000), and this occurs even in juvenile diabetics. Although neutropenia induced by bone marrow suppression during chemotherapy, or immunosuppression induced following transplantation, is also believed to contribute to this type of pathology, zygomycosis was observed in only 13 (0.9 percent) of 1 500 consecutive patients who underwent bone marrow transplantation (Morrison and McGlave 1993). Immune system compromise has also been linked to the development of zygomycosis. Corticosteroid use, in particular, may increase susceptibility to the development of zygomycosis by suppressing the normal inflammatory cell response, and also by inducing a diabetic state. An association has been noted between the occurrence of zygomycosis in patients on dialysis who were receiving deferoxamine/ desferrioxamine for iron or aluminum overload (Daly et al. 1989; Boelaert et al. 1991; Slade and McNab 1991). Rhinocerebral manifestations were reported in 31 percent of 59 dialysis patients who developed zygomycosis due to *Rhizopus* spp. (Boelaert et al. 1991). In an alloxan-induced immunocompromised model of zygomycosis in mice, desferrioxamine iron chelation produced rhinocerebral zygomycosis in those animals that were challenged intrethmoidally with *Rhizopus* spores (Anand et al. 1992), presumably because desferrioxamine, in the iron-chelate form, feroxamine, provides iron to fungi belonging to the Mucorales. Rhino-orbital-cerebral zygomycosis may also supervene in the presence of predisposing factors such as protein-calorie malnutrition and iron overload, intravenous drug abuse, leukemia, aplastic anemia, myelodysplastic syndrome, concurrent diabetes mellitus, and adrenogenital syndrome, and treatment with the immunosuppressive medications necessary to maintain solid organ transplants, such as liver transplants, and in patients receiving donor leukocyte infusions (Boelaert et al. 1991; Weinberg et al. 1993; Nussbaum and Hall 1994; Penalver et al. 1998; Raj et al. 1998; Webb et al. 1998; Stern and Kagan 1999; Ribes et al. 2000). The disease may rarely occur in non-immunocompromised individuals if there is some associated antibiotic use or a breakdown in the mucocutaneous barrier (Bhattacharya et al. 1992; Parthiban et al. 1998; Saltoglu et al. 1998); such patients may respond to treatment better than immunocompromised patients.

In recent years, *Apophysomyces elegans* has been reported as a cause of rhino-orbital-cerebral zygomycosis in immunocompetent hosts (Radner et al. 1995; Brown et al. 1998; Fairley et al. 2000). The disease entity caused by *Apophysomyces elegans* appears to differ from that caused by the more common genera of zygomycetes in that the disease frequently follows traumatic inoculation and/or soil contamination, and tends to occur in warm climates in patients who do not exhibit well-recognized immunological or metabolic abnormalities (the patients are frequently immunocompetent) (Fairley et al. 2000).

Fever has been reported to occur in 44 percent of patients with rhinocerebral zygomycosis, nasal ulceration, or actual necrosis in 38 percent and periorbital or facial edema in 34 percent; other findings (in descending order of frequency of occurrence) include decreased vision, ophthalmoplegia, sinusitis, headache, facial pain,

decreased mental status, leukocytosis, nasal discharge, nasal stuffiness, corneal anesthesia, orbital cellulitis, and proptosis (Yohai et al. 1994). Necrosis of the facial skin and bone with typical black eschar formation, which is usually considered to be pathognomonic of rhinocerebral zygomycosis, is reported by some to be a late manifestation (Ferry and Abedi 1983). Others believe that the classic presentation in a susceptible patient is with unilateral severe headache and facial pain, nasal stuffiness with granular or purulent discharge, facial or eyelid edema, fever, and leukocytosis (Nussbaum and Hall 1994). Bacterial cavernous sinus thrombosis may sometimes mimic rhino-orbital-cerebral zygomycosis, but early visual loss and retinal artery occlusion are believed to suggest a diagnosis of the latter condition, since blindness is a much later finding in the former condition (Fairley et al. 2000).

Orbital findings result from ischemic necrosis of the intraorbital contents and cranial nerves; however, bony involvement is uncommon because of the angioinvasive nature of the fungus. Fungal invasion of the eyeball in patients with rhino-orbital-cerebral zygomycosis is considered to be a poor prognostic index, since all six patients reported up to 1992 with this condition succumbed to the disease (Sponsler et al. 1992). Rhino-orbital-cerebral zygomycosis has also been reported to manifest as a painless orbital apex syndrome without any sign of orbital cellulitis or acute systemic disease (Balch et al. 1997), which may have a good outcome with medical therapy, orbital infarction syndrome (Borruat et al. 1993), bilateral cavernous sinus thrombosis (Van Johnson et al. 1988; Alleyne et al. 1999), isolated pontine infarction (Calli et al. 1999), palatal ulcer (Van der Westhuijzen et al. 1989), chiasmal infarction and sudden blindness (Lee et al. 1996), fever with right-sided hemiparesis, and dysarthria in juvenile diabetics (Adler et al. 1998) and numbness, loss of sensation over the temporal region, loss of vision, and proptosis (Attapattu 1995).

Magnetic resonance imaging (MRI) and computerized tomography (CT) are important aids in establishing an anatomic diagnosis in suspected rhinocerebral fungal infections (Moll et al. 1994; Levin et al. 1996). Reported MRI manifestations of rhino-orbital-cerebral zygomycosis include (Press et al. 1988): sinus and orbital disease followed by deep facial extension, involvement of the basal portions of the cerebral hemispheres, brainstem, and hypothalamus and hyperintense regions (signifying intracerebral inflammation). Limitations in establishing the diagnosis of both cerebral zygomycosis and cavernous sinus thrombosis (Van Johnson et al. 1988) may be overcome by performing sequential CT and MRI studies in a patient suspected of having rhino-orbital-cerebral zygomycosis (Boelaert 1994; Moll et al. 1994). It is also not clear whether there are any specific radiological findings for rhinocerebral zygomycosis. CT nonenhancement of the superior ophthalmic artery and vein may represent a specific sign of orbital zygomycosis (Kilpatrick et al. 1984; Gamba et al. 1986). McLean et al. (1996) documented, by MRI and pathology, the perineural spread of rhinocerebral zygomycosis, following the trigeminal nerve to the pons. While CT and MRI scans help to define the extent of bone and soft-tissue destruction, they are probably more useful in planning surgical intervention than in establishing a diagnosis. MRI scanning may be preferred for the diabetic patient in whom CT contrast agents may be contraindicated.

A prompt and accurate diagnosis of orbital zygomycosis necessitates a high level of clinical suspicion, and good coordination between the clinical and laboratory staff. Multiple biopsies should be taken from necrotic tissue and, if possible, from the nose, paranasal sinuses, and oropharynx as well; abscesses should be aspirated, and lesions on the mucous membranes should be irrigated or scraped (Richardson and Shankland 1999). Swabs are not satisfactory. Once collected, samples should be transported at once to the laboratory since immediate information is required and because zygomycetes are rather fragile organisms that do not survive more than a few hours at refrigerator temperature; if overnight storage is required, samples may be kept in Stuart's transport medium and left at room temperature. The tissue should be minced and not ground, to ensure that any viable fungal elements are not destroyed. Zygomycetes may not be found at the center of the necrotic tissue but, rather, at the edge or proximal to the necrotic material (Richardson and Shankland 1999).

The microscopic demonstration of zygomycetes in clinical material taken from necrotic lesions is more significant than their isolation in culture (Figure 16.10a), since zygomycetous fungi are frequent contaminants in the clinical microbiology laboratory (Ribes et al. 2000). In tissue material stained by hematoxylin-eosin, abundant, large, often bizzare, aseptate or sparsely septate, irregularly branching hyphal elements (Figure 16.8) can be seen invading intact tissue; however, failure to observe such elements does not exclude the diagnosis of such an infection. Although nasal, palatal, and sputum cultures seldom contribute to the diagnosis, isolation of Mucorales from sputum, aspirated material from sinuses, or bronchial washings taken from diabetic or immunosuppressed patients should not be ignored (Richardson and Shankland 1999). The confirmatory identification of the genus or species requires culture of the specimen. Although zygomycetes are not especially fastidious, cultures of necrotic tissue frequently fail to yield fungi even when direct microscopy is positive; therefore, culture media should be heavily inoculated with as much material as possible. Sabouraud glucose neopeptone agar (with an antibacterial such as chloramphenicol or polymyxin B, but no cycloheximide) is adequate; squares of sterile home-made bread (without preservatives) have also been suggested. Mucorales usually grow within 2–5 days (Figure 16.10a), and rapidly fill the petri plate or tube.

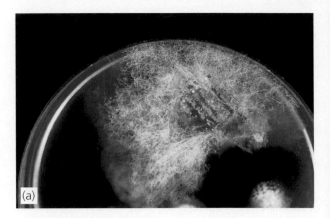

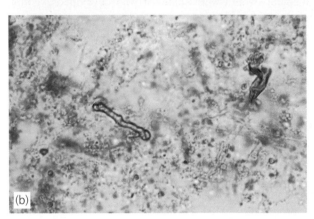

Figure 16.10 (a) *Macroscopic appearance of a* Mucor *sp. (isolated from rhino-orbital zygomycosis) on rose bengal agar.* **(b)** *Photomicrograph of nasal material from a patient with allergic fungal rhinosinusitis, exhibiting sparse fungal hyphae (*Aspergillus fumigatus *recovered in culture). Lactophenol cotton blue stain; magnification ×400*

In addition to the conventional methods of diagnosis, Kaufman et al. (1988) were able to demonstrate antibodies to *Saksenaea vasiformis* by enzyme-linked immunosorbent assay in a patient with rhinocerebral zygomycosis due to this fungus. However, demonstration of antibodies to Mucorales by this serological technique may not be possible if the disease progresses so rapidly that the patient is dead before specific antibodies can be formed.

General principles in the treatment of acute invasive rhinocerebral zygomycosis and other acute invasive orbital mycoses include:

- control of diabetic ketoacidosis or other systemic underlying diseases, and elimination of predisposing factors
- surgical debridement and restoration of sinus drainage
- intravenous amphotericin B (conventional or lipid associated).

In the diabetic patient, there should be prompt correction of acidosis and other metabolic abnormalities, and elimination of predisposing factors.

Surgery consists of wide local excision and debridement of all involved and devitalized oral, nasal, sinus, and orbital tissue while establishing adequate sinus and orbital drainage (Kohn and Hepler 1985); this may be quite mutilating and multiple operations are often required. Infected tissue typically bleeds little due to the vaso-occlusion caused by the Mucorales; therefore, necrotic tissue can be removed until normal bleeding is encountered. However, this may not be possible with extensive infections which extend to the dura or beyond (Yohai et al. 1994). Since rhinocerebral zygomycosis spreads by direct contiguous extension to surrounding tissues, and because the zygomycete hyphae are easily recognized on frozen section biopsies, this disease entity is potentially amenable to complete resection by frozen section guidance; one such technique of frozen section-guided surgical debridement for biopsy-proven rhino-orbital-cerebral zygomycosis, in conjunction with local and systemic amphotericin B, has recently been described (Langford et al. 1997). Serial radiological imaging may also help to identify the extent of disease and response to treatment. In recent years, endoscopic sinus surgery has been tried, alone or in combination with the traditional surgical procedures, to attain the goal of radical resection with less operative morbidity and greater operative accuracy (Jiang and Hsu 1999).

The interval between onset of symptoms and surgery was found to play a key role in the ultimate survival of evaluable patients with rhino-orbital-cerebral zygomycosis reported in the literature between 1970 and 1994; when the interval was 1–6 days, as many as 81 percent survived compared to 42 percent when the interval was 13–30 days (Yohai et al. 1994). In extreme cases, for example in an acutely infected orbit with a blind immobile eye, orbital exenteration, which entails removal of the eyeball together with its extraocular muscles and other soft tissues of the orbit, has to be performed.

Although treatment modalities have not undergone clinical trials, and in spite of the emergence of new antifungals in the therapy of invasive fungal disease, the combination of aggressive surgical debridement and intravenous amphotericin B therapy continues to be the treatment of choice in rhino-orbital-cerebral zygomycosis (O'Keefe et al. 1986; Yohai et al. 1994). An analysis of the case records of 145 patients with rhino-orbital-cerebral zygomycosis reported between 1970 and 1994 revealed that when the interval from onset of symptoms to initiation of amphotericin B therapy was 6 days or less, 76 percent of patients survived, but when the interval was 7–30 days, only 36 percent survived (Yohai et al. 1994). Amphotericin B has been administered by various routes in an effort to improve the outcome of this condition, which normally results in extreme morbidity. Kohn and Hepler (1985) were among the first to use this approach by supplementing the standard therapeutic regimen for rhino-cerebral zygomycosis with daily amphotericin B (1 mg/ml)

irrigation and packing of the involved orbit and sinuses; this is likely to have aided the delivery of amphotericin B to poorly perfused infected and/or necrotic tissue, so that excellent results were obtained in their small series of patients, even though orbital exenteration was not performed. The use of a nasal-sinus catheter (Cohen and Greenberg 1980) or percutaneous catheter (Nussbaum and Hall 1994) for local irrigation, the use of adjunctive local nebulized amphotericin B (Raj et al. 1998) as well as administration of amphotericin B by using the intracavitary/interstitial and cerebrospinal fluid (intraventricular) routes (Adler et al. 1998), in addition to conventional intravenous administration of amphotericin B, have also been tried.

Since the use of intravenous amphotericin B is fraught with numerous side effects, attempts have been made in recent years to prepare new formulations of amphotericin B which maintain the efficacy of the antifungal without its toxicity. One such formulation is amphotericin B encapsulated in liposomes (liposomal amphotericin B). Lipid association significantly decreases the toxicity of amphotericin B, enabling higher dosages to be administered, and appears to enhance delivery to fungi, infected organs, and phagocytes. Treatment comprising liposomal amphotericin B and surgical debridement has been reported useful for a small number of patients with rhinocerebral zygomycosis (Penalver et al. 1998; Saltoglu et al. 1998). Strasser et al. (1996) documented the medical cure of a 60-year-old diabetic woman with rhinocerebral zygomycosis involving the cavernous sinus whose infection responded to medical therapy with another modification of this drug, namely, amphotericin B lipid complex. Yet another modification is amphotericin B colloidal dispersion, a 1:1 complex of cholesteryl sulfate and amphotericin B; administration of this preparation, in combination with surgical debridement, resulted in cure of life-threatening rhinocerebral zygomycosis in two patients, and in a cure of the fungal infection in a third patient (Moses et al. 1998).

Additional therapeutic modalities, including the triazoles itraconazole and fluconazole, have been reported useful in therapy of a small number of patients with rhinocerebral zygomycosis (Sica et al. 1993; Parthiban et al. 1998); however, further clinical studies are necessary to substantiate these preliminary results. The efficacy of granulocyte colony-stimulating factor, in conjunction with fluconazole therapy, was evaluated in the treatment of four patients with histopathologically proven rhinocerebral zygomycosis (Sahin et al. 1996); two patients refractory to fluconazole therapy were treated with liposomal amphotericin B. The improvement in clinical manifestations was closely related to neutrophil recovery, and all patients were alive at the end of therapy; in addition to surgical debridement and antifungal therapy, granulocyte colony-stimulating factor seems to have played a role in their survival. Hyperbaric oxygen has also been described as an additional therapeutic option. Yohai et al. (1994) evaluated the effect of hyperbaric oxygen therapy in patients with a low survival rate, that is, in those with bilateral rhino-cerebral zygomycosis; 22 percent of patients who received standard therapy survived while 83 percent of patients who received standard therapy plus adjunctive hyperbaric oxygen survived. Although the possible adverse effects of hyperbaric oxygen therapy, such as decompression sickness and aeroembolism, need to be kept in mind (Maskin et al. 1989), it has been suggested that hyperbaric oxygen be considered as part of the initial therapy and should be persisted with until evidence of disease regression is observed (Yohai et al. 1994).

In a review of the published case histories of 145 patients who presented with rhinocerebral zygomycosis from 1970 to 1994, delayed diagnosis and treatment, hemiparesis or hemiplegia, bilateral sinus involvement, leukemia, renal disease, and treatment with desferrioxamine were identified as significant factors related to a lower survival rate; the association of facial necrosis with a poor prognosis also appeared to be clinically important (Yohai et al. 1994). The presence of ocular infiltration by fungus may indicate a poor prognosis in rhinocerebral zygomycosis (Sponsler et al. 1992). The mortality of rhinocerebral zygomycosis caused specifically by *Apophysomyces elegans* is currently unknown because of the rarity of diagnosed cases, but it would seem to fall at the more favorable end of the spectrum (Fairley et al. 2000).

ACUTE ORBITAL INFECTIONS DUE TO OTHER FUNGI

Fungi belonging to the order Mucorales are so frequently the causes of acute orbital mycoses that the term 'mucormycosis' has often been used as being synonymous with a fulminant orbital fungal infection. However, species of *Aspergillus* (Levin et al. 1996), *Scedosporium apiospermum* (*Pseudallescheria boydii*) (Anderson et al. 1984), and different genera of dematiaceous filamentous fungi (Jay et al. 1988; Maskin et al. 1989) may also cause this disease entity.

Although orbital aspergillosis usually manifests as a chronic condition, it may sometimes present as acute fulminant disease, where rapid, uncontrollable progression leads to intracranial involvement and sometimes death (Mauriello et al. 1995; Kronish et al. 1996; Johnson et al. 1999; Kusaka et al. 2003); in such cases, the key presenting complaints include abrupt onset of proptosis, ophthalmoplegia, blepharoptosis with precipitous visual loss, debilitating periorbital pain or headache, without inflammatory signs, and sometimes a temporal arteritis (Mauriello et al. 1995; Kronish et al. 1996; Hutnik et al. 1997; Johnson et al. 1999). Acute orbital infections caused by *Scedosporium apiospermum* and dematiaceous fungi are clinically characterized by rapid acute onset, orbital pain, fever, rhinorrhea, and signs of orbital inflammation, such as acute visual loss and total internal and external ophthalmoplegia.

In the treatment of fulminant rhino-orbital infections caused by fungi other than the Mucorales, treatment is essentially as described for orbital zygomycosis; however, antifungals other than amphotericin B may play a prominent role. The dematiaceous fungus *Bipolaris hawaiiensis* was identified as the cause of a devastating bilateral optic neuropathy and orbitopathy following contiguous fungal sinusitis in an apparently immunocompetent young man. Therapy with 3 700 mg of amphotericin B failed, but itraconazole therapy was initiated because of in vitro susceptibility of the fungus to itraconazole, and the detection of fungicidal levels of itraconazole in the sinus mucosa after oral therapy; at the time of reporting, the patient had been on itraconazole for 16 months and appeared stable (Maskin et al. 1989). An acute orbital infection and brain abscess due to *Scedosporium apiospermum* (*Pseudallescheria boydii*) in a 4-year-old boy failed to respond to 6 days of intravenous amphotericin B but was successfully treated with intravenous miconazole and multiple surgical debridements (Anderson et al. 1984). Surgical debridement of the orbit and a 6-week course of intravenous miconazole resulted in reduction of *Scedosporium apiospermum* orbital infection in another patient as well (Nunery et al. 1985).

Chronic invasive orbital infections

The chronic invasive form of fungal sinusitis, which slowly results in rhino-orbital-cerebral disease, manifests as a slowly destructive disease process. It may occur in immunocompromised (Levin et al. 1996) or non-immunocompromised patients (Klapper et al. 1997).

CHRONIC ORBITAL ASPERGILLOSIS

It is not clear whether the entity commonly referred to as 'chronic orbital aspergillosis,' which is described as being more common in hot, humid climates (Deans et al. 1996; Krause and Bullock 1996), refers to the chronic invasive disease (associated with granulomatous inflammation and fibrosis) occurring in immunocompromised persons (Levin et al. 1996), or to chronic invasive disease that occurs in non-immunocompromised patients (Klapper et al. 1997). *Aspergillus* spp. have been implicated in a wide variety of primary ocular orbital conditions. Predisposing factors include alcoholism, high-dose corticosteroid therapy, insulin-dependent diabetes mellitus and AIDS (Mauriello et al. 1995; Kronish et al. 1996; Johnson et al. 1999). Chronic presentations of orbital aspergillosis, which are more common than the acute variety, tend to be more localized and include an aspergilloma extending from one of the periorbital sinuses into the orbit, an aspergilloma in an exenteration socket, a complex dacryocystis, a nerve tumor, and postoperative periorbital swelling (Swoboda and Ullrich 1992; Levin et al. 1996). The presenting complaints in chronic disease may be nonspecific. Chronic sino-orbital aspergillosis may present subacutely in older patients with diabetes mellitus or pulmonary diseases, and in those receiving immunosuppressive drugs. Some presentations of orbital aspergillosis, such as optic nerve involvement, may lead to the use of systemic corticosteroids, delaying diagnosis and, possibly, potentiating the infectious process. In neutropenic, or otherwise immunocompromised patients, a high index of suspicion should be maintained to forestall the emergence of fulminant aspergillosis (Levin et al. 1996). For diagnosis of orbital aspergilloma, aspiration cytology and immunohistochemistry have been described (Oneson et al. 1988).

Management of invasive sino-orbital aspergillosis, as has been reported in patients with AIDS (Vitale et al. 1992; Meyer et al. 1994; Kronish et al. 1996; Johnson et al. 1999), revolves around aggressive surgical debridement and the use of intravenous amphotericin B. In spite of these measures, the prognosis is very poor, and mortality may be as high as 75–80 percent (Kronish et al. 1996; Johnson et al. 1999), primarily due to intracranial extension of the disease process. Most patients succumb to their illness within a few months. However, the survival of a patient with invasive aspergillosis involving the orbit, paranasal sinus, and central nervous system for more than 9 years has recently been reported (Kusaka et al. 2003). The prognosis of this disease depends on the location and duration of the infection and the patient's immunological status, with sphenoidal aspergillosis being particularly aggressive in nature due to its close relation to the skull base. The use of protease inhibitors may prolong survival in some of these patients (Johnson et al. 1999). Fortunately, therapy of the more localized forms of orbital aspergillosis carries a better prognosis. In such cases, conservative orbital debridement (for example, resection of an orbital abscess) with amphotericin B administered intravenously and locally (by means of an indwelling cathether or intraorbital injection), may be effective, especially in patients with reversible immunosuppression and good preoperative visual acuities (Harris and Will 1989; Cahill et al. 1994; Seiff et al. 1999). Rieske et al. (1998) described the use of amphotericin B colloidal dispersion in a child who had sino-orbital aspergillosis. Massry et al. (1996) reported successful resolution of sino-orbital aspergillosis following initiation of itraconazole treatment, without recurrence at 10 months follow-up, in an immunocompetent patient in whom traditional therapeutic modalities (surgical debridement and amphotericin B therapy) had not resulted in resolution; thus, oral itraconazole could be considered as a treatment option in orbital aspergillosis occurring in immunocompetent patients who have recurrent or recalcitrant disease, or in those who cannot tolerate amphotericin B.

CHRONIC RHINOCEREBRAL ZYGOMYCOSIS

In the chronic presentation of rhinocerebral zygomycosis, which is less common than the acute form, the disease is indolent and slowly progressive over weeks to months. Harril et al. (1996), who described two patients with chronic rhinocerebral zygomycosis treated at their institution and reviewed 16 other cases reported in the English-language literature, observed that the median time from onset of symptoms to diagnosis was 7 months; the most common presenting features of chronic rhinocerebral zygomycosis were ophthalmological and included ptosis, proptosis, visual loss, and ophthalmoplegia. This entity was found to occur predominantly in patients with diabetes and ketoacidosis. The incidence of internal carotid artery and cavernous sinus thrombosis was higher in patients with chronic rhinocerebral zygomycosis than in those with the acute disease, although the overall survival rate for the former was 83 percent.

CHRONIC ORBITAL INFECTIONS DUE TO OTHER FUNGI

Other fungi implicated as causes of chronic invasive orbital infection include *Paecilomyces lilacinus*, *Scedosporium apiospermum*, *Bipolaris spicifera*, and *Bipolaris hawaiiensis*, other genera and species of hyaline and dematiaceous fungi, and dimorphic fungi, such as *Blastomyces dermatitidis*, *Coccidioides immitis*, and *Histoplasma capsulatum* var. *duboisii* (Olurin et al. 1969; Streeten et al. 1974; Vida and Moel 1974; Agrawal et al. 1979; Nunery et al. 1985; Jay et al. 1988; Maskin et al. 1989; McGuire et al. 1991; Jacobson et al. 1992). The patients may present with nasal congestion, postnasal drip, and chronic sinus pain, citing frequent bouts of acute sinusitis (Washburn et al. 1988). There may be slowly progressive unilateral proptosis with decreased vision. In *Bipolaris* infections, impairment of extraocular motility and proptosis appear to occur more frequently than visual loss (Jacobson et al. 1992). A mass may be seen originating from an adjacent paranasal sinus, with sclerotic margins, and producing an inflammatory reaction in the orbit (Washburn et al. 1988). Other orbital inflammatory signs may be absent. Histologically, this form is characterized by granulomatous inflammation and tissue invasion. Although the prognosis in this variety of disease is certainly much better than that in acute fulminant disease, intraorbital and intracranial extension may occur, leading to significant morbidity (Klapper et al. 1997). Treatment is surgical debridement and systemic antifungal therapy; antifungals other than amphotericin B may play an important role in such infections.

Chronic localized noninvasive fungal infections

Chronic, localized noninvasive fungal infections or 'fungus balls' tend to occur in normal anatomic cavities (e.g. paranasal sinuses) or cavitations induced by disease (pulmonary cavities due to tuberculosis). An example of such 'fungus balls' is the aspergilloma, which may occur in a paranasal sinus or in a lung cavity (Klapper et al. 1997; Latgé 1999). Histologically, it manifests as a non-invasive, tightly packed fungal mycelium or 'fungus ball' (mycetoma) (Latgé 1999). The normal orbital cavity is filled with the eyeball and ocular adnexa, and surrounding orbital fat. Hence, a fungus ball can occur only where orbital exenteration (removal of the entire contents of the orbit) has been performed. Levin et al. (1996) described the occurrence of *Aspergillus* infection in an exenterated socket. This type of 'fungus ball' may also occur in the orbital prosthesis used to fill an exenterated socket. Oestreicher et al. (1999) described the occurrence of a fungal abscess (subsequently shown by histopathology to be filled with fungal hyphae having the morphology of *Aspergillus* spp.) within an hydroxyapatite orbital implant 58 months after uncomplicated implant surgery; the patient's symptoms resolved following removal of the implant.

Allergic fungal rhinosinusitis with orbital involvement

The name 'allergic *Aspergillus* sinusitis' was proposed in 1983 (Katzenstein et al. 1983) to describe a newly recognized form of chronic sinusitis, but this entity is now called 'allergic fungal sinusitis' or 'allergic fungal rhinosinusitis' since many genera of dematiaceous fungi (*Bipolaris*, *Exserohilum*, *Curvularia*, and *Alternaria*) are now believed to be responsible for many cases (Adam et al. 1986; Brummund et al. 1986; Klapper et al. 1997; Ferguson 1998). This condition occurs in immunocompetent, atopic individuals. Here, the fungus does not invade the surrounding tissue, but serves as an 'allergen,' resulting in the formation of a thick ('peanut butter-like') tenacious, green or brown 'allergic' mucin. The patients may give a history of nasal polyposis or recurrent chronic sinusitis. Symptoms described include nasal obstruction, local pain, rhinorrhea, visual loss, diplopia, proptosis, epiphora, cranial nerve palsies, and facial deformity. Up to 17 percent of patients with allergic fungal rhinosinusitis may experience orbital symptoms (Klapper et al. 1997). Allergic fungal rhinosinusitis is a unique subset of fungal sino-orbital disease which differs from invasive sino-orbital disease in many respects. Key elements to the diagnosis of this condition include a predominantly eosinophilic response of the sinus mucosa, the presence of the characteristic mucoid exudate containing fungal hyphae (Figure 16.10b), peripheral blood eosinophilia, serum precipitins against the fungus involved, elevated serum total and fungus-specific IgE and IgG concentrations, and an immediate hypersensitivity response to the concerned fungal antigen (Brummund et al. 1986). Patients with allergic fungal rhinosinusitis generally have high levels of total IgE and IgE-mediated atopy to multiple fungal and nonfungal antigens; hence, immunotherapy has now been advised for such patients since it appears to significantly alter the clinical course of these patients, sparing them from repeated surgery and the use of systemic corticosteroids (Mabry and Mabry 1998). It has been advised that patients with allergic fungal rhinosinusitis should receive:

- conservative yet thorough surgical exenteration of the lesion by endoscopic techniques
- adjuvant therapy using perioperative systemic corticosteroids and topical corticosteroids as the patients heal and before beginning immunotherapy
- irrigation and debridement as necessary after surgery
- immunotherapy with all relevant fungal and nonfungal antigens to which they are found to be allergic about a month or so following surgery (Ferguson 1998; Mabry and Mabry 1998).

Allergic fungal rhinosinusitis does not require aggressive surgical debridement or intravenous amphotericin B.

MYCOTIC INFECTIONS OF EYELIDS

Although infections of the eyelids are caused most frequently by bacteria (particularly *Staphylococcus* species), fungi may also cause superficial or deep eyelid lesions.

Types

Necrotizing fasciitis of the eyelids and periorbital area due to *Cryptococcus neoformans* was reported in a young man after a trivial trauma by a wooden splinter (Doorenbos-Bot et al. 1990), while an eyelid nodule and ulcerative lesions have been reported to constitute a sentinel lesion of disseminated cryptococcosis in a patient with AIDS (Coccia et al. 1999).

Use of broad-spectrum antibacterials or immunosuppressive agents may predispose to spread of infection from a focus, resulting in the occurrence of an eyelid lesion due to a *Candida* sp. (Alvarez and Tabbara 1996). Ulceration begins at the base of an eyelash; small granulomata appear at its edge, and vesicles and pustules may also be present. In a study of 407 patients with chronic severe ulcerative blepharitis, Huber-Spitzy et al. (1991) noted the occurrence of positive cultures for *Candida* spp. in 47 (12 percent); most of these patients also had atopic dermatitis.

Malassezia spp. (formerly *Pityrosporum orbiculare* and *Pityrosporum ovale*) (Ashbee and Evans 2002) are superficial fungi that may be associated with seborrheic blepharitis and pityriasis versicolor, a chronic mild skin infection sometimes found around the eyebrows and eyelids. When lid scrapings from 40 patients with active seborrheic or mixed seborrheic/staphylococcal blepharitis were subjected to microbiological investigations, yeasts with the morphology of *Malassezia* spp. were detected by direct microscopic examination in the scrapings of 39 of the 40 patients, while fungi were isolated from those of about half the patients (Nelson et al. 1990).

Lesions caused by the dermatophytes begin as erythematous scaly papules that slowly enlarge; healing simultaneously occurs in the central, paler area. Infections caused by species of *Microsporum* and *Trichophyton* result in the hair breaking at the level of the skin surface.

Although the eyelid is generally considered to be the most common site of ocular infection due to *Blastomyces dermatitidis* (Safneck et al. 1990; Slack et al. 1992), Bartley (1995) reported that only one of 79 patients with systemic blastomycosis seen by him had such lesions. The lesions may arise due to contiguous spread from facial lesions, or due to hematogenous dissemination from a pulmonary lesion. Small abscesses may be visible around the eyelashes. These later form granulomatous ulcers with thick crusts and an underlying purplish discoloration of the skin. Healing of the lesions may lead to severe cicatrization and ectropion formation (Slack et al. 1992). Conjunctival lesions generally occur due to contiguous spread from eyelid lesions, but may also occur as separate entities (Slack et al. 1992).

Eyelid lesions, alone or in association with corneal and conjunctival lesions, have been found to occur in more than 50 percent of reported cases of ocular paracoccidioidomycosis (Belfort et al. 1975; Silva et al. 1988; Burnier and Sant' Anna 1997). The palpebral lesion starts as a papule, usually close to the lid border, and grows and ulcerates in the center. The base of the ulceration reveals fine hemorrhagic punctate and elevated, thickened and hardened borders. The lesions evolve toward palpebral coloboma, with loss of the eyelashes.

Rhinosporidiosis of the lid margins is a rare occurrence (Reidy et al. 1997).

Diagnosis

Infections due to dermatophytes may exhibit fluorescence when exposed to a Wood's lamp. The hyphae or yeast cells of fungi causing eyelid lesions can be demonstrated by examination of a 10 percent potassium hydroxide mount or a Gram-stained smear of scrapes from the eyelids.

Therapy

Several therapeutic regimens have been recommended for mycoses of the eyelids (Alvarez and Tabbara 1996;

Behrens-Baumann 1999), but the basis on which these suggestions have been made is not clear.

For infections due to the dermatophytes, terbinafine, itraconazole, or local antifungal drugs (active ingredients being combinations of fatty acids and salicylic acid) probably suffice (Ellis 1985). In one patient, lesions of the eyebrow due to *Trichophyton rubrum* were found to have disappeared almost entirely after 3 weeks of oral itraconazole therapy (Hiruma et al. 1991).

Nelson et al. (1990) reported that when topical 2 percent ketoconazole cream with lid hygiene was used to treat seborrheic blepharitis (due to *Malassezia* spp.) and mixed seborrheic/staphylococcal blepharitis, patients receiving ketoconazole more frequently had normal or markedly improved lids after treatment than did a control group of patients receiving only placebo and lid hygiene.

For infections due to *Candida* spp., topical agents, including ointments of nystatin (100 000 units/g), ketoconazole (2 percent cream), miconazole (2 percent), clotrimazole (1 percent) or econazole (1 percent), and oral ketoconazole (200–400 mg/day) are reportedly effective (Ellis 1985), while oral fluconazole (150 mg) once or twice weekly for 2–3 months has also been reported to be useful for such infections (Alvarez and Tabbara 1996). Oral ketoconazole twice daily (100 mg/day) with topical miconazole ointment for 6 weeks has been recommended for treatment of blepharitis due to *Candida* spp. (Huber-Spitzy et al. 1991).

Lesions of the eyelid in a patient with disseminated infection due to *Cryptococcus neoformans* var. *neoformans* was controlled (although there were occasional recurrences) by a combination of surgical excision and intravenous amphotericin B (Coccia et al. 1999).

Verrucous lesions of the eyelid due to *Blastomyces dermatitidis* were reported to resolve with a combination of antifungals (potassium iodide, intravenous amphotericin B) and surgery (Barr and Gamel 1986); oral fluconazole (150 mg) once or twice weekly for 2–3 months has also been reported to be useful (Alvarez and Tabbara 1996). Itraconazole is an alternative.

Papular lesions of the eyelids due to *Paracoccidioides brasiliensis* were reported to respond to intravenous amphotericin B, alone or in combination with oral ketoconazole, without any surgery being required (Silva et al. 1988).

Lesions of the eyelid due to *Rhinosporidium seeberi*, which are rare, require excision.

Deep infections of the eyelids, which may occur in infections due to the thermally dimorphic fungi or in infections due to species of *Aspergillus*, *Candida*, or *Cryptococcus*, must be treated with drugs administered systemically; intravenous amphotericin B is usually the treatment of choice, although oral ketoconazole, itraconazole, or fluconazole may be considered in milder infections (Ellis 1985; Grant and Clissold 1989; Alvarez and Tabbara 1996). In one patient, preseptal cellulitis due to a *Trichophyton* spp. was described to have completely resolved following two courses of oral itraconazole (100 mg/day) (Velazquez et al. 2002).

MYCOTIC DACRYOCANALICULITIS

Fungi, such as *Alternaria* spp., *Aspergillus fumigatus* and other *Aspergillus* spp., *Candida* spp., dermatophytes, *Fusarium* spp., *Penicillium* spp., *Scopulariopsis* spp., and *Sporothrix schenckii*, have been reported as causes of canaliculitis (Levin et al. 1996; Behrens-Baumann 1999), although the significance of some of these isolates is doubtful since they do not appear to satisfy the criteria described earlier (McGinnis 1980). Most of the clinical features described, as well as the surgical procedures recommended, are for dacryocanaliculitis in general, and are not necessarily specific for fungal infection.

Dacryocanaliculitis usually manifests as persistent unilateral epiphora (watering), with an itching sensation; there may be associated unilateral mucopurulent conjunctivitis. The eyelid is usually red and swollen in the area of the affected canaliculus, with associated unilateral conjunctivitis (conjunctival follicles may be present); reddening and swelling of the canaliculus itself (the opening is dilated and the edges are elevated and inflamed); a mucopurulent discharge and white, yellow, or brown concretions (dacryoliths) in the lacrimal punctum may also be observed (Vecsei et al. 1994). The remainder of the lacrimal passages is patent, and there is no preauricular lymphadenopathy. Local environmental factors in the canaliculus, such as stasis arising out of congenital diverticula, may predispose to these infections.

Pavilack and Frueh (1992) underscored the importance of thorough curettage as the most effective treatment for chronic canaliculitis, although their recommendations were not based on the study of specific actinomycotic or fungal causes of canaliculitis. Essentially, following topical and local anesthesia, the punctum is dilated and a small curette is introduced. If there is extensive ectasia and retention of concretions in diverticuli, a canaliculotomy may be necessary (Behrens-Baumann 1999). Following curettage (Pavilack and Frueh 1992), the canaliculus is thoroughly irrigated to remove any remaining fragments, identify unrecognized pockets of retained debris, and ensure that the distal part of the drainage system is patent.

Mycotic dacryocanaliculitis is reported to be usually satisfactorily treated by topical administration of 5 percent natamycin or by topical application and local syringing of the canaliculi and sac with amphotericin B solution (1.5–8 mg/ml) or nystatin solution (25 000–100 000 units/ml) (Wilson and Ajello 1998; Behrens-Baumann 1999); however, the basis for these recommendations is not clear. If medical therapy fails, surgery (canaliculotomy) is required. All material removed is stained and cultured, and the canaliculus is then syringed with the medications. Silicone intubation may be required for reconstruction of the canaliculus.

In a study on 40 patients with canaliculitis, Vecsei et al. (1994) reported that only 10 percent were cured by medical treatment alone and that 40 percent showed a recurrence; by contrast, 80 percent of patients who underwent canaliculotomy were cured. These results suggest that surgical treatment of canaliculitis, in combination with medical therapy, may yield better results than medical therapy alone.

MYCOTIC DACRYOCYSTITIS

Dacryocystitis, which is an infection of the lacrimal sac (Brook and Frazier 1998), is the most common infection of the entire lacrimal apparatus, and generally arises due to the stasis accompanying obstruction of the nasolacrimal duct; there are several possible causes (Behrens-Baumann 1999). Dacryocystitis may be acute or chronic, but fungi usually do not cause acute dacryocystitis. Chronic dacryocystitis usually arises following partial or complete obstruction at a single site within the lacrimal sac or within the nasolacrimal duct; infection generally follows obstruction, and does not cause it (Kalavathy et al. 1986).

Bacteria (especially aerobic and facultative anaerobic bacteria) may be the etiologic agents of infection in more than 90 percent of patients with acquired dacryocystitis, fungi accounting for only about 5 percent of infections (Kalavathy et al. 1986; Brook and Frazier 1998). However, in congenital dacryocystitis, fungi may account for almost 14 percent of infections (Ghose and Mahajan 1990). The genera of fungi that have been implicated as causes of dacryocystitis include *Acremonium* spp., *Aspergillus* spp., *Candida* sp., *Paecilomyces* spp., *Rhinosporidium seeberi*, dermatophytes, and *Sporothrix schenckii* (Henig et al. 1973; Kalavathy et al. 1986; Ghose and Mahajan 1990; Purgason et al. 1992; Levin et al. 1996; Reidy et al. 1997; Kristinsson and Sigurdsson 1998; Wilson and Ajello 1998; Behrens-Baumann 1999); however, the significance of some of these species as actual causes of dacryocystitis is uncertain, if one analyzes the descriptions per the criteria described earlier (McGinnis 1980). Infections due to *Sporothrix schenckii* and *Acremonium* spp. are reported to manifest as chronic suppurative dacryocystitis, with preauricular and submaxillary lymphadenitis and possible abscess formation. On the other hand, species of *Aspergillus*, *Candida* and *Paecilomyces*, *Rhinosporidium seeberi*, and the dermatophytes may cause chronic granulomatous dacryocystitis (Purgason et al. 1992; Levin et al. 1996; Kalavathy et al. 1998; Behrens-Baumann 1999). With partial or complete obstruction of the nasolacrimal duct, a laminated concretion (dacryolith) may develop in the lacrimal sac.

While epiphora is frequently the only clinical finding in patients with chronic dacryocystitis, there may also be lid edema, conjunctival infection, and a swelling in the medial canthus; pressure over the area usually results in a purulent discharge through the lower punctum (Kalavathy et al. 1986). Kristinsson and Sigurdsson (1998) described a patient in whom *Aspergillus fumigatus* caused plugging of the lacrimal sac, leading to extreme tenderness of the lacrimal sac, epiphora, and discharge from the lacrimal punctum. Krishnan et al. (1986) described the occurrence of a diverticulum of the lacrimal sac in association with rhinosporidiosis while, more recently, Kalavathy et al. (1998) described rhinosporidiosis of the lacrimal sac in two patients in whom bloodstained epiphora (due to the extreme fragility of the lesion) was the presenting complaint.

Management of this condition is by dacryocystectomy (for rhinosporidiosis), where the lacrimal sac is removed in toto, or dacryocystorhinostomy, where the patency of the nasolacrimal duct is restored; dacryoliths, if present, should be surgically removed and dacryocystorhinostomy performed. Chronic mycotic dacryocystitis in six patients was found to resolve completely following dacryocystectomy (Kalavathy et al. 1986), while some patients with congenital mycotic dacryocystitis were reported to respond satisfactorily to topical antifungals, probing, and syringing (Ghose and Mahajan 1990). Dacryocystitis due to *Candida albicans* was reported to resolve completely following surgery, alone or in combination with topical miconazole and natamycin therapy (Purgason et al. 1992). Dacryocystitis following lacrimal sac plugging due to *Aspergillus fumigatus* was relieved by the simple procedure of removing the plug after opening the lacrimal sac; dacryocystorhinostomy was not performed, and the patient was symptom-free one year after the procedure (Kristinsson and Sigurdsson 1998). Recently, minimally invasive, transnasal dacryocystorhinostomy surgery employing endoscopic and laser technologies has been introduced. Postoperative infection could probably be reduced by intraoperative or postoperative antifungal therapy.

MYCOTIC DACRYOADENITIS

Acute fungal infections of the lacrimal gland are rare; zygomycetes can cause a necrotizing dacryoadenitis in the setting of contiguous sino-orbital disease(Behrens-Baumann 1999). Chronic granulomatous dacryoadenitis can be caused by certain filamentous fungi (Wilson and Ajello 1998; Behrens-Baumann 1999). Such an infection is treated by using systemic antifungal agents.

MYCOTIC CONJUNCTIVITIS

Under normal circumstances, fungi are transient inhabitants of the normal conjunctival sac (Williamson et al. 1968; Brinser and Burd 1986). However, the unrestricted use of topical antibacterials or corticosteroids may predispose to fungal infection (Williamson et al. 1968;

Ben Ezra 1994). Acute conjunctivitis due to fungi appears to be uncommon. In a microbiological study of 102 patients with clinically diagnosed acute conjunctivitis (Boralkar et al. 1989), fungi were isolated from only 14 samples, and here also the criteria used to define significant growth of fungi from these 14 samples were not stated. Severe necrotizing granulomatous conjunctivitis due to *Coccidioides immitis* has been described in a patient who had received aggressive treatment with corticosteroids by various routes (Maguire et al. 1994).

The clinical manifestations of mycotic conjunctivitis are described as being dependent on the fungi involved (Ben Ezra 1994; Behrens-Baumann 1999), but the basis for these observations is unclear. The manifestations described include:

- simple acute or subacute superficial epithelial conjunctivitis due to *Candida* spp. (purulent infection), dermatophytic fungi (chronic lesions) and *Malassezia* spp. (catarrhal conjunctivitis)
- nodular conjunctivitis, with associated deep lesions and local lymphadenopathy due to *Sporothrix schenckii*
- follicular conjunctivitis due to *Blastomyces dermatitidis, Coccidioides immitis,* or *Paracoccidioides brasiliensis,* or chronic conjunctivitis with black conjunctival secretions due to *Aspergillus niger.*

Treatment of this condition can be difficult. Local therapy with topical antifungals may suffice if only the conjunctiva is involved, but systemic therapy is also needed if the eyelids are involved (Ellis 1985). In one patient, severe necrotizing granulomatous conjunctivitis due to *Coccidioides immitis* was finally controlled only after aggressive debridement of the affected area and months of topical amphotericin B and oral fluconazole therapy (Maguire et al. 1994).

Conjunctival rhinosporidiosis

Most reported ocular lesions due to rhinosporidiosis have occurred in hot, dry, climatic regions, with the occasional case from temperate zones (Reidy et al. 1997). While nasal lesions predominate in endemic areas, a predominance of ocular lesions is believed to indicate an epidemic (Vukovic et al. 1995). Ocular rhinosporidiosis most frequently manifests as single or multiple polypoidal outgrowths of the palpebral conjunctiva; these are pink or red, granular or lobulated (occasionally flattened), sessile or stalked, and attached to the upper or lower fornix or the tarsal conjunctiva (Ukety et al. 1992; Gaines et al. 1996; Pe'er et al. 1996; Reidy et al. 1997; Shrestha et al. 1998). Conjunctival rhinosporidiosis with associated scleral melting and staphyloma formation, a rare occurrence, was recently reported in three patients in India, with the lesions manifesting as gray–white spherules without polyps (Castelino et al.

2000). The occurrence of rhinosporidiosis of the lid margins, canaliculus, and lacrimal sac has been described earlier.

Rhinosporidial lesions of the eye are usually unilateral and solitary, and cause no discomfort to the patient; however, there may be increased lacrimation, discharge, tenderness of the lids, and photophobia. A clinical diagnosis of ocular rhinosporidiosis is justified if there are similar, extremely friable lesions elsewhere in the body presenting as small, white dot-like structures against a red background (the cysts embedded in the vascular tissue bed). However, a focal lesion on the conjunctiva, eyelid, or sclera needs to be differentiated from a cystic inclusion or adenoma of the various glandular structures, pterygium, pedunculated granuloma due to retained foreign body or end-stage chalazion (Prevost et al. 1980).

Since all attempts to cultivate *Rhinosporidium seeberi* in artificial culture media have been unsuccessful, histopathology is required to establish the diagnosis. The typical histological picture is that of a granuloma with marked inflammatory cell infiltrate (Rippon 1988), but a chronic nongranulomatous type of inflammation has also been described (Pe'er et al. 1996). All stages of the life cycle can be seen in excised tissue, from small trophocytes to large sporangia-containing sporoblasts. Hematoxylin and eosin staining is sufficient to demonstrate the distinctive spherical bodies (spherules or sporangia), which vary in size from 6 to 30 μm (Figure 16.9). The presence of such well-defined spherical bodies of varying size in a rather dense stroma covered by hyperplastic epithelium is a distinctive feature (Rippon 1988); however, these structures need to be differentiated from the spherules of *Coccidioides immitis*. Serological tests have not been found useful for the diagnosis of the condition.

No drug treatment has proven effective for ocular rhinosporidiosis. This condition is treated by surgical excision of the lesions.

MYCOTIC KERATITIS (KERATOMYCOSIS)

Frequency of occurrence

Mycotic keratitis, a suppurative, usually ulcerative, fungal infection of the cornea, may be responsible for more than 50 percent of all cases of ocular mycoses (Srinivasan et al. 1991) and of all patients with culture-proven microbial keratitis (Hagan et al. 1995), especially in tropical and subtropical environments. This condition apparently occurs more frequently in developing countries (e.g. India) than in the developed world (e.g. the USA). In one study in southern India, 139 patients with culture-proven mycotic keratitis were observed in a period of just 3 months (Srinivasan et al. 1997), whereas in the USA, 125 patients with culture-proven mycotic

keratitis were seen over a 10-year period in southern Florida (Rosa et al. 1994) and just 24 such patients over a 9-year period at Philadelphia (Tanure et al. 2000). Although a high incidence of mycotic keratitis can be expected in countries with similar annual rainfall and temperature range, this is not always so and also appears to depend on the extent of urbanization (Houang et al. 2001).

Types

In terms of occurrence, risk factors, and therapeutic approaches, two basic types of this condition are commonly seen, namely, keratitis due to filamentous fungi and keratitis due to yeast-like and related fungi (Jones 1980); a third type, keratitis due to thermally dimorphic fungi such as *Blastomyces dermatitidis* (Rodrigues and Laibson 1973) and *Paracoccidioides brasiliensis* (Silva et al. 1988) occurs rarely. There appears to be a strong geographical influence on the occurrence of the different forms of mycotic keratitis. A review of the data from 39 studies on microbial keratitis or mycotic keratitis reported in the literature between 1976 and 2001 (23 of the studies had been performed in Asia, six in North America and five in Africa) revealed that the proportion of corneal ulcers caused by filamentous fungi tends to increase towards tropical latitudes, whereas in more temperate climates, fungal ulcers appear to be uncommon and to be more frequently associated with *Candida* spp. than filamentous fungi (Leck et al. 2002).

Filamentous fungi are the principal causes of mycotic keratitis in most parts of the world. In 29 of the 39 studies reported between 1976 and 2001, and in another two studies published in 2002 (Gopinathan et al. 2002; Leck et al. 2002), either *Aspergillus* spp. or *Fusarium* spp. were the commonest isolates. The high proportion of corneal infections caused by *Aspergillus* spp. in drier climates may be because spores of *Aspergillus* spp. can tolerate hot, dry weather conditions (Khairallah et al. 1992); however, *Aspergillus* spp. may also predominate in more temperate latitudes (Leck et al. 2002). Dematiaceous fungi, such as *Curvularia* spp., have been reported to be the third most important cause of keratitis in a number of studies (Srinivasan et al. 1997; Garg et al. 2000; Gopinathan et al. 2002; Leck et al. 2002).

Keratitis due to yeasts and yeast-like fungi is most frequently caused by *Candida albicans* (Liesegang and Forster 1980; Tanure et al. 2000); less frequent causes include *Candida parapsilosis*, *Candida tropicalis*, and *Cryptococcus neoformans* var. *neoformans* (see Table 16.3). Since *Candida albicans* is an ubiquitous commensal of mucous membranes in humans with no geographical dominance, keratitis due to this organism tends to occur more frequently in areas where traumatic keratitis is uncommon, but where other predisposing factors are important (Jones 1980). *Candida albicans* was reported to be the commonest fungal species isolated from culture-proven mycotic keratitis in southern California (Ormerod et al. 1987) and in Philadelphia (Tanure et al. 2000), but species of *Candida* accounted for only 12.5 percent of isolates in culture-proven mycotic keratitis in Miami (Rosa et al. 1994). Species of *Candida* and related fungi have been infrequent isolates in most recent studies performed in tropical countries (Dunlop et al. 1994; Gonawardena et al. 1994; Hagan et al. 1995; Srinivasan et al. 1997; Gopinathan et al. 2002; Leck et al. 2002), possibly due to the predominance of livelihoods, such as agriculture, which carry a higher risk for occurrence of trauma-related keratitis caused by filamentous fungi than for keratitis due to *Candida albicans*.

Risk factors

Although the risk factors for microbial keratitis, in general, are well-known (Rosa et al. 1994; Tanure et al. 2000), there have been few attempts to identify specific risk factors for mycotic, as opposed to bacterial, keratitis. In one study in Bangladesh, 35 percent of patients with mycotic keratitis and 52 percent of patients with bacterial keratitis reported antecedent ocular trauma, while dacryocystitis was noted in 12 percent of bacterial and 4 percent of mycotic keratitis (Dunlop et al. 1994). The data derived from a retrospective case–control study in Singapore suggested that mycotic keratitis (principally due to *Fusarium* spp. and *Aspergillus* spp.) was frequently associated with mechanical ocular trauma, whereas bacterial keratitis (principally due to *Pseudomonas aeruginosa*) was more likely to be related to contact lens wear and preexisting ocular diseases; also, antecedent topical corticosteroid therapy (which is traditionally believed to be a specific risk factor for mycotic keratitis) appeared to predispose more frequently to bacterial (38 percent), than to fungal (25 percent), keratitis (Wong et al. 1997).

Certain conditions may predispose more frequently to filamentous fungal keratitis than to keratitis caused by yeast-like fungi. Filamentous fungal keratitis, especially that due to *Fusarium* spp. or *Aspergillus* spp., appears to occur most commonly in healthy young men engaged in agricultural work or outdoor occupations. Trauma is the principal risk factor in 44–55 percent of such patients; various traumatizing agents have been reported, including vegetable matter, mud or dust particles, paddy grain, the swish of a cow's tail, tree branches, and metallic foreign bodies. Prolonged use of topical corticosteroids or antibacterials, diabetes mellitus, preexisting ocular diseases, and contact lens wear, preexisting allergic conjunctivitis or vernal keratoconjunctivitis, and the occupation of onion harvesting, are less frequently encountered factors in filamentous fungal keratitis (Vajpayee et al. 1990; Rosa et al. 1994; Panda et al.

1997; Gupta et al. 1999; Lin et al. 1999; Xie et al. 2001; Gopinathan et al. 2002). Trauma and prior use of corticosteroids have also been noted to be the most frequent risk factors in keratitis due to *Curvularia* spp. (Wilhelmus and Jones 2001). Seasonal variations that have been observed in the incidence of filamentous fungal keratitis, and in the predominant genera of fungi isolated from such cases, have been linked to environmental factors, such as humidity, rainfall, and wind, and also to the harvest (Cuero 1980; Liesegang and Forster 1980; Thomas 1990; Rosa et al. 1994). By contrast, in a study at Philadephia (Tanure et al. 2000) in which *Candida albicans* was the most common isolate (46 percent), the three most common risk factors were found to be chronic ocular surface disease, contact lens wear, and use of topical corticosteroids. A case–control study to compare the relative contribution of different risk factors in determining whether a patient develops keratitis due to filamentous fungi, or that due to yeast or yeast-like fungi, would be of great value.

Diagnosis

A rapid and accurate diagnosis of mycotic keratitis improves the chances of a complete recovery. A detailed clinical history should first be obtained. This should include details of possible predisposing factors (trauma, use of contact lenses), prior therapy with antibacterials, corticosteroids or other compounds, and preexisting ocular disease (allergic conjunctivitis, lagophthalmos). The clinician should then look for ocular or systemic defects that may have predisposed to the keratitis, since these require correction. Symptoms are usually similar to those seen in other types of keratitis, but, perhaps, more prolonged in duration (5–10 days).

Filamentous fungal keratitis, which may involve any area of the cornea, usually presents with firm (sometimes dry) elevated necrotic slough (Figure 16.11a), 'hyphate' lines extending beyond the ulcer edge into the normal cornea, and multifocal granular (or feathery) gray-white 'satellite' stromal infiltrates (Figure 16.11b); an 'immune ring,' minimal cellular infiltration in the adjacent stroma, mild iritis, and an endothelial plaque and hypopyon may also occur. Rosa et al. (1994) observed irregular feathery margins in 62.4 percent, elevated borders (52 percent), dry rough texture (47 percent), satellite lesions (41 percent), and Descemet's folds in 40 percent of 125 patients with culture-proven mycotic keratitis. There may be variations from this pattern depending on the etiologic agent (Wilhelmus and Jones 2001). *Fusarium solani* usually causes a severe infection (Figure 16.11c), with perforation, deep extension, and malignant glaucoma supervening (Verma and Tuft 2002). Keratitis due to *Aspergillus* spp. is generally believed to be less severe than keratitis due to *Fusarium* (Thomas et al. 1988; Thomas and Rajasekaran 1988),

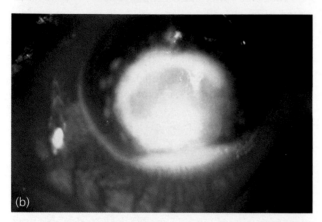

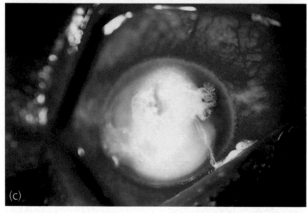

Figure 16.11 (a) *Mycotic keratitis due to* Fusarium solani *exhibiting elevated necrotic slough.* **(b)** *Mycotic keratitis due to* Aspergillus flavus *exhibiting 'satellite' lesions.* **(c)** *Severe mycotic keratitis due to a* Fusarium *sp.*

respond poorly to medical therapy, include *Lasiodiplodia theobromae* (Borderie et al. 1997; Thomas et al. 1998), *Pythium insidiosum* (Virgile et al. 1993; Thianprasit et al. 1996), and *Scedosporium apiospermum* (*Pseudallescheria boydii*) (Zapater and Albesi 1979; Wu et al. 2002). Chronic, severe, filamentous fungal keratitis may resemble bacterial suppuration, and may involve the entire cornea (Klotz et al. 2000). The stromal keratitis caused by *Candida albicans* and related fungi resembles bacterial keratitis, with an overlying epithelial defect, a more discrete infiltrate, and slow progression; such ulcers frequently occur in eyes with preexisting corneal disease, and in areas of exposure, such as inferocentrally, at the junction of the superior two-thirds and inferior one-third of the cornea (O'Day and Burd 1994). Keratitis due to a zygomycete, such as *Rhizopus* spp. (Rosa et al. 1994), or *Absidia corymbifera* (Marshall et al. 1997), which is a rare occurrence, runs a very fulminant course and is unresponsive to medical therapy, so that penetrating keratoplasty is required within a few days of presentation.

The clinical suspicion of mycotic keratitis can sometimes be confirmed by noninvasive means using confocal microscopy, an imaging technique that allows optical sectioning of almost any material, with increased axial and lateral spatial resolution, and better image contrast than conventional slit-lamp microscopy. In clinical keratitis due to an *Aspergillus* sp., fungal hyphae were imaged as high-contrast filaments, 60–400 μm long, and 6 μm in width (Winchester et al. 1997), while in another patient with keratitis due to *Fusarium solani* (Florakis et al. 1997), in vivo scanning slit confocal microscopy helped to first establish the diagnosis, then to demonstrate progression of the disease, and finally to confirm the success of penetrating keratoplasty. This technique has been used, in conjunction with culture, in establishing a diagnosis of mycotic keratitis (Xie et al. 2001; Verma and Tuft 2002). The potential benefits of confocal microscopy in demonstrating the presence of fungal hyphae in vivo within the human cornea by noninvasive means need to be weighed against limitations such as limited resolution of the microscope, difficulty in performing serial examinations, and the lack of a distinctive morphology of some pathogens.

For the microbiological diagnosis of mycotic keratitis, corneal material (as scrapings or biopsy material) is the specimen of choice. Prior to performing a corneal scraping, it is generally recommended that material be collected from the ipsilateral and contralateral lid and conjunctiva (by sterile cotton-tipped or calcium alginate swabs) to ensure that the organisms isolated on the corneal media have not come from the transient commensal fungal flora of the lid margins or conjunctival sac (Jones et al. 1981). However, Sharma et al. (1994) have questioned the rationale for performing such cultures of swabs from the lids and conjunctivae, stating that in their experience, a correlation between

although one group of workers (Vemuganti et al. 2002) noted that a high percentage of patients with *Aspergillus* keratitis ultimately required therapeutic penetrating keratoplasty due to nonresponsiveness to medical therapy. In keratitis due to certain dematiaceous hyphomycetes (species of *Curvularia*, *Bipolaris*, or *Exserohilum*), a persistent, low-grade, smoldering keratitis, with minimal structural alteration and pigmentation, may occur (Garg et al. 2000). Fungi that have been found to cause severe forms of keratitis clinically, which

the corneal and conjunctival isolate occurred in only 3 percent of cases; these workers opined that processing of such samples might, in fact, provide misleading data.

Corneal scrapings are obtained by using an instrument (platinum spatula, Beaver blade, Bard-Parker knife no. 15, blunt cataract knife) to debride material from the base and edges of the ulcerated part of the cornea (Agrawal et al. 1994); this should be done several times to obtain as much material as possible. The blade or spatula may be reused if a sterile medium has been streaked, but must be changed (the spatula can be flamed) if the instrument has made contact with an unsterile slide (Agrawal et al. 1994). Although cotton-tipped swabs do not seem to be a useful means of debriding the necrotic corneal slough, calcium alginate swabs (premoistened with tryptone soy broth), when used for the debridement, may facilitate recovery of fungi in culture (Jacob et al. 1995).

If corneal scrapings do not yield positive results, a corneal biopsy may aid diagnosis since a greater quantum of tissue can be obtained from a greater depth of the cornea. A simple method is to perform free dissection of the corneal lamellae by a sharp surgical knife; here, corneal perforation is a distinct risk. Another method involves removal of the epithelium and necrotic debris overlying the suppurated area and incision of the corneal stroma by a Bard-Parker no. 15 blade and corneal forceps to about one half the corneal thickness (Ishibashi et al. 1987). A corneal trephine may be used to define the precise diameter and depth (0.2–0.3 mm) of corneal tissue that is to be removed (Lee and Green 1990; Kompa et al. 1999). The biopsies may be relatively superficial (keratectomy) or deep. The biopsied tissue may be stained by ink-potassium hydroxide (Ishibashi et al. 1987) or lactophenol cotton blue (Kompa et al. 1999). Several experimental studies have unequivocally highlighted the value of corneal biopsy samples in diagnosis of keratitis due to *Aspergillus fumigatus*, *Candida albicans*, or *Fusarium solani* (Ishibashi and Kaufman 1986b; Ishibashi et al. 1987). The results obtained in a clinical setting are less clearcut. There is anecdotal evidence, in small numbers of patients, to support the use of corneal biopsy samples in diagnosis of mycotic keratitis (Ishibashi and Kaufman 1986b; Brooks and Coster 1993; Rosa et al. 1994; Kompa et al. 1999). However, Lee and Green (1990) reported that only nine of 42 samples submitted from patients with clinically evident infectious ulcerative keratitis were positive for infectious agents (including the presence of fungi in three of the nine positive samples).

While direct microscopic examination of corneal scrapes or corneal biopsy samples permits a rapid presumptive diagnosis of mycotic keratitis, it is generally not considered possible to identify the fungal genus involved. However, the occurrence of adventitious sporulation (presence of conidial structures) in tissue samples, including corneal material, may aid in the differentiation of genera of hyaline filamentous fungi, such as *Acremonium*, *Fusarium*, and *Paecilomyces* (Liu et al. 1998). Examination of a wet preparation (using 10 percent potassium hydroxide, ink-potassium hydroxide, or lactophenol cotton blue), a smear stained by the Gram method (Figure 16.12a) or Giemsa method, and a smear stained by special fungal stains, such as methenamine silver, periodic acid–Schiff, calcofluor white (Figure 16.12b), may yield valuable results. The corneal material should be spread out as thinly as possible on the microscope slides to facilitate easy visualization of the fungal structures.

In culture-proven mycotic keratitis, sensitivities of 72.2 percent (Xie et al. 2001) to 91 percent (Gopinathan et al. 2002) for potassium hydroxide mounts, 31.6 percent (Panda et al. 1997) to 98 percent (Dunlop et al. 1994) for Gram-stained smears, 27 percent (Rosa et al. 1994) to 85 percent (Gopinathan et al. 2002) and 56 percent (Upadhyay et al. 1991) to 86 percent (Forster et al. 1976) for smears stained by Gomori methenamine silver, and 91.4 percent for calcofluor white-stained smears (Gopinathan et al. 2002) have been reported.

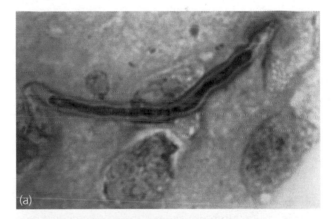

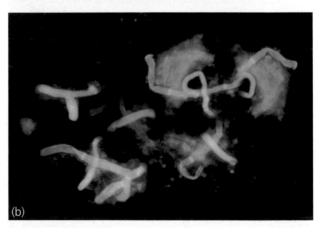

Figure 16.12 (a) *Photomicrograph of fungal hyphae in corneal scrape material. Cytoplasm is stained, cell wall and cross-walls are unstained. Gram stain; magnification ×1 200.*
(b) *Photomicrograph of fungal hyphae in corneal scrape material. Fungal hyphae fluoresce greenish-yellow against a dark background. Calcofluor white stain; magnification ×400*

Lactophenol cotton blue-stained mounts of corneal scrapes (Figure 16.5) have been found to yield positive results in about 78 percent of culture-proven mycotic keratitis (Thomas et al. 1991b; Sharma et al. 1998), while acridine orange-stained smears yielded positive results in 76 percent of clinically suspected cases of mycotic keratitis (Kanungo et al. 1991). A fluorescent Gram-stain technique permitted a rapid presumptive diagnosis of mycotic keratitis in five patients, with the diagnosis being confirmed by culture in all five (Roychoudhury et al. 1997); this stain also detected fungi in the vitreous biopsy of one patient with culture-proven *Aspergillus flavus* endophthalmitis. However, this staining technique has not acquired widespread use. When combined potassium hydroxide–calcofluor white staining was used to detect fungi in corneal scrapings from 114 patients with early keratitis, and 363 patients with advanced keratitis, a sensitivity of 61.1 percent and specificity of 99.0 percent in early keratitis, and a sensitivity of 87.7 percent and specificity of 83.7 percent in advanced keratitis was achieved; while the predictive values were high for potassium hydroxide–calcofluor white in the detection of fungus, this technique yielded false positives in 16.3 percent of patients with advanced keratitis (Sharma et al. 2002). All these results indicate that direct microscopic examination of corneal material permits a rapid presumptive diagnosis of mycotic keratitis, and correlates well with positive results in culture.

CULTURE

A diagnosis of mycotic keratitis is confirmed by isolation of fungi from corneal scrapes or biopsies which are inoculated onto the surface of solid media by making rows of 'C' streaks (two rows from each scraping), with only growth on the 'C' streaks being deemed significant (Figure 16.13), or by immersing the tip of the spatula, loop, or swab into broth (liquid) media. Commonly used culture media include Sabouraud glucose neopeptone agar (Emmons' modification, neutral pH), which is incubated at 25°C, 5 percent blood (preferably sheep blood) agar (25°C and 37°C), brain heart infusion broth (25°C), and thioglycollate broth (25–30°C) (O'Day and Burd 1994). Using these different media, fungi grew out in culture within 2 days in 54 percent, 3 days in 83 percent, and 1 week in 97 percent of patients with mycotic keratitis; 90 percent of scrapings yielded a positive initial culture (Rosa et al. 1994). Chocolate agar (Rosa et al. 1994), cystine tryptone agar (Thomas 1994), and rose bengal agar (P.A. Thomas, unpublished observations) are other media that can be used. Antibacterial compounds, such as chloramphenicol (40 µg/ml) or a penicillin–streptomycin combination, are usually incorporated in the media to suppress bacterial growth, but cycloheximide must never be used since it may suppress most ocular fungal pathogens (O'Day and Burd 1994). While some investigators (Upadhyay et al. 1991)

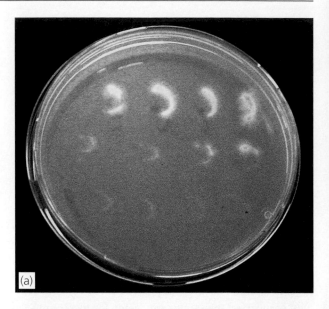

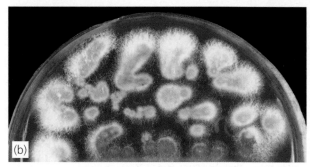

Figure 16.13 (a) *Twenty-four-hour growth of* Fusarium solani *on 'C' streaks of corneal scrape material on Sabouraud glucose neopeptone agar.* **(b)** *Forty-eight-hour growth of* Fusarium solani *on 'C' streaks of corneal scrape material on rose bengal agar.*

advocate the use of liquid-shake cultures to facilitate recovery of ocular fungi, others (Srinivasan et al. 1997) discourage their use since equivocal results may occur. Although fungi are usually recovered in culture within 3–4 days, culture media may need to be kept for up to 4–6 weeks. 'Sham cultures' are maintained to ensure that there is no contamination, either from the environment or media, during sample collection. A fungal strain recovered in culture from corneal material is considered significant if isolated more than once, or in two or more media, or on multiple 'C' streaks of one medium with fungal hyphae or yeast cells detected by direct microscopic examination of the corneal material (Liesegang and Forster 1980; Agrawal et al. 1994).

HISTOPATHOLOGICAL STUDIES

These offer certain advantages over culture, when used in the diagnosis of mycotic keratitis, since contamination is avoided, tissue penetration by fungi can be gauged, and the outcome of surgical procedures can be anticipated (Vemuganti et al. 2002). Some workers (Ishibashi et al. 1987; Rosa et al. 1994) believe that direct examination of corneal biopsies or corneal buttons yields positive results more frequently than fungal cultures of the same specimens; others (Alexandrakis et al. 2000) believe the opposite. Corneal material for histopathology is obtained as a biopsy or button following penetrating keratoplasty. Most fungi can be satisfactorily stained and studied in tissue sections by light microscopy. Sections stained by hematoxylin and eosin demonstrate many salient details, but species of *Fusarium* or *Candida* may not be stained at all. Conversely, while fungi can be easily detected in sections of corneal tissue stained by the Gomori methenamine silver or periodic acid–Schiff stains (Vemuganti et al. 2002), little else can be visualized. Hence, a replicate tissue section stained with hematoxylin and eosin should always be examined before special stains for fungi are used; alternatively, a section stained by Gomori methenamine silver can be counterstained by hematoxylin and eosin to simultaneously demonstrate a mycotic agent and the evoked tissue response (Chandler 1991). The purulent inflammatory cellular reaction (mostly of lymphocytes and plasma cells, with variable involvement of polymorphonuclear leukocytes) is usually less intense in fungal than in bacterial keratitis; filamentous fungi are usually found deep in, and arranged parallel to, the corneal stromal lamellae, while being absent on the surface (Figure 16.14). Stromal abscesses arising from coagulative necrosis, 'satellite' microabscesses, focal necrosis of the corneal stroma, and clusters of acute inflammatory cells may also be seen (Agrawal et al. 1994). At this stage, healing of the epithelium is seen with a coexisting active proliferation of the fungus in the deeper stroma, which may explain why culture of corneal biopsies, where material is obtained from the depth of the stroma, yields results superior to culture of corneal scrapings (Ishibashi and Kaufman 1986b; Brooks and Coster 1993; Rosa et al. 1994; Kompa et al. 1999). Invasion and

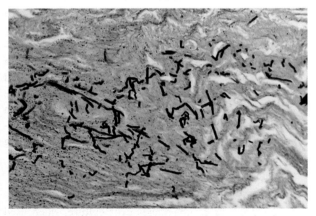

Figure 16.14 *Photomicrograph of a tissue section from a corneal button removed at the time of therapeutic penetrating keratoplasty. Fungal hyphae are seen in the middle and deep layers of the corneal stroma. The hyphae are oriented mostly in a direction parallel to that of the corneal collagen bundles. Gomori methenamine silver stain; magnification ×400*

penetration of an apparently intact Descemet's membrane may occur rarely (Vemuganti et al. 2002), particularly when *Fusarium solani* or *Fusarium* spp. are the infecting organisms (Jones 1975; Kuriakose and Thomas 1991).

Lectins, such as concanavalin A and wheat germ agglutinin, are ubiquitous proteins, found especially in plant seeds that bind specifically to carbohydrates. Fluorescein-conjugated concanavalin A provided consistently bright staining of fungal structures in corneal scrapes from 18 patients with culture-proven mycotic keratitis (Robin et al. 1989). A peroxidase-labeled wheat germ agglutinin staining technique for diagnosis of experimental keratitis due to *Candida albicans*, *Aspergillus fumigatus*, and *Fusarium solani* has recently been described (Garcia et al. 2002); this technique exhibited excellent sensitivities and specificities for all three infections, with a high degree of test–retest and inter-rater concordance between two independent observers. Such a sensitive and specific technique could prove useful in the clinical setting without the need for microscopes with attachments for fluorescence microscopy.

Fluorescence microscopy of a tissue section stained by hematoxylin and eosin revealed the presence of yeast cells of *Blastomyces dermatitidis* in periocular cutaneous lesions that had initially been misdiagnosed as squamous cell carcinoma (Margo and Bombardier 1985). Detection of fungi by direct immunofluorescence of formalin-fixed, paraffin-embedded ocular tissue sections has helped to confirm a presumptive histological diagnosis of ocular infection due to *Blastomyces dermatitidis* (Vida and Moel 1974), *Histoplasma capsulatum* var. *capsulatum* (Macher et al. 1985), *Sporothrix schenckii* (Font and Jakobiec 1976), *Pythium insidiosum* (Mendonza et al. 1996), and a zygomycete (Orgel and Cohen 1989). This technique can be used to detect a variety of dimorphic and hyaline filamentous fungi, even when these are present in small numbers or as atypical forms (Chandler 1991). Moreover, for retrospective studies, tissue sections previously stained by the hematoxylin and eosin, Giemsa, and modified Gram procedures can be decolorized in acid-alcohol and then restained with the specific reagents used for immunofluorescence (Chandler 1991). Performing immunofluorescence on a routine basis requires a microscope with attachments for fluorescence microscopy, antibodies of good quality, and standardization of the reagents and procedures used.

An impression debridement procedure has been described for corneal ulcers, although not specifically for mycotic keratitis (Arora and Singhvi 1994), where cellulose acetate filter paper (of the type used for conjunctival impression cytology) is applied with gentle pressure to the ulcerated part of the cornea; debridement is achieved with no trauma to the cornea since the necrotic material sticks to the filter paper, and this material can then be removed and stained. Degenerated epithelial and inflammatory cells and microorganisms can be seen by microscopic examination of the necrotic material (Arora and Singhvi 1994).

Despite a clear clinical presentation of suppurative keratitis, traditional diagnostic laboratory methods, including microscopy and culture, may yield negative results in 31–50 percent of patients, possibly due to difficulties in obtaining sufficient corneal material from large, late-stage ulcers, due to the risk of perforation, and from early stage, small ulcers from which little material is available (Leck et al. 2002). Self-administration of antimicrobials by patients before seeking medical attention may compound the problem (Dunlop et al. 1994). This has led to an evaluation of the polymerase chain reaction (PCR) as an aid to the diagnosis of mycotic keratitis.

POLYMERASE CHAIN REACTION

This technique, which achieves enzymatic amplification of a specific sequence of DNA, can be used to detect the DNA of fastidious microorganisms which cannot be cultured easily or for which culture is prolonged (for example, in patients who have been on antimicrobial therapy). DNA can be extracted from minute quantities (1 μl) of intraocular fluid (aqueous or vitreous), tears, any fresh ocular tissue, formalin-fixed or paraffin-embedded tissue, and even stained or unstained cytology slides or tissue sections for analysis by the PCR (Rajeev and Biswas 1998). However, this technique can neither be used to monitor a patient's response to treatment, nor to distinguish viable from nonviable organisms. It may thus be difficult to assess the relevance of a positive PCR, especially in locations such as the conjunctival sac, where bacteria, and even fungi, may normally be found; similarly, PCR cannot differentiate between active and latent viral infection (Rajeev and Biswas 1998). A few culture media will suffice to detect and grow the common ocular pathogens, but PCR must be multiplexed for each microorganism that is suspected (the use of panfungal primers may alleviate this problem). PCR can detect only those organisms for which the DNA sequence and primers are known, and cannot provide details of cellular morphology or localization (Rajeev and Biswas 1998).

In ocular fungal infections, PCR has principally been used in the diagnosis of endophthalmitis due to *Candida* spp. (Okhravi et al. 1998; Hidalgo et al. 2000; Jaeger et al. 2000). However, this technique permitted a postmortem diagnosis of *Fusarium* panophthalmitis in one patient (Alexandrakis et al. 1996). When the PCR was applied to the diagnosis of experimental keratitis due to *Fusarium solani*, using primers directed against one part of the *Fusarium* cutinase gene, amplified target DNA sequences were noted in 89 percent of 28 samples from *Fusarium*-infected corneas, whereas only 21 percent of 14 samples from infected corneas were positive by culture. However, one of eight negative control samples

was also positive by PCR (specificity of 88 percent for PCR) while none of the negative controls yielded growth in culture (100 percent specificity for culture) (Alexandrakis et al. 1998). When PCR was applied to the study of corneal scrapes from 30 patients with presumed microbial keratitis, using primers to target the 18S ribosome of fungi, PCR and culture yielded positive results for fungi in samples from 15 patients, and both techniques were negative in seven samples; however, PCR was positive for fungi with a negative culture in seven patients, while PCR was negative for fungi with a positive culture in one patient. Taking a positive fungus culture as the 'gold standard,' the PCR technique had a sensitivity of 94 percent and a specificity of 50 percent in this study (Gaudio et al. 2002). The specificity of this technique for diagnosis of mycotic keratitis requires further evaluation.

ANTIFUNGAL SUSCEPTIBILITY TESTING

Some investigators have reported that antifungal susceptibility testing aids the selection of the appropriate antifungal for successful treatment of mycotic keratitis (Jones 1975; Ishibashi et al. 1984) or orbital infections (Maskin et al. 1989; Massry et al. 1996). Unfortunately, many of these reports have not provided details of the test procedures used, nor have standardized techniques been followed. The future use of reproducible tests conforming to rigorous standards, such as the approved document (M27A) of the National Committee for Clinical Laboratory Standards (1997) for sensitivity testing of yeasts, and a standard method for susceptibility testing of filamentous fungi, especially *Aspergillus* spp., may clarify whether antifungal susceptibility testing is at all useful in guiding therapy of ophthalmic mycoses. Above all, the relationship between in vitro susceptibility data and clinical response to topical antifungal medication needs to be clarified; hitherto, no studies have been performed in this important area.

Management of mycotic keratitis

Mycotic keratitis is managed by medical or surgical means. Medical therapy consists of nonspecific measures, and the use of specific antifungal agents. Cycloplegics are used to relieve the iridocyclitis (anterior uveitis) that usually accompanies mycotic keratitis while broad-spectrum antibacterials may be needed to combat secondary bacterial infection (Rosa et al. 1994; Wong et al. 1997). Where access to specific antifungal drugs is limited, efforts have also been made to evaluate various antiseptics, such as polyhexamethylene biguanide (PHMB), chlorhexidine, and silver sulfadiazine in therapy of mycotic keratitis.

ANTISEPTICS FOR MYCOTIC KERATITIS

PHMB, which exhibits good in vitro activity against bacteria, fungi, and *Acanthamoeba*, has been used as a swimming pool disinfectant, sanitizer, and preservative in topical ophthalmic preparations (Imayasu et al. 1992), and for treatment of *Acanthamoeba* keratitis at concentrations of 0.02–0.053 percent with no adverse effects (Duguid et al. 1997). PHMB 0.02 percent was found to significantly reduce fungal growth in a New Zealand white rabbit model of *Fusarium solani* keratitis, with 58 percent of PHMB-treated eyes exhibiting no growth compared to only 17 percent of placebo-treated eyes (Fiscella et al. 1997).

Chlorhexidine, a bisbiguanide antiseptic which inhibits microbial function by affecting the functioning of the cell membrane, therein leading to a leak of cell electrolytes, has notable bactericidal and amebicidal effects (Seal et al. 1996). When corneal isolates of fungi were tested against chlorhexidine, povidone-iodine, propamidine, PHMB, and econazole by a crude, nonstandardized in vitro method, chlorhexidine exhibited antifungal activity at high concentrations while povidone-iodine was active at all concentrations tested and econazole was the most effective of all (Martin et al. 1995–96). However, in an agar-dilution method of susceptibility testing, chlorhexidine gluconate did not exhibit notable in vitro activity against ocular isolates of *Aspergillus* and *Fusarium* even at concentrations of 64 µg/ml (P. A. Thomas, unpublished observations). In a masked, randomized study on consecutive patients with microscopy-positive (later proven to be culture-positive) mycotic keratitis, 0.2 percent chlorhexidine appeared to be more effective than 5 percent natamycin in patients who did not have prior antifungal treatment (Rehman et al. 1997). In another study on 71 patients in Bangladesh (Rehman et al. 1998), neither 0.2 percent chlorhexidine gluconate nor 2.5 percent natamycin proved effective for healing of severe ulcers at 21 days, while 66.7 percent of nonsevere ulcers treated with chlorhexidine, and 36 percent of those treated with natamycin, had healed at 21 days, a relative efficacy of 1.85 (Rehman et al. 1998). However, since only a 2.5 percent strength natamycin suspension was used, and not the conventional 5 percent concentration, the relative efficacy of chlorhexidine *vis-à-vis* natamycin was probably overestimated.

Silver sulfadiazine functions as an organic base-heavy metal release system by liberating silver. The silver component binds to microbial DNA, preventing unzipping of the helix and therein inhibiting the replication of microorganisms without interfering with epithelial cell regeneration. In a prospective, controlled, randomized, double-masked trial on therapy of mycotic keratitis (Mohan et al. 1988), a higher success rate was achieved with the use of 1 percent silver sulfadiazine ointment (80 percent), without significant ocular or systemic adverse effects, than with 1 percent miconazole ointment (55 percent); the response of *Aspergillus* keratitis was comparable in both groups but silver sulfadiazine was superior to miconazole in patients with *Fusarium* keratitis. However, these excellent results have not been corroborated by other workers.

SPECIFIC ANTIFUNGAL AGENTS

Medical therapy of mycotic keratitis is usually protracted since the effective concentrations achieved by most specific antifungal agents in the cornea, with the possible exception of amphotericin B, only inhibit growth of the fungus, and host defense mechanisms must eradicate the organism (Stern and Buttross 1991). The antifungal agents used in clinical practice can be classed as polyenes, pyrimidines, and azoles.

Polyenes

Amphotericin B exerts its antifungal effect by directly bonding to ergosterol, a sterol unique to fungal cell membranes, therein disrupting the structural integrity of the membrane (O' Brien 1999). This antifungal penetrates the deep corneal stroma after topical application, and its bioavailability is sufficient for susceptible fungi; it also exhibits immunoadjuvant properties (Johns and O'Day 1988). In a model of deep stromal infection due to *Candida albicans* in pigmented rabbits, amphotericin B 0.15 percent and 0.075 percent and natamycin 5 percent exhibited a significant antifungal effect when the cornea was debrided every day; when the epithelium was left intact, the efficacy was much reduced or negligible (O'Day et al. 1984a). In the same model, the efficacy of amphotericin B (0.15 percent and 0.5 percent) appeared unaffected when the antifungal was given with 1 percent prednisolone acetate (O'Day et al. 1984b). Amphotericin B methyl ester was also found to be very efficacious in this model (O'Day et al. 1984c). Since the initial report of Wood and Williford (1976), varying degrees of success have been reported when amphotericin B (0.15 to 0.3 percent) is used to treat clinical mycotic keratitis. Topical hourly applications of amphotericin B (0.15 percent) resulted in resolution of lesions in 56 percent of keratitis due to *Fusarium* and 27 percent of keratitis due to *Aspergillus* in Indian patients (Thomas and Rajasekaran 1988), while use of amphotericin B ointment resulted in resolution of keratitis due to *Aspergillus* spp. and *Fusarium* spp. (Hirose et al. 1997). Keratitis due to *Curvularia brachyspora* resolved with topical amphotericin B therapy alone (Marcus et al. 1992). An analysis of several reports (Wilhelmus and Robinson 1991; Rosa et al. 1994; Hemady 1995; Ainbinder et al. 1998; Lin et al. 1999; Tanure et al. 2000; Djalilian et al. 2001) reveals that 24 of 28 patients with keratitis due to *Candida albicans* and other *Candida* spp. responded to amphotericin B (15 patients to topical amphotericin B alone, one to intravenous amphotericin B alone, and eight to topical amphotericin B in combination with natamycin or systemic azoles). When collagen shields that had been soaked in 0.5 percent amphotericin B for 2 hours at 25°C before application were used in conjunction with amphotericin B 0.25 percent eye drops (applied every 2 hours), cultures from the eyes of three patients with *Aspergillus* keratitis became negative within 15 days of treatment, although two patients subsequently required keratoplasty (Mendicute et al. 1995). Collagen shields, thus prepared and replaced daily, may deliver adequate concentrations of amphotericin B to the cornea with better tolerance than when conventional drops are used, therein improving the prognosis of *Aspergillus* keratitis. Although topical application of amphotericin B (in a desoxycholate vehicle) was found to prevent healing of epithelial defects in rabbit eyes, and to cause severe stromal edema and worsening iridocyclitis (Foster et al. 1981), the 0.15 percent solution prepared from the commercially available intravenous preparation appears to be well-tolerated (Tanure et al. 2000). Peripheral keratitis accompanying nasal rhinosporidiosis was found to resolve following topical administration of amphotericin B, prompting the suggestion that amphotericin B is an effective drug in this condition (Bhomaj et al. 2001). However, the peripheral keratitis in this patient is likely to have resolved due to the diagnostic corneal scraping that was performed, as well as the nasal polypectomy.

Natamycin (pimaricin), the first antifungal specifically developed for topical ophthalmic use, also binds to ergosterol to induce fungal cell membrane damage (O'Brien 1999). This drug is heat stable, permitting heat sterilization to be used, and the 5 percent topical ophthalmic suspension is well-tolerated, effective, and safe, although occasional punctate keratitis has been reported (Johns and O'Day 1988). Subconjunctival injection leads to conjunctival necrosis, and natamycin is not available for systemic administration due to its low aqueous solubility (O'Brien 1999). It has been suggested that natamycin does not penetrate into the deep corneal stroma after topical application. However, radiolabeling studies revealed that 13 topical applications of natamycin (one every 5 minutes) resulted in a drug concentration of approximately 2.5 mg/g of cornea, levels that were substantially higher than those following application of amphotericin B, although the levels achieved were far less in nondebrided corneas with intact epithelium (O'Day et al. 1986). About 2 percent of the total drug in the corneal tissue is bioavailable. Where commercially available, natamycin 5 percent is used as primary therapy for mycotic keratitis, particularly that due to filamentous fungi (Agrawal et al. 1994; Rosa et al. 1994; Panda et al. 1997), but interpretation of data relating to response to therapy has been difficult. In one series in Miami, Florida (Rosa et al. 1994), patients with presumably superficial keratitis due to *Fusarium* spp. received topical natamycin alone (average duration of treatment was 38 days), while those with presumably deep lesions received topical natamycin and systemic antifungals. Details of response to therapy were not provided, but 22 (28 percent) of 79 eyes with *Fusarium* keratitis ultimately required penetrating keratoplasty, and one eye required enucleation. It is uncertain

whether the patients who did not require surgery ultimately responded to therapy with natamycin (with or without systemic antifungals). A recent report from the same institution (Dursun et al. 2003) provided data about 10 cases (out of a total of 159 patients with *Fusarium* keratitis) that progressed to endophthalmitis; combination therapy with oral azoles (fluconazole or ketoconazole) and topical natamycin appeared to be inadequate in severe *Fusarium* keratitis with intraocular spread. In the same center, four patients with keratitis due to *Candida* spp. were reported to have responded to topical natamycin therapy alone (Rosa et al. 1994). In a review of patients with *Curvularia* keratitis in Houston, Texas, it was noted that the corneal lesions resolved with topical natamycin therapy alone in 16 patients; in an additional seven patients, the lesions resolved following medical therapy with topical natamycin and other antifungals, and in another six patients, the keratitis resolved with topical natamycin, other antifungals, and surgery (Wilhelmus and Jones 2001). Variable results have been reported when topical natamycin has been used to treat keratitis due to *Scedosporium apiospermum* (reviewed by Wu et al. 2002).

Pyrimidines

Flucytosine (5-fluorocytosine), a low molecular weight, fluorinated cytosine analog that was first synthesized as a potential antineoplastic agent, exhibits moderate aqueous solubility, a high bioavailability, and achieves anterior chamber levels of 10–40 μg/ml after oral administration of 200 mg/kg body weight/day (Jones 1975). In spite of these advantages, flucytosine has limited use in ocular fungal infections. Since many strains of fungi have rapidly developed resistance to this drug, flucytosine should not be administered alone to treat mycotic keratitis or intraocular infections, even when these are caused by susceptible fungi (O'Brien 1999). Flucytosine also has a limited spectrum of activity against the commonest filamentous fungi implicated in ocular disease (Sutton et al. 1998). Since low intraocular levels of the drug are achieved after topical and subconjunctival administration, the use of oral doses of 50–150 mg/kg body weight/day or a topical 1 percent solution is recommended (O'Brien 1999).

Azoles

As a group, the azoles are potentially as effective as amphotericin B since they have a broad spectrum of activity, are generally less toxic, and can be administered by different routes. Since azoles bind to a cytochrome P450 fungal enzyme involved in the 14α-demethylation of either lanosterol or 25-methylene-dihydrolanosterol, ergosterol synthesis decreases and 14-methylated sterols accumulate, leading to increased membrane permeability, inhibition of growth, and alteration of membrane enzymes, ultimately causing cell death, and also rapid accumulation of chitin over the entire cell wall (Vanden Bossche et al. 1993). All azoles, except for fluconazole, appear to decrease immune cell function, especially that of lymphocytes, thus affecting the degree of tissue damage occurring with the inflammatory reaction, but also affecting the efficacy of the azoles in vivo (Yamaguchi et al. 1993). In view of the limited concentrations of azoles that can be achieved in the eye, these drugs are to be considered as fungistatic in ocular fungal infections (Johns and O'Day 1988).

Clotrimazole is a trityl imidazole of low human toxicity with in vitro activity against species of *Aspergillus*, *Alternaria*, and *Candida*, but not against species of *Fusarium* or *Paecilomyces* (Jones 1975). Clotrimazole is poorly soluble in water, hence parenteral administration is not feasible. Following oral administration, satisfactory blood levels are reached during the first 2 weeks of therapy, but subsequently fall due to rapid induction of hepatic drug-metabolizing enzymes. Hence, clotrimazole is no longer recommended for oral administration (O'Brien 1999). Successful therapy of 16 patients with fungal keratitis (including eight due to various species of *Aspergillus*, and three due to *Candida albicans* and other *Candida* spp.) by using a commercial preparation of clotrimazole (Canesten, Bayer, Germany) was reported in 1976 (Jones et al. 1976). A topical 1 percent solution of clotrimazole in arachis oil was also found useful in therapy of superficial corneal infections (Jones et al. 1979). There are no recent reports of the use of clotrimazole in therapy of mycotic keratitis.

Econazole, a dichloroimidazole that is similar to miconazole in structure and mode of action, has a wide spectrum of activity against filamentous fungi in vitro, but is less effective than miconazole against *Candida* spp. These data may explain why econazole was found effective in treatment of keratitis due to various fungi, including species of *Fusarium* (Jones 1975; Jones et al. 1979; Arora et al. 1983). However, topical administration of a 1 percent econazole solution in polyethylene glycol 400 was not efficacious in experimental deep stromal *Candida albicans* keratitis in pigmented rabbits (O'Day et al. 1984a), irrespective of whether the corneal epithelium was intact or debrided. A recent study in India suggested that 2 percent econazole was as effective as 5 percent natamycin in therapy of filamentous fungal keratitis (Prajna et al. 2003); however, the results were not stratified as to severity of keratitis or the fungus involved.

Miconazole, a phenylethyl imidazole which exhibits a broad spectrum of activity against yeasts and filamentous fungi, is soluble in organic solvents, and may be administered intravenously in normal saline after dissolving in cremophor EL; this vehicle is believed to cause the toxic reactions observed with intravenous miconazole administration (Anderson et al. 1984). Following subconjunctival and topical administration, miconazole was found to penetrate the corneas of rabbit eyes, resulting in high tissue concentrations that were enhanced by epithelial debridement (Foster and Stefanyszyn 1979). In a rabbit model, topical 1 percent miconazole nitrate did not retard the closure of 8.5 mm corneal epithelial defects (Foster et al. 1981). Topical administration of 1 percent miconazole nitrate in polyethylene glycol 400 was not effective in a rabbit model of deep stromal *Candida albicans* keratitis (O'Day et al. 1984a), irrespective of whether the corneal epithelium was intact or debrided. In another study (O'Day et al. 1984b) topical 1 percent prednisolone acetate was found to adversely influence the efficacy of 1 percent miconazole and other antifungals when given in combination (O'Day et al. 1984b). Although intravenous administration of miconazole resulted in undetectable drug concentrations in the cornea and vitreous of rabbit eyes (Foster and Stefanyszyn 1979), intravenous administration in a dose of 600–3 600 mg/day was reported to be effective in the treatment of clinical mycotic keratitis due to *Phialophora gougerotii* (Jones 1975), *Lasiodiplodia theobromae* (Ishibashi and Matsumoto 1984b), *Beauveria bassiana* and *Aspergillus fumigatus* (Ishibashi et al. 1984). Topical and subconjunctival miconazole therapy (Foster 1981) resulted in resolution of all lesions, including endothelial plaques and a descemetocele, in seven patients with keratitis due to *Candida albicans* or *Aspergillus* species; all seven fungal strains were reportedly susceptible in vitro to miconazole. Topical (1 percent) and subconjunctival miconazole therapy, combined with oral ketoconazole, evoked a favorable response in 13 of 20 patients with keratitis due to species of *Fusarium, Curvularia,* or *Candida*; a modification of the regimen produced healing in three more patients, while the lesions progressed in four patients (Fitzsimons and Peters 1986). Since many strains of *Scedosporium apiospermum* (*Pseudallescheria boydii*) are reported to be susceptible to miconazole in vitro (Sutton et al. 1998), miconazole is believed to be an important drug in treatment of clinical keratitis due to this fungus. A recent review of 15 patients with this condition (Wu et al. 2002) appears to endorse this view (only reports where details of treatment regimen and visual outcome were provided were included in the review, and patients with initial scleral involvement were excluded). Four (67 percent) of six individuals who had received miconazole and three (33 percent) of nine persons who had not received miconazole retained form vision (counting fingers or better). It is important to remember that the severity of the keratitis at presentation could be an important determinant of outcome of medical therapy. Topical miconazole therapy is sometimes associated with superficial punctate keratitis which resolves when treatment is withdrawn (Foster 1981).

Ketoconazole, the first successful orally absorbable, broad-spectrum antifungal azole, is a substituted imidazole compound that is currently available as an oral preparation (200 mg) worldwide and as a 1 percent topical ocular medication in India. Following subconjunctival, topical, or oral administration of a 1 percent ketoconazole solution to rabbits with undebrided corneas and an intact epithelium, relatively high concentrations of ketoconazole were achieved in the cornea (44.0±10.1 µg/g) but relatively nondetectable levels in the vitreous; greatly increased corneal drug levels (1 391.5±130.0 µg/g) were achieved, particularly after topical administration and, to a lesser extent, after subconjunctival injection, after debridement of the corneal epithelium (Hemady et al. 1992). Following systemic administration, ketoconazole is highly bound to serum protein, but also has a high tissue distribution (O'Brien 1999). Transscleral iontophoresis (4–6 mA for 15 minutes) resulted in peak ketoconazole concentrations of 10.2 µg/ml in the aqueous after 1 hour (which persisted for about 8 hours) and 0.1 µg/ml in the vitreous; transcorneal iontophoresis (1.5 mA for 15 minutes) resulted in peak corneal concentrations (27.6 µg/ml) and aqueous concentrations (1.4 µg/ml) after 1 hour, which were sustained for 2 hours. These concentrations were significantly higher than those achieved by subconjunctival injections (Grossman and Lee 1989).

Oral doses of ketoconazole greater than 400 mg/day may cause nausea, vomiting, and transient elevations in liver enzymes, which resolve upon withdrawal of the drug; sexual impotence, hair loss, gynecomastia, and oligozoospermia may also occur, due to decreased steroid synthesis (O'Brien 1999). While a 1 percent ketoconazole suspension (prepared in polyethoxylated castor oil) was reported to produce modest pathological changes in the regenerating corneal epithelium in a rabbit model (Foster et al. 1981), topical administration of 1 percent, 2 percent, and 5 percent solutions of ketoconazole in arachis oil did not cause corneal toxicity in rabbit eyes (Oji 1982b). When a topical 2 percent ketoconazole suspension (prepared in a 4.5 percent sterile boric acid solution with hydroxypropylmethyl cellulose added to increase viscosity) was used to treat clinical mycotic keratitis, it did not evoke significant ocular toxicity (Torres et al. 1985).

The results of experimental studies suggest that the response of ocular fungal infections to ketoconazole is species-specific. Topical 1 percent ketoconazole (in arachis oil) applied every hour for 10 hours for 16 days was effective in clearing *Aspergillus flavus* keratitis in a rabbit model (Oji 1982a) whereas oral or topical ketoconazole therapy was not successful in experimental *Aspergillus fumigatus* keratitis (Komadina et al. 1985). Topical 1 percent ketoconazole solution was ineffective in one experimental model of *Candida albicans* keratitis

(O'Day et al. 1983), but topical 2 percent ketoconazole ointment was effective in another model (Ishibashi and Kaufman 1986a). Oral ketoconazole therapy was also effective in experimental *Candida albicans* keratitis (Ishibashi and Matsumoto 1984a).

In a clinical setting, oral ketoconazole therapy was not found effective in one patient with *Aspergillus fumigatus* keratitis, perhaps because of intraocular invasion (Searl et al. 1981), but was effective in a dose of 300 mg/day in two patients with mycotic keratitis, one due to *Fusarium solani* (therapy for 3 weeks), and the other due to an unidentified fungus (therapy for 8 weeks) (Ishibashi 1983). Topical 1 percent ketoconazole, given for 7 weeks, was reported to be effective in the treatment of six patients with mycotic keratitis, principally caused by *Aspergillus* and *Fusarium* (Torres et al. 1985). In these reports, the efficacy of ketoconazole therapy in severe mycotic keratitis caused by different fungi could not be assessed. When 30 patients suffering from mycotic keratitis were treated with oral ketoconazole (200 mg three times per day), there was complete resolution of lesions in all 14 patients with superficial lesions, and six of 16 patients who had deep lesions; none of the patients reported adverse effects to the drug, nor were marked elevations in serum enzyme levels observed (Thomas et al. 1987). In an extension of this study, 10 of 15 patients treated only with a topical 1 percent ketoconazole suspension in distilled water responded, while in 25 of 35 patients treated with topical and oral therapy, a complete or partial response of the keratitis was noted (Rajasekaran et al. 1987a). A comprehensive assessment of the data from both studies indicates that oral therapy alone was effective in 37 percent of patients with deep lesions, topical therapy alone was effective in 20 percent, and combined oral and topical therapy effected complete or partial resolution of the keratitis in 60 percent of patients, including 10 of 30 patients with *Fusarium* keratitis, 11 of 22 patients with *Aspergillus* keratitis, three of five patients with *Penicillium* keratitis, three of four patients with *Candida albicans* keratitis and in all six patients with *Curvularia* keratitis (Rajasekaran et al. 1987a; Thomas et al. 1987). A combination of ketoconazole and amphotericin B therapy and keratoplasty resulted in a favorable outcome of post-traumatic keratitis due to *Scopulariopsis brevicaulis* in a patient (Ragge et al. 1990), whereas treatment with oral ketoconazole could not halt progressive infection due to the thermotolerant mold *Neosartorya fischeri* var. *fischeri*, the teleomorph of *Aspergillus fischerianus* (Coriglione et al. 1990). In south Florida, 25 of 125 patients with mycotic keratitis, seen over a 10-year period, received oral ketoconazole (400 mg/day for a median duration of 2 weeks), in addition to topical 5 percent natamycin, due to severe keratitis, scleritis, and endophthalmitis; nine (36 percent) of these 25 patients did not require a therapeutic penetrating keratoplasty or other major surgical intervention to manage their keratitis (Rosa et al. 1994). All these results suggest that while oral ketoconazole therapy is undoubtedly efficacious in nonsevere mycotic keratitis, its efficacy when deep lesions are present is variable, and also depends on the fungal species involved.

Whether the results of oral ketoconazole therapy for mycotic keratitis have any relevance at all to topical therapy is debatable. Triturated (crushed and suspended) ketoconazole has been recommended for the treatment of mycotic keratitis when commercial antifungal eye drops are not obtainable (Guzek et al. 1998). Ketoconazole and itraconazole tablets were triturated to 20 mg/ml in polyvinyl alcohol, boric acid, olive oil, or balanced salt solution (BSS) and applied topically to de-epithelialized rabbit corneas (one drop/15 minutes for 2 hours). The concentrations of ketoconazole in corneal tissue treated with the drug in BSS, olive oil, polyvinyl alcohol, and boric acid were calculated to be 512, 773, 1 221, and 1 492 µg/g, respectively. Since the vehicle used to triturate antifungals may affect the tissue concentration, the development of effective vehicles may have an impact on the therapy of mycotic keratitis.

Itraconazole, an orally absorbable synthetic dioxolane triazole with a broad spectrum of antifungal activity after oral administration, is larger than fluconazole, very hydrophobic, and poorly soluble in aqueous solution. It is well absorbed orally, especially when given with a meal or formulated in polyethylene glycol, with more than 90 percent being bound to protein in serum, and it is highly concentrated in lipid-rich tissue (Van Cauteren et al. 1987). Although a single 200 mg oral dose results in a peak serum level of about 0.3 µg/ml, which is far below that of ketoconazole or fluconazole (Van Cauteren et al. 1987), the serum level can be increased to 3.5 µg/ml by multiple long-term dosing (200 mg/day orally for 2 weeks) (Carlson et al. 1992). The major drawback of using itraconazole by the oral route for therapy of ocular fungal infections is its poor penetration into the cornea, aqueous humor, and vitreous, when compared to fluconazole and ketoconazole. This was clearly demonstrated in a rabbit model of *Candida* endophthalmitis, even when itraconazole was given in a dose of 80 mg/kg orally (Savani et al. 1987). The ocular pharmacokinetics of a newly developed itraconazole oral solution need to be defined (Harrousseau et al. 2000). Itraconazole is generally well-tolerated after oral administration, the most common complaint being gastrointestinal upset; less frequently observed side effects include hypertriglyceridemia, hypokalemia, edema, decreased libido, and gynecomastia (O'Brien 1999).

Oral itraconazole therapy was found to be effective in the therapy of experimental keratitis due to *Aspergillus* spp. (Van Cutsem and van Gerven 1991). In India (Thomas et al. 1988), 22 of 40 patients with mycotic keratitis (15 with nonsevere and seven with severe keratitis) responded to oral itraconazole (200 mg once daily for a median duration of 17 days); serious adverse reactions did not occur, and all hematological, urine, and biochemical parameters remained normal during and after itraconazole therapy. In a related study (Rajase-karan et al. 1987b), seven of eight nonsevere ulcers and one of eight severe ulcers healed with topical therapy alone, while 17 of 21 nonsevere ulcers and 14 of 33 severe ulcers responded to a combination of oral itraconazole (200 mg once daily) with hourly topical applications of 1 percent itraconazole. To summarize these data, 84 percent of nonsevere ulcers and 57 percent of severe ulcers showed excellent or moderate responses to some form of itraconazole therapy; 76 percent of patients with *Aspergillus* keratitis, 70 percent of those with dematiaceous filamentous fungal keratitis, and 60 percent of patients with *Fusarium* keratitis exhibited excellent or moderate responses to some form of itraconazole therapy (Rajasekaran et al. 1987b; Thomas et al. 1988). In another study, a recurring corneal infection due to *Fonsecaea pedrosoi* was treated by a large penetrating keratoplasty, removal of the involved part of the iris and the entire lens, and a 5-month course of oral itraconazole; this resulted in no recurrence of the infection (Barton et al. 1997).

Bioassay of corneas treated with itraconazole in different vehicles (BSS, polyvinyl alcohol, boric acid, olive oil) demonstrated approximate itraconazole concentrations of 200–250 µg/g tissue (Guzek et al. 1998). Another avenue to pursue is the possible use of itraconazole subconjunctivally. In a novel method for the qualitative evaluation of the pharmacokinetics of subconjunctivally injected antifungals in rabbits, itraconazole (2.5 mg/ml) was found to persist for at least 24 hours in normal and debrided corneas, in contrast to amphotericin B, miconazole, fluconazole, and ketoconazole, which did not persist beyond 4–8 hours (Klippenstein et al. 1993).

Fluconazole, a synthetic bistriazole antifungal compound, is a smaller molecule than itraconazole, soluble in water, and only 10–20 percent protein bound in serum; it has a long half-life and is excreted renally (Savani et al. 1987). In one study (O'Day et al. 1990), oral fluconazole was found to readily penetrate all ocular tissues and fluids of Dutch-belted rabbits, with no difference observed between phakic and aphakic eyes; after a single oral dose of 20 mg/kg, the levels achieved were 13.3±1.4 µg/g (cornea), 7.4±0.3 mg/l (aqueous), 9.8±0.9 mg/l (vitreous), and 5.2±0.4 µg/g (choroid/retina), with the corneal concentrations being highly correlated with serum concentrations. A steady accumulation in both normal corneas, as well as those infected with *Candida albicans*, was noted when given in a twice-daily divided dose; the presence of inflammation induced by fungal infection did not influence corneal uptake (O'Day et al. 1990). In a related experimental study in

Dutch-belted rabbits (O'Day 1990), oral fluconazole therapy resulted in a significant therapeutic effect in keratitis due to *Aspergillus fumigatus* keratitis and that due to *Candida albicans*; pretreatment for 1 day followed by 5 days post-inoculation treatment, led to a significant decrease in isolate recovery and in clinical disease.

Since fluconazole is a stable, water-soluble, low molecular weight bistriazole with high bioavailability and low toxicity, it is potentially useful as a topical ocular agent. In New Zealand white rabbits, topical application of 0.2 percent fluconazole to debrided and nondebrided corneas resulted in peak corneal levels of 8.2 ± 1.2 µg/g (debrided corneas) and 1.6 ± 0.6 µg/g (nondebrided) being noted after 5 minutes, and aqueous humor levels of 9.4 ± 2.3 µg/ml and 1.6 ± 0.6 µg/ml after 15 minutes; the half-life of fluconazole in debrided eyes was 15 minutes and that in nondebrided eyes was 30 minutes (Yee et al. 1997).

In experimental *Candida albicans* keratitis, topical (2 mg/ml) fluconazole, applied to debrided and nondebrided corneas of rabbits, was significantly better than control saline therapy in effecting clinical resolution of corneal lesions and hypopyon, nonprogression to corneal perforation and descemetocele, and eradication of the infecting fungus from the cornea (Behrens-Baumann et al. 1990). In a clinical setting, six consecutive eyes of patients with microbiologically proven *Candida* keratitis with abscess formation, which had failed to respond to topical miconazole or natamycin therapy, responded well to topical fluconazole (20 mg/ml) given for a mean duration of treatment of 22.6 ± 2.3 days (Panda et al. 1996), while oral fluconazole therapy was reported to be useful in therapy of keratitis due to an unspecified fungus (Thakar 1994), and, in association with topical natamycin and intracameral amphotericin B and various surgical measures, to effect the eradication of corneal infection due to *Colletotrichum graminicola* (Ritterband et al. 1997). Although *Candida guilliermondii* is usually susceptible to fluconazole in vitro (Sutton et al. 1998), infectious crystalline keratopathy due to this yeast in a corneal transplant progressed, in spite of 6 weeks of topical amphotericin B, and an additional 6-week course of topical and oral fluconazole (Ainbinder et al. 1998).

SELECTION OF ANTIFUNGAL THERAPY

The selection of an antifungal agent for therapy necessarily depends on its easy availability and on other criteria. If direct microscopic examination of corneal scrapes or corneal biopsies yields definite results that are consistent with the clinical picture, treatment may be initiated; otherwise, therapy may need to be withheld until culture reports become available (Johns and O'Day 1988). Topical natamycin (5 percent), if commercially available, or amphotericin B (0.15 percent) is usually selected as first-line therapy for superficial keratitis (topical econazole is used in many parts of the UK)

whether or not septate hyphae or yeast cells have been seen by direct microscopy; if deep lesions are present, intravenous amphotericin B, subconjunctival or intravenous miconazole, oral ketoconazole, oral itraconazole or oral fluconazole is added to the therapeutic regimen (Rosa et al. 1994; Wong et al. 1997; Tanure et al. 2000). If hyphae have been seen by microscopy, and a filamentous fungus is isolated in culture, natamycin appears to be the treatment of choice when available (Rosa et al. 1994; Tanure et al. 2000); topical 0.15 percent amphotericin B (Johns and O'Day 1988; Wong et al. 1997) is an alternative. If yeasts or pseudohyphae are seen by microscopy, and species of *Candida* or *Cryptococcus* are isolated in culture, topical 0.15 percent amphotericin B appears to be preferred when available (Rosa et al. 1994; Tanure et al. 2000), although natamycin (Rosa et al. 1994; Panda et al. 1996) and topical (1 percent) miconazole (Foster 1981; Panda et al. 1996) are the choices of some other clinicians. In the absence of controlled clinical trials and since the number of patients involved may be very small, the validity of these patterns of prescribing is difficult to assess. However, it may be possible to discern patterns by a detailed analysis of the available literature.

Thomas (2003) performed an analysis of 85 patients with keratitis due to *Fusarium* spp. reported in 15 studies in the literature where details of outcome of therapy were provided. The analysis revealed that more than 70 percent of patients with superficial keratitis due to *Fusarium solani* and other *Fusarium* spp. apparently responded to medical therapy alone (topical amphotericin B alone or in combination with topical natamycin, oral ketoconazole and/or topical ketoconazole, oral itraconazole); although several antifungals were found effective, administration of natamycin may have forestalled surgical intervention. In striking contrast, almost 70 percent of patients with *Fusarium* keratitis with deep lesions did not respond to medical therapy alone, particularly if natamycin was not used, and some form of surgical intervention was necessary. The data pertaining to outcome of therapy was analyzed (Thomas 2003) for 61 patients with *Aspergillus* keratitis (Figure 16.15) reported in 13 studies in the literature. The data suggested that more than 80 percent of patients with superficial keratitis due to *Aspergillus flavus*, *Aspergillus fumigatus*, and other *Aspergillus* spp. responded to medical therapy with a variety of topical (amphotericin B, natamycin, ketoconazole, itraconazole) or systemic (oral ketoconazole, oral itraconazole) antifungals, with surgery not being required; however, in the presence of deep corneal lesions, almost 60 percent of patients did not respond to medical therapy alone, particularly if natamycin was not used, and surgery was required to control the infection.

Details of reponse of keratitis due to *Candida* spp. to therapy were analyzed (Thomas 2003) for 38 patients reported in eight studies. The medical therapy of kera-

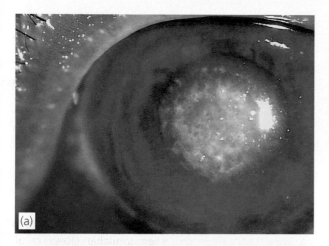

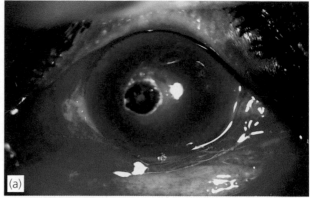

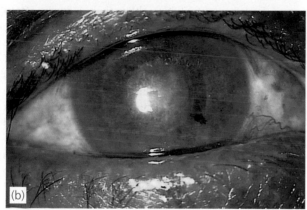

Figure 16.16 (a) *Keratitis due to a* Curvularia *sp. before treatment.* (b) *Keratitis due to a* Curvularia *sp. 1 month after treatment with topical 5 percent natamycin. Lesions have resolved, leaving a faint scar.*

Figure 16.15 (a) *Keratitis due to* Aspergillus flavus *before treatment.* (b) *Keratitis due to* Aspergillus flavus, *1 month after treatment with topical 5 percent natamycin. Lesions have resolved, leaving a faint scar.*

titis due to *Candida albicans* and other *Candida* spp. generally had a favorable prognosis, particularly when topical amphotericin B was used alone or in combination with systemic azoles; topical natamycin and topical fluconazole alone also sufficed in some instances, and the presence of deep lesions appeared not to be a major hurdle.

Data pertaining to 42 patients with keratitis due to *Curvularia* spp. (Figure 16.16) reported in seven studies in the literature were analyzed (Thomas 2003). In 35 (83 percent) of the 42 individuals, the corneal lesions responded to antifungals alone (19 to topical natamycin alone, eight to natamycin and other antifungals, six to oral ketoconazole, and one each to topical miconazole and topical amphotericin B). In an additional three patients, the keratitis resolved with keratectomy and antifungal therapy, while penetrating keratoplasty was required in four patients who did not respond to medical therapy alone. Most of the papers analyzed, however, did not provide details about the severity of the corneal lesions in the patients, which is an important aspect that requires consideration. In one study on dematiaceous fungal keratitis (Garg et al. 2000), antifungal therapy alone (principally natamycin, alone or in combination

with topical azoles) sufficed for resolution of lesions in 88 percent of patients with superficial lesions; however, only 46 percent of patients with deep keratitis responded to antifungal therapy alone (topical antifungals combined with oral ketoconazole), and surgery was required for the other patients. Keratitis due to dematiaceous fungi other than *Curvularia* spp. appears to respond to primary therapy with topical natamycin, oral and/or topical ketoconazole, oral ketoconazole with topical miconazole, topical amphotericin B or oral itraconazole (Thomas et al. 1987, 1988; Rosa et al. 1994; Wong et al. 1997; Garg et al. 2000). However, therapy of keratitis due to *Lasiodiplodia theobromae* is often difficult. A successful outcome of this type of dematiaceous fungal keratitis was reported in patients receiving natamycin ointment and topical or subconjunctival amphotericin B (Rebell and Forster 1976). Intravenous miconazole was also reported to be useful, but this was based on the study of a single patient (Ishibashi and Matsumoto 1984b). Poor results have been reported in patients receiving azoles (Thomas et al. 1991a, 1998; Borderie et al. 1997).

The outcome of therapy of keratitis due to *Scedosporium apiospermum* is varied. A review of 13 cases reported up to 1979 revealed a generally poor outcome, with six of the cases eventually requiring enucleation or evisceration (Zapater and Albesi 1979). At least 14 patients with keratitis due to *Scedosporium apiospermum* have been reported in the literature since 1991 (reviewed by Wu et al. 2002). An analysis of the data pertaining to these patients (Thomas 2003) revealed that medical therapy alone sufficed for resolution of lesions in eight (57 percent) of these 14 patients (three of the 'responders' had keratitis with deep lesions), penetrating keratoplasty was needed in three patients (all had deep keratitis), and evisceration or enucleation was needed for three patients (two of whom had deep corneal lesions). Eight patients (five with deep keratitis and three with keratitis of undetermined severity) received natamycin at some time; corneal lesions resolved with medical therapy alone in three, while penetrating keratoplasty was required in the eyes of three patients, and enucleation had to be done for two patients. Six individuals (two with superficial keratitis, three with deep keratitis and one with keratitis of unknown severity) received miconazole; four resolved with medical therapy alone (two had superficial keratitis), while evisceration/enucleation was needed for two eyes (both with deep keratitis). These data suggest that, currently, keratitis due to *Scedosporium apiospermum* more frequently has a favorable outcome than that reported in the past, with evisceration or enucleation being the final result in 21 percent (compared to 54 percent in patients reported up to 1979). However, if severe keratitis is present, penetrating keratoplasty is required in addition to medical therapy. It is difficult to assess the relative efficacy of miconazole versus natamycin in view of the small numbers of patients involved.

A combination of topical antifungal therapy and keratoplasty appears to provide the most adequate treatment for keratitis due to *Acremonium* spp. (Fincher et al. 1991). A prospective evaluation of the comparative safety and efficacy of topical natamycin and 0.2 percent fluconazole was made in eight patients with filamentous fungal keratitis, including five due to *Acremonium* spp. and two due to *Curvularia* spp. (Rao et al. 1997). Corneal lesions resolved in three of four patients receiving primary natamycin treatment for a mean duration of 20 days (the keratitis worsened in the fourth patient) whereas corneal lesions failed to resolve in all four patients who received topical fluconazole as primary treatment (two subsequently responded to natamycin therapy). This study is important in providing evidence of the efficacy of topical natamycin, and the relative inefficacy of topical fluconazole, in therapy of keratitis due to filamentous fungi.

A common problem reported by all those who have had to treat keratitis due to *Pythium insidiosum* is that it is not sensitive to any of the currently available antifungals; wide surgical excision including penetrating keratoplasty has been advised for such patients, with enucleation or evisceration being required in patients who fail to respond to these measures (Virgile et al. 1993; Imwidthaya 1995; Thianprasit et al. 1996).

Fungal keratitis usually responds slowly over a period of weeks to antifungal therapy. Clinical signs of improvement of a fungal corneal ulcer include:

- diminution of pain
- decrease in size of infiltrate
- disappearance of satellite lesions
- rounding out of the feathery margins of the ulcer
- hyperplastic masses or fibrous sheets in the region of healing fungal lesions (Jones 1975; Johns and O'Day 1988).

Conjunctival chemosis and injection and punctate epithelial keratopathy may indicate toxicity of the antifungal agent being used. Negative scrapings during treatment do not always indicate that the fungus has been eradicated since it may have become deep-seated; hence therapy should be maintained for at least six weeks (Johns and O'Day 1988).

Although a combination of natamycin and ketoconazole was found to be beneficial in an animal model of *Aspergillus fumigatus* keratitis (Komadina et al. 1985), there are believed to be risks of antagonistic effects developing by the combination of certain antifungals such as amphotericin B and miconazole (Johns and O'Day 1988). Thus, methods to enhance the efficacy of existing antifungal agents require careful study.

MEASURES TO SUPPRESS CORNEAL DAMAGE DUE TO MICROBE- OR HOST TISSUE-DERIVED FACTORS

The inflammatory reaction directed against an infecting microorganism may sometimes be so severe as to damage adjacent tissues. To reduce these effects, therapy with corticosteroids and specific antimicrobials has sometimes been contemplated, since this regimen has proven satisfactory in disciform keratitis and central stromal keratitis due to the herpes simplex virus (O'Day 1991). In a recent study, Schreiber et al. (2003) tried to determine the most efficient time point and concentration of topical corticosteroids in *Candida albicans* keratitis treated with fluconazole. Their results suggested that fluconazole plus adjunctive high-dose prednisolone treatment was most effective when administered 9 days after infection. These workers concluded that the delayed application of corticosteroids after treatment with antifungals in patients with mycotic keratitis need not be a contraindication and, in fact, could be beneficial in some instances. However, most available evidence appears to weigh against the use of corticosteroids in the therapy of mycotic keratitis. In an experimental model of keratitis due to *Candida albicans*, the use of a topical corticosteroid was found to reverse the therapeutic effect of all the topical antifungals used, except for amphotericin B (0.5 or 0.15 percent) and to permit a greater replication of organisms, in comparison to untreated controls (O'Day et al. 1984b). Moreover, corticosteroids appear to be necessary to create experimental models of mycotic keratitis (Ishibashi and Kaufman 1986a; O'Day et al. 1991), and have been found to worsen the course of existing, but unrecognized, fungal infection in the cornea (Stern and Buttross 1991; Rosa et al. 1994). This may explain why currently, the use of corticosteroids is definitely contraindicated when a specific microbicidal therapy is not used concurrently, and especially when a fungal pathogen is present. Possible 'inflammatory rebound,' a potentially devastating complication that occurs when corticosteroid therapy is abruptly terminated, also needs to considered, since this could be confused with a worsening of infection (O'Day 1991).

Since corticosteroids cannot be used to decrease the unwanted side effects of the inflammatory response in mycotic keratitis, other molecules are being evaluated. In an animal model of keratitis due to *Candida albicans*, ketorolac (a nonsteroidal anti-inflammatory compound) was found to satisfactorily reduce the tissue necrosis occurring as a result of inflammatory mechanisms, without, however, permitting progression of the infection (Fraser-Smith and Matthews 1987); there do not appear to be published reports regarding the use of this compound in clinical mycotic keratitis. A variety of mechanisms have been found to contribute to corneal damage in sterile (nonmicrobial) corneal ulceration, including:

- activation of corneal collagenase, infiltration by polymorphonuclear leukocytes, and release of free radicals in keratitis due to alkali burns (Burns et al. 1990; Alio et al. 1993, 1995)
- release of lipid mediators (Tjebbes et al. 1993)
- activation of platelet activating factor (Tao et al. 1995).

Whether such mechanisms are relevant in pathogenesis of mycotic keratitis requires further study.

Therapeutic surgery

In the management of mycotic keratitis, surgery may be indicated in those patients who respond poorly, or not at all, to medical therapy, or where perforation or descemetocele formation is likely to occur. Prior to surgery, medical therapy should be administered for as long as possible to render the infecting fungus nonviable and therein to improve the outcome of surgery. The modalities of surgical management include removal of the corneal epithelium and anterior lamellar keratectomy, conjunctival flaps, tissue adhesives, and penetrating keratoplasty. These techniques aid medical therapy by removing infected corneal tissue (therein reducing or eliminating the microbial load) and increasing drug penetration, by bringing in blood vessels (when conjunctival

flaps are formed), by stabilizing the corneal epithelial surface itself or by providing tectonic support to the entire globe when its integrity is threatened by thinning or perforation of the cornea (Agrawal et al. 1994).

In mycotic keratitis, regular debridement of the base of the ulcer helps in elimination of fungi and necrotic material and also facilitates penetration of antifungal drugs such as amphotericin B 0.15 percent and 0.075 percent and natamycin into the corneal stroma (O'Day et al. 1984c). Debridement is usually performed, under topical anesthesia, with a Bard-Parker blade no. 15, ensuring that a margin of 1–2 mm is left at the limbus (Agrawal et al. 1994). Debridement of necrotic material may suffice to induce resolution of dematiaceous fungal keratitis. The procedure of anterior lamellar keratectomy helps to remove the thick mat of fungal filaments on the cornea, therein facilitating increased drug penetration and resolution of the corneal lesions in patients with dematiaceous fungal keratitis (Jones 1975; Garg et al. 2000). Anterior (superficial) stromal corneal infiltrates can also be ablated by the excimer laser for therapeutic purposes. In an experimental model of corneal infection due to a *Fusarium* spp., photoablation by the 193 nm excimer was found useful to eradicate early, localized microbial infections, but advanced infections, with deep stromal involvement and suppuration, could not be eradicated by this technique (Gottsch et al. 1991).

Conjunctival flaps help in achieving a stable conjunctival surface in cases of persistent or recurrent epithelial defects and progressive ulceration; such flaps are especially helpful in chronic peripheral disease, where the flap does not encroach onto the visual axis (Portnoy et al. 1989; Agrawal et al. 1994). Blood vessels present in the flap brought in to cover the ulcerated area help in healing of peripheral fungal corneal ulcers; a superficial lamellar keratectomy should first be done to remove the necrotic stroma, and then a thin conjunctival flap should be anchored over the ulcerated site (Alino et al. 1998). There have been no studies evaluating the specific use of conjunctival flaps in central fungal corneal ulcers.

Permanent or temporary amniotic membrane transplantation may be a useful adjunctive surgical procedure for the management of microbial (including mycotic) keratitis by promoting wound healing and reducing inflammation (Kim et al. 2001). However, a larger number of patients needs to be studied to confirm its use in mycotic keratitis. It is also not clear whether this technique will be useful in patients who have extensive corneal epithelial ulceration and stromal infiltration, and whether the infecting fungus is completely eradicated in this procedure, or whether foci of viable fungi persist in the corneal tissue.

Tissue adhesives (cyanoacrylate 'glue') provide support to a thinned-out cornea, can seal a corneal perforation that is 2 mm or less in size (Forster 1994), and are also bacteriostatic for gram-positive bacteria

(Agrawal et al. 1994). Necrotic stromal and epithelial debris is first removed from the base of the ulcer, following which the adhesive is applied and a bandage contact lens fitted. The adhesive is left in place until it loosens spontaneously, or the bed becomes vascularized, or keratoplasty is performed.

Although penetrating keratoplasty appears to be rarely required for the treatment of active bacterial keratitis, probably due to the availability of specific antibacterial drugs, it may be required in about 15.3–27 percent of cases with mycotic keratitis (Rosa et al. 1994; Garg et al. 2000), particularly when the keratitis is caused by species of *Aspergillus* (Rosa et al. 1994; Vemuganti et al. 2002), *Fusarium* (Vemuganti et al. 2002), *Lasiodiplodia theobromae* (Borderie et al. 1997; Thomas et al. 1998), *Pythium insidiosum* (Imwidthaya 1994), and *Paecilomyces lilacinus* (Gordon and Norton 1985; Okhravi et al. 1997). Mycotic keratitis is reported to be associated with a five- to sixfold higher risk of subsequent perforation and need for penetrating keratoplasty than bacterial keratitis (Wong et al. 1997). The outcome of penetrating keratoplasty (Figure 16.17) also depends on whether the surgery has been performed for a bacterial or fungal corneal ulcer. In one study (Cristol et al. 1996), it was found that the graft tended to opacify much earlier in mycotic keratitis (median duration 4.0 weeks) than in bacterial keratitis (median duration 12.9 weeks). Similarly, the reported success rate for grafts in mycotic keratitis (20–60 percent) appears to be much lower than that in bacterial keratitis (70–75 percent) (Panda et al. 1991; Killingsworth et al. 1993). In one study, 25 percent of grafts performed for mycotic keratitis showed reinfection (Rosa et al. 1994). To decrease the incidence of recurrence, at least 0.5 mm of clear tissue all around the infected area should be excised. Although antifungal therapy should be continued postoperatively, topical corticosteroids should be used with caution (Stern and Buttross 1991). When donor grafts 8 mm or less in diameter were used for penetrating keratoplasty in fungal corneal ulcers, the outcome was better than when larger grafts were used (Killingsworth et al. 1993). Topical cyclosporin A (5 mg/ml) was reported to be useful as a primary or an adjunctive therapy for prevention of allograft rejection in a small number of patients with culture-proven mycotic keratitis (Perry et al. 2002).

In a recently published review of the results obtained in 45 patients with severe mycotic keratitis, in whom therapeutic penetrating keratoplasty was performed using cryopreserved donor corneas, with additional measures being irrigation of the anterior chamber with fluconazole (2 mg/ml), dissection of fibrinoid membrane on the iris and iridectomy, it was noted that the fungal infection was eradicated, and the anatomic integrity was maintained, in 39 eyes (Yao et al. 2003). Thus, cryopreserved donor corneas may be effective substitutes in therapeutic penetrating keratoplasty in severe mycotic keratitis.

MYCOTIC SCLERITIS

Although uncommon, fungal lesions of the sclera are of great importance since they may spread from fungal infections of contiguous structures, such as the cornea or conjunctiva, or may occur secondary to trauma or surgery. Endogenous infections have also been reported. Scleritis arising due to spread of infection from keratitis due to *Absidia corymbifera* (Marshall et al. 1997), *Acremonium* spp., and *Lasiodiplodia theobromae* (Borderie et al. 1997) has been reported. Similarly, scleritis due to fungi (*Aspergillus* spp. or *Sporothrix schenckii*) may follow ocular trauma (Rodriguez-Arres et al. 1995). A unique subset of microbial scleritis following ocular surgical procedures is being increasingly reported; this is dealt with in another section (see later under Fungal ocular infections after ophthalmic surgical procedures).

One patient with *Scedosporium prolificans* corneoscleritis responded to intensive antifungal therapy and aggressive scleral debridement (Kumar et al. 1997), whereas another patient responded poorly to medical therapy (topical natamycin and amphotericin B, oral itraconazole and ketoconazole), necessitating eventual enucleation (Sullivan et al. 1994).

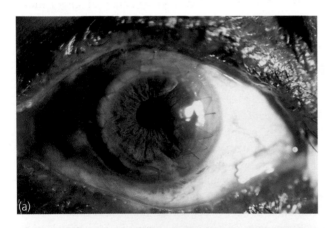

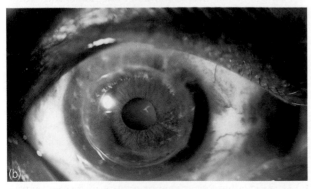

Figure 16.17 (a) *Keratitis due to* Fusarium solani, *1 month after therapeutic penetrating keratoplasty. The corneal graft is clear.* **(b)** *Keratitis due to* Fusarium solani, *4 months after therapeutic penetrating keratoplasty. The corneal graft remains clear.*

The outcome of scleritis due to *Scedosporium apiospermum* is also reported to be varied, with good results being obtained in some patients (Moriarty et al. 1993), and poor results in others (Taravella et al. 1997). Mycotic scleritis due to species of *Candida* or filamentous fungi (except *Fusarium*) may be treated by subconjunctival miconazole (and intravenous miconazole, if necessary). Additional remedies include oral flucytosine, oral ketoconazole, oral itraconazole, oral fluconazole, intravenous amphotericin B, topical miconazole, or topical natamycin. *Sporothrix schenckii* infection has been successfully treated with oral potassium iodide (50 mg/drop), 10 drops thrice daily, slowly increasing to 24 drops thrice daily (Brunette and Stulting 1992). Post-traumatic scleritis due to *Aspergillus fumigatus* was found to worsen in spite of oral fluconazole and topical amphotericin B therapy; cryotherapy and dura mater grafting were then performed, which appeared to control the infection (Rodriguez-Arres et al. 1995). Oral itraconazole therapy brought about resolution of inflammation in scleritis due to *Aspergillus flavus*; the patient's condition had worsened during therapy with oral ketoconazole and topical amphotericin B (Carlson et al. 1992).

INTRAOCULAR MYCOSES

Definitions

Uveitis may be defined as intraocular inflammation (Weinberg 1999). There are several systems classifying the many uveitis entities, one of which is uveitis arising due to infections caused by various microorganisms, including fungi. The intraocular inflammation caused by fungal infection usually develops slowly, is spread to the eye through the bloodstream, and consists of focal or multifocal lesions in the choroid and retina (chorioretinitis). Anterior segment inflammation is variable, but can be severe with hypopyon (Weinberg 1999). When the initial focus of intraocular fungal infection in the choroid and retina extends into the vitreous to produce inflammation, which may involve the entire internal structure of the eye, endophthalmitis results. It may be difficult, from a practical viewpoint, to discuss fungal retinitis separately from fungal endophthalmitis (Hamza et al. 1999). However, intraocular findings have been classified, based on ophthalmoscopic examination in patients with suspected ocular involvement in candidemia, into endophthalmitis (focal chorioretinitis with extension into the vitreous, intravitreal fluff balls, or vitreous haze associated with typical chorioretinal lesions), chorioretinitis (focal, deep, white chorioretinal lesions with no evidence of direct vitreous involvement), and 'nonspecific' fundus lesions (clear vitreous and no focal chorioretinitis, including cotton wool spots, retinal hemorrhages, and Roth spots) (Donahue et al. 1994). Endophthalmitis, in turn, can be divided into endogenous endophthalmitis, which arises from hematogenous spread from a focus of infection elsewhere in the body, and exogenous endophthalmitis, resulting from primary inoculation of the eye following surgery or penetrating trauma.

Etiologic agents

The major causes of intraocular mycoses are species of *Candida* (especially *Candida albicans*), *Aspergillus* (especially *Aspergillus fumigatus*), *Blastomyces dermatitidis*, *Coccidioides immitis*, *Cryptococcus neoformans*, *Histoplasma capsulatum*, *Scedosporium apiospermum* (*Pseudallescheria boydii*), and *Sporothrix schenckii* (Hamza et al. 1999). Species of *Fusarium* (Patel et al. 1994; Goldblum et al. 2000), *Acremonium* (Weissgold et al. 1996), *Paecilomyces* (Pettit et al. 1980; Kozarsky et al. 1984; Scott et al. 2002), zygomycetes (Orgel and Cohen 1989), and *Pneumocystis carinii* (Dugel et al. 1990; Foster et al. 1991) have also been reported to cause such lesions. Although most of these organisms produce endophthalmitis, they may also cause focal (localized) chorioretinitis or a granulomatous lesion in the iris or ciliary body (Weinberg 1999).

Frequency of occurrence

In the 1970s and 1980s, the incidence of endogenous fungal endophthalmitis in patients with candidemia or candidiasis, or both, was reported to range from approximately 10 to 40 percent. A study in the late 1980s reported that ocular lesions developed within 72 hours of the suspected onset of fungemia in approximately 90 percent of nontreated, hospitalized patients with candidemia (Brooks 1989). However, a more recent study (Scherer and Lee 1997) estimated the incidence of chorioretinal findings consistent with early endogenous fungal endophthalmitis to be 2.8 percent. This low incidence was attributed to the fact that 90 percent of the patients were being treated with systemic antifungal medication (mostly amphotericin B and/or oral fluconazole) at the time of the ophthalmological consultation. *Candida* endophthalmitis now also appears to be rare in patients with candidemia, with none of 118 adults with candidemia showing evidence of endophthalmitis, only 9.3 percent exhibiting chorioretinitis, and 20.3 percent nonspecific fundus lesions (Donahue et al. 1994). In a study of 30 hospitalized children with suspected or known systemic infections caused by *Candida* spp., all of whom had risk factors for disseminated candidiasis (broad-spectrum intravenous antibiotics, chronic debilitation with indwelling catheterization, total parenteral nutrition, immunocompromised state), none was found to have either endophthalmitis or chorioretinitis (Donahue et al. 2003). The low rate of ocular involvement in both series, compared with earlier reports, was believed to be due either to earlier treatment with antifungal agents or to the lack of rigorous criteria for endophthalmitis in previous studies.

Predisposing factors

Intravenous drug abusers are at great risk to develop endogenous fungal endophthalmitis, probably because they may use contaminated drugs, syringes, needles, or cotton (Elliott et al. 1979). Contaminated preserved lemon juice used to dissolve heroin was found to be the cause of *Candida albicans* endophthalmitis in intravenous drug abusers in Glasgow (Shankland and Richardson 1989). Vitreous opacities may be a prominent manifestation in such individuals (Aguilar et al. 1979). In a retrospective analysis of clinical and histopathological features in 25 patients who underwent enucleation (13 with morphological features and/or positive culture for *Aspergillus* and 12 with histological evidence and/or positive culture for *Candida*), Rao and Hidayat (2001) noted a history of gastrointestinal surgery, hyperalimentation, or diabetes mellitus in nine of 12 patients with *Candida* endophthalmitis (and none of the patients with aspergillosis), whereas 12 of 13 patients with *Aspergillus* endophthalmitis, and only one of 12 patients with *Candida* endophthalmitis, had received immunosuppressive agents or had undergone organ transplants or valvular cardiac surgery. Thus, it might be possible clinically to suspect *Candida* endophthalmitis in patients who have undergone gastrointestinal surgery, and *Aspergillus* infection in those with drug-induced immune suppression or in patients who have had cardiac surgery. In another retrospective analysis of endogenous fungal endophthalmitis (where *Candida* species were the major causes), in 79 eyes of 46 patients the principal predisposing factors noted were the presence of β-D-glucan in concentrations of at least 20 pg (in 90 percent of patients), intravenous hyperalimentation (87 percent), fever of unknown origin (76 percent), male gender (74 percent), cancer (72 percent), and neutrophils in a count of at least 500/ml (67 percent) (Tanaka et al. 2001). Recent prolonged broad-spectrum antibacterials, indwelling intravenous catheters, pressure-monitoring devices, organ transplantation, and generalized immunosuppression associated with malnutrition, debilitating diseases, iatrogenic neutropenia, or corticosteroid use (Hamza et al. 1999) are other reported predisposing factors.

Diagnosis

Since anterior chamber and/or vitreous inflammation may obscure visualization of the retina, echography may help to determine the anatomic status of the retina, the extent of inflammation, the presence of choroidal detachment, and the presence and location of intraocular foreign bodies. Apart from localizing such foreign bodies, computed tomography may not be very useful, nor are electrophysiological studies (Boldt and Mieler 1996).

Since the diagnosis of exogenous fungal endophthalmitis is ultimately established by demonstrating infectious organisms in the eye, samples of aqueous and vitreous should be obtained prior to instituting therapy. A vitrectomy specimen (where as much as necessary of the vitreous is cut and removed, usually by a vitreous cutter) is preferred to a vitreous aspirate (vitreous tap) sample since the latter may fail to sample the locus of infection (Hamza et al. 1999). Anterior chamber aspirates are also a poor diagnostic technique. Although cultures of vitreous samples are more likely to yield positive results than cultures of aqueous samples, it is recommended that both samples be obtained. A conjunctival swab is of relevance only if there is a leaking filtering bleb. Fungal endophthalmitis can be confirmed by direct microscopic demonstration of fungal hyphae or yeast cells in 10 percent KOH wet mounts, or smears stained by calcofluor white and the Gram method (Brar et al. 2002). Culture is performed on appropriate media. Direct inoculation of vitreous and aqueous samples on the culture media immediately after collection is recommended. To increase the chances of recovering fungi, vitrectomy samples can be concentrated in an ultracentrifuge and then inoculated, or can be passed through cellulose membrane filters which are then applied to the culture media plates.

Fungal chorioretinitis and endogenous fungal endophthalmitis represent exceptions to the rule requiring isolation of the fungus from ocular tissue in order to establish an ocular fungal infection. For example, the presence of a typical fundus lesion, as well as a history of *Candida* spp. being cultured from blood, urine, intravascular cathethers, or other body sources days or weeks prior to the onset of symptoms, permits a presumptive diagnosis of *Candida* retinitis (Hamza et al. 1999). If nonspecific findings are present, or if there is no positive culture from an extraocular site, a diagnostic vitrectomy can be performed (Hamza et al. 1999). Similarly, in intraocular histoplasmosis and coccidioidomycosis, which are commonly associated with characteristic chorioretinal lesions, isolation of the fungus from another anatomic site, or measurement of antibody titers to the fungus, is usually deemed sufficient evidence to establish one of these fungi as the cause of the eye disease (Klotz et al. 2000). Since ophthalmic infections due to *Cryptococcus neoformans* usually occur in conjunction with meningoencephalitis, isolation of cryptococci from blood and/or cerebrospinal fluid is usually sufficient explanation for the associated eye findings (Klotz et al. 2000).

In recent years, the value of DNA-based technology in the diagnosis of fungal endophthalmitis has been evaluated. PCR assay was evaluated in the diagnosis of *Candida* endophthalmitis in four patients (Hidalgo et al. 2000); vitreous cultures were negative in two of the four patients, but characteristic PCR products were generated in all four patient specimens, permitting the diagnosis of *Candida* endophthalmitis in all four. A protocol employing two novel panfungal primers complementary to 18S rRNA sequences of *Candida albicans*, *Aspergillus fumigatus*, and *Fusarium solani*, followed by three nested PCRs utilizing species-specific primers, was applied to ocular material from three patients with suspected endophthalmitis (Jaeger et al. 2000); a positive result for *Candida albicans* DNA and isolation of *Candida albicans* in culture was obtained from one sample and a positive result for *Candida albicans* DNA with a negative culture result was obtained in another sample (the third sample was negative for fungal DNA and was also culture negative). The two patients whose samples were positive for *Candida albicans* DNA were reported to have had clinical signs typical of fungal endophthalmitis and both patients responded to treatment with antifungal agents (Jaeger et al. 2000). Further reports are necessary before conclusions can be drawn.

Treatment

Intravitreal injections are performed in order to improve the intraocular concentration of antifungal drugs, to concentrations greater than can be achieved by systemic preparations, but toxicity may result if the dilutions are not prepared properly, or if the dilutions are injected into air-filled eyes. In the local treatment of endogenous fungal endophthalmitis, intravitreal amphotericin B is the usual recommended initial treatment, while intravitreal miconazole can be considered for those rare fungal cases resistant to amphotericin B (Flynn 2001; Song et al. 2002). Periocular amphotericin B is not commonly used because it has poor intravitreal penetration and usually causes marked conjunctival necrosis. Systemic (intravenous) amphotericin B causes serious and sometimes irreversible toxicity, hence its use is limited to very advanced endogenous fungal endophthalmitis, especially when other nonocular sites are involved (Weishaar et al. 1998; Flynn 2001).

Less toxic azole compounds, for example, fluconazole for *Candida*, can supplement intravitreal therapy and may contribute to a favorable clinical response without the risk of unwanted side effects of systemically administered amphotericin B (Borne et al. 1993; Luttrull et al. 1995; Christmas and Smiddy 1996; Flynn 2001). The possible use of intravitreal itraconazole also requires clarification. Ocular toxicity studies performed in New Zealand rabbits 5 weeks after intravitreal administration of itraconazole 10–100 μg (dissolved in 100 percent DMSO) showed no substantial retinal or histopathological changes in eyes that had been injected with 100 percent DMSO or 10 μg itraconazole, although higher doses caused focal areas of retinal necrosis (Schulman et al. 1991). Based on the results of an experimental study, Yoshizumi and Banihashemi (1988) suggested

that a single intravitreal dose of ketoconazole ($\leqslant$540 μg) in DMSO could be safely used for fungal endophthalmitis. Intravenous administration of fluconazole (5 or 25 mg/kg) in albino rats resulted in aqueous, vitreous, and serum levels (1 hour after administration) of 2.87, 1.72, and 4.6 μg/ml (5 mg/kg) and 14.9, 7.05, and 20.6 μg/ml (25 mg/kg), respectively; the intraocular penetration was moderately enhanced by vitrectomy (Mochizuki et al. 1992). In vitro electroretinograms (ERG) remained unchanged after perfusion with fluconazole (20 μg/ml) while the in vivo ERG and visual evoked potentials were unchanged after daily fluconazole (25 mg/kg) for 8 days, suggesting good safety profile. Following intravenous inoculation of fluconazole 20 mg/kg as a single dose or 20 mg/kg every 12 hours for four doses in nonpigmented rabbits, fluconazole concentrations in the aqueous, vitreous, CSF, and serum were determined by a microbiological assay; the penetration of fluconazole in all the anatomic compartments was found to be >70 percent of that in serum (Mian et al. 1998). Since the CSF and ocular pharmacokinetic parameters closely resemble each other, either could be used as a surrogate for the other (Mian et al. 1998). A biodegradable polymeric scleral implant containing fluconazole was reported to be a promising intravitreal drug delivery system to treat fungal endophthalmitis (Miyamoto et al. 1997). Scleral implants loaded with 10, 20, and 30 percent doses gradually released fluconazole over 4 weeks in vitro while those with 50 percent doses released most of the drug in one week; implants with 30 percent fluconazole that were studied in pigmented rabbits resulted in vitreous concentrations of fluconazole (sustained for 3 weeks) sufficient to inhibit *Candida albicans*. Intravitreal injection of up to 100 μg fluconazole per 0.1 ml of vitreous did not produce biomicroscopic, ophthalmoscopic, electroretinographic, or light microscopic evidence of intraocular toxicity, even 8 days after inoculation (Schulman et al. 1987).

The typically slow replication of fungal elements may contribute to a less favorable response to intravitreal antifungal therapy alone; therefore, pars plana vitrectomy is often selected for initial therapy to obtain a satisfactory specimen for culture and to remove vitreous infiltrates. Vitrectomy for the treatment of fungal endophthalmitis has been advocated in all but very mild or exceptional cases. The indications for vitrectomy in patients with fungal chorioretinitis and endophthalmitis are advanced cases with extensive vitreous involvement and poor response to systemic antifungal therapy (Pettit et al. 1996; Smiddy 1998). The potential benefits of vitrectomy include:

- debulking of inflammatory and infectious material from the vitreous
- acquisition of a larger sample for laboratory study
- potential for concentrating the sample by centrifugation or filtration to give a better yield on culture

- an opportunity for intravitreal injection of antifungals
- removal of the scaffolding for vitreoretinal traction bands and epiretinal membranes that can contribute to late-developing macular pucker and retinal detachment (Pettit et al. 1996).

However, vitrectomy and intravitreal injections are better avoided in cases of neonatal endophthalmitis.

Essman et al. (1997) performed a retrospective analysis of treatment outcomes of 20 eyes (18 patients) with culture-proven endogenous fungal endophthalmitis (17 due to *Candida* species, three due to *Aspergillus* spp.) seen over a 10-year period. After initial examination, 17 of 20 eyes underwent pars plana vitrectomy and 19 of 20 received intravitreal amphotericin B; 16 patients received systemic antifungals (oral ketoconazole, oral fluconazole). Thirteen (76 percent) of 17 eyes with *Candida* infection achieved visual acuities equal to or better than 20/400, while none of the three eyes with *Aspergillus* infection achieved such a visual acuity. Based on these results, these authors recommended pars plana vitrectomy, intravitreal amphotericin B, and administration of appropriate systemic antifungals (oral fluconazole for *Candida* infections) in managing patients with marked vitreous infiltrates due to endogenous fungal endophthalmitis.

Recommendations for postoperative and post-traumatic exogenous fungal endophthalmitis (Pflugfelder et al. 1988) include surgical measures (excision of clinically involved tissue, vitrectomy if there is visible vitreous involvement, retention of the intraocular lens unless there is extensive infiltration around the lens, or recurrent infection) and medical therapy:

- intraocular amphotericin B 5–10 µg in regions of maximal involvement, repeatable
- miconazole 25–50 µg in patients with *Paecilomyces lilacinus* infection or in treatment failure with amphotericin B
- topical (hourly) 5 percent natamycin, or topical (hourly) 0.15 percent and subconjunctival 500–1000 µg amphotericin B or topical (hourly) 1 percent and subconjunctival 5–10 mg miconazole, if there is corneal, scleral, or anterior chamber involvement
- systemic (oral) ketoconazole 400–600 mg/day.

Recommendations for exogenous fungal endophthalmitis developing from fungal keratitis include surgery (penetrating keratoplasty for deep keratitis with retrocorneal or anterior chamber involvement unresponsive to medical therapy, retention of the iris and lens if possible, or iridectomy, lensectomy, and vitrectomy if there is clinical evidence of infiltration of these structures, or if there is recurrence of the endophthalmitis) and medical therapy:

- intravitreal amphotericin B 5–10 µg if the anterior chamber or vitreous is involved
- topical hourly natamycin 5 percent, topical 0.15 percent amphotericin B or topical 1 percent miconazole

- subconjunctival miconazole or amphotericin B
- oral ketoconazole.

The role of intravitreal corticosteroids remains controversial, but they can be considered if appropriate antimicrobial coverage of the causative organism can be assured (Flynn 2001). Interestingly, Majji et al. (1999) reported on the role of intravitreal dexamethasone in the management of 20 patients with exogenous fungal endophthalmitis; all 20 patients had undergone pars plana vitrectomy, and had received intravitreal amphotericin B and oral ketoconazole, and some patients had also received intravitreal dexamethasone. Overall, nine of the 20 patients achieved a final visual acuity of better than counting fingers at 3 metres. The rate of clearance of inflammation was better in eyes that had received intravitreal dexamethasone. Also, a favorable visual outcome was achieved in a higher number of eyes that had received dexamethasone, compared to eyes not receiving dexamethasone; however, this difference was not statistically significant. These authors cautioned that while the results of their retrospective study suggested that corticosteroids may be beneficial in promoting faster clearance of inflammation in fungal endophthalmitis, the sensitivity of the fungi to antifungals, the dose and timing of steroid, and institution of effective antifungal medication prior to the use of corticosteroids were all essential factors which needed to be examined further in a prospective manner.

Specific types of intraocular mycoses due to different etiologic agents

CANDIDA SPECIES

Candida spp. are common causes of disseminated disease in drug abusers, in severely ill hospitalized patients, and in those patients who are immunocompromised (Donahue et al. 1994). Hence, it is not surprising that in patients with endogenous fungal endophthalmitis, the most common causative organism is *Candida* spp. (Flynn 2001). Endogenous *Candida* endophthalmitis differs from endogenous bacterial endophthalmitis in being more often slowly progressive, sometimes capable of successful management by systemic therapy alone, and in having a more favorable visual prognosis (Brod et al. 1990; Christmas and Smiddy 1996; Essman et al. 1997; Flynn 2001). *Candida albicans* is the most common fungal pathogen causing chorioretinitis or endophthalmitis, while other *Candida* spp., including *Candida glabrata* and *Candida parapsilosis*, are less commonly implicated (Parke et al. 1982; Pettit et al. 1996). This greater tendency of *Candida albicans*, compared to other *Candida* spp., to produce chorioretinitis is thought to be because *Candida albicans* rapidly forms germ tubes in serum (whereas other *Candida* spp. do not), which lodge in the choriocapillaries more easily and more frequently than other species, or due to differences in the patterns of phospholipase and protease production among different *Candida* spp. (Moyer and Edwards 1993).

Although a number of factors (mentioned above) may predispose to endogenous fungal endophthalmitis, specific risk factors for endophthalmitis due to *Candida* spp. appear to be major surgery involving the gastrointestinal system, parenteral hyperalimentation, and diabetes mellitus (Rao and Hidayat 2001). *Candida* endophthalmitis has also been reported after induced abortion (Chen et al. 1998) and in association with retinopathy in premature infants (Gago et al. 2002). Cataract and intraocular inflammation, and progressive retinopathy of prematurity and tractional retinal detachment, secondary to *Candida* septicemia were also recently reported in a markedly premature infant (Shah et al. 2000); pars plana vitrectomy and lensectomy revealed lens material infiltrated by *Candida*-like yeast forms and pseudohyphae. Song et al. (2002) reported the occurrence of endogenous fungal retinitis, presumed to be due to a *Candida* sp., in a 3-year-old boy who had acute lymphocytic leukemia and a skin rash proven by culture to be due to a *Candida* sp.

The most characteristic ophthalmoscopic sign of *Candida* chorioretinitis is a creamy-white, well-circumscribed lesion involving the retina and choroid in the posterior pole, which may be single or multiple, unilateral or bilateral; the lesions may exist in a satellite pattern, and retinitis may progress to vitreous involvement. Vitreous opacities are typically yellow–white, may be connected by strands, giving a 'string of pearls' appearance, and may be so severe as to obscure the view of the fundus, thereby making the clinical diagnosis difficult (Aguilar et al. 1979; Hamza et al. 1999). In a study of 12 eyes of 12 patients that had been enucleated due to *Candida* endophthalmitis, it was noted that the initial presentation of *Candida* infection was as a vitritis or vitreoretinitis in seven, panuveitis in three, and endophthalmitis in two patients (Rao and Hidayat 2001). Intraretinal hemorrhages may also occur; in fact, focal retinal necrosis and scarring combined with vitreoretinal membrane formation and contraction are the major causes of permanent visual loss (Aguilar et al. 1979; Barrie 1987; Brod et al. 1990). Choroidal neovascularization is a potential cause of late visual loss in patients who have had *Candida albicans* sepsis and endogenous *Candida albicans* chorioretinitis (Jampol et al. 1996). *Candida parapsilosis* may cause a low-grade postcataract extraction endophthalmitis (Hamza et al. 1999), while multifocal endophthalmitis due to *Candida tropicalis* has been reported as the only initial manifestation of pacemaker endocarditis (Shmuely et al. 1997).

Candida retinitis or endogenous endophthalmitis should be suspected in any patient with one of the known predisposing conditions and who presents with progressive chorioretinitis or posterior uveitis. Criteria for the diagnosis of such an infection are discussed above. In addition, detection of anti-*Candida* antibodies in aspirates of anterior chamber fluid may be useful in establishing a diagnosis (Mathis et al. 1988). In histopathological studies on 12 eyes of *Candida* endophthalmitis, fungi were located in the vitreous alone in seven patients, in the vitreous and retina in two, in the vitreous, retina, subretina, and choroid in two, and in the iris and ciliary body in one (Rao and Hidayat, 2001).

Based on a retrospective analysis of treatment outcomes in 20 eyes of 18 patients, Essman et al. (1997) recommended pars plana vitrectomy, intravitreal amphotericin B, and administration of appropriate systemic antifungals (oral fluconazole for *Candida* infections) in managing patients with marked vitreous infiltrates due to endogenous fungal endophthalmitis. However, other therapeutic approaches have been described for chorioretinitis and endophthalmitis due to *Candida* spp. Laatikainen et al. (1992) described a patient with bilateral *Candida* endophthalmitis who was treated with oral fluconazole, and in whom vitrectomy and intravitreal amphotericin B was given for one eye only; both eyes healed equally well. Borne et al. (1993) described a patient with postoperative endophthalmitis due to *Candida parapsilosis* in whom vitrectomy and intravitreal amphotericin B had failed, whereas oral fluconazole resulted in a successful outcome. Luttrull et al. (1995) described rapid (within 24 hours of initiation of therapy) and dramatic responses to oral fluconazole therapy only (400 mg/day) in one patient with culture-proven and in three patients with presumed endogenous *Candida* endophthalmitis. Christmas and Smiddy (1996) reported that pars plana vitrectomy, oral fluconazole therapy (100 mg once or twice daily for 3–6 months), and subconjunctival gentamicin and dexamethasone were sufficient to bring about clinical resolution of lesions and visual improvement in six eyes of five patients with *Candida* endophthalmitis (four eyes culture-positive for *Candida albicans*, one eye culture-positive for *Candida tropicalis*, and one eye direct microscopy positive but culture-negative) without the need for intravitreal amphotericin B or fluconazole therapy. In one patient, oral antifungals alone had been unsuccessful before pars plana vitrectomy, while in two other patients with asymmetric bilateral vitritis, the second eye had not improved despite 11 days of oral fluconazole therapy. These authors concluded that vitrectomy and oral fluconazole without intravitreals could be recommended for successful management of endogenous *Candida* endophthalmitis. More recently, Mora et al. (2002), based on their experience of management and therapy of fungal endophthalmitis in 17 patients (27 eyes) seen over a 9-year period, concluded that while oral fluconazole was successful in managing mild fungal endophthalmitis, additional vitrectomy (sometimes more than once) and intraocular amphotericin B were necessary for more severe cases. Presumed bilateral endophthalmitis due to *Candida* spp. that developed in an infant with bilateral posterior stage 3 retinopathy of prematurity was successfully eradicated from each eye by vitrectomy with instillation of 5 µg amphotericin B (Gago et al. 2002). In patients who have developed choroidal neovascularization secondary to *Candida* chorioretinitis, laser photocoagulation or surgical excision of the neovascular complex may be of benefit in selected cases (Jampol et al. 1996).

ASPERGILLUS SPECIES

Aspergillus spp. may cause exogenous or endogenous endophthalmitis. Narang et al. (2001), who reviewed data pertaining to 27 patients with exogenous fungal endophthalmitis, principally due to *Aspergillus* spp., noted substantial corneal involvement in 14 eyes; multivariate analysis revealed that this was the single most important risk factor in determining final visual outcome. Tabbara and al-Jabarti (1998) reported an outbreak of ocular aspergillosis (*Aspergillus fumigatus*) after cataract surgery; the outbreak was finally found to be associated with ongoing hospital construction. All five patients developed endophthalmitis, while two patients also had superior limbal, scleral, and corneal infiltration. Endogenous *Aspergillus* endophthalmitis has been reported in patients who have received immunosuppressive agents or who have undergone organ (heart, lung, or liver) transplants or valvular cardiac surgery (Weishaar et al. 1998; Rao and Hidayat 2001) and intravenous drug abusers (Weishaar et al. 1998). Kalina and Campbell (1991) reported endophthalmitis in a patient suffering from chronic lymphocytic leukemia who had presented with acute visual loss and redness in one eye. Although results of stains and cultures of anterior chamber fluid were negative, cultures from antemortem sputum and skin samples were positive for *Aspergillus terreus* while postmortem histological results showed extensive invasion of the posterior vitreous, retina, choroid, and anterior optic nerve by filamentous fungi presumed to be *Aspergillus terreus*. Predisposing systemic conditions may modify the probability of dissemination, risk of mortality, and role for systemic treatment.

The clinical features of endogenous *Aspergillus* endophthalmitis, the ocular presentation, and findings often are characteristic and provide possible diagnostic utility. In a retrospective analysis of ten patients (12 eyes) with culture-proven endogenous *Aspergillus* endophthalmitis, Weishaar et al. (1998) reported that all patients had a 1–3-day history of pain and marked loss of visual acuity in the involved eyes and that varying degrees of vitritis were present in all 12 eyes; in eight of 12 eyes, a central macular chorioretinal inflammatory lesion was present, and an inferior gravitational layering of inflammatory exudate in either or both the subhyaloid and subretinal space. In another retrospective analysis of 13 patients with histopathologically proven *Aspergillus* intraocular infection, the findings at presentation were endophthalmitis in four, panuveitis, acute retinal necrosis, chorioretinitis, or vitritis/vitreoretinitis in two each, and as granulomatous anterior uveitis in one patient (Rao and Hidayat 2001). The severity of retinal involvement in endogenous *Aspergillus* endophthalmitis may range from subretinal or subhyaloid infiltrates to vascular occlusion and full-thickness retinal necrosis; intraretinal hemorrhages are frequent. Rao and Hidayat (2001) reported that in 13 eyes with *Aspergillus* endophthalmitis, fungi were seen in the vitreous, retina, subretina, and choroid in eight, in the vitreous alone in three, and in the vitreous and retina in two. These authors concluded that, histopathologically, the vitreous is the primary focus of infection for *Candida*, whereas it appears that *Aspergillus* grows preferentially along the subretinal pigment epithelium and subretinal space.

Compared with *Candida* spp., *Aspergillus* endophthalmitis is typically more difficult to diagnose because:

- skin test and serology are unreliable
- the diagnostic yield of blood cultures, pulmonary radiographic studies, and echocardiograms is low
- isolation from blood cultures is difficult even in patients who are immunosuppressed with definite dissemination
- systemic manifestations are often lacking in intravenous drug abusers with this infection (Lance et al. 1988; Weishaar et al. 1998).

A diagnostic vitreous aspirate for cytology and culture isolation is usually necessary to identify this etiologic agent (Lance et al. 1988). However, diagnosis of endogenous *Aspergillus* endophthalmitis by anterior chamber or vitreous aspirates alone may be unreliable (Weishaar et al. 1998). According to Rao and Hidayat (2001), since aspergillosis clinically presents with extensive areas of deep retinitis/choroiditis, vitreous biopsy might not always yield positive results. Pars plana vitrectomy specimens assisted by Gram or Giemsa stains appear to have the highest yield of positive cultures for *Aspergillus*; in the series of Weishaar et al. (1998), eight of nine previously untreated eyes had a positive culture from the vitrectomy specimen. Anand et al. (2001) compared PCR against the conventional mycological methods of microscopy and culture for growth of fungi in the diagnosis of *Aspergillus* endophthalmitis in 27 intraocular specimens from 22 patients with suspected fungal endophthalmitis (which were proven to be nonbacterial in origin); 10 patients with non-infective intraocular disorders were the controls. None of the controls was positive by microscopy, culture, or PCR. Among the 27 test samples, four were positive by culture for *Aspergillus* spp. and were also positive by PCR. In addition, PCR detected and identified *Aspergillus* spp. in two culture-negative specimens. The average time required for culture and identification of *Aspergillus* was 10 days, whereas PCR was able to yield results in just 24 hours.

The optimal treatment for endogenous *Aspergillus* endophthalmitis remains controversial. This is in part due to the fact that, in contrast to the outcome of endophthalmitis due to *Candida* spp., the mortality is high in patients with *Aspergillus* endophthalmitis, particularly those who have received organ transplants or who have undergone cardiac surgery (Rao and Hidayat 2001). Moreover, endogenous *Aspergillus* endophthalmitis during systemic aspergillosis usually requires more than systemic treatment to achieve resolution of the ocular infection. The visual outcome in these patients is influenced primarily by the propensity of *Aspergillus* for initial macular choroidal involvement. Based on their results in treatment of endogenous *Aspergillus* endophthalmitis, Weishaar et al. (1998) recommended the following:

- pars plana vitrectomy when vitreous seeding is present to remove the focus of infection and to improve the likelihood of a positive culture, but not necessarily in fellow eyes in which only isolated chorioretinal lesions are present and when the diagnosis can be made from evaluation of the first eye
- amphotericin B (5–10 µg) as intravitreal antifungal treatment
- intravitreal dexamethasone 400 µg (although controversial) to reduce the marked intraocular inflammation in many of these eyes
- intravenous amphotericin B to supplement the effectiveness of intravitreal amphotericin B therapy and to treat proven or clinically suspected systemic aspergillosis
- a well-tolerated and less toxic systemic medication (example, itraconazole) to supplement intravitreal amphotericin B in patients with only endogenous *Aspergillus* endophthalmitis and no systemic manifestations
- repeat intravitreal amphotericin B (5–10 µg) and possibly repeat vitrectomy in patients with persistent vitreous infiltrates and suspected recurrent disease after initial treatment.

Using this regimen to treat 12 eyes of 10 patients with endogenous *Aspergillus* endophthalmitis, final visual acuities of 20/25 to 20/200 were attained in three eyes without central macular involvement; in eight eyes with initial central macular involvement, final visual acuities were 20/400 in three eyes and 5/200 or less in four eyes. Two painful eyes with marked inflammation, hypotony, and retinal detachment were enucleated (Weishaar et al. 1998).

Lance et al. (1988) reported a distinctive case of endogenous endophthalmitis due to *Aspergillus flavus* in an intravenous drug abuser in which the patient recovered useful vision after resolution of the endophthalmitis, which did not recur in spite of extensive retinal and vitreous involvement, since vitrectomy and intravitreal injection of amphotericin B were performed within 48 hours of presentation, followed by intravenous adminis-

tration of amphotericin B several days later. Unfortunately, others have reported less favorable results. Coskuncan et al. (1994) reported on two cases of *Aspergillus flavus* retinitis (proven by recovery of the fungi in culture of vitreous samples) in patients who had undergone bone marrow transplantation 120 days before; the eyes responded poorly to antifungals. Tabbara and al-Jabarti (1998) reported an outbreak of culture-proven exogenous endophthalmitis due to *Aspergillus fumigatus* in five patients who had undergone cataract extraction during hospital construction; severe postoperative uveitis occurred in all five patients 4–15 days postsurgery and failed to subside with topical corticosteroid therapy. One patient received only antifungals (intravitreal and subconjunctival miconazole, intravenous and intravitreal amphotericin B), two patients had pars plana vitrectomy and intravitreal amphotericin B, while one patient had only pars plana vitrectomy. Unfortunately, evisceration or enucleation had to be ultimately performed for all five patients. Brar et al. (2002) reported the occurrence of *Aspergillus niger* endophthalmitis 9 weeks after cataract surgery, with extension to the cornea; the outcome of the infection was unknown since the patients did not return for follow-up evaluation.

DEMATIACEOUS FUNGI

Fungal endophthalmitis has been reported to be caused by dematiaceous fungi such as *Alternaria alternata* (Rummelt et al. 1991), *Curvularia lunata* (Kaushik et al. 2001), *Exophiala jeanselmei* var. *jeanselmei* (Hofling-Lima et al. 1999) following cataract surgery, and by *Lasiodiplodia theobromae* (Borderie et al. 1997) and *Microsphaeropsis olivacea* (Shah et al. 2001) following keratitis. The prognosis in such infections is unsatisfactory. In the cases reported by Kaushik et al. (2001) and Shah et al. (2001), the intraocular part of the infection resolved quickly after intraocular debridement, and intravitreal, topical, and systemic antifungals, but the corneal infection persisted.

Hofling-Lima et al. (1999) reported on two cases of late endophthalmitis caused by *Exophiala jeanselmei* var. *jeanselmei* after cataract surgery. Although intravitreal and anterior chamber amphotericin B were used in both cases, with initial resolution of the lesions, recurrence occurred in both instances, leading to endophthalmitis.

SCEDOSPORIUM APIOSPERMUM (PSEUDALLESCHERIA BOYDII)

Endogenous endophthalmitis due to *Scedosporium apiospermum* (teleomorph *Pseudallescheria boydii*) also appears to carry a poor prognosis. McGuire et al. (1991) reviewed the clinical and other features of 17 patients with ocular infections due to this fungus reported up to 1991. Numerous predisposing factors in these patients included trauma (penetrating injury, burn, or foreign body), a chronic underlying disease (diabetes mellitus or systemic lupus erythematosus), drowning with aspiration, immunosuppression (for renal transplantation), multiple surgical procedures, and cataract surgery. The clinical findings of the 17 infected eyes included ten endophthalmitides, five corneal ulcers, and two orbital cellulitides. *Scedosporium apiospermum* (reported as *Pseudallescheria boydii*)was cultured in 17 of the 17 reported cases; ocular histopathological descriptions reported hyphae and/or conidia in 13 of the 17 cases. Therapeutic surgical procedures included enucleation for five of the 17 eyes, with one instance each of lysis of adhesions and resection of a mycetoma, orbital debridement, evisceration, lensectomy, and keratoplasty; one patient required a sclerotomy, iridectomy, lensectomy, and synechiectomy; two patients underwent a vitrectomy; and two underwent orbital debridement. Antifungal agents were used in 14 of the 17 patients; however, the etiologic diagnosis was delayed in all 17 patients. Seven patients were treated with only amphotericin B; one required enucleation to eradicate the infection, but there were no deaths in this treatment group. Four patients were treated with amphotericin B initially, then treated with miconazole; two underwent enucleation and two died. One patient had partial treatment with amphotericin B, then miconazole, and finally fluconazole; this patient underwent enucleation before death. In this series of ocular infections, four patients (all disseminated disease) died. In an experimental study by the same group of workers, the fungus was found to produce a severe endogenous endophthalmitis in nine of 17 immunosuppressed rabbits following its injection into the carotid artery; three instances of lens/lens capsule destruction were seen in this group (McGuire et al. 1991). These authors concluded that ocular infections due to *Pseudallescheria boydii* (*Scedosporium apiospermum*) mimic both clinically and histologically those caused by *Aspergillus fumigatus*. More recent reports confirm the severity of endogenous endophthalmitis due to *Scedosporium apiospermum*. McKelvie et al. (2001) reported about two patients with acute myeloid leukemia and neutropenia who developed endophthalmitis due to *Scedosporium apiospermum* and who ultimately succumbed to overwhelming fungal septicemia in spite of treatment with amphotericin B alone in one patient, and combined amphotericin B and fluconazole in the other. Luu et al. (2001) reported the development of endophthalmitis due to this fungus (in spite of concurrent intravenous amphotericin B therapy) in a patient who had undergone liver transplantation, and who had a ring-enhancing brain lesion; the inflammation progressed in spite of pars plana vitrectomy and intravitreal amphotericin B, and the patient died of complications due to brain abscess.

COCCIDIOIDES IMMITIS

Intraocular coccidioidomycosis, which may occur in otherwise healthy individuals, usually presents with multiple, yellow–white, juxtapapillary chorioretinal lesions with pigmented borders, but retinal exudates or serous retinal detachment may also occur (Rodenbiker and Ganley 1980). Fluconazole appears to be useful in the treatment of this condition. Luttrull et al. (1995) reported successful treatment of bilateral multifocal choroiditis and unilateral granulomatous iridocyclitis due to *Coccidioides immitis* in a patient who received 600 mg fluconazole daily. Cunningham et al. (1998) described two patients with intraocular coccidioidomycosis, one with a unilateral, granulomatous iridocyclitis with multiple iris nodules and a large vascularized anterior chamber, and the other with papilledema and multifocal choroiditis, in whom the diagnosis was established by detection of *Coccidioides immitis* spherules in skin biopsy samples; the patients were successfully treated with local and systemic amphotericin B and oral fluconazole.

BLASTOMYCES DERMATITIDIS

Intraocular blastomycosis is an unusual occurrence. Choroidal involvement (manifesting as bilateral, multiple, yellow–white posterior fundus lesions) has been reported to occur in association with miliary or disseminated blastomycosis (Sinskey and Anderson 1955; Lewis et al. 1988), while isolated endophthalmitis (Safneck et al. 1990) and panophthalmitis (Font et al. 1967) due to *Blastomyces dermatitidis*, without any systemic manifestations, has also been reported. In a case described by Bond et al. (1982), a patient with biopsy-proven sarcoidosis developed fever, skin lesions and anterior uveitis, vitritis, and a large posterior yellow, elevated choroidal lesion in one eye and three white posterior small choroidal lesions in the other eye. Although a diagnosis of blastomycosis was established by the biopsy of the skin lesion, it was unclear whether the ocular manifestations in this patient were due to blastomycosis or sarcoidosis. Li et al. (1998) reported a patient with concurrent unilateral endophthalmitis and orbital cellulitis proven by histological findings and by culture to be due to *Blastomyces dermatitidis*. Amphotericin B therapy was shown to produce a rapid resolution of choroidal and extraocular lesions in one patient (Lewis et al. 1988), without any recurrence over a 6-month follow-up period. However, in general, the therapy of intraocular blastomycosis appears to be unrewarding, with enucleation of the affected eye having to be performed to control the infection in most instances (Sinskey and Anderson 1955; Font et al. 1967; Safneck et al. 1990; Li et al. 1998). Intraocular dissemination of blastomycosis should be suspected in the differential diagnosis of endophthalmitis in patients with previous or active pulmonary lesions of equivocal nature. Early diagnosis and prompt treatment with antifungal medications are essential.

CRYPTOCOCCUS NEOFORMANS

Intraocular cryptococcosis usually results from cryptococcal septicemia with severe meningeal infection; such sequelae may be seen in patients with AIDS. Cohen and Glasgow (1993) reported the occurrence of sudden, simultaneously bilateral blindness in a patient with AIDS who had cryptococcal meningitis; although the sudden, bilateral visual loss in this patient was ultimately found to have been caused by focal but fulminant necrosis of both optic nerves, the authors believed that the presence of cryptococcal organisms throughout the basal meninges and in the sheaths of both optic nerves suggests that cryptococcosis may produce visual loss by damaging multiple areas of the anterior visual pathway. Agarwal et al. (1991) reported the occurrence of retinitis (presumably due to cryptococcal infection) following disseminated cryptococcosis in a patient who had received a renal allograft. However, Hester et al. (1992) described isolated ocular cryptococcosis in a 62-year-old female immunocompetent patient. Hence, ocular crypto-coccal infection must be suspected, even in the absence of predisposing factors or systemic findings. The diag-nosis of cryptococcal chorioretinitis is a presumptive one in a patient with characteristic fundus lesions, with or without vitritis, and documented cryptococcal meningitis or disseminated cryptococcosis (Klotz et al. 2000). In one patient, transscleral needle biopsy of a subretinal mass was used to establish the diagnosis of subretinal cryptococcosis (Hester et al. 1992). A vitreous tap or biopsy may be done if vitritis is present. Cryptococcal chorioretinitis can be treated with intravenous ampho-tericin B (Hester et al. 1992) or oral fluconazole (Agarwal et al. 1991). Vitrectomy may be needed if chorioretinitis progresses to endophthalmitis.

PARACOCCIDIOIDES BRASILIENSIS

Ocular paracoccidioidomycosis usually manifests as lesions in the eyelid and conjunctiva (Silva et al. 1988). Cases of choroidal granuloma and endophthalmitis have been described, but few have been histopathologically proven (Bonomo et al. 1982). Finamor et al. (2002) recently described the occurrence of ocular and central nervous system paracoccidioidomycosis in a pregnant woman with AIDS. The patient presented with an acute and painful eye, and ophthalmological examination revealed unilateral severe iridocyclitis, severe vitritis, and a granulomatous chorioretinal lesion adjacent to and involving the optic nerve; the lesions evolved after a week to iris neovascularization, neovascular glaucoma, and retinal detachment. The eye was enucleated due to intense pain. Macroscopic study of the enucleated eye showed a total retinal detachment with subretinal exudate and a white mass in the vitreous cavity dis-placing the lens to the anterior chamber. Histopatholo-gical examination of the enucleated eye, as well as a biopsy of an oral lesion, revealed a chronic granuloma-tous inflammation with epithelioid cells, multinucleated giant cells, and several round yeasts exhibiting 'ship's wheel' external budding in sections stained by periodic

acid–Schiff and Grocott methenamine silver stains, an appearance typical of Paracoccidioides brasiliensis. With the diagnosis of disseminated ocular para-coccidioidomycosis, the patient was treated with trime-thoprim-sulfamethoxazole with a satisfactory outcome and reduction in the size of the brain lesion (Finamor et al. 2002).

HISTOPLASMA CAPSULATUM

The 'presumed ocular histoplasmosis syndrome' is char-acterized by the presence of multifocal choroiditis scat-tered throughout the fundus, the peripapillary area, and sometimes in the macular area, with some lesions exhi-biting healing with variable chorioretinal scarring (Katz et al. 1997). This syndrome is not associated with intra-ocular inflammation, and is well tolerated by the eye, unless complications of subretinal neovascularization, such as progressive visual loss, arise (Gonzales et al. 2000). In such instances, vitreoretinal surgical techniques have been used to remove the subfoveal neovascular membranes while preserving the overlying neurosensory retina, and thus preserving central visual acuity (Thomas and Kaplan 1991). Histoplasma capsulatum var. capsu-latum has been isolated from the eyes of patients suffering from this syndrome, suggesting that this fungus is the etiologic agent (Katz et al. 1997), although this is a contentious point. However, Histoplasma capsulatum var. capsulatum has been implicated as the etiologic agent in several other intraocular lesions. Specht et al. (1991) reported on ocular histoplasmosis with retinitis in a patient with AIDS and disseminated histoplasmosis, who had complained of a hazy spot in the vision of his left eye; distinct creamy-white intraretinal and subretinal infiltrates were visualized in both eyes. After the patient succumbed to his illness, light microscopy of his eyes removed at autopsy revealed that the most prominent colonization of ocular tissue by yeasts was in the retinae and optic nerves, with focal choroiditis near the retinal lesions. These lesions showed variable numbers of lymphocytes and histiocytes, and yeasts with a morphology suggestive of Histoplasma capsulatum var. capsulatum, free or phagocytized within cells, in all layers of the retinal lesions, sometimes with extension to the subretinal and subhyaloid spaces. One eye also exhibited necrotizing granulomatous cyclitis with aggre-gates of yeasts. Histoplasmic optic neuritis was reported by both Macher et al. (1985) and Specht et al. (1991), while histoplasmic cyclitis was also noted by Specht et al. (1991). Endophthalmitis due to Histoplasma capsulatum var. capsulatum in a patient with AIDS has also been reported (Gonzales et al. 2000).

SPOROTHRIX SCHENCKII

Risk factors for endophthalmitis due to Sporothrix schenckii include AIDS (Kurosawa et al. 1988) and trauma (Witherspoon et al. 1990); however, this ocular infection may occur even without trauma or systemic infection (Cartwright et al. 1993). Vieira-Dias et al. (1997) reported the occurrence of concomitant ocular and cutaneous sporotrichosis, where the fungus was isolated from skin lesions and the aqueous humor. Endophthalmitis due to *Sporothrix schenckii* usually presents initially as a granulomatous uveitis, which may be treated with corticosteroids, therein leading to progression of the lesion and resulting in frank endophthalmitis (Kurosawa et al. 1988) or scleral perforation (Cartwright et al. 1993). Up to 1990, 17 episodes of endophthalmitis due to *Sporothrix schenckii* had been reported (Witherspoon et al. 1990). Curi et al. (2003) described the occurrence of fluffy vitreous opacities and a retinal granuloma in a young man who presented with disseminated ulcerated skin lesions that were proven by biopsy and by culture to be due to *Sporothrix schenckii*. It is important to remember that although intraocular infection due to *Sporothrix schenckii* is uncommon, it can occur in cases of disseminated sporotrichosis. Systemic therapy with potassium iodide or itraconazole is a successful means to control skin and ocular sporotrichosis.

ACREMONIUM SPECIES

Exogenous endophthalmitis due to *Acremonium* spp. may arise due to contamination from a common environmental source. Weissgold et al. (1996) described four patients who presented after cataract surgery with delayed-onset endophthalmitis caused by *Acremonium kiliense* with in vitro sensitivity to amphotericin B. In all patients, ocular infection was recalcitrant to single-dose intravitreous amphotericin B injection (5 μg) and systemic administration of fluconazole. These patients subsequently responded to vitrectomy followed by additional intravitreous amphotericin B injection. These authors concluded that single-dose administration of intravitreous amphotericin B may be inadequate treatment for fungal endophthalmitis caused by *Acremonium kiliense*, and that vitrectomy with repeated intravitreous administration of amphotericin B may be necessary to eradicate intraocular infection caused by this organism.

ZYGOMYCETES

Orgel and Cohen (1989) described the first case of postoperative zygomycete endophthalmitis that occurred after uncomplicated phacoemulsification and the insertion of a posterior chamber intraocular lens. Diagnosis, in the presence of negative aqueous and vitreous cultures, was confirmed by immunofluorescence staining of an anterior chamber inflammatory mass. Successful treatment of the eye with 20 μg intraocular amphotericin B may have been made possible in part by the fact that the posterior capsule remained intact, keeping the eye bicompartmental. The evolution of fundus changes from edema of the optic disk and peripapillary retina to optic

atrophy and chorioretinal pigmentary changes has been rarely reported in rhino-orbital zygomycosis (Newman and Kline 1997; Kadayifcilar et al. 2001). Unique chorioretinal changes caused by choroidal ischemia secondary to rhino-orbital zygomycosis have also been reported (Kadayifcilar et al. 2001).

FUSARIUM SPECIES

Intraocular lesions due to *Fusarium* spp. may occur in patients with leukemia or as a consequence of the progression of corneal lesions. Patel et al. (1994) reported on a patient with acute lymphocytic leukemia with both endogenous *Fusarium* endophthalmitis and leukemic ocular infiltrates who presented with sudden visual loss in one eye, which exhibited a dense white placoid infiltrate in the macula extending into the vitreous, and with an iris nodule and hypopyon in the other eye. Histopathological studies performed after the patient's death from multisystem failure disclosed panophthalmitis with *Fusarium* invading all the ocular coats in the right eye, while in the left eye there were leukemic infiltrates in the iris, anterior chamber, and trabecular meshwork. These authors postulated that the ocular destructiveness of *Fusarium* spp. may be caused by marked mycotic vascular invasion and occlusion with consequent infarction and necrosis of ocular tissues. Dursun et al. (2003) reported that 10 out of a total of 159 patients with keratitis due to *Fusarium* spp. (seven due to *Fusarium oxysporum*, two to *Fusarium solani*, and one to a *Fusarium* sp.) progressed to endophthalmitis. Cultures of aqueous and intraocular tissues grew *Fusarium* in eight cases, whereas vitreous cultures were positive in two. Nine cases had preexisting risk factors. All patients received oral ketoconazole or fluconazole and topical natamycin 5 percent, and intravitreal amphotericin B injections were also given in two patients. Four patients required a penetrating keratoplasty, enucleation was performed in two patients, two patients required a combination of a penetrating keratoplasty and pars plana vitrectomy, and one patient developed phthisis bulbi.

PAECILOMYCES LILACINUS

Paecilomyces lilacinus is an uncommon cause of exogenous endophthalmitis. However, this form of endophthalmitis has generally been associated with poor visual outcome. In a series of 13 cases of *Paecilomyces lilacinus* endophthalmitis following implantation of contaminated intraocular lenses, eight eyes were enucleated, two had a final vision of no light perception, and one had only light perception; retention of useful vision in the remaining two eyes was attributed to early removal of the lens and inflammatory material in one case and prompt use of intraocular miconazole in the other (Pettit et al. 1980). In another series of patients with postoperative *Paecilomyces lilacinus* endophthalmitis, enucleation was performed in two of the three cases; the salvaged eye was treated with miconazole early in the course of the disease (Kozarsky et al. 1984). In a case report of successful treatment of *Paecilomyces lilacinus* endophthalmitis following cataract extraction, treatment consisted of early vitrectomy, multiple intravitreal injections of amphotericin B and miconazole, intravenous miconazole, and later, oral ketoconazole; the intraocular lens removed 6 weeks after initiation of therapy was found to contain *Paecilomyces lilacinus*. After 8 months, excellent recovery of visual acuity was noted (Levin et al. 1987). Current management of ocular infections due to *Paecilomyces lilacinus* continues to be challenging because of the organism's resistance to amphotericin B, natamycin, and, in some cases, fluconazole (Aguilar et al. 1998; Sutton et al. 1998) and because miconazole and ketoconazole, agents to which *Paecilomyces lilacinus* is usually susceptible, are not usually included in the initial management of suspected fungal endophthalmitis (Scott et al. 2002). The sensitivity profile of *Paecilomyces lilacinus* also differs significantly from that of *Paecilomyces variotii* (Sutton et al. 1998). Therefore, initial treatment with miconazole (intravitreal, 25 µg/0.1 ml) has been suggested as a good option since, to date, miconazole is the only known medication to which *Paecilomyces lilacinus* and *Paecilomyces variotii* are consistently sensitive. Miconazole is currently available in powder form, which can be used to prepare topical and intravitreal preparations. If, however, miconazole is not readily available, a combination of intravitreal amphotericin B and oral ketoconazole (200–400 mg four times daily) has been suggested since *Paecilomyces variotii* is sensitive to amphotericin B and *Paecilomyces lilacinus* is sensitive to ketoconazole (Scott et al. 2002). However, while intravitreal amphotericin B is readily available from most pharmacies, ketoconazole is available only in an oral preparation. Intravitreal administration of fluconazole may be useful since approximately 10 percent of *Paecilomyces lilacinus* organisms are sensitive to fluconazole, which can be prepared in an intravitreal form. In the absence of miconazole, oral ketoconazole and intravitreal fluconazole may be considered (Scott et al. 2002).

PNEUMOCYSTIS CARINII

Patients with AIDS are at risk to develop pulmonary disease due to *Pneumocystis carinii*. Although aerosolized pentamidine may be given as prophylaxis in such patients, this does not protect against extrapulmonary disease. Dugel et al. (1990) and Foster et al. (1991) described the occurrence of choroidal lesions which appeared to be typical of *Pneumocystis carinii* in two and three patients, respectively, who were receiving prophylactic aerosolized pentamidine therapy. The lesions resolved with intravenous pentamidine therapy in four of the patients, while the lesions in the remaining

patient resolved with intravenous trimethoprim and sulfamethoxazole. All these patients did not have clinical or laboratory evidence of *Pneumocystis carinii* infection other than in the eye. The choroidal lesions of *Pneumocystis carinii* manifest as yellow–white to orange spots without vitreous inflammation (Morinelli et al. 1993). Early ophthalmological examination may detect these lesions before they are threatening to sight, and allows systemic therapy to be instituted before widely disseminated infection due to *Pneumocystis carinii* results in a fatal outcome. Intravenous pentamidine therapy appears to control *Pneumocystis carinii* chorioditis but may need to be continued to prevent exacerbation. Presumed *Pneumocystis carinii* choroiditis may serve as a marker for disseminated infection.

FUNGAL OCULAR INFECTIONS AFTER OPHTHALMIC SURGICAL PROCEDURES

Certain surgical procedures are unique to ophthalmology. The most important ophthalmic surgical procedure is cataract extraction, and the most important fungal infection following cataract extraction is fungal endophthalmitis (see earlier under Intraocular mycoses). Fungal infections, such as keratitis and scleritis, have been reported following corneal refractive surgical procedures [radial keratotomy, photorefractive keratectomy, laser-in-situ keratomileusis (LASIK)], keratoplasty (corneal transplantation), pterygium excision, and cataract extraction.

Postoperative infectious keratitis is an uncommon, but serious, complication of radial keratotomy; the use of topical corticosteroids and the presence of corneal incisions are probably risk factors. There have been at least five reported cases of mycotic keratitis following radial keratotomy, two each due to species of *Fusarium* and *Aspergillus*, and one due to *Candida parapsilosis* (Maskin and Alfonso 1992; Gussler et al. 1995; Heidemann et al. 1995; Panda et al. 1998). Three of these responded to antifungals alone; the other two required therapeutic penetrating keratoplasty, wherein the lesions resolved.

Since LASIK combines the precision of excimer laser photoablation with the advantages of an intrastromal procedure that maintains the integrity of Bowman's layer and the overlying corneal epithelium, the risk of infectious keratitis following this procedure should be, theoretically speaking, minimal. However, microbial contamination of the corneal stromal bed may occur during surgery due to the proximity of the eyelids, eyelashes, conjunctiva, and microkeratome. This, combined with the use of topical corticosteroids, unstable epithelium at the edge of the lamellar flap, reduced corneal sensitivity and use of contact lenses, all render these eyes susceptible to infection. Read et al. (2000) reported the first case of mycotic keratitis due to *Acremonium atrogriseum* following LASIK; since then, there have been reports of at least five other patients afflicted with this condition, in which six different species of fungi from five genera (*Aspergillus flavus*, *Aspergillus fumigatus*, *Curvularia* spp., *Fusarium solani*, *Scedosporium apiospermum*) have been implicated (Chung et al. 2000; Sridhar et al. 2000a, b; Kuo et al. 2001; Verma and Tuft 2002). Only two of these patients responded to medical therapy alone, and therapeutic penetrating keratoplasty was ultimately required for the other four. To achieve a favorable outcome in such infections, the fungus involved should first be identified as soon as possible; sampling at the site of infection (from underneath the flap) provides the best chance of obtaining a positive culture. This is followed by irrigation of the stromal bed with antifungal agents, and intensive treatment with specific systemic antifungals.

Mycotic keratitis has been reported to occur following superficial (lamellar) (Panda et al. 1999) or deep (penetrating) keratoplasty (Levenson et al. 1984; Wilhelmus and Robinson 1991; Ainbinder et al. 1998; Benaoudia et al. 1999), following repair of a dehiscent keratoplasty wound (Kisla et al. 2000), and secondary to endophthalmitis occurring after phacoemulsification and intraocular lens implantation surgery (Fekrat et al. 1995; Weissgold et al. 1998). The nine patients involved required surgery (therapeutic penetrating keratoplasty in seven, optical keratoplasty in one, and debridement in one). Four different yeast species (*Candida albicans*, *Candida guilliermondii*, *Candida parapsilosis*, and a *Rhodotorula* sp.) were isolated from the lesions of four patients, and four different species of filamentous fungi (*Exophiala dermatitidis* in two patients, *Acremonium kiliense*, *Beauveria bassiana*, and a presumed *Fusarium* sp. in one patient each) from the others.

Fungal scleritis has been reported to occur in at least 13 patients following various ophthalmic surgical procedures, including excision of pterygia (fleshy conjunctival growths) without or with β-irradiation (Margo et al. 1988; Moriarty et al. 1993; Sullivan et al. 1994; Kumar et al. 1997; Taravella et al. 1997), trabeculectomy (filtering surgery for glaucoma), and cataract extraction (Carlson et al. 1992; Bernauer et al. 1998; Locher et al. 1998). Various modalities of antifungal therapy, as well as surgical debridement, were not found useful in five of these 13 patients, with enucleation eventually having to be performed; *Scedosporium apiospermum* (*Pseudallescheria boydii*) was identified as the cause in two of these patients, *Scedosporium prolificans* in one, and an *Aspergillus* sp. in one. The identification of a *Rhizopus* sp. in the remaining patient is fraught with uncertainty since fungal hyphae were not visualized in the samples collected, and just one colony of the fungus was recovered in culture (Locher et al. 1998). Eight of 13 patients required some form of surgical intervention such as scleral debridement or resection, or removal of plaque, to ensure resolution of the infection; the fungi involved in these eight patients were *Aspergillus flavus* and *Aspergillus* spp. (in five patients), and *Scedosporium*

apiospermum, *Scedosporium prolificans*, and a *Fusarium* sp. (in one patient each) (Carlson et al. 1992; Moriarty et al. 1993; Kumar et al. 1997; Bernauer et al. 1998).

Fungal infection of the self-sealing incision used for cataract surgery was reported in seven patients from different locales in India (Garg et al. 2003). The initial diagnoses at the time of onset of symptoms were keratitis, scleritis, and excessive anterior chamber reaction, but all cases subsequently developed deep keratitis. The diagnosis of fungal infection was established by microscopy and culture of corneal scrapings, corneoscleral biopsies, and anterior chamber aspirates, and the fungi isolated were species of *Aspergillus* in six patients and *Candida albicans* in one patient. The infection resolved with good visual recovery by medical therapy alone in two patients, but progressed to endophthalmitis and complete visual loss in five eyes.

OPHTHALMIC MYCOSES ASSOCIATED WITH AIDS

Infections by opportunistic microorganisms constitute an important ocular manifestation of AIDS, although ocular findings are infrequent in human immunodeficiency virus (HIV)-infected, asymptomatic individuals. Cytomegalovirus retinitis is reported to be the most common intraocular infection in AIDS patients while other opportunistic ocular infections are considerably less common (Jabs et al. 1989; Seal et al. 1998). Various types of ophthalmic mycoses, principally affecting the orbit and intraocular structures, have been reported in association with AIDS.

Frequency

Autopsy findings in 25 patients who died of AIDS revealed opportunistic ocular infections in eight patients, including retinitis due to a *Candida* spp. in one patient and choroiditis due to *Cryptococcus neoformans* var. *neoformans* in one patient (Jabs et al. 1989). Earlier, Schuman and Friedman (1983) had reported the occurrence of retinitis due to *Candida albicans* and *Cryptococcus neoformans* in two of 34 patients with AIDS; bacterial corneal ulceration was noted in 2 percent, but fungal corneal ulceration was not noted at all.

Orbital infections

Opportunistic infections of the orbit due to bacteria, fungi, and protozoa constitute serious complications of HIV infection and frequently lead to ocular morbidity and mortality. Orbital mycoses associated with AIDS have been reported in 15 patients since 1991 (Blatt et al. 1991; Friedberg et al. 1992; Vitale et al. 1992; Meyer et al. 1994; Kronish et al. 1996; Lee et al. 1996; Johnson et al. 1999; Hejny et al. 2001). The orbital mycotic infection resolved or improved in just six of these patients. Surprisingly, there was resolution or improvement following debridement and intravenous amphotericin B therapy in two of the three patients with rhino-orbital zygomycosis, perhaps because these had relatively focal lesions (Blatt et al. 1991; Hejny et al. 2001). There was complete resolution of lesions in the one patient with *Pneumocystis carinii* infection after treatment with trimethoprim and sulfamethoxazole (Friedberg et al. 1992). Eleven patients had infections due to *Aspergillus fumigatus*, and 10 of these were treated with surgery and intravenous amphotericin B. The patient survived and the mycotic infection resolved in only three of these patients who appeared to have relatively focal orbital lesions without intracranial disease; two of the survivors received surgery and amphotericin B (intravenous and local irrigation), while the third was treated by debridement, amphotericin B lipid complex, liposomal amphotericin B, and orbital exenteration (Vitale et al. 1992; Kronish et al. 1996; Johnson et al. 1999). Overall, it appears that orbital aspergillosis in patients with AIDS has a poor outcome.

Optic neuritis

Although there are many causes of optic neutritis in AIDS patients, a fungal etiology has been reported in two patients, the pathogen being *Histoplasma capsulatum* var. *capsulatum* (Specht et al. 1991; Yau et al. 1996); in one of these, the optic neuritis was associated with retinitis and uveitis.

Corneal infections

Fungal corneal infections appear to be infrequent in HIV-infected patients, but tend to be severe and prone to corneal perforation when they do occur and to manifest even when known risk factors for ulcerative keratitis are absent (Hemady 1995). There have been reports of six patients with mycotic keratitis associated with AIDS (in one patient, the diagnosis of AIDS was made postmortem); *Candida albicans* was the fungus implicated in all six patients (Parrish et al. 1987; Hemady 1995). The keratitis resolved in all six with topical amphotericin B (0.15 percent) therapy.

Other mycotic infections of the anterior segment reported in AIDS include limbal nodules (and multifocal choroiditis) in one patient (Muccioli et al. 1995) and an iris inflammatory mass in another (Charles et al. 1992); histopathological studies permitted a presumptive diagnosis of *Cryptococcus neformans* var. *neoformans* infection.

Presumed mycoses of the posterior segment in patients with AIDS include multifocal choroiditis (choroidopathy) and retinitis due to cryptococcosis, histoplasmosis, candidiasis, and *Pneumocystis carinii* infection

(Macher et al. 1985; Jabs et al. 1989; Shami et al. 1991; Morinelli et al. 1993; Muccioli et al. 1995; Verma and Graham 1995). The choroidal lesions of *Pneumocystis carinii* manifest as yellow–white to orange spots without vitreous inflammation (Morinelli et al. 1993). Culture-proven endogenous endophthalmitis due to *Bipolaris hawaiiensis*, a *Fusarium* sp., *Sporothrix schenckii*, and *Histoplasma capsulatum* var. *capsulatum* has also been reported (Kurosawa et al. 1988; Pavan and Margo 1993; Glasgow et al. 1996; Gonzales et al. 2000); complete resolution of lesions was achieved by surgery and amphotericin B and fluconazole therapy only in the patient with the *Bipolaris* infection. Although central nervous system infection with *Cryptococcus neoformans* var. *neoformans* is common in patients with AIDS, actual invasion of the intraocular structures by the fungus appears relatively uncommon. In one study of 80 HIV-seropositive patients with cryptococcal infections, papilledema, visual loss, abducens nerve palsy, and optic atrophy were the important ophthalmic aberrations noted; however, actual invasion of the intraocular structures was an uncommon complication (Kestelyn et al. 1993).

OPHTHALMIC MYCOSES ASSOCIATED WITH OCULAR BIOMATERIALS

Microbial colonization of indwelling and implanted biomedical devices constructed of polymers, silicones, and metals, such as shunts or catheters, can lead to serious, often lethal, infection (Wilson 1996). The organisms responsible for such biomaterial-related infections are usually part of the resident microbial flora at a particular area of the body, therein posing a constant threat. Factors believed to contribute to the mechanisms of infection associated with biomedical devices include:

- intraoperative contamination during surgical implantation or extraluminal migration of organisms, which permits potential pathogens to transcend normal protective barriers
- production of mucoid substances by microorganisms, which facilitates adhesion of colonizing microorganisms, and also protects them from various host defense mechanisms
- the presence of plasma proteins (especially fibronectin) on the surface of the biopolymer, which may promote attachment of staphylococci and *Candida* spp. to the surface (Wilson 1996).

Contact lens plastics and their storage cases, and intraocular lens implants, are the two most important categories of biomaterials used in ophthalmology.

Fungal contamination of intraocular lenses

The occurrence of postoperative inflammation following implantation of an intraocular lens depends on various factors, such as the type and surface quality of the lens, the fixation of the posterior chamber lens, the surgical technique adopted, the use of topical corticosteroids, systemic antibacterials, and nonsteroidal anti-inflammatory drugs as a prophylactic measure, as well as the experience of the surgeon. The use of polymethylmethacrylate single-block lenses in eyes with uveitis is advantageous since the complement cascade is not activated, while surface-modified lenses, such as heparin-coated models, may confer some protection against reactivation of inflammation (Biswas and Kumar 2002). Although improvements in the quality of the lenses and microsurgical techniques have led to a decrease in the frequency of complications from intraocular lenses, lenses that have been implanted intraocularly may still need to be explanted for various reasons. In a cytopathological study of five intraocular lenses that had to be removed due to endophthalmitis, fibrinous material was noted on three, granulomatous inflammation on one, and nongranulomatous inflammation with fungal filaments on the surface on one (Biswas and Kumar 2002); thus, careful examination of the intraocular lens surface in such cases would help in establishing a diagnosis of fungal endophthalmitis (Figure 16.18). In instances where the explantation of a contaminated intraocular lens is contraindicated, drug therapy may be the only alternative. Penk and Pittrow (1999) recently reviewed the role of fluconazole in the long-term suppression of fungal infections (mainly due to *Candida* spp.) in patients with artificial implants; their review included intraocular lenses in nine patients. These authors concluded that in such cases, fluconazole exhibited an excellent safety profile and therapeutic efficacy in doses of 750 mg/day; however, they cautioned that higher doses and lifelong treatment might be indicated in certain instances.

An outbreak of *Candida parapsilosis* endophthalmitis was reported in 13 patients from three states of the USA who had had an intraocular lens implantation between November 1983 and January 1984 (McCray et al. 1986). This outbreak was traced to the introduction in July 1983 of a new brand of BSS as an intraoperative ophthalmic irrigation solution that was subsequently recalled because of intrinsic fungal contamination. A retrospective cohort study revealed that definite exposure to the product was a significant risk factor for *Candida parapsilosis* endophthalmitis, while a retrospective case–control study, including 203 control patients with definite exposure to BSS, suggested that exposure to systemic steroids was an additional risk factor for this type of infection. All 13 patients received topical, intraocular, or systemic antifungal therapy (or a combination of these modalities), and 10 also underwent vitrectomy. No patients had systemic symptoms or complete visual loss. Laboratory investigations showed a 6.7 percent overall contamination of the product with *Candida parapsilosis*. The problem did not recur following recall of the product by the manufacturer.

Thirteen cases of fungal endophthalmitis caused by *Paecilomyces lilacinus* were attributed to implantation of intraocular lenses that had been sterilized in sodium hydroxide and neutralized in sodium bicarbonate (Pettit et al. 1980). Twelve of the 13 patients had received neutralizing solution from one particular lot, and the same fungus that caused the endophthalmitis was cultured from several of the neutralizing solutions from this particular lot. Eight of the 13 eyes eventually required enucleation. Of the remaining five eyes, only two recovered useful vision. This surgically induced epidemic of fungal endophthalmitis clearly shows the major consequences of a breakdown in quality control for any substance or material used intraocularly. Tabbara and al-Jabarti (1998) reported on the occurrence of exogenous endophthalmitis due to *Aspergillus fumigatus* in five patients who had undergone extracapsular cataract surgery with posterior chamber intraocular lens implantation. In these patients, however, the intraocular lens was not at fault but, rather, ongoing hospital construction work.

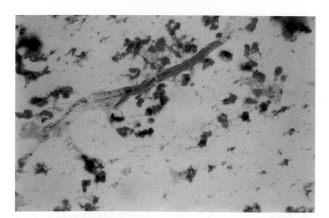

Figure 16.18 *Photomicrograph of fungal hyphae* (Aspergillus fumigatus) *on an intraocular lens that was removed due to endophthalmitis. Lactophenol cotton blue stain; magnification ×400*

Fungal contamination of contact lenses

Contact lens-associated mycotic keratitis appears to be relatively uncommon, when compared to contact lens-associated bacterial or *Acanthamoeba* keratitis, because fungi isolated from the healthy outer eye only transiently colonize this area, and are not normally resident in the outer eye (Srinivasan et al. 1991; Liesegang 1997). When soft lenses are worn continuously, fungal conidia adhere to the lens surface and, under favorable conditions, germinate; fungal hyphae are able to enter the matrix of the soft lens, project through the posterior surface and then penetrate the corneal epithelium, resulting in fungal infection (Simmons et al. 1986). Factors that possibly contribute to such infiltration include fungal enzymatic activity and properties of the lens material, which provide a matrix as well as a source of nutrients for growth of the fungi (Yamamoto et al. 1979). Fungi have been found to penetrate the matrix of extended-wear soft contact lenses, both during normal use (Yamamoto et al. 1979; Berger and Streeten 1981; Wilson and Ahearn 1986) and in laboratory studies (Yamaguchi et al. 1984). When contact lenses with a water content of 45 or 73.5 percent were challenged with *Fusarium solani* or *Aspergillus flavus* on Sabouraud glucose neopeptone agar, it was found that the fungi penetrated both lenses but appeared to grow more vigorously into the lenses with the higher water content, resulting in physical and metabolic degradation of these lenses (Yamaguchi et al. 1984). In another study, it was thought that contamination of 11 extended-wear soft contact lenses by fungi occurred while the lenses were on the eye (Wilson and Ahearn 1986); in two cases, the same fungi isolated from the contaminated lenses were also obtained from corneal ulcers immediately under-lying the fungal contamination on the soft lens. Simmons et al. (1986) reported that filamentous fungi of the genera *Acremonium*, *Aspergillus*, *Alternaria*, *Clados-porium*, *Curvularia*, and *Fusarium* penetrated the matrix of soft contact lenses during normal use and in labora-tory studies. The species of *Acremonium*, *Aspergillus*, and *Fusarium* readily penetrated lenses having a water content of 55 percent or greater following incubation for 120–168 hours; when germinating on lenses, the conidia of these fungi appeared to be often associated with small salt deposits. Dematiaceous fungi appeared to penetrate intraocular lenses very rapidly. Conidia of *Cladosporium cladosporioides* were found to germinate on a new 55 percent water-content lens, either producing flattened appressorium-like cells on the lens surface or pene-trating the lens with a finely coiled and tapered hyphal element within the first 12 hours of incubation; conidia of *Curvularia lunata* were found to germinate and pene-trate into and through a new 55 percent water-content lens within 96 hours. Some hyphae adhered primarily to ridges or crevices or occasionally to small salt deposits on the lenses, but adherence to smooth surfaces was also observed. Growth of the fungal hyphae (which were coiled within the lens matrix) increased with increasing water content of the lens (Simmons et al. 1986). Based on these findings, these authors agreed with the general recommendation that lenses with fungal deposits be replaced, and also advised frequent and regular cleaning and disinfection of extended-wear lenses. They also suggested that activities in moldy habitats, such as raking leaves, cleaning dusty areas, and showering in rooms with moldy plastic curtains or tiles, be completely avoided while wearing extended-wear lenses, or that the lenses be cleaned and disinfected after such activities.

If fungal conidia alight on the surface of a contact lens, they are normally removed by surface cleaning of the lens. If such lenses are worn without proper cleaning for an extended duration, fungi may adhere, and pene-trate the contact lens (Wilson 1996). This may explain why soft lenses for aphakia and therapeutic extended wear have been most frequently implicated in such infections (Wilson and Ahearn 1986; Starr 1987; Wilhelmus et al. 1988).

Fungal infection was reported in four (4 percent) of 90 contact lens wearers, and in four (27 percent) of patients who wore therapeutic bandage-contact lenses (Wilhelmus et al. 1988). Filamentous fungi (*Aspergillus flavus*, *Fusarium dimerum*, *Fusarium* sp.) were also reported to have been isolated from the corneal scrapes of several other patients in South Florida who had contact lens-associated fungal keratitis (Liesegang and Forster 1980; Alfonso et al. 1986). Interestingly, Wilhelmus et al. (1988) observed that filamentous fungi are more likely to be associated with keratitis in persons wearing cosmetic or aphakic lenses, whereas yeasts were more frequently found with therapeutic lens use. Rosa et al. (1994) reported that six of their 125 patients with mycotic keratitis in south Florida wore extended-wear contact lenses; *Fusarium oxysporum* was isolated from four patients and *Candida albicans* and a *Paecilomyces* sp. from one patient each. In one patient who wore a bandage contact lens, keratitis due to *Candida para-psilosis* developed. Liesegang and Forster (1980) had earlier reported the occurrence of fungal keratitis in three patients who wore soft contact lenses, the fungi isolated being *Aspergillus flavus* and *Fusarium dimerum*. *Candida parapsilosis* keratitis, associated with contact lens wear, was reported in an elderly Israeli patient who developed stromal infiltration at the donor/recipient interface 2 years after penetrating keratoplasty, while wearing a 'piggyback' type of contact lens; the infection resolved with amphotericin B and flucytosine (Kremer et al. 1991). Perry et al. (1998) reported the occurrence of a conjunctival mass and keratoconjunctivitis in an immunocompetent patient; detailed examination revealed the presence of a soft contact lens, covered by mucoid material, at the posterior aspect of this mass, and the simple removal of the lens resulted in a resolu-tion of all signs and symptoms, with the contact lens growing *Aspergillus fumigatus*. Keratouveitis associated with the intraocular long-term retention of a contact lens was reported in a 76-year-old female patient (Arthur et al. 2001); a fungus was found to be involved, and this was identified as *Scedosporium prolificans*. However, this identification was found to be erroneous, and the isolate was later reidentified as *Acrophialophora fusis-pora* (Guarro and Gené 2002).

While the occurrence of contact lens-associated bacterial and *Acanthamoeba* keratitis has been corre-lated with the presence of bacteria and *Acanthamoeba* in contact lens cases (Seal et al. 1992; McLaughlin-Borlace et al. 1998), it is unclear whether such a rela-tionship occurs in contact lens-associated fungal kera-titis. However, Wilson et al. (1991) demonstrated the adherence of *Candida albicans* within a biofilm to poly-ethylene contact lens case plastic; this was found to be more resistant to the action of contact lens disinfectants than bacteria. A survey of contact lens cases from 101 asymptomatic daily-wear, cosmetic contact lens wearers in a domiciliary contact lens practice revealed contamination in 82 (81 percent) cases; 77 percent grew bacteria, 24 percent fungi, and 20 percent protozoa (Gray et al. 1995). Keratitis due to *Cryptococcus laurentii* and *Fusarium solani* was recently reported in a diabetic male patient who wore a gas-permeable contact lens (Ritterband et al. 1998); both fungi were isolated from the patient's corneal button, infected toe nails, and contact lens storage case, and enucleation eventually had to be performed.

Overnight soaking of soft lenses in 3 percent hydrogen peroxide (longer than 4 hours), with neutralization in the morning by thiosulfate solution, catalase solution, or catalase tablets is perhaps the safest way to ensure killing of bacteria, *Acanthamoeba*, and fungi (Chandler 1990; Wilson 1996). However, in one study, as many as 75 percent of individuals who used hydrogen peroxide disinfection for their contact lenses were found to have contact lens cases that were contaminated with microbes (Gray et al. 1995); all the contaminating microorganisms were found to possess the enzyme catalase (which breaks down hydrogen peroxide to water and oxygen). Recommendations for contact lens wearers to prevent microbial contamination of the lens and case include regular scrubbing of the interior of contact lens cases to disrupt biofilms, exposure of the contact lens case to very hot water ($\geqslant70°C$), air-drying of the contact lens case between use, use of a two-step system for hydrogen peroxide disinfection, and regular replacement of the contact lens case (Gray et al. 1995; Seal et al. 1998).

Fungal contamination of other ocular biomaterials

Punctal occluders or plugs aid the management of dry eye syndrome. However, since these devices are left in situ for a long duration, nonspecific microbial attachment, surface colonization, and biofilm formation may occur. Fungi such as *Curvularia lunata* may sometimes be implicated (Wilson 1996).

Several alloplastic biomaterials have been used for repair of the orbital floor and to restore an anophthalmic socket. Oestreicher et al. (1999) described a patient who developed an *Aspergillus* abscess within a hydroxyapatite orbital implant 58 months following uncomplicated implant surgery; the symptoms resolved following removal of the implant.

Patients with corneal or scleral defects have been treated with Gore-Tex grafting; Huang et al. (1994) reported the occurrence of fungal contamination in one such graft, which ultimately led some months after graft removal to fungal endophthalmitis and penetrating keratoplasty.

ACKNOWLEDGMENTS

The authors are grateful to J. Kaliamurthy, C.M. Kalavathy, D. Arvind Prasanth, A. Geetha, R.T. Rajagowthamee, and P. Archana Teresa for their abundant help, and to C.A. Nelson Jesudasan for permitting them to utilize the patient and laboratory records and other facilities available at the Institute of Ophthalmology, Joseph Eye Hospital, Tiruchirapalli, in the preparation of this chapter.

REFERENCES

Abboud, I.A. and Hanna, L.S. 1970. Ocular fungus. Report of two cases. *Br J Ophthalmol*, **54**, 477–83.

Adam, R.D., Paquin, M.L., et al. 1986. Phaeohyphomycosis caused by the fungal genera *Bipolaris* and *Exserohilum*: a report of 9 cases and review of the literature. *Medicine (Baltimore)*, **65**, 203–17.

Adisa, V.A. and Fajola, A.O. 1983. Cellulolytic enzymes associated with the fruit rots of *Citrus sinensis* caused by *Aspergillus aculeatus* and *Botryodiplodia theobromae*. *Z Allg Mikrobiol*, **23**, 283–8.

Adler, D.E., Milhorat, T.H. and Miller, J.I. 1998. Treatment of rhinocerebral mucormycosis with intravenous interstitial and cerebrospinal fluid administration of amphotericin B: case report. *Neurosurgery*, **42**, 644–8.

Agarwal, A., Gupta, A., et al. 1991. Retinitis following disseminated cryptococcosis in a renal allograft recipient. Efficacy of oral fluconazole. *Acta Ophthalmol Copenh*, **69**, 402–5.

Agrawal, P.K., Lal, B., et al. 1979. Orbital paecilomycosis due to *Paecilomyces lilacinus* (Thom) Samson. *Sabouraudia*, **17**, 363–9.

Agrawal, V., Biswas, J., et al. 1994. Current perspectives in infectious keratitis. *Indian J Ophthalmol*, **42**, 171–91.

Aguilar, C., Pujol, I., et al. 1998. Antifungal susceptibilities of *Paecilomyces* species. *Antimicrob Agents Chemother*, **42**, 1601–4.

Aguilar, G.L., Blumenkranz, M.S., et al. 1979. *Candida* endophthalmitis after intravenous drug abuse. *Arch Ophthalmol*, **97**, 96–100.

Ainbinder, D.J., Parmley, V.C., et al. 1998. Infectious crystalline keratopathy caused by *Candida guilliermondii*. *Am J Ophthalmol*, **125**, 723–5.

Ajayi, B.G., Osuntokun, B., et al. 1986. Orbital histoplasmosis due to *Histoplasma capsulatum* var. *duboisii*: successful treatment with septrin. *J Trop Med Hyg*, **89**, 179–87.

Ajello, L., Dean, D.F. and Irwin, R.S. 1976. The zygomycete *Saksenaea vasiformis* as a pathogen of humans with a critical review of the etiology of zygomycosis. *Mycologia*, **68**, 52–62.

Albesi, E.J. and Zapater, R.C. 1975. Queratitis por *Aspergillus flavus*. *Arch Oftalmol Buenos Aires*, **50**, 163–8.

Alexandrakis, G., Sears, M. and Gloor, P. 1996. Postmortem diagnosis of *Fusarium* panophthalmitis by the polymerase chain reaction. *Am J Ophthalmol*, **121**, 221–3.

Alexandrakis, G., Jalali, S. and Gloor, P. 1998. Diagnosis of *Fusarium* keratitis in an animal model using the polymerase chain reaction. *Br J Ophthalmol*, **82**, 306–11.

Alexandrakis, G., Haimovici, R., et al. 2000. Corneal biopsy in the management of progressive microbial keratitis. *Am J Ophthalmol*, **129**, 571–6.

Alfonso, E., Mandelbaum, S., et al. 1986. Ulcerative keratitis associated with contact lens wear. *Am J Ophthalmol*, **108**, 64–7.

Alino, A.M., Perry, H.D., et al. 1998. Conjunctival flaps. *Ophthalmology*, **105**, 1120–3.

Alio, J.L., Ayala, M.J., et al. 1993. Treatment of experimental acute corneal inflammation with inhibitors of oxidative metabolism. *Ophthalmic Res*, **25**, 331–6.

Alio, J.L., Artola, A., et al. 1995. Effect of topical antioxidant therapy on experimental infectious keratitis. *Cornea*, **14**, 175–9.

Alleyne, C.H. Jr., Vishteh, A.G., et al. 1999. Long-term survival of a patient with invasive cranial base rhinocerebral mucormycosis treated with combined endovascular, surgical and medical therapies: case report. *Neurosurgery*, **45**, 1461–3.

Alvarez, H. and Tabbara, K.F. 1996. Infections of the eyelids. In: Tabbara, K.F. and Hyndiuk, R.A. (eds), *Infections of the eye*, 2nd edition. Boston, MA: Little, Brown & Co., 559–70.

Anand, A.R., Madhavan, H.N., et al. 2001. Polymerase chain reaction in the diagnosis of Aspergillus endophthalmitis. *Indian J Med Res*, **114**, 133–40.

Anand, V.K., Alemar, G. and Griswold, J.A. Jr 1992. Intracranial complications of mucormycosis: an experimental model and clinical review. *Laryngoscope*, **102**, 656–62.

Anandi, V., Suryawanshi, N.B., et al. 1988. Corneal ulcer caused by *Bipolaris hawaiiensis*. *J Med Vet Mycol*, **26**, 301–6.

Anderson, B., Roberts, S.S., et al. 1959. Mycotic ulcerative keratitis. *Arch Ophthalmol*, **62**, 169–79.

Anderson, R.L., Carroll, T.F., et al. 1984. *Petriellidium (Allescheria) boydii* orbital and brain abscess treated with intravenous miconazole. *Am J Ophthalmol*, **97**, 771–5.

Ando, N. and Takatori, K. 1987. Keratomycosis due to *Alternaria alternata* corneal transplant infection. *Mycopathologia*, **100**, 17–22.

Arora, I. and Singhvi, S. 1994. Impression debridement of corneal lesions. *Ophthalmology*, **101**, 1935–40.

Arora, I., Kulshrestha, O.P. and Upadhyay, S. 1983. Treatment of fungal corneal ulcers with econazole. *Indian J Ophthalmol*, **31**, 1019–21.

Arthur, S., Steed, L.L., et al. 2001. *Scedosporium prolificans* keratouveitis in association with a contact lens retained intraocularly over a long term. *J Clin Microbiol*, **39**, 4579–82.

Ashbee, H.R. and Evans, E.G.V. 2002. Immunology of the diseases associated with *Malassezia* species. *Clin Microbiol Rev*, **15**, 21–57.

Attapattu, M.C. 1995. Acute rhinocerebral mucormycosis caused by *Rhizopus arrhizus* from Sri Lanka. *J Trop Med Hyg*, **98**, 355–8.

Azar, P., Aquavella, J.V. and Smith, R.S. 1975. Keratomycosis due to an *Alternaria* species. *Am J Ophthalmol*, **79**, 881–2.

Badenoch, P.R., Coster, D.J., et al. 2001. *Pythium insidiosum* keratitis confirmed by DNA sequence analysis. *Br J Ophthalmol*, **85**, 502–3.

Balch, K., Phillips, P.H. and Newman, N.J. 1997. Painless orbital apex syndrome from mucormycosis. *J Neuroophthalmol*, **17**, 178–82.

Bansal, R.K., Suseela, A.V. and Gugnani, H.C. 1977. Orbital cyst due to *Histoplasma duboisii*. *Br J Ophthalmol*, **61**, 70–1.

Barr, G.C. and Gamel, J.W. 1986. Blastomycosis of the eyelid. *Arch Ophthalmol*, **104**, 96–7 (letter).

Barrie, T. 1987. The place of elective vitrectomy in the management of patients with candidal endophthalmitis. *Graefe's Arch Clin Exp Ophthalmol*, **225**, 107–13.

Bartley, G.B. 1995. Blastomycosis of the eyelid. *Ophthalmology*, **102**, 2020–3.

Barton, K., Miller, D. and Pflugfelder, S.C. 1997. Corneal chromoblastomycosis. *Cornea*, **16**, 235–9.

Behrens-Baumann, W. 1999. *Mycoses of the eye and its adnexa. Developments in ophthalmology*, vol. 32. Basel: Karger.

Behrens-Baumann, W., Klinge, B. and Ruchel, R. 1990. Topical fluconazole for experimental *Candida* keratitis in rabbits. *Br J Ophthalmol*, **74**, 40–2.

Behrens-Baumann, W., Ruchel, R., et al. 1991. *Candida tropicalis* endophthalmitis following penetrating keratoplasty. *Br J Ophthalmol*, **75**, 565.

Belfort, R. Jr, Fischman, O., et al. 1975. Paracoccidioidomycosis with palpebral and conjunctival involvement. *Mycopathologia*, **56**, 21–4.

Bell, R. and Font, R.L. 1972. Granulomatous anterior uveitis caused by *Coccidioides immitis*. *Am J Ophthalmol*, **74**, 93–8.

Benaoudia, F., Assouline, M., et al. 1999. *Exophiala (Wangiella) dermatitidis* keratitis after keratoplasty. *Med Mycol*, **37**, 53–6.

Ben Ezra, D. 1994. Guidelines on the diagnosis and treatment of conjunctivitis. *Ocular Immunol Inflamm*, **2**, Suppl, S1–S55.

Berger, R.O. and Streeten, B.W. 1981. Fungal growth in aphakic soft contact lens. *Am J Ophthalmol*, **91**, 630–3.

Bernauer, W., Allan, B.D.S. and Dart, J.K.G. 1998. Successful management of *Aspergillus* scleritis by medical and surgical treatment. *Eye*, **12**, 311–16.

Bhattacharya, A.K., Deshpande, A.R., et al. 1992. Rhinocerebral mucormycosis: an unusual case presentation. *J Laryngol Otol*, **106**, 48–9.

Bhomaj, S., Das, J.C., et al. 2001. Rhinosporidiosis and peripheral keratitis. *Ophthalmic Surg Lasers*, **32**, 338–40.

Biswas, J. and Kumar, S.K. 2002. Cytopathology of explanted intraocular lenses and the clinical correlation. *J Cataract Refract Surg*, **28**, 538–43.

Blatt, S.P., Lucey, D.R., et al. 1991. Rhinocerebral zygomycosis in a patient with AIDS. *J Infect Dis*, **164**, 215–16 (letter).

Bloom, P.A., Laidlaw, D.A., et al. 1992. Treatment failure in a case of fungal keratitis caused by *Pseudallescheria boydii*. *Br J Ophthalmol*, **76**, 367–8.

Boelaert, J.R. 1994. Mucormycosis: is there news for the clinician? *J Infect*, **28**, Suppl 1, 1–6.

Boelaert, J.R., Fenvez, A.Z. and Coburn, J.W. 1991. Deferoxamine therapy and mucormycosis in dialysis patients: report of an international registry. *Am J Kidney Dis*, **18**, 660–7.

Boldt, H.C. and Mieler, W.F. 1996. Endophthalmitis. In: Tabbara, K.F. and Hyndiuk, R.A. (eds), *Infections of the eye*, 2nd edition. Boston: Little Brown and Co., 571–94.

Bond, W.I., Sanders, C.V., et al. 1982. Presumed blastomycosis endophthalmitis. *Ann Ophthalmol*, **14**, 1183–8.

Bonomo, P.P., Belfort, R. Jr, et al. 1982. Choroidal granuloma caused by *Paracoccidioides brasiliensis*. A clinical and angiographic study. *Mycopathologia*, **77**, 37–41.

Boralkar, A.N., Dindore, P.R., et al. 1989. Microbiological studies in conjunctivitis. *Indian J Ophthalmol*, **37**, 94–5.

Borderie, V.M., Bourcier, T.M., et al. 1997. Endophthalmitis after *Lasiodiplodia theobromae* corneal abscess. *Graefe's Arch Clin Exp Ophthalmol*, **235**, 259–61.

Borne, M.J., Elliott, J.H. and O'Day, D.M. 1993. Ocular fluconazole treatment of *Candida parapsilosis* endophthalmitis after failed intravitreal amphotericin B. *Arch Ophthalmol*, **111**, 1326–7.

Borruat, F.X., Bogousslavsky, J., et al. 1993. Orbital infarction syndrome. *Ophthalmology*, **100**, 562–8.

Bouchon, C.L., Greer, D.L. and Genre, C.F. 1994. Corneal ulcer due to *Exserohilum longirostratum*. *Am J Clin Pathol*, **101**, 452–5.

Brar, G.S., Ram, J., et al. 2002. *Aspergillus niger* endophthalmitis after cataract surgery. *J Cataract Refract Surg*, **28**, 1882–3.

Brinser, J.H. and Burd, E.M. 1986. Principles of diagnostic ocular microbiology. In: Tabbara, K.F. and Hyndiuk, R.A. (eds), *Infections of the eye*, 1st edition. Boston, MA: Little, Brown & Co., 73–92.

Brod, R.D., Flynn, H.W. Jr., et al. 1990. Endogenous *Candida* endophthalmitis. Management without intravenous amphotericin B. *Ophthalmology*, **97**, 666–74.

Brook, I. and Frazier, E.H. 1998. Aerobic and anaerobic microbiology of dacryocystitis. *Am J Ophthalmol*, **125**, 552–4.

Brooks, J.G. and Coster, D.J. 1993. Non-ulcerative fungal keratitis diagnosed by posterior lamellar biopsy. *Aust N Z J Ophthalmol*, **21**, 115–19.

Brooks, R.G. 1989. Prospective study of *Candida* endophthalmitis in hospitalised patients with candidemia. *Arch Intern Med*, **149**, 2226–8.

Brown, S.R., Shah, I.A. and Grinstead, M. 1998. Rhinocerebral mucormycosis caused by *Apophysomyces elegans*. *Am J Rhinol*, **12**, 289–92.

Brummund, W., Kurup, V.P., et al. 1986. Allergic sino-orbital mycosis. A clinical and immunologic study. *JAMA*, **256**, 3249–53.

Brunette, I. and Stulting, R.D. 1992. *Sporothrix schenckii* scleritis. *Am J Ophthalmol*, **114**, 370–1 (letter).

Bullock, J.D., Jampol, L.M. and Fezza, A.J. 1974. Two cases of orbital phycomycosis with recovery. *Am J Ophthalmol*, **78**, 811–15.

Burnier, S.V. and Sant'Anna, A.E. 1997. Palpebral paracoccidioidomycosis. *Mycopathologia*, **140**, 29–33.

Burns, F.R., Gray, R.D. and Paterson, C.A. 1990. Inhibition of alkali-induced corneal ulceration and perforation by a thiol peptide. *Investig Ophthalmol Vis Sci*, **31**, 107–14.

Cahill, K.V., Hogan, C.D., et al. 1994. Intraorbital injection of amphotericin B for palliative treatment of *Aspergillus* orbital abscess. *Ophthalmic Plast Reconstruct Surg*, **10**, 276–7.

Cailliez, J.C., Séguy, N., et al. 1996. *Pneumocystis carinii*: an atypical fungal micro-organism. *J Med Vet Mycol*, **34**, 227–39.

Calli, C., Savas, R., et al. 1999. Isolated pontine infarction due to rhinocerebral mucormycosis. *Neuroradiology*, **41**, 179–81.

Campbell, C.K., Johnson, E.M., et al. 1996. *Identification of pathogenic fungi*. London: Public Health Laboratory Service.

Carlson, A.N., Foulks, G.N., et al. 1992. Fungal scleritis after cataract surgery. Successful outcome using itraconazole. *Cornea*, **11**, 151–4.

Carroll, D.M. and Franklin, R.M. 1981. Vitreous biopsy in uveitis of unknown cause. *Retina*, **1**, 245–51.

Cartwright, M.J., Promersberger, M. and Stevens, G.A. 1993. *Sporothrix schenckii* endophthalmitis presenting as granulomatous uveitis. *Br J Ophthalmol*, **77**, 61–2.

Castelino, A.M., Rao, S.K., et al. 2000. Conjunctival rhinosporidiosis associated with scleral melting and staphyloma formation: diagnosis and management. *Cornea*, **19**, 30–3.

Castro, L.G., Salebian, A. and Sotto, M.N. 1990. Hyalohyphomycosis by *Paecilomyces lilacinus* in a renal transplant patient and a review of human *Paecilomyces* species infections. *J Med Vet Mycol*, **28**, 15–26.

Champion, C.K. and Johnson, T.M. 1969. Rhino-orbital-cerebral phycomycosis. *Mich Med*, **68**, 807–10.

Chander, J. and Sharma, A. 1994. Prevalence of fungal corneal ulcers in northern India. *Infection*, **22**, 207–9.

Chandler, F.W. 1991. Histologic diagnosis of mycotic diseases. In: Gatti, F., de Vroey, C. and Persi, A. (eds), *Human mycoses in tropical countries. Health Cooperation Papers, No. 13*. Bologna, Italy: Organizzazione per la Cooperazione Sanitaria Internazionale, 235–42.

Chandler, J.W. 1990. Biocompatibility of hydrogen peroxide in soft contact lens disinfection: antimicrobial activity vs. biocompatibility – the balance. *CLAO J*, **16**, Suppl 1, S43–5.

Chang, S.W., Tsai, M.W. and Hu, F.R. 1994. Deep *Alternaria* keratomycosis with intraocular extension. *Am J Ophthalmol*, **117**, 544–5, (letter).

Chapman, F.M., Orr, K.E., et al. 1998. *Candida glabrata* endophthalmitis following penetrating keratoplasty. *Br J Ophthalmol*, **82**, 712–13.

Charles, N.C., Boxrud, C.A. and Small, E.A. 1992. Cryptococcosis of the anterior segment in acquired immune deficiency syndrome. *Ophthalmology*, **99**, 813–16.

Chen, S.J., Chung, Y.M. and Liu, J.H. 1998. Endogenous *Candida* endophthalmitis after induced abortion. *Am J Ophthalmol*, **125**, 873–5.

Chetchotisakd, P., Booma, P., et al. 1991. Rhinocerebral mucormycosis: a report of eleven cases. *Southeast Asian J Trop Med Public Health*, **22**, 268–73.

Chin, G.N., Hyndiuk, R.A., et al. 1975. Keratomycosis in Wisconsin. *Am J Ophthalmol*, **79**, 121–5.

Chitravel, V., Subramanian, S., et al. 1980. Rhinosporidiosis in a south Indian village. *Sabouraudia*, **18**, 241–7.

Cho, B.J. and Lee, Y.B. 2002. Infectious keratitis manifesting as a white plaque on the cornea. *Arch Ophthalmol*, **120**, 1091–3.

Chopin, J.B., Sigler, L., et al. 1997. Keratomycosis in a Percheron cross horse caused by *Cladorrhinum bulbillosum*. *J Med Vet Mycol*, **35**, 53–5.

Christmas, N.J. and Smiddy, W.E. 1996. Vitrectomy and systemic fluconazole for treatment of endogenous fungal endophthalmitis. *Ophthalmic Surg Lasers*, **27**, 1012–18.

Chung, M.S., Goldstein, M.H., et al. 2000. Fungal keratitis after laser *in situ* keratomileusis: a case report. *Cornea*, **19**, 236–7.

Coccia, L., Calista, D. and Boschini, A. 1999. Eyelid nodule: a sentinel lesion of disseminated cryptococcosis in a patient with acquired immunodeficiency syndrome. *Arch Ophthalmol*, **117**, 271–2.

Coetzee, A.S. and de Bruin, G.F. 1974. Mucormycosis case report and review. *S Afr Med J*, **48**, 2486–8.

Cohen, D.B. and Glasgow, B.J. 1993. Bilateral optic nerve cryptococcosis in sudden blindness in patients with acquired immunodeficiency syndrome. *Ophthalmology*, **100**, 1689–94.

Cohen, S.G. and Greenberg, M.S. 1980. Rhinomaxillary mucormycosis in a kidney transplant patient. *Oral Surg, Oral Med Oral Pathol*, **50**, 33–8.

Coriglione, G., Stella, G., et al. 1990. *Neosartorya fischeri var. fischeri* (Wehmer) Malloch and Cain 1972 (anamorph: *Aspergillus fischerianus* Samson and Gams 1985) as a cause of mycotic keratitis. *Eur J Epidemiol*, **6**, 382–5.

Coskuncan, N.M., Jabs, D.A., et al. 1994. The eye in bone marrow transplantation. VI. Retinal complications. *Arch Ophthalmol*, **112**, 372–9.

Cristol, S.M., Alfonso, E.C., et al. 1996. Results of large penetrating keratoplasty in microbial keratitis. *Cornea*, **15**, 571–6.

Cuero, R.G. 1980. Ecological distribution of *Fusarium solani* and its opportunistic action related to mycotic keratitis in Cali, Colombia. *J Clin Microbiol*, **12**, 455–61.

Cunningham, E.T. Jr, Seiff, S.R., et al. 1998. Intraocular coccidioidomycosis diagnosed by skin biopsy. *Arch Ophthalmol*, **116**, 674–7.

Curi, A.L., Felix, S., et al. 2003. Retinal granuloma caused by *Sporothrix schenckii*. *Am J Ophthalmol*, **136**, 205–7.

Custis, P.H., Haller, J.A. and de Juan, E. Jr 1995. An unusual case of cryptococcal endophthalmitis. *Retina*, **15**, 300–4.

Cutler, J.E., Binder, P.S., et al. 1978. Metastatic coccidioidal endophthalmitis. *Arch Ophthalmol*, **96**, 689–91.

Daly, A.L., Velazquez, L.A., et al. 1989. Mucormycosis association with deferoxamine therapy. *Am J Med*, **87**, 468–71.

Dantas, A.M., Yamane, R. and Camara, A.G. 1990. South American blastomycosis: ophthalmic and oculomotor nerve lesions. *Am J Trop Med Hyg*, **43**, 386–8.

Deans, R.M., Harris, G.J. and Gonnering, R.S. 1996. Infections of the orbit. In: Tabbara, K.F. and Hyndiuk, R.A. (eds), *Infections of the eye*, 2nd edition. Boston, MA: Little, Brown & Co., 531–50.

de Garcia, M.C., Arboleda, M.L., et al. 1997. Fungal keratitis caused by *Metarhizium anisopliae* var. *anisopliae*. *J Med Vet Mycol*, **35**, 361–3.

de Hoog, G.S. and Guarro, J. (eds) 1995. *Atlas of clinical fungi*. Baarn, The Netherlands: Centraalbureau voor Schimmelcultures.

de Shazo, R.D., O'Brien, M., et al. 1997. A new classification and diagnostic criteria for invasive fungal sinusitis. *Arch Otolaryngol Head Neck Surg*, **123**, 1181–8.

Diaz-Valle, D., Benitez del Castillo, J.M., et al. 2002. Severe keratomycosis secondary to *Scedosporium apiospermum*. *Cornea*, **21**, 516–18.

Djalilian, A.R., Smith, J.A., et al. 2001. Keratitis caused by *Candida glabrata* in a patient with chronic granulomatous disease. *Am J Ophthalmol*, **132**, 782–3.

D'Mellow, G., Hirst, L.W., et al. 1991. Intralenticular infections. *Ophthalmology*, **98**, 1376–8.

Donahue, S.P., Greven, C.M., et al. 1994. Intraocular candidiasis in patients with candidemia. Clinical implications derived from a prospective multicenter study. *Ophthalmology*, **101**, 1302–9.

Donahue, S.P., Hein, E. and Sinatra, R.B. 2003. Ocular involvement in children with candidemia. *Am J Ophthalmol*, **135**, 886–7.

Doorenbos-Bot, A.C., Hooymans, J.M. and Blanksma, L.J. 1990. Periorbital necrotising fasciitis due to *Cryptococcus neoformans* in a healthy young man. *Doc Ophthalmol*, **75**, 315–20.

Douer, D., Goldschmied-Reouven, A., et al. 1987. Human *Exserohilum* and *Bipolaris* infections: report of *Exserohilum* nasal infection in a neutropenic patient with acute leukaemia and review of the literature. *J Med Vet Mycol*, **25**, 235–41.

Dugel, P.U., Rao, N.A., et al. 1990. *Pneumocystis carinii* choroiditis in patients receiving inhaled pentamidine. *Am J Ophthalmol*, **110**, 113–17.

Duguid, I.G., Dart, J.K., et al. 1997. Outcome of *Acanthamoeba* keratitis treated with polyhexamethylene biguanide and propamidine. *Ophthalmology*, **104**, 1587–92.

Dunlop, A.A., Wright, E.D., et al. 1994. Suppurative corneal ulceration in Bangladesh: a study of 142 cases, examining the microbiological diagnosis, clinical and epidemiological features of bacterial and fungal keratitis. *Aust N Z J Ophthalmol*, **22**, 105–10.

Dursun, D., Fernandez, V., et al. 2003. Advanced fusarium keratitis progressing to endophthalmitis. *Cornea*, **22**, 300–3.

Elliott, J.H., O'Day, D.M., et al. 1979. Mycotic endophthalmitis in drug abusers. *Am J Ophthalmol*, **88**, 66–72.

Ellis, P.P. 1985. *Ocular therapeutics and pharmacology*, 7th edition. St Louis, MO: Mosby.

Engzell, U.G.C. and Jones, A.W. 1973. Rhinosporidiosis in Uganda. *J Laryngol Otol*, **87**, 1217–23.

Ernest, J.T. and Rippon, J.W. 1966. Keratitis due to *Allescheria boydii* (*monosporium apiospermum*). *Am J Ophthalmol*, **62**, 1202–4.

Essman, T.F., Flynn, H.W. Jr, et al. 1997. Treatment outcomes in a 10-year study of endogenous fungal endophthalmitis. *Ophthalmic Surg Lasers*, **28**, 185–94.

Fairley, C., Sullivan, T.J., et al. 2000. Survival after rhino-orbital-cerebral mucormycosis in an immunocompetent patient. *Ophthalmology*, **107**, 555–8.

Fekrat, S., Haller, J.A., et al. 1995. Pseudophakic *Candida parapsilosis* endophthalmitis with a consecutive keratitis. *Cornea*, **14**, 212–16.

Feman, S.S. and Tilford, R.H. 1985. Ocular findings in patients with histoplasmosis. *JAMA*, **253**, 2534–7.

Ferguson, B.J. 1998. What role do systemic corticosteroids, immunotherapy and antifungal drugs play in the therapy of allergic fungal rhinosinusitis? *Arch Otolaryngol Head Neck Surg*, **124**, 1174–8.

Fernandez, V., Dursun, D., et al. 2002. *Colletotrichum* keratitis. *Am J Ophthalmol*, **34**, 435–8.

Ferry, A.P. and Abedi, S. 1983. Diagnosis and management of rhino-orbital cerebral mucormycosis (phycomycosis). *Ophthalmology*, **90**, 1096–104.

Finamor, L.P., Muccioli, C., et al. 2002. Ocular and central nervous system paracoccidioidomycosis in a pregnant woman with acquired immunodeficiency syndrome. *Am J Ophthalmol*, **134**, 456–9.

Fincher, R.-M.E., Fisher, J.F., et al. 1991. Infection due to the fungus *Acremonium* (*Cephalosporium*). *Medicine (Baltimore)*, **70**, 398–409.

Fiscella, R.G., Moshifar, M., et al. 1997. Polyhexamethylene biguanide (PHMB) in the treatment of experimental *Fusarium* keratomycosis. *Cornea*, **16**, 447–9.

Fitzsimons, R. and Peters, A.L. 1986. Miconazole and ketoconazole as a satisfactory first-line treatment for keratomycosis. *Am J Ophthalmol*, **101**, 605–8.

Florakis, G.J., Moazami, G., et al. 1997. Scanning slit confocal microscopy of fungal keratitis. *Arch Ophthalmol*, **115**, 1461–3.

Flynn, H.W. Jr 2001. The clinical challenge of endogenous endophthalmitis. *Retina*, **21**, 572–4.

Font, R.L. and Jakobiec, F.A. 1976. Granulomatous necrotising retinochoroiditis caused by *Sporotrichum schenckii*. Report of a case, including immunofluorescence and electron microscopical studies. *Arch Ophthalmol*, **94**, 1513–19.

Font, R.L., Spaulding, A.G. and Green, W.R. 1967. Endogenous mycotic endophthalmitis caused by *Blastomyces dermatitidis*. Report of a case and review of the literature. *Arch Ophthalmol*, **77**, 217–22.

Font, R.L., Parsons, M.A., et al. 1995. Involvement of the anterior chamber angle structures in disseminated histoplasmosis: report of three cases. *Ger J Ophthalmol*, **4**, 107–15.

Forster, R.K. 1994. Fungal keratitis and conjunctivitis: clinical aspects. In: Smolin, G. and Thoft, R.A. (eds), *The cornea: scientific foundations and clinical practice*, 3rd edition. Boston, MA: Little, Brown & Co., 239–52.

Forster, R.K., Rebell, G. and Stiles, W. 1975a. Recurrent keratitis due to *Acremonium potronii*. *Am J Ophthalmol*, **79**, 126–8.

Forster, R.K., Rebell, G. and Wilson, L.A. 1975b. Dematiaceous fungal keratitis. Clinical isolates and management. *Br J Ophthalmol*, **59**, 372–6.

Forster, R.K., Wirta, M.G., et al. 1976. Methenamine silver nitrate-stained corneal scrapings in keratomycosis. *Am J Ophthalmol*, **82**, 261–5.

Foster, C.S. 1981. Miconazole therapy for keratomycosis. *Am J Ophthalmol*, **91**, 622–9.

Foster, C.S. and Stefanyszyn, M. 1979. Intraocular penetration of miconazole in rabbits. *Arch Ophthalmol*, **97**, 1703–6.

Foster, C.S., Lass, J.H., et al. 1981. Ocular toxicity of topical antifungal agents. *Arch Ophthalmol*, **99**, 1081–4.

Foster, J.B.T., Almeda, E., et al. 1958. Some intraocular and conjunctival effects of amphotericin B in man and in the rabbit. *Arch Ophthalmol*, **60**, 555–64.

Foster, R.E., Lowder, C.Y., et al. 1991. Presumed *Pneumocystis carinii* choroiditis. Unifocal presentation, regression with intravenous pentamidine and choroiditis recurrence. *Ophthalmology*, **98**, 1360–5.

Fraser-Smith, E.B. and Matthews, T.R. 1987. Effect of ketorolac on *Candida albicans* ocular infection in rabbits. *Arch Ophthalmol*, **105**, 264–7.

Friedberg, D.N., Warren, F.A., et al. 1992. *Pneumocystis carinii* of the orbit. *Am J Ophthalmol*, **113**, 595–6 (letter).

Gago, L.C., Capone, A. Jr and Trese, M.T. 2002. Bilateral presumed endogenous candida endophthalmitis and stage 3 retinopathy of prematurity. *Am J Ophthalmol*, **134**, 611–13.

Gaines, J.J. Jr, Clay, J.R., et al. 1996. Rhinosporidiosis: three domestic cases. *South Med J*, **89**, 65–7.

Gamba, J.L., Woodruff, W., et al. 1986. Craniofacial mucormycosis: assessment with CT. *Radiology*, **160**, 207–12.

Garcia, M.L., Herreras, J.M., et al. 2002. Evaluation of lectin staining in the diagnosis of fungal keratitis in an experimental rabbit model. *Mol Vis*, **8**, 10–16.

Garg, P., Gopinathan, U., et al. 2000. Keratomycosis-clinical and microbiologic experience with dematiaceous fungi. *Ophthalmology*, **107**, 574–80.

Garg, P., Mahesh, S., et al. 2003. Fungal infection of sutureless self-sealing incision for cataract surgery. *Ophthalmology*, **110**, 2173–7.

Gaudio, P.A., Gopinathan, U., et al. 2002. Polymerase chain reaction based detection of fungi in infected corneas. *Br J Ophthalmol*, **86**, 755–60.

Gebhard, F., Chastagner, P., et al. 1995. Favorable outcome of orbital nasal sinus mucormycosis complicating the induction treatment of acute lymphoblastic leukemia. *Arch Paediatr*, **2**, 47–51.

Ghose, S. and Mahajan, V.M. 1990. Fungal flora in congenital dacryocystitis. *Indian J Ophthalmol*, **38**, 189–90.

Gingrich, W.D. 1962. Keratomycosis. *J Am Med Assoc*, **179**, 602–8.

Glasgow, B.J., Engstrom, R.E. Jr, et al. 1996. Bilateral endogenous *Fusarium* endophthalmitis associated with the acquired immunodeficiency syndrome. *Arch Ophthalmol*, **114**, 873–7.

Goldblum, D., Frueh, B.E., et al. 2000. Treatment of post-keratitis *Fusarium* endophthalmitis with amphotericin B lipid complex. *Cornea*, **19**, 853–6.

Goldstein, B.G. and Buettner, H. 1983. Histoplasmic endophthalmitis. *Arch Ophthalmol*, **101**, 774–7.

Gonawardena, S.A., Ranasinghe, K.P., et al. 1994. Survey of mycotic and bacterial keratitis in Sri Lanka. *Mycopathologia*, **127**, 77–81.

Gonzales, C.A., Scott, I.U., et al. 2000. Endogenous endophthalmitis caused by *Histoplasma capsulatum* var. *capsulatum*. A case report and literature review. *Ophthalmology*, **107**, 725–9.

Goodwin, W.J. Jr 1985. Orbital complications of ethmoiditis. *Otolaryngol Clin North Am*, **18**, 139–47.

Gopinathan, U., Garg, P., et al. 2002. The epidemiological features and laboratory results of fungal keratitis. A 10-year review at a referral eye care centre in South India. *Cornea*, **21**, 555–9.

Gordon, M.A. and Norton, S.W. 1985. Corneal transplant infection by *Paecilomyces lilacinus*. *Sabouraudia*, **23**, 295–301.

Gottlieb, J.L., McAllister, I.L., et al. 1995. Choroidal blastomycosis. A report of two cases. *Retina*, **15**, 248–52.

Gottsch, J.D., Gilbert, M.L., et al. 1991. Excimer laser ablation of microbial keratitis. *Ophthalmology*, **98**, 146–9.

Grant, S.M. and Clissold, S.P. 1989. Itraconazole. A review of its pharmacodynamic and pharmacokinetic properties, and therapeutic use in superficial and systemic mycoses. *Drugs*, **37**, 310–44.

Gray, T.B., Cursons, R.T., et al. 1995. *Acanthamoeba* bacterial and fungal contamination of contact lens storage cases. *Br J Ophthalmol*, **79**, 601–5.

Greer, D.L., Brahim, C. and Gonzalez, L.A. 1973. Queratitits micotica en Colombia. *Trib Med*, **74**, A15–120.

Grossman, R. and Lee, D.A. 1989. Transscleral and transcorneal iontophoresis of ketoconazole in the rabbit eye. *Ophthalmology*, **96**, 724–9.

Guarro, J. and Gené, J. 1992. *Fusarium* infections. Criteria for the identification of the responsible species. *Mycoses*, **35**, 109–14.

Guarro, J. and Gené, J. 2002. *Acrophialophora fusispora* misidentified as *Scedosporium prolificans*. *J Clin Microbiol*, **40**, 3544.

Guarro, J., Gams, W., et al. 1997. *Acremonium* species: new emerging fungal opportunists – *in vitro* antifungal susceptibilities and review. *Clin Infect Dis*, **25**, 1222–9.

Guarro, J., Akiti, T., et al. 1999. Mycotic keratitis due to *Curvularia senegalensis* and *in vitro* antifungal susceptibilities of *Curvularia* spp. *J Clin Microbiol*, **37**, 4170–3.

Guarro, J., Vieira, L.A., et al. 2000. *Phaeoisaria clematidis* as a cause of keratomycosis. *J Clin Microbiol*, **38**, 2434–7.

Guarro, J., Hofling-Lima, A.L., et al. 2002. Corneal ulcer caused by the new fungal species *Sarcopodium oculorum*. *J Clin Microbiol*, **40**, 3071–5.

Guerra, R., Cavallini, G.M., et al. 1992. *Rhodotorula glutinis* keratitis. *Int Ophthalmol*, **16**, 187–90.

Gugnani, H.C., Talwar, R.S., et al. 1976. Mycotic keratitis in Nigeria. A study of 26 cases. *Br J Ophthalmol*, **60**, 607–13.

Gumbel, H., Ohrloff, C. and Shah, P.M. 1990. The conjunctival flora of HIV-positive patients in an advanced stage. *Fortschr Ophthalmol*, **87**, 382–3.

Gupta, A., Sharma, A., et al. 1999. Mycotic keratitis in non-steroid exposed vernal keratoconjunctivitis. *Acta Ophthalmol Scand*, **77**, 229–31.

Gussler, J.R., Miller, D., et al. 1995. Infection after radial keratotomy. *Am J Ophthalmol*, **119**, 798–9 (letter).

Guzek, J.P., Roosenberg, J.M., et al. 1998. The effect of vehicle on corneal penetration of triturated ketoconazole and itraconazole. *Ophthalmic Surg Lasers*, **29**, 926–9.

Hagan, M., Wright, E., et al. 1995. Causes of suppurative keratitis in Ghana. *Br J Ophthalmol*, **79**, 1024–8.

Hammer, M.E., Harding, S. and Wynn, P. 1983. Post-traumatic fungal endophthalmitis caused by *Exophiala jeanselmei*. *Ann Ophthalmol*, **15**, 853–5.

Hamza, H.S., Loewenstein, A. and Haller, J.A. 1999. Fungal retinitis and endophthalmitis. *Ophthalmol Clin North Am*, **12**, 1, 89–108.

Harril, W.C., Stewart, M.G., et al. 1996. Chronic rhinocerebral mucormycosis. *Laryngoscope*, **106**, 1292–7.

Harris, G.J. and Will, B.R. 1989. Orbital aspergillosis. Conservative debridement and local amphotericin irrigation. *Ophthalmic Plast Reconstruct Surg*, **5**, 207–11.

Harrousseau, J.L., Dekker, A.W., et al. 2000. Itraconazole oral solution for primary prophylaxis of fungal infections in patients with hematological malignancy and profound neutropenia: a randomised, double-blind, double-placebo, multicenter trial comparing itraconazole and amphotericin B. *Antimicrob Agents Chemother*, **44**, 1887–93.

Heidemann, D.G., Dunn, S.P. and Watts, J.C. 1995. *Aspergillus* keratitis after radial keratotomy. *Am J Ophthalmol*, **120**, 254–6.

Hejny, C., Kerrison, J.B., et al. 2001. Rhinoorbital mucormycosis in a patient with the acquired immunodeficiency syndrome (AIDS) and neutropenia. *Am J Ophthalmol*, **132**, 111–12.

Hemady, R.K. 1995. Microbial keratitis in patients infected with the human immunodeficiency virus. *Ophthalmology*, **102**, 1026–30.

Hemady, R.K., Chu, W. and Foster, C.S. 1992. Intraocular penetration of ketoconazole in rabbits. *Cornea*, **11**, 329–33.

Henig, F.E., Lehrer, N., et al. 1973. *Paecilomyces* of the lacrimal sac. *Mykosen*, **16**, 25–8.

Hester, D.E., Kylstra, J.A. and Eifrig, D.E. 1992. Isolated ocular cryptococcosis in an immunocompetent patient. *Ophthalmic Surg*, **23**, 129–31.

Hidalgo, J.A., Alangaden, G.J., et al. 2000. Fungal endophthalmitis diagnosis by detection of *Candida albicans* DNA in intraocular fluid by use of a species-specific polymerase chain reaction assay. *J Infect Dis*, **181**, 1198–201.

Hirose, H., Terasaki, H., et al. 1997. Treatment of fungal corneal ulcers with amphotericin B ointment. *Am J Ophthalmol*, **124**, 836–8.

Hiruma, M., Kawada, A., et al. 1991. Tinea of the eyebrow showing kerion celsi: report of one case. *Cutis*, **48**, 149–50.

Ho, R.H.T., Bernard, P.J. and McClellan, K.A. 1991. *Phialophora mutabilis* keratomycosis. *Am J Ophthalmol*, **112**, 728–9 (letter).

Hofling-Lima, A.L., Freitas, D., et al. 1999. *Exophiala jeanselmei* causing late endophthalmitis after cataract surgery. *Am J Ophthalmol*, **128**, 512–14.

Hornblass, A., Herschorn, B.J., et al. 1984. Orbital abscess. *Surv Ophthalmol*, **29**, 169–78.

Houang, E., Lam, D., et al. 2001. Microbial keratitis in Hong Kong: relationship to climate, environment and contact-lens disinfection. *Trans R Soc Trop Med Hyg*, **95**, 361–7.

Huang, W.J., Hu, F.R. and Chang, S.W. 1994. Clinicopathologic study of Gore-Tex patch graft in corneoscleral surgery. *Cornea*, **13**, 82–6.

Huber-Spitzy, V., Baumgartner, I., et al. 1991. Blepharitis – a diagnostic and therapeutic challenge. *Graefe's Arch Clin Exp Ophthalmol*, **229**, 224–7.

Hutnik, C.M., Nicolle, D.A. and Munoz, D.G. 1997. Orbital aspergillosis. A fatal masquerader. *J Neuroophthalmol*, **17**, 257–61.

Imayasu, M., Moriyama, T., et al. 1992. A quantitative method for LDH, MDH and albumin levels in tears with ocular surface toxicity scored by Draize criteria in rabbit eyes. *CLAO J*, **18**, 260–6.

Imwidthaya, P. 1994. Human pythiosis in Thailand. *Postgrad Med J*, **70**, 558–60.

Imwidthaya, P. 1995. Mycotic keratitis in Thailand. *J Med Vet Mycol*, **33**, 81–2.

Irvine, A.R. Jr 1968. Coccidioidal granuloma of lid. *Trans Am Acad Ophthalmol Otolaryngol*, **72**, 751–4.

Ishibashi, Y. 1983. Oral ketoconazole therapy for keratomycosis. *Am J Ophthalmol*, **95**, 342–5.

Ishibashi, Y. and Kaufman, H.E. 1986a. Topical ketoconazole for experimental *Candida* keratitis in rabbits. *Am J Ophthalmol*, **102**, 522–6.

Ishibashi, Y. and Kaufman, H.E. 1986b. Corneal biopsy in the diagnosis of keratomycosis. *Am J Ophthalmol*, **101**, 288–93.

Ishibashi, Y. and Matsumoto, Y. 1984a. Oral ketoconazole therapy for experimental *Candida albicans* keratitis in rabbits. *Sabouraudia*, **22**, 323–30.

Ishibashi, Y. and Matsumoto, Y. 1984b. Intravenous miconazole in the treatment of keratomycosis. *Am J Ophthalmol*, **97**, 646–7 (letter).

Ishibashi, Y., Matsumoto, Y. and Takei, K. 1984. The effects of intravenous miconazole on fungal keratitis. *Am J Ophthalmol*, **98**, 433–7.

Ishibashi, Y., Kaufman, H.E., et al. 1986. The pathogenicities of *Cylindrocarpon tonkinense* and *Fusarium solani* in the rabbit cornea. *Mycopathologia*, **94**, 145–52.

Ishibashi, Y., Hommura, S. and Matsumoto, Y. 1987. Direct examination vs. culture of biopsy specimens for the diagnosis of keratomycosis. *Am J Ophthalmol*, **103**, 636–40.

Iwen, P.C., Tarantolo, S.R., et al. 2000. Cutaneous infection caused by *Cylindrocarpon lichenicola* in a patient with acute myelogenous leukemia. *J Clin Microbiol*, **38**, 3375–8.

Jabs, D.A., Green, W.R., et al. 1989. Ocular manifestations of acquired immune deficiency syndrome. *Ophthalmology*, **96**, 1092–9.

Jacob, P., Gopinathan, U., et al. 1995. Calcium alginate swabs versus Bard-Parker blade in the diagnosis of microbial keratitis. *Cornea*, **14**, 360–4.

Jacobson, M., Galetta, S.L., et al. 1992. *Bipolaris*-induced orbital cellulitis. *J Clin Neuroophthalmol*, **12**, 250–6.

Jaeger, E.E., Carroll, N.M., et al. 2000. Rapid detection and identification of *Candida, Aspergillus* and *Fusarium* species in ocular samples using nested PCR. *J Clin Microbiol*, **38**, 2902–8.

James, E.A., Orchard, K.P., et al. 1997. Disseminated infection due to *Cylindrocarpon lichenicola* in a patient with acute myeloid leukaemia. *J Infect*, **34**, 65–7.

Jampol, L.M., Sung, J., et al. 1996. Choroidal neovascularization secondary to *Candida albicans* chorioretinitis. *Am J Ophthalmol*, **121**, 643–9.

Jay, W.M., Bradsher, R.W., et al. 1988. Ocular involvement in mycotic sinusitis caused by *Bipolaris*. *Am J Ophthalmol*, **105**, 366–70.

Jiang, R.S. and Hsu, C.Y. 1999. Endoscopic sinus surgery for rhinocerebral mucormycosis. *Am J Rhinol*, **13**, 105–9.

Jimenez, J.F., Young, D.E. and Hough, A.J. Jr 1984. Rhinosporidiosis: a report of two cases from Arkansas. *Am J Clin Pathol*, **82**, 611–15.

Johns, K.J. and O'Day, D.M. 1988. Pharmacologic management of keratomycoses. *Surv Ophthalmol*, **33**, 178–88.

Johnson, T.E., Casiano, R.R., et al. 1999. Sino-orbital aspergillosis in acquired immunodeficiency syndrome. *Arch Ophthalmol*, **117**, 57–64.

Jones, B.R. 1975. Principles in the management of oculomycosis. *Am J Ophthalmol*, **79**, 719–51.

Jones, B.R., Clayton, Y.M., et al. 1976. The place of Canesten in the management of oculomycosis. *Munch Med Wschr*, **118**, Suppl 1, S97–S103.

Jones, B.R., Clayton, Y.M. and Oji, E.O. 1979. Recognition and chemotherapy of oculomycosis. *Postgrad Med J*, **55**, 625–8.

Jones, D.B. 1980. Strategy for the initial management of suspected microbial keratitis. In Barraquer, J.I., Binder, P.S. et al. (eds), *Symposium on medical and surgical diseases of the cornea*. Transactions of the New Orleans Academy of Ophthalmology. St Louis, MO: Mosby, 86–119.

Jones, D.B., Liesegang, T.J. and Robinson, N.M. 1981. Laboratory diagnosis of ocular infections. In: Washington, J.A.III (ed.), *Cumitech 13*. Washington, DC: American Society for Microbiology.

Jones, F.R. and Christensen, G.R. 1974. *Pullularia* corneal ulcer. *Arch Ophthalmol*, **92**, 529–30.

Jones, J., Katz, S.E. and Lubow, M. 1999. *Scedosporium apiospermum* of the orbit. *Arch Ophthalmol*, **117**, 272–3.

Kadayifcilar, S., Gedik, S., et al. 2001. Chorioretinal alterations in mucormycosis. *Eye*, **15**, Part I, 99–102.

Kalavathy, C.M., Thomas, P.A. and Rajasekaran, J. 1986. Spectrum of microbial infection in dacryocystitis. *J Madras State Ophthalmol Assoc*, **24**, 24–7.

Kalavathy, C.M., Thomas, P.A., et al. 1998. Rhinosporidiosis of the lacrimal sac: a report of two cases. *J Tamilnadu Ophthalmic Assoc*, **37**, 11–12.

Kaliamurthy, J., Kalavathy, C.M., et al. 2004. Keratitis due to a coelomycetous fungus: case reports and review of the literature. *Cornea*, **23**, 3–12.

Kalina, P.H. and Campbell, R.J. 1991. *Aspergillus terreus* endophthalmitis in a patient with chronic lymphocytic leukaemia. *Arch Ophthalmol*, **109**, 102–3.

Kamegasawa, A., Viero, R.M., et al. 1988. Protective effect of prior immunization on ocular paracoccidioidomycosis in guinea pigs. *Mycopathologia*, **103**, 35–42.

Kanungo, R., Srinivasan, R. and Rao, R.S. 1991. Acridine orange staining in early diagnosis of mycotic keratitis. *Acta Ophthalmol Copenh*, **69**, 750–3.

Kapoor, S., Sood, G.C., et al. 1976. Lid rhinosporidiosis – simulating a tumour. *Can J Ophthalmol*, **11**, 91–2.

Katz, B.J., Scott, W.E. and Folk, J.C. 1997. Acute histoplasmosis choroiditis in 2 immunocompetent brothers. *Arch Ophthalmol*, **115**, 1470–1.

Katzenstein, A.L., Sale, S.R. and Greenberger, P.A. 1983. Pathologic findings in allergic aspergillus sinusitis. A newly recognised form of sinusitis. *Am J Surg Pathol*, **7**, 439–43.

Kaufman, L., Padhye, A.A. and Parker, S. 1988. Rhinocerebral zygomycosis caused by *Saksenaea vasiformis*. *J Med Vet Mycol*, **26**, 237–41.

Kaushik, S., Ram, J., et al. 2001. *Curvularia lunata* endophthalmitis with secondary keratitis. *Am J Ophthalmol*, **131**, 140–2.

Kestelyn, P., Taelman, H., et al. 1993. Ophthalmic manifestations of infections with *Cryptococcus neoformans* in patients with the acquired immunodeficiency syndrome. *Am J Ophthalmol*, **116**, 721–7.

Khairallah, S.H., Byrne, K.A. and Tabbara, K.F. 1992. Fungal keratitis in Saudi Arabia. *Doc Ophthalmol*, **79**, 269–76.

Khalil, M.K. 1982. Histopathology of presumed ocular histoplasmosis. *Am J Ophthalmol*, **94**, 369–76.

Killingsworth, D.W., Stern, G.A., et al. 1993. Results of therapeutic penetrating keratoplasty. *Ophthalmology*, **100**, 534–41.

Kilpatrick, C., Tress, B. and King, J. 1984. Computed tomography of rhinocerebral mucormycosis. *Neuroradiology*, **26**, 71–3.

Kim, J.S., Kim, J.C., et al. 2001. Amniotic membrane transplantation in infectious corneal ulcer. *Cornea*, **20**, 720–6.

Kirkness, C.M., Seal, D.V., et al. 1991. *Sphaeropsis subglobosa* keratomycosis – first reported case. *Cornea*, **10**, 85–9.

Kiryu, H., Yoshida, S., et al. 1991. Invasion and survival of Fusarium solani in the dexamethasone-treated cornea of rabbits. *J Med Vet Mycol*, **29**, 395–406.

Kisla, T.A., Cu-Unijieng, A., et al. 2000. Medical management of *Beauveria bassiana* keratitis. *Cornea*, **19**, 405–6.

Klapper, S.R., Lee, A.G., et al. 1997. Orbital involvement in allergic fungal sinusitis. *Ophthalmology*, **104**, 2094–100.

Klippenstein, K., O'Day, D.M., et al. 1993. The qualitative evaluation of the pharmacokinetics of subconjunctivally injected antifungal agents in rabbits. *Cornea*, **12**, 512–16.

Klotz, S.A., Penn, C.C., et al. 2000. Fungal and parasitic infections of the eye. *Clin Microbiol Rev*, **13**, 662–85.

Kohn, R. and Hepler, R. 1985. Management of limited rhino-orbital mucormycosis without exenteration. *Ophthalmology*, **92**, 1440–3.

Komadina, T.G., Wilkes, T.D.I., et al. 1985. Treatment of *Aspergillus fumigatus* keratitis in rabbits with oral and topical ketoconazole. *Am J Ophthalmol*, **99**, 476–9.

Kompa, S., Langefeld, S., et al. 1999. Corneal biopsy in keratitis performed with the microtrephine. *Graefe's Arch Clin Exp Ophthalmol*, **237**, 915–19.

Koul, R.L. and Pratap, V.B. 1975. Keratomycosis in Lucknow. *Br J Ophthalmol*, **59**, 47–51.

Kozarsky, A.M., Stulting, R.D., et al. 1984. Penetrating keratoplasty for exogenous *Paecilomyces lilacinus* keratitis followed by post-operative endophthalmitis. *Am J Ophthalmol*, **98**, 552–7.

Krause, D. and Bullock, J.D. 1996. Orbital infections. In: Pepose, J.S., Holland, G.N. and Wilhelmus, K.R. (eds), *Ocular infection and immunity*, 1st edition. St Louis, MO: Mosby Year Book, 1321–40.

Kremer, I., Goldenfeld, M. and Shmueli, D. 1991. Fungal keratitis associated with contact lens wear after penetrating keratoplasty. *Ann Ophthalmol*, **23**, 342–5.

Krishnan, M.M., Kawatra, V.K., et al. 1986. Diverticulum of the lacrimal sac associated with rhinosporidiosis. *Br J Ophthalmol*, **70**, 867–8.

Kristinsson, J.H. and Sigurdsson, H. 1998. Lacrimal sac plugging caused by *Aspergillus fumigatus*. *Acta Ophthalmol Scand*, **76**, 241–2.

Kronish, J.W., Johnson, T.E., et al. 1996. Orbital infections in patients with human immunodeficiency virus infection. *Ophthalmology*, **103**, 1483–92.

Ksiazek, S.M., Morris, D.A., et al. 1994. Fungal panophthalmitis secondary to *Scedosporium apiospermum* (*Pseudallescheria boydii*) keratitis. *Am J Ophthalmol*, **118**, 531–3.

Kumar, B., Crawford, G.J. and Morlet, G.C. 1997. *Scedosporium prolificans* corneoscleritis: a successful outcome. *Aust N Z J Ophthalmol*, **25**, 169–71.

Kuo, I.C., Margolis, T.P., et al. 2001. *Aspergillus fumigatus* keratitis after laser in situ keratomileusis. *Cornea*, **20**, 342–4.

Kuriakose, T. and Thomas, P.A. 1991. Keratomycotic malignant glaucoma. *Indian J Ophthalmol*, **39**, 118–21.

Kurosawa, A., Pollock, S.C., et al. 1988. *Sporothrix schenckii* endophthalmitis in a patient with human immunodeficiency virus infection. *Arch Ophthalmol*, **106**, 376–80.

Kusaka, K., Shimamura, I., et al. 2003. Long term survival of patient with invasive aspergillosis involving orbit, paranasal sinus, and central nervous system. *Br J Ophthalmol*, **87**, 791–2.

Laatikainen, L., Tuominen, M. and von Dickhoff, K. 1992. Treatment of endogenous fungal endophthalmitis with systemic fluconazole with or without vitrectomy. *Am J Ophthalmol*, **113**, 205–7.

Lance, S.E., Friberg, T.R. and Kowalski, R.P. 1988. *Aspergillus flavus* endophthalmitis and retinitis in an intravenous drug abuser. A therapeutic success. *Ophthalmology*, **95**, 947–9.

Langford, J.D., MacCartney, D.L. and Wang, R.C. 1997. Frozen section-guided surgical debridement for management of rhino-orbital mucormycosis. *Am J Ophthalmol*, **124**, 265–7.

Larsen, J.S. 1973. Ultrasonic examinations of foreign bodies in the posterior wall of the eye. *Acta Ophthalmol (Copenh)*, **51**, 861–8.

Latgé, J.P. 1999. *Aspergillus fumigatus* and aspergillosis. *Clin Microbiol Rev*, **12**, 310–50.

Laverde, S., Moncada, L.H., et al. 1973. Mycotic keratitis: 5 cases caused by unusual fungi. *Sabouraudia*, **11**, 119–23.

Leck, A.K., Thomas, P.A., et al. 2002. Etiology of suppurative corneal ulcers in Ghana and south India, and epidemiology of fungal keratitis. *Br J Ophthalmol*, **86**, 1211–15.

Lee, B.L., Grossniklaus, H.E., et al. 1995. *Ovadendron sulphureo-ochraceum* endophthalmitis after cataract surgery. *Am J Ophthalmol*, **119**, 307–12.

Lee, B.L., Holland, G.N. and Glasgow, B.J. 1996. Chiasmal infarction and sudden blindness caused by mucormycosis in AIDS and diabetes mellitus. *Am J Ophthalmol*, **125**, 895–6.

Lee, P. and Green, W.R. 1990. Corneal biopsy. Indications, techniques and a report of a series of 87 cases. *Ophthalmology*, **97**, 718–21.

Leenders, A.C.A.P., Belkum, A.V., et al. 1999. Density and molecular epidemiology of *Aspergillus* in air and relationship to outbreaks of *Aspergillus* infection. *J Clin Microbiol*, **37**, 1752–7.

Legeais, J.M., Blanc, V., et al. 1994. Keratomycoses severes: diagnostic et traitement. *J Fr Ophtalmol*, **17**, 568–73.

Lewis, H., Aaberg, T.M., et al. 1988. Latent disseminated blastomycosis with choroidal involvement. *Arch Ophthalmol*, **106**, 527–30.

Levenson, J.E., Duffin, R.M., et al. 1984. Dematiaceous fungal keratitis following penetrating keratoplasty. *Ophthalmic Surg*, **15**, 578–82.

Levin, L.A., Avery, R., et al. 1996. The spectrum of orbital aspergillosis: a clinicopathological review. *Surv Ophthalmol*, **41**, 142–54.

Levin, P.S., Beebe, W.E. and Abbott, R.L. 1987. Successful treatment of *Paecilomyces lilacinus* endophthalmitis following cataract extraction with intraocular lens implantation. *Ophthalmic Surg*, **18**, 217–19.

Li, S., Perlman, J.I., et al. 1998. Unilateral *Blastomyces dermatitidis* endophthalmitis and orbital cellulitis: a case report and literature review. *Ophthalmology*, **105**, 1466–70.

Liao, W.Q., Shao, J.Z., et al. 1983. *Colletotrichum dematium* caused keratitis. *Chin Med J*, **96**, 391–4.

Liesegang, T.J. 1997. Contact lens-related microbial keratitis. Part I. Epidemiology. *Cornea*, **16**, 125–31.

Liesegang, T.J. and Forster, R.K. 1980. Spectrum of microbial keratitis in South Florida. *Am J Ophthalmol*, **90**, 38–47.

Limathe, J., Feindt, P., et al. 2002. *Cryptococcus neoformans* infection as scleral abscess in a cardiac allograft recipient 6 months after heart transplantation. *Transplant Proc*, **34**, 3252–4.

Lin, S.H., Lin, C.P., et al. 1999. Fungal corneal ulcers of onion harvesters in southern Taiwan. *Occup Environ Med*, **56**, 423–5.

Liu, K., Howell, D.N., et al. 1998. Morphologic criteria for the preliminary identification of *Fusarium*, *Paecilomyces* and *Acremonium* species by histopathology. *Am J Clin Pathol*, **109**, 45–54.

Locher, D.H., Adesina, A., et al. 1998. Post-operative *Rhizopus* scleritis in a diabetic man. *J Cataract Refract Surg*, **24**, 562–5.

Luttrull, J.K., Wan, W.L., et al. 1995. Treatment of ocular fungal infections with oral fluconazole. *Am J Ophthalmol*, **119**, 477–81.

Luu, K.K., Scott, I.U., et al. 2001. Endogenous *Pseudallescheria boydii* endophthalmitis in a patient with ring-enhancing brain lesions. *Ophthalmic Surg Lasers*, **32**, 325–9.

Mabry, R.L. and Mabry, C.S. 1998. Allergic fungal rhinosinusitis: experience with immunotherapy. *Arch Otolaryngol Head Neck Surg*, **124**, 1178.

Macher, A., Rodrigues, M.M., et al. 1985. Disseminated bilateral chorioretinitis due to *Histoplasma capsulatum* in a patient with the acquired immunodeficiency syndrome. *Ophthalmology*, **92**, 1159–64.

Maguire, L.J., Campbell, R.J. and Edson, R.S. 1994. Coccidioidomycosis with necrotising granulomatous conjunctivitis. *Cornea*, **13**, 539–42.

Majji, A.B., Jalali, S., et al. 1999. Role of intravitreal dexamethasone in exogenous fungal endophthalmitis. *Eye*, **13**, Part 5, 660–5.

Mangiaterra, M., Giusiano, G., et al. 2001. Keratomycosis caused by *Cylindrocarpon lichenicola*. *Med Mycol*, **39**, 143–5.

Marcus, L., Vismer, H.F., et al. 1992. Mycotic keratitis caused by *Curvularia brachyspora* (Boedjin). A report of the first case. *Mycopathologia*, **119**, 29–33.

Margo, C.E. and Bombardier, T. 1985. The diagnostic value of fungal autofluorescence. *Surv Ophthalmol*, **29**, 374–6.

Margo, C.E., Polack, F.M. and Hood, C.I. 1988. *Aspergillus* panophthalmitis complicating treatment of pterygium. *Cornea*, **7**, 285–9.

Marshall, D.H., Brownstein, S., et al. 1997. Post-traumatic corneal mucormycosis caused by *Absidia corymbifera*. *Ophthalmology*, **104**, 1107–11.

Martin, M.J., Rahman, M.R., et al. 1995–96. Mycotic keratitis: susceptibility to antiseptic agents. *Int Ophthalmol*, **19**, 299–302.

Maskin, S.L. and Alfonso, E. 1992. Fungal keratitis after radial keratotomy. *Am J Ophthalmol*, **114**, 369–70.

Maskin, S.L., Fetchick, R.J., et al. 1989. *Bipolaris hawaiiensis* – caused phaeohyphomycotic orbitopathy. A devastating fungal sinusitis in an apparently immunocompetent host. *Ophthalmology*, **96**, 175–9.

Massry, G.G., Hornblass, A. and Harrison, W. 1996. Itraconazole in the treatment of orbital aspergillosis. *Ophthalmology*, **103**, 1467–70.

Mathews, M.S. and Kuriakose, T. 1995. Keratitis due to *Cephaliophora irregularis* Thaxter. *J Med Vet Mycol*, **33**, 359–60.

Mathis, A., Malecaze, F., et al. 1988. Immunological analysis of the aqueous humour in *Candida* endophthalmitis. II. Clinical study. *Br J Ophthalmol*, **72**, 313–16.

Matsumoto, T., Masaki, J. and Okabe, T. 1979. *Cylindrocarpon tonkinense* as a cause of keratomycosis. *Trans Br Mycol Soc*, **72**, 503–4.

Matsuzaki, O., Yasuda, M. and Ichinohe, M. 1988. Keratomycosis due to *Glomerella cingulata*. *Rev Iber Micol*, **5**, Suppl 1, 30.

Mauriello, J.A. Jr, Yepez, N., et al. 1995. Invasive rhinosino-orbital aspergillosis with precipitous visual loss. *Can J Ophthalmol*, **30**, 124–30.

McAleer, R., Proudist, J.H. and Cherian, G. 1988. A dimorphic fungus *Arthrographis kalrae*, implicated in two diseases. *Rev Iber Micol*, **5**, Suppl 1, 93.

McCormick, W.F., Schochet, S.S., et al. 1975. Disseminated aspergillosis. *Aspergillus* endophthalmitis, optic nerve infarction, and carotid artery thrombosis. *Arch Pathol*, **99**, 353–9.

McCray, E., Rampell, N., et al. 1986. Outbreak of *Candida parapsilosis* endophthalmitis after cataract extraction and intraocular lens implantation. *J Clin Microbiol*, **24**, 625–8.

McDonnell, P.J., Werblin, T.P., et al. 1984–85. Mycotic keratitis due to *Beauveria alba*. *Cornea*, **3**, 213–16.

McGinnis, M.R. 1980. *Laboratory handbook of medical mycology*. New York: Academic Press.

McGinnis, M.R., Rinaldi, M.G. and Winn, R.E. 1986. Emerging agents of phaeohyphomycosis: pathogenic species of *Bipolaris* and *Exserohilum*. *J Clin Microbiol*, **24**, 250–9.

McGuire, T.W., Bullock, J.D., et al. 1991. Fungal endophthalmitis. An experimental study with a review of 17 human ocular cases. *Arch Ophthalmol*, **109**, 1289–96.

McKelvie, P.A., Wong, E.Y., et al. 2001. *Scedosporium* endophthalmitis: two fatal disseminated cases of *Scedosporium* infection presenting with endophthalmitis. *Clin Exp Ophthalmol*, **29**, 330–4.

McLaughlin-Borlace, L., Stapleton, F., et al. 1998. Bacterial biofilm on contact lenses and lens storage cases in wearers with microbial keratitis. *J Appl Microbiol*, **84**, 827–38.

McLean, F.M., Ginsberg, L.E. and Stanton, C.A. 1996. Perineural spread of rhinocerebral mucormycosis. *AJNR Am J Neuroradiol*, **17**, 114–16.

McQuillen, D.P., Zingman, B.S., et al. 1992. Invasive infections due to *Candida krusei*: report of ten cases of fungaemia that include three cases of endophthalmitis. *Clin Infect Dis*, **14**, 472–8.

Mendicute, J., Ondarra, R., et al. 1995. The use of collagen shields impregnated with amphotericin B to treat *Aspergillus* keratomycosis. *CLAO J*, **21**, 252–5.

Mendonza, L., Ajello, L. and McGinnis, M.R. 1996. Infections caused by the oomycetous pathogen *Pythium insidiosum*. *J Mycol Med*, **6**, 151–64.

Merkur, A.B. and Hodge, W.G. 2002. *Rhodotorula rubra* endophthalmitis in an HIV positive patient. *Br J Ophthalmol*, **86**, 1444–5.

Meyer, R.D., Gaultier, C.R., et al. 1994. Fungal sinusitis in patients with AIDS: report of 4 cases and review of the literature. *Medicine (Baltimore)*, **73**, 69–78.

Mian, U.K., Mayers, M., et al. 1998. Comparison of fluconazole pharmacokinetics in serum, aqueous humour, vitreous humour, and cerebrospinal fluid following a single dose and at steady state. *J Ocular Pharmacol Ther*, **14**, 459–71.

Mikami, R. and Stemmermann, G.N. 1958. Keratomycosis caused by *Fusarium oxysporum*. *Am J Clin Pathol*, **29**, 257–62.

Miloshev, B., Davidson, C.M., et al. 1966. Aspergilloma of paranasal sinuses and orbit in Northern Sudanese. *Lancet*, **1**, 746–7.

Mino de Kaspar, H., Zoulek, G., et al. 1991. Mycotic keratitis in Paraguay. *Mycoses*, **34**, 251–4.

Miyamoto, H., Ogura, Y., et al. 1997. Biodegradable scleral implant for intravitreal controlled release of fluconazole. *Curr Eye Res*, **16**, 930–5.

Mochizuki, K., Yamashita, Y., et al. 1992. Intraocular penetration and effect on the retina of fluconazole. *Lens Eye Tox Res*, **9**, 537–46.

Mohan, M., Gupta, S.K., et al. 1988. Topical silver sulphadiazine: a new drug for ocular keratomycosis. *Br J Ophthalmol*, **72**, 192–5.

Moll, G.W. Jr, Raila, F.A., et al. 1994. Rhinocerebral mucormycosis in IDDM. Sequential magnetic resonance imaging of long-term survival with intensive therapy. *Diabetes Care*, **17**, 1348–53.

Moorthy, R.S., Rao, N.A., et al. 1994. Coccidioidomycosis iridocyclitis. *Ophthalmology*, **101**, 1923–8.

Mora, P., Bovey, E.H. and Guex-Crosier, Y. 2002. Fungal endophthalmitis: management and therapy (a 9 years experience). *Klin Monatsbl Augenheilkd*, **219**, 221–5.

Moriarty, A.P., Crawford, G.J., et al. 1993. Severe corneoscleral infection. A complication of beta irradiation scleral necrosis following pterygium excision. *Arch Ophthalmol*, **111**, 947–51.

Morinelli, E.N., Dugel, P.U., et al. 1993. Infectious multifocal choroiditis in patients with acquired immunodeficiency syndrome. *Ophthalmology*, **100**, 1014–21.

Morrison, V.A. and McGlave, P.B. 1993. Mucormycosis in the BMT population. *Bone Marrow Transplant*, **11**, 383–8.

Moses, A.E., Rahav, G., et al. 1998. Rhinocerebral mucormycosis treated with amphotericin B colloidal dispersion in three patients. *Clin Infect Dis*, **26**, 1430–3.

Moses, J.S., Balachandran, C., et al. 1990. Ocular rhinosporidiosis in Tamilnadu, India. *Mycopathologia*, **111**, 5–8.

Mosier, M.A., Lusk, B., et al. 1977. Fungal endophthalmitis following intraocular lens implantation. *Am J Ophthalmol*, **83**, 1–8.

Moyer, D.V. and Edwards, J.E. Jr 1993. *Candida* endophthalmitis and central nervous system infection. In: Bodey, G.P. (ed.), *Candidiasis pathogenesis, diagnosis and treatment*. New York: Raven Press, 331–5.

Muccioli, C., Belfort, R. Jr, et al. 1995. Limbal and choroidal *Cryptococcus* infection in the acquired immunodeficiency syndrome. *Am J Ophthalmol*, **120**, 539–40.

Muralidhar, S. and Sulthana, C.M. 1995. *Rhodotorula* causing chronic dacryocystitis: a case report. *Indian J Ophthalmol*, **43**, 196–8.

Murdoch, D. and Parr, D. 1997. *Pythium insidiosum* keratitis. *Aust N Z J Ophthalmol*, **25**, 177–9.

Narang, S., Gupta, A., et al. 2001. Fungal endophthalmitis following cataract surgery: clinical presentation, microbiological spectrum, and outcome. *Am J Ophthalmol*, **132**, 609–17.

National Committee for Clinical Laboratory Standards. 1997. *Reference method for broth dilution antifungal susceptibility testing of yeasts. Approved Standard M27A*. Wayne, PA: NCCLS.

Nelson, M.E., Midgley, G. and Blatchford, N.R. 1990. Ketoconazole in the treatment of blepharitis. *Eye*, **4**, 151–9.

Nelson, P.E., Dignani, M.C. and Anaissie, E.J. 1994. Taxonomy, biology and clinical aspects of *Fusarium* species. *Clin Microbiol Rev*, **7**, 479–504.

Neuhann, T. 1976. The treatment of keratomycosis with clotrimazole. *Klin Monatsbl Augenheilkd*, **169**, 459–62.

Newman, R.M. and Kline, L.B. 1997. Evolution of fundus changes in mucormycosis. *J Neuroophthalmol*, **17**, 51–2.

Newmark, E. and Polack, F.M. 1970. *Tetraploa* keratomycosis. *Am J Ophthalmol*, **70**, 1013–15.

Nityananda, K., Sivasubramaniam, P. and Ajello, L. 1962. Mycotic keratitis caused by *Curvularia lunata*: case report. *Sabouraudia*, **2**, 35–9.

Nityananda, K., Sivasubramaniam, P. and Ajello, L. 1964. A case of mycotic keratitis caused by *Curvularia geniculata*. *Arch Ophthalmol*, **71**, 456–8.

Nunery, W.R., Welsh, M.G. and Saylor, R.L. 1985. *Pseudallescheria boydii (Petriellidium boydii)* infection of the orbit. *Ophthalmic Surg*, **16**, 296–300.

Nussbaum, E.S. and Hall, W.A. 1994. Rhinocerebral mucormycosis: changing patterns of disease. *Surg Neurol*, **41**, 152–6.

Nwufo, M.I. and Fajola, A.O. 1988. Production of amylolytic enzymes in culture by *Botryodiplodia theobromae* and *Sclerotium rolfsii* associated with the corn rots of *Colocasia esculenta*. *Acta Microbiol Hung*, **35**, 371–7.

O'Brien, T.P. 1999. Therapy of ocular fungal infections. In O'Brien, T.P. (ed.), *Ocular infections: update on therapy. Ophthalmology Clinics of North America*, **12**(1), 33–50.

O'Day, D.M. 1990. Orally administered antifungal therapy for experimental keratomycosis. *Trans Am Ophthalmol Soc*, **88**, 685–725.

O'Day, D.M. 1991. Corticosteroids: an unresolved debate. *Ophthalmology*, **98**, 845–6 (editorial).

O'Day, D.M. and Burd, E.M. 1994. Fungal keratitis and conjunctivitis. Mycology. In: Smolin, G. and Thoft, R.A. (eds), *The cornea: scientific foundations and clinical practice*, 3rd edition. Boston, MA: Little, Brown & Co., 229–39.

O'Day, D.M., Robinson, R. and Head, W.S. 1983. Efficacy of antifungal agents in the cornea. I. A comparative study. *Investig Ophthalmol Vis Sci*, **24**, 1098–102.

O'Day, D.M., Ray, W.A., et al. 1984a. Influence of the corneal epithelium on the efficacy of topical antifungal agents. *Investig Ophthalmol Vis Sci*, **25**, 855–9.

O'Day, D.M., Ray, W.A., et al. 1984b. Efficacy of antifungal agents in the cornea. II. Influence of corticosteroids. *Investig Ophthalmol Vis Sci*, **25**, 331–5.

O'Day, D.M., Ray, W.A., et al. 1984c. Efficacy of antifungal agents in the cornea. IV. Amphotericin B methyl ester. *Investig Ophthalmol Vis Sci*, **25**, 851–4.

O'Day, D.M., Head, W.S., et al. 1986. Corneal penetration of topical amphotericin B and natamycin. *Curr Eye Res*, **5**, 877–82.

O'Day, D.M., Foulds, G., et al. 1990. Ocular uptake of fluconazole after oral administration. *Arch Ophthalmol*, **108**, 1006–8.

O'Day, D.M., Ray, W.A., et al. 1991. Influence of corticosteroid on experimentally-induced keratomycosis. *Arch Ophthalmol*, **109**, 1601–4.

Oestreicher, J.H., Bashour, M., et al. 1999. *Aspergillus* mycetoma in a secondary hydroxyapatite orbital implant: a case report and literature review. *Ophthalmology*, **106**, 987–91.

Oji, E.O. 1982a. Ketoconazole: a new imidazole has both prophylactic potential and therapeutic efficacy in keratomycosis of rabbits. *Int Ophthalmol*, **5**, 163–7.

Oji, E.O. 1982b. Study of ketoconazole toxicity in rabbit cornea and conjunctiva. *Int Ophthalmol*, **5**, 169–74.

O'Keefe, M., Haining, W.M., et al. 1986. Orbital mucormycosis with survival. *Br J Ophthalmol*, **70**, 634–6.

Okhravi, N., Dart, J.K.G., et al. 1997. *Paecilomyces lilacinus* endophthalmitis with secondary keratitis: a case report and literature review. *Arch Ophthalmol*, **115**, 1320–4.

Okhravi, N., Adamson, P., et al. 1998. Polymerase chain reaction and restriction fragment length polymorphism mediated detection and speciation of *Candida* spp. causing intraocular infection. *Investig Ophthalmol Vis Sci*, **39**, 859–66.

Olurin, O., Lucas, A.O. and Oyediran, A.B. 1969. Orbital histoplasmosis due to *Histoplasma duboisii*. *Am J Ophthalmol*, **68**, 14–18.

Oneson, R.H., Feldman, P.S. and Newman, S.A. 1988. Aspiration cytology and immunohistochemistry of an orbital aspergilloma. *Diagn Cytopathol*, **4**, 59–61.

Orgel, I.K. and Cohen, K.L. 1989. Post-operative zygomycete endophthalmitis. *Ophthalmic Surg*, **20**, 584–7.

Ormerod, L.D., Hertzmark, E., et al. 1987. Epidemiology of microbial keratitis in southern California. A multivariate analysis. *Ophthalmology*, **94**, 1322–33.

Owor, R. and Wamukota, W.M. 1978. Rhinosporidiosis in Uganda: a review of 51 cases. *East African Med J*, **55**, 582–6.

Paiva, C., Batista, A.C. and Gomes, A. 1960. Endoftalmite micotica posoperatoria por *Hyalopus bogolepofii*. *Rev Bras Oftalmol*, **19**, 193–202.

Panda, A., Vajpayee, R.B. and Kumar, T.S. 1991. Critical evaluation of therapeutic keratoplasty in cases of keratomycosis. *Ann Ophthalmol*, **23**, 373–6.

Panda, A., Sharma, N. and Angra, S.K. 1996. Topical fluconazole therapy of *Candida* keratitis. *Cornea*, **15**, 373–5.

Panda, A., Sharma, N., et al. 1997. Mycotic keratitis in children: epidemiologic and microbiologic evaluation. *Cornea*, **16**, 295–9.

Panda, A., Das, G.K., et al. 1998. Corneal infection after radial keratotomy. *J Cataract Refract Surg*, **24**, 331–4.

Panda, A., Pushker, N., et al. 1999. *Rhodotorula* sp. infection in corneal interface following lamellar keratoplasty – a case report. *Acta Ophthalmol Scand*, **77**, 227–8.

Parfrey, N.A. 1986. Improved diagnosis and prognosis of mucormycosis. A clinicopathologic study of 33 cases. *Medicine (Baltimore)*, **65**, 113–23.

Parke, D.W., Jones, D.B. and Gentry, L.D. 1982. Endogenous endophthalmitis in patients with candidemia. *Ophthalmology*, **89**, 789–96.

Parrish, C.M., O'Day, D.M. and Hoyle, T.C. 1987. Spontaneous fungal corneal ulcer as an ocular manifestation of AIDS. *Am J Ophthalmol*, **104**, 302–3.

Parthiban, K., Gnanaguruvelan, S., et al. 1998. Rhinocerebral zygomycosis. *Mycoses*, **41**, 51–3.

Patel, A.S., Hemady, R.K., et al. 1994. Endogenous *Fusarium* endophthalmitis in a patient with acute lymphocytic leukaemia. *Am J Ophthalmol*, **117**, 363–8.

Pavan, P.R. and Margo, C. 1993. Endogenous endophthalmitis caused by *Bipolaris hawaiiensis* in a patient with acquired immunodeficiency syndrome. *Am J Ophthalmol*, **116**, 644–5.

Pavilack, M.A. and Frueh, B.R. 1992. Thorough curettage in the treatment of chronic canaliculitis. *Arch Ophthalmol*, **110**, 200–2.

Pe'er, J., Gnessin, H., et al. 1996. Conjunctival rhinosporidiosis caused by *Rhinosporidium seeberi*. *Arch Pathol Lab Med*, **120**, 854–8.

Penalver, F.J., Romero, R., et al. 1998. Rhinocerebral mucormycosis following donor leukocyte infusion: successful treatment with liposomal amphotericin B and surgical debridement. *Bone Marrow Transplant*, **22**, 817–18.

Penk, A. and Pittrow, L. 1999. Role of fluconazole in the long-term suppressive therapy of fungal infections in patients with artificial implants. *Mycoses*, **42**, Suppl 2, 91–6.

Perlman, E.M. and Binns, L. 1997. Intense photophobia caused by *Arthrographis kalrae* in a contact lens-wearing patient. *Am J Ophthalmol*, **123**, 547–9.

Perry, H.D., Donnenfeld, E.D., et al. 1998. Retained *Aspergillus* contaminated contact lens inducing conjunctival mass and keratoconjunctivitis in an immunocompetent patient. *CLAO J*, **24**, 57–8.

Perry, H.D., Doshi, S.J., et al. 2002. Topical cyclosporin A in the management of therapeutic keratoplasty for mycotic keratitis. *Cornea*, **21**, 161–3.

Perz, M. 1966. *Fusarium nivale* as a cause of corneal mycosis. *Klin Oczna*, **36**, 609–12.

Pettit, T.H., Olson, R.J., et al. 1980. Fungal endophthalmitis following intraocular lens implantation. A surgical epidemic. *Arch Ophthalmol*, **98**, 1025–39.

Pettit, T.H., Edwards, J.E. Jr, et al. 1996. Endogenous fungal endophthalmitis. In: Pepose, J.S., Holland, G.N. and Wilhelmus, K.R. (eds), *Ocular infection and immunity*. St Louis: Mosby, 1262–82.

Pfeifer, J.D., Grand, M.G., et al. 1991. Endogenous *Pseudallescheria boydii* endophthalmitis. Clinicopathologic findings in two cases. *Arch Ophthalmol*, **109**, 1714–17.

Pflugfelder, S.C., Flynn, H.W. Jr, et al. 1988. Exogenous fungal endophthalmitis. *Ophthalmology*, **95**, 19–30.

Pinna, A., Carta, F., et al. 2001. Endogenous *Rhodotorula minuta* and *Candida albicans* endophthalmitis in an injecting drug user. *Br J Ophthalmol*, **85**, 759.

Pitrak, D.L., Koneman, E.W., et al. 1988. *Phialophora richardsiae* infection in humans. *Rev Infect Dis*, **10**, 1195–203.

Polack, F.M., Siverio, C. and Bresky, R.H. 1976. Corneal chromomycosis: double infection by *Phialophora verrucosa* (Medlar) and *Cladosporium cladosporioides* (Frescenius). *Ann Ophthalmol*, **8**, 139–44.

Polenghi, F. and Lasagni, A. 1976. Observations on a case of mycokeratitis and its treatment with BAY b 5097 (Canesten). *Mykosen*, **19**, 223–6.

Portnoy, S.L., Insler, M.S. and Kaufman, H.E. 1989. Surgical management of corneal ulceration and perforation. *Surv Ophthalmol*, **34**, 47–58.

Powers, C.N. 1998. Diagnosis of infectious diseases: a cytopathologist's perspective. *Clin Microbiol Rev*, **11**, 341–65.

Prajna, N.V., John, R.K., et al. 2003. A randomised clinical trial comparing 2% econazole and 5% natamycin for treatment of fungal keratitis. *Br J Ophthalmol*, **87**, 1235–7.

Press, G.A., Weindling, S.M., et al. 1988. Rhinocerebral mucormycosis. MR manifestations. *J Comput Assist Tomogr*, **12**, 744–9.

Prevost, E., Kreutner, A. Jr, et al. 1980. Conjunctival lesion caused by *Rhinosporidium seeberi*. *South Med J*, **73**, 1077–9.

Pulido, J.S., Folberg, R., et al. 1990. *Histoplasma capsulatum* endophthalmitis after cataract extraction. *Ophthalmology*, **97**, 217–20.

Punithalingam, E. 1976. *Phoma oculohominis* sp. nov. from corneal ulcer. *Trans Br Mycol Soc*, **67**, 142–3.

Purgason, P.A., Hornblass, A. and Loeffler, M. 1992. Atypical presentation of fungal dacryocystitis. A report of two cases. *Ophthalmology*, **99**, 1430–2.

Puttana, S.T. 1967. Mycotic infections of the cornea. *J All-India Ophthalmol Soc*, **15**, 11–18.

Radner, A.B., Witt, M.D. and Edwards, J.E. Jr 1995. Acute invasive rhinocerebral zygomycosis in an otherwise healthy patient: case report and review. *Clin Infect Dis*, **20**, 163–6.

Ragge, N., Hart, J.C., et al. 1990. A case of fungal keratitis caused by *Scopulariopsis brevicaulis*: treatment with antifungal agents and penetrating keratoplasty. *Br J Ophthalmol*, **74**, 561–2.

Raj, P., Vella, E.J. and Bickerton, R.C. 1998. Successful treatment of rhinocerebral mucormycosis by a combination of aggressive surgical debridement and the use of systemic liposomal amphotericin B and local therapy with nebulized amphotericin – a case report. *J Laryngol Otol*, **112**, 367–70.

Rajasekaran, J., Thomas, P.A. and Srinivasan, R. 1987a. Ketoconazole in keratomycosis. In: Blodi, F., Brancato, R., et al. (eds), *Acta XXV concilium ophthalmologicum*. Amsterdam, The Netherlands: Kugler Ghedini, 2462–7.

Rajasekaran, J., Thomas, P.A., et al. 1987b. Itraconazole therapy for fungal keratitis. *Indian J Ophthalmol*, **35**, 157–60.

Rajeev, B. and Biswas, J. 1998. Molecular biologic techniques in ophthalmic pathology. *Indian J Ophthalmol*, **46**, 3–13.

Rao, N.A. and Hidayat, A.A. 2001. Endogenous mycotic endophthalmitis. Variations in clinical and histopathologic changes in candidiasis compared with aspergillosis. *Am J Ophthalmol*, **132**, 244–51.

Rao, N.A., Nerenberg, A.V. and Forster, D.J. 1991. *Torulopsis candida* (*Candida famata*) endophthalmitis simulating *Propionibacterium acnes* syndrome. *Arch Ophthalmol*, **109**, 1718–21.

Rao, S.K., Madhavan, H.N., et al. 1997. Fluconazole in filamentous fungal keratitis. *Cornea*, **16**, 700.

Read, R.W., Chuck, R.S., et al. 2000. Traumatic *Acremonium atrogriseum* keratitis following laser-assisted in situ keratomileusis. *Arch Ophthalmol*, **118**, 418–21.

Rebell, G. and Forster, R.K. 1976. *Lasiodiplodia theobromae* as a cause of keratomycosis. *Sabouraudia*, **14**, 155–70.

Rebell, G. and Forster, R.K. 1980. The fungi of keratomycosis. In: Lennette, E.H., Balows, A., et al. (eds), *Manual of clinical microbiology*, 3rd edition. Washington, DC: American Society for Microbiology, 553–61.

Rehman, M.R., Minassian, D.C., et al. 1997. Trial of chlorhexidine gluconate for fungal corneal ulcers. *Ophthalmic Epidemiol*, **4**, 141–9.

Rehman, M.R., Johnson, G.J., et al. 1998. Randomised trial of 0.2% chlorhexidine gluconate and 2.5% natamycin for fungal keratitis in Bangladesh. *Br J Ophthalmol*, **82**, 919–25.

Reidy, J.J., Sudesh, S., et al. 1997. Infection of the conjunctiva by *Rhinosporidium seeberi*. *Surv Ophthalmol*, **41**, 409–13.

Rhem, M.N., Wilhelmus, K.R. and Font, R.L. 1996. Infectious crystalline keratopathy caused by *Candida parapsilosis*. *Cornea*, **15**, 543–5.

Ribes, J.A., Vanover-Sams, C.L. and Baker, D.J. 2000. Zygomycetes in human disease. *Clin Microbiol Rev*, **13**, 236–301.

Richardson, M.D. and Shankland, G.S. 1999. *Rhizopus, Rhizomucor, Absidia* and other agents of systemic and subcutaneous zygomycoses. In: Murray, P.R., Baron, E.J., et al. (eds), *Manual of clinical microbiology*, 7th edition. Washington, DC: American Society for Microbiology, 1242–58.

Rieske, K., Handrick, W., et al. 1998. Therapy of sinuorbital aspergillosis with amphotericin B colloidal dispersion. *Mycoses*, **41**, 287–92.

Rippon, J.W. 1988. *Medical mycology: the pathogenic fungi and the pathogenic actinomycetes*, 3rd edition. Philadelphia, PA: W.B. Saunders.

Rishi, K. and Font, R.L. 2003. Keratitis caused by an unusual fungus, *Phoma* species. *Cornea*, **22**, 166–8.

Ritterband, D.C., Shah, M. and Seedor, J.A. 1997. *Colletotrichum graminicola*: a new corneal pathogen. *Cornea*, **16**, 362–4.

Ritterband, D.C., Seedor, J.A., et al. 1998. A unique case of *Cryptococcus laurentii* keratitis spread by a rigid gas-permeable contact lens in a patient with onychomycosis. *Cornea*, **17**, 115–18.

Robin, J.R., Chan, R., et al. 1989. Fluorescein-conjugated lectin visualization of fungi and acanthamoebae in infectious keratitis. *Ophthalmology*, **96**, 1198–202.

Rodenbiker, H.T. and Ganley, J.P. 1980. Ocular coccidioidomycosis. *Surv Ophthalmol*, **24**, 263–90.

Rodrigues, M.M. and Laibson, P.R. 1973. Exogenous mycotic keratitis caused by *Blastomyces dermatitidis*. *Am J Ophthalmol*, **75**, 782–9.

Rodrigues, M.M. and MacLeod, D. 1975. Exogenous fungal endophthalmitis caused by *Paecilomyces*. *Am J Ophthalmol*, **79**, 687–90.

Rodrigues, M.M., Laibson, P. and Kaplan, W. 1975. Exogenous corneal ulcer caused by *Tritirachium roseum*. *Am J Ophthalmol*, **80**, 804–6.

Rodriguez-Arres, M.T., de Rojas Silva, M.V., et al. 1995. *Aspergillus fumigatus* scleritis. *Acta Ophthalmol Scand*, **73**, 467–9.

Rolon, P.A. 1974. Epidemiology of rhinosporidiosis in the Republic of Paraguay. *Mycopathol Mycol Appl*, **52**, 155–71.

Romano, A., Segal, E. and Ben-Tovim, T. 1973. Epithelial keratitis due to *Rhodotorula*. *Ophthalmologica*, **166**, 353–9.

Roodhooft, J., van Rens, G., et al. 1998. Infectious crystalline keratopathy: a case report. *Bull Soc Belge Ophtalmol*, **268**, 121–6.

Rosa, R.H. Jr, Miller, D., et al. 1994. The changing spectrum of fungal keratitis in South Florida. *Ophthalmology*, **101**, 1005–13.

Rosenfeld, S.I., Jost, B.F., et al. 1994. Persistent *Torulopsis magnoliae* endophthalmitis following cataract extraction. *Ophthalmic Surg*, **25**, 154–6.

Roth, A.M. 1977. *Histoplasma capsulatum* in the presumed ocular histoplasmosis syndrome. *Am J Ophthalmol*, **84**, 293–8.

Roychoudhury, B., Sharma, S., et al. 1997. Fluorescent Gram stain in the microbiologic diagnosis of infectious keratitis and endophthalmitis. *Curr Eye Res*, **16**, 620–3.

Rummelt, V., Ruprecht, K.W., et al. 1991. Chronic *Alternaria alternata* endophthalmitis following intraocular lens implantation. *Arch Ophthalmol*, **109**, 178.

Safneck, J.R., Hogg, G.R. and Napier, L.B. 1990. Endophthalmitis due to *Blastomyces dermatitidis*. *Ophthalmology*, **97**, 212–16.

Sahin, B., Paydas, S., et al. 1996. Role of granulocyte colony-stimulating factor in the treatment of mucormycosis. *Eur J Clin Microbiol Infect Dis*, **15**, 866–9.

Saltoglu, N., Tasova, Y., et al. 1998. Rhinocerebral zygomycosis treated with liposomal amphotericin B and surgery. *Mycoses*, **41**, 45–9.

Samuels, G.J. and Brayford, D. 1993. Phragmosporous *Nectria* species with *Cylindrocarpon* anamorphs. *Sydowia*, **45**, 55–80.

Saracli, M.A., Erdem, V., et al. 2003. *Scedosporium apiospermum* keratitis treated with itraconazole. *Med Mycol*, **41**, 111–14.

Savani, D.V., Perfect, J.R., et al. 1987. Penetration of new azole compounds into the eye and efficacy in experimental *Candida* endophthalmitis. *Antimicrob Agents Chemother*, **31**, 6–10.

Savino, D.F. and Margo, C.E. 1983. Conjunctival rhinosporidiosis: light and electron microscopic study. *Ophthalmology*, **90**, 1482–9.

Scherer, W.J. and Lee, K. 1997. Implications of early systemic therapy on the incidence of endogenous fungal endophthalmitis. *Ophthalmology*, **104**, 1593–8.

Schreiber, W., Olbrisch, A., et al. 2003. Combined topical fluconazole and corticosteroid treatment for experimental *Candida albicans* keratomycosis. *Investig Ophthalmol Vis Sci*, **44**, 2634–43.

Schulman, J.A., Peyman, G., et al. 1987. Toxicity of intravitreal injection of fluconazole in the rabbit. *Can J Ophthalmol*, **22**, 304–6.

Schulman, J.A., Peyman, G.A., et al. 1991. Ocular toxicity of experimental intravitreal itraconazole. *Int Ophthalmol*, **15**, 21–4.

Schuman, J.S. and Friedman, A.H. 1983. Retinal manifestations of the acquired immune deficiency syndrome (AIDS): cytomegalovirus, *Candida albicans*, *Cryptococcus*, toxoplasmosis and *Pneumocystis carinii*. *Trans Ophthalmol Society UK*, **103**, pt.2, 177–90.

Schwarz, J., Salfelder, K. and Viloria, J.E. 1977. *Histoplasma capsulatum* in vessels of the choroid. *Ann Ophthalmol*, **9**, 633–6.

Scott, I.U., Flynn, H.W. Jr, et al. 2002. Exogenous endophthalmitis caused by amphotericin B-resistant *Paecilomyces lilacinus*: treatment options and visual outcomes. *Arch Ophthalmol*, **119**, 916–19.

Seal, D.V., Stapleton, F. and Dart, J. 1992. Possible environmental sources of *Acanthamoeba* sp. in contact lens wearers. *Br J Ophthalmol*, **76**, 424–7.

Seal, D.V., Hay, J., et al. 1996. Successful medical therapy of *Acanthamoeba* keratitis with topical chlorhexidine and propamidine. *Eye*, **10**, 413–21.

Seal, D.V., Bron, J. and Hay, J. 1998. *Ocular infection: investigation and treatment in practice*. London: Martin Dunitz.

Searl, S.S., Udell, I.J., et al. 1981. *Aspergillus* keratitis with intraocular invasion. *Ophthalmology*, **88**, 1244–50.

Segal, E., Romano, A., et al. 1975. *Rhodotorula rubra* – use of eye infection. *Mykosen*, **18**, 107–11.

Seiff, S.R., Choo, P.H. and Carter, S.R. 1999. Role of local amphotericin B therapy for sino-orbital fungal infections. *Ophthalmic Plast Reconstruct Surg*, **15**, 28–31.

Sengupta, P., Bose, J., et al. 1975. Ocular rhinosporidiosis in West Bengal. *J Indian Med Assoc*, **64**, 68–71.

Shah, G.K., Vander, J. and Eagle, R.C. Jr 2000. Intralenticular *Candida* species abscess in a premature infant. *Am J Ophthalmol*, **129**, 390–1.

Shah, C.V., Jones, D.B. and Holz, E.R. 2001. *Microsphaeropsis olivacea* keratitis and consecutive endophthalmitis. *Am J Ophthalmol*, **131**, 142–3.

Shami, M.J., Freeman, W., et al. 1991. A multicentre study of *Pneumocystis* choroidopathy. *Am J Ophthalmol*, **112**, 15–22.

Shankland, G.S. and Richardson, M.D. 1989. Possible role of preserved lemon juice in the epidemiology of *Candida* endophthalmitis in heroin addicts. *Eur J Clin Microbiol Infect Dis*, **8**, 87–9.

Sharma, S., Sankaridurg, P.R. and Ramachandran, L.L. 1994. Is the conjunctival flora a reflection of the pathogenic bacteria causing corneal ulceration? *Investig Ophthalmol Vis Sci*, **35**, Suppl, S1947, Abstract.

Sharma, S., Silverberg, M., et al. 1998. Early diagnosis of mycotic keratitis: predictive value of potassium hydroxide preparation. *Indian J Ophthalmol*, **46**, 31–5.

Sharma, S., Kunimoto, D.Y., et al. 2002. Evaluation of corneal scraping smear examination methods in the diagnosis of bacterial and fungal keratitis. *Cornea*, **21**, 643–7.

Sheikh, H.A., Mahgoub, S. and Badi, K. 1974. Postoperative endophthalmitis due to *Trichosporon cutaneum*. *Br J Ophthalmol*, **58**, 591–4.

Shin, J.Y., Kim, H.M. and Hong, J.W. 2002. Keratitis caused by *Verticillium* species. *Cornea*, **21**, 240–2.

Shmuely, H., Kremer, I., et al. 1997. *Candida tropicalis* multifocal endophthalmitis as the only initial manifestation of pacemaker endocarditis. *Am J Ophthalmol*, **123**, 559–60.

Shrestha, S.P., Hennig, A. and Parija, S.C. 1998. Prevalence of rhinosporidiosis of the eye and its adnexa in Nepal. *Am J Trop Med Hyg*, **59**, 231–4.

Shukla, P.K., Khan, Z.A., et al. 1983. Clinical and experimental keratitis caused by *Colletotrichum* state of *Glomerella cingulata* and *Acrophialophora fusispora*. *Sabouraudia*, **21**, 137–47.

Shukla, P.K., Jain, M., et al. 1989. Mycotic keratitis caused by *Phaeotrichoconis crotalariae*: new report. *Mycoses*, **32**, 230–2.

Sica, S., Morace, G., et al. 1993. Rhinocerebral zygomycosis in acute lymphoblastic leukaemia. *Mycoses*, **36**, 289–91.

Silva, B.M., Mendes, R.P., et al. 1988. Paracoccidioidomycosis: study of six cases with ocular involvement. *Mycopathologia*, **102**, 87–96.

Simmons, R.B., Buffington, J.R. and Ward, M. 1986. Morphology and ultrastructure of fungi in extended-wear soft contact lenses. *J Clin Microbiol*, **24**, 21–5.

Simonsz, H.J. 1983. Keratomycosis caused by *Acremonium recifei*, treated with keratoplasty, miconazole and ketoconazole. *Doc Ophthalmol*, **56**, 131–5.

Singh, S.M., Khan, R., et al. 1989. Clinical and experimental mycotic corneal ulcer caused by *Aspergillus fumigatus* and the effect of oral ketoconazole in the treatment. *Mycopathologia*, **106**, 133–41.

Singh, S.M., Sharma, S. and Chatterjee, P.K. 1990. Clinical and experimental mycotic keratitis caused by *Aspergillus terreus* and the effect of oral oxiconazole in the treatment. *Mycopathologia*, **112**, 127–37.

Sinskey, R.M. and Anderson, W.B. 1955. Miliary blastomycosis with metastatic spread to the posterior uvea of both eyes. *Arch Ophthalmol*, **54**, 602–4.

Slack, J.W., Hyndiuk, R.A., et al. 1992. Blastomycosis of the eyelid and conjunctiva. *Ophthalmic Plast Reconstruct Surg*, **8**, 143–9.

Slade, M.P. and McNab, A.A. 1991. Fatal mucormycosis therapy associated with deferoxamine. *Am J Ophthalmol*, **112**, 594–5.

Slomovic, M.R., Forster, R.K. and Gelender, H. 1985. *Lasiodiplodia theobromae* panophthalmitis. *Can J Ophthalmol*, **20**, 225–8.

Smiddy, W.E. 1998. Treatment outcomes of endogenous fungal endophthalmitis. *Curr Opin Ophthalmol*, **9**, 66–70.

Song, A., Dubovy, S.R., et al. 2002. Endogenous fungal retinitis in a patient with acute lymphocytic leukaemia manifesting as uveitis and optic nerve lesion. *Arch Ophthalmol*, **120**, 1754–6.

Specht, C.S., Mitchell, K.T., et al. 1991. Ocular histoplasmosis with retinitis in a patient with acquired immunodeficiency syndrome. *Ophthalmology*, **98**, 1356–9.

Sponsel, W.E., Graybill, J.R., et al. 2002. Ocular and systemic posaconazole (SCH-56592) treatment of invasive *Fusarium solani* keratitis and endophthalmitis. *Br J Ophthalmol*, **86**, 829–30.

Sponsler, T.A., Sassani, J.W., et al. 1992. Ocular invasion in mucormycosis. *Surv Ophthalmol*, **36**, 345–50.

Sridhar, M.S., Garg, P., et al. 2000a. *Aspergillus flavus* keratitis following laser *in situ* keratomileusis. *Am J Ophthalmol*, **129**, 802–4.

Sridhar, M.S., Garg, P., et al. 2000b. Fungal keratitis after *in situ* keratomileusis. *J Cataract Refract Surg*, **26**, 613–15.

Srimuang, S., Roongruangchui, K., et al. 1996. Immunological diagnosis of tropical ocular diseases: *Toxocara*, *Pythium insidiosum*, *Pseudomonas* (*Burkholderia*) *pseudomallei*, *Mycobacterium chelonei* and *Toxoplasma gondii*. *Int J Tissue Res*, **18**, 23–5.

Srinivasan, M., Gonzales, C.A., et al. 1997. Epidemiology and aetiological diagnosis of corneal ulceration in Madurai, South India. *Br J Ophthalmol*, **81**, 965–71.

Srinivasan, R., Kanungo, R. and Goyal, J.L. 1991. Spectrum of oculomycosis in South India. *Acta Ophthalmol*, **69**, 744–9.

Srivastava, O.P., Lal, B., et al. 1977. Mycotic keratitis due to *Rhizoctonia* sp. *Sabouraudia*, **15**, 125–31.

Starr, M.B. 1987. *Paecilomyces lilacinus* keratitis: two case reports in extended-wear contact lens wearers. *CLAO J*, **13**, 95–101.

Stenson, S., Brookner, A. and Rosenthal, S. 1982. Bilateral endogenous necrotising scleritis due to *Aspergillus oryzae*. *Ann Ophthalmol*, **14**, 67–72.

Stern, G.A. and Buttross, M. 1991. Use of corticosteroids in combination with antimicrobial drugs in the treatment of infectious corneal disease. *Ophthalmology*, **98**, 847–53.

Stern, L.E. and Kagan, R.J. 1999. Rhinocerebral mucormycosis in patients with burns: case report and review of the literature. *J Burn Care Rehab*, **20**, 303–6.

Stone, R.D., Irvine, A.R. and O'Connor, G.R. 1975. *Candida* endophthalmitis: report of an unusual case with isolation of the etiologic agent by vitreous biopsy. *Ann Ophthalmol*, **7**, 757–62.

Strasser, M.D., Kennedy, R.J. and Adam, R.D. 1996. Rhinocerebral mucormycosis. Therapy with amphotericin B lipid complex. *Arch Intern Med*, **156**, 337–9.

Streeten, B.W., Rebuzzi, D.D. and Jones, D.B. 1974. Sporotrichosis of the orbital margin. *Am J Ophthalmol*, **77**, 750–5.

Sullivan, L.J., Snibson, G., et al. 1994. *Scedosporium prolificans* sclerokeratitis. *Aust N Z J Ophthalmol*, **22**, 207–9.

Sunba, M.S. and al-Ali, S.Y. 1989. The histological and the histopathological pattern of conjunctival rhinosporidiosis associated with papilloma virus infection. *Histol Histopathol*, **4**, 257–64.

Sundaram, B.M., Badrinath, S. and Subramanian, S. 1989. Studies on mycotic keratitis. *Mycoses*, **32**, 568–72.

Sutton, D.A., Fothergill, A.W. and Rinaldi, M.G. 1998. *Guide to clinically significant fungi*. Baltimore, MD: Williams & Wilkins.

Swoboda, H. and Ullrich, R. 1992. Aspergilloma in the frontal sinus expanding into the orbit. *J Clin Pathol*, **45**, 629–30.

Tabbara, K.F. and al-Jabarti, A.L. 1998. Hospital construction-associated outbreak of ocular aspergillosis after cataract surgery. *Ophthalmology*, **105**, 522–6.

Tanaka, M., Kobayashi, Y., et al. 2001. Analysis of predisposing clinical and laboratory findings for the development of endogenous fungal endophthalmitis. A retrospective 12-year study of 79 eyes of 46 patients. *Retina*, **21**, 572–4.

Tandon, R.N., Wahab, S. and Srivastava, O.P. 1984. Experimental infection by *Candida krusei* (Cast.) Berkhout isolated from a case of corneal ulcer and its sensitivity to antimycotics. *Mykosen*, **27**, 355–60.

Tanure, M.A., Cohen, E.J., et al. 2000. Spectrum of fungal keratitis at Wills Eye Hospital, Philadelphia, Pennsylvania. *Cornea*, **19**, 307–12.

Tao, Y., Bazan, H.E. and Bazan, N.G. 1995. Platelet-activating factor induces the expression of the metallo-proteinases-1 and -9, but not -2 or -3 in the corneal epithelium. *Investig Ophthalmol Vis Sci*, **36**, 346–54.

Taravella, M.J., Johnson, D.W., et al. 1997. Infectious posterior scleritis caused by *Pseudallescheria boydii*. Clinicopathologic findings. *Ophthalmology*, **104**, 1312–16.

Thakar, M. 1994. Oral fluconazole therapy for keratomycosis. *Acta Ophthalmol Copenh*, **72**, 765–7.

Theodore, F.H., Littman, M.L. and Almeda, E. 1962. Endophthalmitis following cataract extraction due to *Neurospora sitophila*, a so called nonpathogenic fungus. *Am J Ophthalmol*, **53**, 35–9.

Thianprasit, M., Chaiprasert, A. and Imwidthaya, P. 1996. Human pythiosis. *Curr Topics Med Mycol*, **7**, 43–54.

Thomas, M.A. and Kaplan, H.J. 1991. Surgical removal of subfoveal neovascularization in the presumed ocular histoplasmosis syndrome. *Am J Ophthalmol*, **111**, 1 7.

Thomas, P.A. 1990. Fungi in keratitis: detection, susceptibility patterns and pathogenic mechanisms. PhD thesis, Bharathidasan University, Tiruchirappalli, India.

Thomas, P.A. 1994. Mycotic keratitis: an underestimated mycosis. *J Med Vet Mycol*, **32**, 235–54.

Thomas, P.A. 2003. Current perspectives on ophthalmic mycoses. *Clin Microbiol Rev*, **16**, 730–97.

Thomas, P.A. and Rajasekaran, J. 1988. Treatment of *Aspergillus* keratitis with imidazoles and related compounds. In: Vanden Bossche, H., Mackenzie, D.W.R. and Cauwenbergh, G. (eds), *Aspergillus and aspergillosis*. New York: Plenum Press, 267–79.

Thomas, P.A., Abraham, D.J., et al. 1987. Oral ketoconazole in keratomycosis. *Indian J Ophthalmol*, **35**, 197–203.

Thomas, P.A., Abraham, D.J., et al. 1988. Oral itraconazole therapy for mycotic keratitis. *Mycoses*, **31**, 271–9.

Thomas, P.A., Garrison, R.G. and Jansen, T. 1991a. Intrahyphal hyphae in corneal tissue from a case of keratitis due to *Lasiodiplodia theobromae*. *J Med Vet Mycol*, **29**, 263–7.

Thomas, P.A., Kuriakose, T., et al. 1991b. Use of lactophenol cotton blue mounts of corneal scrapings as an aid to the diagnosis of mycotic keratitis. *Diagn Microbiol Infect Dis*, **14**, 219–24.

Thomas, P.A., Jansen, T., et al. 1995. Virulence factors of *Lasiodiplodia theobromae* in fungal keratitis. In: Pasricha, J.K. (ed.), *Indian ophthalmology today*. New Delhi, India: All-India Ophthalmological Society, 3–4.

Thomas, P.A., Kalavathy, C.M. and Devanandan, P. 1998. *Lasiodiplodia theobromae* keratitis – a clinical profile. *J Tamilnadu Ophthalmic Assoc*, **39**, 31–2.

Thygeson, P. and Okumoto, M. 1974. Keratomycosis: a preventable disease. *Trans Am Acad Ophthalmol Otolaryngol*, **78**, OP433–9.

Tjebbes, G.W., van Delft, J.L. and van Haeringen, N.J. 1993. Production of lipid mediators in experimental keratitis of rabbit eye. *J Lipid Mediators*, **8**, 87–93.

Torres, M.A., Mohamed, J., et al. 1985. Topical ketoconazole for fungal keratitis. *Am J Ophthalmol*, **100**, 293–8.

Toth, J., Bausz, M. and Imre, L. 1996. Unilateral *Malassezia furfur* blepharitis after perforating keratoplasty. *Br J Ophthalmol*, **80**, 488.

Toussoun, T.A. and Nelson, P.E. 1976. *Fusarium. A pictorial guide to the identification of Fusarium species according to the taxonomic system of Snyder and Hansen*, 2nd edition. University Park, PA: The Pennsylvania State University Press.

Uecker, F.A. 1994. Ontogeny of the ascoma of *Glomerella cingulata*. *Mycologia*, **86**, 82–8.

Ukety, T.O., Kaimbo, K., et al. 1992. Conjunctival rhinosporidiosis. Report of three cases from Zaire. *Ann Soc Belge Med Trop*, **72**, 219–23.

Upadhyay, M.P., Karmacharya, P.C.D., et al. 1991. Epidemiological characteristics, predisposing factors and etiological diagnosis of corneal ulceration in Nepal. *Am J Ophthalmol*, **111**, 92–9.

Vajpayee, R.B., Gupta, S.K., et al. 1990. Ocular atopy and mycotic keratitis. *Ann Ophthalmol*, **22**, 369–72.

Vajpayee, R.B., Angra, S.K., et al. 1993. Laboratory diagnosis of keratomycosis: comparative evaluation of direct microscopy and culture results. *Ann Ophthalmol*, **25**, 68–71.

Valenton, M.J., Rinaldi, M.G. and Butler, E.E. 1975. A corneal abscess due to the fungus *Botryodiplodia theobromae*. *Can J Ophthalmol*, **10**, 416–18.

Van Cauteren, H., Heykants, J., et al. 1987. Itraconazole: pharmacologic studies in animals and humans. *Rev Infect Dis*, **9**, Suppl 1, S43–6.

Van Cutsem, J. and van Gerven, F. 1991. Activité antifongique in vitro de l' itraconazole sur les champignons filamenteux opportunistes: traitement de la keratomycose et de la penicilliose experimentales. *J Mycol Med*, **1**, 10–15.

Vanden Bossche, H., Marichal, P., et al. 1993. Effects of itraconazole on cytochrome P-450-dependent sterol 14-alpha demethylation and reduction of 3-ketosteroids in *Cryptococcus neoformans*. *Antimicrob Agents Chemother*, **37**, 2101–5.

Van der Westhuijzen, A.J., Grotepass, F.W., et al. 1989. A rapidly fatal palatal ulcer: rhinocerebral mucormycosis. *Oral Surg Oral Med Oral Pathol*, **68**, 32–6.

Van Johnson, E., Kline, L.B., et al. 1988. Bilateral cavernous sinus thrombosis due to mucormycosis. *Arch Ophthalmol*, **106**, 1089–92.

Vecsei, V.P., Huber-Spitzy, V., et al. 1994. Canaliculitis: difficulties in diagnosis, differential diagnosis and comparison between conservative and surgical treatment. *Ophthalmologica*, **208**, 314–17.

Velazquez, A.J., Goldstein, M.H. and Driebe, W.T. 2002. Preseptal cellulitis caused by *Trichophyton* (ringworm). *Cornea*, **21**, 312–14.

Vemuganti, G.K., Garg, P., et al. 2002. Evaluation of agent and host factors in progression of mycotic keratitis. A histologic and microbiologic study of 167 corneal buttons. *Ophthalmology*, **109**, 1538–46.

Verma, S. and Graham, E.M. 1995. *Cryptococcus* presenting as cloudy choroiditis in an AIDS patient. *Br J Ophthalmol*, **79**, 618–19.

Verma, S. and Tuft, S.J. 2002. *Fusarium solani* keratitis following LASIK for myopia. *Br J Ophthalmol*, **86**, 1190–1.

Vida, L. and Moel, S.A. 1974. Systemic North American blastomycosis with orbital involvement. *Am J Ophthalmol*, **77**, 240–2.

Vieira-Dias, D., Sena, C.M., et al. 1997. Ocular and concomitant cutaneous sporotrichosis. *Mycoses*, **40**, 197–201.

Virgile, R., Perry, H.D., et al. 1993. Human infectious corneal ulcer caused by *Pythium insidiosum*. *Cornea*, **12**, 81–3.

Vismer, H.F., Marasas, W.F., et al. 2002. *Fusarium dimerum* as a cause of eye infections. *Med Mycol*, **40**, 399–406.

Vitale, A.T., Spaide, R.F., et al. 1992. Orbital aspergillosis in an immunocompromised host. *Am J Ophthalmol*, **113**, 725–6.

Vukovic, Z., Bobic-Radovanovic, A., et al. 1995. An epidemiological investigation of the first outbreak of rhinosporidiosis in Europe. *J Trop Med Hyg*, **98**, 333–7.

Waddell, K.M., Lucas, S.B. and Downing, R.G. 2000. Case reports and small case series: conjunctival cryptococcosis in the acquired immunodeficiency syndrome. *Arch Ophthalmol*, **118**, 1452–3.

Wagoner, M.D., Badr, I.A. and Hidayat, A.A. 1999. *Chrysosporium parvum* keratomycosis. *Cornea*, **18**, 616–20.

Warren, C.M. 1964. Dangers of steroids in ophthalmology with report of a case of mycotic perforating corneal ulcer. *J Med Assoc Alabama*, **33**, 229–33.

Washburn, R.G., Kennedy, D.W., et al. 1988. Chronic fungal sinusitis in apparently normal hosts. *Medicine (Baltimore)*, **67**, 231–47.

Webb, M., Dowdy, L., et al. 1998. Cerebral mucormycosis after liver transplantation: a case report. *Clin Transplant*, **12**, 596–9.

Weinberg, J.R., Smith, A., et al. 1993. Rhinocerebral mucormycosis, diabetes mellitus and adrenogenital syndrome. *Br J Clin Pract*, **47**, 108–9.

Weinberg, R.S. 1999. Uveitis. Update on therapy. *Ophthalmol Clin North Am*, **12**, 1, 71–81.

Weishaar, P.D., Flynn, H.W. Jr, et al. 1998. Endogenous *Aspergillus* endophthalmitis. Clinical features and treatment outcomes. *Ophthalmology*, **105**, 57–65.

Weissgold, D.J., Maguire, A.M. and Brucker, A.J. 1996. Management of postoperative *Acremonium* endophthalmitis. *Ophthalmology*, **103**, 749–56.

Weissgold, D.J., Orlin, S.E., et al. 1998. Delayed-onset fungal keratitis after endophthalmitis. *Ophthalmology*, **105**, 258–62.

Wilhelmus, K.R. and Jones, D.B. 2001. *Curvularia* keratitis. *Trans Am Ophthalmol Soc*, **99**, 111–32.

Wilhelmus, K.R. and Robinson, N.M. 1991. Infectious crystalline keratopathy caused by *Candida albicans*. *Am J Ophthalmol*, **112**, 322–5.

Wilhelmus, K.R., Robinson, N.M. and Font, R.L. 1988. Fungal keratitis in contact lens wearers. *Am J Ophthalmol*, **106**, 708–14.

Williamson, J., Gordon, A.M., et al. 1968. Fungal flora of the conjunctival sac in health and disease. *Br J Ophthalmol*, **52**, 127–37.

Wilson, L.A. 1996. Biomaterials and ocular infection. In: Pepose, J.S., Holland, G.N. and Wilhelmus, K.R. (eds), *Ocular infection and immunity*, 1st edition. St Louis, MO: Mosby Year Book, 215–31.

Wilson, L.A. and Ahearn, D.G. 1986. Association of fungi with extended wear-soft contact lenses. *Am J Ophthalmol*, **101**, 434–6.

Wilson, L.A. and Ajello, L. 1998. Agents of oculomycosis: fungal infections of the eye. In: Collier, L., Balows, A. and Sussman, M. (eds)*Topley and Wilson's microbiology and microbial infections*, vol. 4. 9th edition. London: Arnold, 525–67.

Wilson, L.A., Sexton, R.R. and Ahearn, D.G. 1966. Keratochromomycosis. *Arch Ophthalmol*, **76**, 811–16.

Wilson, L.A., Sawant, A.D. and Ahearn, D.G. 1991. Comparative efficacies of soft contact lens disinfectant solutions against microbial films in lens cases. *Arch Ophthalmol*, **109**, 1155–7.

Winchester, K., Mathers, W.D. and Sutphin, J.E. 1997. Diagnosis of *Aspergillus* keratitis in vivo with confocal microscopy. *Cornea*, **16**, 27–31.

Witherspoon, C.D., Kuhn, F., et al. 1990. Endophthalmitis due to *Sporothrix schenckii* after penetrating ocular injury. *Ann Ophthalmol*, **22**, 385–8.

Wong, T.-Y., Ng, T.-P., et al. 1997. Risk factors and clinical outcome between fungal and bacterial keratitis. A comparative study. *CLAO J*, **23**, 275–81.

Wood, T.O. and Williford, W. 1976. Treatment of keratomycosis with amphotericin B 0.15%. *Am J Ophthalmol*, **81**, 847–9.

Wright, E.D., Clayton, Y.M., et al. 1990. Keratomycosis caused by *Dichotomophthoropsis nymphearum*. *Mycoses*, **33**, 477–81.

Wu, Z., Ying, H., et al. 2002. Fungal keratitis caused by *Scedosporium apiospermum*. *Cornea*, **21**, 519–23.

Xie, L., Dong, X. and Shi, W. 2001. Treatment of fungal keratitis by penetrating keratoplasty. *Br J Ophthalmol*, **85**, 1070–4.

Yamaguchi, H., Abe, S. and Tokuda, Y. 1993. Immunomodulating activity of antifungal drugs. *Ann N Y Acad Sci*, **685**, 447–57.

Yamaguchi, T., Hubbard, A., et al. 1984. Fungus growth on soft contact lenses with different water contents. *CLAO J*, **10**, 166–71.

Yamamoto, G.K., Pavan-Langston, D., et al. 1979. Fungal invasion of a therapeutic soft contact lens and cornea. *Ann Ophthalmol*, **11**, 1731–5.

Yamamoto, N., Matsumoto, T. and Ishibashi, Y. 2001. Fungal keratitis caused by *Colletotrichum gloeosporioides*. *Cornea*, **20**, 902–3.

Yao, Y.-F., Zhang, Y.-M., et al. 2003. Therapeutic penetrating keratoplasty in severe fungal keratitis using cryopreserved donor corneas. *Br J Ophthalmol*, **87**, 543–7.

Yau, T.H., Rivera-Velazquez, P.M., et al. 1996. Unilateral optic neuritis caused by *Histoplasma capsulatum* in a patient with the acquired immunodeficiency syndrome. *Am J Ophthalmol*, **121**, 324–6.

Yee, R.W., Cheng, C.J., et al. 1997. Ocular penetration and pharmacokinetics of topical fluconazole. *Cornea*, **16**, 64–71.

Yohai, R.A., Bullock, J.D., et al. 1994. Survival factors in rhino-orbital-cerebral mucormycosis. *Surv Ophthalmol*, **39**, 3–22.

Yoshizumi, M.O. and Banihashemi, A.R. 1988. Experimental intravitreal ketoconazole in DMSO. *Retina*, **8**, 210–15.

Zapater, R.C. and Albesi, E.J. 1979. Corneal microsporidiosis. A review and report on a case. *Ophthalmologica*, **178**, 142–7.

Zapater, R.C., de Arrechea, A. and Guevara, V.H. 1972. Queratomicosis porr *Fusarium dimerum*. *Sabouraudia*, **10**, 274–5.

Zapater, R.C., Albesi, E.J. and Garcia, G.H. 1975. Mycotic keratitis by *Drechslera spicifera*. *Sabouraudia*, **13**, 295–8.

Zapater, R.C., Brunzini, M.A., et al. 1976. El genero *Fusarium* como agente etiologico de micosis oculares (presentacion de 7 cases). *Arch Oftalmol Buenos Aires*, **51**, 279–86.

PART IV

SUBCUTANEOUS MYCOSES

Subcutaneous zygomycosis

DAVID H. ELLIS

The Entomophthorales are a group of primitive fungi belonging to the Zygomycetes, with infrequently septate (coenocytic) hyphae in which the sporangium has been reduced to function as a single conidium that is forcibly discharged at maturity. If the conidium does not land on a favorable substrate, it in turn produces a second and smaller conidium, which is again shot away. This cycle may be repeated several times. Members of the Entomophthorales are often parasitic on insects, and other animals and some species have developed very potent proteolytic enzyme systems. Although they are capable of growing saprobically in pure culture, many species require a complex nutrient medium to stimulate sporulation (King 1983). Currently the Entomophthorales are classified into six families with 22 genera containing some 132 species (Humber 1989). However, the order may be heterogeneous as recent molecular data has shown *Basidiobolus* to cluster with some of the Chytridiales, and *Conidiobolus*, *Entomophthora*, and *Zoopthora* with the Mucorales (Nagahama et al. 1995).

Basidiobolus ranarum (*Basidiobolaceae*), *Conidiobolus coronatus* (*Ancylistaceae*), and *Conidiobolus incongruus* are the only species that are known to cause human disease (Table 17.1). *Conidiobolus lamprauges* has also been reported once from a horse (Humber et al. 1989). Infections caused by *B. ranarum* usually present as chronic, painless, indurated subcutaneous masses and have been referred to by a variety of names, such as basidiobolomycosis (Kwon-Chung and Bennett 1992), entomophthoramycosis basidiobolae (Emmons et al. 1977), and subcutaneous zygomycosis (Rippon 1988; Goodman and Rinaldi 1991). *C. coronatus* infections usually occur as chronic granulomatous lesions of the nasal mucosa, often spreading to the contiguous facial skin and have also been called a variety of names, such as conidiobolomycosis, entomophthoramycosis conidiobolae, and rhinoentomophthoromycosis (Emmons et al. 1977; Kwon-Chung and Bennett 1992).

The term 'zygomycosis' has often been used to describe in the broadest sense any infection caused by a member of the Zygomycetes. However, many authors still prefer to use the term 'mucormycosis' to describe mycoses caused by members of the Mucorales and entomophthoromycosis (or entomophthoramycosis – the spelling used seems to depend upon the author) to describe mycoses caused by species belonging to the Entomophthorales. In general, fungi in the order Mucorales cause the more severe forms of disease (see Chapter 33, Systemic zygomycosis), whereas those in the order Entomophthorales cause a more chronic disease of the nasal mucosa and subcutaneous tissue (Goodman and Rinaldi 1991).

ETIOLOGIC AGENTS OF MEDICAL IMPORTANCE

Basidiobolus ranarum Eidem

The genus *Basidiobolus* is distinguished primarily by the morphology and development of forcibly discharged conidia and 'beaked' zygospores, both of which must be observed for reliable identification to genus and species level (King 1983). As all species of *Basidiobolus* lose their ability to sporulate after being maintained in culture for a relatively short time, it is very important to examine freshly isolated strains as soon as possible (Drechsler 1955; King 1979, 1983). This lack of

Table 17.1 *Families and genera of the Entomophthorales of medical importance*

Family	Genus	Key identification features
Ancylistaceae	*Conidiobolus*	Sporophores without subsporangial vesicle; nucleus difficult to observe during mitosis; nucleolus prominent Forcibly discharged multireplicative conidia with prominent papillae and villi
Basidiobolaceae	*Basidiobolus*	Sporophores with subsporangial vesicle; all cells uninucleate Forcibly discharged conidia and zygospore morphology

sporulation in culture, combined with other unsatisfactory criteria for separating the species, such as optimum growth temperature, *Streptomyces*-like odor, and zygospore wall undulation, has led to some confusion in the identification of the pathogenic isolates (Hutchison et al. 1972). In the past, pathogenic isolates have been classified as *B. ranarum* (Emmons et al. 1977), *Basidiobolus meristosporus* (Drechsler 1956; Greer and Friedman 1966; Clark 1968; Coremans-Pelseneer 1973), and *Basidiobolus haptosporus* (Srinivasan and Thirumalachar 1965, 1967; Miller and Pott 1980; Miller and Turnwald 1984). However, recent advances in taxonomy, including antigenic studies (Yangco et al. 1986), restriction analysis of rDNA (Nelson 1989), and isoenzyme banding (Cochrane et al. 1989), indicate that all human pathogenic isolates of *Basidiobolus* belong to *B. ranarum* (McGinnis 1980b; Kwon-Chung and Bennett 1992). *B. ranarum* is also known to produce extracellular protease and lipase, which can be detected in agar media (Okafor et al. 1987).

B. ranarum is commonly present as a saprophyte in soil, decaying fruit and vegetable matter, and in the dung of amphibians and reptiles. The animals show no intestinal lesions (Clark 1968; Nickerson and Hutchinson 1971; Tills 1977; Gugnani and Okafor 1980; Okafor et al. 1984; Zahari and Shipton 1988; Nelson et al. 2002), although cutaneous lesions have been reported in amphibians (Groff et al. 1991; Taylor et al. 1999). Less commonly, it has been recovered from the intestinal tract of insectivorous bats, the dung of macropod species, and from wood lice (Chaturvedi et al. 1984; Speare and Thomas 1985; Zahari and Shipton 1988).

Subcutaneous infections have been well documented in horses (Miller and Pott 1980; Miller and Campbell 1982, 1984; Miller 1985), dogs (Greene et al. 2002), and in humans (Joe and Njo-Imjo 1956; Dasgupta et al. 1976; Mugerwa 1976; Bittencourt et al. 1979, 1991; Davis et al. 1994); gastrointestinal lesions have also been noted in dogs (Miller 1985), and with increasing occurrence in humans (de Agular et al. 1980; Schmidt et al. 1986; Yousef et al. 1999; Khan et al. 2001; Lyon et al. 2001). *B. ranarum* has been reported from tropical Africa, India, Indonesia, and southeast Asia, including northern Australia (Zahari et al. 1990), with human infections being limited to the warmer regions of the world (King 1979, 1983; Gugnani 1999).

Colonies of *B. ranarum* are moderately fast growing at 30°C, flat, yellowish-gray to creamy-gray, glabrous, becoming radially folded and covered by a fine, powdery, white surface mycelium. Note that conidia ejected from the primary colony often germinate to form satellite colonies. Microscopic examination usually shows the presence of large vegetative hyphae (8–20 μm in diameter) forming numerous round (20–50 μm in diameter), smooth, thick-walled zygospores that have two closely appressed beak-like appendages (Figure 17.1). The production of 'beaked' zygospores is diagnostic for the genus. Two types of asexual conidia are formed: primary and secondary. Primary conidia are globose, one celled, solitary, and forcibly discharged from a sporophore. The sporophore has a distinct swollen area just below the spore, which actively participates in the discharge of the spore (Figure 17.1). Secondary (replicative) conidia are clavate, one celled, and passively

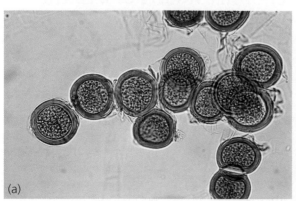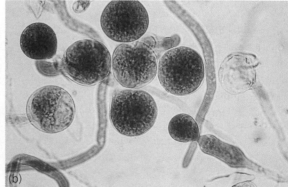

Figure 17.1 Basidiobolus ranarum: **(a)** *zygospores;* **(b)** *conidia and sporophores. Magnification ×495.*

released from a sporophore. These sporophores are not swollen at their bases. The apex of the passively released spore has a knob-like adhesive tip. These spores may function as sporangia, producing several sporangiospores. Note that isolates often lose their sporulating ability with subculture and special media, incorporating glucosamine hydrochloride and casein hydrolysate, may need to be used to stimulate sporulation (Shipton and Zahari 1987). Key features include forcibly discharged conidia and the presence of smooth-walled zygospores with prominent beak-like appendages. (For additional information, see Srinivasan and Thirumalachar 1965, 1967; Greer and Friedman 1966; Dworzack et al. 1978; McGinnis 1980a; King 1983; Rippon 1988; Drouhet and Ravisse 1993; Jong and Dugan 2003.)

Conidiobolus species

The species of the genus *Conidiobolus* produce characteristic multinucleate primary and secondary (replicative) conidia on top of unbranched conidiophores. Each subspherical conidium is discharged as a result of the pressure developed within the conidium, and each bears a more or less prominent papilla after discharge (King 1983). The genus contains 27 species; it has recently undergone taxonomic revision and is now classified in the family *Ancylistaceae* (see King 1976a, b, 1977, 1979; Humber 1989). However, isolates of *C. coronatus* and *C. incongruus* are the only species that are known to cause human disease, although *C. lamprauges* has also been reported once from a horse (Humber et al. 1989). Isolates of the *Conidiobolus* species do not tend to lose the ability to sporulate as readily as isolates of *B. ranarum*. Most strains of *C. coronatus* readily produce villous conidia.

CONIDIOBOLUS CORONATUS (COSTANTIN) BATKO

C. coronatus, synonym *Entomophthora coronata* (Costantin) Kevorkian, has a worldwide, mainly tropical distribution and is commonly present as a saprophyte in soil and on decaying vegetation (King 1983). It has been most frequently isolated in the tropical rain forests of Africa (Fromentin and Ravisse 1977). It is an occasional pathogen of insects worldwide, and has been isolated from dolphins (Sweeney and Migaki 1976), chimpanzees (Roy and Cameron 1972), sheep (Carrigan et al. 1992; Ketterer et al. 1992), llama (Moll et al. 1992; French and Ashworth 1994), and from horses as far north as southern Texas (Emmons and Bridges 1961; French et al. 1985) and as far south as Australia (Hutchins and Johnson 1972; Miller and Campbell 1982). Human infections are usually restricted to the rhinofacial area (Rippon 1988; Costa et al. 1991; Ng et al. 1991; Gugnani 1992; Kwon-Chung and Bennett 1992). However, there are occasional reports of dissemination to other sites (Rinaldi 1989; Jaffey et al. 1990; Walker et al. 1992). All

human infections have been confined to the tropics (King 1983).

There has been much confusion as to the correct name of this species, and it is often listed under an older synonym *Delacroixia coronata* (Cost.) Sacc. & P. Syd. (de Hoog and Guarro 1995). This has been further compounded by recent investigations of sterol content, which suggest separation of *C. coronatus* (as *Delacroixia coronata*) from other fungi in *Entomophthora*, *Basidiobolus*, and *Conidiobolus* (Weete and Gandhi 1997). However, the name *C. coronatus* is now generally accepted and is retained for the purposes of this chapter. With the possible exception of a few human isolates, all strains of *C. coronatus* produce villous conidia, and about 75 percent of the strains isolated from plant debris produce multireplicative conidia (King 1976a, 1979, 1983). Some strains also produce abundant spherical chlamydoconidia submerged in the agar, but zygospores are not known to be formed by this species (King 1983).

Colonies grow rapidly and are flat, cream colored, glabrous, becoming radially folded and covered by a fine, powdery, white surface mycelium and conidiophores. The lid of the Petri dish soon becomes covered with conidia that were forcibly discharged by the conidiophores. The color of the colony may become tan to brown with age. Conidiophores are simple, forming solitary, terminal conidia, which are spherical, 25–45 μm in diameter, multinucleate, single celled, and have a prominent papilla (Figure 17.2). Conidia may also produce hair-like appendages called villi. Conidia germinate to produce either:

- single or multiple hyphal tubes which may also become conidiophores that bear secondary conidia, or
- multiple short conidiophores, each bearing a small secondary conidium.

Key features include forcibly discharged conidia with conspicuous basal papillae and villous hair-like projections that readily distinguish *C. coronatus* from other species. (For additional information see Emmons and Bridges 1961; King 1976a, 1983; McGinnis 1980a; Rippon 1988; Kwon-Chung and Bennett 1992; de Hoog et al. 2000.)

CONIDIOBOLUS INCONGRUUS DRECHSLER

C. incongruus is only rarely isolated as a saprophyte from plant debris or as a pathogen causing human or animal infection. Animal infections have been reported once from a deer (Stephens and Gibson 1997) and from five cases of rhinocerebral and nasal zygomycosis in sheep (Ketterer et al. 1992). Human infections are also infrequent. The first case report was of a 15-month-old boy with indolent pneumonia, mediastinitis, and pericarditis who was successfully treated with amphotericin B (Gilbert et al. 1970; Eckert et al. 1972; King and Jong 1976). A fatal disseminated infection was then reported by Busapakum et al. (1983). Since then a case of

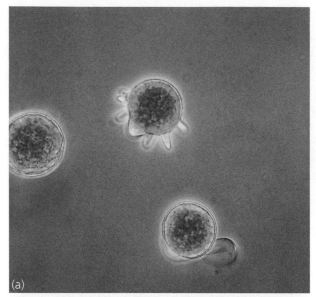

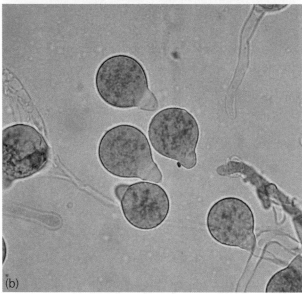

Figure 17.2 Coniodobolus coronatus: **(a, b)** *spherical conidia with hair-like appendages (villi) and prominent papillae. Magnification ×495.*

endocarditis in a cocaine abuser (Jaffey et al. 1990), a fatal disseminated case in a renal transplant patient (Walker et al. 1992), and an orbitofascial infection in a child (al-Hajjar et al. 1996) have also been reported. Another recent human case report described a granulocytic patient with pulmonary and pericardial infection and concluded that *C. incongruus* was an uncommon but highly invasive fungal pathogen that may be resistant to amphotericin B (Walsh et al. 1994). Colony characteristics are similar to *C. coronatus* and are flat, white becoming brown with age, and glabrous becoming radially folded with minimal surface mycelium. Zygospores are yellowish, spherical to elongate (15–25 μm in diameter), thick walled and without beaks. Primary conidia are forcibly discharged, spherical to pyriform (11–37×12–42 μm in diameter) with a tapering, often

sharply pointed, basal papilla. Zygospores and multiplicative conidia must be observed to identify *C. incongruus*. Key features include forcibly discharged conidia with pronounced elongated basal papilla but without the hair-like villous projections that distinguish this species from *C. coronatus*. (For additional information, see King and Jong 1976; King 1977, 1979, 1983; Busapakum et al. 1983; Rippon 1988; Kwon-Chung and Bennett 1992.)

CONIDIOBOLUS LAMPRAUGES DRECHSLER

C. lamprauges is also rare, having been isolated from plant debris and occasionally from insects, and from a granulomatous nasopharyngeal nodule of a horse (Humber et al. 1989). No human infections have been reported. *C. lamprauges*, like *C. coronatus*, produces zygospores but not villose conidia. Colony characteristics are similar to *C. coronatus* and are flat, whitish becoming radially folded with minimal surface mycelium. Zygospores are spherical (12–18 μm in diameter), thick walled and with two protruding lateral appendages or beaks. Conidiophores are hypha-like, irregular, and are not swollen at their apices. Primary conidia are forcibly discharged, densely granular when immature, spherical (13–22 μm in diameter), thin walled, with one to several flat papilla after liberation. Key features include zygospores with two 'beaks,' forcibly discharged conidia with one or more papilla but without the hair-like villous projections that also distinguish this species from *C. coronatus*. (For additional information, see Kwon-Chung and Bennett 1992; de Hoog et al. 2000; Jong and Dugan 2003.)

HOST RESPONSES

Infections caused by entomophthoraceous fungi result from species of 2 genera: *Basidiobolus* and *Conidiobolus*. Infections are chronic, slowly progressive, and generally restricted to the subcutaneous tissue in otherwise healthy individuals. Other characteristics that separate these infections from those caused by mucoraceous fungi are a lack of vascular invasion or infarction and the production of a prolific chronic inflammatory response, often with eosinophils and the Splendore–Hoeppli phenomenon around the hyphae (Kwon-Chung and Bennett 1992).

Entomophthoromycosis caused by *B. ranarum* is a chronic inflammatory or granulomatous disease generally restricted to the subcutaneous tissue of the limbs, chest, back, or buttocks, primarily occurring in children and with a predominance in males (Martinson 1972; Clark 1975; Bittencourt et al. 1979; Rippon 1988; Rinaldi 1989; Kwon-Chung and Bennett 1992; Gugnani 1999). Initially, lesions appear as subcutaneous nodules which develop into massive, firm, indurated, painless swellings which are freely movable over the underlying muscle, but attached to the skin, and which may become hyperpigmented although not ulcerated (Kwon-Chung and

Bennett 1992). The first human case caused by *B. ranarum* was reported in 1956 in Indonesia (Joe and Njo-Imjo 1956) and subsequently cases have been reported in India (Dasgupta et al. 1976), Africa (Mugerwa 1976), South America (Bittencourt et al. 1979), and northern Australia (Davis et al. 1994). The first culture-proven invasive infection by *B. ranarum* (syn. *B. haptosporus*) was reported in the USA (Dworzack et al. 1978). This was an unusual case, with a description of an ulcerative lesion perforating the hard palate of an adult patient. Gastrointestinal infection appears to be an emerging entity (de Agular et al. 1980; Schmidt et al. 1986; Yousef et al. 1999; Khan et al. 2001; Lyon et al. 2001) and poses several diagnostic difficulties. Its clinical presentation is nonspecific, there are no identifiable risk factors, and all age groups are susceptible; as a result many cases have been confused with Crohn's disease (Zavasky et al. 1999). A case of muscle invasion by *B. ranarum* resulting from deep penetration of this fungus has also been described (Kamalam and Thambiah 1984). *B. ranarum* most probably gains entrance to human tissue following traumatic implantation by insect bites, scratches, and minor trauma (Kwon-Chung and Bennett 1992). This may also explain why it is most commonly encountered in children, involving the thighs or buttocks.

Entomophthoromycosis caused by *Conidiobolus* spp. is a chronic inflammatory or granulomatous disease, which is typically restricted to the nasal submucosa and characterized by polyps or palpable restricted subcutaneous masses (Martinson 1972; Herstoff et al. 1978; Winter and Carme 1984; Rippon 1988). Clinical variants, including pulmonary and systemic infections, have also been described (Busapakum et al. 1983). Human infections occur mainly in adults with a predominance in males (80 percent of cases). Most cases have been reported from the tropical rainforest areas of central and west Africa (Okafor et al. 1983; Winter and Carme 1984) and South and Central America (Rippon 1988). Infections usually begin with unilateral involvement of the nasal mucosa. Symptoms include nasal obstruction, drainage, and sinus pain. Subcutaneous nodules develop in the nasal and perinasal regions and progressive generalized facial swelling may occur (Richardson and Warnock 1993). Infections also occur in horses, usually producing extensive nasal polyps (Emmons and Bridges 1961; Miller and Campbell 1982; Miller 1983; French et al. 1985), and other animals (Miller 1985). *C. coronatus* is also a recognized pathogen of termites, other insects, and spiders (King 1979). *Conidiobolus* infections are believed to occur following inhalation of conidia and subsequent tissue invasion of traumatized nasal mucosa (Kwon-Chung and Bennett 1992).

LABORATORY DIAGNOSIS

A tissue biopsy is essential, and swabs are not recommended. Tissue samples should be kept moist with saline or brain–heart infusion broth and be transported to the laboratory as soon as possible. It is essential that both direct microscopy and culture be performed on all specimens. It should also be noted that zygomycetous fungi have coenocytic hyphae, which will often be damaged and become nonviable during the biopsy procedure or by the chopping up or tissue grinding processes in the laboratory. This is why zygomycetous fungi that are clearly visible in direct microscopic or histopathological mounts are often difficult to grow in culture from clinical specimens. Therefore it is important to avoid excessive tissue damage when collecting the specimen and in the laboratory to gently tease the tissue apart and inoculate it directly onto the isolation media.

Direct microscopy and histopathology

Histopathology is the most rapid diagnostic method and the presence in clinical specimens of broad, 5–15 μm wide, infrequently septate, thin-walled hyphae that are surrounded by an eosinophilic sheath, known as the Splendore–Hoeppli phenomenon, is characteristic (Figure 17.3). There are two major histological differences between subcutaneous zygomycosis caused by *Basidiobolus* and *Conidiobolus* (entomophthoromycosis) and zygomycosis caused by members of the Mucorales (mucormycosis). First, in entomophthoromycosis, an eosinophilic sheath surrounds the hyphae and there is a lack of vascular invasion, which is so characteristic of infections caused by the Mucorales. Second, the hyphal elements of Mucorales are sparsely septate in tissue, whereas frequent septation is seen in tissue hyphae of *Basidiobolus* or *Conidiobolus*. Direct microscopy of biopsy tissue may be performed using 10 percent KOH ink or calcofluor mounts. A small piece of tissue may be mounted in KOH on a glass slide, teased apart, and gently heated to dissolve the tissue prior to microscopic

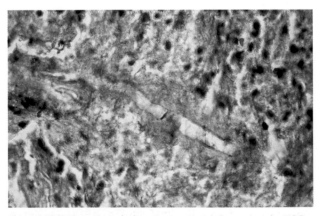

Figure 17.3 *Tissue morphology in entomophthoromycosis: H&E-stained section of infected tissue showing broad, infrequently septate, hyphae surrounded by an eosinophilic sheath (Splenodore–Hoeppli phenomenon), typical of* Basidiobolus ranarum.

examination for the presence of zygomycetous hyphae. Tissue sections should be stained with hematoxylin and eosin (H&E), periodic acid–Schiff (PAS), and Gomori's methenamine silver (GMS) (Grocott 1955) stains. H&E-stained tissue sections are best to observe the distinctive eosinophilic sheath (Splendore–Hoeppli phenomenon) that typically surrounds the hyphae. However, GMS-stained sections are still necessary as often the hyphae stain poorly (Chandler et al. 1980; Rippon 1988; Salfelder 1990).

Culture

Biopsy tissue should be inoculated onto culture media without delay; it should not be kept for an extended time in a refrigerator (Burkitt et al. 1964). Sabouraud's dextrose agar containing antibacterial antibiotics, but without cycloheximide (actidione), is usually used as the primary isolation medium. Strains of *Basidiobolus* also rapidly lose their ability to sporulate after relatively short periods of time in culture. This problem has been partially overcome by the use of media that incorporate glucosamine hydrochloride and casein hydrolysate (Shipton and Zahari 1987). Nevertheless, isolates should be examined and preserved in liquid nitrogen as quickly as possible after isolation (Jong and Dugan 2003). Most isolates of *Basidiobolus* and *Conidiobolus* are fast growing, 2–5 days at 26–28°C, initially whitish, later becoming buff or gray, with a glabrous to waxy or powdery radiate surface. For isolation from environmental sources such as soil, decaying leaves, or the intestinal contents of reptiles, specimens can be shaken in sterile water and the supernatant then pipetted onto a sterile Whatman no. 1 filter paper. The filter paper is then placed on the inside of a petri dish lid covering an agar plate. The petri dish is then incubated upside down to allow the forcible discharged conidia to land on the agar surface. Colonies take about 3 days to appear in positive samples (Rippon 1988).

Immunology and serology

Serological testing for antibodies by immunodiffusion against both *Basidiobolus* and *Conidiobolus* spp. has been developed (Kaufman et al. 1990; Imwidthaya and Srimuang 1992) and has been used on a limited basis for diagnostic purposes, especially when the culture result is negative (Lyon et al. 2001). Exoantigen tests have also been developed for the identification of cultures by immunodiffusion (Yangco et al. 1986; Toriello et al. 1989).

TREATMENT (MANAGEMENT/THERAPY)

In subcutaneous infections caused by *B. ranarum*, the therapy of choice still appears to be saturated potassium iodide solution. The usual dose has been about 30 mg/kg/day, given either as a single dose or divided into three daily doses, which should be given for 6–12 months (Koshi et al. 1972; Mugerwa 1976; Vismer et al. 1980; Kamalam and Thambiah 1982; Drouhet and Ravisse 1993; Nazir et al. 1997). In a few patients, oral ketoconazole (Piens et al. 1985), itraconazole (Gugnani 1999), and fluconazole (Davis et al. 1994; Gugnani et al. 1995) have sometimes been successful, but amphotericin B has seldom been helpful (Khan et al. 2001). Some more recent reports have demonstrated the successful use of combination potassium iodide and ketoconazole (Bittencourt et al. 1991) and itraconazole and fluconazole (Valle et al. 2001). Surgical resection alone is not curative (Kwon-Chung and Bennett 1992). In one series of gastrointestinal infection, all patients were successfully treated by the combination of surgical resection of the infected portions of the gastrointestinal tracts and itraconazole therapy, and this may be the treatment of choice for these patients (Lyon et al. 2001). In vitro antifungal susceptibility testing of nine isolates of *B. ranarum* reported low minimum inhibitory concentrations (MIC) to the azoles fluconazole, itraconazole, ketoconazole, and miconazole (Guarro et al. 1999).

In patients with submucosal infections caused by *Conidiobolus* spp., treatment options have so far been disappointing. Surgical resection of infected tissue is seldom successful and may even hasten spread of infection. Potassium iodide solution, amphotericin B, and trimethoprim–sulfamethoxazole have all been used with mixed success (Martinson 1971; Herstoff et al. 1978; Kwon-Chung and Bennett 1992). Some more recent reports of successful treatment include using fluconazole (Gugnani et al. 1995), or combination therapy with amphotericin B and terbinafine (Foss et al. 1996) and ketoconazole plus potassium iodide (Mukhopadhyay et al. 1995). In vitro resistance to amphotericin B, 5-fluorocytosine, fluconazole, itraconazole, ketoconazole, and miconazole has been reported for seven isolates of *Conidiobolus* (Guarro et al. 1999). In the rare cases of disseminated infections caused by *C. incongruus*, only one of three patients has survived following surgical debridement and amphotericin B treatment (Busapakum et al. 1983; Walsh et al. 1994; Dromer and McGinnis 2003). There is a single patient report of an 18-month-old girl with periorbital cellulites secondary to *C. incongruus* who initially failed amphotericin B therapy was successfully treated by radical resection with antifungal chemotherapy and hyperbaric oxygen (Temple et al. 2001).

REFERENCES

al-Hajjar, S., Perfect, J., et al. 1996. Orbitofascial conidiobolomycosis in a child. *Pediatr Infect Dis J*, **15**, 1130–2.

Bittencourt, A.L., Londero, A.T., et al. 1979. Occurrence of subcutaneous zygomycosis caused by *Basidiobolus haptosporus* in Brazil. *Mycopathologia*, **68**, 101–4.

Bittencourt, A.L., Arruda, S.M., et al. 1991. Basidiobolomycosis: a case report. *Pediatr Dermatol*, **8**, 325–8.

Burkitt, D.P., Wilson, A.M.M. and Felliffe, D.B. 1964. Subcutaneous phycomycosis. *Br Med J*, **1**, 1669–72.

Busapakum, R., Youngchaiyud, U., et al. 1983. Disseminated infection with *Conidiobolus incongruous*. *Sabouraudia*, **21**, 323–30.

Carrigan, M.J., Small, A.C. and Perry, G.H. 1992. Ovine nasal zygomycosis caused by *Conidiobolus incongruus*. *Aust Vet J*, **69**, 237–40.

Chandler, F.W., Kaplan, W. and Ajello, L. 1980. *A colour atlas and textbook of the histopathology of mycotic diseases*. London: Wolfe Medical.

Chaturvedi, V.P., Randhawa, H.S., et al. 1984. Prevalence of *Basidiobolus ranarum* Eidam in the intestinal tract of an insectivorous bat, *Rhinopoma harwickei hardwickei* Gray, in Delhi. *J Vet Med Mycol*, **22**, 185–9.

Clark, B.M. 1968. The epidemiology of phycomycosis. In: Wolstenholme, G.E.W. and Porter, R. (eds), *Systemic mycoses*. London, UK: Churchill, 179–99.

Clark, B.M. 1975. The epidemiology of entomophthoromycosis. In: Al-Doory, Y. (ed.), *The epidemiology of human mycotic diseases*. Springfield, IL: Charles C Thomas, 178–96.

Cochrane, B.J., Brown, J.K., et al. 1989. Genetic studies in the genus *Basidiobolus*. I Isozyme variation among isolates of human and natural populations *Mycologia*, **81**, 504–13.

Coremans-Pelseneer, J. 1973. Isolation of *Basidiobolus meristoporus* from natural sources. *Mycopathol Mycol Appl*, **49**, 173–6.

Costa, A.R., Porto, E., et al. 1991. Rhinofacial zygomycosis caused by *Conidiobolus coronatus*: a case report. *Mycopathologia*, **115**, 1–8.

Dasgupta, L.R., Agarwal, S.C., et al. 1976. Subcutaneous phycomycosis from Pondicherry, south India. *Sabouraudia*, **14**, 123–7.

Davis, S.R., Ellis, D.H., et al. 1994. First human culture-proven Australian case of entomophthoromycosis caused by *Basidiobolus ranarum*. *J Med Vet Mycol*, **32**, 225–30.

de Agular, E., Moraes, W.C. and Londero, A.T. 1980. Gastrointestinal entomophthoramycosis caused by *Basidiobolus haptosporus*. *Mycopathologia*, **72**, 101–5.

de Hoog, G.S. and Guarro, J. 1995. *Atlas of clinical fungi*. Baarn, The Netherlands: Centraalbureau voor Schimmelcultures.

de Hoog, G.S., Guarro, J., et al. 2000. *Atlas of clinical fungi*, 2nd edition. Baarn, The Netherlands: Centraalbureau voor Schimmelcultures.

Drechsler, C. 1955. A southern *Basidiobolus* forming many sporangia from globose and from elongated adhesive conidia. *J Washington Acad Sci*, **45**, 49–56.

Drechsler, C. 1956. Supplementary developmental stages of *Basidiobolus ranarum* and *Basidiobolus haptosporus*. *Mycologia*, **48**, 655–76.

Dromer, F. and McGinnis, M.R. 2003. Zygomycosis. In: Anaissie, E.J., McGinnis, M.R. and Pfaller, M.A. (eds), *Clinical mycology*. Edinburgh: Churchill Livingstone, 297–308.

Drouhet, E. and Ravisse, P. 1993. Entomophthoromycosis. *Curr Top Med Mycol*, **5**, 215–45.

Dworzack, D.L., Pollok, A.S., et al. 1978. Zygomycosis of the maxillary sinus and palate caused by *Basidiobolus haptosporus*. *Arch Intern Med*, **128**, 1274–6.

Eckert, H.L., Khoury, G.H., et al. 1972. Deep *Entomophthora* phycomycotic infection reported for the first time in the United States. *Chest*, **61**, 392–4.

Emmons, C.W. and Bridges, C.H. 1961. *Entomophthora coronata*, the etiologic agent of a phycomycosis of horses. *Mycologia*, **53**, 307–312.

Emmons, C.W., Binford, C.H., et al. 1977. *Medical mycology*, 3rd edition. Philadelphia: Lea & Febiger.

Foss, N.T., Rocha, M.R. and Lima, V.T. 1996. Entomophthoramycosis: therapeutic success by using amphotericin B and terbinafine. *Dermatology*, **193**, 258–60.

French, D.D., Haynes, P.F. and Miller, R.I. 1985. Surgical and medical management of rhinophycomycosis (conidiobolomycosis) in a horse. *J Am Vet Med Assoc.*, **186**, 1105–7.

French, R.A. and Ashworth, C.D. 1994. Zygomycosis caused by *Conidiobolus coronatus* in a llama (*Lama glama*). *Vet Pathol*, **31**, 120–2.

Fromentin, H. and Ravisse, P. 1977. Les entomophthoramycosis tropicales. *Acta Trop*, **34**, 375–94.

Gilbert, E.F., Khoury, G.H. and Pore, R.S. 1970. Histological identification of *Entomophthora* phycomycosis. Deep mycotic infection in an infant. *Arch Pathol*, **90**, 583–7.

Goodman, N.L. and Rinaldi, M.G. 1991. Agents of zygomycosis. In: Balows, A., Hausler, W.J., et al. (eds), *Manual of clinical microbiology*, 5th edition. Washington, DC: American Society for Microbiology.

Greene, C.E., Brockus, C.W., et al. 2002. Infection with *Basidiobolus ranarum* in two dogs. *J Am Vet Med Assoc*, **221**, 528–32.

Greer, D.L. and Friedman, L. 1966. Studies on the genus *Basidiobolus* with reclassification of the species pathogenic to man. *Sabouraudia*, **4**, 231–41.

Grocott, R.G. 1955. A stain for fungi in tissue sections and smears using Gomori's methenamine-silver nitrate technique. *Am J Clin Pathol*, **25**, 975–9.

Groff, J.M., Mughannam, A., et al. 1991. An epizootic of cutaneous zygomycosis in cultured dwarf African clawed frogs (*Hymenochirus curtipes*) due to *Basidiobolus ranarum*. *J Med Vet Mycol*, **29**, 215–23.

Guarro, J., Aguilar, C. and Pujol, I. 1999. In-vitro antifungal susceptibility of *Basidiobolus* and *Conidiobolus* spp. strains. *J Antimicrob Chemother*, **44**, 557–60.

Gugnani, H.C. 1992. Entomophthoromycosis due to *Conidiobolus*. *Eur J Epidemiol*, **8**, 391–6.

Gugnani, H.C. 1999. A review of zygomycosis due to *Basidiobolus ranarum*. *Eur J Epidemiol*, **15**, 923–9.

Gugnani, H.C. and Okafor, J.I. 1980. Mycotic flora of the intestine and other internal organs of certain reptiles and amphibians with special reference to characterization of *Basidiobolus* isolates. *Mykosen*, **23**, 260–8.

Gugnani, H.C., Ezeanolue, B.C., et al. 1995. Fluconazole in the therapy of tropical deep mycosis. *Mycoses*, **38**, 485–8.

Herstoff, J.K., Bogaars, H. and McDonald, C.J. 1978. Rhinophycomycosis entomophthororae. *Arch Dermatol*, **114**, 1674–8.

Humber, R.A. 1989. Synopsis of a revised classification for the Entomophthorales (Zygomycotina). *Mycotaxon*, **34**, 441–60.

Humber, R.A., Brown, C.C. and Kornegay, R.W. 1989. Equine zygomycosis caused by *Conidiobolus lamprogues*. *J Clin Microbiol*, **27**, 573–6.

Hutchins, D.R. and Johnson, K.G. 1972. Phycomycosis in the horse. *Aust Vet J*, **48**, 269–78.

Hutchison, J.A., King, D.S. and Nickerson, M.A. 1972. Studies on temperature requirements, odor production and zygospore wall undulation of the genus *Basidiobolus*. *Mycologia*, **64**, 467–74.

Imwidthaya, P. and Srimuang, S. 1992. Immundodiffusion test for diagnosing basidiobolomycosis. *Mycopathologia*, **118**, 127–31.

Jaffey, P.B., Haque, A.K., et al. 1990. Disseminated *Conidiobolus* infection with endocarditis in a cocaine abuser. *Arch Pathol Lab Med*, **114**, 1276–68.

Joe, L.K. and Njo-Imjo, T.E. 1956. *Basidiobolus ranarum* as a cause of subcutaneous mycosis in Indonesia. *Arch Dermatol*, **74**, 378–83.

Jong, S.C. and Dugan, F.M. 2003. Zygomycetes: the Order Entomophthorales. In: Howard, D.H. (ed.), *Pathogenic fungi in humans and animals*, 2nd edition. New York: Marcel Dekker, 127–39.

Kamalam, A. and Thambiah, A.S. 1982. Basidiobolomycosis following injury. *Mykosen*, **25**, 512–16.

Kamalam, A. and Thambiah, A.S. 1984. Muscle invasion of *Basidiobolus haptosporus*. *Sabouraudia: J Med Vet Mycol*, **22**, 273–7.

Kaufman, L., Mendoza, L. and Standard, P.G. 1990. Immunodiffusion test for serodiagnosing subcutaneous zygomycosis. *J Clin Microbiol*, **28**, 1887–90.

Ketterer, P.J., Kelly, M.A., et al. 1992. Rhinocerebral and nasal zygomycosis in sheep caused by *Conidiobolus incongruus*. *Aust Vet J*, **69**, 85–7.

Khan, Z.U., Khoursheed, M., et al. 2001. *Basidiobolus ranarum* as an etiologic agent of gastrointestinal zygomycosis. *J Clin Microbiol*, **39**, 2360–3.

King, D.S. 1976a. Systematics of *Conidiobolus* (Entomophthorales) using numerical taxonomy. I. Biology and cluster analysis. *Can J Bot*, **54**, 45–6.

King, D.S. 1976b. Systematics of *Conidiobolus* (Entomophthorales) using numerical taxonomy. II. Taxonomic considerations. *Can J Bot*, **54**, 1285–96.

King, D.S. 1977. Systematics of *Conidiobolus* (Entomophthorales) using numerical taxonomy. III. Descriptions of recognized species. *Can J Bot*, **55**, 718–29.

King, D.S. 1979. Systematics of fungi causing entomophthoramycosis. *Mycologia*, **71**, 731–45.

King, D.S. 1983. Entomophthorales. In: Howard, D.H. (ed.), *Fungi pathogenic for humans and animals. Part A Biology*. New York: Marcel Dekker, 61–73.

King, D.S. and Jong, S.C. 1976. Identity of etiological agent of the first deep entomophthoraceous infection of man in the United States. *Mycologia*, **68**, 181–3.

Koshi, G., Kuriem, T., et al. 1972. Subcutaneous phycomycosis caused by *Basidiobolus*: report of three cases. *Sabouraudia*, **10**, 237–43.

Kwon-Chung, K.J. and Bennett, J.W. 1992. *Medical mycology*. Philadelphia: Lea & Febiger.

Lyon, G.M., Smilack, J.D., et al. 2001. Gastrointestinal basidiobolomycosis in Arizona: clinical and epidemiological characteristics and review of the literature. *Clin Infect Dis*, **32**, 1448–55.

Martinson, F.D. 1971. Chronic phycomycosis of the upper respiratory tract, rhinophycomycosis entomophthorae. *Am J Trop Med Hyg*, **20**, 449–55.

Martinson, F.D. 1972. Clinical, epidemiological and therapeutic aspects of entomophthoromycosis. *Ann Soc Belg Med Trop*, **52**, 329–42.

McGinnis, M.R. 1980a. *Laboratory handbook of medical mycology*. London, UK: Academic Press.

McGinnis, M.R. 1980b. Recent taxonomic developments and changes in medical mycology. *Ann Rev Microbiol*, **34**, 109–35.

Miller, R.I. 1983. Investigations into the biology of three phycomycotic agents pathogenic for horses in Australia. *Mycopathologia*, **81**, 23–8.

Miller, R.I. 1985. Gastrointestinal phycomycosis in 63 dogs. *J Am Vet Med Assoc*, **186**, 473–8.

Miller, R.I. and Campbell, R.S.F. 1982. Clinical observations on equine phycomycosis. *Aust Vet J*, **58**, 221–6.

Miller, R.I. and Campbell, R.S.F. 1984. The comparative pathology of equine cutaneous phycomycosis. *Vet Pathol*, **21**, 325–32.

Miller, R. and Pott, B. 1980. Phycomycosis of the horse caused by *Basidiobolus haptosporus*. *Aust Vet J*, **56**, 224–7.

Miller, R.J. and Turnwald, G.H. 1984. Disseminated basidiobolomycosis in a dog. *Vet Pathol*, **21**, 117–19.

Moll, H.D., Schumacher, J. and Hoover, T.R. 1992. Entomophthoramycosis conidiobolae in a llama. *J Am Vet Med Assoc*, **200**, 969–70.

Mugerwa, J.W. 1976. Subcutaneous phycomycosis in Uganda. *Br J Dermatol*, **94**, 539–44.

Mukhopadhyay, D., Ghosh, L.M., et al. 1995. Entomophthoromycosis caused by *Conidiobolus coronatus*: clinicomycological study of a case. *Auris Nasus Larynx*, **22**, 139–42.

Nagahama, T., Sato, H., et al. 1995. Phylogenetic divergence of the entomophthoralean fungi: evidence from nuclear 18S ribosomal RNA gene sequences. *Mycologia*, **87**, 203–9.

Nazir, Z., Hassan, R. and Pervaiz, S. 1997. Invasive retroperitoneal infection due to *Basidiobolus ranarum* with response to potassium iodide: case reports and review of the literature. *Ann Trop Paediatr*, **17**, 161–4.

Nelson, R.T. 1989. Taxonomic relationships among isolates of the genus *Basidiobolus* revealed by restriction analysis of ribosomal DNA. *Mycol Soc Newslett*, **40**, 1.

Nelson, R.T., Cochrane, B.J., et al. 2002. Basidioboliasis in anurans in Florida. *J Wildl Dis*, **38**, 463–7.

Ng, K.H., Chin, C.S., et al. 1991. Nasofacial zygomycosis. *Oral Surg Oral Med Oral Pathol*, **72**, 685–8.

Nickerson, M.A. and Hutchinson, J.A. 1971. The distribution of the fungus *Basidiobolus ranarum* Eidam in fish, amphibians and reptiles. *Am Midl Nat*, **86**, 500–2.

Okafor, B.C., Guignani, H.C. and Jacob, A. 1983. Nasal entomophthoromycosis with laryngeal involvement. *Mykosen*, **26**, 471–4.

Okafor, J.I., Testrake, D., et al. 1984. A *Basidiobolus* sp. and its association with reptiles and amphibians in Southern Florida. *J Vet Med Mycol*, **22**, 47–51.

Okafor, J.I., Gugnani, H.C., et al. 1987. Extracellular enzyme activities by *Basidiobolus* and *Conidiobolus* isolates on solid media. *Mykosen*, **30**, 404–7.

Piens, M.A., Garin, C., et al. 1985. First case of subcutaneous phycomycosis seen in Gabon and treated successfully with ketoconazole. *Bull Soc Pathol Exot Fil*, **78**, 170–8.

Richardson, M.D. and Warnock, D.W. 1993. *Fungal Infection: diagnosis and management*. London: Blackwell Scientific.

Rinaldi, M.G. 1989. Zygomycosis. *Infect Dis Clin North Am*, **3**, 19–41.

Rippon, J.W. 1988. *Medical mycology: the pathogenic fungi and pathogenic actinomycetes*, 3rd edition. Philadelphia: W.B. Saunders.

Roy, A.D. and Cameron, M.H. 1972. Rhinophycomycosis entomophthorae occurring in a chimpanzee in the wild in East Africa. *Am J Trop Med Hyg*, **21**, 234–7.

Salfelder, K. 1990. *Atlas of fungal pathology*. Dordrecht: Kluwer Academic.

Schmidt, J.H., Howard, R.J., et al. 1986. First culture-proven gastrointestinal entomophthoromycosis in the United States: a case report and review of the literature. *Mycopathologia*, **95**, 101–4.

Shipton, W.A. and Zahari, P. 1987. Sporulation media for *Basidiobolus*. *J Med Vet Mycol*, **25**, 323–7.

Speare, R. and Thomas, A.D. 1985. Kangaroos and wallabies as carriers of *Basidiobolus haptosporus*. *Aust Vet J*, **62**, 209–10.

Srinivasan, M. and Thirumalachar, M. 1965. *Basidiobolus* species pathogenic for man. *Sabouraudia*, **4**, 32–4.

Srinivasan, M. and Thirumalachar, M. 1967. Studies on *Basidiobolus* species from India with discussion on some of the characters used in the speciation of the genus. *Mycopathol Mycol*, **33**, 56–64.

Stephens, C.P. and Gibson, J.A. 1997. Disseminated zygomycosis caused by *Conidiobolus incongruus* in a deer. *Aust Vet J*, **75**, 358–9.

Sweeney, J.C. and Migaki, G. 1976. Systemic mycosis in marine animals. *J Am Vet Med Assoc*, **169**, 946–8.

Taylor, S.K., Williams, E.S. and Mills, K.W. 1999. Mortality of captive Canadian toads from *Basidiobolus ranarum* mycotic dermatitis. *J Wildl Dis*, **35**, 64–9.

Temple, M.E., Brady, M.T. and Koranyi, K.I. 2001. Periorbital cellulites secondary to *Conidiobolus incongruus*. *Pharmacotherapy*, **21**, 351–4.

Tills, D.W. 1977. The distribution of the fungus, *Basidiobolus ranarum* Eidam, in fish, amphibians and reptiles of the southern Appalachian Region. *Trans Kansas Acad Sci*, **80**, 75–7.

Toriello, C., Zeron, E., et al. 1989. Immunological separation of Entomophthorales genera. *J Invertebr Pathol*, **53**, 358–60.

Valle, A.C., Wanke, B., et al. 2001. Entomophthoramycosis by *Conidiobolus coronatus*. Report of a case successfully treated with the combination of itraconazole and fluconazole. *Rev Inst Med Trop Sao Paulo*, **43**, 233–6.

Vismer, H.F., De Beer, H.A. and Dreyer, L. 1980. Subcutaneous phycomycosis caused by *Basidiobolus haptosporus* (Dreschsler 1947). *S Afr Med J*, **58**, 644–7.

Walker, S.D., Clark, R.V., et al. 1992. Fatal disseminated *Conidiobolus coronatus* infection in a renal transplant patient. *Am J Clin Pathol*, **98**, 559–64.

Walsh, T.J., Renshaw, G., et al. 1994. Invasive zygomycosis due to *Conidiobolus incogruus*. *Clin infect Dis*, **19**, 423–30.

Weete, J.D. and Gandhi, S.R. 1997. Sterols of the phylum Zygomycota: phylogenetic implications. *Lipids*, **32**, 1309–16.

Winter, C. and Carme, B. 1984. First rhinoentomophthoromycosis observed in Peoples' Republic of Congo (Congo Brazzaville). *Bull Soc Pathol Exot*, **77**, 377–84.

Yangco, B.G., Nettlow, A., et al. 1986. Comparative antigenic studies of species of *Basidiobolus* and other medically important fungi. *J Clin Microbiol*, **23**, 679–82.

Yousef, O.M., Smilack, J.D., et al. 1999. Gastrointestinal basidiobolomycosis. Morphologic findings in a cluster of six cases. *Am J Clin Pathol*, **112**, 610–16.

Zahari, P. and Shipton, W.A. 1988. Growth and sporulation responses of *Basidiobolus* to changes in environmental parameters. *Trans Br Mycol Soc*, **91**, 141–8.

Zahari, P., Hirst, R.G., et al. 1990. The origin and pathogenicity of *Basidiobolus* species in northern Australia. *J Med Vet Mycol*, **28**, 461–8.

Zavasky, D.M., Samowitz, W., et al. 1999. Gastrointestinal zygomycotic infection caused by *Basidiobolus ranarum*: case report and review. *Clin Infect Dis*, **28**, 1244–8.

Chromoblastomycosis

PHILIPPE ESTERRE

Chromoblastomycosis is a non-contagious, chronic, localized fungal infection of cutaneous and subcutaneous tissues caused by several species of pheoid (dematiaceous) fungi. The disease is characterized by the presence in infected tissues of brown, thick-walled, multiseptate, fungal forms known as muriform bodies (synonym: sclerotic cells, Medlar bodies) (Figure 18.1). The etiologic agents of chromoblastomycosis are black molds that are found world-wide growing as saprophytes of wood, other vegetation and soil (Gezuele et al. 1972; Dixon et al. 1980; Iwatsu et al. 1981; Okeke and Gugnani 1986) (Table 18.1). Infection follows traumatic implantation of the etiologic agent beneath the epidermis via penetration by foreign bodies such as wood splinters and tree thorns, or other minor wounds or abrasion. After implantation, the pseudo-yeast cells undergo a morphological change into the distinctive muriform bodies, and the fungus persists in this form within the infected host tissue. Nearly all cases of chromoblastomycosis are accounted for by the species *Fonsecaea pedrosoi*, *Cladophialophora carrionii* (*Cladosporium carrionii*) and *Phialophora verrucosa*. The disease is reported most frequently in individuals living in tropical and subtropical regions, notably the West Indies, Brazil, Northern Venezuela, and Madagascar, but it is cosmopolitan in distribution. The actual prevalence and incidence of chromoblastomycosis are unknown because of sporadic reporting, but in areas where interest and endemicity are high, rates of infection have ranged from 1 per 32 500 population to one per 7 000 population (Ajello 1970). In some villages in Madagascar and Northern Venezuela, the prevalence is as high as 1 per 480 inhabitants (Esterre et al. 1996) or even higher (Rodrigues et al. 1992). Historically, chromoblastomycosis has usually been refractory to treatment, but two recently developed antifungal compounds, itraconazole and terbinafine, appear to hold promise for a greatly improved prognosis in many cases.

HISTORY

The earliest known discovery of the fungal etiology of chromoblastomycosis was made by Alexandrino Pedroso in 1911 in São Paulo, Brazil, when he observed muriform bodies in nodular and ulcerated foot and leg lesions of a patient, and recovered the pheoid mold known today as *Fonsecaea pedrosoi*. Pedroso did not publish his observations, however, until 1920 (Carrion and Silva 1947). In the meantime, Max Rudolph had published, in 1914, his clinical observations from cases encountered in the Brazilian states of Minas Gerais and Goiás. In 1915, Medlar and Lane of Boston became the first investigators to publish clinical observations that were supported by positive culture results. The etiology in their case was *Phialophora verrucosa*.

Subsequently, the nomenclature for infections caused by these fungi became confused. Following the initial clinical reports, chromoblastomycosis was known by several names (McGinnis 1983). The name chromoblastomycosis (*chromo* = coloured; *blasto* = budding)

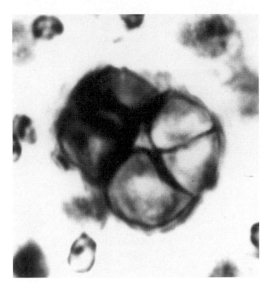

Figure 18.1 *Muriform bodies, pathognomonic for chromoblastomycosis, are subglobose, 5–13 µm in diameter, multiseptate, and dark brown in color as a result of the synthesis of melanin within the cell walls.*

was first proposed in 1922 (Terra et al. 1922). A later proposal, however, reasoned that the term 'chromoblastomycosis' is a misnomer because muriform bodies multiply by splitting apart rather than by budding, and therefore a more appropriate name would be chromomycosis. Further confusion arose from the occasionally indiscriminate use of both these terms to refer to an assortment of infections caused by pheoid fungi, irrespective of whether or not the characteristic muriform bodies were formed in tissue.

To eliminate ambiguity caused by misapplication of these two terms, the name chromoblastomycosis was retained in its original sense, chromomycosis was rejected, and the name pheohyphomycosis was coined to encompass infections caused by pheoid fungi that are clinically and pathologically different from chromoblastomycosis (Ajello et al. 1974; McGinnis 1983). Pheohyphomycosis is characterized by pheoid fungi that grow in infected tissue as pseudohyphae, hyphae, or any combination of these forms, but not as muriform bodies. It is caused by at least 100 fungal species, and has clinical features that generally differ from chromoblastomycosis. Today, the distinctions embodied by the terms 'chromoblastomycosis' and 'pheohyphomycosis' have become widely accepted. As for most biological phenomena, however, efforts to categorize the manifestations of pathogenic fungi have seldom been entirely satisfactory. For example, the two leading agents of chromoblastomycosis, *P. verrucosa* and *F. pedrosoi*, have been encountered in rare cases that can best be characterized as mycetoma and pulmonary pheohyphomycosis, respectively (Morris et al. 1995; Turiansky et al. 1995). Although there is some overlap, there is diagnostic utility in maintaining the two clinical terms, stemming from the strong predictive correlation of

morphology in tissue with etiology, clinical course, and prognosis.

ECOLOGY AND EPIDEMIOLOGY

Virtually all cases of chromoblastomycosis are caused by the species *F. pedrosoi*, *C. carrionii*, and *P. verrucosa*. Additional fungi, including *Fonsecaea compacta* (McGinnis and Schell 1980), *Rhinocladiella aquaspersa* (Borelli 1972; Schell et al. 1983), *Exophiala spinifera* (Barba-Gomez et al. 1992; Padhye et al. 1996), *Cladophialophora arxii* (Tintelnot et al. 1995), *Phaeosclera dematioides* (McGinnis et al. 1985), *Botryomyces caespitosus* (de Hoog and Rubio 1982), *Sporothrix schenckii* var. *luriei* (Padhye et al. 1992) and *Wangiella (Exophiala) dermatitidis* are also capable of forming muriform bodies in infected tissue but, with the exception of *W. dermatitidis*, these species are exceedingly rare as etiologic agents of infection in humans. Infections resulting from *W. dermatitidis* occur occasionally and are typified in tissue by short chains of spherical cells that have 0–1 septa; only very rarely do muriform bodies form, and for this reason most mycologists consider the species to be an agent not of chromoblastomycosis but of pheohyphomycosis.

Worldwide, *F. pedrosoi* causes the great majority of infections. *C. carrionii*, in addition to *F. pedrosoi*, is particularly common in Australia, Southern Madagascar and South Africa. *P. verrucosa* is reported in smaller numbers from various temperate to subtropical climates. The reasons for regional variation in the relative etiologies seemed to be linked to local ecological conditions. For example, investigation of natural and clinical isolates in Australia between 1957 and 1961 has suggested that *C. carrionii* is particularly likely to occur in drier regions that have as little as 500 mm of annual rainfall (Ridley 1961). In 1960, Brygoo and Segretain reached a similar conclusion for Madagascar, finding that *C. carrionii* was predominant in a spiny desert region of 500–600 mm of annual rainfall, whereas *F. pedrosoi* infections predominated in regions having annual rainfall of 2200–3200 mm. These observations were later extended by a retrospective study of 170 clinical isolates in Madagascar from 1970 to 1994 (Esterre et al. 1996). This study found that in the southern dry region, all cases of chromoblastomycosis (representing more than 38 percent of the total cases in Madagascar) were caused by *C. carrionii*, whereas in the evergreen forest region of much greater rainfall only *F. pedrosoi* (62 percent of the total sample) was encountered. Sources of saprobic strains of *C. carrionii* provide additional support for the concept that this species can exist in dry substrata, such as soil in Arizona (USA) (Padhye 1986), fence posts in Australia and Nigeria made from *Eucalyptus cebra* (Ridley 1961) and a *Pinus* sp. (Okeke and Gugnani 1986) and cactacae in northern Venezuela (Richard-Yegres and Yegres 1987). Morphological

Table 18.1 *Representative environmental samples for* Fonsecaea pedrosoi, Phialophora verrucosa, *and* Cladophialophora carrionii

Environmental sample	Isolates/samples (no.)	Reference
F. pedrosoi		
Plant debris, soil	34/328	–
Animal feces, nests, burrows, moss, water, palm fruit, invertebrates, epiphytes	0/294	Gezuele et al. 1972
Rotting wood	1/78	–
Lumber, logs, bark, soil, sawdust, nests	0/99	Iwatsu et al. 1981
Wood piles	2/42	–
Tree stumps, tree trunks, tree bases	6/76	–
Sawdust dump	1/76	–
Pine fenceposts	1/6	–
Beneath trees	2/10	–
Soil, stacked planks	0/16	Okeke and Gugnani 1986
Tree branch bark	1/1	Rubin et al. 1991
Rotten plants, soil	2/98	Montenegro et al. 1996
P. verrucosa		
Plant debris, soil, wasps nests, epiphytes, bird droppings	31/328	–
Animal feces, ant nests, burrows, moss, water, palm fruit, invertebrates	0/297	–
Rotting wood, lumber, sauna bath wood	38/78	Gezuele et al. 1972
Lumber, logs, pine bark, rotting wood, soil	17/157	–
Other bark, sawdust, nests	0/20	Iwatsu et al. 1981
Wooden outdoor furniture, lumber, toolshed *Pinus* sp.	5/10	–
Soil, various woody materials	0/28	Dixon et al. 1980
Wood piles	2/42	–
Stacked planks	2/11	–
Tree trunks, soil, base of tree, fencepost, sawdust dump	0/173	Okeke and Gugnani 1986
Pulp	3/84	Wang 1965
Rotten plants, soil	6/98	Montenegro et al. 1996
C. carrionii		
Soil	NS	Padhye 1986
Pine fenceposts	2/6	–
Stacked planks	3/11	–
Soil	1/5	Okeke and Gugnani 1986
Fenceposts of *Eucalyptus* spp.	2/NS	Ridley 1961
Xerophilous *Prosopis juliflora*	1/25	Richard-Yegres and Yegres 1987
Cactus *Opuntia caribea*	1/7	–

NS, not specified.

identification of muriform-like cells was even made in the medulla of two cactus species (Zeppenfeldt et al. 1994) (see Table 18.1). *P. verrucosa* seems to be the third most common etiologic agent, after *F. pedrosoi* and *C. carrionii*, and has been reported from North, Central, and South America, Asia, and Africa. In one study, it was concluded that isolates of *P. verrucosa* were more likely to be encountered in relatively cooler climates than warmer climates; five of six isolates of *P. verrucosa* among 90 cases of chromoblastomycosis came from temperate regions as opposed to subtropical and tropical regions (Carrion and Silva 1947). It is of interest that the relative proportions of species among clinical isolates do not necessarily agree with relative proportions among

environmental isolates. For example, Fukushiro (1983) found 86 and 1.6 percent of chromoblastomycosis in Japan to be caused by *F. pedrosoi* and *P. verrucosa*, respectively, whereas Iwatsu et al. (1981) were able to culture only one isolate of *F. pedrosoi*, in contrast to 17 isolates of *P. verrucosa* from environmental sources in Japan.

Cutaneous and subcutaneous infections begin when the fungus is introduced by a break in the dermal barrier. In rare instances, splinters that presumably harbored the fungus and initiated the infection were found within infected tissue (Mead and Ridley 1957; Tschen et al. 1984; Connor and Gibson 1985). In one case of chromoblastomycosis that developed following abrasion by the branch of a tree, investigators were able to return to the site 2 months later and isolate the etiologic agent by culturing portions of the branch (Rubin et al. 1991). Although some patients are able to recount specific trauma to the affected site as a possible initiating event, some as recently as 3 weeks before (Gruber et al. 1988), most patients cannot specifically recall any precipitating injury. Presumably, this is because the initiating event was often trivial or unnoticed, and symptoms, which may not occur until much later, can be mild and seemingly of little consequence initially. It is also typical that there is a considerable delay in the medical diagnosis of chromoblastomycosis. For example, in Madagascar, where there is both a relatively high prevalence of infection and a high awareness on the part of local healthcare providers, the average time of diagnosis is 1–4 years after initial symptoms arise (Esterre et al. 1996). Probable factors for delay in medical diagnosis include the gradual and indolent nature of the infection and, in some localities, a limited availability of easily accessible healthcare.

Chromoblastomycosis has been reported in most regions of the world, but is more common in tropical and subtropical settings, as evidenced by Carrion who found that 79 percent of reviewed cases occurred in patients living in such climates (Carrion and Silva 1947). Furthermore, most of these individuals resided in rural areas and were involved in agricultural or related activities. In one rural tropical region of Brazil where mycoses are a relatively frequent finding, chromoblastomycosis has remained one of the most common of the serious mycoses (Talhari et al. 1988). Most patients were between 30 and 55 years of age, and there seems to be no predilection for any particular race, because the regions of high endemicity have included populations of European, African, or Asian origin (Carrion and Silva 1947; Al-Doory 1972; Lavalle 1980; Tsuneto et al. 1989). The role of gender is unclear; most large reports have concluded that infections are more likely to occur in men than in women, but others have found female patients in equal or greater numbers (Carrion and Silva 1947; Campins and Scharyj 1953; Al-Doory 1972; Bansal and Prabhakar 1989; Tsuneto et al. 1989; Hiruma et al. 1993; Esterre et al. 1996). It is probable that these contradictory findings merely reflect the difference in sex ratios within various occupations in different cultures and geographical regions.

Chromoblastomycosis occurs most often in individuals who have no apparent underlying illness or debility. However, some preliminary data suggest that a genetic susceptibility to chromoblastomycosis might exist. In a study of 32 infected people and 77 control individuals, the presence of HLA antigen A29 was linked to a 10-fold increase in susceptibility to chromoblastomycosis (Tsuneto et al. 1989). Other people who are at risk of developing infections are those whose immune system is impaired as a result of organ transplantation or corticosteroid use (Nishimoto et al. 1984; Morales et al. 1985; Wackym et al. 1985; Perry and Sheretz 1987; Greene et al. 1990). In countries where these medical modalities are widely used, such patient populations may constitute potential for an increasing prevalence of chromoblastomycosis. Infections have also been noted in conjunction with malignancy (Foster and Harris 1987; Hiruma et al. 1992) and leprosy (Pavithran 1988; Lacaz et al. 1994), although the significance of these observations is unknown. Chromoblastomycosis has been rarely reported in animals including dogs, horses (Abid et al. 1987), toads (Bube et al. 1992) and frogs, although only some of the reports are well documented. The infection is not known to be transmissible from animals or between humans.

MYCOLOGY

In terms of mycological classification, the phylogeny of the suspected fungi has not been conclusively established and the species are currently placed into form-taxa within a parallel system of classification known as the Deuteromycota. It is suspected that the fungi, which cause chromoblastomycosis, may be related to the family *Herpotrichiellaceae*, order Ophiostomatales, phylum Ascomycota (de Hoog et al. 1995), but no sexual stage has yet been discovered in the lifecycle of any of the causative species. As this classification necessarily employs highly artificial and potentially unreliable criteria, attempts at classifying species usually do not extend above the genus level. However, those species classified in the Deuteromycota and which contain large amounts of melanin in their cell walls, causing the colony color of these species to range from pale gray or brown to black, are often collectively referred to as the black or pheoid fungi. Based on this common feature, they are sometimes ascribed the status of a form-family, the *Dematiaceae* (Matsumoto et al. 1994; Pappagianis and Ajello 1994).

Individual species of these pheoid opportunistic pathogens are identified primarily based upon various features of their microscopic morphology, including characteristics of the conidia (spores) and the cells from which the conidia arise. Some of these species are pleomorphic, meaning that they have the ability to form more than one asexual reproductive morphology in culture. Each of these different morphs is sometimes given a separate latinized name for technical reasons, as permitted by the International Code of Botanical Nomenclature. Thus, the identification process for pleomorphic isolates can be problematic. In such instances, the accepted approach has been to choose the most distinctive, conspicuous and stable morph as the basis of the binomial name (identification) that will be applied to the isolate. Any accompanying morph that is present may be referred to using a separate genus level name.

However, the Code permits taxonomists to disagree about which morph should be chosen as the basis of the binomial. For example, in the case of *F. pedrosoi*, the *Fonsecaea* morph usually predominates, and a *Rhinocladiella* morph is often well represented, whereas a very few examples of a *Phialophora* morph can sometimes be seen as well. For this reason, some taxonomists have chosen to name the fungus *F. pedrosoi* whereas others prefer *Rhinocladiella pedrosoi*. The reader is referred for further discussion of this matter to the literature (McGinnis and Salkin 1993; Schell et al. 1995).

Newer methods of identification are less practical for clinical work, but are considered to be more reflective of phylogeny, and thus have had impact upon the taxonomy of these organisms (Honbo et al. 1984a; de Hoog et al. 1995). For example, results from restriction fragment length polymorphism (RFLP) patterns of mitochondrial DNA (Kawasaki et al. 1999) or large subunit RNA analysis (de Hoog et al. 1995) have led to the confirmation of *F. pedrosoi* as an homogeneous species or the reclassification of *Cladosporium carrionii* as *Cladophialophora carrionii*.

IDENTIFICATION

Fonsecaea pedrosoi

Colonies are olivaceous–black, slow in growth rate, and velvety in texture. Conidiogenous cells are cylindrical, usually with an irregularly swollen apex that is studded with scars from conidial attachment. Conidia measure approximately $(1.5–2.5) \times (3.5–5)$ µm, are one celled and broadly clavate with a flat base. Each conidium usually gives rise to several additional conidia in a repeating process that ultimately forms a complexly branched head consisting of three to four levels of conidia (Figure 18.2). A *Rhinocladiella* morph (Figure 18.3) is usually present, and a *Phialophora* morph (Figure 18.4) can also be present (McGinnis and Schell 1980). *F. pedrosoi* can easily be confused with *Cladosporium* spp. and *Rhinocladiella* spp. (Schell et al. 1995).

Cladophialophora carrionii (Cladosporium carrionii)

Colonies are olivaceous–black, slow in growth rate, and velvety in texture. Conidiophores bear outwardly spreading, sparsely branching, long chains of ellipsoidal, symmetrical, one-celled conidia that measure approximately $(2.2–2.6) \times (4.5–6.0)$ µm (Figure 18.5). A *Phialophora* morph can also be present (Trejos 1954; Honbo et al. 1984a, b). This species can easily be confused with *Cladosporium arxii*, *Cladosporium* spp., *F. pedrosoi*, and *Cladophialophora bantiana* (Schell et al. 1995).

Figure 18.2 Fonsecaea pedrosoi *can produce conidia by multiple mechanisms. Usually, conidia arise from the swollen apex of the conidiophore, and then give rise to additional conidia, resulting in two to four levels of conidia formation. This form is termed the 'Fonsecaea synanamorph,' and is generally considered to be the most distinctive.*

Phialophora verrucosa

Colonies are olivaceous–black, slow in rate of growth, and velvety in texture. Conidiogenous cells are phialides, vase-like in shape, with a prominent collarette (Figure 18.6). Conidia measure approximately $(1.5–2.5) \times (2.5–4.5)$ µm, are one-celled, ellipsoidal, and emerge sequentially from phialides to accumulate in subglobose masses. The species can be confused with other species of *Phialophora* and *Fonsecaea* (Schell et al. 1995).

CLINICAL PRESENTATIONS OF CHROMOBLASTOMYCOSIS

In its various stages, chromoblastomycosis can mimic the following:

- a foreign body granuloma (Wackym et al. 1985)
- a ganglion cyst (Jayalakshmi et al. 1990)
- lupus (Molina Lequizamon et al. 1984)
- leprosy (Banks et al. 1985; Pavithran 1992)
- leishmaniasis or cutaneous tuberculosis (Boudghene-Stambouli and Merad-Boudia 1994)

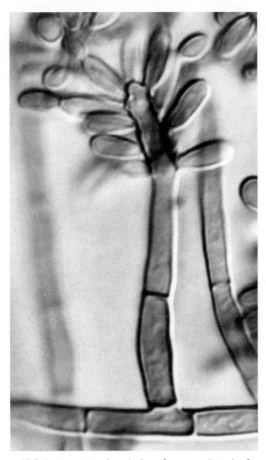

Figure 18.3 Fonsecaea pedrosoi *also often sporulates by forming each conidium sequentially at the continually elongating tip of the conidiophore. This form is termed the 'Rhinocladiella synanamorph' based upon its resemblance to fungi in the genus* Rhinocladiella.

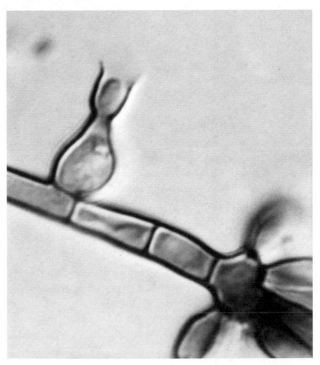

Figure 18.4 Fonsecaea pedrosoi *can sometimes sporulate by forming phialides, where conidia emerge from a cup-like orifice in a sequential process. This is termed the 'Phialophora synanamorph' because of its resemblance to the fungus* Phialophora verrucosa. *In this figure, a portion of the* Fonsecaea *synanamorph is also visible.*

- keratoacanthoma (Wiss et al. 1986)
- cutaneous sarcoidosis (Perry and Sheretz 1987)
- common wart (Jayalakshmi et al. 1990)
- malignancy (Pavlidakey et al. 1986; Jayalakshmi et al. 1990; Hiruma et al. 1992).

Lesions of chromoblastomycosis most often arise on an extremity and usually begin as small scaly papules or nodules which are painless but may be itchy (Fukushiro 1983). Satellite lesions may gradually arise from auto-inoculation as a result of scratching the lesions, or by spread of the fungus through the lymphatic system (Carrion and Silva 1947; Takase et al. 1988; Hiruma et al. 1992). Rash-like areas enlarge and become raised irregular plaques, often scaly or verrucose (Figure 18.7). Lesions in cases of long duration can become large, tumorous and even cauliflower-like in appearance (Figure 18.8). Flat lesions that exhibit central healing and scarring can sometimes be seen. Blood, serous, or caseous material can often be expressed from lesions (Carrion and Silva 1947). Damaged tissue, clotted blood, and fungal cells are expelled through the epidermis by the process of transepithelial elimination and are

deposited in the form of small dark aggregations at the surface of the lesion (Uribe et al. 1989). These are commonly known as black dots, and can be selectively removed and examined microscopically for the diagnostic presence of muriform bodies.

Other prominent features of chromoblastomycosis are epithelial hyperplasia, microabscess formation in the epidermis, and dermal fibrosis. There are data to suggest that intense extracellular matrix cross-linking, associated with the sustained contribution of transforming growth factor beta (TGF-β)-activated fibroblasts, exists in tissues of chromoblastomycosis compared with healthy tissue and tissue from other chronic skin infections. Keloid-like irreversible fibrosis seems to result from lysyl oxidase-mediated pyridinoline (Esterre et al. 1992) and transglutaminase-mediated glutamyl-lysine (Esterre et al. 1998b) collagen cross-linking.

Chromoblastomycosis is slowly progressive and can persist for decades if neglected or not successfully treated. An analysis of cases for a 20-year period in Jamaica found a mean duration of 8 years, and a range of 3 months to 27 years (Bansal and Prabhakar 1989). A persistence up to 30 years has been recorded in the Malagassy focus (Esterre et al. 1996). Untreated lesions can progress to involve an entire limb, although muscle and bone do not become infected. In the absence of adequate management, chromoblastomycosis is a

Figure 18.5 Cladophialophora carrionii *forms conidia in acropetal, occasionally branched, chains.*

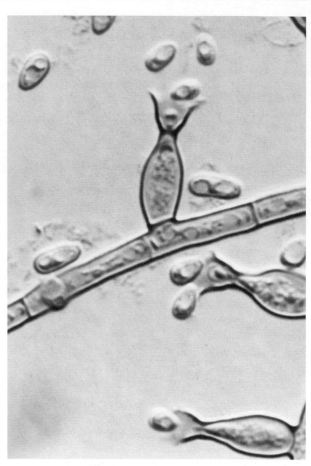

Figure 18.6 Phialophora verrucosa *sequentially forms conidia at the orifice of phialides which bear distinctive cup-like collarettes.*

chronic, progressive, and recalcitrant disease that may eventually become incapacitating. Complications include secondary bacterial infections, lymphostasis, swelling and contraction of the affected limb. For those living in tropical climates, feet and legs are the parts of the body that are most frequently subjected to minor wounds involving soil and vegetation; as a result, the disease most often involves these sites, although any exposed part of the body is vulnerable (Nakamura et al. 1972; Iwatsu et al. 1983; Zaror et al. 1987; Gruber et al. 1988; Gross and Schosser 1991; Hiruma et al. 1992; Esterre et al. 1996 and Figure 18.9). Hematogenous spread and cerebral infections are rare complications, and seem to occur only with *F. pedrosoi* in Japan (Fukushiro 1983). Primary infection of the lung has only been reported twice for *F. pedrosoi* (Zaharopoulos et al. 1988; Morris et al. 1995).

HISTOPATHOLOGY

Muriform bodies are the characteristic tissue form of all agents of chromoblastomycosis. They are pheoid, thick-walled, multiseptate bodies, 5–13 μm in diameter and are typically found at the surface of the lesions (see

Figure 18.1). The host tissue response is a granulomatous reaction with polymorphonuclear neutrophils, sometimes showing ultrastructurally a frustrated phago-cytosis pattern (Esterre et al. 1993), and macrophages infiltrate and formation of numerous microabscesses. Muriform bodies are found singly or in clusters within the microabscesses, and sometimes within giant cells

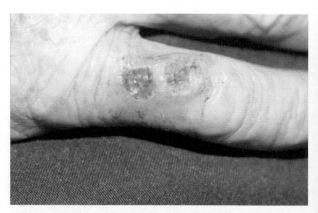

Figure 18.7 Chromoblastomycosis *lesions are often scaly or crusty, and frequently feature small, dark aggregations of clotted blood and fungal cells (right). A portion of the lesion (left) had been biopsied the preceding day.*

Figure 18.8 *Chromoblastomycosis of many years' duration can result in large, verrucous or cauliflower-like lesions. (Courtesy T.G. Mitchell.)*

(Fukushiro 1983; Uribe et al. 1989). Epitheliomatous hyperplasia is generally present. Melanin is produced in the cell walls of the muriform bodies, which results in a dark-brown coloration that can easily be seen in histological sections stained with hematoxylin and eosin. They routinely divide, not by budding as seen in yeasts, but by separation along the interface of double septa. Budding of the muriform bodies is sometimes seen, but is not believed to contribute significantly to multiplication. Muriform bodies can be induced in vitro under defined conditions, and experimental data suggest that the concentration of calcium ions within infected tissue may play a role in regulating in vivo dimorphism (Mendoza et al. 1993). Detecting the presence of muriform bodies establishes a diagnosis, but as all etiologic agents of

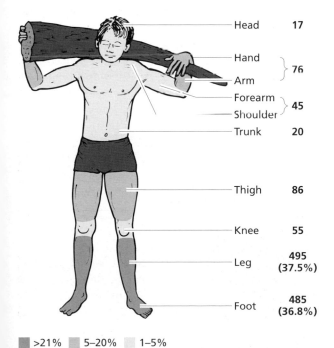

Head	17	
Hand		
Arm	}	76
Forearm		
Shoulder	}	45
Trunk	20	
Thigh	86	
Knee	55	
Leg	495 (37.5%)	
Foot	485 (36.8%)	

■ >21%   ■ 5–20%   ■ 1–5%

Figure 18.9 *Distribution of skin lesions observed on 1 316 cases, during a 40-year (1955–94) epidemiologic survey in Madagascar.*

chromoblastomycosis have the same features in tissue, the particular species involved in a case can be determined only by recovery of the fungus in culture.

IMMUNOLOGY AND SEROLOGY

Relatively little is known about the immunological aspects of chromoblastomycosis, but evidence so far demonstrates that host responses to fungal antigens occur at both cellular and humoral levels. Antibodies against the causative fungi have been detected by various methodologies in a high percentage of patients (Villalba 1988; Villalba and Yegres 1988; Esterre et al. 1997a, 2000), including some asymptomatic cases in endemic areas (Esterre et al. 1997a). Experimentation using animals has suggested that purified rabbit IgG can substantially inhibit growth of *F. pedrosoi* in culture (Ibrahim-Granet et al. 1988), but whether these findings reflect a role for protective immunity in humans against this mycosis is not yet certain. Host cellular response has been experimentally demonstrated in mice. Fungal cell wall lipids coated on to charcoal particles induced significant granulomatous reactions in mice for agents of chromoblastomycosis, as compared with control charcoal particles (Lacaz and Ekizlerian 1985).

In addition, a preparation of fractionated non-lipid cell wall material common to four etiologic agents has been found to induce a granulomatous reaction and death in mice, thereby showing that nonlipid cell wall components, as well as lipids, are capable of causing a cell-mediated host reaction (Lacaz and Fazioli 1985). Data from a recent clinical study of eight patients with chromoblastomycosis, compared with healthy control patients, suggest that individuals with chromoblastomycosis exhibit a normal cellular response to bacterial antigens, but that a partial suppression of cell-mediated immune response to fungal antigens arises at some point during the infection. The authors have suggested that such suppression may play a role in the pathogenicity of these fungi (Fuchs and Pecher 1992). Skin-test antigens have been prepared in research settings, and limited testing has found reactivity among patients diagnosed with chromoblastomycosis. Skin testing of larger population subsets to determine the extent of exposure to the etiologic fungi has not been attempted, but an enzyme-linked immunosorbent assay (ELISA)-based survey has been organized in the endemic eastern region of Madagascar, revealing a 6 percent level of seropositivity in the whole population (Esterre et al. 1997a).

LABORATORY DIAGNOSIS

As chromoblastomycosis can be confused with many other afflictions, it is necessary to establish the diagnosis through laboratory testing before proceeding with specific treatment. Serological testing for chromoblastomycosis is reportedly very sensitive (Esterre et al. 2000), but is not widely used because the necessary specialized reagents are not available commercially. By contrast, microscopy and culture provide highly sensitive means of diagnosis that are both simple and inexpensive. The most rapid diagnostic technique is collection of skin scrapings or a biopsy specimen from the active margin of the lesion and examination microscopically in 10–20 percent KOH for the presence of muriform bodies. So-called black dots, which are often present on the lesion's surface, are most likely to contain the muriform bodies and are excellent specimens (Zaias and Rebell 1973). In unusual cases, superficial crusts from lesions or specimens from pulmonary infection may show septate, pheoid hyphae instead of muriform bodies (Carrion and Silva 1947; Lavalle 1980; Fukushiro 1983; Morris et al. 1995). The addition of the fluorescent reagent calcofluor white to the KOH procedure has greatly increased the sensitivity of microscopic examination for most mycotic infections. However, it is not reliable for detecting heavily melanized fungal cells, such as muriform bodies, and therefore should not be the only method of microscopic examination used in suspected cases of chromoblastomycosis.

When microscopy is negative, culture can establish the diagnosis because of its superior sensitivity as a clinical procedure. As all agents of chromoblastomycosis have the same morphology in tissue, culturing the fungus is the only means of establishing the etiology in a given case, and is a prerequisite if antifungal susceptibility testing is to be performed. Furthermore, because of the current proliferation of newly developed antifungal compounds, it is important to culture and correctly identify causative fungi before instituting therapy so that potential patterns of clinical resistance or susceptibility to individual drugs can eventually become recognized and utilized by the medical community.

TREATMENT

Historically, successful treatment of chromoblastomycosis has been exceedingly difficult. Surgical excision of small areas of infection has provided the most reliable means of cure. Successful surgery necessarily requires removal of a margin of uninfected tissue. In addition to traditional surgical management, success has been reported with cryosurgery using liquid nitrogen in 11 patients infected with *F. pedrosoi* (Pimentel et al. 1989), with CO_2 laser surgery in one case (Kuttner and Siegle 1986) and with Mohs' micrographic surgery in another (Pavlidakey et al. 1986). Both thiabendazole and flucytosine (5-fluorocytosine), an orally administered drug, have led to improvement in several cases where excision was not feasible. As a result of toxicity from metabolites of flucytosine, however, it is necessary to monitor the level of this drug in the patient's serum during the course of therapy.

The first consistently effective antifungal compound for infections caused by pheoid fungi appears to be the new triazole itraconazole. No prospective, controlled trials have yet been conducted for this compound but evidence so far indicates that cures and significant improvements can be expected for many patients, even for some who have failed to respond to other therapies. In one series of 19 patients with inoperable lesions caused by *F. pedrosoi*, duration of their infections ranged from 6 to 36 years, with 14 patients having had previous therapy. All the patients responded to itraconazole therapy, with eight patients achieving clinical and biological cure (Queiroz-Telles et al. 1992). From Natal, South Africa, a report suggests that itraconazole is especially effective for infections with *C. carrionii* (Bayles 1992).

Another new oral compound, terbinafine, has been recently investigated for infections caused by pheoid fungi. An open pilot study in two hospitals in Madagascar, using 500 mg/day during 6–12 months, reported a high clinical and mycological efficiency (about 85 percent) without significant adverse reactions nor relapse during a 2-year follow-up (Esterre et al. 1997b, 1998a). Short treatments (3–4 months) or 250 mg/day dose have been tested with success on recent lesions (Esterre et al. 1997b) and even poorly responsive patients could be cured by a combination therapy with oral itraconazole (200–400 mg/day) and terbinafine (250–750 mg/day) (Gupta et al. 2002). In addition to its fungicidal properties, terbinafine revealed two unexpected and interesting bypass effects at the skin level: an antifibrotic activity (Esterre et al. 1998a; Esterre and Ricard-Blum 2002) and a reversal of immunological anergy (Elewski et al. 2002).

Unfortunately, both itraconazole and terbinafine are expensive at present and therefore of limited affordability. An inexpensive approach that has yielded good results in several cases is treatment with topical heat. Chemical and electric pocket warmers and heat bulbs resulted in cures for 15 of 19 *F. pedrosoi* infections in one reported series (Hiruma et al. 1993). Spontaneous resolution of untreated lesions has been reported but appears to be a rare occurrence (Nishimoto et al. 1984).

ACKNOWLEDGMENTS

This chapter is adapted and expanded from Schell, W.A. 1998. Agents of chromoblastomycosis and sporotrichosis. In: Ajello, L. and Hay, R.J. (eds) *Topley & Wilson's Microbiology and microbial infections*, 9th edn, Volume 4, *Medical Mycology*. London: Edward Arnold, 315–336.

REFERENCES

Abid, H.N., Walter, P.A. and Litchfield, H. 1987. Chromomycosis in a horse. *J Am Vet Med Assoc*, **191**, 711–12.

Ajello, L. 1970. The medical mycological iceberg. *Scientific Publication No. 205*. Washington, DC: Pan American Health Organization, 3–10.

Ajello, L., Georg, L.K., et al. 1974. A case of phaeohyphomycosis caused by a new species of *Phialophora*. *Mycologia*, **66**, 490–8.

Al-Doory, Y. 1972. *Chromomycosis*. Missoula MT: Mountain Press Publishing, 9–50.

Banks, I.S., Palmieri, J.R., et al. 1985. Chromomycosis in Zaire. *Int J Dermatol*, **24**, 302–7.

Bansal, A.S. and Prabhakar, P. 1989. Chromomycosis: a twenty-year analysis of histologically confirmed cases in Jamaica. *Trop Geogr Med*, **41**, 222–6.

Barba-Gomez, J.F., Mayorga, J., et al. 1992. Chromoblastomycosis caused by *Exophiala spinifera*. *J Am Acad Dermatol*, **26**, 367–70.

Bayles, M.A. 1992. Tropical mycoses. *Chemotherapy*, **38**, Suppl 1, 27–34.

Borelli, D. 1972. *Acrotheca aquaspersa* nova species agente de cromomicosis. *Acta Cient Venez*, **23**, 193–6.

Boudghene-Stambouli, O. and Merad-Boudia, A. 1994. Chromomycose: 2 observations. *Ann Dermatol Venereol*, **121**, 37–9.

Brygoo, E.R. and Segretain, G. 1960. Etude clinique épidémiologique et mycologique de la chromoblastomycose a Madagascar. *Bull Soc Pathol Exot*, **53**, 443–75.

Bube, A., Burkhardt, E. and Weiss, R. 1992. Spontaneous chromomycosis in the marine toad (*Bufo marinus*). *J Comp Pathol*, **106**, 73–7.

Campins, H. and Scharyj, M. 1953. Chromoblastomicosis comentarios sobre 24 casos, con estudio clinico, histologico, y micologico. *Gac Med Caracas*, **61**, 127–51.

Carrion, A.L. and Silva, M. 1947. Chromoblastomycosis and its etiologic fungi. *Ann Cryptogam Phytopathol*, **6**, 20–62.

Connor, D.H. and Gibson, D.W. 1985. Association of splinters with chromomycosis and phaeomycotic cyst. *Arch Dermatol*, **121**, 168.

Dixon, D.M., Shadomy, H.J. and Shadomy, S. 1980. Dematiaceous fungal pathogens isolated from nature. *Mycopathologia*, **70**, 153–61.

Elewski, B.E., El Charif, M., et al. 2002. Reactivity to trichophytin antigen in patients with onychomycosis: effect of terbinafine. *J Am Acad Dermatol*, **46**, 371–5.

Esterre, P. and Ricard-Blum, S. 2002. Chromoblastomycosis: new concepts in physiopathology and treatment. *J Mycol Med*, **12**, 21–4.

Esterre, P., Peyrol, S., et al. 1992. Cell-matrix patterns in the cutaneous lesion of chromomycosis. *Pathol Res Pract*, **188**, 894–900.

Esterre, P., Peyrol, S., et al. 1993. Granulomatous reaction and tissue remodelling in the cutaneous lesion of chromomycosis. *Virchows Arch Pathol Anat*, **422**, 285–91.

Esterre, P., Andriantsimahavandy, A., et al. 1996. Forty years of chromoblastomycosis in Madagascar: a review. *Am J Trop Med Hyg*, **55**, 45–7.

Esterre, P., Jahevitra, M., et al. 1997a. Evaluation of the ELISA technique for the diagnosis and the seroepidemiology of chromoblastomycosis. *J Mycol Med*, **7**, 137–41.

Esterre, P., Ratsioharana, M. and Roig, P. 1997b. Potential use of terbinafine in the treatment of chromoblastomycosis. *Rev Contemp Pharmacother*, **8**, 357–61.

Esterre, P., Inzan, C., et al. 1998a. A multicenter trial of terbinafine in patients with chromoblastomycosis: effect on clinical and biological criteria. *J Dermatol Treat*, **9**, 29–34.

Esterre, P., Risteli, L. and Ricard-Blum, S. 1998b. Immunohistochemical study of type I collagen turn-over and of matrix metalloproteinases in chromoblastomycosis before and after treatment by terbinafine. *Pathol Res Pract*, **194**, 847–53.

Esterre, P., Jahevitra, M. and Andriantsimahavandy, A. 2000. Humoral immune response in chromoblastomycosis during and after therapy. *Clin Diagn Labor Immunol*, **7**, 497–500.

Foster, H.M. and Harris, T.J. 1987. Malignant change (squamous carcinoma) in chronic chromoblastomycosis. *Aust N Z J Surg*, **57**, 775–7.

Fuchs, J. and Pecher, S. 1992. Partial suppression of cell mediated immunity in chromoblastomycosis. *Mycopathologia*, **119**, 73–6.

Fukushiro, R. 1983. Chromomycosis in Japan. *Int J Dermatol*, **22**, 221–9.

Gezuele, E., Mackinnon, J.E. and Conti-Diaz, I.A. 1972. The frequent isolation of Phialophora verrucosa from natural sources. *Sabouraudia*, **10**, 266–73.

Greene, J.N., Foulis, P.R. and Yangco, B.G. 1990. Chromomycosis in a steroid-dependent patient with chronic obstructive pulmonary disease. *Am J Med Sci*, **299**, 54–7.

Gross, D.J. and Schosser, R.H. 1991. Chromomycosis of the nose. Chromomycosis. *Arch Dermatol*, **127**, 1831–1832, 1834.

Gruber, B., Rippon, J.W. and Dayal, V.S. 1988. Phaeomycotic cyst (chromoblastomycosis) of the neck. *Arch Otolaryngol Head Neck Surg*, **114**, 1031–2.

Gupta, A.K., Taborda, P.R. and Sanzovo, A.D. 2002. Alternate-week and combination itraconazole and terbinafine therapy for chromoblastomycosis caused by *Fonsecaea pedrosoi* in Brazil. *Medical Mycol*, **40**, 5, 529–34.

Hiruma, M., Ohnishi, Y., et al. 1992. Chromomycosis of the breast. *Int J Dermatol*, **31**, 184–5.

Hiruma, M., Kawada, A., et al. 1993. Hyperthermic treatment of chromomycosis with disposable chemical pocket warmers. Report of a successfully treated case, with a review of the literature. *Mycopathologia*, **122**, 107–14.

Honbo, S., Padhye, A.A. and Ajello, L. 1984a. The relationship of *Cladosporium carronii* to *Cladophialophora ajelloi*. *J Med Vet Mycol*, **22**, 209–18.

Honbo, S., Standard, P.G., et al. 1984b. Antigenic relationships among *Cladosporium* species of medical importance. *J Med Vet Mycol*, **22**, 301–10.

de Hoog, G.S., Rubio, C., et al. 1982. A new dematiaceous fungus from human skin. *Sabouraudia*, **20**, 15–20.

de Hoog, G.S., Guého, E, et al. 1995. Nutritional physiology and taxonomy of human-pathogenic *Cladosporium-Xylohypha* species. *J Med Vet Mycol*, **33**, 339–47.

Ibrahim-Granet, O., de Bievre, C. and Jendoubi, M. 1988. Immunochemical characterisation of antigens and growth inhibition of *Fonsecaea pedrosoi* by species-specific IgG. *J Med Microbiol*, **26**, 217–22.

Iwatsu, T., Miyagi, M. and Okamoto, S. 1981. Isolation of *Phialophora verrucosa* and *Fonsecae apedrosoi* from nature in Japan. *Mycopathologia*, **75**, 149–58.

Iwatsu, T., Tokano, M. and Okamoto, S. 1983. Auriculat chromomycosis. *Arch Dermatol*, **119**, 88–9.

Jayalakshmi, P., Looi, L.M. and Soo-Hoo, T.S. 1990. Chromoblastomycosis in Malaysia. *Mycopathologia*, **109**, 27–31.

Kawasaki, M., Aoki, M., et al. 1999. Molecular epidemiology of *Fonsecaea pedrosoi* using mitochondrial DNA analysis. *Medical Mycol*, **37**, 435–40.

Kuttner, B.J. and Siegle, R.J. 1986. Treatment of chromomycosis with a CO_2 laser. *J Dermatol Surg Oncol*, **12**, 965–8.

Lacaz, C.S. and Ekizlerian, S.M. 1985. Granulomatous reactions induced by lipids extracted from *Fonsecaea pedrosoi*, *Fonsecaea compactum*, *Cladosporium* carronii and *Phialophora verrucosum*. *J Gen Microbiol*, **131**, 187–94.

Lacaz, C.S. and Fazioli, R.A. 1985. Role of the fungal cell wall in the granulomatous response of mice to the agents of chromomycosis. *J Med Microbiol*, **20**, 299–305.

Lacaz, C.S., Silva, A.C., et al. 1994. Chromoblastomycosis associated with leprosy: report of 2 cases. *Rev Patol Trop*, **27**, 241–4.

Lavalle, P. 1980. Chromoblastomycosis in Mexico. *Scientific Publication No. 396*. Washington, DC: Pan American Health Organization, 235–47.

Matsumoto, T., Ajello, L., et al. 1994. Developments in hyalohyphomycosis and phaeohyphomycosis. *J Med Vet Mycol*, **32**, Suppl 1, 329–77.

McGinnis, M.R. 1983. Chromoblastomycosis and phaeohyphomycosis: new concepts, diagnosis and mycology. *Am Acad Dermatol*, **8**, 1–16.

McGinnis, M.R. and Salkin, I.F. 1993. A clinical user perspective of anamorphs and teleomorphs. In: Reynolds, D.R. and Taylor, J.W. (eds), *The fungal holomorph: mitotic, meiotic and pleomorphic speciation in fungal systematics*. Wallingford, UK: CAB International, 87–92.

McGinnis M.R. and Schell W.A., 1980, The genus *Fonsecaea* and its relationship to the genera *Cladosporium*, *Phialophora*, *Ramichloridium* and *Rhinocladiella*. *Scientific Publication No. 396*, Pan American Health Organization, Washington DC, 215–24.

McGinnis, M.R., McKenzie, R.A. and Connole, M.D. 1985. *Phaeosclera dematioides*, a new etiologic agent of phaeohyphomycosis in cattle. *J Med Vet Mycol*, **23**, 133–5.

Mead, M. and Ridley, M.F. 1957. Sporotrichosis and chromoblastomycosis in Queensland. *Med J Aust*, **7**, 192–7.

Mendoza, L., Karuppayil, S.M. and Szaniszlo, P.J. 1993. Calcium regulates in vitro dimorphism in chromoblastomycotic fungi. *Mycoses*, **36**, 157–64.

Molina Lequizamon, E.B., Casas, J.G. and Perini, G.M. 1984. Cromomicosis de la nalga. *Med Cutan Ibero Lat Am*, **12**, 430–8.

Montenegro, M.R., Miyaji, M., et al. 1996. Isolation of fungi from nature in the region of Botucatu, state of Sao Paulo, Brazil, an endemic area of paracoccidioidomycosis. *Mem Inst Oswaldo Cruz*, **91**, 6, 665–70.

Morales, L.A., Gonzalez, Z.A. and Santiago-Delpin, E.A. 1985. Chromoblastomycosis in a renal transplant patient. *Nephron*, **40**, 238–40.

Morris, A., Schell, W.A., et al. 1995. *Fonsecaea pedrosoi* pneumonia and *Emericella nidulan*s cerebral abscesses in a bone marrow transplant patient. *Clin Infect Dis*, **21**, 1346–8.

Nakamura, T., Grant, J.A., et al. 1972. Primary chromoblastomycosis of the nasal septum. *Am J Clin Pathol*, **58**, 365–70.

Nishimoto, K., Yoshimura, S. and Honma, K. 1984. Chromomycosis spontaneously healed. *Int J Dermatol*, **23**, 408–10.

Okeke, C.N. and Gugnani, H.C. 1986. Studies of pathogenic dematiaceous fungi: isolation from natural sources. *Mycopathologia*, **94**, 19–25.

Padhye, A.A. 1986. Identification of the etiologic agents of chromoblastomycosis. *Scientific Publication No. 479*. Washington, DC: Pan American Health Organization, 87.

Padhye, A.A., Kaufman, L., et al. 1992. Fatal pulmonary sporotrichosis caused by *Sporothrix schenckii* var. *luriei* in India. *J Clin Microbiol*, **30**, 2492–4.

Padhye, A.A., Hampton, A.A., et al. 1996. Chromoblastomycosis caused by *Exophiala spinifera*. *Clin Infect Dis*, **22**, 331–5.

Pappagianis, D. and Ajello, L. 1994. Dematiaceous – a mycologic misnomer? *J Med Vet Mycol*, **32**, 319–21.

Pavithran, K. 1988. Chromoblastomycosis in a residual patch of leprosy. *Indian J Lepr*, **60**, 444–7.

Pavithran, K. 1992. Chromoblastomycosis masquerading as tuberculoid leprosy. *Int J Lepr Other Mycobact Dis*, **6012**, 657–8.

Pavlidakey, G.P., Snow, S.N. and Mohs, F.E. 1986. Chromoblastomycosis treated by Mohs micrographic surgery. *J Dermatol Surg Oncol*, **12**, 1073–5.

Perry, A.E. and Sheretz, E.F. 1987. Papules and nodules in a patient with sarcoidosis. Chromoblastomycosis. *Arch Dermatol*, **123**, 520–1.

Pimentel, E.R., Castro, L.G., et al. 1989. Treatment of chromomycosis by cryosurgery with liquid nitrogen: a report on eleven cases. *J Dermatol Surg Oncol*, **15**, 72–7.

Queiroz-Telles, F., Purim, K.S., et al. 1992. Itraconazole in the treatment of chromoblastomycosis due to *Fonsecaea pedrosoi*. *Int J Dermatol*, **31**, 805–12.

Richard-Yegres, N. and Yegres, F. 1987. *Cladosporium carrionii* en vegetacion xerofila: aislamiento en una zona endemica para la cromomicosis en Venezuela. *Dermatol Venezolana*, **25**, 15–18.

Ridley, M.F. 1961. Soil as a source of pathogenic fungi. *Recent advances in botany*, **1**. Toronto: University of Toronto Press, 312–6.

Rodrigues, J.Y., Richard-Yegres, N., et al. 1992. Cromomicosis: susceptibilidad genetica en grupos familiares de la zona endemica en Venezuela. *Acta Cientif Venezol*, **43**, 98–102.

Rubin, H.A., Bruce, S., et al. 1991. Evidence for percutaneous inoculation as the mode of transmission for chromoblastomycosis. *J Am Acad Dermatol*, **25**, 951–4.

Rudolph, M. 1914. Uber die brasilianische 'Figueira'. *Arch Schiffs Tropen-Hygen*, **18**, 498.

Schell, W.A., McGinnis, M.R. and Borelli, D. 1983. *Rhinocladiella aquaspersa*, a new combination for *Acrotheca aquaspersa*. *Mycotaxon*, **17**, 341–8.

Schell, W.A., Pasarell, L., et al. 1995. *Bipolaris*, *Exophiala*, *Scedosporium*, *Sporothrix* and other dematiaceous fungi. In: Murray, P.R. (ed.), *Manual of clinical microbiology*, 6th edition. Washington DC: American Society for Microbiology, 825–46.

Takase, T., Baba, T. and Uyeno, K. 1988. Chromomycosis. A case with a widespread rash, lymph node metastasis and multiple subcutaneous nodules. *Mycoses*, **31**, 343–52.

Talhari, S., Cunha, M.G., et al. 1988. Deep mycoses in Amazon region. *Int J Dermatol*, **27**, 481–4.

Terra, F., Torres, M. and da Fonseca, O. 1922. Novo typo de dermatite verrucosa mycose por *Acrotheca* com associacao de leishmaniosa. *Brazil Med*, **2**, 263–8.

Tintelnot, K., von Hunnius, P., et al. 1995. Systemic mycosis caused by a new *Cladophialophora* species. *J Med Vet Mycol*, **33**, 349–54.

Trejos, A. 1954. *Cladosporium carrionii* n. sp. and the problem of cladosporia isolated from chromoblastomycosis. *Rev Biol Trop*, **2**, 75–112.

Tschen, J.A., Know, J.M., et al. 1984. Chromomycosis. The association of fungal elements and wood splinters. *Arch Dermatol*, **120**, 107–8.

Tsuneto, L.T., Arce-Gomez, B., et al. 1989. HLA-A29 and genetic susceptibility to chromoblastomycosis. *J Med Vet Mycol*, **27**, 181–5.

Turiansky, G.W., Benson, P.M., et al. 1995. *Phialophora verrucosa*: a new cause of mycetoma. *J Am Acad Dermatol*, **32**, 311–15.

Uribe, F., Zuluaga, A.I., et al. 1989. Histopathology of chromoblastomycosis. *Mycopathologia*, **105**, 1–6.

Villalba, E. 1988. Detection of antibodies in the sera of patients with chromoblastomycosis by counter immunoelectrophoresis. I. Preliminary results. *J Med Vet Mycol*, **26**, 73–4.

Villalba, E. and Yegres, J.F. 1988. Detection of circulating antibodies in patients affected by chromoblastomycosis by *Cladosporium carrionii* using double immunodiffusion. *Mycopathologia*, **102**, 17–19.

Wackym, P.A., Gray, G.F. Jr, et al. 1985. Cutaneous chromomycosis in renal transplant recipients. Successful management in two cases. *Arch Intern Med*, **145**, 1036–7.

Wang, C.J.K. 1965. Fungi of pulp and paper. *Technical Publication No. 87*. Syracuse, NY: State University College of Forestry, 115.

Wiss, K., McNeely, M.C. and Solomon, A.R. Jr 1986. Chromoblastomycosis can mimic keratoacanthoma. *Int J Dermatol*, **25**, 385–6.

Zaharopoulos, P., Schnadig, V., et al. 1988. Multiseptate bodies in systemic phaeohyphomycosis diagnosed by fine needle aspiration cytology. *Acta Cytol*, **32**, 885–91.

Zaias, N. and Rebell, G. 1973. A simple and accurate diagnostic method in chromoblastomycosis. *Arch Dermatol*, **108**, 545–6.

Zaror, L., Fischman, O., et al. 1987. A case of primary nasal chromoblastomycosis. *Mykosen*, **30**, 468–71.

Zeppenfeldt, G., Richard-Yegres, N., et al. 1994. *Cladosporium carrionii* hongro dimorfico en cactaceas de la zona endemica para la cromomicosis en Venezuela. *Rev Iberoam Micol*, **11**, 61–3.

Sporotrichosis

ROBERTO ARENAS

INTRODUCTION

Sporotrichosis is a varied disease caused by a single dimorphic fungal species, *Sporothrix schenckii*. It occurs worldwide, growing as a mold in association with dead or senescent plant material. When cells of the mold gain entry to a susceptible animal host, they transform into budding yeast cells and persist as such within the infected tissue. Sporotrichosis most commonly presents as a chronic ulcerative infection of cutaneous and subcutaneous tissue, and tends to spread along channels of the lymphatic system. These cutaneous and lymphocutaneous infections typically are initiated by trauma or minor abrasion from plant material harboring cells of the fungus. Lesions normally develop within 3–12 weeks following inoculation. In unusual instances, the fungus can spread from lesions in subcutaneous tissue to nearby joints, or more rarely to other organs. Pulmonary sporotrichosis is an infrequent syndrome which starts when airborne conidia from the mold, growing as a saprobe upon natural substrata, are inhaled into the alveoli of the lungs. In such cases, an infection can develop that resembles other chronic lung infections such as tuberculosis and histoplasmosis.

Sporotrichosis is a cosmopolitan disease and is probably the most frequently encountered subcutaneous mycosis in the world. Most cases occur among people living in tropical and subtropical regions, and the disease is relatively common in Mexico, Uruguay, and Brazil. In Brazil, the national incidence of sporotrichosis is second only to paracoccidioidomycosis among the deep mycoses (Costa et al. 1994). In Sao Paulo, sporotrichosis has accounted for as much as 0.5 percent of all mycoses for which medical attention was sought (Sampaio et al. 1954). The number of cases in North America has been estimated at 2.4 per million population per year, but because the mycoses have never been designated as notifiable (reportable) diseases, the true incidence of sporotrichosis remains unknown (Ajello 1970; Reingold et al. 1986).

HISTORY

The first case report and documentation of the fungal etiology of sporotrichosis appeared in 1898 in the USA (Schenck 1898). By 1932 about 200 infections had been reported in the USA (Jacobson 1932), and there was subsequently a rapid increase in reported cases in this and other countries (Mackinnon 1948; Goncalves and Peryassu 1954; Sampaio et al. 1954; Mayorga et al. 1978; Lavalle and Mariat 1983; Reingold et al. 1986; Rippon 1988; Talhari et al. 1988). In Japan, for example, 13 cases had been reported as of 1945, but since that time more than 2 500 cases have been dealt with by the Japanese healthcare system (Fukushiro 1984). A retrospective study at one dermatology clinic in Japan found that 0.17 percent (150 patients) of all patients entering the clinic were diagnosed with sporotrichosis (Kusuhara et al. 1988), and similar findings were reported from a second clinic (Itoh et al. 1986; Eisfelder et al. 1993). By contrast. there has been some recent evidence to suggest an overall decline in the prevalence of sporotrichosis in parts of Japan from 1983 onwards, as well as a decrease in the number of infections in Europe, possibly resulting partly from a decreasing number of people being engaged in agriculture (Alberici et al. 1989; Barile et al. 1993; Eisfelder et al. 1993). Epidemiologic data from about 822 cases were recorded in a 37-year period in Jalisco, Mexico (Mayorga et al. 1997) and a review, mainly about epidemiology, microbiology, and treatment, was published in 2001 (De Araujo et al. 2001).

Sporotrichosis continues to be a chronic problem for public health programs in certain rural areas of the world; in addition, there are newer aspects of the disease that are particularly relevant in urban and suburban settings. One of these is the emergence of sporotrichosis within the subpopulation of people who are seropositive for the human immunodeficiency virus (HIV). A second recently recognized aspect is the potential for zoonotic infections, particularly those contracted from the domestic cat *Felis catus*. Several medical therapies are available for the treatment of sporotrichosis, and cure of the lymphocutaneous form of infection can be achieved with relative ease. By contrast, pulmonary, skeletal, and disseminated infections can be very difficult to manage and are life-threatening conditions.

ECOLOGY AND EPIDEMIOLOGY

Sporothrix schenckii is widespread in nature and can be isolated as a saprobe from dead or senescent vegetation, such as thorns, hay, straw, sphagnum moss, and wood. For this reason, sporotrichosis is a recognized occupational hazard for nursery and forestry workers, and is also often associated with farmers, florists, leisure gardeners, and others whose occupation or avocation brings them into frequent and sometimes traumatic contact with plant material or soil. The great majority of infections arise under such circumstances, as reflected by one of the common names for sporotrichosis – 'rose gardener's disease.' Also at risk are people who work as manual laborers in rural areas and who wear little protective clothing or footwear, and thus have increased potential for exposure to the fungus (Forester 1924; Lavalle and Mariat 1983; Itoh et al. 1986).

Children comprise a significant portion of sporotrichosis patients, and presumably acquire most of these infections while playing or helping with chores out of doors (Rudolph 1984; Prose et al. 1986; Frumkin and Tisserand 1989; Yamada et al. 1990). Two reports from Mexico have shown that as many as 10–25 percent of sporotrichosis patients in that country are children (Leon 1964; Solano 1966). Retrospective examination of records at a university in Japan has shown that children under the age of 12 have a particularly high rate of infection. The face is a frequent site of infection in children, as evidenced by a finding that 92 percent of 529 pediatric cases of sporotrichosis included lesions of the face (Fukushiro 1984; Eisfelder et al. 1993).

A report of 238 cases occurring in a relatively remote area of the south central highlands of Peru showed 60 percent in children and the most commonly affected anatomic site was the face. The incidence of sporotrichosis in this region ranged from 48 to 60 cases per 100 000 persons and was highest among children aged 7–14 years, approaching one case per 1 000 persons (Pappas et al. 2000). In Mexico, 10–34 percent of cases are children, the youngest patient has been a child, 2 days old, who acquired sporotrichosis after a rat bite and the oldest a 117-year-old man. Lymphocutaneous forms are observed in 65–82 percent, fixed forms in 20–30 percent and the systemic ones in 2–5 percent (Mayorga et al. 1997; Arenas 2003).

Cluster infections can occur when people are exposed to a common source of inocula. Past examples of small outbreaks included 12 workers in two geographical regions who were infected by hay mulch (Cook et al. 1984), six students who became infected by packing straw (Sanders 1971), four workers who were infected by sphagnum moss at a garden center (Remington et al. 1982), four people who contracted infections from a contaminated camping tent (Campos et al. 1994), and two employees of a meat packing plant who developed pulmonary infection, presumably from a common source (Dewan et al. 1986).

A 32-year-old white man developed lymphocutaneous sporotrichosis at the same site as a self-administered tattoo on the dorsal aspect of the left foot, but he stated that he had mowed the lawn wearing only sandals on the same day that he had tattooed his foot (Bary et al. 1999). Cutaneous sporotrichosis evolved in a 20-year-old woman who underwent electrolysis on the anterior aspect of her neck (Ditmars and Maguina 1998) and a case of *Sporothrix* conjunctivitis was described as an atraumatic infection (Hampton et al. 2002).

In addition, there are numerous reports of multiple people contracting sporotrichosis from an infected cat (Table 19.1).

More rarely, sporotrichosis can occur in epidemic proportions. The largest reported outbreak of sporotrichosis occurred during 1941–44 when at least 2 899 cases were seen among gold miners in the Transvaal region of South Africa (du Toit 1942; Helm and Berman 1947). Of these, 2 441 occurred in several levels of a single mine, although a total of at least nine mines was involved in the outbreak. The fungus was found to be growing upon timbers used to shore the mine shafts, and infections would begin when bare skin was abraded by the colonized timbers. The epidemic was brought under control by application of a fungicidal compound to the timbers (Brown and Weintroub 1947), and formaldehyde fumigation of clothing and living quarters (du Toit 1942). All infections were cured using oral potassium

Table 19.1 *Zoonotic sporotrichosis transmitted by* Felis catus

Vector	No. of infected humans	Associated trauma	Reference
1 cat	3	3	Larsson et al. 1989
5 cats	7	3 of 7 (2 bitten; 1 preexisting scratch)	Dunstan et al. 1986a, b
1 cat	6	None	Read and Sperling 1982
1 cat	2	1 (scratched)	Nusbaum et al. 1983
1 cat	3	1 (scratched)	Schiappacasse et al. 1985
1 cat	1	Scratched	Samorodin and Sina 1984
1 cat	1	None	Reed et al. 1993
1 cat	1	NS	Dunstan et al. 1986a, b
1 cat	3	2 of 3 (bitten, scratched)	Naqvi et al. 1993
1 cat	1	None	Travassos and Lloyd 1980
3 cats	7	None	Samorodin and Sina 1984
5 cats	5	4 of 5 (bitten, scratched)	Zamri-Saad et al. 1990
4 cats	3	1 of 3 (scratched)	Marques et al. 1993
4 cats	4	4 scratched	Fleury et al. 2001
2 cats	13	2 of 13 scratched	Kauffman 1999
117 cats	66	52 contact, 31 (scratched or bitten)	Bastos de Lima-Barros et al. 2001; Schubach et al. 2001
Totals: 149 cats	126[a]	53 of 125	–

a) 89 of 42 people denied incurring trauma from the cats.

iodide daily for 4–5 weeks in most cases (Helm and Berman 1947; Quintal 2000).

The largest single outbreak in the USA occurred in 1988, infecting 84 people in 15 states and spanning a distance of hundreds of kilometers. These individuals, most of whom were forestry workers, became infected from handling sphagnum moss that had been applied to the roots of tree seedlings in order to humidify and protect them during shipment from central nurseries (Coles et al. 1992). Sphagnum is also used by nurseries to protect seedling roots from cold storage conditions during the winter season. Contaminated sphagnum moss, used in horticulture or forestry activities, has caused additional outbreaks (Gastineau et al. 1941; Crevasse and Ellner 1960; Hayes 1960; D'Alessio et al. 1965; Storrs et al. 1969; Powell et al. 1978; Grotte and Younger 1981; Remington et al. 1982; Cote et al. 1988). Nearly all of the sphagnum harvested in the USA grows in bogs in central Wisconsin.

Although sporotrichosis has long been linked in North America to the handling of sphagnum moss, it is still unknown at what point the fungus comes into association with the moss. Environmental sampling from earlier moss-related outbreaks suggested that *S. schenckii* is likely to be found in conjunction with harvested moss that has become wet during storage, rather than with moss that has been continually stored under dry conditions (McDonough et al. 1970; Powell et al. 1978). During investigation of the 1988 outbreak, samples of moss taken at the point of harvest were found to be culture negative for *S. schenckii*. Pond water used at one of the nurseries to dampen the moss before its use as packing material was also culture negative. Additional

testing under laboratory conditions showed that *S. schenckii* did not grow on living sphagnum, but would grow very quickly on dead sphagnum in the presence of adequate humidity and warmth. Thus, it is suspected that *S. schenckii* can be introduced into sphagnum during or after harvest, but that the moss does not pose an infectious hazard as such, until it becomes moistened and warmed to an extent that allows the fungus to multiply (Zhang and Andrews 1993). Similar findings have been reported with regard to hay (Cook et al. 1984).

An outbreak was associated with stored hay or hay bales harvested in a military installation in southwestern Oklahoma. Four patients had maintained hay bales in a house and the fifth patient had visited the house once. As in three previous reports, this contact with hay should be recognized as a risk factor for infection with *S. schenckii* (Dooley et al. 1997).

Although most infections originate when the fungus is introduced by splinters, thorns, abrasions, and other minor wounds, there is evidence to suggest that a wound of the cutaneous barrier is not essential for infection to occur. Experimental results obtained with human subjects in conjunction with the African epidemic showed that one in 10 developed sporotrichosis when a thick suspension of viable conidia was placed onto the skin, covered with gauze, and allowed to incubate (Simson et al. 1947). A 1924 report attributed the source of an infection to the handling of contaminated bandages from lesions of a patient who was being treated for sporotrichosis (Forester 1924). In 1992, there was a compelling incident further suggesting that a wound is not a prerequisite for initiation of cutaneous sporotrichosis. In this instance, infection developed on the finger of a researcher who had used a mortar and pestle to homogenize a strain of *S. schenckii* (Cooper et al. 1992). Molecular typing of both the isolate cultured from the lesion and the stock culture that the researcher had worked with showed that they were the same strain of *S. schenckii*. Molecular typing has been used in additional cases to establish the source of infection (Reed et al. 1993), and to suggest that a person can be simultaneously infected with more than one strain of *S. schenckii* (Kobayashi et al. 1990).

As these accounts imply that sporotrichosis can sometimes develop without a preceding breach of the dermal barrier, the question of contagion warrants consideration. Historically, sporotrichosis has not been considered a transmissible infection even though rare instances of person-to-person spread had been suggested long ago (Forester 1924; Smith 1945). Recent reports, such as that of an infection apparently passed from the arm of a woman to the face of her 3-year-old child, give cause to reconsider this potential aspect of sporotrichosis (Jin et al. 1990). It was a series of animal-acquired infections, however, that provided some of the strongest evidence that, in some circumstances, infections can be contagious and possibly occur through the invasion of apparently intact dermis. The infections in these cases were acquired from domestic cats. In one example, all four family members developed sporotrichosis from their infected pet. When the cat was presented for veterinary care, a veterinary technician and a veterinary student also contracted sporotrichosis soon after handling the animal. Significantly, none of the infected individuals could recall being scratched or bitten by the cat (Read and Sperling 1982).

In another example, a veterinary technician wore gloves and had no wounds while examining an infected cat but 3 weeks later developed lesions of sporotrichosis at the cuff line where glove protection ended (Dunstan et al. 1986b). Infections attributed to cats are known from at least four continents. A recurring feature of interest in these reports is that many patients were certain that they had not been wounded during handling of the infected animals. At a minimum, 43 infections in humans have been attributed to 26 cats, and 26 of 42 people for whom information was available denied incurring trauma from the animal (see Table 19.1).

It has long been known that cats are quite susceptible to sporotrichosis (Barbee et al. 1977). It is not clear how cats become infected, but in one instance eight people contracted infections from two cats that frequented a site where sphagnum moss had been used in landscaping. One possibility is that the fungus might adhere to the feet and claws of the cats, and lesions could arise as the inocula are implanted during fights between animals (Dunstan et al. 1986b). Another possibility is the inclination of cats to rub repeatedly against rough or thorny vegetation, which could initiate lesions of sporotrichosis. A related observation is that lesions in cats can be teeming with cells of the fungus (Werner et al. 1971; Nusbaum et al. 1983; Dunstan et al. 1986b; Marques et al. 1993; Reed et al. 1993). This is in stark contrast to lesions in humans where cells of *S. schenckii* are typically very sparse. It has been suggested that a potentially heavy load of organisms in the lesions of cats may be related to the high number of transmissions to humans and to the lack of evident trauma in some of these cases. In view of these reports, it should become accepted medical practice that patients with sporotrichosis be questioned to ascertain whether pets, particularly cats, may have been the source of infection, and that such animals be investigated as a potential threat to the health of other people.

In southeastern Brazil, four cases of human sporotrichosis transmitted by domestic cats were described. There was a previous history of a cat scratch before the development of the lymphocutaneous lesions. In cats, the lesions were multiple, extensive, necrotic, exudative, and ulcerated. In two necropsied cats, disseminated lymphatic and visceral mycotic infection was observed and histopathology revealed widespread histiocytic reaction with a large number of fungal organisms (Fleury et al. 2001).

In Rio de Janeiro over a period of 12 years (1987–98), 13 cases of human sporotrichosis were recorded. Two of them were associated with the scratch of a sick cat. During the subsequent two years (1998–2000), 66 human, 117 cats and seven dogs with sporotrichosis were diagnosed. Fifty-two human (78.8 percent) reported contact with cats with sporotrichosis, and 31 (47 percent) reported a history of a scratch or bite. This epidemic involving cats, dogs, and human beings is unprecedented in the literature, and may have started insidiously before 1998 (Kauffman 1999; Bastos de Lima-Barros et al. 2001; Schubach et al. 2001). In the study in Brazil, 72 of the cats (61.5 percent) were male with ages between 3 months and 18 years (median 2 years). The majority of the cutaneous lesions were located on head (46.6 percent) and limbs (35.9 percent). Most of the cases were from the underprivileged areas of the city. Nowadays, veterinarians, technicians, and cat owners or caregivers should be considered to be new categories at risk of acquiring the mycosis. The zoonotic potential of feline sporotrichosis is possibly due to the large number of organisms detected in feline tissues (Bastos de Lima-Barros et al. 2001).

Sporotrichosis has also been attributed to wounds inflicted by various other animals. Two well-documented cases occurred in young boys, each of whom was bitten by a captured field mouse (Moore and Davis 1918; Frean et al. 1991). In another rodent-related report, a rancher developed sporotrichosis after killing and handling gophers that had numerous cutaneous lesions (Olson 1912). Sporotrichosis has also been documented in a number of other animals including the armadillo, boar, camel, cattle, chimpanzee, dog, dolphin, donkey, fowl, fox, horse, mule, and rat (Kaplan et al. 1982).

The relationship of sporotrichosis to the armadillo is remarkable. It has been established that *S. schenckii* can be found on vegetation lining the burrows of armadillos (Mackinnon et al. 1969). In Uruguay, two armadillo species, *Dasypus novemcinctus* and *Dasypus septemcinctus*, are hunted as game. For a 16-year period, of 138 cases for which the source of infection could be discerned, 81 percent were attributed to contact with armadillos, their burrows or equipment used in hunting them (Conti-Diaz 1980). Of the remaining infections, wounds from plants or other objects caused 15 percent and 4 percent, respectively. Furthermore, it was shown that infections were most numerous during the peak hunting months, and that very few cases of sporotrichosis were seen during the armadillo breeding season when hunting is traditionally suspended. The indigenous population has long been aware of this association and refers to sporotrichosis as *mulita* (armadillo) sickness (Conti-Diaz 1980). Sporotrichosis has also been found in two wild armadillos (*D. novemcinctus*) captured in the USA for purposes of leprosy research (Kaplan et al. 1982). The widespread exposure of various animals to *S. schenckii* has been demonstrated experimentally by skin-testing surveys. In Sao Paulo, testing of 96 zoo animals with sporotrichin, a skin-test antigen derived from *S. schenckii*, showed a positivity rate of 57 percent among terrestrial animals and 4 percent among arboreal animals (Costa et al. 1994).

Sporotrichosis has also been reported in the Pacific white-sided dolphin (*Lagenorhynchus obliquidens*), and there has been suggestion that it could possibly be associated with fish as well (Migaki et al. 1978). In Guatemala, 53 cases of sporotrichosis were observed during a 3-year period among people living in the Lake Ayarza District (Mayorga et al. 1978). Of 30 patients able to recall injury, six infections were attributed to trauma by woody material, but 24 individuals recounted trauma associated with handling fish taken from the lake. A report from South America relates an infection attributed by the patient to a knife wound sustained while cleaning fish (Beer-Romero et al. 1989).

Based on these types of reports, it seems that the role of animal vectors should probably be given greater attention in the consideration of the epidemiology of sporotrichosis. Outbreaks of the kind associated with sphagnum moss are dramatic, but may prove to be the easiest to prevent or control (Coles et al. 1992). By contrast, the potential for zoonotic infections remains high, and in some settings does not readily lend itself to prevention strategies.

S. schenckii has been also been encountered as a saprobe from a variety of unexpected substrates. It has been recovered from commercial potting soil (Kenyon et al. 1984), found as a food contaminant of meat sausages (Ahearn and Kaplan 1969) and culinary mushrooms (Kazanas and Jackson 1983), and found growing in a container of intravenous fluid at a medical center (Matlow et al. 1985). *S. schenckii* has been recovered from samples taken for routine microbiological monitoring of various surfaces within an indoor swimming pool complex. Although the fungus was detected in moderate numbers from swabs of wall and floor surfaces, no plants or wooden materials, which could have provided a substrate for amplification of the organism, were found in the building (Staib and Grosse 1983). No infections were associated in any of these instances.

The pandemic spread of HIV has brought particular attention to the mycoses, in that most of the acquired immune deficiency syndrome (AIDS)-defining illnesses are fungal infections. Sporotrichosis is now one of the opportunistic infections encountered in very small numbers among AIDS patients (Lipstein-Kresch et al. 1985; Shaw et al. 1989; Heller and Fuhrer 1991; Perfect et al. 1993; Bolao et al. 1994; Donabedian et al. 1994). Only about a dozen cases have been reported to date, but sporotrichosis in HIV-seropositive patients is extremely serious. In patients whose conditions have deteriorated to clinical AIDS, sporotrichosis becomes disseminated and the prognosis is grave. Disseminated infection can involve one or more joints or organ systems, including the central nervous system (CNS). Such infections are problematic not only from the standpoint of therapy but also from the basis of diagnosis. As the AIDS epidemic spreads, there are certain to be more cases of disseminated sporotrichosis. Whether the prevalence of sporotrichosis in this group of people can be expected to increase is unknown.

There appears to be no relationship of gender to sporotrichosis. Although some reports have suggested that patients are more likely to be men, others have shown nearly equal ratios and a few have noted a preponderance of female patients (Conti-Diaz 1980, 1989; Fukushiro 1984; Itoh et al. 1986; Kusuhara et al. 1988). Differences among these reports may result from varying sex distributions for occupations and tasks among the differing regions, and thus might correlate with frequency of exposure to the fungus rather than a gender-based predilection for infection. Statistics from the recent USA outbreak suggest that gender was not a risk factor, and that the greatest predictor of infection in this particular series was the duration of exposure to the fungal-ridden moss (Coles et al. 1992).

In a recent report in Mexico, the frequency was higher in women (62 percent), children, and adolescents under 20 years of age (34 percent) and adults older than 50 years of age (28 percent) (Espinosa-Texis et al. 2001).

MYCOLOGY

It has long been recognized that the genus *Sporothrix* has accommodated a heterogeneous assemblage of morphologically similar species. One species, *Sporothrix cyanescens*, has been found to possess a septal ultrastructure that clearly shows affinity to the basidiomycetes. and so has been transferred to the genus *Cerinosterus* as *Cerinosterus cyanescens* (Smith and Batenburg-van der Vegte 1985). A teleomorph of *S. schenckii* has not been conclusively demonstrated in spite of many attempts at mating various clinical and environmental isolates. Nevertheless, as a result of similarities in morphology of the conidial morph formed by the ascomycete *Ophiostoma (Ceratocystis) stenocera* as compared with conidial formation in clinical isolates of *S. schenckii*, a phylogenetic relationship was surmised (Mariat 1971; de Hoog 1974). By contrast, studies of the polysaccharides and DNA base composition in these two taxa did not support this conclusion (Travassos and Lloyd 1980). Similarly, subsequent mitochondrial DNA typing by restriction fragment length polymorphism (RFLP) analysis showed a great deal of variation among isolates of *S. schenckii*, and this led to the conclusion that *S. schenckii* and *O. stenocera* may not be of the same holomorph (Suzuki et al. 1988). However, recent comparison of 18S rRNA gene sequence characters found that *S. schenckii* (two strains) and *O. stenocera* (one strain) differed at only three sites of the 1 700 nucleotides sequenced (Berbee and Taylor 1992). These findings strongly support the conclusion that *S. schenckii* is, or was at one time, the anamorph of an *Ophiostoma* spp. As a result of the wide variation in morphological and molecular characteristics reported thus far in these fungi, testing of additional strains is warranted. Presently, *S. schenckii* is classified in the mitosporic fungi (deuteromycetes).

S. schenckii synthesizes melanin via the 1,8-dihydoxynaphthalene pentaketide pathway, and this dark pigment has been found exclusively in the conidia, which are the infecting structures of this organism. Cultures of a reddish-brown mutant were identified by thin-layer chromatography, high-performance liquid chromatography, and ultraviolet (UV) spectra and material believed to be melanin was absent in conidial walls of the albino mutant. These results demonstrate that melanin may protect *S. schenckii* against certain oxidative antimicrobial compounds and against attack by macrophages, because melanized cells are less susceptible to oxidative killing in vitro and to phagocytosis by human monocytes and murine macrophages (Romero-Martinez et al. 2000).

IDENTIFICATION

Growth of *S. schenckii* can usually be observed within 3–5 days on culture media incubated at 30°C. Colonies typically are cream-colored initially, smooth and moist, gradually becoming dark brown and filamentous in texture. There is, however, much variation among isolates and even among subcultures of the same isolate. Some remain smooth and exhibit few or no aerial hyphae; others remain white to cream-colored with little or no brown pigmentation, or are brown during all stages of growth. Texture and coloration are also influenced by various culture media. Microscopic features include elongated conidiophores with a swollen, scarred apex which bears one-celled, obovoid (egg-shaped), hyaline to subhyaline conidia that measure approximately $(1.4–2.5) \times (2–4)$ μm; these are sometimes in radial 'flower-like' clusters (Figure 19.1). Isolates also form thick-walled conidia borne laterally along the hyphae, with each conidium being individually attached to the hypha (Figure 19.2).

These conidia are usually subglobose but can be triangular in some isolates. They can be colorless, but usually are dark brown and are chiefly responsible for the dark color of the colony. The various species in the genus *Sporothrix* are similar in appearance, but only *S. schenckii* is known to be pathogenic for humans and other animals. In addition, several other genera bear morphological resemblance to the genus *Sporothrix* (de Hoog 1974; de Hoog et al. 1985). For these reasons, in the absence of positive histology, it is necessary to confirm the identity of a suspected isolate of *S. schenckii*. This is accomplished by proving that, in addition to having both conidial forms as described above, the isolate will grow in a stable yeast-Iike form at 35–37°C (Figure 19.3). It should be noted that environmental, nonpathogenic isolates of *Sporothrix* spp. can grow as a yeast at 35–37°C, but lack the thick-walled, pheoid conidia that are attached individually along the hyphae (Dixon et al. 1991). When cultured at 35–37°C on chocolate agar or other enriched media, most isolates of *S. schenckii* will quickly show partial or complete conversion to a stable yeast-Iike form. Occasional isolates are more difficult to convert and may require extended incubation (up to 3 weeks) or frequent subculturing of the isolate (Kwon-Chung 1979). Some isolates from fixed cutaneous lesions reportedly will grow at 35°C but not at 37°C, and this thermal limitation was also shown through experimentation with mice (Kwon-Chung 1979). Other investigators have been unable to demonstrate this difference in isolates from fixed cutaneous lesions (de Albornoz et al. 1995).

A murine model suggest that the virulence of *S. schenckii* conidia may be determined by their cell wall composition. The more virulent forms of the fungus showed differences in cell wall sugar composition with rhamnose:mannose molar ratios of 1.7:1.0 for cells cultured for 4 days and 1.0:1.7 for conidia cultured for 12 days (Fernandes et al. 1999).

The interaction between *S. schenckii* and several extracellular matrix (ECM) proteins have been studied since adhesion is the first step involved with the dissemination of pathogens in the host. *S. schenckii* can bind to fibronectin, laminin, and type II collagen and also show

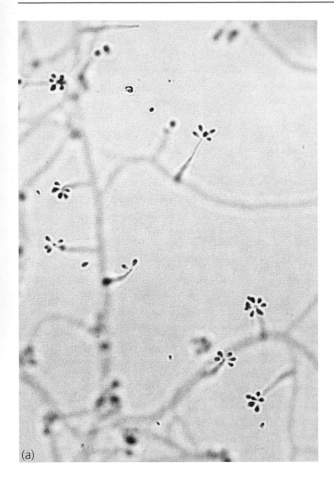

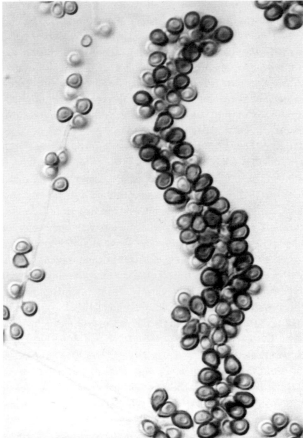

Figure 19.2 S. schenckii *also forms conidia that are thick-walled, dark brown as a result of melanin formation and attached directly to the hyphae, often in a dense sleeve-like manner.*

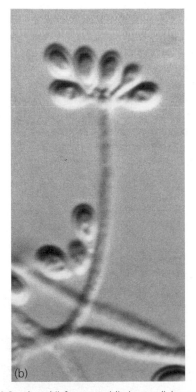

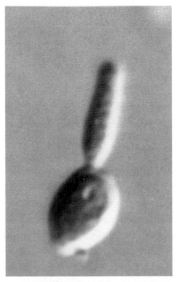

Figure 19.1 S. schenckii *forms conidia in a radial manner, each being attached to the apex of the conidiophore, resulting in a flower-like appearance.*

Figure 19.3 *In vitro conversion of* S. schenckii *from the mold to the yeast phase can be induced using a rich medium incubated at 35–37°C. Buds will often exhibit a characteristic cylindrical or 'cigar' shape.*

differences in binding capacity according to the morphological form of the fungus (Lima et al. 1999).

A new variety, *S. schenckii* var. *luriei*, was described in 1969 based on a case of sporotrichosis in which the fungal morphology manifested in host tissue was significantly different from that of typical sporotrichosis. Three cases have now been reported. In these infections, the fungus exhibited large (10–30 µm), subglobose to elongated, thick-walled (up to 2 µm) cells (Ajello and Kaplan 1969; Alberici et al. 1989; Padhye et al. 1992). These cells each formed a single septum that allowed separation into two separate cells. During separation, a fragment of the cell wall previously shared by the two cells can temporarily remain attached in a bridge-like manner, giving the fungus the distinctive appearance of spectacles. In some instances, a second septum would form, giving the fungus a muriform appearance. In the absence of culture, it is possible that the histological appearance of this new variety could be mistaken for chromoblastomycosis, although its cells do not contain visible amounts of melanin as seen in chromoblastomycosis. Mitochondrial DNA typing has also confirmed that *S. schenckii* var. *luriei* is markedly different from *S. schenckii* var. *schenckii* (Suzuki et al. 1988). In culture, whether as a mold or a yeast, *S. schenckii* var. *luriei* is morphologically indistinguishable from *S. schenckii* var. *schenckii*. There is insufficient experience with this variety to know if there are associated serological, therapeutic, or prognostic differences.

UV exposure of *Sporothrix* strains produces a high frequency of morphological variants and one common feature among them was that they were smaller in size compared to the wild type. These findings suggest that there is a highly sensitive chromatin in the conidial genome that is the target of UV light (Torres-Guerrero and Arenas-López 1998).

CLINICAL PRESENTATIONS OF SPOROTRICHOSIS

Sporotrichosis can be mistaken for numerous skin diseases including pyoderma gangrenosum (Spiers et al. 1986; Liao et al. 1991), rosaceae (Day et al. 1984), sarcoma (Sen et al. 1986) and leishmaniasis (Talhari et al. 1988). A recent field study in Brazil found that most patients with an initial diagnosis of leishmaniasis were infected instead with *S. schenckii* (Dr Aloisio Falqueto, personal communication).

Three cases of coinfection of leishmaniasis with *Sporothrix* spp. in the same lesion have been described in rural areas where people frequently try to remove the scabs using thorns or wooden splinters, thereby facilitating the entry of *S. schenckii*, which is present in the same geographic areas as *Leishmania* spp. (Agudelo et al. 1999)

Confusion with other diseases can lead not only to delay of suitable therapy, but also to the use of anti-inflammatory corticosteroids which can cause marked proliferation of the fungus within the lesions and a worsening of the infection (Bickley et al. 1985; Padhye et al. 1992). In a recent outbreak involving 84 patients, only 11 were correctly diagnosed at their first medical evaluation; 48 cases were misdiagnosed as bacterial infections, six as spider bites and two as cancer (Coles et al. 1992). In another series, the average delay in correct diagnosis was 4 months overall and 25 months for cases of extracutaneous sporotrichosis (Rowe et al. 1989).

The typical clinical presentation of sporotrichosis is that of a chronic cutaneous and subcutaneous infection. In these cases, lesions begin as a firm nodule, which initially is fluctuant but which gradually becomes immovable. The nodule becomes softened and in most cases eventually ulcerates to discharge pus or serous material (Figure 19.4). Satellite lesions will sometimes arise. A review of 235 cases showed that 23 percent were characterized by such fixed cutaneous lesions (Sampaio et al. 1954). Furthermore, facial lesions in children are often of this type (Fukushiro 1984; Eisfelder et al. 1993). In most cases, however, yeast cells from the initial lesion are phagocytosed and carried into the lymphatic system where a line of cutaneous ulcerative lesions then develops in an ascending fashion along lymph channels (Figure 19.5). *S. schenckii* has been isolated from blood, but such a finding is rare (Morgan et al. 1984; Kosinski et al. 1992).

In Mexico, upper limbs are affected in 45–53 percent, the face in 14–21 percent, lower limbs in 18–23 percent, and it is exceptional in the trunk (Arenas 2003). Lymphangitic forms account for 82 percent (Espinosa-Texis et al. 2001).

Complications in sporotrichosis are unusual, but the disease can be very refractory to medical management. Skeletal involvement (Atdjian et al. 1980; Arenas and Latapi 1984; Chang et al. 1984; Kumar et al. 1984; Yacobucci and Santilli 1986; Yao et al. 1986; Janes and Mann 1987; Wilson et al. 1988; Govender et al. 1989) is encountered often enough for sporotrichosis to be considered in the differential diagnosis of patients presenting with chronic joint disease. Reviews have

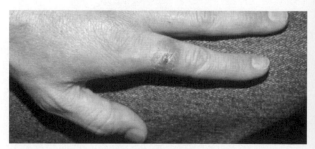

Figure 19.4 *Solitary cutaneous sporotrichosis lesion on the finger of a professional rose gardener*

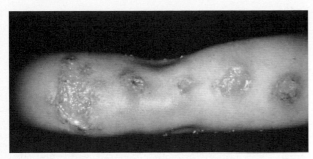

Figure 19.5 *Lymphangitic sporotrichosis ulcers develop in an ascending fashion along channels of the lymph system. (Courtesy T.G. Mitchell)*

noted that 1–14 percent of cases of sporotrichosis involve one or more joints (Rowe et al. 1989; Ortiz and Lefkovitz 1991). Joint infection also has been noted in the absence of accompanying skin lesions (Ortiz and Lefkovitz 1991).

Sporothrix arthritis is a rare disease and should be considered to be a monoarthritis, with only 51 cases reported in the English literature. Most of the patients have been middle-aged men who have a significant alcohol intake (77 percent). Seventeen joints have been affected and in 20 percent the knees were involved (Howell and Toohey 1998)

Dissemination, which can involve the CNS, has historically been relatively rare (Smith et al. 1981; Selman and Hampel 1982; Friedman and Doyle 1983; Morgan et al. 1984; Aronson 1992). However, cases of sporotrichosis now being seen in AIDS patients are usually disseminated, fatal infections (Lipstein Kresch et al. 1985; Kurosawa et al. 1988; Shaw et al. 1989; Heller and Fuhrer 1991; Bolao et al. 1994; Donabedian et al. 1994). Extracutaneous sporotrichosis can also occur in organ transplant recipients who are receiving immunosuppressive therapy for the prevention of organ rejection (Gullberg et al. 1987; Agarwal et al. 1994).

Sixty-three widely distributed cutaneous lesions were observed in a female host with both humoral and cellular immunity within normal limits, but who was under treatment with long-term prednisone (Severo et al. 1999).

Although the typical clinical presentation of sporotrichosis is that of a chronic cutaneous and subcutaneous infection, *S. schenckii* has also caused eye infections (Schell 1986; Rippon 1988), sinusitis, infection of the vocal cords (Smith et al. 1981; Agger and Seager 1985), pyelonephritis (Agarwal et al. 1994) and pulmonary infections. Pulmonary sporotrichosis has been increasingly documented (Rohatgi 1980; Friedman and Doyle 1983; Nusbaum et al. 1983; Pluss and Opal 1986; Pueringer et al. 1986; Heller and Fuhrer 1991) and is suspected of being more common than is generally realized (Lavalle and Mariat 1983; England and Hochholzer 1985, 1987; Gerding 1986; Pluss and Opal 1986; Pueringer et al. 1986; Watts and Chandler 1987; Rippon 1988; Haponik et al. 1989; Farley et al. 1991).

A recent 30-year review at one institution found that 4 percent of sporotrichosis cases were pulmonary infections (Rowe et al. 1989). This form of the disease follows inhalation of conidia that have become airborne from the mold as it grows upon organic substrata in nature. The infection is insidious and sometimes asymptomatic. Symptoms include low-grade fever, weight loss, night sweats, productive cough with or without hemoptysis, and pleuritic pain. Radiological examination of the lungs can reveal adenopathy, nodules or cavitation, usually involving the upper lobes, or diffuse involvement similar to that seen with tuberculosis (Comstock and Woolson 1975; Rohatgi 1980; Pluss and Opal 1986). Ethanol abuse is found in conjunction with some cases of sporotrichosis, especially pulmonary and joint infections, but the reasons for this apparent association are not known.

HISTOPATHOLOGY

Histological examination of lesions reveals a central necrotic region with associated infiltration of neutrophils, macrophages, and giant cells. Yeast cells are usually not seen using routine histological methodology, but can be detected through examination of serially sectioned tissue that has been stained with special fungal stains. This approach reveals areas of host tissue reaction to the fungus and indicates where the yeast cells are likely to be found. The yeast cells of *S. schenckii* are often solitary and quite round in shape, 3–5 μm in diameter, mostly without apparent budding (Figure 19.6). Asteroid bodies (Figure 19.7) and cylindrical 'cigarshaped' buds are often cited as characteristic microscopic features in cases of sporotrichosis, but both these can be difficult to demonstrate in many cases. Asteroid bodies are stellate-shaped conglomerations which develop from host–parasite interactions and can be seen in other diseases in addition to sporotrichosis.

Electron micrography of asteroid bodies obtained from cases of sporotrichosis suggests that materials from disintegrated host cells, particularly polymorphonuclear neutrophils, are deposited upon the surface of the fungal

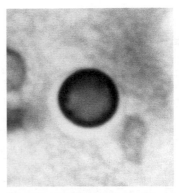

Figure 19.6 *S. schenckii cells in infected tissue are rare in number, round to ellipsoidal, 3–5 μm, yeast-like, with or without obvious budding.*

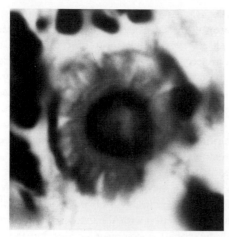

Figure 19.7 *Asteroid bodies consist of degenerate materials from the host immune response that accumulate in a ray-like manner upon the yeast cells of* S. schenckii, *giving them a stellate appearance.*

cell and are subsequently crystallized through an unknown process. The result is a multilayered accretion which features an outer layer of low electron density that appears to contain fragments of cell membrane-Iike material (Hiruma et al. 1991, 1992a). In lesions where the fungus is abundant, cells of *S. schenckii* can be within the macrophages, and sometimes exhibit a staining artifact of a halo or capsule. In these instances, sporotrichosis can be mistaken for *Histoplasma capsulatum* var. *capsulatum* (Nusbaum et al. 1983; Farley et al. 1991; Aronson 1992).

The characteristic infiltrates in three zones (central suppurative, tuberculoid, and peripheral round cell) were found in 27.27 percent, the asteroid bodies in 18 percent, and the yeast cells in 32 percent. The histopathological findings usually offer clues to the diagnosis and in just a few cases when the yeast cells are found, it can be diagnostic (Ruiz-Esmenjaud et al. 1996).

Asteroid bodies and yeast cells with budding, highly suggestive of the disease, were seen in the cytological and histological preparations obtained with fine-needle aspiration (FNA) cytology (Zaharopoulos 1999).

IMMUNOLOGY AND SEROLOGY

Data suggest that exposure to *S. schenckii* is widespread in many populations, and that the host immune response is able to resolve most inoculum challenges. Antigens prepared from *S. schenckii* have permitted small-scale studies of skin-test reactivity to be conducted in areas of both high and low endemicity (Ajello 1970). Results from studies of a prison population and a hospital population in the USA showed delayed hypersensitivity reactions occurring in 11 percent and 10 percent of individuals, respectively. In comparison, people involved in high-risk occupations were found to have a much greater prevalence of skin-test positivity. Of workers tested in a

plant nursery, 33 percent were reactive; for those workers employed for at least 10 years, the rate of positive skin tests was 58 percent. In Brazil, skin testing in a small group from the general population yielded a reactivity rate of 24 percent. These findings suggest that exposure to *S. schenckii* is more common than the number of clinical infections indicates, and that most infections are asymptomatic and self-limiting. Occasional clinical observations of apparent spontaneous remission support this notion for both cutaneous and pulmonary infections (Powell et al. 1978; Bargman 1981; Iwatsu et al. 1985; Pluss and Opal 1986; Pueringer et al. 1986; Ramani et al. 1991).

In 50 patients, a metabolic antigen was used to elicit delayed hypersensitivity skin reaction and delayed hypersensitivity skin reactions were positive in 78 percent and culture in 94 percent (Espinosa-Texis et al. 2001). Immunodiffusion has been positive in 80 percent, complement fixation in 40 percent and also Western blot has been performed (Arenas 2003)

Animal models of sporotrichosis have also provided data which suggest that acquired resistance to infections may occur. Experimentation has indicated that exposure of mice to viable cells of *S. schenckii* can confer increased resistance to a subsequent intravenous challenge with the fungus (Shiraishi et al. 1992). In addition, sensitized spleen cells which were transferred into fresh mice also conferred increased resistance to intravenous challenge. Further testing showed that T-cell-mediated macrophage activation led to increased killing of *S. schenckii* cells from a subsequent challenge, and conversely that pretreatment with a macrophage inhibitor reduced resistance to intravenous challenge. Other studies using the mouse model have also revealed that *S. schenckii* can secrete extracellular proteases, and that mouse antibodies formed against these enzymes correlate with the course of disease progression (Yoshiike et al. 1993). Furthermore, it has been shown, in the mouse model, that the cellular immune response becomes depressed between weeks 4 and 6 of infection, and is a harbinger of worsening disease (Carlos et al. 1992).

The infectivity of this fungus between isolates from human cutaneous and pulmonary sporotrichosis after subcutaneous injection into the footpad showed restricted fungal infection and, after intravenous or intraperitoneal injection of yeast forms, three of four isolates from cutaneous sporotrichosis were unable to establish infection, but the development of systemic infection was observed only with *S. schenckii* isolates obtained from the human lung lesion (Tachibana et al. 1999a)

The cell-mediated immune response against infection with *S. schenckii* has been investigated in vivo and in vitro. There have been few studies reporting the role of cytokines in this infection. In this study it was shown that immune lymph node cells expressed message for interferon-gamma (IFN-γ), tumor necrosis factor-alpha (TNF-α), and interleukin-10 (IL-10) after stimulation with heat-killed *S. schenckii*. These results suggest that acquired immunity against *S. schenckii* is expressed mainly by macrophages activated by CD4+ T cells, exerting protective immunity against *Sporothrix* via enhanced fungicidal activity (Tachibana et al. 1999b)

Differences in human serological response to sporotrichosis have been demonstrated for patients with cutaneous disease compared with patients with extracutaneous (joint, pulmonary, and CNS) disease (Scott and Muchmore 1989). In tests using a single strain of *S. schenckii* for antigen preparation, twice as many antibodies were detected in the sera of patients with extracutaneous infection. It is not clear how this may relate to pathogenicity, but the ability to detect antibodies that are specific for extracutaneous sporotrichosis could potentially provide a useful means for serological diagnosis of this complication.

Antibodies to *S. schenckii* antigens can routinely be detected in most cases of sporotrichosis through the use of complement fixation and latex agglutination tests, but such tests require the use of specialized, labile, and relatively expensive reagents. Therefore, as a practical matter, it is easier and cheaper to diagnose most infections by culture and microscopic examination of infected tissue. In the diagnosis of CNS sporotrichosis, however, detection of antibodies has succeeded in several patients when a culture-based diagnosis had failed. In such cases serological testing offers an invaluable tool (Scott et al. 1987).

A concanavalin A (conA)-binding fraction has been isolated from the *S. schenckii* cell wall (SsCBF) and the immunodominance of the *O*-glycosidically linked oligosaccharides in the reactivity of SsCBF with IgG from patients with sporotrichosis has been observed. These findings suggest that the SsCBF is a species-specific antigenic fraction recognized by human sera and with potential application in the serological diagnosis of sporotrichosis as well as in the epidemiologic survey of the systemic form of this mycosis (Loureiro et al. 2000).

A predominant antigen gp84 has been demonstrated by Western blot with a conA-horseradish peroxidase (HRP) conjugate (Lima and Lopes-Bezerra 1997)

LABORATORY DIAGNOSIS

The quickest means of diagnosing most mycoses is by immediate microscopic examination of infected host material. This approach is usually futile with sporotrichosis because the cells of *S. schenckii* normally occur very sparsely within infected material. However, painstaking histological examination of multiple tissue sections can often establish the diagnosis. Another approach to diagnosis by microscopy is the staining of histological tissue sections with a species-specific fluorescent antibody conjugate which can quickly provide a specific identification of *S. schenckii* (Kaplan and Gonzalez-Ochoa 1963); these fluorescent antibody reagents are, however, available only at certain refer-

ence laboratories and are not commercially offered. The usual form of *S. schenckii* in tissue is that of a yeast, but, in rare cases, superficial crusts from lesions or specimens from pulmonary infection have shown hyphal filaments (Maberry et al. 1966; Lopes et al. 1992).

By indirect immunofluorescence, parasite elements were demonstrated in 50 patients corresponding with a sensitivity and specificity of 100 percent. In this work indirect immunofluorescence was the most effective diagnostic method for sporotrichosis followed by culture, hypersensitivity skin reaction, and histopathology (Espinosa-Texis et al. 2001). It has been found that anti-*Sporothrix* antibodies are specific in sporotrichosis. Immunohistochemistry was performed in 25 paraffin-embedded biopsy specimens of sporotrichosis in which asteroid bodies (AB) were demostrated, ten of them with a positive culture. The AB were extracellular eosinophilic structures, 15–35 μm in diameter, and located within abscesses, with a central yeast. The yeast stains with the anti-*Sporothrix* antibody, while the spicules do not. Visualization of the spicules alone can lead to the demonstration of the AB in adjacent sections, and thus could be a useful clue in the diagnosis of sporotrichosis (Rodríguez and Sarmiento 1998). Immunohistochemistry has been a useful tool as is immunofluorescence when a yeast or AB is present (M. Heurre, personal communication) (Figure 19.8).

The most sensitive means of diagnosing sporotrichosis is recovery of the fungus in culture. Infected material is inoculated onto fungal culture media that are formulated to allow the growth of *S. schenckii* while selectively inhibiting the growth of fungal contaminants and the normal microbiota that could be present. Cutaneous specimens consist of pus or biopsied tissue from the lesion, whereas pulmonary specimens consist of expectorated sputum or material collected by means of a bronchoscope. If meningitis is suspected, cerebrospinal fluid is obtained for culture via a lumbar puncture. As cultures are often negative in cases of CNS infection, serological testing together with culture is recommended

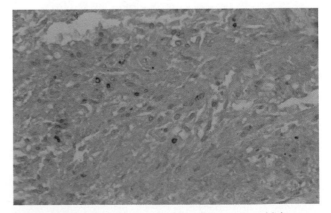

Figure 19.8 *Sensitivity by immunohistochemistry is as high as immunoflorescence for* Sporothrix *yeasts.*

(Scott et al. 1987). Serology should also be considered in any case where sporotrichosis is strongly suspected but cultures are negative.

MOLECULAR BIOLOGY

RFLP analysis of mitochondrial DNA (mtDNA) has been reported to be a useful method of identification, taxonomy, typing, and epidemiology of *S. schenckii*. Originally based on the phylogeny of 14 mtDNA types of *S. schenckii*, types 1–14 were constructed by estimating sequence divergences of mtDNA using restriction patterns with *Hae*III. Based on genetic distance, they have been clustered into two main groups: group A and group B. Isolates of group A exist mainly in North, South, and Central America, including Costa Rica (Ishizaki et al. 1996), and most of the Japanese isolates belong to group B (Takeda et al. 1991; Ishizaki et al. 1998)

S. schenckii strains have been classified into 23 types and clustered into two groups, group A and group B, on the basis of RFLP of mtDNA. The geographical distribution of mtDNA types was very characteristic. All isolates from Spain belonged to group B in contrast to North and South America where group A was predominant (Kawasaki et al. 2000) (Figure 19.9).

mtDNA types based on RFLP patterns with *Hae*III have been investigated also in China. In addition to types 1–23, a new mtDNA type (type 24) was found. Type 24 was divided into two subtypes (24A and 24B) based on RFLP with *Eco*RV. Seventy-seven isolates in China consisted of 58 isolates of type 4 (91 percent), five of type 6, one of type 5, one of type 20, and two of type 24, and classified into group A (types 1–3, 11, 14–19, 22 and 23) and group B (types 12, 13, 20, 21, and 24). Most isolates in China belong to group B and Korean isolates presumably belong to group B (Lin et al. 1999).

As we can see, group B contains most of the Japanese isolates and group A most of the American types. Group A is more branched compared with group B, where most of the types seem to be more closely genetically related and more mtDNA variability is found in group A, probably because the isolates have different geographic origins, although most of them come from the American continent. The molecular evolutionary explanations for the high variability are not clear. It will start to emerge with the molecular study of other genetic characters such as DNA sequences of mitochondrial and nuclear genes. mtDNA diversity was analyzed in 42 clinical isolates from Mexico, Guatemala, and Colombia, and six new types were found. Most of the strains belong to types 14 and 30, the former restricted

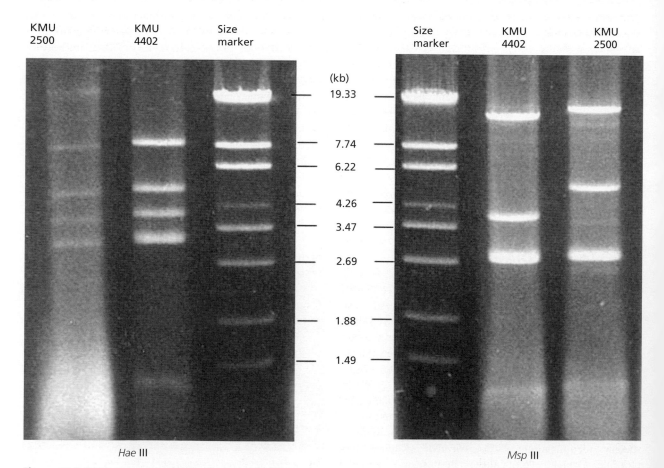

Figure 19.9 *RFLP patterns of mtDNA of* S. schenckii.

to Mexico, whereas the latter was distributed in Mexico, Guatemala, and Colombia. The new types (25–30) were identified in Mexico, Guatemala, and Colombia. These types were clustered in group A (types 26, 28–30) and group B (type 27). It seems that the variability in *S. schenckii* mtDNA is due in part to restriction site changes; however, the variation in size could also indicate length mutations (Mora-Cabrera et al. 2001).

Other clinical isolates were investigated for mtDNA type RFLP patterns with *Nae*III. The 62 isolates in South Africa comprised types 3 (9.7 percent), 4 (1.6 percent), 11 (9.7 percent), 17 (77.4 percent), and 23 (1.6 percent), whereas the 23 Australian isolates comprised types 3 (26 percent), 4 (56.5 percent), 7 (8.6 percent) and 21 (8.6 percent). The isolates in South Africa mainly belong to group A, with the predominance of type 17 (77 percent), and isolates in Australia mainly belong to group B.

The phylogenetic tree of mtDNA types and the world-wide distribution of types 3 and 4, suggest that these two types are ancestral types (or closely related), and that the others diverged from them over the course of continental drift. It may be that the divergence of mtDNA types occurred in South Africa earlier than in Australia (Ishizaki et al. 2000).

Partial cDNA cloning of a putative membrane transporter protein gene expressed in *S. schenckii* is reported and also DNA polymorphism of the isolated gene is demonstrated using polymerase chain reaction single-strand conformation polymorphism (PCR-SSCP) analysis. Also based on mtDNA diversity, *S. schenckii* has been classified into three groups located on the branches of the phylogenetic tree constructed by digestion profiles of mtDNA with restriction enzymes. The correlation of the results of PCR-SSCP analysis with the mtDNA diversity might indicate linkage of the mutation of the membrane transporter protein gene with the evolution of *S. schenckii*, and although the precise mechanism must be elucidated by further investigation, database searches also showed homology with the sugar transporter protein family (Sugita 2000)

Clinical isolates of fixed and lymphocutaneous forms from Mexico, Guatemala, and Colombia, as well as environmental isolates from Mexico, were studied by analyzing their phenotypic characteristics (conidial length, thermotolerance by percent growth inhibition at 35 and 37°C, median lethal dose (LD50), and genotypic characteristics [by random amplified polymorphic DNA (RAPD) analysis polymerase chain reaction (PCR)]. In general, the highest virulence, as determined by measurement of the LD50 for mice, was observed for the Mexican environmental isolates. *Sporothrix* fell into four major groups by hierarchical cluster analysis. A principal component analysis yielded three distinct groups, depending on the geographical origins of the isolates. This study revealed that isolates from Colombia had low thermotolerance at 35 and 37°C and could be associated with superficial skin lesions in patients with fixed clinical forms of sporotrichosis. Distinct patterns dependent on geographical origin were also revealed by RAPD analysis PCR, but these had no relation to the clinical form of the disease (Mesa-Arango et al. 2002).

TREATMENT

Uncomplicated lymphocutaneous sporotrichosis historically has been treated by oral administration of saturated solutions of potassium iodide, an inexpensive treatment that has proved effective in most patients. In some cases, the application of topical heat has been shown to cure lymphocutaneous disease (Hiruma et al. 1992b). A partial explanation for the efficacy of topical heat therapy has been provided by laboratory investigation. When yeast cells of *S. schenckii* were incubated in serum with polymorphonuclear neutrophils at 40°C as compared with 37°C, there was no difference in the rates at which the fungal cells were engulfed by the neutrophils. Once engulfed, however, the rate of killing by the neutrophils was significantly greater at 40°C than at 37°C (Hiruma and Kagawa 1986).

In one study of 58 patients, potassium iodide was equally effective whether administered once or three times daily, although side effects were slightly increased (Cabezas et al. 1996).

Potassium iodide cures approximately 90 percent of immunocompetent patients with sporotrichosis and relapses are rare. A pregnant woman taking excess quantities of iodine may result in possible goiter, hypothyroidism, and fetal death. Potassium iodide is currently a category D drug, meaning that it should not be used in pregnant or breastfeeding women unless the benefits of taking it outweigh the risks. Care must be taken in patients with renal failure and other potential causes of hyperkalemia, or in children with cystic fibrosis as they are susceptible to goitrogenic effects of iodide, or in patients with thyroid disease (Sanchez 2000; Sterling and Heymann 2000). Potassium iodide may cause the Wolff–Chaikoff effect (WCE), the cessation of thyroid hormone synthesis subsequent to the administration of iodide, but also iodine excess may lead to thyrotoxicosis (Jod–Basedow effect) and, rarely, iodine excess may cause an acute thyroiditis manifested by a painful, enlarged thyroid gland (Heymann 2000).

Modern antifungal compounds have been shown to be highly effective in curing lymphocutaneous sporotrichosis. Itraconazole and, to a lesser degree, fluconazole are orally administered drugs that are both effective; however, they are quite expensive and unavailable in many parts of the world (Sharkey-Mathis et al. 1993; Ghodi et al. 2000)

Forty-three Japanese cases of sporotrichosis were treated with itraconazole for 3 months, adult patients received 100 mg/day, whereas young patients under 5 years of age were treated with 50 mg/day; 88 percent have been assessed as cured or improved (Noguchi et al. 1999).

Terbinafine, another of the new antifungals, has been used topically for dermatophytosis, and preliminary results from oral use in cases of sporotrichosis have been promising (Hull and Vismer 1992).

A review article reported that 18 patients with sporotrichosis enrolled in five separate studies were cured by terbinafine at doses of 125–500 mg/day for a median of 12 weeks (range 4–37). They received terbinafine for up to 1 month after clinical resolution of lesions or for a maximum period of up to 6 months (Hay 1999; Pérez 1999). A total of 63 patients with cutaneous or lymphocutaneous sporotrichosis were randomized in a comparative study to evaluate the efficacy and safety of two doses of terbinafine, 28 were treated with 500 mg/day and 35 with 1 000 mg daily during a maximum of 24 weeks. Tolerance was similar in both groups, but 6 cases relapsed in the first group and none in the second. There were two drug adverse events related to the 1000 mg dose group, one case of gastrointestinal hemorrhage and one of progressive abdominal pain (Chapman et al. 2004).

Systemic forms including osteoarticular, pulmonary, and meningeal involvement can be life-threatening and very difficult to treat (Wescott et al. 1999).

Complicated cases of sporotrichosis are those that involve one or more joints, or that have disseminated to other organs, such as the CNS. These cases are very difficult to treat, and require the intravenous administration of amphotericin B. These patients may also need a prolonged subsequent regimen of an oral antifungal azole to prevent reactivation of the infection. In the case of AIDS patients, sporotrichosis (as with other fungal infections) can potentially be suppressed but cannot be cured, so lifelong suppressive antifungal therapy is required.

In patients infected with HIV, sporotrichosis may require systemic administration of amphotericin. The disease is potentially life-threatening. In one case amphotericin therapy was reinstituted but discontinued because of renal complications and amphotericin B lipid complex was then initiated at a dose of 5 mg/kg/day and was tolerated well, resulting in progressive resolution of all lesions. Lesions have recurred, despite the oral administration of itraconazole, fluconazole, or terbinafine. Blood levels of itraconazole are diminished by a number of agents that stimulate its hepatic metabolism, such as carbamazepine and rifampin (Ware et al. 1999) (see Table 19.1).

ACKNOWLEDGMENTS

This chapter is adapted and expanded from Schell, W.A. 1998. Agents of Chromoblastomycosis and Sporotrichosis. In Ajello, L. and Hay, R.J. (eds) *Topley & Wilson's Microbiology and Microbial Infections*, vol. 4, *Medical Mycology*, 9th edn. London: Edward Arnold, pp. 315–36.

REFERENCES

Agarwal, S.K., Tiwari, S.C., et al. 1994. Urinary sporotrichosis in a renal allograft recipient. *Nephron*, **66**, 485.

Agger, W.A. and Seager, G.M. 1985. Granulomas of the vocal cord caused by *Sporothrix schenckii*. *Laryngoscope*, **95**, 595–6.

Agudelo, S.P., Restrepo, S. and Vélez, I.D. 1999. Cutaneous New World leishmaniasis – sporotrichosis coinfection: report of 3 cases. *J Am Acad Dermatol*, **40**, 6 Pt 1, 1002–4.

Ahearn, D.G. and Kaplan, W. 1969. Occurrence of *Sporotrichum schenckii* on a cold-stored meat product. *Am J Epidemiol*, **80**, 116–24.

Ajello L, 1970, The medical mycological iceberg, *Scientific Publication No.205,* Pan American Health Organization, Washington, DC, 3–10.

Ajello, L. and Kaplan, W. 1969. A new variant of *Sporothrix schenckii*. *Mykosen*, **12**, 633–44.

Alberici, F., Paties, C.T., et al. 1989. *Sporothrix schenckii* var *luriei* as the cause of sporotrichosis in Italy. *Eur J Epidemiol*, **5**, 173–7.

Arenas, R. 2003. *Micologia médica ilustrada*. Mexico: Interamericana-McGraw-Hill, 29–37.

Arenas, R. and Latapi, F. 1984. Sporotrichose generalisee. *Bull Soc Pathol Exot Filiales*, **77**, 385–91.

Aronson, N.E. 1992. Disseminated sporotrichosis. *JAMA*, **268**, 2021.

Atdjian, M., Granda, J.L., et al. 1980. Systemic sporotrichosis polytenosynovitis with median and ulnar nerve entrapment. *JAMA*, **243**, 1841–2.

Barbee, W.C., Ewert, A. and Davidson, E.M. 1977. Animal model: sporotrichosis in the domestic cat. *Am J Pathol*, **86**, 281–4.

Bargman, H.B. 1981. Sporotrichosis of the nose with spontaneous cure. *Can Med Assoc J*, **124**, 1027.

Barile, F., Mastrolonardo, M., et al. 1993. Cutaneous sporotrichosis in the period 1978–1992 in the province of Bari, Apulia, Southern Italy. *Mycoses*, **36**, 181–5.

Bary, P., Kuriata, M.A. and Cleaver, L.L. 1999. Lymphocutaneous sporotrichosis. A case report and unconventional source of infection. *Cutis*, **63**, 173–7.

Bastos de Lima-Barros, M., Pacheco-Schubach, T.M., et al. 2001. Sporotrichosis: an emergent zoonosis in Rio de Janeiro. *Mem Inst Oswaldo Cruz*, **96**, 777–9.

Beer-Romero, P., Rodriguez-Ochoa, G., et al. 1989. Sporotrichosis in the Orinoco river basin of Venezuela and Colombia. *Mycopathologia*, **105**, 19–23.

Berbee, M.L. and Taylor, J.W. 1992. 18S ribosomal RNA gene sequence characters place the human pathogen *Sporothrix schenckii* in the genus *Ophiostoma*. *Exp Mycol*, **16**, 87–91.

Bickley, L.K., Berman, I.J. and Hood, A.F. 1985. Fixed cutaneous sporotrichosis: unusual histopathology following intralesional corticosteroid administration. *J Am Acad Dermatol*, **12**, 1007–12.

Bolao, F., Podzamczer, D., et al. 1994. Efficacy of acute phase and maintenance therapy with itraconazole in an AIDS patient with sporotrichosis. *Eur J Clin Microbiol Infect Dis*, **13**, 609–12.

Brown, R.D. and Weintroub, D. 1947. *Sporotrichosis infection on mines of the Witwatersrand.* A *symposium*. Johannesburg: Transvaal Chamber of Mines, 23–7.

Cabezas, C., Bustamante, B., et al. 1996. Treatment of cutaneous sporotrichosis with one daily dose of potassium iodide. *Pediatr Infect Dis*, **15**, 352–4.

Campos, P., Arenas, R. and Coronado, H. 1994. Epidemic cutaneous sporotrichosis. *Int J Dermatol*, **33**, 38–41.

Carlos, I.Z., Sgarbi, D.B., et al. 1992. Detection of cellular immunity with the soluble antigen of the fungus *Sporothrix schenckii* in the systemic form of the disease. *Mycopathologia*, **117**, 139–44.

Chang, A.C., Destouet, J.M. and Murphy, W.A. 1984. Musculoskeletal sporotrichosis. *Skeletal Radiol*, **12**, 23–8.

Chapman, S.W., Pappas, P., et al. 2004. Comparative evaluation of the efficacy and safety of two doses of terbinafine (500 and 1000 mg day) in the treatment of cutaneous or lymphocutaneous sporotrichosis. *Mycoses*, **47**, 62.

Coles, F.B., Schuchat, A., et al. 1992. A multistate outbreak of sporotrichosis associated with sphagnum moss. *Am J Epidemiol*, **136**, 475–87.

Comstock, C. and Woolson, A.H. 1975. Roentgenology of sporotrichosis. *Am J Roentgenol Radium Ther Nucl Med*, **125**, 651–5.

Conti-Diaz, I.A. 1980. Sporotrichosis in Uruguay: epidemiologic and clinical aspects. *Scientific Publication No. 396*. Washington, DC: Pan American Health Organization, 312–21.

Conti-Diaz, I.A. 1989. Epidemiology of sporotrichosis in Latin America. *Mycopathologia*, **108**, 113–16.

Cook, W., Sexton, D.J., et al. 1984. Sporotrichosis among hay-mulching workers – Oklahoma, New Mexico. *Morbid Mortal Wkly Rep*, **33**, 682–3.

Cooper, C.R., Dixon, D.M. and Salkin, I.F. 1992. Laboratory-acquired sporotrichosis. *J Med Vet Mycol*, **30**, 169–71.

Costa, E.O., Diniz, L.S., et al. 1994. Epidemiological study of sporotrichosis and histoplasmosis in captive Latin American wild mammals, Sao Paulo, Brazil. *Mycopathologia*, **125**, 19–22.

Cote, T.R., Kasten, M.J. and England, A.C. 1988. Sporotrichosis in association with Arbor Day activities. *N Engl J Med*, **319**, 1290–1291.

Crevasse, L. and Ellner, P.D. 1960. An outbreak of sporotrichosis in Florida. *JAMA*, **173**, 29–33.

D'Alessio, D.J., Leavens, L.J., et al. 1965. An outbreak of sporotrichosis in Vermont associated with sphagnum moss as the source of infection. *N Engl J Med*, **272**, 1054–8.

Day, T.W., Gibson, G.H. and Guin, J.D. 1984. Rosacea-like sporotrichosis. *Cutis*, **33**, 549–52.

de Albornoz, M.B., Mendoza, M. and de Torres, E.D. 1995. Growth temperatures of isolates of Sporothrix schenckii from disseminated and fixed cutaneous lesions of sporotrichosis. *Mycopathologia*, **95**, 81–3.

De Araujo, T., Marques, A.C. and Kerdel, F. 2001. Sporotrichosis. *Int J Dermatol*, **40**, 737–42.

de Hoog, G.S. 1974. The genera *Blastobotrys, Sporothrix, Calcarisporium* and *Calcarisporiella* gen. nov. *Studies in mycology, No.7*. Baarn: Centraal Bureau voor Schimmelcultures, 12–66.

de Hoog, G.S., Rantio-Lehtimaki, A.H. and Smith, M.T. 1985. *Blastobotrys, Sporothrix* and *Trichosporiella*: generic delimitation, new species, and a *Stephanoascus* teleomorph. *Antonie Van Leeuwenhoek*, **51**, 79–109.

Dewan, N., Bedi, S. and O'Donohue, W.J. Jr 1986. Primary pulmonary sporotrichosis occurring in two meat packers. *Nebr Med J*, **71**, 37–9.

Ditmars, D.M. and Maguina, P. 1998. Neck skin sporotrichosis after electrolysis. *Plast Reconstr Surg*, **101**, 504–6.

Dixon, D.M., Salkin, I.F., et al. 1991. Isolation and characterization of *Sporothrix schenckii* from clinical and environmental sources associated with the largest U.S. epidemic of sporotrichosis. *J Clin Microbiol*, **29**, 1106–13.

Donabedian, H., O'Donnell, E., et al. 1994. Disseminated cutaneous and meningeal sporotrichosis in an AIDS patient. *Diagn Microbiol Infect Dis*, **18**, 111–15.

Dooley, D.P., Bostic, P.S. and Beckius, M.L. 1997. Spook house sporotrichosis. A point-source outbreak of sporotrichosis associated with hay bale props in a Halloween haunted-house. *Arch Intern Med*, **157**, 16, 1885–7.

Dunstan, R.W., Laughan, R.F., et al. 1986a. Feline sporotrichosis: a report of five cases with transmission to humans. *J Am Acad Dermatol*, **15**, 37–45.

Dunstan, R.W., Reimann, K.A. and Langham, R.F. 1986b. Feline sporotrichosis. *J Am Vet Med Assoc*, **189**, 880–3.

du Toit, C.J. 1942. Sporotrichosis on the Witwatersrand. *Proc Transvaal Mine Med Officers' Assoc*, **22**, 241, 111–25.

Eisfelder, M., Okamoto, S. and Toyama, K. 1993. Erlahrungen mit 241 sporotrichose-fallen in Chiba/Japan. *Huatartz*, **44**, 524–8.

England, D.M. and Hochholzer, L. 1985. Primary pulmonary sporotrichosis. Report of eight cases with clinicopathologic review. *Am J Surg Pathol*, **9**, 193–204.

England, D.M. and Hochholzer, L. 1987. *Sporothrix* infection of the lung without cutaneous disease. Primary pulmonary sporotrichosis. *Arch Pathol Lab Med*, **111**, 298–300.

Espinosa-Texis, A., Hernández-Hernández, F., et al. 2001. Estudio de 50 pacientes con esporotricosis. Evaluación clínica y de laboratorio. *Gac Med Mex*, **137**, 2, 111–16.

Farley, M.L., Fagan, M.F., et al. 1991. Presentation of *Sporothrix schenckii* in pulmonary cytology specimens. *Acta Cytol*, **35**, 389–95.

Fernandes, K.S., Mathews, H.L. and Lopes Bezerra, L.M. 1999. Differences in virulence of *Sporothrix schenckii* conidia related to culture conditions and cell-wall components. *J Med Microbiol*, **48**, 2, 195–203.

Fleury, R.N., Taborda, P.R., et al. 2001. Zoonotic sporotrichosis. Transmission to humans by infected domestic cat scratching: report of four cases in Sao Paulo, Brazil. *Int J Dermatol*, **40**, 5, 318–22.

Forester, H.R. 1924. Sporotrichosis. *Am J Med Sci*, **167**, 55–76.

Frean, J.A., Isaacson, M., et al. 1991. Sporotrichosis following a rodent bite. A case report. *Mycopathologia*, **116**, 5–8.

Friedman, S.J. and Doyle, J.A. 1983. Extracutaneous sporotrichosis. *Int J Dermatol*, **22**, 171–6.

Frumkin, A. and Tisserand, M.E. 1989. Sporotrichosis in a father and son. *J Am Acad Dermatol*, **20**, 964–7.

Fukushiro, R. 1984. Epidemiology and ecology of sporotrichosis in Japan. *Zentral Bakteriol Hyg A*, **257**, 228–33.

Gastineau, F.M., Spolyar, L.W. and Haynes, E. 1941. Sporotrichosis: report of six cases among florists. *JAMA*, **117**, 1074–7.

Gerding, D.N. 1986. Treatment of pulmonary sporotrichosis. *Semin Respir Infect*, **1**, 61–5.

Ghodi, S.Z., Shams, S., et al. 2000. Case report. An unusual case of cutaneous sporotrichosis and its response to weekly fluconazole. *Mycoses*, **43**, 75–7.

Goncalves, A.P. and Peryassu, D. 1954. A esporotrichose no Rio de Janeiro (1936–1953). *O Hospital*, **46**, 1–12.

Govender, S., Rasool, M.N. and Ngcelwane, M. 1989. Osseous sporotrichosis. *J Infect*, **19**, 273–6.

Grotte, M. and Younger, B. 1981. Sporotrichosis associated with sphagnum moss exposure. *Arch Pathol Lab Med*, **105**, 50–1.

Gullberg, R.M. and Quintanilla, A. 1987. Sporotrichosis: recurrent cutaneous, articular, and central nervous system infection in a renal transplant recipient. *Rev Infect Dis*, **9**, 369–75.

Hampton, D.E., Adesina, A. and Chodosh, J. 2002. Conjunctival Sporotrichosis in the Absence of Antecedent Trauma. *Cornea*, **21**, 831–3.

Haponik, E.F., Hill, M.K. and Craighead, C.C. 1989. Pulmonary sporotrichosis with massive hemoptysis. *Am J Med Sci*, **297**, 251–3.

Hay, R.J. 1999. Therapeutic potential of terbinafine in subcutaneous and systemic mycoses. *Br J Dermatol*, **141**, Suppl 56, 36–40.

Hayes, W.N. 1960. Sporotrichosis in employees of a tree nursery. *GP*, **22**, 4, 114–15.

Heller, H.M. and Fuhrer, J. 1991. Disseminated sporotrichosis in patients with AIDS: case report and review of the literature. *AIDS*, **5**, 1243–6.

Helm, M.A.F. and Berman, C. 1947. *Sporotrichosis infection on mines of the Witwatersrand*. Johannesburg: Transvaal Chamber of Mines, 59–72.

Heymann, W.R. 2000. Potassium iodide and the Wolff-Chaikoff effect: relevance for the dermatologist. *J Am Acad Dermatol*, **42**, 3, 490–2.

Hiruma, M. and Kagawa, S. 1986. Effects of hyperthermia on phagocytosis and intracellular killing of *Sporothrix schenckii* by polymorphonuclear leukocytes. *Mycopathologia*, **95**, 93–100.

Hiruma, M., Kawada, A. and Ishibashi, A. 1991. Ultrastructure of asteroid bodies in sporotrichosis. *Mycoses*, **34**, 103–107.

Hiruma, M., Kawada, A., et al. 1992a. Tissue response in sporotrichosis: light and electron microscopy studies. *Mycoses*, **35**, 35–41.

Hiruma, M., Kawada, A., et al. 1992b. Hyperthermic treatment of sporotrichosis: experimental use of infrared and far infrared rays. *Mycoses*, **35**, 293–9.

Howell, S.J. and Toohey, J.S. 1998. Sporotrichal arthritis in south central Kansas. *Clin Orthop*, **346**, 207–14.

Hull, P.R. and Vismer, H.F. 1992. Treatment of cutaneous sporotrichosis with terbinafine. *Br J Dermatol*, **126**, Suppl 39, 51–5.

Ishizaki, H., Kawasaki, M., et al. 1996. Mitochondrial DNA analysis of *Sporothrix schenckii* in Costa Rica. *J Med Vet Mycol*, **34**, 71–3.

Ishizaki, H., Kawasaki, M., et al. 1998. Mitochondrial DNA analysis of *Sporothrix schenckii* in North and South America. *Mycopathologia*, **142**, 115–18.

Ishizaki, H., Kawasaki, M., et al. 2000. Mitochondrial DNA analysis of *Sporothrix schenckii* in South Africa and Australia. *Med Mycol*, **38**, 6, 433–6.

Itoh, M., Okamoto, S. and Kariya, H. 1986. Survey of 200 cases of sporotrichosis. *Dermatologica*, **172**, 209–13.

Iwatsu, T., Nishimura, K. and Miyaji, M. 1985. Spontaneous disappearance of cutaneous sporotrichosis. Report of two cases. *Int J Dermatol*, **24**, 524–5.

Jacobson, H.P. 1932. *Fungous diseases: a clinico-mycological text.* London: Baillière, 121–48.

Janes, P.C. and Mann, R.J. 1987. Extracutaneous sporotrichosis. *J Hand Surg*, **12**, 441–5.

Jin, X.Z., Zhang, H.D., et al. 1990. Mother-and-child cases of sporotrichosis infection. *Mycoses*, **33**, 33–6.

Kaplan, W. and Gonzalez-Ochoa, A. 1963. Application of the fluorescent antibody technique to the rapid diagnosis of sporotrichosis. *J Lab Clin Med*, **62**, 835–41.

Kaplan, W., Broderson, J.R. and Pacific, J.N. 1982. Spontaneous systemic sporotrichosis in nine-banded armadillos (*Dasypus novemcinctus*). *Sabouraudia*, **20**, 289–94.

Kauffman, C.A. 1999. Sporotrichosis. *Clin Infect Dis*, **29**, 231–6.

Kawasaki, M., Roberto Arenas, R. et al. 2000. Molecular epidemiology of *Sporthrix schenckii* in Mexico, Brazil and Spain. *14th ISHAM Congress 2000, Buenos Aires, Argentina.* Abstract book. No. 54, p.120.

Kazanas, N. and Jackson, G. 1983. *Sporothrix schenckii* isolated from edible black fungus mushrooms. *J Food Protect*, **46**, 714–16.

Kenyon, E.M., Russell, L.H. and McMurray, D.N. 1984. Isolation of *Sporothrix schenckii* from potting soil. *Mycopathologia*, **87**, 128.

Kobayashi, H., Kawasaki, M., et al. 1990. A case of sporotrichosis caused by two genetically different *Sporothrix schenckii* strains. *Mycopathologia*, **112**, 19–22.

Kosinski, R.M., Axelrod, P., et al. 1992. *Sporothrix schenckii* fungemia without disseminated sporotrichosis. *J Clin Microbiol*, **30**, 501–3.

Kumar, R., van der Smissen, E. and Jorizzo, J. 1984. Systemic sporotrichosis with osteomyelitis. *J Can Assoc Radiol*, **35**, 84.

Kurosawa, A., Pollock, S.C., et al. 1988. *Sporothrix schenckii* endophthalmitis in a patient with human immunodeficiency virus infection. *Arch Ophthalmol*, **106**, 376–80.

Kusuhara, M., Hachisuka, H. and Sasai, Y. 1988. Statistical survey of 150 cases with sporotrichosis. *Mycopathologia*, **102**, 129–33.

Kwon-Chung, K.J. 1979. Comparison of isolates of *Sporothrix schenckii* obtained from fixed cutaneous lesions with isolates from other types of lesions. *J Infect Dis*, **139**, 424–31.

Larsson, C.E., Goncalves, M., et al. 1989. Esporotricosis felina: aspectos clinicos e zoonoticos. *Rev Inst Med Trop Sao Paulo*, **31**, 351–8.

Lavalle, P. and Mariat, F. 1983. Sporotricosis. *Bull Inst Pasteur*, **81**, 295–322.

Leon, L.A. 1964. La esporotricosis en America y su incidencia en la edad infantil. *Medicina*, **44**, 541–59.

Liao, W.Q., Zang, Y.L. and Shao, J.Z. 1991. Sporotrichosis presenting as pyoderma gangrenosum. *Mycopathologia*, **116**, 165–8.

Lima, O.C. and Lopes-Bezerra, L.M. 1997. Identification of a Concanavalin A-binding antigen of the cell surface of *Sporothrix schenckii*. *J Med Vet Mycol*, **35**, 167–72.

Lima, O.C., Figueiredo, C.C., et al. 1999. Adhesion of the human pathogen *Sporothrix schenckii* to several extracellular matrix proteins. *Braz J Med Biol Res*, **32**, 5, 651–7.

Lin, J., Kawasaki, M., et al. 1999. Mitochondrial DNA analysis of *Sporothrix schenckii* clinical isolates froma China. *Mycopathologia*, **148**, 69–72.

Lipstein-Kresch, E., Isenberg, H.D., et al. 1985. Disseminated *Sporothrix schenckii* infection with arthritis in a patient with acquired immunodeficiency syndrome. *J Rheumatol*, **12**, 805–8.

Lopes, J.O., Alves, S.H., et al. 1992. Filamentous forms of *Sporothrix schenckii* in material from human lesions. *J Med Vet Mycol*, **30**, 403–6.

Loureiro, C.V., Penha, Y. and Lopes-Bezerra, L.M. 2000. Concanavalin A-binding cell wall antigens of *Sporothrix schenkii*: a serological study. *Med Mycol*, **38**, 1–7.

Maberry, J.D., Mullins, J.F. and Stone, O.J. 1966. Sporotrichosis with demonstration of hypae in human tissue. *Arch Dermatol*, **93**, 65–67.

Mackinnon, J.E. 1948. The dependency on the weather of the incidence of sporotrichosis. *Mycopathologia*, **4**, 367–74.

Mackinnon, J.E., Conti-Diaz, I.A., et al. 1969. Isolation of *Sporothrix schenckii* from nature and consideration on its pathogenicity and ecology. *Sabouraudia*, **7**, 38–45.

Mariat, F. 1971. Adaption de *Ceratocystis stenoceras* a la vie parasitaire chez l'animal. Etude de l'aquisition d'un pouvoir pathogene comparable a celui de *Sporothrix schenckii*. *Sabouraudia*, **9**, 191–205.

Marques, S.A., Franco, S.R.V.S., et al. 1993. Esporotricose do gato domestico (*Felis catus*): transmissao humana. *Rev Inst Med Trop Sao Paulo*, **35**, 327–30.

Matlow, A.G., Goldman, C.B., et al. 1985. Contamination of intravenous fluid with *Sporothrix schenckii*. *J Infect*, **10**, 169–71.

Mayorga, J.A., Barba-Rubio, J., et al. 1997. Esporotricosis en el Estado de Jalisco, estudio clínico-epidemiológico. *Dermatol Rev Mex*, **41**, 3, 105–8.

Mayorga, R., Caceres, A., et al. 1978. Etude dune zone dendemie sporotrichosique au Guatemala. *Sabouraudia*, **16**, 185–98.

McDonough, E.S., Lewis, A.L. and Meister, M. 1970. *Sporothrix* (*Sporotrichum*) *schenckii* in a nursery barn containing sphagnum. *Public Health Rep*, **85**, 579–86.

Mesa-Arango, A.C., Reyes-Montes, M.R., et al. 2002. Phenotyping and genotyping of *Sporothrix schenckii* isolates according to geographic origin and clinical form of sporotrichosis. *J Clin Microbiol*, **40**, 3004–11.

Migaki, C., Font, R.L. and Kaplan, W. 1978. Sporotrichosis in a Pacific white-sided dolphin (*Lagenorhynchus oliquidens*). *Am J Vet Res*, **39**, 196–9.

Moore, J.J. and Davis, D.J. 1918. Sporotrichosis following mouse bite with certain immunologic data. *J Infect Dis*, **23**, 252–66.

Mora-Cabrera, M., Alonos, R.A. and Ulloa-Arvizu, R. 2001. Análisis of restriction profiles of mitochondrial DNA from *Sporothrix schenckii*. *Med Mycol*, **39**, 439–44.

Morgan, M.A., Cockerill, F.R. III, et al. 1984. Disseminated sporotrichosis with *Sporothrix schenckii* fungemia. *Diagn Microbiol Infect Dis*, **2**, 151–5.

Naqvi, S.H., Becherer, P. and Gudipati, S. 1993. Ketoconazole treatment of a family with zoonotic sporotrichosis. *Scand J Infect Dis*, **25**, 543–5.

Noguchi, H., Hiruma, M. and Kawada, A. 1999. Case report. Sporotrichosis successfully treated with itraconazole in Japan. *Mycoses*, **42**, 571–6 (letter).

Nusbaum, B.P., Gulbas, N. and Horwitz, S.N. 1983. Sporotrichosis acquired from a cat. *J Am Acad Dermatol*, **8**, 386–91.

Olson, G.M. 1912. A case of sporotrichosis in North Dakota: probable infection from gophers. *JAMA*, **59**, 941.

Ortiz, O. and Lefkovitz, Z. 1991. Case report 678. Sporotrichal arthritis. *Skel Radiol*, **20**, 376–8.

Padhye, A.A., Kaufman, L., et al. 1992. Fatal pulmonary sporotrichosis caused by *Sporothrix schenckii* var. *luriei* in India. *J Clin Microbiol*, **30**, 2492–4.

Pappas, P.G., Tellez, I., et al. 2000. Sporotrichosis in Peru: description of an area of hyperendemicity. *Clin Infect Dis*, **30**, 65–70.

Pérez, A. 1999. Terbinafine: broad new spectrum of indications in several subcutaneous and systemic and parasitic diseases. *Mycoses*, **42**, Suppl 2, 11–14.

Perfect, J.R., Schell, W.A. and Rinaldi, M.G. 1993. Uncommon invasive fungal pathogens in the acquired immunodeficiency syndrome. *J Med Vet Mycol*, **31**, 175–9.

Pluss, J.L. and Opal, S.M. 1986. Pulmonary sporotrichosis: review of treatment and outcome. *Medicine*, **65**, 143–53.

Powell, K.E., Taylor, A., et al. 1978. Cutaneous sporotrichosis in forestry workers. *JAMA*, **240**, 232–5.

Prose, N.S., Milburn, P.B. and Papayanopulos, D.M. 1986. Facial sporotrichosis in children. *Pediatr Dermatol*, **3**, 311–14.

Pueringer, R.J., Iber, C., et al. 1986. Spontaneous remission of extensive pulmonary sporotrichosis. *Ann Intern Med*, **104**, 366–7.

Quintal, D. 2000. Sporotrichosis infection on mines of the Witwatersrand. *J Cutan Med Surg*, **4**, 1, 51–4.

Ramani, R., Balachandran, C., et al. 1991. Spontaneous remission of cutaneous sporotrichosis – a case report. *Indian J Pathol Microbiol*, **34**, 288–9.

Read, S.I. and Sperling, L.C. 1982. Feline sporotrichosis. *Arch Dermatol*, **118**, 429–31.

Reed, K.D., Moore, F.M., et al. 1993. Zoonotic transmission of sporotrichosis: case report and review. *Clin Infect Dis*, **16**, 384–7.

Reingold, A.L., Lu, X.D., et al. 1986. Systemic mycoses in the United States, 1980–1982. *J Med Vet Mycol*, **24**, 433–6.

Remington, P.L., Vergeront, J.M., et al. 1982. Sporotrichosis associated with Wisconsin sphagnum moss. *Morbid Mortal Wkly Rep*, **31**, 542–3.

Rippon, J.W. 1988. *Medical mycology. The pathogenic fungi and the pathogenic actinomycetes*, 3rd edition. Philadelphia: W.B. Saunders.

Rodríguez, G. and Sarmiento, L. 1998. The asteroid bodies of sporotrichosis. *Am J Dermatopathol*, **20**, 246–9.

Rohatgi, P.K. 1980. Pulmonary sporotrichosis. *South Med J*, **73**, 1611–17.

Romero-Martinez, R., Wheeler, M., et al. 2000. Biosynthesis and functions of melanin in *Sporothrix schenckii*. *Infect Immun*, **68**, 6, 3696–703.

Rowe, J.G., Amadio, P.C. and Edson, R.S. 1989. Sporotrichosis. *Orthopedics*, **12**, 981–5.

Rudolph, R.I. 1984. Facial sporotrichosis in an infant. *Cutis*, **33**, 171–173, 179.

Ruiz-Esmenjaud, J., Arenas, R. and Vega-Memije, M.E. 1996. Esporotricosis: Estudio histopatológico de 22 casos. *Dermatol Rev Mex*, **40**, 106–12.

Samorodin, C.S. and Sina, B. 1984. Ketoconazole-treated sporotrichosis in a veterinarian. *Cutis*, **33**, 487–8.

Sampaio, S.A.P., Lacaz, C.D. and de Almeida, F.S. 1954. Aspectos clinicos da esporotrichose em Sao Paulo, analise de 235 casos. *Rev Hosp Clin*, **9**, 391–402.

Sanchez, M.R. 2000. Miscellaneous treatments: thalidomide, potassium iodide, levamisole, clofazimine, colchicine and D-penicillamine. *Clin Dermatol*, **18**, 131–45.

Sanders, E. 1971. Cutaneous sporotrichosis. *Arch Intern Med*, **127**, 482–3.

Schell, W.A. 1986. Oculomycosis caused by dematiaceous fungi. *Scientific Publication No.479*. Washington, DC: Pan American Health Organization, 105–9.

Schenck, B.R. 1898. On refractory subcutaneous abscesses caused by a fungus possibly related to the sporotricha. *Johns Hopkins Hosp Rep*, **9**, 286–91.

Schiappacasse, R.H., Colville, J.M., et al. 1985. Sporotrichosis associated with an infected cat. *Cutis*, **35**, 268–70.

Schubach, T., Valle, A., et al. 2001. Isolation of *Sporothrix schenckii* from the nails of domestic cats (*Felis catus*). *Med Mycol*, **39**, 147–9.

Scott, E.N. and Muchmore, H.G. 1989. Immunoblot analysis of antibody responses to *Sporothrix schenckii*. *J Clin Microbiol*, **27**, 300–304.

Scott, E.N., Kaufman, L., et al. 1987. Serologic studies in the diagnosis and management of meningitis due to *Sporothrix schenckii*. *N Engl J Med*, **317**, 935–40.

Selman, S.H. and Hampel, N. 1982. Systemic sporotrichosis: diagnosis through biopsy of epididymal mass. *Urology*, **20**, 620–1.

Sen, S.K., Buford, R.C., et al. 1986. Cutaneous sporotrichosis presenting as soft tissue sarcoma. *J Natl Med Assoc*, **78**, 1099–101.

Severo, L.C., Festugato, M., et al. 1999. Widespread cutaneous lesions due to *Sporothrix schenckii* in a patient under a long-term steroids therapy. *Rev Inst Med Trop Sao Paulo*, **41**, 59–62.

Sharkey-Mathis, P.K., Kauffman, C.A., et al. 1993. Treatment of sporotrichosis with itraconazole. *Am J Med*, **95**, 279–85.

Shaw, J.C., Levinson, W. and Montanaro, A. 1989. Sporotrichosis in the acquired immunodeficiency syndrome. *J Am Acad Dermatol*, **21**, 1145–7.

Shiraishi, A., Nakagaki, K. and Arai, T. 1992. Role of cell-mediated immuity in the resistance to experimental sporotrichosis in mice. *Mycopathologia*, **120**, 15–21.

Simson, F.W., Helm, M.A.F., et al. 1947. *Sporotrichosis infection on mines of the Witwatersrand. A symposium*. Johannesburg: Transvaal Chamber of Mines, 34–40.

Smith, L.M. 1945. Sporotrichosis: report of four clinically atypical cases. *South Med J*, **38**, 505–9.

Smith, M.T. and Batenburg-van der Vegte, W.H. 1985. Ultrastructure of septa in *Blastobotrys* and *Sporothrix*. *Antonie Van Leeuwenhoek*, **51**, 121–8.

Smith, P.W., Loomis, G.W., et al. 1981. Disseminated cutaneous sporotrichosis: three illustrative cases. *Arch Dermatol*, **117**, 143–4.

Solano, E. 1966. Sporotrichosis in children. *Derm Ib Lat Amer*, **1**, 243–5.

Spiers, E.M., Hendrick, S.J., et al. 1986. Sporotrichosis masquerading as pyoderma gangrenosum. *Arch Dermatol*, **122**, 691–4.

Staib, F. and Grosse, G. 1983. Isolation of *Sporothrix schenckii* from the floor of an indoor swimming pool. *Zentral Bakteriol Hyg A*, **177**, 499–506.

Sterling, J.B. and Heymann, W.R. 2000. Potassium iodide in dermatology: a 19th century drug for the 21st century – uses, pharmacology, adverse effects, and contraindications. *J Am Acad Dermatol*, **43**, 691–7.

Storrs, F., Weeks, G., et al. 1969. Sporotrichosis – Oregon. *Morbid Mortal Wkly Rep*, **18**, 170.

Sugita, Y. 2000. Molecular analysis of DNA polymorphism of *Sporothrix schenckii*. *Jpn J Med Mycol*, **41**, 11–15.

Suzuki, K., Kawasaki, M. and Ishizaki, H. 1988. Analysis of restriction profiles of mitochondrial DNA from *Sporothrix schenkii*. *Mycopathologia*, **103**, 147–51.

Tachibana, T., Matsuyama, T. and Mitsuyama, M. 1999a. Characteristic infectivity of *Sporothrix schenckii* to mice depending on routes of infection and inherent fungal pathogenicity. *Med Mycol*, **36**, 21–7.

Tachibana, T., Matsuyama, T. and Mitsuyama, M. 1999b. Involvement of CD4+ T cells and macrophages in acquired protection against infection with *Sporothrix schenckii* in mice. *Med Mycol*, **37**, 6, 397–404.

Takeda, Y., Kawasaki, M. and Ishizaki, H. 1991. Phylogeny and molecular epidemiology of *Sporothrix schenckii* in Japan. *Mycopathologia*, **116**, 9–14.

Talhari, S., Cunha, M.G., et al. 1988. Deppe mycoses in Amazon region. *Int J Dermatol*, **27**, 481–4.

Torres-Guerrero, H. and Arenas-López, G. 1998. UV irradiation induced high frequency of colonial variants with altered morphology in *Sporothrix schenckii*. *Med Mycol*, **36**, 81–7.

Travassos, L.R. and Lloyd, K.O. 1980. *Sporothrix schenckii* and related species of *Ceratocystis*. *Microbiol Rev*, **44**, 683–721.

Ware, A.J., Cockerell, C.J., et al. 1999. Disseminated sporotrichosis with extensive cutaneous involvement in a patient with AIDS. *J Am Acad Dermatol*, **40**, 350–5.

Watts, J.C. and Chandler, F.W. 1987. Primary pulmonary sporotrichosis. *Arch Pathol Lab Med*, **111**, 215–17.

Werner, R.E. Jr, Levine, B.G., et al. 1971. Sporotrichosis in a cat. *J Am Vet Med Assoc*, **159**, 407–12.

Wescott, B.L., Nasser, A. and Jarolim D.R. 1999. *Sporothrix* meningitis. *Nurse Pract*, **24**, 90, 93–4, 97–8.

Wilson, S.D., Grossheim, R., et al. 1988. Case report of synovial sporotrichosis involving both wrists. *J Med Vet Mycol*, **26**, 307–9.

Yacobucci, G.N. and Santilli, M.D. 1986. Sporotrichosis of the knee. A case report. *Orthopedics*, **9**, 387–90.

Yamada, Y., Dekio, S., et al. 1990. A familial occurrence of sporotrichosis. *J Dermatol*, **17**, 255–9.

Yao, J., Penn, R.G. and Ray, S. 1986. Articular sporotrichosis. *Clin Orthop*, **204**, 207–14.

Yoshiike, T., Lei, P.C., et al. 1993. Antibody raised against extracellular proteinases of *Sporothrix schenckii* in *S. schenckii* inoculated hairless mice. *Mycopathologia*, **123**, 69–73.

Zaharopoulos, P. 1999. Fine-needle aspiration cytologic diagnosis of lymphocutaneous sporotrichosis: a case report. *Diagn Cytopathol*, **20**, 74–7.

Zamri-Saad, M., Salmiyah, T.S., et al. 1990. Feline sporotrichosis: an increasingly important zoonotic disease in Malaysia. *Vet Rec*, **127**, 480.

Zhang, X. and Andrews, J.H. 1993. Evidence for growth of *Sporothrix schenckii* on dead but not on living sphagnum moss. *Mycopathologia*, **123**, 87–94.

20

Eumycetomas

RODERICK J. HAY

INTRODUCTION

The ability to form granules in vivo is a property possessed by a variety of microorganisms. Usually these granules are small, visible, microbial aggregates composed of cellular masses which are frequently filamentous. They probably represent the formation in vivo of colonies. However, their small size and the discrete nature of the granules are characteristic. The best examples are the 'grains' of a fungal mycetoma or eumycetoma and the sulphur granules of actinomycosis. A number of fungi produce clusters of hyphae in vivo, and these range in size from large fungus balls, which are easily visible in gross pathological material such as the aspergillomas, to dermatophytes, which can be visualized occasionally by microscopy within foreign body giant cells. Strictly speaking, these are not true granules and the nomenclature will be discussed briefly at the end of this section.

Mycetomas are chronic subcutaneous infections caused by fungi or actinomycetes (Table 20.1), known as eumycetomas and actinomycetomas respectively (Mahgoub and Murray 1973). The organisms are traumatically implanted into the deep dermis or subcutaneous tissue from the natural environment and cause a subcutaneous infection characterized by the formation of large aggregates of fungal or actinomycete filaments known as grains. A dense infiltrate of polymorphonuclear leukocytes and other inflammatory cells accumulates around these structures (Chandler and Watts 1987). This abscess either discharges onto the skin surface through draining sinuses or may involve contiguous structures such as the bone, causing chronic osteomyelitis. Mycetoma is a chronic disease process which causes considerable deformity and disability. It does not often spread beyond the locality of the initial site of infection and is seldom fatal. As there are pronounced differences in the treatment and prognosis of mycetomas resulting from different organisms, in particular between the eumycetomas and actinomycetomas, identification of the causative organisms plays a key part in their management.

The name mycetoma has been applied erroneously to a number of other conditions including those in which a large fungus ball is formed and dermatophyte pseudomycetomas (Ajello et al. 1980). The formation of a fungus ball, exemplified by the disease aspergilloma, usually occurs within a cavity formed as a result of disease. Lung cavities are most often affected although fungus balls may occur in the paranasal sinuses. The fungi find their way to this site either through inhalation or bloodstream spread, but not through implantation. Sinuses affecting the skin are not formed and the pathogenesis of this form of fungal infection is different from that of a true mycetoma. Dermatophytes may produce granules composed of mycelial elements which often develop within giant cells (Ajello et al. 1980). Once again, the pathogenesis is completely different from mycetoma. Infections do not follow implantation of environmental organisms but follow invasion from a site

Table 20.1 *Causes of eumycetoma*

White grains	Black grains
Acremonium falciforme	Curvularia lunata
Acremonium kiliense	Curvularia geniculata
Acremonium recifei	Exophiala jeanselmei
Aspergillus nidulans	Leptosphaeria senegalensis
Aspergillus flavus	Madurella mycetomatis
Chaetospaeronema larense[a]	Madurella grisea
Cylindrocarpon cyanescens[a]	Phialophora parasitica
Fusarium moniliforme	Pyrenochaeta romeroi
Fusarium solani	
Neotestudina rosatii	
Scedosporium apiospermum	

a) Very rare causes of mycetoma.

on the skin, particularly the hair follicle. For this reason granules formed by dermatophytes are sometimes referred to as pseudomycetomas.

HISTORY

Mycetoma, Madura foot, is a disease of great antiquity and was probably known by the ancient Indian writers as padavalmita (foot anthill) (Sran et al. 1972). Changes suggestive of mycetoma have also been found in ancient skeletal remains in Israel (Spigelman and Donoghue 2001). The earliest medical description of the condition in Western literature appears to be that of the German traveller and physician, Kaempfer (1651–1716) and formed part of his doctoral thesis at the University of Leyden in Holland in 1694 (Kaempfer 1694). The thesis, 'Disputatio physica medica inauguralis exhibens decadem observationum', contains a section concerned with a disease called 'Perical' or 'ulcerous hypersarcosis of the feet' which was indigenous to Malabar in southern India. He described a tumorous disease generally affecting the feet which, as a result, would enlarge to three or four times the normal size. A feature of the condition was said to be the presence of oozing and granules. However, it was not fatal and caused considerable deformity often as a result of treatment – generally caustics or the application of heated metal. In 1842, Gill, an English surgeon working in the Madurai dispensary, once again in southern India, was said to have mentioned a condition that may have been mycetoma in a dispensary report (Gill 1842, cited in Carter 1874). The disease was first described in the medical literature by Godfrey (1846), who was the garrison surgeon in Bellary, Madras, India. He described four patients and his paper was published in *The Lancet* entitled 'Diseases of the foot not hitherto described.' Godfrey reported this new condition as 'morbus tuberculosis pedis' and the gross deformity, but non-fatal nature, of the condition was emphasized (Godfrey 1846). In 1859, Eyre drew attention once again to the small characteristic pigmented specks seen within mycetomas and mentioned in Kaempfer's original article (Eyre 1859). Originally these were thought to be caused by blood, but their presence was noted to be a characteristic of this particular infection. Although many cases involved the feet, other sites were also affected; for instance, Minas (1860) first described lesions of the hand and pointed out that the grains in lesions could be either black or pale but that the former were more common. The bone changes were first described by Ballingali (1855).

The causative organisms were first described in 1860 by Carter who suggested that the small bodies were 'fungus particles.' He also proposed that these were fungi of plant origin (Carter 1861). He believed that the white particles or grains were merely degenerate elements of the black form. In order to clarify the etiology of mycetoma, Carter enlisted the help of the Reverend Berkley who isolated a red mold from a lesion. Berkeley named this fungus *Chionyphe carteri*, a new species (Berkley 1862). The true identification of this organism remains unknown as it does not correspond to any fungus recognized as pathogenic today. In 1874, Carter produced a monograph on mycetoma, 'On mycetoma or the fungus disease of India,' using this disease name for the first time (Carter 1874). He also appears to have described the pathological features of an infection caused by the organism now known as *Madurella mycetomatis*. The first isolation of a fungal species recognized today was by Brumpt (1906), who cultured the organism *Madurella mycetomi (-atis)*, from a black grain infection.

The pathogenesis of mycetoma still eluded investigators. For instance, Bidie also carried out detailed studies of the disease in Bombay and studied the grains which he also believed to be caused by fungi that had entered the body as spores (Bidie 1862). However, the most likely explanation for the true portal of entry of organisms came in 1893, when Bocarro found a thorn embedded in the inflammatory mass and proposed that the infection was usually introduced by a minor injury (Bocarro 1893). Although until this time most of the cases originated from India, it soon became clear that cases were being reported from Algeria, the USA, and Sudan.

In 1894, two observations showed that actinomycetes were also causes of this infection. Boyce and Surveyor (1894) observed that the size of filaments within certain grains of mycetomas were fine and more likely to be bacterial in origin. In the same year, Vincent (1894) isolated an organism that he named *Streptothrix madurae* (probably *Actinomadura madurae*) from an Algerian case.

PATHOGENESIS

A unifying theory for the pathogenesis of mycetoma is hampered by the lack of an effective animal model for eumycetomas, although there are experimental models for actinomycete infections such as nocardiosis (mice) and *A. madurae* (goats). However, there is evidence to connect mycetoma with implantation of microorganisms from the environment coupled with an extraordinary series of adaptive mechanisms which are of potential benefit to the survival of the organisms in nature.

Some of the causative fungi have been isolated from the natural environment. For instance, *Leptosphaeria senegalensis* has been identified in the natural environment in acacia thorns (Segretain and Mariat 1969). In addition, it is occasionally possible to see plant material, probably originating from a thorn in some lesions of mycetoma (Cameron et al. 1973). These appear as pieces of lignified tissue containing fungal hyphae. Basset and co-workers removed thorn material from two cases and cultured *L. senegalensis* from one and *Pyrenochaeta romeroi* from the other (Basset et al. 1965).

There is no evidence to suggest that the infection is specifically associated with an underlying host predisposition and, even though some patients appear to have defective T-lymphocyte-mediated responses to the fungi, these have been described in those with extensive disease, suggesting that they may have followed rather than caused the infection (Mahgoub et al. 1978). Other investigators have failed to find any changes in T-lymphocyte function in patients with disease (Bender et al. 1987). There is also nothing to suggest that other less debilitating changes in immune function occur in mycetoma patients, although in one study a high frequency of diabetes mellitus was reported among patients (Hay and Mackenzie 1983). The rare cases reported in immunosuppressed patients may even be atypical with the formation of clusters of hyphae rather than frank grains (Meis et al. 2000) This has also not been mentioned by others (Mahgoub and Murray 1973). By contrast, there is evidence that fungi in mycetomas may develop a number of different adaptive changes in vivo, such as cell wall reduplication or unfolding of the cytoskeleton, or deposition of extra- or intracellular melanin, all of which may affect the ability of the host inflammatory response to destroy the organisms. Cell wall reduplication gives rise to an 'onion skin' appearance around cells in eumycetoma grains caused by different species under electron microscopic examination (Hay and Collins 1983; Wethered et al. 1987). This appears to follow intrahyphal growth of mycelial elements. Adjacent cell walls also fuse, particularly at the rim of the grain where, in addition, there is evidence of deposition of human immunoglobulin. The contribution of the latter to the microstructure of grains is unknown. In the case of *M. mycetomatis*, cells within the grain may be surrounded by electron-dense homogeneous material, fungal melanin. This is very resilient to physical extremes and is difficult to section using a microtome (Findlay and Vismer 1987). Once again it is thought to provide a protection for living cells within the fungal grain.

As yet there is little evidence among the fungal causes of mycetoma of immunomodulation as has been demonstrated with the actinomycete, *Streptomyces somaliensis*, which produces an extracellular protease that affects the ability of macrophages in vitro to kill bacteria (Nasher et al. 1987).

EPIDEMIOLOGY

The epidemiology of the geographical distribution of cases of mycetoma is well documented. As a result of a worldwide postal survey, Mariat (1963) produced evidence that mycetomas were found throughout a range of countries, mainly confined to tropical regions, and that infections outside these areas were rare and usually occurred as imported infections. The countries reporting the highest prevalence of infection were Mexico where *Nocardia brasiliensis* was the dominant organism and Senegal where most infections were caused by fungi. In addition to these countries, India and Sudan have a higher than average prevalence of infection (Mahgoub and Murray 1973; Hay et al. 1992). In these two areas, *M. mycetomatis* is the most common reported cause of mycetoma. *Nocardia* spp. and *M. mycetomatis* were also the most common organisms documented in the Mariat survey, accounting for 32 and 19 percent of the total, respectively. Recently, *M. mycetomatis* DNA has been identified in soil (23 percent) and thorn (5 percent) samples from Sudan but not from the Netherlands (Ahmed et al. 2002). In addition, mycetomas occur throughout central and northern South America, around the Saharan region and Central and South Africa, the Middle East, India, and Pakistan. Infections are less frequently seen further east. The main endemic areas for mycetomas are relatively arid and have a short rainy season, with a constant temperature with little fluctuation with different seasons or at night (Gonzalez-Ochoa 1962). In temperate areas, mycetoma infections are rare and usually caused by fungi rather than actinomycetes. There are reasons for suspecting the accuracy of some of these data. In the UK, for instance, over half the fungal causes of imported mycetomas resulted from pigmented fungi that are sterile and produce nonpigmented grains in vivo (Hay and Mackenzie 1983). They cannot therefore be assigned to a single species or genus. There is evidence from the analysis of the protein profiles of some of these agents, which appear to be very similar in gross morphology, using polyacrylamide gels, that this is a heterogeneous group of fungi (Zaini et al. 1991).

There are variations within the endemic areas although these principally affect the actinomycetes where, for instance, *S. somaliensis* is mainly found in hot and dry climates with less than 50–250 mm of rain per year. Among the fungi, *Pyrenochaeta romeroi* is also found in areas with higher rainfall levels (Mahgoub and Murray 1973). As has been described above, the fungi that cause mycetomas are associated with vegetation such as acacia trees (Segretain and Mariat 1969). This is compatible with the description of the endemic areas because they are mainly seen in regions with low rainfall.

Mycetoma is most often seen in men, the male:female ratio of infections being approximately 4:1. It is mainly a disease of rural workers aged between 16 and 45 years. Although they may not give a history of an initiating trauma, their occupation predisposes to such an event. Women with mycetomas are also usually from rural areas and work in the fields. However, occasionally mycetomas are found in nonmanual workers.

CLINICAL APPEARANCES

The clinical appearances of mycetoma are characteristic (Abbott 1956; Mahgoub and Murray 1973). The incubation period is unknown as most patients do not give a history of a trauma that initiated the lesions. However, in some cases, it is clear that infection started over 8 years after a traumatic injury (Maiti et al. 2002).

The earliest sign of a mycetoma is the appearance of a hard subcutaneous nodule (Figure 20.1). This is painless and usually found on exposed sites such as the arms and legs. More than 95 percent of mycetomas start at these areas. As the lesion enlarges, sinuses appear on the skin surface (Figure 20.2). These present as papules 0.2–0.5 cm in diameter or pustules that discharge their contents and then dry up, leaving a small scar. Sometimes individual lesions are painful before the release of purulent material. At an early stage, pain and tenderness are intermittent and unpredictable. As the lesion increases in size, the whole area becomes hard, swollen, and, in time, limbs become grossly deformed. In advanced mycetomas there is often localized sweating over lesions (R.J. Hay, personal observation). In some

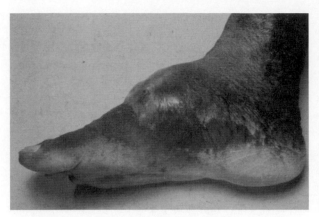

Figure 20.2 *Mycetoma of the foot caused by* Madurella grisea

cases pain becomes prominent in the later stages, but conversely, patients may have very extensive lesions without symptoms.

Occasionally mycetomas may be found on the scalp and thorax, although these are most often caused by actinomycetes, namely *S. somaliensis* and *Nocardia* spp., respectively. It is not possible to distinguish clinically with any degree of certainty between the different causes of mycetoma. The lesions usually remain localized and are locally aggressive. Actinomycetomas usually evolve more rapidly than eumycetomas, but this cannot be relied on as a means of distinguishing between the two types. Very rarely, lesions may spread to lymph nodes, although this is most likely to occur with *S. somaliensis* (El Hassan and Mahgoub 1972). Metastatic mycetomas are exceptionally rare and, on the few occasions where two lesions are seen, these may have been caused by separate infections. Occasionally cystic mycetomas confined to the bone and without draining sinuses have been described (Fahal et al. 1998). The main complications of eumycetomas are deformity which can lead to considerable disability, fungal osteomyelitis, and secondary bacterial infection. The last is not common but may cause an acute exacerbation of the swelling and increased purulent discharge from sinuses combined with pyrexia.

DIFFERENTIAL DIAGNOSIS

Mycetoma must be distinguished from other conditions in which draining sinuses may develop. These include actinomycosis which is usually restricted to sites where the causative organisms are carried, such as the mouth and jaw. The grains in actinomycosis are soft and yellowish. Botryomycosis, a bacterial disease, may also be confused with mycetoma, although there is usually wider internal spread and patients have a pronounced fever. Carcinomas, such as squamous carcinomas, chondrosarcoma, and Kaposi's sarcoma, may have to be distinguished where necessary by biopsy.

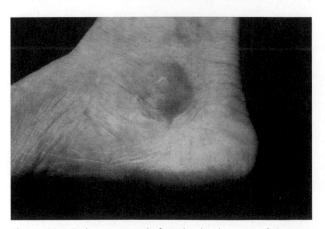

Figure 20.1 *Early mycetoma before the development of sinuses*

Diagnostic imaging

The radiological appearances of mycetomas are well described (Davies 1958; Cockshott 1968). They include soft tissue swelling, endosteal cavitation with small cystic lesions, and periosteal bone erosion (Figure 20.3). New bone growth is often seen in longstanding lesions along the periosteum. Hyperostosis may occur in the skull and other areas and, in advanced lesions, pathological fractures and bone resorption can be seen. Once again it is difficult to distinguish between eumycetomas and actinomycetomas. Cystic lesions are often fewer and larger in eumycetomas. Other methods of detection include bone scanning and echo scanning. Bone scans may be more useful at detecting early bone involvement than x-rays, although not widely available in the endemic areas. Ultrasound scans show the marked tissue and bone swelling with hyperreflective areas (Fahal et al. 1997). However, magnetic resonance imaging provides a clearer picture of both bone and soft tissue distribution (Czechowski et al. 2001)

Laboratory diagnosis

The diagnosis of eumycetoma is confirmed by the demonstration of grains in lesions and their identification. This is important in order to provide information on prognosis and treatment. For instance, it is important to differentiate eumycetomas, which often respond poorly to chemotherapy, from actinomycetomas, which respond well. In addition, there are differences in responses within the eumycetomas. Infections caused by *M. mycetomatis* respond to ketoconazole in about 50 percent of cases. Likewise, it is worth using intravenous miconazole in infections caused by *Scedosporium apiospermum*.

The methods of laboratory diagnosis of granule-forming fungi are no different from those used for other fungal infections: direct microscopy, culture, serology, and histopathology. The key to the diagnostic process is the extraction of grains from lesions. This can be accomplished in one of two ways.

First, an unruptured pustule is identified, cleaned with an alcohol swab and gently pierced with a sterile needle. The edges of the sinus beneath are squeezed, and a small drop of pus and blood is extruded. This is then spread on a glass slide where it is possible to identify small (200 µm to 2 mm) granules that are white/yellow, black, or red. These are then processed for microscopy, culture, and histopathology (see section on Histopathology below). Often it is necessary to open more than one pustule and it is important to scrape beneath the edge of the scab because grains may be trapped there.

Deep biopsy is the second approach. A superficial biopsy removing the mouth of a sinus track is seldom effective because the amount of material is usually insufficient to identify grains, and the chances of the specimen containing a grain are low because the mouth of the sinus merely represents a pathological thoroughfare. A large deep biopsy, which is usually taken under a general anaesthetic, on the other hand, provides a large amount of material from which it may be possible to identify grains. These should be picked out of the tissue for culture and for histopathology – segments containing visible grains should be sent for processing.

Fine needle aspiration can also sometimes provide suitable diagnostic material (El Hag et al. 1996)

Direct microscopy

Grains are mounted in 5–10 percent potassium hydroxide and allowed to stand for 1–2 hours. The material is then gently flattened by pressure on the cover slip. Grains must be distinguished from clusters of neutrophils, but it is possible to make out their structure using this method, as well as the type of pigmentation. Extracellular melanin produced by *M. mycetomatis*, for instance, forms into irregular clusters of pigmentation. The presence of hyphae in grains can be seen under high power objective (Figure 20.4). If broad hyphae are not seen, it is likely that the grain is caused by an actinomycete.

Culture

Provided the appropriate material is selected, culture is usually successful. It is best to use conventional media such as Sabouraud's agar for primary isolation, using plates or tubes both with and without antibiotics such as penicillin and streptomycin (Mahgoub and Murray 1973; Mariat et al. 1977). The rate of growth is highly variable and therefore cultures should be kept for at least 3 weeks before they are discarded as negative.

The criteria for recognition are no different from those used with other fungi. Colonial morphology, patterns of sporulation, pigmentation, and growth rates

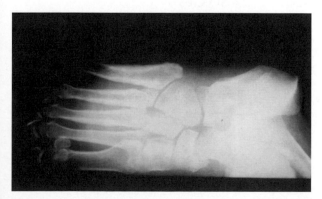

Figure 20.3 *X-ray of eumycetoma of foot caused by a* Fusarium *species*

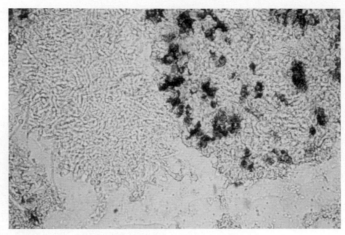

Figure 20.4 *Grains of* Madurella grisea. *Magnification × 300; stain: KOH mount.*

are all important. Media used range from Sabouraud's to Czapek–Dox or Lactrimel (Borelli and Feo 1966). Failure to sporulate is common and a variety of media should be used. The main means of diagnosis is through colonial characteristics. However, some physiological tests have been carried out on the more common mycetoma agents (Table 20.2).

At present there are few studies and no established molecular diagnostic tests although these have been applied to certain organisms, notably *M. mycetomatis* (Lopes et al. 2000). The cultural characteristics of the main organisms are discussed below

ACREMONIUM SPECIES

Most of these fungi implicated in mycetoma were previously described as members of the genus *Cephalosporium*. The main species implicated in mycetoma are *Acremonium kiliense*, *Acremonium falciforme*, and *Acremonium recifei*. Colonies of all three are slow growing and downy. The color is white to pink. On the undersurface there is usually a clear pigmentation that is pink to rose in color. The hyphae are not pigmented. Conidia are borne on aerial hyphae and appear in clusters bound by a gelatinous material. These conidia are single or multicellular and fusiform with a slight curve; some conidia appear as a shallow crescent. They are about 10 × 4 μm in length and width. The different species vary in conidial structure.

A. falciforme is mainly crescentic and nonseptate, but has some two- or three-celled conidia; *A. kiliense* has straight short conidia; *A. recifei* is mainly crescentic and nonseptate.

EXOPHIALA JEANSELMEI

Previously known as *Torula*, *Pullularia*, or *Phialophora jeanselemi*, this fungus forms a distinct colony that changes with time from a predominance of yeast-like cells, which form a glabrous mucoid colony with dark pigmentation, to a downy colony containing a lot of mycelium with aerial hyphae. Conidia are produced by annellation. These are formed and are often seen in clusters at the tip of an annelide. Conidia are oval and about 5 × 1 μm in size (Nielsen et al. 1968).

LEPTOSPHAERIA SPECIES

These fungi can be recognized by their more rapid growth to produce a downy colony with a gray to black reverse pigmentation. With *L. senegalensis*, asci are produced on older colonies and these contain eight oval ascospores that are long, measuring up to 30 μm in length. The ascospores of *Leptosphaeria tompkinsii* are shorter.

MADURELLA SPECIES

M. mycetomatis has other generic synonyms such as *Streptothrix* and *Glenospora*, as well as other *Madurella*

Table 20.2 *Physiological characteristics of common fungal causes of mycetoma*

Species	Starch	Gelatin	Glucose	Galactose	Lactose	Maltose	Sucrose
Scedosporium apiospermum	+	+	+	var	−	−	var
Madurella mycetomatis	+	±	+	+	+	+	−
Madurella grisea	+	−	+	+	−	+	+
Pyrenochaeta romeroi	+	±	+	+	−	+	+
Leptosphaeria senegalensis	+	?	+	+	var	+	+
Exophiala jeanselmei	−	−	+	+	−	+	+

species. It was originally named as *M. mycetomi* but this was subsequently changed to the form used today: *M. mycetomatis*.

The rate of growth of *M. mycetomatis* colonies is very variable, but may start very slowly. Colonies are often glabrous initially and of a white to tan color, with deep folds that develop a down, resulting from hyphal proliferation over the surface. This is cream to gray in color. Diffusible brown pigment can be seen on the undersurface. Conidial formation is best seen on cornmeal agar. *M. mycetomatis* produces either small flask-like phialides which produce small round conidia (2–4 μm in diameter) or elongated pyriform conidia (3–5 μm). In old colonies there are numerous black sclerotia 1 mm in diameter.

Madurella grisea

The colonies are similar to *M. mycetomatis* initially, although there is no diffusible pigment. Once again a small downy cover forms after 2–3 weeks. Colonies are sterile, although on poor media pycnidia form. Recognition is difficult as the fungus is sterile and some concern has been expressed that isolates identified as *M. grisea* may not form a single species. Protein patterns analyzed from extracted fungi by polyacrylamide gels suggest that their protein patterns are not homogeneous. At present there have been no published studies that employed genetic techniques to compare different isolates.

NEOTESTUDINA ROSATII

This was formerly named *Zopfia rosatii*. The organism produces a slow-growing yellow to brown-colored colony which is covered by folds. On certain media such as cornmeal agar older colonies (after 3 weeks) produce asci. The ascostromas are easily distinguished because of their black color. Asci are long, up to 30 μm, and have thick walls; they contain eight two-celled ascospores which are 10 μm in length (Segretain and Destombes 1961).

PYRENOCHAETA SPECIES

P. romeroi produces a dark downy colony with a black to tan undersurface. The colony expands rapidly and produces pycnidia. These have thick outer walls with setae. Pycnidia are 40–100 μm by 50–120 μm in size and contain elliptical and yellowish pycnidiospores which are about 1 μm in diameter. These are produced by phialides in chains. A new species, *Pyrenochaeta mackinnonii*, differs in the size of pycnidia and pycnidiospores (Borelli 1976).

SCEDOSPORIUM APIOSPERMUM

This fungus is the imperfect state of *Pseudallescheria boydii*. It is synonymous with *Monosporium apiospermum*. The colonies are rapidly growing and a lot of fluffy aerial mycelium is produced which is white at first, but with time becomes gray or brown. The underside of the culture is usually gray. Sectoring or contour formation may occur on the colonial surface. Conidia are produced singly or in clusters on anellophores or laterally on conidiogenous hyphae. These conidia are distinctive, large, lemon-shaped and with a yellowish color. They vary from 4–9 to 6–10 μm in size. Some colonies produce asci. The ascospores are lightly pigmented and round to oblong (6 × 6 μm).

The isolation of unusual organisms

The other fungal causes of mycetoma are very rare and have been described on a single or, at the most, two occasions. It is difficult to decide whether an isolated organism is a novel cause of mycetoma or a contaminant. Theoretically, as most of these agents are of low virulence, it is possible for other environmental organisms to be implicated in mycetomas. If the organism is isolated at least twice from grains, and these are compatible with a fungal infection, it is likely that they are the cause of the infection. Such rare organisms include *Pseudochaetosphaeronema larense* (*Chaetosphaeronema larense*) (Borelli and Zamora 1973), *Cylindrocarpon (Phialophora) cyanescens* (Zoutman and Sigler 1991), *Phialophora parasitica* (Pracharktam et al. 2000), and *Pyrenochaeta mackinnonii* (Borelli 1976). Other fungi described as causes of eumycetoma are *Corynespora cassiicola* and *Plenodomus avramii*.

Serodiagnosis

The value of serodiagnosis in the diagnosis of mycetomas is debatable. Certainly it cannot be used as a screening process for suspicious-looking lesions, given that weak positive reactions may occur in otherwise healthy individuals. There are also no commercial systems for the detection of antibody or antigen in mycetoma.

The techniques used have generally depended on either immunodiffusion or counterimmunoelectrophoresis (Murray and Mahgoub 1968; Mariat et al. 1977); there is no enzyme-linked diagnostic system such as an enzyme-linked immunosorbent assay (ELISA), which has been developed specifically for measuring antibodies to mycetoma as a diagnostic measure, even though investigators have used ELISA to study the pathogenesis of infection (Wethered et al. 1988). An immunodiffusion assay has been established in a few laboratories and provides useful help for following the course of treatment where initial reactions are strongly positive. This is more often seen with actinomycetes such as *Nocardia* spp. or *S. somaliensis* (Murray and Mahgoub 1968). Both these groups give strongly positive responses which can be quantitated and followed as a guide to treatment outcome or relapse. Generally, for most eumycetomas, the serological responses are less reliable and their monitoring only occasionally forms a useful part of the management of these infections.

HISTOPATHOLOGY

Provided that the limitations of histopathology in the diagnosis of mycetoma are recognized, this is a useful procedure. It can be performed either on grains extracted from lesions and fixed immediately in formalin or on biopsy material. If grains are used, it is important to ensure that the laboratory staff can also identify the relevant material in the specimen pot, otherwise it will be discarded. For most purposes, recognition is based on the appearances of grains using the hematoxylin and eosin (H&E) stain (Winslow and Steen 1964; Mahgoub and Murray 1973; Mariat et al. 1977; Destombes 1978; Chandler and Watts 1987). Specific fungal stains such as the methenamine silver and periodic acid–Schiff (PAS) stains are useful to ensure that fungi are present, but are seldom helpful for specific diagnosis. Actinomycete filaments also take up silver-based stains.

A scheme for the identification of mycetomas is shown in Table 20.3. The key elements for recognition are the presence of pigmentation and whether it is intramural or extracellular, the size and shape of grains, and their affinity for H&E. Although, with practice, it is possible to separate some of the black grain eumycetomas by their appearances in H&E, this is not the case with pale grain fungi. There is some evidence that infections caused by S. apiospermum show clusters of vesicular expansion of hyphae at the grain periphery and that this is less common with other nonpigmented agents of mycetoma (Figure 20.5). However, this is not completely reliable as a means of distinguishing among different organisms. The definitive diagnosis is made by culture.

Dermatophyte hyphal clusters are often referred to as pseudomycetomas. They seldom produce sinuses, but clusters of fungal hyphae are seen in giant cells in the dermis rather than subcutaneous fat or fascia (Chandler and Watts 1987). This mode of pathogenesis is somewhat similar to that seen with S. somaliensis which also starts as an intracellular infection (Nasher et al. 1987). However, with dermatophytes, the fungal hyphae are discrete and unaltered, in contrast to those seen with other fungi, where there are gross cell wall changes. They also originate from an endogenous infection and are not implanted from the environment. In some cases, pseudomycetoma caused by dermatophytes appears to follow persistent hair shaft infection whereas, in others, it is a manifestation of disseminated subcutaneous disease. Both types are rare.

Electron microscopy does not play a role in the diagnosis of mycetomas although it is useful for studying the pathogenesis of infection. As described above, a characteristic of these infections is the change in cell wall structure with gross thickening and fusion of adjacent cells. The cell walls often show an 'onion skin' appearance, which is probably the result of intrahyphal growth of fungi (Hay and Collins 1983).

Immunofluorescence has not been developed as a means of identification of mycetomas. This is partly the result of autofluorescence of grains which makes recognition difficult. Theoretically there is no reason why nonfluorescent techniques, such as immunoperoxidase staining, should not be used for mycetomas as there is no evidence that there is antigenic variation in vitro versus in vivo that could lead to false-negative reactions. However, reagents for such tests are not available at present. Likewise there is no established place for molecular genetic techniques such as Southern hybridization or restrictive fragment length polymorphism (RFLP) analysis in the diagnosis of these infections.

Table 20.3 Characteristics of the grains of the common eumycetoma agents

Color	Species	Size (µm)	Shape	Appearances in H&E[a]
White (pale)	Scedosporium apiospermum	500–1500	Irregular	Prominent vesicles. Round–oval Eosinophilic fringe
	Neotestudina rosatii	500–1000	Regular oval or round, lobed	Compact with cement-like matrix, basophilic hyphae
	Acremonium or Fusarium spp.	500–2000	Regular oval or round	Dense packing of hyphae which stain poorly
	Aspergillus nidulans	200–1000	Lobed	Pale, poor staining of fungal hyphae, hyaline
Black	Madurella mycetomatis	500–5000	Irregular lobed	Hard, cement-like matrix Hyphae swollen and indistinct
	Madurella grisea	500–2000	Irregular	No cement, swollen hyphae Individual elements distinct
	Leptosphaeria senegalensis	500–2000	Arc-like. Round	Distinct hyphal rim, diffuse central hyphae
	Exophiala jeanselmei	200–1000	Arc-like	Dense hyphal rim, neutrophil invasion common

a) These appearances are features described in H&E stained sections. They are variable.

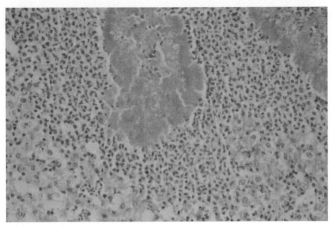

Figure 20.5 *Grains of* Scedosporium apiospermum. *Magnification* × 350; *stain: H&E*

TREATMENT

Treatment of eumycetomas with chemotherapy is difficult and their management often involves deciding from
conservative approaches involving observation, a trial of
chemotherapy, and surgery (Hay et al. 1992).

Conservative management

Eumycetomas are seldom life-threatening and are often
asymptomatic. Where chemotherapy is not available, it
may be preferable to review patients regularly without
active intervention. Relief of pain and dressings to
sinuses may be required from time to time.

Chemotherapy

The range of antifungals that can used with success in
mycetomas is not great. In some cases, responses have
been seen with amphotericin B, either in conventional
formulation or in combination with lipids or liposomes;
ketoconazole may also be effective. Reports of individual successes with itraconazole and terbinafine have
been recorded, although in no case has there been sufficient follow-up to establish cure (Paugam et al. 1997;
Hay 1999). Of all these approaches, more is known
about ketoconazole, which in daily doses of 200 mg has
been recorded as a successful therapy for mycetomas
caused by *M. mycetomatis*. The response rate is variable
though and probably depends on the extent of the infection, but in about half the cases successful results can be
obtained. Therapy is continued for at least 3 months and
12–18 months may be required. In other patients a trial
of therapy is a wise choice as this can be guided by the
in vitro sensitivity pattern [minimal inhibitory concentration (MIC)] of the organism isolated. Although this
approach has some logic, in vitro sensitivity is not a
good guide to cure or remission and it is probably more

realistic to regard this approach as palliative in slowing
enlargement of lesions. As such, it has clinical value
even though it is seldom curative. Griseofulvin is seldom
active against mycetoma agents in vitro and yet it
appears to slow the progress of mycetomas in some
patients. It is possible that this is because it is an active
inhibitor of leukocyte chemotaxis and suppresses the
accumulation of neutrophils in lesions (Yousef and Hay
1987).

There is no published experience on the use of immunological therapy such as cytokines – granulocyte–
macrophage colony-stimulating factors – in mycetoma.

Surgery

The ultimate method of eradicating a fungal mycetoma
is surgical excision (Mahgoub and Murray 1973).
However, it is important to recognize that this has to be
sufficiently radical to ensure an adequate margin of excision, otherwise relapse is inevitable. Local excision of
mycetomas on the foot may cause practical problems if
there is insufficient tissue remaining to allow adequate
weight-bearing support, and pain on standing or walking
may result. In other words, the extent and method used
have to be considered carefully. The definitive curative
process is adequate amputation, again ensuring that the
margins are well clear of infected bone or subcutaneous
tissue. Amputation has to be carefully considered together with the patient, taking into consideration the site,
the resulting disability and its effect on employment
or social life, the availability of good prostheses, and
the discomfort or deformity caused by the lesion. In
many cases this may be the best approach, but it
should be arrived at after proper consideration of all the
alternatives and after informed discussion with the
patient. It may be preferable to accept a measure of
disability with an untreated mycetoma if the alternative
is an operation that would deprive the patient of his or
her livelihood.

APPARENT GRANULE FORMATION BY OTHER FUNGI

In addition to mycetomas, granule-like structures may be formed by a number of different organisms in vivo. Usually this is a pathological phenomenon and not seen normally with infections caused by these organisms. The exception is where mycelial fungi grow within spaces or cavities such as pathological cavities in the lung. Under these conditions a number of different fungal species produce a fungus ball. *Aspergillus* spp. are the most common causes, but other organisms from *Coccidioides immitis* to *Pseudallescheria boydii* have been implicated and it is possible that all mycelial fungi have the propensity to form into balls given the appropriate conditions in vivo. However, two specific and different examples of in vivo granule formation are seen with the dermatophytes and *Scedosporium apiospermum*.

Pseudomycetomas caused by dermatophytes

These have been defined in the Introduction. As their pathogenesis is completely different to that seen with mycetomas, they are termed 'pseudomycetomas' (Ajello et al. 1980). The aggregates are formed of loose collections of dermatophyte mycelium, usually packed into giant cells. This phenomenon has been seen after hair shaft infections caused by both endothrix and ectothrix pathogens; it may also occur in the rare disseminated forms of dermatophytosis.

Strictly speaking, these collections are randomly aggregated hyphae and are not granules at all.

Systemic infections caused by *Pseudallescheria boydii* (*Scedosporium apiospermum*)

This organism is a known cause of mycetoma. Occasionally the rare systemic infections caused by this organism produce a compact aggregate of hyphae like a microcolony. This cannot be distinguished as a 'granule' in gross pathological specimens and is once again best designated as a pseudogranule.

REFERENCES

Abbott, P. 1956. Mycetoma in the Sudan. *Trans R Soc Trop Med Hyg*, **50**, 11–30.

Ahmed, A., Adelmann, D., et al. 2002. Environmental occurrence of *Madurella mycetomatis*, the major agent of human eumycetoma in Sudan. *J Clin Microbiol*, **40**, 1031–6.

Ajello, L., Kaplan, W., and Chandler, F.W. 1980. Dermatophyte mycetomas: fact or fiction? *Proceedings of the Fifth International Conference of Mycoses: superficial, cutaneous and subcutaneous infections*. Washington, DC: PAHO Scientific Publications, **396**, 135–40.

Ballingali, G.R. 1855. An account of a tumour affecting the foot. *Trans Med Phys Soc Bombay (New Series)*, **11**, 272–6.

Basset, A., Camain, R., et al. 1965. Role des epines de mimosacees dans l'innoculation des mycetomes (à propos de deux observations). *Bull Soc Pathol Exot*, **58**, 22–4.

Bender, B.J., Mackey, D., et al. 1987. Mycetoma in Saudi Arabia. *J Trop Med Hyg*, **90**, 51–9.

Berkley, M.J. 1862. The fungus foot of India. *Intelligence*, **10**, 248–59.

Bidie, G. 1862. Notes on morbus pedis entophyticus. *Madras Q J Med Sci*, **4**, 222–7.

Bocarro, J.E. 1893. An analysis of one hundred cases of mycetoma. *Lancet*, **2**, 797–8.

Borelli, D. 1976. *Pyrenochaeta mackinnonii*. Nova species agente de micetoma. *Castellania*, **4**, 227–34.

Borelli, D. and Feo, M. 1966. Diagnóstico morfológico de Candida albicans en os terrenos caseros y tritmel. *Acta Med Venez*, **14**, 448–50.

Borelli, D. and Zamora, R. 1973. *Chaetospaeronema larense* nova species agente de micetoma – presentacion del tipo. *Boletin Mensul Soc Venez Dermatol*, **6**, 17–18.

Boyce, R. and Surveyor, N. 1894. Upon the existence of more than one fungus in madura disease (mycetoma). *Philos Trans R Soc London*, **185**, Part I, 1–14.

Brumpt, E. 1906. Les mycetomes. *Arch Parasitol*, **16**, 489–564.

Cameron, H.M., Gatei, D. and Bremner, A.D. 1973. The deep mycoses of Kenya: a histopathological study 1. Mycetoma. *East Afr Med J*, **50**, 382–95.

Carter, H.V. 1861. On mycetoma or the fungus disease of India including notes of recent cases and new observations on the structure etc of the endophytic growth. *Trans Med Phys Soc Bombay*, **7**, 206–21.

Carter, H.V. 1874. *On mycetoma and fungus diseases of India*. London: J&A Churchill Ltd.

Chandler, F.W. and Watts, J.C. 1987. Mycetoma. In *Pathologic diagnosis of fungal infections*. Chicago: ASCP Press, 251–64.

Cockshott, W.P. 1968. Radiological patterns of deep mycoses. In: Wolstenholme, G.E.W. and Porter, R. (eds), *Systemic mycoses – a CIBA Foundation symposium*. Boston: Little, Brown, 113–25.

Czechowski, J., Nork, M., et al. 2001. MR and other imaging methods in the investigation of mycetomas. *Acta Radiol*, **42**, 24–6.

Davies, P. 1958. The bone changes of madura foot, observations on Ugandan Africans. *Radiology*, **70**, 841–7.

Destombes, P. 1978. Histological diagnosis of mycetoma granules. In *Proceedings of the First International Symposium on Mycetoma*. E.D. Barroeta, Venezuela, 80–94.

El Hag, I.A., Fahal, A.H. and Gasim, E.T. 1996. Fine needle aspiration cytology of mycetoma. *Acta Cytol*, **40**, 461–4.

El Hassan, A.M. and Mahgoub, E.S. 1972. Lymph-node involvement in mycetoma. *Trans R Soc Trop Med Hyg*, **66**, 165–9.

Eyre, E.W. 1859. Account of a peculiar disease, tubercular of the foot. *Indian Ann Med Sci*, **234**, 505–16.

Fahal, A.H., Sheik, H.E., et al. 1997. Ultrasonographic imaging of mycetoma. *Br J Surg*, **84**, 1120–2.

Fahal, A.H., el Hassan, A.M., et al. 1998. Cystic mycetoma: an unusual clinical presentation of *Madurella mycetomatis* infection. *Trans R Soc Trop Med Hyg*, **92**, 66–7.

Findlay, G.H. and Vismer, H.F. 1987. Black grain mycetoma. A study of the chemistry, formation and significance of the tissue grain in *Madurella mycetomi* infection. *Br J Dermatol*, **91**, 297–303.

Gill, R. 1842. *Indian Army Medical Reports*.

Godfrey, J. 1846. Diseases of the foot not hitherto described. *Lancet*, **1**, 593–4.

Gonzalez-Ochoa, A. 1962. Mycetoma caused by *Nocardia brasiliensis* with a note on the isolation of the causative organism from soil. *Lab Invest*, **11**, 1118–23.

Hay, R.J. 1999. Therapeutic potential of terbinafine in subcutaneous and systemic mycoses. *Br J Dermatol*, **141**, Suppl 56, 36–40.

Hay, R.J. and Collins, M.J. 1983. An ultrastructural study of pale eumycetoma grains. *Sabouraudia*, **21**, 261–9.

Hay, R.J. and Mackenzie, D.W.R. 1983. Mycetoma (madura foot) in the United Kingdom – a survey of 44 cases. *Clin Exp Dermatol*, **8**, 553–62.

Hay, R.J., Mahgoub, E.S., et al. 1992. Mycetoma. *J Med Vet Mycol*, **30**, Suppl 1, 41–9.

Kaempfer, E. 1694. Disputatio physica medica inauguralis exhibens decadem observationum exoticarum. Doctoral thesis, University of Leyden.

Lopes, M.M., Freitas, G. and Boiron, P. 2000. Potential utility of random amplified polymorphic DNA (RAPD) and restriction endonuclease assay (REA) as typing systems for Madurella mycetomatis. *Curr Microbiol*, **40**, 1–5.

Mahgoub, E.S. and Murray, I.G. 1973. *Mycetoma*. London: Heinemann Medical Books.

Mahgoub, E.S., Gumaa, S.A. and El Hassan, A.M. 1978. Immunological status of mycetoma patients. *Bull Soc Pathol Exot*, **70**, 48–54.

Maiti, P.K., Ray, A. and Bandyopadhyay, S. 2002. Epidemiological aspects of mycetoma from a retrospective study of 264 cases in West Bengal. *Trop Med Int Health*, **7**, 788–92.

Mariat, F. 1963. Sur la distribution geographique et al repartition des agents des mycetomes. *Bull Soc Pathol Exot*, **56**, 35–45.

Mariat, F., Destombes, P. and Segretain, G. 1977. The mycetomas: clinical features, pathology, etiology and epidemiology. In: Beemer, A.M., Ben-David, A. ct al. (eds), *Contributions to microbiology and immunology*, vol. 4, *Host–parasite relationships in systemic mycoses*: Basel: S. Karger, 1–39.

Meis, J.F., Schouten, R.A., et al. 2000. Atypical presentation of *Madurella mycetomatis* mycetoma in a renal transplant patient. *Transplant Infect Dis*, **2**, 96–8.

Minas, P.A. 1860. Observations in Keerenagraph (tuberculous disease) of the foot. *Indian Ann Med Sci*, **7**, 316–17.

Murray, I.G. and Mahgoub, E.S. 1968. Further studies in the diagnosis of mycetoma by double diffusion in agar. *Sabouraudia*, **6**, 106–10.

Nasher, M., Wethered, D.B., et al. 1987. An ultrastructural study of *Streptomyces somaliensis* grains *in vivo* and *in vitro*. *Am J Hyg Trop Med*, **37**, 174–9.

Nielsen, H.S., Conant, N.F., et al. 1968. Report of a mycetoma due to *Phialophora jeanselmei* and undescribed characteristics of the fungus. *Sabouraudia*, **6**, 330–3.

Paugam, A., Tourte-Schaefer, C., et al. 1997. Clinical cure of fungal madura foot with oral itraconazole. *Cutis*, **60**, 191–3.

Pracharktam, R., Chongtrakool, P., et al. 2000. Mycetoma and phaeohyphomycosis caused by *Phialophora parasitica* in Thailand. *J Med Assoc Thailand*, **83**, Suppl 1, S42–45.

Segretain, G. and Destombes, P. 1961. Description d'un nouvelle agent de maduromycose. *Neotestudina rosatii* n. gen. n. sp. isolé en Afrique. *C R Acad Sci (Paris)*, **253**, 2577–9.

Segretain, G. and Mariat, F. 1969. Recherches sur la presence d'agents de mycetomes dans le sol et sur les epineux du Senegal et de la Mauritanie. *Bull Soc Pathol Exot*, **62**, 194–202.

Spigelman, M. and Donoghue, H.D. 2001. Brief communication: unusual pathological condition in the lower extremities of a skeleton from ancient Israel. *Am J Phys Anthropol*, **114**, 92–3.

Sran, H.S., Narula, I.M.S., et al. 1972. History of mycetoma. *Indian J Hist Med*, **17**, 1–7.

Vincent, M.H. 1894. Etude sur le parasite du pied de Madura. *Ann Inst Pasteur, Paris*, **8**, 129–51.

Wethered, D.B., Markey, M.A., et al. 1987. Ultrastructural and immunogenic changes in the formation of mycetoma grains. *J Med Vet Mycol*, **25**, 39–46.

Wethered, D.B., Markey, M.A., et al. 1988. Humoral immune responses to mycetoma organisms: characterisation of specific antibodies by the use of enzyme-linked immunosorbent assay and immunoblotting. *Trans R Soc Trop Med Hyg*, **82**, 918–23.

Winslow, D.J. and Steen, F.J. 1964. Consideration in the histologic diagnosis of mycetoma. *Am J Clin Pathol*, **42**, 164–9.

Yousef, M.A. and Hay, R.J. 1987. Leukocyte chemotaxis to mycetoma agents. *Trans R Soc Trop Med Hyg*, **81**, 319–21.

Zaini, F., Moore, M.K., et al. 1991. The antigenic composition and protein profiles of eumycetoma agents. *Mycoses*, **34**, 44–9.

Zoutman, D.E. and Sigler, L. 1991. Mycetoma of the foot caused by *Cylindrocarpon destruens*. *J Clin Microbiol*, **29**, 1855–9.

21

Protothecosis

R. SCOTT PORE

ETIOLOGIC AGENTS

The etiologic agents of protothecosis are algae:

- *Prototheca wickerhamii* Tubaki and Soneda 1959
- *Prototheca zopfii* Krüger 1894
- *Chlorella* spp. Beyerinck 1890 may cause infections.

HISTORY

Prototheca spp. were discovered by Zopf (Krüger 1894a, b) and are related to *Chlorella* spp. Species of both genera reproduce only asexually by similar internal formation and subsequent release of spheroidal progeny cells from a parental sporangium. Species of both genera can subsist saprophytically in wastewater (Pore 1983). Comparison of nutritional and cell wall composition data has led to a proposed phylogenetic relationship between the chlorophyll-free *P. wickerhamii* and the photosynthetic *Chlorella protothecoides* Krüger (Pore 1972; Conte and Pore 1973). Confirmation of this relationship came from DNA base composition data (Huss et al. 1988), and sequence similarities between the 16S-like ribosomal RNA (rRNA) and the genetic DNA that separated these two species from the other *Chlorella* spp. tested (Huss and Sogin 1990). On the one hand, *Chlorella* spp. are polyphyletic, but DNA-binding homologies showed heterogeneity among *Prototheca* spp. as well (Huss et al. 1988). Although it has been postulated that *Prototheca* spp. could be nonphotosynthetic mutants of extant *Chlorella* spp. (see Cooke 1968 for references to this mutation theory), there is no evidence that such is

the case. Considerable evolutionary diversity would appear to separate the species of these two genera (Nedelcu 2001); with the lone exception that *C. protothecoides* is the postulated phylogenetic ancestor of *P. wickerhamii* (Ueno et al. 2003). From this summary it may be concluded that *Prototheca* spp. are achlorophyllous descendents of *Chlorella* spp., but evolutionary distinctions and further taxonomic hierarchies are speculative.

The discovery of *P. zopfii* during the late nineteenth century notwithstanding, proof of its infectious potential was not obtained until 1952, when it was shown to cause bovine mastitis (Lerche 1952). *P. zopfii* is now known to be the agent of several different animal infections. *Prototheca* spp. stand as the best, if not the only, examples of algae with infectious capabilities in animals and humans. The index human case of dermal protothecosis of the foot (Davies et al. 1964) may also have been caused by *P. zopfii*; however, nearly all subsequent human infections have been caused by *P. wickerhamii*, a microbe belatedly discovered in studies conducted separately by Phaff, Soneda and Wickerham (Tubaki and Soneda 1959). Finally, a few reports incriminate *Chlorella* spp. themselves as infectious agents (see the section on *Chlorella* spp. below).

EPIDEMIOLOGY

Ecology

P. wickerhamii and *P. zopfii* are ubiquitous inhabitants of wastewater (household and municipal sewage), in which they grow and thrive, ranging in densities up to 10^6 cells/ml (Pore and Boehm 1986). Likewise, they are common in animal waste and agricultural sewage. They also achieve a high population density in some types of tree slime flux (Krüger 1894a, b; Pore 1986). Both species have been isolated in Asia, Europe, North and South America. From these sources they contaminate aquatic systems and food (Pore 1983), and may subsequently be eaten by humans and animals. They pass undigested through the gastrointestinal tract of normal human and mono-gastric animals tested, with little if any multiplication, and are excreted to resume their saprophytic lifecycle (Pore and Shahan 1988). In a few cases, however, sustained colonization of the gut by *P. wickerhamii* was suspected (see section on Intestinal infection below). Canine protothecal enteritis is a recognized syndrome, and a prelude to systemic involvement (section on Canine infection below). Because of the importance of endemic protothecal mastitis in dairy cattle, a question remaining to be answered is, to what degree are ingested *Prototheca* spp. killed by digestion in ruminants?

Epidemiologic studies by Schuster and Blaschke-Hellmessen (1983), Blaschke-Hellmessen et al. (1987), Pore et al. (1987), and Anderson and Walker (1988) have indicated that *P. zopfii* may be isolated routinely from secreted milk, bulk milk collections, and dairy farm environments. Recent evidence that *P. zopfii* could survive milk pasteurization (Melville et al. 1999) is also of epidemiologic concern.

Relative to the high concentration of *P. wickerhamii* in municipal wastewater and the opportunity for occupational exposure, Clark et al. (1985) did not detect elevated protothecal antibody titers in sewage treatment workers tested. Organically enriched water, in general, not just wastewater, has commonly been incriminated as the source of skin and wound infections, the largest category of human protothecosis cases, but the case reports were anecdotal in nature.

Infections

More than 100 human infections have been documented in the medical literature during the previous 50 years (Krcmery 2000) and, even though many cases may have escaped diagnosis or case reporting, human protothecosis is still considered to be a rare disease. Protothecosis is worldwide in occurrence, but its incidence is low and the disease is not thought to be communicable. Animal infections, on the other hand, may be common, as in bovine protothecal mastitis, but in the past they were mainly undiagnosed by veterinary scientists.

HOST RESPONSES

Clinical presentations of protothecosis

CUTANEOUS

Protothecosis, following naturally occurring skin inoculation, may be the best-documented type of infection by *Prototheca* spp. In several cases, aquariums or waste-water were implicated as the source of the infections, but more frequently no such aqueous source was discovered. The primary cutaneous lesion has not been described, but after several weeks or more, a circumscribed or progressive, indurated, papular rash develops as an early manifestation. Inflammation is usually moderate and the primary symptoms may be pruritus and skin blemishes, although eczematoid reactions may occur. Such primary infections may resolve in some patients, although this has not been documented, and all current reports indicate that the disease is progressive over a period of many years; characteristically exemplified by spreading cutaneous nodules (Figure 21.1). Well-illustrated recent reports of cutaneous protothecosis include those of Cho et al. 2002, Piyophirapong et al. 2002 and Chao et al. 2002. The infections are not always benign, but may cause burning irritation (Tyring et al. 1989), pseudoepitheliomas or lymphocutaneous spread of granulomatous lesions for up to 18 years (Yip et al. 1976). Some patients are immunocompromised and may develop acanthosis or ulcerative skin lesions (Wolfe et al. 1976; Dagher et al. 1978; Thianprasit et al. 1983; Tejada and Parker 1994; Polk and Sanders 1997; Chao et al. 2002; Piyophirapong et al. 2002).

The index case of protothecosis was a progressive infection of the epidermis and dermis of the foot of an African rice farmer in Sierra Leone (Davies et al. 1964). It started as a small itchy, weeping, raised lesion, which,

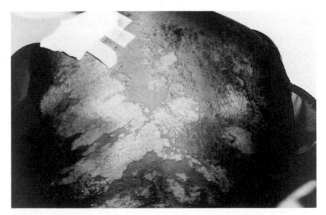

Figure 21.1 *Progressive cutaneous protothecosis of several years' duration. The spreading, raised plaques lacked pigmentation and were culturally and histopathologically positive for* P. wickerhamii. *The case was reported by Venezio et al. (1982). (Photograph courtesy of John Williams and John Phair)*

5 years later, became a well-demarcated, dry, atrophic, depigmented, papular dermatitis. After 2–3 years, complicated by secondary bacterial infections, the foot had the mossy appearance of elephantiasis. Subsequently, the algal pathogen caused femoral lymphadenitis (Davies and Wilkinson 1967). Although this patient was infected by *P. zopfii* (synonym used was *Prototheca segbwema*), almost all subsequent human infections have been attributed to *P. wickerhamii*. Another well-documented case (Venezio et al. 1982) of several years' duration developed from papular lesions into depigmented scaling plaques covering 50 percent of the skin surface (Figure 21.1). This case and two others were considered to be the sole reported cases of 'disseminated cutaneous protothecosis' (Thianprasit et al. 1983; Wirth et al. 1999), and all three patients had severe, defective neutrophil functions.

Cutaneous protothecosis may be a superficial form of wound protothecosis, local trauma has been recalled by patients (Chao et al. 2002). From such wounds papules followed by an extensive spreading granulomatous eruption of the skin can develop (Mars et al. 1971; Mayhall et al. 1976; Tang et al. 1995; Kim et al. 1996). *P. wickerhamii* was also the cause of an ulcerating cutaneous papulopustular lesion on the leg of an immunocompromised patient with metastatic adenocarcinoma of the bones. Like the index case, the algae were localized in small foci within the lesions, but the dermatitis was progressive (Klintworth et al. 1968). Raised, plaque-like, verrucose lesions with ulceration have been reported. Forty-five human case reports of all types of protothecosis were reviewed by Nelson et al. (1987) and 34 additional cases of cutaneous infections have been reviewed by Boyd et al. (1995). Resolution of cutaneous infections after antibiotic therapy left considerable residual scarring and atrophy of the skin (Tyring et al. 1989). In three instances patients had simultaneous cutaneous and systemic infections (Cox et al. 1974; Mohabeer et al. 1997; Marr et al. 1998). An infection of the nasopharyngeal mucosa complicated prolonged endotracheal intubation of an insulin-dependent diabetic individual receiving high-dose systemic dexamethasone (Iacoviello et al. 1992). Ulcerations of the hard palate, esophagus, and tongue were inflamed and necrotic, and contained clusters of algal cells that stained lightly with hematoxylin and eosin (H&E), but intensely with periodic acid–Schiff (PAS) and methylene blue.

WOUND INFECTION

Wounds may also be the cause of deeper infections. Accidental dermal and subdermal traumatic inoculation of *P. wickerhamii* during surgery to relieve carpal tunnel syndrome was thought to be the cause of suppurative tenosynovitis (Pegram et al. 1983; Moyer et al. 1990). Again, almost all such human infections were caused by *P. wickerhamii* as annotated by Nelson et al. (1987),

along with the source of the infections. Cellulitis, with an aspirate of necrotizing fibrinopurulent exudate containing *P. wickerhamii* occurred in a renal transplant recipient who also received immunosuppressants (Mezger et al. 1981). Subcutaneous wound infections are known to involve tendons and bones in humans (Sirikulchayanonta et al. 1989), dogs (unpublished data) and cats (Coloe 1982). Opportunistic wound infections are probably the second most frequent form of human protothecosis. Three cases of protothecal peritonitis secondary to peritoneal dialysis demonstrate the susceptibility of the compromised abdominal cavity opportunistic infection (O'Connor et al. 1986; Gibb et al. 1991; Sands et al. 1991).

OLECRANON BURSITIS

The relatively high incidence of protothecal olecranon bursitis has led to its recognition as a separate clinical entity. Although the circumstances for traumatic implantation were confirmed in many instances, a number of patients inexplicably reported no prior trauma. The bursa evidently provides a favorable environment for *P. wickerhamii*, and the preexistence of inflammatory conditions and the needle delivery of steroid therapeutics may provide the necessary conditions for infection in some of the cases (Cochran et al. 1986). An underlying immunodeficiency is not a necessary condition for an infection, however (Ahbel et al. 1980). Otherwise normal patients were found to have tender, swollen, olecranon bursas. Aspirated bursal contents were serosanguineous and culturally positive for *P. wickerhamii* (Nosanchuk and Greenberg 1973) or, in a single case, for *P. zopfii* (Naryshkin et al. 1987). Protothecal olecranon bursitis was usually less acute than corresponding bacterial infections. Nelson et al. (1987) have compiled many such protothecal olecranon bursitis cases. Finally, in one instance, olecranon bursitis caused by *P. wickerhamii* was antecedent or coincident with systemic infection (Marr et al. 1998).

INTESTINAL INFECTION

Prototheca portoricensis (another synonym for *P. zopfii*) was proposed as the suspected cause of tropical sprue after its isolation from the feces of two Puerto Rican patients (Ashford et al. 1930). This original observation was never again repeated on sprue patients, despite the efforts of several investigators over the intervening years (Klipstein and Schenk 1975), but the account by Ashford and colleagues does stand as an early attempt to incriminate algae as etiologic agents of infections. In fact, *P. wickerhamii* and *P. zopfii*, originating in the diet, may subsequently be isolated from human and animal feces (Pore and Shahan 1988). *P. wickerhamii* (notably not *P. zopfii*) was isolated on prototheca isolation medium (PIM) from the stools of nine of 500 patients tested (unpublished data). Follow-up was not done for all nine patients, but for one of the patients, a nutritionally normal, breast-fed, 1-year-old with nonbloody diarrhea of 2 months' duration, the organism was repeatedly recovered on PIM for the final 2 weeks of the diarrhea. At one point in this case fecal dilution plate counting revealed 25 colony-forming units (c.f.u.) per gram of wet stool. A dietary source of the alga was discounted and the microorganism disappeared from the feces near the end of the diarrheal episodes. This patient did not develop any related adverse health problems during 18 subsequent years. All of the other patients also had episodic chronic diarrhea and two of the patients had colony counts of more than 100 c.f.u. per fecal smear on PIM (100 × 15 mm Petri dish). In no case, however, was a causal relationship of the *P. wickerhamii* with the diarrhea established. One infant had pica, but environmental studies on this patient plus four of the others revealed no ingestion source of *P. wickerhamii*. *P. wickerhamii* was also isolated on PIM from the stools of two Philippine patients, one having acute diarrhea. One culture collection strain of *P. wickerhamii* (CDC B-1280) is labeled 'human intestine.'

An additional case of infant diarrhea, possibly caused by *P. wickerhamii*, has been reported (Casal et al. 1983). In another instance *P. wickerhamii* was repeatedly isolated from the stools of a patient with a well-documented cutaneous prototodecosis, but enteritis was not evident or, if evident, not reported (Venezio et al. 1982). Only one case report exists with strong evidence of intestinal prototodecosis. This patient with underlying chronic mucocutaneous candidiasis developed a 4-year symptomatology including abdominal pain, diarrhea, and a non-necrotizing granulomatous mass partially obstructing the colon. Histological demonstration of prototodecosis was made, but the species was not determined (Raz et al. 1998). Three years later the symptomatology continued and the prototodecosis was related to polyps in the colon and ileum. Two patients with systemic prototodecosis had a history of antecedent diarrhea (Cox et al. 1974; Heney et al. 1991), and intestinal prototodecosis in dogs is not uncommon (see the section on Canine infection below).

SYSTEMIC INFECTION

There are several accounts of human patients with multiorgan prototodecosis. A patient having Hodgkin's disease and multiple predisposing factors for secondary infection, and chronic diarrhea of a year's duration, had a mixed *P. wickerhamii*, *Torulopsis glabrata* sepsis, the source of which was probably the long-term Hickman catheter (Heney et al. 1991). Several case reports are available in which the patients had both cutaneous and systemic prototodecoses (Cox et al. 1974; Mohabeer et al. 1997; Marr et al. 1998). The first was a rare, chronic, disseminated infection involving the gallbladder, surface of the liver, and the duodenum, in an otherwise normal

patient operated upon for symptomatic sclerosing cholangitis (Chan et al. 1990). Two other patients had involvement of the liver, viscera and central nervous system (Matsuda and Matsumota 1992; Takaki et al. 1996). As there was antecedent diarrhea in two cases (Cox et al. 1974; Heney et al. 1991) and a postulated intestinal route of infection or involvement in two others (Chan et al. 1990; Matsuda and Matsumota 1992), these human cases may be compared with the frequently encountered visceral canine prototodecosis, which is associated with bloody diarrhea prior to multiorgan dissemination (see the section on Canine infection below). Finally, one case of prototodecal endocarditis in a premature infant was reported (Buendia et al. 1998).

IMMUNODEFICIENCY, OPPORTUNISM

In at least half of the published prototodecosis cases the patients had a preexisting immune dysfunction or had received immunosuppressive therapy. This was particularly true of the cutaneous and disseminated infections in humans, but has not been as apparent in olecranon bursitis (Woolrich et al. 1994). Corticosteroid treatment appeared to be a significant predisposing factor in several recent cutaneous cases (Chao et al. 2002), and in one otherwise immunocompetent patient, prototodecosis developed at the site of cutaneous corticosteroid injection (Kim et al. 1996); another patient who was corticosteroid dependent developed olecranon bursitis with possible subsequent dissemination (Marr et al. 1998). A myasthenia gravis patient receiving corticosteroids developed algemia (protodeca sepsis), but recovered with amphotericin B and gradual elimination of immunosuppressive therapy (Mohabeer et al. 1997). Thirteen cases of prototodecosis in cancer patients, several with lymphoid malignancies, have been summarized (Torres et al. 2003). *P. wickerhamii* caused meningitis in one human immunodeficiency virus (HIV)-positive patient (Kaminski et al. 1992) and cutaneous prototodecosis in six other AIDS patients (Laeng et al. 1994; Woolrich et al. 1994; Carey et al. 1997; Polk and Sanders 1997; Cunliffe and DiPersio 2000; Piyophirapong et al. 2002). Although other predisposing factors were also present, AIDS was thought to be relevant to the prototodecosis. Regarding dogs, immunodeficiency was found to be a significant precondition in several systemic infections (Thomas and Preston 1990). Experimental mouse infections were facilitated by pretreatment with corticosteroids (Pore 1971) as was a naturally occurring disseminated mastitis in a cow (Taniyama et al. 1994).

Animal infections

BOVINE MASTITIS

After its original description and experimental reproduction in Germany (Lerche 1952), prototodecal bovine mastitis has gained worldwide recognition. Caused by *P. zopfii*, protothecal mastitis is mildly inflammatory in comparison to bacterial mastitis; however, it is invasive, reaching the supramammary lymph nodes, chronic, and results in reduced milk production. Twenty-three of 93 dairy cows in one herd were culled (Frank et al. 1969), and 400 cases were reported in New York State (Mayberry 1984). Ten of 192 infected cows in a Danish herd had decreased milk production (Bodenhoff and Madsen 1978), and five of 130 cows were culled for prototothecosis, after calving, in the UK (Spalton 1985). A report from Hungary lists 223 cases in 32 herds over 2 years (Janosi et al. 2001b). The only pathogen present in three cases (Dion 1979), and in 10 other cases of mastitis (Chengappa et al. 1984), was *P. zopfii*.

The ubiquity of *P. zopfii* in the dairy farm environment (Schuster and Blaschke-Hellmessen 1983; Blaschke-Hellmessen et al. 1987), and its frequency in milk collections (Pore et al. 1987), provided the foundation for epidemiologic studies (also refer to the section on Ecology below) of bovine mastitis in a herd of 263 cows (Anderson and Walker 1988) and a 52-herd survey (Costa et al. 1998). In the later study, *P. zopfii* was the most common mastitis pathogen recovered (41.2 percent), and this included bacteria. In another large herd, 248 cows were culled for protothecal mastitis. The advanced age of the cows and prior antibiotic treatment for bacterial mastitis were found to be significant risk factors (Tenhagen et al. 1999). Two additional risk factors were found to be high capacity machine milking and poor herd management (Corbellini et al. 2001b; Janosi et al. 2001a). A review of protothecal bovine mastitis included a more comprehensive discussion of pathogenesis and possible control measures (Pier et al. 2000).

Despite the postulated decrease in milk production, protothecal mastitis is an underreported dairy cow disease. Factors responsible include:

- the reluctance of dairy farmers to admit to mastitis in their herds
- insufficient information about protothecal mastitis etiology
- the relatively low white blood cell counts in the milk
- the lack of a treatment regimen
- the ease of culling and its economic advantages.

Both mastitic and non-mastitic dairy cows have been found to have elevated levels of anti-*P. zopfii* antibodies (Jensen et al. 1998), and a recently developed enzyme-linked immunosorbent assay (ELISA) test based on elevated serum IgA and IgG, and whey IgA against *P. zopfii* (Roesler and Hensel 2003) presents the opportunity for screening infected herds and advancing epidemiologic research.

CANINE INFECTION

After chronic hemorrhagic diarrhea in a domestic dog that was attributed to *Prototheca* spp., necropsy revealed small, white, granulomatous nodules containing numerous cells of *Prototheca* spp. in the intestinal mucosa and wall, peritoneal cavity, myocardium, kidneys, liver, and spleen (Buyukmihci et al. 1975). Notable in this case was necrotizing granulomatous chorioretinitis leading to blindness. Similarly, Cook et al. (1984) described a dog presenting with blindness and deafness, that had had a previous bloody diarrhea and ultimately developed multiorgan system involvement. In this case, *P. zopfii* was identified by fluorescent antibody stains as it was in a similar third case (Gaunt et al. 1984). In a fourth example, blindness caused by micro-abscesses containing *P. wickerhamii* under the detached retina was preceded by an enterocolitis (Font and Hook 1984). It appears, based on a review of 12 such cases, that hemorrhagic diarrhea and disseminated protothecosis is a distinctive syndrome that can be caused by either *P. wickerhamii* or *P. zopfii* after protothecal colitis in dogs (Migaki et al. 1982). Similar cases were reported and the syndrome reviewed (Rakich and Latimer 1984; Thomas and Preston 1990; Rallis et al. 2002). Attention was drawn to the fact that the syndrome is most common in the Collie breed and the similar features of immunodeficiencies in both canine and human protothecoses. However, not all dogs with disseminated protothecosis had documented antecedent intestinal infections (Tyler et al. 1980), so the syndrome and its epidemiology need further elucidation. Myocardial infection may also be a significant manifestation of systemic canine protothecosis (Moore et al. 1985). To summarize, early recognition of canine protothecosis syndrome should include manifestations of chronic diarrhea and/or acute blindness, this indicating the most opportune time for therapeutic intervention (Hollingsworth 2000).

OTHER ANIMAL INFECTIONS

Kaplan (1978) compiled information on 42 case reports of protothecosis in animals and 18 in humans. He included several cases attributable to *Chlorella* spp. as well (Kaplan et al. 1993). Disseminated protothecosis in cattle has also been described (Migaki et al. 1969), and it may be acute, causing massive necrotic mastitis, kidney infarction, pulmonary artery thrombosis and colitis (Taniyama et al. 1994). In this last instance a cow was simultaneously infected with *P. zopfii*, *Aspergillus* sp. and a zygomycete (Taniyama et al. 1994). Cutaneous protothecosis caused by *P. wickerhamii* in two cats has some of the features of human infections (Dillberger et al. 1988), and a spreading mucocutaneous infection of the nares of a dog has been well documented (Sudman et al. 1973); otherwise, descriptions of cutaneous infections in animals have been rare. One cutaneous dog infection was thought to be the primary manifestation of a case of systemic protothecosis (Ginel et al. 1997).

Experimental infection

P. zopfii was pathogenic for mice after intraperitoneal inoculation (Schiefer and Gedek 1968). Weakly inflammatory, but chronic, infections of the kidney capsule and kidney tubules were the prominent manifestations (Pore 1971), and hydrocortisone treatment enhanced the experimental infection of mice by *P. wickerhamii* (Pore 1971). de Camargo et al. (1980) were less successful and only obtained intratesticular infections in laboratory animals, but Horiuchi and Masuzawa (1995) successfully demonstrated a mouse model for cutaneous protothecosis. In this experimental model *P. wickerhamii* caused epithelioid cell granulomas in BALB/c mice while ICR mice were less vulnerable. Experimental mastitis in cows, after introduction of *P. zopfii* through the teat canal, led to well-documented progressive pyogranulomatous infections (Lerche 1952; Frank et al. 1969; McDonald et al. 1984). The experimental infections mimicked natural infections and *P. zopfii* was subsequently isolated from secreted milk. Schiefer and Gedek (1968) produced mastitis with *Prototheca moriformis* – another potential cause of bovine mastitis (Pore et al. 1987). *P. zopfii* was lethal for immunosuppressed mice in one study (Jensen and Aalbaek 1994). Finally, cells of *Chlorella* spp. persisted in the peritoneal cavity of rats with minimal pathogenesis (Rogers et al. 1980), but even *Prototheca* spp. were not very aggressive pathogens in normal laboratory rodent models of infection.

Immunology

About half the patients with protothecosis have immunodeficiencies prior to or coincident with their protothecosis. These immunodeficiencies differ, but commonly fall within the same category as those for patients with increased susceptiblity to opportunistic fungal infections. One opportunistically predisposed patient was simultaneously infected with *P. wickerhamii* and the fungus *Exophialia jeanselmei* (*Phialophora gougerotti*) (McAnally and Parry 1985), while another patient had *P. wickerhamii* and *T. glabrata* (Heney et al. 1991). A typical example of predisposition was a patient receiving long-term prednisolone for systemic lupus erythematosus who developed cutaneous protothecosis after an injury. The combination of oral prednisolone and cyclophosphamide in a patient with systemic lupus erythematosus was a contributing immunosuppressive factor for another patient with cutaneous protothecosis (Thianprasit et al. 1983). A significant number of protothecosis patients received local corticosteroid injections in soft tissues or the olecranon bursa (Nelson et al. 1987) or in skin lesions (example of two cases; Walsh et al. 1998). Patients with malignancies and other immunosuppressive factors above have been reviewed (Torres et al. 2003). Corticosteroids may predispose patients to opportunistic protothecosis and mycoses by suppressing T-cell-mediated immunity, as do cytotoxic agents and other immunosuppressants. Cutaneous protothecosis has been described more than once in patients receiving combination immunosuppressive therapy to combat graft rejection (Wolfe et al. 1976), as has chemotherapy-induced neutropenia in a patient with acute myelocytic leukemia (Wirth et al. 1999). Corticosteroids, antilymphocyte globulin, and cytotoxic drugs predispose patients to opportunistic bacterial and fungal infections as well as cutaneous abscesses caused by *P. wickerhamii*. There is evidence that *P. wickerhamii* itself may be immunoinhibitory (Perez and Ginel 1997).

Intracellular killing of *P. wickerhamii* by human polymorphonuclear neutrophils (PMNs) in vitro was enhanced in the presence of humoral factors that stimulated oxidative metabolism and degranulation (Phair et al. 1981). Subsequent research linked this finding to a case of protothecosis in a patient who had a specific defect in PMN killing of *P. wickerhamii* (Venezio et al. 1982), but nonspecific acquired immunity in this case was normal. Likewise, a severe oxidative burst defect in peripheral blood neutrophils was found to be contributory to protothecosis in an AIDS patient (Carey et al. 1997). Similarly, a Collie dog with disseminated protothecosis had a serum-mediated defect in PMN chemotaxis (Rakich and Latimer 1984). Natural killer cell activity was suppressed in one patient with cutaneous protothecosis (Tyring et al. 1989). Finally, one patient with disseminated protothecosis, who had developed a high specific IgG titer to *P. wickerhamii*, possessed a normal cell-mediated immune system (Chan et al. 1990), but another with a high serum titre had a blast transformation deficiency specific for the microorganism (Cox et al. 1974). These patient studies indicate that protothecosis, like several of the mycoses, may not be ameliorated by the humoral immune system and indicate that nonspecific leukocyte defects are paramount susceptibility factors.

Treatment

SURGERY

Surgical excision has been a successful means of treating localized cutaneous protothecosis (Thianprasit et al. 1983; Cunliffe and DiPersio 2000), wound protothecosis (Sirikulchayanonta et al. 1989; Tejada and Parker 1994) and prothecal and olecranon bursitis (Ahbel et al. 1980; Vernon and Goldman 1983; Naryshkin et al. 1987), often combined with an antibiotic. Wound infections have responded to débridement plus amphotericin B (Moyer et al. 1990; Walsh et al. 1998), and intrabursal amphotericin B alone was successful (Cochran et al. 1986). Removal of a right atrial mass followed by amphotericin B resolved a case of prothecal endocarditis in a premature infant (Buendia et al. 1998).

AMPHOTERICIN B

Amphotericin B alone has been used successfully to treat cutaneous infections (Mayhall et al. 1976; Thianprasit et al. 1983; Carey et al. 1997; Chao et al. 2002), a mucosal infection of the pharynx (Iacoviello et al. 1992) and a case of algemia, together with the removal of a Hickman catheter (Heney et al. 1991). Amphotericin B, followed by ketoconazole therapy, was successful in a case of systemic human protothecosis (Chan et al. 1990), but failed in a similar case in a dog (Moore et al. 1985). Amphotericin B alone was not efficacious in a case of cutaneous protothecosis (Yip et al. 1976); but amphotericin B has also led to cures (Chao et al. 2002). Combined with transfer factor, amphotericin B was effective in a case of disseminated human protothecosis (Cox et al. 1974). Intestinal protothecosis did not respond to amphotericin B alone (Raz et al. 1998). Tetracycline has had postulated synergistic activity with amphotericin B (Venezio et al. 1982; Tyring et al. 1989; Sands et al. 1991). Aminoglycosides and tetracycline were good inhibitors in susceptibility tests (Shahan and Pore 1991), but they lacked efficacy in a mucocutaneous dog infection (Sudman et al. 1973). Although tetracycline after surgery (Sirikulchayanonta et al. 1989) and oral tetracycline plus topical amphotericin B (Tyring et al. 1989) are reported to be successful therapeutic agents, tetracycline alone has also failed as a treatment for protothecosis (Wolfe et al. 1976). So, in light of increasing success with azoles (see the next section), tetracycline therapy is not promising. In patients with specific neutrophil killing defects, it may be advantageous to use a cidal agent such as amphotericin B (Carey et al. 1997). The same may be true for patients developing chemotherapy-induced neutropenia (Torres et al. 2003), along with amelioration of the neutropenia (Wirth et al. 1999).

AZOLES

Recent reports of successful therapy of cutaneous protothecosis with oral azoles are promising. A case of cutaneous eczematous dermatitis of the eyelid was successfully treated with oral ketoconazole over a 4-month period (Kuo et al. 1987), as was a post-surgical wrist infection (Pegram et al. 1983), but failures of ketoconazole treatment have also been reported (McAnally and Parry 1985; Walsh et al. 1998). Oral fluconazole has proven successful (Follador et al. 2001) as has 200 mg/day oral itraconazole (Boyd et al. 1995; Tang et al. 1995; Okuyama et al. 2001; Cho et al. 2002). Instances of failure of itraconazole (Carey et al. 1997) and fluconazole (Walsh et al. 1998) have also been reported. In one instance, a cutaneous infection that did not respond to itraconazole was successfully resolved with fluconazole (Kim et al. 1996), and another with amphotericin B (Carey et al. 1997). Itraconazole alone was not satisfactory for intestinal protothecosis, but in combination with interferon-gamma (IFN-γ), the patient improved (Raz

et al. 1998). Favorable treatment results were achieved with fluconazole for peritonitis (Gibb et al. 1991).

In a report of five cutaneous cases compromised by steroid treatment, successful resolution of each separate case occurred by using ketoconazole, fluconazole, or itraconazole (Chao et al. 2002). In the other two cases, intravenous amphotericin B or intralesional amphotericin B resulted in cures. In the case of AIDS-related cutaneous infections, oral ketoconazole was not successful, but 2 g intravenous amphotericin B for 10 weeks resulted in resolution that was maintained for 3 years (Piyophirapong et al. 2002).

RESISTANCE

Currently, the management of disseminated protothecosis with conventional antibiotics remains an open question, as exemplified by the patient who received three different azoles and amphotericin B, but remained infected (Takaki et al. 1996). Even though azoles and polyenes may be efficacious in humans, azoles give unreliable susceptibility data, and the development of resistance has been noted (Takaki et al. 1996). *P. wickerhamii* is uniformly sensitive to amphotericin B in in vitro susceptibility tests (Segal et al. 1976), but an exception has been noted (Takaki et al. 1996). Notably, both *P. wickerhamii* and *P. zopfii* are resistant to 5-fluorocytosine (Pore 1973). Likewise, both species may be resistant to fluconazole and itraconazole, at least according to the E-test (Blaschke-Hellmessen 1996; Linares et al. 1998), but this finding is in need of confirmation in light of the successful therapies reported above. Similarly to the situation with yeasts, successful therapy of protothecosis with azoles does not necessarily correspond to conventional minimum inhibitory concentration (MIC) susceptibility testing results, which can be equivocal, yet azoles are often efficacious therapies. Finally, it should be kept in mind that all infections responding to antibiotics were caused by *P. wickerhamii*.

LABORATORY DIAGNOSIS

Isolation and culture

FROM BLOOD, CSF, AND TISSUE

P. wicherhamii and *P. zopfii* grow aerobically on conventional Sabouraud's and blood agar media after 24–48 hours at room temperature up to 35°C. The small colonies are grossly indistinguishable from yeast colonies. As these two pathogenic *Prototheca* spp. may be sparse in clinical specimens, a large, fresh, preferably unrefrigerated specimen should be processed for culture. Notably, *Prototheca* spp. similar to *Cryptococcus* spp., are uniformly inhibited in vitro by common culture medium additives such as cycloheximide, tetracycline

(Mezger et al. 1981), and aminoglycosides (Shahan and Pore 1991).

FROM STOOLS

As *Prototheca* spp. may be overgrown by faster growing microorganisms and hence usually unrecoverable from contaminated specimens, PIM (Table 21.1), which contains phthalate to suppress bacteria and 5-fluorocytosine to suppress yeasts, was found to be a good selective isolation medium for *Prototheca* spp. from contaminated sources (Pore 1973). A subsequent modification included the addition of rose Bengal (4,5,6,7-tetrachloro-2',4',5',7'-tetraiodofluorescein sodium salt) for further suppression of fungal and bacterial growth (Pore 1983). Specimens may be smeared or streaked with a swab and, after 48–72 hours, yeast-like colonies will become visible with the aid of a low-power stereomicroscope. Not all yeast species are suppressed by PIM, so suspicious colonies should eventually be examined at magnifications of 400 × with a compound microscope.

Microbiology

MORPHOLOGY

Colonies have a yeast-like appearance, but microscopically, budding is absent. Reproduction of *Prototheca* spp. is via the formation of a variable number of sporangiospores (commonly two to eight) within a sporangium (Figure 21.2). The sporangiospores are released through a characteristic split in the sporangial wall and usually disperse (Figure 21.3). The reproductive morphology is identical to that of most *Chlorella* spp. and is the basis for a postulated phylogenetic relationship to the algae; however, *Prototheca* spp. lack plastids

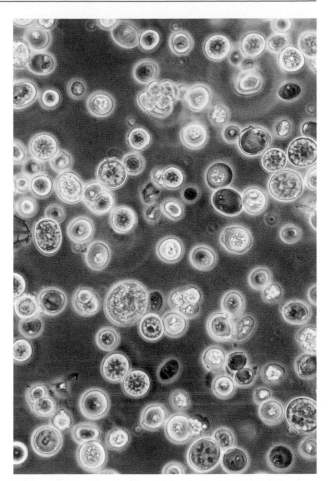

Figure 21.2 P. wickerhamii; *typical stages of the lifecycle in culture. Phase contrast; magnification × 700*

or vestigial plastids as is often erroneously reported (Figure 21.4). Examination of published electron micrographs (Klintworth et al. 1968; Patni and Aaronson 1974; Joshi et al. 1975; Kaplan 1978; Cheville et al. 1984; Font and Hook 1984; Jensen et al. 1998) supports the claim that no leukoplasts or thallocoid structures are present. The sporangium is spheroidal or an enlarged version of the sporangiospore. The sporangiospores are usually spheroidal or ellipsoidal, but reniform and falcate strains occur. The fact that a Dauer cell stage resists staining with methylene blue and rose Bengal is of significance in identification (Pore 1998). Sexual stages are not found (Pore 1985). Three other species – *P. moriformis*, *Prototheca stagnora*, and *Prototheca ulmea* – which have been isolated from nature, but not from animal infections, all have a large capsule that may give a resemblance to *Cryptococcus* spp. (de Camargo and Fischman 1979; Pore 1985, 1986).

PHYSIOLOGY

Chlorophyll is not present in the *Prototheca* spp. Glucose is assimilated oxidatively and the assimilation of a few other carbon substrates has taxonomic significance (Pore 1985). D-Lactic acid is produced from glucose by

Table 21.1 *Prototheca isolation medium (PIM)*

Ingredients	Amount
Distilled water	1 l
Potassium hydrogen phthalate[a]	10 g
Sodium hydroxide	0.9 g
Magnesium sulfate	0.1 g
Potassium phosphate, monobasic	0.2 g
Ammonium chloride	0.3 g
Thiamine	0.001 g
Rose Bengal	0.0025 g
Glucose	10 g
Purified agar	15 g

a) Adjust to pH 5.1 ± 0.1 with NaOH before sterilization; autoclave for 15 min; after sterilization add 5-fluorocytosine (0.25 g) (Pore 1973). The potassium hydrogen phthalate is the unmetabolizable buffer; it also serves to enhance growth in an unknown way and suppresses growth of most bacteria and fungi in contaminated specimens. A modification of this medium (Pore et al. 1987) is an enrichment medium for the recovery of *P. zopfii* from milk and the environment. *C. protothecoides*, but not other species of *Chlorella*, also grows on PIM. Another modification (Pore 1985, 1998) is the filter-sterilized, PIM, minimum essential medium for assimilation and identification to species.

Figure 21.3 P. wickerhamii *releasing sporangiospores from the* *sporangium. Scanning electron micrograph; magnification* × *9 350. (Courtesy of Phillip Allender)*

P. zopfii (Barker 1935; Ueno et al. 2002) and other species also produce acid, but not gas, from glucose (unpublished results). Ueno et al. (2002) do report CO_2 production during fermentation as a function of

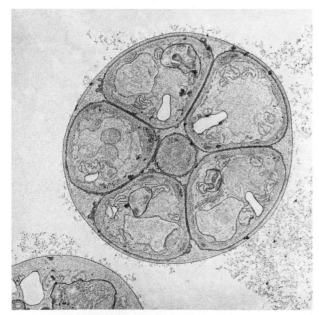

Figure 21.4 P. wickerhamii *sporangium thin section* *demonstrating five sporangiospores. Nuclei, mitochondria, Golgi* *and glycogen storage organelles are present. Transmission* *electron micrograph; magnification* × 16 250

temperature. Ammonium chloride can be assimilated as a sole nitrogen source, but nitrate can not (Pore 1972). Frequently, authors erroneously report starch to be present, but glycogen is the storage carbohydrate (Manners et al. 1973) and starch is absent (Pringsheim 1963). Thiamine (Anderson 1945) and oxygen (Barker 1935) are required for growth. *Prototheca* spp. may store lipids up to 10 percent by weight (Atkinson et al. 1972; Sud and Feingold 1979). The fatty acids 16:0, 18:1, and 18:2 (palmitic, oleic, and linoleic acids) were the predominant extractable lipids of *P. zopfii* grown on glucose or hexadecane (Boehm 1984). Growth on fatty acids (Barker 1935) may well be a reason for the ubiquity of *Prototheca* spp. in fatty acid-enriched wastewater (Pore and Boehm 1986). Assimilation of hexadecane and other hydrocarbons is a notable feature of some strains of *P. zopfii* (Walker and Pore 1978; Pore 1983; Boehm 1984). The cycloartnol-ergosterol biosynthetic pathway of *P. wickerhamii* differs from the lanosterol-ergosterol pathway of yeasts, yet azoles remain effective inhibitors of growth (Mangla and Nes 2000). Ergosterol was not detected in *Prototheca* spp. (unpublished results).

The outermost trilaminar aspects of the cell walls of *Prototheca* spp. resist acid and alkaline hydrolysis and contain neither cellulose (as sometimes erroneously reported) nor chitin (Conte and Pore 1973). Atkinson et al. (1972) proposed that sporopollenin could be the refractory compound in the cell walls responsible for resistance to hydrolysis, but the current status of this proposal is that sporopollenin is not the compound in the outermost nonhydrolysable trilaminar cell wall layers (Puel et al. 1987). Regardless of the resolution of this question, the unique cell wall of *Prototheca* spp. and its role in pathogenesis and ecology is a relevant topic for speculation. The presence or absence of carotenoids, accessory photosynthesis pigments, and the purported monomers of sporopollenin would all help in the resolution of the question. Another potentially significant component of the cell envelope of *P. zopfii* is an immunoreactive linear β(1,6)-galactopyranosyl (Manners et al. 1973; Roy et al. 1981). Significant quantities of similar galactans can be easily extracted and purified from the capsules of *P. zopfii* and *P. moriformis* grown in bioreactors. These viscous, polyanionic galactans contain variable amounts of arabinose and rhamnose (unpublished results) and could have medical or industrial value. Likewise, *P. zopfii* and *P. moriformis* have been shown to produce commercial quantities of L-ascorbic acid (Running et al. 2002).

Identification

Five species of *Prototheca* are currently valid, but only two are documented pathogens. Table 21.2 provides the minimum information needed to differentiate *P. wickerhamii* from *P. zopfii*. For a complete means to

Table 21.2 *Abbreviated key to differentiate the pathogenic species of* Prototheca[a]

Species	Sporangiospore diameter on PIM[b] (μm) [usual results]	Trehalose assimilation on PIM[c] [no exceptions]
P. wickerhamii	<4.3	+
P. zopfii	>5.7	−

a) Based on complete key for the separation of all five species (Pore 1985, 1998).
b) *Prototheca* isolation medium.
c) Modified for assimilation testing (Pore 1985, 1998).

differentiate all five species the reader is referred to Pore (1985, 1986, 1998).

PROTOTHECA WICKERHAMII

P. wickerhamii grows on PIM after 3–5 days as a white yeast-like colony. In many respects the colony resembles *Candida* spp. (Figure 21.5a). The small, usually smooth colonies (Figure 21.5b) characteristically develop a tan pigmentation with advanced age (2–10 weeks) on Sabouraud's medium. All cell stages are spheroidal in culture. Sporangiospores are 2.5–4.5 μm (mean 3.2), sporangia are 7–13 μm (mean 9.4), and Dauer cells are 5.5–8.5 μm (mean 6.5) in diameter. Temperature optima are usually above 25°C and pathogenic strains grow at 37°C. It is the only species that assimilates trehalose (Arnold and Ahearn 1972). The API 20C carbohydrate assimilation (Padhye et al. 1979) and the API ZYM tests (Casal et al. 1985) are of value for clinical identification. A monograph containing a complete taxonomic description was prepared by Pore (1985).

PROTOTHECA ZOPFII

After growing for 3–5 days on PIM, the colonies are whitish with a wrinkled, yeast-like appearance (Figure 21.5b). The colonies are faster growing and larger than those of *P. wickerhamii* and the sporangia, ranging up to 25 μm in diameter, give a 'ground-glass'

appearance when colonies are observed with a stereo-microscope (magnification × 25). Cells are commonly spheroidal or ellipsoidal, waxy, and hydrophobic. Newly released sporangiospores are spheroidal (4.5–7.5, mean 6.5 μm) or ellipsoidal (3–7 × 5–8, mean 5.5 × 6.5 μm). Dauer cells are spheroidal (8.5–14, mean 11 μm) or ellipsoidal (6–11 × 8.5–13, mean 9.5 × 11.5 μm), and sporangia are spheroidal (14–25, mean 17.3 μm) or ellipsoidal (11–20 × 14–23, mean 14.5 × 16.5 μm), but morphology may vary depending on substrate and environmental conditions. Some strains grow at 37°C. *P. zopfii* is the only unencapsulated species capable of growing at pH 5.1 in the presence of acetate (Pore 1985), or assimilating 1-propanol (Arnold and Ahearn 1972). Notably, only 80 percent of the strains tested grew in the presence of clotrimazole 50 μg/ml (Pore 1985), invalidating the use of clotrimazole for identification. No clinically derived strains of *Prototheca* spp. have been described as having a capsule, but spontaneous variants of *P. zopfii* in laboratory culture may produce a polysaccharide capsule; these variants are indistinguishable from *P. moriformis* by conventional morphological and physiological tests (Pore 1985). Several synonyms and varieties have been proposed for the pathogenic *Prototheca* spp., but *P. portoricensis*, *P. segbwema*, and *Portotheca salmonis* have been reduced to synonymy with *P. zopfii* (Pore 1998). Finally, an evaluation of morphological variability in *P. zopfii* was undertaken (Blaschke-Hellmessen et al. 1985).

CHLORELLA SPP.

In one report algae were isolated from cattle with lymphadenitis (Rogers et al. 1980) and in another from sheep liver (Zakia et al. 1989), but identification of genus and species was inconclusive. All other green algal infections were based on isolation or histopathological observations, and identification of species was not carried out. In previous taxonomic studies, identification of species followed the key points of Shihira and Krauss

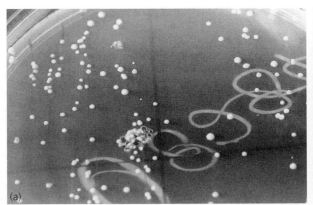
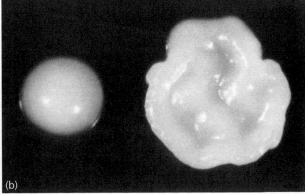

Figure 21.5 (a) *Primary isolation of* P. wickerhamii *from fecal (human) smear after 72 hours on PIM.* **(b)** P. wickerhamii *usually has a smooth, round colony after 72 hours on PIM:* P. zopfii *produces a larger wrinkled colony.*

(1965), who included *C. protothecoides*, the postulated extant ancestor of *P. wickerhamii,* as subsequent studies have shown (Pore 1998; Ueno et al. 2003). Except for the presence or absence of photosynthetic pigments, the physiological similarities, like the morphological similarities, between *C. protothecoides* and *P. wickerhamii* are remarkable. *C. protothecoides* even inhabits wastewater (Pore and Boehm 1986).

Histopathology

PROTOTHECA WICKERHAMII

P. wickerhamii may be rare to abundant in biopsied specimens of infected tissue, and histopathological detection is the usual means of diagnosis of protothecosis. The algal cells may be seen with the aid of H&E, Papanicolaou's, methylene blue, safranin, lactophenol cotton blue, or Gram's stains, but special stains, such as PAS and Gomori's methenamine silver (GMS), facilitate and confirm detection and identification (Figure 21.6). A comprehensive selection of color illustrations of the histopathology has been published

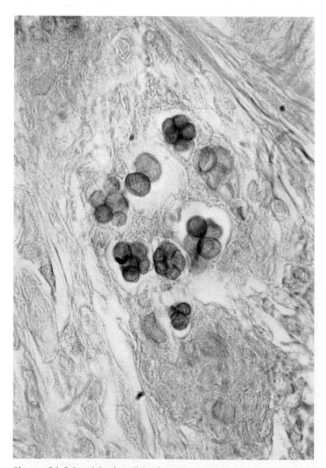

Figure 21.6 P. wickerhamii *in dermis stained by Gridley's silver stain method. In this instance the older sporangial walls are unstained and the sporangiospores alone are strongly stained. Magnification × 700*

(Chandler et al. 1980). Unlike in vitro size data, speciation on the basis of cell size in tissue is not definitive because of a size overlap of the sporangia of the two species, imprecise dating of the age of the algae, and the compounding influence of host tissue effects. Single cells range from 2 to 10 μm in diameter and sporangia from 10 to 30 μm, larger than in vitro sporangia and clearly overlapping the size range for *P. zopfii*. Empty, collapsed sporangia may be common, and could be difficult to differentiate from fungal debris. *P. wickerhamii* is unencapsulated; however, it was reported to stain weakly with mucicarmine (Kuo et al. 1987) – similar to the purported acid fastness; these may be staining artifacts, not observed by all investigators.

Infected epidermal and dermal responses include hyperkeratosis, parakeratosis, acanthosis, and papillomatosis (Davies et al. 1964). Granulomatous inflammatory reactions in the dermis and soft tissues were associated with multinuclear giant cells containing intracellular *P. wickerhamii*, a notable feature, and surrounded by lymphocytes, plasma cells, eosinophils, and a few polymorphonuclear leukocytes (Mars et al. 1971). Intraepidermal to mid-dermal depositions of the algae may be pathognomonic in cutaneous manifestations (Yip et al. 1976). *P. wickerhamii* may cluster in microabscesses, in which case the lesions show central necrosis and caseation. By contrast, other reporters have not always detected notable multinucleated giant cell response nor microabscess formation, but characterize the histopathological response as a mixed inflammatory response (Walsh et al. 1998).

The distinctive internal cleavage of the spheroidal sporangia, described as morula or being mulberry like, is considered pathognomonic for protothecosis; alternatively, sporangia that break open to release two to eight or more sporangiospores into the tissue are also indicative of *P. wickerhamii* (Figure 21.7). The sporangia are usually smaller, and the sporangiospores fewer and larger than those of *Coccidioides immitis* and *Rhinosporidium seeberi*. Juxtaposed algal cells of different sizes may falsely resemble budding, but a distinctive isthmus between two cells is never seen. Reports of internal starch granules and leucoplast lamellae are erroneous; *Prototheca* spp. have neither, although they do have refractive glycogen and lipid granules (unpublished observations).

Immunofluorescent reagents have led to a means for the histopathological identification of protothecosis and the differentiation of *P. wickerhamii* from *P. zopfii* (Sudman and Kaplan 1973). Many case reports would not have been published without the taxonomic corroboration obtainable with these serological reagents, but these reagents have limited availability and staining should be interpreted with caution. The best current method of speciation is still from the morphological and physiological traits of cultured isolates (see Table 21.2).

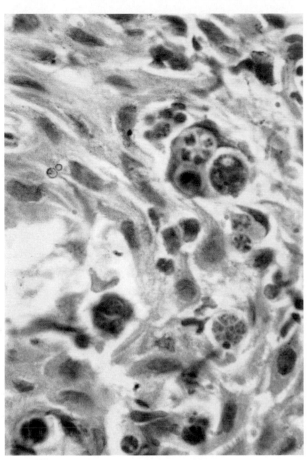

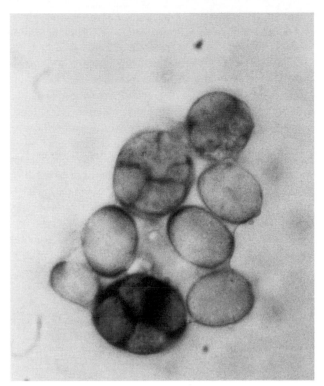

Figure 21.8 P. zopfii *is rare in secreted bovine milk without enrichment culture (see Table 21.1), but it is large and distinct when stained with methylene blue. Magnification × 700*

Figure 21.7 P. wickerhamii *in dermis with little tissue inflammation demonstrated by H&E staining. Several sporangia contain sporangiospores, the best histopathological identifying characteristic. Magnification × 700*

PROTOTHECA ZOPFII

P. zopfii causes chronic progressive pyogranulomatous lesions in the bovine mammary gland and associated lymph nodes. The microorganisms cluster in alveolar lumina and epithelia (Pore et al. 1987). They may also cause acute destruction of blood vessels, acini, and connective tissues in the udder, as well as mildly inflammatory, but highly destructive and necrotic, lesions of other organs (Taniyama et al. 1994; Jensen et al. 1998). The algal cells can be cultured from secreted milk (Figure 21.8). After experimental inoculation via the teat canal, living and degenerated *P. zopfii* cells were found in enlarged macrophages in granulomas, but not in associated neutrophils or epithelial cells (Cheville et al. 1984). Mastitis is almost solely caused by *P. zopfii*, whereas *P. zopfii* or *P. wickerhamii* causes dog infections equally frequently. Only three human infections with *P. zopfii* are known (Davies et al. 1964; Naryshkin et al. 1987; Kwok and Schwartz 1996); another was associated with a case of onycholysis (Magerman et al. 1991).

P. zopfii is unpigmented and the shrunken basophilic cytoplasm (seen by H&E staining) is within a large, unstained, spheroidal cell wall. However, the specific fungal stains, based on GMS or PAS, do stain the cell wall, even though the wall is chitin free. Specific immunohistochemical staining has helped to elucidate the intracellular sites of *P. zopfii* in macrophages and neutrophils from mastitic infections (Jensen et al. 1998; Corbellini et al. 2001a). Although cells resembling fungi may be seen, budding is absent and sporangial reproduction is present instead. All stages in the formation of two to six endospores are frequently seen within a sporangium after thin sectioning. In tissue, the sporangiospores of *P. zopfii*, but not the sporangia, are fewer and larger than those of *C. immitis* and *R. seeberi*. Although *P. zopfii* sporangia and sporangiospores are also larger than those of *P. wickerhamii*, in vitro, no comparable size data are available for the two species in histological material, so differentiation on the basis of size should be done with caution, if at all.

CHLORELLA SPP.

Histopathological identification to species or even genus is not possible among the small unicellular green algae. Green lymph nodes in sheep and cattle that histopathologically contain microorganisms resembling *Prototheca* or *Chlorella* spp. have been interpreted as chlorella infections based on the green appearance of these nodes (Chandler et al. 1980; Rogers et al. 1980; Jones et al. 1983; Connole 1990; Kaplan et al. 1993; Seniw et al. 1997). Reports of green tissues from animals are, however, anecdotal and hard to interpret, and green lymph nodes have been found in a case of bovine protothecosis (Migaki et al. 1969). Chlorophyll is soluble in alcohol and absent from most histopathological preperations. In at least one instance, however, the infectious microorganisms contained electron microscopic evidence of chloroplasts (Kaplan et al. 1993) – definitive evidence of a green alga, although not necessarily *Chlorella* spp., but conclusive evidence that the microorganisms were not *Prototheca* spp. Finally, not all *Chlorella* spp. contain starch granules as is sometimes erroneously reported; *C. prototheroides* is such an example. *C. prototheroides* is an alga that stores glycogen, as does *Prototheca* spp. (unpublished observations). Many other *Chlorella* spp. species do store starch, and such PAS-stainable granules in the cytoplasm may after all prove to be an identification characteristic (Chandler et al. 1980).

REFERENCES

Ahbel, D.E., Alexander, A.H., et al. 1980. Protothecal olecranon bursitis. *J Bone Joint Surg [Am]*, **62A**, 835–6.

Anderson, E.H. 1945. Nature of the growth factor for the colorless alga *Prototheca zopfii. J Gen Physiol*, **28**, 287–96.

Anderson, K.L. and Walker, R.L. 1988. Sources of *Prototheca* spp in a dairy herd environment. *J Am Vet Med Assoc*, **193**, 553–6.

Arnold, P. and Ahearn, D.G. 1972. The systematics of the genus *Prototheca* with a description of a new species *P. filamenta. Mycologia*, **64**, 265–75.

Ashford, B.K., Ciferri, R. and Dalmau, L.M. 1930. A new species of *Prototheca* and a variety of the same isolated from the human intestine. *Archiv für Protistenkunde*, **70**, 619–38.

Atkinson, A.W., Gunning, B.E.S. and John, P.C.L. 1972. Sporopollenin in the cell wall of *Chlorella* and other algae: ultrastructure, chemistry, and incorporation of ^{14}C-acetate, studied in synchronous cultures. *Planta (Berlin)*, **107**, 1–32.

Barker, H.A. 1935. The metabolism of the colorless alga, *Prototheca zopfii* Krüger. *J Cell Comp Physiol*, **7**, 73–93.

Blaschke-Hellmessen, R. 1996. Fluconazole and itraconazole susceptibility testing with clinical yeast isolates and algae of the genus *Prototheca* by means of the Etest. *Mycoses*, **39**, 2, 39–43.

Blaschke-Hellmessen, R., Schuster, H. and Bergman, V. 1985. Differenzierung von Varianten bei *Prototheca zopfii* Krüger 1894. *Arch Exp Vet Med Leipzig*, **39**, 387–97.

Blaschke-Hellmessen, R., Teichmann, G., et al. 1987. Orientierende Untersuchungen zum Nachweis von Antikörpern gegen *Prototheca zopfii* bei Rindern. *Monatsh Vetmed*, **42**, 48–50.

Bodenhoff, J. and Madsen, P.S. 1978. Bovine protothecosis: a brief report of ten cases. *Acta Pathol Microbiol Scand Sect B*, **86**, 51–2.

Boehm, D.F. 1984. Hexadecane utilization by *Prototheca zopfii*. Dissertation, West Virginia University, Morgantown, WV 26506, USA.

Boyd, A.S., Langley, M. and King, L.E. 1995. Cutaneous manifestations of *Prototheca* infections. *J Am Acad Dermatol*, **32**, 758–64.

Buendia, A., Patino, E., et al. 1998. Endocarditis por alga del genero *Prototheca* sp. Saprofito del aqua y de la savia de los arboles? *Arch Inst Cardiol Mex*, **68**, 333–6.

Buyukmihci, N., Rubin, L.F. and DePaoli, A. 1975. Protothecosis with ocular involvement in a dog. *J Am Vet Med Assoc*, **167**, 158–61.

Carey, W.P., Kaykova, Y., et al. 1997. Cutaneous protothecosis in a patient with AIDS and a severe functional neutrophil defect: successful therapy with amphotericin B. *Clin Infect Dis*, **25**, 1265–6.

Casal, M., Zerolo, J., et al. 1983. First human case of possible protothecosis in Spain. *Mycopathologia*, **83**, 19–20.

Casal, M., Linares, M.J. and Morales, M.M. 1985. Enzymatic profile of *Prototheca* species. *Mycopathologia*, **92**, 81–2.

Chan, J.C., Jeffers, L.J., et al. 1990. Visceral protothecosis mimicking sclerosing cholangitis in an immunocompetent host: successful antifungal therapy. *Rev Infect Dis*, **12**, 802–7.

Chandler, F.W., Kaplan, W. and Ajello, L. 1980. *Color atlas and text of the histopathology of mycotic diseases*. Chicago IL: Year Book Medical Publishers, 96–100, 263–70.

Chao, S., Hsu, M.M. and Lee, J.Y. 2002. Cutaneous protothecosis: report of five cases. *Br J Dermatol*, **146**, 688–93.

Chengappa, M.M., Maddux, R.L., et al. 1984. Isolation and identification of yeasts and yeastlike organisms from clinical veterinary sources. *J Clin Microbiol*, **19**, 427–8.

Cheville, N.F., McDonald, J. and Richard, J. 1984. Ultrastructure of *Prototheca zopfii* in bovine granulomatous mastitis. *Vet Pathol*, **21**, 341–8.

Cho, B.K., Ham, S.H., et al. 2002. Cutaneous protothecosis. *Int J Dermatol*, **41**, 304–6.

Clark, C.S., Linnemann, C.C., et al. 1985. Serologic survey of rotavirus, Norwalk agent and *Prototheca wickerhamii* in wastewater workers. *Am J Public Health*, **75**, 83–5.

Cochran, R.K., Pierson, C.L., et al. 1986. Protothecal olecranon bursitis: treatment with intrabursal amphotericin B. *Rev Infect Dis*, **8**, 952–4.

Coloe, P.J. 1982. Protothecosis in a cat. *J Am Vet Med Assoc*, **180**, 78–9.

Connole, M.D. 1990. Review of animal mycoses in Australia. *Mycopathologia*, **111**, 133–64.

Conte, M.V. and Pore, R.S. 1973. Taxonomic implications of *Prototheca* and *Chlorella* cell wall polysaccharide characterization. *Arch Mikrobiol*, **92**, 227–33.

Cook, J.R. Jr, Tyler, D.E., et al. 1984. Disseminated protothecosis causing acute blindness and deafness in a dog. *J Am Vet Med Assoc*, **184**, 1266–72.

Cooke, W.B. 1968. Studies in the genus *Prototheca*. I. Literature review. *J Elisha Mitchell Scientific Society*, **84**, 213–16.

Corbellini, L.G., Driemeier, D. and Cruz, C.E. 2001a. Immunohistochemistry combined with periodic acid-Schiff for bovine mammary gland with prototothecal mastitis. *Biotech Histochem*, **76**, 85–8.

Corbellini, L.G., Driemeier, D., et al. 2001b. Bovine mastitis due to *Prototheca zopfii*: clinical epidemiological and pathological aspects in a Brazilian dairy herd. *Trop Anim Health Prod*, **33**, 463–70.

Costa, E.O., Ribeiro, A.R., et al. 1998. Infectious bovine mastitis caused by environmental organisms. *Zentralbl Veterinärmed [B]*, **45**, 65–71.

Cox, E.G., Wilson, J.D. and Brown, P. 1974. Protothecosis: a case of disseminated algal infection. *Lancet*, **2**, 379–82.

Cunliffe, C.C. and DiPersio, J.R. 2000. Localized cutaneous protothecosis in a patient with acquired immune deficiency syndrome. *Clin Microbiol News*, **22**, 133–5.

Dagher, F.J., Smith, A.G., et al. 1978. Skin protothecosis in a patient with renal allograft. *South Med J*, **71**, 222–4.

Davies, R.R. and Wilkinson, J.L. 1967. Human protothecosis: supplementary studies. *Ann Trop Med Parasitol*, **61**, 112–14.

Davies, R.R., Spencer, H. and Wakelin, P.O. 1964. A case of human protothecosis. *Trans R Soc Trop Med Hyg*, **58**, 448–51.

de Camargo, Z.P. and Fischman, O. 1979. *Prototheca stagnora*, an encapsulated organism. *Sabouraudia*, **17**, 197–200.

de Camargo, Z.P., Fischman, O. and Silva, M.R.R. 1980. Experimental protothecosis in laboratory animals. *Sabouraudia*, **18**, 237–40.

Dillberger, J.E., Homer, B., et al. 1988. Protothecosis in two cats. *J Am Vet Med Assoc*, **192**, 1557–9.

Dion, W.M. 1979. Bovine mastitis due to *Prototheca zopfii. Can Vet J*, **20**, 221–2.

Follador, I., Bittencourt, A., et al. 2001. Cutaneous protothecosis: report of a second Brazilian case. *Rev Inst Med Trop Sao Paulo*, **43**, 287–90.

Font, R.L. and Hook, S.R. 1984. Metastatic prototothecal retinitis in a dog. Electron microscopic observations. *Vet Pathol*, **21**, 61–6.

Frank, N., Ferguson, L.C., et al. 1969. Prototheca, a cause of bovine mastitis. *Am J Vet Res*, **30**, 1785–94.

Gaunt, S.D., McGrath, R.K. and Cox, H.U. 1984. Disseminated protothecosis in a dog. *J Am Vet Med Assoc*, **185**, 906–7.

Gibb, A.P., Aggarwal, R. and Swanson, C.P. 1991. Successful treatment of *Prototheca* peritonitis complicating continuous ambulatory peritoneal dialysis. *J Infect*, **22**, 183–5.

Ginel, P.J., Perez, J., et al. 1997. Cutaneous protothecosis in a dog. *Vet Rec*, **140**, 651–3.

Heney, C., Greeff, M. and Davis, V. 1991. Hickmann catheter-related prothecal algaemia in an immunocompromised child. *J Infect Dis*, **163**, 930–1.

Hollingsworth, S.R. 2000. Canine protothecosis. *Vet Clin North Am Small Anim Pract*, **30**, 1091–101.

Horiuchi, Y. and Masuzawa, M. 1995. Electron microscopic observations of epithelioid cell granulomas experimentally induced by prototheca in the skin of mice. *J Dermatol*, **22**, 643–9.

Huss, V.A.R. and Sogin, M.L. 1990. Phylogenetic position of some *Chlorella* species within the Chlorococcales based upon complete small-subunit ribosomal RNA sequences. *J Mol Evol*, **31**, 432–42.

Huss, V.A.R., Wein, K.H. and Kessler, E. 1988. Deoxyribonucleic acid reassociation in the taxonomy of the genus *Chlorella*. *Arch Microbiol*, **150**, 509–11.

Iacoviello, V.R., DeGirolami, P.C., et al. 1992. Protothecosis complicating prolonged endotracheal intubation: case report and literature review. *Clin Infect Dis*, **15**, 959–67.

Janosi, S., Ratz, F., et al. 2001a. Review of the microbiological, pathological, and clinical aspects of bovine mastitis caused by the alga *Prototheca zopfii*. *Vet Q*, **23**, 58–61.

Janosi, S., Szigeti, G., et al. 2001b. *Prototheca zopfii* mastitis in dairy herds under continental climatic conditions. *Vet Q*, **23**, 80–3.

Jensen, H.E. and Aalbaek, B. 1994. Pathogenicity of yeasts and algae isolated from bovine mastitis secretions. *Mycoses*, **37**, 101–7.

Jensen, H.E., Aalbaek, B., et al. 1998. Bovine mammary protothecosis due to *Prototheca zopfii*. *Med Mycol*, **36**, 89–95.

Jones, J.W., McFadden, H.W., et al. 1983. Green algal infection in a human. *Am J Clin Pathol*, **80**, 102–7.

Joshi, K.R., Gavin, J.B. and Wheeler, E.E. 1975. The ultrastructure of *Prototheca wickerhamii*. *Mycopathologia*, **56**, 9–13.

Kaminski, Z.C., Kapila, R., et al. 1992. Meningitis due to *Prototheca wickerhamii* meningitis in a patient with AIDS. *Clin Infect Dis*, **15**, 704–8.

Kaplan, W. 1978. Protothecosis and infections caused by morphologically similar green algae. *Proceedings of the Fourth International Conference on the Mycoses: Black and White Yeasts*. Scientific Publication no. 356. Washington, DC: Pan American Health Organization, **356**, 218–232.

Kaplan, W., Chandler, F.W., et al. 1993. Disseminated unicellular green algal infection in two sheep in India. *Am J Trop Med Hyg*, **32**, 405–11.

Kim, S.T., Suh, K.S., et al. 1996. Successful treatment with fluconazole of protothecosis developing at the site of an intralesional corticosteroid injection. *Br J Dermatol*, **135**, 803–6.

Klintworth, G.K., Fetter, B.F. and Nielsen, H.S. Jr 1968. Protothecosis, an algal infection: report of a case in man. *J Med Microbiol*, **1**, 211–16.

Klipstein, F.A. and Schenk, E.A. 1975. Prototheca and sprue. *Gastroenterology*, **69**, 1372–3.

Krcmery, V. Jr 2000. Systemic chlorellosis, an emerging infection in humans caused by algae. *Int J Antimicrob Agents*, **15**, 235–7.

Krüger, W. 1894a. Kurz Charakteristik einiger niederer Organismen in Saftflusse der Laubbäume. *Hedwigia*, **33**, 241–66.

Krüger, W. 1894b. Beiträge zur Kenntniss der Organismen des Saftflusses (sog. Schleimflusses) der Laubbäume. *Zopf's Beitr Physiol Morphol Organismen*, **4**, 69–116.

Kuo, T., Hsueh, S., et al. 1987. Cutaneous protothecosis. *Arch Pathol Lab Med*, **111**, 737–40.

Kwok, N. and Schwartz, S.N. 1996. Prototheca sepsis in a long transplant patient. *Clin Microbiol Newsl*, **18**, 183–4.

Laeng, R.H., Egger, C., et al. 1994. Protothecosis in a HIV-positive patient. *Am J Surg Pathol*, **18**, 1261–4.

Lerche, M. 1952. Eine durch Algen (*Prototheca*) herorgerufene Mastitis der Kuh, Berlin. *Münch tierärztl Wochenschr*, **65**, 64–9.

Linares, M.J., Munozfj, J.F., et al. 1998. Study of the susceptibility of yeast isolates of clinical interest to five antifungal agents using the E test. *Rev Esp Quimioter*, **11**, 64–9.

Magerman, K., Gordts, B., et al. 1991. Isolation of *Prototheca zopfii* from a finger. Case report and review of the literature. *Acta Clin Belg*, **46**, 233–6.

Mangla, A.T. and Nes, W.D. 2000. Sterol C-methyl transferase from *Prototheca wickerhamii* mechanism, sterol specificity and inhibition. *Bioorg Med Chem*, **8**, 925–36.

Manners, D.J., Pennie, I.R. and Ryley, J.F. 1973. The molecular structures of a glucan and a galactan synthesised by *Prototheca zopfii*. *Carbohydr Res*, **29**, 63–77.

Marr, K.A., Hirschmann, J.V., et al. 1998. Photo quiz. Protothecosis. *Clin Infect Dis*, **26**, 575, 756–757.

Mars, P.W., Rabson, A.R., et al. 1971. Cutaneous protothecosis. *Br J Dermatol*, **85**, Suppl 7, 76–84.

Matsuda, T. and Matsumoto, T. 1992. Protothecosis: a report of two cases in Japan and a review of the literature. *Eur J Epidemiol*, **8**, 397–406.

Mayberry, D. 1984. Colorless alga can pollute water, cause mastitis. *Agric Res*, **March**, 4–5.

Mayhall, C.G., Miller, C.W., et al. 1976. Cutaneous protothecosis. *Arch Dermatol*, **112**, 1749–52.

McAnally, T. and Parry, E.L. 1985. Cutaneous protothecosis presenting as recurrent chromomycosis. *Arch Dermatol*, **121**, 1066–9.

McDonald, J.S., Richard, J.L. and Cheville, N.F. 1984. Natural and experimental bovine intramammary infection with *Prototheca zopfii*. *Am J Vet Res*, **45**, 592–5.

Melville, P.A., Watanabe, E.T., et al. 1999. Evaluation of the susceptibility of *Prototheca zopfii* to milk pasteurization. *Mycopathologia*, **146**, 79–82.

Mezger, E., Eisses, J.F. and Smith, M.J. 1981. Protothecal cellulitis in a renal transplant patient [abstract]. *Lab Invest*, **44**, 81A.

Migaki, G., Garner, F.M. and Imes, G.D. Jr 1969. Bovine protothecosis, a report of three cases. *Pathol Vet*, **6**, 444–53.

Migaki, G., Font, R.L., et al. 1982. Canine protothecosis: review of the literature and report of an additional case. *J Am Vet Med Assoc*, **181**, 794–7.

Mohabeer, A.J., Kaplan, P.J., et al. 1997. Algaemia due to *Prototheca wickerhamii* in a patient with myasthenia gravis. *J Clin Microbiol*, **35**, 3305–7.

Moore, F.M., Schmidt, G.M., et al. 1985. Unsuccessful treatment of disseminated protothecosis in a dog. *J Am Vet Med Assoc*, **186**, 705–8.

Moyer, R.A., Bush, D.C. and Dennehy, J.J. 1990. *Prototheca wickerhamii* tenosynovitis. *J Rheumatol*, **17**, 701–4.

Naryshkin, S., Frank, I. and Nachamkin, I. 1987. *Prototheca zopfii* isolated from a patient with olecranon bursitis. *Diagn Microbiol Infect Dis*, **6**, 171–4.

Nedelcu, A.M. 2001. Complex patterns of plastid 16S rRNA gene evolution in nonphotosynthetic green algae. *J Mol Evol*, **53**, 670–9.

Nelson, A.M., Neafie, R.C. and Connor, D.H. 1987. Cutaneous protothecosis and chlorellosis, extraordinary 'aquatic-borne' algal infections. *Clin Dermatol*, **5**, 76–87.

Nosanchuk, J.S. and Greenberg, R.D. 1973. Protothecosis of the olecranon bursa caused by achloric algae. *J Clin Pathol*, **59**, 567–73.

O'Connor, J.P., Nimmo, G.R., et al. 1986. Algal peritonitis complicating continuous ambulatory peritoneal dialysis. *Am J Kidney Dis*, **VIII**, 122–3.

Okuyama, Y., Hamaguchi, T., et al. 2001. A human case of protothecosis successfully treated with itraconazole. *Nippon Ishinkin Gakkai Zasshi*, **42**, 143–7.

Padhye, A.A., Baker, J.G. and D'Amato, R.F. 1979. Rapid identification of *Prototheca* species by the API 20C system. *J Clin Microbiol*, **10**, 579–82.

Patni, N.J. and Aaronson, S. 1974. The nutrition, resistance to antibiotics and ultrastructure of *Prototheca wickerhamii*. *J Gen Microbiol*, **83**, 179–82.

Pegram, P.S. Jr, Kerns, F.T., et al. 1983. Successful ketoconazole treatment of protothecosis with ketoconazole-associated hepatotoxicity. *Arch Intern Med*, **143**, 1802–5.

Perez, J. and Ginel, P.J. 1997. Canine cutaneous protothecosis: an immunohistochemical analysis of the inflammatory cellular infiltrate. *J Comp Pathol*, **117**, 83–9.

Phair, J.P., Williams, J.E., et al. 1981. Phagocytosis and algicidal activity of human polymorphonuclear neutrophils against *Prototheca wickerhamii*. *J Infect Dis*, **144**, 72–6.

Pier, A.C., Cabanes, F.J., et al. 2000. Prominent animal mycoses from various regions of the world. *Med Mycol*, **38**, Suppl 1, 47–58.

Piyophirapong, S., Linpiyawan, R., et al. 2002. Cutaneous protothecosis in an AIDS patient. *Br J Dermatol*, **146**, 713–15.

Polk, P. and Sanders, D.Y. 1997. Cutaneous protothecosis in association with the acquired immunodeficiency syndrome. *South Med J*, **90**, 831–2.

Pore, R.S. 1971. Taxonomic status and experimental pathology of *Prototheca* species. *Comptes Rendus des Communications V Congress de la Société Intérnationale de Mycologie Humane et Animale, Paris*. Paris: Institut Pasteur, 63–64.

Pore, R.S. 1972. Nutritional basis for relating *Prototheca* and *Chlorella*. *Can J Microbiol*, **18**, 1175–7.

Pore, R.S. 1973. Selective medium for the isolation of *Prototheca*. *Appl Microbiol*, **26**, 648–9.

Pore, R.S. 1983. *Prototheca* ecology. *Mycopathologia*, **81**, 49–62.

Pore, R.S. 1985. *Prototheca* taxonomy. *Mycopathologia*, **90**, 129–39.

Pore, R.S. 1986. The association of *Prototheca* spp. with slime flux in *Ulmus americana* and other trees. *Mycopathologia*, **94**, 67–73.

Pore, R.S. 1998. *Prototheca* Krüger. In: Kurtzman, C.P. and Fell, J.W. (eds), *The yeasts. A taxonomic study*, 5th edition. Amsterdam: Elsevier, 883–7.

Pore, R.S. and Boehm, D.F. 1986. *Prototheca* (achloric alga) in wastewater. *Water Air Soil Pollut*, **27**, 355–62.

Pore, R.S. and Shahan, T.A. 1988. *Prototheca zopfii*: natural, transient, occurrence in pigs and rats. *Mycopathologia*, **101**, 85–8.

Pore, R.S., Shahan, T.A., et al. 1987. Occurrence of *Prototheca zopfii*, a mastitis pathogen, in milk. *Vet Microbiol*, **15**, 315–23.

Pringsheim, E.G. 1963. *Farblose Algen*. Stuttgart: Gustav Fischer Verlag, 273.

Puel, F., Largeau, C. and Giraud, G. 1987. Occurrence of a resistant biopolymer in the outer walls of the parasitic alga *Prototheca wickerhamii* (Chlorococcales): ultrastructural and chemical studies. *J Phycol*, **23**, 649–56.

Rakich, P.M. and Latimer, K.S. 1984. Altered immune function in a dog with disseminated protothecosis. *J Am Vet Med Assoc*, **185**, 681–3.

Rallis, T.S., Tontis, D., et al. 2002. Prototheceal colitis in a German shepherd dog. *Aust Vet J*, **80**, 406–8.

Raz, R., Rottem, M., et al. 1998. Intestinal protothecosis in a patient with chronic mucocutaneous candidiasis. *Clin Infect Dis*, **27**, 399–400.

Roesler, U. and Hensel, A. 2003. Longitudinal analysis of *Prototheca zopfii*-specific immune responses: correlation with disease progression and carriage in dairy cows. *J Clin Microbiol*, **41**, 1181–6.

Rogers, R.J., Connole, M.D., et al. 1980. Lymphadenitis of cattle due to infection with green algae. *J Comp Pathol*, **90**, 1–9.

Roy, A., Manjula, B.N. and Glaudemans, C.P.J. 1981. The interaction of two polysaccharides containing β1,6-linked galactopyranosyl residues with two monoclonal antigalactan immunoglobulin Fab' fragments. *Mol Immunol*, **18**, 79–84.

Running, J.A., Severson, D.K. and Schneider, K.J. 2002. Extracellular production of L-ascorbic acid by *Chlorella protothecoides*, *Prototheca* species, and mutants of *P. moriformis* during aerobic culturing at low pH. *J Ind Microbiol Biotechnol*, **29**, 93–8.

Sands, M., Poppel, D. and Brown, R. 1991. Peritonitis due to *Prototheca wickerhamii* in a patient undergoing chronic ambulatory peritoneal dialysis. *Rev Infect Dis*, **13**, 376–8.

Schiefer, B. and Gedek, B. 1968. Zum Verhalten von *Prototheca*-Species im Gewebe von Säugetieren. *Berl Münch Tierärztl Wochenschr*, **24**, 485–90.

Schuster, H. and Blaschke-Hellmessen, R. 1983. Zur Epizootiologie der Protothekenmastitis des Rindes-Anzüchtung von Algen der Gattung *Prototheca* aus der Umgebung landwirtschaftlicher Nutztiere. *Monatsh Vetmed*, **38**, 24–9.

Segal, E., Padhye, A.A. and Ajello, L. 1976. Susceptibility of *Prototheca* species to antifungal agents. *Antimicrob Agents Chemother*, **10**, 75–9.

Seniw, C.M., Chandler, F.W. and Connor, D.H. 1997. Chlorellosis. In: Connor, D.H., Chandler, F.W. et al. (eds), *Pathology of infectious diseases*, vol. II. Stamford, CT: Appleton & Lange, 965–9.

Shahan, T.A. and Pore, R.S. 1991. In vitro susceptibility of *Prototheca* spp. to gentamicin. *Antimicrob Agents Chemother*, **35**, 2434–5.

Shihira, O. and Krauss, R.W. 1965. *Chlorella*. Baltimore MD: University of Maryland, Park City Press, 97 pp.

Sirikulchayanonta, V., Visuthikosol, V., et al. 1989. Protothecosis following hand injury a case report. *J Hand Surg*, **14B**, 88–90.

Spalton, D.E. 1985. Bovine mastitis caused by *Prototheca zopfii*; a case study. *Vet Rec*, **116**, 347–9.

Sud, I.J. and Feingold, D.S. 1979. Lipid composition and sensitivity of *Prototheca wickerhamii* to membrane-active antimicrobiol agents. *Antimicrob Agents Chemother*, **16**, 486–90.

Sudman, M.S. and Kaplan, W. 1973. Identification of the *Prototheca* species by immunofluorescence. *Appl Microbiol*, **25**, 981–90.

Sudman, M.S., Majka, J.A. and Kaplan, W. 1973. Primary mucocutaneous protothecosis in a dog. *J Am Vet Med Assoc*, **163**, 1372–4.

Takaki, K., Umeno, O.M., et al. 1996. Chronic prototheca meningitis. *Scand J Infect Dis*, **28**, 321–3.

Tang, W.Y.M., Lo, K.K., et al. 1995. Cutaneous protothecosis: report of a case in Hong Kong. *Br J Dermatol*, **133**, 479–82.

Taniyama, H., Okamoto, F., et al. 1994. Disseminated protothecosis caused by *Prototheca zopfii* in a cow. *Vet Pathol*, **31**, 123–5.

Tejada, D. and Parker, C.M. 1994. Cutaneous erythematous nodular lesion in a crab fisherman. *Arch Dermatol*, **130**, 247–8.

Tenhagen, B.A., Kalbe, P., et al. 1999. Tierindividuelle Risikofaktoren für die Protothekenmastitis des Rindes. *DTW Dtsch Tierarztl Wochenschr*, **106**, 376–80.

Thianprasit, M., Youngchaiyud, U. and Suthipinittharm, P. 1983. Protothecosis: a report of two cases. *Mykosen*, **26**, 455–61.

Thomas, J.B. and Preston, N. 1990. Generalized protothecosis in a collie dog. *Aust Vet J*, **67**, 25–7.

Torres, H.A., Bodey, G.P., et al. 2003. Protothecosis in patients with cancer: case series and literature review. *Clin Microbiol Infect*, **9**, 786–94.

Tubaki, K. and Soneda, M. 1959. Cultural and taxonomic studies on *Prototheca*, Nagaoa. *Mycol J Nagao Inst Tokyo*, **6**, 25–34.

Tyler, D.E., Lorenz, M.D., et al. 1980. Disseminated protothecosis with central nervous system involvement in a dog. *J Am Vet Med Assoc*, **176**, 987–93.

Tyring, S.K., Lee, P.C., et al. 1989. Papular protothecosis of the chest. *Arch Dermatol*, **125**, 1249–52.

Ueno, R., Urano, N., et al. 2002. Isolation, characterization, and fermentative pattern of a novel thermotolerant *Prototheca zopfii* var. *hydrocarbonea* strain producing ethanol an CO_2 from glucose at 40 degrees C. *Arch Microbiol*, **177**, 244–50.

Ueno, R., Urano, N. and Suzuki, M. 2003. Phylogeny of the non-photosynthetic green micro-algal genus *Prototheca* (Trebouxiophyceae, Chlorophyta) and related taxa inferred from SSU and LSU ribosomal DNA partial sequence data. *FEMS Microbiol Lett*, **223**, 275–80.

Venezio, F.R., Lavoo, E., et al. 1982. Progressive cutaneous protothecosis. *Am J Clin Pathol*, **112**, 829–32.

Vernon, S.E. and Goldman, L.S. 1983. Protothecosis in the Southeastern United States. *South Med J*, **76**, 949–50.

Walker, J.D. and Pore, R.S. 1978. Growth of *Prototheca* isolates on n-hexadecane and mixed-hydrocarbon substrate. *Appl Environ Microbiol*, **35**, 694–7.

Walsh, S.V., Johnson, R.A. and Tahan, S.R. 1998. Protothecosis: an unusual cause of chronic subcutaneous and soft tissue infection. *Am J Dermatopathol*, **20**, 379–82.

Wirth, F.A., Passalacqua, J. and Kao, G. 1999. Disseminated cutaneous protothecosis in an immunocompromised host: a case report and literature review. *Cutis*, **63**, 185–8.

Wolfe, I.D., Sacks, H.G., et al. 1976. Cutaneous protothecosis in a patient receiving immunosuppressive therapy. *Arch Dermatol*, **112**, 829–32.

Woolrich, A., Koestenblatt, E., et al. 1994. Cutaneous protothecosis and AIDS. *J Am Acad Dermatol*, **31**, 920–4.

Yip, S.Y., Huang, C. and Clark, W.H. 1976. Protothecosis, an infection by algae: report of a case from Hong Kong. *J Dermatol*, **3**, 309–15.

Zakia, A.M., Osheik, A.A. and Halima, M.O. 1989. Ovine chlorellosis in the Sudan. *Vet Rec*, **125**, 625–6.

Pythiosis

LEONEL MENDOZA

Members of the oomycetous genus *Pythium* are ecologically and physiologically unique. They occur in soil and aquatic habitats worldwide. *Pythium* species are important plant pathogens causing seed decay, pre-emergent and post-emergent damping off, root rot of seedlings, and rot of stored foodstuffs. In addition, some species of *Pythium* have been reported to cause disease in fish. *Pythium insidiosum*, however, is the only member of the genus that has been recognized as a mammalian pathogen. Chandler et al. (1980) coined the term pythiosis to include all clinical and pathological manifestations caused by *P. insidiosum* in mammals. In horses, the disease has been long known as bursatee, leeches, granular dermatitis, hyphomycosis destruens equi, phycomycosis, espundia, summer sores, and swamp cancer. Pythiosis in mammals is characterized by the development of cutaneous, subcutaneous, blood vessel, and intestinal lesions and, less frequently, by the involvement of bones and lungs. If not treated, the disease progresses rapidly, becoming life-threatening.

HISTORY

The first reported cases of equine pythiosis were described in the middle of the last century when veterinarians studied cutaneous granulomas among equines in India. The etiology of the disease was not established.

The first well-documented cases, in which the 'fungal-like' nature of the infection was suggested, were those published by Smith (1884), Fish (1895–96), and Drouin (1896). In Indonesia, de Haan and Hoogkamer (1901), working with several equines with cutaneous granulomas, isolated the etiologic agent of pythiosis for the first time. Their isolate, however, could not be identified because it did not sporulate on their media. These investigators named the disease hyphomycosis destruens. This name was modified later by de Haan (1902) to hyphomycosis destruens equi. In 1924 another Dutch scientist, working with horses in Indonesia, completed the most comprehensive papers on equine pythiosis ever published (Witkamp 1924, 1925). He also isolated a nonsporulating organism from equine cutaneous granulomas and covered several aspects of the disease, including its clinical signs, pathology, microbiology, animal inoculation, diagnosis, treatment, and immunology. Unfortunately, this classic work remained largely unnoticed, in part because of its Dutch language, but also because of the contemporaneous finding that cutaneous granulomas in horses may also be caused by nematode species of the genus *Habronema* (Ransom 1911). Thus, the terms used to describe pythiosis were also indiscriminately used for equine cutaneous habronemiasis.

The early findings on the 'fungal' nature of cutaneous granulomas in equines were obscured by the new hypothesis of equine habronemiasis, and the presumed mycotic etiology was forgotten. Thirty-seven years later, Bridges and Emmons (1961) isolated a sterile, filamentous microorganism from equine cases in Texas and Florida, similar to those that had been studied by de Haan and Hoogkamer and Witkamp earlier in the century. They believed that they had isolated 'a species of *Mortierella*' – a zygomycete. The filamentous isolate was named *Hyphomyces destruens*, based on the disease name previously coined by de Haan and Hoogkamer (1901, 1903). It was not clear, however, if the authors intended to introduce a new binomial for the etiologic agent of cutaneous granulomas in horses. In 1974, Austwick and Copland reported that an organism, isolated from horses in Papua New Guinea afflicted with swamp cancer, formed biflagellate zoospores after transfer to a petri dish of sterile water to which a sterilized decayed piece of rotten maize silage had been added. They concluded that the organism should be classified in the oomycete genus *Pythium* but they failed to provide a scientific name for their isolate.

Later, Ichitani and Amemiya (1980) isolated a filamentous microorganism from a Japanese horse with granular dermatitis. Based on the production of smooth oogonia and aplerotic oospores, they identified their isolate as *Pythium gracile*. In 1987, de Cock et al., working in Costa Rica with numerous isolates of a *Pythium* sp., recovered from horses that had espundia (a regional name used for pythiosis), concluded that these isolates belonged to an undescribed species of *Pythium* and the binomial *P. insidiosum* was validly introduced. They also reported that other isolates from cows, dogs, horses, and humans, including the one described as *P. gracile*, also belonged to the single species *P. insidiosum*. Almost concurrently an isolate, from horses in Australia, was also described as a new species and named *Pythium destruens* by Shipton in 1987. Later this isolate was studied by Mendoza and Marin (1989) and it was found to be indistinguishable from the strains isolated from other pythiosis cases; it thus became a synonym for *P. insidiosum*. Recently, Schurko et al. (2003a, b), using molecular tools, confirmed that *P. insidiosum* is the only etiologic agent in the genus causing pythiosis in mammals.

Since 1961 several new cases of pythiosis have been reported in equines (Habbinga 1967; Hutchins and Johnston 1972; Connole 1973; McMullan et al. 1977; Murray et al. 1978; Miller and Campbell 1982b; Mendoza and Alfaro 1986), cats (Bissonnette et al. 1991; Thomas and Lewis 1998), cattle (Miller et al. 1985; Santurio et al. 1998), dogs (Pavletic et al. 1983; Foil et al. 1984; Thomas and Lewis 1998; Graham et al. 2000), humans (Thianprasit 1986, 1990; Rinaldi et al. 1989; Sathapatayavongs et al. 1989; Chetchotisakd et al. 1992; Triscott et al. 1993; Virgile et al. 1993; Imwidthaya 1994b), and in captive bears and a camel in South Carolina and Florida zoos respectively (personal communications, L. Kaufman and A.A. Padhye, and J.F.X. Wellehan).

TAXONOMY AND MORPHOLOGICAL FEATURES OF *PYTHIUM INSIDIOSUM*

Pythium insidiosum is an organism classified in the kingdom Straminipila, class Oomycetes, order Pythiales, and family *Pythiaceae* (Dick 2001). Although Dick (2001) suggests the term Peronosporomycetes to rename the class Oomycetes in the kingdom Striminipila, for convenience, we will continue using the term Oomycetes throughout this chapter. This was motivated also by his position to treat the straminipilans as members of the kingdom Fungi. In recent years, however, strong phylogenetic evidence has been generated indicating that the straminipilans are not fungi, but protistal microbes very close related to algae and plants (Herr et al. 1999; Baldauf et al. 2000). Patterson in 1989 proposed the term Stramenopila. This name has been widely used. However, Dick (2001) recently modified it to Straminipila. He stated that the term Stramenopila is a bad derivation from the latin 'stramini'= a straw and 'pilus'= a hair. So, the straminipilans are organisms bearing tubular tripartite hairs. Therefore, the right spelling of this term should be Straminipila. Dick (2001) also indicated that the spelling 'straminopile,' appearing in the *Dictionary of fungi* (Hawksworth et al. 1995), was an unintentional misprint corrected in the 2001 edition of the same dictionary (Kirk et al. 2001).

In culture, *P. insidiosum* develops fungal-like sparsely septate hyphae; thus it erroneously has been referred to as a microorganism belonging to the kingdom Fungi. All of its phylogenetic features, however, correlate with the kingdom Straminipila (Baldauf et al. 2000; Martin 2000; Dick 2001). On this basis, *P. insidiosum* should only be referred to as a protist or a parafungal organism. As with other oomycetes, *P. insidiosum* grows relatively well on a variety of media. On cornmeal agar, colonies are colorless to white and submerged, with short aerial filaments, and a finely radiate pattern. The hyphae range between 4 and 10 μm in diameter with perpendicular lateral branches. Cross-septa are only occasionally observed in young hyphae, but they are abundant in old viable hyphae. Appressoria and hyphal swellings, measuring 12–28 μm in diameter, are common in laboratory cultures (Figure 22.1a, b).

Production of zoosporangia is only possible in water cultures incubated at 28–37°C. Zoosporangia increase in number when *P. insidiosum* is placed in contact with pieces of grass leaves in water containing a minimal quantity of different ions (including Ca^{2+}) (Mendoza and Prendas 1988). Early-stage sporangia cannot be differentiated from vegetative hyphae. At maturity, sporangial protoplasm flows into a discharge tube and

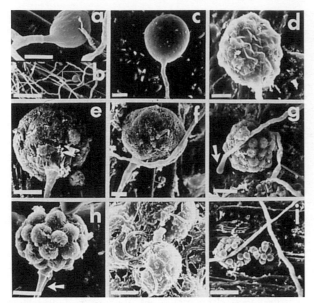

Figure 22.1 *Transmission electron microscopy of* Pythium insidiosum's *different stages of development. The formation of globose vesicles and hyphae on a leaf are the first steps in the colonization of a plant host (a, b). The electron photographs show sporangial formation leading to the release of secondary-type zoospores (c–j). Bar:* **(a, c)** *28 μm,* **(b)** *120 μm,* **(d)** *20 μm,* **(e)** *30 μm,* **(f)** *15 μm,* **(g)** *32 μm,* **(h)** *18 μm,* **(i)** *15 μm,* **(j)** *50 μm. (Reproduced, with permission, from Mendoza et al. 1993)*

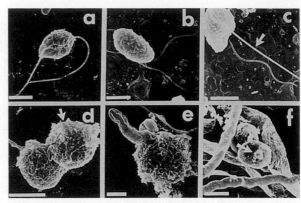

Figure 22.2 *Transmission electron microscopy of* Pythium insidiosum's *motile zoospores (a–c). Encysted zoospores are characterized by their globose morphology (d–f). Arrows indicate the place once occupied by the flagella. An amorphous substance used to bind the zoospores to the host's tissue is evident over the surface of encysted zoospores. Zoospores that encyst away from leaf or animal tissues do not secrete this amorphous material* **(f)**. *Bars:* **(a, b)** *9 μm,* **(c, d, f)** *10 μm,* **(e)** *5 μm. (Reproduced, with permission, from Mendoza et al. 1993)*

forms a vesicle. It is globose and hyaline, and measures 20–60 μm in diameter. Through progressive cleavage, biflagellate zoospores are formed inside the vesicle (Figure 22.1c). The internal developmental process within the undifferentiated vesicle and the release of zoospores takes about 35 minutes. The zoospores mechanically break the vesicle's wall and upon emergence swim for about 20 minutes and then encyst (Figure 22.1j). The zoospores are reniform and of the secondary biflagellate type. The two flagella arise from the deeper part of a lateral groove and seem to emerge from a common point. They are unequal in length. The anterior shorter flagellum is of the tinsel type and is covered with mastigonemas (small hair-like structures). The posterior flagellum is of the whiplash type and lacks mastigonemas (Figure 22.2a–c). After encystment, the zoospore's flagella are detached and it becomes spherical. When chemotactically stimulated by a suitable substrate (plant or animal tissue), the zoospores secrete anamorphous sticky material which covers their surface, and they become attached to the plant's or animal's surface. This substance, which may be a glycoprotein, has been implicated as a potential virulence factor in establishing infection (Mendoza et al. 1993) (Figure 22.2d–f).

In contrast to other species of the genus, only a few strains of *P. insidiosum* have been reported to produce oogonia. Thus, the mechanism implicated in the process of oogonial formation in *P. insidiosum* remains obscure.

When oogonia develop in vitro, they are intercalary, smooth, and subglobose. They have a rigid fertilization tube measuring 23–30 μm in diameter and may have one to three declinous antheridia per oogonium. The antheridia are attached over their entire length to the oogonium (de Cock et al. 1987; Shipton 1987). The oospores are aplerotic, or almost plerotic, and are pressed to one side of the oogonium by the rigid fertilization tube. They measure 20–25 μm in diameter (Figure 22.3k) (de Cock et al. 1987; Shipton 1987). The morphology of the mature oogonia is the basis for speciating members of the genus *Pythium*.

Schurko et al. (2003a, b) using phylogenetic tools, reported that in the analyses done on 23 isolates of *P. insidiosum* from the Americas, Asia, and Australia, all clustered together according to their geographical areas (Figure 22.4). This study showed that *P. insidiosum* is more closely related to each other than to any other *Pythium* species, and quite different from the genera *Phytophthora* and *Lagenidium*. Interestingly, the third cluster of *P. insidiosum* strains was made of isolates from Thailand and the USA. Schurko et al. (2003b) suggested that these unique isolates could belong to new species very close related to *P. insidiosum*. However, they did not treat these cryptic variants as such.

Recently, Grooters (2003) found that clinical cases initially thought to be pythiosis in dogs were not caused by *P. insidiosum*, but by a new species in the genus *Lagenidium*. She based her findings on morphological and molecular studies. In a recent phylogenetic study, however, Schurko et al (2003a, b) described at least three cryptic strains of *P. insidiosum*. These strains formed a sister clade with the other 20 *P. insidiosum* isolates studied by them. They stated that all *P. insidiosum*, and the other *Pythium* spp, were closer to the

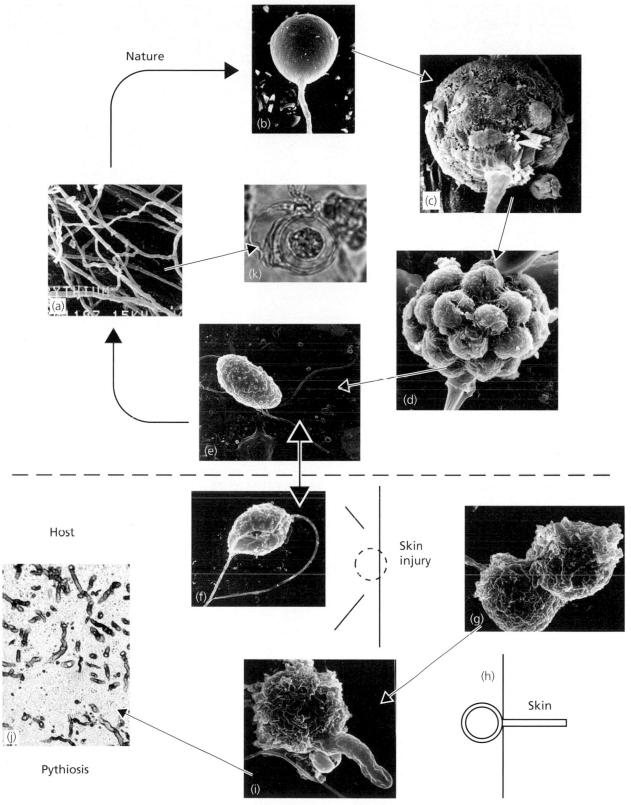

Figure 22.3 *(a–e, k, in upper panel) The life cycle of* Pythium insidiosum *in nature. Plant tissue is first colonized by* **(a)** *hyphae of* P. insidiosum, *and then* **(b–d)** *the differentiation of sporangium into mature stages leads to* **(e)** *zoospore release.* **(f)** *The zoospores swim to locate another plant or may be attracted by injured animal tissue. The encysted zoospores were* **(g,** *dots) attached to tissue by a sticky substance, germinate* **(h),** *invade the host* **(i),** *and cause pythiosis.* **(j)** *The formation of masses, called kunkers, occurs only in horses. The production of oospores occurs in nature and they may serve as* **(k)** *resistant spores.*

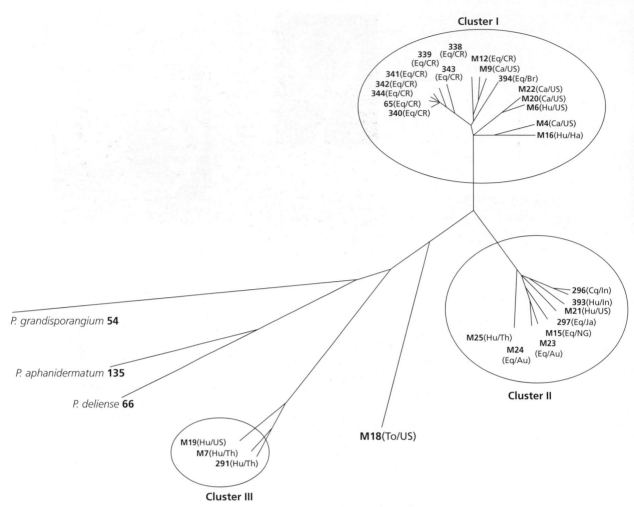

Figure 22.4 *The phylogenetic tree shows an unrooted unweighted pair-group method average (UPGMA) phenogram depicting the clusters formed by the 23 Pythium insidiosum isolates investigated by Schurko el at. (2003a). The numbers and letters in bold identified the strains used to build the tree and in parenthesis the origin and geographical location of the strains. Cluster I contains all the strains from the Americas, cluster II, strains from Asia and cluster III (including M18) has a mixture of Asian and American isolates. (Reproduced, with permission, from Schurko et al. 2003a)*

genus *Lagenidium* and far away from the genus *Phytophthora*. Thus, it could be that the strains studied by Grooters (2003) may be related to the third group described by Schurko et al (2003a, b) rather than being new members of the genus *Lagenidium*. However, this conjecture needs verification.

EPIDEMIOLOGY

Pythium spp. are ecologically versatile microorganisms. They occur in practically all soils and wet environments. They are among the most destructive phytopathogens, inflicting economic losses on a wide variety of crops. However, of all the described species (about 120) only one, *P. insidiosum*, has been implicated as an etiologic agent of mammalian disease. Since the last century, water has been related to cases of cutaneous granulomas in equines. It was noted that, after horses had grazed for a lengthy period in stagnant water, they frequently developed pythiosis. The term 'bursattee,' derived from

the Indian words 'bururs' or 'bursat,' meaning rain or rain sore, was used in India to describe this condition. Moreover, equine pythiosis has been referred to as swamp cancer in some areas of the world. Thus, a relationship between wet habitats and this disease had been suspected from early in its history. In Australia, equine pythiosis is usually observed in late summer and autumn after the formation of large bodies of stagnant water from the remnants of the previous winter's rains and high seasonal temperatures – conditions that favor the rapid growth of *P. insidiosum* (Miller and Campbell 1982b; Shipton 1985). Miller (1983) noted that, under laboratory conditions, the zoospores of *P. insidiosum* are attracted to animal hair and tissue. He used horse hair as bait to isolate *P. insidiosum* from swampy areas in Australia. He speculated that this oomycete lives in wet habitats and that perhaps it may require a plant to complete its life cycle. Although an unidentified water lily was suggested as a possible plant host (Miller 1983), other findings (Mendoza and Prendas 1988; Chaiprasert et al. 1990; Mendoza et al. 1993) indicate that, during its life cycle, *P. insidiosum* may preferentially use grass tissue, but in environments free of gramineae, it may complete its life cycle on other plants.

DISTRIBUTION

Pythiosis is a disease of the temperate, subtropical, and tropical areas of the world. It has been reported in Argentina, Australia, Brazil, Colombia, Costa Rica, Haiti, India, Indonesia, Japan, New Guinea, New Zealand, South Korea, Thailand, the USA, and Venezuela. Verified cases of this disease have not been reported in Europe. A case of cutaneous granulomas in a horse with all the features of pythiosis was, however, published by Drouin (1896) in France. The geographical location and tropical climate of Africa seemingly would make it an ideal region for pythiosis. Nevertheless, cases from that continent have yet to be reported. In the Americas, the disease has been diagnosed in North, Central, and South America, as well as the Caribbean islands.

In the USA, the disease is more common in states near the Gulf of Mexico: Alabama, Florida, Louisiana, Mississippi, and Texas. In addition, sporadic cases have been diagnosed in nearby states such as Georgia, Missouri, North Carolina, South Carolina, Tennessee, and as far off as Illinois, Indiana, New York, and Wisconsin – all temperate regions of the USA near the Canadian border. Cases of pythiosis have not as yet been described from Mexico. In Central America, Costa Rica is the country with the largest number of documented cases of pythiosis in horses. This disease is most prevalent in the Atlantic region. Some cases have been diagnosed along its Pacific coast, however. The disease has been observed in other Central American countries – Guatemala, Nicaragua, and Panama (personal communications), but there are no recorded cases in the literature. In the Caribbean region, a case of human pythiosis was recently recorded in Haiti, so the disease may also be present in the other islands of that tropical region (Virgile et al. 1993). Northern Argentina, Brazil, Colombia, and Venezuela have reported cases of equine pythiosis since 1965. The Pantanal region of Brazil, the world's largest freshwater wetland, may have the highest incidence and prevalence of equine pythiosis, but systematic studies are not available to support this impression (dos Santos and Londero 1974; Santurio et al. 1998; Leal et al. 2001).

Australia, the Pacific islands, and Asia have reported cases of pythiosis since the last century. In Australia, it has been found in Maitland, New South Wales, the coast of Queensland, the Northern Territory, and Western Australia. In Indonesia, the disease occurs in Borneo, Java, and Sumatra. In Japan, pythiosis is known to occur along the southern coast of Kyushu and the Ryukyu Islands. A case of intestinal pythiosis was recently reported in a dog from South Korea, suggesting that the disease could also be present in North Korea and China (Sohn et al. 1996). In Thailand, cases of human pythiosis have been found in the northern and southeastern parts of the country. Figure 22.5 shows the currently known distribution of pythiosis around the world.

LIFE CYCLE OF *PYTHIUM INSIDIOSUM*

P. insidiosum, as with other *Pythium* spp., is an organism of aquatic environments, although it can also be found in soil as a result of its ability to produce resistant spores (de Cock et al. 1987; Mendoza et al. 1993). To maintain its life cycle in nature, *P. insidiosum* requires a low concentration of ions, a pH near neutrality, and a plant host (Shipton et al. 1982; Shipton

Figure 22.5 *Map showing the distribution of pythiosis in the tropical, subtropical, and temperate regions of the world.*

1983, 1985). As for other zoosporic fungal-like organisms (Endo and Colt 1974), *P. insidiosum* probably uses plants to produce sporangia and release zoospores to colonize other plants and expand its ecological niche, but it has not been demonstrated to cause pathology in plants. Early studies on the life cycle of this oomycete have shown that *P. insidiosum* is present in stagnant waters and may possibly require the Australian water lily or other plants to complete its life cycle (Miller 1983). It was also found that its zoospores may play an important role in the propagation of infections among plants and animals. More recently, other investigators (Mendoza et al. 1993) confirmed that *P. insidiosum* has a special tropism for animal and plant tissues, and that, upon release of zoospores, several mechanisms are activated that permit plant and animal tissue invasion. Encysted zoospores are surrounded by an amorphous material that has adhesive properties. This substance attaches the zoospores to the host's surface. They further hypothesized that chemotactic factors may signal zoospores to produce this material. Figure 22.3 illustrates various stages of *P. insidiosum*'s life cycle in detail.

CLINICAL AND PATHOLOGICAL SIGNS OF INFECTIONS CAUSED BY *PYTHIUM INSIDIOSUM*

The pathogenesis of pythiosisis not clear, in part because the disease is not reproducible in natural hosts. Laboratory data obtained from strains of *P. insidiosum* indicated that zoospores have a tropism for animal and plant tissues (Miller 1983; Mendoza et al. 1993). Thus, the zoospores, as motile propagules, in all likelihood, play a role as infecting agents in natural infections. However, reports of pythiosis in animals and humans with no history of contact with stagnant water open the possibility of inoculation by other propagules of *P. insidiosum* (hyphae, resting spores, oogonia) (Rinaldi et al. 1989; Fischer et al. 1994).

When an animal or human enters swampy areas inhabited by *P. insidiosum*, its zoospores would be attracted by any open wound on their skin and encyst on the exposed tissue. On the basis of electron microscopic observations (Mendoza et al. 1993), the zoospores are known to secrete a sticky substance that allows them to make and maintain tight contact with the host during the initial stages of invasion. Stimulated by the temperature of the host's skin, the encysted zoospores develop a germ tube, often directed towards injured tissue (Mendoza et al. 1993), and mechanically penetrate the tissue (see Figure 22.2e). Ravishankar and Davis (2001) experimentally established that *P. insidiosum* cannot penetrate normal skin, but has the potential to successfully penetrate injured skin. Their study strongly supports previous speculations. *P. insidiosum* initiates a cell-mediated immune response in its host, mostly in the form of

eosinophils and a few neutrophils (Miller and Campbell 1983; Mendoza and Alfaro 1986), but the immune response cannot prevent the propagation of *P. insidiosum*. The eosinophils then, in an effort to phagocytose the microorganism, degranulate over the hyphae. New eosinophils fuse around the degranulated eosinophils and are also degranulated in turn. A mass known as a 'kunker' is thus formed. The eosinophilic material (Splendore–Hoeppli-like phenomenon) around *P. insidiosum* hyphae is a main feature of the infection in equines but not in the other species. Kunkers are composed of degranulated eosinophils interlaced with the viable hyphae of *P. insidiosum*. The lower panel of Figure 22.3 summarizes the probable mechanism of infection. The tissue damage observed in acute and chronic cases has been attributed to the release of chemicals from the degranulated eosinophils and mast cells (Miller 1981; Miller and Campbell 1983; Mendoza et al. 2003).

Pythiosis in non-primates

HORSES

Since the beginning of this century, the clinical features of pythiosis in horses have been well described by several authors (de Haan and Hoogkamer 1901, 1903; Witkamp 1924; Bridges and Emmons 1961; McMullan et al. 1977; Miller and Campbell 1982b; Chaffin et al. 1995). There are no reports of predisposition with regard to sex or age of the afflicted animals. Lesions caused by *P. insidiosum* are found on any part of a horse's body, although they are more common on the lower limbs because these are the first areas to come in contact with swampy water. In addition, the lesions in horses can also be found on the thorax, abdomen, neck, shoulders, genitalia, and head. Pythiosis often occurs as a single lesion but unusual cases with multicentric granulomas have been encountered. When lesions develop on the extremities (particularly near the joints), lameness is a frequent finding. In chronic pythiosis, emaciation and secondary infections are also common. There are no reports of animal-to-animal, or animal-to-human transmission of infection.

Lesions are circular in shape, from 5 to 500 mm in diameter (Miller and Campbell 1982b; Mendoza and Alfaro 1986), with a characteristic serosanguineous discharge (Figure 22.6a, b). Intense pruritus is one of the characteristic clinical features of the disease in horses. Ulceration of the tissue is common in large lesions. Smaller ones often show one or more small ulcers. As the disease progresses, small coral-like masses called kunkers, which contain viable hyphae of *P. insidiosum*, are found in the sinuses and surface of the infected tissue. They vary in size and form (10–90 mm in diameter) (Figure 22.7). Metastasis of *P. insidiosum* through lymphatic vessels to regional lymph nodes (Connole 1973; Murray et al. 1978), lungs (Witkamp

Figure 22.6 *Clinical features of cutaneous granulomas caused by* Pythium insidiosum *in two horses (a, shoulder; b, leg). Note the circular shape typical of pythiosis granulomas in horses.*

1924; McMullan et al. 1977; Goad 1984), or bones (Mendoza et al. 1988; Alfaro and Mendoza 1990; Eaton 1993; Neuwirth 1993) is common (Figure 22.8). Equine intestinal pythiosis, resulting from direct inoculation of propagules of *P. insidiosum* in mucous membranes, has been reported (Brown and Roberts 1988; Morton et al. 1991; Purcell et al. 1994).

Histopathologically, in the early stages of pythiosis, abundant microabscesses with an inflammatory reaction, composed mainly of eosinophils, a few neutrophils, lymphocytes, and macrophages, are present in the subcutaneous lesions. In chronic cases, eosinophilic granulomatous tissue with giant cells is often recorded. In central areas of inflammation, sequestered eosinophilic masses (kunkers) are detectable. Sections of the kunkers stained with Gomori's methenamine silver reveal the presence of aseptate hyphae 6–10 μm in diameter along with cellular debris (Figure 22.9).

DOGS

Canine infections were first reported in dogs with cutaneous and gastrointestinal lesions from the Gulf of

Figure 22.8 *Posterior view of a bone in the front leg of a horse affected by* Pythium insidiosum, *with lysis and exostosis of the metacarpal, carpal, and distal radius, and the first phalanx bones.*

Mexico in the USA (Heller et al. 1971; Miller et al. 1983; Pavletic et al. 1983; Foil et al. 1984). Subcutaneous pythiosis lesions have been recorded also on the legs, face, and tail in dogs (Figure 22.10) (Foil et al. 1984; Thomas and Lewis 1998). Systemic pythiosis with involvement of internal organs has also been reported in dogs (Foil et al. 1984; Thomas and Lewis 1998). The pruritic

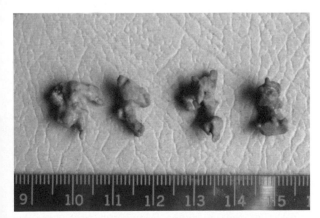

Figure 22.7 *This shows four irregularly shaped, firm, coral-like masses called kunkers which were removed from a cutaneous granulomatous lesion in a horse with pythiosis. (Reproduced, with permission, from Mendoza and Alfaro 1986)*

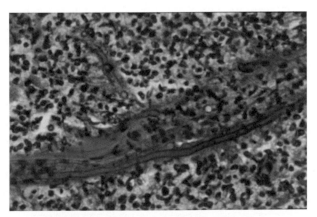

Figure 22.9 *Photomicrograph of a kunker's tissue section, showing the aseptate hyphae of* Pythium insidiosum *and cellular debris. Periodic acid–Schiff stain; magnification × 9 200.*

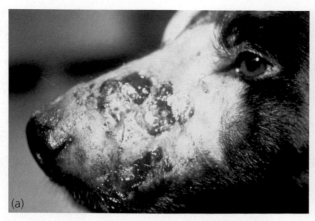
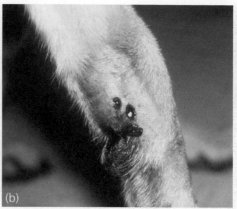

Figure 22.10 (a, b) *Skin lesions caused by* Pythium insidiosum *on two dogs with subcutaneous infection. (Courtesy of Dr Randall C. Thomas)*

skin lesions were denuded of hair and perforated by fistulous sinus tracts discharging a serosanguineous exudate. Histopathological findings included multifocal areas of necrosis with moderate neutrophils and macrophages. Discrete granulomatous infiltrates with eosinophilic material are surrounded by eosinophils, neutrophils, and epithelioid cells. Giant cells have also been reported in chronic cases. Hyphae which measured 4.4–8.6 μm in diameter were detected in these tissues.

Gastrointestinal pythiosis in dogs is characterized by vomiting, weight loss, and sporadic diarrhea (Miller et al. 1983, 1985; Pavletic et al. 1983; Thomas and Lewis 1998). Formation of hard gastrointestinal granulomatous masses, areas of mural thickness, and mucosal ulceration are common in most cases. Lesions can spread to adjacent tissue such as that of the pancreas, uterus, and mesenteric lymph nodes. Histopathologically, the mucosa shows ulceration, atrophy, and epithelial cell hyperplasia. The submucosa is thickened and the muscular mucosa contains focal granulomas. Neutrophils, eosinophils, plasma cells, macrophages, epithelioid cells, and giant cells are observed in the affected tissues. The hyphae of *P. insidiosum* are difficult to detect in hematoxylin–eosin-stained sections, but with Gomori's methenamine silver stain, hyphae between 2.5 and 8.9 μm in diameter are readily detected. Recently, cases of canine pythiosis have been recorded in Australia (English and Frost 1984) and in the USA (Miller et al. 1985).

Grooters (2003) stated that there were clinical cases similar to those observed in canine pythiosis, caused by an undescribed species in the genus *Lagenidium*. She based her findings on histopathological, cultural, and molecular analysis. However, her data have yet to be confirmed by others (see earlier under Taxonomy and morphological features of *Pythium insidiosum*).

Cats and cattle

Few cases of pythiosis in cats (Bissonnette et al. 1991; Thomas and Lewis 1998) and calves (Miller et al. 1985; Santurio et al. 1998) have been described in the literature. Bissonnette et al. (1991) was the first to diagnose a case of pythiosis in a cat with facial swelling, but none of its internal organs was involved (Figure 22.11a). *P. insidiosum* was isolated from the affected tissue. Histopathological features, similar to those recorded in horses and dogs, were observed in this case.

A total of five calves with pythiosis have so far been reported with the disease. Showing multiple focal ulcers and fistulous tracts, draining a watery purulent exudate and with swelling of fetlock joints, they were diagnosed with pythiosis (Miller et al. 1985; Santurio et al. 1998). Miller and colleagues (1985) reported that *P. insidiosum* was isolated from two of the studied calves, whereas *P. insidiosum*-like hyphae were found in the tissue of a third case. Histological sections from that case showed granular encrustations around the hyphae. The perihyphal deposit was composed of granular material similar to that of the Splendore–Hoeppli phenomenon. Similar observations were described by Santurio et al. (1998) in calves from the Pantanal region of Brazil.

More recently, several cases of pythiosis in calves were also described in the State of Apure, Venezuela (Figure 22.11b) (personal communication, Rosa Cristina Perez). Serological testing done in these animals was consistent with *P. insidiosum*. In addition, histopathological and DNA sequencing analysis from the infected tissues confirmed the presence of this pathogen in the infected calves.

Human pythiosis

Most human cases have occurred in Thailand with only a few reports of such cases in Australia, Haiti, New Zealand, and the USA. Two human cases of subcutaneous pythiosis were mentioned by de Cock et al. (1987), but details about those early cases were not available. The first five cases of this disease in humans involved two women and three men in the rural areas of northern Thailand (Sathapatayavongs et al. 1989). Ten

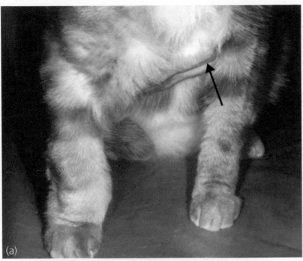

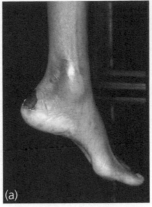

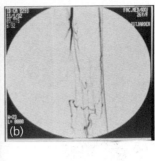

Figure 22.11 (a) *Lesions on a cat from Florida, USA with subcutaneous pythiosis (arrow).* **(b)** *A subcutaneous lesion on the limb of a Venezuelan calf caused by* Pythium insidiosum. *(Courtesy of Dr Randall C. Thomas and Rosa Cristina Perez)*

Figure 22.12 (a) *Dry gangrenous lesion on the left leg caused by* Pythium insidiosum *in a Thai thalassemic patient.* **(b)** *Digital subtraction angiography of the same patient showing occlusion of the left superficial femoral artery. (Courtesty of Dr Ploenchan Chetchotisalkd, Thailand)*

more cases were later reported from the same areas of Thailand (Chetchotisakd et al. 1992; Wanachiwanawin et al. 1993). Most were associated with a hemoglobinopathic syndrome. Clinically, progressive gangrene and pain in the extremities were the main findings. Strikingly, all cases showed large vessel arteritis (Figure 22.12). Angiographic analysis indicated occlusion of the iliac, popliteal, and femoral arteries. Treatment by amputation was successful in seven of the 15 cases. Fatal invasion of the abdominal arteries with occlusion of the infrarenal part of the aorta was also recorded. So far, the majority of cases of human pythiosis have occurred in Thailand, with some 90 new patients reported since the first case in 1987 (personal communication, Theerapong Krajaejun).

Three patients with subcutaneous orbital pythiosis were diagnosed in the USA (Rinaldi et al. 1989). One was a healthy boy in Texas who had accidentally been struck in his right eye. He developed progressive peri-orbital oedema with chemosis, erythema, and periorbital cellulitis (Figure 22.13). *P. insidiosum* was isolated from biopsied tissue. Details of the second case were not discussed. The third case was recorded by Shenep et al. (1998) on a 2-year-old boy from Tennessee. Two additional cases of pythiosis with a similar clinical history were reported in Australia (Triscott et al. 1993). These cases involved an 11-year-old and a 14-year-old boy. Both developed periorbital swellings which rapidly developed into an orbital tumour. The diagnosis was made histopathologically using an immunoperoxidase staining assay (see section on Immunohistochemical assays below). In addition to vessel arteritis and subcutaneous pythiosis in humans, cases of keratitis have also been recorded in Haiti (Virgile et al. 1993), New Zealand (Fraco and Parr 1997), and Thailand (Kunavisarut et al. 1988; Imwidthaya and Methitrairut 1992; Imwidthaya 1995).

Triscott et al. (1993) described in detail the histopathology of subcutaneous pythiosis in humans. They found that the histological features were similar to those of equine pythiosis. Eosinophilic granular masses of about 7 mm in diameter, containing degenerating eosinophils, eosinophilic granules, hyphae, and intact eosinophils at the edge of the granulomatous areas were the main findings. They also reported a mixed inflammatory cell infiltrate comprising eosinophils, lymphocytes, neutrophils, macrophages, and mast cells. As in equine pythiosis, the hyphae of *P. insidiosum* were restricted to the areas of the eosinophilic granular masses. Kunkers, as recorded in equine pythiosis, have not, however, been detected in tissue from human cases.

IMMUNOLOGY

The development of an immune response to *P. insidiosum* antigens during equine infections has been known since early in the century (Witkamp 1924, 1925).

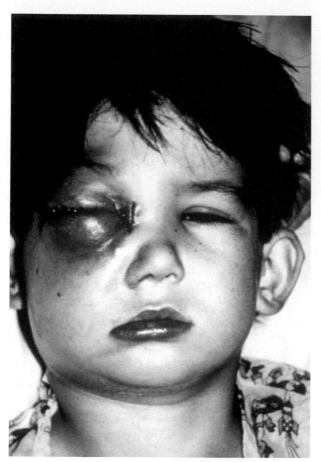

Figure 22.13 *Severe orbital swelling resulting from* Pythium insidiosum *infection in a healthy boy. (Courtesy of Drs Michael G. Rinaldi and Steven Seidemseld, USA)*

In those studies, Witkamp described precipitin and complement fixation antibodies, and a cellular immunity response to a skin test in horses with active pythiosis. Miller and Campbell (1982a), using a similar approach, confirmed this early work. At least three antigens were prepared. A trypsin–hyphal antigen detected one precipitin band by double immunodiffusion (ID) in all horses with the disease. By contrast, only 82 percent of the horses were positive to a complement fixation test. The skin test, using a precipitate protein antigen, was positive in 64 percent of the clinically infected horses and in 31 percent of normal horses inhabiting the enzootic areas. The finding that normal animals reacted positively in a delayed hypersensitivity skin test indicates that subclinical pythiosis may occur in some horses. They also reported detectable levels of *P. insidiosum* antibodies at the day of birth in foals born to mares with active pythiosis, suggesting that their foals received some degree of passive immunity against *P. insidiosum*. Horses with a history of pythiosis but negative to skin tests were considered anergic. Analogous immunological studies in other areas also confirmed these findings (Mendoza and Alfaro 1986; Grooters et al. 2002; Krajaejun et al. 2002).

Immunization (immunotherapy) with antigens derived from *P. insidiosum* was shown to have curative properties in horses afflicted with the disease (Miller 1981; Mendoza and Alfaro 1986; Hensel et al. 2003; Mendoza et al. 2003). Fifty percent of infected horses responded to immunization. Changes from an eosinophilic infiltrate, before immunization, to a mononuclear response (macrophages, T lymphocytes), after immunization, suggested that cellular immunity was playing a major role in the clearance of *P. insidiosum* from immunized horses. Mendoza et al. (2003) suggested that a switching from a T helper 2 (Th2) to a T helper 1 (Th1) response was behind the curative properties of the immunotherapeutic antigens used by these investigators.

Animals in the early stages of pythiosis react positively to a delayed hypersensitivity skin test using a culture filtrate antigen (CFA) and are cured by immunization with precipitated protein antigens. These observations, together with the fact that sera from humans and animals reacted positively in ID and Western blot tests, indicated that cellular and humoral immunity were active in the early stages of the disease. Animals became anergic when the disease reached chronicity (>2 months).

Systemic human pythiosis in Thailand, on the other hand, has been diagnosed mainly in thalassemic patients (Imwidthaya 1994a). This may indicate that the disease usually occurs in debilitated patients. However, there is no such parallel in animals with this disease. Moreover, in the USA and Australia, four subcutaneous human cases of pythiosis were recorded in healthy individuals (Rinaldi et al. 1989; Triscott et al. 1993). Thus, the predisposition to *P. insidiosum* infection in patients with thalassemia hemoglobinopathy syndrome is intriguing and merits further investigation.

LABORATORY DIAGNOSIS

Clinical material from suspected cases of pythiosis should be sent directly to the laboratory for processing. Storage of tissue samples at 4°C during transportation may result in the death of *P. insidiosum*. If the samples cannot be delivered immediately, it is recommended that they should be vigorously washed several times in sterile distilled water, and then transported in sterile distilled water plus antibiotics (ampicillin + streptomycin). The most suitable tissue from which to isolate *P. insidiosum* from horses is the coral-like structures known as kunkers. These masses are removed and repeatedly washed in distilled water. The tissue or kunkers are then cut into small pieces, placed on Sabouraud's dextrose agar (SDA) and incubated at 37°C for 24–48 hours. Parts of the pieces of the kunkers should be placed in 10 percent KOH for direct microscopy examination. Wet mounts are of value for the early detection of *P. insidiosum*, which appear as hyaline sparsely septated hyhae 4.0–9.0 µm in diameter.

Several media have been used for the isolation of *P. insidiosum*. The most common is SDA. *P. insidiosum* grows rapidly at 37°C on that medium. In 24 hours it develops flat colonies that are 20 mm in diameter. Microscopically, however, the hyphae are found to be sterile at this stage. The hyphae are branched at approximately 90° angles and are usually coenocytic, although occasional septa are observed in tissue sections and in old cultures. To induce zoospore production, it is recommended that boiled grass leaves be placed on *P. insidiosum* cultures. After 24 hours of incubation at 37°C, the leaves are immersed in dilute salt solution containing calcium and incubated 2–3 hours at 37°C (Mendoza and Prendas 1988). Zoosporangia containing motile zoospores will be readily observed at the edges of the leaves. Oogonium production is very rare in *P. insidiosum*, so specific identification is based on serological tests and DNA testing (Mendoza et al. 1987; Badenoch et al. 2001; Grooters and Gee 2002; Schurko et al. 2003a).

Histopathology

In tissue sections, the hyphae of *P. insidiosum* in tissue sections are often difficult to differentiate from those found in cases of zygomycosis caused by fungal species of the orders Mucorales and Entomophthorales. In such cases, the diameter of the hyphae may be of help. The hyphae of *P. insidiosum* are between 3.0 and 10.0 μm in diameter, whereas those of *Basidiobolus ranarum* and *Conidiobolus coronatus* are broader, being about 5–15 μm in diameter. However, the isolation of *P. insidiosum* and the use of serological tests are the ultimate basis for the diagnosis of pythiosis. In some cases, clinical data can also be useful. For example, clinically pythiosis and zygomycosis caused by *B. ranarum* occur in the same anatomic areas in horses, so it is often difficult to differentiate between them, although infections caused by *C. coronatus* occur primarily in a horse's nostrils. This clinical difference may be of help, if serological tests and isolation of the etiologic agent are not possible.

DNA-based diagnostic assays

The identification of *P. insidiosum* in the laboratory and histopathological preparations is entirely based on its sexual oogonia and hyphae-like morphological features. However, several fungi have also morphological similarities with the filamentous structures of this oomycete, and the development of oogonia in agar plates is difficult. To overcome these facts, DNA methodologies have recently been used for its diagnosis and identification (Badenoch et al. 2001; Grooters and Gee 2002; Reis et al. 2003). Badenoch et al. (2001) were the first to use DNA sequencing to identify *P. insidiosum* hyphae from

a patient with keratitis. These authors suggested that molecular tools could be important in differentiating *P. insidiosum* from the filamentous fungi. Later, Grooters and Gee (2002) developed nested primers that specifically amplified 105 base pairs from *P. insidiosum*'s genomic DNA. More recently, Reis et al. (2003) obtained several amplicons of *P. insidiosum*'s 18S small-subunit rDNA sequences from three cases of horses with systemic pythiosis. These studies indicated that DNA-based technology is ideal to identify *P. insidiosum* from fixed tissues and cultures, and could be successfully used more frequently in the future.

Serology

Serological methods for diagnosing *P. insidiosum* infections were developed in the beginning of the century (Witkamp 1925). Miller and Campbell (1982a) and Mendoza and Alfaro (1986) based their immunological studies of equine pythiosis on Witkamp's early work. They found that immunodiffusion and complement fixation tests were useful in the diagnosis of pythiosis. Moreover, a skin test was shown to be effective in determining subclinical infections in animals with active disease.

IMMUNODIFFUSION TEST

Two antigens have been used to diagnose *P. insidiosum* infections by immunodiffusion. One was prepared from the hyphae of *P. insidiosum* digested with trypsin (Miller and Campbell 1982a). This trypsin antigen detected one precipitinogen in all animals with active pythiosis. However, its shelf life was short. The other antigen was obtained from concentrated CFAs (Mendoza et al. 1986). It detected three to six precipitins in sera from horses and humans with pythiosis. One of them was the precipitinogen reported earlier in an ID test with trypsin antigens. Healthy individuals were always negative with the ID test. Chronic equine cases of pythiosis showed no bands in the ID test. These animals also did not react to skin tests and were considered anergic. The ID test detects pythiosis in its early stages (3 days), indicating that antibodies against the antigens of *P. insidiosum* are developed early in the course of infection.

No crossreactivity was recorded using the CFA of *P. insidiosum* and sera from patients with zygomycosis or other mycotic and bacterial diseases. The ID test was specific for pythiosis. Thus, the presence of a precipitin band in ID is suggestive of this disease. The number of precipitins in ID does not correlate with the severity of the disease. Horses treated and recovered from the disease showed no precipitin bands after 2 months of successful treatment. Some bands of non-identity were recorded when sera from horses with pythiosis reacted to the antigens of *B. ranarum (haptosporus)* or *C. coronatus* (Kaufman et al. 1990). The CFA proved to be more sensitive, stable and useful, not only in diagnosing the disease in cats, cattle, dogs, horses, and humans, but also in monitoring response to treatment (Mendoza et al. 1986; Imwidthaya and Srimuang 1989; Pracharktam et al. 1991). In addition, complement fixation has been previously used (Miller and Campbell 1982a). However, this test is no longer available.

WESTERN BLOT

Culture-filtrated antigens in the ID test detected at least six precipitinogens. However, ID is too insensitive and also unable to identify major antigenic immunogens. Western blot analysis was introduced to determine which of the antigens of *P. insidiosum* were of importance during infection (Mendoza et al. 1992a). It was found that equine IgG recognized almost all of the cytoplasmic proteins of *P. insidiosum*. Several protein antigens of 28, 30, and 32 kDa) and other immunogens were found to be immunodominant (Figure 22.14). Negative results in Western blot were recorded using sera from healthy horses or sera from horses with various other diseases. Equine immunoglobulin G against the 32, 30, and 28 kDa and other immunodominant antigens was found to persist at least for a year in horses cured by immunotherapy. This suggests that these prominent antigens may also be important as protective immunogens in horses. It was later found that the addition of cytoplasmic antigens, containing the 28, 30, and 32 kDa immunodominant proteins, to the available *Pythium* vaccine enhanced its curative properties (Mendoza et al. 2003). This finding suggested that these proteins may play a major role in the immunotherapy of pythiosis. In equine chronic pythiosis, Western blot fails to detect the cytoplasmic antigens of *P. insidiosum*, indicating that anti-*P. insidiosum* IgG is absent from their sera. This supports the concept that horses with the chronic disease are anergic. IgG from horses with active pythiosis recognized only the 44 kDa cytoplasmic antigen of *C. coronatus*, confirming that the Western blot test is very specific in horses. This test is useful for analyzing immunoglobulin classes during infection. Recently, a single-step immunoblot (SIB) assay to diagnose pythiosis in horses was shown to be sensitive, specific, and easy to perform (Rosa 1993). Studies using human sera in Western blot are not yet available.

ENZYME-LINKED IMMUNOSORBENT ASSAY

The enzyme-linked immunosorbent assay (ELISA) test was originally designed to detect anti-*P. insidiosum* IgG in proved cases of pythiosis with a negative ID (Rosa 1993; Mendoza et al. 1997). More recently, the test was shown to be helpful in cases of a dog and cat with the disease (Mendoza et al. 1997; Grooters et al. 2002). In Thailand, at least another in-house assay was developed using antigens from Thai isolates (Krajaejun et al. 2002). These ELISAs showed high sensitivity and specificity to diagnose pythiosis and were also suitable to monitor the response to treatment, confirming previous observations (Mendoza et al. 1997). In addition, some healthy horses from enzootic regions reacted positively in low titers, suggesting that subclinical infection may occur.

The ELISA developed to detect pythiosis in humans (Krajaejun et al. 2002) and animals (Mendoza et al. 1997; Grooters et al. 2002) both showed 100 percent sensitivity and specificity. These ELISAs were capable of discriminating between sera from apparently healthy individuals or heterologous infections and patients with active pythiosis.

IMMUNOHISTOCHEMICAL ASSAYS

Two tests have been developed for the immunodetection of *P. insidiosum* hyphae in tissue: an immunofluorescence test (Mendoza et al. 1987) and an immunoperoxidase assay (Brown and Roberts 1988; Triscott et al. 1993). The immunofluorescence assay was shown to be entirely specific for *P. insidiosum* in the detection of its hyphae in cat, dog, and human tissues. However, in equine pythiosis, tissue sections of kunkers containing hyphae showed a uniform fluorescence throughout the sections. This was attributed to the exoantigens released by the hyphae of *P. insidiosum* into the matrix of the kunkers. Immunofluorescence has also been used to differentiate *P. insidiosum* from other species of *Pythium*.

The immunoperoxidase assay was performed on the tissue of dogs, horses, and humans with the disease (Brown and Roberts 1988; Triscott et al. 1993; Fischer et al. 1994). *P. insidiosum* in those tissues stained positively after applying anti-IgG peroxidase. Tissue sections

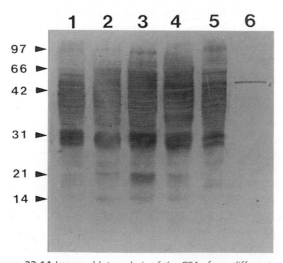

Figure 22.14 *Immunoblot analysis of the CFAs from different strains of* Pythium insidiosum *(lanes 1–5) and* Conidiobolus coronatus *CFAs (lane 6) after reacting with sera from horses with pythiosis. The 28, 30, and 32 kDa proteins are observed as prominent bands. A band of crossreaction was observed against* C. coronatus's *CFA (lane 6). (Reproduced, with permission, from Mendoza et al. 1992a)*

from cases of zygomycosis caused by *B. ranarum* and *C. coronatus* were negative in these assays. Both assays were found to be sensitive and specific.

ANIMAL INOCULATION

All attempts experimentally to reproduce pythiosis in dogs, horses, and mice have failed (Patino-Meza 1988). Early workers, however, found that rabbits were susceptible to infection by the propagules of *P. insidiosum* (Witkamp 1924). Amemiya (1969, 1982), working with strains isolated from horses with granular dermatitis in Japan, confirmed the susceptibility of rabbits to hyphal inoculation. In his work, he found that subcutaneous inoculation of *P. insidiosum* hyphae gave rise to nodules that contained hyphae and inflammatory cells at the injection sites (Figure 22.15). Intravenous injection caused systemic infections in the inoculated rabbits with granulomatous necrotizing masses in the aortas, intestines, livers, and lungs.

Miller and Campbell (1983) evaluated motile zoo spores as inocula in cortisone-treated and non-cortisone-treated rabbits. They found that both groups were extremely susceptible to infection by the zoospores of *P. insidiosum*. All inoculated animals developed necrotizing hepatitis or embolic nephritis, thus demonstrating that *P. insidiosum* does not require a debilitated or immunocompromised host to induce infections. In striking contrast, numerous authors have failed to reproduce the disease in horses and dogs, even after exposing them to motile zoospores by submerging the animals in tanks of seeded water which simulated the conditions encountered in swamps. The predisposing factors required to develop pythiosis in dogs, horses, and humans remain unknown. Mendoza et al. (2003), based on clinical data, however, argued that humans and animals are resistant to pythiosis. They also suggested that perhaps a defect in the immune response is what makes a host susceptible, which in part explains the low occurrence of the infection in the endemic areas.

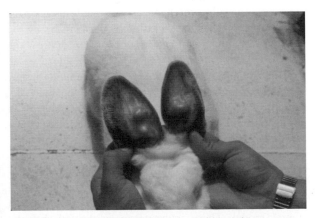

Figure 22.15 *Experimental pythiosis in a rabbit. Subcutaneous nodules and systemic dissemination of* Pythium insidiosum *are the main features of such experimental infections.*

TREATMENT

Treatment of infections caused by *P. insidiosum* in animals and humans is difficult. Three therapeutic methods are, however, often used for this disease: surgery, drugs, and immunotherapy.

Surgery

Radical surgery has been successfully used in cases of equine pythiosis since the last century. It consists of removal of the lesions and their kunkers, followed, in some cases, by cauterization (Habbinga 1967; McMullan et al. 1977). This method is very popular and frequently used by veterinary practitioners, although it is not always successful. In general, the response to surgery is limited. Moreover, lesions on the limbs of equines are not easily treated by this method because of their delicate anatomic structure. Surgical removal of tumor-like lesions in dogs with intestinal pythiosis has been reported, but survival beyond 3 months was extremely rare in the treated dogs (Fischer et al. 1994; Thomas and Lewis 1998). A common shortcoming of surgical treatment is recurrence as a result of incomplete removal of infected tissue.

Radical surgery has also been used in human cases with arteritis caused by *P. insidiosum*. Amputation of the extremities is a drastic procedure of the last resort, which is used to treat patients with severe *Pythium* arteritis. However, this method has proved to be only partially successful because most patients finally died of disseminated abdominal arteritis (Sathapatayavongs et al. 1989; Wanachiwanawin et al. 1993). Early detection of affected vessels using angiography is important in determining the most appropriate amputation site.

Chemotherapy

Two main groups of drugs are commonly used to treat pythiosis: iodide and amphotericin B. Iodides have been used since early in the century in equine pythiosis with contradictory results. Although some authors reported cures with intravenous injections of potassium iodide (0.75 g/45 kg in weight) (Gonzalez et al. 1979) or sodium iodide (1.0 g/15 kg in weight) (Hutchins and Johnston 1972), other investigators reported failures (de Haan and Hoogkamer 1903; Hartsfield 1971; Murray et al. 1978). Similar observations were recorded in humans. Patients with arteritis caused by *P. insidiosum* did not respond to potassium iodide, but patients with subcutaneous pythiosis seem to have responded well to this drug (Thianprasit 1990). The main drawback of iodide and amphotericin B is their toxicity (Murray et al. 1978).

Amphotericin B is the drug of choice in many mycotic infections. As a result of the lack of ergosterols in the cytoplasmic membrane of members of the genus

Pythium, however, one could predict that amphotericin B would be ineffective in infections caused by *P. insidiosum*. Nevertheless, in equine pythiosis, amphotericin B has been used with some success. Eight cases of equine pythiosis were cured out of a total of 10 horses (McMullan et al. 1977) after intravenous injection of amphotericin B (0.38–1.47 mg/kg). Topical application of this drug had little or no effect on equine cases (Miller et al. 1983). The use of amphotericin B has been limited partly as a result of the cost of therapy, the poor rate of success, and its toxic side effects. Amphotericin B in humans with pythiosis gave contradictory results as well. For example, intravenous injections of amphotericin B (0.1 mg/kg) were ineffective in cases with arteritis (Wanachiwanawin et al. 1993) and in one case of subcutaneous infection (620 mg/kg, over a 6-week period) (Rinaldi et al. 1989). It was effective in two subcutaneous pythiosis cases when it was used in combination with 5-fluorocytosine (amphotericin B 0.5 mg/kg/day, 5-fluorocytosine 150 mg/kg/day) (Triscott et al. 1993). A combination of itraconazole and terbinafine saved the life of a boy in Tennessee with subcutaneous pythiosis (Shenep et al. 1998). However, this combination has been also used in dogs and other animals with contradictory results.

Immunotherapy

Immunization of equines with products derived from *P. insidiosum* cultures was reported to have curative properties in Australia (Miller 1981) and Costa Rica (Mendoza and Alfaro 1986). Two antigens, referred to in the literature as Miller's and Mendoza's vaccines (Newton and Ross 1993), have been used for immunization. The Australian vaccine was prepared from sonicated hyphal antigens, whereas the Costa Rican vaccine used precipitated proteins from CFAs. Immunotherapy of horses with pythiosis using Miller's vaccine alone gave a success rate of 53 percent. An increase in the percentage of cured cases was obtained when immunization was followed by surgical removal of cutaneous lesions. Similar results were obtained using Mendoza's vaccine (Figure 22.16). Nevertheless, significant reduction in the swelling, caused by the vaccine at the site of injection, an undesirable side effect of Miller's vaccine, was the main feature of Mendoza's vaccine. Moreover, the vaccine described by Miller was unstable, losing its curative properties after storage at 4°C. By contrast, Mendoza's vaccine was found to be effective even 18 months after its preparation.

Recently Mendoza et al. (2003) introduced new immunotherapeutic immunogens to treat pythiosis in both humans and animals. This new formulation contains exo- and endoproteins extracted from cultures of *P. insidiosum* and was able to cure chronic and acute cases of the disease (Thitithanyanont et al. 1998; Hensel et al. 2003; Mendoza et al. 2003). These studies suggest

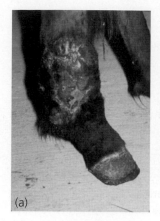

Figure 22.16 (a) *A horse with pythiosis before immunotherapy;* **(b)** *the same horse after successful immunotherapy.* (Reproduced, with permission, from Mendoza and Alfaro 1986)

that hyphal antigens may contain products that are involved in the enhancement of the immunological response to immunization. Horses with the disease that failed to respond to immunization did not develop swellings at the injection sites. All in all, this finding suggests that the response to immunization is directly related not only to the immunostatus of the infected hosts, but also to the type of immunogens used.

Based on histopathological data, the hyphae of this oomycete are always sequestered within kunkers. Thus, it has been postulated that perhaps some antigens are downregulated during infection (Mendoza et al. 2003). Reports on the histological changes that take place after immunotherapy had shown that the initial eosinophilic inflammatory reaction, typical of equine pythiosis, changed to a mononuclear response composed mainly of macrophages and T lymphocytes (perhaps cytotoxic lymphocytes) (Miller 1983; Mendoza and Alfaro 1986). The antigens, used during immunotherapy, apparently trigger a Th1 immune response with macrophages and cytotoxic lymphocytes that eventually clear the organism from tissues. This contention is supported by the fact that cultures of biopsied tissue in immunized horses are always negative for hyphae. This confirms, in part, the observation that the digestion of the kunkers by the mononuclear inflammatory cells seems to destroy the hyphae of *P. insidiosum* (Miller 1981; Mendoza et al. 1992b). Complications associated with immunotherapy are mainly an inflammatory reaction at the injection site, secondary bacterial contamination, and lameness in lesions located on limbs. It was reported that some horses cured by immunization relapsed one year later. This suggests that, if the vaccine has prophylactic properties, it is of short duration. The prophylactic and curative properties of *P. insidiosum* immunogens have been recently evaluated by Santurio and Leal (2003) in an animal model. These investigators confirmed that the antigens derived from this oomycete have, indeed, curative and prophylactic properties.

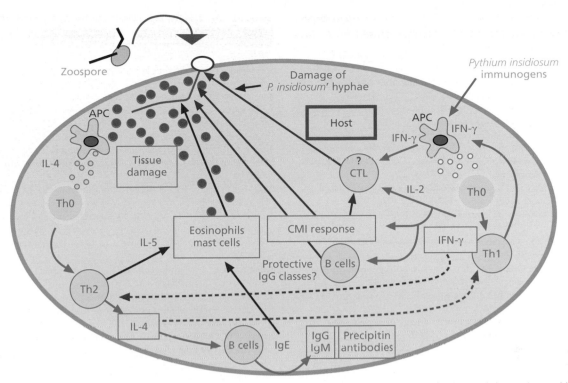

Figure 22.17 *Proposed working hypothesis of the pathogenic mechanisms involved during natural infection and those triggered by immunotherapy. In this model, pythiosis begins when the host contacted a zoospore (or other propagules of the pathogen) through a wound. Pythium insidosum will develop hyphae-like filaments that penetrate the host. Once in the tissues, it will release antigens that will lock the immune system into a Th2 response and which contain typical cellular and other mediators including IL-4 that could downregulate the Th1 response (brown dotted arrow) and cause pythiosi (blue arrow) (left side of the figure). By contrast, after immunotherapy, IFN-γ is immediately triggered, which in turn activate a Th1 immunity and downregulates the Th2 response (pink dotted arrow). The mononuclear response (putative natural killers and cytotoxic lymphocytes, cell-mediated immunity) is believed to be responsible for the killing of the pathogen in the infected tissues (red arrows). APC, antigen presenting cell. (Reproduced, with permission, from Mendoza et al. 2003)*

The mechanism involved in the response of horses to natural infection and to immunotherapy using *P. insidiosum* products has been recently addressed by Mendoza et al. (2003). In their model, *P. insidiosum*, after contact with the host, will release exoantigens that will lock the immune system into a Th2 response (eosinophils, mast cells, IgE, IL-4, IL-5, precipitin, IgG, and IgM). They propose that the degranulation of the eosinophils and mast cells is responsible for the tissue damage observed during natural infection (Figure 22.17). By contrast, after injection with the immunotherapeutic antigens of *P. insidiosum*, the host will trigger a Th1 response with interferon-gamma (IFN-γ) that eventually downregulates the Th2 response, and a prominent mononuclear response. These investigators speculated that the events triggered by the immunotherapeutic antigens ultimately kill the pathogen in the infected tissues, and therefore are more likely responsible for the cure of the infected hosts (Figure 22.17).

REFERENCES

Alfaro, A.A. and Mendoza, L. 1990. Four cases of equine bone lesions caused by *Pythium insidiosum*. *Equine Vet J*, **22**, 295–7.

Amemiya, J. 1969. Isolation of a fungus of the Mortierellaceae from an equine granular dermatitis III. *Bull Fac Agric Kagoshima Univ*, **19**, 31–50.

Amemiya, J. 1982. Granular dermatitis in the horse, caused by *Pythium gracile*. *Bull Fac Agric Kagoshima Univ*, **32**, 141–7.

Austwick, P.K.C. and Copland, J.W. 1974. Swamp cancer. *Nature (Lond)*, **250**, 84.

Badenoch, P.R., Coster, D.J., et al. 2001. *Pythium insidiosum* keratitis confirmed by DNA sequence analysis. *Br J Ophthalmol*, **85**, 502–3.

Baldauf, S.L., Roger, A.J., et al. 2000. A kingdom-level phylogeny of eukaryotes based on combined protein data. *Science*, **290**, 972–7.

Bissonnette, K.W., Sharp, N.J.H., et al. 1991. Nasal and retrobulbar mass in a cat caused by *Pythium insidiosum*. *J Med Vet Mycol*, **29**, 39–44.

Bridges, C.H. and Emmons, C.W. 1961. A phycomycosis of horses caused by *Hyphomyces destruens*. *J Am Vet Med Assoc*, **38**, 579–89.

Brown, C.C. and Roberts, E.D. 1988. Intestinal pythiosis in a horse. *Aust Vet J*, **65**, 88–9.

Chaffin, M.K., Schumacher, J. and McMullan, W.J. 1995. Cutaneous pythiosis in the horse. *Vet Clin North Am: Equine Pract*, **11**, 91–103.

Chaiprasert, A.K., Samerpitak, K., et al. 1990. Induction of zoospore formation in Thai isolates of *Pythium insidiosum*. *Mycoses*, **33**, 317–23.

Chandler, F.W., Kaplan, W. and Ajello, L. 1980. *A color atlas and textbook of the histopathology of mycotic diseases*. Chicago IL: Year Book Medical Publishers, 104–5.

Chetchotisakd, P., Porntaveevuani, O., et al. 1992. Human pythiosis on Srinagarind Hospital: one year experience. *J Med Assoc Thail*, **75**, 248–54.

Connole, M.D. 1973. Equine phycomycosis. *Aust Vet*, **49**, 214–15.

de Cock, W.A.W., Mendoza, L, et al. 1987. *Pythium insidiosum* sp. nov. the etiological agent of pythiosis. *J Clin Microbiol*, **25**, 344–9.

de Haan, J. 1902. Basartige Schimmelkrankheit des pferdes (Hyphomycosis destruens equi). *Zentbl Bakteriol Parasitenkd Infektionskr*, **31**, 758–63.

de Haan, J. and Hoogkamer, L. 1901. Hyphomycosis destruens. *Veeartsenijk Bl v Ned Indie*, **13**, 350–74.

de Haan, J. and Hoogkamer, L. 1903. Hyphomycosis destruens equi. *Archiv Wissenschaft Prakt Tierheilkd*, **29**, 395–410.

Dick, M.W. 2001. *Straminipilous fungi: systematics of the Peronosporomycetes including accounts of the marine straminipilous protist, the plasmodiophorids and similar organisms*. London: Kluwer Academic Publishers.

dos Santos, M.N. and Londero, A.T. 1974. Zigomicosis subcutanea em cavalos. *Pesqui Agropecu Bras*, **9**, 7–8.

Drouin, V. 1896. Sur une nouvelle mycose du cheval. *Rec Med Vet*, **30**, 337–44.

Eaton, S.A. 1993. Osseous involvement by *Pythium insidiosum*. *Compendium*, **15**, 485–8.

Endo, R.M. and Colt, W.M. 1974. Anatomy, cytology and physiology of infections by *Pythium*. *Proc Am Phytopathol Soc*, **17**, 215–23.

English, P.B. and Frost, A.J. 1984. Phycomycosis in a dog. *Aust Vet J*, **61**, 291–2.

Fischer, J.R., Pace, L.W., et al. 1994. Gastrointestinal pythiosis in Missouri dogs: eleven cases. *J Vet Diagn Invest*, **6**, 380–2.

Fish, P.A. 1895–96. Leeches: a histological investigation of two cases of equine mycosis with a historical account of a supposed similar disease, called bursatte, occurring in India. *12th and 13th Annual Report of the Bureau of Animal Industry*, 229–59.

Foil, C.S., Short, B.G., et al. 1984. A report of subcutaneous pythiosis in five dogs and a review of the etiologic agent *Pythium* spp.. *J Am Anim Hosp Assoc*, **20**, 959–66.

Fraco, D.M. and Parr, D. 1997. *Pythium insidiosum* keratitis. *Aust N Z J Ophthalmol*, **25**, 177–9.

Goad, M.E. 1984. Pulmonary pythiosis in a horse. *Vet Pathol*, **21**, 261–2.

Gonzalez, H.E., Threebilcock, P., et al. 1979. Tratamiento de la ficomicosis equine subcutanea empleando yoduro de potasio. *Rev Colomb Agric*, **14**, 115–21.

Graham, J.P., Newell, S.M., et al. 2000. Ultrasonographic features of canine gastrointestinal pythiosis. *Vet Radiol Ultrasound*, **41**, 273–7.

Grooters, A.M. 2003. Pythiosis, lagenidiosis, and zygomycosis in small animals. *Vet Clin Small Anim*, **33**, 695–720.

Grooters, A.M. and Gee, M.K. 2002. Development of a nested polymerase chain reaction assay for the detection and identification of *Pythium insidiosum*. *J Vet Intern Med*, **16**, 147–52.

Grooters, A.M., Leise, B.S., et al. 2002. Development and evaluation of an enzyme-linked immunosorbent assay for the serodiagnosis of pythiosis in dogs. *J Vet Intern Med*, **16**, 142–6.

Habbinga, R. 1967. Phycomycosis in an equine. *Southwest Vet*, **20**, 237–8.

Hartsfield, M. 1971. Phycomycosis in a mare. *Southwest Vet*, **24**, 138–9.

Hawksworth, D.L., Kirk, P.M., et al. 1995. *Ainsworth & Bisby's dictionary of the fungi*, 8th edition. Wallingford: CAB International.

Heller, R.A., Hobson, H.P., et al. 1971. Three cases of phycomycosis in dogs. *Vet Med Small Anim Clin*, **66**, 472–6.

Hensel, P., Greene, C.E., et al. 2003. Immunotherapy for treatment of multicentric cutaneous pythiosis in a dog. *J Am Vet Med Assoc*, **223**, 215–218, 197.

Herr, R.A., Ajello, L., et al. 1999. Phylogenetic analysis of *Rhinosporidium seeberi*'s 18S small-subunit ribosomal DNA groups this pathogen among members of the protoctistan Mesomycetozoa clade. *J Clin Microbiol*, **37**, 2750–4.

Hutchins, D.R. and Johnston, K. 1972. Phycomycosis in the horse. *Aust Vet J*, **48**, 269–78.

Ichitani, T. and Amemiya, J. 1980. *Pythium gracile* isolated from the foci of granular dermatitis in the horse (*Equus caballus*). *Trans Mycol Soc Jpn*, **21**, 263–5.

Imwidthaya, P. 1994a. Systemic fungal infections in Thailand. *J Med Vet Mycol*, **32**, 395–9.

Imwidthaya, P. 1994b. Human pythiosis in Thailand. *Postgrad Med J*, **70**, 558–60.

Imwidthaya, P. 1995. Mycotic keratitis in Thailand. *J Med Vet Mycol*, **33**, 81–2.

Imwidthaya, P. and Methitrairut, A. 1992. *Pythium insidiosum* keratitis. *The 33rd Siriraj Scientific Annual Meeting (Bangkok)*, 537–42.

Imwidthaya, P. and Srimuang, S. 1989. Immunodiffusion test for diagnosing human pythiosis. *Mycopathologia*, **106**, 109–12.

Kaufman, L., Mendoza, L. and Standard, P. 1990. Immunodiffusion test for diagnosing subcutaneous zygomycosis. *J Clin Microbiol*, **28**, 1887–90.

Kirk, P.M., Cannon, P.F., et al. 2001. *Ainsworth & Bisby's dictionary of the fungi*, 9th edition. Wallingford: CAB International.

Krajaejun, T., Kunakorn, M., et al. 2002. Development and evaluation of an in-house enzyme-linked immunosorbent assay for early diagnosis and monitoring of human pythiosis. *Clin Diagn Lab Immunol*, **9**, 378–82.

Kunavisarut, S., Prawinwongwuth, K., et al. 1988. Pythium corneal ulcer: case report. *Thai J Ophthalmol*, **2**, 70–3.

Leal, A.B.M., Leal, A.T., et al. 2001. Pitiose eqüine no Pantanal brasileiro: aspectos clínicos patológicos de casos típicos e atípicos. *Pesq Vet Bras*, **21**, 151–6.

Martin, F.N. 2000. Phylogenetic relationships among some *Pythium* species inferred from sequence analysis of the mitochondrially encoded cytochrome oxidase II gene. *Mycologia*, **92**, 711–27.

McMullan, W.C., Joyce, J.R., et al. 1977. Amphotericin B for the treatment of localized subcutaneous phycomycosis in the horse. *J Am Vet Med Assoc*, **170**, 1293–8.

Mendoza, L. and Alfaro, A.A. 1986. Equine pythiosis in Costa Rica: report of 39 cases. *Mycopathologia*, **94**, 123–9.

Mendoza, L. and Marin, G.M. 1989. Antigenic relationship between *Pythium insidiosum* de Cock et al 1987 and its synonym *Pythium destruens* Shipton 1987. *Mycoses*, **32**, 73–7.

Mendoza, L. and Prendas, J. 1988. A method to obtain rapid zoosporogenesis of *Pythium insidiosum*. *Mycopathologia*, **104**, 59–62.

Mendoza, L., Kaufman, L. and Standard, P. 1986. Immunodiffusion test for diagnosing and monitoring pythiosis in horses. *J Clin Microbiol*, **23**, 813–16.

Mendoza, L., Kaufman, L. and Standard, P. 1987. Antigenic relationship between the animal and human pathogen *Pythium insidiosum* and nonpathogenic *Pythium* spp.. *J Clin Microbiol*, **25**, 2159–62.

Mendoza, L., Alfaro, A.A. and Villalobos, J. 1988. Equine bone lesions in a horse caused by *Pythium insidiosum*. *Med Vet Mycol*, **26**, 5–12.

Mendoza, L., Nicholson, V. and Prescott, J.F. 1992a. Immunoblot analysis of the humoral immune response to *Pythium insidiosum* in horses with pythiosis. *J Clin Microbiol*, **30**, 2980–3.

Mendoza, L., Villalobos, J., et al. 1992b. Evaluation of two vaccines for the treatment of pythiosis in horses. *Mycopathologia*, **119**, 89–95.

Mendoza, L., Hernandez, F. and Ajello, L. 1993. Life cycle of the human and animal oomycete pathogen *Pythium insidiosum*. *J Clin Microbiol*, **31**, 2967–73.

Mendoza, L., Kaufman, L., et al. 1997. Serodiagnosis of *Pythium insidiosum* infections using an enzyme-linked immunodifussion assay. *Clin Diagn Lab Immunol*, **4**, 715–18.

Mendoza, L., Mandy, W. and Glass, R. 2003. An improved *Pythium insidiosum*-vaccine formulation with enhanced immunotherapeutic properties in horses and dogs with pythiosis. *Vaccine*, **21**, 2797–804.

Miller, R. 1981. Treatment of equine phycomycosis by immunotherapy and surgery. *Aust Vet J*, **57**, 377–82.

Miller, R. 1983. Investigation into the biology of the three phycomycotic agents pathogenic for horses in Australia. *Mycopathologia*, **81**, 23–8.

Miller, R.I. and Campbell, R.S. 1982a. Immunological studies on equine phycomycosis. *Aust Vet J*, **58**, 227–31.

Miller, R.I. and Campbell, R.S. 1982b. Clinical observations on equine phycomycosis. *Aust Vet J*, **58**, 221–6.

Miller, R.I. and Campbell, R.S. 1983. Experimental pythiosis in rabbits. *Sabouraudia*, **21**, 331–41.

Miller, R.I., Wold, D., et al. 1983. Complications associated with immunotherapy of equine phycomycosis. *J Am Vet Med Assoc*, **182**, 1227–9.

Miller, R.I., Olcott, B.M. and Archer, M. 1985. Cutaneous pythiosis in beef calves. *J Am Vet Med Assoc*, **186**, 984–6.

Morton, L.D., Morton, D.G., et al. 1991. Chronic eosinophilic enteritis attributed to *Pythium* sp. in a horse. *Vet Pathol*, **288**, 542–4.

Murray, D.R., Ladds, P., et al. 1978. Metastatic phycomycosis in a horse. *J Am Vet Med Assoc*, **172**, 834–6.

Neuwirth, L.B. 1993. Radiographic appearance of lesions associated with equine pythiosis. *Compendium*, **15**, 489–90.

Newton, J.C. and Ross, P.S. 1993. Equine pythiosis: an overview of immunotherapy. *Compendium*, **15**, 491–3.

Patino-Meza, F. 1988. Role of the zoospores of *Pythium insidiosum* in the experimental reproduction of pythiosis in susceptible species. DVM thesis, National University, Heredia, Costa Rica, 1–31.

Patterson, D.J. 1989. Stramenopiles: chromophytes from a protistan perspective. In: Green, G.C., Leadbeater, B.S.C. and Diver, W.L. (eds), *The chromophyte algae, problems and perspectives*. Oxford: Clarendon Press, 357–79.

Pavletic, M.M., Miller, R.I. and Turnwald, G.H. 1983. Intestinal infarction associated with canine phycomycosis. *J Am Anim Hosp Assoc*, **19**, 913–19.

Pracharktam, R., Chantrakool, P.J., et al. 1991. Immunodiffusion test for diagnosis and monitoring of human pythiosis. *Clin Microbiol*, **29**, 2661–2.

Purcell, K.L., Johnson, P.J., et al. 1994. Jejunal obstruction caused by *Pythium insidiosum* granuloma in a mare. *J Am Vet Med Assoc*, **205**, 337–9.

Ransom, B.H. 1911. The life history of a parasitic nematode *Habronema muscaehm*. *Science*, **34**, 690–2.

Ravishankar, J.P. and Davis, C.M. 2001. Mechanics of solid tissue invasion by the mammalian pathogen *Pythium insidiosum*. *Fungal Genet Biol*, **34**, 167–75.

Reis, J.L., de Caravalho, E.C., et al. 2003. Disseminated pythiosis in three horses. *Vet Microbiol*, **96**, 289–95.

Rinaldi, M.G., Seidenfeld, S.M., et al. 1989. *Pythium insidiosum* causes severe disease in a healthy boy. *Mycol Observer*, **9**, 7.

Rosa, P.S. 1993. Development and evaluation of serologic tests to detect pythiosis in horses. MS Thesis, Louisiana State University, Baton Rouge, LA, 1–120.

Santurio, J.M. and Leal, A.T. 2003. Three types of immunotherapics against pythiosis developed and evaluated. *Vaccine*, **21**, 2535–40.

Santurio, J.M., Monteiro, A.B., et al. 1998. Cutaneous pythiosis in calves from the Pantal region of Brazil. *Mycopathologia*, **141**, 123–5.

Sathapatayavongs, B., Ledachaikul, P., et al. 1989. Human pythiosis associated with thalassemia hemoglobinopathy syndrome. *J Infect Dis*, **159**, 274–80.

Schurko, A., Mendoza, L., et al. 2003a. Evidence for geographic clusters: molecular genetic differences among strains of *Pythium insidiosum* from Asia, Australia and the Americas are explored. *Mycologia*, **95**, 200–8.

Schurko, A.M., Mendoza, L., et al. 2003b. A molecular phylogeny of *Pythium insidiosum*. *Mycol Res*, **107**, 537–44.

Shenep, J.L., English, B.K., et al. 1998. Successful medical therapy for deeply invasive facial infection due to *Pythium insidiosum* in a child. *Clin Infect Dis*, **27**, 1388–93.

Shipton, W.A. 1983. Possible relationship of some growth and sporulation responses of *Pythium* to the occurrence of equine phycomycosis. *Trans Br Mycol Soc*, **80**, 13–80.

Shipton, W.A. 1985. Zoospore induction and release in a *Pythium* causing equine phycomycosis. *Trans Br Mycol Soc*, **84**, 147–55.

Shipton, W.A. 1987. *Pythium destruens* sp. nov., an agent of equine pythiosis. *J Med Vet Mycol*, **25**, 137–51.

Shipton, W.A., Miller, R.I. and Lea, I.R. 1982. Cell wall, zoospore and morphological characteristics of Australian isolates of a *Pythium* causing equine phycomycosis. *Trans Br Mycol Soc*, **79**, 15–23.

Smith, F. 1884. The pathology of bursattee. *Vet J*, **19**, 16–17.

Sohn, Y., Kim, D., et al. 1996. Enteric pythiosis in a Jindo dog. *Korean J Vet Res*, **36**, 447–51.

Thianprasit, M. 1986. Fungal infection in Thailand. *Jpn J Dermatol*, **96**, 1343–5.

Thianprasit, M. 1990. Human pythiosis. *Trop Dermatol*, **4**, 1–4.

Thitithanyanont, A., Mendoza, L., et al. 1998. The use of an immunotherapeutic vaccine to treat a life threatening human arteritis infection caused by *Pythium insidiosum*. *Clin Infect Dis*, **27**, 1394–400.

Thomas, R.C. and Lewis, D.T. 1998. Pythiosis in dogs and cats. *Compendium*, **20**, 63–75.

Triscott, J.A., Weedon, D. and Cabana, E. 1993. Human subcutaneous pythiosis. *J Cutan Pathol*, **20**, 267–71.

Virgile, R., Perry, H.D., et al. 1993. Human corneal ulcer caused by *Pythium insidiosum*. *Cornea*, **12**, 81–3.

Wanachiwanawin, W., Thianprasit, M., et al. 1993. Fatal arteritis due to *Pythium insidiosum* infection in patients with thalassaemia. *Trans R Soc Trop Med Hyg*, **8**, 296–8.

Witkamp, J. 1924. Bijdrage tot de kennis van der hyphomycosis destruens. *Ned Ind Bland Diergenaeskd Dierenteelt*, **36**, 229–345.

Witkamp, J. 1925. Het voorkomen van metastasen in de regionaire lymphklieren by hyphomycosis destruens. *Ned Ind Bland Diergenaeskd Dierenteelt*, **37**, 79–102.

23

Lobomycosis

SINÉSIO TALHARI AND ROGER PRADINAUD

This disease was first described in 1931 by Lobo in Recife, Brazil as blastomycosis queloidiforme, following observations in a 52-year-old Amazonian man who had multinodular keloid-like lesions in the lumbar region. Further cases have been identified in Brazil and other countries in the Amazon region, and Central America. Later sporadic and isolated cases have been described in Europe and North America.

The only case described in Europe (France) was of a 24-year-old aquarium employee who presented with Lobo's disease on the dorsum of his hand 3 months after contact with an infected dolphin (Symmers 1983). A single report from the USA was of a patient who probably contracted the infection in Venezuela (Burns et al. 2000). The total number of registered patients reached 464 by December 2002 (Baruzzi et al. 1979; Silva and Brito 1994; Pradinaud and Talhari 1996; Opromolla et al. 1999; Saint-Blancard et al. 2000; Fischer et al. 2002).

The etiologic agent responsible for Lobo's initial case was classified as *Glenosporella loboi* by Fonseca Filho and Area Leão in 1940. Other names that have been proposed include *Glenosporopsis amazonica*, *Paracoccidioides loboi*, *Blastomyces loboi*, *Loboa loboi*, and *Lacazia loboi*.

Lobomycosis has been described in dolphins: *Tursiops truncatus* (bottle-nosed dolphin) in Florida (Migaki et al. 1971; Symmers 1983) and *Sotalia guianensis* in the Surinam river. The disease identification in dolphins has widened the geographical distribution since seven cases have now been reported in Florida and single cases in the Gulf of Mexico, Surinam, and the Gulf of Gascony, off the coast of France.

The human disease affects the tropical zone of the New World and Table 23.1 illustrates the distribution.

There is a clear gender predilection as 90 percent of cases have been reported in men. This is most likely to be related to occupational exposure. However, in the Amazonian Caiabi Indian population, the prevalence in women reaches 32 percent, probably as a consequence of their role in agricultural and forest activities (Baruzzi et al. 1979).

The age range of patients at the time of diagnosis is between 40 and 70 years. Their case histories, however, suggest that clinical manifestations began many years before. Some cases have also been reported in children: a 5-year-old whose lesions evolved from the age of one; there is also a report of disease in a 2-year-old Caiabi Indian.

ETHNIC BACKGROUND

There is no ethnic predominance as all races appear to be equally susceptible. However, lobomycosis is very well documented in the Mato Grosso (Brazil) Caiabi Indians where it was first recognized in 1915. By 1979, 60 cases had been reported in this particular tribe, representing 31 percent of all Brazilian cases (Baruzzi et al. 1979). It has therefore been suggested that this tribe might be genetically predisposed. However, the fact that no new cases have been identified after movement of these tribes to a new habitat in Xingu National

Table 23.1 *Geographical distribution of Lobo's disease*

Country	No. of cases
Brazil	296
Colombia	50
Surinam	34
Costa Rica	21
Venezuela	20
French Guiana	17
Panama	13
Peru	3
Bolivia	3
Ecuador	2
Guyana	2
Mexico	1
Europe: France/Gulf of Gascony	1
USA	1
Total	464

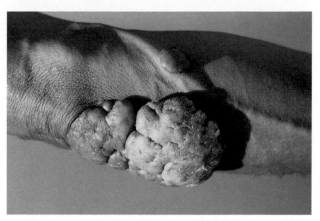

Figure 23.1 *Isolated and confluent keloid-like lesions*

Park points to an environmental or exposure-related etiology.

OCCUPATIONAL EXPOSURE AND INOCULATION

Farmers exposed to harsh conditions and aquatic environments represent most of the reported cases. Other high-risk activities include gold mining, fishing, and hunting.

Patients often report a bite or sting from an arthropod, snake or sting-ray, or otherwise trauma from a cutting instrument. Probably, the fungus is an aquatic saprophyte, which would explain disease in aquatic mammals such as the dolphin. There are no reports of person-to-person spread, even with intimate contact. However, there are reports of voluntary inoculation (Borelli 1968) and accidental surgical inoculation of a patient (Azulay et al. 1975).

CLINICAL FEATURES

Lobomycosis is characterized by variably sized dermal nodules, either lenticular or in plaques, which can attain the size of a small keloid-like cauliflower. They can be either hyper- or hypopigmented, and occasionally achromic. A number of different types of lesions may be seen at various stages of disease progression (Figures 23.1 and 23.2). Infiltrative lesions are often a feature of early disease and can simulate tuberculoid leprosy or burn scars, especially in the Caiabi Indian population (Figure 23.3). Verruciform lesions represent advancing pathology and occur most commonly on the lower extremities and, like other exotic mycoses, can give the 'mousy-foot' appearance. Ulcerative lesions may also represent advanced pathology and on healing result in sclerotic or atrophic scars. Most of the ulcer-ated lesions are caused by trauma. A few cases of dermohypodermic lesions could be referred to as having a gumma-like appearance, but the characteristic softening is absent and therefore these lesions are more typical of nodules. True gummatous lesions are rarely observed. Patients, however, usually present with lesions of various types.

The disease generally does not affect the epidermis and therefore skin lesions usually have a smooth and shiny appearance, especially in the facial region. Fine telangiectasia and sometimes 'black dots,' a feature of chromoblastomycosis and other vegetating dermatoses, can be observed, the 'black dots' corresponding to trans-epidermic elimination of necrotic fungi. Some lesions, especially on the lower extremities, can be keratotic and vegetating. An extreme case was that of a patient whose skin resembled a rock covered with seashells. On the sole of his foot, lesions appeared crater-like.

Any area of the body is potentially susceptible as infection commonly follows trauma. However, the most commonly affected areas include the lower extremities (31 percent), the ears (26 percent), and the upper extremities (20 percent), and the disease is usually unilateral (Figures 23.4 and 23.5).

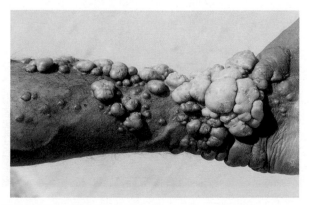

Figure 23.2 *Isolated and confluent papules and nodules*

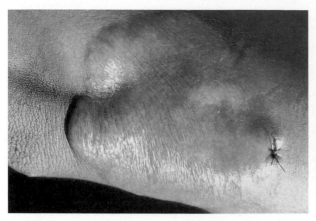

Figure 23.3 *Infiltrative plaque*

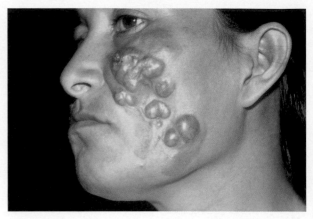

Figure 23.5 *Infiltrative lesions of a woman's face. This is a very unusual location of the disease. (Courtesy of Arival C. Brito, Belém*

Disease varies in extent and local or distant spread of lesions can occur. Local spread probably occurs in part by autoinoculation, producing small satellite nodules at the periphery of the mother lesion. The disease can invade lymph nodes proximal to the affected area but reports, however, vary as to how commonly this occurs, ranging from less than 10 percent (Azulay et al. 1975) to more than 25 percent (Wiersema 1971). Distant skin metastases occur via transdermal and lymphohematological routes. Cutaneous dissemination can result in confluent lesions and plaques (Figures 23.6 and 23.7), with some patients developing hundreds of lesions. There has been a single report of visceral pathology in a native Costa Rican whose lesions evolved over a period of 47 years. He presented with left lower

extremity lymphangitis secondary to left testicular tumoral spread, which necessitated orchidectomy. Histology revealed a multicellular granuloma actively phagocytosing the cells of the fungal pathogen and sparing the scrotum.

The clinical presentation in dolphins is similar to that in humans. They also develop keloidal nodules, either confluent or in plaques, as well as verrucous and vegetating lesions.

EVOLUTION AND PROGNOSIS

Disease onset is usually insidious and therefore difficult to document. The increase in either size or number of lesions is a slow and progressive process, which can continue for up to 40–60 years. The presenting complaint is usually esthetic or functional disability. The disease progresses in a benign fashion and patients usually die of intercurrent illnesses. At present, however, it appears incurable as new lesions usually continue to appear despite treatment either in proximity to, or distant from, the initial lesion. Interestingly, spontaneous regression in three Caiabi Indians has been reported.

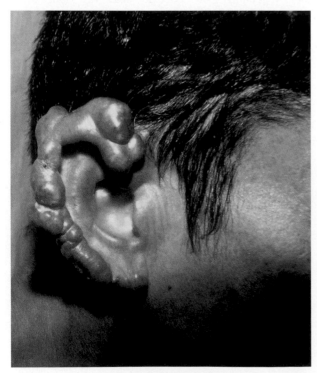

Figure 23.4 *Classic infiltrative nodules of the auricle, on one side only, which are indolent and have evolved over many years.*

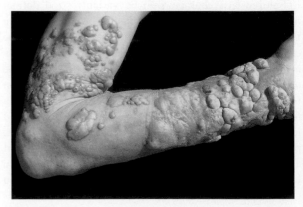

Figure 23.6 *Numerous papular and tumor lesions on the arm and forearm.*

Figure 23.7 *Numerous papules, plaques and tumoral lesions on the foot, leg and thigh.*

DIAGNOSIS

The clinical diagnosis is apparent to physicians who have seen previous cases of lobomycosis and the patient is known to reside in an endemic area. The differential diagnosis includes mainly leprosy, anergic cutaneous leishmaniasis, chromoblastomycosis, paracoccidioidomycosis, and Kaposi's sarcoma.

Diagnosis may also be established by a skin smear or biopsy.

Skin smear

A skin smear permits direct visualization of the parasite, which is abundant in lesions. It is quick and easy and entails scraping a tumoral lesion with a scalpel blade in order to obtain dermal tissue for microscopy. Lobomyces are yeast-like rounded thick-walled cells which range from 5 to 15 μm in diameter and occur in chains of two to ten cells (Figure 23.8). Granules (five to eight) are present in the center and there is usually one that is

larger than the others. They appear mobile due to Brownian motion, especially in a fresh specimen.

In old lesions, one can observe the presence of a single large inert granule. The cells are united by little cell bridges. In most cases, each cell's cytoplasm is clearly demarcated, although, at the end of the chain, one can occasionally see a budding cell, with continuous cytoplasm between the mother and daughter cell.

Histopathology

Histopathology is pathognomonic. The epidermis is usually atrophic, although it can also be hyperplastic, vegetating, and keratotic. Lesions in dolphins are identical to those in human skin (Destombes and Ravisse 1964; Symmers 1978; Arrese et al. 1988; Sesso and Baruzzi 1988; Sesso et al. 1988).

In the dermis a thin band of subepithelial collagen, analogous to lepromatous leprosy's 'Unna band' is usually intact. The remainder of the dermis is occupied by a progressively fibrous, diffuse, inflammatory granuloma. The granuloma is composed of histiocytes and numerous giant cells, which are often clustered in small groups separated by interstitial tissue. The histiocytes, and more often the multinucleated giant cells, can be seen to have numerous phagocytosed thick-walled cells clearly distinguishable by periodic acid-Schiff (PAS), or Gomori's/Grocott's, or Gridley's silver stains (Figures 23.9 and 23.10). Although the cells can sometimes appear empty, they often contain vacuolized inclusions that are birefringent in polarized light. Budding can be observed but hyphae are always absent. These yeast-like cells, which abound in the cytoplasm of giant cells, occasionally appear free floating in the stroma of the granuloma. It is important to note that necrosis and suppuration are never present in the granuloma (unless there is nearby epidermal damage). Histochemical stains reveal polysaccharides in the inflammatory granuloma,

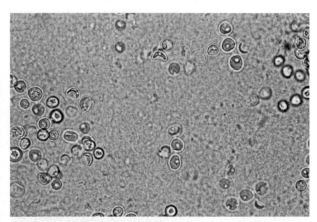

Figure 23.8 *Numerous cells of* P. loboi *in a smear of dermal scrapings: the diagnosis can be confirmed in the office within a few seconds. Multiple buds and chains are connected by tubes; five to eight mobile protoplasmic granules are seen in active fungus.*

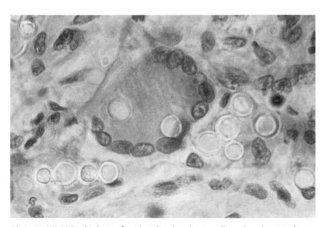

Figure 23.9 *Pathology: foreign body giant cell and polymorphous infiltrate with histiocytes. Parasites are abundant throughout the whole granuloma. Hematoxylin and eosin stained section.*

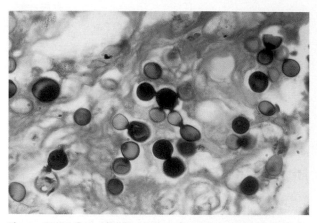

Figure 23.10 *Gomori's–Grocott's stain on the same histological section as seen in Figure 23.9. (Courtesy of A.R. Gradelha)*

which are absent in the deeper areas, especially in the hypodermis. The latter is usually intact.

Disease spread to the lymph nodes produces a similar inflammatory infiltrate with histiocytes and multinucleated giant cells. The cytoplasm of these cells is often xathomatized and contains the same abundance of yeast cells as in the dermis.

Electron microscopic studies demonstrate small spines 1 μm in size on the external glycoprotein membrane which itself measures approximately 0.8 μm. The fungus is multinucleated and central vacuolized inclusion bodies have been identified as triglyceride and polysaccharide glycogenic particles (Furtado et al. 1967; Grimaud et al. 1974).

Culture

There have been many attempts to culture *L. loboi*, but all have been unsuccessful. Sampaio (1974) obtained encouraging results by spreading the cells of *L. loboi* in a 199 TC, containing 5 percent bovine fetal serum penicillin 100 units/ml, amphotericin B μg/2.5 ml, and 2.5 percent phytohemagglutinin at 37°C. The inoculum, which had been taken from the lesion of a patient, was preserved for 5 months in 4°C distilled water.

Animal inoculation has been used with varying success. Armadillo (*Euphractus sexcintus*) inoculation has been more successful and resulted in nodular lesions histologically identical to human pathology (Sampaio and Dias 1977). L. Ajello (unpublished data) has also successfully inoculated the footpads of mice at the Centers for Disease Control and Prevention (CDC) in Atlanta, GA.

IMMUNOLOGY

Patients with Lobo's disease have normal humoral immunity but a partial cellular immune deficiency reflected in absent delayed hypersensitivity responses to dinitrochlorobenzene, streptoccoccus, staphylococcus,

trichophyton and candida antigens (Pecher and Fuchs 1988). In addition, some form of secondary immunodeficiency has been suggested in a dolphin with lobomycosis but to date, lobomycosis in conjunction with human immunodeficiency virus (HIV) or acquired immunodeficiency syndrome (AIDS) has not been reported in humans.

Antigens modified for intradermal use can be used to evaluate the potential of asymptomatic infection due to *L. loboi*. However, these have also given positive results with paracoccidioidomycosis, mycetomas caused by *Nocardia brasiliensis* and in a single case of eczema. This may be due to antigenic relationships between *L. loboi* and other mycological agents which have been demonstrated for *Histoplasma capsulatum* var. *capsultatum* and var. *duboisii*, *Blastomyces dermatitidis*, *Candida albicans*, and the mycelial form of *Coccidioides immitis* (Silva 1978).

When Jorge Lobo described the first case of lobomycosis in 1931, he suggested that the etiologic agent is a fungus similar to *Paracoccidioides brasiliensis*. Taborda et al. (1999) have shown that it is indeed close to *P. brasiliensis* and have designated it *Lacazia loboi*. They suggest that *L. Loboi* is a sister taxon of *P. brasiliensis* as demonstrated on the 18S rDNA phylogenetic tree. However, paracoccidioidin skin tests are negative in cases of lobomycosis.

TREATMENT

The optimal treatment for localized lesions is wide surgical excision ensuring that margins are free of infection to avoid recurrence. It is worth noting that instruments contaminated during an operation can lead to reinfection. Furthermore, in black patients surgical excision carries the risk of inducing true keloids.

The only treatment of choice for disseminated infections is chemotherapy. Clofazamine at doses of 100 and 200 mg daily has been tried in several studies (Silva 1978; Azulay 1995, personal communication; Talhari et al. 1988) as it has shown efficacy with a variety of fungi and actinomycetes (especially the *Nocardia* spp.) and has anti-inflammatory properties in processes involving foreign body granulomas. Of the 16 treated cases in the various studies, four demonstrated 30–80 percent improvement following 2–8 months of therapy. However, compliance to drug therapy, especially when required for prolonged periods, is low as the disease appears to be well-tolerated by patients. However, Niemel (1990, personal communication) reported lack of clinical efficacy of clofazamine in a patient on long-term therapy for *Mycobacterium leprae* who had coincidental *L. loboi* infection. Ketoconazole therapy for lobomycosis has proved to be disappointing.

Of interest is that a dolphin with lobomycosis has been successfully treated with miconazole (Dudok Van Heel 1977). There has been a recent report of successful treatment of a localized plaque lesion on the face with a combination of both clofazamine 100 mg and itraconazole 100 mg daily for a one year period (Fischer et al. 2002). The patient remained disease-free both clinically and histopathogically after a follow-up period of 4 years. Therefore, dual therapy with clofazamine and itraconazole may be a promising treatment for plaque lesions only.

ACKNOWLEDGMENTS

To Mahreen Ameen, Carolina Chrusciak Talhari, and Gottfried Schmer for their help in the preparation of this manuscript.

REFERENCES

Arrese, E.J., Rurangirwa, A. and Pierard, G.E. 1988. Lobomycosis. *Ann Pathol*, **8**, 325–7.

Azulay, R.D., Carnciro, J.A., et al. 1975. Keloidal blastomycosis (Lobós disease) with lymphatic involvement: a case report. *Int J Dermatol*, **15**, 40–2.

Baruzzi, R.G., Lacaz, S. and Souza, P.P.A. 1979. Historia natural da Doença de Jorge Lobo. Ocorrência entre os índios Caiabi (Brasil Central). *Rev Med Trop São Paulo*, **21**, 302–38.

Borelli, D. 1968. Lobomicosis: nomenclatura de su agente. *Med Cutan ILA*, **2**, 151–6.

Burns, R.A., Roy, J.S., et al. 2000. Report of the first human case of lobomycosis in the United States. *J Clin Microbiol*, **38**, 1283–5.

Destombes, P. and Ravisse, P. 1964. Etude histologique de 2 cas guyanais de blastomycose chéloidienne (maladie de J. Lobo). *Bull Soc Path Exot*, **57**, 1018–28.

Dudok Van Heel, W.H. 1977. Successful treatment in a case of lobomycosis (Lobós disease) in *Tursiops truncatus* (Mont) at the Dolfinarium Harderwijk. *Aquatic Mammals*, **5**, 8–15.

Fischer, M., Talhari, A.C., et al. 2002. Lobomykose. Erfolgreiche Therapie mit Clofazimin und Itrakonazol bei einem 46-jährigen Patienten nach 32-jähriger Krankheitsdauer. *Der Hautarzt*, **53**, 677–82.

Fonseca Filho, O. and Area Leão, A.E. 1940. Contribuição para o conhecimento das granulomatoses blastomicoides. O agente etiológico da doença de Jorge Lobo. *Rev Med Cir Brasil*, **48**, 147–58.

Furtado, J.S., Brito, T. and Freymuller, E. 1967. Structure and reproduction of *Paracoccidioides loboi*. *Mycologia*, **59**, 286–94.

Grimaud, J.A., Andrade, L.C. and Lobato, R.M. 1974. Blastomycose chéloidienne (maladie de Jorge Lobo) aspect du parasite en microscopie électronique. *Rev Inst Pasteur de Lyon*, **7**, 95–100.

Lobo, J. 1931. Um caso de blastomicose produzido por uma espécie nova, encontrada em Recife. *Rev Med Pernambuco*, **1**, 763–75.

Migaki, G., Valério, M.G., et al. 1971. Lobo's disease in an Atlantic bottle-nosed dolphin. *J Am Vet Med Assoc*, **159**, 578–82.

Opromolla, D.V., Madeira, S., et al. 1999. Jorge Lobo's disease: experimental inoculation in Swiss mice. *Rev Inst Med Trop Sao Paulo*, **41**, 359–64.

Pecher, S.A. and Fuchs, J. 1988. Cellular immunity in lobomycosis (keloidal blastomycosis). *Allergol Immunopathol (Madr)*, **16**, 413–15.

Pradinaud R. and Talhari, S. 1996. Lobomycose. *Encycl Méd Chir, Maladies Infectieuses, Paris*, 608–A10.

Sampaio, M.M. 1974. A note on the cultivation of the aetiological agent of Jorge lobós disease in 199 TC Medium containing phytohaemagglutinin. *Rev Inst Med Trop São Paulo*, **16**, 121–2.

Sampaio, M.M. and Dias, L.B. 1977. The amardillo *Euphractus sexcinctus* as a suitable animal for experimental studies of Jorge Lobós disease. *Rev Inst Med Trop São Paulo*, **19**, 215–20.

Saint-Blancard, P., Maccari, F., et al. 2000. La lobomycose: une mycose rarement observée en France métropolitaine. *Ann Pathol*, **20**, 241–4.

Sesso, A. and Baruzzi, R.G. 1988. Interaction between macrophage and parasite cells in lobmycosis. The thickened cell wall of *Paracoccidioides loboi* exhibits apertures to the extracellular milieu. *J Sumicrosc Cytol Pathol*, **20**, 537–48.

Sesso, A., Azevedo, R.A. and Baruzzi, R.G. 1988. Lanthanum nitate labelling of the outer cell wall surface of phagocytosed *Paracoccidioides loboi* in human lobomycosis. *J Sumicrosc Cytol Pathol*, **20**, 776–82.

Silva, D. 1978. Traitement de la maladie de Jorge Lobo par la clofazimine (B663). *Bull Soc Pathol Exot*, **71**, 409–12.

Silva, D. and Brito, A. 1994. Formas clínicas não usuais da micose de Lobo. *Ann Bras Dermatol*, **69**, 133–6.

Symmers W.StC. (ed.) 1978. Lobo's disease (keloidal blastomycosis). *Systemic pathology*, 2nd edition. Edinburgh: Churchill Livingstone, 628.

Symmers, W.StC. 1983. A possible case of Lobo's disease acquired in Europe from a bottle-nosed dolphin (*Tursiops truncatus*). *Bull Soc Pathol Exot*, **76**, 777–84.

Taborda, P.R., Taborda, V.A. and McGinnis, R. 1999. *Lacazia loboi* gen. nov., comb. nov., the etiologic agent of lobomycosis. *J Clin Microbiol*, **37**, 2031–3.

Talhari, S., Cunha, M.G.S., et al. 1988. Deep mycoses in the Amazon region. *Int J Dermatol*, **27**, 481–4

Wiersema, J.P. 1971. Lobós disease (keloidal blastomycosis). In: Baker, R.D. (ed.), *Human infection with fungi, actinomycetes and algae*. New York: Springer-Verlag, 577–88.

24

Rhinosporidiosis

SARATH N. ARSECULERATNE AND LEONEL MENDOZA

THE ETIOLOGIC AGENT

'There is a certain air of romance attached to the name "Rhinosporidium"' (Thomas et al. 1956), arising from many synonyms given to its picturesque developmental stages and from its enigmatic nature.

Rhinosporidiosis, the disease that *Rhinosporidium seeberi* causes in humans and animals, presents, as the name implies, with chronic, granulomatous polyps, predominantly in nasal and other respiratory sites. It was first observed by Malbran in Buenos Aires in 1892. Seeber's thesis (1900) first reported the disease, in Argentina. Comprehensive descriptions of the organism (Minchin and Fantham 1905; Beattie 1906; Tirumurti 1914; Ashworth 1923; Karunaratne 1964) and clinical aspects of rhinosporidiosis have been made (Allen and Dave 1936; Karunaratne 1964). While important advances have been made since this chapter was first written in 1998, unresolved enigmas yet remain. Since the organism has yet to be cultured in vitro, the classical Koch's postulates are as yet unfulfilled. Recently, however, using molecular tools, Herr et al. (1999b) found that *R. seeberi* was not a fungus but an unusual protist that shared phylogenetic features with microbes that cause infections in fish.

NOMENCLATURE

The diverse names and descriptions of *R. seeberi* on which tentative classifications were based, were chronologically reviewed by Ashworth (1923) and de Mello (1954) (Table 24.1). The nomenclature of the developmental stages and their contents is given in the section on Light microscopical and ultrastructural features below.

TAXONOMY

The taxonomy of *R. seeberi* (Ashworth 1923; de Mello 1954; Karunaratne 1964) had been, until recently, controversial. Confusion had arisen undoubtedly from the absence of pure isolates of the organism due to its intractability to culture. *R. seeberi* was initially classified as a sporozoon, then as a phycomycete, and when the doubts deepened, it was placed in an orphanage of anomalous fungal and fungal-like organisms, and even as a cyanobacterium. With the advent of phylogenetic analyses, its dignity was restored, by grouping it with *Dermocystidium*, the rosette agent, *Ichthyophonus*, and *Psorospermium* in a new Class – the Mesomycetozoea, formerly Ichthyosporea (Mendoza et al. 2002), a name which signifies the location of this group at the point when the animals and fungi diverged some 2.5 billion years ago.

Table 24.1 *Nomenclatural history of R. seeberi (Wernicke 1903) Seeber 1912*

Year	Event
1892	Malbran observed the organism in a nasal polyp, and regarded it as a sporozoon, but did not publish his findings
1900	Seeber's (1900) thesis described the organism (cysts, spores containing 'sporozoites') in a nasal polyp; he regarded it as a sporozoon allied to the Polysporea of the Coccidia, but did not name the organism
1900	Wernicke (see de Mello 1954) named the organism *Rhinosporidium seeberi*
1903	O'Kinealy (1903) described a nasal polyp and Vaughan described its histology (cyst with a pore, filled with sporules)
1903	Belou (see Ashworth 1923; de Mello 1954) described the organism under the name *Coccidium seeberia* Wernicke 1900
1904	Minchin and Fantham (1905) studied O'Kinealy's (1903) rhinosporidial tissue and named the organism *Rhinosporidium kinealyi*
1912	Seeber (see Ashworth 1923; de Mello 1954) endorsd the generic name *Rhinosporidium* of Minchin and Fantham, and the specific name *seeberi* of Wernicke, and established the identity of *R. kinealyi* with *R. seeberi*
1913	Zschokke (1913) recorded a similar organism in a nasal growth in a horse; he named it *Rhinosporidium equi*
1923	Ashworth (1923) renamed the species *R. seeberi* instead of Belou's *R. seeberia* and considered *R. kinealyi* as identical with *Coccidium seeberia*. He finally validated the designation '*Rhinosporidium seeberi* (Wernicke 1903), emend. Ashworth,' which is now the valid name of the organism. Ashworth also concluded that 'all examples of *Rhinosporidium* recorded from man appear to be referable to one species'
1936	Ciferri et al. (1936) established the identity of *R. seeberi* and *R. equi*, a view endorsed by Ashworth (1923)
1940	Carini (1940) described cysts with spores in skin nodules in frogs (*Hyla rubra*) and created a new genus *Dermosporidium* for the organism which he regarded as having close affinities with, but different from, *R. seeberi*. The relationship of *R. seeberi* to Carini's *Dermosporidium* remained obscure until 1999 (see Herr et al. 1999a)
1958	Vanbreuseghem (1958) included two synonyms: *Rhinosporidium ayyari* (see Allen and Dave 1936) and *Rhinosporidium amazonicum* Aben-Athar 1944
1999	The new classification of *Rhinosporidium seeberi* and its affinities were proposed by Herr et al. (1999b)

Previous attempts to classify *R. seeberi* (see Ashworth 1923) are of interest in the larger context of taxonomy, especially in the use of morphology as a criterion for its taxonomic classification. For instance:

- Minchin and Fantham 1905 – affinities with Neosporidia and on the structure of its simpler 'spore', with the Haplosporidia
- Beattie 1906 – Neosporidia, allied to the Sarcosporidia based on its striated capsule, a framework separating the 'spores', and polar development of the 'spores'
- Ridewood and Fantham 1907 – Order Haplosporidia, Sub-order Polysporulea, based on the numerous 'spores'
- Laveran and Pettit 1910 noted the resemblance of the trout pathogen *Ichthyosporidium*, regarded as a haplosporidian, to *Rhinosporidium*. The organism described by Laveran and Pettit was regarded as a phycomycete, allied to the Chytridineae, by Plehn and Mulsow (1911). The affinity of *R. seeberi* to the Chytridineae was in doubt, because the Chytridineae have flagellated stages

- Ashworth (1923) related *R. seeberi* to the phycomycetes 'as the thallus is formed of a single cell,' and to the Chytridineae on account of the absence of a mycelium and because 'sexual organs are not generally formed so that the spores are usually asexual;' the absence of such flagellation in *R. seeberi* was noted, while from the spores of most Chytridineae, a flagellated stage emerges. He further placed it provisionally in the sub-order Olpidiaceae in which the trophic stage is spherical or elliptical and because 'each such cell becomes transformed into one sporangium only' as in *R. seeberi*
- Vanbreuseghem (1973) failed to identify *R. seeberi* as an alga
- Herr et al. (1999b), through phylogenetic analysis of the organism's 18S SSU rDNA, classified *R. seeberi* in a new clade which they named the Mesomycetozoea. This new group includes fish and amphibian pathogens in the former DRIP clade (*Dermocystidium*, the rosette agent, *Ichthyophonus*, and *Psorospermium*). Cavalier-Smith (1998 quoted by Mendoza et al. 2001a) included the mesomycetozoeans in a class Ichthyosporae but based on the

fact that 'not all members of this group are fish pathogens and that preliminary analysis showed that *R. seeberi* may have chitin synthase genes, render this epithet inappropriate,' Mendoza et al. (2002) amended the term to Mesomycetozoea. The finding of Herr et al. (1999b) on the phylogenetic placement of *R. seeberi* was recently confirmed by Fredricks et al. (2000).

The current classification of *R. seeberi* according to Mendoza et al. (2001a), modifying the proposal of Cavalier-Smith (1998), is as follows (Figure 24.1):

Kingdom – Protozoa
  Sub-kingdom – Neozoa
    Infra-kingdom – Neomonada
      Phylum – Neomonada
        Sub-phylum – Mesomycetozoa
        (formerly Ichthyosporea)
          Class – Mesomycetozoea
            Order – Dermocystida

Family – Rhinosporideaceae
(Mendoza et al. 2002)
  Dermocystidium spp.
  Rhinosporidium seeberi
  Rosette agent
Order – Ichthyophonida
  Family – Ichthyophonae
  (Mendoza et al. 2002)
    Ichthyophonus
    Psorospermium

The taxonomy of *R. seeberi* has finally been resolved (Herr et al. 1999b; Fredricks et al. 2000). These phylogenetic analyses showed that *R. seeberi* is not a fungus but a unique protist that shares morphological and phylogenetic features with several previously orphaned aquatic microbes, located at the divergence between animals and fungi. More information about the mesomycetozoeans could be found in a recent review of this class by Mendoza et al. (2002). In that review, two Orders were recognized – Dermocystida (to which

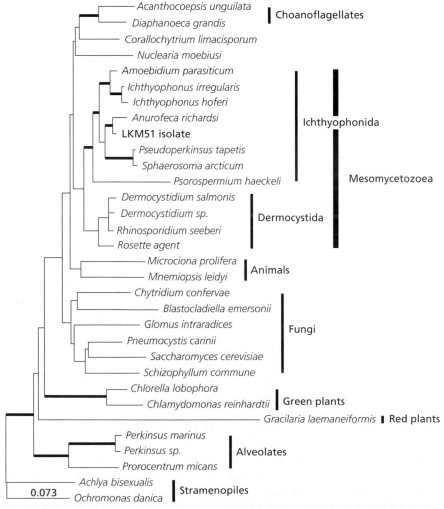

Figure 24.1 *Phylogenetic analysis of* R. seeberi *and other microbes using the 18S SSU rDNA molecule and distances estimated by maximum likelihood. The thickened branches are supported 90 percent of 1000 bootstrapped. The scale for percentage nucleotide substitution is given on the branch to* Ochromonas danica. *In the phylogenetic tree,* R. seeberi *is the sister clade to* Dermocystidium *spp. and closely related to the Rosette Agent in the Order* Dermocystida *(Mendoza et al. 2002).*

R. seeberi belongs) and the Ichthyophonida, with the commonality of several features that *R. seeberi* shares with the other mesomycetozoeans.

BIOLOGY

The absence of a method for the in vitro propagation of *R. seeberi* undoubtedly accounts for the paucity of data on the biology of *R. seeberi*. Vanbreuseghem (1973) commented that African rhinosporidial endospores are smaller than Indian ones, although the organism 'looks similar in sections of very different origins.' We have no evidence (Arseculeratne et al. 2001) that endospores from rhinosporidial tissues from different patients in Sri Lanka show differences in size.

Karunaratne (1964) postulated, from epidemiologic data, that a saprobic form might exist in soil or water; Gori and Scasso (1994) referred to a yeast form in tissues. There is no evidence for an alternate life cycle in any intermediate host.

Random amplified polymorphic DNA (RAPD) analysis of DNA from six strains of *R. seeberi* from patients with differing clinical presentations showed a hetcrogencity in strains of *R. seeberi*, with the six strains falling into three distinct groups on banding patterns (Appuhamy et al. 2002); there was, however, no correlation of the patterns with the clinical picture of the disease. Variations, especially in virulence properties, if they exist, might be relevant to the wide spectrum of clinical features, as well as to the variations scen in histopathology of rhinosporidial granulomata (see Arseculeratne ct al. 2001).

Ecology

It is generally held, from cogent epidemiological evidence, that the main natural habitat of *R. seeberi* is ground water in ponds and lakes, or in soil that is contaminated with such water. This evidence includes:

- a history given by the majority of Asian patients of exposure to these sources
- the predominant occurrence of lesions in nasal and ocular sites
- the reports of Noronha (1933) and of Mandlik (1937) that rhinosporidiosis of the nose was closely associated with the occupation of sand-gathering from river beds
- the outbreaks of rhinosporidiosis in humans who had bathed in the same lake (Vukovic et al. 1995), and in swans inhabiting the same lake (Kennedy et al. 1995)
- the higher prevalence in human males, through occupational exposure to soils and soil dust
- a history of trauma from grass or wood splinters (Karunaratne 1964) sometimes related to agricultural occupations

- the occurrence of rhinosporidiosis in animals with an aquatic habitat, such as river dolphin, geese, swans and ducks
- the frequency of anterior urethral rhinosporidiosis in Muslim males (Ingram 1910; Dhayagude 1941; Gahukamble et al. 1982), which was attributed to infection conveyed by pieces of stone or brick, probably contaminated by soil-borne *R. seeberi*, applied to absorb the last drops of urine.

Many reports on microscopic examination of deposits of stagnant water, silt, manure, and water plants in endemic areas have not yielded confirmation of the hypothesis of the aquatic habitat of *R. seeberi*. Histological studies on samples from aquatic animals such as fish, snails, crabs, and frogs have also yielded negative results. Amphibia exposed to natural waters did not develop rhinosporidiosis (Reddy and Lakshminarayana 1962); deposits from ground waters in which rhinosporidial patients had bathed failed to induce rhinosporidiosis after subcutaneous injection in the following amphibia – *Kaloula taprobanica*, *Tamanella obscura*, *Bufo melanostictus*, *Polypidatus cruciger*, *Rana temporaria* (Arseculeratne 1999, unpublished data).

Deposits of water from a reservoir in Sri Lanka in which many rhinosporidial paients had bathed, however, showcd bodies on periodic acid Schiff (PAS)-stained smears, that were compatible in size and shape with rhinosporidial cndospores, with the presence within these bodies of structures that were also compatible with the electron-dense bodies (EDBs), Figure 24.2). Indirect immunofluorescence tests, however, with antirhinosporidial antibody, gave extensive nonspecific labeling of amorphous, possibly mineral deposits in the water, which made identification of specifically-labeled rhinosporidial bodies difficult. Polymerase chain reaction (PCR) studies on these deposits are in progress in our laboratory.

It should be added that microscopic methods which rely on the morphology of *R. seeberi* as observed in histopathological sections have assumed that the morphology of the tissue stages is similar to that of the saprobic pathogen. There is no evidence, on the other hand, that *R. seeberi* is a dimorphic organism, as some classical fungi are.

Cultural methods applied to material from aquatic habitats have also uniformly proved to have been unsuccessful, undoubtedly due to the uncultivability of the pathogen in vitro.

A reservoir of *R. seeberi* in animals such as horses and cattle, from which transmission to humans is thought to occur, has been postulated (Weller and Riker 1930; Rao 1938), although it is probable that soil or ground water is the common source of infection for these animals, and for humans who work with them.

Reports that the 'causative organism of rhinosporidiosis, viz. *Microcystis aeruginosa*' was isolated from

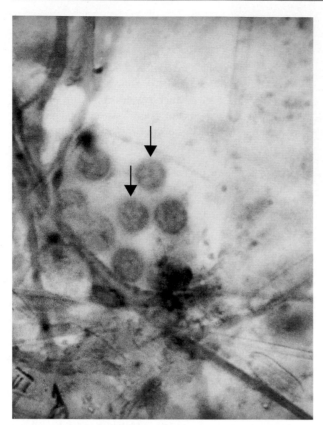

Figure 24.2 *Smear of deposit from reservoir water in which rhinosporidial patients had bathed, showing spherical PAS-positive bodies (arrows) with intracellular structures, compatible in size and shape with rhinosporidial endospores. PAS; initial magnification ×1 000.*

aquatic environments were recently published (Ahluwalia et al. 1997; Ahluwalia 1999). These investigators erroneously concluded that the causative agent of rhinosporidiosis was the aquatic cyanobacterium *Microcystis aeruginosa*, ubiquitously found in stagnant ground waters. *M. aeruginosa* proved to be an effective red-herring on account of:

● its presence in the same type of waters which are the putative habitat of *R. seeberi*
● its presence as 'round bodies', a feature shared with *R. seeberi*
● its isolation from rhinosporidial polyps. It needs to be pointed out that the surface of rhinosporidial polyps, especially in nasal sites, is rough, a feature which allows the retention of contaminant aquatic micro-organisms. Moreover, the sporangia of *R. seeberi* are often invaded by host-cell infiltrates which include phagocytes; these phagocytes could contain *Microcystis* which could be visualized microscopically and be isolated in culture from rhinosporidial polyps
● no genetic or immunological homologies between *Microcystis* and *R. seeberi* were reported by Ahluwalia. Refutations of Ahluwalia's views were made by Arseculeratne (2000b) and by Mendoza et al. (2001b) on the basis of

– non-contamination of the organisms's DNA with human DNA
– absence of control samples from normal people inhabiting the same areas from where the author's water samples were collected
– absence of cultures of endospores and sporangia free of bacteria after multiple washes and filtration
– improper sampling of rhinosporidial tissue
– ultrastructural studies including the morphology of rhinosporidial mitochondria
– inability of *M. aeruginosa* to cause rhinosporidiosis in experimental animals
– the absence of reactivity of this cyanobacterium with human and experimental anti-*R. seeberi* antibody in immunofluorescence tests
– absence of reactions with anti-rhinosporidial sera in immunoblots against *M. aeruginosa* extracts
– the absence of amplification in PCR tests with primers based on the sequence data published by Herr et al. (1999b). A previous report on the culture of a mycelial fungus as the cause of rhinosporidiosis (Krishnamoorthy et al. 1989) had similar flaws.

Light microscopical and ultrastructural features

In parallel with its disputed taxonomy for many decades, a variety of inconsistent and overlapping terms has been applied to the morphological structures of the organism in its developmental stages (Table 24.2). Kennedy et al. (1995) proposed a new terminology for the standardization of the ontogenic nomenclature of the *R. seeberi*; these terms are used in this chapter. It is to be noted that these terms represent stages in a continuum of development of the organism; some of these terms, such as electron-dense body, should be regarded as tentative until more definitive studies, as suggested in the following discussion, are available.

Consensual descriptions of the light-microscopical and ultrastructural features of the different ontogenic stages of *R. seeberi*, as they appear in rhinosporidial tissues of humans and animals, have been made by many authors; rhinosporidial endospores and sporangia in various stages of development are uniformly found in histopathological material from diseased humans (Figure 24.3) and animals.

The 'life cycle' of *R. seeberi*, as deduced from histopathological appearances (Tirumurti 1914; Ashworth 1923; Kutty and Teh 1974; Savino and Margo 1983; Thianprasit and Thagernpol 1989; Kennedy et al. 1995), is represented in Figure 24.4. All developmental stages of *R. seeberi* are readily identified in most hematoxylin and eosin (H&E)-stained histopathological sections, irrespective of the location of the rhinosporidiosis and the species of the host.

Table 24.2 *New terminology of the ontogenic stages of* R. seeberi [a]

Earlier terminology	New terminology
Developmental stages	
Spore, immature spore, endospore, sporoblast, early trophic stage, spherule, conidium, sporule	Immature endospore
Mature spore, sporont, pansporoblast, merozoite, yeast phase in tissue, spore morula	Mature endospore
Trophocyte (early, intermediate, late), trophic stage, trophozoite, precleavage phase, immature sporangium, granular stage, sporocyst, cyst, spherule, sporangium	Juvenile sporangium
Endosporulating stage, sporulation phase, trophic sporangium	Intermediate sporangium
Mature trophocyte, mature cyst, endosporulating stage, post-cleavage phase, post-cleavage sporangium, adult stage	Mature sporangium
Intracytoplasmic structures in juvenile sporangium	
Lipid droplets	Lipid bodies
Multilamellar bodies	Laminated bodies
Intracytoplasmic structures in mature endospore	
Lipid bodies, cytoplasmic vacuoles	Lipid globules
Spherules, electron-dense circular structures, protrusions of cell wall, electron-dense inclusions, germinative bodies, sporozoites, spores, sporules, spherical bodies, spore morulae	Electron-dense bodies

a) Adapted from Kennedy et al. 1995.

THE LAMINATED (MULTILAMELLAR) BODY

Several authors have described these bodies as consisting of concentric rings of electron-dense material around a core of chromatin, and seen in juvenile and intermediate sporangia in human and animal rhinosporidial tissues (Kutty and Teh 1974; Savino and Margo 1983; Thianprasit and Thagernpol 1989; Kennedy et al. 1995). Thianprasit and Thagernpol (1989) regarded the laminated body (LB) as the precursors of the endospores. Apple (1983) found these bodies in mature endospores and suggested that, after discharge from the sporangia, the endospores released the LB and from these arose the juvenile sporangia in continuation of the life cycle of *R. seeberi* (see also Kennedy et al. 1995). Kutty and Gomez (1971) described the LB as being 1–1.5 μm in size, derived from the EDBs, and liberated from the endospores; they concluded that the LB is the earliest infective stage, giving rise to the juvenile sporangium. Further work is necessary to elucidate the ontogenic significance of the LBs.

ENDOSPORES

The endospores are produced by endosporulation in the sporangia, and are thought to be the asexual propagules of *R. seeberi*; there is no evidence of a sexual stage in the life cycle of *R. seeberi*. After nuclear divisions in the juvenile sporangia, endospores are formed by the condensation of the cytoplasm around the nuclei with the formation of cell walls; 'surfaces of contact remain distinguishable as a sort of network between the spores' (Ashworth 1923). The development of endospores from a germinal part of the inner sporangial wall was postulated by Minchin and Fantham (1905) (see also Beattie 1906; Savino and Margo 1983). Ashworth (1923) and Moses et al. (1991) noted the attachment of endospores to the inner aspect of the sporangial wall; Figure 24.5 illustrates the immature endospores and their probable origin from this site. Mature endospores were reported to be in the center of the sporangium (Ashworth 1923;

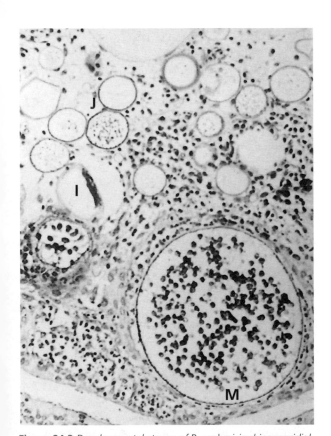

Figure 24.3 *Developmental stages of* R. seeberi *in rhinosporidial tissue. M, mature sporangium, with surrounding cell infiltrate; I, intermediate sporangium; J, juvenile sporangium. H&E; initial magnification ×400*

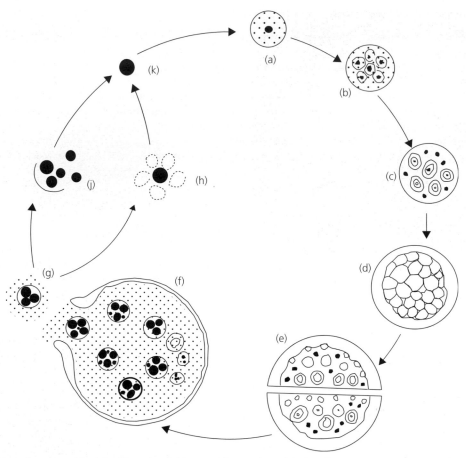

Figure 24.4 *Diagram (not to scale) of the life cycle of* Rhinosporidium seeberi. *(a) Juvenile sporangium;* **(b)** *juvenile sporangium with vesicles containing chromatin;* **(c)** *juvenile sporangium with laminated bodies;* **(d)** *endosporulating immature sporangium with the formation of endospore walls;* **(e)** *immature endospores formed on inner sporangial wall, maturing centripetally (upper half of figure) or formed at the center of the sporangium, maturing centrifugally;* **(f)** *mature sporangium and matrix containing mature endospores with EDBs, departing through annulus and pore;* **(g)** *mature endospore with EDBs;* **(h)** *endospore with disappearing 'nutritive spherules' (Ashworth 1923);* **(j)** *endospore liberated after dissolution of wall of 'spore morula' (Tirumurti 1914);* **(k)** *free endospore*

Shrewsbury 1933; Savino and Margo 1983; Noor-Sunba and al-Ali 1989; Mendoza et al. 1999), indicating a centripetal direction of maturation. Mature endospores may also be seen laterally (Figure 24.6); their location at the pore end of the sporangium (Beattie 1906; Ingram 1910; Tirumurti 1914; Ashworth 1923) is probably in preparation for their discharge through the sporangium's exit pore.

The mature endospores are spherical, measuring approximately 10–12 μm in diameter. Vanbreuseghem (1973) described endospores in Indian rhinosporidial tissue to be larger than those in African tissue. Endospores have a thick wall that is PAS positive (Figure 24.7) and contains chitin (Karunaratne 1964), as do fungi. Ultrastructurally, they were found to have a fibrillar coat (Figure 24.8, Arseculeratne et al. 1999, unpublished data) which, on an analogy with the fimbriae of gram-negative bacteria, could mediate the hemagglutination of red blood cells and adherence to mammalian cells (HEp-2) in culture, described below. The finely granular cytoplasm contains a vesicular nucleus. The most striking microscopical feature is the presence of several spherical bodies of approximately 1–1.5 μm in diameter; these include the EDBs (see Contents of endospores, below). Immature and mature endospores (see Figure 24.6), some of which appear to be degenerate, are present in mature sporangia (see Table 24.3).

Figure 24.5 *Intermediate thick-walled sporangium with formation of new endospores on inner sporangial wall. PAS stain; initial magnification ×400*

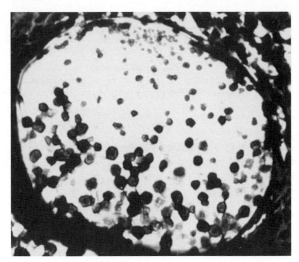

Figure 24.6 *Mature sporangium with atypical walls, containing mature endospores adjacent to the sporangial wall, with immature or degenerate endospores on the opposite side. H&E; initial magnification ×400*

Kutty and Teh (1975) described the endospore as having a prominent Golgi apparatus, endoplasmic reticulum, mitochondria, vacuolated spaces in the cytoplasm with an irregular plasma membrane, and the EDBs.

We have not detected motility of endospores in wet mounts of material from nasal or palpebral rhinosporidiosis; flagella or organs of motility were not seen on scanning electron microphotographs of endospores from different patients (Arseculeratne et al. 2002, unpublished data). The scanning electron microphotographs of Noor-Sunba and al-Ali (1989) also do not reveal flagella or organs of motility on the endospores; nor did Kutty and Teh (1974) note flagella on the sporangia.

Beattie (1906) described free endospores as having residual matricial material from the sporangium, which

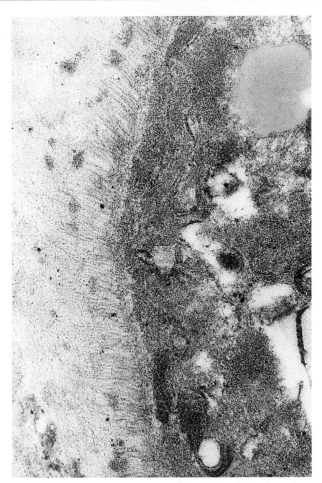

Figure 24.8 *Electron photomicrograph of endospore showing fibrillar coat.*

appears as an envelope around the endospore and trailing off, giving the appearance of a '. . . tail to a comet, the head being the spore-morulae' (endospore). It was pointed out (Arseculeratne 2000a) that in cytodiagnosis of rhinosporidiosis from smears, especially from nasopharyngeal sites, the residual cytoplasm of the epithelial cells (Figure 24.9a) could mimic the tail-like mucoid, matricial material around the endospore (Figure 24.9b). The rhinosporidial endospores with the mucoid covering could be distinguished from epithelial cells by the PAS stain as the nuclei of the latter do not stain magenta with this procedure, whilst the endospore stains strongly positive.

Contents of endospores: the electron-dense body

The most controversial morphological elements of *R. seeberi* are the electron-dense bodies (EDB). They are the most prominent inclusions of the endospores, and are 1–1.5 μm, spherical bodies appearing in clusters (Shrewsbury 1933) of 15–20 EDBs. Kennedy et al. (1995) described their ultrastructural features as 'irregular membrane-bound, vacuolated structures of various

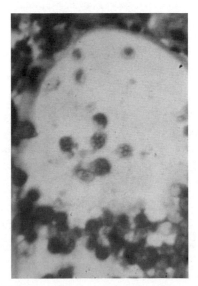

Figure 24.7 *Mature endospores with PAS-positive walls and internal EDBs. PAS stain; initial magnification ×400*

Table 24.3 *Morphological characteristics of rhinosporidial sporangia*

Sporangium	Characteristics
Juvenile	6–10 μm in diameter
	Unilamellar, PAS-positive, mucicarmine and GMS-negative wall
	Fibrillar or granular, PAS-positive, mucicarmine and GMS-negative cytoplasm
	Chromatin present as a single nucleus or as multiple nuclei, each within a vesicle
	Laminated bodies present
	Lipid globules present
Intermediate	100–150 μm in diameter
	Bilamellar wall, inner cellulose, PAS- and mucicarmine-positive
	Outer wall PAS-, mucicarmine- and GMS-negative
	No organized nucleus, lipid globules present
	Mature endospores absent
Mature	100–450 μm in diameter
	Thinner bilamellar outer wall
	Inner wall PAS- and mucicarmine-positive, GMS-variable
	Outer wall PAS-, mucicarmine- and GMS-negative
	Immature and mature endospores present, embedded in matrix
	Mature endospores with lipid globules and electron-dense bodies

GMS, Gomori's methanamine silver; PAS, periodic acid–Schiff.

sizes containing electron-dense granular material.' Their periphery stains deep magenta with the PAS stain (Figure 24.7) and they can be clearly seen in smears or wet mounts of the mucoid discharge on rhinosporidial growths (Figure 24.10). The EDBs have also been seen in the endospores in rhinosporidial tissue of dogs (Easley et al. 1986).

The EDBs have been shown to be yellow-fluorescent with acridine orange, Feulgen-positive and DNAase-sensitive (Vanbreuseghem et al. 1955; Lakshmanan et al. 1978; Savino and Margo 1983; Moses et al. 1991). We noted that these bodies when released by ultrasonic treatment of Percoll-purified endospores or in their native location within endospores, fluoresced yellow

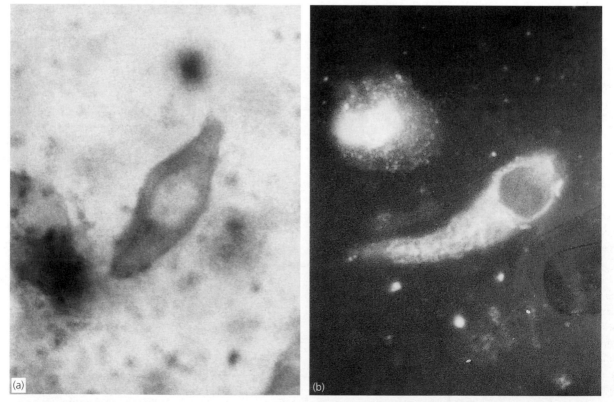

Figure 24.9 (a). *Normal bovine, nasal epithelial cell showing nucleus with residual cytoplasm in elongated fashion. PAS; initial mgnification ×1 000. Compare with* **(b)**. **(b)**. *Smear from human nasal rhinosporidiosis, showing endospore with residual matricial material from sporangium, mimicking an epithelial cell; indirect immunofluorescence with antirhinosporidial antibody. Compare with* **(a)**. *PAS; initial magnification ×1 000*

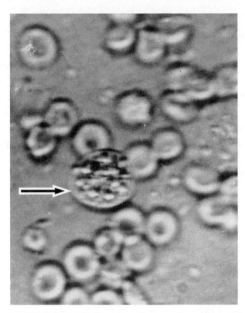

Figure 24.10 *Mature 10–12 µm endospore (arrow) containing EDBs, amidst red blood cells, in unstained wet mount of mucoid discharge from nasal rhinosporidiosis; initial magnification ×1 000*

with acridine orange, suggesting the presence of nucleic acid; acridine orange, however, is also claimed (Chayen and Bitensky 1991) to stain mast cell granules, cartilage matrix (red fluorescence), keratin (red fluorescence), vascular elastic elements (yellow fluorescence), but these elements have not been described in EDBs.

Lakshmanan et al. (1978) concluded from histochemical tests that the EDBs stained deeply with bromophenol blue, indicating that their contents are proteinaceous, as are the walls which also stain positively with Bismark Brown; deep staining of the wall of the EDB with Sudan Black B was interpreted as indicating that the wall of the EDB had lipoprotein. The EDB also stained deeply with methyl green and was Feulgen-positive. These workers concluded that: '. . . the spherule is made up of a lipoprotein coat with a protein matrix and a strongly Feulgen-positive centre' and that the EDB does contain deoxyribonucleic acid; they cite Vanbreuseghem et al. (1955) as also having demonstrated DNA in the EDB. Beattie (1906) had also suggested that the EDB are indeed the 'spores' and describes them as containing a 'well-defined nucleus which reacts to the chromatin stains.' Minchin and Fantham (1905) noted 'a central spot' in the 'refractile granule' (EDB), which stains with hematoxylin and concluded that it is 'a chromatic nucleus.' Teh and Kutty (1975) concurred with this view. Karunaratne (1964) noted that Vanbreuseghem et al. (1955) considered the EDB to be active bodies, the actual spores; this interpretation would then mean that the 'spore' of Ashworth (1923), now named the endospore, is really a collection of the ultimate true spores and hence meriting the term 'spore-morula' used by Beattie (1906) and Tirumurti (1914).

Kutty and Gomez (1971) recognized that '. . . one of the interior "spherical bodies"' shows 'a smaller and darker sphere enclosed within it . . .'; this 'would appear to represent a nucleus with a nucleolus.' This view is compatible with our observation that when the next stage of development, the juvenile sporangium, occurs, only one larger spherical body or very rarely two, appears within the juvenile sporangium (Figure 24.11). If this interpretation is correct, it would imply that this EDB is the ultimate generative, infective unit of *R. seeberi* as these and other authors have concluded. Lakshmanan et al. (1978) came to a similar conclusion: '. . . the spherules could be the initial inciting agent of the disease.'

We have found that endospores from fresh, unfrozen homogenates of rhinosporidial tissues stained deeply purple-blue in the cytoplasm and the EDBs with 3-[4,5-dimethyl thiazol-2-yl]-2-5-diphenyl tetrazolium bromide (MTT) (Figure 24.12; Arseculeratne and Atapattu 2003a). This dye is used in tests of the viability of diverse cells in culture (as in lymphoproliferative assays on lymphocytes – Mosmann 1983), yeasts, and mycelial fungi. The formation of the purple-blue colored formazan from MTT is due to the reduction of MTT by mitochondrial dehydrogenases and this phenomenon in the EDB, while being also shown in the endospore's

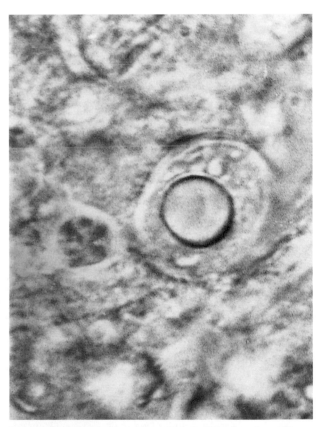

Figure 24.11 *Developing endospore in prejuvenile sporangium stage, showing single body (probably an EDB) which will develop into a juvenile sporangium. Wet mount of endospores in suspension; initial magnification ×1 000.*

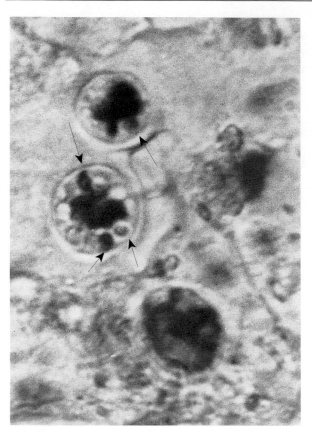

Figure 24.12 *Fresh endospores after incubation with MTT showing deep (purple-blue) staining of cytoplasm, and spherical internal bodies (arrows), probably EDBs. Other bodies of similar shape and size are MTT-negative and could be the protein nutritive spherules of Ashworth (1923) or the lipid bodies of Kennedy et al. (1995); initial magnification ×1 000*

cytoplasm (which contains mitochondria as described by Kutty and Teh 1975), would therefore indicate vital activity in these sites. If this conclusion is valid, then it is further evidence that the EDBs are indeed viable units.

These conclusions on the generative function of the EDB differ from those of Ashworth (1923) that the EDB are proteinaceous nutritive reserves; Shrewsbury (1933), too, disputed Ashworth's view. Easley et al. (1986), on the other hand, agreed with Ashworth (1923) that the EDBs are reserve bodies rather than mere 'protrusions of a morula-like cell wall of the spore' (Bader and Grueber 1970) or germinative bodies (Teh and Kutty 1975).

A basic question therefore, which needs resolution through an experimental model of pathogenicity, is what is the ultimate generative unit of *R. seeberi*? Is it the endospore, as most authors have assumed, or is it the EDB?

Other contents of the endospore

Eosinophilic, lipid globules were described (Kennedy et al. 1995) in the cytoplasm of the mature endospore and were regarded as nutritive reserves; they were absent in immature endospores. Some spherical bodies

that do not stain with MTT (Figure 24.12) could be the protein nutritive bodies postulated by Ashworth (1923) or the lipid bodies described by Kennedy et al. (1995). In correlation with MTT-stained endospores, transmission electron microphotographs (Figure 24.13) also revealed the EDBs that possessed an electron-dense core bounded by a thick electron-dense wall, in contrast to other bodies, perhaps the nutritive (protein or lipid) bodies that were electron-lucent and without a wall.

SPORANGIA

Table 24.3 summarizes the principal morphological features of *R. seeberi*'s sporangia. These, of varying size and shape and in various stages of development, are abundant in most rhinosporidial tissues. Sporangia were classified according to their maturity by Kennedy et al. (1995).

The juvenile sporangium (formerly, the trophocyte)

This developmental stage is the least controversial, and has been identified free in the stroma of infected tissue or even within mature sporangia (Karunaratne 1964). It is a thin-walled sphere of 10–15 μm diameter in its early stages (Figure 24.14), enlarging to about 100 μm (Bader

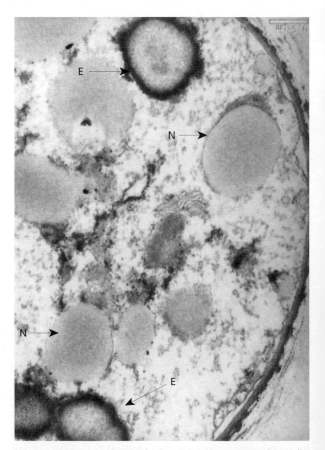

Figure 24.13 *Transmission electron microphotograph of Percoll-purified mature endospore showing EDBs with a relatively thick wall (E) and electron-lucent bodies, perhaps nutritive bodies without such walls (N); initial magnification × 16 000*

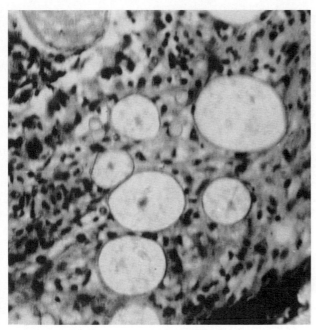

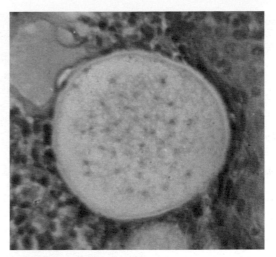

Figure 24.15 *Juvenile sporangium in the stage of nuclear division. H&E; initial magnification ×400*

Figure 24.14 *Thin-walled, early juvenile sporangia, each with a single nucleus and granular cytoplasm. H&E; initial magnification ×400*

and Grueber 1970; Savino and Margo 1983; Kennedy et al. 1995). Bader and Grueber (1970) found its wall to be laminated and that, histochemically, it consisted of a 'chitinous membrane incorporating PAS-positive neutral polysaccharides of a hemicellulose character.' It has variable intracelluar appearances according to its stage of development. Kutty and Teh (1975) recognized early, intermediate, and late stages of the juvenile sporangium.

The chromatin is either organized into a prominent single nucleus (Figure 24.14) with a nucleolus (Kutty and Teh 1975), or may be fragmented and diffuse (Tirumurti 1914). Kennedy et al. (1995) observed 'an aggregate of clumped and granular electron-dense material . . . and an interrupted double layered nuclear membrane . . . as the only nuclear remnants identified.' When the juvenile sporangium is about 50–60 μm in diameter, nuclear division (Figure 24.15) results in the appearance of chromatin within the vesicles (Figure 24.16) (Kutty and Teh 1974; Kennedy et al. 1995).

Kennedy et al. (1995) described ocular rhinosporidiosis in swans in which juvenile sporangia had unilamellar walls ranging from 2.4 to 9.9 μm in thickness and the wall was of electron-dense material which merged with the outer capsule of 0.9–1.5 μm thickness, which consisted of fibrils and hazy electron-dense material. This outer layer was considered in other descriptions as the outer of a bilamellar wall. In later stages, the filamentous structures are replaced by a 'faint encapsulation.' Kutty and Teh (1975) noted the similarity of the radiating structures on the surface to structures on the surface of *Cryptococcus neoformans* and suggested that this appearance was 'additional evidence in support of

its (*R. seeberi*'s) fungal nature . . . .' The pitfalls of using morphological similarities in taxonomy have been pointed out earlier in this discussion in relation to the phenomenon of 'morphological convergence' in fungi.

The granular or fibrillar cytoplasm (Kennedy et al. 1995) contains endoplasmic reticulum, mitochondria, lipid globules, membrane-bound vacuolated structures containing electron-dense material that stained positively with the Feulgen stain, and microbodies that are probably nutritive in nature and are used up as the juvenile sporangium matures. Laminated bodies, with two or more concentric rings surrounding a core, ranging from 1.4 to 2.6 μm were also noted throughout the cytoplasm.

The intermediate sporangium

Intermediate (immature) sporangia (Figure 24.17) are spherical bodies that measure approximately 100–150 μm in diameter. Kennedy et al. (1995) identified a bilamellar wall with an overlying, third capsular layer which thinned as the sporangium matured. Thianprasit and Thagernpol (1989) described this stage to possess a trilaminated wall that included an outer capsule of a 'myriad of curvilinear structures;' the capsule was considered to be the third layer – hence trilamellar. Some descriptions of intermediate sporangia exclude this layer in recognizing the sporangium's wall to be bilamellar. The subcapsular, outer chitinous layer is thicker than the capsular layer and is similar, structurally, to the unilamellar wall of the juvenile sporangium. The new inner layer, probably of cellulose, is electron-lucent and thinner, and is formed when the sporangium is about 100 μm in diameter. The intermediate sporangium contains immature endospores in the cytoplasm that is granular or fibrillar. An intermediate sporangium that is about 140 μm in diameter is said to contain several thousand endospores (Ashworth 1923). Laminated bodies have also been noted in the intermediate sporangia.

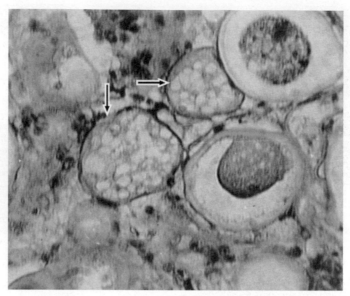

Figure 24.16 *Juvenile sporangia (arrows) with vesicles, after nuclear division. H&E; initial magnification ×400*

The mature sporangium

This spherical body measures approximately 150–400 μm in diameter, and is located in the stroma of the rhinosporidial polyps, as well as intraepithelially (Figure 24.18) and in downgrowths of the squamous epithelium into the subepidermal tissue (Figure 24.19). The mature sporangium has a trilamellar (Savino and Margo 1983) wall that is thinner (1–3 μm) but is similar, tinctorially, to that of the intermediate sporangium. The outer wall contains chitin, whilst the inner wall has cellulose (Ashworth 1923; Shrewsbury 1933; Karunaratne 1964; Jain 1967), and is more translucent. The possession of chitin is a property that is shared with the fungi; the presence of a chitin-synthase gene

in *R. seeberi* was reported by Herr et al. (1999a). Striations of the inner sporangial wall were noted by Beattie (1906) and by Ashworth (1923), although their existence was disputed by Minchin and Fantham (1905), and by Kutty and Teh (1974). A specialized, thickened (15 μm) area of the sporangial wall is the annulus (Figure 24.20) which surrounds the exit pore (Beattie 1906; Ingram 1910; Tirumurti 1914; Ashworth 1923; Karunaratne 1964; Grover 1970; Mendoza et al. 1999), which is the aperture through which the endospores are extruded.

When the sporangium matures, the wall becomes thinner while the annulus surrounding the pore remains thick (Figure 24.20). With maturation, the bilamellar (Savino and Margo 1983; Thianprasit and Thagernpol

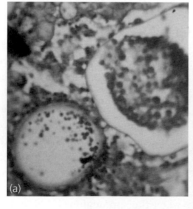

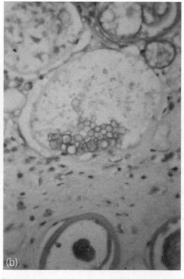

Figure 24.17 (a) *Intermediate (thick-walled) and mature (thin-walled) sporangia in rhinosporidial tissue. PAS stain; initial magnification ×400.* **(b)** *Intermediate, thick-walled sporangium with new formation of endospores on inner wall. PAS; initial magnification ×400*

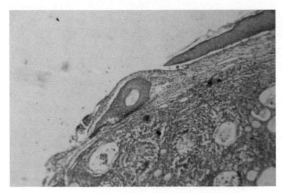

Figure 24.18 *Intraepithelial effete, thick-walled sporangium in subcutaneous rhinosporidiosis. PAS; initial magnification ×200*

1989) wall retains filamentous radiations or 'curvilinear tubular structures' on the outer surface; the 'dense network of threads' in the cell wall was thought to consist of cellulose (Bader and Grueber 1970).

The sporangial walls are PAS-positive as are the walls of the endospores, while also staining with Alcian Blue, indicating the presence of acid mucopolysaccharides; other tinctorial properties were described by Rao (1966) and Jain (1967). PAS-positivity decreases with maturation of the sporangial wall (Jain 1967). Rao (1966) and Jain (1967) considered the presence of acid mucopolysaccharides to confer selective permeability to nutrients, on the sporangial wall. It is relevant that Mendoza et al. (1999) postulated that sporangia apparently absorb nutrients through the walls during the active process of extrusion of the endospores. Extensive histochemical studies on the different developmental stages were also made by Bader and Grueber (1970).

The walls of degenerate sporangia could have very thick, homogeneous walls (Beattie 1906; Karunaratne 1964), or the wall could be absent, and when the sporangium is empty of endospores, it could pose a diagnostic problem in histopathology (Arseculeratne et al.

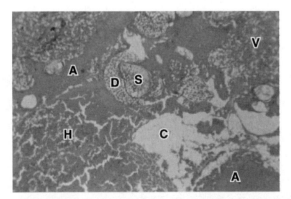

Figure 24.19 *Degenerate sporangium (S) in nasal rhinosporidiosis; it is encircled by ancanthotic squamous epithelium (A) surrounded by dense cell infiltrate (D), in vascular stroma (V) with hemorrhage (H) and cystic spaces (C). H&E; initial magnification ×400*

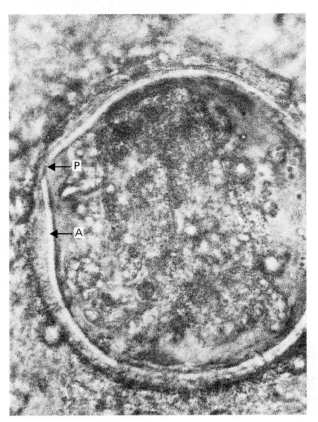

Figure 24.20 *Intermediate sporangium with annulus (A) and pore (P). Wet mount; initial magnification ×400*

2001). Empty, degenerate sporangia may collapse on themselves, giving characteristic linear markings that represent infoldings of the wall. A homogeneous, eosinophilic, vacuolated material sometimes fills the empty sporangia (Figure 24.21; Jimenez et al. 1984). Degenerate sporangia could be seen to have within them infiltrates of host-inflammatory cells (Figure 24.22) and multinucleated giant cells (Figure 24.23).

Herr et al. (1999b) reported that *R. seeberi* possesses mitochondria with flat cristae, a finding that contrasted with that of Fredricks et al. (2000). However, recently Mendoza et al. 2001a) presented new evidence that supports Herr et al.'s (1999b) observations. The presence of mitochondria with flat cristae in *R. seeberi* is also in agreement with the presence of mitochondria with flat cristae in the other members of the Order Dermocystida (the Rosette Agent and *Dermocystidium*).

Ashworth (1923) estimated that a mature sporangium contains about 16 000 endospores, many of which fail to mature and then appear morphologically and tinctorially abnormal.

The mature endospores within a mature sporangium are embedded in an intrasporangial matrix and appear spaced (Figure 24.24). The matricial mucoid material stains with Alcian Blue, suggesting the presence of mucopolysaccharides (Savino and Margo 1983; Thianprasit and Thagernpol 1989) that were considered as

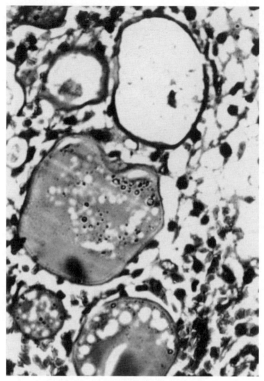

Figure 24.21 *Degenerate, thin-walled sporangia that are empty or contain vacuolated, homogeneous eosinophilic material; nasal rhinosporidiosis. H&E; initial magnification ×400*

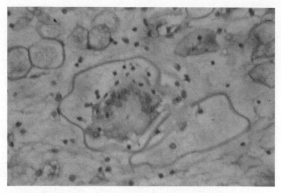

Figure 24.23 *Degenerate sporangium invaded by a multinucleated giant cell. PAS; initial magnification ×400*

mucoid matrix or individually as unenveloped bodies (Figure 24.25). The role of the matrix in the release of endospores is referred to below.

The large, 400–450 μm, subepithelial mature sporangia give the typical strawberry-like appearance to the rhinosporidial polyps, the surface of which is studded with yellow, pinhead-sized spots representing the sporangia.

The sporangia of *R. seeberi* could resemble the spherules of *Coccidiodes immitis* but can be differentiated from the latter by being larger, having thicker walls, and with more numerous endospores and an exit pore. The sporangium of *R. seeberi* also differs in staining with

inhibitors of phagocytosis of *R. seeberi* in the host tissue (Thianprasit and Thagernpol 1989); it was also thought to inhibit the penetration of drugs (Woodard and Hudson 1984). Residual portions of this mucoid material might be represented by an envelope on the endospores that leave the sporangium; this matricial residue might also appear as the 'tail of a comet' (Beattie 1906; Figure 24.9b), and show intense staining in immunofluorescence tests (Atapattu et al. 1999), suggesting the high antigenicity of this material. Smears of crushed polyps reveal the endospores in masses embedded in the

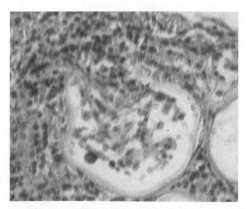

Figure 24.22 *Invasion of an effete sporangium by fibrocellular stromal tissue in a submental rhinosporidial mass in disseminated rhinosporidiosis. H&E; initial magnification ×400*

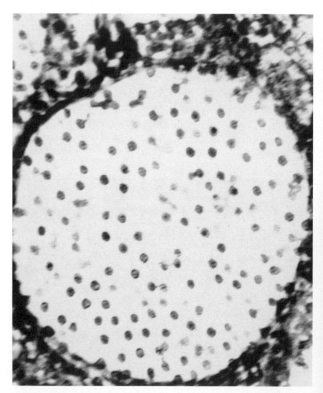

Figure 24.24 *Mature endospores, separated from each other by matrix in mature sporangium; nasal rhinosporidiosis. H&E; initial magnification ×400*

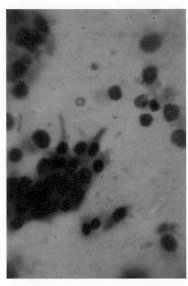

Figure 24.25 Smear of homogenate of nasopharyngeal rhinosporidial growth, showing mature endospores, unenveloped or embedded en masse in homogeneous material which also appears as envelopes, bipolar wisps or 'tails' on individual endospores. Leishman's stain; initial magnification ×1 000.

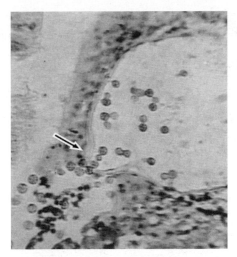

Figure 24.26 Mature sporangium in nasal rhinosporidiosis with mature endospores exiting through the annulus and pore (arrow). H&E; initial magnification ×400

mucicarmine, unlike those of *C. immitis* (Kwon-Chung and Bennett 1992).

Release of endospores from sporangia

Before extrusion, mature endospores are relocated close to the pore. The mature sporangium's pore is surrounded by a thickened annulus of the sporangial wall (Figure 24.20), through which the endospores are extruded into the surrounding tissues to perpetuate the 'life cycle' of *R. seeberi*. de Mello's (1954) review included a figure (Figure 24.23) that illustrated a lid or operculum that closed the pore; a 'film' or 'thin disc' covering the pore was also referred to by Ashworth (1923) and by Karunaratne (1964). A pore might not be visualized in some sections of tissue (see Bader and Grueber 1970); this is attributable to the plane of sectioning of the sporangium falling outside the level of the pore.

It had been been assumed that the escape of endospores resulted from rupture of the sporangium from external tissue pressure (Bader and Grueber 1970) or from within the sporangia (Ashworth 1923; Karunaratne 1964) with the endospores being 'shot-out' from the sporangium after its rupture (Beattie 1906) or after the 'thin film over the pore gives way' (Ashworth 1923). Evidence, also documented on a video recording, was produced for the operation of an active, vital mechanism, probably osmotic, for the extrusion of the endospores (Mendoza et al. 1999). de Mello (1954), too, postulated an osmotic mechanism. The endospores are extruded in an orderly procession (Figure 24.26) and not explosively, *en masse*, through the exit pore.

Isolation and purification of developmental stages

The inability to obtain pure preparations of the developmental stages of *R. seeberi*, free of contamination with tissues of the host or with environmental microorganisms, has been the cause of the lack of unequivocal observations and firm conclusions on the biology and immunology of the pathogen, as well as on immunity mechanisms in diseased hosts. Crude preparations of rhinosporidial bodies, derived from homogenized polyps, have been used after sedimentation, coarse filtration, or pipetting-out from the homogenates. These preparations have probably had contamination from host material and the use of such crude 'antigens' could have led, for instance, to the identification of 'rhinosporidial antigenemia' on immunoprecipitation in gel-diffusion tests (Chitravel et al. 1982). It is more likely that the precipitin reactions that were observed were between contaminating human antigens and the corresponding antibodies from the animal sera which were obtained by immunization against these crude rhinosporidial preparations derived from human tissue.

A method was described (Atapattu et al. 1999) for the isolation, in the pure state, of the different developmental stages of *R. seeberi*, from homogenates of rhinosporidial tissue. This method is based on centrifugation of deposits of tissue homogenates, in density gradients of Percoll (Figure 24.27). The EDBs may be obtained by controlled disintegration of the endospores by ultrasonic or enzymic (chitinase) means. PAS-stained smears of chitinase-treated suspensions of endospores (calcium phosphate buffer at pH 6) showed that the intensity of PAS positivity of the endospore's wall decreased, with better visualization of their EDBs (Arseculeratne 2002, unpublished data).

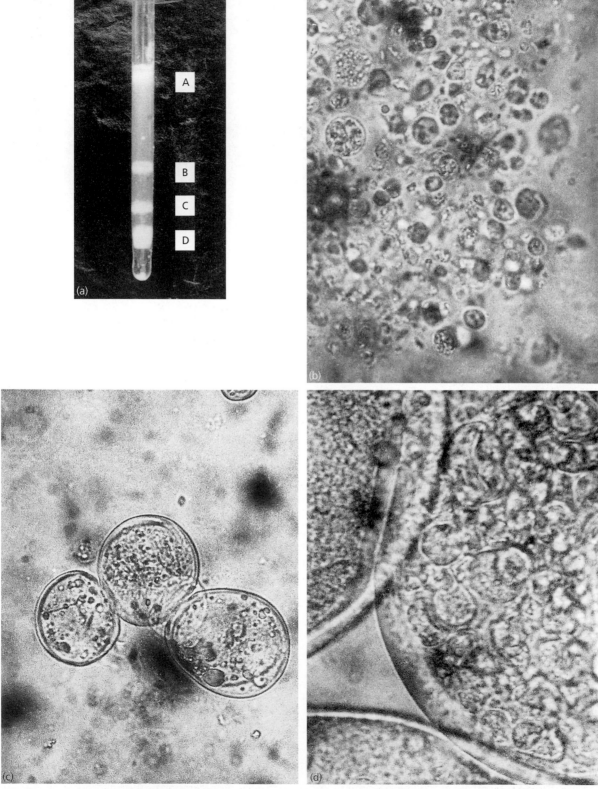

Figure 24.27 (a) *Fractionation of homogenate of rhinosporidial tissue in Percoll of graded density. A, small endospores; B, larger endospores; C, juvenile sporangia; D, intermediate sporangia. Mature sporangia were not visualized, probably after rupture.* **(b)** *Small and large endospores after fractionation in Percoll. Wet mount; initial magnification ×400.* **(c)** *Juvenile sporangia after fractionation in Percoll. Wet mount; initial magnification ×400.* **(d)** *Intermediate sporangia after fractionation in Percoll. Wet mount; initial magnification ×400.*

When aseptic conditions are used in the Percoll purification, the isolated rhinosporidial bodies are free even of bacterial contamination. These preparations would permit investigations, particularly on the immunology and experimental pathogenicity of *R. seeberi*.

VIABILITY OF ENDOSPORES

The question of the viability of the endospores, in homogenates that have been stored at low temperatures or after purification in Percoll, and used for inoculation of the test animals, is a problem that has a bearing on experiments on the pathogenicity of *R. seeberi*. Homogenates of human or animal rhinosporidial tissues have been used by many workers to attempt infection of experimental animals. With congenitally immunodeficient mice (Arseculeratne et al. 2000), the homogenate was mixed with sonicates of Percoll-purified endospores with their ultrasonically liberated EDBs. With the inability to culture *R. seeberi*, in vitro, and the absence of a method for the direct assessment of the viability of these bodies at the time of experimentation, the viability of the endospores and sporangia was assumed in all these experiments. Nigrosin-eosin or nigrosin-neutral red was found to be useful in determination of viability through non-uptake of the dye (Jain 1967). Arseculeratne and Atapattu (2003a) described the use of Evan's blue in a simple test for the assessment of the morphological integrity of *R. seeberi*'s endospores; it is based on their staining by the vital dye Evan's blue (Figure 24.28); unfrozen, fresh endospores were also found to stain deeply (purple-blue) with MTT (Figure 24.12; Arseculeratne and Atapattu 2003a), which is used to assess the viability and hence proliferation of lymphocytes in lymphoproli-

ferative assays for cell-mediated immune competence, and for assessing the viability of fungi. Endospores, after autoclaving, formalinization, and trypsinization, failed to take up the Evan's blue or MTT, whilst endospores from freshly homogenized human rhinosporidial tissue stained deeply, in its cytoplasm and EDBs. *Toxoplasma* behaves in a similar manner in the Sabin–Feldman test (Sabin and Feldman 1948) when dead cells fail to stain with methylene blue. Purified endospores after prolonged storage at –20°C retained their staining with Evan's blue but failed to stain with MTT (Arseculeratne and Atapattu 2003a). Apparently, staining with Evan's blue reveals the morphological integrity of the endospores, especially the EDBs, whilst MTT staining indicates their viability on account of the need for enzymic mechanisms (mitochondrial in eukaryotic cells) to reduce the MTT to the purple-blue colored formazan.

Biological activities of endospores

HEMAGGLUTINATION

Percoll-purified suspensions of endospores from human rhinosporidial tissues were found to cause hemagglutination of red blood cells from humans, a species that is susceptible to infection with *R. seeberi*, and from rats (Figure 24.29), that have not been reported as susceptible; the hemagglutination occurred as well at 37°C as at 4°C, and was reactivated after agitation and disruption of the clumps of hemagglutinated red cells after first incubation (Arseculeratne and Atapattu 2003b).

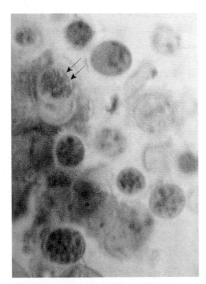

Figure 24.28 *Fresh endospores stained with 0.2 percent Evan's blue in PBS, showing deeply stained bodies probably EDBs, with moderately stained cytoplasm, and unstained spherical bodies (arrows); initial magnification ×1 000.*

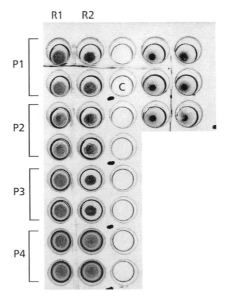

Figure 24.29 *Hemagglutination of rat red blood cells by purified rhinosporidial endospores. C, control red cells in PBS; R1, R2, rats from which red cells were obtained; P1, P2, P3, P4, purified endospores from four rhinosporidial patients.*

ADHERENCE

Endospores were reported to show strong adherence to epithelioid cells from a human rectal tumor, in vitro (Levy et al. 1986). Endospores also showed adherence to HEp-2 cells (derived from human laryngeal carcinoma) in culture in Minimum Essential Medium (MEM) with 10 percent fetal calf serum (Figure 24.30); an intriguing feature of the adherent or free endospores in mixtures with these mammalian cells was the appearance of bodies that were similar in size and shape to the EDBs on the surface of the endospore (Figure 24.31; Arseculeratne and Atapattu 2003b), suggesting that the EDBs were in the process of extrusion from the endospores on contact with the HEp-2 cell. Similar 'protrusions' of the endospore's cell wall were described by Bader and Grueber (1970) and by Teh and Kutty (1975), whilst the latter authors considered the protrusions to contain EDBs. The significance of this phenomenon remains to be elucidated. The HEp-2 cells that showed adherent endospores numbered only about half of the total cells present and a similar phenomenon, attributed to variations in the density of receptors for adherence on the buccal cell-surface, was reported for *Candida albicans* (Sandin et al. 1987).

It was previously pointed out (Atapattu et al. 2000) that the endospore had a fibrillar surface as revealed on transmission electron microscopy (Figure 24.8), which might, on a parallel with the adhesive fimbriae of gram-negative bacilli, be the agent responsible for the adherence to mammalian cells and for hemagglutination.

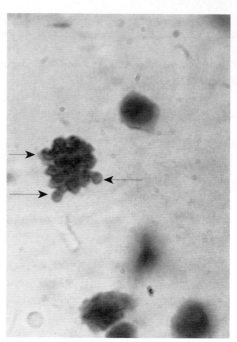

Figure 24.31 *Protruding bodies (arrows), compatible in size and shape with EDBs, attached to endospore in suspension with HEp-2 cells. PAS; initial magnification ×1 000.*

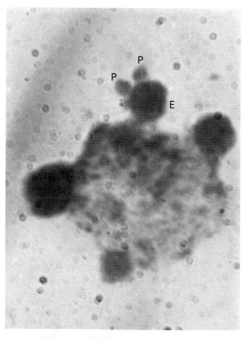

Figure 24.30 *Adherence of purified endospores to HEp-2 cells in MEM with 10 percent fetal calf serum. One endospore (E) has two protrusions (P) suggestive of extruding EDBs. PAS stain; initial magnification ×1 000.*

Sensitivity of *R. seeberi* to antimicrobial drugs and biocides

On account of the inability to culture *R. seeberi* in vitro, conventional tests of sensitivity to antimicrobial drugs and to biocides have not been possible with *R. seeberi*. There have been many agents that were tested clinically. Quinine hydrochloride (Wright 1922), salts of antimony and bismuth, iodine, and pentamidine (Rajam et al. 1955) griseofulvin, amphotericin B, and topical steroids (Jimenez et al. 1984; Ho and Tay 1986), local and systemic antibacterial antibiotics and radiotherapy (Satyanarayana 1966) were clinically ineffective, as were other compounds (see de Mello 1954). Allen and Dave (1936), discussing the treatment of rhinosporidiosis, recorded negative results in therapy with a variety of other compounds. The only agent that is now used in therapy is dapsone (4,4'-diaminodiphenyl sulphone, DDS), and the evidence for its efficacy was obtained from clinical observations and microscopic examination of the pathogen in rhinosporidial tissue from treated patients (discussed under Treatment below). Since leprosy, which also occurs in the same regions of rhinosporidial endemicity (in Sri Lanka), is also treated with this drug, an alternative drug for rhinosporidiosis will be advantageous, in the event of the development of resistance of *Mycobacterium leprae* to dapsone. An additional justification for an alternative drug is that dapsone precipitates hemolytic crises in patients with glucose-6-phosphate dehydrogenase (G6PD) deficiency, which is common in the regions of rhinosporidial endemicity in Sri Lanka.

The development of sporangia from endospores in aqueous suspension under storage (described under in vitro culture, below) was used in an attempt to test the sensitivity of *R. seeberi* to various drugs in vitro (Arseculeratne 1999, unpublished data). Viability of *R. seeberi* was deduced from the formation, over a period of 1–2 weeks, of juvenile sporangia of at least 50 μm in diameter as observed microscopically in flat-bottomed wells of cell culture trays. It was difficult to standardize the optimum conditions for sporangial development from the endospores, and the method was abandoned. A further disincentive to pursue this technique was the long period of observation required; the development of sporangia from endospores, or 'maturation' as Easley et al. (1986) termed the phenomenon, was noted to be upto 35 days while Grover (1970) noted the development within 10 days.

The Evan's blue test of the morphological integrity of endospores has been used to investigate their sensitivity to various biocides and antimicrobial agents (Arseculeratne 2002). The EDBs of untreated endospores stained intensely while the cytoplasm had a lesser intensity of staining (Figure 24.28). The end point used was the loss of staining of the EDBs and the cytoplasm, as in endospores that were treated with autoclaving and formalinizing. Chloroxylenol, chlorhexidine, Betadine, cetrimide, Tincture Iodine, and 10 percent formalin produced marked loss of staining with Evan's blue. Endospores treated with these agents and stained with MTT also showed a similar pattern of loss of staining (Arseculeratne 2002, unpublished observations).

CULTURE

The terms 'growth' and 'development' need definition in discussing the literature on culture of *R. seeberi*. By 'growth,' in this discussion, is meant the increase in size of a given stage, e.g. the enlargement of a sporangium. By 'development' is meant the sequential formation of the ontogenic stages, e.g. the development of an endospore into a juvenile sporangium. 'Development' may also be equated with 'maturation,' a term used by Easley et al. (1986). 'Serial propagation' refers to the repetition of the cycle of development of each stage to the mature sporangium, the release of its endospores and the repetition of the cycle. These differentiations might serve to avoid confusion in deducing from observations on the formation of sporangia from endospores in vitro, that true, cyclical development of the ontogenic stages occurred.

Culture on inanimate media

Numerous attempts are recorded in the literature as not having produced definite evidence of serial propagation of *R. seeberi*, on inanimate culture media (Shrewsbury 1933; Dhayagude 1941; Satyanarayana 1960; Reddy and Lakshminarayana 1962; Kennedy et al. 1995). A wide variety of conventional bacteriological culture media has been used unsuccessfully, sometimes with exotic additives such as horse dung, prune juice, and hydrocele fluid, and incubated under various conditions of temperature and aerobiosis. Claims have, however, been made of successful culture in inanimate media (see Karunaratne 1964), but none of these reports provided unequivocal evidence for serial propagation of *R. seeberi*, in vitro. What appears to have been interpreted as propagation was in fact the maintenance of morphological integrity and continuation of the development of the endospores for a limited period under suitable conditions such as low temperature, and an environment such as that in monolayer cultures of cells. The limited ontogenic development is not indicative of serial propagation of *R. seeberi*.

Other reports of allegedly successful culture of *R. seeberi* in vitro (e.g. the mycelial fungus of Krishnamoorthy et al. 1989) dealt with contamination by environmental microorganisms or by organisms that were confused with *R. seeberi* (e.g. *Microcystis aeruginosa* of Ahluwalia et al. 1997; Ahluwalia 1999).

Grover (1970) suspended material from human rhinosporidial tissues in the synthetic medium TC 199 at 4°C and observed a rapid increase in the number of endospores and the formation of sporangia; the organism was maintained for 4 months. No mycelium was formed. We have made similar observations with Percoll purified endospores (Arseculeratne et al. 1996, unpublished data) in phosphate buffered saline or in tissue culture media, although the development did not proceed beyond the stage of immature sporangia; at no stage did we observe endosporulation in, and endospore-release from, these sporangia. Various conditions, such as temperature (4 °C, room temperature ambient at 28–30 °C), bubbling of air through the suspension, alternative suspending media (distilled water, phosphate-buffered saline pH 7.2, natural ground water) were tested, but the conditions for reproducible ontogenic development are as yet obscure. This limited in vitro development has therefore been unable to provide the basis for a standardizable, reproducible method for the assay of sensitivity of *R. seeberi* to antimicrobial agents.

Growth in cell cultures

Cultures of mammalian cells have also been tested for their ability to support the serial propagation of *R. seeberi*.

Levy et al. (1986) described attempts to grow the organism from canine nasal rhinosporidiosis in monolayers of a cell line from an epithelioid human rectal tumor (HRT); they reported strong adherence of the organism 'within minutes' to the cells. They also claimed to have observed, after 2 days, 'polyp-like structures' consisting of HRT cells and mature sporangia releasing endospores into the medium. The analogy between their polyps in cell cultures, which appeared within 2 days, and the polyps in clinical cases appears doubtful, not only in respect of the time interval of several months required for the development of the natural lesions, but also because the clinical polyps are consequent to the host's tissue response (epithelial proliferation, edema, with cellular infiltration into the fibrovascular stroma) surrounding the multiple sporangia and endospores. These authors regarded the need for contact with mammalian cells for the complete development of the pathogen, and the absence of such cells as a reason for the failure of culture of the pathogen on inanimate culture media. Easley et al. (1986) documented similar results. There was focal proliferation of the tumor cells around the organism, the numbers of which increased; the authors claimed that this 'reproduction' continued for over 3 months in subculture. No independent confirmation of these results has been reported.

Whether these observations were similar to those reported by Grover (1970) is not clear. The development of endospores into sporangia, which was earlier reported by Grover (1970), was also noted by Levy et al. (1986), who added that this development occurred also in the absence of the HRT cells.

We have observed the development of immature sporangia from endospores that were added to cultures of HEp-2 or HeLa cells in MEM with 10 percent fetal calf serum, or even in phosphate buffered saline (PBS) pH 7.2 thawed after storage at −20°C, although the formation of mature sporangia with mature endospores was not observed.

Kennedy et al. (1995), however, did not note any evidence of growth of R. seeberi in fresh rhinosporidial tissue, in culture media with 10 percent fetal bovine serum.

IMMUNOLOGY

Data on the immunology of the developmental stages of R. seeberi were unavailable until recently, attributable to the absence of a method for obtaining pure cultures of the organism in vitro. The method described above for the purification of the developmental stages of R. seeberi from homogenates of rhinosporidial tissue, using graded dilutions of Percoll, now permits the isolation of these stages, free of contamination from host tissue and contaminant microorganisms. EDBs may also be obtained from the endospores by controlled sonic disintegration.

Herr et al. (1999c) used immunoelectron microscopy to demonstrate an antigen that first appears immediately under the mature sporangial wall; this antigen was apparently absent in sporangia of earlier stages of maturation and in endospores. The authors stated that

this was 'the first report in which an antigenic material with a potential role in the immunology of rhinosporidiosis has been detected.' This finding also suggests the occurrence of antigenic variation in R. seeberi that might be a means of immune evasion of this pathogen.

A further finding that is relevant to the pathogenesis of rhinosporidiosis, in particular to its chronicity, recurrence, and dissemination, is that sonic extracts of purified rhinosporidial endospores and sporangia showed evidence on immunoblots of the presence of human immunoglobulins (IgG, IgM, and to a lesser degree IgA) in the rhinosporidial antigens; this suggests that either immunoglobulins were bound to the rhinosporidial antigens or that these antigens had epitopes that cross-reacted with those on human immunoglobulins (Atapattu et al. 2000). Azadeh et al. (1994) reported the presence of IgG and IgA on walls of early and mature sporangia and in endospores; IgM was absent. This phenomenon might also contribute to immune evasion by R. seeberi.

The phenomenon of binding by pathogens of host proteins has also been described in bacteria (Boyle 2000) and in schistosomes (Roitt et al. 1993).

PATHOGENICITY IN EXPERIMENTAL ANIMALS

An experimental animal model of rhinosporidiosis has yet to be established. This lack has contributed to the paucity of information on immunity mechanisms in rhinosporidiosis, some aspects of the developmental 'life cycle' and the in vivo response of R. seeberi to antimicrobial drugs.

Rhinosporidiosis occurs in many species of farm, domestic, and wild animals. Some of these species (cats, goats, horses, bovines), conventional experimental animals (guinea pigs, mice, rabbits, rats), and other animals such as monkeys and snails, and even a human volunteer (Habibi et al. 1944, quoted by Karunaratne 1964) have been inoculated with homogenates or powdered dry material from rhinosporidial tissue, through diverse routes – subconjunctival, subcutaneous, intradermal, inreaperitoneal, intravaginal, intratesticular, as well as by scarification of the skin, buccal and nasal mucosa, and through the gills of fish, the mantle and mouth parts of snails. No sustained or progressive disease occurred (see Karunaratne 1964). The chorioallantoic membrane, yolk sac, and the amniotic cavity of developing chick embryos also failed to support growth of R. seeberi. On a superficial resemblance of the lesions and tissue forms produced by R. seeberi, to those of a Dermosporidium spp. in frogs, Karunaratne and Dissanaike (A.S. Dissanaike, 1995, personal communication) injected homogenized material from human rhinosporidiosis into frogs (Rana temporaria) but no progressive disease ensued. We have also found five genera/species of amphibia to be insusceptible to rhinosporidial infection (Arseculeratne 1999, unpublished data). Since the frog pathogen *Dermocystidium* is now included with *Rhinosporidium* in the proposed, new Order Dermocystida of the Class Mesomycetozoea (Herr et al. 1999b; Mendoza et al. 2002), it is relevant that Carini (1940) failed to produce the disease in the frog *Hyla rubra*, with the frog pathogen.

Negative results were also produced by intraperitoneal injection of suspensions of endospores and sporangia in 5 percent hog gastric mucin or in 100 percent Percoll in mice (Arseculeratne, 2000, unpublished data).

Iatrogenically immunosuppressed animals (steroids, Jain 1967; cheek pouch of hamsters under steroids, Grover 1970; cyclophosphamide in C3H mice, Arseculeratne 1998, unpublished data; x-irradiated, Satyanarayana 1960) also proved to be insusceptible. Congenitally immunodeficient mice also failed to support the growth of *R. seeberi*; Levy et al. (1986) stated that 'aged spores' failed to produce infection in nude mice when the 'spores' were inoculated, on the scarified nasal mucosa, intramuscularly or intraperitoneally. Congenitally immunodeficient SCID and nude mice failed to support the growth of endospores, sporangia and EDBs released from endospores, purified from human rhinosporidial tissue (Arseculeratne et al. 2000).

A uniform finding in all these experiments was that the injected material showed only degenerate sporangia that were subsequently invaded by fibrous tissue, and that the endospores had disappeared in approximately 3 weeks.

The failure to establish infection might have resulted from nonspecific immunity, especially through macrophages, which might have prevented the initial establishment of infection during the first few days after inoculation, rather than from specific adaptive immune responses that would have been operative later.

A second factor that might be involved in the failure of tests of experimental pathogenicity is the viability of the rhinosporidial endospores and sporangia that were obtained from homogenized tissues. Using the MTT dye reduction test described above, we observed that Percoll-purified endospores which had been stored at −20°C for several months, failed to stain with the dye, indicating a reduction of viability (Arseculeratne and Atapattu 2003a). The age of the endospores used as inocula in the experiments reported in the literature has not been stated and the possibility remains that low viability of the endospores contributed to the negative results.

A further factor that might be absent in inocula prepared from rhinosporidial tissue is concomitant microbiota – hydrophilic microorganisms in the putative natural aquatic habitat of *R. seeberi*; it is conceivable that some of these organisms might be synergistic with *R. seeberi* as has been demonstrated of *Wolbachia* spp. with filarial nematodes, lactobacilli with *Trichomonas vaginalis*, and *Staphyococcus aureus* with *Candida albicans*.

RHINOSPORIDIOSIS

Rhinosporidiosis is an infective disease in the sense that the pathogen is detectable in the lesions in rhinosporidiosis. No evidence has, however, been adduced that it is transmissible as there has been no unequivocally documented instance of cross-infection between members of the same family or between animals and humans.

Cases have been overwhelmingly sporadic but two outbreaks have been described. In Serbia an outbreak of ocular and nasal rhinosporidiosis in 17 humans was reported by Vukovic et al. (1995), whilst Kennedy et al. (1995) described an outbreak of ocular and nasal rhinosporidiosis in captive swans in Florida, USA.

EPIDEMIOLOGY

Geographical distribution

Sporadic cases of the disease have been reported from more than 60 countries of different geographical characteristics (Table 24.4), although it occurs predominantly in tropical and subtropical regions. Southern India and Sri Lanka have had the highest incidence; other countries where the disease is endemic include Argentina, Brazil, USA (Texas), and Uganda. Incidence rates, even in hyperendemic countries, vary with region; in India, South India has a high incidence whereas the disease is rare in North India (Paul et al. 1978). In Sri Lanka, the predominant focus is the dry zone. The south central region in the USA is an endemic-enzootic region (Wallin et al. 2001). Karunaratne (1964) noted that ocular rhinosporidiosis (oculosporidiosis) was commoner in South Africa.

In Western temperate and Middle Eastern countries, most cases of human rhinosporidiosis occurred in expatriate Indians who probably acquired the disease in their native lands (Al-Hili 1985; Matusik et al. 1986; Mears and Amerasinghe 1992; Ohgaki et al. 1994). On account of this feature, increasing international movement of persons from countries with endemic rhinosporidiosis might result in a more frequent occurrence of rhinosporidiosis in nonendemic countries; the disease might then merit the term 'an emerging infection.' Yet a few cases of rhinosporidiosis have been reported in persons, living in the West, who have never traveled to endemic areas (Wright 1907; Weller and Riker 1930; Lasser and Smith 1976; Woodard and Hudson 1984; Gaines et al. 1996). The disease in all these cases, irrespective of the country of occurrence, has shown remarkably uniform clinical and pathological features.

Reasons for endemicity

It has also to be explained why the disease is of high endemicity in certain regions of southern India and in

Table 24.4 *The geographical distribution of rhinospiridiosis in humans and animals*

Region	Countries
Africa	Burundi, Cameroons, Chad, Congo, Egypt, Ethiopia, Gabon, Ghana, Guinea, Ivory Coast, Kenya, Liberia, Madagascar, Malwai, Martinique, Mozambique, Nigeria, Rwanda, South Africa, Sudan, Tanzania, Uganda, Zaire, Zambia, Zimbabwe
North America	Canada, Mexico, USA
Central America	Belize, El Salvador
Caribbean Islands	Cuba, Guadeloupe, Tobago, Trinidad
South America	Argentina, Bolivia, Brazil, Chile, Colombia, Ecuador, Paraguay, Uruguay, Venezuela
Asia	Bangladesh, India, Indonesia, Japan, Laos, Malaysia, Pakistan, Papua New Guinea, The Philippines, Sri Lanka, Thailand, Vietnam
Europe	Germany, Great Britain, Italy, Latvia, The Netherlands, Poland, Russia, Serbia, Spain, Switzerland
Middle East	Bahrain, Iran, Israel, Saudi Arabia, Turkey

the dry zone of Sri Lanka. If indeed ground waters are the natural habitat, then the chemical and physical characteristics of these waters need definition. In addition, as pointed out above, other aquatic microorganisms might also be relevant to a possible synergistic action in the establishment of natural rhinosporidiosis.

Seasonal incidence

Differences of opinion exist on the seasonality of the disease. Paul et al. (1978) stated that in Jammu, India, the disease appears to be met with towards the later months of the year (see also Jain 1967). Ratnakar et al. (1992) reported a seasonal variation in ocular and nasal disease. There was no seasonal feature in other reports (Balachandran et al. 1990); indeed, in view of the slow growth of the lesions, a seasonal variation might be obscured. The time of arrival of the patients, especially farmers, at hospital might rather depend on weather and monsoons and on exigencies of occupation.

Incubation period

Rhinosporidial growths are of variable duration, from 8 days to 35 years (Allen and Dave 1936; de Mello 1954), with a shorter duration in ocular than in nasal disease (Karunaratne 1964). When a preceding event such as trauma has been noted, the incubation period has been as short as 6–16 days (Jimenez et al. 1984); more commonly it has been months or a few years. In most cases, it is difficult to estimate the incubation period on account of the slow growth of the polyps and hence clinical manifestation

Differences in race and religion of patients

From Sri Lankan data corrected for population ratios, Karunaratne (1964) recorded differences in the occurrence of nasal and ocular rhinosporidiosis between racial and religious groups; it was noted that nasal and ocular rhinosporidiosis was commoner in the Muslims with the

comment that their religion 'enjoins the thorough washing out of the nose before entering the mosque for prayer, and I am given to understand that this may include the mechanical cleansing of the nostril with the finger,' an ideal scenario for the inoculation of finger-borne *R. seeberi* from the water. Hindus predominated among patients with nasal rhinosporidiosis in India (Agarwal 1966). Penile or urethral rhinosporidiosis was, however, predominant among Muslims (Tirumurti 1914). Kurup (1931) concluded that 'customs, ways, manners and sanitary conditions' were important in its etiology; in his series, the incidence was greatest among the Muslims on the west coast of India.

Occupation and socioeconomic status of patients

Occupationally, most patients, especially in the large Indian series, have been agricultural workers and persons of the lower socioeconomic strata. Of particular interest is the occurrence of rhinosporidiosis in river-sand workers (Noronha 1933; Mandlik 1937) which supports the claim that the pathogen has an aquatic habitat, with sand particles providing mucosal abrasions for its implantation. From Sri Lankan data, Karunaratne (1964) postulated an etiologic link with work that involved contact with mud in paddy fields, although the occurrence of the disease in patients without such an occupation makes this a doubtful link. Caniatti et al. (1998), reporting four dogs with nasal rhinosporidiosis, also suggested mud in rice fields as the source of the pathogen. Indian data do not support that view.

Patients have often been poor, with a low standard of living, and among them farmers and laborers predominated (Allen and Dave 1936; Elles 1941). In India, the disease was found in 'all socio-economic strata and among all communities and religious groups' (Moses et al. 1990); this is of interest on account of the putative aquatic habitat of *R. seeberi* and the suggestion that water in communal pools, which are used for ablutions before religious rites in some communities, is a vehicle of transmission.

Age and sex differences of patients

Most patients in the African, Sri Lankan (Karunaratne 1964) and Indian series (Moses et al. 1988) have been under 25 years of age, with a range from 4 months to 90 years (Sharma et al. 1962), whilst most patients were young adults (15–40 years); this feature is probably linked to occupation.

In patients under 15 years of age, nasal and ocular rhinosporidiosis occurred more frequently in females (Karunaratne 1964). In adults, a male predominance has been noted in most series, African, Sri Lankan, and Indian, although in other series (Moses et al. 1988, 1990) no sex difference was noted in the occurrence of either nasal or ocular disease. In other series, for example of 1 241 patients, 84 percent were males (Kwon-Chung and Bennett 1992). The male preponderance in older age groups was, however, not seen in younger patients. The preponderant male incidence might have an occupational basis, although the male preponderance in animal rhinosporidiosis would argue against an occupational basis for the preponderance of the disease in human males. Karunaratne (1964) noted that nasal rhinosporidiosis was commoner in males whereas ocular rhinosporidiosis was more common in females. Occupational exposure to the postulated water-borne pathogen is insufficient to explain the predominant occurrence in males, in view of the exposure of females, in one series of cases, to the same environment (Wright 1922). A male predominance has been noted in swans (Kennedy et al. 1995), as in humans, although Moses and Balachandran (1987) found no sex difference in 103 farm animals.

Sources of the pathogen

The evidence for ground waters as the natural habitat and source of *R. seeberi* is impressive:

- rhinosporidiosis in patients with a history of having bathed frequently in such waters
- the occurrence of an outbreak in persons who bathed in the same lake (Vukovic et al. 1995)
- the clustering of the disease in river-sand workers (Noronha 1933; Mandlik 1937)
- the occurrence of the disease especially in upper respiratory and ocular sites
- the occurrence of nasal and ocular disease in animals that habitually come in contact with such waters, especially the epizootic disease in swans (Kennedy et al. 1995); see also Ecology, above.

MODE OF INFECTION

Rhinosporidiosis in upper respiratory sites is thought to be initiated by the implantation of endospores from the aquatic habitat into the respiratory mucosa, aided by abrasions caused, for instance, by sand particles in river-sand workers who dive into the water (Mandlik 1937) or by vigorous cleansing of the anterior nares with the fingers (Forsyth 1924; Karunaratne 1964); it is relevant to the initiating role of abrasions caused by sand tha workers who did not enter the water, but merely worked on the river banks, did not show the disease (Jain 1967). Trauma to the nose even in the absence of exposure to stagnant water, has been recorded as an antecedent event (Weller and Riker 1930; Jimenez et al. 1984). 'Autoinoculation' of the mucosa by endospores released from sporangia or after surgery was claimed (Karunaratne 1964) to be responsible for satellite polyps in the adjacent regions. On the same basis, it has been claimed that polyps on the skin result from implantation of endospores by scratching with contaminated fingers, although the rarity of skin polyps in children who might be expected to scratch themselves as often as adults makes this explanation improbable. Polyps on the skin on the other hand could occur through hematogenous dissemination from respiratory sites; lesions on the skin, however, without precedent respiratory rhinosporidiosis have been recorded (Madhavan et al. 1978).

No clear evidence of transmission between humans, even between those living in close contact, has been recorded (Wright 1922); the disease has occurred in members of the same family, although they were not living together (Allen and Dave 1936). Karunaratne (1963) documented three cases among family members who lived together, but such an occurrence could have also been rather the result of exposure to a common source of the pathogen.

Karunaratne (1964) used the term 'transepithelial infection' to refer to the implantation of endospores through minor abrasions in the skin or mucosae, through occupationally acquired injuries, from sand particles, and vegetable matter contaminated with soil; dust laden with the endospores is thought to be the vehicle of infection in ocular rhinosporidiosis. Bader and Grueber (1970) agreed with the concept of transepithelial infection though Ashworth (1923) disputed it.

The mode of infection in ocular rhinosporidiosis is even more obscure; Kaye's (1938) patients gave no history of bathing in ponds, exposure to animals, or trauma. Ocular rhinosporidiosis, as it occurs in dry regions in the Middle Eastern countries, is believed to have been due to the dustborne pathogen. Dust storms were suggested as the possible vehicle although ocular disease has been recorded in wet regions as well (Moses et al. 1990).

From a case of vulval rhinosporidiosis in a woman and penile rhinosporidiosis in her male sexual partner, Symmers (1966) postulated a sexual mode of transmission; Palaniswamy and Bhandari (1983) and Sasidharan et al. (1987) found no evidence for such transmission.

There is no evidence that humans can acquire the disease from animals (Lasser and Smith 1976). Humans and animals probably acquire the pathogen from the common source of ground waters. From the similarity noted between *Ichthyosporidium* and *Rhinosporidium* spp. (Laveran and Pettit 1910), Allen and Dave (1936) surmised that *R. seeberi* might also be a parasite of fish and that animals and humans could acquire the pathogen from inland waters which harbor infected fish. Reddy and Lakshminarayana (1962) found no evidence for the existence of *R. seeberi* in inland fish.

There is evidence for trauma as a predisposing factor; Moses and Balachandran (1987) reported that in 19 cattle with rhinosporidiosis, all had nasal disease at the site of the passage through the nasal septum, of the guiding nose-rope. Trauma followed soon after by exposure to stagnant water, featured in a case of nasal rhinosporidiosis (Atav et al. 1955), and Symmers (1966) documented a case with trauma and simultaneous contact with stagnant water shortly preceding the appearance of the lesion at the traumatized site. The occurrence of rhinosporidiosis at a site of a closed injury that precluded exposure to soil, dust, or water, as with the rhinosporidial mass over the tibia (Figure 24.43, p. 466) could have resulted from bloodborne endospores which extravasated at the traumatized site. Trauma to the nose without exposure to soil or ground water, followed by nasal rhinosporidiosis, has also been reported (Weller and Riker 1930; Jimenez et al. 1984). Finger-borne inoculation in anterior nasal passages has been suggested (Forsyth 1924; Karunaratne 1964), although the development of lesions in the nasopharynx cannot be thus explained.

Penile and anterior urethral rhinosporidiosis in the male is thought to result from the application, by some communities in whom rhinosporidiosis in these sites is commoner, of a stone or piece of brick to absorb the last drops of urine (Kurup 1931; Kutty and Unni 1969).

Predisposition to infection

A history of bathing in ground waters is common in Sri Lankan patients. On the other hand, whilst numerous persons bathe in such waters, only a small proportion of these bathers acquire the disease, suggesting the existence of predisposing host factors. The only host factor that has been investigated is blood group; Kameswaran and Lakshmanan (1999) reported that in India, group O had the highest incidence (70 percent) in rhinosporidial patients followed by group AB, although groups A, B, and O are distributed 'fairly equally,' whilst group AB is rare. Jain (1967) found no association of incidence with blood groups. In our series of 21 Sri Lankan patients, the pattern differed: group O, 43 percent; group A, 29 percent; group B, 19 percent; and group AB, 10 percent; with a population distribution of 43 percent, 21 percent,

26 percent, and 5 percent, respectively. HLA or genetic typing of rhinosporidial patients would be of interest.

The possible existence of host factors that predispose to infection is suggested by the development of rhinosporidiosis in boys but not in girls who lived under the same conditions, using the same swimming baths (Karunaratne 1936).

CLINICAL FEATURES

Sites

The most common (70 percent) sites of solitary growths are the vestibule of the nose (Figure 24.32), the septum, and the inferior and middle turbinates and the floor of the nasal acvity. Bilateral (Figure 24.33) or multiple polyps are less common. Other respiratory sites are the adjacent sinuses, nasopharynx, fauces, soft palate, inner end of the Eustachian tube, tonsil and larynx, and occasionally the trachea and bronchi. Nonrespiratory sites with single or multiple polyps (10 percent) are the skin of the face (Figure 24.34), scalp, body (superficially, subfascially, or intramuscularly), external ear, vulva, vagina, rectum, glans and fossa navicularis of the penis, and the external urethral meatus of males and females. It is claimed (Kutty and Unni 1969) that in anterior urethral rhinosporidiosis in the male, infection is initiated by the rubbing of stones to absorb the last drops of urine by certain communities; the stone could cause abrasions in the epithelium while containing the soil or water-based pathogen,

Pulmonary lesions have been rare and rhinosporidiosis in the anterior buccal cavity has not been reported; this rarity is conceivably associated with continual flushing of the cavity with food and fluids, leaving insufficient contact time for implantation of the organism. Rhinoporidiosis has, however, occurred in the parotid gland

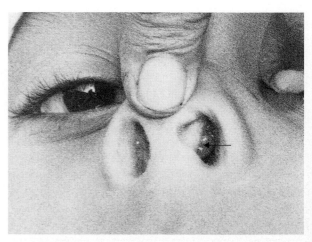

Figure 24.32 *Early, strawberry-like, pedunculated rhinosporidial polyp (arrow) in nasal vestibule. (Courtesy Dr P. Balasooriya, Sri Lanka)*

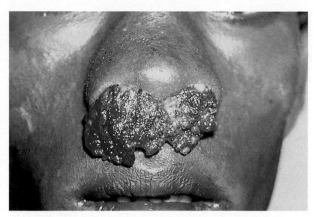

Figure 24.33 *Bilateral rhinosporidial masses obstructing anterior nares, in advanced disease. (Courtesy Dr P. Balasooriya, Sri Lanka)*

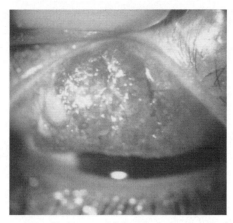

Figure 24.35 *Sessile rhinosporidial mass on upper palpebral conjunctiva, with a fluid-filled vesicle on outer surface of the eyelid. The vesicle disappeared after surgical excision of the rhinosporidial mass. (Courtesy Dr C.R. Seimon, Sri Lanka)*

(Mahadevan 1952), a rare site, infection in which was thought to have arisen through the Stenson's duct from the oral cavity. Approximately 15 percent of all cases of rhinosporidiosis are in ocular sites (oculosporidiosis) with palpebral conjunctival lesions being the commonest (Figure 24.35). Ocular polyps are painless, granular, often sessile masses; they could occur in the limbus, canthus, and lacrimal duct and sac with spread to the nose (Kirkpatrick 1916). With involvement of the bulbar conjunctiva, scleral 'melting' and staphyloma formation are complications (Castellino et al. 2000). The lacrimal apparatus – lacrimal sac, nasolacrimal duct – could also be involved (Kameswaran 1966). Obstruction of the lacrimal duct may lead to suppurative dacryocystitis with secondary infection. The idea that the initial focus of infection is the lacrimal duct is probably invalid as nasal lesions have been on the contralateral side. Ocular cases have occurred predominantly in dusty regions rather than in wet agricultural areas, while also being preceded by injuries to the eye.

Sites in disseminated cases include the brain (Alessandrini 1926 quoted by Karunaratne 1964), bones of limbs

(Nguyen-Van-Ut et al. 1959; Chatterjee et al. 1977; Aravindan et al. 1989; Gokhale et al. 1997; Figure 24.36), the skin, and the abdominal viscera.

Rhinosporidial polyps on nonrespiratory sites of the body, notably the skin and subcutaneous tissues, could occur without primary lesions in respiratory or ocular sites (Madhavan et al. 1978).

Symptoms

The presenting symptoms depend on the site of the disease. In respiratory sites, the commonest symptoms are obstruction to breathing, and bleeding even with minimal trauma of sneezing or nose-blowing. Changes in voice occur in laryngeal disease. Pain is uncommon.

PATHOLOGY

Macroscopic appearances

The macroscopic appearances vary with the site of rhinosporidial polyps. Nasal polyps are typically strawberry-like masses with pinhead-sized, cream-colored spots due to mature subepithelial sporangia. Polyps on the skin, and in respiratory sites, could be sessile or pedunculate and non-infiltrating into deeper tissues; this latter feature makes the radical excision of pedunculate polyps possible with minimal spillage of endospores and uncommon recurrence, while the excision of sessile polyps might be incomplete, with contamination of the adjacent mucosa by endospores leading to recurrence. Most polyps are lobulated and, in the nasopharynx, the lobes appear variegated with smooth, nongranular surfaces or granular and vascular in other areas; histopathological diagnosis is easier with the latter portions of the growth. Polyps with a verrucous surface might

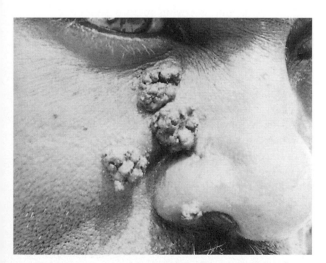

Figure 24.34 *Verrucous rhinosporidial polyps on the face of a man aged 41 years. (Courtesy Dr P. Balasooriya, Sri Lanka)*

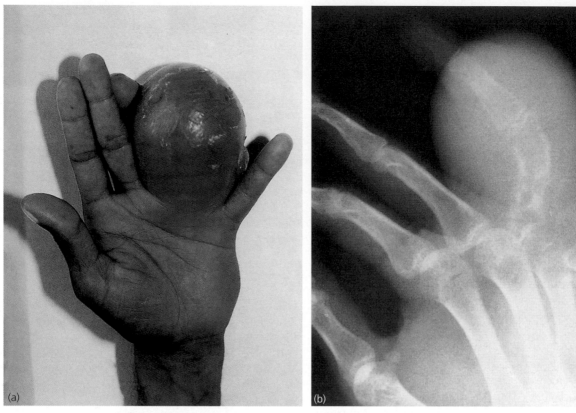

Figure 24.36 (a) *Globular rhinosporidial mass on finger, in a patient with disseminated rhinosporidiosis. (Courtesy Dr S. Gokhale, India).* **(b)** *Radiograph of fingers of patient in* **(a)***, showing destruction and fracture of bones of the affected finger. (Courtesy Dr S. Gokhale, India)*

entrap other microorganisms from the aquatic habitat; this is of importance in sampling for purposes of culture in that contaminating microorganisms which appear in culture might be erroneously incriminated as the cause of rhinosporidiosis (see Krishnamoorthy et al. 1989; Ahluwalia et al. 1997; Ahluwalia 1999).

Polyps in respiratory sites are usually covered with mucus in which numerous endospores may be detected on stained smears. Growths on the skin of the body might undergo ulceration and de-epithelialization, and secondary infection, and then mimic malignant tumors. Pus formation is uncommon. Bones, as in the foot, hand or limbs, which underlie rhinosporidial masses, could show erosion or fractures on radiographs (Chatterjee et al. 1977; Gokhale et al. 1997; Figure 24.36b).

Disseminated rhinosporidiosis is relatively rare, and presents with masses of varying size on the face, trunk, limbs, and in the viscera including the liver, spleen, brain, and kidneys (Rajam et al. 1955; Agrawal et al. 1959). Rhinosporidial endospores and sporangia in these cases have been detected microscopically in the blood vessels. Lymph node enlargement, even in disseminated rhinosporidiosis, is curiously uncommon (Chatterjee et al. 1977), although it is possible that lymph nodes have not been examined in patients with disseminated disease.

Microscopic appearances

The histopathological pattern in human rhinosporidiosis shows wide variation, especially in the stroma and infiltrating cells (Arseculeratne et al. 2001). The various developmental stages of the organism can be seen in a single histological section (Figure 24.3) and their presence, rather than the host's tissue response, affords a definitive diagnosis.

The epithelium overlying mature sporangia is attenuated or hyperplastic with keratinization, and acanthotic with downgrowths that surround the sporangia (see Figure 24.19). This phenomenon is further discussed under Transepidermal elimination below. Proliferation of respiratory (ciliated, columnar) epithelium results in a papillary appearance, with many goblet cells and hypersecretion of mucus (Karunaratne 1964). Growths on the skin may be denuded of epithelium and ulcerated.

Endospores and sporangia (Figure 24.18) have been seen intraepithelially (Noronha 1933; Karunaratne 1964); it is uncertain whether the presence of sporangia, intraepithelially, results from the development of intraepithelial endospores after 'transepithelial infection' (Karunaratne 1964) or from transepidermal elimination (see below), with the organism being 'thrust into the epidermis from the subjacent connective tissue' (Ashworth 1923; Figures 24.37 and 24.38). Endospores

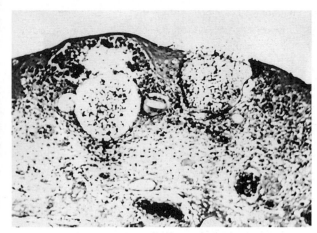

Figure 24.37 *Late stage of 'transepidermal elimination' of sporangia and endospores in ocular rhinosporidiosis. Note thinned epithelium. H&E; initial magnification ×400*

occur in both the mature sporangia as well as free, extrasporangially, in the tissues.

Fibromyxomatous or fibroedematous reactions occur in the subepithelial stroma with few or absent sporangia (Figure 24.39). Tirumurti (1914) noted myxomatous degeneration, especially in nasal polyps. Mild-to-moderate inflammatory cell infiltrates accumulate around a multitude of sporangia (Figure 24.40) and, more prominently, around fee endospores. The cells include lymphocytes, plasma cells, histiocytes, neutrophils (especially around free endospores and with secondary infection), and rarely eosinophils; these are in varying proportions depending on the chronicity of the polyp. The paucity of eosinophils contrasts rhinosporidiosis with opportunistic, mycelial deep mycoses; there is also no eosinophilic precipitate (probably of antigen/immunoglobulin complexes) (Grover 1970), the Splendoré–Hoeppli phenomenon, seen around hyphae of *Entomorphthora coronata* (a cause of nasal polyps in horses) and *Busidiobolus ranarum*, and in other zygomycotic tissues (Williams 1969).

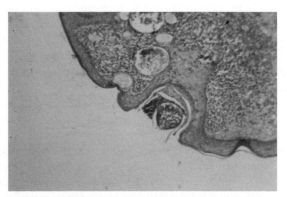

Figure 24.38 *Subcutaneous rhinosporidiosis with down-growth of squamous epithelium encircling a degenerate sporangium, illustrating 'transepidermal elimination' of degenerate sporangia. PAS stain; initial magnification ×200*

Figure 24.39 *Acanthotic squamous epithelium (a), fibromyxomatous stroma (f), with sparse cell infiltration in an area with scarce sporangia, in nasopharyngeal rhinosporidiosis. H&E, initial magnification ×400*

With secondary infection, neutrophils are numerous. Stromal microabscesses may occur (Jimenez et al. 1984). The predominant mononuclear cell in the stroma is the plasma cell (Kaye 1938). In more chronic lesions, lymphocytes and macrophages predominate. Invading connective tissue (Figure 24.22) or multinucleate giant cells (Figure 24.23) may be seen within degenerate or empty sporangia, which may also be calcified. Stromal giant cells occasionally contain juvenile sporangia (Karunaratne 1964). In a rare case of fatal, disseminated visceral rhinosporidiosis, Rajam et al. (1955) found relatively insignificant histopathology or evidence of inflammatory reactions around the sparse sporangia and endospores in the viscera.

Marked vascularity, resulting in focal hemorrhage in the polyp and severe external hemorrhage following surgery or trauma, occurs with the very vascular polyps, especially in nasal rhinosporidiosis (see Figure 24.19). Empty spaces (Figure 24.19) occur in the stroma and are derived from the destruction of tissue elements or from dilated glands or ducts in glandular tissues (Karunaratne 1936). They are commoner in nasal rhinosporidiosis (Beattie 1906). There is minimal fibrosis which is more marked in lesions that are regressing under dapsone therapy (Job et al. 1993).

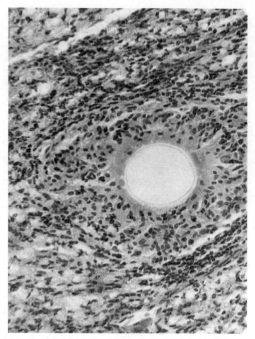

Figure 24.40 *Mononuclear cell infiltration around degenerate sporangium. H&E; initial magnification ×400*

Erosion of bone (Figure 24.36b) or cartilage with perforation of the nasal septum has been recorded (Atav et al. 1955; Chitravel et al. 1990; Gokhale et al. 1997).

TRANSEPIDERMAL (TRANSEPITHELIAL) ELIMINATION OF RHINOSPORIDIAL ENDOSPORES

This term was applied to the presence of rhinosporidial sporangia in subepithelial sites, with their endospores exiting through the breached epithelium (Figures 24.37 and 24.38). In only three reports (Thianprasit and Thagernpol 1989; Azadeh et al. 1994; Kennedy et al. 1995) has this term been used.

In what may be construed as the sequence of this process, the early stages of this phenomenon are characterized by the formation of downgrowths of the deeper layers of the epidermis to surround the mature dermal and subdermal sporangia (Figures 24.19 and 24.38). These downgrowths often appear histologically as 'whorls' (Arseculeratne et al. 2001; see Lever and Schaumberg-Lever 1990) and have also been described as a response to chemical irritants (Lever and Schaumberg-Lever 1990); their presence in rhinosporidial tissues is therefore not a specific reaction in this disease, and it might alternatively be regarded as a nonspecific defense reaction against irritants. This phenomenon was thought to represent a nonspecific defense mechanism of the host for the expulsion of rhinosporidial sporangia and endospores. Thinning of epithelium (Figure 24.37) which overlies the sporangia was thought to result from pressure from the underlying sporangia.

On a parallel with the amphibian skin pathogen *Batrachochytrium* (Daszak et al. 1999), it was postulated

(Arseculeratne et al. 2001) that the phenomenon in rhinosporidiosis could, on the other hand, be the pathogen's mechanism for endospore dispersal. The apparent migration of the mature sporangia towards the surface of the epithelium could be due to the dissolution of the tissues by histolytic enzymes of *R. seeberi* rather than through a mere effect of pressure which forces the sporangium towards the epithelium. Broz and Privora (1952) described a similar appearance in the granulomatous lesions in frogs caused by *Dermocystidium*: 'When the cyst reaches the surface of the skin, it becomes exposed, owing to ulceration of the epithelium, and its membrane ruptures, setting the spores free.' *Dermocystidium* is now included with *Rhinosporidium* in the proposed Class Mesomycetozoea (Herr et al. 1999b). In Figure 24.2 of Wallin et al.'s (2001) paper on rhinosporidiosis in a cat, 'the "nipple-like" projection', of the mature sporangium projected into the nasal cavity; this appearance recalls that described by de Mello (1954) of the sporangial pore that is directed towards the surface of the epithelium.

The conceivable defensive role of the epithelium on the part of the host, together with the apparent migration of the sporangia towards the surface of the epithelium in the postulated pursuance of the release of the endospores on the part of the pathogen, constitute the phenomenon of 'transepidermal (epithelial) elimination' of sporangia and endospores.

MODE OF SPREAD

Rhinosporidial infection spreads locally, regionally, and to distant, anatomically unrelated sites.

From upper respiratory sites, especially the anterior nares where the disease most often begins, local and regional spread occurs in the nasal passages, extending to the nasopharynx, lacrimal apparatus, and paranasal sinuses; in these instances, the spread is probably through autoinoculation (Weller and Riker 1930; Karunaratne 1936) of adjacent sites by endospores released spontaneously from adjacent polyps or after surgery. In Ingram's case (1910), the initial growth in the external urethral meatus spread to involve the glans within a period of 2 years, thus spreading more slowly than in other sites. Dissemination refers to the occurrence of lesions in sites such as the limbs and viscera that are anatomically unrelated or distant from the initial foci in the upper respiratory passages.

Lymphatic spread has not been documented, although postulated (Ashworth 1923) in cutaneous and subcutaneous rhinosporidiosis (Agrawal et al. 1959; Agarwal 1966), although such spread will not explain dissemination to anatomically unrelated sites. None of the numerous reports on rhinosporidiosis has described the occurrence of lymphadenitis. de Mello (1954) stated that 'lymphatic nodes are not affected,' nor did Pumhirun et al. (1983) find enlarged lymph nodes in their case of nasopharyngeal and facial rhinosporidiosis. The presence of enlarged inguinal and epitrochlear lymph nodes in a case of disseminated rhinosporidiosis was noted by Agrawal et al. (1959) though the histology of the nodes was not commented on. Neither Chatterjee et al. (1977) nor Agarwal (1966) who studied 27 cases, found enlargement of lymph nodes in their cases of disseminated disease.

Inguinal lymphadenitis was recently described (Arseculeratne et al. 2001) in a Sri Lankan patient with disseminated disease involving the lower limbs. The criteria on which lymph node involvement was made included the location of the mass in the site of inguinal lymph nodes, the histological presence of a capsule in the mass, trabeculae from the capsule towards the center of the mass, a subcapsular sinus, and masses of lymphocytes which probably represented the cortical follicles (Figure 24.41).

Lymphadenitis in rhinosporidiosis is apparently rare, contrasting with its frequency in systemic mycotic disease; this rarity might be based on mechanisms of immune evasion in rhinosporidiosis which are discussed below.

There is also histological evidence for hematogenous spread of rhinosporidiosis to anatomically distant sites; Rajam et al. (1955) identified endospores and sporangia in the lumen of renal blood vessels, urine, ascitic fluid, and viscera; these authors' case is of interest in that growths were not found in the nose, nasopharynx, or urethra, which are common initial sites of rhinosporidiosis. Sporangia with endospores were also found

within blood vessels in the case of Ho and Tay (1986). Figure 24.34 illustrates facial polyps in a Sri Lankan patient that developed a few years after excision of a nasal polyp; amongst 20 other rhinosporidial masses on the rest of this patient's body was a pedunculated mass on the chest wall (Figure 24.42) and a large subcutaneous mass that appeared at the site of a previous closed injury on the leg (Figure 24.43). Hematogenous spread to these sites could have followed the nasal infection.

Concurrent diseases

Infections or malignancies concurrent with rhinosporidiosis, in the same or different sites or adjacent tissues, have been recorded; in conjunctival rhinosporidiosis with papillomavirus (Noor-Sunba and al-Ali 1989); papovavirus infection (Noor-Sunba et al. 1988); squamous carcinoma of the tongue (Karunaratne 1936), Ahluwalia and Bahadur 1990); leprosy with nasal (Allen and Dave 1936; Rosenbaum and Gan 1948) and cutaneous rhinosporidiosis (Porichha et al. 1986; Jayakumar et al. 1995); syphilis (Satyanarayana 1960); rhinoscleroma (Al-Serhani et al. 1998) in humans; and schistosomiasis in bovines (Rao 1938). Rhinosporidiosis coexisted in 50 percent of bullocks with schistosomiasis of the nose (Datta 1965). Jain (1967) quoted Patnaik that rhinosporidiosis was not coexistent with schistosomiasis; in Sri Lanka, bovine nasal scrapings, from abattoir samples, showed schistosomes in approximately 75 percent of buffaloes and in 40 percent of cattle, while none had concomitant rhinosporidial endospores (Arseculeratne and Rajapakse 2001, unpublished data).

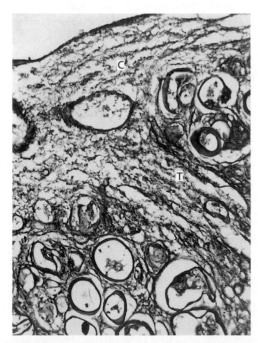

Figure 24.41 *Inguinal lymphadenitis in disseminated rhinosporidiosis affecting lower limbs. C, capsule of lymph node; T, trabeculae from capsule to parenchyma, with degenerate sporangia in node's cortex. Reticulin stain; initial magnification ×400*

Figure 24.42 *Globular, compact, pedunculated rhinosporidial mass (2 cm × 3 cm) on posterior chest wall in a case of disseminated rhinosporidiosis.*

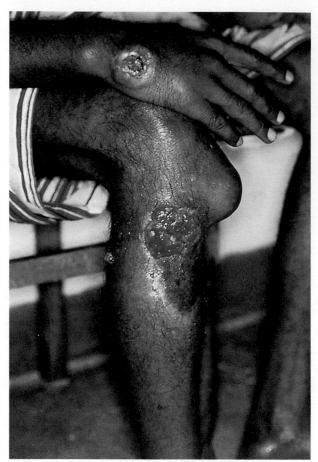

Figure 24.43 *Rhinosporidial mass on the tibia mimicking a malignant tumor, a nonulcerated rhinosporidial lump on the knee, and an ulcerated rhinosporidial mass on the hand, in a patient with disseminated rhinosporidiosis.*

It is not clear whether the concurrence of these diseases with rhinosporidiosis has a causal, sequential, or coincidental basis. In view of the occurrence of suppressor immune reactions in rhinosporidiosis, discussed under Immunity below, local or generalized cell-mediated immunity (CMI) anergy might occur with some of these associated diseases. In cats, rhinosporidiosis was not associated with the feline leukemia and feline immunodeficiency viruses (Wallin et al. 2001). There is no evidence in the voluminous literature that rhinosporidiosis is an opportunistic infection, although Stricker et al. (1999) reported a case of a human immunodeficiency virus (HIV)-positive man who had an unusual infection which these authors claimed to be rhinosporidiosis; their histopathological section photograph, however, does not reveal the different stages of *R. seeberi*'s development which are invariably seen in rhinosporidial tissue (Figure 24.3). In the extensive literature on clinical rhinosporidiosis, from diverse environments, such an association between an identified immunodeficient state and rhinosporidiosis has not been recorded.

IMMUNITY

Information on mechanisms of nonspecific and specific adaptive antirhinosporidial immunity would be necessary to explain some predominant clinical characteristics of the disease – chronicity, spontaneous regression, recurrence after surgery, and occasional dissemination. Such data on mechanisms of antirhinosporidial immunity were scarce until recently, undoubtedly due to the absence of a method for obtaining pure preparations of *R. seeberi*, uncontaminated by host tissue and other microorganisms, and consequently the lack of reagents for suitable immunological tests in diseased hosts as well as in vitro. A further lack was an experimental model of rhinosporidiosis.

Only one report is available on mechanisms of antirhinosporidial humoral immunity (Chitravel et al. 1982) that described the use of crude preparations of endospores and sporangia obtained from human polyps. Homogenates after filtration and centrifugation were used with Freund's complete adjuvant as antigens for the preparation of antirhinosporidial antibodies. These authors were not able to demonstrate antirhinosporidial antibody in patients by double diffusion tests in gels or counterimmunoelectrophoresis (CIE), but reported that rhinosporidial antigen(s) were detected in rhinosporidial patients' sera by gel diffusion tests. Kwon-Chung and Bennett (1992) pointed out that Chitravel's gel precipitin lines could have been due to human antigens that contaminated the crude rhinosporidial antigen preparations and which reacted with antihuman antibody from the experimental animals that were immunized with the crude antigen preparations. If these preparations did not contain mature sporangia, then new rhinosporidial antigens which appear only in maturing sporangia (Herr et al. 1999c), might also have been excluded. Crude preparations of endospores and sporangia were also used by Reddy and Lakshminarayana (1962) in agglutination tests, which, however, gave inconclusive results, as agglutination was caused by sera from normal humans as well as from patients with rhinosporidiosis.

New information on cell-mediated and humoral antirhinosporidial immunity in human rhinosporidial patients and in experimental mice is now available and is reviewed below.

Cell-mediated immune responses

Cell-mediated immune responses (CMI) in rhinosporidiosis was investigated by Reddy and Lakshminarayana (1962) by the 'cutaneous' injection of crude preparations of rhinosporidial endospores. Transient erythematous reactions were produced in normal persons as well as in rhinosporidial patients, and these authors concluded that these reactions were nonspecific.

The only definitive report on cell-mediated immunity in rhinosporidial patients was that of Chitravel et al.

(1981); they used the Leukocyte Migration Inhibition (LMI) test and found that, in patients with disease of less than 9 years' duration, LMI was detectable but that the inhibition decreased when the disease was of longer duration. These authors suggested that immunological 'unresponsiveness or tolerance' might explain this finding. Our data on antirhinosporidial CMIR in mice, discussed below, elucidates Chitravel's results.

CMI responses in human rhinosporidiosis were recently described (de Silva et al. 2001). Two approaches were used:

1 immunohistochemistry with monoclonal antibodies against specific markers on cells in human rhinosporidial tissues
2 in vitro lymphoproliferative responses (LPR) of peripheral blood lymphocytes from rhinosporidial patients to the T-cell mitogen concanavalin A (conA) and to sonic extracts from Percoll-purified rhinosporidial endospores and sporangia.

The cell infiltrate in human rhinosporidial polyps showed similar patterns despite their origin from different clinical presentations and of differing durations. In composition, the infiltrates were mixed; neutrophils were abundant, CD20+ B cells were present in considerable numbers with plasma cells, CD68+ macrophages (Figure 24.44), some of which were foamy, and CD3+ T lymphocytes, were numerous. CD4+ helper T lymphocytes were scarce, while the presence of CD8+ T-cytotoxic/suppressor lymphocytes was marked; CD8+ cytotoxic/suppressor T lymphocytes were found specially around and within mature sporangia (Figure 24.45).

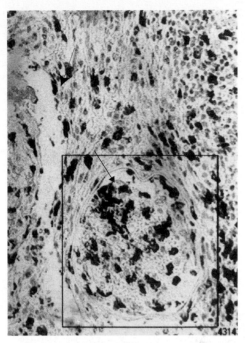

Figure 24.45 *CD8+ T lymphocytes in rhinosporidial stroma and within a mature, degenerate sporangium (arrow), visualized with monoclonal antibodies-Dako Fast-red. (Reproduced with permission from* Mycopathologia, *2001;***52***:59–68)*

There were many TIA1-positive lymphocytes of the cytotoxic subtype located especially around the sporangia. CD56/57+ NK lymphocytes were less numerous than CD8+ T lymphocytes and were also located around mature sporangia.

In LPR assays in vitro (de Silva et al. 2001), rhinosporidial lymphocytes showed stimulatory responses (stimulation indices (SI) >1, Figure 24.46) with conA, but the majority of these samples showed significantly diminished responses to rhinosporidial antigens, indicating that suppressor responses could occur in human rhinosporidiosis. The SI of lymphocytes from control (nonrhinosporidial) persons and the SI of lymphocytes from rhinosporidial patients, to conA showed no significant differences, indicating that the rhinosporidial patients had general adequate cell-mediated immune competence. There was, however, no correlation between the intensity of the depression of the LPR-SI to rhinosporidial antigen and the site, duration of the disease, the number of lesions, or to the presence of dissemination. These results demonstrate that while on the one hand a CMIR does develop in human rhinosporidiosis, suppressor responses, on the other hand, also occur.

In some patients, the peripheral blood lymphocytes showed a significantly lower proliferative response to rhinosporidial antigen than to conA, suggesting that the suppression was antigen-specific; the rhinosporidial antigen did not suppress the response to conA when the two agents were mixed, nor was the antigen toxic to the lymphocytes.

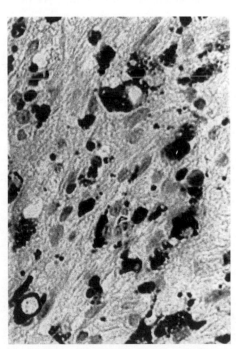

Figure 24.44 *CD68+ macrophages in rhinosporidial stroma, visualized with monoclonal antibodies-Dako Fast-red. (Reproduced with permission from* Mycopathologia, *2001;***52***: 59–68)*

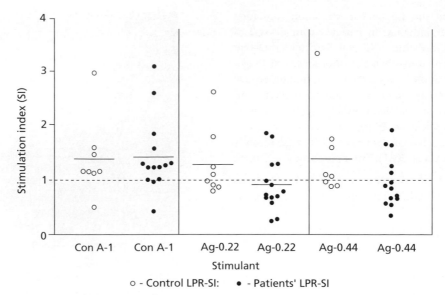

Figure 24.46 *Scatter of LPR-SI of control subjects and rhinosporidial patients to conA and rhinosporidial extract (RhA) as stimulants. Horizontal bars represent mean values. Open circles, control LPR-SI; filled circles, patients' LPR-SI. (Reproduced with permission from* Mycopathologia, *2001;***52***:59–68)*

The delayed-type hypersensitivity (DTH) Mantoux reactivity to PPD-S from *Mycobacterium tuberculosis* was markedly depressed (to approximately 25 percent) in most Sri Lankan patients with rhinosporidiosis (with or without prior bacille Calmette–Guérin (BCG) vaccination), although the Mantoux reactivity was far more pronounced in normal and BCG-vaccinated populations (75 percent); no correlation was found between Mantoux-negativity and the clinical state of rhinosporidiosis – whether respiratory or in other sites, localized or disseminated, or with single or multiple polyps – as with the lymphoproliferative responses in vitro (de Silva et al. 2001). Rhinosporidial antigen prepared from purified, sonicated endospores and sporangia from the polyps of one patient with disseminated disease, injected intradermally in the same patient, gave a negative skin reaction, correlating with the significantly depressed lymphoproliferative response (SI < 1) in this patient.

The CMI anergy demonstrated in rhinosporidial patients and in in vitro tests needs further study for the determination of whether the suppression is a result of general, nonspecific suppression of CMI or is antigen-specific.

CMI responses in experimental animals have hitherto not been examined. CMIR to *R. seeberi* in experimental mice were recently studied by the quantitative footpad response of the DTH, with histopathology of the treated footpads for confirmation of the DTH-type response. Jayasekera et al. (2001) demonstrated that sonicated suspensions of rhinosporidial endospores and sporangia, used for sensitization and challenge, evoked well-marked DTH-type footpad responses of a magnitude similar to that evoked by T-dependent antigens such as sheep red blood cells. The footpads challenged with rhinosporidial extracts showed histopathology

(Figure 24.47) which was typical of DTH reactions in the mouse.

An interesting finding in the mouse experiments was that, in comparison with the results after a single sensitizing dose of rhinosporidial antigen(s), repeated sensitization resulted in the decrease in the intensity of the DTH footpad response; conversely, the humoral immune response was significantly elevated. This phenomenon of immune deviation has been ascribed (Mosmann and Sad

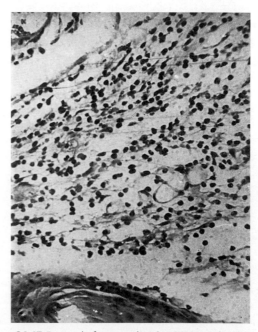

Figure 24.47 *Footpad of mouse showing DTH reaction to challenge following sensitization with extract of rhinosporidial endospores/sporangia, with infiltrate of neutrophils, macrophages and lymphocytes. H&E; initial magnification ×400*

1996) to the switch, on repeated administration of antigens, from activation of CD4+ Th0 cells to the production of CD4+ Th2 cells which encourage antibody production, after an initial production of CD4+ Th1 cells which induce DTH. The ensuing decrease of DTH reactivity with the switch from Th1 to Th2 might contribute to decreased antirhinosporidial cell-mediated immunity and might explain the chronicity, recurrence, and dissemination in clinical rhinosporidiosis. This finding might also explain Chitravel et al.'s observation (1981), that the intensity of LMI waned when the duration of the disease was greater than 9 years.

Humoral immune responses

Jain (1967) used a suspension of 'spores' as antigen, injected subcutaneously in humans, and concluded that '... this disease does not produce any sensitizing antibodies' as no reactions were produced in either one hour or after 3 days.

Reddy and Lakshminarayana (1962) used suspensions of endospores and sporangia from homogenized rhinosporidial tissues, in agglutination tests for experimental antirhinosporidial antibody and observed nonspecific 'clumping.' The same antigen was used in 'cutaneous' tests on patients and normal persons, but no specific reactions were detected.

Chitravel et al. (1982) reported that antirhinosporidial antibody was not detected in patients by immunodiffusion or CIE; they used suspensions of endospores and sporangia, that were ground with glass powder. Precipitin lines were, however, detected in our (Arseculeratne et al. 1999) CIE tests with antigens released by ultrasonic disintegration of Percoll-purified endospores/sporangia, and human rhinosporidial sera, and the possibility remains that Chitravel's antigen preparations had insufficient amounts of soluble antigen. Antirhinosporidial humoral immune responses (HIR) was also detected in experimental mice (Jayasekera et al. 2001) and rabbits (Atapattu et al. 1999) by indirect immunofluorescence tests which used Percoll-purified, sonically disrupted endospores and sporangia as antigen. Rhinosporidial patients' sera, tested by indirect immunofluorescence, had appreciable titers (over 1/320) whereas only two of 30 clinically nonrhinosporidial persons showed low (1/30) titers; these two persons were from an area endemic for rhinosporidiosis and might have been subclinically exposed to the organism with consequent immunization. Antibody titers of over 1/320, especially of IgG and IgM, were also detected in patients' sera in immunodot blot assays on nitrocellulose paper, with specific antihuman immunoglobulin phosphatase-conjugated tracers (Arseculeratne et al. 1999 and unpublished data 2002).

It can be concluded that R. seeberi shows a variety of mechanisms of evasion of immune responses, which

might explain some hitherto enigmatic aspects of rhinosporidiosis, viz. chronicity, recurrence, and dissemination:

- antigenic variation, based on the findings of Herr et al. (1999c)
- immune suppression as demonstrated by de Silva et al. (2001)
- immune distraction: the cell-infiltration patterns in rhinosporidial tissues in some cases indicate that infiltrates occur in some areas of the tissue in which rhinosporidial bodies are absent, suggesting that the infiltrates were in response to free rhinosporidial antigen. If this were the case, the possibility of immune distraction by free antigen in 'mopping up' of antirhinosporidial antibody might conceivably contribute to immune evasion by R. seeberi (Arseculeratne et al. 2001)
- immune deviation: experimental studies on CMIR in mice to R. seeberi (Jayasekera et al. 2001) suggest the occurrence of immune deviation which might further contribute to immune evasion through decreasing antirhinosporidial CMI reactivity, if, indeed, antirhinosporidial CMIR is protective
- binding of host immunoglobulins (Atapattu et al. 2000)
- antigen sequestration: the thick wall of the endospores which contains chitin and cellulose is conceivably impermeable to release of antigens or to immune destruction of the endospores, as perhaps with encysted parasites (Abbas et al. 2000). Selective permeability, however, to smaller molecules probably exists in rhinosporidial endospores, as shown by the entry of Evan's blue and MTT into endospores as described above. Moreover, the endospore of R. seeberi is not comparable with the encysted resting stage of amebae but is a vegetative stage in its life cycle.

It is interesting that several of these mechanisms of evasion of immune responses have also been described in parasitic infections, notably malaria (Goonewardene et al. 1990; see also Kierszenbaum 1994) and in fungal infections (chromomycosis, Fuchs and Pecher 1992; see also Cuff et al. 1986 on *Candida albicans*).

DIAGNOSIS

Urgency in diagnosis of rhinosporidiosis arises from its mimicry of malignant tumors (Figure 24.43).

Clinical clues to diagnosis

The macroscopic features of polyps and the clinical presentation, especially of nasal polyps, facilitate diagnosis. Rhinosporidiosis presents as single, or less frequently multiple, polyps in respiratory or ocular sites. The polyps are strawberry-like, red due to vascularity, with a granular surface that has yellowish, pinhead-sized spots. The polyps are friable and fragile, very vascular, and bleed easily on handling. Lobulation, especially in nasopharyngeal sites, is common. Growths on the skin are often multiple and wart-like.

Differential diagnosis varies with the site. Urethral rhinosporidiosis needs differentiation from caruncles; nasal rhinosporidiosis from hypertrophic rhinitis, allergic polyps, vascular polyps, angiomas and fibromas, carcinoma (Ravi et al. 1992; Lourenco and Costa 1996), granulomatous leishmaniasis, leprosy, rhinoscleroma, and mycotic granulomas. Rhinosporidial lesions on the skin of the body may simulate viral warts, sarcomas, and leprous, tuberculous, or cryptococcal nodules.

Disseminated rhinosporidiosis is not uncommon. A history of nasal polyps and their surgical excision is often given, and the disseminated lesions appear several months or even years after the primary lesions. Growths of diverse appearances – pedunculated or sessile polyps or masses – appear on the body, face, scalp, limbs, and in the viscera (lungs, kidneys, liver, spleen, and brain). Rhinosporidial masses are accompanied by osteolysis and fractures of underlying bones (Figure 24.36b; Gokhale et al. 1997). Lymphadenitis, though hitherto unreported in rhinosporidiosis, was detected in a case of disseminated rhinosporidiosis in Sri Lanka as described above.

The contents of closed, subcutaneous lumps need microscopic examination of needle aspirates. The surface of mucosal polyps is often coated with mucus that contains numerous endospores which afford a convenient method for cytodiagnosis with PAS or silver-stained smears.

DIFFERENTIAL DIAGNOSIS

Rhinosporidiosis mimics, macroscopically, several infective and noninfective diseases. Noninfective diseases include:

- in the rectum: hemorrhoids and polyps
- in the nasal passages: papillary adenocarcinoma, hypertrophic rhinitis, beningn polyps
- in the eye: pterygia, mucocele of the lacrimal sac
- penile and urethral sites: squamous carcinoma, condylomata
- on the body: sarcoma, verruca vulgaris, fibroma, angiofibroma.

Infective diseases include rhinoscleroma, cryptococcosis, chromo(blasto)mycosis, histoplasmosis, infections caused by *Chrysosporium (Emmonsia) parvum* vars. *parvum* and *crescens*, blastomycosis, and paracoccidiodomycosis. The agents of these diseases have distinguishing morphological and tinctorial features and may also be cultured in vitro and in experimental animals, especially mice.

HISTOPATHOLOGY

The definitive diagnosis of rhinosporidiosis is by histopathology on biopsied or resected tissues. A histopathological diagnosis is based on the classical descriptions of rhinosporidial pathology, of Beattie (1906), Tirumurti (1914), Ashworth (1923), and Karunaratne (1964). Most histological sections contain the different developmental stages of the pathogen (Figure 24.4), and these appearances are more specific than the tissue responses – stromal or cell infiltrates – of the host; useful stains are H&E, PAS, mucicarmine, and Gomori's methenamine silver.

Autofluorescence of the sporangia and endospores, in sections of paraffin-embedded tissues, has been noted (Senba and Yamashita 1985; Moses et al. 1991) and merits consideration in tests that involve the use of fluorescent tracers.

Notable features of rhinosporidial histopathology include variations in the host responses:

- within the same specimen
- in different sites of the same and in different hosts, in respect of the density of the developmental stages
- in the stromal and cellular response (Arseculeratne et al. 2001).

Well-marked cell infiltrates, especially of macrophages and lymphocytes, occur around both intact, apparently viable sporangia that contain intact endospores as well as around degenerate, empty sporangia. Cell infiltrates could also be sparse around degenerate or intact sporangia in other regions of the same tissue and in samples from different patients. Kennedy et al. (1995) described similar histopathological variations in rhinosporidial tissues from swans; three patterns were identified:

- a severe, focal or diffuse acute and chronic inflammatory cell infiltrate
- granulomatous inflammation with epithelioid histiocytes and foreign body-type giant cells
- a fibroblastic reaction with paucity of sporangia and an inflammatory cell infiltrate, that was infrequent.

Rhinosporidial endospores and sporangia in disseminated cases involving the kidneys and liver might not be associated with a surrounding inflammatory cell infiltrate, although these bodies in the lung and subcutaneous tissues were found to have been surrounded by well-marked cell infiltrates (Agrawal et al. 1959).

Problems in histopathological diagnosis

Despite the ease of histopathological diagnosis through the ubiquitous developmental stages of *R. seeberi* in most rhinosporidial tissue, there are, however, problematic instances from which arise confusion, misdiagnosis, or a false-negative diagnosis. These instances which were described elsewhere (Arseculeratne et al. 2001) consisted of:

- rhinosporidial tissue in which rhinosporidial bodies were absent despite the presence of marked cell infiltrates while other, more vascular, portions of the polyp, however, had rhinosporidial bodies; this problem occurs especially in the variegated polyps of nasopharyngeal rhinosporidiosis. Selection of typical portions of such polyps for histology might minimize the occurrence of a false-negative diagnosis

- atypical sporangia in which the well-marked, thick wall was absent
- the presence of only fragments of the sporangial wall, while endospores might or might not be present
- the absence of typical rhinosporidial bodies although PAS-positive bodies which are similar to the rhinosporidial endospores might be present indicating incipient or slow-growing rhinosporidiosis.

A noteworthy feature in rhinosporidial histopathology is the absence of the Splendoré–Hoeppli (antibody-mediated) reaction around rhinosporidial bodies. This reaction is often present in (mycelial) mycotic and some bacterial infections. This absence is remarkable in that rhinosporidial patients show high titers of anti-rhinosporidial antibody.

Differential histopathological diagnosis

A problem in differential histopathological diagnosis is that of subcutaneous spherulocystic disease (myospherulosis). Hutt et al. (1971) described it as 'reminiscent of an endosporulating fungus such as rhinosporidiosis;' McClatchie and Bremner (1969) noted it as a 'distant cousin of rhinosporidiosis.' The cysts of myospherulosis remain unstained with Gomori methenamine silver or with mucicarmine and no spores are present inside the cysts. The bodies in myospherulosis were regarded as erythrocytes, altered by exogenous or endogenous fat or lipids, or as derived from malignant cells or macrophages (Chau et al. 2000, see also Ali et al. 2001).

Histopathological differential diagnosis might also consider coccidiodomycosis in which *C. immitis* has mature stages which consist of large, thick-walled bodies which are smaller than the sporangia of *R. seeberi*, but which also contain endospores which are also smaller than those of *R. seeberi* and do not contain EDB.

CYTODIAGNOSIS

Cytodiagnosis is a useful method in diagnosis on aspirates from nonulcerated lumps in extrarespiratory sites (Kamal et al. 1995; Kavishkar et al. 1998), smears of the endospore-laden mucoid covering on the surfaces of polyps in the respiratory passages and from ulcerated polyps elsewhere in the body, washings of sinuses and bronchi, bronchial brushings, or fine-needle aspirates from bones (Pai et al. 1996). With suitable stains such as Gomori methenamine-silver, Papanicolaou's, PAS, Mayer's mucicarmine, and other stains, diagnostic features, especially of the endospores with their unique EDBs, are revealed (Ashworth 1923; Allen and Dave

1936; Kaye 1938; de Mello 1954; Rajam et al. 1955; Thomas et al. 1956; Vanbreuseghem 1958; Khan et al. 1969; Fortin and Meisels 1974; Chaudary et al. 1986; Jaiswal et al. 1992; Van der Coer et al. 1992; Gori and Scasso 1994; Maru et al. 1999). Wet mounts of the mucoid material on rhinosporidial polyps would also show the endospores with their EDBs (Figure 24.10).

In posterior respiratory sites, respiratory (columnar, ciliated) epithelial cells with their large nuclei could mimic endospores with the 'tail of the comet' appearance that is caused by the residue of the intrasporangia, mucoid material; caution was advised (Arseculeratne 2000a) in the identification of endospores in these sites (Figure 24.9). The PAS stain is particularly useful in this differentiation; the epithelial cell's nucleus is PAS-negative whereas the endospore stains markedly magenta.

Allen and Dave (1936) used nasal smears for monitoring antirhinosporidial chemotherapy, although it is noteworthy that the presence of endospores in nasal smears did not always correlate with the presence of nasal growths; smears from nasal passages after resection of the polyps have shown endospores with subsequent recurrence of the growth.

TREATMENT

Spontaneous regression of rhinosporidial polyps has been noted in humans (Forsyth 1924; Allen and Dave 1936; Karunaratne 1964; Anand et al. 1975) and in swans (Kennedy et al. 1995); it is rare, and surgical excision of the polyp, in accessible sites, remains the mainstay of treatment. Cauterization of the base of the excised polyp, especially in sessile growths, is recommended (Desmond 1953; Samaddar and Sen 1990; Ravi et al. 1992) to prevent, or at least minimize, recurrence. Excision with a snare was considered to result in dissemination and recurrence (Khan et al. 1969). Spillage of endospores by mere handling of the polyp or from its cut surface is probably responsible for the recurrence or development of new lesions on adjacent sites; this process was termed 'autoinoculation' (Karunaratne 1964). Satyanarayana (1960) esimated the recurrence rate after surgery at 11 percent, whilst Sasidharan et al. (1987) reported a recurrence rate of 25 percent in penile rhinosporidiosis.

Cryosurgery has also been used. Extensive growths have necessitated amputation, of the penis in disease of the glans, or of the foot or hand. Since rhinosporidial polyps in respiratory sites are very vascular, extensive hemorrhage needs caution (Khan et al. 1969).

Numerous antimicrobial agents – antimony compounds, bismuth, iodine, and pentamidine (Rajam et al. 1955), antifungal agents griseofulvin, amphotericin B, and topical steroids (Jimenez et al. 1984; Ho and Tay 1986), 5-fluorocytosine, potassium iodide, ketoconazole), quinine hydrochloride (Wright 1922), paludrine, pentamidine, antibacterial antibiotics, and deep x-ray therapy

(Satyanarayana 1960) – have not been demonstrated to have any effect. Trivalent or pentavalent compounds of antimony have been reported to have been of some effect and Wright (1922) claimed that ocular rhinosporidiosis responded to 2 percent tartar emetic (antimony potassium tartrate). Allen and Dave (1936) found 'neostibosan' to have been effective, with the disappearance of endospores from nasal secretions. Dapsone is the only drug that has been found, on clinical and microscopic grounds, to have a proven beneficial effect; it minimized recurrence and produced firmer, more fibrous, and less vascular nasal and nasopharyngeal polyps (Nair 1979); recurrence occurred in 28.6 percent of dapsone-treated patients whereas 93 percent of untreated patients required surgery for recurrent disease after a 3-year period. Histologically and ultrastructurally, it caused the arrest of the maturation of the sporangia and promotion of a granulomatous response with fibrosis (Job et al. 1993, see also Krishnan 1979; Venkateswaran et al. 1997). Its therapeutic role is, essentially, as an adjunct to surgery.

Failure of drug therapy was attributed to the impenetrability of the sporangial wall (Woodard and Hudson 1984).

RHINOSPORIDIOSIS IN ANIMALS

Several domesticated species [dogs (Easley et al. 1986), cats (Moisan and Baker 2001; Wallin et al. 2001), parrots], farm species [bovines (buffaloes and cattle), goats (Jain 1967), horses, mules, ducks, geese, swans (Kennedy et al. 1995)], and wild animals [the river dolphin *Inia geoffrensis* (Lipscomb 1998)], and water fowl have shown natural rhinosporidiosis. The only epizootic, of ocular and nasal rhinosporidiosis, occurred in swans (Kennedy et al. 1995). Nasal disease is more common in mammals. Rhinosporidiosis in animals is comparable with the disease in humans in many respects (Karunaratne 1964):

- predominance in males
- occurrence in India, South America, and South Africa
- general characteristics of the growths
- histopathological patterns and their variations.

In swans, Kennedy et al. (1995) described the epithelium as attenuated or acanthotic. The degree of acute or chronic inflammation was mild to moderate and showed three patterns:

1 a severe, focal to diffuse, acute and chronic cell infiltrate
2 a granulomatous reaction with epithelioid histiocytes and foreign body-type multinucleate giant cells commonly found centered around ruptured or collapsed, crescent-shaped sporangia

3 the least common pattern with fibroblasts usually associated with a paucity of sporangia and inflammatory cells.

These patterns are similar to those in human, equine (Rippon 1988), and canine (Easley et al. 1986) rhinosporidiosis.

There are no data on immunological responses in naturally occurring rhinosporidiosis in animals.

REFERENCES

Abbas, A.K., Lichtman, A.H. and Pober, J.S. 2000. *Cellular and molecular immunology.* Philadelphia: W.B. Saunders, 359.

Agarwal, S. 1966. Rhinosporidiosis. *J Indian Med Assoc,* **46**, 8, 442–7.

Agrawal, S., Sharma, K.D. and Shrivastava, J.B. 1959. Generalized rhinosporidiosis with visceral involvement. *AMA Arch Dermatol,* **80**, 22–6.

Ahluwalia, K.B. 1999. Culture of the organism that causes rhinosporidiosis. *J Laryngol Otol,* **113**, 523–8.

Ahluwalia, K.B. and Bahadur, S. 1990. Rhinosporidiosis associated with squamous cell carcinoma in the tongue. *J laryngol Otol,* **104**, 648–50.

Ahluwalia, K.B., Maheswari, N., et al. 1997. Rhinosporidiosis: a study that resolves etiologic controversies. *Am J Rhinol,* **11**, 6, 479–83.

Al-Hili, F. 1985. Rhinosporidiosis in Bahrain, Arabian Gulf. *Mycopathologia,* **89**, 155–9.

Ali, A., Flieder, D., et al. 2001. Rhinosporidiosis: an unusual affication. *Arch Pathol Lab Med,* **125**, 1392–3.

Allen, F.R.W.K. and Dave, M.L. 1936. The treatment of rhinosporidiosis in man based on the study of sixty cases. *Indian Med Gaz,* **71**, 376–95.

Al-Serhani, A.M., Ali, S., et al. 1998. Association of rhinoscleroma with rhinosporidiosis. *Rhinology,* **36**, 43–5.

Anand, C.S., Gupta, S.K. and Srivastava, S. 1975. Rhinosporidiosis. *J Indian Med Assoc,* **64**, 40–2.

Apple, D.J. 1983. 'Papillome' der Conjunktiva bedingt durch Rhinosporidiose. *Fortschr Ophthalmol,* **79**, 571–4.

Appuhamy, S., Atapattu, D.N. et al. 2002. Strain variation in *Rhinosporidium seeberi. Proc Sri Lanka Coll Microbiol,* OP7.

Aravindan, K.P., Visvanathan, M.K., et al. 1989. Rhinosporidioma of bone: a case report. *Indian J Pathol Microbiol,* **32**, 312–13.

Arseculeratne, S.N. 2000a. An update on rhinosporidiosis. *2nd SAARC ENT Congress, Kathmandu, Nepal.*

Arseculeratne, S.N. 2000b. *Microcystis aeruginosa* as the causative organism of rhinosporidiosis. *Mycopathologia,* **151**, 3–4.

Arseculeratne, S.N. 2002. The effect of biocides (antiseptics and disinfectants) on the endospores of *Rhinosporidium seeberi. Proc Sri Lanka College of Microbiologists,* OP8.

Arseculeratne, S.N. and Atapattu, D.N. 2003a. The biological nature of the electron dense bodies of *Rhinosporidium seeberi. Proc 25th Anniv Res Sessions, Kandy Soc Med,* 22.

Arseculeratne, S.N. and Atapattu, D.N. 2003b. Biological activities of the endospores of *Rhinosporidium seeberi. Proc 25th Anniv Sessions, Kandy Soc Med,* 23.

Arseculeratne, S.N., Atapattu, D.N., et al. 1999. The humoral immune response in human rhinosporidiosis. *Proc Kandy Soc Med,* **21**, 19.

Arseculeratne, S.N., Hussein, F.N., et al. 2000. Failure to infect congenitally immunodeficient SCID and NUDE mice with *Rhinosporidium seeberi. Med Mycol,* **38**, 393–5.

Arseculeratne, S.N., Panabokke, R.G., et al. 2001. Lymphadenitis, transepidermal elimination and unusual histopathology in human rhinosporidiosis. *Mycopathologia,* **153**, 57–69.

Ashworth, J.H. 1923. On *Rhinosporidium seeberi* (Wernicke 1903) with special reference to its sporulation and affinities. *Trans R Soc, Edinb,* **53**, 301–42.

Atapattu, D.N., Arseculeratne, S.N., et al. 1999. Purification of the endospores and sporangia of *Rhinosporidium seeberi* on Percoll columns. *Mycopathologia*, **145**, 113–19.

Atapattu, D.N., Arseculeratne, S.N., et al. 2000. Is human IgG bound by endospores and sporangia of *Rhinosporidium seeberi*? – a possible mechanism of immune evasion. *Proc Kandy Soc Med*, **22**, 34.

Atav, T., Goksan, T., et al. 1955. The first case of rhinosporidiosis met with in Turkey. *Annal Otol Rhinol Laryngol*, **64**, 1270–2.

Azadeh, B., Baghoumian, N., et al. 1994. Rhinosporidiosis: immunohistochemical and electron microscopic studies. *J Laryngol Otol*, **108**, 1048–54.

Bader, G. and Grueber, H.L.E. 1970. Histochemical studies of *Rhinosporidium seeberi*. *Virchows Arch Abt A Pathol Anat*, **350**, 76–86.

Balachandran, C., Muthiah, V., et al. 1990. Incidence and clinicopathological studies on rhinosporidiosis in Tamil Nadu. *J Indian Med Assoc*, **88**, 274–5.

Beattie, J.M. 1906. *Rhinosporidium kinealyi*: a sporozoon of the nasal mucous membrane. *J Pathol Bacteriol*, **11**, 270–5.

Boyle, M.D.P. 2000. Immunoglobulin-binding proteins expressed by Gram-positive bacteria. In: Cunningham, M.W. and Fujinami, R.S. (eds), *Effects of microbes on the immune system*. Philadelphia: Lippincott Williams & Wilkins, 195–218.

Broz, O. and Privora, M. 1952. Two skin parasites of *Rana temporaria*: *Dermocystidium ranae* Guyenot & Naville and *Dermosporidium granulosum* N.Sp.. *Parasitology*, **42**, 1& 2, 65–9.

Caniatti, M., Roccabianca, P., et al. 1998. Nasal rhinosporidiosis in four dogs: four cases from Europe and a review of the literature. *Vet Rec*, **142**, 13, 334–8.

Carini, A. 1940. Sobre um parasito Semelhante ao 'Rhinosporidium', Encontrado Em Quistos da Pele de Uma 'Hyla'. *Arq Inst Biol, Sao Paulo*, **11**, 93–6.

Castellino, A.M., Rao, S.K., et al. 2000. Conjunctival rhinosporidiosis associated with scleral melting and staphyloma formation: diagnosis and management. *Cornea*, **19**, 1, 30–3.

Chatterjee, P.K., Khatua, C.R., et al. 1977. Recurrent multiple rhinosporidiosis with osteolytic lesions in hand and foot. A case report. *J Laryngol Otol*, **91**, 729–34.

Chau, K.Y., Pretorius, J.M. and Stewart, A.W. 2000. Myospherulosis in renal cell carcinoma. *Arch Path Lab Med*, **124**, 1476–9.

Chaudhary, S.K., Joshi, J.R., et al. 1986. Diagnosis of rhinosporidiosis by nasal smear examination. *J Indian Med Assoc*, **84**, 274–6.

Chayen, L. and Bitensky, L. 1991. *Practical histochemistry*, 2nd edn. Chichester: John Wiley & Sons, 99.

Chitravel, V., Sundararaj, V., et al. 1981. Cell mediated immune response in human cases of rhinosporidiosis. *Sabouraudia*, **19**, 135–42.

Chitravel, V., Sundararaj, T., et al. 1982. Detection of circulating antigen in patients with rhinosporidiosis. *Sabouraudia*, **20**, 185–91.

Chitravel, V., Sundaram, E.M., et al. 1990. Rhinosporidiosis in man: case reports. *Mycopathologia*, **109**, 11–12.

Ciferri, R., Redaelli, P. and Scatizzi, I. 1936. Unita etiologica della melattia di Seeber (granuloma da *Rhinosporidium seeberi*) accertata con lo studio di materiali originali. *Boll Soc Med-Chir Pavia*, **14**, 723–45.

Cuff, C.F., Rogers, C.M., et al. 1986. Induction of suppressor cells *in vitro* by *Candida albicans*. *Cell Immunol*, **100**, 47–56.

Daszak, P., Berger, L., et al. 1999. Emerging infectious diseases and amphibian population declines. *Emerg Infect Dis*, **5**, 6, 735–48.

Datta, S. 1965. *Rhinosporidium seeberi* – its cultivation and identity. *J Vet Sci Anim Husban*, **35**, 1–17.

de Mello, M.T. 1954. Rhinosporidiosis. In: Simons, R.D.G.P. (ed.), *Medical mycology*. Amsterdam: Elsevier, 368–83.

de Silva, N.R., Huegel, H., et al. 2001. Cell-mediated immune responses (CMIR) in human rhinosporidiosis. *Mycopathologia*, **152**, 59–68.

Desmond, A.F. 1953. A case of multiple rhinosporidiosis. *J Laryngol Otol*, **67**, 51–5.

Dhayagude, R.G. 1941. Unusual rhinosporidial infection in man. *Indian Med Gaz*, **76**, 513–15.

Easley, J.R., Meuten, D.J., et al. 1986. Nasal rhinosporidiosis in the dog. *Vet Pathol*, **23**, 50–6.

Elles, N.B. 1941. *Rhinosporidium seeberi* infection in the eye. *Arch Ophthalmol*, **25**, 969–91.

Forsyth, W.L. 1924. *Rhinosporidium kinealyi*. *Lancet*, **1**, 951–2.

Fortin, R. and Meisels, A. 1974. Rhinosporidiosis. *Acta Cytol*, **18**, 2, 170–3.

Fredricks, D.N., Jolly, J.A., et al. 2000. *Rhinosporidium seeberi*: a human pathogen from a novel group of aquatic protistan parasites. *Emerg Infect Dis*, **6**, 273–82.

Fuchs, J. and Pecher, S. 1992. Partial suppression of cell-mediated immunity in chromoblastomycosis. *Mycopathologia*, **119**, 73–6.

Gahukamble, L.D., John, F., et al. 1982. Rhinosporidiosis of urethra. *Trop Geog Med*, **34**, 266–7.

Gaines, J.J. Jr., Clay, J.R., et al. 1996. Rhinosporidiosis: three domestic cases. *South Med J*, **89**, 1, 65–7.

Gokhale, S., Ohri, V.C., et al. 1997. Subcutaneous and osteolytic rhinosporidiosis. *Indian J Pathol Microbiol*, **40**, 1, 95–8.

Goonewardene, R., Carter, R., et al. 1990. Human T-cell proliferative responses to *Plasmodium vivax* antigens: evidence of immunosuppression following prolonged exposure to endemic malaria. *Eur J Immunol*, **20**, 1387–91.

Gori, S. and Scasso, A. 1994. Cytologic and differential diagnosis of rhinosporidiosis. *Acta Cytol*, **38**, 3, 361–6.

Grover, R. 1970. *Rhinosporidium seeberi*: A preliminary study of the morphology and life cycle. *Sabouraudia*, **7**, 249–51.

Herr, R.A., Ajello, L. et al. 1999a. Chitin synthase Class 2 (*CHS2*) gene from the human and animal pathogen *Rhinosporidium seeberi*. *Proc 99th General Meeting Am Soc Microbiol*, 296.

Herr, R.A., Ajello, L., et al. 1999b. Phylogenetic analysis of *Rhinosporidium seeberi*'s 18S small-subunit ribosomal DNA groups this pathogen among the members of the protoctistan Mesomycetozoa clade. *J Clin Microbiol*, **37**, 9, 2750–4.

Herr, R.A., Mendoza, L., et al. 1999c. Immunolocalization of an endogenous antigenic material of *Rhinosporidium seeberi* expressed only during mature sporangial development. *FEMS Immunol Med Microbiol*, **23**, 205–12.

Ho, M.S. and Tay, B.K. 1986. Disseminated rhinosporidiosis. *Ann Acad Med Singapore*, **15**, 80–3.

Hutt, M.S.R., Fernandes, B.J.J., et al. 1971. Myospherulosis (subcutaneous spherulocystic disease). *Trans R Soc Trop Med Hyg*, **65**, 182–8.

Ingram, A.C. 1910. *Rhinosporidium kinealyi* in unusual situations. *Lancet*, **2**, 726.

Jain, S.N. 1967. Aetiology and incidence of rhinosporidiosis. A preliminary report. *Indian J Otolaryngol*, **19**, 1–21.

Jaiswal, V., Kumar, M., et al. 1992. Cytodiagnosis of rhinosporidiosis. *J Trop Med Hyg*, **95**, 71–2.

Jayakumar, J., Aschhoff, M., et al. 1995. Rhinosporidiosis in leprosy. *Int J Leprosy*, **63**, 3, 448–50.

Jayasekera, S., Arseculeratne, S.N., et al. 2001. Cell-mediated immune responses (CMIR) to *Rhinosporidium seeberi* in mice. *Mycopathologia*, **152**, 69–79.

Jimenez, J.F., Young, D.E., et al. 1984. Rhinosporidiosis. A report of two cases from Arkansas. *Am J Clin Pathol*, **82**, 611–15.

Job, A., Venkateswaran, S., et al. 1993. Medical therapy of rhinosporidiosis with dapsone. *J Laryngol Otol*, **107**, 809–12.

Kamal, M.M., Luley, A.S., et al. 1995. Rhinosporidiosis. Diagnosis by scrape cytology. *Acta Cytol*, **39**, 5, 931–5.

Kameswaran, S. 1966. Surgery in rhinosporidiosis. Experience with 293 cases. *Int Surg*, **46**, 6, 602–5.

Kameswaran, S. and Lakshmanan, M. 1999. In Kameswaran, S. and Kameswaran, M. (eds), *ENT disorders in a tropical environment*. Chennai: MERF Publications, 19–34.

Karunaratne, W.A.E. 1936. The pathology of rhinosporidiosis. *J Pathol Bacteriol*, **42**, 193–202.

Karunaratne, W.A.E. 1963. Rhinosporidiosis in the human female. *Proceedings of the Ceylon Association for the Advancement of Science* **19**, 8.

Karunaratne, W.A.E. 1964. *Rhinosporidiosis in man*. London: The Athlone Press.

Kavishkar, V.S., Naik, L.P., et al. 1998. Fine needle aspiration diagnosis of subcutaneous and osteolytic rhinosporidiosis. *Cytopathology*, **9**, 3, 215–17.

Kaye, H. 1938. A case of rhinosporidiosis on the eye. *Br J Ophthalmol*, **22**, 447–55.

Kennedy, F.A., Buggage, R.R., et al. 1995. Rhinosporidiosis: a description of an unprecedented outbreak in captive swans *(Cygnus* spp.) and a proposal for revision of the ontogenic nomenclature of *Rhinosporidium seeberi. J Med Vet Mycol*, **37**, 157–65.

Khan, A.A., Khaleque, K.A. and Huda, M.N. 1969. Rhinosporidiosis of the nose. *J Laryngol Otol*, **83**, 461–73.

Kierszenbaum, F. (ed.) 1994. *Parasitic infections and the immune system*. San Diego: Academic Press.

Kirkpatrick, H. 1916. Rhinosporidium of the lachrymal sac. *Ophthalmoscope*, **14**, 477–9.

Krishnamoorthy, S., Sreedharan, V.P., et al. 1989. Culture of *Rhinosporidium seeberi*: Preliminary report. *J Laryngol Otol*, **103**, 178–80.

Krishnan, K.N. 1979. Clinical trial of diamnodiphenylsulphone (DDS) in nasal and nasal pharyngeal rhinosporidiosi. *Laryngoscope*, **89**, 291–5.

Kurup, K. 1931. *Rhinosporidium kinealyi* infection. *Indian Med Gaz*, **66**, 239–41.

Kutty, M.K. and Gomez, J.B. 1971. The ultrastructure and life history of *Rhinosporidium seeberi. S E Asian J Trop Med Publ Health*, **2**, 9–16.

Kutty, M.K. and Teh, E.C. 1974. *Rhinosporidium seeberi*: an electron microscopic study of its life cycle. *Pathology*, **6**, 63–70.

Kutty, M.K. and Teh, E.C. 1975. *Rhinosporidium seeberi*. An ultrastructural study of its endosporulation phase and trophocyte phase. *Archiv Pathol*, **99**, 51–4.

Kutty, M.K. and Unni, P.N. 1969. Rhinosporidiosis of the urethra. *Trop Geog Med*, **21**, 338–40.

Kwon-Chung, K.J. and Bennett, J.E. 1992. Rhinosporidiosis. In: Kwon-Chung, K.J. and Bennett, J.E. (eds), *Medical microbiology*. Philadelphia: Lea & Febiger, 695–706.

Lakshmanan, M. and Kameswaran, S. 1978. Development and histochemistry of *Rhinosporidium seeberi*. In: Subramanian, C.V. (ed.), *Taxonomy of fungi. Proceedings of the International Symposium on Taxonomy of Fungi*. Madras: University of Madras.

Lasser, A. and Smith, H.W. 1976. Rhinosporidiosis. *Arch Otolaryngol*, **102**, 308–10.

Laveran, A. and Pettit, A. 1910. Sur une epizootie des truites. *C R Acad Sci Paris*, **151**, 421–3.

Lever, W.F. and Schaumberg-Lever, G. 1990. *Histopathology of the skin*. Lippincott: Philadelphia, 5.

Levy, M.G., Meuten, D.J., et al. 1986. Cultivation of *Rhinosporidium seeberi* in vitro: interaction with epithelial cells. *Science*, **234**, 474–6.

Lipscomb, T.P. 1998. Slide Conference. *Armed Forces Inst Pathol*, **30**, 1.

Lourenco, E.A. and Costa, L.H. 1996. Nasal septal pediculate carcinoma in situ: differential diagnosis. *Rev Paul Med*, **114**, 4, 1216–19.

Madhavan, M., Ratnakar, C., et al. 1978. Rhinosporidial infection of the forehead. *J Postgrad Med*, **24**, 4, 235–6.

Mahadevan, R. 1952. A rare case of parotid salivary cyst due to rhinosporidiosis. *Indian J Surg*, **14**, 271–4.

Mandlik, G.S. 1937. A record of rhinosporidial polypi with some observations on the mode of infection. *Indian Med Gaz*, **72**, 143–7.

Maru, Y.K., Munjal, S., et al. 1999. Brush cytology and its comparison with histopathological examination in cases of diseases of the nose. *J Laryngol Otol*, **113**, 11, 983–7.

Matusik, J., Hira, P.R., et al. 1986. Rhinosporidiosis in Kuwait. *Trop Geog Med*, **38**, 190–2.

McClatchie, S. and Bremner, A.D. 1969. Unusual subcutaneous swellings in African patients. *East Afr Med J*, **46**, 625–33.

Mears, T. and Amerasinghe, C. 1992. View from beneath: pathology in focus, rhinosporidiosis. *J Laryngol Otol*, **106**, 468.

Mendoza, L., Herr, R.A., et al. 1999. In vitro studies on the mechanisms of endospore release by *Rhinosporidium seeberi. Mycopathologia*, **148**, 9–15.

Mendoza, L., Ajello, L., et al. 2001a. The taxonomic status of *Lacazia loboi* and *Rhinosporidium seeberi* has been finally resolved with the use of molecular tools. *Rev Iberoam Micol*, **18**, 95–8.

Mendoza, L., Herr, R.A., et al. 2001b. Causative agent of rhinosporidiosis. Authors' reply. *J Clin Microbiol*, **39**, 1, 413–15.

Mendoza, L., Taylor, J.W., et al. 2002. The Class Mesomycetozoea: a heterogeneous group of microorganisms at the animal-fungal boundary. *Annu Rev Microbiol*, **56**, 315–44.

Minchin, E.A. and Fantham, H.B. 1905. *Rhinosporidium kinealyi* n.g., n.sp., a new sporozoon from the mucous membrane of the septum nasi of man. *Q J Microsc Sci*, **49**, 521–32.

Moisan, P.G. and Baker, S.V. 2001. Rhinosporidiosis in a cat. *J Vet Diag Invest*, **13**, 4, 352–4.

Moses, J.S. and Balachandran, C. 1987. Rhinosporidiosis of bovines in Kanyakumari district, Tamil Nadu India. *Mycopathologia*, **100**, 23–6.

Moses, J.S., Shanmugam, A., et al. 1988. Epidemiology of rhinosporidiosis in Kanyakumari district of Tamil Nadu. *Mycopathologia*, **101**, 177–9.

Moses, J.S., Balachandran, C., et al. 1990. Ocular rhinosporidiosis in Tamil Nadu, India. *Mycopathologia*, **111**, 5–8.

Moses, J.S., Balachandran, C., et al. 1991. *Rhinosporidium seeberi*: light, phase contrast, fluorescent and scanning electron microscopic study. *Mycopathologia*, **114**, 17–20.

Mosmann, T. 1983. Rapid colorimetric assay for cellular growth and survival. Application to proliferation and cytotoxicity assays. *J Immunol Methods*, **5**, 55–63.

Mosmann, T.R. and Sad, S. 1996. The expanding universe of T-cell subsets: Th1, Th2 and more. *Immunol Today*, **17**, 3, 138–46.

Nair, K.K. 1979. Clinical trial of diaminodiphenylsulphone (DDS) in nasal and nasopharyngeal rhinosporidiosis. *Laryngoscope*, **89**, 291–5.

Nguyen-Van-Ut, Nguyen-Van-Ai et al. 1959. Un cas de rhinosporidiose nasal, cutanée et osseuse. *Presse Méd*, **67**, 2073–5.

Noor-Sunba, M.S. and al-Ali, S.Y. 1989. The histological and histopathological pattern of conjunctival rhinosporidiosis associated with papillomavirus infection. *Histol Histopathol*, **4**, 257–64.

Noor-Sunba, M.S., al-Ali, S.Y. and el-Mekki, A.A. 1988. *Rhinosporidium seeberi* and papovavirus infection of the conjunctiva: a clinical and ultrastructural study. *APMIS Suppl*, **3**, 91–3.

Noronha, A.J. 1933. A preliminary note on the prevalence of rhinosporidiosis among sand-workers in Poona, with a brief description of some histological features of the rhinosporidial polypus. *J Trop Med Hyg*, **36**, 115–20.

Ohgaki, T., Ikeda, A., et al. 1994. A case of rhinosporidiosis. *J Otolaryngol Japan*, **97**, 35–40.

O'Kinealy, F. 1903. Localised psorospermosis of the mucous membrane of the septum nasi. *Proc Laryngol Soc Lond*, **10**, 109–12.

Pai, S.A., Naresh, K.N., et al. 1996. Rhinosporidioma of bone: diagnosis by fine needle aspiration. *Acta Cytol*, **40**, 4, 845–6.

Palaniswamy, R. and Bhandari, M. 1983. Rhinosporidiosis of male terminal urethra. *J Urol*, **129**, 598–9.

Paul, J., Khan, A.R., et al. 1978. Rhinosporidiosis in Jammu. *Indian J Pathol Bacteriol*, **21**, 73–7.

Plehn, M., Mulsow, K. 1911. Der erreger der taumelkranheit der Salmoniden. *Centralbl F Bakt 1 Abt Hs*, 63–8.

Porichha, D., Pradhan, S.C., et al. 1986. Cutaneous rhinosporidiosis in a patient of lepromatous leprosy – a case report. *Indian J Leprosy*, **58**, 626–7.

Pumhirun, P., Chuapan, C., et al. 1983. Rhinosporidiosis of nasopharynx and skin. *R Thai Army Med J*, **36**, 6, 367–74.

Rajam, R.V., Viswanathan, G.C., et al. 1955. Rhinosporidiosis: a study with a report of a fatal case with systemic dissemination. *Indian J Surg*, **17**, 269–98.

Rao, M.A.N. 1938. Rhinosporidiosis in bovines in Madras presidency, with a discussion on the probable mode of infection. *Ind J Vet Sci Anim Husb*, **8**, 187–98.

Rao, S.N. 1966. *Rhinosporidium seeberi* – a histochemical study. *Indian J Exp Biol*, **4**, 10–14.

Ratnakar, C., Madhavan, M., et al. 1992. Rhinosporidiosis in Pondicherry. *J Trop Med Hyg*, **95**, 280–3.

Ravi, R., Malikarjuna, V.S., et al. 1992. Rhinosporidiosis mimicking penile malignancy. *Urol Int*, **49**, 224–6.

Reddy, D.G. and Lakshminarayana, C.S. 1962. Investigation into transmission, growth and serology in rhinosporidiosis. *Indian J Med Res*, **50**, 3, 363–70.

Ridewood, W.G. and Fantham, H.B. 1907. On *Neurosporidium cephalodisci*, n.g., n.sp.. *Q J Microsc Sci*, **51**, 93–6.

Rippon, J.W. 1988. *Medical mycology*. Philadelphia PA: W.B. Saunders, 362–72.

Roitt, I.M., Brostoff, J. and Male, D.K. 1993. *Immunology*, 3rd edn. St Louis: Mosby, 16–17.

Rosenbaum, E. and Gan, R. 1948. A case of rhinosporidiosis coexisting with leprosy. *Ann Otol Rhinol Laryngol*, **57**, 223–9.

Sabin, A.B. and Feldman, H.A. 1948. Dyes as microchemical indicators of a new immunity phenomenon affecting a protozoan parasite (*Toxoplasma*). *Science*, **108**, 660–3.

Samaddar, R.R. and Sen, M.K. 1990. Rhinosporidiosis in Bankura. *Indian J Pathol Microbiol*, **33**, 129–36.

Sandin, R.L., Rogers, A.L., et al. 1987. Variations in affinity to *Candida albicans in vitro* among human buccal epithelial cells. *J Med Microbiol*, **24**, 151–5.

Sasidharan, K., Subramonian, P., et al. 1987. Urethral rhinosporidiosis: analysis of 27 cases. *Br J Urol*, **59**, 66–9.

Satyanarayana, C. 1960. Rhinosporidiosis. *Acta Otolaryngol (Stockh)*, **51**, 348–66.

Satyanarayana, C. 1966. Rhinosporidiosis. In: Elbs, M. (ed.), *Clinical surgery*. London: Butterworth, 143–52.

Savino, D.F. and Margo, C.E. 1983. Conjunctival rhinosporidiosis: light and electron microscopic study. *Ophthalmology*, **99**, 1482–9.

Seeber, G.R. 1900. Un nuevo esporozoario parasito del Hombre. Dos casos encontrades en polypos nasales. Unpublished thesis, Universidad Nacional de Buenos Aires.

Senba, M. and Yamashita, H. 1985. Autofluorescence of *Rhinosporidium seeberi*. *Am J Clin Pathol*, **83**, 132.

Sharma, K.D., Junnakar, R.V., et al. 1962. Rhinosporidiosis. *J Indian Med Assoc*, **38**, 640–2.

Shrewsbury, J.F.D. 1933. Rhinosporidiosis. *J Pathol Bacteriol*, **36**, 431–4.

Stricker, J.B., Hurley, D.L., et al. 1999. An unusual infection in a human immunodeficiency virus-positive man: rhinosporidiosis. *Arch Pathol Lab Med*, **123**, 11, 1121–2.

Symmers, W.StC. 1966. Deep-seated fungal infections currently seen in the histopathologic service of a medical school laboratory in Britain. *Am J Clin Pathol*, **46**, 514–37.

Teh, E.C. and Kutty, M.K. 1975. *Rhinosporidium seeberi*: spherules and their significance. *Pathology*, **7**, 133–7.

Thianprasit, M. and Thagernpol, K. 1989. Rhinosporidiosis. *Curr Topics Med Mycol*, **3**, 64–85.

Thomas, T., Gopinath, N., et al. 1956. Rhinosporidiosis of the bronchus. *Br J Surg*, **44**, 316–19.

Tirumurti, T.S. 1914. *Rhinosporidium kinealyi*. *Practitioner*, **93**, 704–19.

Vanbreuseghem, B. 1958. *Mycoses of man and animals*. London: Sir Isaac Pitman & Sons, 211–13.

Vanbreuseghem, B., Thys, A., et al. 1955. Troisieme cas congolais rhinosporidiose. Considerations nouvelles sur la nature des spherules. *Ann Soc Belg Med Trop*, **35**, 225–8.

Vanbreuseghem, R. 1973. Ultrastructure of *Rhinosporidium seeberi*. *Int J Dermatol*, **12**, 20–8.

Van der Coer, J., Marres, H., et al. 1992. Rhinosporidiosis in Europe. *J Laryngol Otol*, **106**, 440–1.

Venkateswaran, S., Date, A., et al. 1997. Light and electron microscopic findings in rhinosporidiosis after dapsone therapy. *Trop Med Int Health*, **2**, 12, 1328–32.

Vukovic, Z., Bobic-Radovanovic, A., et al. 1995. An epidemiological investigation of the first outbreak of rhinosporidiosis in Europe. *J Trop Med Hyg*, **98**, 333–7.

Wallin, L.L., Coleman, G.D., et al. 2001. Rhinosporidiosis in a domestic cat. *Med Mycol*, **39**, 1, 139–41.

Weller, C.V. and Riker, A.D. 1930. *Rhinosporidium seeberi*: Pathological histology and report of the third case from the United States. *Am J Pathol*, **6**, 721–31.

Williams, A.O. 1969. Pathology of phycomycosis due to *Entomophthora* and *Basidiobolus* species. *Arch Pathol*, **87**, 13–20.

Woodard, B. and Hudson, J. 1984. Rhinosporidiosis: ultrastructural study of an infection in South Carolina. *South Med J*, **77**, 1587–8.

Wright, J. 1907. A nasal sporozoon (*Rhinosporidium kinealyi*). *N Y Med J*, **84**, 1149–53.

Wright, R.E. 1922. *Rhinosporidium kinealyi* of the conjunctiva. *Indian Med Gaz*, **57**, 82–3.

Zschokke, E. 1913. Ein *Rhinosporidium* beim pferd. *Schweiz Arch Tierheilkd*, **55**, 641–50.

PART V

SYSTEMIC MYCOSES DUE TO DIMORPHIC FUNGI

25

Blastomycosis

ARTHUR F. DI SALVO AND BRUCE S. KLEIN

ETIOLOGIC AGENT

Blastomyces dermatitidis is the etiologic agent of blastomycosis. This thermally dimorphic, presumably terricolous, fungus causes a primary acute or chronic infection of the respiratory system. In many instances the disease manifests itself as cutaneous lesions, but the fungus may infect almost any organ. *B. dermatitidis* exists in a yeast-like form in tissues or when incubated at 37°C. In nature or when incubated at 25°C on Sabouraud's glucose agar (SGA), it appears as a white, cottony mold. A definitive identification is made by the isolation of the etiologic agent. Prompt diagnosis requires a knowledgable physician and an astute microbiologist.

HISTORY

In 1894, Gilchrist, Assistant Dermatologist at Johns Hopkins Hospital in Baltimore, MD, presented a preliminary report on the first patient with blastomycosis (Gilchrist 1894). Formalinized tissue from the patient was referred to him as a case of scrofuloderma (cutaneous tuberculosis). The clinical appearance, according to the referring physician, was classic for this disease. The cutaneous manifestations of tuberculosis were common at that time and justified the diagnosis. Gilchrist examined the tissue histologically but was unable to find tubercle bacilli. However, he described the 'numerous curious bodies' that he saw in the excised tissue. Gilchrist called this disease process a protozoan

dermatitis but expressed the opinion that the organisms were more probably of plant origin.

By the time he published an extended report on this patient he was convinced that the causative agent was a 'blastomycete,' a general term used at that time to refer to the yeasts (Gilchrist 1896). The histopathological description was thorough and complete. The illustrations in that paper demonstrate all the classic characteristics of this organism; the thick-walled, budding yeast-like cells with a broad base are clearly shown.

In his second report, also published in 1896, Gilchrist, with his colleague Stokes, described the remaining salient features of blastomycosis in another patient (Gilchrist and Stokes 1896). As these physicians had the opportunity to see this second patient, tissue specimens were taken for culture as well as histopathology. The typical unicellular tissue-form cells of *B. dermatitidis* were again described together with a complete histopathological account. The cultures from the diseased tissue grew in both their yeast and mycelial forms. Their observation of hyphae with distal and lateral cells or conidia on the mycelium led them to conclude that this agent was a filamentous fungus and not a true yeast.

The authors also observed that the mycelial form was not seen in tissues. They were inclined to separate *B. dermatitidis* from the other pathogenic 'blastomycetes' which later came to be known as species of *Candida* and *Cryptococcus*. As part of the investigation, they inoculated a dog intravenously with a pure culture of the patient's isolate. From the postmortem examination of this dog, they observed and reported on the gross and microscopic features of bilateral pulmonary blastomycosis.

Extended observations of the isolate from the second patient were published in 1898 (Gilchrist and Stokes 1898). They detailed the budding mechanism, the various tissue stains which aided in the visualization of the organism, and the inflammatory reaction in the tissue. Extensive experiments with the isolate were also performed. They described the septations in the hyphae, observing the morphological variations of the yeast-like cells and mycelium, but did not recognize the thermal cause for the two forms. The isolates were found to be relatively inert biochemically.

In the early 1900s, both pulmonary and cutaneous blastomycosis were frequently misdiagnosed as manifestations of tuberculosis. At that time mycobacterial infections were more prevalent in the general population and there was considerably more medical knowledge of tuberculosis and its etiologic agent. Those clinicians who became familiar with blastomycosis were able to make the correct diagnosis.

During the next 10 years, many case reports of blastomycosis appeared in the literature. Ricketts, later of Rocky Mountain spotted fever fame, published an extensive review of blastomycosis in 1901 (Ricketts 1901). He reviewed the subtle differences between the blastomycetic diseases which were then known: cryptococcosis (*Cryptococcus neoformans*), coccidioidomycosis (*Coccidioides immitis*), and blastomycosis. Ricketts concluded that these were similar diseases with different clinical manifestations and there was enough similarity to state that, if not caused by the same organism, the agents were sufficiently related to be in the same genus. He conceded that thrush (*Candida albicans*) and sporotrichosis (*Sporothrix schenckii*) were different disease entities.

Ricketts described 12 new cases of blastomycosis with a review of their clinical courses, therapy, attempts at serology, and exquisite histopathological details on each patient. He recognized the mycelial and yeast forms of the organism and noted that the mycelial form never appeared in tissue. On one patient, he provided the first histological report of both 'small forms' (4–6 μm) and 'large forms' (20–30 μm) of *B. dermatitidis* in potassium hydroxide (KOH) preparations from fresh tissue. Unfortunately, the cultures from this patient did not grow. Cultures from tissue of another patient, however, were incubated in both a 'brood oven' incubator and the laboratory at room temperature. He described the differences in gross colonial morphology between the two cultures but did not present a microscopic comparison. The brood oven culture was described as having the appearance of 'a heap of earth worms.' He also failed to recognize that temperature was the cause for the two forms.

The thermal basis for dimorphism was first described by Hamburger in 1907 (Hamburger 1907). He specifically reported that the incubator temperature was 37°C and that room temperature varied between 16°C and 24°C. The gross colonial morphology and the microscopic characteristics of both forms were described. He converted the mycelial form of *B. dermatitidis* to the yeast form and vice versa simply by changing the temperature of incubation. Hamburger stated that temperature is the most important factor for this transformation in morphology, and recommended that duplicate cultures of clinical material should be made to accommodate growth at either temperature (see below).

In 1914, Stober published the first attempts to define the ecology and epidemiology of blastomycosis (Stober 1914). From his investigation of the social aspects of several of the patients and his personal inspection of their residence, he made some observations on the probable source of infections. Many patients, Stober reported, lived in close proximity to damp earth, rotted wood, and wet, cardboard-lined walls. He also associated blastomycosis with plumbers who allegedly came in contact with the mold on timbers in cellars where they worked. He grew a white mold from one of the homes but did not provide sufficient mycological features of the isolate to show that this was *B. dermatitidis*. Thus, Stober probably performed the first ecological studies of *B. dermatitidis* and was the first investigator to recognize the ecological association of this mold with the environment. His environmental observations are similar to the associations we make today with the probable natural reservoir of *B. dermatitidis*. This paper also provided a thorough study of the pathological, mycological, and clinical features of systemic blastomycosis. He described the course of the illness in patients as well as the therapy and prognosis of this mycosis.

ECOLOGY

Knowledge of the ecology of the etiologic agent is essential for the investigator to understand the epidemiology of blastomycosis. Although blastomycosis and its specific etiologic agent, *B. dermatitidis*, were first described in 1894, the mystery of the precise ecological niche occupied by this terricolous fungus remains unresolved. The natural reservoir of the three other major systemic fungal pathogens was firmly established by the middle of the twentieth century. *C. immitis* was associated with the deserts of the southwestern USA and northwestern Mexico in 1932 (Stewart and Meyer 1932), *Histoplasma capsulatum* var. *capsulatum* was isolated from infested soil in 1949 (Emmons 1949), and the association of *C. neoformans* with pigeon droppings has been known since 1951 (Emmons 1951). The ecological niche of *B. dermatitidis* cannot be described with certainty but there is sufficient evidence to describe the probable habitat. The first isolation of *B. dermatitidis* from nature was not reported until 1961 (Denton et al. 1961). The fungus was found in a collection of soil, tobacco, and organic debris obtained in a tobacco-drying barn in Lexington, KY, USA. However, it was later found that the outbuilding from which the fungus was isolated had formerly housed a dog with blastomycosis. Although this isolation did not establish the ecological niche of this organism, it did prove the value of the investigational method used to recover the agent and demonstrated that *B. dermatitidis* could survive in certain soil for at least 2 years.

Isolation from nature

There have been only a few credible reports of the isolation of this fungus from a natural habitat (Denton et al. 1961; Denton and Di Salvo 1964, 1979; Bakerspigel et al. 1986; Klein et al. 1986, 1987a; Baumgardner and Paretsky 1999). Each of these isolations was made from material that was rich in organic matter including animal feces, dust, plant fragments, minerals, or insect remains. The environmental samples usually contained soil, or were near the soil. The epigeal substrate from which the fungus has been isolated was moist, shaded from direct sunlight, mixed with organic material, and with a pH of less than 6.0. A more specific ecological association for *B. dermatitidis*, as with *H. capsulatum* var. *capsulatum* and *C. neoformans*, is not yet apparent.

Environmental isolations

In 1964, Denton and Di Salvo reported the first isolation of *B. dermatitidis* from naturally contaminated environmental sites (Denton and Di Salvo 1964). Of 356 samples collected along the flood plain of the Savannah River in Augusta, GA, USA, 10 yielded the fungus. The samples were collected in abandoned sheds which had previously provided shelter to mules, chickens, dogs, or rabbits. Persistent attempts to re-isolate the geophilic fungus from the positive sites were not successful. Only one site yielded the fungus from a sample collected 10 months later (Denton and Di Salvo 1979).

In 1986, Klein et al. reported the isolation of *B. dermatitidis* from a beaver lodge located in Eagle River, WI, USA, in association with an outbreak of blastomycosis among a large number of school children (Klein et al. 1986). This was the first account of an isolate in direct association with human infection and confirmed what was long suspected: the source of infection for blastomycosis was a saprophytic mold in nature. In the same year, Bakerspigel and associates reported the isolation of *B. dermatitidis* from a patient with systemic disease who worked in a shed in Ontario, Canada (Bakerspigel et al. 1986). The shed contained large amounts of diatomaceous earth and the air was often dusty. To control this dust, the area was sprinkled with water on a daily basis. After soil sampling at this worksite, these investigators were successful in isolating *B. dermatitidis* from the environment. In 1987, Klein et al. reported the isolation of *B. dermatitidis* from soil collected at two new sites in Wisconsin where additional outbreaks of blastomycosis had occurred (Klein et al. 1987a). The patients in both these latter outbreaks were involved with the disruption of soil. Thus, it has been shown without a doubt that *B. dermatitidis* is an edaphic organism. In 1999, Baumgardner and Paretsky reported the isolation of *B. dermatitidis* from environmental material. The samples were collected from a woodpile located near a dog kennel, which housed dogs with a confirmed diagnosis of blastomycosis. The significant feature of this isolation is that, for the first time, the organism was isolated from the environment using an in vitro method (Baumgardner and Paretsky 1999).

Water association

Some investigators claim that the ecological niche of *B. dermatitidis* has an association with water. There has been some support for this hypothesis from both epidemiologic and experimental studies. The first isolates of *B. dermatitidis* from nature came from the alluvial flood plain of the Savannah River (Denton and Di Salvo 1964). In 1969, Borelli and his colleagues described the distribution of the fresh water salamanders of the genus *Necturus*, which coincided with the distribution of patients with blastomycosis in the USA. They suggested that these salamanders may play a role in the ecology of *B. dermatitidis* (Borelli et al. 1969). Furcolow et al. published an extensive investigation of the prevalence and incidence of *B. dermatitidis* in dogs and human beings (Furcolow et al. 1970). They reviewed the literature from 1885 until 1968 and found a high prevalence of human and canine cases south of the Ohio River and east of the Mississippi River. From an analysis of these findings they hypothesized an ecological relationship between *B. dermatitidis* and the proximity to water.

The conidia of *B. dermatitidis* are difficult to separate from their conidiophores. McDonough et al. demonstrated that the conidia are readily released from the mycelium in the presence of moisture (McDonough et al. 1976). Subsequently, McDonough and Kuzma, in an epidemiologic study of blastomycosis in Wisconsin, found that many of the human and canine subjects lived near bodies of water. This led them to suggest that foggy weather may contribute to the release of conidia in nature and the infection of susceptible hosts (McDonough and Kuzma 1980).

Archer also studied canine blastomycosis in Wisconsin (Archer 1985). He found that 68 percent of the 200 dogs with this mycosis lived within 500 meters of water. In a similar account from Canada, Harasen and Randall reported the location of eight dogs with blastomycosis in a small town in Saskatchewan (Harasen and Randall 1986). The dogs all resided within two blocks of a creek, which was a resting place for migrating Canada geese

(*Branta canadensis*). Thus, three purported requirements for the ecological niche of *B. dermatitidis* in nature were satisfied (soil, moisture, and organic material). Baker-spigel et al. isolated *B. dermatitidis* from the soil in a workshed in which there was a constant spray of water for dust control. The atmosphere had essentially 100 percent humidity (Bakerspigel et al. 1986). Finally, six of the 13 outbreaks of human blastomycosis occurred along rivers or lakes in the USA.

Geographical distribution

The geographical distribution of histoplasmosis and coccidioidomycosis has been defined by the use of skin test surveys, soil isolations, and serological studies. But these tools have not been available for the investigation of blastomycosis. Neither the skin test nor the serological antigens have sufficient specificity to be reliable, and the methods available to isolate *B. dermatitidis* from natural substrates, utilizing animal inoculation, are cumbersome and have a low sensitivity. The present understanding of the geographical distribution of *B. dermatitidis* comes from published case reports and a few outbreaks which attracted investigation.

These reports and studies usually reflect the interest of clinicians, epidemiologists, and mycologists who publish their experiences, rather than a true reflection of the distribution of *B. dermatitidis*. From the investigations of the outbreaks listed above, it can now be said with some certainty that the desiccation of the substrate, followed by a disturbance of the site by anthropurgic activity or natural phenomena (wind or storms), results in the creation of an aerosol of infectious particles. From

a detailed review of the literature and careful interpretation of the evidence, it appears that autochthonous cases of blastomycosis occur in North America and Africa with a possible focus in central India (Figure 25.1). First thought to be restricted to the North American continent, sporadic case reports of blastomycosis from various parts of the world began to appear in the 1940s. Many of these reports were the result of erroneous identification of the etiologic agent resulting from incomplete studies of the isolate. Other reports did not present sufficient data to validate the claim that these were autochthonous cases. Di Salvo reviewed the original publications of all reports of blastomycosis from outside the USA (Di Salvo 1992a). Each case was re-examined as to the details of diagnosis. Those with insufficient diagnostic information (clinical, cultural, or histological) were discarded. The remaining reports in that publication present the current knowledge of the geographical distribution of *B. dermatitidis*.

With culture and tissue confirmation it became clear that Africa was an endemic area for this arcane mycosis. There have been scattered reports of autochthonous cases of blastomycosis from the Middle East and Europe, but these involved a small number of patients and, in some cases, insufficient information was presented to make a definitive statement of endemicity. It is curious that in North America, the disease occurs almost exclusively east of the 100th meridian which is approximately through the center of the USA (Figure 25.2). There has been a recent report of two patients with blastomycosis from Colorado who were both involved with the relocation of prairie dogs. Although the two patients previously resided in the

Figure 25.1 *Worldwide distribution of patients with documented evidence of blastomycosis. Detailed maps of each of these areas appeared in a previous publication (Di Salvo 1992a). (From Di Salvo 1992a, with permission. Copyright by South Carolina Department of Health and Environmental Control)*

Figure 25.2 *The 100th meridian passes through the center of the USA and Canada. This figure also shows the location of the 13 documented outbreaks of blastomycosis (Di Salvo 1992b). (From Di Salvo 1992b, with permission. Copyright by South Carolina Department of Health and Environmental Control)*

endemic area, circumstantial evidence (temporal onset, substantial disruption of soil, and excess rainfall immediately preceding the anthropurgic activity), despite lack of antibody evidence and negative soil sampling, is convincing that these were autochthonous cases (Hannah et al. 2001). This finding suggests that a previous patient with blastomycosis from Colorado may have also been an autochthonous case of this mycosis instead of reactivation from residency in the endemic area more than 50 years previously (Ehni 1989). The number of cases is extensive in the USA, and moderate in Canada, with only one well-documented patient from Mexico.

There is a need for a simpler, more efficient, in vitro method to isolate or identify the etiologic agent of blastomycosis in the environment. Until this is available much of the ecology of *B. dermatitidis* will remain speculative.

EPIDEMIOLOGY

The infectious particles of *B. dermatitidis* are its mycelial fragments and conidia; the respiratory system is the portal of entry. Particles which are between 0.5 and 5.0 μm can readily pass through the respiratory tract and enter the alveoli of the lungs. The conidia of *B. dermatitidis* vary in size from 2.0 to 10 μm and the mycelial fragments may be even smaller; thus these particles can readily reach the alveoli of humans and other animals. When the host's resistance fails, an infectious process is established. Garrison and Boyd presented in vitro evidence to demonstrate the conversion of the conidia to yeast-like cells (Garrison and Boyd 1978), whereas Denton and Di Salvo demonstrated the infectivity of conidia in vivo (Denton and Di Salvo 1968). The incubation period for blastomycosis is 4–8 weeks after an inhalation exposure and a clinical dermal lesion will be obvious 1–5 weeks after a trauma (Di Salvo 1992b).

Seasonality

A substantial number of publications, which reported the time of year when cases of *B. dermatitidis* occurred, have been reviewed (Di Salvo 1992b). Each of these papers surveyed a large number of human beings or dogs with blastomycosis. From this analysis, there does not appear to be any seasonal occurrence of this mycosis. The evidence is difficult to evaluate because the time of exposure or onset of illness usually cannot be determined with any accuracy. Cutaneous lesions may appear at various intervals after the initial exposure, which is frequently unknown; pulmonary infection may be asymptomatic or unrecognized and there may be prolonged time lapses before a definitive diagnosis is made.

Occupational risk

No association can be made between *B. dermatitidis* and a specific occupation. Ten selected papers published between 1955 and 1988 were reviewed (Di Salvo 1992b). Each paper reported on a large series of patients in which occupation was specified. There were a total of 771 patients with blastomycosis in the analysis. It was obvious that most of the victims of this disease had an occupation or avocation involving outdoor activity. There are many additional single case reports of individuals with an environmental exposure to wood or soil who became infected with *B. dermatitidis*. Thus, it is not the specific occupation that renders a person susceptible to infection with *B. dermatitidis*, but the exposure of an individual to the probable ecological niche of the fungus.

Age, race, and gender

There is no recognized age, race, or gender susceptibility. People of all races and all age groups, male or female, are susceptible to this mycosis. The predominance of patients in the fourth, fifth, and sixth decades of life is consistent with an occupation or avocation that places these individuals at greater exposure to the elusive environmental niche of this fungus. A similar reasoning can be applied to the gender of the patients. Males may be infected more than females, but this bias reflects the workforce more probably exposed to the natural nidus of *B. dermatitidis* or the greater male orientation in the past to outdoor activities. Unlike coccidioidomycosis, race susceptibility for *B. dermatitidis* is not evident. Each report of such predilections merely reflects the population at risk or being studied (Di Salvo 1992b).

Transmission

It is generally accepted that the respiratory system is the portal of entry for all forms of blastomycosis except direct transcutaneous inoculation (Schwarz and Baum 1951). Blastomycosis is not ordinarily transmitted from person to person or animal to human.

With the exception of one report in which two young men working side by side acquired pulmonary disease, there has been no evidence of aerosol transfer (Procknow 1966). The two patients had no other association except at work, thus leading the investigator to believe that there may have been human-to-human transmission. Exposure to the same environmental source of *B. dermatitidis* at the worksite was not absolutely ruled out. There have been two reports with strong evidence to support transplacental passage of *B. dermatitidis* (Watts et al. 1983; Maxson et al. 1992) and two well-documented cases of conjugal transmission of blastomycosis (Farber et al. 1968; Craig et al. 1970; Dyer et al. 1983).

Transcutaneous blastomycosis caused by accidental inoculation through dog bites during examination or treatment is an occupational hazard of veterinarians. There are several well-documented reports of humans becoming infected from animals in this way (Di Salvo 1992b). Two other reports of transmission from dogs to humans were, most probably, a common exposure to a contaminated site of *B. dermatitidis* (Schwartzman et al. 1959; Lieberman 1963). Similarly, pathologists have acquired infections with *B. dermatitidis* through the percutaneous route by self-inoculation during a postmortem examination of a patient with the infection (Di Salvo 1992b).

Skin test

There is currently no reliable skin test available for *B. dermatitidis*. The dermal hypersensitivity to fungal antigens has been a significant tool for delineating the geographical distribution of *H. capsulatum* var. *capsulatum* and *C. immitis*. Information gleaned from these studies has also established the prevalence of the asymptomatic disease caused by these agents. Unfortunately, the delayed-type hypersensitivity skin test antigen for blastomycosis has not been useful. The crude extracts, made from either the mycelial form (blastomycin) or the yeast form of the organism, have had the same deficiencies as the serological tests: a lack of sensitivity and specificity.

The blastomycosis skin test has presently been abandoned by the medical community because it is not suitable for either diagnostic or epidemiologic studies. For these reasons the US Food and Drug Administration (FDA) removed the product from the US market in 1972.

Many scientists have endeavored to develop an improved skin test antigen through the development of purified chemical extracts of specific antigenic fractions, but a successful product has yet to be achieved. Recent studies by Abuodeh and Scalarone have produced encouraging results with regard to the development of a useful skin test antigen (Abuodeh and

Scalarone 1994, 1995). Preparations from a yeast-form isolate were efficacious with respect to their sensitivity to delayed-type hypersensitivity responses in *B. dermatitidis*-sensitized guinea pigs; but there was cross-reaction in animals sensitized with the *capsulatum* variety of *H. capsulatum*. Additional studies with a more purified derivative elicited a more specific reaction (Bono et al. 1995). These investigations may lead to the development of a useful epidemiologic tool in the future.

Animal disease

Next to humans, dogs are the most frequent animal infected with *B. dermatitidis*. Canines, as a surrogate for humans, have served as sentinels for human disease in epidemiologic studies for many years. Menges and his colleagues reported on an extensive study of canine and human blastomycosis in 1965 (Menges et al. 1965). The results of that study indicated that both humans and dogs were exposed to the same natural nidus of *B. dermatitidis* where they inhaled the infectious particles. These investigators also concluded that there was no evidence of animal-to-human transmission. Similar studies have supported these conclusions and have shown that in the endemic areas for blastomycosis, both dogs and humans have a similar geographical distribution of the disease (McDonough and Kuzma 1980; Archer et al. 1987; Baumgardner et al. 1995). Canine blastomycosis has been called the harbinger of human blastomycosis (Sarosi et al. 1979). Legendre has reviewed animal blastomycosis in detail (Legendre 1992).

MYCOLOGY

Nomenclature

The etiologic agent of blastomycosis is the thermal dimorphic species *B. dermatitidis* (Gilchrist and Stokes 1898). It has long been recognized that this binomial is taxonomically incorrect. This specific epithet, first used by Gilchrist and Stokes in 1898, has had several synonyms proposed over the years. All have been rejected (Conant 1939). Although medical mycologists agree that *B. dermatitidis* does not conform to the taxonomic code, the traditional, long-time use of the binomial has justified its continued use.

Identification

An isolate of *B. dermatitidis* is identified by several morphological characteristics. The classic means of identification is the in vitro conversion of this dimorphic fungus from the yeast form to the mycelial form and/or vice versa. Temperature variation is the method used to demonstrate the thermal morphological change. Incubation of yeast colonies on SGA at 25–30°C will yield growth of the mycelial form of the mold in 10–14 days. Conversely, the cultivation of mycelium on an enriched medium, such as brain–heart infusion agar with 5 percent blood (see Chapter 4, Laboratory diagnosis), and incubation at 37°C will yield yeast colonies in 3–5 days.

The culture of *B. dermatitidis* from clinical material requires 2–4 weeks for growth at 25°C on SGA. The typical colony of most isolates of *B. dermatitidis* is a white, aerial, cottony, filamentous mold. The colony may vary from a flat, dull colony of mostly vegetative hyphae to a heaped fungal mass with hyphal tufts. After 2–3 weeks the culture usually becomes more typical in appearance. The mycelial-form colony is white in the early stages of growth but may become tan to light brown as the colony ages. Concentric rings of growth are frequently seen in older colonies, i.e. 4–8 weeks of age (Figure 25.3).

Microscopically, the mycelium and the asexual propagules (conidia) are evident. The hyphae are narrow (6–7 μm wide), hyaline, branching, septate mycelia. New growth occurs at the hyphal tip. The unicellular conidia are formed on thin, lateral conidiophores or directly on the hyphae (sessile). They are slightly oval to round with a smooth wall and vary from 3 to 6 μm in diameter. These structures are readily seen in a slide culture or a wet mount preparation. *B. dermatitidis* cannot, however, be identified solely by its microconidia, which are

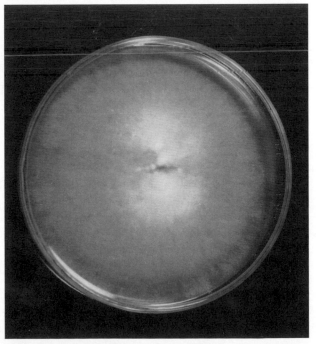

Figure 25.3 Blastomyces dermatitidis *mycelial form colony on Sabouraud's glucose agar. Note the concentric rings of growth. This colony, incubated at 25°C, was about 4 weeks old.*

indistinguishable from those of many other fungal sapro-phytes and pathogens (Figure 25.4).

At 37°C the yeast-like, or tissue, form grows in about 7–10 days when cultured on enriched media (usually brain–heart infusion agar containing 5 percent blood). The colony appears as a buttery-like, cerebri-form, soft colony with a tan color. A wet mount in lactophenol cotton blue can be prepared from the culture. Microscopic examination shows typical yeast cells, 8–12 μm in diameter, with a thick wall (frequently referred to as a 'double wall') and a single bud with a wide base of attachment to the mother cell, characteristic of this species (Figure 25.5). Electron microscopy studies have shown that the cell wall consists of two layers separated by a clear zone (Daniel et al. 1979). The yeast forms contain two to four nuclei per cell (Clemons et al. 1991) while the conidia are uninucleate.

The yeast cells, when found without buds, may be confused by less experienced laboratorians with the immature spherules of *C. immitis*, the yeast forms of *H. capsulatum* var. *duboisii*, *C. neoformans*, or an unbudded yeast-form cell of *Paracoccidioides brasiliensis*.

A defining characteristic of *B. dermatitidis* and several other dimorphic fungi is the fact that their yeast forms will convert into mycelial forms when incubated at 25°C and, conversely, mycelial forms can be changed to the yeast forms when incubated at 37°C. Many strains of *B. dermatitidis* do not readily convert to the yeast form. Several transfers, 3–4 days apart, aid in the transforma-tion, although sometimes the transformation is not complete and transitional forms are seen (Figure 25.6). There are no diagnostic biochemical tests which will aid in the identification of the yeast or mycelial forms of

B. dermatitidis. Morphology is the preliminary or presumptive method of identification and the specific nucleic acid probe is confirmatory.

Teleomorph

Ajellomyces dermatitidis, the ascomycetous or sexual state of this pathogen, was first described by McDo-nough and Lewis (1968) . It is a heterothallic member of the family *Gymnoascaceae*. The perfect state is char-acterized by its globose gymnothecia with spiralled hyphae radiating from their centers. A network of secondary peridial hyphae develops from the spirals. When mature, the tan gymnothecia measure 200–350 μm in diameter. The hyaline, evanescent asci produced in the gymnothecia contain eight smooth, spherical, hyaline, or light-tan ascospores that are uninucleate and measure 1.5–2.0 μm in diameter. It is interesting to note that the teleomorph of *H. capsulatum* is in the same genus as that of *B. dermatitidis*.

African forms

The autochthonous isolates of *B. dermatitidis* from Africa are similar to each other and appear to be distinct from North American isolates (Di Salvo 1992b). It has been suggested that isolates from these two different parts of the world are probably the same species, but different serotypes. McDonough crossed African isolates with North American isolates with the production of cleisothecia that lacked ascospores, thus indicating that the two isolates were closely related but not unequivocally the same species (McDonough 1970). However, four African strains studied by Kwon-Chung

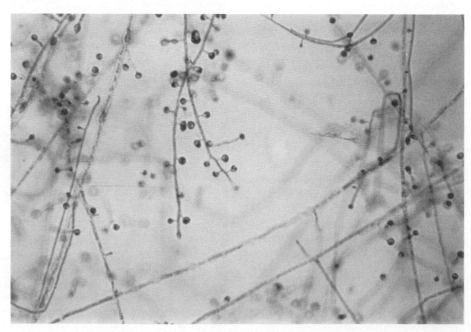

Figure 25.4 *Microscopic view of* Blastomyces dermatitidis *mycelium and conidia. Agar at 25°C, 13 days old. Magnification × 112*

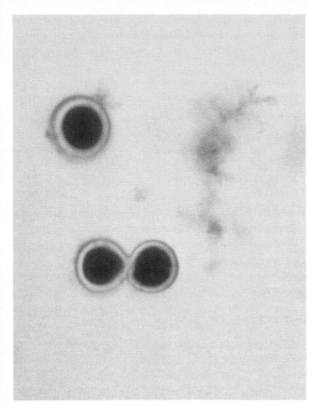

Figure 25.5 *Microscopic view of the yeast form of* Blastomyces dermatitidis *in a methylene blue wet mount. Note the figure-of-eight appearance of two cells of equal size. The broad base of attachment, single bud and 'double contour' wall are obvious. Magnification × 160*

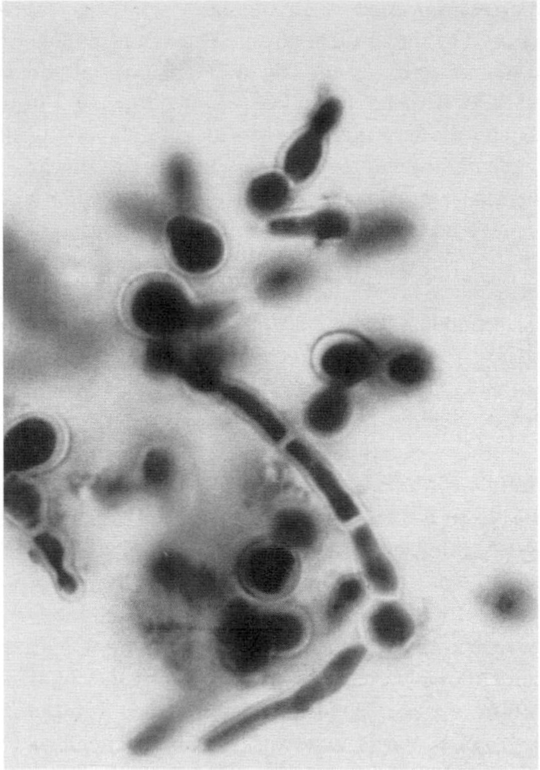

Figure 25.6 *Conversion of the mycelial form of* Blastomyces dermatitidis *to its yeast-form by incubation at 35°C. Note that the conversion was not complete on the first transfer. Some fragments of mycelium are seen intermingled with the yeast forms. At 48 hours; magnification × 160*

did not cross with her North American tester strains (Kwon-Chung 1971).

Sudman and Kaplan investigated the antigenic relationship between North American and African strains of this fungus using fluorescent antibody reagents (Sudman and Kaplan 1974). The conclusion of this study was that, although the strains share certain antigens, the isolates from these two geographical areas were antigenically different. Kaufman et al. (1983) examined the exoantigens of a large number of strains of *B. dermatitidis* from various parts of the world. They reported that the African isolates did not produce the A exoantigen which was evident in all other isolates. They also described gross and microscopic differences in the colonial morphology between African and North American isolates. Some African strains have echinulate conidia as opposed to the smooth-walled conidia of North American strains. These investigators concluded that there are two serotypes of *B. dermatitidis*.

Vermeil et al. (1982), in a comparative study of these geographically distinct strains, reported that the conidia from the African strains were gathered in clusters. They also observed that the mycelial form of the African strains did not convert to the yeast form as readily as the North American strains. Using the enzymatic activity of the yeast-form as a biotyping tool, Summerbell et al.

(1990) failed to show any differences between two African strains and 18 North American strains based on these criteria. The taxonomy and biology of *B. dermatitidis* were recently reviewed by Sugar (1992).

Biochemistry

Comparative studies on the chemical composition of yeast and mycelia have enhanced understanding of the biochemistry of *B. dermatitidis*. Lipids make up about 9 percent of the dry weight of yeast cells (Beck and Hauser 1938). About 75–80 percent of extractable lipids from yeast and mycelia are oleic and linoleic acids. Oleic acid predominates in yeasts, while a higher percentage of linoleic acid is recovered from mycelia (Domer and Hamilton 1971). Studies to localize lipids found that about 2 percent of the yeast cell wall was lipid, while negligible lipid was found in the mycelial cell wall. The lipid content of yeasts is interesting, as several studies have noted differences in lipid content between virulent and avirulent isolates of *B. dermatitidis*. Lipid content directly correlated with virulence of the isolate in mice in one study (Di Salvo and Denton 1963). However, in another study, a spontaneous avirulent mutant had a three- to fourfold increase in total lipid content, primarily composed of an increase in fatty acids, when compared to the virulent strain (Brass et al. 1982).

Other differences in cell wall composition between yeast and mycelia have been studied in an effort to understand dimorphism and virulence. The cell walls of both yeast and mycelia contain chitin and glucan. The glucan in yeast cells is 95 percent α-glucan, while that of mycelia is 60 percent α- and 40 percent β-glucan (Kanetsuna and Carbonell 1971). This shift in predominant type of glucan is similar to *P. brasiliensis* and is thought to be secondary to activation of α-glucan synthase with increased temperature (Kanetsuna et al. 1972). The chitin isolated from *B. dermatitidis* mycelia is similar to chitin obtained from *H. capsulatum* and *P. brasiliensis* (Gow and Gouday 1983). Chitin is a stable cell wall component during dimorphic change. Its role in dimorphism is suggested by electron microscopic studies showing different chitin fibril arrangements in yeast and mycelial cell walls. In yeast, the chitin fibrils are noted to be tightly interwoven in a random orientation, while in mycelia, they are arranged in a more longitudinal manner in many areas of the cell wall (Kanetsuna 1981). Isolation of the enzyme chitin synthetase from the yeast and mycelial forms indicates similar biochemical characteristics, but the yeast form enzyme is in latent form requiring activation by protease, whereas the mycelial form enzyme is in active form (Shearer and Larsh 1985).

The dimorphism of *B. dermatitidis* is primarily temperature-dependent. However, studies into the mechanism of dimorphic transition indicate a more complex regulation other than by temperature alone. Yeasts grown at 37° C can revert to mycelia in the presence of cAMP phosphodiesterase inhibitors, theophylline, or 3-isobutyl-1-methylxanthine. Using electron microscopy to study yeasts at 37°C, cAMP phosphodiesterase is localized to nuclear and mitochondrial membranes and to endoplasmic reticulum and vacuolar areas. By contrast, similar studies of the yeast 24 hours after lowering the incubation temperature to 25°C show decreased enzyme activity of cAMP phosphodiesterase in vacuolar areas. This suggests that increased levels of intracellular cAMP may be required for dimorphic transition to the yeast form (Paris and Garrison 1984). Other physiological changes that occur with mycelial to yeast transition upon change of temperature are related to mitochondrial respiration. Three stages in the transition are described:

1 immediately after an increase in temperature, partial or complete uncoupling of oxidative phosphorylation, leading to decreases in ATP levels, respiration rate, and electron transport components

2 a dormant period of 4–6 days with absent to low rates of respiration

3 a recovery phase marked by increase in respiration, increased ATP, and electron transport components and the appearance of yeast cell morphology.

During stage 2, the presence of cysteine is required for the operation of the sulfhydryl-induced shunt pathway. These stages of form transition are shared by *H. capsulatum* and *P. brasiliensis* (Medoff et al. 1987). Nutritionally dependent dimorphism is also indicated by reported conversion of mycelial to yeast form at 26°C on partially defined media.

HOST RESPONSES

Clinical forms

There are two clinical forms of blastomycosis: cutaneous and systemic or disseminated disease. It was thought that these were two clinically distinct syndromes until 1951 when Schwarz and Baum (1951) clearly demonstrated that the common cutaneous form is largely a manifestation of active or inactive systemic disease.

CUTANEOUS DISEASE

Cutaneous blastomycosis may be a primary transcutaneous disease or secondary to systemic infection. Primary transcutaneous blastomycosis is the result of the direct inoculation of *B. dermatitidis* into the skin, and therefore it is commonly seen as a solitary lesion. This type of disease is unusual and frequently occurs as the result of an accidental inoculation. It is considered an occupational hazard for mycologists, pathologists, and veterinarians (Schwarz 1983). There have been several reports of laboratory accidents which occurred during the course of experimental studies. Accidental self-inoculation during postmortem examinations of patients who have died from blastomycosis has also been reported. Similarly, veterinarians are at risk from transcutaneous inoculation as a result of accidents which occur during postmortem examinations, surgery, or when bitten by animals, usually dogs, infected with *B. dermatitidis*. As the accidents occur when working with the organism or infected tissue, most inoculations occur on the hands (Schwarz 1983). When first noted, the wound site is tender and inflamed. The lesion starts as a papule and promptly develops into a pustule; lymphangitis or regional lymphadenopathy may or may not occur, but the progression of the infection stops at the axilla. There are usually no constitutional symptoms and the course of disease is self-limited. The lesions heal spontaneously in several weeks to several months, and systemic antifungal therapy is not required. The pus from the lesions contains many yeast form cells which are readily visible in KOH wet mounts. This observation, together with knowledge of the previous trauma, is diagnostic. The incubation time for transcutaneous inoculation blastomycosis is from 1 to 5 weeks (Di Salvo 1992b).

Secondary cutaneous blastomycosis occurs as a result of the hematogenous spread of infection from another focus, usually the lungs; therefore the lesions are frequently multiple. Although cutaneous lesions result from disseminated blastomycosis, the initial pulmonary disease may not be obvious and the cutaneous lesions may be the only presenting symptoms. This cutaneous manifestation may occur spontaneously or develop at the site of a subsequent injury. The eliciting trauma may be as slight as a bump to an extremity, where transient circulating blastomycetes localize in the damaged tissue. The skin lesions appear as indolent, ulcerated, or verrucous lesions and may resemble a malignancy. Typically a lesion ulcerates and heals with scarring. Generally there is a sharp edge at the periphery of the lesion where invasive activity is occurring. The boundary frequently consists of minute abscesses which often serve as a good source of clinical material for direct examination or culture. Healing is obvious in the central area. As opposed to primary inoculation blastomycosis, the regional lymph nodes are not involved in secondary cutaneous blastomycosis (Figure 25.7).

SYSTEMIC DISEASE

The systemic disease takes many clinical forms. Acute primary pulmonary blastomycosis probably occurs more frequently than is reported. The symptoms are those of an influenza-like illness with fever, chills, and a productive cough, and chest x-rays will show nonspecific pulmonary disease. The etiology is usually not suspected and, if considered, the patient is closely observed without the administration of therapy. Untreated, acute pulmonary blastomycosis may be self-limited and resolve within a month or progress to chronic pulmonary disease or systemic disease. The

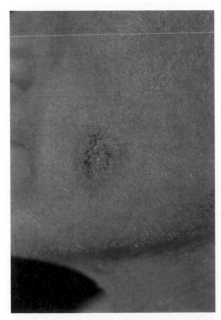

Figure 25.7 *This patient has a cutaneous lesion of secondary blastomycosis. Six months before the appearance of the lesion, he experienced an acute, undiagnosed, pulmonary infection.*

diagnosis of acute primary pulmonary blastomycosis is frequently made retrospectively when the disease progresses to the chronic form of pulmonary infection, secondary cutaneous lesions develop, or the patient is recognized as a member of a cohort exposed at a common point source. The incubation period of blastomycosis acquired by the respiratory route is 4–8 weeks (Di Salvo 1992b).

The most common form of pulmonary blastomycosis is a slowly progressing chronic pneumonia which persists for many months. The symptoms consist of fever, chills, cough, chest pain, hemoptysis, night sweats, and weight loss. The differential diagnosis includes tuberculosis, histoplasmosis, coccidioidomycosis, other non-infectious forms of chronic lung disease, and malignancy. Pleural effusions do occur, but they are an unusual manifestation of blastomycosis (Jay et al. 1977; Kinasewitz et al. 1984; Failla et al. 1995).

Although pulmonary and cutaneous manifestations are the most common form of the disease, there have been many reports of infection of practically every other organ. Systemic blastomycosis may be limited to the lungs or, if disseminated, involve almost any organ in the body. Dissemination may occur by the hematogenous, lymphatic, or direct extension routes (Schwarz and Baum 1951) and the symptoms vary according to the organ systems affected by the disease.

Miliary blastomycosis is an unusual systemic manifestation of this fungous infection. The symptoms consist of a rapid onset of severe dyspnea and respiratory failure due to adult respiratory distress syndrome (ARDS). The sporadic occurrence of this rare form of blastomycosis is considered a medical emergency and the outcome is frequently fatal. An aggressive therapeutic regimen is imperative if the patient is to survive (Shaw et al. 1976; Rippon et al. 1977; Griffith and Campbell 1979; Stelling et al. 1984; Renston et al. 1992; Meyer et al. 1993). It is thought that this clinical entity may represent endogenous reactivation (Berger and Kraman 1981; Stelling et al. 1984).

The male urogenital system seems to be highly susceptible to blastomycosis. This includes the kidney, prostate, testes, epididymis, and seminal vesicles. The prostate is frequently involved whereas the urinary bladder itself is rarely infected. Prostatitis caused by *B. dermatitidis* was extensively reviewed by Schwarz (1982).

Osteomyelitis is a frequent complication of blastomycosis (Denton et al. 1967) and arthritis may be the presenting symptom in disseminated blastomycosis (George et al. 1985; Robert and Kauffman 1988). Blastomycosis of the central nervous system, although not common, does occur. Pleocytosis is usually lymphocytic (Carmody and Tappen 1959; Buechner and Clawson 1967), although a largely polymorphonuclear response has been reported in one patient (Harley et al. 1994). Blastomycosis of the larynx, adrenals, heart, eye, paranasal sinuses, tongue, uterus, ovary, etc., has also been reported. The liver, spleen, and gastrointestinal tract are only rarely involved.

REACTIVATION BLASTOMYCOSIS

It has been suggested that reactivation of latent blastomycosis can occur. Several observations support this concept which is well documented in tuberculosis. Subsequent to the course of benign self-limited disease, in some patients a period of clinical and radiological resolution of the mild pulmonary symptoms has been reported. At some later time, ranging from months to years, the patients exhibit signs of systemic blastomycosis (Laskey and Sarosi 1978). It was reported that one patient, from Colorado, developed fatal progressive pulmonary blastomycosis 40 years after leaving the endemic area (Ehni 1989). However, as a result of a subsequent report of two patients with proven blastomycosis from Colorado, which has been considered outside the endemic area, reactivation disease in that patient should be reconsidered (Hannah et al. 2001). The source of the sequestered organisms, unlike tuberculosis, is not known. Such occurrences reinforce the admonition that a careful patient history is essential when an endemic fungal disease is suspected. There is insufficient evidence to determine if exogenous reinfection occurs.

SUBCLINICAL DISEASE

It is becoming more apparent that there is a form of blastomycosis that is subclinical or an acute, mild, self-limited disease (Denton et al. 1967; Recht et al. 1979; Sarosi et al. 1986; Vaaler et al. 1990). This concept has been reinforced during studies of common source outbreaks of blastomycosis (Sarosi et al. 1974; Centers for Disease Control 1979; Klein et al. 1986). The diagnosis of asymptomatic patients was determined while investigating a cohort of exposed individuals. Many of these individuals recovered spontaneously from their illness. The only evidence of B. dermatitidis infection was detected by antigen-specific immune studies, serological testing, and an abnormal chest x-ray. In some instances, the revelation of a previous history of pulmonary disease, compatible with blastomycosis, was revealed after the diagnosis of systemic blastomycosis was confirmed.

OPPORTUNISTIC INFECTIONS

B. dermatitidis may act as an opportunistic fungus in immunocompromised patients. Sporadic cases of blastomycosis in such patients were reported before 1990. In most patients, the underlying conditions were hematological malignancies and corticosteroid therapy (Schwarz and Salfelder 1977; Recht et al. 1982). Additional patients with blastomycosis, who were compromised by sarcoid, renal transplants, and Hodgkin's disease, have

also been reported (Önal et al. 1976; Chow et al. 1981; Hii et al. 1990; Winquist et al. 1993).

The number of acquired immunodeficiency syndrome (AIDS) patients with blastomycosis is limited when compared with those patients superinfected with the other systemic fungi. Candida spp. and cryptococcal infections are the most prevalent mycoses in AIDS patients. Histoplasmosis is common among AIDS patients in the endemic area for that disease and coccidioidomycosis is frequent among the population where C. immitis is found. There have been only a few reports of AIDS (or other immunocompromised) patients with blastomycosis. The first account of this combination was made by Chiu et al. (1988). The only collective, comparative review was published by Pappas et al. (1992). Their series of 15 AIDS patients with blastomycosis reported that the disease was usually disseminated, survival was short and the central nervous system was frequently involved (Pappas et al. 1992). While only a limited number of such patients have been reported to date (Kitchen et al. 1989; Herd et al. 1990; Fraser et al. 1991; Harding 1991), Pappas et al. have reported that an increasing proportion of blastomycosis patients seen at their medical center (24 percent seen between 1978 and 1991) have underlying impairment of immunity (Pappas et al. 1993). Mortality rates in such patients are high and exceed 30 percent. Maintenance antimycotic therapy may be required for AIDS patients with blastomycosis as it is for those with cryptococcosis.

As with the other endemic mycoses, some AIDS patients with blastomycosis had not lived in an endemic area for several years. These findings support the case for subclinical infection with sequestration of a quiescent nidus of infection. Endogenous reactivation occurs when host defenses are compromised as in the case of these AIDS patients.

It is interesting to note that, in 1914, Stober commented that, in humans, host resistance must be substantially diminished for an infection with B. dermatitidis to occur (Stober 1914). The detailed clinical aspects of blastomycosis were recently reviewed (Causey and Campbell 1992; Pappas and Dismukes 2002).

Radiology

There are no diagnostic radiological features of blastomycosis. Pulmonary infiltrates, consolidation, cavities, miliary and intermediate nodules, and mass-like lesions have been reported, but the distribution of this pathology differs among studies (Pfister et al. 1966; Halvorsen et al. 1984; Sheflin et al. 1990; Brown et al. 1991). Although pathologists claim that pleural effusions are an uncommon finding postmortem (Schwarz and Salfelder 1977), some radiologists claim that 20 percent of their blastomycosis patients had signs (blunting of the costophrenic angle) that were consistent with fluid (Sheflin et al. 1990).

The location of the pulmonary abnormalities also varied with the reports. One or both lungs may be affected. Bilateral disease is more common, although the abnormalities may occur in either the right or left lung. Upper or lower lobes may be involved but more often the disease is limited to the upper lobes.

Pathology

GROSS PATHOLOGY

The gross pathology of the dermal lesions in primary and secondary blastomycosis has been adequately characterized in the clinical description of the disease. The two types of lesions are distinct and an exposure history will support the diagnosis of primary cutaneous blastomycosis.

There is no characteristic gross appearance of organs with blastomycetic lesions. When the patient has had disseminated disease the lesions will vary with the extent of the pathology and the organs involved. The lungs are the most frequently affected organ in blastomycosis. Clinically, radiologically and grossly, the lung lesions frequently resemble carcinoma. Schwarz and Baum (1951) found that the lungs were usually heavy, i.e. 2000 g for both. The cut surface is gray with small pneumonic foci, but the typical cavities, usually seen in other fungal disease or tuberculosis, are not found (Schwarz and Baum 1951).

The bronchial tree may be ulcerated and is often filled with exudate which contains many organisms. Cavitary lesions are not common. Pleuritis is frequent but pleurisy is unusual. Calcified nodules, common in histoplasmosis, are not found and caseation of the lymph nodes is not common. The gross lesions of other organs most closely resemble a tumor upon inspection. Histological examination and culture are required to confirm the diagnosis. Yeast forms of *B. dermatitidis* have been found in blood vessels and their presence is evidence that fungemia does occur (Schwarz and Salfelder 1977).

HISTOPATHOLOGY

Histological stains

The universal histological stain in general use in the pathology laboratory is the hematoxylin and eosin (H&E) stain. This reagent is used on all diagnostic tissue sections. It will stain many, but not all, fungi. The advantage of H&E is its capability to reveal tissue structures and the inflammatory reaction as well as the invading fungus. *B. dermatitidis*, however, is sometimes difficult to visualize in tissue sections stained by this method.

There are three special stains commonly used for *B. dermatitidis* and other fungi. The special stains provide a sharp delineation of the fungus and demonstrate its typical characteristics. Gomori's methenamine–silver nitrate (GMS) stain outlines the fungal cell walls which are stained a deep black whereas the intracellular portions of the yeast cells are rose colored. The tissue background is light green, providing an effective contrast with the organism. This stain is especially useful for the inexperienced pathologist and when there is a paucity of organisms in the infected tissue. A combination of GMS and H&E has the advantages of both stains: readily visualized organisms and the opportunity to evaluate the host tissue reaction. With the periodic acid–Schiff (PAS) stain the fungus is red with a pink or light green background depending on the counterstain. It outlines *B. dermatitidis* distinctly and is preferred by some pathologists for the study of fungal histopathology. The Papanicolaou stain, commonly used in exfoliative cytology, also stains *B. dermatitidis* very well. The translucent cell wall is obvious whereas the cytoplasm is granular and stains a pale pink (Trumbull and Chesney 1981). The yeast forms can be readily visualized with ordinary cervical cytology screening procedures (Dyer et al. 1983).

The direct fluorescent antibody (DFA) test is a serological tool that is useful for the rapid and specific identification of certain fungi. It is specific for *B. dermatitidis*, when evaluated with the morphological characteristics of the suspect organism. DFA is not a stain, but a visible antigen–antibody reaction which fluoresces and outlines the fungus cell (Kaplan and Kaufman 1963). These reagents will stain only the yeast form and therefore this tool is most useful as an adjunct to histopathology. A fluorescence microscope and specific filters are required. Unfortunately, these reagents are not commercially available, so specimens must be sent to a reference laboratory that offers the desired analysis.

Nested polymerase chain reaction (PCR) has been investigated on tissue from dogs infected with *B. dermatitidis* (Bialek et al. 2003). Formalin-fixed, paraffin-embedded tissues were examined for evidence of this organism. PCR amplification of the gene encoding the unique *BAD1* adhesin was found to be as sensitive as, and even more specific than, PCR amplification of 18S rDNA for detection of *B. dermatitidis* in canine tissue. Neither was PCR crossreactivity seen in human tissue from biopsy proven cases of histoplasmosis

Morphology

Morphology is the key to the histological identification of *B. dermatitidis*. Individual cells or budding cells may be seen. The single bud with a wide base and a thick wall ('double contoured') is pathognomonic of blastomycosis. The fungus cells are 8–12 μm in diameter and round or globose in shape. Careful examination may reveal that the yeast cells contain several nuclei, with a retracted cytoplasm within the cell wall. The yeast cells may be free in the tissue or within giant cells. The inflammatory reaction is a mix of polymorphonuclear cells and a granulomatous reaction. Early lesions are usually suppurative with a polymorphonuclear infiltrate. As the inflammatory process proceeds, a granulomatous or mixed response with multinucleate giant cells appears in the lesion. The giant cells contain nuclei both in the centre and the periphery of the cell (Figure 25.8).

Size variation

The microscopist should be aware that both small and large forms of *B. dermatitidis* do occur. Small forms, 2–10 µm in diameter, have been described (Manwaring 1949; Tompkins and Schleifstein 1953; Tuttle et al. 1953; Schwarz and Salfelder 1977). These diminutive cells resemble those of the *capsulatum* variety of *H. capsulatum* in size, but they posses the characteristic broad base of *B. dermatitidis*. *H. capsulatum* yeast forms are usually elliptical and are more apt to be seen within mononuclear or polymorphonuclear leukocytes rather than giant cells. Large forms of *B. dermatitidis* have also been reported. Watts et al. (1990) described a case of pulmonary blastomycosis in which cells of *B. dermatitidis* as large as 40 µm in diameter were found in lung tissue. Walker et al. (2002) described culturally confirmed cutaneous lesions with yeast cells 20–40 µm in diameter. One of 10 wild strains of *B. dermatitidis* isolated from soil also produced large yeast forms. When recovered from experimentally infected mice, most cells from this isolate were 25–30 µm in diameter (Denton and Di Salvo 1964). Histologically, these aberrant forms have frequently been found mixed with yeast cells of normal size and morphology. In those instances, where cultures have been made, the yeast forms grew in the normal size range (W. Kaplan 1988, personal communication). It is interesting to note that Stober described small forms (diameter 3–5 µm) and large forms (diameter 25–30 µm) in 1914 (Stober 1914).

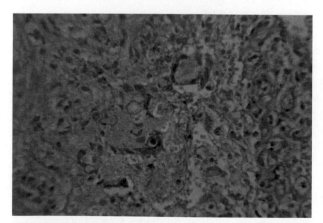

Figure 25.8 *A microscopic view of a tissue section with a bone lesion caused by* Blastomyces dermatitidis. *Note the giant cell containing nuclei and cells of the yeast form. As a result of the large size of the tissue-form of this organism, the yeast invoked a foreign body tissue reaction.*

Mycelial forms

Mycelial forms of *B. dermatitidis* have occasionally been seen in tissue (Hardin and Scott 1974; Kaufmann et al. 1979; Atkinson and McCurley 1983) and bronchial washings (Herd et al. 1990). The latter case was an AIDS patient, and the mixture of tissue and mycelial forms found in this patient was similar to that found in an AIDS patient with coccidioidomycosis. These hyphal forms of *B. dermatitidis* are usually short, septate, and hyaline, with occasional branches which resemble the pseudomycelium of *C. albicans*.

Skin

The common histological finding in cutaneous blastomycosis is pseudoepitheliomatous hyperplasia with squamous cell proliferation and acanthosis. Intraepithelial and dermal microabscesses with budding yeasts are seen. Giant cells may or may not be present and the organisms can be found within or outside the giant cells. In chronic cutaneous blastomycosis eosinophils may be seen.

Other organs

Abscesses with few or numerous organisms may be seen or the yeast-like cells may be dispersed in the lesion. The host inflammatory reaction may be both suppurative and granulomatous. As in the cutaneous lesions, the organisms are frequently found in giant cells. In bones, a granulomatous osteomyelitis is usually seen.

Caveat

For the inexperienced pathologist, it may be difficult to make a definitive diagnosis of *B. dermatitidis* if the typical morphological forms are not present in the tissue sections being examined. This occurs when there are only a few fungal cells present or when budding yeasts are not found. *B. dermatitidis* may then be difficult to distinguish from the *duboisii* variety of *H. capsulatum*, *C. immitis* (immature spherules), *P. brasiliensis* (look for multiple buds, thin walls, narrow attachment of daughter cells), or *C. neoformans* if a large capsule is not present (look for a capsule or use the mucicarmine stain which tints the smallest amount of the polysaccharide capsule). *Candida* spp., and sometimes *S. schenckii*, will appear as round cells. The DFA can resolve many of these problems if no typical forms are present. A comprehensive review of the pathology of blastomycosis can be found in Chandler and Watts (1992).

Host–pathogen interactions

DIMORPHISM AND VIRULENCE

Transition to a yeast form is a requirement for pathogenesis. Medoff et al. (1986) treated *H. capsulatum* mycelia with the sulfhydryl inhibitor *p*-chloromercuriphenylsulfonic acid (PCMS), to prevent conversion to the yeast form, and showed that inoculation with this material no longer caused disease in a murine model of infection. His findings imply that yeast-form-specific factors are essential for pathogenicity of dimorphic fungi. To date, few form-specific genes have been identified in *B. dermatitidis* and other dimorphic fungi. One of the best studied in *B. dermatitidis* is *BAD1* (*Blastomyces* adhesin; formerly WI-1). A number of molecular and genetic tools applied to *B. dermatitidis* have aided the study of this yeast-form-specific virulence factor, as described below. Another yeast-form specific gene of *B. dermatitidis* is *bys 1* (*Blastomyces* yeast-specific), but its function is unknown (Burg and Smith 1994). α-Glucan is an additional differentially expressed product of *B. dermatitidis*. The cell wall glucan content of *B. dermatitidis* yeast is 95 percent α-glucan (Kanetsuna and Carbonell 1971), which is situated in the outer wall. By contrast, the proportion of α- and β-glucan is more evenly distributed in the cell walls of the mold. Hence, these polymers are differentially expressed in the yeast and mold forms of the fungus. In *B. dermatitidis*, as well as in *P. brasiliensis* and chemotype II strains of *H. capsulatum*, the loss of yeast cell wall α-(1,3)-glucan in genetically related variants correlates with the loss of virulence in animal and cell culture models of infection (Hogan et al. 1996).

TOOLS FOR GENETIC MANIPULATION OF *B. DERMATITIDIS*

Study of *B. dermatitidis* pathogenicity has benefited from the application of molecular biological tools (Brandhorst et al. 2002). High-voltage electric pulse was used first for introducing foreign DNA into *B. dermatitidis* (Hogan and Klein 1997). Transformation efficiency by electroporation is low, however (range, 1–42 c.f.u./μg transforming DNA). Where numerous transformants must be screened, a more efficient gene-transfer method is desirable. *Agrobacterium tumefaciens*-mediated gene transfer, developed originally for transformation of plant species, has been adapted for use with *B. dermatitidis* (Sullivan et al. 2002), and routinely generates transformants with a frequency of up to 10^3 per 10^7 per target yeast. Transforming DNA chiefly integrates into a single chromosomal locus. This method of DNA transfer thus offers the advantages of single-site integration, desirable for insertional mutagenesis, and a high yield of transformants.

The *Escherichia coli* hygromycin phosphotransferase gene (*hph*), coding for hygromycin B resistance, serves as the staple dominant marker for transformation. Chlorimuron ethyl resistance, encoded by the *sur* (sulfonylurea resistance) gene from *Magnaporthe grisea*, is another effective dominant selectable marker and has been useful in retransformation of previously manipulated, hygromycin-resistant strains. G418 resistance also

has been employed, yielding three options for antibiotic selection-based transformation (Brandhorst et al. 2002). Auxotrophic strains of *B. dermatitidis* have become available recently, when strains with spontaneous and UV-induced mutations in the *ura5* gene for uracil production were isolated using 5-fluoro-orotic acid selection (Sullivan et al. 2002). Complementation of this auxotrophy offers more efficient selection than antibiotic resistance.

Homologous recombination takes place in *B. dermatitidis*, but its rate relative to the usual mode of illegitimate recombination is uncertain. To knock out the *BAD1* gene by homologous recombination, the *hph* selection marker was placed downstream of a truncated span of *BAD1* promoter. This strategy may have functioned as a promoter trap. Homologous replacements of *BAD1* represented about 1–2 percent of hygromycin B-resistant transformants (Brandhorst et al. 1999). Targeted knockout of genes, while attainable by more conventional techniques, may be enhanced by such a strategy. Reporters for studies of gene regulation have been developed for *B. dermatitidis*. β-Galactosidase activity can be measured in extracts of *B. dermatitidis* transformed with this reporter, and *BAD1*–β-galactosidase reporter fusions have been used to confirm that *BAD1* is transcriptionally regulated (Rooney et al. 2001). A β-glucuronidase reporter is also functional in *B. dermatitidis* (Brandhorst et al. 2002).

IMPORTANCE OF BAD1 IN PATHOGENESIS

Manipulation of the *BAD1* locus is an example of the application of new molecular genetic tools. *BAD1* was successfully deleted from *B. dermatitidis* strain ATCC 26199 by allelic replacement. The knockout strain had diminished capacity to bind macrophages and murine lung tissue, compared to the parent strain, and most importantly, was found to be attenuated in a murine model of blastomycosis (Brandhorst et al. 1999). Binding and virulence defects were rescued by re-expression of *BAD1* in trans in the knockout strain. The indispensable role of *BAD1* in pathogenesis was thus uncovered using the molecular genetic tools now available in *B. dermatitidis*.

Sequence analysis of the *BAD1* gene predicts a product consisting of a short amino-terminal region, a 24-amino acid repeat, present in 30 tandem copies within the core of the protein, and a carboxyl-terminal region that possesses homology to epidermal growth factor (Hogan et al. 1995). The C-terminal EGF domain is important for BAD1 localization on the yeast cell surface (Brandhorst et al. 2003). The tandem repeat of BAD1, which constitutes over 75 percent of the protein's primary structure, mediates binding of macrophage CR3 and CD14 receptors (Newman et al. 1995). BAD1 binding to phagocytes has also been found to modulate host immune responses. For example, BAD1 downregulates the production of tumor necrosis factor-alpha (TNF-α), which is normally stimulated in response to the infection, and critical for host defense (Finkel-Jimenez et al. 2001). A *BAD1* knockout strain evokes the production of TNF-α in vivo in an experimental model of infection and in vitro during co-culture with macrophages and neutrophils, whereas the isogenic, wild-type strain (and purified *BAD1*) curtails TNF-α production, promoting progressive infection. BAD1 modulates TNF-α production partly through inducing transforming growth factor-beta (TGF-β), which antagonizes phagocyte TNF-α responses (Finkel-Jimenez et al. 2002). Hence, in addition to promoting binding of the fungus to cells and tissue, BAD1 enhances pathogenicity by allowing *B. dermatitidis* to evade host immunity by modulating it in a manner that favors fungal survival over host survival.

BAD1 AND FORM TRANSITION

BAD1 is only found on the surface of the yeast form of *B. dermatitidis*, and not on the surfaces of hyphal filaments or conidia that make up the mold form (Rooney et al. 2001). BAD1 is not a prerequisite for the form transition to yeast. Strains from which BAD1 has been deleted undergo form transition in both directions with no hindrance. Given the established role of BAD1 in pathogenesis, these observations help explain how the thermally triggered form transition leads to establishment of a virulent state. Once conversion to yeast has begun, BAD1 rapidly accumulates on the surface of the organism. An analysis of yeast and mycelial morphotypes employing both Northern hybridization and reporter genes fused to the *BAD1* promoter has shown that the form-specific regulation of *BAD1* is tightly controlled at the level of transcription (Rooney et al. 2001). Hence, a virulence gene – *BAD1* – has proven to be under transcriptional control that is tied, either directly or indirectly, to the temperature-induced transition from mold to yeast.

Many African strains of *B. dermatitidis* are naturally devoid of BAD1, as they lack the gene for this protein (Klein et al. 1997). When the BAD1 coding region, under the control of its native promoter, was transferred into an African strain of *B. dermatitidis*, the heterologously expressed BAD1 retained form-specific expression and control at the level of transcription. The same phenomenon was observed when the *BAD1* gene was transferred into *H. capsulatum*, which also lacks a native *BAD1* locus, indicating that the form-specific mechanism of transcriptional control is conserved (Rooney et al. 2001). Sequence analysis of the upstream region of *BAD1* identified regions of homology with the promoter of the *H. capsulatum* yeast-form-specific gene *yps-3*. It seems likely that other, as yet undefined, yeast-form-specific genes may be controlled by the same transcriptional regulatory mechanism.

HOST DEFENSE MECHANISMS

Polymorphonuclear leukocytes (PMN) efficiently phagocytose and kill the conidia of *B. dermatitidis*. By contrast, Drutz and Frey (1985) demonstrated that PMNs do not ingest *B. dermatitidis* yeast cells. Instead they surround and attach to extracellular yeasts. The lack of phagocytosis is attributed to the size of the yeast cells. PMN killing of conidia and yeast is mediated by oxidative mechanisms. Killing of the yeast is thought to involve an extracellular process, whereby PMNs attach to the yeast and empty granule contents in the extracellular space (Diamond and Krzesicki 1978). Brummer and Stevens (1982) have reported that PMNs augment growth of yeast in vivo and in vitro, rather than kill the fungus, fueling controversy about the role of PMNs in host defense.

Alveolar macrophages readily ingest conidia, but have difficulty both ingesting and killing yeast. Peripheral monocytes and especially monocyte-derived macrophages ingest and kill conidia and yeast more efficiently. Macrophage functions are enhanced with cells from immunized patients (Bradsher et al. 1985). Killing of yeast cells by interferon-gamma (IFN-γ)-activated macrophages does not depend on products of the oxidative burst (Brummer and Stevens 1987).

Bradsher described an in vitro test to quantify cell-mediated immunity in patients with blastomycosis (Bradsher and Alford 1981). The *Blastomyces* alkali-soluble water-soluble (ASWS) antigen developed by Cox (Cox and Larsh 1974) was used to measure lymphocyte migration inhibition and lymphocyte transformation. An outbreak of blastomycosis in which 48 patients were identified with acute pulmonary blastomycosis provided an opportunity to evaluate this assay in humans (Klein et al. 1990). Most patients developed specific cellular immunity between 3 months and 21 months after exposure to *B. dermatitidis*. The assay may offer an epidemiologic tool to elucidate the geographical distribution of blastomycosis. A detailed discussion of the immunology of *B. dermatitidis* has been published by Klein (1992).

As early as 1914, results of autogenous vaccines were reported. Stober described three patients (Stober 1914). One patient did well and two improved for a short period of time, then relapsed and died. There has been no recent investigation into the immunization of humans. Mice were shown to be protected against a lethal challenge with *B. dermatitidis* yeast cells by immunization (recovery from nonlethal subcutaneous infection) (Morozumi et al. 1982). Protection was mediated by cells and not serum, supporting the previous studies by Cozad and Chang (1980), which established that cellular immunity mediates protection against *B. dermatitidis*. More recently, Wuthrich et al. (2000) described a genetically engineered attenuated vaccine strain of *B. dermatitidis* that harbors a targeted deletion of *BAD1*. The vaccine confers sterilizing immunity against infection with isogenic and nonisogenic strains in a murine model, and has allowed investigation of the cellular molecular mechanisms of acquired immunity in the murine system. As in prior work, T cells rather than B cells are largely responsible for vaccine immunity, particularly CD4$^+$ T cells. Type 1 cytokines including IFN-γ, TNF-α, and granulocyte-macrophate colony-stimulating factor (GM-CSF) are instrumental in immunity mediated by these cells in normal hosts (Wuthrich et al. 2002). There is sufficient redundancy in immune-deficient hosts so that an absence of one or more of these products can be offset by a compensatory response in one of the others. For example, IFN-γ deficiency can be compensated by TNF-α, and conversely, TNF-α deficiency can be compensated by GM-CSF (Wuthrich et al. 2002). This theme has been extended to studies in CD4$^+$ T-cell-deficient hosts (Wuthrich et al. 2003), where CD8$^+$ T cells subsume the role of T-cell help, producing the type 1 cytokines needed for host defense and sterilizing immunity against *B. dermatitidis* (or *H. capsulatum*). These studies raise the prospect that engineered vaccines can protect against infection with *B. dermatitidis* or related fungi, possibly even in immune-deficient hosts (Casadevall and Pirofski 2003).

LABORATORY DIAGNOSIS

The diagnosis of blastomycosis in a patient frequently requires a combination of several techniques. The clinical signs and symptoms are nonspecific. Some commercial serological tests lack specificity and most lack sensitivity. The histological picture may not be clear as a result of the morphological similarity with other mycotic agents and nonmycotic artifacts. A presumptive diagnosis can frequently be made by integrating patient history, the clinical symptoms, a wet mount examination, and serology. The only absolute laboratory diagnostic criterion for blastomycosis is the isolation and identification of *B. dermatitidis*.

As the fungal etiologic agent is not always apparent on clinical presentation of the patient, universal methods for fungus culture are performed if a mycotic disease is suspected. The techniques for the isolation and identification of mycotic agents are described in detail in Chapter 4, Laboratory diagnosis. Table 4.1 of Chapter 4 describes the appropriate clinical material to be selected for the isolation of *B. dermatitidis*. The application to blastomycosis will be described briefly.

Direct examination

The observation of yeast cells in an adequate clinical sample is a quick way to establish a preliminary diagnosis. However, isolation and culture are essential to confirm the diagnosis. The specimens to be submitted to the laboratory depend on the manifestations of the disease: skin scrapings or pus from skin lesions; sputum in cases of pulmonary disease; and biopsy tissue from any lesion. Urine, even in the absence of urinary tract symptoms, is frequently a good source of the organism in males. The yeasts are most probably liberated from an infected prostate gland. Although hematogenous dissemination is common in blastomycosis, blood cultures have only rarely yielded this fungus (Musial et al. 1987).

Direct microscopic examination of the clinical material can readily be performed. A typical cutaneous lesion shows central healing with microabscesses at the periphery. A pus specimen may be obtained by nicking the top of a microabscess with a scalpel. The exuded purulent material is mixed in a drop of 20 percent KOH on a microscope slide. A coverslip is applied and the slide is heated gently to clear the tissue and purulent material without damage to the fungus. The wet mount is then examined under the microscope. *B. dermatitidis* appears as a round yeast cell with double wall and often with a single bud. Usually the cells are uniform in size, 8–12 μm, readily seen in pus or infected tissue with KOH. The apparent double wall and broad-based bud are characteristic. Bud scars are sometimes seen at the side of the juncture of the mother and daughter cells. Both cells are frequently the same size and appear as a figure-of-eight (see Figure 25.5).

Culture

The preferred medium for the isolation of *B. dermatitidis* is Emmons' modification of SGA. The modification includes the addition of antibiotics to inhibit bacterial contamination, a decrease in glucose from 4.0 percent to 2.0 percent and a final pH of 6.9. The reduced sugar content is more favorable for the recovery of *B. dermatitidis*. Samples submitted for culture are incubated at both 35–37°C and 25–30°C. All cultures should be held 4–6 weeks before discarding as negative.

PRESUMPTIVE DIAGNOSIS

A presumptive identification of *B. dermatitidis* is based on the colonial and microscopic morphology as described above and the conversion of the isolated form to the converse. The dimorphic attribute applies to only a few human pathogens and the structures of one form or the other are sufficiently distinct to the experienced mycologist to allow the identification of the isolate. In recent years several means of verifying the identity of *B. dermatitidis* have been developed: exoantigen, DNA probe, and DFA.

EXOANTIGEN

A simple, specific, and sensitive test to identify the mycelial form of dimorphic fungi is the exoantigen test

(Kaufman et al. 1983). This is a double immunodiffusion reaction in agar performed by using an aqueous extract of a 7–10-day-old mycelial culture as the unknown antigen. This unknown antigen solution is precipitated against a known, standardized antiserum. A sharp, clear precipitin band of identity is formed where the antigen and the homologous antiserum meet. This test, with the proper antiserum, is 100 percent specific for *B. dermatitidis* and can also be applied to the identification of other dimorphic fungi. It is interesting that nonviable mycelial cultures of *B. dermatitidis* can be identified by this method (Di Salvo and Wooten 1982) and *B. dermatitidis* can be specifically identified even when the extracted culture is contaminated (Di Salvo et al. 1981).

DIRECT FLUORESCENT ANTIBODY

The direct fluorescent antibody (DFA) test for *B. dermatitidis* was developed by Kaplan and Kaufman (1963). There are several advantages of this rapid procedure to identify yeast forms of the fungus. The technique may be applied to tissue sections or to a yeast-form culture. Tissue, suspected of being malignant, is often removed at surgery and placed in formalin. Later, when the possibility of a fungous infection is being considered, clinical material for culture is not available. If the organisms in the tissue are not typical because of unusual shapes or size, the morphology may not be sufficient for a definitive diagnosis. Where morphology among yeasts can be very confusing, DFA is specific. When applied to unknown yeast organisms, DFA will fluoresce positive as soon as growth is visible. The stain will identify *B. dermatitidis* whether or not the organisms are viable.

NUCLEIC ACID PROBES

A simplified test using nucleic acid probes has recently been developed for the identification of *B. dermatitidis* cultures. The probes may be applied to either the mycelial or yeast forms of the organism. The single-stranded DNA probe is combined with a chemiluminescence label for detection. If the unknown organism is *B. dermatitidis*, the DNA probe will hybridize with the RNA of the target organism. The labelled DNA/RNA hybrid is then detected for identification of the fungus. In three evaluations of the nucleic acid probe, the sensitivity ranged between 87.8 percent and 100 percent. Two studies reported 100 percent specificity (Scalarone et al. 1992; Stockman et al. 1993). The third evaluation tested 17 strains of *P. brasiliensis*, a morphologically similar dimorphic fungus (Padhye et al. 1994). These investigators found that 10 of the 17 *P. brasiliensis* strains hybridized with the *B. dermatitidis* probe. This crossreaction is not an obstacle if the morphology is distinctive for the suspected organism. The commercially available reagents are expensive, but the test requires less than 2 hours to confirm the identity of the isolate. Any rapid

means of identification is welcome by mycologists because the traditional methods required to isolate and identify dimorphic fungi require 4–6 weeks. In the future, the nested PCR assay for detecting *B. dermatitidis BAD1* sequences, developed by Bialek et al. (2003), may fulfill this criterion.

A comprehensive discussion of specific culture methods for each type of clinical material has been published (Goodman 1992). Detailed diagnostic laboratory procedures for the microbiologist are described in that publication.

Serodiagnosis

As fungal pulmonary infections cannot be etiologically classified clinically, the serological tests for fungal infections are requested, and performed, as a group. Most diagnostic laboratories in the USA perform fungal serology as a battery for the endemic mycoses blastomycosis, histoplasmosis, and coccidioidomycosis.

There are two serological tests in general use which detect antibodies to *B. dermatitidis* and are used for the diagnosis of blastomycosis: the complement fixation (CF) test and the immunodiffusion (ID) precipitin test. The CF test usually requires 2–3 months for antibody to be detected. Both the sensitivity and the specificity of the CF test are poor. However, the results are quantitative and may aid in evaluating the progression of disease or efficacy of therapy. The ID or precipitin test may be positive in up to 80 percent of the patients with blastomycosis, 2–3 weeks after the onset of symptoms, and its specificity is close to 100 percent. A positive test indicates recent or active disease. The enzyme immunoassay (EIA) test is a more recent development in the diagnosis of blastomycosis. In one small trial of a commercial diagnostic kit, the procedure was shown to have a sensitivity of 100 percent and a specificity of 85.6 percent (Sekhon et al. 1995).

An outbreak of blastomycosis among a group of 95 individuals exposed to a point source of *B. dermatitidis* provided an opportunity to study the serology of this disease. A comparison of the CF, ID, and EIA tests on 47 patients meeting the case definition was reported (Klein et al. 1987b). They found these three tests to have a sensitivity of 9, 28, and 77 percent, respectively, and a specificity of 100, 100, and 92 percent. The low sensitivity in this study may be the result of the number of asymptomatic or mild cases of blastomycosis identified in this cohort. The EIA testing was performed at the Centers for Disease Control and Prevention (CDC) using their reagents, not the commercially available kit. As the day of exposure was known with certainty, which is not usually the case in blastomycosis, the time required for the development of antibody could be determined reliably. The conclusion of this study was that, in this instance, the EIA antibody was detected as early as 13 days and peaked at 50–70 days after the onset of symptoms. The ID antibody peaked at the same time but did not appear earlier. Therapy shortened the antibody window, but the three tests did not appear to have any prognostic value. A report by Kaufman (1992) specifically described the present immunodiagnosis of blastomycosis in greater detail.

A noncommercial radioimmunoassay has been reported that incorporated BAD1, which is a target of both antibody and T-cell responses (Klein and Jones 1990). Sensitivity and specificity were 85 and 97 percent, respectively, in 68 patients with blastomycosis, and 73 patients with other systemic mycoses. Follow-up studies demonstrated that BAD1 tandem repeats are the target of the antibody response in patients and that these sequences are present in the A antigen of *B. dermatitidis*. BAD1 and A antigen of *B. dermatitidis* are thus similar antigens, if not the same, although detailed biochemical studies of the antigens demonstrated that carbohydrate on A antigen (lacking on BAD1) accounts for crossreactivity with sera from histoplasmosis patients or rabbit histoplasmosis immune sera (Klein and Jones 1994).

TREATMENT

Not all patients with blastomycosis require antifungal therapy. It is customary to observe the patient with suspected mild pulmonary blastomycosis while the cultures are pending. In many patients, the pneumonia will clear spontaneously. If progression of the disease is obvious, an antimycotic agent should be considered. There are two major classes of antimycotic agents used for blastomycosis: the polyenes and azoles.

The polyene amphotericin B, introduced in the 1960s, is still considered the gold standard of therapy for blastomycosis and other systemic mycoses. This drug must be administered intravenously for several weeks. The initiation of therapy requires hospitalization, close observation of the patient, and monitoring of renal function. Amphotericin B has significant side effects in many patients, but relapses are few after an adequate course of therapy (about 2 g). The mechanism of action of this agent is believed to be an alteration of the permeability of the cell membrane of the yeast cells which allows essential constituents of the fungus to leak from the cell with subsequent death of the fungus. It is still considered the drug of choice by many infectious disease physicians, particularly in patients with severe disease or underlying impairments of immunity.

The azoles cause fungistasis by blocking the enzyme which converts lanosterol to 14-demethyllanosterol. This biochemical sequence is the forerunner of ergosterol synthesis. The triazole, itraconazole, is presently the drug of choice for mild cases of blastomycosis. This oral medication can be administered on an outpatient basis and is less toxic than amphotericin B. Some physicians

still debate whether antimycotic therapy is necessary after surgical excision of pulmonary lesions (Edson and Keys 1981). More conservative clinicians tend to provide the additional antibiotic therapy. For immunocompromised patients, particularly those with AIDS, or in life-threatening infection, amphotericin B is still the drug of choice. A detailed account of the therapy of blastomycosis with various antimycotic agents has recently been published by Bradsher et al. (2003). Additional information on antifungal therapy is presented in Chapter 8, Principles of antifungal therapy.

The recognition of patients infected with blastomycosis requires a detailed patient history and knowledge of the ecology and epidemiology of *B. dermatitidis*.

REFERENCES

Abuodeh, R.O. and Scalarone, G.M. 1994. Comparative studies on the detection of delayed dermal hypersensitivity in experimental animals with lysate and filtrate antigens of *Blastomyces dermatitidis*. *Mycoses*, **37**, 149–53.

Abuodeh, R.O. and Scalarone, G.M. 1995. Induction and detection of delayed dermal hypersensitivity in guinea pigs immunized with *Blastomyces dermatitidis* lysate and filtrate antigens. *J Med Vet Mycol*, **33**, 19–25.

Archer, J.R. 1985. Epidemiology of canine blastomycosis in Wisconsin. Master's thesis, University of Wisconsin, Stevens Point, Wisconsin.

Archer, J.R., Trainer, D.O. and Schell, R.F. 1987. Epidemiologic study of canine blastomycosis in Wisconsin. *J Am Vet Med Assoc*, **190**, 1292–5.

Atkinson, J.B. and McCurley, T.L. 1983. Pulmonary blastomycosis: filamentous forms in an immunocompromised patient with fulminating respiratory failure. *Human Pathol*, **14**, 186–8.

Bakerspigel, A., Kane, J. and Schaus, D. 1986. Isolation of *Blastomyces dermatitidis* from an earthen floor in southwestern Ontario, Canada. *J Clin Microbiol*, **24**, 890–1.

Baumgardner, D.J. and Paretsky, D.P. 1999. The in vitro isolation of *Blastomyces dermatitidis* from a wood pile in North Central Wisconsin, USA. *Med Mycol*, **37**, 163–8.

Baumgardner, D.J., Paretsky, D.P. and Yopp, A.L. 1995. The epidemiology of blastomycosis in dogs: north central Wisconsin, USA. *J Med Vet Mycol*, **33**, 171–6.

Beck, R.L. and Hauser, C.R. 1938. Chemical studies of certain pathogenic fungi. I. The lipids of *Blastomyces dermatitidis*. *J Am Chem Soc*, **60**, 2599–603.

Berger, R. and Kraman, S. 1981. Acute miliary blastomycosis after 'short-course' cortiscosteroid treatment. *Arch Intern Med*, **141**, 1223–5.

Bialek, R., Cirera, A.C., et al. 2003. Nested PCR assays for the detection of *Blastomyces dermatitidis* DNA in paraffin-embedded canine tissue. *J Clin Microbiol*, **41**, 205–8.

Bono, J.L., Legendre, A.M. and Scalarone, G.M. 1995. Detection of antibodies and delayed hypersensitivity with rotofor preparative IEF fractions of *Blastomyces dermatitidis* yeast phase lysate antigen. *J Med Vet Mycol*, **33**, 209–14.

Borelli, D., Marcano, C. and Feo, M. 1969. Actividades de la seccion de micologia medica, Instituto de Medicina Tropical (Caracas), Durante el ano 1968. *Dermatol Venez*, **8**, 887–904.

Bradsher, R.W. and Alford, R.H. 1981. *Blastomyces dermatitidis* antigen-induced lymphocyte reactivity in human blastomycosis. *Infect Immun*, **33**, 485–90.

Bradsher, R.W., Ulmer, W.C., et al. 1985. Intracellular growth and phagocytosis of *Blastomyces dermatitidis* by monocyte-derived macrophages from previously infected and normal subjects. *J Infect Dis*, **151**, 57–64.

Bradsher, R.W., Chapman, S.W. and Pappas, P.G. 2003. Blastomycosis. *Infect Dis Clin N Am*, **17**, 21–40.

Brandhorst, T.T., Wuthrich, M., et al. 1999. Targeted gene disruption reveals an adhesin indispensable for pathogenicity of *Blastomyces dermatitidis*. *J Exp Med*, **189**, 1207–16.

Brandhorst, T.T., Rooney, P., et al. 2002. Genetic manipulation of *Blastomyces dermatitidis* uncovers pathogenic mechanisms. *Trends Microbiol*, **10**, 25.

Brandhorst, T., Wuthrich, M., et al. 2003. A C-terminal EGF-like domain governs BAD1 localization to the yeast surface and fungal adherence to phagocytes, but is dispensable in immune modulation and pathogenicity of *Blastomyces dermatitidis*. *Mol Microbiol*, **48**, 53–65.

Brass, C., Volkman, C.M., et al. 1982. Spontaneous mutant of *Blastomyces dermatitidis* attenuated in virulence for mice. *Curr Microbiol*, **7**, 25–8.

Brown, L.R., Swensen, S.J., et al. 1991. Roentgenologic features of pulmonary blastomycosis. *Mayo Clin Proc*, **66**, 29–38.

Brummer, E. and Stevens, D.A. 1982. Opposite effects of human monocytes, macrophages and polymorphonuclear neutrophils on replication of *B. dermatitidis* in vitro. *Infect Immun*, **36**, 297–303.

Brummer, E. and Stevens, D.A. 1987. Fungicidal mechanisms of activated macrophages: evidence for nonoxidative mechanisms for killing of *B. dermatitidis*. *Infect Immun*, **55**, 3221–4.

Buechner, H.A. and Clawson, C.M. 1967. Blastomycosis of the central nervous system. II. A report of nine cases from the Veterans' Administration Cooperative Study. *Am Rev Respir Dis*, **95**, 820–6.

Burg, E.F. III and Smith, L.H. 1994. Cloning and characterization of *bys1*, a temperature-dependent cDNA specific to the yeast phase of the pathogenic dimorphic fungus *Blastomyces dermatitidis*. *Infect Immun*, **62**, 2521–8.

Carmody, E.J. and Tappen, W. 1959. Blastomycosis meningitis, report of case successfully treated with amphotericin B. *Ann Intern Med*, **51**, 780–91.

Casadevall, A. and Pirofski, L.A. 2003. Exploiting the redundancy in the immune system: vaccines can mediate protection by eliciting 'unnatural' immunity. *J Exp Med*, **197**, 1401–4.

Causey, W.A. and Campbell, G.D. 1992. Clinical aspects of blastomycosis. In: Al-Doory, Y. and Di Salvo, A.F. (eds), *Blastomycosis*. New York: Plenum Medical Book Co., 165–88.

Centers for Disease Control. 1979. Blastomycosis in canoeists – Wisconsin. *MMWR*, **28**, 450–1.

Chandler, F.W. and Watts, J.C. 1992. Pathologic features of blastomycosis. In: Al-Doory, Y. and Di Salvo, A.F. (eds), *Blastomycosis*. New York: Plenum Medical Book Co., 189–220.

Chiu, J., Berman, S. et al. 1988. Disseminated blastomycosis in HIV infected patients (abstract 7209). *Program and Abstracts of the Fourth International Conference on AIDS, Stockholm, Sweden*, 430.

Chow, S., Goldstein, E.J.C. and Brody, N. 1981. North American blastomycosis in an immunosuppressed patient. *Cutis*, **28**, 572–4.

Clemons, K.V., Hurley, S.M., et al. 1991. Variable colonial phenotypic expression and comparison to nuclei number in *Blastomyces dermatitidis*. *J Med Vet Mycol*, **29**, 165–78.

Conant, N.F. 1939. Laboratory studies of *Blastomyces dermatitidis*. *Proc 6th Pacific Scientific Congr*, **5**, 853–61.

Cox, R.A. and Larsh, H.W. 1974. Isolation of skin test-active preparations from yeast-phase cells of Blastomyces dermatitidis. *Infect Immun*, **10**, 42–7.

Cozad, G.C. and Chang, C.T. 1980. Cell-mediated immunoprotection in blastomycosis. *Infect Immun*, **28**, 398–403.

Craig, M.W., Davey, W.N. and Green, M.L. 1970. Conjugal blastomycosis. *Am Rev Respir Dis*, **102**, 86–90.

Daniel, W.C., Nair, S.V. and Bluestein, J. 1979. Light and electron microscopic observations of *Blastomyces dermatitidis* in sputum. *Acta Cytol J*, **23**, 222–6.

Denton, J.F. and Di Salvo, A.F. 1964. Isolation of *Blastomyces dermatitidis* from natural sites at Augusta, Georgia. *Am J Trop Med*, **13**, 716–22.

Denton, J.F. and Di Salvo, A.F. 1968. Respiratory infection of laboratory animals with conidia of *Blastomyces dermatitidis*. *Mycopathol Mycol Appl*, **36**, 129–36.

Denton, J.F. and Di Salvo, A.F. 1979. Additional isolations of *Blastomyces dermatitidis* from natural sites. *Am J Trop Med Hyg*, **28**, 697–700.

Denton, J.F., McDonough, E.S., et al. 1961. Isolation of *Blastomyces dermatitidis* from soil. *Science*, **133**, 1126–7.

Denton, J.F., Di Salvo, A.F. and Hirsch, M.L. 1967. Laboratory acquired North American blastomycosis. *JAMA*, **199**, 935–6.

Diamond, R.D. and Krzesicki, R. 1978. Mechanism of attachment of neutrophils to Candida albicans pseudohyphae in the absence of serum, and subsequent damage to pseudohyphae by microbicidal process of neutrophils in vitro. *J Clin Invest*, **61**, 360–9.

Di Salvo, A.F. and Denton, J.F. 1963. Lipid content of four strains of *Blastomyces dermatitidis* of different mouse virulence. *J Bacteriol*, **85**, 927–31.

Di Salvo, A.F. 1992a. The ecology of *Blastomyces dermatitidis*. In: Al-Doory, Y. and Di Salvo, A.F. (eds), *Blastomycosis*. New York: Plenum Medical Book Co., 43–73.

Di Salvo, A.F. 1992b. The epidemiology of blastomycosis. In: Al-Doory, Y. and Di Salvo, A.F. (eds), *Blastomycosis*. New York: Plenum Medical Book Co., 75–104.

Di Salvo, A.F. and Wooten, A.K. 1982. The identification of non-viable cultures of *Blastomyces dermatitidis* and *Histoplasma capsulatum* by their exoantigens. *Am Soc Microbiol Abstract*, F52.

Di Salvo, A.F., Terreni, A.A. and Wooten, A.K. 1981. Use of the exoantigen test to identify *Blastomyces dermatitidis, Coccidioides immitis* and *Histoplasma capsulatum* in mixed cultures. *Am J Clin Pathol*, **75**, 825–6.

Domer, J.E. and Hamilton, J.G. 1971. The readily extracted lipids of *Histoplasma capsulatum* and *Blastomyces dermatitidis*. *Biochim Biophys Acta*, **231**, 465–78.

Drutz, D.J. and Frey, C.L. 1985. Intracellular and extracellular defenses of human phagocytes against *Blastomyces dermatitidis* conidia and yeasts. *J Lab Clin Med*, **105**, 737–50.

Dyer, M.L., Young, T.L., et al. 1983. Blastomycosis in a Papanicolaou smear. Report of a case with possible venereal transmission. *Acta Cytol (Baltimore)*, **27**, 285–7.

Edson, R.S. and Keys, T.F. 1981. Treatment of primary pulmonary blastomycosis: results of long-term follow-up. *Mayo Clin Proc*, **56**, 683–5.

Ehni, W. 1989. Endogenous reactivation in blastomycosis. *Am J Med*, **86**, 831–2.

Emmons, C.W. 1949. Isolation of *Histoplasma capsulatum* from soil. *Public Health Rep*, **64**, 892–6.

Emmons, C.W. 1951. Isolation of *Cryptococcus neoformans* from soil. *J Bacteriol*, **62**, 685–90.

Failla, P.J., Cerise, F.P., et al. 1995. Blastomycosis: pulmonary and pleural manifestations. *South Med J*, **88**, 405–10.

Farber, E.R., Leary, M.S. and Meadows, T.R. 1968. Endometrial blastomycosis acquired by sexual contact. *Obstet Gynecol*, **32**, 195–9.

Finkel-Jimenez, B., Wuthrich, M., et al. 2001. The WI-1 adhesin blocks phagocyte TNF-alpha production, imparting pathogenicity on *Blastomyces dermatitidis*. *J Immunol*, **166**, 4, 2665–73.

Finkel-Jimenez, B., Wuthrich, M. and Klein, B.S. 2002. BAD1, an essential virulence factor of *Blastomyces dermatitidis*, suppresses host TNF-alpha production through TGF-beta-dependent and -independent mechanisms. *J Immunol*, **168**, 5746–55.

Fraser, V.J., Keath, E.J. and Powderly, W.G. 1991. Two cases of blastomycosis from a common source: use of DNA restriction analysis to identify strains. *J Infect Dis*, **163**, 1378–81.

Furcolow, M.L., Chick, E.W., et al. 1970. Prevalence and incidence studies of human and canine blastomycosis. *Am Rev Respir Dis*, **102**, 60–7.

Garrison, R.G. and Boyd, K.S. 1978. Role of the conidium in dimorphism of *Blastomyces dermatitidis*. *Mycopathologia*, **64**, 29–33.

George, A.F., Hays, J.T. and Graham, B.S. 1985. Blastomycosis presenting as monarticular arthritis. *Arthritis Rheum*, **28**, 516–21.

Gilchrist, T.C. 1894. Protozoan dermatitis. *J Cutan Vener Dis*, **12**, 496.

Gilchrist, T.C. 1896. A case of blastomycotic dermatitis in man. *Johns Hopkins Hosp Rep*, **1**, 269–98.

Gilchrist, T.C. and Stokes, W.R. 1896. The presence of an oidium in the tissues of a case of pseudolupus vulgaris. *Johns Hopkins Hosp Bull*, **7**, 129–33.

Gilchrist, T.C. and Stokes, W.R. 1898. A case of pseudo-lupus vulgaris caused by a blastomyces. *J Exp Med*, **3**, 53–83.

Goodman, N.L. 1992. Diagnosis of blastomycosis. In: Al-Doory, Y. and Di Salvo, A.F. (eds), *Blastomycosis*. New York: Plenum Medical Book Co., 105–21.

Gow, N.A.R. and Gouday, G.W. 1983. Ultrastructure of chitin in hyphae of *Candida albicans* and other dimorphic and mycelial fungi. *Protoplasma*, **115**, 52–8.

Griffith, J.E. and Campbell, G.D. 1979. Acute miliary blastomycosis presenting as fulminating respiratory failure. *Chest*, **75**, 630–2.

Halvorsen, R.A., Duncan, J.D., et al. 1984. Pulmonary blastomycosis: radiologic manifestations. *Radiology*, **150**, 1–5.

Hamburger, W.W. 1907. A comparative study of four strains of organisms isolated from four cases of generalized blastomycosis. *J Infect Dis*, **4**, 201–9.

Hannah, E.L., Bailey, A.M., et al. 2001. Public health response to 2 clinical cases of blastomycosis in Colorado. *Clin Infect Dis*, **32**, e151–153.

Harasen, G.L.G. and Randall, J.W. 1986. Canine blastomycosis in Southern Saskatchewan. *Can Vet J*, **30**, 375–8.

Hardin, H.F. and Scott, D.J. 1974. Blastomycosis: occurrence of filamentous forms in vivo. *Am J Clin Pathol*, **62**, 104–6.

Harding, C.V. 1991. Blastomycosis and opportunistic infections in patients with acquired immunodeficiency syndrome. An autopsy study. *Arch Pathol Lab Med*, **115**, 1133–6.

Harley, W.B., Lomis, M. and Haas, D.W. 1994. Marked polymorphonuclear pleocytosis due to blastomycotic meningitis: case report and review. *Clin Infect Dis*, **18**, 816–18.

Herd, A.M., Greenfield, S.B., et al. 1990. Miliary blastomycosis and HIV infection. *Can Med Assoc J*, **143**, 1329–30.

Hii, J.H., Legault, L., et al. 1990. Successful treatment of systemic blastomycosis with high-dose ketoconazole in a renal transplant recipient. *Am J Kidney Dis*, **15**, 595–7.

Hogan, L.H. and Klein, B.S. 1997. Transforming DNA integrates at multiple sites in the dimorphic fungal pathogen *Blastomyces dermatitidis*. *Gene*, **186**, 219–26.

Hogan, L.H., Josvai, S. and Klein, B.S. 1995. Genomic cloning, characterization, and functional analysis of the major surface adhesin WI-1 on *Blastomyces dermatitidis* yeasts. *J Biol Chem*, **270**, 30725–32.

Hogan, L.H., Klein, B.S. and Levitz, S.M. 1996. Virulence factors of medically important fungi. *Clin Microbiol Rev*, **9**, 469–88.

Jay, S.J., O'Neill, R.P., et al. 1977. Case report. Pleural effusion: a rare manifestation of acute pulmonary blastomycosis. *Am J Med Sci*, **274**, 325–8.

Kanetsuna, F. 1981. Ultrastructural studies on the dimorphism of *Paracoccidiodes brasiliensis*, *Blastomyces dermatitidis* and *Histoplasma capsulatum*. *Sabouraudia*, **19**, 275–86.

Kanetsuna, F. and Carbonell, L.M. 1971. Cell wall composition of the yeastlike and mycelial forms of *Blastomyces dermatitidis*. *J Bacteriol*, **106**, 946–8.

Kanetsuna, F., Carbonell, L.M., et al. 1972. Biochemical studies on the thermal dimorphism of *Paracoccidiodes brasiliensis*. *J Bacteriol*, **110**, 208–18.

Kaplan, W. and Kaufman, L. 1963. Specific fluorescent antiglobulins for the detection and identification of *Blastomyces dermatitidis* yeast-phase cells. *Mycopathologia*, **19**, 173–80.

Kaufman, L. 1992. Immunodiagnosis of blastomycosis. In: Al-Doory, Y. and Di Salvo, A.F. (eds), *Blastomycosis*. New York: Plenum Medical Book Co., 123–32.

Kaufman, L., Standard, P.G., et al. 1983. Detection of two *Blastomyces dermatitidis* serotypes by exoantigen analysis. *J Clin Microbiol*, **18**, 110–14.

Kaufmann, A.F., Kaplan, W. and Kraft, D.E. 1979. Filamentous forms of *Ajellomyces (Blastomyces) dermatitidis* in a dog. *Vet Pathol*, **16**, 271–3.

Kinasewitz, G.T., Penn, R.L. and George, R.B. 1984. The spectrum and significance of pleural disease in blastomycosis. *Chest*, **86**, 580–4.

Kitchen, L.W., Clark, R.A., et al. 1989. Concurrent pulmonary *Blastomyces dermatitidis* and mycobacterium tuberculosis infection in an HIV-1 seropostive man. *J Infect Dis*, **160**, 911–12.

Klein, B.S. 1992. Immunology of blastomycosis. In: Al-Doory, Y. and Di Salvo, A.F. (eds), *Blastomycosis*. New York: Plenum Medical Book Co., 133–63.

Klein, B.S. and Jones, J.M. 1990. Isolation, purification, and radiolabeling of a novel 120-kD surface protein on *Blastomyces dermatitidis* yeasts to detect antibody in infected patients. *J Clin Invest*, **85**, 152–61.

Klein, B.S. and Jones, J.M. 1994. Purification and characterization of the major antigen WI-1 from *Blastomyces dermatitidis* yeasts and immunological comparison with A antigen. *Infect Immun*, **62**, 3890–900.

Klein, B.S., Vergeront, J.M., et al. 1986. Isolation of *B. dermatitidis* in soil associated with a large outbreak of blastomycosis in Wisconsin. *N Engl J Med*, **314**, 529–34.

Klein, B.S., Vergeront, J.M., et al. 1987a. Two outbreaks of blastomycosis along rivers in Wisconsin. *Am Rev Respir Dis*, **136**, 1333–8.

Klein, B.S., Vergeront, J.M., et al. 1987b. Serological tests for blastomycosis: assessments during a large point source outbreak in Wisconsin. *J Infect Dis*, **155**, 262–8.

Klein, B.S., Bradsher, R.W., et al. 1990. Development of long-term specific cellular immunity after acute *Blastomyces dermatitidis* infection: assessments following a large point-source outbreak in Wisconsin. *J Infect Dis*, **161**, 97–101.

Klein, B.S., Aizenstein, B.D. and Hogan, L.H. 1997. African strains of *Blastomyces dermatitidis* that do not express surface adhesin WI-1. *Infect Immun*, **65**, 1505–9.

Kwon-Chung, K.J. 1971. Genetic analysis on the incompatibility system of *Ajellomyces dermatitidis*. *Sabouraudia*, **9**, 231–8.

Laskey, W. and Sarosi, G. 1978. Roentgen appearance of pulmonary blastomycosis. *Radiology*, **126**, 351–7.

Legendre, A.M. 1992. Blastomycosis in animals. In: Al-Doory, Y. and Di Salvo, A.F. (eds), *Blastomycosis*. New York: Plenum Medical Book Co., 249–64.

Lieberman, A. 1963. The case of the fumbled fungus. *J Indiana Med Assoc*, **56**, 1017–22.

Manwaring, J.H. 1949. Unusual forms of *Blastomyces dermatitidis* in human tissues. *Arch Pathol*, **48**, 421–5.

Maxson, S., Miller, S.F., et al. 1992. Perinatal blastomycosis: a review. *Pediatr Infect Dis J*, **11**, 760–3.

McDonough, E.S. 1970. Blastomycosis – epidemiology and biology of its etiologic agent *Ajellomyces dermatitidis*. *Mycopathol Mycol Appl*, **41**, 195–201.

McDonough, E.S. and Kuzma, J.F. 1980. Epidemiological studies on blastomycosis in the State of Wisconsin. *Sabouraudia*, **18**, 173–83.

McDonough, E.S. and Lewis, A.L. 1968. The ascigerous stage of *Blastomyces dermatitidis*. *Mycologia*, **60**, 76–83.

McDonough, E.S., Wisniewski, R.T., et al. 1976. Preliminary studies on conidial liberation of *Blastomyces dermatitidis* and *Histoplasma capsulatum*. *Sabouraudia*, **14**, 199–204.

Medoff, G., Sacco, M., et al. 1986. Irreversible block of the mycelial-to-yeast phase transition of *Histoplasma capsulatum*. *Science*, **231**, 476–9.

Medoff, G., Painter, A. and Kobayashi, G.S. 1987. Mycelial to yeast phase transitions of dimorphic fungi *Blastomyces dermatitidis* and *Paracoccidiodes brasiliensis*. *J Bacteriol*, **169**, 4055–60.

Menges, R.W., Furcolow, M.L., et al. 1965. Clinical and epidemiologic studies on seventy-nine canine blastomycosis in Arkansas. *Am J Epidemiol*, **81**, 164–79.

Meyer, K.C., McManus, E.J. and Maki, D.G. 1993. Overwhelming pulmonary blastomycosis associated with the adult respiratory distress syndrome. *N Engl J Med*, **329**, 1231–6.

Morozumi, P.A., Brummer, E. and Stevens, D.A. 1982. Protection against pulmonary blastomycosis: correlation with cellular and humoral immunity in mice after subcutaneous non-lethal infection. *Infect Immun*, **37**, 670–8.

Musial, C.E., Wilson, W.R., et al. 1987. Recovery of *Blastomyces dermatitidis* from blood of a patient with disseminated blastomycosis. *J Clin Microbiol*, **25**, 1421–3.

Newman, S.L., Chaturvedi, S. and Klein, B.S. 1995. The WI-1 antigen of *Blastomyces dermatitidis* yeasts mediates binding to human macrophage CD11b/CD18 (CR3) and CD14. *J Immunol*, **154**, 753–61.

Önal, E., Lopata, M. and Lourenço, R.V. 1976. Disseminated pulmonary blastomycosis in an immunosuppressed patient. Diagnosis by fiberoptic bronchoscopy. *Am Rev Respir Dis*, **113**, 83–6.

Padhye, A.A., Smith, G., et al. 1994. Comparative evaluation of chemiluminescent DNA probe assays and exoantigen tests for rapid identification of *Blastomyces dermatitidis* and *Coccidioides immitis*. *J Clin Microbiol*, **32**, 867–70.

Pappas, P.G. and Dismukes, W.E. 2002. Blastomycosis: Gilchrist's disease revisited. *Curr Clin Top Infect Dis*, **22**, 61–77.

Pappas, P.G., Pottage, J.C., et al. 1992. Blastomycosis in patients with the acquired immunodeficiency syndrome. *Ann Intern Med*, **116**, 847–53.

Pappas, P.G., Threlkeld, M.G., et al. 1993. Blastomycosis in immunocompromised patients. *Medicine (Baltimore)*, **72**, 311–25.

Paris, S. and Garrison, R.G. 1984. Cyclic adenosine 3′,5′-monsphosphate as a factor in phase morphogenesis of *Blastomyces dermatitidis*. *Mykosen*, **27**, 340–5.

Pfister, A.K., Goodwin, A.W., et al. 1966. Pulmonary blastomycosis: roentgenographic clues to the diagnosis. *South Med J*, **59**, 1441–7.

Procknow, J.J. 1966. Disseminated blastomycosis treated successfully with the polypeptide antifungal agent X-5079C. Evidence for human to human transmission. *Am Rev Respir Dis*, **94**, 761–72.

Recht, L., Philips, J.R., et al. 1979. Self-limited blastomycosis: a report of 13 cases. *Am Rev Respir Dis*, **120**, 1109–12.

Recht, L.D., Davies, S.F., et al. 1982. Blastomycosis in immunosuppressed patients. *Am Rev Respir Dis*, **125**, 359–62.

Renston, J.P., Morgan, J. and Di Marco, A.F. 1992. Disseminated miliary blastomycosis leading to acute respiratory failure in an urban setting. *Chest*, **101**, 1463–5.

Ricketts, H.T. 1901. Oidiomycosis of the skin and its fungi. *J Med Res*, **6**, 373–558.

Rippon, J.W., Zvetina, J.U. and Reyes, C. 1977. Case report: miliary blastomycosis with cerebral involvement. *Mycopathologia*, **60**, 121–5.

Robert, E.M. and Kauffman, C.A. 1988. Blastomycosis presenting as polyarticular septic arthritis. *J Rheumatol*, **15**, 1438–42.

Rooney, P.J., Sullivan, T.D. and Klein, B.S. 2001. Selective expression of the virulence factor BAD1 upon morphogenesis to the pathogenic yeast form of *Blastomyces dermatitidis*: evidence for transcriptional regulation by a conserved mechanism. *Mol Microbiol*, **39**, 875–89.

Sarosi, G.A., Hammerman, K.J., et al. 1974. Clinical features of acute pulmonary blastomycosis. *Arch Dermatol*, **71**, 84–8.

Sarosi, G.A., Eckman, M.R., et al. 1979. Canine blastomycosis as a harbinger of human disease. *Ann Intern Med*, **91**, 733–5.

Sarosi, G.A., Davies, S.F. and Phillips, J.R. 1986. Self-limited blastomycosis: a report of 39 cases. *Semin Respir Infect*, **1**, 40–4.

Scalarone, G.M., Legendre, A.M., et al. 1992. Evaluation of a commercial DNA probe assay for the identification of clinical isolates of *Blastomyces dermatitidis* from dogs. *J Med Vet Mycol*, **30**, 43–9.

Schwartzman, R.M., Fusaro, R.M. and Orkin, M. 1959. Transmission of North American blastomycosis. Possible case of transmission from dog to human. *JAMA*, **171**, 2185–9.

Schwarz, J. 1982. Mycotic prostatitis. *Urology*, **19**, 1–5.

Schwarz, J. 1983. Laboratory infections with fungi. In: Di Salvo, A.F. (ed.), *Occupational mycosis*. Philadelphia: Lea & Febiger, 215–27.

Schwarz, J. and Baum, G.L. 1951. Blastomycosis. *Am J Clin Pathol*, **21**, 999–1029.

Schwarz, J. and Salfelder, K. 1977. Blastomycosis. A review of 152 cases. *Curr Topics Pathol*, **65**, 165–200.

Sekhon, A.S., Kaufman, L., et al. 1995. The value of the Premier enzyme immunoassay for diagnosing *Blastomyces dermatitidis* infections. *J Med Vet Mycol*, **33**, 123–5.

Shaw, G.B., Campbell, G.D. and Busy, J.E. 1976. Miliary blastomycosis. *Am Rev Respir Dis*, **113**, 81.

Shearer, G. and Larsh, H.W. 1985. Chitin synthetase from yeast and mycelial phases of *Blastomyces dermatitidis*. *Mycopathologia*, **90**, 91–6.

Sheflin, J.R., Campbell, J.A. and Thompson, G.P. 1990. Pulmonary blastomycosis: findings on chest radiographs in 63 patients. *Am J Roentgenol*, **154**, 1177–80.

Stelling, C.B., Woodring, J.H., et al. 1984. Miliary pulmonary blastomycosis. *Radiology*, **150**, 7–13.

Stewart, R.A. and Meyer, K.F. 1932. Isolation of *Coccidioides immitis* from soil. *Proc Soc Exp Biol Med*, **29**, 937–8.

Stober, A.M. 1914. Systemic blastomycosis: a report of its pathological, bacteriological and clinical features. *Arch Intern Med*, **13**, 509–56.

Stockman, L., Clark, K.A., et al. 1993. Evaluation of commercially available acridinium ester-labeled chemiluminescent DNA probes for culture identification of *Blastomyces dermatitidis, Coccidioides immitis, Cryptococcus neoformans*, and *Histoplasma capsulatum*. *Clin Microbiol*, **31**, 845–50.

Sudman, M.S. and Kaplan, W. 1974. Antigenic relationship between American and African isolates of *Blastomyces dermatitidis* as determined by immunofluorescence. *Appl Mycol*, **27**, 496–9.

Sugar, A.M. 1992. Taxonomy and biology of *Blastomyces dermatitidis*. In: Al-Doory, Y. and Di Salvo, A.F. (eds), *Blastomycosis*. New York: Plenum Medical Book Co., 9–29.

Sullivan, T., Rooney, P. and Klein, B.S. 2002. *Agrobacterium tumefaciens* transfers T-DNA into single chromosomal sites of dimorphic fungi and yields homokaryotic progeny from multinucleate yeast. *Eukaryot Cell*, **1**, 895.

Summerbell, R.C., Kane, J. and Pincus, D.H. 1990. Enzymatic activity profiling as a potential biotyping method for *Ajellomyces dermatitidis*. *J Clin Microbiol*, **28**, 1054–5.

Tompkins, V. and Schleifstein, J. 1953. Small forms of *Blastomyces dermatitidis* in human tissues. *Arch Pathol*, **55**, 432–5.

Trumbull, M.L. and Chesney, T.Mc. 1981. The cytological diagnosis of pulmonary blastomycosis. *JAMA*, **245**, 836–8.

Tuttle, J.G., Lichtwardt, H.E. and Altshuler, C.H. 1953. Systemic North American blastomycosis. Report of a case with small forms of blastomycetes.. *Am J Pathol*, **23**, 890–7.

Vaaler, A.K., Bradsher, R.W. and Davies, S.F. 1990. Evidence of subclinical blastomycosis in forestry workers in northern Minnesota and northern Wisconsin. *Am J Med*, **89**, 470–6.

Vermeil, C., Bouillard, C., et al. 1982. The echinulate conidia of *B. dermatitidis* Gilchrist and Stokes and the taxonomic status of this species. *Mykosen*, **25**, 251–3.

Walker, K., Skelton, H. and Smith, K. 2002. Cutaneous lesions showing giant yeast forms of *Blastomyces dermatitidis*. *J Cutaneous Pathol*, **29**, 616–18.

Watts, E.A., Gard, P.D. and Tuthill, S.W. 1983. First reported case of intrauterine transmission of blastomycosis. *Pediatr Infect Dis*, **2**, 308–10.

Watts, J.C., Chandler, F.W., et al. 1990. Giant forms of *Blastomyces dermatitidis* in the pulmonary lesions of blastomycosis: potential confusion with *Coccidioides immitis*. *Am J Clin Pathol*, **93**, 575–8.

Winquist, E.W., Walmsley, S.L. and Berinstein, N.L. 1993. Reactivation and dissemination of blastomycosis complicating Hodgkin's disease: a case report and review of the literature. *Am J Hematol*, **43**, 129–32.

Wuthrich, M., Filutowicz, H.I. and Klein, B.S. 2000. Mutation of the WI-1 gene yields an attenuated *Blastomyces dermatitidis* strain that induces host resistance. *J Clin Invest*, **106**, 1381–9.

Wuthrich, M., Filutowicz, H.I., et al. 2002. Requisite elements in vaccine immunity to *Blastomyces dermatitidis*: plasticity uncovers vaccine potential in immune-deficient hosts. *J Immunol*, **169**, 6969–76.

Wuthrich, M., Filutowicz, H.I., et al. 2003. Vaccine immunity to pathogenic fungi overcomes the requirement for CD4 help in exogenous antigen presentation to CD8+ T cells: implications for vaccine development in immune-deficient hosts. *J Exp Med*, **197**, 1405–16.

26

Coccidioidomycosis

DEMOSTHENES PAPPAGIANIS

Coccidioides immitis (a second species, *Coccidioides posadasii* sp.nov., has been proposed by Fisher et al. 2002), is a soil-inhabiting fungus which, usually following inhalation, causes coccidioidomycosis. It usually occurs naturally only in certain areas of the western hemisphere. The disease, in its various manifestations, has also been called coccidioidal granuloma, San Joaquin fever, valley fever, and desert rheumatism. As a result of its persistent geographical distribution, coccidioidomycosis is an 'endemic systemic mycosis.' *C. immitis* undergoes a morphological change from its hyphal/arthroconidial form in nature, or in the usual laboratory culture, to a spherule-endospore form in vivo. It is thus grouped with the dimorphic systemic pathogenic fungi.

Recent reviews of coccidioidomycosis and coccidioides are provided in *Seminars in Respiratory Infections* (Ampel 2001) and by Cole et al. (2004).

Coccidioides species have been included as a 'select agent' of bioterrorism.

ETIOLOGIC AGENT

History

During the nineteenth century, physicians in the central valleys of California described febrile maladies related to disturbance of the soil, which probably included coccidioidomycosis (Thompson 1969). The first definitive written word on the disease emerged in 1892 when Posadas, in Buenos Aires, reported on the disease in an Argentinian soldier. The disease had begun some 4 years earlier, with its first manifestations being cutaneous involvement; a physician, Dr Bengolea, described it as 'mycosis fungoides,' a neoplastic disease which involves the skin, but it became evident that this was a progressive, systemic, and altogether different disease. Posadas had noted microscopically an organism that resembled a protozoan, belonging to the group referred to as coccidia that are characterized by a cyst containing endocystic cells. Two years after the initial report by Posadas, a second case was reported, originating in California (Rixford 1894). In fact, in the next several decades this was the site of occurrence and recognition of most of the cases.

As a result of its resemblance to the protozoan coccidia, the organism was subsequently named *Coccidioides immitis* (coccidia-like, *im*, not; *mitis*, mild). [Interestingly, no additional case was recognized in Argentina until 35 years after Posadas' report (Negroni 1967)]. The fungal nature of the organism was unrecognized until 1900 at which time a filamentous mold, believed for several years to have been a contaminant of cultures obtained from patients' tissues and exudates, was shown by inoculation of laboratory animals to change its morphological form into one resembling a protozoan (Ophüls and Moffitt 1900). This provided proof of the diphasic fungal nature of *C. immitis*.

Significant migration of individuals into the endemic areas of the southwestern USA occurred in the 1930s, as a result of catastrophic ecological and economic changes elsewhere in the USA. With the start of World War II in the early 1940s, military troops were also brought into endemic areas. These two groups of largely susceptible individuals permitted refinement of the knowledge about coccidioidomycosis (Smith et al. 1946a). In 1956, specific antifungal therapy in the form of amphotericin B was introduced for the treatment of coccidioidomycosis

(Fiese 1957). The azoles, introduced in the 1980s, significantly altered the approach to treatment.

Mycology

C. immitis Rixford and Gilchrist, 1896, is diphasic and polymorphic (Figure 26.1). The hyphal arthroconidial form is the saprophytic or saprobic form, found in nature and the usual laboratory cultures, and infrequently in tissues of the infected host (arthroconidia were formerly referred to as arthrospores). Airborne arthroconidia are inhaled and convert to the spherule-endospore ('parasitic') phase in the infected mammalian host. Special cultural conditions permit the production of the spherule-endospore phase in vitro (Converse 1955). The taxonomic placement of *C. immitis* has customarily been in the Fungi Imperfecti (Deuteromycotina) owing to an absence of a demonstrable conventional sexual conjugational stage. However, Burt et al. (1996) have described 'cryptic sex' (which may resemble para sexuality) as providing a means of exchanging genetic information. There are variations in the phenotype of the mycelial phase but not the spherule-endospore phase. Several features have, however, indicated that coccidioides probably fits in the division Ascomycotina

(Ascomycetes), family *Onygenaceae*, and order Onygenales (Currah 1985): the hyphae are septate and have septal pores close to structures that resemble Woronin bodies; thallic arthroconidia are produced from specialized hyphae (i.e. whole hyphal cells become thick walled spore-like structures) which alternate with empty, degenerate 'disjunctor' cells (Sigler and Carmichael 1976) (Figures 26.1 and 26.2). Four pairs of chromosomes are reportedly present.

The percentage of guanine–cytosine (49.1–49.61 percent) was similar in the spherule and hyphal forms of three strains of *C. immitis* [compared with 47.3 percent and 48 percent in *Histoplasma capsulatum* and *Blastomyces dermatitidis*, respectively, both of which are ascomycetes (Pappagianis et al. 1985)]; based on homology/relatedness studies of the 18S ribosomal gene, *C. immitis* was shown to be closely related to *H. capsulatum* and *B. dermatitidis* (as well as *Trichophyton rubrum*), providing affirmation of taxonomic affiliation with ascomycetes (Bowman et al. 1992). The DNA studied by restriction enzyme analysis allowed separation of *C. immitis* into two groups, I and II (Zimmermann et al. 1994). Fisher et al. (2002) concluded that group I corresponded to a new species, *Coccidioides posadasii* ('non-California') and group II, *C. immitis*

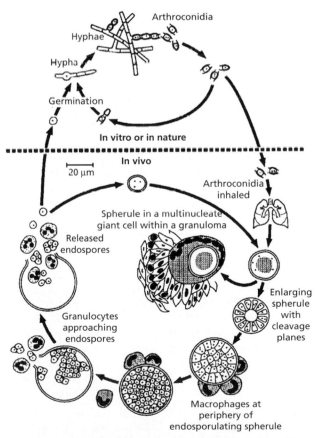

Figure 26.1 *Diagram illustrating the morphological forms of* Coccidioides immitis *in its saprobic form and its 'parasitic' form in vivo. The spherule/endospore form is depicted as evolving after inhalation of the arthroconidia, and the usual cellular reaction to the in vivo forms. (Adapted from Forbus and Bestebreurtje (1946) and others)*

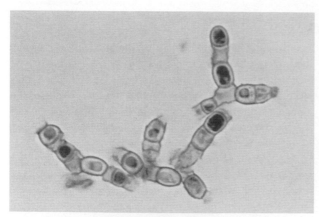

Figure 26.2 *Chains of arthroconidia, some of which are separated by 'disjunctor' cells which undergo degeneration, first of the cytoplasm and later of the nuclei and cell wall.*

('California') species. There is no evidence that disease or its laboratory diagnosis associated with *C. immitis* or *C. posadasii* differs, or that response to therapy differs; and epidemiologic and experimental data indicate that immunity to these putative species is cross-protective (Pappagianis 1999).

The vegetative hyphal form can grow on a variety of media, ranging from those with sparse organic substrates (ammonium acetate suffices as a source of carbon and nitrogen) to the usual rich clinical laboratory culture media: blood agar, Sabouraud's, etc. On suitable media (e.g. 1 percent glucose/0.5 percent yeast extract/2 percent agar) arthroconidia are formed in 5–7 days at 34–37°C. These arthroconidia, usually 2–4×5–6 µm can be barrel-shaped, cylindrical, or oval and, when terminal on a hypha, they can be bullet-shaped. They form in the chain of cells of specialized hyphae developed from short hyphae 1–2 µm in diameter that are borne laterally on vegetative hyphae (2–4 µm in diameter). The arthroconidia, usually multinucleate and having a very refractile inner and outer wall, alternate with degenerating ('disjunctor') cells which share only the outer wall with adjacent arthroconidia (Huppert et al. 1982) (Figures 26.1 and 26.2). This thinner wall readily fractures, leading to disarticulation and, in appropriate currents of air, dispersion of individual or short chains of arthroconidia. When deposited on a suitable (nonviable) substrate with adequate moisture, the arthroconidia will germinate unilaterally or bilaterally to give rise to hyphae which, in turn, can produce arthroconidia. These airborne arthroconidia are highly infectious and it is necessary to be constantly aware of the potential infectious hazard to laboratory workers handling cultures, particularly from cultures more than 3 days old.

When the arthroconidia reach the airways of a human or other species (or are injected parenterally), and in the presence of host inflammatory cells (Baker and Braude 1956) and increased concentrations of CO_2 (Lones and Peacock 1960), the arthroconidia shed an outer wall layer and all but one nucleus, then become round and enlarge to produce a spherule. Nuclear division is followed by formation of cleavage planes of cell wall material which originates in the peripheral cell wall. It is likely that formation of the larger spherule from the smaller endospore involves several coordinated enzyme activities involved in the synthesis and breakdown of cell wall substances: synthases of chitin, glucan, mannan; lysis by chitinase, chitobiase, glucanase, mannosidase and proteases. In about 48 hours in vivo, the progressive growth and segmentation yield a mature spherule 20–150 µm in diameter with variable numbers, from a few to hundreds, of endospores 2–4 µm in diameter (Figure 26.3). These endospores can be released from the mature spherule, although the mechanism of release is uncertain but may involve proteases (Resnick et al. 1987; Yuan and Cole 1987), glucosidase (glucanase) (Kruse and Cole 1992) and chitinase (Johnson and Pappagianis 1992); these act upon the spherule wall and its internal projection around the endospores (Figure 26.4). The chitinase, and possibly also the glucanase and a protease, are important antigens in the induction of antibodies that have a significant role to play in the serodiagnosis of coccidioidomycosis (see section on Immunoserology, below). The released endospores enlarge in vivo and produce endosporulating spherules, which, in turn, can release another generation of endospores; in this way the pathogenic events continue. It is possible to demonstrate a similar formation of endosporulating spherules from an inoculum of arthroconidia or endospores in a synthetic medium (Converse 1955; Levine et al. 1960).

The metabolic activities have been summarized (Pappagianis 1988b). Expression of certain genes and their ultimate protein products (e.g. enzymes such as aldehyde reductase) can be specifically associated with the morphological (i.e. parasitic) spherule-endospore phase (Delgado et al. 2004). Some coccidioidal enzymes have been cloned and expressed by recombinant technology, e.g. chitinase (Zimmermann et al. 1996) and aspartyl protease (Johnson et al. 2000). *C. immitis* is catalase positive in both hyphal and spherule/endospore forms; it grows aerobically and readily uses glucose and other carbohydrates as carbon sources. Under anaerobic conditions, however, some metabolism can be carried out, and spherules produce more CO_2 than is produced by hyphae (Lones and Peacock 1960). The arthroconidia and hyphae can carry out endogenous respiration for many hours in the absence of an added carbon source. Both hyphae and spherules contain enzymes of the tricarboxylate cycle with the exception of α-ketoglutaric dehydrogenase. Isocitric lyase and malate synthase were, however, found in the hyphal form, suggesting that the glyoxylate bypass (conservative metabolism of fatty acids) can be utilized in the absence of 2-oxoglutaric dehydrogenase (Lones 1967). Endospores, but not spherules, use malate as a carbon source.

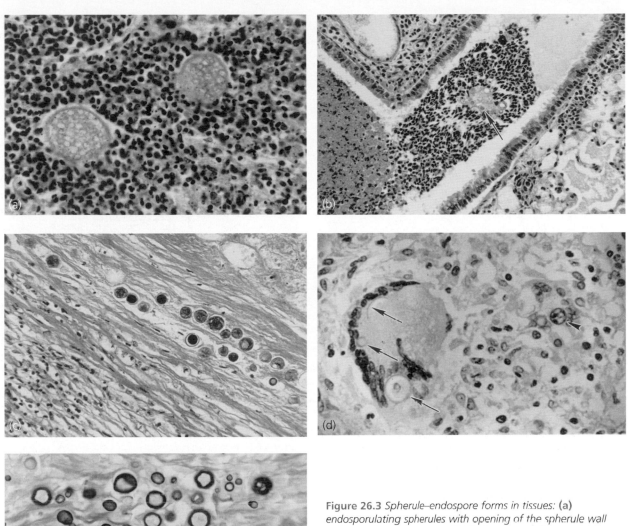

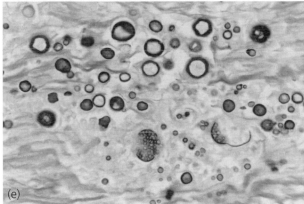

Figure 26.3 *Spherule–endospore forms in tissues:* **(a)** *endosporulating spherules with opening of the spherule wall and granulocytic response.* **(b)** *Endospores and immature spherules with inflammatory exudate in a bronchiole.* **(c)** *Immature spherules in area of chronic inflammation – fibroblasts and collagen fibers, and lymphocytes and other mononuclear cells.* **(d)** *Multinucleated giant cell containing an immature spherule, a fragment of a spherule wall and an endospore (arrows). There is also an immature spherule (arrowhead) in this area of chronic inflammation.* **(e)** *A range of endospore to endosporulating, mature spherules, stained with methenamine silver.*

The hyphae can utilize starch or xylose but the spherule/endospore form does not. Ammonium, nitrate, urea, acetamide, and several amino acids can be utilized as sources of nitrogen.

Ecology/epidemiology/geographic distribution

These aspects have been reviewed (Pappagianis 1988b). Coccidioidomycosis occurs primarily in arid and semi-arid areas of the western hemisphere where *C. immitis* lives in the soil, on the surface, and to a depth of 30 cm (Egeberg and Ely 1956). In the hottest periods of the year the soil temperature at the surface and to a depth of 1 cm can reach 60°C, which is lethal to *C. immitis* within minutes. Disturbance of the soil, with the creation of dust by winds or excavation or earthquakes (Schneider et al. 1997) throws arthroconidia up into the air. These conidia are hydrophobic and have a greater buoyancy with slower settling in air than would be expected for their size, $\sim 2 \times 5$ μm. This allows them to be airborne for up to hundreds of kilometers. Such conidia-laden dust, which leads to infection, is most likely to occur in the hot dry seasons when the soil has become low in moisture.

The relative confinement of *C. immitis* to specific geographic areas, which has no precise explanation, makes coccidioidomycosis an 'endemic mycosis.' The

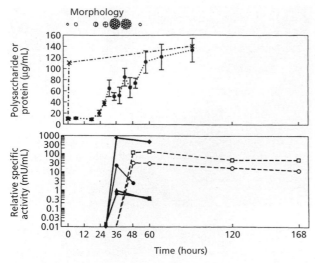

Figure 26.4 *Sequence of appearance of polysaccharide, protein and carbohydrases and proteases during the evolution in vitro of endospore to spherule to endospore. Top: filled circles, protein; crosses, polysaccharide. Bottom: filled circles, chitinase; filled squares, β-1,3-glucanase; filled triangles, β-1,6-glucanase; filled diamonds, β-1,3-glucanase; open circles, protease azocollytic; protease elastase. (From Pappagianis (1989) with permission)*

Figure 26.5 *Geographical distribution of* Coccidioides immitis *sites where coccidioidomycosis is usually acquired. (From Pappagianis (1980) with permission)*

regions fall between 40° North 120° West (Northern California) and 40° South 65° West (Argentina), and include the southwestern USA, California, Arizona, New Mexico, Texas, and small areas of Nevada and Utah, southern and northeastern (Petersen et al. 2004); Mexico; Honduras; Guatemala; Colombia; Venezuela; Paraguay; Argentina and, recently recognized, Brazil (state of Piaui). The name 'San Joaquin fever' or 'valley fever' is derived from a central region of California in which the disease is highly endemic (Figure 26.5).

Many, but not all, of the regions where the fungus is found conform to a bioclimatic area referred to as the Lower Sonoran Life Zone (Maddy 1957). *C. immitis* exists in areas that have relatively low annual rainfall (12.5–50 cm), hot summers (e.g. as in Phoenix, Arizona, a mean temperature 26–32°C) and relatively mild winters, though there is occasional snow at some elevations (900–1000 meters) where *C. immitis* is found. The soil is often sandy and alkaline (Swatek 1970). During the hot, dry season the soil has a high concentration of mineral salts which is better tolerated by *C. immitis* than by other microorganisms in the soil. Despite its survival in hot, dry, desert-like areas, where the relative humidity is usually no lower than 20 percent, it does not survive well at lower, e.g. 10 percent, relative humidity.

Rainfall has two influences: first, rain wets the soil, diminishing the airborne dust and fungal conidia, thereby reducing infections (Smith et al. 1946a); second, rainfall moistening the soil and *C. immitis* contained therein leads to germination of the fungal arthroconida; this supports a new round of vegetative growth, producing another crop of the infectious arthroconidia.

Roughly, the higher the rainfall, the greater the number of cases of the infection that present in the succeeding hot, dry season (Pappagianis 1994). Although there is a greater risk of infection associated with soil-disturbing occupations, overall more cases are observed in persons not occupationally exposed, conforming to the easy movement of airborne arthroconidia.

Increased travel or immigration has led to increased numbers of cases among those entering the endemic areas (Ampel et al. 1998; Pappagianis 1994). Such cases have been observed in Europeans, Canadians, and Asians who have briefly visited the endemic western hemisphere, and in those who live elsewhere in countries with endemic *C. immitis* but who visit the endemic areas. A few infections have occurred in people who have never visited the endemic areas, but who have received or been exposed to exported products from these areas (Symmers 1967).

In addition to humans, a variety of other species have been infected with *C. immitis* (Table 26.1). There is a wide range of responses among animals in which infections have been recognized: cattle, sheep, and swine appear to be resistant to progressive forms of the disease; they are often infected and have mediastinal lymph node involvement while remaining clinically healthy. On the other hand, domestic horses and llamas often have severe disease (Ziemer et al. 1992), though recently subclinical infections have been recognized in some domestic horses (Higgins et al. 2004). Przewalski (non-domesticated) horses have been recognized as having severe infections (Terio et al. 2003). Primates and monkeys have generally had severe disease, although arrest of the disease with anti-fungal therapy has been achieved. Domestic dogs appear to have disease patterns – mild to severe dissemination, especially involving the osteoarticular tissues – as seen in humans. Marine mammals (sea otters, sea lions, and recently a dolphin) which reside in the Pacific Ocean or its bays, are unique among these hosts. How they are infected is uncertain though the arthroconidia can travel long distances in the air, and their buoyancy and hydrophobicity could lead to their flotation on sea water and aerosolization by wind and/or wave action.

HOST RESPONSE

Clinical forms

Infection with *C. immitis* usually follows inhalation of the arthroconidia. Rarely, traumatic introduction of the organism into the skin leads to infection and disease (Wilson et al. 1953). In both instances, the regional lymph nodes are often involved: hilar in pulmonary infection, and inguinal in infection of the lower extre-mities. (Human-to-human transmission of *C. immitis* ordinarily does not occur, but has followed transplanta-tion of organs from individuals with coccidioidomycosis

Table 26.1 *Animal species with naturally acquired coccidioidomycosis*

Animal	Animal
Aardvark[a] (*Orycteropus aferi*)	Lemur, ringtail[a] (*Lemur catta*)
Armadillo, nine banded[b] (*Dasypus novemcinctus*)	Lion, mountain[a] (*Felis concolor*)
Baboon[a] (*Papio* sp.), mandrill (*Mandrillus* sp.)	transvaal[a] (*Panthera leo krugeri*)
Badger[a] (*Taxidea taxus*)	Llama[a] (*Lama glama*)
Bear, Malayan sun[a] (*Ursus malayanus*)	Monkey[a] – tropical American (*Cebus hypoleucus*)
Binturong[a] – a civet (*Arctitis binturong*)	colobus (*Colobus guereza*)
Burro (*Equus assinus*)	guenon DeBrazza (*Cercopithecus neglectus*)
Cat, domestic (*Felis catus*)	sooty mangabey (*Cerocebus atys*)
Cattle, domestic (*Bos taurus*)	spider[c] (*Ateles* sp.)
Cheetah[a] (*Acinonyx jubatus*)	wooley[c] (*Lagothrix* sp.)
Chimpanzee[a] (*Pan troglodytes*)	bonnet macaque (*Macaca radiata*)
Chinchilla[a] (*Chinchilla lanigera*)	Celebes macaque (*Macaca maurus*)
Coyote[a,b] (*Canis latrans*)	lion tail macaque (*Macaca silenus*)
Deer, white-lipped[a] (*Cervus albirostris*)	rhesus (*Macaca mulatta*)
Dog (*Canis familiaris*)	Okapi[a] (*Okapia johnstoni*)
Dog, Cape or African hunting[a] (*Lycaon pictus*)	Onager[a] (*Equus hemionus onager*)
Dolphin, Pacific bottle-nosed[b] (*Tursiops gilli*)	Otter, river[a] (*Lutra canadensis*)
Ferret[a,c] (*Mustela* sp.)	sea[b] (*Enhydra lutris*)
Gazelle (*Gazella thomsonii*)	Rhinoceros, Northern White[a] (*Ceratotherium simum cottoni*)
Genet[a,c] (*Genetta felina*)	black (*Diceros bicornis*)
Gorilla, mountain[a] (*Gorilla gorilla beringeri*)	Rodents[b] – pocket mouse (*Perognathus baileyii,*
lowland[a] (*Gorilla gorilla gorilla*)	*P. penicillatus, P. intermedius*)
	grasshopper mouse (*Onchyomys torridus*)
Horse, domestic (*Equus caballus*)	ground squirrel (*Citellus harrisi*)
Horse[a], Przewalski (*Equus przewalski*)	kangaroo rat (*Dipodomys merriami*)
Human (*Homo sapiens*)	Sea lion[a,b] (*Zalophus californianus*)
Impala[a] (*Aepyceros melampus*)	Sheep (*Ovis aries*)
Jackrabbit[b] (*Lepus californicus*)	Skunk, hog-nose[a,c] (*Mephitis* sp.)
Kangaroo[a] (*Macropus rufus*)	Snake, Sonoran Gopher[b] (*Piticophis melanoleucus affinis*)
wallaroo[a] (*Macropus robustus or erubescens*)	Swine (*Sus scrofa*)
Kiang[a] (*Equus hemionus kiang*)	Tapir[a] (*Tapirus terrestris*)
Kit fox[b] (*Vulpes velox*)	Tiger[a], Bengal and Sumatran (*Panthera tigris*)
Lemur, red ruffed[a] (*Varecia variegata* v. *suber*)	Zebra[a], Grevy's (*Equus grevyi*)

This compilation includes cases detected by Raymond Reed, DVM and Kathryn Orr, DVM.
a) Zoo or captive animals.
b) In the wild.
c) Species name uncertain.

who served as donors (Tripathy et al. 2002; Wright et al. 2003.)

It is estimated that some 60 percent of individuals infected by the respiratory route are asymptomatic or have such a mild illness that medical attention is not sought (Figure 26.6). Their exposure to *C. immitis* is diagnosed from the presence of a positive skin test, or from detection of a chronic pulmonary residual lesion, 'coccidioidoma' or cavity, detected incidentally in a chest radiograph. Occasionally a previously healthy individual may present with hemoptysis or other complaint that can be referred to a cavity.

Of the 40 percent who develop symptoms, most will have an acute febrile illness beginning 7–28 days (average 10–16 days) after exposure to the organism. The clinical presentation has been reviewed in the literature (Galgiani 1993). Chest pain, often pleuritic and sometimes severe, occurs in 75 percent. The following symptoms occur in descending order: cough, fever, malaise, rash, sore throat, headache, arthralgia and/or myalgia. Anorexia, weight loss, and night sweats are encountered in some patients. These signs and symptoms resemble those of many other diseases and have led to designation of coccidioidomycosis as 'flu-like.'

The acute-stage pulmonary lesions, detected by chest radiograph, are also similar to those found in other diseases: small or large single or multiple infiltrates (the latter can resemble metastatic carcinoma); hilar and/or mediastinal adenopathy; transient cavity, pleural, or pleuropericardial effusions; miliary or reticulonodular lesions (each of which is prognostically unfavorable).

Various rashes occur, but erythema nodosum and erythema multiforme are most commonly seen (Quimby et al. 1992).

Of the 40 percent with symptomatic infections, most will recover completely. These patients are immune to exogenous reinfection as are those individuals who have incurred asymptomatic infections. About 10 percent of these patients will, however, be left with a pulmonary residual nodule or cavity, usually detected several months or years later. Infrequently, a chronic progressive fibrocavitary disease resembling apical tuberculosis or histoplasmosis capsulati evolves. Metapulmonary spread of the disease (dissemination) is usually clinically apparent, but can occur silently, although infrequently, to the eyes (Rodenbiker et al. 1981) or kidneys (Petersen et al. 1976). Such dissemination occurs and becomes manifest in the first few weeks to months after onset of primary infection; for example, meningitis became evident about 5 weeks after the initial onset of symptoms (Pappagianis 1988a). Dissemination may not, however, be noted for some years and can occur in the absence of any previously recognized pulmonary focus. In the southwestern USA where coccidioidomycosis is endemic, the overall rate of dissemination in symptomatic cases is 5–7 percent but the rate varies with ethnicity or race. Thus, the primary infection disseminates in one to two per 100 symptomatic adult white males, at a slightly higher rate in Mexicans and Native Americans (Indians) (approximately 5 per 100), but at a markedly higher rate in African-Americans (>20 per 100) and Filipinos (>25 per 100) and other Asians (30 per 100) (Gifford et al. 1936/1937; Smith et al. 1946a, b; Pappagianis 1988b).

Additional risk factors that heighten the risk of dissemination are the following:

- group B blood (Deresinski et al. 1979)
- HLA class II (Louie et al. 1999)

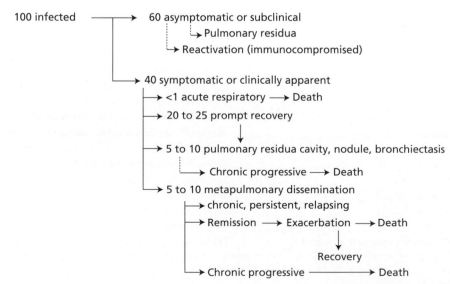

Figure 26.6 *Forms of coccidioidomycosis in humans. Broken lines indicate infrequently occurring phenomena.*

- pregnancy, particularly when acute primary coccidioi-domycosis is acquired during the third trimester (Peterson et al. 1993)
- hematopoietic neoplasms, e.g. lymphoma, AIDS (Deresinski and Stevens 1974; Fish et al. 1990)
- immunosuppression for organ transplant (Cohen et al. 1982; Riley et al. 1994)
- trauma creating a site of lessened resistance (Pappa-gianis 1985).

Dissemination can affect virtually any tissue and organ, although the gastrointestinal mucosal surface (Weisman et al. 1976) and endocardium have rarely been affected. The skin, bones, and meninges are affected (Huntington 1971). The involvement of various tissues and organs in disseminated disease assures that coccidioidomycosis can be confused with a great variety of infections, neoplasia, connective tissue disorders, etc.; in fact, it has become a 'great imitator' similar to syphilis. A history of travel or residence in 'endemic areas' or (rarely) exposure to products exported from such areas, should prompt consideration of coccidioidomycosis in the differential diagnosis. The coexistence of other diseases (infectious, neoplastic, 'connective tissue,' etc.) must not, however, be overlooked despite confirmation of coccidioidomy-cosis (Huth 1994).

Immunology

C. immitis can be a primary pathogen producing disease in normal individuals or opportunistic disease in immu-nocompromised individuals. It is unique among the systemic pathogenic fungi, because its replication in the infected host results in spherules containing tens to hundreds of endospores each of which can, in turn, produce mature endosporulating spherules. Contrast this with the single or few progeny produced by the budding of other systemic pathogens. The mechanism(s) by which *C. immitis* causes disease has not been defined. It may involve toxic action (Gale et al. 1967), enzymatic activity, e.g. proteases (Lupan and Nziramasanga 1986; Resnick et al. 1987; Yuan and Cole 1987; Johnson et al. 2000), and immunological mechanisms: complement-mediated/hypersensitivity/cytokines may contribute to pathogenesis (Magee and Cox 1995).

Mannose is a prominent component of coccidioidal cells (Pappagianis et al. 1961) and may play a role in immune responses via macrophage mannose receptors (Ampel et al. 2004, unpublished).

The immune interactions of host and *C. immitis* have been well reviewed (Cox and Magee 1998). The immunology can be divided roughly into the following categories:

- innate responses of the host
- specific humoral or T-cellular immunity
- induction of protective immunity.

The precise initial events in the interaction between the previously uninfected host and infective arthroco-nidia have not been well defined. The conidia are of such dimensions that at least some of them may elude the upper respiratory filtration and mucociliary escalator to reach the small bronchi, bronchioles, and alveoli. [Bronchoscopic examination had shown that primary endobronchial disease existed in the apparent absence of pneumonitis (alveolitis) (Birsner 1954).] Although the portal of entry is usually respiratory, infrequently infec-tions have resulted from traumatic introduction of arthroconidia through the skin (Wilson et al. 1953). Experimentally, subcutaneous or intraperitoneal injec-tion of conidia is less likely to result in the severe, progressive, and lethal disease than that induced by respiratory challenge.

In the mouse lung, the arthroconidia elicit a response which includes both macrophages and polymorpho-nuclear leukocytes (PML). The presence of PML appears to contribute to conversion of arthroconidia to spherules (Baker and Braude 1956). The outer wall of the arthroco-nidia, before being shed in the evolution to spherules, can, however, have a toxic effect inhibiting phagocytosis and destruction of the arthroconidia by human neutrophil granulocytes (Drutz and Huppert 1983). Mature spher-ules, by virtue of their size, and packets of newly released endospores embedded in a matrix appear to resist phago-cytosis until individual endospores are released from this matrix (Frey and Drutz 1986).

Nevertheless, PMLs may provide an impediment to some morphological forms of *C. immitis* after the activa-tion of complement and production of chemotactic substances in the absence of antibody (Galgiani et al. 1980). This may result after antigen–antibody precipita-tion as in a fibrin clot (Pappagianis and Zimmer 1990), induction of an 'oxidative burst' with production of antimicrobial H_2O_2 and hypochlorous acid (induced by arthroconidia and by immature but not mature spherules) (Galgiani 1995), production of cationic peptides (Segal et al. 1985), or production of lysozyme which is lethal to *C. immitis* spherules in vitro (Collins and Pappagianis 1973). Some studies suggest that enhanced phagocytosis by PMLs may result after immunization (Wegner et al. 1972).

The induction of a *C. immitis*-specific T-lymphocyte response underlies the major protective response to *C. immitis* (reviewed by Cox and Magee 1998). This has been derived from studies in humans and in experi-mental animals. A high rate (~100 percent) of humans with primary acute coccidioidomycosis who recovered had positive skin test (delayed-type hypersensitivity) to coccidioidin compared with only about 70 percent of those with unifocal disseminated (extrapulmonary) disease and about 30 percent with multiple sites of disse-mination (Smith et al. 1948). Anergy was associated with a poor prognosis in those patients with disseminated disease, and may result from the following:

- excessive fungal antigen or of immune complexes
- induction of suppressor cells
- relative increase of Th2 lymphocytes over Th1 (the latter of greater importance in cell-mediated immunity) (Corry et al. 1996)
- depression of the immune response by fungal cell products.

[However, 15–20 percent of patients with chronic coccidioidal pulmonary cavities are nonreactive to skin tests with coccidioidin, but do not have a poor prognosis or a risk of dissemination from their pulmonary lesion (Smith et al. 1948; Winn 1968).]

It has been proposed that part of the protective immune response to *C. immitis* is related to an opsonizing effect of IgM reacting with surface antigen of *C. immitis* (Pappagianis and Zimmer 1990); but the higher IgG titers are associated with a poorer prognosis (see section on Immunoserology, below) It appears that induction of a *C. immitis*-specific T-lymphocyte reponse underlies the major protective induced immune response to *C. immitis*. In experimental animals, several lines of evidence indicate that immune T lymphocytes and lymphokines are crucial in the protection against *C. immitis* (Beaman et al. 1983). The immune T lymphocytes act not by a direct killing action on *C. immitis* but rather through cytokine activation of monocyte/macrophages (Beaman et al. 1983; Slagle et al. 1989; Ampel et al. 1992; Ampel 2003). Experimentally, interferongamma was associated with resistance, interleukins 4 and 10 with susceptibility to *C. immitis* (Magee and Cox 1995; Fierer et al. 1998). Various levels of deficiency of T cell or competency of macrophages may exist in patients who are at higher risk of disseminated coccidioidomycosis. For example, it needs to be determined whether anticoccidioidal enzymatic activities are significantly different in macrophages and this can provide a possible explanation of differences in susceptible versus resistant or immunized hosts.

Resistance (immunity) to second exogenous infection follows subclinical infections (detected by positive skin test) and complete recovery from clinically apparent infection only. Only four cases of second exogenous (re)infection have been documented in >30 000 cases of coccidioidomycosis. Only one was a naturally acquired infection but in an immunocompromised individual. Three others were laboratory acquired (Pappagianis 1999). Endogenous reactivation of a previously arrested, non-disseminated, coccidioidal infection can occur as a result of immunosuppression: iatrogenic as with organ transplantation, from infection with HIV, or through development of a hematopoietic–lymphoreticular malignancy. Reactivation of a previously disseminated coccidioidal infection, even after apparently successful antifungal therapy, is not uncommon (Stevens 1995).

It had been demonstrated by Rixford (Rixford and Gilchrist 1896) that prior infection with *C. immitis* rendered a dog resistant to (parenteral) reinfection. The resistance conferred upon humans by prior infection led to experimental immunization in laboratory animals. In 1951 it was demonstrated by C. E. Smith that injection of the hyphal/arthroconidial form of living avirulent *C. immitis* in mice induced resistance to intraperitoneal challenge. Injection of virulent *C. immitis* in mice or monkeys provided protection against lethal respiratory challenge (Pappagianis et al. 1960). Killed *C. immitis* also afforded protection (Friedman and Smith 1956; Levine et al. 1960). Of the various morphological forms used in nonviable vaccines, the mature spherule (particularly its cell wall) proved to be the most efficacious (Levine et al. 1961). These immunizations did not prevent infection following respiratory challenge but did prevent progressive, lethal disease. After tests for safety of the killed spherule vaccine, tests for protective efficacy in humans showed that the killed spherule vaccine did not prevent clinically apparent coccidioidomycosis (Pappagianis and the Valley Fever Vaccine Study Group 1993). The inadequate response of humans may have resulted from the relatively small dose of antigen that had to be used because of the inflammatory effect of whole spherules injected intramuscularly. It is possible that isolation and/or synthesis of the immunogenic component away from irritating cell wall components, e.g. chitin, may provide a tolerable and effective vaccine for humans.

Some candidate subcellular vaccines that have exhibited protection against respiratory challenge in mice include derivatives of mechanically disrupted spherules (Zimmermann et al. 1998), a soluble proline-rich antigen (PRA) developed by Galgiani and colleagues (Shubitz et al. 2002), which proved to be the same as antigen 2 (Ag2) developed by Cox and colleagues (Zhu et al. 1996); and antigens including coccidioides-specific antigen (CSA) of Cole et al. (2004). Recombinant peptide forms of Ag2/PRA and CSA, when mixed or chemically combined, have the appropriate properties and immunogenicity that may permit them to be tested in nonhuman primates and humans.

Screening immunization of a cDNA library has led to another antigen with protective activity in mice (Ivey et al. 2003).

Histopathological response

Coccidioidal lesions can generally be termed pyogranulomatous, but in fact a variety of histopathological responses can be detected: tuberculosis-like granulomata, caseation necrosis with or without (more often) calcification, hyalinization-fibrosis, fibrinous pleuritis. Endospores attract PML whereas the mature spherules evoke a mononuclear/macrophage response (Figure 26.1). Thus, abscesses and tubercles may be side by side (Huntington 1971). *Coccidioides* in the skin evokes a proliferative ('pseudoepitheliomatous') response in the epidermis. The synovial membranes produce a villous response in coccidioidal arthritis. Bones have lytic lesions (occasionally proliferative) in the human, lytic or proliferative in dogs. In old pulmonary lesions (coccidioidomas), immature (non-endosporulating) spherules are often embedded in fibrocaseous material. In the absence of identifiable *C. immitis*, the granulomatous reaction may resemble that of other diseases such as sarcoidosis. Peripheral (bloodstream) eosenophilia can occur and on occasion the acute coccidioidal pulmonary disease can resemble 'eosenophilic pneumonia' of other etiologies (and disastrous therapy with adrenocorticosteroids may be applied).

In coccidioidal meningitis, the cerebrospinal fluid (CSF) exhibits a pleocytosis which, early on, may have a preponderance of PML and eosenophilia, but usually the lymphocytes predominate. Cerebral arteritis can accompany coccidioidal meningitis.

Treatment

Most patients with coccidioidomycosis could be managed successfully without specific antifungal therapy but, if necessary, the disease can be treated medically or surgically. The decision to treat or not to treat, and the mode of therapy depend on the clinical stage, on risk factors in the hosts, and the type and location of lesions. The variability in the clinical course of the disease, whether treated or untreated, has made evaluation of therapy difficult. More than one therapeutic modality may have to be applied.

MEDICAL THERAPY

Currently available therapy includes preparations of the polyene amphotericin B (AMB), including lipid or liposomal forms, and azoles. The azoles are fungistatic. The AMB under some circumstances can be fungicidal but with clinical concentrations is usually fungistatic. There is a lack of well controlled therapeutic studies which has led to uneven and sometimes uncertain application of therapy. Between the introduction of AMB in the treatment of coccidioidomycosis in humans in 1956 (Fiese 1957), and the introduction of the azoles 25 years later, the usually benign course of primary coccidioidomycosis prompted a conservative approach – to avoid the undesirable side effects of parenteral AMB, particularly nephrotoxicity and adverse effects on the hematopoietic system. Miconazole, also given parenterally, had a short-lived application in disseminated coccidioidomycosis. After the introduction of the azoles that could be taken by mouth [ketoconazole (KTZ), fluconazole (FLU), and itraconazole (ITR)] and were relatively free of undesirable side effects, these oral medications were used initially to treat disseminated disease (Tucker et al. 1990a, b; Diaz et al. 1991; Galgiani 1993; Galgiani et al. 1988, 2000). In some locations, the routine early use of the relatively benign triazoles has become customary in the treatment of suspected or confirmed cases of coccidioi-

domycosis without controlled evaluation of the efficacy at this stage of the disease, i.e. it is not known whether such treatment hastens recovery or prevents complications such as extrapulmonary spread of the disease or development of pulmonary residual lesions. Despite uncertainty regarding the usefulness of treatment in acute primary coccidioidomycosis, the presence of certain threatening conditions (severe intrapulmonary disease, risk factors for dissemination) and unfavorable laboratory values (such as peripheral blood eosinophilia and climbing complement fixation (CF) titer) gives the justification for treatment.

KTZ is now infrequently used in human coccidioidomycosis in the USA. Variconazole, a triazole as are FLU and ITR, has had limited use in coccidioidomycosis but has appeared effective even against meningitis following failure of FLU therapy (Cortez et al. 2003).

Treatment is almost always indicated for patients with demonstrated or surmised disseminated disease, or with rapidly progressive primary pulmonary disease. AMB is indicated for the latter, for widespread extrapulmonary lesions, infected pregnant women (the AMB appears to have little adverse effect on the fetus), and osseous lesions. [The place of the azoles in the treatment of humans with osseous lesions is not well defined (ITR may be preferred), but in the dog KTZ appears to lead to an improvement in osteoarticular disease (Wolf and Pappagianis 1981).] For more indolent disseminated disease, the azoles can be used, and which to use may be guided by the side effects (greater with KTX) or expense (greater with FLU than with KTZ or ITR). However, other properties may also influence the selection: excellent penetration of FLU into the CSF (appropriate for treatment of meningitis); short half-life of KTZ compared with those for FLU and ITR (Tucker et al. 1990a). Significant interaction with other drugs requires close attention when the azoles are used (e.g. with rifampin, cyclosporin, warfarin, and terfenadine). Acquisition of resistance to FLU by *C. immitis* has been suggested but not proven.

Two major problems accompany medical therapy: the long-term therapy of coccidioidomycosis (usually months or years) is very expensive, and there is uncertainty as to when to stop treatment. Relapses have followed seemingly successful treatment, especially with the azoles (Dewsnup et al. 1996).

FUNGICIDAL AGENTS

These are needed, because currently used chemotherapeutic agents have not killed cells in the host. Experimentally, nikkomycin Z, an inhibitor of synthesis of cell wall polysaccharide by chitin synthase, has proved to be curative (Hector et al. 1990) (Figure 26.7).

SURGICAL THERAPY

Surgical approaches are indicated as diagnostic and therapeutic procedures for numerous coccidioidal lesions.

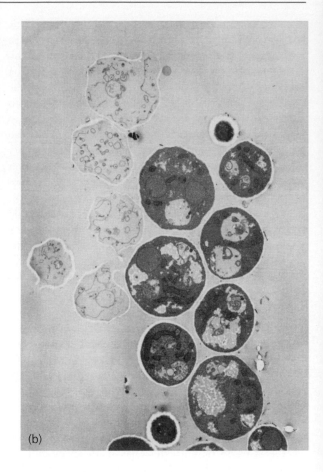

(a)   (b)

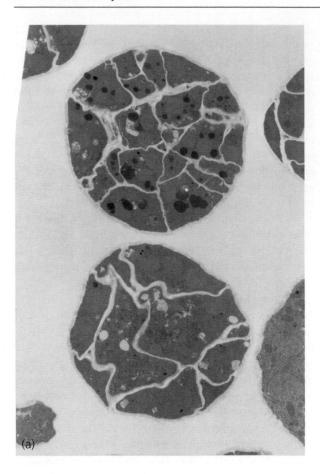

UDP-*N*-acetylglucosamine

(c)   Nikkomycin Z

Figure 26.7 (a) *Transmission electron micrograph of 33-hour-old spherules produced in vitro, untreated* **(a)** *and those exposed to nikkomycin Z 2 μg/mL for 24 hours* **(b)***. Untreated spherules exhibit an intact cell wall and segmentation towards formation of endospores. In the presence of nikkomycin there is a decrease in the size of the spherules, little or no cell wall was formed and the ghosts of some spherules are collapsing. (From Pappagianis (1989) with permission.)* **(c)** *Comparative structure of antifungal nikkomycin Z and UDP-N-acetylglucosamine.*

Biopsy (this may be excisional) of a solitary pulmonary nodule is in the diagnostic category; this type of nodule must be distinguished from neoplastic or other infectious nodules which have not yielded a diagnosis with non-invasive procedures (Forseth et al. 1986). Coccidioidal pulmonary cavities, and their associated complications such as hemorrhage, rupture into the pleural space, or formation of a bronchopleural fistula, may require surgical correction (Cunningham and Einstein 1982).

Drainage of abscesses and empyema, removal of infected bone or joint tissue, and immobilization of a joint by arthrodesis are additional therapeutic surgical procedures (Bried and Galgiani 1986). Stabilization of vertebrae infected with *C. immitis* has been accomplished (Herron et al. 1997).

Surgical placement of prostheses, such as insertion of an Ommaya reservoir and catheter into the lateral ventricle or cisterna magna, may help in the treatment of coccidioidal meningitis (Zealear and Winn 1967).

Surgical procedures are usually carried out with concomitant antifungal medical therapy.

LABORATORY DIAGNOSIS

Direct examination

Microscopic demonstration of *C. immitis*, particularly the endosporulating spherules, provides the most direct and secure method of diagnosis. Detection may be attempted with bronchial washings, exudates, and sputum (the last is often unreliable because of confusing artifacts). Synovia or other body fluids, e.g. pleural, pericardial, or peritoneal, may reveal presence of the organism though the corresponding tissue is more likely to yield *C. immitis* microscopically (or by culture). The endosporulating spherules or other forms can be identified in preparations treated with potassium hydroxide or stained with hematoxylin and eosin (H&E) or Papanicolaou stain, but they may be more readily distinguished when stained by methenamine silver or periodic acid–Schiff (PAS), particularly when the latter is used without a counterstain (Huntington 1971). *C. immitis* can be visualized in gram-stained sputum or body fluids, but this has not been a widely advocated method. Various sized forms of *C. immitis* may be seen, and in some lesions, particularly chronic ones, small immature spherules without endospores may be observed and can resemble *Blastomyces dermatitidis* (Figure 26.3). The presence of chitin in the cell wall permits fluorescent staining of *C. immitis* in sputum or other body fluids using Calcofluor white M2R (Polysciences Inc.) or its equivalent Fluorescent Brightener 28 (Sigma Chemical Co.). Fluorescent antibody staining (of limited availability) has shown a desirable specificity and can be used when the usual histopathological examination has not yielded a definitive answer.

In some tissues, e.g. pulmonary cavities, the hyphal form may be observed but is not definitive for *C. immitis* and further confirmatory studies are required, e.g. conversion of hyphal to endosporulating spherule by inoculation of laboratory animals or special culture media. A wide range of tissue reactions is evoked by *C. immitis* – suppurative, granuloma, pyogranuloma, caseation necrosis with or without calcification, hyaline or fibrotic changes. Thus, even if the characteristic endosporulating spherules are not seen, the diagnosis of coccidioidomycosis should not be rejected based on the nature of the host response.

Culture

C. immitis is hardy and non-exacting in its nutritional requirements, being able to grow in simple media containing salts and ammonium acetate or ammonium lactate as its sources of energy, carbon, and nitrogen. It can grow in media with an acidic pH <3.0 and alkaline pH >9.0). (The tolerance of a high pH may underlie its ability to survive in the alkaline soils of the endemic areas, and in the selective action of NH_4OH added to yeast extract/phosphate medium.) It can grow at 22°C and at 40°C, but its optimum for hyphal growth is about 30°C. In vitro, the conversion from hyphal to spherule/endospore form is enhanced by incubation at 40°C but it can be maintained in the spherule/endospore phase at 37°C. Exposure to 46°C for 2 hours is tolerated but it is killed in 4 minutes at 60°C. Inoculation of sputum, exudates, tissues, onto many of the commonly used culture media will yield satisfactory growth, including: sheep blood agar, brain heart infusion, beef infusion glucose, Sabouraud's glucose agar, glucose/yeast extract agar. The glucose (2 percent)/yeast extract (1 percent) agar (2 percent) provides excellent yields of arthroconidia whereas the peptone-containing Sabouraud's glucose agar can diminish the yield of arthroconidia.

Culture of the blood may yield the organism, but patients with coccidioidal fungemia are frequently very ill and in the preterminal stage of their coccidioidomycosis. Care must be taken to permit recovery of *C. immitis* in specimens that may contain other organisms. Treatment of sputum with alkali or hypochlorite for recovery of *Mycobacterium tuberculosis* can inactivate *C. immitis*. Recovery of *C. immitis* may, however, be enhanced by mucolytic agents such as *N*-acetyl-1-cysteine and dithiothreitol. Competing organisms can be minimized by the addition of the antifungal cycloheximide and antibacterial chloramphenicol. Before diagnostic animal inoculation, specimens can be treated with penicillin G and streptomycin (or comparable aminoglycoside) to reduce possible infection with bacteria. Isolation from the soil can entail similar plating or inoculation of laboratory animals – mice being the most convenient.

Faint grayish-white growth becomes visible in 48 hours on several media; it becomes clearly evident as a cottony to weblike growth in 48–72 hours. Although rarely necessary, it is advisable to retain cultures for 2–3 weeks before discarding them as negative.

Typical gross morphology is exhibited by most isolates of *C. immitis* but atypical colonial growth does occur. For this reason and because the hyphal/arthroconidial form of *C. immitis* resembles other nonpathogenic fungi, it is essential to carry out definitive identification. This can be achieved by demonstrating the spherule/endospore phase after inoculation into laboratory animals or into special culture media.

More recently, a DNA hybrid protection assay (Gen-Probe, Inc.) has proved to be a sensitive and rapid way to identify isolates as *Coccidioides* (Beard et al. 1993). Ribosomal RNA (rRNA) is extracted from the suspect organism and allowed to react with specific DNA from *C. immitis* to which is attached a potentially luminescent acridinium ester. If the rRNA is of coccidioidal origin, it will react with its complementary DNA, and this protects the ester linkage against hydrolysis, thus yielding luminescence. Heterologous rRNA will not interact with the coccidioidal DNA, and the acridinium ester will be hydrolyzed and thus not yield luminescence. The method appears to be quite specific and requires very little of the mycelial growth form.

Another in vitro test which has served for several years as a specific test for *C. immitis* is the exoantigen test (Standard and Kaufman 1977). The suspect culture is grown for several days on a slant culture. An aqueous solution of thimerosal is added to the slant to sterilize the culture and to extract antigen. The antigen-containing supernatant is then tested by immunodiffusion using specific anticoccidioidal antibody to demonstrate the presence of antigen specific for authentic *C. immitis*. A 19 kDa polypeptide with serine protease activity may represent the '*Coccidioides immitis*-specific' antigen (Pan and Cole 1995).

The indispensable condition for morphological identification is demonstration of endosporulating spherules. This can be achieved by intraperitoneal or intranasal inoculation of mice or intratesticular inoculation of guinea pigs. If the animals become overtly ill, tissue with visible lesions can be examined microscopically in a squash preparation or permanent histological sections. If no illness develops, mice should be sacrificed at 7–10 days. If no lesions are evident, subcultures should be made from the tissues, and additional mice sacrificed at 14 and 21 days. Intraperitoneal inoculation usually suffices but some isolates of lesser virulence may only be recoverable after intranasal inoculation.

Immunoserology

Immunological tests have been very useful in epidemiologic studies and in the diagnosis and management of cases of coccidioidomycosis [reviewed by Pappagianis 2001 (in Ampel 2001) and Pappagianis and Zimmer

1990]. They represent sensitive and specific tests which can assist in the diagnosis when there is not ready access to sputum or tissue, and to obviate invasive procedures. In the early application and evaluation of these procedures, conversion to skin test positivity appeared to precede the detection of antibody (Smith et al. 1956). However, more recent studies have indicated that antibody can be detected before the skin test becomes positive. Therefore, a negative skin test should not preclude a diagnosis of coccidioidomycosis nor the use of serological tests (Table 26.2).

Skin testing can be carried out with lysate-culture filtrate of the hyphal phase (coccidioidin) or lysate of the spherule/endospore phase (spherulin) (Levine et al. 1973). Although some individuals yield immediate-type reactions, the delayed cutaneous hypersensitivity reaction (readings at 24–48 hours, therefore earlier than tuberculin reactivity) represents the evidence of past or current infection. Only recent conversion to positivity is of assistance in the diagnosis of an acute primary coccidioidal infection. A reaction in a healthy individual denotes resistance to exogenous reinfection. Cross-reactivity in individuals previously infected with *H. capsulatum* can be observed, but does not represent a serious problem.

Failure to react to the skin test (anergy) may develop in about half the patients with disseminated coccidioidomycosis. However, 15–20 percent of patients with non-disseminated disease, e.g. a single chronic pulmonary cavity or nodule, fail to react (Winn 1968).

In vitro lymphocyte stimulation by coccidioidal antigen is a correlate of skin test responsiveness but has not supplanted the simple intracutaneous test.

In primary non-disseminating coccidioidomycosis, a regular sequential response is noted, indicated by production of two kinds of antibody: a precipitin and a complement-fixing antibody (Figure 26.8). The former was originally detected by a tube precipitin (TP) test (soluble, heat-stable antigen present in coccidioidin yielded a precipitate when mixed with serum of a patient with early, acute coccidioidomycosis) (Smith et al. 1956). It was later shown that the reaction could

Table 26.2 *Application of serologic tests in coccidioidomycosis*

	Detection of	
	IgM	IgG
Tube precipitin (TP)	Yes	No
Latex particle agglutination (LPA)	Yes	No[b]
Immunodiffusion = TP (IDTP)	Yes	No
Complement fixation (CF)	No (rare[a])	Yes
Immunodiffusion = CF (IDCF)	No	Yes
Counterimmunoelectrophoresis (CIE)	Yes	Yes
Enzyme-linked immunoassay (EIA)	Yes	Yes

a) Rarely positive CF in presence of positive IDTP (IgM, and absence of detectable IgG by IDCF).
b) Occasional positive LPA in presence of detectable IgG but not IgM by immunodiffusion.

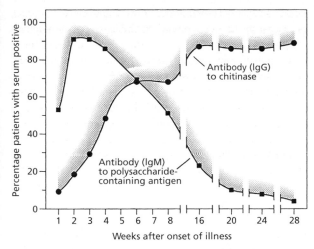

Figure 26.8 *Sequential antibody response in humans infected with* Coccidioides immitis. *The shaded zones represent the greater sensitivity, though not precisely defined, of more recent immunodiffusion and enzyme immunoassay tests for the detection of the IgM or IgG than the results (solid squares and circles) obtained using a TP and CF test by Smith et al. (1956).*

also be detected by agar gel double immunodiffusion (ID). The reaction is referred to as immunodiffusion corresponding to the tube precipitin (IDTP) test; and the responsible antibody was found to be IgM. This antibody could also be detected by antigen-coated latex particles and by enzyme-linked immunoassay (EIA). The antigen that reacts with the IgM appears to be mainly polysaccharide but has also been described in connection with a 120 kDa glucosidase (Kruse and Cole 1992), or a 21 kDa proteinase (Resnick et al. 1987) that can be released into the culture medium (Zimmer and Pappagianis 1989). This IgM antibody is detectable in more than 50 percent of patients with acute coccidioidomycosis by a week after the onset of symptoms. It is subsequently detectable (between 2 and 3 weeks of illness) in more than 90 percent of patients, and gradually fades although, with currently available methods, more than 10 percent of patients still have detectable antibody 6 months after onset of illness. There is no clear correlation between its titer and prognosis. It has been detected in the blood of newborns whose mothers had coccidioidomycosis and the anti-coccidioidal IgM, but this did not denote presence of the disease in the neonate. Presence of the IgM in the CSF has been infrequent but only associated with coccidioidal meningitis.

The antibody appearing later was originally detected by CF (Smith et al. 1956) and its titer clearly correlated with the extent and severity of the disease. It did not yield a precipitate with its corresponding antigen in a liquid milieu, but in a gel (agar, agarose, Gellan, Kelco of Merck & Co., Inc.) the antibody with its corresponding antigen produced a precipitate referred to as immunodiffusion corresponding to complement fixation (IDCF) (Figure 26.9). The antibody can also be detected

by EIA (Pappagianis and Zimmer 1990; Gade et al. 1992). The responsible antigen, a protein of approximately 110 kDa, or 48 kDa under reducing conditions (Zimmer and Pappagianis 1988), is a chitinase (Johnson and Pappagianis 1992), which is released by the maturing then rupturing spherule. [The coccidioidal DNA which encodes for this chitinase has been cloned in *Escherichia coli* which has produced enzymatically and serologically active chitinase (Zimmermann et al. 1996).] The corresponding antibody appears to be primarily IgG. Its titer generally rises with worsening of the disease, and decreases with improvement of the patient. A titer above 1:16 is often associated with meta-pulmonary dissemination, e.g. to the skin, a bone or even the meninges (Smith et al. 1956). With this last, there can be lethal disease even though the CF titer in the serum is less than 1:16. In most patients with meningitis, antibody can also be detected in the CSF but the serum must always be tested along with the CSF

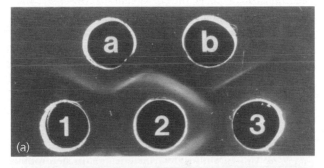

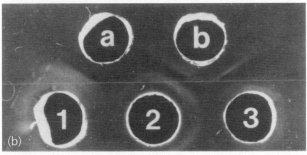

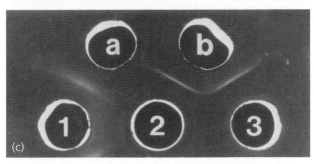

Figure 26.9 *Agar gel immunodiffusion indicating the detection of the usual antibodies (IgM and IgG) in human serum at different stages after infection. Well 2, patient's serum; wells 1 and 3, control IgM- and IgG-positive sera, respectively. Wells a and b contained coccidioidal antigen. (a) Early infection, IgM present; (b) later stage, both IgM and IgG present; (c) even later, only IgG detectable.*

because the CSF of some patients may be seronegative early in the course of meningitis. Testing synovial and other cavity fluids may be helpful in following the disease in corresponding anatomic sites.

Although the serological reactivity may obviate the need for invasive procedures, mediastinoscopy, or bronchoscopy, when possible, and recovery of the organism must be attempted. It is important to remember that in suspected early coccidioidomycosis, the tests for IgM are needed because the coccidioidal IgG may not be detected at that stage.

In the usual patient, serological reactivity is of considerable assistance for diagnosis and prognosis. However, some patients with immunosuppressive conditions (iatrogenic as for organ transplantation, or AIDS) may fail to produce a detectable antibody response (Fish et al. 1990). However, appropriate testing (concentrating serum prior to testing) has reduced seronegativity in AIDS patients with coccidioidomycosis to less than 10 percent.

It has been shown that detection of antigen may be accomplished before the detection of the antibodies described above, and this may provide an additional early serodiagnostic test and one useful in immunocompromised patients (Galgiani 1992).

Histopathology

Gross lesions may provide a clue to the histopathological changes which may involve a pyogranulomatous response with granulocytes, including both neutrophils and eosinophils, macrophages, multinucleate giant cells with palisading epithelioid cells, caseation necrosis, and hyalinization or fibrosis with lymphocytes at the periphery. The granulomatous changes, fibrosis and, occasionally, calcification, may represent chronic lesions, but only the presence of C. immitis distinguishes these from lesions of many other diseases. Vasculitis with perivascular inflammation and endarteritis obliterans has been described in association with coccidioidal meningitis/encephalitis.

The suppurative reaction may be extensive, but microabscesses only a few hundred micrometers in diameter may be found adjacent to granulomas. The former probably represent a response to released endospores, the latter a response to the larger spherules. Occasionally, the surface of mature spherules stained with H&E demonstrates a rough eosinophilic deposit similar to the Splendore–Hoeppli phenomenon associated with other fungi.

In addition to H&E and Papanicolaou's stains, the polysaccharide stains PAS and methenamine silver are useful for detection of C. immitis. Methenamine silver may stain the C. immitis diffusely dark, so that the internal structure, e.g. endospores in a spherule may not be clearly visible. PAS can provide better definition of the spherule/endospore phase. The so-called 'Schiff-positive dust' (Huntington 1957) stained by PAS can occasionally be seen in the cytoplasm of multinucleate giant cells and may represent remains of cell wall polysaccharide, digested autolytically by C. immitis, or by host cell enzymes, e.g. lysozyme (Collins and Pappagianis 1973), or chitinase. Calcofluor, which produces fluorescence with chitin in the cell wall, can be used to detect C. immitis in bronchoalveolar lavage specimens.

REFERENCES

Ampel, N.M. 2001. Coccidioidomycosis. Semin Respir Infect, 16, 229–96.

Ampel, N.M. 2003. Measurement of cellular immunity in human coccidioidomycosis. Mycopathologia, 156, 247–62.

Ampel, N.M., Bejarano, G.C. and Galgiani, J.N. 1992. Killing of Coccidioides immitis by human peripheral blood mononuclear cells. Infect Immun, 60, 4200–4.

Ampel, N.M., Mosley, D.G. and England, B. 1998. Coccidioidoymcosis in Arizona: increase in incidence from 1990 to 1995. Clin Infect Dis, 27, 1528–30.

Baker, O. and Braude, A.I. 1956. A study of stimuli leading to the production of spherules in coccidioidomycosis. J Lab Clin Med, 47, 169–81.

Beaman, L., Benjamini, E. and Pappagianis, D. 1983. Activation of macrophages by lymphokines: enhancement of phagosome-lysosome fusion and killing of Coccidioides immitis. Infect Immun, 39, 1201–7.

Beard, J.S., Benson, P.M. and Skillman, L. 1993. Rapid diagnosis of coccidioidomycosis with a DNA probe to ribosomal RNA. Arch Dermatol, 129, 1589–93.

Birsner, J.W. 1954. The roentgen aspects of five hundred cases of pulmonary coccidioidomycosis. Am J Roentgenol, 72, 556–73.

Bowman, B.H., Taylor, J.W. and White, T.J. 1992. Molecular evolution of the fungi: human pathogens. Mol Biol Evol, 9, 893–904.

Bried, J.M. and Galgiani, J.N. 1986. Coccidioides immitis infections in bones and joints. Clin Orthop Relat Res, 211, 235–43.

Burt, A., Carter, D.A., et al. 1996. Molecular markers reveal cryptic sex in the human pathogen Coccidioides immitis. Proc Natl Acad Sci U S A, 93, 770–3.

Cohen, I.M., Galgiania, J.N., et al. 1982. Coccidioidomycosis in renal replacement therapy. Arch Intern Med, 142, 489–94.

Cole, G.T., Xue, J.-M., et al. 2004. A vaccine against coccidioidomycosis is justified and attainable. Med Mycol, 42, 189–216.

Collins, M.S. and Pappagianis, D. 1973. Effects of lysozyme and chitinase on the spherules of Coccidioides immitis in vitro. Infect Immun, 7, 817–22.

Converse, J.L. 1955. Growth of spherules of Coccidioides immitis in a chemically defined liquid medium. Proc Soc Exp Biol Med, 90, 709–11.

Corry, D.B., Ampel, N.M., et al. 1996. Cytokine production by peripheral blood mononuclear cells in human coccidioidomycosis. J Infect Dis, 174, 440–3.

Cortez, K.J., Walsh, T.J. and Bennett, J.E. 2003. Successful treatment of coccidioidal meningitis with voriconazole. Clin Infect Dis, 36, 1619–22.

Cox, R.A. and Magee, D.M. 1998. Protective immunity in coccidioidomycosis. Res Immunol, 149, 417–28.

Cunningham, R.T. and Einstein, H. 1982. Coccidioidal pulmonary cavities with rupture. J Thorac Cardiovasc Surg, 84, 172–7.

Currah, R.S. 1985. Taxonomy of the Onygenales: Arthrodermataceae, Gymnoascaceae, Myxotrichaceae and Onygenaceae. Mycotaxon, 24, 1–216.

Delgado, N., Hung, C.-Y., et al. 2004. Profiling gene expression in Coccidioides posadesii. Med Mycol, 42, 59–71.

Deresinski, S.C. and Stevens, D.A. 1974. Coccidioidomycosis in compromised hosts. Medicine, 54, 377–95.

Deresinski, S.C., Pappagianis, D. and Stevens, D.A. 1979. Association of ABO blood group and outcome of coccidioidal infection. Sabouraudia, 17, 261–4.

Dewsnup, D.H., Galgiani, J.N., et al. 1996. Is it ever safe to stop azole therapy for *Coccidioides immitis* meningitis? *Ann Intern Med*, **124**, 305–10.

Diaz, M., Puente, R., et al. 1991. Itraconazole in the treatment of coccidioidomycosis. *Chest*, **100**, 682–4.

Drutz, D.J. and Huppert, M. 1983. Coccidioidomycosis: factors affecting the host-parasite interaction. *J Infect Dis*, **147**, 372–90.

Egeberg, R.O. and Ely, A.F. 1956. Coccidioides immitis in the soil of the southern San Joaquin Valley. *Am J Med Sci*, **23**, 151–4.

Fierer, J., Walls, L., et al. 1998. Importance of interleukin-10 in genetic susceptibility of mice to *Coccidioides immitis*. *Infect Immun*, **66**, 4397–402.

Fiese, M.J. 1957. Treatment of disseminated coccidioidomycosis with amphotericin B; report of a case. *Calif Med*, **86**, 119–20.

Fish, D.G., Ampel, N.M., et al. 1990. Coccidioidomycosis during human immunodeficiency virus infection. A review of patients. *Medicine 69*, **77**, 384–91.

Fisher, M.C., Koenig, G.L., et al. 2002. Molecular and phenotypic description of *Coccidioides posadasii* sp. nov., previously recognized as the non-California population of *Coccidioides immitis*. *Mycologia*, **94**, 73–84.

Forbus, W.D. and Bestebreurtje, A.M. 1946. Coccidioidomycosis. *Milit Surg*, **99**, 653–719.

Forseth, J., Rohwedder, J.J., et al. 1986. Experience with needle biopsy for coccidioidal lung nodules. *Arch Intern Med*, **146**, 19 20.

Frey, C.L. and Drutz, D.J. 1986. Influence of fungal surface components on the interaction of *Coccidioides immitis* with polymorphonuclear neutrophils. *J Infect Dis*, **153**, 933–43.

Friedman, L. and Smith, C.E. 1956. Vaccination of mice against *Coccidioides immitis*. *Am Rev Tuberc*, **74**, 245–8.

Gade, W., Ledman, D.W., et al. 1992. Serologic responses to various *Coccidioides* antigen preparations in a new enzyme immunoassay. *J Clin Microbiol*, **30**, 1907–12.

Gale, D., Lockhart, E.A. and Kimbell, E. 1967. Studies of *Coccidioides immitis*: III. Further studies of toxic soluble components of *Coccidioides immitis*. In: Ajello, L. (ed.), *Coccidioidomycosis*. Tucson, AZ: University of Arizona Press, 355–72.

Galgiani, J.N. 1992. Coccidioidomycosis: changes in clinical expression, serological diagnosis, and therapeutic options. *Clin Infect Dis*, **14**, Suppl 1, S100–5.

Galgiani, J.N. 1993. Coccidioidomycosis. *West J Med*, **159**, 153–71.

Galgiani, J.N. 1995. Differences in oxidant release by human polymorphonuclear leukocytes produced by stimulation with different phases of *Coccidioides immitis*. *J Infect Dis*, **172**, 199–203.

Galgiani, J.N., Yam, P., et al. 1980. Complement activation by *Coccidioides immitis in vitro* and clinical studies. *Infect Immun*, **28**, 944–9.

Galgiani, J.N., Stevens, D.A., et al. 1988. Ketoconazole therapy of progressive coccidioidomycosis. *Am J Med*, **84**, 603–10.

Galgiani, J.N., Amepl, N.M., et al. 2000. Practice guidelines for the treatment of coccidioidomycosis. *Clin Infect Dis*, **30**, 658–61.

Gifford, M.A., Buss, W.C. and Douds, R.J. 1936/1937. Data on *Coccidioides* infection, Kern County 1900–1936. Kern County Health Dept Annual Report.

Hector, R.F., Zimmer, B.L. and Pappagianis, D. 1990. Evaluation of nikkomycins X and Z in murine models of coccidioidomycosis, histoplasmosis, and blastomycosis. *Antimicrob Agents Chemother*, **34**, 587–93.

Herron, L.D., Kissel, P. and Smilovitz, D. 1997. Treatment of coccidioidal spinal infection: experience in 16 cases. *J Spinal Disord*, **10**, 215–22.

Higgins, J.C., Leith, G.S. et al. 2004. Seroepidemiologic survey of *Coccidioides immitis* in healthy horses in Arizona. *J Am Vet Med Assoc*, in press.

Huntington, R.W. 1957. Diagnostic and biologic implications of the histopathology of coccidioidomycosis. *Proceedings of the Symposium on Coccidiomycosis. Public Health Service Publication No. 575.* Atlanta: Communicable Disease Center, 34–46.

Huntington, R.W. 1971. Coccidioidomycosis. In: Baker, R. (ed.), *Human infection with fungi, Actinomycetes and algae*. Berlin: Springer-Verlag, 147–210.

Huppert, M., Sun, S.H. and Harrison, J.L. 1982. Morphogenesis throughout saprobic and parasitic cycles of *Coccidioides immitis*. *Mycopathology*, **78**, 107–22.

Huth, R.G. 1994. Concomitant systemic cryptococcosis and coccidioidomycosis in a patient with AIDS. *Clin Infect Dis*, **18**, 262–3.

Ivey, F.D., Magee, D.M., et al. 2003. Identification of a protective antigen of *Coccidioides immitis* by expression library immunization. *Vaccine*, **21**, 4359–67.

Johnson, S.M. and Pappagianis, D. 1992. The coccidioidal complement fixation and immunodiffusion-complement fixation antigen is a chitinase. *Infect Immun*, **60**, 2588–92.

Johnson, S.M., Kerekes, K.M., et al. 2000. Identification and cloning of an aspartyl proteinase from *Coccidioides immitis*. *Gene*, **241**, 213–22.

Kruse, D. and Cole, G.T. 1992. A seroreactive 120-kilodalton β- 1,3-glucanase of *Coccidioides immitis* which may participate in spherule-morphogenesis. *Infect Immun*, **60**, 4350–63.

Levine, H.B., Cobb, J.M. and Smith, C.E. 1960. Immunity to coccidioidomycosis induced in mice by purified spherule, arthrospore and mycelial vaccines. *Trans NY Acad Sci*, **22**, 436–49.

Levine, H.B., Cobb, J.M. and Smith, C.E. 1961. Immunogenicity of spherule-endospore vaccines of *Coccidioides immitis* for mice. *J Immunol*, **87**, 218–27.

Levine, H.B., Gonzalez-Ocho, A. and Ten Eyck, D.R. 1973. Dermal sensitivity to *Coccidioides immitis* a comparison of responses elicited in man by spherulin and coccidioidin. *Am Rev Respir Dis*, **107**, 379–86.

Lones, G.W. 1967. Studies of intermediary metabolism in *Coccidioides immitis*. In: Ajello, L (ed.), *Coccidioidomycosis*. Tucson, AZ: University of Arizona Press, 349–53.

Lones, G.W. and Peacock, C.L. 1960. Role of carbon dioxide in the dimorphism of *Coccidioides immitis*. *J Bacteriol*, **79**, 308–9.

Louie, L., Ng, S., et al. 1999. Influence of host genetics on the severity of coccidioidomycosis. *Emerg Infect Dis*, **5**, 672–80.

Lupan, D.P. and Nziramasanga, P. 1986. Collagenolytic activity of *Coccidioides immitis*. *Infect Immun*, **51**, 360–1.

Maddy, K.T. 1957. Ecological factors of the geographic distribution of *Coccidioides immitis*. *J Am Vet Med Assoc*, **130**, 475–6.

Magee, D.M. and Cox, R.A. 1995. Roles of gamma interferon and interleukin-4 in genetically determined resistance to *Coccidioides immitis*. *Infect Immun*, **63**, 3514–19.

Negroni, P. 1967. Coccidioidomycosis in Argentina. In: Ajello, L. (ed.), *Coccidioidomycosis*. Tucson, AZ: University of Arizona Press, 273–278.

Ophüls, W. and Moffitt, H.C. 1900. A new pathogenic mould (formerly described as a protozoan: *Coccidioides immitis pyogenes*): preliminary report. *Philadelphia Med J*, **5**, 1471–2.

Pan, S. and Cole, G.T. 1995. Molecular and biochemical characterization of a *Coccidioides immitis*-specific antigen. *Infect Immun*, **63**, 3994–4002.

Pappagianis, D. 1980. Epidemiology of coccidioidomycosis. In: Stevens, D.A. (ed.), *Coccidioidomycosis*. New York: Plenum, 63–85.

Pappagianis, D. 1985. The phenomenon of locus minoris resistentiae in coccidioidomycosis. In: Einstein, H.E. and Catanzaro, A. (eds), *Coccidioidomycosis*. Washington, DC: National Foundation for Infectious Diseases, 319–29.

Pappagianis, D. 1988a. Coccidioidomycosis. In: Balows, A., Hausler, W.J. Jr, et al. (eds), *Laboratory diagnosis of infectious diseases*, vol. 1. New York: Springer-Verlag, 600–23.

Pappagianis, D. 1988b. Epidemiology of coccidioidomycosis. In: McGinnis, M. (ed.), *Current topics in medical mycology*, vol. 2. New York: Springer-Verlag, 199–238.

Pappagianis, D. 1989. Enhancement of host resistance by control of fungal growth. *Current topics in infectious diseases and clinical microbiology*, 2nd edition. Wiesbaden: Friedrich Vieweg & Son.

Pappagianis, D. 1994. Marked increase in cases of coccidioidomycosis in California: 1991, 1992 and 1993. *Clin Infect Dis*, **19**, S14–18.

Pappagianis, D. 1999. *Coccidioides immitis* antigen. *J Infect Dis*, **180**, 243–4.

Pappagianis, D. and Zimmer, B.L. 1990. Serology of coccidioidomycosis. *Clin Microbiol Rev*, **3**, 247–68.

Pappagianis, D., Miller, R.L., et al. 1960. Response of monkeys to respiratory challenge following subcutaneous inoculation with *Coccidioides immitis*. *Am Rev Respir Dis*, **82**, 224–50.

Pappagianis, D., Putman, E. and Kobayashi, G.S. 1961. Polysaccharide of *Coccidioides immitis*. *J Bacterol*, **82**, 714–23.

Pappagianis, D., Ornelas, A. and Hector, R.F. 1985. Guanine plus cytosine content of the DNA of *Coccidioides immitis*. *Sabouraudia: J Med Vet Mycol*, **23**, 451–4.

Pappagianis, D. and the Valley Fever Vaccine Study Group. 1993. Evaluation of the protective efficacy of the killed *Coccidioides immitis* spherule vaccine in humans. *Am Rev Respir Dis*, **148**, 656–60.

Petersen, E.A., Friedman, B.A., et al. 1976. Coccidioidouria. Clinical significance. *Ann Intern Med*, **85**, 34–8.

Petersen, L.R., Marshall, S.L., et al. 2004. Coccidioidomycosis among workers at an archeological (sic) site northeastern Utah. *Emerg Infect Dis*, **10**, 637–41.

Peterson, C.M., Schuppert, K., et al. 1993. Coccidioidomycosis and pregnancy. *Obstet Gynecol Surv*, **48**, 149–56.

Posadas, A. 1892. Un nuevo caso de micosis fungoidea con psorospermias. *An Circ Med Argent*, **15**, 585–97.

Quimby, S.R., Connolly, S.M., et al. 1992. Clinicopathologic spectrum of specific cutaneous lesions of disseminated coccidioidomycosis. *J Am Acad Dermatol*, **26**, 79–85.

Resnick, S., Pappagianis, D. and McKerrow, J.H. 1987. Proteinase production by the parasitic cycle of the pathogenic fungus *Coccidioides immitis*. *Infect Immun*, **55**, 2807–15.

Riley, D.K., Galgiani, J.N., et al. 1994. Coccidioidomycosis in bone marrow transplant recipients. *Transplantation*, **56**, 1531–3.

Rixford, E. 1894. Case for diagnosis presented before the San Francisco Medico-Chirurgical Society, March 5, 1894. *Occidental Med Times*, **8**, 326.

Rixford, E. and Gilchrist, T.C. 1896. Two cases of a protozoan (coccidioidal) infection of the skin and other organs. *Johns Hopkins Hosp Rep*, **1**, 209–64.

Rodenbiker, H.T., Ganley, J.P., et al. 1981. Prevalence of chorioretinal scars associated with coccidioidomycosis. *Arch Ophthalmal*, **99**, 71–5.

Schneider, E., Hajjeh, R.A., et al. 1997. A coccidioidomycosis outbreak following the Northridge, California, earthquake. *JAMA*, **277**, 904–8.

Segal, G.P., Lehrer, R.I. and Selsted, M.E. 1985. In vitro effect of phagocyte cationic peptides on *Coccidioides immitis*. *J Infect Dis*, **151**, 890–4.

Shubitz, L., Peng, T., et al. 2002. Protection of mice against *Coccidioides immitis* intranasal infection by vaccination with recombinant antigen 2/PRA. *Infect Immun*, **70**, 3287–9.

Sigler, L. and Carmichael, J.W. 1976. Taxonomy of *Malbranchea* and some other hyphomycetes with arthroconidia. *Mycotoxon*, **4**, 349–488.

Slagle, D.C., Cox, R.A. and Kuraganti, U. 1989. Induction of tumor necrosis factor alpha by spherules of *Coccidioides immitis*. *Infect Immun*, **57**, 1916–21.

Smith, C.E., Beard, R.R., et al. 1946a. Effect of season and dust control on coccidioidomycosis. *JAMA*, **132**, 833–8.

Smith, C.E., Beard, R.R., et al. 1946b. Varieties of coccidioidal infection in relation to the epidemiology and control of the disease. *Am J Pub Health*, **36**, 1394–402.

Smith, C.E., Whiting, E.G., et al. 1948. The use of coccidioidin. *Am Rev Tuberc*, **57**, 330–60.

Smith, C.E., Saito, M.T. and Simons, S.A. 1956. Pattern of 39,500 serologic tests in coccidioidomycosis. *JAMA*, **160**, 546–52.

Standard, P.G. and Kaufman, L. 1977. Immunological procedure for the rapid and specific identification of *Coccidioides immitis* cultures. *J Clin Microbiol*, **5**, 144–53.

Stevens, D.A. 1995. Coccidioidomycosis. *N Engl J Med*, **322**, 1077–82.

Swatek, F.E. 1970. Ecology of *Coccidioides immitis*. *Mycopathol Mycol Appl*, **40**, 3–12.

Symmers, W.StC. 1967. Cases of coccidioidomycosis seen in Britain. In: Ajello, L. (ed.), *Coccidiomycosis*. Tucson, AZ: University of Arizona, 301–5.

Terio, K.A., Stalis, I.H., et al. 2003. Coccidioidomycosis in Przewalski's horses (*Equus przewalski*). *J Zoo Wildl Med*, **34**, 339–45.

Thompson, K. 1969. Irrigation as a menace to health in California. A nineteenth century view. *Geogr Rev*, **59**, 195–214.

Tripathy, U., Yung, G.L., et al. 2002. Donor transfer of pulmonary coccidioidomycosis in lung transplantation. *Ann Thorac Surg*, **73**, 306–8.

Tucker, R.M., Denning, D.W., et al. 1990a. Itraconazole therapy for chronic coccidioidal meningitis. *Ann Intern Med*, **112**, 108–12.

Tucker, R.M., Galgiani, J.N., et al. 1990b. Treatment of coccidioidal meningitis with fluconazole. *Rev Infect Dis*, **12**, Suppl 3, S380–9.

Wegner, T.N., Reed, R.E., et al. 1972. Some evidence for the development of a phagocytic response by polymorphonuclear leukocytes recovered from the venous blood of dogs inoculated with *Coccidioides immitis* or vaccinated with an irradiated spherule vaccine. *Am Rev Respir Dis*, **105**, 845–9.

Weisman, I.M., Moreno, A.J., et al. 1976. Gastrointestinal dissemination of coccidioidomycosis. *Am J Gastroenterol*, **81**, 589–93.

Wilson, J.W., Smith, C.E. and Plunkett, O.A. 1953. Primary cutaneous coccidioidomycosis; the criteria for diagnosis and a report of a case. *Calif Med*, **79**, 233–9.

Winn, W.A. 1968. A long term study of 300 patients with cavitary-abscess lesions of the lung of coccidioidal origin. *Dis Chest*, **54**, Suppl 1, 268–72.

Wolf, A. and Pappagianis, D. 1981. Canine coccidioidomycosis treatment with a new antifungal agent: ketoconazole. *Calif Vet*, **35**, 25–7.

Wright, P.W., Pappagianis, D., et al. 2003. Donor-related coccidioidomycosis in organ transplant recipients. *Clin Infect Dis*, **37**, 1265–9.

Yuan, L. and Cole, G.T. 1987. Isolation and characterization of an extracellular proteinase of *Coccidioides immitis*. *Infect Immun*, **55**, 1970–8.

Zealear, D.S. and Winn, W.A. 1967. The neurosurgical approach in the treatment of coccidioidal meningitis. Report of ten cases. In: Ajello, L. (ed.), *Coccidioidomycosis*. Tucson, AZ: University of Arizona Press, 43–53.

Zhu, Y., Yang, C., et al. 1996. Molecular cloning and characterization of *Coccidioides immitis* antigen 2-cDNA. *Infect Immun*, **64**, 2695–9.

Ziemer, E.L., Pappagianis, D., et al. 1992. Coccidioidomycosis in horses: 15 cases (1975–1984). *J Am Vet Med Assoc*, **201**, 910–16.

Zimmer, B.L. and Pappagianis, D. 1988. Characterization of a soluble protein of *Coccidioides immitis* with activity as an immunodiffusion-complement fixation antigen. *J Clin Microbiol*, **26**, 2250–6.

Zimmer, B.L. and Pappagianis, D. 1989. Immunoaffinity isolation and partial characterization of the *Coccidioides immitis* antigen detected by tube precipitin and immunodiffusion-tube precipitin tests. *J Clin Microbiol*, **27**, 1759–66.

Zimmerman, C.R., Snedker, C.J. and Pappagianis, D. 1994. Characterization of *Coccidioides immitis* isolates by restriction fragment length polymorphisms. *J Clin Microbiol*, **32**, 3040–2.

Zimmermann, C.R., Johnson, S.M., et al. 1996. Cloning and expression of the complement fixation antigen-chitinase of *Coccidioides immitis*. *Infect Immun*, **64**, 4967–75.

Zimmermann, C.R., Johnson, S.M., et al. 1998. Protection against lethal murine coccidioidomycoses by a soluble vaccine from spherules. *Infect Immun*, **66**, 2342–5.

27

Histoplasmosis

RAM P. TEWARI, L. JOSEPH WHEAT, AND L. AJELLO

Histoplasma capsulatum, like Caesar's Gaul, is tripartite. Taxonomically, it has been subdivided into three varieties, each with its own distinctive and defining characteristics. The varieties are validly referred to as *H. capsulatum* var. *capsulatum*, *H. capsulatum* var. *duboisii*, and *H. capsulatum* var. *farciminosum*. As might be expected, each of them causes a distinct type of histoplasmosis, designated respectively as histoplasmosis capsulati, histoplasmosis duboisii, and histoplasmosis farciminosi. They also have a varied geographical distribution. Rather than divide the genus *Histoplasma* into varieties and classes, Kasuga and colleagues propose seven phylogenetic species based on genetic isolation (Kasuga et al. 2003). The *capsulatum* variety is cosmopolitan, being endemic in all the world's continents. The *duboisii* variety has the most limited areas of endemicity, being restricted essentially to the central part of the African continent between the tropics of Cancer and Capricorn, and to the offshore island of Madagascar in the Indian Ocean. *H. capsulatum* var. *farciminosum* is not known to occur or to have been introduced anywhere in the New World. Its endemic areas are currently located in Africa, eastern Europe, the Middle East, Asia (including Japan and the Philippines), and the Far East. All three varieties occur as saprophytic, mitosporic molds (Fungi imperfecti) in their ecological niches in nature or in the laboratory cultures at room temperature (about 25°C). In the tissue of their mammalian hosts, however, or when incubated at 37°C on rich media, they transform into unicellular yeast-like budding organisms.

ETIOLOGIC AGENTS: HISTORICAL BACKGROUND AND NATURAL HISTORY

Histoplasma capsulatum var. capsulatum – Darling 1906

The genus *Histoplasma* was established in 1906 when Darling (1906) described the first case of histoplasmosis capsulati in Panama and named its etiologic agent *H. capsulatum*. He was under the impression that this organism was a protozoan closely related to the species of *Leishmania*, but 'the differences are so marked and the lesions so unusual that I feel the case is a unique one.' The mycotic nature of this 'protozoan' was inferred by da Rocha-Lima of Brazil, who, while in Germany, compared histological sections from Darling's Panamanian human case with sections from east African cases of equine epizootic lymphangitis (histoplasmosis farciminosi) as well as visceral leishmaniasis (kala-azar). The equine cases (*Equus asinus* × *caballus* and *E. caballus*) were known to be caused by the fungus *Cryptococcus farciminosus*, now designated *H. capsulatum* var. *farciminosum*. As a result of these comparative histological studies, da Rocha-Lima concluded that Darling's *H. capsulatum* was a fungus and not a protozoan, based on the intracellular occurrence of the unicellular tissue forms of the organisms and similarities in their morphology and tinctorial reactions (da Rocha-Lima 1912, 1912–13).

It remained for De Monbreun (1934) of Vanderbilt University in Nashville, TN to isolate the *capsulatum* variety of *H. capsulatum* for the first time. He clearly described and depicted the in vitro mycelial and unicellular forms of this mold and proved that it was dimorphic. He also reproduced the critical clinical and histopathological features of Darling's histoplasmosis capsulati in two monkeys (*Macacus rhesus*) after intravenous injection of the yeast form of his historic isolate.

Development of the histoplasmin skin test also was an important milestone in our understanding of histoplasmosis (Palmer 1945). These studies provided the initial evidence that histoplasmosis often caused a benign and usually asymptomatic infection characterized by pulmonary calcification. Subsequently this approach established our understanding of the geographical distribution of histoplasmosis in the USA (Edwards et al. 1969).

The astuteness of De Monbreun as an investigator was revealed when he stated that 'the saprophytic form of *Histoplasma capsulatum* probably exists free in nature.' This probability turned into reality in 1949 when Emmons of the National Institutes of Health in Bethesda, MA reported the isolation from two soil samples collected under the edge of an outbuilding on a Virginia farm where histoplasmosis capsulati had been proven in 7 of 43 trapped brown rats (*Rattus norvegicus*) (Emmons et al. 1947).

The geophilism of *H. capsulatum* var. *capsulatum* and its growth as a saprophyte in soil were irrefutably established by Emmons when he detected and photographed the diagnostically important tuberculate macroconidia of that fungus in the supernatant of suspensions of soil specimens that had been proven to be positive for that mold by passage through laboratory mice (Emmons 1949).

The saprophytic existence of *H. capsulatum* in avian and chiropteran habitats was established respectively by field studies (Zeidberg et al. 1952; Emmons 1959). Although birds are not known to be infected by the *capsulatum* variety (Mengis and Habermann 1955), transitory infections in chickens (*Gallus gallus*) and pigeons (*Columba livia*) were, however, reported by Schwarz et al. (1957) when their birds were inoculated intravenously with 500 000 yeast-form cells/ml of the *capsulatum* variety – a highly improbable type of challenge to occur in nature. The innate resistance of birds, perhaps, may be attributed to their normally high body temperatures of 42–43°C.

Bats of a wide variety of genera (24) and species (33) have been found to be infected by *H. capsulatum* var. *capsulatum* (Schacklette et al. 1962; Ajello et al. 1995, 1996) and one species (*Nyceteris hispida*) by the *duboisii* variety in Nigeria (Gugnani et al. 1994).

Among humans, isolated cases and outbreaks of histoplasmosis capsulati almost invariably can be traced to sites associated with accumulations of bird and bat guanos in many parts of the world (Ajello 1964; Sacks et al. 1986). The *capsulatum* variety of *H. capsulatum* thrives in such soils, especially in caves, chicken coops with dirt floors, attics, and the hollow walls of homes, industrial plants, schools, church steeples, etc., and the roosts of gregarious birds, such as blackbirds, oil birds, pigeons, seagulls, and starlings (Ajello 1964; Smith and Furcolow 1964; Dean et al. 1978).

Disturbances of such sites create aerosols laden with infectious propagules of *H. capsulatum* var. *capsulatum*. When inhaled, these aerosols result in infections varying in degrees of severity depending on the site of the inoculum and the immunological status of the individuals involved.

A possible explanation for the predilection of the *capsulatum* variety for avian bat habitats was suggested by Vining and Weeks (1974). These investigators found significant chemical differences between soils positive and negative for this fungus. The positive soils were richer in nitrogen and phosphorus as well as organic matter and water-holding capacity. Previously, Smith and Furcolow (1964) had found that growth-stimulating substances were present in infusions of starling feces.

Knowing the relationship between avian and chiropteran habitats and the *capsulatum* variety of *H. capsulatum*, decontamination procedures for clearing sites found to be infested by that fungus have been developed (Smith et al. 1964; Tosh et al. 1967). Formalin at a concentration of 3 percent was found to be the most effective agent. Application procedures and costs for decontaminating a 3.23 hectare (7 acre) bird roost site were presented by Bartlett et al. (1982). The total cost of this project was $US75 000.

For a long time *H. capsulatum* was considered to be an asexual mold classified in the Fungi imperfecti or Deuteromycetes grouping where all fungi without known sexual states were classified under a system based on spore groups developed by Saccardo (1899). However, in 1972, Kwon-Chung discovered that the variety *capsulatum* of *H. capsulatum* was heterothallic and described its perfect state. It proved to be an ascomycete and was named *Emmonsiella capsulata* (Kwon-Chung 1972). Isolates of the *capsulatum* variety were found to be one or the other of two mating types designated as (a) and (b) or (+) or (−). When the opposite mating types are paired on appropriate media, cleistothecia are formed that are loosely composed of a core group of ascogenous hyphae surrounded by tightly coiled hyphae. When mature, the cleistothecia contain club-shaped asci ($3–5 \times 10–16$ μm), each of which contains eight globose ascospores measuring 1.5 μm in diameter. Four of the eight ascospores will turn out to be of the (+) mating type and four of the (−) mating type (Kwon-Chung and Bennett 1992, pp. 497–8).

Kwon-Chung (1975) investigated the genetic relationship of the *capsulatum* variety to the *duboisii* variety. It was thus determined that isolates of the *duboisii* variety successfully mated with the appropriate opposite mating types of the *capsulatum* variety giving rise to a teleomorph identical to *E. capsulata*.

The ascospores from the cleistothecia of the *duboisii* matings did not germinate. When injected into mice, however, the ascospores proved to be pathogenic, converting into yeast-form budding cells. The tissues from the experimentally infected mice, when cultured, gave rise to colonies typical of *Histoplasma* (Kwon-Chung and Hill 1981). In 1986, Vincent et al. demonstrated that the restriction enzyme digest pattern of mitochondrial DNA from the *capsulatum* and *duboisii* varieties supported their varietal status.

In 1979, McGinnis and Katz compared herbarium specimens of *Ajellomyces dermatitidis* and *E. capsulata* and concluded that the differences between these two genera were too insignificant to warrant maintenance of the genera as separate entities. As a result, *E. capsulata* was reduced to the status of a synonym of *A. capsulatus*. The species of *Ajellomyces* are classified in the order Onygenales, family *Onygenaceae* of the phylum Ascomycota (Sigler 1993).

Histoplasma capsulatum var. *duboisii* – Vanbreuseghem (1952) – Ciferri 1960

Although much is known about the natural history of the *capsulatum* variety, relatively little is known about the *duboisii* variety in respect to its niche in nature. This variety was first isolated in culture and identified as the cause of a new form of histoplasmosis by Duncan in 1943. He did not, however, publish his findings until 1947 and 1958 (Duncan 1947, 1958). In the meantime, Vanbreuseghem, in a case report by Dubois et al (1952), described the agent as a new species of *Histoplasma* and named it *Histoplasma duboisii*. Duncan, in his 1958 paper, fully described the clinical, histological, and mycological features of the disease that had been contracted by an English mining engineer who had worked for 6 years in the Gold Coast Colony now known as Ghana. Duncan found that the mycelial form of his African isolate was indistinguishable from that of *H. capsulatum*. It was recognizable as being different by the large size of the tissue forms of the pathogen, which measured 8–12 μm and up to 15 μm in the longer axis. These dimensions were in striking contrast to the in vivo measurements of the tissue-form cells of the *capsulatum* variety, which predominantly range in size from 3 to 4 μm in diameter. It is as well to recall, however, that some isolates of the *capsulatum* variety also produce large yeast-like cells of 6–17 μm in diameter in vitro and in vivo, as emphasized by Drouhet and Schwarz (1956). As a matter of fact, these two investigators stated that 'the great variability found in the strains studies does not seem to indicate justification of assembling the African strains with large cells in a distinct species.' Such an opinion was shared by Ciferri in 1960, who named it *H. capsulatum* var. *duboisii*.

Four years later, the disease term 'histoplasmosis duboisii' was coined by Cockshott and Lucas (1964).

They succinctly stated that the 'American isolates and African isolates have negligible differences between their mycelial phases but distinct morphological differences in their parasitic stages whether naturally acquired or experimentally induced infections.' In 1975, Kwon-Chung solidly established the validity of the status of *H. duboisii* as a variety of *H. capsulatum* by successfully mating the tester strains of *A. capsulatus* (as *E. capsulata*), the teleomorphic state of the *capsulatum* variety, with varieties of the *duboisii* variety.

In the disseminated cases of histoplasmosis duboisii, a characteristic clinical picture is presented, revealing bone and subcutaneous abscesses and pleomorphic skin lesions. A definitive diagnosis rests on the demonstration of large, ovoid, budding yeast cells in pus exudate from granulomas and other clinical materials.

The *duboisii* variety of *H. capsulatum* can be readily isolated from clinical material on a variety of media. In its mycelial form grown at 25°C, the isolates cannot be distinguished from the *capsulatum* variety as stated previously. Only by converting the mycelial isolates to their yeast form at 37°C on blood or brain–heart infusion agars can the *duboisii* variety be identified. The diagnostic large-form unicellular cells (12–15 μm in diameter) are formed in significant numbers only in cultures held for several weeks at 37°C.

It has long been surmised that the *duboisii* variety was geophilic, but its isolation from soil was not achieved until 1994. In that year, Gugnani et al. reported the first isolation of that variety from eight of 45 soil specimens collected in an eastern Nigerian bat cave in the state of Anambra. It is noteworthy that one of the eight isolates was recovered by direct plating of a 10^{-5} dilution of a soil suspension on an unspecified agar plate (either Sabouraud's dextrose agar or yeast extract agar supplemented with antibiotics).

In addition, the duboisii variety was recovered from the intestinal contents of one of 35 bats comprising the species *Nycteris hispida* – the hairy slit-faced bat – and *Tadarida* (*Chaerephora*) *pumila* – the lesser mastiff bat. The positive bat was one of the hairy slit-faced bats that are found in most of Africa south of the Sahara (Nowak 1991). Thus, another nonhuman mammalian host of this variety has been added to those previously known: *Papio cynocephalus cynocephalus* – the yellow baboon (Curtois et al. 1955); *Papio cynocephalus anubis* – the olive baboon (Butler and Hubbard 1991); and *Papio cynocephalus papio* – the red baboon (Walker and Spooner 1960).

In a personal communication from P.K.C. Austwick (August 1980), one of us (L.A.) was informed that two aardvarks (*Oryceteropus capensis*), one in the London Zoo and another in the Copenhagen Zoo, were histologically confirmed to have lungs infected by *H. duboisii*. The aforementioned baboons and bat, as well as the soil burrowing aardvarks, may act and serve as sentinels for the presence of the *duboisii* variety in a given territory. This would parallel the role played by badgers (*Meles meles*) in Denmark (Jansen et al. 1992), Germany (Rapp et al. 1992; Grosse et al. 1997), and Switzerland (Burgisser et al. 1961) in revealing endemic foci of the *capsulatum* variety of *H. capsulatum* in these respective countries.

An epizootic of histoplasmosis duboisii in a Texan colony of 3 400 baboons casts a new light on the epidemiology of this disease (Butler et al. 1988; Butler and Hubbard 1991). Twenty-one cases of simian histoplasmosis duboisii developed over a period of 4 years in the primate colony in San Antonio, TX. Eight of the baboons had been born in the USA whereas of 13 wild-caught simians, six (*P. cynocephalus papio*) had been imported from Senegal, a west African country known to be endemic for *H. capsulatum* var. *duboisii*, and the remaining seven (four *P. cynocephalus cynocephalus* and three *P. cynocephalus anubis*) had been captured in Kenya, an east African country with no known endemic cases of histoplasmosis duboisii. The first infection was diagnosed in one of the Senegalese baboons, 18 months after entry into the USA. The Kenyan baboons developed their infections 10–12 years after being colonized in Texas.

In their 1991 report, Butler and Hubbard stated that the incubation period in their baboons was at least 9 months with a possible upper-end incubation period of 18 months. The most likely route of transmission from the Senegalese import was a direct contact brought about by the grooming habits of the baboons. This social habit involves licking of lesions or picking and ingesting material from lesions on other animals.

In humans, however, in analogy with the *capsulatum* variety, histoplasmosis duboisii is acquired via the respiratory route (Gatti and DeVroey 1991) and, as revealed by Gugnani et al. (1994), soil is the natural habitat occupied by this etiologic agent.

Histoplasma capsulatum var. *farciminosum* (Rivolta 1873) Weeks, Padhye et Ajello 1985

The *farciminosum* variety of *H. capsulatum* has the distinction of having been the first of the three varieties to have come to medical attention and to have been identified as the etiologic agent of a widespread endemic disease known variously as the farcy of Africa, the farcy of Naples and, most commonly, epizootic lymphangitis.

The disease had long been confused with glanders, a highly contagious disease of asses, horses, and mules caused by *Pseudomonas mallei*. In 1873, Rivolta of Pisa, Italy, noted the presence of yeast-like organisms in clinical material taken from horses suffering from epizootic lymphangitis, which he considered to be a species of *Cryptococcus*. The yeast cells noted in the pus taken from the lymph nodes of a horse were unicellular,

oval, and budding. In size, these cells measured 2.5–3.5×2.3 μm. Ten years later in 1883, he and Micellone named the organism *C. farciminosum* (Rivolta and Micellone 1883). However, it was not until 1895 that this organism was successfully isolated and cultured by Marcone of Naples. The isolate developed as a sterile mold in a medium composed of horse serum, glucose, glycerin, and sucrose (Marcone 1895).

It remained for Redaelli and Ciferri to recognize that the agent of epizootic lymphangitis was a dimorphic mold and that it was appropriate to transfer the organism to Darling's genus *Histoplasma* on the basis that in tissues the in vivo form cells of *C. farciminosum* were indistinguishable from those of the *capsulatum* variety of *H. capsulatum*. They accordingly transferred Rivolta's *C. farciminosum* to the genus *Histoplasma* as *H. farciminosum* (Ciferri and Redaelli 1934).

In 1985, in studying Egyptian isolates from horses, Weeks et al. induced the production of numerous microconidia and moderate numbers of tuberculate macroconidia. This was accomplished by incubating the cultures in the dark at 30°C for a period of 8–12 weeks on soil-extract agar. The smooth to rough-walled microconidia were pyriform to subglobose and measured 3–5×2.5–3.5 μm. The macroconidia were globose to subglobose (8–12×1–1.5 μm) with thick walls. For the first time, these investigators found that the mycelial cultures of *H. farciminosum* produced micro- and macroconidia that are characteristics of the members of the genus *Histoplasma*. In addition, the close resemblance of the yeast forms of the *capsulatum* and *farciminosum* varieties and the antigenic relationship shown to exist between them by Standard and Kaufman (1976) led to the conclusion that *H. farciminosum* was best treated as another variety of *H. capsulatum*.

Unlike the other two varieties of *Histoplasma*, authenticated cases of histoplasmosis farciminosi have only been known to occur in equines: asses (*Equus asinus*), horses (*E. caballus*), and mules (*E. asinus* × *caballus*). It is pertinent to note that only in the African continent do the three varieties of *H. capsulatum* coexist.

HOST RESPONSE

Human infection with *H. capsulatum* var. *capsulatum* is usually clinically unrecognized, identified only by skin test positivity or pulmonary or splenic calcifications. Symptomatic cases present with a variety of clinical manifestations ranging from an acute pulmonary disease, to a chronic pulmonary infection, or a progressive disseminated form. Among symptomatic cases with no underlying lung disease or immunosuppressive disorders, self-limiting illnesses develop in more than 90 percent of individuals. Chronic pulmonary histoplasmosis develops in those with underlying emphysema, and progressive disseminated disease in those at the extremes of age or who are immunosuppressed. In this section the main clinical manifestations of histoplasmosis capsulati are examined.

Clinical forms

ACUTE SELF-LIMITED HISTOPLASMOSIS CAPSULATI

After low-inoculum exposure, the infection is asymptomatic in at least 90 percent of cases (Wheat and Kauffman 2003). Among symptomatic patients, the most common clinical presentation is acute pulmonary histoplasmosis characterized by fever, cough, and chest pain. Chest x-rays show mediastinal lymphadenopathy with infiltrates (Figure 27.1). Most patients recover in a few weeks, but some experience prolonged fatigue. Heavy exposure causes diffuse pulmonary involvement, which is often accompanied by respiratory failure (Kataria et al. 1981). Patients may also experience symptoms as a result of compression of the mediastinal structures by enlarged lymph nodes (Loyd et al. 1988). Patients with acute histoplasmosis may experience rheumatological syndromes characterized by arthritis or arthralgia, and erythema nodosum or erythema multiforme (Rosenthal et al. 1983). Pericarditis is another inflammatory complication of acute histoplasmosis, and both pericarditis and rheumatological syndromes occur in less than 10 percent of cases (Wheat et al. 1983). Rarely, constrictive pericarditis may develop.

CHRONIC PULMONARY HISTOPLASMOSIS CAPSULATI

Chronic pulmonary histoplasmosis capsulati occurs in individuals with underlying bullous lung disease and is characterized by recurrent pulmonary symptoms, progressive lung infiltrates, fibrosis, and cavitation (Figure 27.2) (Goodwin et al. 1976; Wheat et al. 1984). Most patients experience recurrent illnesses highlighted by progressive radiological worsening and respiratory deterioration. Chest x-rays show cavity enlargement, formation of new cavities, and spread to new areas of the lungs. Rarely, bronchopleural fistula or hematogenous dissemination may develop.

PROGRESSIVE DISSEMINATED HISTOPLASMOSIS CAPSULATI

Hematogenous spread outside the lungs occurs during the acute infection but is rarely recognized clinically. These patients recover with the development of varying degrees of immunity to the *capsulatum* variety of *H. capsulatum*. Progressive disseminated disease occurs in about one in 2 000 acute infections (Sathapatayavongs et al. 1983), mostly in patients who are immunosuppressed or at the extremes of age (younger children and older adults). Common clinical manifestations include fever, weight loss, and respiratory symptoms. Physical examination frequently reveals hepatomegaly and/or splenomegaly. Laboratory tests usually show anemia, leukopenia, thrombocytopenia, and elevated hepatic enzymes and bilirubin. Shock with hepatic, renal, and

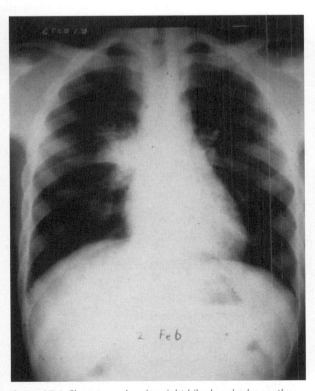

Figure 27.1 *Chest x-ray showing right hilar lymphadenopathy with adjacent infiltrate in acute pulmonary histoplasmosis capsulati*

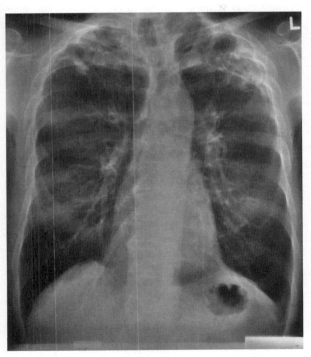

Figure 27.2 *Chest x-ray showing bilateral upper lobe fibrotic infiltrates in a patient with underlying emphysema complicated by chronic pulmonary histoplasmosis capsulati*

respiratory failure and coagulopathy may complicate severe cases (Wheat et al. 1990b). Meningitis, cerebritis, and focal brain or spinal cord lesions occur in 10–20 percent of cases, as either manifestations of widely disseminated infection or isolated findings (Wheat et al. 1990a). Other frequent sites of dissemination include the oral mucosa, gastrointestinal tract, skin, kidneys, and adrenal glands (Figure 27.3). Endocarditis is a rare manifestation of disseminated histoplasmosis. Chest x-rays usually show diffuse infiltrates but are normal in 25–33 percent of cases (Figure 27.4).

MEDIASTINAL FIBROSIS

Mediastinal fibrosis represents an abnormal fibrotic response to past infection and is a rare finding (Goodwin et al. 1972; Loyd et al. 1988; Davis et al. 2001). Recurrent hemoptysis is a common symptom and respiratory failure often ensues. Serological tests may be positive and staining of tissues may show the small tissue-form cells of the *capsulatum* variety of *H. capsulatum* in half to two-thirds of cases, although cultures are usually negative, supporting the belief that fibrosing mediastinitis represents an aberrant inflammatory response rather than a progressive infection. The superior vena cava, airways, pulmonary arteries and veins, and the esophagus are most commonly involved, but any mediastinal structure can be trapped in these fibrotic masses (Goodwin et al. 1972; Loyd et al. 1988). Chest x-rays may be normal or show only mediastinal widening, but computed tomography (CT) reveals restriction of mediastinal structures. Pulmonary angiograms may show obstruction of the pulmonary arteries (Figure 27.5). Calcification is often present. The diagnosis requires demonstration of obstruction of mediastinal structures by CT scan, and cannot be made based on the presence of fibrosis histopathologically in biopsy specimens, as this finding is not specific for fibrosing mediastinitis.

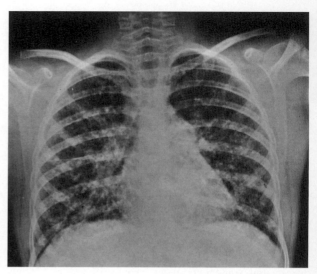

Figure 27.4 *Chest x-ray showing diffuse miliary infiltrate in a patient with disseminated histoplasmosis capsulati*

HISTOPLASMOMA AND BRONCHOLITHIASIS

Rarely, patients may develop a slowly enlarging pulmonary nodule, which has been called an 'enlarging histoplasmoma' (Goodwin and Snell 1969), causing concern about neoplasia. Lesions range in diameter from 8 to 35 mm and enlarge slowly (2 mm/year). Histologically, they are characterized by a necrotic center surrounded by a fibrous-like capsule. Organisms may be seen in the necrotic center but usually cannot be isolated

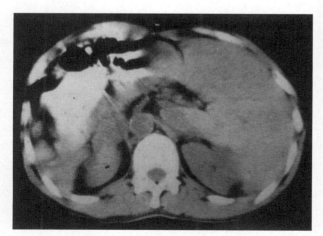

Figure 27.3 *Bilateral adrenal enlargement caused by disseminated histoplasmosis established by fine needle aspiration biopsy showing organisms consistent with* H. capsulatum *var.* capsulatum

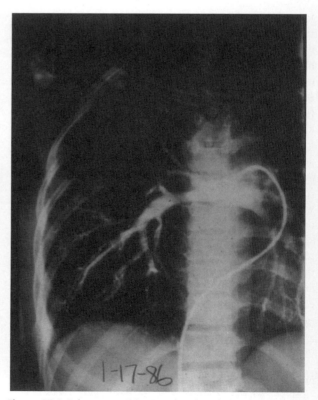

Figure 27.5 *Pulmonary angiogram showing obstruction of right pulmonary artery in a patient with fibrosing mediastinitis*

in culture. As these lesions or mediastinal nodes calcify, they may erode into adjacent bronchi, causing broncholithiasis (Arrigoni et al. 1971). Patients may experience recurrent hemoptysis and expectorate small gravel-like particles of tissue. Bronchial obstruction or tracheoesophageal fistula may also complicate broncholithiasis.

PRESUMED OCULAR HISTOPLASMOSIS CAPSULATI

This distinct clinical entity is usually seen in histoplasmin skin test-positive individuals residing in endemic areas for histoplasmosis capsulati, but scientific evidence that *H. capsulatum* var. *capsulatum* causes this syndrome is not very convincing (Spaeth 1971; Ciulla et al. 2001). However, the eye may also be involved in patients with disseminated histoplasmosis (Specht et al. 1991). Ocular histoplasmosis and presumed ocular histoplasmosis are discussed in Chapter 16, Oculomycosis.

Immunology

Histoplasma capsulatum var. *capsulatum* can be a primary pathogen producing systemic disease in normal individuals or opportunistic infection in immunocompromised hosts. The infection is acquired by inhalation of the infectious propagules of *H. capsulatum* into the respiratory tract. The conversion from mycelial to yeast form is critical because it is the yeast form in which *H. capsulatum* var. *capsulatum* exists as a facultative intracellular parasite of the mononuclear cells in the susceptible host (Schwarz 1981). The virulence factor(s) and the mechanisms(s) by which *H. capsulatum* causes disease are not clearly defined. The disease process may involve an endotoxin-like substance (Salvin 1952), collagenases and elastases (Okeke and Miller 1991), siderophores (Burt 1982), and immunological mechanisms: complement-mediated/hypersensitivity and cytokines may contribute to pathogenesis (review by Howard 1981a; Lane et al. 1994; Zhou et al. 1995). In addition, the ability of live *H. capsulatum* var. *capsulatum* yeast cells to neutralize the acidic environment (pH 5.0) of phagolysosomes, enabling them to survive within macrophages, contributes to the virulence of the organism (Eissenberg et al. 1993). The host's ability successfully to mount a series of complex but interrelated cellular responses in the lung is responsible for the recovery and possible development of resistance to reinfection. The cells of the immune system 'talk' to each other by direct cell–cell interactions, cytokine production, and associated signal transduction. The salient features of the natural (innate) and acquired immunity are discussed in the next section.

NATURAL (INNATE) IMMUNITY

Although the precise initial events in the interaction between the previously uninfected host and infective conidia are not well defined, the innate cellular resistance to histoplasmosis capsulati is mediated by polymorphonuclear leukocytes (PML), macrophages, and natural killer (NK) cells. The inhaled infectious units (conidia and mycelial fragments) of *H. capsulatum* var. *capsulatum* lodge within the terminal bronchioles and alveoli. The organism changes from mycelial to yeast-form cells, the form in which the *capsulatum* variety of *H. capsulatum* exists as a facultative intracellular parasite in the susceptible host (Schwarz 1981). The initial cellular infiltrate is composed predominantly of PMLs, which changes to a predominance of macrophages and lymphocytes with progression of the inflammatory response. The microconidia and yeast cells of the invading pathogen bind to the PMLs via the CD11/CD18 family of integrins (adhesion-promoting glycoproteins) without prior opsonization (Schnur and Newman 1990). After binding, the ingestion of *H. capsulatum* var. *capsulatum* is rapid and maximum phagocytosis occurs within 1 hour. The PMLs from guinea pigs and mice exhibit significant fungicidal activity to both the conidia and yeast cells of *H. capsulatum* in vitro (review by Howard 1981a; Kondoh et al. 1989). The killing of conidia and yeast cells is mediated by the combined effect of hydrogen peroxide, halides, and myeloperoxidase (Howard 1981b).

Although there is some controversy regarding their fungicidal activity, human PMLs exert strong fungistasis against the *capsulatum* variety of *H. capsulatum* (Schnur and Newman 1990; Brummer et al. 1991). Phagocytosis of opsonized yeast cells by PMLs stimulates the respiratory burst with intracellular production of H_2O_2 and O_2^-. This phenomenon appears to be unique to PML–*H. capsulatum* interaction, because O_2^- is not released by monocytes upon phagocytosis of the yeast cells. These findings suggest that PMLs may be more important in early host defenses against *H. capsulatum* than previously recognized.

Normal macrophages recognize, bind, and phagocytose *H. capsulatum* conidia and yeast cells, thereby providing a permissive environment for multiplication of the organism in the yeast form (Figure 27.6). The yeast cells attach to human macrophages via the CD11/CD18 family of integrins – the same receptors by which they bind to PMLs (Schnur and Newman 1990). Once ingested, the *capsulatum* variety of *H. capsulatum* yeast cells confront a hostile environment, which occurs by stimulation of the respiratory burst and phagolysosomal fusion, although the yeast cells survive and grow within normal human and murine macrophages (Figure 27.6). How *H. capsulatum* survives within macrophages is not known. However, the ability of live yeast cells to neutralize the acidic environment (pH 5.0) required for the fungicidal activity of many lysosomal enzymes has been suggested as a possible mechanism (Eissenberg et al. 1993).

There is considerable evidence to indicate that NK cells participate in early host defenses against *H. capsulatum*. var. *capsulatum*. NK cells from lungs and spleens

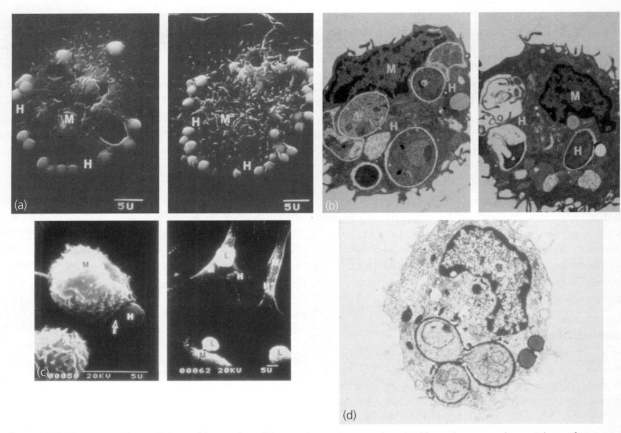

Figure 27.6 *Ultrastructural morphology of interaction of* H. capsulatum *var.* capsulatum *with murine macrophages 1 hour after incubation at 37°C.* **(a)** *Scanning electron micrograph (SEM) of normal (left; magnification ×19 500) immune (right; magnification ×14 300) macrophages showing phagocytosed yeast cells. Note the activated macrophage with ruffled membranes which has ingested more yeast cells.* **(b)** *Transmission electron micrograph (TEM) of normal (left) and immune (right) macrophages with phagocytosed yeast cells. Note the activated macrophage has degraded yeasts (magnification ×62 000).* **(c)** *SEM of the attachment of lymphocytes (L) to peritoneal macrophages (M) from immune mice (left). Two macrophages also contain phagocytosed histoplasma yeast cells (H) (magnification ×10 250) SEM of an activated murine macrophage (M) engulfing yeast cells of* H. capsulatum *(H) (right; magnification ×25 000).* **(d)** *TEM of a macrophage with ingested yeast cells, showing positive acid phosphatase staining (dark areas) around ingested cells indicating phagosome–lysosome fusion (magnification ×62 000)*

of normal mice are cytotoxic to *H. capsulatum* yeast cells in vitro (Yamada et al. 1982; Raman et al. 1989b). The cytotoxicity is enhanced by addition of anti-histoplasma antibody to NK cell–yeast cell cultures, demonstrating antibody-dependent cellular cytotoxicity (ADCC) by NK cells to *H. capsulatum*. Likewise, NK cells isolated from human peripheral blood of both histoplasmin-positive and histoplasmin-negative subjects exhibit natural as well as ADCC to *H. capsulatum* (Tewari et al. 1981; Choudhury et al. 1989). Although the mechanism of NK cell cytotoxicity is not known, the NK cells bind to *H. capsulatum* by microvilli (Figure 27.7), in a similar way to that previously described for binding of NK cells to tumor cell targets (Frey et al. 1982). The binding is followed by exocytosis of NK cell granules and the cytotoxicity is presumably expressed by associated signal transduction.

Additional evidence for involvement of NK cells in innate resistance to histoplasmosis capsulati is provided by NK cell depletion and adoptive transfer experiments. Depletion of NK cells by administration of anti-NK cell

antibody (anti-asialo GM1) enhanced susceptibility of mice to histoplasmosis as assessed by challenge with *H. capsulatum* yeast cells. (Raman et al. 1989a). Furthermore, adoptive transfer of NK cells to cyclophosphamide-treated mice (NK cell deficient) restored the in vitro NK cell activity of their lung and spleen cells and protected them from a subsequent challenge with *H. capsulatum* (Raman et al. 1990).

The yeast-form cells of *H. capsulatum* var. *capsulatum* have a predilection for the reticuloendothelial system and, in fact, the term 'reticuloendotheliosis' was formerly used as a synonym for histoplasmosis capsulati. It is noteworthy that yeast cells were seen in the endothelial cells lining blood and lymph vessels and in alveolar lining of the lungs by Darling (1906), in the first reported human case of histoplasmosis in Panama (review by Schwarz 1981). In an attempt to elucidate the mechanism involved in extrapulmonary dissemination of the organism, an in vitro experimental model has been developed to study the interaction of *H. capsulatum* var. *capsulatum* with human umbilical vein endothelial cells,

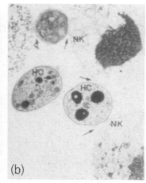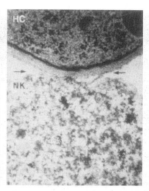

Figure 27.7 *Interaction of* H. capsulatum *var.* capsulatum *with splenic NK cells from mice in culture.* **(a)** *SEM showing binding of two NK cells to a budding yeast cell (left) and one NK cell to two budding yeasts (right) (magnification ×25 000).* **(b)** *Binding of an NK cell to the cell wall of a yeast cell (left; magnification ×77 000) and conjugate formation between NK cells and yeast cells. Note degranulation of NK cells (right; magnification ×51 300)*

mouse pulmonary cells, and mouse heart endothelial cells, in monolayer cell cultures (Bedrosian et al. 1986; Tewari et al. 1986; Hughes et al. 1994). The endothelial cells maintain a cobblestone arrangement and other characteristics described for endothelial cells, but are much larger than macrophages. Both human and mouse endothelial cells adhere to, phagocytose, and kill yeast cells of *H. capsulatum* (Figure 27.8). Unlike macrophages, where maximum phagocytosis is observed within 1 hour, optimum endothelial cell phagocytosis requires 4–6 hours. However, only 25–30 percent of cells in monolayer cultures react with *H. capsulatum*, suggesting a heterogeneous nature for endothelial cells. Ultrastructural studies provide supporting evidence for morphological degradation and killing of yeast cells (Figure 27.8). These findings indicate that endothelial cells may have a critical role in host defenses to histoplasmosis capsulati.

The role of humoral factors in natural resistance to histoplasmosis is not clearly defined. Nevertheless, heat-labile and/or heat-stable serum opsonins enhance phagocytosis of *H. capsulatum* var. *capsulatum* yeast cells by PMLs, but not by macrophages (review by Deepe and Bullock 1992). Furthermore, normal serum inhibits growth of *H. capsulatum* yeast cells and the inhibition is attributed to transferrin, which binds iron and makes it unavailable for nutritional requirement of the organism (Sutcliffe et al. 1980).

ACQUIRED IMMUNITY

H. capsulatum var. *capsulatum* is a highly immunogenic mold and acquired immunity is expressed by induction of a spectrum of cellular as well as humoral immune responses. However, it is generally accepted that the primary host defense mechanism activated in response to *H. capsulatum* is the cell-mediated arm of the immune response. Several lines of evidence support this concept:

- Congenitally athymic (nude) mice lacking T cells are more susceptible to *H. capsulatum* var. *capsulatum* infection (Williams et al. 1978; Miyaji et al. 1981).

- Lymphokines from immune mice or IFN-γ can activate macrophages for enhanced antifungal activity to *H. capsulatum* (Wu-Hsieh and Howard 1984, 1987).

- Adoptive transfer of T cells but not B cells or serum from immunized mice protects naive animals against a lethal challenge with *H. capsulatum* (Tewari et al. 1977, 1978; Williams et al. 1981).

- Transfer of *H. capsulatum*-reactive CD4 cells and to a lesser extent that of CD8 cells confers immunity to naive mice against histoplasma yeast cells (Allendoerfer et al. 1993; Deepe 1994).

- Immunized mice exhibit a delayed-type hypersensitivity (DTH) response to histoplasmin, and lymphocytes proliferate in response to histoplasmin in culture (Tewari et al. 1982; Khardori et al. 1986; Sharma et al. 1986).

How T cells contribute to protective immunity is not well understood. The most logical mechanism by which T cells augment the protective immune response is by release of macrophage-activating lymphokines which, in turn, arm the macrophages for more efficient antifungal activity. Although T cells can produce a number of regulatory cytokines, IFN-γ is the only cytokine that has been shown to arm murine macrophages for antifungal activity against *H. capsulatum* (Wu-Hsieh and Howard 1987). Abrogation of protective immunity by administration of anti-IFN-γ antibody to mice (Zhou et al. 1995) and failure of IFN-γ to knock out mice to control *H. capsulatum* infection (Wu-Hsieh and Howard 1984) further support the role of IFN-γ in histoplasmosis capsulati. In addition to CD4 cells, CD8 and NK cells produce IFN-γ and thus could potentially produce IFN-γ in *H. capsulatum* infection. However, co-culturing of human monocyte-derived macrophages with IFN-γ does not enhance their anti-*H. capsulatum* activity (Fleishmann et al. 1990), but they are activated by interleukin 3 (IL-3), granulocyte–macrophage colony-stimulating factor (GM-CSF), and macrophage colony-stimulating factor (M-CSF) (Newman and

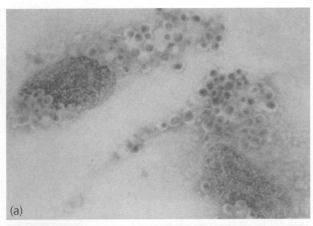

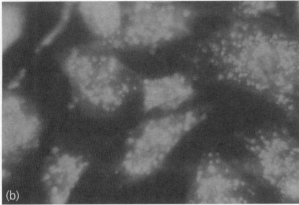

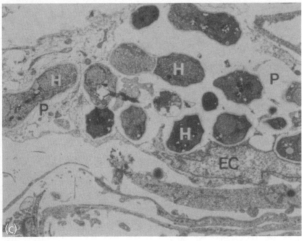

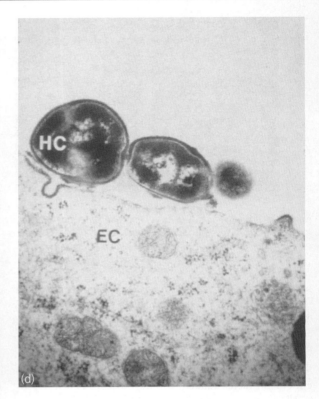

FIG203.8 **Figure 27.8** *Interaction of human umbilical cord endothelial cells (EC) with* H. capsulatum *var.* capsulatum *(HCH) in culture.* **(a)** *An endothelial cell monolayer, 4 hours after incubation with histoplasma yeast cells. Note phagocytosis of yeast cells by EC (Giemsa stain; magnification ×3 900).* **(b)** *Acridine orange-stained EC monolayer showing numerous orange-colored lysosomes (magnification ×3 400).* **(c)** *TEM of a phagocytic EC with numerous H yeast cells in phagosomes (P). Note intact as well as degraded HC within phagosomes (magnification ×12 000).* **(d)** *TEM of an adherent HC yeast cell to an EC. Note morphological alteration in HC even before ingestion (magnification ×34 000)*

Administration of interleukin 12 (IL-12) or neutralizing antibody to interleukin 4 (IL-4) has a protective effect on *H. capsulatum* infection in mice, demonstrating the antagonistic role of these cytokines (Zhou et al. 1995). Treatment of *H. capsulatum*-infected mice with neutralizing antibodies to IL-12, IFN-γ or tumor necrosis factor α (TNF-α), enhances the severity of the disease. Recently treatment with tumor necrosis factor (TNF) inhibitors has been shown to predispose to severe histoplasmosis in man (Wood et al. 2003). Moreover, IL-12 treatment increases the endogenous production of IFN-γ in infected mice, suggesting that the protective effect of IL-12 is primarily mediated through production of IFN-γ. Recently, Zhou et al. (1997) also demonstrated the protective effect of IL-12 against histoplasmosis capsulati in immunodeficient severe combined immunodeficiency disease (SCID) mice. The effect was further increased by combining IL-12 with amphotericin B. The protective effect is associated with increased production of IFN-γ, TNF-α, and nitric oxide from spleen cells.

Macrophages from immunized mice or macrophages activated in vitro have ruffled membranes with high affinity surface receptors and are far more phagocytic than normal macrophages (see Figure 27.6). The activated macrophages also have more lysosomal enzymes and the ability to produce more reactive nitrogen and oxygen intermediates (RNI-NO, NO_2^-, and ROI-H_2O_2, O_2^-). However, it is generally considered that activated macrophages exert fungistatic but not fungicidal activity against the *capsulatum* variety of *H. capsulatum*,

Gootee 1992), indicating differences in signaling mechanisms for macrophage activation between the two species.

although ultrastructural studies indicate morphological degradation of the ingested organisms. Neutralization of the acidic environment of phagolysosomes (pH 5.0) has been proposed as a possible mechanism for survival of *H. capsulatum* within macrophages (Eissenberg et al. 1993).

The opsonization of microconidia and yeast cells with anti-*H. capsulatum* antibodies increases the phagocytic activity of PMLs, but the enhancing effect of antibodies is not observed in phagocytosis of the organism by macrophages (review by Deepe and Bullock 1992). The antibody-dependent cellular cytotoxicity of both murine and human NK cells to *H. capsulatum* yeast cells, and thus antibodies, may have a contributory role in host defenses (Tewari et al. 1982; Choudhury et al. 1989).

The inhalation of a small number of *H. capsulatum* conidia, which frequently occurs in individuals living in an endemic area for histoplasmosis capsulati, results in asymptomatic or mild self-limiting infections with concomitant development of resistance to reinfection. This had also been observed in experimental animals in the early 1950s and provided the major rationale for development of a vaccine for histoplasmosis (review by Salvin 1965). The immunity induced by sublethal infection of experimental animals persisted longer than that induced by killed organisms or fractions thereof. However, only a low level of protection was observed in mice immunized with killed cells and the immunogenicity was attributed to cell wall constituents. Feit and Tewari (1974) demonstrated that immunization of mice with *H. capsulatum* yeast ribosomes or ribosomal proteins elicited a high degree of immunity, protecting up to 90 percent of animals from a lethal challenge. The immunity was comparable to that obtained by a sublethal infection with live yeast cells (Tewari 1975). Furthermore, a cytoplasmic membrane fraction or membrane proteins were as immunogenic as ribosomes (Tewari and LaFemina 1983; Tewari et al. 1989). Recently, Deepe and his associates isolated a glycoprotein (HIS-62) from the cell walls and cell membranes of *H. capsulatum* var. *capsulatum* yeast cells, cloned the gene encoding this antigen, and prepared a recombinant protein that was shown to be a member of the heat-shock protein Hsp60 family. Immunization of mice with the native as well as with the recombinant protein conferred significant protection against a lethal challenge with *H. capsulatum* yeast cells (Gomez et al. 1991, 1995). However, another protein (80 kDa) from *H. capsulatum* var. *capsulatum* with homology to Hsp70 did not provide significant protection. In view of the concern of using heat-shock proteins as a vaccine in humans, four recombinant polypeptides were prepared from Hsp60, but these recombinants did not confer significant protection against histoplasmosis capsulati (Deepe et al. 1996).

There is a strong correlation between protective immunity and delayed hypersensitivity in histoplasmosis.

Lymphocytes from immunized mice proliferate in response to histoplasmin and the proliferative response is used as an in vitro correlate of DTH responses (Tewari et al. 1982; Deepe et al. 1996). Furthermore, there is a strong correlation between footpad reactivity (DTH reaction) to histoplasmin and production of macrophage migration inhibition factor (MIF) when peritoneal exudate cells from immunized mice are cultured with histoplasmin (Tewari et al. 1989). It is noteworthy that both the DTH reaction and MIF production decayed faster than the protective immunity in immunized mice. The immunized mice showed footpad reactivity (DTH) responses to histoplasmin which peaked between 21 and 28 days after immunization, and thereafter showed a gradual decline, reaching the control level at 105 days (Khardori et al. 1986; Sharma et al. 1986). There was, however, no decline in protective immunity for up to 105 days, the maximum observation period used in these studies (Figure 27.9). It is noteworthy that protective immunity persisted longer than the DTH reaction and MIF production, demonstrating a dichotomy between the two manifestations of the immune response in histoplasmosis capsulati.

Treatment

Histoplasmosis capsulati is a systemic fungal infection with a wide spectrum of clinical manifestations in different individuals. The therapy of the disease has to be clinically determined after taking into consideration the immunocompetence of the host, site of infection, and the extent of the disease. Initially, amphotericin B was the only effective drug available for treatment. However, toxicity of the drug limited its use to patients with severe disease. The development of azoles such as ketoconazole and, more recently, itraconazole, which have lower toxicity and can be given orally, has allowed the clinician options for treatment of patients with less severe disease. The current recommendations for treatment are presented in this section.

ACUTE PULMONARY

Patients with extensive acute pulmonary histoplasmosis capsulati benefit from antifungal therapy and those who are hypoxic may also require adjunctive corticosteroid therapy (Table 27.1) (Wheat and Kauffman 2003). Patients with less extensive disease but who remain symptomatic for a month or more may also benefit from therapy. While the benefit of therapy for patients with more severe, diffuse disease seems to be clear, based upon the author's experience and the anecdotal reports of others, the effect in patients with localized disease and persistent symptoms is uncertain. Amphotericin B should be used in severely ill patients requiring hospitalization (Table 27.2). Itraconazole is highly effective in individuals with milder illnesses or as a follow-up

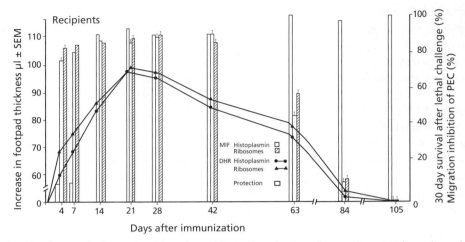

Figure 27.9 *Relationship of protective immunity, footpad reactivity, and production of macrophage MIF by peritoneal exudate cells from mice immunized by sublethal infection with yeast cells of* H. capsulatum *var.* capsulatum *(105 cells, s.c.). Mean ± SEM of five mice per time period for delayed hypersensitivity and MIF and 20 mice per group for protective immunity (10 mice/group per experiment)*

therapy after an initial response to amphotericin B (Wheat et al. 1995).

MEDIASTINAL GRANULOMA

Mediastinal granuloma may produce obstructive symptoms or fistula. Occasional patients with these complications improve after antifungal therapy (Savides et al. 1995), but some may require surgical resection (Gilliland et al. 1984). The benefit of therapy in such cases, however, is difficult to prove. Prophylactic antifungal therapy or surgery to prevent fibrosing mediastinitis is not indicated because progression of granulomatous to fibrosing mediastinitis is rare (Loyd et al. 1988).

RHEUMATOLOGICAL AND PERICARDITIS

Patients with these inflammatory manifestations may respond to aspirin or nonsteroidal anti-inflammatory agents, but also may require treatment with corticosteroids (Rosenthal et al. 1983; Wheat et al. 1983). Antifungal therapy is not required, but should be considered in those receiving corticosteroids at higher than physiological doses for more than 3 weeks, in order to reduce the risk for dissemination during corticosteroid-induced immunosuppression. Rarely, joints and the pericardium are sites of disseminated infection, in which case treatment would be necessary.

Table 27.1 *Summary of indications for treatment of histoplasmosis capsulati*

Treatment indicated	Treatment not indicated
Acute pulmonary, diffuse	Pericarditis
Acute pulmonary, localized with prolonged symptoms	Rheumatological
Chronic pulmonary	Fibrosing mediastinitis
Disseminated	Histoplasmoma
Symptomatic mediastinal granuloma	Presumed ocular histoplasmosis

CHRONIC PULMONARY

Treatment of chronic pulmonary histoplasmosis capsulati improves survival, reduces symptoms, promotes radiological healing, and eradicates *H. capsulatum* from the sputum (Sutliff et al. 1964). Most patients (about 80 percent) respond well to itraconazole 200–400 mg daily (Dismukes et al. 1992). Itraconazole should be continued for at least a year and until maximal clinical and radiological benefits have been achieved. Amphotericin B may be needed initially in patients with severe respiratory insufficiency. Relapse is common (about 25 percent) after discontinuation of treatment, emphasizing the need for prolonged follow-up.

DISSEMINATED

Disseminated histoplasmosis is usually fatal if untreated (Furcolow 1963; Sathapatayavongs et al. 1983). Liposomal amphotericin B induced a more rapid response than the standard deoxycholate formulation, and improved survival in patients with the acquired immunodeficiency syndrome (AIDS) (Johnson et al. 2002). The high cost of the liposomal formulation may preclude its use in all situations, however. Amphotericin B and itraconazole induce a favorable response in 85–90 percent of patients (Wheat et al. 1990b, 1995; Dismukes et al. 1992). Itraconazole should be administered for at least a year in most patients. Relapse is common in patients who are immunosuppressed, including those with AIDS. The level of histoplasma antigen declines during therapy and increases with relapse, providing a useful tool for monitoring the response to therapy (Wheat et al. 1991). Treatment should be continued until antigenemia and antigenuria have resolved or at least reached a stable low level (<2 units). Lifelong maintenance treatment is needed in patients with AIDS who do not achieve a good CD4 cell response to potent antiretroviral therapy. If the CD4 cell count increases to

Table 27.2 *Recommendations for antifungal therapy of histoplasmosis capsulati*

Disease severity	Drug	Dose	Duration
Mild	Itraconazole	200 mg t.i.d. × 3 days or 200 mg q.d or b.i.d.[a]	3–24 months[b]
Moderate or severe	Amphotericin B or liposomal amphotericin B	0.7–1 mg/kg/day or 3 mg/kg/day for liposomal preparation	3 days to 2 weeks based on severity[c]

a) The itraconazole dose should begin as a loading dose followed by a treatment dose of 200 mg once or twice daily. Blood levels should be measured after 1 week of therapy and the dosage should be modified to maintain blood levels between 1 and 10 μg/m .
b) Treatment duration ranges from 3 months in patients with acute pulmonary histoplasmosis to 12–24 months in patients with disseminated or chronic pulmonary histoplasmosis. If treatment is initiated with an amphotericin B formulation, itraconazole may be used to complete the recommended duration of therapy. Patients with AIDS may require lifelong treatment if they fail to achieve CD4 cell improvement to >150 cells/ml in response to potent combination antiretroviral therapy.
c) Amphotericin B dosage is 0.7–1 mg/kg/day whereas the dosage of the liposomal preparation is 3 mg/kg/day.

150 cells/mm^3 or higher, and the antigen levels in urine and serum decline to <2 units, itraconazole may be stopped after one year. Maintenance therapy also may be required in other immunocompromised individuals with persistent immunosuppression, or those who have relapsed after extended courses of amphotericin B or itraconazole.

The response to therapy in patients with meningitis is inferior to that in other types of histoplasmosis: 20–40 percent of patients with meningitis die and up to half of responders relapse after therapy is stopped. The optimal treatment for *Histoplasma* meningitis is unknown, but an aggressive approach is recommended because of the poor outcome. Although the best therapy is still not known, it seems reasonable to begin therapy with liposomal amphotericin B, 3–5 mg/kg/day, for 2–3 months and until CSF cultures are negative and CSF antigen levels are negative, followed by itraconazole, 200 mg twice or three times daily for at least another year. Lifelong maintenance therapy may be needed in patients who relapse or in those whose CSF findings do not return to normal. Patients who fail chronic itraconazole therapy have these options: fluconazole 800 mg daily, voriconazole 200 mg twice daily, and posaconazole.

MEDIASTINAL FIBROSIS

Antifungal treatment usually does not influence the course of mediastinal fibrosis (Goodwin et al. 1972; Loyd et al. 1988; Davis et al. 2001). However, a few patients have shown improvement after treatment with ketoconazole (Urschel et al. 1990). In such cases, differentiation between fibrosing and granulomatous mediastinitis may have been difficult. Thus, a 3-month trial of itraconazole is reasonable, especially in patients with positive serological tests and elevated erythrocyte sedimentation rates, or findings that may be more consistent with granulomatous mediastinitis. It is not likely to be helpful, however, since the pathogenesis is fibrosis rather than infection or inflammation. Resection of the fibrotic tissue may be helpful but has a high operative mortality (about 25 percent) and probably should be reserved for patients with severe and progressive manifestations (Mathisen and Grillo 1992). A nonprogressive course is common, precluding the need for surgical intervention (Goodwin et al. 1972; Loyd et al. 1988).

PRESUMED OCULAR HISTOPLASMOSIS CAPSULATI

Presumed ocular histoplasmosis does not represent an active infection and would not be expected to respond to antifungal therapy, although no prospective trials have been reported (Ciulla et al. 2001).

LABORATORY DIAGNOSIS

Direct examination

Rounded to oval yeast cells measuring 2–4 mm in diameter, characteristic of *H. capsulatum* var. *capsulatum*, may be seen by fungal staining of biopsy specimens from pulmonary, mediastinal, or extrapulmonary tissues. Less commonly, organisms may be observed by fungal staining of sputum, sterile body fluids, or peripheral blood smears. In disseminated histoplasmosis capsulati, the highest yield is from bone marrow (>50 percent) (Figure 27.10). Organisms may be overlooked in patients with histoplasmosis leading to false-negative results, and staining artefacts or other organisms may be mistaken for *H. capsulatum* in patients with other diseases, emphasizing the importance of the skill and experience of the examiner for accurate interpretation of fungal stains. *Candida (Torulopsis) glabrata*, *Blastomyces dermatitidis*, capsule-deficient strains of *Cryptococcus neoformans*, *Penicillium marneffii*, *Toxoplasma gondii*, and *Pneumocystis carinii* may be misidentified in routine clinical laboratories as the tissue-form cells of the *capsulatum* variety of *H. capsulatum*. Mucicarmine staining for capsular material may be used to distinguish *C. neoformans* from *H. capsulatum*, but capsule-deficient strains may be misidentified by this technique. In such cases, melanin staining is positive with *C. neoformans* but negative with *H. capsulatum*.

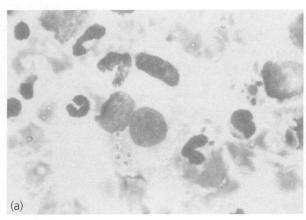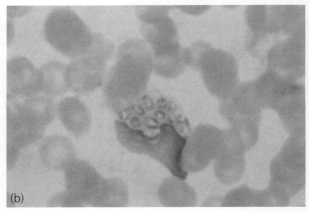

Figure 27.10 (a) *Bone marrow smear from a young girl with disseminated histoplasmosis capsulati showing budding yeast cells (2–4 μm) in a phagocytic monocyte Giemsa stain; magnification ×3 400).* **(b)** *Peripheral blood smear of an AIDS patient with disseminated histoplasmosis capsulati showing numerous yeast cells (2–4 μm) in a monocyte (Giemsa stain; ×3 400)*

Culture

Cultures are most useful in patients with disseminated or chronic pulmonary histoplasmosis, being positive in 50–85 percent of cases. In disseminated disease, the highest culture yield is from bone marrow or blood, being positive in over 75 percent of cases (Sathapatayavongs et al. 1983; Wheat et al. 1990b). Organisms can be found in sputum or bronchoscopy specimens from 60–85 percent of patients with cavitary histoplasmosis if multiple specimens are obtained (Wheat et al. 1984). The sensitivity of culture is only 10–15 percent in patients with other forms of histoplasmosis capsulati (Williams et al. 1994). Slow growth (2–4 weeks) may delay the diagnosis in patients with positive cultures (Williams et al. 1994). The development of specific nucleic acid probes has facilitated the identification of *H. capsulatum* var. *capsulatum* in cultures with atypical morphology (Stockman et al. 1993).

Immunoserology

This section will concentrate on three tests that are commonly used: the immunodiffusion and complement fixation tests for detection of antibodies and an immunoassay for demonstration of histoplasma polysaccharide antigen in body fluids. The precipitin bands (H and M) are demonstrated by the immunodiffusion test using histoplasmin as the antigen. H precipitin bands can be demonstrated in less than 25 percent of patients and clear during the first 6 months after infection, whereas M bands appear in about 75 percent of individuals and may persist for several years after the resolution of infection (see Figure 27.5) (Wheat et al. 1982; Williams et al. 1994). The complement fixation test is performed with both histoplasmin and whole yeast cell antigens. Complement fixation titers of 1:8 or higher are found in most patients with active histoplasmosis capsulati, whereas titers of 1:32 or higher are more

suggestive of active infection. Both the immunodiffusion and complement fixation tests should be performed to obtain the highest sensitivity for diagnosis (Figure 27.11).

Enzyme immunoassay or radioimmunoassay methods for detection of antibodies are more difficult to interpret, because of higher background positivity rates. They have not been validated adequately to recommend their use in place of immunodiffusion and complement fixation methods (Wheat 1993). Antibodies require 4–8 weeks to develop after acute infection and may be negative when the patient is first seen. Furthermore, serological tests may be falsely negative in up to one-third of immunocompromised patients (Williams et al. 1994). False-positive results are caused by antigenic cross reactions with *B. dermatitidis*, *Coccidioides immitis*, and *Paracoccidioides brasiliensis* (Wheat et al. 1986). M precipitin bands and low titers (1:8 or 1:16) of complement-fixing antibodies may require years to clear, causing confusion in patients with other diseases.

Histoplasma polysaccharide antigen is found in the blood, urine, and bronchoalveolar lavage fluid of more than 90 percent of individuals with disseminated histoplasmosis capsulati, and in the urine of 75 percent of those with extensive pneumonitis after heavy acute exposure (Wheat et al. 1986; Wheat and Kauffman 2003). Antigen may be found in the cerebrospinal fluid of 25–50 percent of patients with meningitis caused by histoplasmosis (Figure 27.11). It is important to recognize that lower levels of crossreactions may be seen in patients with histoplasmosis duboisii, blastomycosis, paracoccidioidomycosis, and penicilliosis marneffei. Antigen levels decline during treatment and increase with relapse, providing a tool for monitoring therapy (Wheat et al. 1991). Antigen levels should be monitored every 3–6 months, and at the time of suspected relapse.

Histoplasmin skin test antigen is a culture filtrate of the mycelial form of *H. capsulatum* var. *capsulatum* but

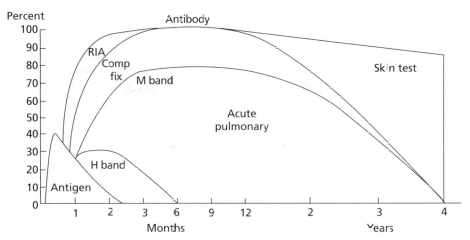

Figure 27.11 *Time course of diagnostic tests for histoplasmosis capsulati in patients with acute pulmonary disease. Antigen refers to histoplasma antigen in urine. H band and M band refer to precipitants measured by immunodiffusion. Comp fix refers to titers of at least 1:8 to either the yeast or mycelial antigen; Skin test refers to histoplasmin skin test reactivity, RIA refers to measurement of antibodies by radioimmunoassay showing that antibodies may become detectable somewhat earlier than by complement fixation of immunodiffusion.*

is no longer available. Histoplasmin skin testing is not recommended for diagnosis, because of high rates of positivity (50–80 percent) in endemic areas and false-positive results in patients with other fungal diseases (Edwards et al. 1969). Furthermore, skin tests may also be falsely negative in more than half of patients with disseminated disease and skin tests boost the levels of anti-histoplasma antibody, causing confusion in the interpretation of serological tests (Campbell and Hill 1964). The skin test is, however, very useful in epidemiologic studies for determining the prevalence of histoplasmosis capsulati.

Molecular diagnostics

Molecular methods have been developed for diagnosis of histoplasmosis. DNA probes are commercially available and widely used in the clinical mycology laboratory, reducing the time for definitive identification of positive cultures. Research is in progress to develop molecular diagnostic methods to detect *H. capsulatum* var. *capsulatum* DNA in tissue sections and body fluids. To date, while studies suggest that these methods will prove useful, none has been validated by comparison to other established procedures. These are promising methods, particularly for diagnosis of specimens containing granulomas but with negative fungal stains, but require more evaluation before they can be recommended for use in the clinical laboratory. Furthermore, how they will compare with antigen detection in body fluids for rapid diagnosis of histoplasmosis requires further investigation. Nevertheless, polymerase chain reaction (PCR) methods are offered at several commercial laboratories, but should not be accepted as accurate, or used for making the diagnosis of histoplasmosis in patients, until they are validated.

Histopathology

The tissue response in histoplasmosis capsulati follows the general course of histopathological reactions described as granulomas. The early lesions in the lung are areas of pneumonia containing a large number of macrophages and lymphocytes with occasional epithelioid cells and multinucleated giant cells. At this stage, tissue-form yeast cells are easily seen within macrophages. Central areas begin to show necrosis early in the evolution of the lesion, and are occasionally accompanied by transudation of fluid into the necrotic areas (Figure 27.12). With the progression of the lesion, tissue architecture is lost and its outer margins exhibit epithelioid cells together with multinucleated giant cells, lymphocytes, and plasma cells (Goodwin et al. 1981). Proliferation of fibroblasts, deposition of collagen around margins of the lesion, and calcification are usually observed in old, localized granulomas. In 'histoplasmoma,' the outer portion of fibrosis may consist of multiple layers of fibrous tissue (Goodwin and Snell 1969).

In cavitary pulmonary histoplasmosis capsulati, the cavity lining is necrotic without a well-localized outer margin of fibrosis. Histoplasma yeasts seen in the cavity lining are pleomorphic, with occasional short hyphal forms. The inflammatory response varies considerably from one area of the lungs to another; some areas have well organized granulomas with whorls of epithelioid cells and necrosis, whereas other areas have caseous necrosis. Demonstration of *H. capsulatum* is very difficult with hematoxylin and eosin (H&E) staining and is almost impossible in areas without necrosis. Gridley's stain is useful for demonstration of yeast cells. The periodic acid–Schiff (PAS) stain can be negative despite the presence of silver-staining yeast cells by Gomori's

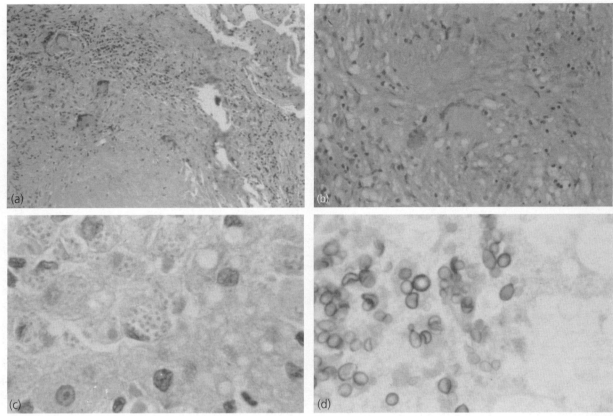

Figure 27.12 *Histopathology of histoplasmosis capsulati.* **(a)** *Cross-section of the lung from a patient with histoplasmosis showing granulomatous inflammatory response consisting of mononuclear cells and giant cells with a central area of necrosis (H&E stain; magnification ×1 200).* **(b)** *Cross-section of a mediastinal lymph node showing caseous granulomatous lymphadenitis resulting from* H. capsulatum *var. capsulatum. Note the epithelioid cells and giant cells around the necrotic centre (H&E stain; magnification ×1 200).* **(c)** *Cross-section of the liver showing numerous intracellular yeast cells in histiocytes with associated mononuclear cell infiltration (H&E stain; magnification ×1 200).* **(d)** *Cross-section of a lung granuloma showing intact and some degraded yeast cells (GMS stain; magnification ×1 200). (Courtesy of Dr Joan Barenfanger, MMC, Springfield, IL)*

methenamine silver (GMS) stain – the preferred technique for recognition of the yeast cells of the three varieties of *H. capsulatum*. The organisms from old lesions may not grow on culture (Figures 27.13 and 27.14).

A critical issue in pulmonary pathology is the distinction of infection from non-infectious granulomas in the lung. In Wegener's granulomatosis, necrotizing vasculitis of blood vessels is prominent, even in areas without necrosis of surrounding tissues. Vessel walls are destroyed and infiltrated with acute and chronic inflammatory cells, mixed with multinucleated giant cells and histiocytes. If one considers only the lungs and not the lesions of Wegener's granulomatosis in the upper airway or kidney, the tissue response can be quite similar to that of chronic pulmonary histoplasmosis capsulati. Lymphomatoid granulomatosis differs in that the vessel necrosis is less prominent, cellular infiltrates are present in all layers of the vessels, and the lesions are predominantly angionecrotic. Lesions of sarcoidosis can also be mistaken for those of histoplasmosis capsulati. Within infectious diseases, tuberculosis most closely resembles histoplasmosis, histopathologically. In cryptococcosis, necrotic areas contain large collections of

yeasts that can be easily recognized by special stains. Furthermore, cryptococcal granulomas have yeast cells throughout the lesion. *C. immitis* is the fungus with an inflammatory response that mimics that of histoplasmosis, but spherules of *C. immitis* are easy to find and are distinctive. Likewise, *B. dermatitidis* yeast cells have thick, double refractile cell walls with a broad-based septum between the mother and daughter cells, and are usually associated with a mixed suppurative and granulomatous tissue reaction.

Involvement of hilar lymph nodes with necrotizing granulomas and calcification is common in pulmonary histoplasmosis (Figure 27.12). Patients who progress to fibrosing mediastinitis have increasing effacement of lymph node architecture with caseous necrosis and dense bands of collagen about the node (Goodwin et al. 1972). With progression of the lesion, fibrotic compression of pulmonary veins, superior vena cava, bronchus or pulmonary arteries can occur. Caseous nodes can adhere and rupture into the esophagus or, rarely, the bronchus. In patients with pericarditis, thickened pericardium shows nonspecific inflammation more often than a granulomatous response or *H. capsulatum*

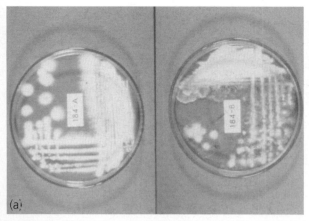

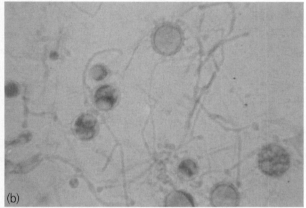

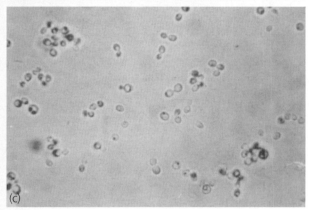

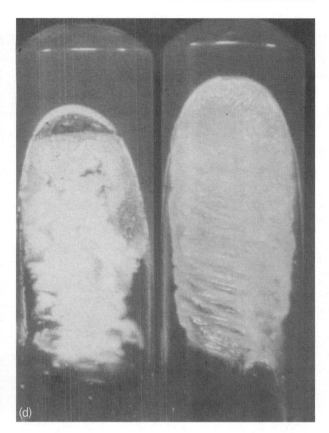

Figure 27.13 *Morphology of* H. capsulatum *var.* capsulatum. **(a)** *Mycelial colonies of brown (left) and albino (right) types at 25°C after 4 weeks (Cotton blue stain).* **(b)** *Microscopic appearance of* H. capsulatum *in mycelial form showing branching septate hyphae with microconidia (2–4 μm) and tuberculate macroconidia (9–15 μm) (Cotton blue stain; magnification ×3 400.)* **(c)** *Yeast form growth of* H. capsulatum *var.* capsulatum *on brain–heart infusion agar slants at 37°C after 48 hours. Morphology of* H. capsulatum *var.* capsulatum. **(d)** *Microscopic appearance of* H. capsulatum *var.* capsulatum *yeast cells (2–5 μm). (Cotton blue stain; magnification ×3 400)*

yeast cells. Occasionally, caseous subcarinal nodes may be attached to the pericardium.

The most prominent lesions of progressive disseminated disease are extrapulmonary. There is a diffuse scattering of infected macrophages throughout the mononuclear phagocytic system (MPS), often referred to as the reticuloendothelial system; granulomas are generally formed poorly, if at all. The organs rich in MPS are most frequently involved. Necrosis is less prominent in lesions outside the lungs because necrosis perhaps reflects an adequate host defense that is compromised in disseminated disease. A notable exception is the adrenal glands where necrosis is a prominent feature. Fungal stains demonstrate a large number of histoplasma yeast cells in lesions with diffuse infiltration by macrophages. In patients with AIDS and other severely immunocompromised hosts, the phagocytic cells are usually engorged with yeast cells, presumably as a result of lack of adequate cytokine production and signaling required for killing of the ingested organisms.

Some extrapulmonary sites have their peculiar lesions. The infected cardiac valves often have large yeast forms or short hyphal cells. The histological characteristics of an embolized vegetation of a peripheral artery can readily be confused with candidal infection. In the CNS, histoplasmosis capsulati may present only as a meningitis or one or more cerebral granulomas. In meningitis, the organisms can be so sparse that histological sections may miss the diagnosis unless cultures of basilar meninges are done.

HISTOPLASMA CAPSULATUM VAR. DUBOISII INFECTION

H. capsulatum var. *duboisii* is endemic to the tropical regions of Africa and the island of Madagascar. It was

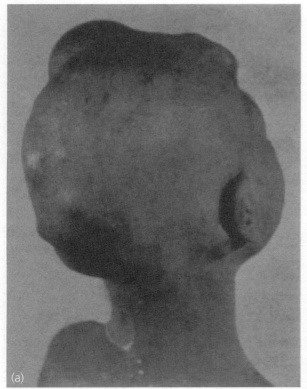

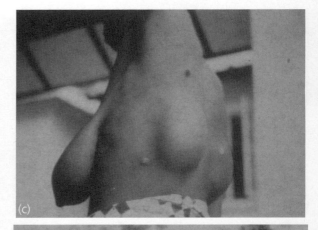

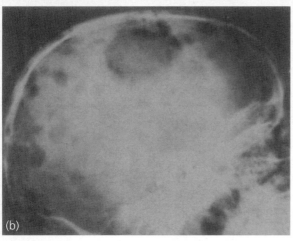

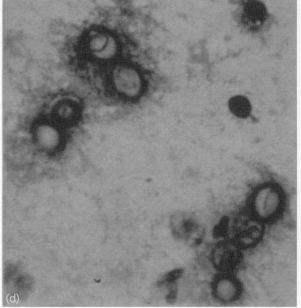

FIG203.14 **Figure 27.14 (a)** *Head of a male child with disseminated form of histoplasmosis duboisii showing multiple subcutaneous abscesses over the bone defects shown in* **(b)**. **(b)** *x-Ray of the skull with multiple osteolytic defects resulting from granulomas in the diploe which have extended to involve both the inner and outer tables of the vault.* **(c)** *A female patient with two large abscesses on the back caused by* H. capsulatum *var. duboisii.* **(d)** *Yeast cells of* H. capsulatum *var.* duboisii *in pus (PAS stain; magnification ×800). (Courtesy of Dr Adatukumbo Lucas, Ibadan Nigeria)*

contrasted to *H. capsulatum* var. *capsulatum* by Houston (1994). Cases have also been diagnosed in travelers who have visited those areas and immigrants from central Africa. Several cases have been reported in people with AIDS. Exposure to bat or bird droppings appears to be a risk factor for acquisition of histoplasmosis duboisii. Isolation of the *duboisii* variety from the environment was conclusively established in 1994 by Gugnani et al.

The clinical manifestations of *H. capsulatum* var. *duboisii* infections differ from those seen in patients infected with the *capsulatum* variety of *H. capsulatum*. Although the infection is acquired by inhalation, the most common clinical manifestations in histoplasmosis duboisii involve skin, bone, and subcutaneous tissues (Cockshott and Lucas 1964). Multiple osteolytic lesions are commonly found in rib, vertebra, femur, humerus, tibia, skull, and wrist (Williams et al. 1971). The granulation tissue from a vertebral lesion may cause compression of the spinal cord, resulting in paraplegia. Skin lesions may be papular, nodular, ulcerative, circinate, and eczematoid. Formation of subcutaneous abscesses from underlying bone lesions are common in histoplasmosis duboisii. With the enlargement of nodular or papular lesions, the centre ulcerates. Patients with only skin or bone lesions may have an indolent disease course, and may sometimes recover without any treatment.

If present, the pulmonary lesions of histoplasmosis duboisii, in chest radiographs, may have a diffuse reticular pattern extending throughout both lung fields or localized with or without involvement of mediastinal lymph nodes. These radiographic findings are similar to that observed in *H. capsulatum* var. *capsulatum* infections. The patients with a mild form of the disease presumably recover without any treatment.

In patients with the disseminated form of histoplasmosis duboisii, multiple lesions are usually present in the liver, spleen, lymph nodes, bone marrow, and other visceral organs. Fever, anemia, and loss of weight are common accompaniments. Such patients follow a progressive, wasting course and their disease may prove fatal if untreated. At autopsy, numerous nodular lesions are found throughout the reticuloendothelial system. This feature is remarkably similar to that observed in patients with histoplasmosis capsulati (Wheat et al. 1981).

Microscopic examinations of tissue sections show numerous organisms in areas of granulomatous inflammation (Williams et al. 1971). The giant cells contain large, thick-walled yeast cells that are larger than those of the *capsulatum* variety. An acute inflammatory response may be seen in ulcerated skin lesions and in soft tissue abscesses. The well-organized granulomata with their caseous centres and surrounding epitheloid cells, which are commonly associated with pulmonary histoplasmosis capsulati, are rarely observed in histoplasmosis duboisii.

Diagnosis is made by fungal stains that reveal ovoid, thick-walled yeasts measuring 12–15 µm in diameter that are distinct from those of *H. capsulatum* var. *capsulatum* (2–5 µm in diameter). Morphologically, *H. capsulatum* var. *duboisii* and var. *capsulatum* are identical in culture; however, they can be distinguished by differences in the size of their yeast-form cells in experimentally infected animals. Treatment of histoplasmosis duboisii is similar to that of *H. capsulatum* var. *capsulatum* infections. Both amphotericin B and itraconazole are effective (Eichmann and Schar 1996) in HIV-positive as well as in HIV-negative individuals (Figure 27.14).

HISTOPLASMA CAPSULATUM VAR. FARCIMINOSUM

As has been emphasized, histoplasmosis farciminosi is an equine disease. Clinically, it is a virulent disease notable for the suppuration of the superficial and subcutaneous lymphatic vessels, especially in the neck and limbs of the infected animals. Other parts of the body are more rarely involved – flanks, back. In general, the initial lesions are most frequently associated with those parts of the body which are most apt to be subject to wounds by kicks, contusions, and harness galls. The ulcerations are painful and usually pruritic. The infected lymphatic vessels become dilated and form a linear series of abscesses which in time rupture and discharge serosanguineous pus for several weeks. Noncutaneous lesions may occur in the mucous membranes and

proliferate along the nasal septum to the pharynx, larynx, and trachea. Ocular involvement may also occur, especially on males. Such infections are characterized by watery discharge from one or both eyes with blepharitis that may lead to the closing of the affected eyes (Singh 1966). Dissemination of histoplasmosis farciminosi is essentially through the lymphatic ducts. Involvement of the internal organs has not been documented.

Diagnosis is based on the detection of the tissue-form cells of the dimorphic *H. capsulatum* var. *farciminosi*. These unicellular hyaline cells are 2.5–3.5×2.0–3.0 µm. They are best detected in pus smears on tissue sections through staining with the combined GMS–H&E stain or the PAS stain. The yeast-like cells are found to be intracellular within phagocytes. As noted previously, they are indistinguishable from the tissue-form cells of the *capsulatum* variety of *Histoplasma* in respect of size, form, and staining reactions. Confirmation of isolating the etiologic agent can be achieved by the use of a soil extract medium as reported by Weeks et al. (1985) and by converting the mycelial form to its yeast-like form by growing the isolates at 37°C in brain–heart infusion agar with or without 5 percent sheep blood.

Treatment is empirical as no consistent effective treatment has been developed. Surgery for extirpation of early lesions has been effective in some instances. Recently administration of amphotericin B has been recommended (Chermette and Bussieras 1993). For further information on histoplasmosis farciminosi, the readers are referred to the 1980 publication of Chandler, Kaplan and Ajello.

REFERENCES

Ajello, L. 1964. Relationship of *Histoplasma capsulatum* to avian habitats. *Public Health Rep*, **79**, 266–70.

Ajello, L., Fadaye, A., et al. 1995. Occurrence of *Penicillium marneffei* infections among bamboo rats in Thailand. *Mycopathologia*, **131**, 1–8.

Ajello, L., Fadaye, A., et al. 1996. Erratum: occurrence of *Penicillium marneffei* infections among bamboo rats in Thailand. *Mycopathologia*, **135**, 195–7.

Allendoerfer, R., Magee, D.M., et al. 1993. Transfer of protective immunity in murine histoplasmosis by a CD4+ T-cell clone. *Infect Immun*, **61**, 714–18.

Arrigoni, M.G. Bernatz, P.E. and Donoghue, F.E. 1971. Bronchlithiasis. *J Thorac Cardiovasc Surg*, **62**, 231–7.

Bartlett, P.C., Weeks, R.J. and Ajello, L. 1982. Decontamination of a *Histoplasma* infected bird roost in Illinois. *Arch Environ Health*, **37**, 221–3.

Bedrosian, N., Gatchel, S.L. and Tewari, R.P. 1986. Interactions of *Histoplasma capsulatum* with murine lung endothelial cells in culture. *Abstracts of 26th ICAAC Annual Meeting, Washington DC*, Abstract no. 830.

Brummer, E., Kurita, N., et al. 1991. Fungistatic activity of human neutrophils against *Histoplasma capsulatum*: correlation with phagocytosis. *J Infect Dis*, **164**, 158–62.

Burgisser, H., Frankhauser, R., et al. 1961. Mykose bei einem Dachs in der Schweiz: histologisch Histoplasmose. *Pathol Microbiol Basel*, **24**, 794–802.

Burt, W.R. 1982. Identification of coprogen B and its breakdown products from *Histoplasma capsulatum*. *Infect Immun*, **35**, 990–6.

Butler, T.M. and Hubbard, G.B. 1991. An epizootic of *Histoplasma duboisii* (African histoplasmosis) in an American baboon colony. *Lab Animal Sci*, **41**, 407–10.

Butler, T.M., Gleiser, C.A., et al. 1988. Case of disseminated African histoplasmosis in a baboon. *J Med Primatol*, **17**, 153–61.

Campbell, C.C. and Hill, G.B. 1964. Further studies on the development of complement-fixing antibodies and precipitins in healthy histoplasmin-sensitive persons following a single histoplasmin skin test. *Am Rev Respir Dis*, **90**, 927–34.

Chandler, F.W., Kaplan, W. and Ajello, L. 1980. Histoplasmosis farciminosi. In *A colour atlas and textbook of the histopathology of mycotic diseases*. London: Wolfe Medical Publications, 70–2.

Chermette, R. and Bussieras, J. 1993. Lymphoangite épizootique des Equides. *Parasitologie vétérinaire mycologie*. Service de Parasitologie, Ecole National Vétérinaire d'Alfort, 94–7.

Choudhury, C., Raman, C., et al. 1989. Restoration of natural killer cell activity and immunocompetence against *Histoplasma capsulatum* in hairy cell leukemia following treatment with recombinant alpha interferon. *Curr Ther Res*, **45**, 179.

Ciferri, R. 1960. Le istoplasmosi. In *Manuale di micologia medica*, vol. 2. Pavia: Casa Editrice Renzo Corina, 339–42.

Ciferri, R. and Redaelli, P. 1934. Sulla posizione sistematica dell'agente patogeno del farcino equino. *Bull Istit Sieroterapico Milanese*, **13**, 1–8.

Ciulla, T., Piper, H., et al. 2001. Presumed ocular histoplasmosis syndrome: update on epidemiology, pathogenesis, and photodynamic, antiangiogenic, and surgical therapies. *Curr Opin Ophthalmol*, **12**, 442–9.

Cockshott, W.P. and Lucas, A.O. 1964. Histoplasmosis duboisii. *Q J Med*, **33**, 223–8.

Curtois, G., Segretain, G., et al. 1955. Mycose cutanée à corps levuriformes observée chez singes africains en captivité. *Ann Inst Pasteur*, **89**, 124–7.

Darling, S.T. 1906. A protozoan general infection producing pseudotubercles in the lungs and focal necroses in the liver, spleen and lymph nodes. *JAMA*, **46**, 1283–5.

da Rocha-Lima, H. 1912. Histoplasmosis und epizootic lymphangitis. *Arch Schiffs Tropenhyg*, **16**, 79–85.

da Rocha-Lima, H. 1912–13. Beitrag zur Kentnis der Blastoimykosen – Lymphangitis epizootica und Histoplasmosis. *Zentralbl Bakteriol*, **67**, 233–49.

Davis, A., Pierson, D. and Loyd, J.E. 2001. Mediastinal fibrosis. *Semin Respir Infect*, **16**, 119–30.

Dean, A.G., Bates, J.H., et al. 1978. An outbreak of histoplasmosis at an Arkansas courthouse with five cases of probable reinfection. *Am J Epidemiol*, **108**, 34–46.

Deepe, G.S. 1994. Role of CD8+ T cells in host resistance to systemic infection with *Histoplasma capsulatum* in mice. *J Immunol*, **152**, 3491–500.

Deepe, G.S. Jr and Bullock, W.E. 1992. *Inflammation: basic principles and clinical correlates*. New York: Raven Press.

Deepe, G.S. Jr, Gibbons, R., et al. 1996. A protective domain of heat-shock protein 60 from *Histoplasma capsulatum*. *J Infect Dis*, **174**, 828–34.

De Monbreun, W.A. 1934. The cultivation and cultural characteristics of Darling's *Histoplasma capsulatum*. *Am J Trop Med*, **14**, 93–125.

Dismukes, W.E., Bradsher, R.W. Jr, et al. 1992. Itraconazole therapy for blastomycosis and histoplasmosis. *Am J Med*, **93**, 489–97.

Drouhet, E. and Schwarz, J. 1956. Croissance et morphogenèse *Histoplasma*. I. Etude comparative des phases mycelienne et levure de 18 souches d'*Histoplasma capsulatum* d'origine amércaine et Africaine. *Ann Inst Pasteur*, **90**, 144–60.

Dubois, A., Janssens, P.G. and Brutsaert, P. 1952. Un cas d'histoplasmose Africaine. Avec une note mycologique sur *Histoplasma duboisii* n sp por R. Vanbreuseghems. *Am Soc Belg Med Trop*, **32**, 569–84.

Duncan, J.T. 1947. A unique form of *Histoplasma*. *Trans R Soc Trop Med Hyg*, **40**, 364–5.

Duncan, J.T. 1958. Tropical African histoplasmosis. *Trans R Soc Trop Med Hyg*, **52**, 468–74.

Edwards, L.B., Acquaviva, F.A., et al. 1969. An atlas of sensitivity to tuberculin, PPD-B and histoplasmin in the United States. *Am Rev Respir Dis*, **99**, 1–18.

Eichmann, A. and Schar, G. 1996. African histoplasmosis in a patient with HIV-2 infection. *Schweiz Med Wochenschr*, **126**, 765–9.

Eissenberg, L.G., Goldman, W.E. and Schlesinger, P.H. 1993. *Histoplasma capsulatum* modulates acidification of phagolysosomes. *J Exp Med*, **177**, 1605–11.

Emmons, C.W. 1949. Isolation of *Histoplasma capsulatum* from soil. *Public Health Rep*, **64**, 892–6.

Emmons, C.W. 1959. Association of bats with histoplasmosis. *Public Health Rep*, **73**, 590–5.

Emmons, C.W., Bell, J.A. and Olson, B.J. 1947. Naturally occurring histoplasmosis in *Mus musculus* and *Rattus norvegicus*. *Public Health Rep*, **62**, 1642–6.

Feit, C. and Tewari, R.P. 1974. Immunogenicity of ribosomal preparations from yeast cells of *Histoplasma capsulatum*. *Infect Immun*, **10**, 1091.

Fleischmann, J., Wu-Hsieh, B. and Howard, D.H. 1990. The intracellular fate of *Histoplasma capsulatum* in human macrophages is unaffected by recombinant human interferon-γ. *J Infect Dis*, **161**, 143–5.

Frey, T., Petty, H.R. and McConell, H.M. 1982. Electron microscopic study of natural killer cells – tumor cells conjugates. *Proc Natl Acad Sci U S A*, **79**, 5317–21.

Furcolow, M.L. 1963. Comparison of treated and untreated severe histoplasmosis. *JAMA*, **183**, 121–7.

Gatti, F. and DeVroey, C. 1991. African histoplasmosis. In: Gatti, F., DeVroey, C. and Parsi, A. (eds), *Human mycoses in tropical countries*, 2nd edition. Bologna, Italy: Associazione Italiana 'Amici di Raoul Follereau', 211–18.

Gilliland, M.D., Scott, L.D. and Walker, W.E. 1984. Esophageal obstruction caused by mediastinal histoplasmosis: beneficial results of operation. *Surgery*, **95**, 59–62.

Gomez, F.J., Gomez, A.M. and Deepe, G.S. Jr 1991. Protective efficacy of a 62-kilodalton antigen, HIS-62, from the cell wall and cell membrane of *Histoplasma capsulatum* yeast cells. *Infect Immun*, **59**, 4459–64.

Gomez, F.J., Allendoerfer, R. and Deepe, G.S. Jr 1995. Vaccination with recombinant heat shock protein 60 from *Histoplasma capsulatum* protects mice against pulmonary histoplasmosis. *Infect Immun*, **63**, 2587–95.

Goodwin, R.A. Jr and Snell, J.D. Jr 1969. The enlarging histoplasmoma: concept of a tumor-like phenomenon encompassing the tuberculoma and coccidioidoma. *Am Rev Respir Dis*, **100**, 1–12.

Goodwin, R.A., Nickell, J.A. and des Prez, R.M. 1972. Mediastinal fibrosis complicating healed primary histoplasmosis and tuberculosis. *Medicine*, **51**, 227–46.

Goodwin, R.A., Owens, F.T., et al. 1976. Chronic pulmonary histoplasmosis. *Medicine (Baltimore)*, **55**, 413–52.

Goodwin, R.A., Loyd, J.E. and des Prez, R.M. 1981. Histoplasmosis in normal hosts. *Medicine*, **60**, 231–66.

Grosse, G., Staib, F., et al. 1997. Pathological and epidemiological aspects of skin lesions in histoplasmosis. Observations in an AIDS patient and badgers outside endemic areas of histoplasmosis. *Zentralbl Bakteriol*, **285**, 531–9.

Gugnani, H.C., Muotoc, F., et al. 1994. Natural focus of *Histoplasma capsulatum* var. *duboisii* in a bat cave. *Mycopathologia*, **127**, 151–7.

Houston, S. 1994. Histoplasmosis and pulmonary involvement in the tropics. *Trop Respir Med*, **49**, 598–601.

Howard, D.H. 1981a. *Immunology of human infection*. New York: Plenum, 475–94.

Howard, D.H. 1981b. Comparative sensitivity of *Histoplasma capsulatum* conidiospores and blastospores to oxidative antifungal systems. *Infect Immun*, **32**, 381–7.

Hughes, R.M., Von Behren, L.A. and Tewari, R.P. 1994. Interactions of heart endothelial cells (EC) with *Histoplasma capsulatum* (HC). *XII International Society for Human and Animal Mycology, Adelaide, Australia*, Abstract no. P036.

Jansen, H.E., Bloch, B., et al. 1992. Disseminated histoplasmosis in a badger (*Meles meles*). *Acta Microbiol Immunol Scand*, **100**, 586–92.

Johnson, P.C., Wheat, L.J., et al. 2002. Safety and efficacy of liposomal amphotericin B compared with conventional amphotericin B for induction therapy of histoplasmosis in patients with AIDS. *Ann Intern Med*, **137**, 105–9.

Kasuga, T., White, T.J., et al. 2003. Phylogeography of the fungal pathogen *Histoplasma capsulatum*. *Mol Ecol*, **12**, 3383–401.

Kataria, Y.P., Campbell, P.B. and Burlingham, B.T. 1981. Acute pulmonary histoplasmosis presenting as adult respiratory distress syndrome: effect of therapy on clinical and laboratory features. *South Med J*, **74**, 534–7.

Khardori, N., Von Behren, L., et al. 1986. Cellular mediators of anti-*Histoplasma* immunity: I. Protective immunity and cellular changes in spleens of mice immunized by sublethal infection with yeast cells of *Histoplasma capsulatum*. *Mykosen*, **29**, 103–15.

Kondoh, Y., Raman, C. and Tewari, R.P. 1989. Antifungal activity of murine circulating and peritoneal neutrophils to *Histoplasma capsulatum*. *Abstracts of the Annual Meeting of American Society for Microbiology, New Orleans LA*, Abstract no. F-70.

Kwon-Chung, K.J. 1972. *Emmonsiella capsulata*: perfect state of *Histoplasma capsulatum*. *Science*, **177**, 368–9.

Kwon-Chung, K.J. 1975. Perfect state (*Emmonsiella capsulata*) of the fungus causing large form African histoplasmosis. *Mycologia*, **67**, 980–90.

Kwon-Chung, K.J. and Bennett, J.E. 1992. *Medical mycology*. Philadelphia, PA: Lea & Febiger.

Kwon-Chung, K.J. and Hill, W.B. 1981. Virulence of the two mating types of *Emmonsiella capsulata* and the mating experiments with *Emmonsiella capsulata* and *Emmonsiella capsulata* var. *duboisii*. In: Vanbreuseghem, R. and DeVroey, C.H. (eds), *Sexuality and pathogenicity of fungi*. New York: Masson, 48–56.

Lane, T.E., Wu-Hsieh, B.A. and Howard, D.H. 1994. Antihistoplasma effect of activated mouse splenic macrophages involves production of reactive nitrogen intermediates. *Infect Immun*, **62**, 1940–5.

Loyd, J.E., Tillman, B.F., et al. 1988. Mediastinal fibrosis complicating histoplasmosis. *Medicine*, **67**, 295–310.

Marcone, G. 1895. La saccaramicosi degli equini (farcino d'Africa, linfangite epizootica, farcino criptococcio ecc). *Atti 1st Incorragiemento Napoli*, **8**, 1–19.

Mathisen, D.J. and Grillo, H.C. 1992. Clinical manifestation of mediastinal fibrosis and histoplasmosis. *Soc Thorac Surg*, **54**, 1053–8.

McGinnis, M.R. and Katz, B. 1979. *Ajellomyces* and its synonym *Emmonsiella*. *Mycotaxon*, **8**, 157–164, 564.

Mengis, R.M. and Habermann, R. 1955. Experimental avian histoplasmosis. *Am J Vet Med*, **16**, 314–20.

Miyaji, M., Chandler, F.W. and Ajello, L. 1981. Experimental histoplasmosis capsulati in athymic nude mice. *Mycopathologia*, **75**, 139–48.

Newman, S.L. and Gootee, L. 1992. Colony stimulating factors activate human macrophages to inhibit intracellular growth of *Histoplasma capsulatum* yeasts. *Infect Immun*, **60**, 4593–7.

Nowak, R.M. 1991. *Walker's mammals of the world*, 5th edition, vol. 1. Baltimore, MD: The Johns Hopkins Press, 247.

Okeke, C.N. and Miller, J. 1991. In vitro production of extracellular elastolytic proteinase by *Histoplasma capsulatum* var *duboisii* and *Histoplasma capsulatum* var *capsulatum* in the yeast phase. *Mycoses*, **34**, 461–7.

Palmer, C.E. 1945. Nontuberculous pulmonary calcification and sensitivity to histoplasmin. *Public Health Rep*, **60**, 513–20.

Raman, C., Kondoh, Y. et al. 1989a. Impaired clearance of *Histoplasma capsulatum* in mice depleted of NK cells. *Abstracts of the Annual Meeting of FASEB, New Orleans LA*, Abstract no. 808.

Raman, C., McConnachie, P.R. et al. 1989b. Characterization of murine lung natural killer (NK) cell activity to Histoplasma capsulatum. Abstracts of the 27th ICAAC Annual Meeting New York, Abstract no. 578.

Raman, C., Gourakonti, N. et al. 1990. Adoptive transfer of NK cells restores resistance to histoplasmosis in cyclophosphamide-treated mice. *Abstracts of the FASEB Annual Meeting, Washington DC*, Abstract no. 4405.

Rapp, J., Lofqvist, A., et al. 1992. Sprosspilze als ursache von hautgranulomen beim dachs in Süddeutschland. *Tierartzl Umschau*, **47**, 451–2.

Rivolta, S. 1873. *Dei Parassiti Vegetali con Introduzione allo Studio delle Malattie Parassitarie e delle Alterazioni dell'Alimento degli Animali Domestici*. Turin, Italy: Tipografi Giulio Speirani F Figli, 246–52, 524–5.

Rivolta, S. and Micellone, I. 1883. Del farcino criptococcio. *Giorn Anat Tatol Anim Domest*. **15**, 143–62.

Rosenthal, J., Brandt, K.D., et al. 1983. Rheumatologic manifestations of histoplasmosis in the recent Indianapolis epidemic. *Arthritis Rheum*, **26**, 1065–70.

Saccardo, P.A. 1899. *Sylloge fungorum omnium hucusque cognitorium*. Pavia, Italy: Saccardo.

Sacks, J.J., Ajello, L. and Crockett, L.K. 1986. An outbreak and review of cave associated histoplasmosis capsulati. *J Med Vet Mycol*, **24**, 313–27.

Salvin, S.B. 1952. Endotoxin in pathogenic fungi. *J Immunol*, **69**, 89–99.

Salvin, S.B. 1965. Constituents of the cell wall of the yeast-phase of *Histoplasma capsulatum*. *Am Rev Respir Dis*, **92**, 119–25.

Sathapatayavongs, B., Batteiger, B.E., et al. 1983. Clinical and laboratory features of disseminated histoplasmosis during two large urban outbreaks. *Medicine (Baltimore)*, **62**, 263–70.

Savides, T.J., Gress, F.G., et al. 1995. Dysphagia due to mediastinal granulomas: diagnosis with endoscopic ultrasonography. *Gastroenterology*, **109**, 366–73.

Schacklette, M.H., Diercks, F.H. and Gale, N.B. 1962. *Histoplasma capsulatum* recovered from bat tissues. *Science*, **135**, 1135.

Schnur, R.A. and Newman, S.L. 1990. The respiratory burst response to *Histoplasma capsulatum* by human neutrophils. Evidence for intracellular trapping of superoxide anion. *J Immunol*, **144**, 4765–72.

Schwarz, J. 1981. *Histoplasmosis*. New York: Praeger.

Schwarz, J, Baum, G.L., et al. 1957. Successful infections of pigeons and chickens with *Histoplasma capsulatum*. *Mycopathol Mycol Appl*, **8**, 189–93.

Sharma, D., Khadori, N., et al. 1986. Cellular mediators of anti-*Histoplasma* immunity: II Protective immunity and delayed hypersensitivity in mice immunized by sublethal infection with yeast cells of *Histoplasma capsulatum*. *Mykosen*, **29**, 116–26.

Sigler, L. 1993. Perspective on Onygenales and their anamorphs by a traditional taxonomist. In: Reynolds, D.R. and Taylor, J.W. (eds), *The fungal holomorph: mitotic, meiotic and pleomorphic speciation in fungal systematics*. Wallingford: CAB International, 161–8.

Singh, T. 1966. Studies on epizootic lymphangitis. Study of clinical cases and experimental transmission. *Indian J Vet Sci*, **36**, 45–59.

Smith, C.D and Furcolow, M.L. 1964. The demonstration of growth stimulating substances for *Histoplasma capsulatum* and *Blastomyces dermatitidis* in infusions of starling (*Sturnus vulgaris*) manure. *Mycopathol Mycol Appl*, **22**, 73–80.

Smith, C.D., Furcolow, M.L. and Tosh, F.E. 1964. Attempts to eliminate *Histoplasma capsulatum* from soil. *Am J Hyg*, **79**, 170–80.

Spaeth, G.L. 1971. *Histoplasmosis*. Springfield IL: Charles C Thomas, 221–30.

Specht, C.S., Mitchell, K.T., et al. 1991. Ocular histoplasmosis with retinitis in a patient with acquired immune deficiency syndrome. *Ophthalmology*, **98**, 1356–9.

Standard, P. and Kaufman, L. 1976. Specific immunological text for the rapid identification of members of the genus *Histoplasma*. *J Clin Microbiol*, **3**, 191–9.

Stockman, L., Clark, K.A., et al. 1993. Evaluation of commercially available acridinium ester-labeled chemiluminescent DNA probes for culture identification of *Blastomyces dermatitidis*, *Coccidioides immitis*, *Cryptococcus neoformans* and *Histoplasma capsulatum*. *J Clin Microbiol*, **31**, 845–50.

Sutcliffe, M.C., Savage, A. and Alford, R.H. 1980. Transferrin-dependent growth inhibition of yeast-phase *Histoplasma capsulatum* by human serum and lymph. *J Infect Dis*, **142**, 209–19.

Sutliff, W.D., Andrews, C.E., et al. 1964. Histoplasmosis cooperative study: Veterans Administration–Armed Forces Cooperative study on histoplasmosis. *Am Rev Respir Dis*, **89**, 641–50.

Tewari, R.P. 1975. *The immune system and infectious diseases*. Basel, Switzerland: S Karger, 441.

Tewari, R.P. and LaFemina, R. 1983. Immunogenicity of subcellular fractions from yeast cells of *Histoplasma capsulatum*. *Jpn J Med Mycol*, **24**, 15.

Tewari, R.P., Sharma, D., et al. 1977. Adoptive transfer of immunity from mice immunized with ribosomes or live yeast cells of *Histoplasma capsulatum*. *Infect Immun*, **15**, 789–95.

Tewari, R.P., Sharma, D.K. and Mathur, A. 1978. Significance of thymus-derived lymphocytes in immunity elicited by immunization with ribosomes or live yeast cells of *Histoplasma capsulatum*. *J Infect Dis*, **138**, 605–13.

Tewari, R.P., Mkwananzi, J.B. et al. 1981. Natural and antibody-dependent cellular cytotoxicity (ADCC) of human peripheral blood mononuclear cells (PBMCC) to yeast cells of *Histoplasma capsulatum* (HC). *Abstracts of the Annual Meeting of the American Society for Microbiology, Dallas TX*, Abstract no. F-66.

Tewari, R.P., Khardori, N., et al. 1982. Blastogenic responses of lymphocytes from mice immunized by sublethal infection with yeast cells of *Histoplasma capsulatum*. *Infect Immun*, **36**, 1013–18.

Tewari, R.P., Gatchel, S.L. et al. 1986. Phagocytosis of *Histoplasma capsulatum* by human endothelial cells in culture. *Abstracts of the XIV International Congress of Microbiology, Manchester, England*, Abstract no. PM9-1.

Tewari, R.P., Kohler, R.B. and Wheat, L.J. 1989. *Fungal antigens, isolation, purification and detection*. New York: Plenum, 431.

Tosh, F.E., Weeks, R.J., et al. 1967. The use of formalin to kill Histoplasma capsulatum at an epidemic site. *Am J Epidemiol*, **85**, 259–65.

Urschel, H.C. Jr, Razzuk, M.A., et al. 1990. Sclerosing mediastinitis: improved management with histoplasmosis titer and ketoconazole. *Ann Thorac Surg*, **50**, 215–21.

Vincent, R.D., Goewet, R., et al. 1986. Classification of *Histoplasma capsulatum* isolated by restriction fragment polymorphisms. *J Bacteriol*, **165**, 813–18.

Vining, L.K. and Weeks, R.J. 1974. A preliminary chemical and physical comparison of blackbird-starling roost soils which do or do not contain *Histoplasma capsulatum*. *Mycopathol Mycol Appl*, **54**, 541–8.

Walker, J. and Spooner, E.T.C. 1960. Natural infection of the African baboon (*Papio papio*) with the large-cell form of *Histoplasma*. *J Pathol Bacteriol*, **80**, 436–9.

Weeks, R.J., Padhye, A.A. and Ajello, L. 1985. *Histoplasma capsulatum* variety *farciminosum*: a new combination for histoplasma farciminosum. *Mycologia*, **77**, 964–70.

Wheat, J., Hafner, R., et al. 1995. Itraconazole treatment of disseminated histoplasmosis in patients with the acquired immunodeficiency syndrome. *Am J Med*, **98**, 336–42.

Wheat, L.J. 1993. *Fungal diseases of the lung*, 2nd edition. New York: Raven Press, 29–38.

Wheat, L.J. and Kauffman, C.A. 2003. Histoplasmosis. *Infect Dis Clin North Am*, **17**, 1–19, vii.

Wheat, L.J., Slama, T., et al. 1981. A large urban outbreak of histoplamosis: clinical features. *Ann Intern Med*, **94**, 331–7.

Wheat, L.J., French, M.L., et al. 1982. The diagnostic laboratory tests for histoplasmosis: analysis of experience in a large urban outbreak. *Ann Intern Med*, **97**, 680–5.

Wheat, L.J., Stein, L., et al. 1983. Pericarditis as a manifestation of histoplasmosis during two large urban outbreaks. *Medicine*, **62**, 110–19.

Wheat, L.J., Wass, J., et al. 1984. Cavitary histoplasmosis occurring during two large urban outbreaks: analysis of clinical, epidemiologic, roentgenographic, and laboratory features. *Medicine*, **63**, 201–9.

Wheat, L.J., French, M.L., et al. 1986. Evaluation of cross-reactions in *Histoplasma capsulatum* serologic tests. *J Clin Microbiol*, **23**, 493–9.

Wheat, L.J., Batteiger, B.E. and Sathapatayavongs, B. 1990a. *Histoplasma capsulatum* infections of the central nervous system: a clinical review. *Medicine*, **69**, 244–60.

Wheat, L.J., Connolly-Stringfield, P.A., et al. 1990b. Disseminated histoplasmosis in the acquired immune deficiency syndrome: clinical findings, diagnosis and treatment, and review of the literature. *Medicine*, **69**, 361–74.

Wheat, L.J., Connolly-Stringfield, P.A., et al. 1991. Histoplasmosis relapse in patients with AIDS: detection using *Histoplasma capsulatum* variety *capsulatum* antigen levels. *Ann Intern Med*, **115**, 936–41.

Williams, A.O., Lawson, E.A. and Lucas, A.O. 1971. African histoplasmosis due to *Histoplasma duboisii*. *Arch Pathol*, **92**, 306–18.

Williams, B., Fojtasek, M., et al. 1994. Diagnosis of histoplasmosis by antigen detection during an outbreak in Indianapolis, IN. *Arch Pathol Lab Med*, **118**, 1205–8.

Williams, D.M., Graybill, J.R. and Drutz, D.J. 1978. *Histoplasma capsulatum* infection in nude mice. *Infect Immun*, **21**, 973–7.

Williams, D.M., Graybill, J.R. and Drutz, D.J. 1981. Adoptive transfer of immunity to *Histoplasma capsulatum* in athymic nude mice. *Sabouraudia*, **19**, 39–48.

Wood, K.L., Hage, C.A., et al. 2003. Histoplasmosis after treatment with anti-TNF-α therapy. *Am J Respir Crit Care Med*, **167**, 1279–82.

Wu-Hsieh, B. and Howard, D.H. 1984. Inhibition of growth of *Histoplasma capsulatum* by lymphokine-stimulated macrophages. *J Immunol*, **132**, 2593–7.

Wu-Hsieh, B.A. and Howard, D.H. 1987. Inhibition of the intracellular growth of *Histoplasma capsulatum* by recombinant murine gamma interferon. *Infect Immun*, **55**, 1014–16.

Yamada, T., Khardori, N. and Tewari, R.P. 1982. Natural and antibody-dependent cellular cytotoxicity (ADCC) of macrophages and lymphocytes from normal and immune mice for yeast cells of *Histoplasma capsulatum*. *Abstracts of the Annual Meeting of the American Society for Microbiology, Atlanta GA*, Abstract no. F-9.

Zeidberg, L.D., Ajello, L., et al. 1952. Isolation of *Histoplasma capsulatum* from soil. *Am J Public Health*, **42**, 930–5.

Zhou, P., Sieve, M.C., et al. 1995. IL-12 protects mice infected with *Histoplasma capsulatum* through induction of IFNγ. *J Immunol*, **155**, 785–95.

Zhou, P., Sieve, M.C., et al. 1997. Interleukin-12 modulates the protective immune response in SCID mice infected with *Histoplasma capsulatum*. *Infect Immun*, **65**, 936–42.

28

Paracoccidioidomycosis

GIL BENARD AND MARCELLO FRANCO

ETIOLOGIC AGENT

History

Paracoccidioidomycosis, a term officially recognized in 1971 in Medellín, Colombia, during a meeting of mycologists, is a deep mycosis which may be manifested as an overt disease or occur as an asymptomatic form. The mycosis is of great interest for Latin American countries and is caused by a dimorphic fungus *Paracoccidioides brasiliensis*, which still has not revealed its perfect, sexual phase (Lacaz 1994a).

In 1908, in the city of São Paulo, Brazil, Adolpho Lutz (1855–1940), one of the greatest scientists in Brazilian medicine, described this morbid entity for the first time in two patients, who presented with extensive oropharyngeal lesions associated with cervical lymphadenitis. Lutz was able to isolate the microorganism in culture; the parasite's colony showed gross resemblance to the fur of a white rat, grew as a filamentous microorganism in various media at room temperature, and had a dimorphic nature. As Lutz never saw endogenous sporulation, he considered the fungus to be different from *Coccidioides immitis*, a fungal agent previously described in Argentina (Lutz 1908).

Between 1909 and 1912, Alfonso Splendore (1871–1953), a distinguished Italian bacteriologist, observed new cases of the disease, gave a more complete description of its causative agent and of the clinico-histopathological features of the disease, and assigned the causative agent to the *Zymonema* genus under the name *Zymonema brasiliense* (Splendore 1912).

Later on, the disease was baptized 'Brazilian blastomycosis' and soon after, 'South American blastomycosis' due to the observation of cases in other South American countries.

In 1923, Floriano Paulo de Almeida (1898–1977), a researcher working at the University of São Paulo Medical School, confirmed that the agent differed from that of coccidioidomycosis, and, in 1930, after a series of systematic studies, established a new genus within the kingdon of fungi, i.e. *Paracoccidioides*. He revalidated the name of the species created by Splendore and officially assigned the name of the new fungus, *Paracoccidioides brasiliensis*, according to the rule established by Linneus (Campos and Almeida 1927).

Synonymy of the disease then proliferated, with the coining of terms such as paracoccidioidic granuloma, paracoccidioidic granulomatosis, malignant ganglionic granuloma of blastomycetic origin, neotropical blastomycoid granulomatosis, adenomycosis, Lutz disease, and Lutz–Splendore–Almeida disease.

As a result of its geographical distribution, the mycosis was also designated as South American blastomycosis. After the First Pan-American Symposium on Paracoccidioidomycosis, held in Medellin, Colombia, in 1971, the term paracoccidioidomycosis became well established, and is nowadays the denomination universally accepted (PAHO, 1972).

Mycology

The etiologic agent of paracoccidioidomycosis, *P. brasiliensis*, is a dimorphic fungus only known in its asexual state (anamorph). The dimorphism depends primarily on the temperature.

In its parasitic phase and in culture at 35–37°C, the growth is yeast-like. The fungal elements are rounded cells up to 5–40 μm in diameter, with a well-defined refringent apparently double wall and reproduction by single or multiple buds, which are connected to the parent cells by a narrow bridge. Sometimes, in the lesions, the parasite presents as small diminute yeast cells, measuring about 3–5 μm (Lacaz 1994b).

The type of reproduction of *P. brasiliensis* is characterized by the formation of chromatin masses in the cytoplasm, which may cross already existing small openings in the cell walls, or newly formed ones, dragging the cytoplasmatic mass and the enveloping membrane (Lacaz 1994b).

The variability in size and number of the daughter cells and their connection to the parent cells are characteristic. The large cells, with multiple exosporulation, which in cross-section resembles a pilot's wheel, are pathognomic.

In culture at room temperature, the growth is mycelial. The colonies do not grow rapidly and start to develop after 20–30 days of incubation: they become velvety and cracked, and may show glabrous, cerebriform, and brownish features. Under the microscope, the colonies show fine, septated, interlaced, mycelial hyaline hyphae, 0.8–2.5 μm in diameter. Round to subspherical, 20–50 μm in diameter, terminal, and intercalary chlamydoconidia may be observed (Lacaz 1994b) (Figure 28.1).

In the parasitic state of *P. brasiliensis*, the blastoconidium has a blastic development which is characterized by the blowing-out of a conidiogenous cell or of a fertile hypha to form a conidium. The electron microscopy studies suggested that both walls of the conidiogenous cell are used in the formation of the conidial wall, which means that the conidia can be considered as holoblastic. Regarding the order of arrangement and the position of the conidium, the terms used for *P. brasiliensis* will be botryose, asynchronous (Lacaz 1994b).

It should be mentioned that *P. brasiliensis* is a eukaryotic cell or eukaryote, showing a cell wall made up by chitin as visualized by periodic acid–Schiff (PAS). This wall can be well detected by electron microscopy.

The molecular aspects of *P. brasiliensis* dimorphism, diagnosis, epidemiology, taxonomy, and genetics have recently been extensively reviewed (San-Blas et al. 2002).

Molecular tools have been used to discover genes with differential expression according to the fungal phase. It is important to stress that glucose polymers are arranged mainly as α-1,3-glucan in the cell wall of the pathogenic yeast phase (95 percent) of the fungus whereas β-1,3-glucan is the only glucose polymer found in the mycelial cell wall (San-Blas 1979). These features have a remarkable importance in pathogenicity and dimorphism of *P. brasiliensis*.

The synthesis in vitro of β-glucan by β-1,3-glucan synthase requires UDP-glucose as the preferred nucleotide precursor. So far, only one related *P. brasiliensis* gene, *FKSP61*, has been cloned and sequenced (Pereira et al. 2000).

In addition, two other DNA fragments have been shown to correspond to distinct glucan synthase genes, which indicates that genes encoding glucan synthases appear to comprise multigene families.

On the other hand, no α-1,3-glucan synthase gene has been reported in *P. brasiliensis*. This polysaccharide is organized as the outer capsule in the yeast phase of the fungus and replaces almost entirely the β-glucan that comprises the neutral polysaccharide of the vegetative mycelial phase, behaving like a virulence factor.

Chitin represents a major structural component of the cell wall with functions in morphogenesis and virulence. Five chitins synthase genes, representing different classes of enzyme, are active in *P. brasiliensis* (Niño-Vega et al. 2000b).

The best studied proteins and genes related to fungal morphogenesis include the heat shock proteins (Hsp). The Hsp 70 gene isolated in *P. brasiliensis* is differentially expressed during transition from the mycelial to yeast form (Silva et al. 1999).

The synthesis of polyamines is a metabolic process that appears to be involved in the dimorphic process; this process is catalyzed by ornithine decarboxylase (ODC). The *PbrODC* gene has been cloned and sequenced in *P. brasiliensis*.

Several *P. brasiliensis* DNA sequences of potential diagnostic use have been reported, such as a species-specific 110 bp DNA fragment, a species-specific 14-base DNA probe originated from the 28S ribosomal gene of the fungus, a specific 418 bp fragment from the 5.8S rDNA flanking internal transcribed spacer regions, and primers derived from the sequence of the gene coding for the immunodominant gp43 antigen (Goldani et al. 1995; Sandhu et al. 1997; Imai et al. 2000).

The molecular methods used for identifying *P. brasiliensis* are also used for epidemiologic screening and in molecular taxonomic studies. Random amplified polymorphic DNA (RAPD) analysis of *P. brasiliensis* strains have indicated that DNA variation correlates with geographical areas but not with pathological findings (Calcagno et al. 1998). RAPD analysis also achieved discrimination that reflected degree of virulence. Altogether these data suggest that *P. brasiliensis* may consist of several genetically distinct groups under a single morphological species (Molinari-Madlum et al. 1999).

The integration of data between morphological features, virulence, and DNA variation might eventually lead to a more appropriate species differentiation concept in the fungus.

Molecular methods have also been useful to redefine the taxonomy of *P. brasiliensis*. Phylogenetic comparison based on large subunit (28S) ribosomal rDNA sequences has placed the fungus as belonging in the order Onygenales, family *Onygenaceae* (phylum Ascomycota) (Leclerc et al. 1994).

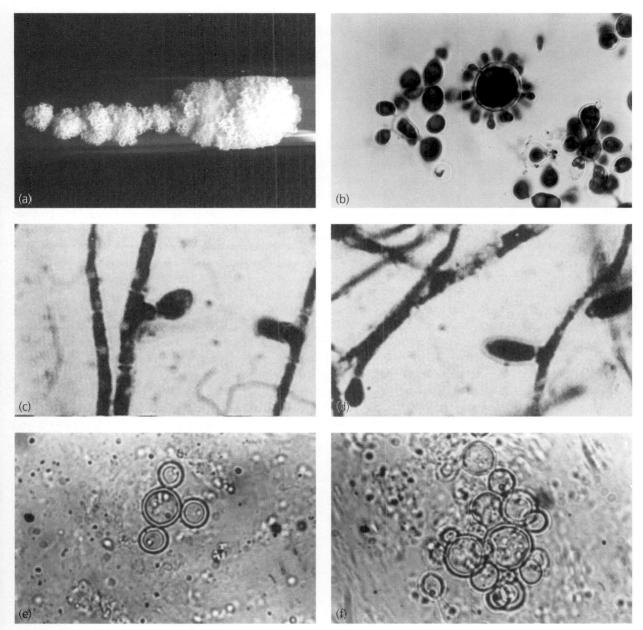

Figure 28.1 (a) *Macroscopic aspect of a* Paracoccidiodes brasiliensis *culture at 37°C on Fava Netto's medium.* **(b)** *Yeast-like forms on lactophenol culture grown at 35°C.* **(c, d)** *Mycelial form grown at 25°C on Sabouraud agar, showing fertile hyphae producing terminal conidia.* **(e, f)** *Direct examination: KOH mount of sputum showing yeast-form-like cells.*

Four or five chromosomes of variable molecular weights have been reported in *P. brasiliensis* depending on the samples and the techniques used for karyotyping. The data so far indicate that the fungus has chromosomal polymorphism, due to genetic translocations or large-scale deletions, as is the case for other pathogenic fungi. The phenomenon may play a role in promoting the genetic variability and accelerated evolution of different isolates (Montoya et al. 1999). The genome size of *P. brasiliensis* is around 45.7–60.9 Mbp and the nuclei of the yeast cells may be diploid (Cano et al. 1998).

The *P. brasiliensis* genome project is under way. There has been constructed a *P. brasiliensis* DNA library from the yeast pathogenic forms and sequenced 13 490 Expressed Sequence Tags (EST) and clusters. Sequence analyses of these clusters have already identified several virulence factors (Goldman 2002). In addition, a similar approach is aiming at identifying expressed genes in mycelium and yeast forms. Seven genes that code for members of the Hsp family have been identified which could potentially be involved in the dimorphic and thermoregulated transition of the fungus. Genes coding to multidrug resistance proteins have also been identified. These issues will only be further addressed as gene-disruption and/or RNA interference approaches become available for the pathogen (Felipe et al. 2002).

Ecology and epidemiology

TRANSMISSION AND ECOLOGICAL NICHE

Paracoccidioidomycosis is acquired most commonly by inhalation of propagules from the filamentous phase of the infecting agent present in the nature (Borelli 1963–64). The true prevalence and incidence of paracoccidioidomycosis are not precisely known. Many factors contribute to this situation. The mycosis is not a notifiable disease in the countries where it is endemic. Cases are sporadic; neither epidemic outbreaks nor interhuman transmission have been reported to date. Familiar incidence is rare. Moreover, the ecological niche of the fungus is not yet solved (Restrepo et al. 2001). The long time of incubation, sometimes decades, in the most common form of the disease, the chronic form, makes the delimitation of the areas where the disease was probably acquired difficult. Acute/subacute form cases, more rarely seen, make this task somewhat easier. The presence of a significant percentage of children under 14 years of age with the acute/subacute form within a circumscribed region would imply that it is an endemic region, because such children have less frequently moved away from their living place as compared to adult patients The regions considered endemic in Latin America are despicted in Figure 28.1.

In the face of these difficulties, Borelli (1963–64) created the concept of 'reservárea' for the place in endemic areas where man acquires the infection, and consequently where the fungus has its habitat, in opposition to 'endemic area,' which refers to the place where the disease is diagnosed.

Previous estimates of incidence were based on the number of cases per year diagnosed in traditional mycoses services and encompassed mostly the period from 1940 to 1980. Based on these data, Londero and Ramos (1990) had estimated an annual incidence of one to three cases per 100 000 habitants in highly endemic areas, while Franco et al. (1989) estimated an annual incidence of four new cases per million habitants. However, this may not hold valid because of the increasing urbanization of the population in the Latin American countries in the last decades. A recent study analyzed 3 181 deaths from paracoccicioidomycosis in Brazil, based on 16 years of sequential data. The mean annual mortality rate was 1.45 per million inhabitants, representing the eighth most common cause of death from predominantly chronic or recurrent infectious and parasitic diseases; the data justified classifying the systemic mycosis as a major health problem in Brazil (Coutinho et al. 2002).

The *P. brasiliensis* habitat still is a puzzle and has been reviewed recently (Restrepo et al. 2001). It is presumed that the fungus lives in its saprophytic filamentous phase in humid, protein-rich soils, in areas with small temperature variations, where it would produce the arthroconidia and aleurionoconidia. Once man disturbs this habitat, these small infectious propagules, which are less than 5 μm in diameter, would disperse in the air, be inhaled, and reach the terminal bronchi and alveoli. Thus the strong link that exists between agricultural activity and paracoccidioidomycosis infection and disease. Recently, the role played by deforestation in increasing both exposure and incidence rates has been documented in Brazil in Amerindians and in children (Coimbra et al. 1994; Rios-Gonçalves et al. 1998; Fonseca et al. 1999). Unexpectedly, however, there is only a small number of successful recoveries of *P. brasiliensis* from the soil (Franco et al. 2000). One possible reason comes from a recent study on the action of agricultural pesticides on *P. brasiliensis*, showing that its in vitro growth was inhibited by concentrations of pesticides that can usually be found in agricultural soils (Ono et al. 2001).

Borelli (1963–64) has also previously suggested that the fungus survives in cavities in the underground, associated with rodents' holes and armadillo's burrows. Although there is still no information regarding rodents, the hypothesis is consistent with the recent findings of fungal isolation from the nine bands armadillos (*Dasypus novemcinctus*) (Naiff et al. 1986; Cadavid and Restrepo 1993; Bagagli et al. 1998; Silva-Vergara and Martinez 1999). Moreover, skin surveys of domesticated animals such as cows, horses, and sheep revealed a high proportion of positive responses to paracoccidioidin (Costa and Fava Netto 1978). Serological studies in dogs showed high percentages of positive tests, mainly in dogs from rural areas as compared to urban animals (Mós and Fava Netto 1974; Ono et al. 2001). Up to very recently, the enigma of why *P. brasiliensis* could not be found in autopsies of dogs highly exposed to the fungus as demonstrated by high titers of specific antibodies was explained in principle by the assumption that dogs are extremely resistant to paracoccidioidomycosis, as opposed to coccidioidomycosis. However, a recently proven case of paracoccidioidomycosis in a dog will reshape this discussion (Ricci et al. 2002).

A careful study in Colombia by Restrepo's group attempted to characterize the ecological factors that prevail in the 'reservárea.' Altitude from 1 000 to 1 499 meters above sea level, rainfall from 2 000 to 2 999 mm, presence of humid forests and of coffee and tobacco crops were significantly associated with the disease (Calle et al. 2001). In Brazil more than 800 cases have been notified in the last 6 years in an area at the border of the amazonic region that had experienced an intense migration process and had become a new agricultural frontier (Lima and Durlacher 2003).

AGE AND GENDER

The disease predominates in men aged 30–50 years. It decreases gradually over 60 and abruptly in the young.

This is because the chronic form of the disease is the most frequent presentation of the mycosis, the acute/subacute form representing usually less than 10 percent of the cases. Males are afflicted with much greater frequency than females, with a ratio 14:1 or more, depending on the patient series. The reason for the lower incidence in women was elegantly explained by observations showing that *P. brasiliensis* mycelia and conidia possess an estrogen receptor and that estrogens inhibit the mycelia/conidia to yeast transformation (Restrepo et al. 1997; Aristizábal et al. 1998). These data fit well with the finding that the acute/subacute form of the disease affects equally both genders before puberty, but preferentially men after puberty (Silvestre et al. 1997), and with the knowledge that infection is acquired at an early age with no gender differences (Londero et al. 1987). In the review published by Londero et al. (1996), which included children up to 14 years of age, there was a slight predominance of males in the older group. In endemic areas, skin test surveys showed that positive reactions start before completion of the first decade of life, reaching their peak by 15–19 years of age, and only decrease after the age of 50, probably due to immunological senescence. However, paracoccidioidomycosis disease has occasionally been seen in fertile adult women; in these cases a search for immunosuppressive-associated conditions must be undertaken, with some reports of a disseminated disease in pregnant women (Martinez et al. 1993).

OCCUPATION AND RACE

Almost half of the reported cases in adults occur in individuals whose occupations require extensive exposure to the soil. Among the agricultural activities, work in coffee plantations has been traditionally linked to the mycosis (Naiff et al. 1988; Coimbra et al. 1994). In a recently published epidemiologic survey of 584 cases from an endemic region in São Paulo State, Brazil, bricklaying and masonry were, after agricultural activities, the second most important risk factors, which suggests that digging and excavating may also favor the encounter with fungus from under the ground, as first pointed out by Borelli in 1972 (Blotta et al. 1999). No ethnicity susceptibility has been documented. Although a few studies had previously suggested some link between HLA and disease susceptibility, the influence of genetic traits on the mycosis has thus far not been elucidated (Dias et al. 2000).

GEOGRAPHICAL DISTRIBUTION

The mycosis was limited to continental Latin America until 1987, when cases were reported in the Caribbean islands of Trinidad, Grenada, and Guadeloupe (Restrepo-Moreno 1994). Paracoccidioidomycosis is distributed heterogeneously in the endemic region; areas of high endemicity may be localized next to areas of very low incidence. The highest numbers of cases have been reported from Brazil, Colombia, and Venezuela (Restrepo 1985); 7 000 of the approximately 10 000 patients reported from that period were natives of Brazil (Figure 28.2). Cases diagnosed outside Latin America referred to individuals who, with no exception, once visited or lived in endemic areas from Latin America up to 10–40 years before (Ajello and Polonelli 1985). Within endemic areas. 10–50 percent of healthy adult individuals react to paracoccidioidin, suggesting previous exposure to *P. brasiliensis*. Caution in the interpretation of these skin test surveys comes from studies showing that concomitant histoplasmin reactivity significantly increases the rate of paracoccidioidin reactivity when crude paracoccidioidal antigens are used. The use of the specific gp43 antigen has significantly reduced the percentage of skin-reactive persons in areas where *Histoplasma capsulatum* is endemic (Kalmar et al. 2004). Recent molecular biology studies have indicated that there are at least five different groups of *P. brasiliensis* strains and that they correspond closely with the borders of the various endemic areas and countries (Niño-Vega et al. 2000a).

OTHER FACTORS

Alcoholism was shown to be a predisposing factor to paracoccidioidomycosis. It also probably accounts for a worse prognosis for this infection, at least because of the poorer compliance to treatment of such patients (Martinez and Moya 1992). Undernourishment is another factor commonly associated with paracoccidioidomycosis (Restrepo and Benard 2004).

HOST RESPONSES

Most authors consider that the infection takes place through the inhalation of fungal propagules and that the lungs are the primary focus of the lesions. In most cases, the initial pulmonary infection does not cause noticeable symptoms. Once the conidia reach the terminal bronchi or alveolar spaces, the conidia to yeast transformation takes place (McEwen et al. 1987). In some cases, the fungus disseminates by the lymphohematogenous route, producing distant subclinical and quiescent foci. Apparently, the local defenses are capable of controlling fungal spread, but some viable yeast cells may remain silent in such foci in the pulmonary/mediastinal lymph nodes or even elsewhere. Rarely, the initial pulmonary infection overcomes the host's immune defenses, causing an acute/subacute disease, the juvenile type, with predominant involvement of the reticuloendothelial system. Most often, the quiescent foci remain so throughout life, as demonstrated by the high number of subclinical infections versus the low disease incidence of this mycosis in endemic regions (Wanke and Londero 1994). Nevertheless, the most common clinical presentation of the disease is the chronic form, or adult-type disease,

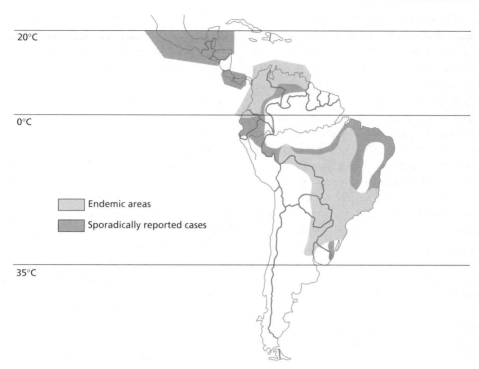

Figure 28.2 *Geographical distribution of paracoccidioidomycosis.*

believed to result from fungal reactivation in these foci, which not infrequently occurs when the patient has already left the endemic area, as demonstrated by the patients who developed the disease many years after they have migrated to countries outside Latin America (Figure 28.3).

Actually, paracoccidioidomycosis, either in the acute/subacute or chronic form, is more frequently a disseminated disease, even though clinical manifestations appear to be restricted to a sole organ (Yamaga et al. 2003). However, determination of the degree of dissemination is certainly influenced by the availability of diagnostic procedures. Thus, as a polymorphic disorder that frequently involves more than one organ system, any topographic classification is ineffective. The currently accepted classification of the mycosis takes into consideration not only the organs involved but also the host's immune condition and the disease's

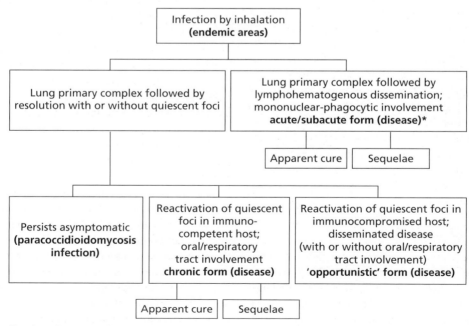

Figure 28.3 *Classification of paracoccidioidomycosis.*

natural history (Franco et al. 1987). The different outcomes that may result from the *P. brasiliensis*–host interaction are shown in Figure 28.1. Those individuals who remain infected are categorized as having only the subclinical form. The disease that develops soon after the initial infection is classified as the acute/subacute form. The acute/subacute pattern predominates in children and young adults. According to the severity of the process, acute/subacute patients are assigned to two subgroups, severe and moderate. The chronic progressive form of paracoccidioidomycosis, more common in adult men over 30 years of age, may remain localized in the lungs, the unifocal disease, or may disseminate from its primary foci, the multifocal disease. The chronic form can be mild, moderate, or severe (Franco et al. 1987).

Clinical forms

PARACOCCIDIOIDOMYCOSIS INFECTION

P. brasiliensis-infected persons would correspond to up to 10 percent of the population in countries where paracoccidioidomycosis is endemic, according to previous skin-test surveys (Wanke and Londero 1994). However, there are only sparse reports on the encounter of the fungus in quiescent foci, usually calcified lymph nodes, accidentally or in autopsies of individuals from these countries who died of unrelated causes (reviewed in Restrepo 2000). There are also anedoctal case reports of a pulmonary regressive form (Restrepo et al. 1976; Londero 1986). The study of the morphology and viability of the dormant yeast cells from a patient's residual lesion demonstrated a large proportion of aberrant cells, but there were still 5 percent viable cells (Restrepo 2000). Analysis of some immunological parameters from a small number of infected individuals revealed a Th1-type immune profile that distinguishes them from patients with this disease or cured individuals (Mamoni et al. 2002). Another example is a recent serological survey of blood donors from the countryside of a Brazilian endemic region, employing a more sensitive test, which disclosed that 12 percent of the putatively healthy donors produce specific antibodies (Botteon et al. 2002). The reasons why in some of these infected persons the fungus reactivates at some moment and provokes the chronic form of the disease are not known. A thorough clinical and immunopathological study of persons living in endemic areas is still lacking and is needed before any definitive statement on the concept of paracoccidioidomycosis infection can stand.

ACUTE/SUBACUTE FORM OF THE DISEASE

The time elapsed from infection to the onset of symptoms in the acute/subacute form is not precisely known, but has been estimated to be short, a few months. The acute form involves preferentially the reticuloendothelial system. Fever, malaise, weight loss, and emaciation may be seen in the acute/subacute form. Actually, only a minor proportion of the cases reported have had lung symptomatology. However, this involvement may be underestimated because the severe extrapulmonary lesions tend to minimize the more discrete respiratory manifestations (Londero 1986). Frequently, the mycosis is misdiagnosed as ganglionar tuberculosis or certain lymphomatous disorders. However, induced sputum or other maneuvers may reveal the characteristic *P. brasiliensis* yeast cells (Restrepo et al. 1989). New imaging methods, such as gallium and computerized tomography, have allowed detection of incipient or discrete interstitial pulmonary lesions not revealed by plain radiographs (Funari et al. 1999; Yamaga et al. 2003).

Mucosal lesions are rather exceptional, but skin involvement is more common and tends to be multiple, in contrast with the chronic form (Restrepo and Benard 2004). The lesions represent hematogenous spread of the fungus. In this case, they may appear as ulcerated or ulcerovegetative lesions, papules, crust-covered ulcers, usually at the same stage of development. Sporadic reports of septic shock due to septicemia by *P. brasiliensis* show that the fungus can be bloodborne.

Almost every patient with acute/subacute paracoccidioidomycosis exhibits involvement of the superficial and/or deep lymph node chains. Lymph nodes vary in size, number, consistency, and location; with time, they liquefy, forming abscesses or fistulae (Figure 28.4). The spleen and liver are frequently involved. Splenic lesions are nodular or miliary. Gross hepatic lesions may not be apparent, but histopathological examination regularly reveals fungal invasion of this organ (Teixeira et al. 1978). Abdominal lymph node involvement is also common. Hypertrophied lymph nodes, usually generalized but particularly periaortic, around the hepatic hilum and in the retroperitoneum, can be detected by radiological examination. Patients may complain of abdominal masses, lymph node enlargement, diarrhea, vomiting, abdominal distension and/or pain, and ascites. Coalescent masses may become palpable and may result in extrinsic compression of adjacent structures, leading to jaundice (by compression of the biliary duct), pancreatitis, or intestinal obstruction (Martinez et al. 1979a).

The intestinal mucosa may be affected and the involvement may be secondary to blockade of the regional lymphatic flow, with retrograde progression of *P. brasiliensis* to the mucosa, a process resulting in mycotic enteritis (Martinez et al. 1979b). In this case, the submucosal inflammatory process is granulomatous, fungal cells are visualized, and the intestinal changes may vary from dilated loops, edema, congestion, nodule formation to multiple mucosal ulcers (Martinez et al. 1979a).

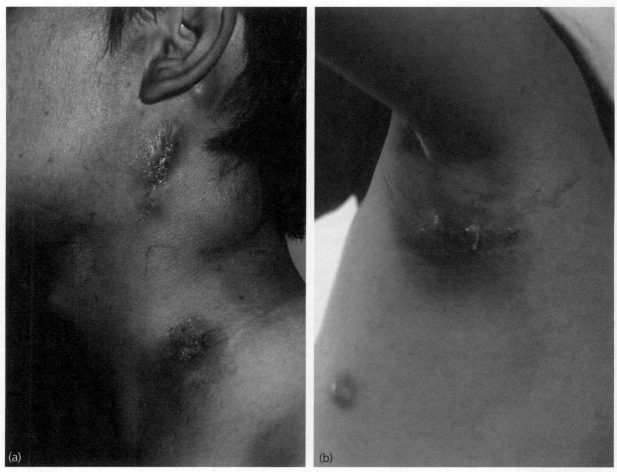

Figure 28.4 *Subacute form of paracoccidioidomycosis in:* **(a)** *a young adult man: enlarged cervical suppurated lymph nodes;* **(b)** *a girl: suppurated axilar lymph node.*

Recent studies have also shown bone marrow infiltration mainly, but not exclusively, in the acute/subacute form of the disease (Resende 2000). The histological pattern is variable but appropriate staining always displays fungal cells. Bone marrow invasion is frequently associated with marked eosinophilia (Shikanai-Yasuda et al. 1992a). Bone and joint lesions are frequent, and appear closely related to bone marrow infiltration. The lytic lesions mainly locate at the diaphyseal or metaphyseal–epiphyseal regions of the long bones, but ribs, skull, phalanges, and vertebral lytic lesions have also been documented. A pathological fracture may occasionally occur.

CHRONIC FORM

Most often, chronic-form patients present a history of a few to several months of cough, either productive or not, weight loss, anorexia. Fever may be absent or present in low grade and nocturnal, as in tuberculosis, which is the main differential diagnosis. Occasionally, these respiratory symptoms are mild and the patient seeks medical care only when lymphohematogenous dissemination results in lymph node involvement or distant cutaneous lesions. Plain films taken during admission reveal the

primary, pulmonary focus. The pulmonary involvement may be present as micronodular or miliary, nodular, infiltrative or interstitial, cavitary, fibrotic, or mixed patterns (Valle et al. 1992) (Figure 28.5). Emphysematous areas, pleural thickening, and enlarged hilar and mediastinal lymph nodes can also be observed. Right ventricular hypertrophy may be found in cases of long duration. Lung involvement is probably secondary to a chronic lymphangitic process due to the fungus itself and to the host's response represented by formation of granulomas and fibrosis, the latter predominating at the peri-hilar region (Tuder et al. 1985). This aspect correlates with the butterfly-like (peri-hilar) micronodular and interstitial infiltration observed on plain films.

Alternatively, in some patients the main complaints may be dysphagia and hoarseness. Generally they are associated with lung infiltrates on the x-ray, and in these cases the decrease in the nutritional status is more prominent due to the patient's impairment in swallowing. The lesions in the oropharyngeal and laryngeal mucosa may be infiltrative, ulcerated, nodular, or vegetative, and usually have a granulomatous aspect (Sant'Anna et al. 1999) (Figure 28.6). The base of the ulcerated lesions is usually covered by small abscesses

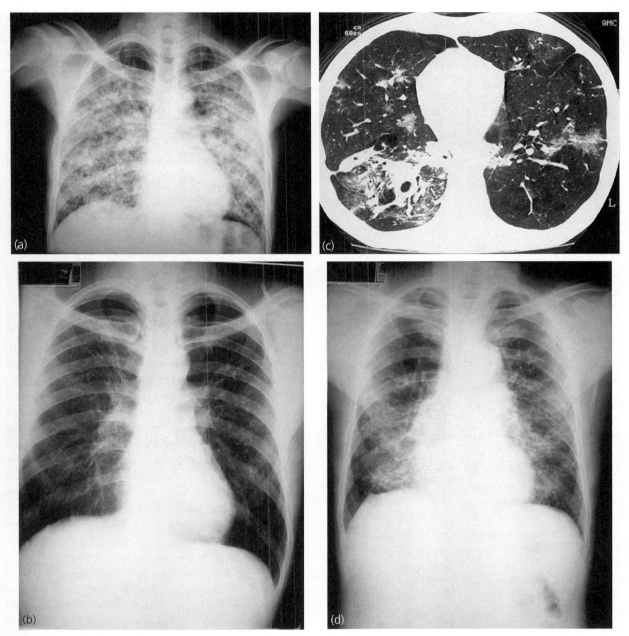

Figure 28.5 *Pulmonary involvement in the chronic form paracoccidioidomycosis:* **(a)** *extensive pneumonic involvement of the lungs and* **(b)** *the same patient, after 2 years of treatment, with discrete fibrotic, mainly perihilar, lines;* **(c)** *computed tomography showing diffuse pleuropathy with bronchocentric nodules, ground-grass opacities, cavities of several sizes, airspace consolidations, and centrilobular opacities (<1 mm) associated with bronchiectasis and parenchymal distortion.* **(d)** *Micronodular and reticular bilateral infiltrates predominating in the middle and lower fields.*

(the mulberry-like lesions) that probably represent fungal dissemination through the lymphatic system, as they are usually accompanied by regional lymph node involvement (Castro et al. 1999a). Cutaneous lesions are mostly represented by contiguous involvement of the periorificial mucosal lesions or draining lymph nodes or may represent hematogenous spread in patients with long-lasting disease. The latter, thus, have no preferential localization. The adrenals are often involved in patients with the chronic form of the mycosis, but only a small proportion of them will suffer from adrenal hypofunction or insufficiency (Addison's disease) (Del Negro

et al. 1980). The glands containing multiple, granulomatous foci, and diffuse necrosis may be seen in the most severe cases. Recently, silent CNS granulomatous-like lesions attributable to paracoccidioidomycosis have been reported in a small survey of patients using magnetic resonance imaging; the clinical importance of this finding awaits confirmation (Castro et al. 1999b). On the other hand, granulomatous involvement of the spinal cord has been reported in a sizable number of patients and is a severe condition because it is frequently associated with significant, often irreversible, neurological impairment.

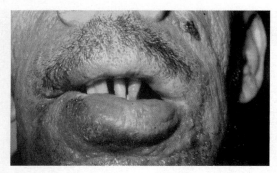

Figure 28.6 *Chronic form paracoccidioidomycosis in an adult man: a lips lesion with the moriform stomatitis aspect and a facial skin ulcer.*

'OPPORTUNISTIC' FORM

Paracoccidioidomycosis, as an apparently opportunistic infection in a patient with cancer, was first documented by Lacaz et al. in 1948. The association with hematological malignancies, predominantly, or other types of cancer, generally resulted in a disseminated form of the mycosis with clinical features of both the acute/subacute and chronic forms (Marques and Shikanai-Yasuda 1994). Unless the clinician is armed with a high degree of suspicion, the diagnosis of the mycosis is delayed. The same applies to AIDS patients. This subject has been recently reviewed (Benard and Duarte 2000). It is believed that in most instances it represents reactivation from quiescent foci as a result of the failure of the immune surveillance in blocking the parasite multiplication. The two reasons for the lower than expected number of cases of this mycosis in patients with the acquired immunodeficiency syndrome (AIDS) are supposedly that the two infections have non-overlapping epidemiologic features (e.g. urban versus rural) and that many patients infected with the human immunodeficiency virus (HIV) are on continuous antifungal regimens which, although directed at other fungal infections, also have a suppressive effect on *P. brasiliensis*. Paracoccidioidomycosis mainly afflicts those patients with low (<200 cells/µl) CD4 counts. Nonetheless, it has seldom been the cause of death. As with other fungal opportunistic infections, the mycosis responds adequately to the usual antifungal treatments.

SEQUELAE

Mycologically, successfully treated patients may persist with high morbidity. In the acute/subacute form, disabsorptive syndromes may ensue due to the digestive involvement that resolves by intense fibrosis and obstruction of the abdominal lymphatics, aggravating the nutritional status of the patients, resulting in immunodeficiency and opportunistic infections with high mortality rates (Shikanai-Yasuda et al. 1992b; Benard et al. 1996). Patients may also present with relapsing ascites or chylothorax. In the adult form, fibrosis, either pulmonary, tracheal, labial, or adrenal, leading to, respectively,

emphysema, permanent tracheostomy, microstomy, and Addison's disease, are the cause of serious problems in patients who respond to therapy (Mendes et al. 1994). Thus, despite the recent advances in the treatment, these sequelae still preclude the complete recovery in several patients.

Immunology

The initial events of the host–parasite interaction are unknown due to our inability to detect the precise moment when the initial infection occurs. The data collected up to now indicate that the host–parasite interaction in paracoccidioidomycosis is complex, and both arms of the immune system participate in the host's response. On the one hand, the patients have a polyclonal activation of the humoral system with high serum concentrations of specific IgA, IgG isotypes, and IgE (Yarzábal et al. 1980; Biagioni et al. 1984; Baida et al. 1999; Biselli et al. 2001; Mamoni et al. 2002). However a protective role for antibodies has not yet been demonstrated. On the other hand, there is hyporesponsiveness of the patients' T-cell-mediated immunity characterized by decreased lymphocyte proliferation and absence of hypersensitivity skin-test responses, the intensity of which parallels the severity of the disease and correlates inversely with the amount of circulating anti-*P. brasiliensis* antibodies (Musatti et al. 1976; Motta et al. 1985). This imbalance was particularly noted when antigen-specific responses were analyzed (Benard et al. 1996, 1997; Oliveira et al. 2002). Several studies showed that a marked imbalance in the cytokine secretion pattern underlies this picture: lack or strongly decreased secretion of Th1-type cytokines (IL-2, IL-12, and IFN-γ) associated with normal or increased secretion of Th2 cytokines (IL-10, IL-5, and IL-4) (Karhawi et al. 2000; Benard et al. 2001; Oliveira et al. 2002). Previous studies on animal models of paracoccidioidomycosis had already shown that Th2-type responses dominated over Th1-type responses (Kashino et al. 2000). Recently, the role of IL-10 and IL-12 in the regulation of the host's immune responses has been emphasized (Romano et al. 2002). In situ studies of patients' biopsy lesions have shown overexpression of TGF-β, a cytokine with strong immunosuppressive activity (Neworal et al. 2001; Parisi-Fortes et al. 2001). Furthermore, high levels of specific antibodies of the IgE and IgG4 (switch dependent upon IL-4) isotypes were found in patients with the acute/subacute form of the disease as well as in chronic cases with severely disseminated disease, denoting the Th2 control of the humoral immune response in these patients (Biselli et al. 2001; Mamoni et al. 2002). Another aspect in favor of Th2 control was the observation of peripheral and bone marrow eosinophilia in severe cases, an abnormality linked to increased production of IL-5 (Oliveira et al. 2002). Thus, apparently in the acute/subacute disease, the immune response fails to control the proliferation of the fungus from the beginning of the infectious process due to a profoundly downregulated Th1 response concomitant to upregulated Th2 immune response. In chronic less severe cases, the initial infection is apparently controlled, which allows a better balance of the immune response, with lower antigenic challenge and a less marked Th2 immune response. Apoptosis of the antigen-specific T cells has also been linked to the depressed cell-mediated immune response (Cacere et al. 2002). Patients with apparent cure demonstrate strong lymphocyte proliferative responses as well as positive skin tests, probably in consequence of the recovery of the fungal antigen-induced Th1 type reactivity (Benard et al. 1996; Karhawi et al. 2000).

The cytokine imbalance also affects the macrophages, which represent a crucial cell in the host defense mechanism against *P. brasiliensis*. It has been shown that the intracellular killing of *P. brasiliensis* yeast cells and conidia by this cell occurs only when previous activation has taken place (Gonzalez et al. 2000), which in humans has been shown to be mediated by the synergistic action of TNF-α and IFN-γ (Anjos et al. 2002). The main fungal antigen, gp43, appears to directly affect macrophage functions (Benard et al. 2001; Popi et al. 2002). Polymorphonuclear neutrophils (PMN) also participate in the immune response to the fungus, but their role is less well understood. PMNs are found in some paracoccidioidal lesions, specially in those representing acute inflammation or microabscesses (Sandoval et al. 1996); in experimental models they appear early in the inflammatory process and may play a role in the control of the initial infection (Burger et al. 1996). PMNs can also be activated by proinflammatory cytokines to have enhanced anti-*P. brasiliensis* activity (Kurita et al. 2000; Rodrigues et al. 2001) .

Treatment

The disease caused by *P. brasiliensis* can be managed successfully with oral antifungal drugs. Many drugs are available, and the list is still increasing. Historically, sulfa drugs were introduced in 1940 (Ribeiro 1940), followed by intravenous amphotericin B, in 1958 (Lacaz and Sampaio 1958). The use of azole derivatives (miconazole, soon discarded, ketoconazole, itraconazole, fluconazole, and more recently, voriconazole) was introduced in the 1970s. The treatment is divided into an attack phase and maintenance treatment. Two different drugs can be used for each phase, or the same drug can be used for the two phases but at different dosages (Mendes et al. 1994). Among the reasons for this two-phase prolonged treatment is the chronic progressive nature of the disease and the knowledge that all currently available drugs, albeit with apparently

different efficacy, are only fungistatic. In parallel, measures in support of the general condition are required, because cellular immune mechanisms may be affected by malnutrition, concomitant infections, or by the disease process itself, and they are crucial to controlling the fungus dissemination.

Sulfadiazine is used at 60–100 mg/kg daily doses divided in four equal parts to a maximum daily dose of 6 g. Water intake and urine alkalinization (usually by bicarbonated water intake) are recommended to prevent crystalluria and tubular deposits of sulfadiazine. Duration of the attack treatment is dictated by the patient's response, once clinical and serological improvement has been attained; in the chronic form, 2–6 months are required; in acute/subacute cases, the duration of treatment should be longer. Slow-acting sulfa drugs (sulfamethoxypyridazine or sulfadimethoxine), 1 g/day, can be provided as maintenance treatment (Mendes et al. 1994). Sulfadiazine is usually recommended for mild to moderate cases, but our own experience is that it can be used for severely ill acute/subacute cases with good results. It can be used for patients who have been treated initially with amphotericin B. Sulfadiazine has been replaced in many services by trimethoprim-sulfamethoxazole (80 mg of trimethoprim and 400 mg of sulfamethoxazole per tablet, two tablets at 8- or 12-hour intervals) due to its easier posology and availability. This combination has the advantage of permitting alternative parenteral administration whenever necessary. Maintenance treatment in these cases can be achieved by using half the dose of the attack treatment or by using slow-acting sulfas.

Amphotericin B is quite effective, but it should be reserved for severely disseminated cases or in whom gastrointestinal involvement may impair drug absorption. A total dose of 1–2 g is usually required to achieve substantial clinical improvement. There are no sufficient data regarding effectiveness of the new amphotericin B lipid formulations in paracoccidioidomycosis.

As to the imidazole derivatives, patients can be treated with 200–400 mg ketoconazole once daily for periods of 6–12 months and low relapse rates have been documented; however, other studies indicated higher relapses rates and the need for 18 months or more of treatment to achieve good results (Del Negro 1982; Restrepo et al. 1983; Marcondes et al. 1984; Dillon et al. 1985; Marques et al. 1985). These discrepant results may be related to differences in severity of disease among the studies and to problems in drug absorption. Indeed, caution should be urged regarding its absorption, which is dependent on acid gastric pH. Poor absorption has been described with concomitant use of antacids, or when the drug is taken with meals (Mendes et al. 1994). Liver toxicity as well as testosterone synthesis blockade, clinically manifested or not by gynecomastia or sexual impotence, should also be evaluated, especially during prolonged therapy.

With the advent of the new triazole, itraconazole, which was shown to be less toxic and more potent in vitro, ketoconazole tends to be used less. Itraconazole is administered in 100 mg capsules that should be given with a meal; a 100–200 mg regimen for adults, and half dosage for children, daily for 6 months, have been shown to be effective in reducing all active lesions in the large majority (98 percent) of patients, including acute/subacute and chronic forms (Mendes et al. 1994). It is important to remember that both ketoconazole and itraconazole penetrate the brain barrier poorly, thus CNS lesions should be carefully discarded. Instead of amphotericin B, we have used itraconazole, at higher doses (300–400 mg adults, 100–200 mg children) for some severe cases with good results, provided there is no involvement of the digestive tract that could hinder its absorption. Some other authors (Marques 2002; Shikanai-Yasuda et al. 2002a) suggest that patients should be kept on maintenance therapy with slow-acting sulfas after clinical and mycological cure to prevent relapses, while other investigators argue that this is not required (Telles 2002; Tobón et al. 2002). Side effects have been few and include transient elevation of hepatic enzymes (Tobón et al. 1995). A randomized trial with sulfadiazine, ketoconazole, and itraconazole for the treatment of patients with moderate severity disease failed to show higher efficacy of one regimen over the others. However, long-term follow-up was not evaluated (Shikanai-Yasuda et al. 2002b). Experience with fluconazole is far less extensive than with itraconazole, and due to its higher cost is still used less. It may be useful in severely ill patients who must be treated intravenously. There is only one report on the successful use of terbinafine in an adult patient (Ollague et al. 2000).

New perspectives in paracoccidioidomycosis treatment include the new itraconazole formulations (oral solution and intravenous formulations) that will probably further extend its indications. A multicentric prospective trial on voriconazole, 400 mg daily for 6 months, showed encouraging results (F.Q. Telles Filho, M.A. Shikanai-Yasuda and L.Z. Goldani, 2003, unpublished observations). This drug will probably be an excellent option due to its efficacy and low incidence of side effects. Nevertheless, some challenges in the treatment of patients with this mycosis still remain. Neither the clinical and laboratory criteria that allow the transition from the attack to the maintenance treatment phases nor the parameters that could be used to end treatment thus assuring that the patient is definitively cured have been fully standardized. The gradual decrease in the titers of the serological tests currently available is one of the laboratory parameters most frequently used; restoration of some cellular immunity responses, such as reactivity to the paracoccidioidin skin test, is also used. Newer methods, e.g. antigenemia detection and molecular biology approaches, will probably be useful.

LABORATORY DIAGNOSIS

Direct examination

The direct microscopic observation of the characteristic *P. brasiliensis* multiple-budding yeast forms in 10 percent potassium hydroxide preparations of clinical specimens is the most recommended way for diagnosing the mycosis (Lacaz et al. 2002). Usually the sources are sputum, secretion, or scraps/debris from ulcerated skin and mucosal lesions, draining material from suppurating lymph nodes or abscesses, bronchioalveolar fluid, etc. Bronchioalveolar, articular, cerebrospinal fluid or other fluids must be centrifuged before examination. For bone marrow aspirates, Giemsa staining is recommended. When the first examination of the sputum is negative, it can be homogenized with an equal volume of 4 percent sodium hydroxide or *N*-acetyl-L-cysteine and then centrifuged again. Visualization of the multiple-budding yeast cells (pilot's wheel) with the well-defined refringent double wall is mandatory, because the *P. brasiliensis* yeast cells may range from a few microns to 30–40 μm, and when small forms predominate they may be mistaken for *Blastomyces dermatitidis*, capsule-deficient *Cryptococcus neoformans*, *H. capsulatum* yeast cells, endospores, and small *C. immitis* empty spherules or even *Pneumocystis carinii* (Silletti et al. 1996).

More recently, several molecular biology diagnostic tests based on polymerase chain reaction (PCR) are being implemented which will probably open a new field in the rapid and specific diagnosis of paracoccidioidomycosis (Motoyama et al. 2000; Lindsley et al. 2001; Sano et al. 2001). For example, the gene for gp43 from 10 *P. brasiliensis* yeast cells/ml of sputum samples could be amplified, providing sufficient accuracy for the diagnosis of the mycosis (Gomes et al. 2000).

Cultures

Cultures from every patient's specimen should be obtained whenever possible. However, they are far less useful in diagnosis compared to direct examination since other more rapidly growing microorganisms such as bacteria, yeasts (especially *Candida*), and contaminant molds eventually present in the samples inhibit their growth. Isolation should be attempted by the use of modified Sabouraud agar, yeast extract agar plus antibiotics and cycloheximide, brain–heart infusion (BHI) agar plus blood and antibiotics (without cycloheximide), or tryptic soy agar incubated at 37°C and room temperature. Cultures should be observed for up to 4 weeks. As the mycelial form is not characteristic, subcultures in BHI, Fava Netto's agar, or peptone–yeast extract–glucose agar, for up to 14 days at 35–37°C, allow yeast conversion and the definitive diagnosis (Lacaz et al.

2002). In cases of severe disseminated disease, mainly those with the acute/subacute form, blood cultures may become positive (Negroni 1993).

Serology

Patients with paracoccidioidomycosis not only have a polyclonal B-cell activation but also produce high amounts of anti-*P. brasiliensis* antibodies, which are long-lasting and generally correlate with the severity of the disease. Once having started the treatment, serological titers tend to fall, accompanying the clinical improvement. Many serological tests have been developed and extensively studied. They are highly useful for both diagnosis (especially in patients with only deep-seated lesions) and treatment follow-up of patients. Agar gel immunodiffusion (ID), complement fixation (CF), and counterimmunoelectrophoresis (CIE) tests have been employed extensively (Mendes-Giannini et al. 1994). The CF tends to be used less often due to its difficult and time-consuming technique. Also available are indirect immunofluorescence, indirect hemagglutination, enzyme-linked immunosorbent assay (ELISA), and the dot blot immunobinding and Western blotting (Mendes-Giannini et al. 1984, 1992; Taborda and Camargo 1993, 1994; Martins et al. 1997). Some of the newer tests have employed purified antigens, which are well characterized and more specific. The use of these purified antigens has resulted in improvement in the efficiency of the serological tests presently available. However, they are still not easily available for most diagnostic centers. The best characterized and most used is the 43 kDa glycoprotein (Puccia and Travassos 1991). ID, CF, and CIE, which employ crude antigens derived from the yeast phase, can detect ⩾95 percent of active cases. Most authorities recommend the use of two different tests, since none of these low-cost, widely used tests *per se* guarantees more than 90 percent sensitivity when employed alone. ID, for example, is very simple and specific. A conjoint effort of several diagnostic centers assessing the value of this test using a reference crude antigen (Ag7) demonstrated it to be 84.3 percent sensitive and 98.9 percent specific (Camargo 1994). Patients with very localized, benign disease may present negative results with these tests, but when tested with more sensitive techniques, e.g. Western blot or gp43-ELISA, the diagnosis is usually reached. Crossreactivity is a less common problem, but may occur between the patients' sera and *Histoplasma* and *Lacazia loboi* antigens (Puccia and Travassos 1991).

The value of serological tests in the follow-up of patients under treatment has also been addressed by several centers. Serial serological evaluation, usually 3 months apart, together with the clinical, mycological, and radiological surveys, can guide the treatment schedule, and two negative results in two of the serological tests routinely employed is an additional criterion that helps in the difficult decision of therapy discontinuation (Restrepo and Benard 2004). The ID, CF, CIE, ELISA, and Western blot tests were shown to be able to document the decline in antibody levels that parallels the clinical improvement: serum antibodies became undetectable approximately 2 years after cessation of therapy, depending on the clinical form (Fava Netto 1961; Ferreira-da-Cruz et al. 1990; Mendes-Giannini et al. 1994; Bueno et al. 1997; Martins et al. 1997; Del Negro et al. 2000). Thus, in this circumstance, rising antibody levels mean a relapse. Nonetheless, a considerable number of patients maintain, sometimes all life long, the so-called 'healing titers,' i.e. the persistence of low serological titers in the tests. This is more frequent in patients with the acute or chronic multifocal form disease (Del Negro et al. 2000). The debate on the clinical relevance of this phenomenon, whether or not it represents the persistence of residual active lesions, thus eventually predicting future relapses, is inconclusive. Part of this problem has been lately addressed by the search of tests for antigenemia detection. Although not yet completely standardized, preliminary results of antigenemia detection in blood and urine reveal that this is a promising tool (Gómez et al. 1998; Salina et al. 1998). It will also be very helpful in the diagnosis of immunocompromised patients, specially with AIDS, in whom the search of antibody response is not a completely reliable diagnostic tool. Antigenemia follow-up studies would also permit a more precise determination of improvement as antigen load decreases with treatment response.

Histopathology

The histopathological features of paracoccidioidomycosis are similar to those of other systemic mycoses and it is essentially characterized by a granulomatous inflammation (Montenegro and Franco 1994).

It was first asserted by Motta that granulomatous inflammation response is assumed to be the most evolved form of the host tissue reactivity against the invading fungal cells. *P. brasiliensis* granuloma generally centers around one or more fungal cells and is made up of giant cells and epithelioid cells. Polymorphonuclear leukocytes may be observed close to the fungi in the central area; surrounding the epithelioid cells, there is a halo of mononuclear cells and frequently of eosinophils. Fungal antigens tend to be trapped inside epithelioid granulomata. In addition to central suppuration, *P. brasiliensis* granulomata may show central necrosis of the coagulation type.

This type of inflammation represents an immune-specific tissue response of the host in an attempt to destroy, block, and circumscribe the parasite and to prevent its multiplication. Accordingly the two clinicohistopathological forms of the disease are classified as

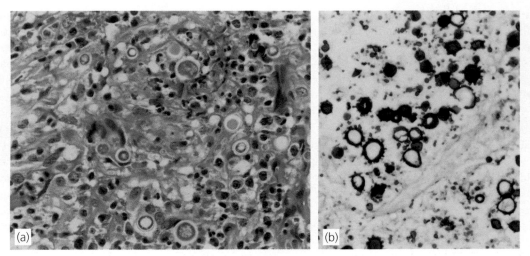

Figure 28.7 **(a)** *Epithelioid granulomatous inflammation in lymph node with numerous fungal cells in hematoxylin and eosin.* **(b)** *The pathognomic multiple-budding yeast cells are Gomori methenamine silver stain strongly positive.*

the hyperergic and anergic poles. The anergic pole, as seen in cases associated with AIDS, is characterized by disseminated infection, with a mixed suppurative and looser granulomatous inflammation, showing extensive areas of necrosis and large numbers of fungal cells. On the other hand, the hyperergic pole is defined by a more benign, localized infection with persistent cellular immune responses and histopathology showing compact epithelioid granuloma with few fungi (Figures 28.7 and 28.8).

Skin and mucous membrane lesions usually exhibit pseudoepitheliomatous hyperplasia and intraepithelial microabscesses. Particularly in ulcerated skin lesions or ruptured lymph nodes, the granulomatous inflammation is associated with a mixed pyogenic component with the formation of intraepidermic microabscesses.

Due to the chronic nature of the infection and the dynamics of the host–parasite relationship, the histopathology of the disease may show areas of extremely active disease characterized by pyogenic reaction and loose granulomata, rich in budding fungal cells, intermingled with areas with compact granulomas, rare fungal cells, and variable degrees of fibrosis. This mixed aspect can be observed either in lymph nodes, skin, or pulmonary lesions, suggesting that the disease evolves through localized new bouts of fungal multiplication and tissue invasion, whereas the adjacent older lesions are on their way to fibrotic resolution. Computer tomographic imaging of the lung confirmed this aspect by depicting areas with alveolar condensation alongside fibrotic and emphysematous zones.

As the morphology of the parasitic forms of *P. brasiliensis* is pathognomonic, the diagnosis of paracoccidioidomycosis is frequently first established by the histopathology examination of a biopsy and further confirmed either by culture or serology. If the parasite is

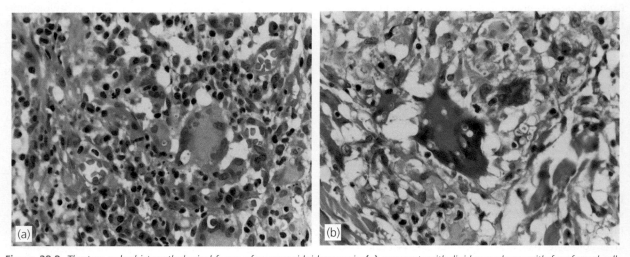

Figure 28.8 *The two polar histopathological forms of paracoccidoidomycosis:* **(a)** *compact epithelioid granuloma with few fungal cells and* **(b)** *loose, exudative granulomatous inflammation.*

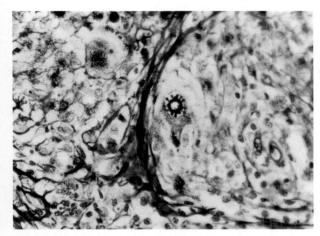

Figure 28.9 *The pathognomic pilot-wheel yeast form of* P. brasiliensis *in Grocott methenamine silver stain.*

abundant, it may be identified by hematoxylin and eosin stain and confirmed by a special fungal stain such as Grocott silver methenamine (Figure 28.9). When the disease is chronic, most of the fungal cells are found inside the macrophages, but free yeast cells predominate in disseminated cases. Pauciparasitic forms of the *P. brasiliensis* granuloma may occur, which shows a sarcoidic pattern with rare yeast forms. In these cases, the diagnosis may rely on epidemiologic data plus molecular diagnostic tools such as PCR on paraffin-embedded tissue. On the other hand, some cases may show predominance of diminute parasitic yeast forms which may be taken by those of *H. capsulatum*. Diagnosis then should rely on culture or on the use of immunohistochemistry with specific polyclonal or monoclonal anti-*P. brasiliensis* antibody.

Residual lesions of the pulmonary primary complex of the disease have rarely been described (Londero 1986; Angulo-Ortega and Pollak 1971) and they may present as fibrogranulomatous, fibrocaseous, and calcified lesions in the lungs and/or in the regional lymph nodes.

In old inactive fibrotic lesions, the remaining fungi have the morphology of nonviable cells and require a long time to multiply in microaerophilic conditions, which could explain the long latency period needed for the mycosis to become manifested in the chronic form patients (Restrepo 2000).

Immunohistochemical techniques and monoclonal antibodies to T-lymphocyte subsets have been used to characterize the tegumentary and lymphatic granulomatous lesions from patients with paracocciciodomycosis. In the granuloma, T cells formed a peripheral mantle around the centrally aggregated macrophages. The majority of the lymphocytes were T-helper cells with few suppressor cells, indicating that those cells were actively involved in the pathogenesis of the lesions and in disease control. The phagocytic histiocyte is lys+ S-100−, whereas the antigen-presenting histiocyte is lys− S-100+. In the granulomata, the histiocyte population was predominantly lys+ with few S-100− cells, a pattern

similar to that of other immunogenic granulomata. It is possible that the interplay between antigen-presenting cells and lymphocytes in *P. brasiliensis* granuloma would result in the release of T-cell stimulating factors, such as IL-2. Lymphokines released by activated lymphocytes would attract, fix, and activate macrophages in the inflammatory foci. The activated macrophages would then show enhanced killing of *P. brasiliensis*, secrete cytokines, and further differentiate into epithelioid cells. Altogether, the effector cells in the *P. brasiliensis* granuloma are: natural killer cells, lymphocytes, T-cytotoxic cells, activated neutrophils and eosinophils, activated macrophages, epithelioid cells, giant cells, and B-lymphocytes (Moscardi-Bacchi et al. 1989).

In accordance with these data, studies on protective immunization in paracocciciodomycosis have indicated a positive correlation between elevated cellular immune responses and restriction of infection and in the numbers of fungal cells in granulomata (Meira et al. 1996).

REFERENCES

Ajello, L. and Polonelli, L. 1985. Imported paracoccidioidomycosis: a public health problem in non-endemic area. *Eur J Epidemiol*, **1**, 160–5.

Angulo-Ortega, A. and Pollak, L. 1971. Paracoccidioidomycosis. The pathologic anatomy of mycoses. In: Backer, R.D. (ed.), *Human infection with fungi, actinomycetes and algae*. Berlin: Springer-Verlag, 507–76.

Anjos, A.R., Calvi, S.A., et al. 2002. Role of *Paracoccidioides brasiliensis* cell wall fraction containing beta-glucan in tumor necrosis factor-alpha production by human monocytes: correlation with fungicidal activity. *Med Mycol*, **40**, 377–82.

Aristizábal, B.H., Clemons, K.V., et al. 1998. Morphological transition of *Paracoccidioides brasiliensis* conidia to yeast cells: in vivo inhibition in females. *Infect Immun*, **66**, 5587–91.

Bagagli, E., Sano, A., et al. 1998. Isolation of *Paracoccidioides brasiliensis* from armadillos (*Dasypus noveminctus*) captured in an endemic area of paracoccidioidomycosis. *Am J Trop Med Hyg*, **58**, 505–12.

Baida, H., Biselli, P.J., et al. 1999. Differential antibody isotype expression to the major *Paracoccidioides brasiliensis* antigen in juvenile and adult form paracoccidioidomycosis. *Microb Infect*, **1**, 273–8.

Benard, G. and Duarte, A.J.S. 2000. Paracoccidioidomycosis: a model for evaluation of the effects of human immunodeficiency virus infection on the natural history of endemic tropical diseases. *Clin Infect Dis*, **31**, 1032–9.

Benard, G., Hong, M.A., et al. 1996. Antigen-specific immunosuppression in paracoccidioidomycosis. *Am J Trop Med Hyg*, **54**, 7–12.

Benard, G. Mendes-Giannini, M.J., et al. 1997. Immunosuppression in paracoccidioidomycosis: T cell hyporesponsiveness to two *Paracoccidioides brasiliensis* glycoproteins that elicit strong humoral immune response. *J Infect Dis*, **175**, 1263–7.

Benard, G., Romano, C.C., et al. 2001. Imbalance of IL-2, IFN-gamma and IL-10 secretion in the immunosuppression associated with human paracoccidioidomycosis. *Cytokine*, **13**, 248–52.

Biagioni, L., Souza, M.J. et al. 1984. Serology of paracoccidioidomycosis. II. Correlation between class-specific antibodies and clinical forms of the disease. *Trans R Soc Trop Med Hyg*, **78**, 617–21.

Biselli, P.J., Juvenale, M., et al. 2001. IgE antibody response to the main antigenic component of *Paracoccidioides brasiliensis* in patients with paracoccidioidomycosis. *Med Mycol*, **39**, 475–8.

Blotta, M.H., Mamoni, R.L., et al. 1999. Endemic regions of paracoccidioidomycosis in Brazil: a clinical and epidemiologic study of 584 cases in the southeast region. *Am J Trop Med Hyg*, **61**, 390–4.

Borelli, D. 1963–1964. Concepto de reservárea. La reducida reservárea de la paracoccidioidomicosis. *Dermatol Venez*, **4**, 71.

Botteon, F.A., Camargo, Z.P., et al. 2002. *Paracoccidioides brasiliensis*-reactive antibodies in Brazilian blood donors. *Med Mycol*, **40**, 387–91.

Bueno, J.P., Mendes-Giannini, M.J., et al. 1997. IgG, IgM and IgA antibody response for the diagnosis and follow-up of paracoccidioidomycosis: comparison of counterimmunoelectrophoresis and complement fixation. *J Med Vet Mycol*, **35**, 213–17.

Burger, E., Miyaji, M., et al. 1996. Histopathology of paracoccidioidomycotic infection in athymic and euthymic mice: a sequential study. *Am J Trop Med Hyg*, **55**, 235–42.

Cacere, C.R., Romano, C.C., et al. 2002. The role of apoptosis in the antigen-specific T cell hyporesponsiveness of paracoccidioidomycosis patients. *Clin Immunol*, **105**, 215–22.

Cadavid, D. and Restrepo, A. 1993. Factors associated with *Paracoccidioides brasiliensis* infection among permanent residents of three endemic areas in Colombia. *Epidemiol Infect*, **111**, 121–3.

Calcagno, A.M., Niño-Vega, G., et al. 1998. Geographic discrimination of *Paracoccidioides brasiliensis* strains by randomly amplified polymorphic DNA análisis. *J Clin Microbiol*, **36**, 1733–6.

Calle, D., Rosero, D.S., et al. 2001. Paracoccidioidomycosis in Colombia: an ecological study. *Epidemiol Infect*, **126**, 309–15.

Camargo, Z.P. 1994. Report on the activities of the committee on paracoccidioidomycosis serodiagnosis (PS) in 1993. *Mycoses Newsl*, **65**, 2–3.

Campos, E.S. and Almeida, F.P. 1927. Contribuição para o estudo das 'Blastomycoses' (granulomas coccidioides) observadas em São Paulo. *An Fac Med São Paulo*, **2**, 203.

Cano, M.I.N., Cisalpino, P.S., et al. 1998. Eletrophoretic karyotypes and genome sizing of the pathogenic fungus *Paracoccidioides brasiliensis*. *J Clin Microbiol*, **36**, 742–7.

Castro, C.C., Benard, G., et al. 1999a. MRI of head and neck paracoccidioidomycosis. *Br J Radiol*, **72**, 717–22.

Castro CC, Yamaga LI et al. 1999b. MRI of central nervous system paracoccidioidomycosis. *Proceedings of the Seventh Scientific Meeting of the International Society for Magnetic Resonance Imaging*, Philadelphia, Pennsylvania, 907.

Coimbra, C.E.A., Wanke, B., et al. 1994. Paracoccidioidin and histoplasmin sensitivity in Tupí-Mondé Amerindian populations from Brazilian Amazonia. *Ann Trop Med Parasitol*, **88**, 197–207.

Costa, E.O. and Fava Netto, C. 1978. Contribution to the epidemiology of paracoccidioidomycosis and histoplasmosis in the state of Sao Paulo, Brazil. Paracoccidioidin and histoplasmin intradermic tests in domestic animals. *Sabouraudia*, **16**, 93–101.

Coutinho, Z.F., Silva, D., et al. 2002. Paracoccidioidomycosis mortality in Brazil (1980–1995). *Cad Saúde Pública*, **18**, 1441–54.

Del Negro, G. 1982. Ketoconazole in paracoccidioidomycosis. A long-term therapy study with prolonged folllow-up. *Rev Inst Med Trop São Paulo*, **24**, 27–39.

Del Negro, G., Melo, E.H., et al. 1980. Limited adrenal reserve in paracoccidioidomycosis: cortisol and aldosterone responses to 1-24 ACTH. *Clin Endocrinol*, **13**, 553–59.

Del Negro, G.M.B., Pereira, C.N., et al. 2000. Evaluation of tests for antibody response in the follow-up of patients with acute and chronic forms of paracoccidioidomycosis. *J Med Microbiol*, **49**, 37–46.

Dias, M.F., Pereira, A.C., et al. 2000. The role of HLA antigens in the development of paracoccidioidomycosis. *Eur Acad Dermatol Venereol*, **14**, 166–71.

Dillon, N.L., Habermann, M.C., et al. 1985. Ketoconazole. Tratamento da paracoccidioidomicose no período de 2 anos. *An Bras Dermatol*, **60**, 45–8.

Fava Netto, C. 1961. Contribuição para o estudo imunológica da blastomicose de Lutz. *Rev Inst Adolfo Lutz*, **21**, 99–194.

Felipe, M.S.S. and the Pbgenome Network. 2002. Functional and differential genome project of *Paracoccidioides brasiliensis*. *Annu Rev Biomed Sci*, **special issue**, 18.

Ferreira-da-Cruz, M.F., Francesconi-do-Vale, A.C., et al. 1990. Study of antibodies in paracoccidioidomycosis: follow-up of patients during and after treatment. *J Med Vet Mycol*, **28**, 151–7.

Fonseca, E.R., Pardal, P.P. and Severo, L.C. 1999. Paracoccidioidomicose em crianças em Belém do Pará. *Rev Soc Bras Med Trop*, **32**, 31–3.

Franco, M., Mendes, R.P., et al. 1987. Paracoccidioidomycosis, a recently proposed classification of its clinical forms. *Rev Soc Bras Med Trop*, **20**, 129–32.

Franco, M., Mendes, R.P., et al. 1989. Paracoccidioidomycosis. *Baillière's Clin Trop Commun Dis*, **4**, 185.

Franco, M., Bagagli, E., et al. 2000. A critical analysis of isolation of *Paracoccidioides brasiliensis* from soil. *Med Mycol*, **38**, 185–91.

Funari, M., Kavakama, J., et al. 1999. Chronic pulmonary paracoccidioidomycosis (South American blastomycosis): high-resolution CT findings in 41 patients. *Am J Roentgenol*, **173**, 59–64.

Goldani, L.Z., Maia, A.L. and Sugar, A.M. 1995. Cloning and nucleotide sequence of a specific DNA fragment from *Paracoccidioides brasiliensis*. *J Clin Microbiol*, **36**, 3960–6.

Goldman, G.H. 2002. The *Paracoccidioides brasiliensis* EST genome project. *Annu Rev Biomed Sci*, **special issue**, 18.

Gomes, G.M., Cisalpino, P.S., et al. 2000. PCR for diagnosis of paracoccidioidomycosis. *J Clin Microbiol*, **38**, 3478–80.

Gómez, B.L., Figueroa, J.I., et al. 1998. Antigenemia in patients with paracoccidioidomycosis: detection of the 87 kDa determinant during and after antifungal therapy. *J Clin Microbiol*, **36**, 3309–16.

Gonzalez, A., Gregory, W., et al. 2000. Nitric oxide participation in the fungicidal mechanism of gamma interferon-activated murine macrophages against *P. brasiliensis* conidia. *Infect Immun*, **68**, 2546–52.

Imai, T., Sano, A., et al. 2000. A new PCR primer for the identification of *Paracoccidioides brasiliensis* based on rRNA sequences coding the internal transcribed spacers (ITS) and 5.8S regions. *Med Mycol*, **38**, 323–6.

Kalmar, E.M.N., Alencar, F.E.C., et al. 2004. Paracoccidioidomycosis: an epidemiological survey in a pediatric population from the Brazilian Amazon using skin tests. *Am J Trop Med Hyg*, **71**, 82–6.

Karhawi, A.S., Colombo, A.L. and Salomão, R. 2000. Production of IFN-gamma is impaired in patients with paracoccidioidomycosis during active disease and is restored after clinical remission. *Med Mycol*, **38**, 225–9.

Kashino, S.S., Fazioli, R.A., et al. 2000. Resistance to *Paracoccidioides brasiliensis* infection is linked to a preferential Th1 immune response, whereas susceptibility is associated with absence of IFN-gamma production. *J Interferon Cytokine Res*, **20**, 89–97.

Kurita, N., Oarada, M., et al. 2000. Effect of cytokines on antifungal activity of human polymorphonuclear leucocytes against yeast cells of *Paracoccidioides brasiliensis*. *Med Mycol*, **38**, 177–82.

Lacaz, C.S. 1994a. Historical evolution of the knowledge on paracoccidioidomycosis and its etiologic agent, *Paracoccidioides brasiliensis*. In: Franco, M., Lacaz, C.S., et al. (eds), *Paracoccidioidomycosis*. Boca Raton, FL: CRC Press, 1–11.

Lacaz, C.S. 1994b. *Paracoccidioides brasiliensis*: morphology, evolutionary cycle, maintenance during saprophytic life, biology, virulence, taxonomy. In: Franco, M., Lacaz, C.S., et al. (eds), *Paracoccidioidomycosis*. Boca Raton, FL: CRC Press, 13–25.

Lacaz, C.S. and Sampaio, S.A.P. 1958. Tratamento da blastomicose sul americana com anfotericina B – resultados preliminares. *Rev Paul Med*, **52**, 443.

Lacaz, C.S., Faria, J.L. and Moura, R.A.A. 1948. Blastomicose sulamericana associada a moléstia de Hodgkin. *Hospital (Rio de Janeiro)*, **34**, 313.

Lacaz, C.S., Porto, E., et al. 2002. Paracoccidioidomicose. In: Lacaz, C.S., Porto, E., et al. (eds), *Micologia médica*, 9th edition. Sao Paulo, Brasil: Sarvier, 629–729.

Leclerc, M.C., Phillipe, H. and Guého, E. 1994. Phylogenic of dermatophytes and dimorphic fungi based on large subunit ribosomal RNA sequence comparisons. *J Med Vet Mycol*, **32**, 331–41.

Lima, S.M.D. and Durlacher, R.R. 2003. Situação da paracoccidioidomicose em Rondônia. Presented at *XXXIX Proceedings of the Congress of the Brazilian Society of Tropical Medicine*, Belém, Pará, Brazil, February.

Lindsley, M.D., Hurst, S.F., et al. 2001. Rapid identification of dimorphic and yeast-like fungal pathogens using specific DNA probes. *J Clin Microbiol*, **39**, 3505–11.

Londero, A.T. 1986. Paracoccidioidomicose. Patogenia, formas clínicas, manifestações pulmonares e diagnóstico. *J Pneumol*, **12**, 41–57.

Londero, A.T. and Ramos, C.D. 1990. Paracoccicioidomicose. Estudo clínico e micológico de 260 casos observados no interior do Estado do Rio Grande do Sul. *J Pneumol*, **16**, 129–32.

Londero, A.T., Rios-Gonçalves, A.J., et al. 1987. Paracoccidioidomicose disseminada 'infanto-juvenil' em adolescentes. *Arq Bras Med*, **61**, 5–12.

Londero, T.A., Rios-Gonçalves, A.J., et al. 1996. Paracoccidioidomycosis in Brazilian children. A critical review (1911–1994). *Arq Bras Med*, **70**, 197–203.

Lutz, A. 1908. Uma mycose pseudococcidica localizada na bocca e observada no Brazil. Contribuição ao conhecimento das hyphoblastomycoses americanas. *Brasil Med*, **22**, 121–124, 141.

Mamoni, R.L., Nouer, S.A., et al. 2002. Enhanced production of specific IgG4, IgE, IgA and TGF-beta in sera from patients with the juvenile form of paracoccidioidomycosis. *Med Mycol*, **40**, 153–9.

Marcondes, J., Meira, D.A., et al. 1984. Avaliação do tratamento da paracoccidioidomicose com o ketoconazole. *Rev Inst Med Trop São Paulo*, **26**, 113–21.

Marques, S.A. 2002. Treatment of paracoccidioidomycosis with itraconazole. *Annu Rev Biomed Sci*, **special issue**, 24–5.

Marques, S.A. and Shikanai-Yasuda, M.A. 1994. Paracoccidioidomycosis associated with immunosuppression, AIDS, and cancer. In: Franco, M., Lacaz, C.S., et al. (eds), *Paracoccidioidomycosis*. Boca Raton, FL: CRC Press, 393–405.

Marques, S.A., Dillon, N.L., et al. 1985. Paracoccidioidomycosis: a comparative study of the evolutionary serologic, clinical and radiologic results for patients treated with ketoconazole or amphotericin B plus sulfonamides. *Mycopathologia*, **89**, 19–25.

Martinez, R. and Moya, M.J. 1992. The relationship between paracoccidioidomycosis and alcoholism. *Rev Saúde Pública*, **26**, 12–16.

Martinez, R., Meneghelli, U.G., et al. 1979a. O comprometimento gastrintestinal na blastomicose sul-americana (paracoccidioidomicose). I. Estudo clínico, radiológico e histopatológico. *Rev Ass Med Bras*, **25**, 31–4.

Martinez, R., Meneghelli, U.G., et al. 1979b. O comprometimento gastrintestinal na blastomicose sul-americana (paracoccidioidomicose) II. Estudo funcional do intestino delgado. *Rev Ass Med Bras*, **25**, 70–3.

Martinez, R., Figueiredo, J.F.C. et al. 1993, Paracoccidioidomicose durante a gestação. *Proceedings of the XXXIII Congress of the Brazilian Society of Tropical Medicine*, Brazil, 133.

Martins, R., Marques, S., et al. 1997. Serological follow-up of patients with paracoccidioidomycosis treated with itraconazole using Dot-blot, ELISA and western-blot. *Rev Inst Med Trop Sao Paulo*, **39**, 261–9.

McEwen, J.G., Bedoya, V., et al. 1987. Experimental paracoccidioidomycosis induced by the inhalation of conidia. *J Med Vet Mycol*, **25**, 165–75.

Meira, D.A., Pereira, P.C., et al. 1996. The use of glucan as immunostimulant in the treatment of paracoccidioidomycosis. *Am J Trop Med Hyg*, **55**, 496–503.

Mendes, R.P., Negroni, R. and Arechavala, A. 1994. Treatment and control of cure. In: Franco, M., Lacaz, C.S., et al. (eds), *Paracoccidioidomycosis*. Boca Raton, FL: CRC Press, 373–92.

Mendes-Giannini, M.J., Camargo, M.E., et al. 1984. Immunoenzymatic absorption test for serodiagnosis of paracoccidioidomycosis. *J Clin Microbiol*, **20**, 103–8.

Mendes-Giannini, M.J., Shikanai-Yasuda, M.A., et al. 1992. Imunochemical study of *Paracoccidioides brasiliensis* antigens by Western blotting. *Rev Arg Micol*. **15**, 45.

Mendes-Giannini, M.J.S., del Negro, G.B. and Siqueira, A.M. 1994. Serodiagnosis. In: Franco, M., Lacaz, C.S., et al. (eds), *Paracoccidioidomycosis*. Boca Raton, FL: CRC Press, 345–63.

Molinari-Madlum, E.E.W.I., Felipe, M.S.S. and Soares, C.M.A. 1999. Virulence of *Paracoccidioides brasiliensis* isolates can be correlated to groups defined by random amplified polymophic DNA analysis. *Med Mycol*, **37**, 269–76.

Montenegro. M.R. and Franco, M. 1994. Pathology. In: Franco, M., Lacaz, C.S., et al. (eds), *Paracoccidioidomycosis*. Boca Raton, FL: CRC Press, 131–50.

Montoya. A.E., Alvarez, A.L., et al. 1999. Electrophoretic karyotype of environmental isolates of *Paracoccidioides brasiliensis. Med Mycol*, **37**, 219–22.

Mós, E.N. and Fava Netto, C. 1974. Contribution to the study of paracoccidioidomycosis. II. Possible epidemiological role of dogs. Anatomopathological and serological study. *Rev Inst Med Trop São Paulo*, **16**, 4, 232–7.

Moscardi-Bacchi, M., Soares, A., et al. 1989. In situ localization of T lymphocytes subsets in human paracoccidioidomycosis. *J Med Vet Mycol*, **27**, 149.

Motoyama, A.B., Venancio, E.J., et al. 2000. Molecular identification of *Paracoccidioides brasiliensis* by PCR amplification of ribosomal DNA. *J Clin Microbiol*, **38**, 8, 3106–9.

Motta, N.G.S., Rezkallah-Iwasso, M.T., et al. 1985. Correlation between cell-mediated immunity and clinical forms of paracoccidioidomycosis. *Trans R Soc Trop Med Hyg*, **79**, 765–72.

Musatti, C.C., Rezkallah-Iwasso, M.T., et al. 1976. In vivo and in vitro evaluation of cell-mediated immunity in patients with paracoccidioidomycosis. *Cell Immunol*, **24**, 365–78.

Naiff, R.D., Ferreira, L.C.L., et al. 1986. Paracoccidioidomicose enzoótica em tatus (*Dasypus novemcinctus*) no Estado do Pará. *Rev Inst Med Trop São Paulo*, **28**, 19–27.

Naiff, R.D, Barret. T.V., et al. 1988. Encuesta epidemiológica de histoplasmosis, paracoccidioidomicosis y leishmaniasis mediante pruebas cutáneas. *Bol Of Sanit Panam*, **104**, 35–50.

Negroni, R. 1993. Paracoccidioidomycosis (South American blastomycosis, Lutz's mycosis). *Int J Dermatol*, **32**, 847–59.

Neworal, E.P.M., Altemani, A., et al. 2001. Immunocytochemical localization of inducible nitric oxide synthase and transforming growth factor beta 1 in oral mucosa of patients with paracoccidioidomycosis. *Rev Soc Bras Med Trop*, **34**, 141.

Niño-Vega, G.A., Calgagno, A.M., et al. 2000a. RFLP analysis reveals marked geographical isolation between strains of *Paracoccidioides brasiliensis. Med Mycol*, **38**, 437–41.

Niño-Vega, G.A., Munro, C.A., et al. 2000b. Differential expression of chitin synthase genes during temperature-induced dimorphic transition in *Paracoccidioides brasiliensis. Med Mycol*, **38**, 31–9.

Oliveira, S.J., Mamoni, R.L., et al. 2002. Cytokines and lymphocyte proliferation in juvenile and adult forms of paracoccidioidomycosis: comparison with infected and non-infected controls. *Microbes Infect*, **4**, 139–44.

Ollague, J.M., Zurita, A.M. and Calero, G. 2000. Paracoccidioidomycosis (South American blastomycosis) successfully treated with terbinafine: first case report. *Br J Dermatol*, **143**, 188–91.

Ono, M.A., Bracarense, A.P., et al. 2001. Canine paracoccidioidomycosis: a seroepidemiologic study. *Med Mycol*, **39**, 277–82.

PAHO. 1972. Paracoccidioidomycosis. *Scientific Publication 254*.

Parisi-Fortes, M.R., Kurokawa, C.S., et al. 2001. Detection of TGF-beta 1 in lesions of patients with paracoccidioidomycosis. *Rev Soc Bras Med Trop*. **34**. 129.

Pereira, M., Felipe, M.S.S., et al. 2000. Molecular cloning and characterization of a glucan synthase from the human pathogenic fungus *Paracoccidioides brasiliensis*. *Yeast*, **16**, 451–62.

Popi, F.A., Daniel Lopes, J. and Mariano, M. 2002. GP43 from *Paracoccidioides brasiliensis* inhibits macrophage functions. An evasion mechanism of the fungus. *Cell Immunol*, **218**, 87–94.

Puccia, R. and Travassos, L.R. 1991. 43-kilodalton glycoprotein from *Paracoccidioides brasiliensis*. Immunochemical reactions with sera from patients with paracoccidioidomycosis, histoplasmosis and Jorge Lobo's disease. *J Clin Microbiol*, **29**, 1610–15.

Resende, L.S.R. 2000. Mielopatia infiltrativa por paracoccidioidomicose. PhD thesis, Faculdade de Medicina de Botucatu, UNESP, São Paulo State, Brazil.

Restrepo, A. 1985. The ecology of *Paracoccidioides brasiliensis*: a puzzle still unsolved. *J Med Vet Mycol*, **23**, 323–34.

Restrepo, A. 2000. Morphological aspects of *Paracoccidioides brasiliensis* in lymph nodes: implications for the prolonged latency of paracoccidioidomycosis. *Med Mycol*, **38**, 317–22.

Restrepo, A., Robledo, M., et al. 1976. The gamut of paracoccidioidomycosis. *Am J Med*, **61**, 33–42.

Restrepo, A., Gomez, I., et al. 1983. Treatment of paracoccidioidomycosis with ketoconazole: a three-year experience. *Am J Med*, **74**, 48–52.

Restrepo, A., Trujillo, M. and Gómez, I. 1989. Inapparent lung involvement in patients with the subacute juvenile type of paracoccidioidomycosis. *Rev Inst Med Trop São Paulo*, **31**, 18–22.

Restrepo, A., Salazar, M.E., et al. 1997. Hormonal influences in the host-interplay with *Paracoccidioides brasiliensis*. In: Stevens, D.A., Vanden Bosche, H. and Odds, F. (eds), *Topics on fungal infections*. Bethesda, MD: National Foundation for Infectious Diseases, 125–33.

Restrepo, A., McEwen, J.G. and Castañeda, E. 2001. The habitat of *Paracoccidioides brasiliensis*: How far from solving the riddle? *Med Mycol*, **39**, 233–41.

Restrepo, A.M. and Benard, G. 2004. Paracoccidioidomycosis. In: Feigin, R.D. and Cherry, J.D. (eds), *Textbook of pediatric infectious diseases*, 4th edition. Houston, TX: Harcourt Health Sciences.

Restrepo-Moreno, A. 1994. Ecology of *Paracoccidioides brasiliensis*. In: Franco, M., Lacaz, C.S., et al. (eds), *Paracoccidioidomycosis*. Boca Raton, FL: CRC Press, 121–30.

Ribeiro, D.O. 1940. Nova terpêutica para a blastomicose. *Publ Med (São Paulo)*, **12**, 36–54.

Ricci, G., Silva, I.D.C.G., et al. 2002. Canine paracoccidioidomycosis: report of the first case of the literature. *Annu Rev Biomed Sci*, **special issue**, 80.

Rios-Gonçalves, A.J., Londero, A.T., et al. 1998. Paracoccidioidomycosis in children in the state of Rio de Janeiro (Brazil). Geographic distribution and the study of a 'reservarea'. *Rev Inst Med Trop São Paulo*, **40**, 11–13.

Rodrigues, D.R., Calvi, A.S., et al. 2001. The role of cytokines in the killing of high virulent strain of *P. brasiliensis* by human PMN cells. *Rev Soc Bras Med Trop*, **34**, 155.

Romano, C.C., Mendes-Giannini, M.J., et al. 2002. IL-12 and neutralization of endogenous IL-10 revert the in vitro antigen-specific cellular immunosuppression of paracoccidioidomycosis patients. *Cytokine*, **18**, 149–57.

Salina, M.A., Shikanai-Yasuda, M.A., et al. 1998. Detection of circulating *Paracoccidioides brasiliensis* antigen in urine of paracoccidioidomycosis patients before and during treatment. *J Clin Microbiol*, **36**, 1723–8.

San-Blas, G. 1979. Biosynthesis of glucans by subcellular fractions of *Paracoccidioides brasiliensis*. *Exp Mycol*, **3**, 249–58.

San-Blas, G., Niño-Vega, G. and Iturriaga, T. 2002. *Paracoccidioides brasiliensis* and paracoccidioidomycosis: molecular approaches to morphogenesis, diagnosis, epidemiology, taxonomy and genetics. *Med Mycol*, **40**, 225–42.

Sandhu, G.S., Aleff, R.A., et al. 1997. Molecular detection and identification of *Paracoccidioides brasiliensis*. *J Clin Microbiol*, **35**, 1894–6.

Sandoval, M., de Brito, T., et al. 1996. Antigen distribution in mucocutaneous biopsies of human paracoccidioidomycosis. *Int J Surg Pathol*, **3**, 181–8.

Sano, A., Yokoyama, K., et al. 2001. Detection of gp43 and ITS1-5.8S-ITS2 ribosomal RNA genes of *Paracoccidioides brasiliensis* in paraffin-embedded tissue. *Nippon Ishinkin Gakkai Zasshi*, **42**, 23–7.

Sant'Anna, G.D., Mauri, M., et al. 1999. Laryngeal manifestations of paracoccidioidomycosis (South American blastomycosis). *Arch Otolaryngol Head Neck Surg*, **125**, 1375–8.

Shikanai-Yasuda, M.A., Higaki, Y., et al. 1992a. Comprometimento da medula óssea e eosinofilia na paracoccidiodiomicose. *Rev Inst Med Trop São Paulo*, **34**, 85–90.

Shikanai-Yasuda, M.A., Segurado, A.A.C., et al. 1992b. Immunodeficiency secondary to juvenile paracoccidioidomycosis. *Mycopathologia*, **120**, 23–8.

Shikanai-Yasuda, M.A., Benard, G., et al. 2002a. Randomized trial with itraconazole, ketoconazole and sulfadiazine in paracoccidioidomycosis. *Med Mycol*, **40**, 411–17.

Shikanai-Yasuda, M.A., Yoshida, M., et al. 2002b. Treatment of paracoccidioidomycosis patients: duration of treatment, maintenance therapy, relapses and sequels. *Annu Rev Biomed Sci*, **special issue**, 27.

Silletti, R.P., Glezerov, V. and Schwartz, I.S. 1996. Pulmonary paracoccidioidomycosis misdiagnosed as *Pneumocystis* pneumonia in an immunocompromised host. *J Clin Microbiol*, **34**, 2328–30.

Silva, S.P., Borges-Walmsley, M.I., et al. 1999. Differential expression of an *hsp*70 gene during transition from the mycelial to the infective yeast form of the human pathogenic fungus *Paracoccidioides brasiliensis*. *Mol Microbiol*, **31**, 1039–50.

Silva-Vergara, M.L. and Martinez, R. 1999. Role of the armadillo *Dasypus novemcinctus* in the epidemiology of paracoccidioidomycosis. *Mycopathologia*, **144**, 131–3.

Silvestre, M.T., Nishioka, A.S. et al. 1997. Forma juvenil da paracoccidioidomycose em adultos: estudo de oito casos. In: *Proceedinngs of the XXXIII Congress of the Brazilian Society of Tropical Medicine*, Brazil, 93.

Splendore, A. 1912. Blastomicosi-sporotricosi e rapporti con processi affini. *Proc Congr Int Dermatol Sifilog, Roma*, 1.

Taborda, C.P. and Camargo, Z.P. 1993. Diagnosis of paracoccidioidomycosis by passive haemagglutination assay of antibody using a purified and specific antigen-gp43. *J Med Vet Mycol*, **31**, 155–60.

Taborda, C.P. and Camargo, Z.P. 1994. Diagnosis of paracoccidioidomycosis by dot immunobinding assay for antibody detection using the purified and specific antigen gp43. *J Clin Microbiol*, **32**, 554–6.

Teixeira, F., Gayotto, L.C. and de Brito, T. 1978. Morphological patterns of the liver in South American blastomycosis. *Histopathology*, **2**, 231–7.

Telles, F.Q. 2002. Treatment of paracoccidioidomycosis patients: new insights and approaches. *An Rev Biomed Sci*, **special issue**, 26–7.

Tobón, A.M., Gómez, I., et al. 1995. Seguimiento post-terapia en pacientes con paracoccidioidomicosis tratados con itraconazol. *Rev Colombiana Neumol*, **7**, 74–8.

Tobón, A.M., Agudelo, C.A., et al. 2002. Paracoccidioidomycosis and the influence of itraconazole treatment on lung abnormalities: extended follow up observations. *An Rev Biomed Sci*, **special issue**, 25–6.

Tuder, R.M., El Ibrahim, R., et al. 1985. Pathology of the pulmonary paracoccidioidomycosis. *Mycopathologia*, **92**, 179–88.

Valle, A.C.F., Guimarães, R.R., et al. 1992. Aspectos radiológicos torácicos na paracoccidioidomicose. *Rev Inst Med Trop São Paulo*, **34**, 107–15.

Wanke, B. and Londero, A.T. 1994. Epidemiology and paracoccidioidomycosis infection. In: Franco, M., Lacaz, C.S., et al. (eds), *Paracoccidioidomycosis*. Boca Raton, FL: CRC Press, 109–20.

Yamaga, L.I., Benard, G., et al. 2003. The role of gallium-67 scan in defining extent of disease in an endemic deep mycosis, paracoccidioidomycosis. A predominantly multifocal disease. *Eur J Nucl Med Mol Imaging*, **30**, 888–94.

Yarzábal, L., Dessaint, J.P., et al. 1980. Demonstration and qualification of IgE antibodies against *Paracoccidioides brasiliensis* in paracoccidioidomycosis. *Int Arch Allergy Immunol*, **62**, 346–51.

29

Penicilliosis

MARIA ANNA VIVIANI AND NONGNUCH VANITTANAKOM

ETIOLOGIC AGENT

Penicillium marneffei is the only species known to be dimorphic among the over 200 described species of the genus *Penicillium*. This fungus is currently only known to be endemic in southeast Asia. It can cause infections of the reticuloendothelial system in humans and lower animals.

Infections have been reported in apparently immunocompetent subjects but, more often, in immunocompromised patients undergoing corticosteroid treatment or affected by such underlying diseases as tuberculosis, Hodgkin's lymphoma and the acquired immunodeficiency syndrome (AIDS). The interest of medical mycologists in this mycosis has increased since it appeared in association with human immunodeficiency virus (HIV) infections in patients traveling or living in southeast Asia. As a result of the recent enormous increase of HIV infection in that region, *P. marneffei* has become one of the principal new emerging fungal pathogens. The disease that it causes is also known as penicilliosis marneffei.

HISTORY

P. marneffei was first described by Capponi et al. in 1956 as an etiological agent infecting the reticuloendothelial system of Chinese bamboo rats (*Rhizomys sinensis*). It was isolated by Capponi from the hepatic lesions of one of the Chinese bamboo rats maintained in captivity for experimental infections at the Pasteur Institute of Indo-

china at Dalat (now Vietnam). The organism was identified as a new species by Segretain (1959a), who named the fungus *Penicillium marneffei* in honor of Hubert Marneffe, then Director of the Pasteur Institute of Indochina. In the same year, Segretain (1959b) also described the first human infection that was caused by the accidental puncture of his finger by a needle used to inoculate hamsters. Nine days later, a small nodule appeared at the site, followed by lymphangitis and axillary adenopathy.

The first spontaneous infection in a human was reported in 1973 in an American minister with Hodgkin's disease, who had traveled in southeast Asia one year before (Di Salvo et al. 1973). *P. marneffei* was incidentally found in an isolated splenic abscess when the patient underwent splenectomy for management of a lymphoma. The fungus was isolated from the spleen and observed in histological sections.

A second imported case in the USA was described in a 59-year-old man who had traveled extensively in the Far East. The disease, a focal pulmonary infection, was recognized following pneumonectomy one month before death (Pautler et al. 1984).

In the same period penicilliosis marneffei was reported in southeast Asia in subjects living in Thailand (Jayanetra et al. 1984), Hong Kong (So et al. 1985; Yuen et al. 1986), and the Guangxi region of southern China (Deng and Connor 1985; Wei et al. 1985). Most of the cases in the Guangxi region were observed between 1964 and 1983 and had been originally misdiagnosed as histoplasmosis capsulati (Deng and Connor 1985).

During the period 1988–89, disseminated *P. marneffei* infections began to be observed in AIDS patients who had traveled to or were living in southeast Asia (Ancelle et al. 1988; Peto et al. 1988; Piehl et al. 1988; Coen et al. 1989; Sathapatayavongs et al. 1989; Stern et al. 1989). The recent rapid rise in the number of reported cases of penicilliosis marneffei coincides with the explosive epidemic of AIDS in northern Thailand where this mycosis is now estimated to be the third most common HIV-related opportunistic infection after tuberculosis and cryptococcosis.

With the aim of identifying the natural habitat of *P. marneffei* and its mode of transmission, several studies on wild bamboo rats have been carried out in different regions of southeast Asia where most patients with auto-chthonous infection have been observed (Deng et al. 1986, 1988; Wei et al. 1987; Li et al. 1989; Ajello et al. 1995; Chariyalertsak et al. 1996b).

EPIDEMIOLOGY

Ecology

The natural habitat of *P. marneffei* is probably soil. Since the first discovery of *P. marneffei* as an agent of disease of captive wild Chinese bamboo rats (*Rhizomys sinensis*) in Vietnam (Capponi et al. 1956), these rodents have been erroneously regarded by some investigators as a potential source of human infection.

Epidemiologic investigations were therefore focused on wild rodents in the Asian areas where the disease was suspected to be endemic (Deng et al. 1986, 1988; Wei et al. 1987; Li et al. 1989; Ajello et al. 1995; Chariyalertsak et al. 1996b). The animals were trapped and portions of their internal organs cultured. Table 29.1 lists the bamboo rat species trapped in different Asian countries and found positive for *P. marneffei*. The most studied animals are hoary bamboo rats (*Rhizomys pruinosus*), which have been found to be apparently healthy carriers of the fungus (Figure 29.1). Among the internal

Figure 29.1 *The hoary bamboo rat Rhizomys pruinosus* **(a)** *(reproduced, with permission, from Nowak 1991) and the large bamboo rat (Rhizomys sumatrensis)* **(b)** *(Chariyalertsak et al. 1996b).*

organs, the lungs had the highest positivity (89–83 percent), followed by the liver (83–33 percent), spleen (78 percent), mesenteric lymph nodes (39 percent) and pancreas (33 percent) (Deng et al. 1986; Ajello et al. 1995).

The isolation of the fungus from four species of bamboo rats in China, Thailand, and Vietnam confirmed the value of these wild rodents as epidemiologic markers. However, there is no evidence that the disease is transmitted to humans by the bamboo rat because

Table 29.1 *Bamboo rat species trapped in different southeast Asian areas and found positive for* Penicillium marneffei

Bamboo rat species	Positive/examined No. (%)	Country	References
Rhizomys sinensis (Chinese bamboo rat)	1/1	Vietnam	Capponi et al. 1956
	2/2	Southeast China	Deng et al. 1988
Rhizomys pruinosus (hoary bamboo rat)	114/179 (64)	Southeast China	Wei et al. 1987
	37/41 (90)	Southeast China	Deng et al. 1988
	15/16 (94)	Southeast China	Li et al. 1989
	6/8 (75)	Thailand	Ajello et al. 1995
Cannomys badius (bay bamboo rat)	6/31 (19)	Thailand	Ajello et al. 1995
	3/61 (5)[a]	Thailand	Chariyalertsak et al. 1996b
Rhizomys sumatrensis (large bamboo rat)	13/14 (93)	Thailand	Chariyalertsak et al. 1996b

a) *P. marneffei* were isolated from three of 10 reddish-brown *C. badius*. All 51 grayish-black *C. badius* were negative for *P. marneffei*.

these animals live far away from humans in remote mountain areas where few people reside (Deng et al. 1986). In addition, few people are in contact with, or consume, them. Both humans and bamboo rats are probably infected from a common environmental source. *P. marneffei* was isolated from soil collected from the burrows of hoary bamboo rats in China by Deng et al. in 1988 and in Thailand by Chariyalertsak et al. (1996b) from the burrow of *R. sumatrensis*. The modified flotation method (Vanittanakom et al. 1995) combined with mouse inoculation was applied in the isolation in Thailand.

By the analysis of 550 patients with disseminated *P. marneffei* infection diagnosed at Chiang Mai University Hospital in northern Thailand between 1991 and 1994, a consistent significant seasonal variation in the incidence of penicilliosis marneffei was observed. The infection was significantly more frequent in the rainy season (May to October) than in the dry season (Chariyalertsak et al. 1996a).

A case-control study performed in the same region did not suggest bamboo rats as a reservoir for infection in humans, nor could a significant association be demonstrated between *P. marneffei* infection and exposure to bamboo, a plant commonly growing in rural areas in northern Thailand and often eaten by Thais (Chariyalertsak et al. 1997). This study identified an age range of 16–30 years and occupational or other exposure to soil, especially during the rainy season, as factors independently associated with an increased risk for *P. marneffei* infection.

Infection in humans

The route of acquisition of *P. marneffei* has not yet been definitively established, but a respiratory portal of entry would be consistent with infections caused by other dimorphic fungal pathogens that produce conidia in their saprophytic habitats in nature. The attachment of *P. marneffei* conidia to the bronchoalveolar epithelium is facilitated by the recognition of extracellular matrix proteins, laminin and fibronectin, by a protein present on the surface of the conidia (Hamilton et al. 1999).

Studies of bamboo rats and human infections with *P. marneffei* have indicated that this organism is endemic in Thailand, the Guangxi region of China, Hong Kong, Vietnam, and Indonesia (Ajello et al. 1995). It is also suspected to be endemic in neighboring countries such as Cambodia, Laos, Malaysia, Myanmar (Burma), and, as recently reported, the Manipur State of India (Singh et al. 1999) and Taiwan (Hsueh et al. 2000). A specific skin test would be extremely useful to establish the precise geographical distribution of *P. marneffei*.

Penicilliosis marneffei has been associated with living in rural areas and with predisposing factors in the human hosts, such as cell-mediated immunodeficiency caused by lymphoproliferative disorders, corticosteroid treatment, and, most importantly, HIV infection (Drouhet 1993).

This fungal disease was a rare event in southeast Asia until the end of the 1980s (Table 29.2) when it began to be reported in some patients affected with AIDS who were living or had been traveling in the Asian endemic regions (Ancelle et al. 1988; Peto et al. 1988; Piehl et al. 1988; Coen et al. 1989; Sathapatayavongs et al. 1989; Stern et al. 1989). But it was from 1991, with the spread of the AIDS pandemic across the endemic areas, that *P. marneffei* emerged as a significant pathogen among HIV-infected Asian natives, especially in Thailand where the mycosis is accounting for 15–20 percent of all AIDS-related illness (Supparatpinyo et al. 1998). The disease was designated an indicator of AIDS by the Thai Ministry of Public Health (Patamasucon 1992). The possibility of spread of the fungus to the nearby nonendemic countries, as suggested in a report from Taiwan (Hsueh et al. 2000), is worrying. Penicilliosis marneffei will remain a significant public health problem in those endemic Asian countries where the incidence of AIDS is foreseen to continue its dramatic rise. At times, this fungal infection has been reported in Europe and in other nonendemic countries (Table 29.3). Some of the patients were native of, or had been living in, southeast Asia, but most of them had just been visiting the endemic areas for a short period of time, suggesting that exposure to the fungal propagules is probably high and the infection may be easily acquired. A non-Asian case in a Congolese HIV-positive physician suspected to have acquired penicilliosis during a training course at the Pasteur Institute in Paris (France), although not working directly with the fungus, was considered suggestive of the highly infective nature of *P. marneffei* airborne conidia (Hilmarsdottir et al. 1994). A second case in a patient native of Ghana and living in Germany, who had never been to Asia and for whom the possible source of contamination has not been found, arouses the suspicion that other natural habitats outside Asia might exist (Lo et al. 2000). Attempts to identify different genotypes of strains isolated from human and natural sources have been undertaken (Vanittanakom et al. 1996; Hsueh et al. 2000; Imwidthaya et al. 2000; Trewatcharegon et al. 2001). Further studies, however, are needed to analyze the existence of multiple genotypes of *P. marneffei* and their geographical variations.

From the history of heterochthonous cases, it appears that the period between infection in the endemic areas and manifestation of clinically active disease is extremely variable, ranging from a few weeks to 11 years, which suggests reactivation in the presence of immunosuppression (Peto et al. 1988; Jones and See 1992).

Penicilliosis marneffei occurs mainly in adults; however, cases have also been reported in children with or without AIDS, living in endemic areas (Jayanetra et al. 1984; Deng and Connor 1985; Yuen et al. 1986;

Table 29.2 Penicillium marneffei *infections from southeast Asia as of Nov 2002*

Country where diagnosed	No. of cases Non-AIDS	References	No. of cases AIDS	References
Cambodia			2	Bailloud et al. 2002
China	8	Deng and Connor 1985	1	Liao et al. 2002
	1	Wei et al. 1985		
	1	Li et al. 1985		
	4	Deng et al. 1988		
	3	Li et al. 1991		
	5	Deng and Ma, 1995		
Hong Kong	1	So et al. 1985	2	Tsang et al. 1991
	1	Yuen et al. 1986	2	Tsui et al. 1992
	1	Tsang et al. 1988	1	Ko 1994
	1	Chan et al. 1989	1	Chang et al. 1998
	2	Chan and Woo 1990	19	Wong and Lee 1998
	1	Lo et al. 1995		
	1	Wong et al. 2001c		
India			4	Singh et al. 1999
			46	Ranjana et al. 2002
Malaysia	1	Saadiah et al. 1999	1	Rokiah et al. 1995
Taiwan	1	Wang et al. 1989	1	Liu et al. 1994
	8	Hsueh et al. 2000	1	Chang et al. 1995
			1	Chiang et al. 1998
			1	Hung et al. 1998
			16	Hsueh et al. 2000
Thailand	5	Jayanetra et al. 1984	5443	HIV/AIDS report, Thailand, 2002; Sathapatayavongs et al. 1989; Chiewchanvit et al. 1991; Thongcharoen et al. 1992; Supparatpinyo et al. 1992,1994; Sirisanthana and Sirisanthana 1993,1995; Louthrenoo et al. 1994; Srison et al. 1995; Vanittanakom and Sirisanthana 1997; Kantipong et al. 1998; Ukarapol et al. 1998; Chokephaibulkit et al. 2001
	5	Supparatpinyo et al. 1992		
	1	Supparatpinyo et al. 1994		

Table 29.3 Penicillium marneffei *infections diagnosed in nonendemic areas, reported as of Nov 2002*

Country where diagnosed	No. of cases	References	Country where diagnosed	No. of cases	References
Australia	1	Jones and See 1992	Germany	1	Sobottka et al. 1996
	1	Heath et al. 1995		1	Rimek et al. 1999
Canada	1[a]	Sekhon et al. 1994		1	Lo et al. 2000
	1[a]	Manion et al. 1991	Italy	1	Coen et al. 1989
Japan	1	Mohri et al. 2000	Sweden	1	Julander and Petrini 1997
			Switzerland	1	Kronauer et al. 1993
Belgium	2	Depraetere et al. 1998		2	Borradori et al. 1994
France	1	Ancelle et al. 1988		2	Garbino et al. 2001
	1	Stern et al. 1989	The Netherlands	1	Hulshof et al. 1990
	2	Hilmarsdottir et al. 1993		2	Kok et al. 1994
	1	de Truchis et al. 1991	UK	1	Peto et al. 1988
	1	Hilmarsdottir et al. 1994		1	McShane et al. 1998
	2	Grise et al. 1997		1	Vilar et al. 2000
	1	Lachaud et al. 1998		1	Bateman et al. 2002
	1	Miegeville et al. 1998	USA	1[a]	Di Salvo et al. 1973
	1	Valeyrie et al. 1999		1[a]	Pautler et al. 1984
	1	Rosenthal et al. 2000		1[a]	Piehl et al. 1988
				1	Nord et al. 1998

a) Non-AIDS patients.

Deng et al. 1988; Sirisanthana and Sirisanthana 1993; Kwan et al. 1997). *P. marneffei* infection appears to be an important problem in children who acquired HIV infection perinatally (Sirisanthana and Sirisanthana 1995). The clinical presentation of this mycosis in children is reported to be similar to the infection in adults.

MYCOLOGY

Morphology and physiology

P. marneffei grows in culture at 25°C on Sabouraud's dextrose agar as a mycelial fungus typical of the genus, producing rapidly growing greenish–yellow sporulating colonies, with a pink or red center and dark green edges. A characteristic brick-red pigment is released into the medium. On potato dextrose agar or Sabouraud's dextrose agar the pigment appears after 24–72 hours of incubation at 25–30°C (Figure 29.2). Little or no pigment is produced at 35–37°C. At this higher temperature colonies are glabrous, off-white, and yeast-like.

Microscopically, the mycelial form of growth is characterized by sinuous smooth-walled conidiophores (diameter 1.5–2 μm, length 70–100 up to 175 μm), borne on aerial hyphae, which have terminal, typically divaricate, verticils, either symmetrical or asymmetrical and seldom ramified (Figure 29.3). Metulae (7–11 × 2.5–3.5 μm) originate from the apex of the conidiophore in groups of four to five. Sometimes one of the metulae is much longer than the others and is usually septate. The

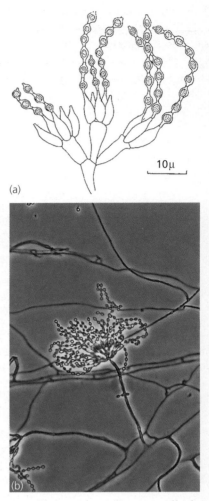

(a)

(b)

Figure 29.3 *Conidiophore of* Penicillium marneffei *from Segretain's drawing of the type strain* **(a)** *and microscopy of the mold form of this fungus (lactophenol cotton blue mount)* **(b)**. *Magnification,* × 600

(a)

(b)

Figure 29.2 *Colony of* Penicillium marneffei *on* **(a)** *Sabouraud's dextrose agar and* **(b)** *potato dextrose agar after 10 days of incubation at 28°C*

phialides, arranged in verticils of four to six, are the same size as the metulae, and have a flask-like shape gradually tapering to a slender collar. Conidia are ellipsoidal to globose (2–3 μm), smooth walled, often with prominent disjunctors, and arranged in short disordered chains. The terminal conidia of the conidial chains are sometimes larger than the ones beneath them, which is a characteristic of *P. marneffei* (Corda's phenomenon) (Segretain 1959b).

P. marneffei requires an organic source of nitrogen for mycelial growth in a well-defined, synthetic, chemical medium. Casein hydrolysate, peptone, and asparagine are utilized, whereas $NaNO_3$ and $(NH_4)_2PO_4$ are not. Glucose (optimal growth), lactose, xylose, maltose, levulose, and mantel are used as carbon sources. Sucrose is weakly used (Segretain 1959a). The fungus is sensitive to cycloheximide. *P. marneffei* has proved difficult to maintain in culture, and pigment production may be reduced by prolonged storage. More extensive study of the biochemical properties of *P. marneffei* and their possible use in strain biotyping has been reported (Wong et al.

2001a). In addition, some enzymatic activities were examined in *P. marneffei* human isolates (Youngchim et al. 1999). Both mycelial and yeast form expressed several interesting enzymes, for examples, acid phosphatase, esterases, and lipases, which might be implicated with virulence in this fungus.

The fungus is characterized by its thermal dimorphism. When grown at 37°C on 5 percent sheep blood agar, wort agar, or in liquid synthetic media containing maltose or glucose and amino acids, *P. marneffei* undergoes a transformation to a yeast-like single-cell form, also named arthroconidium.

Hyphae first become shorter, develop more septa and branches, and cease to produce conidia. After 2 weeks there is a gradual shift to spherical or ellipsoidal yeast-like cells which are 2–6 μm in diameter and divide by transverse septation (fission) in a somewhat similar fashion to *Schizosaccharomyces* spp.

The in vitro yeast-like form resembles that found in rodents and humans. Globose, ovoid, and elongated yeast-like cells measuring 2–3 × 2–6.5 μm are seen free and within macrophages. Extracellular, elongated, and sausage-shaped cells up to 8–13 μm long, and more rarely short hyphae, no longer than 20 μm, are also observed. The multiplying cells usually have a single central septum and occasionally two.

The cells of *P. marneffei* are poorly stained by hematoxylin and eosin (H&E). Their cell wall, however, is well stained by periodic acid–Schiff (PAS) and Gomori's methenamine silver (GMS) stains. Initially, *P. marneffei* yeast forms were mistaken for *Histoplasma capsulatum* var. *capsulatum* yeast cells (Deng and Connor 1985). The principal differences between the tissue forms of these two dimorphic fungi are presented in Figures 29.4 and 29.5.

The ultrastructural characteristics of *P. marneffei* in golden hamsters were described by Drouhet and colleagues who showed the evolution of the fungal cell in liver macrophages (Drouhet et al. 1988). The various stages of the multiplication process (formation of the septum and division by fission) were recognized (Figure 29.6). The yeast-like cells were often included in phagolysosomes, and some were surrounded by concentric calcified layers as in an onion bulb. This aspect of the host defense reaction has also been observed in animal models and human infections caused by *Histoplasma capsulatum* var. *capsulatum*.

Identification

In the first mycological description of *P. marneffei* (Segretain 1959a), the new species was classified in the genus *Penicillium*, section Asymmetrica, subsection Divaricata, according to Raper and Thom's classification (Raper and Thom 1949). Subsequently, however, *P. marneffei* was assigned to the subgenus Biverticillium of

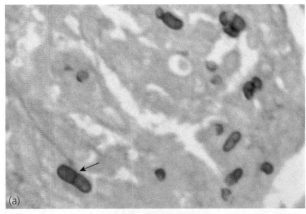

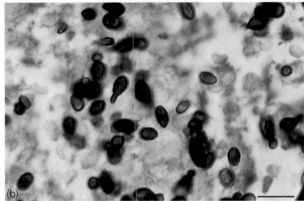

Figure 29.4 (a) *Oval, round, and sausage-shaped cells of* Penicillium marneffei, *some with septum (arrows), in a tissue section.* **(b)** *Uniform round and ovoid budding cells of* Histoplasma capsulatum var. capsulatum *in a tissue section. GMS stain; bar, 10 μm.*

Pitt's taxonomy because of its frequently biverticillate conidiophores and on the basis of its poor growth on 25 percent glycerol nitrate agar (Pitt 1979). Phylogenetic analysis also confirmed that *P. marneffei* is closely related to the *Penicillium* species in the subgenus Biverticillium, and the sexual *Talaromyces* species with an asexual biverticillate penicillium state (LoBuglio and Taylor 1995).

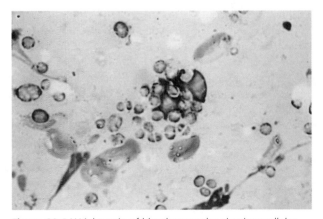

Figure 29.5 *Wright stain of blood smear showing intracellular* Penicillium marneffei yeasts with septa, magnification × 1 500

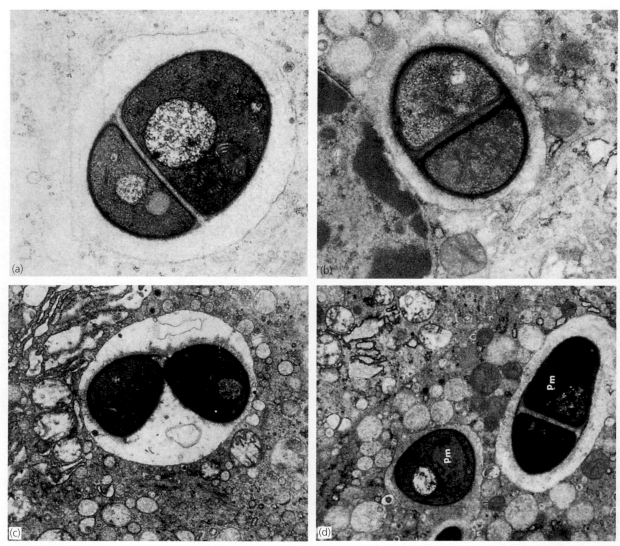

Figure 29.6 *Ultrastructure of yeast-like cells of* Penicillium marneffei *at various stages of division by fission within vacuoles of macrophages:* **(a, b)** *magnification* ×126 500; **(c, d)** *magnification* ×73 600. (Reproduced, with permission, from Drouhet 1993)

The above-described macro- and micromorphology of *P. marneffei* is relatively characteristic, and identification of the fungus is not considered difficult. A reddish pigment is, however, produced by other *Penicillium* species from which *P. marneffei* must be differentiated (Table 29.4).

Another approach to the differentiation of *P. marneffei* from other penicillia consists of the use of an exoantigen test (Sekhon et al. 1989). The exoantigen extract from a slant culture of *P. marneffei* is tested by a reverse microimmunodiffusion procedure against rabbit anti-*P. marneffei* reference serum in the presence of appropriate reference antigens. The production of two precipitin lines of the same identity as the *P. marneffei* reference system confirms identification.

A significant criterion in the identification of this fungus is based on the observation that *P. marneffei* is the only species of the *Penicillium* genus which, on enriched media at 37°C, converts into a schizogenous yeast form.

HOST RESPONSES

Clinical presentation

The clinical presentation of penicilliosis marneffei is similar in patients with and without HIV infection. Predominant signs and symptoms are chills, persistent fever, lymphadenopathy, hepatosplenomegaly, leukocytosis, anemia, persistent cough, weakness, and weight loss. Skin lesions and subcutaneous abscesses have often been reported. This clinical presentation is relatively nonspecific. It may be suggestive of tuberculosis or other systemic mycoses. Patients with AIDS present similar manifestations caused by HIV as well as other opportunistic infections or lymphoma.

P. marneffei usually causes an invasive disseminated disease, characterized in AIDS patients by a more rapid onset and more severe symptoms. Focal infections have been described rarely. The spleen, lung, or nasopharynx

Table 29.4 *Main characteristics of Penicillium spp. producing reddish diffusible pigments*

Penicillium spp.	Penicillium subgenus	Penicilli	Conidia	Pigment
P. citrinum	Furcatum sect. Furcatum	Bi-(mono-ter) verticillate divaricate	Smooth or very finely roughened	Yellow–brown to reddish brown
P. janthinellum	Furcatum sect. Divaricatum	Bi-(mono-ter) verticillate strongly divaricate	Smooth to finely roughened	Orange–reddish to vinaceous-purple
P. marneffei	Biverticillatum	Bi-(ter) verticillate divaricate	Smooth	Brick-red
P. purpurogenum	Biverticillatum	Biverticillate symmetrical appressed	Smooth, finely roughened or verrucose	Red to purple–red
P. rubrum	Biverticillatum	Biverticillate symmetrical appressed	Smooth	Red to purple–red

was reported as the only involved organ in HIV-negative patients (Di Salvo et al. 1973; Pautler et al. 1984; Deng et al. 1988; Chan et al. 1989). Among the AIDS population, one patient was reported to have a localized *P. marneffei* infection of the lung, which was diagnosed in the very early stage by a normal chest x-ray (Romaña et al. 1989).

Data on the extent of organ involvement are incomplete because several cases have been poorly described. An analytical overview of the clinical aspects of the first 88 cases, half of them in AIDS patients, has been published (Drouhet 1993). Pulmonary manifestations, mostly persistent cough and dyspnea, and occasionally chest pain and hemoptysis, associated with broncho-pneumonia, pulmonary abscess or pulmonary infiltrates, have been reported in 70 percent of HIV-negative and 86 percent of HIV-positive patients. In some cases, however, respiratory involvement has been found in the presence of a normal chest x-ray. In addition to lymphadenopathy in 45–56 percent of patients and hepatosplenomegaly in 43–50 percent, lesions have frequently been observed in other body sites such as the meninges, kidney, pericardium, and digestive tract – expressions of hematogenous spread of the infection.

Cutaneous and subcutaneous involvement has been reported in 68–71 percent of patients with or without AIDS (Supparatpinyo et al. 1994; Drouhet and Dupont 1995). Papules, rashes, acne-like pustules, and nodules are usually multiple, and located on the face, trunk, and extremities. Some of the papules have central necrotic umbilications resembling molluscum contagiosum (Figure 29.7). A more rapid appearance of multiple cutaneous lesions is observed in AIDS patients.

Bone and joint infections have also occurred in patients with disseminated disease, affected or not with AIDS (Jayanetra et al. 1984; Deng and Connor 1985; Chan and Woo 1990; Li et al. 1991; Louthrenoo et al. 1994; Deng and Ma 1995; Pun and Fang 2000). Arthritis affected both the large peripheral joints and the small finger joints, and osteomyelitis affected the flat bones of the skull, ribs, and the long bones. Most of the patients had multiple osteolytic lesions associated with swelling of the soft tissue and pain, particularly of the joints.

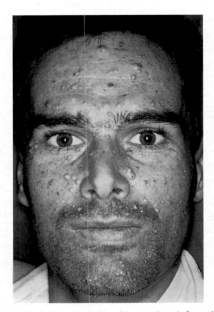

Figure 29.7 *Skin lesions in HIV-positive patient infected with* Penicillium marneffei. *Some of the papules have a central umbilication resembling lesions of molluscum contagiosum. (Reproduced, with permission, from Kronauer et al. 1993)*

Histopathology

The host response to *P. marneffei* invasion mainly involves the mononuclear phagocytic cells of the lung and the reticuloendothelial system, and resembles the response in acute histoplasmosis capsulati. In the early lesions, *P. marneffei* yeast-like cells are engulfed by histiocytes where they proliferate within the phagocyte's vacuoles. As the lesions progress, necrosis develops in the central areas, and fungal cells are released into the cytoplasm. Infiltration of neutrophils is followed by abscess formation. A granulomatous reaction takes place, especially in the organs of the reticuloendothelial system, with epithelioid cells, lymphocytes, plasma cells, and giant cells surrounding the necrotic area. Slowly evolving granulomas in the lung may lead to fibrosis and cavitation. Calcifications, such as those seen in experimental and human pulmonary histoplasmosis capsulati, have not been observed in human patients with penicilliosis marneffei, but have been reported in liver macrophages of experimentally infected golden hamsters (Drouhet et al. 1988). The fungal cells were surrounded by concentric calcified layers like onion bulbs, which represented a host defense response.

A totally different histopathological pattern – an anergic and necrotizing reaction involving mainly the lung, liver, and skin – is associated with compromised immunity of the host. The reaction is characterized by necrosis surrounded by histiocytes, which are distended by the engulfed proliferating fungi and are unable to form phagocytic vacuoles. This histopathological feature indicates a progressive disseminated infection.

In anergic as well as in pyogranulomatous lesions, special stains such as PAS or GMS are essential to distinguish penicilliosis marneffei from tuberculosis and other systemic mycoses as the fungi stain poorly in H&E preparations.

Immunology

The inhaled *P. marneffei* conidia are known to recognize fibronectin and to bind laminin of the bronchoalveolar epithelium via a sialic acid-containing receptor, avoiding entrapment by the mucus and removal from the respiratory tract (Hamilton et al. 1999). This adherence represents the first step of the host–parasite interaction. After attachment to the bronchoalveolar epithelium, *P. marneffei* conidia are easily phagocytized by macrophages.

The response of phagocytic cells to *P. marneffei* was studied with a macrophage culture system, using murine J774 macrophages and culture conditions suitable to reproduce the transformation of the fungus from the conidial to the characteristic in vivo yeast-like form. The results showed that following phagocytosis of *P. marneffei* conidia, nonstimulated murine macrophages are damaged by the fungus, which, growing intracellularly, undergoes transformation into a yeast-like form. By contrast, macrophages stimulated with interferon-gamma (IFN-γ) (50 U/ml) and lipopolysaccharide at 1 μg/ml significantly inhibit the intracellular growth of the conidia and kill the fungus. This correlates with the amount of nitric oxide produced by stimulated macrophages, suggesting that the L-arginine-dependent nitric oxide pathway has an important role in the murine host defense against *P. marneffei* (Cogliati et al. 1997). Killing of intracellular *P. marneffei* conidia was also shown to occur in IFN-γ-lipopolysaccharide-activated human THP1 cells (Taramelli et al. 2000). An increase in pH within the phagocytic vacuole, by threatening the *P. marneffei*-infected human THP1 and mouse J774 macrophages with chloroquine, induced macrophage killing of this fungus (Taramelli et al. 2001). The increase in intravacuolar pH may directly reduce fungus growth or may inhibit pH-dependent yeast virulence

factors. It has been shown that both mycelial and yeast form of *P. marneffei* expressed acid phosphatase activity (Youngchim et al. 1999). This enzyme has been considered one of the virulence factors for intracellular pathogen (Baca et al. 1993).

Other polymorphonuclear (PMN)-activating cytokines have been proven to enhance PMN inhibitory effects on germination and morphological changes of *P. marneffei*, but only granulocyte-macrophage colony stimulating factor (GM-CSF), granulocyte colony stimulating factor (G-CSF), and IFN-γ enhance PMN activity from being fungistatic to fungicidal (Kudeken et al. 1999). The killing mechanism of GM-CSF-stimulated neutrophils on *P. marneffei* was not mediated by a superoxide-dependent mechanism, but through exocytosis of granular enzymes which were largely heat-labile (Kudeken et al. 2000).

The ability of unopsonized *P. marneffei* to parasitize mononuclear phagocytes without stimulating the production of tumor necrosis factor (TNF)-alpha, observed in vitro using human leukocytes, may be critical for host defenses as it enhances the virulence of this intracellular parasite (Rongrungruang and Levitz 1999).

Iron availability was reported to affect the immunity to, and the pathogenicity of, *P. marneffei*, as iron overloading significantly reduces the antifungal activity of macrophages, whereas exogenous iron enhances, and iron chelators inhibit, the extracellular growth of *P. marneffei* (Taramelli et al. 2000).

It was shown that the immunocompetent host generates a cell-mediated response against *P. marneffei* (Deng et al. 1988) and that the failure of a CD4+ T-cell-dependent immunity in AIDS patients contributes to the development of disseminated systemic infection (Viviani and Tortorano 1990; Supparatpinyo et al. 1992).

The role of T lymphocytes in host defenses against *P. marneffei* was studied in mice experimentally depleted of CD4+ or CD8+ T cells (Viviani et al. 1993a). The results suggest that CD4+ T cells and, to a lesser extent CD8+ T cells, play a role in the protective host response against the fungus. *P. marneffei* strains differed in their virulence for mice. Depending on this virulence, CD4+ T-cell-mediated protective immunity acts to slow or halt multiplication of the fungus in the viscera. It is therefore likely that AIDS patients are susceptible to systemic *P. marneffei* infection because they are unable to generate a CD4+ T-cell response. Furthermore, CD8+ T cells, which are spared by the virus, may be insufficient to eliminate the organisms in developing granulomas. Further evidence that cell-mediated immunity plays a central role in a host defense mechanism against *P. marneffei* infection was provided in work involving experimentally infected mice (Kudeken et al. 1996).

In the mouse model, CD4+ T cells play an important role in eradicating *P. marneffei* from infected sites (Kudeken et al. 1997). Interleukin-12 (IL-12) protects mice against this infection. The induction of IL-12 synthesis by human peripheral blood mononuclear cells is promoted by a phosphoprotein, osteopontin (OPN), after stimulation with *P. marneffei* (Koguchi et al. 2002). Furthermore, the OPN production by monocytes is regulated by GM-CSF. It is suggested that the possible mechanism is mediated by mannoprotein in the OPN production upon stimulation by *P. marneffei*. The mannoprotein (Mp1p) is known as an abundant cell wall antigen of the yeast cells of this fungus (Cao et al. 1998a). Thus, the OPN may promote a Th1 response through the mannoprotein-induced IL-12 secretion by human macrophages stimulated with *P. marneffei* and contribute to host defense against this pathogen. In addition, IFN-γ renders murine macrophages highly active in killing *P. marneffei* yeast cells by promoting the release of nitric oxide (Kudeken et al. 1998).

LABORATORY DIAGNOSIS

The high mortality rate in the first reported cases of *P. marneffei* infection is attributable to a delay in diagnosis because the mycosis was often misdiagnosed as tuberculosis and treated accordingly. The recently improved survival rate can be ascribed to clinicians' and microbiologists' increased awareness of the disease.

Diagnosis rarely requires invasive procedures even though biopsy of lymph nodes, lung, liver, and bone marrow provides specimens rich in fungal organisms trapped by the reticuloendothelial cells. Biopsy is recommended in the presence of skin nodules or subcutaneous abscesses. Skin lesions should be sought carefully because they may facilitate diagnosis. Biopsy samples must always be processed for both microscopy and culture. Other laboratory tools including polymerase chain reaction (PCR) and methods for detecting specific antibody against and antigen of the fungus in clinical specimens have been developed.

Microscopic examination

A variety of clinical specimens is appropriate to diagnose this condition. The clinical specimens include bone marrow aspirates, blood, lymph node biopsies, skin biopsies, skin scrapings, sputum, bronchoalveolar lavage pellets, pleural fluid, liver biopsies, cerebrospinal fluid, pharyngeal ulcer scrapings, palatal papule scrapings, urine, kidney, pericardium, stomach or intestine, and stools (Vanittanakom and Sirisanthana 1997). Rapid bedside diagnosis can be made by microscopic examination of Wright's stained bone marrow aspirates and/or touch smear of skin biopsy or lymph node biopsy specimens (Supparatpinyo et al. 1994). The yeast cells of *P. marneffei* could be seen in the peripheral blood smear of a patient with fulminant infection (Supparatpinyo and Sirisanthana 1994).

P. marneffei can be seen in histopathological sections stained with H&E, GMS, or PAS. They must be discriminated from *Leishmania donovani*, *Pneumocystis carinii*, and *Histoplasma capsulatum* var. *capsulatum*, which also frequently cause disease in patients with AIDS *P. marneffei* superficially resembles the *H. capsulatum* var. *capsulatum* in tissue when yeast-like cells are clustered within macrophages or histiocytes, but the two organisms multiply differently. *P. marneffei* replicates by fission, instead of budding, and the considerable variation in size and shape of the extracellular yeasts (enlarged, occasionally septate, tubular or sausage shaped) makes *P. marneffei* easily distinguishable from the budding tissue-form cells of *H. capsulatum* (see Figures 29.4 and 29.5).

A specific indirect fluorescent antibody reagent was developed to facilitate the rapid detection and identification of *P. marneffei* in histological sections (Kaufman et al. 1995). This antiglobulin preparation has proved to be specific because it failed to react with the yeast and hyphal forms of *H. capsulatum* and with a variety of other fungi, including *Aspergillus* spp. Specific IgM monoclonal antibodies were also produced from mice immunized with *P. marneffei* mycelial culture filtrate and proven strongly reactive in immunofluorescence staining of the fungus in tissue samples (Trewatcharegon et al. 2000).

Cultures

P. marneffei can be cultured from various clinical specimens. Bone marrow culture is the most sensitive (100 percent), followed by skin biopsy culture (90 percent), and blood culture (76 percent) (Supparatpinyo et al. 1994). A blood culture method, the lysis centrifugation system (Isolator System, Wampole Laboratories, Cranbury, NJ), has proved highly effective for the isolation of *P. marneffei*, especially in AIDS patients who frequently have concomitant bacteremia.

P. marneffei grows easily in routine culture media containing antibacterial agents. The fungus is inhibited by cycloheximide, which has therefore to be excluded from Sabouraud's dextrose agar. Inoculated Petri dishes should be incubated at 25°C and 37°C to demonstrate thermal dimorphism. The fungus grows in a mycelial phase at 25°C after 3–5 days of incubation. The mycelial phase produces red pigment which diffuses into the agar (Figure 29.2). Aged colonies are deep red. Microscopic examination shows typical structure of the genus *Penicillium* (see Figure 29.3). At 35–37°C on Sabouraud's agar or on brain–heart infusion agar, colonies are glabrous, off-white and yeast like, and the diffusible red pigment may be poorly produced (Figure 29.8, Figure 29.9). Pigment may also be masked by the color of media such as blood or chocolate agar. Microscopically, yeast cells of *P. marneffei* are

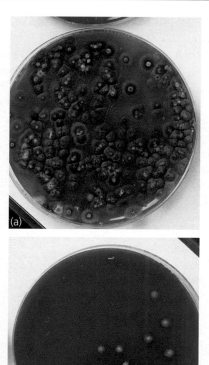

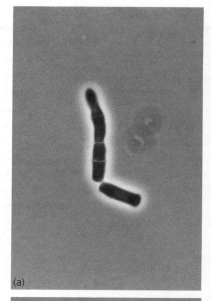

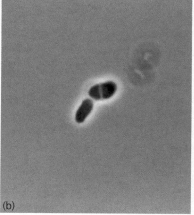

Figure 29.8 *Colonies of* Penicillium marneffei *grown from blood (Isolator system blood culture) on* **(a)** *Sabouraud's dextrose agar and* **(b)** *blood agar, after 7 days at 37°C.*

globose to oval with single septum, or arthroconidia-like cells.

Serodiagnosis

ANTIBODY DETECTION

Several studies were carried out to detect reliable reagents for antibody and antigen detection in body fluids. Although antigen is more likely to be detected than antibodies in patients with impaired immunity, antibodies to *P. marneffei* were shown to be produced by AIDS patients and may be useful for monitoring the efficacy of treatment. A microimmunodiffusion test, using a mycelial culture filtrate antigen of *P. marneffei* (Sekhon et al. 1982), has been used to monitor the serological response of an AIDS patient with penicilliosis marneffei (Viviani et al. 1993b). Sera drawn early in the course of the disease gave positive antibody reactions whereas sera taken 3–5 months after the start of therapy were negative. In another study of the immunodiffusion test, using yeast culture filtrate antigen, only two of 17 patients infected with *P. marneffei* were positive antibody reactions in their serum specimens (Kaufman et al. 1996). An indirect fluorescent antibody test for detecting IgG antibody, using germinating conidia and yeast cells

Figure 29.9 *Yeast cells of* Penicillium marneffei *cultured at 37°C showing arthroconidia-like cells; magnification ×1 500 (Chariyalertsak et al. 1996b).*

as antigens (Yuen et al. 1994), could detect the IgG titer of 160 or more in patients infected with *P. marneffei*.

Thereafter, several tests to detect antibodies to *P. marneffei* have been developed using crude antigen preparation from yeast phase and mold phase, purified proteins from yeast phase, and recombinant antigen. An immunoblot analysis of sera from 33 *P. marneffei*-infected AIDS patients, 29 non-infected AIDS patients, 25 AIDS patients from nonendemic areas, and 84 healthy individuals revealed the specificity of secreted yeast phase proteins of 54 and 50 kDa in the diagnosis of penicilliosis marneffei (Vanittanakom et al. 1997). Other diagnostic antigens of *P. marneffei* included antigens with relative molecular masses of 61 kDa (Jeavons et al. 1998) and 38 kDa (Chongtrakool et al. 1997), which were prepared from purified cytoplasmic yeast antigen and from acetone-precipitated culture filtrate of mold form, respectively. In addition, a gene (*MP1*) encoding a highly antigenic cell wall mannoprotein (Mp1p) has been cloned and the purified recombinant antigenic mannoprotein produced and tested in an enzyme-linked immunosorbent assay (ELISA) for antibody detection (Cao et al. 1998a, b). The sensitivity of these antigenic preparations is difficult to compare as different methods have been used and different patient populations studied. Recently, recombinant Mp1p protein vaccine was used for generation of protective immune responses against *P. marneffei* infection using a mouse model, and compared with *MP1* DNA vaccine delivered by live-attenuated *Salmonella typhimurium* (Wong et al. 2002). The intramuscular *MP1* DNA vaccine offered the best protection after challenge with *P. marneffei* yeast cells intravenously.

The scanty antibody response of immunosuppressed patients or the presence of antibodies in patients previously exposed to the fungus may limit the usefulness of the antibody detection. Tests for antigen detection would be of major clinical value, mainly to diagnose promptly the infection in patients without skin lesions, to identify AIDS patients with an initial infection and to predict relapse.

ANTIGEN DETECTION

In order to assess the presence of antigen in body fluids of patients affected with *P. marneffei*, rabbit polyclonal antibodies against arthroconidial filtrate (or yeast phase) were used in an immunodiffusion and a latex agglutination (LA) test and, despite the crude antigen preparation used to raise the antibody reagent, the tests showed a surprisingly high specificity, 100 percent (Kaufman et al. 1996). Encouraging results were also obtained with the antibody reagent against the recombinant mannoprotein Mp1p, that was used to test sera from AIDS and non-AIDS patients with penicilliosis in an ELISA assay with a specificity and sensitivity of 100 and 65 percent, respectively (Cao et al. 1999). The combined use of antigen and antibody tests was proven to increase the diagnostic sensitivity from 76 to 82 percent (Kaufman et al. 1996) and from 65 to 88 percent (Cao et al. 1999). The ELISA system using the recombinant Mp1p was further tested with sera from eight *P. marneffei*-infected HIV-positive patients and seven *P. marneffei*-infected HIV-negative patients (Wong et al. 2001b). Serum antigen titers were found to be higher in HIV-positive patients, whereas serum antibody titers were found to be higher in HIV-negative patients. A urinary antigen detection assay was also studied using a rabbit hyperimmune IgG preparation which, at a cutoff titer of 1:40, had a diagnostic sensitivity of 97 percent and specificity of 98 percent (Desakorn et al. 1999). A dot blot ELISA and a LA test were developed further for detecting *P. marneffei* antigenuria by using the same polyclonal antibody and compared with the ELISA (Desakorn et al. 2002). All tests appeared to be highly specific. The sensitivities of the tests were as follows: dot blot ELISA, 94.6 percent; ELISA, 97.3 percent; and LA test, 100 percent. The LA test will be evaluated for the detection of antigenemia and for the treatment follow-up.

PCR

Oligonucleotide primers for selective amplification of *P. marneffei* DNA were designed from the nuclear ribosomal DNA (rDNA) internal transcribed spacer region (LoBuglio and Taylor 1995). Using the nested PCR, the test was 100 percent successful in amplifying *P. marneffei* DNA. Thereafter, nucleotide sequence from the 18S rDNA of this fungus was determined (Vanittanakom et al. 1998). An oligonucleotide probe was designed and proved to be specific for *P. marneffei* in the PCR-hybridization reaction, regardless of whether the fungus was isolated from humans or natural habitats. This PCR-hybridization technique could be used to detect *P. marneffei* DNA in EDTA-blood samples collected from AIDS patients with penicilliosis. Recently, new PCR primers were designed based on the 18S rDNA sequence, and were used in a single and nested PCR methods for the rapid identification of *P. marneffei* (Vanittanakom et al. 2002). Very young culture of *P. marneffei* (2-day-old filamentous colony, 2 mm in diameter) could be performed by this assay. The application of this method for early diagnosis of the disease needs to be studied further. Another method which was developed to provide rapid identification of dimorphic fungi including *P. marneffei* was a PCR-enzyme immunoassay (PCR-EIA) method (Lindsley et al. 2001). The DNA was amplified by using universal fungal primers which was directed to the conserved regions of rDNA. The PCR amplicons were then detected colorimetrically in an enzyme immunoassay format for specific fungal species. The probes developed in this test were found to be highly specific and should prove to be useful in differentiating these organisms in the clinical setting.

THERAPY

P. marneffei infection is fatal in the absence of treatment, as documented by a fatality rate of 91.3 percent in non-AIDS patients and 100 percent in AIDS patients (Drouhet 1993; Supparatpinyo et al. 1993). In addition, a high rate (50 percent) of relapse is reported after successful primary therapy (Supparatpinyo et al. 1998).

In vitro studies have shown that most *P. marneffei* strains are very sensitive to the antifungals used in systemic treatment, mostly to amphotericin B, ketoconazole, itraconazole, voriconazole, and terbinafine (Sekhon et al. 1992; Drouhet 1993; Supparatpinyo et al. 1993; Boon-Long et al. 1996; McGinnis et al. 1997, 2000).

The antifungal drugs mostly used in the treatment of *P. marneffei* infection are amphotericin B, itraconazole, and ketoconazole. At present, the treatment regimen recommended in Thailand consists of a 2-week course of intravenous amphotericin B, followed by itraconazole (400 mg/day) for 10 weeks, given on an outpatient basis (Sirisanthana et al. 1998). The short course of amphotericin B was designed to lower the incidence of adverse drug effects and to reduce the length of hospital stay. As *P. marneffei* is highly susceptible to itraconazole, patients with mild to moderately severe infection can be given itraconazole 400 mg/day as primary therapy (Supparatpinyo et al. 1998). Secondary prophylaxis with itraconazole 200 mg/day is well tolerated and should be the standard care for patients with AIDS successfully treated for *P. marneffei* infection (Supparatpinyo et al. 1998). Itraconazole (200 mg/day) has been proved to be safe and effective as primary prophylaxis to prevent *P. marneffei* infection in patients with advanced HIV infection, especially those with CD4+ lymphocyte counts of <100 cells/μl (Chariyalertsak et al. 2002).

FUNGAL GENETICS

P. marneffei is a dimorphic fungus which displays a temperature-dependent dimorphism. In host cells or at 37°C, the fungal hyphae undergo phase transition via the arthroconidiation to form fission yeast cells. The genetic factors that express during phase transition may be involved in the molecular mechanisms of fungal cell morphogenesis and pathogenesis. A homolog of the *Aspergillus nidulans abaA* gene has been cloned from *P. marneffei* and shown that it is involved in both asexual conidiation at 25°C and dimorphic hyphal-yeast switching (Borneman et al. 2000).

Another gene, *cflA*, which is a homolog of the *Saccharomyces cerevisiae cdc42*, has been cloned from *P. marneffei* (Boyce et al. 2001). *cflA* is required for polarized growth of the vegetative mycelium at 25°C and for correct cellular morphology and separation of yeast cells at 37°C. However, it does not appear to be involved in development or in dimorphic switching. Furthermore, a *gasA* gene encoding a Gα subunit has been cloned from *P. marneffei*, and investigated as a key regulator of asexual development (Zuber et al. 2002). By contrast, *gasA* mutants have no apparent defect in dimorphic switching or yeast-like growth. Borneman et al. (2002) has cloned and characterized an APSES protein-encoding gene from *P. marneffei* that has a high degree of similarity to *Aspergillus nidulans stuA*. This gene is required for metula and phialide formation of *P. marneffei* but is not required for dimorphic growth.

A search for phase-specific genes in *P. marneffei* has been investigated by using differential display technique (Cooper and Haycocks 2000). Many of the genes whose expression displays during mold-to-arthroconidium transition are related to energy metabolism. Expression of the gene which encodes malate synthase is upregulated in the arthroconidial or yeast phase as compared to the mold phase (Cooper and Haycocks 2000). The authors

plan to assess the role of those genes in the virulence of *P. marneffei* in murine model by creating specific gene-disruption mutants.

REFERENCES

Ajello, L., Padhye, A.A., et al. 1995. Occurrence of *Penicillium marneffei* infections among wild bamboo rats in Thailand. *Mycopathologia*, **131**, 1–8.

Ancelle, T., Dupouy-Camet, J., et al. 1988. Un cas de pénicilliose disséminée à *Penicillium marneffei* chez un malade atteint d'un syndrome imunodéficitaire acquis. *Presse Méd*, **17**, 1095–6.

Baca, O.G., Roman, M.J., et al. 1993. Acid phosphatase activity in *Coxiella burnetii*: a possible virulence factor. *Infect Immun*, **61**, 4232–9.

Bailloud, R., Sumanak, M., et al. 2002. Premiers cas d'infection *Penicillium marneffei* identifis chez l'immunodprim au Cambodge. *J Mycol Med* (Paris), **12**, 138–42.

Bateman, A.C., Jones, G.R., et al. 2002. Massive hepatosplenomegaly caused by *Penicillium marneffei* associated with human immunodeficiency virus infection in a Thai patient. *J Clin Pathol*, **55**, 143–4.

Boon-Long, L., Mekha, N., et al. 1996. *In vitro* antifungal activity of the new triazole D0 870 against *Penicillium marneffei* compared with that of amphotericin B, fluconazole, itraconazole, miconazole and flucytosine. *Mycoses*, **39**, 453–6.

Borneman, A.R., Hynes, M.J. and Andrianopoulos, A. 2000. The *abaA* homologue of *Penicillium marneffei* participates in two developmental programmes: conidiation and dimorphic growth. *Mol Microbiol*, **38**, 1034–47.

Borneman, A.R., Hynes, M.J. and Andrianopoulos, A. 2002. A basic helix-loop-helix protein with similarity to the fungal morphological regulators, Phd1p, Efg1p and StuA, controls conidiation but not dimorphic growth in *Penicillium marneffei*. *Mol Microbiol*, **44**, 621–31.

Borradori, L., Schmit, J.C., et al. 1994. *Penicilliosis marneffei* infection in AIDS. *J Am Acad Dermatol*, **31**, 843–6.

Boyce, K.J., Hynes, M.J. and Andrianopoulos, A. 2001. The *CDC42* homologue of the dimorphic fungus *Penicillium marneffei* is required for correct cell polarization during growth but not development. *J Bacteriol*, **183**, 3447–57.

Cao, L., Chan, C.M., et al. 1998a. *MP1* encodes an abundant and highly antigenic cell wall mannoprotein in the pathogenic fungus *Penicillium marneffei*. *Infect Immun*, **66**, 966–73.

Cao, L., Chen, D.L., et al. 1998b. Detection of specific antibodies to an antigenic mannoprotein for diagnosis of *Penicillium marneffei* penicilliosis. *J Clin Microbiol*, **36**, 3028–31.

Cao, L., Chan, K.M., et al. 1999. Detection of cell wall mannoprotein Mp1p in culture supernatants of *Penicillium marneffei* and in sera of penicilliosis patients. *J Clin Microbiol*, **37**, 981–6.

Capponi, M., Sureau, P. and Segretain, G. 1956. Pénicillose de *Rhizomys sinensis*. *Bull Soc Pathol Exot*, **49**, 418–21.

Chan, J.K.C., Tsang, D.N.C. and Wong, D.K.K. 1989. *Penicillium marneffei* in bronchoalveolar lavage fluid. *Acta Cytol*, **33**, 523–6.

Chan, Y.F. and Woo, K.C. 1990. *Penicillium marneffei* osteomyelitis. *J Bone Joint Surg [Br]*, **72**, 500–3.

Chang, C.C., Liao, S.T., et al. 1995. Disseminated *Penicillium marneffei* infection in a patient with acquired immunodeficiency syndrome. *J Formos Med Assoc*, **94**, 572–5.

Chang, K.C., Chan, C.K., et al. 1998. *Penicillium marneffei* infection and solitary pulmonary nodule. *Hong Kong Med J*, **4**, 59–62.

Chariyalertsak, S., Sirisanthana, T., et al. 1996a. Seasonal variation of disseminated *Penicillium marneffei* infections in Northern Thailand: a clue to the reservoir. *J Infect Dis*, **173**, 1490–3.

Chariyalertsak, S., Vanittanakom, P., et al. 1996b. *Rhizomys sumatrensis* and *Cannomys badius*, new natural animal hosts for *Penicillium marneffei*. *J Med Vet Mycol*, **34**, 105–10.

Chariyalertsak, S., Sirisanthana, T., et al. 1997. Case-control study of risk factors for *Penicillium marneffei* infection in human immunodeficiency virus-infected patients in Northern Thailand. *Clin Infect Dis*, **24**, 1080–6.

Chariyalertsak, S., Supparatpinyo, K., et al. 2002. A controlled trial of itraconazole as primary prophylaxis for systemic fungal infections in patients with advanced human immunodeficiency virus infection in Thailand. *Clin Infect Dis*, **34**, 277–84.

Chiang, C.T., Leu, H.S., et al. 1998. *Penicillium marneffei* fungemia in an AIDS patient: the first case report in Taiwan. *Changgeng Yi Xue Za Zhi*, **21**, 206–10.

Chiewchanvit, S., Mahanupab, P., et al. 1991. Cutaneous manifestations of disseminated *Penicillium marneffei* mycosis in five HIV-infected patients. *Mycoses*, **34**, 245–9.

Chokephaibulkit, K., Veerakul, G., et al. 2001. Penicilliosis-associated hemophagocytic syndrome in a human immunodeficiency virus-infected child: the first case report in children. *J Med Assoc Thai*, **84**, 426–9.

Chongtrakool, P., Chaiyaroj, S.C., et al. 1997. Immunoreactivity of a 38-kilodalton *Penicillium marneffei* antigen with human immunodeficiency virus-positive sera. *J Clin Microbiol*, **35**, 2220–3.

Coen, M., Viviani, M.A. et al. 1989. Disseminated infection due to *Penicillium marneffei* in a HIV positive patient. *Abstracts of the 5th International Conference on AIDS*. Ottawa: International Development Research Centre, abstract MBP 94.

Cogliati, M., Roverselli, A., et al. 1997. Development of an in vitro macrophage system to assess *Penicillium marneffei* growth and susceptibility to nitric oxide. *Infect Immun*, **65**, 279–84.

Cooper, C.R. and Haycocks, N.G. 2000. *Penicillium marneffei*: an insurgent species among the Penicillia. *J Eukaryot Microbiol*, **47**, 24–8.

Deng, Z.L. and Connor, D.H. 1985. Progressive disseminated penicilliosis caused by *Penicillium marneffei*. Report of eight cases and differentiation of the causative organism from *Histoplasma capsulatum*. *Am J Clin Pathol*, **84**, 323–7.

Deng, Z.L. and Ma, Y. 1995. Disseminated penicilliosis marneffei complicated with osteolytic lesions. *J Mycol Méd (Paris)*, **5**, 44–9.

Deng, Z.L., Ma, Y. and Ajello, L. 1986. Human penicilliosis marneffei and its relation to the bamboo rat (*Rhizomys pruinosus*). *J Med Vet Mycol*, **24**, 383–9.

Deng, Z.L., Ribas, J.L., et al. 1988. Infections caused by *Penicillium marneffei* in China and Southeast Asia: review of eighteen published cases and report of four more Chinese cases. *Rev Infect Dis*, **10**, 640–52.

Depraetere, K., Colebunders, R., et al. 1998. Two imported cases of *Penicillium marneffei* infection in Belgium. *Acta Clin Belg*, **53**, 255–8.

Desakorn, V., Smith, M.D., et al. 1999. Diagnosis of *Penicillium marneffei* infection by quantitation of urinary antigen by using an enzyme immunoassay. *J Clin Microbiol*, **37**, 117–21.

Desakorn, V., Simpson, A.J.H., et al. 2002. Development and evaluation of rapid urinary antigen detection tests for diagnosis of penicilliosis marneffei. *J Clin Microbiol*, **40**, 3179–83.

de Truchis, P., Bounioux, M.E. et al. 1991. Septicémie à *Penicillium marneffei* au cours du SIDA. *Abstracts of the Eleventh Interdisciplinary Meeting on Anti-Infections Chemotherapy*. Paris: Société Française de Microbiologie, Société de Pathologie Infectieuse de Langue Française, 89, Abstract 35/C4.

Di Salvo, A.F., Fickling, A.M. and Ajello, L. 1973. Infection caused by *Penicillium marneffei*: Description of first natural infection in man. *Am J Clin Pathol*, **59**, 259–63.

Drouhet, E. 1993. Penicilliosis due to *Penicillium marneffei*: A new emerging systemic mycosis in AIDS patients travelling or living in Southeast Asia. Review of 44 cases reported in HIV infected patients during the last 5 years compared to 44 cases of non AIDS patients reported over 20 years. *J Mycol Méd (Paris)*, **4**, 195–224.

Drouhet, E. and Dupont, B. 1995. Infection à *Penicillium marneffei*: mycose systémique à manifestations cutanées associée au SIDA. *J Mycol Méd (Paris)*, **5**, 21–34.

Drouhet, E., Ravisse, P., et al. 1988. Étude mycologique, ultrastructurale et expérimentale sur *Penicillium marneffei* isolé d'une pénicilliose disséminée chez un SIDA. *Bull Soc Fr Mycol Méd*, **17**, 77–82.

Garbino, J., Kolarova, L., et al. 2001. Fungemia in HIV-infected patients: a 12-year study in a tertiary care hospital. *AIDS Patients Care STDS*, **15**, 407–10.

Grise, G., Aouar, M., et al. 1997. Infection à *Penicillium marneffei*: une pathologie à connaître. *Ann Biol Clin*, **55**, 241–2.

Hamilton, A.J., Jeavons, L., et al. 1999. Recognition of fibronectin by *Penicillium marneffei* conidia via a sialic acid-dependent process and its relationship to the interaction between conidia and laminin. *Infect Immun*, **67**, 5200–5.

Heath, T.C., Patel, A., et al. 1995. Disseminated *Penicillium marneffei* presenting illness of advanced HIV infection: a clinicopathological review, illustrated by a case report. *Pathology*, **27**, 101–5.

Hilmarsdottir, I., Meynard, J.L., et al. 1993. Disseminated *Penicillium marneffei* infection associated with human immunodeficiency virus: a report of two cases and a review of 35 published cases. *J AIDS*, **6**, 466–71.

Hilmarscottir, I., Coutellier, A., et al. 1994. A French case of laboratory-acquired disseminated *Penicillium marneffei* infection in a patient with AIDS. *Clin Infect Dis*, **19**, 357–8 (letter).

HIV/AIDS epidemiology report as of July 2002. Division of Epidemiology, Ministry of Public Health, Thailand. http://epid.moph.go.th/epi31.html

Hsueh, P.R., Teng, L.J., et al. 2000. Molecular evidence for strain dissemination of *Penicillium marneffei*. An emerging pathogen in Taiwan. *J Infect Dis*, **181**, 1706–12.

Hulshof, C.M.J., van Zanten, R.A.A., et al. 1990. *Penicillium marneffei* infection in an AIDS patient. *Eur J Clin Microbiol Infect Dis*, **9**, 370.

Hung, C.C., Hsueh, P.R., et al. 1998. Bacteremia and fungemia in patients with advanced human immunodeficiency virus (HIV) infection in Taiwan. *J Formos Med Assoc*, **97**, 690–7.

Imwidthaya, P., Thipsuvan, K., et al. 2000. *Penicillium marneffei*: types and drug susceptibility. *Mycopathologia*, **149**, 109–15.

Jayanetra, P., Nitiyanant, P., et al. 1984. Penicilliosis marneffei in Thailand: report of five human cases. *Am J Trop Med Hyg*, **33**, 637–44.

Jeavons, L., Hamilton, A.J., et al. 1998. Identification and purification of specific *Penicillium marneffei* antigens and their recognition by human immune sera. *J Clin Microbiol*, **36**, 949–54.

Jones, P.D. and See, J. 1992. *Penicillium marneffei* infection in patients infected with human immunodeficiency virus: late presentation in an area of nonendemicity. *Clin Infect Dis*, **15**, 744 (letter).

Julander, I. and Petrini, B. 1997. *Penicillium marneffei* infection in a Swedish HIV-infected immunodeficient narcotic addict. *Scand J Infect Dis*, **29**, 320–2.

Kantipong, P., Panich, V., et al. 1998. Hepatic penicilliosis in patients without skin lesions. *Clin Infect Dis*, **26**, 1215–17.

Kaufman, L., Standard, P.G., et al. 1995. Development of specific fluorescent-antibody test for tissue form of *Penicillium marneffei*. *J Clin Microbiol*, **33**, 2136–8.

Kaufman, L., Standard, P.G., et al. 1996. Diagnostic antigenemia tests for penicilliosis marneffei. *J Clin Microbiol*, **34**, 2503–5.

Ko, K.F. 1994. Retropharyngeal abscess caused by *Penicillium marneffei*: an unusual cause of upper airway obstruction. *Otolaryngol Head Neck Surg*, **111**, 445–6.

Koguchi, Y., Kawakami, K., et al. 2002. *Penicillium marneffei* causes osteopontin-mediated production of interleukin-12 by peripheral blood mononuclear cells. *Infect Immun*, **70**, 1042–8.

Kok, I., Veenstra, J., et al. 1994. Disseminated *Penicillium marneffei* infection as an imported disease in HIV-1 infected patients. Description of two cases and a review of the literature. *Neth J Med*, **44**, 18–22.

Kronauer, C.M., Schär, G., et al. 1993. Die HIV-assoziierte *Penicillium-marneffei*-infektion. *Schweiz Med Wochenschr*, **123**, 385–90.

Kudeken, N., Kawakami, K., et al. 1996. Cell-mediated immunity in host resistance against infection caused by *Penicillium marneffei*. *J Med Vet Mycol*, **34**, 371–8.

Kudeken, N., Kawakami, K. and Saito, A. 1997. CD4+ T cell-mediated fatal hyperinflammatory reactions in mice infected with *Penicillium marneffei. Clin Exp Immunol*, **107**, 468–73.

Kudeken, N., Kawakami, K. and Saito, A. 1998. Different susceptibilities of yeasts and conidia of *Penicillium marneffei* to nitric oxide-mediated fungicidal activity of murine macrophages. *Clin Exp Immunol*, **112**, 287–93.

Kudeken, N., Kawakami, K. and Saito, A. 1999. Cytokine-induced fungicidal activity of human polymorphonuclear leukocytes against *Penicillium marneffei. FEMS Immunol Med Microbiol*, **26**, 115–24.

Kudeken, N., Kawakami, K. and Saito, A. 2000. Mechanism of the *in vitro* fungicidal effects of human neutrophils against *Penicillium marneffei* induced by granulocyte-macrophage colony-stimulating factor (GM-CSF). *Clin Exp Immunol*, **119**, 472–8.

Kwan, E.Y., Lau, Y.L., et al. 1997. *Penicillium marneffei* infection in a non-HIV infected child. *J Paediatr Child Health*, **33**, 267–71.

Lachaud, L., Gayvallet-Montredon, N., et al. 1998. Nouveau cas de péniciliose disséminée à *Penicillium marneffei* chez une patiente sidéenne. *J Mycol Méd (Paris)*, **8**, 211–12 (letter).

Li, J.C., Pan, L.Q. and Wu, S.X. 1989. Mycologic investigation on *Rhizomys pruinosus senex* in Guangxi as natural carrier with *Penicillium marneffei. Chung Hua I Hsueh Tsa Chih*, **102**, 477–85.

Li, J.S., Pan, L.Q., et al. 1985. A case report on *Penicillium marneffei. J Clin Dermatol (China)*, **14**, 24–6.

Li, J.S., Pan, L.Q., et al. 1991. Disseminated penicilliosis marneffei in China. Report of three cases. *Chin Med J*, **104**, 247–51.

Liao, X., Ran, Y., et al. 2002. Disseminated *Penicillium marneffei* infection associated with AIDS, report of a case. *Zhonghua Yi Xue Za Zhi*, **82**, 325–9.

Lindsley, M.D., Hurst, S.F., et al. 2001. Rapid identification of dimorphic and yeast-like fungal pathogens using specific DNA probes. *J Clin Microbiol*, **39**, 3505–11.

Liu, M.T., Wong, C.K. and Fung, C.P. 1994. Disseminated *Penicillium marneffei* infection with cutaneous lesions in an HIV-positive patient. *Br J Dermatol*, **131**, 280–3.

Lo, C.Y., Chan, D.T., et al. 1995. *Penicillium marneffei* infection in a patient with SLE. *Lupus*, **4**, 229–31.

Lo, Y., Tintelnot, K., et al. 2000. Disseminated *Penicillium marneffei* infection in an African AIDS patient. *Trans R Soc Trop Med Hyg*, **94**, 187.

LoBuglio, K.F. and Taylor, J.W. 1995. Phylogeny and PCR identification of the human pathogenic fungus *Penicillium marneffei. J Clin Microbiol*, **33**, 85–9.

Louthrenoo, W., Thamprasert, K. and Sirisanthana, T. 1994. Osteoarticular penicilliosis marneffei. A report of eight cases and review of the literature. *Br J Rheumatol*, **33**, 1145–50.

Manion, D.J., Auclair, F. and Saginur, R. 1991. *Penicillium marneffei* mycosis: the first Canadian case report. *Abstracts of 59th Conjoint Meeting on Infectious Diseases*. Québec: Canadian Association for Clinical Microbiology and Infectious Diseases, abstract C-3.

McGinnis, M.R., Pasarell, L., et al. 1997. In vitro evaluation of voriconazole against some clinically important fungi. *Antimicrob Agents Chemother*, **41**, 1832–4.

McGinnis, M.R., Nordoff, N.G., et al. 2000. In vitro comparison of terbinafine and itraconazole against *Penicillium marneffei. Antimicrob Agents Chemother*, **44**, 1407–8.

McShane, H., Tang, C.M. and Conlon, C.P. 1998. Disseminated *Penicillium marneffei* infection presenting as a right upper lobe mass in an HIV positive patient. *Thorax*, **53**, 905–6.

Miegeville, M., Leautez, S., et al. 1998. Pneumopathie à *Penicillium marneffei* chez une patiente VIH positive d'origine Thaïlandaise. Premier cas nantais. *J Mycol Méd (Paris)*, **8**, 159–62.

Mohri, S., Yoshikawa, K., et al. 2000. A case of *Penicillium marneffei* infection in an AIDS patient: the first case in Japan. *Nippon Ishinkin Gakkai Zasshi*, **41**, 23–6.

Nord, J., Karter, D. and LaBombardi, V. 1998. An AIDS patient with fever and pancytopenia. *Int J Infect Dis*, **2**, 173–5.

Nowak, R.M. 1991. *Walker's mammals of the world*, 5th edition, Baltimore: The Johns Hopkins University Press.

Patamasucon, P. 1992. AIDS in Thailand. *J Infect Dis Antimicrob Agents*, **9**, 35–6.

Pautler, K.B., Padhye, A.A. and Ajello, L. 1984. Imported penicilliosis marneffei in the United States: report of a second human infection. *Sabouraudia*, **22**, 433–8.

Peto, T.E.A., Bull, R., et al. 1988. Systemic mycosis due to *Penicillium marneffei* in a patient with antibody to human immunodeficiency virus. *J Infect*, **16**, 285–90.

Piehl, M.R., Kaplan, R.L. and Haber, M.H. 1988. Disseminated penicilliosis in a patient with acquired immunodeficiency syndrome. *Arch Pathol Lab Med*, **112**, 1262–4.

Pitt, J.I. 1979. *The genus Penicillium and its teleomorphic states Eupenicillium and Talaromyces*. London: Academic Press.

Pun, T.S. and Fang, D. 2000. A case of *Penicillium marneffei* osteomyelitis involving the axial skeleton. *Hong Kong Med J*, **6**, 231–3.

Ranjana, K.H., Priyokumar, K., et al. 2002. Disseminated *Penicillium marneffei* infection among HIV-infected patients with Manipur State, India. *J Infect*, **45**, 268–71.

Raper, K.B. and Thom, C. 1949. *A manual of the Penicillia*. Baltimore, MA: Williams & Wilkins.

Rimek, D., Zimmermann, T., et al. 1999. Disseminated *Penicillium marneffei* infection in an HIV-positive female from Thailand in Germany. *Mycoses*, **42**, 25–8.

Rokiah, I., Ng, K.P. and Soo Hoo, T.S. 1995. *Penicillium marneffei* infection in an AIDS patient – a first case report from Malaysia. *Med J Malaysia*, **50**, 101–4.

Romaña, C.A., Stern, M., et al. 1989. Pénicilliose pulmonaire à *Penicillium marneffei* chez un patient atteint d'un syndrome immunodéficitaire acquis. Deuxième cas français. *Bull Soc Fr Mycol Méd*, **18**, 311–16.

Rongrungruang, Y. and Levitz, S.M. 1999. Interactions of *Penicillium marneffei* with human leukocytes in vitro. *Infect Immun*, **67**, 4732–6.

Rosenthal, E., Marty, P., et al. 2000. Infection à *Penicillium marneffei* évoquant une leishmaniose viscérale chez un patient infecté par le VIH. *Presse Méd*, **29**, 363–4.

Saadiah, S., Jeffrey, A.H. and Mohamed, A.L. 1999. *Penicillium marneffei* infection in a non AIDS patient: first case report from Malaysia. *Med J Malaysia*, **54**, 264–6.

Sathapatayavongs, B., Damrongkitchaiporn, S., et al. 1989. Disseminated penicilliosis associated with HIV infection. *J Infect*, **19**, 84–5 (letter).

Segretain, G. 1959a. Description d'une nouvelle espèce de penicillium: *Penicillium marneffei* n. sp. *Bull Soc Mycol Fr*, **75**, 412–16.

Segretain, G. 1959b. *Penicillium marneffei* n. sp., agent d'une mycose du système réticulo-endothélial. *Mycopathol Mycol Appl*, **11**, 327–53.

Sekhon, A.S., Li, J.S.K. and Garg, A.K. 1982. Penicillosis marneffei: serological and exoantigen studies. *Mycopathologia*, **77**, 51–7.

Sekhon, A.S., Garg, A.K., et al. 1989. Antigenic relationship of *Penicillium marneffei* to *P. primulinum. J Med Vet Mycol*, **27**, 105–12.

Sekhon, A.S., Padhye, A.A. and Garg, A.K. 1992. In vitro sensitivity of *Penicillium marneffei* and *Pythium insidiosum* to various antifungal agents. *Eur J Epidemiol*, **8**, 427–32.

Sekhon, A.S., Stein, L., et al. 1994. Pulmonary penicillosis marneffei: report of the first imported case in Canada. *Mycopathologia*, **128**, 3–7.

Singh, P.N., Ranjana, K., et al. 1999. Indigenous disseminated *Penicillium marneffei* infection in the state of Manipur, India: report of four autochthonous cases. *J Clin Microbiol*, **37**, 2699–702.

Sirisanthana, T., Supparatpinyo, K., et al. 1998. Amphotericin B and itraconazole for treatment of disseminated *Penicillium marneffei* infection in human immunodeficiency virus-infected patients. *Clin Infect Dis*, **26**, 1107–10.

Sirisanthana, V. and Sirisanthana, T. 1993. *Penicillium marneffei* infection in children infected with human immunodeficiency virus. *Pediatr Infect Dis J*, **12**, 1021–5.

Sirisanthana, V. and Sirisanthana, T. 1995. Disseminated *Penicillium marneffei* infection in human immunodeficiency virus-infected children. *Pediatr Infect Dis J*, **14**, 935–40.

So, S.Y., Chau, P.Y., et al. 1985. A case of invasive penicilliosis in Hong Kong with immunologic evaluation. *Am Rev Respir Dis*, **131**, 662–5.

Sobottka, I., Albrecht, H., et al. 1996. Systemic *Penicillium marneffei* infection in a German AIDS patient. *Eur J Clin Microbiol Infect Dis*, **15**, 256–9 (letter).

Srison, D., Thisyakorn, U., et al. 1995. Perinatal HIV infection in Thailand. *Southeast Asian J Trop Med Public Health*, **26**, 559–63.

Stern, M., Romaña, C.A., et al. 1989. Pénicilliose pulmonaire à *Penicillium marneffei* chez un malade atteint d'un syndrome immunodéficitaire acquis. *Presse Méd*, **18**, 2087.

Supparatpinyo, K. and Sirisanthana, T. 1994. Disseminated *Penicillium marneffei* infection diagnosed on examination of a peripheral blood smear of a patient with human immunodeficiency virus infection. *Clin Infect Dis*, **18**, 246–7.

Supparatpinyo, K., Chiewchanvit, S., et al. 1992. *Penicillium marneffei* infection in patients infected with human immunodeficiency virus. *Clin Infect Dis*, **14**, 871–4.

Supparatpinyo, K., Nelson, K.E., et al. 1993. Response to antifungal therapy by human immunodeficiency virus-infected patients with disseminated *Penicillium marneffei* infections and in vitro susceptibilities of isolates from clinical specimens. *Antimicrob Agents Chemother*, **37**, 2407–11.

Supparatpinyo, K., Khamwan, C., et al. 1994. Disseminated *Penicillium marneffei* infection in Southeast Asia. *Lancet*, **344**, 110–13.

Supparatpinyo, K., Perriens, J., et al. 1998. A controlled trial of itraconazole to prevent relapse of *Penicillium marneffei* infection in patients infected with the human immunodeficiency virus. *N Engl J Med*, **339**, 1739–43.

Taramelli, D., Brambilla, S., et al. 2000. Effects of iron on extracellular and intracellular growth of *Penicillium marneffei*. *Infect Immun*, **68**, 1724–6.

Taramelli, D., Tognazioli, C., et al. 2001. Inhibition of intramacrophage growth of *Penicillium marneffei* by 4-aminoquinolines. *Antimicrob Agents Chemother*, **45**, 1450–5.

Thongcharoen, P., Vithayasai, P., et al. 1992. Opportunistic infections in AIDS/HIV infected patients in Thailand. *Thai AIDS J*, **4**, 117–22.

Trewatcharegon, S., Chaiyaroj, S.C., et al. 2000. Production and characterization of monoclonal antibodies reactive with the mycelial and yeast phases of *Penicillium marneffei*. *Med Mycol*, **38**, 91–6.

Trewatcharegon, S., Sirisinha, S., et al. 2001. Molecular typing of *Penicillium marneffei* isolates from Thailand by NotI macrorestriction and pulsed-field gel electrophoresis. *J Clin Microbiol*, **39**, 4544–8.

Tsang, D.N.C., Chan, J.K.C., et al. 1988. *Penicillium marneffei* infection: an underdiagnosed disease? *Histopathology*, **13**, 311–18.

Tsang, D.N.C., Li, P.K.C., et al. 1991. *Penicillium marneffei*: another pathogen to consider in patients infected with human immunodeficiency virus. *Rev Infect Dis*, **13**, 766–7 (letter).

Tsui, W.N., Ma, K.F. and Tsang, D.N.C. 1992. Disseminated *Penicillium marneffei* infection in HIV-infected subject. *Histopathology*, **20**, 287–93.

Ukarapol, N., Sirisanthana, V. and Wongsawasdi, L. 1998. *Penicillium marneffei* mesenteric lymphadenitis in human immunodeficiency virus-infected children. *J Med Assoc Thai*, **81**, 637–40.

Valeyrie, L., Botterel, F., et al. 1999. Prolonged fever revealing disseminated infection due to *Penicillium marneffei* in a French HIV-seropositive patient. *AIDS*, **13**, 731–2 (letter).

Vanittanakom, N. and Sirisanthana, T. 1997. *Penicillium marneffei* infection in patients infected with human immunodeficiency virus. *Curr Top Med Mycol*, **8**, 35–42.

Vanittanakom, N., Mekaprateep, M., et al. 1995. Efficiency of the flotation method in the isolation of *Penicillium marneffei* from seeded soil. *J Med Vet Mycol*, **33**, 271–3.

Vanittanakom, N., Cooper, C.R., et al. 1996. Restriction endonuclease analysis of *Penicillium marneffei*. *J Clin Microbiol*, **34**, 1834–6.

Vanittanakom, N., Mekaprateep, M., et al. 1997. Western immunoblot analysis of protein antigens of *Penicillium marneffei*. *J Med Vet Mycol*, **35**, 123–31.

Vanittanakom, N., Merz, W.G., et al. 1998. Specific identification of *Penicillium marneffei* by a polymerase chain reaction/hybridization technique. *Med Mycol*, **36**, 169–75.

Vanittanakom, N., Vanittanakom, P. and Hay, R.J. 2002. Rapid identification of *Penicillium marneffei* by PCR-based detection of specific sequences on the rRNA gene. *J Clin Microbiol*, **40**, 1739–42.

Vilar, F.J., Hunt, R., et al. 2000. Disseminated *Penicillium marneffei* in a patient infected with human immunodeficiency virus. *Int J STD AIDS*, **11**, 126–8.

Viviani, M.A. and Tortorano, A.M. 1990. Unusual mycoses in AIDS patients. In: Vanden Bosche, H., Mackenzie, D.W.R., et al. (eds), *Mycoses in AIDS patients*. New York: Plenum Press, 147–53.

Viviani, M.A., Hill, J.O. and Dixon, D.M. 1993a. *Penicillium marneffei*: dimorphism and treatment. In: Vanden Bossche, H., Odds, F. and Kerridge, D. (eds), *Dimorphic fungi in biology and medicine*. New York: Plenum Press, 413–22.

Viviani, M.A., Tortorano, A.M., et al. 1993b. Treatment and serological studies of an italian case of penicilliosis marneffei contracted in Thailand by a drug addict infected with the human immunodeficiency virus. *Eur J Epidemiol*, **9**, 79–85.

Wang, I.L., Yeh, H.P., et al. 1989. Penicilliosis caused by *Penicillium marneffei*. A case report. *Derm Sinica*, **7**, 19–22.

Wei, X.G., Zhou, L.T., et al. 1985. Report of the first case of penicilliosis marneffei in China. *Natl Med J China*, **65**, 533–4.

Wei, X.G., Ling, Y.M., et al. 1987. Study of 179 bamboo rats carrying *Penicillium marneffei*. *China J Zoonoses*, **3**, 34–5 (in Chinese).

Wong, K.H. and Lee, S.S. 1998. Comparing the first and second hundred AIDS cases in Hong Kong. *Singapore Med J*, **39**, 236–40.

Wong, L.P., Woo, P.C., et al. 2002. DNA immunization using a secreted cell wall antigen Mp1p is protective against *Penicillium marneffei* infection. *Vaccine*, **20**, 2878–86.

Wong, S.S., Ho, T.Y.C., et al. 2001a. Biotyping of *Penicillium marneffei* reveals concentration-dependent growth inhibition by galactose. *J Clin Microbiol*, **39**, 1416–21.

Wong, S.S., Wong, K.H., et al. 2001b. Differences in clinical and laboratory diagnostic characteristics of penicilliosis marneffei in human immunodeficiency virus (HIV)- and non-HIV-infected patients. *J Clin Microbiol*, **39**, 4535–40.

Wong, S.S., Woo, P.C. and Yuen, K.Y. 2001c. *Candida tropicalis* and *Penicillium marneffei* mixed fungemia in a patient with Waldenstrom's macroglobulinaemia. *Eur J Clin Microbiol Infect Dis*, **20**, 132–5.

Youngchim, S., Vanittanakom, N. and Hamilton, A.J. 1999. Analysis of the enzymatic activity of mycelial and yeast phases of *Penicillium marneffei*. *Med Mycol*, **37**, 445–50.

Yuen, K., Wong, S.S., et al. 1994. Serodiagnosis of *Penicillium marneffei* infection. *Lancet*, **344**, 444–5

Yuen, W.C., Chan, Y.F., et al. 1986. Chronic lymphadenopathy caused by *Penicillium marneffei*: a condition mimicking tuberculous lymphadenopathy. *Br J Surg*, **73**, 1007–8.

Zuber, S., Hynes, M.J. and Andrianopoulos, A. 2002. G-protein signaling mediates asexual development at 25°C but has no effect on yeast-like growth at 37°C in the dimorphic fungus *Penicillium marneffei*. *Eukaryot Cell*, **1**, 440–7.

PART VI

SYSTEMIC MYCOSES CAUSED BY OPPORTUNISTIC FUNGI

30

Candidiasis

ESTHER SEGAL AND DANIEL ELAD

The genus *Candida* comprises about 200 species, of which close to 20 have been associated with pathology in humans or animals (Meyer et al. 1998; de Hoog et al. 2000), including species, such as *Candida famata* or *Candida inconspicua*, which have been added more recently to the list of potential pathogens. This chapter focuses only on the major pathogenic species:

- *Candida albicans*
- *Candida dubliniensis*
- *Candida glabrata*
- *Candida guilliermondii* and its teleomorph *Pichia guilliermondii*
- *Candida kefyr* and its teleomorph *Kluyveromyces marxianus*
- *Candida krusei* and its teleomorph *Issatchenkia orientalis*
- *Candida lusitaniae* and its teleomorph *Clavispora lusitaniae*
- *Candida parapsilosis*
- *Candida tropicalis*

HISTORIC NOTES

Based on detailed historic reviews of the genus *Candida* (Odds 1988; Rippon 1988; Kwon-Chung and Bennett 1992), the following significant data regarding the genus and the diseases caused by it may be outlined.

The first known description of *Candida* infection, oral candidiasis (thrush) in two patients with other underlying disease, may be found in Hippocrates' 'Epidemics'

from the fourth century BC. The first descriptions of thrush in modern medicine were made by Rosen von Rosenstein in 1771 and by Underwood in 1784, who identified the infection as a pediatric problem and accordingly described it in books dealing with such entities. Veron in 1935 suggested that the gastrointestinal tract of newborns might become infected during birth. Although Lagenbeck, in 1839, described the fungus in a case of oral thrush observed in a patient suffering from typhus, he misidentified it as the causative agent of the underlying disease. The correct association between oral thrush and the fungus was made only 3 years later, in 1842, by Gruby who classified the microorganism as *Sporotrichum*.

During the following decades various pathological conditions were shown to be associated with yeasts. The fungus was isolated by Bennett in 1844 in the sputum of a tuberculotic patient, by Wilkinson in 1849 from vaginal candidiasis, by Robin in 1853 from a systemic infection and by Zenker in 1861 from a brain infection in a debilitated patient, in whom the fungus spread hematogenously from an oral infection. In 1875, Hausemann established the possibility of infant infection during birth by demonstrating the analogy between the causative agent of oral and vaginal thrush. Additional pathological entities caused by *Candida* were described at the beginning of the twentieth century: onychomycosis by Dubendorfer in 1904, dermatitis by Jacobi in 1907, chronic mucocutaneous candidiasis by Forbes in 1923, and cystitis by Rafin in 1910. Later, in 1928, Conner described osteomyelitis, in 1940 Joachim and Polayes described endocarditis, and in 1943 Suthin noted the association between pathology of the endocrine system and *Candida* infections. Castellani, in 1912, while describing 'tea tasters' cough,' was probably the first to suggest the possibility that *Candida* species other than *C. albicans* may be involved in pathological processes.

The nomenclature of the yeast isolated from the patients, changed often. Robin in 1853 named it *Oidium albicans*, Quinquad in 1868 *Syringospora robinii*, and Reess in 1875 *Saccharomyces albicans*. Mycological studies by Grawitz, published in 1877, described the various morphological forms of *Candida*. The first binomial to gain wide acceptance over a long period, and which is sometimes still used, albeit wrongly, was *Monilia albicans*, which was suggested by Zopf in 1890. The criteria defined by Saccardo for the genus *Monilia* permitted the inclusion of certain fungi isolated from rotting vegetation into the genus. Berkhout, in 1923, after recognizing the differences between *Monilia* spp. isolated from rotting plants and fruit and those isolated from medical cases, established the genus *Candida* to accommodate the latter. This was accepted as the official name of the genus by the Eighth Botanical Congress in Paris in 1954.

Two major medical events have revived the interest in fungal diseases in general and *Candida* infections in particular. The first was the introduction of antibacterial drugs in the second half of the twentieth century. These drugs, especially those having a broad spectrum of activity, may act as predisposing factors for mycotic infections by causing an imbalance of the host's natural microflora in favor of fungi, upon which they have no inhibitory activity. The second event was the increase in the prevalence of immunosuppressed patients during the last few decades, as a result of chemotherapy or disease [acquired immunodeficiency syndrome (AIDS)], which led to a parallel increase in the incidence of *Candida* infections in general, and the less pathogenic non-*C. albicans* species in particular.

These events, in conjunction with the development of more sophisticated biological research techniques, led to a highly intensive study of the pathogenic mechanisms of *Candida* infection, such as adhesion, production of proteinases, and interaction with the defenses of the host and, consequently, to new concepts for diagnosis and management. Another result of the intense interest in *Candida* was the increased demand for more efficient drugs with fewer side effects. Recent reports indicating the possibility of innate or acquired resistance of a number of *Candida* spp. to several antimycotic drugs (see section on Treatment below) has emphasized the importance of in vitro antimycotic susceptibility testing and has prompted efforts toward the standardization of these procedures and their interpretation.

The extensive use of antimycotic drugs, particularly azoles, for prolonged therapeutic courses has led to changes in the relative prevalence of various *Candida* species, with a decrease in the proportion of *C. albicans* as the etiologic agent of candididiasis and an increase in the proportion of non-*albicans* species such as *C. glabrata* or *C. krusei* (Krcmery and Barnes 2002).

An additional recent aspect is the development of molecular technologies in the use of taxonomy, resulting in changes of the status of certain *Candida* spp. and the recognition of new species such as *C. dubliniensis* by Sullivan and colleagues in 1995 (Sullivan et al. 1995; Gutierrez et al. 2002). This species was discovered in the course of an epidemiologic survey of *Candida* in human immunodeficiency virus (HIV)-positive patients. It was identified initially as an 'atypical *Candida albicans*' strain that did not hybridize with a *C. albicans* specific probe (Sullivan et al. 1993). Subsequently, it was recognized as a separate taxon (Sullivan et al. 1995).

TAXONOMIC COMMENTS

Most yeast species dealt with in this chapter often went through name changes, usually following the discovery of synonymy among species, discovery of teleomorphic stages or nonvalidity of binomials. For a comprehensive treatise on the taxonomic aspects of yeasts the reader is referred to Kurtzman and Fell (1998) and de Hoog et al. (2000).

The more recent changes in nomenclature are described below, some of them based on molecular biological methods which have been applied in search of a better understanding of taxonomic relationships between yeast species. The analysis of small ribosomal subunit sequences has shown (Barns et al. 1991; Hendriks et al. 1991) that *C. albicans*, *C. tropicalis*, *C. parapsilosis*, and *Candida viswanatii* form one subgroup with a more distant connection to *C. guilliermondii*. *C. kefyr* and *C. glabrata* are located on two different but interconnected branches, while *C. lusitaniae* and *C. krusei* are on a different branch each. These findings are in agreement with those of the coenzyme Q analysis (Yamada and Kondo 1972) which assesses the number of isoprene units per ubiquinone molecule, according to which *C. albicans* and *C. tropicalis* and *C. guilliermondii* share the same system (Q9), *C. kefyr* and *C. glabrata* share the Q6 system, whereas *C. lusitaniae* and *C. krusei* have Q8 and Q7 systems, respectively.

Candida species, exhibiting a teleomorphic stage, are considered ascomycetous fungi (Kurtzman 1993). Meyer et al. (1998) divided the *Candida* species on the basis of physiological characteristics into 12 groups. The pathogenic *Candida* species dealt with in this chapter are included in group VI. They do not assimilate inositol, erythritol, and nitrate and grow at 40°C.

Candida albicans

The most significant change in the taxonomy of this species was the acceptance of the present binomium in place of *Monilia albicans* (see section on Historic notes below). The synonymy between *C. albicans*, *C. langeronii*, and *C. clausenii* has been confirmed by molecular biological methods (Mahrous et al. 1992; Wickes et al. 1992) as has been that of *Candida stellatoidea* (Kwon-Chung and Bennett 1992).

Two serotypes, A and B, based on differences between the mannan component of the cell wall have been described (Hasenclever and Mitchell 1961). Serotype A was found to be antigenically related to *C. tropicalis*, whereas serotype B was related to *C. stellatoidea*, currently integrated into the *C. albicans* species. Additional attempts of subspeciation of *C. albicans* have been carried out, including various biotyping and chemotyping methods as well as susceptibility to 'killer factors' (Polonelli et al. 1985). Recent efforts are focused on typing of *C. albicans* isolates by restriction enzyme profiles, karyotyping, and other molecular biological techniques (see section on Genetics below).

The use of molecular technologies yielded new insights on the possible sexual mating of *C. albicans*, identifying a mating type-like locus (Hull and Johnson 1999; Hull et al. 2000; Magee and Magee 2000).

Candida dubliniensis

Phylogenetic trees have shown that *C. dubliniensis* is a discrete taxon within the *Candida* genus (Gutierrez et al. 2002). *C. dubliniensis* differs in its DNA fingerprint pattern from *C. albicans* (Sullivan et al. 1993, 1995). In addition, pulse field gel electrophoresis showed that *C. dubliniensis* has 10 or more chromosomes whereas *C. albicans* has seven to eight. *C. dubliniensis* differs by 2.3 percent from *C. albicans* in the V3 variable region of the large ribosomal RNA genes.

Candida glabrata

The taxonomic affiliation of this species is the most controversial among those described in this chapter as this yeast was classified in the genus *Torulopsis* (Kwon-Chung and Bennett 1992). The principal point of argument was whether the ability or inability to produce pseudohyphae/hyphae is a cogent reason to differentiate between two genera (*Candida* and *Torulopsis*) or not. This controversy was complicated by additional arguments on whether the name *Torulopsis* was validly published before *Candida* and the importance of keeping the two genera divided to avoid confusion among the mycologists (Odds 1988). It is currently generally accepted that this yeast should be assigned to the genus *Candida*.

Candida guilliermondii

Two varieties of *C. guilliermondii* have been described: *C. guilliermondii* var. *guilliermondii* and *C. guillier-*

mondii var. *membranaefaciens*. Both variants have teleomorphs *Pichia guilliermondii* and *Pichia ohmeri*, respectively (Kurtzman 1998a).

Candida kefyr

C. kefyr is the anamorph of *Kluyveromyces marxianus*. *C. kefyr* was formerly designated as *Candida pseudotropicalis*, considerd to be the anamorph of *Kluyveromyces fragilis*. The two teleomorphs and their anamorphs were found to be conspecific (Kwon-Chung and Bennett 1992) and since *K. marxianus* and *C. kefyr* have precedence on *K. fragilis* and *C. pseudotropicalis*, respectively, they have been accepted.

Candida krusei

C. krusei is the anamorph of *Issatchenkia orientalis*. *C. krusei* differs from other pathogenic yeast species in a number of characteristics, such as its ability to grow on vitamin-free media (Odds 1988), the formation of flat, spreading colonies on Sabouraud's agar, cell wall mannan composition (Kogan et al. 1988), coenzyme Q numbers (Yamada and Kondo 1972) and other properties recently reviewed (Samaranayake and Samaranayake 1994).

Candida lusitaniae

C. lusitaniae is the anamorph of the ascomycetous yeast *Clavispora lusitaniae*. Genetic variability among clinical isolates was described (Merz et al. 1992). The involvement of *C. lusitaniae* as an opportunistic pathogen in humans was initially reported by Holzschu et al. (1979).

Candida parapsilosis

Nakase et al. (1979) showed that *C. parapsilosis* isolates could be divided into form I (arabinose positive) and form II (arabinose negative). The latter was shown subsequently to consist of anascosporogenous strains of *Loderomyces elongisporus* (Hamajima et al. 1987).

Candida tropicalis

The synonymy between *C. tropicalis* and *C. paratropicalis* has been confirmed by molecular biological methods (Wickes et al. 1992).

MYCOLOGICAL CHARACTERISTICS

Morphology

With a few exceptions (described below), macroscopic and microscopic cultural characteristics of the yeasts dealt with in this chapter are not specific enough to be a basis for differentiation. Colonies cultured on glucose peptone agar (GPA), incubated at 25°C for 3 days, have a diameter of 2–3 mm, are white to cream colored, smooth or umbonate, and may become wrinkled after further incubation, as well as dull to glistening. Intraspecific variability of these characteristics is significant enough to preclude their use as criteria for species identification. In spite of the basic morphological similarity between various *Candida* spp., a number of characteristics, which are species specific, are evident on certain media, such as corn meal agar (CMA).

CANDIDA ALBICANS

On CMA, after 3 days of incubation, *C. albicans* will produce true- and pseudomycelium, grape-like groups of blastospores clustered at the septa and, typically for this species, at the end of the hyphae or short lateral branches, chlamydospores.

When incubated for 2 hours in 10 percent serum at 37°C, typical cell elongations – germ tubes – are formed. They differ from other, similar outgrowths produced by several other species by the absence of a constriction at the tube's base.

An interesting phenomenon is the 'phenotypic switching' described initially by Slutsky et al. (1985). It was demonstrated that strains of *C. albicans* possess the ability to undergo phenotypic switching expressed in colony morphology variations, such as white–opaque (Slutsky et al. 1987), traits associated with virulence (see section on Virulence factors below). This phenomenon was extensively investigated, including various genetic aspects (Soll 2002).

CANDIDA DUBLINIENSIS

C. dubliniensis is morphologically very similar to *C. albicans*. The main similarities are germ tube and chlamydospore production, while the main differences are the color produced when grown on ChromAgar medium and it does not grow well, or not at all, at 45°C (Gutierrez et al. 2002). Additional, subtler differences may be noted (Al-Mosaid et al. 2001).

CANDIDA GLABRATA

A characteristic of this species is the absence of hyphae or pseudohyphae on CMA.

CANDIDA GUILLIERMONDII (ANAMORPH) AND PICHIA GUILLIERMONDII (TELEOMORPH)

On CMA, formation of pseudohyphae may vary according to the strain from abundant to sparse. Blastospores may develop in small chains or in clusters. True hyphae are not produced.

P. guilliermondii is heterothallic. On 5 percent malt extract agar, asci containing one to four hat-shaped ascospores, which are liberated shortly after their generation, are formed. The majority of *C. guillier-*

mondii natural isolates, however, do not produce the teleomorph (Kurtzman 1998a).

CANDIDA KEFYR (ANAMORPH) AND KLUYVEROMYCES MARXIANUS (TELEOMORPH)

Colonies of this species, cultured on GPA and incubated for 3 days at 25°C, are relatively small: 1–2 mm. On CMA, in addition to pseudohyphae, blastospores may be observed. Occasionally the blastospores may be elongated and reach a length of up to 16 µm, a characteristic they share with *C. krusei*.

Kluyveromyces ascii contain one to four reniform ascospores, which tend to agglutinate after their release (Lachance 1998).

CANDIDA KRUSEI (ANAMORPH) AND ISSATCHENKIA ORIENTALIS (TELEOMORPH)

Colonies of *C. krusei* cultured on GPA become flat-topped after 5–7 days of incubation. Most blastospores are elongated and may reach 25 µm in length after a week. In Sabouraud's broth, *C. krusei* grows on top of the medium as a thick pellicle that may be observed after 2–3 days' incubation at 25°C (Figure 30.1).

Figure 30.1 C. krusei, C. tropicalis, *and* C. albicans, *in Sabouraud's dextrose broth. Note the large pellicle of* C. krusei *(left) and small pellicle of* C. tropicalis *(center).*

Issatchenkia orientalis The taxonomic connection between *I. orientalis* and *C. krusei* has been confirmed only by comparison of genetic material (Wickes et al. 1992). Asci of *I. orientalis* contain one, occasionally two, smooth, spherical ascospores (Kurtzman 1998b).

CANDIDA LUSITANIAE (ANAMORPH) AND CLAVISPORA LUSITANIAE (TELEOMORPH)

On CMA, blastospores in verticillate chains are born on branching, often curved, pseudohyphae.

Clavispora lusitaniae is heterothallic. One to four ascospores, sometimes containing an oil droplet, are contained in each ascus. The ascospores are clavate (hence the name *Clavispora*), a characteristic unique to the genus. The spores might be slightly echinulate, a characteristic sometimes seen only by electron microscope (EM) examination. The type strain of *C. obtusa* (CBS 1944), which unlike *C. lusitaniae* does not assimilate lactate or ferment galactose, produced *Cl. lusitaniae* teleomorphs when mated with α mating types of *C. lusitaniae* (Rodriguez de Miranda 1979).

CANDIDA PARAPSILOSIS

On CMA, after 3 days incubation, long, regularly branching pseudohyphae with a 'pine forest' aspect are formed. Production of pseudomycelium may be, however, slight in some strains.

CANDIDA TROPICALIS

C. tropicalis grows on the surface of Sabouraud's dextrose broth as a thin pellicle which becomes evident after 2 days of incubation at 25°C (Figure 30.1).

Physiology

Candida spp. metabolize glucose via the hexose monophosphate pathway under aerobic conditions (assimilation), or via the Embden–Meyerhof pathway in anaerobiosis (fermentation). Additional metabolic mechanisms such as mitochondrial oxidative phosphorylation and the Krebs cycle and protein synthesis by 80S ribosomes (composed by a 60S and a 38S subunit) do not differ from those described in other eukaryotic cells (Braude 1986).

The enzymatic apparatus of *Candida* spp. is complex and has been extensively researched and reviewed (Odds 1988). The practical importance of *Candida* enzymes is significant since a number of them, such as proteases and those involved in sterol synthesis, may be directly involved in pathogenesis, or serve as target for antimycotic drugs, respectively.

Among catabolites secreted by some *Candida* spp. are acids, fatty acids, and alcohols, including ethanol.

Growth temperature seems to have an important influence upon morphogenesis of several *Candida* spp. While temperatures around 25°C promote primarily the formation of chlamydospores in *C. albicans*, higher temperatures, such as those present in potential hosts (around 37°C but sometimes reaching 43°C), will promote the formation of pseudohyphae (Braude 1986). Not all *Candida* spp. are able to grow at 37°C and higher temperatures, and consequently such capabilities are considered important pathogenic factors that separate potentially pathogenic strains from environmental saprophytes.

Genetics

A significant amount of data regarding the genomes, including genome size, number of chromosomes, and allocation of genes to specific chromosomes, of medically important *Candida* spp. has been accumulated during the last decade as recently summarized by Magee and Chibana (2002). Since *C. albicans* was studied more intensively than other species, this section focuses primarily on this microorganism. A significant recent development in *Candida* genetics is the *Candida albicans* Genome Project, targeted at the recognition of the fungus' genes (www-sequence.stanford.edu/group/candida).

The medically important *Candida* species have been traditionally assigned to the Deuteromycetes as only an anamorph state was known. Over the years, however, an ascomycetous teleomorph state was discovered for a number of *Candida* species (see section on Taxonomic comments below). *C. albicans* was considered as a species with no recognized sexual state. However, recent studies indicated that mating might occur in this species as well (Hull and Johnson 1999; Hull et al. 2000; Magee and Magee 2000). The *Candida* spp. dealt with in this chapter are considered diploid, except for *C. glabrata* (Doi et al. 1992; Kitada et al. 1995). Classical genetic studies, particularly those of *C. albicans*, were hampered due to its diploid nature and the lack of sexual reproduction. Nevertheless, molecular-genetic techniques enabled investigators to carry out studies revealing significant information regarding the genetics of *C. albicans*, and, to some extent, of other *Candida* spp. as well.

The principal findings related to *C. albicans* regarding the number of chromosomes, isolation of various mutations including site-directed mutagenesis, identification and sequencing of various genes, as well as allocation of part of them to specific chromosomes, DNA amplification, genetic cloning and gene expression are as follows.

The DNA content of *C. albicans* (nuclear and mitochondrial) was estimated to be in the range of 40 pg/cell (Riggsby et al. 1982). As cytological methods for demonstration of chromosomes are difficult, most karyotyping studies are based on electrophoretic methods (Magee and Magee 1987).

An important characteristic of *C. albicans* and other *Candida* spp. is the electrophoretic karyotype variability, which led to inconsistent results regarding the number of chromosomes (Magee and Magee 1987; Magee et al.

1988; Lasker et al. 1989; Wickes et al. 1991; Altboum 1994). The current notion is that *C. albicans* has eight pairs of chromosomes (Chibana et al. 2000). A physical mapping of chromosome 7 has been reported by Chibana et al. (1998).

Various types of mutants of *C. albicans* have been isolated spontaneously or following treatment with chemical or physical mutagens (Kirsch et al. 1990; Altboum 1994). These include auxotrophic strains, strains resistant to various substances, or temperature-sensitive and respiratory-deficient mutants expressed as 'petite' phenotype (Figure 30.2). The most frequently induced mutations were in genes involved in the biosynthesis of amino acids and nucleic acids expressed phenotypically by auxotrophic strains. Various techniques, including the widely applied method of complementation of homologous mutations in *Saccharomyces cerevisiae* (Malathi et al. 1994; Rosenbluh et al. 1985) have been used to isolate *C. albicans* genes from genomic libraries. This led to the isolation of a significant number of genes, some of which have been sequenced and/or located on chromosomes (Magee et al. 1988; Lasker et al. 1989; Rustchenko-Bulgac et al. 1990). Isolation of defined mutants and the corresponding genes led to development of transformation methods that resulted in gene expression. An important development in the area of mutagenesis was the introduction by Fonzi and Irwin (1993) of the targeted mutagenesis method, allowing the isolation and identification of specific mutants and the corresponding genes.

Genetic-molecular studies of other species besides *C. albicans* focused primarily on karyotyping (Asakura et al. 1991; Lott et al. 1993; Doi et al. 1994) or use of molecular techniques for epidemiological typing (Carruba et al. 1991; Doebbeling et al. 1991, 1993; Branchini et al. 1994) or phylogeny (Hendriks et al. 1991; Kurtzman 1993). Isolation of mutants (Gleeson et al. 1990) and transformation in *C. tropicalis* (Sanglard and Fiechter 1992), as well as isolation and sequencing of the *SAP* gene (Togni et al. 1991; Ganesan et al.

1991) in this species have been reported. In addition, cloning of genes encoding for acid proteases in *C. parapsilosis* (de Viragh et al. 1993) and genetic manipulation in *C. glabrata* (Zhou et al. 1994) have also been reported. Furthermore, efforts have been focused on use of molecular methods, primarily polymerase chain reaction (PCR), for diagnostic purposes attempting identification and detection of various species (Lehmann et al. 1992; Burgener Kairuz et al. 1994; Maiwald et al. 1994).

ECOLOGY AND EPIDEMIOLOGY

Candida spp. are currently considered the fourth most common cause of hospital acquired systemic infections in North America, accounting for 10 percent of all bloodstream infections (Richardson and Warnock 1997; Sullivan and Coleman 2002), and which may lead to mortality rates of up to 35 percent (Wenzel 1995). Although most *Candida* spp. have been isolated from animal and other environmental sources (Mok et al. 1984; Mok and de Carvalho 1985; de Hoog et al. 2000), human infections are usually endogenous. *Candida* spp. can be isolated from various sites on or in the human body of 30–50 percent of asymptomatic individuals (Richardson and Warnock 1997), primarily the gastrointestinal (GI) tract, the oropharynx, the vagina, and the skin (Mok and da Silva 1984; de Hoog et al. 2000). Variations regarding the presence of *Candida* spp. in healthy individuals may be a function of various factors such as climate, age, or diet of the surveyed population. As summarized by Odds (1988), the range of *Candida* isolations from the oral cavity is 1.9–41.4 percent, the GI tract 0–55 percent and the vagina 2.2–68 percent. The same source (Odds 1988) indicates that mean values of data from a number of epidemiologic surveys reveal that *C. tropicalis* and *C. glabrata* are the second most frequent species isolated, from the oropharynx in the case of the former, and from the vagina and the GI tract the latter.

C. albicans is generally considered the major pathogen among the *Candida* species. Although an increase in the prevalence of non-*albicans* species has been noted during the last decade (Hazen 1995; Krcmery and Barnes 2002; Nguyen et al. 1996) it is still usually the most frequently isolated species (Edmond et al. 1999; Pfaller and Diekema 2002), albeit the differences have become less substantial. Recent surveys indicate (Krcmery and Barnes 2002; Pfaller 2003) that besides *C. albicans*, the four most common *Candida* spp. isolated from candidemia patients include *C. parapsilosis*, *C. tropicalis*, *C. glabrata*, and *C. krusei*. About a dozen more rarely isolated additional non-*albicans Candida* spp., such as *C. guillermondii*, *C. kefyr* or *C. lusitaniae*, have been reported as causes of candidiasis (Krcmery and Barnes 2002; Pfaller and Diekema 2002). Particu-

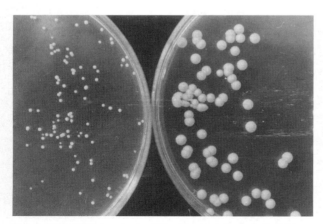

Figure 30.2 *'Petite' mutants of* C. albicans: *left, mutant; right, parent strain*

larly noteworthy is the identification of a novel *Candida* species – *C. dubliniensis* – as a pathogen (see section on *Candida dubliniensis* below).

Analogy between two isolates based primarily on phenotypic morphological and physiological characteristics is often insufficient. Therefore, efforts have been made during recent years to apply molecular biological techniques (Pfaller 1992b; Sullivan and Coleman 2002), such as pulsed-field gel electrophoresis (Pfaller et al. 1998), randomly amplified polymorphic DNA analysis (Lehmann et al. 1992; Lin and Lehmann 1995; Holmberg and Feroze 1996), and restriction fragment length polymorphism (Joly et al. 1996, 1999; Lockhart et al. 1997), to epidemiological studies of various *Candida* spp. This approach is exemplified in a recent study of Singh et al. (2002), who by using contour-clamped homogenous electrical field analysis in a survey on vaginitis caused by *C. krusei*, identified two genotypes (Singh et al. 2002).

It is generally accepted that the incidence of yeast infections has increased during the last decades (Pfaller and Wenzel 1992). Several reasons have been mentioned as predisposing to this increase, such as immunosuppression (pathological and iatrogenic) and increase in the use of broad-spectrum antibacterials, to mention but two. However, other factors, such as higher awareness and, more importantly, improved and easily accessible diagnostic tools, may have contributed significantly to the observed epidemiologic changes.

In addition to the overall increase of *Candida* infections, epidemiologic shifts in the relative prevalence of the different *Candida* spp. and emergence of new pathogens has occurred (Hazen 1995; Pfaller and Diekema 2002). Among the possible factors contributing to these shifts, an important role has been attributed to antifungal drugs. It is believed that the introduction of various antimycotic agents into extensive use has resulted in the selection of specific species that are inherently less susceptible to the specific drug (most known examples are *C. glabrata* and *C. krusei* exhibiting resistance to fluconazole). In addition, more recently, other factors, such as demographic parameters or even specific genetic susceptibility, have been suggested as contributing to the change in the relative proportions of the different *Candida* spp. involved in human pathology (Pfaller 2003).

In view of the scope of this chapter, the following sections deal only briefly with epidemiologic aspects of the major pathogenic *Candida* spp., and the reader interested in more detailed information is referred to specific literature.

Candida albicans

C. albicans is isolated more rarely than other *Candida* spp. from environmental sources. This might be a result of this species' relatively high adaptation to a parasitic

way of life and consequent loss of some of the capacities possessed by other species. Two serotypes have been recognized: serotype A is more prevalent than serotype B among clinical isolates (Auger et al. 1979). Studies with *C. albicans* isolated from immunosuppressed patients (Brawner et al. 1992) indicated a higher incidence of serotype B. These finding may, however, differ according to the geographical area. Interestingly, a correlation between susceptibility to flucytosine and serotype has been noted: while serotype A isolates are susceptible, serotype B isolates often exhibit resistance (Auger et al. 1979).

Outbreaks of *Candida* infections traced back to hospital personnel or contaminated materials, in risk populations such as neonates (Shian et al. 1993; el-Mohandes et al. 1994) or coronary bypass patients (Isenberg and Tucci 1989) have been reported.

Candida dubliniensis

The initial isolation of *C. dubliniensis* was from the oral cavity (Sullivan et al. 1995) and subsequently from other sites including vagina (Coleman et al. 1997), urine (Polacheck et al. 2000), and blood (Meis et al. 1999). Although the original isolates were from HIV-positive patients, later this yeast was cultured from non-HIV carriers receiving chemotherapy, following bone marrow transplantation (Meis et al. 1999), or broad-spectrum antibacterial treatment (Polacheck et al. 2000). It was also reported recently by Jabra-Rizk et al. (2001) that *C. dubliniensis* was isolated from healthy children.

It is interesting that *C. dubliniensis* may develop resistance to fluconazole and this may possibly have contributed to the emergence of this species in HIV carriers/AIDS patients (Sullivan et al. 1999).

Candida guilliermondii

C. guilliermondii has been isolated from various environmental sources such as swimming pools (Aho and Hirn 1981), soil, sand, and seawater (Boiron et al. 1983), as well as amphibians (Mok and de Carvalho 1985), wild and domestic birds (Mangiarotti et al. 1993), and mammals, including the human cutaneous surface. In addition, *C. guilliermondii* has been isolated from the hands of hospital personnel (Strausbaugh et al. 1994). In another report of nosocomial infections caused by this yeast in a neonatal intensive care unit, the source of infection was the heparin solution used to flush 'butterfly needles' used for blood sampling (Yagupsky et al. 1991).

Candida glabrata

This species is one of the most prevalent among those isolated from the oral cavity (Stenderup 1990) or vagina of healthy individuals (Odds 1988). The prevalence of this species among those causing human disease has increased significantly (Borg von-Zepelin et al. 1999; Pfaller and Diekema 2002), becoming an important cause of candidemia. In addition, it is a major pathogen involved in candidal vaginitis (Sobel 1998) or denture stomatitis (Cumming et al. 1990). It is generally believed that increased prevalence of this species as a pathogen is due to selective pressure resulting from its resistance to fluconazole (Richardson and Warnock 1997). In addition, it has been recently suggested that other factors, such as age or geographical area, may also play a role (Pfaller 2003).

Candida kefyr

This species has been involved only rarely in human pathology. Several cases of disseminated yeast infections caused by C. kefyr have been described and this species was isolated sporadically from vaginal, urinary, ear, and GI infections (Davis 1986).

Candida krusei

C. krusei has been isolated from various environmental sources (Do Carmo Sousa 1969) but only rarely from the mucosal surfaces of healthy carriers (Odds 1988). The involvement of C. krusei in pathological processes has increased in the last few years and may continue to do so, considering its resistance to fluconazole, widely used as an antimycotic prophylactic drug. In fact, a comparison between the incidence of disseminated C. krusei in 1989 and 1990, the year in which fluconazole prophylaxis was introduced, showed that the latter practice led to an increase of such cases while no such change was observed for C. albicans and C. tropicalis cases (Wingard et al. 1991). Indeed, C. krusei became an important cause of difficult to treat cases of candidemia (Pfaller 2003). Among mucocutaneous Candida infections it is an important cause of vaginitis (Singh et al. 2002).

Candida parapsilosis

C. parapsilosis has a wide distribution in nature and has been found to be part of the normal human skin flora especially subungual spaces (Weems 1992).

C. parapsilosis has been reported as a contaminant of high concentration glucose solutions and prosthetic material (Branchini et al. 1994). A number of nosocomial infections with this yeast species have been reported. It was isolated from cases of sepsis in an intensive care nursery (Shian et al. 1993; el-Mohandes et al. 1994), it was reported to be prevalent in patients with indwelling devices or those fed parenterally (Herrero et al. 1992; Martino et al. 1993; Krcmery and Barnes 2002), possibly as a result of its tendency to form biofilms on catheter tips (Kuhn et al. 2002). Nosocomial

infections were connected with the presence of C. parapsilosis on hospital equipment, personnel hands, and pigeon guano on window sills (Greaves et al. 1992; Oie and Kamiya 1992; Sanchez et al. 1993), and infection of high-risk patients connected with such contamination (Greaves et al. 1992; Sanchez et al. 1993). Cases of endophtalmitis caused by C. parapsilosis following lens implants has been reported (Kauffman et al. 1993).

Candida tropicalis

Odds (1988) reports that according to some surveys, C. tropicalis is a relatively prevalent pathogenic yeast species in the oral cavity of asymptomatic carriers. C. tropicalis has been shown to be involved in various human pathologies such as fungemia (Pfaller 2003) post-surgery infections, osteomyelitis (Ferra et al. 1994), endocarditis (Shmuely et al. 1997; Gerritsen et al. 1998). In addition to endogenous infections, which are the most common, cases of fungemia in a neonatal intensive care unit have been connected to the presence of the fungus on the hands of personnel (Finkelstein et al. 1993).

Candida lusitaniae

C. lusitaniae was first described as part of the normal microbial flora of warm-blooded animals by van Uden and do Carmo-Sousa (Sanchez et al. 1992). C. lusitaniae has been isolated from various animal sources (pigs, mastitis milk, bird droppings), alimentary products (citrus peel juice), and clinical specimens (sputum, feces, urine) (Merz 1984).

The prevalence of C. lusitaniae among yeast species isolated from clinical cases is relatively low. During a 15-month study, only 0.64 percent of the isolated yeast species were identified as C. lusitaniae (Merz 1984). Nosocomial colonization of the digestive or urinary tracts has been reported and the possibility of transmission by hands of hospital personnel suggested (Sanchez et al. 1992; Strausbaugh et al. 1994). Restriction enzyme analysis of C. lusitaniae isolates cultured in the same study showed that although most patients carry the same strain throughout the period of colonization, occasionally, two different strains may be present in different body sites (Sanchez et al. 1992).

The distribution of the two mating types of Cl. lusitaniae is unequal: a ratio of 1:6 between the 'a' and the 'α' mating types has been reported.

PATHOGENICITY AND PATHOGENESIS

Virulence factors

In addition to previously available information regarding virulence properties of Candida spp. (Cutler 1991; Odds 1994; Segal 1994b), newer studies with this subject have been summarized in several review articles in recent literature (Staib et al. 2000; Calderone and Fonzi 2001; Haynes 2001; Navarro-Garcia et al. 2001; Soll 2002). While many of the reviews describe investigations concentrating on specific virulence attributes, such as adherence to host tissues (Cotter and Kavanagh 2000; Sundstrom 2002), proteinases (De Bernardis et al. 2001; Monod and Borg-von Zepelin 2002), phospholipases (Ghannoum 2000), or biofilm formation (Waldvogel and Bisno 2000; Ramage et al. 2001), the most intriguing facet of the recent studies is the focus on isolation and/or expression of specific virulence genes (Staib et al. 2000; Hoyer 2001; Navarro-Garcia et al. 2001). Although the majority of these reviews report on research involving *C. albicans*, studies on additional *Candida* spp., such as *C. glabrata*, *C. tropicalis*, *C. parapsilosis*, or *C. dubliniensis* have also been reviewed (Haynes 2001; Gutierrez et al. 2002).

The virulence traits of *Candida* spp., particularly of the major pathogen, *C. albicans*, to which possible roles in pathogenesis of candidiasis have been attributed, are believed to be associated primarily, with:

- the ability of the fungus to bind (attach, adhere) to host tissue as an initial step in the recognition and interaction with the host. Special attention has been given more recently to an associated phenomenon – biofilm formation, as a significant factor in the pathogenesis of candidiasis, affecting adversely the host's response to infection and causing difficulties in therapy
- the production of specific enzymes that could facilitate tissue penetration and invasion, such as secretory aspartyl proteinases (SAPs) and phospholipases
- yeast–hyphal morphogenetic transformation, which could also facilitate penetration and, in addition, assist the microbe to evade the host's defense system
- phenotypic switching
- various immunomodulatory effects of fungal determinants, which could contribute to reduced activity of the host's defense system.

Although the above-mentioned mechanisms of virulence are supported by evidence provided by various experimental studies and epidemiologic–clinical observations, it should be emphasized that assessment of microbial virulence under actual in vivo conditions is a complex issue, since virulence involves a combination of traits, while in experimental systems a single characteristic is assessed under given conditions.

The ability of *Candida* to attach to host tissues has been extensively studied both in vitro and in vivo, the latter mostly in experimental animal models, and reviewed in several articles (Segal 1987a, 1994b; Calderone and Braun 1991; Kennedy et al. 1992; Segal and Sandovsky-Losica 1995, 1997). As stated above, the more recent studies and reviews summarizing those studies (Cotter and Kavanagh 2000; Sundstrom 2002)

concentrate on genetic aspects related to adhesion or association with the host (Hoyer 2001). As it is outside the scope of the present text to exhaust in detail the subject, only the principles and the major conclusions drawn from the various studies are detailed.

In vitro systems showed that *Candida* can adhere to a variety of substrates, such as exfoliated human epithelial cells (buccal, vaginal, dermal) (Segal and Sandovsky-Losica 1995), human tissue lines (endothelial, Hela) (Hostetter 1994a, b) animal tissue segments (oral, gastro-intestinal) (Segal and Sandovsky-Losica 1995) or inert surfaces (various polymers used for indwelling medical devices) (Waldvogel and Bisno 2000). These experiments revealed that a gradient in the level of adherence could be noted, which correlates with the pathogenicity of the *Candida* spp. in man and animals. Thus, *C. albicans* was, generally, the most adherent *Candida* species. However, intraspecies variation expressed by different adhesion ability of various isolates of the species was also noted (Samaranayake et al. 1994). In addition, it was shown that mutants with reduced adherence exhibited also decreased pathogenicity in vivo (Roth-Ben Arie et al. 2001). Adhesion to tissues was found to be dependent on various environmental conditions affecting the fungus, and influenced by various host factors, such as fungal surface hydrophobicity, growth medium, and growth conditions, or hormonal and immune status of the host, respectively (Kennedy 1988).

Animal experimental models and ex vivo systems have confirmed many of the data derived from the in vitro systems. They showed that the attachment of the fungus to the host's tissues initiates colonization and the infectious process. This concept was the basis for attempts to prevent development of infection by intercepting the binding of the fungus to host's tissues (Segal and Sandovsky-Losica 1997). Such an approach could be considered as a novel prophylaxis modality for prevention of disease.

Studies aimed at identifying the fungal surface molecules involved in the binding process led to the recognition of several putative adhesins: mannan (Calderone and Braun 1991), the mannoproteins – primarily their proteinous moiety (Douglas 1987) – and chitin (Segal and Sandovsky-Losica 1995, 1997). Miyakawa et al. (1992) suggested that antigen 6 of the mannan might be associated with adhesin activity. Different adhesins may, possibly, exist on different *Candida* species (Bendel and Hostetter 1993). The counterpart receptor molecule on the host cell is, depending on the type of cell, apparently, fucosyl, glucosamine, fibronectin or arginine-glycine-asparagine (RGD) (Hostetter 1994a, b).

More recent studies reported several additional adhesins in *C. albicans*, such as Alsp and Hwp1p. Hoyer's group described a family of surface glycoproteins, the Alsp (Hoyer 2001), that is involved in fungal binding and may act as adhesins, thus being considered virulence attributes, playing a role in pathogenesis. It was also found that at least nine genes encode these molecules. Another probable adhesin, a cell wall protein as well, the Hwp1p, was described (Staab and Sundstrom 1998; Tsuchimori et al. 2002). It was shown in these studies that mutants deficient in these adhesins were less virulent. Gale et al. (1998) have demonstrated the linkage of adhesion, hyphal growth, and virulence of *C. albicans* to the *INT1* gene. Int1p, the integrin-like protein, is considered a *C. albicans* virulence factor (Hostetter 2000).

Regarding other *Candida* spp., Epa1p, a cell wall protein acting as an adhesin in *C. glabrata*, was reported (Cormack et al. 1999). *ALS* genes have been reported for *C. tropicalis* and *C. dubliniensis* (Hoyer et al. 2001).

C. albicans also possesses surface components (e.g. iC3b), which are associated with binding to fibrinogen and complement (Calderone and Braun 1991). This interaction and speculated molecular mimicry mechanisms of mammalian integrins by *C. albicans* surface proteins (Hostetter 1994a) may affect the binding to neutrophils and thereby phagocytosis, which in turn will affect the host's defense system, and should thus be considered as additional factors playing a role in the pathogenesis of candidiasis.

As sequelae of attachment, microorganisms can grow in colonial communities and produce the so-called 'biofilms,' instead of single entities – the 'planktonic forms.' Biofilm production was shown for *Candida* spp. (Kumamoto 2002), particularly for *C. albicans* (Chandra et al. 2001). The biofilms contain extracellular materials (ECM), composed of proteins, carbohydrates, and other substances (Baillie and Douglas 2000). Other significant characteristics of biofilms are the poorer response of the pathogen in the biofilm to antimicrobials (Ramage et al. 2002) and the difficulties of the hosts' defense systems to cope with the microbe, resulting in difficulties of eradication of infection. *Candida* biofilm formation has important clinical ramifications, as biofilms are formed on inert surfaces of medical devices (Sanchez-Sousa et al. 2001), such as catheters, artificial dentures, prosthetic valves, the use of which is known to be associated with increased risk for development of systemic infections.

The factors involved in fungal tissue penetration and invasion, which may follow the initial step of interaction between the fungus and host cell, are related to the morphogenetic change from yeast to hyphae, and to production of hydrolytic enzymes (Odds 1994). Among the latter, the SAPs (Ruchel et al. 1992) and phospholipases (Ogawa et al. 1992) seem to be the major enzymes playing a significant role in pathogenesis of candidiasis (Ghannoum 2000; Hube 2000).

SAPs can be found following adhesion of *C. albicans* to epithelial cells. As in the case of adhesion, SAP production was exhibited, primarily, by pathogenic *Candida* spp. Specifically, while enzyme activity was revealed by *C. albicans*, *C. tropicalis*, and *C. parapsilosis* (Sono et al. 1992; Fusek et al. 1994) other *Candida* spp. exhibit it only partially. Enzymes were detected in vivo

during infection (Ruchel et al. 1991) and in association with phagocytosis (Borg and Ruchel 1990). SAP-deficient mutants were less pathogenic in inducing experimental infection and unable to invade tissue (Kwon-Chung et al. 1985; Ross et al. 1990). It is believed that SAP production may facilitate invasion through degradation of keratin and collagen.

Currently, the SAP family consists of nine enzymes, for which genes have been isolated (Hube and Naglik 2001), and for which correlations with specific clinical forms have been described. Thus, *SAP2* has been associated with vaginal candidiasis, while other SAPs (e.g. *SAP1–SAP6*) have been associated with systemic infection (Hube et al. 1997; Sanglard et al. 1997). A most recent review by Monod and Borg-von Zepelin (2002) states that 10 genes of *C. albicans* SAPs have been cloned thus far. Of interest are the studies in AIDS patients undergoing antiretroviral treatment with inhibitors of aspartic proteinases, that revealed a decrease in incidence of oropharyngeal candidiasis in these patients (Egger et al. 1997), possibly due to inhibition of production of Sap protein, involved in the fungal adherence to the mucosal surface (Borg-von Zepelin et al. 1999).

Phospholipases have been shown to be fungal virulence factors (Ghannoum 2000; Hube 2000; Monod and Borg-von Zepelin 2002; Sundstrom 2002). Phospholipases have been shown for *C. albicans* and other *Candida* spp., such as *C. glabrata*, *C. parapsilosis*, *C. tropicalis*, or *C. krusei* (Ghannoum 2000). As in the case of the SAP enzymes, a role in the adherence of *Candida* to host cells and tissue penetration has been attributed to phospholipase production (Ghannoum 2000). Two genes encoding phospholipases in *C. albicans* have been cloned (Leidich et al. 1998; Sugiyama et al. 1999) and disruption of the genes affected the fungal virulence (Leidich et al. 1998). It is interesting in this context to mention the study of Hube and colleagues (Hube 2000) reporting on ten genes encoding lipases that, the authors suggest, may have a role in the microorganism's virulence.

The importance of the transition to the hyphal form as a virulence factor, recognized for many years, has been corroborated more recently by molecular studies (Madhani and Fink 1998). The transformation into the hyphal form is noted during active infection (Odds 1988; Edwards 2000). It is believed that phospholipases concentrated at the hyphal tip may be related to the greater invasiveness of this form as compared to the yeast. In addition, the hyphae, being larger than the yeast form, are more resistant to phagocytosis (Diamond and Krzesicki 1978), and thus the morphological change contributes to increased pathogenic potential of the fungus. An interesting recent observation reported by Sundstrom (2002) indicates the presence of the adhesin Hwp1 only on the surfaces of germ tubes and mature hyphae, and not on that of the other morphological forms of *C. albicans*. Furthermore, it was also shown by the same group of investigators (Sundstrom et al. 2002) that this cell wall component contributed to the yeast's virulence in murine oropharyngeal candidiasis. Recent molecular studies on *C. albicans* dimorphism (Liu 2001) revealed the association between specific genes involved in the morphological transition to hyphae and the fungus' virulence. A *Candida CLA4* mutant, with reduced yeast to hyphae transition ability, was found to have a lower virulence in mice (Madhani and Fink 1998). Cyclic AMP has been shown to be involved in blastospore-hypha transition; thus the cAMP signaling pathway plays a role in pathogenesis (Bahn and Sundstrom 2001).

Phenotypic switching associated with morphological colony changes from smooth to rough or fuzzy type, white to opaque, and other morphological changes is a phenomenon described first in the 1980s (Slutsky et al. 1985, 1987). Later studies associated phenotypic switching with differences in antigenicity (Anderson et al. 1990), pathogenic potential (Jones et al. 1994), and indicated genetic regulation (Perez-Martin et al. 1999). Phenotypic switching is considered as a potential virulence characteristic, as it may express the fungal plasticity, contributing to its adaptability to various anatomic sites of the human body and thereby possibly enabling the variety of clinical entities it causes, as well as play a role in the transition from commensalisms to pathogenicity (Soll 2002). Furthermore, phenotypic switching may assist the microbe to evade the host's defense systems, in analogy to the bacterial 'phase transition' phenomenon.

Several investigators (Cassone 1989; Domer and Garner 1989; Garner et al. 1990; Domer et al. 1992) described immunomodulatory activity of fungal surface components. An immunosuppressive effect of *C. albicans* in experimental models was noted, which was attributed particularly to manann (Domer 1989; Domer and Carrow 1989).

Should such a mechanism or a similar one occur in clinical situations, this could be added to the arsenal of candidal pathogenic traits contributing to the pathogenesis of the infection.

This section can be concluded by the belief that the current technology of transcript profiling using DNA arrays, which are already commercially available, may lead in the future to identification of additional virulence traits and, most importantly, possibly those expressed in vivo.

Pathogenesis of candidiasis

In discussing pathogenesis of candidiasis, it has to be kept in mind that *Candida* is a human commensal, so that the infectious source is mostly endogenous (Kennedy 1989). The GI tract is considered a major reservoir for candidiasis, e.g. in vaginitis or diaper rash. Moreover, from the GI tract the fungus can invade the bloodstream following damage to the GI mucosa, such as induced by anticancer treatment (irradiation or chemotherapy) (Cunningham et al. 1985) or major surgery, and spread hematogenously into various organs, causing deep-seated/disseminated infection. It is believed that *Candida* may also be able to cross the intact GI mucosa by a process termed 'persorption' (Krause et al. 1969) following fungal overgrowth, which can result from changes in the normal balance of the microbial flora due to extensive antibiotic treatment. Although previous experimental studies, such as those reported by Sandovsky-Losica et al. (1992), have already shown that systemic candidiasis originating from the GI tract follows adhesion to its mucosa (Sandovsky-Losica and Segal 1990; Sandovsky-Losica et al. 1992), more recent investigations have revealed additional aspects of this interaction. Wiesner et al. (2002), using enterocyte cultures, reported that the *INT1* gene was involved in adherence of *C. albicans* to enterocytes. Moreover, this involvement was more marked for blastospores than the hyphal form (Wiesner et al. 2002). However, *Candida* can be introduced from exogenous sources as well (Edwards 2000). These may include introduction through various catheters and lines, or other indwelling/prosthetic medical devices (Khatib and Clark 1995). This route is of particular importance in the development of deep-seated and systemic candidiasis as most of those therapeutic modalities are used primarily in compromised hosts whose defense systems are unable to combat the introduced pathogen. An interesting observation was made by Rangel Frausto et al. (1994) regarding nosocomial transmission of yeasts. The survival and transmission of *Candida* spp. on and by human volunteers has been assessed and it has been shown that, although the half-life of the microorganisms was relatively low, more than 10^5 colony-forming units of the yeast remained on the hands of the volunteers after 45 minutes (Rangel Frausto et al. 1994).

Once *Candida* entered the bloodstream, whether from exogenous or endogenous source, the microorganisms have to adhere to the endothelial surface of the blood vessels, before dissemination into tissues (Klotz 1992). This process involves endocytosis, endothelial invasion, and response to the invading pathogen expressed in production of proinflammatory cytokines, such as tumor necrosis factor-alpha (TNF-α) or interleukin (IL)-6 (Orozco et al. 2000). It was also noted that germ tubes are better endocytosed than blastospores (Phan et al. 2000).

It is noteworthy that proinflammatory cytokines have been shown recently to be involved in the pathogenesis of candidiasis (Brieland et al. 2001). In a model using immunocompetent mice, Brieland and colleagues compared the pathogenesis of intravenous (IV)-induced infection by *C. albicans* versus that of *C. glabrata*, analyzing organ colonization and cytokine production. They found that while kidneys of *C. glabrata* infected animals were colonized, the animals nevertheless, survived, contrary to those infected by *C. albicans*, that were colonized as well and succumbed to the infection. The authors suggest that the difference is associated primarily with a more rapid induction of the proinflammatory cytokines in response to *C. glabrata* than to *C. albicans*.

Person-to-person transmission is not a predominant mechanism of pathogenesis in candidiasis, and is expressed primarily in oral thrush of newborns acquired during birth from their mothers affected by vaginal infection, or noted more rarely in sexual transmission from vaginitis patients to their male partners.

The normal innate mammalian defense system against *Candida* infection includes intact dermal surfaces, intact mucosa, unspecific humoral factors, and the immune system – humoral and cellular (Greenfield 1992; Levitz 1992).

The first line of defense, which protects, primarily, before the mucocutaneous forms of candidiasis, is the unbreached skin and mucosa (Edwards 2000). As detailed under the section on Clinical entities below, risk factors, which increase the susceptibility to these types of *Candida* infections, include maceration of skin, changes of microbial flora, trauma to skin/mucosa, hormonal changes affecting skin integrity or local defense and/or excess of carbohydrates on surfaces, as seen in diabetics. It is believed that the skin, in addition to its function as a barrier, possesses an inflammatory–immunological activity as well (Wagner and Sohnle 1995). The latter may involve the specific skin cells – epidermal Langerhans cells and keratinocytes – which may function as antigen-presenting cells, engage in phagocytosis or/and produce various cytokines. As indicated above, breaching the intactness of the skin (e.g. through intravenous catheters) or of the GI mucosa can lead also to introduction of the organism into the bloodstream, either from exogenous or endogenous (GI tract) sources, respectively. Thus, damage to the integuments may be considered as a risk factor for the development of systemic candidiasis. Additional nonspecific, nonimmune factors to which a possible role in defense has been attributed include iron-binding proteins, such as transferrin or lactoferrin (Nikawa et al. 1993).

The second line of defense, which comes into action following fungal penetration, includes phagocytic and candidacidal activity of polymorphonuclear (PMN) (primarily, neutrophils) and mononuclear (primarily, monocytes) cells (Marodi et al. 1991a, b; Greenfield 1992; Levitz 1992). These processes involve myeloperoxidase and superoxide production, or cationic proteins. PMN activity is of crucial significance in protection against the deep-seated forms of *Candida* infections, as can be judged by epidemiologic data that indicate that an increased rate of these infections is associated with neutropenia or defective PMNs (see sections on Clinical entities and Immunity below).

As to specific activity of the immune system, despite numerous experimental studies and clinical observations there is no clearcut concept of the activity of the various elements of the system and the mechanisms underlying that activity. T lymphocytes involved in cell-mediated immune response seem to be of significance, particularly in defense against mucocutaneous forms of candidiasis (Domer and Carrow 1989). Patients with T-cell deficiencies, such as AIDS patients, are especially prone to these infections. The role of these cells in resistance to systemic candidiasis is more controversial (see discussion in the section on Immunity below). Opsonizing antibodies (IgG) seem to contribute to phagocytosis and thereby to defense. An additional humoral element of the immune system involved in defense is, apparently, the complement cascade (Gelfand et al. 1978).

ANTIGENICITY, IMMUNE RESPONSE, AND IMMUNITY

The immune response of the mammalian host to *Candida* infection, the antigens eliciting this response, and the eventual resulting immunity, as well as the defense mechanisms involved in the process, constitute complex issues which have attracted extensive research over the years, research which has not always led to unequivocal conclusions. Nevertheless, certain pertinent generalizations can still be drawn.

Antigenicity

In discussing the topic of antigenicity, one should bear in mind that the definition of the term should be in the frame of a specific context; whether:

- in relation to antigens which elicit immune responses in general
- as antigens which elicit a response during infection, a response which can be measured and used as a diagnostic tool
- as antigens which confer resistance to the infection and may be considered 'protective antigens.' As in many microbial systems, the antigens eliciting a given response or having a protective role may be different.

In defining the antigenicity of *Candida*, primarily that of *C. albicans*, most studies, particularly earlier investigations, concentrated on the cell wall. Those studies indicated that the mannoprotein complex, particularly the polysaccharidic moiety of the *Candida* cell wall, is the major antigenic component. It was postulated that the backbone structure of the mannan confers the antigenic properties to the whole group, and the specific antigenicity of a given species or variety is apparently due to the different side chains of the mannan molecule (Reiss 1986; Reiss et al. 1992). This hypothesis was the background of extensive experimental studies of Tsuchiya et al. (1965), and over the years of other investigators (Miyakawa et al. 1986) as well. These studies, using agglutination assays with anti-*Candida* sera, culminated in the development of an antigenic profile of *Candida* spp. of medical interest. The presence of heat-labile and heat-stable antigens, which could be common or specific to a particular species or several species, was shown. A commercial kit based on this principle enables us to differentiate among the most commonly encountered pathogenic *Candida* spp. Limited serotyping of *C. albicans* into two serotypes, serotype A and serotype B, as described by the studies of Hasenclever and Mitchell (1961), was also based on a similar principle. Summaries of studies conducted in the last decade on *Candida* spp. cell wall, such as that of Chauhan et al. (2002), contribute to a more detailed knowledge of its composition and molecular structure. It became evident that, besides the traditionally recognized polysaccharides, other components such as chitinase and glycolipids can be considered important cell wall constituents. Glycosyl-phosphatidylinositol (GPI) was found to play a role in anchoring proteins to the cell wall polymers (Fonzi 1999). In addition to cell wall composition and structure, studies were also aimed at characterizing cell wall antigens. Ponton described several types of antigens expressed either on the surface of germ tubes, blastoconidia, or both (Ponton et al. 1993, 2002). La Valle et al. (2000) studied the 65 kDa mannoprotein of *C. albicans* and found it to be a major immunodominant antigen. The group of Poulain concentrated on phospholipomannans (Poulain et al. 2002; Trinel et al. 2002). They showed that *C. albicans* can synthesize specific β-1,2-oligomannosides which have antigenic properties.

If antigens are to be used as reagents for diagnostic purposes, it is of importance to define the aim. If the aim is to differentiate between colonization of superficial mucocutaneous surfaces and deep-seated organ involvement, it would seem reasonable not to employ *Candida* cell wall antigens. Since *Candida* is an ubiquitous microorganism and human commensal, it elicits immune responses in the host to the fungal cell wall to which there is constant exposure. On the other hand, a reasonable approach would be to search for antigens of the *Candida* cytosol, which are released during infection and elicit a measurable immune response. This rationale was the basis for development of immunodiagnostic tests in which the reagent was a cytoplasmic *Candida* extract, standardized by its protein content (Reiss et al. 2002). More defined antigens, mostly of proteinaceous nature, to which the above-mentioned characteristics have been attributed, have also been suggested. These include the enolase, a protein with catabolic enzymatic activity (Franklin et al. 1990); a 47–48 kDa protein described by Matthews et al. (1988), who believe that it has some

protective role in AIDS patients and which could be identical with the enolase, and more recently La Valle et al. (1995) described a 70–80 kDa heat shock protein (HSP) with immunogenic activity, and which the authors suggest as a possible reagent in diagnosis. Undefined nonprotein antigens, probably a glycoprotein, and the polysaccharide mannan, have been believed to be released during systemic *Candida* infections and consequently tests aimed at the detection of these antigens for diagnostic purposes were devised (the Ramco and Pastorex kits, respectively).

It should be noted that some of the antigens mentioned above, such as the 65 kDa protein or the oligomannosides, may be considered as potential reagents with diagnostic value as well (La Valle et al. 2000; Sendid et al. 2002).

The identification of antigens, the response to which would induce protection to the infection, is a subject even more complicated and controversial than the identification of antigens of diagnostic relevance. This is so despite experimental attempts, as mentioned later, to induce protection against experimental infection by immunization with various *Candida* preparations. These have included whole live organisms, live-attenuated *Candida* cells and those killed by different methods or noncellular preparations (Segal 1987b, 1991). Among the latter can be included antigens prepared by the author with *Candida* ribosomal fractions, which induced, in animal models, significant protection against systemic candidiasis (Segal 1989). The protection is, apparently, due primarily to the ribosomal-protein(s) moiety. Interestingly, specific ribosomal proteins, the L7/L12 of *Brucella* spp., have been shown to be immunogenic and to elicit a delayed-type hypersensitivity (DTH) response (Bachrach et al. 1994).

It is noteworthy that a recent study of Mizutani et al. (2000) suggested that a membrane fraction of *C. albicans* may act as an antigen which confers CD4+ T-cell-mediated resistance to systemic candidiasis in mice. The 65 kDa mannoprotein described by La Valle et al. (2000) was also able to induce a cell-mediated immune (CMI) response (La Valle et al. 2000). A more recent study of Dromer et al. (2002) reported on the ability of synthetic analogs of β-1,2-oligomannosides of *C. albicans* to prevent intestinal colonization by the fungus.

Immune response

Exposure to *Candida* species stimulates both humoral and CMI responses. Anti-*Candida* antibodies can be detected in experimental animal infections, in humans with naturally acquired infections, as well as in healthy individuals harboring the fungus in a carrier state. The antibodies represent the different immunoglobulin types – IgG, IgM, IgA or IgE – albeit not all immunoglobulin types can be detected in all the clinical entities, under all conditions, and in different body fluids.

IgG and IgM are generally found in sera of patients with deep-seated forms of candidiasis, except if the patients are highly immunosuppressed and unable to mount an immune response. IgG and IgM antibodies can also be found in the mucocutaneous forms of candidiasis, as well as in asymptomatic carriers.

Anti-*Candida* IgE antibodies have been found in serum and in other body fluids in situations associated with allergy. In addition to measuring IgE antibodies in serum and other body fluids by serological tests, these antibodies can be revealed by skin testing with a *Candida* antigen. Skin inoculation with a *Candida* antigen may reveal an immediate type hypersensitivity (ITH) reaction resulting in the appearance of a flare-like erythema at the inoculation site 20–30 minutes after the test inoculation. The ITH response will be, generally, found in allergic states. An interesting observation was made by Witkin (Witkin et al. 1989) who noted the presence of IgE antibodies in women with vaginitis, indicating a possible role or association of ITH in the pathogenesis of vaginal infection.

IgA antibodies have been found in sera and vaginal secretions of patients with candidal vulvovaginitis (Schonheyder et al. 1983; Sobel 1985) and in vaginal fluids from experimentally infected rats (Polonelli et al. 1994; De Bernardis et al. 2002). However, the differences in the titers of IgA antibodies in patients in comparison to those of healthy individuals, or in other conditions with candidal colonization, are not discriminatory enough to permit the use of detection of this type of antibodies as a reliable diagnostic tool for candidal vaginitis.

As to CMI, a *Candida* infection or an asymptomatic colonization stimulates the response of the cellular immune system (Domer and Carrow 1989; Baum 1994). The CMI response can be detected by in vitro assays as well as by in vivo measurements. The current concept of the chain of events culminating in the expression of CMI postulates that PMNs produce IL-12, which stimulates Th1 cells to release interferon-gamma (IFN-γ) and IL-2, which activate phagocytes. In vivo measurements include skin testing with a candidal antigen (candidin). The CMI response thus measured will be expressed in a DTH reaction, revealing an indurated erythema at the site of the test-antigen inoculation, apparent about 48 hours postinoculation. As *Candida* do colonize healthy individuals, the skin test for detection of a DTH response will be positive in most of the tested individuals and cannot, therefore, serve as a tool for diagnosis of candidiasis. It can, however, indicate that the individual tested possesses an adequate cell-mediated immune system in order to mount the response to the antigen with which he/she was tested. Indeed, the *Candida* skin test is used to evaluate the CMI competence of patients. A negative candidin test will generally point to the inadequacy of the patient's immune system to mount a DTH response, and will suggest that the patient is immunodeficient or immunosuppressed. However, as there are different responses to different *Candida* antigens, this test must be interpreted with caution. CMI responses can be, as mentioned earlier, detected by several in vitro assays. The most commonly used assay is the lymphocyte transformation/proliferation test, which measures the in vitro stimulation of T lymphocytes with *Candida* antigens, indicating the sensitization of the CMI system by the fungus. This assay or other in vitro assays for detection of CMI response to *Candida*, similar to the skin test for in vivo evaluation of DTH, do not have marked value for diagnosis of candidiasis. However, as in the case of the skin test, the in vitro assays are useful for evaluation of the cell-mediated immune status of the individual. Moreover, the use of more than a single evaluation method, as for example a skin test and the in vitro lymphocyte transformation assay or the macrophage migration inhibition assay, may enable one to define more specifically the putative immune defect of the tested patient. This was demonstrated by Kirkpatrick (1989) in a series of studies of patients with chronic mucocutaneous candidiasis (CMC) in whom a differentiation of the individual immune defect(s) has been shown, revealing, for example, incompetence in antigen recognition versus inability to mount an immune response. The differentiation of different T-lymphocyte subsets (e.g. CD4, CD8) may also be of clinical relevance in evaluating the patients' immune status and possible susceptibility to *Candida* infection.

Immunity

Immunity to *Candida* infection, including the innate and adaptive defense system, is also a topic that has attracted considerable research interest. As can be judged by recent reviews (Romani 2002a, b) detailed, in-depth information on defense systems against candidiasis is emerging from studies undertaken during the last decade. Nevertheless, it is still difficult to draw clear and unequivocal conclusions. This might be due to the complexity of the topic and, partially, to the fact that most experimental evidence has been derived from animal models, among which divergences exist, and which do not necessarily correspond to human situations.

The contemporary view as to the innate versus adaptive immunity is that both should be looked upon as an integrated system. Immunity to *Candida* is currently believed to depend on various cells, including polymorphonuclear and mononuclear leukocytes, keratinocytes, and dendritic, epithelial, and endothelial cells. In addition, humoral elements, including antibodies, cytokines, and complement, are involved (Romani and Kaufmann 1998).

Among the T cells, the Th1 subset is the predominant cell population responsible for protection (Romani 1999). However, an important role in the immune response should be attributed to the Th2 subset as well.

It is believed that immunity is affected by the subtle interactions between the two subsets and their cytokines/chemokines. The current paradigm regarding events occurring following exposure to *C. albicans* yeasts includes the production, by PMNs, dendritic or other cells, of cytokines, which in turn induce Th1 cells to secrete cytokines activating effector cells. Among the cytokines, IL-2, IL-12, TNF-α, IFN-γ were reported to be of significance in protection against *Candida*. However, in superficial infections, local immunity at different anatomic sites apparently plays a significant role in immune response and defense to *Candida* (De Bernardis and Boccanera 2002).

MUCOSAL AND CUTANEOUS CANDIDIASIS

CMI plays a major role in the defense against mucosal and cutaneous *Candida* involvement. Indeed, CMI deficiencies as noted in AIDS or CMC patients result in increased susceptibility for development of cutaneous, nail, or mucosal candidiasis (Edwards 2000). As stated above, investigations of underlying mechanisms of CMI (Fukazawa et al. 1994) revealed the involvement of CD4[+] T lymphocytes, various cytokines, such as IFN-γ, IL-2 and 12, granulocyte-macrophage colony-stimulating factor (GM-CSF) (Smith et al. 1990; Watanabe et al. 1991; Romani and Kaufmann 1998; Romani 1999). Macrophages may act as effectors exerting their candidacidal effect following activation by IFN-γ produced by the Th1 CD4+ subset. It is also believed that oxygen radicals or cationic proteins or peptides released from the activated cells may be associated with the antifungal effect. In addition to CMI, antibodies have been shown to have a role in immunity as deduced from studies on experimental vaginitis in laboratory animals (De Bernardis and Boccanera 2002). A study by Fidel et al. (1995) pointed out, interestingly, that systemic CMI seems not to have a protective role in vaginitis (De Bernardis et al. 2002). Thus, apparently, defense against specific forms of superficial candidiasis may depend on different immune mechanisms, emphasizing the possibility of immunological compartmentalization.

DEEP-SEATED CANDIDIASIS

Immunity to deep-seated/systemic candidiasis does involve T-cell CMI as well; however, additional cellular and humoral elements seem to contribute significantly. This conclusion can be derived from the clinical–epidemiologic observations as to the individuals particularly susceptible to these forms of *Candida* infection. While, as indicated above, individuals with defects in the CMI system are especially prone to mucosal and cutaneous candidiasis, the other candidiasis forms are present mostly in neutropenic patients (Edwards 2000). It is, thus, evident that the PMNs are of importance in defense against these *Candida* infections. The role of granulocytopenia as a risk factor for systemic candidiasis

was also demonstrated in experimental animal models, in which a neutropenic state was induced, revealing increased susceptibility to *Candida*. It is interesting that some studies (Fukazawa et al. 1994) have shown the involvement of both the Th1 and Th2 subsets of the CD4+ T cells in systemic *Candida* infection. A complex and not fully elucidated interaction between the two subsets takes place.

Extensive research has been devoted to the elucidation of the intricate role of cytokines in immunity to systemic *Candida* infections. The studies indicated that a major role might be attributed to IL-12, TNF-α (Mencacci et al. 1998), transforming growth factor beta (TGF-β) (Spaccapelo and Romani 1995) and IFN-γ (Cenci and Mencacci 1998) that are required for Th1 response induction (Mencacci et al. 1998). It should also be noted that different cytokines are induced by different *Candida* antigens (Cassone and De Bernardis 1998). Of interest are recent findings of Brieland et al. (2001). These researchers investigated the role of proinflammatory and anti-inflammatory cytokines in murine systemic candidiasis and showed that TNF-α played a significant role in protection against infection.

Phagocytosis and killing of *Candida* involve complement, antibodies, and various cytokines, such as IFN-γ or tumor necrosis factor (TNF) (Diamond et al. 1991) and granulocyte colony-stimulating factors (Kullberg and Anaissie 1998; Kullberg et al. 1998). Differences in the activity of phagocytes against the hyphal versus the yeast form have been described indicating increased efficacy of PMNs towards the blastospores (Romani and Bistoni 2002). It was also shown that phagocytosis of *C. albicans* yeasts by PMNs involve a mannose receptor. PMNs are known for their ability to kill *Candida* via oxidative burst, enzymatic activities such as myeloperoxidase, and defensins (Domer and Carrow 1989; Greenfield 1992). Intracellular killing of the phagocytized fungus seems to be more efficient in the presence of opsonins. Yeast cells surviving in murine macrophage may develop a germ tube, perforate the phagosome, and cause the cell's destruction (Kaposzta et al. 1999). Studies have shown the presence of specific receptor sites on *C. albicans* for CR2 and CR3 of the complement (Calderone et al. 1994; Hostetter 1994b), which could be involved in phagocytosis.

It has been postulated by Matthews (Matthews et al. 1988) that anti-*Candida* antibodies found in sera of AIDS patients directed against a 47 kDa candidal antigen might have a protective role (see the following section on Immunization below).

Immunization

Experimental animal models have indicated that sublethal *Candida* infections may induce partial immunity to reinfection (Giger et al. 1978; Oblack and Holder 1979), an observation that indicated that obtainment of adaptive immunity could be feasible. This was the rationale for various attempts to induce protection against *Candida* infections by immunization with a fungal antigen. Various fungal preparations, including live, attenuated or killed organisms, subcellular components, and defined antigens have been explored as possible immunogens (Segal 1994a). Various routes and procedures of immunization have been investigated as well. It is not practical, nor is it purposeful, within the scope of this chapter to detail all these studies, and the specifically interested reader should refer to relevant bibliography.

It should be noted that although, despite the efforts, no clinically available anti-*Candida* vaccine of proven efficacy has yet been obtained, some recent developments seem promising. These are related, particularly, to the possibility of using anti-*Candida* antibodies in immunotherapy and/or immunoprophylaxis of systemic candidiasis, based on the observations of Matthews and colleagues (Matthews et al. 1988) indicating a possible role for anti-*Candida* antibodies in resistance to systemic candidiasis. A specific *Candida* antigen, the heat shock protein 90 (HSP90), was identified as an immunodominant antigen that elicits protective antibodies (Matthews et al. 1991, 1995). Use of molecular engineering technology enabled production of a human antibody, which is being evaluated in human clinical trials (Matthews et al. 2000; Matthews and Burnie 2001; Rigg et al. 2001) (under the designation 'Mycograb') as an adjunct to antifungal therapy and possible immunoprophylaxis in compromised patients (Matthews and Burnie 2001). Other investigators have also suggested the usefulness of anti-*Candida* antibodies (Cutler et al. 2002). Cutler and colleagues demonstrated that antibodies to a mannan–protein complex, encapsulated in liposomes, protected mice against challenge with the fungus.

Various attempts to immunize against candidiasis have been undertaken for several decades and have been summarized previously (Segal 1987b, 1991). Among these were studies on the ribosomal fraction of *C. albicans* as immunogen (Segal 1987b, 1991), revealing their protectivity. Based on this background, Levy et al. (1989) found that supplementing *C. albicans* ribosomes with a cell wall fraction from *Klebsiella pneumoniae*, an adjuvant permissible for human use, augmented the efficacy of the vaccine.

The group of Segal (Eckstein et al. 1997) investigated the use of liposomal particles as carriers of the *C. albicans* ribosomes. The efficacy of ribosomes incorporated into liposomes, with and without addition of lipid A, was compared to that of ribosomes, and incomplete Freund's adjuvant (IFA), revealing that the liposomal preparations were as effective as ribosomes and IFA. Thus, *C. albicans* ribosomal particles have the potential for human use.

Glyceraldehyde-3-phosphate-dehydrogenase (GAPDH) was found to be a surface antigen (Gil-Navarro et al. 1997) that binds to fibronectin and laminin (Gozalbo et al. 1998) and is expressed in vitro and in infected tissues (Gil et al. 1999). The possibility to generate 'humanized' antibodies, using *Candida* GAPDH as an immunogen, has been recently suggested (Priel et al. 2002).

PATHOLOGY AND HISTOPATHOLOGY

Many *Candida* spp., notably *C. albicans*, are part of the normal microflora of various mucosal surfaces. Consequently, pathology caused by these yeasts is a result of one or more underlying predisposing factors which:

- impair, to various degrees, the immune response to these microorganisms (metabolic diseases, AIDS, immunosuppressive chemotherapy)
- produce an imbalance, in favor of the fungal microflora (antibacterial drugs), or
- damage the integrity of the integument (surgery, intravenous catheters).

The macroscopic and microscopic appearance of lesions caused by *Candida* spp. is primarily influenced by the interaction of three factors:

1 the site of infection
2 the pathogenicity of the infecting microorganism and to a lesser extent its species
3 the competence of the host's immune system.

Consequently, pathological and histopathological features of *Candida* infections are highly variable: the same microorganism may cause dissimilar pathological lesions in patients whose immune system is intact and in those who are in a state of immunosuppression. Nevertheless, certain common pathological characteristics of *Candida* infection may be described (Salfelder 1990; Luna and Tortoledo 1993)

Superficial infections

Superficial infections result from invasion of the superficial layers of skin and/or mucosae by the microorganism. Macroscopically these infections are characterized by the formation of a grayish plaque, surrounded by edema, which, at histopathological examination, consists of the infecting microorganism, neutrophils, and cell debris.

In deeper ulcerative lesions of the mucosae, such as those of the digestive system (especially the esophagus) often observed in immunosuppressed patients, the infecting microorganism may be present in the submucosa and may reach blood vessels and be disseminated hematogenously. The cellular reaction to the infection is limited or absent, and necrosis, typically accompanied by hemorrhage, are the salient microscopic features of *Candida* infection in these patients.

A special form of superficial infection, chronic mucocutaneous candidiasis, is characterized by the formation of warty proliferation of the skin, which results from hyperkeratosis and epithelial hyperplasia. Granulomatous tissue may be formed in the subcutis.

Deep infections

Several parenchymatous organs may be involved. These infections are characterized by microabscesses, macroscopically similar to miliary tuberculosis in which the microorganism may be individuated as blastospores and/or pseudohyphae (the latter absent in *C. glabrata* infections), neutrophils, and mononuclear cells, and a necrotic center. Granulomata with giant cells and lymphocytes may be formed in chronic infections.

Although various organs (such as liver, spleen, and brain) (Figure 30.3) may be affected in disseminated *Candida* infections, kidneys seem to be target organs of *C. albicans*, presenting nephritis, pyelonephritis, and, more rarely, papillary necrosis. Infection may be ascending or hematogenous. The former is a frequent sequel to prolonged catheterization and antibacterial therapy. It is characterized by small white lesions, especially in the medullar part of the kidney. Histopathological examination of these lesions reveals the presence of the infecting agent in the tubules, invading the surrounding parenchyma through the tubule's wall. When the spread of the infecting agent is hematogenous, the lesions are evenly distributed between cortex and medulla, lesions in the lower urinary tract are typically absent, and, microscopically, fungal elements may be noticed in the glomeruli.

Candida endocarditis, often a sequel to intravascular catheterization or prosthetic valve implantation, is characterized by vegetations on the surface of the endocardium, especially the mitral valves. The composition of these vegetations may vary and they consist of blood clots, cell debris, and fungal elements. Microabscesses may develop in the myocardium and evolve, in chronic cases, to granulomata.

Liver and spleen lesions are typically larger than those observed in other organs and may be detected by computerized tomography or other related techniques. Microbscesses characterize the acute stage of the infection, whereas granulomatous lesions without many organisms may develop in the chronic forms.

CLINICAL ENTITIES

Candida species can cause a range of clinical forms, from superficial manifestations involving skin, nails, and mucosal surfaces to deep-seated infections involving various internal organs and to disseminated disease (Odds 1988; Rippon 1988; Kwon-Chung and Bennett 1992; Bodey 1993; Edwards 2000). For the sake of simplicity, clinical entities can be subdivided into two large groups:

1 mucocutaneous candidiasis
2 deep-seated candidiasis.

Mucocutaneous candidiasis

Mucocutaneous candidiasis can be further subdivided into:

● cutaneous infections
● nail infections
● mucosal infections,

each group including a number of specific clinical forms.

It is evident that such a division is, to a certain degree, artificial, since in certain clinical forms, more than a single system can be involved. For example, patients suffering from CMC can present clinical manifestations on the glabrous skin, on mucosal surfaces, and in the nails. This may also apply, to some degree, to nail infection, which can include also the surrounding skin area (paronychia).

CUTANEOUS INFECTIONS

Based on the dermatological classifications (Goslen and Kobayashi 1993), the major syndromes considered here are:

● candidal intertrigo
● interdigital candidiasis
● perianal (diaper) rash
● candidids
● CMC.

As tabulated by Odds (1988), *C. albicans* is the most commonly isolated species from most forms of cutaneous involvement. However, other pathogenic *Candida* spp., such as *C. tropicalis*, *C. guilliermondii*, *C. kefyr*, *C. krusei* and *C. parapsilosis* are also known to be involved. If not otherwise specifically stated, the descriptions given below apply to *C. albicans*, as well as to other *Candida* spp.

Candidal intertrigo

Candida intertrigo is believed to be the most common clinical form of the cutaneous infection, as *Candida* spp. readily colonize skin folds, particularly in moist and macerated sites (Hymes and Duvic 1993; Edwards 2000). These may include ear folds or the submammary, crural, and axillary areas. The lesions are erythematous, with vesicles and pustules, in combination with pruritus.

These infections are frequently associated with specific predisposing conditions (Odds 1988; Wagner and Sohnle 1995), such as obesity, diabetes mellitus (Montes 1992), various endocrine disturbances, HIV infection

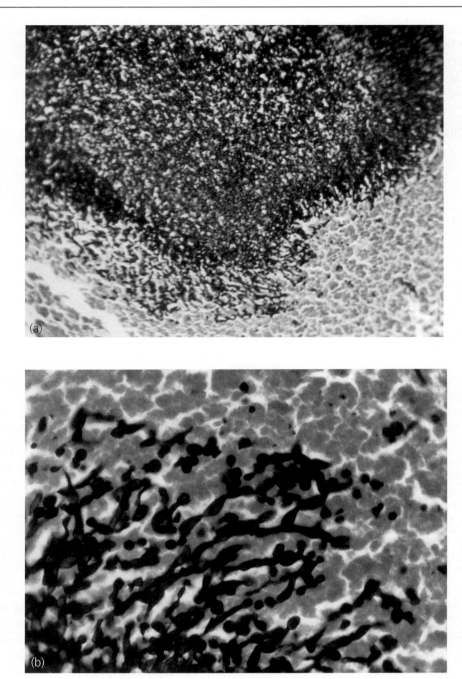

Figure 30.3 *Hepatic candidiasis:* **(a)** *candidal lesion;* **(b)** *candida mycelial and yeast elements. GMS. (Courtesy Professor Salfelder, Venuzuela)*

(Odds et al. 1990) or various iatrogenic factors including antibacterial and steroid therapy.

Interdigital candidasis

Among the intertriginous areas affected by *Candida*, the skin folds between the fingers of the hands should be particularly mentioned. In this form the skin folds between the fingers are macerated and itching. This condition is associated with excessive exposure to moisture, and thus seen particularly among dishwashers, bartenders, and fruit cannery workers. A similar erosive rash may occur between the toes in hot climates.

Perianal (diaper) rash

Another clinical form associated with *Candida* is that involving infants wearing diapers, in whom rashes are seen in the perianal area and on the buttocks. It should, however, be pointed out that diaper rash is not associated exclusively with *Candida*, and it is possible that the organisms are secondary to a preexisting inflammatory condition.

Candidids

A less commonly seen clinical form involves lesions mediated by the host's immune response to the infection – the candidids (Rippon 1988). These may be considered allergic responses, similar, albeit less frequent, to that noted in dermatophyte infections (the dermatophytids). Generally, with the treatment of the *Candida* infection, the candidid reaction clears.

Chronic mucocutaneous candidiasis

This paragraph will refer only to the cutaneous manifestations of chronic mucocutaneous candidiasis (CMC), while the other manifestations (mucosal and nail), which can also be associated with this form, as stated above, are covered later under the appropriate sections.

CMC is a relatively rare condition, however extensively studied, particularly by Kirkpatrick (1989, 1993), since it is considered the most severe clinical entity of the superficial candidiasis forms, and associated with a set of specific predisposing factors. Most of the cases are known to be caused by *C. albicans*.

The predisposing factors include, generally, congenital immunological or endocrinological defects. The immunological defects are mostly in CMI, which can be either at the stage of general or specific antigen recognition, and/or response to the antigen. This can be expressed in defects of T-lymphocyte activation and/or cytokine production leading to impairment of macrophage activation.

Clinically, the entity is characterized by the presence of persistent lesions, with high rate of recurrence, starting in early childhood and possibly persisting throughout the individual's lifetime. Lesions can be seen on various skin sites, and are not limited particularly to skin folds as in the other cutaneous forms of candidiasis. CMC can appear in a generalized form or can be localized and assume a form of hyperkeratotic lesions – 'Candida granuloma.'

An interesting feature of CMC, from which conclusions as to the immune-defense mechanisms in candidiasis have been drawn, relates to the phenomenon that CMC patients, who may suffer from extensive cutaneous, nail, and/or mucosal infections, do not develop deep-seated or systemic candidiasis. It should be added in this context that AIDS patients, who frequently also suffer from severe and repeated mucocutaneous candidiasis, and may thus be considered in a similar category as CMC patients (Filler and Edwards 1993), do not generally develop systemic candidiasis.

This observation led to the hypothesis that, apparently, the immune-defense mechanism in candidiasis differs for the mucocutaneous versus the deep-seated forms (Domer and Lehrer 1993). It is thus believed that, while the resistance to the mucocutaneous forms is based, primarily, on intact and adequately functioning CMI, particularly, that associated with T-cell activity, resistance to deep-seated and systemic candidiasis involves additional factors. The CMC condition, in addition to serving as a relevant 'in vivo model' for providing valuable insights on the immune-defense mechanisms in candidiasis, was also one of the few entities in which various immunomodulating substances, such as transfer factor (Kirkpatrick et al. 1976; Mackie 1979) or thymosin (Wara and Amman 1978), have been evaluated as possible therapeutic approaches.

NAIL INFECTIONS

Under this section two syndromes are discussed: *Candida* onychomycosis and paronychia (Haneke 1986; Hymes and Duvic 1993).

Paronychia

Paronychia is an infection of the nail folds. It is a condition characterized by an inflammation with painful, erythematous, and swollen nail folds. It generally appears around the nails of the hands and less frequently on those of the feet. Nail plate involvement can follow, but not necessarily. *C. albicans* is the major etiologic agent; however, *C. parapsilosis* and *C. guilliermondii* are also recognized agents of this syndrome. Secondary bacterial infections are a known complication.

Paronychia is often seen in patients who also suffer from erosio interdigitalis. Similar to erosio interdigitalis, it is associated with excessive water immersion and is therefore seen in a similar population segment: dishwashers, bartenders, and fruit cannery workers. It can thus be considered, to some degree, an occupational disease. CMC patients often suffer from paronychia. Paronychia is more frequent among women than among men.

Onychomycosis

As in paronychia, involvement of the nail itself by *Candida* is more often seen in nails of the hands. The affected nails may become discolored, eroded, brittle, detached from the nail bed, and painful. Generally there is less accumulation of debris under the nail plate than in dermatophyte onychomycosis; however, the condition is more painful. In most instances the infection begins with the nail fold involvement, although as indicated above, not all the paronychia cases do lead to onychia. Candida onychomycosis is caused by the same *Candida* species as paronychia, and is more frequent among women. It is seen in CMC patients and in immunosuppressed patients or those with symptomatic peripheral vascular disease.

MUCOSAL INFECTIONS

Involvement of mucosal surfaces by *Candida* is considered the most frequent clinical manifestation of candidiasis. Although the term mucosal surfaces encompasses in the strict sense of the word mucosal membranes of various organs, what is generally understood by the term mucosal candidiasis is involvement, primarily, of the oral and vaginal mucosa. Thus, the following section covers *Candida* infections of these two mucosal sites, while additional mucosal manifestations will be dealt elsewhere, under the appropriate sections.

Oral candidiasis

Candidiasis of the oral mucosa (Samaranayake and MacFarlane 1990; Lynch 1994; Cannon et al. 1995) is a disease recognized since antiquity, which has gained renewed significance more recently as an infection frequently seen in AIDS patients and in other conditions. As cited by Samaranayake and Macfarlane (1990) in their monograph on oral candidiasis, the renewed interest in this infection is expressed in the large number of scientific papers published in the last two decades.

Although *C. albicans* is, as in other types of *Candida* infection, the most frequently isolated etiological agent, additional *Candida* spp., such as *C. glabrata*, *C. tropicalis*, *C. parapsilosis*, and *C. guilliermondii*, are also implicated in oral candidiasis (Bodey 1993). It should be emphasized, however, that these species can also be isolated from the oral cavity of normal individuals without obvious pathology. This observation, which demonstrates the complexity of the diagnosis, is relevant also to other anatomic sites, such as the vagina.

Oral candidiasis can be clinically classified into several different forms:

- acute pseudomembranous candidiasis
- acute atrophic candidiasis
- chronic atrophic candidiasis
- chronic hyperplastic candidiasis
- angular cheilitis.

Oral candidiasis of several clinical forms, mostly of chronic nature, is seen in patients with CMC. It can assume the pseudomembranous or hyperplastic form, with or without angular cheilitis. It can also, similar to the situation in AIDS patients, spread to the esophageal mucosa.

Acute pseudomembranous and acute atrophic candidiasis

Acute pseudomembranous candidiasis, known also under the term oral thrush, is manifested by white-gray lesions on the gums, tongue, or oral mucosa, which can appear as single lesions or as confluent large plaques. The lesions, particularly when covering larger areas, may be painful, and disturb intake of food. They may spread to the mucosa of the esophagus, as seen in a significant number of AIDS patients, and cause dysphagia. In addition to AIDS or cancer patients and otherwise debilitated individuals, the infection is generally present in

infants, particularly those born to mothers suffering from vaginal candidiasis, and in the elderly.

Acute atrophic candidiasis is characterized by painful, erythematous mucosa, particularly on the tongue, which can be associated with loss of the tongue papillae, affecting food intake. Acute atrophic candidiasis may follow the acute pseudomembranous form with the disappearance of the pseudomembranous lesions. It can appear as a sequelae of antibiotic treatment.

Chronic atrophic and hyperplastic candidiasis

Chronic atrophic oral candidiasis, known also as denture stomatitis, is frequently seen in elderly individuals (Cumming et al. 1990), particularly those wearing dentures, and is reported in a very high proportion of this group. It is believed to be the most common clinical form of oral mucosa involvement. Clinically it is characterized by erythema and/or edema of the mucosa under the dentures, and generally, contrary to acute atrophic candidiasis, is not painful.

The chronic hyperplastic form, termed also *Candida* leukoplakia, is a much rarer condition. It presents with white plaques, which can appear on various sites of the buccal mucosa, and contrary to the acute pseudomembranous form, can not be removed. The importance of this clinical entity lies in the association with possible transformation into a malignant state.

Angular cheilitis

Clinically this form is characterized by erythema and fissures at the folds of the corners of the mouth. It may accompany other clinical forms of oral candidiasis, such as denture stomatitis or oral thrush. When accompanying denture stomatitis its appearance is attributed to unfitting dentures, which create the appropriate physical conditions for development of this type of infection. It is also believed that angular cheilitis may be associated with avitaminosis.

VAGINAL CANDIDIASIS

Candidal vulvovaginitis (Odds 1988; Sobel 1992a, b, 1993; Edwards 2000) is a common female infection, primarily during the fecund period. It is believed to be the most frequent or second most frequent vaginal infection, depending on geographical area, and its incidence is estimated in the range of 5–20 percent. The incidence increases in particular groups, such as pregnant or diabetic women, those using oral contraceptives or who are post antibiotic treatment. The predisposing factors believed to be associated with this specific condition can be associated primarily with the host and also with some virulence attributes of the pathogen. The former may include hormonal effects as seen in pregnancy, immunological or nutritional factors (Witkin et al. 1983, 1986, 1989). The latter may be correlated with certain fungal factors, such as increased adherence ability (Segal et al. 1984).

Besides *C. albicans*, *C. glabrata* and *C. tropicalis* are the most frequently isolated *Candida* spp. (Horowitz et al. 1992) both from vulvovaginitis patients and from healthy carriers. Again, as already stated, here too the problem of microbiological diagnosis is demonstrated.

Clinically the syndrome includes complaints of vulvovaginal pruritus and discharge, which can be thick and 'curd-like' or thin. The infection may be associated with erythema of the vulvovaginal mucosa and possibly also of the perianal area. The lesions on the mucosal surface are basically adherent plaques. The condition may cause discomfort and pain during intercourse.

An important feature of candidal vulvovaginitis is the recurrence of the infection, although the majority of vulvovaginitis cases will respond to treatment. Chronic, recurrent vaginal candidiasis poses a medical problem and was quite extensively studied, as to possible risk or predisposing factors, and fungal reservoir (Sobel 1992b).

Sexual transmission to a male partner is known (Sobel 1993). In the male the infection presents, generally, as a balanitis with lesions and erythema on the penis.

Deep-seated candidiasis

The term deep-seated candidiasis (Odds 1988; Bodey 1993; Edwards 2000) refers to infection of visceral organs and possibly to multiple-organ or disseminated disease. Basically, any organ or system can be affected, thus the clinical entities may include: candidiasis of the gastrointestinal tract, respiratory system, central nervous system, renal and urinary tract, cardiovascular system; hepatosplenic candidiasis; hematogenous disseminated disease; ocular infections; and a variety of other specific manifestations. As it is impossible to cover all the clinical manifestations, the description here is limited to the major clinical entities.

CANDIDIASIS OF THE GASTROINTESTINAL TRACT
Esophagitis

This syndrome (Musial et al. 1988) includes painful dysphagia and possibly chest pain. Additional symptoms may include nausea and/or vomiting. White patches, which resemble those of oral candidiasis, can be noted by endoscopy on the esophageal mucosa. Esophagitis may be associated with presence of oral candidiasis, but may also present as a separate clinical entity without oral involvement.

The frequency of this syndrome increased since the appearance of AIDS. It is estimated that 10–30 percent of AIDS patients with oral candidiasis may suffer also from candidal esophagitis. The infection is also seen in cancer patients following anticancer therapy.

Gastrointestinal candidiasis

As stated earlier, *Candida* spp., primarily *C. albicans*, *C. glabrata*, or *C. tropicalis*, colonize the gastrointestinal (GI) tract in a significant proportion of normal individuals (Stone et al. 1973; Bolivar and Bodey 1985; Odds 1988) as judged by fungal isolation from feces. An increase in the percentage of colonized individuals is noted among hospitalized patients (Kusne et al. 1994).

Clinical involvement of mucosal surfaces of the stomach and/or small and large intestine, with mucosal white plaques and ulcerations are found primarily in cancer patients. A large survey conducted by Eras et al. (1972) on cancer patients, including postmortem studies, revealed involvement of stomach and intestines.

As already mentioned, *Candida* GI colonization and infection are believed to play a role in the pathogenesis of development of disseminated candidiasis. The GI tract can serve as a reservoir for the fungus, from where it can spread, particularly if there is a breach in the intactness of the mucosal lining. In patients with malignancies, there are several factors contributing to an increased risk of dissemination. These include increased *Candida* GI colonization, as these patients are generally treated with antibiotics, which change the GI flora in favor of *Candida*, and impairment of the GI mucosa, mostly due to anticancer therapy. Furthermore, such individuals may also have deficiencies in the activity of the immune system, both due to the neoplastic process itself and as possible sequelae of the anticancer treatment, therefore being unable to overcome the infectious agent once it was introduced. Thus, GI colonization and infection in cancer patients or in other categories of debilitated individuals, such as organ transplant patients (Kusne et al. 1994), is of particular significance and should be monitored and, if possible, controlled.

The possibility that *Candida* colonization may cause diarrhea is speculative, although reports in this regard are found in the literature (Talwar et al. 1990; Enweani et al. 1994; Levine et al. 1995).

Candidiasis of the liver, spleen, and other organs

Hepatosplenic candidiasis, which can be considered also as part of a disseminated infection, became a recognized clinical entity recently (Haron et al. 1987; Thaler et al. 1988; Edwards 2000). It is seen primarily in leukemics, with the hepatosplenic involvement becoming apparent generally during recovery from the neutropenic state, while the infection started during the patient's neutropenic period. This clinical entity is difficult to diagnose and manage. Diagnosis can be assisted by computed tomography (CT) scans demonstrating lesions in the liver and the spleen, and by biopsied tissues revealing the presence of fungal elements. Most cases are caused by *C. albicans*; however, infections by other species have also been described (Pines et al. 1994).

Other organs of the GI system affected by *Candida* include the gallbladder and pancreas. In this connection peritonitis due to *Candida* should also be mentioned, which occurs particularly in patients on peritoneal dialysis (Yuen et al. 1992; Amici et al. 1994).

CANDIDIASIS OF THE RESPIRATORY SYSTEM

Respiratory *Candida* infections involving the lungs or bronchial system appear predominantly in patients with underlying primary diseases (Masur et al. 1977; Gueteau et al. 1991). Bronchopneumonia can originate from hematogenous spread of the fungus as part of a disseminated infection or from introduction of the pathogen into the lung. Diagnosis is difficult as *Candida* can be found in the sputum of individuals without candidiasis, so that it is pertinent to demonstrate the presence of the fungus in bronchopulmonary tissues.

CANDIDIASIS OF THE CARDIOVASCULAR SYSTEM

Candida spp. can cause clinical manifestations in various organs of the cardiovascular system (Musial et al. 1988; Kwon-Chung and Bennett 1992; Edwards 2000): the pericardium, myocardium or endocardium, with endocarditis being the best-known clinical entity.

Endocarditis (Hallum and Williams 1993; Wilson H.A. et al. 1993) is seen primarily in intravenous drug addicts and in individuals with impaired or prosthetic heart valves. It was also described in patients following various cardiac surgery procedures or as sequelae of anti-cancer therapy. Due to the increase in drug addiction and frequency of cardial surgery, an increase in the incidence of *Candida* endocarditis was noted.

Endocarditis can be caused by *C. albicans* and other species, such as *C. parapsilosis* (Cancelas et al. 1994) and *C. tropicalis*. Interestingly, non-*albicans Candida* infection is seen mostly in heroin addicts. It is possible that the infection originates from contaminated heroin. In the other susceptible groups, such as those with impaired or prosthetic valve patients, the infection may originate from introduction of the fungus during the surgical procedure, with the prosthetic or impaired valve serving as a suitable substrate to which *Candida* can readily adhere and form the typical vegetations.

Clinically, candidal endocarditis is difficult to differentiate from that caused by bacteria. It may, however, have a more prolonged onset; e.g. in post-surgery patients it may become apparent months later. The patients present with fever, heart murmurs, possible blood vessel obstruction, splenomegaly, and typical endophthalmitis. Pathologically, *Candida* endocarditis is characterized by large vegetations, from which emboli may be released.

RENAL AND URINARY TRACT CANDIDIASIS

Lower urinary tract infection

Candida lower urinary tract infection (UTI) (Musial et al. 1988; Gentry and Price 1993) is quite frequently seen in association with indwelling catheters, and may originate from the GI tract or genital flora. It is, thus, more frequently seen in females. The infection is also found in diabetics.

The symptoms are not unique and the course of the infection can be mild or more severe. An interesting clinical feature is the possible formation of fungal masses, 'fungus balls' (Scerpella and Alhalel 1994), which may cause obstruction and impair normal urine flow.

A basic problem associated with the diagnosis of this entity is the significance of candiduria in respect of discriminating infection from colonization due to contamination with flora of adjacent anatomic sites. Quantitative assessment of *Candida* colony-forming units (CFU.) in urine may be helpful, but does not always provide a definitive unequivocal answer.

Renal infection

Renal candidiasis (Edwards 2000), can, theoretically, originate from hematogenous *Candida* dissemination or as an ascending UTI. In reality, however, the latter is quite a rare event, and most of the cases of renal involvement come from hematogenous spread.

Pathologically, as mentioned in the relevant section, renal candidiasis is characterized by microabscess formation, mostly evident in the cortex of the kidneys. Similar pathology is noted in experimental *Candida* infections in animal models. The kidneys are a target organ for the fungus in infections induced by different ways of inoculation (intraperitoneal or intravenous).

CENTRAL NERVOUS SYSTEM CANDIDIASIS

Central nervous system (CNS) involvement by *Candida* (Walsh et al. 1985) is uncommon and is limited predominantly to *C. albicans*. The most susceptible individuals are AIDS patients and premature infants (Fakes 1984; Baley et al. 1986; Hughes et al. 1993; Edwards 2000). The CNS is generally seen as part of disseminated candidiasis involving in most cases primarily the meninges, although abscess formation in brain tissue was also reported.

Candida meningitis patients may present with symptoms similar to those seen in meningitis of other etiology, including various neurological abnormalities. *Candida* organisms may be detected microscopically in direct smears from the cerebrospinal fluid (CSF). In addition, abnormal protein and/or sugar values may be found in the CSF.

DISSEMINATED CANDIDIASIS AND CANDIDEMIA

Disseminated candidiasis can be defined as a multiorgan infection including possible candidemia, although blood cultures do not always yield the fungus. As already stated in the previous sections, disseminated candidiasis may include involvement of the CNS, kidneys, heart, eyes, or other organs and systems. As previously indicated, hepatosplenic candidiasis, although dealt with in an other section, is a specific clinical manifestation of disseminated infection.

Disseminated candidiasis caused by *C. albicans* and other species is associated with debilitation. It is seen in cancer patients (Musial et al. 1988), particularly those with acute leukemia (D'Antonio et al. 1992), in patients post-surgery, particularly GI and cardiac surgery, in transplant patients, particularly bone marrow transplants, in premature infants (Walter et al. 1990), burn patients (Green et al. 1994), and drug addicts.

Candidemia may present with the unspecific symptoms of a septic state, including fever. Presence of cutaneous and ocular involvement may be helpful diagnostic criteria. Candidemia patients may present (Edwards 2000) with nodular lesions on the skin and typical white lesions in the retina. These criteria are of particular importance, as cultural evidence of fungal etiology may be lacking.

Ocular *Candida* infections

Candida species can affect both the outer and inner eye (Langenhaeck and Zeyen 1993; Romano and Madjarov 1994). Infection may originate from hematogenous dissemination or from direct fungal introduction, the first, generally resulting in inner-eye infection, and the latter in clinical manifestations of the outer parts of the eye. Both categories of eye involvement are caused by *C. albicans* and some other *Candida* species, such as *C. parapsilosis*, *C. krusei*, and *C. glabrata*.

Outer-eye infections could also be considered in the section on superficial mucocutaneous candidiasis. These include infections of the conjunctiva (conjunctivitis), cornea (keratititis), eyelids (blepharitis), or lacrimal duct (lacrimal canaliculitis). Such infections may be associated with some ocular trauma, such as surgery, injury, or even the use of contact lenses. *Candida* infections may also follow other ocular infections, such as herpes simplex, or appear as sequelae of corticosteroid or antibacterial treatment.

Endophthalmitis, as indicated above, is generally a result of hematogenous fungal spread (McQuillen et al. 1992; Cohen and Montgomerie 1993), although it may also result from an exogenous source. The infection can cause loss of vision and is characterized by presence of typical white cotton-like lesions, which, as indicated, are an important diagnostic criterion for disseminated candidiasis.

ANIMAL DISEASE

Candida spp. may constitute part of the normal microflora in various anatomic sites, such as the digestive and genital tracts, skin, and ears (Yeruham et al. 1999; Duarte et al. 2001), of animals. Consequently, their involvement in pathological processes is secondary, often as a sequel to prolonged antibacterial chemotherapy, stress, and substandard management practices such as overcrowding and bad nutrition. Immune insufficiencies, of primary importance as predisposing factors to human mycotic infections, seem to have a more limited significance in animals. Nevertheless, the possibility of correlation between hereditary immunodeficiency in foals (McClure et al. 1985) and *Parvovirus*-induced immunosuppression in dogs (Anderson and Pidgeon 1987) has been suggested.

Most *Candida* infections, which have been described in animals, can be grouped into three principal categories as follows.

Candidiasis of the digestive tract

Esophagogastric candidiasis has been reported in various animal species. *C. albicans* and other non-specified *Candida* spp. have been isolated from pigs suffering from ulcers related to altered keratinization of the esophagogastric tract more frequently than from healthy animals (Kadel et al. 1969). The ability of *C. albicans* to utilize keratin as sole nitrogen source has been suggested as an explanation for this observation (Kapica and Blank 1957). Other possible predisposing factors to esophagogastric candidiasis in swine may be bad management, overcrowding, and antibacterial chemotherapy.

Other animals from which *Candida* spp. were isolated from esophagogastric lesions are foals (Gross and Mayhew 1983) and calves (Cross et al. 1970; Elad 1993). *C. glabrata* was found to colonize the abomasums of preweaned calves, possibly being involved in neonatal calf diarrhea (Elad et al. 1998). Colonization was more frequent and shedding more intensive in calves fed with milk replacers than in those fed dam's milk (Elad et al. 2002).

Candida spp., especially *C. albicans*, may be part of the normal flora of the avian digestive tract. However, they may be also involved as primary pathogens or as secondary invaders of damaged mucosa (such as is often the case in force-fed geese), in infections of various parts of this system, especially the crop. These infections may unsettle the function of the digestive tract and cause vomiting, increase in crop emptying time, and impaction. Antibacterial chemotherapy is assumed to play an important role as a predisposing factor to avian candidiasis by changing the microbial balance. The outcome depends primarily on the immune competence of the host (Ritchie et al. 1994), influenced by environmental factors such as stress and captivity. *Candida* infections of birds, in sites other than the digestive tract, have been reported sporadically (Ritchie et al. 1994). In addition to *C. albicans*, other *Candida* spp. have been reported to be involved in avian candidiasis (Moretti et al. 2000), albeit more rarely.

Candidiasis of the urogenital tract

Candida spp. have been reported to be part of the normal flora of the genital tract of various animals (Chengappa et al. 1984), but have been associated with cases of bovine abortion (Foley and Schlafer 1987) and reduced fertility in cattle (Panangala et al. 1978; Sinha et al. 1980; Sutka 1983; Foley and Schlafer 1987; Tainturier et al. 1993) and horses (Abou-Gabal et al. 1977). The isolation of *Candida* spp. from aborted bovine fetuses is rare relative to other pathogens, including hyphomycetes, but has a worldwide distribution (Foley and Schlafer 1987). *C. tropicalis*, *C. albicans*, *C. krusei*, and *C. parapsilosis* are among the species isolated from such cases.

C. albicans as the causetive agent of cloaca and vent infections of geese and turkeys has been reported (Beemer et al. 1973; Marius-Gestin et al. 1987).

Candida mastitis

Candidal mastitis is usually self-limiting with spontaneous recovery and the infection has usually no systemic consequences. *Candida* spp. have been isolated from milk sampled from cattle with and without signs of mastitis. Most cases are sporadic but outbreaks have been reported (Elad et al. 1995). These were usually associated with intramammary instillation of contaminated antibacterial preparations, usually home made under nonhygienic conditions (Kirk et al. 1986). However, heavy environmental contamination with the yeast, in conjunction with insufficient milking hygiene, have been suspected of precipitating such outbreaks (Elad et al. 1995). All the yeast species dealt with in this chapter (and many others) have been isolated from cases of mycotic mastitis. *C. krusei* seems, however, to be more prevalent than other species in many reports. This might be the consequence of this species' relatively high environmental prevalence, especially in silage, where it may thrive during certain stages of anaerobiosis development, by using lactic acid as carbon source (Burmeister and Hartman 1966).

In addition, unique yeast infections, such as pyothorax (Fales et al. 1984) and urocystitis (Fulton and Walker 1992) in two cats, enteritis in a dog (Anderson and Pidgeon 1987), and keratitis in a horse (Desbrosse 1994), all caused by *C. albicans*, have been reported. In wild animals, *C. albicans* has been isolated from marine mammals (Sweeney et al. 1976), wild birds, marsupials, and reptiles.

LABORATORY DIAGNOSIS

Laboratory diagnosis of candidiasis depends upon the nature of the different clinical entities, specifically, whether they are mucocutaneous or deep-seated forms. Thus, the discussion of laboratory diagnosis will be in reference to the two groups of clinical manifestations.

Mucocutaneous candidiasis

Laboratory diagnosis of the different clinical forms of mucocutaneous candidiasis, which follows the clinician's diagnosis of suspected *Candida* involvement of the skin, nails, or mucosal surfaces, includes, generally, two steps:

1 direct examination of pathological specimen to demonstrate fungal presence
2 isolation of the fungus in culture and its identification.

DIRECT EXAMINATION
Skin and nail infections

Clinical specimens from diseased skin or nails can be collected by scraping the affected area, or, in some instances, by the use of swabs; material from nails can also be obtained by cutting the infected nail (Merz and Roberts 2003).

The preferred method for direct examination of clinical specimens from cutaneous and nail candidiasis is the wet mount technique. Specimens have, in most cases, to be treated with a keratinolytic substance, generally 10–30 percent KOH, which facilitates microscopic examination of the specimen. It is also helpful to add Parker's ink or the lactophenol cotton blue stain, which enables easier demonstration of the fungal elements. A valuable development in this regard involves the use of fluorochromes, such as calcofluor white or analogous products, with affinity for chitin and glucan, which makes the demonstration of fungal elements with a fluorescent microscope relatively simple, not requiring specific expertise (occasionally, nonspecific fluorescence, e.g. cotton fibers, may occur). Stained mounts, such as Gram staining of dermal and nail specimens, or histopathological sections are, generally, not recommended for routine use.

Direct microscopic examination of infected material is expected to reveal the presence of budding yeasts, pseudohyphae, and/or hyphae. The finding is indicative of candidiasis, but does not result in most cases in a definite speciation. The latter can be obtained only by isolation of the fungus in culture and its identification.

Mucosal infections

Specimens from patients with vaginal and oral infections can be collected by swabs (Merz and Roberts 2003) and kept, preferably, in transport medium before being processed in the laboratory. Specimens from oral lesions can also be obtained by scraping.

Processing of the specimen does not generally require treatment with a keratinolytic substance. Both wet and fixed mounts can be used. Wet mounts can be unstained, prepared in saline or water, or stained with lactophenol cotton blue or calcofluor white. Fixed mounts can be stained by Gram, methylene blue or Giemsa. As in the case of skin and nail infections, histopathological sections are not necessary for routine diagnosis of candidal vulvovaginitis or oral candidiasis.

Microscopic examination of specimens from vulvovaginitis or oral candidiasis will demonstrate the presence of budding yeasts, pseudohyphae, and/or hyphae. Demonstration of hyphal elements in direct microscopic mounts is important for the diagnosis, as both the vagina and the mouth are colonized also normally by *Candida* spp., and it is believed that the presence of the hyphal elements noted directly in the clinical specimen, in addition to the yeasts, is an indicator of infection (Figure 30.4). However, it should be taken into account that *C. glabrata*, a significant non-*albicans* species involved in vaginal infections, does not produce hyphae or pseudohyphae in clinical specimens.

In mucosal infections as well, microscopic examinations serve only as indications of candidiasis, without defining the exact etiology, which can be established only by culture. However, in this context, the significance of the direct examination for diagnosis should be emphasized. Since *Candida* spp. can be found in the mouth and vagina of healthy individuals, positive cultures without direct demonstration of the fungus in the clinical specimen are not always sufficient for diagnosis.

CULTURE

Isolation of *Candida*

The routine medium used for isolation of fungi in culture (Hazen and Howell 2003) from mucocutaneous infections is Sabouraud dextrose agar (SDA) supplemented with antibiotics (chlomphenicol, gentamicin, and/or tetracycline) to prevent bacterial overgrowth. It is also recommended to use an additional isolation medium – SDA to which cycloheximide (Actidione) has been added as well, a medium on which contamination by airborne molds can be limited. It is, however, not recommended to use the latter medium (commercially known as Mycosel or as Mycobiotic agar) as a single isolation medium, since some *Candida* are sensitive to cycloheximide and will not grow.

Candida from mucocutaneous sources are, generally, relatively easily isolated in culture. Cultures can be incubated at 28°C or/and at 37°C, and *Candida* colonies will be apparent within 2–3 days; in some cases growth will be noted already after 24 hours, but may take in others more than 3 days.

Identification and speciation

The description in this section is also relevant to the parallel one in diagnosis of deep-seated *Candida* infections.

It is advisable to obtain pure cultures prior to the identification process and to check by microscopy that the colonies reveal yeasts. Identification of *Candida* and

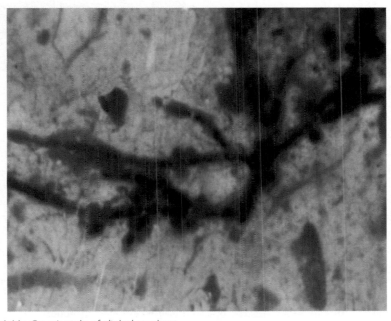

Figure 30.4 Candida *vaginitis: Gram's stain of clinical specimen*

its speciation is, generally, based on the following categories of criteria: macroscopic/microscopic morphology and physiological/biochemical characteristics; serology for antigenic identification of *Candida* spp. or serotypes can also be applied, although this is not done routinely.

Use of morphology for identification

Macroscopic morphology of *Candida* spp. (Kwon-Chung and Bennett 1992; de Hoog et al. 2000) on routine isolation medium (SDA), is rather similar. *Candida* spp. appear on SDA as smooth, in some species (e.g. *C. krusei*) more dry, cream colonies. Although fine differences between the species can be noted, they are, generally, very difficult to interpret and influenced by subjective judgment, and thus are not believed to be a valid criterion for routine *Candida* speciation.

A number of investigators (Quindos et al. 1992) have undertaken to develop specific media, which would enable, already at the isolation stage, investigators to differentiate, on the basis of macroscopic morphology, the most commonly encountered pathogenic *Candida* spp. The systems were, however, not specific enough or too difficult to standardize for a reliable method. An important development was introduced by Odds and Bernaerts 1994, consisting of the CHROMagar system, which uses chromogenic substances. It is based on the reaction between specific enzymes of the different species and chromogenic substrates, which results in formation of differently colored colonies. This system recommended by the producers for isolation and identification, permits, based on the color of the colonies, fast presumptive identification of *C. albicans*, *C. krusei*, *C. tropicalis* (Figure 30.5), and others. Of particular notice is *C. dubliniensis*, which grows on this medium in a darker shade of green than *C. albicans*, a characteristic that is helpful in differentiating between the two species.

Microscopic morphology of the culture on SDA isolation medium is also not reliable for species identification. The *Candida* spp. will appear as budding yeasts, and although some differences in the size and shape (oval or more elongated) do exist between the species (see section on Mycological characteristics below), they are too subtle to serve as a reliable, repeatable tool for routine diagnosis.

The commonly used differential medium both for genus identification as well as speciation, is the CMA plate (Dalmau plate) supplemented with Tween 80 or rice agar (de Hoog et al. 2000). Subcultures on CMA plates, made from the SDA isolation medium by furrowing the CMA plates, and incubated for 2–5 days at 28°C, will reveal, upon direct (inverted microscope) or mount microscopic examination, the presence of pseudo- and true hyphae, a characteristic of the genus *Candida* (except for *C. glabrata*, see section on Taxonomic comments, below). In addition, microscopic examination of CMA plates will also reveal species-

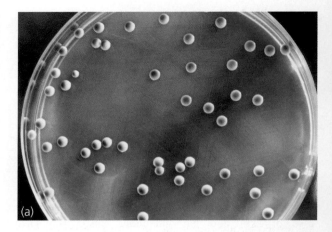

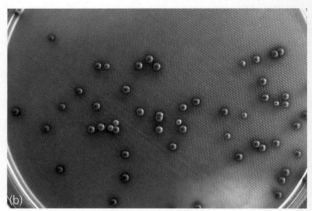

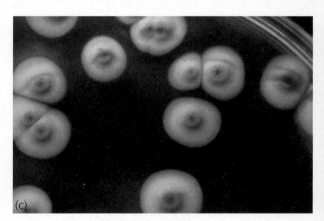

Figure 30.5 Candida *species on CHROMagar* Candida: **(a)** C. albicans; **(b)** C. tropicalis; **(c)** C. krusei

specific characteristics, such as chlamydospores in *C. albicans* and *C. dubliniensis*, and the typical morphology of the different species (see description under section on Mycological characteristics, below). Both unstained and stained wet or fixed mounts, respectively, can be used to demonstrate the microscopic morphology formed on the CMA plates.

In the context of morphology, as a basis for species identification, the so-called 'germ tube test' (Merz and Roberts 2003) should be mentioned. The principle of this test is the ability of *C. albicans* or *C. dubliniensis* blastospores to produce, under defined conditions (generally, in serum incubated at 37°C for 2 hours), germ tubes. The formation of germ tubes and chlamydospores is indicative of these two species and therefore offers reliable and easily performed tests for routine diagnosis.

Physiological/biochemical characterization

The genus *Candida* and the different *Candida* spp., similar to other yeasts, can be characterized by the pattern of utilization of specific carbohydrate and nitrogen substances (see section on Mycological characteristics below). As mentioned earlier, *Candida* spp. can utilize carbohydrates both oxidatively (assimilation) and anaerobically (fermentation). Yeasts possessing the ability to ferment a given carbohydrate do also assimilate that substance, but not necessarily vice versa. Thus, biochemical identification of *Candida* spp. is based primarily on assimilation and fermentation tests.

Various techniques have been developed for assessing the assimilation patterns. The classical Wickerham and Burton (1948) method assessed assimilation by determining the ability of a given yeast isolate to grow in a set of defined minimal liquid media supplemented with different carbohydrates. This method, which, although precise, was laborious and time consuming and therefore not suitable for routine use, was replaced in the clinical laboratory by the auxanographic technique (Hazen and Howell 2003). This technique employs minimal media agar plates on which paper disks impregnated with different carbohydrates are placed, and the growth ability of the yeast around a specific disk is an indication of the yeast's ability to assimilate that carbohydrate. This method is simpler and quicker, and therefore more adapted for routine use. Modifications of this principle are used by various commercial kits, such as the API 20C, API ID32C (Figure 30.6), Vitek, Minitek, Uni-Yeast-Tek, and others. These commercial systems are easy to use and some are automated. A number of studies focused on comparison of the various methods (Fenn et al. 1994; Riddle et al. 1994). Other methods, such as MicroScan, have also been described (Land et al. 1991; St Germain and Beauchesne 1991).

Fermentation tests are more difficult to perform. The classical tests involved liquid media supplemented with different carbohydrates, a color indicator to assess pH changes to measure acid formation, and a tool to assess gas production (generally an inverted small tube, the Durham tube), a criterion of fermentation. There are several modifications for assessment of gas production, such as use of semisolid media or a wax layer on top of the liquid medium. Most of the commercial kit systems do not use fermentation assays but rely on assimilation tests.

Other identification methods

Unlike diagnosis of mucocutaneous candidiasis, in the deep-seated forms isolation of the fungus may be problematic, its identification, once isolated, is relatively straightforward by the described methods. Nevertheless, additional methodologies for identification have been described.

Candida spp. can also be identified serologically, using specific antisera. A commercially available set of sera (Iatron System), prepared against *Candida* cell wall antigens (see section on Antigenicity below) enables quick species identification of the commonly encountered pathogenic *Candida* spp. by simple slide agglutination. However, serological identification is not used as a standard routine method in clinical laboratories for diagnosis of candidiasis.

Other non-culture identification methods may include molecular techniques (Buchheidt et al. 2000; Alexander 2002; Sullivan and Coleman 2002) such as the use of

Figure 30.6 Candida *spp. in API 32C: upper panel,* C. albicans; *lower panel,* C. glabrata

electrophoretic patterns of DNA, RNA profiling, restriction enzyme analysis (Velegraki et al. 1999), and PCR (Niesters et al. 1993; Maiwald et al. 1994; Santos et al. 1994). Common molecular probes used by a number of investigator groups for identification of *Candida* at genus or species level are ribosomal DNA primers, including the large or small rRNA subunits (Haynes and Westerneng 1996). Additional probes described include SAP DNA (Flahaut et al. 1998), the *ERG11* gene involved in ergosterol biosynthesis (Posteraro et al. 2000), and others.

However, once again, these too do not yet, generally, constitute part of routine diagnosis.

Deep-seated candidiasis

Although the basic principles involved in the diagnosis of deep-seated and disseminated candidiasis are similar to those of the mucocutaneous candidiasis, the diagnosis of the deep-seated entities is more difficult and complicated, including various problems at the different steps of the process. While laboratory diagnosis of mucocutaneous candidiasis is based on direct examination of clinical specimen, followed by isolation and identification of *Candida* spp., that of the deep-seated forms involves in addition also use of immunodiagnostic and other noncultural methods.

DIRECT EXAMINATION

Due to the nature of the different clinical forms, collection of clinical specimens may be problematic. In some clinical forms it is difficult to obtain a sample from the infected organ or the specimen can be collected only by invasive methods. In others, such as in the case of samples from the respiratory tract, the demonstration of *Candida* in sputum, a sample which is easily obtainable, poses a problem in interpretation, as *Candida* is part of the oral flora; thus to have meaningful results, it is necessary to obtain samples by the invasive procedures of bronchial lavage or biopsy. Some clinical specimens, such as CSF, may require procedures of concentration due to the possibility of paucity of the pathogen in the clinical sample (Konneman et al. 1997). However, if samples from normally sterile sites are obtained, demonstration in such samples of the presence of yeasts and/or hyphal elements, indicative of *Candida*, is important for the establishment of diagnosis.

Direct demonstration of *Candida* in specimens from deep-seated candidiasis can be done both by wet and fixed mounts, aided by calcofluor white. Tissue biopsies can be prepared and stained by histopathological techniques (Salfelder 1990), using the periodic acid–Schiff (PAS) or the Gomori methamine silver (GMS) stain. An additional method using an immunohistochemical technique was described by Marcilla et al. (1999). These investigators used monoclonal antibodies (3H8), in identification of *C. albicans* (Marcilla et al. 1999).

CULTURE

Isolation of *Candida* species

A basic problem is the difficulty in isolating *Candida* in culture from specimens of deep-seated candidiasis, particularly if the infection is localized in an internal organ. The specimen that is most easily obtainable, which leads to the highest isolation rates, is blood cultures. However, even these are characterized by a relatively high proportion of failure to isolate the fungus (Jones 1990; Pfaller 1992a; Reiss and Morrison 1993). The most recommended methods for blood cultures include the lysis-centrifugation isolator system (Wilson M.L. et al. 1993; Kirkley et al. 1994), which apparently yields the highest rates of positive cultures (Telenti and Roberts 1989), the Bactec radiometric or nonradiometric systems (Yagupsky et al. 1990), and use of biphasic media (Richardson and Carlson 2002). Other specimens from normally sterile sites with lesser rates of positive cultures may include CSF, tissue biopsies, or bronchial washings. Such specimens are inoculated on Sabouraud's agar with antibiotic but without cycloheximide. Growth conditions are similar to those for specimens from mucocutaneous candidiasis. However, it is advisable to wait much longer periods (up to 3 weeks) before a decision is made that a culture is negative.

Identification of *Candida* species

Identification of the *Candida* isolated is done as in the case of mucocutaneous candidiasis. It is based, primarily, on macroscopic and microscopic morphology, and on physiological/biochemical characteristics of the isolates.

Immunodiagnosis

The basis for immunodiagnosis is the patient's immune response to *Candida*, as expressed by presence of antibodies or CMI, and/or presence of microbial antigens in the patient's body fluids. Contrary to the mucocutaneous forms of candidiasis, in which the use of direct examination of the clinical material and culture is quite satisfactory for diagnosis, in the deep-seated entities, these methods, as indicated above, are in many instances not sufficient. Thus, the search for additional approaches, such as the use of immunodiagnostic techniques or other non-culture methods, is warranted, and indeed this led to the development of some diagnostic procedures, primarily in the area of immunodiagnosis (de Repentigny 1992; Reiss et al. 2002).

Immunodiagnosis of candidiasis is based, primarily, on detection of the humoral immune response as expressed in antibody production, and detection of fungal antigens. As a consequence two types of assays are involved in immunodiagnosis:

1 tests for detection of antibodies, primarily in serum
2 tests for detection of the fungal antigen(s) in body fluids, again, primarily in serum.

Tests for detection of antibodies

The classical serology included assays for detection of serum antibodies by various modifications of the agglutination test. However, in view of the ubiquitous and commensal nature of *Candida* spp., antibodies to *Candida* detected by the agglutination technique, which generally represent antimannan antibodies, can be found in healthy individuals or in those with superficial infections (Lew 1989). Thus, this test is not discriminatory enough to be useful in the diagnosis of deep-seated candidiasis. By contrast, tests determining the presence of antibodies raised to *Candida* internal antigens, which, it is assumed, are released into the patient's body fluids, primarily during invasive infection, could be more discriminatory and therefore more useful for diagnosis. In addition, it should be considered that immunocompromised patients, who are the principal candidates at risk to develop systemic infections, have difficulties in mounting an antibody response at detectable levels.

Tests aimed at detecting antibodies to *Candida* antigens released and/or expressed during deep-seated infections, include the gel immunodiffusion (ID) or counterimmunoelectrophoresis (CIE) techniques to detect antibodies to *Candida* cytoplasmic antigens, which are standardized by their protein content (Reiss et al. 2002). Other systems used include the enzyme-linked immunosorbent assay (ELISA) or latex-agglutination (LA) test. Detection of antibodies to defined antigens believed to circulate in body fluids during infection, such as enolase (Mitsutake et al. 1994), heat shock proteins (47 kDa – breakdown of HSP90) Walsh and Chanock (1997), and more recently, specific mannosides (Poulain et al. 2002), although not widely used, has also been evaluated. Low sensitivity is the major problem of using antibody-detection methods in diagnosis, which may be either the result of the use of more discriminatory, less sensitive tests, such as the ID or the poor immune status of a significant proportion of the candidiasis patients.

Although attempts to use techniques that demonstrate the different immunoglobulin types for diagnostic purposes have been described (Fegeller et al. 1990), the presence of IgM anti-*Candida* antibodies as an indicator of a recent infection has, apparently, contrary to other microbial infections, no proven validity. Thus, the differentiation between types of immunoglobulins has limited diagnostic value in candidiasis and, consequently, is generally not carried out for this purpose. Moreover, determination of the specific immunoglobulin type cannot be used effectively for differentiating between the mucocutaneous *Candida* infections and the deep-seated candidiasis forms. The possibility of using antibody detection as a valid diagnostic means is dependent on the specific serological test, which measures the response to a specific candidal antigen rather than a specific immunoglobulin type. It should be emphasized, however, that precipitins detected by the immunodiffusion or counterelectrophoresis tests represent generally IgG antibodies, while other tests, such as hemagglutination, used in some laboratories, can detect the different immunoglobulin types (Fegeller et al. 1990).

Tests for detection of antigens

Detection of *Candida* antigens in body fluids is an important diagnostic tool, particularly in the immunocompromised patients who have difficulties in mounting antibodies at a detectable level (Walsh and Chanock 1997).

Although various techniques have been described (de Bernardis et al. 1993; Reboli 1993), currently the most widely used tests include LA and ELISA (Fujita and Hashimoto 1992). Some of these are commercially available, such as the Cand-Tec or Pastorex LA systems or the lately introduced Platelia system, a modified ELISA method. Antigen is generally detected in serum; however, it can be found also in other body fluids, such as urine (Ferreira et al. 1990). The antigens detected by the different systems include mannan (Pastorex LA), an undefined glycoprotein (Cand-Tec LA), a 47 kDa protein (HSP), enolase (ELISA), and specific mannosides (Platelia) (Walsh and Chanock 1997; Sendid et al. 2002). Generally, the antigen-detection tests are more significant for diagnosis as they detect active infection. Although as a rule they are sensitive tests, they still may not be sensitive enough to detect all the cases of antigenemia, particularly since some of the antigens are cleared quickly from the bloodstream. Another source for false-negative results may come from cases where the infective agent is one of the rarer *Candida* spp., due to the specificity of the antigenic determinants, which are absent or not recognized by the detecting reagent, which is generally an anti-*C. albicans* serum. It has been suggested that concurrent determination of both antigens and antibodies in clinical samples may increase the sensitivity of the diagnosis of systemic candidiasis (Sendid et al. 2002).

CMI as evaluated in vivo by a skin test to measure DTH to candidal antigen, or by in vitro assays such as lymphocyte transformation, are valid tools in assessing the immune competence of the patient, but do not contribute much to diagnosis of candidiasis.

In summary, although the immunodiagnostic techniques are associated with problems of standardization and interpretation, and cannot be used as the sole methods for laboratory diagnosis, they are, nevertheless, a helpful adjunct in the complex diagnosis of deep-seated candidiasis. The use of a combination of antigen-detection and antibody-detection tests is helpful in diagnosing the infection in both the immunocompetent and the non-immunocompetent patient.

Other nonculture methods

Since, as indicated above, in spite of the existence of culture and immunodiagnostic techniques, the diagnosis of deep-seated candidiasis is still problematic, the search for more sensitive and/or more specific methods continued. These included the possibility of detecting non-antigenic fungal metabolites released during infection into the patients' body fluids and, more recently, detection of fungal nucleic acids by sensitive molecular techniques.

Detection of fungal metabolites

Most studies in this area concentrated on detection of arabinitol as a candidal metabolite which is released during infection into the patient's body fluids, and which could possess a diagnostic value (Reiss and Morrison 1993). As arabinitol was first described by Kiehn and Bernard in 1979 as a fungal metabolite detected in the serum of patients with disseminated candidiasis, the techniques to detect this metabolite are in general applicable for serum, although arabinitol can be found also in urine. Arabinitol is released by most of the pathogenic *Candida* species, except *C. krusei* and *C. glabrata* (and not by *Cryptococcus neoformans*).

The techniques to detect arabinitol include the original method by which the metabolite was first detected, gas-liquid chromatography (GLC), and the more recent techniques involving enzymatic-fluorometric measurements (Reiss and Morrison 1993; Switchenko et al. 1994). Moreover, a kit for the latter technique has been described (Tokunaga et al. 1992). Serum arabinitol levels can also be elevated in patients with renal insufficiency and some other clinical conditions. In order to overcome this problem, determination of the serum arabinitol/creatinine ratio is advised.

Another *Candida* metabolite, suggested by Miyazaki et al. (1995) to be useful for diagnosis, is the cell wall component (1→3)-beta-D-glucan for which commercial kits already exist (Hossain et al. 1997). It should, however, be remembered that glucan is found in other fungi as well, therefore its value as a diagnostic tool is debatable (Verweij et al. 1998; Chryssanthou et al. 1999).

Molecular biology techniques

These techniques are based on detection of candidal DNA in patients' body fluids involving the use of specific DNA probes. This approach became more promising as a possible diagnostic tool with the development of DNA amplification techniques, particularly PCR. Recent review articles suggest the usefulness of such methodologies in diagnosis (Reiss et al. 2000; Turin et al. 2000).

Various possible probes have been described, such as the actin gene, the gene encoding cytochrome P450 14-lanosterol demethylase (Burgener Kairuz et al. 1994), the chitin synthetase gene (Jordan 1994), SAP genes (Flahaut et al. 1998), mitochondrial DNA or the candidal DNA repetitive elements (CARE). A widely used probe consists of the rRNA gene complex (Maiwald et al. 1994; Kappe et al. 1998). Some of these probes are species specific while others have wider specificity (Niesters et al. 1993). Although most of the studies aim for fungal detection in serum, some involve other clinical specimens such as urine or tracheal aspirate.

As already indicated in the section on Other identification methods above, such DNA probes can also be used for species identification after these have been isolated in culture. An example of such an approach is a study by Weissman et al. (1995) who used the benomyl resistance gene for *C. albicans* molecular identification.

Successful use of molecular techniques for detection of the fungal DNA in clinical specimens could possibly provide a sensitive means for early diagnosis. However, presently, these techniques are still mostly at the investigatory stage, and do not provide easily applicable methods for routine clinical use.

TREATMENT

Susceptibility testing

Susceptibility testing of fungi to antimycotic drugs, including that of yeasts, was until recently performed only rarely. This was a result of several factors:

- the cases necessitating systemic antifungal treatment were relatively rare
- the number of antifungal drugs was limited
- the incidence of resistance to the drugs was rare
- most importantly, correlation between in vitro and in vivo results was questionable.

These factors have changed: a marked increase in the incidence of mycotic diseases, especially disseminated infections, has been noted, the number of available drugs has increased, and continues to increase, and an innate or acquired resistance of certain yeast species to several drugs has been reported. Thus, collaborative efforts have been initiated, which have culminated in a proposal of a standard for broth dilution susceptibility testing of yeasts (but so far not yeast forms of dimorphic fungi) published as National Committee for Clinical Laboratory Standards (NCCLS) document M27-A2 (National Committee for Clinical Laboratory Standards 2002). The acceptance and use of these standards may, in analogy with similar standards for antibacterial drugs, lead to a better predictive value of in vitro susceptibility testing of antimycotic drugs.

The performance of in vitro fungal susceptibility testing poses considerable difficulties, including the test's procedure (i.e. necessity of adjusting the medium to the drug, estimation of inoculum size) and particularly their interpretation (Cook et al. 1990; Odds 1993). Several additional problems make reliable antifungal susceptibility testing complicated. Among these are:

- the fact that different morphological cell forms of the same fungus may show different susceptibility to the same drug (azoles seem to be more efficient in inhibiting hyphal forms than blastospores of *C. albicans* (Fromtling 1988; Walsh and Pizzo 1988)
- low water solubility and diffusibility of the drugs as well as their instability (especially polyenes)
- the determination of endpoint minimum inhibitory concentration (MIC) in liquid media, is complicated for some drugs, especially 5-fluorocytosine and the azoles, by the 'trailing' phenomenon: a certain turbidity is noted in the broth above the MIC, indicating a partial growth inhibition (Espinel-Ingroff et al. 1992; Fromtling et al. 1993). To overcome this problem, it has been suggested that it is best to ignore the slight turbidity above the MIC or to use a 1:5 dilution of the drug-free control culture as a reference of 80 percent growth inhibition (Espinel-Ingroff et al. 1992).

Several methods may be implemented in the testing of fungal susceptibility, but only one, the macrodilution technique, has been standardized so far, and is detailed in the NCCLS publication.

Other methods include various microdilution techniques, the results of which, according to preliminary investigations (Espinel-Ingroff et al. 1992, 1995; Tiballi et al. 1995; To et al. 1995), do not diverge significantly from those of the macrodilution method. Tests on solid media include the agar dilution and agar diffusion methods. The diffusion method may be implemented by using drug-impregnated disks or strips, the latter giving an indication of the MIC. This is the principle of the commercially available Etest (Figure 30.7). In a comparison between results obtained by the Etest and the NCCLS standard method, differences for certain species or specific drugs were shown and the comparison varied, according to the drug tested, between 71 percent and 84 percent (Colombo et al. 1995).

In spite of these advances in standardization of antifungal susceptibility test methods, the most troublesome problem remains the predictive value of the results for therapy. In fact, discrepancies between in vitro results and therapy success still afflict antibacterial susceptibility tests, more than a decade after their standardization, making them useful primarily for the indication of drugs to which the tested microorganism is resistant and thus probably useless for therapeutic purposes. The increase in amphotericin B (Dick et al. 1980; Hadfield et al. 1987) and/or azole-resistant (see section on Prophylaxis below) clinical isolates of yeasts, suggests that indication of possible resistance will presumably also be the primary function of antimycotic susceptibility tests.

Therapy

The therapy of candidiasis, similar to laboratory diagnosis, depends on the type of clinical manifestation and is different for the mucocutaneous forms as opposed to the deep-seated entities. Thus, the topic of therapy will be discussed in the light of the two groups of clinical forms (Richardson and Warnock 1997; Edwards 2000).

THERAPY OF MUCOCUTANEOUS CANDIDIASIS

Patients suffering from cutaneous or mucosal *Candida* infections can in principle be successfully treated topically. The specific drugs include, primarily, the polyene nystatin, and the azoles miconazole, clotrimazole, and ketoconazole (Georgopapadakou and Walsh 1994). The polyene amphotericin B in a formulation for topical administration, and additional azole compounds, such as econazole, sulconazole, and terconazole, have also been used for treatment of mucocutaneous candidiasis (Richardson and Warnock 1997; Edwards 2000; Stevens and Bennett 2000).

Nystatin (Brajtburg et al. 1990) is poorly soluble in water, and the specific formulation may vary according to the clinical entity. Thus, nystatin can be administered as a suspension or gel for treatment of oral candidiasis, and it can also be applied in the form of a cream or suppository, for dermal or vaginal infections, respectively. The duration of treatment may also vary according to the clinical entity. So for nystatin, suppositories of 100 000 units/ml, a 14-day course, is recommended in the treatment of vaginitis (Bennett 1995), while the treatment of oral infection by nystatin suspension of similar concentration may require 2–3 weeks (Richardson and Warnock 1997). For the treatment of cutaneous candidiasis, such as candidal intertrigo or diaper rash, nystatin may be used as topical preparation in combination with anti-inflammatory agents. *Candida* nail infections do not, generally, respond to topical nystatin treatment, although such treatment may be helpful for the accompanying paronychia. Nystatin, as other polyenes, acts on the fungal cytoplasmic membrane through complexing with the ergosterol in the membrane. In the form of topical preparations it is nontoxic and usually devoid of side effects, except for the bitter taste of the suspension used in treatment of oral infection.

The azoles (Fromtling 1988) miconazole, clotrimazole, and ketoconazole can also be used in different formulations, such as lotion, cream, suppository, powder, or spray. They can be applied for the treatment of both cutaneous and mucosal involvement, but are not useful in treating nail infections. The advantage of the azole compounds over nystatin is that the necessary treatment duration is shorter. So a treatment regimen for candidal vulvovaginitis with miconazole cream would require a 7-day use, and for some azoles (e.g. tioconazole) even a

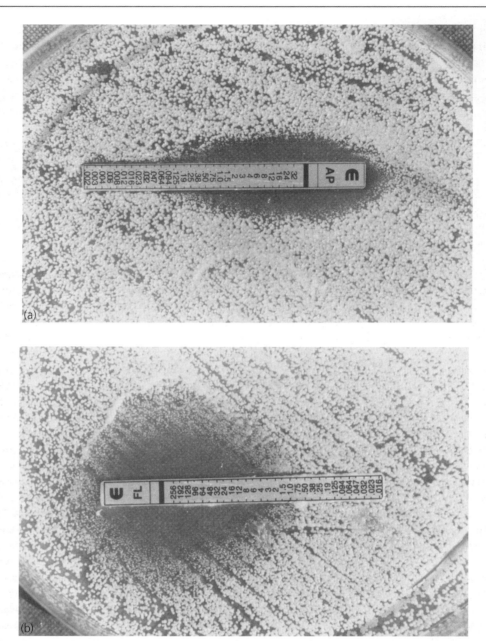

Figure 30.7 C. albicans *susceptibility testing by Etest:* **(a)** *amphotericin B (AP);* **(b)** *fluconazole (FL). Note the partial inhibition zone around the fluconazole strip.*

single-day treatment could be sufficient in comparison to a 14-day course for nystatin. An additional advantage of the azoles is their broad antifungal spectrum, which permits the use of these drugs even without laboratory confirmation of the candidal etiology of the infection. The azoles, as the polyenes, act on the fungal cytoplasmic membrane, however, through another mechanism – interference in the biosynthesis of ergosterol. Similar to the polyenes, they too, basically lack side effects when used as topical formulations.

Topical treatment will not suffice in nail infections, in CMC patients, in AIDS patients with *Candida* infections, or in cases of recurrent vulvovaginitis. These may require systemic oral therapy. The oral administration of the drugs can replace or be an adjunct to the topical treatment. The major antifungals administered orally for treatment of mucocutaneous *Candida* infections are primarily the triazoles: fluconazole and itraconazole (Bennett 1995). The use of these drugs is currently preferred over that of the oral preparation (tablets) of ketoconazole, as the use of the latter (see below) is associated with more adverse side effects and is not recommended for the treatment of vulvovaginitis (Stevens and Bennett 2000).

Tablets of fluconazole (Zervos and Meunier 1993) have been extensively used for therapy of oral and esophageal candidiasis in AIDS patients (Dupont and Drouhet 1988). Administration of 50–100 or 100–200 mg/day for 2 weeks is effective in the treatment of oral or esophageal candidiasis, respectively. However, the problem in these patients is the high rate of relapses of the infection. Thus, fluconazole has been used for prolonged periods, as a prophylactic means of preventing relapses of the infection (Just-Nubling et al. 1991). Long-term administration and prophylactic use of fluconazole are also applied for the management of CMC and prevention of recurrences in cases of recurrent vaginal infections, respectively.

Oral use of fluconazole is basically not associated with major side effects (nausea or abdominal pain have been noted in a small percentage of individuals; Bennett 1995). However, prolonged administration of the drug can lead to induction of resistance and was expressed in the appearance of fluconazole-resistant *C. albicans* isolates (Sanguineti et al. 1993; Newman et al. 1994). In addition, fluconazole was also found to be inherently ineffective against certain *Candida* spp., such as *C. krusei* and *C. glabrata* (Hitchcock et al. 1993; Richardson and Warnock 1997). As a consequence, *Candida* speciation seems to be important in regard to therapy and it would be advisable to perform sensitivity testing prior to commencement of treatment.

Itraconazole (Sporanox) has a similar profile of use: AIDS patients, CMC, or vaginitis (Rees and Philips 1992). It has also been applied in treatment of *Candida* nail infections (Hay et al. 1988). As with fluconazole, itraconazole use is apparently devoid of significant adverse reactions. Cross-resistance with fluconazole has been reported (Richardson and Warnock 1997).

As indicated above, the application of the two triazoles is currently preferred over the oral use of ketoconazole (Nizoral), the first azole for systemic administration, due to possible side effects associated with the use of this drug. Administration of ketoconazole can be, albeit rarely, associated with severe hepatotoxicity (Lewis et al. 1984). Ketoconazole use can also interfere with steroid hormone synthesis and be expressed in reduced levels of testosterone (Hanger et al. 1988). Milder, but much more frequent, side effects include nausea and vomiting, at higher doses of the drug. Another problem associated with ketoconazole use is the relatively low absorption of the drug in AIDS patients due to lower gastric acidity in these patients (Richardson and Warnock 1997).

Nystatin can also be used in the form of tablets. Since, as indicated above, nystatin is poorly soluble in water, it is not well absorbed by the GI mucosa. Nystatin tablets have been used as an adjunct to topical treatment in the therapy of recurrent vaginal candidiasis. The rationale of this therapeutic approach is that oral use of nystatin may lead to clearance or reduction of the *Candida* load in the GI tract, and thereby eliminate a possible reservoir for relapse of the vaginal infection.

In CMC patients, who tend to develop relapses, which are explained by their underlying immune defect that renders them susceptible to the infection, an additional therapeutic approach has been investigated. Attempts have been made to potentiate the immunoreactivity of such patients by use of immunomodulatory agents (Kirkpatrick 1993), such as 'transfer factor,' levamisole, or thymosine. The most extensively studied immunomodulator was the transfer factor, a leukocyte extract, which was able to transfer to CMC patients the ability to mount a DTH response, an ability that they lacked.

Additional antimycotic drugs, such as the allylamine terbinafine (Perea et al. 2002), have been shown to have in vitro activity against some *Candida* spp., but have not been widely adopted for clinical use to treat infections with the fungus.

DEEP-SEATED CANDIDIASIS

It is evident that therapy of the deep-seated forms of *Candida* infections requires systemic administration of the antimycotics. The gold standard of antifungals suitable for treatment of systemic mycoses was, and still is, amphotericin B (AMB) (Sarosi 1990). AMB (Fungizone) has a wide antifungal spectrum, it is effective against molds, yeasts, and dimorphic fungi (Bennett 1995). Although, as a rule, resistance to AMB is not a phenomenon restricting the use of this agent, reports on resistant strains of *C. tropicalis* and *C. lusitaniae* are found in the literature (Richardson and Warnock 1997). Resistance of *C. lusitaniae* to AMB may emerge during treatment (Pappagianis et al. 1979; Merz 1984).

AMB is a polyene that acts like other polyenes, through binding to the ergosterol in the fungal cytoplasmic membrane (Georgopapadakou and Walsh 1994), which leads to increased membrane permeability, and subsequently to fungal cell death. As other polyenes, it is water insoluble and consequently not well absorbed from the GI tract and not applicable for oral use, but has to be administered parenterally. The clinically applicable parenteral AMB formulation consists of a lyophilized powder of 50 mg AMB, 41 mg sodium deoxycholate and 25 mg of sodium phosphate, made up in water supplemented with 5 percent of glucose. This formulation is given in the form of a slow (several hours) IV infusion (Edwards 2000).

The dose and duration of AMB treatment of deep-seated candidiasis is dependent upon the specific clinical entity, and the category of patients involved: non-immunosuppressed, granulocytopenic, or otherwise immunosuppressed. Nevertheless, a protocol has been adopted, which in general terms can be outlined as involving treatment with 0.5–1 mg/kg/day of AMB for a duration of 6–10 weeks. In view of the possible side effects associated with AMB therapy, it is recommended to initiate the treatment at a low dose, and monitor the patient's response before dose increase.

Treatment with AMB is associated with considerable side effects. These may include acute reactions and cumulative effects. The acute reactions may include, fever, chills, hypotension, nausea, and vomiting. These side effects can be ameliorated or prevented by some supportive measures, such as use of steroids and/or Ibuprofen. The major risk of AMB treatment is its cumulative effect – the nephrotoxicity (Sabra and Branch 1990). AMB therapy causes renal impairment in most patients. Thus, AMB treatment requires careful monitoring of renal functions. Although the renal impairment can be reversible several months after cessation of treatment, irreversible damage does also occur. Other, less common, not immediate, effects may include anemia.

The problem of AMB therapy associated with its significant toxicity seems to be overcome to some extent by newer formulations involving AMB–lipid complexes (Janoff 1990; Meunier 1989; Fielding 1991). Three major formulations of this type are currently available: AMB with dimyristoyl phosphatidyl choline and dimyristoyl phosphatidyl glycerol (ABLC, Abelcet), AMB with cholesterol sulfate (ABCD, Amphocil), and a liposomal AMB preparation (AmBisome) (Chopra et al. 1991). The AMB–lipid complexes are characterized by lower nephrotoxicity. They can, therefore, be administered at significantly higher doses (5 mg/kg/day), doses which are not tolerated with the traditional AMB formulation. The major drawback of these formulations is their current extremely high cost, which hampers their use on a wider scale. It should be added, in this context, that trials with mixtures of AMB and Intralipid (Chavanet et al. 1992; Caillot et al. 1993; Nath et al. 1999), which significantly lower the cost of the treatment, have been carried out.

Disseminated candidiasis can also be treated by a combination of AMB and 5-fluorocytosine (5-FC) (Polak 1987). 5-FC (flucytosine, Ancobon) is a fluorinated pyrimidine, which acts as an antimetabolite (Diasio et al. 1978). Following deamination it converts into 5-fluorouracil (5-FU), which is incorporated into RNA resulting, subsequently, in impaired protein synthesis. Additional conversion of 5-FU into 5-fluorodeoxyuridilic acid, which can act as an inhibitor of thymidilate synthetase, may result in inhibition of DNA synthesis.

5-FC is water soluble and absorbed quite well by the GI mucosa following oral administration (tablets, capsules); it is usually given at a dose of 150 mg/kg/day (Bennett 1995). It is also well distributed in the body fluids, reaching significant levels in CSF. In patients with no underlying major renal dysfunction, 5-FC treatment leads rarely to side effects, which may include nausea, vomiting, and diarrhea. However, more serious side effects, such as leukopenia, thrombocytopenia, and hepatotoxicity, are also known, particularly among patients with underlying renal or hematological impairments. In such patients careful monitoring is required. The major disadvantage of 5-FC treatment is the frequent occurrence of resistant strains during the course of treatment, and even the presence of primary resistant strains (Stiller et al. 1982) being a recognized phenomenon. Thus, laboratory assessment of the fungal isolate's sensitivity to 5-FC prior to commencement and during the course of treatment, as well as careful monitoring of serum levels of the drug, is recommended.

5-FC is generally not administered as a single therapeutic agent, but rather given in combination with AMB. It is believed that the combined treatment may enable the use of lower AMB doses, thereby reducing possible side effects, without compromising its therapeutic efficacy. The combined treatment is particularly recommended in Candida meningitis, due to the high passage of 5-FC through the blood–brain barrier.

The triazoles fluconazole and itraconazole have been considered as possible alternatives to AMB, or additional agents in the treatment of deep-seated candidiasis. The triazoles act as the other azoles (imidazoles) by interfering in the biosynthesis of ergosterol, through inhibition of lanosterol demethylation, involving binding to cytochrome P450 enzymes. The advantage of the triazoles over the imidazoles is that they affect less the synthesis of human sterols (Hanger et al. 1988) and, therefore, are less associated with side effects. As a consequence, use of triazoles would seem preferable over the imidazole ketoconazole, which is also suitable for systemic (oral) use and could be considered for treatment of deep-seated candidiasis. (Kauffman et al. 1991, 1993; Meunier et al. 1992).

Fluconazole is active against most Candida spp. However, resistant C. albicans strains have been isolated, primarily from AIDS patients under prolonged fluconazole treatment (Richardson and Warnock 1997). Of clinical significance is also the phenomenon of certain species, such as C. krusei and C. glabrata, which reveal initial insensitivity to the drug. Epidemiologic data point to an increase in incidence of systemic candidiasis due to these species (Hazen 1995; Nguyen et al. 1996; Krcmery and Barnes 2002), which may possibly be the result of selective pressure of fluconazole use.

Fluconazole is available in two forms, tablets (50, 100, 200 mg) for oral use and a formulation for IV use. Fluconazole is quite soluble and well absorbed from the GI tract after oral administration (Bennett 1995). It is also well distributed in the human body sites, including the CNS; its level in the CSF can reach 70 percent of that in the serum. Its use is rarely followed by serious side effects, but nausea, vomiting, and diarrhea have been recorded in a small percentage of treated individuals.

Fluconazole dose and treatment duration depend on the specific clinical entity and take into consideration the patient involved. Generally, doses of 200–400 mg/day (higher doses of up to 800 mg/day have also been reported) for 6–8 weeks are used. Of the deep-seated Candida infections, fluconazole has been used extensively and successfully in treatment of esophagitis; it has also been applied in treatment of endocarditis, hepatosplenic candidiasis or in candidemia of non-immunosupressed patients (Bennett 1995).

Itraconazole can be used orally in the form of capsules and, more recently, as solution and parenterally by IV administration. It is less well absorbed from the GI tract than fluconazole, and crosses the blood–brain barrier less well. Cross-resistance with fluconazole does, apparently, exist, and reports on resistance of *C. glabrata* have been published (Dermoumi 1994). Itraconazole has been used in treatment of *Candida* esophagitis. In view of the recent introduction of the IV formulation, its role in other deep-seated *Candida* infections is not yet clear.

It is noteworthy that voriconazole, the most novel licensed triazole, has activity against non-*albicans* species resistant to fluconazole, such as *C. glabrata* and *C. krusei*.

Another more recent antifungal drug approved for use in treatment of candidemia is the lipopeptide (echinocandin) caspofungin. Caspofungin is a semisynthetic compound acting as a glucan synthesis inhibitor. It is administered intravenously and is the first antifungal drug active on the fungal cell wall. Combinations of caspofungin and azole are currently being evaluated by in vitro studies (Roling et al. 2002).

At present, most clinicians feel that AMB therapy, particularly that using AMB–lipid complexes, is still preferable for treatment of systemic candidiasis in severely ill immunosuppressed patients.

Immunotherapy as an adjunct to antimicrobial treatment has also been undertaken. This included the use of cytokines, such as for example, granulocyte colony-stimulating factor (G-CSF). However, as these attempts were sporadic and limited, it is difficult to draw conclusions as to the efficacy of this approach (Stevens 1998; Edwards 2000).

Prophylaxis

Vaccination is considered a most effective measure for prevention of microbial diseases. The last decade has seen several attempts to prevent *Candida* infections in humans by employing anti-*Candida* antibodies (Rigg et al. 2001). However, this approach is still experimental and has not yet reached the stage of a commercial vaccine. Thus, other alternatives, primarily the use of antifungals on a prophylactic basis in patients at risk, seem to be a reasonable approach. Since one of the principal aims in the earlier studies was to reduce the fungal load, particularly in the GI tract, which is a major source for *Candida*, and thereby reduce the risk of infection, nystatin was used in these attempts.

Later studies, with few exceptions (Damjanovic et al. 1993), focused on the triazoles (Thunnissen et al. 1991; Wingard et al. 1991, 1993; Goodman et al. 1992; Cesaro et al. 1993; Winston et al. 1993; Chandrasekar and Gatny 1994), mostly fluconazole (Thunnissen et al. 1991). The population groups in whom prophylactic attempts have been undertaken include, primarily, leukemia patients (Winston et al. 1993), neutropenic

individuals (Cesaro et al. 1993), bone marrow transplants (Wingard et al. 1991), and neonates in intensive care units (Damjanovic et al. 1993), all of which are considered at high risk for the development of candidiasis.

These studies indicated that prophylactic use of antifungals could reduce *Candida* colonization (Chandraseker and Gatny) or GI carriage (Damjanovic et al. 1993), and possibly also infection (Goodman et al. 1992). However, they also pointed to the important observation that the possible preventive effect was mostly noted in regard to *C. albicans* and not to the non-*albicans* species: *C. krusei* (Wingard et al. 1991; Goodman et al. 1992; Winston et al. 1993), *C. guilliermondii*, *C. parapsilosis* (Cesaro et al. 1993), and *C. glabrata* (Wingard et al. 1993; Chandrasekar and Gatny 1994). Thus, an increase in candidiasis due to non-*albicans* species has been noted in at-risk patients (Hazen 1995; Nguyen et al. 1996; Kremery and Barnes 2002).

It is noteworthy that, while the use of nystatin would have an effect solely in preventing *Candida* infections, the use of triazoles is expected to have a broader spectrum of activity. While this approach still continues, studies comparing the efficacy and toxicity of different compounds have been published (Walsh 2002). Walsh and colleagues compared voriconazole and liposomal AMB in neutropenic patients and reached the conclusion that 'voriconazole is a suitable alternative to amphotericin B preparations for empirical therapy in patients with neutropenia.' Another study by Wolff et al. (2000), which compared fluconazole with low-dose AMB in bone marrow transplants, found both antifungals have been effective in prophylaxis, with better tolerance of the triazole.

A more recent study of Bow et al. (2002) undertook a meta-analysis including several azoles and AMB, reaching the conclusion that antifungal prophylaxis reduced morbidity and mortality caused by fungi.

It has recently been indicated (Sobel 2003) that the echinocandins, being characterized by their action on the fungal cell wall, show promise for prophylactic use in high-risk patients.

In addition, other possible prophylactic measures that have been suggested include a more cautious use of venous catheters, a risk factor for candidemia (Nucci 2003), use of immunomodulatory means to boost the immune system such as granulocyte transfusions (Van Burik 2003) or cytokines (e.g. G-CSF) (Rodriguez-Adrian et al. 1998; Stevens 1998).

REFERENCES

Abou-Gabal, M., Hogle, R.M. and West, J.K. 1977. Pyometra in a mare caused by *Candida rugosa*. *J Am Vet Assoc*, **170**, 177–8.
Aho, R. and Hirn, P. 1981. A survey of fungi and some indicator bacteria in chlorinated water of indoor public swimming pools. *Zentralbl Bakteriol Mikrobial Hyg 1B*, **173**, 242–9.

Alexander, B.D. 2002. Diagnosis of fungal infection: new technologies for the mycology laboratory. *Transpl Infect Dis*, **4**, Suppl 3, 32–7.

Al-Mosaid, A., Sullivan, D. and Sulkin, I.F. 2001. Differentiation of *Candida dubliniensis* from *Candida albicans* on Staib agar and caffeic acid-ferric citrate agar. *J Clin Microbiol*, **39**, 323–7.

Altboum, Z. 1994. Genetic studies in *Candida albicans*. In: Segal, E and Baum, G.L. (eds), *Pathogenic yeasts and yeast infections*. Boca Raton, FL: CRC Press, 33–48.

Amici, G., Grandesso, S., et al. 1994. Fungal peritonitis in peritoneal dialysis: critical review of six cases. *Adv Perit Dial*, **10**, 169–73.

Anderson, J., Mihalik, R. and Soll, D.R. 1990. Ultrastructure and antigenicity of the unique cell wall pimple of the Candida opaque phenotype. *J Bacteriol*, **172**, 224–35.

Anderson, P.G. and Pidgeon, G. 1987. Candidiasis in a dog with parvoviral infection. *J Am Anim Hosp Assoc*, **23**, 27–30.

Asakura, K., Iwaguchi, S., et al. 1991. Electrophoretic karyotypes of clinically isolated yeasts of *Candida albicans* and *C. glabrata*. *J Gen Microbiol*, **137**, 2531–8.

Auger, P., Dumas, C. and Joly, J. 1979. A study of 666 strains of *Candida albicans*: correlation between serotype and susceptibility to 5-fluorocytosine. *J Infect Dis*, **139**, 590–4.

Bachrach, G., Banai, M., et al. 1994. Brucella ribosomal protein L7/L12 is a major component in the antigenicity of Brucellin INRA for delayed-type hypersensitivity in *Brucella* sensitized guinea pigs. *Infect Immun,: 5361-5366*, **62**, 5361–6.

Bahn, Y.S. and Sundstrom, P. 2001. CAP1, an adenylate cyclase-associated protein gene, regulates bud-hypha transitions, filamentous growth, and cyclic AMP levels and is required for virulence of *Candida albicans*. *J Bacteriol*, **183**, 3211–23.

Baillie, G.S. and Douglas, L.J. 2000. Matrix polymers of Candida biofilms and their possible role in biofilm resistance to antifungal agents. *J Antimicrob Chemother*, **46**, 397–403.

Baley, J.E., Kligman, R.M., et al. 1986. Fungal colonization in the very low birth weight infant. *Pediatrics*, **78**, 225–32.

Barns, S.M., Lane, D.J., et al. 1991. Evolutionary relationships among pathogenic *Candida* species and relatives. *J Bacteriol*, **173**, ?2250–2255.

Baum, G.L. 1994. Epidemiology, pathogenesis and immunology. In: Segal, E. and Baum, G.L. (eds), *Pathogenic yeasts and yeast infections*. Boca Raton, FL: CRC Press, 89–92.

Beemer, A.M., Kutin, E.S. and Katz, Z. 1973. Epidemic venereal disease due to *Candida albicans* in geese in Israel. *Avian Dis*, **17**, 639–49.

Bendel, C.M. and Hostetter, M.K. 1993. Distinct mechanisms of epithelial adhesion for *Candida albicans* and *Candida tropicalis*. Identification of the participating ligands and development of inhibitory peptides. *J Clin Invest*, **92**, 1840–9.

Bennett, J.E. 1995. Antifungal agents. In: Mandell, G.L., Bennett, J.E. and Dolin, R. (eds), *Principles and practice of infectious diseases*, 4th edition. New York: Churchill Livingstone, 401–10.

Bodey, G.P. (ed.). 1993. *Candidiasis: pathogenesis, diagnosis, and treatment*, 2nd edition. New York: Raven Press.

Boiron, P., Agis, F. and Nguyen, V.H. 1983. A study of the yeast flora of medical interest on Saint Anne beach in Guadeloupe. *Bull Soc Pathol Exot Fil*, **76**, 351–6.

Bolivar, R. and Bodey, G.P. 1985. Candidiasis of the intestinal tract. In: Bodey, G.P. and Fainstein, V. (eds), *Candidiasis*. New York: Raven Press, 181–201.

Borg, M. and Ruchel, R. 1990. Demonstration of fungal proteinase during phagocytosis of *Candida albicans* and *Candida tropicalis*. *J Med Vet Mycol*, **28**, 3–14.

Borg-von Zepelin, M., Meyer, I., et al. 1999. HIV-Protease inhibitors reduce cell adherence of *Candida albicans* strains by inhibition of yeast secreted aspartic proteases. *J Invest Dermatol*, **113**, 747–51.

Bow, E.J., Laverdiere, M., et al. 2002. Antifungal prophylaxis for severely neutropenic chemotherapy recipients: a meta analysis of randomized-controlled clinical trials. *Cancer*, **94**, 3230–46.

Brajtburg, J., Powderly, W.G., et al. 1990. Amphotericin B: delivery system. *Antimicrob Agents Chemother*, **34**, 381–4.

Branchini, M.L., Pfaller, M.A., et al. 1994. Genotypic variation and slime production among blood and catheter isolates of *Candida parapsilosis*. *J Clin Microbiol*, **32**, 452–6.

Braude, I.A. 1986. Candida. In: Braude, A.I., Davis, C.E. and Fiere, J. (eds), *Infectious diseases and medical microbiology*. Philadelphia: W.B. Saunders, 571–7.

Brawner, D.L., Anderson, G.L. and Yuen, K.Y. 1992. Serotype prevalence of *Candida albicans* from blood culture isolates. *J Clin Microbiol*, **30**, 149–53.

Brieland, J., Essig, D., et al. 2001. Comparison of pathogenesis and host immune responses to *Candida glabrata* and *Candida albicans* in systemically infected immunocompetent mice. *Infect Immun*, **69**, 5046–55.

Buchheidt, D., Skladny, H., et al. 2000. Systemic infections with *Candida* sp. and *Aspergillus* sp. in immunocompromised patients with hematological malignancies: Current serological and molecular diagnostic methods. *Chemotherapy*, **46**, 219–28.

Burgener-Kairuz, P., Zuber, J.P., et al. 1994. Rapid detection and identification of *Candida albicans* and *Torulopsis (Candida) glabrata* in clinical specimens by species-specific nested PCR amplification of a cytochrome P-450 lanosterol-alpha-demethylase (L1A1) gene fragment. *J Clin Microbiol*, **32**, 1902–7.

Burmeister, H.R. and Hartman, P.A. 1966. Yeasts in ensiled high moisture corn. *Appl Microbiol*, **14**, 35–8.

Caillot, D., Casasnovas, O., et al. 1993. Efficacy and tolerance of an amphotericin B lipid (Intralipid) emulsion in the treatment of candidemia in neutropenic patients. *J Antimicrob Ther*, **31**, 161–3.

Calderone, R.A. and Braun, P. 1991. Adherence and receptor relationship of *Candida albicans*. *Microbiol Rev*, **55**, 1–20.

Calderone, R.A. and Fonzi, W.A. 2001. Virulence factors of *Candida albicans*. *Trends Microbiol*, **9**, 327–35.

Calderone, R., Diamond, R., et al. 1994. Host cell-fungal cell interaction. *J Med Vet Mycol*, **32**, Suppl 1, 151–68.

Cancelas, J.A., Lopez, J., et al. 1994. Native valve endocarditis due to *Candida parapsilosis*: a late complication after bone marrow transplantation-related fungemia. *Bone Marrow Transplant*, **13**, 333–4.

Cannon, R.D., Holmes, A.R., et al. 1995. Oral *Candida*: clearance, colonization, or candidiasis. *J Dent Res*, **74**, 1152–61.

Carruba, G., Pontieri, E., et al. 1991. DNA fingerprinting and electrophoretic karyotype of environmental and clinical isolates of *Candida parapsilosis*. *J Clin Microbiol*, **29**, 916–22.

Cassone, A. 1989. Cell wall of *Candida albicans*: its function and impact on the host. *Curr Top Med Mycol*, **3**, 248–314.

Cassone, A. and De Bernardis, F. 1998. Immunogenic and protective *Candida albicans* constituents. *Res Immunol*, **149**, 289–99.

Cenci, E. and Mencacci, A. 1998. IFN-gamma is required for IL-12 responsiveness in mice with *Candida albicans* infection. *J Immunol*, **161**, 3543–50.

Cesaro, S., Rossetti, F., et al. 1993. Fluconazole prophylaxis and *Candida* fungemia in neutropenic children with malignancies. *Haematologica*, **78**, 249–51.

Chandra, J., Kuhn, D.M., et al. 2001. Biofilm formation by the fungal pathogen *Candida albicans*: development, architecture, and drug resistance. *J Bacteriol*, **183**, 5385–94.

Chandrasekar, P.H. and Gatny, C.M. 1994. The effect of fluconazole prophylaxis on fungal colonization in neutropenic cancer patients. Bone Marrow Transplantation Team. *J Antimicrob Chemother*, **33**, 309–18.

Chauhan, N., Li, D., et al. 2002. The cell wall of *Candida* spp.. In: Calderone, R. (ed.), *Candida and candidiasis*. Washington, DC: ASM Press, 159–75.

Chavanet, P.Y., Garry, I., et al. 1992. Trial of glucose versus fat emulsion in preparation of amphotericin for use in HIV infected patients with candidiasis. *Br Med J*, **305**, 921–5.

Chengappa, M.M., Maddux, R.L., et al. 1984. Isolation and identification of of yeast and yeast-like organisms from clinical veterinary sources. *J Clin Microbiol*, **19**, 427–8.

Chibana, H., Magee, B.B., et al. 1998. A physical map of chromosome 7 of *Candida albicans*. *Genetics*, **149**, 1739–52.

Chibana, H., Beckerman, J.L. and Magee, P.T. 2000. Fine-resolution physical mapping of genomic diversity in *Candida albicans*. *Genome Res*, **10**, 1865–77.

Chopra, R., Blair, S., et al. 1991. Liposomal amphotericin B (Ambisome) in the treatment of fungal infections in neutropenic patients. *J Antimicrob Chemother*, **28**, Suppl B, 93–104.

Chryssanthou, E., Klingspor, L., et al. 1999. PCR and other non-culture methods for diagnosis of invasive candida infections in allogenic bone marrow and solid organ transplant recipients. *Mycoses*, **42**, 239–47.

Cohen, M. and Montgomerie, J.Z. 1993. Hematogenous endophthalmitis due to *Candida tropicalis*: report of two cases and review. *Clin Infect Dis*, **17**, 270–2.

Coleman, D.C., Sullivan, D.J., et al. 1997. Candidiasis: the emergence of a novel species, *Candida dubliniensis*. *AIDS*, **11**, 557–67.

Colombo, A.L., Barchiesi, F., et al. 1995. Comparison of Etest and National Committee for Clinical Laboratory Standards broth macrodilution method for azole antifungal susceptibility testing. *J Clin Microbiol*, **33**, 535–40.

Cook, R.A., McIntyre, K.A. and Galgiani, J.N. 1990. Effects of incubation temperature, inoculum size, and medium on agreement of macro- and microdilution broth susceptibility test results for yeasts. *Antimicrob Agents Chemother*, **34**, 1542–5.

Cormack, B.P., Ghori, N. and Falkow, S. 1999. An adhesin of the yeast pathogen *Candida glabrata* mediating adherence to human epithelial cells. *Science*, **285**, 578–82.

Cotter, G. and Kavanagh, K. 2000. Adherence mechanisms of *Candida albicans*. *Br J Biomed Sci*, **57**, 241–9.

Cross, R.F., Moorhead, P.D. and Jones, J.E. 1970. *Candida albicans* infection of the forestomachs of a calf. *J Am Vet Med Assoc*, **157**, 1525–30.

Cumming, C.G., Wight, C., et al. 1990. Denture stomatitis in the elderly. *Oral Microbiol Immunol*, **5**, 82–5.

Cunningham, D., Morgan, R.J., et al. 1985. Functional and structural changes of the human proximal small intestine after cytotoxic therapy. *J Clin Pathol*, **38**, 265–70.

Cutler, J.E. 1991. Putative virulence factors of *Candida albicans*. *Annu Rev Microbiol*, **45**, 187–218.

Cutler, J.E., Granger, B.L. and Hahn, Y. 2002. Immunoprotection against candidiasis. In: Calderone, R.A. (ed.), *Candida and candidiasis*. Washington, DC: ASM Press, 243–58.

Damjanovic, V., Connolly, C.M., et al. 1993. Selective decontamination with nystatin for control of a *Candida* outbreak in a neonatal intensive care unit. *J Hosp Infect*, **24**, 245–59.

D'Antonio, D., Fioritoni, G., et al. 1992. Hepatosplenic infection caused by *Candida parapsilosis* in patients with acute leukemia. *Mycoses*, **35**, 311–13.

Davis, C. 1986. CMI description sheets: set 88. *Mycopathologia*, **96**, 171–3.

De Bernardis, F. and Boccanera, M. 2002. Mucosal infection and immunity in candidiasis. In: Calderone, R.A. and Cihlar, R.L. (eds), *Fungal pathogenesis: principles and clinical applications*. New York: Marcel Dekker, 461–82.

De Bernardis, F., Girmenia, C., et al. 1993. Use of a monoclonal antibody in a dot immunobinding assay for detection of a circulating mannoprotein of *Candida* spp., in neutropenic patients with invasive candidiasis. *J Clin Microbiol*, **31**, 3142–6.

De Bernardis, F., Sullivan, P.A. and Cassone, A. 2001. Aspartyl proteinases of *Candida albicans* and their role in pathogenicity. *Med Mycol*, **39**, 303–13.

De Bernardis, F., Boccanera, M., et al. 2002. Intravaginal and intranasal immunizations are equally effective in inducing vaginal antibodies and conferring protection against vaginal candidiasis. *Infect Immun*, **70**, 2725–9.

de Hoog, G.S., Guarro, J. et al. (eds). 2000. Ascomycetous yeasts genus: *Candida*. In *Atlas of clinical fungi*, 2nd edition. Delft: Centraalbureau voor Schimmelcultures, 180–226.

de Repentigny, L. 1992. Serodiagnosis of candidiasis, aspergillosis and cryptococcosis. *Clin Infect Dis*, **14**, Suppl 1, S11–22.

Dermoumi, H. 1994. In vitro susceptibility of fungal isolates of clinically important specimens to itraconazole, fluconazole and amphotericin B. *Chemotherapy*, **40**, 92–8.

Desbrosse, A.M. 1994. Keratite mycose a *Candida albicans*. *Pratique Vet Equine*, **26**, 63–4.

de Viragh, P.A., Sanglard, D., et al. 1993. Cloning and sequencing of two *Candida parapsilosis* genes encoding acid proteases. *J Gen Microbiol*, **139**, 335–42.

Diamond, R.D. and Krzesicki, R. 1978. Mechanism of attachment of neutrophils to *Candida albicans* pseudohyphae in the absence of serum and subsequent damage to pseudohyphae by microbicidal process of neutrophils in vitro. *J Clin Invest*, **61**, 360–9.

Diamond, R.D., Lyman, C.A. and Wysang, D.A. 1991. Disparate effect of interferon-gamma and tumor necrosis factor-alpha on early respiratory burst and fungicidal responses to *Candida albicans* hyphae in vitro. *J Clin Invest*, **87**, 711–20.

Diasio, R.B., Bennett, J.E. and Myers, C.E. 1978. Mode of action of 5-flucrocytosine. *Biochem Pharmacol*, **27**, 703–7.

Dick, J.D., Merz, W.G. and Saral, R. 1980. Incidence of polyene resistant yeasts recovered from clinical specimens. *Antimicrob Agents Chemother*, **18**, 158–63.

Do Carmo Sousa, L. 1969. Distribution of yeasts in nature. In: Rose, A.H. and Harrison, J.H. (eds), , *The yeasts*, vol. 1. . London: Academic Press, 79–105.

Doebbeling, B.N. Hollis, R.J. et al. 1991. Restriction fragment analysis of a *Candida tropicalis* outbreak of sternal wound infections. *J Clin Microbiol*, **29**, 1268–70.

Doebbeling, B.N., Lehmann, P.F., et al. 1993. Comparison of pulsed-field gel electrophoresis with isoenzyme profiles as a typing system for *Candida tropicalis*. *Clin Infect Dis*, **16**, 377–83.

Doi, M., Homma, M., et al. 1992. Estimation of chromosome number and size by pulsed-field gel electrophoresis (PFGE) in medically important *Candida* species. *J Gen Microbiol*, **138**, 2243–51.

Doi, M., Mizuguchi, I., et al. 1994. Electrophoretic karyotypes of *Candida* yeasts recurrently isolated from single patients. *Microbiol Immunol*, **38**, 19–23.

Domer, J.E. 1989. Candida cell wall mannan: a polysaccharide with diverse immunologic properties. *CRC Crit Rev Microbiol*, **17**, 31–51.

Domer, J.E. and Carrow, E.W. 1989. Candidiasis. In: Cox, R.A. (ed.), *Immunology of the fungal diseases*. Boca Raton, FL: CRC Press, 57–92.

Domer, J.E. and Garner, R.E. 1989. Immunomodulation in response to *Candida*. *Immunol Ser*, **47**, 293–317.

Domer, J.E. and Lehrer, R.I. 1993. Introduction to *Candida*: systemic candidiasis. In: Murphy, J.W., Friedman, H. and Bendinelli, M. (eds), *Fungal infections and immune responses*. New York: Plenum Press, 49–116.

Domer, J.E., Murphy, J.W., et al. 1992. Immunomodulation in the mycoses *J Med Vet Mycol*, **30**, Suppl 1, 157–66.

Douglas, J. 1987. Adhesion of *Candida albicans* to epithelial surfaces. *CRC Crit Rev Microbiol*, **15**, 27–43.

Dromer, F., Chevalier, R., et al. 2002. Synthetic analogues of beta-1,2 oligomannosides prevent intestinal colonization by the pathogenic yeast *Candida albicans*. *Antimicrob Agents Chemother*, **46**, 3869–76.

Duarte, E.R., Resende, J.C., et al. 2001. Prevalence of yeasts and mycelial fungi in bovine parasitic otitis in the State of Minas Gerais, Brazil. *J Vet Med B Infect Dis Vet Public Health*, **48**, 631–5.

Dupont, B. and Drouhet, E. 1988. Fluconazole in the management of oropharingeal candidosis in in a predominantly HIV antibody-positive group of patients. *J Med Vet Mycol*, **26**, 67–71.

Eckstein, M., Barenholz, Y., et al. 1997. Liposomes containing *Candida albicans* ribosomes as a prophylactic vaccine against disseminated candidiasis in mice. *Vaccine*, **15**, 220–4.

Edmond, M.B., Wallace, S.E., et al. 1999. Nosocomial blood stream infections in United States hospitals: a three-year analysis. *Clin Infect Dis*, **29**, 239–44.

Edwards, J.E. 2000. *Candida* species. In: Mandell, G.L., Bennett, J.E. and Dolin, R. (eds), *Principles and practice of infectious diseases*, 5th edition. Philadelphia: Churchill Livingstone, 2656–74.

Egger, M., Hirschel, B., et al. 1997. Impact of new antiretroviral combination therapies in HIV infected patients in Switzerland: prospective multicentre study. Swiss HIV Cohort Study. *Br Med J*, **315**, 1194–9.

Elad, D. 1993. Epidemiology and diagnostic organization of veterinary mycology in Israel. *Microbiol Med*, **8**, 405–8.

Elad, D., Shpigel, N.Y., et al. 1995. Feed contamination with *Candida krusei* as a probable source of mycotic mastitis in dairy cows. *J Am Vet Med Assoc*, **207**, 620–2.

Elad, D., Brenner, J., et al. 1998. Yeasts in the gastrointestinal tract of preweaned calves and possible involvement of *Candida glabrata* in neonatal calf diarrhea. *Mycopathologia*, **141**, 7–14.

Elad, D., Brenner, J., et al. 2002. Influence of diet on the shedding of *C. glabrata* by experimentally infected preweaned calves. *Vet J*, **164**, 275–9.

el-Mohandes, A.E., Johnson-Robbins, L., et al. 1994. Incidence of *Candida parapsilosis* colonization in an intensive care nursery population and its association with invasive fungal disease. *Pediatr Infect Dis J*, **13**, 520–4.

Enweani, I.B., Obi, C.L. and Jokpeyibo, M. 1994. Prevalence of *Candida* species in Nigerian children with diarrhoea. *J Diarrhoeal Dis Res*, **12**, 133–5.

Eras, P., Goldstein, M.J. and Sherlock, P. 1972. Candida infection of the gastrointestinal tract. *Medicine*, **51**, 367–79.

Espinel-Ingroff, A., Kish, C.W. Jr, et al. 1992. Collaborative comparison of broth macrodilution and microdilution antifungal susceptibility tests. *J Clin Microbiol*, **30**, 3138–45.

Espinel-Ingroff, A., Rodriguez-Tudela, J.L. and Martinez-Suarez, J.V. 1995. Comparison of two alternative microdilution procedures with the National Committee for Clinical Laboratory Standards Reference Macrodilution Method M27-P for in vitro testing of fluconazole-resistant and -susceptible isolates of *Candida albicans*. *J Clin Microbiol*, **33**, 3154–8.

Fakes, R.G. 1984. Systemic *Candida* infections in infants in intensive care nurseries: high incidence of central nervous system involvement. *J Pediatr*, **105**, 616–22.

Fales, W., Stockham, S. and Lattimer, J. 1984. Pyothorax caused by *Candida albicans* in a cat. *J Am Vet Med Assoc*, **185**, 311–12.

Fegeller, W., Kappe, R., et al. 1990. *Diagnostishes Vorgehen bei Verdacht auf Endomykosen*. Basel: Editiones 'Roche'.

Fenn, J.P., Segal, H., et al. 1994. Comparison of updated Vitek Yeast Biochemical Card and API20C yeast identification systems. *J Clin Microbiol*, **32**, 184–7.

Ferra, C., Doebbling, B.N., et al. 1994. *Candida tropicalis* vertebral osteomyelitis: a late sequela of fungemia. *Clin Infect Dis*, **19**, 697–703.

Ferreira, R.P., Yu, B., et al. 1990. Detection of *Candida* antigenuria in disseminated candidiasis by immunoblotting. *J Clin Microbiol*, **28**, 1075–8.

Fidel, P.L., Cuthright, J. and Sobel, J. 1995. Effects of systemic cell mediated immunity on vaginal candidiasis in mice resistant and susceptible to *Candida albicans* infections. *Infect Immun*, **63**, 4191–4.

Fielding, R.M. 1991. Liposomal drug delivery, advantages and limitations from a clinical pharmacokinetic and therapeutic perspective. *Clin Pharmacokinetics*, **21**, 1155–64.

Filler, S.G. and Edwards, J.E. Jr 1993. Chronic mucocutaneous candidiasis. In: Murphy, J.W., Friedman, H. and Bendinelli, M. (eds), *Fungal infections and immune responses*. New York: Plenum Press, 117–34.

Finkelstein, R., Reinhertz, G., et al. 1993. Outbreak of *Candida tropicalis* fungemia in a neonatal intensive care unit. *Infect Control Hosp Epidemiol*, **14**, 587–90.

Flahaut, M., Sanglard, D., et al. 1998. Rapid Detection of *Candida albicans* in clinical samples by DNA amplification of common regions from *C. albicans*-secreted aspartic proteinase genes. *J Clin Microbiol*, **36**, 395–401.

Foley, G.L. and Schlafer, D.H. 1987. Candida abortion in cattle. *Vet Pathol*, **24**, 532–6.

Fonzi, W.A. 1999. PHR1 and PHR2 of *Candida albicans* encode putative glycosidases required for proper cross-linking of beta-1-3- and beta-1-6-glucans. *J Bacteriol*, **181**, 7071–9.

Fonzi, W.A. and Irwin, M.Y. 1993. Isogenic strain construction and gene mapping in *Candida albicans*. *Genetics*, **134**, 717–28.

Franklin, K.M., Warmington, J.R., et al. 1990. An immunodominant antigen of *Candida albicans* shows homology to the enzyme enolase. *Immunol Cell Biol*, **68**, 173–8.

Fromtling, R. 1988. Overview of medically important antifungal azole derivatives. *Clin Microbiol Rev*, **1**, 187–217.

Fromtling, R.A., Galgiani, J.N., et al. 1993. Multicenter evaluation of a broth macrodilution antifungal susceptibility test for yeasts. *Antimicrob Agents Chemother*, **37**, 39–45.

Fujita, S. and Hashimoto, T. 1992. Detection of serum *Candida* antigens by enzyme-linked immunosorbent assay and a latex agglutination test with anti-*Candida albicans* and anti-*Candida krusei* antibodies. *J Clin Microbiol*, **30**, 3132–7.

Fukazawa, Y., Cassone, A., et al. 1994. Mechanisms of cell-mediated immunity in fungal infection. *J Med Vet Mycol*, **32**, Suppl 1, 123–31.

Fulton, R.B. and Walker, R.D. 1992. *Candida albicans* urocystitis in a cat. *J Am Vet Med Assoc*, **200**, 524–6.

Fusek, M., Smith, E.A., et al. 1994. Extracellular aspartic proteinases from *Candida albicans, Candida tropicalis*, and *Candida parapsilosis* yeasts differ substantially in their specificities. *Biochemistry*, **33**, 9791–9.

Gale, C.A., Bendel, C.M., et al. 1998. Linkage of adhesion, filamentous growth, and virulence in *Candida albicans* to a single gene, *INT1*. *Science*, **279**, 1355–8.

Ganesan, K., Banerjee, A. and Datta, A. 1991. Molecular cloning of the secretory acid proteinase gene from *Candida albicans* and its use as a species-specific probe. *Infect Immun*, **59**, 2972–7.

Garner, R.E., Childress, A.M., et al. 1990. Characterization of *Candida albicans* mannan-induced, mannan-specific delayed hypersensitivity suppressor cells. *Infect Immun*, **58**, 2613–20.

Gelfand, J.A., Hurley, D.L., et al. 1978. Role of complement in defense against experimental disseminated candidiasis. *J Infect Dis*, **139**, 9–16.

Gentry, L.O. and Price, M.F. 1993. Urinary and peritoneal *Candida* infections. In: Bodey, G.P. (ed.), *Candidiasis: pathogenesis, diagnosis, and treatment*, 2nd edition. New York: Raven Press, 249–60.

Georgopapadakou, N.H. and Walsh, T.J. 1994. Human mycoses: drugs and targets for emerging pathogens. *Science*, **264**, 371–3.

Gerritsen, J., van Dissel, J.T., et al. 1998. *Candida tropicalis* endocarditis. *Circulation*, **98**, 90–1.

Ghannoum, M.A. 2000. Potential role of phospholipases in virulence and fungal pathogenesis. *Clin Microbiol Rev*, **13**, 122–43.

Giger, D.K., Domer, J.E. and Mcquity, J.T. 1978. Experimental murine candidiasis: patholological and immune response to cutaneous inoculation with *Candida albicans*. *Infect Immun*, **19**, 499–509.

Gil, M.L., Villamon, E., et al. 1999. Clinical strains of *Candida albicans* express the surface antigen glyceraldehyde 3-phosphate dehydrogenase in vitro and in infected tissues. *FEMS Immunol Med Microbiol*, **23**, 229–34.

Gil-Navarro, I., Gil, M.L., et al. 1997. The glycolytic enzyme glyceraldehyde-3-phosphate dehydrogenase of *Candida albicans* is a surface antigen. *J Bacteriol*, **179**, 4992–9.

Gleeson, M.A., Haas, L.O. and Cregg, J.M. 1990. Isolation of *Candida tropicalis* auxotrophic mutants. *Appl Environ Microbiol*, **56**, 2562–4.

Goodman, J.L., Winston, D.J., et al. 1992. A controlled trial of fluconazole to prevent fungal infections in patients undergoing bone marrow transplantation. *N Engl J Med*, **326**, 845–51, (see comments).

Goslen, B.J. and Kobayashi, G.S. 1993. Mycological infections. In: Fitzpatrick, T.B. (ed.), *Dermatology in general medicine*, 3rd edition. New York: McGraw-Hill, 2193–248.

Gozalbo, D., Gil-Navarro, I., et al. 1998. The cell wall associated glyceraldehydes 3-phosphate dehydrogenase of *Candida albicans* is also fibronectin and laminin binding protein. *Infect Immun*, **66**, 2052–9.

Greaves, I., Kane, K., et al. 1992. Pigeons and peritonitis? *Nephrol Dial Transplant*, **7**, 967–9.

Green, D., Still, J.M. Jr and Law, E.J. 1994. *Candida parapsilosis* sepsis in patients with burns: report of six cases. *J Burn Care Rehabil*, **15**, 240–3.

Greenfield, R.A. 1992. Host defense system interaction with *Candida*. *J Med Vet Mycol*, **30**, 89–104.

Gross, T.L. and Mayhew, I.G. 1983. Gastroesophageal ulceration and candidiasis in foals. *J Am Vet Med Assoc*, **182**, 1370–3.

Gueteau, N., Darras, A., et al. 1991. Bilateral *Torulopsis glabrata* pneumonia. *Rev Pneumol Clin*, **47**, 137–9.

Gutierrez, J., Morales, P., et al. 2002. *Candida dubliniensis*, a new fungal pathogen. *J Basic Microbiol*, **42**, 207–27.

Hadfield, T.L., Smith, M.B. and Winn, R.E. 1987. Mycoses caused by *Candida lusitaniae*. *Rev Infect Dis*, **9**, 1006–12.

Hallum, J.L. and Williams, T.W. 1993. Candida endocarditis. In: Bodey, G.P. (ed.), *Candidiasis: pathogenesis, diagnosis, and treatment*, 2nd edition. New York: Raven Press, 357–69.

Hamajima, K., Nishikawa, A., et al. 1987. Deoxyribonucleic acid base composition and its homology between two forms of *Candida parapsilosis* and *Loderomyces elongisporus*. *Gen Appl Microbiol*, **33**, 299–302.

Haneke, E. 1986. Differential diagnosis of mycotic nail diseases. In: Hay, R.J. (ed.), *Advances in topical antifungal therapy*. Berlin: Springer-Verlag, 94–101.

Hanger, D.P., Jewos, S. and Shaw, J.T.B. 1988. Fluconazole and testosterone: in vivo and in vitro studies. *Antimicrob Agents Chemother*, **32**, 646–8.

Haron, E., Feld, R. and Tuffnel, P. 1987. Hepatic candidiasis: an increasing problem in immunocompromised host. *Am J Med*, **83**, 17–26.

Hasenclever, H.F. and Mitchell, W.O. 1961. Antigenic studies of *Candida*. I. Observations of two antigenic groups in *Candida albicans*. *J Bacteriol*, **82**, 570–3.

Hay, R.J., Clayton, Y.M., et al. 1988. An evaluation of itraconazole in the management of onychomycosis. *Br J Dermatol*, **119**, 359–66.

Haynes, K. 2001. Virulence in *Candida* species. *Trends Microbiol*, **9**, 591–6.

Haynes, K.A. and Westerneng, T.J. 1996. Rapid identification of *Candida albicans*, *C. glabrata*, *C. parapsilosis* and *C. krusei* by species-specific PCR of large subunit ribosomal DNA. *J Med Microbiol*, **44**, 390–6.

Hazen, K.C. 1995. New and emerging yeast pathogens. *Clin Microbiol Rev*, **8**, 462–78.

Hazen, K.C. and Howell, S.A. 2003. Candida, Cryptococcus, and other yeasts of medical importance. In: Murray, P.R., Baron, E.J., et al. (eds), *Manual of Clinical Microbiology*, 8th edn. Washington DC: ASM Press, 1693–711.

Hendriks, L., Goris, A., et al. 1991. Phylogenetic analysis of five medically important *Candida* species as deduced on the basis of small ribosomal subunit RNA sequences. *J Gen Microbiol*, **137**, 1223–30.

Herrero, J.A., Lumbreras, C., et al. 1992. Nosocomial fungemia caused by *Candida parapsilosis*. *Enferm Infecc Microbiol Clin*, **10**, 520–4.

Hitchcock, C.A., Pye, G.W., et al. 1993. Fluconazole resistance in *Candida glabrata*. *Antimicrob Agents Chemother*, **37**, 1962–5.

Holmberg, K. and Feroze, F. 1996. Evaluation of an optimized system for random amplified polymorphic DNA (RAPD) analysis for genotypic mapping of *Candida albicans* strains. *J Clin Lab Anal*, **10**, 59–69.

Holzschu, D.L., Presley, H.L., et al. 1979. Identification of *C. lusitaniae* as opportunistic yeast in humans. *J Clin Microbiol*, **10**, 202–5.

Horowitz, B.J., Giaquinta, D. and Ito, S. 1992. Evolving pathogens in vulvovaginal candidiasis: implications for patient care. *J Clin Pharmacol*, **32**, 248–55.

Hossain, M.A., Miyazaki, T., et al. 1997. Comparison between Wako-WB003 and Fungitech G tests for detection of (1,3)-beta-D-glucan in systemic mycoses. *J Clin Lab Anal*, **11**, 73–7.

Hostetter, M.K. 1994a. Interaction of *Candida albicans* with eukaryotic cells. *ASM News*, **60**, 370–4.

Hostetter, M.K. 1994b. Adhesins and ligands involved in the interaction of *Candida* spp. with epithelial and endothelial surfaces. *Clin Microbiol Rev*, **7**, 29–42.

Hostetter, M.K. 2000. RGD mediated adhesion in fungal pathogens of humans, plants and insects. *Curr Opin Micobiol*, **3**, 344–8.

Hoyer, L.L. 2001. The ALS gene family of *Candida albicans*. *Trends Microbiol*, **9**, 176–80.

Hoyer, L.L., Fundyga, R., et al. 2001. Characterization of agglutinin-like sequence genes from non-*albicans Candida* and phylogenetic analysis of the ALS family. *Genetics*, **157**, 1555–67.

Hube, B. 2000. Extracellular proteinases of human pathogenic fungi. *Contrib Microbiol*, **5**, 126–37.

Hube, B. and Naglik, J. 2001. *Candida albicans* proteinases: resolving the mystery of a gene family. *Microbiology*, **147**, 1997–2005.

Hube, B. Sanglard, D., et al. 1997. Disruption of each of the secreted aspartyl proteinase genes SAP1, SAP2 and SAP3 of *Candida albicans* attenuates virulence. *Infect Immun*, **65**, 3529–38.

Hughes, P.A., Lepow, M.L. and Hill, H.R. 1993. Neonatal candidiasis. In: Bodey, G.P. (ed.), *Candidiasis: pathogenesis, diagnosis, and treatment*, 2nd edition. New York: Raven Press, 261–77.

Hull, C.M. and Johnson, A.D. 1999. Identification of a mating type-like locus in the asexual pathogenic yeast *Candida albicans*. *Science*, **285**, 1271–5.

Hull, C.M., Raisner, R.M. and Johnson, A.D. 2000. Evidence for mating of the 'asexual' yeast *Candida albicans* in a mammalian host. *Science*, **289**, 307–10.

Hymes, S.R. and Duvic, M. 1993. Cutaneous candidiasis. In: Bodey, G.P. (ed.), *Candidiasis: pathogensis, diagnosis, and treatment*, 2nd edition. New York: Raven Press, 159–66.

Isenberg, H.D. and Tucci, V. 1989. Single-source outbreak of *Candida tropicalis* complicating coronary bypass surgery. *J Clin Microbiol*, **27**, 242–68.

Jabra-Rizk, M.A., Falkler, W.A. Jr, et al. 2001. Prevalence of yeast among children in Nigeria and the United States. *Oral Microbiol Immunol*, **16**, 383–5.

Janoff, A.S. 1990. Liposomes and lipid structures as carriers of amphotericin B. *Eur J Clin Microbiol Infect Dis*, **9**, 146–50.

Joly, S., Pujol, C., et al. 1996. Development of two species-specific fingerprinting probes for broad computer-assisted epidemiological studies of *Candida tropicalis*. *J Clin Microbiol*, **34**, 3063–71.

Joly, S., Pujol, C., et al. 1999. Development and characterization of complex DNA fingerprinting probes for the infectious yeast *Candida dublinieasis*. *J Clin Microbiol*, **37**, 1035–44.

Jones, J.M. 1990. Laboratory diagnosis of invasive candidiasis. *Clin Microbiol Rev*, **3**, 379–88.

Jones, S., White, G. and Hunter, P.R. 1994. Increased phenotypic switching in strains of *Candida albicans* associated with invasive infections. *J Clin Microbiol*, **32**, 2869–70.

Jordan, J.A. 1994. PCR identification of four medically important *Candida* species by using a single primer pair. *J Clin Microbiol*, **32**, 2962–7.

Just-Nubling, G., Gentschew, G., et al. 1991. Fluconazole prophylaxis of recurrent oral candidiasis in HIV-positive patients. *Eur J Clin Microbiol Infect Dis*, **10**, 917–21.

Kadel, W.L., Kelley, D.C. and Coles, E.H. 1969. Survey of yeastlike fungi and tissue changes in esophagogastric region of stomachs in swine. *Am J Vet Res*, **30**, 401–8.

Kapica, L. and Blank, F. 1957. Growth of *Candida albicans* on keratin as sole source of nitrogen. *Dermatologica*, **115**, 81–105.

Kaposzta, R., Marodi, L., et al. 1999. Rapid recruitment of late endosomes and lysosomes in mouse macrophages ingesting *Candida albicans*. *J Cell Sci*, **112**, 3237–48.

Kappe, R., Okeke, C.N., et al. 1998. Molecular probes for the detection of pathogenic fungi in the presence of human tissue. *J Med Microbiol*, **47**, 811–20.

Kauffman, C.A., Bradley, S.F., et al. 1991. Hepatosplenic candidiasis: succsessful treatment with fluconazole. *Am J Med*, **91**, 137–41.

Kauffman, C.A., Badley, S.F. and Vine, A.K. 1993. Candida endophthalmitis associated with intraocular lens implantation: efficacy of fluconazole therapy. *Mycoses*, **36**, 13–17.

Kennedy, M.J. 1988. Adhesion and association mechanisms of *Candida albicans*. *Curr Top Med Mycol*, **2**, 123–32.

Kennedy, M.J. 1989. Regulation of *Candida albicans* populations in the gastrointestinal tract: mechanisms and significance in GI and systemic candidiasis. *Curr Top Med Mycol*, **3**, 315–402.

Kennedy, M.J., Calderone, R.A., et al. 1992. Molecular basis of *Candida albicans* adhesion. *J Med Vet Mycol*, **30**, Suppl 1, 95–122.

Khatib, R. and Clark, J.A. 1995. Relevance of culturing *Candida* species from intravascular catheters. *J Clin Microbiol*, **33**, 1635–7.

Kiehn, T.E. and Bernard, E.M. 1979. Candidiasis: detection by gas liquid chromatography of D-arabinitol, a fungal metabolite, in human serum. *Science*, **206**, 577–80.

Kirk, J.H., Bartlett, P.C. and Newman, J.P. 1986. Candida mastitis in a dairy herd. *Compend Con Educ Pract Vet*, **8**, F150–2.

Kirkley, B.A., Easley, K.A. and Washington, J.A. 1994. Controlled clinical evaluation of Isolator and ESP aerobic blood culture systems for detection of bloodstream infections. *J Clin Microbiol*, **32**, 1547–9.

Kirkpatrick, C.H. 1989. Chronic mucocutaneous candidiasis. *Eur J Clin Microbiol Infect Dis*, **8**, 448–56.

Kirkpatrick, C.H. 1993. Chronic mucocutaneous candidiasis. In: Bodey, G.P. (ed.), *Candidiasis: pathogenesis, diagnosis and treatment*, 2nd edition. New York: Raven Press, 167–84.

Kirkpatrick, C.H., Ottenson, E.A., et al. 1976. Reconstitution of defective cellular immunity with foetal thymus and dialysable transfer factor: long term studies in a patient with chronic mucocutaneous candidiasis. *Clin Exp Immunol*, **23**, 414–28.

Kirsch, D.R., Kelly, R. and Kurtz, M.B. (eds) 1990. *The genetics of Candida*. Boca Raton, FL: CRC Press.

Kitada, K., Yamaguchi, E. and Arisawa, M. 1995. Cloning of the *Candida glabrata TRP1* and *HIS3* genes, and construction of their disruptant strains by sequential integrative transformation. *Gene*, **165**, 203–6.

Klotz, S.A. 1992. Fungal adherence to the vascular compartment; a critical step in the pathogenesis of disseminated candidiasis. *Clin Infect Dis*, **14**, 340–7.

Kogan, G., Pavliak, V., et al. 1988. Novel structure of the cellular mannan of the pathogenic yeast *Candida krusei*. *Carbohydr Res*, **184**, 171–82.

Konneman, E.W., Stephen, D.A. et al. 1997. *Color atlas and textbook of diagnostic microbiology*, 5th edition. Philadelphia: Lippincott, 983–1069.

Krause, W., Matheis, H. and Wulf, K. 1969. Fungemia and funguria after oral administration of *Candida albicans*. *Lancet*, **1**, 598–9.

Krcmery, V. and Barnes, A.J. 2002. Non-albicans *Candida* spp. causing fungaemia: pathogenicity and antifungal resistance. *J Hosp Infect*, **50**, 243–60.

Kuhn, D.M., Chandra, J., et al. 2002. Comparison of biofilms formed by *Candida albicans* and *Candida parapsilosis* on bioprosthetic surfaces. *Infect Immun*, **70**, 878–88.

Kullberg, B.J. and Anaissie, E.J. 1998. Cytokines as therapy for opportunistic fungal infections. *Res Immunol*, **149**, 478–88.

Kullberg, B.J., Netea, M.G., et al. 1998. Recombinant murine granulocyte colony-stimulating factor protects against acute disseminated *Candida albicans* infection in nonneutropenic mice. *J Infect Dis*, **177**, 175–81.

Kumamoto, C.A. 2002. Candida biofilms. *Curr Opin Microbiol*, **5**, 608–11.

Kurtzman, C.P. 1993. Systematics of the ascomycetous yeasts assessed from ribosomal RNA sequence divergence. *Antonie van Leeuwenhoek*, **63**, 165–74.

Kurtzman, C.P. 1998a. *Pichia* E.C.Hansen emend. Kurtzman. In: Kurtzman, C.P. and Fell, J.W. (eds), *The yeasts, a taxonomic study*, 4th edition. Amsterdam: Elsevier, 273–352.

Kurtzman, C.P. 1998b. *Issatchenkia* Kudryavtsev emend. Kurtzman, Smiley and Johnson. In: Kurtzman, C.P. and Fell, J.W. (eds), *The yeasts, a taxonomic study*, 4th edition. Amsterdam: Elsevier, 221–6.

Kurtzman, C.P. and Fell, J.W. (eds) 1998. *The yeasts, a taxonomic study*, 4th edition. Amsterdam: Elsevier.

Kusne, S., Tobin, D., et al. 1994. *Candida* carriage in the alimentary tract of liver transplant candidates. *Transplantation*, **57**, 398–402.

Kwon-Chung, K.J. and Bennett, J.E. 1992. *Medical mycology*. Philadelphia: Lea & Febiger, 81–104, 768–770.

Kwon-Chung, K.J., Lehman, D., et al. 1985. Genetic evidence for role of extracellular proteinase in virulence of *Candida albicans*. *Infect Immun*, **49**, 571–5.

Lachance, M.A. 1998. *Kluyveromyces* van der Walt emend. van der Walt. In: Kurtzman, C.P. and Fell, J.W. (eds), *The yeasts, a taxonomic study*, 4th edition. Amsterdam: Elsevier, 227–47.

Land, G.A., Salkin, I.F., et al. 1991. Evaluation of the Baxter-MicroScan 4-hour enzyme-based yeast identification system. *J Clin Microbiol*, **29**, 718–22.

Langenhaeck, M. and Zeyen, T. 1993. Fungal endophthalmitis: a report of two cases. *Bull Soc Belge Ophtalmol*, **250**, 63–6.

Lasker, B.A., Carle, G.F., et al. 1989. Comparison of the separation of *Candida albicans* chromosomes sized DNA by pulsed field gel electrophoresis techniques. *Nucleic Acids Res*, **17**, 3783–93.

La Valle, R., Bromuro, C., et al. 1995. Molecular cloning and expression of 70-kilodalton heat shock protein of *Candida albicans*. *Infect Immun*, **63**, 4039–45.

La Valle, R., Sandini, S., et al. 2000. Generation of a recombinant 65-kilodalton mannoprotein, a major antigen target of cell-mediated immune response to *Candida albicans*. *Infect Immun*, **68**, 6777–84.

Lehmann, P.F., Lin, D. and Lasker, B.A. 1992. Genotypic identification and characterization of species and strains within the genus *Candida* by using random amplified polymorphic DNA. *Clin Microbiol*, **30**, 3249–54.

Leidich, S.D., Ibrahim, A.S., et al. 1998. Cloning and disruption of *caPLB1*, a phospholipase B gene involved in the pathogenicity of *Candida albicans*. *J Biol Chem*, **273**, 26078–86.

Levine, J., Dykoski, R.K. and Janoff, E.N. 1995. Candida associated diarrhea: a syndrome in search of credibility. *Clin Infect Dis*, **21**, 881–6.

Levitz, S.M. 1992. Overview of host defenses in fungal infections. *Clin Infect Dis*, **14**, Suppl 1, S37–42.

Levy, D.A., Bohbot, J.M., et al. 1989. Phase II study of D.651, an oral vaccine designed to prevent recurrences of vulvovaginal candidiasis. *Vaccine*, **7**, 4, 337–40.

Lew, M.A. 1989. Diagnosis of systemic candidiasis. *Annu Rev Med*, **40**, 87–97.

Lewis, J.H., Zimmerman, H.J., et al. 1984. Hepatic injury associated with ketoconazole therapy. *Gastroenterology*, **86**, 503–13.

Lin, D. and Lehmann, P.F. 1995. Random amplified polymorphic DNA for strain delineation within *Candida tropicalis*. *J Med Vet Mycol*, **33**, 241–6.

Liu, H. 2001. Transcriptional control of dimorphism in *Candida albicans*. *Curr Opin Microbiol*, **4**, 728–35.

Lockhart, S.R., Joly, S., et al. 1997. Development and verification of fingerprinting probes for *Candida glabrata*. *Microbiology*, **143**, 3733–46.

Lott, T.J., Kuykendall, R.J., et al. 1993. Genomic heterogeneity in the yeast *Candida parapsilosis*. *Curr Genet*, **23**, 463–7.

Luna, M.A. and Tortoledo, M.E. 1993. Histologic identification and pathologic patterns of disease caused by *Candida*. In: Bodey, G.P. (ed.), *Candidiasis: pathogenesis, diagnosis, and treatment*, 2nd edition. New York: Raven Press, 21–42.

Lynch, D.P. 1994. Oral candidiasis: history, classification, and clinical presentation. *Oral Surg Oral Med Oral Pathol*, **78**, 189–93.

Mackie, R.M. 1979. Mucocutaneous candidiasis responsive to transfer factor therapy. *J R Soc Med*, **72**, 926–7.

Madhani, H.D. and Fink, G.R. 1998. The control of filamentous differentiation and virulence in fungi. *Trends Cell Biol*, **8**, 348–53.

Magee, B.B. and Magee, P.T. 1987. Electrophoretic karyotypes and chromosome numbers in *Candida* species. *J Gen Microbiol*, **133**, 425–30.

Magee, B.B. and Magee, P.T. 2000. Induction of mating in *Candida albicans* by construction of MTLa and MTLalpha strains. *Science*, **289**, 310–13.

Magee, B.B., Koltin, Y., et al. 1988. Assignment of cloned genes to the seven electrophoretically separated *Candida albicans* chromosomes. *Mol Cell Biol*, **8**, 4721–6.

Magee, P.T. and Chibana, H. 2002. The genomes of *Candida albicans* and other *Candida* species. In: Calderone, R.A. (ed.), *Candida and candidiasis*. Washington, DC: ASM Press, 293–306.

Mahrous, M., Sawant, A.D., et al. 1992. DNA relatedness, karyotyping and gene probing of *Candida tropicalis*, *Candida albicans* and its synonyms *Candida stellatoidea* and *Candida claussenii*. *Eur J Epidemiol*, **8**, 444–51.

Maiwald, M., Kappe, R. and Sonntag, H.G. 1994. Rapid presumptive identification of medically relevant yeasts to the species level by polymerase chain reaction and restriction enzyme analysis. *J Med Vet Mycol*, **32**, 115–22.

Malathi, K., Ganesan, K. and Datta, A. 1994. Identification of a putative transcription factor in *Candida albicans* that can complement the mating defect of *Saccharomyces cerevisiae ste12* mutants. *J Biol Chem*, **269**, 22945–51.

Mangiarotti, A.M., Caretta, G., et al. 1993. Fungi isolated from feathers and feces of hens in a poultry factory in Monferrat. *Bol Mycol*, **8**, 91–8.

Marcilla, A., Monteagudo, C., et al. 1999. Monoclonal antibody 3H8: a useful tool in the dianosis of candidiasis. *Microbiology*, **145**, 695–701.

Marius-Gestin, V., Thibault, E., et al. 1987. Etiology of the venereal disease of the gander. *Rec Med Vet*, **163**, 645–54.

Marodi, L., Forehand, J.R. and Johnston, R.B. Jr 1991a. Mechanisms of host defense against *Candida* species. II. Biochemical basis for the killing of Candida by mononuclear phagocytes. *J Immunol*, **146**, 2790–4.

Marodi, L., Korchak, H.M. and Johnston, R.B. Jr 1991b. Mechanisms of host defense against *Candida* species. I. Phagocytosis by monocytes and monocyte-derived macrophages. *J Immunol*, **146**, 2783–9.

Martino, P., Girmenia, C., et al. 1993. Fungemia in patients with leukemia. *Am J Med Sci*, **306**, 225–32.

Masur, H., Rosen, P.P. and Armstrong, D. 1977. Pulmonary disease caused by *Candida* species. *Am J Med*, **63**, 914–25.

Matthews, R. and Burnie, J. 2001. Antifungal antibodies: a new approach to the treatment of systemic candidiasis. *Curr Opin Investig Drugs*, **2**, 472–6.

Matthews, R., Burnie, J., et al. 1988. Candida and AIDS: evidence for protective antibodies. *Lancet*, **2**, 263–6.

Matthews, R.C., Burnie, J.P., et al. 1991. Autoantibody to heat-shock protein 90 can mediate protection against systemic candidosis. *Immunology*, **74**, 20–4.

Matthews, R., Hodgetts, S. and Burnie, J. 1995. Preliminary assessment of a human recombinant antibody fragment to hsp90 in murine invasive candidiasis. *J Infect Dis*, **171**, 1668–71.

Matthews, R.C., Burnie, J.P. et al. 2000. Human recombinant antibody to hsp90 in the treatment of disseminated candidiasis. ASM Abstract No. 063.

McClure, J.J., Addison, D. and Miller, R.I. 1985. Immunodeficiency manifested by oral candidiasis and bacterial septicemia in foals. *J Am Vet Med Assoc*, **186**, 1195–7.

McQuillen, D.P., Zingman, B.S., et al. 1992. Invasive infections due to *Candida krusei*: report of ten cases of fungemia that include three cases of endophthalmitis. *Clin Infect Dis*, **14**, 472–8.

Meis, J.F., Ruhnke, M., et al. 1999. *Candida dubliniensis* candidemia in patients with chemotherapy-induced neutropenia and bone marrow transplantation. *Emerg Infect Dis*, **5**, 150–3.

Mencacci, A., Cenci, E., et al. 1998. Specific and non-specific immunity to *Candida albicans*; a lesson from genetically modified animals. *Res Immunol*, **149**, 352–61.

Merz, W.G. 1984. *Candida lusitaniae*: frquency of recovery, colonization, infection and amphotericin B resistance. *J Clin Microbiol*, **20**, 1194–5.

Merz, W.G. and Roberts, G.D. 2003. Algorithms for detection and identification of fungi. In: Murray, P.R. and Baron, E.J. et al. (eds), *Manual of clinical microbiology*, 8th edition. Washington, DC: ASM Press, 668–85.

Merz, W.G., Khazan, U., et al. 1992. Strain delineation and epidemiology of *Candida (Clavispora) lusitaniae*. *J Clin Microbiol*, **30**, 449–54.

Meunier, F. 1989. New methods for delivery of antifungal agents. *Rev Infect Dis*, **11**, Suppl, S1605–12.

Meunier, F., Aoun, M. and Bitar, N. 1992. Candidemia in immunocompromised patients. *Clin Infect Dis*, **14**, Suppl 1, S120–5.

Meyer, S.A., Payne, R.W. and Yarrow, D. 1998. Candida Berkhout. In: Kurtzman, C.P. and Fell, J.W. (eds), *The yeasts, a taxonomic study*, 4th edition. Amsterdam: Elsevier, 454–573.

Mitsutake, K., Kohno, S., et al. 1994. Detection of *Candida* enolase antibody in patients with candidiasis. *J Clin Lab Anal*, **8**, 207–10.

Miyakawa, Y., Kagaya, K., et al. 1986. Production and characterization of agglutinating monoclonal antibodies against predominant antigenic factors for *Candida albicans*. *J Clin Microbiol*, **23**, 881–6.

Miyakawa, Y., Kuribayashi, T., et al. 1992. Role of specific determinants in mannan of *Candida albicans* serotype A in adherence to human buccal epithelial cells. *Infect Immun*, **60**, 2493–9.

Miyazaki, T., Kohno, S., et al. 1995. Plasma (1→3)-beta-D-glucan and fungal antigenemia in patients with candidemia, aspergillosis, and cryptococcosis. *J Clin Microbiol*, **33**, 3115–18.

Mizutani, S., Endo, M., et al. 2000. CD4$^+$-T-cell-mediated resistance to systemic murine candidiasis induced by a membrane fraction of *Candida albicans*. *Antimicrob Agents Chemother*, **44**, 2653–8.

Mok, W.Y. and da Silva, M.S.B. 1984. Mycoflora of the human dermal surfaces. *Can J Microbiol*, **30**, 1205–9.

Mok, W.Y. and de Carvalho, C.M. 1985. Association of anurans with pathogenic fungi. *Mycopathologia*, **92**, 37–43.

Mok, W.Y., Luizao, R.C., et al. 1984. Ecology of pathogenic yeasts in Amazonian soil. *Appl Environ Microbiol*, **47**, 390–4.

Monod, M. and Borg-von Zepelin, M. 2002. Secreted proteinases and other virulence mechanisms of *Candida albicans*. *Chem Immunol*, **81**, 114–28.

Montes, L.F. 1992. Candidiasis. In Moschella, S.L. and Harley, H.J. (eds), *Dermatology*, 3rd edition. Philadelphia: W.B. Saunders, 913–23.

Moretti, A., Piergili Fioretti, D., et al. 2000. Isolation of *Candida rugosa* from turkeys. *J Vet Med B Infect Dis Vet Public Health*, **47**, 433–9.

Musial, C.E., Cockerill, F.R. and Roberts, G.D. 1988. Fungal infections of the immunocompromised host: clinical and laboratory aspects. *Clin Microbiol Rev*, **1**, 349–64.

Nakase, T., Komagata, K. and Fukazawa, Y. 1979. A comparative taxonomic study on two forms of *Candida parapsilosis* (Ashford) Langeron et Talice. *J Gen Appl Microbiol*, **25**, 375–86.

Nath, C.E., Shaw, P.J., et al. 1999. Amphotericin B in children with malignant disease: a comparison of the toxicities and pharmacokinetics of amphotericin B administered in dextrose versus lipid emulsion. *Antimicrob Agents Chemother*, **43**, 1417–23.

National Committee for Clinical Laboratory Standards. 2002. *Reference method for broth dilution susceptibility testing of yeasts; approved standard*, 2nd edition. NCCLS document M27-A2. Villanova, PA: National Committee for Clinical Laboratory Standards.

Navarro-Garcia, F., Sanchez, M., et al. 2001. Virulence genes in the pathogenic yeast *Candida albicans*. *FEMS Microbiol Rev*, **25**, 245–68.

Newman, S.L., Flanigan, T.P., et al. 1994. Clinically significant mucosal candidiasis resistant to fluconazole treatment in patients with AIDS. *Clin Infect Dis*, **19**, 684–6.

Nguyen, M.H., Peacock, J.E. Jr, et al. 1996. The changing face of candidemia: emergence of non-*Candida albicans* species and antifungal resistance. *Am J Med*, **100**, 617–23.

Niesters, H.G., Goessens, W.H., et al. 1993. Rapid, polymerase chain reaction based identification assays for *Candida* species. *J Clin Microbiol*, **31**, 904–10.

Nikawa, H., Samaranayake, L.P., et al. 1993. The fungicidal effect of human lactoferrin on *Candida albicans* and *Candida krusei*. *Arch Oral Biol*, **38**, 1057–63.

Nucci, M. 2003. Should all venous catheters be removed in all patients with hematogenous candidiasis? Review of evidence. *Focus on fungal infections 13*, Maui, Hawaii, March 19–21, pp. 41–42.

Oblack, D.L. and Holder, I.A. 1979. Active immunization against muscle damage mediated by *Candida albicans*. *J Med Microbiol*, **12**, 503–6.

Odds, F.C. 1988. *Candida and candidosis*, 2nd edition. London: Baillière Tindall.

Odds, F.C. 1993. Effects of temperature on anti-*Candida* activities of antifungal antibiotics. *Antimicrob Agents Chemother*, **37**, 685–91.

Odds, F.C. 1994. Candida species and virulence. *ASM News*, **60**, 31–8.

Odds, F.C. and Bernaerts, R.I.A. 1994. CHROMagar Candida, a new differential isolation medium for presumptive identification of clinically important *Candida* species. *J Clin Microbiol*, **32**, 1923–9.

Odds, F.C., Schmid, J. and Sell, D.R. 1990. Epidemiology of *Candida* infection in AIDS. In: van den Bossche, H., Drouhet, E., et al. (eds), *Mycoses in AIDS patients*. New York: Plenum Press, 67–74.

Ogawa, H., Nozawa, Y., et al. 1992. Fungal enzymes in pathogenesis of fungal infections. *J Med Vet Mycol*, **30**, Suppl 1, 189–96.

Oie, S. and Kamiya, A. 1992. Microbial contamination of brushes used for preoperative shaving. *J Hosp Infect*, **21**, 103–10.

Orozco, A.S., Zhou, X. and Filler, S.G. 2000. Mechanisms of the pro-inflammatory response of endothelial cells to *Candida albicans* infection. *Infect Immun*, **68**, 1134–41.

Panangala, V.S., Fish, N.A. and Barnum, D.A. 1978. Microflora of the cervico-vaginal mucus of repeat breeder cows. *Can Vet J*, **19**, 226–8.

Pappagianis, D., Collins, M.S., et al. 1979. Development of resistance to amphotericin B in *Candida lusitaniae* infecting a human. *Antimicrobial Agents Chemoter*, **16**, 123–6.

Perea, S., Gonzales, G., et al. 2002. In vitro activities of terbinafine in combination with fluconazole, itraconazole, voriconazole and posaconazole against clinical isolates of *Candida glabrata* with decreased susceptibility to azoles. *J Clin Microbiol*, **40**, 1831–3.

Perez-Martin, J., Uria, J.A. and Johnson, A.D. 1999. Phenotypic switching in *Candida albicans* is controlled by a SIR2 gene. *EMBO J*, **18**, 2580–92.

Pfaller, M.A. 1992a. Laboratory aids in diagnosis invasive candidiasis. *Mycopathologia*, **120**, 65–72.

Pfaller, M.A. 1992b. The use of molecular techniques for epidemiologic typing of *Candida* species. *Curr Top Med Mycol*, **4**, 43–63.

Pfaller, M.A. 2003. Nosocomial fungal infections: a look at emerging pathogens and threats to the future. *Focus on fungal infections 13*, Maui, Hawaii, March 19–21, 2003.

Pfaller, M.A. and Diekema, D.J. 2002. The role of sentinel surveyance of candidemia: trends in species distribution and antifungal susceptibility. *J Clin Microbiol*, **40**, 3551–7.

Pfaller, M. and Wenzel, R. 1992. Impact of the changing epidemiology of fungal infections in the 1990s. *Eur J Clin Microbiol Infect Dis*, **11**, 287–91.

Pfaller, M.A., Messer, S.A., et al. 1998. National epidemiology of mycoses survey: a multi center study of strain variation and antifungal susceptibility among isolates of *Candida* species. *Diagn Microbiol Infect Dis*, **31**, 289–96.

Phan, Q.T., Belanger, P.H. and Filler, S.G. 2000. Role of hyphal formation in interactions of *Candida albicans* with endothelial cells. *Infect Immun*, **68**, 3485–90.

Pines, E., Malbec, D., et al. 1994. Hepatic candidiasis caused by *Candida glabrata*. *Ann Gastroenterol Hepatol Paris*, **30**, 208–11.

Polacheck, I., Strahilevitz, J., et al. 2000. Recovery of *Candida dubliniensis* from non-human immunodeficiency virus-infected patients in Israel. *J Clin Microbiol*, **38**, 170–4.

Polak, A. 1987. 5-Fluorocytosine and combinations. *Ann Biol Clin (Paris)*, **45**, 669–72.

Polonelli, L., Castagnola, M., et al. 1985. Use of killer toxin for computer aided differentiation of *Candida albicans* strains. *Mycopathologia*, **91**, 175–9.

Polonelli, L., De Bernardis, F., et al. 1994. Idiotypic intravaginal vaccination to protect against candidal vaginitis by secretory, yeast killer toxin-like anti-idiotypic antibodies. *J Immunol*, **152**, 3175–3182.

Ponton, J., Marot-Leblond, A., et al. 1993. Characterization of *Candida albicans* cell wall antigens with monoclonal antibodies. *Infect Immun*, **61**, 4842–7.

Ponton, J., Moragues, M.D. and Quindos, G. 2002. Non-culture-based diagnostics. In: Calderone, R. (ed.), *Candida and candidiasis*. Washington, DC: ASM Press, 395–425.

Posteraro, B., Sanguinetti, M., et al. 2000. Reverse cross blot hybridization assay for rapid detection of PCR-amplified DNA from *Candida* species, *Cryptococcus neoformans*, and *Saccharomyces cerevisiae* in clinical samples. *J Clin Microbiol*, **38**, 1609–14.

Poulain, D., Slomianny, C., et al. 2002. Contribution of phospholipomannan to the surface expression of beta-1,2-oligomannosides in *Candida albicans* and its presence in cell wall extracts. *Infect Immun*, **70**, 4323–8.

Priel, S., Brosh, N. et al. 2002. Generation of human monoclonal antibodies specific to *Candida albicans* in a human/mouse radiation chimera: the Trimera system. *HUMS Congress*, Abstract, p. 16.

Quindos, G., Fernandez Rodriguez, M., et al. 1992. Colony morphotype on Sabouraud-triphenyltetrazolium agar: a simple and inexpensive method for *Candida* subspecies discrimination. *J Clin Microbiol*, **30**, 2748–52.

Ramage, G., Wickes, B.L. and Lopez-Ribot, J.L. 2001. Biofilms of *Candida albicans* and their associated resistance to antifungal agents. *Am Clin Lab*, **20**, 42–4.

Ramage, G., Bachmann, S., et al. 2002. Investigation of multidrug efflux pumps in relation to fluconazole resistance in *Candida albicans* biofilms. *J Antimicrob Chemother*, **49**, 973–80.

Rangel Frausto, M.S., Houston, A.K., et al. 1994. An experimental model for study of *Candida* survival and transmission in human volunteers. *Eur J Clin Microbiol Infect Dis*, **13**, 590–5.

Reboli, A.C. 1993. Diagnosis of invasive candidiasis by a dot immunobinding assay for *Candida* antigen detection. *J Clin Microbiol*, **31**, 518–23.

Rees, T. and Philips, R. 1992. Multicenter comparison of one-day oral therapy with fluconazole or itraconazole in vaginal candidiasis. *Int J Gynecol Obstet*, **37**, 33–8.

Reiss, E. 1986. *Candida albicans*. In *Molecular immunology of mycotic and actinomycotic infections*. New York: Elsevier, 191–250.

Reiss, E. and Morrisson, C.J. 1993. Nonculture methods for diagnosis of disseminated candidiasis. *Clin Microbiol Rev*, **6**, 311–23.

Reiss, E., Hearn, V.M., et al. 1992. Structure and function of the fungal cell wall. *J Med Vet Mycol*, **30**, Suppl 1, 143–56.

Reiss, E., Obayashi, T., et al. 2000. Non-culture based diagnostic tests for mycotic infections. *Med Mycol*, **38**, Suppl 1, 147–50.

Reiss, E., Kaufman, L., et al. 2002. Clinical immunomycology. In: Rose, N.R., Hamilton, R.G. and Detrick, B. (eds), *Manual of clinical laboratory immunology*. Washington, DC: ASM Press, 559–83.

Richardson, M.D. and Carlson, P. 2002. Culture and non-culture based diagnostics of *Candida* species. In: Calderone, R.A. and Cihlar, R.L. (eds), *Fungal pathogenesis: principles and clinical applications*. New York: Marcel Dekker, 387–94.

Richardson, M.D. and Warnock, D.W. 1997. *Fungal infection: diagnosis and management*. Oxford: Blackwell Science, 20–58; 78–93; 131–148.

Riddle, D.L., Giger, O., et al. 1994. Clinical comparison of the Baxter MicroScan Yeast Identification Panel and the Vitek Yeast Biochemical Card. *Am J Clin Pathol*, **101**, 438–42.

Rigg, G.P., Matthews, R., et al. 2001. Antibodies in the treatment of disseminated candidosis. *Mycoses*, **44**, Supp. 1, 63–4.

Riggsby, W.S., Torres-Bauza, L.J., et al. 1982. DNA content, kinetic complexity and ploidy question in *Candida albicans*. *Mol Cell Biol*, **2**, 853–62.

Rippon, J.W. 1988. Candidiasis and the pathogenic yeasts. In *Medical mycology*, 3rd edition. Philadelphia: W.B. Saunders, 536–81.

Ritchie, B.W., Harrison, G.J. and Harrison, L.R. 1994. *Avian medicine: principles and applications*. Lake Worth, FL: Wingers Publishing Inc., 998–9.

Rodriguez de Miranda, L. 1979. *Clavispora*, a new yeast genus of the *Saccharomycetales*. *Antonie van Leeuwenhoek*, **45**, 479–83.

Rodriguez-Adrian, L.J., Grazziutti, M.L., et al. 1998. The potential role of cytokine therapy for fungal infections in patients with cancer: is recovery from neutropenia all that is needed? *Clin Infect Dis*, **26**, 1270–8.

Roling, E.E., Klepser, M.E., et al. 2002. Antifungal activities of fluconazole, caspofungin (MK0991), and anidulafungin (LY303366) alone and in combination against *Candida* spp. and *Cryptococcus neoformans* via time-kill methods. *Diagn Microbiol Infect Dis*, **43**, 13–17.

Romani, L. 1999. Immunity to *Candida albicans*: Th1, Th2 cells and beyond. *Curr Opin Microbiol*, **2**, 363–7.

Romani, L. 2002a. Innate immunity against fungal pathogens. In: Calderone, R.A. and Cihlar, R.L. (eds), *Fungal pathogenesis: principles and clinical applications*. New York: Marcel Dekker, 401–32.

Romani, L. 2002b. Immunology of invasive candidiasis. In: Calderone, R.A. (ed.), *Candida and candidiasis*. Washington DC: ASM Press, 223–42.

Romani, L. and Bistoni, F. 2002. Systemic immunity in candidiasis. In: Calderone, R.A. and Cihlar, R.L. (eds), *Fungal pathogenesis: principles and clinical applications*. New York: Marcel Dekker, 483–514.

Romani, L. and Kaufmann, S.H. 1998. Immunity to fungi: editorial review. *Res Immunol*, **149**, 277–81.

Romano, A. and Madjarov, B. 1994. Ocular candidiasis. In: Segal, E. and Baum, G.L. (eds), *Pathogenic yeasts and yeast infections*. Boca Raton, FL: CRC Press, 103–11.

Rosenbluh, A., Mevarech, M., et al. 1985. Isolation of genes from *Candida albicans* by complementation in *Saccharomyces cerevisiae*. *Mol Gen Genet*, **200**, 500–2.

Ross, I.K., de Bernardis, F., et al. 1990. The secreted aspartate proteinase of *Candida albicans*: physiology of secretion and virulence of a proteinase deficient mutant. *J Gen Microbiol*, **136**, 687–94.

Roth-Ben Arie, Z., Altboum, Z., et al. 2001. Loss and regain of virulence of a *Candida albicans* mutant as assessed in systemic murine candidiasis. *J Med Mycol*, **11**, 117–22.

Ruchel, R., Zimmermann, F., et al. 1991. Candidiasis visualised by proteinase-directed immunofluorescence. *Virchows Arch A Pathol Anat Histopathol*, **419**, 199–202.

Ruchel, R., de Bernardis, F., et al. 1992. Candida acid proteinases. *J Med Vet Mycol*, **30**, Suppl 1, 123–32.

Rustchenko-Bulgac, E.P., Sherman, F. and Hicks, J.B. 1990. Chromosomal rearrangements associated with morphological mutants provide a means for genetic variation of *Candida albicans*. *J Bacteriol*, **172**, 1276–83.

Sabra, R. and Branch, R.A. 1990. Amphotericin B nephrotoxicity. *Drug Saf*, **5**, 94–108.

Salfelder, K. 1990. Candidiasis. In *Atlas of fungal pathology*. Dordrecht: Kluwer Academic Publishers, 28–37.

Samaranayake, L.P. and MacFarlane, T.W. (eds) 1990. *Oral candidosis*. London: Butterworth.

Samaranayake, Y.H. and Samaranayake, L.P. 1994. *Candida krusei*: biology, epidemiology, pathogenicity and clinical manifestations of an emerging pathogen. *J Med Microbiol*, **41**, 295–310.

Samaranayake, Y.H., Wu, P.C., et al. 1994. Adhesion and colonisation of *Candida krusei* on host surfaces. *J Med Microbiol*, **41**, 250–8.

Sanchez, V., Vazquez, J.A., et al. 1992. Epidemiology of nosocomial acquisition of *Candida lusitaniae*. *J Clin Microbiol*, **30**, 3005–8.

Sanchez, V., Vazquez, J.A., et al. 1993. Nosocomial acquisition of *Candida parapsilosis*: an epidemiologic study. *Am J Med*, **94**, 577–82.

Sanchez-Sousa, A., Tarrago, D., et al. 2001. Adherence to polystyrene of clinically relevant isolates of *Candida* species. *Clin Microbiol Infect*, **7**, 379–82

Sandovsky-Losica, H. and Segal, E. 1990. Interaction of *Candida albicans* with murine gastrointestinal mucosa from methotrexate and 5-fluorouracyl treated animals: in vitro adhesion prevention. *J Med Vet Mycol*, **28**, 274–87.

Sandovsky-Losica, H., Bar-Nea, L. and Segal, E. 1992. Fatal systemic candidiasis of gastrointestinal origin. An experimental model in mice compromised by anti-cancer treatment. *J Med Vet Mycol*, **30**, 219–31.

Sanglard, D. and Fiechter, A. 1992. DNA transformations of *Candida tropicalis* with replicating and integrative vectors. *Yeast*, **8**, 1065–75.

Sanglard, D., Hube, B., et al. 1997. A triple deletion of the secreted aspartyl proteinase genes SAP4, SAP5 and SAP6 of *Candida albicans* causes attenuated virulence. *Infect Immun*, **65**, 3539–46.

Sanguinetti, A., Carmichael, J.K. and Campbell, K. 1993. Fluconazole resistant *Candida albicans* after long-term suppressive therapy. *Arch Intern Med*, **153**, 1122–4.

Santos, M.A., el Adlouni, C., et al. 1994. Transfer RNA profiling: a new method for the identification of pathogenic *Candida* species. *Yeast*, **10**, 625–36.

Sarosi, G.A. 1990. Amphotericin B: still the gold standard for antifungal therapy. *Postgad Med*, **88**, 151–66.

Scerpella, E.G. and Alhalel, R. 1994. An unusual cause of acute renal failure: bilateral ureteral obstruction due to *Candida tropicalis* fungus balls. *Clin Infect Dis*, **18**, 440–2.

Schonheyder, H., Johansen, J.A., et al. 1983. IgA and IgG serum antibodies to *Candida albicans* in women of childbearing age. *Sabouraudia*, **21**, 223–31.

Segal, E. 1987a. Pathogenesis of mycoses: role of adhesion to host surfaces. *Microbiol Sci*, **4**, 344–7.

Segal, E. 1987b. Vaccines against fungal infections. *CRC Crit Rev Microbiol*, **14**, 229–70.

Segal, E. 1989. Fungal ribosomal vaccines, I. *Histoplasma* and *Candida* vaccines. *Mycopathologia*, **105**, 45–8.

Segal, E. 1991. Immunizations against fungal disease in man and animals. In: Arora, D.K., Mukerji, K.G. and Ajello, L. (eds), *Applied mycology*, vol. 2. New York: Marcel Dekker, 341–68.

Segal, E. 1994a. Vaccination as an expression of immunity to *Candida*. In: Segal, E. and Baum, G.L. (eds), *Pathogenic yeasts and yeast infections*. Boca Raton, FL: CRC Press, 79–86.

Segal, E. 1994b. Virulence factors. In: Segal, E. and Baum, G.L. (eds), *Pathogenic yeasts and yeast infections*. Boca Raton FL: CRC Press, 49–60.

Segal, E. and Sandovsky-Losica, H. 1995. Interaction of *Candida* with mammalian tissues in vitro and in vivo. *Methods Enzymol*, **253**, 439–52.

Segal, E. and Sandovsky-Losica, H. 1997. Basis for *Candida albicans* adhesion and penetration. In: Jacobs, P.H. and Nall, L. (eds), *Fungal disease*. New York: Marcel Dekker, 321–34.

Segal, E., Sroka, A. and Schechter, A. 1984. Correlative relationship between adherence of *Candida albicans* to human vaginal epithelial cells in vitro and vaginitis. *J Med Vet Mycol*, **22**, 191–200.

Sendid, B., Poirot, J.L., et al. 2002. Combined detection of mannanaemia and anti mannan antibodies as a strategy for the diagnosis of systemic infection caused by pathogenic *Candida* species. *J Med Microbiol*, **51**, 433–42.

Shian, W.J., Chi, C.S., et al. 1993. Candidemia in the neonatal intensive care unit. *Acta Paediatr Sin*, **34**, 349–55.

Shmuely, H., Kremer, I., et al. 1997. *Candida tropicalis* multifocal endophtalmitis as the only initial manifestation of pacemaker endocarditis. *Am J Ophthalmol*, **123**, 559–60.

Singh, S., Sobel, J.D., et al. 2002. Vaginitis due to *Candida krusei*: epidemiology, clinical aspects, and therapy. *Clin Infect Dis*, **35**, 1066–70.

Sinha, B.K., Sharma, T.S. and Mehrotra, V.K. 1980. Fungi isolated from the genital tract of infertile cows and buffaloes in India. *Vet Rec*, **106**, 177–8.

Slutsky, B., Buffo, J. and Soll, D.R. 1985. High frequency switching of colony morphology in *Candida albicans*. *Science*, **230**, 666–9.

Slutsky, B., Staebel, M., et al. 1987. 'White-opaque transition': a second high frequency switching system in *Candida albicans*. *J Bacteriol*, **169**, 189–97.

Smith, P.D., Lamerson, C.L., et al. 1990. Granulocyte-macrophage colony-stimulating factor augments human monocyte fungicidal activity to *Candida albicans*. *J Infect Dis*, **161**, 999–1005.

Sobel, J.D. 1985. Epidemiology and pathogenesis of recurrent vulvovaginal candidiasis. *Am J Obstet Gynecol*, **152**, 924–35.

Sobel, J.D. 1992a. Pathogenesis and treatment of recurrent vulvovaginal candidiasis. *Clin Infect Dis*, **14**, Suppl 1, S148–53.

Sobel, J.D. 1992b. Vulvovaginitis. *Dermatol Clin*, **10**, 339–59.

Sobel, J.D. 1993. Genital candidiasis. In: Bodey, G.P. (ed.), *Candidiasis: pathogenesis, diagnosis, and treatment*, 2nd edition. New York: Raven Press, 225–47.

Sobel, J.D. 1998. Vulvovaginitis due to *Candida glabrata*. An emerging problem. *Mycoses*, **41**, Suppl 2, 18–22.

Sobel, J.D. 2003. New treatment options for invasive *Candida* infections. *Focus on fungal infections 13*. Maui, Hawaii, March 19–21.

Soll, D.R. 2002. Phenotypic switching. In: Calderone, R.A. (ed.), *Candida and candidiasis*. Washington, DC: ASM Press, 123–43.

Sono, E., Masuda, T., et al. 1992. Comparison of secretory acid proteinases from *Candida tropicalis*, *C. parapsilosis* and *C. albicans*. *Microbiol Immunol*, **36**, 1099–104.

Spaccapelo, R. and Romani, L. 1995. TGF-beta is important in determining the in vivo patterns of susceptibility or resistance in mice infected with *Candida albicans*. *J Immunol*, **155**, 1349–60.

Staab, J.F. and Sundstrom, P. 1998. Genetic organization and sequence analysis of the hypha-specific cell wall protein gene *HWP1* of *Candida albicans*. *Yeast*, **14**, 681–6.

Staib, P., Kretschmar, M., et al. 2000. Expression of virulence genes in *Candida albicans*. *Adv Exp Med Biol*, **485**, 167–76.

Stenderup, A. 1990. Oral mycology. *Acta Odontol Scand*, **48**, 3–10.

St Germain, G. and Beauchesne, D. 1991. Evaluation of the MicroScan Rapid Yeast Identification panel. *J Clin Microbiol*, **29**, 2296–9.

Stevens, D.A. 1998. Combination immunotherapy and antifungal chemotherapy. *Clin Infect Dis*, **26**, 1266–9.

Stevens, D.A. and Bennett, J.E. 2000. Antifungal agents. In: Mandell, G.L., Bennett, J.E. and Dolin, R. (eds), *Principles and practice of infectious diseases*, 5th edition. New York: Churchill Livingstone, 448–59.

Stiller, R.L., Bennett, J.E., et al. 1982. Susceptibility to 5-fluorocytosine and prevalence of serotype in 402 *Candida albicans* isolates in United States. *Antimicrob Agents Chemother*, **22**, 482–7.

Stone, H.H., Geheber, C.E., et al. 1973. Alimentary tract colonization by *Candida albicans*. *J Surg Res*, **14**, 273–6.

Strausbaugh, L.J., Sewell, D.L., et al. 1994. High frequency of yeast carriage on hands of hospital personnel. *J Clin Microbiol*, **32**, 2299–300.

Sugiyama, Y., Nakashima, S., et al. 1999. Molecular cloning of a second phospholipase B gene, *caPLB2* from *Candida albicans*. *Med Mycol*, **37**, 61–7.

Sullivan, D.J. and Coleman, D.C. 2002. Molecular typing and epidemiology of *Candida* spp. and other important human fungal pathogens. In: Calderone, R.A. and Cihlar, R.L. (eds) *Fungal pathogenesis, principles and clinical applications*. New York: Marcel Dekker, 717–37.

Sullivan, D., Bennett, D., et al. 1993. Oligonucleotide fingerprinting of isolates of *Candida* species other than *Candida albicans*, and of atypical *Candida* species from human immunodeficiency virus positive and AIDS patients. *J Clin Microbiol*, **31**, 2124–33.

Sullivan, D.J., Westerneng, T.J., et al. 1995. *Candida dubliniensis* sp. nov., phenotypic and molecular characterization of a novel species associated with oral candidosis in HIV-infected individuals. *Microbiology*, **141**, 1507–21.

Sullivan, D.J., Moran, G., et al. 1999. *Candida dubliniensis*, an update. *Rev Iberoamer Mycol*, **16**, 72–6.

Sundstrom, P. 2002. Adhesion in *Candida* spp.. *Cell Microbiol*, **4**, 461–9.

Sundstrom, P., Balish, E. and Allen, C.M. 2002. Essential role of the *Candida albicans* transglutaminase substrate, hyphal wall protein 1, in lethal oroesophageal candidiasis in immunodeficient mice. *J Infect Dis*, **185**, 521–30.

Sutka, P. 1983. *Candida guillermondii* var. *guillermondii* infection in cows slaughtered because of infertility and udder inflammation. *Acta Vet Yug*, **33**, 287–97.

Sweeney, J.C., Migaki, G., et al. 1976. Systemic mycoses in marine animals. *J Am Vet Med Assoc*, **169**, 946–8.

Switchenko, A.C., Miyada, C.G., et al. 1994. An automated enzymatic method for measurement of D-arabinitol, a metabolite of pathogenic *Candida* species. *J Clin Microbiol*, **32**, 92–7.

Tainturier, D., Fieni, F., et al. 1993. Candidose et troubles de la reproduction chez la vache. *Rev Med Vet*, **144**, 411–14.

Talwar, P., Chakrabarti, A., et al. 1990. Fungal diarrhoea: association of different fungi and seasonal variation in their incidence. *Mycopathologia*, **110**, 101–5, (see comments).

Telenti, A. and Roberts, G.D. 1989. Fungal blood cultures. *Eur J Clin Microbiol Infect Dis*, **8**, 825–31.

Thaler, M., Pastakia, B., et al. 1988. Hepatic candidiasis in cancer patients: the evolving picture of the syndrome. *Ann Intern Med*, **108**, 88–100.

Thunnissen, P.L., Sizoo, W. and Hendriks, W.D. 1991. Safety and efficacy of itraconazole in prevention of fungal infections in neutropenic patients. *Neth J Med*, **39**, 84–91.

Tiballi, R.N., He, X., et al. 1995. Use of colorimetric system for yeast susceptibility testing. *J Clin Microbiol*, **33**, 915–17.

To, W.K., Fothergill, A.W. and Rinaldi, M.G. 1995. Comparative evaluation of macrodilution and Alamar colorimetric microdilution broth methods for antifungal susceptibility testing of yeast isolates. *J Clin Microbiol*, **33**, 2660–4.

Togni, G., Sanglard, D., et al. 1991. Isolation and nucleotide sequence of the extracellular acid protease gene (ACP) from the yeast *Candida tropicalis*. *FEBS Lett*, **286**, 181–5.

Tokunaga, S., Ohkawa, M., et al. 1992. Clinical significance of measurement of serum D-arabinitol levels in candiduria patients. *Urol Int*, **48**, 195–9.

Trinel, P.A., Maes, E., et al. 2002. *Candida albicans* phospholipomannan, a new member of the fungal mannose inositol phosphoceramide family. *J Biol Chem*, **277**, 7260–71.

Tsuchimori, N., Sharkey, L.L., et al. 2002. Reduced virulence of HWP1-deficient mutants of Candida albicans and their interactions with host cells. *Infect Immun*, **68**, 1997–2002.

Tsuchiya, T., Fukazawa, Y. and Kawakita, S. 1965. Significance of serological studies on yeasts. *Mycopathol Mycol Appl*, **26**, 1–15.

Turin, L., Riva, F., et al. 2000. Fast, simple and highly sensitive double-rounded polymerase chain reaction assay to detect medically relevant fungi in dermatological specimens. *Eur J Clin Invest*, **30**, 511–18.

Van Burik, J.A.H. 2003. Granulocyte transfusions as treatment or prophylaxis for fungal infections. *Focus on fungal infections 13*. Maui, Hawaii, 45–46.

Velegraki, A., Kambouris, M.E., et al. 1999. Identification of medically significant fungal genera by polymerase chain reaction followed by restriction enzyme analysis. *FEMS Immunol Med Microbiol*, **23**, 303–12.

Verweij, P.E., Poulain, D., et al. 1998. Current trends in detection of antigenemia, metabolites and cell wall markers for the diagnosis and therapeutic monitoring of fungal infections. *Med Mycol*, **36**, Suppl 1, 146–55.

Wagner, D.K. and Sohnle, P.G. 1995. Cutaneous defenses against dermatophytes and yeasts. *Clin Microbiol Rev*, **8**, 317–35.

Waldvogel, F.A. and Bisno, A.L. (eds) 2000. *Infections associated with indwelling medical devices*, 3rd edition. Washington, DC: ASM Press.

Walsh, T.J. 2002. Echinocandins – an advance in the primary treatment of invasive candidiasis. *N Engl J Med*, **347**, 2070–2.

Walsh, T.J. and Chanock, S.J. 1997. Laboratory diagnosis of invasive candidiasis: a rationale for complementary use of culture and nonculture-based detection systems. *Int J Infect Dis*, **1**, Suppl 1, S11–19.

Walsh, T.J. and Pizzo, A. 1988. Treatment of systemic fungal infections: recent progress and current problems. *Eur J Clin Microbiol Infect Dis*, **7**, 460–75.

Walsh, T.J., Hier, D.B. and Caplan, L.R. 1985. Fungal infections of the central nervous system: comparative analysis of risk factors and clinical signs in 57 patients. *Neurology*, **35**, 1654–7.

Walsh, T.J., Pappas, P., et al. 2002. Voriconazole compared with liposomal amphotericin B for empirical antifungal therapy in patients with neutropenia and persistent fever. *N Engl J Med*, **346**, 225–34.

Walter, E.B. Jr, Gringras, J.L. and McKinney, R.E. Jr 1990. Systemic *Torulopsis glabrata* infection in a neonate. *South Med J*, **83**, 837–8.

Wara, D.W. and Amman, A.J. 1978. Thymosin treatment of children with primary immunodefficiency disease. *Transplant Proc*, **10**, 203–9.

Watanabe, K., Kagaya, K., et al. 1991. Mechanism for candidacidal activity in macrophages activated by recombinant gamma-interferon. *Infect Immun*, **59**, 521–8.

Weems, J.J. Jr 1992. *Candida parapsilosis*: epidemiology, pathogenicity, clinical manifestations, and antimicrobial susceptibility. *Clin Infect Dis*, **14**, 756–66, (see comments).

Weissman, Z., Berdicevsky, I. and Cavari, B. 1995. Molecular identification of *Candida albicans*. *J Med Vet Mycol*, **33**, 205–7.

Wenzel, R.P. 1995. Nosocomial candidemia: risk factors and attributable mortality. *Clin Infect Dis*, **20**, 1531–4.

Wickerham, L.J. and Burton, K.A. 1948. Carbon assimilation tests for the classification of yeasts. *J Bacteriol*, **56**, 363–71.

Wickes, B.L., Staudinger, J., et al. 1991. Physical and genetic mapping of *Candida albicans*: several genes previously assigned to chromosome 1 map tochromosome R, the rDNA containing linkage group. *Infect Immun*, **59**, 2480–4.

Wickes, B.L., Hicks, J.B., et al. 1992. The molecular analysis of synonymy among medically important yeasts within the genus *Candida*. *J Gen Microbiol*, **138**, 901–7.

Wiesner, S.M., Bendel, C.M., et al. 2002. Adherence of yeast and filamentous forms of *Candida albicans* to cultured enterocytes. *Crit Care Med*, **30**, 677–83.

Wilson, H.A. Jr, Downes, T.R., et al. 1993. Candida endocarditis: a treatable form of pacemaker infection. *Chest*, **103**, 283–4.

Wilson, M.L. Davis, T.E., et al. 1993. Controlled comparison of the BACTEC high-blood-volume fungal medium, BACTEC Plus 26 aerobic blood culture bottle, and 10-milliliter isolator blood culture system for detection of fungemia and bacteremia. *J Clin Microbiol*, **31**, 865–71, (see comments).

Wingard, J.R., Merz, W.G., et al. 1991. Increase in *Candida krusei* infection among patients with bone marrow transplantation and neutropenia treated prophylactically with fluconazole. *N Engl J Med*, **325**, 1274–7.

Wingard, J.R., Merz, W.G., et al. 1993. Association of *Torulopsis glabrata* infections with fluconazole prophylaxis in neutropenic bone marrow transplant patients. *Antimicrob Agents Chemother*, **37**, 1847–9.

Winston, D.J., Chandrasekar, P.H., et al. 1993. Fluconazole prophylaxis of fungal infections in patients with acute leukemia. Results of a randomized placebo-controlled, double-blind, multicenter trial. *Ann Intern Med*, **118**, 495–503, (see comments).

Witkin, S.S., Yu, I.R. and Ledger, W.J. 1983. Inhibition of *Candida albicans* induced lymphocyte proliferation by lymphocytes and sera from women with recurrent vaginitis. *Am J Obstet Gynecol*, **147**, 809–11

Witkin, S.S., Hirsch, J. and Ledger, W.J. 1986. A macrophage defect in women with recurrent *Candida* vaginitis and its reversal by prostaglandin inhibitors. *Am J Obst Gynecol*, **155**, 790–5.

Witkin, S.S., Jeremias, J. and Ledger, W.J. 1989. Vaginal eosinophilia and IgE antibodies to *Candida albicans* in women with recurrent vaginitis. *J Med Vet Mycol*, **27**, 57–8.

Wolff, S.N., Fay, J., et al. 2000. Fluconazole vs low-dose amphotericin B for the prevention of fungal infections in patients undergoing bone marrow transplantation: a study of the North American Marrow Transplant Group. *Bone Marrow Transplant*, **25**, 853–9.

Yagupsky, P., Nolte, F.S. and Menegus, M.A. 1990. Enhanced detection of *Candida* in blood cultures with the BACTEC 460 system by use of the aerobic-hypertonic (8B) medium. *Epidemiol Infect*, **105**, 553–8.

Yagupsky, P., Dagan, R., et al. 1991. Pseudooutbreak of *Candida guillermondii* fungemia in a neonatal intensive care unit. *Pediatr Infect Dis J*, **10**, 928–32.

Yamada, Y. and Kondo, K. 1972. Taxonomic of the coenzyme Q system in yeasts and yeast-like fungi. In: Iwata, K. (ed.), *Yeasts and yeast-like organisms in medical science. Proceedings of the Tokyo International Special Symposium on Yeasts*. Tokyo: University of Tokyo Press, 61–9.

Yeruham, I., Elad, D. and Liberboim M. 1999. Clinical and microbiological study of an otitis media outbreak in calves in a dairy herd. *Zentralbl Veterinarmed B*, **46**, 145–50.

Yuen, K.Y., Seto, W.H., et al. 1992. An outbreak of *Candida tropicalis* peritonitis in patients on intermittent peritoneal dialysis. *J Hosp Infect*, **22**, 65–72.

Zervos, M. and Meunier, F. 1993. Fluconazole (Diflucan): a review. *Int J Antimicrob Agents*, **3**, 147–70.

Zhou, P., Szczypka, M.S., et al. 1994. A system for gene cloning and manipulation in the yeast *Candida glabrata*. *Gene*, **142**, 135–40.

31

Serious infections caused by uncommon yeasts

JANINE R. MAENZA AND WILLIAM G. MERZ

There are more than 500 species in almost 50 genera of yeast, or yeast-like organisms, that have been described in the world's literature. This vegetative morphological form must convey advantages to the organism, since yeast have arisen in several different branches in the evolution of fungi. This is exemplified by many ascomycetous and basidiomycetous yeast genera.

Members of the asexual genus *Candida*, and one of the sexual basidiomycetous *Cryptococcus* spp., *Cryptococcus neoformans*, cause most serious human yeast infections. In addition, however, there are less common yeast genera that may also cause human infection. These include members of the genera *Blastoschizomyces*, *Hansenula*, *Malassezia*, *Rhodotorula*, *Saccharomyces*, *Sporobolomyces*, *Trichosporon*, and *Ustilago*.

Most members of these genera have caused fungemia with or without organ invasion in immunocompromised patient populations. Although uncommon, these infections are an important clinical entity as they are often difficult to treat, due to both the compromised nature of the patient and the refractoriness of the organisms to standard antifungal regimens.

SPECTRUM OF CLINICAL DISEASE

Most of these organisms are associated with environmental sources and are less frequently encountered as colonizers of humans. The major exception is the genus *Malassezia*, which is a very common component of the normal skin flora of humans and many animal species. The other genera are not usual components of our normal flora, but may be found as transient colonizers. Persistent colonization, when it occurs, is seen more often in hospitalized or immunocompromised patients. Colonization may then serve as a precursor to infection.

Given the relative rarity of these infections most reported information is in the form of case reports and case series. Derived from these reports, Table 31.1 summarizes the types of infections caused by individual genera. The most common infection is fungemia, but there are also reports of solid organ involvement with most of these yeast. The host characteristics of patient populations who are more at risk for infection are described in Table 31.2. As apparent from this table, most of these infections occur in immunocompromised patient populations.

Blastoschizomyces

Blastoschizomyces capitatus, formerly included in the genus *Trichosporon*, is known to be an environmental organism. Nevertheless, an observational report that represents the largest collected series of infection with this organism failed to reveal a common environmental exposure (Martino et al. 1990b). In this study, 20 patients with evidence of *B. capitatus* colonization or infection were identified, yet no environmental source was found, nor was the isolation of *B. capitatus* from environmental cultures related to cases of colonization or infection (Martino et al. 1990b). There was also no evidence for a common source of transmission through food, equipment, or personnel. Given the rarity of clinical disease with *B. capitatus*, risk factors for the development of infection are not clearly defined, but notably the single case series (Martino et al. 1990b) and multiple case reports (Martino et al. 1990b; Girmenia et al. 1991;

Table 31.1 *Types/sites of infection*[a]

Type or site of infection	*Blastoschizomyces*	*Hansenula*	*Malassezia*	*Rhodotorula*	*Saccharomyces*	*Sporobolomyces*	*Trichosporon*	*Ustilago*
Fungemia	++	++	++	++	++	−	++	+
Pneumonia	+	++	+	−	+	−	++	−
Peritonitis	−	−	+	+	−	−	+	−
Meningitis	+	−	−	+	−	−	−	−
Brain abscess	+	−	−	−	−	−	+	−
Endophthalmitis	−	−	−	+	−	−	+	−
Endocarditis	+	+	−	+	−	−	+	−
Hepatosplenic infection	+	−	−	−	+	−	+	−
Lymphadenitis	−	+	−	−	−	+	−	−
Bone marrow infiltration	−	−	−	−	−	+	−	−
Osteomyelitis	+	−	−	+	−	−	+	−
Renal/urinary infection	+	+	−	−	−	−	+	−

+ Reported only in isolated case reports.
++ Reported in larger case series or observational studies.
a) *Hanseniospora* is not included because it has not been shown to cause infection (see section on Microbiology below).

D'Antonio et al. 1994; Pagano et al. 1996) describe the development of disease in patients with hematological malignancies, the vast majority of whom were neutropenic at the time of development of infection.

Twelve of the 20 patients in Martino's case series had *B. capitatus* infection, rather than simply colonization. Each of these 12 presented with fever in the setting of fungemia or solid organ invasion. Sites of solid organ involvement included lung, liver, spleen, kidney. central nervous system, and heart (Martino et al. 1990b). Lung involvement was manifest by productive cough and radiographic studies showing infiltrates (which appeared to be mycetomas in some) (Martino et al. 1990a, b). Hepatic, renal, and/or splenic involvement appeared as focal, nodular lesions seen by imaging studies, and laboratory evaluation demonstrating abnormal liver function tests (Martino et al. 1990a). Central nervous system involvement was manifest clinically by neurological deficits and radiographically by the appearance of focal intracerebral lesions. Myocarditis and endocarditis were both diagnosed at autopsy in one patient with symptomatic heart failure (Martino et al. 1990a). In contrast to the patients with infection, those with transient colonization in this series had *B. capitatus* in stool or urine cultures with no evidence for invasive disease.

In additional reports, other types of infection have included osteomyelitis and diskitis, which developed after *B. capitatus* fungemia (D'Antonio et al. 1994), and

Table 31.2 *Defined characteristics of susceptible patient populations*

Population	*Blastoschizomyces*	*Hansenula*	*Malassezia*	*Rhodotorula*	*Saccharomyces*	*Trichosporon*	*Ustilago*
Neonate	−	−	+	−	−	+	−
Malignancy	+	+	−	+	+	+	−
Central venous catheter	−	+	+	+	+	+	+
Immunosuppressive medications	−	+	+	+	−	+	−
Burns	−	−	−	+	−	+	−
Dialysis	−	−	−	+	−	−	−
Surgery	−	+	−	+	+	−	−
HIV infection	−	−	−	−	+	−	−

meningitis (Girmenia et al. 1991). Both these infections were in children who had received bone marrow transplants for acute leukemia. An adult leukemia patient with hepatosplenic *B. capitatus* infection (DeMaio and Colman 2000) and a pediatric leukemia patient with *B. capitatus* osteomyelitis (Cheung et al. 1999) have also been reported.

Diagnosis of *B. capitatus* infection is usually made by culture of blood, or other affected body site. Biopsy may be needed to show invasive disease; e.g. in patients with pulmonary symptoms and positive sputum cultures, lung biopsy may be used to show histological evidence of invasion.

Treatment of disseminated *B. capitatus* infection requires systemic antifungal therapy. Fluconazole and flucytosine have good in vitro activity (Vendetti et al. 1991). In clinical practice, amphotericin B with or without flucytosine has frequently been used (Martino et al. 1990a, b; D'Antonio et al. 1994). Fluconazole was used as treatment in the case of meningitis: symptoms resolved and the organism was eradicated from the cerebrospinal fluid. Three months after fluconazole was discontinued, however, the patient was found on autopsy to have persistent evidence of meningeal invasion with *B. capitatus* (Girmenia et al. 1991). After a course of amphotericin B, itraconazole was used as continuation treatment in the child with osteomyelitis (D'Antonio et al. 1994). A crucial aspect of treatment is also likely to be resolution of neutropenia; one case report describes a lack of response to antifungal therapy until the patient was treated with granulocyte-macrophage colony-stimulating factor (GM-CSF) (Pagano et al. 1996), another describes the effective use of interferon-gamma (IFN-γ) for hepatosplenic infection after the failure of amphotericin B alone (DeMaio and Colman 2000).

Hansenula

The more commonly recognized *Hansenula* species, *Hansenula anomala*, is an environmental colonizer that may be isolated from organic substances including vegetables, fruit, and soil. Less common is *Hansenula polymorpha* which may also be found colonizing environmental sources (Lodder 1970; McGinnis et al. 1980).

The first reported case of human infection with *H. anomala* was of a child with interstitial pneumonia (Csillag et al. 1953). Subsequent cases include additional patients with pulmonary involvement (Csillag and Brandstein 1954; Kane et al. 2002), neonatal ventriculitis (Murphy et al. 1986), endocarditis (Nohinek et al. 1987), and urinary tract infection (Qadri et al. 1988). Most reported cases have described fungemia without end-organ involvement (Milstoc and Siddiqui 1986; Murphy et al. 1986; Klein et al. 1988; Yamada et al. 1995; Ma et al. 2000; Wong et al. 2000). *H. anomala* has also been reported as the actual etiologic agent in cases of oral 'candidiasis' (Cameron et al. 1993).

Risk factors for *H. anomala* infection appear to be similar to those described for other opportunistic fungal infections. These include central venous catheters, use of broad-spectrum antibacterials and hyperalimentation, immunosuppression/neutropenia, and surgery (Klein et al. 1988; Thuler et al. 1997).

Diagnosis of *Hansenula* infection is usually by isolation of the organism in culture of the infected site. Treatment of systemic *H. anomala* infection should include removal of the central venous catheter, if one is present, and use of a systemic antifungal agent. The optimal agent and dose have not been established. Parenteral amphotericin B is commonly used, and in vitro studies demonstrate susceptibility to this agent (Klein et al. 1988; Goss et al. 1994). There is, however, one report of amphotericin B failure, with subsequent successful sterilization of blood cultures by use of ketoconazole (Wong et al. 2000). Lipid complex amphotericin was successfully used to treat *H. anomala* in one patient with pneumonia, fungemia, and acute renal failure (Kane et al. 2002). The utility of fluconazole is unclear since there are reports of its successful use (Hirasaki et al. 1992; Goss et al. 1994; Yamada et al. 1995), but also reported instances of in vitro fluconazole resistance (Yamada et al. 1995), clinical failure of fluconazole treatment (Yamada et al. 1995), of the development of *H. anomala* infection in a patient already taking fluconazole for treatment of a *Candida* urinary tract infection (Alter and Farley 1994), and of breakthrough fungemia with *H. anomala* despite fluconazole prophylaxis (Krcmery et al. 1998).

There is only one clearly documented case of tissue infection with *H. polymorpha*: a child with chronic granulomatous disease developed hilar lymphadenopathy (McGinnis et al. 1980). *H. polymorpha* was recovered from cultures of hilar, posterior mediastinal, and paratracheal lymph nodes. The patient was successfully treated with amphotericin B.

Malassezia

Among the *Malassezia* species, *Malassezia furfur* is the most commonly identified cause of human infection. Infections with *Malassezia pachydermatis* are less common. There are a few case reports of *Malassezia sympodialis* infection. The most common types of *Malassezia* infection are superficial cutaneous and folliculitis (see Chapter 12, Superficial diseases caused by *Malassezia* species).

The major type of serious infection caused by *M. furfur* is central venous catheter-related fungemia. Neonates, especially those receiving lipid infusions (Sizun et al. 1994) and immunosuppressed patients, both children and adults (Barber et al. 1993; Morrison and Weisdorf 2000) are the patient populations most at risk for infection. Early reports suggested that *M. furfur* fungemia was almost always associated with the use of parenteral lipid infusions, but later reports describe the infection in patients without this risk factor (Myers et al. 1992; Barber et al. 1993). The clinical findings of *M. furfur* fungemia are nonspecific and the same as those seen with other types of bloodstream infection: fever, leukocytosis, thrombocytopenia (Barber et al. 1993).

In many reports of *M. furfur* infection, the organism is isolated only from blood cultures drawn through the central venous catheter (i.e. peripheral blood cultures are negative), and solid organ involvement is absent. There are, however, cases reported of pneumonia (Redline and Dahms 1981; Richet et al. 1989), peritonitis (Wallace et al. 1979; Gidding et al. 1989), and an intra-cardiac mass (Schleman et al. 2000) caused by *M. furfur*.

Although venous catheter colonization clearly occurs, all *M. furfur* fungemia should be treated, since prediction of catheter colonization versus true systemic infection is not possible. As for the other uncommon yeast infections, treatment of serious *M. furfur* infections is not standardized, but usually involves removal of the central venous catheter and use of systemic antifungal therapy (Marcon and Powell 1992; Morrison and Weisdorf 2000). Intravenous amphotericin B is the most commonly utilized antifungal agent.

M. sympodialis is a human colonizer and is felt to be a pathogen despite a paucity of reports of human infection. The strongest evidence for a pathogenic role comes from the isolation of the organism from the scalp of a patient with the acquired immunodeficiency syndrome (AIDS) with a focal skin infection (Simmons and Gueho 1990), as a central venous catheter-related infection in a surgical-oncology patient (Kikuchi et al. 2001), and as the cause of malignant otitis externa in a man with diabetes mellitus (Chai et al. 2000). It is also possible that *M. sympodialis* is under-recognized as a cause of infection since most laboratories do not distinguish it from *M. furfur*.

M. pachydermatis is also recognized as a cause of human infection. Essentially all reported cases are of fungemia in neonates (Marcon and Powell 1992; Lautenbach et al. 1998; Chryssanthou et al. 2001). There is also at least one reported occurrence of fungemia in an immunosupressed adult with a central venous catheter. As with *M. furfur* and other yeasts, central venous catheters and the administration of intravenous lipids may be risk factors for infection (Welbel et al. 1994). Removal of the venous catheter and administration of amphotericin B are both components of treatment of this infection as well. There are also reports of susceptibility to fluconazole and itraconazole, and resistance to flucytosine (Chryssanthou et al. 2001).

Rhodotorula

Rhodotorula species have been isolated from many environmental sources including soil, water, fruit juice, milk products, shower curtains, toothbrushes, and fiber-optic bronchoscopy equipment (Kwon-Chung and Bennet 1992; Whitlock et al. 1992; Warren and Hazen 1995). Infections due to the genus are rare, but are most frequently due to *Rhodotorula rubra*. Infections due to *Rhodotorula minuta* (Gregory and Haller 1992; Goldani et al. 1995; Pinna et al. 2001; Cutrona et al. 2002) and *Rhodotorula glutinis* (Pien et al. 1980; Casolari et al. 1992; Gaytan-Martinez et al. 2000; Dorey et al. 2002; Hsueh et al. 2003) have also been reported.

As with the other genera, the most commonly described infection due to *Rhodotorula* is fungemia. There are also reports of endocarditis (Naveh et al. 1975), meningitis (Pore and Chen 1976; Ahmed et al. 1998; Huttova et al. 1998), ventriculitis (Donald and Sharp 1988), peritonitis (Eisenberg et al. 1983; Wong et al. 1988; Pennington et al. 1995), and eye infections (Casolari et al. 1992; Gregory and Haller 1992; Guerra et al. 1992; Muralidhar and Sulthana 1995).

Rhodotorula fungemia most commonly occurs in patients with central venous catheters (Kiehn et al. 1992; Alliot et al. 2000; Kiraz 2000). Risk factors for the development of this infection are the same as those described for other opportunistic fungal bloodstream infections (immunosuppression, neutropenia, broad-spectrum antibacterials, hyperalimentation, burns, and surgery) (Marinová et al. 1994), but may also include endocarditis (Pien et al. 1980). The clinical presentation of *Rhodotorula* fungemia is nonspecific, with findings that may be seen in any bloodstream infection: fever, chills, tachycardia, and hypotension (Pien et al. 1980). Of note, many reports of *Rhodotorula* fungemia describe bloodstream infections that are polymicrobial.

Since *Rhodotorula* infections other than fungemia are rare, few general conclusions can be drawn about their clinical presentation. *Rhodotorula* peritonitis has only been reported in patients undergoing peritoneal dialysis and is notable for a propensity to lead to peritoneal fibrosis (Donald and Sharp 1988). Eye infections due to *Rhodotorula* species have included chronic postoperative endophthalmitis (caused by *R. minuta*) (Gregory and Haller 1992), keratitis (*R. glutinis*) (Casolari et al. 1992; Guerra et al. 1992), and dacryocystitis (Muralidhar and Sulthana 1995).

When deciding on treatment, it must be recognized that *Rhodotorula* may be an environmental contaminant. In one series where *Rhodotorula* was isolated from blood cultures (lysis centrifugation method) in 36 patients, only 23 were felt to have clinically significant fungemia. In the remainder there was only one colony of yeast isolated on one culture plate, and no clinical evidence of infection (Kiehn et al. 1992). In vitro susceptibility testing has shown that *Rhodotorula* is sensitive to flucytosine, moderately sensitive to amphotericin B, miconazole, ketoconazole, and itraconazole, and often resistant to fluconazole (Kiehn et al. 1992; Marinová et al. 1994; Galan-Sanchez et al. 1999). Treatment of fungemia has historically included systemic anti-fungal treatment and/or removal of a central venous catheter. Current recommendations for treatment are not standardized, but usually involve removal of the central catheter when feasible (Pien et al. 1980) and the use of amphotericin B at doses of 0.7 mg/kg/day (Kiehn et al. 1992). It may also be reasonable to add flucytosine to this regimen. *Rhodotorula* peritonitis has often been treated with intraperitoneal amphotericin B (Eisenberg et al. 1983; Wong et al. 1988), but this may exacerbate the risk of peritoneal fibrosis, and therefore intravenous amphotericin B may be preferable.

Saccharomyces

Saccharomyces cerevisiae is commonly known as baker's yeast or brewer's yeast, and is used in baking bread or making beer. Human colonization may occur at any mucosal surface, but happens more commonly in patients with underlying illnesses (Greer and Gemoets 1943; Kiehn et al. 1980; Aucott et al. 1990). A recent case series showed that increasing length of hospitalization was also a risk factor for greater likelihood of colonization with *S. cerevisae* (Salonen et al. 2000).

Clinical infections most commonly occur at the sites of mucosal colonization (thrush, esophagitis, vulvovaginitis), but invasive infections have also been reported. Overall, these infections occur in patients with significantly compromised immune systems. *S. cerevisiae* is thus a weak pathogen at best, but one that can clearly cause invasive opportunistic disease in an immunocompromised host. The specific underlying diseases associated with invasive *S. cerevisiae* infection include malignancy (Aucott et al. 1990), AIDS (Sethi and Mandell 1988; Doyle et al. 1990), myelodysplastic syndromes (Aucott et al. 1990; Oriol et al. 1993), burns (Eschette and West 1980), and rheumatoid arthritis (Feld et al. 1982). Clinical characteristics that may also be associated with disease are those described for other opportunistic fungal infections [invasive procedures/surgery (Dougherty and Simmons 1982; Chertow et al. 1991), hyperalimentation (Aucott et al. 1990), broad-spectrum antibacterial therapy (Aucott et al. 1990), and the presence of central venous catheters (Cimolai et al. 1987)]. Isolated fungemia is most common, but solid organ involvement may occur. Documented sites of deep infection include the lungs (Aucott et al. 1990), liver (Aucott et al. 1990), and joints (Feld et al. 1982). Empyema (Chertow et al. 1991) and peritonitis (Dougherty and Simmons 1982) have also been reported. There are also several case reports of *S. cerevisiae* fungemia after therapeutic use of *Saccharomyces boulardii* (Cesaro et al. 2000; Perapoch et al. 2000)

Clinical features of *S. cerevisiae* infection are not specific to the organism, but instead are determined by the site of infection. Blood cultures are diagnostic of fungemia, but since the organism is a common colonizer of mucosal surfaces, histological confirmation of tissue invasion is often necessary in order to diagnose solid organ involvement (Aucott et al. 1990).

Successful treatment of invasive *S. cerevisiae* infection with oral azoles has been noted in case reports (Dougherty and Simmons 1982; Cairoli et al. 1995), but there are also reports of azole-resistant *S. cerevisiae* (Salonen et al. 2000), and, in general, fungemia and/or solid-organ infection should be treated with parenteral amphotericin B.

OTHER *SACCHAROMYCES* SPECIES

S. boulardii is a *Saccharomyces* species, not usually considered a pathogen, that has been used in the treatment of bacterial diarrhea. In this setting, there have been reported cases of fungemia due to *S. boulardii* (Zunic et al. 1991; Pletincx et al. 1995; Hennequin et al. 2000; Rijnders et al. 2000; Lherm et al. 2002; Riquelme et al. 2003). Treatment should involve discontinuation of *S. boulardii* and institution of antifungal therapy.

Sporobolomyces

Sporobolomyces is usually recovered only from soil, occasionally from other environmental sources (leaves, bark, grasses, fruit) and rarely as a cause of human infection. Case reports describe one instance of mycetoma in which *Sporobolomyces roseus* was isolated (Janke 1954), one patient with dermatitis due to *Sporobolomyces holsaticus* (Bergman and Kauffman 1984), and several infections that appear due to *Sporobolomyces salmonicolor*. The reports of *S. salmonicolor* in which the organism is acting as pathogen have included one report of a positive culture from a removed nasal polyp (Dunnette et al. 1986), lymphadenitis (Plazas et al. 1994) and bone marrow involvement in AIDS patients (Morris et al. 1991), and a prosthetic cranioplasty infection (Morrow 1994). In addition, there is one reported case of extrinsic allergic alveolitis being caused by *Sporobolomyces* spp. in a veterinary student who developed fever, cough, and dyspnea when exposed to straw from which the organism was isolated: the patient had serum precipitans against *Sporobolomyces* and a positive intradermal skin test (Cockcroft et al. 1983). Furthermore, there is a description of an outbreak of asthma after a thunderstorm during which time there had been an increase in the number of *Sporobolomyces* spores in the air, suggesting the organism may have contributed to this event (Packe and Ayres 1985).

Given the rarity of clinical infection due to *Sporobolomyces* spp., there is clearly no standard therapy. Treatment which has been used successfully includes amphotericin B (Morrow 1994), amphotericin B followed by ketoconazole (Morris et al. 1991), and amphotericin B followed by fluconazole (Plazas et al. 1994).

Trichosporon

Trichosporon beigelii is a part of normal soil flora, and is occasionally found as colonizer of the oropharynx, skin, and nails. The organism was initially noted to cause superficial infections of the hair shaft (see Chapter 12 Superficial diseases caused by *Malassezia* species), then recognized as a pathogen capable of causing invasive disease (Watson and Kallichurum 1970). Most reported cases of disseminated infection have been in patients who are immunosuppressed in the setting of hematologic or solid organ malignancy, or solid organ transplantation (Sarfati et al. 1983; Murray-Leisure et al. 1986; Ness et al. 1989; Mirimoto et al. 1994; Tashiro et al. 1994). A series from Japan described 43 patients identified with disseminated *T. beigelii* infection: 37 (86 percent) of these patients had an underlying hematological malignancy. The majority of these patients (26 of 43) had been profoundly neutropenic (ANC <100/mm^3) prior to the development of infection (Tashiro et al. 1994). Notably, nearly all the patients in this series without a hematological malignancy were receiving systemic corticosteroids. Other populations in which *T. beigelii* fungemia has been described include premature infants (Giacoia 1992; Fisher et al. 1993; Singh et al. 1999), burn patients (Hajjeh and Blumberg 1995), and patients with the human immunodeficiency virus (HIV) infection (Anuradha et al. 2000; Girija et al. 2001).

Both a colonized gastrointestinal tract and central venous catheters are considered potential portals of entry for this infection. Pulmonary involvement is, however, the most common site of end-organ disease (Tashiro et al. 1995). Chest x-rays may show diffuse interstitial infiltrates or patchy reticulonodular involvement (Hoy et al. 1986; Tashiro et al. 1995). Signs and symptoms are similar to those of other fungal pneumonias in immunosuppressed patients: persistent fever in the face of antibacterial therapy associated with dyspnea, cough, and bloody sputum production (Tashiro et al. 1995). Other organs that are less commonly involved in disseminated *T. beigelii* infection include the brain (Watson and Kallichurum 1970), eyes (Sheikh et al. 1974), heart (Sidarous et al. 1994), liver (Haupt et al. 1983), spleen (Bhansali et al. 1986), bone (Goodman et al. 2002), and peritoneum (Anuradha et al. 2000). It is also important to note, however, that *T. beigelii* funguria may be a benign finding, not associated with invasive infection, in renal transplant patients (Lussier et al. 2000). There is also one report of a breast implant infection with *T. beigelii* in an immunocompetent patient (Reddy et al. 2002).

Cutaneous findings in association with deep *T. beigelii* infection have included macules, papules, vesicopustules, and nodules which may be localized to the extremities or found dispersed over the entire body (Piérard et al.

1992; Hsiao et al. 1994). Clinically, these lesions may appear similar to those seen in disseminated candidiasis. Cellulitis may also occur (Libertin et al. 1983). Skin biopsy with culture and/or histological studies may be necessary to clarify the diagnosis.

Disseminated trichosporon infection is often diagnosed by blood cultures. In the face of end-organ disease, diagnosis may also be made by biopsy of the affected site with culture or histopathological studies. Histology most frequently shows both yeast forms and hyphal elements that are larger than those seen in *Candida* infections (up to 10 μm). An example is illustrated by renal involvement in a neutropenic patient with leukemia (Figure 31.1). Arthrospores may be seen in some infections. Another distinguishing histological feature is that *T. beigelii* is commonly found arranged in a radial pattern (Tashiro et al. 1994). Immunoperoxidase staining has also been used as a diagnostic modality since discrimination between *Candida* and *T. beigelii* in tissue sections may occasionally be difficult (Tashiro et al. 1994). Of note, serum from patients with disseminated *T. beigelii* infection may show cross-reactivity with the latex agglutination test commonly used to diagnose infection with *Cryptococcus neoformans*.

Although intravenous amphotericin B is used as standard treatment of disseminated trichosporonosis, there are many reports of therapeutic failures (Walsh et al. 1990; Fisher et al. 1993; Tashiro et al. 1994). In one series, nine of 11 patients with trichosporon fungemia died of systemic fungal infection despite therapy with amphotericin B (Krcmery et al. 1999). In vitro susceptibility testing has demonstrated that although *T. beigelii* is inhibited by standard concentrations of amphotericin B, the levels required for the lethal effects on the organism are much higher. For example, using seven clinical isolates of *T. beigelii*, Walsh demonstrated that the organism was inhibited by standard concentrations of amphotericin B (< 2.0 μg/ml). With both a macrodilution technique and timed kill studies, however, they found that the majority of strains were not killed unless

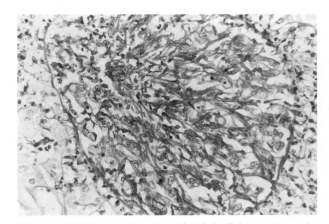

Figure 31.1 *Kidney involvement with* T. beigelii *in a neutropenic patient with leukemia (periodic acid–Schiff (PAS) 300×)*

extremely high concentrations of amphotericin B (> 20 µg/ml) were used (Walsh et al. 1990). Thus, amphotericin B is a static agent at attainable concentrations, making treatment difficult in patients who lack neutrophils to combat the infection. Other therapeutic options which have been considered include: the use of high-dose amphotericin B, combination treatment with amphotericin B and flucytosine or rifampin or fluconazole, liposomal amphotericin B, and systemic azoles (Walsh et al. 1990; Anaissie et al. 1992; Hajjeh and Blumberg 1995; Canales et al. 1998; Cawley et al. 2000). Although there are in vitro data showing synergism between terbinafine and azoles against *T. beigelii* (Ryder 1999), to our knowledge there are no clinical reports of its use.

OTHER *TRICHOSPORON* SPP.

In the past, species other than *T. beigelii* were only rarely reported as the cause of clinical infections. The taxonomy of this genus has now been redefined (see section on Microbiology below). *Trichosporon capitatum*, *Trichosporon fermentans*, and *Trichosporon penicillatum* are now classified as *Blastoschizomyces* and *Geotrichum* species. In addition, newly defined species, including *Trichosporon asahii*, *Trichosporon inkin*, *Trichosporon louberi*, *Trichosporon mucoides*, and *Trichosporon pullulans*, are being reported with increased frequency. There may be an impact on human infection since virulence may vary among these species. There are several case reports of invasive *T. pullulans* infection: one of pneumonia and the remainder of intravascular catheter-related fungemia in neutropenic patients (Shigehara et al. 1991; Kunová et al. 1995; Krcmery et al. 1999). Patients with *T. pullulans* fungemia have developed the infection while receiving ketoconazole and itraconazole antifungal prophylaxis (Kunová et al. 1995, 1996). There are also reports of bone marrow transplant patients receiving fluconazole prophylaxis who have nevertheless acquired serious *Trichosporon* infections (Moretti-Branchini et al. 2001). In one patient *T. asahii* caused fatal fungemia, despite the use of amphotericin. In the other, *T. inkin* caused a vascular catheter infection. One case report also describes a *T. inkin* lung abscess (Piwoz et al. 2000). Table 31.3 summarizes these sites of infection, and other case reports of infection with trichosporon species other than *T. beigelii* (Abliz et al. 2002; Fournier et al. 2002;

Gokahmetoglu et al. 2002; Meyer et al. 2002; Crowther et al. 2003; Nettles et al. 2003; Padhye et al. 2003).

Ustilago

Ustilago species are known as pathogens of corn and other grains. In corn, they are also known to induce tumor formation (Banuett 1995). *Ustilago maydis* has been described as an allergen causing asthma and *Ustilago esculenta* has been reported as a cause of hypersensitivity pneumonitis (Yoshida et al. 1996). There is one report of a central venous catheter infection with *Ustilago*. This infection occurred in a man with a chronic indwelling venous catheter who visited a farm frequently. The organism was isolated from the catheter, and subsequently from blood cultures 3 days after the catheter was removed. The infection resolved without use of an antifungal agent, but susceptibility testing showed that amphotericin B, fluconazole, and itraconazole should each have been effective (Patel et al. 1995).

MICROBIOLOGY

In order to understand the phylogenetic position of a yeast species, the sexual, teleomorph, stage needs to be demonstrated. This may be difficult since some species are heterothallic, requiring two genetically different, but compatible, strains for sexual reproduction to occur. In addition, there are data that suggest that some fungi, including yeast species, have lost their ability to reproduce sexually.

For this discussion, the less common yeast will be divided by their known or implied teleomorph classification. The ascomycetous yeast include members of the genera *Hansenula*, *Haseniospora*, and *Saccharomyces*. The basidiomycetous yeast include species of *Blastoschizomyces*, *Malassezia*, *Rhodotorula*, *Sporobolomyces*, *Trichosporon*, and *Ustilago*. In the absence of asci/ascospore or basidia/basidiospore production, other characteristics can be used to define their phylogenetic positions. This separation may be very important, as there are data that some cell wall active antifungals may not inhibit cell wall synthesis of all classes of fungi. Simple tests that can be performed in a clinical laboratory include testing for urease activity and determining staining with DBB. Methods such as DNA reassociation

Table 31.3 *Sites of* Trichosporon *infection*

Site	T. asahii	T. inkin	T. louberi	T. mucoides	T. pullulans
Pulmonary	–	+	–	–	+
Bloodstream/disseminated	+	+	–	+	+
Renal/urinary	–	–	+	–	–
Peritoneum	–	+	–	–	–
Liver/spleen	+	–	–	–	–

Table 31.4 *Characteristics of ascomycetous (non-Candida) yeast recovered from clinical specimens*

Characteristic	*Hansenula* spp.	*Hanseniospora* spp.	*Saccharomyces* spp.
Budding	Multilateral with narrow attachments	Bipolar, elongated daughter cells	Multilateral
Hyphae	± Pseudohyphae, rarely true	Absent to extensive	± Pseudohyphae
Ascospores	Hat-shaped	Spherical to hat-shaped	Spherical
Temperature	± 37°C	± 37°C	± 37°C
Fermentation	± (species dependent)	+	−
Nitrate utilization	+	−	−

studies, G/C content and rRNA/DNA sequencing are also available in research laboratories to elucidate phylogenetic relationships.

Ascomycetous yeast

Ascomycetous yeasts that have been recovered from clinical specimens, and may lead to human infection, include species of *Hansenula*, *Haseniospora*, and *Saccharomyces*. Characteristics of these yeasts are shown in Table 31.4. All three genera are urease negative and DBB negative.

HANSENULA

Although there are at least 30 species of *Hansenula* that have been described, the two most frequently encountered are *H. anomala* (*Candida pelliculosa*) and *H. polymorpha*. Key characteristics are the lack of urease, presence of hat-shaped ascospores, ability to utilize nitrate, compatible carbohydrate assimilation patterns and both species are fermenters. (Kurtzman 1984; Warren and Hazen 1995). Clinically relevant *Hansenula* spp. are shown in Table 31.5.

HANSENIOSPORA

There are at least eight recognized species of *Hanseniospora*, but only three may be encountered in clinical specimens (*Hanseniospora uvarum*, *Hanseniospora valbyensis*, and *Hanseniospora guilliermondii*) (Table 31.6). To date, none is known to have caused actual infection. Identification is based on ascospore production, lack of urease and nitrate utilization, fermentation, and compatible carbohydrate assimilation patterns (Smith 1984).

SACCHAROMYCES

Pseudohyphae or chains of budding yeast cells may be present. Asci and ascospores (usually four) are seen in many strains. There are seven species within this genus of which *S. cerevisiae* is the one associated with humans. Identification of this species can be made by carbohydrate assimilation patterns and microscopic morphology (ascospores can be seen in approximately 60 percent of strains). The genus cannot utilize nitrates, and all species are capable of fermenting sugars (Yarrow 1984; Warren and Hazen 1995).

Basidiomycetous yeast

Members of six genera of basidiomycetous yeast have been reported to cause human infection. Characteristics of these yeasts are shown in Table 31.7. Most are positive for urease activity and DBB staining.

BLASTOSCHIZOMYCES

Blastoschizomyces is a basidiomycetous genus with a single species, *B. capitatus*. This species was originally classified in the genus *Trichosporon*, which shares many similarities. Identification of *B. capitatus* is established by morphology with the production of true hyphae and annelloconidia, lack of urease activity, growth up to 42–45°C, growth in the presence of cycloheximide, and compatible carbohydrate assimilation patterns (Salkin et al. 1985; Warren and Hazen 1995).

MALASSEZIA

All *Malassezia* species but one (*M. pachydermatis*) require exogenous fatty acids for growth. Identification to the level of genus may be made based on the char-

Table 31.5 *Clinically relevant* Hansenula *species*

Characteristic	*H. anomala*	*H. polymorpha*
Ascospores	1–4, hat-shaped	1–4, less hat-shaped, more hemispherical
Hyphae	Pseudohyphae may be produced, but no true hyphae	Neither pseudo- or true hyphae
Growth at 37°C	±	+
Specific carbohydrate assimilation patterns	+	+
Fermentation	+	+

Table 31.6 *Clinically relevant* Hanseniospora *species*

Characteristic	H. uvarum	H. valbyensis	H. guilliermondii
Ascospores	1–2, spherical	2, rarely 4, hat- to helmet-shaped	1–4, mostly 4, hat- to helmet-shaped
Hyphae	Pseudohyphae, ± extensive	Pseudohyphae, ± extensive	Pseudohyphae, ± extensive
Growth at 37°C	–	–	+
Specific carbohydrate assimilation patterns	+	+	+
Fermentation	+	+	+

acteristics in Table 31.7; more detailed information for speciation is provided in Chapter 12, Superficial diseases caused by *Malassezia* species).

RHODOTORULA

Rhodotorula is a common yeast in environments associated with water. A key characteristic is the production of carotenoid pigments that convey an orange/pink color to the colonies. No *Rhodotorula* species produce true hyphae. Identification is based on the pigment, urease activity, nonfermenter, lack of utilization of inositol, and nitrate utilization. Some strains may produce capsular-type material. There at least eight species; some are the anamorph stage of the genus *Rhodosporidium*. Of the eight, *R. rubra*, *R. glutinis*, and *R. minuta* are encountered. They can be speciated by the characteristics noted in Table 31.8 (Fell et al. 1984; Warren and Hazen 1995).

SPOROBOLOMYCES

Sporobolomyces is another pigmented, basidiomycetous yeast genus. Key characteristics of the genus include the production of carotenoid pigments, presence of budding, true hyphae, forcibly discharged ballistospores, presence of urease activity, and the inability to ferment carbohydrates. There are at least seven species and perhaps the anamorph stage of *Sporidiobolus*. Of the seven species, *S. holsaticus*, *S. roseus*, and *S. salmonicolor* have been recovered from clinical specimens. Separation of these two species is based on morphology, temperature

optima, and carbohydrate assimilation patterns. See Table 31.9 (Fell and Statzell Tallman 1984).

TRICHOSPORON

Trichosporon is a genus characterized by the production of septate hyphae, abundant arthrospores, budding cells (that may be difficult to discern), and an inability to ferment carbohydrates. Most also express urease activity. *T. beigelii* (or *T. cutaneum*, which is the correct taxonomic name) is the most common species in this genus associated with humans, as both a colonizer and as a pathogen. Three species associated with humans have been re-evaluated and transferred to other genera; *T. capitatum* has been transferred to the genus *Blastoschizomyces*, and *T. penicillatum* and *T. fermentans* have been transferred to the genus *Geotrichum*. Most other species are environmental and rarely associated with humans.

In the past, the asexual genus *Trichosporon* encompassed a very heterogenous group of species that were the anamorph stages of both ascomyceterous and basidiomycetous fungi. Over the last 20 years, this genus has received significant taxonomic evaluation and revision. This genus now is considered a basidiomycetous yeast genus based on cell wall structure, positive DBB staining, septal pore morphology, urease production, similar rRNA sequences, and similarity of a complex polysaccharide antigen cross-reactive with the capsular antigen of *Cryptococcus neoformans* (Kreger-van Rij 1984).

Table 31.7 *Characteristics of basidiomycetous (non-cryptococccal) yeast recovered from clinical specimens*

Characteristic	Blastoschizomyces	Malassezia	Rhodotorula	Sporobolomyces	Trichospsoron	Ustilago
Colony pigmentation	–	–	+	+	–	–
Budding	May be present (anellospores)	Unipolar	Multilateral	Multilateral	May be difficult to observe	Present
Hyphae	True	±	Usually absent	True and pseudo	True	–
Other spores	Arthrospores	–	–	± ballistospores	Arthrospores	Absent
Temperature	up to 45°C	37°C	± 37°C	± 37°C	± 37°C	–
Lipid-dependent	No	All, except *M. pachydermatis*[a]	No	No	No	No
Urease	–	+	+	+	Most positive	No
Fermenter	No	No	No	No	No	

a) *M. pachydermatis* is lipophilic, but not lipid-dependent.

Table 31.8 *Clinically relevant* Rhodotorula *species*

Characteristic	R. rubra	R. glutinis	R. minuta
Hyphae	Rudimentary, pseudohyphae may be produced	Usually absent, pseudohyphae if present	Absent
Growth at 37°C	±	±	±
Nitrate utilization	−	+	−
Specific carbohydrate assimilation patterns	+	+	+

Table 31.9 *Clinically relevant* Sporobolomyces *species*

Characteristic	S. holsaticus	S. roseus	S. salmonicolor
Hyphae	Extensive true hyphae, rare pseudohyphae	Absent	Absent–present, pseudo or true
Ballistospores	Usually produced	usually abundant	Usually produced
Growth at 37°C	−	−	±
Nitrate utilization	+	+	+
Specific carbohydrate assimilation patterns	+	+	+

Identification of *Trichosporon* spp., especially *T. cutaneum* (*beigelii*), currently requires morphological and biochemical studies (for details, see Chapter 12, Superficial diseases caused by *Malassezia* species). Morphologically, abundant true hyphae are produced within 24–72 hours that give rise to arthrospores which are initially rectangular in shape. Blastospore production is usually more difficult to observe. They may be produced along the true hyphae, or in chains from pseudohyphae, or they may even bud off from the arthrospores. Most species including *T. cutaneum* (*beigelii*) are urease positive, do not ferment carbohydrates, but assimilate many carbohydrates. Assimilation patterns are commonly used for identification of members of this genus.

Gueho et al. 1992 have performed an extensive evaluation of 101 strains representing the full range of species recovered from humans, animals, and environmental sources. Characteristics used for assessment included morphology, ultrastructure morphology, physiological parameters, ubiquinone systems, G/C content of DNA, DNA/DNA reassociation percentages, and partial sequences of 26S rRNA. A total of 19 taxa were delineated within this genus using data from all of the characteristics. Members of six of the 19 taxa were associated with humans: *T. asahii*, *T. asteroides*, *T. cutaneum*, *T. mucoides*, *T. ovoides*, and *T. inken*. Fortunately, speciation within this genus can be accomplished using temperature studies, assimilation reactions, and cyloheximide susceptibility or resistance. DNA-related studies by Sugita et al. 1995 confirmed the studies of Gueho et al. 1992. Human infections caused by *T. cutaneum* (*beigelii*) were caused by at least four different DNA-based species. When DNA association percentages of 10 clinical strains of *Trichosporon* were determined, seven of the eight isolates recovered from blood or urine specimens were *T. asahii* and one was

T. ovoides. Of the two strains recovered from superficial infections, one was *T. cutaneum* and one was *T. montevideense*. Clinical studies with identification to the levels of these taxa will be needed to determine whether there are differences in virulence or treatment associated with any of these new taxa.

USTILAGO

Ustilago is a yeast-like phase of an important plant disease called the smuts. They are recognized by their elongated budding cells.

CONCLUSIONS

With growing populations of immunocompromised patients, it is likely that the uncommon yeast described in this chapter and other species will become ever more important. Further changes in epidemiology and spectrum of disease may be noted in upcoming years with the increasing use of extended-spectrum azoles and cell wall acting echinocandins. Accurate recognition of the causative organisms of specific infections, continued reporting of efficacy of different therapeutic options, and the use of standardized in vitro susceptibility testing should allow ongoing refinements in the treatment of the patients affected by such infections.

REFERENCES

Abliz, P., Fukushima, K., et al. 2002. Identification of the first isolates of *Trichosporon asahii* var asahii from disseminated trichosporonosis in China. *Diagn Microbiol Infect Dis*, **44**, 17–22.

Ahmed, A., Aggarwal, M., et al. 1998. A fatal case of *Rhodotorula* meningitis in AIDS. *Med Health Rhode Island*, **81**, 1, 22–3.

Alliot, C., Desablens, B., et al. 2000. Opportunistic infection with *Rhodotorula* in cancer patients treated by chemotherapy: two case reports. *Clin Oncol (R Coll Radiol)*, **12**, 2, 115–17.

Alter, S.J. and Farley, J. 1994. Development of *Hansenula anomala* infection in a child receiving fluconazole therapy. *Pediatr Infect Dis J*, **13**, 2, 158–9.

Anaissie, E., Gokaslan, A., et al. 1992. Azole therapy for trichosporonosis: clinical evaluation of eight patients, experimental therapy for murine infection, and review. *Clin Infect Dis*, **15**, 781–7.

Anuradha, S., Chatterjee, A., et al. 2000. *Trichosporon beigelii* peritonitis in a HIV-positive patient on continuous ambulatory peritoneal dialysis. *J Assoc Physicians India*, **48**, 10, 1022–4.

Aucott, J.N., Fayen, J., et al. 1990. Invasive infection with *Saccharomyces cerevisiae*: report of three cases and review. *Rev Infect Dis*, **12**, 3, 406–11.

Banuett, F. 1995. Genetics of *Ustilago maydis*, a fungal pathogen that induces tumors in maize. *Annu Rev Genet*, **29**, 179–208.

Barber, G.R., Brown, A.E., et al. 1993. Catheter-related *Malassezia furfur* fungemia in immunocompromised patients. *Am J Med*, **95**, 365–70.

Bergman, A.G. and Kauffman, C.A. 1984. Dermatitis due to *Sporobolomyces* infection. *Arch Dermatol*, **120**, 1059–60.

Bhansali, S., Karanes, C., et al. 1986. Successful treatment of disseminated *Trichosporon beigelii* (*cutaneum*) infection with associated splenic involvement. *Cancer*, **58**, 1630–2.

Cairoli, R., Marenco, P., et al. 1995. *Saccharomyces cerevisiae* fungemia with granulomas in the bone marrow in a patient undergoing BMT. *Bone Marrow Transplant*, **15**, 785–6.

Cameron, M.L., Schell, W.A., et al. 1993. Correlation of in vitro fluconazole resistance of *Candida* isolates in relation to therapy and symptoms of individuals seropositive for human immunodeficiency virus type 1. *Antimicrob Agents Chemother*, **37**, 11, 2449–53.

Canales, M.A., Sevilla, J., et al. 1998. Successful treatment of *Trichosporon beigelii* pneumonia with itraconazole. *Clin Infect Dis*, **26**, 999–1000.

Casolari, C., Nanetti, A., et al. 1992. Keratomycosis with an unusual etiology (*Rhodotorula glutinis*): a case report. *Microbiologica*, **15**, 1, 83–7.

Cawley, M.J., Braxton, G.R., et al. 2000. *Trichosporon beigelii* infection: experience in a regional burn center. *Burns*, **26**, 5, 483–6.

Cesaro, S., Chinello, P., et al. 2000. *Saccharomyces cerevisiae* fungemia in a neutropenic patient treated with *Saccharomyces boulardii*. *Support Care Cancer*, **8**, 6, 504–5.

Chai, F.C., Auret, K., et al. 2000. Malignant otitis externa caused by *Malassezia sympodialis*. *Head Neck*, **22**, 1, 87–9.

Chertow, G.M., Marcantonio, E.R. and Wells, R.G. 1991. *Saccharomyces cerevisiae* empyema in a patient with esophago-pleural fistula complicating variceal sclerotherapy. *Chest*, **99**, 1518–19.

Cheung, M.Y., Chiu, N.C., et al. 1999. Mandibular osteomyelitis caused by *Blastoschizomyces capitatus* in a child with acute myelogenous leukemia. *J Formos Med Assoc*, **98**, 11, 787–9.

Chryssanthou, E., Broberger, U. and Petrini, B. 2001. *Malassezia pachydermatis* fungaemia in a neonatal intensive care unit. *Acta Paediatr*, **90**, 3, 323–7.

Cimolai, N., Gill, M.J. and Church, D. 1987. *Saccharomyces cerevisiae* fungemia: case report and review of the literature. *Diagn Microbiol Infect Dis*, **8**, 113–17.

Cockcroft, D.W., Berscheid, B.A., et al. 1983. *Sporobolomyces*: a possible cause of extrinsic allergic alveolitis. *J Allergy Clin Immunol*, **72**, 305–9.

Crowther, K.S., Webb, A.T. and McWhitney, P.H. 2003. *Trichosporon inkin* peritonitis in a patient on continuous ambulatory peritoneal dialysis returning from the Caribbean. *Clin Nephrol*, **59**, 69–70.

Csillag, A. and Brandstein, L. 1954. The role of *Blastomyces* in the aetiology of interstitial plasmocytic pneumonia of the premature infant. *Acta Microbiol Hung*, **2**, 179–90.

Csillag, A., Brandstein, L., et al. 1953. Adatok a koraszulottkori interstitialis pneumonia koroktanahoz. *Orv Hetil*, **94**, 1303–4.

Cutrona, A.F., Shah, M., et al. 2002. *Rhodotorula minuta*: an unusual fungal infection in hip-joint prosthesis. *Am J Orthop*, **31**, 3, 137–40.

D'Antonio, D., Piccolomini, R., et al. 1994. Osteomyelitis and intervertebral discitis caused by *Blastoschizomyces capitatus* in a patient with acute leukemia. *J Clin Microbiol*, **32**, 1, 224–7.

DeMaio, J. and Colman, L. 2000. The use of adjuvant interferon-gamma therapy for hepatosplenic *Blastoschizomyces capitatus* infection in a patient with leukemia. *Clin Infect Dis*, **31**, 3, 822–4.

Donald, F.E. and Sharp, J.F. 1988. *Rhodotorula rubra* ventriculitis. *J Infect*, **16**, 187–91.

Dorey, M.W., Brownstein, S., et al. 2002. *Rhodotorula glutinis* endophthalmitis. *Can J Ophthalmol*, **37**, 3, 416–18.

Dougherty, S.H. and Simmons, R.L. 1982. Postoperative peritonitis caused by *Saccharomyces cerevisiae*. *Arch Surg*, **117**, 2, 248 (Letter).

Doyle, M.G., Pickering, L.K., et al. 1990. *Saccharomyces cerevisiae* infection in a patient with acquired immunodeficiency syndrome. *Pediatr Infect Dis J*, **9**, 850–1.

Dunnette, S.L., Hall, M.M., et al. 1986. Microbiologic analyses of nasal polyp tissue. *J Allergy Clin Immunol*, **78**, 102–8.

Eisenberg, E.S., Alpert, B.E., et al. 1983. *Rhodotorula rubra* peritonitis in patients undergoing continuous ambulatory peritoneal dialysis. *Am J Med*, **75**, 349–52.

Eschette, M.L. and West, B.C. 1980. *Saccharomyces cerevisiae* septicemia. *Arch Intern Med*, **140**, 1539.

Feld, R., Fornasier, V.L., et al. 1982. Septic arthritis due to *Saccharomyces* species in a patient with chronic rheumatoid arthritis. *J Rheumatol*, **9**, 637–40.

Fell, J.W. and Statzell Tallman, A. 1984. Genus 13. *Sporobolomyces Kluyver et van Niel*. In: Kreger-van Rij, N.J.W. (ed.), *The yeasts, a taxonomic study*, 3rd revised edition. Amsterdam, Netherlands: Elsevier Science Publications, 911–20.

Fell, J.W., Statzell Tallman, A.S. and Ahearn, D.G. 1984. Genus 10. *Rhodotorula Hansen*. In: Kreger-van Rij, N.J.W. (ed.), *The yeasts, a taxonomic study*, 3rd revised edition. Amsterdam, Netherlands: Elsevier Science Publications, 893–905.

Fisher, D.J., Christy, C., et al. 1993. Neonatal *Trichosporon beigelii* infection: report of a cluster of cases in a neonatal intensive care unit. *Pediatr Infect Dis J*, **12**, 149–55.

Fournier, S., Pavageau, W., et al. 2002. Use of voriconazole to successfully treat disseminated *Trichosporon asahii* infection in a patient with acute myeloid leukaemia. *Eur J Clin Microbiol Infect Dis*, **21**, 892–6.

Galan-Sanchez, F., Garcia-Martos, P., et al. 1999. Microbiological characteristics and susceptibility patterns of strains of rhodotorula isolated from clinical samples. *Mycopathologia*, **145**, 3, 109–12.

Gaytan-Martinez, J., Mateos-Garcia, E., et al. 2000. Microbiological findings in febrile neutropenia. *Arch Med Res*, **31**, 4, 388–92.

Giacoia, G.P. 1992. *Trichosporon beigelii*: a potential cause of sepsis in premature infants. *South Med J*, **85**, 12, 1247–8.

Gidding, H., Hawes, L. and Dwyer, B. 1989. The isolation of *Malassezia furfur* from an episode of peritonitis. *Med J Aust*, **151**, 603 (Letter).

Girija, T., Kumari, R., et al. 2001. Pneumonia due to *Trichosporon beigelii* in HIV-positive patient – a case report. *Indian J Pathol Microbiol*, **44**, 3, 379–80.

Girmenia, C., Micozzi, A., et al. 1991. Fluconazole treatment of *Blastoschizomyces capitatus* meningitis in an allogeneic bone marrow recipient. *Eur J Clin Microbiol Infect Dis*, **10**, 752–6.

Gokahmetoglu, S., Nedret Koe, A., et al. 2002. Case reports. *Trichosporon mucoides* infection in three premature newborns. *Mycoses*, **45**, 123–5.

Goldani, L.Z., Craven, D.E. and Sugar, A.M. 1995. Central venous catheter infection with *Rhodotorula minuta* in a patient with AIDS taking suppressive doses of fluconazole. *J Med Vet Mycol*, **33**, 4, 267–70.

Goodman, D., Pamer, E., et al. 2002. Breakthrough trichosporonosis in a bone marrow transplant recipient receiving caspofungin acetate. *Clin Infect Dis*, **35**, E35–36.

Goss, G., Grigg, A., et al. 1994. *Hansenula anomala* infection after bone marrow transplantation. *Bone Marrow Transplant*, **14**, 995–7.

Greer, A.E. and Gemoets, H.N. 1943. The coexistence of pathogenic fungi in certain chronic pulmonary diseases: with especial reference to pulmonary tuberculosis. *Dis Chest*, **9**, 212–24.

Gregory, J.K. and Haller, J.A. 1992. Chronic postoperative *Rhodotorula* endophthalmitis. *Arch Ophthalmol*, **110**, 1686–7.

Gueho, E., Smith, M.T., et al. 1992. Contributions to a revision of the genus *Trichosporon*. *Antonie van Leeuwenhoek*, **61**, 289–316.

Guerra, R., Cavallini, G.M., et al. 1992. *Rhodotorula glutinis* keratitis. *Int Ophthalmol*, **16**, 3, 187–90.

Hajjeh, R.A. and Blumberg, H.M. 1995. Bloodstream infection due to *Trichosporon beigelii* in a burn patient: case report and review of therapy. *Clin Infect Dis*, **20**, 913–16.

Haupt, H.M., Merz, W.G., et al. 1983. Colonization and infection with *Trichosporon* species in the immunosuppressed host. *J Infect Dis*, **147**, 2, 199–203.

Hennequin, C., Kauffmann-Lacroix, C., et al. 2000. Possible role of catheters in *Saccharomyces boulardii* fungemia. *Eur J Clin Microbiol Infect Dis*, **19**, 1, 16–20.

Hirasaki, S., Ijichi, T., et al. 1992. Fungemia caused by *Hansenula anomala*: successful treatment with fluconazole. *Int Med*, **31**, 622–4.

Hoy, J., Hsu, K., et al. 1986. *Trichosporon beigelii* infection: a review. *Rev Infect Dis*, **8**, 959–67.

Hsiao, G.H., Chang, C.C., et al. 1994. *Trichosporon beigelii* fungemia with cutaneous dissemination: a case report and literature review. *Acta Derm Venereol (Stockh)*, **74**, 481–2 (letter).

Hsueh, P.R., Teng, L.J., Ho, S.W. and Luh, K.T. 2003. Catheter-related sepsis due to *Rhodotorula glutinis*. *J Clin Microbiol*, **41**, 2, 857–9.

Hurtova, M., Kralinsky, K., et al. 1998. Prospective study of nosocomial fungal meningitis in children – report of 10 cases. *Scand J Infect Dis*, **30**, 5, 485–7.

Janke, A. 1954. *Sporobolomyces roseus* var. *Madurae* var. nov. und die Beziehungen zwischen den Genera *Sporbolomyces* und *Rhodotorula*. *Zentralbl Bakteriol Parasitenkd*, **161**, 514–20.

Kane, S.L., Dasta, J.F. and Cook, C.H. 2002. Amphotericin B lipid complex for *Hansenula anomala* pneumonia. *Ann Pharmacother*, **36**, 1, 59–62.

Kiehn, T.E., Edwards, F.F. and Armstrong, D. 1980. The prevalence of yeasts in clinical specimens from cancer patients. *Am J Clin Pathol*, **73**, 518–21.

Kiehn, T.E., Gorey, E., et al. 1992. Sepsis due to *Rhodotorula* related to use of indwelling central venous catheters. *Clin Infect Dis*, **14**, 841–6.

Kikuchi, K., Fujishiro, Y., et al. 2001. A case of central venous catheter-related infection with *Malassezia sympodialis*. *Nippon Ishinkin Gakkai Zasshi*, **42**, 4, 220–2.

Kiraz, N., Gulbas, Z. and Akgun, Y. 2000. Case report. *Rhodotorula rubra* fungamia due to use of indwelling venous catheters. *Mycoses*, **43**, 5, 209–10 (letter).

Klein, A.S., Tortora, G.T., et al. 1988. *Hansenula anomala*: a new fungal pathogen: two case reports and a review of the literature. *Arch Intern Med*, **148**, 1210–13.

Kremery, V., Oravcova, E., et al. 1998. Nosocomial breakthrough fungaemia during antifungal prophylaxis or empirical antifungal therapy in 41 cancer patients receiving antineoplastic chemotherapy: analysis of aetiology risk factors and outcome. *J Antimicrob Chemother*, **41**, 3, 373–80.

Kremery, V., Mateicka, F., et al. 1999. Hematogenous trichosporonosis in cancer patients: report of 12 cases including 5 during prophylaxis with itraconazole. *Support Care Cancer*, **7**, 39–43.

Kreger-van Rij, N.J.W. 1984. Genus 16: *Trichosporon Behrend*. In: Kreger-van Rij, N.J.W. (ed.), *The yeasts, a taxonomic study*, 3rd revised edition. Amsterdam, Netherlands: Elsevier Science Publications.

Kuncvá, A., Sorkovská, D., et al. 1995. Report of catheter-associated *Trichosporon pullulans* break-through fungemia in a cancer patient. *Eur J Clin Microbiol Infect Dis*, **14**, 729–30, (Letter).

Kunová, A., Godal, J., et al. 1996. Fatal *Trichosporon pullulans* break-through fungemia in cancer patients: report of three patients who failed on prophylaxis with itraconazole. *Infection*, **24**, 3, 273–4.

Kurtzman, C.P. 1984. Genus 11. *Hansenula H. et P. Sydow*. In: Kreger-van Rij, N.J.W. (ed.), *The yeasts, a taxonomic study*, 3rd revised

edition. Amsterdam, Netherlands: Elsevier Science Publications, 165–213.

Kwon-Chung, K.J. and Bennet, J.E. 1992. Infections due to *Trichosporon* and other miscellaneous yeast-like fungi. In: Kwon-Chung, K.J. and Bennets, J.E. (eds), *Medical mycology*. Philadelphia: Lea & Febiger.

Lautenbach, E., Nachamkin, I. and Schuster. M. 1998. *Malassezia pachydermatis* infections. *N Engl J Med*, **339**, 4, 270–1.

Lherm, T., Monet, C., et al. 2002. Seven cases of fungemia with *Saccharomyces boulardii* in critically ill patients. *Intensive Care Med*, **28**, 6, 797–801.

Libertin, C.R., Davies, N.J., et al. 1983. Invasive disease caused by *Trichosporon beigelii*. *Mayo Clinic Proc*, **58**, 684–6.

Lodder, J. (ed.). 1970. *The yeasts: a taxonomic study*, 2nd edition. Amsterdam, Netherlands: North-Holland Publishing Co.

Lussier, N., Laverdiere, M., et al. 2000. *Trichosporon beigelii* funguria in renal transplant recipients. *Clin Infect Dis*, **31**, 5, 1299–301.

Ma, J.S., Chen, P.Y., et al. 2000. Neonatal fungemia caused by *Hansenula anomala*: a case report. *J Microbiol Infect*, **33**, 4, 267–70.

Marcon, M.J. and Powell, D.A. 1992. Human infections due to *Malassezia* spp.. *Clin Microbiol Rev*, **5**, 2, 101–19.

Marinová, I., Szabadosová, V., et al. 1994. *Rhodotorula* spp. fungemia in an immunocompromised boy after neurosurgery successfully treated with miconazole and 5-flucytosine: case report and review of the literature. *Chemotherapy*, **40**, 287–9.

Martino, P., Girmenia, C., et al. 1990a. Spontaneous pneumothorax complicating pulmonary mycetoma in patients with acute leukemia. *Rev Infect Dis*, **12**, 4, 611–17.

Martino, P., Venditti, M., et al. 1990b. *Blastoschizomyces capitatus*: an emerging cause of invasive fungal disease in leukemia patients. *Rev Infect Dis*, **12**, 4, 570–82.

McGinnis, M.R., Walker, D.H. and Folds, J.D. 1980. *Hansenula polymorpha* infection in a child with chronic granulomatous disease. *Arch Pathol Lab Med*, **104**, 290–2.

Meyer, M.H., Letscher-Bru, V., et al. 2002. Chronic disseminated *Trichosporon asahii* infection in a leukemic child. *Clin Infect Dis*, **35**, E22–25.

Milstoc, M. and Siddiqui, N.A. 1986. Fungemia due to *Hansenula anomala*. *NY State J Med*, **86**, 541–2.

Mirimoto S., Shimaziki, C., et al. 1994. *Trichosporon cutaneum* fungemia in patients with acute myeloblastic leukemia and measurement of serum D-arabinitol, *Candida* antigen (CAND-TEC), and β-D-glucan. *Ann Hematol*, **68**, 159–61.

Moretti-Branchini, M.L., Fukushima, K., et al. 2001. *Trichosporon* species infection in bone marrow transplanted patients. *Diagn Microbiol Infect Dis*, **39**, 3, 161–4.

Morris, J.T., Beckius, M. and McAllister, C.K. 1991. *Sporobolomyces* infection in an AIDS patient. *J Infect Dis*, **164**, 623–4.

Morrison, V.A. and Weisdorf, D.J. 2000. The spectrum of *Malassezia* infections in the bone marrow transplant population. *Bone Marrow Transplant*, **26**, 6, 645–8.

Morrow, J.D. 1994. Prosthetic cranioplasty infection due to *Sporobolomyces*. *J Tenn Med Assoc*, **87**, 11, 466–7.

Muralidhar, S. and Sulthana, C.M. 1995. *Rhodotorula* causing chronic dacryocystitis: a case report. *Indian J Ophthalmol*, **43**, 4, 196–8.

Murphy, N., Buchanan, C.R., et al. 1986. Infection and colonization of neonates by *Hansenula anomala*. *Lancet*, **2**, 291–3.

Murray-Leisure, K.A., Aber, R.C., et al. 1986. Disseminated *Trichosporon beigelii* (*cutaneum*) infection in an African heart recipient. *JAMA*, **256**, 2995–8.

Myers, J.W., Smith, R.J., et al. 1992. Fungemia due to *Malassezia furfur* in patients without the usual risk factors. *Clin Infect Dis*, **14**, 620–1.

Naveh, Y., Friedman, A., et al. 1975. Endocarditis caused by *Rhodotorula* successfully treated with 5-fluorocytosine. *Br Heart J*, **37**, 101–4.

Ness, M.J., Markin, R.S., et al. 1989. Disseminated *Trichosporon beigelii* infection after orthotopic liver transplantation. *Am J Clin Pathol*, **92**, 119–23

Nettles, R.E., Nichols, L.S., et al. 2003. Successful treatment of *Trichosporon mucoides* infection with fluconazole in a heart and kidney transplant recipient. *Clin Infect Dis*, **36**, E63–66.

Nohinek, B., Zee-Cheng, C.-S., et al. 1987. Infective endocarditis of a bicuspid aortic valve caused by *Hansenula anomala*. *Am J Med*, **82**, 165–8.

Oriol, A., Ribera, J.M., et al. 1993. *Saccharomyces cerevisiae* septicemia in a patient with myelodysplastic syndrome. *Am J Hematol*, **43**, 4, 325–6, (Letter).

Packe, G.E. and Ayres, J.G. 1985. Asthma outbreak during a thunderstorm. *Lancet*, **2**, 8448, 199–204.

Padhye, A.A., Verghese, S., et al. 2003. *Trichosporon louberi* infection in a patient with adult polycystic kidney disease. *J Clin Microbiol*, **41**, 479–82.

Pagano, L., Morace, G., et al. 1996. Adjuvant therapy with rhGM-CSF for the treatment of *Blastoschizomyces capitatus* systemic infection in a patient with acute myeloid leukemia. *Ann Hematol*, **73**, 1, 33–4.

Patel, R., Roberts, G.D., et al. 1995. Central venous catheter infection due to *Ustilago* species. *Clin Infect Dis*, **21**, 4, 1043–4.

Pennington, J.C., Hauer, K. and Miller, W. 1995. *Rhodotorula rubra* peritonitis in an HIV+ patient on CAPD. *Del Med J*, **67**, 3, 184.

Perapoch, J., Planes, A.M., et al. 2000. Fungemia with *Saccharomyces cerevisiae* in two newborns, only one of whom had been treated with ultra-levura. *Eur J Clin Microbiol Infect Dis*, **19**, 6, 468–70.

Pien, F.D., Thompson, R.L., et al. 1980. *Rhodotorula* septicemia: two cases and a review of the literature. *Mayo Clin Proc*, **55**, 258–60.

Piérard, G.E., Read, D., et al. 1992. Cutaneous manifestations in systemic trichosporonosis. *Clin Exp Dermatol*, **17**, 79–82.

Pinna, A., Carta, F., et al. 2001. Endogenous *Rhodotorula minuta* and *Candida albicans* endophthalmitis in an injecting drug user. *Br J Ophthalmol*, **85**, 6, 759.

Piwoz, J.A., Stadtmauer, G.J., et al. 2000. *Trichosporon inkin* lung abscesses presenting as a penetrating chest wall mass. *Pediatr Infect Dis*, **19**, 10, 1025–7.

Plazas, J., Portilla, J., et al. 1994. *Sporobolomyces salmonicolor* lymphadenitis in an AIDS patient: pathogen or passenger? *AIDS*, **8**, 387–8.

Pletincx, M., Legein, J. and Vandenplas, Y. 1995. Fungemia with *Saccharomyces boulardii* in a 1-year-old girl with protracted diarrhea. *J Pediatr Gastroenterol Nutr*, **21**, 113–15.

Pore, R.S. and Chen, J. 1976. Meningitis caused by *Rhodotorula*. *Sabouraudia*, **14**, 331–5.

Qadri, S.M.H., Dayel, F.A., et al. 1988. Urinary tract infection caused by *Hansenula anomala*. *Mycopathologia*, **104**, 99–101.

Reddy, B.T., Torres, H.A. and Kontoyiannis, D.P. 2002. Breast implant infection caused by *Trichosporon beigelii*. *Scand J Infect Dis*, **34**, 43–4.

Redline, R.W. and Dahms, B.B. 1981. *Malassezia* pulmonary vasculitis in an infant on long-term intralipid therapy. *N Engl J Med*, **305**, 23, 1395–8.

Richet, H.M., McNeil, M.M., et al. 1989. Cluster of *Malassezia furfur* pulmonary infections in infants in a neonatal intensive-care unit. *J Clin Microbiol*, **27**, 6, 1197–200.

Rijnders, B.J., Van Wijngaerden, E., et al. 2000. *Saccharomyces* fungemia complicating *Saccharomyces boulardii* treatment in a non-immunocompromised host. *Intensive Care Med*, **26**, 6, 825.

Riquelme, A.J., Calvo, M.A., et al. 2003. *Saccharomyces cerevisiae* fungemia after *Saccharomyces boulardii* treatment in immunocompromised patients. *J Clin Gastroenterol*, **36**, 1, 41–3.

Ryder, N.S. 1999. Activity of terbinafine against serious fungal infections. *Mycoses*, **42**, Suppl 2, 115–19.

Salkin, I.F., Gordon, M.A., et al. 1985. *Blastoschizomyces capitatus*, a new combination. *Mycotaxon*, **22**, 373–80.

Salonen, J.H., Richardson, M.D., et al. 2000. Fungal colonization of haematological patients receiving cytotoxic chemotherapy: emergence of azole-resistant *Saccharomyces cerevisiae*. *J Hosp Infect*, **45**, 4, 293–301.

Sarfati, C., et al. 1983. Septicémie mortelle à *Trichosporon cutaneum* chez un sujet immunodéprimé. *Bull Soc Fr Mycol Med*, **12**, 287–9.

Schleman, K.A., Tullis, G. and Blum, R. 2000. Intracardiac mass complicating *Malassezia furfur* fungemia. *Chest*, **118**, 6, 1828–9.

Sethi, N. and Mandell, W. 1988. *Saccharomyces* fungemia in a patient with AIDS. *NY State J Med*, **88**, 278–9.

Sheikh, H.A., Mahgoub, S. and Badi, K. 1974. Postoperative endophthalmitis due to *Trichosporon cutaneum*. *Br J Ophthalmol*, **58**, 591–4.

Shigehara, K., Tahakshi, K., et al. 1991. A case of *Trichosporon pullulans* infection of the lung in patient with adult T cell leukemia. *Jpn J Med*, **30**, 135.

Sidarous, M.G., Reilly, M.V.O. and Cherubin, C.E. 1994. A case of *Trichosporon beigelii* endocarditis years after aortic valve replacement. *Clin Cardiol*, **8**, 215–19.

Simmons, R.B. and Gueho, E. 1990. A new species of *Malassezia*. *Mycol Res*, **94**, 8, 1146–9.

Singh, K., Chakrabarti, A., et al. 1999. Yeast colonisation and fungaemia in preterm neonates in a tertiary care centre. *Indian J Med Res*, **110**, 169–73.

Sizun, J., Karangwa, A., et al. 1994. *Malassezia furfur*-related colonization and infection of central venous catheters. *Intensive Care Med*, **20**, 496–9.

Smith, M.T. 1984. Genus 10. *Hanseniospora Zikes*. In: Kreger-van Rij, N.J.W. (ed.), *The yeasts, a taxonomic study*, 3rd revised edition. Amsterdam, Netherlands: Elsevier Science Publications, 154–64.

Sugita, T., Nishikawa, A., et al. 1995. Taxonomic position of deep-seated, mucosa-associated, and superficial isolates of *Trichosporon cutaneum* from trichosporonosis patients. *J Clin Microbiol*, **33**, 1368–70.

Tashiro, T., Nagai, H., et al. 1994. Disseminated *Trichosporon beigelii* infection in patients with malignant diseases: immunohistochemical study and review. *Eur J Clin Microbiol Infect Dis*, **13**, 3, 218–24.

Tashiro, T., Nagai, H., et al. 1995. *Trichosporon beigelii* pneumonia in patients with hematologic malignancies. *Chest*, **108**, 190–5.

Thuler, L.C., Faivichenco, S., et al. 1997. Fungaemia caused by *Hansenula anomala* – an outbreak in a cancer hospital. *Mycoses*, **40**, 5-6, 193–6.

Venditti, M., Posteraro, B., et al. 1991. In-vitro comparative activity of fluconazole and other antifungal agents against *Blastoschizomyces capitatus*. *J Chemother*, **3**, 1, 13–15.

Wallace, M., Bagnall, H., et al. 1979. Isolation of lipophilic yeasts in 'sterile' peritonitis. *Lancet*, **2**, 956.

Walsh, T.J., Melcher, G.P., et al. 1990. *Trichosporon beigelii*, a emerging pathogen resistant to amphotericin B. *J Clin Microbiol*, **28**, 7, 1616–22.

Warren, N.G. and Hazen, K.C. 1995. *Candida, Cryptococcus*, and other yeasts. In: Murray, P.R., Baron, E.J., et al. (eds), *Manual of clinical microbiology*, 6th edition. Washington DC: ASM Press, 723–37.

Watson, K.C. and Kallichurum, S. 1970. Brain abscess due to *Trichosporon cutaneum*. *J Med Microbiol*, **3**, 191–3.

Welbel, S.F., McNeil, M.M., et al. 1994. Nosocomial *Malassezia pachydermatis* bloodstream infections in a neonatal intensive care unit. *Pediatr Infect Dis J*, **13**, 104–8.

Whitlock, W.L., Dietrich, R.A., et al. 1992. *Rhodotorula rubra* contamination in fiberoptic bronchoscopy. *Chest*, **102**, 5, 1516–19.

Wong, A.R., Ibrahim, H., et al. 2000. *Hansenula anomala* infection in a neonate. *J Paediatr Child Health*, **36**, 6, 609–10.

Wong, V., Ross, L., et al. 1988. *Rhodotorula rubra* peritonitis in a child undergoing intermittent cycling peritoneal dialysis. *J Infect Dis*, **157**, 2, 393–4.

Yamada, S., Maruoka, T., et al. 1995. Catheter-related infections by *Hansenula anomala* in children. *Scand J Infect Dis*, **27**, 85–7.

Yarrow, D. 1984. Genus 22. *Saccharomyces Meyen* ex Riess. In: Kreger-van Rij, N.J.W. (ed.), *The yeasts, a taxonomic study*, 3rd revised edition. Amsterdam, Netherlands: Elsevier Science Publications, 379–95.

Yoshida, K., Suga, M., et al. 1996. Hypersensitivity pneumonitis induced by a smut fungus *Ustilago esculenta*. *Thorax*, **51**, 6, 650–1.

Zunic, P., Lacotte, J., et al. 1991. Fungémie à *Saccharomyces boulardii*. *Therapie*, **45**, 498–9.

32

Cryptococcosis

JOHN R. PERFECT AND GARY M. COX

Cryptococcus neoformans is an encapsulated hetero-basidiomycetous fungus that has become a major human pathogen and a common infection in certain immuno-compromised hosts (Perfect and Casadevall 2002). Cryptococcosis, the disease resulting from infection with *C. neoformans*, varies from a localized skin lesion or asymptomatic colonization of the respiratory tree to a widely disseminated life-threatening infection, which may infect all organs of the body. However, *C. neoformans* has a special propensity for invading the central nervous system and cryptococcal meningoencephalitis is the primary clinical presentation for the life-threatening stage of this infection.

HISTORY

Historically, human cryptococcosis was first described in 1894 when Busse and Buschke independently reported on the same case of a 31-year-old woman with a history of enlarged lymphatic glands who had developed a large ulcer over her tibia (Buschke 1895). Busse observed yeast-like forms in histological sections of the lesion and he was able to culture the yeast, which he initially referred to as *Saccharomyces*. The patient died and was found at autopsy to have multiple abscesses in the lungs, spleen, kidneys, bones, and skin. No mention of symptoms attributable to the central nervous system (CNS) or examination of the CNS was made in this patient. The first isolation of *Cryptococcus* from the environment was reported in 1894 when Sanfelice

isolated the yeast from peaches and named it *Saccharomyces neoformans* (Sanfelice 1894). A few more reports of isolation of this fungus from humans and animals occurred, and then in 1901 Vuillemin renamed the yeast *Cryptococcus hominis* to distinguish it from *Saccharomyces* spp. because it did not form ascospores (Vuillemin 1901).

The first published case of cryptococcal meningitis was diagnosed in a 29-year-old woman with leptomeningeal involvement described by Verse in 1914 (Verse 1914). Two cases of meningitis were then described in 1916 by Stoddard and Cutler and, on complete pathological inspection of the CNS tissue obtained postmortem, these investigators observed yeast forms with surrounding areas of clearing in the tissue (Stoddard and Cutler 1916). This finding was initially misinterpreted as evidence of tissue histolysis and they renamed the yeast *Torula histolytica*, and the syndrome was referred to clinically as torulosis. There was confusion over the fact that one yeast was now considered under three different genera (*Saccharomyces*, *Cryptococcus*, and *Torula*) until 1935 when Benham published a comprehensive study of yeasts that contained 22 cryptococcal strains including the original Busse–Buschke strain (Benham 1935). Based on morphology, fermentation, and serological studies, she concluded that all the cryptococcal isolates should be considered as one species and she designated the name as *Cryptococcus hominis*. Unfortunately, the term 'torulosis' continued to be used for another quarter of a century despite Benham's proposals to call the disease cryptococcosis. *C. neoformans* is now considered the most valid name based on priority since Sanfelice first proposed the species name in 1894, and for the last 40 years *C. neoformans* has been the name used to designate the yeast causing cryptococcosis. In 1976, Kwon-Chung discovered and characterized the sexual stage of this basidiomycete and the teleomorph was named *Filobasidiella neoformans* (Kwon-Chung 1976). Finally, the entire genome of *C. neoformans* was sequenced in 2003.

Cryptococcosis was considered a clinical rarity in the early 1900s. In Littman and Zimmerman's textbook on cryptococcosis, published in 1956, they had found just over 300 cases of cryptococcosis in the world's medical literature before 1955 (Littman and Zimmerman 1956). The number of cases of cryptococcosis has dramatically increased since the 1950s. This increase results partly from greater awareness of *C. neoformans* as a pathogen, but mostly the increase parallels the ever-enlarging population of immunocompromised individuals as a result of aggressive chemotherapies, organ transplantation, and the acquired immunodeficiency syndrome (AIDS). In many medical centers throughout the world, *C. neoformans* is the most commonly identified pathogen obtained from cerebrospinal fluid (CSF) cultures in adults.

MYCOLOGY

There are 19 species within the genus *Cryptococcus*, and they have been found on every continent and in or on a variety of terrains and animals. All members of the genus are nonfermentative, assimilate inositol, and generally produce urease. Although occasional cases of other cryptococci such as *Cryptococcus laurentii* and *Cryptococcus albidus* are alleged to cause human infections (Lynch et al. 1981; Horowitz et al. 1993), *C. neoformans* should be considered the only cryptococcal species that is routinely pathogenic for humans. Isolation of other cryptococcal species from clinical specimens should generally require both culture and histological proof of tissue invasion before attributing disease to them.

Serotypes

C. neoformans can be subclassified into four serotypes and three varieties. The serotypes are based on variation in capsular epitopes and thus different capsular aggluti-

nation reactions (Ikeda et al. 1982). These serotypes are designated as A, B, C, and D (Evans 1950; Evans and Kessel 1951; Vogel 1966) and there are diploid or hybrid strains designated AD. There are specific polyclonal or monoclonal antibodies that can recognize each of the serotypes. Prior to the molecular age of *C. neoformans*, serotype A and D strains were included under var. *neoformans* but today are separated into two varieties, *neoformans* and *grubii*, and the serotype B and C strains were considered under var. *gattii*. The varieties differ in their biochemical properties (Table 32.1) such as the ability of var. *gattii* strains to use glycine or proline as sole sources of nitrogen whereas var. *neoformans/grubii* strains cannot (Dufait et al. 1987). The var. *gattii* strains are also resistant to the chemical canavanine, whereas var. *neoformans/grubii* strains are usually sensitive (Polacheck and Kwon-Chung 1986). The ability to utilize glycine and the resistance to canavanine were exploited to distinguish var. *gattii* from var. *neoformans/grubii* strains. A medium called CGB contains canavanine (C) and has glycine (G) as the sole nitrogen and carbon source. When variety *gattii* strains are grown on CGB agar, glycine is metabolized and results in the formation of ammonia which alkalinizes the medium and converts the bromothyol blue (B). The variety *neoformans/grubii* strains usually cannot metabolize glycine and those rare isolates that can are almost always inhibited by the canavanine that is also present in the medium. Urease activity is another difference between the two varieties in that var. *gattii* urease is sensitive to dicationic chelating agents such as EDTA, but var. *neoformans/grubii* urease is resistant. This difference can be important if isolates are grown on media containing EDTA and then subjected to a rapid urease test. The test can be falsely negative for var. *gattii* strains. There have only been rare reported cases of *C. neoformans* infections with urease-negative strains (Ruane et al. 1988; Bava et al. 1993). With recent DNA typing methods and other genetic features, some mycologists have even proposed that var. *gattii* be actually reclassified as a separate species and named *Cryptococcus bacillospora*.

The only reliable differences in morphology between these varieties are in the teleomorph stage (see section on Life cycle and genetics below). There is no credible way to distinguish the yeast forms of the varieties based on morphology. However, var. *gattii* basidiospores tend to appear more bacilliform or rod-shaped oval when compared with var. *neoformans/grubii* basidiospores.

Table 32.1 *Comparison of* C. neoformans *var.* neoformans *with var.* gattii

Variety	Serotypes	Canavanine resistance	Glycine as sole nitrogen source?	Shape of basidiospores
neoformans	A and D	No	No	Round
gattii	B and C	Yes	Yes	Elliptical

There are significant differences in the ecology and epidemiology between the two varieties and these will be further discussed in those sections.

Through a variety of genetic analyses, it has been demonstrated that the serotype A and D strains have significantly diverged from each other. It is estimated from population genetic studies that these serotypes have separated from each other over 18 million years ago (Xu et al. 2002). To account for significant genetic diversity, there is the proposal to separate these serotypes into two varieties (Franzot et al. 1999) which we use in this chapter. Serotype A was proposed to be named var. *grubii* and serotype D would be var. *neoformans*. The varietal status for the serotype AD strains is uncertain, but there is evidence that most of these strains represent diploids and/or hybrids. These findings show that these varieties still have the ability to mate with each other. As previously mentioned, Kwon-Chung identified the sexual stages of *C. neoformans* and these findings allowed for the completion of the fungal life cycle. Furthermore, *C. neoformans* can also be divided into eight distinct molecular genotypes: VN 1–4 for serotype A and VG 1–4 for serotypes B and C (Meyer et al. 2003). Serotype D isolates have proven more difficult to cluster into distinct genotypes.

Life cycle and genetics

The life cycle of *C. neoformans* involves two distinct forms: asexual and sexual. First, in the asexual stage *C. neoformans* exists as yeast cells and reproduces by simple budding (Figure 32.1). These haploid, unicellular yeasts are the only forms of *C. neoformans* that have been recovered from the environment and human infections. These yeast-like forms exist in one of two mating types, 'a' or 'alpha', and when yeasts of opposite mating type are combined under certain conditions they can undergo conjugation to produce the perfect (sexual) state (Kwon-Chung 1975). Generally, these matings occur in the laboratory on nonenriched media such as V8 juice or hay-infusion agar, and temperatures must be held under 37°C. In fact, room temperature, minimal water exposure, and environmental $P\mathrm{CO_2}$ concentrations are the best conditions for mating to occur. These observations suggest that there are regulatory controls for mating that respond to certain environmental signals such as nitrogen starvation and temperature. Recent work has begun to dissect the genes and pathways in *C. neoformans* that control this bipolar mating system (Hull and Heitman 2002). The mating type loci for several serotypes have been sequenced and annotated. Genes for pheromones and their receptors have been identified and the signaling networks that control mating have been determined and studied.

Conjugation between the 'a' and 'alpha' mating strains results in the formation of the teleomorph (sexual) stage which consists of dikaryotic hyphae that bear true clamp connections (Figure 32.2). Some of these hyphae develop specialized terminal structures called basidia, and the formation of these structures is the basis of classifying the teleomorphs in the phylum Basidiomycota. Furthermore, careful study of the hyphae, basidia, and spore chains showed them to be similar to members of the genus *Filobasidium*. Meiosis occurs at the terminal portion of the basidium and uninucleate basidiospores are formed. These basidiospores bud off at the ends of the basidia in a basipetal fashion to form four chains of spores. These spores are initially unencapsulated but can quickly develop capsules when released from the basidia onto agar and begin budding as yeasts and thus complete the sexual life cycle. The perfect state of *C. neoformans* has been observed only in the laboratory and never conclusively shown to occur in nature. In fact, it is unknown to what extent, if any, sexual reproduction presently occurs in nature. Some investigators have hypothesized that the basidiospores produced by the perfect state may be the infectious propagule for humans because they are ideally suited in size (1–2 μm) for inhalation into the alveoli (Ajanee et al. 1996). Strains of the appropriate mating types of serotype A or D can mate with each other (A × A, A × D or D × D), and the resulting teleomorph is called *Filobasidiella neoformans* var. *neoformans* (Kwon-Chung 1975). Likewise, serotype B and C strains can mate with each other (B × B, B × C or C × C), and the teleomorphs in this case are called *Filobasidiella neoformans* var. *bacillispora* (Kwon-Chung 1975). The teleomorphs of the two varieties can be distinguished on morphological grounds on the basis of their basidiospores. The var. *neoformans* produces chains of spherical basidiospores, whereas the var. *bacillispora* produces oval, bacilliform basidiospores (Figure 32.2).

When the mating types of *C. neoformans* strains recovered from either the environment or patient samples have been determined, there has always been a large excess of the 'alpha' mating type over 'a' mating type strains

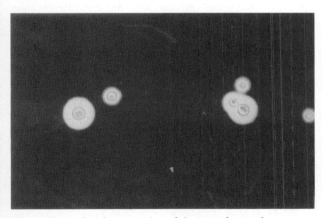

Figure 32.1 *India ink preparation of the yeast forms of* C. neoformans. *The India ink particles serve to outline the large polysaccharide capsules of the yeasts, making them appear as clear areas surrounding the yeasts. Note the one mother yeast cell with the attached daughter cell that is budding.*

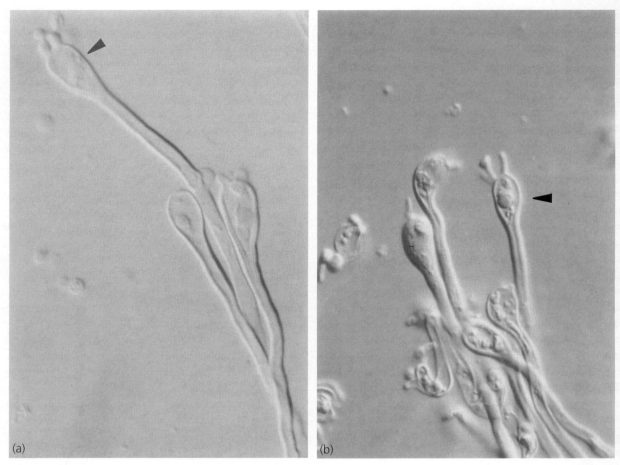

Figure 32.2 *The teleomorph stages of* **(a)** C. neoformans *var.* neoformans *and* **(b)** C. neoformans *var* gattii. *This stage results from mating between two isolates of the opposite mating type, and is characterized by hyphae containing nuclei from each parent (dikaryotic hyphae). At the end of some hyphae, specialized structures called basidia (arrow) can form. Meiosis occurs in the basidia, and the resulting haploid basidiospores bud off at the apex. Note the shape of the basidiospores of each variety. The basidiospores of* var. neoformans *are round, whereas those of var.* gattii *are elliptical.* **(a)** *Courtesy of Wiley Schell;* **(b)** *courtesy of Michael McGinnis*

(i.e. >95 percent) (Kwon-Chung and Bennett 1978). There has been some evidence for recombination in nature among strains, but the reason for this biased difference in mating type remains unclear. One possibility for predominance of the 'alpha' mating type comes from Wickes et al. (1996), in which they found that several haploid 'alpha' mating type strains were able to form hyphal structures on starvation media. The haploid fruiting formed by these strains produced basidia and basidiospores that were indistinguishable from those produced by hyphae resulting from matings. This ability to form basidiospores by haploid fruiting was independent of serotype, and was only seen in alpha mating type strains. However, in a recent investigation MAT 'a' strains have been observed to produce haploid fruiting (Tscharke et al. 2003), but MAT 'alpha' strains seem to more commonly perform this morphological change.

Growth and Identification

C. neoformans can be grown on a variety of agars, and develops into white, mucoid colonies that usually become visible to the naked eye within 48 hours of incubation. The mucoid appearance of the colonies is caused by the yeast cell's production of a well-defined polysaccharide capsule. The optimal growth temperature of most cryptococcol strains ranges between 30 and 35°C, and the maximally tolerated temperature is approximately 40°C. *C. neoformans* var. *grubii* generally is more thermotolerant than the other two varieties (Martinez et al. 2001). On rich media at 30°C the doubling time for most cryptococcal strains is between 2.5 and 6 hours. The yeast forms are round to oval with diameters between 2.5 and 10 μm and hyphae are never seen with the yeast forms in nutrient media, although pseudohyphal and hyphal forms can sometimes be found when the yeast cells are stressed by either harsh environmental conditions such as elevated temperatures or under nutrient-poor conditions.

In the clinical microbiology laboratory, *C. neoformans* can be readily differentiated from other yeasts on the basis of morphology and biochemical tests. The finding of a mucoid colony on an agar plate is usually the first clue to the presence of *C. neoformans* from a clinical specimen. This suspicion is further heightened by observing encapsulated, budding yeasts on an India ink preparation of the colony (see Figure 32.1). The rapid urease test is another presumptive means for identifying the yeasts as *C. neoformans*. Virtually all cryptococci (both *neoformans* and non-*neoformans* species) have the enzyme urease and thus can hydrolyze urea to ammonia, and increase the ambient pH. This property can be detected by using various broths containing urea and a pH marker. As cryptococci have large amounts of urease, positive results using such tests can be seen within 15 minutes (Zimmer and Roberts 1979). The identification of *C. neoformans* specifically can be confirmed by using various biochemical tests, and most clinical laboratories use a battery of biochemical tests contained in commercially available kits. Among the cryptococci, *C. neoformans* is the only one that possesses prominent laccase activity, and testing for its presence is another means of accurate identification. This feature can be detected by culturing the isolated yeasts on special agars such as niger seed agar, birdseed agar, caffeic acid, or dopamine agar, and examining for the growth of brown or black colonies (melanin-positive). These assays are particularly helpful when attempting to identify cryptococci from environmental sources with contamination by other fungi and bacteria. However, most clinical microbiology laboratories do not use these special agars because of the expense of keeping them, and instead depend on the biochemical testing. The yeast forms of cryptococci can be identified in histopathological specimens by using mucicarmine or alcian blue staining of histopathological specimens to demonstrate that the yeast forms possess a polysaccharide capsule (Figure 32.3). There may be strains with a variety of capsule sizes in tissue but only rarely is a true hypocapsular strain detected in tissue. Finally, a Fontana–Masson stain, although not specific for melanin, can demonstrate production of melanin by these yeasts in tissue.

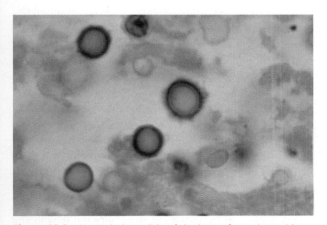

Figure 32.3 *Histopathology slide of the lung of a patient with cryptococcal pneumonia. The yeasts are easily identified in tissue and the bright red staining around the yeast body demonstrates how this mucicarmine stain identifies the polysaccharide capsule of* C. *neoformans.*

Capsule

The polysaccharide capsule that surrounds *C. neoformans* is easily observed when yeast cells from the site of infection are examined under the microscope in a suspension of India ink (the ink particles serve to outline the capsule, and make it appear as a clear area or halo surrounding the yeasts) (see Figure 32.1). Capsule thickness can vary, and in some isolates the capsule can account for more than 50 percent of the diameter of the yeast cells. The capsule is an important virulence determinant for *C. neoformans*, and it is considered to be one of the 3 'classic' virulence factors of cryptococci (the other two are melanin production and growth at 37°C which are discussed in the sections on Melanin production and Growth at 37°C below). Mutant cryptococci that are either hypocapsular or acapsular are consistently less virulent in animal models than encapsulated strains (Chang and Kwon-Chung 1994), and are more easily taken up by white blood cells in in vitro assays (Bulmer et al. 1967; Kozel and Cazin 1971; Fromtling et al. 1982; Kozel and Gotschlich 1982). Infections caused by capsule-free or poorly encapsulated strains are associated with less severe disease (Farmer and Komorowski 1973; Levinson et al. 1974; Milchgrub et al. 1990). The major polysaccharide present in the capsule is glucuronoxylomannan. It is the variable structure of this polysaccharide that accounts for the different serotypes. The glucuronoxylomannans of the serotypes differ in the degree of mannosyl substitution and in the molar ratios of mannose, xylose, and glucuronic acid (Turner and Cherniak 1991). This capsule material can be released from the yeast and extrudes into various tissues such as blood and CSF. It is this antigen that is detected by the cryptococcal antigen test which is so effective for diagnosis in the clinics (see section on Meningoencephalitis below).

Several physiological conditions have been linked to the growth and regulation of the capsule in *C. neoformans*. The $P\text{CO}_2$ concentration of the surrounding environment, ferric iron availability, and presence of serum can dramatically influence capsular size (Granger et al. 1985; Vartivarian et al. 1993; Zargoza et al. 2003). Cryptococcal polysaccharide synthesis and capsule size are increased additively by physiological concentrations of CO_2, limitation of ferric iron, and addition of serum. Most *C. neoformans* strains appear to produce large capsules at physiological $P\text{CO}_2$ concentrations and small capsules when exposed to environmental $P\text{CO}_2$ concentrations. This environmental regulation of the capsule appears to be congruent with the yeast's needs as a pathogen. For example, a small capsule is required for deposition in the host's airways from their environmental source, but a large capsule is needed for protection against the host's defenses. As *C. neoformans* does not produce hydroxamate siderophores for iron acquisition (Jacobson and Petro 1987), but responds to iron in the environment, it is likely that there are specific iron receptors on the surface of *C. neoformans*; iron is extremely important to the growth of *C. neoformans* (Jacobson and Vartivarian 1992). Several genes that are essential for capsule production and regulation have recently been identified (Chang and Kwon-Chung 1994, 1999; Moyrand et al. 2002). It is also clear that there are signaling pathways such as the cAMP pathway that control the expression of capsule production (Alspaugh et al. 1997) and the capsule appears to attach to the cell wall through glucan linkages (Reese and Doering 2003). Understanding the formation of this unique structure is underway, but further insights will require continued focus on molecular and biochemical aspects of the formation and maintenance of this imposing physical structure.

The impact of the capsular polysaccharide on host immunity can be profound. For instance, it has been shown to have the following effects on immunity:

- antiphagocytosis
- depletes complement
- decreases antibody responsiveness
- inhibits leukocyte migration
- dysregulates cytokine secretion
- produces brain edema
- enhances human immunodificiency virus (HIV) infection
- interferes with antigen presentation
- produces selectin and tumor necrosis factor receptor loss
- creates highly negative charge around cells
- extrudes itself into the intracellular environment with potential local toxicity for cellular organelles (Casadevall and Perfect 1998).

Since a major component of the yeasts' growth in the host is intracellular (Feldmesser et al. 2001), the negative impact of the polysaccharide extruded into the interior of a phagocytic host cell is likely to be profound. The capsule is a major virulence factor through its multipotential ability to abrogate a successful host immune challenge.

Melanin production

The presence of a functional laccase(s) enzyme in *C. neoformans* is unique among other members of this genus. This enzyme participates in the conversion of diphenolic compounds into melanin. The production of this pigment by a simple biochemical pathway is both a means for identifying *C. neoformans* in the laboratory and also a major virulence factor. It is thought that this enzyme is bound on the inner aspect of the cytoplasmic membrane, and the primary gene encoding for this enzyme has been identified and characterized (Polacheck et al. 1982). A wide variety of diphenolic

substrates can be utilized by the cryptococcal laccase, including catecholamines such as dopamine and norepinephrine. The ability to produce melanin by a variety of pathogenic fungi has been touted as a potential protectant from host cell damage. Cryptococcal mutants lacking laccase activity have been found to be attenuated in animal models of infection (Salas et al. 1996) and are also killed by the epinephrine oxidative system in the presence of a transition metal ion and hydrogen peroxide compared to the more resistant laccase-positive cells (Kwon-Chung et al. 1982; Rhodes et al. 1982; Polacheck et al. 1990). The melanin produced by degradation of these catecholamines accumulates in the cell wall and may provide protection against a variety of host oxidative stresses (Jacobson and Emery 1991). Importantly, the ability to form melanin in the host has been demonstrated (Nosanchuk et al. 2000a; Rosas et al. 2000). The ability to degrade naturally occurring catecholamines such as norepinephrine to an antioxidant to protect the yeast when it is in the catecholamine-rich CNS may partially explain its propensity to infect the central nervous system. Besides melanin's antioxidant properties, it may also provide other protections for the yeast including:

- cell wall support or integrity
- alterations in cell wall charge
- interference with T-cell response
- abrogation of antibody-mediated phagocytosis
- protection from temperature changes and antifungal agents (Casadevall and Perfect 1998).

Growth at 37°C

A very basic trait for all pathogenic fungi is their ability to grow well at 37°C. In fact, mutants of *C. neoformans* that cannot grow well at 37°C are avirulent even when they possess capsules and laccase activity (Kwon-Chung et al. 1982). Cryptococcal strains differ in their rates of growth at 37°C and their virulence is directly related to their rate of growth at physiological temperatures. Very high mammalian body temperatures of 39–40°C will significantly slow the growth rate of most *C. neoformans* strains and at this temperature the yeasts will actually begin to produce intracellular vacuolization with aberrant budding patterns and pseudohyphal structures. The varieties *gattii* and *neoformans* appear to be more sensitive to high temperatures than var. *grubii*, and at 40°C most strains of var. *gattii* and var. *neoformans* will lose viability within 24 hours (Martinez et al. 2001). Furthermore, it is clear that high-temperature growth is under genetic control with the isolation of a series of temperature-sensitive mutants. Recent work, beginning with the findings of Odom et al., showing that calcineurin in *C. neoformans* was important to high-temperature growth (Odom et al. 1997), has begun to unravel the molecular networks and specific genes necessary for *C. neoformans* to grow well at mammalian body temperatures (Alspaugh et al. 2000; Erickson et al. 2001). This includes both the signaling pathways of calcineurin and *RAS* and includes biochemical pathways such as the stress protectant sugar trehalose and certain vacuolar functions (Erickson et al. 2001).

ECOLOGY

C. neoformans was first identified in nature by Sanfelice in 1894 from samples of peach juice (Sanfelice 1894). Although fruits are currently not thought to be a significant environmental source for this fungus, certain species of trees are a potential reservoir or habitat for it. *C. neoformans* var. *grubii* can also be found in soil contaminated by bird guano. Its ability to survive within scavenging predators of the soil has led to studies of *C. neoformans* growth in ameba (Steenbergen et al. 2001), dictosteylium (Steenbergen et al. 2003), and nematodes such as *Caenorhabditis elegans* (Mylonakis et al. 2002).

C. neoformans var. *neoformans* and var. *grubii*

Emmons isolated *C. neoformans* from soil, pigeon droppings, and pigeon nests in the 1950s (Emmons 1951, 1955). Since those initial reports, this fungus has been found in soil samples from around the world. The soils that most commonly yield *C. neoformans* var. *neoformans* and var. *grubii* have been frequented by birds, especially pigeons and chickens. For instance, this fungus has commonly been isolated from roosting sites of pigeons in which their droppings had accumulated over many years. Guano from other birds such as canaries, parrots, and turkeys has also yielded *C. neoformans* var. *neoformans* and var. *grubii*. It is not known why there is a frequent association of pigeons with *C. neoformans* because pigeons rarely develop disease with *C. neoformans* in nature. This probably results from the fact that the normal body temperature for a pigeon is above 40°C and most cryptococci are inhibited by these elevated temperatures (Littman and Borok 1968). Pigeons can, however, harbor the yeast as saprophytes in their gastrointestinal tracts. In fact, when force-fed *C. neoformans*, pigeons had viable yeasts cultured from their crops for more than 2 months (Swinne-Desgain 1976). Thus birds could act as a vector for spread of the fungus in the environment.

This fungus can metabolize low-molecular-weight nitrogenous compounds, such as creatinine, which are present in bird guano, and it has been postulated that there is a natural selection for *C. neoformans* to inhabit and grow in this niche. The exact role of pigeon guano in the pathogenesis of human infections is obscure. Only rarely can one elicit a history of intense pigeon exposure from patients with cryptococcosis (Fessel 1993), and

there have never been outbreaks of the disease traced back to pigeon-roosting areas. However, there is a case report of infection in an immunosuppressed patient linked to a strain found in a pet bird's cage (Nosanchuk et al. 2000b).

C. neoformans var. *neoformans* and var. *grubii* have occasionally been isolated from rotting vegetation and wood of trees (Pal and Mehrotra 1985). A report from Brazil demonstrated that var. *grubii* isolates could be cultured from areas of wood decay existing on living trees (Lazera et al. 1996). It is probable that decaying wood of certain trees is one natural ecological niche for *C. neoformans* var. *neoformans* and var. *grubii*.

C. neoformans var. *gattii*

As opposed to var. *neoformans* and var. *grubii*, var. *gattii* has never been cultured from bird guano. There have, however, been several environmental niches identified for var. *gattii* strains. Ellis and Pfeiffer were able to culture *C. neoformans* var. *gattii* from vegetation associated with river red gum trees (*Eucalyptus camaldulensis*) and forest red gum trees (*Eucalyptus tereticornis*) in Australia (Ellis and Pfeiffer 1990; Pfeiffer and Ellis 1992). These trees were exported from Australia to various parts of the world, and it was reasoned that var. *gattii* may have been exported along with the trees. These investigators were able to culture a strain of var. *gattii* from a river red gum tree in San Francisco, CA (Pfeiffer and Ellis 1991). There may also be an association with var. *gattii* release of spores with the flowering of the trees since cultures of air samples taken from under flowering trees can yield cryptococci, whereas samples taken when the trees are not in flower do not. Areas of the world that have river red gum trees in the environment also have a fair number of var. *gattii* cryptococcal strains involved in their cases of cryptococcosis. However, a recent significant outbreak of cryptococcosis on Vancouver Island in British Columbia demonstrated that other trees such as fir and oak will also contain var. *gattii* strains (Stephen et al. 2002) and possibly become a vector for human disease.

EPIDEMIOLOGY

Cryptococcosis is a worldwide infection which only rarely causes disseminated disease in healthy individuals. Most patients with cryptococcosis have evidence of some underlying immunocompromising condition. The most common of these underlying conditions worldwide is AIDS, followed by prolonged treatment with corticosteroids, organ transplantation, malignancies, and sarcoidosis (Pappas et al. 2001; Perfect and Casadevall 2002). However, recent estimates are that if patients with HIV infection are excluded, approximately 20 percent of patients with cryptococcosis will present with no apparent underlying disease or risk factor (Pappas et al. 2001).

Cryptococcosis is not a routinely reportable disease in the USA, so accurate nationwide estimates of prevalence are difficult to determine. However, in 1992, before the availability of highly active antiretroviral therapy (HAART), an incidence of 4.9 cases of cryptococcosis per 100 000 population was reported in both San Francisco, CA and Atlanta, GA (Pinner et al. 1993). Between 1992 and 1994 in four large metropolitan areas within the USA, the incidence of infection ranged between 0.2 to 0.9 cases/100 000 population. With widespread use of fluconazole and the introduction of HAART, this lower rate of infection has continued in the US cities. Cryptococcosis in AIDS within developed countries has generally focused on a group of patients with less access to medical care (Mirza et al. 2002). Prior to HAART it was estimated that 6–10 percent of all patients with AIDS in the USA, western Europe, and Australia were diagnosed with cryptococcosis, making it the fourth most common opportunistic infection in this population (Kovacs et al. 1985; Powderly 1993). The rate of cryptococcal meningitis in patients with AIDS is even higher in sub-Saharan Africa where rates of 15–30 percent have been encountered (Clumeck et al. 1984; Van de Perre et al. 1984; Powderly 1993), and HAART has yet to make an impact on the incidence of cryptococcal disease in many of these countries. It can be anticipated that the numbers of cases of cryptococcal meningitis in the world will continue to mirror the worldwide control of the HIV epidemic. In addition, in developed countries more individuals are undergoing organ transplantation and aggressive cancer chemotherapeutic regimens and thus the pool of at-risk patients continues to enlarge.

Interestingly, there are only rare reports of outbreaks of cryptococcosis traced to a common source such as the recent Vancouver Island outbreak (Stephen et al. 2002), and this is unlike the occasional known outbreaks of infection from an environmental source with the classic dimorphic fungi. There have been no documented cases of spread of infection from person to person through routine contact, but there are several cases of cryptococcosis produced by transplantation of infected tissue (Beyt and Waltman 1978; Kanj et al. 1996). C. neoformans occasionally has been introduced into a host through direct trauma with a fomite (Glaser and Garden 1985; Casadevall et al. 1994), although inhalation into the lung is the most common route of initial infection. Cryptococcosis appears to be primarily a disease of adults because it is uncommon to produce disease in children, even in those with AIDS. The reason for the reduced incidence of disease in children is unclear and it occurs despite the fact that studies have shown that children are exposed to this fungus at a young age (Chen et al. 1999; Goldman et al. 2001).

The frequency of serotypes or varieties of C. neoformans strains that have been found in cases of cryptococcosis varies according to geographical location and whether the patient has HIV infection as a predisposing condition. Kwon-Chung and Bennett did an extensive review of 725 clinical isolates of C. neoformans obtained from around the world before the AIDS epidemic (Kwon-Chung and Bennett 1984). These investigators found that, overall, var. neoformans and grubii were far more common than var. gattii, and accounted for more than 80 percent of all isolates, and of these isolates 80 percent were var. grubii. The var. gattii isolates were almost exclusively found in tropical and subtropical areas such as southern California, Hawaii, Brazil, Australia, Southeast Asia, and Central Africa. This geographical restriction is most probably the result of the ecological relationship of var. gattii with certain trees such as eucalyptus and firs. Serotype D strains (var. neoformans) predominately come from Europe, especially Denmark, Germany, Italy, and Switzerland. Isolates from the UK consisted of 87 percent serotype A and 13 percent serotypes B and C, whereas isolates from continental USA, excluding southern California, were 81 percent serotype A, 7 percent serotype D, and 6 percent serotypes B and C (some had no identifiable serotype)(-Bennett et al. 1977; Friedman 1983; Kwon-Chung and Bennett 1984). Before the AIDS epidemic, the serotypes of cryptococcal strains isolated from patients living in southern California had a higher proportion of serotypes B and C when compared with the rest of the USA. It is important to keep in mind that these isolates were all obtained before the AIDS epidemic. The vast majority of cryptococcal isolates from patients with AIDS have been var. neoformans or var. grubii, regardless of geographical location. This is most strikingly demonstrated by data from Central Africa. Over 90 percent of isolates obtained before 1970 from Central Africa were var. gattii. However, among the many cryptococcal strains obtained from patients with AIDS in Central Africa, there has been only one var. gattii strain identified (Kapend'a et al. 1987). The reason(s) why patients with AIDS are more likely to be infected by var. neoformans or var. grubii strains is not known. It is hypothesized that differences in infection patterns are the result of certain epidemiologic factors of exposure. For instance, HIV remains a predominantly urban infection; patients with AIDS may not be exposed to the rural environmental niches of var. gattii strains. However, it may also be possible that var. neoformans or var. grubii strains more commonly produce dormant infections that are reactivated when immunity is impaired during HIV infection compared with var. gattii in which disease may occur primarily during initial infection.

PATHOPHYSIOLOGY AND IMMUNOLOGY

The exact pathophysiology of cryptococcosis is not known for certain, but a reasonable scheme starts with the host coming into contact with cryptococci from the environment through inhalation of small yeasts or basidiospores, with their subsequent deposition in the pulmonary alveoli. It is advantageous for the infecting propagules to be as small as possible because this would favor deposition into the lung, and it is for this reason that some investigators have postulated that the infectious propagules of *C. neoformans* are the basidiospores produced during the sexual cycle or with haploid fruiting. Basidiospores are smaller than the yeast cells obtained in clinical specimens and appear to be the ideal size for inhalation. With identification of the haploid fruiting process, spores can be made in the environment without the need for mating. However, there have been small poorly encapsulated yeasts of sizes consistent with basidiospores isolated from the environment (Neilson et al. 1977; Ruiz and Bulmer 1980), and it must be stressed that the basidiospore structures have yet to be directly observed in nature.

Once the yeasts or spores arrive at the alveoli, the earliest innate immune response elicited is composed of alveolar macrophages, which is then followed by the arrival of polymorphonuclear cells, activated macrophages, and lymphocytes. Data from both experimental animal and human experiments show that the protective components of the immune system for cryptococcosis are primarily cell-mediated, and are composed of activated alveolar macrophages, natural killer (NK) cells and both CD4+ and CD8+ lymphocytes. Several studies have indicated that CD4+ and CD8– cells are especially important in containing the spread of cryptococci from the site of initial infection in the lung to the central nervous system (Hill and Harmsen 1991; Mody et al. 1993, 1994).

Patients with intact immune systems can mount an effective, coordinated response to cryptococcal infections, and the result is a granulomatous-type inflammatory lesion which is primarily driven by the Th1 cytokine responses. A series of studies have identified the positive features of IL-2, IL-12, and interferongamma for an effective immune response against *C. neoformans*. A Th2 response is associated with an ineffective inflammatory reaction. Most primary infections of the lungs are asymptomatic in immunocompetent individuals because they are readily contained by the immune system. As cryptococci are widespread in nature and yet cryptococcosis is a rare event, it is assumed that subclinical primary infections are common in the general population. Unfortunately, there is not a standardized skin test to prove this hypothesis. Besides the host's immune status, there may be some contribution of inoculum size or particular virulence of infecting strain as to whether or not immunocompetent individuals become symptomatic during primary infection.

Elegant postmortem studies have shown the existence of pulmonary foci and hilar lymph nodes containing yeasts during autopsies in individuals who had no antecedent respiratory complaints (Baker and Haugen 1955; Baker 1976). These focal areas of quiescent infection have the potential to disseminate should the host's immune system become impaired. The widespread and acute presentation of cryptococcosis in patients with AIDS shows that the lack of an effective cell-mediated immune response with deterioration in numbers and functions of CD4+ cells results in a rapid dissemination of yeasts throughout the body with a higher fungal burden in tissues compared to apparently immunocompetent individuals. Disseminated cryptococcosis in the immunocompromised population may represent either reactivation of latent infection or in some cases a primary infection with immediate disease due to an initial ineffective immune response.

Although the cell-mediated arm of the immune system is the primary host reaction in responding to invading cryptococci, antibodies directed against *C. neoformans* can be protective. For instance, passive immunization of mice with monoclonal antibodies directed against the capsule can protect and modulate infection (Mukherjee et al. 1992, 1993, 1994). The basis for the humoral protection may be the result of opsonization or decreased levels of glucuronoxylomannan in tissue. As a result of the protection afforded by antibodies in animal models of infection, there remains hope of developing a vaccine for use in people at high risk for cryptococcosis or the use of serotherapy in the treatment of active disease. Both polyclonal and monoclonal anticryptococcal antibodies have been infused and evaluated in patients, but further studies are needed to define their clinical effectiveness.

C. neoformans appears to have some type of neurotropism in humans because of its common propensity to cause meningoencephalitis. The reasons for this unique presentation are not known. There is some speculation that because of its ability to use catecholamines as substrates for melaninogenesis, *C. neoformans* has a survival advantage in the CNS where these products occur in large amounts. It is also possible that this yeast has special surface properties or has the ability to reside within host cells and traverse across the blood–brain barrier within them. Once the yeast is in the CNS compartment, the relatively immunologically sequestered environment of the CNS allows it to grow easily.

CLINICAL PRESENTATION

Meningoencephalitis

Cryptococcal meningoencephalitis is the most frequently encountered manifestation of cryptococcosis. The term 'meningoencephalitis' is more appropriate than meningitis because pathological inspection of human cases shows that the underlying brain parenchyma is generally involved, although radiographically it may not be apparent. The clinical presentation of this infection in AIDS versus non-AIDS patients is slightly different (Table 32.2). This clinical difference results from the more severe immunosuppression generally observed in patients with AIDS. It is a manifestation of a higher burden of yeasts and a reduced host inflammatory response. When cryptococcal meningoencephalitis is diagnosed in the setting of HIV infection, it is almost always in patients with CD4 lymphocyte counts lower than 100 cells/mm^3. The features of CNS cryptococcosis in patients with AIDS collected from several different series are shown in Table 32.3 (Kovacs et al. 1985; Eng et al. 1986; Zuger et al. 1986; Chuck and Sande 1989; Clark et al. 1990a; Saag et al. 1992). This infection was the AIDS-defining illness for 60 percent of the patients in these series. The onset of symptoms was usually over a period of 1–2 weeks, and the three most common symptoms were headache, fever, and malaise, seen in 83, 75, and 68 percent of patients, respectively. Symptoms such as stiff neck, photophobia, and vomiting were seen only in a minority of patients. The initial physical examination revealed altered mentation in 24 percent of the patients and focal neurological deficits in 6 percent (Table 32.3). Rarely, patients with cryptococcal meningitis and severe immunosuppression can present in a fulminant fashion with rapidly developing coma and death.

By contrast, patients without AIDS usually have symptoms for a longer period of time. Those less common patients with no obvious underlying risk factors can have symptoms lasting for months before a diagnosis is made. However, most patients present with signs and symptoms of subacute meningoencephalitis. Headaches, fever, lethargy, coma, personality changes, and memory loss can develop over weeks. However, there can be tremendous variability in the clinical presentation. Some patients may present with severe headaches for only a few days, intermittent headaches for months, or no headaches at all. The presence of a headache may, in fact, be an important positive prognostic factor by allowing an earlier diagnosis (Dismukes et al. 1987).

It is generally thought that there is no difference in the clinical presentation of patients with cryptococcal meningoencephalitis based on the serotype of the infecting strain. However, the possibility of strain influence should not be entirely neglected. Reviews of cryptococcosis in Australia compared the clinical features of patients with infections caused by the varieties (Speed and Dunt 1995). In contrast to the patients with infection resulting from var. *neoformans* or var. *grubii* strains, the patients with var. *gattii* were more likely to have neurological complications at presentation such as hydrocephalus and cranial nerve deficits. These differences might be explained by the possibility that var. *gattii* strains are generally less virulent than var. *neoformans* or var. *grubii*. Thus, less virulent cryptococcal strains might allow a more robust immune response by the host. The active inflammatory response may account for the insidious presentation and focal neurological signs. It is again interesting to note that in several large reviews of *C. neoformans* there were few infections caused by var. *gattii* strains in highly immunosuppressed patients.

Lung

Despite the probability that most pulmonary infections with cryptococci are asymptomatic, hundreds of cases of symptomatic pulmonary cryptococcosis in apparently immunocompetent hosts have been described in the medical literature (Kent and Layton 1962; Newberry et al. 1967; Tynes et al. 1968; Warr et al. 1968; Hammerman et al. 1973; Rosenheim and Schwarz 1975; Paillas et al. 1982; Henson and Hill 1984; Cohen 1985; McDonnell and Hutchins 1985; Diamond and Levitz 1988). Campbell's comprehensive review of the English medical literature up to 1965 remains the most complete description of primary pulmonary cryptococcosis (Campbell 1966). This review suggests that primary pulmonary cryptococcosis in the immunocompetent host may present asymptomatically in 32 percent, i.e. abnormal findings on a chest x-ray taken for other reasons. However, most patients described in this review presented with symptoms including cough (54 percent), chest pain (46 percent), sputum production (32 percent), weight loss and fever (26 percent each), and hemoptysis (18 percent). Other presentations including dyspnea, night sweats, and superior vena cava obstruction are rarer (Menon and Rajamani 1976; Lehmann et al. 1984). Patients may simply present with airway colonization (Duperval et al. 1977), but this colonized state is frequently observed in patients with underlying chronic lung disease.

Table 32.2 *Comparison of cryptococcal meningoencephalitis in AIDS versus non-AIDS patients*

Findings	Patients with AIDS	Patients without AIDS
Duration of symptoms	Usually <2 weeks	Usually <2 weeks
Positive India ink of CSF	Approx. 75%	Approx 50%
CSF antigen titer (1:1,024	Common	Rare
Serum antigen positive	Common	Common
CSF white blood cells (20/μl	Very common	Rare
Extraneural involvement	Common	Rare
Opening pressure (200 mm	Common	Less common

Table 32.3 *Clinical features of cryptococcal meningoencephalitis in patients with AIDS*

Clinical feature	Frequency among patients with AIDS (%)
Headache	83
Fever	75
Malaise	68
Vomiting	42
Meningismus	32
Altered mentation	24
Photophobia	19
Focal neurological deficit	6

Immunocompromised patients with cryptococcal pneumonia may have a more rapid symptomatic clinical course than that seen in immunocompetent hosts. Although *C. neoformans* typically enters through the lungs, in the immunocompromised host it tends to disseminate rapidly or reactivate from a primary focus, eventually to establish infection within the CNS; hence these patients may present with a meningitis syndrome rather than a pulmonary syndrome. However, overwhelming cryptococcal pneumonia with adult respiratory distress syndrome can also occur on initial presentation (Kent and Layton 1962; Henson and Hill 1984; Murray et al. 1988). These immunocompromised patients include those with HIV infection, cirrhosis, diabetes, Cushing's syndrome, sarcoidosis, chronic leukemia, lymphoma, those undergoing treatment with glucocorticoids, and solid organ transplant recipients (Kerkering et al., 1981; Blanc et al. 1982; Kramer et al. 1983; Christoph 1990; Cameron et al. 1991). Kerkering et al., in a classic retrospective review of pulmonary cryptococcosis, described 41 patients with pulmonary infection, 34 of whom had an underlying immunocompromising condition other than HIV (Kerkering et al., 1981). Twenty-nine of these patients developed disseminated disease, but only one of the patients with disseminated disease was identified as immunocompetent. Unlike immunocompetent hosts who may have an inapparent pneumonitis, 83 percent of the immunosuppressed patients had constitutional symptoms. The most common presenting symptoms were:

- fever 63 percent
- malaise 61 percent
- chest pain 44 percent
- weight loss 37 percent
- dyspnea 27 percent
- night sweats 24 percent
- cough 17 percent
- hemoptysis and headache 7 percent each.

Subsequent to the diagnosis of pulmonary cryptococcosis, dissemination to the meninges occurred in 25 patients within 2–20 weeks of the pulmonary presentation. All the patients had abnormal chest x-rays. The most common findings were alveolar or interstitial infiltrates, followed by single or multiple coin lesions, masses, cavitary lesions, and pleural effusions. In this review the outcome of infection clearly supported the recommendation that immunodeficient hosts with pulmonary cryptococcosis require antifungal therapy (Kerkering et al. 1981).

Pulmonary cryptococcosis is less common than meningitis as a presenting complaint in patients with AIDS, but is well described. Pulmonary cryptococcosis in patients with HIV infection has slightly different manifestations than in other types of immunocompromised hosts (Wasser and Talavera 1987; Khardori et al. 1988; Chechani and Kamholz 1990; Clark et al. 1990b; Cameron et al. 1991). In one series of pulmonary manifestations of AIDS, *C. neoformans* was the implicated pathogen in approximately 10 percent of the cases (Suster et al. 1986). Almost all patients with AIDS present with symptoms, including:

- fever 81 percent
- cough 63 percent
- dyspnea 50 percent
- weight loss 47 percent
- headache 41 percent
- occasionally pleuritic chest pain and hemoptysis.

Dissemination of cryptococci, particularly to the meninges or blood, occurred in 94 percent of the patients with pulmonary disease, based on the results from two studies (Wasser and Talavera 1987; Cameron et al. 1991). Physical examination findings were not stated in most reports, but may include lymphadenopathy, rales, tachypnea, and splenomegaly (Cameron et al. 1991). Concomitant second infections with other opportunistic pathogens, including *Pneumocystis jiroveci*, *Mycobacterium avium-intracellulare*, cytomegalovirus, and *Histoplasma capsulatum* var. *capsulatum*, can occur in conjunction with pulmonary cryptococcosis (Chechani and Kamholz 1990; Clark et al. 1990b; Cameron et al. 1991). In addition, *C. neoformans* pneumonia may occur as a consequence of adjunctive steroid therapy for *P. jiroveci* (Lambertus et al. 1990) and its clinical presentation can be confused in AIDS patients with pneumocystis pneumonia.

Skin

It has been known for over 50 years that *C. neoformans* infection can produce a variety of skin lesions. Littman and Zimmerman (1956), in their classic monograph, described the observations of Cawley, Grekin, and Curtis (Cawley et al. 1950) on cutaneous cryptococcosis. It was noted that skin lesions could appear as acneiform lesions, purpura, papules, vesicles, nodules, tumors, abscesses, ulcers, superficial granulomas, plaques resembling ecchymoses, and sinus tracts. Recent case reports in AIDS patients have shown an even more expanded variety of skin presentations to include herpetiformis- (Borton and Wintroub 1984) or molluscum contagiosum-like lesions (Concus et al. 1988). Furthermore, cellulitis (Gauder 1977; Mayers et al. 1981) around an intravenous catheter, which may appear similar to a bacterial infection, can be caused by *C. neoformans* in patients at high risk for infection (Sarosi et al. 1971; Schupbach et al. 1976). The impressive array of cutaneous presentations of infection emphasizes the particular need for skin biopsy and proper histopathology of all new skin lesions for diagnosis in immunocompromised patients. These sentinel lesions may be the first presentation of disseminated cryptococcosis, and the identification of cutaneous cryptococcosis represents disseminated disease in most cases. However, direct inoculation of *C. neoformans* into the skin with production of primary infection has been described (Neuville et al. 2003) and it has been linked to both direct trauma with contaminated fomites and/or laboratory or clinical accidents (Glaser and Garden 1985; Casadevall et al. 1994). There are some cryptococcal strains which appear to have a tropism for skin in animal models and *C. neoformans* var. *neoformans* appears to infect the skin more commonly than the other varieties. Finally, in solid organ transplant recipients with the use of immunosuppressives like tacrolimus, some institutions have begun to observe more infections with *Cryptococcus* limited to the skin (Singh et al. 1997).

Prostate

The prostate gland is known to be a site for cryptococcal infection (Braman 1981). Infection at this site typically does not cause symptoms of prostatitis, but these yeasts have been isolated from prostate tissue and blood after urological procedures (Plunkett et al. 1981). *C. neoformans* has been described to cause a penile ulcer (Perfect and Seaworth 1985) and a vulvar lesion (Blocker et al. 1987), but there have been no reports of conjugal spread of this infection. The significance of genitourinary infection for clinical disease may reside in the prostate's potential as a protected sanctuary for this yeast during drug treatment. In careful follow-up of patients with AIDS-associated cryptococcosis, it has frequently been found that routine urine or seminal fluid cultures for cryptococci after prostatic massage were positive at the end of an extended course of therapy (Larsen et al. 1989; Staib et al. 1989). However, it is not certain that the prostatic foci of infection are the reasons why some patients with cryptococcosis relapse after therapy is stopped.

Eye

In patients with disseminated cryptococcosis, ocular involvement with *C. neoformans* is not a rare event (Blachie et al. 1985). Before HIV infection, one series found ocular signs and symptoms in approximately 45 percent of patients (Okun and Butler 1964). From ocular palsies to involvement of the retina (Crump et al. 1992), which can also be simultaneously infected with other pathogens such as HIV or cytomegalovirus (Doft and Curtin 1982), ocular cryptococcosis is a sight-threatening infection. In fact, most cases of cryptococcal endophthalmitis lead to severe visual loss with only an occasional case successfully managed (Denning et al. 1991a). In about a quarter of cases, eye involvement may present before the diagnosis of cryptococcal meningoencephalitis (Crump et al. 1992).

There have been recent descriptions of catastrophic visual loss in patients without evidence of cryptococcal endophthalmitis (Johnston et al. 1992; Rex et al. 1993). Funduscopic examination is either normal or demonstrates papilledema. Investigators have suggested that two pathogenic mechanisms exist for this complication. One group has rapid visual loss in a period as short as 12 hours and a clinical syndrome suggestive of optic neuritis with the optic nerve possibly infiltrated with yeasts. Few therapeutic maneuvers have been successful for this form of visual loss. The second group presents with slow visual loss which typically begins later in therapy and slowly progresses over weeks to months. In this group, symptoms may be related to increased intracranial pressure and treatment of these pressure conditions with shunts, medications, or optic nerve fenestrations may halt progression of the visual loss.

The eye may also be the entry site for *C. neoformans* infection. For example, one case of a corneal transplant transmitting this yeast from the donor to recipient was reported with eventual development of meningitis in the patient (Beyt and Waltman 1978); a second case of cryptococcal keratitis after a keratoplasty has been described (Perry and Donnenfeld 1990); and, finally, the observations that meningitis was only identified after eye involvement have been documented in some cases (Granger et al. 1986; Birkmann and Bennett 1988). All these findings suggest the possibility that trauma to the eye with *C. neoformans* contaminated debris could be a portal of entry for *C. neoformans* in some clinical cases of infection.

DIAGNOSIS

Meningoencephalitis

The clinical diagnosis of cryptococcal meningoencephalitis can be very difficult given the subacute onset of symptoms and the nonspecific presentation that many patients can have. The most important point for the diagnosis is to appreciate the signs and symptoms referable to the CNS in high-risk patients, and to consider *C. neoformans* as a possible pathogen in cases of apparently immunocompetent individuals presenting with subacute to chronic meningitis.

A lumbar puncture is necessary for the definitive diagnosis of cryptococcal meningoencephalitis. The opening pressure may be quite elevated, especially in patients with AIDS-associated infections. About two-thirds of patients with AIDS-associated cryptococcal meningoencephalitis described in the literature had initial spinal taps with opening pressures greater than 200 mmH$_2$O. Examination of the CSF with an India ink preparation will show the typical encapsulated yeast forms in more than 75 percent of patients with AIDS and approximately 50 percent of non-AIDS cases. White blood cell counts of the CSF are characteristically low in AIDS-associated cases (<20 cells/mm^3) and somewhat higher in non-AIDS cases (50–350 cells/mm^3) with a mononuclear cell predominance. Protein and glucose levels in the CSF are usually only slightly abnormal in AIDS-associated cases but cases of cryptococcal meningitis can present with very elevated protein levels and prominent hypoglycorrhachia. On the other hand, the CSF glucose and protein levels, as well as the cell counts, can all be completely normal in some severely immunosuppressed patients with cryptococcal meningoencephalitis.

The diagnosis of meningoencephalitis is firmly established by culture of the yeasts from the CSF. In patients with AIDS-associated infections, CSF cultures are almost always positive because of the high burden of organisms associated with severe immune suppression. In non-AIDS patients, 90 percent will have positive cultures and this rate can be further increased by the culturing of large volumes of CSF (15–20 ml). For those cases where the India ink examination is negative, the cryptococcal polysaccharide antigen assay in the CSF is an important adjunct to diagnosis. Results of the polysaccharide antigen test may be obtained soon after the lumbar puncture is performed and can suggest the presence of the infection before the cultures become positive in the laboratory. The antigen test is both very sensitive and specific (>90 percent). In all patients, the antigen test is positive in more than 90 percent of cases, but false-negative results are slightly more common in non-immunosuppressed patients. On the other hand, false-positive tests can result from infections caused by the fungus *Trichosporon beigelii* or possibly other bacteria (McManus and Jones 1985; Westernik et al. 1987; Chanock et al. 1995). Testing for cryptococcal polysaccharide antigen in serum can also be performed and, in patients with AIDS, the sensitivity of using serum for the antigen test is useful as a screening test in patients who do not initially receive a lumbar puncture in geographical areas with a high incidence of cryptococcal disease. The antigen test can be quantified by testing serial dilutions of CSF or serum. In general, the higher the antigen titer, the higher the burden of yeasts and this finding may help in assessing prognosis. Patients with high polysaccharide antigen titers at the beginning or end of therapy are at higher risk for treatment failure. However, the polysaccharide antigen titers cannot be precisely used to follow therapy and to predict relapse in patients (Powderly et al.

1994) and the use of polysaccharide antigen titers to make treatment decisions can be faulty.

On some occasions, the diagnosis of cryptococcal meningoencephalitis is made only after *C. neoformans* is recovered from another body site. This is especially true in patients with AIDS who have a much higher likelihood of having extraneural disease involving the respiratory tract, urinary tract, or bloodstream at the same time as a CNS infection. There are occasional cases in which a blood culture that was sent during the workup of fever of uncertain etiology in a patient with advanced HIV infection returned positive for *C. neoformans*. Then a subsequent lumbar puncture confirmed the diagnosis of a CNS infection. Blood cultures are positive for cryptococci in about two-thirds of AIDS-associated cases of meningoencephalitis, but occasionally a blood culture is positive without evidence of CNS involvement. However, all patients with a positive serum cryptococcal antigen or blood culture should have a lumbar puncture performed to examine for CNS infection and, in patients that are severely immunosuppressed, these findings should support the diagnosis of disseminated infection and treatment administered for invasive disease.

Radiological imaging of the brain by computed tomography (CT) or magnetic resonance imaging (MRI) is helpful when there is a concern about performing a lumbar puncture in a patient with focal neurological signs or symptoms. It is also useful when it shows significant hydrocephalus because this is an indication for placement of a ventricular shunt. Mass lesions can be seen in about 10 percent of patients with cryptococcal meningoencephalitis. Caution should be exercised in patients with AIDS and mass lesions because lymphomas and cerebral toxoplasmosis can coexist with cryptococcal meningitis. The mass lesions caused by cryptococci may not always be true abscesses or granuloma, but may be gelatinous, cyst-like areas resulting from collection of yeasts and antigen with sparse granulomatous inflammation. It is also important to note that, in following patients with cryptococcomas, these brain lesions may worsen radiographically despite effective therapy and this likely reflects an improved inflammatory reaction during elimination of infection (Hospenthal and Bennett 2000).

Pneumonitis

Diagnosis of infection in immunocompetent hosts has been made antemortem by lung biopsy, for histopathology and/or culture, cytopathology, sputum cultures, and serum antigen testing, in a patient with an abnormal chest x-ray (Smith et al. 1976; Kauffman et al. 1981; Fouret et al. 1985; Davies and Sarosi 1987; Murata et al. 1989). Several publications on cryptococcal pneumonia have reported patients with asymptomatic colonization of the respiratory tree by *C. neoformans* (Tynes et al. 1968; Warr et al. 1968; Hammerman et al. 1973; Subramanian et al. 1982). This condition generally occurs in patients with an underlying lung disease such as chronic obstructive pulmonary disease. Hence, a sputum culture positive for *C. neoformans* in a patient, with a lack of clinical symptoms or chest x-ray findings of an infiltrative process, needs to be interpreted carefully and may not represent clinical disease. In the normal host, pulmonary *C. neoformans* generally does not disseminate outside the lung, and thus blood, urine, or CSF cultures, and serum or CSF cryptococcal polysaccharide antigens, will be negative. Nevertheless, all patients with cryptococcal pneumonia and/or pulmonary colonization should be evaluated for disseminated disease (Kerkering et al. 1981; Lehmann et al. 1984). Clinicians might consider following patients without symptoms, with no apparent immunocompromising conditions and a negative serum polysaccharide antigen titer, without checking for CNS involvement since their chance of meningitis is very low. All others should be considered for lumbar puncture to rule out meningitis.

Radiographically, *C. neoformans* pneumonia in the normal host may present with well-defined, noncalcified, single or multiple lung nodules, indistinct to mass-like infiltrates, hilar and mediastinal lymphadenopathy, occasionally pleural effusions, and, more rarely, cavitation (Hunt et al. 1976; Young et al. 1980; Zlupko et al. 1980; Paillas et al. 1982; Feigin 1983; McAllister et al. 1984; Murata et al. 1989). Single or multiple peripheral nodules are the most common radiological findings (Figure 32.4), and many of these infections are discovered when the nodules are aspirated or removed to rule out malignancy (Khoury et al. 1984). These infiltrates may resolve without any specific therapy.

Diagnostic studies in HIV-infected patients with cryptococcal pneumonitis should include arterial P_{O_2}, chest x-ray, cultures, and serum cryptococcal polysaccharide antigen detection. Mild-to-moderate hypoxemia can be found, although both normal P_{O_2} and profound hypoxemia have been documented with cryptococcal pneumonia (Clark et al. 1990b; Cameron et al. 1991). Adult respiratory distress syndrome has also occurred during this infection (Perla et al. 1985; Murray et al. 1988; Similowski et al. 1989). Chest radiographs most often reveal interstitial infiltrates, either focal or diffuse, and lymphadenopathy; unlike immunocompetent and other types of immunocompromised hosts, nodular and alveolar infiltrates are more unusual (Miller and Edelman 1990; Clark et al. 1990b; Cameron et al. 1991). Large masses and pleural effusions are also unusual in AIDS patients (Cameron et al. 1991). As the most common radiological picture is one of interstitial infiltrates, *C. neoformans* pneumonia can easily be confused with *P. jiroveci* pneumonia, a common cause of interstitial infiltrates in patients with AIDS (Loerinc et al. 1988). Cryptococcal serum antigen detection can be extremely helpful while awaiting culture results. As patients with AIDS frequently have disseminated cryptococcosis, positive serum or CSF cryptococcal antigen tests are also frequently detected (Clark et al. 1990b; Cameron et al. 1991).

MANAGEMENT

Meningoencephalitis

For an invasive mycosis, the clinical management of cryptococcosis has been relatively well studied and there is substantial evidence-based data to make therapeutic recommendations. In fact, a group of experts in medical mycology have developed guidelines for the therapy of cryptococcosis (Saag et al. 2000). Despite the studies, there remains a wide range of options and consequences

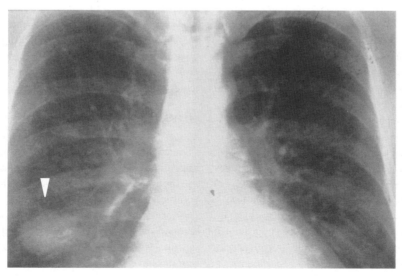

Figure 32.4 *A chest x-ray from a patient with an asymptomatic nodule in the right lower lung field (arrow). The nodule was removed because of suspicion for cancer, and was found to be caused by infection with* C. neoformans. *Most cases of pulmonary cryptococcosis are asymptomatic, and it is not unusual to find nodules secondary to pulmonary cryptococcosis on chest x-rays taken for other reasons.*

in the management of cryptococcosis in many cases algorithms cannot be followed.

Amphotericin B therapy converted cryptococcal meningitis from a uniform fatal infection to one that is curable. Amphotericin B therapies prior to the AIDS epidemic reported therapeutic successes in 60–70 percent of patients (Sarosi et al. 1969). Recent studies have suggested that higher daily doses of amphotericin B may be more effective (deLalla et al. 1995) and with the use of lipid formulations of amphotericin B, such as AmBisome at 4 mg/kg/day, the use of a polyene as a single agent results in a reasonable treatment outcome (Leenders et al. 1997).

In early studies, flucytosine was used alone in the treatment of cryptococcal meningitis (Utz et al. 1972). However, enthusiasm for flucytosine as a single agent was quickly discouraged because of the rapid development of drug resistance. Fluconazole has outstanding CNS pharmacokinetics and has been used extensively in the treatment of cryptococcal meningitis (Stern et al. 1988; Berry et al. 1992; Saag et al. 1992; Yamaguchi et al. 1996). Because of concern about its slow CSF sterilization, its use has tended to occur after the burden of yeast has been reduced with other drug regimens. Despite its poor penetration into CSF, itraconazole has been used in the successful management of cryptococcal meningitis (Denning et al. 1989; Viviani et al. 1989). New extended-spectrum azoles such as voriconazole can treat cases of cryptococcosis (Perfect et al. 2003), but it is not certain how these azoles will improve on fluconazole treatment. The echinocandins, a new class of antifungal agents, do not appear to be effective in treatment of cryptococcal infections.

The most recent strategy for the management of cryptococcal meningitis uses a combination of antifungal drugs in sequence. Years of study have shown that conventional amphotericin B and flucytosine together are excellent for the initial management of infection (Utz et al. 1975; Bennett et al. 1979; Dismukes et al. 1987; Schmitt et al. 1988; Larsen et al. 1990; DeGans et al. 1992). The addition of flucytosine to amphotericin B helps sterilize the CSF quicker than amphotericin B alone or combinations of amphotericin B and fluconazole. It has also been associated with fewer cases of relapse compared to a single agent (Saag et al. 1999). A large comparative study in AIDS patients for the management of cryptococcal meningitis has also been adopted in the treatment of non-AIDS patients with cryptococcal meningitis (van der Horst et al. 1997; Pappas et al. 2001). The regimen used in this study was amphotericin B at 0.7 mg/kg/day and flucytosine 100 mg/kg/day for at least 2 weeks of induction therapy and then switching to oral fluconazole at 400–800 mg/day alone for 8 weeks. Ideally, a lumbar puncture would be performed after 2 weeks of therapy with extension of the amphotericin B therapy if the CSF cultures are still postive. After these 2.5–3 months of therapy, the fluconazole at 200–400 mg/day is continued for 6 months to a year in non-AIDS patients and in AIDS patients suppression therapy had been recommended for a lifetime prior to HAART. In the suppressive stage of treatment, there are data that suggest that the use of fluconazole may be better than itraconazole (Saag et al. 1999). With the advent of HAART and thus improvement in immunity with antiretroviral therapies, the need for lifelong suppression in patients who have reduced their viral load, increased their CD4 cells ($\geqslant$200 cells/ml), and been free of clinical cryptococcal disease for 1–2 years is being questioned. Recent studies have supported stopping fluconazole in this patient population where the underlying disease is being controlled with HAART (Martinez et al. 2000; Vibhagool et al. 2003).

The issue of discontinuing chronic suppressive therapy was evaluated by a retrospective observational study of 100 patients with a definitive diagnosis of cryptococcal meningitis who subsequently had CD4 counts above 100 cells/μL while receiving HAART (Mussini et al. 2004). Prior to discontinuation of chronic suppressive antifungal therapy, the median duration of HAART in this patient cohort was 26 months, the median duration of maintenance therapy was 33 months, and the median time the CD4 count had been above 100 cells/μL was one month. Forty-one patients had a positive serum cryptococcal antigen at the time maintenance therapy was discontinued. Over a median follow-up of 28 months, there were two relapses of cryptococcal meningitis and two cases of extrameningeal cryptococcosis. The relapse in one patient occurred at a CD4 count of 46 cells/μL, but the other three patients all had negative serum cryptococcal antigen test results at the time therapy was discontinued, and at the time of relapse or extrameningeal disease had undetectable HIV viral loads and CD4 counts above 100 cells/μL (two above 400 cells/μL). These data demonstrate that it may be possible to withdraw chronic suppressive therapy in select patients, but that there remains a small, but significant, risk for recurrence of cryptococcosis.

Site of infection

Unlike cryptococcal meningitis, pulmonary cryptococcosis has not been prospectively studied and most data come from retrospective examination of patient outcomes. Azoles such as fluconazole at 200–400 mg/day for 3–6 months are generally safe and effective (Dromer et al. 1996; Yamaguchi et al. 1996; Pappas et al. 2001). In fact, fluconazole can be used in the treatment of both symptomatic and asymptomatic patients with *C. neoformans* isolated from the lung or sputum. In seriously ill patients, the use of amphotericin B or the combination of amphotericin B plus flucytosine may be more attractive. Surgery has been recommended for a large CNS lesion (>3 cm) and lung abscesses with *C. neoformans* var. *gattii*, but this strategy should only be followed on an individual case basis. Treatment of other sites of cryptococcal infection should be determined by their location and host factors. Amphotericin B-containing regimens should be considered as induction therapy when there is a high burden of yeasts such as during severe immune depression, but otherwise fluconazole is a reasonable choice.

Immune status of host

Every attempt to improve immunity of the host during treatment needs to be made. On a practical basis this may mean an attempt to reduce immunosuppressive agents, for instance, reduction of prednisone doses to 20 mg/day or less per day. Treatment with HAART can help reconstitute the immune cells of patients with HIV infection and should be encouraged in the management of cryptococcal meningitis, but the timing of a HAART regimen in relationship to stage of treatment remains uncertain. For instance, it is clear that a reconstitution syndrome can occur in the central nervous system or other sites when patients with cryptococcosis are started on HAART (Jenny-Avital and Abadi 2002). This syndrome is marked by evidence of sterile inflammation at this site of infection and this could add to new symptoms in the patient. In patients with increased intracranial pressure, the immune reconstitution may actually worsen symptoms and immune reconstitution syndrome may actually improve with the use of corticosteroid therapy to reduce symptomatic inflammation. Better guidance into what is the best time for HAART therapy in patients with cryptococcosis is needed, but it may be reasonable to start HAART after several months of antifungal therapy to control the initial yeast burden of infection.

Suppressive therapy

Relapse rates in AIDS patients with cryptococcal meningitis before HAART approached 50–60 percent when therapy was stopped within the first 6 months. Fluconazole suppressive therapy with 200 mg/day in AIDS patients has dramatically reduced these numbers to less than 5 percent (Bozette et al. 1991) and it is probably better than intermittent intravenous amphotericin B and/or daily suppression with itraconazole (Powderly et al. 1992). Since non-AIDS patients also relapse (10–15 percent) in the first 6 months after therapy is stopped, this suppressive strategy is similar to the AIDS patients' strategy with 6–12 months of fluconazole after induction therapy being used. In AIDS patients, suppressive therapy has been used for a lifetime, but recent data have studied the discontinuation of drugs after 1–2 years in those with a satisfactory response to HAART (Martinez et al. 2000; Vibhagool et al. 2003).

Role of intracranial pressure

Patients with a high burden of yeasts and polysaccharide antigen can develop increased intracranial pressure before, during, or after treatment (Denning et al. 1991b; Graybill et al. 2000). In this syndrome of early increased intracranial pressure, the patient can suddenly develop systemic hypertension, decreased sensorium, cranial neuropathies, visual loss, worsening headaches, and mental confusion. It is suggested that this elevated pressure is related to reduced CSF resorption, brain edema, and CSF outflow resistance. This increased pressure may respond to CSF withdrawal techniques (lumbar–peritoneal shunts, repeated lumbar taps) (Van Gemert and Vermeulen 1991), but the preciseness of management for this life-threatening complication remains unclear. On the other hand, development of classic noncommunicating hydrocephalus weeks or months after treatment is probably related to inflammation and scar tissue from the original infection and should respond to a ventriculoperitoneal (VP) shunt. A VP shunt can be placed successfully while cryptococcal meningitis is being effectively treated. However, shunts should not be placed until adequate antifungal treatment has begun (Yadav et al. 1988; Ingram et al. 1993).

Antigen titers

CSF polysaccharide antigen titers generally report the burden of yeasts at initial diagnosis, and high antigen titers do predict poor prognosis. However, in an individual patient the reliability of following antigen titers to guide therapy is unclear (Powderly et al. 1994), and thus it is difficult to use titer changes during therapy to make clinical decisions.

Cytokines and specific antibodies

Disseminated cryptococcosis generally occurs during an episode of immunosuppression. Granulocyte-macrophage colony-stimulating factor (GM-CSF), macrophage colony-stimulating factor (M-CSF), IL-2, IL-12, and interferon-gamma can positively interact with host cells to inhibit or kill cryptococci. Both GM-CSF and interferon-gamma have been examined in open human trials, but it remains uncertain whether these biological modifiers will have a positive impact on the outcome of this infection. Interferon-gamma treatment has been studied prospectively; there was a trend toward a faster reduction in yeast counts in the CSF but no impact on outcome. Specific monoclonal antibodies and serum therapy have a positive effect on the prevention and treatment of experimental cryptococcosis. Early safety studies with a monoclonal antibody have been completed in humans, but efficacy data with the use of antibodies are lacking. It remains important to continue examination on the impact of biological enhancers and strategies for management of infection since there is still a 10–25 percent acute mortality rate for cryptococcal meningitis during early therapies even in medically advanced countries.

PROGNOSTIC FACTORS

The most important prognostic factor for a patient with cryptococcosis is the nature of their underlying illnesses. Certainly, those patients with underlying malignancies or AIDS have a much poorer prognosis than those patients with no apparent risk factors for cryptococcosis. In one comparison of patients with cryptococcal meningoencephalitis, patients with malignancies as an underlying risk factor actually had a much shorter median survival rate than patients with AIDS (2 versus 9 months) (White et al. 1992), and this observation was made before the use of HAART, which even further improves the survival of patients with AIDS and cryptococcosis.

In patients without AIDS, several clinical features have been identified which correlate with treatment failure (Diamond and Bennett 1974; Bennett et al. 1979; Dismukes et al. 1987). Patients who have died during therapy have been more likely to have presented with:

- positive India ink examinations of the CSF
- CSF white blood cell counts of less than 20 cells/mm^3
- initial CSF or serum cryptococcal antigen titers of more than 1:32
- extraneural sites of infection
- high opening pressures (>350 mmH$_2$O) on lumbar puncture (Diamond and Bennett 1974).

Patients who relapsed after treatment have been characterized by:

- persistently low CSF glucose concentrations after 4 weeks of therapy
- low initial CSF white blood count, $\geqslant$ post-treatment CSF, or serum antigen titers of 1:8 or more
- treatment with corticosteroids with at least 20 mg prednisone or its equivalent daily after completion of therapy (Diamond and Bennett 1974).

In patients without AIDS as an underlying risk factor, the following factors predicted a favorable response:

- normal mental status, headache, CSF white blood count over 20 cells/mm^3
- absence of another underlying disease
- negative CSF India ink preparations (Dismukes et al. 1987).

Although these observations were made more than 15–25 years ago, they still likely identify patients with a high burden of yeasts, poor mental status, difficult-to-treat underlying disease on presentation, and risk for failure of treatment.

In patients with AIDS as a risk factor, significant pretreatment predictors of death during therapy include:

- abnormal mental status
- a CSF antigen titer of more than 1:1 024
- a CSF white blood count of less than 20 cells/mm^3 (Saag et al. 1992).

Diastolic hypertension has also been associated with earlier death in these patients (Fan-Havard et al. 1992), and this finding is probably reflective of increased intracranial pressure. The AIDS patient with cryptococcal meningitis at highest risk for complications has:

- high burden of yeasts
- poor inflammatory response
- increased intracranial pressure $\geqslant$350 mmH$_2$O.

PREVENTION

Trying to prevent cryptococcosis in the general population is an unrealistic task given its rarity. However, in certain high-risk populations, antifungal prophylaxis is a consideration. Some data have been collected from patients with AIDS who were given chronic fluconazole for prophylaxis (Nightingale et al. 1992). Patients receiving fluconazole 200 mg/day had cryptococcosis rates of 1 percent over 35 months, compared with patients given topical clotrimazole who had rates of 7 percent (Powderly et al. 1995). However, general recommendations for fluconazole prophylaxis in patients with AIDS can be made only after very careful consideration of costs, impact on survival, prevalence of the infection, and the potential development of azole resistance in other fungal species. It is rarely considered today in the era of HAART.

Immunization is an attractive form of preventing cryptococcosis. A cryptococcal glucuronoxylomannan tetanus toxoid conjugate vaccine was developed, and it appeared to be highly immunogenic in murine models (Devi et al. 1991). There are now several C. neoformans immunogenic proteins in which genes have been cloned and recombinant proteins made. Thus progress toward vaccine development is being made. Trials in humans have not yet been done with any product, but it seems that, for patients at highest risk for cryptococcosis, any vaccine will have to be administered at a time when their immune systems are competent enough to make a protective response. Without a standardized skin test, it is more difficult to determine who is at risk and the magnitude of potential need.

REFERENCES

Ajanee, N., Alam, M., et al. 1996. Brain abscess caused by Wangiella dermatitidis: case report. Clin Infect Dis, 23, 197–8.

Alspaugh, J.A., Perfect, J.R. and Heitman, J. 1997. Cryptococcus neoformans mating and virulence are regulated by the G-protein gamma subunit GPA1 and cAMP. Genes Dev, 11, 3206–17.

Alspaugh, J.A., Cavallo, L.M., et al. 2000. RAS1 regulates filamentation, mating, and growth at high temperature of Cryptococcus neoformans. Mol Microbiol, 36, 352–65.

Baker, R.D. 1976. The primary pulmonary lymph node complex of cryptococcosis. Am J Clin Pathol, 65, 83–92.

Baker, R.D. and Haugen, R.K. 1955. Tissue changes and tissue diagnosis of cryptococcosis: a study of 26 cases. *Am J Pathol*, **25**, 14.

Bava, A.J., Negroni, R. and Bianchi, M. 1993. Cryptococcosis produced by a urease negative strain of *Cryptococcus neoformans*. *J Med Vet Mycol*, **31**, 87–9.

Benham, R.W. 1935. Cryptococci, their identification by morphology and serology. *J Infect Dis*, **57**, 255–74.

Bennett, J.E., Kwon-Chung, K.J. and Howard, D.H. 1977. Epidemiology differences among serotypes of Cryptococcus neoformans. *Am J Epidemiol*, **105**, 582–6.

Bennett, J.E., Dismukes, W., et al. 1979. A comparison of amphotericin B alone and combined with flucytosine in the treatment of cryptococcal meningitis. *N Engl J Med*, **301**, 126–31.

Berry, A.J., Rinaldi, M.G. and Graybill, J.R. 1992. Use of high dose fluconazole as salvage therapy for cryptococcal meningitis in patients with AIDS. *Antimicrob Agents Chemother*, **36**, 690–2.

Beyt, B.E. and Waltman, S.R. 1978. Cryptococcal endophthalmitis after corneal transplantation. *N Engl J Med*, **298**, 825–6.

Birkmann, L.W. and Bennett, D.R. 1988. Meningoencephalitis following evaluation for cryptococcal endophthalmitis. *Ann Neurol*, **4**, 476–7.

Blachie, J.D., Danta, G., et al. 1985. Ophthalmological complications of cryptococcal meningitis. *Clin Exp Neurol*, **21**, 263–70.

Blanc, M., Leuenberger, P. and Favez, G. 1982. Cryptococcose pulmonaire invasive isole. Presentation d'un cas. *Schweiz Med Wochenschr J Suisse Med*, **112**, 421–4.

Blocker, K.S., Weeks, J.A. and Noble, R.C. 1987. Cutaneous cryptococcal infection presenting as vulvar lesion. *Genitourin Med*, **63**, 341–3.

Borton, L.K. and Wintroub, B.U. 1984. Disseminated cryptococcosis presenting as herpetiform lesions in a homosexual man with acquired immunodeficiency syndrome. *J Am Acad Dermatol*, **10**, 387–90.

Bozette, S.A., Larsen, R.A., et al. 1991. A placebo-controlled trial of maintenance therapy with fluconazole after treatment of cryptococcal meningitis in the acquired immunodeficiency syndrome. *N Engl J Med*, **324**, 580–4.

Braman, R.T. 1981. Cryptococcosis (*Torulopsis*) of prostate. *Urology*, **17**, 284–6.

Bulmer, G.S., Sans, M.D. and Gunn, C.M. 1967. *Cryptococcus neoformans*. I. Nonencapsulated mutants. *J Bacteriol*, **94**, 1475–9.

Buschke, A. 1895. Ueber eine durch coccidien hemorgerufene krankheit des menschen. *Dtsch Med Wochenschr*, **21**, 14.

Cameron, M.L., Bartlett, J.A., et al. 1991. Manifestations of pulmonary cryptococcosis in patients with acquired immunodeficiency syndrome. *Rev Infect Dis*, **13**, 64–7.

Campbell, G.D. 1966. Primary pulmonary cryptococcosis. *Am Rev Respir Dis*, **94**, 236–43.

Casadevall, A. and Perfect, J.R. 1998. *Cryptococcus neoformans*. Washington, DC: ASM Press.

Casadevall, A.J., Mukherjee, J., et al. 1994. Management of *Cryptococcus neoformans* contaminated needle injuries. *Clin Infect Dis*, **19**, 951–3.

Cawley, E.P., Grekin, R.H. and Curtis, A.C. 1950. Torulosis, a review of the cutaneous and adjoining mucous membrane manifestations. *J Invest Dermatol*, **14**, 327–41.

Chang, Y.C. and Kwon-Chung, K.J. 1994. Complementation of a capsule-deficiency mutation of *Cryptococcus neoformans* restores its virulence. *Mol Cell Biol*, **14**, 4912–19.

Chang, Y.C. and Kwon-Chung, K.J. 1999. Isolation, characterization, and localization of a capsule-associated gene, *CAP10*, of *Cryptococcus neoformans*. *J Bacteriol*, **181**, 5636–43.

Chanock, S.J., Toltzis, P. and Wilson, C. 1995. Cross-reactivity between *Stomatococcus mucilaginosus* and latex agglutination for cryptococcal antigen. *Lancet*, **342**, 1119–20.

Chechani, V. and Kamholz, S.L. 1990. Pulmonary manifestations of disseminated cryptococcosis in patients with AIDS. *Chest*, **98**, 1060–6.

Chen, L.C., Goldman, D.L. and Doering, T.L. 1999. Antibody response to *Cryptococcus neoformans* proteins in rodents and humans. *Infect Immun*, **67**, 2218–24.

Christoph, I. 1990. Pulmonary *Cryptococcus neoformans* and disseminated *Nocardia brasiliensis* in an immunocompromised host. *North Carolina Med J*, **51**, 219–20.

Chuck, S.L. and Sande, M.A. 1989. Infections with *Cryptococcus neoformans* in the acquired immunodeficiency syndrome. *N Engl J Med*, **321**, 794–9.

Clark, R.A., Greer, D., et al. 1990a. Spectrum of *Cryptococcus neoformans* infection in 68 patients infected with acquired immunodeficiency virus. *Rev Infect Dis*, **12**, 768–77.

Clark, R.A., Greer, D.L., et al. 1990b. *Cryptococcus neoformans* pulmonary infection in HIV-1-infected patients. *J Acquir Immune Defic Syndr*, **3**, 480–5.

Clumeck, N., Sonnet, J., et al. 1984. Acquired immunodeficiency syndrome in African patients. *N Engl J Med*, **310**, 492–7.

Cohen, I. 1985. Isolated pulmonary cryptococcosis in a young adolescent. *Pediatr Infect Dis J*, **4**, 416–19.

Concus, A.P., Helfand, R.F., et al. 1988. Cutaneous cryptococcosis mimicking molluscum contagiosum in a patient with AIDS. *J Infect Dis*, **158**, 897–8.

Crump, J.R., Elner, S.G., et al. 1992. Cryptococcal endophthalmitis: case report and review. *Clin Infect Dis*, **14**, 1069–73.

Davies, S.F. and Sarosi, G.A. 1987. Role of serodiagnostic tests and skin tests in the diagnosis of fungal disease. *Clin Chest Med*, **8**, 135–46.

DeGans, J., Portegies, P. and Tiessens, G. 1992. Itraconazole compared with amphotericin B plus flucytosine in AIDS patients with cryptococcal meningitis. *AIDS*, **6**, 185–90.

deLalla, F., Pellizzer, G. and Vaglia, A. 1995. Amphotericin B as primary therapy for cryptococcosis in patients with AIDS: reliability of relatively high doses administered over a relatively short period. *Clin Infect Dis*, **20**, 263–6.

Denning, D.W., Tucker, R.M., et al. 1989. Itraconazole therapy for cryptococcal meningitis and cryptococcosis. *Arch Intern Med*, **149**, 2301–8.

Denning, D.W., Armstrong, R.W., et al. 1991a. Endophthalmitis in a patient with disseminated cryptococcosis and AIDS who was treated with itraconazole. *Rev Infect Dis*, **13**, 1126–30.

Denning, D.W., Armstrong, R.W., et al. 1991b. Elevated cerebrospinal fluid pressures in patients with cryptococcal meningitis and acquired immunodeficiency syndrome. *Am J Med*, **91**, 267–72.

Devi, S.J., Scheerson, R., et al. 1991. *Cryptococcus neoformans* serotype A glucuronoxylomannan protein conjugate vaccines: synthesis, characterization, and immunogenicity. *Infect Immun*, **59**, 3700–7.

Diamond, R.D. and Bennett, J.E. 1974. Prognostic factors in cryptococcal meningitis. A study of 111 cases. *Ann Intern Med*, **80**, 176–81.

Diamond, R.D. and Levitz, S.M. 1988. *Cryptococcus neoformans* pneumonia. In: Pennington, J.E. (ed.), *Respiratory infections: diagnosis and management*. New York: Raven Press, 457–71.

Dismukes, W.E., Cloud, G., et al. 1987. Treatment of cryptococcal meningitis with combination amphotericin B and flucytosine for four as compared with six weeks. *N Engl J Med*, **317**, 334–41.

Doft, B.H. and Curtin, V.T. 1982. Combined ocular infection with cytomegalovirus and cryptococcosis. *Arch Ophthalmol*, **100**, 1800–3.

Dromer, F., Mathoulin, S., et al. 1996. Comparison of the efficacy of amphotericin B and fluconazole in the treatment of cryptococcosis in human immunodeficiency virus-negative patients: retrospective analysis of 83 cases. *Clin Infect Dis*, **22**, Suppl 2, s154–60.

Dufait, R., Velho, R. and de Vroey, C. 1987. Rapid identification of the two varieties of *Cryptococcus neoformans* by D-proline assimilation. *Mykosen*, **30**, 483.

Duperval, R., Hermans, P.E., et al. 1977. Cryptococcosis, with emphasis on the significance of isolation of *Cryptococcus neoformans* from the respiratory tract. *Chest*, **72**, 13–19.

Ellis, D.H. and Pfeiffer, T.J. 1990. Natural habitat of *Cryptococcus neoformans* var *gattii*. *J Clin Microbiol*, **28**, 1642–4.

Emmons, C.W. 1951. Isolation of *Cryptococcus neoformans* from soil. *J Bacteriol*, **62**, 685–90.

Emmons, C.W. 1955. Saprophytic sources of *Cryptococcus neoformans* associated with the pigeon. *Am J Hyg*, **62**, 227–32.

Eng, R.H., Bishburg, E. and Smith, S.M. 1986. Cryptococcal infections in patients with acquired immune deficiency syndrome. *Am J Med*, **81**, 19–23.

Erickson, T., Liu, L., et al. 2001. Multiple virulence factors of *Cryptococcus neoformans* are dependent on VPH1. *Mol Microbiol*, **42**, 1121–31.

Evans, E.E. 1950. The antigenic composition of *Cryptococcus neoformans*. I. A serologic classification by means of the capsular agglutinations. *J Immunol*, **64**, 423–30.

Evans, E.E. and Kessel, J.F. 1951. The antigenic composition of *Cryptococcus neoformans*. II. Serologic studies with the capsular polysaccharide. *J Immunol*, **67**, 109–14.

Fan-Havard, P., Yamaguchi, E., et al. 1992. Diastolic hypertension in AIDS patients with cryptococcal meningitis. *Am J Med*, **93**, 347–8.

Farmer, S.G. and Komorowski, R.A. 1973. Histologic response to capsule-deficient *Cryptococcus neoformans*. *Arch Pathol*, **96**, 383–7.

Feigin, D.S. 1983. Pulmonary cryptococcosis: radiologic-pathologic correlates of its three forms. *Am J Radiol*, **141**, 1263–72.

Feldmesser, M., Tucker, S.C. and Casadevall, A. 2001. Intracellular parasitism of macrophages by *Cryptococcus neoformans*. *Trends Microbiol*, **9**, 273–8.

Fessel, W.J. 1993. Cryptococcal meningitis after unusual exposures to birds. *N Engl J Med*, **328**, 1354–5.

Fouret, P., Roux, P., et al. 1985. Diagnostic rapide d'une cryptococcose pulmonaire par examen cytologique du liquide de lavage broncho-alveolaire (LBA). *Arch Anat Cytol Pathol (Paris)*, **33**, 90–2.

Franzot, S.P., Salkin, I.F. and Casadevall, A. 1999. *Cryptococcus neoformans* var. *grubii*: separate variety status for *Cryptococcus neoformans* serotype A isolates. *J Clin Microbiol*, **37**, 838–40.

Friedman, G.D. 1983. The rarity of cryptococcosis in Northern California: the 10-year experience of a large defined population. *Am J Epidemiol*, **117**, 230–4.

Fromtling, R.A., Shadomy, H.J. and Jacobson, E.S. 1982. Decreased virulence in stable, acapsular mutants of *Cryptococcus neoformans*. *Mycopathologia*, **79**, 23–9.

Gauder, J.P. 1977. Cryptococcal cellulitis. *J Am Med Assoc*, **237**, 672–3.

Glaser, J.B. and Garden, A. 1985. Inoculation of cryptococcosis without transmission of the acquired immunodeficiency syndrome. *N Engl J Med*, **313**, 264.

Goldman, D.L., Khine, H. and Abadi, J. 2001. Serologic evidence for cryptococcus infection in early childhood. *Pediatrics*, **107**, 66.

Granger, D.L., Perfect, J.R. and Durack, D.T. 1985. Virulence of *Cryptococcus neoformans*: regulation of capsule synthesis by carbon dioxide. *J Clin Invest*, **76**, 508–16.

Granger, D.L., Perfect, J.R. and Durack, D.T. 1986. Macrophage-mediated fungistasis: requirement for a macromolecular component in serum. *J Immunol*, **137**, 693–701.

Graybill, J.R., Sobel, J., et al. 2000. Diagnosis and management of increased intracranial pressure in patients with AIDS and cryptococcal meningitis. *Clin Infect Dis*, **30**, 47–54.

Hammerman, K.J., Powell, K.E., et al. 1973. Pulmonary cryptococcosis: clinical forms and treatment. A Center for Disease Control Cooperative Mycoses Study. *Am Rev Respir Dis*, **108**, 1116–23.

Henson, D.J. and Hill, A.R. 1984. Cryptococcal pneumonia: a fulminant presentation. *Am J Med*, **228**, 221.

Hill, J.O. and Harmsen, A.G. 1991. Intrapulmonary growth and dissemination of an avirulent strain of *Cryptococcus neoformans* in mice depleted of CD4+ or CD8+ T cells. *J Exp Med*, **173**, 755–8.

Horowitz, I.D., Blumberg, E.A. and Krevolin, L. 1993. *Cryptococcus albidus* and mucormycosis empyema in a patient receiving hemodialysis. *South Med J*, **86**, 1070–2.

Hospenthal, D. and Bennett, J.E. 2000. Persistence of cryptococcomas on neuroimaging. *Clin Infect Dis*, **31**, 1303–6.

Hull, C. and Heitman, J. 2002. Genetics of *Cryptococcus neoformans*. *Ann Rev Genet*, **36**, 557–615.

Hunt, K.K. Jr., Enquist, R.W. and Bowen, T.E. 1976. Multiple pulmonary nodules with central cavitation. *Chest*, **69**, 529–30.

Ikeda, R., Shinoda, T., et al. 1982. Antigenic characterization of *Cryptococcus neoformans* serotypes and its application to serotyping of clinical isolates. *J Clin Microbiol*, **36**, 22–9.

Ingram, C.W., Haywood, H.B., et al. 1993. Cryptococcal ventricular peritoneal shunt infection: clinical and epidemiological evaluation of two closely associated cases. *Infect Immun*, **14**, 719–22.

Jacobson, E.S. and Emery, H.S. 1991. Catecholamine uptake, melanization, and oxygen toxicity in *Cryptococcus neoformans*. *J Bacteriol*, **173**, 401–3.

Jacobson, E.S. and Petro, M.J. 1987. Extracellular iron chelation in *Cryptococcus neoformans*. *J Med Vet Mycol*, **25**, 415–18.

Jacobson, E.S. and Vartivarian, S.E. 1992. Iron assimilation in *Cryptococcus neoformans*. *J Med Vet Mycol*, **30**, 443–50.

Jenny-Avital, E.R. and Abadi, M. 2002. Immune reconstitution cryptococcosis after initiation of successful highly active antiretroviral therapy. *Infect Immun*, **35**, 128–33.

Johnston, S.R., Corbett, E.L., et al. 1992. Raised intracranial pressure and visual complications in AIDS patients with cryptococcal meningitis. *J Infect*, **24**, 185–9.

Kanj, S.S., Welty-Wolf, K., et al. 1996. Fungal infections in lung and heart-lung transplant recipients, report of 9 cases and review of the literature. *Medicine*, **75**, 142–56.

Kapend'a, K., Komichelo, K., et al. 1987. Meningitis due to *Cryptococcus neoformans* biovar *gattii* in a Zairean AIDS patient. *Eur J Clin Microbiol*, **6**, 320–1.

Kauffman, C.A., Bergman, A.G., et al. 1981. Detection of cryptococcal antigen. Comparison of two latex agglutination tests. *Am J Clin Pathol*, **75**, 106–9.

Kent, T.H. and Layton, J.M. 1962. Massive pulmonary cryptococcosis. *Am J Clin Pathol*, **38**, 596–604.

Kerkering, T.M., Duma, R.J. and Shadomy, S. 1981. The evolution of pulmonary cryptococcosis. Clinical implications from a study of 41 patients with and without compromising host factors. *Ann Intern Med*, **94**, 611–16.

Khardori, N., Butt, F. and Rolston, K.V.I. 1988. Pulmonary cryptococcosis in AIDS. *Chest*, **93**, 1319–20.

Khoury, M.B., Godwin, J.D., et al. 1984. Thoracic cryptococcosis: immunologic competence and radiologic appearance. *Am J Radiol*, **141**, 893–6.

Kovacs, J.A., Kovacs, A.A., et al. 1985. Cryptococcosis in the acquired immunodeficiency syndrome. *Ann Intern Med*, **103**, 533–8.

Kozel, T.R. and Cazin, J. 1971. Nonencapsulated variant of *Cryptococcus neoformans*. I. Virulence studies and characterization of soluble polysaccharides. *Infect Immun*, **3**, 287–94.

Kozel, T.R. and Gotschlich, E.C. 1982. The capsule of *Cryptococcus neoformans* passively inhibits phagocytosis of the yeast by macrophages. *J Immunol*, **129**, 1675–80.

Kramer, M., Corrado, M.L., et al. 1983. Pulmonary cryptococcosis and Cushing's syndrome. *Ann Intern Med*, **143**, 2179–80.

Kwon-Chung, K.J. 1975. A new genus, *Filobasidiella*, the perfect state of *Cryptococcus neoformans*. *Mycologia*, **67**, 1197–200.

Kwon-Chung, K.J. 1976. Morphogenesis of *Filobasidiella neoformans*, the sexual state of *Cryptococcus neoformans*. *Mycologia*, **68**, 821–833.

Kwon-Chung, K.J. and Bennett, J.E. 1978. Distribution of 'alpha' and 'a' mating types of *Cryptococcus neoformans* among natural and clinical isolates. *Am J Med*, **108**, 337–40.

Kwon-Chung, K.J. and Bennett, J.E. 1984. Epidemiologic differences between the two varieties of *Cryptococcus neoformans*. *Am J Epidemiol*, **120**, 123–40.

Kwon-Chung, K.J., Polacheck, I. and Popkin, T.J. 1982. Melanin-lacking mutants of *Cryptococcus neoformans* and their virulence for mice. *J Bacteriol*, **150**, 1414–21.

Lambertus, M.W., Goetz, M.B., et al. 1990. Complications of corticosteroid therapy in patients with the acquired immunodeficiency syndrome and *Pneumocystis carinii* pneumonia. *Chest*, **98**, 38–43.

Larsen, R.A., Bozzette, S., et al. 1989. Persistent *Cryptococcus neoformans* infection of the prostate after successful treatment of meningitis. *Ann Intern Med*, **111**, 125–8.

Larsen, R.A., Leal, M.A.E. and Chan, L.S. 1990. Fluconazole compared with amphotericin B plus flucytosine for cryptococcal meningitis in AIDS. *Ann Intern Med*, **113**, 183–7.

Lazera, M.S., Pires, F.D.A., et al. 1996. Natural habitat of *Cryptococcus neoformans* var. *neoformans* in decaying wood forming hollows in living trees. *J Med Vet Mycol*, **34**, 127–31.

Leenders, A.C., Reiss, P., et al. 1997. Liposomal amphotericin B (AmBisome) compared with amphotericin B followed by oral fluconazole in the treatment of AIDS-associated cryptococcal meningitis. *AIDS*, **11**, 1463–71.

Lehmann, P.F., Morgan, R.J. and Freimer, E.H. 1984. Infection with *Cryptococcus neoformans* var. *gattii* leading to a pulmonary cryptococcoma and meningitis. *J Infect*, **9**, 301–6.

Levinson, D.J., Silcox, D.C., et al. 1974. Septic arthritis due to nonencapsulated *Cryptococcus neoformans* with co-existing sarcoidosis. *Arthritis Rheum*, **17**, 1037–47.

Littman, M.L. and Borok, R. 1968. Relation of the pigeon to cryptococcosis: natural carrier state, heat resistance and survival of *Cryptococcus neoformans*. *Mycopathologia*, **35**, 329–45.

Littman, M.L. and Zimmerman, L.E. 1956. Cryptococcosis, torulosis or European blastomycosis. *Cryptococcosis*. New York: Grune and Stratton, 38–46.

Loerinc, A.M., Bottone, E.J., et al. 1988. Primary cryptococcal pneumonia mimicking *Pneumocystis carinii* pneumonia in a patient with AIDS. *Mt Sinai J Med*, **55**, 181–6.

Lynch, J.P. III, Schaberg, D.R., et al. 1981. *Cryptococcus laurentii* lung abscess. *Am Rev Respir Dis*, **123**, 135–8.

Martinez, E., Garcia-Viejo, M.A. and Marcos, M.A. 2000. Discontinuation of secondary prophylaxis for cryptococcal meningitis in HIV-infected patients responding to highly active antiretroviral therapy. *AIDS*, **14**, 2615.

Martinez, L.R., Garcia-Rivera, J. and Casadevall, A. 2001. *Cryptococcus neoformans* var. *neoformans* (serotype D) strains are more susceptible to heat than *C. neoformans* var. *grubii* (serotype A strains). *J Clin Microbiol*, **39**, 3365–7.

Mayers, D.L., Martone, W.J. and Mandell, G.L. 1981. Cutaneous cryptococcosis mimicking gram-positive cellulitis in a renal transplant patient. *South Med J*, **74**, 1032.

McAllister, K., Ognibene, A.J., et al. 1984. Cryptococcal pleuro-pulmonary disease: infection of the pleural fluid in the absence of disseminated cryptococcosis. Case report. *Milit Med*, **149**, 684–6.

McDonnell, J.M. and Hutchins, G.M. 1985. Pulmonary cryptococcosis. *Hum Pathol*, **16**, 120–8.

McManus, E.J. and Jones, J.M. 1985. Detection of a *Trichosporon beigelii* antigen cross-reactive with *Cryptococcus neoformans* capsular polysaccharides in serum from a patient with disseminated trichosporon infection. *J Clin Microbiol*, **21**, 681–5.

Menon, A. and Rajamani, R. 1976. Giant 'cryptococcoma' of the lung. *Br J Dis Chest*, **70**, 269–72.

Meyer, W., Castaneda, A., et al. 2003. Molecular typing of Ibero American *Cryptococcus neoformans* isolates. *Emerg Infect Dis*, **9**, 189–95.

Milchgrub, S., Visconti, E. and Avellini, J. 1990. Granulomatous prostatitis induced by capsule-deficient cryptococcal infection. *J Urol*, **143**, 365–6.

Miller, W.T. and Edelman, J.M. 1990. Cryptococcal pulmonary infection in patients with AIDS: radiographic appearance. *Radiology*, **175**, 725–8.

Mirza, S., Phelan, M., et al. 2002. The changing epidemiology of cryptococcosis: an update from population-based active surveillance in 2 large metropolitan areas. *Clin Infect Dis*, **36**, 789-794, 1992–2000.

Mody, C.H., Chen, G.H., et al. 1993. Depletion of murine CD8+ T cells *in vivo* decreases pulmonary clearance of a moderately

virulent stain of *Cryptococcus neoformans*. *J Lab Clin Med*, **121**, 765–73.

Mody, C.H. and Paine, R. III 1994. CD8 cells play a critical role in delayed type hypertensivity to intact *Cryptococcus neoformans*. *J Immunol*, **152**, 3970–9.

Moyrand, F., Klaproth, B., et al. 2002. Isolation and characterization of capsule structure mutant strains of *Cryptococcus neoformans*. *Mol Microbiol*, **45**, 837–49.

Mukherjee, J., Sharff, M.D. and Casadevall, A. 1992. Protective murine monoclonal antibodies to *Cryptococcus neoformans*. *Infect Immun*, **60**, 4534–41.

Mukherjee, J., Pirofski, L.A., et al. 1993. Antibody-mediated protection in mice with lethal intracerebral *Cryptococcus neoformans* infection. *Proc Natl Acad Sci U S A*, **90**, 3636–40.

Mukherjee, J., Zuckier, L.S., et al. 1994. Therapeutic efficacy of monoclonal antibodies to *Cryptococcus neoformans* glucuronoxylomannan alone and in combination with amphotericin B. *Antimicrob Agents Chemother*, **38**, 580–7.

Murata, K., Khan, A. and Herman, P.G. 1989. Pulmonary parenchymal disease: evaluation with high-resolution CT. *Radiology*, **170**, 629–35.

Murray, R.J., Becker, P., et al. 1988. Recovery from cryptococcemia and the adult respiratory distress syndrome in the acquired immunodeficiency syndrome. *Chest*, **93**, 1304–7.

Mussini, C., Pezzotti, P., et al. 2004. Discontinuation of maintenance therapy for cryptococcal meningitis in patients with AIDS treated with highly active antiretroviral therapy: an international observational study. *Clin Infect Dis*, **38**, 565.

Mylonakis, E., Ausubel, F.M., et al. 2002. Killing of *Caenorhabditis elegans* by *Cryptococcus neoformans* as a model of yeast pathogenesis. *Proc Natl Acad Sci U S A*, **99**, 15675–80.

Neilson, J.B., Fromtling, R.A. and Bulmer, G.S. 1977. *Cryptococcus neoformans*: size range of infectious particles from aerosolized soil. *Infect Immun*, **17**, 634–8.

Neuville, S., Dromer, F., et al. 2003. Primary cryptococcosis: a distinct clinical entity. *Clin Infect Dis*, **36**, 347.

Newberry, W.M. Jr, Walter, J.E., Chandler, J.W. Jr and Tosh, F.E. 1967. Epidemiologic study of *Cryptococcus neoformans*. *Ann Intern Med*, **67**, 724–32.

Nightingale, S.D., Cal, S.X., et al. 1992. Primary prophylaxis with fluconazole against systemic fungal infections in HIV-positive patients. *AIDS*, **6**, 191–4.

Nosanchuk, J.D., Rosas, A.L. and Lee, S.C. 2000a. Melanisation of *Cryptococcus neoformans* in human brain tissue. *Lancet*, **355**, 2049–50.

Nosanchuk, J.D., Shoham, S., et al. 2000b. Evidence for zoonotic transmission of *Cryptococcus neoformans* from a pet cockatoo to an immunocompromised patient. *Ann Intern Med*, **132**, 205–8.

Odom, A., Muir, S., et al. 1997. Calcineurin is required for virulence of *Cryptococcus neoformans*. *EMBO J*, **16**, 2576–89.

Okun, E. and Butler, W.T. 1964. Ophthalmologic complications of cryptococcal meningitis. *Arch Ophthalmol*, **71**, 52–7.

Paillas, J., Sinico, M., et al. 1982. Un cas de cryptococcose pulmonaire pseudo-tumorale. *Sem Hop Paris*, **58**, 2402–4.

Pal, M. and Mehrotra, B.S. 1985. Studies on the isolation of *Cryptococcus neoformans* from fruits and vegetables. *Mykosen*, **28**, 200–5.

Pappas, P.G., Perfect, J.R., et al. 2001. Cryptococcosis in HIV-negative patients in the era of effective azole therapy. *Clin Infect Dis*, **33**, 690–9.

Perfect, J.R. and Casadevall, A. 2002. Cryptococcosis. *Infect Dis Clin North Am*, **16**, 837–74.

Perfect, J.R. and Seaworth, B. 1985. Penile cryptococcosis with a review of mycotic infections of the penis. *Urology*, **25**, 528–31.

Perfect, J.R., Marr, K.A., et al. 2003. Voriconazole treatment for less common, emerging or refractory fungal infections. *Clin Infect Dis*, **36**, 1122–31.

Perla, E.N., Maayan, S., et al. 1985. Disseminated cryptococcosis presenting as the adult respiratory distress syndrome. *N Y State J Med*, **85**, 704–6.

Perry, H.D. and Donnenfeld, E.D. 1990. Cryptococcal keratitis after keratoplasty. *Am J Ophthalmol*, **110**, 320–1.

Pfeiffer, T.J. and Ellis, D.H. 1991. Environmental isolation of *Cryptococcus neoformans gattii* from California. *J Infect Dis*, **163**, 929–30.

Pfeiffer, T.J. and Ellis, D.H. 1992. Environmental isolation of *Cryptococcus neoformans* var. *gattii* from *Eucalyptus tereticornis*. *J Med Vet Mycol*, **30**, 407–8.

Pirner, R.W., Shihata, N. et al. 1993. Population-based active surveillance for cryptococcal disease. *Thirty-third Interscience Conference on Antimicrobial Agents and Chemotherapy, New Orleans, 1993*, abst. 1326.

Plunkett, J.M., Turner, B.I. and Tallent, M.B. 1981. Cryptococcal septicemia associated with urologic instrumentation in a renal allograft recipient. *J Urol*, **125**, 241–2.

Polacheck, I. and Kwon-Chung, K.J. 1986. Canavanine resistance in *Cryptococcus neoformans*. *Antimicrob Agents Chemother*, **29**, 468–73.

Polacheck, I., Hearing, V.J. and Kwon-Chung, K.J. 1982. Biochemical studies of phenoloxidase and utilization of catecholamines in *Cryptococcus neoformans*. *J Bacteriol*, **150**, 1212–20.

Polacheck, I., Platt, Y. and Aronovitch, J. 1990. Catecholamines and virulence of *Cryptococcus neoformans*. *Infect Immun*, **58**, 2919–2922.

Powderly, W.G. 1993. Cryptococcal meningitis and AIDS. *Clin Infect Dis*, **17**, 837–42.

Powderly, W.G., Saag, M.S., et al. 1992. A controlled trial of fluconazole or amphotericin B to prevent relapse of cryptococcal meningitis in patients with the acquired immunodeficiency syndrome. *N Engl J Med*, **326**, 793–8.

Powderly, W.G., Cloud, G.A., et al. 1994. Measurement of cryptococcal antigen in serum and cerebrospinal fluid: value in the management of AIDS-associated cryptococcal meningitis. *Clin Infect Dis*, **18**, 789–92.

Powderly, W.G., Finkelstein, D.M., et al. 1995. A randomized trial comparing fluconazole with clotrimazole troches for the prevention of fungal infections in patients with advanced human immunodeficiency virus infection. *N Engl J Med*, **332**, 700–5.

Reese, A.J. and Doering, T.L. 2003. Cell wall alpha -1,3-glucan is required to anchor the *Cryptococcus neoformans* capsule. *Mol Microbiol*, **50**, 1401–9.

Rex, J.H., Larsen, R.A., et al. 1993. Catastrophic visceral loss due to *Cryptococcus neoformans* meningitis. *Medicine*, **72**, 207–24.

Rhodes, J.C., Polacheck, I. and Kwon-Chung, K.J. 1982. Phenoloxidase activity and virulence in isogenic strains of *Cryptococcus neoformans*. *Infect Immun*, **36**, 1175–84.

Rosas, A.L., Nosanchuk, J.D. and Feldmesser, M. 2000. Synthesis of polymerized melanin by *Cryptococcus neoformans* in infected rodents. *Infect Immun*, **68**, 2845–53.

Rosenheim, S.H. and Schwarz, J. 1975. Cavitary pulmonary cryptococcosis complicated by aspergilloma. *Am Rev Respir Dis*, **111**, 549–53.

Ruane, P.J., Walker, L.J. and George, W.L. 1988. Disseminated infection caused by urease-negative *Cryptococcus neoformans*. *J Clin Microbiol*, **26**, 2224–6.

Ruiz, A. and Bulmer, G.S. 1980. Particle size of airborne *Cryptococcus neoformans* in a tower. *Appl Environ Microbiol*, **41**, 1225–9.

Saag, M.S., Powderly, W.G., et al. 1992. Comparison of amphotericin B with fluconazole in the treatment of acute AIDS-associated cryptococcal meningitis. *N Engl J Med*, **326**, 83–9.

Saag, M.S., Cloud, G.A., et al. 1999. A comparison of itraconazole versus fluconazole as maintenance therapy for AIDS-associated cryptococcal meningitis. *Clin Infect Dis*, **28**, 291–6.

Saag, M.S., Graybill, J.R., et al. 2000. Practice guidelines for the management of cryptococcal disease. Infectious Disease Society of America. *Clin Infect Dis*, **30**, 710–18.

Salas, S.D., Bennett, J.E., et al. 1996. Effect of the laccase gene, CNLAC1, on virulence of *Cryptococcus neoformans*. *J Exp Med*, **184**, 377–86.

Sanfelice, F. 1894. Contributo alla morfologia e biolgia dei blastomiceti che si sviluppano nei succhi di alcuni frutti. *Ann d'igiene*, **4**, 463–495.

Sarosi, G.A., Parker, J.D., et al. 1969. Amphotericin B in cryptococcal meningitis: long-term results of treatment. *Ann Intern Med*, **71**, 1079–87.

Sarosi, G.A., Silberfarb, P.M. and Tosh, F.E. 1971. Cutaneous cryptococcosis. *Arch Dermatol*, **104**, 1–3.

Schmitt, H., Bernard, E., et al. 1988. Aerosol amphotericin B is effective for prophylaxis and therapy in a rat model of pulmonary aspergillosis. *Antimicrob Agents Chemother*, **32**, 1676–9.

Schupbach, C.W., Wheeler, C.E. and Briggaman, R.A. 1976. Cutaneous manifestations of disseminated cryptococcosis. *Arch Dermatol*, **112**, 1734–44.

Similowski, T., Datry, A., et al. 1989. AIDS-associated cryptococcosis causing adult respiratory distress syndrome. *Respir Med*, **83**, 513–515.

Singh, N., Gayowski, T., et al. 1997. Clinical spectrum of invasive cryptococcosis in liver transplant recipients receiving tacrolimus. *Clin Transplant*, **11**, 66–70.

Smith, F.S., Gibson, P., et al. 1976. Pulmonary resection for localized lesions of cryptococcosis (torulosis): a review of eight cases. *Thorax*, **31**, 121–6.

Speed, B. and Dunt, D. 1995. Clinical and host differences between infections with the two varieties of *Cryptococcus neoformans*. *Clin Infect Dis*, **21**, 28–34.

Staib, F., Seibold, M., et al. 1989. *Cryptococcus neoformans* in the semina fluid of an AIDS patient. A contribution to the clinical course of cryptococcosis. *Mycoses*, **32**, 171–80.

Steenbergen, J.N., Shuman, H.A. and Casadevall, A. 2001. *Cryptococcus neoformans* interactions with amoebae suggest an explanation for its virulence and intracellular pathogenic strategy in macrophages. *Proc Natl Acad Sci U S A*, **98**, 15245–50.

Steenbergen, J.N., Shuman, H.A. and Casadevall, A. 2003. *Cryptococcus neoformans* is enhanced after growth in the genetically malleable host, *Dictyostelium discoideum*. *Infect Immun*, **71**, 4862–72.

Stephen, C., Lester, S., et al. 2002. Multispecies outbreak cryptococcosis on Southern Vancouver Island, British Columbia. *Can J Vet Res*, **43**, 792–4.

Stern, J.J., Hartman, B.J., et al. 1988. Oral fluconazole therapy for patients with the acquired immunodeficiency syndrome and cryptococcosis: experience with 22 patients. *Am J Med*, **85**, 477–80.

Stoddard, J.L. and Cutler, E.C. (1916) Torula Infection in Man. *Rockefeller Institute for Medical Research, Monograph No. 6*, 1–98.

Subramanian, S., Kherdekar, S.S., et al. 1982. Lipoid pneumonia with *Cryptococcus neoformans* colonisation. *Thorax*, **37**, 319–20.

Suster, B., Akerman, M., et al. 1986. Pulmonary manifestations of AIDS: review of 106 episodes. *Radiology*, **161**, 86–93.

Swinne-Desgain, D. 1976. *Cryptococcus neoformans* in the crops of pigeons following its experimental administration. *Sabouraudia*, **14**, 313–17.

Tscharke, R.L., Lazera, M.S., et al. 2003. Haploid fruiting in *Cryptococcus neoformans* is not mating type alpha-specific. *Fungal Genet Biol*, **39**, 230–7.

Turner, S.H. and Cherniak, R. 1991. Glucuronoxylomannan of *Cryptococcus neoformans* serotype B: structural analysis by gas-liquid chromatography mass spectrometry and 13C-nuclear magnetic resonance spectroscopy. *Carbohydr Res*, **211**, 103–16.

Tynes, B., Mason, K.N., et al. 1968. Variant forms of pulmonary cryptococcosis. *Ann Intern Med*, **69**, 1117–25.

Utz, J.P., Shadomy, S. and McGehee, R.F. 1972. Flucytosine. *N Engl J Med*, **286**, 777–8.

Utz, J.P., Garrigues, I.L., et al. 1975. Therapy of cryptococcosis with a combination of flucytosine and amphotericin B. *J Infect Dis*, **132**, 368–73.

Van de Perre, P., Lepage, P. and Kestelyn, P. 1984. Acquired immunodeficiency syndrome in Rwanda. *Lancet*, **2**, 62–5.

van der Horst, C., Saag, M.S., et al. 1997. Treatment of cryptococcal meningitis associated with the acquired immunodeficiency syndrome. *N Engl J Med*, **337**, 15–21.

Van Gemert, H.M. and Vermeulen, M. 1991. Treatment of impaired consciousness with lumbar punctures in a patient with cryptococcal meningitis and AIDS. *Clin Neurol Neurosurg*, **93**, 257–258.

Vartivarian, S.E., Anaissie, E.J., et al. 1993. Regulation of cryptococcal capsular polysaccharide by iron. *J Infect Dis*, **167**, 186–90.

Verse, M. 1914. Uber einen Fall von generalisierter Blastomykose beim menschen. *Verh Dtsch Pathol Ges*, **17**, 275–8.

Vibhagool, A., Sungkanuparph, S., et al. 2003. Discontinuation of secondary prophylaxis for cryptococcal meningitis in human immunodefiiency virus-infected patients treated with highly active antiretroviral therapy: a prospective, multicenter randomized study. *Clin Infect Dis*, **36**, 1329–31.

Viviani, M.A., Tortorano, A.M., et al. 1989. Experience with itraconazole in cryptococcosis and aspergillosis. *J Infect*, **18**, 151–65.

Vogel, R.A. 1966. The indirect fluorescent antibody test for the detection of antibody in human cryptococcal disease. *J Infect Dis*, **116**, 573–80.

Vuillemin, P. 1901. Les blastomycetes pathogenes. *Rev Gen Sci Pures Appl*, **12**, 732–51.

Warr, W., Bates, J.H. and Stone, A. 1968. The spectrum of pulmonary cryptococcosis. *Ann Intern Med*, **69**, 1109–16.

Wasser, L. and Talavera, W. 1987. Pulmonary cryptococcosis in AIDS. *Chest*, **92**, 692–5.

Westernik, M.A., Amsterdam, D., et al. 1987. Septicemia due to DF-2. Cause of a false-positive cryptococcal latex agglutination result. *Am J Med*, **83**, 155–8.

White, M., Cirrincione, C., et al. 1992. Cryptococcal meningitis with AIDS and patients with neoplastic disease. *J Infect Dis*, **165**, 960–6.

Wickes, B.L., Mayorga, M.E., et al. 1996. Dimorphism and haploid fruiting in *Cryptococcus neoformans* association with the alpha-mating type. *Proc Natl Acad Sci U S A*, **93**, 7327–31.

Xu, J., Vilgalys, R. and Mitchell, T.G. 2002. Multiple gene genealogies reveal recent dispersion and hybridization in the human fungus, *Cryptococcus neoformans*. *Mol Ecol*, **9**, 1471–81.

Yadav, Y.R., Perfect, J.R. and Friedman, A. 1988. Successful treatment of cryptococcal ventriculo-atrial shunt infection with systemic therapy alone. *Neurosurgery*, **23**, 317–22.

Yamaguchi, H., Ikemoto, H., et al. 1996. Fluconazole monotherapy for cryptococcosis in non-AIDS patients. *Eur J Clin Microbiol Infect Dis*, **15**, 787–92.

Young, E.J., Hirsh, D.D., et al. 1980. Pleural effusions due to *Cryptococcus neoformans*: a review of the literature and report of two cases with cryptococcal antigen determinations. *Am Rev Respir Dis*, **121**, 743–6.

Zargoza, O., Fries, B.C. and Casadevall, A. 2003. Induction of capsule growth in *Cryptococcus neoformans* by mammalian serum and CO_2. *Infect Immun*, **71**, 6155–64.

Zimmer, B.L. and Roberts, G.D. 1979. Rapid selective urease test for presumptive identification of *Cryptococcus neoformans*. *J Clin Microbiol*, **10**, 380–1.

Zlupko, G.M., Fochler, F.J. and Goldschmidt, Z.H. 1980. Pulmonary cryptococcosis presenting with multiple pulmonary nodules. *Chest*, **77**, 575.

Zuger, A., Louie, E., et al. 1986. Cryptococcal disease in patients with acquired immunodeficiency syndrome. Diagnostic features and outcome of treatment. *Ann Intern Med*, **104**, 234–40.

33

Systemic zygomycosis

DAVID H. ELLIS

INTRODUCTION

The Zygomycetes are a class of relatively primitive, fast growing, widely distributed, terrestrial fungi which are largely saprobic on plant debris and in soil. Many species are common environmental contaminants, often causing food spoilage; some are sources of food in the Orient; and a few are pathogens of plants, insects, and, more rarely, of humans. Several species are thermophilic, capable of growing at well over 40°C. Hyphae are coenocytic and mostly aseptate, making them clearly distinguishable from those of other filamentous fungi. Although the absence of hyphal cross-walls or septa facilitates rapid growth and sporulation by allowing the uninhibited absorption and translocation of nutrients, together with cellular organelles between growth sites, it is also a significant disadvantage in that hyphal elements are prone to physical damage and subsequent death. By comparison, when septate hyphae, which are typically found in higher fungi belonging to the Basidiomycetes and Ascomycetes, are damaged, the pores between adjacent compartments are plugged, thus preventing death of the whole hyphal strand.

Thus, in human cases of zygomycosis, reliable culture identification of the causative fungus is made in less than 50 percent of the cases reported in the literature (Scholer et al. 1983). Asexual reproduction is by nonmotile sporangiospores, by modified sporangial units functioning as conidia, or by true conidia (Hesseltine and Ellis 1973). Sexual reproduction is isogamous through the formation of zygospores. However, most isolates appear to be heterothallic, requiring mating by two distinct strains for the formation of zygospores.

Thus, in pure culture the sexual or zygosporic state is of limited taxonomic value, although recent mating experiments, leading to the formation of zygospores, have been used to confirm the identity of three zygomycetous fungi isolated from clinical specimens (Weitzman et al. 1995). Occasionally, resting spores known as azygospores, which develop from a single gametangium without any evidence of sexual fusion, may also be observed in some species (Benjamin 1979). At present, the classification and identification of zygomycetous fungi are based primarily on the asexual or anamorphic state, which is characterized by the production of sporangia and sporangiospores.

Sporangiospores are distinguished from other asexual fungal spores, in that they are cleaved out of the cytoplasmic contents of the sporangium without the wall of the sporangium being directly involved in the formation of the spore wall (Benjamin 1979). Other asexual reproductive structures occasionally observed in some species are chlamydoconidia and yeast-like blastoconidia. Chlamydoconidia are thick-walled resting structures formed endogenously within the hyphae. Yeast-like growth usually occurs only under specific culture conditions, especially anaerobiosis or high glucose concentrations, and has been reported in species from a number of genera, including *Cokeromyces*, *Mucor*, and *Rhizomucor* (Benjamin 1979). As a result of the above characteristics, the Zygomycetes, as a class, have long been considered a rather natural group. This grouping has been supported by the composition of the cell wall which is chitosan–chitin in nature (Hesseltine and Ellis 1973) and by molecular evidence (Kwon-Chung 1994). More recent molecular phylogeny studies have provided strong support for the monophyly of the Mucorales, with the exception of *Mortierella* spp., which appear to be misclassified within the Mucorales (Voigt et al. 1999). Accordingly a new order, the Mortierellales, has been proposed to accommodate *Mortierella* (Cavalier-Smith 1998; Voigt and Wostemeyer 2001). Based on the 18S gene tree topology, *Absidia corymbifera* and *Rhizomucor variabilis* also appear to be misplaced taxonomically. *A. corymbifera* is strongly supported as a sister group of the *Rhizomucor miehei–Rhizomucor pusillus* clade, while *R. variabilis* is nested within *Mucor* (Voigt et al. 1999). To date, some 870 species of the Zygomycetes have been described (Kirk et al. 2001). Until 1979, the Zygomycetes were divided into two orders: the Mucorales and the Entomophthorales (Onions et al. 1981). Hesseltine and Ellis (1973) gave a comprehensive key to the groups and genera and accepted 14 families in the Mucorales. Based on phylogenetic relationships, Benjamin (1979), O'Donnell (1979), and Jefferies (1985) produced a new classification of the Zygomycetes at ordinal level, creating seven orders and 30 families. Currently 10 orders and 32 families are recognized (Kirk et al. 2001); with three orders containing genera and species of medical importance: the Mucorales, Mortierellales, and Entomophthorales. The term 'zygomycosis' is used in this chapter to describe any infection caused by a member of the Mucorales or Mortierellales. However, it should be noted that some authors still prefer to use the term 'mucormycosis' to describe mycoses caused by members of the Mucorales and Mortierellales and entomophthoromycosis (or entomophthoramycosis – the spelling used seems to depend upon the author) to describe mycoses caused by species belonging to the Entomophthorales. In general, fungi in the order Mucorales cause the more severe forms of disease, whereas those in the order Entomophthorales cause disease of the nasal mucosa and subcutaneous tissue (Goodman and Rinaldi 1991). Recent reviews of the Zygomycetes of medical interest have been made by Ribes et al. (2000), Schipper and Stalpers (2003), and Dromer and McGinnis (2003).

Mucorales

The Mucorales (commonly called bread or pin molds) consist of fungi that reproduce asexually by means of nonmotile sporangiospores borne in sporangia (multispored), merosporangia (spores aligned in rows), sporangiola (containing few spores), or as one-spored sporangia or conidia. Sporangia generally have a central columella, which may extend below the sporangium to form a funnel-shaped apophysis. Several species also develop stolons and rhizoids.

Most of the species known to cause human or animal infections belong to a few genera within the family *Mucoraceae*, and are readily recognizable by their rapid growth rate, filling a Petri dish within 3–5 days, and their characteristic appearance, colonies usually being loosely floccose and of a gray or brownish-gray color (Table 33.1). These include members of the genera *Absidia*, *Apophysomyces*, *Mucor*, *Rhizomucor*, and *Rhizopus*. *Rhizopus oryzae* (*arrhizus*) is the most frequent infectious agent reported, followed by *Rhizopus microsporus* var. *rhizopodiformis*, *Absidia corymbifera*, and *Rhizomucor pusillus* (Scholer et al. 1983). These four species account for more than 80 percent of culture-proven cases of zygomycosis. Other mucoraceous species reported less frequently include *R. microsporus* var. *microsporus*, *Mucor ramosisimus*, *Mucor circinelloides*, *Mucor indicus*, and *Apophysomyces elegans*. Members of several other genera belonging to other families, such as *Cunninghamella bertholletiae* (*Cunninghamellaceae*), and *Saksenaea vasiformis* (*Saksenaeceae*), are also occasionally reported.

Mortierellales

This is a newly created order to accommodate a single family, the *Mortierellaceae* (Cavalier-Smith 1998). All structures are generally more delicate than in the Mucorales and the sporangia have no columellae and are split into numerous, few, or single spores. *Mortierella*

Table 33.1 *Families and genera of the Mucorales of medical importance*

Family	Genera	Key identification features
Mucoraceae	*Absidia, Apophysomyces, Mucor, Rhizomucor, Rhizopus*	Multispored sporangia with columella which may or may not show an apophysis. Sporangia may be globose, pyriform, or ellipsoidal in shape
Cunninghamellaceae	*Cunninghamella*	Single-spored, globose to ovoid, echinulate sporangiola (conidia) borne on globose to clavate fertile vesicles
Saksenaeaceae	*Saksenaea*	Flask-shaped sporangia with columella and simple, darkly pigmented rhizoids
Syncephalastraceae	*Syncephalastrum*	Cylindrical merosporangia with spores in one row borne on globose vesicles
Thamnidiaceae	*Cokeromyces*	Dimorphic with yeast growth at 37°C, sporangiola with few spores borne on long recurved pedicles arising from a fertile vesicle

wolfii is probably the only species pathogenic to animals (Zycha et al. 1969; Domsch et al. 1980; de Hoog and Guarro 1995), although Scholz and Meyer (1965) have reported a case of pulmonary infection in cattle caused by *Mortierella polycephala*.

ETIOLOGIC AGENTS OF MEDICAL IMPORTANCE

By definition, all pathogenic zygomycotic species will grow at 37°C, with the possible exception of the *M. circinelloides* group. Thus occasional reports of infections caused by species unable to grow at temperatures above 35°C must be regarded as questionable. Usually laboratory incubation temperatures range between 26 and 35°C; however, growth temperature studies at 40 or 45°C can also be helpful. Most isolates are heterothallic, i.e. zygospores are absent, so identification of those species is based primarily on sporangial morphology. This includes the arrangement and number of sporangiospores, shape, color, presence or absence of columellae and apophyses, as well as the arrangement of the sporangiophores and the presence or absence of rhizoids. Tease mounts are best, using a drop of 95 percent alcohol as a wetting agent to reduce air bubbles. There are very few molecular techniques currently in use for the identification of zygomycetes, most studies have been directed at phylogeny. (For detailed descriptions of species, keys to taxa, and additional information, see Hesseltine and Ellis 1964, 1966; Ellis and Hesseltine 1965, 1966; Nottebrock et al. 1974; O'Donnell 1979; Domsch et al. 1980; McGinnis 1980; Onions et al. 1981; Scholer et al. 1983; Samson et al. 1995; Goodman and Rinaldi 1991; Ribes et al. 2000; Schipper and Stalpers 2003.)

Absidia corymbifera (Cohn) Saccardo & Trotter

The genus *Absidia* is characterized by a differentiation of the hyphae into arched stolons bearing more or less verticillate sporangiophores in the raised part (internode), and rhizoids formed at the point of contact with the substrate (at the node). This feature separates the species of *Absidia* from those of the genus *Rhizopus*, where the sporangia arise from the nodes and are, therefore, found opposite the rhizoids. The sporangia are relatively small, globose, pyriform, or pear shaped, and are supported by a characteristic funnel-shaped apophysis. This distinguishes the genus *Absidia* from the genera *Mucor* and *Rhizomucor*, which have large, globose sporangia without an apophysis. *Absidia* currently contains 21 mostly soil-borne species (Zycha et al. 1969; Domsch et al. 1980; Schipper 1990; Schipper and Stalpers 2003). *A. corymbifera* is the only species of *Absidia* known to cause disease in humans and animals.

Absidia ramosa, previously reported as a pathogenic species, has now been reduced to synonymy with *A. corymbifera* (Nottebrock et al. 1974; Scholer et al. 1983).

A. corymbifera (synonym: *A. ramosa*) is a relatively rare agent of human zygomycosis (Kwon-Chung and Bennett 1992); it is more often reported as an animal pathogen (Smith 1989; Jensen et al. 1989, 1990; Jensen 1992; Knudtson and Kirkbride 1992; Guillot et al. 2000). However, it was probably responsible for the first reported case of human pulmonary zygomycosis (Furbringer 1876) and it is occasionally reported from this site (the lungs) (Murphy and Bornstein 1950; Darja and Davy 1963; Lake et al. 1988; Leleu et al. 1999). However, *A. corymbifera* is most often associated with cutaneous infections in both immunocompetent patients, often following localized trauma (Roger et al. 1989; Marshall et al. 1997; Scalise et al. 1999; Seguin et al. 1999; Narain et al. 2001; Thami et al. 2003), and immuno-supressed patients, including bone marrow transplant (Hagensee et al. 1994; Jantunen et al. 1996; Leong et al. 1997; Leleu et al. 1999; Paterson et al. 2000), idiopathic aplastic anemia (Cloughley et al. 2002), neonates (Amin et al. 1998; Buchta et al. 2003), and AIDS patients (Chavanet et al. 1990; Hopwood et al. 1992). Lopes et al. (1995) described an unusual case of cutaneous infection from a leukemic patient, which probably resulted from latent osteomyelitis. It has also been reported as the causative agent of meningitis that followed a penetrating head injury (Mackenzie et al. 1988), renal infection in intravenous drug-using AIDS patients (Smith et al. 1989; Torres-Rodriguez et al. 1993), and rhinocerebral infection in neutropenic patients (Manso et al. 1994; Ryan et al. 2001). *A. corymbifera* has a worldwide distribution mostly in association with soil and decaying plant debris (Domsch et al. 1980).

Colonies are fast growing, floccose, white at first, becoming pale gray with age, and up to 1.5 cm high. Sporangiophores are hyaline to faintly pigmented, simple or sometimes branched, arising solitary from the stolons, in groups of three, or in whorls of up to seven. Rhizoids are very sparingly produced and may be difficult to find without the aid of a dissecting microscope to examine the colony on the agar surface. Sporangia are small (10–40 μm in diameter) and are typically pyriform in shape with a characteristic conical columella and pronounced apophysis, often with a short projection at the top (Figure 33.1). Sporangiospores vary from subglobose to oblong–ellipsoidal (3.0–7.0×2.5–4.5 μm) and are hyaline to light-gray and smooth walled. Zygospores formed by compatible mating strains are reddish-brown, thick walled (60–100×45–80 μm) and have flat projections and one to three equatorial ridges. Isolates are thiamine dependent and have an optimum growth temperature of 35°C, with a maximum of 48–52°C. Key laboratory diagnostic features are growth at 40°C and the formation of small pyriform sporangia with characteristic conical columellae and pronounced apophyses.

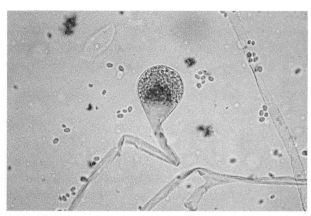

Figure 33.1 Absidia corymbifera *showing a typical pyriform-shaped sporangium with a conical columella and pronounced apophysis. Magnification ×330*

(For additional taxonomic descriptions, see Ellis and Hesseltine 1965; Hesseltine and Ellis 1966; Nottebrock et al. 1974; O'Donnell 1979; Domsch et al. 1980; McGinnis 1980; Scholer et al. 1983; Schipper 1990; Samson et al. 1995; de Hoog et al. 2000; Schipper and Stalpers 2003.)

Apophysomyces elegans Misra, Srivastava & Latas

The genus *Apophysomyces* was erected by Misra et al. (1979) to accommodate a new mucoraceous fungus with prominent funnel-shaped apophyses, which they named *A. elegans*. It is readily distinguishable from other zygomycetes of medical importance, especially the morphologically similar, strongly apophysate pathogen *Absidia corymbifera*. *A. elegans* has sporangiophores with distinctive funnel- or bell-shaped apophyses and hemispherical columellae, a conspicuous pigmented subapical thickening which constricts the lumen of the sporangiophore below the apophysis, and distinctive foot cells. However, *A. elegans* may resemble *S. vasiformis* in gross colony morphology and in its failure to sporulate on routinely used media (Ellis and Ajello 1982; Padhye and Ajello 1988).

A. elegans is an emerging human pathogen of apparently immunocompetent patients and is usually associated with invasive soft tissue infections complicated by burns or wounds contaminated by soil (Winn et al. 1982; Wieden et al. 1985; Newton et al. 1987; Cooter et al. 1990; Huffnagle et al. 1992; McGinnis et al. 1993; Weinberg et al. 1993; Holland 1997; Burrell et al. 1998; Kimura et al. 1999; Page et al. 2001; Blair et al. 2002). Other reports include infections of the kidney and bladder (Lawrence et al. 1986; Okhuysen et al. 1994; Naguib et al. 1995), abdominal wall (Lakshmi et al. 1993), rhinocerebral infection (Brown et al. 1998; Garcia-Covarrubias et al. 2001), and osteomyelitis (Eaton et al. 1994; Meis et al. 1994). Interestingly, most cases occur in apparently immunocompetent patients,

indicating that traumatic implantation of the fungus through the skin is the most important predisposing factor. *A. elegans* is a soil fungus with a tropical to subtropical distribution. It was initially isolated from soil samples collected from a mango orchard in northern India (Misra et al. 1979) and subsequently from soil in northern Australia (Cooter et al. 1990).

Colonies are fast growing, white becoming creamy white to buff with age, downy with no reverse pigment, and are composed of broad, sparsely septate (coenocytic) hyphae typical of a zygomycetous fungus. Sporangiophores are unbranched, straight or curved, slightly tapering towards the apex, up to 200 μm long, 3–5 μm in width near the apophysis, and hyaline when young but developing, as the culture ages, a sepia to brown pigmentation and a conspicuous subapical thickening 10–16 μm below the apophysis (Figure 33.2). Sporangiophores arise at right angles from the aerial hyphae and often have a septate basal segment resembling the 'foot cell' commonly seen in *Aspergillus* spp. Rhizoids are thin walled, subhyaline, and predominantly unbranched. Sporangia are multispored, small (20–50 μm in diameter), typically pyriform in shape, hyaline at first, sepia colored when mature, columellate and strongly apophysate (Figure 33.2). Columellae are hemispherical in shape and the apophyses are distinctively funnel or

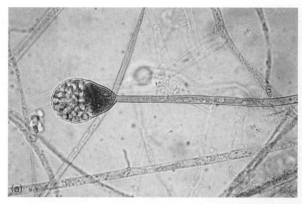
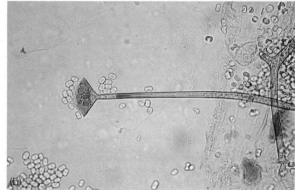

Figure 33.2 Apophysomyces elegans*: (a) young sporangium;* **(b)** *mature sporangium showing distinctive funnel-shaped apophyses and a conspicuous pigmented subapical thickening below the apophysis. Magnification ×495*

bell shaped. Sporangiospores are smooth walled, mostly oblong, occasionally subglobose (3–4×5–6 um), and subhyaline to sepia *en masse*. Good growth occurs at 26, 37 and 42°C. Key laboratory diagnostic features include rapid growth at 42°C (thermotolerant), and the formation on nutrient-deficient media of characteristic 'cocktail glass' sporangia. (For additional taxonomic descriptions, see Misra et al. 1979; Ellis and Ajello 1982; Wieden et al. 1985; Lawrence et al. 1986; Padhye and Ajello 1988; Cooter et al. 1990; McGinnis et al. 1993; de Hoog et al. 2000.)

Cokeromyces recurvatus Poitras

C. recurvatus is the type species of the genus *Cokeromyces* and is usually found in soil or isolated from rodent or lizard dung (Shanor et al. 1950; Benny and Benjamin 1976). It has been reported on several occasions from humans as a colonizer of the vagina (Rippon and Dolan 1979; McGough et al. 1990; Kwon-Chung and Bennett 1992; Kemna et al. 1994) and once from the urinary bladder (Axelrod et al. 1987). *C. recurvatus* has also been reported as causing severe diarrhea in bone marrow transplant recipients (Alvarez et al. 1995; Tsai et al. 1997) and pleuritis and peritonitis in a patient with a ruptured peptic ulcer (Munipalli et al. 1996). Its validity as a human pathogen is, however, doubtful because there has not been any evidence of tissue invasion in any cases. *C. recurvatus* is one of the few zygomycetous fungi that exhibits thermal dimorphism by growing at 37°C in a yeast-like form characterized by the development of large, spherical cells, bearing one to many buds, similar in appearance to those of *Paracoccidioides brasiliensis*, but growing as a mold when incubated at lower temperatures.

Colonies grow moderately at 25°C and are flat, radially wrinkled, tan to brown, becoming grayish with the production of sporangiospores. Sporangiophores are mostly unbranched, 9 µm wide and produce terminal vesicles up to 30 µm in diameter; rhizoids are absent. Sporangiola have few spores, and are globose to obovoid, with a columella, but no apophysis; they are borne terminally on long, recurved, and contorted pedicles which arise from the fertile vesicle formed at the apex of the sporangiophore. Sporangiospores are smooth walled and ovoid to ellipsoid in shape (2.5×4.5 µm). Zygospores are produced in abundance (homothallic) and are dark brown, globose (33.5–54.5 µm in diameter) with rough walls resulting from sharply pointed projections. Isolates readily convert from mold-like to yeast-like colonies when incubated at 35–37°C on enriched media such as brain heart infusion agar with 5 percent sheep blood. *C. recurvatus* has a maximum growth temperature of 42°C, is thiamine dependent, has a positive assimilation of nitrate, negative assimilation of sucrose, negative fermentation of glucose and is sensitive to cycloheximide (Kemna et al. 1994). Key features

include dimorphic growth as a yeast at 37°C, sporangiola borne terminally on long recurved stalks, and the development of abundant zygospores at 25°C. (For further information see Zycha et al. 1969; Goodman and Rinaldi 1991; Kwon-Chung and Bennett 1992; Kemna et al. 1994; de Hoog et al. 2000.)

Cunninghamella bertholletiae Stadel

The genus *Cunninghamella* is characterized by white to gray, rapidly growing colonies, producing erect, straight, branching sporangiophores. These sporangiophores end in globose or pyriform vesicles from which several one-celled, globose to ovoid, echinulate or smooth-walled sporangiola develop on swollen denticles. Chlamydoconidia and zygospores may also be present. *Cunninghamella* species are mainly soil fungi of the Mediterranean and subtropical zones, and they are only rarely isolated in temperate regions. The genus now contains seven species, with *C. bertholletiae*, the only one known to cause disease in humans and animals (Domsch et al. 1980).

Once again, there has been some confusion as to the correct name of this zygomycete. Many medical mycologists (McGinnis 1980; Weitzman 1984; Rippon 1988) preferred the name *C. bertholletiae* because of the thermophilic nature of the human isolates that grow at temperatures as high as 45°C. However, Samson (1969) and Domsch et al. (1980) preferred the name *Cunninghamella elegans* and Lunn and Shipton (1983) went further and reduced *C. elegans* (*C. bertholletiae*) to a variety of *Cunninghamella echinulata*, i.e. *C. echinulata* var. *elegans*. However, *C. bertholletiae* is currently the most acceptable name; *C. elegans* differs by having purely gray colonies and by not growing at temperatures above 40°C (Weitzman and Crist 1979; de Hoog et al. 2000).

C. bertholletiae is a common soil fungus found throughout the temperate regions of the world. It is a rare cause of zygomycosis in humans, often associated with traumatic implantation (Boyce et al. 1981) and immunosuppression (Robinson et al. 1990). Pulmonary infections are being reported with increasing frequency in cancer patients (Kwon-Chung et al. 1975; Kiehn et al. 1979; Ventura et al. 1986; Reed et al. 1988; Maloisel et al. 1991; Cohen-Abbo et al. 1993; Dermoumi 1993; Kontoyianis et al. 1994; Mazade et al. 1998; Rickerts et al. 2000). Reports of pulmonary infection in immunocompetent patients are rare (Zeilender et al. 1990). Disseminated disease has been reported by Sands et al. (1985) and Iwatsu et al. (1990), and rare cases of rhino-cerebral infection (Brennan et al. 1983) and sinusitis (Ng et al. 1994) have also been reported. Infections have also been reported in association with desferrioxamine therapy (Rex et al. 1988; Sane et al. 1989; Maloisel et al. 1991), transplant recipients (Kolbeck et al. 1985; Nimmo et al. 1988; Darrisaw et al. 2000; Garey et al. 2001; Zhang et al. 2002), and in AIDS patients (Mostaza et al. 1989).

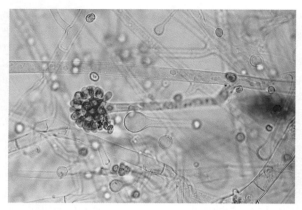

Figure 33.3 Cunninghamella bertholletiae: *microscopic morphology showing the formation of single-celled sporangiolas from a swollen vesicle. Magnification* ×495

Colonies are very fast growing, white at first, but becoming rather dark gray and powdery with development. Sporangiophores are up to 20 μm wide, straight, with verticillate or solitary branches. Vesicles are subglobose to pyriform, with the terminal ones being up to 40 μm and the lateral ones 10–30 μm in diameter. Sporangiola are globose (7–11 μm diameter), ellipsoidal (9–13×6–10 μm), verrucose or short echinulate, hyaline singly but brownish *en masse* (Figure 33.3). Temperature is at optimum 25–30°C and has a maximum up to 50°C. Key features include growth at 40°C and the presence of one-celled, globose to ovoid, echinulate 'conidia' borne on terminal or lateral globose to clavate fertile vesicles. (For additional information see Samson 1969; Weitzman and Crist 1979, 1980; Domsch et al. 1980; McGinnis 1980; Lunn and Shipton 1983; Weitzman 1984; Sands et al. 1985; Iwatsu et al. 1990; de Hoog et al. 2000.)

Mortierella wolfii Mehrotra & Baijal

The genus *Mortierella* has now been placed in a separate order, the Mortierellales (Cavalier-Smith 1998), and is characterized by gray to yellowish-grey, rapidly growing colonies, often spreading in overlapping 'waves' or lobes, which produce small delicate sporangia lacking a columella, on simple or branched sporangiophores (Figure 33.4). Sporangia may be one- or multispored and a collarette is typically left after the sporangial wall dissolves. Sporangiospores are one celled, globose to ellipsoidal, and chlamydoconidia may also be present. *Mortierella* spp. are common soil fungi and the genus contains about 90 recognized species; however, *M. wolfii* is probably the only species pathogenic to animals (Zycha et al. 1969; Domsch et al. 1980; de Hoog et al. 2000), although Scholz and Meyer (1965) have reported a case of pulmonary infection in cattle caused by *Mortierella polycephala*.

M. wolfii is an important causal agent of bovine mycotic abortion, pneumonia, and systemic mycosis in Australia, Europe, New Zealand, and the USA (di

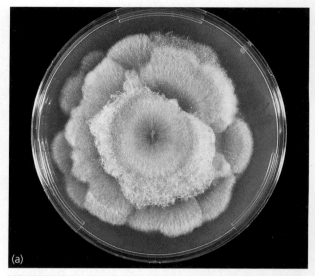

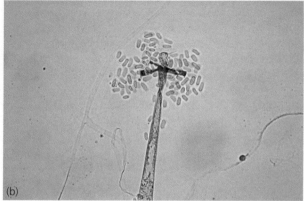

Figure 33.4 Mortierella wolfii: **(a)** *culture showing typical lobed growth pattern;* **(b)** *sporangium with acrotonous (terminal) branches. Magnification* ×330

Menna et al. 1972; Carter et al. 1973; Wohlgemuth and Knudtson 1977; MacDonald and Corbel 1981; Neilan et al. 1982; Komoda et al. 1988; Smith 1989; Knudtson and Kirkbride 1992; Done et al. 1994; Johnson et al. 1994; Uzal et al. 1999). Confirmed human infections have not been documented. *M. wolfii* has also been isolated from soil, rotten silage, hay, and coal spoil tips (Mehrotra and Baijal 1963; Austwick 1965). Seviour et al. (1987) have described a specialized medium, silage extract agar, to stimulate sporulation of isolates of *M. wolfii*.

Cultures are fast growing, white to grayish-white, downy, often with a broadly zonate or lobed (rosette-like) surface appearance, and no reverse pigment. Sporangiophores are typically erect, delicate, 80–250 μm in height, 6–20 μm wide at the base, arising from rhizoids or bulbous swellings on the substrate hyphae, and terminating with a compact cluster of short acrotonous (terminal) branches. Sporangia are usually 15–48 μm in diameter, with transparent walls, and a conspicuous collarette is usually present after dehiscence of the sporangiospores (Figure 33.4). Columellae are generally lacking and sporangiospores are single-celled,

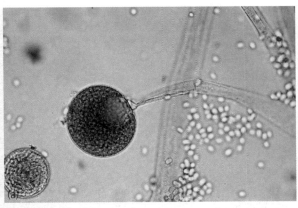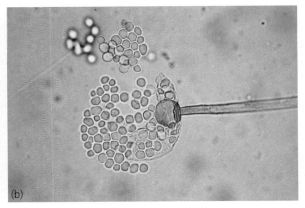

Figure 33.5 Mucor *species:* **(a, b)** *sporangia, columella and sporangiospores. Magnification* ×495

short–cylindrical, 6–10×3–5 µm, with a double membrane. Chlamydoconidia with or without blunt appendages (ameba-like) may be present. Zygospores have not been observed. Isolates grow well at 40–42°C, with a maximum of 48°C. Key features include rapid growth at 40°C (thermotolerant) and characteristic delicate acrotonous branching sporangia without columellae. (For additional information, see Mehrotra and Baijal 1963; Zycha et al. 1969; Domsch et al. 1980; Seviour et al. 1987; de Hoog et al. 2000.)

Mucor species

The genus *Mucor* can be differentiated from *Absidia*, *Rhizomucor*, and *Rhizopus* by the absence of stolons and rhizoids. Colonies are very fast growing, cottony to fluffy, white to yellow, becoming dark gray, with the development of sporangia. Sporangiophores are erect, simple or branched, forming large (60–300 µm in diameter), terminal, globose to spherical, multispored sporangia, without apophyses and with well-developed subtending columellae (Figure 33.5). A conspicuous collarette (remnants of the sporangial wall) is usually visible at the base of the columella after sporangiospore dispersal. Sporangiospores are hyaline, gray or brownish, globose to ellipsoidal, and smooth walled or finely ornamented. Chlamydoconidia and zygospores may also be present. The genus contains about 50 recognized taxa, many of which have widespread occurrence and are of considerable economic importance (Zycha et al. 1969;

Schipper 1978a; Domsch et al. 1980). However, only a few thermotolerant species are of medical importance and human infections are only rarely reported (Table 33.2). Most infections reported list *M. circinelloides* and similar species such as *M. indicus* (*M. rouxii*), *Mucor ramosissimus*, and *Mucor amphibiorum* as the causative agents. However, *Mucor hiemalis* and *Mucor racemosus* have also been reported as infectious agents, although their inability to grow at temperatures above 32°C raises doubt as to their validity as human pathogens and their pathogenic role may be limited to cutaneous infections (Scholer et al. 1983; Goodman and Rinaldi 1991; Kwon-Chung and Bennett 1992; de Hoog et al. 2000). Key laboratory diagnostic features are the presence of large, spherical, non-apophysate sporangia with pronounced columellae and conspicuous collarettes after sporangiospore dispersal. (For additional information, see Schipper 1978a; Domsch et al. 1980; McGinnis 1980; Onions et al. 1981; Scholer et al. 1983; Rippon 1988; Goodman and Rinaldi 1991; Samson et al. 1995; de Hoog et al. 2000; Schipper and Stalpers 2003.)

MUCOR AMPHIBIORUM SCHIPPER

This species has been reported as the causative agent of zygomycosis in several amphibia (Frank et al. 1974; Schipper 1978a; Speare et al. 1994, 1997; Berger et al. 1997) and in free-living platypuses (*Ornithorhynchus anitinus*) (Obendorf et al. 1993; Connolly et al. 2000). Human infections have not been reported. Colonies are

Table 33.2 *Maximum temperature of growth for the reported pathogenic species of* Mucor

Species	Maximum temperature (°C)	Pathogenicity
M. amphibiorum	36	Animals, principally amphibians
M. circinelloides	36–40	Animals, occasionally humans
M. hiemalis	30	Questionable cutaneous infections only
M. indicus	42	Humans and animals
M. racemosus	32	Questionable
M. ramosissimus	36	Humans and animals

grayish-brown, slightly aromatic, and do not grow at 37°C (maximum temperature for growth is 36°C). Sporangiophores are hyaline, erect, and mostly unbranched, rarely sympodially branched. Sporangia are dark brown, up to 75 μm in diameter, and are slightly flattened with a diffluent membrane. Columellae are subglobose to ellipsoidal or pyriform, up to 60×50 μm, with small collars. Sporangiospores are smooth walled, spherical, and 3.5–5.5 μm in diameter. Zygospores, when formed by compatible mating types, are spherical to slightly compressed, up to 70×60 μm in diameter, with stellate projections. *M. amphibiorum* is distinguished by poor ramification of the sporangiophores and by globose sporangiospores. Ethanol and nitrates are not assimilated (Schipper 1978a; Scholer et al. 1983; de Hoog et al. 2000).

MUCOR CIRCINELLOIDES VAR. TIEGH

M. circinelloides is a common and variable species that includes four forms: *circinelloides*, *lusitanicus*, *griseocyanus*, and *janssenii* (Schipper 1978a; Scholer et al. 1983). It has been reported as a rare cause of cutaneous infections in humans (Fetchick et al. 1986; Berenguer et al. 1988; Wang et al. 1990; Fingeroth et al. 1994; Chandra and Woodgyer 2002) and from mycoses in cattle and swine (Morquer et al. 1965), birds (Porges et al. 1935; Marjankova et al. 1978), and a platypus (Stewart et al. 1999). Colonies are floccose, pale grayish-brown, and grow poorly at 37°C (maximum growth temperature 36–40°C). Sporangiophores are hyaline and mostly sympodially branched with long branches erect and shorter branches becoming circinate (recurved). Sporangia are spherical, varying from 20 to 80 μm in diameter, with small sporangia often having a persistent sporangial wall. Columellae are spherical to ellipsoidal and are up to 50 μm in diameter. Sporangiospores are hyaline, smooth walled, ellipsoidal, and 4.5–7×3.5–5 μm in size. Chlamydoconidia are generally absent. Zygospores are only produced in crosses of compatible mating types and are reddish-brown to dark brown, spherical with stellate spines, up to 100 μm in diameter and have equal to slightly unequal suspensor cells. *M. circinelloides* differs from other species of *Mucor* in its formation of short circinated (coiled), branched sporangiophores bearing brown sporangia and its ability to assimilate ethanol and nitrates (Schipper 1976; Scholer et al. 1983; Samson et al. 1995; de Hoog et al. 2000; Schipper and Stalpers 2003).

MUCOR HIEMALIS WEHMER

This species has been reported from a few cases of human cutaneous infection (Neame and Rayner 1960; Costa et al. 1990; Prevoo et al. 1991). However, with a maximum growth temperature of 30°C its pathogenicity is questionable. It is most commonly recovered from soil (Schipper 1973). Colonies are grayish-ochraceous in color. Sporangiophores are hyaline, erect, mostly unbranched, with occasional sympodial branching. Sporangia are yellow to dark brown, up to 80 μm in diameter and have a diffluent membrane. Columellae are ellipsoidal, 30×38 μm in diameter. Sporangiospores are hyaline, smooth walled, ellipsoidal, sometimes flattened on one side, and 5.5–9×2.5–5.5 μm in size. Oidia may be present in the substrate hyphae, chlamydoconidia are uncommon, and zygospores are not formed in pure culture. *M. hiemalis* is similar to *M. circinelloides* but produces larger sporangiospores, usually has unbranched sporangiophores, and is unable to grow at 37°C (Schipper 1976; Domsch et al. 1980; Samson et al. 1995; de Hoog et al. 2000; Schipper and Stalpers 2003).

MUCOR INDICUS LENDNER

M. indicus, usually referred to in the medical literature by its synonym *M. rouxii* (Calmette)Wehmer *sensu* Webmer, has been reported from human gastric (Douvin et al. 1975; Borg et al. 1990) and pulmonary (Krasinski et al. 1985) infections and from a case of necrotizing fasciitis (Mata-Essayag et al. 2001). Oliver et al. (1996) reported an interesting case of an hepatic infection in a bone marrow transplant recipient who had ingested contaminated naturopathic medicine. Colonies are characteristically deep yellow, aromatic, and have a maximum growth temperature of 42°C. Sporangiophores are hyaline to yellowish, erect or, rarely, circinate, and repeatedly sympodially branched, with long branches. Sporangia are yellow to brown, up to 75 μm in diameter, with diffluent membranes. Columellae are subglobose to pyriform, often with truncate bases, up to 40 μm high. Sporangiospores are smooth walled, subglobose to ellipsoidal, and 4–5 μm in diameter. Chlamydoconidia are produced in abundance, especially in the light. Zygospores, when formed by crosses of compatible mating strains, are black, spherical, up to 100 μm in diameter, with stellate spines and unequal suspensor cells. *M. indicus* differs from other species of *Mucor* by its characteristic deep-yellow colony color, growth at over 40°C, assimilating ethanol, but not nitrate, and being thiamine dependent (Schipper 1978a; de Hoog et al. 2000; Schipper and Stalpers 2003).

MUCOR RACEMOSUS FR.

This species has been infrequently reported as a causative agent of animal and human zygomycosis (Scholer et al. 1983). With a maximum growth temperature of only 32°C, however, its pathogenicity is questionable. Colonies are light grayish-brown in color. Sporangiophores are hyaline, erect, and show both sympodial and monopodial branching. Sporangia are brownish, up to 80–90 μm in diameter, and the membrane is mostly diffluent but persistent in smaller sporangia. Columellae are subglobose to pyriform, often with truncated bases, light brown with collars. Sporangiospores are hyaline to pale brown, smooth walled, subglobose to broadly ellipsoidal, 8–10 μm in diameter. Chlamydoconidia are formed abundantly, but zygospores are not produced in pure culture. *M. racemosus* differs from other species of *Mucor* by the formation of chlamydoconidia in sporangiophores, its inability to grow at 37°C, and its positive assimilation of sucrose (Schipper 1976; Samson et al. 1995; de Hoog et al. 2000).

MUCOR RAMOSISSIMUS SAMUTSEVICH

M. ramosissimus has been isolated from mucocutaneous (Vignale et al. 1964), cutaneous (Weitzman et al. 1993), and rhinocerebral (Bullock et al. 1974) infections and from a case of septic arthritis in a neonate (Sharma et al. 1994). Colonial growth is restricted; it is grayish and does not grow at 37°C (maximum temperature for growth is 36°C). Sporangiophores are hyaline, slightly roughened, tapering towards the apex, and are erect with repeated sympodial branching. Sporangia are gray to black, globose or somewhat flattened, up to 80 μm in diameter and have very persistent sporangial walls. Columellae are applanate (flattened), up to 40–50 μm in size, and are often absent in smaller sporangia. Sporangiospores are faintly brown, smooth walled, subglobose to broadly ellipsoidal, 5–8×4.5–6 μm in size. Oidia may be present in the substrate hyphae, chlamydoconidia and zygospores are absent. Assimilation of ethanol is negative and that of nitrate is positive. *M. ramosissimus* differs from other species of *Mucor* by its low, restricted growth on any medium, extremely persistent sporangial walls, columellae that are applanate or absent in smaller sporangia (often resembling *Mortierella* species), short sporangiophores that repeatedly branch sympodially as many as 12 times, and the occurrence of racket-shaped enlargements in the sporangiophores (Hesseltine and Ellis 1964; Schipper 1976; Scholer et al. 1983; de Hoog et al. 2000; Schipper and Stalpers 2003).

Rhizomucor species

The genus *Rhizomucor* is distinguished from *Mucor* by the presence of stolons and poorly developed rhizoids at the base of the sporangiophores and by the thermophilic behavior of its three species: *R. miehei*, *R. pusillus*, and *R. tauricus*. All three of these species are potential human and animal pathogens and were originally classified in the genus. *R. variabilis* as described by de Hoog et al. (2000) is not thermophilic and is probably a degenerate culture of *M. hiemalis* (Voigt et al. 1999). *R. pusillus* is cosmopolitan and both *R. miehei* and *R. pusillus* have been reported as pathogens to humans and animals, the latter to the greater extent (Cooney and Emerson 1964; Hesseltine and Ellis 1973; Schipper 1978b; Scholer et al. 1983; de Hoog et al. 2000; Schipper and Stalpers 2003).

RHIZOMUCOR MIEHEI (COONEY AND EMERSON) SCHIPPER

R. miehei (synonym: *Mucor miehei* Lindt) has been reported as a rare cause of bovine mastitis (Scholer et al. 1983) and is similar in most respects to *R. pusillus* (described in the next section). However, all strains are homothallic, forming numerous zygospores, which are reddish-brown to blackish-brown, globose to slightly compressed, up to 50 μm in diameter, with stellate warts and equal suspensor cells. Colony color is a dirty gray rather than brown, and sporangia have spiny walls, are up to 50–60 μm in diameter, with columellae rarely larger than 30 μm in diameter. Growth is stimulated by thiamine, with no assimilation of sucrose and maximum temperature of growth is 54–58°C. Key features include growth at 45°C, the formation of numerous zygospores, a dirty gray culture color and a partial growth requirement for thiamine. (For further information, see Cooney and Emerson 1964; Schipper 1978b; Scholer et al. 1983; de Hoog et al. 2000.)

RHIZOMUCOR PUSILLUS (LINDT) SCHIPPER

R. pusillus (synonym: *Mucor pusillus* Lindt) is a rare human pathogen. It has been reported from cases of pulmonary (Nicod et al. 1952; Zagoria et al. 1985; Latif et al. 1997; Bjorkholm et al. 2001; Ma et al. 2001), disseminated (Meyer et al. 1973; Kramer et al. 1977; Severo et al. 1991; Gonzalez et al. 1997), and cutaneous types of infection (Wickline et al. 1989). Erdos et al. (1972) have documented a case of endocarditis and St-Germain et al. (1993) reported three cases of pulmonary and one of disseminated infection in patients with leukemia. It is more often associated with animal disease, especially bovine abortion (Smith 1989). *R. pusillus* has a worldwide distribution and is commonly associated with compost heaps. This thermophilic zygomycete is readily recognizable by its characteristic compact, low growing (2–3 mm high), gray to grayish-brown colored mycelium, and by the development of typical, sympodially branched, hyaline to yellow–brown sporangiophores (8–15 μm in diameter), always with a septum below the sporangium. Sporangia are brown or gray, globose (40–60 μm in diameter), each possessing an oval or pear-shaped columella (20–30 μm), often with a collarette (Figure 33.6). Sporangiospores are hyaline, smooth walled, globose to subglobose, occasionally oval (3–5 μm), and are often mixed with crystalline remnants of the sporangial wall. Chlamydoconidia are absent. Zygospores are rough walled, reddish-brown to black, 45–65 μm in diameter, and may be produced throughout the aerial hyphae in matings between compatible isolates. Temperature growth range is: minimum 20–27°C; optimum 35–55°C; maximum 55–60°C. There is positive assimilation of sucrose and no thiamine

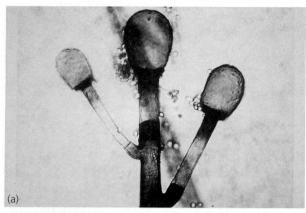

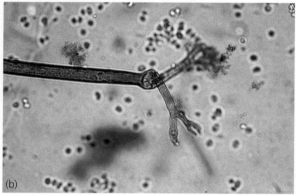

Figure 33.6 Rhizomucor pusillus: **(a)** *branched sporangiophores and columellae;* **(b)** *primitive rhizoids. Magnification:* **(a)** *×330 and* **(b)** *×495*

dependence. Key features include growth at 45°C, poorly developed stolons and rhizoids, branching sporangiophores with a septum below the sporangium, dark-colored sporangia without apophyses, and smooth-walled globose to subglobose sporangiospores. (For additional information, see Cooney and Emerson 1964; Schipper 1978b; Domsch et al. 1980; Ellis and Keane 1981; McGinnis 1980; Scholer et al. 1983; de Hoog et al. 2000; Schipper and Stalplers 2003.)

Rhizopus species

The genus *Rhizopus* is characterized by the presence of stolons and darkly pigmented (pheoid) rhizoids, the formation of sporangiophores singly or in groups from nodes directly above the rhizoids, and apophysate, columellate, multispored, generally globose sporangia. After spore release, the apophyses and columella often collapse to form an umbrella-like structure. Sporangiospores are globose to ovoid, one celled, hyaline to brown, and striate in many species. Colonies are fast growing and cover an agar surface with a dense cottony growth which is at first white, becoming gray or yellowish-brown with sporulation. Species of *Rhizopus* are often associated with soil, plant material, fruit, and similar substrates. Many species grow well at high temperatures, some are used in food fermentations,

whereas others are pathogenic to humans and animals. (For descriptions of species, keys to taxa, and additional information, see Domsch et al. 1980; McGinnis 1980; Onions et al. 1981; Scholer et al. 1983; Schipper 1984; Schipper and Stalpers 1984, 2003; Ellis 1985, 1986; Rippon 1988; Kwon-Chung and Bennett 1992; Samson et al. 1995; de Hoog et al. 2000.)

In the past, numerous attempts have been made to clarify the species concepts of the genus *Rhizopus* (Inui et al. 1965; Zycha et al. 1969; Hesseltine and Ellis 1973). Three excellent revisions with easy-to-use keys have been produced by Schipper (1984), Ellis (1985, 1986), and Schipper and Stalpers (2003). Basically, three groups have been recognized: the 'stolonifer' group, the 'oryzae' group, and the 'microsporus' group. The GC values of the three groups have been defined by Frye and Reinhardt (1993), and temperature growth studies at 30, 36, and 45°C are characteristic for each of the groups.

The 'stolonifer' group has sporangia up to 275 μm in diameter and grows at 30°C, but has a maximum growth temperature of 36°C. Species in this group include *Rhizopus sexualis* and *Rhizopus stolonifer*. The latter has been unconvincingly implicated in human infection (Ferry and Abedi 1983), although with a maximum growth temperature of only 32°C its pathogenicity is thus questionable.

The 'oryzae' group has been reduced to a single species that is able to grow at 40°C but not at 45°C, and has sporangia not exceeding 240 μm in diameter. There is no doubt that *R. oryzae* and *Rhizopus arrhizus* are synonymous, the contentious issue being which species name to use (Schipper 1984; Ellis 1985; Frye and Reinhardt 1993). Briefly, *R. arrhizus* was described before *R. oryzae* and some authors would therefore give that name priority (Ellis 1985, 1986; Rippon 1988; Goodman and Rinaldi 1991). However, other authors consider the original identity of *R. arrhizus* as somewhat doubtful and at best an extreme or atypical form of *R. oryzae*, because it shows applanate columellae that are very unusual in *R. oryzae* (Domsch et al. 1980; Scholer et al. 1983; Schipper 1984; Kwon-Chung and Bennett 1992). The taxonomic treatment of Schipper (Schipper 1984; Schipper and Stalpers 2003) is used in this chapter; however, the synonym *R. arrhizus* is commonly used in the medical literature. *R. oryzae* is an important human pathogen.

The 'microsporus' group has simple rhizoids, and smaller sporangia up to 100 μm in diameter and grows at both 40 and 45°C. This group contains four species: *Rhizopus homothallicus*, *Rhizopus azygosporus*, *Rhizopus schipperae*, and *Rhizopus microsporus*, with the latter subdivided into three varieties, namely *R. microsporus* var. *microsporus*, *R. microsporus* var. *oligosporus*, and *R. microsporus* var. *rhizopodiformis* (Table 33.3). All are thermophilic and *R. microsporus* is a well-recognized pathogen of humans and animals.

Table 33.3 *Differentiation of pathogenic* Rhizopus microsporus *isolates*

Species	Growth at 45°C	Growth at 50°C	Main species characteristics
R. azygosporus	Good	No	Abundant azygospores
R. microsporus var. microsporus	Good	No	Sporangiospores angular to ellipsoidal and distinctly striate, up to 5–6 μm diameter
R. microsporus var. oligosporus	Restricted	No	Sporangiospores globose, up to 9 μm diameter or more, heterogeneous
R. microsporus var. rhizopodiformis	Good	Good	Sporangiospores globose rarely over 5 μm in diameter minutely spinulose
R. schipperae	Good	No	Abundant chlamydoconidia and restricted sporulation

RHIZOPUS AZYGOSPORUS YUAN & JONG

R. azygosporus is closely related to *R. microsporus* (Yuan and Jong 1984) and has been reported as the causative agent of three fatal cases of gastrointestinal infection in premature babies (Woodward et al. 1992; Schipper et al. 1996). Previously, this fungus was only known from its type culture, which has been isolated from tempeh, a solid fermented soybean food from Indonesia (Yuan and Jong 1984). Colonies are whitish to gray–black, producing pale brown simple rhizoids. Sporangiophores are brownish, up to 350 μm high, and 6–14 μm wide. Sporangia are grayish-black, spherical, and 50–100 μm in diameter. Columellae are subglobose to globose. Sporangiospores are ovoid to ellipsoidal, 4–5 to 6–7 μm in diameter with faint striations. Azygospores are pale to dark brown, spherical to subglobose, 30–70 μm in diameter, with coarse conical projections. All strains produce abundant azygospores in unmated isolates as a species characteristic. There is good growth at 45°C with a maximum of 46–48°C. (For further information, see Yuan and Jong 1984; Schipper et al. 1996; de Hoog et al. 2000.)

RHIZOPUS MICROSPORUS VAR. MICROSPORUS

R. microsporus var. *microsporus* is a rare cause of human infection (Kerr et al. 1988; Kwon-Chung and Bennett 1992). An unusual case of cellulitis in the leg of a diabetic patient was reported by West et al. (1995) and a cutaneous case in a patient with idiopathic thrombopenic purpura was reported by Kobayashi et al. (2001). Colonies are pale brownish-gray producing simple rhizoids. Sporangiophores are brownish, up to 400 μm high and 10 μm wide, but most are smaller and are produced in pairs. Sporangia are grayish-black, spherical, up to 80 μm in diameter. Columellae are subglobose to globose to conical. Sporangiospores are angular to broadly ellipsoidal to lemon-shaped, quite equal in size, up to 5–6 μm in diameter and are distinctly striate. Zygospores, formed by crosses of compatible mating strains, are dark red–brown, spherical, up to 100 μm in diameter, with stellate projections and unequal suspensor cells. There is good growth at 45°C, with a

maximum of 46–48°C. (For further information, see Scholer et al. 1983; Schipper and Stalpers 1984, 2003.)

RHIZOPUS MICROSPORUS V. TIEGH. VAR. OLIGOSPORUS (SAITO) SCHIPPER AND STALPERS

R. microsporus var. *oligosporus* is a rare cause of human zygomycosis (Tintelnot and Nitsche 1989). Colonies are pale yellowish-brown to gray and sporulation is often poor. Rhizoids are subhyaline and simple. Sporangiophores are brownish, up to 300 μm high, and 15 μm wide, with one to three produced together. Sporangia are black, spherical, up to 100 μm in diameter. Columellae are subglobose to somewhat conical. Sporangiospores are subglobose to globose, up to 9 μm in diameter, almost smooth, with larger spores often irregular in shape (Figure 33.7). Chlamydoconidia are abundant, hyaline, single or in chains, spherical, ellipsoidal, or cylindrical, 7–35 μm in diameter. Zygospores are not known. There is growth at 45°C with a maximum of 46–48°C. (For further information, see Scholer et al. 1983; Schipper and Stalpers 1984; de Hoog and Guarro 1995; Samson et al. 1995.)

RHIZOPUS MICROSPORUS V. TIEGH VAR. RHIZOPODIFORMIS (COHN) SCHIPPER AND STALPERS

R. microsporus var. *rhizopodiformis* is the second most frequently isolated zygomycete, accounting for between 10 and 15 percent of reported human cases (Scholer et al. 1983; Kwon-Chung and Bennett 1992). It is most commonly isolated from cutaneous infections with several reports citing the use of contaminated Elastoplast bandages or wooden tongue depressors as the environmental source of the fungus (Baker et al. 1962; Ellis 1978; Gartenberg et al. 1978; Bottone et al. 1979; Everett et al. 1979; Mead et al. 1979; Sheldon and Johnson 1979; Myskowski et al. 1983; West et al. 1983; Parfrey 1986; Norden et al. 1991; Paparello et al. 1992; Mitchell et al. 1996; Verweij et al. 1997; Holzel et al. 1998). Gastrointestinal infections are also commonly reported (Neame and Rayner 1960; Waller et al. 1993; Kimura et al. 1995), but disseminated and rhinocerebral forms of infection are rare (Rosenberger et al. 1983;

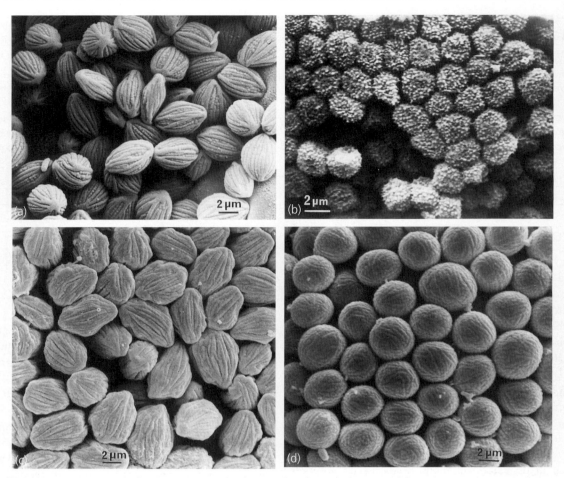

Figure 33.7 Rhizopus *spp.: sporangiospore ornamentation by scanning electron microscopy.* **(a)** R. microsporus *var.* microsporus; **(b)** R. microsporus *var.* rhizopodiformis; **(c)** R. oryzae *(*R. arrhizus*).* **(d)** R. microsporus *var.* oligosporus

Rangel-Guerra et al. 1985; Iqbal and Scheer 1986; Levy et al. 1986). *R. microsporus* var. *rhizopodiformis* is also an important pathogen in animals, especially in swine and bovine abortion (Dion and Sandford 1985; Smith 1989). Colonies are dark grayish-brown, up to 10 mm high with simple rhizoids. Sporangiophores are brownish, up to 500 μm high and 8 μm wide, with one to four produced together. Sporangia are bluish to grayish-black, spherical and up to 100 μm in diameter. Columellae are pyriform, comprising 80 percent of the sporangium. Sporangiospores are subglobose to globose, quite equal in size, up to 6 μm in diameter, and minutely spinulose (Figure 33.7). Zygospores, when formed by crosses of compatible mating strains, are reddish-brown, spherical, up to 100 μm in diameter, with stellate projections and unequal suspensor cells. There is good growth at 45°C with a maximum of 50–52°C (Bottone et al. 1979; Scholer et al. 1983; Schipper and Stalpers 1984, 2003; Ellis 1986; Polonelli et al. 1988; Waller et al. 1993; de Hoog et al. 2000).

RHIZOPUS ORYZAE WENT & PRINSEN GEERLIGS

R. oryzae (synonym: *R. arrhizus*) is the most common causative agent of zygomycosis, accounting for some 60 percent of the reported culture positive cases, and nearly 90 percent of the rhinocerebral form of infection (Scholer et al. 1983; Hyatt et al. 1992; Kwon-Chung and Bennett 1992; Hofman et al. 1993; Attapattu 1995; Adler et al. 1998; Romano et al. 2002). Pulmonary infections (Muhm et al. 1996; Yokoi et al. 1999; Vincent et al. 2000) and primary cutaneous infections (Johnson et al. 1993; Linder et al. 1998; Song et al. 1999) have also been reported. Cerebral infections have occasionally been diagnosed in leukemia patients (McGinnis 1980) and drug addicts (Oliveri et al. 1988; Fong et al. 1990). Sanchez et al. (1994) reviewed infections in AIDS patients and Winkler et al. (1996) reported a case of gastric mucormycosis in a renal transplant patient. This species has rarely been reported from animals (Smith 1989; Perelman and Kuttin 1992). *R. oryzae* (*arrhizus*) has a worldwide distribution with a high prevalence in tropical and subtropical regions. It has been isolated from many substrates, including a wide variety of soils, decaying vegetation, foodstuffs, and animal and bird dung. *R. oryzae* is often used in the production of fermented foods and alcoholic beverages in Indonesia, China, and Japan. However, it also produces the ergot alkaloid agroclavine which is toxic to humans and animals (Domsch et al. 1980; Samson et al. 1995).

Colonies are very fast growing at 25°C, reaching 5–8 mm in height, with some tendency to collapse, white and cottony at first, becoming brownish-gray to blackish-gray depending on the amount of sporulation. Sporangiophores are up to 1 500 µm in length and 18 µm in width, smooth walled, nonseptate, simple or branched, arising from stolons opposite rhizoids usually in groups of three or more (Figure 33.8). Sporangia are globose, often with a flattened base, grayish-black, powdery in appearance, up to 175 µm in diameter, and with many spores. Columella and apophysis together are globose,

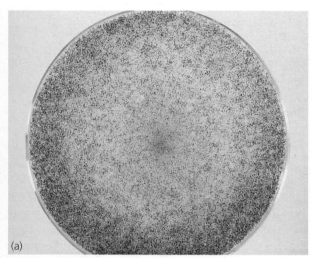

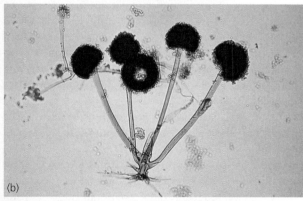

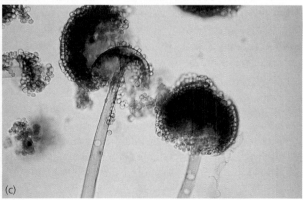

Figure 33.8 Rhizopus oryzae (arrhizus): **(a)** *culture showing abundant pin-head sporangia;* **(b)** *sporangiophores, rhizoids, and sporangia. Magnification ×80.* **(c)** *Sporangium with collapsed columell and sporangiospores. Magnification ×330*

subglobose or oval, up to 130 µm in height, and soon collapse to an umbrella-like form after spore release (Figure 33.8). Sporangiospores are angular, subglobose to ellipsoidal, with ridges on the surface, and up to 8 µm in length. Chlamydoconidia may be formed singly or in chains and are hyaline, smooth walled, spherical to ovoid, and 10–35 µm in diameter. Zygospores, when formed by crosses of compatible mating strains, are red to brown, spherical or laterally flattened, 60–140 µm in diameter, with flat projections and unequal suspensor cells. There is no growth at 45°C, but good growth at 40°C. Key features include growth at 40°C, sporangiophores often more than 1 mm in height, rhizoids with secondary branching, and sporangia between 100 and 240 µm in diameter (Domsch et al. 1980; Scholer et al. 1983; Schipper 1984; Ellis 1985; Oliveri et al. 1988; Samson et al. 1995; de Hoog et al. 2000).

RHIZOPUS SCHIPPERAE WEITZMAN, MCGOUGH, RINALDI & DELLA-LATTA

R. schipperae belongs to the *R. microsporus* complex and is closest to *R. microsporus* var. *microsporus* (Weitzman et al. 1996). It was recently isolated from the bronchial wash and lung specimens from a patient with myeloma (Weitzman et al. 1996) and from paranasal and gastrointestinal specimens from a patient suffering heatstroke (Anstead et al. 1999). Distinguishing characteristics include the production of sporangia on a limited range of media, best on Czapek Dox agar [Oxoid (CM97), Basingstoke, Hants, UK], with clusters of up to 10 sporangiophores arising from each bundle of rhizoids and the formation of abundant, large, mostly rounded chlamydoconidia on all media. Colonies on Czapek Dox agar are grayish-white producing hyaline to pale brown simple rhizoids. Sporangiophores are brownish, 100–460 µm in length, 5–15 µm wide and may be produced singly, in pairs, or in aggregates of up to 10. Sporangia are grayish-black, spherical, and are up to 80 µm in diameter. Columellae are subglobose to subglobose–conical. Sporangiospores are subglobose to ovoid, 5–6 to 6–7 µm in diameter with faint striations. Chlamydoconidia are abundant, terminal or intercalary, globose, oval to ellipsoidal, up to 20 µm in diameter. Zygospores are unknown. Growth is on potato dextrose agar at 45°C, but not at 48–50°C, with optimal development observed at 30–35°C (Weitzman et al. 1996).

RHIZOPUS STOLONIFER (EHRENB:FR) VUILL.

R. stolonifer has been implicated only once in human infection (Ferry and Abedi 1983), and with a maximum growth temperature of only 32°C its pathogenic role is questionable. Colonies are whitish with black spots of sporangia and dark sporangiophores; they produce well-developed complex rhizoids. Sporangiophores are brownish, up to 2 000 µm high and 20 µm wide, and are usually produced in groups of one to three. Sporangia are blackish, spherical, up to 275 µm in diameter.

Columellae are conical and up to 140 μm in height. Sporangiospores are angular to globose to ellipsoidal, up to 13 μm in length, and distinctly striate. Zygospores, when formed by crosses of compatible mating strains, are black, spherical, up to 200 μm in diameter, with stellate projections and unequal suspensor cells. There is no growth at 33°C, but good growth and sporulation at 15–30°C. (For additional information, see Domsch et al. 1980; Onions et al. 1981; Schipper 1984; Samson et al. 1995; de Hoog et al. 2000.)

Saksenaea vasiformis Saksena

The genus *Saksenaea* is characterized by the formation of flask-shaped sporangia with columellae and simple, darkly pigmented rhizoids. *S. vasiformis* is the only known species and appears to have a worldwide distribution in association with soil (Saksena 1953; Hodges 1962; Ajello et al. 1976). *S. vasiformis* is an emerging human pathogen (Holland 1997) that is most often associated with cutaneous or subcutaneous lesions after trauma (Oberle and Penn 1983; Ellis and Kaminski 1985; Parker et al. 1986; Pritchard et al. 1986; Padhye et al. 1988; Goldschmied-Reouven et al. 1989; Tauphaichitr et al. 1990; Mathews et al. 1993; Bearer et al. 1994; Lye et al. 1996; Chakrabarti et al. 1997; Wilson et al. 1998). It has also been associated with rhinocerebral infection (Kaufman et al. 1988), cranial infection (Ajello et al. 1976; Dean et al. 1977), osteomyelitis (Pierce et al. 1987), necrotizing cellulitis (Patino et al. 1984), and disseminated types of infection (Torrell et al. 1981; Hay et al. 1983; Solano et al. 2000). *S. vasiformis* has also been reported from a case of bovine cranial zygomycosis (Hill et al. 1992) and from cases of invasive infection in a captive killer whale and a number of dolphins (Robeck and Dalton 2002).

Laboratory identification of this fungus may be difficult or delayed because of the mold's failure to sporulate on primary isolation media or on subsequent subculture onto potato dextrose agar. Sporulation may be stimulated by the use of nutrient-deficient media, such as cornmeal–glucose–sucrose–yeast extract agar or Czapek Dox agar, or by using the agar block method described by Ellis and Ajello (1982), Ellis and Kaminski (1985), and Padhye and Ajello (1988).

Colonies are fast growing, downy, white with no reverse pigment, and are made up of broad, nonseptate hyphae typical of a zygomycetous fungus. Sporangia are typically flask shaped with a distinct spherical venter and long neck, arising singly or in pairs from dichotomously branched, darkly pigmented rhizoids (Figure 33.9). Columellae are prominent and dome shaped. Sporangiospores are small, oblong, 1–2×3–4 μm in size, and are discharged through the neck after the dissolution of an apical mucilaginous plug. Key features include the formation of unique flask-shaped sporangia and failure to sporulate on primary isolation media. (For additional information, see Saksena 1953;

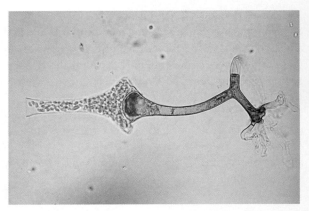

Figure 33.9 Saksenaea vasiformis *showing a distinct flask-shaped sporangium. Magnification ×495*

Ellis and Hesseltine 1966; Chien et al. 1992; Kwon-Chung and Bennett 1992; de Hoog et al. 2000.)

Syncephalastrum racemosum

The genus *Syncephalastrum* is characterized by the formation of cylindrical merosporangia on a terminal swelling of the sporangiophore. Sporangiospores are arranged in a single row within the merosporangia. *S. racemosum* is the type species of the genus and a potential human pathogen; however, well-documented cases are lacking. Kamalam and Thambiah (1980) reported a case of cutaneous infection and Turner (1964) reported one of bovine mycotic abortion. It is often isolated from soil and dung in tropical and subtropical regions, and it can be a persistent laboratory contaminant. The sporangiophore and merosporangia of the *Syncephalastrum* spp. can also be mistaken for an *Aspergillus* sp. if the isolate is not looked at carefully.

Colonies are very fast growing, cottony to fluffy, white to light gray, becoming dark gray with the development of sporangia. Sporangiophores are erect, stolon-like, often producing adventitious rhizoids, and show sympodial branching (racemose branching) producing curved lateral branches. The main stalk and branches form terminal, globose to ovoid vesicles which bear finger-like merosporangia directly over their entire surface (Figure 33.10). At maturity, merosporangia are thin walled, evanescent, and contain globose to ovoid, 5–10(18) μm, smooth-walled sporangiospores (merospores). Optimum growth temperature is 20–40°C. Key features include the production of sympodially branching sporangiophores with terminal vesicles bearing merosporangia. (For additional information, see Domsch et al. 1980; McGinnis 1980; Onions et al. 1981; Samson et al. 1995; de Hoog et al. 2000.)

HOST RESPONSES

In general, members of the Mucorales cause the more severe forms of zygomycosis, *Mortierella* (Mortierellales)

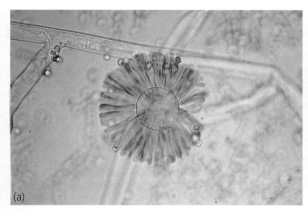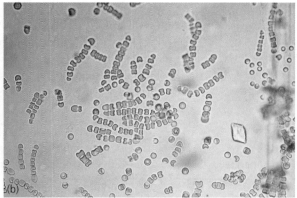

Figure 33.10 Syncephalastrum racemosum: **(a)** *terminal vesicle and merosporangia;* **(b)** *merospores. Magnification* ×495

is not a human pathogen, and the Entomophthorales cause more chronic disease of the nasal mucosa and subcutaneous tissue (see Chapter 17, Subcutaneous Zygomycosis). Infections are rare in the immunocompetent patient and usually present as cutaneous lesions after the traumatic implantation of fungal elements through the skin. Lesions usually remain localized around the initial site of the inoculation and respond well to local débridement and antifungal therapy. However, zygomycosis in the debilitated patient is the most acute and fulminate fungal infection known. The disease typically involves the rhinofaciocranial area, lungs, gastrointestinal tract, skin, or, less commonly, other organ systems (Ribes et al. 2000). It is often associated with acidotic diabetes, starvation, severe burns, intravenous drug abuse, and other diseases such as leukemia and lymphoma, immunosuppressive therapy, or the use of cytotoxins and corticosteroids, therapy with desferrioxamine (an iron-chelating agent for the treatment of iron overload), and other major trauma. Infections are increasingly being reported from HIV-positive patients (Teira et al. 1993; Nagy-Agren et al. 1995; van den Saffele and Boelaert 1996). Symptoms usually develop rapidly depending on the site of infection, the immune status of the host, and degree of pathology. The infecting fungi, primarily *R. oryzae*, have a predilection for invading vessels of the arterial system, causing embolization and subsequent necrosis of surrounding tissue. A rapid diagnosis is extremely important if management and therapy are to be successful (see reviews by Rippon 1988; Goodman and Rinaldi 1991; Kwon-Chung and Bennett 1992; Ribes et al. 2000; Dromer and McGinnis 2003).

Rhinocerebral zygomycosis

Rhinocerebral infection is the most common form of the disease and is usually associated with acute, uncontrolled diabetes mellitus or acidosis (Abramson et al. 1967; Baker 1971; Rippon and Dolan 1979; Kline 1985; Reich et al. 1985; Oakley et al. 1986; Huddle et al. 1987; Ryan-

Poirier et al. 1988; Bessler et al. 1994; Hopkins and Treloar 1997; Adler et al. 1998; Dokmetas et al. 2002; Romano et al. 2002). Other predisposing factors include steroid-induced hyperglycemia, especially in patients with leukemia and lymphoma, renal transplant recipients, and those receiving concomitant treatment with corticosteroids and azathioprine, and in patients with chronic alcoholism (Meyer et al. 1972; Meyer and Armstrong 1973; Hammer et al. 1975; Lyon et al. 1979; Blitzer et al. 1980; Fisher et al. 1980; Kilpatrick et al. 1983; Carbone et al. 1985; Rippon 1988; Sica et al. 1993; Nussbaum and Hall 1994; Alloway et al. 1995; Schuster and Stern 1995; Shpitzer et al. 1995; Ryan et al. 2001). Rare cases of rhinocerebral infection have also been reported in two elderly, nonketotic diabetic patients with periorbital cellulitis and blindness (O'Brien and McKelvie 1994), a patient with burns (Stern and Kagan 1999), and in an otherwise healthy patient (Radner et al. 1995). Rhinocerebral zygomycosis is usually an acute, rapidly progressive, fulminant disease, typified by local tissue necrosis progressing to sinusitis with invasion of the arterial walls, and leads to invasion of the periorbital tissue and brain. Several species have been reported as causative agents, but most human cases are caused by *R. oryzae* (Carbone et al. 1985; Rangel-Guerra et al. 1985; Parfrey 1986; Rinaldi 1989; Ribes et al. 2000). Infections usually begin in the upper turbinates or paranasal sinuses after the inhalation of sporangiospores, and may involve the orbit, palate, face, nose, or brain (Abidi et al. 1984; Morduchowicz et al. 1986; Kwon-Chung and Bennett 1992). Initial symptoms include fever, unilateral headache, nasal or sinus congestion or pain, and a serosanguineous nasal discharge. However, patients are rarely seen during early stages of infection and most present in a comatose state or with symptoms of advanced disease, such as periorbital or perinasal swelling, and a thick, bloody nasal discharge. Other symptoms include ophthalmoplegia with ptosis and proptosis, decreased visual acuity, and facial paralysis. Abnormal mental status often indicates brain involvement. If the fungus invades the palate, a black, necrotic, plaque-like lesion or fistula is often formed (Lehrer et al. 1980;

Table 33.4 *Agents of human zygomycosis and clinical manifestations*

Clinical manifestation	Pathogenic species reported
Rhinocerebral	*R. oryzae* (most common), *A. corymbifera*, *C. bertholletiae*, *S. vasiformis*, *A. elegans*, *M. ramosissimus*
Pulmonary	*R. oryzae* (most common), *A. corymbifera*, *C. bertholletiae*, *R. pusillus*, *R. schipperae*, *M. indicus*
Gastrointestinal	*A. corymbifera* (most common), *R. oryzae*, *R. azygosporus*, *R. microsporus* var. *rhizopodiformis*, *M. indicus*, *A. corymbifera*
Cutaneous	*R. microsporus* var. *rhizopodiformis*, *R. oryzae*, *R. pusillus* (most common), *A. corymbifera*, *C. bertholletiae*, *A. elegans*, *S. vasiformis*, *R. microsporus* var. *microsporus*, *R. microsporus* var. *oligosporus*, *M. circinelloides*, *M. hiemalis*, *M. ramosissimus*, *S. racemosum*
Disseminated	*R. oryzae*, *R. pusillus*, *A. corymbifera*, *C. bertholletiae*, *S. vasiformis*, *A. elegans*

Bigby et al. 1986; Kotzamanoglou et al. 1988; Rippon 1988; Goodman and Rinaldi 1991; Kwon-Chung and Bennett 1992; Ribes et al. 2000; Dromer and McGinnis 2003) (Table 33.4).

Pulmonary zygomycosis

This form of disease is progressive and usually fatal within 2–3 weeks. Infections result by inhalation of sporangiospores into the bronchioles and alveoli, leading to pulmonary infraction and necrosis with cavitation. Hematogenous dissemination to other organs, particularly the brain, often occurs (Meyer et al. 1972; Carbone et al. 1985; Rangel-Guerra et al. 1985; Bigby et al. 1986; Parfrey 1986; Rothstein and Simon 1986; Lake et al. 1988; Rippon 1988; Rinaldi 1989; Goodman and Rinaldi 1991; Kwon-Chung and Bennett 1992; Ribes et al. 2000). Predisposing conditions include: hematological malignancies, lymphoma and leukemia, or severe neutropenia, treatment with cytotoxins and corticosteroids, desferrioxamine therapy, uncontrolled diabetes, organ transplantation, and AIDS (Winston 1965; Schwartz et al. 1982; Hoffman 1987; Chavanet et al. 1990; St-Germain et al. 1993; Piliero and Deresiewicz 1995; Rickerts et al. 2000; Vincent et al. 2000). Clinical symptoms include unremitting fever and a rapidly progressive pneumonia. Hemoptysis and pleuritic chest pain may also be present. The progressive development of lung infiltrates despite broad-spectrum antibacterial treatment is a common radiological finding (Bartram et al. 1973; Skahan et al. 1991; Richardson and Warnock 1993). *R. oryzae* is the most common causative agent, followed by *A. corymbifera*, *C. bertholletiae*, and *R. pusillus* (Ventura et al. 1986; Reed et al. 1988; Rippon 1988; Ribes et al. 2000).

Gastrointestinal zygomycosis

This is a rare entity, usually associated with severe malnutrition, particularly in children, and gastrointestinal diseases which disrupt the integrity of the mucosa. Primary infections probably result after the ingestion of fungal elements and usually present as necrotic ulcers (Watson 1957; Neame and Rayner 1960; Isaacson and Levin 1961; Michalik et al. 1980; Agha et al.

1985; Carbone et al. 1985; Parfrey 1986; Gordon et al. 1988; Ismail et al. 1990; Skahan et al. 1991; Thomson et al. 1991; Mooney and Wagner 1993; Reimund and Ramos 1994; Vadeboncoeur et al. 1994; Hughes et al. 1995; Kimura et al. 1995; Singh et al. 1995; Winkler et al. 1996). Invasive gastrointestinal zygomycosis has been reported in liver transplant patients (Mazza et al. 1999; Vera et al. 2002), a renal transplant patient (Tinmouth et al. 2001), heart–lung and heart transplantation (Knoop et al. 1998), and in neonates (Amin et al. 1998; Nissen et al. 1999). An unusual case was reported from a patient with emphysematous gastritis (Cherney et al. 1999) and another in an immunocompetent host following surgery (Carr et al. 1999). Lesions are most common in the stomach, colon, and ileum (Richardson and Warnock 1993). Symptoms vary depending on the site and extent of disease.

Nonspecific abdominal pain and hematemesis are typical. Necrotic ulcers develop and peritonitis follows if intestinal perforation occurs. Intestinal infections are usually fatal within 2–3 weeks as a result of bowel infarction, sepsis, or hemorrhagic shock (Lyon et al. 1979; Rippon 1988; Rinaldi 1989; Richardson and Warnock 1993). *A. corymbifera* is probably the most frequently isolated mucoraceous causative agent, but *R. azygosporus* has been reported as the causative agent in three cases of gastrointestinal infection in premature babies (Woodward et al. 1992; Schipper et al. 1996). A rare case of colonization of gastric cancerous ulcer by *R. microsporus* var. *rhizopodiformis* has also been reported by Kimura et al. (1995).

Cutaneous zygomycosis

Clinical manifestations of cutaneous zygomycosis vary considerably depending on the status of the host and the portal of entry. Cutaneous lesions are typically single, nonspecific, and include plaques, pustules, ulcerations, deep abscesses, and ragged necrotic patches. There is often a mixed suppurative and necrotizing inflammatory reaction in the dermis and subcutaneous tissue (Goodman and Rinaldi 1991). Primary cutaneous zygomycosis is usually caused by the traumatic implantation of fungal elements through the skin, especially in patients with extensive burns, diabetes, or steroid-induced hyperglycemia, and in cases of major trauma resulting from automobile accidents or machinery (Wilson et al. 1976; Tomford et al. 1980; Sands et al. 1985; Venezio et al. 1985; Parfrey 1986; Johnson et al. 1987; Rippon 1988; Vainrub et al. 1988; Rinaldi 1989; Tintelnot and Nitsche 1989; Skahan et al. 1991; Adam et al. 1994; Hicks et al. 1995; Liao et al. 1995; Lopes et al. 1995; Seguin et al. 1999; Song et al. 1999; Chandra and Woodgyer 2002). Cases have arisen at insulin injection sites, spider bites, entry sites of intravenous or peritoneal catheters, or operative wounds (Kwon-Chung and Bennett 1992; Baraia et al. 1995; Leong et al. 1997). In patients with burns, the initial clinical signs include fever, swelling, and changes in the appearance of the burn wound leading to severe underlying necrosis and infarction (Bruck et al. 1971; Cooter et al. 1990). Necrotizing cutaneous lesions have occurred in patients who have had contaminated surgical dressings or Elastoplast bandages applied to their skin (Gartenberg et al. 1978; Dennis et al. 1980; Patterson et al. 1986; White et al. 1986; Kerr et al. 1988; Lakshmi et al. 1993). An outbreak in neonates has been reported, caused by the use of wooden tongue depressors as splints to immobilize limbs (Mitchell et al. 1996; Tonks 1996). Cutaneous infections have also been reported from adhesive tape used to secure endotracheal tubes (Dickinson et al. 1998; Alsuwaida 2002) and from a snapdragon skin patch test (Blair et al. 2002). Reports of primary cutaneous infections in neonates appear to be increasing over time (du Plessis et al. 1997; Amin et al. 1998; Linder et al. 1998; Oh and Notrica 2002; Buchta et al. 2003). Many species have been reported as causative agents with *R. microsporus* var. *rhizopodiformis*, *R. oryzae*, and *R. pusillus* being the most common (Rippon 1988). *C. bertholletiae* has also been reported as causing a cutaneoarticular infection in an AIDS patient (Mostaza et al. 1989).

Most of the reported infections with *S. vasiformis* and *A. elegans* have been localized to the site of trauma in a previously normal host (Kwon-Chung and Bennett 1992). In the immunocompetent patient, infections usually remain localized around the site of the initial trauma and respond well to local débridement and antifungal therapy (Song et al. 1999; Kimura et al. 1999); however, in the immunosuppressed patient, infections may rapidly disseminate and require much more aggressive management. Cutaneous and subcutaneous zygomycosis may also be the result of hematogenous spread or direct invasion from other organs, and usually indicates a very poor prognosis (Goodman and Rinaldi 1991; Ribes et al. 2000).

Disseminated zygomycosis

Systemic infections may originate from any of the above, but are usually seen in neutropenic patients with pulmonary infection and less commonly from the gastrointestinal tract, burns, or other cutaneous lesions (Ingram et al. 1989; Richardson and Warnock 1993). Recent reports have also indicated an association between desferrioxamine use and disseminated zygomycosis, particularly in dialysis patients (Windus et al. 1987; Boelaert et al. 1988; Rex et al. 1988; Ryan-Poirier et al. 1988; Sane et al. 1989; Goodman and Rinaldi 1991; Prokopowicz et al. 1994). The most common site of spread is the brain, but metastatic necrotic lesions have also been found in the spleen, heart, and other organs (Tuder 1985; Iqbal and Scheer 1986; Bosken et al. 1987; Kalayjian et al. 1988; Rippon 1988; Rozich et al. 1988; Rinaldi 1989; Severo et al. 1991; Teira et al. 1993; Fortun et al. 1995; Paterson et al. 2000; Eucker et al. 2001). Infections are usually diagnosed postmortem unless there are metastatic cutaneous lesions. Cerebral infection after hematogenous dissemination results in abscess formation and infarction, and is thus distinct from the rhinocerebral form of infection (Richardson and Warnock 1993). Patients usually present with sudden onset of focal neurological deficits or coma. Traumatic implantation leading to brain abscess has been reported in intravenous drug abusers and AIDS patients (Pierce et al. 1982; Lawrence et al. 1986; Parfrey 1986; Woods and Hanna 1986; Kasantikul et al. 1987; Cuadrado et al. 1988; Oliveri et al. 1988; Smith et al. 1989; Stave et al. 1989; Fong et al. 1990; Hopkins et al. 1994). Sporadic reports of other miscellaneous types of zygomycosis include endocarditis (Virmani et al. 1982; Zhang et al. 2002), osteomyelitis (Echols et al. 1979; Pierce et al. 1987; Eaton et al. 1994; Meis et al. 1994), renal infections (Langston et al. 1973; Low et al. 1974; Levy and Bia 1995; Melnick et al. 1995; Utas et al. 1995; Gonzalez et al. 1997), and a case of fatal penile necrosis in a patient with undiagnosed diabetes (Williams et al. 1995).

LABORATORY DIAGNOSIS

As zygomycosis is such an aggressive and rapidly fatal infection, it is essential to make an early diagnosis, although at times this may be very difficult. The diagnosis is usually made by demonstrating the characteristic ribbon-like, aseptate hyphae in tissue and culture of the causative agent. Despite being recognized as common laboratory contaminants, zygomycetes are infrequently isolated in the clinical laboratory. Therefore, in patients with any of the above predisposing conditions, especially diabetes or immunosuppression and/or clinical symptoms, the isolation of any zygomycete fungus should be regarded as potentially significant. Obviously, in patients without predisposing conditions, the isolation of a zygomycete from a nonsterile site, such as skin or sputum, must be interpreted with caution, especially in the absence of direct microscopic detection in clinical specimens.

Clinical material

Clinical material may include:

- skin biopsies from cutaneous lesions
- sputum and needle biopsies from pulmonary lesions
- nasal discharges, scrapings, and aspirates from sinuses in patients with rhinocerebral lesions
- biopsy tissue from patients with gastrointestinal and/or disseminated disease.

However, as most specimens are collected from abscesses and necrotic tissue, a biopsy usually provides the best specimen; swabs are not recommended. Ideally, tissue specimens should include both normal tissue, and the center and edge of the lesion. Tissue samples should be kept moist with saline or brain–heart infusion broth and be transported to the laboratory as soon as possible. It is essential that both direct microscopy and culture be performed on all specimens. It should also be noted that zygomycetous fungi have coenocytic hyphae, which will often be damaged and become nonviable during the biopsy procedure (especially scrapings and aspirates), or by the chopping up or tissue grinding processes in the laboratory. This is why zygomycetous fungi which are clearly visible in direct microscopic or histopathological mounts are often difficult to grow in culture from clinical specimens. If on clinical and/or radiological evidence zygomycosis is suspected, excessive tissue damage when collecting the specimen should be avoided if possible and in the laboratory the tissue should be gently teased apart and inoculated directly onto the isolation media.

Direct microscopy and histopathology

Microscopy is the most rapid diagnostic method and the presence in clinical specimens of broad, 10–15 µm wide, infrequently septate, thin-walled hyphae, which often show focal bulbous dilatations and irregular branching, is diagnostic for zygomycosis (Figure 33.11). Scrapings, sputum, and exudates may be examined using 10–20 percent KOH and Parker ink or Calcofluor mounts. The demonstration of zygomycetous hyphae on direct microscopy from such nonsterile sites is essential in order to determine the significance of a positive culture. Tissue sections should be stained with hematoxylin and eosin (H&E), Gomori's methenamine silver (GMS) (Grocott 1955), and periodic acid–Schiff (PAS) stains. Staining reactions may be variable; however, unlike *Aspergillus* spp. mycelium, zygomycete hyphae are usually clearly visible in H&E-stained sections. Nevertheless, a GMS stain should always be performed (Chandler et al. 1980; Rippon 1988). Apart from the presence of the characteristic ribbon-like hyphae, the main histopathological features of zygomycosis are necrosis with acute and chronic infiltrates and involvement of blood vessels. Blood vessel invasion is quite frequent, resulting in

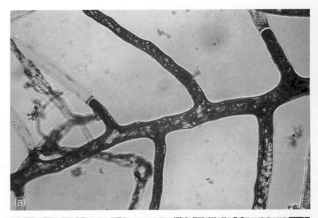

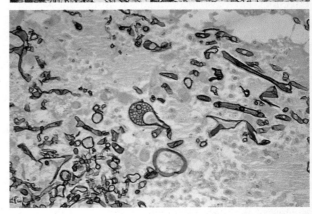

Figure 33.11 *Tissue morphology in zygomycosis:* **(a)** *lactophenol cotton blue mount showing distinctive wide infrequently septate hyphae;* **(b)** *H&E-stained section of infected tissue showing broad, infrequently septate, thin-walled hyphae with focal bulbous dilatations and irregular branching, typical for those species belonging to the Mucorales;* **(c)** *GMS-stained tissue section showing typical zygomycotic hyphae and by chance a sporangium of* Absidia corymbifera. *Magnification ×330*

thrombi, infarction, and necrosis of blood vessels. Occasionally, sporangia with well-delineated columellae and sporangiospores may be seen in tissue sections of nasal and pulmonary tissue that are well aerated (Chandler et al. 1980; Salfelder 1990; Ribes et al. 2000).

Culture

Nutritionally, members of the Mucorales are not particularly fastidious and will grow on most routine fungal culture media provided they do not contain cyclohex-imide (actidione). However, the use of antibacterial anti-biotics such as chloramphenicol and gentamicin in primary isolation media is recommended to prevent bacterial contamination. Media containing dextrose as a carbon source and ammonium or peptone as nitrogen sources are preferred (Scholer et al. 1983). Sabouraud's dextrose agar containing antibiotics is usually used as the primary isolation medium, with potato dextrose agar frequently used for culture maintenance and identification purposes. Thiamine is the only growth factor required by some species, such as A. corymbifera; however, sufficient amounts are usually present in agar containing peptone (Scholer et al. 1983). Most mucorac-eous species sporulate profusely and are fast growing, often pushing the lid off the petri dish in a few days. However, laboratory identification of some species, especially A. elegans and S. vasiformis, may be difficult or delayed because of their failure to sporulate on primary isolation media or on subsequent subculture onto potato dextrose agar. Sporulation may be stimu-lated by the use of nutrient deficient media, such as cornmeal–glucose–sucrose–yeast extract agar, Czapek Dox agar, or by using the agar block method described by Ellis and Ajello (1982) and Ellis and Kaminski (1985).

Immunology and serology

Several studies have investigated the ability of members of the Mucorales to cause infection in normal, cortisone-treated and alloxan-induced diabetic animals (Waldorf and Diamond 1989). Diabetic animals develop a rapidly fatal infection when challenged with R. oryzae, R. pusillus, R. microsporus var. oligosporus, and A. corym-bifera (Reinhardt et al. 1981). Intranasal or itrasinus inoculation of R. oryzae into streptozotocin-induced diabetic mice results in typical pulmonary or rhinocerebral infections (Waldorf et al. 1984b). Unlike other models of zygomycosis (Waldorf et al. 1982), these diabetic models lead to induction of specific enhanced susceptibility to infection by Rhizopus spp., resembling the predisposition to zygomycosis which occurs in uncontrolled human diabetic patients (Waldorf and Diamond 1989). Rhizopus spp. spore germination occurs in tissues of diabetic or cortisone-treated mice after intranasal inoculation, but not in normal animals (Lund-borg and Holma 1972; Smith 1976; Waldorf et al. 1982, 1984a; Waldorf and Diamond 1984). Thus, prevention of zygomycosis appears to require inhibition of spor-angiospore germination by the bronchoalveolar macro-phage, thereby preventing conversion of the fungus to its hyphal form, although resident bronchoalveolar macro-phages are unable to kill R. oryzae spores (Lundborg and Holma 1972). The invasive hyphal elements seen in tissue are too large to be ingested by phagocytic cells, and it would appear that neutrophils have an important

role to play in killing hyphae of R. oryzae (Diamond and Clark 1982; Schaffner et al. 1986). Isolates of R. oryzae also produce an alkaline protease with proteolytic activity within the normal physiological pH range. However, further studies are needed to examine the role of enzymes and toxins in the virulence of these fungi (Waldorf and Diamond 1989). In general, however, most data suggest that pathogenicity of the zygomycetes depends primarily on the susceptibility of the host rather than on the pathogenic potential of the fungus.

The development of serological assays for the early diagnosis of zygomycosis has proved to be very diffi-cult. Sera from proven cases are difficult to obtain because infections are infrequently diagnosed. Only weak antibody responses are detected in humans, prob-ably related to the depressed immunological status of the patients and the rapid progressive course of infec-tion, which often leads to death before significant production of specific antibodies. However, some laboratories have developed 'in house' immunodiffusion and ELISA tests for the detection of antibodies to zygomycetes (Jones and Kaufman 1978; Waldorf et al. 1982; Lawrence et al. 1986; Wysong and Waldorf 1987; Kaufman et al. 1989, 1990). Exoantigen tests for the rapid identification of nonsporulatng strains of A. elegans and S. vasiformis have been developed (Lombardi et al. 1989). The development of more sensitive DNA-based diagnostic methods, antigen detec-tion, or specific serological procedures may prove to be more valuable in the future (Dromer and McGinnis 2003).

TREATMENT (MANAGEMENT/THERAPY)

The successful management of infections caused by mucoraceous zygomycetes requires an early diagnosis, control or reversal of any predisposing factors or under-lying disease, antifungal therapy, and aggressive surgical débridement which may have to be repeated until all infected necrotic tissue is removed.

Amphotericin B is the drug of choice and full-dose therapy of 1.0 or 1.5 mg/kg/day is necessary. Lower dosages of 0.8–1.0 mg/kg/day and/or alternate day therapy may be considered after the patient has been stabilized with no new areas of necrosis developing (Andriole and Bodey 1994). Some patients may require a total dose of up to 4 g (Sugar 1992). Liposomal ampho-tericin B is increasingly being used, as it is much better tolerated than conventional amphotericin B and doses as high as 3–5 mg/kg/day may be given (Grauer et al. 1993; Munckhof et al. 1993; Berenguer et al. 1994; Lim et al. 1994; Ng and Denning 1995; Strasser et al. 1996; Saltoglu et al. 1998; Bjorkholm et al. 2001; Cagatay et al. 2001; Cloughley et al. 2002; Ma et al. 2001; Handzel et al. 2003; Kofteridis et al. 2003). Amphotericin B is generally continued for 8–10 weeks. To date other antifungal agents such as 5-fluorocytosine, the azoles, or the echinocandins have had no role in the management of zygomycosis. However, recent in vitro data (Sun et al. 2002a, b; Dannaoui et al. 2003) and in vivo data for the new triazole agent posaconazole have demonstrated promising activity against the zygomycetes (Greenberg et al. 2003; Tobon et al. 2003). Antifungal susceptibility testing of these fungi is also not reliable and has an uncertain place in guiding therapeutic decisions.

In the diabetic patient prompt correction of acidosis is essential and with early diagnosis and treatment, 50–85 percent of patients with rhinocerebral zygomycosis can be cured (Pillsbury and Fischer 1977; Meyers et al. 1979; Lehrer et al. 1980). The best prognosis is among those without extension into the brain or the internal carotid artery. The most critical decisions all concern balancing the extent of surgery between that necessary to control progressive disease and that which causes unnecessary loss of the eye or gaping operative wounds (Kwon-Chung and Bennett 1992). By contrast, few leukemia patients with the infection recover (Richardson and Warnock 1993).

Treatment of pulmonary, gastrointestinal, or disseminated zygomycosis has been successful too rarely to judge appropriate therapy, but intravenous amphotericin B and decreased immunosuppression appear important in survival (Kwon-Chung and Bennett 1992). In normal patients with localized cutaneous lesions surgical débridement is often sufficient alone, although intravenous treatment with amphotericin B is also indicated. Adjunctive therapies using G-CSF (Sahin et al. 1996; Gonzalez et al. 1997; Leong et al. 1997; Ma et al. 2001; Slavin et al. 2002), GM-CSF (Leleu et al. 1999; Garcia-Diaz et al. 2001; Ma et al. 2001; Mileshkin et al. 2001; MacKenzie et al. 2002), interferon-gamma (Okhuysen et al. 1994), and hyperbaric oxygen have been used with increasing frequency (Ferguson et al. 1988; Okhuysen et al. 1994; Kajs-Wyllie 1995; Bentur et al. 1998; Gonzalez et al. 2002).

REFERENCES

Abidi, E., Sismanis, A., et al. 1984. Twenty-five years experience treating cerebro-rhino orbital mucormycosis. *Laryngoscope*, **94**, 1060–2.

Abramson, E., Wilson, D. and Arky, R.A. 1967. Rhinocerebral phycomycosis in association with diabetic ketoacidosis: report of two cases and a review of clinical and experimental experience with amphotericin B therapy. *Ann Intern Med*, **66**, 735–42.

Adam, R.D., Hunter, G., et al. 1994. Mucormycosis: emerging prominence of cutaneous infections. *Clin Infect Dis*, **19**, 67–76.

Adler, D.E., Milhorat, T.H. and Miller, J.I. 1998. Treatment of rhinocerebral mucormycosis with intravenous interstitial, and cerebrospinal fluid administration of amphotericin B: case report. *Neurosurgery*, **42**, 644–8.

Agha, F.P., Lee, H.H., et al. 1985. Mucormycoma of the colon: early diagnosis and successful management. *Am J Roentgenol Radium Ther Nucl Med*, **145**, 739–41.

Ajello, L., Dean, D.F. and Irwin, R.S. 1976. The zygomycete *Saksenaea vasiformis* as a pathogen of humans with a critical review of the etiology of zygomycosis. *Mycologia*, **68**, 52–62.

Alloway, J.A., Buchsbaum, R.M., et al. 1995. Mucormycosis in a patient with sarcoidosis. *Sarcoidosis*, **12**, 143–6.

Alsuwaida, K. 2002. Primary cutaneous mucormycosis complicating the use of adhesive tape to secure the endotracheal tube. *Can J Anaesth*, **49**, 880–2.

Alvarez, O.A., Maples, J.A., et al. 1995. Severe diarrhea due to *Cokeromyces recurvatus* in a bone marrow transplant recipient. *Am J Gastroenterol*, **90**, 1350–1.

Amin, S.B., Ryan, R.M., et al. 1998. *Absidia corymbifera* infections in neonates. *Clin Infect Dis*, **26**, 990–2.

Andriole, V.T. and Bodey, G.P. 1994. *Pocket guide to systemic antifungal therapy*. New Jersey: Scientific Therapeutics Information Inc.

Anstead, G.M., Sutton, D.A., et al. 1999. Disseminated zygomycosis due to *Rhizopus schipperae* after heatstroke. *J Clin Microbiol*, **37**, 2656–62.

Attapattu, M.C. 1995. Acute rhinocerebral mucormycosis caused by *Rhizopus arrhizus* from Sir Lanka. *J Trop Med Hyg*, **98**, 355–8.

Austwick, P.K.C. 1965. Environmental aspects of *Mortierella wolfii* infections in cattle. *NZ J Agric Res*, **19**, 25–33.

Axelrod, P., Kwon-Chung, K.J., et al. 1987. Chronic cystitis due to *Cokeromyces recurvatus*: a case report. *J Infect Dis*, **155**, 1062–4.

Baker, R.D. 1971. Mucormycosis (opportunistic phycomycosis). In: Baker, R.D. (ed.), *Human infection with fungi, actinomycetes and algae*. New York: Springer-Verlag, 832–918.

Baker, R.D., Seabury, J.H. and Schneidau, J.H. 1962. Subcutaneous and cutaneous mucormycosis and subcutaneous phycomycosis. *Lab Invest*, **11**, 1091–102.

Baraia, J., Munoz, P., et al. 1995. Cutaneous mucormycosis in a heart transplant patient associated with a peripheral catheter. *Eur J Clin Microbiol Infect Dis*, **14**, 813–15.

Bartram, R.J. Jr, Watnick, M. and Herman, P.B. 1973. Roentgenographic findings in pulmonary mucormycosis. *Am J Roentgenol Radium Ther Nucl Med*, **117**, 810–15.

Bearer, E.A., Nelson, P.R., et al. 1994. Cutaneous zygomycosis caused by *Saksenaea vasiformis* in a diabetic patient. *J Clin Microbiol*, **32**, 1823–4.

Benjamin, R.K. 1979. Zygomycetes and their spores. In: Kendrick, B. (ed.), *The whole fungus*, Vol. 2. Ottawa: National Museums of Canada, 573–621.

Benny, G.L. and Benjamin, R.K. 1976. Observations on Thamnidiaceae (Mucorales). II *Chaetocladium, Cokeromyces, Mycotypha, Phascolomyces. Aliso*, **8**, 391–424.

Bentur, Y., Shupak, A., et al. 1998. Hyperbaric oxygen therapy for cutaneous/soft-tissue zygomycosis complicating diabetes mellitus. *Plast Reconstr Surg*, **102**, 822–4.

Berenguer, J., Moreno, S., et al. 1988. Mucormycosis, clinical manifestations, diagnosis, treatment and outcome of 12 cases. *Rev Iber Micol*, **5**, Suppl 1, 100.

Berenguer, J., Munoz, P., et al. 1994. Treatment of deep mycoses with liposomal amphotericin B. *Eur J Clin Microbiol Infect Dis*, **13**, 504–7.

Berger, L., Speare, R. and Humphrey, J. 1997. Mucormycosis in a free-ranging green tree frog from Australia. *J Wildl Dis*, **33**, 903–7.

Bessler, S.C., Hailemariam, S. and Gammert, C. 1994. Rhino-orbital mucormycosis: surgical aspects. *ORL J Otorhinolaryngol Relat Spec*, **56**, 244–6.

Bigby, T.D., Serota, M.L., et al. 1986. Clinical spectrum of pulmonary mucormycosis. *Chest*, **89**, 435–9.

Bjorkholm, M., Runarsson, G., et al. 2001. Liposomal amphotericin B and surgery in the successful treatment of invasive pulmonary Mucormycosis in a patient with acute T-lymphoblastic leukemia. *Scand J Infect Dis*, **33**, 316–19.

Blair, J.E., Fredrikson, L.J., et al. 2002. Locally invasive cutaneous *Apophysomyces elegans* infection acquired from snapdragon patch test. *Mayo Clin Proc*, **77**, 717–20.

Blitzer, A., Lawson, W., et al. 1980. Patient survival factors in paranasal sinus mucormycosis. *Laryngoscope*, **90**, 635–48.

Boelaert, J.R., van Roost, G.F., et al. 1988. The role of desferrioxamine in dialysis-associated mucormycosis: report of three cases and review of the literature. *Clin Nephrol*, **29**, 261–6.

Borg, F., Kuijper, E.J. and van der Leilie, H. 1990. Fatal mucormycosis presenting as an appendiceal mass with metastatic spread to the liver during chemotherapy-induced granulocytopenia. *Scand J Infect Dis*, **22**, 499–501.

Bosken, C.H., Szporn, A.H. and Kleinerman, J. 1987. Superior vena cava syndrome due to mucormycosis in a patient with lymphoma. *Mt Sinai J Med*, **54**, 508–11.

Bottone, E.J., Weitzman, I. and Hanna, B.A. 1979. *Rhizopus rhizopodiformis*: emerging etiological agent of mucormycosis. *J Clin Microbiol*, **9**, 530–7.

Boyce, J.M., Lawson, L.A., et al. 1981. *Cunninghamella bertholletiae* wound infection of probable nosocomial origin. *South Med J*, **74**, 1132–5.

Brennan, O.R., Crain, B.J., et al. 1983. *Cunninghamella*: a newly recognized cause of rhinocerebral mucormycosis. *Am J Clin Pathol*, **80**, 98–102.

Brown, S.R., Shar, I.A. and Grinstead, M. 1998. Rhinocerebral mucormycosis caused by *Apophysomyces elegans. Am J Rhinol*, **12**, 239–92.

Bruck, H.M., Nash, G., et al. 1971. Opportunistic fungal infection of the burn wound with phycomycetes and *Aspergillus*: a clinical-pathologic review. *Arch Surg*, **102**, 476–82.

Buchta, V., Kalous, P., et al. 2003. Primary cutaneous *Absidia corymbifera* infection in a premature newborn. *Infection*, **31**, 57–9.

Bullock, J.D., Jampol, L.M. and Fezza, A.J. 1974. Two cases of orbital phycomycosis with recovery. *Am J Opthalmol*, **78**, 811–15.

Burrell, S.R., Ostlie, D.J., et al. 1998. *Apophysomyces elegans* infection associated with catus spine injury in an immunocompetent pediatric patient. *Pediatr Infect Dis*, **17**, 663–4.

Cagatay, A.A., Oncu, S.S., et al. 2001. Rhinocerebral mucormycosis treated with 32 gram liposomal amphotericin B and incomplete surgery: case report. *BMC Infect Dis*, **1**, 22.

Carbone, K.M., Pennington, L.R., et al. 1985. Mucormycosis in renal transplant patients: a report of two cases and a review of the literature. *Q J Med*, **57**, 825–31.

Carr, E.J., Scott, P. and Gardon, J.D. 1999. Fatal gastrointestinal mucormycosis that invaded the postoperative abdominal wall wound in an immunocompetent host. *Clin Infect Dis*, **29**, 956–7.

Carter, M.E., Cordes, D.O., et al. 1973. Fungi isolated from bovine mycotic abortion and pneumonia with special reference to *Mortierella wolfii. Res Vet Sci*, **14**, 201–6.

Cavalier-Smith, T. 1998. A revised six-kingdom system of life. *Biol Rev Camb Philos Soc*, **73**, 203–66.

Chakrabarti, A., Kumar, P., et al. 1997. Primary cutaneous zygomycosis due to *Saksenaea vasiformis* and *Apophysomyces elegans. Clin Infect Dis*, **24**, 580–3.

Chandler, F.W., Kaplan, W. and Ajello, L. 1980. *A colour atlas and textbook of the histopathology of mycotic diseases*. London: Wolfe Medical.

Chandra, S. and Woodgyer, A. 2002. Primary cutaneous zygomycosis due to *Mucor circinelloides. Australas J Dermatol*, **43**, 39–42.

Chavanet, P., Lefranc, T., et al. 1990. Unusual cause of pharyngeal ulcerations in AIDS. *Lancet*, **8711**, 383–4.

Cherney, C.L., Chutuape, A. and Fikrig, M.K. 1999. Fatal invasive gastric mucormycosis occurring with emphysematous gastritis: case report and literature review. *Am J Gastroenterol*, **94**, 252–6.

Chien, C.-Y., Bhat, D.J. and Kendrick, W.B. 1992. Mycological observations on *Saksenaea vasiformis* (Saksenaceae, Mucorales). *Trans Mycol Soc Jpn*, **33**, 443–8.

Cloughley, R., Kelehan, J., et al. 2002. Soft tissue infection with *Absidia corymbifera* in a patient with idiopathic aplastic anemia. *J Clin Microbiol*, **40**, 725–7.

Cohen-Abbo, A., Bozeman, P.M. and Patrick, C.P. 1993. *Cunninghamella* infections: review and report of two cases of *Cunninghamella* pneumonia in immunocompromised children. *Clin Infect Dis*, **17**, 173–7.

Connolly, J.H., Canfield, P.J. and Obendorf, D.L. 2000. Gross, histological and immunohistochemical features of mucormycosis in the platypus. *J Comp Pathol*, **123**, 36–46.

Cooney, D.G. and Emerson, R. 1964. *Thermophilic fungi*. San Francisco, CA: WH Freeman.

Cooter, R.D., Lim, I.S., et al. 1990. Burn wound zygomycosis caused by *Apophysomyces elegans. J Clin Microbiol*, **28**, 2151–3.

Costa, A.R., Porto, E. and Tayah, M. 1990. Subcutaneous mucormycosis caused by *Mucor hiemalis* Wehmer f. Luteus (Linnemann) Schipper. *Mycoses*, **33**, 241–6.

Cuadrado, L.M., Guerrero, A. and Garcia Asenjo, J.A. 1988. Cerebral mucormycosis in two cases of acquired immunodeficiency syndrome. *Arch Neurol*, **45**, 109–11.

Dannaoui, E., Meis, J.F., et al. 2003. Activity of posaconazole in treatment of experimental disseminated zygomycosis. *Antimicrob Agents Chemother*, **47**, 3647–50.

Darja, M. and Davy, M.I. 1963. Pulmonary mucormycosis with cultural identification. *Can Med Assoc J*, **89**, 1235–8.

Darrisaw, L., Hanson, G., et al. 2000. *Cunninghamella* infection post bone marrow transplant: case report and review of the literature. *Bone Marrow Transplant*, **25**, 1213–16.

Dean, D.F., Ajello, L., et al. 1977. Cranial zygomycosis caused by *Saksenaea vasiformis. J Neurosurg*, **46**, 97–103.

de Hoog, G.S. and Guarro, J. 1995. *Atlas of clinical fungi*. Baarn, The Netherlands: Centraalbureau voor Schimmelcultures.

de Hoog, G.S., Guarro, J., et al. 2000. *Atlas of clinical fungi*, 2nd edition. Baarn, The Netherlands: Centraalbureau voor Schimmelcultures.

Dennis, J.E., Rhodes, K.H., et al. 1980. Nosocomial *Rhizopus* infection (zygomycosis) in children. *J Pediatr*, **96**, 924–8.

Dermourni, H. 1993. A rare zygomycosis due to *Cunninghamella bertholletiae. Mycoses*, **36**, 293–4.

Diamond, R.D. and Clark, R.A. 1982. Damage to *Aspergillus fumigatus* and *Rhizopus oryzae* hyphae by oxidative and nonoxidative microbicidal products of human neutrophils in vitro. *Infect Immun*, **38**, 487–95.

Dickinson, M., Kalayanamit, T., et al. 1998. Cutaneous zygomycosis (mucormycosis) complicating endotracheal intubation: diagnosis and successful treatment. *Chest*, **114**, 340–2.

di Menna, M.E., Carter, M.E. and Cordes, D.O. 1972. The identification of *Mortierella wolfii* isolated from cases of abortion and pneumonia in cattle and a search for its infection source. *Res Vet Sci*, **13**, 439–442.

Dion, W.L. and Sandford, S.E. 1985. Isolation of *Rhizopus rhizopodiformis* from a case of mucormycosis in a pig. *Mycopathologia*, **89**, 127–8.

Dokmetas, H.S., Canbay, E., et al. 2002. Diabetic ketoacidosis and rhino-orbital mucormycosis. *Diabetes Res Clin Pract*, **57**, 139–42.

Domsch, K.H., Gams, W. and Anderson, T.-H. 1980. *Compendium of soil fungi*. London: Academic Press.

Done, S.H., Sharp, M.W. and Lupson, G.R. 1994. Isolation of *Mortierella wolfii* from bovine lung. *Vet Rec*, **134**, 194.

Douvin, D., Lefichoux, Y. and Hugent, C. 1975. Phycomycose gastrique. *Arch Pathol*, **23**, 133–8.

Dromer, F. and McGinnis, M.R. 2003. Zygomycosis. In: Anaissie, E.J., McGinnis, M.R. and Pfaller, M.A. (eds), *Clinical mycology*. Edinburgh: Churchill Livingstone, 297–308.

du Plessis, P.J., Wentzel, L.F., et al. 1997. Zygomycotic necrotizing cellulites in a premature infant. *Dermatology*, **195**, 179–81.

Eaton, M.E., Padhye, A.A., et al. 1994. Osteomyelitis of the sternum caused by *Apophysomyces elegans. J Clin Microbiol*, **32**, 2827–2828.

Echols, R.M., Selinger, D.G., et al. 1979. *Rhizopus* osteomyelitis: a case report and a review. *Am J Med*, **66**, 141–5.

Ellis, D.H. and Kaminski, G.W. 1985. Laboratory identification of *Saksenaea vasiformis*: a rare cause of zygomycosis in Australia. *J Med Vet Mycol*, **23**, 137–40.

Ellis, D.H. and Keane, P.J. 1981. Thermophilic fungi isolated from some Australian soils. *Aust J Bot*, **29**, 689–704.

Ellis, J.J. 1978. Follow-up on *Rhizopus* infections associated with Elastoplast bandages (United States). *Minnesota Morbid Mortal Week Rep*, **27**, 243–4.

Ellis, J.J. 1985. Species and varieties in the *Rhizopus arrhizus-Rhizopus oryzae* group as indicated by their DNA complementarity. *Mycologia*, **77**, 243–7.

Ellis, J.J. 1986. Species and varieties in the *Rhizopus microsporus* group as indicated by their DNA complementarity. *Mycologia*, **78**, 508–10.

Ellis, J.J. and Ajello, L. 1982. An unusual source of *Apophysomyces elegans* and a method for stimulating sporulation of *Saksenaea vasiformis*. *Mycologia*, **74**, 144–5.

Ellis, J.J. and Hesseltine, C.W. 1965. The genus *Absidia*: globose spored species. *Mycologia*, **57**, 222–35.

Ellis, J.J. and Hesseltine, C.W. 1966. Species of *Absidia* with ovoid sporangiospores II. *Sabouraudia*, **5**, 59–77.

Erdos, M.S., Butt, K. and Weinstein, L. 1972. Mucormycotic endocarditis of the pulmonary valve. *J A M A*, **222**, 951–3.

Eucker, J., Sezer, O., et al. 2001. Mucormycosis. *Mycoses*, **44**, 253–60.

Everett, E.D., Pearson, S. and Rogers, W. 1979. *Rhizopus* surgical wound infection associated with elasticized adhesive tape dressings. *Arch Surg*, **114**, 738–9.

Ferguson, B.J., Mitchell, T.G., et al. 1988. Adjunctive hyperbaric oxygen for treatment of rhinocerebral mucormycosis. *Rev Infect Dis*, **10**, 551–9.

Ferry, A.P. and Abedi, S. 1983. Diagnosis and management of rhino-orbitocerebral mucormycosis (phycomycosis). *Ophthalmology*, **90**, 1096–104.

Fetchick, R.J., Rinaldi, M.G., Sun, S.H. 1986. Zygomycosis due to *Mucor circinelloides*, a rare agent of human fungal disease: clinical and mycological aspects. *Bact Proc*, 1986, F-42.

Fingeroth, J.D., Roth, R.S., et al. 1994. Zygomycosis due to *Mucor circinelloides* in a neutropenic patient receiving chemotherapy for acute myelogenous leukemia. *Clin Infect Dis*, **19**, 135–7.

Fisher, J., Tuazon, C.U. and Geelhoed, G.W. 1980. Mucormycosis in transplant patients. *Am Surg*, **46**, 315–22.

Fong, K.M., Seneviratne, E.M. and McCormack, J.G. 1990. Mucor cerebral abscess associated with intravenous drug abuse. *Aust N Z J Med*, **20**, 74–7.

Fortun, J., Cobo, J., et al. 1995. Post traumatic cranial mucormycosis in an immunocompetent patient. *J Oral Maxillofac Surg*, **53**, 1099–102.

Frank, W., Roester, U. and Scholer, H.J. 1974. Sphaerulen-Bilddung bie einer *Mucor*-spezies in inneren Organen von Amphibien. *Zentralbl Bakt Parasitenkd Abt 1*, **226**, 405–17.

Frye, C.B. and Reinhardt, J. 1993. Characterization of groups of the zygomycete genus *Rhizopus*. *Mycopathologia*, **124**, 139–47.

Furbringer, P. 1876. Beobachtungen uber Lungenmycose beim Menschen. *Virchows Arch Pathol Anat Physiol Klin Med*, **66**, 330–6.

Garcia-Covarrubias, L., Bartlett, R., et al. 2001. Rhino-orbitocerebral mucormycosis attributable to *Apophysomyces elegans* in an immunocompetent individual: case report and review of the literature. *J Trauma*, **50**, 353–7.

Garcia-Diaz, J.B., Palau, L. and Pankey, G.A. 2001. Resolution of rhinocerebral zygomycosis associated with adjuvant administration of granulocyte-macrophage colony-stimulating factor. *Clin Infect Dis*, **32**, 145–50.

Garey, K.W., Pendland, S.L., et al. 2001. *Cunninghamella bertholletiae* infection in a bone marrow transplant patient: amphotericin lung penetration, MIC determinations, and review of the literature. *Pharmacotherapy*, **21**, 855–60.

Gartenberg, G., Bottone, E.J., et al. 1978. Hospital acquired mucormycosis (*Rhizopus rhizopodiformis*) of skin and subcutaneous tissue. *N Engl J Med*, **299**, 1115–18.

Goldschmied-Reouven, A., Shvoron, A., et al. 1989. *Saksenaea vasiformis* infection in a burn wound. *J Med Vet Mycol*, **27**, 427–9.

Gonzalez, C.E., Couriel, D.R. and Walsh, T.J. 1997. Disseminated zygomycosis in a neutropenic patient: successful treatment with amphotericin B lipid complex and granulocyte colony-stimulating factor. *Clin Infect Dis*, **24**, 192–6.

Gonzalez, C.E., Rinaldi, M.G. and Sugar, A.M. 2002. Zygomycosis. *Infect Dis Clin North Am*, **16**, 895–914.

Goodman, N.L. and Rinaldi, M.G. 1991. Agents of zygomycosis. In: Balows, A., Hausler, W.J., et al. (eds), *Manual of clinical microbiology*, 5th edition. Washington, DC: American Society for Microbiology, 674–92.

Gordon, G., Indeck, M., et al. 1988. Injury from silage wagon accident complicated by mucormycosis. *J Trauma*, **28**, 866–7.

Grauer, M.E., Bokemeyer, C., et al. 1993. Successful treatment of *Mucor* pneumonia in a patient with relapsed lymphoblastic leukemia after bone marrow transplantation. *Bone Marrow Transplant*, **12**, 421.

Greenberg, R.N., Anstead, G. et al. 2003. Posaconazole experience in the treatment of zygomycosis. *43rd ICAAC Abstracts, American Society for Microbiology, September 2003*, 476.

Grocott, R.G. 1955. A stain for fungi in tissue sections and smears using Gomori's methenamine-silver nitrate technique. *Am J Clin Pathol*, **25**, 975–9.

Guillot, J., Collobert, C., et al. 2000. Two cases of equine mucormycosis caused by *Absidia corymbifera*. *Equine Vet J*, **32**, 453–6.

Hammer, G.S., Bottone, E.J. and Hirschman, S.Z. 1975. Mucormycosis in a transplant recipient. *Am J Clin Pathol*, **64**, 389–98.

Hagensee, M.E., Bauwens, J.E., et al. 1994. Brain abscess following marrow transplantation: experience at the Fred Hutchinson cancer research center, 1984–1992. *Clin Infect Dis*, **19**, 402–8.

Handzel, O., Landau, Z. and Halperin, D. 2003. Liposomal amphotericin B treatment for rhinocerebral mucormycosis: how much is enough? *Rhinology*, **41**, 184–6.

Hay, R.J., Campbell, C.K., et al. 1983. Disseminated zygomycosis (mucormycosis) caused by *Saksenaea vasiformis*. *J Infect*, **7**, 162–5.

Hesseltine, C.W. and Ellis, J.J. 1964. An interesting case of *Mucor*, *M. ramosissimus*. *Sabouraudia*, **3**, 151–4.

Hesseltine, C.W. and Ellis, J.J. 1966. Species of *Absidia* with ovoid sporangiospores I. *Mycologia*, **58**, 761–85.

Hesseltine, C.W. and Ellis, J.J. 1973. Mucorales. In: Ainsworth, G.G., Sparrow, F.K. and Sussman, A.S. (eds), *The fungi IVB*. New York: Academic Press, 187–217.

Hicks, W.L. Jr, Nowels, K. and Troxel, J. 1995. Primary cutaneous mucormycosis. *Am J Otolaryngol*, **16**, 265–8.

Hill, B.D., Black, P.F., et al. 1992. Bovine cranial zygomycosis caused by *Saksenaea vasiformis*. *Aust Vet J*, **69**, 173–4.

Hodges, C.S. 1962. Fungi isolated from southern forest tree nursery soils. *Mycologia*, **54**, 221–9.

Hoffman, R.M. 1987. Chronic endobronchial mucormycosis. *Chest*, **91**, 469.

Hofman, P., Gari-Toussaint, M., et al. 1993. Rhino-orbito-cerebral mucormycosis caused by *Rhizopus oryzae*. A typical case in a cirrhotic patient. *Ann Pathol*, **13**, 180–3.

Holland, J. 1997. Emerging zygomycosis of humans: *Saksenaea vasiformis* and *Apophysomyces elegans*. *Curr Top Med Mycol*, **8**, 27–34.

Holzel, H., Macqueen, S., et al. 1998. *Rhizopus microsporus* in wooden tongue depressors: a major threat or minor inconvenience? *J Hosp Infect*, **38**, 113–18.

Hopkins, M.A. and Treloar, D.M. 1997. Mucormycosis in diabetes. *Am J Crit Care*, **6**, 363–7.

Hopkins, R.J., Rothman, M., et al. 1994. Cerebral mucormycosis associated with intravenous drug use: three case reports and review. *Clin Infect Dis*, **19**, 1133–7.

Hopwood, V., Hicks, D.A., et al. 1992. Primary cutaneous zygomycosis due to *Absidia corymbifera* in a patient with AIDS. *J Med Vet Mycol*, **30**, 399–402.

Huddle, K.R.L., Hale, M.J., et al. 1987. Rhinocerebral mucormycosis in diabetic keto-acidosis. *South Afr Med J*, **72**, 713–14.

Huffnagle, K.E., Southern, P.M. Jr, et al. 1992. *Apophysomyces elegans* as an agent of zygomycosis in a patient following trauma. *J Med Vet Mycol*, **30**, 83–6.

Hughes, C., Driver, S.J. and Alexander, K.A. 1995. Successful treatment of abdominal wall *Rhizopus* necrotizing cellulitis in a preterm infant. *Pediatr Infect Dis J*, **14**, 336.

Hyatt, D.S., Young, Y.M., et al. 1992. Rhinocerebral mucormycosis following bone marrow transplantation. *J Infect*, **24**, 67–71.

Ingram, C.W., Sennesh, J., et al. 1989. Disseminated zygomycosis: report of four cases and review. *Rev Infect Dis*, **11**, 741–54.

Inui, T., Takeda, Y. and Iiuka, H. 1965. Taxonomical studies on genus *Rhizopus*. *J Gen Appl Microbiol*, **11**, Suppl, 1–21.

Iqbal, S.M. and Scheer, R.I. 1986. Myocardial mucormycosis with emboli in a hemodialysis patient. *Am J Kidney Dis*, **8**, 455–8.

Isaacson, C. and Levin, S.E. 1961. Gastrointestinal mucormycosis in infancy. *S Afr J Med*, **35**, 581–4.

Ismail, M.H., Hodkinson, H.J., et al. 1990. Gastric mucormycosis. *Trop Gastroenterol*, **11**, 103–5.

Iwatsu, T., Udagawa, S., et al. 1990. *Cunninghamelia bertholletiae* recovered from human disseminated zygomycosis in Japan. *Trans Mycol Soc Jpn*, **31**, 259–70.

Jantunen, E., Kolho, E., et al. 1996. Case report. Invasive cutaneous mucormycosis caused by *Absidia corymbifera* after allogenic bone marrow transplantation. *Bone Marrow Transplant*, **18**, 229–30.

Jefferies, P. 1985. Mycoparasitism. *Bot J Linn Soc*, **91**, 135–50.

Jensen, H.E. 1992. Murine subcutaneous granulomatous zygomycosis induced by *Absidia corymbifera*. *Mycoses*, **35**, 261–8.

Jensen, H.E., Jorgensen, J.B. and Schonheyder, H. 1989. Pulmonary mycosis in farmed deer: allergic zygomycosis and invasive aspergillosis. *J Med Vet Mycol*, **27**, 329–34.

Jensen, H.E., Schonheyder, H. and Jorgensen, J.B. 1990. Intestinal and pulmonary mycotic lymphadenitis in cattle. *J Comp Pathol*, **102**, 345–55.

Johnson, A.S., Ranson, M., et al. 1993. Cutaneous infection with *Rhizopus oryzae* and *Aspergillus niger* following bone marrow transplantation. *J Hosp Infect*, **25**, 293–6.

Johnson, C.T., Lupson, G.R. and Lawrence, K.E. 1994. The bovine placentome in bacterial and mycotic abortions. *Vet Rec*, **134**, 263–6.

Johnson, P.C., Satterwhite, T.K., et al. 1987. Primary cutaneous mucormycosis in trauma patients. *J Trauma*, **27**, 437–41.

Jones, K.D. and Kaufman, L. 1978. Development and evaluation of an immunodiffusion test for diagnosis of systemic zygomycosis (mucormycosis): preliminary report. *J Clin Microbiol*, **7**, 97–103.

Kajs-Wyllie, M. 1995. Hyperbaric oxygen therapy for rhinocerebral fungal infection. *J Neurosci Nurs*, **27**, 174–81.

Kalayjian, R.C., Herzig, R.H., et al. 1988. Thrombosis of the aorta caused by mucormycosis. *South Med J*, **81**, 1180–2.

Kamalam, A. and Thambiah, A.S. 1980. Cutaneous infection by *Syncephalastrum*. *Sabouraudia*, **18**, 19–20.

Kasantikul, V., Shuangshoti, S. and Taecholarn, C. 1987. Primary phycomycosis of the brain in heroin addicts. *Surg Neurol*, **28**, 468–72.

Kaufman, L., Padhye, A.A. and Parker, S. 1988. Rhinocerebral zygomycosis caused by *Saksenaea vasiformis*. *J Med Vet Mycol*, **26**, 237–41.

Kaufman, L., Turner, L.F. and McLaughlin, D.W. 1989. Indirect enzyme-linked immunosorbent assay for zygomycosis. *J Clin Microbiol*, **27**, 1979–82.

Kaufman, L., Mendoza, L. and Standard, P.G. 1990. Immunodiffusion test for serodiagnosing subcutaneous zygomycosis. *J Clin Microbiol*, **28**, 1887–90.

Kemna, M.E., Neri, R.C., et al. 1994. *Cokeromyces recurvatus*, a mucoraceous zygomycete rarely isolated in clinical laboratories. *J Clin Microbiol*, **32**, 843–5.

Kerr, P.G., Turner, H., et al. 1988. Zygomycosis requiring amputation of the hand: an isolated case in a patient receiving haemodialysis. *Med J Aust*, **148**, 258–9.

Kiehn, T.E., Edwards, F., et al. 1979. Pneumonia caused by *Cunninghamella bertholletiae* complicating chronic lymphatic leukemia. *J Clin Microbiol*, **10**, 374–9.

Kilpatrick, C.J., Speer, A.G., et al. 1983. Rhinocerebral mucormycosis. *Med J Aust*, **1**, 308–10.

Kimura, M., Udagawa, S., et al. 1995. Isolation of *Rhizopus microsporus* var. *rhizopodiformis* in the ulcer of human gastric carcinoma. *J Med Vet Mycol*, **33**, 137–9.

Kimura, M., Smith, M.B. and McGinnis, M.R. 1999. Zygomycosis due to *Apophysomyces elegans*: report of 2 cases and review of the literature. *Arch Pathol Lab Med*, **123**, 386–90.

Kirk, P.M., Cannon, P.F., et al. 2001. *Ainsworth & Bisby's dictionary of the fungi*, 9th edition. London: CAB International Mycological Institute.

Kline, M.W. 1985. Mucormycosis in children: review of the literature and report of cases. *Pediatr Infect Dis*, **4**, 672–5.

Knoop, C., Antoine, M., et al. 1998. Gastric perforation due to mucormycosis after heart-lung and heart transplantation. *Transplantation*, **15**, 932–5.

Knudtson, W.U. and Kirkbride, C.A. 1992. Fungi associated with bovine abortion in the northern plains states (USA). *J Vet Diagn Invest*, **4**, 181–5.

Kobayashi, M., Hiruma, M., et al. 2001. Cutaneous zygomycosis: a case report and review of the literature. *Mycoses*, **44**, 311–15.

Kofteridis, D.P., Karabekios, S., et al. 2003. Successful treatment of rhinocerebral mucormycosis with liposomal amphotericin B and surgery in two diabetic patients with renal dysfunction. *J Chemother*, **15**, 282–6.

Kolbeck, P.C., Makhoul, R.G., et al. 1985. Widely disseminated *Cunninghamella* mucormycosis in an adult renal transplant recipient. Case report and review of literature. *Am J Clin Pathol*, **83**, 747–53.

Komoda, M., Itoi, Y., et al. 1988. An infection of cow with *Mortierella wolfii*. *Mycopathologia*, **101**, 89–93.

Kontoyianis, D.P., Vartivarian, S., et al. 1994. Infections due to *Cunninghamella bertholletiae* in patients with cancer: report of three cases and review. *Clin Infect Dis*, **18**, 925–8.

Kotzamanoglou, K., Tzanakakis, G., et al. 1988. Orbital cellulitis due to mucormycosis. *Graefe's Arch Clin Exp Ophthalmol*, **266**, 539–41.

Kramer, B.S., Hernandez, A.D., et al. 1977. Cutaneous infarction. Manifestation of disseminated mucormycosis. *Arch Dermatol*, **113**, 1075–6.

Krasinski, K., Holzman, R.S., et al. 1985. Nosocomial fungal infection during hospital renovation. *Infect Control*, **6**, 278–82.

Kwon-Chung, K.J. 1994. Phylogenetic spectrum of fungi that are pathogenic to humans. *Clin Infect Dis*, **19**, Suppl 1, S1–7.

Kwon-Chung, K.J. and Bennett, J.W. 1992. *Medical mycology*. Philadelphia: Lea & Febiger.

Kwon-Chung, K.J., Young, R.C. and Orlando, M. 1975. Pulmonary mucormycosis caused by *Cunninghamella elegans* in a patient with chronic myelogenous leukemia. *Am J Clin Pathol*, **64**, 544–8.

Lake, F.R., McAleer, R. and Tribe, A.E. 1988. Pulmonary mucormycosis without underlying systemic disease. *Med J Aust*, **149**, 323–6.

Lakshmi, V., Sudha Rani, T., et al. 1993. Zygomycotic necrotizing fasciitis caused by *Apophysomyces elegans*. *J Clin Microbiol*, **31**, 1368–9.

Langston, C., Roberts, D.A., et al. 1973. Renal phycomycosis. *J Urol*, **109**, 941–4.

Latif, S., Saffarian, N., et al. 1997. Pulmonary mucormycosis in diabetic renal allograft recipients. *Am J Kidney Dis*, **29**, 461–4.

Lawrence, R.M., Snodgrass, W.T., et al. 1986. Systemic zygomycosis caused by *Apophysomyces elegans*. *J Med Vet Mycol*, **24**, 57–65.

Lehrer, R.I., Howard, D.H., et al. 1980. Mucormycosis. *Ann Intern Med*, **93**, 93–108.

Leleu, X., Sendid, B., et al. 1999. Combined antifungal therapy and surgical resection as treatment of pulmonary zygomycosis in allogeneic bone marrow transplantation. *Bone Marrow Transplant*, **24**, 417–20.

Leong, K.W., Crowley, B., et al. 1997. Cutaneous mucormycosis due to *Absidia corymbifera* occurring after bone marrow transplantation. *Bone Marrow Transplant*, **19**, 513–15.

Levy, E. and Bia, M.J. 1995. Isolated renal mucormycosis: case report and review. *J Am Soc Nephrol*, **5**, 2014–19.

Levy, S.A., Schmitt, K.W. and Kaufman, L. 1986. Systemic zygomycosis diagnosed by fine needle aspiration and confirmed with enzyme immunoassay. *Chest*, **90**, 146–8.

Liao, W.Q., Yao, Z.R., et al. 1995. Pyoderma gangraenosum caused by *Rhizopus arrhizus*. *Mycoses*, **38**, 75–7.

Lim, K.K., Potts, M.J., et al. 1994. Another case report of rhinocerebral mucormycosis treated with liposomal amphotericin B and surgery. *Clin Infect Dis*, **18**, 653–4.

Linder, N., Keller, N., et al. 1998. Primary cutaneous mucormycosis in a premature infant: case report and review of the literature. *Am J Perinatol*, **15**, 35–8.

Lombardi, G., Padhye, A.A., et al. 1989. Exoantigen tests for the rapid and specific identification of *Apophysomyces elegans* and *Saksenaea vasiformis*. *J Med Vet Mycol*, **27**, 113–20.

Lopes, J.O., Pereira, D.V., et al. 1995. Cutaneous zygomycosis caused by *Absidia corymbifera* in a leukemic patient. *Mycopathologia*, **130**, 89–92.

Low, A.I., Tulloch, A.G.S. and England, E.J. 1974. Phycomycosis of the kidney associated with a transient immune defect and treated with clotrimazole. *J Urol*, **111**, 732–4.

Lundborg, M. and Holma, B. 1972. *In vitro* phagocytosis of fungal spores by rabbit lung macrophages. *Sabouraudia*, **10**, 152–6.

Lunn, J.A. and Shipton, W.A. 1983. Re-evaluation of taxonomic criteria in *Cunninghamella*. *Trans Br Mycol Soc*, **81**, 303–12.

Lye, G.R., Wood, G. and Nimmo, G. 1996. Subcutaneous zygomycosis due to *Saksenaea vasiformis*: rapid isolate identification using a modified sporulation technique. *Pathology*, **28**, 364–5.

Lyon, D.T., Schubert, T.T., et al. 1979. Phycomycosis of the gastrointestinal tract. *Am J Gastroenterol*, **72**, 379–94.

Ma, S., Seymour, J.F., et al. 2001. Cure of pulmonary *Rhizomucor pusillus* infection in a patient with hairy-cell leukemia: role of liposomal amphotericin B and GM-CSF. *Leuk Lymphoma*, **42**, 1393–1399.

MacDonald, S.M. and Corbel, M.J. 1981. *Mortierella wolfii* infection in cattle in Britain. *Vet Rec*, **109**, 419–21.

Mackenzie, D.W.R., Soothill, J.F. and Millar, J.H.D. 1988. Meningitis caused by *Absidia corymbifera*. *J Infect*, **17**, 241–8.

MacKenzie, K.M., Baumgarten, K.L., et al. 2002. Innovative medical management with resection for successful treatment of pulmonary mucormycosis despite diagnostic delay. *J LA State Med Soc*, **154**, 82–5.

Maloisel, F., Dufour, P., et al. 1991. *Cunninghamella bertholletiae*: an uncommon agent of opportunistic fungal infection. Case report and review. *Nouv Rev Fr Hematol*, **33**, 311–15.

Manso, E., Montillo, M., et al. 1994. Rhinocerebral mucormycosis caused by *Absidia corymbifera*: an unusual localization in a neutropenic patient. *J Mycol Med*, **4**, 104–7.

Marjankova, K., Krivanek, K. and Zajicek, J. 1978. Mass occurrence of necrotic inflammation of the penis in ganders caused by phycomycetes. *Mycopathologia*, **66**, 21–6.

Marshall, D.H., Brownstein, S., et al. 1997. Post-traumatic corneal mucormycosis caused by *Absidia corymbifera*. *Ophthalmology*, **104**, 1107–11.

Mata-Essayag, S., Magaldi, S., et al. 2001. *Mucor indicus* necrotizing fasciitis. *Int J Dermatol*, **40**, 406–8.

Mathews, M.S., Mukundan, U., et al. 1993. Subcutaneous zygomycosis caused by *Saksenaea vasiformis* in India. A case report and review of the literature. *J Mycol Med*, **3**, 95–8.

Mazade, M.A., Margolin, J.F., et al. 1998. Survival from pulmonary infection with *Cunninghamella bertholletiae*: case report and review of the literature. *Pediatr Infect Dis J*, **17**, 835–9.

Mazza, D., Gugenheim, J., et al. 1999. Gastrointestinal mucormycosis and liver transplantation; a case report and review of the literature. *Transpl Int*, **12**, 297–8.

McGinnis, M.R. 1980. *Laboratory handbook of medical mycology.* London, UK: Academic Press.

McGinnis, M.R., Midez, J., et al. 1993. Necrotizing fasciitis caused by *Apophysomyces elegans*. *J Mycol Med*, **3**, 175–9.

McGough, D.A., Fothergill, A.W. and Rinaldi, M.G. 1990. *Cokeromyces recurvatus* Poitras, a distinctive zygomycete and potential pathogen: criteria for identification. *Clin Microbiol Newsl*, **12**, 113–17.

Mead, J.H., Lupton, G.P., et al. 1979. Cutaneous *Rhizopus* infection. Occurrence as a postoperative complication associated with elasticized adhesive dressing. *J A M A*, **242**, 272–4.

Mehrotra, B.S. and Baijal, U. 1963. Species of *Mortierella* from India – III. *Mycopath Mycol Appl*, **20**, 50–4.

Meis, J.F., Kullberg, B.J., et al. 1994. Severe osteomyelitis due to the zygomycete *Apophysomyces elegans*. *J Clin Microbiol*, **32**, 3078–81.

Melnick, J.Z., Latimer, J., et al. 1995. Systemic mucormycosis complicating acute renal failure: case report and a review of the literature. *Renal Fail*, **17**, 619–27.

Meyer, R.D. and Armstrong, D. 1973. Mucormycosis changing status. *CRC Crit Rev Clin Lab Sci*, **4**, 421–51.

Meyer, R.D., Rosen, P. and Armstrong, D. 1972. Phycomycosis complicating leukemia and lymphoma. *Ann Intern Med*, **77**, 871–9.

Meyer, R.D., Kaplan, M.H., et al. 1973. Cutaneous lesions in disseminated mucormycosis. *J A M A*, **225**, 737–8.

Meyers, B.R., Wormser, G., et al. 1979. Rhinocerebral mucormycosis: premortem diagnosis and therapy. *Arch Intern Med*, **139**, 557–60.

Michalik, D.M., Cooney, D.R., et al. 1980. Gastrointestinal mucormycosis in infants and children: a cause of gangrenous intestinal cellulitis and perforation. *J Pediatr Surg*, **15**, 320–4.

Mileshkin, L., Slavin, M., et al. 2001. Successful treatment of rhinocerebral zygomycosis using liposomal nystatin. *Leuk Lymphoma*, **42**, 1119–23.

Misra, P.C., Srivastava, K.J. and Latas, K. 1979. *Apophysomyces*, a new genus of the Mucorales. *Mycotaxon*, **8**, 377–82.

Mitchell, S.J., Gray, J. and Morgan, M.E. 1996. Nosocomial infection with *Rhizopus microsporus* in preterm infants: association with wooden tongue depressors. *Lancet*, **348**, 441–3.

Mooney, J.E. and Wagner, A. 1993. Mucormycosis of the gastrointestinal tract in children: report of a case and review of the literature. *Pediatr Infect Dis J*, **12**, 872–6.

Morduchowicz, G., Shmueli, D., et al. 1986. Rhinocerebral mucormycosis in renal transplant recipients: report of three cases and review of the literature. *Rev Infect Dis*, **8**, 441–6.

Morquer, R., Lombard, C., et al. 1965. Pouvoir pathogene des Mucorales dans le regne animal. Une nouvelle mycose chez les bovides et les porcins. *CR Acad Sci Paris*, **260**, 6173–6.

Mostaza, J.M., Barrado, F.J., et al. 1989. Cutaneorticular mucormycosis due to *Cunninghamella bertholletiae* in a patient with AIDS. *Rev Infect Dis*, **11**, 316–18.

Muhm, M., Zuckerman, A. and Prokesch, R. 1996. Early onset of pulmonary mucormycosis with pulmonary vein thrombosis in a heart transplant recipient. *Transplantation*, **62**, 1185–7.

Munckhof, W., Jones, R., et al. 1993. Cure of *Rhizopus* sinusitis in a liver transplant recipient with liposomal amphotericin B. *Clin Infect Dis*, **16**, 183.

Munipalli, B., Rinaldi, M.G. and Greenberg, S.B. 1996. *Cokeromyces recurvatus* isolated from pleural and peritoneal fluid: case report. *J Clin Microbiol*, **34**, 2601–3.

Murphy, J.D. and Bornstein, S. 1950. Mucormycosis of the lung. *Ann Intern Med*, **33**, 442–53.

Myskowski, P.L., Brown, A.E. and Dinsmore, R. 1983. Mucormycosis following bone marrow transplantation. *J Am Acad Dermatol*, **9**, 111–15.

Naguib, M.T., Huycke, M.M., et al. 1995. *Apophysomyces elegans* infection in a renal transplant recipient. *Am J Kidney Dis*, **26**, 381–4.

Nagy-Agren, S.E., Chu, P., et al. 1995. Zygomycosis (mucormycosis) and HIV infection: report of three cases and review. *J Acquir Immune Defic Syndr Hum Retrovirol*, **1**, 441–9.

Narain, S., Mitra, M., et al. 2001. Post-traumatic fungal keratitis caused by *Absidia corymbifera*, with successful medical treatment. *Eye*, **15**, 352–3.

Neame, P. and Rayner, D. 1960. Mucormycosis. A report of twenty-two cases. *Arch Pathol Chicago*, **70**, 261–8.

Neilan, M.C., McCausland, I.P. and Maslen, M. 1982. Mycotic pneumonia, placentitis and neonatal encephalitis in dairy cattle caused by *Mortierella wolfii*. *Aust Vet J*, **59**, 48–9.

Newton, W.D., Cramer, F.S. and Norwood, S.H. 1987. Necrotizing fasciitis from invasive phycomycetes. *Crit Care Med*, **15**, 331–2.

Ng, T.T. and Denning, D.W. 1995. Liposomal amphotericin B (AmBisome) therapy in invasive fungal infections. Evaluation of United Kingdom compassionate use data. *Arch Intern Med*, **155**, 1093–8.

Ng, T.T., Campbell, C.K., et al. 1994. Successful treatment of sinusitis caused by *Cunninghamella bertholletiae*. *Clin Infect Dis*, **19**, 313–16.

Nicod, J.L., Fleury, C. and Schlegel, J. 1952. Mycose pulmonaire double a *Aspergillus fumigatus* Fres. et a *Mucor pusillus* Lindt. *Schweiz Z Allg Pathol Bakteriol*, **15**, 307–21.

Nimmo, G.R., Whiting, R.F. and Strong, R.W. 1988. Disseminated mucormycosis due to *Cunninghamella bertholletiae* in a liver transplant recipient. *Postgrad Med J*, **64**, 82–4.

Nissen, M.D., Jana, A.K., et al. 1999. Neonatal gastrointestinal mucormycosis mimicking necrotizing enterocolitis. *Acta Paediatr*, **88**, 1290–3.

Norden, G., Bjorck, S., et al. 1991. Cure of zygomycosis caused by a lipase-producing *Rhizopus rhizopodiformis* strain in a renal transplant patient. *Scand J Infect Dis*, **23**, 377–82.

Nottebrock, H., Scholer, H.J. and Wall, M. 1974. Taxonomy and identification of mucormycosis causing fungi. 1. Synonymity of *Absidia ramosa* with *A. corymbifera*. *Sabouraudia*, **12**, 64–74.

Nussbaum, E.S. and Hall, W.A. 1994. Rhinocerebral mucormycosis: changing patterns of disease. *Surg Neurol*, **41**, 152–6.

Oakley, L.A., Fisher, J.F. and Dennison, J.H. 1986. Bread mold infection in diabetes. *Postgrad Med*, **80**, 93–102.

Obendorf, D.L., Peel, B.F. and Munday, B.L.O. 1993. *Mucor amphibiorum* infection in platypus (*Ornithorhynchus anatinus*) from Tasmania. *J Wildlife Dis*, **29**, 485–7.

Oberle, A.C. and Penn, R.L. 1983. Nosocomial invasive *Saksenaea vasiformis* infection. *Am J Clin Pathol*, **80**, 885–8.

O'Brien, T.J. and McKelvie, P. 1994. Rhinocerebral mucormycosis presenting as periorbital cellulitis with blindness: report of 2 cases. *Clin Exp Neurol*, **31**, 68–78.

O'Donnell, K.L. 1979. Zygomycetes in culture. *Palfrey contributions in botany 2*. Athens, USA: University of Georgia.

Oh, D. and Notrica, D. 2002. Primary cutaneous mucormycosis in infants and neonate: case report and review of the literature. *J Pediatr Surg*, **37**, 1607–711.

Okhuysen, P.C., Rex, J.H., et al. 1994. Successful treatment of extensive post-traumatic soft tissue and renal infections due to *Apophysomyces elegans*. *Clin Infect Dis*, **19**, 329–31.

Oliver, M.R., van Voohis, W.C., et al. 1996. Hepatic mucormycosis in a bone marrow transplant recipient who ingested naturopathic medicine. *Clin Infect Dis*, **22**, 521–4.

Oliveri, S., Cammarata, E., et al. 1988. *Rhizopus arrhizus* in Italy as the causative agent of primary cerebral zygomycosis in a drug addict. *Eur J Epidemiol*, **4**, 284–8.

Onions, A.H.S., Allsopp, D. and Eggins, H.C.W. 1981. *Smith's introduction to industrial mycology*. London: Edward Arnold.

Padhye, A.A. and Ajello, L. 1988. Simple method of inducing sporulation by *Apophysomyces elegans* and *Saksenaea vasiformis*. *J Clin Microbiol*, **26**, 1861–3.

Padhye, A.A., Koshi, G., et al. 1988. First case of subcutaneous zygomycosis caused by *Saksenaea vasiformis* in India. *Diagn Microbiol Infect Dis*, **9**, 69–77.

Page, R., Gardam, D.J. and Heath, C.H. 2001. Severe cutaneous mucormycosis (zygomycosis) due to *Apophysomyces elegans*. *Aust N Z J Surg*, **71**, 184–6.

Paparello, S.F., Parry, R.L., et al. 1992. Hospital-acquired wound mucormycosis. *Clin Infect Dis*, **14**, 350–2.

Parfrey, N. 1986. Improved diagnosis and prognosis of mucormycosis. A clinico-pathologic study of 33 cases. *Medicine*, **65**, 113–23.

Parker, C., Kaminski, G. and Hill, D. 1986. Zygomycosis in a tattoo, caused by *Saksenaea vasiformis*. *Aust J Derm*, **27**, 107–11.

Paterson, P.J., Marshall, S.R., et al. 2000. Fatal invasive cerebral *Absidia corymbifera* infection following bone marrow transplantation. *Bone Marrow Transplant*, **26**, 701–3.

Patino, J.F., Mora, R., et al. 1984. Mucormycosis: a fatal case by *Saksenaea vasiformis*. *World J Surg*, **8**, 419–22.

Patterson, J.E., Barden, G.E. and Bia, F.J. 1986. Hospital-acquired gangrenous mucormycosis. *Yale J Biol Med*, **59**, 453–9.

Perelman, B. and Kuttin, E. 1992. Zygomycosis in ostriches. *Avian Pathol*, **21**, 675–80.

Pierce, P.F., Solomon, S.L., et al. 1982. Zygomycetes brain abscesses in narcotic addicts with serological diagnosis. *J A M A*, **248**, 2881–92.

Pierce, P.F., Wood, M.B., et al. 1987. *Saksenaea vasiformis* osteomyelitis. *J Clin Microbiol*, **25**, 933–5.

Piliero, P.J. and Deresiewicz, R.L. 1995. Pulmonary zygomycosis after allogeneic bone marrow transplantation. *South Med J*, **88**, 1149–52.

Pillsbury, H.C. and Fischer, N.D. 1977. Rhinocerebral mucormycosis. *Arch Otolaryngology*, **103**, 600–4.

Polonelli, L., Dettori, G., et al. 1988. Antigenic studies on *Rhizopus microsporus*: *Rh. rhizopodiformis*, progeny and intermediates (*Rh. chinensis*). *Antonie Leeuwenhoek*, **54**, 5–17.

Porges, N., Muller, J.F. and Lockwood, L.B. 1935. A *Mucor* found in fowl. *Mycologia*, **27**, 330–1.

Prevoo, R.L.M.A., Starink, T.M. and de Haan, P. 1991. Primary cutaneous mucormycosis in a healthy young girl. Report of a case caused by *Mucor hiemalis* Wehmer. *J Am Acad Dermatol*, **24**, 882–5.

Pritchard, R.C., Muir, D.B., et al. 1986. Subcutaneous zygomycosis due to *Saksenaea vasiformis* in an infant. *Med J Aust*, **145**, 630–1.

Prokopowicz, G.P., Bradley, S.F. and Kauffman, C.A. 1994. Indolent zygomycosis associated with deferoxamine chelation therapy. *Mycoses*, **37**, 427–31.

Radner, A.B., Witt, M.D. and Edwards, J.E. 1995. Acute invasive rhinocerebral zygomycosis in an otherwise healthy patient: case report and review. *Clin Infect Dis*, **20**, 163–6.

Rangel-Guerra, R., Martinez, H.R. and Saenz, C. 1985. Mucormycosis, report of 11 cases. *Arch Neurol*, **42**, 578–81.

Reed, A.E., Body, B.A., et al. 1988. *Cunninghamella bertholletiae* and *Pneumocystis carinii* pneumonia as a fatal complication of chronic lymphocytic leukemia. *Hum Pathol*, **19**, 1470–2.

Reich, H. Behr, W. and Barnert, J. 1985. Rhinocerebral mucormycosis in a diabetic ketoacidotic patient. *J Neurol*, **232**, 115–17.

Reimund, E. and Ramos, A. 1994. Disseminated neonatal gastrointestinal mucormycosis: a case report and review of the literature. *Pediatr Pathol*, **14**, 385–9.

Reinhardt, D.J., Licata, I., et al. 1981. Experimental cerebral zygomycosis in alloxan-diabetic rabbits: variation in virulence among zygomycetes. *Sabouraudia*, **19**, 245–55.

Rex, J.H., Ginsberg, A.M., et al. 1988. *Cunninghamella bertholletiae* infection associated with deferoxamine therapy. *Rev Infect Dis*, **10**, 1187–94.

Ribes, J.A, Vanover-Sams, C.L. and Baker, D.J. 2000. Zygomycetes in human disease. *Clin Microbiol Rev*, **13**, 236–301.

Richardson, M.D. and Warnock, D.W. 1993. *Fungal Infection: diagnosis and management*. London: Blackwell Scientific.

Rickerts, V., Bohme, A., et al. 2000. Cluster of pulmonary infections caused by *Cunninghamella bertholletiae* in immunocompromised patients. *Clin Infect Dis*, **31**, 910–13.

Rinaldi, M.G. 1989. Zygomycosis. *Infect Dis Clin North Am*, **3**, 19–41.

Rippon, J.W. 1988. *Medical mycology. the pathogenic fungi and pathogenic actinomycetes*, 3rd edition. Philadelphia: W.B. Saunders.

Rippon, J.W. and Dolan, C.T. 1979. Colonization of the vagina by fungi of the genus *Mucor*. *Clin Microbiol Newsl*, **11**, 4–5.

Robeck, T.R. and Dalton, L.M. 2002. *Saksenaea vasiformis* and *Apophysomyces elegans* zygomycotic infections in bottlenose dolphins (*Tursiops truncates*), a killer whale (*Orcinus orca*), and pacific white-sided dolphins (*Lagenorhynchus obliquidens*). *J Zoo Wildl Med*, **33**, 356–66.

Robinson, B.E., Stark, M.T., et al. 1990. *Cunninghamella bertholletiae*: an unusual agent of zygomycosis. *South Med J*, **83**, 1088–91.

Roger, H., Biat, I., et al. 1989. *Absidia corymbifera* cutaneous zygomycosis (mucormycosis), gangrenosum-like ecthyma in a non-immunosuppressed non-diabetic patient. Treatment with ketoconazole. *Ann Dermatol Venerol*, **116**, 844–6.

Romano, C., Miracco, C., et al. 2002. Case report: fatal rhinocerebral zygomycosis due to *Rhizopus oryzae*. *Mycoses*, **45**, 45–9.

Rosenberger, R.S., West, B.C. and King, J.W. 1983. Case report – survival from sino-orbital mucormycosis due to *Rhizopus rhizopodoformis*. *Am J Med Sci*, **286**, 25–30.

Rothstein, R.D. and Simon, G.L. 1986. Subacute pulmonary mucormycosis. *J Med Vet Mycol*, **24**, 391–4.

Rozich, J., Holley, H.P., et al. 1988. Cauda equina syndrome secondary to disseminated zygomycosis. *J A M A*, **260**, 3638–40.

Ryan, M., Yeo, S., et al. 2001. Rhinocerebral zygomycosis in childhood acute lymphoblastic leukaemia. *Eur J Pediatr*, **160**, 235–8.

Ryan-Poirier, K., Eiseman, R.M., et al. 1988. Post-traumatic cutaneous mucormycosis in diabetes mellitus. *Clin Pediatr*, **27**, 609–12.

Sahin, B., Paydas, S., et al. 1996. Role of granulocyte colony-stimulating factor in the treatment of mucormycosis. *Eur J Clin Microbiol Infect Dis*, **15**, 866–9.

Saksena, S.B. 1953. A new genus of Mucorales. *Mycologia*, **45**, 426–36.

Salfelder, K. 1990. *Atlas of fungal pathology*. Dordrecht: Kluwer Academic.

Saltoglu, N., Tasova, Y., et al. 1998. Rhinocerebral zygomycosis treated with liposomal amphotericin B and surgery. *Mycoses*, **41**, 45–9.

Samson, R.A. 1969. Revision of the genus *Cunninghamella* (Fungi Mucorales). *Proc K Ned Akad Wet Ser C*, **72**, 332–5.

Samson, R.A., Hoekstra, E.S., et al. 1995. *Introduction to food-borne fungi*. Baarn, The Netherlands: Centraalbureau voor Schimmelcultures.

Sanchez, M.R., Ponge-Wilson, I., et al. 1994. Zygomycosis in HIV infection. *J Am Acad Dermatol*, **30**, 904–8.

Sands, J.M., Macher, A.M., et al. 1985. Disseminated infection caused by *Cunninghamella bertholletiae* in a patient with beta-thalassemia. Case report and review of the literature. *Ann Intern Med*, **102**, 59–63.

Sane, A., Manzi, S., et al. 1989. Deferoxamine treatment as a risk factor for zygomycete infection. *J Infect Dis*, **159**, 151–2.

Scalise, A., Barchiesi, F., et al. 1999. Infection due to *Absidia corymbifera* in a patient with a massive crush trauma of the foot. *J Infect*, **38**, 191–2.

Schaffner, A., Davis, C.E., et al. 1986. In vitro susceptibility of fungi to killing by neutrophil granulocytes discriminates between primary pathogenicity and opportunism. *J Clin Invest*, **78**, 511–24.

Schipper, M.A.A. 1973. A study on variability in *Mucor hiemalis* and related species. *Stud Mycol*, **4**, 1–40.

Schipper, M.A.A. 1976. On *Mucor circinelloides*, *Mucor racemosus* and related species. *Stud Mycol*, **12**, 1–40.

Schipper, M.A.A. 1978a. On certain species of *Mucor* with a key to all accepted species. *Stud Mycol*, **17**, 1–52.

Schipper, M.A.A. 1978b. On the genera *Rhizomucor* and *Parasitella*. *Stud Mycol*, **17**, 53–71.

Schipper, M.A.A. 1984. A revision of the genus *Rhizopus* 1. The *Rhizopus stolonifer* group and *Rhizopus oryzae*. *Stud Mycol*, **25**, 1–19.

Schipper, M.A.A. 1990. Notes on Mucorales – I. Observations on *Absidia*. *Persoonia*, **14**, 133–48.

Schipper, M.A.A. and Stalpers, J.A. 1984. A revision of the genus *Rhizopus* II. The *Rhizopus microsporus* group. *Stud Mycol*, **25**, 30–4.

Schipper, M.A.A. and Stalpers, J.A. 2003. Zygomycetes: the Order Mucorales. In: Howard, D.H. (ed.), *Pathogenic fungi in humans and animals*, 2nd edition. New York: Marcel Dekker, 67–125.

Schipper, M.A.A., Maslen, M.M., et al. 1996. Human infection by *Rhizopus azygosporus* and the occurrence of azygospores in Zygomycetes. *J Med Vet Mycol*, **34**, 199–203.

Scholer, H.J., Müller, E. and Schipper, M.A.A. 1983. Mucorales. In: Howard, D.H. (ed.), *Fungi pathogenic for humans and animals, Part A Biology*. New York: Marcel Dekker, 9–59.

Scholz, H.D. and Meyer, L. 1965. *Mortierella polycephela* as a cause of pulmonary mycosis in cattle. *Berl MunchTierarzl Wochenshr*, **78**, 27–30.

Schuster, M.G. and Stern, J. 1995. Zygomycosis orbital apex syndrome in association with a solitary lung carcinoma. *J Med Vet Mycol*, **33**, 73–5.

Schwartz, J.R., Nagle, M.G., et al. 1982. Mucormycosis of the trachea: an unusual cause of acute upper airway obstruction. *Chest*, **81**, 653–4.

Seguin, P., Musellec, H., et al. 1999. Post-traumatic course complicated by cutaneous infection with *Absidia corymbifera*. *Eur J Clin Microbial Infect Dis*, **18**, 737–9.

Severo, L.C., Job, F. and Mattos, T.C. 1991. Systemic zygomycosis: nosocomial infection by *Rhizomucor pusillus*. *Mycopathologia*, **113**, 79–80.

Seviour, R.J., Cooper, A.L. and Skilbeck, N.W. 1987. Identification of *Mortierella wolfii*, a causative agent of mycotic abortion in cattle. *J Med Vet Mycol*, **25**, 115–23.

Shanor, L.S., Poitras, A.W. and Benjamin, R.K. 1950. A new genus in the Choanophoraceae. *Mycologia*, **42**, 271–8.

Sharma, R., Prem, R.R., et al. 1994. Disseminated septic arthritis due to *Mucor ramosissimus* in a premature infant a rare fungal infection and its successful management. *Pediatr Res*, **35**, 303A.

Sheldon, D.L. and Johnson, W.C. 1979. Cutaneous mycomycosis. Two documented cases of suspected nosocomial cause. *J A M A*, **241**, 1032–4.

Shpitzer, T., Stern, Y., et al. 1995. Mucormycosis: experience with 10 patients. *Clin Otolaryngol*, **20**, 374–9.

Sica, S., Morace, G., et al. 1993. Rhinocerebral zygomycosis in acute lymphoblastic leukaemia. *Mycoses*, **36**, 289–91.

Singh, N., Gayowski, T., et al. 1995. Invasive gastrointestinal zygomycosis in a liver transplant recipient: case report and review of zygomycosis in solid organ transplant recipients. *Clin Infect Dis*, **20**, 617–20.

Skahan, K.J., Wong, B. and Armstrong, D. 1991. Clinical manifestations and management of mucormycosis in the compromised patient. In: Warnock, D.W. and Richardson, M.D. (eds), *Fungal infections in the compromised patient*. Chichester: John Wiley & Sons, 153–90.

Slavin, M.A., Kannan, K., et al. 2002. Successful allogeneic stem cell transplant after invasive pulmonary zygomycosis. *Leuk Lymphoma*, **43**, 437–9.

Smith, A.G., Bustamante, C.I. and Gilmor, G.D. 1989. Zygomycosis (absidiomycosis) in AIDS patient. Absidiomycosis in AIDS. *Mycopathologia*, **105**, 7–10.

Smith, J.M.B. 1976. *In vivo* development of spores of *Absidia ramosa*. *Sabouraudia*, **14**, 11–15.

Smith, J.M.B. 1989. *Opportunistic mycoses of man and other animals*. London: CAB International Mycological Institute.

Solano, T., Atkins, B., et al. 2000. Disseminated mucormycosis due to *Saksenaea vasiformis* in an immunocompetent adult. *Clin Infect Dis*, **30**, 942–3.

Song, W.K., Park, H.J., et al. 1999. Primary cutaneous mucormycosis in a trauma patient. *J Dermatol*, **26**, 825–8.

Speare, R., Thomas, A.D., et al. 1994. *Mucor amphibiorum* in the toad, *Bufo marinus*, in Australia. *J Wildl Dis*, **30**, 399–407.

Speare, R., Berger, I., et al. 1997. Pathology of mucormycosis of cane toads in Australia. *J Wildl Dis*, **33**, 105–11.

Stave, G.M., Heimberger, T. and Kerkering, T.M. 1989. Zygomycosis of the basal ganglia in intravenous drug users. *Am J Med*, **86**, 115–17.

Stern, L.E. and Kagan, R.J. 1999. Rhinocerebral mucormycosis in patients with burns: case report and review of the literature. *J Burn Care Rehibil*, **20**, 303–6.

Stewart, N.J., Munday, B.L. and Hawkesford, T. 1999. Isolation of *Mucor circinelloides* from a case of ulcerative mycosis of platypus (*Ornithorhynchus anatinus*), and a comparison of the response of *Mucor circinelloides* and *Mucor amphibiorum* to different culture temperatures. *Med Mycol*, **37**, 201–6.

St-Germain, G., Robert, A., et al. 1993. Infection due to *Rhizomucor pusillus*: report of four cases in patients with leukemia and review. *Clin Infect Dis*, **16**, 640–5.

Strasser, M.D., Kennedy, R.J. and Adam, R.D. 1996. Rhinocerebral mucormycosis. Therapy with amphotericin B lipid complex. *Arch Intern Med*, **156**, 337–9.

Sugar, A.M. 1992. Mucormycosis. *Clin Infect Dis*, **14**, Suppl 1, 126–9.

Sun, Q.N., Fothergill, A.W., et al. 2002a. In vitro activities of posaconazole, itraconazole, voriconazole, amphotericin B, and fluconazole against 37 clinical isolates of zygomycetes. *Antimicrob Agents Chemother*, **46**, 1581–92.

Sun, Q.N., Najvar, L.K., et al. 2002b. In vivo activity of posaconazole against Mucor spp. in an immunosuppressed-mouse model. *Antimicrob Agents Chemother*, **46**, 2310–12.

Tauphaichitr, V.S., Chaiprasert, A., et al. 1990. Subcutaneous mucormycosis caused by *Saksenaea vasiformis* in a thallassaemic child: first case report in Thailand. *Mycoses*, **33**, 303–9.

Terra, R., Trinidad, J.M., et al. 1993. Zygomycosis of the spleen in a patient with the acquired immunodeficiency syndrome. *Mycoses*, **36**, 437–9.

Thami, G.P., Kaur, S., et al. 2003. Post-surgical zygomycotic necrotizing subcutaneous infection caused by *Absidia corymbifera*. *Clin Exp Dermatol*, **28**, 251–3.

Thomson, S.R., Bade, P.G., et al. 1991. Gastrointestinal mucormycosis. *Br J Surg*, **78**, 952–4.

Tinmouth, J., Baker, J. and Gardiner, G. 2001. Gastrointestinal mucormycosis in a renal transplant patient. *Can J Gastroenterol*, **15**, 269–71.

Tintelnot, K. and Nitsche, B. 1989. *Rhizopus oligosporus* as a cause of mucormycosis in man. *Mycoses*, **32**, 115–18.

Tobon, A.M., Arango, M., et al. 2003. Mucormycosis (zygomycosis) in a heart-kidney transplant recipient: recovery after posaconazole therapy. *Clin Infect Dis*, **36**, 1488–91.

Tomford, J.W., Whittlesey, D., et al. 1980. Invasive primary cutaneous phycomycosis in diabetic leg ulcers. *Arch Surg*, **115**, 770–1.

Tonks, A. 1996. Fatal outbreak traced to wooden tongue depressors. *Br Med J*, **312**, 1186.

Torrell, J., Cooper, B.H. and Helgeson, N.G.P. 1981. Disseminated *Saksenaea vasiformis* infection. *Am J Clin Pathol*, **76**, 116–21.

Torres-Rodriguez, J.M., Lowinger, M., et al. 1993. Renal infection due to *Absidia corymbifera* in an AIDS patient. *Mycoses*, **36**, 255–8.

Tsai, T.W., Hammond, M., et al. 1997. *Cokeromyces recurvatus* infection in a bone marrow transplant recipient. *Bone Marrow Transplant*, **19**, 301–3.

Tuder, R.M. 1985. Myocardial infarct in disseminated mucormycosis: case report with special emphasis on the pathogenic mechanisms. *Mycopathologia*, **89**, 81–8.

Turner, P.D. 1964. *Syncephalastrum* associated with bovine mycotic abortion. *Nature*, **204**, 309.

Utas, C., Unluhizarci, K., et al. 1995. Acute renal failure associated with rhinosinus-orbital mucormycosis infection in a patient with diabetic nephropathy. *Nephron*, **71**, 235.

Uzal, F.A., Connole, M.D., et al. 1999. *Mortierella wolfii* isolation from the liver of a cow in Australia. *Vet Rec*, **145**, 260–1.

Vadeboncoeur, C., Walton, J.M., et al. 1994. Gastrointestinal mucormycosis causing an acute abdomen in the immunocompromised pediatric patient – three cases. *J Pediatr Surg*, **29**, 1248–9.

Vainrub, B., Macareno, A. and Mandel, S. 1988. Wound zygomycosis (mucormycosis) in otherwise healthy adults. *Am J Med*, **84**, 546–8.

van den Saffele, J.K. and Boelaert, J.R. 1996. Zygomycosis in HIV positive patients: a review of the literature. *Mycoses*, **39**, 77–84.

Venezio, F.R., Sexton, D.J., et al. 1985. Mucormycosis after open fracture injury. *South Med J*, **78**, 1516–17.

Ventura, G.J., Kantarjian, H.M., et al. 1986. Pneumonia with *Cunninghamella* species in patients with hematological malignancies. *Cancer*, **58**, 1534–6.

Vera, A., Hubscher, S.G., et al. 2002. Invasive gastrointestinal zygomycosis in a liver transplant recipient: case report. *Transplantation*, **15**, 145–7.

Verweij, P.E., Voss, A. and Donnelly, J.P. 1997. Wooden sticks as the source of a pseudoepidemic of infection with *Rhizopus microsporus* var. *rhizopodiformis* among immunocompromised patients. *J Clin Microbiol*, **35**, 2422–3.

Vignale, R., Mackinnon, J.E., et al. 1964. Chronic destructive mucocutaneous phycomycosis in man. *Sabouraudia*, **3**, 143–7.

Vincent, L., Biron, F., et al. 2000. Pulmonary mucormycosis in a diabetic patient. *Ann Med Interne*, **151**, 669–72.

Virman, R., Connor, D.H. and McAllister, H.A. 1982. Cardiac mucormycosis. A report of five patients and review of 14 previously reported cases. *Am J Clin Pathol*, **78**, 42–7.

Voigt, K. and Wostemeyer, J. 2001. Phylogeny and origin of 82 zygomycetes from all 54 genera of the Mucorales and Mortierellales based on combined analysis of actin and translocation elongation factor EF-1α genes. *Gene*, **70**, 113–20.

Voigt, K., Cigelnik, E. and O'Donnell, K. 1999. Phylogeny and PCR identification of clinically important zygomycetes based on nuclear ribosomal-DNA sequence data. *J Clin Microbiol*, **37**, 3957–64.

Waldorf, A.R. and Diamond, R.D. 1984. Cerebral mucormycosis in diabetic mice after intrasinus challenge. *Infect Immun*, **44**, 194–5.

Waldorf, A.R. and Diamond, R.D. 1989. Aspergillosis and mucormycosis. In: Cox, R.A. (ed.), *Immunology of the fungal diseases*. Boca Raton, FL: CRC Press, 29–55.

Waldorf, A.R., Halde, C. and Vedros, N.A. 1982. Murine model of pulmonary mucormycosis in cortisone treated mice. *Sabouraudia*, **20**, 217–24.

Waldorf, A.R., Peter, L. and Polak, A. 1984a. Mucormycotic infection in mice following prolonged incubation of spores in vivo and the role of spore agglutinating antibodies on spore germination. *Sabouraudia*, **22**, 101–8.

Waldorf, A.R., Ruderman, N. and Diamond, R.D. 1984b. Specific susceptibility to mucormycosis in murine diabetes and bronchoalveolar macrophage defence against *Rhizopus*. *J Clin Invest*, **74**, 150–60.

Waller, J., Woehl-Jaegle, M., et al. 1993. Mucormycose abdomiale nosocomiale a *Rhizopus rhizopodiformis* chez un transplante hepatiqu. Revue de la litterature. *J Mycol Med*, **3**, 180–6.

Wang, J.J., Satch, H., et al. 1990. A case of cutaneous mucormycosis in Shanghai, China. *Mycoses*, **33**, 311–15.

Watson, K.C. 1957. Gastric perforation due to the fungus *Mucor* in a child with kwashiorkor. *S Afr J Med*, **31**, 99–101.

Weinberg, W.G., Wade, B.H., et al. 1993. Invasive infection due to *Apophysomyces elegans* in immunocompetent hosts. *Clin Infect Dis*, **17**, 881–4.

Weitzman, I. 1984. The case for *Cunninghamella elegans*, *C. bertholletiae* and *C. echinulata* as separate species. *Trans Br Mycol Soc*, **83**, 527–8.

Weitzman, I. and Crist, M.Y. 1979. Studies with clinical isolates of *Cunninghamella*. I. Mating behaviour. *Mycologia*, **71**, 1024–33.

Weitzman, I. and Crist, M.Y. 1980. Studies with clinical isolates of *Cunninghamella*. II. Physiological and morphological studies. *Mycologia*, **72**, 661–9.

Weitzman, I., Della-Latta, P., et al. 1993. *Mucor ramosissimus* Samutsevitsch isolated from a thigh lesion. *J Clin Microbiol*, **31**, 2523–5.

Weitzman, I., Whittier, S., et al. 1995. Zygospores: the last word in identification of rare or atypical zygomycetes isolated from clinical specimens. *J Clin Microbiol*, **33**, 781–3.

Weitzman, I., McGough, D.A., et al. 1996. *Rhizopus schipperae* sp. nov. a new agent of zygomycosis. *Mycotaxon*, **59**, 217–25.

West, B.C., Kwon-Chung, K.J., et al. 1983. Inguinal abscess caused by *Rhizopus rhizopodiformis*: successful treatment with surgery and amphotericin B. *J Clin Microbiol*, **18**, 1384–7.

West, B.C., Oberle, A.D. and Kwon-Chung, K.J. 1995. Mucormycosis caused by *Rhizopus microsporus* var. *microsporus*: cellulitis in a leg of a diabetic patient cured by amputation. *J Clin Microbiol*, **33**, 3341–4.

White, C.B., Barcia, P.J. and Bass, J.W. 1986. Neonatal zygomycotic necrotizing cellulitis. *Pediatrics*, **78**, 100–2.

Wickline, C.L., Cornitius, T.C. and Butler, T. 1989. Cellulitis caused by *Rhizomucor pusillus* in a diabetic patient receiving continuous insulin infusion pump therapy. *South Med J*, **82**, 1432–4.

Wieden, M.A., Steinbronn, K.K., et al. 1985. Zygomycosis caused by *Apophysomyces elegans*. *J Clin Microbiol*, **22**, 522–6.

Williams, J.C., Schned, A.R., et al. 1995. Fatal genitourinary mucormycosis in a patient with undiagnosed diabetes. *Clin Infect Dis*, **21**, 682–4.

Wilson, C.B., Siber, G.R., et al. 1976. Phycomycotic gangrenous cellulitis: a report of two cases and a review of the literature. *Arch Surg*, **111**, 532–8.

Wilson, M., Robson, J., et al. 1998. *Saksenaea vasiformis* breast abscess related to gardening injury. *Aust N ZJ Med*, **28**, 845–6.

Windus, D.W., Stokes, T.J., et al. 1987. Fatal *Rhizopus* infections in hemodialysis patients receiving deferoxamine. *Ann Intern Med*, **107**, 678–80.

Winkler, S., Susani, S., et al. 1996. Gastric mucormycosis due to *Rhizopus oryzae* in a renal transplant recipient. *J Clin Microbiol*, **86**, 546–9.

Winn, R.E., Ramsey, P.D. and Adams, E.D. Jr 1982. Traumatic mucormycosis secondary to *Apophysomyces elegans*, a new genus of the Mucorales. *Clin Res*, **30**, 382A.

Winston, R.M. 1965. Phycomycosis of the bronchus. *J Clin Pathol*, **18**, 729–31.

Wohlgemuth, K. and Knudtson, W.V. 1977. Abortion associated with *Mortierella wolfii* in cattle. *J Am Vet Med Assoc*, **171**, 437–9.

Woods, K.R. and Hanna, B.J. 1986. Brain stem mucormycosis in a narcotic addict with eventual recovery. *Am J Med*, **80**, 126–8.

Woodward, A., McTigue, C., et al. 1992. Mucormycosis of the neonatal gut: a new disease or a variant of necrotizing entercolitis? *J Pediatr Surg*, **27**, 737–40.

Wysong, D.R. and Waldorf, A.R. 1987. Electrophoretic and immunoblot analysis of *Rhizopus arrhizus* antigens. *J Clin Microbiol*, **25**, 358–863.

Yokoi, S., Iizasa, T., et al. 1999. Case report. Localized pulmonary zygomycosis without pre-existing immunocompromised status. *Mycoses*, **42**, 675–7.

Yuan, G.F. and Jong, S.C. 1984. A new obligate azygosporic species of *Rhizopus*. *Mycotaxon*, **20**, 397–400.

Zagoria, R.J., Choplin, R.H. and Karstaedt, N. 1985. Pulmonary gangrene as a complication of mucormycosis. *A J R Am J Roentgenol*, **144**, 1195–6.

Zeilender, S., Drenning, D., et al. 1990. Fatal *Cunninghamella bertholletiae* infection in an immunocompetent patient. *Chest*, **97**, 1482–1483.

Zhang, R., Zhang, J.W. and Szerlip, H.M. 2002. Endocarditis and hemorrhagic stroke caused by *Cunningham bertholletiae* infection after kidney transplantation. *Am J Kidney Dis*, **40**, 842–6.

Zycha, H., Siepmann, R. and Linnemann, G. 1969. *Mucorales, eine Beschreibung aller Gattungen und Arten dieser Pilzgruppe*. Lehre: Verlag von J. Cramer.

Aspergillosis

MALCOLM D. RICHARDSON

The accumulation of data on many aspects of *Aspergillus* spp. and aspergillosis justifies a text of major proportions. Many monographs, reviews, and journal supplements have more than adequately covered the organisms and the complete disease area (Raper and Thom 1965; Rinaldi 1983; Al-Doory and Wagner 1985; Walsh and Pizzo 1988; Cohen 1991; Barnes and Denning 1993; Brakhage et al. 1999; Latgé 1999). More recent overviews of aspergillosis can be found in Marr et al. (2002), Richardson and Warnock (2003), Wiederhold et al. (2003), and a supplement of *Clinical Infectious Diseases* (Advances against aspergillosis 2003). Arguably the most comprehensive resource is the *Aspergillus* website (www.aspergillus.man.ac.uk). The content of this website is updated regularly. The aim of this chapter is to present the reader with a summary of the organisms and an overview of the major clinical manifestations in order to present a contemporary account of the pathogenesis of the disease, new diagnostic procedures, and new perspectives on treatment.

BRIEF DESCRIPTION OF THE GENUS

Aspergillus is a very large genus containing over 185 species to which humans are constantly exposed. Only a small number of these species have, however, been associated with disease. Of these, over 95 percent of all infections are caused by three species: *Aspergillus fumigatus*, *Aspergillus flavus*, and *Aspergillus niger*. Several more species have been reported in association with aspergillosis cases, including *Aspergillus nidulans*, *Aspergillus terreus*, *Aspergillus oryzae*, *Aspergillus ustus*, and *Aspergillus versicolor*.

Of the documented species of *Aspergillus*, *A. fumigatus* causes the large majority of cases of both invasive and non-invasive aspergillosis. Indeed, the allergic forms of the disease appear to be almost exclusively caused by this organism. Both aspergilloma and invasive aspergillosis are also caused by *A. flavus* and *A. niger*. More recently, cases of invasive aspergillosis caused by *A. terreus* and *A. nidulans* have been reported.

The fungi classified in the genus *Aspergillus* are anamorphic (asexual) filamentous organisms which reproduce by means of asexual spores termed conidia. Teleomorphic (sexual) forms of many aspergilli have been described. The aspergilli produce conidia in a basipetal fashion, which results in a chain of asexual conidia (the youngest conidium at the base and the oldest at the tip of the chain). The conidiogenous cell is termed a phialide. The base of the conidiophore, where it originates from the parent vegetative hypha, is termed a foot cell.

The conidiophore is hyphal-like and enlarges at its apex to form a swollen vesicle. Phialides may arise directly from the vesicle (uniseriate) or from sterile cells called metulae (biseriate); in some species, such as *A. flavus*, both conditions may exist in the same head. In the past phialides were incorrectly referred to as sterigmata, a term that should be restricted to structures formed by Basidiomycetes.

A number of reference sources on the mycology of the aspergilli culminated in the publication in 1965 of a manual by Raper and Thom (1965) in which the total number of known species and varieties were classified into 18 groups, 132 species, and 18 varieties. Since that time numerous new species and varieties have been described.

The identification of species of *Aspergillus* is not easy. Several excellent guides are available to aid in the identification of the common species of medically important aspergilli (Raper and Thom 1965; de Hoog et al. 2000). The genus *Aspergillus* is characterized by septate hyphae from which arise nonseptate conidiophores; although these conidiophores may be rough or smooth, hyaline or pigmented, they all terminate in a swollen vesicle. One or two rows of phialides are formed on the surface of the vesicle. Chains of round to oval, variously pigmented conidia are produced from the tips of the phialides. The shape of the vesicle and the arrangement of the phialides thereon determine the shape of the conidial head, from columnar to radiate, a key feature in speciating members of the genus. Finally, pigmentation of the conidia is the main factor determining the color of the colony. Following growth on appropriate media, observation of these macroscopic and microscopic features is usually sufficient to identify the organism.

The restricted list of pathogenic species of the genus suggests that these organisms may share some properties not found in other species which confer on them an intrinsic pathogenic advantage. One obvious feature that these human pathogens must share is the ability to grow efficiently at 37°C. The production of proteolytic enzymes is another property or virulence mechanism that has been investigated in the pathogenesis of invasive and allergic disease. The ability to produce elastase has been correlated with mouse virulence in strains of *A. fumigatus*. Likewise, isolates from cases of invasive aspergillosis in humans all showed the ability to digest elastin, regardless of the species. Finally, antigens with protease activity have been isolated from *Aspergillus* spp. and have been shown to react with sera from patients with allergic aspergillosis, aspergilloma, and invasive disease.

Most of the species of *Aspergillus* reproduce asexually and are classified with the Deuteromycetes (Fungi Imperfecti). However, some species, in addition to their asexual reproduction, have the characteristic of reproducing sexually by the formation of ascospores, which typifies Ascomycetes. However, for the purpose of classifying all fungi which reproduce asexually within one genus, similar to that for *Aspergillus* spp., and to avoid confusion by initiating new genera, most mycologists agree that the generic name *Aspergillus* should be applied to all species of these fungi, regardless of whether an ascosporic stage has been observed. It has been suggested that finding the sexual stage in a fungus merely completes its characterization but does not justify its inclusion in another genus; therefore, the names of the ascosporic species (the sexual stage) should be used as synonyms for certain *Aspergillus* spp., whenever this is applicable.

Accurate identification of any species of *Aspergillus* requires the isolation of a pure culture and its examination on a culture medium of known composition. Differences in the nature and composition of the culture media produce marked changes in growth characteristics and colony color, as well as in the dimensions and morphology of the microscopic components, which may lead to an inaccurate identification. Even though aspergilli can grow on almost every microbiological culture medium (liquid, solid, or semisolid), various authors have proposed standardized and reproducible formulae for media which in their experience have provided uniform cultures over long periods and which are of value for comparative studies. These media include Czapek Dox agar which is widely used as a routine medium for comparative studies and was used extensively by Raper and Thom (1965) and their coworkers. A medium used extensively for primary isolation from clinical specimens is malt extract agar on which most aspergilli sporulate freely.

In identifying aspergilli, it is important to keep in mind that there is variation among strains within species, as well as among species within a genus. Therefore, characteristics of the various groups of aspergilli may overlap. Within each strain, species, or group there are certain characteristics which can be identified.

There are several key features which are frequently consistent enough to be used to speciate individual isolates. As a result of the overlapping of criteria for identification of certain species, however, they may be classified within more than one group. The criteria in Table 34.1 are useful when identifying unknown isolates to species level.

The 18 groups of aspergilli classified by Raper and Thom (1965) are: *Aspergillus clavatus*, *Aspergillus glaucus*, *Aspergillus ornatus*, *Aspergillus cervinus*, *Aspergillus restrictus*, *A. fumigatus*, *Aspergillus ochraceus*, *A. niger*, *Aspergillus candidus*, *A. flavus*, *Aspergillus wentii*, *Aspergillus cremeus*, *Aspergillus sparsus*, *A. versicolor*, *A. nidulans*, *Aspergillus ustus*, *Aspergillus flavipes*, and *A. terreus*.

Although more than 180 species are found within the genus, three species, *A. fumigatus*, *A. flavus*, and *A. terreus*, account for most cases of invasive aspergillosis. The ability to distinguish between the various clinically relevant species has diagnostic value, as certain species are associated with higher mortality and increased virulence and vary in their resistance to antifungal therapy. A molecular method has been developed to identify *Aspergillus* at the species level using the 18S and 28S rRNA genes for primer binding sites (Henry et al. 2000). The contiguous internal transcribed spacer (ITS) regions, ITS 1-5.8S-ITS 2, from reference strains and clinical isolates of aspergilli were amplified, sequenced, and compared with non-reference strains in GenBank.

Table 34.1 *Morphological criteria for the identification of Aspergillus spp.*

Criteria
Growth condition
Medium, temperature, light, age and source
Colony characteristics
Rate of growth
Pattern of growth
Color and texture
Basal mycelium
Surface mycelium
Abundance and arrangement of heads, sclerotia, cleistothecia
Reverse color
Conidial stage
Heads
Manner in which borne
Color
Form
Overall dimension
Vesicle
Shape
Dimension
Color
Fertile area
Phialides
Development
Arrangement – uniseriate or biseriate
Primaries – dimension and color
Secondaries – dimension and color
Conidiophore
Length
Diameter
Wall characteristics
Conidia
Dimension
Wall characteristics
Color
Hülle cells and other elements
Shape
Dimension
Sclerotia or sclerotium-like structures
Form and structure
Dimension
Color
Ascosporic stage
Cleistothecia
Origin
Form and structure
Quantity present
Dimension
Color
Asci
Shape
Dimension

(Continued over)

Table 34.1 *Morphological criteria for the identification of Aspergillus spp. (Continued)*

Criteria
Ascospores
Pattern
Dimension
Color

Adapted from Raper and Thom (1955).

ITS amplicons from *Aspergillus* spp. ranged in size from 565 to 613 bp. Comparison of reference strains and GenBank sequences demonstrated that both ITS 1 and ITS 2 regions were needed for accurate identification of *Aspergillus* at the species level. Intraspecies variation among clinical isolates and reference strains was minimal. Other pathogenic molds demonstrated less than 89 percent similarity with *Aspergillus* ITS 1 and 2 sequences. A blind study of 11 clinical isolates was performed and each was correctly identified. The clinical usefulness of this rapid approach to identification has still to be determined. Further reviews of the major pathogenic species of aspergilli can be found in de Hoog et al. 2000; Brakhage and Langfelder 2002; Denning 2000; Varga and Tóth 2003.

MORPHOLOGY

Description of colony appearance

Colonies of aspergilli may be black, brown, yellow, red, white, green, or other colors depending on the species and the growth conditions (Figures 34.1 and 34.2). The colony color always depends on the color of the microscopic components of the fungus, e.g. the color of the vegetative hyphae, the conidial heads, and the sexual structures if they are present. In addition to the color of the aerial parts of the fungal colonies, there may be pigmentation of the underlying medium, which may be different from the

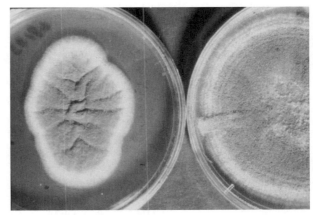

Figure 34.1 A. fumigatus *(right)* and A. flavus *(left)* cultures

Figure 34.2 A. niger *culture*

color of the aerial parts of the colonies. Therefore, at any one time colonies may have one or more shades.

Even though most species of aspergilli are fast-growing fungi, there is a great deal of variation in their rate of growth. Growth rate is an important characteristic for identification of the species, so colony diameter at a certain age, under standard conditions, is a useful feature to note in the identification.

The marginal appearance of the colonies is another important characteristic. Margins may appear as heavy and sharply delineated, thin and diffuse, smooth and entire, irregularly lobed, submerged, or aerial. Also, the texture of the colonial surface may be velvety, floccose, or granular, with or without zonation. Zonation is usually expressed by alternating production of either conidial heads and sclerotia or conidial heads and cleistothecia (depending on the species).

The mycelium

The mycelium of the *Aspergillus* spp. is similar to that of most other fungi. It is well developed, with branching hyaline and septate hyphae. The hyphae may be thin or dense, and light or heavily sporulating. The cells of the hyphae are usually multinucleated. The mycelium can produce copious levels of enzymes and some produce mycotoxins. The mycelial form of *Aspergillus* spp. is characterized by vigorous growth and an abundant production of conidia carried on long, erect conidiophores. These arise from a specialized cell within the vegetative hyphae known as a 'foot cell'.

The conidial head

The conidial head of aspergilli is conventionally considered to comprise the conidia, the phialides, the vesicle, and the conidiophore, which arise from the foot cell (Figures 34.3, 34.4, 34.5, 34.6 and 34.7). In most species, the shape, the size, and the color of the conidial heads within the same colony show no variation.

In the early stages of fungal growth, certain cells in the submerged or aerial parts of the vegetative mycelium enlarge and form a heavy wall. These foot cells will form a branch, which is always produced at a right angle to the hyphal cells. Not all aspergilli have distinct foot cells. The branch formed, which develops into a conidiophore, terminates in a swollen head known as a vesicle. The length of the conidiophore (between the foot cell and the vesicle) and the nature of its wall (smooth,

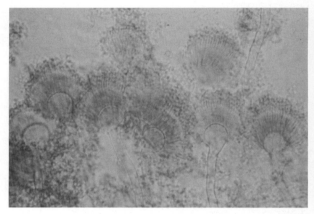

Figure 34.3 A. fumigatus *sporing head*

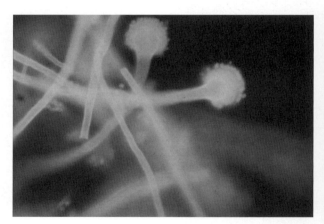

Figure 34.4 A. fumigatus *sporing heads stained with Calcofluor white and viewed under the fluorescence microscope*

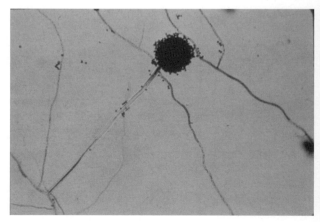

Figure 34.5 A. niger *sporing head*

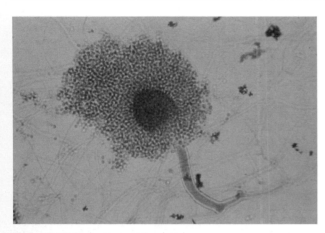

Figure 34.6 A. clavatus *sporing head*

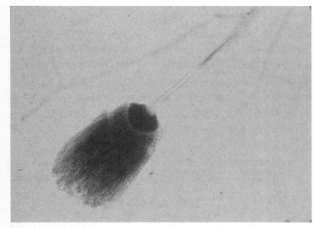

Figure 34.7 A. terreus *sporing head*

rough, echinulate, or pitted) vary from one species to another and are considered important characteristics of the species. Also, the conidiophore may be septate or nonseptate, in addition to being uniform throughout in length or greater in diameter at the base of the foot cell. Certain species, such as *A. glaucus*, may show branched conidiophores.

The vesicles vary in shape and size. They may be globose, hemispherical, elliptical, or elongated and clavate. They may have a very thin or a relatively thick wall. They are mostly hyaline, but in a few species they may appear pigmented. The lumen of the vesicle is usually a part of the conidiophore without any septum separating the vesicle from the conidiophore. These vesicles are usually borne upright on the conidiophore, with a few exceptions where they are formed at an angle to its main axis.

Depending on the species, certain areas of the vesicle surface become fertile and give rise to a layer of conidium-producing cells, which are called 'phialides'. These phialides, in some species, will cover the entire surface of the vesicle, whereas in others they may cover the upper half or three-quarters. They may vary in color in different species, from hyaline to lightly or darkly pigmented. The phialides are usually cylindrical, and are

produced in different sizes and shapes, although they are usually uniform within the same culture of each species. They are produced perpendicular to the point of origin on the vesicle surface.

The cylindrical body of the phialides narrows at the apex to form a conidium-producing tube. However, depending on the species, the phialides may be formed in one layer (primary phialides), or each of these primary ones will bear one or more phialides (secondary phialides) in one linear surface. The size, shape, and length of the secondary phialides are usually different from those of primary origin. When there are secondary phialides, however, they will be the ones that narrow to form a conidium-producing tube. In some species, some of the primary phialides have been found to be septate.

Nuclear division starts within the conidium-producing structures of the phialides as a preliminary stage of conidia formation. One of the new nuclei moves upwards to the conidium-producing tube; this is followed by the formation of a septum separating the new nucleus from the rest of the cylinder. Another division follows within the cylinder and one of the newly formed nuclei moves upwards towards the tube, pushing the first nucleus (which has become a conidium by this time) out of the tube, but remaining attached to its tip. As this process continues, newly formed conidia push previously formed conidia out of the tube, thereby forming a chain of conidia at the tip of each phialide. The conidia remain attached to each other within each chain by a connective bridge, which is usually not visible with a light microscope. There are variations among the various species in the flexibility of these bridges, which affects the ease with which conidia are freed from each other.

Conidia are usually globose with a rough surface and are found in various sizes; they are mostly uninucleate. However, some have been found to contain up to 12 nuclei. The length of conidial chains, their density of packing and their orientation around the conidial heads vary from species to species and are considered part of the characteristics of the species and group. The individual color of the conidia (which may be hyaline), the collective color of the conidial mass, as well as the color of the aerial hyphae, are species characteristics that give the shade observed in the colonies.

Sexual reproduction

Many species of *Aspergillus* can reproduce sexually. Sexual reproduction in the ascosporic aspergilli is mainly by the formation of ascospores within asci. The mature ascospores are usually either in the shape of a double convex lens or two symmetrical valves, with variations among the different species. In members of the *A. nidulans* group, the color of the ascospores varies from red–brown to purple–red or violet; in the *A. ornatus* group, individual species display shades of red–brown at maturity; otherwise the ascospores are usually hyaline.

The usual number of ascospores per ascus is eight in most species.

The cleistothecium (ascocarp), where the asci are formed, varies in size, color, shape, and appearance in different species. They may appear naked or surrounded by a layer of specialized chlamydospores known as Hülle cells. These cells (which have no known function) are usually borne in hyphae as terminal or intercalary cells similar to chlamydospores. They may be globose (as in A. nidulans), subglobose, or elliptical in shape with very thick walls, which may occupy the whole cell body.

THE CAUSAL ORGANISMS AND THEIR HABITAT

Most human infections are caused by A. fumigatus, but A. flavus, A. nidulans, A. niger, and A. terreus have also been implicated. These molds are widespread in the environment. They are common soil inhabitants and are also found in large numbers in dust and decomposing organic matter. Their spores are often found in the outside air (Goodley et al. 1994; Anderson et al. 1996).

A. fumigatus is thermotolerant, able to grow over a temperature range from below 20°C up to 50°C, and grows well at over 40°C. It abounds in vegetable matter decomposing in warm environments, such as self-heating hay and composts. Strains of A. fumigatus can be distinguished from one another by analysis of polymorphic DNA markers (Aufauvre-Brown et al. 1992; van Belkum et al. 1993; de Hoog et al., 2000, and reviewed in Warnock et al. 2001). This approach has allowed environmental strains to be associated with particular disease entities (Symoens et al. 1993; Buffington et al. 1994; Girardin et al. 1994a, b; Tang et al. 1994; Mondon et al. 1995).

Inhalation of conidia leads to a variety of disease patterns. Atopic subjects may develop asthma. In non-atopic subjects, the presence of damaged lung tissue may result in the growth of a fungus ball (aspergilloma). Inhalation of massive doses of the conidia may lead to alveolitis. Aspergillosis also occurs in many species of birds and mammals, domesticated and wild, infected primarily by inhalation and starting as a pulmonary disease, although sometimes involving other organs. A. fumigatus is also one of several molds implicated in bovine mycotic abortion, probably initiated by inhalation, but with alimentary infection from spores ingested in moldy fodder as another possible route.

Nosocomial outbreaks of aspergillosis have become a well-recognized complication of construction work in or near hospital wards in which neutropenic patients are housed. In several reported outbreaks, building works adjacent to the unit in which patients were accommodated led to contamination of the air. In other outbreaks, the ventilation system for the unit drew contaminated air from neighboring building sites, or became contaminated in some other way (Goodley et al.

1994; Anderson et al. 1996; Cornet et al. 1999; Thio et al. 2000, and reviewed in Warnock et al. 2001).

The factors influencing the recovery of Aspergillus from the air and a comparison of the instruments available for sampling air are reviewed by Morris et al. 2000.

CLINICAL MANIFESTATIONS

Conidia of A. fumigatus can, because of their small size (<5 μm) and aerodynamic properties, bypass the upper respiratory tract defenses and may reach distal regions of the lung (Amitani et al. 1992, 1995). In this region the host defenses rely on phagocytic cells to remove the conidia efficiently.

An understanding of the spectrum of clinical manifestations seen in aspergillosis is also dependent on an insight into the ways in which Aspergillus spp. can transform from a saprophyte to a parasite. The ability of A. fumigatus to invade living tissue is dependent upon a number of virulence attributes (reviewed in Bouchara et al. 1995; Latgé 2001). Several reports suggest that the virulence of A. fumigatus is associated with the production of secreted proteases and elastases (Monod et al. 1993; Frosco et al. 1994; Moser et al. 1994; Reichard et al. 1994). Extensive colonization of lung tissue by A. fumigatus has suggested a key role for fungal proteases during infection. However, work by Jaton-Ogay et al. (1994) demonstrated that a serine alkaline protease (ALP) was not essential for invasion of lung tissue when ALP-negative mutants were shown to cause comparable mortality and to invade lung tissue in immunocompromised mice, as well as the parent strain (Monod et al. 1993; Tang et al. 1993). These data contrast with the work of Kolattukudy et al. (1993) who observed that an ALP-deficient mutant obtained by nitrosoguanidine treatment was less virulent than a wild-type strain of A. fumigatus. As proteases do not appear to play a role in the establishment of infection, other hypotheses for virulence of A. fumigatus have been put forward. One proposal (Jaton-Ogay et al. 1994) is that A. fumigatus has an ecological niche in the bronchi characterized by nutrient and temperature conditions favorable to its saprophytic growth; this results because, although this fungus accounts for only a small proportion of the airborne mold spores, it is the one isolated from lung and sputum most frequently. Consequently, invasive aspergillosis may occur when the host phagocytic defenses have been weakened by immunosuppression.

Inhalation of the conidia of Aspergillus spp. can give rise to a number of different clinical forms of aspergillosis, depending on the immunological status of the host (Cohen 1991; Soubani and Chandrasekar 2002; Richardson and Warnock 2003). However, there are no currently accepted classification schemes so the terminology used here may not always correspond to that adopted by other texts. Moreover, some clinical entities defy precise classification and the pathological features at the time of diagnosis may not be known. In non-compromised individuals, *Aspergillus* spp. can act as potent allergens or cause localized infection of the lungs or sinuses.

In neutropenic patients, there is widespread growth of the fungus in the lungs and dissemination to other organs often follows. This condition is usually fatal, even if diagnosed during life and treated. It must, however, be emphasized that, with early diagnosis and treatment, a small but significant number of patients is cured.

The following sections deal, in turn, with the clinical manifestations of aspergillosis in the various organ and tissue systems, pointing out, where possible, those features that may assist in the early and definitive recognition of the disease.

Allergic aspergillosis

Allergic bronchopulmonary aspergillosis (ABPA) is characterized by recurrent pyrexia, cough, wheezing, sputum plugs containing aspergilli, and recurrent pulmonary infiltrates (Patterson et al. 1982; Ikemoto 1992; Greenberger 2002a, b, 2003; Moss 2002). This is an uncommon condition, most often seen in atopic individuals who develop bronchial allergic reactions (asthma) following inhalation of *Aspergillus* spores (Singh et al. 2003). Mucus plugs then form in the bronchi, leading to atelectasis. The illness may be mild, but it is an episodic condition and can often progress to bronchiectasis and fibrosis. ABPA is a specific disease seen in about 5 percent of asthmatic individuals; the pathology is thought to be a type III and IV reaction to inhaled conidia of *A. fumigatus*. Dual skin reactions (immediate and late) and IgG and IgE antibodies are also recognized. ABPA is thought to result from type I and III, and perhaps type IV hypersensitivity reactions to antigens released from the fungus colonizing the bronchial tree.

Against this pathological and immunological background a number of major criteria for the diagnosis of ABPA are recognized:

- asthma (reversible airway obstruction)
- eosinophilia of sputum and blood
- recurrent pulmonary infiltrates
- allergy to antigens of *Aspergillus* spp. by skin test (immediate weal–flare reaction, type I) and late reaction (Arthus, or type III).

Other criteria are:

- *Aspergillus* spp. in sputum
- raised total IgE and specific IgE and IgG in serum
- history of recurrent fever or pneumonia
- history of coughing up plugs.

The clinical course of ABPA is characterized by frequent exacerbations of cough, wheezing, and fever, with eosinophilia of sputum and blood. Allergy or infec-

tion or both may trigger the exacerbations. A significant number of patients with ABPA have clinically and immunologically demonstrable atopy. The incidence of atopy is particularly high in those patients with early-onset asthma (first decade of life).

The expectoration of sputum containing brownish mucus plugs is another feature of ABPA. Microscopy shows that the plugs are made up of eosinophils and septate hyphae embedded in mucus and fibrinous debris. Culture of the sputum may yield *A. fumigatus*. However, this is not pathognomonic, because aspergilli may occasionally be isolated from the sputum of normal people.

The most frequent symptoms seen in ABPA include fever, intractable asthma, productive cough, malaise, and weight loss. Expectoration of brown eosinophilic mucus plugs containing *Aspergillus* mycelia is common.

The radiological findings range from small, fleeting, unilateral, or bilateral infiltrates with ill-defined margins (often in the upper lobes) and hilar or paratracheal lymph node enlargement, to chronic consolidation and lobar contractions. Bronchiectasis represents the later stage of the disease. Bronchial obstruction secondary to mucus plugs often produces radiological signs and is believed to be the first step in the development of inflammation and infection that eventually lead to fibrosis. Correlation is poor between radiological and clinical findings.

The clinical features of ABPA appear to result from reaginic (IgE) and precipitating (IgG) antibodies (Brummund et al. 1987; Marchant et al. 1994). Specific IgE is believed to cause the asthma, eosinophilia, and immediate skin reaction, whereas precipitating antibodies are believed to cause the pulmonary infiltrations, damage to the bronchial wall, and late skin reaction. Serum IgE is significantly elevated in ABPA but not in other hypersensitivity lung diseases caused by, for example, inhaled organic dusts.

Reviews by Cockrill and Hales (1999), Wark et al. (2001), Chetty (2003), and Kauffman (2003) summarize the role of cellular immunity in ABPA and give very clear guidelines regarding diagnosis and treatment.

ALLERGIC BRONCHOPULMONARY ASPERGILLOSIS AND CYSTIC FIBROSIS

An association between ABPA and cystic fibrosis was first reported by Mearns et al. (1965). ABPA occurs with an incidence of 10–11 percent in patients with cystic fibrosis. Nearly one-half of cystic fibrosis patients have a positive immediate weal and flare skin test to *A fumigatus* and *A. fumigatus*-specific IgE (Marchant et al. 1994) and IgG without evidence of ABPA (Knutsen et al. 1994). *Aspergillus* allergy appears to be coincident with *Pseudomonas aeruginosa* colonization; hence it is difficult to distinguish the effects of either organism. The diagnosis of ABPA is, however, clearly important from a clinical viewpoint because of its association with severe proximal bronchiectasis and a far more rapid decline in lung function in cystic fibrosis.

A greater understanding of the association between ABPA and cystic fibrosis is provided by the identification of mutations within the cystic fibrosis transmembrane conductance regulator (CFTR) gene in patients with ABPA but without clinical criteria for cystic fibrosis. Miller and colleagues analysed the entire coding region in 11 individuals who met criteria for the diagnosis of ABPA but had normal sweat electrolyte tests (Miller et al. 1996). Of the 11 patients, six had minor mutations of the CFTR gene, suggesting that CFTR plays an etiologic role in a subset of patients with ABPA. Recurrence of ABPA after lung transplant has also been described in a patient with cystic fibrosis (Fitzsimmons et al. 1997). An exhaustive review of ABPA and cystic fibrosis has been published recently (Stevens et al. 2003).

ASPERGILLUS SPP. AS FUNGAL ALLERGENS

Aspergillus spp. antigens have been studied thoroughly as allergens because *Aspergillus* conidia were among the first to be recognized as important aeroallergens (Kauffman et al. 1984).

Two antigens (18 and 20 kDa) with potential for the diagnosis of allergy have been purified from *A. fumigatus* by chromatography (reviewed in Ikemoto 1992). The 18 kDa antigen was isolated from the mycelium of ten different strains of *A. fumigatus*. Monospecific polyclonal rabbit antiserum directed against this antigen reacted with antigens of 26, 28, 44, and 46 kDa as well (Latgé et al. 1991). The characteristics of the 18 kDa antigen suggest that it is similar to Ag-2 or Ag-10 described previously from counterimmunoelectrophoresis patterns. The 20 kDa allergen is a glycoprotein that appears to be different from antigens Ag-3, Ag-5, Ag-7, and Ag-13. Another glycoprotein allergen, designated gp55, was recently isolated; its allergenic activity was sensitive to protease but not to deglycosylation. On the basis of the amino-terminal protein sequence, this appears to be a novel protein.

The amino acid sequence of the 18 kDa protein (Asp fI) has been determined partially (Burnie and Matthews 1991). Asp fI mRNA was detected in *A. fumigatus* but not in seven other *Aspergillus* spp. Asp fI shows extensive sequence homology (95 percent) to mitogillin (a cytotoxin) produced by *A. restrictus*. A cross-inhibition radioimmunoassay with a murine monoclonal antibody and human IgG and IgE antibodies revealed that Asp fI and mitogillin are indistinguishable antigenically. The simultaneous toxicity and allergenicity of Asp fI have significant implications for understanding the etiology of ABPA. Asp fI cDNA from *A. fumigatus* has now been cloned, and the recombinant allergen has been expressed and used in skin test trials. A recombinant *A. fumigatus* protein (65 kDa) expressed from a cDNA

clone has also been shown to bind IgE from ABPA patients. On the basis of cDNA and deduced amino acid sequence homology to Hsp90 from other organisms, this allergen is apparently a heat-shock protein of the Hsp90 family (Burnie and Matthews 1991). The presence of non-allergenic homologs provides a significant opportunity to study the basis of allergen epitopes by focusing on the differences between these proteins, i.e. Hsp90 from *A. fumigatus* (allergenic) and Hsp90 from humans (non-allergenic). These important achievements provide a significant aid to standardizing *Aspergillus* spp. extracts and diagnosing *Aspergillus*-related diseases. However, optimal methods for the preparation of relevant antigens have still to be defined (Little et al. 1993). For detailed descriptions of those *Aspergillus* antigens approved by the allergen nomenclature committee see Banerjee and Kurup (2003) and Kurup et al. (2000).

Infection of the paranasal sinuses

Aspergillosis of the sinuses includes a number of diseases ranging from a benign non-invasive form to an aggressively invasive type. To rationalize a number of classification schemes, Talbot et al. (1991) have defined four basic types: saprophytic aspergillus colonization of a previously abnormal sinus, allergic aspergillus sinusitis, subacute or chronic invasive aspergillus sinusitis, and fulminant invasive aspergillus sinusitis. These clinical entities have in many ways pulmonary counterparts: aspergilloma, ABPA, chronic necrotizing pulmonary aspergillosis, and invasive pulmonary aspergillosis, respectively (Chang et al. 1992; De Carpentier et al. 1994; Schubert 2001; Dhiwakar et al. 2003).

Aspergillosis is currently the most common fungal infection of the paranasal sinuses (see reviews by Blitzer and Lawson 1993, Drakos 1993, and Richardson and Warnock 2003). Most patients who have developed aspergillus sinusitis have no underlying disease, although invasive rhinosinusitis has been seen in patients with acute leukemia (Talbot et al. 1991). In 20 cases of orbital aspergillosis, only one patient had diabetes. Thirty-seven patients with sino-orbital aspergillosis presented with no systemic disease. There have been only occasional reports of sinus aspergillosis arising in diabetes and leukemia patients (Peterson and Schimpff 1989). The disease runs a more fulminant course in immunocompromised patients; the mortality rate was reported as 100 percent in bone marrow transplant recipients. There is a more recent association of aspergillus sinusitis in patients with nasal allergies.

The nose and paranasal sinuses have local factors that may promote fungal infection, including nasal polyps, recurrent bacterial infections and chronic rhinitis with stagnation of nasal secretions. Some authors have suggested that occlusion of the nasal ostia of the sinuses creates an anaerobic environment that may promote fungal pathogenicity. Other reports have challenged this concept, citing aspergillus infections in such well-aerated regions as the nose, bronchi, and external ear. Other underlying factors include prolonged antibiotic therapy for sinusitis and the greater use of antibiotics and immunosuppressive agents.

Two different forms of sinusitis due to *Aspergillus* spp. have been recognized. Acute sinusitis is a life-threatening condition encountered in immunocompromised patients. The clinical presentation is similar to that of rhinocerebral mucormycosis. The presenting symptoms include fever, nasal discharge, headache, and facial pain. Necrotic lesions develop on the hard palate or nasal turbinates and disfiguring destruction of facial tissue may occur. The infection can spread into the orbit and brain, causing thrombosis and infarction. Paranasal aspergillus granuloma formation is a slowly progressive condition. It is most common in the tropics, where *A. flavus* is the most common cause, although cases have been reported from temperate climates. Affected individuals usually complain of longstanding symptoms of nasal obstruction and headache, suggesting chronic sinusitis, but are otherwise normal. Patients present with unilateral facial pain and headache, or with facial swelling and proptosis. The swelling is firm but not usually tender. In the later stages of this condition, upward spread of the fibrosing paranasal granuloma results in focal cerebral or orbital infection. The typical radiological finding is a dense filling defect within the maxillary or ethmoid sinuses with erosion of the surrounding bone. This can be confirmed by computed tomography (CT) or magnetic resonance imaging (MRI). A third form of aspergillus sinusitis, termed allergic fungal sinusitis (AFS), has recently been described. Up to 7 percent of patients requiring sinus surgery may have AFS.

Aspergillus sinusitis is a worldwide disease; the largest number of cases occur in hot dry climates. It may be that a hot, dry, dusty climate produces chronic nasal inflammation, allowing an ingrowth and tissue damage by the fungus and its metabolites, followed by the immunological reaction of the host to the fungal antigens. In hot, dry environments aspergillus infection has a more virulent course.

Generally, only one sinus is involved with *Aspergillus* spp., most commonly the maxillary sinus. The others, in decreasing order of frequency, are the ethmoid, sphenoid, and frontal sinuses. In a review of 103 cases (cited by Blitzer and Lawson 1993), 67 patients were found to have solitary sinus involvement (46 maxillary, eight sphenoidal, eight nasal, three ethmoidal, two frontal). Aspergillosis of the sphenoid sinus can be confusing, with reported stone formation and cavernous sinus syndrome; it can also mimic a pituitary tumor. When the disease becomes invasive, it usually spreads from the maxillary sinus to the ethmoid and then extends to the orbit and nasal cavity. Most cases of orbital aspergillosis originate in the maxillary and ethmoidal sinuses. Nasal, postnasal, sinus, and orbital symptoms are found in

upper respiratory tract infections with *Aspergillus* spp. The nasal symptoms are associated with a mucoid or mucopurulent discharge, swollen nasal mucosa, enlarged turbinates, and occasional polyps or facial pain. In some cases, the secretions are gelatinous, containing necrotic tissue. Irrigation of the maxillary antrum may be difficult, and viscous or greasy material may appear in the lavage fluid. Repeated antral irrigations and antibiotics do not provide symptomatic or radiological improvement. Surgical access to the antrum reveals a thickened or polyploid antral lining, granulation tissue, and necrotic or gelatinous material, often in a cheese-like mass of brownish or greenish material. The administration of systemic corticosteroids seems to be a major factor in predisposing humans to aspergillus infection. Aspergillosis of the nose and paranasal sinuses can be classified into two types: invasive and non-invasive. In the non-invasive form, the patient complains of a nasal obstruction, rhinorrhea, and a sensation of fullness of the face. The sinus involved is opaque radiologically, without invasion. These cases appear to respond well to Caldwell–Luc procedures with an excellent prognosis. In patients with the invasive form, the disease involves the nose, orbit, palate, turbinates, and skin. Patients have pain, proptosis, and blurred vision. Radiologically, there is bone destruction, usually involving the floor of the medial orbital wall. The erosion may involve the basisphenoid and pituitary fossa, the temporal bone and pterygopalatine fossa, the alveolus with loosening of teeth, and the frontal bone. CT scanning has been very useful in defining the full extent of the disease (Ashdown et al. 1994; Krennmair et al. 1994).

Aspergillus sinusitis often produces a mixture of high- and low-density areas within the sinuses. Bone windows allow a very accurate assessment of possible invasion. MRI can be used for better definition of the high and low signals produced by the mycetoma. Despite surgical therapy, recurrences occur and multiple surgery is necessary to eradicate the disease. Direct extension or hematogenous spread from an infected sinus may cause vascular invasion. The invasive form also involves bone or meninges. In one series there were 17 deaths among 103 patients, all of which were caused by intracranial extension. Interestingly, in almost all these cases, there was no known underlying disease. Invasive sinus aspergillosis has also been reported to be associated with facial nerve paralysis, optic neuritis, arteritis, and rupture of the carotid artery. Involvement of the skull base can simulate a malignancy. Histopathologically, epithelium is ulcerated, with a dense inflammatory infiltrate of lymphocytes, plasma cells, and neutrophils. There may be bone necrosis. The granulomas have been classified as proliferative with pseudotubercles in a fibrous stroma, necrotizing with large areas of edematous necrosis, and mixed.

The diagnosis of paranasal sinusitis is nonspecific and often confusing. The differential diagnosis includes bacterial sinusitis, malignant tumors, tuberculosis, syphilis, osteomyelitis, Wegener's granulomatosis, and rhinoscleroma. Treatment for the non-invasive form consists of sinusotomy and curettage of all diseased and necrotic tissue. This is usually curative on its own. The invasive form, however, requires radical surgical débridement and intravenous amphotericin B. Azole antifungals may have a role in treatment. Early work with itraconazole suggests that it may also have a role. Despite these measures, however, multiple recurrences requiring several procedures are the rule.

Initially, AFS was attributed to *Aspergillus* spp.; it is now clear, however, that many environmental fungi are associated with the disease (Buzina et al. 2003). *Bipolaris* spp. are the most common of this group, but *Exserohilum*, *Curvularia*, and *Alternaria* spp. have also been identified by culturing tissue. The typical patient is an immunocompetent, atopic, young adult with a longstanding history of allergic rhinitis, nasal congestion, headache, nasal polyposis, asthma, and/or recurrent sinusitis. There is a male preponderance of 2 or 3:1. The patient may have had several sinus operations over a period of years. Visual changes may have occurred, and even facial deformity has been reported as a consequence of the expansile nature of the alterations of the sinus. This form of sinusitis is initiated by hypersensitivity to fungal antigens. Immunologically, the serum IgE is increased, and the patient demonstrates a positive cutaneous skin test to the fungus. Elevated serum fungal allergen-specific IgE and IgG precipitins can be demonstrated. A non-contrast CT scan will frequently demonstrate multiple sinus involvement, with areas of high signal within the sinus. This is almost pathognomonic. Bone expansion is commonplace. Bone destruction is reported 30–50 percent of the time. At surgery, dense, inspissated, rubbery, green–brown, allergic mucoid material, and hyperplastic mucosa and polyps are noted. This mucus contains an increased number of eosinophils and Charcot–Leyden crystals, epithelial cells, and cellular debris. Areas of bone may include posterior sphenoid, ethmoid septa, laminar papyracea, and the medial antral wall. The dura, however, remains intact.

Treatment is conservative surgical drainage. Endonasal approaches to the ethmoid, sphenoid, and frontal sinuses, and the Caldwell–Luc approach to the maxillary sinuses, are reasonable choices. Systemic antifungals should be avoided unless there is definite evidence of tissue invasion or there is orbital or intracranial invasion.

The diagnosis and incidence of AFS has been revised recently by Ponikau and colleagues (1999) and Buzina et al. (2003). The objective of these studies was to re-evaluate the current criteria for diagnosing AFS and to determine its incidence in patients with chronic rhinosinusitis (CRS). This prospective study evaluated the incidence of AFS in 210 consecutive patients with CRS with or without polyposis, of whom 101 were treated surgically. A novel method of patient sampling is described where each nostril was lavaged with saline. The histopathological examination of mucin for fungi and eosinophils is emphasized. Fungal cultures of nasal secretions were positive in 202 (96 percent) of 210 consecutive CRS patients. Allergic mucin was found in 97 (96 percent) of 101 consecutive surgical cases of CRS. Allergic fungal sinusitis was diagnosed in 94 (93 percent) of 101 consecutive surgical cases with CRS, based on histopathological findings and culture results. IgE-mediated hypersensitivity to fungal allergens was not evident in the majority of AFS patients. The data from this report indicate that the diagnostic criteria for AFS are present in the majority of patients with CRS with or without polyposis. The authors propose that because the presence of eosinophils in the allergic mucin, and not a type 1 hypersensitivity, is the common feature in the pathophysiology of AFS, the clinical entity AFS should be renamed eosinophilic fungal rhinosinusitis. Molecular techniques have also been applied (Willinger et al. 2003).

Aspergilloma

This is the most familiar of the localized infections produced by *Aspergillus* spp. Typically, aspergillomas develop in old tuberculous cavities, but any cause of lung necrosis acts as a suitable environment. Fungus ball (aspergilloma) formation usually occurs in patients with residual lung cavities following tuberculosis, sarcoidosis, bronchiectasis, pneumoconiosis, and ankylosing spondylitis, or where there is a neoplasm of the lungs. Hemoptysis is the only serious complication. Fungus balls are usually located in the upper lobes. Less frequently, they occur in the apical segments of the lower lobes. Spontaneous lysis has been reported to occur in up to 10 percent of cases. Patients are often asymptomatic, but may present with chronic cough, malaise, and weight loss. Hemoptysis is the most common symptom, occurring in 50–80 percent of cases, and can, on occasion, be massive and life threatening (Jewkes et al. 1983). Chest x-rays will reveal a characteristic oval or round mass with a radiolucent halo or crescent of air over the superior aspect (Figure 34.8). The mass can often be shown to move as the patient changes position (Figure 34.9). CT will help to delineate the lesion.

Chronic pulmonary aspergillosis

A number of clinical entities have been described, including chronic necrotizing pulmonary aspergillosis (CNPA), semi-invasive aspergillosis, chronic invasive pulmonary aspergillosis, symptomatic pulmonary aspergilloma, and aspergillus pseudotuberculosis. This distinction between subacute invasive pulmonary aspergillosis, CNPA, and aspergilloma has not been rigorously defined, and an overlap in clinical and radiological features bwtween these different entities probably exists (Denning et al. 2003). In the course of their clinical practice, Denning and colleagues have observed patients

(CFPA). In these patients pleural involvement has been seen, either as direct invasion of the pleural cavity or as fibrosis. The third proposed category comprises those patients where there is progressive enlargement of a single cavity, usually with a thin wall, in some cases to substantial dimensions, occurring slowly over months or rapidly in weeks. This last group of patients appear to have slowly progressive invasive aspergillosis, a disease entity similar to or identical to the previously described CNPA. These patients appear to differ slightly from those with CCPA and CFPA because they usually have minor or moderate degrees of immune dysfunction, such as diabetes or corticosteroid use. The authors propose that this condition be called subacute invasive pulmonary aspergillosis, interchangeable with CNPA (Denning et al. 2003).

This condition usually occurs in middle-aged or older men with chronic or previously treated lung disease such as tuberculosis (reviewed by Binder et al. 1982). Pleural spread has been reported, but dissemination beyond the lung does not occur. The most frequent symptoms include fever, productive cough, malaise, and weight loss, often lasting for months before diagnosis. The radiological findings include a chronic progressive infiltrate representing parenchymal necrosis involving the upper lobes or the superior segment of the lower lobes. Cavitation is common and about 50 percent of patients develop single or multiple fungus balls.

Infection of the central nervous system

It is much more common for cerebral aspergillosis to occur after hematogenous dissemination of infection from the lungs than as a result of direct spread from the nasal sinuses. The brain is involved in about 10 percent of cases of disseminated aspergillosis, but cerebral infection is seldom diagnosed during life (Haran and Chandy 1993; Ashdown et al. 1994).

The symptoms of cerebral aspergillosis are gradual in onset. Confusion, behavioral alterations, and reduced consciousness in a neutropenic patient should suggest the diagnosis. Multiple brain lesions with infarction caused by cerebral arterial thrombosis often result in focal neurological signs, fits, and raised cerebrospinal fluid (CSF) pressure.

The CSF findings are normal in 50 percent of cases. In the remainder, the protein concentration may be raised, but the glucose concentration is usually normal. On occasion, a marked pleocytosis is seen. It is most unusual to recover the fungus from the CSF.

CT is often helpful in locating the lesions, but the findings are nonspecific. Meningitis is a most unusual manifestation of CNS aspergillosis.

Invasive aspergillosis

Various forms of invasive infection have increased markedly in number over the past few years (Anaissie et al.

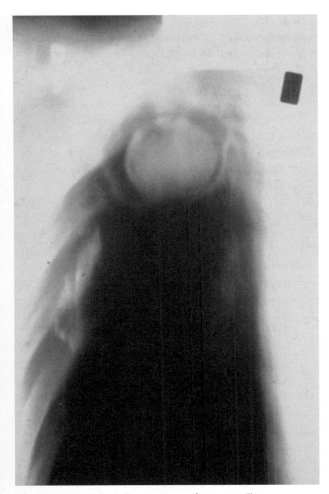

Figure 34.8 Radiological appearance of an aspergilloma

who have chronic pulmonary disease undoubtedly caused by A. fumigatus, with clinical features often different from that described in the older literature. Three distinct radiological patterns have been proposed. The first is characterized by the formation and expansion of multiple cavities, some containing fungus balls, which has been termed chronic cavitary pulmonary aspergillosis (CCPA). In some cases, this progresses to marked and extensive pulmonary fibrosis, which has been termed chronic fibrosing pulmonary aspergillosis

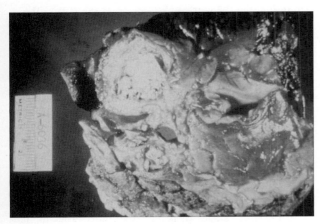

Figure 34.9 Gross pathological appearance of an aspergilloma

1989; Walsh 1990; Cohen et al. 1993; Denning 1994b; Williamson et al. 1999; Dimopoulos et al. 2003; Richardson and Warnock 2003). Although the lung is the most common site of infection, aspergillosis can disseminate to virtually any body site, and indeed the true extent of this spread is often only apparent postmortem. Invasive aspergillosis is almost always seen in the setting of the immunocompromised host and is often fatal, even if diagnosed during life and treated. Those at risk include neutropenic patients with hematological malignancies, transplant recipients, and children with chronic granulomatous disease. There is widespread growth of the fungus in the lung tissue, resulting in hemorrhagic infarction. Hematogenous dissemination to other organs often follows.

The increase in the incidence of invasive aspergillosis is exemplified in a report of infections in 60 consecutive adults undergoing unrelated donor bone marrow transplantation over a 7-year period (Williamson et al. 1999). T-cell depletion was employed in 93 percent. There was a high incidence of invasive aspergillosis (17 cases), despite the use of itraconazole or amphotericin B prophylaxis. Ten of these cases occurred beyond 100 days. Two patients (11 percent) with invasive aspergillosis survived. Invasive aspergillosis was significantly associated with bacteremia and multiple viral infections.

The most common presentation in the neutropenic patient is an unremitting fever (higher than 38°C) which fails to respond to broad-spectrum antibacterial treatment. Pleuritic chest pain is not unusual. Cough may be present, but sputum production is usually minimal. Hemoptysis is uncommon.

The radiological findings are nonspecific, but the earliest lesions are single or multiple opacities. These usually progress to diffuse bilateral consolidation, or cavitation, or large wedge-shaped peripheral lesions, representing hemorrhagic infarction. These last suggest aspergillosis and their detection is sufficient justification for treatment. In 10 percent of patients with proven aspergillosis, the chest x-ray has been normal within a week of death. CT often reveals nodular lesions in patients with normal chest x-rays.

Expectoration of necrotic tissue from an infarcted lesion can leave behind what looks like a fungus ball in the lung. However, this should not be confused with the classic benign condition seen in the non-immunocompromised patient.

Hematogenous dissemination of infection from the lungs to the brain, gastrointestinal tract, and other organs occurs in up to 30 percent of patients.

The risk factors associated with invasive aspergillosis are in general similar to those for disseminated candidosis, with an even higher emphasis on prolonged neutropenia and corticosteroid treatment, particularly in combination with other immunosuppressive drugs as used in organ and bone marrow transplant recipients.

The degree and duration of granulocytopenia together are the most important predisposing host factor. The risk of developing an aspergillus infection seems to be low during the first 2 weeks of neutropenia, but after the second and third week the risk increases dramatically. Cell-mediated immunodeficiency, as in patients undergoing bone marrow transplantation or with corticosteroid therapy, is an additional risk factor.

Pulmonary CT may detect nodular lesions, sometimes with a surrounding zone of intermittent attenuation, in patients with a normal chest x-ray, and has therefore been considered extremely valuable (McWhinney et al. 1993; Gotway et al. 2002).

Specific diagnostic aspects of invasive aspergillus infections in allogenic bone marrow transplant recipients have been described also by Jantunen and colleagues (Jantunen et al. 2000). The charts of 22 consecutive patients with invasive aspergillosis transplanted in 1989–95 were reviewed. Invasive aspergillosis was diagnosed 69–466 days (median 131 days) after bone marrow transplantation. In 16 patients (73 percent), a definite or probable diagnosis of invasive aspergillosis was made during life. Respiratory symptoms were the presenting feature in half of the patients, followed by neurological symptoms (27 percent). Chest x-ray revealed single or multiple nodular lesions in ten patients; cavitation was observed in five patients. Tissue biopsy was the most common method of diagnosis (nine patients: lungs, six; liver, one; subcutaneous tissue, one; brain, one). Five cases of invasive aspergillosis were detected by nine guided fine-needle lung biopsies in eight patients and without complications. Bronchoalveolar lavage (BAL) was performed in 14 patients, with findings suggestive of invasive pulmonary aspergillosis in eight cases. Lungs were the most common organ affected (90 percent), followed by the central nervous system (CNS) (41 percent). More recent reviews and clinical surveys include those by Marr et al. (2002), Guermazi et al. (2003), and Wiederhold et al. (2003).

Acute invasive pulmonary aspergillosis

This is the most common form of invasive aspergillus infection and carries a very high mortality rate, in some studies about 95 percent (Gerson et al. 1984; Weinberger et al. 1992; Gentile et al. 1993; Ribrag et al. 1993; Richard et al. 1993; Saugier-Veber et al. 1993; Walmsley et al. 1993, and reviewed by Marr et al. 2002, Richardson and Warnock 2003, and Wiederhold et al. 2003). This type of infection may present in several forms. A bronchopneumonia with solitary or multiple infiltrates, fever, and lack of response to treatment with antibiotics is one common manifestation of acute pulmonary aspergillosis. Invasive aspergillosis of the upper airways has been described (Logan et al. 1994). Also a lobar pneumonia which resembles a bacterial infection may occur.

In the early stages of acute aspergillus bronchopneumonia, the small, patchy infiltrates may be undetectable with ordinary chest x-rays, and in these cases CT has proved valuable for the detection of these lesions (Palmer et al. 1991; Aquino et al. 1994; Blum et al. 1994). Even in cases where the aspergillus infection presents as nonspecific infiltrates on conventional chest x-rays, a CT scan may give very valuable clues to the diagnosis by revealing a wedge-shaped, peripheral halo zone of intermediate CT attenuation surrounding a nodular infiltrate (Taccone et al. 1993). This halo is the CT correlate of hemorrhage and edema surrounding an infarct caused by thrombosis (Figure 34.10).

Bronchopneumonic aspergillus infiltrates later give rise to abscess formation, and not infrequently cavitary lesions may develop, often responding with neutrophil recovery (Pai et al. 1994). CT is also valuable for the demonstration of such cavitary lesions. These latter lesions may sometimes give rise to a secondary aspergilloma. Sudden life-threatening hemoptysis may supervene.

Another major manifestation of acute pulmonary aspergillosis is hemorrhagic pulmonary infarction caused by invasion and thrombosis of a large pulmonary artery. The classic clinical manifestation is sudden onset of fever, and a pleural friction rub, suggesting pulmonary thromboembolism. If uncontrolled, the aspergillus infection may extend by invasive growth into neighboring organs and structures, e.g. the mediastinum, ribs, vertebrae, esophagus, and pericardium (Ribrag et al. 1993).

The clinical manifestations described above in acute pulmonary aspergillosis are not diagnostic for this infection (Saugier-Veber et al. 1993). The diagnosis can only be made with certainty by histopathological demonstra-

tion of the fungus in a fine-needle lung biopsy, but this is not always possible in acutely ill patients, not infrequently with severe hemorrhagic diathesis. Also the tissue obtained may show only necrotic debris from infarcted areas without the characteristic aspergillus hyphae. Often, therefore, the diagnosis may only be strongly suspected by judicious interpretation of the clinical and radiological manifestations, in the presence of the most important risk factors for invasive aspergillus infection. In this setting the presence of *Aspergillus* spp. in tracheobronchial secretions should not be ignored and ascribed to contamination or colonization. It increases the likelihood of pulmonary aspergillosis.

Aspergillus can more readily be found in sinus washings and biopsies from necrotic lesions in the sinuses. The use of surveillance cultures is not common practice in most hospitals but may be useful. The value of antigen and antibody detection in the serological diagnosis and prognosis of invasive aspergillosis is discussed below. The value of imaging in the diagnosis of invasive aspergillosis is highlighted in reviews by Worthy et al. (1997) and Berger (1998).

Systematic CT scanning appears to allow earlier diagnosis of invasive aspergillosis. A number of studies have shown that the halo sign was highly indicative of invasive pulmonary aspergillosis in neutropenic patients. The image occurs early in the disease and predicts aspergillosis before typical cavitation. The experience of one center highlights the value of CT scanning in the early diagnosis of invasive aspergillosis. Caillot and colleagues have systematically performed CT scanning over the past 7 years in febrile neutropenic patients with pulmonary x-ray infiltrates (Caillot et al. 1997). This approach allowed the investigators to detect suggestive CT halo signs in 92 percent of 23 histologically proven and 14 highly probable invasive aspergillosis patients, compared with 13 percent before the period of analysis. The mean time to invasive pulmonary aspergillosis diagnosis was reduced dramatically from 7 ± 5.5 to 1.9 ± 1.5 days. Other centers have reported similar findings (Franquet et al. 2001).

Aspergillus fungemia

Aspergillus fungemia is encountered infrequently. In most published studies of fungemia, it is seldom mentioned. The interpretation of these data is difficult because media contamination, yielding false-positive results of cultures, often occurs. Therefore, determining the importance of aspergillus fungemia in the immunocompromised patient is particularly difficult. A review of the literature of aspergillus fungemia by Duthie and Denning (1995) suggests that both clinical and laboratory criteria have to be critically appraised before assigning the isolation of *Aspergillus* spp. from blood into one of the following: a genuine case of fungemia or, where the organism has been isolated from the blood but in an atypical clinical setting, a case not compatible

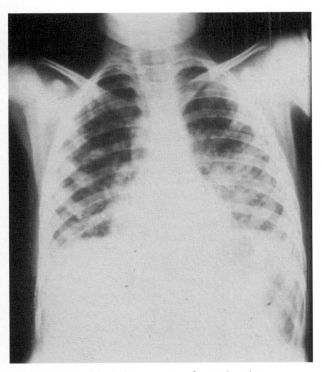

Figure 34.10 *Radological appearance of acute invasive aspergillosis*

with disseminated infection (pseudofungemia). In their review of 30 cases of genuine fungemia and 34 cases of pseudofungemia, and the inclusion of two further cases of genuine infection from their own practice, a number of features emerge. A wide range of media and blood culture systems was used to isolate *Aspergillus* spp. The median time to positive blood culture was 8.5 days (range 1–27 days) in the genuine cases. Genuine aspergillus fungemia was observed more often after cardiac surgery or during neutropenia than in other settings. Other patients at risk for aspergillus fungemia were similar to those at risk for invasive aspergillosis, including patients with the acquired immunodeficiency syndrome (AIDS). Of 19 patients who were treated, seven (44 percent) survived. In the group of patients with aspergillus pseudofungemia, there were no deaths, and cultures of additional specimens from the same patient were not positive. The conclusion to be drawn from the numbers of disseminated aspergillosis is that aspergillus fungemia occurs at a far greater rate than is normally detected. Development of an optimal blood culture system may be justified for the increasing numbers of high-risk patients. Many modifications of classic blood culture systems for optimizing the isolation of fungi have been described, but very few of these have resulted in an increased isolation rate of *Aspergillus* spp. It is possible that agitation of some of the static systems may enhance the recovery of hyphal elements of this organism (Duthie and Denning 1995).

Cerebral aspergillosis

The most important organ affected in disseminated aspergillosis, apart from the lungs, is the brain (invasive aspergillosis infection), because of both its frequency and its staggering mortality. The clue to the interpretation of this clinical manifestation is the striking tendency for aspergillus hyphae to invade cerebral vessels, causing thrombosis and infarction. Accordingly, the typical manifestations of cerebral aspergillosis are focal neurological deficits including hemiparesis, focal seizures, and cranial nerve palsy. In addition, progressive signs of CNS depression may be seen. Solitary or multiple abscesses may form in an infarcted area, and a CT or MRI scan will show abnormalities consistent with infarctions or abscesses. Although the clinical and radiological manifestations of cerebral aspergillosis are not diagnostic, the presence of CNS manifestations as described in a febrile, immunocompromised patient with pulmonary infiltrates must be considered highly suggestive of disseminated aspergillosis.

Ocular infections

Three forms of ocular infection with *Aspergillus* spp. have been recognized. The traumatic implantation of the fungus into the eye may result in a corneal ulcer which can progress to perforation. Endophthalmitis is an uncommon condition, but it has been described in drug abusers, patients with endocarditis, and organ transplant recipients. It can arise following ocular trauma or hematogenous spread of the fungus. Hematogenous spread is more usual in immunocompromised patients. The symptoms include ocular pain and impaired vision.

Orbital aspergillosis can develop as an extension from infection of the paranasal sinuses. The presenting symptoms include orbital pain, proptosis, and loss of vision. In 25 percent of cases the infection spreads into the brain and causes death (Sugata et al. 1994).

Endocarditis and myocarditis

Aspergillus endocarditis tends to occur in patients undergoing open heart surgery, although it has also been described as a complication of parenteral nutrition and drug addiction. The aortic and mitral valves are the most frequent sites of infection (Motte et al. 1993). It often gives rise to large friable vegetations and large emboli are common.

The symptoms and clinical signs are similar to those of bacterial endocarditis, with prolonged fever and abnormal heart murmurs. More specific features include large friable vegetations. Emboli that obstruct major arteries, particularly those of the brain, occur in about 80 percent of cases.

Myocardial infection with abscess formation or mural vegetations may occur as a result of hematogenous dissemination. Myocarditis has been reported in about 15 percent of patients dying with disseminated aspergillosis. It can result in nonspecific ECG abnormalities or congestive heart failure.

Osteomyelitis

Aspergillus osteomyelitis is an uncommon condition, but children with chronic granulomatous disease seem to be at particular risk. In these children, spread from an adjacent lung lesion is usual, and the ribs and spine are the most common sites of aspergillus infection. In adults, the spine is also a very common site of infection, but hematogenous spread of the fungus may be more common. Paraplegia can occur.

Otomycosis

Otomycosis is the name given to growth of *Aspergillus* spp., usually *A. niger* or *A. fumigatus*, within the external auditory canal (Paulose et al. 1989). Patients present with decreased hearing, itching, pain, or discharge from the canal. Otoscopy reveals greenish or black fuzzy growth on the cerumen or debris in the auditory canal. Treatment usually consists of careful cleaning of the canal and the application of topical nystatin suspension or ointment morning and evening for 2–3 weeks. Imidazole creams such as econazole nitrate also give excellent results. The course is chronic with acute episodes, especially in summer, and intermittent remissions. With antifungal treatment the prognosis is good.

Aspergillus spp. may invade the external auditory canal of immunocompromised patients, extending into contiguous bone or even brain.

Skin infections

Two different forms of cutaneous aspergillosis have been reported in immunocompromised patients (Carlisle 1978; McCarthy et al. 1986; Allo et al. 1987). Cutaneous lesions may arise at catheter insertion sites and as the source of a subsequent disseminated infection. The lesions begin as erythematous to violaceous, edematous, indurated plaques which evolve into necrotic ulcers covered with a black eschar.

In about 5 percent of patients with aspergillosis, hematogenous spread of infection gives rise to cutaneous lesions. These may be single or multiple, well-circumscribed, maculopapular lesions which become pustular. They evolve into ulcers with distinct borders covered by a black eschar. The lesions enlarge and may become confluent.

Primary cutaneous aspergillosis requires direct inoculation and is usually seen in immunocompromised hosts (reviewed in Groll et al. 1998b; Walsh 1998). Primary cutaneous infection has been reported at intravenous catheter sites, under occluded skin, in surgical wounds and burns, and in damaged skin, for example, venisection sites and pyoderma gangrenosum. Other cases have been reported where patients were nursed on a spinal rotorest bed designed to obviate the need for turning patients with unstable spinal injuries. The bed therefore rotates to alter sites of pressure, but the skin remains occluded with the bed sheets. Another case of primary cutaneous aspergillosis has been reported in a tetraplegic patient where there were erythematous and violaceous papules and plaques studded with pustules (Galimberti et al. 1998). It is anticipated that as the numbers of immunocompromised hosts continue to increase, the numbers of primary cutaneous aspergillosis will continue to increase in parallel with a rise in cases of invasive pulmonary aspergillosis.

Infections of the gastrointestinal tract

Gastrointestinal tract infection has been detected in 40–50 percent of patients dying with disseminated infection. The esophagus is the most frequent site of involvement, but intestinal ulcers also occur and these often result in bleeding or perforation.

Hepatic and splenic infection

This has been seen in up to 30 percent of patients with disseminated aspergillosis. The symptoms include liver tenderness, abdominal pain, and jaundice, but many patients are asymptomatic. CT scans will reveal numerous, small, radiolucent lesions scattered throughout the liver. Modest elevations in alkaline phosphatase or bilirubin concentrations can often be detected.

Aspergillosis in solid organ transplants

The incidence of invasive aspergillosis varies according to the organ transplanted, with renal transplant recipients being least at risk. Lung transplantation for cystic fibrosis carries a high risk of invasive aspergillosis (Bag 2003). In some units preoperative colonization remains a contraindication to transplantation. In early studies in lung transplant recipients, 27.5 percent developed a fungal infection postoperatively, and invasive aspergillosis accounted for 9.8 percent. Other reviews show the following pattern of incidence: liver recipients, 1–4.5 percent; kidney recipients, 0.5–2.2 percent; lung or lung–heart recipients, 18 percent (Panackal et al. 2003).

Other case series indicate that the incidence of invasive aspergillosis in liver transplant recipients appears to be 1.5–14.7 percent and the time period of highest risk post-transplant has been reported to be between 14 and 100 days. The predictors for invasive fungal infections complicating orthotopic liver transplantation (OLT) include severe renal insufficiency, prolonged operative time (>11 h), retransplantation, and colonization within 3 days of OLT. If a number of these predictors are present then the incidence of invasive fungal infection can be as high as 67 percent.

The risk factors for invasive aspergillosis in solid organ transplant recipients include: pulse steroid administration, the use of OKT3, antibiotic usage, organ failure, retransplantation, thrombocytopenia, and hospital construction. Recent data support the association between cytomegalovirus (CMV) and fungal infections in liver transplant recipients. CMV infection has been found to be present in 28.5 percent of cases of invasive aspergillosis. The use of antilymphocyte globulin, antithymocyte globulin, OKT3 monoclonal antibody, and the number of episodes of acute rejection treated are all associated with an increased risk of fungal infections.

Aspergillosis in solid organ transplant recipients present nonspecifically and their signs and symptoms often overlap with those of other infectious and noninfectious processes. In any infection diagnosed in a solid organ transplant recipient, a careful search should be made for metastatic infection, especially of the skin, the skeletal system, and the CNS. Pulmonary symptoms predominate – including nonproductive cough, pleuritic chest pain, dyspnea and low-grade fever. Chest radiology may suggest a patchy pneumonia, cavitary lung disease, or a pulmonary embolus; the chest x-ray may also be normal. From the lungs, *Aspergillus* spp. may disseminate to almost any organ, including the brain, liver, spleen, kidney, thyroid, heart, blood vessels, bone and joints, among others. *Aspergillus* spp. may also invade the paranasal sinuses, gastrointestinal tract, or skin; rarely, it may gain entry through an intravenous catheter. *Aspergillus* spp. may cause peritonitis in renal transplant recipients on continuous ambulatory peritoneal dialysis (CAPD), or in liver transplant recipients with intra-abdominal abscesses. Endophthalmitis may occur, usually in conjunction with endocarditis. Other unusual presentations of aspergillus infections in solid organ transplant recipients include tracheobronchitis, with infection limited to the anastomotic site and large airways in heart–lung and lung transplant recipients, and wound infections.

LUNG TRANSPLANTATION

Aspergillus infections, documented in 6–8 percent of lung transplant recipients, remain among the most significant opportunistic infections after lung transplantation (Kubak 2002). Lung transplant recipients have a unique predisposition for and clinical manifestations of aspergillus infection. Direct communication of the transplanted organ with the environment, impairment in local host defenses (e.g. mucociliary clearance and cough reflex), disruption of lymphatic drainage, ischemic airway injury, altered alveolar phagocytic function, and an overall greater requirement of immunosuppression render lung transplant recipients uniquely susceptible to aspergillus airway colonization and invasive disease. Isolated tracheobronchitis and bronchial anastomotic infections are entities distinct from aspergillus pneumonia. In one study, although the early mortality of patients with bronchial anastomotic aspergillus infections did not differ significantly from patients without these infections, their long-term survival was reduced.

Aspergillus infections in lung transplant recipients have been reported largely as institutional experiences with few patients or case reports. Consequently, the risk factors, role of aspergillus airway colonization, and variables influencing outcome have not been well defined.

A recent review has identified a total of 159 cases of aspergillus infections in 40 published reports (Singh and Husain 2003). Of these, 87 were individually detailed cases, and 72 were summarized in reports containing 2–14 cases. The type of lung transplantation was reported in 66 patients. Overall, 51.5 percent (34 of 66) of the patients with aspergillus infections were single lung transplant recipients, 34.9 percent (23 of 66) were bilateral lung transplant recipients, and 13.6 percent (9 of 66) were heart–lung transplant recipients. Patients ranged in age from 19 to 64 years (median 48 years), and 64 percent were male.

The median incidence of aspergillus infection was 6.2 percent and ranged from 2.2 to 20 percent. Aspergillus infections occurred a median of 3.2 months and up to 5 years after lung transplantation. Twenty-six percent of all infections occurred within 1 month, 51 percent within

3 months, and 72 percent within 6 months of lung transplantation. Sixteen percent of the infections occurred after 6 months, and only 12 percent were documented after 12 months of transplantation.

Time of onset of infections varied significantly for different types of lung transplant recipients. The median time to onset after transplantation was 0.7 months for heart–lung, 3.9 months for bilateral lung, and 5 months for single lung transplant recipients ($P=0.046$). Single lung transplant recipients developed aspergillus infections significantly later after transplantation than all other patients (median 4.9 months versus 2.1 months, $P=0.019$). Sixty-eight percent (15 of 21) of the single lung as compared with 35 percent (9 of 26) of the bilateral lung transplant recipients developed aspergillus infection at >3 months after transplantation ($P=0.021$). Time of onset also differed for various types of aspergillus infections. Invasive pulmonary or disseminated aspergillosis occurred significantly later than tracheobronchitis ($P=0.03$). Of the aspergillus infections occurring within 3 months of transplantation, 75 percent (21 of 28) were tracheobronchitis or bronchial anastomotic infections, 18 percent (5 of 28) were invasive pulmonary infections, and 7 percent (2 of 28) were disseminated invasive infections ($P=0.06$). Other variables (e.g. age, chronic obstructive pulmonary disease as underlying lung disease, antilymphocyte globulin use for induction, use of antifungal prophylaxis, prior rejection or CMV infection) did not differ for patients with early- versus late-onset aspergillus infections.

Of 78 lung transplant recipients, 29 (37 percent) had tracheobronchitis, 25 (32 percent) had invasive pulmonary aspergillosis, and 16 (20 percent) had bronchial anastomotic infection. Only 10 percent (8 of 78) of the patients had disseminated invasive aspergillosis. Of these, 4 percent (3 of 78) had multiple sites of dissemination and 5 percent (5 of 78) had a single extrapulmonary site of infection documented; these included osteomyelitis in two patients (involving pelvis in one and vertebrae in the other), and thoracic wound infection, endophthalmitis, and retroperitoneal abscess in one patient each. Eight percent (3 of 36) of the patients with invasive aspergillosis were diagnosed only at autopsy.

Fifty-eight percent (19 of 33) of the single lung as compared with 45 percent (14 of 31) of all other patients had invasive aspergillosis ($P=0.3$). Single lung transplant recipients were older ($P=0.006$) and more likely to have chronic obstructive pulmonary disease as an underlying illness ($P=0.05$). Of the patients with invasive pulmonary aspergillosis, 67 percent (16 of 24) were single lung, 21 percent (5 of 24) bilateral lung, and 13 percent (3 of 24) heart–lung transplant recipients.

Whether invasive aspergillosis first occurred in the native or transplanted lung could be ascertained in ten patients. In nine of these, invasive aspergillosis was first documented in the native lung. The underlying lung disease in these patients was chronic obstructive pulmonary disease in five, 1-antitrypsin deficiency in three, idiopathic pulmonary veno-occlusive disease in one, and unknown in one patient. Invasive aspergillosis occurred a median of 7 months post-transplant in these patients (range 18 days to 5 years). Seven of these 10 patients died.

In total, 91 percent of the aspergillus infections were due to *A. fumigatus*, 2 percent (1 of 40) to *A. flavus*, 2 percent (1 of 40) to *A. niger*, and 5 percent (2 of 40) were mixed infections due to two *Aspergillus* spp. Non-*A. fumigatus* spp. were associated with 10 percent (3 of 29) of tracheobronchial infection episodes, but were never associated with invasive aspergillus infections. Only 15 percent (4 of 27) of the aspergillus infections were accompanied by fever. Fever was significantly more likely to be documented in patients with invasive as compared with localized tracheobronchial aspergillus infections; 50 percent (2 of 4) of the patients with disseminated aspergillosis, 20 percent (2 of 10) of those with invasive pulmonary aspergillosis, and 0 percent (0 of 13) of those with tracheobronchial or bronchial anastomotic infections were febrile ($P=0.046$). Prior rejection episodes were documented in 50 percent, and use of antilymphocyte preparation or OKT3 in 38 percent of the lung transplant recipients with invasive pulmonary aspergillosis. The radiographic appearance of pulmonary aspergillus infections was a focal consolidation in 40 percent (4 of 10), cavitary lesions in 30 percent (3 of 10), and nodules or mass-like lesions in 30 percent (3 of 10).

Data on aspergillus colonization could be assessed in 395 patients in 25 studies that had numerator and denominator data available. *A. fumigatus* was the most common species cultured (63 percent, 174 of 276), followed by *A. niger* (12 percent, 34 of 276), *A. versicolor* (9 percent, 26 of 276), *A. flavus* (8 percent, 21 of 276), *A. nidulans* (4 percent, 10 of 276), *A. glaucus* (3 percent, 7 of 276) and *A. terreus* (1 percent, 4 of 276). In 60.9 percent (14 of 23) of the patients, the first positive culture for *Aspergillus* was detected within 3 months of transplantation, in 4.3 percent (1 of 23) between 3 and 6 months, in 26.1 percent (6 of 23) between 6 and 12 months, and in 8.7 percent (2 of 23) at >12 months after transplantation. Of the patients with an airway culture positive for *Aspergillus*, 7.4 percent (15 of 207) had tracheobronchitis or bronchial anastomotic infection, and 20.1 percent (41 of 202) had invasive aspergillosis upon first detection of *Aspergillus*. An additional 5.7 percent (11 of 193) of the patients with a positive airway culture progressed to aspergillus infection at a later time point. These included 3.1 percent (6 of 193) with progression to tracheobronchitis, and 2.6 percent (5 of 193) with progression to invasive aspergillosis. The mortality rate was 62.5 percent (5 of 8) in those with progression due to infection and 1 percent (1 of 105) in those who remained colonized.

Overall mortality in lung transplant recipients with aspergillus infections was 52 percent (59 of 113) and varied significantly with site of infection, type of lung transplantation, time of onset of infection, and antifungal therapy employed. Mortality rate was 23.7 percent (9 of 38) for patients with tracheobronchial or bronchial anastomotic infections, 81.9 percent (18 of 22) for patients with invasive pulmonary aspergillosis, 67 percent (2 of 3) for those with disseminated invasive infection, and 33 percent (1 of 3) for patients with only an extrapulmonary site of infection ($P=0.0001$). Mortality for single lung transplant recipients (60 percent, 18 of 30) was significantly higher than for bilateral lung (40 percent, 6 of 15) or heart–lung transplant recipients (56 percent, 3 of 9; $P=0.03$). Patients with late-onset aspergillus infections had significantly higher mortality than those with early-onset infections (57 percent, 12 of 23 versus 28 percent, 7 of 25; $P=0.045$). However, when only the patients with invasive aspergillosis were analyzed, the mortality rate did not differ for those with late (73 percent, 8 of 11) versus early-onset aspergillus infections (71 percent, 5 of 7). Age (44.7 versus 45.5 years), rejection (76 percent, 16 of 21 versus 47.1 percent, 8 of 17), or antilymphocyte globulin use (34.6 percent, 9 of 26 versus 34.5 percent, 10 of 29) did not differ significantly in patients with aspergillus infection who lived versus those who died, respectively.

Of lung transplant recipients with invasive aspergillosis, 18 received amphotericin B, five received a liposomal formulation of amphotericin B, and one patient received amphotericin B plus itraconazole. Mortality was significantly higher in patients with invasive aspergillosis who received amphotericin B (83 percent, 15 of 18) as compared with those who received a liposomal formulation of amphotericin B or amphotericin B plus itraconazole (33 percent, 2 of 6; $P=0.038$). Mortality rate, however, did not differ for patients with tracheobronchitis or bronchial anastomotic infections who received amphotericin B (33 percent, 2 of 17) as compared with the other regimens (19.3 percent, 5 of 26, $P >0.20$). Only four patients underwent pneumonectomy for the treatment of invasive aspergillosis; all were single lung transplant recipients and three of four of these patients died. Of the four patients who underwent pneumonectomy, one each had received amphotericin B, a liposomal formulation of amphotericin B, and amphotericin B plus itraconazole as antifungal therapy; the antifungal therapy was unavailable in one patient.

Although previous reports have alluded to a higher incidence of aspergillus infections in single lung transplant recipients, the report by Singh and Husain (2003) has highlighted the unique characteristics and overall greater impact of aspergillus infections after single lung transplantation. Single lung transplant recipients were significantly older, more likely to have chronic obstructive pulmonary disease as an indication for lung transplantation, or have developed aspergillus infections later after transplantation, tended to have a higher incidence of invasive aspergillosis, and had higher mortality than all other lung transplant recipients with aspergillus infections.

In the vast majority of the single lung transplant recipients, invasive aspergillosis was documented in the native lung, suggesting that the native lung may harbor a nidus for *Aspergillus* and serve as a source of infection in these patients. It is known that single lung transplant recipients are more likely to have chronic obstructive pulmonary disease, a condition which predisposes to airway colonization with *Aspergillus*. A molecular epidemiologic study in lung transplant recipients that used DNA primers for strain typing documented that, although the clinical strain of a single lung transplant recipient was identical to the one collected at home, the isolates in other lung transplant recipients were deemed to be more likely of nosocomial origin. Thus, aspergillus infections in single lung transplant recipients likely represent reactivation of a preexisting focus. The higher mortality (despite pneumonectomy) in these patients may be related to the higher incidence of invasive pulmonary as compared with tracheobronchial infections.

Characterized by endobronchial lesions ranging from erythema to ulceration and pseudomembrane formation, tracheobronchitis is an entity observed specifically in lung transplant recipients. Lesions in the vicinity of or involving the anastomotic site can result in bronchopleural fistula and fatal hemorrhage. Bronchopleural fistulas were documented in 4.4 percent (2 of 45) of the lung transplant recipients with tracheobronchial aspergillus infections. A majority of the patients in whom tracheobronchitis or bronchial anastomotic infections occurred were bilateral lung transplant recipients. Bilateral lung and right lung transplant recipients in a previous study were documented to have a higher incidence of bronchial anastomotic infection. It was proposed that bilateral lung transplant recipients may be more likely to have greater impairment of cough reflex and mucociliary clearance, leading to a higher risk of colonization and subsequent infection of the anastomotic sites. Also, bilateral lung transplant recipients undergo longer operations and may have a higher risk of ischemia in the area of anastomosis.

Two observations relevant to clinical and radiographic manifestations of aspergillus infections in lung transplant recipients deserve mention. Fever was documented in only 15 percent of the patients with aspergillus infections. Patients with invasive pulmonary or extrapulmonary aspergillosis were more likely to be febrile than those with tracheobronchitis or bronchial anastomotic infections. Second, aspergillus lung infections frequently lacked a characteristic radiographic appearance and presented most often with focal areas of patchy consolidation or infiltrate. Nodular lesions were documented in 30 percent of the patients with pulmonary aspergillosis in our study and in 27 percent in a previous report. Characteristic CT findings (e.g. halo sign) were distinctly unusual.

An optimal antifungal prophylactic strategy in lung transplant recipients has not been defined. The antifungal agent employed and the duration of prophylaxis remain controversial in these patients. Aerosolized amphotericin B and itraconazole are currently the most frequently employed antifungal prophylactic agents in lung transplant recipients. Given the variability in the route of administration, dosages, and duration of prophylaxis, the efficacy of either agent cannot be discerned incontrovertibly. However, the analysis described here allows the following conclusions that may be relevant in the approach toward antifungal prophylaxis or designing antifungal prophylactic trials. Tracheobronchial or anastomotic infections were the most frequently occurring infections within 3 months; the median time to onset was 2.7 months. Antifungal prophylaxis, if employed, may be limited to 3 months in most patients. Such a strategy utilizing an 'effective' antifungal agent could potentially prevent 62 percent of the trancheobronchial, 36 percent of the invasive pulmonary, and 50 percent of the disseminated infections. Invasive pulmonary aspergillosis, however, occurred a median of 5.5 months after transplantation. Single lung transplant recipients were significantly more likely to develop aspergillus infections after 3 months, and the native lung may have been the source in these patients. Thus, a longer duration of prophylaxis, employing a systemic antifungal agent able to achieve adequate pulmonary levels, may be preferable in these patients.

In summary, the review has highlighted an overall greater morbidity and poorer outcome in single lung transplant recipients with aspergillus infections. The single lung transplant recipients with aspergillus infections were older, more likely to have chronic obstructive pulmonary disease as an underlying illness and to have developed aspergillus infections later after transplantation, and tended to have a higher incidence of invasive aspergillosis than other lung transplant recipients. Overall mortality in lung transplant recipients with aspergillus infections was 52 percent; single lung transplant recipients and patients with late-onset aspergillus infections had significantly higher mortality. The unique epidemiologic characteristics and variables influencing outcome delineated in the review have implications relevant for the management and for designing antifungal prophylactic studies in lung transplant recipients in the future.

HEART TRANSPLANTATION

Most systemic fungal infections in heart transplant recipients are caused by *Aspergillus*, second in incidence to CMV. The infection manifestation is pneumonia occurring around 35 days post-transplantation. In one study in 844 heart transplant recipients between the years 1980 and 1998, there were 21 definite cases of invasive pulmonary aspergillosis and four probable. The median time to onset was 46 days. The presentation and diagnostic features included fever/cough, single or multiple pulmonary nodules, abnormal radiological findings (nodules), and a positive culture. Neutropenia was not present (Montoya et al. 2003).

Further information on aspergillosis in solid organ transplant recipients can be found in Kubak (2002) and Singh (2003).

Invasive aspergillus infection in AIDS

Invasive aspergillus infections appear to be rather uncommon in individuals with AIDS, although the incidence appears to be increasing (Minamoto et al. 1992; Lortholary et al. 1993; Khoo and Denning 1994). The rarity of invasive aspergillosis during the first few years of the AIDS epidemic led to its exclusion as an AIDS-defining illness. The low incidence of this infection in patients with AIDS is highlighted by the report from the Memorial Sloan Kettering Cancer Center, New York (Pursell et al. 1992). Of the 972 patients with AIDS who were observed over a 10-year period, Aspergillus spp. were isolated from various respiratory sites from 45 patients before death. Invasive aspergillosis was documented postmortem in four of these patients and was strongly suspected in an additional patient in whom a postmortem examination was not performed. Traditional risk factors for the development of invasive disease (neutropenia, hematological malignancy, and/or corticosteroid use) were present in all the patients with invasive aspergillosis. A review of the literature up to 1992 reveals reports of 13 cases of invasive aspergillosis in patients with AIDS.

In a report from France (Daleine et al. 1993) where respiratory secretions from 614 patients positive for the human immunodeficiency virus (HIV) over a 2-year period were analyzed retrospectively, it was found that the prevalence of Aspergillus spp. isolated was 2.7 percent in BAL specimens (21 of 757), 15.1 percent in sputum specimens (3 of 53), 12.6 percent in bronchial aspirates (11 of 87), 7 percent in protected brush specimens (2 of 28) and 16.6 percent in lung biopsy specimens (4 of 24). A total of 20 patients (rate = 3.14 percent) had lung specimens positive for Aspergillus spp.

Patients with advanced HIV infection certainly have profound underlying immunosuppression and are exposed to aspergillus conidia, so it would appear that these infections should be more common. One explanation for the low prevalence may be the fact that neutrophils and macrophages are the main immunological defenses against Aspergillus spp. and that these components of the host defense system are more or less spared in HIV infection compared with the defects in T lymphocytes. However, in vitro, neutrophils from children with HIV appear to show impaired ability to kill hyphae of A. fumigatus (Roildes et al. 1993a).

One explanation may be that these infections are now being diagnosed more frequently than in the past

because of less reluctance to perform bronchoscopies and lung biopsies. Other explanations may relate to an increased prevalence of risk factors for invasive aspergillosis in the HIV-infected population. One consideration is whether advanced HIV infection itself is a risk factor, and whether certain pulmonary infections, such as tuberculosis or Pneumocystis carinii pneumonia, may lead to pulmonary damage that predisposes to invasive aspergillosis. Aerosolized pentamidine may also be a risk factor, because most patients with invasive aspergillosis reported in the literature had received this form of therapy. Finally, it should be considered whether prolonged treatment with certain antifungal agents having little activity against Aspergillus spp., and thus possibly leading to colonization, is an additional risk factor. It must be emphasized, however, that, as yet, there is little to support the establishment of anything but neutropenia and steroid treatment as risk factors for invasive aspergillosis infection in people infected with HIV.

The clinical presentation of invasive aspergillosis in the HIV-infected population ranges from isolated cutaneous involvement to widely disseminated disease. Most patients reported thus far have advanced HIV infection, with CD4 counts below $50/mm^3$ and histories of other opportunistic infections. Pulmonary infections are the most common manifestation of invasive aspergillosis in patients with AIDS (Miller et al. 1994). The lungs are involved as the sole site in about 60–80 percent of all cases, and lung involvement with extrapulmonary disease occurs in about 25 percent. Extrapulmonary infection in the absence of lung involvement has been seen in about 10–15 percent of cases.

Extrapulmonary invasive aspergillosis has been seen in virtually all organs of the body in patients with AIDS (Kempner et al. 1993). The most common site has been the CNS (Woods and Goldsmith 1990); focal abscesses, meningitis, and spinal cord lesions have all been described. Only a few patients with CNS lesions have been intravenous drug users; most probably result from hematogenous spread, presumably from the lungs, but there is an occasional case in which it seems likely that spread from the sphenoid sinus occurred, with resulting basilar meningitis. The heart has been the next most commonly involved extrapulmonary organ; many pathological descriptions have shown multiple myocardial abscesses. Gastrointestinal aspergillosis has been described in one patient, with scattered ulcers extending from the stomach to the colon. Abscesses of the kidney, liver, spleen, and peripancreatic spaces have all been reported in the literature; most were diagnosed postmortem.

Invasive pulmonary aspergillosis in neonates

Only a few cases of invasive aspergillosis have been reported in neonates and infants <3 months of age. A case report from Frankfurt, Germany reviews 43 cases of invasive aspergillosis during the period 1955–96 (Groll et al. 1998a). Eleven of the 44 patients had primary cutaneous aspergillosis, 10 had invasive pulmonary aspergillosis, and 14 had disseminated disease. Most infections were nosocomial in origin. Prematurity (43 percent), proven chronic granulomatous disease (14 percent), and a complex of diarrhea, dehydration, malnutrition, and invasive bacterial infection (23 percent) accounted for the majority of underlying conditions. At least 41 percent of the patients had received corticosteroid therapy before diagnosis, but only one patient had been neutropenic. Among patients who received medical and/or surgical treatment, outcome was relatively favorable, with an overall survival rate of 73 percent. From this review it is evident that invasive aspergillosis may occur in neonates and young infants. The current recommendation regarding treatment is high-dose amphotericin B and appropriate surgical interventions.

A North American experience of aspergillosis in children is illustrative (Abbasi et al. 1999). Sixty-six patients with culture-documented aspergillus infection were identified from a 34-year period. The most common underlying diagnosis was leukemia. Risk factors included neutropenia, immunosuppression, and prior antibiotic therapy. On the basis of clinical presentation, 23 patients were believed to have disseminated disease and 43 to have localized disease. The lung was the most frequently affected organ. Overall mortality was 85 percent within the first year of diagnosis. Patients who presented with disease in sites other than the lungs fared better than patients with initial pulmonary involvement. A further North American experience is reviewed by Wright et al. (2003).

EPIDEMIOLOGY OF ASPERGILLOSIS

Aspergillus spp. grow well on virtually any organic debris. They are, therefore, found naturally in leaves, grains, soil, and composts. Compost is a particularly rich source for the thermotolerant species. *A. fumigatus* is often the predominant fungal member of the compost microbiota, with more than 10^8 conidia per gram of compost material. Likewise, numbers of airborne conidia in such places as hay barns have been estimated to be as high as $10^9/m^3$. These numbers are exceptional, however, and the number of conidia in outdoor air probably rarely exceeds $200/m^3$. Interestingly, indoor air usually contains larger numbers of aspergillus conidia than outdoor air sampled over the same period. This has led hospital epidemiologists to advocate the use of high-efficiency filtration units to reduce to undetectable levels the number of airborne particles in the rooms of patients at risk. Various hospital routines may increase the level of contamination. Bursts of airborne conidia have, for example, been detected during cleaning procedures

(Rhame et al. 1984). In general, however, with the exception of those individuals with occupational exposure, everyday exposure to conidia of *Aspergillus* spp. may be of little concern to most of the population. Nosocomial outbreaks of invasive aspergillosis are often attributed, correctly or incorrectly, to an environmental point source, such as a contaminated air vent or a demolition project (Walsh and Pizzo 1988; Walsh and Dixon 1989). The interrelationship between isolates of *A. fumigatus* from outside air with strains implicated in clusters of invasive disease has been analyzed by various molecular techniques (Loudon et al. 1993, 1994; Symoens et al. 1993; Girardin et al. 1994b; Goodley et al. 1994; Tang et al. 1994; Kennedy et al. 1995; Anderson et al. 1996).

HOST DEFENSE

Host defense to invasive aspergillosis is primarily an innate mechanism. By contrast, in the allergic forms of aspergillosis, the acquired responses appear to contribute more to the pathology than to the control of disease. In immunocompromised patients, various defects in neutrophil function have been recognized in, for example, patients with chronic granulomatous disease or patients with prolonged neutropenia induced by cytotoxic drugs. Elucidation of the host defense mechanisms in aspergillosis has been the focus of many studies (reviewed by Rhodes 1993, Kauffman and Tomee 2002).

Innate mechanisms

COLONIZATION OF THE RESPIRATORY MUCOSA

Initial interaction with host cells and colonization of mucosal surfaces are now considered crucial steps in the establishment of infection. Ligand-mediated adherence may be a component of this process. For example, fibrinogen, which plays a key role in inflammatory reactions and in the coagulation pathway, appears to be a good candidate for mediating adherence. Various studies have shown that, among filamentous fungi belonging to different groups (opportunistic fungi, strictly saprophytic or phytopathogenic fungi, and dermatophyte or related species), only the pathogenic aspergilli, particularly *A. fumigatus*, bind crude fibrinogen (reviewed by Bouchara et al. 1988, 1995, and Annaix et al. 1992). The binding was detected mainly at the surface of the conidia. Further studies demonstrated that the site of maximum fibrinogen binding was on the outer conidial surface but that binding sites also lay in the inner part of the wall and in cytoplasmic granules (Annaix et al. 1992). Other virulence determinants of *A. fumigatus*, relevant in vivo, may include proteases and elastase enzymes (Kolattukudy et al. 1993; Jaton-Ogay et al. 1994; Bouchara et al. 1995).

Aspergillus spp., unlike many other fungal species, can colonize the respiratory mucosa. This can occur in the normal airway, and the individual may become sensitized, causing ABPA (see section on Allergic aspergillosis below). More commonly, the fungus colonizes damaged airways, such as in patients with healed tuberculosis, bronchiectasis, and cystic fibrosis (see section on Allergic aspergillosis below). Although the type III hypersensitivity reaction to aspergillus antigens in the bronchial wall in ABPA and the importance of the neutrophil in host defense against *Aspergillus* spp. are well appreciated (see section on Immune response below), the mechanisms of colonization of the respiratory mucosa are poorly understood. Amitani and colleagues (1992) have shown that culture filtrates of most clinical isolates of *A. fumigatus* studied slow ciliary beat frequency and damage human respiratory epithelium in vitro. These changes appeared to occur concurrently. The characteristics of a factor purified from a culture filtrate of one clinical isolate of *A. fumigatus* has been described (Amitani et al. 1995). The ciliary-inhibitory activity was heat labile and associated with both high- and low-molecular-weight factors. As a result of purification of the low-molecular-weight activity, one such component was identified as gliotoxin, a known metabolite of *A. fumigatus*. It is suggested that *A. fumigatus* produces a number of biologically active substances which slow ciliary beating and damage epithelium and which may influence colonization of the airways.

INTERACTION WITH ALVEOLAR MACROPHAGES AND POLYMORPHONUCLEAR LEUKOCYTES

Numerous reviews of the acquisition and pathogenesis of invasive aspergillosis have identified common features which appear to be responsible for placing patients at risk for these infections (Holden et al. 1994). Neutropenia and broad-spectrum antibiotics and corticosteroids are consistently identified as key predisposing factors. Likewise, the recognition that patients with chronic granulomatous disease, in which phagocytic cells fail to produce the respiratory burst, are susceptible to invasive aspergillosis adds further weight to the concept that the presence of normally functioning phagocytes is the key factor in defense against invasive disease. It has been shown, in a murine model, that both neutrophil and monocyte function have to be ablated for invasive infection to develop. Bronchoalveolar macrophages are believed to serve as the first line of defense and to be responsible for the ingestion and intracellular killing of inhaled conidia (Henwick et al. 1993). Those conidia escaping the macrophages and going on to germinate and form hyphae are the target of attack by the granulocytes, which adhere to the hyphae, degranulate, and kill the fungus.

Our knowledge of the role that phagocytic cells play in the eradication of *A. fumigatus* has been advanced only recently. The published studies on the ability of phagocytic cells to kill conidia of *A. fumigatus* have produced divergent and conflicting results, largely as a result of the many different in vitro assays of phagocytosis and intracellular killing being used (Ibrahim-Granet et al. 2003).

One important aspect of host defense to *Aspergillus* in relation to phagocytosis in the alveoli may be the role of surfactant proteins. Surfactant protein A (SP-A) and surfactant protein D (SP-D) are thought to play important roles in pulmonary defense. These proteins are members of the C-type lectin superfamily, which also includes serum mannose-binding protein, conglutinin, and collectin 43. SP-A and SP-D are synthesized by type II cells and Clara cells in the lung and share many structural features. The host defense role played by SP-A and SP-D very much depends on the target organs and specific surfactant protein involved. Recent reports have shown that these proteins bind to and enhance killing and clearance of the organism in vitro (Madan et al. 1997; Allen et al. 1999). It is suggested that SP-D specifically plays a major role in the recognition of *Aspergillus* conidia in alveolar fluid (Allen et al. 1999).

ANTIPHAGOCYTIC ACTIVITY OF *ASPERGILLUS*-DERIVED PRODUCTS

It has been shown over the past 10 years that spores and spore diffusates of *A. fumigatus* suppress the production of superoxide anion and hydrogen peroxide by rodent phagocytic cells and that diffusates of *A. fumigatus* spores inhibit phagocytosis of antibody-coated radiolabeled sheep red blood cells by mouse peritoneal exudate cells. Similar experiments with human cells have shown that spore diffusates of *A. fumigatus* reduce the migration of human polymorphonuclear leukocytes (PMNL) and decrease the capacity of mouse peritoneal exudate cells to spread on glass. In addition, it has been shown that human pulmonary macrophages can bind and kill *A. fumigatus* conidia. More recently, Murayama and colleagues (1998) showed that *Aspergillus*-derived products had a suppressive effect on PMNL-mediated damage of *A. fumigatus* hyphae. The conclusion drawn from these studies is that *A. fumigatus* produces a variety of substances, some of which may suppress antifungal (anti-*A. fumigatus*) activity of human alveolar macrophages and PMNLs.

RECENT PROGRESS IN AN UNDERSTANDING OF HOST DEFENSE MECHANISMS AGAINST INVASIVE ASPERGILLOSIS

The fact that invasive aspergillosis disproportionately afflicts immunocompromised patients indicates the critical importance of the immune status of the host in this infection, but the defense mechanisms against this pathogen remain incompletely understood. Recent work suggests that the chemokine ligand monocyte chemotactic protein-1, also designated CC chemokine ligand-2 (MCP-1/CCL2), may be necessary for effective host defense against invasive aspergillosis (Morrison et al.

2003). It was found that there was a rapid and marked induction of MCP-1/CCL2 in the lungs of neutropenic mice with invasive aspergillosis. Neutralizing MCP-1/CCL2 resulted in twofold greater mortality and greater than threefold increase in pathogen burden in the lungs. Neutralization of MCP-1/CCL2 also resulted in reduced recruitment of natural killer (NK) cells to the lungs at early time points, but did not affect the number of other leukocyte effector cells in the lungs. Ab-mediated depletion of NK cells similarly resulted in impaired defenses against the infection, resulting in a greater than twofold increase in mortality and impaired clearance of the pathogen from the lungs. This study suggests that MCP-1/CCL2-mediated recruitment of NK cells to the lungs may be a critical early host defense mechanism in invasive aspergillosis and indicates that NK cells may be an important and previously unrecognized effector cell in this infection.

Another area of innate host defenses against *Aspergillus* are defensins. Defensins and other antimicrobial peptides act in the innate defense of epithelial surfaces. Human β-defensin 1 (hBD-1) has recently been shown to be expressed in airway epithelial cells and so has been implicated as a primary component of antibacterial activity in human lung. β-Defensins have been implicated in host defense of human airway epithelia. Antibacterial activity associated with low-molecular-weight, salt-sensitive, and heat-stable factors has been detected in the surface fluid of primary cultures from normal airway epithelia, and expression of hBD-1 messenger RNA has been detected in human lung. Hence, hBD-1 has been implicated as a major component of host defense in human airways. Furthermore, inactivation of hBD-1 in the lungs of cystic fibrosis patients may be an important cause for the onset of chronic lung infection. It is tempting to speculate that β-defensins have a role to play in the pathogenesis of pulmonary aspergillosis. Research in this area is continuing.

Modulation of innate mechanisms

There has been some concern expressed regarding the immunosuppressive effects of cyclosporin A on the function of peripheral blood phagocytes. Cyclosporin A has been reported to induce immunosuppressive effects on the fungicidal activity of murine polymorphonuclear monocytes against blastoconidia of *Candida albicans* in vivo and has rendered the animals susceptible to subsequent challenge with this organism (Vecchiarelli et al. 1989; Roildes et al. 1993b, c). Roildes et al. (1994a, b) addressed this issue by looking at the effects of cyclosporin A on phagocytic defenses against *A. fumigatus* in vitro and ex vivo. Cyclosporin A at therapeutically relevant concentrations did not suppress antifungal activity of phagocytes except that of circulating monocytes. However, cyclosporin A did appear to induce significant immunosuppression of phagocytes' antifungal function

at relatively high concentrations in vitro, especially when combined with corticosteroids. This observation has potential implications for patients receiving these immunosuppressive agents.

Treatment with recombinant human interferon-γ is known to augment the ability of monocytes and neutrophils from patients with chronic granulomatous disease (CGD) to generate superoxide. This observation prompted Rex et al. (1990) to look at the effect of interferon-γ on neutrophils, from patients on interferon-γ therapy for CGD, in their ability to damage aspergillus hyphae. Neutrophils from patients receiving interferon-γ produced significantly more damage to hyphal elements than those from the placebo group. The authors conclude that interferon-γ therapy improves the ability of CGD neutrophils to damage *A. fumigatus* hyphae in an in vitro assay.

Immune response

In colonizing and in invasive forms of aspergillosis, the study of the immune response has centered on the humoral response as an aid to determining the diagnosis or prognosis of patients. In ABPA, studies have attempted to explain the pathology of the disease, which appears to be caused primarily by that very response. In all studies on the humoral response to aspergillus antigens, progress has been hampered by the lack of common, reference antigens which are capable of detecting responses to all common species and by which results obtained in multiple centers may be compared.

HUMORAL RESPONSE IN ALLERGIC ASPERGILLOSIS

Patients with extrinsic allergic alveolitis or hypersensitivity pneumonitis develop IgG precipitin antibodies. The pulmonary pathology appears to be mediated by immune complex deposition, perhaps in conjunction with cell-mediated mechanisms. By contrast, IgE appears to play a major role in the pathophysiology of ABPA. The patients have elevated total IgE, which rises and falls with exacerbations and remission of clinical symptoms. Specific anti-aspergillus IgE is also demonstrable in ABPA patients and IgG anti-aspergillus precipitin antibodies are also produced by ABPA patients, with almost 100 percent having positive serological tests for these antibodies (Brummund et al. 1987). Pulmonary immune complex deposition, secondary to elevated IgG antibodies, may account for much of the inflammatory component seen during active phases of ABPA.

Significantly elevated levels of specific *A. fumigatus* IgG have been also been detected in cystic fibrosis patients with ABPA (Murali et al. 1994).

HUMORAL RESPONSE IN ASPERGILLOMA

Patients with aspergilloma, almost without exception, mount a strong IgG response to aspergillus antigens, judging by the multiple precipitin bands formed on agar gel double diffusion or counterimmunoelectrophoresis to extracts of the infecting fungus. The response does not appear to provide any protection to the patient, however. In cases of aspergilloma in which the fungus ball has formed secondary to cavitation of an invasive lesion, the ability of patients to produce antibodies is governed more by their degree of immunosuppression than by the type of disease.

HUMORAL RESPONSE IN INVASIVE ASPERGILLOSIS

As a result of their profound degrees of immunosuppression, patients with invasive aspergillosis mount little to no humoral response to the fungus when the standard serological tests are used. However, the use of purified antigens in immunoblotting procedures appears to detect antibody in immunocompromised individuals (Hearn et al. 1995). Antibody responses in these patients do not appear to confer any degree of protection in invasive aspergillosis.

CELLULAR RESPONSE IN ALLERGIC ASPERGILLOSIS

The presence of bronchocentric granulomatosis as a pathological hallmark of ABPA indicates that there must be a cell-mediated component in the response of these patients to *Aspergillus* spp. Lymphocyte proliferation in response to antigens of *A. fumigatus* has been demonstrated in some ABPA patients. Likewise, fractionation of antigenic extracts of *A. fumigatus* yielded components that elicited positive-type (type IV) skin reactivity in, and proliferation of lymphocytes from, sensitized guinea pigs.

CELLULAR RESPONSE IN ASPERGILLOMA

There is little evidence to support the presence of a cell-mediated response in aspergilloma patients. In general, the pathology of aspergillomas is one in which an inflammatory response, either acute or chronic, is minimal. Host defense against this form of infection has not been well studied.

CELLULAR RESPONSE IN INVASIVE ASPERGILLOSIS

The fungicidal activity of neutrophils against hyphae involves the release of granular contents on the surface of the organism. Both macrophages and neutrophils use oxidative and non-oxidative fungicidal mechanisms including myeloperoxidase and myeloperoxidase-independent oxidants. The granular constituents of neutrophils have all been implicated as antifungal agents and include: lactoferrin, lysozyme, cathepsin G, azarocidin, and defensins. Interferon-γ and granulocyte-macrophage colony-stimulating factors (GM-CSF) both augment oxidant release and fungicidal activity. Specific antibodies or T lymphocytes do not appear to be of major importance in host protection against *Aspergillus* spp. Complement facilitates neutrophil damage to hyphae and monocyte killing of conidia.

The role of T cells in invasive disease has been investigated in experimental systems, with little consensus as to its importance. For example, nu/nu mice fare no worse than their nu/+ littermates when infected with conidia of *A. fumigatus*. Nu/+ can be immunized protectively against the fungus, however, and the protective immunity can be transferred to nu/nu mice recipients with adherent peritoneal cells from the immunized nu/+ mice. This suggests a role for T cells and macrophages in the acquisition of immunity to aspergillosis, but the issue is still open to investigation. In humans, reviews of cases of invasive aspergillosis have not suggested that T-cell defects were primary predisposing factors for disease but rather that granulocytopenia is the risk factor most commonly associated with infection. Additional recent evidence on the incidence of invasive aspergillosis in patients with AIDS also suggests that, even in the face of profound loss of CD4+ T cells, this fungal infection rarely occurs (see above), except when patients become neutropenic, secondary to azidothymidine (AZT) therapy.

Immune regulation in aspergillosis

Regulation of the immune response to aspergillosis has been little studied. Of interest are the descriptions of metabolic products of the fungus that are immunomodulatory in vitro. One of the best characterized of these is gliotoxin, produced in vitro by *A. fumigatus*. Gliotoxin is a potent inhibitor of macrophage phagocytosis and also appears to inhibit the ability of spleen cells to induce major histocompatibility complex (MHC)-restricted cytotoxic cells. *A. fumigatus* also produces phospholipid compounds that are active in inhibiting the activation of complement via the alternative pathway. These factors have the potential to influence both the innate and immune responses that provide adequate host defense to *Aspergillus* spp. in the normal host. It is not known whether gliotoxin or the complement inhibitor is, however, produced in biologically significant quantities in vivo.

DIAGNOSIS

Establishing the diagnosis of aspergillosis in a compromised patient is difficult because the clinical presentation is nonspecific and the fungus is seldom isolated from blood or other body fluids, or from sputum. Interpretation of serological test results is also difficult because failure to detect precipitins in a compromised individual does not mean that aspergillosis is not present (Saugier-Veber et al. 1993). Nor is the detection of circulating antigen a consistent finding in such patients (Jones and McLintock 2003; McLintock and Jones 2004).

The problems associated with the diagnosis of invasive aspergillosis are highlighted by the diagnostic criteria that have to be fulfilled before patients are eligible for inclusion in clinical trials of new antifungal agents. Very often these criteria are either presumptive or definitive depending on whether tissue diagnosis has been achieved. The following patient selection criteria are required by the European Organization for Research in the Treatment of Cancer (EORTC):

- persistence of fever: $\geqslant 38°C$ for 3–5 days despite broad-spectrum antibiotics, unless biopsy-proven aspergillosis is documented before day 5 of fever
- current negative blood cultures for bacterial pathogens
- no presumptive diagnosis of infections caused by viruses, acid-fast bacilli, legionellae, Q fever, psittacosis, mycoplasmas, chlamydia pneumonia, and *Pneumocystis* sp.
- appropriate physical symptoms and signs suggestive of invasive aspergillosis such as:
  pulmonary: cough, dyspnea, hemoptysis, crepitation, hypoxia, chest pain
  sinus: headache, nasal discharge, facial swelling, tenderness, cellulitis
  liver: may be asymptomatic, discomfort in the right upper quadrant, jaundice, hepatomegaly, abnormal liver function tests
  CNS: headache, focal fits, focal neurological deficit, confusion, altered conscious level, abnormal CSF findings
  other sites: these include bone, kidney, skin, pleura, eyes, spine, external ear, endocarditis, pericardium, joints, adrenal, peritoneal, gastrointestinal tract, lymph node, and thyroid, as appropriate
- appropriate radiological investigations:
  pulmonary: pulmonary infiltrates either nonspecific or suggestive (nodules, cavities)
  sinus: mucosal thickening, opacification/clouded sinus spaces, fluid levels, bone destruction
  liver: nodules/abscesses demonstrated by computed tomography, ultrasonography, or MRI
  spleen: nodules/abscesses demonstrated by computed tomography, ultrasonography, or MRI
  CNS: nodules/abscesses demonstrated by computed tomography, ultrasonography, or MRI
  other sites: as appropriate
- positive identification of *Aspergillus* spp. either by histology or by culture from an appropriate site (i.e. involved organ):
  pulmonary: bronchial washing/bronchoalveolar lavage/tracheal aspirate/anterior nares/transbronchial biopsy/radiologically guided fine-needle aspiration (FNA)/open lung biopsy
  sinus: nasal eschars/sinus aspirate/sinus biopsy
  liver: FNA/laparotomy/laparoscopic guided biopsy
  spleen: FNA/laparotomy/laparoscopic guided biopsy
  CNS: FNA/stereotactic biopsy, CSF
  other sites: appropriate sampling.

A positive blood culture (highly unlikely in aspergillosis) in the presence of radiological and clinical features can be used as an alternative mycological confirmation.

Entry into clinical trials often depends on direct tissue evidence but obviously some patients will be too ill or considered at high risk (thrombocytopenia) for invasive procedures, e.g. open lung biopsy.

Definitively diagnosed invasive aspergillosis requires all of the above criteria to be satisfied and *Aspergillus* spp. should be isolated from the organ/site involved.

Microscopy

Microscopic examination of sputum preparations is often helpful in the diagnosis of allergic aspergillosis, because abundant septate mycelia with characteristic dichotomous branching are usually seen.

Microscopic examination of sputum is seldom helpful in patients with suspected invasive aspergillosis, but examination of BAL specimens is often rewarding (Delvenne et al. 1993) (Figure 34.11). Typical mycelium may also be detected in wet preparations of necrotic material from cutaneous lesions or sinus washings, but isolation of the etiologic agent in culture is essential to confirm the diagnosis.

The most reliable method for the diagnosis of acute invasive aspergillosis is the examination of stained tissue sections. The detection of nonpigmented, septate filaments which show repeated dichotomous branching is characteristic of aspergillus infection (Figures 34.12, 34.13 and 34.14). However, other less common organisms, such as *Fusarium* spp. and *Scedosporium apiospermum*, appear similar. More precise identification can sometimes be achieved with immunochemical staining methods. One of the main limitations in using this approach has been crossreactivity between species, and even monoclonal antibodies may suffer the same problems. In one report it has been shown that murine monoclonal antibodies specifically recognize cytoplasmic antigens of *A. fumigatus*, *A. flavus*, and *A. niger*

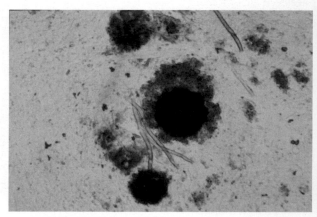

Figure 34.11 *Direct microscopy of sputum from a patient with allergic bronchopulmonary aspergillosis caused by* A. niger

Figure 34.12 *Histopathological appearance of an aspergilloma*

(Fenelon et al. 1999). These antibodies were used to identify *Aspergillus* spp. in frozen sections of tissue from cases of invasive aspergillosis by immunofluorescence, and in paraffin-embedded clinical specimens by immunofluorescence and immunoperoxidase staining.

Optical brighteners, such as Calcofluor white, which label glucan and chitin in the fungal cell wall, have been

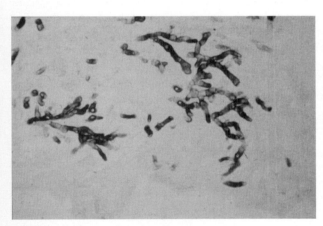

Figure 34.13 *Histopathological appearance of an* A. fumigatus *infection of the paranasal sinuses*

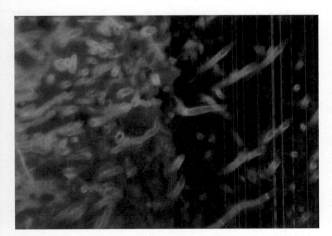

Figure 34.14 *Postmortem tissue from a case of upper airway aspergillosis stained with Calcofluor white. This technique provides a very rapid staining method for biopsy specimens.*

used to highlight fungal elements in a variety of clinical specimens for the past 20 years. Further refinements to this technique for the recognition of aspergillus hyphae in respiratory secretions have been reported by Ruchel and Schaffrinski (1999).

Culture

The definitive diagnosis of aspergillosis depends upon the isolation of the etiologic agent in culture. The fungus may be recovered from sputum specimens from patients with allergic aspergillosis, but cultures from patients with other forms of aspergillosis are less successful. Moreover, because *Aspergillus* spp. are commonly found in the air, their isolation must be interpreted with caution (Nalesnik et al. 1980). Their isolation from sputum is more convincing if multiple colonies are obtained on a plate, or the same fungus is recovered on more than one occasion. Often the number of colonies isolated is small which may result from the filamentous character of the organism. Positive culture may also be a sign of transient exposure to inhaled spores. If sputum cannot be obtained from an immunocompromised patient with a lung infiltrate, alveolar lavage specimens should be obtained. Isolation of an *Aspergillus* sp. from such specimens is often indicative of infection, but is positive in less than 60 percent of cases. *Aspergillus* spp. may be recovered from sputum or bronchial lavage specimens, especially in patients with diffuse pulmonary infiltrates, whereas recovery of fungus from patients with focal lesions is more difficult. A positive culture is indicative of infection but may also merely represent colonization (Delvenne et al. 1993). However, the isolation of *A. fumigatus* from respiratory tract specimens in heart transplant recipients is highly predictive of invasive aspergillosis (Munoz et al. 2003). It has been estimated that around 40 percent of neutropenic patients from whom *Aspergillus* spp. are isolated do not have invasive disease (McWhinney et al. 1993).

Aspergillus spp. are seldom recovered from blood, urine, or CSF specimens, although cultures of blood have been positive in occasional patients with endocarditis. More often, however, their isolation is the result of contamination. It has not been established whether lysis-centrifugation is any more useful than traditional blood culture methods in the diagnosis of aspergillosis.

The diagnosis of aspergillus sinusitis is less difficult to establish than infection of other sites. The fungus can usually be isolated from sinus washings or biopsies of the necrotic lesions in the nose or palate.

Skin tests

Skin tests with *A. fumigatus* antigen are useful in the diagnosis of allergic aspergillosis. Patients with uncomplicated/extrinsic asthma caused by *Aspergillus* spp. give an immediate type I reaction. Those with allergic bronchopulmonary aspergillosis give an immediate type I reaction and many also give a delayed type IV reaction.

Serological tests

Many potential systems for the immunodiagnosis of aspergillosis have been described (Kurup and Kumar 1991; Barnes 1993; Kappe and Seeliger 1993). Those based on detection of antibody to the organism have been very successful in allergic aspergillosis and aspergilloma, and those used for the detection of fungal antigen have great potential for the diagnosis of invasive aspergillosis (Manso et al. 1994). To understand the potentials and limitations of serology in aspergillosis, it is necessary to review the extensive literature on the antigens of *Aspergillus* spp. and how this knowledge has been applied to the design of immunoassays for improved diagnosis.

ANTIGENS OF *ASPERGILLUS* SPP.

A major problem that has prevented the development of standardized serological tests is the vast heterogeneity of antigen types and the lack of understanding of their relevance in vivo. Most of the antigens identified in crude extracts from *A. fumigatus* grown in vitro lack specificity (Hearn 1992; Latgé et al. 1993). For example, Western blot experiments have demonstrated that circulating antibodies from sera of either aspergilloma patients or non-infected individuals bind to most antigens secreted in glucose–peptone culture medium (Latgé et al. 1993). In this study, the levels of anti-*A. fumigatus* antibodies were 10–100 times higher in aspergilloma patients than in control individuals. Furthermore, antigens of *A. fumigatus* crossreact with other fungi (Kumar and Kurup 1993).

The presence of detectable levels of anti-*Aspergillus* spp. antibodies in every individual is thought to result from the continuous inhalation of conidia from the atmosphere. However, normal individuals inhale only conidia whereas the development of aspergillosis results from mycelial growth in lung tissue. Although conidial and mycelial antigens are very similar, growth-phase-specific antigens have been demonstrated in *Aspergillus* spp. (Kauffman et al. 1984). In addition, expression of different mycelial antigens in vitro is highly dependent on the nutrient composition of the culture medium (Moutaouakil et al. 1993). These findings suggest the existence of antigens, released during mycelial growth in vivo, which can only be recognized by antibodies in patients infected with *Aspergillus* spp. To identify such antigens, Sarfati et al. (1995) probed immunoblots of extracts from kidneys of mice infected intravenously with *A. fumigatus* with sera from patients with aspergilloma. A limited number of antigens were detected with molecular masses of 31, 36, 56, 84, 88, and 200 kDa. A 88 kDa antigen secreted by *A. fumigatus* has been

described by Kobayashi et al. (1993). Antigens of this type have been characterized as belonging to two classes:

1 galactomannan- and/or galactofuran-containing glycoproteins
2 antigenic proteins without the galactofuran epitope.

These studies have confirmed that antigens bearing a galactofuranosyl epitope are present both in vivo and in vitro. An exhaustive review of *Aspergillus* antigens can be found in Latgé (1999).

TESTS FOR ANTIBODIES TO *ASPERGILLUS* SPP.

Detection of specific precipitating antibodies to *Aspergillus* spp. by double diffusion, counterimmunoelectrophoresis or enzyme-linked immunosorbent assay (ELISA) has provided the basis for the most frequently used serological tests for the diagnosis of aspergillosis (Hopwood and Evans 1991; Barnes 1993). Some of the tests are available commercially from a number of sources and are technically simple to perform and to interpret. The presence of one or more weak precipitin bands is one of the diagnostic criteria accepted for the diagnosis of ABPA. Published studies would suggest that 70–100 percent of patients with ABPA are positive for IgG-precipitating antibodies directed against *Aspergillus* spp., depending on the antigen used and whether or not serum was concentrated before testing. Furthermore, specific IgE is a reliable indicator of ABPA in cystic fibrosis (Marchant et al. 1994). In addition, rates of positivity in patients with extrinsic allergic alveolitis are also very high. Precipitins are present in 98–99 percent of aspergilloma patients, and multiple bands, often three or more, are routinely present. In patients with invasive aspergillosis, however, the experience with detecting precipitating antibodies has been disappointing. Immunoblotting is a sensitive technique that has been used to detect antibodies in sera from patients with invasive aspergillosis and patients with ABPA. Although the correlation between the sizes of the recognized antigens has not been exact, it has great research potential for determining which antigens may be the most appropriate for future testing. This information, in conjunction with ELISA data, may allow wider availability of a more sensitive and specific test for aspergillosis.

More recently, a number of highly sensitive methods for the detection of very low levels of antibodies in serum samples have been described. It has been reported that all serum specimens from patients with invasive aspergillosis contain antibodies to a 58 kDa concanavalin A-binding antigen identified by sodium dodecylsulfate-polyacrylamide gel electrophoresis (SDS-PAGE) and immunoblotting. However, the specificity of this antigen has not been well documented (Fratamico and Buckley 1991). Both antigen and antibody levels have been monitored successfully in a number of cases of invasive aspergillosis (Manso et al. 1994; Hearn et al. 1995). With these highly sensitive methods for antibody detection, a problem of specificity arises because IgG antibodies to aspergillus antigens can be detected in a proportion of healthy individuals.

Immunoblotting in conjunction with SDS-PAGE has been used to detect serological responses to aspergillus antigens. Piechura and coworkers (1985) have used two-dimensional electrophoresis and isoelectrofocusing to analyze complex mixtures of aspergillus antigens. Hearn et al. (1995) have extended these studies by using immunoblotting to detect specific IgG antibodies to aspergillus antigens in serum specimens from patients with different types of aspergillosis. Different *A. fumigatus* extracts, including culture filtrates, surface components, and mycelial fractions, were tested. All these preparations were shown to be highly reactive antigenically when tested previously by precipitin and ELISA procedures. In particular, patients with invasive aspergillosis were capable of mounting a response to culture filtrates, surface washes, and mycelial extracts of *A. fumigatus* (a total of 12 antigenic fractions); 11 were reactive with serum specimens from patients with aspergilloma. Eight of these antigens showed good responses with serum specimens from patients with ABPA, which were used to assess the sensitivity of IgG detection. No measurable activity was detected in 18 negative control serum specimens, whereas 11 of 13 patients with proven, highly probable, or probable cases of invasive aspergillosis had anti-*Aspergillus* spp. IgG to multiple antigenic preparations. Patients with invasive aspergillosis who were capable of mounting a substantial humoral response to aspergillus antigens gave an antibody profile with five antigenic preparations, which seemed to be characteristic of the disease. The authors concluded that this approach was highly sensitive and may allow the selection of fractions that are both highly antigenic and specific for the detection of antibody to aspergillus antigens. This study also indicates that the use of a spectrum of antigenic molecules is advisable, given the variability observed in the immune responses of individual patients.

DETECTION OF ASPERGILLUS ANTIGEN

Antibody production in the immunocompromised host with invasive aspergillosis is invariably difficult to detect. Therefore, methods have been sought for diagnosing this infection which would rely on the measurement of a cell component of fungal origin and thereby be independent of the host's ability to respond (Richardson 1987; Barnes 1993). Aspergillus galactomannan circulates during infection and appears in the urine, presumably after clearance by a receptor-mediated process by Kupffer's cells in the liver. In addition to galactomannan, at least seven other *Aspergillus*-related antigens have been detected, by immunoblotting, in the urine of patients with invasive disease (Ansborg et al. 1994).

Various immunoassay formats have been designed to detect antigen, either free or in immune complexes in serum, bronchoalveolar fluid, or urine. In two of the largest studies on patient samples, a radioimmunoassay for antigen detection showed 74 percent sensitivity and 90 percent specificity, and either of two ELISA procedures used to measure antigenemia or antigenuria, with a rat monoclonal antibody (EB-42) specific for galactomannan, showed greater than 95 percent sensitivity and specificity. The value of granulocyte-macrophage (GM) detection by the Pastorex *Aspergillus* latex agglutination (LA) test (Sanofi Diagnostics Pasteur, Marnes-La-Coquette, France) has been evaluated by a number of groups (Haynes and Rogers 1994; Hopwood et al. 1995; Verweij et al. 1995a, b, c) and the test showed sensitivities of up to 95 percent with serum samples from patients with a high index of suspicion for invasive aspergillosis. The latex test was also found to yield positive results earlier than conventional microbiological procedures for 68 percent of patients with proven invasive aspergillosis. However, these observations have not been confirmed by others (Warnock et al. 1991), and a sensitivity as low as 38 percent has been reported (Warnock et al. 1991; Hopwood et al. 1995). In the Verweij et al. (1995b) study, the LA test yielded positive results only during advanced stages of infection in most patients with suspected invasive aspergillosis and did not contribute to an early diagnosis. This study showed that an ELISA using a monoclonal antibody to GM detected GM in serum up to 5 days earlier than the LA test did. Although both LA and ELISA failed to detect one proven infection, the ELISA detected GM in two additional patients for whom the LA test continued to yield negative results. Moreover, GM was detected in more serum samples by ELISA than by LA. This suggests that monitoring sequential serum samples from high-risk patients during neutropenia may allow the diagnosis of invasive aspergillosis to be made at an earlier stage of infection. Furthermore, GM has been detected by ELISA in BAL fluid, but the antigen may appear sooner in serum (Verweij et al. 1995a). Galactofuran detection also has been described and may be an additional marker of invasive aspergillosis (Stynen et al. 1995).

One problem long associated with the increased sensitivity of immunoassays such as ELISA or LA has been the occurrence of false-positive results (Kurup and Kumar 1991; Warnock et al. 1991; Kappe and Schulze-Berg 1993). Therefore, to identify a genuine elevation of antigen level in serum, positive ELISA results should be found for at least two consecutive serum samples.

The studies by Verweij and colleagues (1995a, b, c) are encouraging and form the basis for a number of recommendations for the immunodiagnosis of invasive aspergillosis:

- antigen detection in serum at regular intervals by the double sandwich ELISA may allow the early diagnosis of invasive aspergillosis in immunocompromised patients

- twice-weekly collection and testing of serum samples from a patient during periods of neutropenia should be sufficient to detect an increase in the GM in serum early in the course of infection
- where there is a positive ELISA result, serum samples should be tested in order to exclude the possibility of a false-positive result
- confirmation of suspected invasive aspergillosis should be obtained by chest x-ray, computed tomography, or BAL.

Although antigen detection for the early diagnosis of invasive aspergillosis holds great promise for the future, determination of its full diagnostic potential awaits widespread availability and evaluation. In addition, greater effort has to be made in the search for immunodominant antigens.

Another approach to the diagnosis of aspergillosis that is independent of a host immunological response is the detection of fungal metabolites in the body fluids of patients. The presence of high levels of D-mannitol in the serum and tissues of experimentally infected animals has been shown to correlate with the presence and extent of invasive aspergillosis. Likewise, the presence of oxalic acid in BAL fluid has been proposed as a presumptive marker for aspergillosis.

Further information regarding the Platelia *Aspergillus* sandwich ELISA has been reported. Galactomannan has been detected by this test in BAL fluid specimens from patients with invasive aspergillosis and an excellent correlation has been found between serum and BAL fluid results (Verweij et al. 1995a). The detection of galactomannan in both BAL fluid and serum from a patient suspected of having invasive aspergillosis seems to provide strong evidence for the presence of this disease. The concentration of galactomannan in serum has been shown to correspond to the tissue burden of *Aspergillus* (Stynen et al. 1995), and there is some evidence that the course of the antigen titer is closely related to clinical outcome (Rohrlich et al. 1996; Verweij et al. 1997). Indeed, the amount of circulating galactomannan decreased in six of ten pediatric patients during antifungal treatment (Rohrlich et al. 1996) and in patients with invasive aspergillosis following bone marrow transplantation who responded to antifungal therapy (Patterson et al. 1995; Verweij et al. 1997).

Until now the sandwich-ELISA has been evaluated predominantly with serum samples from patients receiving treatment for hematological malignancies. The presence of a well-defined period of increased risk for invasive aspergillosis means that the sandwich ELISA can be used to screen patients at the time of risk of acquiring this infection. This approach may allow patients with invasive aspergillosis to be identified promptly and to benefit from early antifungal treatment.

The combination of antigen testing by ELISA and early CT scanning has been shown to improve diagnosis and outcome in immunocompromised patients (Caillot et al. 1997). Twenty-three histologically proven and 14 highly probable cases of invasive aspergillosis in 37 hematology patients were analyzed retrospectively. BAL fluid was positive in 22 (69 percent) of 32 cases. When BAL fluid was tested for antigen, the *Aspergillus* ELISA test was positive in 83 percent of cases. Starting from October 1991, early thoracic CT scans were routinely performed in febrile neutropenic patients with pulmonary infiltrates on chest radiograph. This approach allowed the investigators to recognize suggestive CT halo signs in 93 percent of patients, compared with 13 percent before this date. The mean time to a diagnosis of invasive pulmonary aspergillosis was reduced from 7 to 1.9 days.

The ELISA for detection of galactomannan becomes positive at an early stage of infection. Early detection is probably the most important feature of this assay, because the detection of antigenemia dictates the initiation of therapy. Recent studies, reviews, and recommendations highlight the usefulness of the galactomannan assay (Maertens et al. 1999, 2002; Denning 2000; Dupont et al. 2000). However, a number of important issues have to be addressed, for example false-negative reactivity, influence of host factors, and the degree of angioinvasion. Furthermore, some reports disagree with the consensus view (reviewed by Jones and McLintock 2003).

Other ELISA formats for the detection of *Aspergillus* antigen have been designed. Chumpitazi and colleagues developed an inhibition enzyme immunoassay (inhibition-EIA) to monitor for the occurrence of invasive aspergillosis in sera from 45 immunocompromised patients (Chumpitazi et al. 2000). The test used rabbit polyclonal antibodies and a mixture of components from *A. fumigatus*, containing three predominant antigens with molecular masses of 18 000, 33 000, and 56 000 Da. Circulating antigens were found in five of seven proven cases of invasive aspergillosis due to *A. fumigatus*. In two of the five positive cases, antigenemia was detected with inhibition-EIA earlier than with x-ray or other biological methods. No antigens were detected in the sera from two patients with proven invasive aspergillosis due to *A. flavus* and *A. terreus*, nor in the sera from four patients with probable invasive aspergillosis. Circulating antigens were not detected in the control group, composed of 30 healthy adult blood donors. Four of the 32 at-risk patients examined, although they displayed no definite evidence of invasive aspergillosis, gave a positive result in this test. The sensitivity, specificity, and positive predictive value of inhibition-EIA were 71.4, 94.4, and 71.2 percent, respectively. The data were compared with those obtained by an LA assay of galactomannan that was positive in only one patient with probable invasive aspergillosis. The higher sensitivity obtained by inhibition-EIA may well be as a result of its ability to detect circulating antigens other than galactomannan in the sera of immunocompromised patients with invasive aspergillosis. It would be useful for the inhibition ELISA to be compared with the Platelia ELISA for galactomannan.

DETECTION OF *ASPERGILLUS* DNA BY POLYMERASE CHAIN REACTION (PCR)

The limitations of antibody detection and the problems of sensitivity associated with antigen detection have prompted the evaluation of the PCR for the diagnosis of invasive aspergillosis (Alexander 2002; Kawazu et al. 2003). The main advantages of PCR appear to be that it detects low burdens of fungal genetic material and warns of the presence of possible invasive aspergillosis. A number of PCR techniques have been developed to detect either individual species or general primer-mediated methods to detect filamentous fungi in general (Melchers et al. 1994; Montone and Litzky 1995). Various clinical specimens have been analyzed by these methods, including sputum, whole blood, and BAL fluid. Early reports described methods to amplify the gene for *A. fumigatus* 18 kDa ribonucleotoxin (Reddy et al. 1993). Subsequently, a PCR, based on universally conserved sequences within fungal large rDNA, including that of *A. fumigatus*, was described (Haynes et al. 1995). Primers to sequences of large subunit rDNA genes, which are universally conserved within the fungal kingdom, were capable of amplifying DNA from 43 strains representing 20 species (12 strains) of medically important fungi (Haynes et al. 1995). Sequence analysis of the products from *A. fumigatus* allowed the design of specific primers which only amplified homologous DNA. This approach allowed the detection and identification of *A. fumigatus* within 8 hours from simulated specimens.

The sensitivity of PCR is an important issue. The lower limit of detection reported for *A. fumigatus* corresponds to 10–100 c.f.u. per sample (Reddy et al. 1993; Spreadbury et al. 1993; Tang et al. 1993). Of equal concern is the question of false positives. In a competitive PCR assay applied to BAL samples, this problem was highlighted (Bretagne et al. 1995). Here a competitive, internal control was incorporated into a PCR designed for the detection of *Aspergillus* sp. DNA in BAL samples. For this purpose, a 1 kb mitochondrial DNA fragment of *A. fumigatus* was sequenced. The primers used allowed amplification of *A. fumigatus*, *A. flavus*, *A. terreus*, and *A. niger* DNAs. but not DNA of other fungi and yeasts. BAL samples from 55 consecutively enrolled patients were tested. Of 28 immunocompromised patients, six were PCR positive, three died of invasive pulmonary aspergillosis and their BAL cultures yielded *A. fumigatus*, and three were culture negative and did not develop invasive pulmonary asper-

gillosis. Of 15 HIV-positive patients and nine immunocompetent patients, five and four, respectively, were both PCR positive and culture negative, and none developed aspergillosis. Therefore, in this evaluation, PCR confirmed invasive pulmonary aspergillosis in three patients but gave positive results for 25 percent (12 of 49) of the patients who did not have evidence of aspergillosis. The authors conclude that the predictive value of PCR-positive results seems low for patients at risk for aspergillosis and that the risk of contamination of reaction buffers or biological samples with aspergillus conidia was high and was a major consideration if the potential diagnostic benefit of PCR was going to be realized.

There have been a number of reports on the use of PCR for the detection and identification of *Aspergillus* spp. Einsele and colleagues reported an oligonucleotide primer pair consisting of a consensus sequence for a variety of fungal pathogens, whilst the species-specific probes used for species identification were derived from a comparison of the sequences of the 18S rRNA genes of *Aspergillus* and *Candida* spp. (Einsele et al. 1997).

Verweij and coworkers sought to compare the sensitivity and utility of the double-sandwich ELISA for *Aspergillus* galactomannan with that of a PCR-based system in BAL fluid samples from 19 patients who were treated for hematological malignancies and suspected of having invasive pulmonary aspergillosis (Verweij et al. 1995a). ELISA was also performed with serum samples. All patients had fever and pulmonary infiltrates on the chest radiograph on the day that the BAL fluid was obtained. *Aspergillus* spp. were detected by PCR or ELISA in five of seven patients who had radiological evidence of invasive aspergillosis.

While most PCR systems for detection of aspergillosis use whole blood or BAL as a source, the possibility of using serum was investigated by Yamakami et al. (1996). Employing a nested PCR method, these investigators worked with two sets of oligonucleotide primers derived from the sequence of the variable regions V7–V9 of the 18S rRNA genes of *A. fumigatus*. The DNA fragment of interest was detected in the serum of mice with disseminated aspergillosis and in patients with invasive aspergillosis.

Walsh and colleagues reported the use of single-strand conformational polymorphism (SSCP) as a technique for detecting and delineating differences between fungal species and genera (Walsh et al. 1995). These investigators used a 197 bp fragment amplified from the 18S rRNA gene, which is common to all medically important fungi. After amplification, the fragments were denatured and run on an acrylamide-glycerol gel at room temperature or 4°C for 4.5 or 4 hours. The SSCP patterns of major and minor bands at room temperature permitted a distinction to be made between strains of *A. fumigatus* and *A. flavus*. The SSCP banding pattern among different strains of the same *Aspergillus* sp. was also consistent.

A variety of PCR protocols for human samples have been published, including PCR assays, methods detecting one species, members of a fungal family, or several species. As the incidence of fungal species in various infections is relatively low and identification of the infecting agent to the species level is required to guide appropriate treatment, the only efficient and economic approach may be a single protocol that is able to detect and identify many species. However, the protocols detecting several species use labor-intensive blotting protocols and sequential hybridizations with various radiolabeled probes for differentiation of species, which make these approaches impractical for routine laboratory use. In a recent study, a PCR method is described that is based on the sequence variation of the fungal ITS region (Hendolin et al. 2000). The method includes multiplex liquid hydrization with species-specific probes, and it employs nonradioactive and automated PCR product detection on a fluorescent automated DNA sequencer. Nonhybridized products are easily recognized on the sequencer and they can be used for identification by sequencing. The method has been applied to the analysis of tissue samples from a range of clinical specimens from deep sites. Total DNA was extracted from deep-tissue samples from 13 patients with suspected or proven fungal infection, and from nasal polyp tissue from seven patients with CRS. Broad-range primers were used to amplify the fungal ITS region (ITS 1-5.8S rRNA-ITS 2). Eight common fungal pathogens were identified by multiplex liquid hybridization with species-specific probes. The nonhybridized products were identified by sequencing either directly, or after segregation by cloning to allow assessment of mixed fungal populations. Twenty (95 percent) of 21 tissue specimens yielded PCR products, whereas culture resulted in only eight (38 percent) positive specimens. All culture or direct-microscopy positive specimens were PCR positive. Sequencing the nonhybridized PCR products identified an infecting agent in seven specimens, and sequences of unknown fungal origin were detected in three specimens. *Phoma glomerata* was detected by PCR in two CRS patients but culture was negative.

Another development has been the report of a two-step PCR assay that specifically amplifies a region of the 18S rRNA gene that is highly conserved in *Aspergillus* spp. (Skladny et al. 1999). A number of primers with the least homology to equivalent human or *Candida* gene sequences were screened for the pairs that gave the highest sensitivity and specificity. No crossreaction with numerous fungal and bacterial pathogens was observed. The assay allowed direct and rapid detection of down to 10 fg of *Aspergillus* DNA corresponding to 1–5 c.f.u./ml blood. A total of 315 blood and BAL samples from 140 subjects, including 93 patients at risk for invasive disease, were screened. The authors found 100 percent correlation between positive histology, culture, or high-resolution CT findings and PCR results. The test specificity was 89 percent. The method appears to be simple, specific, rapid, and inexpensive.

Other PCR formats have been adapted for the detection of *Aspergillus* DNA. A nested PCR test targeting the large ribosomal subunit genes of *Aspergillus* spp. was evaluated retrospectively on 175 serum samples from 37 bone marrow transplant recipients, 70 percent of whom received grafts from unrelated donors (Williamson et al. 2000). Six patients had proven infection, seven had probable infection, and three had possible infection, using the revised EORTC case definitions. These 16 patients were all PCR positive (57 out of 93 samples tested). Two additional patients who did not fulfill current diagnostic criteria, but in whom invasive aspergillosis was thought clinically probable, were also PCR positive (five of nine samples). Invasive aspergillosis was unlikely in the remaining 19 patients, four of whom were PCR positive on a single occasion (four of 70 samples). Three samples were inhibitory to PCR. The sensitivity of PCR in diagnosing patients with invasive aspergillosis was 100 percent, specificity was 79 percent, and positive predictive value was 80 percent, using the criterion of a single positive result. If two positive results were required, these values became 81 percent, 100 percent, and 100 percent, respectively. The median duration of infection documented by PCR was 36 days (range 3–248 days) in 17 of 18 patients (94 percent) who did not survive. Positive PCR results predated the institution of antifungal therapy in two-thirds of patients. Four patients became PCR positive during pretransplant conditioning therapy.

Many of these PCR formats are labor intensive and require a minimum of one working day to perform. The Light Cycler technique combines rapid in vitro amplification of DNA in glass capillaries with real-time species determination and quantification of DNA load. Loeffler and colleagues have established a quantitative PCR protocol for *A. fumigatus*, as well as for *C. albicans* (Loeffler et al. 2000). The sensitivity of the assay was comparable to those of previously described PCR protocols (5 c.f.u./ml). Specific detection of *A. fumigatus* could be achieved. The assay showed a high reproducibility of 96–99 percent. The assay was linear in a range between 10^1 and 10^4 aspergillus conidia. The Light Cycler allowed quantification of the fungal loads in clinical specimens from patients with hematological malignancies and histologically proven invasive fungal infections. In blood specimens spiked with aspergillus conidia, a sensitivity of 5 c.f.u./ml was demonstrated. All samples were also found to be PCR positive by PCR-ELISA analysis.

DETECTION OF CELL WALL β-1,3-D-GLUCAN

The cell wall of aspergillus hyphae, and the cell walls of other pathogenic fungi consist of mannans and glucans. The detection of circulating β-1,3-D-glucan is another investigative strategy for diagnosis of invasive aspergillosis.

The plasma concentration of β-1,3-glucan has been measured at the time of routine cultures in febrile episodes (Obayashi et al. 1995). With a plasma cutoff value of 20 pg/ml, 37 of 41 episodes of proven fungal infections (confirmed at postmortem or by microbiological methods), including aspergillosis, were detected. All of the 59 episodes of nonfungal infections, tumor fever, or collagen-vascular diseases had concentrations below the cutoff value (specificity 100 percent). Of 102 episodes of fever of unknown origin, 26 had plasma glucan concentrations of more than 20 pg/ml. Of these, 102 cases are taken as nonfungal infections, the positive predictive value of the test was estimated as 59 (37 of 63), the negative predictive value as 97 percent (135 of 139), and the efficacy as 85 percent (172 of 202). Although a positive result does not indicate the specific cause of the detected fungal infection, this approach is very encouraging and warrants more extensive investigation in selected patient populations. More recent studies have confirmed the usefulness of this approach (Hossain et al. 1997). The test has now been reformatted as an ELISA (Glucatell – Associates of Cape Cod Inc.). Ostrosky-Zeichner and coworkers reported in an abstract from a multicenter trial evaluating the usefulness of the Glucatell assay as an aid to diagnosis of invasive fungal infections in humans (ICAAC 2003: www.iccac.org). The study evaluated the assay in 170 healthy controls and 163 patients with either proven or probable invasive fungal infections; the assay was able to provide a very high specificity and a high predictive value. This assay may be a useful diagnostic adjunct to cultures in the earlier diagnosis of invasive fungal infections.

CONCLUSIONS

Considering the differences in patient populations, conditions of specimen collection, and controls, no comparisons can be legitimately drawn between different antigen and PCR detection systems. The fact that these approaches have been found consistently to enhance the diagnosis of invasive aspergillosis is very encouraging. The recent approval of the Platelia *Aspergillus* galactomannan ELISA by the FDA in the United States should facilitate further understanding. A more comprehensive review of recent developments in the serological and immunological diagnosis of invasive aspergillosis is given by Yeo and Wong (2002), Jones and McLintock (2003), and McLintock and Jones (2004).

HISTOLOGICAL FEATURES

An extremely important characteristic of the histopathology of invasive aspergillosis is the striking tendency of the fungal hyphae to invade large and small arteries and veins, causing inflammation, thrombosis, and infarction. A variety of clinical manifestations and signs of organ dysfunction reflects this angiotropic behavior of *Aspergillus* spp. Aspergillus hyphae can be seen in thrombosed vessels and necrotic tissues, with regular septation and dichotomous branching at about 45°, advancing in the same direction. Aspergillus hyphae stain poorly with hematoxylin and eosin, and are best highlighted with Gomori's methenamine silver. A varying admixture of inflammatory cells is seen depending on the immune status of the patient.

Prognostic markers in aspergillosis

In allergic forms of aspergillosis, antibody titers may be of some prognostic value. It has been suggested that precipitins may decrease with successful corticosteroid therapy. IgE levels are, however, the most useful parameter to follow in the treatment of ABPA patients. The total IgE often rises before a clinical relapse, and the duration of therapy may be judged based on the fall in IgE level.

Tests for aspergillus precipitins are often helpful in the diagnosis of the different forms of aspergillosis that can occur in the noncompromised patient. Precipitins can be detected in up to 70 percent of patients with allergic aspergillosis and over 90 percent of patients with aspergillomas.

The precipitin test is also useful for the diagnosis of chronic necrotizing aspergillosis of the lung and other invasive forms of aspergillus infection, such as endocarditis, provided that the patient is not immunosuppressed.

The detection of precipitins in a neutropenic patient with unresponsive fever or a lung infiltrate is often sufficient to prompt the initiation of therapy, but it must be stressed that a positive test result is not proof of infection. Nor does a negative precipitin test result preclude the diagnosis of aspergillosis in an immunocompromised patient, because such individuals are often incapable of mounting a detectable serological response.

Tests for the detection of circulating aspergillus antigen in blood and urine offer an alternative means of diagnosing aspergillosis in the immunocompromised patient. Changing titers of galactomannan may predict treatment outcomes or indicate the progression of disease. However, aspergillus galactomannan is rapidly cleared from the circulation and frequent sampling is required for optimal detection of antigen.

Levels of antigenemia and antigenuria may correlate with the clinical course in invasive aspergillosis. Data in animal studies and some from patient series would suggest that antigen levels rise as the clinical condition worsens. Also, some studies have shown that efficacious antifungal therapy causes antigen levels to fall; however, this has not been confirmed in all studies. At this time, it is probably premature to derive any universal correlates for antigen levels in this disease.

In conclusion, rapid serological and molecular diagnostic methods facilitate the early diagnosis of invasive fungal infection and would appear to be most useful when used prospectively to screen high-risk patients. However, in order to determine the optimal approach to treatment it is essential that these tests are incorporated into management strategies and their impact on incidence of invasive fungal infection and clinical outcome evaluated in further clinical trials (Jones and McLintock 2003).

Future directions of diagnostic techniques

In the field of medical mycology, DNA-Chip technology appears to be the most promising molecular technique and has several potential applications, including:

- pathogen identification either of single species or part of a panel (as for example fungi causing fungemia)
- detection of antifungal resistance alleles
- virulence factors in bacteria; fungal typing in cases of nosocomial infection or for molecular epidemiologic purposes
- host–pathogen interaction with gene expression arrays.

It is likely that the panel strategy will be increasingly applied to the diagnosis of systemic fungal infections. Currently, most of the applications of DNA-Chip technology have been in the field of cancer, human genetics, or the monitoring of cellular gene expression. A DNA-Chip or DNA-Microarray is a small surface, usually glass or silica, which is covered by hundreds to hundreds of thousands of different oligonucleotides. Each oligonucleotide is represented on the Chip as millions of copies. These Chips are used in hybridization reactions to detect nucleic acids generated from a sample by an amplification technique. We are only at the beginning of the story of the use of DNA-Chips in the field of microbial diagnostics. Small microarrays have been used to identify bacteria in blood cultures using the 23S ribosomal gene. Very few fungal applications have been described. A DNA-Chip test requires approximately 6 hours to be performed. First DNA or RNA must be extracted from the sample using standard existing procedures. Amplification is then made with PCR, reverse transcriptase polymerase chain reaction (RT-PCR), or nucleic acid sequence based amplification (NASBA). The amplification products are cleaved to fragments of approximately bases and labelled with a fluorescent marker. The labelled products are then hybridized on the chip and the chip is thoroughly washed. Finally, using a confocal laser reader, the fluorescence, which is proportional to the amount of hybridized material on an individual oligonucleotide, is measured on each printed spot. One particular technique that is currently being evaluated for fungal diagnostics is arrayed primer extension (APEX).

This is an integrated system with multiplex primer extension on a DNA array, fluorescence imaging, and data analysis. Fungal DNA is amplified by PCR, digested enzymatically, and annealed to the immobilized primers, which promote sites for template-dependent DNA polymerase extension reactions using four fluorescently labeled dideoxynucleotides. This and other methodologies will hopefully be developed into commercially available tests for the mycology laboratory.

MANAGEMENT

A number of new reviews and monographs have been published which should be consulted for detailed aspects and recommendations regarding the treatment of diseases caused by *Aspergillus* (Chiller and Stevens 2000; Stevens et al. 2000a; De Marie 2000; Judson and Stevens 2001; Reichenberger et al. 2002; Advances against aspergillosis 2003; Richardson and Jones 2003; Richardson and Warnock 2003; Anonymous 2004). Further information can be found on the *Aspergillus* website: www.aspergillus.man.ac.uk. A detailed account of the clinical pharmacology of all drugs currently used and those being developed for treating aspergillosis has been published by Groll et al. 1998b.

Allergic aspergillosis

Mild disease may not require treatment. Prednisone is the drug of choice because it is effective in reducing symptoms, improving chest x-rays, and clearing positive sputum cultures. The usual dosage regimen is 1.0 mg/kg/day until x-rays are clear, then 0.5 mg/kg/day for 2 weeks. The same dose is given at intervals of 48 hours for another 3–6 months, and then the dose is tapered off over another 3 months. The initial regimen should be resumed if the condition recurs. Bronchodilators and postural drainage may help to prevent mucus plugging. Treatment with antifungal drugs is not known to be helpful (Ikemoto 1992). However, more recent reports suggest that antifungal treatment may be helpful (Leon and Craig 1999; Salez et al. 1999; Chiu 2003; Stevens et al. 2000b, 2003).

Aspergilloma (fungus ball) of the lung

Surgical resection is indicated if massive or recurrent hemoptysis should occur. On occasion, segmental or wedge resection will suffice, but lobectomy is usually required to ensure complete eradication of the disease (Jewkes et al. 1983).

If surgical intervention is contraindicated, endobronchial instillation or percutaneous injection of amphotericin B may be helpful (Jewkes et al. 1983; Yamada et al. 1993). The optimum dosage has not been determined, but 10–20 mg amphotericin B in 10–20 ml distilled water instilled two or three times per week for about 6 weeks has proved successful. Larger doses (40–50 mg) have been instilled into lung cavities using percutaneous catheters. Itraconazole does not appear to be effective (Campbell et al. 1991).

The treatment of mild-to-moderate bleeding and asymptomatic patients remains controversial, but observation without intervention may be the best approach to management. Recent reports suggest that antifungal treament may be useful in the management of aspergilloma (Kaestel et al. 1999; Ikemoto 2000; Kawamura et al. 2000; Regnard et al. 2000).

Chronic necrotizing aspergillosis of the lung

Treatment with an antifungal drug, such as amphotericin B, is often the first step in management, but surgical resection of necrotic lung and surrounding infiltrated tissue may also be required. The long-term prognosis is poor.

Infection of the paranasal sinuses

In some cases of paranasal aspergillus granuloma, surgical removal of infected material, with drainage and aeration, is curative. Often, however, the condition will recur, necessitating further surgical intervention. The long-term results are generally poor. Postoperative treatment with itraconazole appears promising as a means of preventing relapse. The drug should be given at a dosage of 200 mg/day for at least 6 weeks.

Neutropenic patients with acute aspergillus sinusitis require surgical débridement and treatment with amphotericin B (1.0 mg/kg/day).

Endophthalmitis

Patients with aspergillus endophthalmitis should be treated with intravenous amphotericin B (1.0 mg/kg/day). Surgical débridement and intravitreal instillation of amphotericin B (5 μg doses two or three times) may also be required.

Endocarditis

Aspergillus endocarditis requires aggressive medical and surgical treatment. Treatment with high-dose amphotericin B (1.0 mg/kg/day) alone is ineffective. Infected tissue and prostheses must be removed.

Osteomyelitis

Surgical débridement of necrotic tissue is important in the management of aspergillus osteomyelitis. Most patients with vertebral osteomyelitis undergo simple

débridement as part of their initial diagnostic procedure. Later procedures include radical débridement with bone grafting. Both medical and surgical treatment are required if ribs are infected.

Treatment with itraconazole (400 mg/day) has proved successful in several patients with aspergillus osteomyelitis.

Cutaneous aspergillosis

High-dose amphotericin B (1 mg/kg/day) is the treatment of choice. Débridement of cutaneous lesions that arise at catheter insertion sites should be delayed until the neutrophil count has recovered.

Acute invasive aspergillosis

The successful management of acute invasive aspergillosis in the neutropenic patient depends on the prompt initiation of antifungal treatment (within 96 hours of the onset of infection) (Denning and Stevens 1990; Denning 1994a; Keating et al. 1994). The prognosis is poor if the neutrophil count does not recover.

The drug of choice for the treatment of disseminated aspergillosis is amphotericin B. There are numerous regimens for the administration of this drug, but there is widespread agreement that in neutropenic patients it is important to give the full dose of amphotericin B from the outset. High doses must be used (at least 1.0 mg/kg/day).

The optimum duration of treatment has not been established, but amphotericin B should be continued at least until the neutrophil count is more than 0.5×10^9/l. Thereafter treatment should be continued until symptoms resolve and relevant radiological abnormalities (on x-rays and CT scans) disappear. The shortcomings of current methods of diagnosis often require clinicians to proceed to amphotericin B treatment without waiting for formal proof that a neutropenic patient who has persistent fever (>72–96 hours duration), and is unresponsive to antibacterial drugs, has aspergillosis. Empirical treatment should be initiated with the usual test dose (1 mg) of amphotericin B. If possible, the full therapeutic dosage level should be reached within the first 24 hours of treatment (Fraser and Denning 1993).

Neutropenic patients who recover from aspergillosis may suffer from reactivation of the infection during subsequent periods of immunosuppression. One solution to this problem is to begin empirical treatment with amphotericin B (1 mg/kg/day) not less than 48 hours before antileukemic treatment is commenced. The drug should then be discontinued until the neutrophil count has recovered.

During the 1980s, several investigators incorporated amphotericin B into lipid vehicles and showed a reduction in the toxic adverse effects without loss in the effectiveness of amphotericin B. Initially, such formulations were produced locally and used at single hospitals. In more recent years, lipid complexes of amphotericin B have become commercially available. These formulations include:

- the liposome AmBisome (Gilead Sciences, Foster City, CA, USA)
- a lipid vesicle – amphotericin B colloidal dispersion: Amphotec, Amphocil (InterMune, Brisbane, CA, USA)
- an amphotericin B–lipid complex: Abelcet (Enzon, Bridgewater, NJ, USA).

All of these preparations differ in size, structure, and pharmacokinetics, and, to a certain extent, in the clinical efficacy in the treatment of invasive aspergillosis (reviewed by Tollemar and Ringden 1995 and Barrett et al. 2003).

The treatment of invasive aspergillosis in bone marrow transplant patients has been reviewed recently by Marr et al. 2002; Richardson and Warnock 2003; Wiederhold et al. 2003, and treatment guidelines are published as *Therapeutic guidelines in systemic fungal infections* (Richardson and Jones 2003) and on Clinical Mycology Online (www.clinical-mycology.com).

AMBISOME (LIPOSOMAL AMPHOTERICIN B)

Liposomal amphotericin B (AmBisome) is well tolerated and doses as high as 3–5 mg/kg/day have been administered without significant side effects. Some 10 000 patients have been treated worldwide over the past 6 years. Administration of the drug in this form has sometimes eradicated aspergillus infection in neutropenic patients, and it should be considered in patients who have failed to respond to the conventional parenteral formulation, or who have developed side effects that would otherwise necessitate discontinuation of the drug.

A representative study highlighting the use of AmBisome in invasive aspergillosis is that by Mills et al. (1994). In a single-center study, the diagnosis of aspergillosis was proved in 21 of 116 patients with signs indicative of invasive aspergillosis (with four patients diagnosed postmortem). Of these 21 patients, 13 (62 percent) obtained complete or excellent partial resolution. Of the 17 patients with aspergillosis confirmed in life, the response was 77 percent (13 of 17). In eight patients, liposomal amphotericin B was discontinued when the neutrophil count was below 1×10^9/l, without recrudescence of the infection. Complete hematological remission was confirmed in nine responders on recovery from chemotherapy; three were found to have persistent disease and two had recurrent aspergillosis following subsequent chemotherapy. Of the eight nonresponders, four had refractory malignancy and four died before assessment was possible. In a further report, Ng and Denning (1995) detail the results from 34 centers in the

UK of 17 cases of definite aspergillosis treated with AmBisome. The overall response rate was 59 percent (10 of 17). Mean ($\pm$SD) cumulative doses of AmBisome received by the responders and those who failed to respond were 2 081$\pm$1 189 mg and 1 207$\pm$839, respectively. Mean durations of treatment were 17$\pm$8 days and 7$\pm$4 days, respectively. All six patients who received therapy for more than 14 days responded, whereas 75 percent of those (three of four) who were treated for 5 days or less died. The average maximum daily dose received/tolerated by the group as a whole was 3$\pm$1 mg/kg/day (range 1–5 mg/kg/day).

AmBisome was used as 'salvage' therapy in seven patients who received conventional amphotericin B as first-line treatment but either failed to respond or developed renal insufficiency as a result of treatment. The response rate among this group of patients was 43 percent (three of seven). Of ten of those who did not receive prior therapy with conventional amphotericin B, eight (80 percent) responded to AmBisome therapy.

The data concerning the efficacy of AmBisome in the treatment of mycologically proven cases of invasive aspergillosis, in the analysis by Ng and Denning (1995) of compassionate use, are similar to those reviewed by Tollemar and Ringden (1995) (17 of 29 or 60 percent cure/improvement). This survey emphasizes the point that those patients who responded to AmBisome tended to have received a larger (cumulative) dose for a longer period of time than those in whom treatment failed. Among the patients who had proven invasive aspergillosis, those in whom AmBisome was used as salvage therapy had a less favorable prognosis than those who had never been treated with conventional amphotericin B. AmBisome (up to 5 mg/kg/day) was well tolerated in this study, even in one patient who had previously had a severe drug reaction after conventional amphotericin B therapy.

A notable feature of AmBisome is the very low incidence of acute infusion-related adverse reactions; in fact, where such reactions do occur patients appear to tolerate subsequent doses much more readily. There is no requirement for test dosing, slow escalation, or premedication.

A number of recent clinical evaluations and reviews have reinforced the earlier perception of AmBisome being a safe and effective treatment for invasive aspergillosis (Ribaud and Gluckman 1996; Coukell and Brogden 1998; Ringden et al. 1998). An illustrative study is that of Leenders and colleagues who reported a comparison between AmBisome 5 mg/kg/day with conventional amphotericin B 1 mg/kg/day (Leenders et al. 1998). Among patients with suspected or documented pulmonary aspergillosis, 11 (42 percent) patients on AmBisome and six (21 percent) patients on conventional amphotericin B had a complete response. Other trials have been reported by Andström et al. (1996) and Clark et al. (1998).

More recent information can be found in Marr et al. (2002) and Wiederhold et al. (2003). High-dose AmBisome has become an attractive therapeutic option for the management of invasive aspergillosis, although further studies are needed to assess this approach (Walsh et al. 2001; Martin et al. 2003).

AMPHOCIL/AMPHOTEC

Amphocil [amphotericin B colloidal dispersion (ABCD)] is formed from equimolar amounts of amphotericin B and cholesterol sulfate; it has a disk-like form with a mean size of 122 nm and is rapidly taken up by the liver. Despite the lack of published studies, ABCD has been used in a number of centers and the data presented in abstract form (Oppenheim et al. 1995). The effectiveness of ABCD has been studied in a number of animal models of infection. In a rabbit model of invasive aspergillosis, conventional amphotericin B, at a dosage of 1.5–4.5 mg/kg/day, was more effective in eradicating hepatic and renal infection than ABCD at dosages from 1.5 to 4.5 mg/kg/day (reviewed by Warnock 1995). Infection persisted in the lungs even when animals were treated with a lethal 4.5 mg/kg dose of conventional amphotericin B. However, ABCD could be administered at a much higher dosage (15 mg/kg/day) and this eradicated the infection. In a second report involving a rabbit model of invasive aspergillosis (reviewed by Warnock 1995), animals treated with ABCD had better survival rates (33 percent and 70 percent at dosages of 1 and 5 mg/kg/day) than those treated with conventional amphotericin B (14 percent at 1 mg/kg/day).

The clinical data are still too limited to allow firm conclusions to be drawn, but published information on the compassionate use of ABCD in patients with invasive aspergillosis compares favorably with that for conventional amphotericin B, but not with AmBisome. The collective global experience has been reviewed by Barrett et al. (2003). Of patients at 45 centers, 168 were entered into an open-label, non-comparative evaluation of the efficacy and safety of ABCD in confirmed systemic mycoses. To be included in the study, patients had either to have a systemic fungal infection that had failed to respond to conventional amphotericin B or to be unsuitable for conventional amphotericin B because of nephrotoxicity from conventional amphotericin B or other drugs or because of preexisting renal impairment. A total of 97 patients were evaluable for clinical efficacy of whom 33 percent had invasive aspergillosis. Thirty-four per cent of cases of aspergillosis responded. In a separate analysis of 21 patients with proven fungal infections treated with Amphocil in two UK hospitals, a clinical response was seen in eight of 13 patients with aspergillosis.

Further US experience is reported by White et al. (1997). Eighty-two patients with proven or probable aspergillosis who were treated with Amphocil were compared retrospectively with 261 patients with aspergillosis who were treated with conventional amphotericin B. Response rates (48.8 percent) and survival rates (50 percent) among Amphocil-treated patients were higher than those (23.4 percent and 28.4 percent, respectively) among amphotericin B-treated patients. This study suggested also that in the treatment of aspergillosis, Amphocil caused fewer nephrotoxic effects than conventional amphotericin B and that the efficacy of Amphocil was at least comparable with that of amphotericin B.

More recent information about the use of Amphocil in aspergillosis can be found in Noskin et al. 1999, Roland 1999, and Barrett et al., 2003.

ABELCET

Abelcet [amphotericin B lipid complex (ABLC)] consists of amphotericin B complexed with two lipids, dimyristoylphosphatidylcholine and dimyristoylphosphatidylglycerol, in a 1:1 drug:lipid molar ratio. In animal models, ABLC has been shown to be at least as effective as amphotericin B and substantially less toxic.

Two key studies represent the current clinical experience with ABLC. As part of a 225 patient study where ABLC was used at 5.0 mg/kg/day, complete response or improvement was achieved in 60 percent (43 of 72) of aspergillosis cases (reviewed by Barrett et al. 2003). Among the different patterns of invasive aspergillosis, the response rate was greatest for sinus aspergillosis (83 percent), followed by pulmonary aspergillosis (58 percent) and extrapulmonary aspergillosis (52 percent). Extrapulmonary aspergillosis included single-organ infection ($n=14$) and disseminated infection ($n=13$). The response to ABLC in single-organ extrapulmonary aspergillosis showed a trend towards a more favorable response in nine of 14 cases (64 percent) versus five of 13 (38 percent) in disseminated aspergillosis. There was also a trend towards a more favorable response in all nondisseminated aspergillosis (38 of 59 or 64 percent) versus disseminated aspergillosis cases (five of 13 or 38 percent). The authors note that responses were obtained in patients with disseminated aspergillosis unresponsive to or intolerant of conventional amphotericin B, and that this response was encouraging considering that the mortality rate in this disease approaches 100 percent.

The second major study of the use of ABLC in aspergillosis is where the efficacy, nephrotoxicity, and mortality of patients with invasive aspergillosis treated with ABLC (5 mg/kg/day) in an emergency program were compared with those of a historical control group of patients treated with conventional amphotericin B (reviewed by Barrett et al. 2003). The patients selected ($n=151$) were those who had failed to respond to antifungal therapy, and had nephrotoxicity or preexisting renal insufficiency. The historical control patients ($n=122$) had received 500 mg or more of conventional amphotericin B or had nephrotoxicity. All patients met the definitions for definite or probable aspergillosis. A distinction between the different disease patterns of aspergillosis is not made. ABLC-treated patients showed greater probability of survival (P=0.0233) (median of 52 versus 31 days) and higher response rates (40 percent versus 23 percent, P=0.002) than patients treated with conventional amphotericin B. Among patients with serum creatinine of 221 μmol/l or more when ABLC treatment began, a significant decrease from baseline levels was observed between week 2 and week 5 (P=0.004). The authors conclude from this study that ABLC appears to be associated with at least equal, and possibly better, efficacy and clearly less toxicity compared with conventional amphotericin B in the treatment of disseminated aspergillosis.

Data on ABLC are limited, although the formulation seems to have an acute toxicity profile (pyrexia, chills) very similar to that of amphotericin B. As clinical experience with this drug is restricted to a few centers, it will be some time before more definitive statements can be made regarding overall toxicity.

A number of studies have confirmed the usefulness of ABLC in the treatment of severe invasive candidosis and as second-line treatment of invasive aspergillosis, cryptococcosis in HIV patients, and miscellaneous fungal infections (Oravcova et al. 1995; Franklin et al. 1997; Mehta et al. 1997). The drug appears to be cleared rapidly from the serum. High tissue levels, especially in the lung, have been recorded in animals, followed by rapid clearance of the drug from the lung area over 24 hours. Human data regarding accumulation of ABLC in lung tissue appears to be based on postmortem tissue from one patient. The initial clinical trials were conducted in the USA (reviewed by Lister 1996). The size and quality of these studies varied considerably and the conclusions that can be drawn are limited in the absence of confirmatory studies.

The recommended dosage of ABLC is 5 mg/kg and this should be infused over a period of 2 hours for at least 2 weeks. Response to treatment times is highly variable. This formulation of amphotericin B has been administered to individual patients for up to 11 months, to a cumulative dosage of 50 g without significant toxic side effects.

Very few studies evaluating the effectiveness of ABLC in invasive aspergillosis have been described. Mehta et al. (1997) report that five of seven evaluable patients with aspergillus pneumonia responded to ABLC (daily dose of 5 mg/kg).

A further study regarding the safety and efficacy of ABLC has recently been reported by Walsh et al. (1998). ABLC was evaluated in 556 cases of invasive fungal infection treated through an open-label, single-patient, emergency-use study of patients who were refractory to or intolerant of conventional antifungal therapy. This large clinical trial demonstrated a significant improvement in renal function following the initiation of therapy with ABLC, particularly in patients with amphotericin B-induced nephrotoxicity or primary renal dysfunction. Among the 291 mycologically confirmed cases evaluable for therapeutic response, there was a complete or partial response to ABLC in 55 of 130 cases of aspergillosis. Among the different patterns of invasive aspergillosis, the response rate was greatest for single-organ extrapulmonary aspergillosis and for sinus aspergillosis, followed by pulmonary and disseminated aspergillosis.

As clinical experience with this drug is restricted to a few centers, it will be some time before more definitive statements can be made regarding overall efficacy and toxicity.

There is some concern regarding the incidence and severity of toxicity of ABLC. Ringden et al. (1998) reported their experience with ABLC for proven or suspected invasive fungal infection in 19 patients with a variety of hematological conditions. ABLC was discontinued in 14 of the 19 patients due to side effects. These included renal insufficiency, increase in bilirubin, erythema, increased alanine aminotransferase (ALAT), fever and chills, severe vomiting, and a variety of other drug-related effects. The authors express the opinion that ABLC is far more toxic than AmBisome. In their experience, AmBisome was discontinued in only six of one series of 187 patients. However, it is pointed out that the dose of AmBisome used was much lower in most patients.

Further experience with Abelcet is reported by Boyle and Swenson (1999), Linden et al. (1999), Martino et al. (1999a, b), Walsh et al. (1999b), and reviewed by Barrett et al. (2003).

ITRACONAZOLE

Itraconazole is comparable in vitro to amphotericin B in efficacy. Itraconazole capsules are tolerated with an acceptable level of toxicity. Treatment with itraconazole (400 mg/day) has sometimes proved successful in neutropenic individuals with invasive aspergillus infection (Jennings and Hardin 1993). However, absorption of the drug from the gastrointestinal tract can be a problem and blood concentrations must be measured at regular intervals. Recently, there have been several reports on the successful treatment of invasive pulmonary aspergillosis in neutropenic patients (reviewed by Beyer et al. 1994a; Denning et al. 1994). The accumulated data on itraconazole are highlighted by a small randomized trial where itraconazole (capsules) 400 mg/day was compared with amphotericin B at a dose of 0.6 mg/kg/day in 32 patients with suspected or proven fungal infection (reviewed by Beyer et al. 1994b). Invasive pulmonary aspergillosis was suspected or documented in 13 of 32 (41 percent) of the patients. The overall response rate was similar, with 63 percent in the itraconazole arm and

56 percent in the amphotericin B arm, but all three fatalities with documented invasive pulmonary aspergillosis were treated with amphotericin B. Patients with candida infections responded better to amphotericin B, whereas patients with aspergillus infections had a better response to itraconazole, although the response was not significant. The median treatment duration of 20 days with itraconazole and 13 days with amphotericin B was short, and the responses to conventional amphotericin B and itraconazole were associated with neutrophil recovery in most patients. Itraconazole has also been successfully used in the treatment of invasive pulmonary aspergillosis in patients who did not respond to amphotericin B (see Beyer et al. 1994b). The use of oral itraconazole plus intranasal amphotericin B for prophylaxis of invasive aspergillosis has been described (Todeschini et al. 1993).

More recent studies reaffirm the value of itraconazole in the treatment of invasive aspergillosis, especially where an early diagnosis is achieved. An illustrative study is that of Caillot et al. (1997) where systematic use of CT added to the early diagnosis of invasive aspergillosis. Twenty-six of 37 patients with proven or highly probable invasive aspergillosis were improved or cured by antifungal treatment. Twenty-two patients received itraconazole (median dose 400 mg/day) for a median length of 302 days (range 103–1 185). No severe adverse reactions were seen. The plasma levels of itraconazole were determined in 16 patients, and in 14 of them the residual levels of itraconazole were greater than 0.6 µg/ml during the first 20 days of treatment. In 13 cases, amphotericin B (median dose 1.3 mg/kg/day) was combined with itraconazole for a median duration of 9 days (range 5–20).

There is limited experience with the new oral solution formulation of itraconazole as first-line treatment of invasive aspergillosis – 400–600 mg/day for 4 days then 200 mg twice daily. An alternative approach appears to be the use of high-dose itraconazole solution (400–600 mg) following 7–14 days of AmBisome.

An intravenous formulation of itraconazole is currently being evaluated and has been licensed in the UK, the USA, and The Netherlands. Data are available only in abstract form at the moment. Further studies are being conducted to determine the efficacy of this formulation in the treatment of aspergillosis and as empirical therapy for fever of unknown origin. The current dosing schedule of the intravenous formulation is as follows:

- Day 1 and 2: 1-hour infusion 200 mg twice daily
- From day 3 onwards: a single 1-hour infusion 200 mg each day. Safety of use for periods longer than 14 days has not been established.

Recent reports on all formulations of itraconazole include: Popp et al. (1999); Prentice et al. (1999); Persat et al. (2000); Caillot (2003). Additional information and guidelines can be found on the *Aspergillus* website (www.aspergillus.man.ac.uk), and in *Therapeutic guidelines in systemic fungal infection* (Richardson and Jones 2003). Recent findings on itraconazole are found in a *Journal of Antimicrobial Chemotherapy* supplement (in press).

VORICONAZOLE

A new azole antifungal, voriconazole (Pfizer), formulated in both oral and intravenous forms, is currently being evaluated in clinical studies. It exhibits potent antifungal activity against clinical isolates of *Aspergillus* spp. in vitro. It appears to be more active in vitro compared to amphotericin B (Clancy and Nguyen 1998). For *A. fumigatus* the minimum inhibitory concentration (MIC) has been shown to be in the range 0.03–0.5 µg/ml and the MIC_{90} to be 0.25 µg/ml. In experimental aspergillosis in mice, oral voriconazole 30 mg/kg body weight per day appears to significantly delay or prevent mortality (Murphy et al. 1997). A number of clinical evaluations have demonstrated the effectiveness of voriconazole in invasive aspergillosis (initial clinical experience reviewed by De Pauw and Meis 1998). Caillot et al. (1997) described three patients who were cured and one who improved with voriconazole treatment (median dose 6 mg/kg/day; median duration 63 days). One patient failed to respond to a 26-day treatment course with voriconazole.

Further animal data have been reported by Chandrasekar et al. (2000). Their work compared the efficacies of amphotericin B and voriconazole against pulmonary aspergillosis in a guinea pig model. Voriconazole-treated animals had significantly better survival rates and a decreased fungal burden in the lungs compared with controls. Although no statistical difference was seen between the efficacies of voriconazole and amphotericin B, a trend favoring voriconazole was noted.

Currently, voriconazole is licensed for the treatment of patients aged 2 years or over with invasive aspergillosis. It may be administered either by mouth or intravenously, depending on the condition of the patient. It also appears to be active against other filamentous fungi besides *Aspergillus* (Perfect et al. 2003).

A review of the clinical efficacy of voriconazole in invasive aspergillosis can be found in Muijsers et al. (2002), and the *Drugs and Therapeutic Bulletin* (Anonymous 2004). The conclusion is that voriconazole appears to be a reasonable choice as a second-line therapy, not least because it offers the potential for intravenous or oral treatment of a broad spectrum of fungal infections, in both adults and children. However, there is no convincing published evidence to justify the claim that voriconazole is superior to amphotericin B at increasing survival rates in patients with invasive aspergillosis (Anonymous 2004).

CASPOFUNGIN

Caspofungin is a synthetic lipopeptide. As one of the echinocandins it inhibits the synthesis of β-1,3-D-glucan, an essential component of fungal but not mammalian cells. Its range of activity is restricted to *Aspergillus* and *Candida* spp. Caspofungin is licensed for the treatment of invasive aspergillosis in adults who have failed to tolerate or respond to conventional or lipid formulations of amphotericin B and/or itraconazole, and for the treatment of candidosis in non-neutropenic adults.

Published evidence on the efficacy of caspofungin in invasive aspergillosis is limited to an uncontrolled retrospective study, published only in abstract form (Maertens et al. 2002). This involved 83 patients with presumed invasive aspergillosis who had either failed to tolerate or not responded to unspecified antifungal therapy, and were given intravenous caspofungin for up to 162 days. A favorable response (i.e. resolution of, or improvement in, attributable clinical and radiological or bronchoscopic features) was reported in 45 percent of patients.

In conclusion, caspofungin appears to be better tolerated than conventional amphotericin B but there is little published evidence to justify its use in invasive aspergillosis (Anonymous 2004).

IN VIVO EFFECTS OF COLONY-STIMULATING FACTORS ON NEUTROPHIL FUNCTION AND INVASIVE ASPERGILLOSIS

The assessment of the effects of colony-stimulating factors on neutrophil function is contradictory, largely because of the diversity of methods used and the growth phase of the organism used. In vitro, granulocyte colony-stimulating factor (G-CSF) appears to be a weaker stimulus to the neutrophil metabolic burst than GM-CSF, tumor necrosis factor (TNF), or various interleukins (IL). Phagocytosis and fungicidal activity appear to be enhanced by exposure of normal effector cells to G-CSF, GM-CSF, or IL-8 (Roildes et al. 1993b, c; Richardson and Patel 1995). Pre-exposure to concentrations of IL-8 achievable in vivo significantly enhances phagocytic ingestion of *A. fumigatus* conidia (Richardson and Patel 1995). Furthermore, IL-8-primed neutrophils showed enhanced phagocytic activity to the chemotactic peptide *N*-formyl-methionyl-leucyl-phenylalanine.

CLINICAL TRIALS OF COLONY-STIMULATING FACTORS IN INVASIVE ASPERGILLOSIS

Closely paralleling preclinical trials in experimental infections, a number of studies have shown that macrophage colony-stimulating factor (M-CSF), G-CSF, and GM-CSF can reduce the severity and duration of chemotherapy-induced neutropenia; these growth factors are a useful adjunct to antifungal therapy (Spielberger et al. 1993). Undoubtedly, there will soon be other

investigations of the benefits and risks of using colony-stimulating factors to enhance neutrophil production and upregulate their function. Furthermore, it is anticipated that this will produce a rapid growth in our understanding of the role of neutrophils in response to invasive aspergillosis. A limited number of studies have demonstrated the safe concomitant use of amphotericin B, granulocyte transfusions, and GM-CSF. Concern has been raised about the use of GM-CSF during serious fungal infection because this cytokine can inhibit neutrophil migration into the inflammatory zone measured in a standardized skin chamber assay.

The use of cytokines, together with standard and new formulations of amphotericin B, should be considered for granulocytopenic patients with such invasive fungal infections as invasive aspergillosis, particularly when there has been no response to antifungal therapy alone, and when bone marrow recovery is not expected for at least 10–14 days (Iwen et al. 1993). In a series of cases reported by Bodey et al. (1993), there were two patients with aspergillosis who were treated with GM-CSF and amphotericin B. A patient with underlying breast cancer, who had received a bone marrow transplant, contracted an aspergillus pneumonia. This patient only partially responded. A second patient with acute myeloid leukemia and *A. flavus* sinopulmonary infection failed to respond to combination treatment. The dose range of GM-CSF used in this series was 100–759 μg/m^2/day, based on initial phase 1 trials. This was considered to be too high because a number of patients developed capillary leak syndrome. It is advocated that doses in the range 15–30 μg/m^2 be used in future trials. However, cytokines should be used cautiously in the setting of persistent acute myeloid leukemia because the issues surrounding such treatment remain unresolved at the present time.

EMPIRICAL TREATMENT OF PRESUMED INFECTION

Persistent fever suggests an occult fungal infection. The likelihood of this increases with the number of preceding febrile episodes. It has been shown that 44 percent of patients suffering their fourth bout of fever had a fungal infection as the cause.

Patients with prolonged neutropenia (more than 7–10 days) or receiving high doses of glucocorticoids during neutropenia, and a high probability of invasive aspergillosis should be given empirical systemic antifungal therapy. This recommendation is based on numerous reviews and clinical studies. Such patients, with fever persisting despite antibiotic therapy or presenting at the onset of a febrile episode with findings suggesting the possibility of an invasive infection, should be started on antifungals. In this situation amphotericin B must be used. If possible, the full therapeutic dosage level (1.0–1.5 mg/kg/day of the conventional formulation) should be reached within the first 24 hours of treatment. There is no need for gradual escalation of dosage, nor is there evidence to support the clinical prejudice that a lower dose can be used in suspected invasive disease.

Should the conventional formulation be contra-indicated, one of the lipid-complexed formulations should be used instead. For example, in a Europe-wide randomized, double-blind comparative trial of the lipo-somal formulation of amphotericin B (AmBisome) versus conventional amphotericin B in the empirical treatment of both pediatric and adult febrile neutropenic patients, AmBisome (3 mg/kg/day) was as effective as conventional amphotericin B in the prevention of proven treatment-emergent fungal infections (Prentice et al. 1997). Of significance was the reduction in overall drug-related toxicity by two- to sixfold in patients treated with AmBisome compared with conventional amphotericin B. Severe drug-related adverse reactions were almost absent in patients treated with AmBisome. Nephrotoxicity, in patients not receiving concomitant nephrotoxic agents, was seen only in 3 percent of patients treated with AmBisome, compared with 23 percent of patients on conventional amphotericin B. Time to develop nephrotoxicity was longer in patients on AmBisome compared with conventional amphotericin B.

A significantly larger study, which formed the basis of a successful new drug application to the Federal Drug Administration (FDA) for the use of AmBisome in empirical therapy of presumed fungal infections in febrile, neutropenic patients has been reported (Walsh et al. 1999a). This study was a multicenter, double-blind trial that compared AmBisome (3 mg/kg/day) with conventional amphotericin B (0.6 mg/kg/day). Nearly 700 adult and pediatric patients with neutropenia and pyrexia of unknown origin (PUO) were randomized to receive either conventional amphotericin B or AmBi-some. Therapeutic success, measured by a combination of factors including resolution of fever, absence of emer-gent fungal infection, and patient survival for at least 7 days after therapy, was equivalent between the two groups. There were significantly fewer proven treatment-emergent fungal infections in patients treated with AmBisome than in patients treated with conventional amphotericin B. Similar to the Prentice et al. (1997) report, there was a significant reduction in the frequency of infusion-related fever and in the development of nephrotoxicity. The conclusions that can be drawn from the Walsh et al. (1999a) study are that AmBisome was equivalent to conventional amphotericin B for empirical antifungal therapy in neutropenic patients, but superior in reducing proven treatment-emergent fungal infec-tions, nephrotoxicity, and infusion-related toxicity.

The duration of treatment will differ from patient to patient. If the patient responds and a diagnosis of fungal infection is established, a full course of treatment should be given. More often, however, the patient responds and/or the neutrophil count recovers, but a firm diag-nosis is not obtained. In this situation it is reasonable to discontinue amphotericin B when the neutrophil count goes above $1 \times 10^9/l$, the fever and other symptoms and signs resolve, and relevant radiological abnormalities return to normal.

Neutropenic patients who recover from invasive aspergillosis may suffer from reactivation of the infec-tion during subsequent periods of immunosuppression. One solution to this problem is to begin empirical treat-ment with amphotericin B (1 mg/kg/day) not less than 48 hours before antileukemic treatment is commenced. This drug should be continued until the neutrophil count has recovered.

It is important to remember that empirical antifungal therapy does not preclude further investigation for other occult causes of fever (drug-induced fever, viral infec-tion, possibly *Pneumocystis*, rarely parasites or myco-bacteria). In addition, a change in fever pattern, including abatement of previously continuous fevers, which frequently follows initiation of amphotericin B therapy, cannot be taken as a guarantee of a favorable response to empirical antifungal therapy. Furthermore, efforts to document fungal infection during empirical therapy in order to adapt antifungal therapy to specific agents and their clinical symptoms should not be aban-doned.

Recent reports, reviews, and guidelines regarding empirical treatment of fever in neutropenia include: Hamacher et al. (1999); Roland (1999); Wingard et al. (1999); Prentice et al. (2000); Marr (2002); Richardson and Jones (2003).

A recent review of voriconazole and caspofungin in the *Drugs and Therapeutics Bulletin* (Anonymous 2004) recommends that neither caspofungin nor voriconazole be used for empirical treatment of fever in patients with neutropenia.

TREATMENT OF INVASIVE ASPERGILLOSIS IN NEONATES

High dose (1.0–1.5 mg/kg/day) amphotericin B remains the cornerstone of treatment for both suspected and proven invasive aspergillosis. In neonates, the pharmaco-kinetics of amphotericin B are characterized by extreme variability between individuals and a comparatively lower clearance rate. However, no correlations between plasma concentrations and pharmacological effects have ever been established, and there is no evidence that these pharmacokinetic characteristics result in any clin-ical consequences. Indeed, several case series indicate that amphotericin B is usually tolerated without nephro-toxicity at daily doses of up to 1 mg/kg/day, even in very low-birthweight infants, and daily dosages of up to 1.5 mg/kg have been safely administered. It is recom-mended that a starting dosage of at least 1 mg/kg daily be used, in view of the high mortality associated with invasive aspergillosis. Because it has the potential to cause cardiac arrhythmias, amphotericin B should be infused under careful monitoring, in particular in the presence of hyperkalemia and/or renal impairment.

Liposomal amphotericin B (AmBisome) has been tolerated at dosages of up to 5 mg/kg/day in more than 50 term and preterm infants (Groll et al. 1998b; Scarcella et al. 1998; Weitkamp et al. 1998). Published data regarding the efficacy of either Abelcet or Amphocil are lacking. The collective opinion is that AmBisome should be the salvage agent of choice in neonates refractory to or intolerant of conventional amphotericin B. A starting dose of 5 mg/kg/day is recommended for proven infections, with dosage adjustment only to limit toxicity (Groll et al. 1998b). Where there is fever of unknown origin without conclusive evidence of invasive aspergillosis, then empirical/pre-emptive AmBisome should be given at a rate of 3 mg/kg/day.

Itraconazole as the oral solution may be indicated in a stable patient with residual lesions. The drug is absorbed by preterm neonates. The recent licensing of intravenous itraconazole offers the opportunity to treat the condition parenterally.

COMBINATION THERAPY

Published data on the use of the new antifungals caspofungin or voriconazole in combination with other antifungal drugs is limited to case reports or case series. Further investigation of the efficacy and safety of caspofungin and voriconazole in combination therapy is needed (for recent reviews see: Lewis and Kontoyiannis 2001; Antoniadou and Kontoyiannis 2003; Steinbach et al. 2003).

PROPHYLAXIS

Old and new approaches to prophylaxis include aerosolized amphotericin B, low-dose intravenous amphotericin B, liposomal amphotericin B, and oral itraconazole (for general reviews see Prentice et al. 2000; Singh 2000; Cornely et al. 2003; Richardson and Warnock 2003). Guidelines can be found in Richardson and Jones (2003).

Amphotericin B inhalation

Aerosolized amphotericin B appeared to be promising in some of the early clinical trials, which have included patients undergoing bone marrow transplantation, and the incidence of aspergillus infections in some series dramatically decreased (compared with historical controls) (Jorgensen et al. 1989; Connealy et al. 1990; Jeffery et al. 1991; Myers et al. 1992; Beyer et al. 1993, 1994a, b). Unfortunately, these results have not been confirmed. A more recent study shows that, in contrast, aerosolized amphotericin B was poorly tolerated (Erjavec et al. 1997). Here, the value of inhaled amphotericin B against invasive pulmonary aspergillosis was

evaluated in 61 neutropenic episodes in 42 patients treated for a hematological malignancy. Each patient was assigned to receive amphotericin B in doses escalating to 10 mg thrice daily but only 20 (48 percent) patients managed to complete the schedule regimen. One patient tolerated the full dose initially, but had to discontinue treatment when dyspnea developed as a result of pneumonia and acute respiratory distress. Another 22 patients (52 percent) experienced side effects, including eight (19 percent) who reported mild coughing and dyspnea but who tolerated the full dose, and three (7 percent) patients whose dose was reduced to 5 mg thrice daily. Another six (14 percent) patients could tolerate only 5 mg thrice daily, and five (12 percent) others stopped treatment because of intolerance. The study indicated that elderly patients and those with a history of chronic pulmonary obstructive disease were more likely to develop side effects during inhalation. Twelve (28 percent) patients developed proven or possible invasive fungal infection but no correlation was established between infection and the total amount of amphotericin B inhaled. The investigators concluded that prophylaxis with inhaled amphotericin B does not appear useful in preventing invasive pulmonary aspergillosis in neutropenic patients and is tolerated poorly. The type of nebulizer would appear to be a key issue in the delivery of amphotericin B aerosols throughout the respiratory tract. Many types of nebulizers exist. The nebulizer used by Erjavec and colleagues was capable of producing particles of 0.5–5.5 μm, sufficiently small to reach most components of the respiratory tree including the terminal bronchioles and alveolar sacs. However, nothing is known regarding the level of amphotericin B that is achieved in the pneumocytes of the alveolar epithelium that lines the alveolar ducts and sacs.

The question of safety and tolerance of inhaled amphotericin B has been investigated as part of a study to evaluate the benefit of aerosolized amphotericin B as prophylaxis against invasive aspergillosis (Dubois et al. 1995). The investigators looked specifically at oxygen saturation levels, peak flow values, and symptoms of patients given amphotericin B. Data are presented on a series of 18 patients and 132 amphotericin B administrations. Four (22 percent) of the patients stopped treatments because of nausea and vomiting which were believed to be due to the inhaled amphotericin B. For the remaining patients, no treatment was stopped because of symptoms or physiological changes caused by amphotericin B, although there were nine instances of clinically significant bronchospasm as defined by a drop in peak flow of 20 percent or more, nine clinically relevant increases in cough, and three clinically relevant increases in dyspnea. The findings from this study led the authors to draw the following conclusions. First, most prophylactic amphotericin B was well tolerated and did not cause any clinically or physiologically significant changes. Second, inhaled amphotericin B might in some cases cause nausea and vomiting, which would lead to discontinuation of therapy. Third, asthmatic subjects are more likely to experience significant drops in peak flow values. The authors advocate that peak flow should be monitored and bronchodilators should be used prior to amphotericin B inhalation for known asthmatic subjects. Fourth, an occasional patient with no known history of asthma will consistently experience adverse symptoms with amphotericin B inhalation. It is clear from this study that whilst side effects do occur, neither the frequency nor severity of reactions appears to preclude the use of inhaled amphotericin B in the majority of patients. However, the specific benefit as prophylaxis against pulmonary aspergillosis has still to be resolved. It is possible that nebulized liposomal amphotericin B (AmBisome) may provide greater protection. In terms of drug levels in alveolar sacs and lung tissue of mice, aerosolization and delivery of liposomal amphotericin B was shown to be far superior to that of conventional amphotericin B and that liposomal amphotericin B was more effective than conventional amphotericin B when given as a prophylactic aerosol in clearing mouse lungs of *Aspergillus*, although this was very much inoculum concentration dependent (Allen et al. 1994).

Oral itraconazole prophylaxis (see below) combined with intranasal amphotericin B is another approach (Todeschini et al. 1993). Here, itraconazole capsules (200 mg/day) combined with intranasal amphotericin B (10 mg/day) was prescribed as prophylaxis in 164 patients. The incidence of proven aspergillosis was lower, as was the mortality rate. However, more experience is needed before this drug combination can be recommended for daily clinical practice, especially as the oral solution formulation of itraconazole will undoubtedly supersede the capsule form for routine prophylaxis.

Low-dose amphotericin B

The value of low-dose amphotericin B (0.15–0.25 mg/kg/day) as prophylaxis against *Aspergillus* spp. in patients undergoing allogeneic bone marrow transplantation has been explored in a number of studies (Rousey et al. 1991; Perfect et al. 1992). The data do not suggest a benefit compared with historical controls. For example, in a placebo-controlled prospective study involving 188 patients who underwent autologous bone marrow transplantation, amphotericin B (0.1 mg/kg daily) did not prevent aspergillus infection compared to the placebo group. One aspergillus infection was documented in each group. Prophylaxis was discontinued once an individual developed a persistent fever for 5 days despite treatment with broad-spectrum antibiotics. Suspected fungal infections were treated empirically with amphotericin B 0.6 mg/kg/day. In this study amphotericin B at 0.1 mg/kg/day did not prevent the emergence of aspergillosis compared with placebo.

The conclusions drawn from these earlier studies of the benefit of low-dose amphotericin B as prophylaxis against invasive aspergillosis are not clear. Overall, low-dose amphotericin B does not appear to provide a protective effect compared with placebo, but there does appear to be a possible effect in decreasing mortality.

Liposomal amphotericin B

The proven efficacy and improved toxicity profile of liposomal amphotericin B (AmBisome) prompted Tollemar and colleagues to evaluate the role of this lipid formulation of amphotericin B in preventing fungal infections after bone marrow transplantation (Tollemar et al. 1993). Thirty-six patients received the drug, at a dosage of 1 mg/kg for between 1 and 11 weeks, and 40 patients received no treatment. Fungal colonization decreased in the AmBisome group whereas it increased in the placebo group. By the end of prophylaxis, eight of 24 (33 percent) patients who received AmBisome were colonized compared with 18 of 29 (62 percent) patients in the placebo group. Presumed fungal infection occurred in five of 36 patients given AmBisome. None of these infections was fatal. In the control group, seven patients developed presumed fungal infection and two died. Proven fungal infection occurred in one patient receiving AmBisome (*Candida guilliermondii*) compared with three patients receiving placebo (*C. guilliermondii*, two; *C. albicans*, one). The study showed possible benefits regarding the reduction of proven fungal infection in the patients given AmBisome prophylactically. A beneficial effect of prophylaxis was found as none of the patients with a suspected fungal infection in the AmBisome group died with proven fungal infection compared with two placebo-treated patients.

The value of AmBisome as a prophylactic agent has been examined further by Kelsey et al. (1999) where AmBisome 2 mg/kg was compared with placebo in patients undergoing chemotherapy or bone marrow transplantation. Prophylaxis began on day 1 of chemotherapy and continued until neutrophils regenerated or infection was suspected. Of 161 evaluable patients, 74 received AmBisome and 87 received placebo. Proven fungal infections developed in no patients on AmBisome and in three patients on placebo. Suspected fungal infections requiring intervention with systemic antifungal therapy (usually amphotericin B) occurred in 31 patients (42 percent) on AmBisome and in 40 on placebo (46 percent)

Itraconazole

Itraconazole is an orally active triazole with a wide spectrum of activity which has been successfully used in the prevention of aspergillus infection.

Itraconazole has been compared with ketoconazole and nystatin in neutropenic patients in retrospective analyses, and was found to be superior (reviewed by Beyer et al. 1994b). In a placebo-controlled, randomized trial of itraconazole prophylaxis, 400 mg/day reduced the overall incidence of proven fungal infection to nine of 83 neutropenic episodes (11 percent), compared with 15 of 84 episodes (18 percent) in the placebo arm (Beyer et al. 1994b). However, it should be noted that this effect was mainly the result of a reduction of systemic candida infections and was not statistically significant. The incidence of suspected or proven cases of invasive pulmonary aspergillosis was similar in both the prophylactic itraconazole (five of 83 or 6 percent) and the placebo (four of 84 or 5 percent) arms.

The absorption of itraconazole from the gastrointestinal tract varies within a wide range and is largely unpredictable, so itraconazole doses higher than 400 mg or therapeutic drug monitoring might be necessary for effective antifungal prophylaxis with this drug. There are a number of drugs that interfere with the absorption of itraconazole, among which are agents such as antacids that are commonly used in neutropenic patients. Further trials including larger numbers of patients are needed to make a definitive statement on the effectiveness of itraconazole for the prophylaxis of invasive pulmonary aspergillosis. To facilitate the prophylactic use of this agent, the availability of an itraconazole formulation with improved and more predictable bioavailability would clearly be beneficial.

ITRACONAZOLE ORAL SOLUTION

The known problems associated with the bioavailability of the capsule form in neutropenic patients have prompted the development of an oral solution of itraconazole. This bioavailablilty problem is thought to arise because gastrointestinal mucositis (following intensive cancer chemotherapy or radiotherapy) can make drug dissolution and absorption erratic, leading to abnormally low blood levels and lack of drug efficacy in some cases. It appears that this formulation has improved pharmacokinetics and more predictable bioavailability in neutropenic patients (Prentice et al. 1994; Michallet et al. 1998). In this study trough serum concentrations of itraconazole were measured weekly during prophylaxis with itraconazole liquid, 2.5 mg/kg b.i.d. in neutropenic patients undergoing treatment for hematological malignancies. From week 2 of prophylaxis, at least 84 percent of patients had itraconazole plasma concentrations higher than 250 ng/ml; this concentration is considered the minimum level required for antifungal efficacy. These pharmacokinetic findings have been confirmed by a much larger study where the value of prophylactic itraconazole solution was compared with fluconazole suspension (see below).

There has been a considerable expansion in the literature on the use of the itraconazole oral solution as prophylaxis against aspergillosis (see, for example, Foot et al. 1999; Glasmacher et al. 1999a, b; Kibbler 1999; Menichetti et al. 1999; Harrousseau et al. 2000; and

reviewed in a *Journal of Antimicrobial Chemotherapy* supplement (in press).

ITRACONAZOLE LIQUID COMPARED WITH ORAL AMPHOTERICIN B PROPHYLAXIS

A randomized, double-blind, double-dummy study compared the efficacy of itraconazole liquid with oral amphotericin B capsules in patients with hematological malignancy and profound neutropenia (Harrousseau et al. 2000). The incidence of fungal infections was compared between groups and comparisons were also made of the frequency and time and initiation of intravenous amphotericin B rescue therapy. A total of 557 patients were randomized to receive either itraconazole liquid, 2.5 mg/kg body weight b.i.d. (*n*=281) or oral amphotericin B, 2×250 mg capsules q.i.d (*n*=276). Prophylaxis was initiated with chemotherapy and continued until the neutrophil count recovered or another study point was reached, up to a maximum of 8 weeks. Although not reaching statistical significance, there was a trend in favor of itraconazole liquid for proven invasive aspergillosis (five versus nine cases) as well as overall documented deep fungal infections including aspergillosis (eight versus 13 cases). One of eight patients died in the itraconazole group in contrast to five of 13 in the amphotericin B group. In this study plasma levels of itraconazole were measured and adequate levels (>250 ng/ml) were attained in at least 80 percent of patients 10–14 days after the start of treatment. After 4 weeks of prophylaxis, these levels were reached in 96 percent of patients.

ITRACONAZOLE LIQUID COMPARED WITH FLUCONAZOLE PROPHYLAXIS

Itraconazole in its liquid suspension form (5 mg/kg) has been compared with fluconazole suspension (100 mg) in a randomized trial where either itraconazole or fluconazole was given from the start of cytoreduction to neutrophil recovery in 581 neutropenic episodes in 445 adults receiving a bone marrow transplant or chemotherapy for hematological malignancy (Morgenstern et al. 1999). The dose of fluconazole suspension was chosen based on accepted practice at the time the study commenced. There were more proven cases of invasive aspergillosis in patients given fluconazole prophylaxis compared with itraconazole (four in the fluconazole arm, none in the itraconazole arm), demonstrating the superiority of itraconazole in suppressing invasive aspergillosis in immunocompromised patients. However, there were more gastrointestinal disturbances associated with itraconazole. These are most likely due to the cyclodextrin vehicle used in the solution formulation.

SAFETY OF ITRACONAZOLE LIQUID FOR ANTIFUNGAL PROPHYLAXIS

Safety information from data-on-file studies and summaries in abstract form is available for about 1 000 patients.

Adverse experiences observed frequently in these patients included nausea, vomiting, mucositis, diarrhea, and fever. Biochemical abnormalities frequently occurring for all treatment groups included decreases in calcium, potassium, and uric acid and increases in urea and bilirubin. The incidence and nature of adverse events reported in these studies to be definitely related to prophylactic drug treatment were similar for itraconazole liquid, oral amphotericin B, and placebo and were mainly gastrointestinal in nature.

ITRACONAZOLE ORAL SOLUTION AND CHILDREN

Itraconazole oral solution has been evaluated in neutropenic children undergoing an allogeneic bone marrow transplant (Foot et al. 1999). In an open study of antifungal prophylaxis in 103 children, patients received 5.0 mg/kg itraconazole oral solution per day. Prophylaxis was started at least 7 days before the onset of neutropenia and continued until neutrophil recovery. No proven systemic fungal infections occurred during the study but 26 patients received intravenous amphotericin B for antibiotic-unresponsive PUO. One patient received amphotericin B for mycologically confirmed esophageal candidosis. Three patients developed suspected oral candidosis but none was mycologically proven and no treatment was given. Serious adverse events occurred in 27 patients. The most common adverse events attributable to itraconazole were vomiting, abnormal liver function, and abdominal pain. This study indicates that itraconazole oral solution may be used as antifungal prophylaxis in neutropenic children.

PREVENTION

Prevention of aspergillosis is relatively difficult. Simple precautions – such as eliminating potted plants from patients' rooms and using barriers during hospital construction – are recommended. The use of high-efficacy particulate air (HEPA) filters appears to be the only currently effective means of decreasing the incidence of aspergillus infection.

The principles of environmental control of nosocomial aspergillosis are complex given that even HEPA units are not completely effective in preventing disease. Fungal exposure would be more precisely studied using a personal air sampler for the patient, but there is no fungal sampler currently available which can be used in this way, and there are also severe technical limitations on the duration of the sampling time of available fungal samplers (Morris et al. 2000; Thio et al. 2000). Alternatively, a systematic program of longitudinal patient and environmental surveillance may predict cases of invasive aspergillosis. Indeed, there appears to be a correlation between the recovery of *Aspergillus* spp. from the nose and mouth of patients in an open hematology ward and an elevated number of conidia in the air (Richardson et al. 2000)

The relationship between aspergillosis in predisposed patients and building work is also complex. Hospitals are buildings of continuous change and adaptation, so construction is an inevitable prospect which may extend throughout the year. Whether or not this activity is complicated by an outbreak of infection in the susceptible patients nearby, or is a risk related directly to the amount of disruption or some other factor, is unknown. In one study, an increase in the number of patients with invasive aspergillosis could not be explained by an increase in the number of aspergillus conidia in the outside air (Leenders et al. 1998). Genotyping of the isolates from the unit where the survey was carried out showed that clonally related isolates were persistently present for more than 1 year. Clinical isolates of *A. fumigatus* obtained during the outbreak period were different from the clones normally persistent in that environment.

Currently, the environmental mycology of most outbreaks of nosocomial aspergillosis is poorly defined. However, the development of molecular biology techniques more directly applicable to *Aspergillus* spp. may help resolve some of these difficulties.

Aspergillus spp. have a major reservoir in organic debris, bird droppings, dust, and building material. The principal approach in the prevention of aspergillosis is to minimize patients' exposure to aspergillus conidia by filtering air or initiating some form of patient isolation (Sherertz et al. 1987; Rhame 1991; Hay 1993). Further steps consist of elimination of obvious sources of aspergilli, such as removing plants from the surrounding environment of a patient. In some instances surface disinfection with copper-8-quinolinolate has been reported to be effective. Susceptible patients should not be treated in areas where there is construction or demolition activity, and if such activities are under way, measures should be instigated to seal these sites to prevent air exchange with the patients' environment.

Certain foodstuffs, such as cereals, nuts, and spices, e.g. ground black pepper, have been found to be contaminated with aspergilli and should not be offered to patients at risk of developing invasive pulmonary aspergillosis (reviewed by Beyer et al. 1994b). Although outbreaks of invasive aspergillosis have been associated with construction within or around a hospital, the precise source of the fungus is occasionally difficult to trace with certainty (Opal 1986; Iwen et al. 1994; Anderson et al. 1996). There have been few studies that have prospectively examined the aeromycology in and around a hospital during major building alterations and then compared these findings with samples from patients and the incidence of invasive aspergillosis (reviewed by Beyer et al. 1994b).

In a seminal study (Goodley et al. 1994), the authors took advantage of the opportunities that arose during widespread building operations around their hospital where several groups of patients seemed at risk of fungal infection – in wards for renal transplantation, bone marrow transplantation, oncology, and intensive care. Air samples were taken in these wards (by SAS Sampler, pbi International, Milan, Italy) and various outdoor sites around the hospital, at specific sites throughout the hospital, sequentially throughout the year as well as in particular areas during periods of construction activity. Nasal swabs were also taken from patients for comparison with the air sampling results. The most commonly isolated fungal species was *A. fumigatus*. Nasal swabs were positive in 12 of 188 samples – 11 *A. fumigatus* and one *Aspergillus sydowi*. Most of the air samples cultured less than 10 c.f.u./m^3 throughout the year. A peak of higher counts occurred in March (190 c.f.u./m^3, confirmed at various sites) which could not be explained either by building work or by meteorology. Eight of the positive nasal swabs were obtained during March; three cases of invasive aspergillosis developed through the year and did not seem to be related to the spell of higher spore counts. One of the buildings was demolished, but there was no significant rise in spore counts and no change in the background pattern of fungal isolation in the wards or the corridors. Air sampling was repeated over the following year when a peak was recorded in June at 90 c.f.u./m^3, and very similar low levels throughout the rest of the year. The authors' interpretation of the results was that, because cases of invasive aspergillosis seemed to develop at low spore levels, then all highly susceptible patients should have protective isolation (HEPA ventilation and sterile management procedures). Routine nasal swab sampling was not proposed as an alternative to air sampling. Avoidance measures appear to be appropriate if minimal exposure is the only component necessary to induce invasive aspergillosis in transplant recipients.

The debate concerning the value of laminar air flow (LAF) rooms in protecting high-risk patients from *Aspergillus* has continued. Wald and coworkers conducted a retrospective analysis of aspergillosis in bone marrow transplant recipients (Wald et al. 1997). One of the conclusions of this study was that the use of LAF rooms is protective against early, but not late, aspergillosis, during the transplant course. The conclusions in this study were questioned by Viscoli (1998), who was of the opinion, based on personal experience and the existing literature, that even if it was shown that LAF rooms were associated with a reduced risk of early aspergillosis, it is unlikely that this benefit resulted from the use of the LAF room system as a whole, but rather by the air filtration through the HEPA filters. The opinion was offered that the use of LAF rooms in bone marrow transplant recipients should not be encouraged without convincing supporting data.

A preventative strategy has been proposed by a consortium of The Mycologic Study Group of the French Society of Hematology and The Research Group on Fungal Infections, and The European Group for Research on Biotypes and Genotypes of *Aspergillus* Network (Gangneux et al. 2002). The salient points of this strategy are as follows:

- Air-control measures remain crucial for the reduction of environmental dissemination of fungal conidia. The consortium's premise is that the nosocomial origin of aspergillosis has been convincingly demonstrated in epidemic situations. A number of investigations have shown that the concentration of *Aspergillus* spp. and other fungi in the air in a hematology unit correlated with the incidence of invasive aspergillosis in nonepidemic situations. Thus, it is strongly recommended that patients who are at high risk of infection benefit from the measures put forth by the Centers for Disease Control and Prevention, that is, the use of HEPA filtration, use of LAF systems, high rates of room-air change, use of positive pressure, and use of well-sealed rooms.

- Air-control efficiency must be monitored. Several years ago, a regular monitoring of environmental fungal contamination (for *Aspergillus* and other airborne fungal species) was started, with two major goals: to detect increases in conidia density and to assess air-control efficiency. Environmental monitoring requires the following:
  - air sampling with an efficient biocollector
    surface sampling with contact petri dishes or swabs, which is a simple and efficient monitoring method that can detect minor contamination, even when concomitant air samples test negative for fungi
  - sampling of patient rooms that are equipped with HEPA filters, with or without LAF, and sampling of all parts of the ward that are provided with air filtration and positive pressure (particularly corridors)
  - use of guidelines for patient management and cleaning procedures in protective environments, which should be adapted according to the results of monitoring.

- Determination of the baseline values of air and surface concentrations of *Aspergillus* spp. and other fungal conidia is essential for valuable assessment of any further increase in fungal contamination. Given the genetic diversity of *A. fumigatus* isolates, and given the current limitations of molecular typing methods to localize the fungal source or to date the infection, environmental sampling is not, in the authors' opinion, simply a means of comparing environmental isolates with clinical isolates.

- The sources and routes of conidia transmission are unclear. It has been suggested that aspergillosis is waterborne, and that nosocomial aspergillosis can be airborne from water sources in the hospital setting.

However, this is not a unversal finding. It is apparent that the environmental risk of invasive aspergillosis linked to water should be interpreted according to the local situation, namely, the source of water (underground versus surface water). In addition, because humidity favors fungal growth, high fungal densities near water sources may simply reflect the presence of conidia in the air or on surfaces and indicate the need for new cleaning procedures. Furthermore, the biotope of *A. fumigatus* differs from that of *Fusarium* and *Acremonium* spp.

Alternative sources of conidia inhalation should not be neglected. These sources include the clothing of visitors and medical staff as well as personal and medical materials. Therefore, specific preventive measures may be appropriate. The gastrointestinal route of infection is rarely considered; however, a high rate of food contamination by filamentous fungi, such as *Aspergillus* spp. and zygomycetes, has been reported. The existence of isolated gastrointestinal filamentous fungal infection without pulmonary or disseminated infection supports the hypothesis that, in addition to the risk of conidia inhalation, contact with food or water can lead to conidia absorption.

The authors' conclusions and recommendations in this consensus document are many-fold. Aspergillosis mainly occurs in neutropenic patients. Air-control measures are presently the most effective way of significantly reducing the incidence of nosocomial aspergillosis. Although they are expensive and only partially effective, these measures should not be called into question, as shown by their protective effect against early infections after bone marrow transplantation. However, the efficiency of air-control measures should be continuously monitored by regular measurement of environmental levels of fungal conidia. This approach is strongly recommended in units that use air-control measures, as are specific investigations of cases of aspergillus infection. However, there should be more investigation and control of alternative sources of contamination. Prevention of delayed acquired aspergillosis is even more difficult: control of environmental sources of contamination and follow-up of aspergillus colonization are difficult outside the hospital setting, and, until now, there has been no demonstration that any antifungal drug or cytokine can significantly prevent or reduce the risk of aspergillosis in patients who are at risk of infection.

REFERENCES

Abbasi, S., Shenep, J.L. and Hughes, W.T. 1999. Aspergillosis in children with cancer: a 34-year experience. *Clin Infect Dis*, **29**, 1210–19.

Advances against aspergillosis. *Clin Infect Dis*, 2003; **37**, Suppl 3, S155–S292.

Al-Doory, Y. and Wagner, G.E. 1985. *Aspergillosis*. Springfield, IL: Charles C Thomas, 274.

Alexander, B.D. 2002. Diagnosis of fungal infection: new technologies for the mycology laboratory. *Transpl Infect Dis*, **4**, Suppl 3, 32–7.

Allen, S.D., Sorensen, K.N.. et al. 1994. Prophylactic efficacy of aerosolized liposomal (AmBisome) and non-liposomal (Fungizone) amphotericin B in murine pulmonary aspergillosis. *J Antimicrob Chemother*, **34**, 1001–13.

Allen, M.J., Harbeck, R., et al. 1999. Binding of rat and human surfactant proteins A and D to *Aspergillus fumigatus* conidia. *Infect Immun*, **67**, 4563–9.

Allo, M.D.. Miller, J., et al. 1987. Primary cutaneous aspergillosis associated with Hickman intravenous catheters. *N Engl J Med*, **317**, 1105–8.

Amitani, R., Sato, A., et al. 1992. Effects of *Aspergillus* species culture filtrates on human respiratory ciliated epithelium *in vitro*. *Am Rev Respir Dis*, **145**, A548.

Amitani, R., Taylor, G., et al. 1995. Purification and characterization of factors produced by *Aspergillus fumigatus* which affect human ciliated respiratory epithelium. *Infect Immun*, **63**, 3266–71.

Anaissie, E.J., Bodey, G.P., et al. 1989. New spectrum of fungal infections in patients with cancer. *Rev Infect Dis*, **11**, 369–78.

Anderson, K.. Morris, G., et al. 1996. Aspergillosis in immunocompromised paecriatric patients: associations with building hygiene, design and indoor air. *Thorax*, **51**, 256–61.

Andström, E.E., Ringden, O., et al. 1996. Safety and efficacy of liposomal amphotericin B in allogeneic bone marrow transplant recipients. *Mycoses*, **39**, 185–93.

Annaix, V., Bouchara, J.-P., et al. 1992. Specific binding of human fibrinogen fragment D to *Aspergillus fumigatus* conidia. *Infect Immun*, **60**, 1747–55.

Anonymous. 2004. Caspofungin and voriconazole for fungal infections. *Drugs Ther Bull*, **42**, 5–8.

Ansorg, R., von Heinegg, E.H. and Rath, P.M. 1994. *Aspergillus* antigenuria compared to artigenemia in bone marrow transplant recipients. *Eur J Clin Microbiol Infect Dis*, **13**, 582–9.

Antoniadou. A. and Kontoyiannis, D.P. 2003. Status of combination therapy for refractory mycoses. *Curr Opin Infect Dis*, **16**, 539–45.

Aquino, S.L., Kee, S.T., et al. 1994. Pulmonary aspergillosis: imaging findings with pathologic correlation. *Am J Roentgenol*, **163**, 811–15.

Ashdown, B.C., Tien, R.D. and Felsberg, G.J. 1994. Aspergillosis of the brain and paranasal sinuses in immunocompromised patients: CT and MR imaging findings. *Am J Roentgenol*, **162**, 155–9.

Aufauvre-Brown, A., Cohen, J., et al. 1992. Use of randomly amplified polymorphic DNA markers to distinguish isolates of *Aspergillus fumigatus*. *J Clin Microbiol*, **30**, 2991–3.

Bag, R. 2003. Fungal pneumonias in transplant recipients. *Curr Opin Pulm Med*, **9**, 193–8.

Banerjee, B. and Kurup, V.P. 2003. Molecular biology of *Aspergillus* allergens. *Front Biosci*, **8**, s128–39.

Barnes, A.J. 1993. *Aspergillus* infection: does serodiagnosis work? *Serodiagn Immunother Infect Dis*, **5**, 135–8.

Barnes, A.J. and Denning, D.W. 1993. *Aspergillus*: sigificance as a pathogen. *Rev Med Microbiol*, **4**, 176–80.

Barrett, J.P., Vardulaki, K.A., et al. 2003. A systematic review of the antifungal effectiveness and tolerability of amphotericin B formulations. *Clin Ther*, **25**, 1293–320.

Berger, L.A. 1998. Imaging in the diagnosis of infections in immunocompromised patients. *Curr Opin Infect Dis*, **11**, 431–6.

Beyer, J., Barzen, G., et al. 1993. Aerosol amphotericin B for prevention of invasive pulmonary aspergillosis. *Antimicrob Agents Chemother*, **37**, 1367–9.

Beyer, J., Schwartz, S., et al. 1994a. Use of amphotericin B aerosols for the prevention of pulmonary aspergillosis. *Infection*, **22**, 143–8.

Beyer, J., Schwartz, S., et al. 1994b. Strategies in prevention of invasive pulmonary aspergillosis in immunosuppressed or neutropenic patients. *Antimicrob Agents Chemother*, **38**, 911–17.

Binder, R.E., Faling, L.J., et al. 1982. Chronic necrotising pulmonary aspergillosis: a discrete clinical entity. *Medicine*, **61**, 109–24.

Blitzer, A. and Lawson, W. 1993. Fungal infections of the nose and paranasal sinuses, Part 1. *Otolaryngol Clin North Am*, **26**, 1007–35.

Blum, U., Windfuhr, M., et al. 1994. Invasive pulmonary aspergillosis. MRI, CT, and plain radiographic findings and their contribution for early diagnosis. *Chest*, **106**, 1156–61.

Bodey, G.P., Anaissie, E., et al. 1993. Role of granulocyte-macrophage colony-stimulating factor as adjuvant therapy for fungal infection in patients with cancer. *Clin Infect Dis*, **17**, 705–7.

Boyle, J.A. and Swenson, C.E. 1999. ABELCET treatment. *J Clin Pharmacol*, **39**, 427–8.

Bouchara, J.-P., Bouali, G., et al. 1988. Binding of fibrinogen to the pathogenic *Aspergillus* species. *J Med Vet Mycol*, **26**, 327–34.

Bouchara, J.-P., Tronchin, G., et al. 1995. The search for virulence determinants in *Aspergillus fumigatus*. *Trends Microbiol*, **3**, 327–30.

Brakhage, A.A. and Langfelder, K. 2002. Menacing mould: the molecular biology of *Aspergillus fumigatus*. *Annu Rev Microbiol*, **56**, 433–55.

Brakhage, A.A., Jahn, B. and Schmidt, A. 1999. *Aspergillus fumigatus, biology, clinical aspects and molecular approaches to pathogenicity*. Basel: Karger.

Bretagne, S., Costa, J.-M., et al. 1995. Detection of *Aspergillus* species DNA in bronchoalveolar lavage samples by competitive PCR. *J Clin Microbiol*, **33**, 1164–8.

Brummund, W., Resnick, A., et al. 1987. *Aspergillus fumigatus*-specific antibodies in allergic bronchopulmonary aspergillosis and aspergilloma: evidence for a polyclonal antibody response. *J Clin Microbiol*, **25**, 5–9.

Buffington, J., Reporter, R., et al. 1994. Investigation of an epidemic of invasive aspergillosis: utility of molecular typing with the use of random amplified polymorphic DNA probes. *Pediatr Infect Dis J*, **12**, 386–93.

Burnie, J.P. and Matthews, R.C. 1991. Heat shock protein 88 and *Aspergillus* infection. *J Clin Microbiol*, **29**, 2099–106.

Buzina, W., Braun, H., et al. 2003. Fungal biodiversity – as found in nasal mucus. *Med Mycol*, **41**, 149–61.

Caillot, D. 2003. Intravenous itraconazole followed by oral itraconazole for the treatment of amphotericin-B-refractory invasive pulmonary aspergillosis. *Acta Haematol*, **109**, 111–18.

Caillot, D., Casasnovas, O., et al. 1997. Improved management of invasive pulmonary aspergillosis in neutropenic patients using early thoracic computed tomographic scan and surgery. *J Clin Oncol*, **15**, 139–47.

Campbell, J.H., Winter, J., et al. 1991. The treatment of pulmonary aspergilloma with itraconazole. *Thorax*, **46**, 839–41.

Carlisle, J.R. 1978. Primary cutaneous aspergillosis in a leukemic child. *Arch Dermatol*, **114**, 78–80.

Chandrasekar, P.H., Cutright, J. and Manavathu, E. 2000. Efficacy of voriconazole against invasive pulmonary aspergillosis in a guinea-pig model. *J Antimicrob Chemother*, **45**, 673–6.

Chang, T., Teng, M.M.H., et al. 1992. Aspergillosis of the paranasal sinuses. *Neuroradiology*, **34**, 520–3.

Chetty, A. 2003. Pathology of allergic bronchopulmonary aspergillosis. *Front Biosci*, **8**, E110–14.

Chiller, T.M. and Stevens, D.A. 2000. Treatment strategies for *Aspergillus* infections. *Drug Resist Update*, **3**, 89–97.

Chiu, A. 2003. The treatment of allergic bronchopulmonary aspergillosis. *Front Biosci*, **8**, S243–5.

Chumpitazi, B.F.F., Pinel, C., et al. 2000. *Aspergillus fumigatus* antigen detection in sera from patients at risk for invasive aspergillosis. *J Clin Microbiol*, **38**, 438–43.

Clancy, C.J. and Nguyen, M.H. 1998. In vitro efficacy and fungicidal activity of voriconazole against *Aspergillus* and *Fusarium* species. *Eur J Clin Microbiol Infect Dis*, **17**, 573–5.

Clark, A.D., McKendrick, S., et al. 1998. A comparative analysis of lipid-complexed and liposomal amphotericin B preparations in haematological oncology. *Br J Haematol*, **103**, 198–204.

Cockrill, B.A. and Hales, C.A. 1999. Allergic bronchopulmonary aspergillosis. *Annu Rev Med*, **50**, 303–16.

Cohen, J. 1991. Clinical manifestations and management of aspergillosis in the compromised patient. In: Warnock, D.W. and Richardson, M.D. (eds), *Fungal infection in the compromised patient*, 2nd edition. Chichester: John Wiley & Sons, 117–52.

Cohen, J., Denning, D.W. and Viviani, M.A. 1993. Epidemiology of invasive fungal infection in European cancer centres. *Eur J Clin Microbiol Infect Dis*, **12**, 392–3.

Connealy, E., Cafferkey, M.T., et al. 1990. Nebulized amphotericin B as prophylaxis against invasive aspergillosis in granulocytopenic patients. *Bone Marrow Transplant*, **5**, 403–6.

Cornely, O.A., Bohme, A. and Buchheidt, D. 2003. Prophylaxis of invasive fungal infections in patients with hematological malignancies and solid tumors – guidelines of the Infectious Diseases Working Party (AGIHO) of the German Society of Hematology and Oncology (DGHO). *Ann Hematol*, **82**, Suppl 2, S186–200.

Cornet, M., Levy, V., et al. 1999. Efficacy of prevention by high-efficacy particulate air filtration or laminar airflow against *Aspergillus* airborne contamination during hospital renovation. *Infect Control Hosp Epidemiol*, **20**, 508–13.

Coukell, A.J. and Brogden, R.N. 1998. Liposomal amphotericin B: therapeutic use in the management of fungal infections and visceral leishmaniasis. *Drugs*, **55**, 585–612.

Daleine, G., Salmon, D., et al. 1993. Prevalence of *Aspergillus* species in bronchopulmonary specimens from patients infected with the human-immunodeficiency virus: pathogenic role. *Pathol Biol (Paris)*, **41**, 237–41.

De Carpentier, J.P., Ramamurthy, L., et al. 1994. An algorithmic approach to *Aspergillus* sinusitis. *J Laryngol Otol*, **108**, 314–18.

de Hoog, G.S., Guarro, J., et al. 2000. *Atlas of clinical fungi*, 2nd edition. Baarn, The Netherlands: Centraalbureau voor Schimmelcultures.

Delvenne, P., Arrese, J.E., et al. 1993. Detection of cytomegalovirus, *Pneumocystis carinii* and *Aspergillus* species in bronchoalveolar fluid – a comparison of techniques. *Am J Clin Pathol*, **100**, 414–18.

De Marie, S. 2000. New developments in the diagnosis and management of invasive fungal infections. *Haematologia*, **85**, 88–93.

Denning, D.W. 1994a. Treatment of invasive aspergillosis. *J Infect*, **28**, Suppl 1, 25–33.

Denning, D.W. 1994b. Invasive aspergillosis in immunocompromised patients. *Curr Opin Infect Dis*, **7**, 456–62.

Denning, D.W. 2000. Early diagnosis of invasive aspergillosis. *Lancet*, **355**, 423–4.

Denning, D.W. and Stevens, D.A. 1990. Antifungal and surgical treatment of invasive aspergillosis: review of 2121 published cases. *Rev Infect Dis*, **12**, 1147–201.

Denning, D.W., Hostetler, J.S., et al. 1994. NIAID Mycoses Study Group multicenter trial of oral itraconazole therapy for invasive aspergillosis. *Am J Med*, **97**, 135–44.

Denning, D.W., Riniotis, K., et al. 2003. Chronic cavitary and fibrosing pulmonary and pleural aspergillosis: case series, proposed nomenclature change, and review. *Clin Infect Dis*, **37**, Suppl 3, S265–80.

De Pauw, B.E. and Meis, J.F.G.M. 1998. Progress in fighting systemic fungal infections in haematology neoplasia. *Support Care Cancer*, **6**, 31–8.

Dhiwakar, M., Thakar, A., et al. 2003. Preoperative diagnosis of allergic fungal sinusitis. *Laryngoscope*, **113**, 688–94.

Dimopoulos, G., Piagnerelli, M., et al. 2003. Disseminated aspergillosis in intensive care unit patients: an autopsy study. *J Chemother*, **15**, 71–5.

Drakos, P.E. 1993. Invasive fungal sinusitis in patients undergoing bone marrow transplantation. *Bone Marrow Transplant*, **12**, 203–8.

Dubois, J., Bartter, T., et al. 1995. The physiologic effects of inhaled amphotericin B. *Chest*, **108**, 750–3.

Dupont, B., Richardson, M., et al. 2000. Invasive aspergillosis. *Med Mycol*, **38**, Suppl 1, 215–24.

Duthie, R. and Denning, D.W. 1995. *Aspergillus* fungemia: report of two cases and review. *Clin Infect Dis*, **20**, 598–605.

Einsele, H., Hebart, H., et al. 1997. Detection and identification of fungal pathogens in blood by using molecular probes. *J Clin Microbiol*, **35**, 1353–60.

Erjavec, Z., Woolthuis, G.M.H., et al. 1997. Tolerance and efficacy of amphotericin B inhalations for prevention of invasive pulmonary aspergillosis in haematological patients. *Eur J Clin Microbiol Infect Dis*, **16**, 364–8.

Fenelon, L.E., Hamilton, A.J., et al. 1999. Production of specific monoclonal antibodies to *Aspergillus* species and their use in immunohistochemical identification of aspergillosis. *J Clin Microbiol*, **37**, 1221–3.

Fitzsimmons, E.J., Aris, R. and Patterson, R. 1997. Recurrence of allergic bronchopulmonary aspergillosis in the post-transplant lungs of a cystic fibrosis patient. *Chest*, **112**, 281–2.

Foot, A.B.M., Veys, P.A. and Gibson, B.E.S. 1999. Itraconazole oral solution as antifungal prophylaxis in children undergoing stem cell transplantation or intensive chemotherapy for haematological disorders. *Bone Marrow Transplant*, **24**, 1089–93.

Franklin, I.M., Mehta, J. and Root, T. 1997. The use of amphotericin B lipid complex. *J Antimicrob Chemother*, **39**, 288–90.

Franquet, T., Muller, N.L., et al. 2001. Spectrum of pulmonary aspergillosis: histologic, clinical, and radiologic findings. *Radiographics*, **21**, 825–37.

Fraser, I. and Denning, D.W. 1993. Empiric amphotericin B therapy. The need for a reappraisal. *Blood Rev*, **7**, 208–14.

Fratamico, P.M. and Buckley, H.R. 1991. Identification and characterization of an immunodominant 58-kilodalton antigen of *Aspergillus fumigatus* recognised by sera of patients with invasive aspergillosis. *Infect Immun*, **59**, 309–15.

Frosco, M.-B., Chase, T. and Macmillan, J.D. 1994. The effect of elastase-specific monoclonal and polyclonal antibodies on the virulence of *Aspergillus fumigatus* in immunocompromised mice. *Mycopathologia*, **125**, 65–76.

Galimberti, R., Kowalczuk, A., et al. 1998. Cutaneous aspergillosis: a report of six cases. *Br J Dermatol*, **139**, 522–6.

Gangneux, J.P., Bretagne, S. and Cordonnier, C. 2002. Prevention of nosocomial fungal infection: the French approach. *Clin Infect Dis*, **35**, 343–6.

Gentile, G., Micozzi, A., et al. 1993. Pneumonia in allogeneic and autologous bone marrow recipients. *Chest*, **104**, 371–5.

Gerson, S.L., Talbot, G.H., et al. 1984. Prolonged granulocytopenia: the major risk factor for invasive pulmonary aspergillosis in patients with prolonged leukemia. *Ann Intern Med*, **100**, 345–51.

Girardin, H., Sarfati, J., et al. 1994a. Use of DNA moderately repetitive sequence to type *Aspergillus fumigatus* isolates from aspergillus patients. *J Infect Dis*, **169**, 683–5.

Girardin, H., Sarfati, J., et al. 1994b. Molecular epidemiology of nosocomial invasive aspergillosis. *J Clin Microbiol*, **32**, 684–90.

Glasmacher, A., Hahn, C., et al. 1999a. Fungal surveillance cultures during antifungal prophylaxis with itraconazole in neutropenic patients with acute leukaemia. *Mycoses*, **42**, 395–402.

Glasmacher, A., Hahn, C., et al. 1999b. Itraconazole trough concentrations in antifungal prophylaxis with six different dosing regimens using hydroxypropyl-beta-cyclodextrin oral solution or coated-pellet capsules. *Mycoses*, **42**, 591–600.

Goodley, J.M., Clayton, Y.M. and Hay, R.J. 1994. Environmental sampling for aspergilli during building construction on a hospital site. *J Hosp Infect*, **26**, 27–35.

Gotway, M.B., Dawn, S.K., et al. 2002. The radiologic spectrum of pulmonary *Aspergillus* infections. *J Comput Assist Tomogr*, **26**, 159–73.

Greenberger, P.A. 2002a. Allergic bronchopulmonary aspergillosis. *J Allergy Clin Immunol*, **110**, 685–92.

Greenberger, P.A. 2002b. Allergic bronchopulmonary aspergillosis, allergic fungal sinusitis, and hypersensitivity pneumonitis. *Clin Allergy Immunol*, **16**, 449–68.

Greenberger, P.A. 2003. Clinical aspects of allergic bronchopulmonary aspergillosis. *Front Biosci*, **8**, S119–27.

Groll, A.H., Jaeger, G., et al. 1998a. Invasive pulmonary aspergillosis in a critically ill neonate: case report and review of invasive aspergillosis during the first 3 months of life. *Clin Infect Dis*, **27**, 437–52.

Groll, A.H., Piscitelli, S.C. and Walsh, T.J. 1998b. Clinical pharmacology of systemic antifungal agents: a comprehensive review of agents in clinical use, current investigational compounds, and putative targets for antifungal drug development. *Adv Pharmacol*, **44**, 343–500.

Guermaz, A., Gluckman, E., et al. 2003. Invasive central nervous system aspergillosis in bone marrow transplantation recipients: an overview. *Eur Radiol*, **13**, 377–88.

Hamacher, J., Spiliopoulos, A., et al. 1999. Pre-emptive therapy with azole in lung transplant patients. Geneva Lung Transplantation Group. *Eur Respir J*, **13**, 180–6.

Haran, R.P. and Chandy, M.J. 1993. Intracranial *Aspergillus* granuloma. *Br J Neurosurg*, **7**, 383–8.

Harrousseau, J.L., Dekker, A.W., et al. 2000. Itraconazole oral solution for primary prophylaxis of fungal infections in patients with hematological malignancy and profound neutropenia: a randomized, double-blind, double-placebo, multicentre trial comparing itraconazole and amphotericin B. *Antimicrob Agents Chemother*, **44**, 1887–93.

Hay, R.J. 1993. The prevention of invasive aspergillosis – a realistic goal? *J Antimicrob Chemother*, **32**, 515–17.

Haynes, K. and Rogers, T.R. 1994. Retrospective evaluation of a latex agglutination test for diagnosis of invasive aspergillosis in immunocompromised patients. *Eur J Clin Microbiol Infect Dis*, **13**, 670–4.

Haynes, K., Westerneng, T.J., et al. 1995. Rapid detection and identification of pathogenic fungi by polymerase chain reaction amplification of large subunit ribosomal DNA. *J Med Vet Mycol*, **33**, 319–25.

Hearn, V. 1992. Antigenicity of *Aspergillus* species. *J Med Vet Mycol*, **30**, 11–25.

Hearn, V., Pinel, C., et al. 1995. Antibody detection in invasive aspergillosis by analytical electrofocusing and immunoblotting methods. *J Clin Microbiol*, **33**, 982–6.

Hendolin, P.H., Paulin, L., et al. 2000. Panfungal PCR and multiplex liquid hybricization for detection of fungi in tissue specimens. *J Clin Microbiol*, **38**, 4186–92.

Henry, T., Iwen, P.C. and Hinrichs, S.H. 2000. Identification of *Aspergillus* species using internal transcribed spacer regions 1 and 2. *J Clin Microbiol*, **38**, 1510–15.

Henwick, S., Hetherington, S.V. and Patrick, C.C. 1993. Complement binding to *Aspergillus* conidia correlates with pathogenicity. *J Lab Clin Med*, **122**, 27–35.

Holden, D.W., Tang, C.M. and Smith, J.M. 1994. Molecular genetics of *Aspergillus* pathogenicity. *Antonie van Leeuwenhoek*, **65**, 251–5.

Hopwood, V. and Evans, E.G.V. 1991. Serological tests in the diagnosis and prognosis of fungal infection in the compromised patient. In: Warnock, D.W. and Richardson, M.D. (eds), *Fungal infection in the compromised patient*, 2nd edition. Chichester: John Wiley & Sons, 311–53.

Hopwood, V., Johnson, E.M., et al. 1995. Use of the Pastorex *Aspergillus* antigen latex agglutination test for the diagnosis of invasive aspergillosis. *J Clin Pathol*, **48**, 210–13.

Hossain, M.A., Miyazaki, T., et al. 1997. Comparison between Wako-WB003 and Fungitec G tests for detection of (1-3)-beta-D-glucan in systemic mycoses. *J Clin Lab Anal*, **11**, 73–7.

Ibrahim-Granet, O., Philippe, B., et al. 2003. Phagocytosis and intracellular fate of *Aspergillus fumigatus* conidia in alveolar macrophages. *Infect Immun*, **71**, 891–903.

Ikemoto, H. 1992. Bronchopulmonary aspergillosis: diagnostic and therapeutic considerations. In: Borgers, M., Hay, R. and Rinaldi, M.G. (eds), *Current topics in medical mycology*, 4th edition. New York: Springer-Verlag, 64–87.

Ikemoto, H. 2000. Medical treatment of pulmonary aspergilloma. *Intern Med*, **39**, 191–2.

Iwen, P.C., Reed, E.C., et al. 1993. Nosocomial invasive aspergillosis in lymphoma patients treated with bone marrow or peripheral stem cell transplants. *Infect Control Hosp Epidemiol*, **14**, 131–9.

Iwen, P.C., Davis, J.C., et al. 1994. Airborne fungal spore monitoring in a protective environment during hospital construction, and correlation with an outbreak of invasive aspergillosis. *Infect Control Hosp Epidemiol*, **15**, 303–6.

Jantunen, E., Piilonen, A. and Volin, L. 2000. Diagnostic aspects of invasive Aspergillus infections in allogeneic BMT recipients. *Bone Marrow Transplant*, **25**, 867–71.

Jaton-Ogay, K., Paris, S., et al. 1994. Cloning and disruption of the gene encoding an extracellular metalloprotease of *Aspergillus fumigatus*. *Mol Microbiol*, **14**, 917–28.

Jeffery, G.M., Beard, M.E.J., et al. 1991. Intranasal amphotericin B reduces the frequency of invasive aspergillosis in neutropenic patients. *Am J Med*, **90**, 685–92.

Jennings, T.S. and Hardin, T.C. 1993. Treatment of aspergillosis with itraconazole. *Ann Pharmacother*, **27**, 1206–11.

Jewkes, J., Kay, P.H., et al. 1983. Pulmonary aspergilloma: analysis of prognosis in relation to haemoptysis and survey of treatment. *Thorax*, **38**, 572–8.

Jones, B.L. and McLintock, L.A. 2003. Impact of diagnostic markers on early antifungal therapy. *Curr Opin Infect Dis*, **16**, 521–6.

Jorgensen, C., Dreyfus, F., et al. 1989. Failure of amphotericin B spray to prevent aspergillosis in granulocytopenic patients. *Nouv Rev Fr Hematol*, **31**, 327–8.

Judson, M.A. and Stevens, D.A. 2001. The treatment of pulmonary aspergilloma. *Curr Opin Investig Drugs*, **2**, 1375–7.

Kaestel, M., Meyer, W., et al. 1999. Pulmonary aspergilloma – clinical findings and surgical treatment. *Thorac Cardiovasc Surg*, **47**, 340–5.

Kappe, R. and Schulze-Berg, A. 1993. New cause for false-positive results with the Pastorex *Aspergillus* antigen latex agglutination test. *J Clin Microbiol*, **31**, 2489–90.

Kappe, R. and Seeliger, H.P.R. 1993. Serodiagnosis of deep-seated fungal infection. *Curr Top Med Mycol*, **5**, 247–80.

Kauffman, H.F. 2003. Immunopathogenesis of allergic bronchopulmonary aspergillosis and airway remodelling. *Front Biosci*, **8**, E190–6.

Kauffman, H.F. and Tomee, J.F. 2002. Defense mechanisms of the airways against Aspergillus fumigatus: role in invasive aspergillosis. *Chem Immunol*, **81**, 94–113.

Kauffman, H.F., van der Heide, S., et al. 1984. The allergenic and antigenic properties of spore extracts of *Aspergillus fumigatus*: a comparative study of spore extracts with mycelium and culture filtrate extracts. *J Allergy Clin Immunol*, **73**, 567–73.

Kawamura, S., Maesaki, S., et al. 2000. Clinical evaluation of 61 patients with pulmonary aspergilloma. *Intern Med*, **39**, 209–12.

Kawazu, M., Kanda, Y., et al. 2003. Rapid diagnosis of invasive pulmonary aspergillosis by quantitative polymerase chain reaction using bronchial lavage fluid. *Am J Hematol*, **72**, 27–30.

Keating, J.J., Rogers, T., et al. 1994. Management of pulmonary aspergillosis in AIDS: an emerging clinical problem. *J Clin Pathol*, **47**, 805–9.

Kelsey, S.M., Goldman, J.M., et al. 1999. Liposomal amphotericin (AmBisome) in the prophylaxis of fungal infections in neutropenic patients: a randomized, double-blind, placebo-controlled study. *Bone Marrow Transplant*, **23**, 163–8.

Kempner, C.A., Hostetler, J.S., et al. 1993. Ulcerative and plaque-like tracheobronchitis due to infection with *Aspergillus* in patients with AIDS. *Clin Infect Dis*, **17**, 344–52.

Kennedy, H.F., Michie, J.R. and Richardson, M.D. 1995. Air sampling for *Aspergillus* species during building activity in a paediatric hospital ward. *J Hosp Infect*, **31**, 322–5.

Khoo, S.H. and Denning, D.W. 1994. Invasive aspergillosis in patients with AIDS. *Clin Infect Dis*, **19**, Suppl 1, S41–8.

Kibbler, C.C. 1999. Antifungal prophylaxis with itraconazole oral solution in neutropenic patients. *Mycoses*, **43**, 2, 121–4.

Knutsen, A.P., Mueller, K.R., et al. 1994. Serum anti-*Aspergillus fumigatus* antibodies by immunoblot and ELISA in cystic fibrosis with allergic bronchopulmonary aspergillosis. *J Allergy Clin Immunol*, **93**, 926–31.

Kobayashi, H., Debeaupuis, J.P., et al. 1993. An 88-kilodalton antigen secreted by *Aspergillus fumigatus*. *Infect Immun*, **61**, 4767–71.

Kolattukudy, P.E., Lee, J.D., et al. 1993. Evidence for possible involvement of an elastolytic serine protease in aspergillosis. *Infect Immun*, **61**, 2357–68.

Krennmair, G., Lenglinger, F. and Muller-Schelken, H. 1994. Computed tomography (CT) in the diagnosis of sinus aspergillosis. *J Craniomaxillofac Surg*, **22**, 120–5.

Kubak, B.M. 2002. Fungal infection in lung transplantation. *Transpl Infect Dis*, **4**, Suppl 3, 24–31.

Kumar, A. and Kurup, V.P. 1993. Murine monoclonal antibodies to glycoprotein antigens of *Aspergillus fumigatus* show cross-reactivity with other fungi. *Allergy Proc*, **14**, 189–93.

Kurup, V.P. and Kumar, A. 1991. Immunodiagnosis of aspergillosis. *Clin Microbiol Rev*, **4**, 439–56.

Kurup, V.P., Shen, H.D. and Banerjee, B. 2000. Respiratory fungal allergy. *Microbes Infect*, **2**, 1101–10.

Latgé, J.-P. 1999. *Aspergillus fumigatus* and aspergillosis. *Clin Microbiol Rev*, **12**, 310–50.

Latgé, J.-P. 2001. The pathobiology of *Aspergillus fumigatus*. *Trends Microbiol*, **9**, 382–9.

Latgé, J.-P., Moutaouakil, M., et al. 1991. The 18-kilodalton antigen secreted by *Aspergillus fumigatus*. *Infect Immun*, **59**, 2586–94.

Latgé, J.-P., Debeaupuis, J.P., et al. 1993. Cell wall antigens in *Aspergillus fumigatus*. *Arch Med Res*, **24**, 269–74.

Leenders, A.C.A., Daenen, S., et al. 1998. Liposomal amphotericin B compared with amphotericin B deoxycholate in the treatment of documented and suspected neutropenia-associated invasive fungal infections. *Br J Haematol*, **103**, 205–12.

Leon, E.E. and Craig, T.J. 1999. Antifungals in the treatment of allergic bronchopulmonary aspergillosis. *Ann Allergy Asthma Immunol*, **82**, 511–16.

Lewis, R.E. and Kontoyiannis, D.P. 2001. Rationale for combination therapy. *Pharmacotherapy*, **21**, 149S–64S.

Linden, P., Lee, L. and Walsh, T.J. 1999. Retrospective analysis of the dosage of amphotericin B lipid complex for treatment of invasive fungal infections. *Pharmacotherapy*, **19**, 1261–8.

Lister, J. 1996. Amphotericin B lipid complex (Abelcet®) in the treatment of invasive mycoses: the North American experience. *Eur J Haematol*, **56**, Suppl 57, 18–23.

Little, S.A., Longbottom, J.L. and Warner, J.O. 1993. Optimized preparation of *Aspergillus fumigatus* extracts for allergy diagnosis. *Clin Exp Allergy*, **23**, 835–42.

Loeffler, J., Henke, N., et al. 2000. Quantification of fungal DNA by using fluorescence resonance energy transfer and the Light Cycler system. *J Clin Microbiol*, **38**, 586–90.

Logan, P.M., Primack, S.L., et al. 1994. Invasive aspergillosis of the airways: radiographic, CT, and pathologic findings. *Radiology*, **193**, 383–8.

Lortholary, O., Meyohas, M.C., et al. 1993. Invasive aspergillosis in patients with acquired immunodeficiency syndrome: report of 33 cases. *Am J Med*, **95**, 177–87.

Loudon, K.W., Burnie, J.P., et al. 1993. Application of polymerase chain reaction to fingerprinting *Aspergillus fumigatus* by random amplification of polymeric DNA. *J Clin Microbiol*, **31**, 1117–21.

Loudon, K.W., Coke, A.P., et al. 1994. Invasive aspergillosis: clusters and sources? *J Med Vet Mycol*, **32**, 217–24.

Madan, T., Eggleton, P., et al. 1997. Binding of pulmonary surfactant proteins A and D to *Aspergillus fumigatus* conidia enhances phagocytosis and killing by human neutrophils and alveolar macrophages. *Infect Immun*, **65**, 3171–9.

Maertens, J., Verhaegen, J., et al. 1999. Autopsy-controlled prospective evaluation of serial screening for circulating galactomannan by a sandwich enzyme-linked immunosorbent assay for haematological patients at risk for invasive aspergillosis. *J Clin Microbiol*, **37**, 3223–8.

Maertens, J., Van Eldere, J., et al. 2002. Use of circulating galactomannan screening for early diagnosis of invasive aspergillosis in allogeneic stem cell transplant recipients. *J Infect Dis*, **186**, 1297–306.

Manso, E., Montillo, G., et al. 1994. Value of antigen and antibody detection in the serological diagnosis of invasive aspergillosis in patients with hematological malignancies. *Eur J Clin Microbiol Infect Dis*, **13**, 756–60.

Marchant, J.L., Warner, O. and Bush, A. 1994. Rise in total IgE as an indicator of allergic bronchopulmonary aspergillosis in cystic fibrosis. *Thorax*, **49**, 1002–5.

Marr, K.A. 2002. Empirical antifungal therapy – new options, new tradeoffs. *N Engl J Med*, **346**, 278–80.

Marr, K.A., Patterson, T. and Denning, D. 2002. Aspergillosis. Pathogenesis, clinical manifestations, and therapy. *Infect Dis Clin North Am*, **16**, 875–94.

Martin, M.T., Gavalda, J., et al. 2003. Efficacy of high doses of liposomal amphotericin B in the treatment of experimental aspergillosis. *J Antimicrob Chemother*, **52**, 1032–4.

Martino, R., Subira, M., et al. 1999a. Low-dose amphotericin B lipid complex for the treatment of persistent fever of unknown origin in patients with haematologic malignancies and prolonged neutropenia. *Chemotherapy*, **45**, 205–12.

Martino, R., Subira, M., et al. 1999b. Amphotericin B lipid complex at 3 mg/kg/day for treatment of invasive fungal infections in adults with haematological malignancies. *J Antimicrob Chemother*, **44**, 569–72.

McCarthy, J.M., Flam, M., et al. 1986. Outbreak of primary cutaneous aspergillosis related to intravenous arm boards. *J Pediatr*, **108**, 721–4.

McLintock, L. and Jones, B.L. 2004. Advances in the molecular and serological diagnosis of invasive fungal infection in haemato-oncology patients. *Br J Haematol*, **126**, 289–97.

McWhinney, P.H.M., Kibbler, C.C., et al. 1993. Progress in the diagnosis and management of aspergillosis in bone marrow transplantation: 13 years experience. *Clin Infect Dis*, **17**, 397–404.

Mearns, M., Young, W. and Batten, J. 1965. Transient pulmonary infiltration in cystic fibrosis due to allergic aspergillosis. *Thorax*, **20**, 385–92.

Mehta, J., Chu, P., et al. 1997. Amphotericin B lipid complex for the treatment of presumed or confirmed fungal infections in immunocompromised patients with haematolgical malignancies. *Bone Marrow Transplant*, **20**, 39–43.

Melchers, W.J.G., Verweij, P.E., et al. 1994. General primer-mediated PCR for detection of *Aspergillus* species. *J Clin Microbiol*, **32**, 1710–17.

Menichetti, F., Del Favero, A., et al. 1999. Itraconazole oral solution as prophylaxis for fungal infections in neutropenic patients with hematologic malignancies: a randomised, placebo-controlled, double-blind, multicenter trial. *Clin Infect Dis*, **28**, 250–5.

Michallet, M., Persat, F., et al. 1998. Pharmacokinetics of itraconazole oral solution in allogeneic bone marrow transplant patients receiving total body irradiation. *Bone Marrow Transplant*, **21**, 1239–43.

Miller, P.W., Hamosh, A., et al. 1996. Cystic fibrosis transmembrane conductance regulator (CFTR) gene mutations in allergic bronchopulmonary aspergillosis. *J Exp Med*, **59**, 45–51.

Miller, W.T., Sais, G.J., et al. 1994. Pulmonary aspergillosis in patients with AIDS. Clinical and radiographic correlations. *Chest*, **105**, 37–44.

Mills, W., Chopra, R., et al. 1994. Liposomal amphotericin B in the treatment of fungal infections in neutropenic patients: a single-centre experience of 133 episodes in 116 patients. *Br J Haematol*, **86**, 754–60.

Minamoto, G., Barlam, T. and Van der Els, N. 1992. Invasive aspergillosis in patients with AIDS. *Clin Infect Dis*, **14**, 66–74.

Mondon, P., Thélu, J., et al. 1995. Virulence of *Aspergillus fumigatus* strains investigated by random amplified polymorphic DNA analysis. *J Med Microbiol*, **42**, 299–303.

Monod, M., Paris, S., et al. 1993. Virulence of alkaline protease-deficient mutants of *Aspergillus fumigatus*. *FEMS Microbiol Lett*, **106**, 39–46.

Montone, K.T. and Litzky, L.A. 1995. Rapid method for detection of *Aspergillus* 5S ribosomal RNA using a genus-specific oligonucleotide probe. *Am J Clin Pathol*, **103**, 48–51.

Montoya, J.G., Chaparro, S.V., et al. 2003. Invasive aspergillosis in the setting of cardiac transplantation. *Clin Infect Dis*, **37**, Suppl 3, S281–92.

Morgenstern, G.R., Prentice, A.G., et al. 1999. A randomized controlled trial of itraconazole versus fluconazole for the prevention of fungal infections in patients with haematological malignancies. *Br J Haematol*, **105**, 901–11.

Morris, G., Kokki, M.H., et al. 2000. Sampling of *Aspergillus* spores in air. *J Hosp Infect*, **441**, 81–92.

Morrison, B.E., Park, S.J., et al. 2003. Chemokine-mediated recruitment of NK cells is a critical host defense mechanism in invasive aspergillosis. *J Clin Invest*, **112**, 1862–70.

Moser, M., Menz, G., et al. 1994. Recombinant expression and antigenic properties of a 32-kilodalton extracellular alkaline protease, representing a possible virulence factor from *Aspergillus fumigatus*. *Infect Immun*, **62**, 936–42.

Moss, R.B. 2002. Allergic bronchopulmonary aspergillosis. *Clin Rev Allergy Immunol*, **23**, 87–104.

Motte, S., Bellens, B., et al. 1993. Vascular graft infection caused by *Aspergillus* species: case report and review of the literature. *J Vasc Surg*, **17**, 607–12.

Moutaouakil, M., Monod, M., et al. 1993. Identification of the 33-kDa alkaline protease of *Aspergillus fumigatus* in vitro and in vivo. *J Med Vet Mycol*, **39**, 393–9.

Muijsers, R.B., Goa, K.L. and Scott, L.J. 2002. Voriconazole: in the treatment of invasive aspergillosis. *Drugs*, **62**, 2655–64.

Munoz, P., Alcala, L., et al. 2003. The isolation of *Aspergillus fumigatus* from respiratory tract specimens in heart transplant recipients is highly predictive of invasive aspergillosis. *Transplantation*, **75**, 326–9.

Murali, P.S., Pathial, K., et al. 1994. Immune responses to *Aspergillus fumigatus* and *Pseudomonas aeruginosa* antigens in cystic fibrosis and allergic bronchopulmonary aspergillosis. *Chest*, **106**, 513–19.

Murayama, A., Amitani, R., et al. 1998. Effects of *Aspergillus fumigatus* culture filtrate on antifungal activity of human phagocytes in vitro. *Thorax*, **53**, 975–8.

Murphy, M., Bernard, E.M., et al. 1997. Activity of voriconazole (UK-109, 496) against clinical isolates of *Aspergillus* species and its effectiveness in an experimental model of invasive pulmonary aspergillosis. *Antimicrob Agents Chemother*, **41**, 696–8.

Myers, S.E., Devine, S.M., et al. 1992. A pilot study of prophylactic aerolized amphotericin B in patients at risk for prolonged neutropenia. *Leuk Lymphoma*, **8**, 229–33.

Nalesnik, M.A., Myerowitz, R.L., et al. 1980. Significance of *Aspergillus* species isolated from respiratory secretions in the diagnosis of invasive pulmonary aspergillosis. *J Clin Microbiol*, **11**, 370–6.

Ng, T.T.C. and Denning, D.W. 1995. Liposomal amphotericin B (AmBisome) therapy in invasive fungal infections. Evaluation of United Kingdom compassionate use data. *Ann Intern Med*, **155**, 1093–8.

Noskin, G., Pietrelli, L., et al. 1999. Treatment of invasive fungal infections with amphotericin B colloidal dispersion in bone marrow transplant recipients. *Bone Marrow Transplant*, **23**, 697–703.

Obayashi, T., Yoshida, M., et al. 1995. Plasma (1-3)-β-D-glucan measurement in diagnosis of invasive deep mycosis and fungal febrile episodes. *Lancet*, **345**, 17–20.

Opal, S.M. 1986. Efficacy of infection control measures during a nosocomial outbreak of disseminated aspergillosis associated with hospital construction. *J Infect Dis*, **153**, 634–7.

Oppenheim, B.A., Herbrecht, R. and Kusne, S. 1995. The safety and efficacy of amphotericin B colloidal dispersion in the treatment of invasive mycoses. *Clin Infect Dis*, **21**, 1145–53.

Oravcova, E., Mistrik, M., et al. 1995. Amphotericin B lipid complex to treat invasive fungal infections in cancer patients: report of efficacy and safety in 20 patients. *Chemotherapy*, **41**, 473–6.

Pai, U., Blinkhorn, R.J. and Tomashefski, J.F. 1994. Invasive cavitary pulmonary aspergillosis in patients with cancer: a clinicopathologic study. *Hum Pathol*, **25**, 293–303.

Palmer, L.B., Greenberg, H.E. and Schiff, M. 1991. Corticosteroid treatment as a risk factor for invasive aspergillosis in patients with lung disease. *Thorax*, **46**, 15–20.

Panackal, A.A., Dahlman, A., et al. 2003. Outbreak of invasive aspergillosis among renal transplant recipients. *Transplantation*, **75**, 1050–3.

Patterson, R., Greenberger, P., et al. 1982. Allergic bronchopulmonary aspergillosis: staying as an aid to management. *Ann Intern Med*, **96**, 286–91.

Patterson, T.F., Miniter, P., et al. 1995. *Aspergillus* antigen detection in the diagnosis of invasive aspergillosis. *J Infect Dis*, **171**, 1553–8.

Paulose, K.O., Al-Khalifa, S., et al. 1989. Mycotic infection of the ear (otomycosis): a prospective study. *J Laryngol Otol*, **103**, 30–5.

Perfect, J.R., Klotman, M.E., et al. 1992. Prophylactic intravenous amphotericin B in neutropenic autologous bone marrow transplant recipients. *J Infect Dis*, **165**, 891–7.

Perfect, J.R., Marr, K.A., et al. 2003. Voriconazole treatment for less-common, emerging, or refractory fungal infections. *Clin Infect Dis*, **36**, 1122–31.

Persat, F., Schwartzbrod, P.E., et al. 2000. Abnormalities in liver enzymes during simultaneous therapy with itraconazole and amphotericin B in leukaemic patients. *J Antimicrob Chemother*, **45**, 928–9.

Peterson, D.E. and Schimpff, S.C. 1989. *Aspergillus* sinusitis in neutropenic patients with cancer: a review. *Biomed Pharmacother*, **43**, 307–12.

Piechura, J.E., Huang, C.T., et al. 1985. Antigens of *Aspergillus fumigatus*. III. Comparative immunochemical analysis of clinically relevant aspergilli and related fungal taxa. *Clin Exp Immunol*, **59**, 716–24.

Ponikau, J.U., Sherris, D.A., et al. 1999. The diagnosis and incidence of allergic fungal sinusitis. *Mayo Clin Proc*, **74**, 877–84.

Popp, A.L., White, M.H., et al. 1999. Amphotericin B with and without itraconazole for invasive aspergillosis: a three-year retrospective study. *Int J Infect Dis*, **3**, 157–60.

Prentice, A.G., Warnock, D.W., et al. 1994. Multiple dose pharmacokinetics of an oral solution of itraconazole in autologous bone marrow transplant recipients. *J Antimicrob Chemother*, **34**, 247–52.

Prentice, H.G., Hann, I.M., et al. 1997. A randomized comparison of liposomal versus conventional amphotericin B for the treatment of pyrexia of unknown origin in neutropenic patients. *Br J Haematol*, **98**, 711–18.

Prentice, H.G., Caillot, D., et al. 1999. Oral and intravenous itraconazole for systemic fungal infections in neutropenic haematological patients: meeting report. *Acta Haematol*, **101**, 56–62.

Prentice, H.G., Kibbler, C.C. and Prentice, A.G. 2000. Towards a targeted, risk based, antifungal strategy in neutropenic patients. *Br J Haematol*, **110**, 273–84.

Pursell, K.J., Telzak, E.E., et al. 1992. *Aspergillus* species colonization and invasive disease in patients with AIDS. *Clin Infect Dis*, **14**, 141–8.

Raper, K.B. and Thom, D.I. 1965. *The genus Aspergillus*. Baltimore, MA: Williams & Wilkins Co., 686.

Reddy, L.V., Kumar, A., et al. 1993. Specific amplification of *Aspergillus fumigatus* DNA by polymerase chain reaction. *Mol Cell Probes*, **7**, 121–6.

Regnard, J.F., Icard, P., et al. 2000. Aspergilloma: a series of 89 surgical cases. *Ann Thorac Surg*, **69**, 893–903.

Reichard, U., Eiffert, H. and Ruchel, R. 1994. Purification and characterization of an extracellular aspartic proteinase from *Aspergillus fumigatus*. *J Med Vet Mycol*, **32**, 427–36.

Reichenberger, F., Habicht, J.M., et al. 2002. Diagnosis and treatment of invasive pulmonary aspergillosis in neutropenic patients. *Eur Respir J*, **19**, 743–55.

Rex, J.H., Bennett, J.E., et al. 1990. In vivo interferon-γ therapy augments the in vitro ability of chronic granulomatous disease neutrophils to damage *Aspergillus* hyphae. *J Infect Dis*, **163**, 849–52.

Rhame, F.S. 1991. Prevention of nosocomial aspergillosis. *J Hosp Infect*, **18**, Suppl A, 466–72.

Rhame, F.S., Striefel, A.J., et al. 1984. Extrinsic risk factors for pneumonia in the patient at high risk of infection. *Am J Med*, **76**, Suppl 5A, 42–5.

Rhodes, J.C. 1993. Aspergillosis. In: Murphy, J.W. (ed.), *Fungal infections and immune responses*. New York: Plenum Press, 359–77.

Ribaud, P. and Gluckman, E. 1996. Critical review of lipid amphotericin B formulations as treatment of invasive aspergillosis in patients with hematological disorders or bone marrow transplantation (BMT). *J Mycol Med*, **6**, 17–20.

Ribrag, V., Dreyfus, F., et al. 1993. Prognostic factors of invasive pulmonary aspergillosis in leukemic patients. *Leuk Lymphoma*, **10**, 317–21.

Richard, C., Romón, I., et al. 1993. Invasive pulmonary aspergillosis prior to BMT in acute leukemia patients does not predict a poor outcome. *Bone Marrow Transplant*, **12**, 237–41.

Richardson, M.D. 1987. *Aspergillus* antigenaemia in the diagnosis of invasive aspergillosis. *Serodiagn Immunother*, **1**, 313–15.

Richardson, M.D. and Jones, B.L. 2003. *Therapeutic guidelines in systemic fungal infections*, 3rd edition. London: Current Medical Literature.

Richardson, M.D. and Patel, M. 1995. Stimulation of neutrophil phagocytosis of *Aspergillus fumigatus* conidia by interleukin-8 and N-formylmethionyl-leucylphenylalanine. *J Med Vet Mycol*, **33**, 99–104.

Richardson, M.D., Rennie, S., et al. 2000. Fungal surveillance of an open haematology ward. *J Hosp Infect*, **45**, 288–92.

Richardson, M.D. and Warnock, D.W. 2003. *Fungal infection: diagnosis and management*, 3rd edition. Oxford: Blackwell Publishing.

Rinaldi, M.G. 1983. Invasive aspergillosis. *Rev Infect Dis*, **5**, 1061–77.

Ringden, O., Jonsson, V., et al. 1998. Severe and common side-effects of amphotericin B lipid complex (Abelcet). *Bone Marrow Transplant*, **22**, 733–4.

Rohrlich, P., Sarfati, J., et al. 1996. Prospective sandwich enzyme-linked immunosorbent assay for serum galactomannan: early predictive value and clinical use in invasive aspergillosis. *Pediatr Infect Dis J*, **15**, 232–7.

Roildes, E., Holmes, A., et al. 1993a. Impairment of neutrophil antifungal activity against hyphae of *Aspergillus fumigatus* in children infected with human immunodeficiency virus. *J Infect Dis*, **167**, 905–11.

Roildes, E., Uhlig, K., et al. 1993b. Enhancement of oxidative response and damage caused by human neutrophils to *Aspergillus fumigatus* hyphae by granulocyte colony-stimulating factor and γ-interferon. *Infect Immun*, **61**, 1185–93.

Roildes, E., Uhlig, E., et al. 1993c. Prevention of cortisone-induced suppression of human polymorphonuclear leukocyte-induced damage of *Aspergillus fumigatus* hyphae by granulocyte colony-stimulating factor and γ-interferon. *Infect Immun*, **61**, 4870–7.

Roildes, E., Holmes, A., et al. 1994a. Antifungal activity of elutriated human monocytes against Aspergillus fumigatus hyphae: enhancement by granulocyte-macrophage colony stimulating factor and interferon-γ. *J Infect Dis*, **170**, 894–9.

Roildes, E., Robinson, T., et al. In vitro and ex vivo effects of cyclosporin A on phagocytic host defenses against *Aspergillus fumigatus*. *Antimicrob Agents Chemother*, **38**, 2883–8.

Roland, W.E. 1999. Amphotericin B colloidal dispersion versus amphotericin B in the empirical treatment of fever and neutropenia. *Clin Infect Dis*, **28**, 935–6.

Rousey, S.R., Russler, S., et al. 1991. Low-dose amphotericin B prophylaxis against invasive *Aspergillus* infections in allogeneic marrow transplantation. *Am J Med*, **91**, 484–92.

Ruchel, R. and Schaffrinski, M. 1999. Versatile fluorescent staining of fungi in clinical specimens by using the optical brightner Blankophor. *J Clin Microbiol*, **37**, 2694–6.

Salez, F., Brichet, A., et al. 1999. Effects of itraconazole therapy in allergic bronchopulmonary aspergillosis. *Chest*, **116**, 1665–8.

Sarfati, J., Boucias, D.G. and Latgé, J.-P. 1995. Antigens of *Aspergillus fumigatus* produced in vivo. *J Med Vet Mycol*, **33**, 9–14.

Saugier-Veber, P., Devergie, A., et al. 1993. Epidemiology and diagnosis of invasive pulmonary aspergillosis in bone marrow transplant patients: results of a 5 year retrospective study. *Bone Marrow Transplant*, **12**, 121–4.

Scarcella, A., Pasquariello, M.B., et al. 1998. Liposomal amphotericin B for neonatal fungal infections. *Pediatr Infect Dis*, **17**, 146–8.

Sherertz, R.J., Belani, A., et al. 1987. Impact of air filtration on nosocomial aspergillus infections. *Am J Med*, **8**, 709–18.

Schubert, M.S. 2001. Fungal rhinosinusitis: diagnosis and therapy. *Curr Allergy Asthma Rep*, **2001**, 268–76.

Singh, B.P., Banerjee, B. and Kurup, V.P. 2003. *Aspergillus* antigens associated with allergic bronchopulmonary aspergillosis. *Front Biosci*, **8**, S102–9.

Singh, N. 2000. The current management of infectious diseases in the liver transplant recipient. *Clin Liver Dis*, **4**, 657–73.

Singh, N. 2003. Fungal infections in the recipients of solid organ transplantation. *Infect Dis Clin N Am*, **17**, 113–34.

Singh, N. and Husain, S.J. 2003. *Aspergillus* infections after lung transplantation: clinical differences in type of transplant and implications for management. *Heart Lung Transplant*, **22**, 258–66.

Skladny, H., Buchheidt, D., et al. 1999. Specific detection of *Aspergillus* species in blood and bronchoalveolar lavage samples of immunocompromised patients by two-step PCR. *J Clin Microbiol*, **37**, 3865–71.

Soubani, A.O. and Chandrasekar, P.H. 2002. The clinical spectrum of pulmonary aspergillosis. *Chest*, **121**, 1988–99.

Spielberger, R.T., Falleroni, M.J., et al. 1993. Concomitant amphotericin B therapy, granulocyte transfusions and GM-CSF administration for disseminated infection with Fusarium in a granulocytopenic patient. *Clin Infect Dis*, **16**, 528–30.

Spreadbury, C., Holden, D., et al. 1993. Detection of *Aspergillus fumigatus* by polymerase chain reaction. *J Clin Microbiol*, **31**, 615–21.

Steinbach, W.J., Stevens, D.A. and Denning, D.W. 2003. Combination and sequential antifungal therapy for invasive aspergillosis: review of published in vitro and in vivo interactions and 6281 clinical cases from 1966 to 2001. *Clin Infect Dis*, **37**, Suppl 3, S188–224.

Stevens, D.A., Kan, V.L., et al. 2000a. Practice guidelines for diseases caused by *Aspergillus*. *Clin Infect Dis*, **30**, 696–709.

Stevens, D.A., Schwartz, H.J., et al. 2000b. A randomized trial of itraconazole in allergic bronchopulmonary aspergillosis. *N Engl J Med*, **342**, 756–62.

Stevens, D.A., Moss, R.B., et al. 2003. Allergic bronchopulmonary aspergillosis in cystic fibrosis – state of the art: Cystic Fibrosis Foundation Consensus Conference. *Clin Infect Dis*, **37**, Suppl 3, S225–64.

Stynen, D., Goris, J., et al. 1995. A new sensitive sandwich enzyme-linked immunosorbent assay to detect galactofuran in patients with invasive aspergillosis. *J Clin Microbiol*, **33**, 497–500.

Sugata, T., Myoken, Y., et al. 1994. Invasive oral aspergillosis in immunocompromised patients with leukemia. *J Oral Maxillofac Surg*, **52**, 382–6.

Symoens, F., Viviani, M.A. and Nolard, N. 1993. Typing by immunoblot of *Aspergillus fumigatus* from nosocomial infections. *Mycoses*, **36**, 229–37.

Taccone, A., Occhi, M., et al. 1993. CT of invasive pulmonary aspergillosis in children with cancer. *Pediatr Radiol*, **23**, 177–80.

Talbot, G.H., Huang, A. and Provencher, M. 1991. Invasive aspergillus rhinosinusitis in patients with acute leukemia. *Rev Infect Dis*, **13**, 219–32.

Tang, C.M., Holden, D.W., et al. 1993. The detection of Aspergillus species by the polymerase chain reaction and its evaluation in bronchoalveolar lavage fluid. *Am Rev Respir Dis*, **148**, 1313–17.

Tang, C.M., Cohen, J., et al. 1994. Molecular epidemiological study of invasive pulmonary aspergillosis in a renal transplant unit. *Eur J Clin Microbiol Infect Dis*, **13**, 318–21.

Thio, C.L., Smith, D., et al. 2000. Refinements of environmental assessment during an outbreak investigation of invasive aspergillosis in a leukemia and bone marrow transplant unit. *Infect Control Hosp Epidemiol*, **21**, 18–23.

Todeschini, G., Murari, C., et al. 1993. Oral itraconazole plus nasal amphotericin B for prophylaxis of invasive aspergillosis in patients with hematological malignancies. *Eur J Clin Microbiol Infect Dis*, **12**, 614–18.

Tollemar, J. and Ringden, O. 1995. Lipid formulations of amphotericin B. Less toxicity but at what economic cost? *Drug Saf*, **13**, 207–18.

Tollemar, J., Ringden, O., et al. 1993. Randomized double-blind study of liposomal amphotericin B (AmBisome) prophylaxis of invasive fungal infections in bone marrow transplant recipients. *Bone Marrow Transplant*, **12**, 577–82.

van Belkum, A., Quint, W.G.V., et al. 1993. Typing of *Aspergillus* species and *Aspergillus fumigatus* isolates by interrepeat polymerase chain reaction. *J Clin Microbiol*, **31**, 2502–5.

Varga, J. and Tóth, B. 2003. Genetic variablity and reproductive mode of *Aspergillus fumigatus*. *Infect Genet Evol*, **3**, 3–17.

Vecchiarelli, A., Cenci, E., et al. 1989. Immunosuppressive effect of cyclosporin A on resistance to systemic infection with *Candida albicans*. *J Med Microbiol*, **30**, 183–92.

Verweij, P.E., Latgé, J.-P., et al. 1995a. Comparison of antigen detection and PCR assay using bronchoalveolar lavage fluid for diagnosing invasive pulmonary aspergillosis in patients receiving treatment for haematological malignancies. *J Clin Microbiol*, **33**, 3150–3.

Verweij, P.E., Rijs, A.J.M.M., et al. 1995b. Clinical evaluation and reproducibility of the Pastorex *Aspergillus* antigen latex agglutination test for diagnosing invasive aspergillosis. *J Clin Pathol*, **48**, 474–6.

Verweij, P.E., Stynen, D., et al. 1995c. Sandwich enzyme-linked immunosorbent assay compared with Pastorex latex agglutination test for diagnosing invasive aspergillosis in immunocompromised patients. *J Clin Microbiol*, **33**, 1912–14.

Verweij, P.E., Dompeling, E.C., et al. 1997. Serial monitoring of *Aspergillus* antigen in the early diagnosis of invasive aspergillosis. Preliminary investigations with two examples. *Infection*, **25**, 86–9.

Viscoli, C. 1998. Prevention of aspergillosis in bone marrow transplantation. *J Infect Dis*, **177**, 1775–6.

Wald, A., Leisering, W. and van Burik, J.A. 1997. Epidemiology of *Aspergillus* infections in a large cohort of patients undergoing bone marrow transplantation. *J Infect Dis*, **175**, 1459–66.

Walmsley, S., Devi, S., et al. 1993. Invasive *Aspergillus* infections in a pediatric hospital: a ten-year review. *Pediatr Infect Dis J*, **12**, 673–82.

Walsh, T.J. 1990. Invasive pulmonary aspergillosis in patients with neoplastic diseases. *Semin Respir Infect*, **5**, 111–22.

Walsh, T.J. 1998. Editorial response: primary cutaneous aspergillosis – an emerging infection among immunocompromised patients. *Clin Infect Dis*, **27**, 453–7.

Walsh, T.J. and Dixon, D.M. 1989. Nosocomial aspergillosis: environmental microbiology, hospital epidemiology, diagnosis and treatment. *Eur J Epidemiol*, **5**, 131–42.

Walsh, T.J. and Pizzo, P.A. 1988. Nosocomial fungal infections: a classification for hospital-acquired fungal infections and mycoses arising from endogenous flora or reactivation. *Annu Rev Microbiol*, **42**, 517–45.

Walsh, T.J., Francesconi, A., et al. 1995. PCR and single-stranded conformational polymorphism for recognition of medically important fungi. *J Clin Microbiol*, **33**, 3216–20.

Walsh, T.J., Hiemenz, J.W., et al. 1998. Amphotericin B lipid complex for invasive fungal infection: analysis of safety and efficacy in 556 cases. *Clin Infect Dis*, **26**, 1383–96.

Walsh, T.J., Finberg, R.W., et al. 1999a. Liposomal amphotericin B for empirical therapy in patients with persistent fever and neutropenia. *N Engl J Med*, **340**, 764–71.

Walsh, T.J., Seibel, N.L., et al. 1999b. Amphotericin B lipid complex in pediatric patients with invasive fungal infections. *Pediatr Infect J*, **18**, 702–8.

Walsh, T.J., Goodman, J.L., et al. 2001. Safety, tolerance, and pharmacokinetics of high-dose liposomal amphotericin B (AmBisome) in patients infected with Aspergillus species and other filamentous fungi: maximum tolerated dose study. *Antimicrob Agents Chemother*, **45**, 3487–96.

Wark, P., Wilson, A.W. and Gibson, P.G. 2001. Azoles for allergic bronchopulmonary aspergillosis associated with asthma. *Cochrane Database Syst Rev*, CD001108.

Warnock, D.W. 1995. Lipid-associated amphotericin B: a new development in the treatment of invasive fungal infection. *Rev Iberoam Micol*, **12**, 18–21.

Warnock, D.W., Foot, A.B.M., et al. 1991. *Aspergillus* antigen latex test for diagnosis of invasive aspergillosis. *Lancet*, **338**, 1023–4.

Warnock, D.W., Hajjeh, R.A. and Lasker, B.A. 2001. Epidemiology and prevention of invasive aspergillosis. *Curr Infect Dis Rep*, **3**, 507–16.

Weinberger, M., Elattor, I., et al. 1992. Patterns of infection in patients with aplastic anemia and the emergence of aspergillus as a major cause of death. *Medicine (Baltimore)*, **71**, 24–43.

Weitkamp, J.H., Poets, C.F., et al. 1998. *Candida* infection in very low birth-weight infants: outcome and nephrotoxicity of treatment with liposomal amphotericin B (AmBisome). *Infection*, **26**, 11–15.

White, M.H., Anaissie, E.J., et al. 1997. Amphotericin B colloidal dispersion vs. amphotericin B as therapy for invasive aspergillosis. *Clin Infect Dis*, **24**, 635–42.

Wiederhold, N.P., Lewis, R.E. and Kontoyiannis, D.P. 2003. Invasive aspergillosis in patients with hematologic malignancies. *Pharmacotherapy*, **23**, 1592–610.

Williamson, E.C., Millar, M.R., et al. 1999. Infections in adults undergoing unrelated donor bone marrow transplantation. *Br J Haematol*, **104**, 560–8.

Williamson, E.C., Leeming, J.P., et al. 2000. Diagnosis of invasive aspergillosis in bone marrow transplant recipients by polymerase chain reaction. *Br J Haematol*, **108**, 132–9.

Willinger, B., Obradovic, A., et al. 2003. Detection and identification of fungi from fungus balls of the maxillary sinus by molecular techniques. *J Clin Microbiol*, **41**, 581–5.

Wingard, J.R. et al. 1999. A randomised, double-blind safety study of AmBisome and Abelcet in febrile neutropenic patients. *Focus on fungal infection 9*, San Diego, 17–19 March 1999. Poster session and Abstract no. 015.

Woods, G.L. and Goldsmith, J.C. 1990. *Aspergillus* infection of the central nervous system in patients with acquired immunodeficiency syndrome. *Arch Neurol*, **47**, 181–4.

Worthy, S.A., Flint, M.D. and Muller, N.L. 1997. Pulmonary complications after bone marrow transplantation: high-resolution CT and pathologic findings. *RadioGraphics*, **17**, 1359–71.

Wright, J.A., Bradfield, S.M., et al. 2003. Prolonged survival after invasive aspergillosis: a single institution review of 11 cases. *J Pediatr Hematol*, **25**, 286–91.

Yamada, H., Kohno, S., et al. 1993. Topical treatment of pulmonary aspergilloma by antifungals. Relationship between duration of the disease and efficacy of therapy. *Chest*, **103**, 1421–5.

Yamakami, Y., Hashimoto, A., et al. 1996. PCR detection of DNA specific for *Aspergillus* species in serum of patients with invasive aspergillosis. *J Clin Microbiol*, **34**, 2464–8.

Yeo, S.F. and Wong, B. 2002. Current status of nonculture methods for diagnosis of invasive fungal infections. *Clin Microbiol Rev*, **15**, 465–84.

35

Deep phaeohyphomycosis

CHESTER R. COOPER, JR

The incidence of fungal infections in humans has been rising for the past several decades due to a number of diverse reasons, some of which include social and geopolitical factors (Dixon et al. 1996; Cooper 2002). The primary dynamic leading to this marked increase in fungal infections has been the expanding population of immunocompromised individuals. Moreover, the etiologic agents in many of these cases are misplaced saprotrophs, i.e. opportunistic pathogens. Such organisms tend to be benign towards humans and animals, rarely causing disease. These fungi exist in nature by gathering nutrition via consuming organic debris or through parasitic relationships with living plants. Yet, under the proper conditions, these opportunists can readily exploit a dysfunctional mammalian immune system to cause severe cutaneous, subcutaneous, and deep-seated disease.

One group of opportunistic fungal pathogens is characterized by the single shared characteristic of possessing a darkly pigmented appearance. The pigment has been demonstrated to be a melanin produced via a pentaketide pathway (Wheeler and Bell 1988). Traditionally, the dark color of these fungi caused them to be designated as 'dematiaceous'. However, the use of the word 'dematiaceous' to describe melanized fungi has been argued to be epistemologically incorrect (Pappagianis and Ajello 1994). Subsequently, the term 'phaeoid' has been used to replace the prior designation. Nonetheless, the descriptive word 'dematiaceous' has been so entrenched within the mycological literature that it is still used today in synonymy with the term 'phaeoid'. In this chapter, both expressions will be used

interchangeably. Despite the semantic arguments, an important point needs to be emphasized with regard to the use of either designation – both have no real taxonomic significance. Rather, these terms refer to a heterogeneous collection of generally unrelated fungi that share the common phenotypic trait of being darkly colored under normal laboratory conditions.

The phaeoid (dematiaceous) fungi are a curious group of organisms capable of causing a wide array of clinical pathologies ranging from the superficial to the deep-seated infections (Perfect et al. 2003; Sanche et al. 2003). The focus of this chapter is placed upon those molds that cause the latter type in humans and animals. Such infections have been designated as visceral, systemic, or disseminated phaeohyphomycosis. In this chapter, these terms are reserved to describe a particular type of phaeohyphomycosis. The more general phrase, deep-seated phaeohyphomycosis (DSP), is proffered in their place. As discussed below, DSP can be divided into disseminated phaeohyphomycosis as well as a second clinical entity, cerebral phaeohyphomycosis. While sharing many features, these two categories of disease caused by dematiaceous fungi are disparate in several epidemiologic, etiologic, and clinical aspects. These differences are presented, as are the general diagnostic and intervention methods for both types of DSP. Also, the potential relationship of certain biological features to virulence is discussed. However, a comprehensive description regarding the mycology and taxonomy of the etiologic agents of DSP is not presented. For this, as well as a fuller discourse pertaining to superficial, cutaneous, and subcutaneous diseases caused by pigmented fungi, the reader is referred to Chapters 11, White piedra, black piedra and tinea nigra; and 18, Chromoblastomycosis in this volume as well as several recent reviews (Kwon-Chung and Bennett 1992; Matsumoto and Ajello 1998; Brandt and Warnock 2003; Perfect et al. 2003; Sanche et al. 2003; Schell et al. 2003).

PHAEOHYPHOMYCOSIS – HISTOPATHOLOGICAL AND CLINICAL CONCEPTS

Dematiaceous fungi are etiologic agents responsible for a number of diverse diseases in humans and animals (Table 35.1). Though darkly pigmented in culture, not all these fungi appear so in vivo. In addition, studies have indicated that some fungal pathogens traditionally considered not to be phaeoid (e.g. *Cryptococcus neoformans*) do apparently either make minute amounts of melanin or have the ability to do so in vivo (Kimura and McGinnis 1998). This has led to a range of difficulties and confusion in the diagnosis of these diseases and their corresponding causative agents (Kwon-Chung and Bennett 1992).

To help resolve these problems, Ajello and colleagues proposed the terms 'phaeohyphomycosis' and 'hyalohyphomycosis' to distinguish between those fungal infections caused by phaeoid and hyaline molds, respectively, and which exhibit a predominant filamentous morphology in vivo (Ajello et al. 1974). In particular, these medical mycologists envisioned phaeohyphomycosis to encompass cutaneous, subcutaneous, and systemic infections in which the etiologic agent 'develops in the host tissues in the form of dark-walled dematiaceous septate mycelial elements.' Thus, this designation defined the infection based upon its histopathology rather than its clinical presentation. However, the original definition was limited to hyphomycetous fungi, i.e. a class of molds taxonomically placed within the form-division Fungi Imperfecti [Deuteromycota; asexual (mitosporic) fungi]. As new cases of phaeohyphomycosis were reported, it became clear that not all were hyphomycetes. Subsequently, Ajello amended the definition of phaeohyphomycosis to include infections caused by

fungi belonging not only to the form-class Hyphomycetes, but also those grouped within the form-class Coelomycetes (form-division Fungi Imperfecti) and division Ascomycota [fungi that form sexually derived spores (meiospores) within an ascus] (Ajello 1981).

As conceived, phaeohyphomycosis was intended to exclude the well-established disease states that also featured the formation of phaeoid hyphae in tissue, e.g. tinea nigra, chromoblastomycosis, sporotrichosis, and eumycotic mycetoma. These latter diseases are distinct, separate entities having their own unique clinical and histopathological presentation in addition to etiologic differences. The reader is referred to the appropriate chapters within this book (Chapters 11, White piedra, black piedra and tinea nigra; 18, Chromoblastomycosis; 19, Sporotrichosis; and 20, Eumycetomas, respectively) for a more in-depth discussion of these disease states.

As a cosmopolitan disease, phaeohyphomycosis mainly afflicts adults among whom many are immunocompromised. Underlying conditions that contribute to infection by dematiaceous fungi include tuberculosis, diabetes, cancer, tissue transplantation, and use of corticosteroids or other immunosuppressive drugs, surgery, and a number of other conditions including acquired immunodeficiency syndrome (AIDS) due to the human immunodeficiency virus (HIV) (Table 35.2).

Many cases of phaeohyphomycosis appear in the perfectly healthy and may have no evident exposure history. These non-life-threatening infections include focal-subcutaenous, mycotic keratitis, and paranasal sinus infections. Subcutaneous phaeohyphomycosis is a common site of infection and is generally thought to occur as a result of traumatic implantation of fungal material from contaminated plants or soil. The most common outcome of this type of inoculation is a localized cyst or abscess that appears primarily as a single, discrete and asymptomatic or mildly painful nodule. From this entity, granulomatous reactions may occur and the disease may on occasion progress to produce verrucose plaques reminiscent of chromoblastomycosis.

Table 35.1 *Fungal infections with phaeoid etiologies*

Infection
Black piedra
Chromoblastomycosis
Eumycotic mycetoma
Phaeohyphomycosis
Pseudallescheriasis
Sporotrichosis
Tinea nigra
Zygomycosis

Table 35.2 *Risk factors associated with phaeohyphomycosis*

Risk factor
Bone marrow transplantation
Cancer
Cardiac surgery
Chronic ambulatory peritoneal dialysis
Corticosteroid therapy
Diabetes
Fungal sinusitis
HIV infection
Intravenous drug use
Neutropenia
Solid organ transplantation
Trauma to skin or soft tissues
Tuberculosis

Mycotic keratitis also is a common site for this fungal infection where asexual spores of these fungi that are common in our environment can contaminate a corneal abrasion following trauma. Paranasal sinus infections are also frequently diagnosed. They present with sinuses impacted with a phaeohyphomycotic "fungoma" seen in biopsy. Frequently, there is an allergic host response to the fungus as evidenced by the presence of eosinophils. Rarely is this type of sinus infection highly invasive. These non-life-threatening infections are caused by the same genera/species that cause deep-seated infections implying the importance of the host in the dynamics of the phaeoid fungus/host interactions.

By comparison, DSP is rare. Recently, Revankar and colleagues (Revankar et al. 2002, 2004), as well as Filizzola et al. (Filizzola et al. 2003), published reviews focused on DSP. Case reports from 1996 through 2002 were obtained from citations listed in the Medline database (National Library of Medicine, Bethesda, MD). Additional cases of DSP were reviewed from other references cited in these reports. Though not fully comprehensive, since non-English language literature was generally excluded, these reviews clearly separate DSP into two broad groups of infections. In one category of DSP, infection is disseminated throughout the body and may involve the central nervous system (CNS). Most individuals affected with disseminated disease tend to have decreased host immunity or possess other risk factors. A second distinct type of DSP is generally restricted to the CNS. Curiously, about half of the victims of cerebral phaeohyphomycosis do not appear to possess underlying conditions or risk factors predisposing them to infection. Both types of DSP are discussed in greater detail below.

It is significant to note, however, that DSP appears to involve only a subset of fungal species listed as etiologic agents of phaeohyphomycosis (Matsumoto et al. 1994; Matsumoto and Ajello 1998). Of the more than 100 species among the 60 genera associated with infection by dematiaceous fungi, about a third are known to cause DSP (Table 35.3). [The specific epithets presented in Table 35.3 and elsewhere in the text represent the currently accepted nomenclature for a particular fungal species. Previously employed binomials for fungal species causing phaeohyphomycosis, as well as a detailed mycological description of these etiologic agents, can be readily found in several publications (Kwon-Chung and Bennett 1992; Matsumoto and Ajello 1998; Perfect et al. 2003; Sanche et al. 2003; Schell et al. 2003).] This list should not be considered comprehensive since new cases of DSP are likely to be reported that involve species not listed. Moreover, not all medical mycologists may agree with the inclusion of certain fungi as agents of phaeohyphomycosis (see discussion below). Nonetheless, the fungal species presented here typify those that have been adequately documented as the major etiologic agents of DSP.

Table 35.3 *Fungal species associated with DSP[a]*

Fungal species
Acrophialophora
Alternaria infectoria
Amium leproinum
Aureobasidium pullulans
Bipolaris
B. australiensis
B. hawaiiensis
B. specifera
Bipolaris spp.
Chaetomium
C. atrobrunneum
C. globosum
C. perlucidum
C. strumarium
Cladophialophora
C. bantiana
C. devriesii
C. modesta
Cladophialophora spp.
Cladosporium cladosporoides
Curvularia
C. geniculata
C. lunata
C. pallescens
Exophiala
E. castellanii
E. jeanselmei
E. mansonii
Exserchilum rostratrum
Fonsecaea pedrosoi
Hormonema dematioides
Lecythophora mutabilis
Microascus cirrosus
Mycelophthora thermophila
Nodulisporium spp.
Ochroconis gallopavum
Phaeoacremonium parasiticum
Phialemonium curvatum
Phialophora richardsiae
Rhamichloridium obovoideum
Rhinocladiella atrovirens
Scopularioposis
S. brumptii
Scopulariopsis spp.
Scytalidium dimidiatum
Wangiella dermatitidis

a) Data collected from Revankar et al. (2002, 2004); Barron et al. (2003); Filizzola et al. (2003); Teixeira et al. (2003).

The list of phaeohyphomycotic agents given in Table 35.3 is notable in its exclusion of *Scedosporium* spp. [*Scedosporium apiospermum* (anamorph of *Pseudallescheria boydii*) and *Scedosporium prolificans* (formerly *Scedosporium inflatum*)] as agents of DSP.

The inclusion of these fungi as a cause of phaeohyphomycosis has been a source of controversy and needless consternation (Matsumoto et al. 1994). Some investigators persist in embracing *Scedosporium* spp. in the etiology of phaeohyphomycosis (Revankar et al. 2002, 2004; Schell et al. 2003). Others seemingly do not agree with this position, yet continue to include this organism in the description of possible causative agents (Matsumoto and Ajello 1998; Perfect et al. 2003). Moreover, some recent publications clearly state that *Scedosporium* is an agent of hyalohyphomycosis (McGinnis and Ajello 1998; Tadros et al. 1998; Walsh 1998; Torres and Kontoyiannis 2003) or the form-genus is conspicuously omitted as a phaeohyphomycotic agent (Sanche et al. 2003). The confusion is further compounded by the fact that several authors of the above-cited publications have taken both positions.

Perhaps the source of this chaos lies in the misapplication of the true, intended definition of phaeohyphomycosis. As noted above, this clinical entity is based upon the histopathological observation of 'dark-walled dematiaceous (i.e. phaeoid) septate mycelial elements' in diseased tissue (Ajello et al. 1974; Ajello 1981). As dematiaceous fungi, *Scedosporium* spp. do form a melanin-like color in culture (Schell et al. 2003). These species also stain positive for melanin in vivo, using the Fontana–Masson silver stain, which is specific for such pigments (Wood and Russel-Bell 1983). The histopathology of tissues infected with *Scedosporium* spp. is also similar to phaeohyphomycosis, but more strongly resembles the dichotomous hyphal branching seen in aspergillosis (Chandler and Watts 1987; Cooper and Salkin 1993). However, the hyphae observed in these tissues are hyaline except when aggregates are formed such as in a pulmonary 'fungoma' (fungus ball). Brown conidia are sometimes observed in the latter. Yet, a fungoma due to *Scedosporium* clearly represents a separate clinical presentation that tends not to be invasive (Cooper and Salkin 1993). Therefore, despite the occasional presence of pigmentation, inclusion of *Scedosporium* spp. as etiologic agents of phaeohyphomycosis is inconsistent with the present definition of this disease state. Unless the description of this term is amended formally, clinicians should be aware that not all phaeoid fungi are agents of phaeohyphomycosis. Indeed, some melanized fungi cause infections better attributed to markedly different, classical presentations (e.g. sporotrichosis, chromoblastomycosis, etc.). At the risk of eliminating more fungi listed in Table 35.3, if a particular mold species does not consistently form dark-colored hyphae in vivo, it should be considered an agent of hyalohyphomycosis. Such is the case with *Scedosporium* spp. and is a position more in line with the accepted definitions first proposed by Ajello and colleagues (Ajello et al. 1974; Ajello 1981).

This same argument, however, might be used against other dematiaceous fungi generally acknowledged by medical mycologists to be phaeohyphomycotic agents. For example, species of *Bipolaris*, as well as other fungi, do not always produce pigmented fungal elements in tissue. In one such case involving *Bipolaris spicifera*, the clinical isolate did form dark-colored hyphae in experimentally infected mice (Yoshimori et al. 1982). Given the ever-increasing spectrum of pathogenic dematiaceous fungi and the inconsistent nature of pigment production in vivo, perhaps the definitions of phaeohyphomycosis and hyalohyphomycosis need to be re-evaluated. However, except in the clinical presentation of fungoma, *Scedosporium* spp. do not produce visible melanin in vivo despite the large number of reports of disease involving these species. Collectively, these cases certainly outnumber those of phaeohyphomycosis caused by the most frequently encountered dematiaceous fungi in which melanin is consistently visible in tissues.

CLINICAL MANIFESTATIONS OF DSP

DSP can be divided into the separate clinical presentations of systemic disease and cerebral involvement. These different manifestations share the same histopathological concept – the presence of darkly pigmented hyphae in vivo. In broad terms, many of the symptoms, treatment outcomes, and types of patients are common to both. Yet, each does have unique aspects that distinguish one from the other. The following succinct descriptions compare both the mutual and exclusive aspects of disseminated and cerebral phaeohyphomycosis.

Disseminated phaeohyphomycosis

Compared to cutaneous and subcutaneous forms of phaeohyphomycosis, there are relatively few reports of disseminated disease by phaeoid fungi in the medical literature. Although infrequent in morbidity, disseminated phaeohyphomycosis deserves special consideration as a separate disease type given its level of mortality (Revankar et al. 2002). Disseminated disease results by the spread of the dematiaceous pathogen from a previously colonized or infected body site (Kwon-Chung and Bennett 1992; Brandt and Warnock 2003; Perfect et al. 2003; Sanche et al. 2003). Trauma or surgery is typically a cause of cutaneous or subcutaneous infection sites. In addition, dissemination can originate from the lungs or infected sinus tissue following inhalation of fungal cells or conidia. Under the appropriate circumstances, direct extension from the sinus or hematogenous spread of the fungus to one or more distant sites of the body can result in visceral infections of the heart and heart valves, brain, joints, bone, kidney, liver, lymphatics, pancreas, and other organs. Cases of dialysis-associated peritonitis have also been shown to be caused by dematiaceous fungi.

Recently, Revankar and co-workers (Revankar et al. 2002) reviewed 72 case studies of disseminated phaeohyphomycosis published in the English language literature between 1996 and 2001. The authors considered each of these cases valid based upon the following criteria:

- the presence of a clinical syndrome consistent with disseminated disease
- histopathological evidence of a true infection
- the recovery and identification of a dematiaceous fungus.

Included in this review were 30 infections caused by *S. prolificans*. As previously noted, infection by this fungus is better termed hyalohyphomycosis. Therefore, the 30 cases cited by these investigators are excluded from the following discussion. The remaining 42 infections, though drawn from a selected pool of reports, do represent a general overview of disseminated disease caused by dematiaceous fungi.

Most of the 27 species cited by Revankar et al. (2002) as etiologic agents are described in single case reports. This reflects the uncommon nature of disseminated phaeohyphomycosis. Of the 42 patients studied, 35 (83 percent) possessed some risk factor predisposing them to infection, including two cases of AIDS. In addition, some of the apparently normal individuals exhibited eosinophila. This suggests that certain patients have developed a poor TH2 response, thereby hindering cell-mediated immunity and inadvertently promoting dissemination of the infection. Clinically, the most common symptoms among the 42 affected individuals included fever, skin manifestations, respiratory complaints, and neurological problems. Sepsis and gastrointestinal symptoms were also recorded. With regard to the sites of infection, half the cases involved the lungs. Other significantly affected organs included the brain, heart/heart valve, kidney, liver, and lymph nodes. In some instances, the spleen and pancreas were involved. Death resulted in 29 (70 percent) of the 42 cases, including five patients who did not receive antifungal therapy. The remaining 13 patients received surgical intervention, antifungal therapy, or both. Amphotericin B alone or in combination with other drugs and/or surgery was used in 10 cases. However, specific therapeutic recommendations based upon the outcome of these treatment regimens are not possible given the general resistance of dematiaceous fungi to amphotericin B as well as the number of diverse species implicated in these infections. Yet, the results do make it clear that no single drug or combination therapy improved mortality rates.

Cerebral phaeohyphomycosis

Cerebral infections by phaeoid fungi represent a specialized type of DSP. For instance, most of the commonly encountered dematiaceous fungi causing cerebral phaeohyphomycosis produce mainly brain abscesses as opposed to disseminated disease (Kwon-Chung and Bennett 1992; Filizzola et al. 2003; Sanche et al. 2003; Revankar et al. 2004). These fungi include *Cladophialophora bantiana*, *Ochronconis gallopavum*, *Ramichloridium obovoideum*, *Wangiella dermatitidis*, *Curvularia* spp., and *Bipolaris* spp. Studies of human infections and in animal models have shown with certainty that *C. bantiana* is neurotropic, i.e. it has a propensity to cause infection of the CNS (Dixon et al. 1987a, b). The remaining major etiologic agents also possess neurotropic tendencies (Walsh et al. 1987). Moreover, whereas disseminated phaeohyphomycosis mainly afflicts individuals having immunodeficiencies or other risk factors, nearly half of patients with cerebral phaeohyphomycosis are immunocompetent. There does appear to be some symptomatic overlap in that disseminated phaeohyphomycosis can also involve the CNS. However, the latter type of phaeohyphomycosis is usually accompanied by disease of multiple organs and typically originates by hematogenous spread from a primary focal infection. Cerebral phaeohyphomycosis, particularly in apparently normal hosts, usually does not involve other organs. Hence, these observations and the limited etiologic spectrum lend special consideration to cerebral phaeohyphomycosis as a separate type of DSP.

CNS infections by phaeoid fungi are distributed worldwide. Presumably, most infections that lead solely to brain abscess or meningitis occur initially from the inhalation of fungal propagules. Following colonization in the lung, the fungus is transferred via a hematogenous route to the CNS. Alternatively, chronic fungal sinusitis may serve as the cause of CNS invasion (Sanche et al. 2003; Schell et al. 2003). Phaeohyphomycotic agents commonly associated with this condition include species of *Bipolaris*, *Exserohilum*, *Curvularia*, *Alternaria*, and *Cladosporium*.

Clinical symptoms of cerebral phaeohyphomycosis include headache, fever, seizures, and neurological deficits. In contrast to bacterial infections of the CNS, these symptoms do not always occur together. Sudden onset of blindness may also occur mainly in those infections by *Bipolaris* or *Curvularia* due to compression of the optic nerve (Walsh 1998). Again, immunocompetent patients tend not to have signs of organ infection other than the CNS.

Abscesses in cerebral phaeohyphomycosis may occur singly or as multiple masses (Perfect et al. 2003; Sanche et al. 2003). The frontal lobe is the most common site for abscess formation, but other areas of the brain can be affected. Another type of presentation, either alone or in conjunction with abscess development, is meningitis. Reports of dematiaceous fungi in CSF, though very rare, have been documented.

Two independent surveys of the English-language literature from 1996 through 2002 have been published that describe authentic cases of cerebral phaeohyphomycosis. One survey focused solely on infection in the immunocompetent host (Filizzola et al. 2003), whereas the other was broader in scope (Revankar et al. 2004). Though not fully comprehensive, these two reviews provide a well-documented synopsis of the current status of cerebral phaeohyphomycosis. Collectively, both study groups cited 109 cases of cerebral phaeohyphomycosis. Each report was considered bona fide if it met specified clinical, histopathological, and mycological criteria that define phaeohyphomycosis. However, a single case due to the hyalohyphomycotic agent *S. prolificans* was incorrectly included (Revankar et al. 2004). Therefore, among the 108 valid case reports, a total of 25 phaeoid fungal species were identified. The number of infected males was three times that of females. Of all the patients afflicted with cerebral phaeohyphomycosis, 55 percent were immunocompetent. Only a single individual was HIV positive among those predisposed to infection. Overall mortality was 73 percent amid the 97 patients for whom an outcome was reported. Nearly the same mortality was reported for both apparently normal and at-risk individuals (74 and 71 percent, respectively). When cases caused by selected etiologic agents are considered, some marked differences in mortality exist (Table 35.4). With regard to *C. bantiana*, which was responsible for 44 percent of the cases reported, overall survival of the 42 patients followed was 29 percent. Interestingly, compromised individuals survived at a higher rate (45 percent) than normal hosts (23 percent) (Filizzola et al. 2003; Revankar et al. 2004). Though not quite significantly different ($P=0.06$), and the underlying reasons unclear, the limited data set does seem to suggest that immunocompetent persons with cerebral phaeohyphomycosis due to *C. bantiana* face a greater mortality risk. Comparable overall levels of mortality resulted from infections with *Bipolaris* spp. and *O. gallopavum*, though much fewer cases have been observed. By contrast, virtually all cases of cerebral phaeohyphomycosis involving *R. obovoideum*, *W. dermatitidis*, *Fonsecaea pedrosoi*, and *Chaetomium* spp. were fatal regardless of host immune status.

HISTOPATHOLOGY AND DIAGNOSIS OF DSP

Phaeohyphomycotic lesions exhibit distinct histopathologies depending upon the clinical form of the disease (McGinnis 1983; Chandler and Watts 1987). Moreover, the pathological features are similar regardless of the etiologic agent. In DSP, lesions of a given organ may occur as single or multiple abscesses that broadly resemble the subcutaneous cystic form of phaeohyphomycosis. Alternatively, infection may consist of an inflammatory infiltrate with the presence of necrosis. When formed, abscesses are usually circumscribed and encapsulated granulomas, the walls of which comprise epithelial histiocytes and giant cells surrounded by connective tissue. A pyogenic inflammatory response is common as well. The centers of the granulomas contain necrotic debris, fibrin, and degenerated polymorphonuclear cells.

Within the necrotic debris and inflammatory infiltrates, as well as in the giant cells, fungal elements can be observed. The morphology of these fungi may consist of septated hyphae (2–6 μm in diameter), pseudohyphae (moniliform hyphae), budding yeasts, enlarged subglobose cells (up to 25 μm in diameter), chlamydoconidia, or spherical cells divided internally in a single plane. The latter resemble the muriform (sclerotic) cells characteristically found in chromoblastomycosis (McGinnis 1983; Chandler and Watts 1987), but differ in having thinner cell walls and a single planate septum. Lesions may contain one or more of the above structures. The brown pigment in the walls of these fungal cells may be visible in hemaotoxylin and eosin-stained tissue sections. This is not always the case, however. For example, the fungal elements of the species of *Bipolaris*, *Curvularia*, and *Alternaria* often appear hyaline in vivo although these fungi do form melanin in culture. In such circumstances, the Fontana–Mason stain will help reveal if a potential dematiaceous fungus is present (Wood and Russel-Bell 1983). Still, fungi other than those causing phaeohyphomycosis (and the similar clinical entity, chromoblastomycosis) also stain positive using this method (Kimura and McGinnis 1998).

Table 35.4 *Survival rates of cerebral phaeohyphomycosis patients infected with selected etiologic agents*[a]

| Fungal species | Number of cases | Number of deaths per total number of patients | | Outcome unknown | Overall death rate (%) |
		Normal host	Compromised host		
C. bantiana	48 (44%)	24/31	6/11	6	71
R. obovoideum	13 (12%)	4/4	5/5	4	100
Bipolaris spp.	9 (8%)	2/4	3/4	1	63
F. pedrosoi	5 (5%)	1/1	2/3	1	75
Chaetomium spp.	5 (5%)	0/0	5/5	0	100
O. gallopavum	5 (5%)	0/0	3/5	0	60
W. dermatitidis	4 (3%)	4/4	0/0	0	100

a) Data taken from Filizzola et al. (2003) and Revankar et al. (2004).

A presumptive diagnosis of DSP can be made if brown fungal elements are observed in vivo and are accompanied by a clinical presentation consistent with the disease. However, such observations do not indicate which pathogen is involved. It is critical to the therapeutic intervention in cases of systemic and cerebral phaeohyphomycosis that the etiologic agent be identified. Because phaeoid fungi can occur as contaminants, documentation that an infection is present requires that cultures be obtained from otherwise sterile body sites or from tissue in which the histopathology is consistent with DSP. For a fuller treatment regarding culture and identification methods for other major species of melanized fungi that cause phaeohyphomycosis, several resources can be consulted (McGinnis 1980; Dixon and Polak-Wyss 1991; Matsumoto and Ajello 1998; Sanche et al. 2003; Schell et al. 2003). In addition to identifying isolates by morphological characteristics, other phenotypic measurements that can be used include nitrate and carbohydrate assimilation tests, exoantigen tests, proteolytic activity, growth temperature response, etc. Generally, the biochemical and exoantigen methods have limited applications in the identification of dematiaceous fungi.

It is important to note that there is a paucity of molecular methods to identify dematiaceous fungi. Virtually all are based upon the identification or amplification of specific DNA from ribosomal or internal transcribed spacer regions (Abliz et al. 2003, 2004a, b; Hall et al. 2004). One commercially available method, the MicroSeq D2 large-subunit rDNA sequencing kit, shows some promise. Of the 80 phaeoid fungi tested, representing 36 separate species, 52 (65 percent) yielded identifications consistent with those derived by classical phenotypic methods (Hall et al. 2004). Also, this molecular-based identification method appears equivalent to the phenotypic procedures in terms of costs. However, more isolates of the same and different species need to be tested to determine the true efficacy of this kit. Despite the generally good results with this kit, though, it remains essential for the near future in medical mycology that requisite laboratory skills in the identification of fungi by more traditional phenotypic means continue to be developed and maintained.

TREATMENT OF DSP

The mortality rate of individuals having DSP, even if treated, is grimly high compared to that of persons receiving therapy for other forms of phaeohyphomycosis. Moreover, because publications on DSP have reported individual cases of disease and treatment, if any, there is relatively little information to recommend global treatment strategies. The situation is further exacerbated by the fact that these case reports collectively describe a diverse spectrum of species rather than a series of frequently encountered pathogens. It seems that each case of DSP should be approached therapeutically based upon its own particular circumstances. Nonetheless, the collective evidence discussed below seems to suggest some general guidelines for treatment of DSP as well as hope that newer antifungal agents will aid in the cure of infections caused by the phaeoid fungi.

Treatment of disseminated phaeohyphomycosis by surgical means is typically impractical. Prior to the development of effective azole and echinocandin drugs, chemotherapy instituted using amphotericin B, flucytosine, or a combination of both yielded mixed results (Revankar et al. 2002). The advent of azole and echinocandin classes of drugs, however, has added to the available antifungal armamentarium. The so-called older azole drugs, such as itraconazole, are very effective against dematiaceous fungi in vitro (McGinnis and Pasarell 1998b; McGinnis et al. 1998). In one landmark study, nine of 17 patients with phaeohyphomycosis showed a marked positive response to itraconazole therapy (Sharkey et al. 1990). In addition, many of the newer azoles show good promise both in vitro (Espinel-Ingroff 1998a, b; McGinnis and Pasarell 1998a) and in vivo (Graybill et al. 2004; Negroni et al. 2004). In a mouse model of disseminated phaeohyphomycosis due to W. dermatitidis, posaconazole reduced tissue burden and prolonged survival. Additionally, posaconazole provided a complete cure to a woman with a 12-year case of relapsing phaeohyphomycosis due to Exophiala spinifera that disseminated upon pregnancy. Though small in number, these experiences suggest that therapeutic regimens employing azoles alone may be effective in the treatment of systemic phaeohyphomycosis.

Cerebral phaeohyphomycosis should be treated aggressively given the rather poor clinical outcome in both normal patients and those at risk (Filizzola et al. 2003; Perfect et al. 2003; Revankar et al. 2004). For single brain abscesses, surgical debridement may be essential for survival. Even partial removal may facilitate recovery, particularly when combined with antifungal therapy. When surgery is contraindicated, a concerted therapeutic approach must be taken. Yet, treatment failures occur even with high doses of amphotericin B and flucytosine. This may be due in part to the innate or acquired resistance of dematiaceous fungi to these drugs. As with disseminated phaeohyphomycosis, the newer azoles and echinocandins show great potential in treating cerebral infections. Murine models of cerebral phaeohyphomycosis due to either R. obovoideum (Al-Abdely et al. 2000) or C. bantiana (cited in Sanche et al. 2003) indicated that posaconazole was superior to amphotericin B and itraconazole in prolonging survival and reducing fungal counts in the brain. In addition, Trinh and colleagues (Trinh et al. 2003) recently reported that voriconazole and caspofungin markedly delayed the progression in a premature infant of a fatal cerebral infection caused by C. bantiana.

In summary, the treatment options for DSP are generally limited due to the efficacy of available antifungal agents and the limitations of surgical interventions. The newer antifungals, especially posaconazole, appear to hold promise for future advances in curing these infections. Perhaps a more concerted effort to clinically assess the azoles in the treatment of DSP needs to be undertaken, and in particular comparing the in vitro susceptibility of the different causative agents with patient outcome. It seems critical that such 'real world' data be collected and analyzed in order to determine if the promise of the newer antifungals will be fulfilled. In addition, another avenue ripe for exploration is immune modulation. A complete understanding of the immune response to DSP is lacking. More information in this area is needed before potential protocols can be established to enhance or target a host's immune system towards prohibiting and eliminating infections by dematiaceous fungi.

BIOLOGICAL ASPECTS RELATED TO PATHOGENESIS

Given the wide array of phylogenetically diverse fungi causing DSP, it is unlikely that there will be a great number of universal themes pertaining to the underlying molecular mechanisms responsible for virulence. In fact, as in most pathogenic fungi, virulence is not dependent upon one or two factors acting alone, but rather two or more modalities acting in conjunction with each other. Recent reviews have emphasized these same points (van Burik and Magee 2001; Cole 2003).

The dematiaceous fungi do, however, share a few common themes associated with virulence. These factors are best exemplified by *W. dermatitidis*, which can serve as a paradigm for most phaeoid fungi (Szaniszlo 2002). This fungus is polymorphic and is a cause of subcutaneous and deep-seated phaeohyphomycosis. Moreover, under the appropriate culture conditions, or through the use of particular morphological mutants, *W. dermatitidis* will produce muriform-like cells that resemble the 'sclerotic body' characteristic of chromoblastomycosis Hence, this fungus may serve as a model to study this disease as well. Two particular attributes of *W. dermatitidis* that contribute to virulence are briefly summarized below.

Perhaps one of the best-studied virulence factors in *W. dermatitidis* is melanin. The genetics and biochemistry of the pentaketide biosynthetic pathway leading to melanin production has been extensively characterized in this pathogen (Cooper and Szaniszlo 1997; Feng et al. 2001; Szaniszlo 2002). Prior to the cloning of a key gene in this pathway, melanin-deficient mutants of *W. dermatitidis* were shown to be significantly less virulent in a murine infection model than the pigment wild-type strain (Dixon et al. 1987b, 1989a, 1992). Subsequently, a genetically derived null mutant, defective in the gene

encoding polyketide synthase (*WdPKS1*), was shown to be similarly less virulent (Feng et al. 2001). The proposed mechanism by which melanin functions in pathogenesis relates to experimental findings that this pigment acts as an antioxidant, thereby promoting survival of the oxidative bursts produced by phagocytes (Jacobson and Tinnell 1993; Schnitzler et al. 1999; Jacobson 2000). Melanin may also bind host hydrolytic enzymes that target the fungal plasma membrane. These experimental results have dramatic implications for other dematiaceous fungi given that melanin production in most of these species involves the same biochemical pathway (Wheeler and Bell 1988). Indeed, melanins have been shown to be important virulence factors in a phylogenetically diverse number of fungi pathogenic for humans, animals, and plants (Langfelder et al. 2003).

Chitin synthase represents another virulence factor in *W. dermatitidis* and, by extension, the phaeoid fungi in general. Chitin is a crucial cell-wall polymer that plays a significant role in the morphogenesis of nearly all fungi. This polysaccharide and the enzymes involved in its biosynthesis have been extensively studied in *W. dermatitidis* (Szaniszlo 2002). Null mutants were derived in five different chitin synthase genes. In addition, mutants bearing various combinations of these mutations were generated. Collectively, these mutations produced a wide range of morphological and growth phenotypes. When selected mutants were incubated at temperatures equivalent to that found in infection, cellular development was significantly altered or stopped. Moreover, certain mutants able to grow well at these temperatures and not exhibiting observable cellular abnormalities were less virulent than the wild-type strain in both immunocompetent and immunocompromised mouse models of infection (Wang et al. 1999, 2001; Liu et al. 2004). In short, chitin represents not only a target for the treatment of disease caused by *W. dermatitidis* and like fungi, but it also serves as yet another factor involved in pathogenesis.

SUMMARY AND FUTURE DIRECTIONS

DSP is characterized by the clinical presentation of disseminated and cerebral infections. Each has its own unique characteristics, although they share some similar histopathological features. They are both caused by a wide variety of fungal species having high levels of mortality even though most patients receive surgical and/or chemotherapeutic treatment. In the future, it is essential for medical mycologists, research scientists, and clinicians to make a concerted effort to develop additional intervention strategies employing newer antifungal agents that seem to show promise in early trials. Such strategies may focus upon the discovery of potential virulence factors and exploiting any unique fungus-specific features they present. Incorporated into these efforts should be the development of modern and more efficient molecular diagnostics for DSP, although current skills in the more traditional areas of fungal culture and identification must be maintained as well as instilled in future generations of medical mycologists.

REFERENCES

Abliz, P., Fukushima, K., et al. 2003. Rapid identification of the genus *Fonsecaea* by PCR with specific oligonucleotide primers. *J Clin Microbiol*, **41**, 873–6.

Abliz, P., Fukushima, K., et al. 2004a. Identification of pathogenic dematiaceous fungi and related taxa based on large subunit ribosomal DNA D1/D2 domain sequence analysis. *FEMS Immunol Med Microbiol*, **40**, 41–9.

Abliz, P., Fukushima, K., et al. 2004b. Specific oligonucleotide primers for identification of *Cladophialophora carrionii*, a causative agent of chromoblastomycosis. *J Clin Microbiol*, **42**, 404–7.

Ajello, L. 1981. The gamut of human infections caused by dematiaceous fungi. *Jpn J Med Mycol*, **22**, 1–5.

Ajello, L., Georg, L.K., et al. 1974. A case of phaeohyphomycosis caused by a new species of *Phialophora*. *Mycologia*, **66**, 490–8.

Al-Abdely, H.M., Najvar, L., et al. 2000. SCH 56592, amphotericin B, or itraconazole therapy of experimental murine cerebral phaeohyphomycosis due to *Ramichloridium obovoideum* ("*Ramichloridium mackenziei*"). *Antimicrob Agents Chemother*, **44**, 1159–62.

Barron, M.A., Sutton, D.A., et al. 2003. Invasive mycotic infections caused by *Chaetomium perlucidum*, a new agent of cerebral phaeohyphomycosis. *J Clin Microbiol*, **41**, 5302–7.

Brandt, M.E. and Warnock, D.W. 2003. Epidemiology, clinical manifestations, and therapy of infections caused by dematiaceous fungi. *J Chemother*, **15**, Suppl 2, 36–47.

Chandler, F.W. and Watts, J.C. 1987. *Pathologic diagnosis of fungal infections*. Chicago: ASCP Press.

Cole, G.T. 2003. Fungal pathogenesis. In: Anaissie, E.J., McGinnis, M.R. and Pfaller, M.A. (eds), *Clinical mycology*. Philadelphia: Churchill Livingston, 20–45.

Cooper, C.R. Jr 2002. New and emerging pathogens: what is a lab to do? In: Calderone, R.A. and Cihlar, R.L. (eds), *Fungal pathogenesis: principles and clinical applications*. New York: Marcel Dekker, 751–7.

Cooper, C.R. Jr and Salkin, I.F. 1993. Pseudallescheriasis. In: Murphy, J.W., Friedman, H. and Bendinelli, M. (eds), *Fungal infections and immune responses*. New York: Plenum Press, 335–58.

Cooper, C.R. Jr and Szaniszlo, P.J. 1997. Melanin as a virulence factor in dematiaceous pathogenic fungi. In: Bossche. H.V., Stevens, D.A. and Odds, F.C. (eds), *Host–fungus interplay*. Bethesda, MD: National Foundation for Infectious Diseases, 81–93.

Dixon, D.M. and Polak-Wyss, A. 1991. The medically important dematiaceous fungi and their identification. *Mycoses*, **34**, 1–18.

Dixon, D.M., Merz, W.G., et al. 1987a. Experimental central nervous system phaeohyphomycosis following intranasal inoculation of *Xylohypha bantiana* in cortisone-treated mice. *Mycopathologia*, **100**, 145–53.

Dixon, D.M., Polak, A. and Szaniszlo, P.J. 1987b. Pathogenicity and virulence of wild-type and melanin-deficient *Wangiella dermatitidis*. *J Med Vet Mycol*, **25**, 97–106.

Dixon, D.M., Polak, A. and Conner, G.W. 1989a. Mel-mutants of *Wangiella dermatitidis* in mice: evaluation of multiple mouse and fungal strains. *J Med Vet Mycol*, **27**, 335–41.

Dixon, D.M., Walsh, T.J., et al. 1989b. Infections due to *Xylohypha bantiana* (*Cladosporium trichoides*). *Rev Infect Dis*, **11**, 515–25.

Dixon, D.M., Migliozzi, J., et al. 1992. Melanized and non-melanized multicellular form mutants of *Wangiella dermatitidis* in mice: mortality and histopathology studies. *Mycoses*, **35**, 17–21.

Dixon, D.M., McNeil, M.M., et al. 1996. Fungal infections: a growing threat. *Public Health Rep*, **111**, 226–35.

Espinel-Ingroff, A. 1998a. Comparison of in vitro activities of the new triazole SCH56592 and the echinocandins MK-0991 (L-743,872) and LY303366 against opportunistic filamentous and dimorphic fungi and yeasts. *J Clin Microbiol*, **36**, 2950–6.

Espinel-Ingroff, A. 1998b. In vitro activity of the new triazole voriconazole (UK-109 496) against opportunistic filamentous and dimorphic fungi and common and emerging yeast pathogens. *J Clin Microbiol*, **36**, 198–202.

Feng, B., Wang, X., et al. 2001. Molecular cloning and characterization of WdPKS1, a gene involved in dihydroxynaphthalene melanin biosynthesis and virulence in *Wangiella (Exophiala) dermatitidis*. *Infect Immun*, **69**, 1781–94.

Filizzola, M.J., Martinez, F. and Rauf, S.J. 2003. Phaeohyphomycosis of the central nervous system in immunocompetent hosts: report of a case and review of the literature. *Int J Infect Dis*, **7**, 282–6.

Graybill, J.R., Najvar, L.K., et al. 2004. Posaconazole therapy of disseminated phaeohyphomycosis in a murine model. *Antimicrob Agents Chemother*, **48**, 2288–91.

Hall, L., Wohlfiel, S. and Roberts, G.D. 2004. Experience with the MicroSeq D2 large-subunit ribosomal DNA sequencing kit for identification of filamentous fungi encountered in the clinical laboratory. *J Clin Microbiol*, **42**, 622–6.

Jacobson, E.S. 2000. Pathogenic roles for fungal melanins. *Clin Microbiol Rev*, **13**, 708–17.

Jacobson, E.S. and Tinnell, S.B. 1993. Antioxidant function of fungal melanin. *J Bacteriol*, **175**, 7102–4.

Kimura, M. and McGinnis, M.R. 1998. Fontana-Masson-stained tissue from culture-proven mycoses. *Arch Pathol Lab Med*, **122**, 1107–11.

Kwon-Chung, K.J. and Bennett, J.E. 1992. *Medical mycology*. Philadelphia: Lea & Febiger.

Langfelder, K., Streibel, M., et al. 2003. Biosynthesis of fungal melanins and their importance for human pathogenic fungi. *Fungal Genet Biol*, **38**, 143–58.

Liu, H., Kauffman, S., et al. 2004. *Wangiella (Exophiala) dermatitidis* WdChs5p, a class V chitin synthase, is essential for sustained cell growth at temperature of infection. *Eukaryot Cell*, **3**, 40–51.

Matsumoto, T. and Ajello, L. 1998. Agents of phaeohyphomycosis. In: Ajello, L. and Hay, R.J. (eds), *Medical mycology. Topley and Wilson's microbiology and microbial infections*, Vol. 4. London: Arnold, 503–24.

Matsumoto, T., Ajello, L., et al. 1994. Developments in hyalohyphomycosis and phaeohyphomycosis. *J Med Vet Mycol*, **32**, 329–49.

McGinnis, M.R. 1980. *Laboratory handbook of medical mycology*. New York: Academic Press.

McGinnis, M.R. 1983. Chromoblastomycosis and phaeohyphomycosis: new concepts, diagnosis, and mycology. *J Am Acad Dermatol*, **8**, 1–16.

McGinnis, M.R. and Ajello, L. 1998. Conceptual basis for hyalohyphomycosis. In: Ajello, L. and Hay, R.J. (eds), *Medical mycology. Topley and Wilson's microbiology and microbial infections*, Vol. 4. London: Arnold, 499–502.

McGinnis, M.R. and Pasarell, L. 1998a. In vitro evaluation of terbinafine and itraconazole against dematiaceous fungi. *Med Mycol*, **36**, 243–6.

McGinnis, M.R. and Pasarell, L. 1998b. In vitro testing of susceptibilities of filamentous ascomycetes to voriconazole, itraconazole, and amphotericin B, with consideration of phylogenetic implications. *J Clin Microbiol*, **36**, 2353–5.

McGinnis, M.R., Pasarell, L., et al. 1998. In vitro activity of voriconazole against selected fungi. *Med Mycol*, **36**, 239–42.

Negroni, R., Helou, S.H., et al. 2004. Case study: posaconazole treatment of disseminated phaeohyphomycosis due to *Exophiala spinifera*. *Clin Infect Dis*, **38**, e15–20.

Pappagianis, D. and Ajello, L. 1994. Dematiaceous – a mycologic misnomer? *J Med Vet Mycol*, **32**, 319–21.

Perfect, J.R., Schell, W.A. and Cox, G.M. 2003. Phaeohyphomycosis. In: Dismukes, W.E., Pappas, P.G. and Sobel, J.D. (eds), *Clinical mycology*. New York: Oxford University Press, 271–82.

Revankar, S.G., Patterson, J.E., et al. 2002. Disseminated phaeohyphomycosis: review of an emerging mycosis. *Clin Infect Dis*, **34**, 467–76.

Revankar, S.G., Sutton, D.A. and Rinaldi, M.G. 2004. Primary central nervous system phaeohyphomycosis: a review of 101 cases. *Clin Infect Dis*, **38**, 206–16.

Sanche, S.E., Sutton, D.A. and Rinaldi, M.G. 2003. Dematiaceous fungi. In: Anaissie, E.J., McGinnis, M.R. and Pfaller, M.A. (eds), *Clinical mycology*. Philadelphia: Churchill Livingstone, 325–51.

Schell, W.A., Salkin, I.F. and McGinnis, M.R. 2003. *Bipolaris, Exophiala, Scedosporium, Sporothrix*, and other dematiaceous fungi. In: Murray, P., Baron, E.J., et al. (eds), *Manual of clinical microbiology*. Washington, DC: American Society for Microbiology, 1820–47.

Schnitzler, N., Peltroche-Llacsahuanga, H., et al. 1999. Effect of melanin and carotenoids of *Exophiala (Wangiella) dermatitidis* on phagocytosis, oxidative burst, and killing by human neutrophils. *Infect Immun*, **67**, 94–101.

Sharkey, P.K., Graybill, J.R., et al. 1990. Itraconazole treatment of phaeohyphomycosis. *J Am Acad Dermatol*, **23**, 577–86.

Szaniszlo, P.J. 2002. Molecular genetic studies of the model dematiaceous pathogen *Wangiella dermatitidis*. *Int J Med Microbiol*, **292**, 381–90.

Tadros, T.S., Workowski, K.A., et al. 1998. Pathology of hyalohyphomycosis caused by *Scedosporium apiospermum* (*Pseudallescheria boydii*): an emerging mycosis. *Hum Pathol*, **29**, 1266–72.

Teixeira, A.B., Trabasso, P., et al. 2003. Phaeohyphomycosis caused by *Chaetomium globosum* in an allogeneic bone marrow transplant recipient. *Mycopathologia*, **156**, 309–12.

Torres, H.A. and Kontoyiannis, D.P. 2003. Hyalohyphomycosis (other than aspergillosis and penicilliosis). In: Dismukes, W.E., Pappas, P.G. and Sobel, J.D. (eds), *Clinical mycology*. New York: Oxford University Press, 252–70.

Trinh, J.V., Steinbach, W.J., et al. 2003. Cerebral phaeohyphomycosis in an immunodeficient child treated medically with combination antifungal therapy. *Med Mycol*, **41**, 339–45.

van Burik, J.A. and Magee, P.T. 2001. Aspects of fungal pathogenesis in humans. *Annu Rev Microbiol*, **55**, 743–72.

Walsh, T.J. 1998. Emerging fungal pathogens: evolving challenges to immunocompromised patients. In: Scheld, W.M., Armstrong, D. and Hughes, J.M. (eds), *Emerging infections 1*. Washington, DC: ASM Press, 221–32.

Walsh, T.J., Dixon, D.M., et al. 1987. Comparative histopathology of *Dactylaria constricta, Fonsecaea pedrosoi, Wangiella dermatitidis* and *Xylohypha bantiana* in experimental phaeohyphomycosis of the central nervous system. *Mykosen*, **30**, 215–25.

Wang, Z., Zheng, L., et al. 1999. WdChs4p, a homolog of chitin synthase 3 in *Saccharomyces cerevisiae*, alone cannot support growth of *Wangiella (Exophiala) dermatitidis* at the temperature of infection. *Infect Immun*, **67**, 6619–30.

Wang, Z., Zheng, L., et al. 2001. WdChs2p, a class I chitin synthase, together with WdChs3p (class III) contributes to virulence in *Wangiella (Exophiala) dermatitidis*. *Infect Immun*, **69**, 7517–26.

Wheeler, M.H. and Bell, A.A. 1988. Melanins and their importance in pathogenic fungi. *Curr Top Med Mycol*, **2**, 338–87.

Wood, C. and Russel-Bell, B. 1983. Characterization of pigmented fungi by melanin staining. *Am J Dermatopathol*, **5**, 77–81.

Yoshimori, R.N., Moore, R.A., et al. 1982. Phaeohyphomycosis of brain: granulomatous encephalitis caused by *Drechslera spicifera*. *Am J Clin Pathol*, **77**, 363–70.

36

Deep hyalohyphomycosis

CAROLINE B. MOORE AND DAVID W. DENNING

THE CONCEPT OF HYALOHYPHOMYCOSIS

Definition

The clinical nomenclature 'hyalohyphomycosis,' first used by Ajello (1982), serves as an umbrella term grouping together infections caused by molds that lack melanin in their cell walls (known as hyaline, non-dematiaceous or non-pheoid), which do not have other specific, well-established names, such as aspergillosis. This terminology allows for the inclusion of future non-melanized filamentous etiologic agents or disease states, thus avoiding the need to create a multitude of unnecessary names. Furthermore, it is resistant to the many taxonomic changes that inevitably happen (Matsumoto et al. 1994).

Causative organisms

Currently known agents of hyalohyphomycosis are many and varied, presently totaling more than 20 genera and 70 species (Table 36.1). Some of these species are well documented in causing disease (Table 36.2); others have been encountered less often.

The hyalohyphomycetes comprise a heterogeneous group characterized by production of colorless septate hyphae in host tissue. Hyphae are parallel-walled, typically showing irregular branching at both 45° and 90°. Hyphal diameter can vary substantially (2–8 μm), but is more slender than that of the Zygomycetes. Colony color can also differ widely, but is never darkly pigmented.

They are opportunistic pathogens, normally found as saprophytes in soil and vegetative matter, and frequently identified as plant pathogens. The most common agents of disease are species of *Fusarium*, *Scedosporium*, *Acremonium*, and *Paecilomyces*.

Identification of hyaline filamentous fungi is not always straightforward, and this chapter does not intend to provide an exhaustive text on the subject. Key characteristics of common genera are provided. However, the reader is referred to excellent guides on the identification of medically important species (Sutton et al. 1998; de Hoog et al. 2000; Larone 2002). Additional resources can be found at www.doctorfungus.org and www.clinical-mycology.com. Expert help may also prove advantageous.

Categories of disease

Hyaline hyphomycetes are capable of causing a wide spectrum of disease that can be further divided into superficial, deep tissue, and disseminated infections. This chapter aims to concentrate on deep-seated infection, and as such, does not give detail on other forms of disease, including those of the eye (see Chapter 16, Oculomycosis).

Deep infections may involve the lungs, sinuses, heart, liver, spleen, kidney, bones, or central nervous system (Table 36.2), and are commonly acquired through the respiratory tract, gastrointestinal tract (e.g. after major surgery), or blood vessels (e.g. catheter-related). Skin breakdown or toenail onychomycosis can be the origin of infection in invasive fusarial disease (Boutati and Anaissie 1997). Major trauma may also result in

Table 36.1 *Currently known agents of hyalohyphomycosis*[a]

Genera and species (synonym)	Genera and species (synonym)	Genera and species (synonym)
Acremonium	**Fusarium**	**Scedosporium**
A. alabamense	F. anthophilum	S. apiospermum
A. atrogriseum	F. aquaeductuum	S. prolificans[b]
A. blochii	F. chlamydosporum	**Schizophyllum**
A. curvulum	F. dimerum	S. commune
A. falciforme	F. equiseti	**Scopulariopsis**
A. kiliense	F. napiforme	S. acremonium
A. potronii	F. nygamai	S. asperula
A. recifei	F. oxysporum	S. brevicaulis
A. roseogriseum	F. polyphialidicum	S. brumptii[b]
A. strictum	F. proliferatum	S. candida
Aphanoascus	F. sacchari	S. fusca
A. fulvescens	F. semitectum (F. incarnatum)	S. koningii
Beauveria	F. solani	**Scytalidium**
B. bassiana	F. subglutinans	S. hyalinum
B. brongniartii	F. tabacinum	**Trichoderma**
Cephaliophora	F. verticillioides (F. moniliforme)	T. citrinoviride
C. irregularis	**Myriodontium**	T. harzianum
Chrysonilia	M. keratinophilum	T. koningii
C. sitophila	**Neocosmospora**	T. longibrachiatum
Chrysosporium	N. vasinfecta	T. pseudokoningii
C. zonatum	**Ovadendron**	T. viride
Coprinus	O. sulphureo-ochraceum	**Tritirachium**
C. cinereus	**Paecilomyces**	T. oryzae
Cylindrocarpon	P. crustaceus	T. roseum
C. cyanescens	P. javanicus	**Tubercularia**
C. destructans	P. lilacinus	T. vulgaris
C. lichenicola	P. marquandii	**Verticillium**
C. vaginae	P. taitungiacus	V. serrae
Engyodontium	P. variotii	**Volutella**
E. album	P. viridis	V. cinerescens

a) Species are often unidentified in published reports.
b) Generally regarded as dematiaceous.

implantation. Infection can remain localized in deep tissue and organs, or can disseminate hematogenously or via the lymphatic system. Secondary dissemination to the skin is more frequently seen with some of the fungi described here than with disseminated aspergillosis.

Patients at risk

Recent advances in medical procedures, therapy, and education have been extremely successful in increasing survival in patients, whom, historically, would have had

Table 36.2 *Common manifestations of deep infection caused by principal agents of hyalohyphomycosis*

	Dissem.	Skin	Blood	Lung	Brain	Sinus	Heart	Liver/spleen	Kidney	Bones/joints
Fusarium spp.	✓✓	✓✓	✓✓	✓	✓	✓	✓	✗	✗	✓
Scedosporium apiospermum	✓	✓	✓	✓✓	✓✓	✓	✓	✗	✓	✓✓
Scedosporium prolificans[a]	✓✓	✓	✓✓	✓✓	✓✓	✓	✓	✓	✓	✓✓
Acremonium spp.	✓	✓✓	✓✓	✓	✓	✗	✓	✗	✗	✓✓
Paecilomyces spp.	✓	✓✓	✓	✓	✓	✓	✓	✗	✓	✓
Scopulariopsis spp.	✓✓	✓✓	✗	✓	✓	✓	✓✓	✗	✗	✗

✓✓ common site of infection of this species; ✓ less commonly reported; ✗ rarely or not reported to date; Dissem., propensity for disseminated infection.
a) Generally regarded as dematiaceous; included to enable comparison between species.

limited treatment options and a poor prognosis. As a direct result, the incidence of infection with opportunistic fungal pathogens, such as hyaline molds, has increased substantially worldwide. Whilst overall a relatively small number of infections are encountered, associated mortality is disproportionately high. Persistent neutropenia is probably the key risk factor for infection with hyaline molds. Highly susceptible patients include those with underlying hematological malignancies, bone marrow transplant recipients, and solid organ transplantation. Interestingly, hyalohyphomycosis in individuals infected with the human immunodeficiency virus (HIV) appears to be relatively uncommon. HIV specifically targets the T-lymphocyte population whereas hyaline mold pathogens are primarily granulocyte-controlled infections (Matsumoto et al. 1994).

Immunocompetent individuals may also be infected by these pathogens. Common predisposing factors include near-drowning or penetrating trauma (Walsh et al. 2004).

DIAGNOSIS

Early diagnosis is important for a successful outcome in invasive hyalohyphomycosis; this, however, can be problematic. Such agents of disease are commonly encountered as laboratory contaminants. Furthermore, isolation of the fungus, particularly from respiratory tract sites, may simply represent colonization, which may or may not progress to infection depending primarily on the immunological status of the host and the virulence of the organism. Consequently, a positive culture from a nonsterile site, such as sputum, must be interpreted with caution. Nevertheless, since these infections can be rapidly devastating, isolation of the fungus must not be automatically dismissed without investigation, particularly in the immunosuppressed host. Definitive diagnosis requires the histological detection of hyphae in normally sterile body fluids and/or tissue sections, together with a positive culture whose morphology matches that seen by microscopy. However, biopsies may be contraindicated in some patients due to severe thrombocytopenia (Hennequin et al. 2002).

Microscopy

Direct microscopy may be helpful in preliminary diagnosis. It will rarely indicate which fungus is responsible, but should distinguish the zygomycetes and hyalohyphomycetes. The test is rapid and the diagnostic yield may be higher than culture alone (Denning et al. 2003). If both microscopy and culture are positive, the likelihood of invasive disease is much higher. Sputum, bronchial washings, and aspirates should be examined using 10 percent potassium hydroxide and Parker ink (1:1), Gram stain, or, preferably, a fluorescent chitin dye such as Calcofluor (to increase sensitivity). Body fluids and exudates should be concentrated by centrifugation and the sediment similarly examined microscopically.

Histology

Hyphae are difficult to detect in hematoxylin and eosin (H&E)-stained sections due to poor staining, and any delay waiting for the results of subsequent specialized fungal stains can have fatal consequences. Therefore, tissue biopsies should be stained with a fungal stain in parallel with standard stains. Periodic acid–Schiff (where cell walls will be red) or a silver stain such as Gomori methenamine silver or Grocott stain (which will stain cell walls black) is recommended. Absence of melanin in the cell wall can usually be confirmed with a melanin stain such as the Fontana–Masson stain, although unstained or H&E-stained specimens should reveal melanin pigment, if present. In histological preparations, it can be almost impossible to distinguish between the various genera of hyaline molds, and agents of hyalohyphomycosis may be misidentified as *Aspergillus* spp. Moreover, histological appearance of fungi may be altered after antifungal therapy (Walts 2001; Denning et al. 2003). Immunohistological staining and in situ hybridization have both been used on tissue sections to differentiate *Fusarium* and *Scedosporium* spp. from other hyalohyphomycetes, including *Aspergillus* spp. However, antibodies to hyaline hyphomycetes, other than *Aspergillus*, are not commercially available (Kaufman et al. 1997; Hayden et al. 2003).

Culture

Since direct microscopy or histopathology may not be reliable means of distinguishing between the various genera of hyaline molds, all specimens should be cultured using a primary isolation medium, such as Sabouraud glucose agar. Cycloheximide should be avoided since many fungal species are sensitive to this compound. Furthermore, prior antifungal therapy could reduce the yield or sensitivity of culture. On occasion, isolates may appear colonially atypical or fail to sporulate on primary isolation media. Subculturing on to less nutritionally rich agars, such as half-strength cornmeal or potato dextrose agar, may help induce sporulation. Some organisms grow rather slowly; therefore, it would be prudent to incubate cultures for 4–6 weeks, generally at 30°C.

Definitive identification of the etiologic agent can then be accomplished using macroscopic and microscopic morphology. Identification is crucial for guiding antifungal therapy since intrinsic resistance to one or more drugs has been noted in many genera (Table 36.3). Acquired resistance is less well characterized. The National Committee for Clinical Laboratory Standards (NCCLS)(National Committee for Clinical Laboratory

Table 36.3 *Summary of in vitro susceptibility patterns of common hyalohyphomycetes to licensed and investigational systemic antifungal agents*

	FCZ	ITZ	KCZ	MCZ	VOR	POS	AMB	5FC	CASP	MICA	ANID
Fusarium spp.	−	−	−	−	+	+/−	+	−	−	−	−
Scedosporium apiospermum	−	+	+/−	+	+	+	−	−	+/−	+/−	+/−
Scedosporium prolificans[a]	−	−	−	−	+/−	−	−	−	+/−	+/−	+/−
Acremonium spp.	−	−	−	−	+	+	+	−	+	?	+/−
Paecilomyces lilacinus	−	−	−	−	+	+	−	−	−	+	?
Paecilomyces variotii	−	+	+	+	+	+	+	+	+	+	?
Scopulariopsis spp.	−	−	+/−	+/−	+/−	?	−	−	+	?	?

AMB, amphotericin B; ANID, anidulafungin; CASP, caspofungin; 5FC, flucytosine; FCZ, fluconazole; ITZ, itraconazole; KCZ, ketoconazole; MCZ, miconazole; MICA, micafungin; POS, posaconazole; VOR, voriconazole.
+, susceptible; −, not susceptible; +/−, some isolates susceptible, limited susceptibility; ?, unknown.
a) Generally regarded as dematiaceous; included to enable comparison between species.

Standards 2002) has developed a standardized method, M38-A, for susceptibility testing of these organisms. The E-test method (AB Biodisk, Sweden) has also been investigated, achieving 70–100 percent agreement depending on drug and organism, with the M38 method (Pfaller et al. 2000; Espinel-Ingroff 2001). However, due to the relative infrequency of such organisms, these tests are unlikely to be clinically validated for some time. Such testing should be performed by a specialist laboratory.

Other tests

In contrast to other mold infections, including aspergillosis, *Fusarium*, *Scedosporium*, *Acremonium*, and *Paecilomyces* can be isolated relatively easily from routine blood cultures. This is most likely due to the expression of unicellular structures in infected tissues (adventitious forms), which are thought to facilitate hematological dissemination more easily than hyphal elements (Boutati and Anaissie 1997; Pontón et al. 2000). Furthermore, use of a specific fungal medium may provide a reduction in detection time compared to standard aerobic medium (Hennequin et al. 2002).

To date, no commercially available serologically based methods are available to aid in the diagnosis of any specific hyalohyphomycosis. Assays exist for the detection of 1,3-β-D-glucan, a component of most fungal cell walls, including the hyaline molds. Whilst these may be useful in diagnosis of fungal infection, they are unable to distinguish between different fungal genera (Obayashi et al. 1995). A number of molecular tests, generally polymerase chain reaction (PCR)-based, have been developed both for diagnosis of systemic mycoses from clinical samples, and for identification of fungal cultures, including *Fusarium* and *Scedosporium* spp. (Hue et al. 1999; Chen et al. 2002; Pryce et al. 2003).

PREVENTION

Nosocomial hyalohyphomycete infections have been reported. Potential reservoirs include hospital water supplies, soil of potted plants, and nearby landscaping or construction work. Contaminated drinking water or foodstuffs such as nuts and spices may also be hazardous (Summerbell et al. 1989; Berenguer et al. 1997; Anaissie et al. 2001; Shaw 2002; Dignani and Anaissie 2004). Of particular note, *Paecilomyces* spp. are resistant to standard sterilization methods and outbreaks have been associated with contaminated skin lotion (Orth et al. 1996). Every attempt should be made to ensure immunocompromised patients avoid exposure to potential environmental sources.

Fluconazole prophylaxis is routinely used in immunocompromised patients, especially bone marrow transplant recipients, and has been successful in reducing the frequency of *Candida* infections. However, a higher prevalence of mold infections at autopsy has been recorded in marrow transplant patients, possibly due to increased early post-transplantation survival (van Burik et al. 1998). Moreover, empirical treatment in febrile neutropenic patients is now generally regarded as accepted practice since therapeutic delay can result in increased morbidity and mortality. The drug of choice is usually amphotericin B. Many agents of hyalohyphomycosis are clinically resistant to this agent (Table 36.3), and invasive infection may develop regardless of therapy (Merz et al. 1988). Consequently, a high index of suspicion should be maintained amongst neutropenic patients receiving antifungal prophylaxis or empiric therapy.

PRINCIPAL HYALINE MOLDS CAUSING DEEP INFECTIONS

Fusarium species

BRIEF DESCRIPTION OF THE GENUS

Members of the genus *Fusarium* are anamorphic (asexual) filamentous organisms that propagate by producing conidia (asexual spores). Sexual states (teleomorphs) have been described in some of the fusaria.

Conidiogenous cells (termed phialides) are formed on aerial septate hyphae or short, branched conidiophores that are often in dense clusters (sporodochia). Microconidia and macroconidia are typically produced, although microconidia may be absent in some species. Some species also contain blastoconidia. In addition, hyaline or pale chlamydospores, produced intercalary or terminally, singly or in chains, may be seen in some species. In comparison to plant-dwelling strains, clinical isolates often exhibit a reduced morphology.

MORPHOLOGY

For the majority of species, colonies grow rapidly (40–80 mm within 7–10 days) although some species such as *Fusarium dimerum* are rather slower growing (less than 20 mm in 7–10 days). Initially colonies may be white but darken with age to shades of cream, purple, pink, red, or yellow (Figure 36.1). Similarly, the reverse may be pale but can darken to purple, red, green, blue, or brown as the culture matures. Sporodochia may not always be seen in culture. If present, they are often cream, orange, or tan. Colony texture can vary, but is usually floccose, woolly, or cottony.

Large smooth sickle- or canoe-shaped macroconidia (5–73×1.5–6 μm) with one to five transverse septa (Figure 36.1) are produced in basipetal fashion (growing from the base) from monophialides or sporodochia, accumulating in slimy masses. These conidia are thick-walled with a basal foot cell and an often pointed apical end. Macroconidia, typically used in identifying the genus and species, may only be apparent in the later stages of growth; thus care must be taken not to confuse young cultures, where only microconidia are evident, with *Acremonium* spp.

Microconidia are produced also in basipetal series on mono- or polyphialides, accumulating in small slimy heads or, sometimes, chains (Figure 36.1). They are generally unicellular, although two- or three-celled varieties are occasionally observed, and variable in size and shape (4–26×1.5–11 μm, smooth, oval, subspherical or pyriform, straight or curved with a rounded or flat base) (Nelson et al. 1994; Sutton et al. 1998; de Hoog et al. 2000).

Molecular methods, such as 28S rRNA gene sequencing, have been used to identify *Fusarium* spp. (Hennequin et al. 1999).

EPIDEMIOLOGY

Fusarium spp. are frequently found as saprophytes on organic debris and in soil. They are commonly implicated as plant pathogens, and indeed may cause economically devastating disease. Conversely, *Fusarium venenatum* (previously misidentified as *Fusarium graminearum*) is a commercially useful organism, being used for mycoprotein production for human consumption (Quorn).

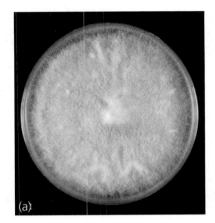

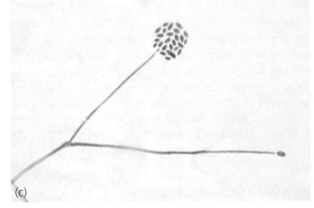

Figure 36.1 (a) *Culture of* Fusarium *sp. (Sabouraud glucose agar, 30°C, 7 days) and* **(b)** *microscopic appearance showing macroconidia and* **(c)** *microconidia. © Salford Royal Hospitals NHS Trust*

Some species are more commonly found in tropical or temperate climates, whilst others are more geographically widespread, including desert and arctic areas. Despite worldwide distribution, most cases of invasive fusariosis have been reported from the United States and Southern Europe, particularly France. Moreover, *Fusarium* spp. are more abundant in the air during the rainy summer–autumn months, coinciding with increased infection rates (Nelson et al. 1994; Boutati and Anaissie 1997; Girmenia et al. 2000).

More than 50 species of *Fusarium* have been described, the taxonomy of which is confusing and under revision. Only a small number of species, however, have been documented as causing human infection (Table 36.1). The most virulent appears to be *Fusarium solani*, which, not surprisingly, is also most commonly isolated (in approximately 50 percent of cases). *Fusarium oxysporum* and *Fusarium verticillioides* (*moniliforme*) are also commonly encountered. Other rarer species include *Fusarium proliferatum* and *F. dimerum* (Mayayo et al. 1999; Pontón et al. 2000).

Fusarium spp. are the most common cause of fungal keratitis and were previously considered to cause only corneal, skin, and nail infections (not discussed in this chapter). However, they are now recognized as having the ability to cause disseminated disease and, indeed, in some centers, have emerged as the most common non-*Aspergillus* mold pathogen (Anaissie et al. 1988; Marr et al. 2002). Disseminated fusarial infection was first described in 1973 in a child with acute leukemia (Cho et al. 1973). One US institution recorded a significant increase from 0.5 to 3.8 cases per year from 1975 to 1995 (Boutati and Anaissie 1997). Incidence rates in other centers worldwide have remained more stable (5.1 and 6.3 mean cases per year for the periods 1981–90 and 1991–96, respectively) (Girmenia et al. 2000).

Fusarium spp. are able to produce several mycotoxins. Alimentary toxic aleukia is one of the most important mycotoxicoses. During the 1930s to 1950s, this disease was widespread in Siberia due to overwintering cereal grains on the field. Upon ingestion of contaminated food, mycotoxins cause tissue breakdown and suppress cellular and humoral immunity, resulting in bone marrow failure and potentially fatal disease (Nelson et al. 1994). It is currently unknown whether such toxins are expressed in human tissue during disseminated infection. Adherence factors may also be important in catheter-related infections (Pontón et al. 2000; Dignani and Anaissie 2004).

RISK FACTORS AND CLINICAL MANIFESTATIONS

Fusarium spp. have been documented as causing infections in almost any organ, including mycetoma, keratitis, pulmonary infections, endocarditis, catheter infections, brain abscess, and fungemia (Table 36.2). Skin and soft tissue infections may be indolent, locally invasive, or manifestations of systemic disease. Toenail onychomycosis and skin lesions can both be a source of dissemination in immunosuppressed individuals; thus a thorough skin and nail examination should be performed in patients prior to commencement of immunosuppressive therapy. Endogenous endophthalmitis may also be an indication of disseminated infection (Castagnola et al. 1993; Velasco et al. 1995; Boutati and Anaissie 1997; Camin et al. 1999; Nucci et al. 2003).

Deep infections within the immunocompetent patient setting usually remain localized, and include continuous ambulatory peritoneal dialysis (CAPD)-associated peri-

tonitis, or septic arthritis and osteomyelitis, generally as a result of trauma or surgery. Disseminated infection has occurred rarely in immunocompetent patients; risk factors include those with extensive burns (Guarro and Gené 1995; Dignani and Anaissie 2004).

Risk factors for invasive disease include persistent neutropenia and graft-versus-host disease (GVHD) (Walsh et al. 2004). Patients with underlying hematological malignancies are most susceptible, with more than 90 percent of reported cases of fusariosis occurring in this population, half of which were acute leukemic patients (Pontón et al. 2000). Hematopoietic stem cell transplant recipients may be at particular risk, especially those who have received grafts with human leukocyte antigen-mismatched or unrelated donors or have undergone multiple transplants (Marr et al. 2002; Nucci 2003). Fusarial infection in solid organ transplant recipients appears to be less prevalent, with only eight cases reported until 2003, five in renal transplant recipients (Nucci 2003).

Invasive disease initially presents as persistent fever (>90 percent) refractory to broad-spectrum antibiotics (and perhaps also to antifungal drugs) in a profoundly neutropenic patient, possibly with pulmonary infiltrates and/or sinusitis (70–80 percent), similar to invasive aspergillosis. However, unlike aspergillosis (<10 percent of cases), metastatic cutaneous lesions may be seen in 70–90 percent of cases, sometimes preceding fungemia (1–10 days). Moreover, skin lesions are characteristic, evolving from subcutaneous nodules to erythematous lesions, later becoming necrotic, and may provide the sole diagnostic sign in >50 percent of patients. In further contrast to aspergillosis, routine blood cultures are positive in 40–60 percent of cases. Antemortem diagnosis is achieved in >90 percent of cases (Merz et al. 1988; Martino et al. 1994; Boutati and Anaissie 1997; Nucci and Anaissie 2002; Dignani and Anaissie 2004).

TREATMENT AND PROGNOSIS

In patients with catheter-associated infection, such as those with CAPD peritonitis, catheter removal is required, in addition to systemic antifungal treatment. Localized infection in the solid organ transplant recipient may respond to surgical resection, together with long-term antifungal therapy. Favorable response rates may be achieved with such localized infections. Nonetheless, timely treatment is crucial if disseminated infection is to be avoided (Velasco et al. 1995).

The overall mortality rate associated with invasive fusarial infection remains very high (52–70 percent). Persistent neutropenia and corticosteroid administration are significant prognostic factors. In those who do not recover from neutropenia, the death rate will be almost 100 percent, despite antifungal therapy. Attributable mortality in solid organ transplant recipients with fusariosis appears to be lower, possibly because these infections tend to be more localized (Groll and Walsh 2001; Nucci et al. 2003; Dignani and Anaissie 2004).

Optimal therapy has yet to be established. Since outcome is poor unless the neutrophil count increases, efforts to modulate the host response by administration of growth factor [granulocyte colony-stimulating factor (G-CSF) or granulocyte-macrophage colony-stimulating factor (GM-CSF)] and/or colony-stimulating factor (CSF)-stimulated granulocyte transfusion and, if possible, concomitant reduction of immune suppressive therapy may be beneficial (Spielberger et al. 1993; Boutati and Anaissie 1997; Hennequin et al. 1997).

Fusarium spp., particularly *F. solani*, are resistant to most of the established antifungal drugs (Table 36.3). Amphotericin B shows some activity in vitro, although in vivo data may be contradictory (Pujol et al. 1997; Arikan et al. 1999; Guarro et al. 1999; Ortoneda et al. 2002a). Nevertheless, satisfactory response, or at least control of the infection allowing recovery of immune function, may sometimes be achieved with high doses of conventional or, preferably, liposomal amphotericin B (Boutati and Anaissie 1997; Letscher-Bru et al. 2002).

Voriconazole has also been used to effectively treat invasive *Fusarium* infection, and recently was approved for such indications (Consigny et al. 2003; Vincent et al. 2003). This efficacy is supported by in vitro and in vivo data, although *F. solani* may be less susceptible (Clancy and Nguyen 1998; Arikan et al. 1999; Paphitou et al. 2002; Graybill et al. 2003).

Posaconazole shows modest in vitro and in vivo activity (Lozano-Chiu et al. 1999; Paphitou et al. 2002), whilst *Fusarium* spp. are intrinsically resistant to echinocandin agents (Pfaller et al. 1998; Tawara et al. 2000; Arikan et al. 2001).

Combination therapy, often amphotericin B with an azole agent, has shown promising activity in vitro and has been used clinically with some degree of success (Durand-Joly et al. 2003; Ortoneda et al. 2004).

Scedosporium species

BRIEF DESCRIPTION OF THE GENUS

The genus *Scedosporium* contains two medically important species, *Scedosporium apiospermum* and *Scedosporium prolificans*. *S. prolificans* (formerly *Scedosporium inflatum*) is regarded as dematiaceous (darkly pigmented), but has been included within this chapter to ease comparison between the two species.

S. apiospermum is the asexual state of teleomorph *Pseudallescheria boydii*, which is homothallic (sexual reproduction requiring only a single organism). The synanamorph *Graphium eumorphum* may also exist in some strains.

The sexual form of *S. prolificans* has been described as belonging to the genus *Petriella*.

MORPHOLOGY

Colonies of *S. apiospermum* spread rapidly (40 mm within 7–10 days), initially white but becoming dark gray or brown, with a hairy, woolly, or cottony texture (Figure 36.2). A pale reverse is initially seen, developing brown or black areas.

The teleomorph *P. boydii* produces spherical brown fruiting bodies, 140–200 µm in diameter, known as cleistothecia, which contain asci and ascospores. However, it is rare for clinical strains to produce these sexual structures; prolonged incubation may be required to induce their formation. In the asexual form, cylindrical conidiogenous cells (termed annellides) arise from branching septate hyphae (2–4 µm in diameter) and produce slimy masses of conidia. Conidia (6–12×3.5–6 µm) are unicellular, smooth, and oval with a flat base, initially colorless or pale becoming brown and thick walled. Conidia are sometimes produced directly on the hyphae (sessile). In the *Graphium* state, conidiophores are sometimes bundled together, forming synnemata, producing clusters of conidia at their apical tip (Figure 36.2). Both asexual forms may exist in a single isolate.

Colonies of *S. prolificans* also grow rapidly (20 mm within 7–10 days), initially with a moist, almost yeast-like texture, later developing tufts of aerial hyphae. The

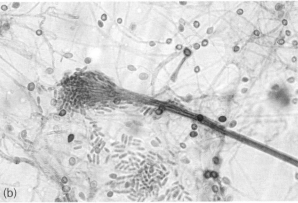

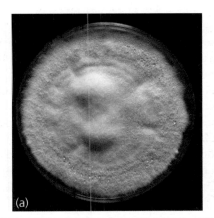

Figure 36.2 (a) *Culture of* Scedosporium apiospermum *(Sabouraud glucose agar, 30°C, 14 days) and* **(b)** *microscopic appearance showing synnemata (*Graphium *state).* © *Salford Royal Hospitals NHS Trust*

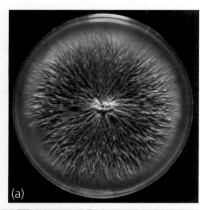

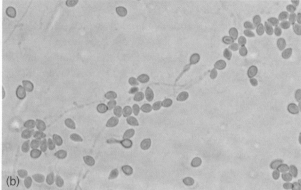

Figure 36.3 (a) *Culture of* Scedosporium prolificans *(Sabouraud glucose agar, 30°C, 14 days) and* **(b)** *microscopic appearance.* © *Salford Royal Hospitals NHS Trust*

colony color may be olive-gray or black with a gray to black reverse, more darkly pigmented than that of *S. apiospermum* (Figure 36.3).

Conidiogenous cells of *S. prolificans* are characteristically flask-shaped with an inflated base, often with an elongated tip (Figure 36.3). Single-celled, smooth, oval, flat-based, olive-brown conidia (3–7×2.5 μm) accumulate in slimy heads. Sessile conidia, which are darker and rounder, may also be seen (de Hoog et al. 2000).

S. apiospermum and *S. prolificans* also differ in their physiology: unlike *S. prolificans*, *S. apiospermum* is able to assimilate ribitol, xylitol, and L-arabinitol. Furthermore, *S. apiospermum* can grow on cycloheximide-containing media (e.g. Mycosel), whereas *S. prolificans* is inhibited by the compound. Molecular methods have also been used to discriminate between the two species (Wedde et al. 1998).

EPIDEMIOLOGY

Scedosporium spp. occur worldwide. *S. apiospermum* is frequently found as a saprophyte in soil, sewage, manure, and contaminated water. The natural habitat of *S. prolificans* appears to be soil, while it has been isolated from the hospital environment during building works. Both species have historically been important causes of subcutaneous infection in immunocompetent individuals (not discussed here), normally following

minor trauma such as splinters or thorns. They are, however, becoming increasingly implicated in invasive disease. In vivo work suggests that *S. prolificans* is more virulent than *S. apiospermum*. Interestingly, most cases of disseminated *S. prolificans* infection have been recorded in Spain and Australia; the reasons for this distribution remain unclear (Cano et al. 1992; Berenguer et al. 1997; Bouza and Muñoz 2004).

RISK FACTORS AND CLINICAL MANIFESTATIONS

Scedosporium infection may involve almost any organ (Table 36.2). Reports of *S. apiospermum* infection include invasive pulmonary disease, sinusitis, brain abscess, endocarditis, fungemia, mycetoma, and peritonitis (Jabado et al. 1998; Nesky et al. 2000; Castiglioni et al. 2002; Raj and Frost 2002; Bouza and Muñoz 2004). *S. prolificans* has also caused renal and splenic abscess, in addition to pulmonary infections, endocarditis, fungemia, and many other manifestations (Marin et al. 1991; Carreter de Granda et al. 2001; Simarro et al. 2001; Bouza and Muñoz 2004).

For both species, manifestations in immunocompetent individuals are usually localized, and include subcutaneous infections, septic arthritis, osteomyelitis, and sinusitis. Pulmonary infection and endocarditis in the normal host have also been reported (Sobottka et al. 1999; Greig et al. 2001; Tirado-Miranda et al. 2001; Levine et al. 2002). The main risk factors are surgery or trauma, although intravenous drug use has also been reported as such (Berenguer et al. 1997; Bouza and Muñoz 2004). *S. apiospermum* has caused fatal pulmonary, cerebral, and disseminated infection in near-drowning victims in polluted water and sewage (Dworzack et al. 1989). In addition, either species may colonize the airway of cystic fibrosis patients, triggering allergic disease in some (Cimon et al. 2000).

Patients at high risk of disseminated disease include those receiving corticosteroids or with prolonged neutropenia. Indeed, at least 90 percent of cases are associated with persistent neutropenia. Those with hematological malignancy, particularly acute leukemia, or hematopoietic stem cell transplant recipients are especially at risk (Berenguer et al. 1997; Revankar et al. 2002). The incidence of *S. apiospermum* infection in solid organ transplant recipients is approximately 0.1 percent, with lung transplant recipients most at risk (Castiglioni et al. 2002).

Clinical signs of invasive infection are typically similar to those produced by *Aspergillus* or *Fusarium*: unremitting fever (>90 percent) unresponsive to broad-spectrum antibiotics (and possibly also to antifungal agents) in a persistently neutropenic patient, followed by pulmonary involvement (approximately 75 percent). Indeed, with both *Scedosporium* spp. the lung is the most common site of infection in immunocompromised patients. Frequent dissemination to other organs is seen, with a particular propensity for the central nervous system.

Neurological symptoms may be noted (approximately 30–40 percent for *S. prolificans*, possibly even higher with *S. apiospermum*), in addition to widespread metastatic cutaneous lesions (25 percent), which typically appear as nodules, possibly becoming necrotic. Metastatic eye involvement may also occur (approximately 10 percent). Dissemination is often seen in solid organ transplant patients, in contrast to fusarial infection in such patients (Berenguer et al. 1997; Bouza and Muñoz 2004). Positive blood cultures are obtained in approximately 80 percent of *S. prolificans* infections (Revankar et al. 2002), whereas the isolation rate is much lower for *S. apiospermum*. Occasional case reports suggest that indwelling central venous catheters may also serve as a portal of entry, particularly with *S. prolificans* (Berenguer et al. 1997).

TREATMENT AND PROGNOSIS

Successful outcomes are often achieved in immunocompetent patients since disease tends to be more localized. A maximal dose of antifungal treatment in combination with surgical debridement and/or drainage, if appropriate, is usually the therapy of choice (Wood et al. 1992; Berenguer et al. 1997).

On the contrary, in the immunocompromised host, *Scedosporium* usually results in a rapidly disseminated infection that is almost universally lethal (85–100 percent) (Wood et al. 1992; Berenguer et al. 1997; Castiglioni et al. 2002; Marr et al. 2002; Revankar et al. 2002). These data indicate that resolution of neutropenia is absolutely crucial for survival, particularly since these organisms are resistant to many antifungal drugs (Table 36.3).

Therapeutic choices are very limited. As with *Fusarium* infections, administration of hematopoietic growth factors may help with neutropenic resolution. Immunosuppressants should also be reduced if clinically justifiable (Idigoras et al. 2001).

S. prolificans is intrinsically resistant to amphotericin B. Most *S. apiospermum* isolates are resistant in vitro and treatment failures have been widely reported (Ellis 2002; Meletiadis et al. 2002). Indeed, *Scedosporium* infection can often develop during amphotericin B treatment (Berenguer et al. 1997; Safdar et al. 2002). Interestingly, in vitro studies have shown that antifungal drugs, including amphotericin B, cooperate with polymorphonuclear leukocytes to cause increased damage to *Scedosporium* hyphae (Gil-Lamaignere et al. 2002). Moreover, combination therapy of liposomal amphotericin B and G-CSF prolonged survival in an in vivo disseminated *S. prolificans* model (Ortoneda et al. 2002b).

Itraconazole and miconazole both show some activity in vitro and have been used clinically with some degree of success against *S. apiospermum*, but rarely so with *S. prolificans* infections. Intravenous miconazole, however, has toxic side effects (Nomdedeu et al. 1993; Cuenca-

Estrella et al. 1999; Meletiadis et al. 2002; Bouza and Muñoz 2004).

Voriconazole was recently indicated for *Scedosporium* infections, and numerous case studies have demonstrated efficacy against *S. apiospermum* (Girmenia et al. 1998; Muñoz et al. 2000; Nesky et al. 2000; Poza et al. 2000). In vitro data confirm activity against *S. apiospermum*, and to a lesser extent, *S. prolificans* (Cuenca-Estrella et al. 1999; Carrillo and Guarro 2001).

Posaconazole activity against *S. prolificans* is poor. However, in vitro and in vivo responses against *S. apiospermum* have been noted (Espinel-Ingroff 1998; Carrillo and Guarro 2001; Meletiadis et al. 2002; González et al. 2003). In addition, Mellinghoff et al. (2002) report successful posaconazole treatment of *S. apiospermum* brain abscesses in a leukemic patient, after failure of itraconazole, amphotericin B, and ketoconazole.

S. apiospermum and *S. prolificans* appear moderately susceptible in vitro to the echinocandin agents, but no clinical data exist (Espinel-Ingroff 1998; Uchida et al. 2000).

Combination antifungal therapy may provide additional therapeutic options. Voriconazole in combination with terbinafine has been used successfully to treat disseminated *S. prolificans* infection (Howden et al. 2003). In vitro studies confirm synergy with azole/terbinafine combinations (Meletiadis et al. 2003). In addition, liposomal amphotericin B combined with itraconazole produced a favorable outcome in *S. apiospermum* infection, but not *S. prolificans* (Barbaric and Shaw 2001).

Other species

Many other species, particularly *Acremonium*, *Paecilomyces*, and *Scopulariopsis*, have been documented rarely as causing deep tissue or disseminated infection (Table 36.1).

MORPHOLOGY

The growth rate of *Acremonium* spp. is moderate (10–30 mm in 7 days), generally slower than *Fusarium* spp. Colonies are flat, occasionally raised in the center, glabrous to velvety and may be pink, yellow, or white with a colorless, pinkish, or pale brown reverse. Microconidia ($3–10 \times 1.5–3$ μm) accumulate at the tips of narrow tapering phialides (de Hoog et al. 2000).

Paecilomyces spp. grow rapidly (30–50 mm in 7 days), initially white, becoming brownish or violet, with a powdery to floccose texture. Conidiophores with characteristic tapering phialides and chains of conidia are observed (de Hoog et al. 2000).

Colonies of *Scopulariopsis* spp. are typically white, becoming light or dark brown, with a cream to brown reverse. They grow moderately quickly (30 mm in 7 days) with a powdery to velvety texture. Conidiophores, annellides, and often rough-walled conidia arranged in chains are seen (de Hoog et al. 2000).

EPIDEMIOLOGY

Acremonium, *Paecilomyces*, and *Scopulariopsis* spp. inhabit soil and plant material; some species of the latter two can also be isolated from insects. All three genera occur worldwide. They may be encountered as contaminants, but are also capable of causing infection, especially in the immunocompromised host.

Each genus contains a number of species, although only a few have been reported as human pathogens (Table 36.1). Most *Acremonium* infections have been caused by *Acremonium kiliense* or *A. strictum*. Other species include *Acremonium alabamense*, *Acremonium falciforme*, and *Acremonium recifei*. There are two medically prevalent species of *Paecilomyces*, *Paecilomyces lilacinus* and *Paecilomyces variotii*. Other species include *Paecilomyces javanicus* and *Paecilomyces marquandii*. *Scopulariopsis* infections are primarily caused by *Scopulariopsis brevicaulis*, although other species have been noted, including *Scopulariopsis acremonium*, and *Scopulariopsis candida* (Herbrecht et al. 2002; Steinbach et al. 2004; Walsh et al. 2004).

RISK FACTORS AND CLINICAL MANIFESTATIONS

Acremonium, *Paecilomyces*, and *Scopulariopsis* spp. have been documented as causing a wide spectrum of disease (Table 36.2).

In the immunocompetent patient, ocular infection (not discussed in this chapter) is the main clinical manifestation of *Acremonium* and *Paecilomyces* spp., typically following traumatic or surgical injury. *Acremonium* is one of the main causative agents of mycetoma, whereas the most common presentation of *Scopulariopsis* infection is onychomycosis, generally of the toenail (neither discussed) (Guarro et al. 1997; Chan-Tack et al. 1999; Steinbach et al. 2004). Nonetheless, *Acremonium* and *Paecilomyces* have been reported as causing peritonitis, endocarditis, and sinusitis (Guarro et al. 1997; Heitmann et al. 1997; Aguilar et al. 1998; Kovac et al. 1998; Nayak et al. 2000; Manzano-Gayosso et al. 2003). Endocarditis caused by *Scopulariopsis* has also been documented (Chen-Scarabelli and Scarabelli 2003).

In the immunocompromised individual, the spectrum of disease includes fungemia and pulmonary and cerebral infections (Boltansky et al. 1984; Guarro et al. 1997; Sillevis Smitt et al. 1997; Sellier et al. 2000; Nedret Koç et al. 2002; Kantarcioğlu et al. 2003; Steinbach et al. 2004). Similar to *Fusarium* and *Scedosporium* spp., predisposing factors for such invasive disease includes persistent neutropenia and corticosteroid use. Hematopoietic stem cell and organ transplant recipients are most at risk (Guarro et al. 1997; Steinbach et al. 2004; Walsh et al. 2004). Furthermore, clinical presentation is also comparable to other invasive hyalohyphomycoses. Cutaneous

lesions may be the first indication of infection, although blood cultures are often positive during *Acremonium* and *Paecilomyces* infections (Guarro et al. 1997; Pontón et al. 2000; Steinbach et al. 2004).

TREATMENT AND PROGNOSIS

Since these infections occur rarely, few clinical data exist to confirm optimal treatment. Localized infection may be resolved by surgery and antifungal therapy. On the contrary, complete recovery from disseminated infection depends almost entirely on recovery of the neutrophil count. Choice of antifungal drug effective against these agents is limited (Table 36.3); hence combination therapy has generally been utilized.

Amphotericin B, often in combination with an azole, has typically been used against *Acremonium* infection. In vitro activity is seen with some isolates, although clinical response is variable. Higher doses may improve the success rate (Guarro et al. 1997; Groll and Walsh 2001). Voriconazole and posaconazole are active in vitro against some species. More importantly, clinical successes have been reported in patients with hematological malignancies who had failed amphotericin B therapy (Herbrecht et al. 2002; Mattei et al. 2003). Of the limited in vitro data available, caspofungin appears active against *Acremonium* spp., though no clinical data exist (Espinel-Ingroff 1998; Pfaller et al. 1998).

Species identification of *Paecilomyces* is important due to variations in antifungal susceptibility (Table 36.3). *P. lilacinus* is resistant in vitro to amphotericin B and itraconazole, whilst *P. variotii* is susceptible (Radford et al. 1997; Aguilar et al. 1998; Espinel-Ingroff et al. 2001). Amphotericin B, either alone or in combination with other agents, has been the most widely used regimen against *Paecilomyces* infections. However, a 40 percent failure rate is seen (Aguilar et al. 1998). Voriconazole and posaconazole are active in vitro against some *Paecilomyces* spp. (Radford et al. 1997; Wildfeuer et al. 1998; Espinel-Ingroff et al. 2001; Uchida et al. 2001). Furthermore, voriconazole has been used with partial success against *P. lilacinus* skin and soft tissue infections in immunocompromised individuals who had failed amphotericin B and/or itraconazole therapy (Hilmarsdóttir et al. 2000; Martin et al. 2002). Caspofungin appears active in vitro against *P. variotii*, although *P. lilacinus* is resistant (Del Poeta et al. 1997). Interestingly, caspofungin in combination with itraconazole was used successfully against progressive cutaneous *P. lilacinus* infection (Safdar 2002). Micafungin shows good activity in vitro against both *P. variotii* and *P. lilacinus*, although this has not been substantiated clinically (Uchida et al. 2000). Terbinafine has been used successfully in a soft tissue infection caused by *P. lilacinus* in a heart transplant recipient (Clark 1999).

Historically, amphotericin B has been the therapy of choice in invasive *Scopulariopsis* infections, but is often ineffective (approximately 60 percent), correlating with in vitro resistance (Aguilar et al. 1999; Steinbach et al. 2004). Itraconazole also shows poor in vitro activity, although a few case studies report success (Radford et al. 1997; Espinel-Ingroff et al. 2001; Steinbach et al. 2004). Variable in vitro responses to voriconazole and clinical failure of voriconazole combined with caspofungin have been reported (Wildfeuer et al. 1998; Espinel-Ingroff et al. 2001; Cuenca-Estrella et al. 2003; Steinbach et al. 2004). In vitro and clinical response to caspofungin was documented against *Scopulariopsis* endocarditis (Chen-Scarabelli and Scarabelli 2003). Recurrent subcutaneous infection caused by *S. brevicaulis* in a liver transplant recipient was controlled with surgery and long-term terbinafine therapy (Sellier et al. 2000).

REFERENCES

Aguilar, C., Pujol, I., et al. 1998. Antifungal susceptibilities of *Paecilomyces* species. *Antimicrob Agents Chemother*, **42**, 1601–4.

Aguilar, C., Pujol, I. and Guarro, J. 1999. In vitro antifungal susceptibilities of *Scopulariopsis* isolates. *Antimicrob Agents Chemother*, **43**, 1520–2.

Ajello, L. 1982. Hyalohyphomycosis: a disease entity whose time has come. In: Newsletter. *Med Mycol Soc N Y*, **20**, 3–5.

Anaissie, E.J., Kantarjian, H., et al. 1988. The emerging role of *Fusarium* infections in patients with cancer. *Medicine (Baltimore)*, **67**, 77–83.

Anaissie, E.J., Kuchar, R.T., et al. 2001. Fusariosis associated with pathogenic *Fusarium* species colonization of a hospital water system: a new paradigm for the epidemiology of opportunistic mold infections. *Clin Infect Dis*, **33**, 1871–8.

Arikan, S., Lozano-Chiu, M., et al. 1999. Microdilution susceptibility testing of amphotericin B, itraconazole, and voriconazole against clinical isolates of *Aspergillus* and *Fusarium* species. *J Clin Microbiol*, **37**, 3946–51.

Arikan, S., Lozano-Chiu, M., et al. 2001. In vitro susceptibility testing methods for caspofungin against *Aspergillus* and *Fusarium* isolates. *Antimicrob Agents Chemother*, **45**, 327–30.

Barbaric, D. and Shaw, P.J. 2001. *Scedosporium* infection in immunocompromised patients: successful use of liposomal amphotericin B and itraconazole. *Med Pediatr Oncol*, **37**, 122–5.

Berenguer, J., Rodriguez-Tudela, J.L., et al. 1997. Deep infections caused by *Scedosporium prolificans*: a report on 16 cases in Spain and a review of the literature. *Medicine (Baltimore)*, **76**, 256–65.

Boltansky, H., Kwon-Chung, K.J., et al. 1984. *Acremonium strictum*-related pulmonary infection in a patient with chronic granulomatous disease. *J Infect Dis*, **149**, 653.

Boutati, E.I. and Anaissie, E.J. 1997. *Fusarium*, a significantly emerging pathogen in patients with hematologic malignancy: ten years' experience at a cancer center and implications for management. *Blood*, **90**, 999–1008.

Bouza, E. and Muñoz, P. 2004. Invasive infections caused by *Blastoschizomyces capitatus* and *Scedosporium* spp. *Clin Microbiol Infect*, **10**, Suppl 1, 76–85.

Camin, A.-M., Michelet, C., et al. 1999. Endocarditis due to *Fusarium dimerum* four years after coronary artery bypass grafting. *Clin Infect Dis*, **28**, 150.

Cano, J., Guarro, J., et al. 1992. Experimental infection with *Scedosporium inflatum*. *J Med Vet Mycol*, **30**, 413–20.

Carreter de Granda, M.E., Richard, C., et al. 2001. Endocarditis caused by *Scedosporium prolificans* after autologous peripheral blood stem cell transplantation. *Eur J Clin Microbiol Infect Dis*, **20**, 215–17.

Carrillo, A.J. and Guarro, J. 2001. In vitro activities of four novel triazoles against *Scedosporium* spp. *Antimicrob Agents Chemother*, **45**, 2151–5.

Castagnola, E., Garaventa, A., et al. 1993. Survival after fungemia due to *Fusarium moniliforme* in a child with neuroblastoma. *Eur J Clin Microbiol Infect Dis*, **12**, 308–9.

Castiglioni, B., Sutton, D.A., et al. 2002. *Pseudallescheria boydii* (anamorph *Scedosporium apiospermum*) infection in solid organ transplant recipients in a tertiary medical center and review of the literature. *Medicine (Baltimore)*, **81**, 333–48.

Chan-Tack, K.M., Thio, C.L., et al. 1999. *Paecilomyces lilacinus* fungemia in an adult bone marrow transplant recipient. *Med Mycol*, **37**, 57–60.

Chen, S.C.A., Halliday, C.L. and Meyer, W. 2002. A review of nucleic acid-based diagnostic tests for systemic mycoses with an emphasis on polymerase chain reaction-based assays. *Med Mycol*, **40**, 333–357.

Chen-Scarabelli, C. and Scarabelli, T.M. 2003. Fungal endocarditis due to *Scopulariopsis*. *Ann Intern Med*, **139**, E–795.

Cho, C.T., Vats, T.S., et al. 1973. *Fusarium solani* infection during treatment for acute leukemia. *J Pediatr*, **83**, 1028–31.

Cimon, B., Carrere, J., et al. 2000. Clinical significance of *Scedosporium apiospermum* in patients with cystic fibrosis. *Eur J Clin Microbiol Infect Dis*, **19**, 53–6.

Clancy, C.J. and Nguyen, H.M. 1998. In vitro efficacy and fungicidal activity of voriconazole against *Aspergillus* and *Fusarium* species. *Eur J Clin Microbiol Infect Dis*, **17**, 573–5.

Clark, N.M. 1999. *Paecilomyces lilacinus* infection in a heart transplant recipient and successful treatment with terbinafine. *Clin Infect Dis*, **28**, 1169–70.

Consigny, S., Dhedin, N., et al 2003. Successful voriconazole treatment of disseminated *Fusarium* infection in an immunocompromised patient. *Clin Infect Dis*, **37**, 311–13.

Cuenca-Estrella, M., Ruiz-Díez, B., et al. 1999. Comparative *in-vitro* activity of voriconazole (UK-109,496) and six other antifungal agents against clinical isolates of *Scedosporium prolificans* and *Scedosporium apiospermum*. *Antimicrob Agents Chemother*, **43**, 149–51.

Cuenca-Estrella, M., Gomez-Lopez, A., et al. 2003. *Scopulariopsis brevicaulis*, a fungal pathogen resistant to broad-spectrum antifungal agents. *Antimicrob Agents Chemother*, **47**, 2339–41.

de Hoog, G.S., Guarro, J., et al. 2000. *Atlas of clinical fungi*. Utrecht: Centraalbureau voor Schimmelcultures.

Del Poeta, M., Schell, W. and Perfect, J.R. 1997. In vitro antifungal activity of pneumocandin L-743,872 against a variety of clinically important molds. *Antimicrob Agents Chemother*, **41**, 1835–6.

Denning, D.W., Kibbler, C.C. and Barnes, R.A. 2003. British Society for Medical Mycology proposed standards of care for patients with invasive fungal infections. *Lancet Infect Dis*, **4**, 230–40.

Dignani, M.C. and Anaissie, E. 2004. Human fusariosis. *Clin Microbiol Infect*, **1)**, Suppl 1, 67–75.

Durand-Joly, L., Alfandari, S., et al. 2003. Successful outcome of disseminated *Fusarium* infection with skin localization treated with voriconazole and amphotericin B-lipid complex in a patient with acute leukemia. *J Clin Microbiol*, **41**, 4898–900.

Dworzack, D.L., Clark, R.B., et al. 1989. *Pseudallescheria boydii* brain abscess: association with near drowning and efficacy of high-dose, prolonged miconazole therapy in patients with multiple abscesses. *Medicine (Baltimore)*, **68**, 218–24.

Ellis, D. 2002. Amphotericin B: spectrum and resistance. *J Antimicrob Chemother*, **49**, Suppl S1, 7–10.

Espinel-Ingroff, A. 1998. Comparison of *in vitro* activities of the new triazole SCH-56592 and the echinocandins MK-0991 (L-743,872) and LY303366 against opportunistic filamentous and dimorphic fungi and yeasts. *J Clin Microbiol*, **36**, 2950–6.

Espinel-Ingroff, A. 2001. Comparison of the E-test with the NCCLS M38-P method for antifungal susceptibility testing of common and emerging pathogenic filamentous fungi. *J Clin Microbiol*, **39**, 1360–7.

Espinel-Ingroff, A., Boyle, K. and Sheehan, D.J. 2001. In vitro antifungal activities of voriconazole and reference agents as determined by NCCLS methods: review of the literature. *Mycopathologia*, **150**, 101–15.

Gil-Lamaignere, C., Roilides, E., et al. 2002. Amphotericin B lipid complex exerts additive antifungal activity in combination with polymorphonuclear leukocytes against *Scedosporium prolificans* and *Scedosporium apiospermum*. *J Antimicrob Chemother*, **50**, 1027–30.

Girmenia, C., Luzi, G., et al. 1998. Use of voriconazole in treatment of *Scedosporium apiospermum* infection: case report. *J Clin Microbiol*, **36**, 1436–8.

Girmenia, C., Pagano, L., et al. 2000. The epidemiology of fusariosis in patients with haematological disease. *Br J Haematol*, **111**, 272–6.

González, G.M., Tijerina, R., et al. 2003. Activity of posaconazole against *Pseudallescheria boydii*: in vitro and in vivo assays. *Antimicrob Agents Chemother*, **47**, 1436–8.

Graybill, J.R., Najvar, L.K., et al. 2003. Improving the mouse model for studying the efficacy of voriconazole. *J Antimicrob Chemother*, **51**, 1373–6.

Greig, J.R., Khan, M.A., et al. 2001. Pulmonary infection with *Scedosporium prolificans* in an immunocompetent individual. *J Infect*, **43**, 15–17.

Groll, A.H. and Walsh, T.J. 2001. Uncommon opportunistic fungi: new nosocomial threats. *Clin Microbiol Infect*, **7**, Suppl 2, 8–24.

Guarro, J. and Gené, J. 1995. Opportunistic fusarial infections in humans. *Eur J Clin Microbiol Infect Dis*, **14**, 741–54.

Guarro, J., Gams, W., et al. 1997. *Acremonium* species: new emerging fungal opportunists–in vitro antifungal susceptibilities and review. *Clin Infect Dis*, **25**, 1222–9.

Guarro, J., Pujol, I. and Mayayo, E. 1999. In vitro and in vivo experimental activities of antifungal agents against *Fusarium solani*. *Antimicrob Agents Chemother*, **43**, 1256–7.

Hayden, R.T., Isotalo, P.A., et al. 2003. In situ hybridization for the differentiation of *Aspergillus*, *Fusarium* and *Pseudallescheria* species in tissue section. *Diagn Mol Pathol*, **12**, 21–6.

Heitmann, L., Cometta, A., et al. 1997. Right-sided pacemaker-related endocarditis due to *Acremonium* species. *Clin Infect Dis*, **25**, 158–60.

Hennequin, C., Lavarde, V., et al. 1997. Invasive *Fusarium* infections: a retrospective survey of 31 cases. *J Med Vet Mycol*, **35**, 107–14.

Hennequin, C., Abachin, E., et al. 1999. Identification of *Fusarium* species involved in human infections by 28S rRNA gene sequencing. *J Clin Microbiol*, **37**, 3586–9.

Hennequin, C., Ranaivoarimalala, C., et al. 2002. Comparison of aerobic standard medium with specific fungal medium for detecting *Fusarium* spp. in blood cultures. *Eur J Clin Microbiol Infect Dis*, **21**, 748–50.

Herbrecht, R., Letscher-Bru, V., et al. 2002. *Acremonium strictum* pulmonary infection in a leukemic patient successfully treated with posaconazole after failure of amphotericin B. *Eur J Clin Microbiol Infect Dis*, **21**, 814–17.

Hilmarsdóttir, I., Thorsteinsson, S.B., et al. 2000. Cutaneous infection caused by *Paecilomyces lilacinus* in a renal transplant patient: treatment with voriconazole. *Scand J Infect Dis*, **32**, 331–2.

Howden, B.P., Slavin, M.A., et al. 2003. Successful control of disseminated *Scedosporium prolificans* infection with a combination of voriconazole and terbinafine. *Eur J Clin Microbiol Infect Dis*, **22**, 111–13.

Hue, F.-X., Huerre, M., et al. 1999. Specific detection of *Fusarium* species in blood and tissues by a PCR technique. *J Clin Microbiol*, **37**, 2434–8.

Idigoras, P., Pérez-Trallero, E., et al. 2001. Disseminated infection and colonization by *Scedosporium prolificans*: a review of 18 cases, 1990–1999. *Clin Infect Dis*, **32**, e158–65.

Jabado, N., Casanova, J.-L., et al. 1998. Invasive pulmonary infection due to *Scedosporium apiospermum* in two children with chronic granulomatous disease. *Clin Infect Dis*, **27**, 1437–41.

Kantarcioğlu, A.S., Hatemi, G., et al. 2003. *Paecilomyces variotii* central nervous system infection in a patient with cancer. *Mycoses*, **46**, 45–50.

Kaufman, L., Standard, P.G., et al. 1997. Immunohistologic identification of *Aspergillus* spp. and other hyaline fungi by using polyclonal fluorescent antibodies. *J Clin Microbiol*, **35**, 2206–9.

Kovac, D., Lindic, J., et al. 1998. Treatment of severe *Paecilomyces variotii* peritonitis in a patient on continuous ambulatory peritoneal dialysis. *Nephrol Dial Transplant*, **13**, 2943–6.

Larone, D.H. 2002. *Medically important fungi: a guide to identification*. Washington, DC: ASM Press.

Letscher-Bru, V., Campos, F., et al. 2002. Successful outcome of treatment of a disseminated infection due to *Fusarium dimerum* in a leukemia patient. *J Clin Microbiol*, **40**, 1100–2.

Levine, N.B., Kurokawa, R., et al. 2002. An immunocompetent patient with primary *Scedosporium apiospermum* vertebral osteomyelitis. *J Spinal Disord Tech*, **15**, 425–30.

Lozano-Chiu, M., Arikan, S., et al. 1999. Treatment of murine fusariosis with SCH 56592. *Antimicrob Agents Chemother*, **43**, 589–91.

Manzano-Gayosso, P., Hernández-Hernández, F., et al. 2003. Fungal peritonitis in 15 patients on continuous ambulatory peritoneal dialysis (CAPD). *Mycoses*, **46**, 425–9.

Marin, J., Sanz, M.A., et al. 1991. Disseminated *Scedosporium inflatum* infection in a patient with acute myeloblastic leukemia. *Eur J Clin Microbiol Infect Dis*, **10**, 759–61.

Marr, K.A., Carter, R.A., et al. 2002. Epidemiology and outcome of mold infections in hematopoietic stem cell transplant recipients. *Clin Infect Dis*, **34**, 909–17.

Martin, C.A., Roberts, S. and Greenberg, R.N. 2002. Voriconazole treatment of disseminated *Paecilomyces* infection in a patient with acquired immunodeficiency syndrome. *Clin Infect Dis*, **35**, e78–81.

Martino, P., Gastaldi, R., et al. 1994. Clinical patterns of *Fusarium* infections in immunocompromised patients. *J Infect*, **28**, Suppl 1, 7–15.

Matsumoto, T., Ajello, L., et al. 1994. Developments in hyalohyphomycosis and phaeohyphomycosis. *J Med Vet Mycol*, **32**, Suppl 1, 329–49.

Mattei, D., Mordini, N., et al. 2003. Successful treatment of *Acremonium* fungemia with voriconazole. *Mycoses*, **46**, 511–14.

Mayayo, E., Pujol, I. and Guarro, J. 1999. Experimental pathogenicity of four opportunistic *Fusarium* species in a murine model. *J Med Microbiol*, **48**, 363–6.

Meletiadis, J., Meis, J.F.G.M., et al. 2002. In vitro activities of new and conventional antifungal agents against clinical *Scedosporium* isolates. *Antimicrob Agents Chemother*, **46**, 62–8.

Meletiadis, J., Mouton, J.W., et al. 2003. In vitro drug interaction modeling of combinations of azoles with terbinafine against clinical *Scedosporium prolificans* isolates. *Antimicrob Agents Chemother*, **47**, 106–17.

Mellinghoff, I.K., Winston, D.J., et al. 2002. Treatment of *Scedosporium apiospermum* brain abscesses with posaconazole. *Clin Infect Dis*, **34**, 1648–50.

Merz, W.G., Karp, J.E., et al. 1988. Diagnosis and successful treatment of fusariosis in the compromised host. *J Infect Dis*, **158**, 1046–55.

Muñoz, P., Marín, M., et al. 2000. Successful outcome of *Scedosporium apiospermum* disseminated infection treated with voriconazole in a patient receiving corticosteroid therapy. *Clin Infect Dis*, **31**, 1499–501.

National Committee for Clinical Laboratory Standards. 2002. *Reference method for broth dilution antifungal susceptibility testing of filamentous fungi*. Approved standard. Document M38-A. Wayne, PA: National Committee for Clinical Laboratory Standards.

Nayak, D.R., Balakrishnan, R., et al. 2000. *Paecilomyces* fungus infection of the paranasal sinuses. *Int J Pediatr Otorhinolaryngol*, **52**, 183–7.

Nedret Koç, A., Erdem, F. and Patiroğlu, T. 2002. Case report. *Acremonium falciforme* fungemia in a patient with acute leukaemia. *Mycoses*, **45**, 202–3.

Nelson, P.E., Dignani, M.C. and Anaissie, E.J. 1994. Taxonomy, biology, and clinical aspects of *Fusarium* species. *Clin Microbiol Rev*, **7**, 479–504.

Nesky, M.A., McDougal, E.C. and Peacock, J.E. Jr 2000. *Pseudallescheria boydii* brain abscess successfully treated with voriconazole and surgical drainage: case report and literature review of central nervous system pseudallescheriasis. *Clin Infect Dis*, **31**, 673–7.

Nomdedeu, J., Brunet, S., et al. 1993. Successful treatment of pneumonia due to *Scedosporium apiospermum* with itraconazole. *Clin Infect Dis*, **16**, 731–3.

Nucci, M. 2003. Emerging moulds: *Fusarium*, *Scedosporium* and Zygomycetes in transplant recipients. *Curr Opin Infect Dis*, **16**, 607–12.

Nucci, M. and Anaissie, E. 2002. Cutaneous infection by *Fusarium* species in healthy and immunocompromised hosts: implications for diagnosis and management. *Clin Infect Dis*, **35**, 909–20.

Nucci, M., Anaissie, E.J., et al. 2003. Outcome predictors of 84 patients with hematologic malignancies and *Fusarium* infection. *Cancer*, **98**, 315–19.

Obayashi, T., Yoshida, M., et al. 1995. Plasma $(1\rightarrow3)$-β-D-glucan measurement in diagnosis of invasive deep mycosis and fungal febrile episodes. *Lancet*, **345**, 17–20.

Orth, B., Frei, R., et al. 1996. Outbreak of invasive mycoses caused by *Paecilomyces lilacinus* from a contaminated skin lotion. *Ann Intern Med*, **125**, 799–806.

Ortoneda, M., Capilla, J., et al. 2002a. Efficacy of liposomal amphotericin B in treatment of systemic murine fusariosis. *Antimicrob Agents Chemother*, **46**, 2273–5.

Ortoneda, M., Capilla, J., et al. 2002b. Liposomal amphotericin B and granulocyte colony-stimulating factor therapy in a murine model of invasive infection by *Scedosporium prolificans*. *J Antimicrob Chemother*, **49**, 525–9.

Ortoneda, M., Capilla, J., et al. 2004. In vitro interactions of licensed and novel antifungal drugs against *Fusarium* spp. *Diagn Microbiol Infect Dis*, **48**, 69–71.

Paphitou, N.I., Ostrosky-Zeichner, L., et al. 2002. In vitro activities of investigational triazoles against *Fusarium* species: effects of inoculum size and incubation time on broth microdilution susceptibility test results. *Antimicrob Agents Chemother*, **46**, 3298–3300.

Pfaller, M.A., Marco, F., et al. 1998. *In vitro* activity of two echinocandin derivatives, LY303366 and MK-0991 (L-743,792), against clinical isolates of *Aspergillus*, *Fusarium*, *Rhizopus*, and other filamentous fungi. *Diagn Microbiol Infect Dis*, **30**, 251–5.

Pfaller, M.A., Messer, S.A., et al. 2000. In vitro susceptibility testing of filamentous fungi: comparison of Etest and reference microdilution methods for determining itraconazole MICs. *J Clin Microbiol*, **38**, 3359–61.

Pontón, J., Rüchel, R., et al. 2000. Emerging pathogens. *Med Mycol*, **38**, Suppl 1, 225–36.

Poza, G., Montoya, J., et al. 2000. Meningitis caused by *Pseudallescheria boydii* treated with voriconazole. *Clin Infect Dis*, **30**, 981–2.

Pryce, T.M., Palladino, S., et al. 2003. Rapid identification of fungi by sequencing the ITS1 and ITS2 regions using an automated capillary electrophoresis system. *Med Mycol*, **41**, 369–81.

Pujol, I., Guarro, J., et al. 1997. In-vitro antifungal susceptibility of clinical and environmental *Fusarium* spp. strains. *J Antimicrob Chemother*, **39**, 163–7.

Radford, S.A., Johnson, E.M. and Warnock, D.W. 1997. In vitro studies of activity of voriconazole (UK-109,496), a new triazole antifungal agent, against emerging and less-common mold pathogens. *Antimicrob Agents Chemother*, **41**, 841–3.

Raj, R. and Frost, A.E. 2002. *Scedosporium apiospermum* fungemia in a lung transplant recipient. *Chest*, **121**, 1714–16.

Revankar, S.G., Patterson, J.E., et al. 2002. Disseminated phaeohyphomycosis: review of an emerging mycosis. *Clin Infect Dis*, **34**, 467–76.

Safdar, A. 2002. Progressive cutaneous hyalohyphomycosis due to *Paecilomyces lilacinus*: rapid response to treatment with caspofungin and itraconazole. *Clin Infect Dis*, **34**, 1415–17.

Safdar, A., Papadopoulos, E.B. and Young, J.W. 2002. Breakthrough *Scedosporium apiospermum* (*Pseudallescheria boydii*) brain abscess during therapy for invasive pulmonary aspergillosis following high-risk allogeneic hematopoietic stem cell transplantation. Scedosporiasis and recent advances in antifungal therapy. *Transplant Infect Dis*, **4**, 212–17.

Sellier, P., Monsuez, J.-J., et al. 2000. Recurrent subcutaneous infection due to *Scopulariopsis brevicaulis* in a liver transplant recipient. *Clin Infect Dis*, **30**, 820–3.

Shaw, P.J. 2002. Suspected infection in children with cancer. *J Antimicrob Chemother*, **49**, Suppl S1, 63–7.

Sillevis Smitt, J.H., Leusen, J.H.W., et al. 1997. Chronic bullous disease of childhood and a *Paecilomyces* lung infection in chronic granulomatous disease. *Arch Dis Childh*, **77**, 150–2.

Simarro, E., Marín, F., et al. 2001. Fungemia due to *Scedosporium prolificans*: a description of two cases with fatal outcome. *Clin Microbiol Infect*, **7**, 645–7.

Sobottka, I., Deneke, J., et al. 1999. Fatal native valve endocarditis due to *Scedosporium apiospermum* (*Pseudallescheria boydii*) following trauma. *Eur J Clin Microbiol Infect Dis*, **18**, 387–9.

Spielberger, R.T., Falleroni, M.J., et al. 1993. Concomitant amphotericin B therapy, granulocyte transfusions and GM-CSF administration for disseminated infection with *Fusarium* in a granulocytopenic patient. *Clin Infec Dis*, **16**, 528–30.

Steinbach, W.J., Schell, W.A., et al. 2004. Fatal *Scopulariopsis brevicaulis* infection in a paediatric stem-cell transplant patient treated with voriconazole and caspofungin and a review of *Scopulariopsis* infections in immunocompromised patients. *J Infect*, **48**, 112–16.

Summerbell, R.C., Krajden, S. and Kane, J. 1989. Potted plants in hospitals as reservoirs of pathogenic fungi. *Mycopathologia*, **106**, 13–22.

Sutton, D.A., Fothergill, A.W. and Rinaldi, M.G. (eds) 1998. *Guide to significant fungi*. Baltimore: Williams and Wilkins.

Tawara, S., Ikeda, F., et al. 2000. In vitro activities of a new lipopeptide antifungal agent, FK463, against a variety of clinically important fungi. *Antimicrob Agents Chemother*, **44**, 57–62.

Tirado-Miranda, R., Solera-Santos, J., et al. 2001. Septic arthritis due to *Scedosporium apiospermum*: case report and review. *J Infect*, **43**, 210–12.

Uchida, K., Nishiyama, Y., et al. 2000. In vitro antifungal activity of a novel lipopeptide antifungal agent, FK463, against various fungal pathogens. *J Antibiot*, **53**, 1175–81.

Uchida, K., Yokota, N. and Yamaguchi, H. 2001. In vitro antifungal activity of posaconazole against various pathogenic fungi. *Int J Antimicrob Agents*, **18**, 167–72.

van Burik, J.-A.H., Leisenring, W., et al. 1998. The effect of prophylactic fluconazole on the clinical spectrum of fungal diseases in bone marrow transplant recipients with special attention to hepatic candidiasis: an autopsy study of 355 patients. *Medicine (Baltimore)*, **77**, 246–54.

Velasco, E., Martins, C.A. and Nucci, M. 1995. Successful treatment of catheter-related fusarial infection in immunocompromised children. *Eur J Clin Microbiol Infect Dis*, **14**, 697–9.

Vincent, A.L., Cabrero, J., et al. 2003. Successful voriconazole therapy of disseminated *Fusarium solani* in the brain of a neutropenic cancer patient. *Cancer Control*, **10**, 414–19.

Walsh, T.J., Groll, A., et al. 2004. Infections due to emerging and uncommon medically important fungal pathogens. *Clin Microbiol Infect*, **10**, Suppl 1, 48–66.

Walts, A.E. 2001. *Pseudallescheria*: an underdiagnosed fungus? *Diagn Cytopathol*, **25**, 153–7.

Wedde, M., Müller, D., et al. 1998. PCR-based identification of clinically relevant *Pseudallescheria*/*Scedosporium* strains. *Med Mycol*, **36**, 61–7.

Wildfeuer, A., Seidl, H.P., et al. 1998. In vitro evaluation of voriconazole against clinical isolates of yeasts, moulds and dermatophytes in comparison with itraconazole, ketoconazole, amphotericin B and griseofulvin. *Mycoses*, **41**, 309–19.

Wood, G.M., McCormack, J.G., et al. 1992. Clinical features of human infection with *Scedosporium inflatum*. *Clin Infect Dis*, **14**, 1027–33.

37

Pneumocystis pneumonia

MELANIE T. CUSHION

A NOTE ABOUT NOMENCLATURE

Since the publication of the previous (9th) edition of this book in 1998, distinct species within the genus *Pneumocystis* have been described. The species which infects human beings was described as *Pneumocystis jirovecii*, while *Pneumocystis carinii* was retained as the name of the species first identified in rats (Frenkel 1999). The species infecting mice was named *Pneumocystis murina* (Keely et al. 2004) and another species found in rats was given the binomial, *Pneumocystis wakefieldiae* (Cushion et al. 2004a). *P. jirovecii* as the name for the human species is controversial (Stringer et al. 2002, 2003; Hughes 2003) and its validity will be decided by the International Botanical Congress in 2005. In this chapter, *Pneumocystis carinii* is used in the section on History below to refer to organisms from any host. Throughout the remaining sections, care is taken to use the correct species assignations when known. The term '*Pneumocystis*' is used when referring to the organisms in general or when the host of origin is not known.

HISTORY

Pneumocystis carinii was first described in Brazil by Carlos Chagas in 1909, while studying infections with *Trypanosoma cruzi* in humans and animal models (Chagas 1909). He interpreted sickle and rounded forms within cyst forms of the organism as male and female stages of trypanosoma. An Italian investigator who agreed with this description, Antonio Carini, sent slides from the lungs of rats infected by *Trypanosoma lewisi* to associates in Paris. Dr and Mrs Delanoë at the Pasteur Institute reviewed the slides and proved conclusively in 1912 that the organism was not a trypanosome, but an entirely different species of protozoan parasite (Delanoë and Delanoë 1912). Since the original diagnostic slides were contributed by Dr Carini, the organism species name bore his surname while the genus recognized both the tropism for the lung (pneumo-) and the characteristic developmental form (-cyst) of the microbe.

Throughout the next three decades, little notice was paid to these organisms, although their presence was observed in the lungs of several mammalian species. Epidemics of the pneumonia were first reported in the 1940s and were especially virulent in European orphanages that housed poorly nourished and sickly children in crowded conditions following World War II (Gajdusek 1957). The pneumonia was characterized by a prominent plasma cell infiltrate in the lung interstitium and the presence of an alveolar exudate described as 'honeycombed' upon histological examination. In 1942, van der Meer and Brug unambiguously reported the presence of the organisms in the lungs of humans (van der Meer and Brug 1942). In 1951, Vanek established the etiologic agent of interstitial plasma cell pneumonia as *Pneumocystis* by demonstration of the organisms in alveolar exudate of children dying of the pneumonia (Vanek 1951). By 1955, the first cases of infection were reported in the USA, although the American pediatricians were reluctant to accept the etiologic agent as *Pneumocystis* (Gajdusek 1957).

Improved conditions associated with economic recovery in Europe led to a decline in cases of the pneumonia by the 1960s. In 1958, Ivady and Paldy reported the successful treatment of infants with interstitial plasma cell pneumonia (Ivady and Paldy 1958). Previously, this family of cationic diamidine compounds was routinely used for therapy of African trypanosomiasis, or 'sleeping sickness.' This drug became generally accepted as the treatment of choice for *Pneumocystis* pneumonia until the 1970s when it was supplanted by the less toxic and highly efficacious dual regimen of trimethoprim-sulfamethoxazole (TMP-SMX). Extending the previous work of Dutz (1970) and Frenkel (Frenkel et al. 1966), Hughes et al. showed that this combination was highly effective for both prophylaxis and therapy of the infection (Hughes et al. 1974a).

During the 1960s and 1970s, the major target populations for *Pneumocystis* pneumonia were children with primary immunodeficiency diseases and patients of all ages receiving corticosteroids and other immunosuppressive therapy to prevent rejection of organ transplants and for the treatment of cancer. Most patients were responsive to TMP-SMX and its widespread use was associated with a dramatic decline in the number of cases of *Pneumocystis* pneumonia at most major medical centers in the USA.

Pneumocystis pneumonia (PCP) once again became a focus of medical attention during the 1980s as the most common opportunistic infection of individuals with the acquired immunodeficiency syndrome (AIDS) (Mills 1986). Unlike the previous subpopulations infected with *Pneumocystis*, AIDS patients experienced a more subtle clinical presentation, slow response to treatment, recurrent infections, and a high rate of adverse reactions to TMP-SMX.

During the early years of the AIDS epidemic, only a rudimentary knowledge of the organism's basic biology, structural and antigenic characteristics, and molecular genetics existed. Despite a lack of continuous in vitro propagation system and undefined life cycle, almost three decades of concerted basic and applied research has resulted in a better understanding of the family of organisms once known singularly as *Pneumocystis carinii*. Sequence analysis of the 16S-like ribosomal RNA gene as well as protein-encoding genes revealed members of the *Pneumocystis* genus to be fungi rather than protozoons as once thought. In response, pharmaceutical companies began development of novel anti-*Pneumocystis* compounds targeted to unique fungal processes, such as cell wall synthesis (Schmatz et al. 1990, 1995). Molecular and antigenic studies support species distinctions for organisms residing in different mammalian hosts, and in some cases, for different species of *Pneumocystis* within the same mammalian host. The process of naming and describing individual species has begun.

Although the incidence has declined significantly from 1992 through 1997 for each of 15 of the 26 specific AIDS-defining opportunistic infections (OI) including PCP ($P<0.05$), PCP remains as the most common AIDS-defining OI to occur first, in 36 percent of individuals infected with the human immunodeficiency virus (HIV) (Jones et al. 1999). PCP was the most common AIDS-defining OI (274 cases per 1 000 person-years), and the most common AIDS-defining OI to have occurred during the course of AIDS (53 percent of persons who died with AIDS had PCP diagnosed at some time during their course of AIDS).

The administration of highly active anti-retroviral therapy (HAART) targeting the HIV infection reduced dramatically the incidence and mortality of *P. jirovecii*. However, a growing body of evidence indicates that *P. jirovecii* populations are evolving resistance to current therapies, with specific mutations in the target genes of sulfamethoxazole (dihydropteroate synthase) (Beard et al. 2000; Huang et al. 2000; Navin et al. 2001) and atovaquone (cytochrome bc_1) (Walker et al. 1998; Armstrong et al. 2000; Kazanjian et al. 2000, 2001, 2004). Eradication of *Pneumocystis* with the currently available therapies is unlikely. Concomitant with the lack of knowledge of the infection's transmission and life cycle, history predicts a re-emergence of PCP. In this chapter, the current state of knowledge in many areas of *Pneumocystis* investigation will be reviewed. Care was taken to include key historical studies as well as the most recent reports on each subject.

THE ORGANISM

Pneumocystis are eukaryotic fungal organisms found in low number in the lungs of mammalian hosts with intact immune systems and in more abundance in homeotherms with compromised immune status. Most mammalian species have been reported to harbor organisms with morphology ascribed to *Pneumocystis*, although marine mammals and some exotic species have not been adequately evaluated. *Pneumocystis* has not been detected in surveys of poikilotherms or birds, although antibodies were detected in one study measuring the serological responses of chickens to rat-derived *Pneumocystis* (Settnes et al. 1994).

Morphology and life cycle

The nomenclature associated with descriptions of the life cycle stages and morphological features of *Pneumocystis* bears many remnants of its previous classification as a protozoan parasite. In this chapter, terms more suitable for a fungus are introduced (Cushion et al. 1997b; Ruffolo 1994). Because *Pneumocystis* has been refractory to all attempts to propagate it in vitro, histochemical and ultrastructural analyses of organisms found in human and rat lungs have significantly contributed to the current understanding of its life cycle and morphology.

MORPHOLOGY

The morphology of *Pneumocystis* found in the lungs of various mammalian hosts is quite similar, with only subtle ultrastructural differences reported between rabbit-, mouse-, and rat-derived organisms (Dei-Cas et al. 1994; Nielsen et al. 1998). Most investigators have relied on the immunosuppressed rat model of PCP for morphological analyses because of the ease of obtaining tissue for study.

Three developmental forms of *Pneumocystis* are generally recognized: the trophic stage, precyst, and cyst. A filamentous phase has not been identified and attempts to detect an environmental cycle in which amplification takes place have been unsuccessful. Proposed life cycles have relied exclusively on microscopic observations of developmental forms found in infected mammalian lungs.

The trophic form

The trophic form of *Pneumocystis* is the smallest of its life cycle stages. These forms range in size from 1.5 to 5.0 μm and are ellipsoid in shape when viewed by light microscopy of unfixed preparations, but can appear ameboid in appearance after fixation and analysis by electron microscopy (Ruffolo 1994) (Figure 37.1a). Trophic forms are thought to reproduce by binary fission, with some evidence of this process provided by electron microscopy (Campbell 1972; Richardson et al. 1989; Yoshida 1989). This stage of the organism has been subdivided by some investigators into 'small' and 'large' categories, the primary characteristics of each category being:

- overall size
- the presence or absence of tubular extensions on the cell surface
- size of nucleus
- number of glycogen granules and cellular inclusions.

Since the smaller trophic form is presumed to be the direct product of excystation (spore release), it is reduced in size (1.5–2.0 μm) and contains fewer inclusions. The larger trophic form is thought to represent the vegetative stage and thus has more cellular inclusions and a larger nucleus as it proceeds towards mitotic division. Cytoplasmic organelles include a nucleus (~0.5–1.0 μm) bounded by a typical nuclear envelope with pores, which at times is associated with a nucleolus; usually a single mitochondrion with lamellar cristae which can be branched or irregular in shape (Palluault et al. 1991; Ruffolo et al. 1989); rough endoplasmic reticulum usually contiguous with the nucleus; ribosomes; occasional vacuoles; and glycogen granules. Smooth endoplasmic reticulum and Golgi-like vesicles were described in *Pneumocystis* from neonatal rabbits (Palluault et al. 1990).

Freeze fracture and transmission electron microscopy have revealed that the trophic forms are circumscribed by two distinct unit membranes separated by a thin electron-lucent space (Campbell 1972; De Stefano et al. 1990b). A dense matrix associated with the outermost membrane contains polysaccharides, the predominant surface antigen of the organism [the major surface glycoproteins (MSG)] and several classes of host-derived molecules, e.g. cardiolipin, fibronectin. The entire complex spans 20–50 nm. The lack of staining with methenamine silver (used for diagnosis of fungal infections and PCP) by this form of the organism is probably due to the almost nonexistent electron-lucent layer thought to be composed in part of β-glucan (De Stefano et al. 1990a). Usually present only on the surface of large trophic forms is a feature called 'tubular expansions or extensions' (Yoshida 1989; Dei-Cas et al. 1991). Ultrastructural studies show them as long, thin, coiled, membranous structures filled with electron-dense granular material, the function of which is not known but suggested to be a means of adherence to host cells or for uptake of nutrients (Figure 37.1c, d).

The observation that some trophic forms are ameboid in appearance has led some investigators to assume that at least this stage is motile (Campbell 1972; Haselton et al. 1981). By contrast, studies with live organisms have not supported this contention (Shiota 1984; Cushion et al. 1988), and the presence of cytoskeletal elements indicative of motility (e.g. microtubules, microfilaments supporting the membrane) has not been detected in trophic forms.

The precyst

The precyst (sporocyte) is recognized as an intermediate stage of the sexual phase of reproduction leading to cyst development. It is presumed that a mating event first occurs to provide a zygote which initiates sporogenesis. The zygotic nucleus undergoes meiosis and a subsequent mitosis within the sporocyte. Sporocyte development has been characterized as early, intermediate, and late. An early sporocyte has an oval shape, a cell wall of 40–50 nm containing the two unit membranes with a widening electron-lucent space, and contains several mitochondria which may be clumped around the diploid nucleus (Figure 37.2a, b). Ultrastructural studies of this early stage have revealed the only morphological evidence for sexual reproduction of *Pneumocystis*, the presence of synaptonemal complexes, indicating a meiotic prophase (Matsumoto and Yoshida 1984). During the intermediate sporocyte stage, the zygotic nucleus meiotically divides to yield four nuclei, while the cell becomes more spherical and the electron-lucent space between the two membranes continues to enlarge. Cytoplasmic inclusions characteristic of this stage include spindle microtubules, which appear to arise from a dense body called the nucleus-associated organelle (NAO) (Matsumoto and Yoshida 1984). The

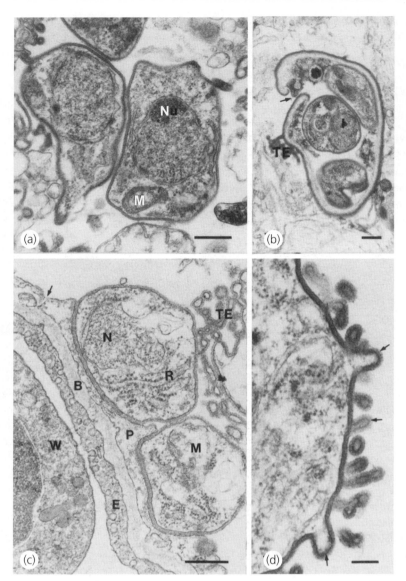

Figure 37.1 *Trophic forms of P. carinii.* **(a)** *The small trophic forms lack tubular extensions. A prominent nucleolus (Nu) is adjacent to the nuclear envelope of the trophic form on the right. A mitochondrion (M) and endoplasmic reticulum (R) can be observed in the same trophic form.* **(b)** *Mature intracystic bodies (ICB) (spores) are thought to be released through ruptured cyst (spore case) walls (arrow). A mature ICB (I) has the same structure as the small trophic form. Tubular extensions (TE) are also present.* **(c)** *Small trophic forms adhere tightly to the type I pneumocyte (P). The alveolar endothelium is discontinuous (arrow). Tubular extensions (TE) are evident on the alveolar lumenal side of the trophic forms. The type I pneumocyte lies adjacent to the basement membrane (B) which separates the endothelial cells (E) that line the capillary. A white blood cell (W) appears in the capillary lumen. A nucleus (N) is also present.* **(d)** *The surface of the trophic form consists of two unit membranes and a thick external glycocalyx. Tubular extensions show continuity with the cell surface (arrows). The bar lines represent 0.5 μm in Figures 37.1, 37.2 and 37.3, except in Figure 37.1* **(d)** *and Figure 37.2* **(b)**, *where it represents 0.2 μm. Figures 37.1, 37.2 and 37.3 are modifications of those appearing in Schrager et al. (1993).*

late sporocyte is defined by its spherical shape and prominent three-layer cell wall (80–120 nm). This form contains a total of eight nuclei resulting from mitotic division of the products of meiosis, aggregated mitochondria, and the presence of unit membranes within the cytoplasm, apparently for compartmentalization of the organelles into spores and completion of sporogenesis.

The cyst stage

The cyst stage (spore case, ascus) is the end product of sporogenesis. This form is spherical, 5–8.0 μm in

diameter, and has a thick cell wall composed of the three layers (Figure 37.2c). Eight spores are commonly found within the case. *Pneumocystis* asci are of the protunicate type, characterized by a thin delicate wall and passive discharge of their ascospores. The other two types of asci found in the ascomycetes, unitunicate and bitunicate, actively discharge their spores by forcible ejection (Liu and Hall 2004). Biochemical analyses and staining reactions provide strong evidence for the presence of substantial amounts of glucan in the spore case cell wall and there is some evidence that chitin or at least, *N*-acetylglucosamine may also be

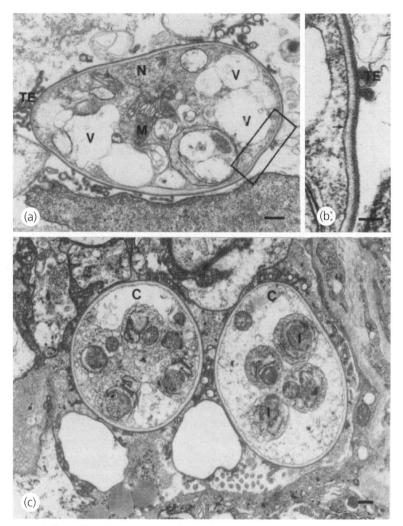

Figure 37.2 *Precyst and cysts (sporocytes and spore cases).* **(a)** *Aggegration of mitochondria (M) are a characteristic of precysts (sporocytes). Lamellar cristae can be seen in these organelles. Vacuoles (V), tubular extensions (TE), and a nucleus (N) are also present.* **(b)** *Higher magnification of the sporocyte surface within the rectangle in* **(a)** *shows the plasma membrane (arrow) and a thickened cell wall consisting of an inner electron-lucent region and an outer electron-dense layer in which the second unit membrane is enmeshed. Tubular extensions are reduced in number in precysts (sporocytes) and cysts (spore cases–I.C).* **(c)** *It is presumed that as the spore cases mature, nuclei and cytoplasm are segmented into ICB/spores (I) within the spore cases.*

present (Garner et al. 1991; De Stefano et al. 1992). The spores are spherical, 1–2 μm in diameter, contain the same organelles of the small trophic forms, and are bounded by two unit membranes. One irregularity of the spore case cell wall can be observed at one of the poles, where a thickening of the electron-lucent space is apparent by electron microscopy and by light microscopy of methenamine silver stained organisms or interference contrast microscopy of unstained organisms (Ruffolo et al. 1986). Membranous particles, mitochondria, and unincorporated cytoplasm are often present at this location.

The process of spore release is not well documented, except by Shiota (1984) who described the excystation of spherical and elongated intracystic bodies using light microscopic techniques. Prior to release, the spores have been reported to exhibit a variety of shapes, such as:

- spherical, which imparts a rosette array within the spore case
- large and irregular shapes that seem to be tightly packed into the case
- elongated banana-shaped spores that often exhibit dense cytoplasm
- ellipsoidal forms that sometimes appear as if attached to stalks.

Contractile-like movement was observed in some light microscopic studies of the ellipsoidal and elongated shaped spores (Shiota 1984; Cushion et al. 1988) and subsequently documented by videomicroscopy (Newsome et al. 1991). Neither the function of the various shapes nor the significance of the movement is understood. Empty spore cases are frequently observed by light and electron microscopy and assume a crescent shape. In some, a rent in the wall is evident (Figure 37.3). Within

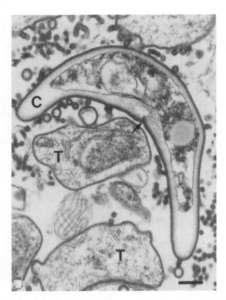

Figure 37.3 *Collapsed cyst (spore case). A cresent-shaped collapsed spore case (C). A trophic form (T) is also shown. The spores have been released, presumably through a rent in the cell wall (arrow). The residual cytoplasm has begun to degenerate.*

the collapsed thick wall, a remaining spore may be observed as well as residual cytoplasm.

In addition to the archetypal thick-walled form, a thin-walled cyst has been described by some investigators (Vossen et al. 1978; Matsumoto and Yoshida 1986). This stage contains large irregularly shaped 'daughter forms' that fill the entire cell which is delimited by a cell wall lacking the inner electron-lucent layer.

LIFE CYCLE

The intrapulmonary life cycle of *Pneumocystis* likely involves an asexual phase of cell division by the apparent haploid trophic form and sexual phase (sporogenesis) leading to the production of a thick-walled reproductive cyst (spore case, ascus) containing eight intracystic bodies (spores) (Figure 37.4). An additional asexual cycle leading to formation of a thin-walled cyst (endogeny) has been proposed (Vossen et al. 1978; Matsumoto and Yoshida 1986).

Several life cycles of *Pneumocystis* have been described. All relied upon organism morphologies observed in infected lungs by techniques of light and electron microscopy and not by direct observations made from replicating organisms (Walzer et al. 1989). Thus, the complete life cycle of *Pneumocystis* is not known, but has been deduced from static images and based on the life cycles of other protists, especially parasitic protozoa, but more recently, fungi (Cushion 2004b). It is hoped that a clearer understanding of the fungal nature of *Pneumocystis*, coincident with improvement in ex vivo culture conditions, will lead to definition of its life cycle.

The asexual replication of the trophic forms appears to resemble binary fission, rather than budding, a process used by many but not all yeasts. *Schizosaccharomyces pombe* is one yeast that divides by binary fission and also has a sexual phase initiated by morphologically indistinguishable mating types. *S. pombe* is phylogenetically related to *Pneumocystis* (see below) and it is conceivable that they share some similarities between their life cycles.

In 1984, Matsumoto and Yoshida provided ultrastructural evidence for sexual reproduction (Matsumoto and Yoshida 1984). Synaptonemal complexes, zipper-like conformations which occur during pairing of homologous chromosomes, were observed in a *Pneumocystis* life cycle stage the authors called an 'early precyst.' If sexual reproduction does take place, cyst development would necessarily follow a mating event. Ultrastructural observation of binucleate trophic forms has been interpreted as resulting from conjugation (Yoshida 1989), but could just as easily represent binary fission. The trophic forms/trophozoites involved in this process have been postulated to be 'isogametes' because of the lack of morphological characteristics to distinguish male or female forms. However, isogametes are usually associated with a protozoan life cycle. Rather, the trophic forms may be more akin to the mating types of yeast which are, in general, morphologically indistinct, but are defined by pheromones and receptors encoded by mating type-specific genes (Herskowitz 1989). For example, in the ascomycete family, *Saccharomyces cerevisiae* has two mating types, 'α' and '**a**,' and *S. pombe* has 'minus (−)' and 'plus (+).' Since the mating process in yeast is usually initiated by an environmental stress, e.g. nitrogen starvation, perhaps sporogenesis of *Pneumocystis* is induced by unfavorable conditions in the mammalian lung brought about by an increasing organism burden or depletion of nutrients.

Molecular genetic techniques have led to significant advances in our understanding of the *Pneumocystis* life cycle. Like that in yeast, the sexual cycle likely involves coordination of mating types. The process is initiated by attachment of the pheromone from one mating type to the other's receptor. In yeast, the receptor *STE2* in **a** cells is recognized by α-factor and the receptor *STE3* of α cells is recognized by **a**-factor. The complementary binding results in signaling to the interior of the cells via a heterotrimeric G protein complex, then to a mitogen-activated kinase cascade. Activation of Ste12p stimulates the cells into expression of mating type genes, cell cycle arrest, fusion of mating types, and nuclear fusion (Herskowitz 1989). Within the genome of *P. carinii* lies homologs to *STE3*, *STE2*, *STE11*, *STE12*, *STE20*, and Gpa1(Smulian et al. 1996, 2001), suggesting that the organisms do indeed mate and can transduce this signal via the appropriate signal cascade. The *P. carinii* Ste12 homolog (which also has homology to the Ste11 of *S. pombe*) was shown to be phosphorylated by a map kinase of *P. carinii* (PCM) (Vohra et al. 2003b). A homolog of the **a** mating

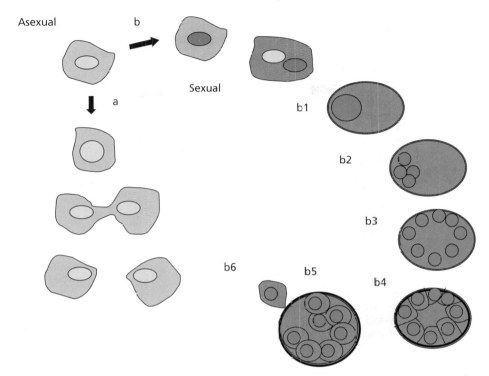

Figure 37.4 *Proposed life cycle of* P. carinii. *Trophic forms of opposite mating type (shown as blue or pink) can undergo asexual binary fission* **(a)** *or enter into sexual reproduction via mating* **(b)**. *Asexual phase* **(a)**. *Trophic forms replicate by mitosis where the haploid nucleus replicates, cytoplasmic volume increases, and the cell splits by simple cell division. No budding has been observed in the life cycle of* Pneumocystis. *Sexual phase. Haploid trophic forms (mating types, blue and pink) conjugate to form a diploid zygote (early phase sporocyte)* **(b1)** *which undergoes meiosis* **(b2)** *and subsequent mitosis* **(b3)** *to form eight haploid nuclei (late phase sporocyte). The spores are formed by compartmentalization of nuclei and cytoplasmic organelles, e.g. mitochondria* **(b4)**. *Release is thought to occur through a rent in the cell wall* **(b5)**. *The ascospores then emerge and become the vegetative stage, or trophic forms, which presumably can enter either phase of division* **(b6)**.

factor was found flanking the *STE3* gene of *P. carinii* (Smulian et al. 2001). The linkage of an **a**-factor pheromone with an **a**-factor receptor has not been reported for any other member of the fungi. Despite the lack of in vitro system and thus, any molecular tools to evaluate function, investigators in the field have relied upon expression in heterologous systems like yeast and *S. pombe*, with generally good rates of success (Thomas et al. 1998; Fox and Smulian 1999; Vohra et al. 2003a). This approach will be invaluable for definition of the complete replication cycle and other pathways of *P. carinii*.

Putative gene homologs that function in mitotic and meiotic replication in other fungal systems have been identified through the *Pneumocystis* Genome Project (Porollo et al. 2004) and by investigators in the field (Thomas et al. 1996, 2001; Gustafson et al. 1999; Vohra et al. 2003b). A partial summary of the homologous genes involved in cell division and replication from the Expressed Sequence Tag (EST) portion of the Genome Project has been reported (Cushion 2004b).

HOST RANGE

Early surveys of wild and domestic animal populations used microscopic methods for detection of *Pneumocystis*. Presence of organisms consistent with the

morphology of the cyst form of *Pneumocystis* in the lungs of mammals was used as the diagnostic criteria. Later studies have relied upon the polymerase chain reaction (PCR) with primers targeting the mitochondrial large subunit (mtLSU) of the ribosomal RNA in most cases. Both standard and nested PCR techniques have been used. Table 37.1 summarizes the majority of these reports. The organisms were detected in phylogenetically disparate classes of mammals spanning rodents, canines, felines, ungulates, and primates. A recent study using nested PCR surveyed a large number of diverse species of nonhuman primates and found evidence supporting the coevolution of the *Pneumocystis* species with their distinct host (Hugot et al. 2003). A striking finding of this study was the genetic diversity of *Pneumocystis* populations as well as their prevalence within the primates (20–100 percent). *Pneumocystis* was also detected with high frequency in wild shrews, ranging from 30–70 percent of the animals sampled (Laakkonen and Henttonen 1995; Laakkonen and Soveri 1995; Bishop et al. 1997). The presence of *Pneumocystis* has not been reported in any of the poikilotherms, although a detailed survey of such animals is absent from the literature. Studies of members of the Aves kingdom have not found evidence of *Pneumocystis* using both microscopic and molecular genetic methods, although some antibodies to rat-derived *Pneumocystis*

Table 37.1 Pneumocystis *detected in mammals*

Animal	Reference	Country	Prevalence (if available)
Wild rats	Yoshida and Ikai 1979	Japan	
	Shimizu et al. 1985	Japan	20.5% (8/39)
	Settnes and Lodal 1980	Denmark	22% (33/150)
Misc. rodents	Lainson and Shaw 1975	Brazil	
	Kucera et al. 1971	Czech Rep	
	Laakkonen et al. 2001a	USA	
Ferret	Stokes et al. 1987	USA	
Shrew	Laakkonen and Henttonen 1995	Finland	
	Laakkonen and Soveri 1995		
	Laakkonen et al. 2001a, b	USA	
Sloth	Lainson and Shaw 1975	Brazil	
Hare	Blazek and Pokorny 1963	Czech Rep	
	Poelma 1972	Netherlands	17% (75/437)
Rabbit	Sheldon 1959	USA	
	Mata 1959	Mexico	
	Soulez et al. 1988	France	
Dog	Carini and Maciel 1916	Italy	
	Farrow et al. 1972	Australia	
	Copland 1974	New Guinea/Australia	
	McCully et al. 1979	South Africa	
	Davalos 1963	Mexico	2.7% (3/108)
	Zavala and Rosado 1972	Mexico	6.3% (4/63)
	Botha and van Rensburg 1979	South Africa	
	Yoshida and Ikai 1979	Japan	15% (2/13)
	Settnes and Hasselager 1984	Denmark	
	Sedlmeier and Dahme 1955	Germany	
Cat	Settnes and Hasselager 1984	Denmark	4% (3/75)
	Zavala and Rosado 1972	Mexico	12.6% (10/79)
Goat	McConnell et al. 1971	South Africa	
Swine	Kucera et al. 1968	Czech Rep	
	Seibold and Munnell 1977	USA	
	Jecny 1973	Russia	
	Bille-Hansen et al. 1990	Denmark	
Cattle, sheep	Jecny 1973	Russia	
Horse (foals)	Shively et al. 1973, 1974	USA	
(CID)	Perryman et al. 1978	USA	
	Ainsworth et al. 1993	USA	100% (5/5)
	Peters et al. 1994	UK	
	Ewing et al. 1994	USA	100% (3/3)
	Perron LePage et al. 1999	France	
	Jensen et al. 2001	Denmark	100% (6/6)
Non-human primates			
Marmosets	Richter et al. 1978	USA	11.3% (50/441)
	Hugot et al. 2003	France	28.6% (10/35)
Owl monkeys	Long et al. 1975	USA	
	Chandler et al. 1976	USA	
Owl-faced monkey	Hugot et al. 2003	France	100% (1/1)
White-nosed monkey	Hugot et al. 2003	France	100% (1/1)
Squirrel monkey	Hugot et al. 2003	France	20% (1/5)
Swamp monkey	Hugot et al. 2003	France	100% (1/1)
Chimpanzee	Chandler et al. 1976	USA	
Macaque	Matsumoto et al. 1987	Japan	
	Norris et al. 2003 (SIV+)	USA	100% (11/11)
	Hugot et al. 2003	France	40% (4/10)
	Guillot et al. 2004	France	
Tamarins	Hugot et al. 2003	France	
		French Guyana	75% (9/12)
Lemurs	Hugot et al. 2003	France	100% (2/2)

(Continued over)

Table 37.1 Pneumocystis detected in mammals (Continued)

Animal	Reference	Country	Prevalence (if available)
Sakis	Hugot et al. 2003	French Guyana	100% (2/2)
Zoo animals	Poelma 1975	Netherlands	
Red kangaroo			
Tree shrew			
Senegal galago			
Demidoff's galago			
Brown howler monkey			
Woolly monkey			
Long-haired spider monkey			
White-eared marmoset			
Chimpanzee			
Three-toed sloth			
Palm squirrel			
Red panda			
Fennec fox			
Tree hyrax			
Large-toothed hyrax			

were detected in the sera from some birds (Settnes et al. 1994). Temperature may play a role in the organism's proclivity for mammals, as birds maintain a higher basal temperature than mammals, at approximately 41°C, and poikilotherms have variable temperatures only slightly higher than the ambient temperature. As discussed later in this chapter, the organism populations harbored by the different species of mammals are genetically and phenotypically distinct, warranting the formal process of naming of these *Pneumocystis* species.

TAXONOMY AND PHYLOGENY

Since the publication of the previous, 9th edition of this text in 1998, there has been a general acceptance within the scientific community that members of the genus *Pneumocystis* are bonafide fungi, albeit 'atypical' fungi. Individual gene sequence comparisons and the fungal homologies of a large number of putative genes produced by the *Pneumocystis* Genome Project have contributed in large part to this consensus. With the question of 'What are they, exactly?' (Stringer 1996) apparently settled, the issue of describing distinct species within the genus has become the focus of attention. In this section, a summary of the controversy regarding the fungal versus protozoan nature of *Pneumocystis* will be presented. The newly named species within the genus will be highlighted as well as the studies and observations that lead to these distinctions.

Beginning with its first identification as a life cycle stage of a trypanosome by Carlos Chagas in 1909, the taxonomy and phylogenetic classification of *Pneumocystis* has been fraught with controversy. The Delanöes believed it to be a protozoan and that assertion was generally accepted by the rather uninterested scientific community. After *P. carinii* was found to be the

etiologic agent of epidemic interstitial plasmacellular pneumonia in the 1950s, a renewed effort to classify it was initiated. Application of the tools available at that time, including vitro culture, microscopic analysis, histological staining, and animal inoculation experiments, resulted in two divided opinions; that *P. carinii* was a protozoan or that it was a fungus (Gajdusek 1957).

Ultrastructural studies of the 1960s and 1970s reached 'conclusions' similar to those of the earlier microscopists, that *Pneumocystis* was a protozoon or a yeast, based on the 'characteristic' cellular structures of the cyst and trophic forms. Exploiting the immunosuppressed rat model of pneumocystosis established by Frenkel in 1966, investigators were able to compare organisms obtained from rats with those from humans. In 1967, Barton and Campbell concluded that the organisms identified as *Pneumocystis* in rat lungs most likely were protozoons based on the presence of a protozoan-like pellicle and predicted motility as indicated by morphological features identified as 'pseudopodia' in the trophozoites (Barton and Campbell 1967, 1969). In 1972, Campbell confirmed the structural similarities of rat and human-derived *Pneumocystis* and again underscored the protozoan nature of the organism, adding the presence of filopodia as an additional feature supporting this contention (Campbell 1972). In contrast, Vavra and Kucera in 1970 reached the conclusion that *Pneumocystis* was a yeast after extensive ultrastructural analyses of rat and human organisms (Vavra and Kucera 1970). They found the morphology of the intracystic bodies within the cyst similar to ascospores within the ascus of yeasts like *Saccharomyces*. In addition, the organization of the cell membrane and wall were likened to those observed in *Rhodotorula* or *Coccidioides immitis*, and the lamellar cristae of the mitochondria reminiscent of the yeast organelle. Placement in the Ascomycetes was tentatively proposed with verification once a culture system was found.

More recent ultrastructural studies echo the same conundrum. In 1984, Matsumoto and Yoshida not only provided evidence for meiosis as a means of reproduction for *Pneumocystis* (the presence of synaptonemal complexes) but proposed a life cycle with striking similarity to that of a protozoan parasite with isogamous gametes (Matsumoto and Yoshida 1984). Studies by Ruffolo and colleagues (Ruffolo et al. 1989) reported the morphology of the cristae in *Pneumocystis* mitochondria to be lamellar, rather than vesicular, and the process of daughter cell formation more like that of a fungus. There are no morphological features of *Pneumocystis* entirely consistent with either a yeast or protozoan classification. Ultrastructural analysis of the cyst wall by De Stefano et al. (1990b) confirmed an earlier finding of Bommer (1962) which described two unit membranes, a plasma membrane delimiting the daughter forms as well as an outer membrane enmeshed within the dense carbohydrate matrix of the cyst wall. A two-membrane outer layer is more reminiscent of protozoons than any common fungus. Most investigators recognized the limitations of ultrastructural analyses for taxonomic purposes, especially in the case of *Pneumocystis* which cannot be cultured in vitro to provide further verification of their observations.

Compositional analyses of the lipids and carbohydrates of *Pneumocystis* have not yet yielded any signature molecules that would define its taxonomic placement. Rather, these studies confirmed the atypical nature of this family of organisms. Instead of finding ergosterol as the main structural sterol, as is the case with most yeast, cholesterol was the bulk sterol in *P. carinii* (Kaneshiro et al. 1989). Kaneshiro et al. have suggested that cholesterol (C27), is probably totally scavenged from the host. By contrast, another study provided evidence that cholesterol could be synthesized, at least in part, by the organism (Zhou et al. 2002). Gas chromatography-mass spectrometry (GC-MS) analyses of the monosaccharide fragments of the carbohydrates of the cyst wall and cytoplasmic contents of *P. carinii* identified glucose as the major constituent; mannose and galactose were present in equal ratios; *N*-acetyl-D-glucosamine was found in lesser amounts, while trace amounts of ribose and sialic acid were present. Sugars were representative of about 8 percent of the cyst wall components, a somewhat lower amount than most fungi (De Stefano et al. 1990a). In contrast to the rat-derived carbohydrate analyses, high-pressure liquid chromatography (HPLC) analyses conducted by Lundgren et al. found mannose, glucose, galactose, and glucosamine to be present in *P. jirovecii* at equimolar concentrations (Lundgren et al. 1991). The presence of chitin was inferred by lectin binding (Garner et al. 1991), but enzymes involved in chitin biosynthesis have not yet been detected in its genome.

Molecular biological analyses

There is overwhelming genetic evidence supporting the fungal identity of *Pneumocystis*. Whereas consideration of morphological characteristics, biochemical composition, or drug susceptibilities produced no definitive phylogenetic association, the use of molecular genetic tools initiated in 1988 provides a distinct answer to the question of the phylogenetic placement of this complex of organisms. *Pneumocystis* have been placed in the Phylum Ascomycota; the Subphylum Taphrinomycotina (syn. Archiascomycotina); Class Pneumocystidomycetes; Order Pneumocystidales; and Family *Pneumocystidaceae* (Eriksson and Winka 1998). The Taphrinomycotina are a paraphyletic assemblage of basal taxa and the identity of the closest extant relative of *Pneumocystis* remains unsettled, with phylogenetic results varying by gene sequence used and method for construction. The morphologies and biochemical characteristics of the fungi within this group are highly diverse and include such members as the fission yeast *S. pombe*, the plant pathogen *Taphrina deformans*, and *Neolecta vitellina*, the only member with a fruiting body structure (Liu and Hall 2004).

SMALL SUBUNIT RRNA SEQUENCES

Sequence comparison of rRNA genes is the most widely used method for studying phylogenetic relationships among prokaryotic and eukaryotic microbes (Vandamme et al. 1996). Both the large and small nuclear ribosomal RNA genes contain regions that are conserved and evolve at a slower pace as well as other regions that are more rapidly evolving and have increased sequence variation. Comparisons of sequences at the more conserved regions are used to predict more inclusive relationships such as kingdom, while the variable regions are used to define the closer associations, e.g. genus, species. Two laboratories reported the fungal nature of *P. carinii* based on comparisons of the DNA gene sequence of the small subunit ribosomal RNA (16S) gene (Edman et al. 1988) and of cDNA from the same gene of *P. carinii* (Stringer et al. 1989). Construction of several phylogenetic trees based on 16S-like sequence infer a relationship of *P. carinii* with the ascomycetes *S. cerevisiae*, *S. pombe* or *T. deformans* (Gargas et al. 1995) (Figure 37.5).

5S RNA SEQUENCE

Sequencing of the 5S RNA gene showed a relationship with the 'Rhizopoda/Myxomycota/Zygomycota group' (Watanabe et al. 1989). The 16S-like genes have replaced the 5S genes for phylogenetic analyses because of associated evolutionary clock anomalies and ambiguities due to the short 120 bp sequence length.

MITOCHONDRIAL GENE SEQUENCES

The use of the mitochondrial genome as an evolutionary clock has been questioned because of their supposed

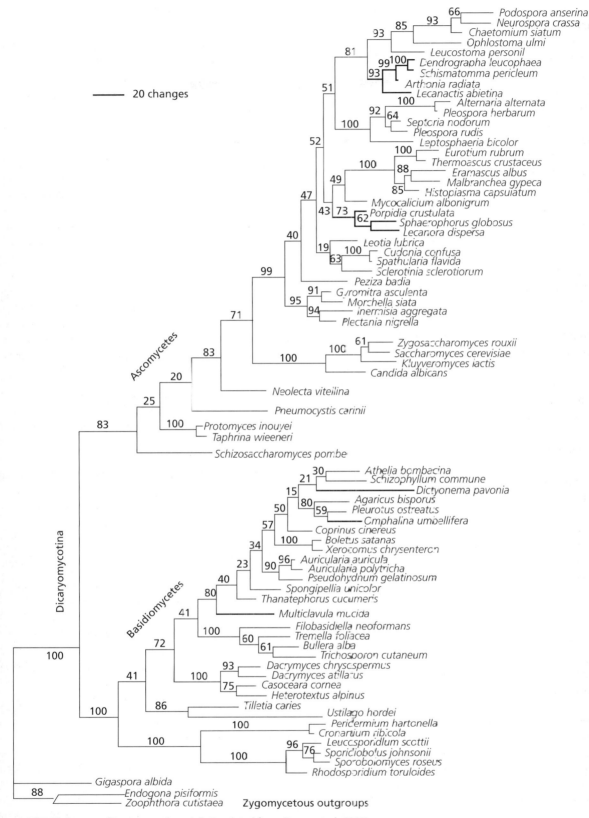

Figure 37.5 *Phylogeny of* Pneumocystis carinii. *Reprinted from Gargas et al. 1995*

bacterial origins and endosymbiotic lifestyles that probably arose more than once throughout microbial evolution. Nonetheless, comparison of several *Pneumocystis* mitochondrial genes supports the fungal identity of the organism. Comparison of the putative amino acid sequences of the entire genes or portions of the mitochondrial cytochrome oxidase II, apocytochrome *b*, *NADH 1, NADH 2, NADH 3, NADH 6* genes from

P. carinii with like genes from three fungi and two protozoans showed greater homology to the fungi than to the protozoans (Pixley et al. 1991). Comparison of the mitochondrial small subunit rRNA secondary structures of representative members of the fungi and protozoa with *Pneumocystis* sequence supported its fungal identity. Wakefield et al. (1992) employed a strategy whereby oligonucleotide primers were targeted to a sequence specific for a portion of the large subunit of the mitochondrial RNA of *P. carinii*. A series of fungi representing genera of Ascomycota and Basidiomycota were subjected to the PCR under conditions of reduced stringency. Members of the ustomycetous red yeast fungi were found to produce amplicons with these primers. Comparisons of gene sequences from these basidiomycetes with those of *P. carinii* inferred a relationship among these organisms. However, placement of *Pneumocystis* among the basidiomycetes has not been upheld by recent data from the *Pneumocystis* Genome Project (pgp.cchmc.org) or by comparisons of other genes (Liu and Hall 2004). Rather, BLAST analyses (Altschul et al. 1990) of cDNA and genomic sequences of *P. carinii* show that the greatest majority of putative genes have homology to the ascomycetes, *S. pombe*, *S. cerevisiae* and *Neurospora crassa* (Cushion et al. 2004b).

ELONGATION FACTOR 3

Identification of the *Pneumocystis* gene for translation elongation factor 3 (EF3) provided strong evidence for a fungal ancestry (Ypma-Wong et al. 1992). This factor is essential for protein synthesis in fungi but not in other eukaryotes. Alignment of the deduced amino acid sequence showed 57 percent identity with *S. cerevisiae* and *Candida albicans* EF3 proteins, while comparison of the *S. cerevisiae* and *C. albicans* amino acid sequences showed a 78 percent identity. These data suggest that *Pneumocystis*, although a member of the fungi, is significantly divergent from other fungi.

RNA POLYMERASE II

Slowly evolving genes that encode proteins are being routinely used to infer phylogeny. These genes are usually those in replication, transcription, and translation of genetic material and did not arise from horizontal transfer (Sicheritz-Ponten and Andersson 2001). One such gene is that encoding the second subunit of the nuclear DNA-dependent RNA polymerase II (*RPB2*). This and other protein-encoding genes have been found to resolve internal branches that were unresolved in rDNA trees (Liu and Hall 2004). Analyses of the *RPB2* gene from *P. carinii*, and those from 60 other fungi including seven basiodiomycetes as outgroups, supported the small subunit rRNA trees in its placement in the Taphrinomycotina, but differed in its closest fungal affinities. Maximum parsimony with bootstrap values greater than 40 percent showed a closer affinity with *S. pombe*, while Bayesian inference with a posterior probability of 97 percent placed *P. carinii* with the plant pathogen *T. deformans* (Liu and Hall 2004).

COMPARISONS OF OTHER GENE SEQUENCES

Several other protein-encoding genes from *Pneumocystis* have been sequenced and compared with like genes from protozoans and fungi. Edlind et al. found the β-tubulin gene to be more homologous with the genes of filamentous fungi (89–91 percent) such as *N. crassa* than with those of yeasts such as *S. pombe* (Edlind et al. 1992). The thymidylate synthase (*TS*) and dihydrofolate reductase (*DHFR*) genes in protozoa are fused and encode a bifunctional protein (Garret and Coderre 1984). In the yeasts and mammals evaluated thus far, these genes are separate. In 1989, it was shown that *Pneumocystis* encodes these proteins with separate genes found on separate chromosomes (Edman et al. 1989a, b). Construction of phylogenetic trees with limited numbers of fungal and protozoal members using *TS* and *DHFR* sequences supported the association of *Pneumocystis* genes with ascomycetes and basidiomycetes. Similar studies with transcription factor IID and a P-type cation-translocating ATPase gene (Stringer 1993) also supported the fungal identity of *Pneumocystis*.

GENETIC DIVERSITY OF *PNEUMOCYSTIS* FOUND IN DIFFERENT MAMMALIAN HOSTS

It has long been appreciated that there are differences among the antigenic reactivities of *Pneumocystis* obtained from the lungs of different mammalian hosts (Graves et al. 1994). Molecular genetic evidence supports these initial findings. Comparison of the srRNA and mitochondrial large ribosomal subunit RNA, and *TS* gene sequences between rat, human, mouse, ferret, and pig *Pneumocystis* shows differences similar to those observed between different yeast species (Keely et al. 2004). Likewise, the sequences of genes from the multigene family encoding the MSG of *Pneumocystis* are quite distinct, supporting the observed antigenic reactivities of the expressed proteins (Gigliotti et al. 1993a; Kovacs et al. 1993; Stringer et al. 1993; Wada et al. 1993; Keely et al. 1994; Sunkin et al. 1994). Comparison of the sequence of a portion of the 5-enolpyruvyl shikimate phosphate synthase, a gene in the *arom* locus, among rat-, human-, mouse-, rabbit-, and ferret-derived *Pneumocystis* showed divergence at the nucleotide level ranging from 7 to 22 percent and at the deduced amino acid sequence level of 7–26 percent (Banerji et al. 1995). The human and rat *Pneumocystis* α-tubulin gene were 92 percent identical in their predicted amino acid sequence and 26 percent divergent in the nucleotide sequence (Stringer 1993). Table 37.2 illustrates some of these differences.

Table 37.2 *Pairwise distances (percent) of six* Pneumocystis *genes*

Species compared	SODA[a]		DHPS		DHFR		mtrRNA		AROM		TS	
	nt	aa	nt	aa	nt	aa	LSU	SSU	nt	aa	nt	aa
P. murina/P. carinii	16	4	6	6	17	19	8	10	7	7	6	5
P. murina/P. jirovecii	27	23	15	19	31	34	20	18	17	19	21	17
P. carinii/P. jirovecii	33	25	16	18	36	39	24	23	18	20	22	14
P. jirovecii/primate[a]	10	2	10	11	19	21	15	12	NA	NA	NA	NA

Modified from Keely et al. 2004.

AROM (5-enolpyruvylshikimate-3-phosphate synthase); DHFR (dihydrofolate reductase); DHPS (dihydropteroate synthase); mtrRNA (mitochondrial large subunit rRNA); NA, No sequence available; SODA (manganese co-factored superoxide dismutase); TS (thymidylate synthase)

a) *P. jirovecii* was compared to non-human primate-derived *Pneumocystis* from owl-monkey and macaque.

Electrophoretic karyotypes from rat, mouse, human, and ferret *Pneumocystis* have been shown to be distinct (Stringer and Cushion 1998; Weinberg and Bartlett 1991) (see Figure 37.6). The rat *Pneumocystis* profiles from *P. carinii* and *P. wakefieldiae* ranged between 700 and 300 kb with 13–15 bands, with estimated total genome sizes of 8.2 and 7.7 Mb, respectively. *P. murina* from mouse contained 15–16 bands within a shorter range of 600–350 kb but with a similar estimated genome size of 8.2 Mb. The human *P. jirovecii* karyotype of 12 bands ranged in size from 800 to 350 kb with a unique band at approximately 800 kb and an estimated genome size of 7.0 Mb. *Pneumocystis* from ferret produced the most distinct karyotype, with about 13 bands separating between 525 and 700 kb and one large band at ~850 kb with the largest estimated genome size for *Pneumocystis* species to date, 11.0 Mb.

Supporting the concept of species specificity were the negative results from several animal studies in which *Pneumocystis* obtained from one host was inoculated or injected into different immunosuppressed recipient hosts. For example, Gigliotti et al. attempted to inoculate severe combined immunodeficiency disease (SCID) mice with *Pneumocystis* derived from immunosuppressed ferrets and tracked the infection with *Pneumocystis*-specific molecular markers (Gigliotti et al. 1993b). No cross-infection could be detected.

GENETIC DIVERSITY WITHIN A SINGLE HOST

Rat

The presence of two genetically distinct *Pneumocystis* residing in the same infected rat lung was described in 1992 by detection of divergent 5.8S and 26S rRNA sequences (Liu et al. 1992). That these two 'sequovars' were from two discrete genomes was reported in 1993 using a combination of pulsed-field gel electrophoresis (PFGE), Southern hybridizations, and gene sequence comparisons (Cushion et al. 1993). The karyotpe profiles of the two populations, termed 'prototype' and 'variant' at that time, produced by PFGE were distinct (Figure 37.6) and hybridization with eight single-copy gene probes to the Southern blotted profiles resulted in localization to chromosomes of different sizes (Cushion

et al. 2004a). Of note, 13 patterns of karyotypic profiles have been identified from rats infected with the 'prototype' population (Rebholz and Cushion 2001) while only a single profile has emerged for the 'variant' organism. The most common profile of 'prototype,' form 1, is illustrated in Figure 37.6. The form 1 genome was chosen as the first *Pneumocystis* to be sequenced (Cushion and Arnold 1997). Both populations were often found within the same rat colony (identified by PFGE profiles), and many of the infections (within the same rat lung) were composed of both organisms (Cushion et al. 1993).

The small subunit rRNA sequences of these two *Pneumocystis* populations were 6.6 percent divergent in a highly variable region of the gene. The sequence differences in each population have been reported at several other loci and ranged from 4 to 31 percent (Table 37.3). Such levels of divergence are associated with species differences among other fungal genera. Calculations based on the rates of change of the small subunit rRNA genes and the internal transcribed spacer (ITS) regions of the nuclear ribosomal locus estimate that these two species diverged 22–15 million years ago (Cushion et al. 2004a).

The MSG that are on the surface of all *Pneumocystis* organisms are encoded by a multigene family. MSG genes from each of the *Pneumocystis* populations in rats were found to be genetically distinct, as evidenced by the lack of hybridization of a probe from prototype to variant *Pneumocystis* populations separated by contour clamped homogeneous electrical field (CHEF) (Cushion et al. 1993) and by sequence comparisons that showed 35 percent divergence (Cushion et al. 2004a). Moreover, the genetic elements associated with regulation of the transcription of MSG genes from each of these populations were found to be quite different (Schaffzin and Stringer 2000). These genetic differences were verified at a protein level with immunoblotting techniques (Vasquez et al. 1996).

Other animals

Two *Pneumocystis* populations coinfecting ferrets have been found to vary in sequence at the *arom* locus by a similar level of divergence (15 percent) observed between *P. carinii* and *P. wakefieldiae* (Banerji et al. 1994). These

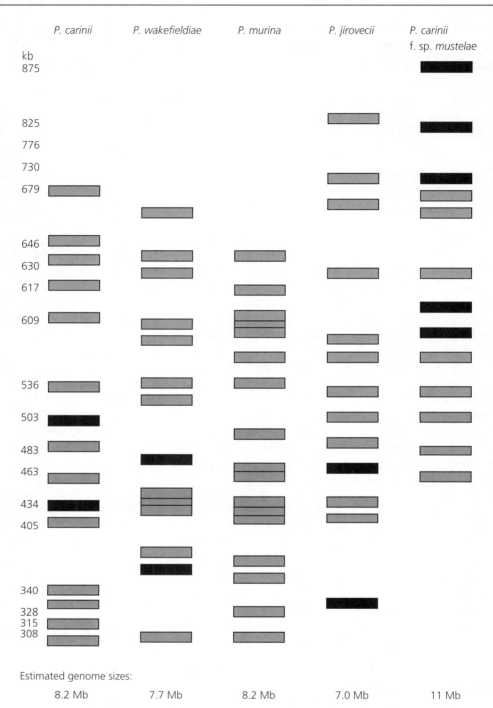

Figure 37.6 *Schematic of electrophoretic karyotypes from different species and putative species of* Pneumocystis. *Profiles were based on the migration of chromosome-sized bands using the pulsed field technique of CHEF. Sizes of chromosomes were calculated by linear regression using concatenated lambda phage DNA. Reprinted with permission from Cushion et al. 2004b, p. 165*

two sequences were 17 and 18 percent divergent from the *P. carinii* and *P. wakefieldiae arom* gene sequences.

PCR amplification directed to the mtLSU of *Pneumocystis* was used to amplify DNA purified from the lung tissue of a female Cavalier King Charles Spaniel diagnosed with PCP. Two sequences of about 350 bp each had 86.9 percent identity to each other and identity with other *Pneumocystis* species that ranged from 87 to 73 percent (English et al. 2001). The 13 percent divergence between the two sequences is similar to differences observed

between the two rat *Pneumocystis* species (Cushion et al. 2004a), and indicates that these may be from two distinct genomes infecting the dog. Further studies of the organisms from the ferret and dog will require additional sequence data to validate these potential species.

Human *Pneumocystis*

The level of genetic divergence among *Pneumocystis* organisms infecting humans appears to be less than

Table 37.3 *Percentage divergence between DNA sequences of* P. carinii *and* P. wakefieldiae

DNA sequence	Percent divergence	Reference
Thymidylate synthase	6	Keely et al. 1994
TATA binding protein (TBP)	9	Sunkin and Stringer 1996
α-Subunit of guanine nucleotide binding protein	17.5	Smulian et al. 1996
BiP precursor	17	Stedman and Buck 1996
Heat-shock protein 70	18	Stedman et al. 1998
Mitochondrial large subunit rRNA	14	Sinclair et al. 1991
Nuclear small subunit rRNA	6.6	Edman et al. 1988; Stringer et al. 1989; Liu and Leibowitz 1993
26S rRNA	5	Liu et al. 1992
Internal transcribed spacer regions I and II	30, 31	Ortiz-Rivera et al. 1994, 1995
5.8S	4	Liu et al. 1992

that between the two species found in rats (Nimri et al. 2002). However, significantly more genotypes of *P. jirovecii* have been identified based on differences of a few nucleotides, but only one karyotypic profile has been reported for human *Pneumocystis* (Stringer et al. 1993).

A number of different genetic loci have been used to type human infections, but most studies employ the ITS of the nuclear rRNA genes, the mtLSU, and more recently, the *DHPS* gene, the target of sulfamethoxazole (Beard 2004). Use of these markers has been used to:

- show that humans can be infected by a single *P. jirovecii* genotype or mixed genotypes
- explore the distribution of genotypes among geographic locations and association of genotypes with underlying disease states
- detect the presence of potential mutations associated with drug resistance.

The ITS regions of the nuclear rRNA operon have produced the most diversity thus far. This locus in *Pneumocystis* consists of genes encoding the 18S, 5.8S, and 26S subunits. The ITS1 and ITS2 regions are located between the 18S and 5.8S, and the 5.8S and 26S genes, respectively (Liu and Leibowitz 1993). These regions evolve more rapidly than the genes that they flank and are useful for detection of differences among closely related populations of organisms (Chen et al. 2000). To date, 15 types of ITS1 sequences (types A through O) and 14 types of ITS2 sequences (a to n) have been reported, for a total (in combination) of more than 60 genotypes (Lee et al. 1998; Nimri et al. 2002). Despite dramatic geographic differences, ranging across nine countries and spanning four continents and numbering greater than 300 specimens, the most common genotype of *P. jirovecii* was 'Eg' followed by type 'Ne' (Lee et al. 1998; Nimri et al. 2002).

Four genotypes occur in the *P. jirovecii* mtLSU loci with two polymorphisms each at nucleotides 85 and 248 (Beard et al. 2000). Genotypes 1 (85:C/248:C) and 2 (85:A/248:C) were those most frequently detected in a

study evaluating geographic variation among *P. jirovecii* populations, while genotypes 3 (85:T/248:C) and 4 (85:C/248:T) were found in less than 10 percent of the 324 samples analyzed (Beard et al. 2000).

Four distinct genotypes are also found in the *P. jirovecii DHPS* gene, at locations in which the same mutations conferred resistance to sulfa-based drugs in other pathogens (Skold 2000), namely (165:A/171:C), genotype 1; (165:G/171:C), genotype 2; (165:A/171:T), genotype 3; and (165:G/171:T). All these mutations are nonsynonymous, resulting in amino acid substitutions. Genotype 1 encodes a threonine at amino acid position 55 and a proline at position 57; genotype 2 has an alanine at position 55 and a proline at 57; genotype 3, a threonine at 55 and a serine at 57, and genotype 4 contains an alanine at 55 and a serine at 57. Combination of the the four genotypes of the mtLSU and the *DHPS* predicts 16 possible combinations, of which 14 have been observed (Beard et al. 2000).

New *Pneumocystis* nomenclature

In 1976, Frenkel proposed that a distinction be made between the organisms obtained from humans and those from rats on the basis of inability to cross-infect and due to serological reactivities (Frenkel 1976). The suggested names, *Pneumocystis jiroveci* for the human organism and *Pneumocystis carinii* for the one found in rats, did not gain wide acceptance.

The cumulative evidence provided by genetic data and cross-infection studies led a large group of *Pneumocystis* investigators to reconsider the issue of nomenclature at the 3rd International Workshop on *Pneumocystis*, Cleveland, Ohio, June 1994. It was the consensus of the group that sufficient evidence for speciating the organisms had not been obtained and a provisional system of nomenclature was proposed until other requirements as outlined by the Code were met to initiate species distinctions. In sexually reproducing organisms, individuals within a species are interfertile and reproductively isolated from individuals of other species. As the reproductive capacity of *Pneumocystis* is not fully understood and the organism cannot be cultured or cloned, taxonomic hierarchy relying on historical methods may not be forthcoming. It was proposed that the organisms found in each mammalian species be given a tripartite name based on the host of origin and in accordance with the International Botanical Code of Nomenclature for 'physiological variants' (The Pneumocystis Workshop 1994; Greuter 2000). For example, the name provisionally given to *Pneumocystis* from mouse was *Pneumocystis carinii* f. sp. *muris*. Occasionally these names may still be found in the literature.

In 1999, Frenkel began the formal process required for describing new species of *Pneumcoystis* (Frenkel 1999). He retained his originally proposed name for the species found in humans, *P. jirovecii*, and the name for the most common organism found in rats as *P. carinii*. Subsequently, the other species found in rats was described and named *P. wakefieldiae* (Cushion et al. 2004a), followed by description of the species in mouse as *P. murina* (Keely et al. 2004). There are now a total of four formally described *Pneumocystis* species in the fungal phylum Ascomycota, subphylum Taphrinomycotina (O.E. Eriksson and Winka 1997), Order Pneumocystidales (O.E. Erikss. 1994), Class Pneumocystidomycetes (sensu O.E. Erikss. and Winka 1997), Family Pneumocystidaceae (O.E. Erikss. 1994), Genus *Pneumocystis* (Delanoë and Delanoë 1912) (Table 37.4).

These species were described using a combination of molecular and phenotypic characteristics following a phylogenetic species concept (Taylor et al. 2000) rather than standard methods of species using mating crosses. Phylogenetic analyses with neighbor-joining trees applied to five *Pneumocystis* taxa, seven *Taphrina* taxa, *S. pombe*, and *S. cerevisiae* separated the archiascomycetes (*Taphrina*) cluster from the hemiascomycete cluster (*S. pombe*, *S. cerevisiae*), with the *Pneumocystis* cluster between the two, strongly supporting monophyly for the three groups (Keely et al. 2004). Within the *Pneumocystis* group, *P. murina* forms a monophyletic clade with the other two species found in rodents, *P. carinii* and *P. wakefieldiae*, with a higher affinity for *P. carinii* (Figure 37.7). Based on estimates using the rDNA and *DHFR* genes as well as taking into consideration the divergence of the hosts, it appears that *P. murina* has been living in the mouse for about 30–40 million years (Keely et al. 2004), while the *Pneumocystis* group probably diversified about 100 million years ago.

IN VITRO PROPAGATION

The single most important need in *Pneumocystis* research is the establishment of a continuous in vitro propagation system. Since the 1940s a multitude of investigators have attempted to culture primary isolates from mammalian hosts onto various artificial substrates, monolayer and cell tissue cultures, and organ cultures,

Table 37.4 *The taxonomic hierarchy of* Pneumocystis

Level	Taxa
Kingdom	Fungi
Phylum	Ascomycota
Subphylum	Taphrinomycotina
Order	Pneumocystidales
Class	Pneumocystidomycetes
Family	*Pneumocystidaceae*
Genus	*Pneumocystis*
Species	*jirovecii, carinii, wakefieldiae, murina*

without success. Even in this new century, the ex vivo culture of any *Pneumocystis* species defies the efforts of scientists. At best, the organism is able to replicate five- to tenfold after inoculation from the host, with only a few passages of diminishing numbers (Cushion 1989). The lack of in vitro propagation has hampered basic biological studies and also more sophisticated molecular genetic techniques. Many approaches for in vitro growth have been tried, some of which are summarized here. Other reviews are recommended to the interested reader for further details (Cushion 1989; Sloand et al. 1993; Armstrong and Cushion 1994b).

Monolayer-based systems

A relatively large number of cell lines and conditions have been explored with little improvement over the tenfold increase in number of *Pneumocystis* most frequently reported (Sloand et al. 1993). Laboratories currently using tissue culture systems routinely employ the human lung epithelioid A549 cell line, the human lung fibroblast line, HEL, or the mink lung cell line, Mv 1 Lu. Growth rates are variable within a laboratory and among laboratories. The use of monolayer-based systems for in vitro drug screening remains problematic. A viability stain must be included in the organism counts over the short-term culture period; dead *Pneumocystis* can stain with Giemsa/Diff-Quik and lead to misinterpretation of a drug's effects. Certain drugs detrimentally influence the viability and metabolic processes of the monolayer cells, leading to an incorrect conclusion of efficacy for *Pneumocystis*. Evaluation of the uptake of radiolabeled precursors is also confounded by the presence of monolayer cells. Monolayer cell systems have been useful for studying the mechanisms of *Pneumocystis* adhesion (Limper et al. 1991, 1993) and should provide additional information in this area of research and other host cell–pathogen interactions.

Cell-free systems

In the late 1980s and early 1990s, a number of cell-free systems were evaluated for growth and drug-screening capabilities. An increased incorporation of [^{35}S]methionine

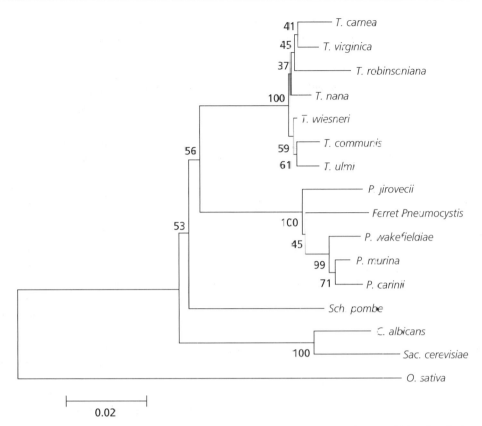

Figure 37.7 *Phylogenetic tree depicting the relationships between Pneumocystis species and among higher fungi. The tree was constructed using the neighbor-joining method, pairwise deletion, and Kimura two-parameter (K2P) in MEGA version 2.1 software (Kumar et al. 2001). One thousand bootstrap replications were performed and the values (expressed as percentages) are shown at the nodes. Bar indicates K2P distance. Reprinted with permission from Keely et al (2004)*

into *Pneumocystis*-specific proteins over time and a five- to sevenfold increase in organism nuclei were observed in organisms maintained in a neopeptone-based medium (Cushion and Ebbets 1990). *Pneumocystis* from nude mice increased fourfold in a primary culture containing Dulbecco's MEM medium supplemented with fetal bovine serum and reducing agents (Tegoshi and Yoshida 1989). Four serial passages of declining growth were observed. Comley and Sterling (1994) used rat *Pneumocystis* to demonstrate incorporation of *para*-aminobenzoic acid (pABA) and de novo synthesis of folates in an RPMI-1640 based medium. In this system, *Pneumocystis* were most viable in the first 24 hours of ex vivo maintenance as shown by pABA uptake, but organisms that stained with Giemsa persisted throughout the 3-day culture period. Exposure to pentamidine at high levels (100 µM) resulted in a dramatic decrease in radiolabel incorporation, but not in organism number.

A cell-free system using a collagen-coated porous membrane support matrix that could maintain the continuous cultivation of *Pneumocystis* from rats was described in 1999 (Merali et al. 1999). The key ingredients included *S*-adenosylmethionine, an iron component in the form of ferric pyrophosphate, *N*-acetylglucosamine, pABA, putrescine, and horse serum. Temperature of incubation was lower than previously

reported, 31°C, and the pH was higher at 8.8. The doubling times ranged from 19 to 65 hours, depending on the inocula density. The addition of *S*-adenosylmethionine was thought to be an essential component and it was later reported that *P. carinii* was auxotrophic for this compound (Merali et al. 2000). Although enthusiasm for the method was initially very high, other laboratories were unable to repeat the success of the originating laboratory (Larsen et al. 2002; Cushion and Walzer 2004).

In 1994, a cell-free system for the maintenance of *Pneumocystis* from rats using an ATP bioluminescent assay was reported (Chen and Cushion 1994a). Organisms cultured in an RPMI-based system with fetal bovine serum and nutritional additives were responsive to the standard anti-*Pneumocystis* compounds, TMP-SMX and pentamidine, and could increase modestly (three- to fivefold) over a 7-day period. Refinement of the system using a multisample luminometer permitted large numbers of compounds to be screened per year, resulting in a profile of responses to more than 60 compounds representing several classes of drugs (Cushion et al. 1997a). More recent improvements to the system include standardization of inocula by ATP content, karyotypic profile, and use of cryopreserved organisms (Collins and Cushion 2001). The latter modification reduced reliance upon availability of organisms from the immunosuppressed animal model and permitted a series of assays to be performed with a single characterized 'lot' of organisms.

BIOCHEMISTRY AND METABOLISM

The lack of a continuous culture system has compromised metabolic studies in two significant ways. First, the veracity of the incorporation of radiolabeled precursor molecules or the identification of biochemical constituents can be difficult to interpret due to the presence of contaminating host cells that may take up the same compounds or contribute to the biomolecular pools. Second, the interpretation of results in a suboptimal in vitro environment must acknowledge the tenuous state of organisms after isolation from the mammalian lung and subsequent placement in less than adequate culture conditions. Nonetheless, investigators have persevered and the results of these biochemical analyses are summarized here.

The presence of gene homologs representing many metabolic processes has been identified by sequencing of the *P. carinii* genome and should help discern the actual metabolic capacities of these organisms(Cushion 2004a, b; Porollo et al. 2004). In this section, the potential gene homologs identified by the *Pneumocystis* Genome Project or by other laboratories belonging to metabolic cycles that have been explored by standard biochemical approaches are noted. For a more comprehensive summary of potential metabolic cycles in *Pneumocystis*, the reader is directed to the following resources: Cushion 2004a, b; Porollo et al. 2004.

General and intermediate metabolism

Early studies using standard biochemical approaches provided evidence for an operational glycolytic pathway in *Pneumocystis* (Pesanti and Cox 1981; Mazer et al. 1987; Pesanti 1989). Recent genomic data lend strong support for a typical glycolytic pathway in *P. carinii*. Homologs to fungal enzymes in the pathway found in its genome include hexokinase, phosphofructokinase, fructose-biphosphate aldolase, triose phosphate isomerase, phosphoglycerate kinase, pyruvate kinase (Porollo et al. 2004), and enolase (Fox and Smulian 2001).

Several lines of investigations indicate that *Pneumocystis* respires aerobically. The presence of intact mitochondria observed by light microscopy and at the ultrastructural level, cytochemical and electrophoretic techniques, as well as incorporation studies predict a functional tricarboxylic acid (TCA) cycle, oxidative phosphorylation, and an active electron transport chain (Pesanti and Cox 1981; Mazer et al. 1987; Pesanti 1989). Cytochrome activities were predicted by inhibition studies with cyanide (Pesanti 1984; Chen and Cushion 1994b), hydroxynaphthoquinones (Dohn and Frame 1994;

Hughes et al. 1990), and antimycin A (Chen and Cushion 1994a). The synthesis of ATP was verified by direct analysis using bioluminescence detection (Chen and Cushion 1994a). Gene sequences isolated from mitochondrial DNA of *Pneumocystis* indicate that *Pneumocystis* contains the genes necessary for mitochondrial processes (Pixley et al. 1991; Cushion and Smulian 2001). *Pneumocystis* has been shown to use oxygen at a low rate, perhaps due to unfavorable assay conditions (Pesanti 1984).

All the genes necessary for a functional citric acid cycle in *P. carinii* have been identified in genomic and cDNA sequences by homology to fungal genes (Porollo et al. 2004) and include pyruvate dehydrogenase, citrate synthetase, aconitase, isocitrate dehydrogenase, alpha ketoglutarate, succinyl-coA synthetase, succinate dehydrogenase, fumarase, and malate dehydrogenase. Likewise, the various subunits of the complexes necessary for oxidative phosphorylation are present in its mitochondrial genome, including seven of the NADH dehydrogenase (complex I); the succinate dehydrogenase of complex II; cytochrome *c*–coenzyme Q oxidoreductase of complex III; cytochrome oxidase of complex IV; and the ATP synthase subunits (Porollo et al. 2004).

Sterol metabolism

Sterols are thought to be used for two purposes in fungal cells, in the maintenance of cell architecture and function of membranes and as a requirement in cellular processes: 'metabolic sterols' (Benveniste 2002). Yeast sterols are distinct from those in animal cells, with ergosterol and 24(28)-dehydroergosterol predominating in yeast membranes.

Kaneshiro and colleagues performed direct biochemical analyses that confirmed cholesterol, rather than ergosterol, was found as the most abundant bulk sterol in *Pneumocystis* preparations (Kaneshiro et al. 1989). The activity of a key enzyme in the biosynthesis of mevolonate, 3-hydroxy-3-methylglutaryl coenzyme A reductase, was detected by biochemical methods and could be inhibited by lovastatin (Kaneshiro et al. 1994a, b). Incorporation of radiolabeled squalene and mevolonate into specific *Pneumocystis* sterols provided evidence that the organism was capable of its own sterol biosynthesis (Kaneshiro et al. 1994a). Incorporation studies with radiolabeled serine, ethanolamine, and choline provided evidence that the organism could metabolize phospholipid head groups and de novo synthesize sphingolipids (Florin-Christensen et al. 1995). Furlong et al. (1994) provided chromatographic and mass spectrometry data indicating the presence of phytosterols in *Pneumocystis*-infected rat lungs and verified the earlier findings of the lack of ergosterol and cholesterol as the predominate bulk sterol.

Inhibitor studies provided evidence that at least a portion of the sterol pathway of *P. carinii* may be operational. Terbinafine, an inhibitor of squalene epoxidase, was shown to be effective against *Pneumocystis* infections in vivo and to inhibit organism growth in vitro (Contini et al. 1994, 1996; Cirioni et al. 1995). Likewise, Urbina et al. (1997) could inhibit sterol C-24 alkylation in vitro using targeted inhibitors and subsequently decrease proliferation of the organisms in short-term culture as well as inhibit the biosynthesis of its sterols. An in vitro study utilizing the bioluminescent ATP assay was conducted to probe several enzymatic steps of the sterol biosynthetic pathway in *P. carinii* (Kaneshiro et al. 1999b). Coincident with these inhibition studies, several orthologs of the genes in this pathway were identified in gene databases generated by the *Pneumocystis* Genome Project (Porollo et al. 2004). Significantly, the inhibitory functions probed by the ATP assay could be associated with the presence of the gene in the *Pneumocystis* genome. Thus, inhibitors targeted to the enzymatic steps driven by *erg1* (squalene epoxidase), *erg7* (oxidosqualene-lanosterol cyclase), and *erg6* [S-adenosyl-L-methionine:sterol C-24 methyltransferase (SAM:SMT)], resulted in decreased organism viability. The lanosterol demethylase gene of *P. carinii* was recently sequenced (Morales et al. 2003) and its presence could explain the inhibition shown by treatment with proprietary inhibitors GR 40317A, GR 42539X, and GR 40665X (Kaneshiro et al. 2000) but not with standard azoles in clinical use. The precise structure of the *P. carinii erg11* gene homolog awaits further study, but such structural information should reveal the specific properties that render it refractory to standard azoles.

Orthologs to HMG-CoA reductase 1 and 2 (Hmg1, 2, *S. cerevisiae*) genes have recently been detected in the *P. carinii* genome (Porollo et al. 2004). Although the compounds targeting this step were ineffective at reducing the ATP pools of the *P. carinii* in this study, more recent studies performed in our laboratory showed strong inhibition of ATP with the HMG-CoA reductase inhibitor, simvastatin (Collins et al. 2003). Curiously, a homolog to *erg4*, which encodes the enzyme that catalyzes the last step in ergosterol biosynthesis [the sterol C-24(28) reductase], was identified in the *P. carinii* genome (Porollo et al. 2004). Whether this genomic sequence is functional, and what role it plays, if any, in ergosterol biosynthesis, is not known at this time. Members of the genus *Pneumocystis* may not be typical fungal organisms due to the lack of ergosterol as their major sterol, but these organisms still have a functional sterol biosynthetic pathway that can be exploited for drug development.

A rare C_{32} sterol, the C-24-alkylated lanosterol derivative 24(Z)-ethylidinelanost-8,24(28)-3β-ol, was detected in large amounts in *P. jirovecii* (Kaneshiro et al. 1999a). Given the trivial name, pneumocysterol, this compound was reported to comprise almost 50 percent of the organism-specific sterols in some samples derived from human lungs. Although present in other *Pneumocystis*

species pneumocysterol appears to be most abundant in human organisms.

As a direct result of the *Pneumocystis* Genome Project, genes associated with sterol biosynthesis encoding *erg7* (AF285825) and *erg6* (AY032981) (Kaneshiro et al. 2001a; Zhou et al. 2002), were sequenced. The *erg6* gene of *P. carinii* was cloned, expressed in *Escherichia coli*, and was able to complement the null mutant in *S. cerevisiae* (Kaneshiro et al. 2001a). The sterol-binding motif of the *P. carinii* SAM:SMT product was novel and reported to have high-affinity binding properties of lanosterol. Furthermore, the enzyme was able to transfer methyl groups from *S*-adenosyl-L-methionine (SAM) to the C-24 position of both lanosterol and 24-methylenelanosterol, thus defining it as an enzyme capable of two methyl group transfers (Kaneshiro et al. 2001a). The oxidosqualene cyclase (Erg7p) encoded by the *ERG7* gene converts oxidosqualene to lanosterol, the first cyclic component of sterol biosynthesis.

The isoprenoid biosynthetic pathway leading to the production of squalene and other compounds includes a branch giving rise to the polyprenyl chain of coenzyme Q (ubiquinone). Coenzyme Q homologs function in the mitochondrial electron transport chain and offer a potential drug target in some parasite–host situations. Kaneshiro et al. (1994b) identified the homolog CoQ_{10} as the predominant species in *Pneumocystis*-infected lung homogenates and further showed that the ubiquinone could be de novo synthesized by incorporation of radiolabeled mevalonic acid into the molecule. Most higher animals contain CoQ_{10} as the sole endogenous ubiquinone, except in the rat, where CoQ_9 is the major homolog (Ellis 1994). Coenzyme Q analogs that function as antiparasitic drugs include the 8-aminoquinolones and hydroxynapthoquinones. Atovaquone, a hydroxynapthoquinone which is used for the treatment of moderate PCP, is an analog of coenzyme Q and blocks mitochondrial electron transport by binding to cytochrome *b* of the bc_1 complex (Gutteridge 1991). In addition, atovaquone was recently reported to inhibit *P. carinii* coenzyme Q biosynthesis (Cushion et al. 2000; Kaneshiro et al. 2001b).

FATTY ACIDS

Pneumocystis lives in an environment in which it is bathed in lipids comprising mainly saturated fatty acids. Purified preparations of *Pneumocystis* have been shown to take up the radiolabeled and fluorescently labeled fatty acids oleic, linoleic, palmitic, stearic acids, and phospholipids (Paulsrud and Queener 1994; Sleight et al. 1994). An epoxy fatty acid, *cis*-9,10-epoxyoctadecanoic acid (epoxy stearate), was identified in purified *Pneumocystis* preparations (Ellis et al. 1994). This unusual fatty acid has been detected in rust fungi where it is a major component. It is interesting to note that like *Pneumocystis*, the rust fungi also lack ergosterol.

Transport

Little is known of the transport or efflux systems of *Pneumocystis*, but these functions are likely important in determining the efficacy of therapeutic compounds. Studies to identify mechanisms of amino acid and glucose uptake were recently conducted (Basselin-Eiweida et al. 2001). Simple diffusion was found as the mechanism for translocation of aspartic acid. By contrast, Michaelis–Menten first-order kinetics, indicating the presence of carrier-mediated mechanisms, was exhibited by the neutral amino acids glutamine, leucine, and serine; the basic amino acid arginine; and the aromatic amino acid tyrosine. Leucine, glutamine, and serine used the same carrier system, as illustrated by competition assays. Exogenous glucose was necessary for efficient amino acid uptake, while other sugars, e.g. galactose, fructose, were less effective. Neutral, basic, and aromatic acids alike were taken up by facilitated diffusion. This conclusion was drawn from observations regarding the effects on uptake with reduced sodium or potassium ions; the lack of inhibition of transport by oubain and valinomycin; and the lack of effect when inhibitors to ATP synthesis and maintenance of the electrochemical ion gradients and membrane potentials were added to the assay. Transport of arginine appeared to be facilitated by both a high-affinity carrier and a low-affinity carrier. In addition, *P. carinii* may possess a separate carrier for aromatic amino acids that may also transport neutral amino acids.

Transport of glucose, and the inhibitor, 2-deoxyglucose, utilized two facilitated transport systems (Basselin-Eiweida et al. 2001). The high-affinity transporter was inhibited by protein synthesis inhibitors, suggesting active transport. The properties of the low-affinity system indicated it was likely to operate by facilitated diffusion, probably in environments of high exogenous glucose levels. Importantly, these investigators noted the dramatically slow uptake rate of both amino acids and glucose by *P. carinii*. They conjectured that *P. carinii* may be capable of synthesizing most of these compounds, negating the requirements for high-affinity transporters or such mechanisms.

Polyamines

Polyamines are low molecular weight polycations thought to play an important role in the regulation of cell growth and differentiation (McCann et al. 1986). *Pneumocystis* appear to have functional polyamine biosynthesis (Merali and Clarkson 1996). However, they are unable to regulate polyamine catabolism as a result of uncontrolled activity of the key enzyme spermidine/spermine-N^1-acetyltransferase (SSAT) (Merali 1999). This results in a net loss of polyamines from the cell and back conversion to smaller polyamines which could make it vulnerable to compounds that interfere with polyamine biosynthesis. α-Difluoromethylornithine (DFMO) is an irreversible inhibitor of the first and rate-limiting step in polyamine biosynthesis, the conversion of ornithine to putrescine by the enzyme ornithine decarboxylase. This compound was effective in reducing *Pneumocystis* infection in immunosuppressed rats in one laboratory (Clarkson et al. 1990) but not in another (Walzer 1994). Ornithine decarboxylase activity was not detected in an early study by Pesanti et al. (1988), but later, Clarkson et al. (1990) reported measurable activity. To date, a fungal homolog for the gene encoding ornithine decarboxylase has not been identified in the EST library or the genome of *P. carinii*, but the gene encoding an inhibitor of the enzyme ornithine decarboxylase antizyme has been detected in the EST and genomic libraries (Ivanov et al. 2000; Porollo et al. 2004).

S-Adenosylmethionine (AdoMet) is a key molecule in both methylation reactions and polyamine biosynthesis. Most AdoMet is used for transmethylation reactions which transfer the *N*-methyl group of the methionine group to other molecules such as proteins, lipids, or DNA which forms new molecules in some cases (e.g. lecithin in the case of complex lipids), or regenerates the methionine. The remaining AdoMet (~2–5 percent) enters into polyamine biosynthesis after decarboxylation (Merali et al. 2000). AdoMet synthetase transfers the adenosyl group of ATP to methionine to form AdoMet and release of P_i and PP_i. Recently, it was reported that the long-term culture of *P. carinii* was dependent upon exogenous supplementation of AdoMet (Merali et al. 1999). Further studies by these investigators described the lack of AdoMet synthetase activity by *P. carinii* in an in vitro system and characterized two transporters, one high affinity and one low affinity, that were capable of transporting exogenous AdoMet into the cells (Merali et al. 2000). They concluded that *P. carinii* was auxotrophic for AdoMet, a finding unprecedented in the fungal kingdom. Supporting their discovery was the apparent reduction of plasma AdoMet concentration in *P. carinii*-infected rats that was correlated with infection burden (Merali et al. 2000). Translation of the effect to the clinical setting confirmed the reduction of plasma AdoMet in patients suffering from PCP, but not in those with bacterial pneumonia, tuberculosis, or cryptococcal meningitis (Skelly et al. 2003). These findings are very intriguing and demand further investigation. On the one hand, replication of the success reported by the authors of these studies using a culture supplemented with AdoMet has not been reproduced in other laboratories (Cushion and Beck 2001; Larsen et al. 2002) and the presence of a fungal gene homolog encoding AdoMet synthetase in the genome of *P. carinii* also seems to contradict the assertion that the organism does not have the capacity to synthesize AdoMet (Porollo et al. 2004).

However, expression of the gene has not been verified and the clinical and experimental findings of plasma AdoMet depletion in infected hosts lends support for the salvage of this compound by these organisms.

Replication

Thermal denaturation studies and gene sequence data show the genome of rat-derived *Pneumocystis* to be adenine and thymine rich (Stringer and Cushion 1998). These initial findings have been supported by the sequencing of over 3 000 genes by the *Pneumocystis* Genome Project. The current estimate of genome-wide A+T content is ~70 percent.

The ploidy of the *Pneumocystis* genome has not been definitively defined. Cytofluorometric evaluations of cyst and trophic forms using the DNA-intercalating stain 4′,6-diamidino-2-phenylindole (DAPI) indicated that 90 percent of the trophic forms contained the same DNA content as those within the cyst (Yamada et al. 1986). Later studies using a number of nuclear intercalating dyes and image analysis confirmed the apparent haploid state of the trophic forms and ascospores within the cysts (Wyder et al. 1994); however, there was some variation observed in the nuclear content, suggesting that there may be other states of ploidy. By contrast, two-dimensional separation of apparently homologous chromosomes by pulsed-field gel-separated chromosomes predicted a diploid nucleus (Cornillot et al. 2002).

There are conflicting data on the incorporation of precursors into *Pneumocystis* nucleic acids. One early study found organisms sustained in a balanced salt solution could incorporate [^{3}H]uridine into TCA-precipitable material, although the uptake was not inhibited by the anti-*Pneumocystis* drug pentamidine (Pesanti and Cox 1981). These investigators reported no uptake of radiolabeled thymidine or hypoxanthine. By contrast, organisms incubated in a low pH minimal medium were able to incorporate [^{3}H]uridine, uracil, thymidine, and hypoxanthine as well as [^{35}S]methionine (Cushion and Ebbets 1990). More recently, the ability of *Pneumocystis* to incorporate 5-bromodeoxyuridine in the lungs of rats and mice was evaluated to determine whether a salvage pathway for thymidine was operational (Vestereng and Kovacs 2004). No incorporation was observed by either *P. murina* or *P. carinii*, although host cells readily took up the analog, suggesting that if *Pneumocystis* possesses a thymidine salvage pathway it was not active during infection in the mammalian host.

Analysis of a partial EST database from *P. carinii* revealed about 11 percent of the genes identified as fungal homologs were associated with the transcription process, including initiation factors, regulators, repressors, terminators, chromatin remodeling proteins, and histone proteins and regulators (Cushion et al. 2004b; Porollo et al. 2004). In addition, sequencing of its genome revealed a multitude of fungal homologs to

DNA repair genes such as *rad2*, *rad8*, *rad15*, *rad16*, *rad18*, *rad26*, *rad32*, and *rad50*; repair helicases *rhp26*, *rhp54*, and *rhp57*; and *MSH2* (Porollo et al. 2004).

Carbohydrates

Besides the chromoatographic analyses of monomeric sugars in the cell wall of *Pneumocystis*, described above, other studies suggest the presence of more complex carbohydrate polymers. The lytic action of the enzyme preparation, Zymolyase, which contains as its major component β-1,3-glucan laminaripentaohydrolase, and the staining of the cyst wall with aniline blue suggest the organism maintains a glucan layer, at least in the cyst form. The gene encoding the *P. carinii* β-1,3-glucan synthetase gene *Gsc1* was cloned and shown to mediate the polymerization of uridine 5′-diphosphoglucose into β-1,3-glucan (Kottom and Limper 1999, 2000). The glucan in the cyst wall of *Pneumocystis* was implicated as a factor in the detrimental inflammatory response in the lungs of its mammalian hosts (Hahn et al. 2003). Compounds targeting glucan synthesis were found to arrest the replication of cysts in the lungs of treated animals and offer an attractive alternative target for chemotherapy of infection (Schmatz et al. 1990, 1995). In addition to β-glucan, direct synthesis of α-1,4-glucan in short-term culture was shown by biochemical methods, GC-MS, and uptake experiments (Williams et al. 1991). α-Glucan in the cell wall plays a key role in fission yeast morphogenesis, maintenance of the characteristic cylindrical shape, and separation of the mother and daughter cell after mitotic replication (Konomi et al. 2003).

PNEUMOCYSTIS INFECTION

Incidence of PCP

In the pre-AIDS era, defined as those decades prior to 1980, the number of reported cases of PCP in humans was fewer than 100 (Walzer et al. 1974; Kovacs et al. 2001). In the decades that followed, a dramatic increase in cases was observed due to the onset of the AIDS epidemic in 1982. In 1991, the incidence of PCP peaked at 20 000 cases, with a subsequent decline due to the recommendation for prophylaxis of HIV-positive patients with a CD4 cell count of less than 200/μL from a Public Health Service task (1989). This decline was even more dramatic after 1995 when widespread HAART was instituted, targeting the HIV virus and permitting immune reconstitution of patients (Kaplan et al. 2000). After 1995, PCP declined at an average of 21.5 percent per year (Kovacs et al. 2001). In spite of such dramatic decreases in PCP cases, this pneumonia still remains the most common life-threatening infection associated with HIV-infected persons. Of significant concern is the increasing incidence of mutations in genes targeted by the mainstays of therapy, TMP-SMX and other secondary therapies such as atovaquone.

Description of clinical features

RISK FACTORS

A decrease in cell-mediated immunity is the single most important risk factor for the development of PCP. Reduction in the cellular immune response can be potentiated by:

- administration of immunosuppressive agents, such as corticosteroids or cancer chemotherapeutic agents
- congenital conditions like the severe combined immunodeficiency disease syndrome (SCIDS)
- starvation or malnutrition
- most recently, infection with the human immunodeficiency virus (HIV).

Quantification of the number of circulating CD4+ lymphocytes has long been recognized as an indicator of the level of immunosuppression and of the risk of developing pneumocystosis (Masur et al. 1989; Phair et al. 1990). Coordination of large center studies (Multicenter AIDS Cohort Study and Pulmonary Complications of HIV Infection Study) determined that the risk for *P. carinii* pneumonia sharply increases when the CD4+ lymphocyte counts reach 200/μL in individuals not on a prophylactic regimen, while in those with less than 200/μL, the risk becomes five times greater as compared with patients maintaining counts above 200/μL.

Although depression of cell-mediated immunity is clearly a risk factor in the development of PCP, there are clinical and experimental data implicating a role for humoral immunity. Patients with only B-lymphocyte abnormalities and mice with B-cell defects have been reported to develop PCP (Saulsbury et al. 1979; Rao and Gelfand 1983; Marcotte et al. 1996) while passive antibody administration aids in the clearance of the infection in experimental animal models (Gigliotti and Hughes 1988; Roths and Sidman 1993; Gigliotti et al. 2002).

CLINICAL PRESENTATION
Adults

PCP in adults commonly presents with increasing dyspnea, a nonproductive cough, inability to breathe deeply, substernal tightness, and often with night sweats. A fever of 38.5°C and tachypnea are often encountered. Rarely, there is sputum production or hemoptysis. Physical examination of the lungs may find little deviation from normal, but various degrees of respiratory distress, small respiratory volumes and fine basilar rales may be present. AIDS patients have a more insidious progression to clinical disease than patients with immune depression unrelated to HIV, approximately 25 days versus 5–10 days (Dohn and Frame 1994).

The chest radiograph may appear normal or indicative of severe disease. Subsequent follow-up of normal radiographs by high-resolution computed tomography (CT) scan of the chest will show a ground glass attenuation if PCP is present (Gruden et al. 1997). Diffuse, symmetric interstitial infiltrates are most commonly encountered, while focal infiltrates, lobar consolidations, cavities, and nodules are less common. In early infection, the infiltrates may be widely distributed but consolidation and air bronchograms develop with progression of the disease. Increased incidence of apical infiltrates and pneumothoraces have been associated with aerosolized prophylactic pentamidine administration (Stansell and Hopewell 1995). Pneumatoceles have been observed with increasing frequency in cases of prolonged PCP (Stansell and Hopewell 1995). The degree and severity of abnormalities on the chest radiograph are prognostic and can be correlated with higher mortality.

Impaired oxygenation is the primary abnormality detected by laboratory methods (Walzer 2004b). Many patients exhibit a widening of the alveolar–arterial oxygen gradient (A–a/DO_2) proportional to the severity of disease and respiratory alkalosis (Forrest et al. 1999). However, a significant number of patients can present with a normal (A–a)DO_2 gradient at rest. Other physiological changes include hypoxemia, impaired diffusing capacity, alterations in lung compliance, total lung capacity, and vital capacity (Walzer 2004b). Analyses of bronchoalveolar lavage fluids (BALF) have revealed a generalized fall in surfactant phospholipids and phosphatidylcholine, but an increase in surfactant proteins (SP) A and D and sphingomyelin has been observed (Phelps and Rose 1991; Hoffman et al. 1992; Limper et al. 1994; Mason et al. 1998; Martin and Wright 1999). SP-A appears to inhibit phagocytosis of *Pneumocystis* (Koziel et al. 1998b) in some studies, while in others, it was shown to mediate binding (McCormack et al. 1997). SP-D has also been shown to bind to the organism's surface glycoproteins and facilitate adherence to macrophages (Limper et al. 1995; O'Riordan et al. 1995).

Children

The presentation in children usually involves cyanosis, flaring of the nasal alae, mild cough usually without fever, and intercostal retractions in severe disease (Walzer 2004b). In contrast to increased risk of the pneumonia in adults with CD4+ counts of 200/μL or below, 27 percent of children 1 year of age or less who developed *Pneumocystis* had CD4+ counts greater than 1000/μL (Anonymous 1995), due to the higher overall CD4 count in children.

EXTRAPULMONARY PNEUMOCYSTOSIS

Prior to the AIDS era, reports of extrapulmonary spread of *Pneumocystis* were rare (Telzak et al. 1990; Ng et al.

1997). Although an incidence of 1–3 percent of extrapulmonary cases has been reported in the post-AIDS era, it is likely there are a significant number of cases that go unreported.

Most extrapulmonary *Pneumocystis* is detected as an incidental finding during postmortem examination. It is rarely considered as a cause of the clinical manifestations associated with the HIV syndrome. Extrapulmonary spread of *Pneumocystis* is found most commonly in patients in the late stages of AIDS who have not received prophylactic treatment or only aerosolized pentamidine (Walzer 2004b).

The frequency of involvement in organs other than the lungs or body sites in a series of 52 patients, 40 of whom were HIV positive and 12 who were not infected, was lymph nodes (44 percent); spleen (33 percent); liver (31 percent); bone marrow (33 percent); adrenal glands (19 percent); gastrointestinal tract (19 percent); genitourinary tract (17 percent); thyroid (17 percent); heart (15 percent); pancreas (13 percent); eyes (12 percent); ears (12 percent); skin (4 percent); and other sites (2–3 percent) (Telzak and Armstrong 1994). The incidence of *Pneumocystis* in extrapulmonary sites in HIV-positive patients was evenly distributed at about 33 percent among the lymph nodes, spleen, liver, and bone marrow. By contrast, the preferential site for extrapulmonary *Pneumocystis* in non-AIDS patients was the lymph nodes with 75 percent of the 12 patients showing involvement. Infection of multiple extrapulmonary sites has been associated with a rapidly fatal outcome. Areas of focal involvement present as retinal cotton wool spots; polyploid lesions in the external auditory canal; thyroid mass; pancytopenia from bone marrow necrosis; pleural effusion; and numerous hypodense low attenuation lesions in the spleen or other organs on CT scans. Involvement of pleural and peritoneal spaces often resulted in calcification of these surfaces. Histopathological findings of fine needle aspirates or biopsies of these lesions showed areas of necrosis with exudative foamy material. Staining with methenamine silver or fluorescently labeled monoclonal antibodies revealed characteristic *Pneumocystis* morphologies (Walzer 2004b).

PCP IN PATIENTS WITHOUT UNDERLYING IMMUNOSUPPRESSION

PCP has been rarely reported in individuals with intact immune systems. However, two reports describe the infection in the elderly. In the first, a 67-year-old man presented with tachypnea, chills, and fever (Ludmerer and Kissane 1991). The chest radiograph showed bilateral pulmonary infiltrates and a biopsy confirmed the diagnosis of PCP. No evidence of an underlying immunosuppression was detected in laboratory tests. In the second report, a cluster of five elderly patients (ages 66–78 years) presented during a 3-month period at the New York Hospital–Cornell Medical Center with somewhat atypical symptoms of *P. carinii* pneumonia (Jacobs et al.

1991). Only two had fever and the chest radiographs revealed focal, lobar infiltrates or pleural effusions rather than bilateral involvement. No obvious immune impairment was detected, although it was suggested that subtle decreases in immune function associated with age may predispose the elderly population to this infection.

Autopsy studies conducted on individuals without predisposing immune suppression showed little evidence for infection with *P. jirovecii*, either by standard histological methods (Cushion 2004c) or by use of PCR (Peters et al. 1992). However, limited sample population numbers and the small amounts of lung tissue analyzed could explain these negative results.

More recent studies employing nested PCR for detection of *P. jirovecii* in bronchoalveolar fluids or oropharyngeal washes have reported positive amplicons from patients without underlying disease as well as from patients with a wide variety of disease states not including infection with HIV (reviewed by Cushion 2004c). The accumulating evidence from these studies suggests that colonization of humans by *P. jirovecii* is not a rare event and some subpopulations may serve as reservoirs of the infection (discussed in the section on Epidemiology below).

Pathology and pathogenesis

In the advanced diffuse form of pneumocystosis, all lobes of the lung are involved. They manifest a rubbery consistency, are bulky and densely consolidated. The cut surfaces appear yellowish to reddish gray, are granular and dry with destruction of air spaces. In less severe disease, there are foci of consolidation and areas of atelectasis admixed with compensatory hyperinflation. The nodular form of pneumocystosis is rare and typified by solitary and multiple nodules with some cavitation (Walzer et al. 1989).

Histopathological examination reveals two characteristic traits

- alveolar interstitial thickening
- a frothy eosinophilic honeycombed exudate in the lumens (Figure 37.8a).

The interstitial thickening is due to hyperplasia and hypertrophy of the type II pneumocytes, interstitial edema, mononuclear cell infiltration, and in some cases, mild fibrosis (Walzer et al. 1989). The cellular infiltrates are limited in the alveolar septa but more apparent in the perivascular and peribronchial regions. The honeycomb exudate usually contains organisms when stained with methenamine silver to visualize the cyst form. Organisms cannot be visualized with hematoxylin and eosin. As the disease progresses, there may be hyaline membrane formation accompanied by interstitial fibrosis and edema. Histopathological assessment of the disease in AIDS and non-AIDS patients revealed a higher organism burden but less tissue damage in the former

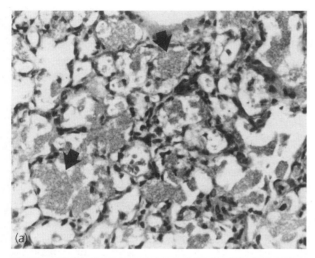

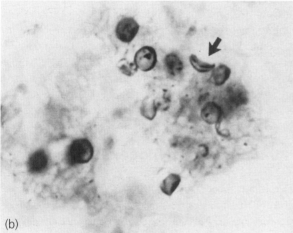

Figure 37.8 *Diagnostic histopathology.* **(a)** *Hematoxylin and eosin-stained lung sections. Note the characteristic honeycombed exudate, a hallmark of fulminant* P. carinii *pneumonia. (Original magnification 100×)* **(b)** *Cup-shape morphology of cysts (spore cases) visualized by Grocott's methenamine silver staining. (Original magnification 250×) Photographs courtesy of Paul Steele, MD, University of Cincinnati College of Medicine, Department of Pathology*

patient group (Dohn and Frame 1994). In some patients, the alveolar exudate may not be present or microcalcifications, granulomas, or cavitary lesions instead may appear. Ultrastructural studies have detected alveolar-capillary leakage and type I cell damage associated with attachment and presence of the organism (Yoneda and Walzer 1980).

Diagnosis

PCP should be considered in any immunocompromised patient who presents with fever, respiratory symptoms, or infiltrates on chest radiogram. Because these symptoms can be associated with a multitude of infectious and non-infectious conditions, definitive diagnosis must be made by morphological or histopathological demonstration of the organism. An algorithm for evaluation of patients with suspected PCP has been published by the

National Institutes of Health in 2001 (Kovacs et al. 2001) and remains a helpful guide in establishing a diagnosis and for initiation of treatment (Figure 37.9). This algorithm includes screening with a non-invasive sampling method, induced sputum. It is likely an updated algorithm will be issued in which alternative non-invasive techniques such as oral washes or oral swabs will be usd for DNA extraction and subsequent amplification by PCR.

Patients with a clinical presentation consistent with PCP and with a chest radiograph demonstrating reticular or granular infiltrate can be placed on therapy immediately or the diagnosis confirmed with a procedure to collect respiratory secretions for microscopic evaluation. Those with a negative chest radiograph with a high degree of suspicion should undergo additional tests.

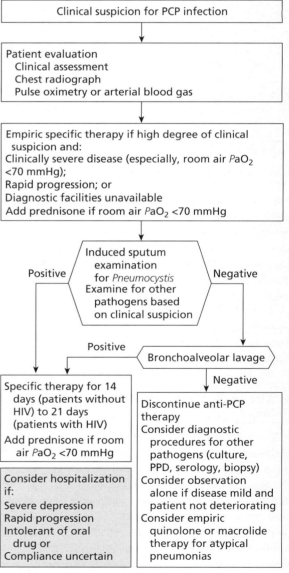

Figure 37.9 *Algorithm for evaluation of patients with suspected* Pneumocystis *pneumonia. Reprinted with permission from Kovacs et al. (2001)*

High-resolution CT scan of the chest or pulmonary function can be used to identify patients without the infection (Huang 2004; Walzer 2004b).

METHOD OF SAMPLE COLLECTION

The diagnosis of pneumocystosis relies on efficient sampling of the infected area. The organism can be detected in a variety of respiratory and other samples including induced sputum, BALF, tracheal aspirate fluid, tissue obtained by transbronchial biopsy, bronchial brushes, pleural fluid, oropharyngeal washes, and swabs. Generally, samples collected from HIV-infected patients provide a higher diagnostic yield than non-HIV-infected patients because of their increased organism burden.

Open thorax lung biopsy and transbronchial biopsy

Open thorax lung biopsy is the most invasive of the sampling procedures and requires the use of an operating room and general anesthesia. Fiberoptic bronchoscopy is the most common invasive procedure for collection of diagnostic tissue, but like open thorax lung biopsy, is rarely performed today. These procedures may be indicated after nondiagnostic bronchoscopy, suspicion of an infection concomitant with pneumocystosis, or for diagnosis of Kaposi's sarcoma.

Bronchoalveolar lavage (BAL)

BAL alone is more efficacious than brushings or washings, carries little morbidity, and has become the method of collection in most institutions. Sensitivity can be increased if multiple lobes are sampled and combined (Baughman 1994; Baughman et al. 1994; Walzer 2004b). Twenty to thirty milliliters of the lavage fluid is collected and concentrated by centrifugation. The pellet is smeared onto glass slides for fixation followed by histochemical staining.

Induced sputum

Deep inhalation of a 3 percent sodium chloride solution by the patient results in osmotic accumulation of fluid in and irritation of the respiratory passages, which can improve the organism yield, especially in AIDS patients. Depending on the institution, the patient base, and technical expertise, the diagnostic yield can range from 50 to 90 percent (Kroe et al. 1997; Caliendo et al. 1998). There are many fine technical points that must be considered during the procedure to provide the best diagnostic yield, including the use of toothpaste, which should be avoided, and the use of mucolytic agents to liquefy the sample (Cushion 2003). Although the procedure is non-invasive and relatively inexpensive, drawbacks to its use are the lack of information concerning other infections or disease processes that may be present

in the lung and low diagnostic yield in non-AIDS patients.

Nasopharyngeal aspirates

Recent studies with neonates have increased the popularity of this sampling method (Vargas et al. 2001, 2003). Detection of *P. jirovecii* from these samples relies mostly on amplification of its DNA by PCR, but in some cases organisms have been detected (Vargas et al. 2001). An advantage of the procedure is that these samples are commonly used for diagnosis of repiratory viruses and can concomitantly be evaluated for the presence of *P. jirovecii*. The aspirates are collected with a suction catheter after instillation of 1–1.5 mL of nonbacteriostatic saline (pH 7.0) in each nostril. The sample can then be treated with a mucolytic agent and subsequently stained or prepared for the PCR by DNA extraction

MICROSCOPIC IDENTIFICATION

A variety of stains can be employed for the detection of *Pneumocystis*. Variations of the methenamine silver stain (e.g. Grocott's, Gomori's) are the most widely used for diagnosis with tissue sections, but can also be used with body fluids such as BALF. A positive reaction results in a distinctive black cup-shaped morphology of the cysts against green-colored host cell architecture (Figure 37.8b) (Cushion 2003). In some reactions, the cysts will appear with a darkened structure resembling a 'double comma,' or 'parentheses' within the cyst. There is no budding associated with *Pneumocystis*, and this feature can be used to differentiate *Pneumocystis* infection from other respiratory fungal diseases such as histoplasmosis. Only the cyst form of the organism can be visualized with these stains, which is a drawback to their use. If the trophic form predominates in an infection, the organism burden can be greatly underestimated. The silver stain and other stains that selectively stain the wall of *Pneumocystis* cysts (e.g. toluidine blue O, cresyl echt violet) are favored among pathologists because they provide a sharp contrast between background host cells and organism. The Papanicolaou stain, frequently used for staining cytopathological specimens, can demonstrate the foamy exudate surrounding the organisms, although the organisms stain poorly. Likewise, the gram stain produces a negative reaction with poorly defined organism morphology.

Most laboratories now use rapid variants of the Giemsa stain for accelerated diagnoses. The most rapid modification of the methenamine silver stain takes approximately 30 minutes to perform, while the rapid Giemsa-like stains (e.g. Diff-Quik[R], LeukoStat[R]) take 30 seconds (Cushion et al. 1985). With the Giemsa stains, all forms of the organism are stained, which permits rapid assessment of the total organism burden. The nuclei appear reddish-purple in contrast to the light blue cytoplasm. The cyst wall excludes the dyes and appears as a circumscribed clear zone around the cyst contents. This stain not only permits visualization of all forms of *Pneumocystis*, but also permits assessment of the sample quality by demonstration of host cells such as alveolar macrophages and neutrophils.

Commercial kits employing fluoresceinated monoclonal antibodies against surface antigens of both cyst and trophic forms are widely used for diagnosis. Although laboratories must be equipped with epifluorescent microscopes, the reaction is specific for the organism and usually can be readily visualized against the background host cells. Depending on the monoclonal antibody supplied with the kit, staining may target only the cyst form or may target all forms of the organism. Since trophic forms are more numerous than cyst forms, it is recommended that laboratories use kits with antibodies directed to all forms of the organism, such as 2G2 or 3F6, which are available from DAKO Corp., Carpinteria, CA and Chemicon International, Inc. (Kovacs et al. 2001). Fluorescein isothionate is the typical fluorophore conjugated to the antibody in a direct assay, or to the detection antibody in an indirect assay. Organisms appear a brilliant apple green in color. Immunohistochemical techniques employing antibodies specific for *Pneumocystis* have been used with success with tissue sections, although silver stains can often provide the same information. Immunohistochemistry, however, provides confirmation of diagnosis in unusual settings such as extrapulmonary pneumocystosis.

CULTURE

Detection of *Pneumocystis* by amplification using artificial media or tissue culture is not an option since there is no continuous in vitro culture system for *Pneumocystis*.

SERUM TESTS

Serum antibody detection has had little success in the diagnosis of pneumocystosis in immunocompromised patient populations (Walzer and Cushion 1989). The high prevalence of circulating anti-*P. carinii* antibodies in the general population has confounded attempts to establish a serodiagnostic test. Prior exposure to the organism cannot be distinguished from active disease.

Other serological tests sometimes used for diagnosis are similarly nonspecific. Serum lactic dehydrogenase (LDH) levels are often elevated in PCP and decline with successful treatment. However, similar increases can be associated with other diseases (Benson et al. 1991) and thus, LDH levels must be used in concert with other diagnostic modalities.

OTHER LABORATORY TESTS

Several other diagnostic tools have been used for diagnosis of PCP. Arterial blood gas (ABG) is an important laboratory test for PCP patients, due to the frequency of impaired gas exchange. Typically, patients with PCP will have a decreased arterial oxygen tension with an increased arterial–alveolar gradient, and respiratory alkalosis. The ABG can be used to direct decisions regarding patient admissions, administration of adjunctive corticosteroids, and to assess PCP therapy (Huang 2004). Gallium scans and diethylenetriamine pentaacetic scans are imaging techniques reported to be abnormal in patients with PCP, but again, are nonspecific for PCP (Huang 2004).

Pulmonary function tests are useful in those patients where there is a clinical suspicion for PCP, but in which there is a normal chest radiograph (Huang 2004). Patients with PCP will often demonstrate a restrictive ventilatory defect with decreased lung volumes and increased airflows. Measurement of the diffusing capacity for carbon monoxide, DL_{CO}, is a sensitive indicator of alveolar-capillary block (Sankary et al. 1988). In patients with PCP, this is decreased on average from 89 to 100 percent (Coleman et al. 1984; Hopewell and Luce 1985; Shaw et al. 1988; Camus et al. 1993). Although such decreases are not definitive, a normal DL_{CO} would strongly suggest the absence of PCP.

MOLECULAR METHODS

Amplification of *Pneumocystis* DNA by PCR offers a dramatically increased level of detection and the ability to genotype the infecting organism populations. Although it is not yet routinely used in the clinical laboratory, it will likely eclipse standard microscopic methods because of its specificity, sensitivity, and relative low cost. In addition to diagnosis, PCR is routinely used for investigations of colonization in normal populations and those with underlying disease states (Helweg-Larsen et al. 2002; Huang et al. 2003; Morris et al. 2004).

P. jirovecii is routinely detected in DNA extracted from human samples by PCR using primers targeting the mtLSU (Wakefield et al. 1990b). These primers are: pAZ102-E (5′-GATGGCTGTTTCCAAGCCCA-3′) and pAZ102-H (5′-GTGTACGTTGCAAAGTACTC-3′) and result in a product of 346 bp. The mtLSU gene is present in multiple copies per organism and provides the most sensitive target for detection. The reaction can be a direct one, or nested, using a second set of primers directed to a region within the product of pAZ102-E and pAZ102-H (Wakefield et al. 1990a). The internal primers are: pAZ102-X (5′-GTGAAATA-CAAATCGGACTAGG-3′) and pAZ102-Y (5′-TCAC-TTAATATTAATTGGGGAGC-3′), which produce a 267 bp product (Vargas et al. 2000). The nested approach is used for samples with anticipated low template levels.

TREATMENT

Trimethoprim-sulfamethoxazole (TMP-SMX)

The combination therapy of TMP-SMX is the therapy of choice for all forms of pneumocystosis (Walzer 2004b). TMP-SMX, which inhibits the folic acid pathway at two distinct steps, is administered orally or intravenously in a dose of 15–20 mg/kg/day trimethoprim (TMP) and 75–100 mg/kg/day sulfamethoxazole (SMX) in three or four divided doses for 21 days in HIV-infected patients and for 14 days in non-infected patients (Walzer 2004b). The parenteral preparation is warranted for gravely ill patients or those with gastrointestinal disturbances. Serum levels of 5–8 μg/mL in adults and 3–5 μg/mL TMP in children and 100–150 μg/mL SMX in both are recommended for maximum efficacy, although there is some question as to the value and practicality of this practice (Walzer 2004b). A significant number of AIDS patients (50–80 percent) experience adverse reactions to this regimen, with gastrointestinal complaints and skin rashes being the most common problems (Stein and Stevens 1995). Other adverse reactions include fever, neutropenia, thrombocytopenia, nausea and vomiting, pancreatitis, nephritis, hyperkalemia, metabolic acidosis, anaphylactoid reaction, and hepatic disturbances (Walzer 2004b). It is thought that the adverse reactions are due to the sulfonamide component, but TMP has been implicated as the cause of skin rashes and hyperkalemia (Velazquez et al. 1993).

Pentamidine

Two parenteral drugs, the cationic diamidine pentamidine isethionate and the dihydrofolate synthase inhibitor trimetrexate, are considered the major alternatives to oral TMP-SMX for moderate to severe pneumocystosis in hospitalized patients (Barry and Johnson 2001).

Pentamidine is administered parenterally as a single daily dose of 4 mg/kg/day or, in some patients, a lower dose of 3 mg/kg/day has been shown to be effective with fewer side effects (Conte et al. 1990). About 50 percent of patients experience adverse effects including local reaction at the site of administration; systemic effects including dizziness, flushing, and cardiac arrhythmias. Hypotension occurs in a significant number of patients and requires vigorous supportive therapy. Nephrotoxicity is a frequent side effect but is usually reversible. The risk of hyperglycemia is related to pentamidine dose, length of administration, and serum levels exceeding 100 ng/mL (Waskin et al. 1988). Other side effects reported were pancreatitis, dysglycemia, hypocalcemia, hypomagnesemia, neutropenia, and hepatic disturbances (Walzer 2004b). Administration of pentamidine via the aerosolized route at 300 mg every month via Respiragard II nebulizer system efficiently delivers high doses of the drug to infected areas, but has been shown to be less effective than other drugs in the treatment of pneumocystosis, probably because it is not well distributed to all lobes of the lungs (Kovacs et al. 2001). It is used primarily as a prophylactic measure.

Trimetrexate

Trimetrexate, a lipid-soluble derivative of methotrexate, was shown to be less effective than TMP-SMX in a controlled trial of hospitalized patients, but better tolerated (Sattler et al. 1994). Trimetrexate is delivered intravenously as a single daily dose of 45 mg/m^2. Bone marrow suppression is the primary side effect associated with this therapy, but administration of folinic acid (20 mg/m^2 every 6 hours) has been found to ameliorate or prevent toxicity. Other side effects reported have been skin rash, peripheral neuropathy, and liver function abnormalities (Walzer 2004b).

Alternatives

Several alternative regimens have been recommended for patients who cannot tolerate or who fail TMP-SMX. These include dapsone at 100 mg daily in combination with TMP taken orally at 320 mg every 8 hours. Atovaquone, a hydroxynaphthoquinone, is another alternative which is administered at 1 500 mg orally per day. Although not as effective as TMP-SMX, it is better tolerated. Dapsone and its combinations appear to be the most efficacious alternative therapies, although some patients still manifest reactions to dapsone which is structurally similar to sulfonamides.

The combination of primaquine and clindamycin has been shown to be most effective against mild to moderate pneumocystosis (Black et al. 1994) but is contraindicated in glucose-6-phosphate dehydrogenase (G6PD) deficiency in which hemolysis has been reported as a major adverse reaction (Korraa and Saadeh 1996). For this purpose, atovaquone has been recommended for patients with G6PD deficiency and who are intolerant to TMP-SMX or pentamidine. The recommended doses are clindamycin at 600 mg (Walzer 2004b) or 300–450 mg (Kovacs et al. 2001) every 6 hours intravenously, with 15–30 mg of primaquine per day orally. Treatment may be initiated with intravenous clindamycin, then switched to oral administration.

Corticosteroids

Adjunctive corticosteroid treatment within the first 72 hours of anti-*Pneumocystis* of moderate to severe pneumocystosis has resulted in improved survival for many patients (Masur 1992). This important finding led to the recommendation by a panel of experts that steroids be added to the treatment of all patients with moderate to severe pneumocystosis (i.e. arterial oxygen pressure of <70 mmHg, alveolar arterial gradient of >35 mmHg) in the following dose regimen: prednisone 40 mg orally twice daily, days 1–5; 40 mg once daily, days 6–10; and 20 mg once daily, days 11–20 (Anonymous 1990). Steroids are thought to reduce the host inflammatory response, which may then reduce the decline in oxygenation often associated with the first few days of therapy. The principal side effects associated with steroid administration were oral candidiasis, exacerbation of herpes simplex infections, and metabolic effects, such as hypoglycemia.

Mutations in *DHPS* and cytochrome *b*. Is resistance occurring?

TMP-SMX is a combination of two antifolates: trimethoprim (TMP), which inhibits the dihydrofolate reductase enzyme (*DHFR*), and sulfamethoxazole (SMX), which inhibits the activity of dihydropteroate synthase (*DHPS*), in folic acid synthesis. *DHFR* reduces dihydrofolic acid to tetrahydrofolic acid, which can be converted to the kinds of tetrahydrofolate cofactors used in 1-carbon transfer chemistry. *DHPS* catalyzes the condensation of pABA and 6-hydroxymethyl-7,8-dihydropterin pyrophosphate. Since many pathogens, inlcuding *Pneumocystis*, must make their own folates and their mammalian hosts do not, this combination has a high therapeutic index.

Mutations in the dihydropteroate synthase gene (*DHPS*), the target of the sulfa component of TMP-SMX, have been identified in over 50 percent of the *P. jirovecii* samples in surveyed geographic areas (Beard et al. 2000). These point mutations have been associated with sulfa and sulfone resistance in bacterial pathogens (Lopez et al. 1987; Fermer and Swedberg 1997; Kai et al. 1999) and malaria (Brooks et al. 1995). In 1997, the first evidence for such mutations in *P. jirovecii* was identified (Lane et al. 1997). The nucleotide polymorphisms all resulted in changes in the amino acid sequence of the enzyme protein, suggesting evolutionary selection was occurring, perhaps in response to drug pressure. These changes were at amino acids 55 and 57. The pterin substrate binds to threonine at site 55 and arginine at site 56. The mutations at position 55 usually substitute alanine for the threonine, which lacks the hydroxyl group needed for binding. The wild-type proline at site 57 is replaced with serine, which may influence the three-dimensional position of Arg56, critical for binding to pABA and sulfa, and the pterin moiety.

Subsequent to these first observations, several studies have been published showing the dramatic spread of the mutant *DHPS* genotypes among *P. jirovecii* infecting human populations around the globe (Helweg-Larsen et al. 1999; Santos et al. 1999; Beard et al. 2000; Huang et al. 2000; Nahimana et al. 2003; Calderon et al. 2004b; Kazanjian et al. 2004; Montes-Cano et al. 2004) and the association of the mutant genotypes with sulfa prophylaxis and potential resistance (Ma et al. 1999; Armstrong et al. 2000; Huang et al. 2000; Kazanjian et al. 2000; Calderon et al. 2004b). In each study, significantly more *DHPS* mutations occurred in sulfa-treated patients than in non-treated study groups. The appearance of the *DHPS* mutations is a relatively recent event, with no evidence for the polymorphisms prior to 1993 (Kazanjian et al. 1998, 2000).

Although the presence of the mutant *DHPS* genotypes and their association with prior sulfa exposure is clear, less clear is how these mutations respond to TMP-SMX therapy. Studies with contrasting results showed a significant decline in survival of patients with the mutations or treatment failure with a sulfa-containing drug (Helweg-Larsen et al. 1999; Nahimana et al. 2003) while other studies have shown no effect in patient outcome (Huang et al. 2000; Kazanjian et al. 2000). Such study outcomes can be confounded by such variables as infections with mixed genotypes, for example. Direct evidence showing a correlation of the *DHPS* gene of *P. carinii* containing the mutations and increased resistance to sulfa was recently demonstrated in a heterologous yeast system, providing evidence that it is likely these mutations will contribute to some decrease in sensitivity to these compounds (Iliades et al. 2004; Meneau et al. 2004).

Last, the encroachment of *P. jirovecii DHPS* mutations into human populations is not relegated to those infected with HIV, rather these mutations are now being reported in association with other types of underlying diseases such as chronic bronchitis (Calderon et al. 2004a). It is apparent that *P. jirovecii DHPS* mutants now enjoy widespread prevalence among the general human population. Whether these mutations will confer a greater survival rate with a higher level of resistance to the sulfa compounds or increased virulence remains to be seen. It should be noted though, that *P. jirovecii* is now being reported in association with a variety of underlying disease states and the issue has been raised as to whether its presence represents a comorbidity factor (De La Horra et al. 2004; Morris et al. 2004).

Atovaquone is a second-line therapeutic for mild to moderate PCP and is used as a prophylactic agent as well. As a ubiquinone mimic, atovaquone binds to cytochrome *b* in the mitochondria, affecting mitochondrial electron transport and parallel processes such as ATP and pyrimidine biosynthesis. Some evidence indicates it may also inhibit ubiquinone synthesis in *Pneumocystis* (Kaneshiro 2001). Mutations in the cytochrome *b* gene of *P. jirovecii* in many patient isolates have been reported (Walker et al. 1998; Kazanjian et al. 2001) and their association with atovaquone exposure documented (Kazanjian et al. 2001). This finding is not altogether surprising since resistance develops in almost all patients receiving atovaquone as monotherapy for the treatment of malaria (Looareesuwan et al. 1996). Unlike the prevalence of two major mutations in *DHPS*, several polymorphisms have been identified in the cytochrome *b* gene of *P. jirovecii*. Like similar studies conducted with the *DHPS* gene, mutations in the cytochrome *b* gene of *Pneumocystis* expressed in a yeast system was shown to confer a resistant phenotype in vitro (Hill et al. 2003; Kessl et al. 2004).

Despite strong evidence supporting the emergence of new *P. jirovecii* genotypes induced by selective drug pressure, there is no reason as yet to change treatment or prophylactic regimens, and dosing schedules as outlined above remain the recommended procedures(Kovacs et al. 2001).

EPIDEMIOLOGY

As the populations at risk for HIV infection have changed since the 1980s, so too have the demographics of PCP. The medically underserved African American and female populations in the USA are now considered at risk (Justice et al. 2001; Kilbourne et al. 2001; Rabeneck et al. 2001). In these populations, PCP remains a leading opportunistic infection (Centers for Disease Control and Prevention 1999; Fisk et al. 2003). In countries considered to be 'developing' (i.e. Africa, Asia, the Phillipines, and Central and South America), the number of those infected with HIV has increased dramatically, sometimes infecting a third of the population in some sub-Saharan countries. Although PCP rarely occurred in African adults in the first decade of the AIDS pandemic, the incidence has risen dramatically as it has in other developing countries, reaching a 25–80 percent coinfection rate with tuberculosis and a mortality rate ranging from 20 to 80 percent (Fisk et al. 2003).

Moreover, PCP is occurring more frequently in other patient groups that in some cases were rarely affected before, including transplantation recipients, especially in lung transplantation patients (Kubak 2002); patients receiving high dose corticosteroids for brain neoplasms, inflammatory or collagen-vascular disorders, and especially patients with Wegener's granulomatosis (Russian and Levine 2001); in association with asthma (Kuitert and Harrison 1991; Abernathy-Carver et al. 1994; Sy et al. 1995; Dziedziczko and Banach-Wawrzenczyk 1997; Contini et al. 1998); chronic obstructive pulmonary disease (Palange et al. 1994); cystic fibrosis (Royce and Blumberg 2000; Sing et al. 2001); Epstein–Barr virus infection (Stiller et al. 1992); lupus erythematosus (Wainstein et al. 1993; Nguyen et al. 1995; Ward and Donald 1999); rheumatoid arthritis (Oien et al. 1995; Ohosone et al. 1997; Rolland et al. 1998); thyroiditis

(Guttler et al. 1993), and ulcerative colitis (Khatchatourian and Seaton 1997; Quan et al. 1997).

The complete life cycle of any *Pneumocystis* species is poorly understood and this has confounded epidemiological studies. The current concepts of the life cycle are discussed in the section on The organism below, but it should be emphasized that neither the route of exit from the lung nor the process it uses to travel to the alveoli is known. There is good evidence that the infection is acquired by an airborne route (Walzer et al. 1977; Hughes 1982), but the agent of transmission has not been identified. In fact, the transmission and acquisition phases are almost completely unknown. Although recent studies have detected *Pneumocystis* DNA in neonatal human and rat populations, an assessment of potential reservoirs in the general population is only in its infancy. Significantly, there is no information about replication within the non-immunocompromised mammalian lung nor its transmission or acquisition among intact hosts.

Seroepidemiologic surveys have shown that the majority of adults in the USA and Europe have anti-*P. carinii* antibodies and healthy children seroconvert at 4 years of age or less (Pifer et al. 1978; Peglow et al. 1990). The primary infection was once thought to be asymptomatic or with generalized symptoms that may be mistaken for a mild viral infection. However, recent studies have shown the high prevalence of *P. jirovecii* in neonates with respiratory symptoms (Vargas et al. 2001), calling into question this widespread assumption. Moreover, reports of the association of *P. jirovecii* in cases of sudden infant death syndrome (SIDS) (Vargas et al. 1999; Morgan et al. 2001; Chabe et al. 2004) also suggest that the organism may be more virulent than once thought. Serological studies have shown that exposure to *P. carinii* is worldwide, although the seroreactivities to specific organism antigens varied among geographic regions (Smulian et al. 1993a).

The evidence on hand supports transmission by immunocompetent individuals or those with subclinical infections. First, more than 75 percent of the human populations surveyed in the USA and Europe are seropositive to *Pneumocystis* antigens by 4 years of age, implying that the organisms are ubiquitous, or at least plentiful, and exposure is a common occurrence. The sheer number of *Pneumocystis*-seropositive immunocompetent individuals argues against the fulminant PCP-infected host as the primary source of infection, as these patients are rare (in relation to the general public) and often hospitalized or have limited access to the general population. More recent data strongly support the concept of host–host transmission. A geographic study of *P. jirovecii* genotypes showed that the genotype with the double mutation in *DHPS* was the most prevalent in the populations surveyed, replacing the wild-type (Beard et al. 2000). The fact that these mutations were not observed before 1993 (Kazanjian et al. 1998, 2000) but had spread over the globe throughout the proceeding

10 years (Miller et al. 2003; Kazanjian et al. 2004; Montes-Cano et al. 2004) argues for a selective pressure and host-dependent spread. Direct evidence is provided by the rat animal model of PCP. Infection with *Pneumocystis* is widespread in the healthy non-immunosuppressed rat populations available from commercial vendors (Icenhour et al. 2001), arguing for highly efficient transmission and maintenance of the infection within these healthy populations. The transmission of the organism from mother to offspring occurs within 1–2 hours after birth in rats (Icenhour et al. 2002) and animals within an infected colony remain positive for *P. carinii* amplicons throughout their residence.

Studies now being conducted in human populations, both immunosuppressed and non-immunsuppressed, point to certain reservoirs of infection in the general population, including patients with underlying disease states besides HIV or overt immunosuppression; neonates; and possibly pregnant women. This is an active and concerted area of research and basic epidemiologic principles for *Pneumocystis* infection should emerge within the next few years.

Asymptomatic colonization

Probable asymptomatic colonization has been shown for both healthy and HIV-infected individuals. Nested or single round PCR amplification targeting the small subunit and large subunit mtLSU of *P. jirovecii* showed 35 of the 47 BAL samples from HIV-infected patients contained *P. jirovecii*- specific DNA; 18 of these came from patients who did not have clinical evidence of *P. jirovecii* pneumonia, suggesting asymptomatic colonization occurred in these patients. In a larger study, 17 of 93 (18 percent) consecutive BALF samples from HIV-negative patients undergoing diagnostic bronchoscopy were found to be positive for the presence of *P. jirovecii* by PCR (Maskell et al. 2003). Analysis of potential risk factors showed only glucocorticoid use [and not lung cancer or chronic obstructive pulmonary disease (COPD)] was associated with the presence of *P. jirovecii*.

The state of pregnancy may facilitate asymptomatic colonization. A study of 33 pregnant healthy women in their third trimester versus 28 healthy women within 15 days of their menstrual cycle showed the presence of *P. jirovecii* DNA in five (15.5 percent) of deep nasal swabs from the pregnant women as compared to none in the non-pregnant women (Vargas et al. 2003). Evidence for the early acquisition of *Pneumocystis* in neonates with the mother as a potential source of organisms is accumulating in both human studies (Miller et al. 2002) and animal studies (Icenhour et al. 2002).

Other factors appear to favor colonization with *P. jirovecii*. In a large study of 367 patients with suspected bacterial pneumonia, the presence of *P. jirovecii* (detected by the same PCR methods discussed above) was statistically associated with old age, concurrent disease, and steroid treatment (Helweg-Larsen et al. 2002).

HOST RESPONSES

Impaired cellular immune function has long been held to be the primary predisposing factor for development of PCP with the CD4 cell playing a central role. The risk of developing PCP in HIV-infected patients increases with a decline in the number of circulating CD4 cells. It is formally recommended that chemoprophylaxis be instituted when the CD4 count in adults falls below 200/mm³ (Centers for Disease Control and Prevention 2002). However, the decrease in CD4 cells rarely occurs without perturbation of other immune responses and there is now considerable evidence that defects in humoral immunity also contribute to the development of PCP.

Antigens

Immunoblotting studies have identified the major antigens of *Pneumocystis* from humans and animal models that elicit serum IgG antibody responses (Walzer 2004a). There are two distinct antigenic regions that are similar across species lines, the MSGs that migrate between 95 and 120 kDa and a broad-based band at 35–55 kDa that likely contains many discrete antigens.

Each *Pneumocystis* organism encodes about 100 different MSG genes distributed among each of its chromosomes as detected by hybridization of Southern blotted electrophoretic karyotypes (Stringer and Keely 2001). These antigens were localized to the surface of the organism (Kovacs et al. 1989) and have been shown to evoke strong humoral and cell-mediated responses (Smulian and Theus 1994) and to interact with host proteins on lung cells such as fibronectin, vitronectin, and SP-A (Pottratz and Martin 1990a; Limper et al. 1991, 1993; Pottratz et al. 1991; Zimmerman et al. 1992) and the mannose receptor on macrophages (Ezekowitz et al. 1991). It is thought that the MSGs function in a variety of ways:

- by promoting adherence to host lung cells and to other *Pneumocystis* organisms
- by complexing with host molecules to avoid the host immune responses
- by modulating the species of surface MSGs to circumvent the immune response
- perhaps by functioning as adhesions during the mating response (Cushion 2004b).

A sequence common to nearly all MSG mRNAs, the upstream conserved sequence (UCS), has been identified and is present in a single copy per organism genome and resides on a single chromosome (Wada et al. 1995; Sunkin and Stringer 1996). It is thought that the MSG genes are transcribed only when adjacent to the UCS and the genes are translocated from subtelomeric regions on each chromosome via reciprocal exchange, gene conversion, or a recombination event. The UCS encodes a signal peptide thought to aid in chaperoning the molecule to the endoplasmic reticulum and golgi for processing and ultimately the cell surface where they are the most abundant surface protein.

The MSGs are heavily glycosylated, highly immunogenic, and have been shown to contain protective B- and T-cell epitopes (Gigliotti and Hughes 1988; Roths and Sidman 1992, 1993; Gigliotti et al. 2002). Immunizations with native or recombinant MSG have resulted in conflicting results; some report a protective effect, while others found no such protection (Theus and Walzer 1997; Gigliotti et al. 1998; Theus et al. 1998; Lasbury et al. 1999). Besides the multigene family of MSGs, other multiply repeated genes are present at the telomeric ends of the chromosomes (Schaffzin et al. 1999; Keely et al. 2001; Ambrose et al. 2004). These include the PRT family of genes that encode kexin-like subtilisin proteases that are likely involved in processing the signal peptide on the nascent MSG and the MSG-related proteins (MSR) that encode proteins similar but distinct to the MSG. These genes are grouped in arrays at the telomeric ends that vary in number of each gene family member present (Keely et al. 2001). The PRT genes exist as a multiple gene family in Pneumocystis infecting rats, but only as a single copy in P. murina (Lee et al. 2000) and P. jirovecii (Kutty and Kovacs 2003), illustrating that there are significant differences among the species of Pneumocystis.

The other major antigen group includes proteins that migrate as a broad band between 35 and 45 kDa in P. jirovecii (Walzer and Linke 1987) and 45–55 kDa in P. carinii and P. wakefieldiae (Vasquez et al. 1996). A gene encoding an antigen within the 45–55 kDa region of rat-derived Pneumocystis (p55) was cloned and shown to be a single copy gene which hybridized to a chromosome band of about 670 kbp on southern blotted electrophoretic karyotypes (Smulian et al. 1992). The function of this antigen is unclear but has been shown to elicit a cell-mediated response in rats previously exposed to Pneumocystis (Smulian et al. 1993b; Theus et al. 1994). An immunodominant epitope was mapped to the 3′ portion of the molecule which contains a 7 amino acid repeat motif. The cells responding to p55 were shown to be CD4+ cells which secreted a Th1 cytokine pattern. The corresponding p55 gene from P. murina was recently cloned and characterized (Ma et al. 2003) and shown to be in single copy as well.

Humoral Immunity

The importance of the humoral arm of the immune response in the control of Pneumocystis infection has been shown by several experimental animal studies and indicated by specific immunodeficiencies. Pneumocystis infection has been reported in B-cell-deficient mice and other animals (Marcotte et al. 1996; Walzer et al. 1997; Lobetti 2000; Macy et al. 2000) and in patients with hypogammaglobulinemia (Walzer et al. 1989; Esolen et al. 1992). Passive immunotherapy with monoclonal antibodies or hyperimmune sera raised against Pneumocystis decreased the organism burden in immunosuppressed animal models (Gigliotti and Hughes 1988; Roths and Sidman 1993). Immunization by intratracheal administration of organisms in CD4-depleted mice was shown to be protective against subsequent Pneumocystis challenge possibly by stimulating serum antibodies (Harmsen et al. 1995; Garvy et al. 1997b; Gigliotti et al. 1998; Pascale et al. 1999). Development of spontaneous PCP in colonies of mice with only B-cell defects suggests a role for the humoral branch in the control of the infection (Marcotte et al. 1996). It is clear that macrophages avidly bind and degrade the organism and addition of antisera and complement enhances the rapid internalization and degradation (Smulian and Theus 1994).

Several studies have shown that the organism elicits a strong humoral response. Early studies used the indirect fluorescent antibody assay (IFA), the enzyme-linked immunosorbent assay (ELISA), and complement fixation assay with whole or fractionated Pneumocystis organisms or crude antigen preparations to show that most of the human population had antibodies to Pneumocystis (Walzer 2004a). Later studies using immunoblotting techniques expanded these studies to evaluate antibodies to specific antigens (Buhl et al. 1993; Lundgren et al. 1993a, b; Smulian et al. 1993a; Chen et al. 1994; Elvin et al. 1994; Smulian and Walzer 1994). Such studies showed that exposure to P. jirovecii in humans occurs early in life. By age 3 years, most children have developed antibodies to the organism (Meuwissen et al. 1977; Pifer et al. 1978; Peglow et al. 1990; Vargas et al. 2001; Daly et al. 2002). A study which examined 680 serum samples from healthy adults in several different areas of the world showed 76 percent reacted with the P. jirovecii antigens by immunoblotting techniques (Smulian et al. 1993a). Most (60 percent) of the reactivity was observed in the area of 35–45 kDa while only about 30 percent of the sera reacted at ~95 kDa, the area at which the MSG family migrates. Of note, the frequency of antibodies to the MSG region exhibited significant geographic variation, suggesting that different strains of P. jirovecii may be predominant in a given area. The high prevalence of serum antibodies and the ubiquitous nature of Pneumocystis have confounded efforts to develop serodiagnostic assays. Most serological surveys have shown the frequence and/or level of antibodies to Pneumocystis among HIV-infected patients and other immunocompromised hosts were similar to those in healthy controls (Smulian and Walzer 1994). The frequent presence of antibodies in most human populations has contributed to the belief that humoral immunity contributes little to the host defense against *Pneumocystis* infection.

The MSG family of genes expressed in part or as the whole recombinant protein have been used to probe the immune response in humans. One laboratory expressed a highly conserved region of the carboxy terminus of an MSG gene that was recognized by all serum samples, and thus could not discriminate among patients with active disease, previous episodes, or healthy controls (Mei et al. 1998). More recently, three overlapping fragments that spanned the entire human MSG gene were generated: *MSGA*, *MSGB*, and *MSGC* (Daly et al. 2002). Recognition of *MSGA* and *MSGB* was significantly lower in HIV-positive patients than normal healthy controls. *MSGC* was recognized by patients with previous episodes of PCP. Responses did not vary by geographic location. These data are important because they show for the first time that different regions of the MSG molecule can identify previously unrecognized differences among populations. Serological response may yet play a role in defining disease state.

Cellular immune response

Evidence for the critical role of T cells in resistance to *Pneumocystis* infection has come from observations of the susceptibility of hosts with T-cell deficiencies to pneumocystosis and directly through reconstitution experiments. Infection with HIV, congenital syndromes with T-cell defects, and administration of drugs to suppress the immune system are all factors that place a mammalian host at risk for pneumocystosis. Except for syndromes with specific T-cell defects, the other immunosuppressive states include a more global immune depression. However, adoptive transfer studies, cell depletion, and reconstitution experiments in rodents and the use of knockout mice models have provided good evidence that the primary defect for development of PCP is CD4 cell dysfunction. CD4 cells also interact with B cells and other cells via the CD40–CD40 ligand pathway. Disruption of these interactions resulted in development of PCP in young children with the hyper-IgM syndrome (Levy et al. 1997) and in impaired clearance of *P. murina* from the lungs of mice (Wiley and Harmsen 1995).

Compelling evidence for the critical role of CD4 cells in resolution of pneumocystosis comes from adoptive transfer experiments in which splenic T cells from normal healthy animals or those exposed to *Pneumocystis* were transferred to animals depleted of CD4 cells. The recipient animals went on to clear or substantively reduce their *Pneumocystis* burden (Graves et al. 1994). Chronic administration of anti-CD4 antibodies to BALB/c mice permitted the development of pneumocystosis following inoculation of *Pneumocystis* organisms

(Shellito et al. 1990). SCID mice were able to resolve the infection after transfer of splenic cells from genetically matched healthy donors (Harmsen and Stankiewicz 1991; Roths and Sidman 1993). However, when the splenic populations were divided into CD8+ and CD4+ populations, transfer of the former cell population was not effective in decreasing the organism burden (Roths and Sidman 1992). When the SCID mice that were reconstituted with splenic populations were then treated with anti-CD4 antibody, the mice developed *Pneumocystis* infection, while those receiving anti-CD8 antibodies were able to resist infection. *Pneumocystis*-infected SCID mice infused with splenic cells depleted of B cells could not resolve the infections (Harmsen and Stankiewicz 1991), suggesting the CD4 cell was not the only immune effector cell involved in mechanisms of resistance. These studies provide strong evidence for the critical role of CD4 cells in resistance and recovery from PCP but suggest a concerted effort by several immune compartments.

CD8 cells accrue in the lungs of experimental animals with pneumocystosis and can be found in the BALF from human patients infected with *P. jirovecii*. CD8 cells have been shown to contribute to host defenses in experimental mouse models (Beck et al. 1996; Beck and Harmsen 1998). BALB/c mice depleted of CD4 lymphocytes, CD8 lymphocytes, or both CD4 and CD8 lymphocytes, and infected with *P. murina*, showed moderate infection in mice depleted of CD4 lymphocytes while mice depleted of CD8+ lymphocytes cleared the inoculum. This result would seem to show that CD8 lymphocytes are unnecessary for defense when CD4 lymphocytes are available. However, mice depleted of both CD4 and CD8 lymphocytes developed PCP with a significantly higher organism burden than mice depleted of CD4 lymphocytes alone. Thus, CD8 lymphocytes function in the defense against *Pneumocystis* in vivo in the context of depleted populations of CD4 lymphocytes. CD8 taken from the lungs of the CD4-depleted mice proliferated in response to *P. murina* antigen and produced interferon-gamma (IFN-γ) in vitro. Thus the mechanism of host defense by CD8 cells likely involves IFN-γ.

Alveolar macrophages serve as the first line of defense against *Pneumocystis* and clear the organism from the alveoli (Koziel et al. 1994; Limper et al. 1997). The nature of the binding occurs via the mannose macrophage receptor (Ezekowitz et al. 1991) and this binding may be enhanced by host-derived molecules (Pottratz and Martin 1990a, b; Limper et al. 1993) or via the gp120 molecule on the organism's surface (Pottratz et al. 1991). More recent studies showed that additional components were necessary for the macrophage: *Pneumocystis* interaction including β-glucan (Vassallo et al. 1999), which resulted in stimulation of tumor necrosis factor alpha (TNF-α) release; extracellular matrix (Fishman et al. 1991); surfactant proteins (Limper et al.

1995; O'Riordan et al. 1995; McCormack et al. 1997); and Fc receptors (Ezekowitz et al. 1991). Hidalgo et al. (1992) were able to demonstrate induction of the oxidative burst in alveolar macrophages isolated from rats and in a cell line (NR8383) and that the evolution of hydrogen peroxide could be enhanced by complement. The release of nitric oxide by mouse alveolar macrophages was elevated in CD4-depleted *Pneumocystis*-infected mice, but the levels did not correlate with either clearance or severity of infection (Shellito et al. 1996). Urokinase-type plasminogen activator, a protease that facilitates recruitment of inflammatory cells, was shown to participate in the clearance of *Pneumocystis* from the lung (Beck et al. 1999). It is interesting to note that *P. carinii* contains a gene that encodes enolase, a function of which is to activate plasminogen (Fox and Smulian 2001), which may serve to regulate host plasminogen activator. Infection with HIV impairs the mannose receptor-mediated binding and phagocytosis of *Pneumocystis* and also changes the cytokine response (Kandil et al. 1994; Koziel et al. 1998a).

The modulation of cytokine profiles by *Pneumocystis* has primarily been studied in animal models. The proinflammatory cytokines TNF-α, and interleukin (IL)-1 play important roles in host defense against this infection, especially in the early stages (Chen et al. 1992a; Garvy et al. 1997a; Kolls et al. 1997; Limper 1997). The role of IL-1 in the clearance of infection was evaluated in the SCID mouse model (Chen et al. 1992b). Reconstituted SCID mice were able to clear the infection and IL-1 activity in the lung homogenate supernates correlated with an increase during infection and a reduction to baseline after clearance. Antibody to IL-1 could inhibit organism clearance, implicating this cytokine as an important host defense against *Pneumocystis* infection. It is interesting to note that secretion of IL-1 by human macrophages from patients with HIV is significantly reduced, perhaps mediating in part the susceptibility to the infection (Smulian and Theus 1994). Activated CD4 cells produce IFN-γ which has been shown to mediate activation of macrophages. Administration of aerosolized IFN-γ to CD4-depleted mice infected with *Pneumocystis* results in a dramatic reduction in organism burden (Beck et al. 1991), suggesting a role in organism clearance. IFN-γ may mediate this response through induction of nitric oxide synthesis in macrophages (Ishimine et al. 1995) or by other means such as TNF-α release. Investigations into the molecular mechanisms of the roles of some cytokines have produced some puzzling results. In one report, deletion of IFN-γ or TNF-α receptor genes did not adversely affect clearance of *Pneumocystis* infection, while elimination of both receptor genes resulted in severe disease (Rudmann et al. 1998). Some cytokines appear to be produced in response to *Pneumocystis* infection (IL-6), but their role, if any, in host defense remains unclear (Chen et al. 1993), while others seem to modulate the host inflamma-

tory response, such as IL-10 (Ruan et al. 2002; Qureshi et al. 2003) or have no apparent role (IL-4, G-CSF) (Garvy et al. 1997b).

The roles of the costimulatory molecules CD2 and CD28 were investigated after mice with the double knockouts spontaneously developed *P. murina* pneumonia (Beck et al. 2003). Mice inoculated with *P. murina* could develop the pneumonia with the loss of CD28 alone, but the organisms eventually cleared. However, mice deficient in both CD2 and CD28 accumulated CD8+ T cells in their lungs in response to infection and demonstrated markedly reduced specific antibody titers. Cytokine profiles suggested IL-10 and IL-15 may be important for the response to *P. murina*.

Although it may seem apparent that a host response to *Pneumocystis* would be a positive effect, increasing evidence implicates a detrimental role in some cases. Immune reconstitution has been associated with marked cellular inflammatory response, rise in proinflammatory cytokines and chemokines, decreased lung compliance, and impaired oxygenation (Chen et al. 1992c; Wright et al. 1997, 1999b). Adoptive transfer of CD4 cells leads to an early hyperinflammatory response, even in the context of profound immunosuppression by corticosteroids (Roths and Sidman 1992; Theus et al. 1995, 1998). CD4 cells expressing the CD25 phenotype appear to be the cell type mediating this response (Hori et al. 2002). A role for the CD8 cell in the pathological inflammatory response has also been implicated. Immunocompetent mice depleted of CD4 cells exposed to *P. murina* develop the infection with a concomitant influx of CD8 cells and increase in neutrophil numbers, experience altered alveolar-capillary permeability, surfactant dysfunction, and impaired respiration (Wright et al. 1997, 1999a, b, 2001). Depletion of the CD8 cells in these CD4-depleted mice dramatically reduced the associated pathology. In rats, the early hyperinflammatory response after adoptive transfer of sensitized CD4 cells could be ameliorated by addition of CD8 cells (Theus et al. 1995; Theus and Walzer 1997). It appears that both cell types contribute to the pathology of *Pneumocystis* infection, and modulation of this response will require further knowledge of the immune circuitry and host immune status.

ANIMAL MODELS OF INFECTION

Animal models of PCP have played an instrumental role in the scientific investigation of this organism. Without a continuous supply of organisms furnished by an in vitro culture system, researchers have relied on animal models for a source of organisms and for assessment of drug responses. The observations, made in the 1950s and 1960s, that the pneumonia could be provoked in rats and rabbits by corticosteroid administration (Weller 1956; Sheldon 1959; Frenkel et al. 1966) led to the current models used today. This subject is reviewed thoroughly by Armstrong and Cushion (1994a). The animal models of pneumocystosis have been shown to be strikingly similar to human pneumocystosis in pathology, organism morphology, and most of the basic biology known to date. Drug responses are similar with only few exceptions (Walzer 1994).

There are basically two types of animal models of pneumocystosis. The first type is a provocation of organisms resident within the lung or acquired from ambient air by chronic administration (e.g. 6–8 weeks) of corticosteroids or another immunosuppressant. The immunosuppressants can be administered in the drinking water (e.g. dexamethasone 1 mg/L) or by injection (e.g. Depo-Medrol, 4 mg/kg/IM or SC). This immunosuppressive regimen is associated with wasting, increased debility, and marked involution of all lymphoid tissues (Walzer et al. 1984a, b). Lymphopenia in the rat model was found to result in inverted CD4/CD8 cells, similar to that in AIDS patients. Oftentimes, an antibiotic (e.g. tetracycline) is concurrently administered to prevent secondary bacterial infections resulting from the immunosuppressed state of the animals. Administration of a low-protein diet (8 percent) has been reported to exacerbate the infection in the rat model (Hughes et al. 1974b). It is presumed that organisms within the lung begin to replicate or replicate without clearance once the immune system is sufficiently suppressed. This has been shown by experiments in which rats are received into barrier facilities that exclude all contact with ambient air, are immunosuppressed, and develop the infection (Cushion et al. 1993). The only source of organisms are those within the lungs of the animals. Rats receive the infectious propagules through an airborne mode, as shown by development of the infection in immunosuppressed *Pneumocystis*-naive animals placed in the same room (but not in contact) with *Pneumocystis*-infected rats (Walzer et al. 1977; Hughes 1982).

The second approach to establishing *Pneumocystis* in animal models is via inoculation. A surgical procedure in which the organisms are instilled into the trachea of lightly anesthetized, immunosuppressed rats (Bartlett et al. 1987), or mice (Bartlett et al. 1991), or instillation via a feeding cannula (Boylan and Current 1992) is used for inoculation. An immunosuppressant regimen as described above is required for replication of the inoculated organisms.

Both methods are useful for studying different aspects of the infection. The 'naturally' acquired infection permits an evaluation of the pathological processes that more closely mimic the actual disease progression. The inoculated model affords standardization in drug evaluations, a consistent supply of organisms, and the ability to use a genetically defined population.

Several species of animals have been used for the study of pneumocystosis by employing either one or both of these techniques. Rats and mice are the models used by most researchers. The ferret model had been used because of its larger size, permitting BAL, but now is rarely employed for studies (Armstrong and Cushion 1994a).

Other approaches to reducing the immune response are gaining favor, especially those designed to preferentially deplete the CD4 cell populations. CD4 cell depletion has been accomplished by administration of anti-CD4 monoclonal antibodies administered on a weekly basis or by implantation of a hybridoma which secretes monoclonal antibodies to this population (Shellito et al. 1990; McFadden et al. 1994; Thullen et al. 2003).

Animals rendered immunodeficient by genetic defects have been used on a more limited basis, but afford a model with fewer manipulations than the previously described ones. Many colonies of immunodeficient mice or rats are often infected with *Pneumocystis*. Homozygous scid/scid mice are especially prone to the infection as well as nude mice and rats (Roths et al. 1990; Sidman and Roths 1994). Neonate rabbits can be naturally infected with *Pneumocystis* without administration of immunosuppressants during their first few weeks of life (Soulez et al. 1988). This model has primarily been used for ultrastructural studies.

CONCLUSIONS

The placement of *Pneumocystis* in the Kingdom Fungi is now generally accepted. The genus has been placed within the Ascomycetes within its own family and order. Four species have been described: *P. jirovecii* which infects humans (Frenkel 1999); *P. carinii* and *P. wakefieldiae* from rats (Frenkel 1999; Cushion et al. 2004a); and from mice, *P. murina* (Keely et al. 2004).

Experimental models and human epidemiologic studies have implicated neonates as potential reservoirs of infection, as well as members of the general population that may be only slightly immunocompromised. Association of *P. jirovecii* with respiratory symptoms and with SIDS calls into question the concept of primary infection as a benign event. The association of *P. jirovecii* with underlying disease states such as COPD suggests its presence may be a comorbidity factor.

P. jirovecii has evolved mutations in the *DHPS* gene which have been linked to sulfa prophylaxis. The *DHPS* mutations in *P. jirovecii* have been reported in many different countries and continents, evidencing the rapid spread of these strains. Mutations in the cytochrome *b* gene, the target of atovaquone, have been reported, raising the specter of drug-resistant organisms.

The *Pneumocystis* Genome Project and efforts of individual investigators have contributed to a better understanding of the metabolic capacity of these organisms and the molecular processes of replication. Genes associated with mating, meiosis, cell cycle, sterol metabolism, heme biosynthesis, and energy production have been identified. Identification of unique cycles or genes should provide new strategies and desperately needed drug targets.

REFERENCES

Abernathy-Carver, K.J., Fan, L.L., et al. 1994. *Legionella* and *Pneumocystis* pneumonias in asthmatic children on high doses of systemic steroids. *Pediatr Pulmonol*, **18**, 135–8

Ainsworth, D.M., Weldon, A.D., et al. 1993. Recognition of *Pneumocystis carinii* in foals with respiratory distress. *Equine Vet J*, **25**, 103–8.

Altschul, S.F., Gish, W., et al. 1990. Basic local alignment search tool. *J Mol Biol*, **215**, 403–10.

Ambrose, H.E., Keely, S.P., et al. 2004. Expression and complexity of the PRT1 multigene family of *Pneumocystis carinii*. *Microbiology*, **150**, Pt 2, 293–300.

Anonymous. 1990. Consensus statement on the use of corticosteroids as adjunctive therapy for pneumocystis pneumonia in the acquired immunodeficiency syndrome. The National Institutes of Health-University of California Expert Panel for Corticosteroids as Adjunctive Therapy for Pneumocystis Pneumonia. *N Engl J Med*, **323**, 1500–4.

Anonymous. 1995. 1995 revised guidelines for prophylaxis against *Pneumocystis carinii* pneumonia for children infected with or perinatally exposed to human immunodeficiency virus. National Pediatric and Family HIV Resource Center and National Center for Infectious Diseases, Centers for Disease Control and Prevention. *MMWR Morb Mortal Wkly Rep*, **44**, RR-4, 1–11.

Armstrong, M.Y.K. and Cushion, M.T. 1994a. Animal models. In: Walzer, P.D. (ed.), *Pneumocystis carinii pneumonia*, 2nd edition. New York: Marcel Dekker, 181–222.

Armstrong, M.Y.K. and Cushion, M.T. 1994b. In vitro cultivation. In: Walzer, P.D. (ed.), *Pneumocystsi carinii pneumonia*, 2nd edition. New York: Marcel Dekker, 3–24.

Armstrong, W., Meshnick, S. and Kazanjian, P. 2000. *Pneumocystis carinii* mutations associated with sulfa and sulfone prophylaxis failures in immunocompromised patients. *Microbes Infect*, **2**, 61–7.

Banerji, S., Lugli, E.B. and Wakefield, A.E. 1994. Identification of two genetically distinct strains of *Pneumocystis carinii* in infected ferret lungs. *J Eukaryot Microbiol*, **41**, 73S.

Banerji, S., Lugli, E.B., et al. 1995. Analysis of genetic diversity at the arom locus in isolates of *Pneumocystis carinii*. *J Eukaryot Microbiol*, **42**, 675–9.

Barry, S.M. and Johnson, M.A. 2001. *Pneumocystis carinii* pneumonia: a review of current issues in diagnosis and management. *HIV Med*, **2**, 123–32.

Bartlett, M.S., Queener, S.F., et al. 1987. Improved rat model for studying *Pneumocystis carinii* pneumonia. *J Clin Microbiol*, **25**, 480–4.

Bartlett, M.S., Queener, S.F., et al. 1991. Inoculated mouse model of *Pneumocystis carinii* pneumonia. *J Protozool*, **38**, 130S–1S.

Barton, E.G.J. and Campbell, W.G.J. 1967. Further observations on the ultrastructure of pneumocystis. *Arch Pathol*, **83** 527–34.

Barton, E.G.J. and Campbell, W.G.J. 1969. *Pneumocystis carinii* in lungs of rats treated with cortisone acetate. Ultrastructural observations relating to the life cycle. *Am J Pathol*, **54**, 209–36.

Basselin-Eiweida, M., Qiu, Y.H., et al. 2001. Mechanisms of amino acid and glucose uptake by *Pneumocystis carinii*. *J Eukaryot Microbiol*, **Suppl**, 155S–6S.

Baughman, R.P. 1994. Use of bronchoscopy in the diagnosis of infection in the immunocompromised host. *Thorax*, **49**, 3–7.

Baughman, R.P., Dohn, M.N. and Frame, P.T. 1994. The continuing utility of bronchoalveolar lavage to diagnose opportunistic infection in AIDS patients. *Am J Med*, **97**, 515–22.

Beard, C.B. 2004. Strain typing methods and molecular epidemiology of *Pneumocystis* pneumonia. *Emerg Infect Dis*, **10**, 1729–35.

Beard, C.B., Carter, J.L., et al. 2000. Genetic variation in *Pneumocystis carinii* isolates from different geographic regions: implications for transmission. *Emerg Infect Dis*, **6**, 265–72.

Beck, J.M. and Harmsen, A.G. 1998. Lymphocytes in host defense against Pneumocystis carinii. *Semin Respir Infect*, **13**, 330–8.

Beck, J.M., Liggitt, H.D., et al. 1991. Reduction in intensity of *Pneumocystis carinii* pneumonia in mice by aerosol administration of gamma interferon. *Infect Immun*, **59**, 3859–62.

Beck, J.M., Newbury, R.L., et al. 1996. Role of CD8+ lymphocytes in host defense against *Pneumocystis carinii* in mice. *J Lab Clin Med*, **128**, 477–87.

Beck, J.M., Preston, A.M. and Gyetko, M.R. 1999. Urokinase-type plasminogen activator in inflammatory cell recruitment and host defense against *Pneumocystis carinii* in mice. *Infect Immun*, **67**, 879–84.

Beck, J.M., Blackmon, M.B., et al. 2003. T cell costimulatory molecule function determines susceptibility to infection with *Pneumocystis carinii* in mice. *J Immunol*, **171**, 1969–77.

Benson, C.A., Spear, J., et al. 1991. Combined APACHE II score and serum lactate dehydrogenase as predictors of in-hospital mortality caused by first episode *Pneumocystis carinii* pneumonia in patients with acquired immunodeficiency syndrome. *Am Rev Respir Dis*, **144**, 319–23.

Benveniste, P. 2002. Sterol metabolism. In *The Arabidopsis book*. Rockville, MD: American Society of Plant Biologists, 1–31.

Bille-Hansen, V., Jorsal, S.E. and Henriksen, S.A. 1990. *Pneumocystis carinii* pneumonia in Danish piglets. *Vet Rec*, **127**, 407–8.

Bishop, R., Gurnell, J., et al. 1997. Detection of *Pneumocystis* DNA in the lungs of several species of wild mammal. *J Eukaryot Microbiol*, **44**, 57S.

Black, J.R., Feinberg, J., et al. 1994. Clindamycin and primaquine therapy for mild-to-moderate episodes of *Pneumocystis carinii* pneumonia in patients with AIDS: AIDS Clinical Trials Group 044. *Clin Infect Dis*, **18**, 905–13.

Blazek, K. and Pokorny, B. 1963. Pneumocystic pneumonia in the field hare (*Lepus europaeus* 1778). In: Ludvik, J. (ed.), *Progress in Protozoology*. New York: Academic Press.

Bommer, W. 1962. *Pneumocystis carinii* from human lungs under electron microscope. *Am J Dis Child*, **104**, 657–61.

Botha, W.S. and van Rensburg, I.B.J. 1979. Pneumocystosis: a chronic respiratory distress syndrome in the dog. *J S Afr Vet Assoc*, **50**, 173–9.

Boylan, C.J. and Current, W.L. 1992. Improved rat model of *Pneumocystis carinii* pneumonia: induced laboratory infections in *Pneumocystis*-free animals. *Infect Immun*, **60**, 1589–97.

Brooks, D.R., Wang, P., et al. 1995. Sequence variation of the hydroxymethyldihydropterin pyrophosphokinase: dihydropteroate synthase gene in lines of the human malaria parasite. *Eur J Biochem*, **224**, 397–405.

Buhl, L., Settnes, O.P. and Andersen, P.L. 1993. Antibodies to *Pneumocystis carinii* in Danish blood donors and AIDS patients with and without *Pneumocystis carinii* pneumonia. *APMIS*, **101**, 707–10.

Calderon, E., de la Horra, C., et al. 2004a. *Pneumocystis jiroveci* isolates with dihydropteroate synthase mutations in patients with chronic bronchitis. *Eur J Clin Microbiol Infect Dis*, **23**, 545–9.

Calderon, E., de la Horra, C., et al. 2004b. Genotypic resistance to sulfamide drugs among patients with *Pneumocystis jiroveci* pneumonia. *Med Clin (Barc)*, **122**, 617–19.

Caliendo, A.M., Hewitt, P.L., et al. 1998. Performance of a PCR assay for detection of *Pneumocystis carinii* from respiratory specimens. *J Clin Microbiol*, **36**, 979–82.

Campbell, W.G.J. 1972. Ultrastructure of *Pneumocystis* in human lung. Life cycle in human pneumocystosis. *Arch Pathol*, **93**, 312–24.

Camus, F., de Picciotto, C., et al. 1993. Pulmonary function tests in HIV-infected patients. *AIDS*, **7**, 1075–9.

Carini, A. and Maciel, J. 1916. Über *Pneumocystis carinii*. *Zentbl Bakteriol Parasitenkd Infektionskr*, **77**, 46.

Centers for Disease Control and Prevention. 1999. Surveillance for AIDS-defining opportunistic illnesses 1992–1997. *MMWR Morb Mortal Wkly Rep*, **48**, SS-2.

Centers for Disease Control and Prevention. 2002. Guidelines for preventing opportunistic infections among HIV-infected persons – 2002 recommendations of the US Public Health Service and the Infectious Diseases Society of America. *MMWR*, **51**, RR-8, 1–52.

Chabe, M., Vargas, S.L., et al. 2004. Molecular typing of *Pneumocystis jirovecii* found in formalin-fixed paraffin-embedded lung tissue sections from sudden infant death victims, 150. *Microbiology*, **150**, Pt 5, 1167–72.

Chagas, C. 1909. Nova tripanozomiaze humana. *Mem Inst Oswaldo Cruz*, **1**, 159–218.

Chandler, F.W., McClure, H.M., et al. 1976. Pulmonary pneumocystosis in nonhuman primates. *Arch Pathol Lab Med*, **100**, 163–7.

Chen, F. and Cushion, M.T. 1994a. Use of an ATP bioluminescent assay to evaluate viability of *Pneumocystis carinii* from rats. *J Clin Microbiol*, **32**, 2791–800.

Chen, F. and Cushion, M.T. 1994b. Use of fluorescent probes to investigate the metabolic state of *Pneumocystis carinii* mitochondria. *J Eukaryot Microbiol*, **41**, 79S.

Chen, W., Havell, E.A. and Harmsen, A.G. 1992a. Importance of endogenous tumor necrosis factor alpha and gamma interferon in host resistance against *Pneumocystis carinii* infection. *Infect Immun*, **60**, 1279–84.

Chen, W., Havell, E.A., et al. 1992b. Interleukin 1: an important mediator of host resistance against *Pneumocystis carinii*. *J Exp Med*, **176**, 713–18.

Chen, W., Mills, J.W. and Harmsen, A.G. 1992c. Development and resolution of *Pneumocystis carinii* pneumonia in severe combined immunodeficient mice: a morphological study of host inflammatory responses. *Int J Exp Pathol*, **73**, 709–20.

Chen, W., Havell, E.A., et al. 1993. Interleukin-6 production in a murine model of *Pneumocystis carinii* pneumonia: relation to resistance and inflammatory response. *Infect Immun*, **61**, 97–102.

Chen, Y., Mei, Q., et al. 1994. Laboratory and clinical study on *Pneumocystis carinii*. II. Immunoblot analysis for detection of anti-*Pneumocystis carinii* antibodies in the sera from healthy blood donors. *Zhongguo Ji Sheng Chong Xue Yu Ji Sheng Chong Bing Za Zhi*, **12**, 115–18.

Chen, Y.C., Eisner, J.D., et al. 2000. Identification of medically important yeasts using PCR-based detection of DNA sequence polymorphisms in the internal transcribed spacer 2 region of the rRNA genes. *J Clin Microbiol*, **38**, 2302–10.

Cirioni, O., Giacometti, A., et al. 1995. In-vitro activity of terbinafine, atovaquone and co-trimoxazole against *Pneumocystis carinii*. *J Antimicrob Chemother*, **36**, 740–2.

Clarkson, A.B.J., Saric, M. and Grady, R.W. 1990. Deferoxamine and eflornithine (DL-alpha-difluoromethylornithine) in a rat model of *Pneumocystis carinii* pneumonia. *Antimicrob Agents Chemother*, **34**, 1833–5.

Coleman, D.L., Dodek, P.M., et al. 1984. Correlation between serial pulmonary function tests and fiberoptic bronchoscopy in patients with *Pneumocystis carinii* pneumonia and the acquired immune deficiency syndrome. *Am Rev Respir Dis*, **129**, 491–3.

Collins, M.S. and Cushion, M.T. 2001. Standardization of an in vitro drug screening assay by use of cryopreserved and characterized *Pneumocystis carinii* populations. *J Eukaryot Microbiol*, **Suppl**, 178S–9S.

Collins, M.S., Bansil, S. and Cushion, M.T. 2003. Expression profiling of the responses of *Pneumocystis carinii* to drug treatment using DNA macroarrays. *J Eukaryot Microbiol*, **50**, Suppl, 605–6.

Comley, J.C. and Sterling, A.M. 1994. Artificial infections of *Pneumocystis carinii* in the SCID mouse and their use in the in vivo evaluation of antipneumocystis drugs. *J Eukaryot Microbiol*, **41**, 540–6.

Conte, J.E.J., Chernoff, D., et al. 1990. Intravenous or inhaled pentamidine for treating *Pneumocystis carinii* pneumonia in AIDS. A randomized trial. *Ann Intern Med*, **113**, 203–9.

Contini, C., Manganaro, M., et al. 1994. Activity of terbinafine against Pneumocystis carinii in vitro and its efficacy in the treatment of experimental pneumonia. *J Antimicrob Chemother*, **34**, 727–35.

Contini, C., Colombo, D., et al. 1996. Employment of terbinafine against Pneumocystis carinii infection in rat models. *Br J Dermatol*, **134**, Suppl 46, 30–2, discussion 40, 30–32.

Contini, C., Villa, M.P., et al. 1998. Detection of *Pneumocystis carinii* among children with chronic respiratory disorders in the absence of HIV infection and immunodeficiency. *J Med Microbiol*, **47**, 329–33.

Copland, J.W. 1974. Canine pneumonia caused by *Pneumocystis carinii*. *Aust Vet J*, **50**, 515–18.

Cornillot, E., Keller, B., et al. 2002. Fine analysis of the *Pneumocystis carinii* f. sp. carinii genome by two-dimensional pulsed-field gel electrophoresis. *Gene*, **293**, 87–95.

Cushion, M.T. 1989. In vitro studies of *Pneumocystis carinii*. *J Protozool*, **36**, 45–52.

Cushion, M.T. 2003. *Pneumocystis*. In: Murray, P.R., Barron, E.J., et al. (eds), *Manual of clinical microbiology*, 8th edition. Washington, DC: ASM Press, 1712–25.

Cushion, M.T. 2004a. Comparative genomics of *Pneumocystis carinii* with other protists: implications for life style. *J Eukaryot Microbiol*, **51**, 30–7.

Cushion, M.T. 2004b. *Pneumocystis*: unraveling the cloak of obscurity. *Trends Microbiol*, **12**, 243–9.

Cushion, M.T. 2004c. Transmission and epidemiology. In: Cushion, M.T. and Walzer, P.D. (eds), *Pneumocystis pneumonia*, 3rd edition. New York: Marcel Dekker, 141–62.

Cushion, M.T. and Arnold, J. 1997. Proposal for a *Pneumocystis* genome project. *J Eukaryot Microbiol*, **44**, 7S.

Cushion, M.T. and Beck, J.M. 2001. Summary of *Pneumocystis* research presented at the 7th International Workshop on Opportunistic Protists. *J Eukaryot Microbiol*, **Suppl**, 101S–5S.

Cushion, M.T. and Ebbets, D. 1990. Growth and metabolism of *Pneumocystis carinii* in axenic culture. *J Clin Microbiol*, **28**, 1385–94.

Cushion, M.T. and Smulian, A.G. 2001. The *Pneumocystis* genome project: update and issues. *J Eukaryot Microbiol*, **Suppl**, 182S–3S.

Cushion, M.T. and Walzer, P.D. 2004. In vitro and in vivo testing of new compounds. In: Cushion, M.T. and Walzer, P.D. (eds), *Pneumocystis pneumonia*. New York: Marcel Dekker, 641–94.

Cushion, M.T., Ruffolo, J.J., et al. 1985. *Pneumocystis carinii*: growth variables and estimates in the A549 and WI-38 VA13 human cell lines. *Exp Parasitol*, **60**, 43–54.

Cushion, M.T., Ruffolo, J.J. and Walzer, P.D. 1988. Analysis of the developmental stages of *Pneumocystis carinii*, in vitro. *Lab Invest*, **58**, 324–31.

Cushion, M.T., Kaselis, M., et al. 1993. Genetic stability and diversity of *Pneumocystis carinii* infecting rat colonies. *Infect Immun*, **61**, 4801–13.

Cushion, M.T., Chen, F. and Kloepfer, N. 1997a. A cytotoxicity assay for evaluation of candidate anti-*Pneumocystis carinii* agents. *Antimicrob Agents Chemother*, **41**, 379–84.

Cushion, M.T., Walzer, P.D., et al. 1997b. Terminology for the life cycle of *Pneumocystis carinii*. *Infect Immun*, **65**, 4365.

Cushion, M.T., Collins, M., et al. 2000. Effects of atovaquone and diospyrin-based drugs on the cellular ATP of *Pneumocystis carinii* f. sp. *carinii*. *Antimicrob Agents Chemother*, **44**, 713–19.

Cushion, M.T., Keely, S. and Stringer, J.R. 2004a. Molecular and phenotypic description of *Pneumocystis wakefieldiae* sp. nov., a new species in rats. *Mycologia*, **96**, 429–38.

Cushion, M.T., Slaven, B.E. and Smulian, A.G. 2004b. *Pneumocystis* genome project and genomic organization. In: Walzer, P.D. and Cushion, M.T. (eds), *Pneumocystis pneumonia*, 3rd edition. New York: Marcel Dekker, 163–82.

Daly, K.R., Fichtenbaum, C.J., et al. 2002. Serologic responses to epitopes of the major surface glycoprotein of *Pneumocystis jiroveci* differ in human immunodeficiency virus-infected and uninfected persons. *J Infect Dis*, **186**, 644–51.

Davalos, A. 1963. Infeccion latente producida por *Pneumocystis carinii* en animales domesticos de la ciudad de Mexico. *Salud Publ Mex*, **5**, 975–7.

Dei-Cas, E., Jackson, H., et al. 1991. Ultrastructural observations on the attachment of *Pneumocystis carinii* in vitro. *J Protozool*, **38**, 205S–7S.

Dei-Cas, E., Mazars, E., et al. 1994. Ultrastructural, genomic, isoenzymatic and biological features make it possible to distinguish rabbit *Pneumocystis* from other mammal *Pneumocystis* strains. *J Eukaryot Microbiol*, **41**, 84S.

De La Horra, C., Varela, J.M., et al. 2004. Association between human-*Pneumocystis* infection and small-cell lung carcinoma. *Eur J Clin Invest*, **34**, 229–35.

Delanoë, P. and Delanoë, Mme. 1912. Sur les rapports des kystes des Carinii des poumon des rats avec le *Trypanosoma lewisii*. *C R Acad Sci (Paris)*, **155**, 658–60.

De Stefano, J.A., Cushion, M.T., et al. 1990a. Analysis of *Pneumocystis carinii* cyst wall. II. Sugar composition. *J Protozool*, **37**, 436–41.

De Stefano, J.A., Cushion, M.T., et al. 1990b. Analysis of *Pneumocystis carinii* cyst wall. I. Evidence for an outer surface membrane. *J Protozool*, **37**, 428–35.

De Stefano, J.A., Trinkle, L.S., et al. 1992. Flow cytometric analyses of lectin binding to *Pneumocystis carinii* surface carbohydrates. *J Parasitol*, **78**, 271–80.

Dohn, M.N. and Frame, P.T. 1994. Clinical manifestations in adults. In: Walzer, P.D. (ed.), *Pneumocystis carinii pneumonia*, 2nd edition. New York: Marcel Dekker, 331–59.

Dutz, W. 1970. *Pneumocystis carinii* pneumonia. *Pathol Annu*, **5**, 309–41.

Dziedziczko, A. and Banach-Wawrzenczyk, E. 1997. Pneumocystis pneumonia in patients with steroid-dependent asthma. *Pneumonol Alergol Pol*, **65**, 399–405.

Edlind, T.D., Bartlett, M.S., et al. 1992. The beta-tubulin gene from rat and human isolates of *Pneumocystis carinii*. *Mol Microbiol*, **6**, 3365–73.

Edman, J.C., Kovacs, J.A., et al. 1988. Ribosomal RNA sequence shows *Pneumocystis carinii* to be a member of the fungi. *Nature*, **334**, 519–22.

Edman, J.C., Edman, U., et al. 1989a. Isolation and expression of the *Pneumocystis carinii* dihydrofolate reductase gene. *Proc Natl Acad Sci U S A*, **86**, 8625–9.

Edman, U., Edman, J.C., et al. 1989b. Isolation and expression of the *Pneumocystis carinii* thymidylate synthase gene. *Proc Natl Acad Sci U S A*, **86**, 6503–7.

Ellis, J. 1994. Coenzyme Q homologs in parasitic protozoa as targets for chemotherapeutic attack. *Parasitol Today*, **43**, 165–70.

Ellis, J.E., Reilly, M.H. and Kaneshiro, E.S. 1994. Identification of an epoxy fatty acid in *Pneumocystis carinii* lipids. *J Eukaryot Microbiol*, **41**, 87S.

Elvin, K., Bjorkman, A., et al. 1994. Seroreactivity to *Pneumocystis carinii* in patients with AIDS versus other immunosuppressed patients. *Scand J Infect Dis*, **26**, 33–40.

English, K., Peters, S.E., et al. 2001. DNA analysis of *Pneumocystis* infecting a Cavalier King Charles spaniel. *J Eukaryot Microbiol*, **Suppl**, 106S.

Eriksson, O.E. and Winka, K. 1998. Families and higher taxa of Ascomycota.. *Myconet*, **1**, 17–24.

Esolen, L.M., Fasano, M.B., et al. 1992. *Pneumocystis carinii* osteomyelitis in a patient with common variable immunodeficiency. *N Engl J Med*, **326**, 999–1001.

Ewing, P.J., Cowell, R.L., et al. 1994. *Pneumocystis carinii* pneumonia in foals. *J Am Vet Med Assoc*, **204**, 929–33.

Ezekowitz, R.A., Williams, D.J., et al. 1991. Uptake of *Pneumocystis carinii* mediated by the macrophage mannose receptor. *Nature*, **351**, 155–8.

Farrow, B.R.H., Watson, A.D., et al. 1972. *Pneumocystis* pneumonia in the dog. *J Comp Pathol*, **82**, 447–53.

Fermer, C. and Swedberg, G. 1997. Adaptation to sulfonamide resistance in *Neisseria meningitidis* may have required compensatory changes to retain enzyme function: kinetic analysis of dihydropteroate synthases

from *N. meningitidis* expressed in a knockout mutant of *Escherichia coli*. *J Bacteriol*, **179**, 831–7.

Fishman, J.A., Samia, J.A., et al. 1991. The effects of extracellular matrix (ECM) proteins on the attachment of *Pneumocystis carinii* to lung cell lines in vitro. *J Protozool*, **38**, 34S–7S.

Fisk, D.T., Meshnick, S. and Kazanjian, P.H. 2003. *Pneumocystis carinii* pneumonia in patients in the developing world who have acquired immunodeficiency syndrome. *Clin Infect Dis*, **36**, 70–8.

Florin-Christensen, M., Florin-Christensen, J. and Kaneshiro, E.S. 1995. Uptake and metabolism of L-serine by *Pneumocystis carinii carinii*. *J Eukaryot Microbiol*, **42**, 669–75.

Forrest, D.M., Zala, C., et al. 1999. Determinants of short- and long-term outcome in patients with respiratory failure caused by AIDS-related *Pneumocystis carinii* pneumonia. *Arch Intern Med*, **159**, 741–7.

Fox, D. and Smulian, A.G. 1999. Mitogen-activated protein kinase Mkp1 of *Pneumocystis carinii* complements the slt2Δ defect in the cell integrity pathway of *Saccharomyces cerevisiae*. *Mol Microbiol*, **34**, 451–62.

Fox, D. and Smulian, A.G. 2001. Plasminogen-binding activity of enolase in the opportunistic pathogen *Pneumocystis carinii*. *Med Mycol*, **39**, 495–507.

Frenkel, J.K. 1976. *Pneumocystis jiroveci* n. sp. from man: morphology, physiology, and immunology in relation to pathology. *Natl Cancer Inst Monogr*, **43**, 13–30.

Frenkel, J.K. 1999. *Pneumocystis* pneumonia, an immunodeficiency-dependent disease (IDD): a critical historical overview. *J Eukaryot Microbiol*, **46**, 89S–92S.

Frenkel, J.K., Good, J.T. and Shultz, J.A. 1966. Latent *Pneumocystis* infection of rats, relapse, and chemotherapy. *Lab Invest*, **15**, 1559–77.

Furlong, S.T., Samia, J.A., et al. 1994. Phytosterols are present in *Pneumocystis carinii*. *Antimicrob Agents Chemother*, **38**, 2534–40.

Gajdusek, D.C. 1957. *Pneumocystis carinii* – etiologic agent of interstitial plasma cell pneumonia of premature and young infants. *Pediatrics*, **19**, 543–64.

Gargas, A., DePriest, P.T., et al. 1995. Multiple origins of lichen symbioses in fungi suggested by SSU rDNA phylogeny. *Science*, **268**, 1492–5.

Garner, R.E., Walker, A.N. and Horst, M.N. 1991. Morphologic and biochemical studies of chitin expression in *Pneumocystis carinii*. *J Protozool*, **38**, 12S–4S.

Garret, C.E. and Coderre, J.A. 1984. A bifunctional thymidylate synthase-dihydrofolate reductase in protozoa. *Mol Biochem Parasitol*, **11**, 257–65.

Garvy, B.A., Ezekowitz, R.A. and Harmsen, A.G. 1997a. Role of gamma interferon in the host immune and inflammatory responses to *Pneumocystis carinii* infection. *Infect Immun*, **65**, 373–9.

Garvy, B.A., Wiley, J.A., et al. 1997b. Protection against *Pneumocystis carinii* pneumonia by antibodies generated from either T helper 1 or T helper 2 responses. *Infect Immun*, **65**, 5052–6.

Gigliotti, F. and Hughes, W.T. 1988. Passive immunoprophylaxis with specific monoclonal antibody confers partial protection against *Pneumocystis carinii* pneumonitis in animal models. *J Clin Invest*, **81**, 1666–8.

Gigliotti, F., Haidaris, P.J., et al. 1993a. Further evidence of host species-specific variation in antigens of *Pneumocystis carinii* using the polymerase chain reaction. *J Infect Dis*, **168**, 191–4.

Gigliotti, F., Harmsen, A.G., et al. 1993b. *Pneumocystis carinii* is not universally transmissible between mammalian species. *Infect Immun*, **61**, 2886–90.

Gigliotti, F., Wiley, J.A. and Harmsen, A.G. 1998. Immunization with *Pneumocystis carinii* gpA is immunogenic but not protective in a mouse model of *P. carinii* pneumonia. *Infect Immun*, **66**, 3179–82.

Gigliotti, F., Haidaris, C.G., et al. 2002. Passive intranasal monoclonal antibody prophylaxis against murine *Pneumocystis carinii* pneumonia. *Infect Immun*, **70**, 1069–74.

Graves, D., Smulian, A.G. and Walzer, P.D. 1994. Humoral and cellular immunity. In: Walzer, P.D. (ed.), *Pneumocystis carinii pneumonia*, 2nd edition. New York: Marcel Dekker, 267–88.

Greuter, W. (ed.) 2000. *International Code of Botanical Nomenclature. 16th International Botanical Congress 1999*, St. Louis Code edition. Konigstein, Germany: Lubrecht & Cramer Ltd.

Gruden, J.F., Huang, L., et al. 1997. High-resolution CT in the evaluation of clinically suspected *Pneumocystis carinii* pneumonia in AIDS patients with normal, equivocal, or nonspecific radiographic findings. *AJR Am J Roentgenol*, **169**, 967–75.

Guillot, J., Demanche, C., et al. 2004. Phylogenetic relationships among *Pneumocystis* from Asian macaques inferred from mitochondrial rRNA sequences. *Mol Phylogenet Evol*, **31**, 988–96.

Gustafson, M.P., Limper, A.H. and Leof, E.B. 1999. Characterization of the Cdc25 phosphatase in *Pneumocystis carinii*. *J Eukaryot Microbiol*, **46**, 129S.

Gutteridge, W.E. 1991. 566C80, an antimalarial hydroxynaphthoquinone with broad spectrum: experimental activity against opportunistic parasitic infections of AIDS patients. *J Protozool*, **38**, 141S–3S.

Guttler, R., Singer, P.A., et al. 1993. *Pneumocystis carinii* thyroiditis. Report of three cases and review of the literature. *Arch Intern Med*, **153**, 393–6.

Hahn, P.Y., Evans, S.E., et al. 2003. *Pneumocystis carinii* cell wall beta-glucan induces release of macrophage inflammatory protein-2 from alveolar epithelial cells via a lactosylceramide-mediated mechanism. *J Biol Chem*, **278**, 2043–50.

Harmsen, A.G. and Stankiewicz, M. 1991. T cells are not sufficient for resistance to *Pneumocystis carinii* pneumonia in mice. *J Protozool*, **38**, 44S–5S.

Harmsen, A.G., Chen, W. and Gigliotti, F. 1995. Active immunity to *Pneumocystis carinii* reinfection in T-cell-depleted mice. *Infect Immun*, **63**, 2391–5.

Haselton, P.S., Curry, A. and Rankin, E.M. 1981. *Pneumocystis carinii* pneumonia: a light microscopical and ultrastructural study. *J Clin Pathol*, **34**, 1138–46.

Helweg-Larsen, J., Benfield, T.L., et al. 1999. Effects of mutations in *Pneumocystis carinii* dihydropteroate synthase gene on outcome of AIDS-associated *P. carinii* pneumonia. *Lancet*, **354**, 1347–51.

Helweg-Larsen, J., Jensen, J.S., et al. 2002. Detection of *Pneumocystis* DNA in samples from patients suspected of bacterial pneumonia – a case-control study. *BMC Infect Dis*, **2**, 28, 1–6.

Herskowitz, I. 1989. A regulatory hierarchy for cell specialization in yeast. *Nature*, **342**, 749–57.

Hidalgo, H.A., Helmke, R.J., et al. 1992. *Pneumocystis carinii* induces an oxidative burst in alveolar macrophages. *Infect Immun*, **60**, 1–7.

Hill, P., Kessl, J., et al. 2003. Recapitulation in *Saccharomyces cerevisiae* of cytochrome *b* mutations conferring resistance to atovaquone in *Pneumocystis jiroveci*. *Antimicrob Agents Chemother*, **47**, 2725–31.

Hoffman, A.G., Lawrence, M.G., et al. 1992. Reduction of pulmonary surfactant in patients with human immunodeficiency virus infection and *Pneumocystis carinii* pneumonia. *Chest*, **102**, 1730–6.

Hopewell, P.C. and Luce, J.M. 1985. Pulmonary involvement in the acquired immunodeficiency syndrome. *Chest*, **87**, 104–12.

Hori, S., Carvalho, T.L. and Demengeot, J. 2002. CD25+CD4+ regulatory T cells suppress CD4+ T cell-mediated pulmonary hyperinflammation driven by *Pneumocystis carinii* in immunodeficient mice. *Eur J Immunol*, **32**, 1282–91.

Huang, L. 2004. Clinical presentation and diagnosis of *Pneumocystis* pneumonia in HIV-infected patients. In: Cushion, M.T. and Walzer, P.D. (eds), *Pneumocystis pneumonia*, 3rd edition. New York: Marcel Dekker, 349–406.

Huang, L., Beard, C.B., et al. 2000. Sulfa or sulfone prophylaxis and geographic region predict mutations in the *Pneumocystis carinii* dihydropteroate synthase gene. *J Infect Dis*, **182**, 1192–8.

Huang, L., Crothers, K., et al. 2003. *Pneumocystis* colonization in HIV-infected patients. *J Eukaryot Microbiol*, **50**, Suppl, 616–17.

Hughes, W.T. 1982. Natural mode of acquisition for de novo infection with *Pneumocystis carinii*. *J Infect Dis*, **145**, 842–8.

Hughes, W.T. 2003. *Pneumocystis carinii* vs. *Pneumocystis jiroveci*: Another misnomer (response to Stringer et al.). *Emerg Infect Dis*, **9**, 276–7.

Hughes, W.T., McNabb, P.C., et al. 1974a. Efficacy of trimethoprim and sulfamethoxazole in the prevention and treatment of *Pneumocystis carinii* pneumonitis. *Antimicrob Agents Chemother*, **5**, 289–93.

Hughes, W.T., Price, R.A., et al. 1974b. Protein-calorie malnutrition. A host determinant for *Pneumocystis carinii* infection. *Am J Dis Child*, **128**, 44–52.

Hughes, W.T., Gray, V.L., et al. 1990. Efficacy of a hydroxynaphthoquinone, 566C80, in experimental *Pneumocystis carinii* pneumonitis. *Antimicrob Agents Chemother*, **34**, 225–8.

Hugot, J.P., Demanche, C., et al. 2003. Phylogenetic systematics and evolution of primate-derived *Pneumocystis* based on mitochondrial or nuclear DNA sequence comparison. *Syst Biol*, **52**, 735–44.

Icenhour, C.R., Rebholz, S.L., et al. 2001. Widespread occurrence of *Pneumocystis carinii* in commercial rat colonies detected using targeted PCR and oral swabs. *J Clin Microbiol*, **39**, 3437–41.

Icenhour, C.R., Rebholz, S.L., et al. 2002. Early acquisition of *Pneumocystis carinii* in neonatal rats as evidenced by PCR and oral swabs. *Eukaryot Cell*, **1**, 414–19.

Iliades, P., Meshnick, S.R. and Macreadie, I.G. 2004. Dihydropteroate synthase mutations in *Pneumocystis jiroveci* can affect sulfamethoxazole resistance in a *Saccharomyces cerevisiae* model. *Antimicrob Agents Chemother*, **48**, 2617–23.

Ishimine, T., Kawakami, K., et al. 1995. Analysis of cellular response and gamma interferon synthesis in bronchoalveolar lavage fluid and lung homogenate of mice infected with *Pneumocystis carinii*. *Microbiol Immunol*, **39**, 49–58.

Ivady, G. and Paldy, L. 1958. Ein neves Behandlungsverfahren der interstitiellen plasmazelligen Pneumonie Fruhgeborener mit funfwertigen Stibium und aromatischen Diamidinen. *Monatsschr Kinderheilkd*, **106**, 10–15.

Ivanov, I.P., Gesteland, R.F. and Atkins, J.F. 2000. Antizyme expression: a subversion of triplet decoding, which is remarkably conserved by evolution, is a sensor for an autoregulatory circuit. *Nucleic Acids Res*, **28**, 3185–96.

Jacobs, J.L., Libby, D.M., et al. 1991. A cluster of *Pneumocystis carinii* pneumonia in adults without predisposing illnesses. *N Engl J Med*, **324**, 246–50.

Jecny, V. 1973. *Pneumocystis carinii* findings in the offspring of various farm animals. *Cesk Epidemiol Mikrobiol Immunol*, **22**, 135–40.

Jensen, T.K., Boye, M. and Bille-Hansen, V. 2001. Application of fluorescent in situ hybridization for specific diagnosis of *Pneumocystis carinii* pneumonia in foals and pigs. *Vet Pathol*, **38**, 269–74.

Jones, J.L., Hanson, D.L., et al. 1999. Surveillance for AIDS-defining opportunistic illnesses, 1992–1997. CDC Surveillance Summaries. *MMWR*, **48**, SS-2, 1–22.

Justice, A.C., Landefeld, C.S., et al. 2001. Justification for a new cohort study of people aging with and without HIV infection. *J Clin Epidemiol*, **54**, Suppl 1, S3–8.

Kai, M., Matsuoka, M., et al. 1999. Diaminodiphenylsulfone resistance of *Mycobacterium leprae* due to mutations in the dihydropteroate synthase gene. *FEMS Microbiol Lett*, **177**, 231–5.

Kandil, O., Fishman, J.A., et al. 1994. Human immunodeficiency virus type 1 infection of human macrophages modulates the cytokine response to *Pneumocystis carinii*. *Infect Immun*, **62**, 644–50.

Kaneshiro, E.S. 2001. Are cytochrome *b* gene mutations the only cause of atovaquone resistance in *Pneumocystis*? *Drug Resist Updat*, **4**, 322–9.

Kaneshiro, E.S., Cushion, M.T., et al. 1989. Analyses of *Pneumocystis* fatty acids. *J Protozool*, **36**, 69S–72S.

Kaneshiro, E.S., Ellis, J.E., et al. 1994a. Evidence for the presence of metabolic sterols in *Pneumocystis*: identification and initial characterization of *Pneumocystis carinii* sterols. *J Eukaryot Microbiol*, **41**, 78–85.

Kaneshiro, E.S., Ellis, J.E., et al. 1994b. Isoprenoid metabolism in *Pneumocystis carinii*. *J Eukaryot Microbiol*, **41**, 93S.

Kaneshiro, E.S., Amit, Z., et al. 1999a. Pneumocysterol [(24Z)-ethylidenelanost-8-en-3beta-ol], a rare sterol detected in the opportunistic pathogen *Pneumocystis carinii hominis*: structural identity and chemical synthesis. *Proc Natl Acad Sci U S A*, **96**, 97–102.

Kaneshiro, E.S., Collins, M. and Cushion, M.T. 1999b. Effects of sterol inhibitors on the ATP content of *Pneumocystis carinii*. *J Eukaryot Microbiol*, **46**, 142S–3S.

Kaneshiro, E.S., Collins, M.S. and Cushion, M.T. 2000. Inhibitors of sterol biosynthesis and amphotericin B reduce the viability of *Pneumocystis carinii* f. sp. *carinii*. *Antimicrob Agents Chemother*, **44**, 1630–8.

Kaneshiro, E.S., Rosenfeld, J.A., et al. 2001a. *Pneumocystis carinii* erg6 gene: sequencing and expression of recombinant SAM:sterol methyltransferase in heterologous systems. *J Eukaryot Microbiol*, **Suppl**, 144S–6S.

Kaneshiro, E.S., Sul, D., et al. 2001b. *Pneumocystis carinii* synthesizes four ubiquinone homologs: inhibition by atovaquone and bupravaquone but not by stigmatellin. *J Eukaryot Microbiol*, **Suppl**, 172S–3S.

Kaplan, J.E., Hanson, D., et al. 2000. Epidemiology of human immunodeficiency virus-associated opportunistic infections in the United States in the era of highly active antiretroviral therapy. *Clin Infect Dis*, **30**, Suppl 1, S5–S14.

Kazanjian, P., Locke, A.B., et al. 1998. *Pneumocystis carinii* mutations associated with sulfa and sulfone prophylaxis failures in AIDS patients. *AIDS*, **12**, 873–8.

Kazanjian, P., Armstrong, W., et al. 2000. *Pneumocystis carinii* mutations are associated with duration of sulfa or sulfone prophylaxis exposure in AIDS patients. *J Infect Dis*, **182**, 551–7.

Kazanjian, P., Armstrong, W., et al. 2001. *Pneumocystis carinii* cytochrome *b* mutations are associated with atovaquone exposure in patients with AIDS. *J Infect Dis*, **183**, 819–22.

Kazanjian, P.H., Fisk, D., et al. 2004. Increase in prevalence of *Pneumocystis carinii* mutations in patients with AIDS and *P. carinii* pneumonia, in the United States and China. *J Infect Dis*, **189**, 1684–7.

Keely, S., Pai, H.J., et al. 1994. *Pneumocystis* species inferred from analysis of multiple genes. *J Eukaryot Microbiol*, **41**, 94S.

Keely, S.P., Wakefield, A.E., et al. 2001. Detailed structure of *Pneumocystis carinii* chromosome ends. *J Eukaryot Microbiol*, **Suppl**, S118–120S.

Keely, S.P., Fischer, J.M., et al. 2004. Phylogenetic identification of *Pneumocystis murina* sp. nov., a new species in laboratory mice. *Microbiology*, **150**, Pt 5, 1153–65.

Kessl, J.J., Hill, P., et al. 2004. Molecular basis for atovaquone resistance in *Pneumocystis jirovecii* modeled in the cytochrome bc1 complex of *Saccharomyces cerevisiae*. *J Biol Chem*, **279**, 2817–24.

Khatchatourian, M. and Seaton, T.L. 1997. An unusual complication of immunosuppressive therapy in inflammatory bowel disease. *Am J Gastroenterol*, **92**, 1558–60.

Kilbourne, A.M., Justice, A.C., et al. 2001. General medical and psychiatric comorbidity among HIV-infected veterans in the post-HAART era. *J Clin Epidemiol*, **54**, Suppl 1, S22–8.

Kolls, J.K., Lei, D., et al. 1997. Exacerbation of murine *Pneumocystis carinii* infection by adenoviral-mediated gene transfer of a TNF inhibitor. *Am J Respir Cell Mol Biol*, **16**, 112–18.

Konomi, M., Fujimoto, K., et al. 2003. Characterization and behaviour of alpha-glucan synthase in *Schizosaccharomyces pombe* as revealed by electron microscopy. *Yeast*, **20**, 427–38.

Korraa, H. and Saadeh, C. 1996. Options in the management of pneumonia caused by *Pneumocystis carinii* in patients with acquired immune deficiency syndrome and intolerance to trimethoprim/sulfamethoxazole. *South Med J*, **89**, 272–7.

Kottom, T.J. and Limper, A.H. 1999. Assembly of cell wall glucans by *Pneumocystis carinii*: characterization of the Gsc-1 subunit mediating beta-glucan synthesis. *J Eukaryot Microbiol*, **46**, 131S.

Kottom, T.J. and Limper, A.H. 2000. Cell wall assembly by *Pneumocystis carinii*. Evidence for a unique gsc-1 subunit mediating beta-1 3-glucan deposition. *J Biol Chem*, **275**, 40628–34.

Kovacs, J.A. Halpern, J.L., et al. 1989. Monoclonal antibodies to *Pneumocystis carinii*: identification of specific antigens and characterization of antigenic differences between rat and human isolates. *J Infect Dis*, **159**, 60–70.

Kovacs, J.A. Powell, F., et al. 1993. Multiple genes encode the major surface glycoprotein of *Pneumocystis carinii*. *J Biol Chem*, **268**, 6034–40.

Kovacs, J.A. Gill, V.J., et al. 2001. New insights into transmission, diagnosis, and drug treatment of *Pneumocystis carinii* pneumonia. *JAMA*, **286**, 2450–60.

Koziel, H., C'Riordan, D., et al. 1994. Alveolar macrophage interaction with *Pneumocystis carinii*. *Immunol Ser*, **60**, 417–36.

Koziel, H., Eichbaum, Q., et al. 1998a. Reduced binding and phagocytosis of *Pneumocystis carinii* by alveolar macrophages from persons infected with HIV-1 correlates with mannose receptor downregulation. *J Clin Invest*, **102**, 1332–44

Koziel, H., Phelps, D.S., et al. 1998b. Surfactant protein-A reduces binding and phagocytosis of pneumocystis carinii by human alveolar macrophages in vitro. *Am J Respir Cell Mol Biol*, **18**, 834–43.

Kroe, D.M., Kirsch, C.M. and Jensen, W.A. 1997. Diagnostic strategies for *Pneumocystis carinii* pneumonia. *Semin Respir Infect*, **12**, 70–8.

Kubak, B.M. 2002. Fungal infection in lung transplantation. *Transpl Infect Dis*, **4**, Suppl 3, 24–31.

Kucera, K., Schlesinger, L. and Kadlee, A. 1968. Pneumocystosis in pigs. *Folia Parasitol*, **17**, 75.

Kucera, K., Vanek, J. and Jirovec, O. 1971. Pneumozystose. In *Leitfaden der Zooanthroponosen*. Berlin: VEB Verlad Volk und Gesundheit, 279.

Kuitert, L.M. and Harrison, A.C. 1991. *Pneumocystis carinii* pneumonia as a complication of methotrexate treatment of asthma. *Thorax*, **46**, 936–7.

Kumar, S., Tamura, K., et al. 2001. MEGA2: molecular evolutionary genetics analysis software. *Bioinformatics*, **17**, 1244–5.

Kutty, G. and Kovacs, J.A. 2003. A single-copy gene encodes Kex1, a serine endoprotease of *Pneumocystis jiroveci*. *Infect Immun*, **71**, 571–4.

Laakkonen, J. and Henttonen, H. 1995. *Pneumocystis carinii* in arvicoline rodents: seasonal, interspecific and geographic differences. *Can J Zool*, **73**, 961–6.

Laakkonen, J. and Soveri, T. 1995. Characterization of *Pneumocystis carinii* infection in *Sorex araneus* from southern Finland. *J Wildl Dis*, **31**, 228–32.

Laakkonen, J., Fisher, R.N. and Case, T.J. 2001a. Pneumocystosis in wild small animals from California. *J Wildl Dis*, **37**, 408–12.

Laakkonen, J., Fisher, R.N. and Case, T.J. 2001b. Spatial analysis on the occurrence of *Pneumocystis carinii* in the shrew *Notiosorex crawfordi* in fragmented landscape in southern California. *J Eukaryot Microbiol*, **Suppl**, 111S–2S.

Lainson, R. and Shaw, J.J. 1975. Pneumocystis and histoplasma infection in wild animals from the Amazon region of Brazil. *Trans R Soc Trop Med Hyg*, **69**, 505–8.

Lane, B.R., Ast, J.C., et al. 1997. Dihydropteroate synthase polymorphisms in *Pneumocystis carinii*. *J Infect Dis*, **175**, 482–5.

Larsen, H.H., Kovacs, J.A., et al. 2002. Development of a rapid real-time PCR assay for quantitation of *Pneumocystis carinii* f. sp. *carinii*. *J Clin Microbiol*, **40**, 2989–93.

Lasbury, M.E., Angus, C.W., et al. 1999. Recombinant major surface glycoprotein of *Pneumocystis carinii* elicits a specific immune response but is not protective in immunosuppressed rats. *J Eukaryot Microbiol*, **46**, 136S–7S

Lee, C.H., Helweg-Larsen, J., et al. 1998. Update on *Pneumocystis carinii* f. sp. *hominis* typing based on nucleotide sequence variations in internal transcribed spacer regions of rRNA genes. *J Clin Microbiol*, **36**, 734–41.

Lee, L.H., Gigliotti, F., et al. 2000. Molecular characterization of KEX1, a kexin-like protease in mouse *Pneumocystis carinii*. *Gene*, **242**, 141–50.

Levy, J., Espanol-Boren, T., et al. 1997. Clinical spectrum of X-linked hyper-IgM syndrome. *J Pediatr*, **131**, Pt 1, 47–54.

Limper, A.H. 1997. Tumor necrosis factor alpha-mediated host defense against *Pneumocystis carinii*. *Am J Respir Cell Mol Biol*, **16**, 110–11.

Limper, A.H., Pottratz, S.T. and Martin, W.J. 1991. Modulation of *Pneumocystis carinii* adherence to cultured lung cells by a mannose-dependent mechanism. *J Lab Clin Med*, **118**, 492–9.

Limper, A.H., Standing, J.E., et al. 1993. Vitronectin binds to *Pneumocystis carinii* and mediates organism attachment to cultured lung epithelial cells. *Infect Immun*, **61**, 4302–9.

Limper, A.H., O'Riordan, D.M., et al. 1994. Accumulation of surfactant protein D in the lung during *Pneumocystis carinii* pneumonia. *J Eukaryot Microbiol*, **41**, 98S.

Limper, A.H., Crouch, E.C., et al. 1995. Surfactant protein-D modulates interaction of *Pneumocystis carinii* with alveolar macrophages. *J Lab Clin Med*, **126**, 416–22.

Limper, A.H., Hoyte, J.S. and Standing, J.E. 1997. The role of alveolar macrophages in *Pneumocystis carinii* degradation and clearance from the lung. *J Clin Invest*, **99**, 2110–17.

Liu, Y. and Leibowitz, M.J. 1993. Variation and in vitro splicing of group I introns in rRNA genes of *Pneumocystis carinii*. *Nucleic Acids Res*, **21**, 2415–21.

Liu, Y., Rocourt, M., et al. 1992. Sequence and variability of the 5.8S and 26S rRNA genes of *Pneumocystis carinii*. *Nucleic Acids Res*, **20**, 3763–72.

Liu, Y.J. and Hall, B.D. 2004. Body plan evolution of ascomycetes, as inferred from an RNA polymerase II phylogeny. *Proc Natl Acad Sci U S A*, **101**, 4507–12.

Lobetti, R. 2000. Common variable immunodeficiency in miniature dachshunds affected with *Pneumocystis carinii* pneumonia. *J Vet Diagn Invest*, **12**, 39–45.

Long, G.G., White, J.D. and Stookey, J.L. 1975. *Pneumocystis carinii* infection in splenectomized owl monkeys. *J Am Vet Assoc*, **167**, 651–4.

Looareesuwan, S., Viravan, C., et al. 1996. Clinical studies of atovaquone, alone or in combination with other antimalarial drugs, for treatment of acute uncomplicated malaria in Thailand. *Am J Trop Med Hyg*, **54**, 62–6.

Lopez, P., Espinosa, M., et al. 1987. Sulfonamide resistance in *Streptococcus pneumoniae*: DNA sequence of the gene encoding dihydropteroate synthase and characterization of the enzyme. *J Bacteriol*, **169**, 4320–6.

Ludmerer, K.M. and Kissane, J.M. 1991. Pulmonary infiltrates in a 67 year old man. *Am J Med*, **90**, 509–15.

Lundgren, B., Lipschik, G.Y. and Kovacs, J.A. 1991. Purification and characterization of a major human *Pneumocystis carinii* surface antigen. *J Clin Invest*, **87**, 163–70.

Lundgren, B., Kovacs, J.A., et al. 1993a. IgM response to a human *Pneumocystis carinii* surface antigen in HIV-infected patients with pulmonary symptoms. *Scand J Infect Dis*, **25**, 515–20.

Lundgren, B., Lebech, M., et al. 1993b. Antibody response to a major human *Pneumocystis carinii* surface antigen in patients without evidence of immunosuppression and in patients with suspected atypical pneumonia. *Eur J Clin Microbiol Infect Dis*, **12**, 105–9.

Ma, L., Borio, L., et al. 1999. *Pneumocystis carinii* dihydropteroate synthase but not dihydrofolate reductase gene mutations correlate with prior trimethoprim-sulfamethoxazole or dapsone use. *J Infect Dis*, **180**, 1969–78.

Ma, L., Kutty, G., et al. 2003. Characterization of variants of the gene encoding the p55 antigen in *Pneumocystis* from rats and mice. *J Med Microbiol*, **52**, Pt 11, 955–60.

Macy, J.D.J., Weir, E.C., et al. 2000. Dual infection with *Pneumocystis carinii* and *Pasteurella pneumotropica* in B cell-deficient mice: diagnosis and therapy. *Comp Med*, **50**, 49–55.

Marcotte, H., Levesque, D., et al. 1996. *Pneumocystis carinii* infection in transgenic B cell-deficient mice. *J Infect Dis*, **173**, 1034–7.

Martin, W.J. and Wright, J.R. 1999. *Pneumocystis carinii* pneumonia and pulmonary surfactant. *J Lab Clin Med*, **133**, 406–7.

Maskell, N.A., Waine, D.J., et al. 2003. Asymptomatic carriage of *Pneumocystis jiroveci* in subjects undergoing bronchoscopy: a prospective study. *Thorax*, **58**, 594–7.

Mason, R.J., Greene, K. and Voelker, D.R. 1998. Surfactant protein A and surfactant protein D in health and disease. *Am J Physiol*, **275**, Pt 1, L1–L13.

Masur, H. 1992. Prevention and treatment of pneumocystis pneumonia. *N Engl J Med*, **327**, 1853–60.

Masur, H., Ognibene, F.P., et al. 1989. CD4 counts as predictors of opportunistic pneumonias in human immunodeficiency virus (HIV) infection. *Ann Intern Med*, **111**, 223–31.

Mata, A.D. 1959. Infeccion latente producida por *Pneumocystis carinii* en couejo de la enidad de Mexico. *Rev Inst Salubr Enferm Trop*, **19**, 357.

Matsumoto, Y. and Yoshida, Y. 1984. Sporogony in *Pneumocystis carinii*: synaptonemal complexes and meiotic nuclear divisions observed in precysts. *J Protozool*, **31**, 420–8.

Matsumoto, Y. and Yoshida, Y. 1986. Advances in *Pneumocystis* biology. *Parasitol Today*, **2**, 137–42.

Matsumoto, Y., Yamada, T., et al. 1987. Pneumocystis infection in macaque monkeys: *Macaca fuscata fuscata* and *Macaca fascicularis*. *Parasitol Res*, **73**, 324–7.

Mazer, M.A., Kovacs, J.A., et al. 1987. Histoenzymological study of selected dehydrogenase enzymes in *Pneumocystis carinii*. *Infect Immun*, **55**, 727–30.

McCann, P.P., Bacchi, C.J., et al. 1986. Inhibition of polyamine biosynthesis by alpha-difluoromethylornithine in African trypanosomes and *Pneumocystis carinii* as a basis of chemotherapy: biochemical and clinical aspects. *Am J Trop Med Hyg*, **35**, 1153–6.

McConnell, E.E., Basson, P.A. and Pienaar, J.G. 1971. Pneumocystosis in a domestic goat. *Onderstepoort J Vet Res*, **38**, 117–24.

McCormack, F.X., Festa, A.L., et al. 1997. The carbohydrate recognition domain of surfactant protein A mediates binding to the major surface glycoprotein of *Pneumocystis carinii*. *Biochemistry*, **36**, 8092–9.

McCully, R.M., Loyd, J. and Kuys, D. 1979. Canine pneumocystis pneumonia. *J S Afr Vet Assoc*, **50**, 207–9.

McFadden, D.C., Powles, M.A., et al. 1994. Use of anti-CD4+ hybridoma cells to induce *Pneumocystis carinii* in mice. *Infect Immun*, **62**, 4887–92.

Mei, Q., Turner, R.E., et al. 1998. Characterization of major surface glycoprotein genes of human *Pneumocystis carinii* and high-level expression of a conserved region. *Infect Immun*, **66**, 4268–73.

Meneau, I., Sanglard, D., et al. 2004. *Pneumocystis jiroveci* dihydropteroate synthase polymorphisms confer resistance to sulfadoxine and sulfanilamide in *Saccharomyces cerevisiae*. *Antimicrob Agents Chemother*, **48**, 2610–16.

Merali, S. 1999. *Pneumocystis carinii* polyamine catabolism. *J Biol Chem*, **274**, 21017–22.

Merali, S. and Clarkson, A.B.J. 1996. Polyamine content of *Pneumocystis carinii* and response to the ornithine decarboxylase inhibitor DL-alpha-difluoromethylornithine. *Antimicrob Agents Chemother*, **40**, 973–8.

Merali, S., Frevert, U., et al. 1999. Continuous axenic cultivation of *Pneumocystis carinii*. *Proc Natl Acad Sci U S A*, **96**, 2402–7.

Merali, S., Vargas, D., et al. 2000. *S*-Adenosylmethionine and *Pneumocystis carinii*. *J Biol Chem*, **275**, 14958–63.

Meuwissen, J.H., Tauber, I., et al. 1977. Parasitologic and serologic observations of infection with *Pneumocystis* in humans. *J Infect Dis*, **136**, 43–9.

Miller, R.F., Ambrose, H.E., et al. 2002. Probable mother-to-infant transmission of *Pneumocystis carinii* f. sp. *hominis* infection. *J Clin Microbiol*, **40**, 1555–7.

Miller, R.F., Lindley, A.R., et al. 2003. Genotypes of *Pneumocystis jiroveci* isolates obtained in Harare, Zimbabwe and London, United Kingdom. *Antimicrob Agents Chemother*, **47**, 3979–81.

Mills, J. 1986. *Pneumocystis carinii* and *Toxoplasma gondii* infections in patients with AIDS. *Rev Infect Dis*, **8**, 1001–11.

Montes-Cano, M.A., de la Horra, C., et al. 2004. *Pneumocystis jiroveci* genotypes in the Spanish population. *Clin Infect Dis*, **39**, 123–8.

Morales, I.J., Vohra, P.K., et al. 2003. Characterization of a Lanosterol 14-[alpha]-demethylase from *Pneumocystis carinii*. *Am J Respir Cell Mol Biol*, **29**, 232–8.

Morgan, D.J., Vargas, S.L., et al. 2001. Identification of *Pneumocystis carinii* in the lungs of infants dying of sudden infant death syndrome. *Pediatr Infect Dis J*, **20**, 306–9.

Morris, A., Sciurba, F.C., et al. 2004. Association of chronic obstructive pulmonary disease severity and *Pneumocystis* colonization. *Am J Respir Crit Care Med*, **170**, 408–13.

Nahimana, A., Rabodonirina, M., et al. 2003. Association between a specific *Pneumocystis jiroveci* dihydropteroate synthase mutation and failure of pyrimethamine/sulfadoxine prophylaxis in human immunodeficiency virus-positive and -negative patients. *J Infect Dis*, **188**, 1017–23.

Navin, T.R., Beard, C.B., et al. 2001. Effect of mutations in *Pneumocystis carinii* dihydropteroate synthase gene on outcome of *P. carinii* pneumonia in patients with HIV-1: a prospective study. *Lancet*, **358**, 545–9.

Newsome, A.L., Durkin, M.M., et al. 1991. Videomicroscopic recording of *Pneumocystis carinii* motion. *J Protozool*, **38**, 207S–8S.

Ng, V.L., Yajko, D.M. and Hadley, W.K. 1997. Extrapulmonary pneumocystosis. *Clin Microbiol Rev*, **10**, 401–18.

Nguyen, T.B., Galezowski, N., et al. 1995. *Pneumocystis carinii* infection disclosing untreated systemic lupus erythematosus. *Rev Med Interne*, **16**, 146–9.

Nielsen, M.H., Settnes, O.P., et al. 1998. Different ultrastructural morphology of *Pneumocystis carinii* derived from mice, rats, and rabbits. *APMIS*, **106**, 771–9.

Nimri, L.F., Moura, I.N., et al. 2002. Genetic diversity of *Pneumocystis carinii* f. sp. *hominis* based on variations in nucleotide sequences of internal transcribed spacers of rRNA genes. *J Clin Microbiol*, **40**, 1146–51.

Norris, K.A., Wildschutte, H., et al. 2003. Genetic variation at the mitochondrial large-subunit rRNA locus of *Pneumocystis* isolates from simian immunodeficiency virus-infected rhesus macaques. *Clin Diagn Lab Immunol*, **10**, 1037–42.

Ohosone, Y., Okano, Y., et al. 1997. Clinical characteristics of patients with rheumatoid arthritis and methotrexate induced pneumonitis. *J Rheumatol*, **24**, 2299–303.

Oien, K.A., Black, A., et al. 1995. *Pneumocystis carinii* pneumonia in a patient with rheumatoid arthritis, not on immunosuppressive therapy and in the absence of human immunodeficiency virus infection. *Br J Rheumatol*, **34**, 677–9.

O'Riordan, D.M., Standing, J.E., et al. 1995. Surfactant protein D interacts with *Pneumocystis carinii* and mediates organism adherence to alveolar macrophages. *J Clin Invest*, **95**, 2699–710.

Ortiz-Rivera, M., Liu, Y. and Felder, R. 1994. Sequence differences in the rRNA ITS regions identify specific isolates of *P. carinii*. *J Eukaryot Microbiol*, **41**, 107S.

Ortiz-Rivera, M., Liu, Y., et al. 1995. Comparison of coding and spacer region sequences of chromosomal rRNA-coding genes of two sequevars of *Pneumocystis carinii*. *J Eukaryot Microbiol*, **42**, 44–9.

Palange, P., Serra, P., et al. 1994. *Pneumocystis carinii* pneumonia in a patient with chronic obstructive pulmonary disease but no evident immunoincompetence. *Clin Infect Dis*, **19**, 543–4.

Palluault, F., Dei-Cas, E., et al. 1990. Golgi complex and lysosomes in rabbit derived *Pneumocystis carinii*. *Biol Cell*, **70**, 73–82.

Palluault, F., Pietrzyk, B., et al. 1991. Three-dimensional reconstruction of rabbit-derived *Pneumocystis carinii* from serial-thin sections. I: Trophozoite. *J Protozool*, **38**, 402–7.

Pascale, J.M., Shaw, M.M., et al. 1999. Intranasal immunization confers protection against murine *Pneumocystis carinii* lung infection. *Infect Immun*, **67**, 805–9.

Paulsrud, J.R. and Queener, S.F. 1994. Incorporation of fatty acids and amino acids by cultured *Pneumocystis carinii*. *J Eukaryot Microbiol*, **41**, 633–8.

Peglow, S.L., Smulian, A.G., et al. 1990. Serologic responses to *Pneumocystis carinii* antigens in health and disease. *J Infect Dis*, **161**, 296–306.

Perron Lepage, M.F., Gerber, V. and Suter, M.M. 1999. A case of interstitial pneumonia associated with *Pneumocystis carinii* in a foal. *Vet Pathol*, **36**, 621–4.

Perryman, L.E., McGuire, T.C. and Crawford, T.B. 1978. Maintenance of foals with combined immunodeficiency: causes and control of secondary infections. *Am J Vet Res*, **39**, 1043–7.

Pesanti, E.L. 1984. *Pneumocystis carinii*: oxygen uptake, antioxidant enzymes, and susceptibility to oxygen-mediated damage. *Infect Immun*. **44**, 7–11.

Pesanti, E.L. 1989. Enzymes of *Pneumocystis carinii*: electrophoretic mobility on starch gels. *J Protozool*, **36**, 2S–3S.

Pesanti, E.L. and Cox, C. 1981. Metabolic and synthetic activities of *Pneumocystis carinii* in vitro. *Infect Immun*, **34**, 908–14.

Pesanti, E.L., Bartlett, M.S. and Smith, J.W. 1988. Lack of detectable activity of ornithine decarboxylase in *Pneumocystis carinii*. *J Infect Dis*, **158**, 1137–8.

Peters, S.E., Wakefield, A.E., et al. 1992. A search for *Pneumocystis carinii* in post-mortem lungs by DNA amplification. *J Pathol*, **166**, 195–8.

Peters, S.E., Wakefield, A.E., et al. 1994. *Pneumocystis carinii* pneumonia in thoroughbred foals: identification of a genetically distinct organism by DNA amplification. *J Clin Microbiol*, **32**, 213–16.

Phair, J., Munoz, A., et al. 1990. The risk of *Pneumocystis carinii* pneumonia among men infected with human immunodeficiency virus type 1. Multicenter AIDS Cohort Study Group. *N Engl J Med*, **322**, 161–5.

Phelps, D.S. and Rose, R.M. 1991. Increased recovery of surfactant protein A in AIDS-related pneumonia. *Am Rev Respir Dis*, **143**, Pt 1, 1072–5.

Pifer, L.L., Hughes, W.T., et al. 1978. *Pneumocystis carinii* infection: evidence for high prevalence in normal and immunosuppressed children. *Pediatrics*, **61**, 35–41.

Pixley, F.J., Wakefield, A.E., et al. 1991. Mitochondrial gene sequences show fungal homology for *Pneumocystis carinii*. *Mol Microbiol*, **5**, 1347–51

Poelma, F.G. 1972. *Pneumocystis carinii* in hares, *Lepus europaeus* Pallas, in the Netherlands. *Z Parasitenkd*, **40**. 195–202.

Poelma, F.G. 1975. *Pneumocystis carinii* infections in zoo animals *Z Parasitenk*, **46**. 61–8.

Porollo, A., Slaven, B.E. et al. 2004. The *Pneumocystis* Genome Project Website http://pgp.cchmc.org.

Pottratz, S.T. and Martin, W.J. 1990a. Mechanism of *Pneumocystis carinii* attachment to cultured rat alveolar macrophages. *J Clin Invest*, **86**, 1678–83.

Pottratz, S.T. and Martin, W.J. 1990b. Role of fibronectin in *Pneumocystis carinii* attachment to cultured lung cells. *J Clin Invest*, **85**, 351–6.

Pottratz, S.T., Paulsrud, J., et al. 1991. *Pneumocystis carinii* attachment to cultured lung cells by pneumocystis gp 120, a fibronectin binding protein. *J Clin Invest*, **88**, 403–7.

Quan, V.A., Saunders, B.P., et al. 1997. Cyclosporin treatment for ulcerative colitis complicated by fatal *Pneumocystis carinii* pneumonia. *Br Med J*, **314**, 363–4.

Qureshi, M.H., Harmsen, A.G. and Garvy, B.A. 2003. IL-10 modulates host responses and lung damage induced by *Pneumocystis carinii* infection. *J Immunol*, **170**, 1002–9.

Rabeneck, L., Menke, T., et al. 2001. Using the national registry of HIV-infected veterans in research: lessons for the development of disease registries. *J Clin Epidemiol*, **54**, 1195–203.

Rao, C.P. and Gelfand, E.W. 1983. *Pneumocystis carinii* pneumonitis in patients with hypogammaglobulinemia and intact T cell immunity. *J Pediatr*, **103**, 410–12.

Rebholz, S.L. and Cushion, M.T. 2001. Three new karyotype forms of *Pneumocystis carinii* f. sp. *carinii* identified by contoured clamped homogeneous electrical field (CHEF) electrophoresis. *J Eukaryot Microbiol*, **Suppl**, 109S–10S.

Richardson, J.D., Queener, S.F., et al. 1989. Binary fission of *Pneumocystis carinii* trophozoites grown in vitro. *J Protozool*, **36**, 27S–9S.

Richter, C.B., Humanson, G.L. and Godbold, J.H. 1978. Endemic *Pneumocystis carinii* in a marmoset colony. *J Comp Pathol*, **88**, 171.

Rolland, Y., Cantagrel, A., et al. 1998. *Pneumocystis carinii* pneumopathy in rheumatoid polyarthritis treated by methotrexate in a patient with pulmonary asbestosis. *Rev Med Interne*, **19**, 581–3.

Roths, J.B. and Sidman, C.L. 1992. Both immunity and hyperresponsiveness to *Pneumocystis carinii* result from transfer of CD4+ but not CD8+ T cells into severe combined immunodeficiency mice. *J Clin Invest*, **90**, 673–8.

Roths, J.B. and Sidman, C.L. 1993. Single and combined humoral and cell-mediated immunotherapy of *Pneumocystis carinii* pneumonia in immunodeficient SCID mice. *Infect Immun*, **61**, 1641–9.

Roths, J.B., Marshall, J.D., et al. 1990. Spontaneous *Pneumocystis carinii* pneumonia in immunodeficient mutant SCID mice. Natural history and pathobiology. *Am J Pathol*, **136**, 1173–86.

Royce, F.H. and Blumberg, D.A. 2000. *Pneumocystis carinii* isolated from lung lavage fluid in an infant with cystic fibrosis. *Pediatr Pulmonol*, **29**, 235–8.

Ruan, S., Tate, C., et al. 2002. Local delivery of the viral interleukin-10 gene suppresses tissue inflammation in murine *Pneumocystis carinii* infection. *Infect Immun*, **70**, 6107–13.

Rudmann, D.G., Preston, A.M., et al. 1998. Susceptibility to *Pneumocystis carinii* in mice is dependent on simultaneous deletion of IFN-gamma and type 1 and 2 TNF receptor genes. *J Immunol*, **161**, 360–6.

Ruffolo, J.J. 1994. *Pneumocystis carinii* cell structure. In: Walzer, P.D. (ed.), *Pneumocystis carinii pneumonia*, 2nd edition. New York: Marcel Dekker, 25–43.

Ruffolo, J.J., Cushion, M.T. and Walzer, P.D. 1986. Techniques for examining *Pneumocystis carinii* in fresh specimens. *J Clin Microbiol*, **23**, 17–21.

Ruffolo, J.J., Cushion, M.T. and Walzer, P.D. 1989. Ultrastructural observations on life cycle stages of *Pneumocystis carinii*. *J Protozool*, **36**, 53S–4S.

Russian, D.A. and Levine, S.J. 2001. *Pneumocystis carinii* pneumonia in patients without HIV infection. *Am J Med Sci*, **321**, 56–65.

Sankary, R.M., Turner, J., et al. 1988. Alveolar-capillary block in patients with AIDS and *Pneumocystis carinii* pneumonia. *Am Rev Respir Dis*, **137**, 443–9.

Santos, L.D., Lacube, P., et al. 1999. Contribution of dihydropteroate synthase gene typing for *Pneumocystis carinii* f. sp. *hominis* epidemiology. *J Eukaryot Microbiol*, **46**, 133S–4S.

Sattler, F.R., Frame, P., et al. 1994. Trimetrexate with leucovorin versus trimethoprim-sulfamethoxazole for moderate to severe episodes of *Pneumocystis carinii* pneumonia in patients with AIDS: a prospective, controlled multicenter investigation of the AIDS Clinical Trials Group Protocol 029/031. *J Infect Dis*, **170**, 165–72.

Saulsbury, F.T., Bernstein, M.T. and Winkelstein, J.A. 1979. *Pneumocystis carinii* pneumonia as the presenting infection in congenital hypogammaglobulinemia. *J Pediatr*, **95**, 559–61.

Schaffzin, J.K. and Stringer, J.R. 2000. The major surface glycoprotein expression sites of two special forms of rat *Pneumocystis carinii* differ in structure. *J Infect Dis*, **181**, 1729–39.

Schaffzin, J.K., Sunkin, S.M. and Stringer, J.R. 1999. A new family of *Pneumocystis carinii* genes related to those encoding the major surface glycoprotein. *Curr Genet*, **35**, 134–43.

Schmatz, D.M., Romancheck, M.A., et al. 1990. Treatment of *Pneumocystis carinii* pneumonia with 1,3-beta-glucan synthesis inhibitors. *Proc Natl Acad Sci U S A*, **87**, 5950–4.

Schmatz, D.M., Powles, M.A., et al. 1995. New semisynthetic pneumocandins with improved efficacies against *Pneumocystis carinii* in the rat. *Antimicrob Agents Chemother*, **39**, 1320–3.

Schrager, L.K., Vermund, S.H. and Langreth, S.G. 1993. *Pneumocystis carinii*. In: Krier, J.P. (ed.), *Parasitic protozoa*, 2nd edition. San Diego: Academic Press, 227–97.

Sedlmeier, H. and Dahme, E. 1955. *Pneumocystis carinii* Infektion beim Hund. *Zentbl Allg Pathol*, **93**, 150–5.

Seibold, H.R. and Munnell, J.F. 1977. *Pneumocystosis carinii*, in a pig. *Vet Pathol*, **14**, 89–91.

Settnes, O.P. and Hasselager, E. 1984. Occurrence of *Pneumocystis carinii* Delanoë and Delanoë, 1912 in dogs and cats. *Nord Ved Med*, **36**, 179–81.

Settnes, O.P. and Lodal, J. 1980. Prevalence of *Pneumocystis carinii* Delanoë and Delanoë, 1912 in rodents in Denmark. *Nord Ved Med*, **32**, 17–27.

Settnes, O.P., Nielsen, P.B., et al. 1994. A survey of birds in Denmark for the presence of *Pneumocystis carinii*. *Avian Dis*, **38**, 1–10.

Shaw, R.J., Roussak, C., et al. 1988. Lung function abnormalities in patients infected with the human immunodeficiency virus with and without overt pneumonitis. *Thorax*, **43**, 436–40.

Sheldon, W.H. 1959. Experimental pulmonary *Pneumocystis carinii* infection in rabbits. *J Exp Med*, **110**, 147–60.

Shellito, J., Suzara, V.V., et al. 1990. A new model of *Pneumocystis carinii* infection in mice selectively depleted of helper T lymphocytes. *J Clin Invest*, **85**, 1686–93.

Shellito, J.E., Kolls, J.K., et al. 1996. Nitric oxide and host defense against *Pneumocystis carinii* infection in a mouse model. *J Infect Dis*, **173**, 432–9.

Shimizu, A., Kumura, F. and Kumura, S. 1985. Occurrence of *Pneumocystis carinii* in animals in Japan. *Jpn J Vet Sci*, **47**, 309–11.

Shiota, T. 1984. Morphology and development of *Pneumocystis carinii* observed by phase-contrast and semiultrathin section light microscopy. *Jpn J Parasitol*, **33**, 443–55.

Shively, J.N., Dellers, R.W., et al. 1973. *Pneumocystis carinii* pneumonia in two foals. *J Am Vet Med Assoc*, **162**, 648–52.

Shively, J.N., Moe, K.K. and Dellers, R.W. 1974. Fine structure of spontaneous *Pneumocystis carinii* pulmonary infection in foals. *Cornell Vet*, **64**, 72–88.

Sicheritz-Ponten, T. and Andersson, S.G. 2001. A phylogenomic approach to microbial evolution. *Nucleic Acids Res*, **29**, 545–52.

Sidman, C.L. and Roths, J.B. 1994. New animal models for *Pneumocystis carinii* research: immunodeficient mice. In: Walzer, P.D. (ed.), *Pneumocystis carinii pneumonia*, 2nd edition. New York: Marcel Dekker, 223–36.

Sinclair, K., Wakefield, A.E., et al. 1991. *Pneumocystis carinii* organisms derived from rat and human hosts are genetically distinct. *Mol Biochem Parasitol*, **45**, 183–4.

Sing, A., Geiger, A.M., et al. 2001. *Pneumocystis carinii* carriage among cystic fibrosis patients, as detected by nested PCR. *J Clin Microbiol*, **39**, 2717–18.

Skelly, M., Hoffman, J., et al. 2003. *S*-Adenosylmethionine concentrations in diagnosis of *Pneumocystis carinii* pneumonia. *Lancet*, **361**, 1267–8.

Skold, O. 2000. Sulfonamide resistance: mechanisms and trends. *Drug Resist Updat*, **3**, 155–60.

Sleight, R.G., Mehta, M.A. and Kaneshiro, E.S. 1994. Uptake and metabolism of fluorescent lipid analogs by *Pneumocystis carinii*. *J Eukaryot Microbiol*, **41**, 111S.

Sloand, E., Laughon, B., et al. 1993. The challenge of *Pneumocystis carinii* culture. *J Eukaryot Microbiol*, **40**, 188–95.

Smulian, A.G. and Theus, S.A. 1994. Cellular immune response in *Pneumocystis carinii* infection. *Parasitol Today*, **10**, 229–31.

Smulian, A.G. and Walzer, P.D. 1994. Serological studies of *Pneumocystis carinii* infection. In: Walzer, P.D. (ed.), *Pneumocystis carinii pneumonia*. New York: Marcel Dekker, 141–54.

Smulian, A.G., Stringer, J.R., et al. 1992. Isolation and characterization of a recombinant antigen of *Pneumocystis carinii*. *Infect Immun*, **60**, 907–15.

Smulian, A.G., Sullivan, D.W., et al. 1993a. Geographic variation in the humoral response to *Pneumocystis carinii*. *J Infect Dis*, **167**, 1243–7.

Smulian, A.G., Theus, S.A., et al. 1993b. A 55 kDa antigen of *Pneumocystis carinii*: analysis of the cellular immune response and characterization of the gene. *Mol Microbiol*, **7**, 745–53.

Smulian, A.G., Ryan, M., et al. 1996. Signal transduction in *Pneumocystis carinii*: characterization of the genes (*pcg1*) encoding the alpha subunit of the G protein (PCG1) of *Pneumocystis carinii carinii* and *Pneumocystis carinii ratti*. *Infect Immun*, **64**, 691–701.

Smulian, A.G., Sesterhenn, T., et al. 2001. The ste3 pheromone receptor gene of *Pneumocystis carinii* is surrounded by a cluster of signal transduction genes. *Genetics*, **157**, 991–1002.

Soulez, B., Dei-Cas, E. and Camus, D. 1988. The rabbit experimental host of *Pneumocystis carinii*. *Ann Parasitol Hum Comp*, **63**, 5–15.

Stansell, J.D. and Hopewell, P.C. 1995. *Pneumocystis carinii* pneumonia: risk factors, clinical presentation and natural history. In: Sattler, F.R. and Walzer, P.D. (eds), *Pneumocystis carinii*. Philadelphia: Baillière Tindall, 449–60.

Stedman, T.T. and Buck, G.A. 1996. Identification, characterization, and expression of the BiP endoplasmic reticulum resident chaperonins in *Pneumocystis carinii*. *Infect Immun*, **64**, 4463–71.

Stedman, T.T., Butler, D.R. and Buck, G.A. 1998. The HSP70 gene family in *Pneumocystis carinii*: molecular and phylogenetic characterization of cytoplasmic members. *J Eukaryot Microbiol*, **45**, 589–99.

Stein, D.S. and Stevens, R.C. 1995. Treatment associated toxicities. In: Sattler, F.R. and Walzer, P.D. (eds), *Pneumocystis carinii*. London: Baillière Tindall, 505–30.

Stiller, R.A., Paradis, I.L. and Dauber, J.H. 1992. Subclinical pneumonitis due to *Pneumocystis carinii* in a young adult with elevated antibody titers to Epstein–Barr virus. *J Infect Dis*, **166**, 926–30.

Stokes, D.C., Gigliotti, F., et al. 1987. Experimental *Pneumocystis carinii* pneumonia in the ferret. *Br J Exp Pathol*, **68**, 267–76.

Stringer, J.R. 1993. The identity of *Pneumocystis carinii*: not a single protozoan, but a diverse group of exotic fungi. *Infect Agents Dis*, **2**, 109–17.

Stringer, J.R. 1996. *Pneumocystis carinii*: what is it, exactly? *Clin Microbiol Rev*, **9**, 489–98.

Stringer, J.R. and Cushion, M.T. 1998. The genome of *Pneumocystis carinii*. *FEMS Immunol Med Microbiol*, **22**, 15–26.

Stringer, J.R. and Keely, S.P. 2001. Genetics of surface antigen expression in *Pneumocystis carinii*. *Infect Immun*, **69**, 627–39.

Stringer, J.R., Stringer, S.L., et al. 1993. Molecular genetic distinction of *Pneumocystis carinii* from rats and humans. *J Eukaryot Microbiol*, **40**, 733–41.

Stringer, J.R., Beard, C.B., et al. 2002. A new name (*Pneumocystis jiroveci*) for *Pneumocystis* from humans. *Emerg Infect Dis*, **8**, 891–6.

Stringer, J.R., Beard, C.B., et al. 2003. A new name (*Pneumocystis jiroveci*) for *Pneumocystis* from humans (response to Hughes). *Emerg Infect Dis*, **9**, 277–9.

Stringer, S.L., Stringer, J.R., et al. 1989. *Pneumocystis carinii*: sequence from ribosomal RNA implies a close relationship with fungi. *Exp Parasitol*, **68**, 450–61.

Sunkin, S.M. and Stringer, J.R. 1996. Translocation of surface antigen genes to a unique telomeric expression site in *Pneumocystis carinii*. *Mol Microbiol*, **19**, 283–95.

Sunkin, S.M., Stringer, S.L. and Stringer, J.R. 1994. A tandem repeat of rat-derived Pneumocystis carinii genes encoding the major surface glycoprotein. *J Eukaryot Microbiol*, **41**, 292–300.

Sy, M.L., Chin, T.W. and Nussbaum, E. 1995. *Pneumocystis carinii* pneumonia associated with inhaled corticosteroids in an immunocompetent child with asthma. *J Pediatr*, **127**, 1000–2.

Taylor, J.W., Jacobson, D.J., et al. 2000. Phylogenetic species recognition and species concepts in fungi. *Fungal Genet Biol*, **31**, 21–32.

Tegoshi, T. and Yoshida, Y. 1989. New system of in vitro cultivation of *Pneumocystis carinii* without feeder cells. *J Protozool*, **36**, 29S–31S.

Telzak, E.E. and Armstrong, D. 1994. Extrapulmonary infection and other unusual manifestations of *Pneumocystis carinii*. In: Walzer, P.D. (ed.), *Pneumocystis carinii pneumonia*, 2nd edition. New York: Marcel Dekker, 361–80.

Telzak, E.E., Cote, R.J., et al. 1990. Extrapulmonary *Pneumocystis carinii* infections. *Rev Infect Dis*, **12**, 380–6.

The Pneumocystis Workshop. 1994. Revised nomenclature for *Pneumocystis carinii*. *J Eukaryot Microbiol*, **41**, 121s–2s.

Theus, S.A. and Walzer, P.D. 1997. Adoptive transfer of specific lymphocyte populations sensitized to the major surface glycoprotein of *Pneumocystis carinii* decreases organism burden while increasing survival rate in the rat. *J Eukaryot Microbiol*, **44**, 23S–4S.

Theus, S.A., Sullivan, D.W., et al. 1994. Cellular responses to a 55-kilodalton recombinant *Pneumocystis carinii* antigen. *Infect Immun*, **62**, 3479–84.

Theus, S.A., Andrews. R.P., et al. 1995. Adoptive transfer of lymphocytes sensitized to the major surface glycoprotein of *Pneumocystis carinii* confers protection in the rat. *J Clin Invest*, **95**, 2587–93.

Theus, S.A., Smulian, A.G., et al. 1998. Immunization with the major surface glycoprotein of *Pneumocystis carinii* elicits a protective response. *Vaccine*, **16**, 1149–57.

Thomas, C.F., Anders. R.A., et al. 1998. *Pneumocystis carinii* contains a functional cell-division-cycle Cdc2 homologue. *Am J Respir Cell Mol Biol*, **18**, 297–306.

Thomas, C.F., Park, J.G., et al. 2001. Analysis of a pheromone receptor and MAP kinase suggest a sexual replicative cycle in *Pneumocystis carinii*. *J Eukaryot Microbiol*. **Suppl**, 141S.

Thomas, C.F.J., Gustafson, M., et al 1996. Identification of a cell division cycle (cdc2) homologue in *Pneumocystis carinii*. *J Eukaryot Microbiol*, **43**, 11S.

Thullen, T.D., Daly, K.R., et al. 2003. New rat model of *Pneumocystis* pneumonia induced by anti-CD4+ T-lymphocyte antibodies. *Infect Immun*, **71**, 6292–7.

Urbina, J.A., Visbal, G., et al. 1997. Inhibitors of delta24 (25) sterol methyltransferase block sterol synthesis and cell proliferation in *Pneumocystis carinii*. *Antimicrob Agents Chemother*, **41**, 1428–32.

van der Meer, M.G. and Brug, S.L. 1942. Infection a *Pneumocystis* chez l'homme et chez les animaux. *Ann Soc Belge Med Trop*, **22**, 301–9.

Vandamme, P., Pot, B., et al. 1996. Polyphasic taxonomy, a consensus approach to bacterial systematics. *Microbiol Rev*, **60**, 407–38.

Vanek, J. 1951. Atypical interstitial pneumonia of infants produced by *Pneumocystis carinii*. *Casop Lek Cesk*, **90**, 1121–30.

Vargas, S.L., Ponce, C.A., et al. 1999. Association of primary *Pneumocystis carinii* infection and sudden infant death syndrome. *Clin Infect Dis*, **29**, 1489–93.

Vargas, S.L., Ponce, C.A., et al. 2000. Transmission of *Pneumocystis carinii* DNA from a patient with *Pneumocystis carinii* pneumonia to immunocompetent contact health care workers. *J Clin Microbiol*, **38**, 1536–8.

Vargas, S.L., Hughes, W.T., et al. 2001. Search for primary infection by *Pneumocystis carinii* in a cohort of normal, healthy infants. *Clin Infect Dis*, **32**, 855–61.

Vargas, S.L., Ponce, C.A., et al. 2003. Pregnancy and asymptomatic carriage of *Pneumocystis jiroveci*. *Emerg Infect Dis*, **9**, 605–6.

Vasquez, J., Smulian, A.G., et al. 1996. Antigenic differences associated with genetically distinct *Pneumocystis carinii* from rats. *Infect Immun*, **64**, 290–7.

Vassallo, R., Standing, J. and Limper, A.H. 1999. Beta-glucan from *Pneumocystis carinii* stimulates TNF alpha release from alveolar macrophages. *J Eukaryot Microbiol*, **46**, 145S.

Vavra, J. and Kucera, K. 1970. *Pneumocystis carinii* Delanoë, its ultrastructure and ultrastructural affinities. *J Protozool*, **17**, 463–83.

Velazquez, H., Perazella, M.A., et al. 1993. Renal mechanism of trimethoprim-induced hyperkalemia. *Ann Intern Med*, **119**, 296–301.

Vestereng, V.H. and Kovacs, J.A. 2004. Inability of *Pneumocystis* organisms to incorporate bromodeoxyuridine suggests the absence of a salvage pathway for thymidine. *Microbiology*, **150**, Pt 5, 1179–82.

Vohra, P.K., Kottom, T.J., et al. 2003a. *Pneumocystis carinii* BCK1 complements the *Saccharomyces cerevisiae* cell wall integrity pathway. *J Eukaryot Microbiol*, **50**, Suppl, 676–7.

Vohra, P.K., Puri, V., et al. 2003b. *Pneumocystis carinii* STE11, an HMG-box protein, is phosphorylated by the mitogen activated protein kinase PCM. *Gene*, **312**, 173–9.

Vossen, M.E., Beckers, P.J., et al. 1978. Developmental biology of *Pneumocystis carinii*, and alternative view on the life cycle of the parasite. *Z Parasitenkd*, **55**, 101–18.

Wada, M., Kitada, K., et al. 1993. cDNA sequence diversity and genomic clusters of major surface glycoprotein genes of *Pneumocystis carinii*. *J Infect Dis*, **168**, 979–85.

Wada, M., Sunkin, S.M., et al. 1995. Antigenic variation by positional control of major surface glycoprotein gene expression in *Pneumocystis carinii*. *J Infect Dis*, **171**, 1563–8.

Wainstein, E., Neira, O. and Guzman, L. 1993. Lupus erythematosus disseminatus and *Pneumocystis carinii* pneumonia. *Rev Med Chil*, **121**, 1422–5.

Wakefield, A.E., Pixley, F.J., et al. 1990a. Amplification of mitochondrial ribosomal RNA sequences from *Pneumocystis carinii* DNA of rat and human origin. *Mol Biochem Parasitol*, **43**, 69–76.

Wakefield, A.E., Pixley, F.J., et al. 1990b. Detection of *Pneumocystis carinii* with DNA amplification. *Lancet*, **336**, 451–3.

Wakefield, A.E., Peters, S.E., et al. 1992. *Pneumocystis carinii* shows DNA homology with the ustomycetous red yeast fungi. *Mol Microbiol*, **6**, 1903–11.

Walker, D.J., Wakefield, A.E., et al. 1998. Sequence polymorphisms in the *Pneumocystis carinii* cytochrome b gene and their association with atovaquone prophylaxis failure. *J Infect Dis*, **178**, 1767–75.

Walzer, P.D. 1994. Development of new anti-*Pneumocystis carinii* drugs: cumulative experience at a single institution. In: Walzer, P.D. (ed.), *Pneumocystis carinii pneumonia*, 2nd edition. New York: Marcel Dekker, 511–44.

Walzer, P.D. 2004a. Immunological features of *Pneumocystis* infection in humans. In: Cushion, M.T. and Walzer, P.D. (eds), *Pneumocystis pneumonia*, 3rd edition. New York: Marcel Dekker, 451–77.

Walzer, P.D. 2004b. *Pneumocystis* species. In: Mandell, G.L., Bennet, J.E. and Dolin, R. (eds), *Principles and practice of infectious diseases*, 6th edition. New York: Churchill Livingstone, 3080–94.

Walzer, P.D. and Cushion, M.T. 1989. Immunobiology of *Pneumocystis carinii*. *Pathol Immunopathol Res*, **8**, 127–40.

Walzer, P.D. and Linke, M.J. 1987. A comparison of the antigenic characteristics of rat and human *Pneumocystis carinii* by immunoblotting. *J Immunol*, **138**, 2257–65.

Walzer, P.D., Perl, D.P., et al. 1974. *Pneumocystis carinii* pneumonia in the United States. Epidemiologic, diagnostic, and clinical features. *Ann Intern Med*, **80**, 83–93.

Walzer, P.D., Schnelle, V., et al. 1977. Nude mouse: a new experimental model for *Pneumocystis carinii* infection. *Science*, **197**, 177–9.

Walzer, P.D., LaBine, M., et al. 1984a. Lymphocyte changes during chronic administration of and withdrawal from corticosteroids: relation to *Pneumocystis carinii* pneumonia. *J Immunol*, **133**, 2502–8.

Walzer, P.D., LaBine, M., et al. 1984b. Predisposing factors in *Pneumocystis carinii* pneumonia: effects of tetracycline, protein malnutrition, and corticosteroids on hosts. *Infect Immun*, **46**, 747–53.

Walzer, P.D., Kim, C.K. and Cushion, M.T. 1989. *Pneumocystis carinii*. In: Walzer, P.D. and Genta, R.M. (eds), *Parasitic infections in the immunocompromised host*. New York: Marcel Dekker, 83–178.

Walzer, P.D., Runck, J., et al. 1997. Immunodeficient and immunosuppressed mice as models to test anti-*Pneumocystis carinii* drugs. *Antimicrob Agents Chemother*, **41**, 251–8.

Ward, M.M. and Donald, F. 1999. *Pneumocystis carinii* pneumonia in patients with connective tissue diseases: the role of hospital experience in diagnosis and mortality. *Arthritis Rheum*, **42**, 780–9.

Waskin, H., Stehr-Green, J.K., et al. 1988. Risk factors for hypoglycemia associated with pentamidine therapy for *Pneumocystis* pneumonia. *JAMA*, **260**, 345–7.

Watanabe, J., Hori, H., et al. 1989. 5S ribosomal RNA sequence of Pneumocystis carinii and its phylogenetic association with Rhizopoda/Myxomycota/Zygomycota group. *J Protozool*, **36**, 16S–8S.

Weinberg, G.A. and Bartlett, M.S. 1991. Comparison of pulsed field gel electrophoresis karyotypes of *Pneumocystis carinii* derived from rat lung, cell culture, and ferret lung. *J Protozool*, **38**, 64S–5S.

Weller, R. 1956. Weitere untersuchungen uber experimentele rattenpneumocystose in hinblick auf die interstitelle Pneumonie der Fruhgeborenen. *Z Kinderheilkd*, **78**, 166–76.

Wiley, J.A. and Harmsen, A.G. 1995. CD40 ligand is required for resolution of *Pneumocystis carinii* pneumonia in mice. *J Immunol*, **155**, 3525–9.

Williams, D.J., Radding, J.A., et al. 1991. Glucan synthesis in *Pneumocystis carinii*. *J Protozool*, **38**, 427–37.

Wright, T.W., Johnston, C.J., et al. 1997. Analysis of cytokine mRNA profiles in the lungs of *Pneumocystis carinii*-infected mice. *Am J Respir Cell Mol Biol*, **17**, 491–500.

Wright, T.W., Gigliotti, F., et al. 1999a. Immune-mediated inflammation directly impairs pulmonary function, contributing to the pathogenesis of *Pneumocystis carinii* pneumonia. *J Clin Invest*, **104**, 1307–17.

Wright, T.W., Johnston, C.J., et al. 1999b. Chemokine gene expression during *Pneumocystis carinii*-driven pulmonary inflammation. *Infect Immun*, **67**, 3452–60.

Wright, T.W., Notter, R.H., et al. 2001. Pulmonary inflammation disrupts surfactant function during *Pneumocystis carinii* pneumonia. *Infect Immun*, **69**, 758–64.

Wyder, M.A., Rasch, E.M. and Kaneshiro, E.S. 1994. Assessment of *Pneumocystis carinii* DNA content. *J Eukaryot Microbiol*, **41**, 120S.

Yamada, M., Matsumoto, Y., et al. 1986. Demonstration and determination of DNA in *Pneumocystis carinii* by fluorescence microscopy with 4', 6-diamidino-2-phenylindole (DAPI). *Zentbl Bakteriol Mikrobiol Hyg [A]*, **262**, 240–6.

Yoneda, K. and Walzer, P.D. 1980. Interaction of *Pneumocystis carinii* with host lungs: an ultrastructural study. *Infect Immun*, **29**, 692–703.

Yoshida, Y. 1989. Ultrastructural studies of *Pneumocystis carinii*. *J Protozool*, **36**, 53–60.

Yoshida, Y. and Ikai, T. 1979. *Pneumocystis carinii* pneumonia: epidemiology in Japan, and cyst concentration method. *Zentbl Bakteriol Parasitenkd Infektionskr Hyg Abt Orig Reihe A*, **244**, 405.

Ypma-Wong, M.F., Fonzi, W.A. and Sypherd, P.S. 1992. Fungus-specific translation elongation factor 3 gene present in *Pneumocystis carinii*. *Infect Immun*, **60**, 4140–5.

Zavala, J. and Rosado, R. 1972. *Pneumocystis carinii* en animales domesticos de la ciudad de Merida, Yucatan. *Salud Publ Mex*, **14**, 103–6.

Zhou, W., Nguyen, T.T., et al. 2002. Evidence for multiple sterol methyl transferase pathways in *Pneumocystis carinii*. *Lipids*, **37**, 1177–86.

Zimmerman, P.E., Voelker, D.R., et al. 1992. 120-kD surface glycoprotein of *Pneumocystis carinii* is a ligand for surfactant protein A. *J Clin Invest*, **89**, 143–9.

PART VII

UNUSUAL DISEASE PRESENTATIONS BY FUNGI

38

Adiaspiromycosis and other infections caused by *Emmonsia* species

LYNNE SIGLER

The mycosis caused by *Emmonsia* species is called adiaspiromycosis. The disease name is based on the cells found in tissue which have been termed adiaspores. Adiaspores are formed in vivo when inhaled conidia, which initially measure about 4 μm diameter, enlarge to form thick-walled spherule-like cells reaching 15–300 μm diameter. The size of the adiaspores is dependent upon the species involved, with those of *Emmonsia crescens* being larger than those of *Emmonsia parva* (Table 38.1). Although adiaspores of *Emmonsia* spp. may resemble the spherules containing endospores produced by *Coccidioides immitis* (Table 38.1), adiaspores do not form internal spores, nor do they further replicate. Infections usually occur in animals, and many species of burrowing animals, including rodents, small carnivores, insectivores, marsupials, and edentates, throughout the world, have been found to be infected. Adiaspiromycosis in humans usually involves the lung, is often self-limited, benign and asymptomatic, with rare cases progressing to a fatal outcome. Diagnosis is based most commonly on radiological and histopathological findings. Available evidence suggests that *Emmonsia* spp. are weak pathogens and that pulmonary impairment and elicitation of symptoms is dependent upon the number of conidia inhaled. *Emmonsia* spp. are only rarely reported as the cause of extrapulmonary infection; one such report involved *Emmonsia pasteuriana*, described as a new pathogen in 1998.

ETIOLOGIC AGENTS

The genus *Emmonsia* currently includes three species: *E. parva* (Emmons & Ashburn), Ciferri and Montemartini (1959), *E. crescens* Emmons and Jellison (1960), and *E. pasteuriana* Drouhet, Gueho & Gori (Drouhet et al. 1998). The taxonomic history of *Emmonsia* spp. has been complicated. Mycologists have disagreed about the appropriate genus for the species, and about their relationship to the dimorphic pathogens. Recently, classical mating experiments and DNA sequence comparisons have provided evidence to confirm that the closest relative to *Emmonsia* spp. is *Blastomyces dermatitidis* Gilchrist & Stokes. Molecular techniques also helped to classify an *Emmonsia*-like isolate that differed in producing yeast-like cells, rather than adiaspores, in vivo. *E. pasteuriana* is known thus far only from a single case of disseminated cutaneous infection in a patient with the acquired immunodeficiency syndrome (AIDS). Other species formerly assigned to *Emmonsia*, i.e. *Emmonsia brasiliensis* and *Emmonsia ciferrina*, have been reclassified (Padhye and Carmichael 1968).

ETYMOLOGY

The term adiaspore comes from the Greek verb *speirein* (*a* – negative, *dia* – through, *speirein* – to scatter or

Table 38.1 *Comparison of* Emmonsia spp.[a] *with* Blastomyces dermatitidis, Paracoccidioides brasiliensis, Histoplasma capsulatum, *and* Coccidioides immitis

	Ecology and distribution	Mycelial form	Teleomorph	Parasitic form in vivo	Confirmatory tests	Figure showing morphology in vitro and in vivo
Emmonsia crescens	Many animal hosts (about 120 species of rodents, lagomorphs, carnivores, insectivores, herbivores, marsupials), rarely soil. Worldwide except Africa, Australia	Narrow hyphae forming solitary sessile conidia and 1–3 conidia at the ends of narrow, often slightly swollen stalks. Conidia subglobose, often slightly broader than long, 2.5–4 μm long and 3–5 μm wide, smooth to finely roughened	*Ajellomyces crescens*	Oval or globose adiaspores, 50–500 μm, commonly 300 μm in diameter, walls 10–70 μm thick	Maximum growth temperature 37°C (lower for some strains), forming adiaspores measuring 20–140 μm. Exoantigen tests may show nonspecific precipitin lines against *B. dermatitidis* and *H. capsulatum* antisera	
Emmonsia parva	Few animal hosts (about 21 species of rodents, lagomorphs, carnivores, and marsupials), rarely soil. Geographically restricted (parts of USA, Kenya, Europe, Australia)	Narrow hyphae forming solitary sessile conidia and 1–3 conidia at the ends of narrow, often slightly swollen stalks. Conidia subglobose, often slightly broader than long, 2.5–4 μm long and 3–5 μm wide, smooth to finely roughened	Not known; relationship to *Ajellomyces* inferred	Oval or globose adiaspores, 10–40 μm in diameter, walls thin	Maximum growth temperature 40°C, forming adiaspores measuring 8–20 μm diameter. Exoantigen tests may show nonspecific precipitin lines against *B. dermatitidis* and *H. capsulatum* antisera	Similar to *E. crescens*, but smaller adiaspores
Emmonsia pasteuriana	Known only from human host in Italy	Narrow hyphae forming solitary sessile conidia and 1–3 (sometimes up to 8) conidia at the ends of narrow stalks, often swollen distally to form a vesicle-like structure. Conidia subglobose to oval, often slightly broader than long, 2–5.5 μm long by 2–3.5 μm. Conidia subglobose, smooth to finely roughened or tuberculate	Not known; relationship to *Ajellomyces* inferred	Small and large yeast cells; small cells oval to pyriform (2–4 μm), often intracellular; occasionally larger, thick-walled (8–15 μm); all with narrow-based buds; budding mostly unipolar, sometimes bipolar	Converts to yeast form at 37°C on enriched media; no adiaspores formed	
Blastomyces dermatitidis	River and forest soil, beaver dams, habitat poorly known. North America in areas along Mississippi and Ohio River valleys, eastern and central Canada, rarely Africa, Asia and Middle East	Narrow hyphae forming solitary sessile conidia and 1, uncommonly 2, conidia at the ends of narrow, usually unswollen stalks. Conidia mostly pyriform, 4–5 μm long and 3–4 μm wide	*Ajellomyces dermatitidis*	Large yeast cells, 10–12 μm in diameter, double contoured wall and single broad-based bud	Converts to yeast form at 37°C on enriched media; yeast cells large, thick-walled with broad-based buds. DNA probe and exoantigen test	
Paracoccidioides brasiliensis	Rarely isolated from soils (coffee plantation, rain forest), armadillos, habitat considered to be forest areas of moderate climate, medium to high annual rainfall, in Central and South America	Narrow hyphae often remaining sterile; occasionally forming lateral or terminal pyriform conidia or intercalary arthroconidia that are slightly swollen (2–5.5 μm long and 2–3.5 μm wide)	Not known; relationship to *Ajellomyces* inferred	Thick-walled central cell up to 30 μm in diameter, bearing 1 to many daughter cells of varying size and joined to mother cell by narrow pedicel, occasionally occurring in short chains	Converts to yeast stage at 37°C on enriched media. DNA probe and exoantigen test	
Histoplasma capsulatum var. *capsulatum*	Soil associated with dung of birds and bats. North America in areas along the Mississippi, Missouri, and Ohio River valleys of the USA, eastern Canada and Mexico, rarely in Central and South America, Asia and Europe	Narrow hyphae forming solitary sessile smooth or roughened, microconidia and tuberculate macroconidia at the ends of unswollen stalks. Microconidia pyriform, 4–5 μm long and 3–4 μm wide. Macroconidia broadly pyriform, 8–17 μm long and 8–13 μm wide or globose, 8–13 μm diameter	*Ajellomyces capsulatus*	Intracellular, rarely extracellular, small budding cells, 2–5 μm in diameter	Converts to yeast form at 37°C on enriched media; yeast cells small, oval, with narrow-based buds. DNA probe and exoantigen test	
Coccidioides immitis	Soil of arid regions, animal burrows, has been found in mixed infections with *E. parva* in lungs of rodents and other burrowing animals. Geographically restricted to arid regions of southwest USA, Mexico, and parts of Central and South America	Hyphae forming alternate arthroconidia. Arthroconidia cylindrical 3–8 μm long and 3–5 μm wide	Not known; placement within the *Onygenaceae* inferred	Spherules 30–60 (–80) μm with walls 2 μm thick and containing endospores 2–5 μm in diameter	DNA probe and exoantigen test	

a) Modified from Sigler 1998, 2003 and based on data from Sigler 1996; figures copyright L. Sigler.

disperse), for the spherule which enlarges from the inhaled conidium (Emmons and Jellison 1960).

HISTORY

The first observation of infected animals and the suggestion of a fungal etiology are generally attributed to Kirschenblatt (Kirschenblatt 1939; Jellison 1969; Schwarz 1978; Rippon 1988) who, in 1939, applied the name *Rhinosporidium pulmonale* to the cyst-like bodies observed in lungs of several rodents. However, the name has not been used because it was not accompanied by the required Latin diagnosis or confirmed as applicable to a fungus by culture. The first valid name for the fungus dates from 1942. The earliest record of infection dates from a rodent specimen deposited in 1845 in a Swedish museum (Jellison 1969).

Infection in North America was first recorded in animals in Arizona. During the 1930s and 1940s, mycologists conducted surveys to define the natural habitats and endemic areas of the dimorphic systemic pathogens. These involved isolation of the fungi from soil, dung, or tissues of trapped birds and animals, and analysis of human exposure by measuring reactivity to antigens through skin tests. In 1942, Emmons reported that soil and desert rodents were natural reservoirs for *Coccidioides immitis* in regions of Arizona. Later that year, Emmons and Ashburn (1942) reported that some animals exhibited the presence of a second fungus forming spherule-like structures measuring up to 20 μm in diameter but lacking internal spores. Lungs of some animals contained both these cells and spherules of *C. immitis*. Thinking that their fungus had features typical of a zygomycete, Emmons and Ashburn (1942) named it *Haplosporangium parvum* and distinguished it from other species of the genus by the smaller size of the solitary sporangia. Some mycologists then considered *C. immitis* also to be a zygomycete because of the internal cleavage of the spherules into endospores (Emmons 1942; Baker et al. 1943). Congruences between *C. immitis* and *H. parvum*, including similarities in tissue morphologies, animals with dual infections, and skin test crossreactivity, led Emmons and Ashburn to speculate on a possible genetic relationship between them, even though *H. parvum* more closely resembled *B. dermatitidis* and *Histoplasma capsulatum* in cultural features.

Five years later, Dowding (1947) observed similar but much larger cells within lungs of rodents captured over a wide range in Alberta, Canada. The cells reached diameters of 300 μm and were often aggregated into pearly white nodules visible by gross examination. Cultures of the Alberta isolates grew poorly at 37°C, but conidia enlarged to form swollen cells as much as ten times greater in diameter and with walls up to 10 μm thick; on return to room temperature, these enlarged cells germinated to form hyphae. Dowding disagreed that *H. parvum* was most closely related to *C. immitis*

because of the different type of conidia formed (i.e. arthroconidia in *C. immitis* versus solitary ones in *H. parvum*) and the lack of internal spores within spherules. Dowding also suggested that *H. parvum* was more closely related to *B. dermatitidis*, *H. capsulatum*, and *Paracoccidioides brasiliensis* and that these fungi, including *C. immitis*, formed a 'natural group.' She was the first to suggest a close biological relationship among the dimorphic pathogens. The placement of *H. parvum* in the zygomycete genus *Haplosporangium* was also challenged by Carmichael (1951). He suggested that the species was aligned most closely with *B. dermatitidis*; it formed conidia of the type that he later called 'aleuries' or aleurioconidia, and had affinities to members of the ascomycete family *Gymnoascaceae* (Ascomycota) (Carmichael 1951, 1962).

Ciferri and Montemartini (1959) made the necessary transfer of *H. parvum* to a new hyphomycete genus *Emmonsia* with *E. parva* as the type species. Emmons and Jellison (1960) described *E. crescens* for morphologically similar isolates that formed larger cells in vivo, for which they coined the term 'adiaspore.' They distinguished *E. crescens* from *E. parva* by the larger multinucleate (rather than uninucleate) adiaspores and by the lower maximum growth temperature. However, mycologists have disagreed on whether these differences were sufficient to recognize distinct species or varieties, and on the most appropriate generic name for the agents of adiaspiromycosis. For example, Carmichael (1962) reevaluated the genus *Chrysosporium* and included in it many fungi producing aleurioconidia, including *B. dermatitidis* and the species of *Emmonsia*. While some mycologists accepted his proposed name *Chrysosporium parvum* with two varieties, *C. parvum* var. *parvum* and *C. parvum* var. *crescens* (McGinnis 1980; Boisseau-Lebreuil 1981; Rippon 1988; McGinnis and Rinaldi 1995), others argued for use of the genus name *Emmonsia* with the varieties *E. parva* var. *parva* and *E. parva* var. *crescens* (von Arx 1973; van Oorschot 1980; Kwon-Chung and Bennett 1992).

Recent evidence confirms that *Emmonsia* is the appropriate name for the agents of adiaspiromycosis, that the taxa should be considered species rather than varieties, and that there is a close relationship between *Emmonsia* spp. and *B. dermatitidis*. *Emmonsia* spp. and *B. dermatitidis* could be accommodated in the same genus but have been retained in separate genera mainly due to the differences in their parasitic forms (i.e. adiaspores versus yeast cells) and the clinical syndromes that they elicit. That distinction becomes less clear with the finding that the recently described *E. pasteuriana* demonstrates budding cells. Teleomorphs (meiotic or sexual stages) of *E. crescens* and *B. dermatitidis* are in the ascomycete genus *Ajellomyces*, as is that of *H. capsulatum*. The genus *Ajellomyces* is currently classified in the order Onygenales of the Ascomycota. Anamorphs (mitosporic or asexual stages) of onygenalean fungi are commonly aleurioconidia or arthroconidia that are released by fracture or dissolution of all or part of the supporting hyphal cell (Carmichael 1962; Sigler and Carmichael 1976; Sigler 1997, 2003). For further discussion on relationships between *Emmonsia* spp. and the dimorphic fungi, see section on Teleomorphs and relationships below.

EPIDEMIOLOGY

Ecology, distribution, and animal infection

Even though *E. parva* was discovered first, *E. crescens* appears to be the primary agent of adiaspiromycosis in both animals and humans. However, it is difficult to verify the species involved in some literature reports due to uncertainty caused by nomenclatural controversy. The species can be verified only when size of the adiaspores observed in tissues is given. *E. crescens* has been reported worldwide from rodents, insectivores, carnivores, lagomorphs, and marsupials. Approximately 118 animal hosts were identified in detailed records compiled up to the 1970s and sporadic cases of animal infection have been reported subsequently (Dvorák 1968; Jellison 1969; Doby et al. 1970; Dvorák et al. 1973). *E. parva* is geographically more restricted and associated with fewer animal hosts. It has been recorded from about 21 species of rodents, lagomorphs, carnivores, and marsupials in the USA, Kenya, (former) Czechoslovakia, the former Soviet Union, and Australia (Jellison 1969; Doby et al. 1970; Krivanec and Mason 1980; Mason and Gauhwin 1982). Factors affecting geographical distribution are largely unknown. In North America, *E. crescens*-infected animals have been recorded from 13 states and four Canadian provinces, whereas *E. parva*-infected animals have been found in Arizona, Texas, Montana, and Utah. The original Arizona survey revealed a high prevalence of *E. parva* infection (33 percent of 303 animals) (Emmons and Ashburn 1942), but a survey of almost 1 800 animals collected over several years in (former) Czechoslovakia yielded 260 infected animals, of which only four showed infection with *E. parva* (Krivanec 1977). *E. crescens* has been found in New Zealand in a marsupial introduced from Australia (Johnstone et al. 1993), but it appears to be absent in neighboring Australia (Connole 1990). There, *E. parva*-like adiaspores have been observed in wombats, but culture yielded atypical nonsporulating isolates (Krivanec and Mason 1980; Mason and Gauhwin 1982). There is some basis to suggest that species could be introduced through transport of infected animals (Peterson and Sigler 1998).

Similarly to *B. dermatitidis*, the habitat for *Emmonsia* spp. is not well known, but animals are presumed to acquire infection by inhalation exposure to conidia in soil of wooded areas. Long-term studies in the Czech Republic have revealed that the highest prevalence of infection (33 percent) occurs in rodents from wooded windbreaks separating cultivated fields, compared with rodents from the fields themselves or from small plots of deciduous woods (Hubálek et al. 1995). Infection is absent in rodents that do not build their nests in soil (Hubálek 1999). Carnivores may play a role in distributing the organism by liberating adiaspores trapped in the tissues of their prey (Krivanec et al. 1975b; Krivanec and Otcenášek 1977). Viable adiaspores of *E. crescens* survive passage through the digestive tracts of mice, birds, and carnivores (Krivanec et al. 1975a; Krivanec and Otcenášek 1977; Hejtmanek 1985), as evidenced by the isolation of the fungus from the dung of coyotes for which rodents are a major dietary constituent (Sigler and Flis 1998). There are only a few records of direct isolation from soil and inanimate materials including moldy straw and bird's nest, all from Alberta, Canada (Sigler and Flis 1998; Peterson and Sigler 1998). Indirect isolation from soil occurred on at least one occasion when an *E. parva*-like isolate was recovered following experimental inoculation of soil into mice during a search for *H. capsulatum* in Missouri (Menges and Habermann 1954; Jellison 1969). However, molecular analysis revealed that the Missouri isolate is genetically distinct (Peterson and Sigler 1998).

In addition to habitat, factors affecting prevalence of infection are species and age of animal, season, geographical area, and altitude (Doby et al. 1971; Hubálek et al. 1995, 1998; Hubálek 1999). *E. crescens* infection is higher in mustelid carnivores, ranging from 30 to 76 percent of animals, than in small rodents and insectivores (1–50 percent) and large carnivores (<1 percent) (Doby et al. 1971; Krivanec et al. 1975a, b; Krivanec and Otcenášek 1977); however, the numbers of carnivores surveyed have been small. An 11-year survey of 10 081 small mammals from the Czech Republic found adiaspiromycosis in 13 percent of animals and a higher prevalence in rodents (13 percent of juveniles; 20 percent of adults) than in insectivores (2 percent of juveniles; 4 percent of adults) (Hubálek 1999). Infection is higher in adult rodents than in juveniles, but there is no difference in infection rates between adults according to sex (Hubálek et al. 1995). Dvorák et al. (1967, 1969) found seasonal variation with the highest incidence in spring. Hubálek et al. (1993) found the highest incidence in spring for voles and in winter for mice, reflecting differences in the animals' digging and burrowing activities. They also suggested that the peak of fungal proliferation may occur in spring when mean soil temperature, 5 cm below the surface, ranges from 3.4 to 12.2°C.

There have been few reports of illness or death of animals naturally infected with *Emmonsia* spp., but severely emaciated animals have been found postmortem to be heavily infected (Jellison 1969; Simpson and Gavie-Widen 2000). Hubálek et al. (1995)

determined the mean intensity of infection of *E. crescens* as 177.2 adiaspores per juvenile and 212.5 per adult rodent, with the number of adiaspores ranging from one to 4 618, and adiaspore size varying between 46 and 451 μm (mean 239.4 μm in adults). Lung tissues of some of these animals showed adiaspores of two distinct sizes, suggesting that they had been reinfected. Rare reports concern domestic or farm animals including two dogs (Al-Doory et al. 1971; Koller et al. 1976), a goat (Koller and Helfer 1978), and a horse (Pusterla et al. 2002). Findings were incidental in the goat and one dog, but the horse and a second dog demonstrated advanced pulmonary infection. In the case of the dog, which also had degenerative arthritis, organisms were found only in the lung and lymph nodes at necropsy (Al-Doory et al. 1971). The isolated fungus was inoculated into mice intraperitoneally, resulting in pulmonary lesions and recovery of the same fungus in culture. These findings are more suggestive of blastomycosis, known to be common in dogs, but the cells in tissue were described as irregular in size and nonbudding.

Experimental infection has been induced in dogs, guinea pigs, mice, rabbits, rats, and sheep, but not birds, by inoculating conidial suspensions, mainly of *E. crescens*, via intranasal, intraperitoneal, subcutaneous, or intravenous routes (Jellison 1969). Intravenous inoculation induces adiaspore development in all soft tissues, especially in the liver, lungs, and spleen. Golden hamsters were found to be less susceptible than other animals to developing adiaspiromycosis (Boisseau-Lebreuil 1975), and infectivity as well as persistence of adiaspores was greater after subcutaneous inoculation with *E. crescens* isolates than with *E. parva* isolates (Carmichael 1951). Time since injection, the strain used, the resistance of the animal, and the inoculation route are factors generally affecting the development of the adiaspores (Carmichael 1951; Dvořák et al. 1973; Slais 1976; Boisseau-Lebreuil 1977). Enlargement of conidia to adiaspores reaching 50–100 μm occurs within 2 weeks of intraperitoneal inoculation with *E. crescens* (Dvořák et al. 1973; Krivanec et al. 1974; Hejtmanek 1985). Experimental infection was obtained for *E. pasteuriana* only in golden hamsters following intratesticular inoculation; other routes of inoculation in the hamster, mice, and guinea pigs were unable to induce infection (Drouhet et al. 1998; Drouhet and Huerre 1999).

Human adiaspiromycosis

The number of recorded human infections is low, with approximately 50 cases of pulmonary or disseminated adiaspiromycosis reported from Central and South America, Europe, USA, and North Africa (Kwon-Chung and Bennett 1992; England and Hochholzer 1993; Sigler 1998). As with animals, infection occurs following inhalation exposure to conidia. Eleven patients whose case histories were reviewed by England and Hochholzer (1993) were adult white males from Argentina, Brazil, Germany, Guatemala, Spain, and the USA. Four of these individuals may have acquired the infection following inhalation of dust through farming, woodworking, or bird-handling activities. Eighteen cases of infection have occurred in Brazil in adult males between the ages of 26 and 62 (Barbas Filho et al. 1990; de Almeida Barbosa et al. 1997; Martins et al. 1997; dos Santos et al. 1997; Moraes et al. 2001). Three of these patients developed acute pulmonary infection following cleaning or renovation of buildings or the handling of rodent and bat carcasses or nesting material (Barbas Filho et al. 1990; de Almeida Barbosa et al. 1997). Soil dust from a garden shop was thought to have been the source of exposure for a 2-year-old Finnish girl (Nuorva et al. 1997). Of 14 case histories reviewed by Schwarz (1978) from the Czech Republic, France, Guatemala, Honduras, Russia, and Venezuela, four involved women and four involved boys aged 10 and 11 years. Most cases of adiaspiromycosis occur in non-AIDS patients. Underlying conditions have included tuberculosis, yellow fever, cancer of the bladder, diffuse lymphoma, chronic bronchitis, steroid-dependent emphysema, atherosclerosis, aortic atherosclerotic aneurysm, and myocardial infarction (Kodousek 1974; Schwarz 1978; Bambirra and Nogueira 1983; England and Hochholzer 1993; Moraes et al. 2001).

Pulmonary and disseminated infection have been reported in a few patients with AIDS (Echavarria et al. 1993; Gori et al. 1998; Turner et al. 1999). The histopathological presentations in these patients are dissimilar to classical adiaspiromycosis, suggesting that different *Emmonsia* spp. may be involved (Drouhet et al. 1998; Gori et al. 1998). The unusual features of these cases are discussed in the section on Infections in AIDS patients below.

HOST RESPONSES

Clinical manifestations of adiaspiromycosis

Adiaspiromycosis in humans generally involves the lungs and is often self-limited, benign, and asymptomatic. Functional disability and onset of symptoms is dependent upon the number of conidia inhaled (Barbas Filho et al. 1990). A heavy or repeated exposure may result in acute pulmonary disease with bilateral granulomatous lesions (de Almeida Barbosa et al. 1997). Patients are either asymptomatic or present with nonproductive cough, dyspnea on exertion, and low-grade fever; less common findings are hemoptysis, pain, chills, malaise, weight loss, night sweats, and crackles in the lungs (Barbas Filho et al. 1990; England and Hochholzer 1993;

de Almeida Barbosa et al. 1997). Almost all cases of human pulmonary adiaspiromycosis involve *E. crescens* (sometimes reported as *E. parva* var. *crescens*) (Schwarz 1978; Kwon-Chung and Bennett 1992; England and Hochholzer 1993; de Almeida Barbosa et al. 1997).

Diagnosis of infection is based most commonly on radiological (Figure 38.1) and histopathological findings without confirmation by culture. Diagnosis may be complicated by symptoms that mimic those of other fungal diseases, tuberculosis, sarcoidosis, Hodgkin's disease, or farmer's lung, or the co-occurrence with mycobacterial or other infections (England and Hochholzer 1993; de Almeida Barbosa et al. 1997). This may lead to failure to perform fungal culture, but even if adiaspiromycosis is suspected, fungal culture of sputum, bronchial washings, or bronchoalveolar lavage is often

unsuccessful. Adiaspores may be discovered incidentally. In the first case of human infection, a single adiaspore was found in a nodule from a patient who had aspergillosis and cystic disease of the lung (Doby-Dubois et al. 1964). If numerous, adiaspores may be present in multiple granulomas involving both lungs (Schwarz 1978; England and Hochholzer 1993). Three forms of infection have been recognized (Kodousek et al. 1971; Vojtek et al. 1972; Schwarz 1978; England and Hochholzer 1993; de Almeida Barbosa 1997). These include:

1 a solitary or localized granuloma
2 a sporadically disseminated form
3 an acute, densely disseminated type with clinical symptoms.

The most prevalent findings in chest x-rays or computed tomography scans are diffuse bilateral reticulonodular infiltrates with areas of honeycombing (Barbas Filho et al. 1990; Peres et al. 1992; England and Hochholzer 1993). Most patients recover spontaneously or with antifungal therapy and/or prednisone, but in a few cases, the infection progresses to a fatal outcome (Peres et al. 1992; England and Hochholzer 1993; de Almeida Barbosa et al. 1997). Extrapulmonary infection is uncommon. Adiaspores of *E. crescens* were observed in two sections taken from an appendix of an 11-year-old boy with bronchitis (Kodousek 1974).

Infections in AIDS patients

Three patients with AIDS have been described recently as having infection caused by *Emmonsia* spp. In all cases, the fungus was grown in culture. This is in contrast to the failure to culture the fungus from non-AIDS patients with pulmonary adiaspiromycosis. Unusual findings in the case described by Echavarria et al. (1993) were dissemination involving bone, bone marrow, and lungs and extensive osteomyelitis. The fungus was identified as *E. parva* based on the smaller adiaspores, measuring between 25 and 35 μm in diameter. Although grown from pus, sputum, and bone marrow, the fungus was later lost to further investigation as to its relationship with other *Emmonsia* spp. Turner et al. (1999) described pulmonary adiaspiromycosis in a male with advanced HIV infection who was negative for both *Pneumocystis carinii* and mycobacterial infection. The fungus was grown from a bronchoalveolar lavage and a protected brush specimen and identified as *E. parva* by formation of adiaspores at 37°C. Fungal elements were reported to be present in the specimens, but their nature was not described nor were they illustrated in the report. The third case involved a female who developed disseminated cutaneous mycosis; histopathology revealed small yeast-like cells rather than adiaspores (Drouhet et al. 1998; Gori et al. 1998)

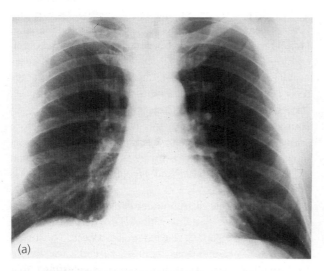

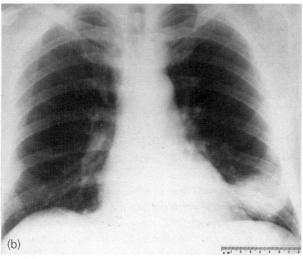

Figure 38.1 *Progressive radiological findings in a 43-year-old man with localized adiaspiromycosis.* **(a)** *Minimal infiltrate in the lower lobe of the left lung;* **(b)** *mass-like lesion 4 months later. (From the Armed Forces Institute of Pathology Photograph, Neg. Nos 76-924-1 and 76-924-3, respectively. Reproduced with the permission of Dr D.M. England from England and Hochholzer 1993. Work of the US Government. Not subject to copyright in the USA)*

(Table 38.1). The fungus in culture demonstrated morphological features compatible with *Emmonsia*, but the isolate was found to be genetically distinct and described as a new species, *E. pasteuriana* (Drouhet et al. 1998).

Unconfirmed reports concerning *Emmonsia* species

Some reports which have been attributed to *Emmonsia* spp. are doubtful. In cases of skin (Thammayya and Sanyal 1973; Kamalam and Thambiah 1979) and corneal infection (Wagoner et al. 1999), the fungal elements in tissue are incompatible with those of *Emmonsia* spp. In two reports of deep infections (Stillwell et al. 1984; Warwick et al. 1991), one of endocarditis (Toshniwal et al. 1986), and one of sinusitis (Levy et al. 1991), the isolated organism was identified in each report only as *Chrysosporium* species; however, the discussions in each report implicate *E. parva* (as *Chrysosporium parvum*) and the reports have been cited by subsequent authors under that etiology. None of these cases can be confirmed as adiaspiromycosis. The fungus causing sinusitis described by Levy et al. (1991) is suspected to be the basidiomycete *Schizophyllum commune* (Sigler and Verweij 2003).

Histopathology

In histopathological examination of biopsy specimens, the adiaspores of *E. crescens* are generally found within granulomas, with each granuloma containing one or occasionally two adiaspores (Figure 38.2). In a severe infection, the pleural surface may appear grossly as grayish-white and thickened and marked by numerous discrete white nodules measuring up to 1 mm in size, with the intervening pulmonary tissue unchanged (Watts et al. 1975). Adiaspores appear as round or oval cells,

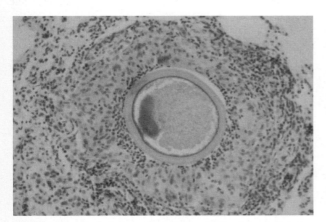

Figure 38.2 *Histological section of lung of striped skunk showing inflammatory reaction around adiaspore of E. crescens. Hematoxylin and eosin; note thick fungal cell wall. (Reproduced, with permission, from Albassam et al. 1986)*

ranging from 50 to 500 μm in diameter (commonly <300 μm), either lacking internal contents or appearing granular, and having walls between 10 and 70 μm thick (Kodousek 1974; Chandler et al. 1980; Hejtmanek 1985; England and Hochholzer 1993). The wall appears either trilaminar, staining positive with Gomori's methenamine silver, Gridley, and periodic acid–Schiff stains, or bilaminar when stained with hematoxylin and eosin (Watts et al. 1975). Adiaspores may be surrounded by a few macrophages and lymphocytes or by a granulomatous inflammatory reaction composed of epithelioid cells, macrophages, lymphocytes, plasma cells, scattered neutrophils, and fibrous tissue; eosinophils are uncommon (Albassam et al. 1986; England and Hochholzer 1993). Multinucleated giant cells may be seen in some granulomas. England and Hochholzer (1993) noted that patients with more acute infections of short duration (<2 months) presented with necrotic and suppurative granulomas, having adiaspores that averaged 100 μm and showed a paucity of T cells, B cells, and KP-1 macrophages on immunohistochemistry, whereas patients with infections of longer duration had non-necrotizing granulomas containing larger adiaspores (average diameter 155 μm) and demonstrated good immunoreactivity for T cells and KP-1-positive macrophages. Adiaspores in stages of collapse or being broken down by inflammatory cells can sometimes be observed (Albassam et al. 1986; England and Hochholzer 1993). The in vivo structures of other fungi, including the spherules of *C. immitis* or nonbudding yeast cells of *B. dermatitidis* (Table 38.1) and even *Curvularia* species in allergic bronchopulmonary disease, as well as parasites and plant cells, have been suggested as possible confounding factors in the differential diagnosis of adiaspiromycosis (England and Hochholzer 1993). Spherules of *C. immitis* are smaller on average (30–60 μm), have walls about 2 μm thick, and contain endospores 2–5 μm in diameter.

In naturally infected animals, adiaspores occur mostly in lungs within solitary or multiple non-necrotic, sometimes coalescing, granulomas; rarely, adiaspores are extrapulmonary and involve the lymph nodes (Albassam et al. 1986). In severely infected animals, pearl-white or grayish-white nodules measuring up to 1 mm in diameter can be seen on gross examination (Figure 38.3) (Dowding 1947; Albassam et al. 1986). As with human infections, adiaspores may be found incidentally when only a few are present and the animal is asymptomatic. Large numbers of adiaspores have been found in lungs of trapped rodents with higher numbers in adults compared with juveniles (mean 212.5 versus 177.2) (Hubálek et al. 1995). In the first case of disseminated pulmonary disease in a horse, biopsy examination revealed multiple granulomas composed of an outer layer of epitheloid macrophages and occasional giant cells, and an inner neutrophilic focus surrounding spherical adiaspores of *E. crescens*, 30–80 μm in diameter; at

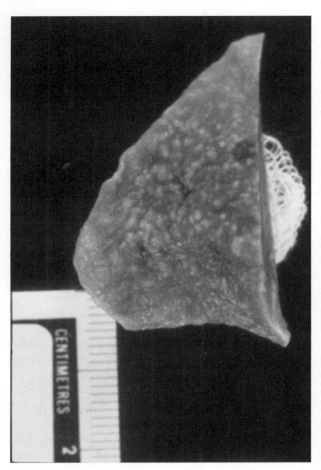

Figure 38.3 *Formalin-fixed lung of mature striped skunk* (Mephitis mephitis) *showing multiple nodules caused by* Emmonsia crescens. *(Reproduced, with permission, from Albassam et al. 1986)*

postmortem, the lung parenchyma showed diffuse infiltration with 1–2 mm diameter miliary, white nodules (Pusterla et al. 2002). Different inflammatory responses and the absence of reaction around some adiaspores have been explained as arising from repeat exposure, the time involved in maturation of adiaspores or varying antigenicity of the adiaspore wall (Albassam et al. 1986). In animals experimentally infected by intraperitoneal inoculation (Figure 38.4), between one and several adiaspores develop within granulomas in the liver, spleen, kidney, peritoneum, or other organs (Hejtmanek 1985).

Cytochemistry and structure of adiaspores and hyphae

The adiaspores of *E. crescens* have a complex wall structure, especially when immature (England and Hochholzer 1993) (Figure 38.5). Cytochemistry has revealed three layers, of which the outer is thinnest and contains the proteins, arginine, tyrosine, and tryptophan, and traces of hydrophilic lipids. The inner wall is thickest and consists of neutral mucosubstances (Hejtmanek

1985). Using transmission electron microscopy, the three layers are almost amorphous or appear as fibrils of different electron densities (Watts et al. 1975; Hejtmanek 1985; Albassam et al. 1986). The innermost layer consists of finer fibrils than the middle layer, while the outer layer appears as irregular heterogeneous fibrils of higher electron density (Watts et al. 1975; Albassam et al. 1986). The middle zone has been described as a fenestrated layer composed of discontinuous dense fibrillar lamellae (Watts et al. 1975). Neutral and acid mucins have been located in the inner and middle layers by colloidal iron and mucicarmine stains (England and Hochholzer 1993). Adiaspores developing on agar demonstrate the hydrolytic enzymes, alkaline and acid phosphatases, nonspecific esterase, aminopeptidase, and respiratory enzymatic activity, including glucose-6-phosphate dehydrogenase, diaphorase, and peroxidase (Kodousek and Hejtmanek 1971; Hejtmanek 1985). Similar enzymatic activity has been reported for hyphae that have cell walls containing chitin, septa with simple pores and Woronin bodies, mitochondria, glycogen granules, vacuoles, lipid bodies, granular endoplasmic reticulum, and numerous nuclei in the hyphal tips (Hejtmanek 1985).

There has been disagreement about the internal structure of adiaspores and about their propensity to divide internally or to replicate by budding. Some authors have indicated that the adiaspores of *E. parva* are uninucleate whereas those of *E. crescens* are multinucleate, and that this factor is associated with unipolar or multipolar budding (Emmons and Jellison 1960; Dvořák et al. 1973; Hejtmanek 1985). Others have found adiaspores to be commonly empty (England and Hochholzer 1993) or to contain lipid vacuoles, mitochondria, and ribosomes with (Hejtmanek et al. 1974) or without (Albassam et al. 1986) nuclei. A form of 'budding' could be induced in the adiaspores of *E. crescens* by changing in vitro growth conditions of immature adiaspores (Emmons 1964), but most investigators now agree that, in the host, adiaspores demonstrate neither budding nor internal replication.

IMMUNODIAGNOSIS

There is no reliable serological test for the diagnosis of adiaspiromycosis. Environmental exposure to *Emmonsia* spp. was first detected by positive skin-test reactions to antigens prepared from culture extracts in a group of school children, some of whom also demonstrated reactivity to coccidioidin (Emmons and Ashburn 1942). Delayed-type hypersensitivity reactions to antigens of *E. crescens* and precipitating antibodies in sera can be demonstrated in humans with infections (Tomšíková et al. 1970, 1974; Simecek et al. 1982). Antigenic activity has been found to vary among strains and to be greater in the parasitic stage (Tomšíková et al. 1972, 1976, 1978). Both strain- and species-specific antigens were

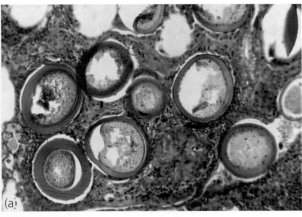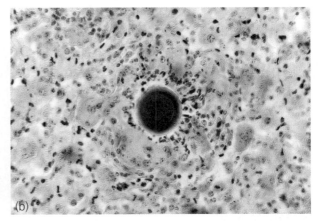

Figure 38.4 *Histological sections of mouse lung after experimental inoculation showing* **(a)** *numerous thick-walled adiaspores of* E. crescens *and* **(b)** *solitary thin-walled adiaspore of* E. parva.

detected by immunodiffusion tests, using antisera raised against partially purified cytoplasmic antigen (Cano and Taylor 1974). A passive cutaneous anaphylaxis test in young guinea pigs detected two possible cases of human infection among donors of 83 sera collected from trappers and mammologists who handled a variety of animals in their work (Jellison et al. 1962). Immunization with specific antiserum achieved some inhibition of adiaspore development in experimentally infected animals (Tomšíková et al. 1982).

Immunological crossreactivity has been demonstrated between *Emmonsia* spp. and *B. dermatitidis* and

H. capsulatum (Tomšíková et al. 1970; Al-Doory et al. 1971; Sekhon et al. 1986), but not with *C. immitis* (Pusterla et al. 2002). Nonspecific precipitin lines are produced in exoantigen tests between *E. parva* antisera and the A antigen of *B. dermatitidis* or the H and M antigens of *H. capsulatum*, but the exoantigen is considered reliable in differentiating among the species (Sekhon et al. 1986; Kwon-Chung and Bennett 1992; Padhye et al. 1994). A commercially available DNA probe for *B. dermatitidis* was found to show crossreactivity with *P. brasiliensis* but not with *E. parva* (Padhye et al. 1994). In exoantigen tests, no crossreactivity was found between *E. pasteuriana* and *B. dermatitidis*, *H. capsulatum*, or *C. immitis* (Gori et al. 1998).

LABORATORY DIAGNOSIS

Direct examination and isolation

Diagnosis of adiaspiromycosis is generally made by observation of characteristic adiaspores by histopathology. Based on the larger size of the adiaspores, most reports concern *E. crescens*. The fungus is rarely cultured from specimens in cases of human infection. The differential clinical diagnosis is complicated by mimicking of other syndromes (including other fungal infections, tuberculosis, sarcoidosis, idiopathic granulomatous disease, etc.) and this may lead to failure to perform fungal culture. However, even if adiaspiromycosis is suspected, fungal culture of sputum, bronchial washings, or bronchoalveolar lavage is often negative (Barbas Filho et al. 1990; England and Hochholzer 1993).

Most *Emmonsia* isolates have been recovered from dissected animal lung tissue. In the case of a horse with disseminated pulmonary adiaspiromycosis, culture was negative from lung tissue and from transtracheal

Figure 38.5 *Electron micrograph of an adiaspore in lung of striped skunk showing thick outer wall, large central lipid vacuoles, glycogen clumps, and outer mitochondria. Uranyl acetate and lead citrate; bar, 10 μm. (Reproduced, with permission, from Albassam et al. 1986)*

aspirate, but etiology of the fungus as *E. crescens* was confirmed by the adiaspores measuring 30–80 µm and by polymerase chain reaction (PCR) assay of the tissue (Pusterla et al. 2002). The sequence of the PCR-amplified DNA fragment from the lung was found to be identical to an LSU-rDNA sequence on deposit in GenBank for an *E. crescens* isolate (accession number AF038335, Peterson and Sigler 1998). *Emmonsia*-like isolates are occasionally cultured from human respiratory or cutaneous specimens, but the significance of these isolates is difficult to evaluate when information on results of direct examination is lacking (Peterson and Sigler 1998). The fungus is occasionally isolated from other substrates including soil, dung, or straw.

By contrast, culture is usually successful from specimens of patients with AIDS. In the osteomyelitic infection attributed to *E. parva*, direct examination of pus showed oval to elongated thick-walled cells lacking internal contents; the fungus was isolated from pus, sputum, and bone marrow (Echavarria et al. 1993). The fungus was grown from a respiratory specimen of a male patient with pulmonary infection and advanced AIDS (Turner et al. 1999). *E. pasteuriana* was isolated with some difficulty from a cutaneous lesion in the Italian patient with AIDS (Gori et al. 1998). Although disseminated cutaneous lesions were present, these were all culture negative except for the last biopsy taken just before death.

Mycology

The colonial and microscopic features of *Emmonsia* spp. resemble those of their closest relative, *B. dermatitidis*. *Emmonsia* spp. differ in their propensity to form one to three conidia at the ends of narrow, often slightly swollen stalks, in the conidia being flattened and as broad as they are long, and in their growth responses at higher temperature (Table 38.1). Conidia are aleurioconidia, i.e. solitary conidia released by lytic disintegration of part of the supporting hyphal cell (Carmichael 1962; Sigler, 1996, 1997, 2003). The colonial features of *Emmonsia* spp. are shown in Figures 38.6 and 38.7 and compared with a typical colony of *B. dermatitidis* shown in Figure 38.8. Microscopically, *Emmonsia* spp. appear similar to each other (Figure 38.9) (Table 38.1). Conidia are single celled and formed on the sides of the hyphae (sessile) or at the ends of a short narrow stalks which are either nonswollen or slightly swollen at the end closest to the conidium. Commonly, two or three conidia are arranged at the tips of spine-like pegs arising by secondary proliferation from the swollen cell. Conidia appear slightly flattened and slightly broader than long (subglobose or oval) or are pyriform and have narrow basal scars. Conidia of *E. crescens* and *E. parva* have smooth walls

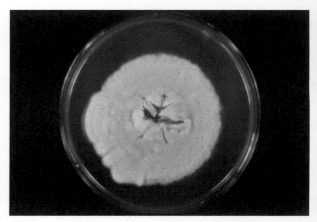

Figure 38.6 *Velvety, buff colored colony of* E. crescens *on phytone yeast extract agar after 28 days at 25°C (UAMH 7365). Note that colonies demonstrate variation in color and texture.*

and measure 2.5–4 µm long and 3–5 µm wide, but they may swell in age and become finely roughened. Up to eight conidia are formed in *E. pasteuriana* and they measure 2–5.5 µm long by 2–3.5 µm wide; conidia are smooth to finely roughened or appear tuberculate by scanning electron microscopy.

Emmonsia spp. grow well on routine mycological media including Sabouraud's glucose agar without cycloheximide, but conidiation is enhanced on media such as potato dextrose agar (PDA). *Emmonsia* spp. are slightly to moderately inhibited on media containing cycloheximide at a concentration of 400 µg/ml; *E. crescens* is less tolerant than *E. parva* (Sigler 1996). Isolates of both species vary in growth rates, morphology, and color (Otcenášek and Zlatanov 1975; Sigler 1996). On PDA, colonies reach diameters of 4.5–8 cm after 4 weeks at 25°C and may be glabrous (smooth or bald), woolly, velvety, or coarsely powdery. Colonies are flat, raised in the center, or radiately folded and the color varies from white to yellowish-white, buff, or tan (Sigler 1996). Two main colony

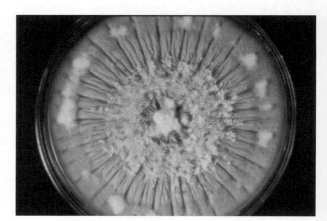

Figure 38.7 *Glabrous to hairy yellowish-white colony of* Emmonsia parva *on phytone yeast extract agar after 28 days at 25°C (UAMH 134). Note that colonies demonstrate variation in color and texture.*

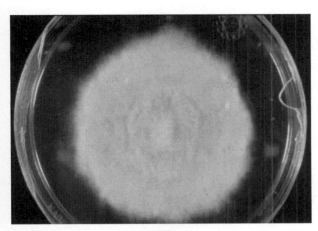

Figure 38.8 *Buff colored colony of* B. cermatitidis *on phytone yeast extract agar after 28 days at 25°C (UAMH 5438). (Compare with Figure 38.6)*

types may be observed, but intermediate forms are common (colony colors are described according to the color charts of Kornerup and Wanscher 1978). One type is yellowish-white to orange-white (4A2/5B3), with a densely woolly center dotted with exudate droplets, and having a broad glabrous margin. The reverse of the colony is grayish-brown (6D3). A second type is pale orange to grayish-orange (5A3/B5), with a coarsely powdery center and an irregular margin. Exudate droplets are present or absent and the reverse is reddish-brown (8E4). Colonies of *E. pasteuriana* are similar in being yellowish-white (4A2), densely woolly, furrowed, or zonate with powdery or glabrous sectors.

Emmonsia spp. are distinguished by their growth responses at higher temperature. In vitro, isolates of *E. crescens* are almost completely inhibited at 37°C and some strains show restricted growth even at 30°C (Boisseau-Lebreuil 1981; Sigler 1996). At 35–37°C,

conidia in the residual inoculum develop into globose swollen cells or adiaspores which measure 20–140 μm in diameter (average 59 μm) (Figure 38.10). These cells are thinner walled than those formed in vivo and have sometimes been given other names (e.g. chlamydospores or trophocytes) (Carmichael 1962; Dvorák et al. 1973). Maturing adiaspores germinate from many sites if transferred to culture media at lower temperatures (25°C) (Emmons 1964; Boisseau-Lebreuil 1970; Hejtmanek 1976). At 37°C, isolates of *E. parva* form colonies ranging from 9 to 77 mm in diameter after 21 days but are almost completely inhibited at 40°C (Sigler 1996). Adiaspores formed at this temperature are smaller than those of *E. crescens* and range from 8 to 20 μm in diameter (mean 12 μm). Complete conversion of hyphae to adiaspores is rarely achieved. When grown on enriched media such as brain heart infusion agar at 37°C, *E. pasteuriana* converts to a yeast phase consisting of oval to lemon-shaped, small (2–4 μm) yeast-like cells having narrow-based buds (Table 38.1) but conversion is incomplete and hyphal forms are present.

Teleomorphs and relationships

Recently, classical mating experiments and DNA sequence comparisons have provided evidence that *Emmonsia* spp. are anamorphs (mitotic, asexual stages) of *Ajellomyces* spp. and that their closest relative is *B. dermatitidis*. Species of *Ajellomyces* are heterothallic and teleomorphs (meiotic, sexual stages) form when compatible isolates are mated. Ascocarps are composed of branched, anastomosing, pale-brown hyphae, in which individual cells are somewhat diamond shaped, and terminate in coiled appendages. Ascospores are small (<1–1.5 μm diameter) and covered with minute spiny projections. The teleomorph of *B. dermatitidis* is *Ajellomyces dermatitidis* (McDonough and Lewis 1967, 1968); that of *H. capsulatum* was described first as *Emmonsiella capsulata* but later renamed as *Ajellomyces capsu-*

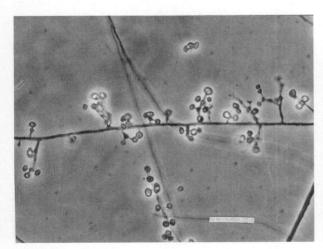

Figure 38.9 *Microscopic morphology of* E. crescens *(UAMH 7365). In Emmonsia species, one to three conidia are borne at the ends of stalks which are commonly slightly swollen. Bar, 20 μm. (Reproduced, with permission, from Sigler 1996)*

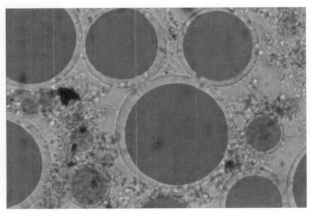

Figure 38.10 *Adiaspores of* Emmonsia crescens *formed in vitro on phytone yeast extract agar at 37°C (UAMH 7365) and stained with acid fuchsin in lactic acid.*

latus (Kwon-Chung 1972; McGinnis and Katz 1979). Matings among 12 strains of *E. crescens* resulted in a teleomorph described as *Ajellomyces crescens* (Sigler 1996) (Figures 38.11–38.14). However, no fertility was seen among mated isolates of *E. parva* or when isolates were paired with those of *E. crescens*.

Phylogenetic trees derived from comparison of DNA sequences obtained from the ribosomal rRNA gene confirm the relationship among these species. *Emmonsia* spp. and *B. dermatitidis*, together with *H. capsulatum*, *Paracoccidioides brasiliensis*, and *Lacazia loboi* (not known in culture), are placed as members of the monophyletic *Ajellomyces* clade (Bowman and Taylor 1993; Leclerc et al. 1994; Drouhet et al. 1998; Peterson and Sigler 1998; Herr et al. 2001; Untereiner et al. 2002; Sigler 2003). Although *Ajellomyces* was classified in the family *Onygenaceae*, order Onygenales (Currah 1985), phylogenetic analyses of DNA sequence data have shown the *Ajellomyces* clade to be a lineage distinct from the *Onygenaceae* (Sugiyama and Mikawa 2001; Untereiner et al., 2002). The new family *Ajellomyceta-ceae* has been proposed for this lineage (Untereiner et al. 2004). Although the *Ajellomyces* clade is well supported, the relationships within it are not clearly resolved. *P. brasiliensis* is in the clade, but its position is not well defined. *C. immitis* is placed within the *Onygenaceae* and apart from *Ajellomyces* species; its closest relative is a nonpathogenic species *Uncinocarpus reesii* (*Malbranchea* anamorph) (Sigler and Carmichael 1976; Bowman and Taylor 1993; Pan et al. 1994; Sugiyama and Mikawa 2001).

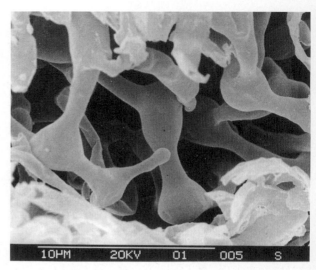

Figure 38.12 *Scanning electron micrograph showing the swollen, diamond-shaped cells of the hyphae composing the ascocarp (UAMH 128 × 129). (Reproduced, with permission, from Sigler 1996)*

Phylogenetic analysis of *Emmonsia* isolates based on LSU-ITS rDNA sequences placed *E. crescens* isolates into two groups according to their geographical origin and found greater than expected variation among isolates identified as *E. parva*, leading to the suggestion that additional species exist, and that some human associated isolates may represent undescribed species (Peterson and Sigler 1998). *E. parva* isolates originally obtained by Emmons from rodents in Arizona were found to be genetically closer to *B. dermatitidis* than to *E. crescens*. Guého et al. (1997), comparing LSU rDNA sequences, found almost the same level of

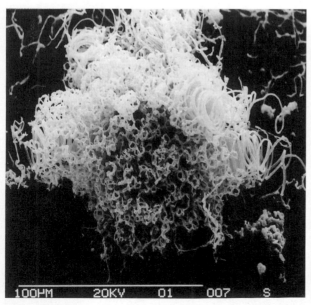

Figure 38.11 *Scanning electron micrograph of an ascocarp of* Ajellomyces crescens, *the sexual stage of* Emmonsia crescens, *obtained by mating sexually compatible strains. Note the helically coiled appendages arising from the ascocarp (UAMH 127 × 349). (Reproduced, with permission, from Sigler 1996)*

Figure 38.13 *Scanning electron micrograph of ascospore of* Ajellomyces crescens. *Note that ascospores are tiny measuring 1– 1.5 μm in diameter and have short spiny projections (UAMH 127 × 349). (Reproduced, with permission, from Sigler 1996)*

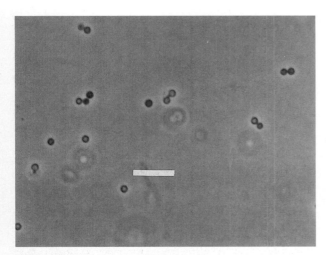

Figure 38.14 *Ascospores of* Ajellomyces crescens *examined by light microscopy (UAMH 128 × 349). Bar, 10 μm. (Reproduced, with permission, from Sigler 1996)*

genetic diversity between African and North American isolates of *B. dermatitidis* as between the latter and *E. parva*. A later study showed that *E. pasteuriana*, which demonstrates a yeast form in vivo, is closer to *E. crescens* than to *B. dermatitidis* (Drouhet et al. 1998). The close relationship between *Blastomyces* and *Emmonsia*, and the finding that some human-associated *Emmonsia* spp. produce budding cells in vivo, suggests that these anamorph taxa should be combined. Similar arguments have been made previously and species of both genera were at one time placed in *Chrysosporium*. An impediment to combining the species in *Blastomyces* comes from the fact that the *Blastomyces* Gilchrist & Stokes is illegitimate according to the Botanical Code. To have this medically important genus retain priority over *Emmonsia*, it must first be conserved (Peterson and Sigler 1998).

Differential diagnosis

Diagnosis of adiaspiromycosis is generally based on radiological presentations and the finding of characteristic adiaspores in tissue. In culture, isolates of *Emmonsia* spp. resemble those of *B. dermatitidis*, but cultures are rarely established from patients with pulmonary adiaspiromycosis. *E. crescens* and *E. parva* do not convert to a yeast form, and their strongly inhibited growth and formation of adiaspores in culture at 37°C or 40°C, respectively, may aid in identification (Sigler 1996). *B. dermatitidis* isolates, by comparison, convert to the yeast form typical for the species (large thick-walled cells with broad-based buds) and are more tolerant of cycloheximide (Table 38.1). *B. dermatitidis* isolates can be confirmed by the DNA probe or exoantigen test (Table 38.1). Microscopically, conidia of *B. dermatitidis* are usually solitary at the ends of stalks with rare proliferation to form another conidium, in

contrast to those of *Emmonsia* spp. which often demonstrate secondary proliferation (Figure 38.9).

TREATMENT

The optimum therapeutic protocol for patients with pulmonary adiaspiromycosis is not clear, but association of a steroid with an antifungal drug appears to aid recovery (de Almeida Barbosa et al. 1997). These authors reviewed the outcome of therapy for three Brazilian patients and case histories for 16 other patients. Two of three Brazilian patients improved without therapy. A third did not respond to itraconazole alone but recovered when prednisone was added to the regimen. A few patients have shown improvement after therapy with steroids alone (Kodousek et al. 1971; Vojtek et al. 1971), amphotericin B alone (de Almeida Barbosa et al. 1997) or in combination with flucytosine (Quilici et al. 1977) or prednisone (de Almeida Barbosa et al. 1997), ketoconazole (Severo et al. 1989; Moraes et al. 2001), and antituberculosis agents (Simecek et al. 1982). Improvement in one case followed therapy with thiobendazole, although in vitro susceptibility tests demonstrated that *E. crescens* was resistant at concentrations up to 128 μg/ml, whereas *E. parva* was sensitive at 1 μg/ml (Barbas Filho et al. 1990). Mycelial stages of both *E. crescens* and *E. parva* are sensitive to imidazoles by in vitro susceptibility tests (Otcenášek and Ditrich 1983). Although *E. crescens* appears less sensitive to amphotericin B in vitro (minimal inhibitory concentration 4 μg/ml) (Otcenášek and Hubálek 1968), enlargement of conidia into adiaspores in experimentally infected mice was prevented by treatment with amphotericin B (Otcenášek and Dvorák 1967).

Infection in two AIDS patients responded to therapy with amphotericin (Echavarria et al. 1993) or amphotericin followed by fluconazole (Turner et al. 1999). However, a third patient whose CD4 counts were very low, died after a short course of therapy with amphotericin (Gori et al. 1998).

REFERENCES

Albassam, M.A., Bhatnagar, R., et al. 1986. Adiaspiromycosis in striped skunks in Alberta, Canada. *J Wildl Dis*, **22**, 13–18.

Al-Doory, Y., Vice, T.E. and Mainster, M.E. 1971. Adiaspiromycosis in a dog. *J Am Vet Med Assoc*, **159**, 87–90.

Baker, E.E., Mrak, E.M. and Smith, C.E. 1943. The morphology, taxonomy, and distribution of *Coccidioides immitis* Rixford and Gilchrist 1896. *Farlowia*, **1**, 199–244.

Bambirra, E.A. and Nogueira, N.M.F. 1983. Human pulmonary granulomas caused by *Chrysosporium parvum* var. *crescens* (*Emmonsia crescens*). *Am J Trop Med Hyg*, **32**, 1184–5.

Barbas Filho, J.V., Amato, M.B.P., et al. 1990. Respiratory failure caused by adiaspiromycosis. *Chest*, **97**, 1171–5.

Boisseau-Lebreuil, M.T. 1970. Particularités de l'infestation adiaspiromycosique par *Emmonsia crescens* Emmons et Jellison, 1960 chez *Pithymys subterraneus* (De Selys-Longchamps 1836) et *Mustela nivalis nivalis* L. 1766. *Mycopathol Mycol Appl*, **40**, 183–91.

Boisseau-Lebreuil, M.T. 1975. Sensibilité comparée de divers animaux de laboratoire à l'infection par instillations nasales de la phase saprophytique d'*Emmonsia crescens* Emmons & Jellison, 1960: fréquence et intensité du parasitisme, réactions histopathologiques. *Mycopathologia*, **56**, 143–8.

Boisseau-Lebreuil, M.T. 1977. Different behavior of 22 strains of *Emmonsia* Ciferri and Montemartini 1959, a monilial fungus, in the lungs of laboratory mice in comparison with their parasitic morphology in vitro. *Mycopathologia*, **61**, 85–91.

Boisseau-Lebreuil, M.T. 1981. *Emmonsia parva* (Emmons et Ashburn 1942) Ciferri et Montemartini, 1959 = *Chrysosporium parvum* (Emmons et Ashburn) comb. nov.; *Emmonsia crescens* Emmons et Jellison, 1960 = *Chrysosporium parvum* var. *crescens*. *Mycopathologia*, **73**, 135–41.

Bowman, B.H. and Taylor, J.W. 1993. Molecular phylogeny of pathogenic and non-pathogenic Onygenales. In: Reynolds, D.R. and Taylor, J.W. (eds), *The fungal holomorph: mitotic, meiotic and pleomorphic speciation in fungal systematics*. Wallingford: CAB International, 169–78.

Cano, R.J. and Taylor, J.J. 1974. Experimental adiaspiromycosis in rabbits: host response to *Chrysosporium parvum* and *C. parvum* var. *crescens* antigens. *Sabouraudia*, **12**, 54–63.

Carmichael, J.W. 1951. The pulmonary fungus *Haplosporangium parvum*. II. Strain and generic relationships. *Mycologia*, **43**, 605–24.

Carmichael, J.W. 1962. *Chrysosporium* and some other aleuriosporic hyphomycetes. *Can J Bot*, **40**, 1137–74.

Chandler, F.W., Kaplan, W. and Ajello, L. 1980. *Histopathology of mycotic diseases*. Chicago, IL: Year Book Medical Publishers.

Ciferri, R. and Montemartini, A. 1959. Taxonomy of *Haplosporangium parvum*. *Mycopathologia*, **10**, 303–16.

Connole, M.D. 1990. Review of animal mycoses in Australia. *Mycopathologia*, **111**, 133–64.

Currah, R.S. 1985. Taxonomy of the Onygenales: Arthrodermataceae, Gymnoascaceae, Myxotrichaceae and Onygenaceae. *Mycotaxon*, **24**, 1–216.

de Almeida Barbosa, A., Moreira Lemos, A.C. and Severo, L.C. 1997. Acute pulmonary adiaspiromycosis. Report of three cases and a review of 16 other cases collected from the literature. *Rev Iberoam Micol*, **14**, 177–80.

Doby, J.M., Boisseau-Lebreuil, M.T. and Beaucournu, C. 1970. Nouveaux hôtes pour *Emmonsia crescens* Emmons et Jellison, agent de l'adiaspiromycose. *Bull Soc Pathol Exot Fil*, **63**, 324–34.

Doby, J.M., Boisseau-Lebreuil, M.T. and Rault, B. 1971. L'adiaspiromycose par *Emmonsia crescens* chez les petits mammifères sauvages en France. *Mycopathol Mycol Appl*, **44**, 107–15.

Doby-Dubois, M., Chevrel, M.L., et al. 1964. Premier cas humain d'adiaspiromycose, par *Emmonsia crescens*, Emmons et Jellison. *Bull Soc Pathol Exot*, **57**, 240–4.

dos Santos, V.M., Santana, J.H., et al. 1997. Disseminated pulmonary adiaspiromycosis. A case report. *Rev Soc Bras Med Trop*, **30**, 397–400.

Dowding, E.S. 1947. The pulmonary fungus, *Haplosporangium parvum*, and its relationship with some human pathogens. *Can J Res E*, **25**, 195–206.

Drouhet, E. and Huerre, M. 1999. Yeast tissue phase of *Emmonsia pasteuriana* inoculated in golden hamster by intratesticular way. *Mycoses*, **52**, 11–18.

Drouhet, E., Guého, E., et al. 1998. Mycological, ultrastructural and experimental aspects of a new dimorphic fungus *Emmonsia pasteuriana* sp. nov. isolated from a cutaneous disseminated mycosis in AIDS. *J Mycol Med*, **8**, 64–77.

Dvorák, J. 1968. The current geographical distribution of adiaspiromycosis caused by *Emmonsia crescens* Emmons et Jellison 1960. *Mykosen*, **11**, 227–32.

Dvorák, J., Otcenásek, M. and Prokopic, J. 1967. Seasonal incidence of adiaspores of *Emmonsia crescens* Emmons et Jellison 1960 in wildly living animals. *Mycopathol Mycol Appl*, **31**, 71–3.

Dvorák, J., Otcenásek, M. and Prokopic, J. 1969. The spring peak of adiaspiromycosis due to *Emmonsia crescens* Emmons et Jellison 1960. *Sabouraudia*, **7**, 12–14.

Dvorák, J., Otcenásek, M. and Rosický, B. 1973. Adiaspiromycosis caused by *Emmonsia crescens* Emmons and Jellison 1960. *Acta Univ Palacki Olomuc Fac Med*, **70**, 1–120.

Echavarria, E., Cano, E.L. and Restrepo, A. 1993. Disseminated adiaspiromycosis in a patient with AIDS. *J Med Vet Mycol*, **31**, 91–7.

Emmons, C.W. 1942. Coccidioidomycosis. *Mycologia*, **34**, 452–63.

Emmons, C.W. 1964. Budding in *Emmonsia crescens*. *Mycologia*, **56**, 415–19.

Emmons, C.W. and Ashburn, L.L. 1942. The isolation of *Haplosporangium parvum* n. sp. and *Coccidioides immitis* from wild rodents. Their relationship to coccidioidomycosis. *Public Health Rep*, **57**, 1715–27.

Emmons, C.W. and Jellison, W.L. 1960. *Emmonsia crescens* sp. n. and adiaspiromycosis (haplomycosis) in mammals. *Ann NY Acad Sci*, **89**, 91–101.

England, D.M. and Hochholzer, L. 1993. Adiaspiromycosis: an unusual fungal infection of the lung. *Am J Surg Pathol*, **17**, 876–86.

Gori, S., Drouhet, E., et al. 1998. Cutaneous disseminated mycosis in a patient with AIDS due to a new dimorphic fungus. *J Mycol Med*, **8**, 57–63.

Guého, E., Leclerc, M.C., et al. 1997. Molecular taxonomy and epidemiology of *Blastomyces* and *Histoplasma* species. *Mycoses*, **40**, 69–81.

Hejtmanek, M. 1976. Conversion of adiasporic and mycelial stages of *Emmonsia crescens* examined in the scanning electron microscope. *Folia Parasitol (Praha)*, **23**, 165–7.

Hejtmanek, M. 1985. Dimorphism in *Chrysosporium parvum*. In: Szaniszlo, P. (ed.), *Fungal dimorphism with emphasis on fungi pathogenic for humans*. New York: Plenum Press, 237–61.

Hejtmanek, M., Necas, O. and Havelkova, M. 1974. The ultrastructure of *Emmonsia crescens* studied by the technique of freeze-etching. *Mycopathol Mycol Appl*, **54**, 79–83.

Herr, R.A., Tarcha, E.J., et al. 2001. Phylogenetic analysis of *Lacazia loboi* places this previously uncharacterized pathogen with the dimorphic Onygenales. *J Clin Microbiol*, **39**, 309–14.

Hubálek, Z. 1999. Emmonsiosis of wild rodents and insectivores in Czechland. *J Wildl Dis*, **35**, 243–9.

Hubálek, Z., Zejda, J., et al. 1993. Seasonality of rodent adiaspiromycosis in a lowland forest. *J Med Vet Mycol*, **31**, 359–66.

Hubálek, Z., Nesvadbová, J. and Rychnovský, B. 1995. A heterogeneous distribution of *Emmonsia parva* var. *crescens* in an agro-system. *J Med Vet Mycol*, **33**, 197–200.

Hubálek, Z., Nesvadbová, J. and Halouzka, J. 1998. Emmonsiosis of rodents in an agroecosystem. *Med Mycol*, **36**, 387–90.

Jellison, W.L. 1969. *Adiaspiromycosis (haplomycosis)*. Missoula MT: Mountain Press, 1–99.

Jellison, W.L., Glesne, L. and Owen, C. 1962. The passive cutaneous anaphylaxis (PCA) test in adiaspiromycosis. *Mycologia*, **54**, 466–75.

Johnstone, A.C., Hussein, H.M. and Woodgyer, A. 1993. Adiaspiromycosis in suspected cases of pulmonary tuberculosis in the common bushtail possum (*Trichosurus vulpecula*). *N Z Vet J*, **41**, 175–8.

Kamalam, A. and Thambiah, A.S. 1979. Adiaspiromycosis of human skin caused by *Emmonsia crescens*. *Sabouraudia*, **17**, 377–81.

Kirschenblatt, J.D. 1939. A new parasite of the lungs in rodents. *C R (Doklady) Acad Sci (USSR)*, **23**, 406–8, not seen.

Kodousek, R. 1974. Adiaspiromycosis. *Acta Univ Palacki Olomuc Fac Med*, **70**, 6–68.

Kodousek, R. and Hejtmanek, M. 1971. Enzymatic activity of the spherules of the adiaspiromycosis causing *Emmonsia crescens* [Emmons & Jellison 1960]. *Acta Histochem*, **41**, 349–52.

Kodousek, R., Vortel, V., et al. 1971. Pulmonary adiaspiromycosis in man caused by *Emmonsia crescens*: report of a unique case. *Am J Clin Pathol*, **56**, 394–9.

Koller, L.D. and Helfer, D.M. 1978. Adiaspiromycosis in the lungs of a goat. *J Am Vet Med Assoc*, **173**, 80–1.

Koller, L.D., Patton, N.M. and Whitsett, D.K. 1976. Adiaspiromycosis in the lungs of a dog. *J Am Vet Med Assoc*, **169**, 1316–17.

Kornerup, A. and Wanscher, J.H. 1978. *Methuen handbook of color*, 3rd edition. London: Methuen, 1–252.

Krivanec, K. 1977. Adiaspiromycosis in Czechoslovakian mammals. *Sabouraudia*, **15**, 221–3.

Krivanec, K. and Mason, R.W. 1980. Adiaspiromycosis in Tasmanian wombats? *Mycopathologia*, **71**, 125–6.

Krivanec, K. and Otcenášek, M. 1977. Importance of free living mustelid carnivores in circulation of adiaspiromycosis. *Mycopathologia*, **60**, 139–44.

Krivanec, K., Otcenášek, M. and Prokopic, J. 1974. Experimental adiaspiromycosis of the common vole (*Microtus arvalis*) and other small wild mammals after intraperitoneal inoculation. *Mycopathol Mycol Appl*, **53**, 133–40.

Krivanec, K., Otcenášek, M. and Rosický, B. 1975a. The role of polecats of the genus *Putorius* Cuvier, 1817 in natural foci of adiaspiromycosis. *Folia Parasitol (Praha)*, **22**, 245–9.

Krivanec, K., Otcenášek, M. and Slais, J. 1975b. Adiaspiromycosis in large free living carnivores. *Mycopathologia*, **58**. 21–5.

Kwon-Chung, K.J. 1972. *Emmonsiella capsulata*: perfect state of *Histoplasma capsulatum*. *Science*, **177**, 368–9.

Kwon-Chung, K.J. and Bennett, J.E. 1992. *Medical mycology*. Malvern, PA: Lea & Febiger, 1–866.

Leclerc, M.C., Philippe, H. and Guého, E. 1994. Phylogeny of dermatophytes and dimorphic fungi based on large subunit ribosomal RNA sequence comparison. *J Med Vet Mycol*, **32**, 331–41.

Levy, F.E., Larson, J.T., et al. 1991. Invasive *Chrysosporium* infection of the nose and paranasal sinuses in an immunocompromised host. *Otolaryngol Head Neck Surg*, **104**, 384–8.

Mason, R.W. and Gauhwin, M. 1982. Adiaspiromycosis in south Australian hairy-nosed wombats. *J Wildl Dis*, **18**, 3–8.

Martins, R.L., Santos, C.G., et al. 1997. Human adiaspiromycosis. A report of a case treated with ketoconazole. *Rev Soc Bras Med Trop*, **30**, 507–9.

McDonough, E.S. and Lewis, A.L. 1967. *Blastomyces dermatitidis*: production of the sexual stage. *Science*, **156**, 528–9.

McDonough, E.S. and Lewis, A.L. 1968. The ascigerous stage of *Blastomyces dermatitidis*. *Mycologia*, **60**, 76–83.

McGinnis, M.R. 1980. Recent taxonomic developments and changes in medical mycology. *Annu Rev Microbiol*, **34**, 109–35.

McGinnis, M.R. and Katz, B. 1979. *Ajellomyces* and its synonym *Emmonsiella*. *Mycotaxon*, **8**, 157–64.

McGinnis, M.R. and Rinaldi, M.G. 1995. Selected medically important fungi and some common synonyms and obsolete names. *Clin Infect Dis*, **21**, 277–8.

Menges, R.W. and Habermann, R.T. 1954. Isolation of *Haplosporangium parvum* from soil and results of experimental inoculations. *Am J Hyg*, **60**, 106–16.

Moraes, M.A.P., Gomes, M.I. and Souze Vianna, L.M. 2001. Adiaspiromicose pulmonar: achado casual em paciente falecido de febre amarela. *Rev Soc Bras Med Trop*, **34**, 83–5.

Nuorva, K., Pitkanen, R., et al. 1997. Pulmonary adiaspiromycosis in a two year old girl. *J Clin Pathol*, **50**, 82–5.

Otcenášek, M. and Ditrich, O. 1983. Comparative susceptibility of the agents of adiaspiromycosis to imidazole derivatives in vitro. *Sabouraudia*, **21**, 239–42.

Otcenášek, M. and Dvorák, J. 1967. The effect of fungizone on experimental adiaspiromycosis of laboratory mice. *Mycopathol Mycol Appl*, **33**, 349–52.

Otcenášek, M. and Hubálek, Z. 1968. In vitro antifungal activity of amphotericin B against *Emmonsia crescens*. *Sabouraudia*, **6**, 255–9.

Otcenášek, M. and Zlatanov, Z. 1975. Natural variability in the mycelial form of *Emmonsia crescens*. *Mycopathol*, **55**, 97–104.

Padhye, A.A. and Carmichael. J.W. 1968. *Emmonsia brasiliensis* and *Emmonsia ciferrina* are *Chrysosporium pruinosum*. *Mycologia*, **60**, 445–7.

Padhye, A.A., Smith, G., et al. 1994. Comparative evaluation of chemiluminescent DNA probe assays and exoantigen tests for rapid identification of *Blastomyces dermatitidis* and *Coccidioides immitis*. *J Clin Microbiol*, **32**, 867–70.

Pan, S., Sigler, L. and Cole, G.T. 1994. Evidence for a phylogenetic connection between *Coccidioides immitis* and *Uncinocarpus reesii* (Onygenaceae). *Microbiology*, **140**, 1481–94.

Peres, L.C., Figueiredo, F., et al. 1992. Fulminant disseminated pulmonary adiaspiromycosis in humans. *Am J Trop Med Hyg*, **46**, 146–50.

Peterson, S.W. and Sigler, L. 1998. Molecular genetic variation in *Emmonsia crescens* and *E. parva*, etiologic agents of adiaspiromycosis, and their phylogenetic relationship to *Blastomyces dermatitidis* (*Ajellomyces dermatitidis*) and other systemic fungal pathogens. *J Clin Microbiol*, **36**, 2918–25.

Pusterla, N., Pesavento, P.A., et al. 2002. Disseminated pulmonary adiaspiromycosis caused by *Emmonsia crescens* in a horse. *Equine Vet J*, **34**, 749–52.

Quilici, M., Orsini, A., et al. 1977. Adiasporomycose pulmonaire disseminée (a propos d'une observation). *Arch Anat Cytol Pathol*, **25**, 227–34.

Rippon, J.W. 1988. Adiaspiromycosis. *Medical mycology: the pathogenic fungi and pathogenic actinomycetes*, 3rd edition. Philadelphia, PA: W.B. Saunders, 718–21.

Schwarz, J. 1978. Adiaspiromycosis. *Pathol Ann*, **13**, 41–53.

Sekhon, A.S., Standard, P.G., et al. 1986. Reliability of exoantigens for differentiating *Blastomyces dermatitidis* and *Histoplasma capsulatum* from *Chrysosporium* and *Geomyces* species. *Diagn Microbiol Infect Dis*, **4**, 215–21.

Severo, L.C., Geyer, G.R., et al. 1989. Adiaspiromycosis treated successfully with ketoconazole. *J Med Vet Mycol*, **27**, 265–8.

Sigler, L. 1996. *Ajellomyces crescens* sp. nov., taxonomy of *Emmonsia* spp., and relatedness with *Blastomyces dermatitidis* (teleomorph *Ajellomyces dermatitidis*). *J Med Vet Mycol*, **34**, 303–14.

Sigler, L. 1997. *Chrysosporium* and molds resembling dermatophytes. In: Summerbell, R.C., Kane, J., et al. (eds), *Handbook of dermatophytes. A clinical guide and laboratory manual of dermatophytes and other filamentous fungi from skin, hair and nails*. Belmont, CA: Star Publishing Co, 261–311.

Sigler, L. 1998. Agents of adiaspiromycosis. In: Ajello, L. and Hay, R. (eds), *Topley & Wilson's microbiology and microbial infections*, vol. 4. 9th edition. London: Edward Arnold, 571–83.

Sigler, L. 2003. Ascomycetes: the Onygenaceae and other fungi from the order Onygenales. In: Howard, D.H. (ed.), *Pathogenic fungi in humans and animals*. New York: Marcel Dekker, 195–236.

Sigler, L. and Carmichael, J.W. 1976. Taxonomy of *Malbranchea* and some other hyphomycetes with arthroconidia. *Mycotaxon*, **4**, 349–488.

Sigler, L. and Flis, A. 1998. *Catalogue of the University of Alberta Microfungus Collection and Herbarium*. Edmonton: University of Alberta Devonian Botanic Garden, 1–213.

Sigler, L. and Verweij, P. 2003. *Aspergillus, Fusarium* and other opportunistic moniliaceous fungi. In: Murray, P.R., Baron, E.J., et al. (eds), *Manual of clinical microbiology*, 8th edition. Washington, DC: American Society for Microbiology, 1726–60.

Simecek, C., Slais, J., et al. 1982. Disseminated pulmonary adiaspiromycosis identified in bioptic material from the lung. *Z Erkrank Atm Org*, **159**, 282–8.

Simpson, V.R. and Gavie-Widen, D. 2000. Fatal adiaspiromycosis in a wild Eurasian otter (*Lutra lutra*). *Vet Rec*, **147**, 239–41.

Slais, J. 1976. Histopathological changes and the genesis of adiaspiromycosis in mice infected intraperitoneally with *Emmonsia crescens* Emmons et Jellison 1960. *Folia Parasitol (Praha)*, **23**, 373–81.

Stillwell, W.T., Rubin, B.D. and Axelrod, J.L. 1984. *Chrysosporium*, a new causative agent in osteomyelitis. *Orthopedics*, **184**, 190–2.

Sugiyama, M. and Mikawa, T. 2001. Phylogenetic analysis of the non-pathogenic genus *Spiromastix* (*Onygenaceae*) and related onygenalean taxa based on large subunit ribosomal DNA sequences. *Mycoscience*, **42**, 413–21.

Thammayya, A. and Sanyal, M. 1973. *Chrysosporium parvum* from eczematous lesions in man. *Bull Calcutta School Trop Med*, **21**, 23–4.

Tomšíková, A., Zavazal, V., et al. 1970. Reactivity of humans to the antigen of *Emmonsia crescens* Emmons et Jellison 1960. *Mykosen*, **13**, 491–8.

Tomšíková, A., Dvorák, J., et al. 1972. Antigenic activity of *Emmonsia crescens* Emmons et Jellison 1960. *Acta Univ Palacki Olomuc Fac Med*, **63**, 51–7.

Tomšíková, A., Sterba, J., et al. 1974. On the possibility of diagnosing lung adiaspiromycosis. *Mykosen*, **17**, 349–58.

Tomšíková, A., Sterba, J. and Novackova, D. 1976. Antigenic characteristics of *Emmonsia crescens* Emmons et Jellison 1960. *Mykosen*, **19**, 407–18.

Tomšíková, A., Hejtmanek, M. and Novackova, D. 1978. Antigenic activity of *Emmonsia crescens* mutants. *Mycopathologia*, **66**, 83–90.

Tomšíková, A., Slais, J., et al. 1982. Contribution to active and passive immunisation in deep-seated mycoses. *Chrysosporium parvum* var. *crescens* as a model. *Mykosen*, **25**, 393–403.

Toshniwal, R., Goodman, S., et al. 1986. Endocarditis due to *Chrysosporium* species: a disease of medical progress? *J Infect Dis*, **153**, 638–9.

Turner, D., Burke, M., et al. 1999. Pulmonary adiaspiromycosis in a patient with acquired immunodeficiency syndrome. *Eur J Clin Microbiol Infect Dis*, **18**, 893–5.

Untereiner, W.A., Scott, J.A., et al. 2002. Phylogeny of *Ajellomyces*, *Polytolypa* and *Spiromastix* (Onygenaceae) inferred from rDNA sequence and non-molecular data. *Stud Mycol*, **47**, 25–35.

Untereiner, W.A., Scott, J.A., et al. 2004. The *Ajellomycetaceae*, a new family of vertebrate-associated *Onygenales*. *Mycologia*, **96**, 811–200.

van Oorschot, C.A.N. 1980. A revision of *Chrysosporium* and allied genera. *Stud Mycol*, **20**, 1–89.

Vojtek, V., Kodousek, R., et al. 1971. Disseminated adiaspiromycosis of the lungs in an 11-year-old boy caused by *Emmonsia crescens*. *Rev Czech Med*, **17**, 164–9.

Vojtek, V., Serý, Z. and Berková, I. 1972. Clinical pattern and early result achieved in a case of pulmonary adiasporomycosis of miliary disseminated type. *Acta Univ Palacki Olomuc Fac Med*, **63**, 23–30.

von Arx , J.A. 1973. Further observations on *Sporotrichum* and some similar fungi. *Persoonia*, **7**, 127–30.

Wagoner, M.D., Badr, I.A. and Hidayat, A.A. 1999. *Chrysosporium parvum* keratomycosis. *Cornea*, **18**, 616–20.

Warwick, P., Ferrieri, B., et al. 1991. Presumptive invasive *Chrysosporium* infection in a bone marrow transplant recipient. *Bone Marrow Transplant*, **8**, 319–22.

Watts, J.C., Callaway, C.S., et al. 1975. Human pulmonary adiaspiromycosis. *Arch Pathol*, **99**, 11–15.

Index

Notes

(Fig.) and (Tab.) refer to figures and tables respectively. *vs.* indicates a comparison or differential diagnosis.
To save space in the index, the following abbreviations have been used:

CNS -central nervous system
EIA - enzyme immunoassay
ELISA - enzyme-linked immunosorbent assay
Ig - immunoglobulin
PCP - pneumocystis pneumonia
PCR - polymerase chain reaction

Complete table of contents for *Topley & Wilson's Microbiology and Microbial Infections*

VIROLOGY, VOLUMES 1 AND 2

BACTERIOLOGY, VOLUMES 1 AND 2

MEDICAL MYCOLOGY

PARASITOLOGY

IMMUNOLOGY